THE WORLD OF BIOLOGY

FOURTH EDITION

P. WILLIAM DAVIS

Hillsborough Community College

ELDRA PEARL SOLOMON

University of South Florida
Center for Research in Behavioral Medicine
and Health Psychology

LINDA R. BERG

University of Maryland

SAUNDERS COLLEGE PUBLISHING

PHILADELPHIA   FT. WORTH   CHICAGO   SAN FRANCISCO
MONTREAL   TORONTO   LONDON   SYDNEY   TOKYO

Text Typeface: Baskerville
Compositor: York Graphic Services
Acquisitions Editor: Julie Levin Alexander
Developmental Editors: Raymond Tschoepe, Gabrielle Goodman, and Martha Colgan
Managing Editor and Project Editor: Carol Field
Copy Editors: Becca Gruliow and Diane Lamsback
Manager of Art and Design: Carol Bleistine
Art Assistant: Doris Bruey
Text Designer: Edward A. Butler
Cover Designer: Lawrence R. Didona
Text Artwork: J & R Technical Services
Layout Artists: Edward A. Butler and Dorothy Chattin
Photo Researcher: Robin Bonner
Director of EDP: Tim Frelick

Cover: The ecosystem of the Smokey Mountain National Park (*background*), includes producers such as hepatica (*left*), consumers represented by the mountain lion kitten (*center*), and decomposers such as amanita (*right*). The interaction of these organisms maintains the delicate biological cycle of life. (Credits: Smokey Mountains, David Muench; hepatica and amanita, Skip Moody/ M. L. Dembinsky, Jr.; and mountain lion, Judd Cooney/ Phototake)

Frontmatter photo credits: Title page, VIREO/Doug Wechsler; page iii, Ed Reschke; page iv, Susan Blanchet; page v, Dwight R. Kuhn; page xix, David Phillips/Visuals Unlimited; page x, Stock Imagery; page xiii, E. R. Degginger; page xiv, Frans Lanting; page xviii, Runk/Schoenberger, from Grant Heilman.

Printed in the United States of America

THE WORLD OF BIOLOGY

0-03-030253-6 (ISBN)

Library of Congress Catalog Card Number: 89-043490

0123 063 987654321

For Freda, Nathan, Seth, Amy, Belicia, Mical, Karla, and Jennifer
. . . and for their generation

Preface

THE WORLD OF BIOLOGY explores the diverse life forms that inhabit Planet Earth, their interdependence, and their interactions with the environment. The ever-expanding knowledge afforded by research in the biological sciences provides us with the tools necessary to understand the marvelous complexity and precise function of life from the molecular and cellular levels to its ecological organization in the ecosphere. One of our principal goals in preparing this book is to convey to the student an appreciation of and sense of excitement about modern biological science.

■ THE EVOLUTION OF THE WORLD OF BIOLOGY

THE WORLD OF BIOLOGY has evolved through its four editions in synchrony with the evolution of the biological sciences and biological education. Every effort has been made to update its content so that this book lies on the cutting edge of modern biological science. THE WORLD OF BIOLOGY focuses on the genetic, evolutionary, and ecological relationships of life on earth. The interdependence of all living things is emphasized as a theme throughout the book.

The social and environmental emphasis that has been one of the trademarks of THE WORLD OF BIOLOGY has been retained in this new edition. The impact of human population growth and of often thoughtless human activity on the ecosphere is brought to the student's attention wherever appropriate. Deliberate attempts are made at consciousness-raising.

Author Team

The author team has been expanded to include botanist Dr. Linda Berg who has rewritten the botany and evolution sections of the book. Coverage of plant biology and of evolution has been expanded and given increased emphasis.

Conceptual Approach

THE WORLD OF BIOLOGY, fourth edition, continues to emphasize biological principles rather than specific facts. All of the basic facts of biology are here, but they are integrated to form whole concepts. In THE WORLD OF BIOLOGY, we take the time and space to integrate the facts so that the student understands how these details fit together to explain cells, organisms, and ecosystems, and how these relate to one another.

The authors develop the conceptual approach in the fourth edition by (1) presenting key concepts at the beginning of each chapter; (2) writing headings and subheadings within each chapter in the form of conceptual statements; and (3) introducing fundamental biological principles in Chapter 1 and then reintroducing them numerous times throughout the book.

"Window On The Animal Cell" And "Window On The Plant Cell"

Another new feature in the fourth edition are two series of acetate overlays one of the plant, the other of the animal cell. Accurately rendered, state-of-the-art drawings are presented as progressively deeper views of the plant and animal cell structures. These overlays will help students learn and remember the parts of the cell and at the same time help make learning fun.

Learning Aids

In addition to the interesting, conversational reading style for which this book is known, the numerous learning aids that have been the hallmark of THE WORLD OF BIOLOGY have been expanded and refined.

1. **Chapter outlines** in the form of concept statements reflect the headings within each chapter and provide students with an overview of the material covered.
2. **Learning objectives** tell students how to demonstrate mastery of the material.
3. **Key concepts** summarize the main principles presented in each chapter.
4. **Concept-statement headings** introduce each section, informing students what key idea will be discussed.
5. Numerous **tables,** many of them illustrated, summarize and organize material presented in the text.
6. Carefully rendered **illustrations,** over half of them new in this edition, support concepts covered in the text.

7. **Focus boxes,** such as Focus on AIDS, spark student interest, present applications of concepts discussed, and familiarize students with the directions and methods of research in modern biology.

8. **Chapter summaries** in outline form at the end of each chapter provide a quick review of the material presented.

9. **Post Tests** provide an opportunity to evaluate the students' mastery of the material; answers are provided at the back of the book.

10. **Review questions** focus on important concepts and applications.

11. **Readings** provide references for further learning.

12. New **terms** are in boldface, permitting easy identification, and providing emphasis.

13. In this edition, a separate **glossary** is provided, facilitating rapid location of definitions.

▣ THE ORGANIZATION OF THE WORLD OF BIOLOGY, FOURTH EDITION

Part 1 Basic Concepts Of Biology: An Introduction

Chapter 1, The World of Biology, introduces several major concepts of biology including the fundamental similarities of all living things; the organization of life on individual and ecological levels; the evolution of life on our planet; the diversity of life and how biologists classify living things; the interdependence of all living things, and of human impact on the ecosphere.

Written at the request of instructors using the third edition, the new Chapter 2 lays the foundation for understanding how science works. This chapter is an introduction to the methods and practice of science, and conveys the importance of science in the modern world.

Part 2 The Organization Of Life

Chapters 3 and 4 provide the basic tools of chemistry needed to understand biology, and introduce the biological molecules that interact in the construction and maintenance of living things. Chapters 5 and 6 focus on the structure and functions of cells and their membranes. Chapter 5 has been enhanced by the addition of the animal and plant cell "Windows"—new and unique learning tools for mastering cell structure and understanding the three dimensional nature of the cell.

Part 3 Energy Flow Through The World Of Life

Chapters 7, 8, and 9 focus on the energy transactions involved in life processes. New art has been developed to help clarify the processes of cellular respiration and photosynthesis. The chapter on photosynthesis has been entirely rewritten by botanist Linda Berg who presents this elegant and complex process in clear, straightforward terms.

Part 4 The Continuity Of Life: Cell Division And Genetics

The genetics unit has been expanded and reorganized. The molecular basis of inheritance is presented first (Chapters 10–12) as the foundation for understanding the transmission of traits from one generation to the next. In Chapter 13 the processes of mitosis and meiosis are explained and contrasted, and Chapters 14 and 15 present patterns of inheritance. Many important principles of inheritance are made relevant for the student by using human applications. Chapter 16, Genetic Frontiers: Recombinant DNA and Genetic Engineering, presents concepts on the cutting edge of genetic research.

Part 5 Evolution

The unit on evolution has been rewritten and expanded with increased coverage for this edition and is presented earlier in the book than in the last edition. Chapter 17 introduces evolution by natural selection, and presents various types of evidence used to support evolution. In Chapter 18 and throughout the book, the concept that evolution occurs in populations, not individuals, is emphasized. Chapter 19 describes the evolution of new species and describes macroevolution as involving changes in the kinds of species over evolutionary time. Chapter 20 surveys geological time from the origin of life on earth to the present and discusses classification.

Part 6 Diversity

This unit has been greatly expanded, so that in place of the three diversity chapters in the last edition there are now five—Chapters 21–25. An evolutionary frame-

work is used to present the various groups of organisms. Chapter 22 on fungal life and Chapter 23 on plant life have been entirely rewritten and expanded by Dr. Linda Berg. Sections on primate evolution and human cultural evolution are included in Chapter 25, Animal Life: Chordates.

Part 7 Plant Structure And Life Processes

Chapters 26 through 30 have been reorganized and rewritten with added depth and detail by botanist Linda Berg. The newly rendered art work has been designed by an experienced botanical illustrator. The coverage of genetic and environmental controls of differentiation in Chapter 26 is an important addition not found in competing texts. The relationship between leaf structure and function is emphasized in Chapter 27. In Chapter 28 the discussion of stem and root structure is integrated with the mechanisms of transport in xylem and phloem, and mineral nutrition. Chapter 29 discusses reproduction in flowering plants including asexual reproduction, flowers, fruits, and seeds. Chapter 30 focuses on plant hormones and responses.

Part 8 Animal Structure And Life Processes

Chapter 31 describes animal tissues; Chapters 32 through 42 go on to present each animal organ system. Each chapter begins by comparing how different animal groups carry on digestion, gas exchange, internal transport or whatever process is being discussed. Then, the human adaptations for carrying on the process are considered. The unit ends with a discussion of development in Chapter 43.

Part 9 Behavior And Ecology

The ecology unit has been reorganized in this edition and much of it has been rewritten. Chapters 44 and 45 present the concepts of animal behavior. Chapters 46 through 49 provide the foundations of ecology with the final chapter focusing on human ecology.

■ SUPPLEMENTS

To further facilitate learning and teaching, a supplement package has been carefully designed for the student and instructor. Included are the **Student Resource Manual, Instructor's Resource Manual, Test Bank** and **Computerized Test Bank, Lecture Outline on Disk, BIO/XL™** electronic study guide, **Laboratory Manual** and **Instructor's Laboratory Manual,** a set of **overhead transparencies** or **35mm slides** of 200 selected color illustrations, and a set of 99 electron micrograph overhead transparencies.

■ THE ECOLOGY OF THE WORLD OF BIOLOGY

The development and production of this new edition of THE WORLD OF BIOLOGY was a process involving interaction and cooperation among the author team and between the authors and many individuals in our home and professional environments. We appreciate the valuable input and support from family, friends, editors, colleagues, and students. We thank our families and friends for their understanding, support, and encouragement as we struggled through many revisions and deadlines. We especially thank Mical Solomon for sharing his computer expertise, Amy Solomon and Belicia Efros for their help in researching and proofreading, Kathleen M. Heide for her input and support, Rabbi and Freda Brod for their encouragement, and Alan Berg for his support and understanding.

The Editorial Environment

Preparing this book has been hard work, but working with the outstanding editorial and production staff at Saunders has also been enjoyable. We thank our Publisher Elizabeth Widdicombe for her support and help. After Aquisitions Editor Ed Murphy expertly launched the project, Julie Alexander took over at the helm and guided the book through the channels of development and production. We appreciate the willingness of Liz and Julie to stand by us and maintain our good humor through the intense work.

Our Developmental Editor, Ray Tschoepe, worked with us and made a unique contribution by sharing his artistic talents to reconceptualize and draw much of the art. Gabrielle Goodman gave us valuable input on the text and coordinated many aspects of the project. We thank Photography Editor Robin Bonner for helping us find the wonderful photographs that enhance the text. Martha Colgan, the text Developmental Editor, did an outstanding job of helping us sort through and make decisions about sometimes conflicting suggestions from numerous reviewers.

We greatly appreciate our Project Editor Carol Field, who contributed her expertise to guiding the project through the pitfalls of production. We thank Art Director Carol Bleistine for coordinating the art program and design. All of these dedicated professionals and many others at Saunders provided the skill and attention needed to produce the fourth edition of THE WORLD OF BIOLOGY. We thank them for their help and support throughout this project.

The Professional Environment—Reviewers

Our colleagues and students have provided valuable input by sharing their responses to the third edition of THE WORLD OF BIOLOGY with us. We thank them and ask again for their comments and suggestions as they use this new edition. We can be reached through our editors at Saunders College Publishing.

We here express our thanks to the many professors and researchers who have read the manuscript during various stages of its preparation and provided us with valuable suggestions for improving it. Their input has contributed greatly to our final product. They are as follows:

Karen Anderson, Hofstra University
Richard Blazer, Parkland College
J. D. Brammer, North Dakota State University
Carol Crafts, Providence College
Kate Denniston, Towson State University
Jean DeSaix, University of North Carolina—Chapel Hill
Judith Goodenough, University of Massachusetts—Amherst
Robert Hamilton, Loyola University of Chicago
Lazlo Hanzelly, Northern Illinois University
George Hudock, Indiana University
John D. Jackson, North Hennepin Community College
Margaret Kort, Southwest Baptist University
Chuck Kugler, Radford University
C. S. Lee, University of Texas—Austin
Jon Maki, Eastern Kentucky State University
Diana Martin, Rutgers University
James Miller, Delaware Valley College
Ronald Quinn, California State Polytechnic University
Russell Skavaril, Ohio State University
James Sorenson, Radford University
Ronald Sterner, University of Texas—Arlington
Robin Tyser, University of Wisconsin—La Crosse
Elizabeth Waldorf, Mississippi Gulf Coast Community College
Elton Woodward, Daytona Beach Community College

P.W.D.
E.P.S.
L.R.B.

Contents Overview

PART 1 BASIC CONCEPTS OF BIOLOGY: AN INTRODUCTION ... 1

1 The World of Biology ... 2
2 The Method of Science ... 23

PART 2 THE ORGANIZATION OF LIFE ... 41

3 Chemical Principles: Atoms, Molecules, and Reactions ... 42
4 The Chemistry of Life: Organic Compounds ... 60
5 Cell Structure and Function ... 79
6 Biological Membranes ... 107

PART 3 ENERGY FLOW THROUGH THE WORLD OF LIFE ... 133

7 The Energy of Life ... 134
8 Energy-Releasing Pathways ... 152
9 Capturing Energy: Photosynthesis ... 169

PART 4 THE CONTINUITY OF LIFE: CELL DIVISION AND GENETICS ... 185

10 DNA: The Molecular Basis of Inheritance ... 186
11 Gene Function: RNA and Protein Synthesis ... 198
12 Gene Regulation ... 212
13 Chromosomes and Cell Division ... 223
14 Patterns of Inheritance ... 241
15 Some Topics in Human Genetics ... 253
16 Genetic Frontiers: Recombinant DNA and Genetic Engineering ... 265

PART 5 EVOLUTION ... 281

17 Darwin and Natural Selection ... 282
18 Population Genetics ... 297
19 Speciation and Macroevolution ... 310
20 The Origin, History, and Diversity of Life ... 325

PART 6 DIVERSITY ... 343

21 Monerans and Protists ... 344
22 Fungal Life ... 364

23 Plant Life ... 381
24 Animal Life: Non-Chordate Invertebrates ... 402
25 Animal Life: The Chordates ... 433

PART 7 PLANT STRUCTURE AND LIFE PROCESSES ... 455

26 Plant Development and Growth ... 456
27 Leaf Structure and Function ... 475
28 Stems and Roots ... 489
29 Reproduction in Flowering Plants ... 507
30 Plant Hormones and Responses ... 523

PART 8 ANIMAL STRUCTURE AND LIFE PROCESSES ... 537

31 Animal Tissues, Organs, and Organ Systems ... 538
32 Protection, Support, and Movement: Skin, Skeleton, and Muscle ... 557
33 Responsiveness: Neural Control ... 577
34 Responsiveness: Nervous Systems ... 593
35 Sense Organs ... 613
36 Internal Transport ... 636
37 Internal Defense ... 660
38 Gas Exchange ... 681
39 Processing Food ... 698
40 Osmoregulation and Disposal of Metabolic Wastes ... 725
41 Endocrine Regulation ... 742
42 Reproduction ... 761
43 Development ... 789

PART 9 BEHAVIOR AND ECOLOGY ... 815

44 Animal Behavior ... 816
45 Social Behavior ... 833
46 Communities and Population Ecology ... 847
47 The Ecosphere ... 866
48 Life Zones ... 883
49 Human Ecology ... 907
Appendix ... A-1
Glossary ... G-1
Post-Test Answers ... ANS-1
Index ... I-1

Contents

PART 1 BASIC CONCEPTS OF BIOLOGY: AN INTRODUCTION   1

1 The World of Biology   2
Life can be defined in terms of the characteristics shared by living things   4
  Living things are composed of cells   4
  Living things grow and develop   4
  Metabolism includes the chemical processes essential to growth, repair, and reproduction   5
  Organisms work to maintain an appropriate internal environment   5
  Movement is a basic property of cells   6
  Living things respond to stimuli   8
  DNA transmits information from one generation to the next   8
  A species has the ability to evolve and adapt to its environment   10
The theory of evolution has become one of the great unifying concepts in biology   10
  Evolution proceeds by natural selection   10
  Species evolve in response to changes in the environment   11
  An evolutionary perspective is apparent in almost every aspect of biology   12
Biological organization is hierarchical   12
  We can identify an organizational hierarchy within the organism   12
  We can also identify organization on the ecological level   14
Millions of kinds of organisms inhabit our planet   16
  Biologists use a binomial system of nomenclature   17
  Taxonomic classification is hierarchical   19
  Most biologists recognize five kingdoms   19
Focus on society and the ecosphere   20
Focus on evolution in action: the case of the peppered moth   17

2 The Method of Science   23
Science is a means of investigation   24
Science seeks to understand the universe   24
  Science began in ancient times   25
Conclusions should be based on adequate evidence   28
  Deductive reasoning applies general principles   28
  Inductive reasoning discovers general principles   29
  Evidence must be interpreted   29
The scientific method tests induction   30
  Recognizing and stating problems is probably the most crucial step   31
  A hypothesis is a proposed explanation or generalization   32
  A prediction is a logical consequence of a hypothesis   32
  Predictions are tested by observation   32
  Hypotheses must be falsifiable   34

Principles are well-supported hypotheses   36
Science and technology interact   36
Science is a human activity   36
  Science is costly   37
  Scientific findings must be published   37
  Scientists must guard the truth   37
  Science has ethical dimensions   38
Focus on Piltdown man   38

PART 2 THE ORGANIZATION OF LIFE   41

3 Chemical Principles: Atoms, Molecules, and Reactions   42
All materials are made up of basic substances called elements   43
  Atoms are the basic particles of elements   43
  Atoms combine in definite proportions   46
Chemical formulas describe the composition and structure of elements and compounds   46
Elements and compounds interact in regular ways that may be expressed as chemical equations   48
Atoms combine by forming chemical bonds   48
  Atoms can share electrons to form covalent bonds   49
  Atoms can gain or lose electrons to form ionic bonds   50
  Hydrogen bonds are weak attractions involving partially charged hydrogen atoms and electronegative atoms   52
  Van der Waals and hydrophobic attractions are often important weak forces in large organic molecules   52
Oxidation involves the loss of electrons or hydrogen; reduction involves the gain of electrons or hydrogen   52
Inorganic compounds are relatively simple compounds without carbon backbones   53
  Water has many unique properties essential to living things   53
  Acids produce hydrogen ions; bases dissociate in water to produce hydroxide ions   55
  Salts form from acids and bases and dissociate into anions and cations   56
  Buffers minimize pH change   57
Focus on isotopes   47

4 The Chemistry of Life: Organic Compounds   60
Carbohydrates include the sugars, starches, and cellulose   61
  Monosaccharides are single-unit sugars   61
  Disaccharides are two-unit sugars   64
  Polysaccharides are large, multiple-unit carbohydrates   65
Lipids are fats or fat-like substances   66
  Neutral fats are composed of glycerol and fatty acids   68
  Phospholipids are important components of cell membranes   68
  Steroids are lipids whose molecules contain four rings of carbon atoms   68

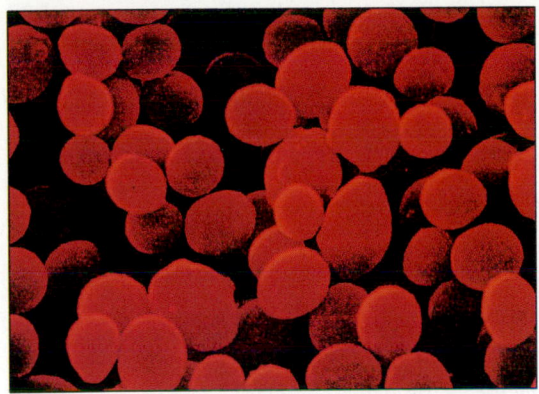

Cells can ingest materials by endocytosis and eject them by exocytosis 126
Focus on splitting the lipid bilayer **113**

PART 3 ENERGY FLOW THROUGH THE WORLD OF LIFE   **133**

7 The Energy of Life   **134**
Biological work requires energy **135**
Energy can be described as potential or kinetic 136
Heat is a form of energy that can be conveniently measured 136
Two laws of thermodynamics govern energy transformations **137**
The first law of thermodynamics holds that the quantity of energy in the universe does not change 137
The second law of thermodynamics holds that without an energy input systems tend to become increasingly disorganized 138
Metabolic reactions involve energy transformations **139**
Chemical reactions may be exergonic or endergonic 139
Chemical reactions are reversible 140
Reversible reactions reach a state of equilibrium 141
Endergonic and exergonic reactions can be coupled in living systems 141
Adenosine triphosphate (ATP) is the energy currency of the cell **142**
Enzymes are chemical regulators **143**
Catalysts lower the activation energy necessary to initiate a chemical reaction 144
If the name of a compound ends in -ase it is an enzyme 144
Enzymes are very efficient catalysts 144
Enzymes are specific 145
Enzymes work by forming enzyme-substrate complexes 145
Many enzymes require cofactors 146
Enzymes often work in teams 147
The cell regulates enzymatic activity 147
Enzymes work best at appropriate temperature and pH 147
Enzymes can be inhibited by certain chemical agents 148

8 Energy-Releasing Pathways   **152**
Fuel molecules can be degraded via aerobic or anaerobic pathways **153**
Catabolic pathways utilize oxidation-reduction reactions **153**
Hydrogen, along with some of its energy, can be transferred to hydrogen acceptor molecules 154
Cellular respiration is a redox process 154
Cellular respiration has four phases **155**
Glycolysis requires an energy investment 156
Pyruvate can be used to make acetyl coenzyme A 158
The citric acid cycle completes the breakdown of the fuel molecule 159
Most of the ATP is produced by the electron transport system and chemiosmotic phosphorylation 160
Cellular respiration is regulated by the amount of ADP and phosphate available **162**
The energy yield from one molecule of glucose may amount to 38 ATPs **163**

Proteins are macromolecules formed from linked amino acids **68**
There are many kinds of proteins 70
Amino acids contain a carboxyl and an amino group 71
Proteins are structured on several levels of organization 73
Nucleic acids consist of nucleotide subunits **75**
Windows on the animal cell **78**

5 Cell Structure and Function   **79**
The cell is the basic unit of life **80**
Cells have many common characteristics **80**
Prokaryotic cells are fundamentally different from eukaryotic cells **81**
Plant cells differ from animal cells **85**
The eukaryotic cell's interior contains numerous structures **88**
Some organelles are specialized for manufacturing products 89
Other organelles function in cell metabolism and energy use 94
Still other structures support the cell, connect it with other cells, and help it move 97
Vesicles and vacuoles are cavities within cells 99
The cell nucleus contains genetic information 99
Focus on viewing the cell **86**

6 Biological Membranes   **107**
Every cell has an outer membrane **108**
Biological membranes are phospholipid bilayers **108**
The membrane bilayer consists mainly of phospholipids 109
Cell membranes contain proteins and glycoproteins 110
Windows on the plant cell **110**
The cytoskeleton has membrane connections 112
Microvilli increase the surface area of some cells 112
Cell walls occur in four of the five kingdoms 113
Cells are joined by specialized structures **116**
Materials pass through membranes in several ways **117**
Diffusion is passive molecular movement 120
In facilitated diffusion molecules pass through special channels 121
Osmosis poses special problems for cells 122
Active transport uses energy to concentrate selected materials 125

Other nutrients can be used as energy sources 163

When oxygen is not available many cells utilize anaerobic pathways 163

In fermentation the final electron acceptor is an organic compound 164

Anaerobic metabolism is inefficient 166

9 Capturing Energy: Photosynthesis 169

Light exhibits properties of both waves and particles 170

Chlorophyll is a pigment that absorbs light 172

Plants use light energy to make sugar 173

In light-dependent reactions, light energy is used to make the high-energy compounds, ATP and NADPH + H⁺ 174

The fixation of carbon dioxide into organic compounds like sugar occurs during light-independent reactions 179

Focus on photosynthetic efficiency 179

PART 4 THE CONTINUITY OF LIFE: CELL DIVISION AND GENETICS 185

10 DNA: The Molecular Basis of Inheritance 186

The nucleus contains hereditary information 187

Experiments with the giant *Acetabularia* cell demonstrated the central role of the nucleus 187

The nucleus controls the cell through a messenger substance 188

The nucleus alone determines the shape of the *Acetabularia* cell 189

DNA can transfer genetic information from one cell to another 189

Viral nucleic acid contains genetic information to produce new viruses 190

A variety of methods have uncovered the shape and structure of the DNA molecule 191

X-ray diffraction provided clues to the shape of the DNA molecule 191

Watson and Crick demonstrated that DNA is a double helix 191

DNA is composed of nucleotide subunits 193

DNA replicates semiconservatively 195

11 Gene Function: RNA and Protein Synthesis 198

Proteins are central to an organism's identity 199

The genetic code links gene structure to protein structure 199

RNA encodes and expresses the genetic message stored in DNA 200

RNA is made by transcription 201

In RNA processing, introns are removed from the new molecule 203

In translation, protein is made using the genetic message encoded in mRNA 203

Ribosomal RNA forms part of the ribosome 203

Transfer RNA carries amino acids to the ribosome 204

Messenger RNA codons are the instructions the ribosome uses to construct a protein 204

Special signals direct the ribosome to start and stop adding amino acids 207

Changes can occur in some stages of genetic expression 209

Mutations are random, permanent changes in DNA 209

Factors in the environment can produce mutations 210

12 Gene Regulation 212

Gene regulation in prokaryotes primarily involves control of transcription 213

Operons are units of gene expression in prokaryotes 213

The shape of the bacterial chromosome has little or no effect on gene expression 216

There are multiple levels of gene regulation in eukaryotes 217

Gene control is important in cell growth and differentiation 217

Most cells of a multicellular eukaryote have the same genes 218

Chromosome structure is an important part of transcriptional regulation in eukaryotes 219

13 Chromosomes and Cell Division 223

All eukaryotes have chromosomes, but chromosome number depends on the species 224

The life cycle of the eukaryotic cell has definite stages 225

The cell spends most of its life in interphase 226

Mitosis is the process of nuclear division 226

The cell cycle is controllable 230

Some diploid cells produce haploid cells through meiosis 231

Meiosis differs extensively from mitosis 232

Meiosis I is the first main stage of meiosis 233

In meiosis II, chromatids separate 235

The cells that result from meiosis mature and form gametes or spores 235

Sex is determined by chromosome inheritance in animals 235

Chromosome defects can produce inherited disease 237

14 Patterns of Inheritance 241

Gregor Mendel founded the science of genetics in the 19th century 242

Inheritance is governed by laws of chance 244

Genes occur in pairs called alleles and are inherited as parts of chromosomes 244

The outcome of a simple cross may be predicted by possible combinations of a single pair of alleles 245

Organisms may have both members of a pair of genes alike, or unlike each other 246

Some genes are always expressed, but others are expressed only when homozygous 246
Dominance can be incomplete 246
Only one allele of each pair goes to each gamete **248**
Genes located on different chromosomes are inherited independently 250

15 Some Topics in Human Genetics **253**
The sex of the offspring determines the expression of genes on the X chromosome **255**
Genetic linkage can be deduced from the way genes are inherited **256**
Some traits are governed by more than one pair of genes **257**
Blood type is an inherited trait **257**
Blood type reflects blood cell surface antigens 258
The antigens of a blood cell are governed by a set of multiple alleles 258
The three ABO alleles produce four blood types 259
The Rh blood factors are potentially hazardous to the newborn 259
Genetic disease is widespread and important **261**
Marriage between blood relatives increases the chance of genetic disease 261
Some genetic disorders are dominant 261
Genetics raises ethical questions 261

16 Genetic Frontiers: Recombinant DNA and Genetic Engineering **265**
Recombinant DNA techniques involve manipulating specific genes in the cells of living organisms, thereby causing them to produce new or unusual substances **266**
First a specific region of DNA is isolated from the rest of the cell's DNA 268
Recombinant DNA molecules are constructed using DNA from two different sources 268
The recombinant DNA molecules are transferred into a new host organism's cell 269
The cells that have taken up the gene of interest are identified 270
The gene is expressed inside the new cell 271
Gene insertion in eukaryotes is technically challenging, but feasible 271
Genetic engineering holds great promise for the future **272**
It may be possible to repair human genetic defects 272
Genetically engineered animals may boost agricultural production 274
Scientists are using recombinant DNA techniques to improve plants 276
Concern over potential danger caused by genetic engineering has led to the formation of policies on utilizing recombinant DNA techniques **278**
Focus on probing for genetic disease **273**
Focus on molecular genetics and cancer **275**

PART 5 EVOLUTION **281**

17 Darwin and Natural Selection **282**
The concept of evolution originated before Charles Darwin **283**
Darwin gathered much evidence for the theory that organisms evolve by natural selection **284**

Many types of scientific evidence support evolution **287**
Comparative anatomy of different organisms reveals evidence of evolution 287
Developmental biology provides more clues to the evolution of species 289
Fossils provide still more clues about evolution 290
Biogeography, the distribution of organisms on the Earth, also supports evolution 291
Comparisons of the biochemistry and molecular biology of different organisms offer evidence of evolution 293

18 Population Genetics **297**
The Hardy-Weinberg law explains gene frequencies in the gene pool of a population that is not evolving **298**
Evolution occurs when there are changes in gene frequencies in a gene pool **299**
Mutation increases variation in the gene pool, thereby causing changes in gene frequencies 299
Genetic drift causes changes in gene frequencies by random, or chance, events 300
Gene flow, which changes the amount of variation in the gene pool, is caused by the differential migration of organisms 301
Natural selection changes gene frequencies in a way that leads to adaptation to the environment 302
Natural selection increases the fitness of a species for the environment in which it lives **303**
Stabilizing selection selects against phenotype extremes and favors phenotypes near the mean, or average 305
Directional selection results in the change from one phenotype to another 307
Disruptive selection separates the population into several distinct phenotypes 307
Focus on evolution of the Africanized honeybee **303**

19 Speciation and Macroevolution **310**
Different species have various mechanisms to achieve reproductive isolation from one another **312**
Prezygotic isolating mechanisms interfere with mating 312
Postzygotic isolating mechanisms prevent successful reproduction if mating occurs 313
The key to speciation is the development of reproductive isolating mechanisms **314**
Allopatric speciation occurs through the effects of long physical isolation and different selective pressures 314
Sympatric speciation results from the divergence of two populations in the same physical location 315
Macroevolution involves changes in the *kinds* of species over evolutionary time **316**
Evolution is gradual, occurs in spurts, or is a combination of both processes **318**
Extinction of species is an important aspect of evolution **320**
Focus on the pace of evolution in trilobites **312**

20 The Origin, History, and Diversity of Life **325**
Early earth provided the conditions for chemical evolution **326**
Before cells existed, organic molecules formed on primitive Earth 326

The first cells probably assembled from organic
  molecules   328
Eukaryotic cells evolved after prokaryotic cells   329
**The fossil record provides us with clues to the history of
life**   **329**
Evidence of living cells may be found in Precambrian
  times   329
An incredible diversity of life forms evolved during the
  Paleozoic era   330
The dinosaurs and other reptiles dominated the
  Mesozoic era   331
The Cenozoic era is known as the Age of Mammals   333
**Organisms are classified using the binomial system of
nomenclature**   **335**
Many biologists recognize five kingdoms   **336**
Protists do not fit together well in one kingdom   338
Viruses cannot be assigned to any of the five kingdoms   339
Focus on continental drift   **336**

PART 6   DIVERSITY   **343**

21   Monerans and Protists   **344**
**Monerans are prokaryotes, the simplest and oldest known
cells**   **345**
Prokaryotes differ structurally from eukaryotes   345
Bacteria differ in their needs for oxygen   347
Eubacteria have diverse methods of obtaining food   347
The archaebacteria differ from eubacteria   348
Bacteria reproduce by fission   348
Bacteria can exchange DNA   348
Bacteria survive harsh conditions through dormancy
  and endospore formation   350
Studies of bacteria have increased our knowledge of all
  cells   350
Protists are eukaryotes and are mostly unicellular   **350**
Protozoa are animal-like protists   351
Algae are plant-like protists   353
Slime molds and water molds are fungus-like protists   358
The simple protists are considered the earliest
  eukaryotes   361

22   Fungal Life   **364**
**Fungi make an important contribution to the ecological
balance of our world**   **365**
Most fungi are filamentous in structure   **366**
Fungi are classified into four divisions   **368**
Zygomycetes reproduce sexually by forming zygospores   368
Ascomycetes (sac fungi) reproduce sexually by forming
  ascospores   369
Basidiomycetes (club fungi) reproduce sexually by
  forming basidiospores   372
Deuteromycetes (imperfect fungi) do not reproduce
  sexually   374
**Lichens are dual organisms composed of a fungus and an
alga or cyanobacterium**   **374**
Fungi are economically important   **376**
Fungi provide food for humans   376
Fungi produce useful chemicals   377
Fungi cause many important diseases of plants   377
Fungi cause certain diseases of animals   378

23   Plant Life   **381**
**Complex photosynthetic organisms are placed in the
kingdom Plantae**   **383**
The mosses and other bryophytes are nonvascular plants   **384**
Seedless vascular plants include the ferns and their allies   **386**
**The production of seeds represents a major evolutionary
advancement**   **391**
The gymnosperms are the "naked seed" plants   391
The flowering plants produce flowers, fruits, and seeds   395
Focus on ancient plants and coal formation   **390**
Focus on pollen and hay fever   **394**

24   Animal Life: Non-Chordate Invertebrates   **402**
Animals are multicellular heterotrophs   **403**
Most animal groups inhabit salt water   404
Animals can be grouped according to body structure
  or pattern of development   404
We can describe the animal body using standard
  directions and body planes   407
Sponges have specialized cells but no true tissues   **407**
**Cnidarians (hydras, jellyfish, and corals) have tissue
layers but no true organs**   **409**
Cnidarians have two body plans: a polyp and a medusa
  form   411
The *Hydra* has a solitary, carnivorous life-style   411
**Flatworms have three tissue layers, a head region, and
organs**   **412**
Planarians are free-living carnivores   412
Flukes have suckers and other adaptations for their
  parasitic life-style   413
Tapeworms are strikingly specialized for their parasitic
  life-style   414
**Proboscis worms are the simplest animals to have organ
systems**   **416**
**The roundworms (nematodes) have organ systems but
lack circulatory structures**   **417**
**Most mollusks are covered by a shell and have an open
circulatory system**   **418**
There are four principal classes of mollusks   418
The clam has all the organ systems typical of complex
  animals   419
The annelids have segmented bodies   **420**
**Arthropods have jointed appendages and an exoskeleton
of chitin**   **423**
The arthropods can be assigned to three subphyla   425
The insects have adapted to almost every available
  ecological niche   425
The grasshopper has a representative insect life-style   426
The echinoderms are spiny-skinned animals of the sea   **428**
Focus on adaptations of squid and octopods   **421**

25   Animal Life: The Chordates   **433**
**The tunicates have a protective covering made of
cellulose**   **434**
**The lancelets may be similar to primitive chordates that
gave rise to vertebrates**   **435**
**The success of the vertebrates is linked to the evolution
of a few key adaptations**   **435**
The jawless fish are the most primitive vertebrates   436
The cartilaginous fish include the sharks, rays, and
  skates   437

The bony fish are the most numerous vertebrates 437
The amphibians were the first successful land vertebrates 438
Reptiles were the dominant land animals for almost 200 million years 439
Birds are adapted for flight 441
Mammals evolved from reptiles 442
Primates evolved from shrewlike mammals **445**
The fossil record suggests general trends in hominid evolution 445
Humans undergo cultural evolution 450

PART 7 PLANT STRUCTURE AND LIFE PROCESSES **455**

26 Plant Development and Growth **456**
Embryonic development in plants follows an orderly and predictable path **457**
A number of external and internal factors affect seed germination **458**
Plants exhibit localized growth after seed germination **458**
Primary growth takes place at apical meristems 460
Secondary growth takes place at lateral meristems 461
Cells and tissues differentiate in the developing plant body **462**
The ground tissue system is composed of parenchyma, collenchyma, and sclerenchyma 464
The vascular system is composed of xylem and phloem 465
Epidermis and periderm comprise the dermal tissue system 467
Roots, stems, leaves, flowers and fruits make up the plant body **468**
Differentiation in plants is under both genetic and environmental control **470**
Focus on some experimental methods in embryogenesis **463**

27 Leaf Structure and Function **475**
Epidermis, mesophyll, xylem, and phloem are the major tissues of the leaf **476**
There is a relationship between structure and function in leaves **480**
The potassium ion mechanism explains how stomates open and close 481
Leaves lose water by transpiration and guttation 482
Shedding leaves allows plants in temperate climates to survive winter **484**
Leaves with functions other than photosynthesis exhibit modifications in structure **485**
Focus on comparative plant anatomy **480**
Focus on photosynthesis in desert plants **483**

28 Stems and Roots **489**
Herbaceous monocots and dicots can be distinguished by the arrangement of vascular bundles in their stems **490**
Gymnosperms and certain dicots have stems with secondary growth **492**
There are structural differences between primary roots and primary stems **492**
Considerable variation occurs in the internal arrangement of tissues in dicot and monocot roots 493

There is a relationship between structure and function in primary roots 498
Gymnosperms and certain dicots have roots with secondary growth **499**
Transport in plants occurs in the xylem and phloem **499**
Water and minerals are translocated in the xylem 499
Sugar is translocated in the phloem 500
Roots obtain most of the naturally occurring minerals that are found in plant tissues **501**
Sixteen elements are essential for plant growth 501
Both organic and inorganic fertilizers can replace certain key elements if they are missing from the soil 504
Focus on tree ring analysis **497**
Focus on commercial hydroponics **500**

29 Reproduction in Flowering Plants **507**
Asexual reproduction in flowering plants may involve modified stems, leaves, or roots **508**
Sexual reproduction in flowering plants involves flowers, fruits, and seeds **510**
Many mechanisms have evolved that accomplish pollination 510
Fruits are mature, ripened ovaries 513
Fruit and seed dispersal is highly varied in flowering plants 514
Environmental cues may induce flowering in plants **515**
Flowering may be initiated by light 516
Temperature may also affect reproduction 520
Focus on localizing phytochrome in plant cells **517**

30 Plant Hormones and Responses **523**
Changes in turgor can induce plant movements **524**
A biological clock influences many plant responses **524**
A tropism is plant growth in response to an external stimulus **525**
Hormones regulate plant growth and development **526**
Charles Darwin first provided evidence for the existence of auxin 526
Gibberellins were first discovered in a fungus 531
Cytokinins promote cell division 533
Ethylene is the only gaseous plant hormone 534
Abscisic acid promotes dormancy in higher plants 534

**PART 8 ANIMAL STRUCTURE AND LIFE
PROCESSES** **537**

31 Animal Tissues, Organs, and Organ Systems **538**
Animals are multicellular **539**
**Tissues of a complex animal are adapted to carry out
specific functions** **539**
Epithelial tissue covers the body and lines its cavities 540
Connective tissue joins and supports other body
structures 540
Muscle tissue contracts, permitting movement 549
Nervous tissue receives stimuli and transmits
information 549
**An organ consists of more than one type of tissue and
performs one or more functions** **550**
Ten organ systems make up the complex animal organism **550**
Focus on neoplasms: unwelcome tissues **548**

**32 Protection, Support, and Movement: Skin,
Skeleton, and Muscle** **557**
**The animal body is covered and lined with a protective
epithelial tissue** **559**
The protective epithelial covering of invertebrates may
function in secretion or gas exchange 559
The vertebrate skin is an important organ system that
performs diverse functions 560
**Skeletons are important in locomotion, protection, and
support of the animal body** **561**
In hydrostatic skeletons, body fluids are used to
transmit force 561
External skeletons of mollusks and arthropods are
nonliving shells 562
Internal skeletons are living tissue capable of growth 563
**Muscle is the contractile tissue that allows movement in
complex animals** **567**
All animals have the ability to move 567
A vertebrate muscle consists of hundreds of muscle
fibers wrapped in connective tissue 567

Muscle contraction occurs when actin and myosin
filaments slide past each other 569
ATP powers muscle contraction 570
Skeletal muscle action depends on muscle pairs
working antagonistically 572
Smooth, cardiac, and skeletal muscle are each
specialized for particular types of responses 573

33 Responsiveness: Neural Control **577**
**Neural response depends on reception, transmission,
integration, and response by muscles or glands** **578**
**The cell types of the nervous system are neurons and
glial cells** **578**
Glial cells support and protect neurons 578
A typical neuron consists of a cell body, dendrites, and
an axon 578
**Neurons convey information by transmitting rapidly
moving electrical impulses** **581**
The resting potential is the electrical potential
difference across the plasma membrane of the
neuron 581
An excitatory stimulus can cause a local depolarization
of the membrane 583
The action potential is a wave of depolarization that
moves down the axon 583
Movement of potassium ions out of the axon
repolarizes the membrane 583
Saltatory conduction occurs in myelinated neurons 584
Information must be transmitted across synapses **584**
Axons release neurotransmitters that affect
postsynaptic neurons 585
Many neurotransmitters have been identified 586
Neurons are one-way streets 587
Myelination and large axon diameter increase rate of
impulse transmission 587
**Neural integration is the process of averaging all of the
incoming information and determining whether an
action potential will be generated** **587**
The reflex arc is an example of a simple neural pathway **587**
**Complex neural pathways are possible because neurons
associate in a variety of ways** **589**

34 Responsiveness: Nervous Systems **593**
All invertebrates except sponges have nervous systems **594**
Some invertebrates have nerve nets or radial nervous
systems 594
Most invertebrates have bilateral nervous systems 594
**Key features of the vertebrate nervous system are the
hollow, dorsal nerve cord and well-developed brain** **596**
**The evolution of the vertebrate brain is marked by
increasing complexity especially of the cerebrum and
cerebellum** **596**
The hindbrain develops into medulla, pons, and
cerebellum 597
The midbrain is most prominent in fish and
amphibians 598
The forebrain gives rise to the thalamus,
hypothalamus, and cerebrum 598
**The human central nervous system is the most complex
biological mechanism known** **599**

The spinal cord transmits impulses to and from the brain and controls many reflex activities 599

The largest, most prominent part of the human brain is the cerebrum 599

The limbic system affects emotional aspects of behavior 600

The reticular activating system is an arousal system 603

When signals from the RAS slow, a person may fall asleep 604

Learning involves the storage of information in the nervous system and its retrieval on demand 604

Experience affects the brain 605

The peripheral nervous system includes somatic and autonomic systems 605

The somatic system helps the body adjust to the external environment 605

The autonomic system helps maintain homeostasis in response to changes in the internal environment 606

Many mood drugs change the levels of neurotransmitters in the brain 607

Focus on dopamine and motor function 597

Focus on cerebral dominance 604

Focus on alcohol abuse 607

Focus on crack cocaine 609

35 Sense Organs 613

Different types of sense organs respond to different types of energy 614

Sense organs work by producing receptor potentials 614

Sensation depends on transmission of a "coded" message 616

Receptors adapt to stimuli 618

Mechanoreceptors respond to touch, pressure, gravity, stretch, or movement 618

Touch receptors are located in the skin 618

Proprioceptors help coordinate muscle movement 619

Lateral line organs supplement vision in fish 620

Many invertebrates have gravity receptors called statocysts 620

The labyrinth of the vertebrate ear is an organ of equilibrium 620

Auditory receptors are located in the cochlea 622

Chemoreceptors detect taste and smell 623

Taste buds are the organs of taste in humans 623

The olfactory epithelium is responsible for the sense of smell 625

Thermoreceptors are sensitive to heat 625

Electroreceptors detect electrical currents in water 625

Photoreceptors use pigments to absorb light 626

Eyespots, simple eyes, and compound eyes are found among invertebrates 626

Vertebrate eyes form sharp images 629

Focus on defects in vision 628

Focus on pain perception 617

36 Internal Transport 636

Some invertebrates have no circulatory system 637

Some invertebrates have an open circulatory system 637

Many invertebrates have a closed circulatory system 638

Vertebrates have a closed circulatory system adapted to carry out a variety of functions 639

Vertebrate blood consists of plasma, blood cells, and platelets 639

Plasma is the fluid component of blood 640

Red blood cells transport oxygen 640

White blood cells defend the body against disease organisms 641

Platelets function in blood clotting 642

Vertebrates have three main types of blood vessels 643

The evolution of the vertebrate heart culminated in a four-chambered heart and double-circuit circulation 644

The fish heart has a single atrium and single ventricle 644

Amphibians have a three-chambered heart 645

The reptilian heart consists of two atria and two ventricles 645

Birds and mammals have a four-chambered heart 645

The structure of the human heart is marvelously adapted for pumping blood 646

Arterial pulse results from the alternate expansion and recoil of an artery 648

Blood pressure depends on blood flow and resistance to blood flow 649

Blood pressure is highest in arteries 650

Blood pressure is carefully regulated 651

In mammals blood is pumped through a pulmonary and a systemic circulation 651

The pulmonary circulation functions to oxygenate the blood 653

The systemic circulation delivers blood to all of the tissues 653

The lymphatic system is an accessory circulatory system 655

The lymphatic system consists of lymphatic vessels and lymph tissue 655

The lymphatic system plays an important role in fluid homeostasis 656

Focus on the electrical activity of the heart 649

Focus on cardiovascular disease 652

37 Internal Defense 660

Internal defense depends on the ability to distinguish between self and nonself 661

Invertebrates have internal defense mechanisms that are mainly nonspecific 661

Vertebrates can launch both nonspecific and specific immune responses 661

Nonspecific defense mechanisms include mechanical and chemical barriers against pathogens 662

Specific defense mechanisms include antibody-mediated immunity and cell-mediated immunity 664

A secondary immune response is more rapid than a primary response 671

Active immunity follows exposure to antigens 672

Passive immunity is borrowed immunity 672

Normally the body effectively defends itself against cancer 673

Graft rejection is an immune response against transplanted tissue 673

Certain sites in the body are immunologically privileged 675

In an autoimmune disease the body attacks its own tissues 675

Allergic reactions are inappropriate immune responses 675

Focus on AIDS 676

38 Gas Exchange — 681

Respiratory structures are adapted for gas exchange in air or water — 682

Animals have evolved several different adaptations for gas exchange — 682

The body surface may be adapted for gas exchange — 683

Tracheal tubes are an adaptation for gas exchange in arthropods — 683

Many aquatic animals exchange gases through gills — 683

Terrestrial vertebrates exchange gases through lungs — 686

The human respiratory system is typical of air-breathing vertebrates — 687

The airway conducts air into the lungs — 688

Gas exchange occurs in the lungs — 688

Ventilation is accomplished by breathing — 688

Breathing is regulated by respiratory centers in the brain — 689

Gas exchange takes place in the air sacs — 690

Oxygen is transported in combination with hemoglobin — 691

Carbon dioxide is transported mainly as bicarbonate ions — 692

Physiological adaptation to changes in pressure takes time — 693

The effect of breathing dirty air is respiratory insult — 694

A variety of defense mechanisms protect the lungs — 694

Continued respiratory insult leads to respiratory disease — 694

Focus on choking — 693

Focus on cardiopulmonary resuscitation (CPR) — 691

Focus on facts about smoking — 695

39 Processing Food — 698

In the animal kingdom we find all kinds of dinner jackets — 700

Some invertebrates have incomplete digestive systems — 700

Most invertebrates and all vertebrates have complete digestive systems — 701

The human digestive system has highly specialized structures for processing food — 703

Food begins its journey inside the mouth — 703

The pharynx and esophagus conduct food to the stomach — 704

Food is mechanically and enzymatically digested in the stomach — 704

Most enzymatic digestion takes place inside the small intestine — 705

The liver secretes bile which mechanically digests fats — 708

The pancreas secretes digestive enzymes — 708

Nerves and hormones regulate digestion — 708

Absorption takes place mainly through the villi of the small intestine — 709

The large intestine eliminates wastes — 709

Adequate amounts of required nutrients are necessary to support metabolic processes — 710

Carbohydrates are a major energy source in the human diet — 711

Lipids are used as an energy source and to make needed biological molecules — 712

Proteins serve as enzymes and are essential structural components of cells — 714

Vitamins are organic compounds essential for normal metabolism — 715

Minerals are inorganic nutrients required by cells — 715

Energy metabolism is balanced when energy input equals energy output — 716

Obesity is a serious nutritional problem — 717

Malnutrition can cause serious health problems — 721

Focus on peptic ulcers — 705

Focus on vegetarian diets — 715

40 Osmoregulation and Disposal of Metabolic Wastes — 725

Excretory systems help maintain homeostasis — 726

The principal metabolic waste products are water, carbon dioxide, and nitrogenous wastes — 726

Invertebrates have solved problems of osmoregulation and metabolic waste disposal in a number of ways — 726

Nephridial organs are tubules specialized for osmoregulation and/or excretion in some invertebrates — 727

Antennal glands are important in osmoregulation in crustaceans — 729

Malpighian tubules are an important adaptation for conserving water in insects — 729

The kidney is the key vertebrate organ of osmoregulation and excretion — 729

Osmoregulation is a continuous challenge for aquatic vertebrates — 729

The mammalian kidney is vital in maintaining homeostasis — 730

Focus on kidney disease, dialysis, and transplant — 735

41 Endocrine Regulation — 742

Hormone secretion is regulated by negative feedback mechanisms — 743

Hormones combine with specific receptor proteins in target cells — 744

Some hormones activate genes — 744

Some hormones work through second messengers — 744

In invertebrates hormones regulate growth, metabolism, reproduction, molting, and pigmentation — 745

In vertebrates hormones help regulate growth, reproduction, and many aspects of metabolism — 746

Endocrine disorders may involve hyposecretion or hypersecretion — 746

Nervous regulation and endocrine regulation are integrated by the hypothalamus — 746

The posterior lobe of the pituitary gland releases two hormones — 747

The anterior lobe of the pituitary gland regulates growth and several other endocrine glands — 750

Thyroid hormones stimulate metabolic rate — 752

The parathyroid glands regulate calcium concentration — 753

The islets of the pancreas regulate glucose concentration — 753

The adrenal glands help the body adapt to stress — 756

42 Reproduction — 761

Asexual reproduction is common among some animal groups — 762

Sexual reproduction is the most common type of animal reproduction — 762

Animals have evolved interesting reproductive variations — 763

Metagenesis is characteristic of some animal groups   763
In parthenogenesis there is no fertilization   763
In hermaphroditism one individual produces sperm and eggs   764
Human reproduction: The male provides sperm   **765**
The testes produce sperm   765
A series of ducts transport sperm   766
Semen is produced by the accessory glands   767
The penis transfers sperm to the female   767
Reproductive hormones promote sperm production and maintain masculinity   767
Human reproduction: The female produces ova and incubates the embryo   **768**
The ovaries produce ova and sex hormones   768
The uterine tubes transport the ovum   772
The uterus incubates the embryo   772
The vagina receives sperm   772
External genital structures are the vulva   773
The breasts function in lactation   773
The menstrual cycle is regulated by hormones   774
Sexual response involves physiological changes   **777**
Fertilization is the fusion of sperm and egg to produce a zygote   **779**
Infertility is the inability to achieve conception   **780**
Birth control methods allow individuals to choose   **780**
Oral contraceptives prevent ovulation   780
Usage of the intrauterine device (IUD) has declined   782
Other common contraceptive methods include the diaphragm and condom   783
Sterilization renders an individual incapable of producing offspring   783
There are three types of abortion   **784**
Sexually transmitted diseases are spread by sexual contact   **785**
Focus on breast cancer   **774**
Focus on novel origins   **782**

43 Development   **789**
Development is a balanced combination of several processes   **791**
Fertilization restores the diploid number of chromosomes   **791**
The first steps in fertilization are contact and recognition   792
Sperm entry is regulated   792
Sperm and egg pronuclei fuse   792
Fertilization activates the egg   792

The zygote contains instructions for producing a complete individual   **793**
During cleavage the zygote divides, forming many cells   **793**
Cleavage provides building blocks for development   793
The amount of yolk determines the pattern of cleavage   793
The germ layers form during gastrulation   **796**
Each germ layer has a specific fate   796
The pattern of gastrulation varies somewhat   796
Organogenesis begins with the development of the nervous system   **797**
Developmental processes are carefully regulated   **797**
Cytoplasmic factors influence the course of development   797
Cells can influence each other with inducers   799
Nongenetic and genetic factors interact   800
Extraembryonic membranes and placenta protect and nourish the embryo   **801**
The chorion and amnion enclose the embryo   801
The allantois functions in waste disposal   802
The yolk sac encloses the yolk   802
The placenta is an organ of exchange   802
Human prenatal development requires about 266 days   **802**
Development begins in the uterine tube   803
Twins develop when cells separate   805
Implantation begins on about the seventh day   805
Differentiation and organ development begin during gastrulation   805
The fetus continues to develop during the second and third trimesters   807
The birth process may be divided into three stages   807
The neonate must adapt to its new environment   809
Environmental factors affect the embryo   **809**
The human life cycle extends from conception to death   **811**
The aging process is marked by a decrease in homeostatic response to stress   **811**

PART 9 BEHAVIOR AND ECOLOGY   **815**

44 Animal Behavior   **816**
Behavior fits life-style   **817**
Behavior is adaptive   **818**
Biological rhythms anticipate environmental changes   818
Behavior capacity is inherited   **820**
Behavior is modified by learning   **823**
In classical conditioning, a reflex becomes associated with a new stimulus   823
In instrumental conditioning, spontaneous behavior is reinforced   824
Some innate behaviors may be perfected by instrumental conditioning   824
Imprinting is a form of learning that occurs only during a critical period   825
Habituation enables the organism to ignore irrelevant stimuli   825
Insight learning adapts recalled events to new situations   826
Learning abilities are biased   826
Instincts are unlearned behavior patterns   **827**
Innate behavior is often triggered by specific stimuli   827
Sign stimuli can trigger inappropriate behavior   828
Behavior develops   829

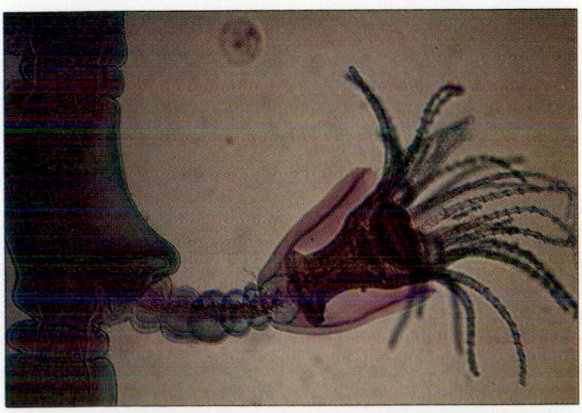

Migration is triggered by environmental changes   829
Migrating animals must navigate correctly   831

45   Social Behavior   833
Sociality requires communication   **835**
Animals communicate in a wide variety of ways   835
Pheromones are chemical signs used in communication   835
Animals often arrange themselves in dominance hierarchies   **836**
Dominance hierarchies suppress aggression   836
Dominance results from many causes   836
Many animals defend territory   **837**
Sexual behavior is usually social   **838**
Pair bonds establish reproductive cooperation   839
Many organisms care for their young   839
Play is often practice behavior   **840**
Highly organized societies occur among insects and vertebrates   **840**
The social insects include some hymenopterans and termites   840
Vertebrate societies tend to be relatively flexible   842
Kin selection could produce altruistic behavior   **843**
Sociobiology attempts to explain altruism by kin selection   **844**

46   Communities and Population Ecology   847
Organisms live together in communities and ecosystems   **848**
Organisms play specific roles in energy transfer   **848**
Producers make food from simple chemicals   849
Consumers obtain food from other organisms   849
Decomposers recycle materials from corpses and wastes   849
Substances are recycled within communities   **851**
Carbon is recycled via carbon dioxide   851
Bacteria are essential to the nitrogen cycle   851
The phosphorus cycle is determined by insolubility and a lack of gaseous compounds   852
Organisms are adapted to characteristic ecological niches   **854**
Populations of organisms make up communities   **856**
Organisms vary in the number of young they can produce   856
Life spans vary among different species   856
A population tends to grow   857
Some limiting factors reduce population regardless of size   858
Some limiting factors dynamically regulate population growth   858
Energy and substances travel through food chains and webs   **858**
Food relationships are best described as webs   858
Food pyramids reflect the biomass, energy, and numeric relationships of trophic levels   858
Ecosystems vary in productivity   859
Communities vary in species diversity   **860**
Communities develop   **861**
Sometimes communities must start with a lifeless habitat   861
Communities sometimes develop where a predecessor community already exists   862
Why does succession occur?   863
Why is the climax community stable?   863
Focus on microcosms   **861**

47   The Ecosphere   866
Life is a thin film on the surface of the earth   **867**
The sun warms the earth   868
Atmospheric circulation is driven by solar input   869
Air pollution affects global patterns of heat transfer   871
The Coriolis effect perturbs the movement of air and water masses   873
Water dominates the surface of the earth   **874**
Water has a very high specific heat   874
Water changes its phase readily and is continuously recycled by the hydrologic cycle   874
Vast oceanic currents redistribute marine water masses   875
Climate and precipitation result from patterns of air and water movement   **877**
Microclimates can differ markedly from overall climate   878
Focus on the last winter   **880**

48   Life Zones   883
Organisms have unique geographic distributions   **884**
Major land life zones are called biomes   **884**
Tundra is the northernmost land biome   886
The boreal forest is dominated by conifers   888
Temperate coniferous forests occur in western North America   888
Temperate biomes vary with precipitation   888
Tropical life zones also vary with precipitation   891
Aquatic life zones occupy most of the earth's surface   **895**
Aquatic organisms fall into three ecological categories   895
Freshwater habitats include streams, lakes, and ponds   896
The marine zone is the largest aquatic life zone   897
Plankton forms the basis of most marine food chains   903
Succession occurs in aquatic habitats   904
Life zones interact   **904**
Focus on coral reefs   **900**

49   Human Ecology   907
Agriculture produces artificial communities   **908**
Agricultural communities have unique characteristics   908
New systems of agriculture may permit increased production   910
Forestry is similar to agriculture in many ways   910
Pesticides are employed for high-productivity agriculture despite the ecological damage that they produce   911
Pollution results from improper waste disposal   **913**
Water pollution is both a threat to public health and an ecological issue   914
Much water pollution robs aquatic habitats of oxygen   914
Air pollution results mainly from combustion   914
Radioactive pollutants are long lived   917
Solid waste destroys wetlands and wildlife and can produce other kinds of pollution   918
Energy is the most basic need of technological civilization   **919**
Wild plants and animals are now becoming extinct at an unprecedented rate   **920**
Human population growth consumes resources and produces pollution   **922**
The human population is growing exponentially   922
Overpopulation is not just a third-world problem   924
What is to be done?   **924**

PART 1

An Introduction to Biology:
Some Basic Concepts

Living things are composed of cells • Metabolism and homeostasis • Perpetuation of species •
DNA transmits information • Adaptations enhance survival • The evolutionary perspective

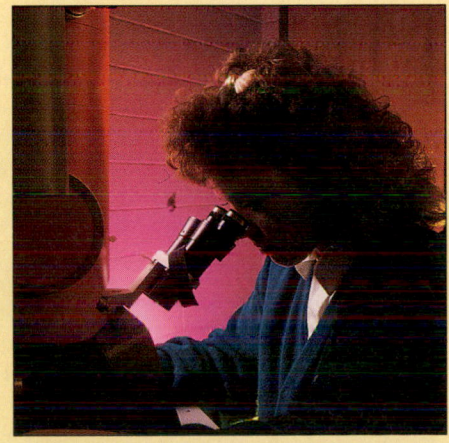

Ecological organization • Society and the ecosphere • Science is a means of investigation •
Understanding the universe • Recognizing and stating problems • Science and technology interact

1
The World of Biology

OUTLINE

I. Life can be defined in terms of the characteristics shared by living things
   A. Living things are composed of cells
   B. Living things grow and develop
   C. Metabolism includes the chemical processes essential to growth, repair, and reproduction
   D. Organisms work to maintain an appropriate internal environment
   E. Movement is a basic property of cells
   F. Living things respond to stimuli
   G. DNA transmits information from one generation to the next
   H. A species has the ability to evolve and adapt to its environment
II. The theory of evolution has become one of the great unifying concepts in biology
   A. Evolution proceeds by natural selection
   B. Species evolve in response to changes in the environment
   C. An evolutionary perspective is apparent in almost every aspect of biology
III. Biological organization is hierarchical
   A. We can identify an organizational hierarchy within the organism
   B. We can also identify organization on the ecological level
      1. Producers make their own food
      2. Consumers depend on producers for food
      3. Decomposers break down wastes
      4. Natural ecosystems survive through diversity
IV. Millions of kinds of organisms inhabit our planet
   A. Biologists use a binomial system of nomenclature
   B. Taxonomic classification is hierarchical
   C. Most biologists recognize five kingdoms

Focus on society and the ecosphere
Focus on evolution in action: the case of the peppered moth

Blossom of a pear-apple tree (Dwight Kuhn)

LEARNING OBJECTIVES

After you have studied this chapter you should be able to:

1. Define biology and discuss its applications to human life and society.
2. Distinguish between living and nonliving things by describing the features that characterize living things.
3. Relate metabolism and homeostasis and give specific examples of these life processes.
4. Construct a hierarchy of biological organization including individual and ecological levels.
5. Compare the roles of producers, consumers, and decomposers, and cite examples of their interdependence.
6. Apply the theory of natural selection to any given adaptation, postulating a plausible explanation of how the adaptation may have evolved.
7. Demonstrate use of the binomial system of nomenclature using several specific examples.
8. Classify an organism such as a human according to kingdom, phylum or division, class, order, family, genus, and species.
9. Contrast the five kingdoms of living organisms and cite examples of each group.

 The world of biology encompasses all of the living things that inhabit our planet, from minute bacteria to giant whales and redwood trees. This broad science examines the diversity, structure, and internal processes of living organisms and extends to their origins, relationships with one another, and interaction with the environment (Fig. 1–1). **Biology,** then, is the study of life. The word itself is derived from two Greek word parts, *bio,* "life," and *logos,* "the study of."

This book emphasizes the interdependence of living things and examines our own interactions with other organisms and with the environment. Early human beings were a harmonious part of the biological world, for their activities had little impact upon the environment. As human society has become increasingly technological, however, our activities have exerted a significant and often damaging effect upon our planet. The expanding human population, coupled with increasing consumption of natural resources, has transformed the earth. Chemicals from industries and modern agricultural practices have spread throughout the soil, water, and atmosphere and threaten to disrupt the delicate network of life upon earth's surface. As a result, we face many critical problems. Of all the sciences, biology is perhaps of the greatest interest to us, for a knowledge of the principles of biology may be the key to human survival on planet earth.

Today biologists are working to improve environmental quality, increase the world food supply, iden-

FIGURE 1–1   Modern biology examines the world of life in all its details and interactions. Biologists are concerned with the physical characteristics of each organism, how its body functions, its behavior, and its interaction with other living things in its environment. Among the various biological levels they study are communities of organisms such as the tidepool community shown here. (Frans Lanting)

tify factors that contribute to health and longevity, and conquer killers like heart disease, cancer, and AIDS (Fig. 1–2). Their work requires the support of an informed citizenry. This book provides you with tools that will enable you to become a biologically literate citizen.

The work of biologists is so broad that it affects us at many levels of our lives. Understanding the principles of biology can help us deal more intelligently with

FIGURE 1–2   Biologists work to improve the quality of life. (a) Under sterile conditions, this biologist checks bottles containing cultures of human cells. He is doing research on interferon, a substance released by cells when they are invaded by viruses. Interferon stimulates noninfected cells to produce an antiviral protein and has a variety of other actions that help defend the body against disease. For these reasons, biologists are studying its potential clinical usefulness in treating viral diseases and perhaps cancer. **(b) Chick embryos are used to test the effects of drugs and various other substances on development. (c) This field biologist is releasing beetles that will destroy an agricultural pest (the tansy ragwort). As biologists develop more effective methods of biological control, the amount of harmful pesticides used can be decreased.** (a, Science VU-NCI/Visuals Unlimited; b, E.R. Degginger; c, Doug Wechsler)

(a)

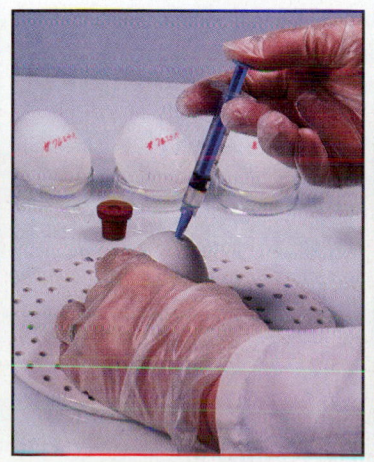

(b)

(c)

a wide range of routine concerns. Health care, nutrition, dieting, smoking, the use of drugs, and the care of domestic plants and animals are a few of the topics that may be of immediate interest. You may be intrigued by specific areas of biological research such as animal communication, biological control of agricultural pests, killer bees, or genetic engineering. The study of biology will also expand your awareness of the millions of diverse life forms that share our planet and your appreciation for the exquisite precision and complexity of living processes and systems.

KEY CONCEPTS

☐ Living systems consist of cells; they grow, move, reproduce, carry on self-regulated metabolism, respond to stimuli, and adapt to environmental changes.

☐ The theory of evolution, one of the great unifying concepts of biology, explains how organisms are related and how they adapt to changes in their environment.

☐ We can identify a hierarchy (ranking) of biological organization on both individual and ecological levels.

☐ Each type of organism is classified in one of five kingdoms and is assigned to a genus and species.

☐ Human beings are part of a delicately balanced environmental system.

☐ LIFE CAN BE DEFINED IN TERMS OF THE CHARACTERISTICS SHARED BY LIVING THINGS

We have defined biology as the the study of life, but what is life? Even biologists have difficulty in defining such terms as *life, living,* and *alive.* The living things that inhabit our planet are so diverse that it is difficult to lump them together with a simple definition. There are, however, certain characteristics and activities that a human being has in common with an earthworm, a tree, or even a single-celled amoeba. Taken together, these features constitute life. All living things—more formally referred to as living systems or organisms—consist of one or more cells, grow, carry on self-regulated metabolism, move, respond to stimuli, reproduce, and adapt to environmental changes.[1] In this section we explore these characteristics in some detail (Fig. 1–3).

---

[1] The viruses lack several of these characteristics, and many biologists do not consider them living things; others describe the viruses as being on the threshold of life.

FIGURE 1–3 **This scorpion fish looks very much like the rock on which it rests on the ocean floor. Blending so well with its background gives the fish an advantage in making dinner of any small organism that unwarily swims by. What characteristics enable us to distinguish the scorpion fish as living and the rock as inanimate?** (Charles Seaborn)

Living Things Are Composed Of Cells

We recognize each different kind of organism by its characteristic appearance and structure. Although living things may vary greatly in size and appearance, all are composed of basic building blocks called cells (see windows on Plant Cell and Animal Cell). The **cell** is the simplest unit of living matter that can carry on all the activities necessary for life. Some of the simplest organisms, such as bacteria or certain algae, consist of a single cell. In contrast, the body of a human or an oak tree is made of trillions of cells (Fig. 1–4). The life processes of such complex organisms depend on the coordinated functions of their component cells.

Living Things Grow And Develop

Some nonliving things appear to grow. A snowball rolling down a hill becomes larger as snow gathers around it, and a stream may swell with accumulating rain water. These inanimate objects increase in size by adding pre-existing materials.

In contrast, living systems grow by taking in raw materials from the environment and refashioning them into their own specific types of substances. Biological growth, then, usually proceeds from the inside out. In biological growth there is an increase in the amount of living substance. This increase results from an increase in the *size* of the individual cells, by an increase in the *number* of cells, or by both (Fig. 1–5).

Some organisms—most trees, for example—continue to grow indefinitely. In contrast, most animals have a growth period that ends when a characteristic size is reached in adulthood. A remarkable feature

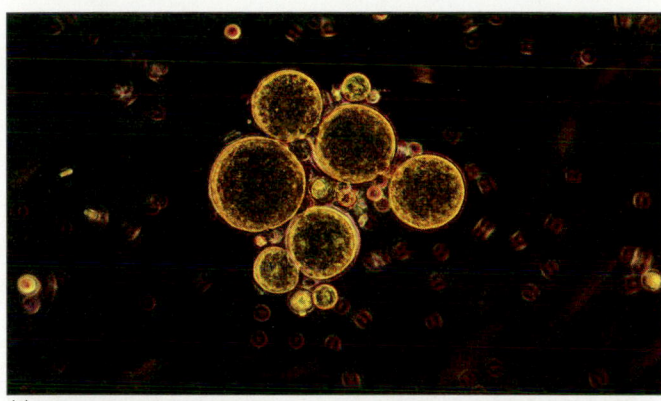

(a)

(b)

FIGURE 1–4 Single-celled and multicellular organisms. (a) Single-celled organisms are generally smaller and less specialized than multicellular organisms. This coccoid green alga, common on tree trunks, carries on all of its life functions within its one cell. Several individual algae are grouped together here, forming a colony. (b) These hybrid water lilies (Nymphaea "Attraction") are much more complex. The body parts of each of these plants contains specialized cells that carry on certain tasks. The leaves contain cells that carry on photosynthesis; the flower, reproduction; and the roots, absorption of nutrients. (a, Ed Reschke; b, E.R. Degginger)

FIGURE 1–5 Biological growth involves the refashioning of raw materials to construct the organism as determined by its genes. This Layson albatross chick will grow to adult size by using nutrients from food to increase the size and number of its cells. (Frans Lanting)

of the growth process is that each part of the organism continues to function as it grows.

Living things undergo development as well as growth. **Development** includes all the changes that take place during the life of an organism from conception to death. Typically, an organism begins life as a fertilized egg, which then grows and develops specialized structures and form.

Metabolism Includes The Chemical Processes Essential To Growth, Repair, And Reproduction

To grow and maintain itself, an organism must be able to convert food materials into living cells. The complex chemical reactions that transform nutrients from food into components needed to build new parts requires the expenditure of energy. In the chemical process known as **cellular respiration** the needed energy is lib-

erated from nutrients. Many other chemical reactions and energy transformations maintain the routine operations of cells. The chemical activities and energy transformations essential to growth, repair, and reproduction are termed **metabolism** (Fig. 1–6). Metabolic reactions are constantly occurring in every living system. When they cease, the organism dies.

Organisms Work To Maintain An Appropriate Internal Environment

Living systems need the appropriate machinery for carrying on metabolic activities. But the machinery alone is insufficient. Metabolic activities must also be self-regulated so as to maintain a balanced state within the organism. The organism must "know" when to synthesize what or when more nutrients or energy are required. On the other hand, it must not produce too much of any specific substance. When enough of a product has been made, the synthesizing mechanisms must be turned off. The organism must also be capable of adjusting its metabolism in response to changes in its external environment. The automatic tendency to

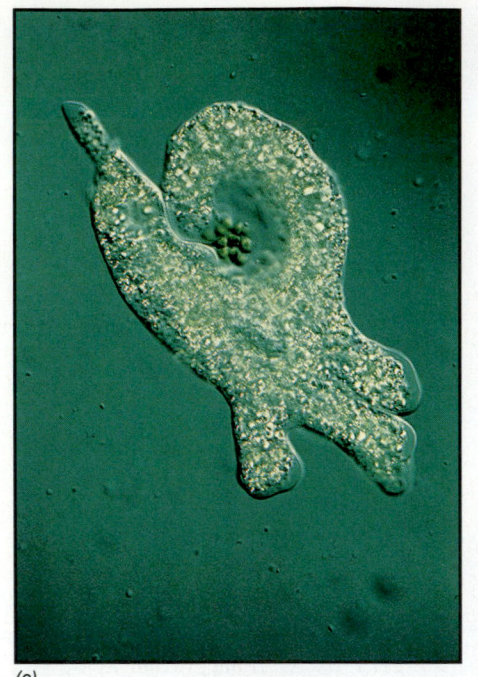

(a)

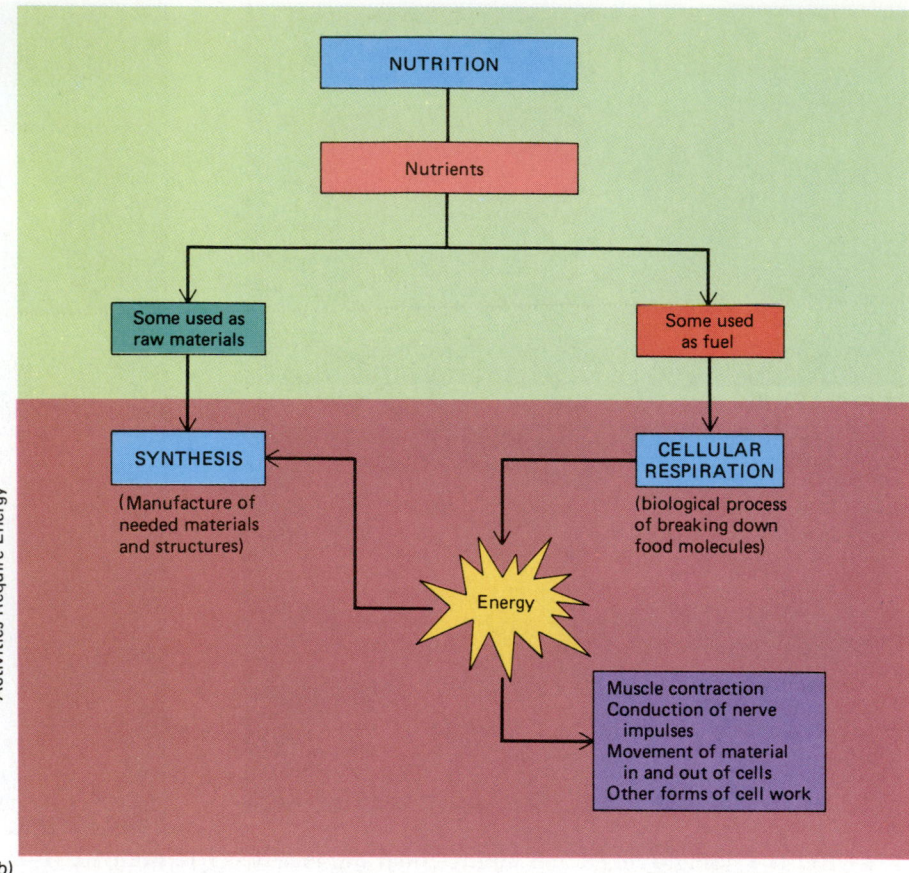

Activities Require Energy

(b)

FIGURE 1–6   Metabolic reactions occur continuously in every living cell.
(*a*) **Like all living things, this one-celled amoeba must take in nutrients and oxygen in order to stay alive. The green structures within the cell are food particles. The amoeba captured them by flowing around and engulfing them.**
(*b*) **Relationships of some metabolic activities. Some of the nutrients from food are used to synthesize needed materials and cell parts; other nutrients are used as fuel for cellular respiration, a process that captures energy stored in food. This energy is needed for synthesis and for other forms of cellular work. Cellular respiration also requires oxygen, which is provided by the process of gas exchange. Wastes from the cells, such as carbon dioxide and water, must be excreted from the body.** (*a*, M. Abbey, Visuals Unlimited)

maintain an appropriate internal environment is called **homeostasis,** and the mechanisms designed to accomplish this task are known as **homeostatic mechanisms.**

Regulation of body temperature is a good example of a homeostatic mechanism (Fig. 1–7). When body temperature rises above its normal 37° Celsius (C; that is, 98.6° Fahrenheit, F), the increase is sensed by a "thermostat" composed of specialized cells in the brain. This temperature-regulating center sends nerve impulses to the sweat glands in the skin. Sweat production then increases. Evaporation of sweat from the skin requires heat, so as sweat evaporates, body temperature is lowered. At the same time, capillaries (tiny blood vessels) in the skin dilate (expand), permitting the blood to carry body heat to the surface more efficiently. The heat radiates from the surface.

When body temperature falls below normal, messages from the "thermostat" in the brain cause blood vessels in the skin to constrict. This reduces heat loss. Heat may also be generated by the muscular contractions we call shivering.

Movement Is A Basic Property Of Cells

The living material of cells is itself in continuous motion. Although many living things do not carry on locomotion (move from one place to another), internal movement is characteristic of all life. A tree cannot pull up its roots and walk away, but it moves as it grows, its buds open, food is transported, and it responds to changes in its environment. Complex animals, such

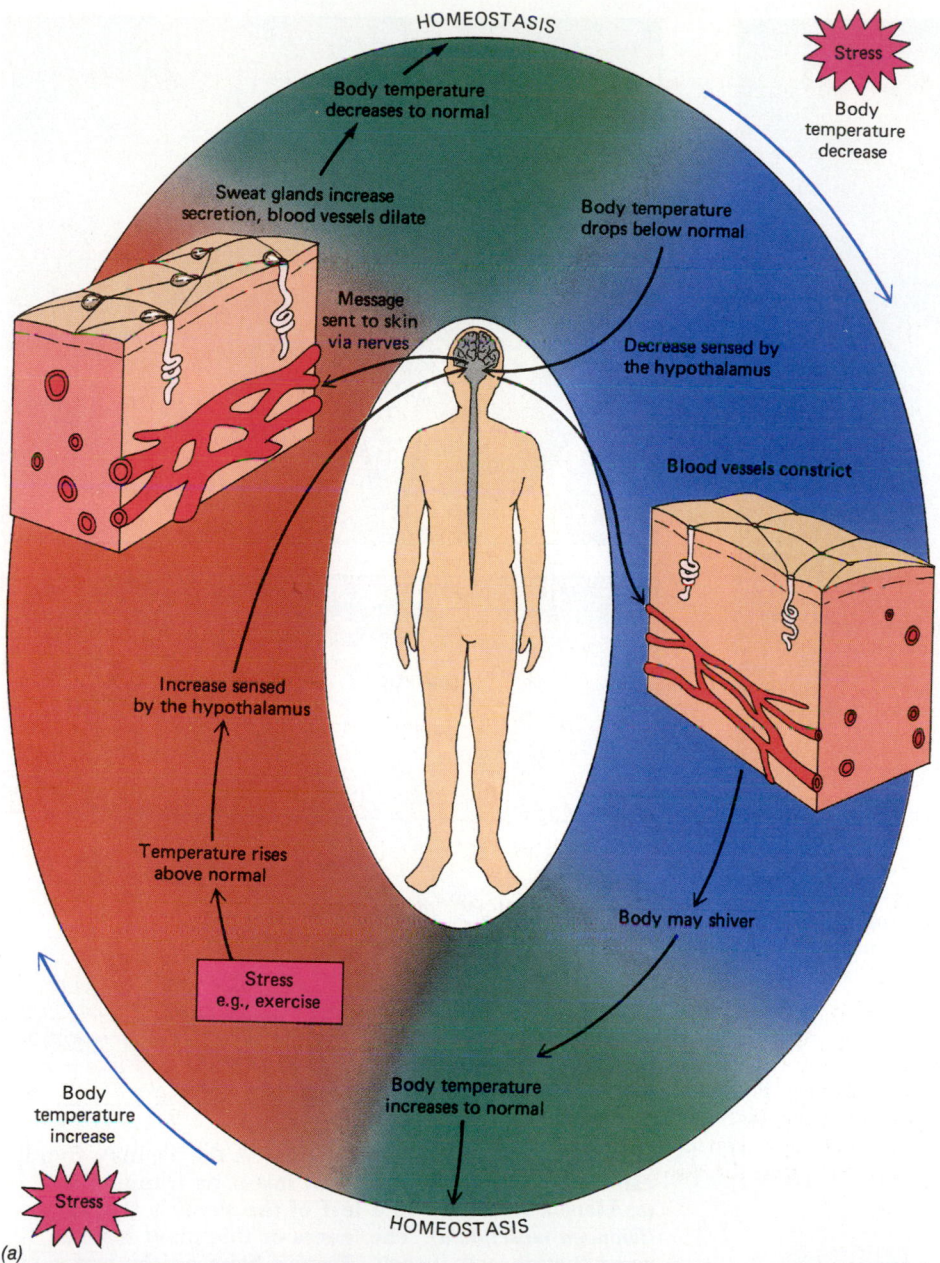

(a)

(b)

FIGURE 1-7 (a) Regulation of body temperature in the human by homeostatic mechanisms. An increase in body temperature above the normal range stimulates cells in the "thermostat" of the brain to send messages to sweat glands and small blood vessels (capillaries) in the skin. Increased circulation of blood in the skin and increased sweating are mechanisms that help the body to get rid of excess heat. When body temperature falls below the normal range, blood vessels in the skin constrict so that less heat is carried to the body surface. Shivering, in which muscle contractions generate heat, may also occur. (b) The sunning behavior of this marine iguana, *Amblyrhynchus cristatus*, a native of the Galapagos Islands, is homeostatic. The animal positions itself to maximize the heat it receives from the sun, thus increasing its body temperature. (b, Animals Animals © 1990, Breck P. Kent)

FIGURE 1–8   A Mediterranean chameleon *(Chamaeleo chamaeleon)* **rapidly darts its unusual tongue into a flower to capture an unwary insect. Not all movement in the biological world is as dramatic or as difficult to photograph as what you see here. This photograph was taken at night using a high-speed strobe.** (Animals Animals © 1990, Stephen Dalton)

FIGURE 1–9   A few plants, such as the Venus's flytrap, can respond to the touch of an insect by trapping it. (a) Here a fly lights on a leaf of the Venus's flytrap *(Dionaea musciplula)*. **The leaves of this plant have a scent that attracts insects. Trigger hairs on the leaf surface detect the presence of the insect, and the leaf, hinged along its midrib, folds. (b) The edges come together and hairs interlock, preventing the fly's escape. The leaf then secretes enzymes that kill and digest the insect.** (Grant Heilman)

as insects and tigers, possess groups of muscles that make complicated, purposeful movements possible (Fig. 1–8).

Living Things Respond To Stimuli

Living things actively respond both to *stimuli* (changes) in the external environment and to changes inside themselves. When in need of nourishment, a one-celled organism, such as an amoeba, reacts positively to food in its watery surroundings by flowing toward it and engulfing it. Most organisms respond to changes in temperature, pressure, or sound; changes in light intensity; and changes in the chemical composition of their surroundings.

The responsiveness of plants is often not as obvi-ous as that of animals. However, most of us have at some time observed how plants grow toward light, roots tend to grow toward water, and vines wrap around solid objects. A few plants, such as the Venus flytrap of the Carolina swamps, are sensitive to touch and can trap insects (Fig. 1–9).

DNA Transmits Information From One Generation To The Next

Although the life spans of various organisms range from minutes to centuries, they are always limited. The death of an individual, or even of a generation of indi-viduals, does not mark the end of that type of organ-ism, however. Perpetuation of each type of organism is provided for by the process of **reproduction,** and a biologically successful individual reproduces before it dies. Reproduction involves the transmission of infor-

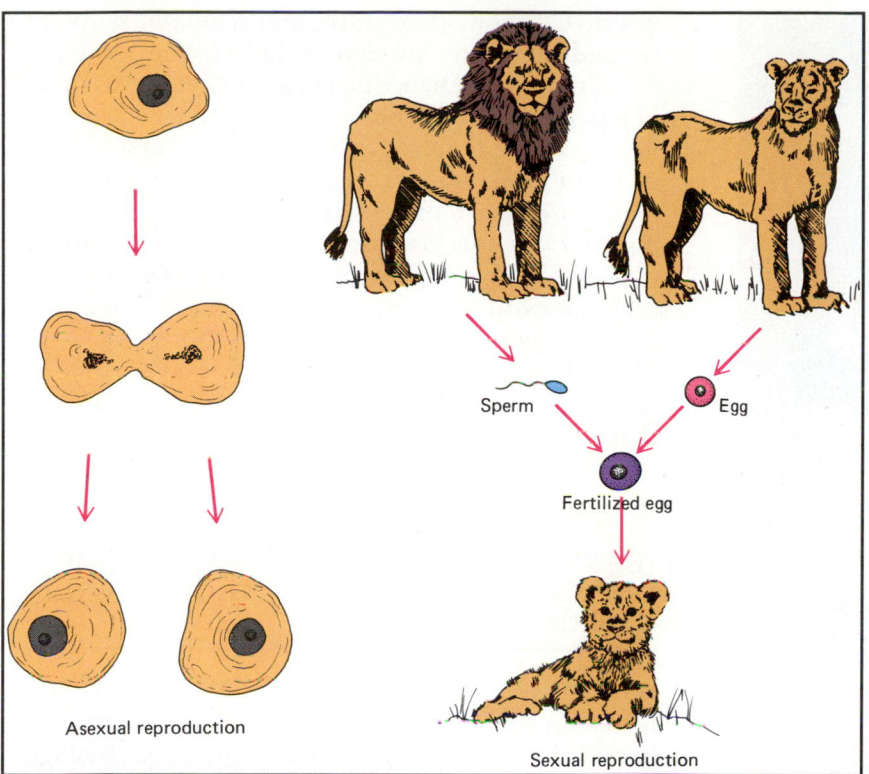

Asexual reproduction

Sexual reproduction

(a & c)

FIGURE 1–10 Approaches to reproduction. (*a*) **In asexual reproduction one individual gives rise to two or more offspring—all identical to the parent.** (*b*) **Asexual reproduction in** *Paramecium (Paramecium caudatum),* **a unicellular organism (a ciliate protozoan).** (*c*) **In sexual reproduction two parents each contribute a sex cell; these join to give rise to the offspring, which is a combination of both parents.** (*d*) **A pair of tropical flies mating.** (*b,* A.M. Siegelman/Visuals Unlimited; *d,* L.E. Gilbert, University of Texas at Austin/BPS)

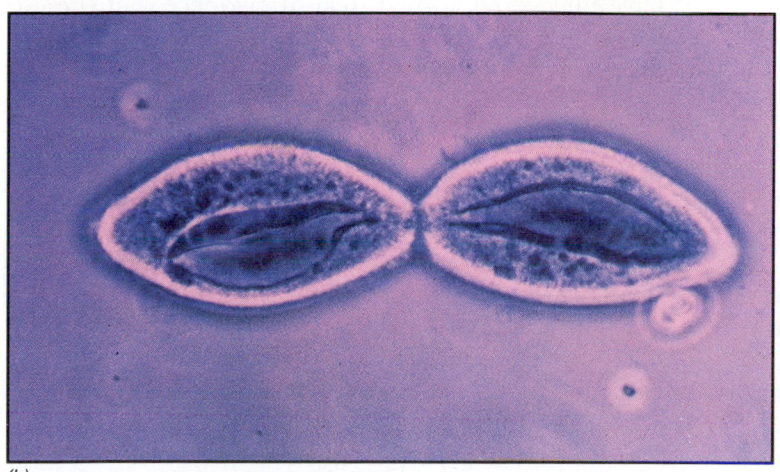

(b)

(d)

mation by the remarkable hereditary material known as **deoxyribonucleic acid (DNA).** DNA codes the information governing the structure and function of the organism. For example, DNA ensures that cows always give rise to calves—never to cats or rose bushes.

In simple organisms, such as the amoeba, reproduction may be asexual, in which a single parent divides (or buds), giving rise to offspring (Fig. 1–10). When an amoeba has grown to a certain size, it reproduces by dividing in half to form two new amoebas. Before it divides, an amoeba makes a duplicate copy of its DNA and one complete set of DNA is distributed to each new cell. Because its DNA is identical, each new amoeba is similar to the parent cell. Unless eaten or destroyed by environmental conditions, such as pollution, an amoeba does not die. It becomes a part of the new generation.

In complex organisms, sexual reproduction is carried out by certain specialized cells. Usually two types of reproductive cells fuse to form a fertilized egg. Generally, though not always, the sexual process involves two individuals, male and female.

Sexual reproduction is biologically important because it increases variation in a species. Each offspring is not a duplicate of a single parent but is, instead, the product of DNA contributed by both the mother and the father (Fig. 1–11). Variety is the raw material for the vital processes of evolution.

FIGURE 1–11   This female bengal tiger is normally colored, but her offspring is white (albinistic) due to an inherited inability to produce normal pigments. The father, not shown, is also normally pigmented. The parents of this white tiger are both genetic carriers for the white trait. The white animal is genetically pure for this trait. Sexual reproduction makes it possible for offspring to differ genetically, and therefore in appearance, from their parents. Only about 50 white tigers are known to exist in the world, all descendants of one white male captured in India in 1951. (Photographed by the authors at Busch Gardens, Tampa)

A Species Has The Ability To Evolve And Adapt To Its Environment

A **species** is a group of organisms having structural, functional, and developmental similarities that breed with one another under natural conditions and produce fertile offspring. Members of one species do not breed with members of another species under natural conditions.

The ability of a species to **evolve,** or change over time, enables it to survive in a changing world. As a species evolves, it develops structural, physiological, and behavioral traits that enable its members to more

effectively grow, reproduce, and maintain homeostasis. Individuals do not evolve; only species evolve.

Traits that enhance an organism's ability to survive in a particular environment are referred to as **adaptations.** Woodpeckers have structural adaptations—powerful neck muscles, beaks fitted for chiseling, and long chisel-like tongues—that enable them to secure insects from tree trunks. Cactus plants have adaptations needed for surviving in dry areas. Every biologically successful organism may be viewed as an impressive collection of coordinated adaptations (Fig. 1–12). Most adaptations evolve over long periods of time and involve many generations; they are the result of evolutionary processes.

■ THE THEORY OF EVOLUTION HAS BECOME ONE OF THE GREAT UNIFYING CONCEPTS OF BIOLOGY

A major focus of biology is on how the millions of life forms on our planet are related to one another and how each kind of organism has come to be. Biologists have uncovered a great deal of evidence that the life forms that exist today arose from earlier species by a process of evolution.

Evolution Proceeds By Natural Selection

In his book *The Origin of Species,* published in 1859, Charles Darwin synthesized many new findings in geology and biology and outlined a comprehensive theory that has helped shape the nature of biological science to the present day. Darwin presented a wealth of evidence that the present forms of life on earth descended with modifications from previously existing forms—his **theory of organic evolution.** His book raised a storm of controversy in both religion and sci-

**FIGURE 1–12   Some diverse adaptations. (*a*) The leaves of the sundew, modified for trapping insects, contain digestive enzymes. This sundew is in the process of devouring a fly. (*b*) A tool-using woodpecker finch (*Camarhyncus pallidus*). The behavior of this species from the Galapagos Islands is adapted for feeding on insects in crevices within the bark of trees. Lacking a long beak, the finch uses twigs or cactus spines as tools to dig out its prey. (*a,* E.R. Degginger; *b,* Miguel Castro, Photo Researchers)

(a)   (b)

FIGURE 1–13 Strings of eggs of the American Toad (*Bufo americanus*) wrapped about Elodea plants. Many more eggs are produced than can possibly develop into adult toads. Random events may be largely responsible for determining which of these developing organisms will hatch, reach adulthood, and reproduce. However, certain traits that each organism might have will also contribute to its probability for success in its environment. Although not all organisms are as prolific as the toad, the generalization that more organisms are born than survive is true throughout the living world. (Runk/Schoenberger from Grant Heilman)

FIGURE 1–14 A successful organism is well adapted to its environment. The long neck of the giraffe is an adaptation for reaching leaves high on trees. (E.R. Degginger)

ence, some of which still lingers. It also generated a great wave of scientific research and observation that has provided much additional evidence that evolution is responsible for the great diversity of organisms present on our planet.

Darwin based his theory of **natural selection** on several observations, which are presented here in contemporary form. (1) Many more organisms are born than survive into adulthood and reproduce (Fig. 1–13). (2) There is variation among the offspring of a population. For example, there may be differences in size, body structure, and color. (3) Individuals compete for survival. Resources in any environment are limited, and members of a population compete with one another for necessities like food, sunlight, and space. (4) Individuals that are best adapted to the environment live to reproduce, passing on their genetic "recipe" for survival. The tendency for individuals with traits best suited for their environment to reproduce in greater numbers than individuals which lack those traits is known as **differential reproduction.** Thus, the best adapted individuals of a species leave (on average) more offspring than do other individuals. Because of this differential reproduction, a greater proportion of the species becomes adapted to the prevailing environmental conditions. Nature *selects* the best adapted organisms for survival.

Although Darwin did not know about DNA or understand the mechanisms of inheritance, we now know that the variation that exists among organisms is genetic. The source of this variation is **mutations,** chemical changes in the DNA that persist and can be inherited.

Species Evolve In Response To Changes In The Environment

If the environment stayed the same, organisms would not change. However, the environment changes continuously, and if species are to survive, they must change with it.

If every organism were exactly like every other individual of its population, any change in the environment might be disastrous to all, and the population would become extinct. Differences among individuals —initiated by random mutation, spread by sexual reproduction, and molded by natural selection—enable populations to adapt to an ever-changing environment. Those traits that promote survival become more widely distributed in the population. Over long periods of time, as organisms continue to change in response to changes in the environment, members of the population begin to look increasingly unlike their ancestors.

The long neck of the giraffe is an adaptation for reaching leaves on trees (Fig. 1–14). The antelope-like ancestors of modern-day giraffes did not have elongated necks. As with other traits, there was a bell-shaped distribution of neck heights (i.e., the length of most giraffe necks was about the same, but some giraffes had relatively short necks and others had relatively long necks). Giraffes with the shortest necks could not compete effectively for the leaves on the trees of the African veldt; they were less likely to survive to reproduce. Giraffes with the longest necks were best able to compete for food. These giraffes survived and reproduced, passing on their genes (DNA) for long necks. Through hundreds of generations, the giraffe neck became longer and longer.

An Evolutionary Perspective Is Apparent In Almost Every Aspect Of Biology

The details and applications of the theory of evolution will be discussed in Chapters 17 through 20. However, in this first chapter, it is important to emphasize that although evolution is itself a subdiscipline of biology, some element of an evolutionary perspective is present in almost every specialized field within biology.

Darwin's theory of evolution has proved to be one of the great unifying concepts of biology. Biologists in almost every subdiscipline try to understand the features and functions of organisms and their constituent cells and parts by considering them in light of the long, continuing process of evolution. Additionally, biologists are constantly checking for verification of the evolutionary relationships among different organisms and often reinterpret these in the light of new evidence. Biology today is more than a science of describing and naming organisms and life processes. Biologists are concerned not only with the existence of structural similarities, but also with what these similarities (and differences) tell us about how organisms are related to one another and how things have come to be as they are.

■ BIOLOGICAL ORGANIZATION IS HIERARCHICAL

Order is the hallmark of life. Whether we study an individual organism or the world of life as a whole, its exquisite organization is evident. Several levels of biological organization are shown in Fig. 1–15.

We Can Identify An Organizational Hierarchy Within The Organism

What levels of organization can we identify in the individual organism? The **chemical level** is the simplest level of organization. There we find the basic particles of all matter, the minute atoms and combinations of atoms called molecules. An **atom** is the smallest amount of a chemical element (fundamental substance) that retains the characteristic properties of that element. For example, an atom of iron is the smallest possible unit of iron, and an atom of sodium is the smallest unit of sodium. (As we will see in Chapter 2, even atoms consist of subatomic particles and have a very definite organization.) Atoms combine chemically to form **molecules.** For example, two atoms of hydrogen combine with one atom of oxygen to form a molecule of water.

Atoms and molecules associate with one another to form the specialized structures that make up the **cellular level.** Each cell consists of a discrete body of jelly-like **cytoplasm** surrounded by a **plasma membrane.** Cytoplasm is a complex mixture of water and many other chemical substances. Some of these substances are organized into specialized structures called **organelles,** which perform specific functions inside the cell. Most types of cells contain a large organelle, the **nucleus,** that houses its DNA (Fig. 1–16).

There are many different types of cells. Each type is specialized to perform specific functions. In most multicellular organisms, including human beings, similar cells associate to form **tissues,** such as muscle tissue or nervous tissue. In turn, various tissues are arranged into functional structures called **organs,** such as the brain or heart. Each group of biological functions, such as circulation, is performed by a coordinated group of tissues and organs, known as an **organ system.** The nervous and circulatory systems are examples of organ systems. Working together with far greater precision and complexity than the most complicated machine created by human beings, the organ systems make up the complex living **organism.**

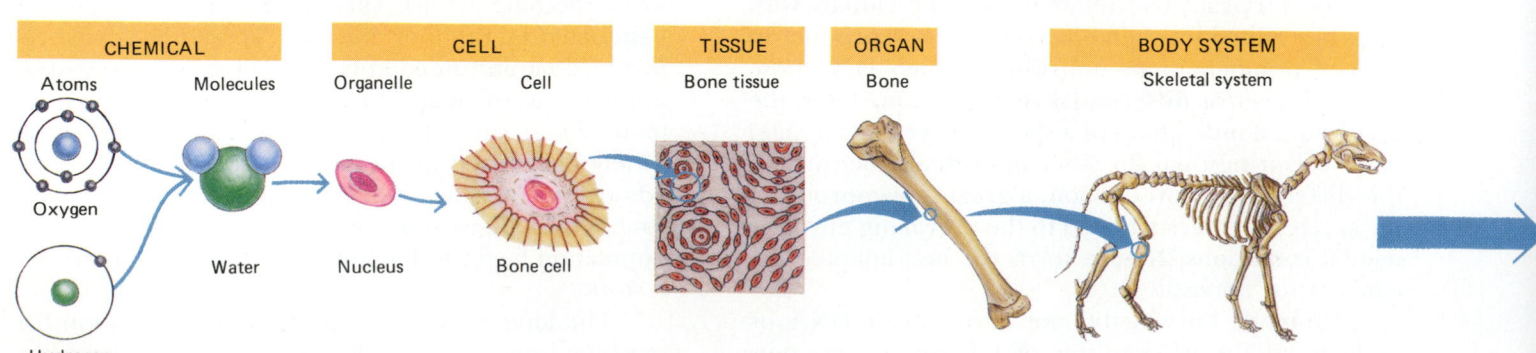

CHEMICAL		CELL		TISSUE	ORGAN	BODY SYSTEM
Atoms	Molecules	Organelle	Cell	Bone tissue	Bone	Skeletal system
Oxygen		Nucleus	Bone cell			
Hydrogen	Water					

Ecosphere
Planet earth and all of its inhabitants

ECOSYSTEM
Wolves, other organisms + nonliving environment

COMMUNITY
Wolves + trees + rabbits, etc

POPULATION
Wolf pack

ORGANISM
Wolf

FIGURE 1–15 Levels of biological organization.

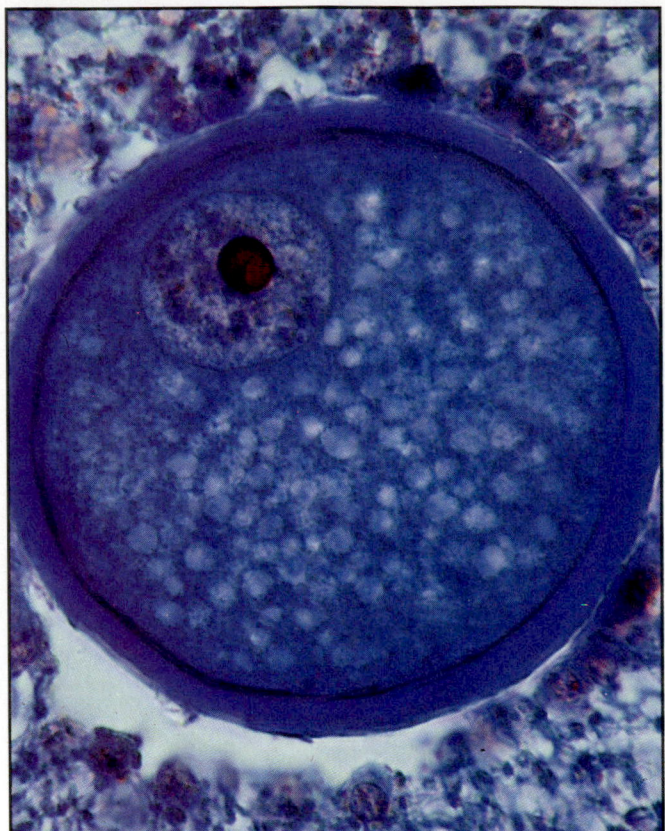

FIGURE 1–16 When fertilized by a sperm cell, an egg cell, such as the one shown here, can give rise to all of the cells of an organism. This egg cell, from the ovary of a mammal, has been magnified more than 250 times. The nucleus is visible as a circular lavendar area at the upper left of the cell. (Ed Reschke)

We Can Also Identify Organization On The Ecological Level

Organisms interact to form several higher categories of organization. The members of a species are not uniformly or even randomly scattered over the earth. They are found in groupings called **populations** that inhabit a particular region (Fig. 1–17). Various populations of different species occupying the same area interact with one another to form **communities.** A community may be composed of hundreds or even thousands of different types of organisms. An **ecosystem** consists of the community plus its nonliving environment. An ecosystem is a community that is self-sufficient in an ecological sense. The study of how organisms of a community interact with one another and relate to their nonliving environment is called **ecology** (derived from the Greek *oikos,* meaning "house" or "dwelling").

An ecosystem generally consists of three varieties of organisms—producers, consumers, and decomposers—and has a physical environment appropriate for their survival.

(a)

(b)

FIGURE 1–17 Ecological organization. (*a*) A population of penguins. (*b*) A characteristic of ecosystems is the presence of various organisms with specialized life-styles. In this scene from the Nyorobgoro crater in Africa, two populations—zebra and wildebeest—graze together on a population of grass. (*a,* W.J.L. Sladen/ Vireo; *b,* Frans Lanting)

By biological tradition, the major natural communities, such as the Great Plains or the arctic tundra, are considered ecosystems, for they appear to be somewhat self-sufficient. But the largest ecosystem of all, and ultimately the only self-sustaining one known, is planet earth with all its inhabitants—the **ecosphere.**[2]

Producers Make Their Own Food

Producers, or **autotrophs** (self-nourishing organisms), can produce their own food from simple raw materials. Algae, plants, and certain bacteria are producers. Most

---

[2] The terms *ecosphere* and *biosphere* are both used in this book. Although sometimes used interchangeably, they have slightly different meanings. The term ecosphere emphasizes the living world as an ecosystem, whereas **biosphere** refers merely to the totality of life on our planet.

autotrophs carry on **photosynthesis** using sunlight as an energy source.

During photosynthesis the energy from sunlight is used to make complex food molecules from carbon dioxide and water. The light energy is transformed into chemical energy, which is stored in the food molecules produced. Oxygen, which is required not only by plant cells but also by the cells of most other organisms, is produced as a by-product of photosynthesis.

Carbon dioxide + Water
+ Energy (from sunlight) $\longrightarrow$ Food + Oxygen

Consumers Depend On Producers For Food

Animals, including human beings, are **consumers.** Consumers, as well as decomposers, are **heterotrophs,** organisms that are dependent upon producers for food—and for energy and oxygen as well (Fig. 1–18). However, these organisms also contribute to the balance of the ecosystem. Like all living things (including producers), consumers and decomposers obtain energy by breaking down food molecules originally produced during photosynthesis. When food molecules are broken down during this process, called **cellular respiration,** their stored energy is made available for life processes.

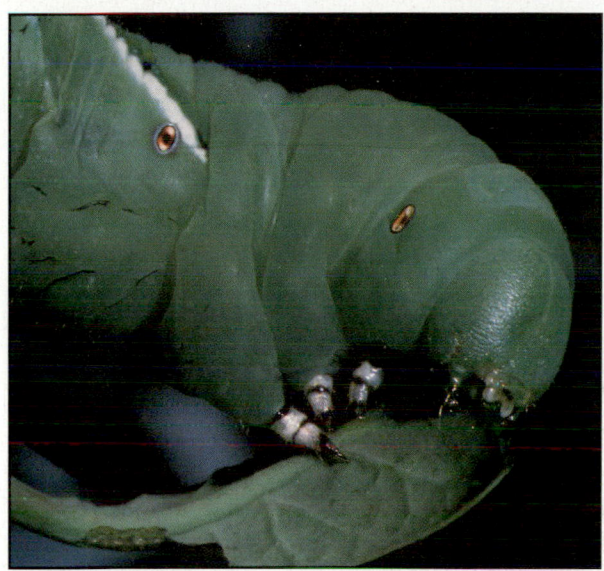

(a)

(b)

(c)

FIGURE 1–18 Consumers. (*a*) **A tomato hornworm larva, *Manduca sexta*, feeding on a plant. All animals ultimately depend upon producers for their source of energy. Insects often do great damage to agricultural crops, thereby placing them in direct competition with humans for food resources. (*b*) An example of a consumer feeding upon another consumer. Shown here is a fishing spider feeding on a small fish. Rapid movements and a venomous sting make the spider an extremely effective hunter. (*c*) In this coral reef scene from the Philippines, a clownfish, *Amphiprion percula*, is shown living in the tentacles of a sea anemone. The tentacles of the sea anemone contain poison used to capture small marine animals; however, a mucus coating on the clownfish prevents the anemone from stinging it. Thus the clownfish is afforded protection from its enemies and also has the opportunity to obtain a meal that the anemone had intended for itself.** (*a*, Peter J. Bryant, UC-Irvine/BPS; *b*, Robert Noonan, Photo Researchers, Inc.; *c*, Charles Seaborn)

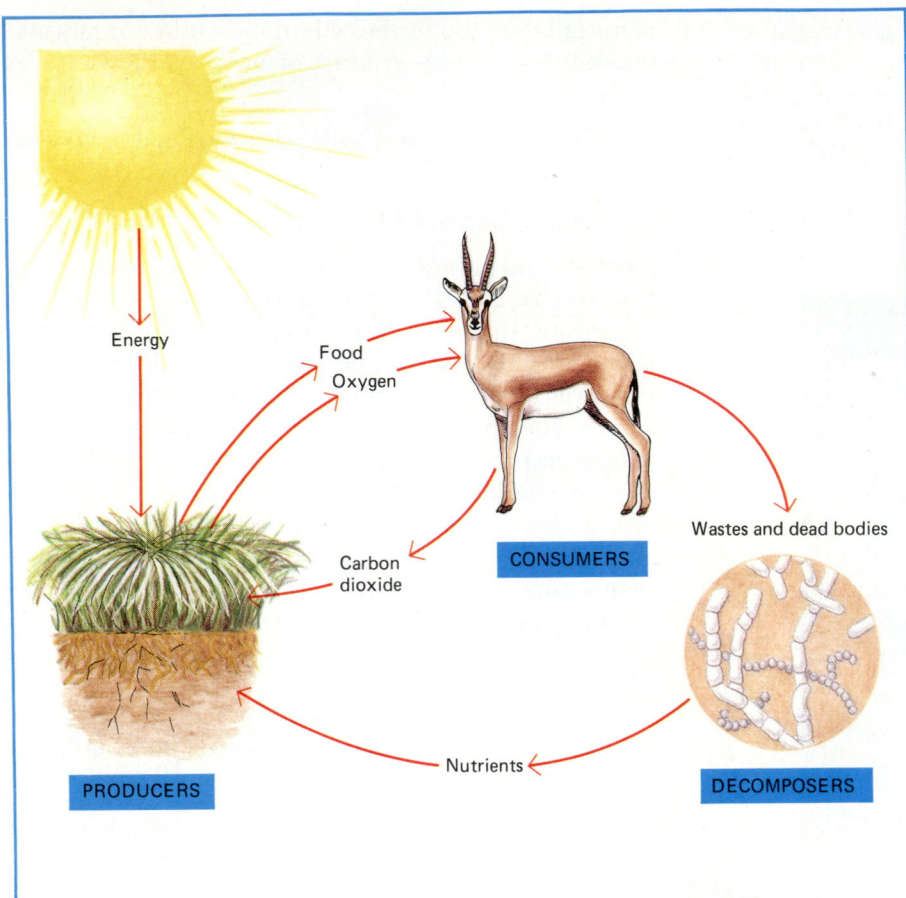

FIGURE 1–19 Interdependence of producers, consumers, and decomposers. Producers provide oxygen and food containing energy and nutrients for consumers. In turn, consumers provide the carbon dioxide needed for photosynthesis by producers. Decomposers break down wastes and dead bodies so that minerals are recycled.

Food + Oxygen $\longrightarrow$
    Carbon dioxide + Water + Energy

Gas exchange between producers (autotrophs) and consumers (heterotrophs) by way of the nonliving environment helps maintain the life-sustaining mixture of gases in the atmosphere (Fig. 1–19).

Decomposers Break Down Wastes

Decomposers—the bacteria and fungi—are an important component of an ecosystem, because they break down wastes and the bodies of dead organisms, making their components available for reuse. If decomposers did not exist, nutrients would become locked up in the dead bodies of plants and animals, and the supply of elements required by living systems would soon be exhausted.

Natural Ecosystems Survive Through Diversity

A familiar example of an ecosystem is a balanced aquarium. Plants or algae serve as producers, providing oxygen, food, and energy for the consumers, the fish. Bacteria and fungi function as decomposers, breaking down waste products and dead bodies so that nutrients within them can be recycled. As long as there is a continuing input of light energy for the plants, such an artificial system may persist for months or even years. Eventually, though, it will collapse.

A natural ecosystem, such as a pond, is far more stable than an aquarium and likely to last much longer. Stability is characteristic of such natural ecosystems, for unlike the aquarium, they consist of a vast diversity of organisms. Chemicals and energy have a multitude of alternative pathways within the system. Should one type of organism decline in numbers or die out, blocking one pathway, other organisms may take its place and the system as a whole can survive. Diversity is nature's grand tactic of survival.

■ MILLIONS OF KINDS OF ORGANISMS INHABIT OUR PLANET

The variety of living organisms that inhabit the ecosphere challenges the imagination. If we are to study their interrelationships, we need to make some sense

FOCUS ON   Evolution in Action: the Case of the Peppered Moth

An interesting case of evolution in action has been documented in England since 1850. The tree trunks in a certain region of England were once white because of a type of fungus, a lichen, that grew on them. The common peppered moth was beautifully adapted for landing upon these white tree trunks, since its light color blended with the trunks and protected it from predaceous birds (see figure). At that time a black moth was an oddity.

Then human beings changed the environment. They built industries that polluted the air with soot, killing the lichens and coloring the tree trunks black. The light-colored moths became easy prey to the birds. Now the black moths blended with the dark trunks and escaped the sharp eyes of predators. In these new surroundings, the dark moths were better adapted and were selected for survival. Eventually, more than 90% of the peppered moths in the industrial areas of England were dark. Interestingly, with recent efforts to control air pollution, there has been an increase in the popula-

tion of the light-colored moths.

Adaptation of the peppered moth was studied in the 1950s by breeding hundreds of male moths, then marking them with a spot of paint under their wings, and releasing them in both rural and industrial areas. After a period of time, the survivors were recaptured by attracting them with light or females. This study confirmed that significantly more dark moths survived in industrial areas and more light forms survived in rural areas.

Dark and light peppered moths. Which is most likely to become dinner for the bird? (John D. Cunningham/Visuals Unlimited)

of this diversity. To facilitate effective communication with one another, biologists have developed a formal system of classifying and naming organisms. The science of classifying and naming organisms is known as **taxonomy,** and the biologists who specialize in classification are **taxonomists.**

Biologists Use A Binomial System Of Nomenclature

In the 18th century Carolus Linnaeus, a Swedish botanist, developed a system of classification that has survived with some modification to the present. The basic unit of classification is the **species.** Closely related species may be grouped together in the next higher unit of classification, the **genus** (plural, genera).

The Linnaean system is referred to as the **binomial system of nomenclature** because each organism is assigned a two-part name. The first part of the name designates the genus, and the second part, the *specific epithet.* This is a descriptive word expressing some quality of the organism. The specific epithet is always used together with the full or abbreviated generic name preceding it. For example, the dog, *Canis familiaris (sometimes abbreviated C. familiaris),* and the timber wolf, *Canis lupus (C. lupus),* belong to the same genus. The cat, *Felis domestica,* belongs to a different genus. Note that each organism has its own scientific name; the first part of the name indicates the genus and is capitalized, and the second part of the name indicates the specific epithet and is not capitalized. Both names are italicized.

FIGURE 1–20 A survey of life. (a) This bacterium, *Micrococcus*, a member of kingdom Monera has been magnified several hundred times in this photomicrograph. (b) *Paramecium*, a protozoan, belongs to kingdom Protista. Although one-celled, *Paramecium* is a complex organism. Note its hair-like projections, called cilia, which are used in locomotion. (c) The mushroom (*Mycena* sp.) is a member of kingdom Fungi. This mushroom was found on a tree trunk in the Coast Range Forest. (d) A member of the plant kingdom, the red passion flower (*Passiflora vitifolia*), was photographed in Panama. (e) African lions, *Panthera leo*, among the fiercest of animals, are also among the most sociable. They live peaceably in prides (groups) consisting of as many as 35 lions. (a, David M. Phillips, Visuals Unlimited; b, M. Abbey, Visuals Unlimited; c, Frans Lanting; d, David Cavagnaro; e, courtesy of Busch Gardens, Tampa)

TABLE 1–1
Classification of the Domestic Cat, the Human Being, and Corn

Category	Classification of Cat	Classification of Human	Classification of White Oak
Kingdom	Animalia	Animalia	Plantae
Phylum (Division)	Chordata	Chordata	Magnoliophyta
Subphylum (Subdivision)	Vertebrata	Vertebrata	None
Class	Mammalia	Mammalia	Magnoliopsida
Order	Carnivora	Primates	Fagales
Family	Felidae	Hominidae	Fagaceae
Genus	*Felis*	*Homo*	*Quercus*
Species	*catus*	*sapiens*	*alba*

Taxonomic Classification Is Hierarchical

Just as species may be grouped together in a common genus, a number of related genera constitute a **family.** In turn, families may be grouped into **orders,** orders into **classes,** and classes into **phyla** (for animals) or **divisions** (for plants or fungi). For example, the family Canidae includes all dog-like carnivores (animals that eat mainly meat). This family includes 12 genera and about 34 living species. Family Canidae, along with family Ursidae (bears), Family Felidae (cat-like animals), and several other families that eat mainly meat, is placed in order Carnivora (Table 1–1). Order Carnivora, order Primates (the order to which humans belong), order Rodentia (rodents), order Cetacea (whales), and several other orders belong to class Mammalia (mammals). Class Mammalia, class Aves (birds), class Reptilia (reptiles), and three other classes are grouped together as subphylum Vertebrata. The vertebrates belong to kingdom Animalia.

Most Biologists Recognize Five Kingdoms

Since the time of Aristotle, biologists have divided the living world into two kingdoms, Plantae and Animalia. After microscopes were developed it became increasingly obvious that many organisms could not easily be assigned to either the plant or animal kingdom.

According to the system of classification used in this book, every organism (except the viruses) is assigned to one of five kingdoms: Monera, Protista, Fungi, Plantae, or Animalia (see Chapter 20). The members of kingdom **Plantae,** the plants, and of kingdom **Animalia,** the animals, are the organisms most familiar to us. The single-celled bacteria and cyanobacteria (blue-green algae) belong to kingdom **Monera** (Fig. 1–20). They differ from all other organisms in that they lack a membrane around the nucleus and also lack other cellular organelles.

Kingdom **Protista** consists of protozoa, algae, water molds, and slime molds. These organisms are single-celled or simple multicellular organisms. Kingdom **Fungi** is composed of the molds and yeasts. These organisms serve as decomposers, generally absorbing nutrients from dead leaves and other organic matter in the soil.

■ CHAPTER SUMMARY

I. Life may be defined in terms of the characteristics shared by living things.
  A. All living things are composed of cells.
  B. Living things grow and develop. Growth involves an increase in size and number of cells.
  C. Metabolism includes all the chemical activities and energy transformations essential to nutrition, growth and repair, and conversion of energy to useful forms.
  D. Metabolic activities are self-regulated so as to maintain an appropriate internal environment; the automatic tendency to maintain a steady state is called homeostasis.
  E. Although not all living things carry on locomotion, all exhibit movement.
  F. Living things respond to stimuli in their internal and external environments.
  G. Reproduction involves the transmission of DNA from one generation to the next.
    1. Reproduction may be asexual, in which the offspring receive a copy of the single parent's DNA and so are identical to the parent.
    2. Reproduction may be sexual, in which the offspring reflect the characteristics of two parents, but are not identical to either parent.

(Summary continues on page 21.)

Our technological society exerts tremendous impact on the delicately balanced ecosphere. We are rapidly depleting resources such as oil, coal, minerals, timber, drinking water, and soil. Those of us who live in developed nations make up about 25% of the world population, but we consume nearly 80% of the earth's resources. With only 5% of the earth's population, humans in the United States consume about 35% of the world's raw materials.

When we stress one aspect of an ecosystem, we often trigger a chain of events that magnifies the impact. The changes that occur affect the life support systems of our planet. Many forms of air and water pollution destroy plants and algae needed to provide oxygen and food (see figure). These producers serve as the nutritional base for animals that eat and in turn are eaten in the food webs that make up our ecosphere.

When we burn fossil fuels, carbon dioxide is released. The large amount of carbon dioxide released by industry is leading to a greenhouse effect in which more heat is retained in the lower levels of the atmosphere. Some scientists are concerned that the climate of the entire earth has already been affected and that we are experiencing a gradual increase in temperature. If the global temperature rises by even a few degrees, polar ice caps and glaciers may melt, raising the sea level 1 to 3 meters, and many coastal regions would be flooded.

Another concern is depletion of the ozone layer (in the stratosphere) that surrounds the earth and absorbs ultraviolet radiation from the sun. Chlorofluorocarbons (freons), used as coolants in refrigerators and air conditioners and as the propellant in aerosol cans, has been shown to destroy ozone. As more ultraviolet radiation reaches the earth, growth of algae and plants may be affected, and cases of human skin cancer will increase. It has been estimated that for each 1% decrease in ozone concentration, the incidence of skin cancer will increase 2 to 5%. Already the incidence of melanoma, a type of skin cancer, is increasing at the rate of 3.4% a year. After years of debate, the United Nations Environmental Program negotiated a tentative agreement to hold production of freon at 1986 levels beginning in 1990. Do you think such action is justified? Is it sufficient? Thousands of public policy decisions such as this one are currently being debated.

In the early 1970s there was widespread awareness of environmental problems, and public concern was responsible for passage of legislation and other actions directed at solving them. Unfortunately, just as styles of clothing change, so do intellectual fashions, and the environmental movement enjoyed only a brief heyday. The need for concern and action, however, is even greater today, for the problem, unlike the fad, shows no signs of disappearing. For example, a significant number of the chemicals we dump into the environment return to haunt us within our own bodies. Each of us now carries strontium-90 and lead in our bones, mercury in our blood and other tissues, DDT in our fatty tissues, asbestos and other particulate matter in our lungs, and an unhealthy concentration of carbon monoxide in our blood.

Any chance of dealing with environmental problems depends upon our degree of commitment, and enlightened commitment demands sound knowledge. However overpopulated the earth may be, energetic, knowledgeable, and committed people are in short supply. Many aspects of biology are intertwined with the environmental predicament. The principles presented in this book should provide the foundation necessary to understand the critical relationship between society and the ecosphere. The preservation of life itself depends on it.

(a)

(b)

(c)

Biologists study the effects of human activities on the environment. (*a*) **Fishing line and other types of plastic carelessly strewn about the environment prove lethal to thousands of animals each year. This sea turtle has become hopelessly entangled in a fishing net.** (*b*) **Runoff of fertilizers from nearby farmland has caused a buildup of nutrients in this pond. As a consequence, there has been an explosive growth of protists (*Euglena rubra*). The growth looks like red scum on hot days, then turns green in early evening. Among its many effects, this explosive growth of organisms prevents sunlight from reaching plants and algae below, leading to their death. Their decomposition reduces the concentration of oxygen in the water, resulting in fish kills.** (*c*) **Massive erosion caused by deforestation.** (*a*, Center for Environmental Education; *b*, Runk Schoenberger from Grant Heilman; *c*, Frans Lanting)

H. Through the process of evolution, living systems acquire adaptations that enable them to survive in changing environments.

II. The theory of evolution explains how species of organisms arose from earlier species and how species are related.

   A. Darwin based his theory of natural selection on the following observations: (1) More offspring are produced than survive and reproduce. (2) Variation exists among offspring of a population. (3) Individuals compete for resources. (4) Individuals which are best adapted to the environment survive and reproduce.

   B. The source of variation in a population is mutation, and the spread of variation through a population is enhanced by sexual reproduction.

   C. Species continuously evolve in response to changes in their environment.

III. There is a hierarchy of biological organization.

   A. In a complex organism there is a chemical level of organization consisting of atoms and molecules, a cellular level consisting of organelles and cells, a tissue level, organ level, and organ system level. The organ systems working together make up the organism.

   B. The basic unit of ecological organization is the population. The various populations that inhabit an area form a community; a community and its physical environment make up an ecosystem. Planet earth and all its inhabitants constitute the ecosphere.

   1. An ecosystem is a balanced community of producers, consumers, and decomposers living within an appropriate nonliving environment.

   2. During photosynthesis producers capture energy from sunlight and transform it into chemical energy. All living cells obtain energy by breaking down molecules originally produced during photosynthesis; this process is called cellular respiration.

IV. Taxonomy is the science of classifying and naming organisms.

   A. Biologists use a binomial system of nomenclature in which each type of organism is assigned a two-part name designating genus and specific epithet.

   B. Taxonomic classification is hierarchical: levels include species, genus, family, order, class, phyla or division, and kingdom.

   C. Biologists classify organisms in one of five kingdoms—Monera (bacteria and cyanobacteria), Protista (protozoa, algae, water molds and slime molds), Fungi, Plantae, and Animalia.

V. Human beings are part of a delicately balanced environmental system which is threatened by human activity. Increased, enlightened commitment to the task of solving environmental problems is needed.

◼ POST-TEST

1. The term *biology* literally means _____.
2. The chemical and energy transformations that take place within an organism are referred to as its _____.
3. The automatic tendency to maintain a steady state is called _____.
4. Reproduction involves the transmission of information by the hereditary material known as _____.
5. Traits that enhance an organism's ability to survive in its environment are called _____.
6. Living things are composed of one or more _____.
7. Various tissues are organized into functional structures called _____.
8. Various populations interact with one another and with the environment to form _____.
9. _____ is the study of how organisms of a community relate to one another and to the environment.
10. During _____ producers use the energy of sunlight to make food from carbon dioxide and water.
11. In cellular respiration food is broken down in the presence of oxygen, yielding carbon dioxide, _____, and _____.
12. Bacteria and fungi function ecologically as _____.
13. Variation among members of a population can be traced to changes in DNA called _____.
14. Darwin's theory of evolution emphasizes natural _____.
15. Protozoa and algae belong to the kingdom _____.
16. Several closely related species may be assigned to a common _____.

◼ REVIEW QUESTIONS

1. A child might argue that an automobile is alive. After all, it drinks water, guzzles gasoline, moves, and even responds to certain types of stimuli. How would you explain that a car is inanimate?

2. How does growth of a living organism differ from "growth" of a snowball as it rolls downhill?
3. Give an example of: (a) homeostasis. (b) metabolism. (c) adaptation.
4. Contrast producers, consumers, and decomposers. What would happen if all the decomposers on earth were suddenly to disappear?
5. Compare photosynthesis with cellular respiration.
6. Organisms A and B are classified in the same species; organisms X and Y are assigned to the same genus but not to the same species. Which pair of animals would have the most characteristics in common?
7. In which kingdom would you classify a bacterium? an elephant?
8. How can the study of biology help you to function as a more enlightened citizen?
9. Polar bears have very thick coats. Apply the theory of natural selection to explain how this adaptation may have evolved.

■ RECOMMENDED READINGS

Cloud, P. "The Biosphere," *Scientific American,* September 1983, 176–189. A fascinating discussion of the relationship between microbial, animal, and plant life on earth and the physical environment.

McMahon, T.A., and J.T. Bonner. *On Size and Life.* San Francisco, W.H. Freeman and Company, 1984. How the design of living things is influenced by their size.

O'Leary, P.R., P.W. Walsh, and R.K. Ham. "Managing Solid Waste," *Scientific American,* December 1988, Vol. 259, No. 36–42. An overview of a timely environmental problem that will require personal and public effort to solve.

Wiesner, J.B. "On Science Advice To The President," *Scientific American,* January 1989, Vol. 260, No. 1, 34–39. A high-quality science-advisory system is needed to advise the President on matters from science education to sending humans to Mars.

2

The Method of Science

OUTLINE

   I. Science is a means of investigation
  II. Science seeks to understand the universe
     A. Science began in ancient times
 III. Conclusions should be based on adequate evidence
     A. Deductive reasoning applies general principles
     B. Inductive reasoning discovers general principles
     C. Evidence must be interpreted
       1. Experimental samples that are not typical may be misleading
       2. Experimenter bias can render conclusions doubtful
 IV. The scientific method tests induction
     A. Recognizing and stating problems is probably the most crucial step
     B. A hypothesis is a proposed explanation or generalization
     C. A prediction is a logical consequence of a hypothesis
     D. Predictions are tested by observation
       1. Experiments generate observations
       2. An investigation of diabetes serves as an example of an experiment
     E. Hypotheses must be falsifiable
       1. True predictions can result from coincidence
       2. Historical hypotheses are difficult to falsify
  V. Principles are well-supported hypotheses
 VI. Science and technology interact
VII. Science is a human activity
     A. Science is costly
     B. Scientific findings must be published
     C. Scientists must guard the truth
     D. Science has ethical dimensions
Focus on Piltdown Man

LEARNING OBJECTIVES

After you have studied this chapter you should be able to:

1. Explain the importance of a knowledge of science in a college education.
2. Describe and distinguish between the inductive and deductive modes of thought and summarize their characteristics.
3. Analyze an experiment and analyze its features in terms of the scientific method.
4. Distinguish critically between well and poorly posed hypotheses.
5. Compare and give examples of the concepts of hypothesis, theory, and principle.
6. Describe the relationship between science and objective truth.
7. Outline the ethical dimensions of the scientific method, summarizing the chief areas in which ethical problems are likely to occur in the practice of science.

 A biology major or a premedical student knows why he or she is reading this book—almost certainly a large part of it is professionally useful or will be necessary to understand other required courses. A business or history major also knows why he or she is reading this book. A course in science is required for graduation, and biology is one way to meet that requirement.

There may be other reasons, but it is likely that most users of WORLD OF BIOLOGY fit into the latter category and may have a less than totally positive attitude about science. Is there really any good reason to ask a budding commercial artist to learn about science? We hope to show you that there is. For now, we will put it this way: A degree implies an education, and anyone who claims to be educated must know something of science, because the modern world simply cannot be understood apart from the influence of science.

tions, physics formulas, mineral types, or the Latin names of trees. Granted, one of the tasks of science is the collection of information. But a scientist is much more than an encyclopedist. He or she organizes information into a coherent structure, infers general principles on the basis of that information, tests these principles, and if they prove valid, uses these principles to reinterpret the significance of old information and to help gather new. As new principles and data emerge, they form the basis of new technology.

Technology transforms the world materially; the principles themselves transform it intellectually. Biology is an excellent example—a specimen science that can teach us much about science itself but whose information can also be used in a vast number of ways. In other words, the essence of a science isn't just its information content, though information must indeed be mastered. But to understand not just biology but science itself you must learn how science operates.

KEY CONCEPTS

☐ Science as a philosophy and method has heavily influenced the modern world and is one of the most important factors in our culture.

☐ Science embodies more than the systematic collection and arrangement of data. It is a self-correcting means to discover new facts and to formulate new principles.

☐ Science involves the careful use of inductive and deductive logic in the formulation of hypotheses and the testing of predictions by observation.

☐ To be useful, hypotheses must be both testable and falsifiable.

☐ The scientific method embodies certain ethical principles and, in addition, science is subject to the same ethical and moral norms as other areas of human endeavor.

☐ SCIENCE IS A MEANS OF INVESTIGATION

If you are reading this book inside a building, look about the room. Apart from any living things it may contain (for example, your roommate), little or nothing will meet your eye that could have been constructed if there were no such thing as science. Indeed almost every detail of your life that is different from what it would have been in a 12th-century university has resulted, in whole or in part, from the influence of science and its expression as technology.

But even in our day, not everyone is clear about just what science *is*. Science is not just chemical equa-

☐ SCIENCE SEEKS TO UNDERSTAND THE UNIVERSE

All science proceeds on the basis of two assumptions and a method. The assumptions are perfectly reasonable: There is a universe that is independent of the human mind, but the human mind can understand that universe to some extent. It takes longer to describe the method, but one of the things that distinguishes it is its goal. Science seeks to reduce the complexity of the world to general descriptions, or principles, that can be used to solve problems or suggest new insights. Science is able to uncover ever more facets of reality, to lead us to an expanded appreciation of reality, and sometimes even to tell us things that cannot be directly pictured by the mind.

There are important areas of human life that are not scientific, which is not to say that they are in any way invalid, of course. It is difficult, if not impossible, to reduce the arts, for example, to a set of principles, and perhaps we should not try to do so. Yet science could be applied even to the arts. Musical instruments can be improved or designed using physics and electronics, for instance, and it is likely that methods of teaching the arts could be investigated by scientific means.

But to return to science itself, from the earliest known times of all civilizations, people have tried to bring order into the complex apparent chaos of the world around them, partly because it is human to be curious. Also, however, understanding the world is likely to lead to greater control over the forces of nature.

FIGURE 2–1 The School of Athens (Raphael) was a gathering place for philosophers of the day, some of which were among the founders of what we today call natural science.

Science Began In Ancient Times

The first known stirrings of what we would call science date from the philosophers of ancient Greece. Many of their ideas, such as Aristotle's system of logic, are still in use today.

Socrates, Plato, and their followers (Fig. 2–1), who were among the more influential Greek philosophers, believed in the existence of ideas or ideals. A Platonic ideal was a perfect pattern of which all earthly examples are reflections.

For Aristotle, who followed Plato, an idea was not a perfect pattern but a generalization or concept constructed by the mind to create an overall definition for specific observations. Today we call this synthetic process of generalization **induction.**

Aristotle also recognized another basic thought process, **deduction,** which operates in the opposite direction, reasoning from generalities to specifics.

Although much information was gathered in ancient times, what was largely lacking for long over a thousand years was a logical framework into which scientific facts might be fitted.

In the 13th century an English scholastic philosopher, Roger Bacon (Fig. 2–2), re-emphasized the role of induction in reason. He taught that once a subject of discourse had been identified, facts bearing upon it should be collected until they fell into a pattern suggesting a new insight. To determine whether the insight was valid or not, a logical consequence should then be examined to see if it would be consistent with other known or discoverable data. If there were no contradiction, the insight, or hypothesis, should be tentatively accepted.

The first great generalization of biology to be discovered was **biogenesis,** the idea that life always arises from life; living things have living parents. Perhaps

FIGURE 2–2 Roger Bacon, a scholastic philosopher who clarified the nature of inductive thought. (Bettmann Archives)

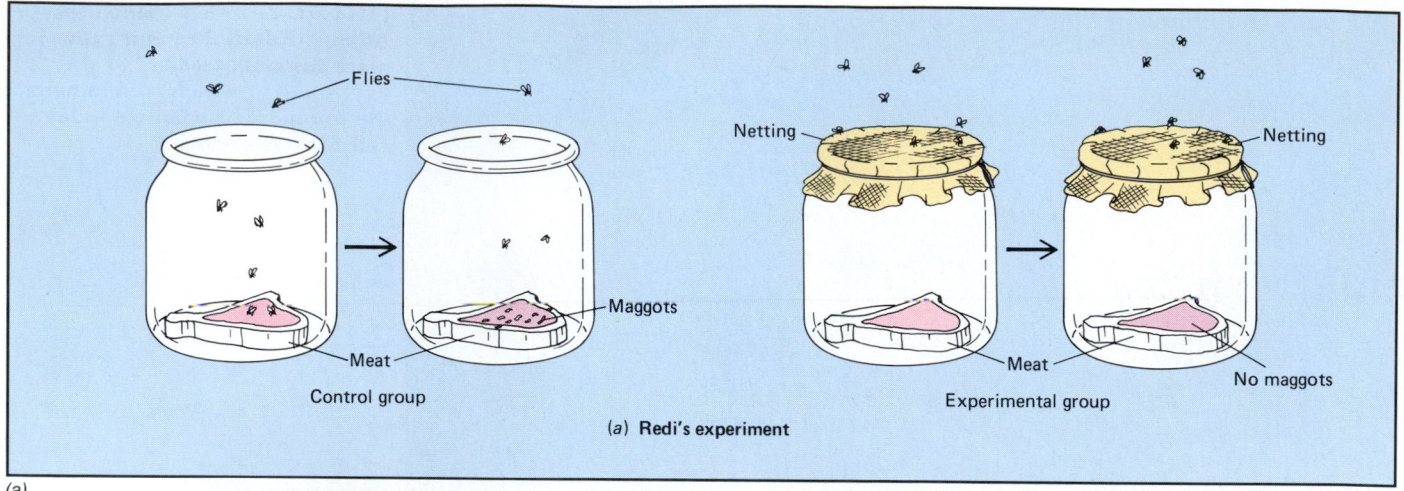

(b)

FIGURE 2–3 (*a*) **In Redi's experiment, flies failed to appear in the screened jars. (*b*) Francesco Redi, a pioneer investigator of biogenesis.** (Bettmann Archives)

surprisingly, it had long been believed that living things could spring from nonliving matter, especially decaying matter. Some ancient Greeks, for instance, thought that barbarians (by which they meant anyone who was not a Greek) could arise spontaneously from carrion, muck, or other noisome substances.

Such beliefs persisted through the Middle Ages and later. According to one treatise, mice could be produced by placing rags and a handful of grain in a crock. This assemblage was then to be placed in the thatch of a cottage roof. Before too long, a family of mice should appear in the crock. In the unlikely event that someone might actually have wanted mice, and tried the experiment, no doubt mice were obtained. Of course, spontaneous generation was not the only possible explanation.

Scientists must consider all possibilities. That is one of the things that distinguishes science from superstition—scientists should have an attitude of skepticism

not just toward the views of others but also toward their own. The skeptical inquirer must consciously seek alternative, possibly more reasonable, explanations of the data and test them if possible. Devotees of superstition are convinced by the first result that seems to confirm their own views and hope that evidence to the contrary will not arise.

One way to exclude alternative explanations is the use of a **control**—a second experiment or series of observations that differ from the main experiment in a known way. We could distinguish the **spontaneous generation** of mice from the usual method of making more mice by just placing a screen across the mouth of the crock. The Italian physician Francesco Redi (1621–1697) performed just such a controlled experiment with flies. He began by observing the natural course of decay of dead animals. Flies visited the corpses, and in due course, maggots appeared, which eventually changed (the biological term is "metamorphosed") into flies themselves. This suggested that the flies produced originated as eggs or larvae deposited by other flies (Fig. 2–3).

If containers of decaying meat were sealed so that flies could not enter, no flies developed. Yet some observers still felt that spontaneous generation might nevertheless be true, that it might require some force or substance from the air, which the seals had excluded along with flies. If the sealed containers are considered controls, they actually differed from the unsealed ones in *two* ways: They had no flies and no air. To make sure that the mere presence of air could not cause spontaneous generation, Redi replaced the seals with fine cotton mesh. Air could move freely through this, but flies could not. The results were the same—maggots were produced only when adult flies had access to the decaying meat. The appearance of flies had been considered

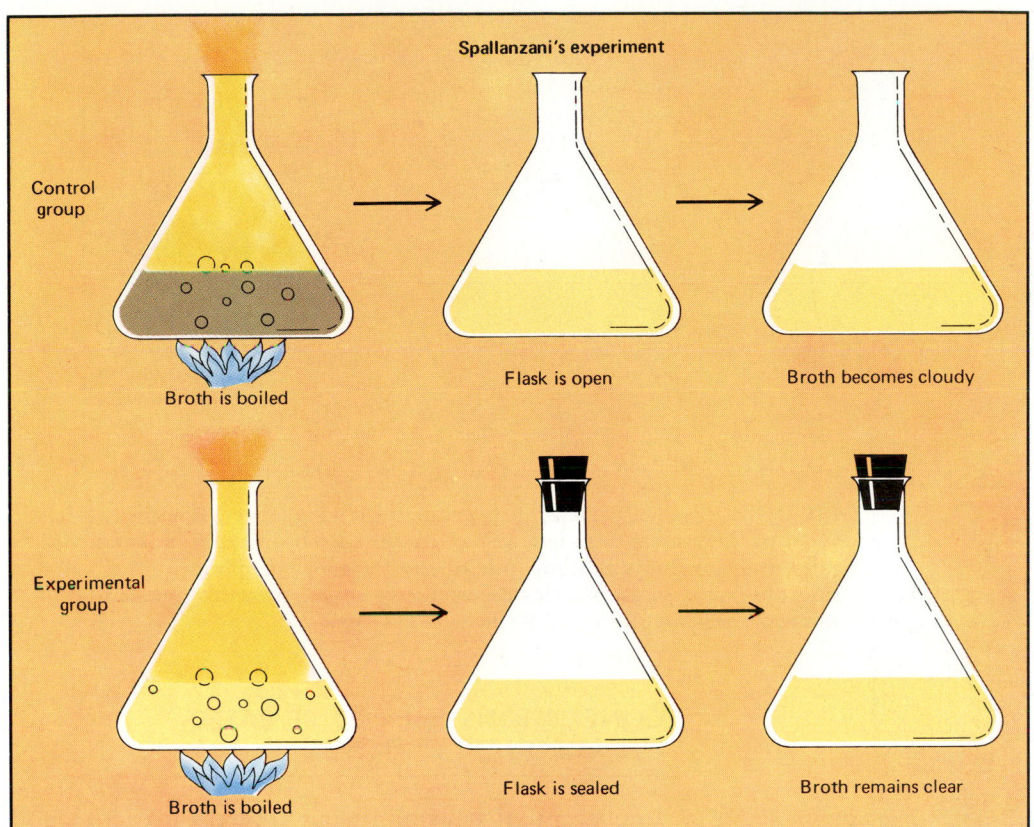

Spallanzani's experiment

Control group

Broth is boiled

Flask is open

Broth becomes cloudy

Experimental group

Broth is boiled

Flask is sealed

Broth remains clear

FIGURE 2–4 A sterilized, sealed flask does not grow microorganisms.

the classic example of spontaneous generation. By showing that flies were not actually produced by spontaneous generation, Redi convinced most informed people of the day that other insects were not produced spontaneously either.

Notice that without the accompanying control container, Redi could not have been certain that it was the mesh that made the difference. Perhaps some other factor—the season or something that was unusual about the dead animals—might have been the reason that no maggots appeared.

At least flies and even their eggs can be seen with the naked eye; microbes cannot. Many people believed that simple microscopic organisms arose spontaneously. Yeasts and other fungi, bacteria, and even protists (animal-like or plant-like single-celled organisms such as amoebas) swiftly swarm in decaying matter. No screen could possibly exclude them; neither could one always be sure that they were not initially present in decaying matter. Therefore an approach like Redi's could not be used to disprove the spontaneous generation of microbes.

The French microscopist Louis Joblot (1645–1723) made a tea or infusion of dried hay, boiling the liquid to destroy any organisms that might have been on or in the hay. He then divided the infusion into two parts, keeping one in a sealed container and one in an open container. Only the open container developed microbial inhabitants. This approach was further developed by the Italian physiologist Lazaro Spallanzani (1729–1799) using several kinds of media and very careful procedures for sealing the vessels (Fig. 2–4). In properly boiled, sealed vessels, microbial growth never took place.

Some raised the objection that in the sealed vessels there was no oxygen. What if spontaneous generation would take place only if oxygen or some other unknown life-giving substance in the air were available? It was hard to answer this objection, because any treatment one might give air to kill its microbes (bubbling it through lye, for instance, or passing it through a red-hot tube) might also destroy any "vital substance" the air might contain.

The matter was given its final resolution by the experiments of two French investigators, Felix Pouchet (1800–1872) and Louis Pasteur (1822–1895). Pouchet discovered a "screen" that could exclude microbes without concieveably damaging or excluding air—ordinary cotton wool. When cotton wool was first sterilized by heat and then used as a plug in the neck of a flask of boiled medium, no microbes grew inside.

Pasteur's studies were even more convincing. He drew out the neck of a flask containing boiled culture medium into a long, curved shape in a flame. Gradual

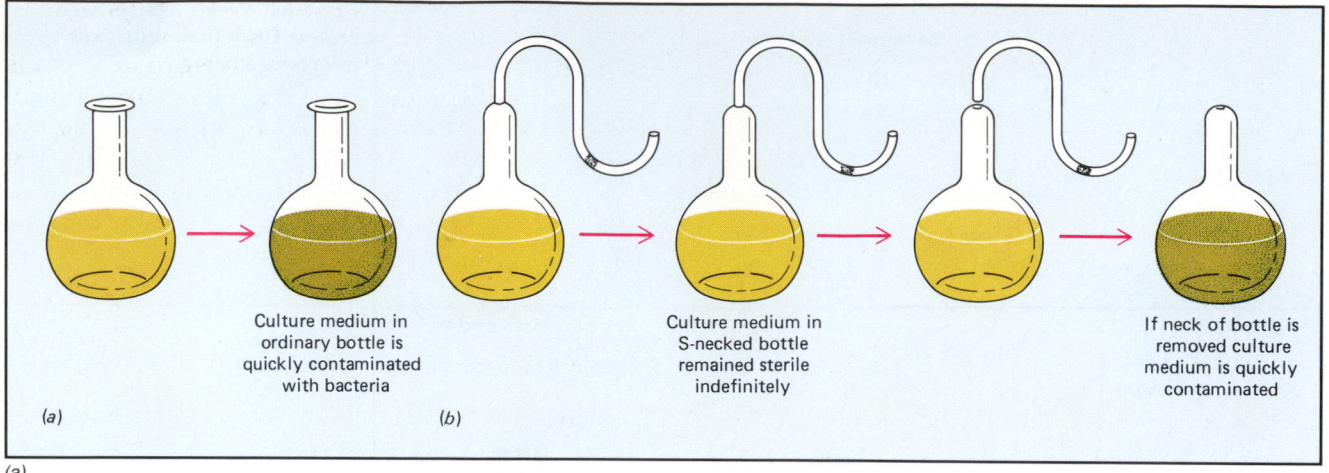

(a)

(b)

FIGURE 2–5 (*a*) **Pasteur's experiment. The sterilized flask was left open to the atmosphere but was so designed that microorganisms could not enter.** (*b*) **Louis Pasteur, one of the founders of modern microbiology, whose studies discredited the spontaneous generation of microorganisms.** (Biological Photo Service)

cooling allowed the air to enter slowly; any microorganisms were trapped in the curve of the neck. No life grew in the flask unless the medium inside came in contact with the region of the neck that held the microbes. If that were to happen, the flask swarmed with life (Fig. 2–5). A more perfect control could scarcely be imagined; the control flask was the *same* as the experimental flask. The air in the flask could not have been damaged in any way, yet when seed (or "germ") microbes were kept out, none developed.

The principle of biogenesis, that all living things had living ancestors, was a major scientific advance, with far-reaching practical applications. We take it for granted that canned food will not spoil and that disease microbes will not spontaneously originate in the body but must travel by contagion from a definite source, but it took generations of careful work to lay the groundwork for all this. Notice that a new *principle* was the outcome, that this principle could be used to predict more facts, and that some of those facts were of great practical importance.

■ CONCLUSIONS SHOULD BE BASED ON ADEQUATE EVIDENCE

Implicit in virtually all of our thought processes and underlying them is the **law of rationality.** Simply stated, it is that conclusions should be based on adequate evidence or that conclusions should agree with the facts, whether they are already known, or whether they are yet to be discovered. By implication, good conclusions should serve as a basis from which new facts can be predicted and should assist in their discovery.

Deductive Reasoning Applies General Principles

Once a general principle is understood, many of its consequences can be inferred with confidence, using a form of logic called **deduction.** Deduction operates from generalities to specifics. The classic form of deductive reasoning, the **syllogism,** is a three-part statement which begins with supplied information, the premises, and draws conclusions based on their common elements. Consider the following syllogism:

All birds have wings.	first premise	(a general rule)
Sparrows are birds.	second premise	(a specific example)
Therefore, sparrows have wings.	conclusion	(how the premises are related)

This example is a *valid* argument. That is, the conclusion that sparrows have wings follows *inevitably* from the information given. No other conclusion is possible.

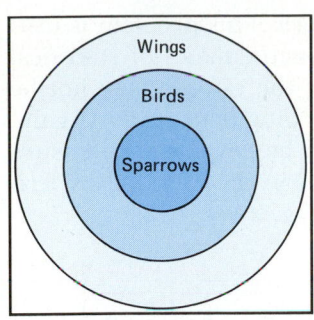

FIGURE 2–6 Why birds have wings. (A diagrammatic example of a syllogism, the classic form of deductive reasoning.)

A diagram of the syllogism would look like Figure 2–6, whose concentric circles represent the premises. Notice that sparrows are a subset of birds, which in turn are a subset of organisms with wings. There is no way that sparrows can be in the category of birds without also being in the category of winged animals. As deduction proceeds, its statements become less general, so that the conclusion contains nothing that was not present in the premises.

But consider also the following example:

> All birds have wings.
> All sparrows have wings.
> Therefore, all birds are sparrows.

It is immediately evident that something is wrong on the basis of general experience. This leads us to examine our reasoning, which we find can be diagrammed in more than one way and is therefore unreliable and invalid. That in itself does not make it untrue. If every kind of bird other than a sparrow were to become extinct, for example, all birds would indeed be sparrows.

If invalid arguments can be true, or untrue conclusions follow from valid arguments, what good is deduction? It actually has great value, if its limitations are appreciated. If the premises of deduction are true, *and* our use of deduction is valid, the conclusions *must* be true, as surely as two plus two equals four.

Syllogisms are simple examples of deduction. Much deduction is far more complex. Given the general principles of arithmetic, plus a record of deposits and checks written, the amount of one's current bank balance is foreordained. One does not know that figure, though, until hours of work have been done. Deduction, by making the relationship of known data explicit, allows us to evaluate many of its implications that we may not have previously realized.

Inductive Reasoning Discovers General Principles

No matter what undiscovered relationships may exist among the facts and principles that we already know, science's fundamental business is discovering new general principles. This is done by using a technique that is the opposite of deduction: **induction.** As we have already seen, inductive reasoning begins with specific examples and seeks to draw a conclusion or discover a unifying rule or general principle on the basis of those examples. It begins by organizing raw data into manageable categories and answering the question: What do these facts have in common? It continues by seeking a common *explanation* for the facts. Then this explanation is brought into some logical relationship with the rest of the known facts and validated explanations of science.

Table 2–1 sums up the characteristics of both inductive and deductive reasoning.

Evidence Must Be Interpreted

Finding facts is one thing, but deciding their significance is another. In one of the minor engagements of the long battle over spontaneous generation, an English clergyman, John T. Needham (1713–1781), boiled hay infusion and sealed it in flasks. Microorganisms

TABLE 2–1				
Method	Progression	Conclusion (Information Content)	Truth of Conclusion	Major Uses
Deduction	General principle to specific conclusion	Decreased	True if premises are true	Discovering explicit relationships among facts
Induction	Specific observations to generalized conclusions	Increased	True or false, even if observations are accurate	Synthesis of data; discovering new general principles

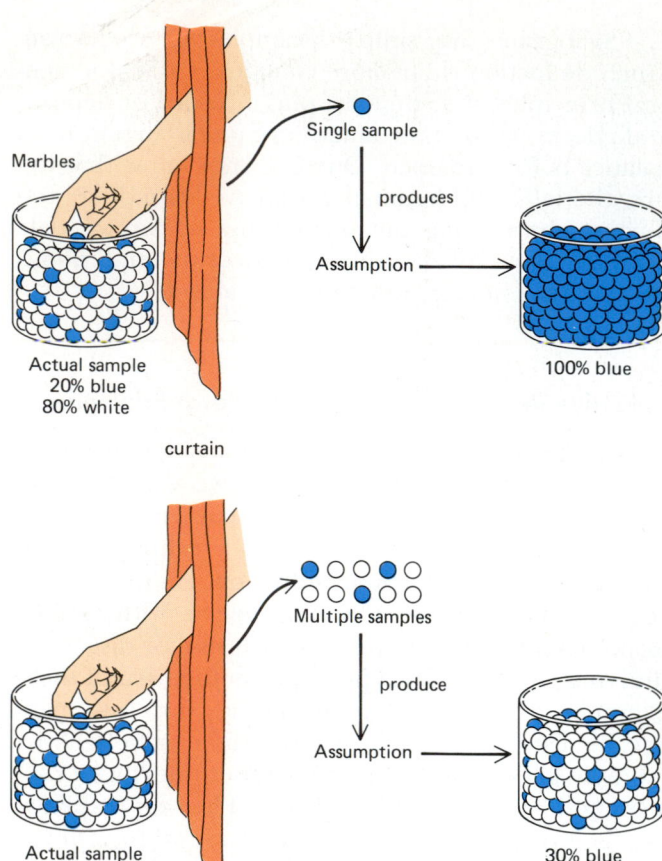

Marbles

Single sample

Actual sample
20% blue
80% white

produces

Assumption

100% blue

curtain

Multiple samples

produce

Assumption

Actual sample
20% blue
80% white

30% blue
70% white

FIGURE 2–7   The greater the number of samples we take of an unknown, the more likely we are to make valid assumptions about it.

grew in the infusion, apparently having originated there by spontaneous generation. These results baffled supporters of biogenesis. Needham and his party contended that spontaneous generation had been demonstrated, but eventually *it was shown by experiment* that hay contains the spores of bacteria that are resistant to ordinary boiling and can survive it. There was nothing wrong with Needham's evidence itself; it was, however, incomplete and could be properly interpreted only in the light of further evidence. It is not usually easy to determine whether evidence is incomplete, and even if this seems to be the case, the additional facts that are needed are often not obtainable right away.

Experimental Samples That Are Not Typical May Be Misleading

One reason for this is **sampling error.** Since *all* cases of what is being studied cannot be observed, we must be content with a sample of them. Yet how can we know whether that sample is truly representative of the class we are studying? In the first place, the sample may be too small, so that it is likely to be different from the

class because of random factors. This problem is usually soluble by applying the mathematics of statistical analysis. In the second place, the sample may not be typical of the group that we intend to study. Again, there are techniques that can be employed to ensure that there is no consistent bias in the way that experimental samples are chosen (Fig. 2–7).

Experimenter Bias Can Render Conclusions Doubtful

We may not have taken every relevant factor into account in choosing the sample, so that our choice may have introduced **bias** into the sample. In one famous case, a telephone poll was taken to determine the results of a forthcoming election. At the time, telephones were owned mainly by fairly affluent people, but since they were not the only ones who voted, the election did not turn out as predicted.

Bias also affects the kinds of conclusions drawn even from excellent, representative data. Bias involves interpreting all data in terms of the hypothesis we wish to prove, ignoring or "explaining away" contrary evidence. Needham and his followers had to dismiss a great mass of observations that contradicted spontaneous generation, while focusing exclusively on one that *seemed* to favor it. More skepticism of his own work and a more open mind might have secured for him a far higher place in the history of science. His opponents were skeptical but not really biased. Needham's experiments were repeated and intensely studied until an explanation was found. True bias would have rejected them out of hand.

■ THE SCIENTIFIC METHOD TESTS INDUCTION

One may propose explanations, but how can they be known to be true? Some means is needed to check the results of induction against reality. The **scientific method** is really the only means we have of doing so. Let us consider more formally how the scientific method was applied to the biogenesis question:

STEPS IN THE SCIENTIFIC METHOD	ILLUSTRATION
1. Recognize and state problem	1. How do living things originate?
2. Make observations or collect data	2. (a) In all known cases, living things can be shown to have living parents. (b) Regardless of other factors, in all known cases, microorganisms do not

FIGURE 2–8 Did life originate on the early earth under conditions that now rarely or never recur? (E.R. Degginger)

3. Propose hypothesis

4. Make prediction to be tested

5. Make observations to check prediction

6. Conclusion

develop in environments from which they are excluded.

3. Living things come only from living things (biogenesis).

4. Microorganisms originate from pre-existing microorganisms.

5. When properly treated so as to kill spores, hay infusions do not develop microbes unless innoculated with them.

6. Biogenesis is confirmed.

Notice that confirming biogenesis in this instance means that its status as a general principle is vindicated, but we must understand that even so, it is not completely secure. It could not stand as a true universal rule or law if so much as one counter-example could be demonstrated. We would then have to modify it to state: "Usually, life originates only from life." But isn't that really the case even without a known counter-example? Given long enough—millions of years, perhaps—and the right set of conditions (even if we cannot say exactly what they might have been), most biologists believe that life indeed originated spontaneously on the face of the earth (Fig. 2–8). So even for biogenesis, a mental reservation must be made. In front of even the most seemingly secure law of science there is an invisible and unspoken "perhaps."

Recognizing And Stating Problems Is Probably The Most Crucial Step

Significant discoveries are usually made by those who are in the habit of looking sharply at nature and whose minds have been prepared—usually by some preliminary idea of what they are looking for and by a subconscious collection of miscellaneous information, which eventually begins to fall in place and suggest a pattern. There is much of chance and luck involved at this stage, but as Louis Pasteur said, "Chance favors the prepared mind." On the one hand, the minds of those expert in a field are best prepared by knowledge of that subject area. On the other hand, familiarity with and long immersion in conventional ways of thinking may make it difficult to view a subject area in a new light. Some of the greatest scientific advances seem obvious in retrospect, and are, in a sense, quite simple. But they must have been extraordinarily difficult to think of in the first place, because they were unprecedented. The scientific method, in other words, is more than just a cookbook exercise. The personal qualities of the scientist, particularly his or her creativity and originality, determine how the method is used.

In 1928 the British bacteriologist Alexander Fleming noticed that one of his bacterial cultures had become invaded by a blue mold. He almost discarded it, but before he did, he noticed that the area contaminated by the mold was surrounded by a zone where bacterial colonies did not grow well. His culture looked something like the one shown in Figure 2–9a.

The bacteria were disease organisms of the genus *Staphylococcus*. Anything that could kill them was interesting! Fleming saved the mold, a variety of *Penicillium*

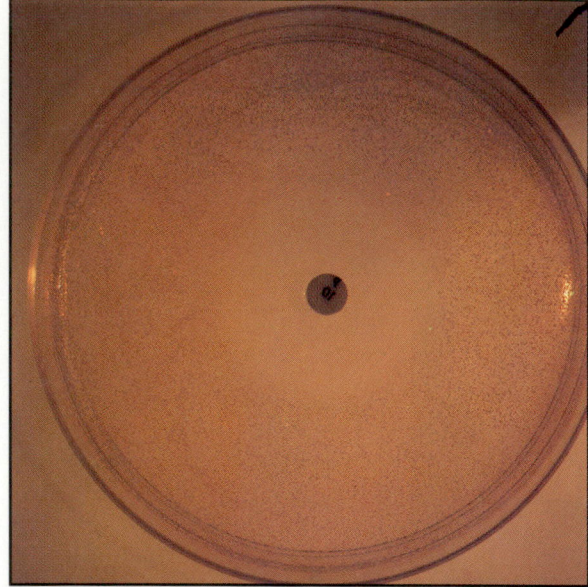

(a)

(b)

FIGURE 2–9 (*a*) **Fleming's culture. Notice the zone of inhibited bacterial growth around the colony of mold.** (*b*) **Sir Alexander Fleming, whose chance discovery led to the discovery of the antibiotic penicillin.** (*a*, Walter Dawn; *b*, Bettmann Archives)

(blue bread mold). What could be responsible? It was subsequently discovered that the mold produced a substance which slowed bacterial growth but which was usually harmless to laboratory animals and human beings—penicillin, one of the first antibiotics.

We may wonder how often the same thing happened to other bacteriologists who failed to notice the key fact and just threw away their contaminated cultures. The evidence is that it sometimes did.[1] Yet, evi-

---

[1] The student is urged to consult a letter to the editor of *Science* (Feb. 28, 1975), by A. C. Hilding, describing this occurrence, and "Missed Chance," *Scientific American,* Nov. 1978, p. 90. Though Fleming recognized the potential practical benefit of penicillin, he did not vigorously promote it, and it was more than ten years before the drug was put to significant use.

dently, not often. Only a few of the many strains of *Penicillium* produce sufficient quantities of penicillin, the staphylococci must be actively multiplying when the antibiotic is administered, and the temperature must be just right. Fleming benefited from chance, but his mind was prepared to observe it and his pen to publish it. It was left to others, however, to apply it.

A Hypothesis Is A Proposed Explanation Or Generalization

If a problem is recognized only by a prepared mind, a hypothesis is generated only by a creative one. In the early stages of an investigation, a scientist usually thinks of many possible explanations and hopes that the right one is among them. He or she then decides which, if any of them, could and should be subjected to experimental test. Why not test them all? Time and money are usually important to the conduct of research, so we must establish priority among the hypotheses so as to decide which to test first. Fortunately, some guidelines do exist. A good hypothesis should be:

1. Reasonably consistent with all well-established facts, or more consistent with them than competing hypotheses.
2. Capable of being tested, that is, it should generate definite predictions, whether the results are positive or negative. Test results should also be repeatable by independent observers.
3. Simpler than competing hypotheses.

A Prediction Is A Logical Consequence Of A Hypothesis

Since a hypothesis is an abstract idea, there is no way to test it directly. But hypotheses should suggest certain logical consequences, that is, observable things that cannot be false if the hypothesis is true. On the other hand, if the hypothesis is, in fact, false, *other* definite predictions should disclose that. In the specialized sense of the word that is employed here, then, a **prediction** is a deductive logical consequence of a hypothesis. It does *not* have to be a future event.

Predictions Are Tested By Observation

Einstein predicted, on the basis of his general theory of relativity, that light would be deflected by a strong field of gravity such as that possessed by the sun (Fig. 2–10). Yet the only way this prediction could be tested was to measure precisely the apparent position of a star whose image was close to the disk of the sun and compare it to its actual mapped position. The sun is so

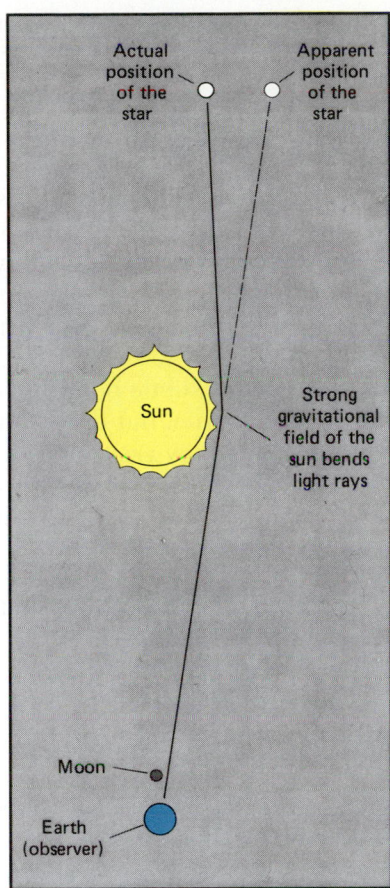

FIGURE 2–10 **Strong gravity changes the pathway of light, changing the apparent position of a star. But you can only see this at a very special time.**

bright that such an observation is usually impractical, but during a rare total eclipse of the sun by the moon, stars can be seen in the sky near the darkened solar disk. When they were precisely surveyed, the star images were found to be displaced from their known position to about the extent that Einstein had predicted. If Einstein's hypothesis had been incorrect, the stars should have appeared in their previously charted and surveyed positions.

Experiments Generate Observations

Notice, incidentally, that the astronomers performed no experiments. One cannot manipulate the sun and moon! They had to go to a great deal of trouble, however, to make their observations and almost did not succeed in doing so (it was a cloudy day). That is one reason why scientists employ experimentation whenever possible. An experiment can generate the needed observations at a convenient time and place, can generate them repeatedly, and may make possible some observations of occurrences that never would take place by themselves.

Also, the experimental approach allows us to be more certain that what we observe really is related to the hypothesis we are testing. Suppose we observe that a species of cricket chirps more frequently as the air temperature rises. How can we know that something other than temperature but related to it—humidity, for instance—is not the real reason? One simple way would be to capture a representative sample of crickets and place them in a room whose lighting, temperature, humidity, and anything else we can think of can be completely controlled. *Then* we could vary these factors one at a time and observe their effect, if any. We would probably discover that variations in humidity and in light similar to natural fluctuations of these factors have no effect on the chirping, but that temperature changes do. In actual experiments similar to this, it has been shown that the frequency with which some tree crickets chirp can be related to the temperature by a mathematical formula. This removes any reasonable doubt about the cause of the changes in chirping.

An Investigation Of Diabetes Serves As An Example Of An Experiment

In July 1983 a research report published in the *New England Journal of Medicine* challenged traditional medical practice of treating diabetics.[1] (Diabetes mellitus is a disorder in which carbohydrates are not utilized effectively by the cells, resulting in abnormally high blood sugar levels.) For many years diabetics have been instructed to avoid dietary intake of sucrose (table sugar). They gave up all types of desserts and avoided sucrose as a sweetener. This practice had been accepted so widely for so long that few researchers ever tested the concept.

The research team set up an experiment to test the hypothesis that diabetics should not eat sugar. They found that there was no significant difference in blood sugar levels when sucrose was included in *moderate* amounts in *nutritionally balanced* meals. This experiment was clearly a controlled situation that produced observations or results relating to a hypothesis.

The research team that tested the responses of diabetics to sugar set up a complex experiment in which 12 patients with Type I (insulin-requiring) diabetes and 10 patients with Type II (non-insulin-requiring) diabetes were fed a breakfast of common foods on five mornings. On three days the breakfasts included simple sugars—glucose, fructose, or sucrose. On two days the breakfasts included complex carbohydrates—potato starch and wheat starch. The breakfasts includ-

[1]John P. Bantle, et al., "Postprandial Glucose and Insulin Responses to Meals Containing Different Carbohydrates in Normal and Diabetic Subjects," *The New England Journal of Medicine*, Vol. 309, No. 1, July 7, 1983.

(a)

(b)

FIGURE 2–11 **(a) Spiny anteater, *Tachyglossus*. (b) Duck-billed platypus. These animals are warm blooded, furry, nurse their young, and nevertheless lay eggs.** (*a*, Tom Stack; *b*, Biological Photo Service)

ing the complex carbohydrates served as controls. From these breakfasts the researchers could gather data indicating what the blood level responses were when the diabetics were not challenged with simple sugars. The blood sugar level patterns resulting from the experimental breakfasts could then be compared with the controls. In addition, ten healthy subjects were included in the experiments as an additional control to provide data regarding the responses of healthy subjects to each type of breakfast.

In any experiment it is important that sufficient numbers of subjects be tested. If the research team testing the diabetics had used only one or two subjects, the experiment would not be considered statistically significant: The subjects selected may not have been typical of most diabetics. Because a total of 22 diabetics (and ten healthy subjects) were included in the study, the results probably reflect the responses that would be made by most diabetics. Many members of the medical profession, however, may be skeptical of the results because only 22 diabetics were tested. It is likely that this experiment will be repeated, perhaps many times, using larger numbers of subjects and with many varia-

tions before physicians change the treatment plans for their diabetic patients.

Hypotheses Must Be Falsifiable

A hypothesis is a tentative statement and so must be formulated in such a way that if it is supported its uncertainty can be reduced; or if it is false it can be shown to be false. Unfortunately, hypotheses sometimes are formulated in such a way that one cannot predict the results, especially the results to be expected if the hypothesis is *false*. To make a subtle point, such hypotheses are not necessarily untrue, but one cannot be even reasonably confident that they are true. If one believes in such a proposition, it must be on grounds other than scientific ones. To appreciate the reason, consider the following syllogism:

All female mammals bear live young (hypothesis).
Fido is a mammal.
Therefore, Fido (if female) should bear live young (prediction).

When Fido has a litter of puppies, this seems to vindicate the hypothesis. Yet it does not really prove the hypothesis, even so. It is *consistent* with the hypothesis and increases its credibility. Before the Southern Hemisphere was explored, most individuals would probably have believed that hypothesis without question, because any known furry, milk-giving creature did, in fact, bear live young. But it was discovered that egg-laying creatures that had fur and gave milk lived in Australia and some nearby islands (Fig. 2–11).

The hypothesis, as stated, was false no matter how many times it had previously been supported. A hypothesis cannot really be proved true, but in theory (though not necessarily in practice) it can be proved false. As a result, biologists had to either consider the platypus and spiny anteater as nonmammals or had to broaden their definition of mammals to include them (they chose the latter).

True Predictions Can Result From Coincidence

A hypothesis is not true just because some of its predictions (the ones we happen to have thought of or have thus far been able to test) have come true. After all, they could be true by coincidence. In particular, *failure* to observe a prediction does not make a hypothesis false, but it does not show that the hypothesis is true, either.

Historical Hypotheses Are Difficult To Falsify

Falsification is particularly difficult for historical hypotheses. Dinosaurs are extinct. There is a lot we do not know about them, including some very basic infor-

(a)

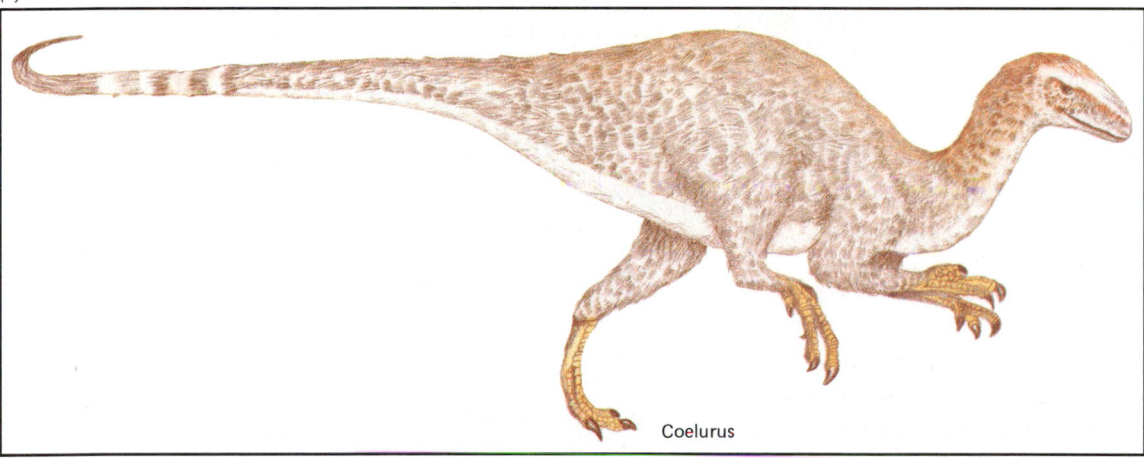

Coelurus

(b)

FIGURE 2–12 (*a*) Dinosaurs as originally conceived: cold-blooded, awkward, sluggish creatures barely able to muster the energy to eat one another! (*b*) A more modern concept. This warm-blooded dinosaur is shown with a conjectural coat of feathers which, if present, would have conserved body heat. Such a dinosaur would have been as active as a modern bird or mammal.

mation. For instance, were dinosaurs warm-blooded, or did their body temperature vary passively with that of their surroundings?

The issue is warmly debated. Because dinosaurs were reptiles (as judged by features of their skeletons) 19th-century scientists assumed almost without question that they, like modern reptiles, did not regulate their body temperature from within. Like a modern lizard or alligator, a dinosaur was believed to have slow life processes and to have required much less food than a bird or mammal of comparable size. Such an ectothermic (cold-blooded) dinosaur would have spent most of its time at rest and would have sunk into a torpor when the temperature of its surroundings dropped. In contrast, endothermic (warm-blooded) animals maintain a constant high body temperature even in cold climates, but to do so they require a great

deal of food, much of which their bodies "burn" as fuel. Also, their high body temperature allows them to be active for long periods of time, unlike most modern reptiles, which spend much of their time resting (Fig. 2–12).

On the basis of our experience today, though, we can make certain predictions about the dinosaurs that would be true if they really were endothermic instead of ectothermic. For instance, their skeletons should be adapted to a fast-moving mode of life (like an ostrich's or an elephant's) rather than to a largely sluggish existence like that of a modern alligator.

Studies with modern predators and their prey disclose that there are fewer predator organisms than prey organisms; the ratio of predators to prey is much less than one. If predators are ectothermic they require less food, so more of them can be supported by a given number of prey organisms. If predators are endothermic, though, they require more food and fewer of them can exist. The predator to prey ratio of dinosaurs appears to resemble that observable today among endothermic animals. The microscopic structure of the bone, when preserved, should resemble that of modern endothermic animals rather than that of modern reptiles. Finally, if they were endothermic, at least some dinosaurs should have had heat-conserving adaptations such as blubber, fur, or feathers, some evidence of which might still survive. One would not expect such features among ectothermic organisms.

Every one of these predictions has been realized, in the view of many biologists, even the last one (ancient pterodactyls—flying reptiles—have been discovered whose downy fur covering has been preserved in their fossils). Not everyone is convinced, however. Those contrary-minded hold that the evidence is open to more than one interpretation. For instance, we don't really have enough surviving dinosaur skeletons of all types to come to statistically assured conclusions about the ratio of predators to prey. Furthermore, pterodactyls were not really dinosaurs. The discussion continues, because it is interesting and enjoyable and shows no sign of being settled. To conclusively falsify the hypothesis of endothermy we need to do one impossible thing: take the temperature of a number of different species of live dinosaurs.

PRINCIPLES ARE WELL-SUPPORTED HYPOTHESES

It is important for researchers to communicate their results to one another, so that all their observations can be brought together and compared. In this way the equivalent of lifetimes of study devoted to a question can be accumulated in a relatively brief time. This is one of the most important reasons for the existence of learned journals and societies.

A hypothesis supported by a large body of observations and experiments becomes a **theory.** A good theory relates facts that previously appeared to be unrelated and that could not be explained on common ground. A good theory grows as additional facts become known.

The germ theory of infectious disease, for example, incorporates biogenesis, microscopy, mathematics, chemistry, and much else. It does not merely state that infectious disease is caused by microbes but tells us much about how such disease spreads and can be prevented or cured. Yet at the time of Pasteur and Koch, the germ theory of disease was no more than an elaborate hypothesis. It has been promoted to the status of theory by a kind of social concensus. Yet no committee met to decide this.

A theory that, over a long period of time, has yielded true predictions and is thus almost universally accepted, is referred to as a scientific **principle,** or **law.** The term "law" is usually reserved for a principle judged to be of great basic importance, for instance, the law of gravity or biogenesis.

Not every hypothesis has the potential of becoming a law. There is no doubt, for instance, that many blood-sucking parasites have anticoagulants in their saliva, but no one would call this trivial generalization a law, however true it may be.

SCIENCE AND TECHNOLOGY INTERACT

The discovery of scientific principles should lead to an increased control of nature. This is probably the chief justification for science for most practical-minded people. Oddly, though, discoveries that proved to have the greatest practical value often came from purely abstract research, the discovery of the principle of the transistor being one such example. The reason for this is that really basic discoveries are likely to be those of which we have little inkling ahead of time. Refining the applications of already known material is like investing in blue-chip stocks. The risk is smaller, but so is the likely return.

Our modern knowledge of the cell's structure could not have been obtained without the use of the electron microscope, first developed in the late 1930s with what was then the latest in electronic technology (Fig. 2–13). The 16th- and 17th-century microscopists such as Leeuwenhoek and Hooke could not have seen cells at all without the use of lens-making techniques that were then state-of-the-art.

Almost everything we know about photosynthesis and, to a somewhat lesser degree, cellular respiration, has been discovered using radioactively labelled chemical substances. Without nuclear technology, we would probably still be wondering whether the hereditary material of the cell was protein, nucleic acid, or something else. The lensmakers, electron-tube designers, and nuclear physicists who made modern biology possible probably never dreamt of the further scientific advances that would occur because of *their* scientific advances! Science feeds on itself, particularly when information from different areas of study is allowed to interact.

SCIENCE IS A HUMAN ACTIVITY

Science is a human activity, which must be practiced by people in a social environment. The very concepts that enter into hypothesis formation result from the gen-

FIGURE 2–13 Modern scanning electron microscope. Science and technology tend to promote one another. Advances in the physical sciences produced the electron microscope, which in turn led to further advances in the physical and biological sciences. Most of our current knowledge of cell structure has been gained by the use of the electron microscope. (E. R. Degginger)

eral social and intellectual mileu into which a scientist is born. Like everything else, science costs money, and research must be funded. Also, so that results may be evaluated by others and incorporated into their own work, research results must be published. Where will the necessary money come from? Who will print the results?

Science Is Costly

Most scientific research today is done in the setting of a university or research institute, itself often university-affiliated. Usually, these are not profit-making institutions and are not in a position to directly support research. Universities do serve as pipelines through which research funds flow, but these funds generally are provided by foundations or government agencies. It is in some ways regrettable, but the days of the back-yard or basement researcher seem to be gone. Research today typically requires very expensive equipment and often a number of assistants whose salaries must be paid. Sometimes the researcher's own salary is paid out of research funds. Those funds are crucial; little or nothing can be accomplished without them. Yet they are limited.

To obtain financing, researchers usually must compose formal proposals, sometimes of almost book length, to be submitted to a funding agency. Writing these proposals is itself a specialized skill—universities often employ personnel who help researchers to do so or offer formal courses on how to go about it. Then once the grant application has been sent in, the funding agency must try to decide whether the research proposal has merit. This is usually done by a process of peer review in which a board of scientists analyzes the proposal and tries to estimate its significance and chance of success.

Scientific Findings Must Be Published

Once a study is complete, its results must be published. There are several reasons for this. In the first place, experiments must be described in sufficient detail to be independently performed by others. This permits objective observers to detect errors or bias in the original study and helps to guard against the occasional odd result that occurs as a result of random or uncontrolled factors as well as those that may even be tainted by conscious dishonesty on the part of the original researcher.

Second, the methods employed in one research project can often be modified for use by someone else in a different project. Publication helps to avoid the need to "re-invent the wheel." Third, publication is the most effective way to disseminate knowledge and encourage the cross-fertilization of ideas that is necessary for creative science.

Fourth, publication is usually necessary in order to obtain renewal of a research grant, promotion, or even the retention of one's position. "Successful" research is often the chief criterion employed by universities to decide who shall be advanced or will obtain academic tenure.

Professional meetings encourage personal communication and discussion and usually provide for the preliminary presentation of research studies by illustrated lecture or other means. These meetings can be surprisingly lively, with spirited debate between rival teams of researchers not uncommon.

Scientists Must Guard The Truth

We may well wish that government officials, business people, journalists, and even students were always honest. In the world as it actually is, though, this is simply not the case. Important as honesty is in all walks of life, it is particularly important in science. In the long run, fortunately, science tends to be uniquely self-correcting and self-purifying by the consistent use of the scientific method itself. Sooner or later, someone's experimental results are bound to cast doubt on dishonest data.

Still, consider the great (though hopefully temporary) damage that is done whenever an unprincipled or even desperate researcher (whose career might depend on publication of a research study) knowingly disseminates false data. If the deception is not uncovered, it may become part of the generally accepted knowledge used by peer reviewers to determine funding or worthiness of publication. Excellent work may be denied a hearing due to the prior general accept-

FOCUS ON   Piltdown Man

By the early 1900s, many fossil remains of early humans had been found. "Heidelberg man," known from a huge lower jaw found in Germany, plus "Java man" and "Peking man" found later, were of the species *Homo erectus*, small-brained but clearly human. Numerous large-brained fossils known as "Neanderthal man" had been found in Germany and France. Though these humans had apelike features—jutting jaws, heavy brow ridges, low sloping foreheads—we now include them in our own species, *Homo sapiens*. Yet none of these were intermediate between humans and the apelike animals from which humans had supposedly evolved. A "missing link" was in need of discovery.

In 1912, Charles Dawson, an amateur fossil collector, announced that he had found a remarkable fossil, a human skull fragment, in an ancient river bed in a place called Piltdown, England. This discovery, he claimed, at long last filled the evolutionary gap between apes and humans.

Attempts to reconstruct the complete skull were complicated by the fragmentary nature of the remains. Unlike apes and other fossil humans

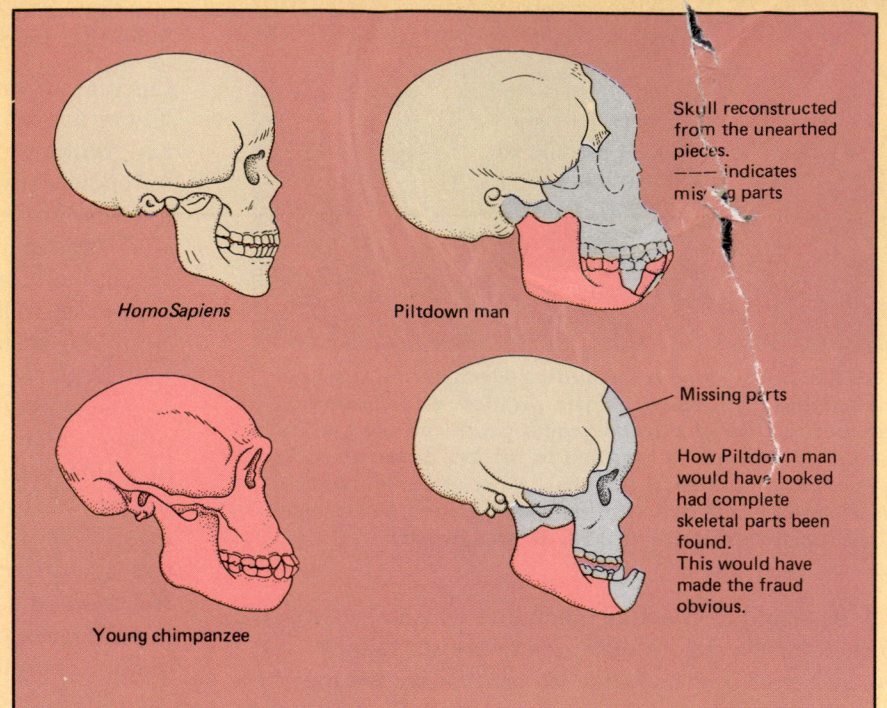

HomoSapiens

Piltdown man

Skull reconstructed from the unearthed pieces.
----- indicates missing parts

Young chimpanzee

Missing parts

How Piltdown man would have looked had complete skeletal parts been found.
This would have made the fraud obvious.

then known, the Piltdown skull had no great brow ridge; the high forehead and the apparently large brain were also characteristic of modern humans. The facial bones were missing, but considering the size of the jaw the face must have sloped sharply forward as did that of Neanderthal man. The jaw was apelike, but the part that would connect with the skull was missing, as were the chin region and teeth. If these missing parts were reconstructed, the total picture was that of a creature with a human skull and an apelike jaw, a mixture of traits that one

ance of inferior research. Other researchers might devote many thousands of dollars or hours of precious professional labor to futile lines of research inspired by untrue reports. Important policies may be wrongly decided on the basis of bad data furnished by prestigious but dishonest researchers. Perhaps worst of all is the combination of a true hypothesis and data purporting to support it that has been exposed as deliberate untruth. Researchers may refuse to take such a hypothesis seriously enough to re-investigate it because it has acquired a distasteful association.

Science Has Ethical Dimensions

We have seen that scientific investigation depends on commitment to such practical ideals as truthfulness and the obligation to communicate results. Since these are really part of the scientific method, they are the **intrinsic ethics** of science. However, there are also **extrinsic ethics** whose violation might not destroy the value of the data and ideas that result from a scientific investigation, but which must be taken into account for reasons of decency and humanity.

This topic is a very large and difficult one. Consider issues such as the use of fetal tissue obtained by abortion in research, the ethics of human or even animal experimentation, military weapons research, and applications of genetic engineering. Many such ethical issues will be raised throughout this book. As most scientists would hasten to agree, no occupation, profession, or field of study is isolated from social concerns, including science.

might well expect to see in a missing link.

Some raised doubts, however. The skull and lower jaw had not been found together and could not be proved to belong to the same individual. The jaw, in fact, could have been that of a young ape. If so, this fact would be evident from the canine teeth, which are much larger and longer in apes than in humans. To prove that the skull and jaw were from the same creature, the missing jaw parts, or at least some of the canine teeth, had to be found.

Not long after this need was voiced, a canine tooth came to light. Though similar to that of a modern chimpanzee, this tooth was smaller and shorter, intermediate, in a way, between those of a human and an ape. Yet since it was found separately, it could not be proved to belong to the jaw.

Another question persisted. Geological evidence suggested that Piltdown man was too young to be an intermediate: The much older Java and Neanderthal fossils were already more human-looking than Piltdown man. Two years after Dawson's death in 1916, it was announced that he had discovered a second Piltdown

specimen, which closely resembled the first. With this, most skeptics became believers, though the place of Piltdown man in human evolution remained unclear.

Meanwhile, investigators in southern Africa had unearthed the human-like *Australopithecus*, apparently much older than any known human fossils. These creatures had chimpanzee-sized brains, long down-slanting faces, and large jaws with essentially human teeth. Seemingly the right age, they were as good a missing link as Piltdown man, though quite different. As more australopithecine bones were found, more and more people doubted Piltdown man.

In 1949, the British Museum tested the Piltdown fossils for their fluorine content, a method of dating fossils developed in the nineteenth century. The results showed that the bones were of various ages and from various sources. Then, in 1952, scientists examined the Piltdown jaw and found that someone had altered the molars by filing them. As for the canine, not only had it been filed, but a cavity had been filled with a rubbery substance. Artificial staining and painting had concealed the tam-

pering. It became clear that the skull was a modern human skull, and the jaw, so apelike, was indeed a chimpanzee's jaw.

It is by no means certain that Dawson was responsible for the clever hoax. Someone else might well have planted the bogus fossils for him to find. Indeed, the whole affair may have been a practical joke that got out of hand. Yet there are lessons we can learn from the story.

Piltdown man appeared to be such a perfect fulfillment of evolutionary prediction that scholars, convinced that such a creature would one day be found, were predisposed to swallow the fraud whole. This postponed for years their recognition of the importance of *Australopithecus*. Decades passed before scientists themselves finally detected the fraud by testing their hypotheses. Piltdown man reminds us that the scientific method depends not only on logical thinking but also on an attitude of open-minded skepticism combined with the strictest honesty.

■ CHAPTER SUMMARY

I. Science is a systematic attempt to investigate the universe and organize the knowledge thus gained into a logical structure.
II. The usual method of science can be summarized thus:
    A. Recognize and state the problem.
    B. Collect information and data bearing upon the problem.
    C. Formulate a hypothesis.
    D. Produce a prediction on the basis of the hypothesis.
    E. Design and perform experiments or otherwise obtain observations that may test the hypothesis by their consistency with its predictions.
    F. Formulate a conclusion.
III. Conclusions should be based upon adequate evidence; logic is a means of drawing conclusions that are consistent with the evidence used to support them. Deductive and inductive modes of logic may be distinguished:
    A. Inductive and deductive logic are both employed in the scientific method.
    B. Deduction adds nothing new to knowledge although it can make relationships among data apparent.
    C. Induction produces new knowledge and generalizations but is error-prone.
    D. The scientific method tests the proposals of induction by deducing their consequences and testing these by observation and experiment.
IV. "Good" hypotheses are reasonably consistent with all known facts, generate definite predictions sub-

ject to test, and are simpler in comparison with competing hypotheses.

A. Total disproof of a hypothesis is rarely possible, and even well-supported hypotheses retain some uncertainty.

B. Unfalsifiable hypotheses are impossible to disprove.

C. Theories are usually broadly conceived, logically coherent, and very well supported. Principles are theories that have been almost universally accepted. Laws are principles judged to be of great importance.

V. Science is a social activity, greatly influenced by the current social mileu and subject to ethical and other constraints.

■ POST-TEST

1. According to the law of rationality, conclusions should be based upon _____ _____.

2. In deductive reasoning, one proceeds from general principles to _____ _____.

3. When properly carried out, the conclusions of deduction follow inevitably from the _____.

4. Yet the conclusions of deduction contain _____ _____ information than that which is implied by the premises.

5. Induction begins with specific examples or facts and seeks to discover a _____ rule or _____ principle.

6. A hypothesis is formed by the process of _____.

7. Thus, a hypothesis must be a _____ explanation.

8. A good hypothesis must be subject to _____.

9. A _____ is one or several well-confirmed hypotheses: it relates facts that previously appeared to be unrelated.

10. An experiment may be viewed as a means of generating appropriate _____ whereby a hypothesis may be tested.

11. Most funding for scientific research originates from _____ sources.

■ REVIEW QUESTIONS

1. Contrast induction and deduction.

2. What is the biggest difference between a well-confirmed hypothesis and a theory? A law?

3. What are the advantages of experimentation in science?

4. What is a control and why is it important?

5. Medical and behavioral researchers sometimes deliberately avoid knowing which persons have received a particular treatment. Why might this be desirable?

6. What are the intrinsic ethics of science? What are some extrinsic ethics?

7. Does society have the right to govern scientific research? If so, why, and what kinds of research should be so governed? How should it be done?

8. Give a new example of an unfalsifiable hypothesis.

9. What parts of the scientific method are inductive? Deductive?

10. Give four examples of ethical problems in recent scientific research as reported in the media.

11. An endless series of studies has been performed on the relationship between smoking, cancer, and numerous other diseases. All responsible scientists familiar with the issue are convinced that smoking and other tobacco use constitute a very serious health menace. Yet tobacco companies continue to contest these findings. Part of the difficulty has been that some kind of case, however implausible, could be made that "something else," alcohol consumption by smokers, for instance, could have been responsible for the observed facts rather than smoking itself. Propose an *experiment* that could settle the issue beyond any doubt. Would there be ethical objections to such an experiment?

■ READINGS

Baker, J., and G. Allen: *Hypothesis, Prediction and Implication in Biology*, Reading, Mass., Addison-Wesley, 1968. A good source for the beginning student who does not wish to become buried a tome.

Borg, W.R., and M.D. Gall: *Educational Research: An Introduction*, Fifth ed. New York, Longman, 1989. The first chapter contains an extensive discussion of the philosophy of science.

Crossland, J. "Human Experimentation," *Environment*, May 1974, p. 18. The ethics of drug testing using human subjects.

Gardner, E.: *History of Biology*, Minneapolis, Burgess, 1960. A useful summary of the development of the scientific method as applied to biology.

Morris, I. "Is Science really scientific?" *Science Journal*, Dec. 1966, p. 76. In our opinion, the finest short summary of modern scientific philosophy in print.

Wilson, D.: *In Search of Penicillin*, New York, Knopf, 1976. A thorough and fascinating account of the role played by Fleming and others in the discovery and development of penicillin.

P A R T 2

The Organization of Life

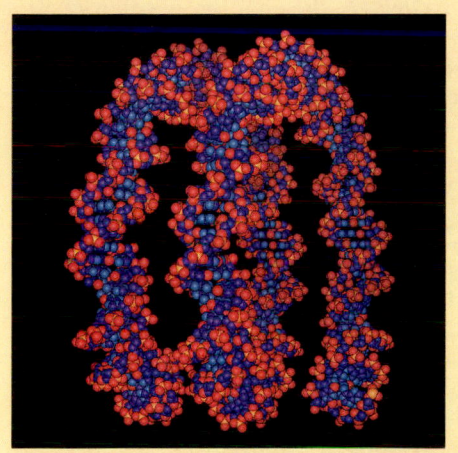 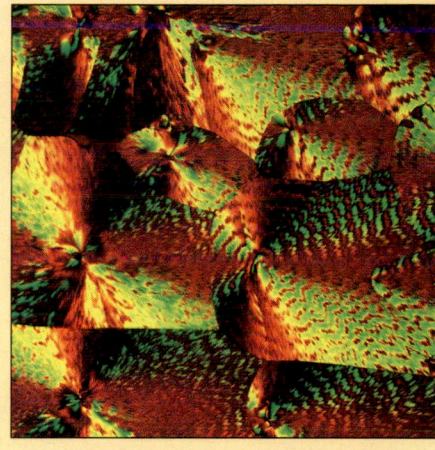 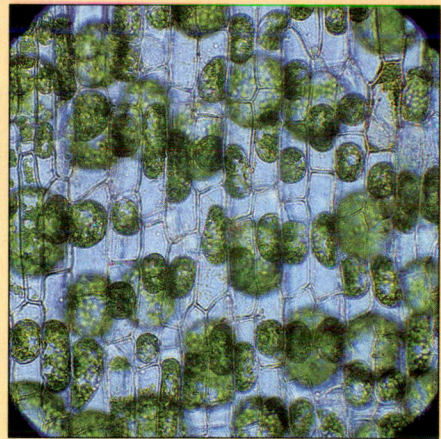

Atoms and molecules • Inorganic compounds are relatively simple • Cell junctions • Diffusion and osmosis • Cell shape and movement • Plant and animal cells • Organic compounds contain carbon

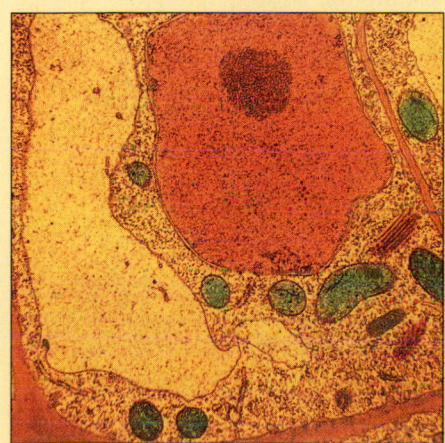  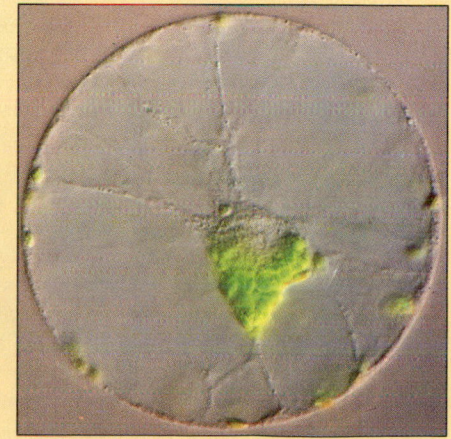

Cell is basic unit of life • Power houses of the cell • Energy traps and storage sacs • Every cell has an outer membrane • Manufacture of cellular products • The nucleus contains genetic information

3

Chemical Principles: Atoms, Molecules, and Reactions

OUTLINE

I. All materials are made up of basic substances called elements
   A. Atoms are the basic particles of elements
      1. Atoms are composed of subatomic particles
      2. Atomic electrons are regularly arranged
   B. Atoms combine in definite proportions
II. Chemical formulas describe the composition and structure of elements and compounds
III. Chemical equations describe reactions among elements and compounds
IV. Atoms combine by forming chemical bonds
   A. Atoms can share electrons to form covalent bonds
   B. Atoms can gain or lose electrons to form ionic bonds
   C. Hydrogen bonds are weak attractions involving partially charged hydrogen atoms and electronegative atoms
   D. Van der Waals and hydrophobic attractions are often important weak forces in large organic molecules
V. Oxidation involves the loss of electrons or hydrogen; reduction involves the gain of electrons or hydrogen
VI. Inorganic compounds are relatively simple compounds without carbon backbones
   A. Water has many unique properties essential to living things
      1. Water molecules are polar
      2. Water is an excellent solvent
      3. Water molecules are mutually attracted to themselves and some other substances by both cohesive and adhesive forces
      4. Water tends to maintain a stable temperature
   B. Acids produce hydrogen ions; bases dissociate in water to produce hydroxide ions
   C. Salts form from acids and bases and dissociate into anions and cations
   D. Buffers minimize pH change

Focus on Isotopes

LEARNING OBJECTIVES

After reading this chapter you should be able to:

1. Diagram the basic structure of the atom in accordance with the conventions presented in this chapter, showing the position of protons, neutrons, and electrons.
2. Identify the biologically significant elements as identified in this chapter by their chemical symbols and summarize the main functions of each in living organisms.
3. Interpret simple chemical formulas, structural formulas, and equations.
4. Define the term *electron orbital* and relate orbitals to energy levels; relate the number of valence electrons to the chemical properties of the element.
5. Distinguish between the types of chemical bonds that join atoms to form ionic and covalent compounds, and give the characteristics of each type.
6. Contrast oxidation and reduction and explain how these processes are linked.
7. Distinguish between inorganic and organic compounds and identify biologically important inorganic compounds.
8. Discuss the properties of water molecules and their importance to living things.
9. Compare *acids* and *bases*, use the pH scale in describing the hydrogen ion concentration in living systems, and describe how buffers help minimize changes in pH.

Computer model of superhelix of DNA.
(University of California Lawrence Livermore Laboratory/BPS)

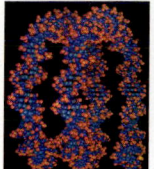

Despite their great diversity, the chemical composition and metabolic processes of all living things are remarkably similar. At one time it was believed that there was a unique substance in their chemical composition that distinguished living organisms from inanimate matter. The search for this substance proved fruitless. Instead, biologists came to understand that it is the specific organization and precise interaction of its chemical components that characterize life. Living things are chemical machines—not just machines in which chemical reactions occur, but machines that are themselves immensely complicated chemical systems. To understand life processes, then, one *must* know the basic principles of chemistry. Only then can one expect to be able to understand biology as it exists in these last years of the 20th century. This and the next chapter are provided because not all users of this book have taken chemistry courses, and those who have may need to update or review their knowledge so that it may be applied throughout the book.

KEY CONCEPTS

☐ All material substances are made up of elements which cannot themselves be broken down by any chemical means. Elements and indeed all materials are composed of tiny particles called atoms.

☐ Electrons are regularly arranged about the nucleus of the atom in ways that may be thought of as potential locations or as energy levels.

☐ The outermost electrons of an atom determine its ability to combine with other atoms in chemical compounds. The ability of an atom to combine with other atoms or groups of atoms depends on its valence electrons.

☐ Atoms combine with one another in a variety of ways, some involving the sharing of electrons, some involving differences in electrical charge, but all involving electrons.

☐ Water, a unique compound, has many physical and chemical properties related to the polar nature of its molecules and their tendency to hydrogen-bond with one another.

☐ Acids, alkalis, and salts are common ionic compounds. Acids yield hydrogen ions when dissolved in water; bases combine with hydrogen ions.

☐ ALL MATERIALS ARE MADE UP OF BASIC SUBSTANCES CALLED ELEMENTS

All matter, living and nonliving alike, is composed of chemical elements, substances that cannot be broken down into simpler substances by chemical reactions. The matter of the universe is composed of 92 naturally occurring elements, ranging from hydrogen, the lightest, to uranium, the heaviest. In addition to the naturally occurring elements, about 17 elements heavier than uranium have been made by bombarding elements with subatomic particles in devices known as particle accelerators.

For convenience we will consider weight and mass as equal, although this is not always true. Mass does not depend upon the force of gravity, but weight does. Thus a person on the moon has the same mass as a person on earth, but, because of the lower gravity, his or her body weight is less.

About 98% of an organism's mass is composed of only six elements—oxygen, carbon, hydrogen, nitrogen, calcium, and phosphorus. Approximately 14 other elements are consistently present in living things, but in smaller quantities. Some of these, such as iodine and copper, are known as **trace elements** because they are present in such minute amounts. Table 3–1 lists the elements that make up a living organism and explains why each is important.

Instead of writing out the name of each element, chemists use a system of abbreviations called chemical symbols—usually the first one or two letters of the English or Latin name of the element. For example, O is the symbol for oxygen, C for carbon, Cl for chlorine, N for nitrogen, and Na for sodium (its Latin name is *natrium*). Chemical symbols for the elements found in living organisms are given in Table 3–1.

Atoms Are The Basic Particles Of Elements

Imagine a bit of gold being divided into smaller and smaller pieces. The smallest possible particle of gold that could be obtained would be an atom of gold. The **atom** is the smallest subdivision of an element that retains the characteristic chemical properties of that element. The subdivision of any kind of matter ultimately yields atoms. This is true no matter what physical state matter may assume—solid, liquid, or gas. Atoms are almost unimaginably small, much smaller than the tiniest particle visible under a light microscope. By special scanning electron microscopy, with magnification as many as 5 million times, researchers have been able to photograph some of the larger atoms, such as those of uranium.

TABLE 3–1
Elements that Make Up the Human Body

Name	Chemical Symbol	Approximate Composition by Mass (%)	Importance or Function
Oxygen	O	65	Required for cellular respiration; present in most organic compounds; component of water
Carbon	C	18	Backbone of organic molecules; can form four bonds with other atoms
Hydrogen	H	10	Present in most organic compounds; component of water
Nitrogen	N	3	Component of all proteins and nucleic acids
Calcium	Ca	1.5	Structural component of bones and teeth; important in muscle contraction, conduction of nerve impulses, and blood clotting
Phosphorus	P	1	Component of nucleic acids; structural component of bone; important in energy transfer
Potassium	K	0.4	Principal positive ion (cation) within cells; important in nerve function; affects muscle contraction
Sulfur	S	0.3	Component of most proteins
Sodium	Na	0.2	Principal positive ion in interstitial (tissue) fluid; important in fluid balance; essential for conduction of nerve impulses
Magnesium	Mg	0.1	Needed in blood and other body tissues
Chlorine	Cl	0.1	Principal negative ion (anion) of interstitial fluid; important in fluid balance; component of sodium chloride
Iron	Fe	Trace amount	Component of hemoglobin and myoglobin; component of certain enzymes
Iodine	I	Trace amount	Component of thyroid hormones

Other elements found in very small amounts in the body include manganese (Mn), copper (Cu), zinc (Zn), cobalt (Co), fluorine (F), molybdenum (Mo), selenium (Se), and a few others. They are called trace elements.

Atoms Are Composed Of Subatomic Particles

An atom is composed of smaller components called subatomic particles. For our purposes we need consider only three types: protons, neutrons, and electrons. **Protons** have a positive electric charge; **neutrons** are uncharged particles with about the same mass as protons. Protons and neutrons make up almost all the mass of an atom and are concentrated in the atomic nucleus. **Electrons** have a negative electrical charge and an extremely small mass (only about 1/1800 of the mass of a proton). The electrons, as we will see, spin about in the space surrounding the atomic nucleus (Fig. 3–1).

Each kind of element has a fixed number of pro-

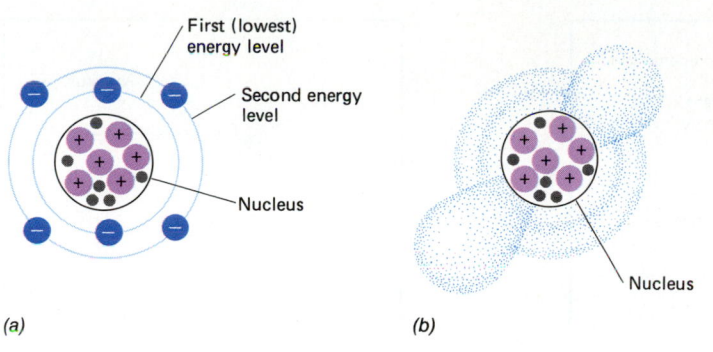

First (lowest) energy level

Second energy level

Nucleus

Nucleus

(a)

(b)

Electron

+ Proton

• Neutron

FIGURE 3–1 Two ways of representing an atom. (*a*) Bohr model of an atom. Although the Bohr model is not an accurate way to depict electron configuration, it is commonly used because of its simplicity and convenience. (*b*) Electron clouds surrounding an atom. Dots represent the probability of the electron's being in that particular location at any given moment.

tons in the atomic nucleus. This number, called the **atomic number,** is written as a subscript to the left of the chemical symbol. Thus $_1H$ and $_8O$ indicate that the hydrogen nucleus contains one proton and the oxygen nucleus has eight protons. It is the atomic number, the number of protons in the nucleus, that determines the chemical identity of the atom. The total number of protons plus neutrons in the nucleus is termed the **mass number** and is indicated by a superscript to the left of the chemical symbol. The common form of oxygen atom, with eight protons and eight neutrons in its nucleus, has an atomic number of 8 and a mass number of 16. It is indicated by the symbol $^{16}_8O$ (see Focus on Isotopes).

Atomic Electrons Are Regularly Arranged

When an atom is uncombined, it contains the same number of electrons as protons. Some kinds of chemical combinations and certain other circumstances change the number of electrons, but chemical reactions do not affect anything in the atomic nucleus. (Although nuclear reactions can do so, we may ignore this reservation for the time being, since living things are not nuclear-powered!). Because electrons and protons have equal though opposite charges, an uncombined atom is electrically neutral as a whole.

The way electrons are arranged around an atomic nucleus is referred to as the atom's **electron configuration.** Knowing the locations of electrons enables chemists to predict how atoms can combine to form different types of chemical compounds.

Because electrons are negatively charged, they are attracted to the positively charged protons in the atomic nucleus (opposite charges tend to attract). At the same time, electrons repel one another (like charges tend to repel one another). These considerations help determine the locations of electrons.

An atom may have several **energy levels,** or **electron shells,** where electrons are located. The lowest energy level is the one closest to the nucleus. Only two electrons can occupy this energy level. The second energy level can accommodate a maximum of eight elec-

trons. Although the third and outer shells can each contain more than eight electrons, they are most stable when only eight are present. We may consider the first shell complete when it contains two electrons, and every other shell complete when it contains eight electrons. The atomic structures of some elements that are important in biological systems—carbon, hydrogen, oxygen, nitrogen, sodium, and chlorine—are shown in Figure 3–2.

Although the simple diagrams, called Bohr models, of electron configuration shown here are helpful in understanding atomic structure, they are highly oversimplified. The energy levels correspond roughly to physical locations, called **orbitals.** There may be several orbitals within a given energy level. Electron orbitals may be represented by spherical, dumbbell-shaped, or more complex three-dimensional coordinates (Fig. 3–3).

Electrons are thought to whirl around the nucleus in an unpredictable manner, now close to it, now farther way. Orbitals represent the places where electrons are most probably found. In fact, one way of illustrating an atom is to represent each occupied orbital as an electron cloud. It is not that each orbital is crowded with a huge number of electrons. The intent of such a diagram is to give an impression, by means of the density of the shaded areas, of the likelihood that an electron is present there at a given time.

The distance of the electrons from the atomic nucleus depends upon their respective energy levels. The specific orbital that an electron occupies is determined by the total energy of the system and the number of electrons available to the atom. An electron can be moved to an orbital farther from the nucleus by providing it with more energy, or an electron can give up energy and sink back to a lower energy level in an orbital nearer the central nucleus. Energy is required to move a negatively charged electron further away from the positively charged nucleus.

When energy is added to the system, an electron can jump from one level to the next, *but it cannot stop in the space in between.* To move an electron from one level to the next, the atom must absorb a discrete packet of

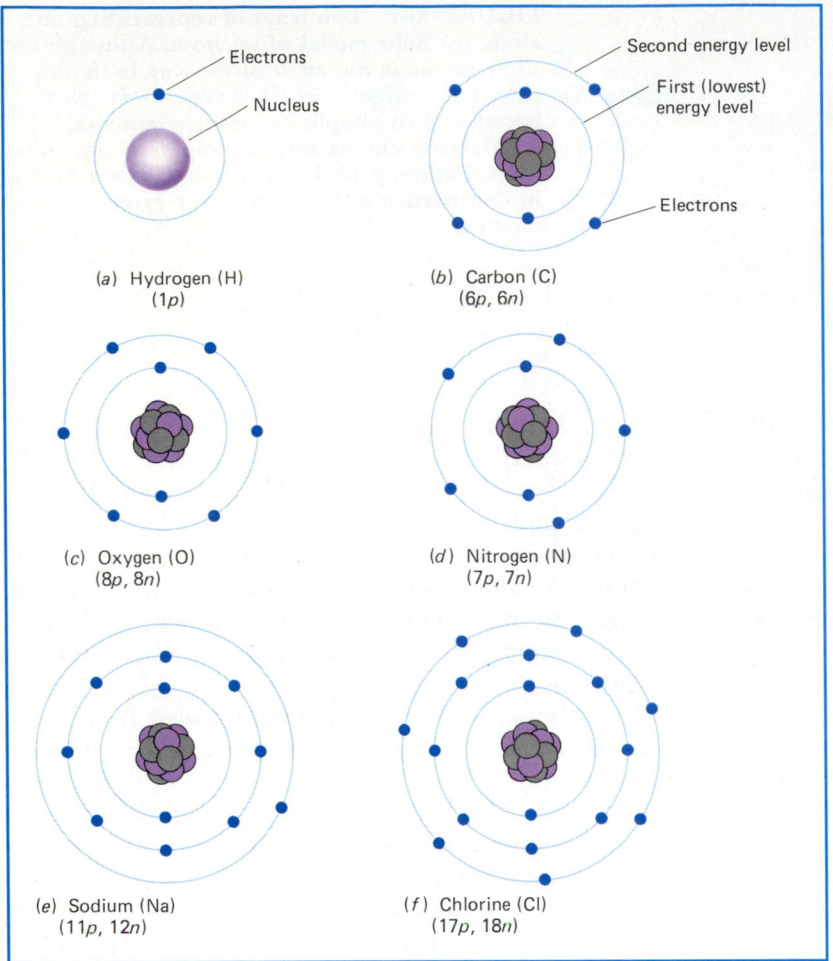

(a) Hydrogen (H)
(1p)

(b) Carbon (C)
(6p, 6n)

(c) Oxygen (O)
(8p, 8n)

(d) Nitrogen (N)
(7p, 7n)

(e) Sodium (Na)
(11p, 12n)

(f) Chlorine (Cl)
(17p, 18n)

**FIGURE 3–2   Bohr models of some biologically important atoms.
(a) Hydrogen. (b) Carbon. (c) Oxygen. (d) Nitrogen. (e) Sodium.
(f) Chlorine. Each circle represents an energy level, or electron shell.
Electrons are represented by dots on the circles; p, proton; n, neutron.**

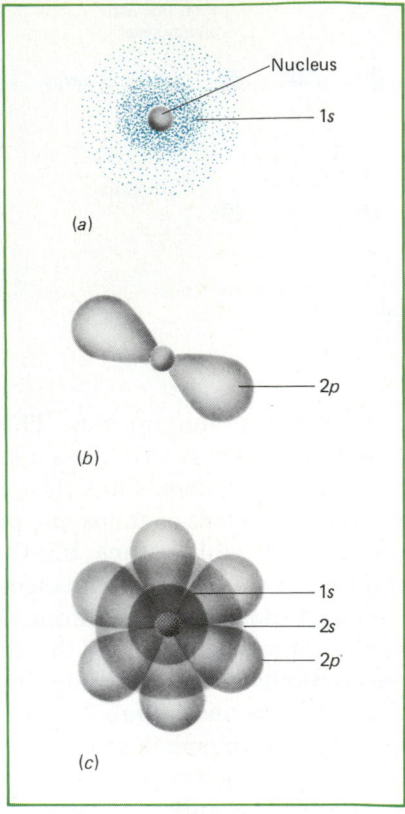

**FIGURE 3–3   Electron clouds or
orbitals. The first, or lowest,
energy level (a) can contain at
most two electrons. (b) An atom
with two energy levels. Electron
clouds or orbitals represent
potential electron locations.
(c) The more complex electron
structure of a carbon atom.**

energy known as a **quantum,** which contains just the
right amount of energy needed for the transition—no
more and no less. The term *quantum jump* is used in
everyday language to indicate a sudden discontinuous
move from one level to another.

Atoms Combine In Definite Proportions

Two or more atoms may combine chemically to form a
molecule. When two atoms of oxygen combine, a molecule of oxygen is formed. That is, in fact, the usual
state of gaseous oxygen and of many other gaseous
elements. Such double molecules consisting of the
same element are called diatomic molecules. *Different*
kinds of atoms can combine to form chemical compounds. A **chemical compound** is a substance that consists of two or more different elements combined in a

fixed ratio. Water is a chemical compound consisting
of two atoms of hydrogen combined with one atom of
oxygen.

■ CHEMICAL FORMULAS DESCRIBE THE COMPOSITION AND STRUCTURE OF ELEMENTS AND COMPOUNDS

A **chemical formula** is a shorthand method for describing the chemical composition of a molecule.
Chemical symbols are used to indicate the types of
atoms in the molecule, and subscript numbers are used
to indicate the number of each type of atom present.
The chemical formula for diatomic molecular oxygen,
O_2, tells us that this molecule consists of two atoms of
oxygen. This formula distinguishes it from another

(*text continues on page 48*)

FOCUS ON   Isotopes

Atoms of the same element containing the same number of protons but different numbers of neutrons have different mass numbers and are called **isotopes.** The three isotopes of hydrogen, $_1^1H$, $_1^2H$, and $_1^3H$, contain zero, one, and two neutrons, respectively. Elements usually occur in nature as a mixture of isotopes.

All isotopes of a given element have essentially the same chemical characteristics. Some isotopes with excess neutrons are unstable and tend to break down, or decay, to a more stable isotope (usually of a different element). Such isotopes are termed **radioisotopes,** since they emit high-energy radiation when they decay.

Radioisotopes like ^3H (tritium)

and ^{14}C have been extremely valuable research tools in biology and are useful in medicine for both diagnosis and treatment. Despite the difference in the number of neutrons, the body treats all isotopes of a given element the same chemically. The reactions of a chemical—a fat, a hormone, a drug—can be followed in the body (see figure) by tagging the substance with a radioisotope, such as carbon-14 or tritium. This is done by replacing one of the nonradioactive atoms in a small sample of molecules of that substance with its radioactive equivalent. Radioactivity can be detected wherever the radioactive atoms, or molecules containing them, may go. For example, the active component in marijuana (tetrahydro-

cannabinol) has been tagged and administered intravenously. By measuring the amount of radioactivity in the blood and urine at successive intervals, experimenters determined that this compound remains in the blood and products of its metabolism remain in the urine for several weeks.

Because radiation from radioisotopes can interfere with cell division, such isotopes have been used in the treatment of cancer (a disease characterized by rapid cell division). Radioisotopes are also used to test thyroid gland function, measure the rate of red blood cell production, and study many other aspects of body chemistry.

(a)

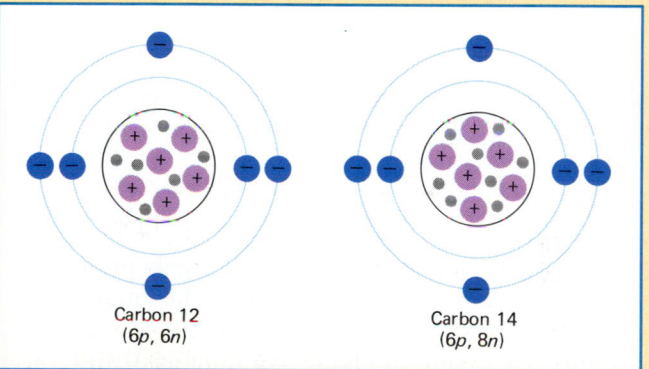

Carbon 12
(6p, 6n)

Carbon 14
(6p, 8n)

(b)

(a) Anthropologists use radioisotope content to date and study fossils. The skeleton of this 11th-century inhabitant of a South African Iron Age village posed an anthropological puzzle. Physically, the man's skeleton was different from those of the other villagers, suggesting that he was not a native of the area. When the skeleton was analyzed for isotopes, its carbon-12 to carbon-13 (b) ratio was found to be similar to that of other skeletons from the same village. Since different kinds of plants incorporate different proportions of isotopes into the food produced from them, this similarity of isotope content indicates that these individuals all ate the same foods. Thus, anthropologists concluded that this man had probably spent most of his life in the village after migrating there from a distant region. (Nikolaas J. van der Merwe, *American Scientist* 70:596–606, 1982)

form of oxygen, ozone, which has three atoms and is written O_3. Hydrogen gas is normally H_2, chlorine gas is Cl_2, and so on. The chemical formula for water, H_2O, indicates that each molecule consists of two atoms of hydrogen and one atom of oxygen. (Note that when a single atom of one type is present, it is not necessary to write 1; one does *not* write H_2O_1.)

Another type of formula is the **structural formula,** which shows not only the types and numbers of atoms in a molecule but also their arrangement. In any type of chemical compound the atoms are always arranged in the same way. From the chemical formula for water, H_2O, you could only guess whether the atoms were arranged H—H—O or H—O—H. The structural formula, H—O—H, settles the matter, indicating that the two hydrogen atoms are attached to the oxygen atom (though it does not give any information on the angles of their attachment or the exact geometry of the water molecule). In water this arrangement is the only one chemically possible. However, there are other substances that consist of the same atoms yet have different chemical properties due to alternative arrangements. Such compounds are known as **structural isomers.** The sugars glucose and fructose are examples (see Chapter 4 for their structural formulas).

■ ELEMENTS AND COMPOUNDS INTERACT IN REGULAR WAYS THAT MAY BE EXPRESSED AS CHEMICAL EQUATIONS

During any moment in the life of an organism, be it an earthworm or a pine tree, many complex, highly organized chemical reactions are taking place. The chemical reactions that occur between atoms and molecules —for instance, between methane (swamp gas) and oxygen—can be described on paper by means of **chemical equations:**

$$CH_4 \quad + \quad 2O_2 \quad \longrightarrow \quad CO_2 \quad + 2H_2O$$
Methane + Oxygen $\longrightarrow$ Carbon dioxide + Water

Methane, produced during some decay processes, is broken down in this reaction.

In a chemical equation the **reactants** (the substances that participate in the reaction) are written on the left side of the equation, and the **products** (the substances formed by the reaction) are written on the right side. The arrow means **yields** and indicates the direction in which the reaction tends to proceed. The number preceding a chemical symbol or formula indicates the number of atoms or molecules reacting. Thus $2O_2$ means two molecules of oxygen and $2H_2O$ means two molecules of water. The absence of a number indicates that only one atom or molecule is present. Thus the equation can be translated into ordinary language as,

"One molecule of methane reacts with two molecules of oxygen to yield one molecule of carbon dioxide and two molecules of water."

In some cases the reaction will proceed in the reverse direction as well as forward; at **equilibrium** a certain amount of the product continuously breaks up to form the reactants, and the rate of the forward reaction equals the rate of the reverse action. Reversible reactions are indicated by double arrows. The reaction:

$$N_2 \quad + \quad 3H_2 \quad \longrightarrow \quad 2NH_3$$
Nitrogen Hydrogen Ammonia

is not readily reversible. But the reaction between ammonia gas and water forms a solution of ammonium hydroxide. If we heat the ammonium hydroxide solution, or just leave the bottle that contains it open, ammonia and water form again readily:

$$NH_3 \quad + H_2O \quad \Longleftrightarrow \quad NH_4OH$$
Ammonia Water Ammonium hydroxide

■ ATOMS COMBINE BY FORMING CHEMICAL BONDS

The chemical properties of an element are determined primarily by the number and arrangement of electrons in the *outermost* energy level (electron shell). In a few elements, called the **noble gases,** the outermost shell is filled. These elements are chemically inert, meaning that they will not readily combine with other elements. Two such elements are helium with two electrons (a complete inner shell) and neon with ten electrons (a complete inner shell of two and a complete second shell of eight). The electrons in the outermost energy level of an atom are referred to as **valence electrons.** The valence electrons are chiefly responsible for the chemical activity of an atom. When the outer shell of an atom contains fewer than eight electrons, the atom tends to lose, gain, or share electrons to achieve an outer shell of eight (the exceptions are zero or two in the case of the lightest elements, hydrogen and helium).

The elements in a given compound are always present in a certain proportion by mass. This reflects the fact that atoms are attached to each other by chemical bonds in a precise way to form the compound. A **chemical bond** is the attractive force that holds two atoms together. Each bond represents a certain amount of potential chemical energy. The atoms of each element form a specific number of bonds with the atoms of other elements—a number dictated by the number of valence electrons. The two principal types of chemical bonds are covalent bonds and ionic bonds.

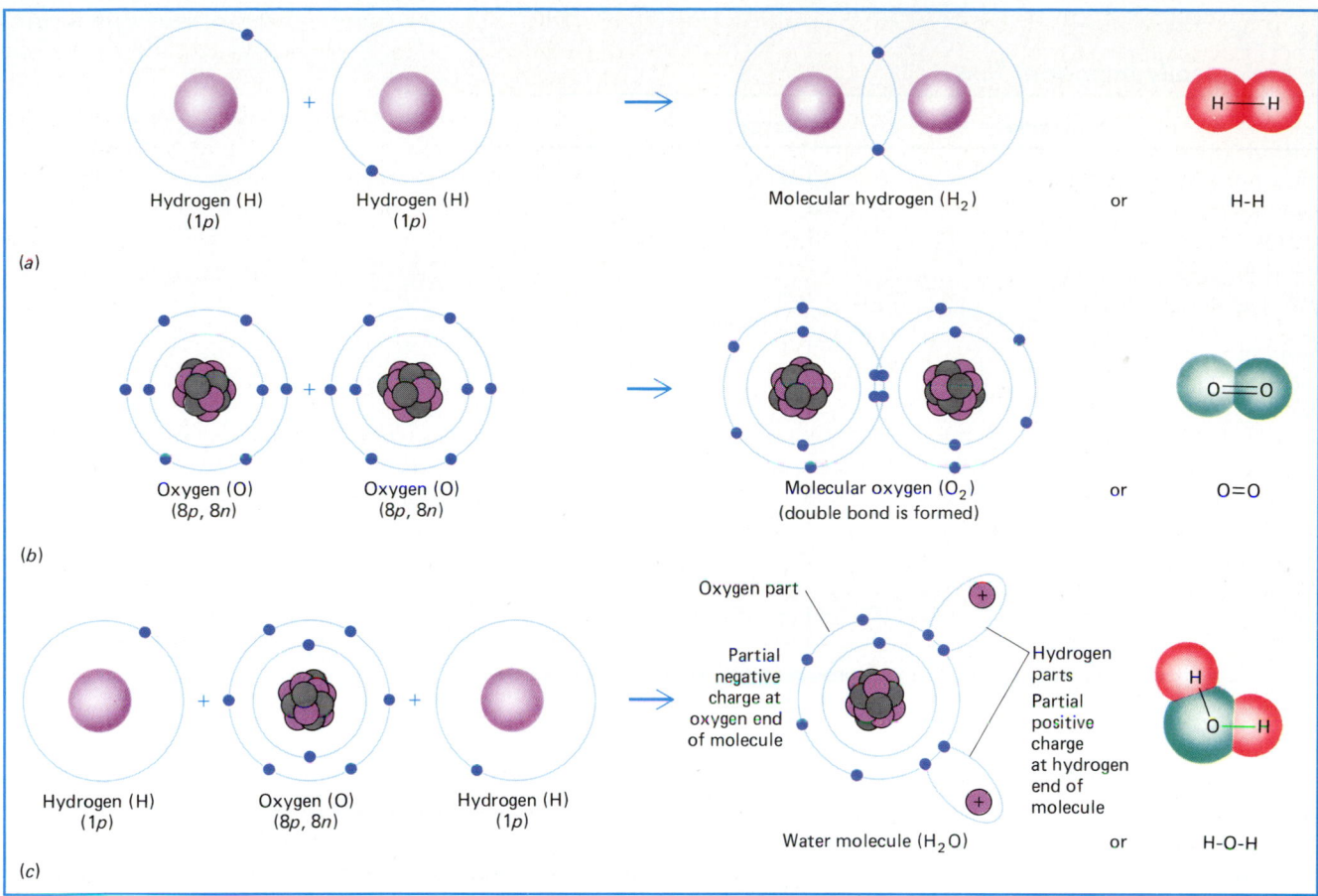

FIGURE 3–4 Formation of covalent compounds. (*a*) Two hydrogen atoms achieve stability by sharing electrons, thereby forming a molecule of hydrogen. The structural formula shown on the right is a simpler way of representing molecular hydrogen. The straight line between the hydrogen atoms represents a single covalent bond. (*b*) Two oxygen atoms share two pairs of electrons to form molecular oxygen. Note the double bond. (*c*) When two hydrogen atoms share electrons with an oxygen atom, the result is a molecule of water. Note that the electrons tend to stay closer to the nucleus of the oxygen atom than to the hydrogen nuclei. This results in a partial negative charge on the oxygen portion of the molecule, and a partial positive charge at the hydrogen end. Although the water molecule as a whole is electrically neutral, it is a polar covalent compound.

Atoms Can Share Electrons To Form Covalent Bonds

Covalent bonds involve the sharing of electrons between atoms. The more precise definition of a molecule is a combination of two or more atoms joined by *covalent* chemical bonds. A simple example of a covalent bond is the one joining two hydrogen atoms in a molecule of hydrogen gas, H_2 (Fig. 3–4). Each atom of hydrogen has one electron, but two electrons are required to complete the first energy level. Each hydrogen atom has the same capacity to attract electrons, so one does not donate an electron to the other. Instead, the two hydrogen atoms share their single electrons so that each of the two electrons is attracted simultaneously to the two protons in the two hydrogen nuclei.

The two electrons are thus under the influence of *both* atomic nuclei, and they join the two atoms together.

The carbon atom has four electrons in its outer energy level. These four electrons are available for covalent bonding. When one carbon and four hydrogen atoms share electrons, a molecule of methane (CH_4) is formed. Each line represents a single pair of shared electrons.

$$H-\overset{\displaystyle H}{\underset{\displaystyle H}{C}}-H$$

Each atom shares its outer-level electrons with the others, thereby completing the first energy level of

TABLE 3–2
Some Biologically Important Ions

Name	Formula	Charge
Sodium	Na^+	1 +
Potassium	K^+	1 +
Hydrogen	H^+	1 +
Magnesium	Mg^{2+}	2 +
Calcium	Ca^{2+}	2 +
Iron	Fe^{2+} or Fe^{3+}	2 + [iron(II)] or 3+ [iron(III)]
Ammonium	NH_4^+	1 +
Chloride	Cl^-	1 −
Iodide	I^-	1 −
Carbonate	CO_3^{2-}	2 −
Bicarbonate	HCO_3^-	1 −
Phosphate	PO_4^{3-}	3 −
Acetate	CH_3COO^-	1 −
Sulfate	SO_4^{2-}	2 −
Hydroxide	OH^-	1 −
Nitrate	NO_3^-	1 −
Nitrite	NO_2^-	1 −

each hydrogen atom and the second energy level of the carbon atom.

The nitrogen atom has five electrons in its outer shell. When a nitrogen atom shares electrons with three hydrogen atoms, a molecule of ammonia, NH_3, is formed:

$$H-N-H$$
$$|$$
$$H$$

When one electron pair is shared between two atoms, the covalent bond is referred to as a **single bond.** Two shared pairs of electrons comprise a **double bond,** shown in diagrams as a pair of lines resembling an equal sign. Double bonds are especially common in compounds of carbon when carbon atoms are linked to one another.

$$\begin{array}{cc} H & H \\ | & | \\ H-C & =C-H \end{array}$$

Two oxygen atoms may achieve stability by forming covalent bonds with one another. Each oxygen atom has six electrons in its outer shell. To become stable, the two atoms share two pairs of electrons, forming a molecule of oxygen gas. Note that the covalent bond formed is a double bond. Some atoms even form triple bonds with one another, sharing three pairs of electrons.

Electronegativity is a measure of an atom's attraction for the electrons that are shared in covalent bonds. A covalent bond between atoms of different electronegativities is an unequal sharing that results in an unsymmetrical molecular charge or polarity. Such a bond is known as a **polar covalent bond.** In polar covalent bonds, electrons are much closer to one atom than to the other. The presence or absence of polarity has important consequences for the structure and properties of biological membranes. Some covalent bonds have no polarity. In these nonpolar covalent bonds, electrons are shared equally, as in the hydrogen molecule.

Atoms Can Gain Or Lose Electrons To Form Ionic Bonds

In the ionic bond, electrons are pulled completely or almost completely from one atom to the other. The number of protons in the nucleus remains unchanged, so the loss or gain of electrons produces an atom with a net positive or negative charge. Such electrically charged atoms are termed **ions.** Atoms with one, two, or three electrons in the outer shell generally donate electrons to other atoms. These atoms, then, become positively charged because of the excess of protons in the nucleus. For example, sodium has one valence electron. It tends to donate this electron, becoming a sodium ion (Na^+) with a one plus charge. Calcium tends to lose its two valence electrons to become a calcium ion (Ca^{2+}) with a two plus charge. Positively charged ions are termed **cations** (see Table 3–2).

Atoms with five, six, or seven electrons in their outer shell tend to *gain* electrons from other atoms and become negatively charged **anions** (e.g., Cl^-, chloride ion). Charged particles, both anions and cations, play many important roles in biological systems, such as the transmission of nerve impulses and the contraction of

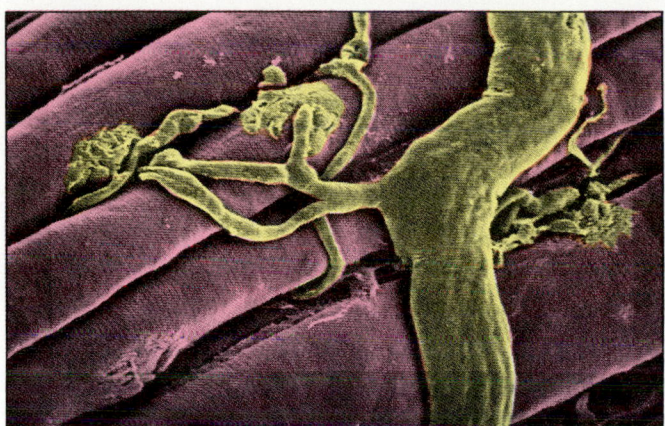

FIGURE 3–5 Sodium, potassium, and chlorine ions are among the ions essential in the conduction of a nerve impulse. This scanning electron micrograph shows a nerve fiber communicating with several muscle cells (approximately ×900). The nerve fiber transmits impulses to the muscle cells, stimulating them to contract. The muscle cells are rich in calcium ions, which are essential for muscle contraction. (From Desaki, J., "Vascular autonomic plexuses and skeletal neuromuscular junctions: A scanning electron microscopic study," *Biomedical Research Supplement,* 139–143, 1981)

muscles (Fig. 3–5). Incidentally, a good way to remember which kind of ion is which is to think of anion as a kind of abbreviation for "*a negative ion.*"

An **ionic bond** is the force of electrical attraction between two positively charged ions. When held together by ionic bonds, oppositely charged ions form an **ionic compound.** Sodium chloride is a good example of an ionic compound. A sodium atom, with atomic number 11, has two electrons in its inner shell, eight in the second, and one in the third. A sodium atom cannot fill its third shell by obtaining seven electrons from other atoms because it would then have a vast excess of negative charge. Instead, it gives up the single electron in its third shell to some electron acceptor, leaving the second shell as the complete outer shell (Fig. 3–6). A chlorine atom, atomic number 17, has two electrons in its inner shell, eight in the second, and seven in the third. The chlorine atom achieves a complete outer shell not by losing the seven electrons in its third shell (for it would then have a vast positive charge) but by accepting an electron from an electron donor, such as sodium, to complete its third shell.

When an electron donor, such as sodium, meets an electron acceptor, such as chlorine, the electron may be transferred completely from the donor to the acceptor. The sodium ion now has 11 protons in its nucleus and 10 electrons around the nucleus, giving it a net charge of 1^+. The chlorine ion has 17 protons in its nucleus and 18 electrons around the nucleus, so it has a net charge of 1^-. These ions attract each other as a result of their opposite charges. They are held together by this electrical attraction to form sodium chloride (NaCl), common table salt. This transfer of one or more electrons from one atom to another and the binding together of two ions of opposite charge (the ionic bond) results in the formation of an ionic compound. Ionic bonds occur between electron donors and electron acceptors.

Whether an ionic compound is in solid form or is dissolved in water, its constituent particles (ions) do not share electrons. Because of this, the term *molecule* does not adequately explain the properties of ionic compounds such as NaCl. Chemists simply refer to them as compounds.

FIGURE 3–6 Formation of an ionic compound. Sodium donates its single valence electron to chlorine, which has seven electrons in its outer energy level. With this additional electron, chlorine completes its outer energy level. The two atoms are now electrically charged ions. They are attracted to one another by their unlike electrical charges, forming the ionic compound sodium chloride. The force of attraction holding these ions together is called an ionic bond.

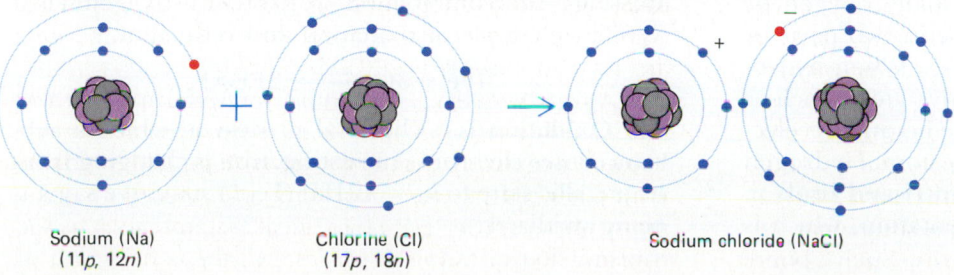

Sodium (Na)   Chlorine (Cl)   Sodium chloride (NaCl)
(11*p*, 12*n*)   (17*p*, 18*n*)

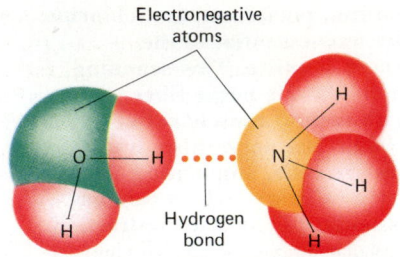

Electronegative atoms

Hydrogen bond

FIGURE 3–7   How hydrogen bonds form. In a hydrogen bond, a weakly charged hydrogen atom that is connected to a negatively charged atom by a polar covalent bond (as in the water molecule shown here) is attracted to the negatively charged side of another polar molecule, here ammonia. Hydrogen bonds are very important in large organic molecules that possess many hydrogen atoms.

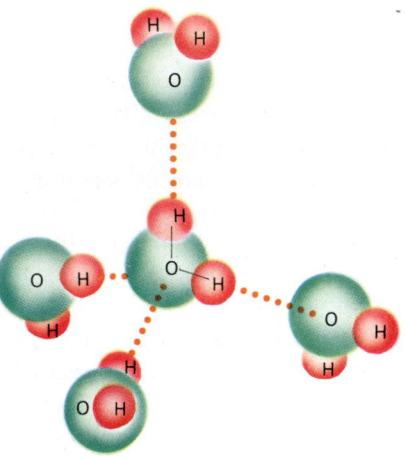

FIGURE 3–8   Hydrogen bonding of water molecules. Each water molecule tends to form hydrogen bonds with four neighboring water molecules. The hydrogen bonds are indicated by dotted lines. The covalent bonds between the hydrogen and oxygen atoms are represented by solid lines.

Hydrogen Bonds Are Weak Attractions Involving Partially Charged Hydrogen Atoms

Another type of bond that is important in biological systems is the hydrogen bond (Fig. 3–7). This type of bond is formed when a hydrogen atom is attracted to two other atoms, one of which is usually oxygen or nitrogen.

Hydrogen bonds tend to form between a hydrogen atom covalently bonded to oxygen or nitrogen and some other electronegative atom, usually oxygen or nitrogen. The atoms involved may be in two parts of the same molecule or in two molecules. When hydrogen is combined with an electronegative atom, such as oxygen, it has a partial positive charge because its electron is positioned closer to the oxygen atom. Hydrogen bonds are weak and are readily formed and broken. They have a specific length and orientation, which is

very important in their role in helping to determine the three-dimensional structure of large molecules, such as nucleic acids and proteins. Though relatively weak individually, the large numbers of hydrogen bonds that occur in some of the very big molecules such as DNA and protein (Chapter 4) are together very effective in stabilizing and giving shape to these molecules. Hydrogen bonds are also very important in determining the properties of water (Fig. 3–8).

van der Waals And Hydrophobic Attractions Are Often Important Weak Forces In Large Organic Molecules

The so-called **van der Waals force** is another weak but important form of attraction especially noteworthy in large organic molecules such as proteins. The van der Waals force results from the fact that since the position of any one electron in an atom is unpredictable, the distribution of all the electrons is a matter of chance. Chance will place all or most of the electrons on one side of an atom on occasion, so that the atom itself exhibits a temporary polarity. If another nearby atom just happens to possess the opposite polarity at the same time, the two will be attracted to another very briefly but repeatedly. This attraction is so weak and uncertain that it is effective only at distances short on even an atomic scale. Weak though they are, van der Waals forces do help to determine the shape of immense protein and DNA molecules.

Hydrophobic interaction is a relationship among molecules that will be more fully explained in Chapters 4 and 6. For now it suffices to say that some substances or molecular parts are repelled by water, which is what "hydrophobic" really means. Under some circumstances this repulsion by water can drive hydrophobic molecules together. Hydrophobic interaction can determine the location and orientation of molecules in a cell, and, along with hydrogen bonding and van der Waals forces, also helps to determine the shapes of protein and nucleic acid molecules.

☐ OXIDATION INVOLVES THE LOSS OF ELECTRONS OR HYDROGEN; REDUCTION INVOLVES THE GAIN OF ELECTRONS OR HYDROGEN

Rusting—the combination of iron with oxygen—is a familiar example of oxidation and reduction:

$$4Fe + 3O_2 \longrightarrow 2FE_2O_3$$

Oxidation is a chemical process in which a substance loses electrons. In rusting, iron is changing from its metallic state to its iron(III) (Fe_3^+) state. We say it is being oxidized:

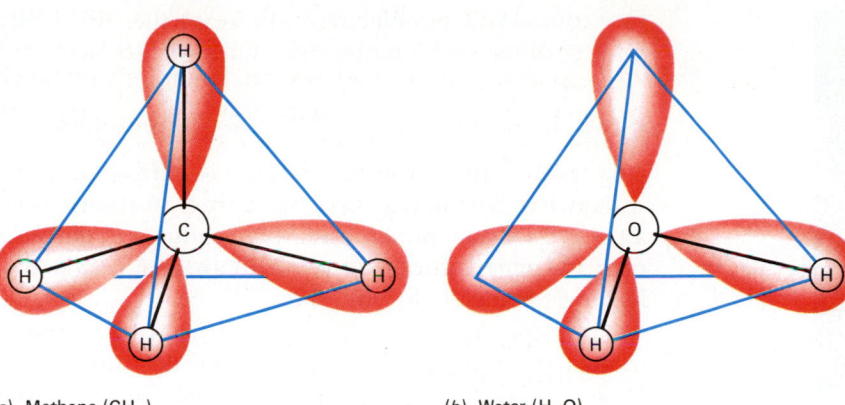

(a) Methane (CH_4)   (b) Water (H_2O)

FIGURE 3–9 Organic and inorganic compounds. (*a*) Methane and (*b*) water. Even a conventional structural formula does not completely express the layout of such simple compounds. These three-dimensional diagrams reveal these molecules as having shapes that must be packed into space.

$$4Fe \longrightarrow 4FE^{3+} + 12e^-$$

The e^- stands for electron; the + sign stands for an electron *deficit*. (When there is loss of an electron, the atom acquires a positive charge from the excess of one proton. The loss of two electrons produces an atom with a double positive charge, and so on.) At the same time, oxygen is changing from its molecular state to its charged state:

$$3O_2 + 12e^- \longrightarrow 6O^{2-}$$

Oxygen accepts the electrons removed from the iron and is said to be reduced. **Reduction** is a chemical process in which a substance gains electrons. An oxidation cannot take place without a reduction because electrons have to go somewhere; free electrons are not usually found in nature. Some substance must accept the electrons that are lost. On the other hand, reduction cannot occur without a corresponding oxidation. Oxidation-reduction reactions are sometimes called **redox reactions.**

Electrons are not easy to remove from covalent compounds unless an entire atom is removed. In living cells oxidation almost always involves the removal of a hydrogen atom from a compound. Reduction often involves a gain in hydrogen atoms. As will be discussed in later chapters, redox reactions are an essential part of photosynthesis, cellular respiration, and other aspects of metabolism.

■ INORGANIC COMPOUNDS ARE RELATIVELY SIMPLE COMPOUNDS WITHOUT CARBON BACKBONES

Chemical compounds can be divided into two broad groups—inorganic and organic (Fig. 3–9). Inorganic compounds are relatively small, simple substances. Organic compounds are usually large and complex and always contain carbon; usually many carbon atoms

are bonded to one another within these compounds to form a kind of molecular backbone. Organic compounds are the focus of Chapter 4. Among the biologically important groups of inorganic compounds are water, simple acids and bases, and salts. Organisms depend upon appropriate amounts of these substances for fluid balance, acid-base balance, and many cell activities.

Water Has Many Unique Properties Essential To Living Things

Water is an essential ingredient of life. It accounts for about 80% of the weight of an average active cell. In fact, the human body is about 70% water by weight. Many organisms make their homes in lakes, rivers, or the sea, and the cells of terrestrial organisms are bathed in body fluids composed largely of water.

Water Molecules Are Polar

The physical and chemical properties of water permit life to exist on our planet. Water molecules are polar; that is, they bear a partial positive and a partial negative charge. The water molecules in liquid water and in ice are held together in part by hydrogen bonds. The hydrogen atom of one water molecule, with its partial positive charge, is attracted to the oxygen atom of a neighboring water molecule, with its partial negative charge, forming a hydrogen bond. Each water molecule can form hydrogen bonds with a maximum of four neighboring water molecules.

Water Is An Excellent Solvent

Because its molecules are polar, water is an excellent solvent, a liquid capable of dissolving many substances, particularly polar substances. For example, the ions of a salt, such as sodium chloride, are held together by strong ionic bonds; in fact, they form a three-dimen-

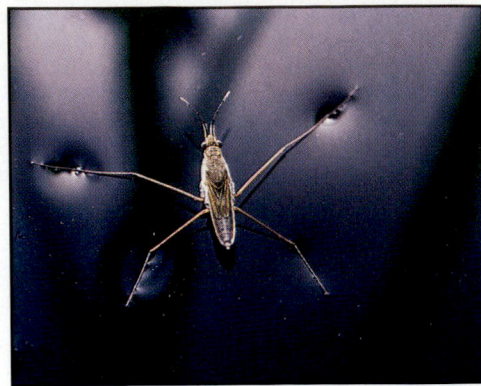

FIGURE 3–10 **A water strider on the surface of a pond. Fine hairs at the ends of its legs spread its weight over a large area, allowing the body of the insect to be supported by the surface tension of the water. Water striders also use the surface tension of the water for communication, setting up patterns of ripples that other water striders can sense. If you look closely, you can see that the water strider has captured a prey.** (Animals Animals © 1988)

sional structure, a crystal lattice. Considerable energy is required to pull the positively and negatively charged ions apart. However, when the sodium chloride is placed in water, the strong electrical attractions between the polar water molecules and Na^+ and Cl^- ions pull ions out of their positions. This results in the formation of a solution of dissociated Na^+ and Cl^- ions kept apart by shells (hydration spheres) of surrounding polar water molecules. Because of its solvent properties and its tendency to cause the ionization of compounds in solution, water is important in facilitating chemical reactions.

Water Molecules Are Mutually Attracted To Themselves And Some Other Substances By Both Cohesive And Adhesive Forces

Water exhibits both cohesive and adhesive forces. Water molecules have a very strong tendency to stick to each other; that is, they are *cohesive*. This is due to the hydrogen bonds among the molecules. Water has a high degree of surface tension because of the cohesiveness of its molecules; its molecules have a much greater attraction for other water molecules than for molecules in the air. Thus water molecules at the surface crowd together, producing a strong layer as they are pulled downward by the attraction of other water molecules beneath them (Fig. 3–10).

Water molecules also stick to many other kinds of substances (that is, those substances that have charged groups of atoms or molecules on their surfaces). These *adhesive* forces explain how water makes things wet.

Adhesive and cohesive forces account for the tendency, termed **capillary action,** of water to rise in very fine tubes. Water molecules adhere to the tube walls, pulling other water molecules along with them.

Water Tends To Maintain A Stable Temperature

The temperature of water changes less drastically and usually less frequently than the temperature of almost all other substances. This ability to minimize temperature changes comes from the hydrogen bonds that hold water molecules together.

In any substance, the physical basis of temperature is the random motion of molecules and atoms. When a substance heats up, its particles move faster. Thus, for the temperature of a substance to be raised, heat energy must be added to make the molecules move faster, that is, to increase their kinetic energy (energy of motion). A measure known as **specific heat** is the amount of heat energy required to raise the temperature of one gram of a substance by 1°C. The hydrogen bonds holding water molecules together tend to restrict their motion, so that it takes more heat energy to raise the temperature of water than it would for a more ordinary substance that lacks these intermolecular hydrogen bonds. Thus, water has a very high specific heat, or put differently, it takes a lot of heat gain or loss to raise or lower the temperature of water.

Because so much heat input (or heat loss) is required to raise (or lower) the temperature of water, the oceans and other large bodies of water have relatively constant temperatures. Thus the aquatic environment provides the multitude of organisms that inhabit it with a relatively constant environmental temperature. The water within all organisms, even those that live on land, contributes to a relatively constant internal temperature. This is important because metabolic reactions can take place only within a relatively narrow temperature range.

Because its molecules are held together by hydrogen bonds, water also has a high heat of vaporization, another property that helps stabilize temperature. More than 500 calories are required to change 1 gram of liquid water into 1 gram of water vapor. A **calorie** is a unit of heat energy (defined as 4.184 joules). Because of the heat of vaporization, we can rid ourselves of excess heat by the evaporation of sweat, and plants can remain cool in the midday heat by evaporating water from their surfaces.

Although most substances become more and more dense as the temperature decreases, water reaches its maximum density at 4°C and then begins to expand again as the temperature decreases. Hydrogen bonds become more rigid and ordered, and ice floats upon the denser cold water (Fig. 3–11). This important property of water explains why lakes and ponds freeze from the surface down rather than from the bottom up. The sheet of ice that forms at the pond surface

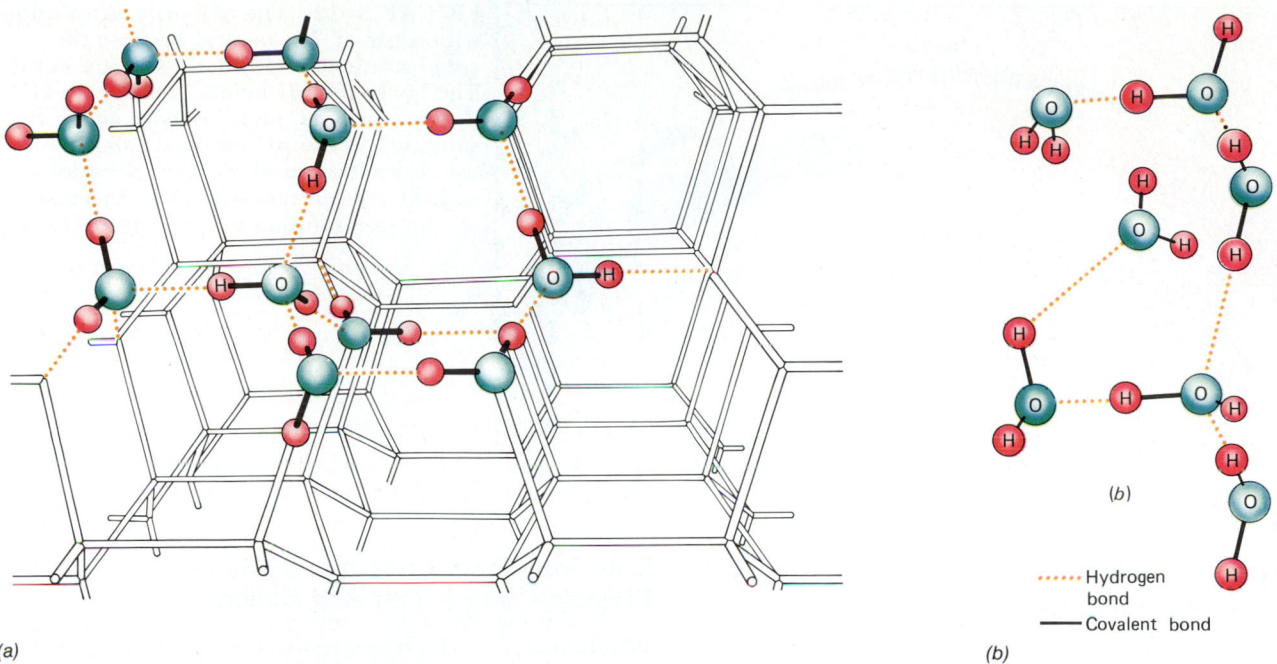

(a)

(b)

······ Hydrogen
       bond
——— Covalent bond

(b)

(c)

insulates the water below from the wintry chill so that it is less likely to freeze. Further, organisms that inhabit northern lakes and ponds are able to carry on their life activities despite the frigid winter and the surface ice sheet.

FIGURE 3–11 The hydrogen bonding in ice compared with that in liquid water. (*a*) Note the regular, evenly distanced hydrogen bonds in the superstructure of ice. (*b*) When ice melts, the hydrogen bonds occur less consistently and are of unequal length, and the crystal structure collapses. (*c*) Icebergs, Tracy Arm, Alaska. Because water expands as it freezes, ice is one of the very few substances that is lighter in its solid than its liquid form. Thus ice will float on water instead of **accumulating on the bottom.** (Frans Lanting)

Acids Produce Hydrogen Ions; Bases Dissociate In Water And Release Hydroxide Ions

An **acid** is a compound that ionizes in solution to yield hydrogen ions (H^+)[1] (that is, protons) and an anion. Acids turn blue litmus paper red and have a sour taste. Hydrochloric acid (HCl) and sulfuric acid (H_2SO_4) are examples of inorganic acids. The strength of an acid depends upon the degree to which it ionizes in water, releasing hydrogen ions. Thus HCl is a very strong acid because most of its molecules dissociate, producing hydrogen and chloride ions.

$$HCl \xrightarrow{\text{in } H_2O} H^+ + Cl^-$$

Hydrochloric   Hydrogen   Chloride
acid   ion   ion

Most **bases** are substances that yield a hydroxide ion (OH^-) and a cation when dissolved in water. Bases turn red litmus paper blue and feel slippery to the touch. Sodium hydroxide (NaOH) and aqueous am-

---

[1] The H^+ immediately combines with an electronegative region of a water molecule, forming a hydronium ion (H_3O^+).

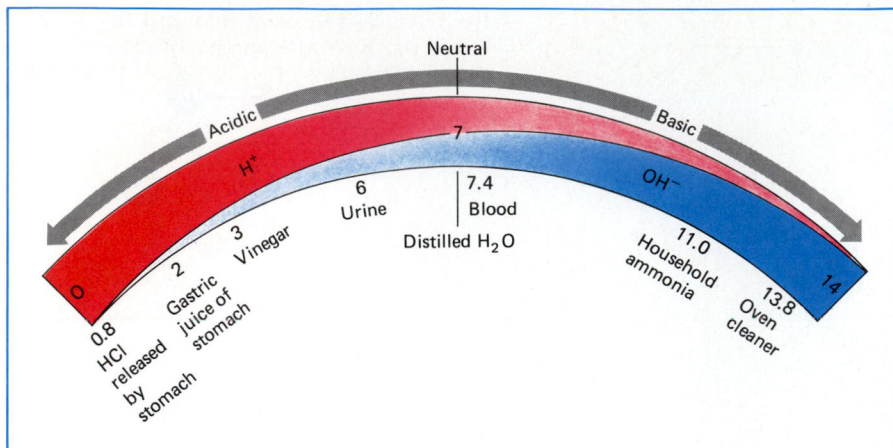

FIGURE 3–12 The pH scale. A solution with a pH of 7 is neutral because the concentrations of H^+ and OH^- are equal. The lower the pH below 7, the more H^+ ions are present, and the more acidic the solution. As the pH increases above 7, the concentration of H^+ ions decreases and the concentration of OH^- increases, making the solution more alkaline (basic).

monia (NH_4OH) are inorganic bases. Bases react with hydrogen ions and remove them from solution.

$$NaOH \longrightarrow Na^+ + OH^-$$

Sodium   Sodium   Hydroxide
hydroxide   ion   ion

Since the concentration of hydrogen or hydroxide ions is usually small, it is convenient to express the degree of acidity or alkalinity in a solution in terms of **pH,** formally defined as the logarithm of the reciprocal of the hydrogen ion concentration, $\log (1/[H^+])$.[1] The pH scale is logarithmic, extending from 0, the pH of a strong acid, such as HCl, to 14, the pH of a strong base, such as NaOH (Fig. 3–12). The pH of pure water is 7, neither acidic nor alkaline, but neutral. Even though water does ionize slightly, the concentrations of H^+ ions and OH^- ions are exactly equal, and each of them has a concentration of 10^{-7}, which is the reason we can say that water has a pH of 7. Solutions with a pH of *less* than 7 are acidic and contain more H^+ ions than OH^- ions. Solutions with a pH greater than 7 are alkaline, or basic, and contain more OH^- ions than H^+ ions.

Because the scale is logarithmic (to base 10), a solution with a pH of 6 has a hydrogen ion concentration that is ten times greater than a solution with a pH of 7 and is much more acidic. A pH of 5 represents another tenfold increase, and so a solution with a pH of 4 is 10×10 or 100 times more acidic than a solution with a pH of 6. The contents of most animal and plant cells are neither strongly acidic nor alkaline but are an essentially neutral or very slightly basic mixture of acidic and basic substances. Most life cannot exist if the pH of the cell changes very much (Fig. 3–13). (The pH of living cells ordinarily ranges around the value of about 7.2 to 7.4. Human blood has a normal pH of 7.4.

---

[1][ ] are symbols indicating concentration.

Salts Form From Acids And Bases And Dissociate Into Anions And Cations

When an acid and a base are mixed together, the H^+ of the acid unites with the OH^- of the base to form a molecule of water. The remainder of the acid (anion) combines with the remainder of the base (cation) to form a salt. This type of reaction is called **acid-base neutralization.** Hydrochloric acid reacts with sodium hydroxide to form water and sodium chloride:

$$HCl + NaOH \longrightarrow H_2O + NaCl$$

Hydrochloric   Sodium   Water   Sodium
acid   hydroxide   chloride

A **salt** can be defined as a compound in which the hydrogen atom of an acid is replaced by some other cation. A salt contains a cation other than H^+ and an anion other than OH^-. Sodium chloride, NaCl, is a compound in which the hydrogen ion of HCl has been replaced by the cation, Na^+.

When a salt, an acid, or a base is dissolved in water, its constituent ions separate (Fig. 3–14). Because these charged particles can conduct an electrical current, these substances are called **electrolytes.** Sugars, alcohols, and many other substances do not ionize when dissolved in water; they do not conduct an electrical current and are termed **non-electrolytes.**

The cells and extracellular fluids (such as sap or blood) of plants and animals contain a variety of dissolved salts. They are a source of many important mineral ions. Such ions are essential for fluid balance, acid-base balance, nerve and muscle function in animals, blood clotting, bone formation, and many other aspects of body function. Although the concentration of salts in the cells and body fluids of plants and animals is small, these salts are of great importance for normal cell function. The concentrations of the respective cations and anions are kept remarkably constant under

FIGURE 3–13   Acid rain and fog have contributed to widespread forest destruction. (J. McDonald/Visuals Unlimited)

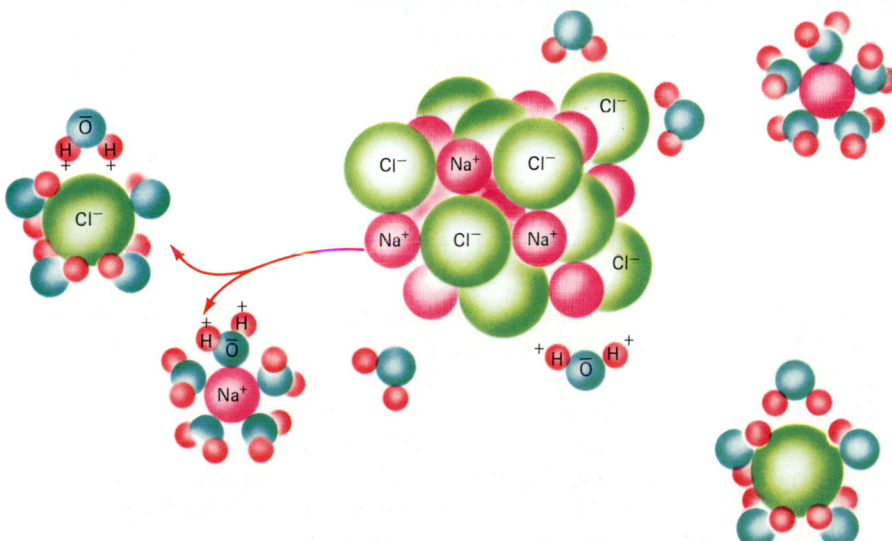

FIGURE 3–14   The crystal lattice of a salt like NaCl is held together by the strong ionic bonds between the Na$^+$ and Cl$^-$ ions, and considerable energy is required to pull the ions apart. When NaCl is added to water, the negative ends of the water molecules are attracted to the positive sodium ions and tend to pull them away from the chlorine ions. At the same time, the positive ends of the water molecules are attracted to the negative chlorine ions, separating them from the sodium ions. When dissolved, each sodium and chlorine atom is surrounded by water molecules electrically attracted to it.

normal conditions. Any marked change results in impaired cellular functions and ultimately in death.

Buffers Minimize pH change

Many homeostatic mechanisms operate to maintain appropriate pH levels. For example, the pH of human blood is about 7.4 and must be maintained within very narrow limits. Should the blood become too acidic (for instance as a result of respiratory disorders or metabolic diseases such as diabetes mellitus), coma and death may result. Excessive alkalinity, on the other hand, can result in overexcitability of the nervous system and

even convulsions. A **buffer** is a substance or combination of substances that minimizes changes in pH when acids or bases are added. The buffer either accepts or releases hydrogen ions. A buffer consists of a weak acid and a salt of that acid, or a weak base and a salt of that base.

One of the most common buffering systems and one that is important in human blood is carbonic acid and the bicarbonate ion. Bicarbonate ions are formed in the body as follows:

$$CO_2 + H_2O \rightleftharpoons H_2CO_3 \rightleftharpoons H^+ + HCO_3^-$$

| Carbon dioxide | Water | | Carbonic acid | | | Bicarbonate ion |

As indicated by the arrows, the reactions are reversible. When excess hydrogen ions are present in blood or other body fluids, bicarbonate ions combine with them to form carbonic acid, a weak acid.

$$H^+ + HCO_3^- \longrightarrow \underset{\text{Carbonic acid}}{H_2CO_3}$$

In this way a strong acid can be converted to a weak acid. The carbonic acid is unstable and quickly breaks down into carbon dioxide and water.

Buffers also work to reduce pH when excessive numbers of hydroxide ions are present. A buffer may release hydrogen ions, which combine with the hydroxide ions to form water.

$$OH^- + 2H^+ + CO_3 \longrightarrow {}^-HCO_3 + H_2O$$

Naturally occurring bodies of water, especially sea water, often contains substances that serve as buffers. For instance, the water of streams with limestone bottoms often contains calcium-based minerals that help to remove the hydrogen ions that seep into them from decaying vegetation or acid rain. Streams that have granite beds, on the other hand, contain different and less alkaline minerals and are much less able to tolerate acid precipitation.

One measure that has been attempted to restore acid-damaged waters to health involves *neutralization,* counteraction of an inappropriate pH by the addition of a substance which has the opposite effect on pH. In streams with granite beds, an alkali is required. This is supplied by adding lime, calcium hydroxide ($Ca(OH)_2$). Suppose the offending acid were sulfuric acid, derived from the burning of high-sulfur fuel in electrical generation plants:

$$\underset{\text{Hydrogen}}{2H^+} + \underset{\text{Sulfate}}{SO_4^{2-}} + \underset{\text{Calcium hydroxide}}{Ca(OH)_2} \longrightarrow$$

$$\underset{\text{Calcium sulfate}}{Ca_2SO_4} + \underset{\text{Water}}{H_2O}$$

This disposes of the hydrogen ion, leaving water in its place. It also removes the sulfate, because calcium sulfate is quite insoluble. The lime is a fairly strong alkali, however, and must be added carefully so as not to raise the pH of the water too much. That is the disadvantage of neutralization as opposed to buffering: the alkali or acid that is used to neutralize an inappropriate pH may produce an undesirable pH itself. Body fluid buffering agents are themselves usually near neutrality.

■ CHAPTER SUMMARY

I. The atom is the smallest unit of a chemical element that retains the characteristic properties of that element.
  A. An atom consists of subatomic particles, including protons, neutrons, and electrons.
  B. Electrons are found in orbitals located within energy levels. The electrons in the outermost energy level of an atom are valence electrons; they help determine the chemical properties of the atom.

II. A chemical compound consists of two or more elements combined in a fixed ratio.
  A. The composition of a compound may be described by a chemical formula, such as H_2O, or by a structural formula, such as HOH.
  B. The chemical reactions that occur between atoms and molecules can be described by means of chemical equations.

III. The atoms of a chemical compound are attached to one another by bonds.
  A. In a covalent bond, atoms share electrons.
  B. An ionic bond is formed when one atom donates an electron to another, each atom thereby becoming charged and attracted to the other because of these electrical charges.
  C. Hydrogen bonds are weak chemical bonds formed between hydrogen atoms and an electronegative atom, usually oxygen or nitrogen. Molecules of water form hydrogen bonds with one another.

IV. When a substance is oxidized, it loses electrons; when a substance is reduced, it gains electrons. In living systems, oxidation-reduction reactions usually involve the loss and gain of hydrogen.

V. Among the biologically important inorganic compounds are water, acids, bases, and salts.
  A. Water is an essential component of living things and is necessary for many vital activities, especially as a solvent and for temperature control and stabilization.
  B. Acids ionize in solution, yielding hydrogen ions; bases usually ionize, yielding hydroxide ions. The pH scale extends from 0 to 14, with 7 considered neutral. As the pH decreases below 7, the solution becomes more acidic. As a solution becomes more basic, its pH increases above 7.
  C. A salt is a compound in which the hydrogen of an acid is replaced by some other cation.
  D. Buffering, which usually involves a weak acid plus a salt of that acid, helps to maintain appropriate pH.

◼ POST-TEST

1. A(n) _____ is the smallest amount of an element that retains the chemical properties of the element.
2. Isotopes are atoms of the same element that differ in their number of _____.
3. Within energy levels electrons may be found within specific _____.
4. What is the composition of a compound with the formula C_2H_6O? _____.
5. Atoms with one to three valence electrons generally behave as _____.
6. The type of bond in which atoms share electrons is a _____ bond.
7. A chemical process in which a substance gains electrons is referred to as _____.
8. A compound that ionizes in solution to yield hydrogen ions and an anion is a(n) _____.
9. A solution with a pH of 8 is best described as _____.
10. A substance that resists changes in pH is a _____.
11. The tendency for water to rise in very fine tubes is called _____; it is due to _____ and _____ forces.
12. Water molecules in liquid water and ice are held together by _____ bonds.

◼ REVIEW QUESTIONS

1. Distinguish between (a) an atom and an element, (b) a molecule and a compound, and (c) an atom and an ion.
2. How do isotopes of the same element differ? What is a radioisotope?
3. How do valence electrons help determine the chemical properties of an atom?
4. Compare ionic and covalent bonds and give specific examples of each.
5. Write a chemical equation depicting the ionization of (a) sodium hydroxide (NaOH), and (b) hydrochloric acid (HCl).
6. What properties of water make it an essential component of living matter?
7. How would a solution with a pH of 5 differ from one with a pH of 9? of 7?
8. Why are buffers important in living organisms? Give a specific example of how a buffer system works.
9. Differentiate clearly among acids, bases, and salts. What are the functions of salts in living organisms?
10. Why do oxidation and reduction occur simultaneously?
11. Describe a reversible reaction that is at equilibrium.
12. What are hydrogen bonds? What is their significance?
13. A person with an upset stomach may take sodium bicarbonate ($NaHCO_3$) to treat the excess acidity. Since stomach acid is hydrochloric acid (HCl), this produces water and carbon dioxide gas (hence the burp that follows!). But before carbon dioxide is emitted, a weak acid, carbonic acid, is produced as an intermediate. Since all that is done is to replace one acid with another, why does sodium bicarbonate reduce excess stomach acidity?

◼ READINGS

Atkins, P. W. *Molecules*. New York, W. H. Freeman and Company (Scientific American Books), 1987. A fascinating and beautifully illustrated collection of different kinds of substances, especially organic substances.

Baker, J. J. W., and G. E. Allen. *Matter, Energy, and Life.* Reading, Mass., Addison-Wesley, 1981. A presentation of the principles of thermodynamics and their application to studies of living systems. A difficult subject clarified.

Frieden, E. "The Chemical Elements of Life." *Scientific American*, 1972, 53–64. A discussion of the biological actions of various elements.

4

The Chemistry of Life: Organic Compounds

OUTLINE

I. Carbohydrates include the sugars, starches, and cellulose
   A. Monosaccharides are single-unit sugars
   B. Disaccharides are two-unit sugars
   C. Polysaccharides are large, multiple-unit carbohydrates
II. Lipids are fats or fat-like substances
   A. Neutral fats are composed of glycerol and fatty acids
   B. Phospholipids are important components of cell membranes
   C. Steroids are lipids whose molecules contain four rings of carbon atoms
III. Proteins are macromolecules formed from linked amino acids
   A. There are many kinds of proteins
   B. Amino acids are simple organic compounds containing a carboxyl and an amino group
   C. Proteins are structured on several levels of organization
IV. Nucleic acids consist of nucleotide subunits

LEARNING OBJECTIVES

After you have studied this chapter you should be able to:
1. Compare the major groups of organic compounds—carbohydrates, lipids, proteins, and nucleic acids—with respect to their chemical composition and function.
2. Distinguish among monosaccharides, disaccharides, and polysaccharides, giving examples of each.
3. Distinguish among neutral fats, phospholipids, and steroids, giving the biologic functions of each group.
4. Describe the functions and chemical structure of proteins.
5. Describe the chemical structure of nucleotides and nucleic acid and explain the importance of these compounds in living organisms.

Crystals of cholesterol, an organic compound essential to the life of vertebrate animals. (G. Musil/Visuals Unlimited)

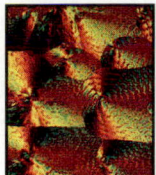 In a way, the chemistry of the carbon atom is the chemistry of life itself. With four electrons in its outer energy level, the carbon atom can share electrons with other atoms, *including other carbon atoms,* forming four covalent bonds. This ability of carbon atoms yields an immense variety of **organic compounds,** complex compounds that can contain carbon atoms. Organic compounds are the main structural components of cells and tissues, the participants in and regulators of thousands of metabolic reactions, and the fuel molecules of living systems.

In all organic compounds the main chain of atoms that makes up the principal axis of the molecule consists of carbon atoms covalently bonded to each other (Fig. 4–1; Table 4–1). Thus carbon atoms share electrons with other carbon atoms to form chains of varying lengths. These chains may be unbranched or branched, or the carbon atoms may join to form rings. Adjacent carbon atoms may form single bonds, or by sharing additional pairs of electrons, they may form double (C=C) or triple (C≡C) bonds.

In this chapter we discuss some of the major groups of organic compounds that are important in living organisms, including carbohydrates, lipids, proteins, and nucleic acids.

KEY CONCEPTS

- Four main groups of organic compounds important in living things are carbohydrates, lipids, proteins, and nucleic acids.

- Organic compounds are made possible by the ability of carbon atoms to combine with one another so as to form straight, branching, and ring-like chains.

- Carbohydrates are molecules of various sizes built of single-unit sugars linked together to form biopolymers.

- Lipids are high-energy fat-like substances usually composed of fatty acids and glycerol. Phospholipids comprise the bilayer of cellular membranes, and steroids often function as hormones.

- Proteins are macromolecules consisting of linked amino acids. Some proteins are enzymes, catalysts which govern the total chemical life of the organism.

- The information needed to construct proteins is stored in nucleic acids, usually deoxyribonucleic acid (DNA), a macromolecular substance composed of nucleotide units.

■ CARBOHYDRATES INCLUDE THE SUGARS, STARCHES, AND CELLULOSE

Familiar to us as sugars and starches, **carbohydrates** serve as fuel molecules and are also important structural components, especially in plant cells. Carbohydrates contain carbon, hydrogen, and oxygen atoms in a ratio of approximately one carbon to two hydrogens to one oxygen $(CH_2O)_n$. The term carbohydrate, meaning "hydrate (water) of carbon," stems from an approximately 2 to 1 ratio of hydrogen to oxygen, the same ratio found in water (H_2O). Carbohydrates may be classified as monosaccharides, disaccharides, or polysaccharides.

Monosaccharides Are Single-Unit Sugars

Monosaccharides are simple sugars that usually contain from three to six carbon atoms. **Glucose** (also called dextrose) and fructose are examples of monosaccharides. Each is composed of a single hexose unit (consists of six carbon atoms) with the formula $C_6H_{12}O_6$. Such compounds as glucose and fructose, which have identical molecular formulas but different arrangements of atoms, are termed **isomers.** Because of their different arrangements of atoms, the two sugars have different chemical properties.

Glucose, often referred to as blood sugar, is the most abundant hexose in the bodies of humans and other animals. Its concentration is kept at a homeostatic level in the blood, and glucose is utilized by the cells as a fuel molecule. The **pentoses** (five-carbon sugars) **ribose** and **deoxyribose** are components of nucleotides and nucleic acids.

Molecules are not the simple two-dimensional structures that are usually depicted on a printed page. In fact, the properties of each compound depend in part on its three-dimensional structure. In solution, molecules of glucose and other monosaccharides can take up more than one three-dimensional form. In fact, in solution they exist mainly as rings (Fig. 4–2).

In some compounds certain atoms can be arranged in more than one position in space around the carbon atom to which they are bonded. Such **stereoisomers** differ from one another only in some geometric or three-dimensional way so that a right-handed form and a left-handed mirror image form both exist. Only the right-handed form of glucose (sometimes called "dextrose" from the Latin word for "right") is usable by most organisms. Since it is sweet to the taste but cannot be utilized by our body cells, the left-handed form of glucose may some day be used as an artificial sweetener in dietetic foods.

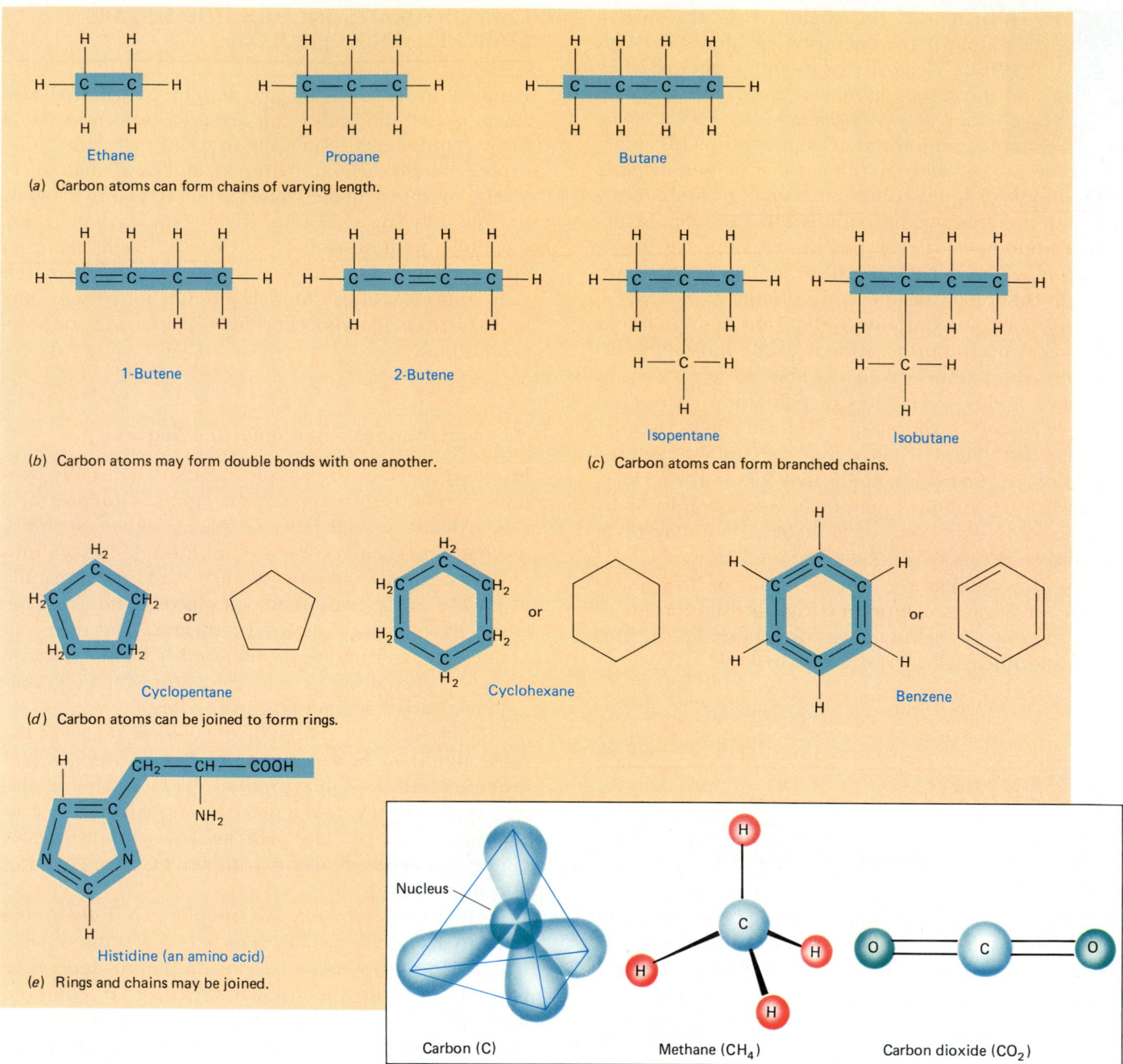

(a) Carbon atoms can form chains of varying length.

Ethane   Propane   Butane

(b) Carbon atoms may form double bonds with one another.

1-Butene   2-Butene

(c) Carbon atoms can form branched chains.

Isopentane   Isobutane

(d) Carbon atoms can be joined to form rings.

Cyclopentane   or   Cyclohexane   or   Benzene

(e) Rings and chains may be joined.

Histidine (an amino acid)

Nucleus

Carbon (C)   Methane (CH_4)   Carbon dioxide (CO_2)

FIGURE 4–1   Some simple organic compounds. Note that each carbon atom has four covalent bonds. (a) Some simple hydrocarbons. These compounds consist only of carbon and hydrogen. (b) When a hydrocarbon is less than completely saturated with hydrogen, adjacent carbon atoms may share two pairs of electrons, forming double bonds with one another. Double bonds act like a pair of nails joining two pieces of wood; atoms so joined cannot rotate around one another. (c) Branched hydrocarbons. There is no reason why carbons cannot be attached to a chain at its side, forming subsidiary chains of their own. (d) Carbon atoms can join to one another to form rings, and (e) side chains can attach to the rings. (f) An even better impression of the arrangement of atoms within an organic molecule can be gained from a space-filling picture or model that shows its three-dimensional shape.

TABLE 4–1
Some of The Groups of Biologically Important Organic Compounds

Class of Compound	Component Elements	Description	How to Recognize	Principal Function in Living Systems
Carbohydrates	C, H, O	Contain approximately 1 C : 2 H : 1 O (but make allowance for loss of oxygen and hydrogen when sugar units are linked)	Count the carbons, hydrogens, and oxygens.	Cellular fuel; energy storage; structural component of plant cell walls; component of other compounds such as nucleic acids and glycoproteins
		1. Monosaccharides (simple sugars)—mainly five-carbon (pentose) molecules like ribose or six-carbon (hexose) molecules such as glucose and fructose	Look for the ring shapes: hexose or pentose	Cellular fuel; components of other compounds
		2. Disaccharides—two sugar units linked by a glycosidic bond, e.g., maltose, sucrose	Count sugar units.	Components of other compounds
		3. Polysaccharides—many sugar units linked by glycosidic bonds, e.g., glycogen, cellulose	Count sugar units.	Energy storage; structural components of plant cell walls
Lipids	C, H, O	Contain less oxygen relative to carbon and hydrogen than do carbohydrates.		Energy storage; cellular fuel, structural components of cells; thermal insulation
		1. Neutral fats. Combination of glycerol with one to three fatty acids. Monacylglycerol contains one fatty acid; diacylglycerol contains two fatty acids; triacylglycerol contains three fatty acids. If fatty acids contain double carbon-to-carbon linkages (C=C), they are unsaturated; otherwise they are saturated.	Look for glycerol at one end of molecule.	Cellular fuel; energy storage
	C, H, O, N, P	2. Phospholipids. Composed of glycerol attached to one or two fatty acids and to an organic base containing phosphorus.	Look for glycerol and side chain containing phosphorus and nitrogen.	Components of cell membranes
		3. Steroids. Complex molecules containing carbon atoms arranged in four interlocking rings (three rings contain six carbon atoms each and the fourth ring contains five)	Look for four interlocking rings:	Some are hormones; others include cholesterol, bile salts, vitamin D. Components of cell membranes.
Proteins	C, H, O, N, usually S	One or more polypeptides (chains of amino acids) coiled or folded in characteristic shapes	Look for amino acid units joined by C—N bonds.	Serve as enzymes; structural components; muscle proteins; hemoglobin

(continued)

TABLE 4–1 (continued)
Some of The Groups of Biologically Important Organic Compounds

Class of Compound	Component Elements	Description	How to Recognize	Principal Function in Living Systems
Nucleic acids	C, H, O, N, P	Backbone composed of alternating pentose and phosphate groups, from which nitrogenous bases project. DNA contains the sugar deoxyribose and the bases guanine, cytosine, adenine, and thymine. RNA contains the sugar ribose, and the bases guanine, cytosine, adenine, and uracil. Each molecular subunit, called a *nucleotide*, consists of a pentose, a phosphate, and a nitrogenous base.	Look for a pentose-phosphate backbone. DNA forms a double helix.	Storage, transmission, and expression of genetic information

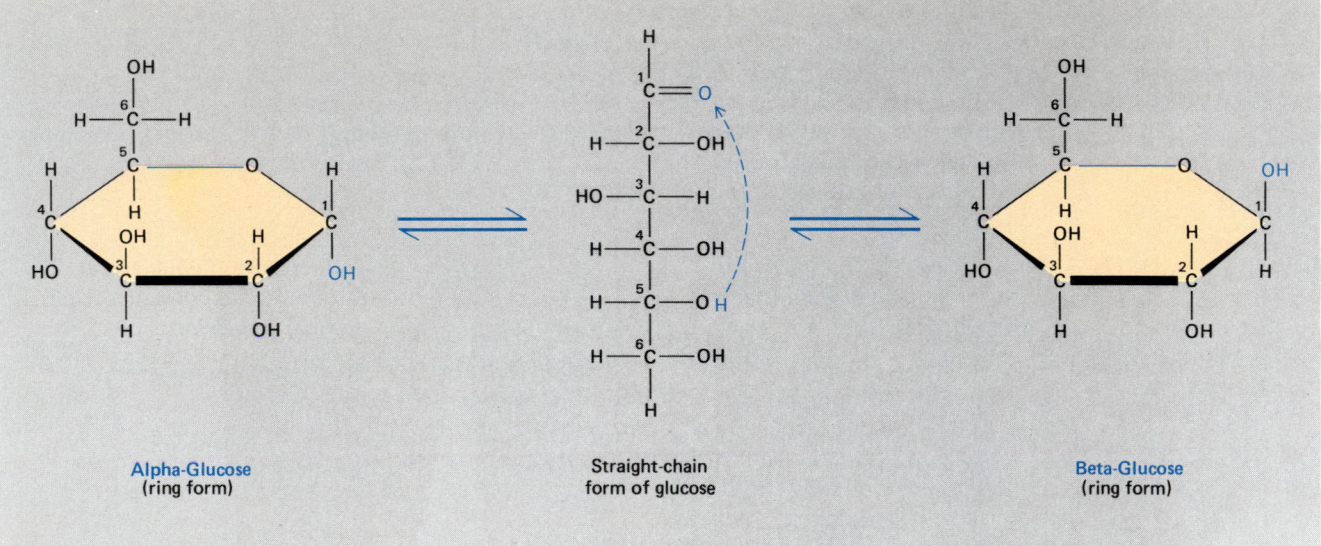

Alpha-Glucose
(ring form)

Straight-chain
form of glucose

Beta-Glucose
(ring form)

FIGURE 4–2 Glucose can exist in three common forms. The alpha and beta varieties can change into one another after first rearranging themselves in the straight-line form. All three configurations coexist in our body fluids. In alpha-glucose the —OH of carbon-1 is below the ring; in beta-glucose, the —OH of carbon-1 is above the ring. The thick, tapered bonds in the lower portion of each ring indicate that the molecule is a three-dimensional structure. The thickest portion of the bond is interpreted as being the part of the molecule "nearest" the viewer.

Disaccharides Are Two-Unit Sugars

A **disaccharide** (two sugars) is a carbohydrate that can be broken into its monosaccharide subunits. The disaccharide **maltose** (malt sugar) consists of two chemically combined glucose units. **Sucrose,** the sugar we use to sweeten our foods, consists of a glucose unit combined with a fructose unit. **Lactose** (the sugar present in milk) is composed of one molecule of glucose and one of galactose, another hexose monosaccharide. The covalent bond that joins two monosaccharide units is called a **glycosidic bond.**

During digestion maltose is cleaved (degraded) to form two molecules of glucose:

$$\text{Maltose} + \text{Water} \longrightarrow \text{Glucose} + \text{Glucose}$$

Similarly, sucrose is cleaved during digestion to form glucose and fructose:

$$\text{Sucrose} + \text{Water} \longrightarrow \text{Glucose} + \text{Fructose}$$

Structural formulas for the compounds in this reaction are shown in Figure 4–3. Because water is added during the cleavage of a disaccharide, this type

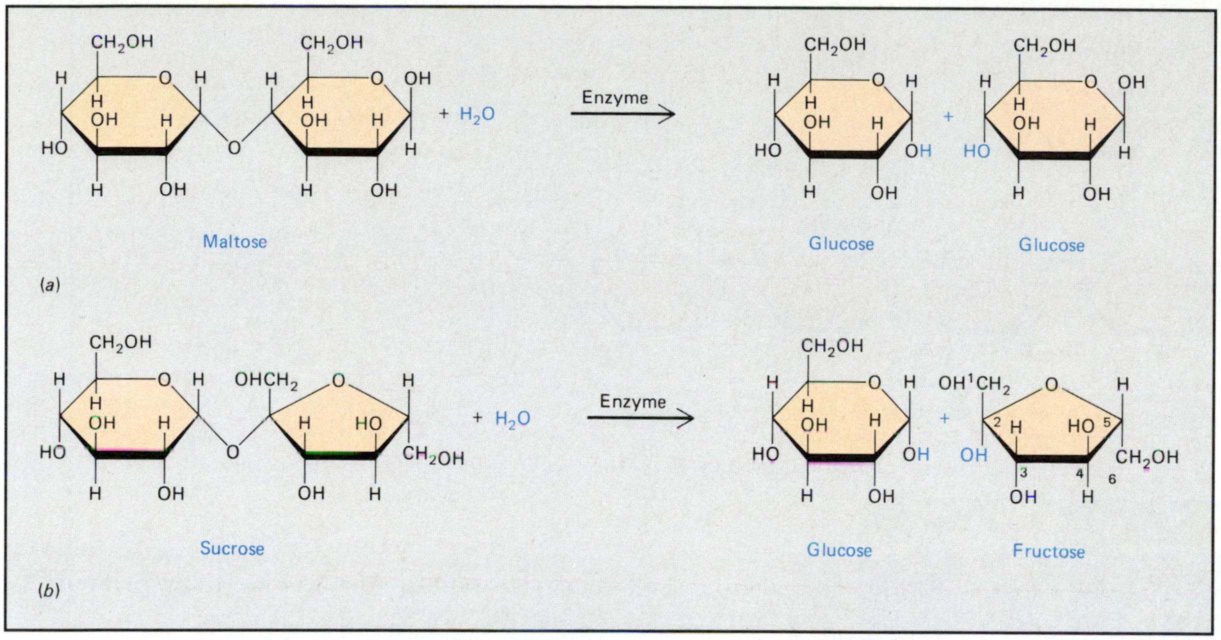

FIGURE 4–3 A disaccharide **can** be cleaved to yield two monosaccharide units. (*a*) Maltose may be broken down (as it is during digestion) to form two molecules of glucose. This is a hydrolysis reaction that requires the addition of water. (*b*) Sucrose can be cleaved to yield a molecule of glucose and a molecule of fructose.

of reaction is called a **hydrolysis** ("water-splitting") reaction.

If a disaccharide can be broken down by the chemical addition of water, as it were, one might think that it could be formed by the chemical removal of water from the two monosaccharide molecules that comprise the disaccharide. In effect, this is indeed what is done when a sugar beet makes sucrose from fructose and glucose in the first place. However, it is an indirect and slightly more complicated process than the simple linking of the two monosaccharides. Basically, sucrose contains more energy than glucose plus fructose, so to make sucrose from glucose and fructose, energy must somehow be injected. This is done by a series of reactions that are beyond the scope of this book. Notice that if two monosaccharides could be linked with one another, three could be fitted together, or four, or almost any number. Thus, monosaccharides are linked to one another to make not only disaccharides, but very large polysaccharides.

Polysaccharides Are Large, Multiple-Unit Carbohydrates

The most abundant carbohydrates are **polysaccharides** (many sugars), such as starches, glycogen, cellulose, and the chitin of insect shells. A polysaccharide is a single long chain, or a branched chain, consisting of repeating units of a simple sugar, usually glucose. Such immense molecules are known as **macromolecules** and also, if their subunits are identical or similar, as **biopolymers.**[1] The precise number of sugar units varies, but typically, thousands of units may be present in a single molecule of polysaccharide. Because they are composed of different stereoisomers of glucose, or because the glucose units are arranged differently, these polysaccharides have different properties.

Starch is the typical storage form of carbohydrate in plants, whereas **glycogen** (sometimes referred to as animal starch) is the form in which glucose is stored in animal tissues (Fig. 4–4). Glycogen is a highly branched polysaccharide, more water-soluble than plant starch. Glucose cannot be stored as such because its small, readily soluble molecules would leak out of the cells. The larger, less soluble starch and glycogen molecules do not readily pass through the cell membrane. Thus, instead of storing simple sugars, cells store the more complex polysaccharides, such as glycogen, which can readily be broken down into simple sugars.

Carbohydrates are the most abundant group of organic compounds on earth, and **cellulose** is the most

---

[1] A single-unit substance is a monomer; a two-unit substance is a dimer; and a many-unit substance is known as a polymer. Many or most plastics are polymers, so to distinguish the polymers found in living things from other polymers we employ the term "biopolymer."

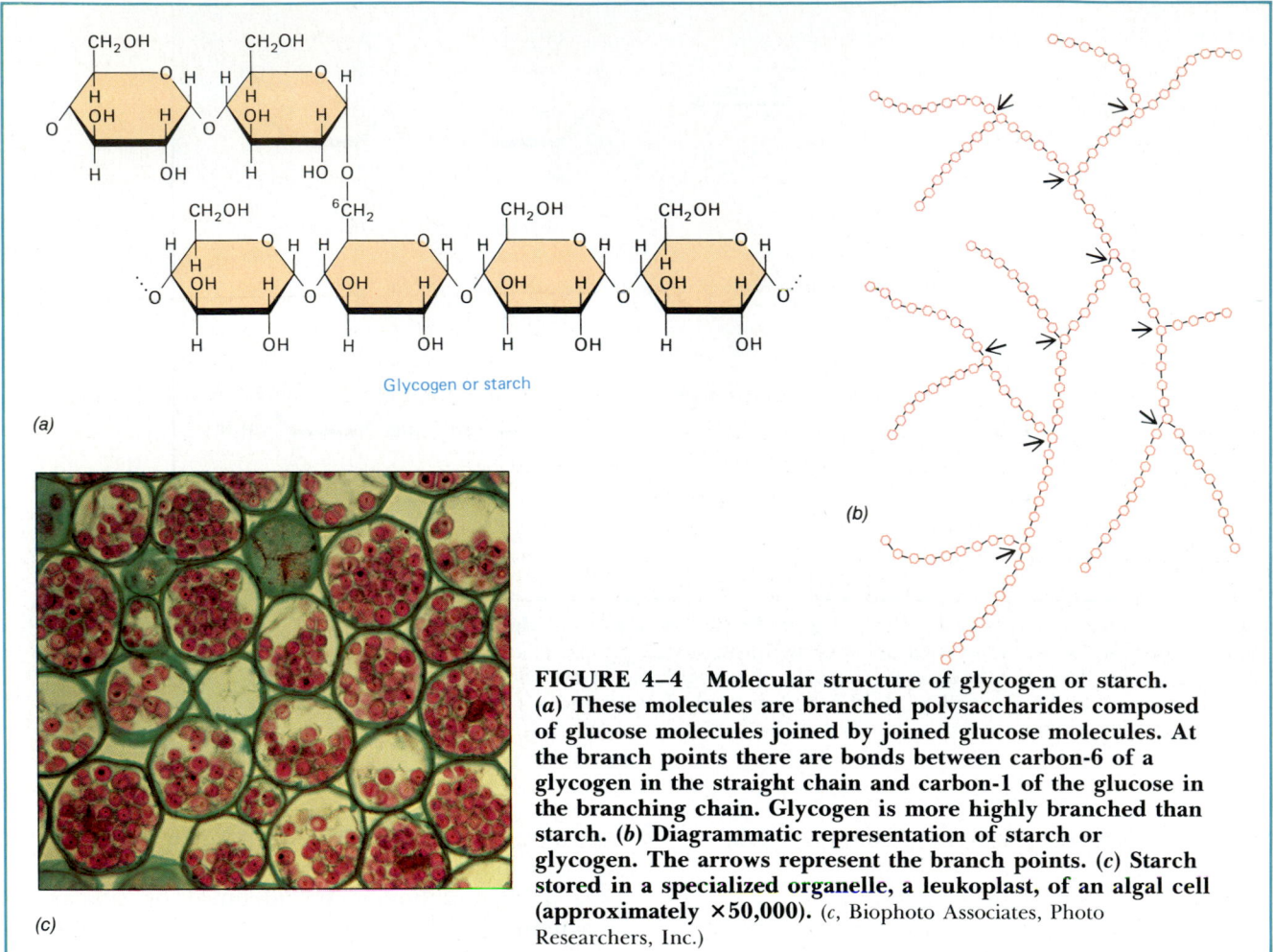

(a)

Glycogen or starch

(b)

(c)

FIGURE 4–4   Molecular structure of glycogen or starch. (a) These molecules are branched polysaccharides composed of glucose molecules joined by joined glucose molecules. At the branch points there are bonds between carbon-6 of a glycogen in the straight chain and carbon-1 of the glucose in the branching chain. Glycogen is more highly branched than starch. **(b)** Diagrammatic representation of starch or glycogen. The arrows represent the branch points. **(c)** Starch stored in a specialized organelle, a leukoplast, of an algal cell **(approximately ×50,000).** (c, Biophoto Associates, Photo Researchers, Inc.)

abundant carbohydrate, accounting for 50% or more of all the carbon in plants. Wood is about half cellulose, and cotton is at least 90% cellulose. Plant cells are surrounded by a strong supporting cell wall consisting mainly of cellulose. Cellulose is an insoluble polysaccharide composed of many glucose molecules joined together (Fig. 4–5).

The bonds joining the sugar units in cellulose are different from those in starch and are not split by the enzyme that cleaves the bonds in starch. For this reason only a few animals are able to digest cellulose. Human beings cannot—that is why plant fiber, which consists mainly of cellulose, adds bulk to the intestinal contents and aids bowel function. Animals whose diet is high in cellulose, such as cattle, rabbits, or termites, usually harbor microorganisms in the digestive tract that can produce enzymes capable of digesting cellulose.

The amino sugars **glucosamine** and **galactosamine** are modified carbohydrates, compounds in which a hydroxyl group (OH) of a monosaccharide is replaced by an amino group (NH_2; see the section on proteins). Glucosamine is the molecular unit found in chitin; galactosamine is found in cartilage. **Chitin,** a tough modified polysaccharide, is the main component of the external skeletons of insects, crayfish, and other arthropod animals.

◼ LIPIDS ARE FATS OR FAT-LIKE SUBSTANCES

Lipids are a heterogeneous group of compounds that have a greasy or oily consistency and are relatively insoluble in water. Like carbohydrates, lipids are composed of carbon, hydrogen, and oxygen atoms, but they have relatively less oxygen in proportion to the carbon and hydrogen than carbohydrates do. Among the groups of lipids especially significant biologically are the neutral fats, phospholipids, steroids, carotenoids (orange and yellow plant pigments), and waxes. Lipids are important biological fuels; they serve as

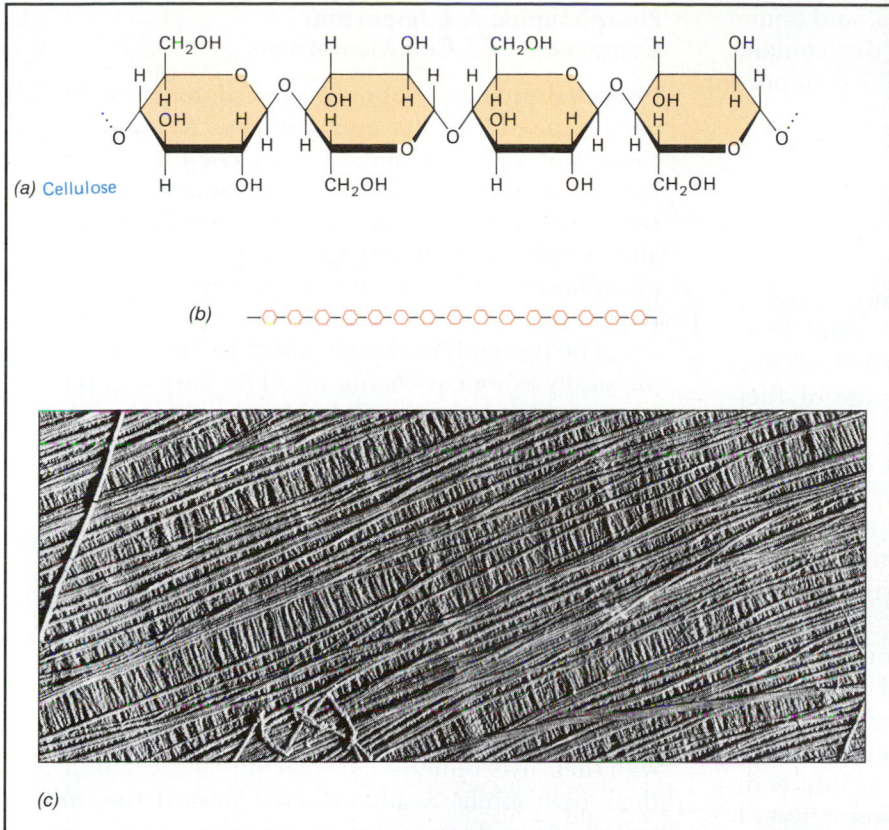

(a) Cellulose

(b)

(c)

(d)

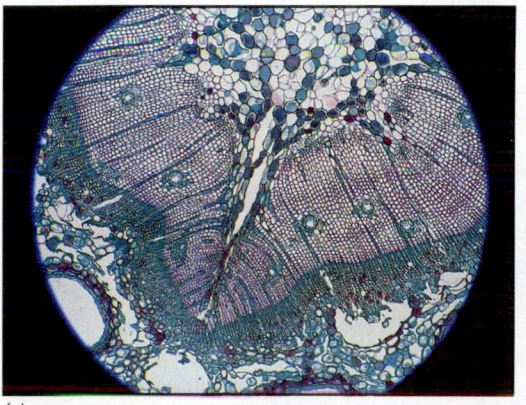

(e)

(f)

FIGURE 4–5 **The structure of cellulose. (*a*) The cellulose molecule is an unbranched polysaccharide composed of approximately 10,000 glucose units. (*b*) A more diagrammatic representation of cellulose structure. Each hexagon represents a glucose molecule bound to the adjacent glucose molecule. (*c*) An electron micrograph of cellulose fibers from the cell wall of a marine alga (approximately ×24,000). (*d*) Tendrils of pumpkin vine. The complex polysaccharide cellulose occurs in plant cell walls. By growing at different rates on two sides of a stalk, the walls of plant cells produce the slow movements of plants, such as the ability of tendrils such as these to attach themselves to supporting objects. (*e*) The multitude of cells in a cross-section of a 1-year-old pine stem. The cell walls are clearly visible as a fine network. (*f*) Clear tunicates, *Clavelin picta*, photographed in the Virgin Islands.** (*c*, Omikron, Photo Researchers, Inc., *e*, Grant Heilman; *f*, Robin Lewis, Coastal Creations)

structural components of cell membranes, and some are essential hormones. Glycolipids, lipids that contain water-soluble carbohydrate components, are important in interactions among cells.

Neutral Fats Are Composed Of Glycerol And Fatty Acids

The most abundant group of lipids in living things is the **neutral fats.** These compounds yield more than twice as much energy per gram as do carbohydrates and are an economical form for the storage of fuel energy. Fats also can serve as insulation (as in blubber), to produce body contours (as they do in human beings), or even as support tissue (they hold the kidneys in place). Carbohydrates and proteins can be transformed by enzymes into fats and stored within the cells of the adipose (fat) tissue that we accumulate so readily (Chapters 31 and 39).

A fat consists of glycerol joined to one, two, or three molecules of a fatty acid. **Glycerol** is a three-carbon alcohol[1] that contains three —OH groups (Fig. 4–6). A **fatty acid** is a long, straight chain of carbon atoms with a carboxyl group (—COOH) at one end. About 30 varieties of fatty acids are commonly found in animal lipids. They typically have an even number of carbon atoms. For example, butyric acid, present in rancid butter, has four carbon atoms, and oleic acid, the most widely distributed fatty acid in nature, has 18 carbon atoms.

Saturated fatty acids contain the maximum possible number of hydrogen atoms, while **unsaturated fatty acids** contain carbon atoms that are doubly bonded with one another and are not fully saturated with hydrogens. Fatty acids with several double bonds are called **polyunsaturated fatty acids.** Fats containing unsaturated fatty acids are the oils (such as corn oil, olive oil, peanut oil, and many others), and most of them are liquid at room temperature. Saturated fats are usually solids; butter and animal fat are examples. At least two fatty acids (linoleic and linolenic) cannot be manufactured by the human body and are therefore **essential nutrients,** which must be included in the diet. Saturated fats have been linked to circulatory disorders (Chapter 39).

When a glycerol molecule combines chemically (by an indirect process) with one fatty acid, a **monoacylglycerol** (sometimes called monoglyceride) is formed. When two fatty acids combine with a glycerol, a **diacylglycerol** (or diglyceride) is formed, and when three fatty acids combine with one glycerol molecule, a **triacylglycerol** (or triglyceride) is formed.

---

[1]For our purposes an alcohol may be defined as a compound in which an —OH group is bonded to a carbon atom.

Phospholipids Are Important Components Of Cell Membranes

Phospholipids are important constituents of the cell membranes of plants and animals. A phospholipid consists of a glycerol molecule attached to one or two fatty acids, and the glycerol is also bonded to phosphorus, which is part of an organic base. Phospholipids also usually contain nitrogen (Fig. 4–7). (Note that phosphorus and nitrogen are absent from neutral fats.)

The two ends of the phospholipid molecule differ physically as well as chemically. The fatty acid portion of the molecule is **hydrophobic** (water-hating) and is not soluble in water. However, the portion composed of glycerol and the organic base is ionized and readily water-soluble. This end of the molecule is said to be **hydrophilic** (water-loving). The polarity of these lipid molecules causes them to take up a certain configuration in the presence of water, with their hydrophilic water-soluble heads facing outward toward the surrounding water. The hydrophobic tails face in the opposite direction. The cell membrane is a lipid bilayer composed of two layers of phospholipid molecules with their hydrophobic tails meeting in the middle and their hydrophilic heads oriented toward the outside and inside surfaces of the membrane.

Steroids Are Lipids Whose Molecules Contain Four Rings Of Carbon Atoms

Although steroids are classified as lipids, their structure is quite different from that of other lipids. A **steroid** molecule contains carbon atoms arranged in four interlocking rings; three of the rings contain six carbon atoms and the fourth contains five (Fig. 4–8). The length and structure of the side chains that extend from these rings distinguish one steroid from another.

Among the steroids of biological importance are vitamin D, cholesterol, bile salts, male and female sex hormones (estrogen, progesterone, and testosterone), and the hormones secreted by the adrenal cortex. Cholesterol is a structural component of animal cell membranes. Steroid hormones regulate certain aspects of metabolism in a variety of animals, including vertebrates, insects, and crabs. Steroids are discussed further in Chapter 41.

■ PROTEINS ARE MACROMOLECULES FORMED FROM LINKED AMINO ACIDS

The macromolecules called **proteins** serve as important structural components of cells and tissues, so growth and repair, as well as maintenance of the or-

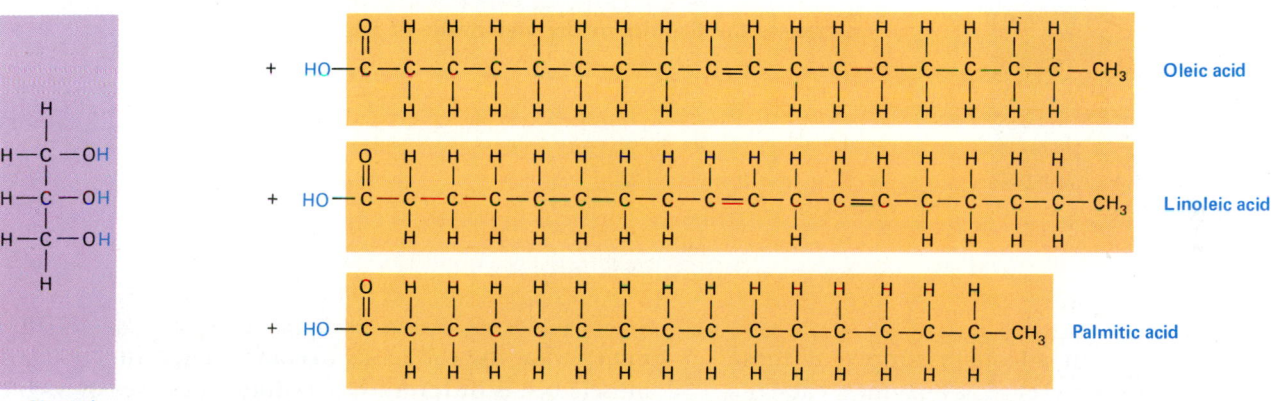

Glycerol   **Fatty acid**

(a)

A triacylglycerol

+ 3H₂O → (Enzyme)

PRODUCTS

Glycerol

Oleic acid

Linoleic acid

Palmitic acid

(b)

(c)

FIGURE 4–6 Neutral fats. (*a*) Structure of glycerol and of a fatty acid. The carboxyl (—COOH) group is present in all fatty acids. The R represents the remainder of the molecule, which varies with each type of fatty acid. (*b*) Hydrolysis of a triacylglycerol yields glycerol plus three fatty acids. (*c*) Honeybees on a brood comb. The comb is composed of wax secreted by special abdominal glands of the bees. It is a compound consisting of fatty acids and alcohols, and although it is classified as a lipid, it can be digested by very few animals.

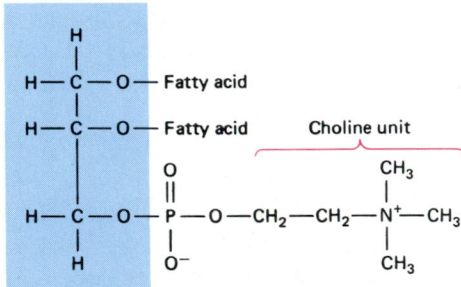

Phosphatidic acid

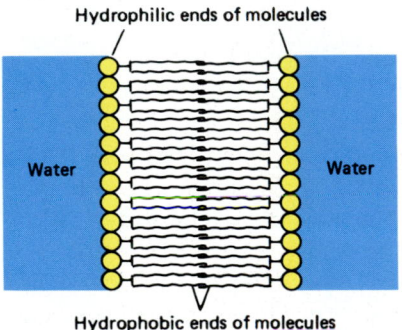

Lecithin

(a)

Hydrophilic ends of molecules

Water   Water

Hydrophobic ends of molecules

(b)

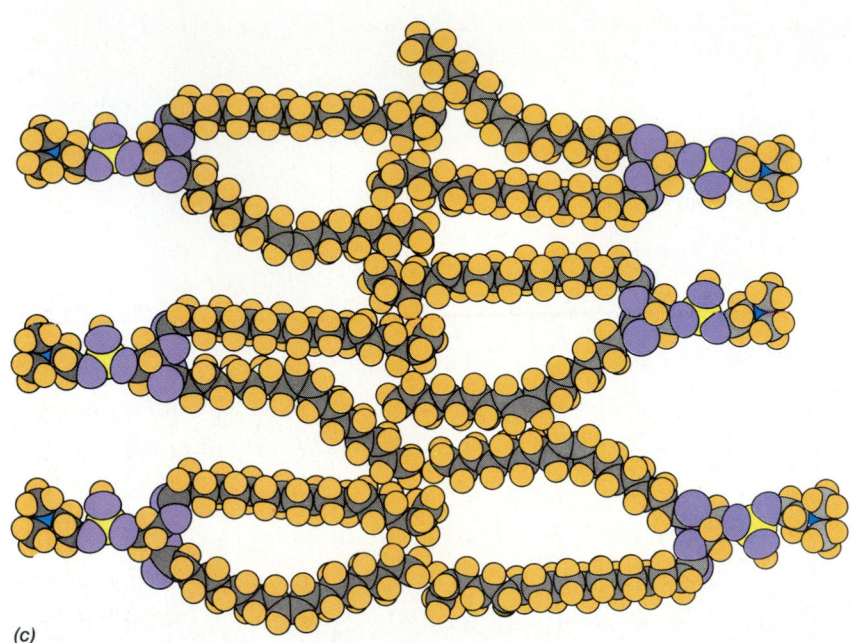

(c)

FIGURE 4–7   Phospholipids. (a) Many phospholipids are derivatives of phosphatidic acid, a compound consisting of glycerol chemically combined with two fatty acids and a phosphate group. Lecithin is a phospholipid found in cell membranes. It forms when phosphatidic acid combines with the compound choline. (b) A lipid bilayer, such as is found in cell membranes. (c) Close view of some of the phospholipids in a cell membrane. Notice that irregular shapes resulting from unsaturated fatty acid chains prevent close packing of the phospholipid molecules. This increases the fluidity of the membrane of which they are a part.

ganism, depend upon an adequate supply of these compounds. Some proteins serve as **enzymes,** catalysts that regulate the thousands of different chemical reactions that take place in a living system.

There Are Many Kinds Of Proteins

There is practically no end to the variety of proteins found within an organism's body. To simplify the task of understanding them, a number of ways to classify proteins have been developed. Three common methods are by function, by composition, and according to their solubility.

The protein constituents of a cell determine its function. Each kind of cell (e.g., muscle, bone, blood, nerve) has characteristic types, distributions, and amounts of protein. The protein composition deter-

mines what the cell looks like and how it functions. A muscle cell is different from other cell types by virtue of its content of the contractile proteins myosin and actin, which are largely responsible for its microscopic appearance as well as its ability to contract. The protein hemoglobin, found in red blood cells, is responsible for the specialized function of oxygen transport.

Most proteins are species-specific; that is, they vary slightly in each species, so the protein complement (as determined by the instructions encoded in the nucleic acids) is also mainly responsible for differences among species. Thus the types and distributions of proteins in the cells of a dog vary somewhat from those in the cells of a fox or a coyote. The degree of difference in the proteins of two species is thought to depend upon and to reflect evolutionary relationships, so that organisms less closely related by evolution have proteins that differ more markedly than those of

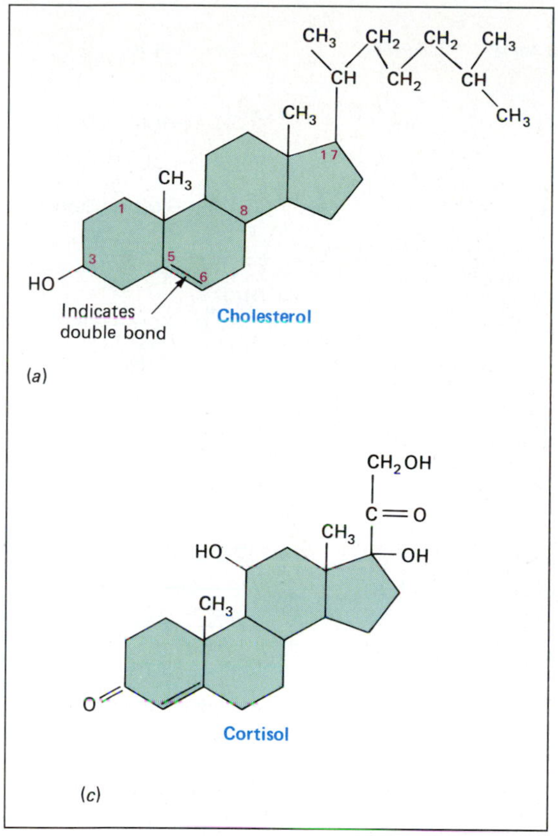

FIGURE 4-8   Steroids. All steroids have the basic skeleton of four interlocking rings of carbon atoms. Note that a carbon atom is present at each point in each ring. Each of the first three rings contains six carbon atoms, and the fourth ring contains five. For simplicity, hydrogen atoms have not been drawn within the ring structures.

closely related forms. Some proteins differ slightly even among individuals of the same species, so each individual is biochemically unique. Only genetically identical organisms—identical twins, clones of asexually reproducing organisms, or members of closely inbred strains of organisms—have identical proteins.

Amino Acids Contain A Carboxyl And An Amino Group

A basic knowledge of protein chemistry is essential for understanding nutrition as well as other aspects of metabolism. Proteins are composed of carbon, hydrogen, oxygen, nitrogen, and usually sulfur. Atoms of these elements are arranged into molecular subunits called **amino acids.** More than 20 kinds of amino acids are commonly found in proteins. All contain an **amino group** ($-NH_2$) and a **carboxyl group** ($-COOH$) bonded to the same carbon atom, but they differ in their side chains, abbreviated as "R" groups. **Glycine,** the simplest amino acid, has a hydrogen atom as its R group or side chain; alanine has a methyl ($-CH_3$) group (Fig. 4–9).

Amino acids combine chemically with one another by bonding the carboxyl carbon of one molecule to the amino nitrogen of another. This covalent bond linking two amino acids is referred to as a **peptide bond.** When two amino acids combine, a **dipeptide** is formed; a longer chain of amino acids is a **polypeptide.** Amino acids are linked by the action of cellular structures called **ribosomes** using a complex process that will be more fully discussed later on. For now, the important point to understand is that, unlike polysaccharides, proteins are macromolecules in which the subunits

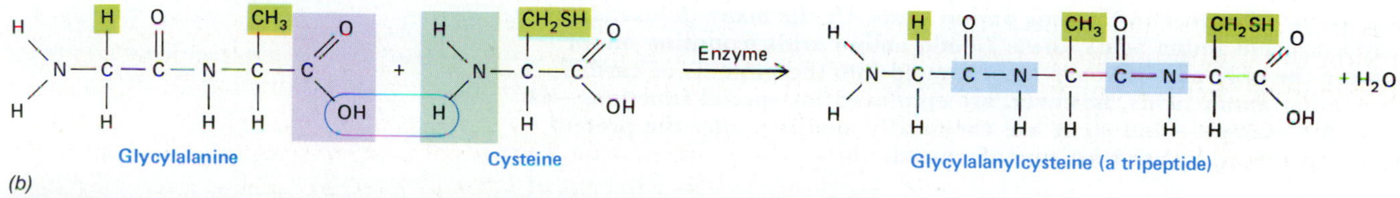

FIGURE 4-9   (a) Formation of a dipeptide. Two amino acids combine chemically to form a dipeptide. Water is produced as a by-product during this reaction. (b) A third amino acid is added to the dipeptide to form a chain of three amino acids (a tripeptide or small polypeptide). The bond between two amino acids is a peptide bond.

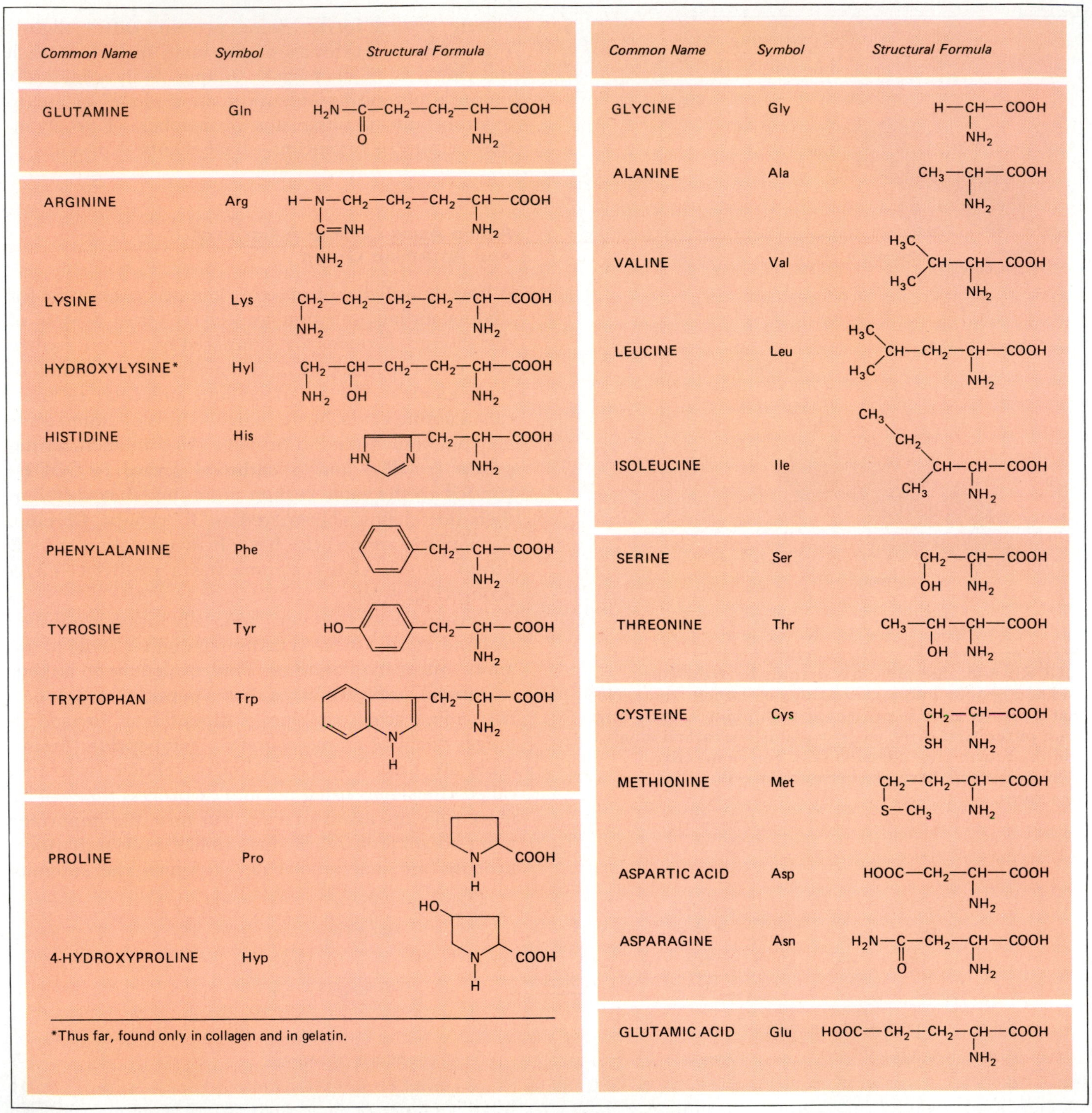

Common Name	Symbol	Structural Formula
GLUTAMINE	Gln	$H_2N-\underset{O}{C}-CH_2-CH_2-\underset{NH_2}{CH}-COOH$
ARGININE	Arg	$H-N-CH_2-CH_2-CH_2-\underset{NH_2}{CH}-COOH$, $\;C=NH$, $\;NH_2$
LYSINE	Lys	$\underset{NH_2}{CH_2}-CH_2-CH_2-CH_2-\underset{NH_2}{CH}-COOH$
HYDROXYLYSINE*	Hyl	$\underset{NH_2}{CH_2}-\underset{OH}{CH}-CH_2-CH_2-\underset{NH_2}{CH}-COOH$
HISTIDINE	His	(imidazole ring)$-CH_2-\underset{NH_2}{CH}-COOH$
PHENYLALANINE	Phe	(benzene ring)$-CH_2-\underset{NH_2}{CH}-COOH$
TYROSINE	Tyr	$HO-$(benzene ring)$-CH_2-\underset{NH_2}{CH}-COOH$
TRYPTOPHAN	Trp	(indole ring)$-CH_2-\underset{NH_2}{CH}-COOH$
PROLINE	Pro	(pyrrolidine ring)$-COOH$
4-HYDROXYPROLINE	Hyp	(HO–pyrrolidine ring)$-COOH$

*Thus far, found only in collagen and in gelatin.

Common Name	Symbol	Structural Formula
GLYCINE	Gly	$H-\underset{NH_2}{CH}-COOH$
ALANINE	Ala	$CH_3-\underset{NH_2}{CH}-COOH$
VALINE	Val	$\overset{H_3C}{\underset{H_3C}{}}CH-\underset{NH_2}{CH}-COOH$
LEUCINE	Leu	$\overset{H_3C}{\underset{H_3C}{}}CH-CH_2-\underset{NH_2}{CH}-COOH$
ISOLEUCINE	Ile	$\underset{CH_3}{\overset{CH_3-CH_2}{}}CH-\underset{NH_2}{CH}-COOH$
SERINE	Ser	$\underset{OH}{CH_2}-\underset{NH_2}{CH}-COOH$
THREONINE	Thr	$CH_3-\underset{OH}{CH}-\underset{NH_2}{CH}-COOH$
CYSTEINE	Cys	$\underset{SH}{CH_2}-\underset{NH_2}{CH}-COOH$
METHIONINE	Met	$CH_2-CH_2-\underset{NH_2}{CH}-COOH$, $\;S-CH_3$
ASPARTIC ACID	Asp	$HOOC-CH_2-\underset{NH_2}{CH}-COOH$
ASPARAGINE	Asn	$H_2N-\underset{O}{C}-CH_2-\underset{NH_2}{CH}-COOH$
GLUTAMIC ACID	Glu	$HOOC-CH_2-CH_2-\underset{NH_2}{CH}-COOH$

FIGURE 4–10   The protein-forming amino acids. Of the many thousands of chemically possible amino acids, these 20-odd amino acids (counting minor variants) are the only ones that are incorporated into the proteins of earthly life. Some other amino acids, however, are employed for special functions—as hormones, for example—and some are chemically modified *after* the protein of which they are a part has been manufactured.

vary along the length of the molecule. This means that the amino acids cannot be joined at random but in accordance with a definite plan or blueprint; proteins do not just happen—they are built. The 20 types of amino acids (Fig. 4–10) found in biological proteins may be thought of as letters of a protein alphabet. Each pro-

tein is a word made up of amino acid letters. To make a given protein, all the necessary amino acids must be available.

With some exceptions, plants can synthesize all their needed amino acids from simpler substances. The cells of humans and animals in general can manu-

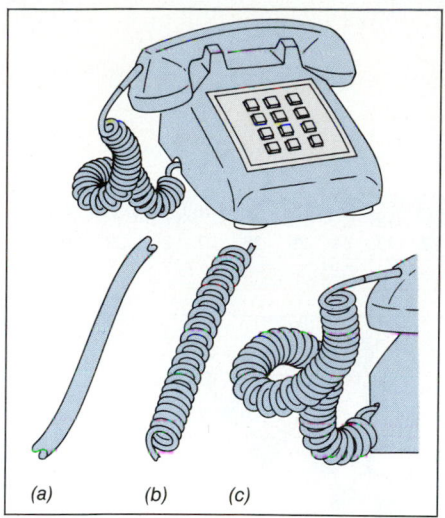

FIGURE 4–11 A telephone handset cord illustrates levels of structure in proteins and nucleic acids. (*a*) Primary structure; (*b*) secondary structure; (*c*) tertiary structure.

(a)   (b)   (c)

facture some, but not all, of the various kinds of biologically significant amino acids if the proper raw materials are available. Amino acids that animals cannot synthesize but must obtain in the diet are known as **essential amino acids.** Animals differ in their biosynthetic capacities; what is an essential amino acid for one species may not be essential in the diet of another.

Proteins Have Several Levels Of Organization

Think about a telephone handset cord (Fig. 4–11). The cord itself is long, thin, and flexible. That is its basic shape or primary structure. To avoid clumsy tangling, it is made in such a way that it coils itself into a long, spring-like secondary shape, which is its secondary structure. Despite that, the cord is usually further twisted or kinked into an unplanned tertiary shape, its tertiary structure.

Several such levels of organization can also be distinguished in the protein molecule. The sequence of amino acids in a polypeptide chain constitutes its **primary structure** (Fig. 4–12). It is this sequence, as we shall see later, that is directly specified by the instructions in a gene.

Because the chemical bonds between and within amino acids in a protein molecule are somewhat flexible, it can be formed into secondary and higher-order shapes. The **secondary structure** of protein molecules involves the coiling of the peptide chain into a helix or some other regular **conformation** (shape). Peptide chains ordinarily do not lie flat or coil randomly, but undergo coiling to yield a specific three-dimensional structure. A common secondary structure in protein molecules is known as the **alpha-helix.** This involves the formation of spiral coils of the polypeptide chain. The alpha-helix is a very uniform geometric structure with 3.6 amino acids occupying each turn of the helix. The helical structure is determined and maintained by the formation of hydrogen bonds between nonadja-

cent amino acids in successive turns of the spiral coil. Other shapes are also possible, though the alpha-helix is probably the most common.

The **tertiary structure** of a protein molecule involves the folding of the alpha-helix upon itself, in folds that impart a specific overall structure to the protein molecule. Hydrogen bonds and other weak bonds between one part of the peptide chain and another part help hold the folds in place. Disulfide bonds (S—S) between certain amino acids and other covalent bonds may also be important in maintaining the tertiary structure of many proteins. The biological activity of the protein depends in large part on the specific tertiary structure of the molecule, held together by these bonds. Almost everyone has fried an egg and observed the change in the "white" (actually an albumin protein) from a transparent, thick liquid to an opaque, rubbery solid. When a protein is heated or treated with any of a number of chemicals, the tertiary structure becomes disordered and the coiled peptide chains unfold to give a more random conformation. This unfolding is accompanied by a loss of the biological activity of the protein, for example, its ability to act as an enzyme. This change is termed **denaturation** of the protein. The structures of proteins form under exact conditions at the time they are produced and usually cannot be re-established after denaturization. For this reason denaturization is usually not reversable. You cannot "un-fry" an egg.

Proteins composed of two or more polypeptide chains have a **quaternary structure.** This refers to the combination of two or more like or unlike peptide chain subunits, each with its own primary, secondary, and tertiary structures, to form the biologically active protein molecule. Hemoglobin, the protein in red blood cells that is responsible for oxygen transport, consists of 574 amino acids arranged in four polypeptide chains—two identical alpha chains and two identical beta chains.

(*text continues on page 75*)

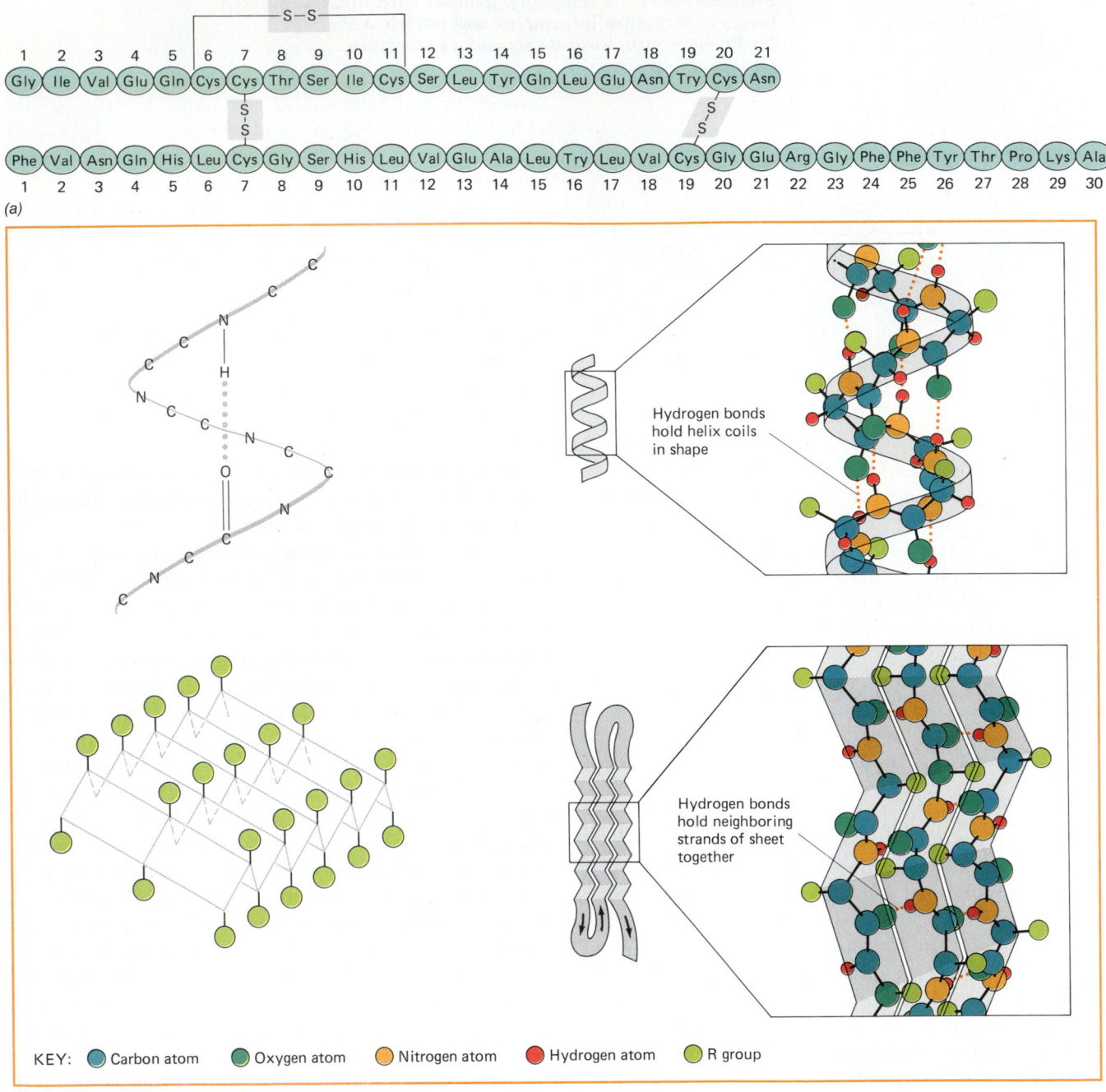

(a)

(b)

KEY: ● Carbon atom ● Oxygen atom ● Nitrogen atom ● Hydrogen atom ● R group

FIGURE 4–12 Protein structure. (*a*) The primary structure of the two polypeptide chains that make up the protein insulin. The primary structure is the linear sequence of amino acids. Each oval in the diagram represents an amino acid. The letters inside the ovals are symbols for the names of the amino acids. Insulin is a very small protein. (*b*) The secondary structure of proteins is commonly an alpha helix. The folds in the helix are held together mainly by hydrogen bonds between oxygen and hydrogen atoms. In some proteins, such as the silk protein fibroin, the backbone of the polypeptide chain is stretched out into a zigzag structure. (*c*) The tertiary structure results from the coiling and folding of the alpha helix (or other secondary structure) into an overall globular or other shape. Hydrogen bonds, bonds between sulfur atoms, and other attractions between atoms are among the forces that hold the parts of the molecule in the designated shape. (*d*) Proteins that consist of more than one polypeptide subunit assume a final quaternary shape. Hemoglobin, a globular-shaped protein containing four polypeptide subunits, is illustrated here. Its quaternary structure consists of the final shape in which the subunits combine. In hemoglobin each polypeptide encloses an iron-containing structure (shown as green discs). (*e*) The silk of a spider's web is composed of a protein with what is known as a "pleated sheet" tertiary structure. To conserve protein, many spiders eat their own webs when they become torn or worn. (Skip Moody/Warren Dembinsky, Jr., Photography Associates)

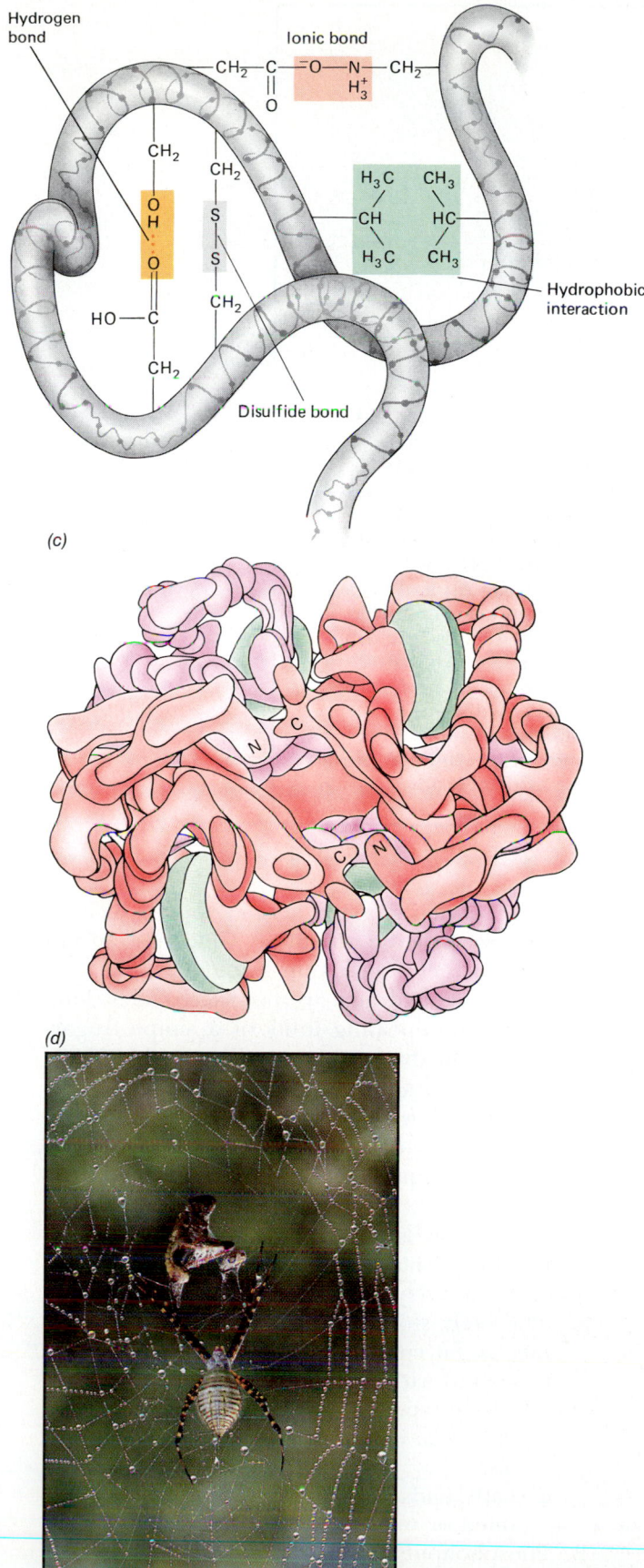

(c)

(d)

(e)

■ NUCLEIC ACIDS CONSIST OF NUCLEOTIDE SUBUNITS

Nucleic acids, like proteins, are large, complex molecules. They were first isolated by Friedrich Miescher in 1870 from the nuclei of pus cells (actually white blood cells), and their name stems from the fact that they are acidic and were first identified in cell nuclei. There are two classes of nucleic acids, **ribonucleic acids (RNA)** and **deoxyribonucleic acids (DNA);** the different kinds of RNA and DNA vary in some of their structural components and in their metabolic functions. DNA contains the instructions for making all the proteins needed by the organism. This information content constitutes the genes themselves, the hereditary material of the cell. The various kinds of RNA function in the process of protein synthesis.

Nucleic acids are composed of **nucleotides,** molecular units that consist of (1) a five-carbon sugar, either ribose or deoxyribose, (2) a phosphate group, and (3) a nitrogenous base that may be either a double-ringed purine or a single-ringed pyrimidine (Fig. 4–13). DNA contains the purines adenine (A) and guanine (G) and the pyrimidines cytosine (C) and thymine (T), together with the sugar deoxyribose and phosphate. RNA contains the purines adenine and guanine and the pyrimidines cytosine and uracil (U), together with the sugar ribose and phosphate.

The molecules of nucleic acids are made of linear chains of nucleotides, each attached to the next by bonds between the sugar molecule of one and the phosphate group of the next. As we will see in our discussion of the genetic code (Chapter 10), the specific information of the nucleic acid is coded in the unique sequence of the four kinds of nucleotides present in the chain. Despite the fact that nucleic acids are inherently simpler than proteins, the information necessary to make specific proteins is stored in and expressed by nucleic acids. The amino acids of proteins are specified by unique sequences of nucleotides.

Besides the importance of nucleotides as subunits of nucleic acids, a number of them serve other vital functions in living cells. **Adenosine triphosphate (ATP),** composed of adenine, ribose, and three phosphates, is of major importance as the energy currency of all cells. The two terminal phosphate groups are joined to the nucleotide by special "energy-rich" bonds, indicated by the ~P symbol. These are energy-rich bonds in the sense that much free energy is released when the bonds are hydrolyzed. The biologically useful energy of these bonds can be transferred to other molecules. Most of the chemical energy of the cell is stored in these high-energy phosphate bonds of ATP, ready to be released when the phosphate group is transferred to another molecule.

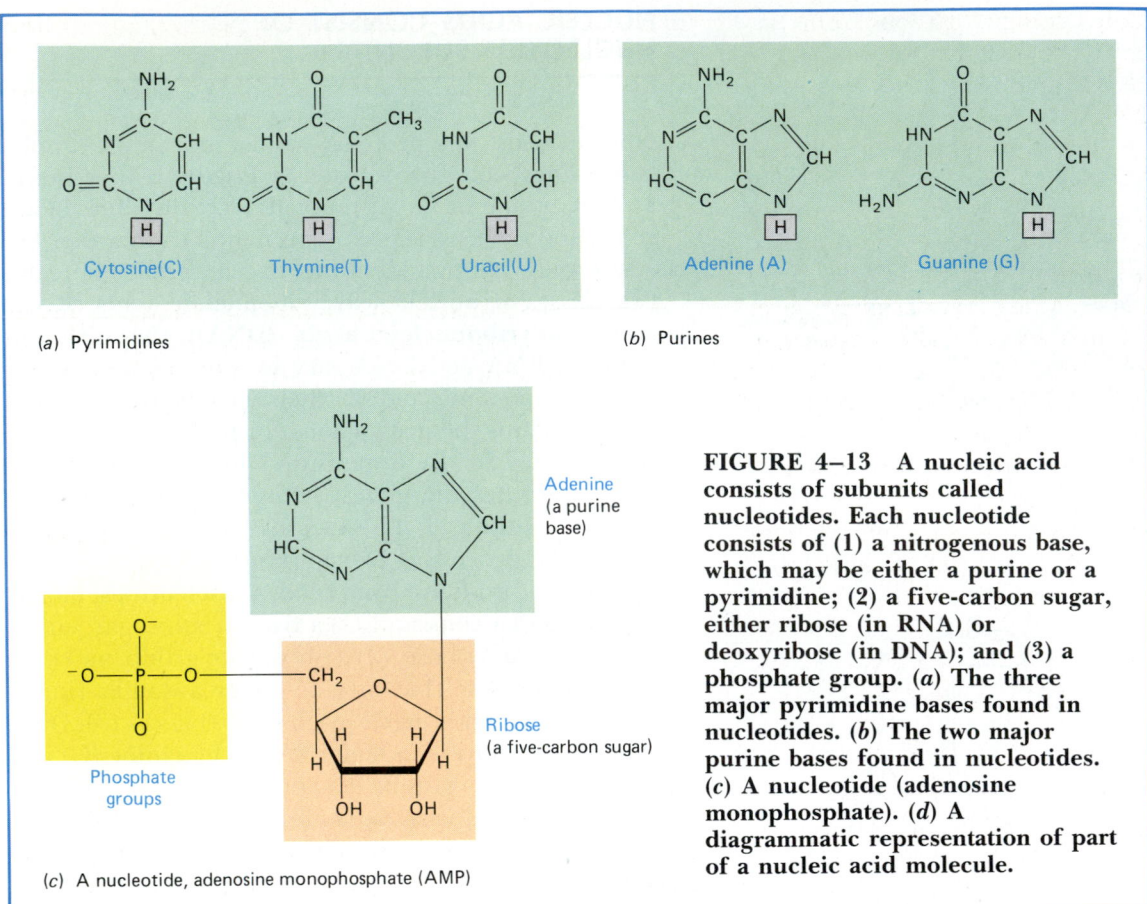

(a) Pyrimidines

(b) Purines

(c) A nucleotide, adenosine monophosphate (AMP)

FIGURE 4–13 A nucleic acid consists of subunits called nucleotides. Each nucleotide consists of (1) a nitrogenous base, which may be either a purine or a pyrimidine; (2) a five-carbon sugar, either ribose (in RNA) or deoxyribose (in DNA); and (3) a phosphate group. (a) The three major pyrimidine bases found in nucleotides. (b) The two major purine bases found in nucleotides. (c) A nucleotide (adenosine monophosphate). (d) A diagrammatic representation of part of a nucleic acid molecule.

A nucleotide may be converted by enzymes called cyclases to a cyclic form. ATP, for example, is converted to cyclic adenosine monophosphate (**cyclic AMP**) by the enzyme adenylate cyclase. **Cyclic nucleotides** play important roles in mediating the effects of peptide or protein hormones and in regulating various aspects of cellular function.

Cells contain several dinucleotides that are of great importance in metabolic processes. For example, nicotinamide adenine dinucleotide (NAD^+) is very important as a primary electron and hydrogen acceptor and donor in biological oxidations and reductions within cells.

■ CHAPTER SUMMARY

I. The major groups of organic compounds are carbohydrates, lipids, proteins, and nucleic acids.

II. Carbohydrates contain carbon, hydrogen, and oxygen in a ratio of approximately 1 carbon to 2 hydrogens to 1 oxygen. Sugars, starches, and cellulose are typical carbohydrates.

A. Monosaccharides are simple sugars such as glucose, fructose, or ribose. Glucose is an important fuel molecule in living cells.

B. Most carbohydrates are polysaccharides, long chains of repeating units of a simple sugar.
1. Carbohydrates are typically stored in plants as starch and in animals as glycogen.
2. The cell walls of plant cells are composed mainly of the polysaccharide cellulose.

III. Lipids are composed of carbon, hydrogen, and oxygen but have relatively less oxygen in proportion to carbon and hydrogen than do carbohydrates. Lipids have a greasy or oily consistency and are relatively insoluble in water.

A. The body stores fuel in the form of neutral fats. A fat consists of a molecule of glycerol combined with one to three fatty acids.
1. Three types of neutral fats are monoacylglycerols, diacyglycerols, and triacylglycerols.
2. Fatty acids, and therefore fats, can be saturated or unsaturated.

B. Phospholipids are structural components of cell membranes.

C. Steroid molecules contain carbon atoms arranged in four interlocking rings. Cholesterol, bile salts, vitamin D, and certain hormones are important steroids.

IV. Proteins are large, complex molecules made of simpler components termed amino acids that are joined by peptide bonds. They are composed of carbon, hydrogen, oxygen, nitrogen, and sulfur.

A. Proteins are important structural components of cells and tissues. Many serve as enzymes or as hormones, and they may also be used as fuel.

B. Four levels of organization can be distinguished in protein molecules.

1. Primary structure is the sequence of amino acids in the peptide chain.
2. Secondary structure refers to the coiling of the peptide chains in a helix or some other regular conformation.
3. Tertiary structure is the folding of the chain upon itself.
4. Quaternary structure is the spatial relationship of the combination of two or more peptide chains.

V. The nucleic acids DNA and RNA store information that governs the structure and function of the organism. Nucleic acids are composed of carbon, hydrogen, oxygen, nitrogen, and phosphorus.

A. Nucleic acids are composed of long chains of nucleotide units, each composed of (1) a nitrogenous base (a purine or a pyrimidine); (2) a five-carbon sugar (ribose or deoxyribose); and (3) a phosphate group.

B. ATP is a nucleotide of special significance in energy metabolism and in the control of cellular function. NAD^+ is an electron and hydrogen acceptor in biological oxidations.

■ POST-TEST

Select the most appropriate term from column B for each entry in column A.

Column A
_____ **1.** a monosaccharide
_____ **2.** a steroid
_____ **3.** a nucleic acid
_____ **4.** oleic acid

Column B
a. cellulose
b. DNA
c. glucose
d. cholesterol
e. none of the preceding

_____ **5.** important constituent of cell membranes
_____ **6.** subunits of proteins
_____ **7.** energy currency of cell
_____ **8.** component of fatty acids

a. ATP
b. glycerol
c. phospholipids
d. amino acids
e. none of the preceding

9. Peptide bonds are found linking _____ .
10. The primary structure of a protein refers to the _____ .

11. _____ is an important component of the cell walls of plant cells.
12. Animals store glucose in the form of _____ .

■ REVIEW QUESTIONS

1. What property of carbon makes it so important in living organisms?
2. Contrast a monosaccharide, such as glucose, with a polysaccharide, such as starch.
3. Why are each of the following biologically important? (a) steroids; (b) phospholipids; (c) polysaccharides; (d) nucleic acids; (e) amino acids.
4. Draw a structural formula of a simple amino acid and identify the carboxyl and amino groups.
5. There are thousands of types of proteins. How does one protein differ chemically from another?
6. Compare proteins with nucleic acids.
7. Why are neutral fats important? What are the molecular components of a neutral fat?

RECOMMENDED READINGS

Bettelheim, F.A., and J. March. *Introduction to General, Organic and Biochemistry*. Philadelphia, Saunders College Publishing, 1990. A readable reference for those who would like to know more about basic chemistry.

Caplan, A. "Cartilage," *Scientific American*, November 1984, 84–94. This fundamental skeletal tissue has unique properties of strength and resilience which now can be explained in terms of the molecular structure of the chemical constituents of the tissue. An excellent example of the chemical basis of biology.

Sharon, N. "Carbohydrates," *Scientific American*, November 1980, 90–116. A discussion of the chains of sugar units that are the most abundant component of living cells and of their roles in normal biological processes and disease.

(Also consult the readings for Chapters 3, 5, and 6.)

5
Cell Structure and Function

OUTLINE

I. The cell is the basic unit of life
II. Cells have many common characteristics
III. Prokaryotes are fundamentally different from those of eukaryotic organisms
IV. Plant cells differ from animal cells
V. The eukaryotic cell's interior contains numerous structures
   A. Some organelles are specialized for manufacturing products
      1. The ER produces protein and other substances
      2. The rough ER contains ribosomes
      3. The Golgi complex is a packaging plant
   B. Other organelles function in cell metabolism and energy use
      1. Lysosomes are digestive organelles
      2. Microbodies contain a variety of enzymes
      3. Mitochondria are the power houses of the cell
      4. Plastids are plant-cell energy traps and storage sacs
   C. Still other structures support the cell, connect it with other cells, and help it move
      1. Cytoskeletal elements give cells their shape and allow movement
      2. Centrioles function in animal mitosis
      3. Cells swim or move materials with cilia and flagella
   D. Vesicles and vacuoles are cavities within cells
   E. The cell nucleus contains genetic information
      1. The nuclear envelope surrounds the nucleus
      2. The nucleolus assembles ribosomes
      3. Chromosomes are packages of hereditary information

Focus on studying cells in a variety of ways

LEARNING OBJECTIVES

After reading this chapter you should be able to:

1. Justify that the cell is considered the basic unit of life and state the cell theory.
2. Describe the general characteristics of cells, for example, size range, and shape.
3. Identify methods by which scientists study cells.
4. Contrast prokaryotic and eukaryotic cells; contrast plant and animal cells.
5. Draw and label a diagram of a prokaryotic cell, a plant cell, and an animal cell. Describe and list the functions of the principal cell organelles.
6. Distinguish between smooth and rough endoplasmic reticulum and describe the functional relationship between ribosomes and endoplasmic reticulum.
7. Describe how the Golgi complex packages secretions and manufactures lysosomes.
8. State the function of the mitochondria.
9. Compare microtubules and microfilaments.
10. Describe the structure and summarize the proposed function of the eukaryotic flagellum or cilium.
11. Describe the structure and function of the cell nucleus.

Cells of *Elodea*, an aquatic plant, that have been exposed to a concentrated salt solution. Each cell has shrunk away from its wall due to loss of water by osmosis. (Runk/Schoenburger from Grant Heilman)

Although some living systems consist of only one cell and others of several billion, even the most complex organism begins life as a single cell, the fertilized egg. In most multicellular organisms, including human beings, this cell divides to form two cells, and each new cell divides again and again, eventually forming the complex tissues, organs, and systems of the developed organism. Like the bricks of a building, cells are the building blocks of the organism. (Refer often in your study of this chapter to Windows on the Plant and Animal Cells, which are bound into this book.)

KEY CONCEPTS

☐ The cell is the basic unit of life.

☐ Cells are adapted for life as independent organisms or specifically adapted for their roles in the body of a multicellular organism.

☐ Cells must usually be viewed by microscope or studied by techniques of indirect chemical or physical analysis. The electron microscope makes high magnifications practical by means of its superior resolving power.

☐ There are fundamentally two kinds of cells—those of eukaryotes, which are equipped with nuclear envelopes and other elaborate internal membranes, and those of prokaryotes, which have no nuclear membranes and usually no other membrane organelles.

☐ The eukaryotic cell contains mitochondria which release energy for cellular use, a nucleus which contains genetic information, and rough and smooth endoplasmic reticulum which synthesize proteins and lipids. Some eukaryotic cells contain chloroplasts which function in photosynthesis and a cell wall that imparts shape.

☐ The eukaryotic cell has a cytoskeleton made of microtubules and microfilaments, vacuoles that hold fluid, a cell wall that imparts shape, lysosomes that digest old organelles, and a Golgi complex that produces lysosomes and that packages and/or stores cell secretions.

☐ THE CELL IS THE BASIC UNIT OF LIFE

The **cell** is the smallest unit of living material capable of carrying on all the activities necessary for life. We might say that it is the smallest structure with a complete metabolism because it has all of the physical and chemical components needed for its own maintenance and growth. When provided with essential nutrients and an appropriate environment, cells can be kept alive in laboratory glassware for many years. No cell part is capable of such survival by itself.

One of the basic generalizations of biology is the **cell theory.** In 1838 and 1839 two German biologists, Matthias Schleiden (a botanist) and Theodor Schwann (a zoologist), proposed that all living things are made up of cells and cell products and that the cell is the basic unit of living organisms. The cell theory was extended in 1855 by Rudolf Virchow, who stated that new cells come into existence only by the division of previously existing cells. Cells cannot arise by spontaneous generation from nonliving matter. In 1880 August Weismann pointed out the corollary of this, that all the cells living today can trace their ancestry back to ancient times (Fig. 5–1).

☐ CELLS HAVE MANY COMMON CHARACTERISTICS

A cell consists of jelly-like **cytoplasm** composed mostly of water (70 to 90%) surrounded by a **plasma membrane,** a physical boundary that separates the cell from the outside environment. Most cells contain a nucleus which contains the deoxyribonucleic acid (DNA), the chemically coded instructions for synthesizing all the proteins needed for growth, repair, and reproduction. Many other **organelles,** internal cellular structures with specialized functions, are suspended within the cytoplasm.

Almost all cells are microscopic. An "average" cell measures about 10 micrometers (1/2500 inch) in diameter. This means that if you could line up about 2500 typical cells end to end, the resulting cellular parade would measure less than 3 centimeters (only about 1 inch).

Although they have many features in common, cells vary widely in size and appearance (Fig. 5–2). In fact, the size and shape of a cell are related to its specific function. Epithelial cells, which are specialized to cover body surfaces, look like tiny building blocks. Nerve cells have long extensions that receive or transmit messages long distances through the body. An extension of the sciatic nerve, for example, may extend from spinal cord to foot. Although such a nerve cell may be more than a meter long, its diameter is so tiny that no part of it can be seen without the aid of a microscope. Certain white blood cells in the body resemble unicellular amebas in their ability to change shape as they flow along from one location to another. Plant cells often have large, fluid-filled structures called vacuoles, and these cells may also contain chloroplasts. The largest cells are birds eggs, which consist largely of yolk that provides nourishment for the developing bird.

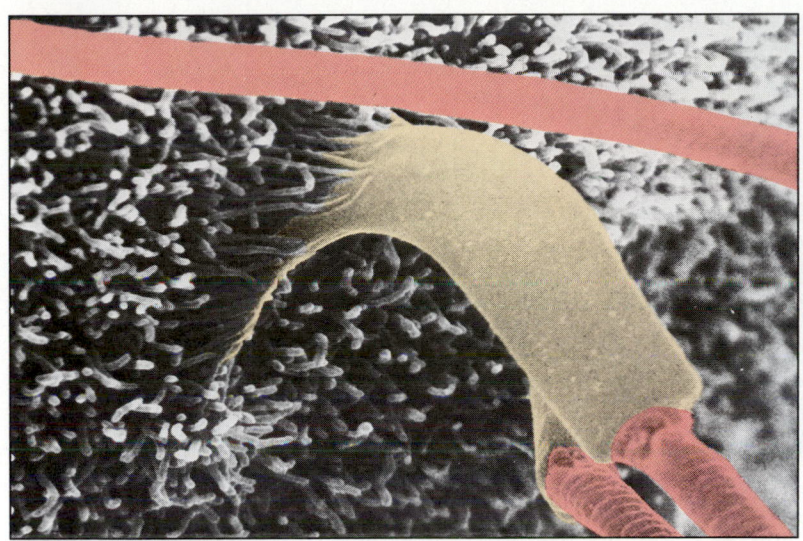

(a)

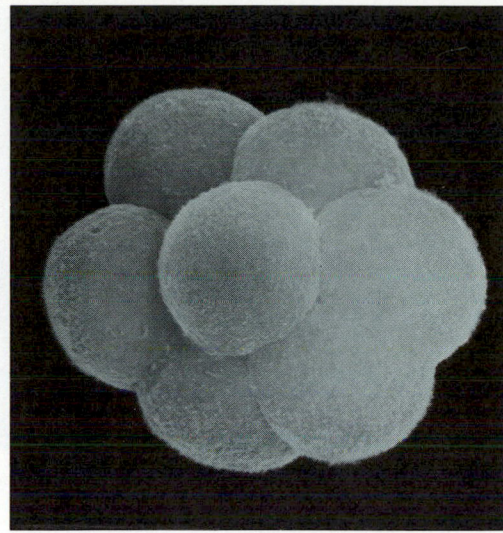

(b)

FIGURE 5–1 Every cell comes from a pre-existing cell, according to the cell theory, and here we see the drama enacted whereby a new individual comes into existence by the union of two cells. (*a*) Shaped like a scimitar, the head of a hamster sperm is enveloped by finger-like microvilli of the egg surface into which it gradually sinks. When these two reproductive cells have united, the resulting fertilized egg, still a single cell, soon divides, and does so again and again. (*b*) A cluster of cells results, each of which continues to divide while the new individual takes form. Though the organism may be composed of billions of cells, the ancestry of each cell can be traced back to the fertilized egg. (*a*, Drs. David M. Phillips and Ryuzo Yanagimachi, *Development, Growth and Differentiation* 24, 1982; *b*, Drs. Yehuda Ben-Shaul, Karen Atzt, and Dorothea Bennett)

■ PROKARYOTIC CELLS ARE FUNDAMENTALLY DIFFERENT FROM EUKARYOTIC CELLS

Members of kingdom Monera—the bacteria and cyanobacteria—consist of prokaryotic cells, which are fundamentally different from the eukaryotic cells of all other organisms. **Prokaryotic cells** are simpler and generally smaller than eukaryotic cells and are thought to have evolved first. The word prokaryotic means "before the nucleus," and these cells have no distinct nucleus. Prokaryotic cells often do have one or more nuclear regions, sometimes referred to as **nucleoids,** in which DNA is concentrated (Fig. 5–3). However, a nucleoid is not surrounded by a membrane.

Prokaryotic cells also lack most other membrane-bound organelles typical of eukaryotic cells. In some prokaryotic cells the plasma membrane is folded inward to form a complex of internal membranes (the **mesosome**) along which the reactions of cellular respiration are thought to take place. Prokaryotic cells that carry out photosynthesis contain the green pigment chlorophyll associated with flat, sheet-like membranes called **lamellae,** but these lamellae are not distinct organelles. Finally, since prokaryotes only possess a small fraction of the genetic information of a typical eukaryote cell, their genetic material is more simply organized and these cells divide by a simple form of binary fission rather than the elaborate process of mitosis characteristic of the eukaryotic cell.

As implied by the term *eukaryotic,* which means "good nucleus," or "true nucleus," **eukaryotic cells** do have distinct nuclei surrounded by nuclear envelopes. They also have many types of membrane-bound organelles that partition the cytoplasm into compartments.

Cells are so small that one might well wonder how we know so much about what goes on inside them. Actually, a variety of instruments and techniques can be used to supplement the naked human eye (see Figure 5–4).

The biologist's most important tool for studying the internal structure of cells has been the microscope. Anton van Leeuwenhoek (1632 to 1723) is credited with developing some of the earliest microscopes and with leaving written records of the structures he studied. In 1665 Robert Hooke (1635 to 1703) examined a slice of cork with the aid of a crude, homemade microscope. Because the tiny compartments he saw reminded him of the little rooms, or cells, of a monastery, he called them cells. What Hooke saw were the

(*Text continues on page 83*)

(a)

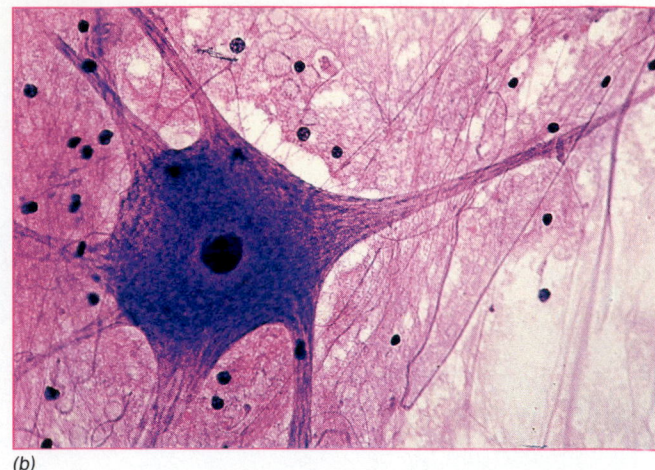

(b)

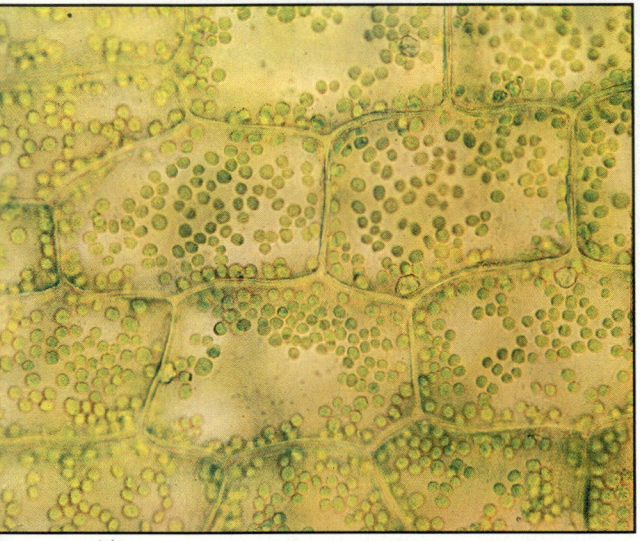

(c)

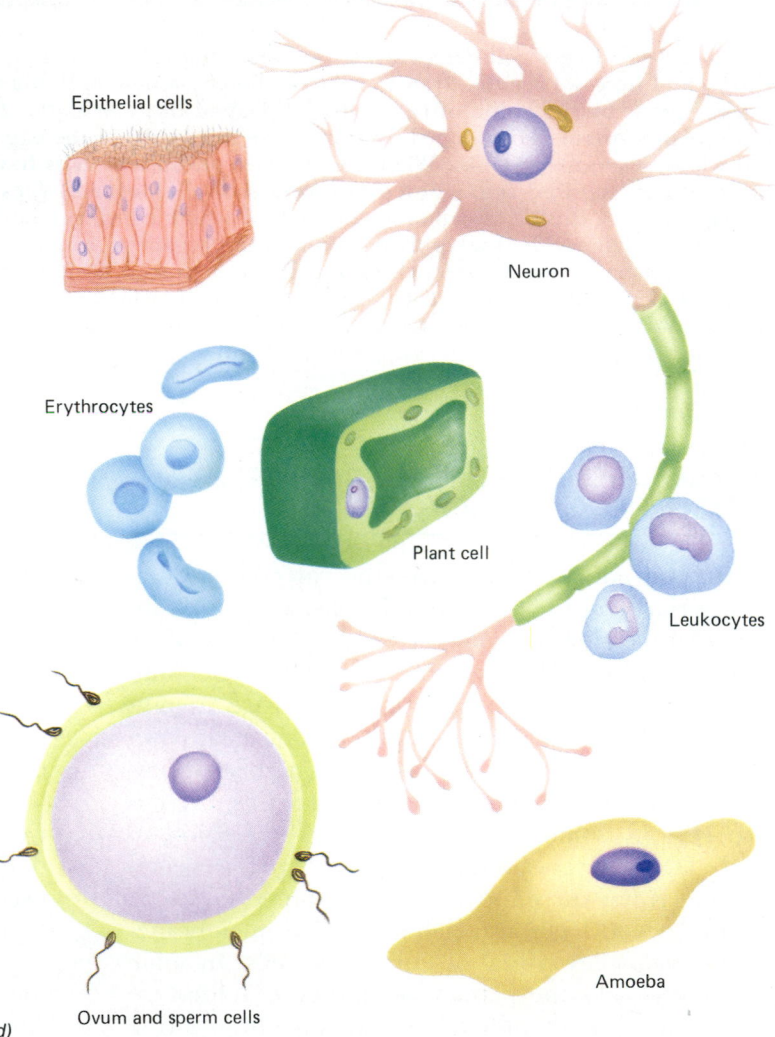

Epithelial cells

Neuron

Erythrocytes

Plant cell

Leukocytes

Ovum and sperm cells

Amoeba

(d)

FIGURE 5–2 Although all the cells of higher organisms share the same basic plan, there is a great variety of specific cellular body plans. Some cells are organisms in their own right, and some make up the bodies of multicellular organisms. (*a*) Mixed algae. These protists are single-cell or colonial, in some cases joined end to end to form long thread-like colonies. Such plant-like protists have cellulose cell walls. (*b*) Nerve cell. Animal cells lack cell walls but, like other cells, are surrounded by plasma membranes. In a multicellular animal each cell has a function to perform and is specialized for that function, in this case, the transmission of impulses. (*c*) Elodea leaf. This aquatic plant has roughly rectangular cells in which the cell walls and chloroplasts are visible. The location of the nucleus can be made out in one or two of them as a rounded area surrounded by chloroplasts. Most cell parts can be made out only if stained. (*d*) Various cells. Most of these are differentiated cells, specialized for a particular body function. Even the ovum is specialized in a sense; it gives rise to an embryo. The amoeba, an organism in its own right, must carry out all the functions in a single cell which are performed by specialized cells in a multicellular organism. (*a*, Dwight R. Kuhn; *b*, Ed Reschke; *c*, Runk/Schoeberger from Grant Heilman)

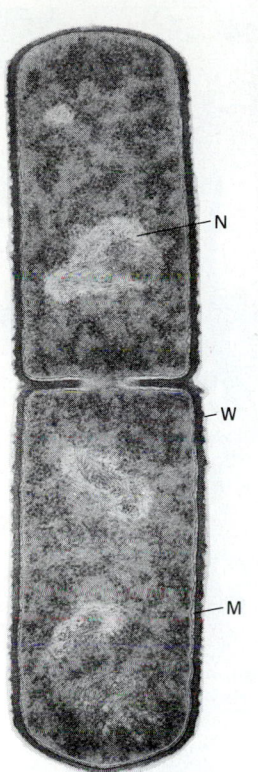

FIGURE 5–3 **The structure of a dividing prokaryotic cell. An electron micrograph of the bacterium,** *Bacillus subtilis.* **This cell has a prominent cell wall (W) surrounding the plasma membrane (M). The nuclear regions (N) are clearly visible.** (Courtesy of A. Ryter)

cell walls of dead cork cells (see Figure 5–5). In later observations he described cell contents, but it was not until two centuries later that scientists realized that the important part of the cell is its contents and not its outer walls or membranes. During the 1800s scientists studied various cells and observed a variety of intracellular structures. They refer to these structures as **organelles** (little organs) because they were thought to perform special jobs within the cell, just as our organs perform specific jobs within our bodies.

The ordinary compound light microscope, the kind used by students in most college laboratories, was responsible for the discovery of most cell structures. The **light microscope,** gradually improved since Hooke's time, uses visible light as the source of illumination. During the last four decades the development of the electron microscope has enabled researchers to study the fine detail, called the **ultrastructure,** of cells. The **electron microscope** floods the specimen being studied with a beam of electrons rather than with light waves.

Magnification is the ratio of the size of the image to the size of the specimen. Whereas the ordinary light microscope can magnify a structure about 1000 times, the electron microscope can magnify it 250,000 times or more (Fig. 5–6).

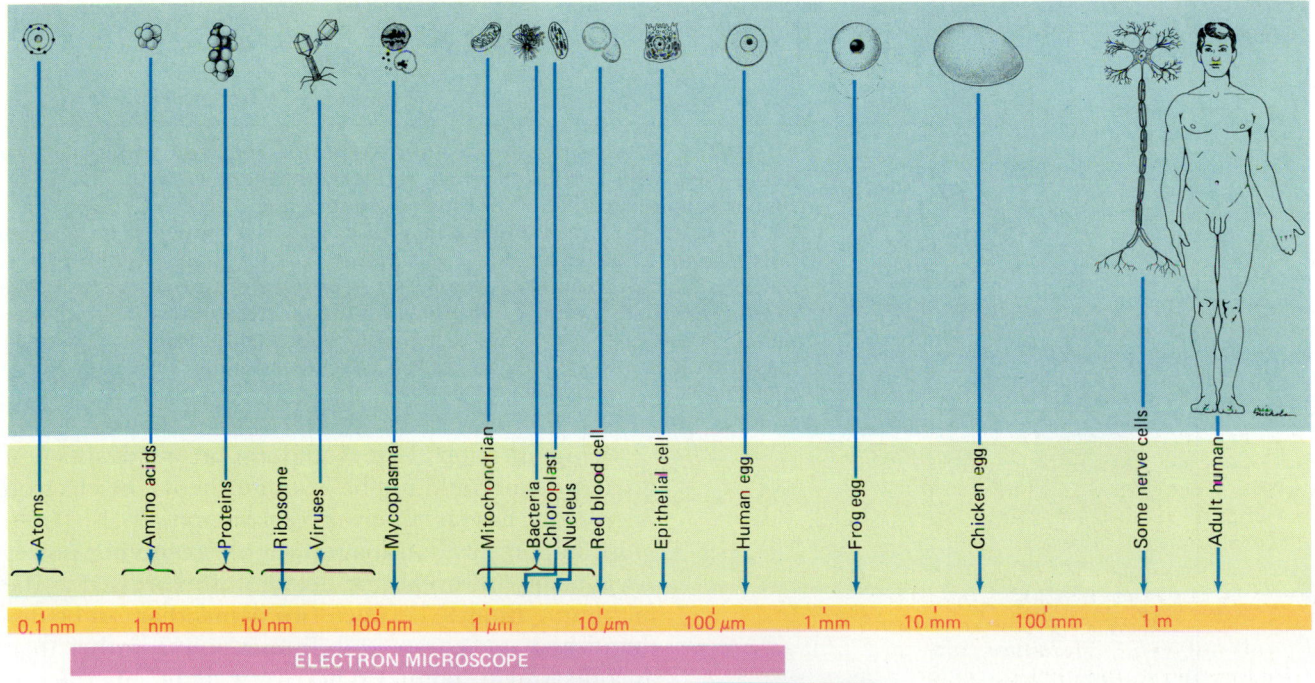

FIGURE 5–4 **Living things are composed of many progressively smaller parts. Most levels of biological organization are imperceptable to the unaided senses, so that to study them we must make use of a variety of instruments and indirect techniques. Modern technology serves biological science by making a variety of methods available that probably could not even have been** understood in principle by scholars of the past. As a result, by the time you have completed this course, you will know vastly more about biology than anyone living earlier than 1900. Yet it is the labors of those scholars that have bestowed this great treasury of knowledge upon us.

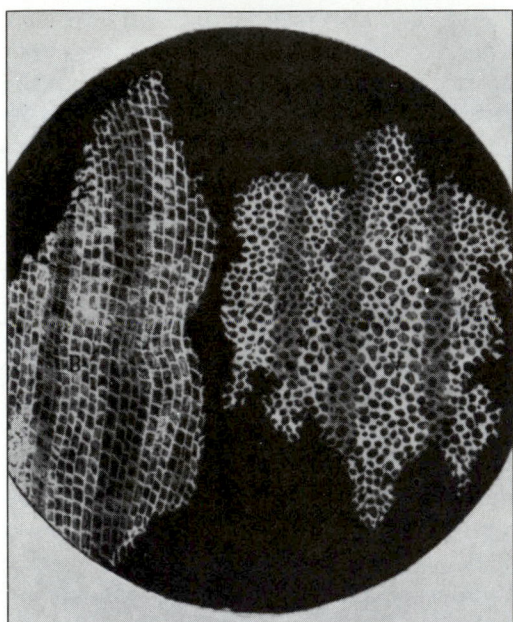

FIGURE 5–5 (*a*) Drawing by Robert Hooke of the microscopic structure of a thin slice of cork. Hooke was the first to describe cells—his observations were based on the cell walls of these dead cork cells. (From the book *Micrographia*, published in 1665, in which Hooke described many of the objects he had viewed using the compound microscope he constructed.)

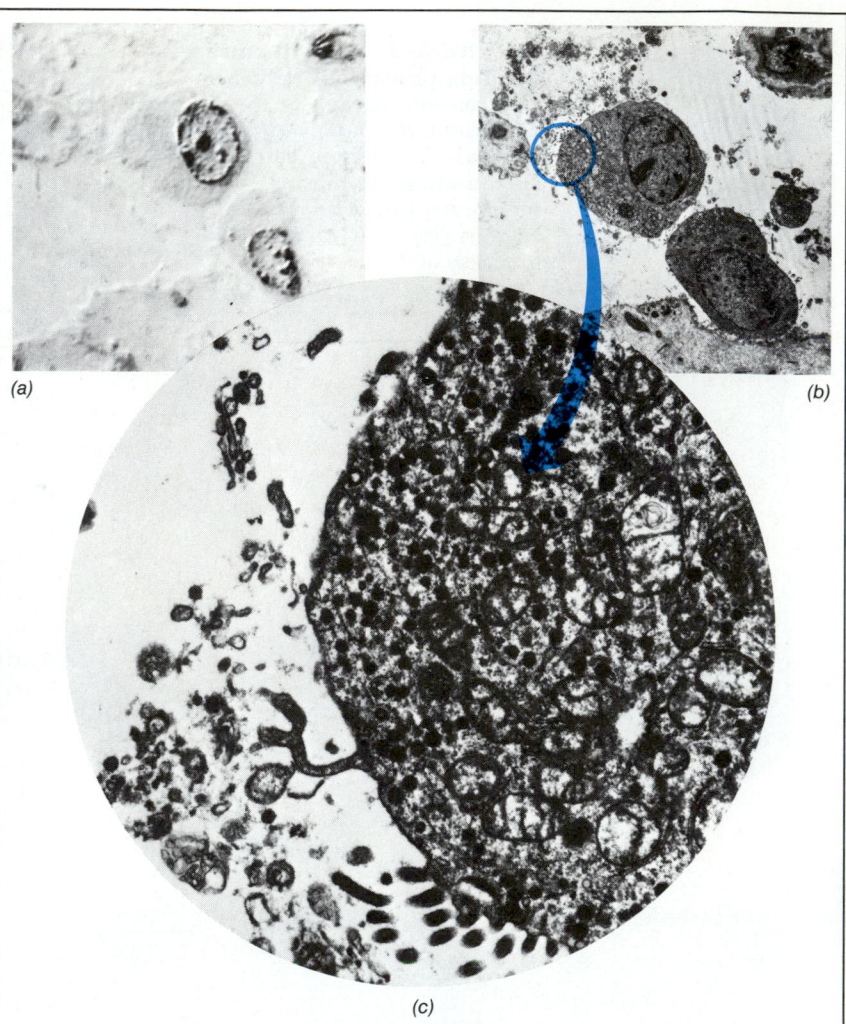

FIGURE 5–7 Comparison of a photograph taken with a modern light microscope with two taken using an electron microscope. (*a*) Lung cancer cells magnified 1800 times as seen using a light microscope. (*b*) The same cells, at the same magnification, seen through an electron microscope. The clearer detail is a result of the greater resolving power of the electron microscope. (*c*) A portion of one of the cells seen in (*b*) magnified about 17,000 times by the electron microscope. Note the black granules and other detail not visible at the lower magnification. (Courtesy of Zeiss. The electron micrographs were taken with Zeiss EM 9S-2 by Dr. Harry Carter.)

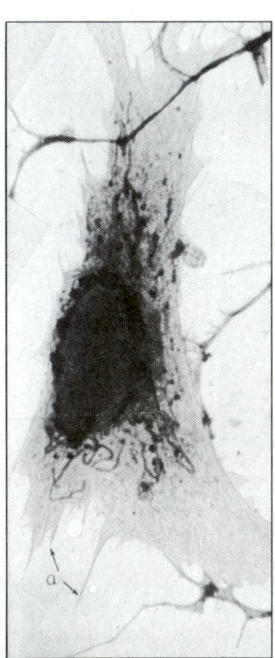

FIGURE 5–6 The first published electron micrograph of a cell was taken by Keith R. Porter, Albert Claude, and Ernest J. Fullam. It was published in the *Journal of Experimental Medicine* in 1945.

Magnification alone would not be an advantage if greater detail could not be distinguished. The electron microscope has far superior resolving power also. Even more important than magnification, **resolving power** is the ability to reveal fine detail; it is expressed as the minimum distance between two points that can be distinguished as separate and distinct points, rather than as one single point. Whereas a light microscope equipped with very fine lenses can resolve objects about 500 times better than the unaided human eye, the electron microscope has a resolving power of more than 10,000 times that of the human eye (Fig. 5–7).

Cell structure has also been studied extensively by physical and chemical methods (Fig. 5–8). Cells can be broken apart and then centrifuged (spun) at high speeds to separate the cellular components. Using a

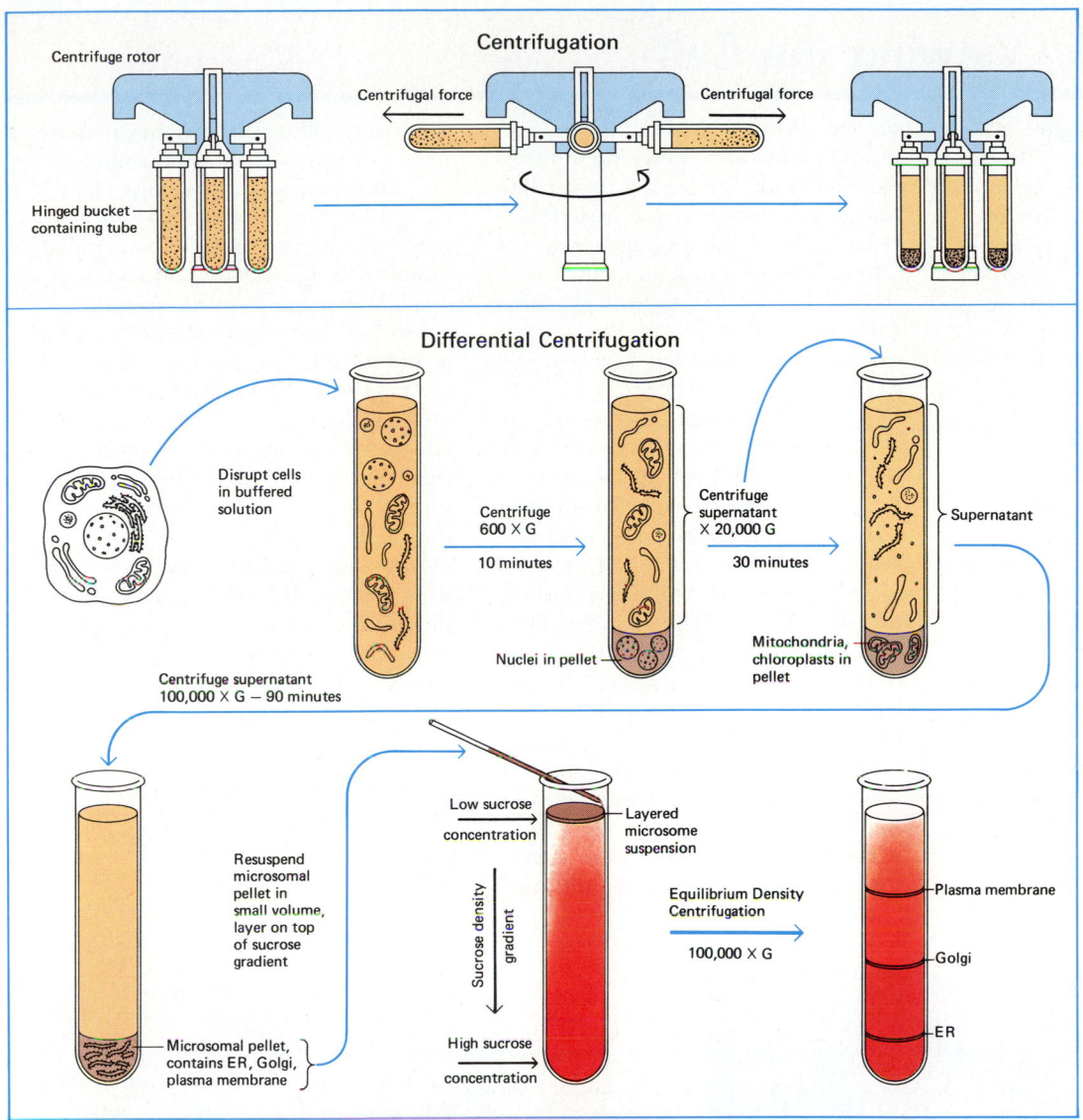

FIGURE 5–8   Before cells can be studied, the tissue they form must be ground up, as often must the cells themselves. The resulting slurry is treated to complete the disruption of the cells, then centrifuged and resuspended in special solutions of precisely controlled density. Careful centrifugation can separate many of the cell parts, which can then be subjected to chemical or other analysis.

variety of techniques, organelles can be separated from one another on the basis of size or density. Once separated from its surrounding cellular components, a given type of organelle can be analyzed biochemically to determine its composition. Another technique used to determine the function of an organelle is to monitor its activities in a test tube under controlled conditions.

■ PLANT CELLS DIFFER FROM ANIMAL CELLS

Plant cells (see acetate insert) differ from animal cells in several ways: (1) Although all cells are limited by plasma membranes, plant cells are also surrounded by stiff **cell walls** of cellulose, which limits any change in

position or shape; (2) plant cells contain **plastids,** membrane-bound structures that produce and store food material, the most familiar and abundant of which are **chloroplasts;** (3) most plant cells have one large or several small, conspicuous compartments, called **vacuoles,** used for storing nutrients and waste products; and (4) certain organelles such as centrioles and lysosomes are absent in plant cells.

The plant cell is often more complex than that of an animal due to the fact that it contains chloroplasts and other plastids (Fig. 5–9). Most of what is present in an animal cell is also found in plant cells. This raises the intriguing possibility that future biological engineering may make it possible to permanently implant

(*Text continues on page 88*)

FOCUS ON   Viewing the Cell

Viewing Cells with the Light Microscope

The light microscope uses visible light as the source of illumination (Fig. A). It has a resolving power of about 200 nanometers (0.2 micrometer). Objects can be magnified about 1000 times with good resolution using the light microscope. Beyond that magnification, structures may appear larger, but they are not clearer.

To be viewed with the light microscope, specimens must be very thin. Single-celled organisms—such as amebas and some cells from multicellular organisms, like sperm cells—can be viewed in the living state. However, tissues are too thick and must be sectioned (cut) into very thin slices. They are generally preserved and then stained with special dyes. When using an ordinary light microscope (such as those you will probably be using in the biology laboratory), we can distinguish certain parts of the cell only when they are

stained differently from other parts. Finding ways to produce such differential staining has been a preoccupation of microscopy for at least 150 years and is still important today. Often, however, staining is only possible if the cells are dead, and even when this is not the case the stain may alter cell function. It is best to work with unstained cells, if possible.

In ordinary light microscopy the absorption of light by the specimen is about the same for all cell parts. But both the **phase-contrast microscope** and the **interference microscope** permit us to readily view parts of living cells that would ordinarily not absorb much light (Fig. B). The technique converts small differences in the way specimens refract (bend) light to much greater apparent differences, causing parts of the specimen to appear brighter than others. In the **dark-field microscope** only the light scattered from edges or particles in the specimen can enter the microscope lenses that produce

the image. Thus the cell shows up as bright against a dark background.

Fluorescence microscopy (Fig. C) is a kind of staining that can be made even more specific than traditional methods. If an animal such as a rabbit is injected with a particular substance that is suspected of playing a role in cells, its immune system will often produce specialized protein molecules, called antibodies, that will attach to that substance whenever they contact it. If these antibodies are added to a preparation of cells, they will attach to the cells containing the suspected material. Certain dyes are then added which attach to the antibodies, and ultraviolet light is shined on the preparation. Cells that have the dye then light up brilliantly, or fluoresce.

Viewing Cells with the Electron Microscope

The electron microscope uses a beam of electrons as a source of illumination instead of light (Fig. D). The electron beams have much shorter wavelengths than visible light. Because electrons have electrical charges, a magnetic field can be used to direct them. Two types of electron microscopes in common use are the transmission electron microscope and the scanning electron microscope. A photograph taken with an electron microscope is called an **electron micrograph (EM).**

Electrons Pass Through the Object Being Viewed with the Transmission Electron Microscope

In the transmission electron microscope a beam of electrons is transmitted through the specimen and falls upon a photographic plate or a fluorescent screen. Before it can be viewed (Fig. E), the specimen must be embedded in plastic and cut in ultrathin sections so that the beam of electrons can pass through it. This is a disadvantage because live specimens cannot be viewed. However, the transmission electron microscope has been invaluable for studying details of internal cell structure.

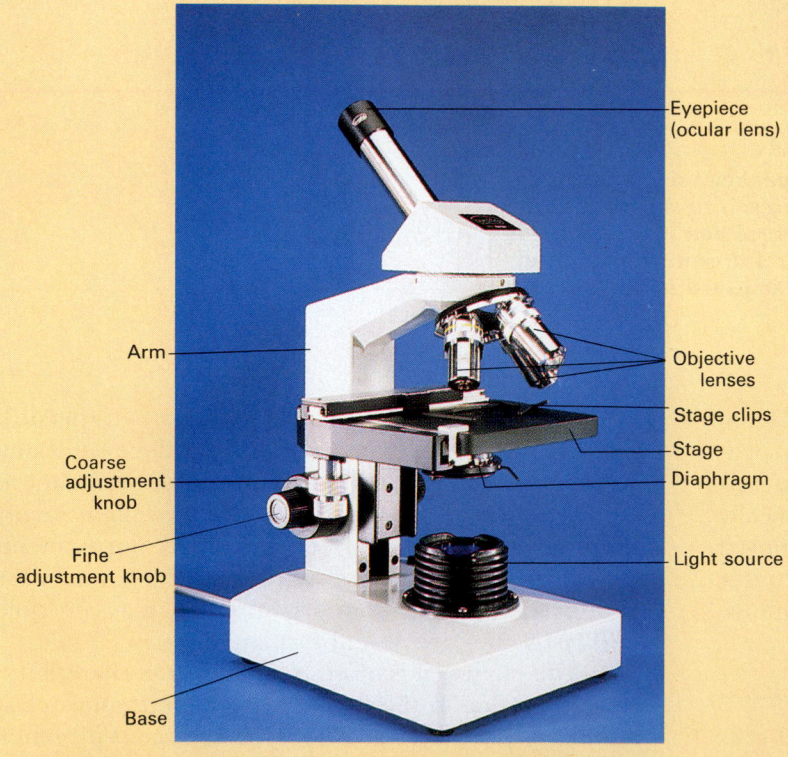

Eyepiece (ocular lens)

Arm

Objective lenses

Stage clips

Stage

Diaphragm

Coarse adjustment knob

Fine adjustment knob

Light source

Base

A   Parts of the light microscope. (Courtesy of the American Optical Corporation)

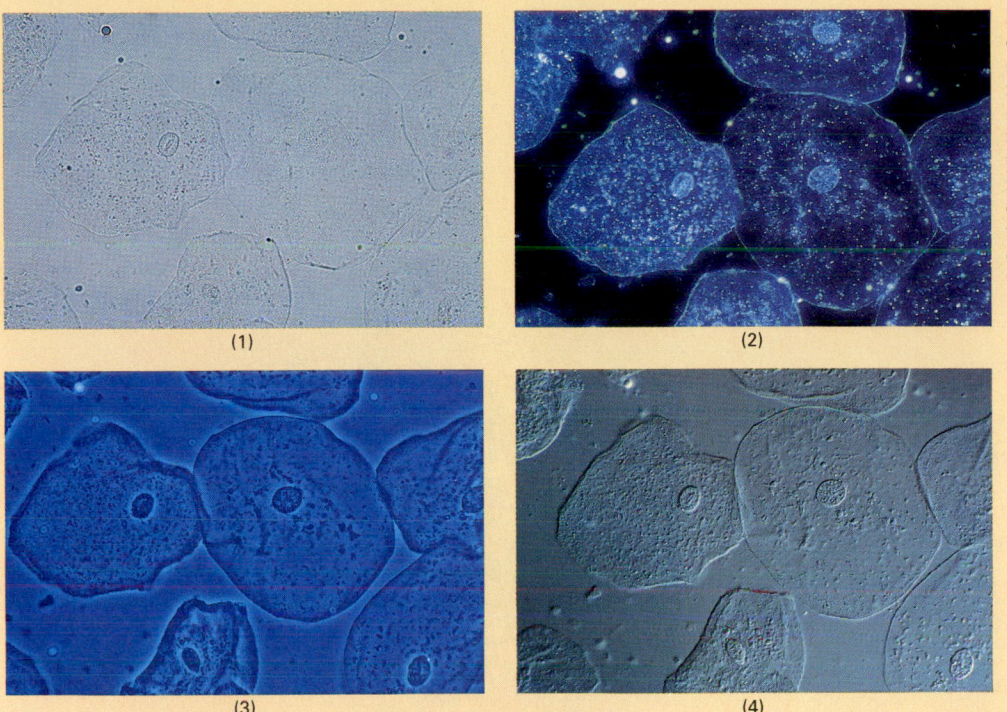

(1)　(2)

(3)　(4)

B (1) through (4), epithelial cells using (1) bright field (transmitted light), (2) dark field, (3) phase contrast, and (4) Nomarski differential interference microscopy. The phase contrast and differential interference microscopes enhance detail by increasing the differences in optical density in different regions of the cells. (Jim Solliday/Biological Photo Service)

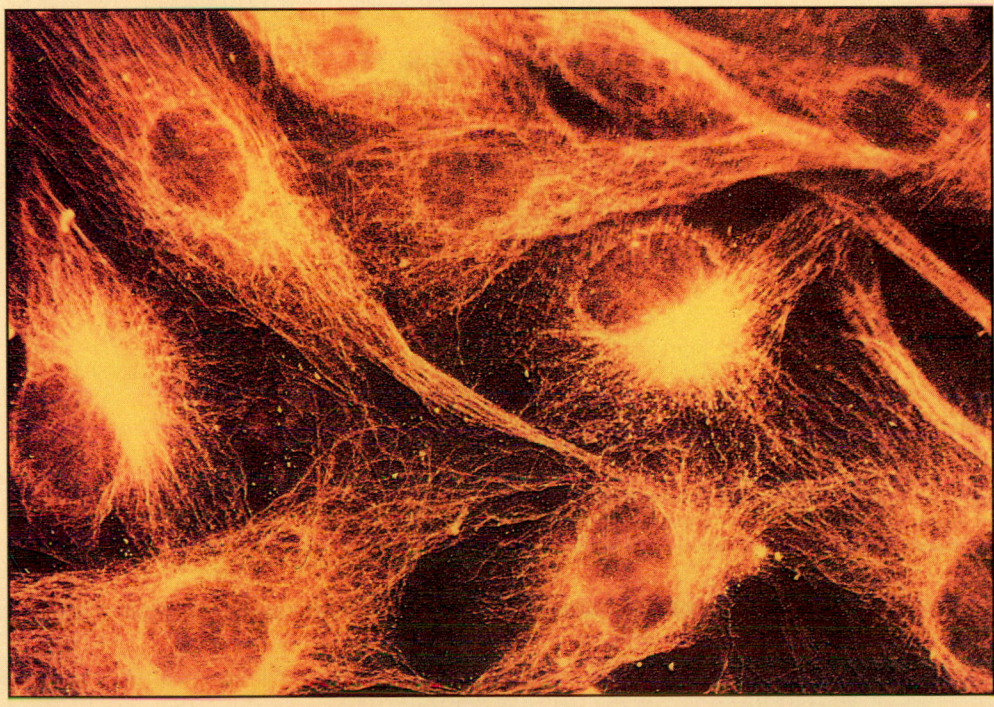

C Microtubules in embryonic (fibroblast) cells stained by a special technique that causes them to fluoresce. (K. G. Murti/ Visuals Unlimited)

The Scanning Electron Microscope Produces an Illusion of Depth

In **scanning electron microscopy** (Fig. F) the specimen is coated with a metal, often gold. The electron beam does not pass through the specimen. Instead, it is directed along the surface of the specimen. The contour of the specimen causes variations in the angle with which the beam strikes the various points of the specimen. This leads to variations in the intensity with which secondary electrons are emitted from the specimen's surface. A recording of the emission from the specimen provides a picture of the three-dimensional nature of the specimen.

Scanning electron microscopy provides information about the

(Text continues on page 89)

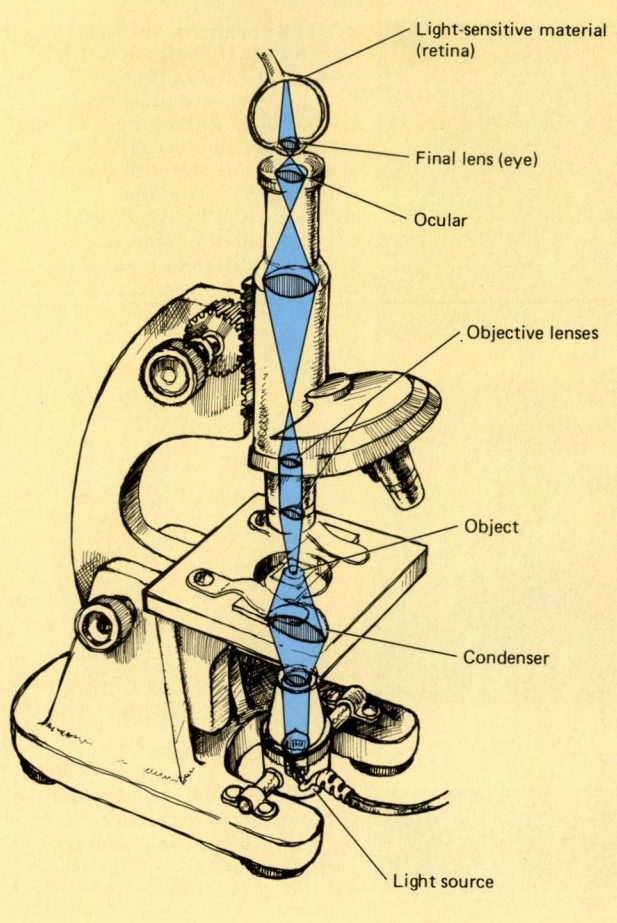

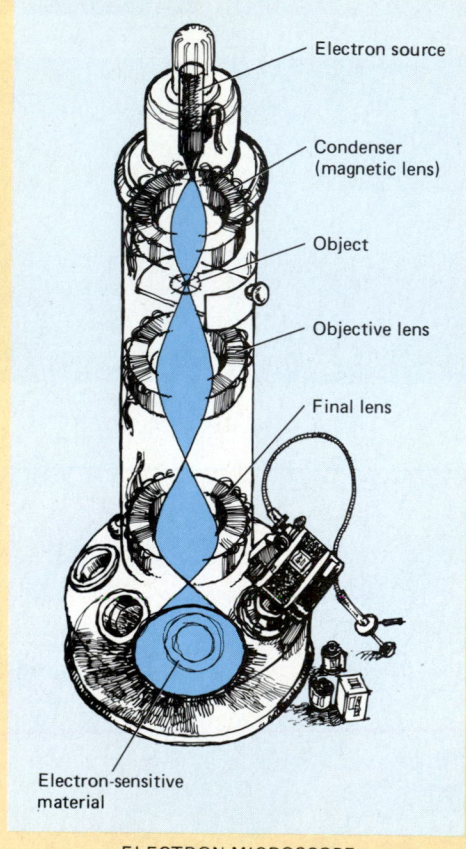

ELECTRON MICROSCOPE

D   A comparison of a light microscope with an electron microscope. In each kind of instrument, light rays or an electron beam are focused by the condenser lens onto the specimen. The objective lens forms a first magnified image of the specimen, which is further magnified by the projector lens onto a ground glass screen in the light microscope or onto a fluorescent screen in the electron microscope. The lenses in the electron microscope are actually magnets that bend the beam of electrons.

chloroplasts in animal cells. In fact, this has already been done in cells raised in tissue culture, though temporarily.

▪ THE EUKARYOTIC CELL'S INTERIOR CONTAINS NUMEROUS STRUCTURES

Early biologists believed that the cell consisted of a homogeneous jelly, which they referred to as protoplasm. With the electron microscope and other modern research tools, perception of the world within the cell has been greatly expanded, and we no longer view the cellular contents as a substance. We now know that the cell is a highly organized, amazingly complex structure (see acetate insert and Fig. 5–10). It has its own control center, internal transportation system, power plants, factories for making needed materials, packaging plants, and even a "self-destruct" system. Today the word protoplasm, if used at all, is used in a very general way. Specifically, the portion of protoplasm outside the nucleus is called **cytoplasm,** and the corresponding jelly-like material within the nucleus is termed **nucleoplasm.** Within the fluid component of

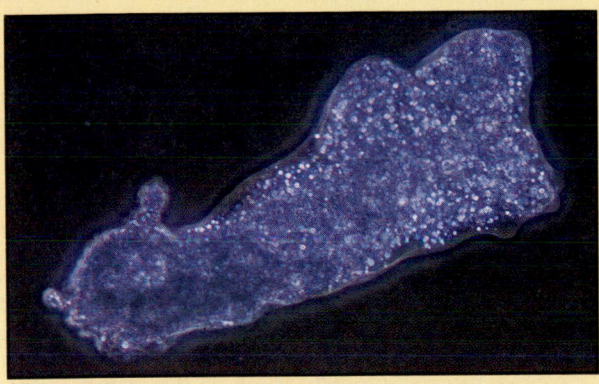
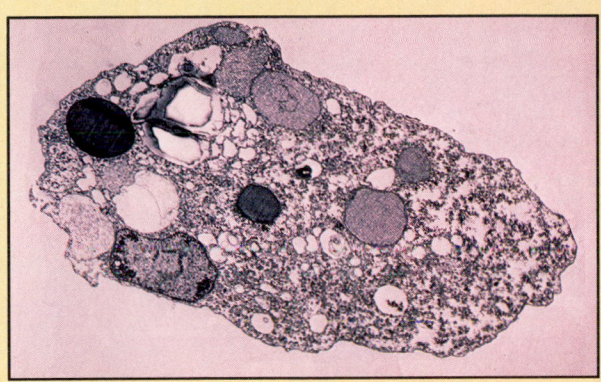

(1)   (2)

**E (1) Light micrograph of *Amoeba proteus*, a sarcodinian (approximately ×120).
(2) Transmission electron micrograph of *Entamoeba histolytica*, a parasitic
sarcodinian. The dark bodies are recently ingested red blood cells (approximately
×550).** (Walker England, Photo Researchers, Inc.; *b*, Science Photo Library, Photo Researchers,
Inc.)

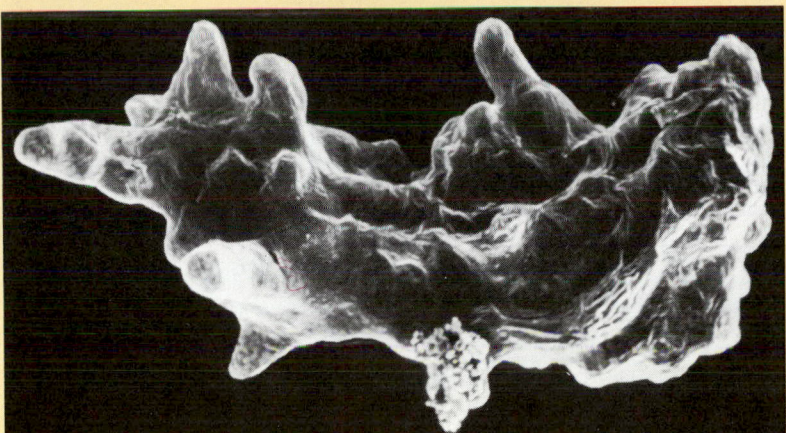

**F Scanning electron micrograph of *Amoeba
proteus*, a sarcodinian, giving a remarkably clear
picture of the surface of the animal.** (Courtesy of Dr.
Eugene Small)

shape and surface of the specimen
that could not be gained from transmission electron microscopy. Scanning electrons provide striking views
of a specimen's surface but cannot
be used to study internal structures
without special preparation. As is the
case with other forms of electron
microscopy, the scanning electron
microscope cannot usually be used to
view living specimens.

New Kinds of Microscopy May Pave the Way for New Advances in the Understanding of Cell Structure and Function

There is no reason to think that microscope technology is at a standstill.
New techniques involving gamma
rays and even sound waves hold
promise for examining *living* cells at
magnifications equal to or even
greater than those achieved by elec-

tron microscopes. This will enable us
to observe changes in the cell as they
actually occur and can be expected
to settle many questions about the
development of cellular structures
that are controversial today.

the cytoplasm and nucleoplasm the various organelles
are suspended.

Some Organelles Are Specialized For Manufacturing Products

Since cells must grow, replace worn out parts, and repair themselves, they must be efficient chemical factories, producing the complex substances of which they
are made. If the cell is part of the body of some multicellular organism, it may produce some product to be

exported for use by other cells. As we saw in Chapter 4,
proteins are the key to the chemical life of the cell because they serve as enzymes which regulate the production of other substances. We will begin by looking
at where proteins are made.

The ER Produces Protein And Other Substances

A maze of internal membranes (Fig. 5–10) extends
throughout the cytoplasm of many eukaryotic cells,
forming an extensive complex of branching tubules,
the **endoplasmic reticulum (ER).** The ER appears to

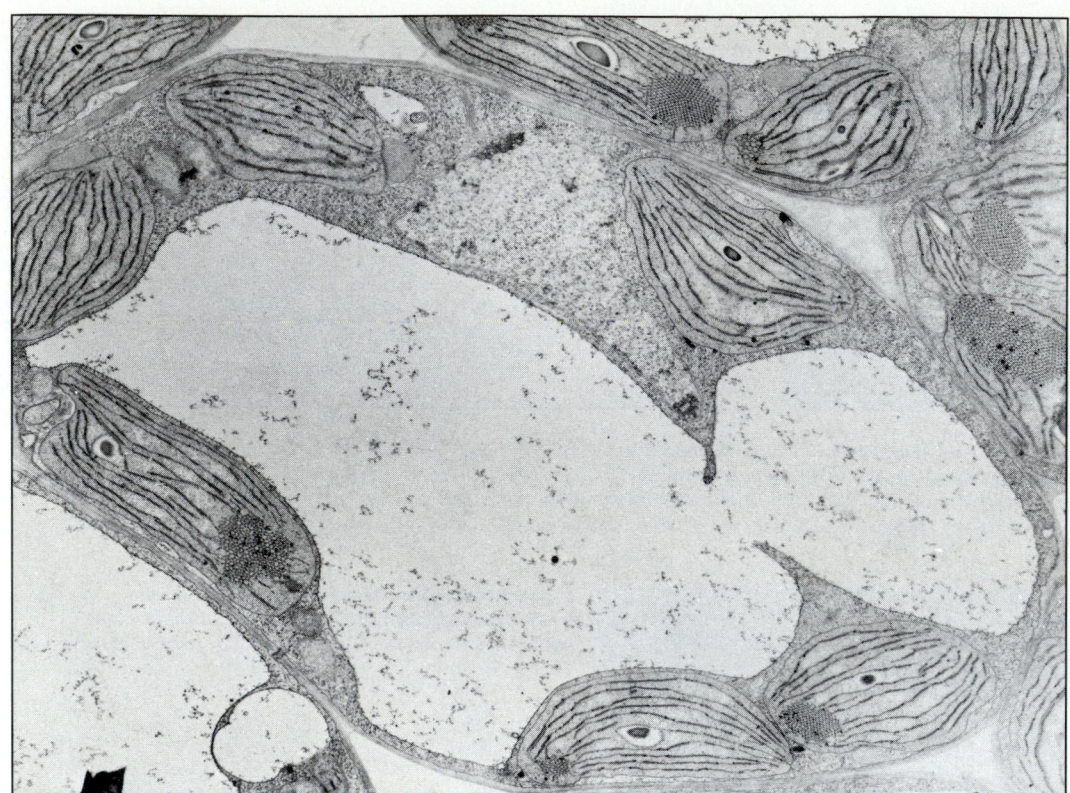

(a)

FIGURE 5-9   The structure of a plant cell, as revealed by the transmission electron microscope. (a) Electron micrograph of a cell from the leaf of a young bean plant, *Phaseolus vulgaris* (magnified ×14,000). (b) A drawing highlighting the structures of the plant cell. Prolamellar bodies are membranous regions typically seen in developing chloroplasts. The structure of this cell would be different were it taken from some other section of the plant, such as the stem or root.
(*a*, Courtesy of Kenneth Miller, Brown University)

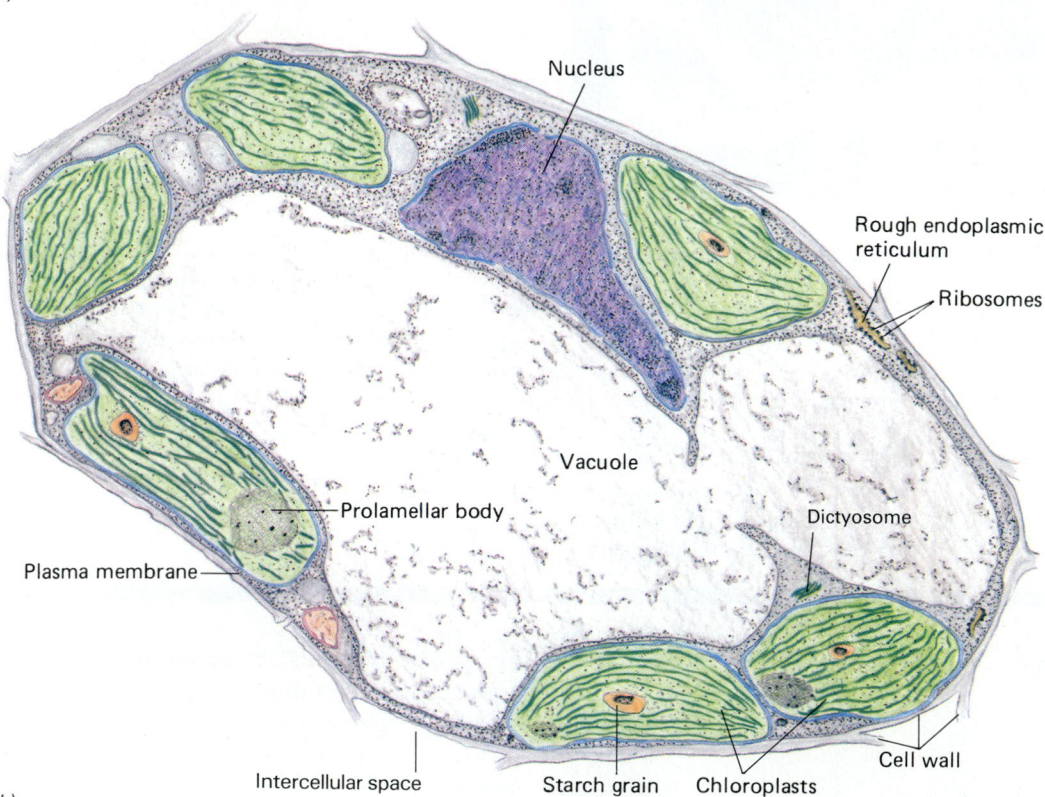

(b)

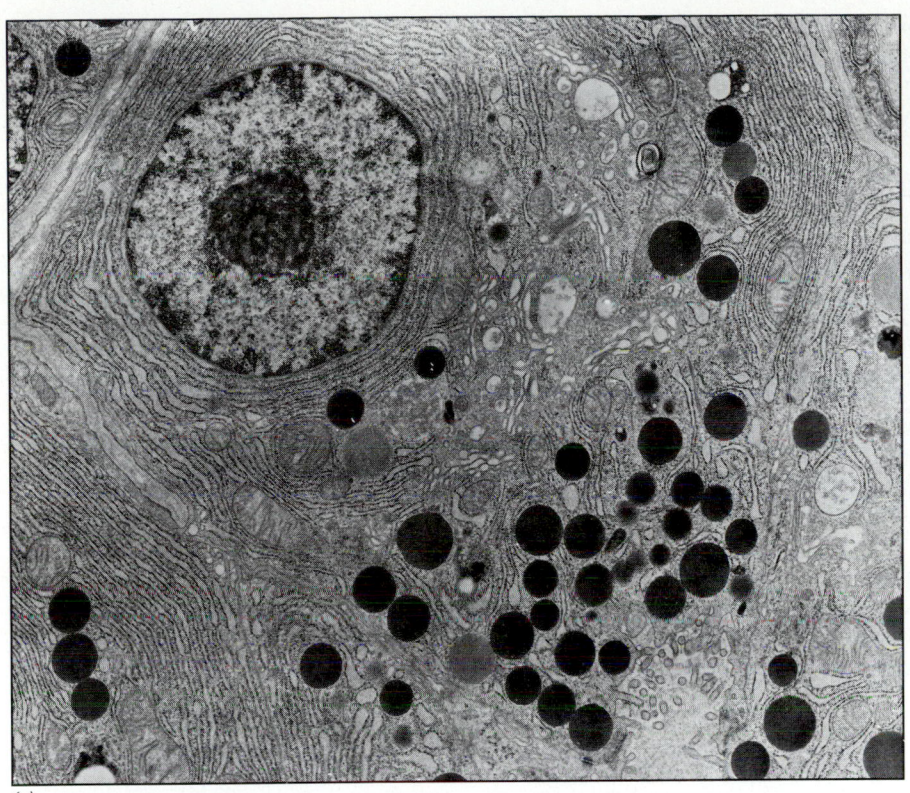

(a)

FIGURE 5–10 The structure of an animal cell as revealed by the scanning electron microscope. (*a*) Electron micrograph of a human pancreas cell (magnified ×16,000). Most of the structures of a typical animal cell are present here; however, like most of the cells of a complex, multicellular animal, this cell has certain features that permit it to carry out a specialized function. The larger, circular dark bodies within the cell are zymogen granules containing inactive enzymes. Released from storage and activated, these enzymes facilitate reactions such as the breaking down of peptide bonds during the digestion of proteins. (*b*) A drawing based on the electron micrograph emphasizes the important structures of the cell. Desmosomes, important structures in maintaining adhesion between cells, are discussed in **Chapter 6.** (*a*, Courtesy of Dr. Susumu Ito, Harvard Medical School)

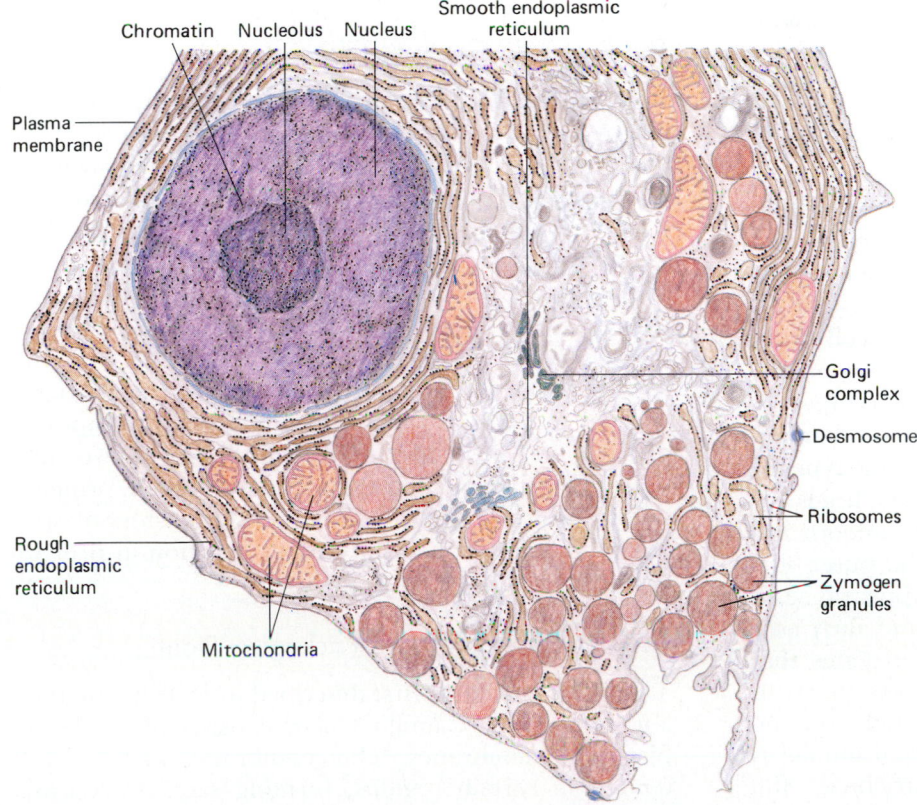

(b)

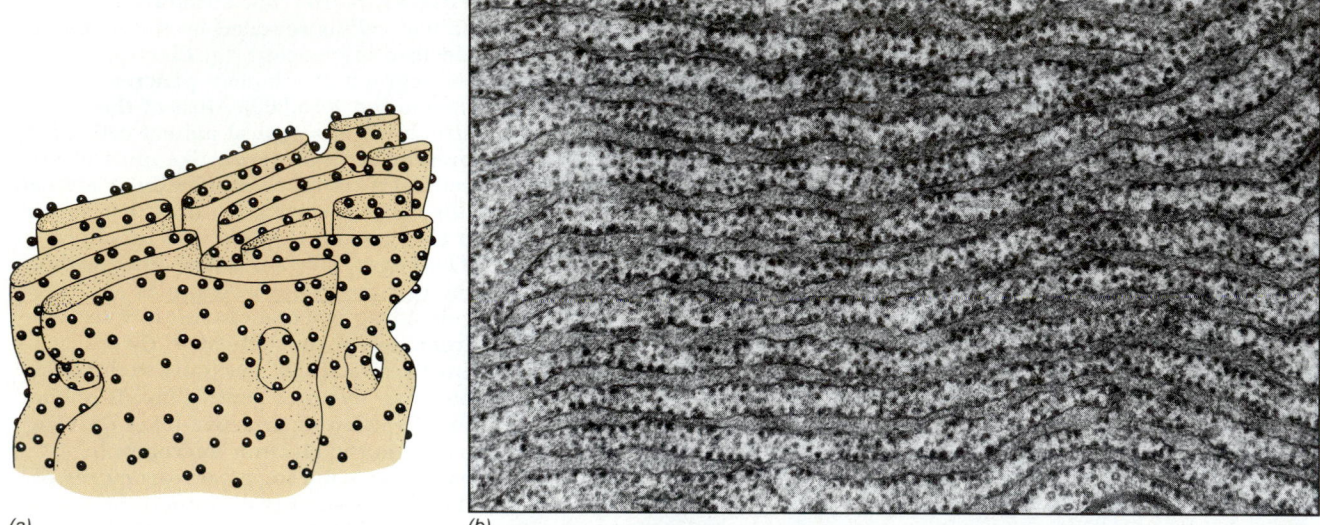

(a)   *(b)*

**FIGURE 5–11   Rough endoplasmic reticulum. (*a*) Diagram of rough ER.
(*b*) Electron micrograph of the rough ER from a secretory cell of the sea
anemone *Metridium*. This form of ER consists of parallel arrays of broad, flat
sacs or cisternae. The outer surface of their limiting membranes is studded
with ribosomes (approximately ×70,000).** (Courtesy of Dr. E. Anderson)

be continuous with the outer membrane surrounding
the nucleus. In electron micrographs, such as
Figure 5–11(*b*), the ER may appear to be discontinuous. This is because such photographs are taken of a
thin slice through the cell, but the ER is continuous in
three dimensions.

The membranes of the ER (Fig. 5–11) divide the
cytoplasm into a multitude of compartments in which
different types of reactions may occur. In fact, the
membranes of the ER contain a variety of enzymes and
serve as a framework of surfaces on which many reactions take place. The ER also functions as a system for
transporting materials from one part of the cell to another and perhaps to the outside environment as well.
Expanded regions of the ER may serve as temporary
storage areas for certain substances. Still another important function of the ER is synthesis of some types of
compounds, particularly proteins and some lipids.

Two types of ER can be distinguished, smooth and
rough. **Smooth ER,** so called because the outer surfaces of its membranes have a smooth appearance, is
the main site of phospholipid, steroid, and fatty acid
metabolism. Smooth ER also contains enzymes that
detoxify harmful chemicals, breaking them down to
water-soluble substances that can be excreted from the
body. For example, when an experimental animal is
injected with the barbiturate phenobarbital, the
amount of smooth ER in the liver cells increases over a
period of several days. In addition, enzymes known to
break down phenobarbital increase in concentration

within the smooth ER membranes when this drug is
administered over a period of time.

The Rough ER Contains Ribosomes

Rough ER has a granular appearance due to the presence of dark particles called ribosomes that stud its
outer walls. **Ribosomes** are the site of protein synthesis. Rough ER is especially extensive in cells that synthesize proteins for export from the cell. Not all ribosomes are attached to the ER. Some float freely in the
cytoplasm or are attached to the cytoskeleton (discussed later). These assemble proteins for use inside
the cell.

Ribosomes are found in all kinds of cells, from
bacteria to complex plant and animal cells. Composed
of RNA and protein, a ribosome consists of two subunits that are combined to form the complete protein-
synthesizing ribosome. In many cells, clusters of several ribosomes termed **polysomes** function in protein
synthesis.

The Golgi Complex Is A Packaging Plant

The **Golgi complex,** first described in 1898 by the Italian investigator Camillo Golgi, consists of stacks of
plate-like membranes. The membranes may be distended in certain regions, forming sacs, or vesicles,
filled with cell products (Fig. 5–12). In animal cells the
Golgi complex may be located at one side of the nucleus.

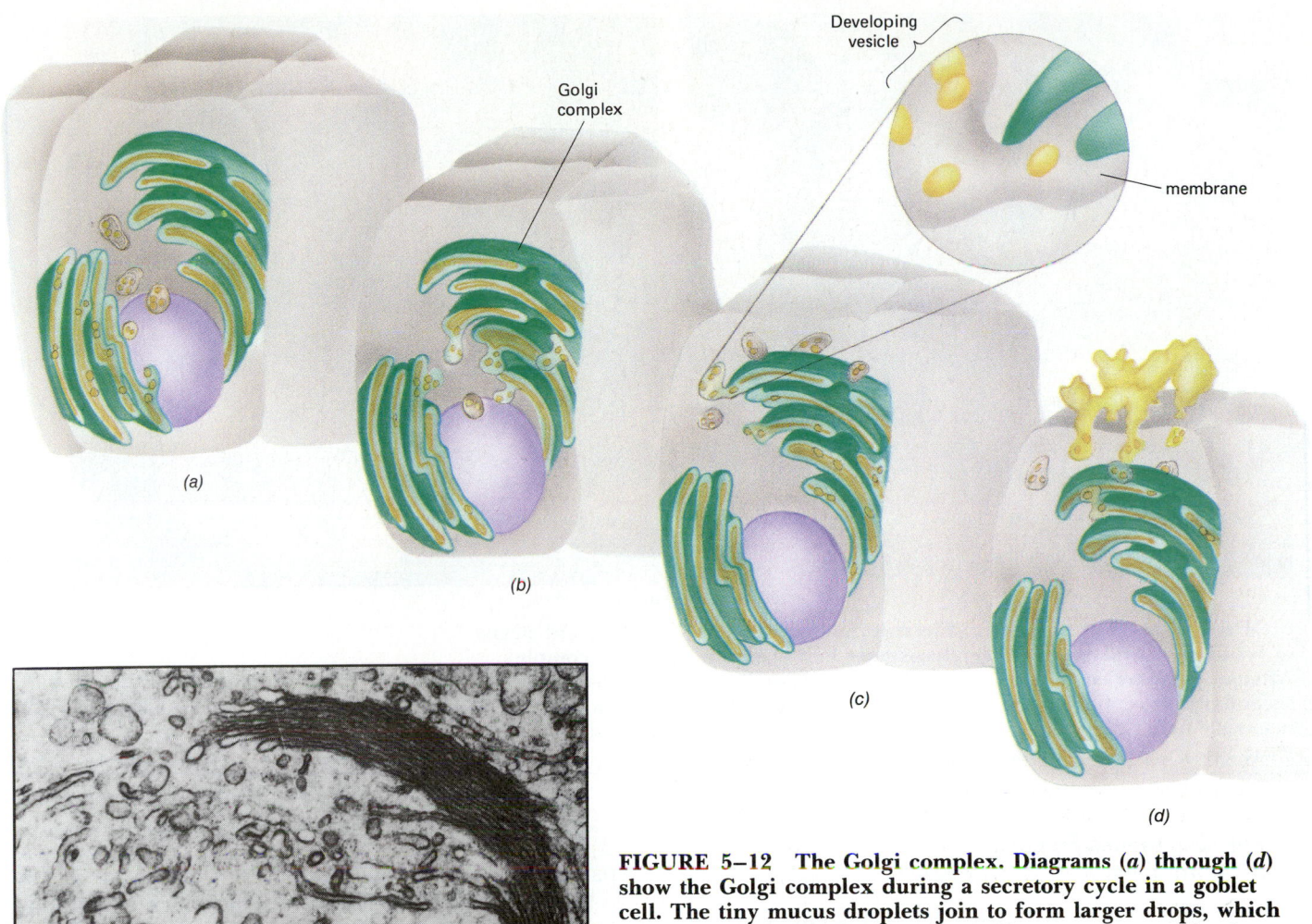

Developing
vesicle

Golgi
complex

membrane

(a)

(b)

(c)

(d)

(e)

FIGURE 5–12   The Golgi complex. Diagrams (a) through (d) show the Golgi complex during a secretory cycle in a goblet cell. The tiny mucus droplets join to form larger drops, which are then released from the cell. (e) Electron micrograph of a section through the Golgi complex from a sperm cell of a ram. (e, Don Fawcett, Photo Researchers, Inc.)

The Golgi complex functions partly as a packaging plant and is most highly developed in cells that are specialized to secrete products. Proteins manufactured along the rough ER pass to the Golgi complex. They are released from the ER in sealed-off little packets of membrane, or **vesicles.** These vesicles fuse with the older membranes of the Golgi complex. Within the Golgi complex the proteins are modified in various ways. Often some carbohydrate component is added to form a **glycoprotein.** (Glycoproteins are proteins with complex branched-chain polysaccharides attached to a number of amino acids.) The Golgi complex of plant cells produces polysaccharides used to construct the cell wall.

Vesicles containing the modified product are released from the Golgi complex. In cells that secrete substances, such as gland cells, the vesicles fuse with the plasma membrane and release their contents to the exterior of the cell.

Although it is especially prominent in cells specialized to secrete products, the Golgi complex also performs an important function in nonsecreting cells. It packages intracellular digestive enzymes in the little organelles called lysosomes.

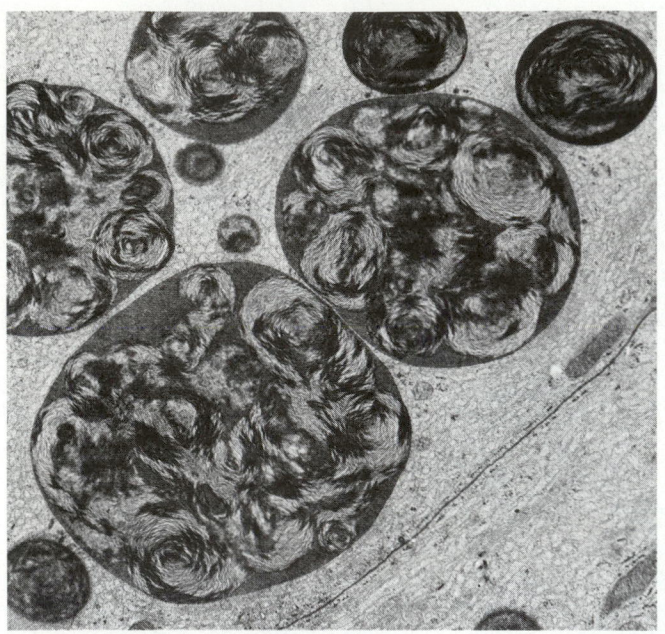

(a)

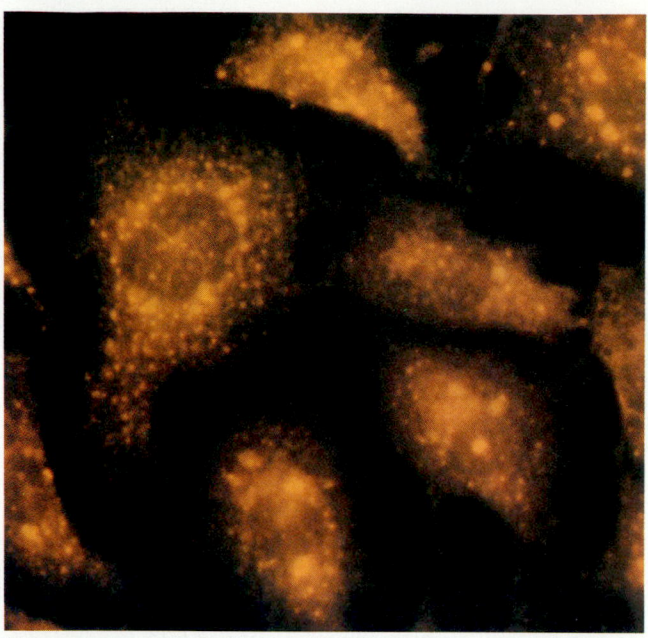

(b)

FIGURE 5–13 Lysosomes. (a) Electron micrograph showing different stages of lysosome formation. Primary lysosomes bud off from the Golgi complex. After a lysosome encounters material to be digested, it is known as a secondary lysosome. The secondary lysosomes shown here contain various materials being digested. (b) Distribution of lysosomes in cells. (a, Don Fawcett, Photo Researchers Inc.; b, courtesy of Dr. Paul Gallup)

Other Organelles Function In Cell Metabolism And Energy Use

All cells must have some way of liberating energy from chemical compounds and of putting that energy to use. Plant and some bacterial cells make their own organic compounds, but animal cells must obtain them from their environment. Often they must first digest organic compounds into simple compounds that can be absorbed and incorporated into the chemical pathways of their metabolism. A multicellular organism may have a digestive system to provide preliminary treatment and breakdown of food, but no organism has a central dynamo to provide energy. Each cell must package its own energy even if it need not provide its own food.

Lysosomes Are Digestive Organelles

Digestion requires enzymes. Small vesicles called **lysosomes** that contain digestive enzymes are released from the Golgi complex and dispersed throughout the cytoplasm (Fig. 5–13). With their potent enzymes, lysosomes break down complex molecules including carbohydrates, fats, proteins, and nucleic acids that originate both within and outside the cell. About 40 different enzymes have been identified in lysosomes.

When a white blood cell ingests a bacterium or some other scavenger cell ingests debris or dead cells, the foreign matter is enclosed in a vesicle composed of

part of the plasma membrane. One or more of the cell's lysosomes then fuse with the vesicle containing the foreign matter. The lysosome pours its powerful digestive enzymes into the vesicle, destroying the material within it. These enzymes can digest almost any type of large molecule found in cells, including the cell that produces the lysosomes.

One very important function of lysosomes is the digestion and breakdown of old cellular components, which otherwise tend to accumulate and interfere with proper cell function. The lysosomal membrane itself is able to resist the digestive action of these powerful enzymes, but this is thought to require a continuous expenditure of energy.

In a cell that is short of fuel, lysosomes may break down organelles so that their component molecules can be used as fuel. Sometimes this digestion occurs in a special food vacuole, but more often it takes place in the lysosome itself.

When a cell dies, the lysosome membrane breaks down, releasing digestive enzymes into the cytoplasm, where they break down the cell itself. This "self-destruct" system accounts for the rapid deterioration of many cells following death.

Some forms of tissue damage as well as the aging process may be related to leaky lysosomes. Rheumatoid arthritis is thought to result in part from damage done to cartilage cells in the joints by enzymes that have been

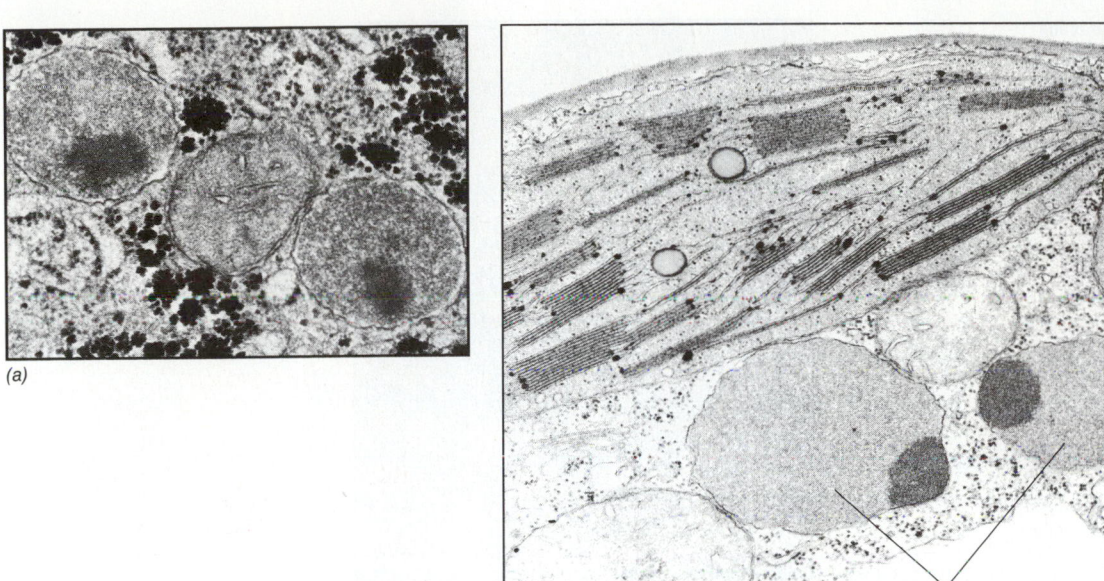

(a)

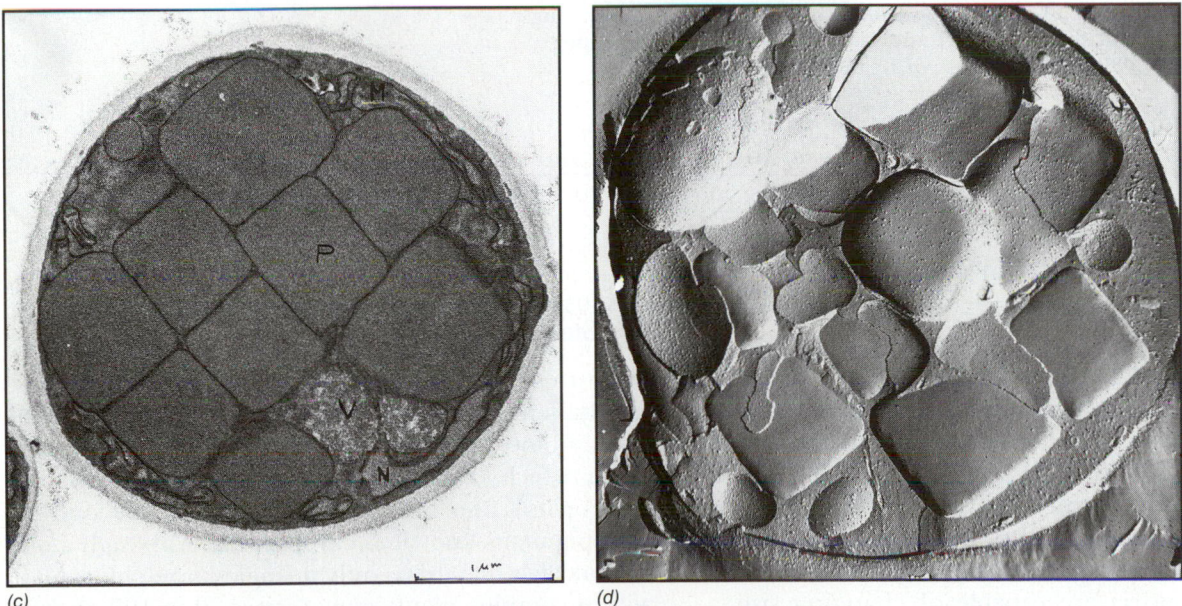

Peroxisomes

(b)

(c)

(d)

FIGURE 5–14   (*a*) **Transmission electron micrograph of a rat liver cell showing two circular peroxisomes with a section of a mitochondrion (*center*). Each peroxisome shows a region of crystalline material that may consist of oxidizing enzymes (approximately ×66,000). (*b*) Peroxisomes in a green plant cell. (*c*) Peroxisomes (marked P) in the cytoplasm of a cell. The cubical shape probably results from crystallization of the enzyme they contain. (*d*) This freeze-fractured preparation of the cytoplasm of a similar cell shows the cavities occupied by peroxisomes.** (*a*, Courtesy of Drs. Christian deDuve and Helen Shio, Rockefeller University; *b*, BPS; *c* and *d*, Dr. Marten Veenhuis)

released from lysosomes. Cortisone-type drugs, which are used as anti-inflammatory agents, stabilize lysosome membranes so that leakage of damaging enzymes is reduced.

Microbodies Contain A Variety Of Enzymes

Microbodies are membrane-bound organelles that contain a variety of oxidative enzymes rather than digestive enzymes. These structures contain enzymes that promote an assortment of metabolic reactions, such as the conversion of fats to carbohydrates. During some of these reactions, hydrogen peroxide, a substance toxic to the cell, is produced. **Peroxisomes,** the types of microbodies (Fig. 5–14) in which these reactions occur, also contain enzymes that split hydrogen

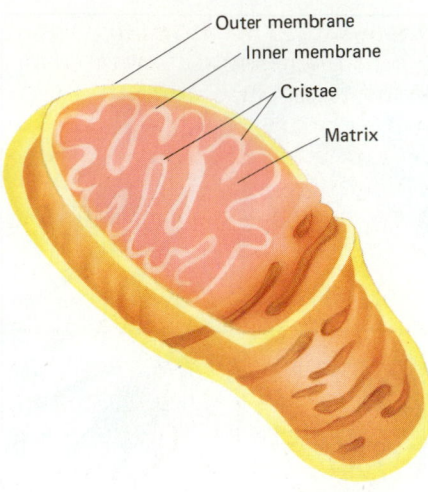

Outer membrane
Inner membrane
Cristae
Matrix

(a)

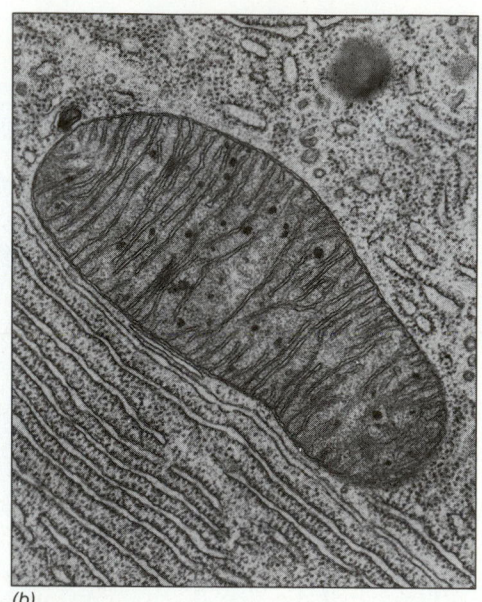

(b)

FIGURE 5–15   The mitochondrion. (*a*) Diagram of a mitochondrion cut open to show the cristae. (*b*) Electron micrograph of a typical mitochondrion from the pancreas of a bat shows the cristae and matrix. Note the extensive rough endoplasmic reticulum at the lower left and some lysosomes at the upper right (approximately ×80,000). (Courtesy of Keith R. Porter)

peroxide, rendering it harmless. In mammals such as ourselves, peroxisomes occur mainly or solely in the cells of the kidney or liver.

Mitochondria Are The Power Houses Of The Cell

Often referred to as the power houses of the cell, the **mitochondria** (singular, *mitochondrion*) are the site of most of the chemical reactions that convert the chemical energy present in organic compounds to another form of energy, ATP (Fig. 5–15). More than 1000 mitochondria have been counted in a single liver cell, but the number varies among cell types. In general they are more abundant in cells that are very active metabolically. Mitochondria are capable of changing size and shape, of fusing with other mitochondria to form larger structures, or of splitting to form smaller ones. They may appear as spheres, rods, sausages, threads, spirals, or irregular shapes.

The mitochondrion is bounded by a double membrane. Both the outer and inner membranes consist of lipid bilayers (discussed in the next chapter) in which a variety of protein molecules are embedded. The outer layer of the mitochondrial membrane is smooth, but the inner layer is folded. Its folds, called **cristae,** project into the interior of the mitochondrion. Cristae serve to increase the available surface area, and some of the enzymes needed for cellular respiration are organized along these folds. Other enzymes are found in the semifluid material within the inner compartment, called the **matrix.** Each mitochondrion has a small amount of DNA, enough to code for about 15 proteins, and also contains ribosomes. Some proteins are indeed synthesized within these organelles.

Plastids Are Plant-Cell Energy Traps And Storage Sacs

Structures known as **plastids** produce and store food materials in algae and plant cells. **Chloroplasts,** the most common type, contain the green pigment **chlorophyll,** which traps light energy for photosynthesis. Chloroplasts also contain a variety of yellow and orange pigments known as **carotenoids.** Although a unicellular alga may have only a single large chloroplast, cells of complex plants may possess 20 to 100 of these organelles.

Though chloroplasts of single-celled organisms occur in a variety of shapes, those of many algae and true plant cells are typically disc-shaped structures bounded by an inner and outer membrane (Fig. 5–16). Inside the chloroplast there are parallel, plate-like membranes called **lamellae.** At various points along the lamellae the membranes are expanded to form disc-like structures called **thylakoids.** Chlorophyll is present within the thylakoids, and the light-dependent reactions of photosynthesis (those involved in capturing the energy of sunlight) take place within these structures. The thylakoids are arranged in stacks called **grana** (singular, *granum*). Like mitochondria, chloroplasts contain DNA and ribosomes and manufacture some proteins.

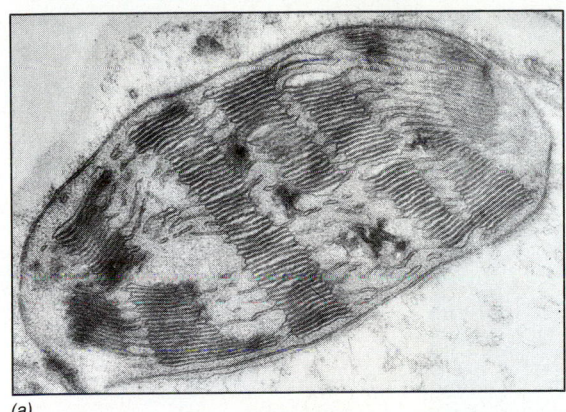

(a)

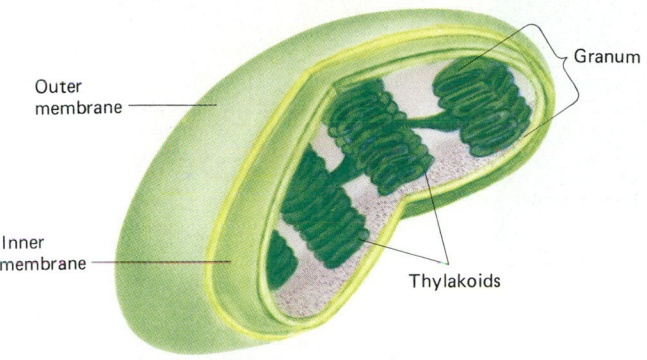

(b)

FIGURE 5–16   The chloroplast. (a) Electron micrograph of a chloroplast from a leaf of the tobacco plant, *Nicotiana rustica*, showing its fine structure. The thylakoids are arranged in stacks called grana (approximately ×30,000). (b) An inside view of a chloroplast showing the arrangement of the thylakoids. The membranous interconnections between thylakoids can only be suggested in this diagram. (a, Courtesy of Dr. E. T. Weier)

Chloroplasts are also able to grow and divide to form daughter chloroplasts and sometimes differentiate from plastid precursors that lack the characteristic chloroplast structure. More details of chloroplast structure and function will be discussed in Chapter 9.

Plant cells also contain colorless plastids termed **leukoplasts,** which serve as centers for the storage of starch and other materials. Starch is commonly deposited in leukoplasts present in storage roots and stems and in seeds. A third type of plastid, the **chromoplast,** contains pigments that give color to flowers, autumn leaves, and ripe fruit.

Still Other Structures Support The Cell, Connect It With Other Cells, And Help It Move

A cell must be more than a microscopic gob of jelly if it, or the organism of which it may form a part, is to have a definite shape and structure. The plastic, semiliquid membranes we have been discussing are not mechanically suitable for use as cellular struts, girders, or cables. But a complex network of threadlike and tubelike structures form the **cytoskeleton** which gives shape to cells. Some of these structures hold cells together to form tissues, and others produce movement of the cell, body fluids, or the entire organism.

Cytoskeletal Elements Give Cells Their Shape And Allow Movement

Microtubules give strength to the cell and produce movement within it and outside of it. They are very small, hollow cylinders composed mainly of proteins called **tubulins** (Fig. 5–17). Microtubules can grow by the addition of more tubulin subunits or can shorten by disassembly into subunits. Microtubules located just beneath the plasma membrane determine the shape of many cells.

Microtubules play a role in some types of cell movements. For example, they evidently help move chromosomes during nuclear division. In nerve cells microtubules are important in the rapid transport of substances down the axon (the elongated extension of a nerve cell). Microtubules are the main structural components of cilia and flagella, structures that are discussed in a later section.

Some thread-like microstructures have no central cavity or lumen. Solid cytoplasmic **microfilaments** and **intermediate filaments** play additional roles in cell structure and movement. Microfilaments are composed of the protein **actin** (or actin-like substances) and often **myosin,** both important in muscle contraction. Microfilaments can rapidly assemble and disassemble. Microfilaments are essential in cellular movement, such as the flowing of cytoplasm that enables a cell to move from one place to another. They are also involved in cytoplasmic streaming, which results in the continuous motion of organelles characteristic of many cells. Some microfilaments hold cells together to form tissues.

Intermediate filaments are stable, tough fibers made of polypeptides. They are thought to help strengthen the cytoskeleton and are abundant in parts of a cell that are subject to mechanical stress.

Centrioles Function In Animal Mitosis

Many types of cells possess a pair of tiny **centrioles,** organelles that function in nuclear division (Chapter 13) in animal and some protist cells. A **centriole** is a hollow rod composed of nine sets of three microtubules (Fig. 5–18). Centrioles are located within a dense area of cytoplasm, the **centrosome,** which is generally located near the nucleus.

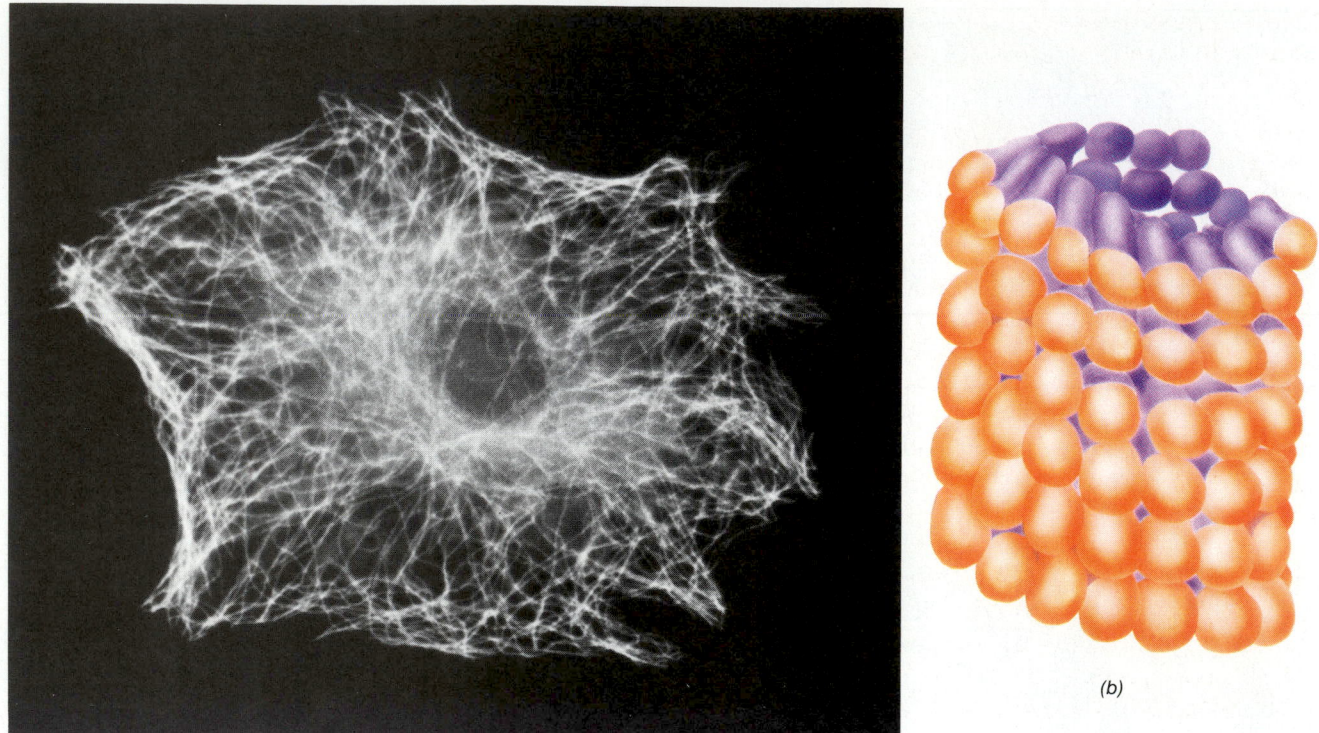

(a)

(b)

FIGURE 5–17 Microtubules. (a) Electron micrograph of a human cell showing the extensive distribution of microtubules throughout the cell. This cell was stained with fluorescent antitubulin, permitting the tubulin that makes up the microtubules to be viewed. (b) Structure of a microtubule. Part of the microtubule has been split in half lengthwise. Note the spiral arrangement of the component units and their two-lobed "barbell" shape.
(a, Courtesy of Drs. Keigi Fujiwara, Hugh Randolph Byers, and Elena McBeath, Department of Anatomy, Harvard Medical School)

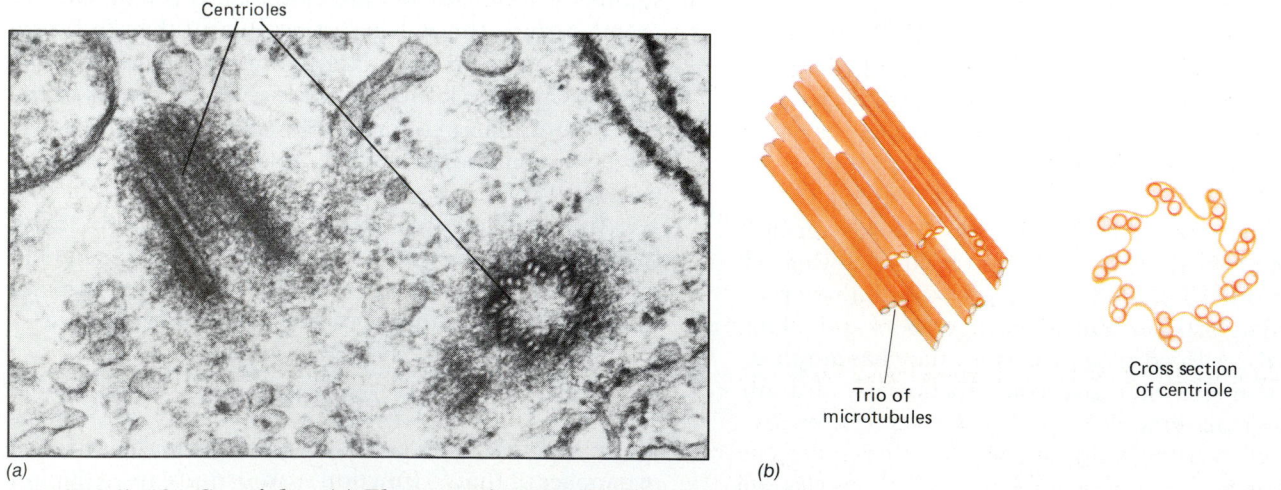

Centrioles

Trio of
microtubules

Cross section
of centriole

(a)

(b)

FIGURE 5–18 Centrioles. (a) Electron micrograph of a pair of centrioles from monkey endothelial cells. Note that one centriole has been cut longitudinally and one transversely. (b) A line drawing of the centrioles.
(a, B.F. King, School of Medicine, University of California, Davis/BPS)

Cells Swim Or Move Materials With Cilia And Flagella

Microtubules are important in cell movement. Many cells have movable, whiplike structures projecting from their free surface that exhibit a beating motion. If a cell has one, or only a few, of these appendages and they are relatively long in proportion to the size of the cell, they are termed **flagella.** If a cell has many short appendages, they are called **cilia.** Flagella and cilia are very similar in structure, and both function mainly in movement—either in moving the cell along through a watery environment or in moving liquids and particles across the surface of a layer of cells.

Flagella or cilia are commonly found on one-celled and small multicellular organisms and on the sperm cells of animals and some plants. They are the principal means of locomotion of such cells. Cilia commonly occur on the cells lining the internal ducts of animals; their beating assists in moving materials through these passageways (e.g., through the respiratory passageways). The description that follows is intended to refer only to the flagella and cilia of eukaryotes. Prokaryote flagella are fundamentally different, and will be discussed in Chapter 21.

Each cilium or flagellum consists of a slender, cylindrical stalk covered by an extension of the plasma membrane. The core of the stalk contains a group of microtubules (Fig. 5–19). Nine pairs of microtubules are arranged around the circumference, and two single microtubules are located in the center. This **9 + 2 arrangement** is characteristic of all eukaryotic cilia and flagella whether they occur on a clam's gill, a gingko tree sperm, or in the human respiratory tract. Microtubules are important in the movement of the cilium or flagellum, which are thought to be activated by a complex sliding of microtubules past one another that bends the entire cilium or flagellum.

At the base of the stalk of a cilium or flagellum there is a **basal body,** which has a 9 × 3 structure similar to that of a centriole (in some single-celled organisms the flagellar basal body and the centriole appear to be one and the same). The basal body is the structure from which the stalk arises. Cilia develop as the cell matures (Fig. 5–20).

Vesicles And Vacuoles Are Cavities Within Cells

Cells have a need for internal cavities that can be used for storage, excretion, digestion, and other functions. We have already described several types of vesicles, for example, those that are released from the Golgi complex filled with a secretory product. The term **vesicle** is used to describe many kinds of small membranous bags found within the cell.

Although the terms vesicle and vacuole are sometimes used interchangeably, vacuoles are actually larger structures, sometimes produced by the merging of many vesicles. A **vacuole** may be defined as a bubble-like space that is usually filled with watery fluid and bordered by a membrane. Present in many types of cells, vacuoles are most common in plant cells and the cells of protists. Most protozoa have food vacuoles containing food undergoing digestion (Fig. 5–21), and many have contractile vacuoles, which remove excess water from the cell.

More than half the volume of a plant cell may be occupied by a large central vacuole containing stored food, salts, pigments, and wastes. The vacuole also functions to maintain the hydrostatic pressure or turgor of the plant cell. Plants lack waste disposal systems and often utilize vacuoles as storage places for toxic materials. Such waste products often aggregate and form small crystals.

The Cell Nucleus Contains Genetic Information

If we were designing a machine that could repair and duplicate itself the way a cell does, we would find it necessary to include a complete set of plans, specifications, and directions somewhere in it, as well as some kind of internal shop or factory for carrying them out. We have looked at the microscopic dynamos and forges of the cell, but where are the plans deposited?

Of all the structures within the cell, the nucleus is usually the most prominent (see acetate insert and Fig. 5–22). Perhaps for this reason investigators guessed that the nucleus served as the control center for the cell even before techniques to prove that hypothesis were available. During recent years many experiments have been performed that have confirmed the vital role of the nucleus. In one such experiment the researcher surgically removed the nucleus from a living amoeba. The enucleated amoeba was unable to eat or grow, and it died after a few weeks. However, if after a day or two an enucleated amoeba was given a new nucleus, it made a complete recovery. These and other experiments show that the nucleus is essential to the well-being of the cell.

The Nuclear Envelope Surrounds The Nucleus

The nucleus is separated from the surrounding cytoplasm by a **nuclear envelope,** a double membrane that regulates the flow of materials into and out of the nucleus. The two membranes of the nuclear envelope are fused at intervals to form **nuclear pores**—channels of complex structure through which the interior of the nucleus communicates with the cytoplasm. However,

(*Text continues on page 102*)

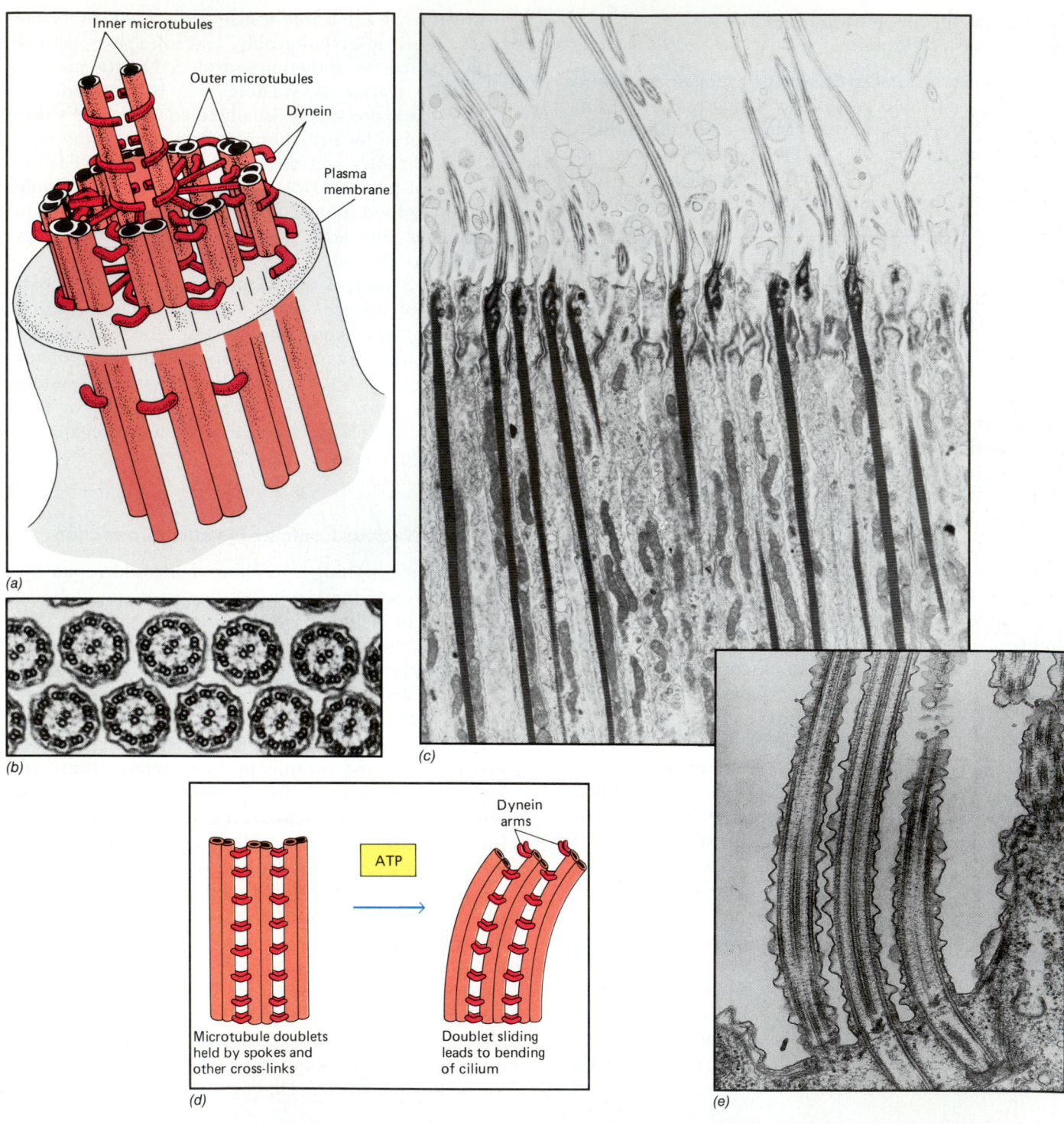

FIGURE 5–19   Cilia. (*a*) Structure of a cilium.
(*b*) **Electron micrograph of cross sections through cilia
showing the 9 + 2 arrangement of microtubules.**
(*c*) **Electron micrograph of the bases of the cilia that
cover the gills of the primitive chordate *Amphioxus*.
Notice the striated rootlets that penetrate deep into the
cell and the long worm-like mitochondria arranged along
them. Presumably the mitochondria provide the energy
for the ciliary contraction (approximately ×12,000).**
(*d*) **How cilia and flagella are believed to function. The**
**arms and spokes within the flagellum enable the
microtubules to exert force on one another, thus bending
the organelle and causing it to move. (*e*) Electron
micrograph of a longitudinal section of the cilia of the
protist *Tetrahymena*, an organism often employed in
genetic research. Some of the interior microtubules may
be clearly seen. In the original photograph some of the
connections between the microtubules also are evident.
(×73,000) (*b*, Omikron, Photo Researchers, Inc; *c*, Dr. M.C.
Holley; *e*, W.L. Dentler/BPS)**

(a)

(b)

(c)

(d)

FIGURE 5–20 How cilia grow. (*a*) A carpet of cilia lining the trachea of a rat and (*b*) one of the millions of cells bearing that cilia at an early stage in its life. The new cilia project like spines from the cell surface. In (*c*) they are much longer, and in (*d*) they form a pattern like a crown or flower on the top of the cell. In the center of the radiating cilia a group of finger-like microvilli (Chapter 6) are apparent. (*a, b,* and *c,* courtesy of Dr. Ulf Nordin, *Acta Otolaryngol,* 94, 1982; *d,* courtesy of Drs. James A. Papp and Joseph T. Martin, *American Journal of Anatomy,* 169, 1984)

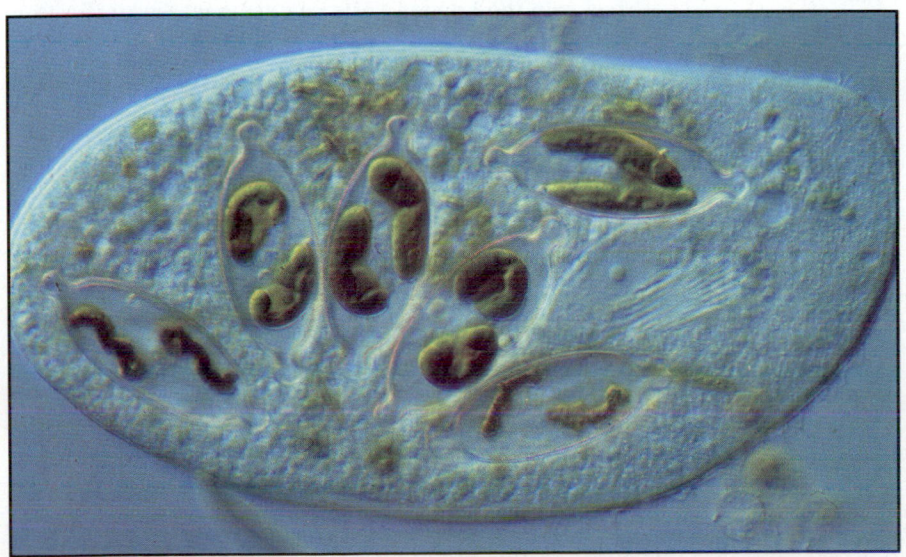

FIGURE 5–21 The protozoan *Chilodonella.* Inside its body are vacuoles containing ingested diatoms (small, photosynthetic, plant-like protists). From the number of diatoms scattered about its insides, one might judge that *Chilodonella* has a rather voracious appetite (approximately ×150). (Walker England, Photo Researchers, Inc.)

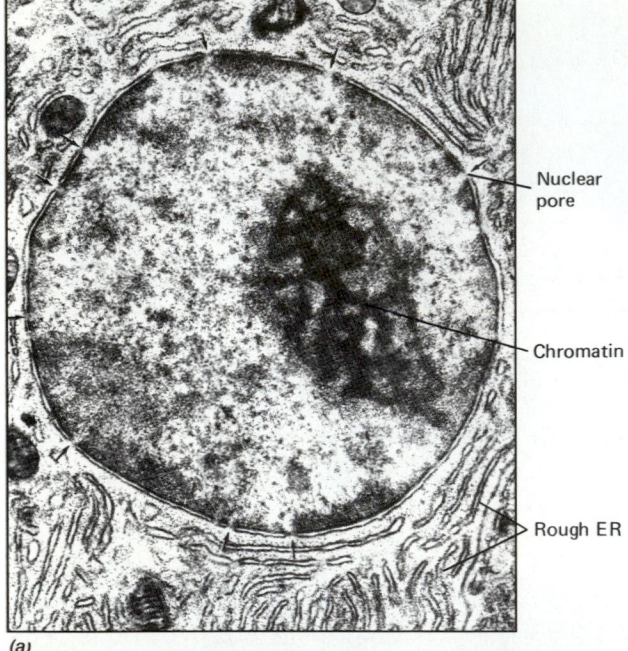

Nuclear pore

Chromatin

Rough ER

(a)

(b)

FIGURE 5–22 **The cell nucleus** *(a)* **Electron micrograph showing the nucleus of a pancreatic cell. Note the two membranes that form the nuclear envelope. Arrows indicate nuclear pores (approximately ×40,000). *(b)* Scanning electron micrograph of the surface of the nuclear envelope (approximately ×60,000). The outermost of the two nuclear envelopes is shown above, but it has been broken partly to expose part of the inner membrane, shown below.** *(a,* Don Fawcett; *b,* courtesy of Dr. Daniel Branton, University of California, Berkeley)

FIGURE 5–23 **Scanning electron micrograph of a chromosome from a hamster cell. Just prior to division, the loose threads of DNA that make up the chromosomes shrivel into the knotted coils you see here (magnification ×15,000).** (Courtesy of Drs. Susanne M. Gollin and Wayne Wray, Kleberg Cytogenetics Laboratory, Department of Medicine, Baylor College, and Biology Department, The Johns Hopkins School of Medicine)

passage of materials through the pores is thought to be selective, that is, only some materials are permitted to pass through the pores.

The Nucleolus Assembles Ribosomes

The nucleus may contain one or more large, prominent **nucleoli** (singular, **nucleolus**). The nucleolus (Fig. 5–22*a*) is not walled off from the rest of the nucleus by any kind of membrane. Rather, it is a dense region rich in ribonucleic acid (RNA) and protein—materials used to manufacture ribosomes. The nucleolus is the assembly site for ribosomes.

A nucleolus develops around a certain region of one chromosome, the **nucleolar organizer,** which contains the genes for synthesizing **ribosomal RNA** (the type of RNA found in ribosomes). During nuclear division, the nucleolus usually becomes disorganized and its components disperse. When division is completed, the nucleolus reorganizes.

Chromosomes Are Packages Of Hereditary Information

In a cell that is not dividing, an irregular network of strands termed **chromatin** is evident in the nucleus. The chromatin consists of protein and uncoiled DNA that may be actively synthesizing substances needed by the cell. When a cell begins the process of nuclear division (mitosis), the chromatin coils and condenses into discrete rod-shaped bodies, the **chromosomes** (Fig. 5–23). (You might think of the condensing of chromatin to form a discrete chromosome as being somewhat like taking two intertwined Slinky toys that have been stretched out as far as they can extend, then releasing them slowly to recoil into a small, tightly wound double helix.)

Each chromosome contains several thousand genes arranged in a specific linear order; the genes, in turn, are composed of the nucleic acid DNA. Chemically coded within the DNA of the genes are instructions for producing all the proteins needed by the cell. These proteins determine what the cell will look like and what functions it will perform. The chromosomes serve as a chemical cookbook for the cell, whereas the genes might be compared to the individual recipes. When condensed, the chromosomes may be compared to a closed cookbook: The recipes are all inside but the pages cannot be read. When the chromosomes elongate, forming chromatin, the book is open and the instructions can be followed. Chromosomes will be discussed in more detail in Chapter 13.

Table 5–1 summarizes the types of organelles typical of eukaryotic cells. Remember that some organelles are found only in specific kinds of cells. For example, chloroplasts are found only in cells that carry on photosynthesis, and cilia or flagella usually occur only in cells that swim or where some kind of fluid motion is produced.

TABLE 5–1
Eukaryotic Cell Structures and Their Functions

Structure	Description	Function
The Membrane System of the Cell		
Plasma membrane (cell membrane)	Membrane boundary of living cell	Surrounds the cytoplasm; regulates movement of materials in and out of cell; helps maintain cell shape; communicates with other cells
Endoplasmic reticulum (ER)	Network of internal membranes extending through cytoplasm	Synthetic site of membrane lipids and many membrane proteins; origin of intracellular transport vesicles carrying proteins to be secreted
Smooth	Lacks ribosomes on outer surface	Lipid biosynthesis; drug detoxification
Rough	Ribosomes stud outer surface	Manufacture of many proteins destined for secretion or for incorporation into membranes
Ribosomes	Granules composed of RNA and protein; some attached to ER, some free in cytoplasm	Synthesize polypeptides
Golgi complex	Stacks of flattened membrane sacs	Modifies substances, packages secreted compounds, sorts other substances to vacuoles and other organelles
Lysosomes	Membranous sacs (in animals)	Contain enzymes to break down ingested materials, secretions, wastes
Vacuoles	Membranous sacs (mostly in plants, fungi, algae)	Transport and store ingested materials, wastes, water
Microbodies (e.g., peroxisomes)	Membranous sacs containing a variety of enzymes	Sites of many diverse metabolic reactions

(Table continues on next page)

TABLE 5–1 (continued)
Eukaryotic Cell Structures and Their Functions

Structure	Description	Function
Energy-Transducing Organelles		
Mitochondria	Sacs consisting of two membranes; inner membrane is folded to form cristae	Site of most reactions of cellular respiration; transformation of energy originating from glucose or lipids into ATP
Plastids	System of three membranes; chloroplasts contain chlorophyll in internal thylakoid membranes	Chlorophyll captures light energy; ATP and other energy-rich compounds are formed and then used to convert CO_2 to glucose
The Cytoskeleton		
Microtubules	Hollow tubes made of subunits of tubulin protein	Provide structural support; have role in cell and organelle movement and cell division; components of cilia, flagella, centrioles
Microfilaments	Solid, rodlike structures consisting of actin protein	Provide structural support; play role in cell and organelle movement and cell division
Centrioles	Pair of hollow cylinders located near center of cell; each centriole consists of nine microtubule triplets (9×3 structure)	Mitotic spindle forms between centrioles during animal cell division; may anchor and organize microtubule formation in animal cells; absent in plants
Cilia	Relatively short, projections extending from surface of cell covered by plasma membrane; made of two central and nine peripheral microtubules ($9 + 2$ structure)	Movement of some single-celled organisms; used to move materials on surface of some tissues
Flagella	Long, projections made of two central and nine peripheral microtubules ($9 + 2$ structure); extend from surface of cell; are covered by plasma membrane	Cellular locomotion by sperm cells and some single-celled organisms
The Cell Nucleus		
Nucleus	Large structure surrounded by double membrane; contains nucleolus and chromosomes	Control center of cell
Nucleolus	Granular body within nucleus; consists of RNA and protein	Site of ribosomal RNA synthesis; ribosome, subunit assembly
Chromosomes	Composed of a complex of DNA and protein known as chromatin; visible as rodlike structures when the cell divides	Contain genes (units of hereditary information that govern structure and activity of cell)

■ CHAPTER SUMMARY

I. The cell is considered the basic unit of life because it is the smallest self-sufficient unit of living material and because, as stated in the cell theory, organisms are composed of cells and their products.

II. Most cells are microscopic, but their size and shape vary according to their function.

III. Biologists have learned much about cellular structure by studying cells with light and electron microscopes and by using chemical techniques.

IV. Prokaryotic cells lack a nucleus with a distinct nuclear membrane and other membranous organelles. Eukaryotic cells have distinct membrane-bound nuclei and a variety of other organelles. Plant cells differ from animal cells in that they possess rigid cell walls, plastids, and large vacuoles; plant cells lack centrioles.

V. The cell is bounded by a plasma membrane, and most eukaryotic cells have elaborate organelles specialized to perform specific intracellular functions.

A. The endoplasmic reticulum (ER) is a system of internal membranes that transport and store materials within the cell and divide the cytoplasm into compartments.

1. The smooth ER is the site of phospholipid, steroid, and fatty acid metabolism.
2. The rough ER is studded along its outer surface with ribosomes that manufacture proteins.

B. The Golgi complex modifies substances that are produced in the ER and packages some for export from the cell. It also produces lysosomes.

C. Lysosomes function in intracellular digestion; peroxisomes contain enzymes that break down hydrogen peroxide.

D. Mitochondria are the power houses of the cell; the cristae of the inner membrane contain enzymes needed for cellular respiration.

E. Cells of algae and plants contain plastids: chloroplasts that contain chlorophyll, pigment-filled chromoplasts, and colorless leukoplasts.

F. Microtubules and microfilaments help maintain the shape of the cell and play a role in cellular movement. These structures form the cytoskeleton.

G. Centrioles, cilia, and flagella are composed of microtubules. Cilia and flagella move cells through the surrounding medium or move the surrounding medium past the cells.

H. In eukaryotic cells the nucleus, control center of the cell, is bounded by a double-layered nuclear envelope.

1. The nucleolus functions in the assembly of ribosomes.
2. When a cell begins to divide, chromatin condenses, forming long chromosomes. Chromatin and chromosomes are composed of DNA and proteins and contain the hereditary units, the genes.

▢ POST-TEST

1. The ability of a microscope to reveal fine detail is known as _____.
2. The sites of protein synthesis in the cell are the _____.
3. Fatty acid and phospholipid metabolism take place along the _____.
4. The _____ modifies proteins and packages cellular secretions.
5. Many of the reactions of cellular respiration take place within the _____.
6. _____ are plastids that contain chlorophyll.
7. In eukaryotic cells cilia and flagella are composed of _____ in a 9 + 2 arrangement.
8. _____ are important in cell movement, such as cytoplasmic streaming.
9. The control center of the cell is the _____;
it is bounded by a _____ _____.
10. In a cell that is not dividing, the uncoiled DNA and protein are evident as _____; when the cell begins to divide, this material condenses to form discrete _____.
11. Each chromosome contains several hundred _____ arranged in a specific linear order.
12. In addition to a plasma membrane, plant cells are bounded by a _____ _____.
13. Many plant cells have a large central _____.
14. Prokaryotic cells lack a _____ as well as most other organelles that occur in eukaryotic cells.
15. In multicellular organisms, the size and shape of a cell is related to the type of _____ it performs.

▢ REVIEW QUESTIONS

1. Label the diagram of the cell. See Window on the Plant Cell for the correct labels.
2. Imagine that a mutant cell was produced that lacked mitochondria. What would be its fate? Why?
3. Explain the following statement: All living cells can trace their ancestry back to ancient times. Trace a muscle cell in your own body back to a cell of a grandparent.
4. Compare mitochondria with chloroplasts.
5. A cell from a gastric (stomach) gland secretes the enzyme pepsin. Trace in sequence its production, transport, packaging, and release from the cell, naming each of the organelles involved.
6. Why are lysosomes sometimes referred to as the self-destruct system of the cell? Do you think this name is justified? Why?

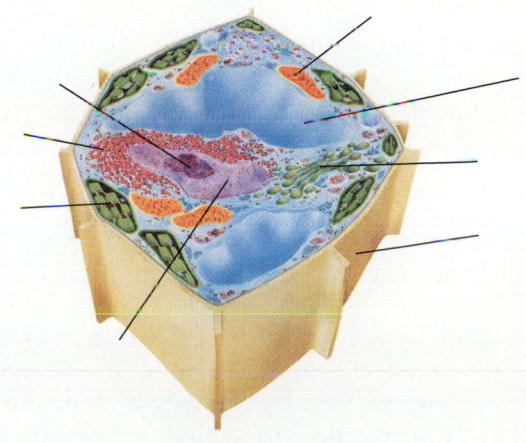

7. What is the relationship between chromatin and chromosomes? Between chromosomes and genes?
8. Draw a typical animal cell and a typical plant cell and label as many structures as you can in each. What are the fundamental differences between the two cell types?
9. How does a bacterial cell differ from an animal cell?
10. Define (a) magnification and (b) resolving power. What are the advantages of the electron microscope over the ordinary light microscope?
11. What do centrioles, cilia, and flagella have in common?
12. Which do you think are greater: the differences between plant and animal cells, or the differences between prokaryote and eukaryotic cells? Why?

■ RECOMMENDED READINGS

(Also consult reading lists in Chapters 6 through 10)

Alberts, B., et al.: *Molecular Biology of the Cell,* second ed. New York, Garland Publishing, 1989. A thorough presentation of cell structure and function. A well-illustrated and easy-to-read reference textbook.

Avers, J. *Cell Biology,* 2nd ed. New York, Van Nostrand, 1986. A fine presentation of the details of cell structure and function.

de Duve, C. *A Guided Tour of The Living Cell.* San Francisco, W.H. Freeman Company, 1985 (2 vols). The discoverer of the lysosome discusses every organelle.

Fawcett, D.W. *The Cell,* 2nd ed. Philadelphia, W.B. Saunders Company, 1981. A study of cell structure through an exciting collection of electron micrographs.

Grivell, L.A. "Mitochondrial DNA," *Scientific American,* March 1983, 78–88. A discussion of the genetic system of mitochondria.

Hoagland, M. *The Roots of Life: A Layman's Guide to Genes, Evolution and the Ways of Cells.* New York, Avon, 1979. A paperback written for the general public by a working scientist.

Holtzman, E., and A.B. Novikoff. *Cells and Organelles,* 3rd ed. Philadelphia, Saunders College Publishing, 1984. An integrated approach to the structural, biochemical, and physiological aspects of the cell.

Lazarides, E., and J.P. Revel. "The Molecular Basis of Cell Movements," *Scientific American,* May 1978, 100–112. The role of microfilaments in cell movement.

Margulis, L. *Symbiosis in Cell Evolution.* San Francisco, W.H. Freeman, 1981. A fascinating summary—written from a position of strong personal advocacy—of the hypothesis that the cell organelles of eukaryotic organisms evolved from prokaryotes.

6

Biological Membranes

OUTLINE

I. Every cell has an outer membrane
II. Biological membranes are phospholipid bilayers
  A. The membrane bilayer consists mainly of phospholipids
  B. Cell membranes contain proteins and glycoproteins
  C. The cytoskeleton has membrane connections
  D. Microvilli increase the surface area of some cells
  E. Cell walls occur in four of the five kingdoms
III. Cells are joined by specialized structures
IV. Materials pass through membranes in several ways
  A. Diffusion is passive molecular movement
  B. In facilitated diffusion molecules pass through special channels
  C. Osmosis poses special problems for cells
  D. Active transport uses energy to concentrate selected materials
  E. Cells can ingest materials by endocytosis and eject them by exocytosis
Focus on splitting the lipid bilayer

LEARNING OBJECTIVES

After you have read this chapter you should be able to:

1. Discuss the importance of the plasma membrane to the cell, describing the various functions it performs.
2. Describe the currently accepted model for the structure of the plasma membrane.
3. Describe the functions of membrane proteins.
4. Describe microvilli and give their function.
5. Describe the cell wall and its function.
6. Contrast desmosomes, tight junctions, and gap junctions.
7. Contrast the physical with the physiological processes by which materials are transported across cell membranes.
8. Solve simple problems involving osmosis. For example, predict whether cells will swell or shrink under various osmotic conditions.
9. Describe and compare exocytosis with endocytosis.

Transmission electron micrograph of the root tip meristem of mouse-ear cress *(Arabidopsis thaliana)*. (Biophoto Associates/Science Source/Photo Researchers, Inc.)

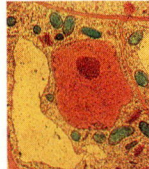

Any living cell must have an interior, separated from the outside world, in which its life processes can take place. A cell must preserve a fundamental distinction between interior and exterior, between itself and everything that is "not itself." Every cell accomplishes this self-definition by means of an external membrane, the **plasma membrane,** that surrounds it (see Windows on the Plant and Animal Cells).

The plasma membrane does not function only as a barrier. In addition to keeping foreign substances out, the cell must also bring necessary materials in, materials that it needs for life and growth. It must also be able to excrete waste products. Thus, the covering of the cell must be versatile and dynamic. The plasma membrane is so thin that·it cannot be seen with a light microscope. Its true complexity of structure and function is only now beginning to be understood.

The plasma membrane is only one of a multitude of membranes that each eukaryote cell possesses. In fact, many of the organelles discussed in the preceding chapter are made of specialized membranes which function together to carry out most of the life functions of the cell. Membranes are the machinery of cellular life; we are made of membranes. In this chapter we focus on the plasma membrane. However, all biological membranes are basically similar.

KEY CONCEPTS

■ Every cell is surrounded by a plasma membrane that is a phospholipid bilayer.

■ Proteins are associated with the plasma membrane; their hydrophobic portions may be embedded in the membrane.

■ The proteins of biological membranes perform a vast number of specialized functions; they are vital in maintaining the cell's homeostasis.

■ Materials can pass through membranes nonselectively by diffusion (by osmosis in the case of water) or selectively by facilitated diffusion. These processes do not consume cellular energy. Materials move from regions of high concentration to regions of low concentration, or "down" a concentration gradient.

■ Materials can also move through membranes by active transport, which does consume cellular energy. In active transport, materials are moved from regions of low concentration to regions of high concentration; that is, active transport concentrates materials.

■ Cells can ingest materials of large molecular size or even solid particles by an inpocketing of the plasma membrane called endocytosis. Large particles or molecules can be ejected by exocytosis.

■ EVERY CELL HAS AN OUTER MEMBRANE

The plasma membrane is far from being an inanimate wall. Quite the contrary, it is a complex structure that permits a multitude of interactions between the cell and its environment. Among its many functions we can list the following:

1. *The plasma membrane regulates the passage of materials into and out of the cell.* Each cell differs from its surroundings in physical properties and chemical composition. If the plasma membrane were completely permeable, substances would pass freely into and out of the cell. The cell would then reflect the chemical composition of the surrounding medium, with disastrous results, for it could no longer function as a cell. This does not happen because the plasma membrane is **selectively permeable,** which means that it can prevent the entrance of certain substances while permitting, even facilitating, the entrance of others, such as needed nutrients. The plasma membrane also prevents the loss of certain substances from the cell but facilitates the exit of specific products and wastes. This ability to regulate passage of materials enables the cell to maintain a fairly constant set of internal conditions despite changes in the external environment.

2. *The plasma membrane receives information that permits the cell to sense changes in its environment and to respond to them.* Receptor proteins in the plasma membrane receive chemical messages from other cells. Hormones, growth factors, and neurotransmitters (chemicals released by nerve cells) are among the substances that combine with such receptors. When a compound combines with a receptor, the membrane is stimulated to send a signal into the cell, resulting in some type of behavioral, physiological, or structural response.

3. *The plasma membrane maintains structural and chemical relationships with neighboring cells and with the organism as a whole.* Certain proteins in the plasma membrane permit cells to recognize one another, to adhere to each other when appropriate, and to exchange materials. It is often the plasma membrane that initially responds to hormones or other stimuli that integrate it with the other cells and larger life of the organism of which it is a part.

4. *The plasma membrane protects the cell, may function in secretion of substances, may be involved in cell movement, and in some cells is important in transmitting impulses.*

■ BIOLOGICAL MEMBRANES ARE PHOSPHOLIPID BILAYERS

The plasma membrane is so thin, about 6 to 10 nanometers, that it can be seen only with the electron micro-

(a)

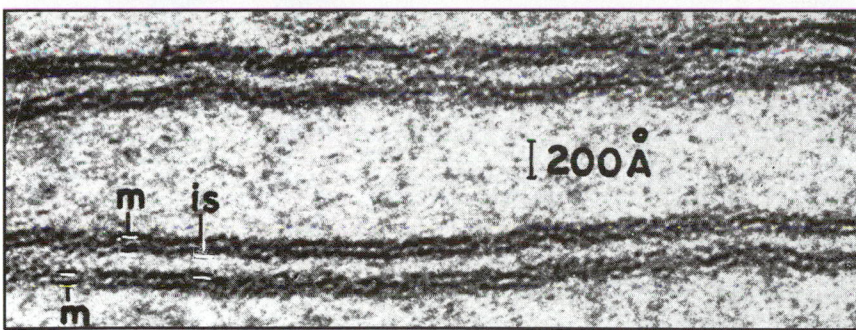

(b)

FIGURE 6–1 The plasma membrane.
(a) The plasma membrane is the interface between the cell and its environment.
(b) Electron micrograph of four regions of a plasma membrane (approximately ×240,000). In each of the four membranes, the dark lines represent the hydrophilic heads of its lipid molecules, whereas the light zone represents the hydrophobic tails. *m*, membrane; *is*, intercellular space.
(*a*, Omikron, Photo Researchers, Inc.)

scope (Fig. 6–1; though it appears to be visible with a light microscope, this is actually an optical illusion). The currently accepted hypothesis of the structure of plasma membranes and other biological membranes is the **fluid mosaic model.** Despite the use of the word "model," there is almost no doubt of its accuracy. We now understand that the plasma membrane and all cellular membranes consist of a rather fluid lipid bilayer (a double layer of lipid) in which a variety of proteins are embedded. Each of the two layers of which the membrane is composed is known as a **leaflet.** It can be important to distinguish between the two leaflets.

The Membrane Bilayer Consists Mainly Of Phospholipids

The lipid components of the plasma membrane include phospholipids, glycolipids, and cholesterol. All of these have an important structural feature in common. They are asymmetrical, elongated molecules that have one highly polar, **hydrophilic** (water-loving) portion and one nonpolar, **hydrophobic** (water-hating) portion. Most of the lipids in cell membranes are **phospholipids,** in which each molecule has a hydrophilic head group (the portion containing phosphate) and two hydrophobic tails, which are usually fatty acid

chains (Fig. 6–2). These fatty acids vary, especially in their degree of saturation. The double bonds found within an unsaturated fatty acid are incapable of rotation and are angled. Thus, as you can see by examining the diagram, double bonds produce bent phospholipid tails.

In the plasma membrane the nonpolar, hydrophobic fatty acid chains (the tails) of the phospholipids meet and overlap in the middle, while the polar, hydrophilic heads are directed toward the outside of the membrane. The tails of the phospholipid molecules pack loosely, and the fatty acid chains are in constant motion. This results in a fluid state of the membrane that is essential for its function.

As shown in Figure 6–1, when viewed with the electron microscope, the plasma membrane is seen as two dark lines separated by an intermediate light zone. The dark lines represent the hydrophilic heads of the lipids, and the light zone is produced by the hydrophobic tails. Diagrams cannot ordinarily show these molecules in detail, so by convention the hydrophilic head groups are represented by circles and the hydrophobic tails by two wavy lines (Fig. 6–3).

The formation of a **lipid bilayer** from phospholipids is a rapid, spontaneous process. The driving force for this process is the hydrophobic interactions of the hydrocarbon chains. Even cells of terrestrial organ-

FIGURE 6–2 (*a*) **A typical phospholipid consists of a molecule of glycerol united with a hydrophilic "head" such as choline (outlined in color) and two fatty acids that make up a double "tail." (*b*) A more realistic depiction of a phospholipid molecule. Notice how the geometry of the molecule determines the space it occupies. (*c*) Since water occurs inside *and* outside a cell, the tails of the phospholipid molecules are forced together to form a bilayer, with the heads facing to the outside. Saturated fatty acids pack tightly together in the center of the bilayer; unsaturated fatty acids pack loosely.**

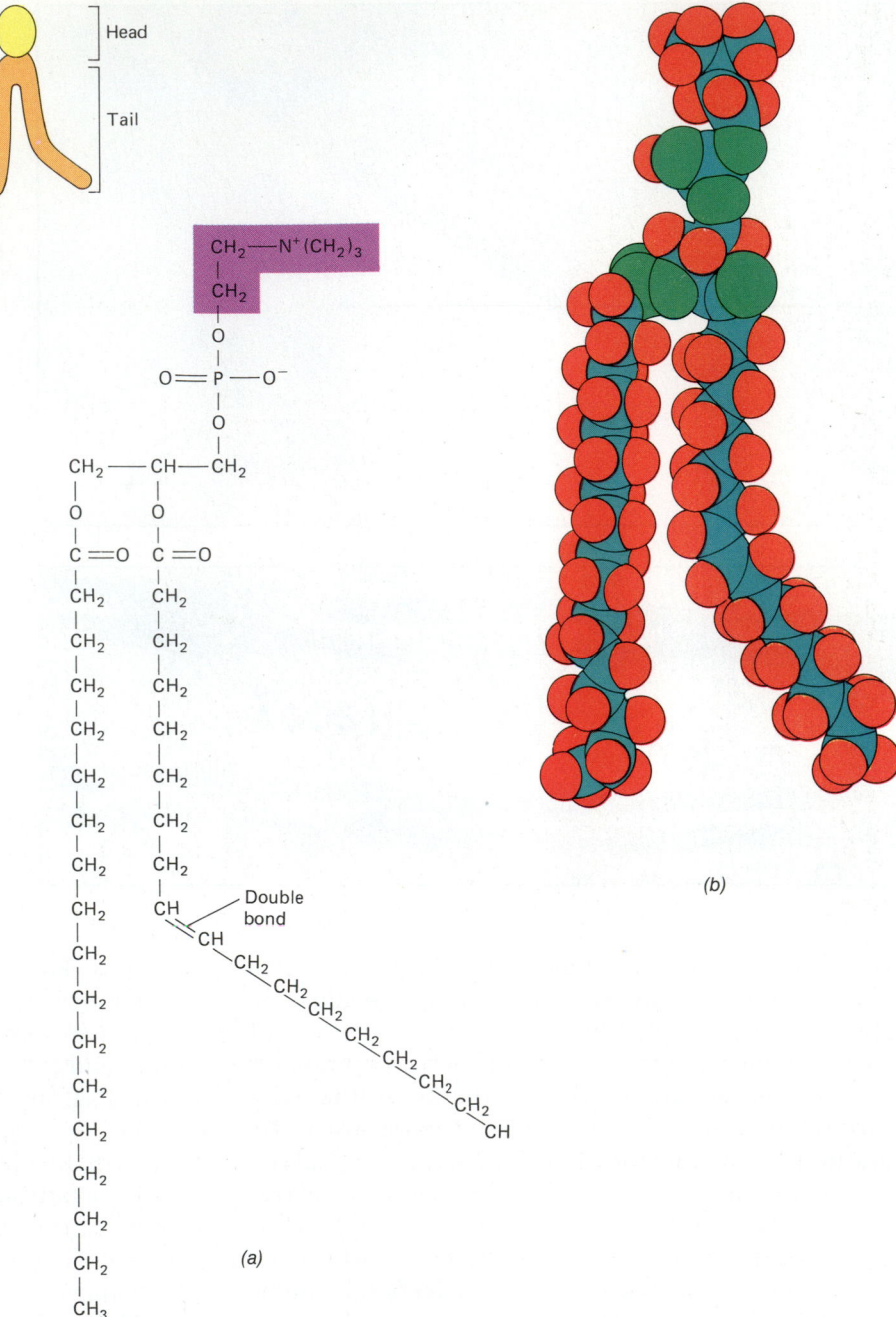

(a)

(b)

isms dwell in a watery medium—the various body fluids, such as tissue fluid. When these lipids are placed in a watery medium, the polar heads show an affinity for water, but the hydrocarbon tails avoid it. Thus the molecules spontaneously become oriented to form a bilayer in a watery environment. No covalent bonds link adjacent lipid molecules to one another. Only the hydrophobic forces hold the molecules in place. Still, the bilayer is strong enough to be a membrane.

Lipid bilayers are self-sealing as well as self-assembling. If a hole is made in the membrane, as long as the hole is not too large, the tails of the phospholipid molecules in the bilayer will be forced together by repulsion of the surrounding water, thus repairing the small hole.

The lipid bilayer is very impermeable to ions and polar molecules. Water is an exception and is able to pass in and out of the cell through the lipid bilayer with ease.

Cell Membranes Contain Proteins And Glycoproteins

The lipids of cellular membranes serve as a barrier to polar molecules and ions, whereas the membrane proteins carry out specific functions of the membrane,

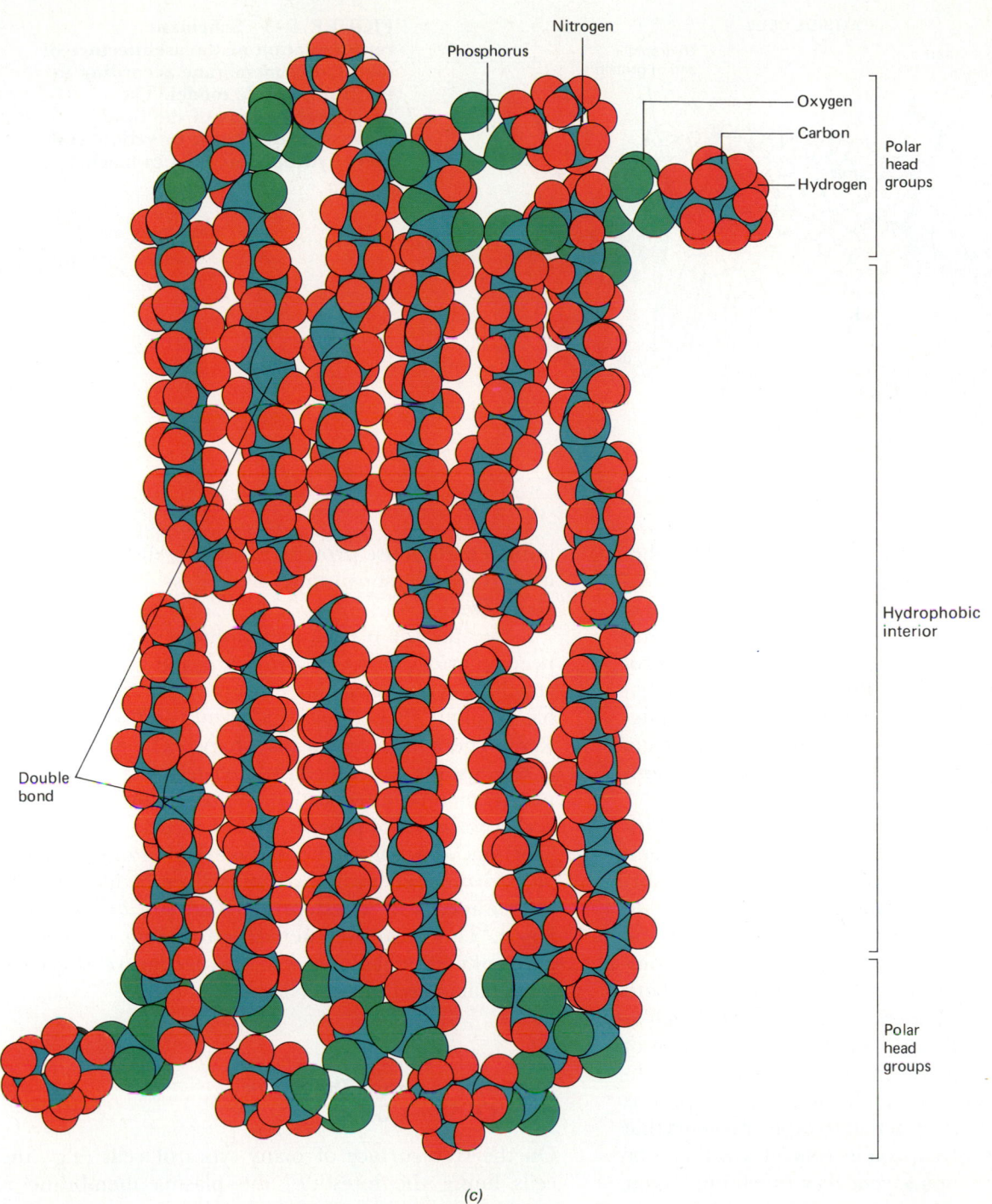

(c)

such as chemical transport, energy transfer, and transmission of messages. Thus the proteins are the valves, cogwheels, and gates of the membrane machinery. Since all biological membranes are quite similar (though not exactly the same) in their phospholipid makeup, membranes with different functions contain different proteins. Also, the chemical composition and molecular arrangement of membranes can change with varying conditions.

Proteins must be chemically compatible with the lipids of a membrane in order to be associated with it.

Proteins, as you know, are long chains of amino acids. Some amino acids, once they have been incorporated into a protein, are hydrophobic. The parts of a protein that are composed of such amino acids are therefore wholly or partly hydrophobic themselves. Such proteins are attracted to and become embedded in the phospholipid bilayer. The hydrophilic parts of a protein protrude into the water inside or outside of the cell.

If a protein has one hydrophilic end and one hydrophobic end, it usually occurs in just one of the two

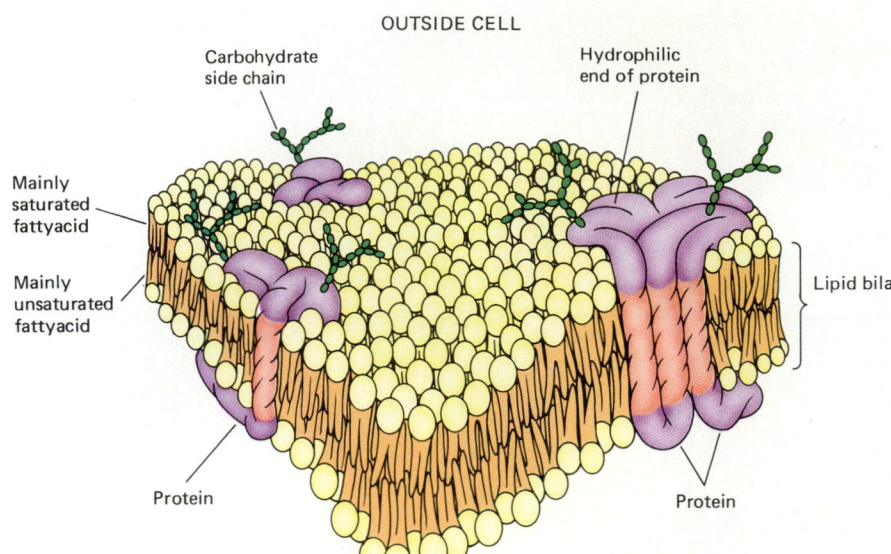

OUTSIDE CELL

Carbohydrate side chain

Hydrophilic end of protein

Mainly saturated fattyacid

Mainly unsaturated fattyacid

Lipid bilayer

Protein

Protein

FIGURE 6–3 Schematic representation of the architecture of the plasma membrane according to the fluid mosaic model. The hydrophilic heads of the lipid molecules are shown in yellow and their hydrophobic tails in black.

leaflets of the bilayer. If only the middle is hydrophobic, it bridges the entire bilayer, with the two hydrophilic ends protruding on both sides. In some remarkable cases, a large and complex protein threads in and out of the membrane numerous times (Fig. 6–4). Proteins that have regions which are inserted into the hydrophobic regions of the lipid bilayer are called **integral proteins.** The other type of membrane protein, **peripheral proteins,** usually bind to exposed regions of integral proteins.

Most of the plasma membrane proteins are associated with the inner, cytoplasmic surface of the membrane. The proteins that protrude from the outer surface are mainly **glycoproteins,** that is, proteins to which carbohydrates are attached. Glycoproteins on the plasma membrane appear to serve as the cell's communication system with its outer environment, both with other cells and with messenger molecules, such as hormones, and are very significant in the operation of the immune system. Some glycoproteins are peripheral proteins and are not embedded in the plasma membrane at all but are attached to other proteins that are (Fig. 6–5). The glycoprotein coat of a cell can be very extensive, forming a **glycocalyx** (meaning "sugar coating"), which coats the surface of many cells.

The Cytoskeleton Has Membrane Connections

As we mentioned previously, the plasma membrane is quite fluid; it has to be, because many of its proteins function by changing their shape or otherwise moving. Are membrane proteins fixed in position or are they free to move from place to place? To settle this question, experimenters fused human and mouse cells in tissue culture (Fig. 6–6). The resulting large cell had human and mouse proteins randomly distributed over its surface, demonstrating that these proteins move about freely.

Because the plasma membrane is so fluid, the cell needs a cytoskeleton to give it a shape and allow it to attach to other cells. For example, the red blood cell has a unique shape, that of a biconcave disc (Fig. 6–7). Researchers have shown that shape to be produced by a fibrous protein (spectrin) located just under the plasma membrane. Spectrin, in turn, is attached to several other kinds of protein, including ankyrin. Ankyrin "anchors" spectrin to the plasma membrane by means of hydrophobic proteins that are embedded in the membrane itself. These proteins are not free to move about. The complex internal scaffolding of spectrin and ankyrin makes possible the biconcave shape of mammalian red blood cells.

Microvilli Increase The Surface Area Of Some Cells

On the free surface of many types of cells (e.g., the cells lining the intestine) the plasma membrane is marked by numerous tiny evaginations (outpocketings) known as **microvilli** (Fig. 6–8). Microvilli enormously increase the surface area of the cell that is available for absorption of materials from the cell's environment. Each microvillus has a central cytoskeletal thread composed of **actin,** a protein also found in muscle cells, where it functions in contraction and movement. Microvilli can extend and retract due to polymerization and depolymerization of their actin fibers. Thus, the number of microvilli can rapidly increase or decrease in response to environmental conditions or to changes in the metabolic needs of the cell.

FOCUS ON   Splitting the Lipid Bilayer

High-resolution electron micrographs of plasma membranes or other cell membranes typically show a three-layered structure of lines in a dark–light–dark pattern. The two dark lines correspond to the polar heads of the lipid bilayer, and the light area between them corresponds to the hydrophobic region of the fatty acid chains. The plasma membranes of animal, plant, and microbial cells and the membranes of a great many subcellular organelles all appear to have this three-layered structure (Fig. A).

Cells can be rapidly frozen in liquid nitrogen and then fractured with a microtome knife. A small amount of ice is evaporated from the fracture surface ("etched") and then a small amount of platinum or gold is deposited on that surface, forming a replica of the fractured surface. When the metallic coating is examined with the transmission electron microscope, one can see that the fracture typically splits each plasma membrane into two half-membranes in the middle of the lipid bilayer (see figure). By this means, termed **freeze fracture** or **freeze etch,** one can examine the interior of the split membrane. The two faces are not identical. The face nearer the cytoplasm contains many particles, and the face near the outside of the cell contains many pits (see Fig. B). The particles are integral proteins (proteins that extend into the lipid bilayer), and the pits are spaces where the proteins had been.

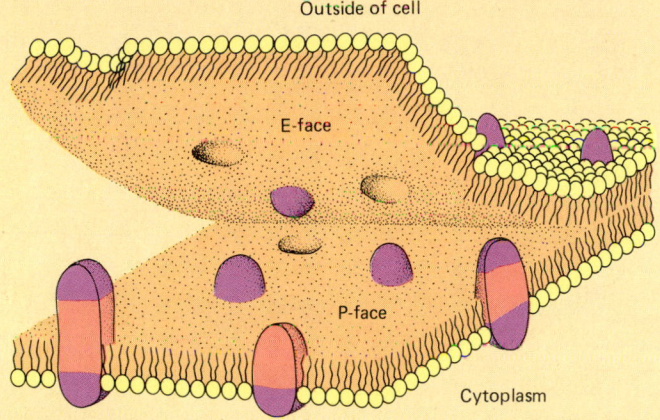

A In the freeze-fracture method, tissue specimens are often broken along the hydrophobic interior of the lipid bilayer in such a way that the two monolayers of the plasma membrane are separated and can be photographed separately. (1) The inner monolayer is an outwardly directed half-membrane (the so-called P-face) that presents the majority of globular proteins, seen here to project from it. (2) The outer monolayer is an inwardly directed half-membrane (the so-called E-face). Although relatively smooth, it nevertheless shows some protein particles.

B Freeze fracture made of the plasma membrane from a cell of the eye of a monkey. Notice the greater number of proteins present in the P-face of the membrane. (Don Fawcett, Photo Researchers, Inc.)

Cell Walls Occur In Four Of The Five Kingdoms

Cells of plants, certain protists, bacteria, and fungi have thick cell walls that lie just outside their plasma membranes (see Window on the Plant Cell). The **cell wall,** consisting of carbohydrate, is secreted by the cell to the outside of the plasma membrane. It protects the cell, gives it its characteristic shape, and often contributes to the shape and strength of the organism of which the cell may be a part (Fig. 6–9).

The rigid plant cell wall is pierced in many places by tiny pores through which water and dissolved materials can pass. The cell wall helps to give shape to plant cells and, perhaps more importantly, serves to contain the high internal pressure of plant cells, somewhat as the outer casing of a tire contains the internal pressure of its air. When, for any reason, the internal pressure of plant cells is reduced, the plant tissue becomes limp, and wilting results since the membrane and cell contents are no longer pressing against the cell walls.

Cellulose, a large carbohydrate, is the main component of the plant cell wall and is usually present in the form of long, thread-like fibers. The cell wall has been compared to reinforced concrete in which the cel-

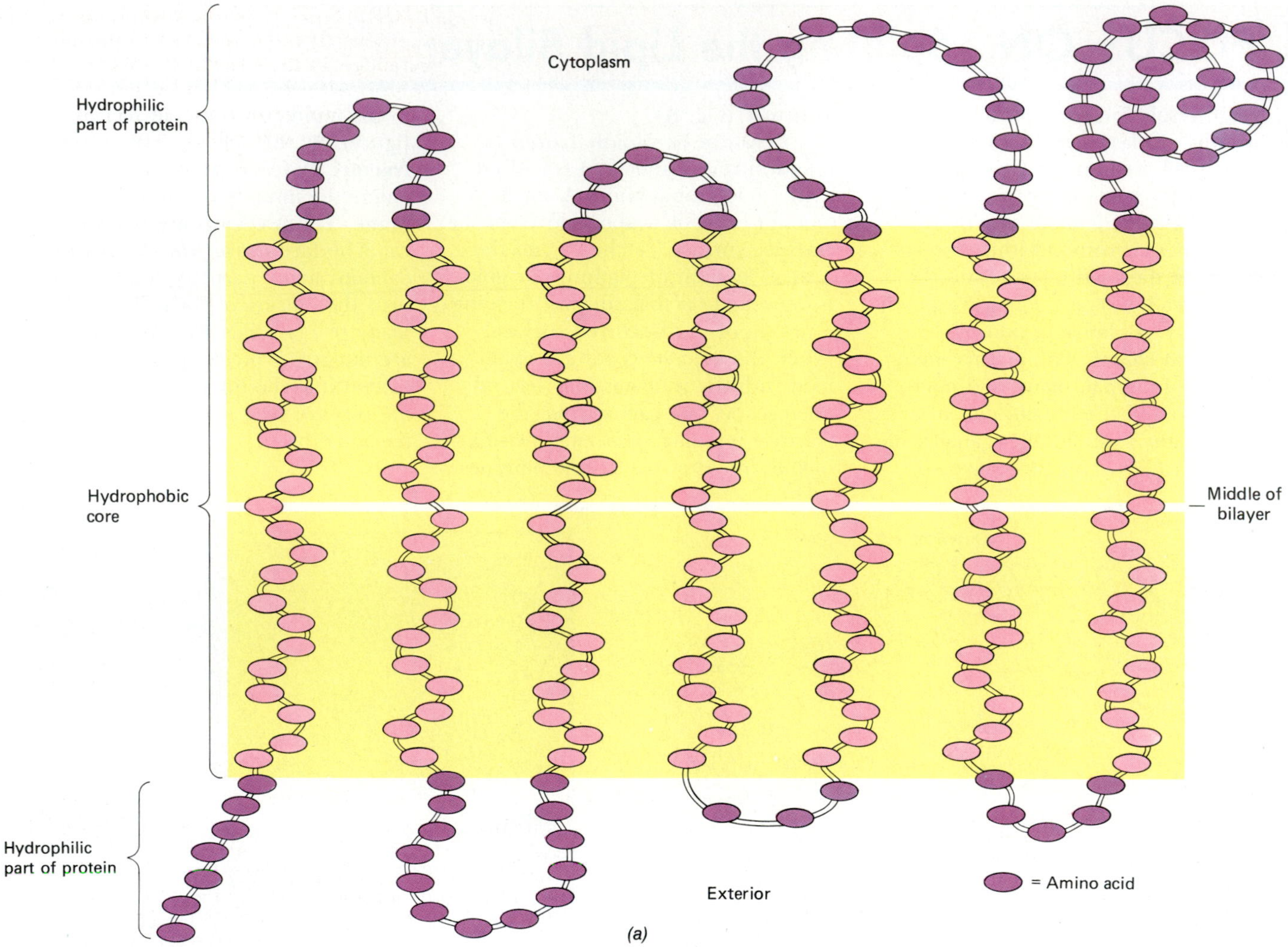

Hydrophilic part of protein

Cytoplasm

Hydrophobic core

Middle of bilayer

Hydrophilic part of protein

Exterior

= Amino acid

(a)

Cytoplasm

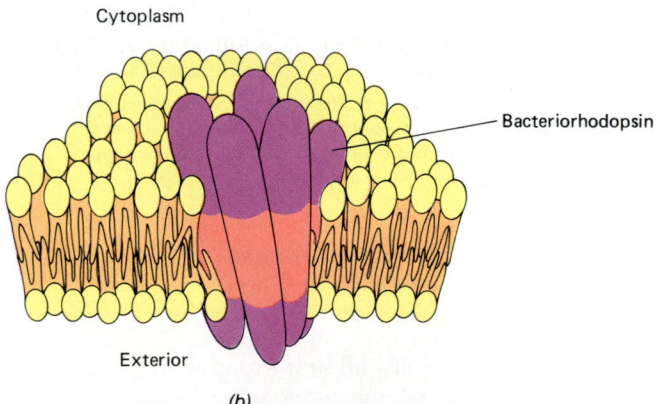

Bacteriorhodopsin

Exterior

(b)

FIGURE 6–4 *(a)* **Bacteriorhodopsin is a protein found in the plasma membrane of certain purple bacteria, in which it plays a crucial role in photosynthesis. It takes the form of a long alpha helix with seven subdivisions. The hydrophobic portions of the molecule are embedded in the membrane, with the hydrophilic parts protruding toward the inside or outside of the cell. The seven subdivisions of the molecule assume a cylindrical configuration as shown in** *(b)***.**

lulose fibers are like steel rods and the matrix material around them is the concrete. Other compounds present in the plant cell wall include **pectins** (jelling agents used in the preparation of jams and jellies) and **lignin,** a substance that provides rigidity. Cells whose main function is support contain a great deal of lignin in their walls. In fact, wood consists mainly of cell walls in which cellulose is reinforced by lignin. Early in the development of the plant cell, the primary cell wall is

quite flexible to allow for growth. A more rigid secondary cell wall is deposited subsequently underneath the primary cell wall.

Fungi and algae (now included in the protist kingdom) also have cell walls. Fungal cell walls are composed mainly of the polysaccharide chitin (which also occurs in the external skeletons of insects and other arthropod animals). Algal cell walls are usually composed of cellulose. Some animal-like protists have a shell-like **pellicle,** really a thickened plasma membrane, which serves some of the functions of a cell wall. True animal cells do not possess cell walls, though an aberrant group of animals, the tunicates (Chapter 25) do have a sort of cellulose skeleton. It is likely that ei-

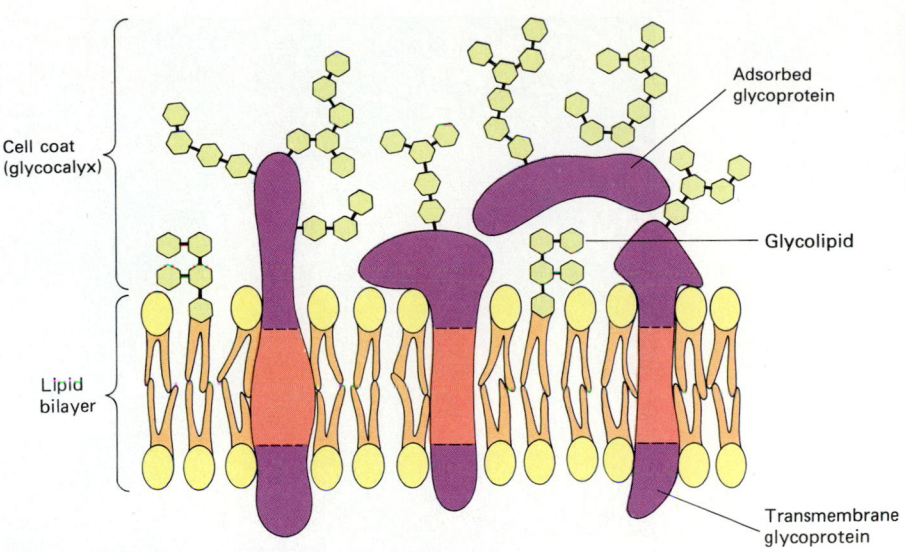

Cell coat (glycocalyx)

Lipid bilayer

Adsorbed glycoprotein

Glycolipid

Transmembrane glycoprotein

FIGURE 6–5 How the carbohydrates of the glycocalyx are attached to proteins embedded in the plasma membrane. Some glycoproteins are said to be adsorbed, which means loosely attached to the surface of the cell rather than embedded in the plasma membrane.

FIGURE 6–6 Fusion of human and mouse cells results in the random distribution of the proteins of both species on the surface of the resulting cell. This shows that most cellular surface proteins are not fixed in position and that the plasma membrane is essentially fluid.

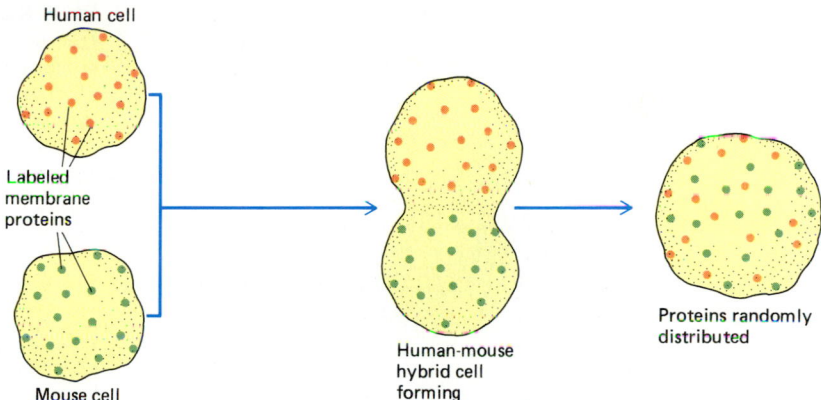

Human cell

Labeled membrane proteins

Mouse cell

Human-mouse hybrid cell forming

Proteins randomly distributed

FIGURE 6–7 The red blood cell has a cytoskeleton located just under its plasma membrane that is made of the protein spectrin. Spectrin is attached to the plasma membrane by means of another protein, ankyrin, whose hydrophobic parts are embedded in the plasma membrane itself. This arrangement helps to produce the characteristic shape of the red blood cell.

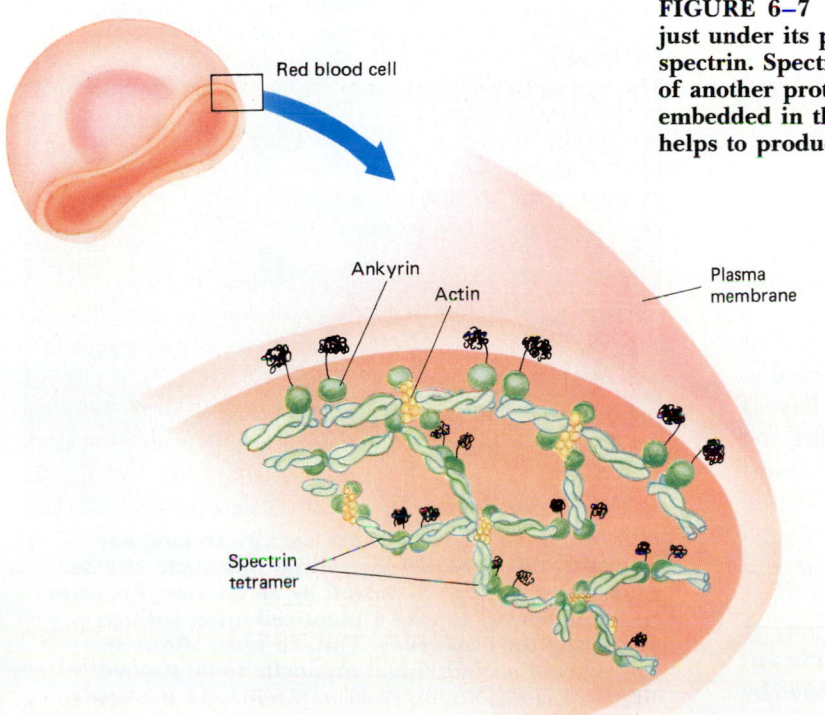

Red blood cell

Ankyrin

Actin

Plasma membrane

Spectrin tetramer

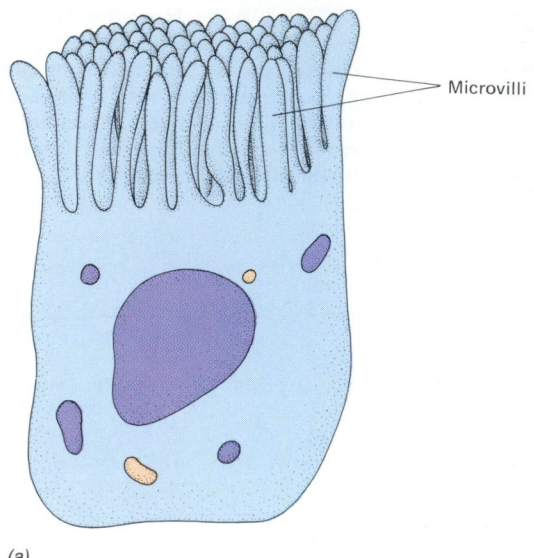

(a)

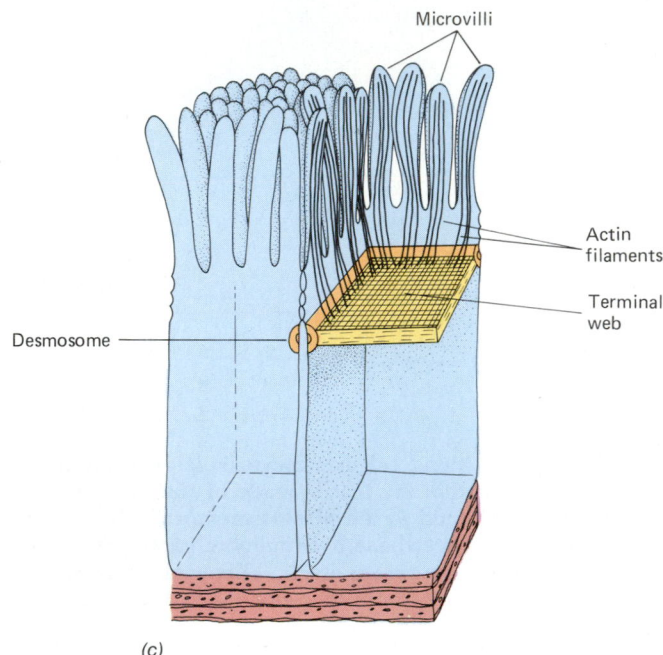

(c)

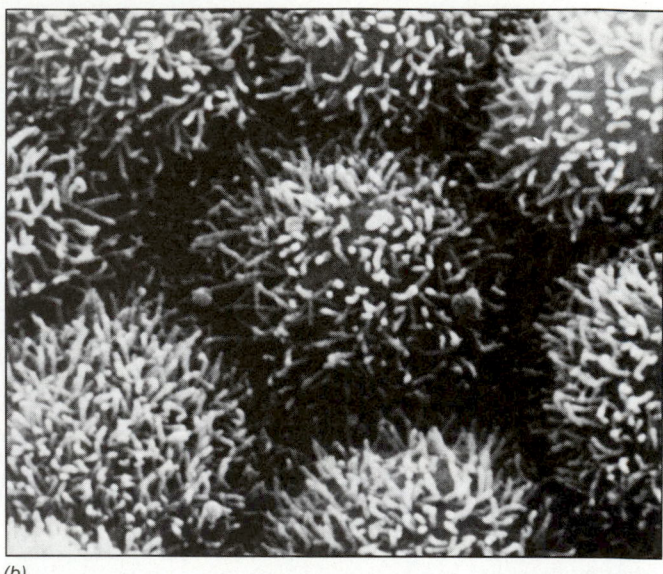

(b)

FIGURE 6–8   Microvilli. (*a*) Microvilli, present on the free surfaces of many kinds of cells, greatly increase the surface area for absorption of materials. (*b*) Scanning electron micrograph of ovarian epithelial cells from a mouse, showing microvilli (approximately ×65,000). (*c*) How actin filaments within the cell can shorten microvilli. (*b*, courtesy of Dr. E. Anderson, Harvard Medical School)

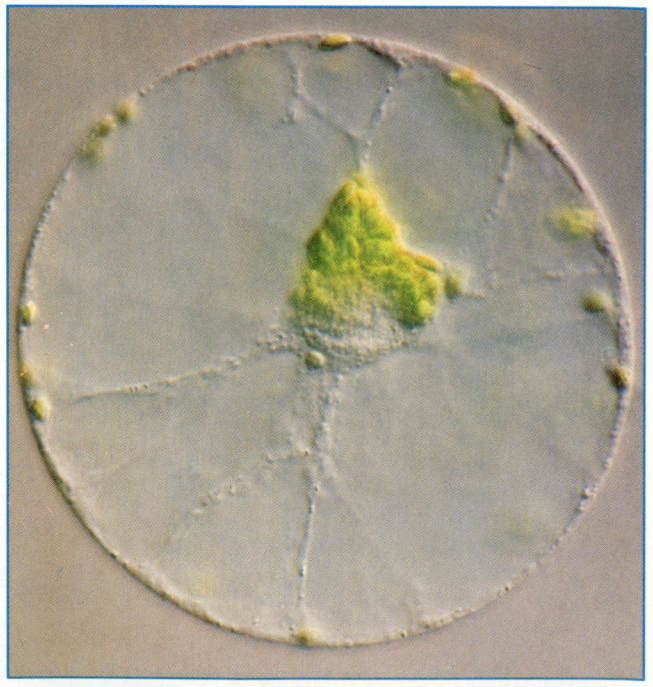

FIGURE 6–9   Plant cells are usually rectangular. However, this *Amaryllis* protoplast was made circular when its walls were dissolved by an enzyme. Enzymes enable biologists to take a plant cell apart without disrupting the organelles. This, in turn, allows the functions of an individual organelle to be studied (approximately ×600). (William J. Marin, Jr., Brookhaven National Laboratory)

ther cellulose or chitin is the most common organic compound on the face of the earth.

The cell walls of prokaryotes are not composed of cellulose. They contain a unique complex carbohydrate known as **peptidoglycan.** Many bacteria have a slimy polysaccharide capsule that lies just outside the cell wall and serves as an additional protective layer.

■ CELLS ARE JOINED BY SPECIALIZED STRUCTURES

Some types of cells communicate directly with one another. Adjacent plant cells may actually be joined by extensions of cytoplasm that pass through openings in

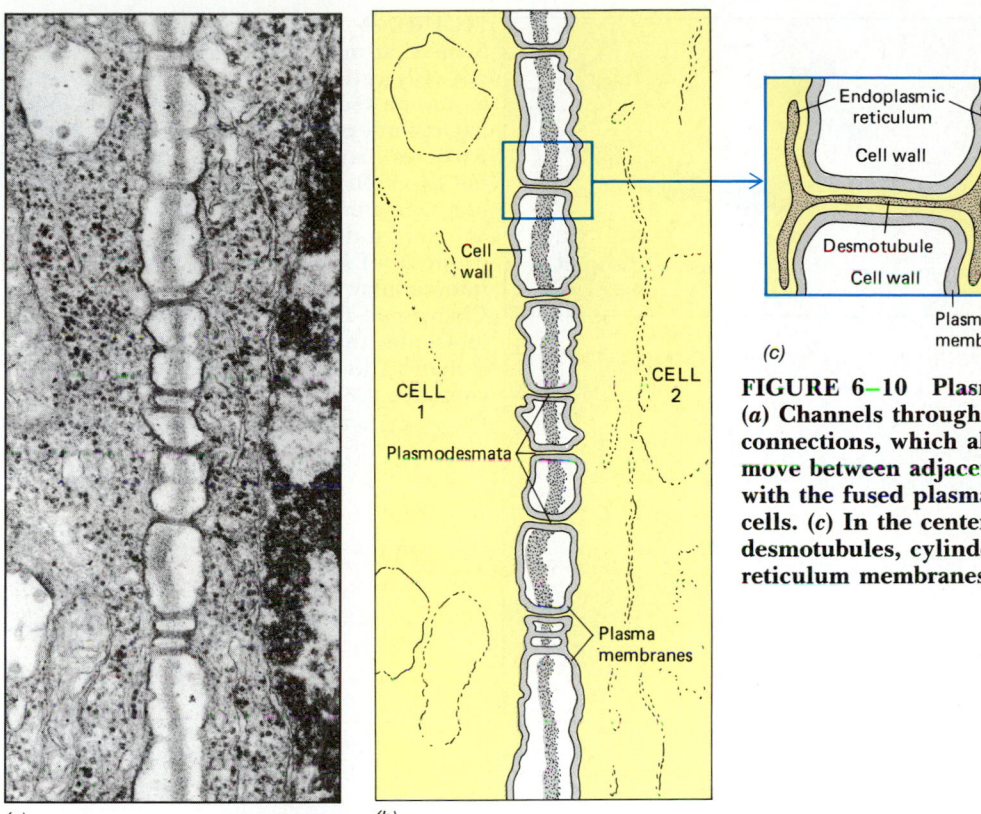

(c)

FIGURE 6–10   Plasmodesmata in plant cell walls.
(*a*) **Channels through plant cell walls form cytoplasmic connections, which allow water and small molecules to move between adjacent cells.** (*b*) **The channels are lined with the fused plasma membranes of the two adjacent cells.** (*c*) **In the center of most plasmodesmata are desmotubules, cylinders formed from endoplasmic reticulum membranes from both cells.**

the cell walls and plasma membranes. These cytoplasmic extensions, called **plasmodesmata,** provide a pathway for the passage of water, ions, nutrients, and other materials from one cell to another (Fig. 6–10).

Serving a somewhat similar function as plant cell plasmodesmata, **gap junctions** connect the cytoplasmic compartments of animal cells (Fig. 6–11). When a marker substance is injected into one of the cells connected by gap junctions, the marker passes rapidly into the adjacent cell. Gap junctions are thought also to provide for electrical communication between certain types of cells. In some species the gap junction (Chapter 33) permits the transmission of an electrical impulse from one cell to the next, perhaps in the form of a surge of ions. Such an arrangement is found in the electrical organs of the electric eel and the electric catfish. Cardiac muscle cells are also connected by gap junctions that permit the rapid transmission of neural impulses from one cell to the next so that all the muscle fibers in the ventricles of the heart can contract one after the other (almost simultaneously) in a rapid, regular fashion.

In tissues where cells are in close contact with one another, specialization of the plasma membrane may provide for the adhesion of neighboring cells. Adjacent epithelial cells, especially those of the upper layer of the skin, are anchored together by button-like plaques, called **desmosomes,** on the two opposing cell surfaces (Fig. 6–12). Even with the desmosomes, a tiny intercellular space about 24 nanometers wide separates the two opposing cell surfaces. The two cells are held together by protein filaments that cross the intercellular space between the desmosomes.

In certain tissues, cells are held together by **tight junctions,** connections so tight that materials cannot pass between the cells. In a tight junction the two plasma membranes are often fused, and there is no intercellular space (Fig. 6–13). Cells connected by tight junctions constitute a continuous barrier and are found where a sharp physical separation between two body compartments is essential. A classic example is the "blood-brain barrier," which involves the lining of the blood vessels in the brain. The cells lining these blood vessels are joined by tight junctions whose role is essentially protective. They block the passage of certain molecules from the blood into the brain tissue.

■ MATERIALS PASS THROUGH MEMBRANES IN SEVERAL WAYS

Whether a membrane will permit the molecules of any given substance to pass through it depends on the structure of the membrane and the size and charge of the molecules. A membrane is said to be **permeable** to a given substance if it permits the substance to pass through and **impermeable** if it prevents passage of the substance. A **selectively permeable** membrane allows some, but not other, substances to pass through it. Re-

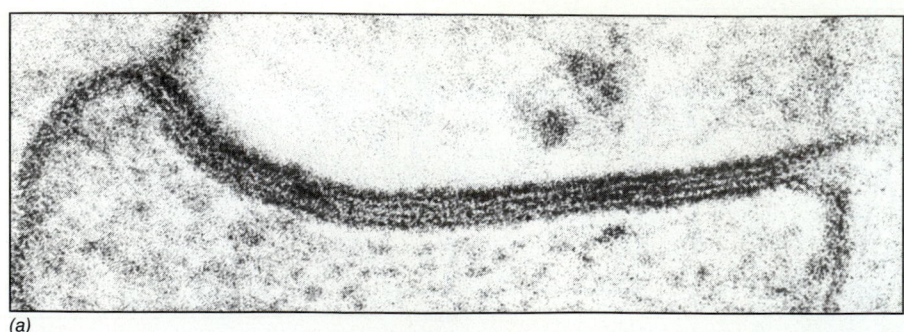

(a)

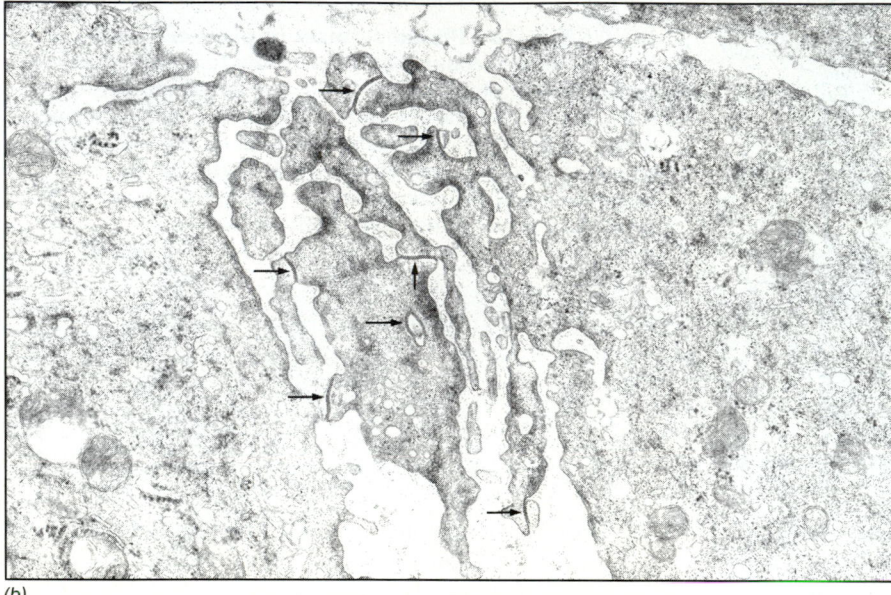

(b)

FIGURE 6–11   (a) A gap junction. Somewhat misnamed, gap junctions join not only cells but their interiors as well, as can be seen in (b). Each cell involved has cylinders consisting of six protein particles that serve as pipes connecting the two cells. (c) Gap junctions sometimes permit materials to pass from one cell to another, and sometimes they do not. One proposal of how they may function as intercellular valves is shown here. Changes in the shape of the protein molecules that comprise them may cause a channel to form or close between the subunits. (Courtesy of Puri and Garfield, from *Biology of Reproduction* 27: 967–975, 1982)

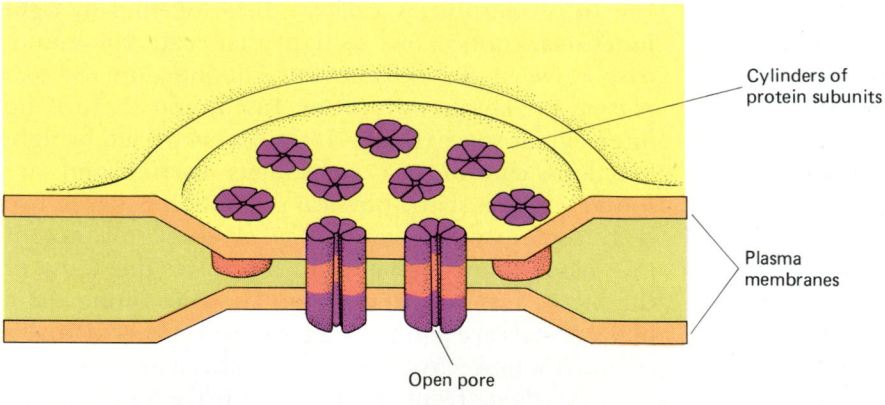

Cylinders of
protein subunits

Plasma
membranes

Open pore

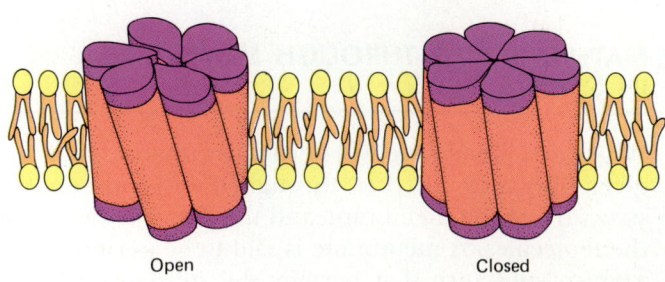

Open   Closed

(c)

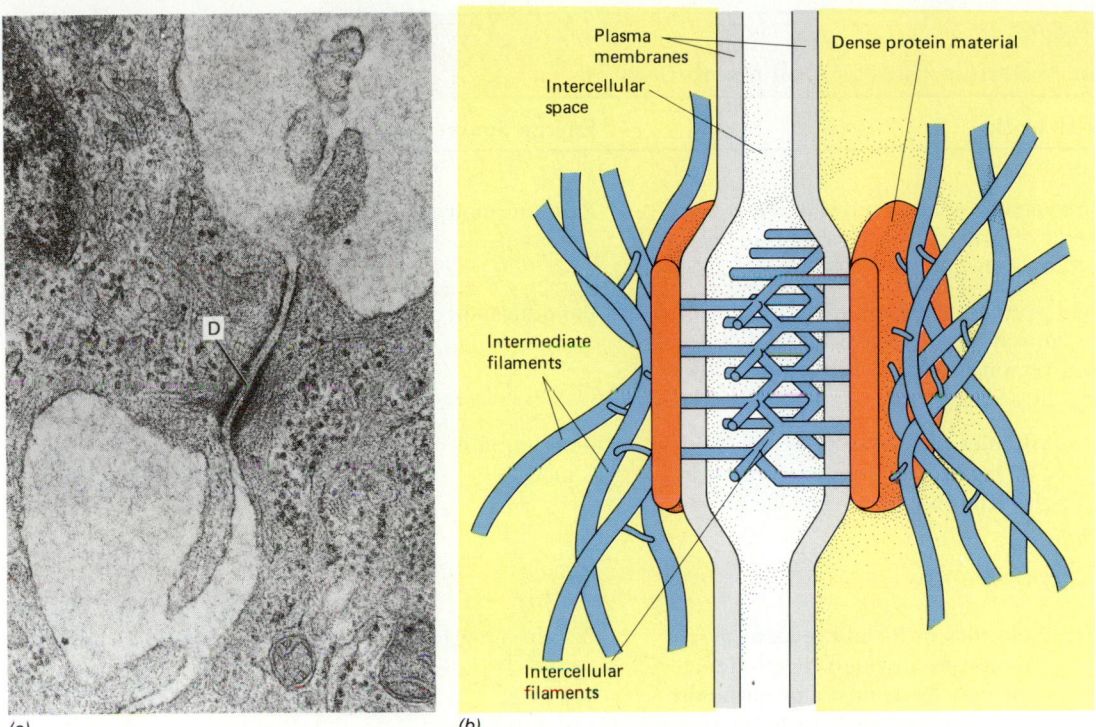

(a)   (b)

FIGURE 6–12   (*a*) **Electron micrograph showing a desmosome (*D*) between two cells of the ovarian epithelium of a rabbit (approximately ×70,000). (*b*) Structure of a desmosome. Cytoskeletal fibers in the two joined cells attach to the button of dense protein material, which in turn clings to a similar button on the adjacent cell via an interlacing mass of external filaments.**

(*a*, Courtesy of Dr. Everett Anderson, *Journal of Morphology* 150:136–166, 1976)

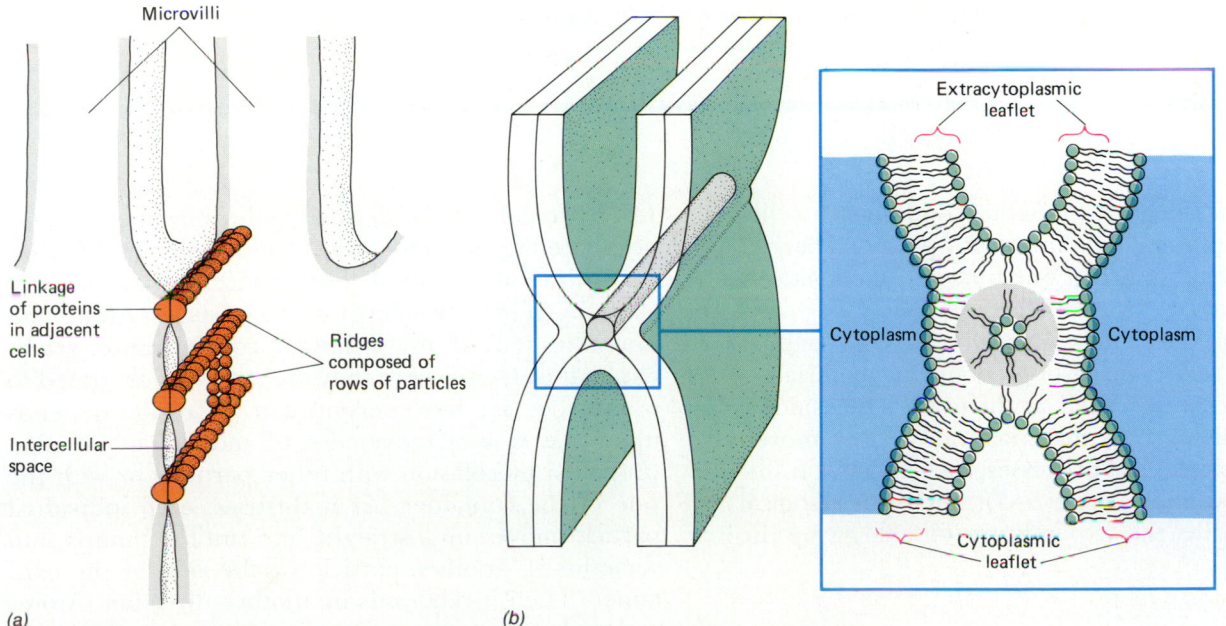

(a)   (b)

FIGURE 6–13   Structure of a tight junction. (*a*) In some tight junctions, adjacent cells are held together by rows of protein particles. (*b*) In many tight junctions the plasma membranes of adjacent cells actually fuse. If the tight junctions are extensive there is no way that materials can pass between cells so united.

TABLE 6-1
Mechanisms For Moving Materials Through Cell Membranes

Process	How It Works	Energy Source	Example
Physical Process			
Diffusion	Net movement of molecules (or ions) from region of greater concentration to region of lower concentration	Random molecular motion	Movement of oxygen in tissue fluid
Facilitated diffusion	Carrier protein in plasma membrane accelerates movement of relatively large molecules from region of higher to region of lower concentration	Random molecular motion	Movement of glucose into cells
Osmosis	Water molecules diffuse from region of higher to region of lower concentration through differentially permeable membrane	Random molecular motion	Water enters red blood cell placed in distilled water
Physiological Process			
Active transport	Protein molecules in membrane transport ions or molecules through membrane; movement may be against concentration gradient (i.e., from region of lower to region of higher concentration)	Cellular energy	Pumping of sodium out of cell against concentration gradient
Endocytosis			
Phagocytosis	Plasma membrane encircles particle and brings it into cell by forming vacuole around it	Cellular energy	White blood cells ingest bacteria
Pinocytosis	Plasma membrane takes in fluid droplets by forming vesicles around them	Cellular energy	Cell takes in needed solute dissolved in fluid
Exocytosis	Plasma membrane ejects materials; vesicle filled with material fuses with cell membrane	Cellular energy	Secretion of mucus

sponding to varying environmental conditions or cellular needs, the plasma membrane may present a barrier to a particular substance at one time and then actively promote its passage at another.

Some materials move passively through cellular membranes—as they do through nonliving materials—by physical processes, such as diffusion and osmosis. However, in living cells materials can also be moved actively by physiological processes, such as active transport, or endocytosis (Table 6–1). Such physiological processes require the expenditure of energy by the cell.

Diffusion Is Passive Molecular Movement

Some substances pass into or out of cells by simple physical diffusion. All molecules in liquids and gases tend to move (diffuse) in all directions until they are spread evenly throughout the available space (Fig. 6–

14). **Diffusion** may be defined as the movement of particles (atoms, ions, molecules) from a region of higher concentration to one of lower concentration brought about by the kinetic energy of the particles. Atoms and molecules tend to move down a **concentration gradient,** that is, from where they are more concentrated to where they are less concentrated. Diffusion depends upon the *random* movement of individual particles, propelled by collision with other particles or with the side of the container. As it diffuses, each individual particle moves in a straight line until it bumps into something—another particle or the side of the container. Then it rebounds in another direction. Atoms and molecules continue to move even when they have become uniformly distributed throughout a given space. However, as fast as some particles travel in one direction, others travel in the opposite direction, so that on the whole all the particles remain uniformly distributed; thus, a dynamic equilibrium exists.

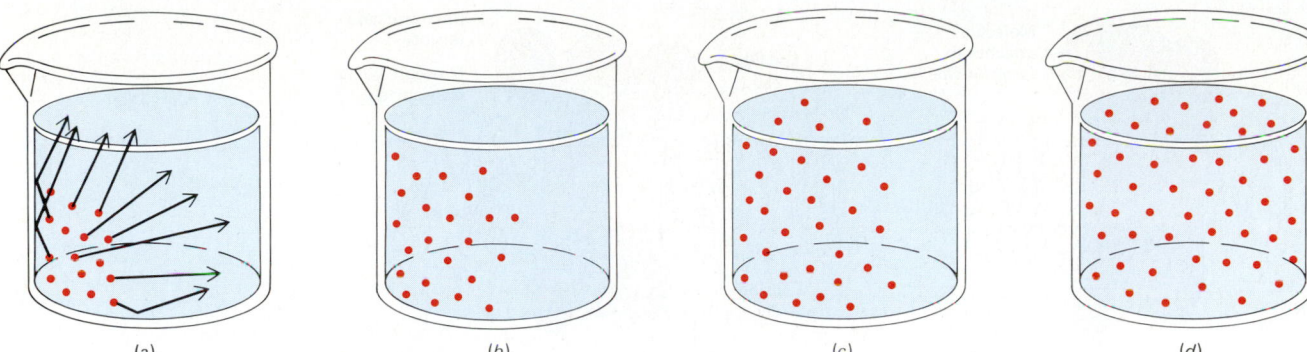

FIGURE 6–14   The process of diffusion. When a small lump of sugar is dropped into a beaker of water, its molecules dissolve (a) and begin to diffuse (b and c). Over a long period of time, diffusion results in an even distribution of sugar molecules throughout the water (d).

The rate of diffusion is a function of the size and shape of the atoms or molecules, the temperature, and the state of matter (that is, whether it is a solid, liquid, or gas). Small particles move faster than larger ones at the same temperature. Heat energy causes molecules to move more rapidly, so as temperature rises, the rate of diffusion increases. Atoms and molecules diffuse independently of each other within the same solution; eventually, all become uniformly distributed.

You might demonstrate the diffusion of gases by opening a bottle of ammonia on a front-row desk of your classroom. Students in the second row would begin to smell ammonia within a few moments because some molecules of ammonia would have left the bottle and begun to diffuse through the air. Some time later the odor would be evident throughout the room. If the room was closed and there were no air currents, the molecules would eventually distribute themselves evenly throughout the room.

Diffusion is important in living systems. A large variety of substances are distributed throughout the cytoplasm by diffusion, and this process is also responsible for moving a great many substances in and out of the cell across the plasma membrane. Oxygen, carbon dioxide, water, and numerous other small ions and molecules can diffuse into or out of the cell readily. However, molecules over a certain size cannot diffuse through the membrane, and molecules that differ fundamentally from the lipid molecules of the membrane (i.e., polar or hydrophilic molecules) in general cannot diffuse through it.

FIGURE 6–15   Tiny granules are visible embedded in the plasma membranes of the microvilli of one of the cells from the lining of the small intestine of a mouse. Some of the granules are thought to be proteins involved in the transport of digested food into the cell. (Courtesy of Drs. Tohru Arima and Torno Yamamoto, *Cell Tissue Research* 233, 1983)

that they are not admitted or not permitted to leave. It's not hard to understand why substances might not get through the plasma membrane easily. It has been hard to explain how some of them can get through it at all! For instance, the question of how hydrophilic compounds, such as glucose and amino acids, can pass through a hydrophobic lipid bilayer had long puzzled biologists. Electron micrographs have revealed membrane proteins prominently attached to the membranes of absorptive cells (Fig. 6–15) that evidently are involved.

In Facilitated Diffusion Molecules Pass Through Special Channels

The concentration of a great many materials inside the cell is different from their concentrations outside the cell. In some cases, this results merely from the fact

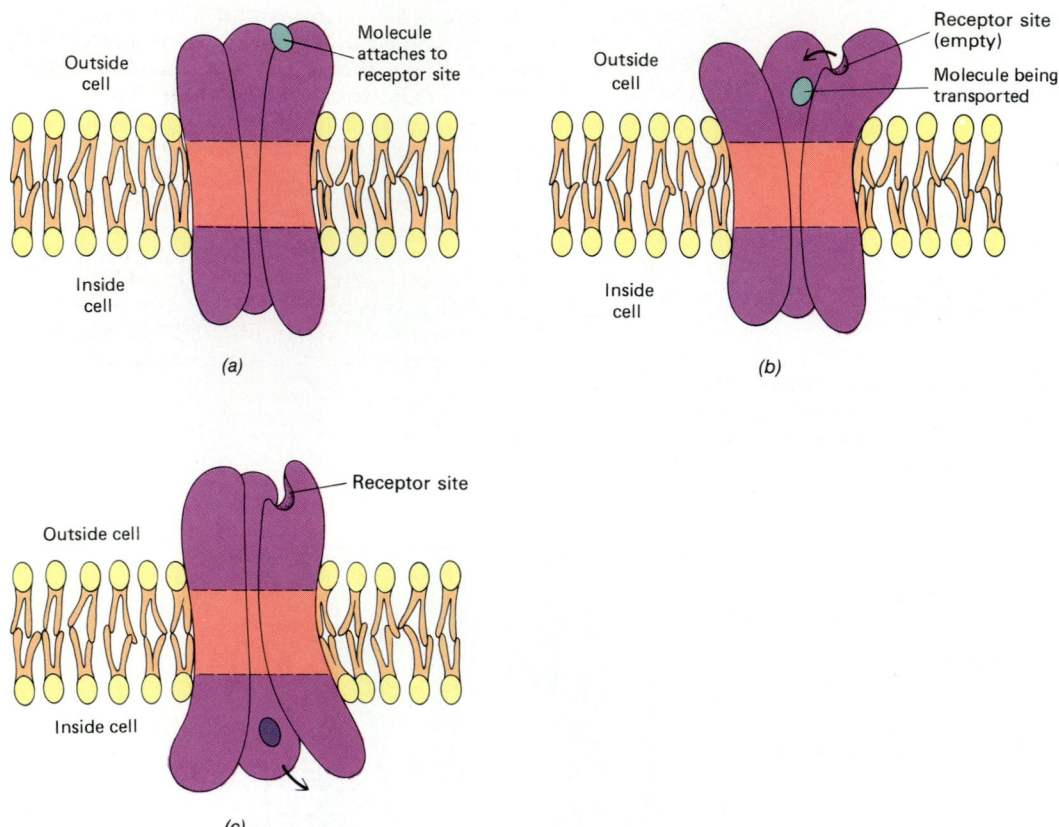

FIGURE 6–16   Facilitated diffusion. (*a*) According to current theory, a group of transport proteins forms a fluid-filled pore through which hydrophilic molecules can pass. When the molecule being transported binds to a receptor site on one of these proteins, it is thought that the transport protein changes shape. (*b*) The molecule is then forced through the channel and ejected on the other side (*c*).

These membrane proteins, called **carrier proteins** or **transport proteins,** permit facilitated diffusion—diffusion that must occur through a special passageway in the membrane. In **facilitated diffusion,** carrier proteins in the plasma membrane assist the passage of small solute molecules through the membrane. The proteins serve as a passive conveyor belt that permits the substance to pass in either direction down a concentration gradient. As with simple diffusion, molecules move from a region of greater concentration to a region of lesser concentration.

Carrier proteins probably extend entirely through the membrane. Four or more carrier protein molecules grouped in contact with each other would provide a water-filled channel about 1 nanometer in diameter between them, through which hydrophilic molecules could pass (Fig. 6–16). A receptor site on the protein specifically binds to one kind of molecule—glucose, for example. When the molecule being transported binds with the carrier protein, the protein is thought to change shape. As a result, the molecule being transported travels through the channel and is ejected on the other side. In this case the movement of the molecule results because the same molecules are

more highly concentrated on the outside of the cell than on the inside, much as in simple diffusion. The difference here is that the receptor site on the carrier protein accepts only certain substances whose molecules have the right shape. This kind of diffusion is therefore selective.

Like simple diffusion, facilitated diffusion is a passive process that consumes no energy other than that inherent in the concentration gradient. Thus facilitated diffusion, like simple diffusion, cannot concentrate materials. Unlike simple diffusion, however, it is controllable, since only certain molecules can pass through the protein channels. This is a very important property, as we will see when we consider the sodium channels that make the operation of nerve, muscle, and many gland cells possible.

Osmosis Poses Special Problems For Cells

Osmosis is a special kind of diffusion—the diffusion of water molecules across a differentially permeable membrane from a region where water molecules are more concentrated to one where they are less concen-

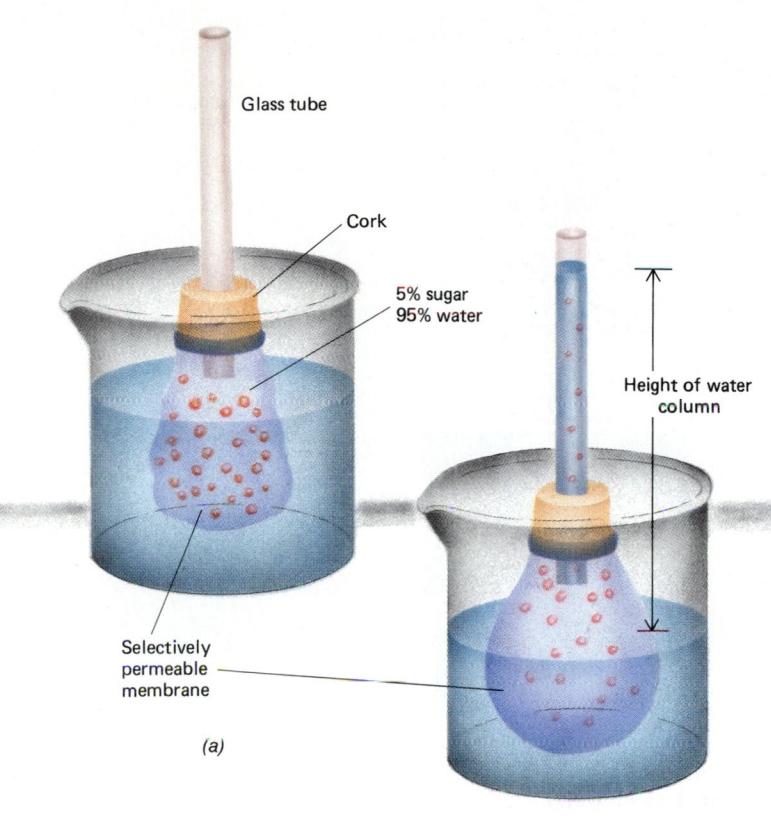

Glass tube

Cork

5% sugar
95% water

Height of water
column

Selectively
permeable
membrane

(a)

(b)

FIGURE 6–17 Osmosis. (*a*) A 5% sugar solution is placed in a sac made of a selectively permeable membrane and suspended in water. The sac is attached to a glass tube. The membrane permits passage of the water molecules but not of the larger sugar molecules. Therefore, the water molecules pass across the membrane *into* the sac. (The water can be thought of as less concentrated inside the sac than outside it.) This causes the column of liquid in the glass tube to rise. (*b*) When equilibrium is reached, the pressure of the column of water just equals, and is a measure of, the osmotic pressure of the sugar solution.

TABLE 6–2
Osmotic Terminology

Solute Concentration In Solution A	Solute Concentration In Solution B	Tonicity	Solute Diffusion	Solvent Diffusion
Greater	Less	A hypertonic to B B hypotonic to A	A to B	B to A
Less	Greater	B hypertonic to A A hypotonic to B	B to A	A to B
Equal	Equal	Isotonic	Equal	Equal

trated. Plasma membranes selectively regulate the passage of most solutes (dissolved materials such as sugar and salts), but water is able to move rather freely in and out of the cell. When living cells are placed in a solution that has a solute concentration equal to that in the cells, the water molecule concentration is also equal, and therefore water molecules move in and out of cells at the same rate, establishing a dynamic equilibrium. The net movement of the water molecules is zero (Fig. 6–17). Such a solution is described as **isotonic** to the cells, that is, of equal solute concentration.

Cells—especially of single-celled organisms—may find themselves in solutions that are of greater or lesser solute concentration relative to the solute con-

centration within the cytoplasm. If the solution has a greater solute concentration, it is said to be **hypertonic** (above strength) to that of the cell, whereas if it has a lesser concentration, it is **hypotonic** (under strength) compared with the cell. Note that the terms hypertonic and hypotonic are relative to each other (Table 6–2). A 5% solution is hypertonic to a 2% solution but hypotonic to a 10% solution.

Suppose that we place some living cells in distilled water that contains 100% water molecules and that will certainly be hypotonic to the cell. If the total number of solute molecules in the cells amounted to 1% of the total molecules present, then water molecules would account for only 99% of the total. Since water mole-

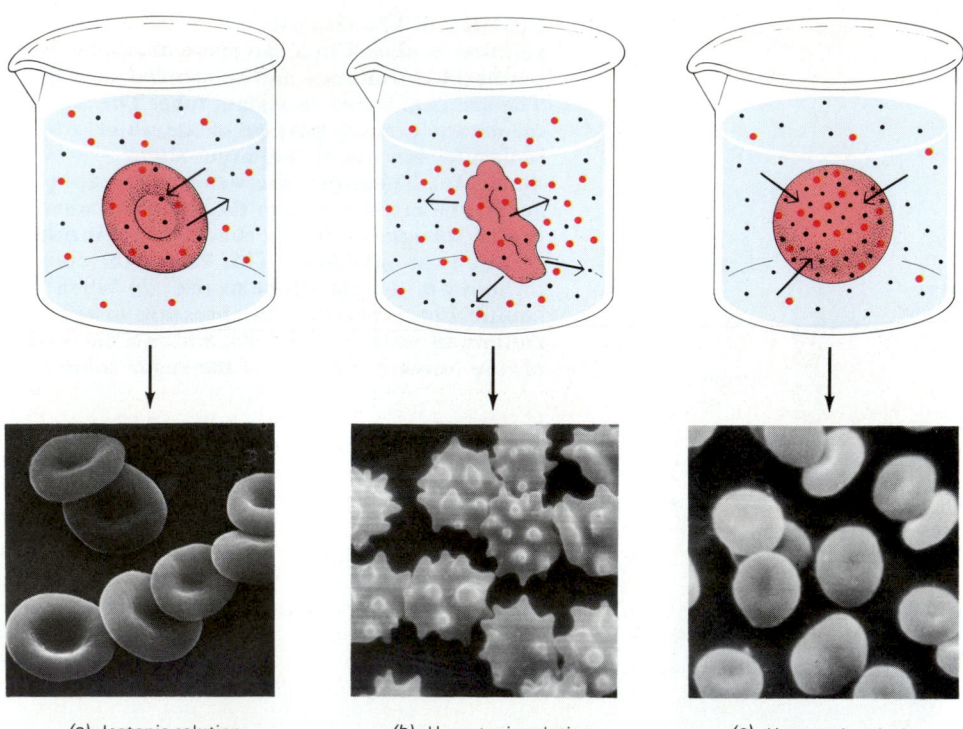

(a) Isotonic solution   *(b)* Hypertonic solution   *(c)* Hypotonic solution

FIGURE 6–18   Osmosis and the living cell. (*a*) **A cell is placed in an isotonic solution. Because the concentration of solutes (and thus of water molecules) is the same in the solution as in the cell, the net movement of water molecules is zero.** (*b*) **A cell is placed in a hypertonic solution. This solution has a greater solute concentration (thus a lower water concentration) than the cell and therefore exerts an osmotic pressure on the cell. This results in a net movement of water molecules out of the cell, causing the cell to dehydrate, shrink, and perhaps die.** (*c*) **A cell is placed in a hypotonic solution. The solution has a lower solute (and thus a greater water) concentration than the cell. The cell contents thus exert an osmotic pressure on the solution, drawing water molecules inward. There is a net diffusion of water molecules into the cell, causing the cell to swell and perhaps even to burst.** (Micrographs of human red blood cells courtesy of Dr. R.F. Baker, University of Southern California Medical School)

cules, like solute molecules, tend to move from a region where they are more concentrated to a region where they are less concentrated, they diffuse inward across the plasma membrane (Fig. 6–18). Although the solute molecules have a tendency to diffuse in the opposite direction, the plasma membrane prevents them from "leaking out" to any great extent. Instead, they may be thought of as being trapped within the cell and exerting an osmotic pressure upon the less-concentrated solution on the other side of the membrane. So much water may enter the cell that it swells and bursts.

On the other hand, when cells are placed in a solution that is hypertonic to them, water tends to flow out of them. The cells may become dehydrated, shrink, and die. Can you explain this in terms of the relative concentrations of water molecules in the two solutions? Remember, when a solution contains *more* solute molecules, it has proportionately *fewer* water molecules, so the solvent and solute concentrations are reciprocally

related. Also remember that water passes freely through the plasma membrane and moves from regions of high concentration to regions of low concentration.

Although the excretory systems of multicellular terrestrial animals like ourselves usually are able to maintain their body fluids in an isotonic state, organisms such as freshwater protists (and other freshwater organisms) have a continuous osmotic problem. Their watery surroundings are hypotonic to them, and water tends to pass into their cells. Some of these organisms possess an adaptation that solves the problem—a **contractile vacuole** that takes up the excess water and pumps it out of the cell (Fig. 6–19). Just how this is done is still not fully understood, but it is believed that contractile fibers pull on the membranes surrounding the vacuole, lowering the pressure inside it. The resulting differential in hydrostatic pressure would force pure water through the differentially permeable vacu-

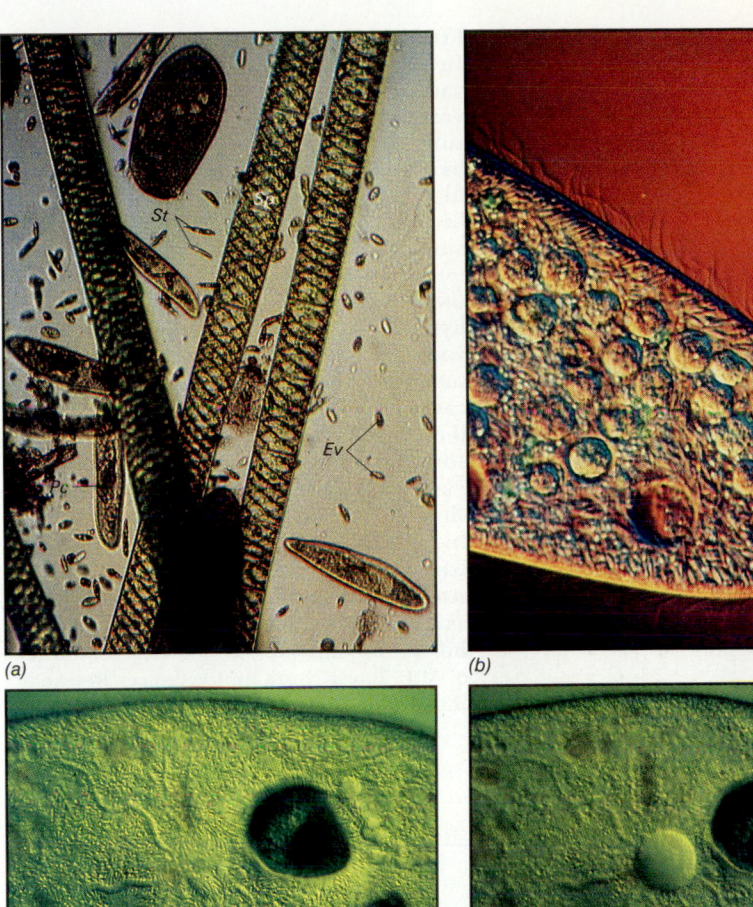

(a)

(b)

(c)

(d)

FIGURE 6–19 (a) **Some life in a drop of pond water. A sample of unappetizing, greenish pond water can be a thing of beauty swarming with life under the microscope. Strands of filamentous green algae are commonly seen, along with animal-like protists, small multicellular organisms, and bacteria. In this picture (×20) we see** *Paramecium caudatum (Pc)*, *Stentor (St)*, *Euglena viridis (Ev)*, **and** *Spirogyra (Sp)*. (b) **Contractile vacuole of a** *Paramecium*. **The vacuole is the cavity near the end of the microorganism. This structure functions somewhat like a cellular kidney, collecting water from the cytoplasm and periodically ejecting it. This adaptation permits the organism to live in hypotonic, freshwater habitats.** (c, d) **Contractile vacuole function in** *Paramecium;* (c) **empty vacuole;** (d) **vacuole filled with fluid about to be ejected.** (*a*, Runk/Schoenberger from Grant Heilman; *b*, M. Abbey/Visuals Unlimited; *c, d,* Thomas Eisner)

olar membrane into the vacuole despite the tendency of osmosis to cause it to flow in the opposite direction. Then the vacuole is squeezed so that it empties to the exterior.

Plant cells are also adapted to the hypotonic water that often bathes their roots. Their rigid cell walls enable them to withstand, without bursting, pressure exerted by the water that seeps in, filling their central vacuoles. As this internal pressure, called **turgor pressure,** increases, it forces water molecules back out of the cell (Fig. 6–20). The turgor pressure levels off when the outward passage of water molecules equals the rate of inward movement of water molecules. When conditions become dry and the plant cell does not have enough water, the central vacuole decreases in size and the cell shrinks. Thus the plant wilts.

Active Transport Uses Energy To Concentrate Selected Materials

A cell does require certain substances—potassium ions, for example—in greater concentration than they are present in the cell's surroundings. The potassium concentration is about 35 times greater inside the cell than outside. Other substances—for example, sodium ions—are more concentrated in the environment than could be tolerated inside the cell. Yet sodium and potassium ions are roughly the same size. By itself, diffusion cannot account for the concentration differences of these two ions inside and outside the cell. Indeed, given the opportunity to do so, diffusion would quickly eliminate such differences in solute concentration and the cell would die. In **active transport** the cell moves materials from a region of lower concentration to a region of higher concentration (the opposite of diffusion or osmosis). Working uphill this way against a concentration gradient requires energy. Expenditure of energy in active transport is an example of how even cells that appear to be resting are actually performing work just to remain alive.

Active transport proteins may operate in a fashion similar to the way carrier proteins work in facilitated diffusion, except that in active transport, metabolic energy is involved. It is supplied by the molecule **adenosine triphosphate (ATP),** the cell's energy-carrying molecule. Sodium-potassium pumps in the plasma

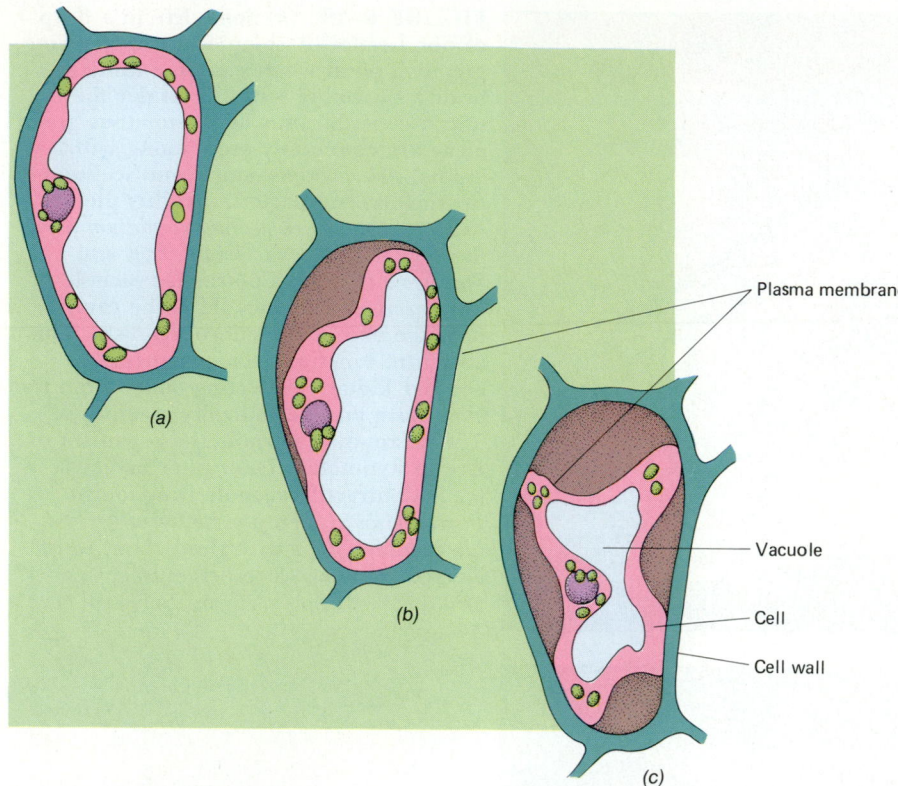

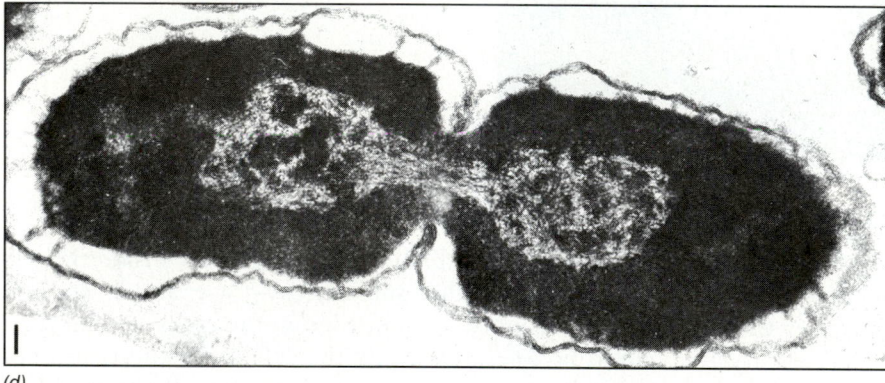

FIGURE 6–20 (*a*) **A turgid plant cell. In a hypotonic solution the central vacuole fills with water because the vacuole contents are hypertonic to the surrounding fluid. The cell is kept from bursting by the tough cell wall, but the high internal pressure causes the cell to become turgid, in a manner like an automobile tire full of compressed air.** (*b*) **When conditions become dry or if the surrounding solution becomes hypertonic, water leaves the central vacuole. First this causes loss of turgidity (wilting), followed in extreme cases by an actual decrease in size of the central vacuole so that** (*c*) **the cell shrinks.** (*d*) **Section through a dividing** *Escherichia coli*, **a bacterium that has shrunk (plasmolyzed) due to a hypertonic environment. Although some bacteria normally live in very salty surroundings, hypertonic solutions such as pickling brines or syrups effectively inhibit most bacterial growth.** (Courtesy of M.E. Bayer, Institute for Cancer Research/ BPS)

membrane produce the differential concentrations of sodium and potassium ions. Look at Figure 6–21 as you follow this explanation:

1 and 2. When a sodium ion inside the cell binds to the receptor site, a molecule of ATP transfers its energy to the pump proteins.

3. The resulting change in protein shape ejects the sodium ion.

4. Another receptor site accepts a potassium ion from outside the cell.

5 and 6. The potassium ion is released into the cell, which depletes the energy stored in the transport protein.

1. Another sodium ion binds to the receptor site and the cycle repeats.

Cells Can Ingest Materials By Endocytosis And Eject Them By Exocytosis

In diffusion and in active transport, individual small molecules and ions pass through the plasma membrane. Larger materials—very large molecules, particles of food, or even whole cells—must sometimes be moved in or out of the cell. Such cellular work requires the cell to expend energy. In **exocytosis** (Fig. 6–22), a cell ejects waste products or specific secretion products, such as mucus or hormones. Generally, the material to be ejected is enclosed within a membrane, and the vesicle thus formed fuses with the plasma membrane. The vesicle then opens at the point of fusion, and the enclosed material is released to the exterior without the loss of other cell contents.

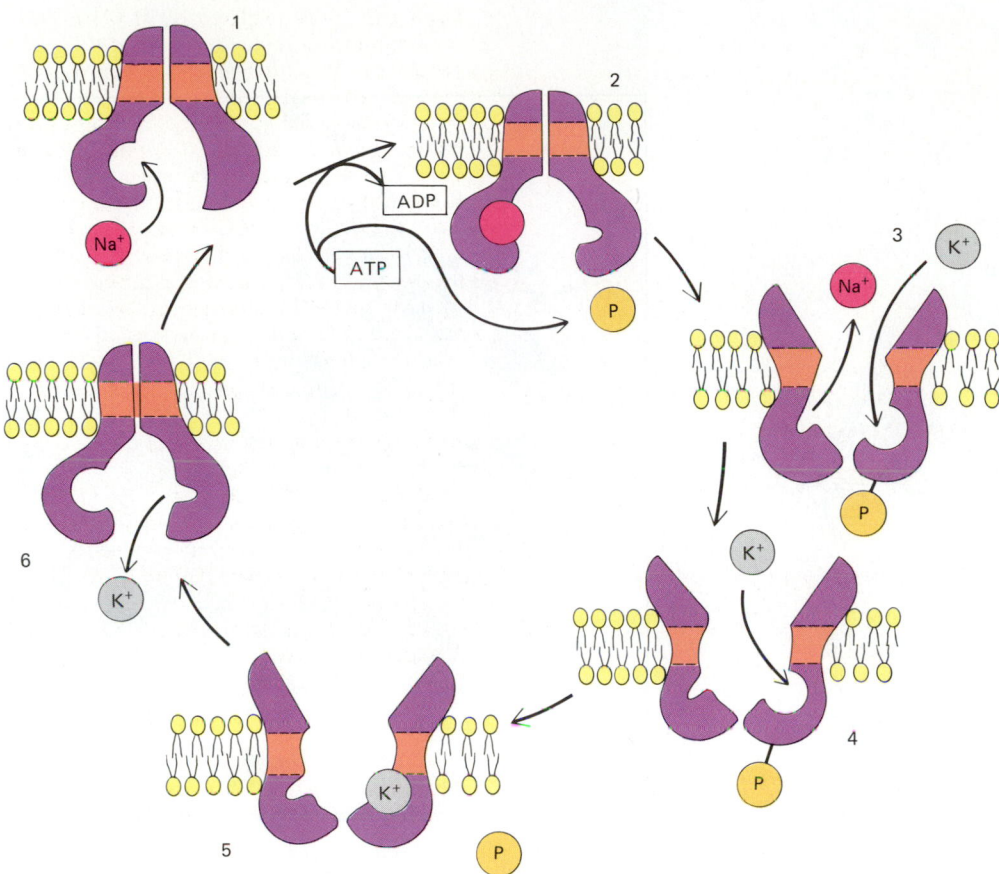

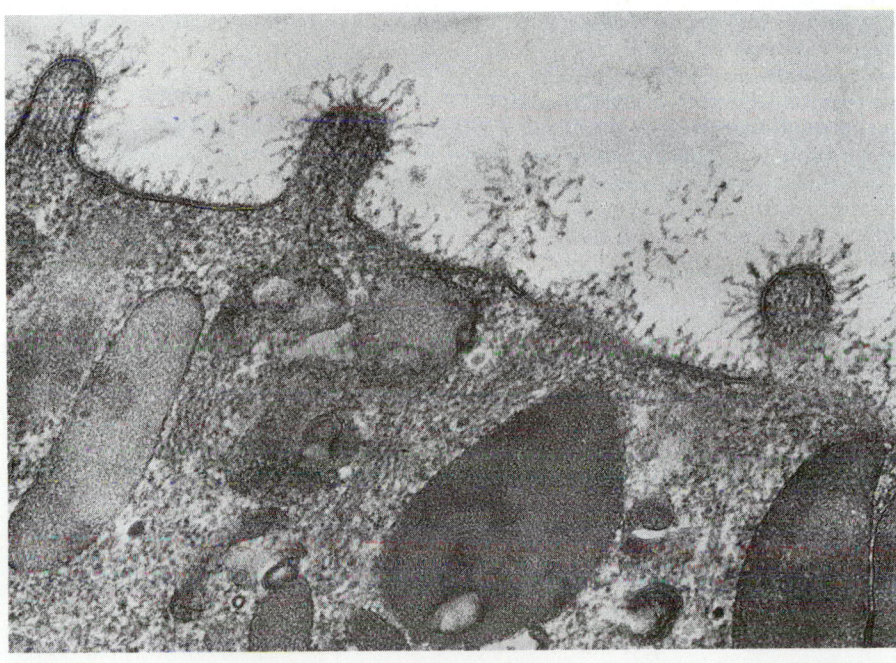

FIGURE 6–22 Exocytosis. A high-magnification electron micrograph of the upper surface of a secreting cell (approximately ×125,000). Secretion granules can be seen in the cytoplasm approaching the plasma membrane. The filaments projecting diffusely from the cell surface are of unknown significance but may be proteins. (J.F. Gennaro, Photo Researchers, Inc.)

In **endocytosis** materials are taken into the cell. In one type of endocytosis, termed **phagocytosis** ("cell eating"), the cell ingests large solid particles, such as bacteria or food (Fig. 6–23). Folds of the plasma membrane extend outward and enclose the particle to be ingested, forming a vacuole around it. The vacuole, still attached to the plasma membrane, bulges into the cell interior. The membrane then tightens like a draw-

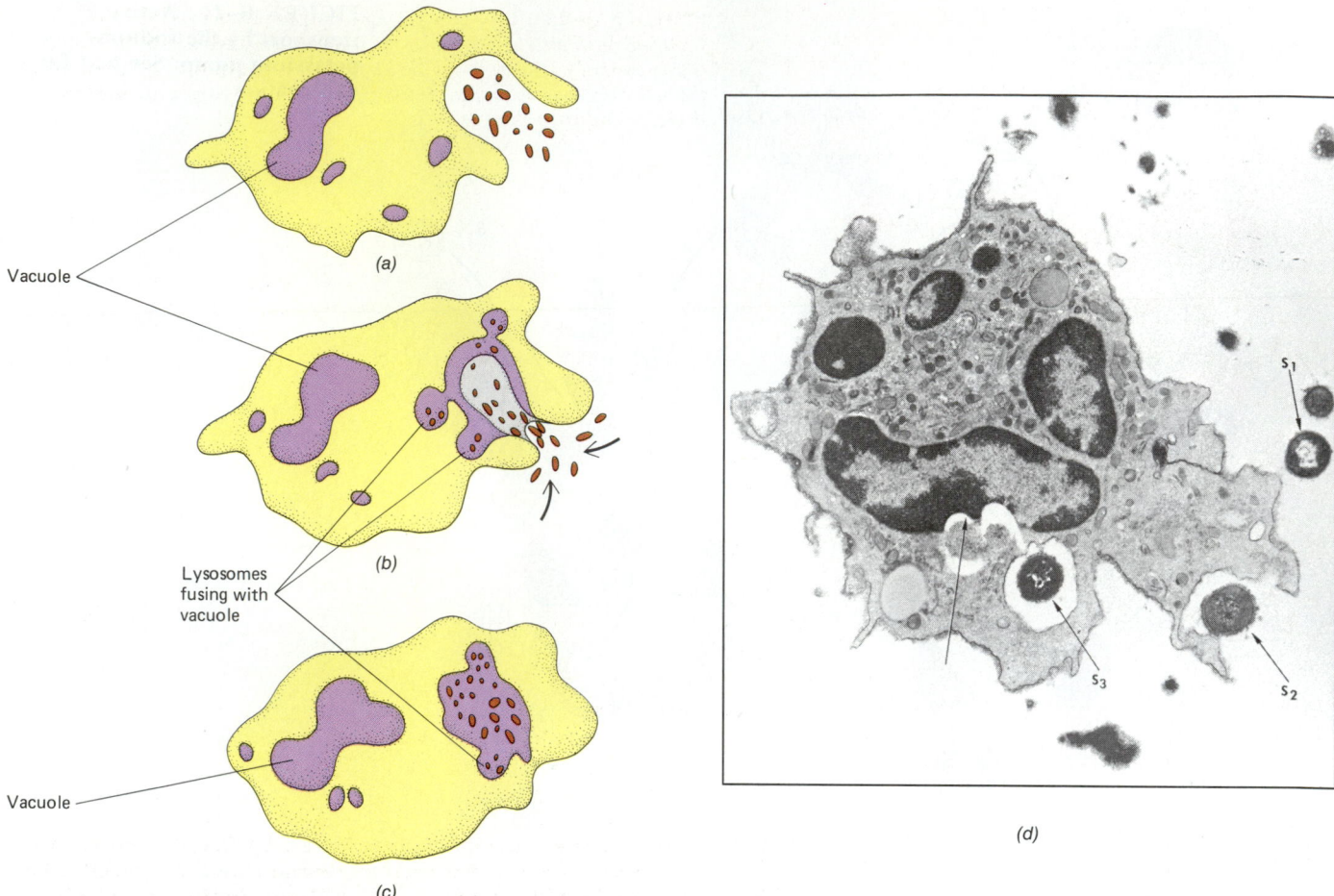

Vacuole

(a)

Lysosomes
fusing with
vacuole

(b)

Vacuole

(c)

(d)

FIGURE 6–23 Phagocytosis. (a) In phagocytosis the cell ingests large solid particles, such as bacteria. Folds of the plasma membrane surround the particle to be ingested, forming a small vacuole around it. (b) This vacuole then pinches off inside the cell. (c) Lysosomes may fuse with the vacuole and pour their potent digestive enzymes onto the digested material. (d) A white blood cell in the presence of *S. pyogenes* (approximately ×23,000). One bacterium (*S1*) is free, one bacterium (*S2*) is being phagocytized, and a third bacterium (*S3*) has been phagocytized and is seen within a vacuole (phagosome). Note that near the vacuole (see *arrow*) the white blood cell's own nucleus has been partly digested. (J.G. Hadley, Battelle/BPS)

string purse and fuses together, leaving the vacuole floating freely in the cytoplasm. The contents are then further broken down by fusing with lysosomes, which exposes their contents to digestive enzymes.

In **pinocytosis** ("cell drinking") the cell takes in dissolved materials. Tiny droplets of fluid are trapped by microvilli, folds of the plasma membrane. These folds pinch off into the cytoplasm as tiny vesicles of fluid (Fig. 6–24). The contents of these vesicles are slowly transferred into the cytoplasm, and the vesicles themselves may become smaller and smaller until they appear to vanish.

Endocytosis, exocytosis, and fusions of organelle membranes within the cell are made possible by the

bilayer structure of these membranes (Fig. 6–26). Because hydrophobic interactions force the lipid bilayer together, the hole in the plasma membrane that endocytosis and exocytosis must leave is quickly repaired (Fig. 6–26).

This chapter has emphasized the surface membrane of the cell. As you know, the eukaryotic cell possesses a multitude of membranes in addition to the plasma membrane, each with specialized and often very complex functions. The wonderfully complex membranes of the mitochondria and chloroplasts generate ATP and release and store energy. The membranes of the rough endoplasmic reticulum are absolutely essential for efficient synthesis of most proteins.

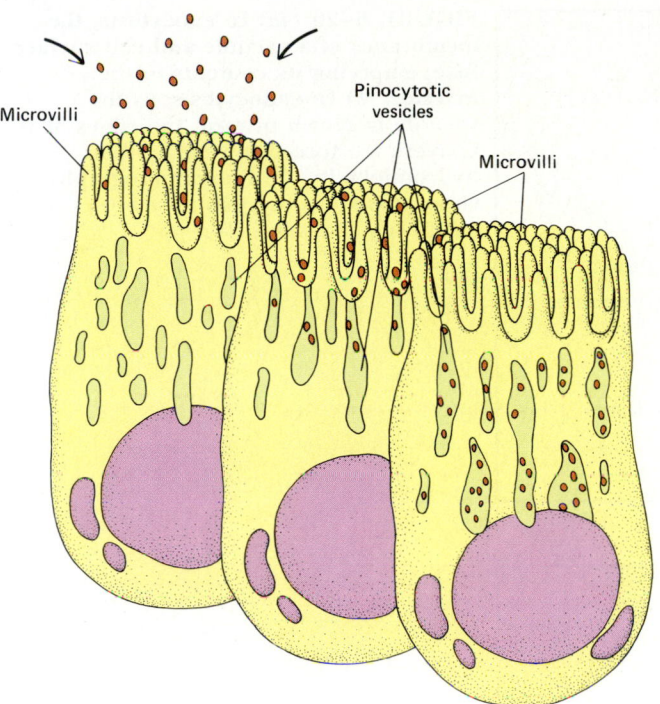

Microvilli

Pinocytotic
vesicles

Microvilli

FIGURE 6–24 In pinocytosis tiny particles of fluid are trapped by folds of the plasma membrane, which then pinch off into the cytoplasm as little vesicles of fluid. The content of these vesicles is slowly transferred to the cytoplasm across their membrane linings.

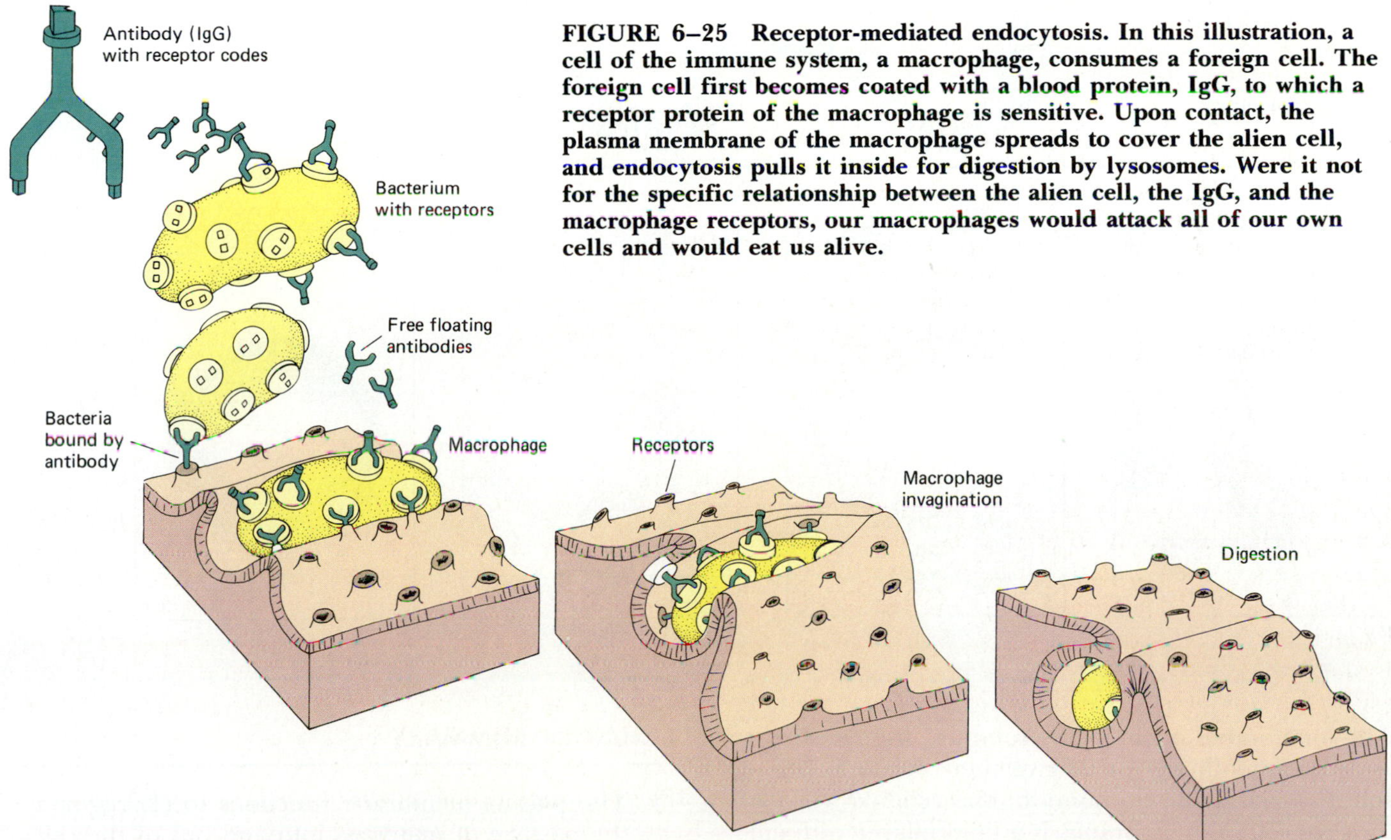

Antibody (IgG)
with receptor codes

Bacterium
with receptors

Free floating
antibodies

Bacteria
bound by
antibody

Macrophage

Receptors

Macrophage
invagination

Digestion

FIGURE 6–25 Receptor-mediated endocytosis. In this illustration, a cell of the immune system, a macrophage, consumes a foreign cell. The foreign cell first becomes coated with a blood protein, IgG, to which a receptor protein of the macrophage is sensitive. Upon contact, the plasma membrane of the macrophage spreads to cover the alien cell, and endocytosis pulls it inside for digestion by lysosomes. Were it not for the specific relationship between the alien cell, the IgG, and the macrophage receptors, our macrophages would attack all of our own cells and would eat us alive.

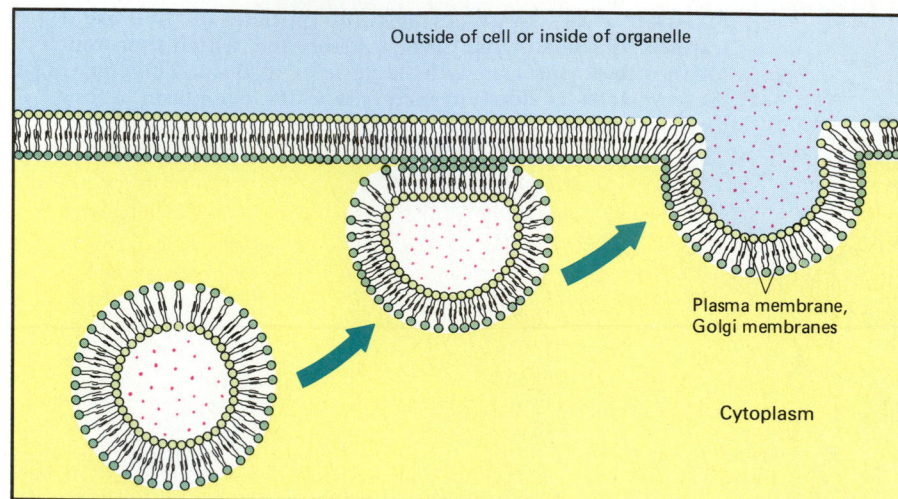

(a) Exocytosis, endomembrane transport

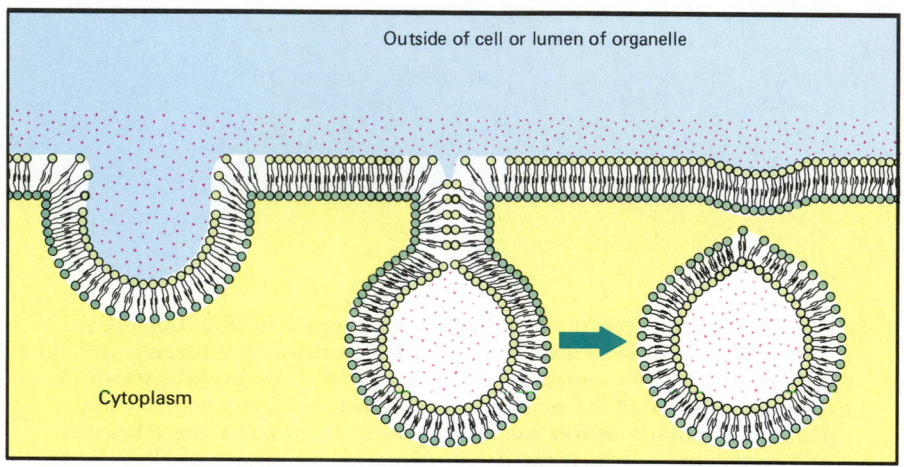

(b) Endocytosis

FIGURE 6-26 *(a)* **In exocytosis, the membranes of a vacuole and cell surface fuse, emptying its contents to the exterior.** *(b)* **In endocytosis, as the vacuole is drawn inward, the edges of the bilayers are forced to fuse by their hydrophobicity.** *(c)* **Photomicrographs of endocytosis in progress.** (*c*, courtesy of M.M. Perry and A.B. Gilbert, *J. Cell. Sci.*)

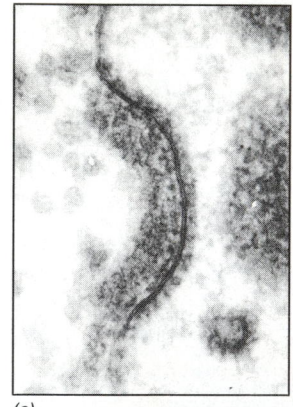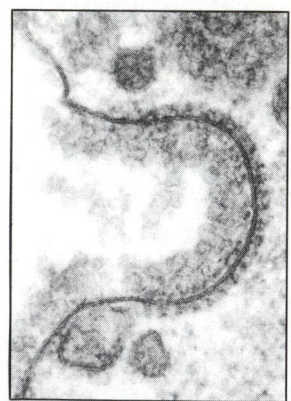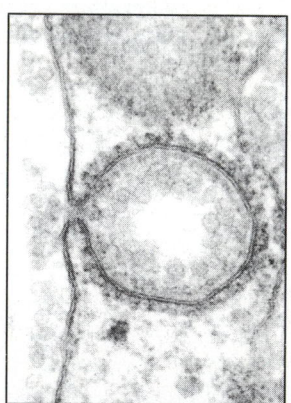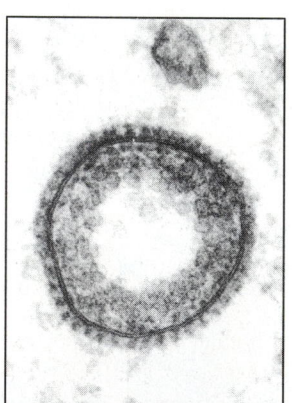

(c)

The membranes of the Golgi complex, the lysosomes, and the smooth ER each have their own specialized role to play in the economy of the cell. Yet each of these biological membranes is a lipid bilayer with embedded proteins, not fundamentally different from the plasma membrane.

▨ CHAPTER SUMMARY

I. The plasma membrane functions to (1) regulate the passage of materials into and out of the cell, (2) receive information that permits the cell to sense changes in its environment and respond to them, (3) communicate with other cells, and (4) protect the cell.

II. The plasma membrane consists of a fluid lipid bilayer in which a variety of proteins are embedded.

   A. The nonpolar, hydrophobic fatty acid chains of the membrane phospholipids meet and overlap in the middle, while the polar, hydrophilic heads are directed toward the outside of the membrane.

   B. Glycoproteins on the plasma membrane communicate with other cells. Other proteins may function in transport of materials through the membrane, in energy transfer, or in transmission of messages.

   C. Microvilli increase the surface area of the cell for absorption of materials from the environment.

III. Cell walls are found outside the plasma membranes of plant cells, algae, bacteria, and fungi. They consist of carbohydrates secreted by the cell. Cell walls protect the cell and give it a characteristic shape.

IV. Adjacent cells are connected and held together by specializations of the plasma membrane, such as desmosomes, tight junctions, and gap junctions.

V. The selectively permeable plasma membrane allows the passage of some, but not other, substances.

   A. Some ions and molecules pass through the plasma membrane by simple diffusion; others pass by facilitated diffusion, in which a carrier protein helps move a molecule through the membrane.

   B. Osmosis is a kind of diffusion in which molecules of water pass through a differentially permeable membrane from a region where water is more concentrated to a region where water is less concentrated.

   C. In active transport the cell expends energy to move ions or molecules against a concentration gradient.

   D. In endocytosis (phagocytosis and pinocytosis) materials, such as food, may be moved into the cell; a portion of the plasma membrane envelops the material, enclosing it in a vacuole that is then released inside the cell. In exocytosis the cell ejects waste products or secretes substances, such as mucus.

◻ POST-TEST

1. A membrane that permits passage of some materials but not of others is described as _____ _____.

2. The lipid components of the plasma membrane have a polar, _____ portion and a nonpolar, _____ portion.

3. In an electron micrograph the light zone of the plasma membrane represents the _____ of the lipid molecules.

4. _____ increase the surface area of the cell for absorption of materials.

5. The main component of the plant cell wall is _____.

6. _____ are button-like plaques that hold epithelial cells tightly together.

7. Cardiac muscle cells are connected by _____, which permit rapid transmission of impulses from one cell to the next.

8. Red dye poured into a beaker of water spreads throughout the water. This is an example of _____.

9. The diffusion of water through a differentially permeable membrane from a region of greater concentration to a region of lesser concentration of water molecules is termed _____.

10. A solution with a greater solute concentration than a tissue is said to be _____ to the tissue.

11. Cells will neither swell nor shrink if they are placed in _____ solutions.

12. Waste products may be ejected from a cell through the process of _____.

13. A white blood cell engulfs a bacterium by the process called _____.

14. In pinocytosis a cell takes in _____.

15. In _____ _____, the cell moves ions from a region of lower concentration to a region of higher concentration.

16. Protists in a hypotonic medium can get rid of excess water by means of their _____ _____.

◻ REVIEW QUESTIONS

1. Imagine a cell with a cell wall but no plasma membrane. In what ways would it be handicapped? Could it live?

2. The plasma membrane has been described as a "fluid mosaic." Is this a good description? Why?

3. The plasma membrane has also been described as having "protein icebergs in a lipid sea." Explain why this is a

good description or why it is not.

4. Why is it advantageous for cells lining the digestive tract to be equipped with microvilli?

5. The blood-brain barrier prevents the passage of certain materials from the blood into the brain tissue. What type of junctions would be found holding the cells of the barrier together? Why?

6. Why do you think the glycocalyx might be important in the function of the immune system?

7. What problems would a cell face if its plasma membrane were permeable rather than differentially permeable?

8. A 0.9% sodium chloride solution is isotonic to red blood cells. A laboratory technician accidentally places a sample of red blood cells in a 1.8% sodium chloride solution. What happens? Explain.

9. Why do carrot and celery sticks become limp after a time? How could they be made crisp once more? Explain in terms of turgor pressure.

10. A saltwater amoeba transferred to fresh water forms a contractile vacuole. In what way is this adaptive? Explain. If placed in salt water would a freshwater amoeba form a contractile vacuole?

11. Carrier proteins are utilized in both facilitated diffusion and active transport, yet these processes are basically very different. Contrast them.

12. Consider the diagram on the right, and describe all the events of endocytosis, secretion, and exocytosis that are numbered. This will provide a review of both this and much of the preceding chapter.

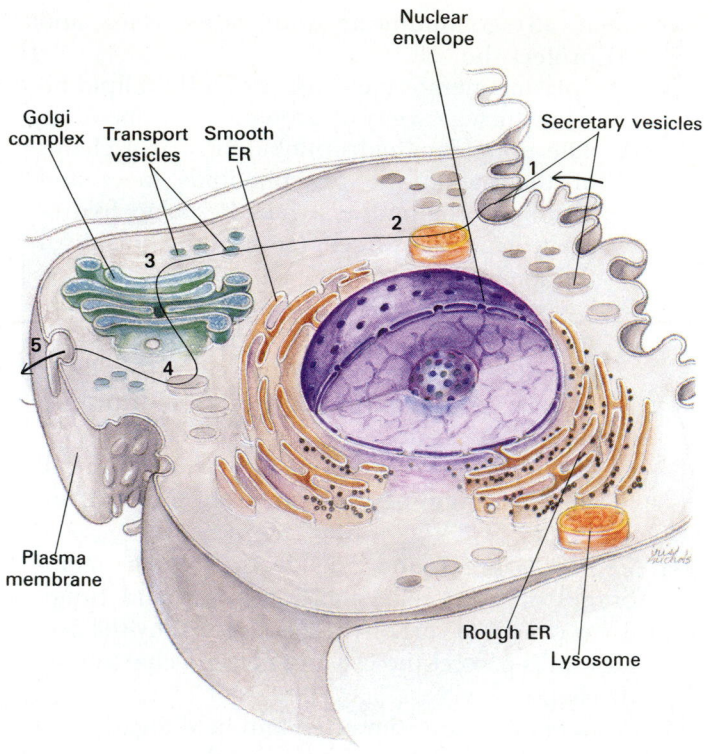

■ READINGS

Bretscher, M.S. "The Molecules of the Cell Membrane," *Scientific American*, October 1985. A good discussion of the relationships of membrane molecules.

Holtzman, E., and A.B. Novikoff. *Cells and Organelles,* 3rd ed. Philadelphia, Saunders College Publishing, 1984. An integrated approach to the structural, biochemical, and physiological aspects of the cell.

Unwin, N., and R. Henderson. "The Structure of Proteins in Biological Membranes," *Scientific American*, February 1984, 78–94. A discussion of the configurations of membrane proteins that permit them to be embedded in lipids yet function in the watery medium that surrounds the membrane.

P A R T 3

Energy Flow Through the World of Life

Energy for biological work • Metabolic reactions require energy transformations • Glucose degrades to carbon dioxide and water • Life on earth depends on photosynthesis • Laws of thermodynamics • Citric acid cycle

Cells break down organic compounds • Aerobic and anaerobic pathways • Enzymes work in teams • ATP is the energy currency of the cell • Photosystems and energy • Chlorophyll absorbs light

7
The Energy of Life

OUTLINE

I. Biological work requires energy
   A. Energy can be described as potential or kinetic
   B. Heat is a form of energy that can be conveniently measured
II. Two laws of thermodynamics govern energy transformations
   A. The first law of thermodynamics holds that the quantity of energy in the universe does not change
   B. The second law of thermodynamics holds that without an energy input systems tend to become increasingly disorganized
III. Metabolic reactions involve energy transformations
   A. Chemical reactions may be exergonic or endergonic
   B. Chemical reactions are reversible
   C. Reversible reactions reach a state of equilibrium
   D. Endergonic and exergonic reactions can be coupled in living systems
IV. Adenosine triphosphate (ATP) is the energy currency of the cell
V. Enzymes are chemical regulators
   A. Catalysts lower the activation energy necessary to initiate a chemical reaction
   B. If the name of a compound ends in -ase it is an enzyme
   C. Enzymes are very efficient catalysts
   D. Enzymes are specific
   E. Enzymes work by forming enzyme-substrate complexes
   F. Many enzymes require cofactors
   G. Enzymes often work in teams
   H. The cell regulates enzymatic activity
   I. Enzymes work best at appropriate temperature and pH
   J. Enzymes can be inhibited by certain chemical agents

LEARNING OBJECTIVES

After you have studied this chapter you should be able to:
1. Contrast potential and kinetic energy and identify different forms in which energy can exist.
2. Apply the first and second laws of thermodynamics to living organisms and to the ecosphere.
3. Describe the energy dynamics of a reaction that is in equilibrium.
4. Distinguish between endergonic and exergonic reactions and explain how they may be coupled so that the second law of thermodynamics is not violated.
5. Relate the chemical structure of ATP to its role in cellular metabolism.
6. Summarize the role of enzymes as chemical regulators and describe how they work.
7. Identify factors such as pH and temperature that influence enzymatic activity.
8. Compare the action and effects of the various types of enzyme inhibitors.

Life processes require energy. Flight requires an especially large amount. Red-winged blackbirds, Klamath Basin, California. (Frans Lanting)

Every activity of a living cell or organism requires energy. Movement, transport of materials, synthesis of needed compounds, and manufacture of new cells are a few of the life processes that require energy (Fig. 7–1). Recall from Chapter 1 that all of the chemical and energy transformations that occur within the living organism are referred to as its *metabolism.*

An organism has no way of creating new energy. It can only use up energy. Thus, an organism is described as having a **one-way flow of energy.** Neither can an organism recycle the energy that it has used. If an organism could recycle energy, it would be a **closed system.** The energy would be used again and again, never leaving the organism. Since energy is lost, however, an organism is an **open system.**

For these reasons, life depends upon a continuous input of energy. Producers (e.g., plants) trap light energy from the sun during photosynthesis and incorporate some of that energy in the chemical bonds of carbohydrates and other organic compounds. Then, a portion of that chemical energy can be transferred to the consumers which eat the producers and to the decomposers which feed on them all, sooner or later (Fig. 7–2). There is a one-way flow of energy through any individual organism and through the ecosphere.

In the one-way energy flow through the open system of a living organism, energy is captured, temporarily stored, and then used to perform biological work. This work is required in order for the organism to maintain its highly organized state. During these processes some energy is converted to heat and dispersed into the environment. Because the organism cannot reuse this energy, it must continually obtain fresh supplies of energy. But just how is this accomplished? In this chapter we will focus on some principles of energy capture, storage, transfer, and use. In the following two chapters we will explore some of the main metabolic pathways utilized by cells in their continuous quest for energy.

KEY CONCEPTS

☐ Living things depend on a continuous input of energy to counteract the universal trend toward increasing disorder. They maintain their high level of organization by using energy to maintain an appropriate internal environment, to synthesize compounds needed for growth and repair, and to move.

☐ Life processes involve energy transactions, and in each transaction some energy is dissipated as heat and then is no longer available in a form useful to the cell.

FIGURE 7–1 This bobcat is expending energy in an effort to capture the snowshoe hare. If caught and eaten, the hare will provide nutrients containing energy for future activity. For its part, the hare is expending a great deal of energy in its effort to escape becoming an energy source for the bobcat. (Sharon Cummings/Marvin L. Dembinsky, Jr., Photography Associates)

☐ ATP is the energy currency of the cell; energy is temporarily stored within chemical bonds of this compound.

☐ An enzyme is a biological catalyst; enzymes increase the rate of chemical reactions.

☐ BIOLOGICAL WORK REQUIRES ENERGY

At this very moment you are expending considerable amounts of energy to maintain your breathing, concentrate urine in your kidneys, digest food, circulate your blood, and maintain countless other metabolic activities. As you read these words your brain is using enough energy to light a 75-watt bulb. These are all forms of biological work, and they require energy.

We can define **matter** as anything that has mass and takes up space, whether a subatomic particle or a planet. **Energy** is the ability to produce a change in the state or motion of matter. For example, it takes energy (in the form of heat) to change ice to water, that is, to change the state of water from solid to liquid. It takes energy to produce the increase in the motion of water molecules that is responsible for this change in state. For our purposes, **energy** may be simply defined as the capacity to do work.

Energy can exist in several different forms. These include heat, electrical, mechanical, chemical, sound, and radiant energy (the energy of electromagnetic waves, such as radio waves, visible light, x-rays, and gamma rays).

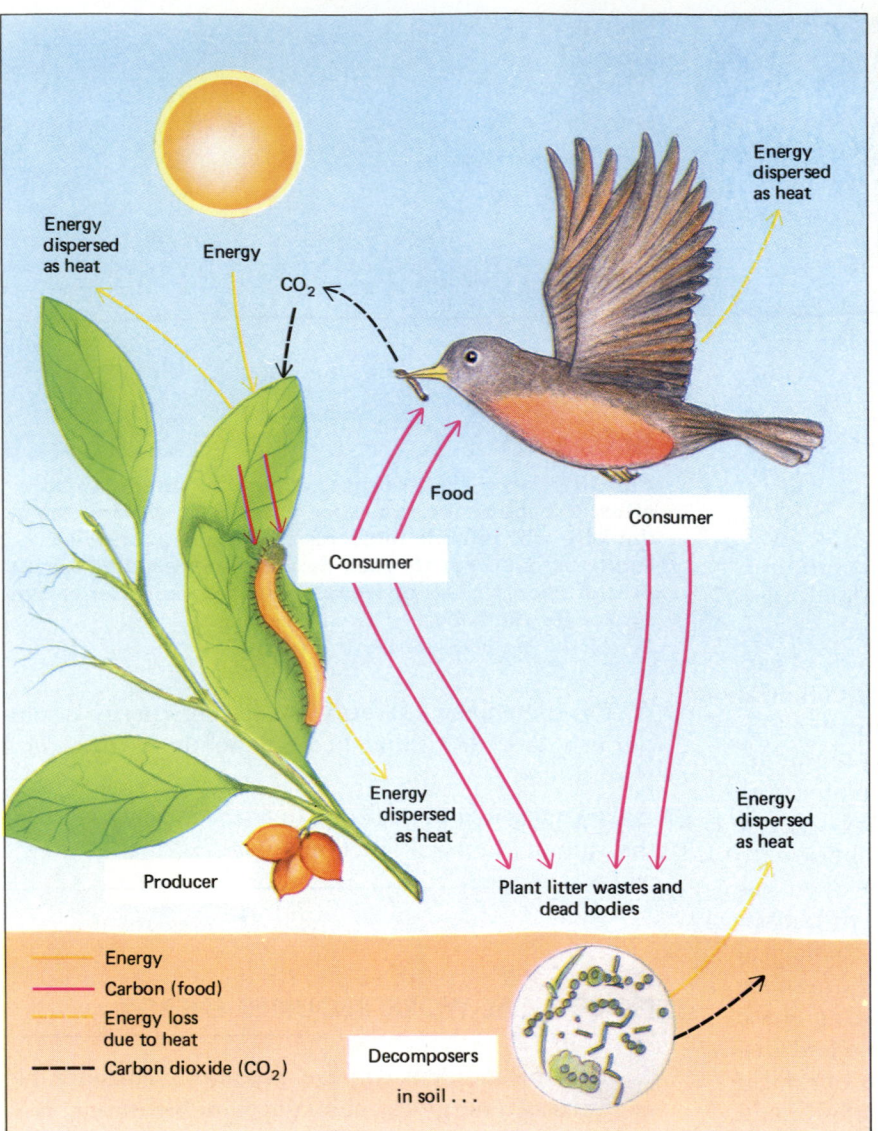

FIGURE 7–2 Flow of energy through the ecosphere. Carbon and many other chemical elements are continuously recycled. In contrast, energy cannot be recycled. During every energy transaction some energy is dispersed as heat. For this reason a constant energy input from the sun is required to keep the ecosphere in operation. Producers trap solar energy which powers our "Spaceship Earth."

Energy Can Be Described As Potential Or Kinetic

Potential energy is stored energy; it has the capacity to do work, owing to its position or state. In contrast, **kinetic energy** is the energy of motion. A boulder at the top of a hill has potential energy because of its position. As it rolls downhill the potential energy is converted to kinetic energy. It would require the input of energy to push the boulder back up the hill and restore the potential energy of its position at the top.

Another example of the conversion of potential energy to kinetic energy is the release of a drawn bow (Fig. 7–3). The tension in the bow and string represents stored energy; when the string is released, this potential energy is released so that the motion of the bow propels the arrow. It would require the input of additional energy to draw the bow once again and restore the potential energy. Most of the actions of a complex organism involve a complex series of energy transformations. For example, to prepare for a running event, athletes eat foods that build up their reserves of glycogen. Potential energy is present in the chemical bonds of the glycogen. During the event, the athlete's body continuously converts the energy stored in the glycogen into the kinetic energy used to run the race.

Heat Is A Form Of Energy That Can Be Conveniently Measured

To study energy transformations, scientists must be able to measure energy. How is this done? **Heat** is a form of energy that can be conveniently measured because all other forms of energy can be converted into heat. In fact the study of energy and its transforma-

POTENTIAL   KINETIC

FIGURE 7–3 Potential energy is stored energy. Kinetic energy is the energy of motion. Shooting a bow and arrow illustrates some of the energy transformations that take place in living systems. When the archer's muscles contract, the muscular force is transmitted to the bow and it is moved. Potential energy is stored in the drawn bow and is released as the arrow speeds toward its target. Energy is neither created nor destroyed, but in each of these transformations the remaining total useful energy is less than in any of the steps that preceded it.

tions has been named **thermodynamics,** that is, heat dynamics.

Although several units may be used in measuring energy, the most widely used in biological systems is the **kilocalorie (kcal).** A kilocalorie is the amount of heat required to raise the temperature of a kilogram of water from 14.5°C to 15.5°C. Nutritionists use the kcal in measuring the potential energy of foods and usually refer to it as a **Calorie** (with a capital C).

◼ TWO LAWS OF THERMODYNAMICS GOVERN ENERGY TRANSFORMATIONS

All of the activities of our universe—from the life and death of cells to the life and death of stars—are governed by two laws of energy, the laws of thermodynamics.

The First Law Of Thermodynamics Holds That The Quantity Of Energy In The Universe Does Not Change

According to the first law of thermodynamics, known also as the **law of conservation of energy,** the total amount of energy in the universe remains constant. The universe is a closed system when it comes to energy. As far as we know, the energy present when it formed some 20 billion years ago is all the universe can ever have. This energy can not be added to or subtracted from.

Energy cannot be created or destroyed during ordinary chemical or physical processes. However, energy can be transferred and changed from one form to another. Although organisms can neither make nor destroy energy, they can capture some from the environment and use it for their own needs. Organisms also transform energy from one form to another. During photosynthesis plant cells transform light energy to electrical energy and then to chemical energy stored in chemical bonds. Some of that chemical energy may later be transformed by some animal that eats the plant to the mechanical energy of muscle contraction or to some other needed form.

As each energy transformation takes place, some energy is converted to heat energy and dissipated to the environment. Because this energy can never again be used by the organism, it is "lost" from the biological point of view. However, it is not really "lost" from a thermodynamic point of view because it can be ac-

FIGURE 7–4   As a bird flies, some of its energy is lost as heat. That energy is gained by the surroundings, so the net energy change is zero. Hovering booby. (Frans Lanting, © 1984)

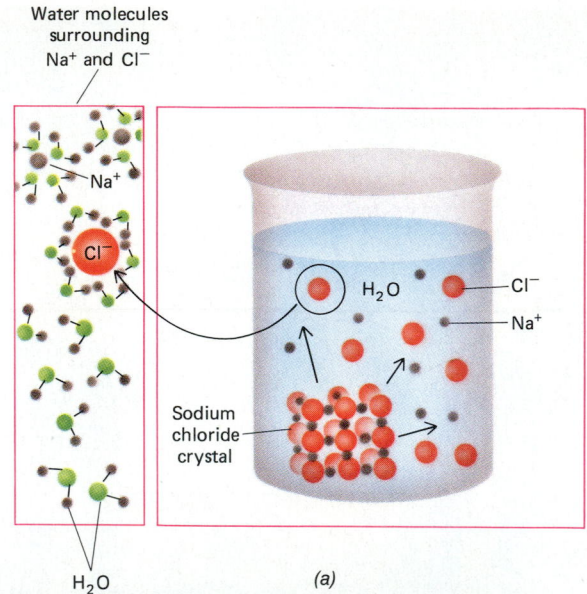

(a)

(b)

FIGURE 7–5   Entropy. (*a*) As particles leave a crystal to go into solution, they become more disordered. The entropy of this system increases during the process. (*b*) Imagine that you shake a beaker containing two colors of marbles. A disordered arrangement (*left*) is much more probable than an ordered arrangement (*right*) in which all marbles of the same color are found together. (*b*, Charles Winters)

counted for in the surrounding physical environment.

Biologists generally study a particular object, or *system*—for example, a cell, plant, animal, or ecosystem. The rest of the universe is referred to as the *surroundings*. The total energy of any system and its surroundings remains constant. As a system undergoes a change, it may absorb energy from its surroundings or deliver energy into its surroundings (Fig. 7–4). The difference in the energy content of the system in its initial and final states must be equalled by a corresponding change in the energy content of the surroundings.

The Second Law Of Thermodynamics Holds That Without An Energy Input Systems Tend To Become Increasingly Disorganized

Although the total amount of energy in the universe remains constant, the energy available to do work is decreasing with time. This is because useful forms of energy are continuously degraded to heat, which is a less useful form of energy. In almost all energy transformations, there is a loss of some energy in the form of heat to the surroundings. The energy lost from the object is no longer available to do work. And when heat is dissipated evenly into the surroundings, it is likewise no longer able to do work.

Heat is the energy of the random motion of molecules and is the most *disorganized* form of energy. Heat can be made to do work only when there is a temperature difference that results in heat flowing from a warmer region to a cooler region. Temperature is the same throughout a living cell, so heat cannot be used to do biological work.

Entropy may be defined as a measure of randomness or disorder, or, more precisely, a measure of the amount of energy that is unavailable to do work. According to the **second law of thermodynamics** the entropy of the universe is continuously increasing. The universe as a whole is a closed system; no new energy is created in it. Physical and chemical processes proceed in such a way that the entropy of every system increases. Because entropy in the universe is continuously increasing, eventually, some billions of years in the future, all energy will be random and uniform in

FIGURE 7–6 Consumers and decomposers must eat in order to replace the energy they continuously lose as they carry on life processes. Baby gorilla eating plant. (Sharon Cummings/Marvin L. Dembinsky, Jr., Photography Associates)

distribution. With only this useless form, no energy will be able to flow, and no work will be able to be performed. The universe will have run down.

It is important to understand that the second law of thermodynamics is consistent with the first law. The total amount of energy in the universe is not decreasing with time; but the energy available to do work is being degraded to random molecular motion. All energy will be "useless" (which is another way of stating the second law).

Because of the second law of thermodynamics, no process requiring energy is ever 100% efficient. Much of the energy is dispersed as heat, so that there is an increase in entropy. Cellular energy utilization is about 55% efficient, with the other 45% of the energy being lost as heat. Such biological processes are actually quite efficient compared to most machines made by human beings. For example, a gasoline engine is only about 17% efficient.

Living systems are highly organized, yet order is an extremely unlikely state (Fig. 7–5). If you drop a crystal vase, the fragments of glass will not spontaneously jump back into place to reconstruct the vase. In fact, it would require a tremendous input of energy for you to gather every small shard of glass and fit these fragments together to mend the vase. Out of the multitude of possible ways in which the pieces can be assembled, only one represents the highly ordered form that is the vase.

Because organisms are highly organized, they are very unstable. In fact, life is a constant struggle against the second law of thermodynamics. Survival of individual organisms, as well as of ecosystems, depends upon continuous energy input. Thus, producers must carry on photosynthesis, and consumers and decomposers must eat the products of photosynthesis (Fig. 7–6). Hence, life on earth depends on a constant input of radiant energy from the sun.

■ METABOLIC REACTIONS INVOLVE ENERGY TRANSFORMATIONS

Metabolic processes involve changes in amount and type of energy. **Free energy** is energy that is available to do work under conditions of constant temperature and pressure. For this reason free energy is the aspect of thermodynamics of greatest interest in biology. The force that drives all processes in living and nonliving systems is the tendency of the system to reach the condition of maximal entropy. All physical and chemical processes proceed with a decline in free energy until they reach an equilibrium in which the free energy of the system is at a minimum and the entropy is at a maximum. Free energy is useful energy in biological systems; entropy is a state of degraded, useless energy.

Chemical Reactions May Be Exergonic Or Endergonic

A **chemical reaction** is a change involving the molecular structure of one or more substances; matter is changed from one substance which has characteristic properties to another with new properties. During the reaction energy is released or absorbed. For example,

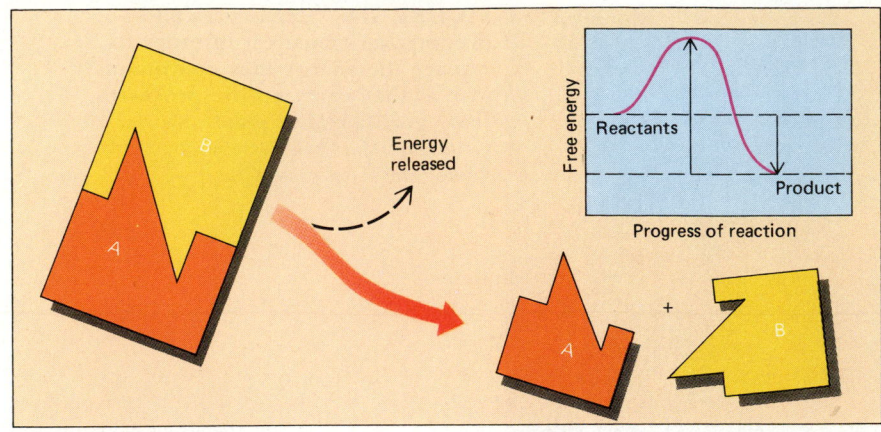

(a) Exergonic

FIGURE 7–7   Energy changes in exergonic and endergonic reactions. (*a*) In exergonic reactions free energy is released, and the product has less energy than the reactants. (*b*) In endergonic reactions there is a net input of free energy, so the products contain more energy than the reactants. An endergonic reaction can occur only when energy lost from some other system is fed into the reaction. Note that even the exergonic reaction requires some input of energy to get started. This initial investment energy is termed *activation energy.*

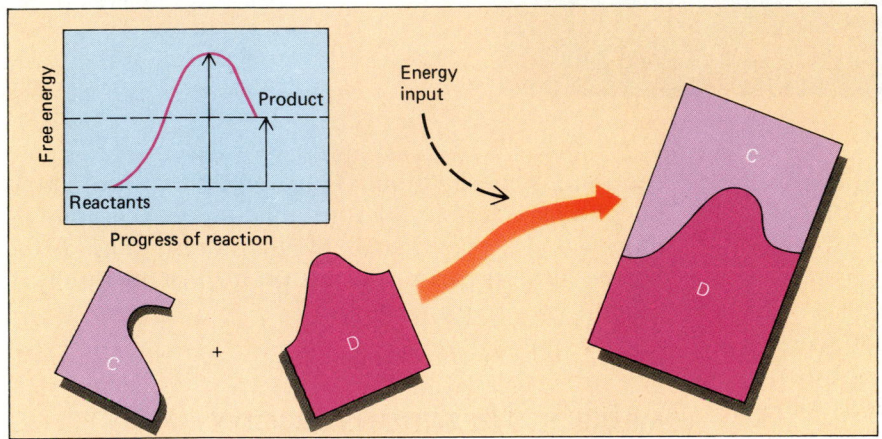

(b) Endergonic

hydrochloric acid, HCl, reacts with the base, sodium hydroxide, NaOH, to yield water, H_2O, and the salt, sodium chloride, NaCl. In the process energy is released as heat:

$$HCl + NaOH \longrightarrow H_2O + NaCl + Energy\ (heat)$$

The chemical properties of HCl and NaOH are very different from those of H_2O and NaCl. Note that the number of atoms of a given element in the products is just equal to the number of atoms of that element in the reactants. Atoms are neither destroyed nor created in a chemical reaction, but simply change partners, in something like a complex atomic square dance. Likewise, energy is neither created nor destroyed but can be released if the reactants are in a more energetic state than the products, as in our example.

Chemical reactions that release free energy are referred to as **exergonic reactions** (Fig. 7–7). Because energy is released, the products of the reaction contain less energy than the reactants. Exergonic reactions tend to proceed spontaneously; some occur rapidly, whereas others occur over a long period of time. Be-

cause exergonic reactions release heat, they are *exothermic* (heat-releasing) reactions (Fig. 7–8).

Reactions that require a net input of free energy are not spontaneous and are described as **endergonic reactions.** In these reactions, the products contain more energy than the reactants. Because endergonic reactions absorb heat, they are *endothermic* reactions.

Chemical Reactions Are Reversible

In most metabolic reactions, there is little free energy difference between reactants and products. As a result, as long as external energy is available, most of the reactions that occur within living cells are theoretically reversible. In fact, reversibility is characteristic of many biochemical reactions. Reversibility allows cells to control their release of free energy according to their needs. It also permits cells to resynthesize their large biological molecules for continued use in metabolic processes.

Reversibility is indicated by a double arrow, $\rightleftarrows$.

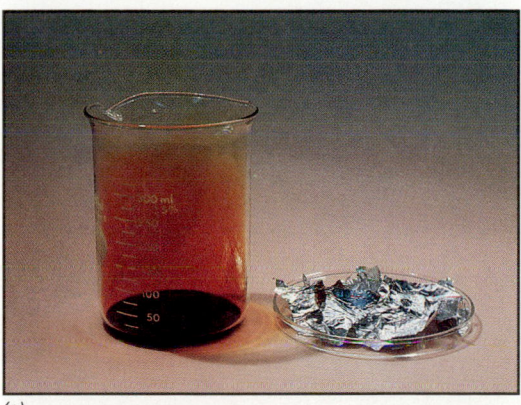

FIGURE 7–8   An exothermic reaction. (*a*) The element bromine is in the beaker, and the element aluminum is on the watch glass. (*b*) When the aluminum is added to the bromine, the reaction proceeds so vigorously that the aluminum melts and glows white hot. (Charles Winters)

Whether a reaction will occur and whether it will proceed from right to left or left to right depends upon factors such as the energy relations of the several chemicals involved, their relative concentrations, and their solubility.

Reversible Reactions Reach A State Of Equilibrium

Imagine that over a 10-year period the population of a city remains the same. Although some new folks have moved into town, others have moved out or perhaps died. However, the net change in the population is zero. We might say that the population in this city is in a state of dynamic equilibrium. In a **dynamic equilibrium** the rate of change in one direction is exactly equal to the rate of change in the opposite direction.

Consider a reaction in which there is only a small free energy difference between reactants and products in terms of the numbers of each type of molecule involved. At the beginning of a reaction only the reactant molecules may be present. These molecules move about and collide with one another with sufficient en-

ergy to react. As more and more product molecules are produced, there are fewer and fewer reactant molecules left. As the number of product molecules increases they collide more frequently and some have sufficient energy to initiate the reverse reaction. The reaction proceeds in both directions simultaneously and reaches an equilibrium in which the rate of the reverse reaction is about the same as the rate of the forward reaction (Fig. 7–9).

When a reaction is at equilibrium, the free energy difference between the products and reactants of a chemical reaction is zero. Any change, such as a change in temperature or pressure, that affects the reacting system may cause the equilibrium to shift. The reaction may then proceed in the opposite direction until once again the free energy difference is zero, and a new equilibrium has been established.

When there is little free energy difference on the two sides of a chemical equation, the direction of a reaction will be determined mainly by the concentrations of the reactants and products. The reaction tends to proceed in the direction that will minimize the difference in concentration between the substances on the two sides of the equation. If one of them, say the product, is continuously removed as it is formed, the reaction will indeed proceed to completion or until all of the reactant is used up. In the laboratory, reactions may be directed to completion by deliberate removal of product as soon as it forms. A cell may "remove" a product by directing it down some chemical pathway as soon as it is formed.

Endergonic And Exergonic Reactions Can Be Coupled In Living Systems

Many metabolic reactions—protein synthesis, for example—in a living organism are endergonic (requiring an input of energy to proceed). The energy to drive these reactions is often energy released from exergonic reactions. Thus, endergonic and exergonic reactions are often coupled to one another. The thermodynamically favorable exergonic reaction provides the energy required to drive the thermodynamically unfavorable endergonic reaction. As we will discuss in the next section, exergonic and endergonic reactions are often linked to ATP synthesis and breakdown.

How does a living organism, from the time it is "born" until the time it dies, employ outside energy inputs to compensate for its continuous loss of free energy? Two factors make these energy inputs available. First, organisms are part of a large universe with a vast reserve of free energy. Second, organisms have within themselves special structures, enzymes, and genetic information needed to direct their life processes and maintain an ordered state.

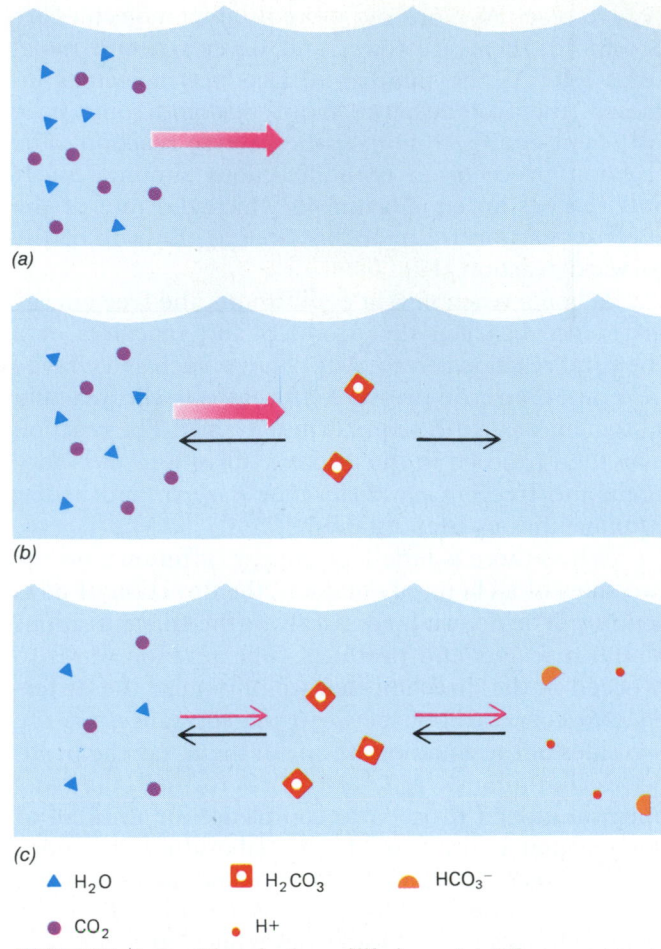

(a)

(b)

(c)

▲ H_2O ▢ H_2CO_3 ◖ HCO_3^-

● CO_2 • H^+

FIGURE 7–9   Chemical equilibrium. (*a*) When reactant molecules are in high concentration, a chemical reaction proceeds strongly in a forward direction. (*b*) As product forms, the reaction proceeds in both directions. (*c*) At equilibrium, the rates of the forward and reverse reactions are the same.

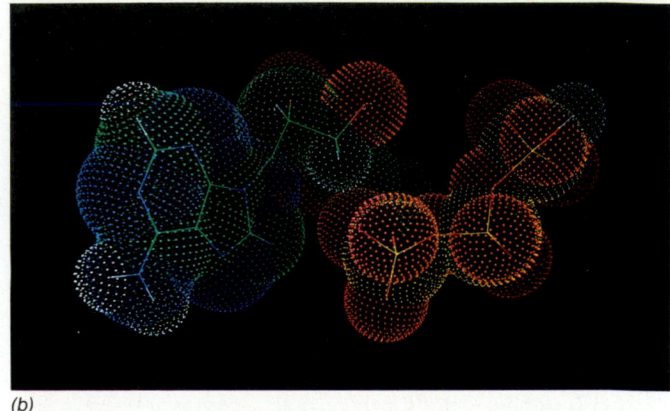

(a)

(b)

FIGURE 7–10   ATP, the energy currency of all living things. (*a*) Chemical structure of ATP. (*b*) A computer-generated model of ATP. The red balls are oxygen atoms; blue = nitrogen, green = carbon, yellow = phosphate, white = hydrogen. Note the hydrogen atom attached to the last oxygen in the triphosphate group. At different pHs, this and other oxygen atoms might be bonded with hydrogen or be present in ionized form. (*b*, Courtesy of Computer Graphics Laboratory, University of California, San Francisco)

☐ ADENOSINE TRIPHOSPHATE (ATP) IS THE ENERGY CURRENCY OF THE CELL

In all living cells energy is temporarily packaged within a remarkable chemical compound called **adenosine triphosphate,** or **ATP.** This compound stores large amounts of energy for very short periods of time. When you work you earn money, and so you might say that your energy is symbolically stored in the money you earn. In the same way the energy of the cell may be stored in ATP. When you have earned extra money you might deposit some in the bank. Similarly the cell deposits energy as lipid in fat cells or as glycogen in liver and muscle. Moreover just as you dare not make less money than you spend, so too the cell must avoid energy bankruptcy which would mean its death. Finally, just as you (alas) do not keep what you make very long, so too the cell is forever spending its ATP.

ATP is a nucleoside triphosphate, consisting of three main parts (Fig. 7–10): (1) a nitrogen-containing base, **adenine,** which also occurs in RNA and DNA; (2) **ribose,** a 5-carbon sugar; and (3) **three inorganic phosphate groups** identifiable as phosphorus atoms surrounded by oxygen atoms. Inorganic phosphate can be designated P_i. Notice that the phosphate groups are attached to the end of the molecule in a series, rather like three passenger cars behind a locomotive. The couplings, that is the chemical bonds attaching the last two phosphates, also resemble those of a train in that they are readily attached and detached. In diagrams, energy-rich bonds are designated by wavy lines.

The bonds linking the phosphate groups of ATP are unstable; they can be broken by **hydrolysis** (water is added as the molecule is split). All three phosphate groups are negatively charged and tend to repel one another. As a result the phosphate bonds, although energy-rich, are relatively weak and are easily broken. When ATP is hydrolyzed the energy released is used to

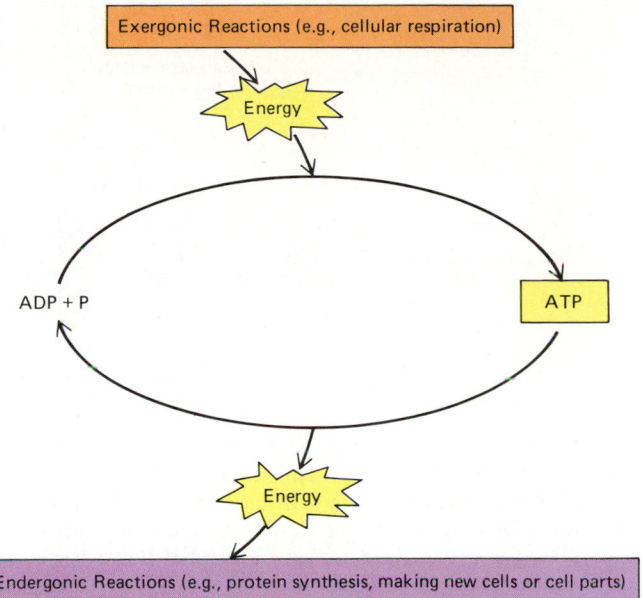

FIGURE 7–11   ATP is an important link between endergonic and exergonic reactions in living cells.

FIGURE 7–12   The chemical energy of ATP is converted to light energy in the light organs of this deep-sea angler fish, *Caulophryne jordani*. The energy transformation actually occurs within luminous bacteria that inhabit the light organs of the fish. (Peter David, Seaphot)

form new bonds that are stronger than the phosphate bonds. Energy is released during the reaction due to the chemical change to a more stable condition.

When the phosphate at the very end of the molecule (the third phosphate in the chain) is removed, the remaining molecule is called **adenosine diphosphate,** or simply **ADP.** This is an exergonic (energy-releasing) reaction:

$$ATP + H_2O \longrightarrow ADP + P + Energy$$

When the two terminal phosphate groups (called a **pyrophosphate group**) are removed, the molecule that remains is **adenosine monophosphate (AMP).**

The energy from the hydrolysis of ATP is coupled to endergonic processes within the cell with the help of specific enzymes. This process involves the transfer of a phosphate group from ATP to some other compound. Addition of a phosphate group to a molecule is referred to as **phosphorylation** (Fig. 7–11).

When a phosphate is attached to AMP it becomes ADP, and when a phosphate is added to ADP, ATP is produced. Note that these reactions are readily reversible:

$$AMP + P_i + Energy \rightleftharpoons ADP$$
$$ADP + P_i + Energy \rightleftharpoons ATP$$

As these equations indicate, energy is required to add a phosphate to either the AMP or the ADP molecule. Conversely, since energy can neither be created nor destroyed, the energy is released or transferred to another molecule when the phosphate is detached. Thus, ATP is an important link between exergonic (en-

ergy-releasing) and endergonic (energy-requiring) reactions. Because ATP can temporarily store energy, it allows exergonic and endergonic reactions to be separated in time or place inside the cell.

ATP is formed from ADP and inorganic phosphate when nutrients are oxidized or when the radiant energy of sunlight is trapped in photosynthesis. Energy released from exergonic reactions is packaged in ATP molecules for use in endergonic reactions (Fig. 7–11). This energy may be used to produce fats or glycogen, molecules stockpiled for long-term energy storage.

The cell contains a pool of ADP, ATP, and phosphate in a state of equilibrium. However, large quantities of ATP cannot be stockpiled in the cell. In fact, recent studies suggest that a bacterial cell has no more than a one-second supply of ATP. Thus ATP molecules are used almost as quickly as they are produced. A human at rest uses about 45 kilograms of ATP each day, but the amount present in the body at any given moment is less than 1 gram. Every second in every cell an estimated ten million molecules of ATP are made from ADP and phosphate and an equal number are hydrolyzed, yielding their energy to whichever life processes may require them (Fig. 7–12).

■ ENZYMES ARE CHEMICAL REGULATORS

The principles of thermodynamics help us predict whether a reaction can occur but tell us nothing about the speed of the reaction. If a glucose solution is sealed in a bottle free of bacteria and molds it keeps indefinitely. In order to break it down we must subject it to high temperature, strong acids, or bases. Living cells can neither wait for centuries nor use extreme means to break down glucose or any other type of molecule. Thus, in cells chemical reactions are brought about by molecules called **enzymes,** substances which affect the speed of a chemical reaction without being consumed by the reaction. Enzymes allow reactions to occur at physiological temperatures and pH. (The temperature of cells is too low to allow chemical reactions to proceed at high enough rates. With enzymes, they proceed at rates high enough to support life.) Most enzymes are proteins.

Cells require a slow, steady release of energy and they must be able to regulate that release to meet metabolic energy requirements. Accordingly, fuel molecules (such as glucose) are slowly and methodically oxidized during the 30 or so reactions of cellular respiration so that energy is released in small amounts. Cellular metabolism generally proceeds by a series of steps, so that a given molecule may go through many chemical transformations before it reaches some final state. And then the seemingly completed molecule may enter yet another chemical pathway so as to be totally transformed or consumed in the course of energy production. The changing needs of the cell require a system of flexible chemical control. The key elements of this control system are the remarkable enzymes.

Catalysts Lower The Activation Energy Necessary To Initiate A Chemical Reaction

In order to form new chemical bonds, existing ones must first be broken. The process of breaking these existing bonds represents a barrier to the reaction. The energy required to overcome this barrier and start the reaction going is called **activation energy.** A **catalyst** lowers the activation energy necessary to initiate a chemical reaction. An enzyme is a biological catalyst.

In a population of molecules of any kind, some have a relatively high energy content, others have a lower energy content, and the energy content of the entire population of molecules conforms to a bell-shaped curve of normal distribution. Only those molecules with a relatively high energy content are likely to react to form the product. To make the reaction proceed more quickly we must raise the energy content of more of the molecules so that the activation barrier is overcome. An enzyme lowers the activation energy of the reaction and allows a larger fraction of the population of molecules to react at any one time (Fig. 7–13). The enzyme does this by forming an unstable intermediate complex with the **substrate,** the substance upon which it operates. Then, the **enzyme-substrate** com-

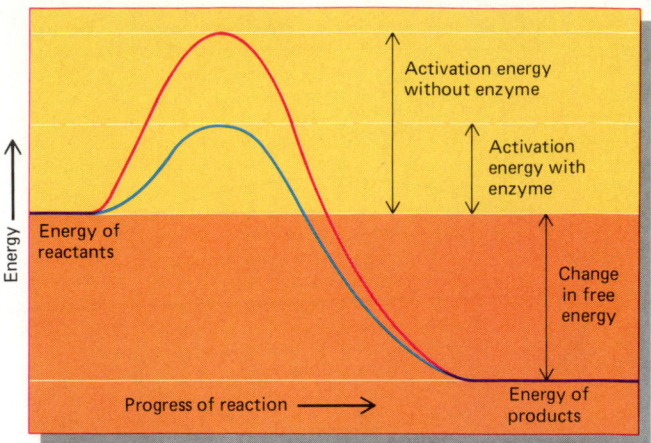

FIGURE 7–13   An enzyme speeds up a chemical reaction by lowering its activation energy. A catalyzed reaction (*blue curve*) proceeds more quickly than an uncatalyzed reaction (*red curve*) because it has a lower barrier of activation to overcome.

plex breaks down, forming the product and freeing the enzyme to react with a new molecule of reactant.

An enzyme can only promote, or speed up, a chemical reaction that could be made to proceed without it (although at a much slower rate). There is nothing in the action of a catalyst of any kind that could change the operation of the second law of thermodynamics, so enzymes do not influence the direction of a chemical reaction or the final concentrations of the molecules involved. They simply speed up reactions.

If The Name Of A Compound Ends In -ase It Is An Enzyme

Enzymes are usually named by adding the suffix *-ase* to the name of the substance acted upon. For example, sucrose is split by the enzyme *sucrase* to give glucose and fructose. There are group names for enzymes that catalyze similar reactions. *Lipases* break down triacylglycerols into glycerol and fatty acids; *proteinases* break the peptide bonds between amino acids in proteins; and *dehydrogenases* transfer hydrogens from one compound to another.

Enzymes Are Very Efficient Catalysts

The catalytic ability of some enzymes is truly phenomenal. For example, one molecule of the iron-containing enzyme **catalase** brings about the decomposition of 5,000,000 molecules of hydrogen peroxide (H_2O_2) per minute at 0°C. Hydrogen peroxide is a substance that is toxic, or poisonous, to cells if it is allowed to accumulate, but which is produced as a by-product in a num-

FIGURE 7–14 A bombadier beetle uses the catalyzed decomposition of hydrogen peroxide as a defense mechanism. The oxygen gas formed in the reaction forces out water and other chemicals with explosive force. Since the reaction is very exothermic, the water comes out as steam. (Thomas Eisner, Daniel Aneshansley)

ber of enzyme reactions. Catalase protects the cell by destroying the peroxide.

Hydrogen peroxide can be split by iron atoms alone but only at a very slow rate. It would take 300 years for an iron atom to split the same number of molecules of H_2O_2 that a molecule of catalase (containing one iron atom) splits in 1 second! The bombadier beetle has evolved a most interesting adaptation using this reaction (Fig. 7–14).

Enzymes Are Specific

Most enzymes are highly specific and catalyze only a few closely related chemical reactions, or in many cases only one particular reaction. For example, the enzyme *urease*, which decomposes urea to ammonia and carbon

dioxide, attacks no other substrate. The enzyme sucrase splits only sucrose; it does not act on maltose or lactose. Although peroxidase decomposes several different peroxides, including hydrogen peroxide, it does not break down other types of oxides.

A few enzymes are specific only in requiring that the substrate have a certain kind of chemical bond. The lipase secreted by the pancreas will split the ester bonds connecting the glycerol and fatty acids of a wide variety of fats.

Enzymes Work By Forming Enzyme-Substrate Complexes

Enzymes form temporary chemical compounds with their substrates. These complexes then break up, releasing the product and regenerating the original enzyme molecule for reuse:

Enzyme + Substrate 1 + Substrate 2 $\longrightarrow$
Enzyme-substrate complex

Enzyme-substrate complex $\longrightarrow$
Enzyme + Product(s)

Note that the enzyme itself is not permanently altered or consumed by the reaction.

Why does the enzyme-substrate complex break up into different chemical products than those that participated in its formation? As shown in Figure 7–15, each enzyme has one or more regions called **active sites** which in the case of a few enzymes have been shown to be actual indentations in the enzyme molecule. These active sites are located close to one another on the enzyme's surface. During the course of a reaction, substrate molecules occupying these sites are temporarily brought close together and so can react with one another.

FIGURE 7–15 Lock-and-key mechanism for enzyme action. The substrates fit the active sites of the enzyme molecule much as keys fit locks. However, in this model the lock acts on the key rather than the other way around. Note that when the products separate from the enzyme, the enzyme is free to catalyze production of additional products. The enzyme is not permanently changed by the reaction. In recent years the lock-and-key concept of enzyme action has undergone some modification, as shown in Figure 7–16.

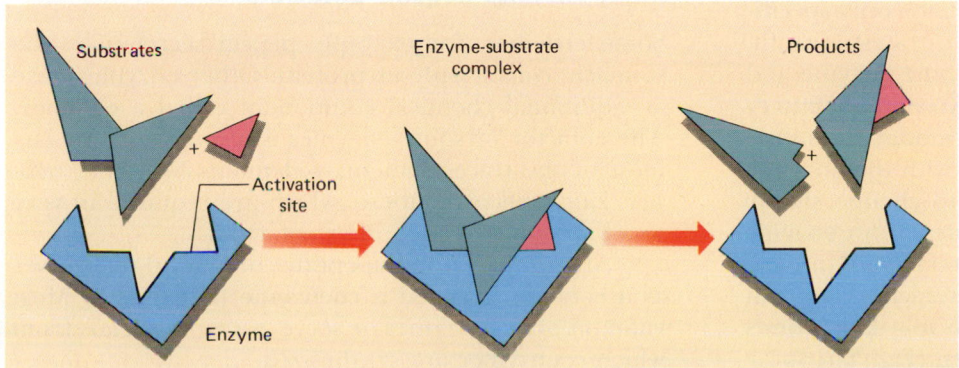

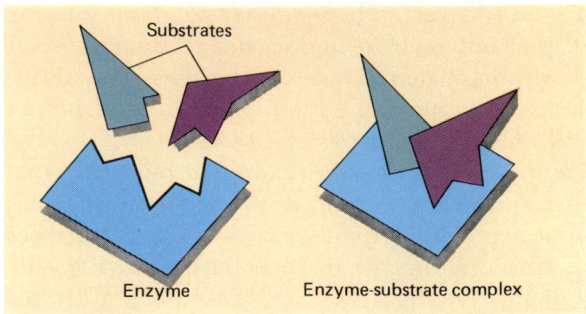

(a) Lock and key model

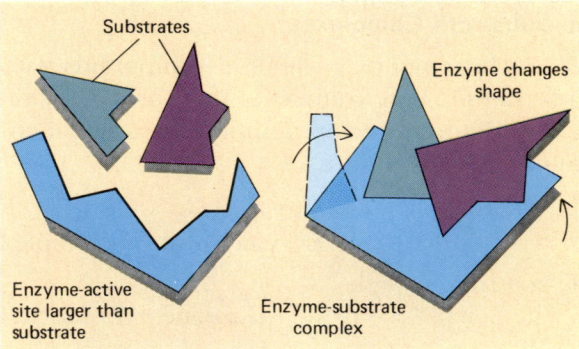

(b) Induced fit model

FIGURE 7–16 Comparison of models of enzyme action. (*a*) The lock-and-key model. (*b*) The induced-fit model. Chemical reactions are favored when substrate molecules get close enough to one another to react, when they are presented to each other in the right orientation, and when their existing chemical bonds are strained. Enzymes often do all three of these things. The straining of the substrate's bonds is accomplished by the apparent fact that most active sites are a bit bigger than the size of their substrate molecules. Accordingly, when a fit is forced upon them, the active site exerts a kind of pull on the substrate, helping to pull it apart. To be sure, the fit of the enzyme and substrate must not be too poor, or they will have no affinity for one another.

When the enzyme and substrate bind together, the shape of the enzyme molecule is thought to change slightly. This produces strain in critical bonds in the substrate molecules so that the bonds break. The new chemical compound thus formed has little affinity for the enzyme and moves away from it. An enzyme can be thought of as a molecular lock into which only specifically shaped molecular keys—the substrates—can fit.

Unlike a lock and key, however, the enzyme and its substrate seem not to be exactly complementary shapes. According to the **induced-fit model** of enzyme action, when the substrate combines with the enzyme, it may cause the enzyme molecule to change shape. This is possible because the active sites of an enzyme are not rigid. The change in shape results in an optimum fit for the substrate-enzyme interaction and can put strain on the substrate. This stress may help bonds to break, thus promoting the reaction (Fig. 7–16).

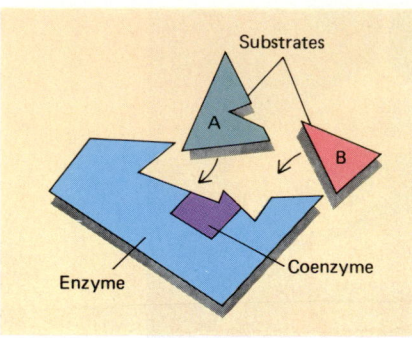

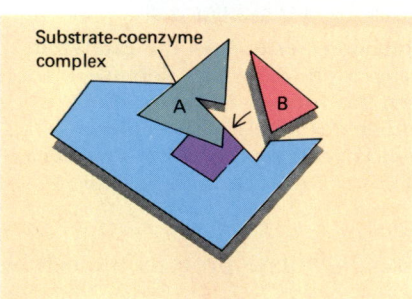

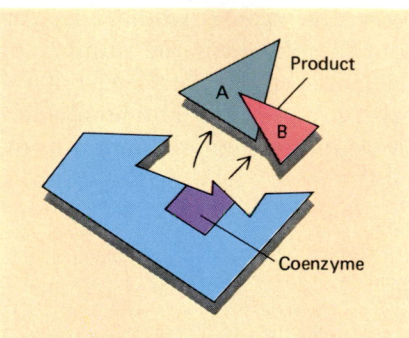

FIGURE 7–17 Coenzyme action. Some enzymes are not able to attach directly to the substrates whose chemical reactions they catalyze. Such enzymes employ accessory coenzymes to serve as adaptors, which facilitate the attachment of one or more substrates to the enzyme's active sites. One substrate combines first with the coenzyme to form a coenzyme-substrate complex. Then the coenzyme-substrate complex combines with the second substrate. This complex yields the products and releases the coenzyme.

Many Enzymes Require Cofactors

Some enzymes, for example pepsin secreted by the stomach, consist only of protein. Other enzymes have an additional chemical component called a **cofactor.** The cofactor of some enzymes is a metal ion. In fact most of the trace elements—elements like iron, copper, zinc, and manganese, which are required in very small amounts—function as cofactors.

An organic nonpolypeptide compound that serves as a cofactor is called a **coenzyme** (Fig. 7–17). Most vitamins are coenzymes or serve as raw materials from which coenzymes are synthesized.

Enzymes Often Work In Teams

Enzymes usually work in teams to catalyze a chain of reactions. The product of one enzyme-controlled reaction may serve as the substrate for the next. We can picture the inside of a cell as a factory with many different assembly lines (and disassembly lines) operating simultaneously. An assembly line is composed of a number of enzymes. Each enzyme carries out one step, such as changing molecule A into molecule B, and then passes it along to the next enzyme, which converts molecule B into molecule C and so on.

$$A \xrightarrow{\text{Enzyme 1}} B \xrightarrow{\text{Enzyme 2}} C$$

From germinating barley seeds, one can extract two enzymes that will convert starch to glucose. The first, amylase, hydrolyzes starch to maltose, and the second, maltase, splits maltose to glucose. Eleven different enzymes, working consecutively, are required to convert glucose to lactic acid. The same series of 11 enzymes is found in human cells, in green leaves, and in bacteria.

The Cell Regulates Enzymatic Activity

Enzymes regulate the chemistry of the cell, but what controls the enzymes? One mechanism of enzyme control simply depends upon the amount of enzyme produced. The synthesis of each type of enzyme is directed by a specific gene. The gene, in turn, may be switched on by a signal from a hormone or by some other type of cellular product. When the gene is switched on, the enzyme is synthesized. Then the amount of enzyme present influences the rate of the reaction.

If the pH and the temperature are kept constant and if an excess of substrate is present, the rate of an enzymatic reaction is directly proportional to the *concentration* of enzyme present up to a maximum value. If the enzyme concentration, pH, and temperature are kept constant, the initial rate of an enzymatic reaction is proportional to the concentration of substrate present, up to a limiting value.

The product of one enzymatic reaction may control the activity of another enzyme, especially in a complex sequence of enzymatic reactions. For example, in the following system

$$A \xrightarrow{\text{Enz. 1}} B \xrightarrow{\text{Enz. 2}} C \xrightarrow{\text{Enz. 3}} D \xrightarrow{\text{Enz. 4}} E$$

each step is catalyzed by a different enzyme. The final product, E, may inhibit the activity of Enzyme 1. When the concentration of E is low, the sequence of reactions proceeds rapidly. However, as the concentration of E increases, it serves as a signal for enzyme 1 to slow down and eventually to stop functioning. Inhibition of

Enzyme 1 stops this entire sequence of reactions. This type of enzyme regulation in which formation of a product inhibits an earlier reaction in the sequence is called **feedback control.**

Another important method of enzymatic control depends upon the activation of enzyme molecules that are present in an inactive form in the cytoplasm. In the inactive form the active sites of the enzyme are inappropriately shaped so that the substrates do not fit. Among the factors that influence the shape (conformation) of the enzyme are acidity and alkalinity and the concentration of certain salts.

Some enzymes, known as **allosteric** enzymes, possess a receptor site, called an **allosteric site,** located on some region of the enzyme molecule other than the active site. (The word *allosteric* means "another space.") Most allosteric enzymes contain more than one polypeptide chain, each with its own active site.

If a substance binds covalently to a site other than the active site, it may stimulate enzyme action or inhibit it. The substance that binds to the allosteric enzyme is referred to as a **regulator,** or modulator.

The enzyme *protein kinase* is an allosteric enzyme with only one polypeptide chain; both the active site and the allosteric site are located on this single chain. Another protein molecule serves as a regulator that binds reversibly to the allosteric site. This regulator inactivates the enzyme by altering the shape of the active site. In the body, protein kinase is in this inactive form most of the time (Fig. 7–18). When protein kinase activity is needed, the compound cyclic AMP contacts the enzyme-regulator complex and removes the regulator. This action changes the shape of the enzyme so that the substrate can combine with the active site:

Enzyme + Regulator $\longrightarrow$

Enzyme-regulator complex

Enzyme-regulator complex + cAMP $\longrightarrow$

Active enzyme + regulator-cAMP

Enzymes Work Best At Appropriate Temperature And pH

Enzymes generally work best under certain narrowly defined conditions referred to as *optima*. These include appropriate temperature, pH, and salt concentration (Fig. 7–19). For example, pepsin, the protein-digesting enzyme of the stomach, works best at the strongly acid pH of 2. In contrast the starch-digesting enzyme amylase in saliva and pancreatic juice has a pH optimum of 8.5 (slightly alkaline). Strong acids or bases irreversibly inactivate most enzymes by permanently changing their molecular conformation (shape).

Enzymatic reactions occur very slowly or not at all at low temperatures, but activity resumes when the

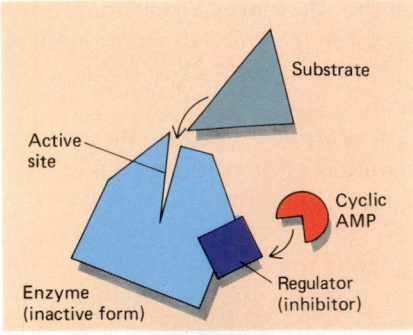

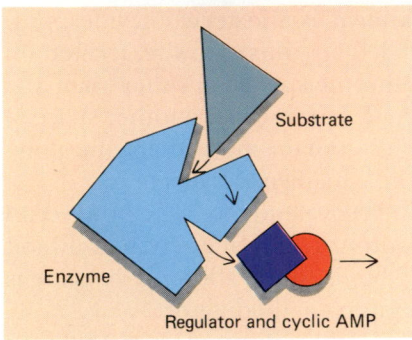

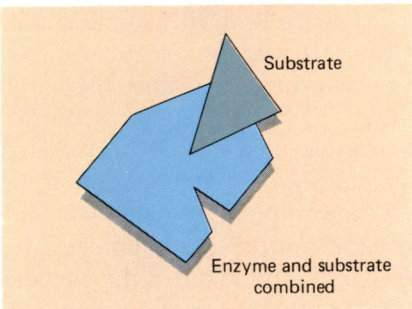

FIGURE 7–18 Allosteric enzyme function. The enzyme protein kinase is inhibited by a regulator protein that binds reversibly to the allosteric site. When the enzyme is in this inactive form, the shape of the active site is modified so that substrate cannot combine with it. When protein kinase activity is needed, cyclic AMP removes the regulator, thus activating the enzyme. The shape of the enzyme changes, allowing the substrate to combine with the active site.

temperature is raised to normal level. The rates of most enzyme-regulated reactions increase with rising temperature within limits. Temperatures greater than 50°C to 60°C rapidly inactivate most enzymes by denaturing the protein (altering its secondary and tertiary structures). Most organisms are killed by even brief exposure to high temperature.

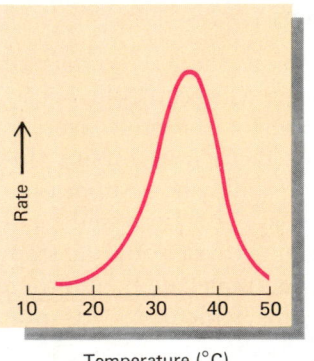 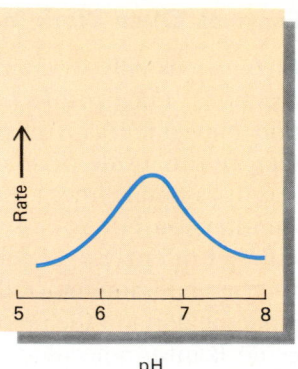

FIGURE 7–19 The effect of temperature (*a*) and pH (*b*) on the rate of enzyme-catalyzed reactions. Substrate and enzyme concentrations are constant.

Enzymes Can Be Inhibited By Certain Chemical Agents

Enzymes can be *inhibited* or even destroyed by certain chemical agents referred to as **inhibitors.** Enzyme inhibition may be reversible or irreversible. **Reversible inhibitors** can be competitive or noncompetitive. In **competitive inhibition,** the inhibitor competes with the normal substrate for the active site of the enzyme (Fig. 7–20). A competitive inhibitor usually is chemically similar to the normal substrate and so fits the active site and binds with the enzyme. However, it is not similar enough to the normal substrate to effectively take its place. The enzyme cannot act upon it to form reaction products. A competitive inhibitor occupies the active site only temporarily and does not damage the enzyme irreversibly.

In **noncompetitive inhibition** the inhibitor binds with the enzyme at a site other than the active site. Such an inhibitor renders the enzyme inactive by altering its shape. Many important noncompetitive inhibitors are metabolic substances that help regulate enzyme activity by combining reversibly with the enzyme.

Many poisons are **irreversible inhibitors** that permanently inactivate or even destroy the enzyme. Nerve gases are irreversible inhibitors that poison the enzyme cholinesterase, essential in the normal function of nerves and muscles. A number of insecticides and drugs are irreversible inhibitors. The antibiotic penicillin and its chemical relatives inhibit a bacterial enzyme necessary for bacterial cell wall construction. Unable to produce new cell walls, susceptible bacteria are unable to multiply effectively (Fig. 7–21). Since human body cells do not possess cell walls (and so do not employ the susceptible enzyme), penicillin is harmless to humans, except for the occasional allergic patient.

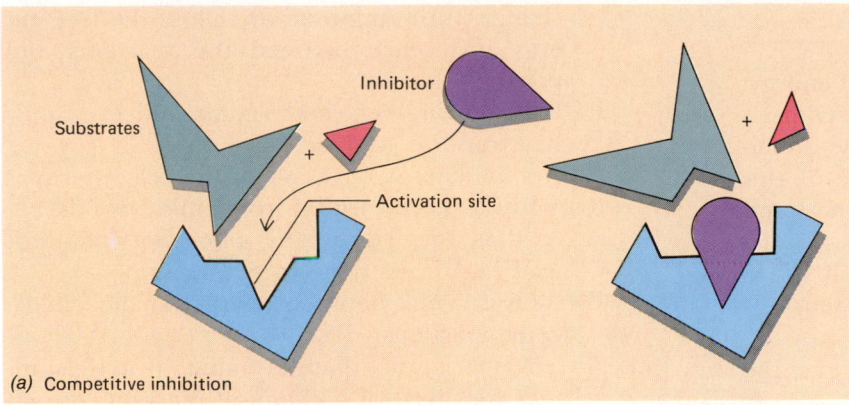

(a) Competitive inhibition

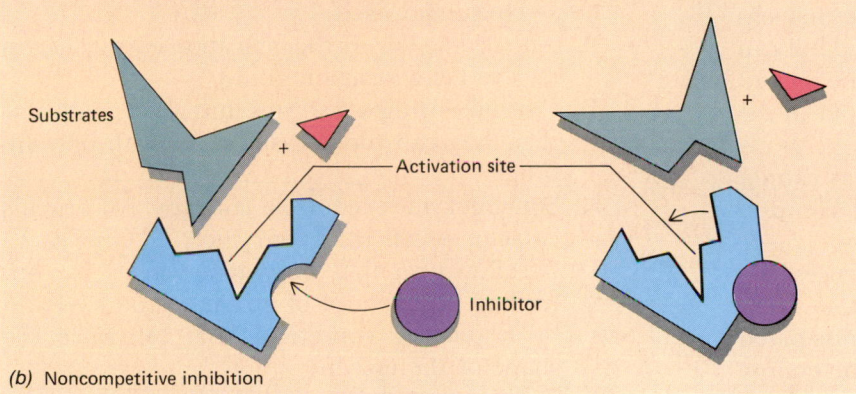

(b) Noncompetitive inhibition

FIGURE 7–20 Competitive and noncompetitive inhibition. (*a*) In competitive inhibition the inhibitor competes with the normal substrate for the active site of the enzyme. A competitive inhibitor occupies the active site only temporarily. (*b*) In noncompetitive inhibition, the inhibitor binds with the enzyme at a site other than the active site, altering the shape of the enzyme and so inactivating it. Noncompetitive inhibition may be reversible. Allosteric action, used by cells to control enzyme action, is a somewhat similar process (Fig. 7–18).

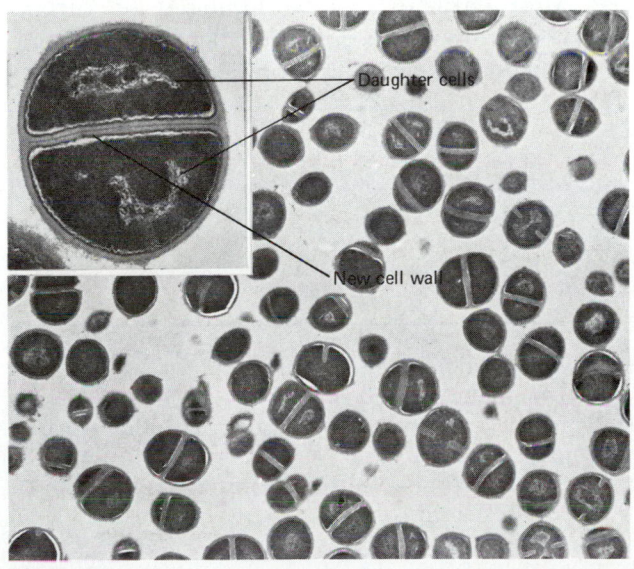

(a)

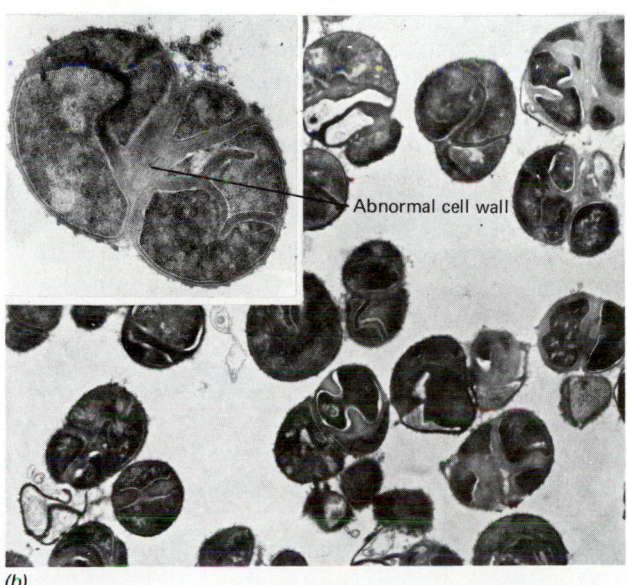

(b)

FIGURE 7–21 Antibiotic damage to bacterial cell walls. Penicillin is an irreversible enzyme inhibitor. (*a*) Normal bacteria. The insert shows the new cell wall laid down between daughter cells of a dividing bacterium. (*b*) Penicillin has damaged these bacterial cell walls. The inserts are magnified approximately × 54,000. (Courtesy of Victor Lorian and Barbara Atkinson, with permission of the *Journal of Clinical Pathology*)

CHAPTER SUMMARY

I. Life depends upon a continuous input of energy. Producers capture energy during photosynthesis and incorporate some of that energy in the chemical bonds of organic compounds. Some of this energy can then be transferred to consumers and decomposers.

II. Energy may be defined as the capacity to do work.
   A. Potential energy is stored energy; kinetic energy is energy of motion.
   B. A common unit used to measure energy is the kilocalorie.

III. The first law of thermodynamics states that energy can neither be created nor destroyed but can be transferred and changed in form. The second law of thermodynamics states that disorder in the universe is continuously increasing.
   A. The first law explains why organisms cannot produce energy but must borrow it continuously from somewhere else.
   B. The second law explains why no process requiring energy is ever 100% efficient; in every energy transaction some energy is dissipated as heat. The term *entropy* refers to the energy no longer available to do work.

IV. Metabolic reactions involve energy transformations.
   A. Reactions that release free energy are exergonic reactions; endergonic reactions require a net input of free energy.
   B. In an equilibrium the rate of change in one direction is exactly the same as the rate of change in the opposite direction; the free energy difference between the reactants and products is zero.
   C. In the living cell, endergonic and exergonic reactions are coupled.

V. ATP is the energy currency of the cell; energy is temporarily stored within its chemical bonds.
   A. ATP is formed by the phosphorylation of ADP, a process that requires energy.
   B. ATP is a link between exergonic and endergonic reactions.

VI. An enzyme is an organic catalyst; it greatly increases the speed of a chemical reaction without being consumed itself.
   A. An enzyme lowers the activation energy necessary to get a reaction going.
   B. Enzymes bring substrates into close contact so that they can more easily react with one another.
   C. Some enzymes require a cofactor. An organic cofactor is called a coenzyme.
   D. A cell can regulate enzymatic activity by controlling the amount of enzyme produced and by regulating conditions that influence the shape of the enzyme.
   E. Enzymes work best at specific temperatures and pH.
   F. Most enzymes can be inhibited by certain chemical substances. Reversible inhibition may be competitive or noncompetitive.

POST-TEST

1. _____ is the ability to do work.
2. Stored energy is referred to as _____ _____.
3. The study of energy is called _____.
4. Energy that can be used to do work is referred to as _____ energy.
5. _____ reactions require a net input of free energy.
6. In a dynamic _____ the rate of change in one direction is equal to the rate of change in the opposite direction.
7. The number of phosphate groups in ATP is _____.
8. Phosphorylation of ADP requires _____; the compound formed is _____.
9. A biological catalyst that regulates a chemical reaction is a(n) _____.
10. The energy required to break existing bonds and get a reaction started is known as _____ _____.
11. The name of an enzyme generally ends in -_____.
12. An organic compound that serves as a cofactor is called a(n) _____.
13. In _____ inhibition the inhibitor renders the enzyme inactive by altering its shape.

■ REVIEW QUESTIONS

1. Trace the various forms that energy takes as it flows through the ecosphere—from sunlight to the heat released during muscle contraction.
2. Give three examples of (a) potential energy; (b) kinetic energy.
3. When a consumer eats a producer it cannot obtain for itself all of the energy captured by the producer during photosynthesis. Why not? Explain in terms of the second law of thermodynamics.
4. Imagine that you could redesign your body so that you could create all the energy you require. What advantages would this ability confer on you?
5. Why is it adaptive that endergonic and exergonic reactions can be coupled?
6. What is activation energy? How does an enzyme affect activation energy?
7. Explain the lock and key analogy of enzymatic action.
8. How are enzymes affected by (a) pH; (b) temperature; (c) inhibitors?
9. In what way does ATP serve as a link between exergonic and endergonic reactions? From what is ATP synthesized?

■ READINGS

Atkins, P.W. *The Second Law.* San Francisco, W.H. Freeman and Company, 1984. A basic, understandable introduction to thermodynamics with an extensive section denoted to its biological implications.

Becker, W.M., *The World of the Cell.* Menlo Park, California, Benjamin Cummings Company, 1986. Chapter 5 contains a detailed account of the flow of energy in the cell.

Cloud, P. "The Biosphere," *Scientific American,* September 1983, 176–189. A fascinating discussion of the relationship between microbial, animal, and plant life on earth and the physical environment—and the energy transfer pathways that knit it all together.

Harold, F., *The Vital Force: A Study of Bioenergetics.* San Francisco, W.H. Freeman & Company, 1986. An extensive examination of what is currently known about how organisms utilize energy.

Westheimer, F., "Why Nature Chose Phosphates," *Science* 235:1173, 1987. An appraisal of the suitability of phosphates for energy reactions in organisms.

8
Energy-Releasing Pathways

OUTLINE

I. Fuel molecules can be degraded via aerobic or anaerobic pathways
II. Catabolic pathways utilize oxidation-reduction reactions
  A. Hydrogen, along with some of its energy, can be transferred to hydrogen acceptor molecules
  B. Cellular respiration is a redox process
III. Cellular respiration has four phases
  A. Glycolysis requires an energy investment
  B. Pyruvate can be used to make acetyl coenzyme A
  C. The citric acid cycle completes the breakdown of the fuel molecule
  D. Most of the ATP is produced by the electron transport system and chemiosmotic phosphorylation
    1. Electron transport provides energy for pumping protons across the inner mitochondrial membrane
    2. Electron transport and ATP synthesis are coupled by a proton gradient
IV. Cellular respiration is regulated by the amount of ADP and phosphate available
V. The energy yield from one molecule of glucose may amount to 38 ATPs
VI. Other nutrients can be used as energy sources
VII. When oxygen is not available many cells utilize anaerobic pathways
  A. In fermentation the final electron acceptor is an organic compound
  B. Anaerobic metabolism is inefficient

LEARNING OBJECTIVES

After you have studied this chapter you should be able to:

1. Write a general equation illustrating hydrogen and electron transfer from a substrate to a hydrogen acceptor such as NAD^+.
2. Write a summary reaction for cellular respiration, giving the origin and fate of each substance involved.
3. Give a brief overview of the four phases of cellular respiration, and indicate where the reactions of each phase take place in the cell.
4. Summarize the events of glycolysis, identifying the key organic compounds formed and the number of carbon atoms in each; indicate the number of ATP molecules used and produced and the transactions in which hydrogen transfer occurs.
5. Summarize the events of the citric acid cycle; indicate the fate of carbon-oxygen segments and of hydrogens removed from the fuel molecule.
6. Summarize the operation of the electron transport system.
7. Describe chemiosmotic phosphorylation, explaining how a gradient of protons is established across the inner mitochondrial membrane, and the process by which the proton gradient drives ATP synthesis.
8. Indicate how the products of protein and fat metabolism feed into the same metabolic pathways that oxidize glucose.
9. Contrast aerobic and anaerobic pathways used by cells to extract energy from nutrients in terms of ATP formation, final hydrogen acceptor, and end products.
10. Compare alcoholic and lactic acid fermentation in terms of final hydrogen acceptor and end products.

Giant panda. (Sharon Cummings/Marvin Dembinsky, Jr., Photography Associates)

Every living cell must extract energy from the organic food molecules it manufactures or captures from the environment (Fig. 8–1). In humans and other complex animals, food is broken down by the digestive system. For example, proteins are split into their component amino acids, carbohydrates are digested to simple sugars such as glucose, and fats are split into glycerol and fatty acids. These nutrients are then absorbed into the blood and transported to all of the cells. Within each cell, these nutrients are oxidized (broken down by combining with oxygen or removing hydrogen) and some of their chemical energy is transferred to adenosine triphosphate (ATP) for later use in cellular work.

The process of splitting larger molecules to smaller ones is an aspect of metabolism referred to as **catabolism,** and the individual reactions involved are called **catabolic reactions.** Cells use three different catabolic pathways to extract energy from nutrients: cellular respiration, anaerobic respiration, and fermentation.

KEY CONCEPTS

☐ During cellular respiration, cells break down organic compounds and gradually release their stored energy. The energy released is used to make ATP.

☐ During aerobic respiration, an organic compound such as glucose can be completely degraded to carbon dioxide and water.

☐ In the absence of oxygen, anaerobic pathways may be used. The fuel molecule can be only partially broken down, and only a small amount of energy is released and packaged in ATP.

☐ FUEL MOLECULES CAN BE DEGRADED VIA AEROBIC OR ANAEROBIC PATHWAYS

The type of environment a cell inhabits determines which catabolic pathway is used to break down nutrients. Cells that live in environments where oxygen is plentiful utilize the very efficient aerobic pathway. An **aerobic** pathway requires molecular oxygen (O_2). Cells that inhabit the soil or polluted waters where oxygen is in short supply have become adapted to utilize less efficient **anaerobic** pathways that do not require oxygen (Fig. 8–2).

Most cells extract energy from fuel molecules—glucose, fatty acids, and other organic compounds—using aerobic cellular respiration. This process involves a long sequence of 30 or more chemical reac-

FIGURE 8–1 Some of the organic nutrients this python obtains from the impala it is eating will be oxidized during cellular respiration. The energy released is packaged in ATP and ultimately used to carry on life processes. (Stan Osolinski/Marvin L. Dembinsky, Jr., Photography Associates)

tions, each regulated by a specific enzyme. Energy is released efficiently in small amounts.

During complete aerobic respiration nutrients are catabolized to carbon dioxide and water. One of the most common pathways of aerobic cellular respiration involves the breakdown of the common nutrient glucose. The overall reaction for the aerobic catabolism of glucose can be summarized as follows:

$$C_6H_{12}O_6 + 6O_2 + 6H_2O \longrightarrow 6CO_2 + 12H_2O + Energy$$

☐ CATABOLIC PATHWAYS UTILIZE OXIDATION-REDUCTION REACTIONS

Energy processing by cells involves the transfer of energy through the flow of electrons and protons. Recall from Chapter 3 that **oxidation** is a chemical process in

FIGURE 8–2 Organisms such as this tetanus bacteria, that live in oxygen-deficient environments have anaerobic pathways for acquiring energy. (Biophoto Associates)

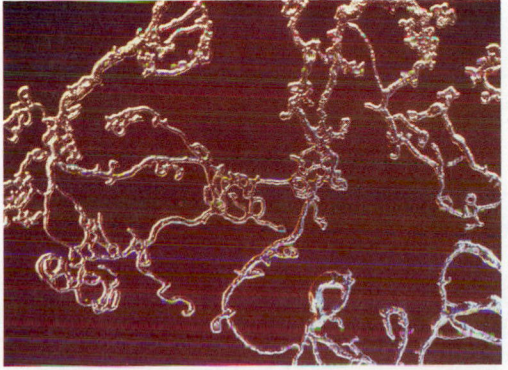

which a substance loses electrons. **Reduction** is a chemical process in which a substance gains electrons. Oxidation must be accompanied by reduction because electrons cannot exist in the free state in living cells. As quickly as they are released, electrons are accepted by another atom or molecule. The oxidized molecule gives up energy, and the reduced molecule receives energy. Oxidation-reduction, or simply, **redox reactions,** are characteristic of many metabolic processes, including cellular respiration and photosynthesis. Generally, there is a sequence of oxidation-reduction reactions that take place as hydrogen or its electrons are transferred from one compound to another. A generalized example of an oxidation-reduction reaction follows.

$$\overset{\text{\small Oxidation}}{\overbrace{Ae^- + B \longrightarrow A + \underset{\text{\small Reduction}}{\underbrace{Be^-}}}}$$

The e^- stands for electron. Note that compound A is *oxidized* as it gives up electrons to compound B. As compound B accepts electrons it is *reduced.*

Hydrogen, Along With Some Of Its Energy, Can Be Transferred To Hydrogen Acceptor Molecules

Electrons are difficult to remove from covalent compounds unless an entire atom is removed. In the cell, oxidation almost always involves the removal of a hydrogen atom (and its single electron) from a compound. Reduction often involves a gain in hydrogen atoms (and thus a gain in electrons).

When hydrogen atoms are removed from an organic compound, they take with them some of the energy that had been stored in their chemical bonds. The hydrogen, along with its energy, is transferred to a hydrogen acceptor molecule, which is generally a coenzyme (an organic, nonpolypeptide component of an enzyme). One of the most common hydrogen acceptor coenzymes in cells is **nicotinamide adenine dinucleotide,** more conveniently referred to as **NAD.**

NAD^+ can temporarily package large amounts of free energy. Here is a generalized equation showing the transfer of hydrogen from a compound we will call X to NAD^+:

$$\overset{\text{\small Oxidation}}{\overbrace{XH_2 + NAD^+ \longrightarrow X + \underset{\text{\small Reduction}}{\underbrace{NAD\!-\!H + H^+}}}}$$

NAD^+ is reduced when it combines with hydrogen. Before it is reduced NAD^+ is an ion with a net charge of $+1$. When hydrogen is added, the charge is neutralized. The reduced form of the compound, NADH, is electrically neutral. (The electron from each hydrogen combines with the NAD^+.)

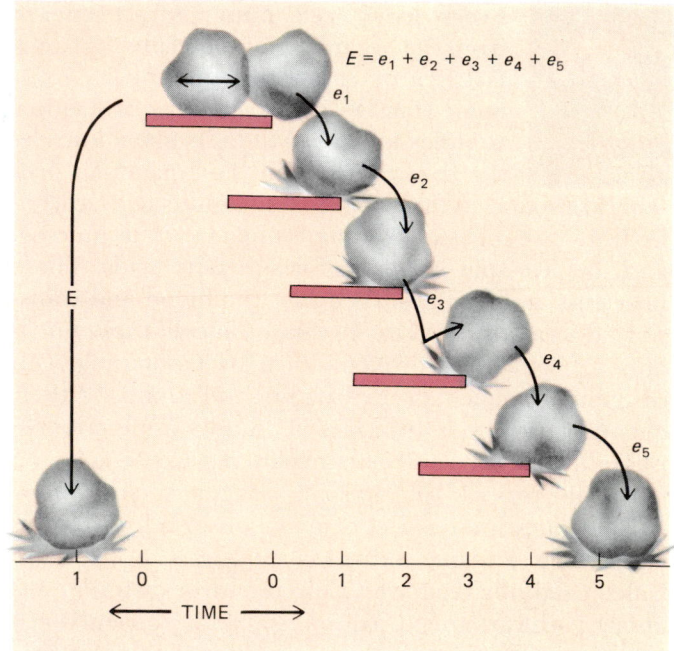

FIGURE 8–3   The total energy released by a falling object is the same whether the object is released all at once or in a series of steps. Similarly, in cellular metabolism the energy of an electron liberated from an organic compound is the same whether it is released all at once, or, as is actually the case, gradually as it passes through a series of electron acceptors. Such chains of electron acceptors permit the slow, controlled extraction of energy from fuel molecules.

Some of the energy stored in the chemical bonds holding the hydrogens to molecule X has been transferred by this reaction to the NADH. This energy can now be used to drive some metabolic process, or it can be transferred through a complex series of reactions to ATP.

Other important hydrogen or electron acceptor compounds are **flavin adenine dinucleotide (FAD)** and the cytochromes. FAD is a nucleotide that accepts hydrogens and their electrons. **Cytochromes** are proteins that contain iron.

Electron acceptors are often arranged in chains. As electrons are transferred from one acceptor to the next along the chain, energy is slowly released. Electron transport chains provide a mechanism for the cell to slowly and efficiently capture energy from fuel molecules (Fig. 8–3).

Cellular Respiration Is A Redox Process

Cellular respiration is a redox process in which hydrogen is transferred from glucose to oxygen. In a long series of reactions, glucose is oxidized and oxygen is reduced.

$$\overbrace{C_6H_{12}O_6 + 6O_2 + 6H_2O \longrightarrow 6CO_2 + 12H_2O + E}$$

During this process potential energy of electrons is decreased and chemical energy (E) is released. This energy can then be used for ATP synthesis.

☐ CELLULAR RESPIRATION HAS FOUR PHASES

The chemical reactions of cellular respiration can be grouped into four phases (Figs. 8–4 and 8–5; Table 8–1): (1) glycolysis, (2) formation of acetyl coenzyme A, (3) citric acid cycle, and (4) electron transport and ATP synthesis.

1. **Glycolysis.** During glycolysis (literally, "splitting sugar") the cell breaks down the six-carbon glucose molecule to two three-carbon molecules of pyruvate.[1] During this sequence of reactions hydrogen is removed from the fuel molecule and ATP is produced. The hydrogens combine with NAD$^+$ to form NADH.

   The reactions of glycolysis take place in the cytoplasm. Glycolysis does not require oxygen and so technically this pathway is not a part of aerobic cellular respiration.

2. **Formation of acetyl coenzyme A.** Pyruvate is degraded to a two-carbon fuel molecule and combines with coenzyme A, forming acetyl coenzyme A. The carbon removed from pyruvate is released as carbon dioxide. Hydrogens are also removed and combine with NAD$^+$ to produce more NADH. This pathway prepares the fuel molecule to enter the citric acid cycle.

3. **Citric acid cycle.** The two-carbon fuel molecule enters the cycle. Two carbon-oxygen segments are released as carbon dioxide. A small amount of ATP is produced, but most importantly, the remaining hydrogens are removed from the fuel molecule and picked up by NAD$^+$ and FAD.

4. **Electron transport and associated chemiosmotic ATP synthesis.** The NADH and FADH$_2$ produced in the preceding phases transfer the hydrogens they have accepted to a chain of hydrogen and electron acceptor molecules. These compounds are located within the inner membrane of the mitochondria. As they are passed along the chain of acceptor molecules the hydrogens are separated from their electrons. When the electron is removed from a hydro-

---

[1] Pyruvate and many other compounds in glycolysis and the citric acid cycle exist as ions at the pH found in the cell. They sometimes associate with H$^+$ to form acids; for example, pyruvate forms pyruvic acid. In some textbooks these compounds are presented in the acid form.

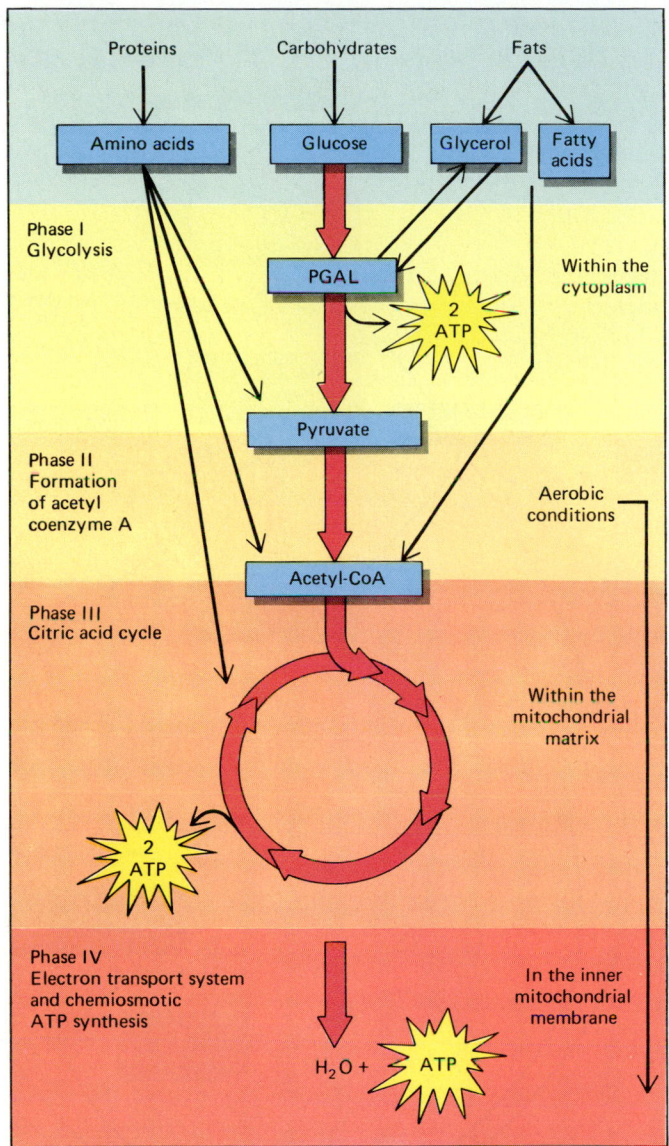

FIGURE 8–4 Four main phases in cellular respiration are (I) glycolysis, (II) formation of acetyl coenzyme A, (III) the citric acid cycle, and (IV) the electron transport system with its associated chemiosmotic ATP synthesis. Glycolysis occurs in the cytoplasm; the remaining phases take place in the mitochondria.

gen atom, only a proton remains. The protons (H$^+$) are pumped across the inner mitochondrial membrane from the matrix of the mitochondrion to the intermembrane space. The difference in concentration of protons between the matrix and the intermembrane space is referred to as a **proton gradient;** this gradient represents potential energy.

The protons can flow back into the matrix only through special channels in the inner membrane. These channels occur within the enzyme **ATP synthetase.** When the protons move through this channel they move down an energy gradient. Energy is released and is used to drive the synthesis of ATP. Thus,

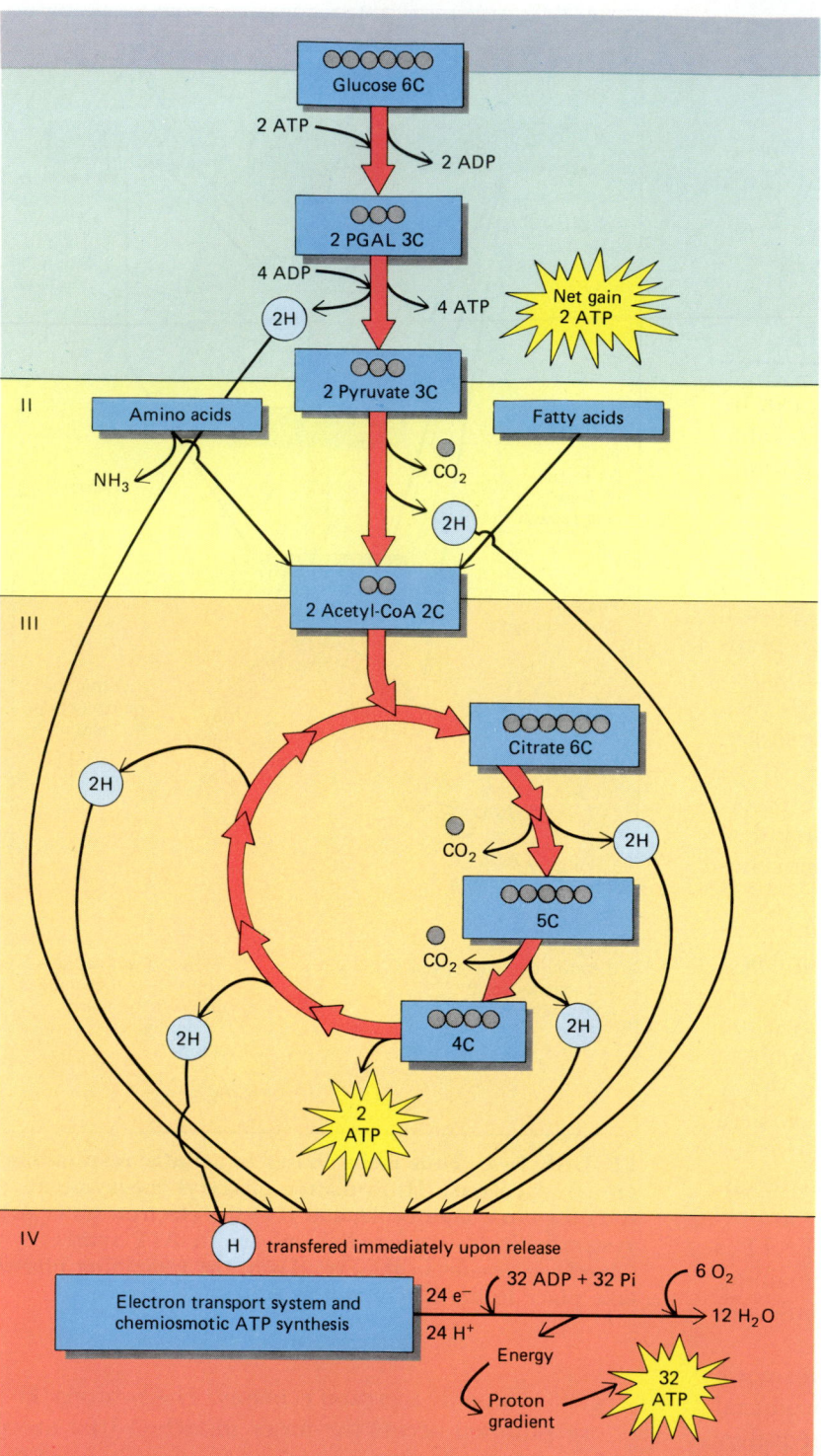

FIGURE 8–5 An overview of cellular respiration. Note that as the fuel molecule is degraded, hydrogen is removed from it, and carbon-oxygen segments are released as carbon dioxide. Hydrogen is transferred to NAD^+ and then passes through the electron acceptors of the electron transport system. The chemical energy released is used to make ATP by chemiosmosis.

this proton gradient across the inner mitochondrial membrane couples **phosphorylation** (addition of phosphate to ADP to form ATP) with oxidation. These processes will be discussed in greater detail later in the chapter.

Glycolysis Requires An Energy Investment

The ingredients necessary for **glycolysis**—glucose, NAD^+, ATP, ADP, and phosphates—float freely in the cytoplasm. The energy of two ATPs must be invested in order to produce four ATPs. This process results in

TABLE 8–1
Summary of Cellular Respiration

Phase	Summary	Needed materials	End products
Glycolysis (takes place in cytoplasm)	Series of about ten reactions during which glucose is degraded to pyruvate; net profit of 2 ATPs; hydrogens released; can proceed anaerobically	Glucose, 2 ATPs, ADP, P_i, NAD^+	Pyruvate, ATP, NADH
Formation of acetyl CoA (takes place in mitochondria)	Pyruvate is degraded and combined with coenzyme A to form acetyl CoA; CO_2 is released	Pyruvate, coenzyme A, NAD^+	Acetyl CoA, NADH, CO_2
Citric acid cycle (takes place in mitochondria)	Series of reactions in which fuel molecule (part of acetyl CoA) is degraded to hydrogen and carbon dioxide	Acetyl CoA, H_2O hydrogen acceptors (e.g., NAD^+), ADP, P_i	CO_2, NADH, $FADH_2$, CoA, ATP
Electron transport and chemiosmosis (take place in mitochondria)	Chain of several electron transport molecules; hydrogens (or their electrons) are passed along chain; energy released is used to generate proton gradient across inner mitochondrial membrane; as protons move through ATP synthetase in membrane, ATP is synthesized; for each pair of hydrogens that enters chain, a maximum of 3 ATPs can be synthesized; aerobic	Hydrogen (in NADH, $FADH_2$), ADP, P_i, oxygen	ATP, water

a net gain of two ATPs. Glycolysis does not require oxygen and can proceed under aerobic or anaerobic conditions.

The reactions of glycolysis are shown in Figure 8–6. In a highly simplified explanation of glycolysis we can divide the process into two stages. The first stage corresponds to the first five steps of glycolysis illustrated in Figure 8–6. During this stage two phosphates from two ATPs are added to the glucose molecule. Addition of phosphate to a molecule is termed **phosphorylation.** The glucose molecule is split, forming two molecules of the three-carbon compound **glyceraldehyde-3-phosphate,** or **PGAL.** This transformation requires energy, and the cell has invested two molecules of ATP in order to initiate the process.

We may summarize this portion of glycolysis as follows:

Glucose + 2ATP $\longrightarrow\!\!\longrightarrow$ 2PGAL + 2ADP + 2Pi
6-C   3-C

Several arrows are used to indicate that the equation summarizes a sequence of several reactions.

In the second phase of glycolysis (steps 6 through 10 in Fig. 8–6), PGAL is oxidized with the removal of two hydrogen atoms, and certain other atoms are rearranged so that each molecule of PGAL is transformed into a molecule of **pyruvate.** During these reactions enough chemical energy is released from the sugar molecule to produce four ATP molecules.

2PGAL + 4ADP + $4P_i$ $\longrightarrow\!\!\longrightarrow$ 2 Pyruvate + 4H + 4ATP
3-C   3-C

In the first phase of glycolysis two molecules of ATP were used to initiate the process, but in the second phase four molecules of ATP are produced. Thus glycolysis yields a net energy profit of two ATP molecules.

One hydrogen atom removed from each PGAL immediately combines with the hydrogen carrier molecule, NAD^+. The other hydrogen ion is found in the surrounding medium. However, both electrons are transferred to the NAD^+.

NAD^+ + 2H $\longrightarrow$ NADH + H^+
Oxidized   Reduced

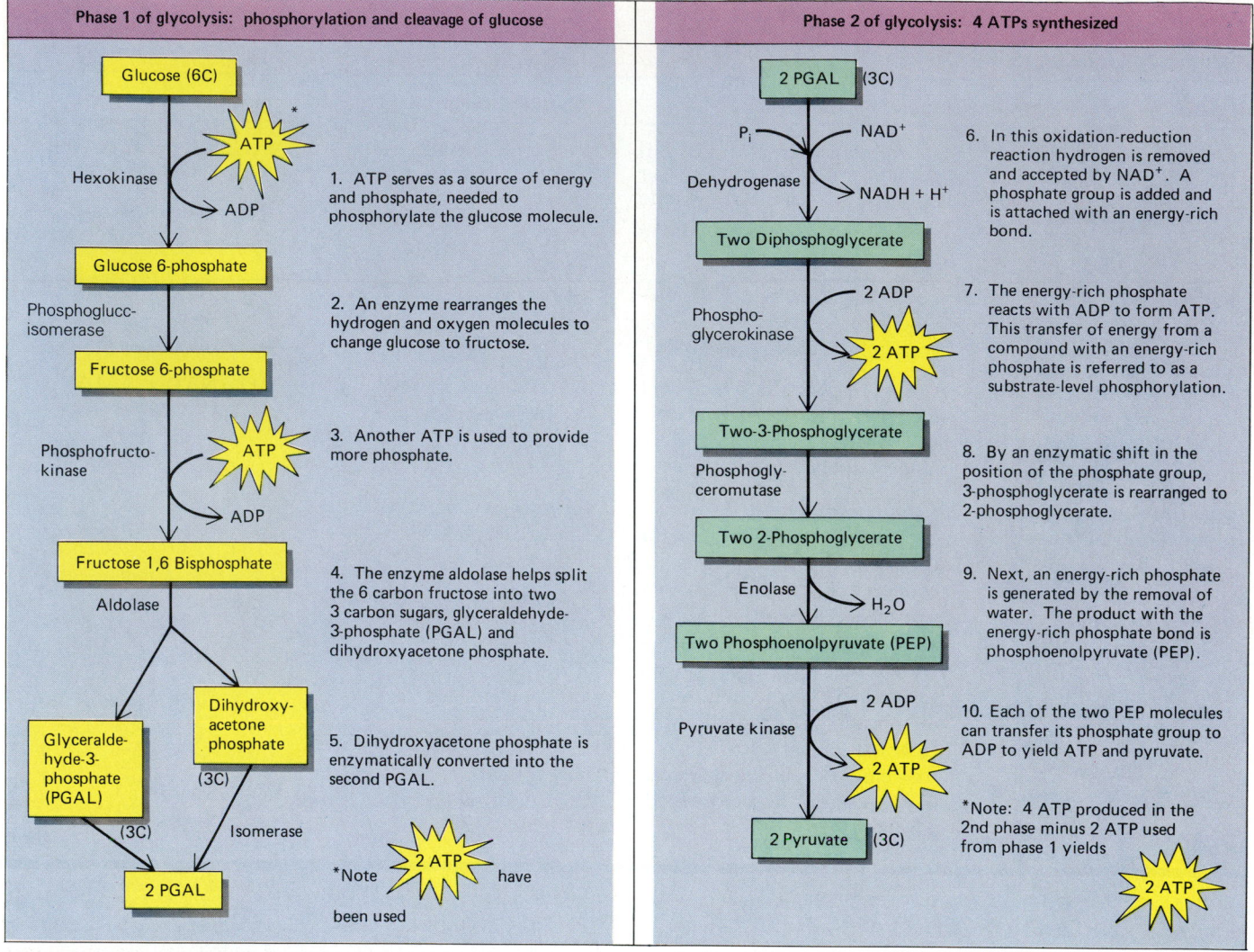

FIGURE 8–6   Glycolysis. Each reaction is catalyzed by a specific enzyme. Note that we can calculate a net yield of two ATPs by subtracting the number of ATPs used in the process from those produced.

The hydrogen atoms and their associated electrons are next given to the electron transport system or to anaerobic pathways. These pathways will be discussed in later sections.

Pyruvate Can Be Used To Make Acetyl Coenzyme A

Pyruvate, the end product of glycolysis, contains most of the energy present in the original glucose molecule. When oxygen is available, pyruvate molecules can be completely degraded during the reactions of the citric acid cycle. Pyruvate molecules move into the mitochondria where all subsequent reactions of cellular respiration take place.

Once a pyruvate molecule enters the mitochondrion it is converted to the two-carbon compound acetyl coenzyme A, or simply, **acetyl CoA.** A large, complex enzyme catalyzes the reaction. First, a carbon-oxygen segment (carboxyl group) is removed and released as carbon dioxide. Then, the two-carbon fragment remaining is oxidized; the hydrogens removed are accepted by NAD^+. Finally, the oxidized fragment, an acetyl group, is attached to coenzyme A. This coenzyme contains a sulfur atom which attaches to the acetyl group by an unstable bond. The overall reaction for the formation of acetyl CoA is

$$2 \text{ Pyruvate} + 2NAD^+ + 2CoA \longrightarrow$$
$$2 \text{ Acetyl CoA} + 2NADH + 2H^+ + 2CO_2$$

Coenzyme A is manufactured in the cell from one of the B vitamins, **pantothenic acid.** Fats and amino acids can also be degraded to become acetyl CoA. The two-carbon acetyl portion of acetyl CoA enters the citric acid cycle.

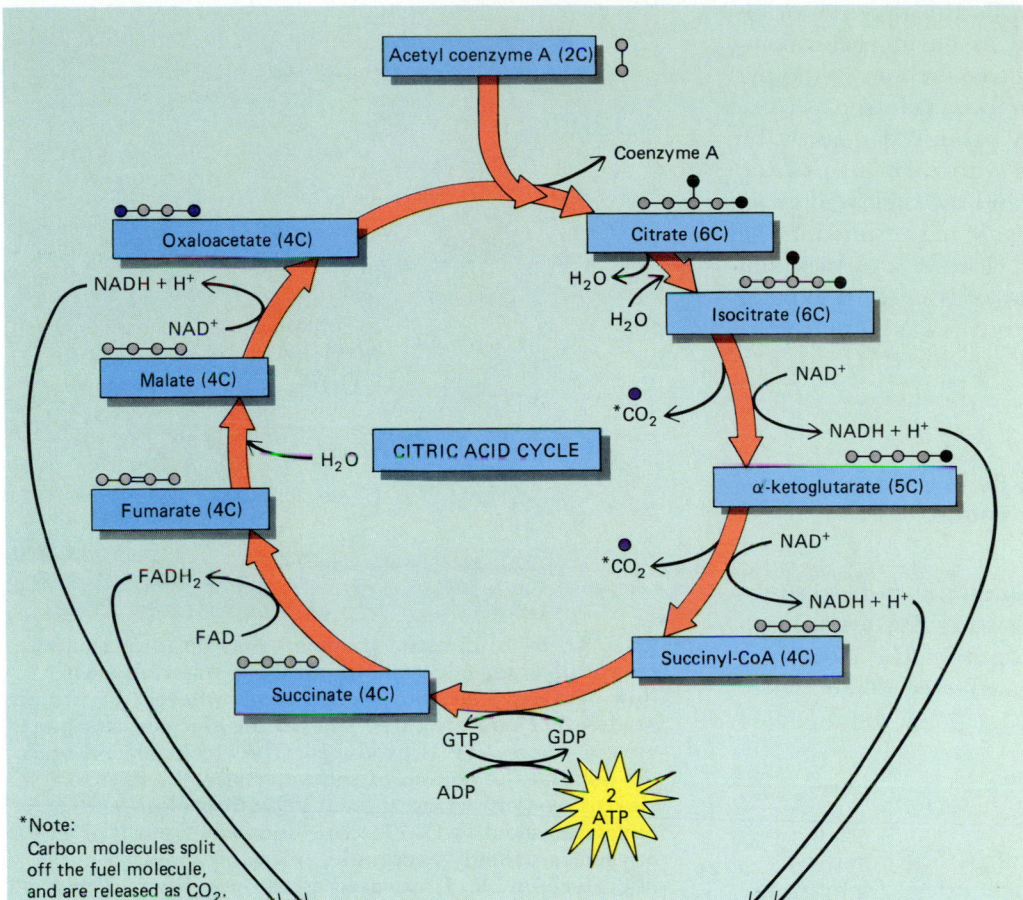

FIGURE 8–7 The citric acid cycle. During this series of reactions, acetyl coenzyme A, produced from glucose and other organic compounds, is oxidized to yield carbon dioxide and hydrogen. The hydrogen is immediately transferred to NAD^+ or FAD and is fed into the electron transport chain.

Note that the original glucose molecule has now been oxidized to two acetyl groups and two CO_2 molecules. Hydrogens have been removed and accepted by NAD^+, reducing it to NADH + H^+. Two NADH molecules are formed during glycolysis and two during the oxidation of pyruvate.

The Citric Acid Cycle Completes The Breakdown Of The Fuel Molecule

The **citric acid cycle** is also known as the **tricarboxylic acid (TCA) cycle** or **Krebs cycle** after Sir Hans Krebs, who worked it out in the 1930s. This cycle is the final common pathway for the oxidation of pyruvate, fatty acids, and the carbon chains of amino acids. The cycle, which takes place in the mitochondria, consists of the eight steps illustrated in Figure 8–7. Each reaction is catalyzed by a specific enzyme.

The first reaction of the cycle occurs when acetyl CoA transfers its two-carbon acetyl group to the four-carbon compound **oxaloacetate,** forming **citrate,** a six-carbon compound.

Oxaloacetate + Acetyl CoA $\longrightarrow$ Citrate
4-C   2-C   6-C

The citrate then goes through a series of chemical transformations losing first one, then another carboxyl group (—COOH) as CO_2. Most of the energy made available by the oxidative steps of the cycle is transferred as energy-rich electrons to NAD^+. For each acetyl group that enters the citric acid cycle, three molecules of NAD^+ are reduced to NADH. In the conversion of succinate to fumarate, electrons are transferred to the electron acceptor FAD, rather than to NAD^+.

In the course of the cycle, two molecules of CO_2 and eight hydrogen atoms (eight protons and eight electrons) are removed. The CO_2 produced accounts for the two carbon atoms that entered the citric acid cycle. You may wonder why more hydrogen is generated by these reactions than entered the cycle with the fuel molecule. These hydrogens come from water molecules that are added during the reactions of the cycle. Extracting hydrogen from water requires energy. The needed energy is provided by the disruption of the chemical bonds of the fuel molecule. Some of this energy is in effect stored in the hydrogen that is generated.

Because two pyruvate molecules are produced from each glucose molecule, the cycle must turn twice to process each glucose. At the end of a complete cycle a four-carbon oxaloacetate is all that is left, and the cycle is ready for another turn. By this time the original pyruvate may be regarded as having been completely consumed because the three carbons in the pyruvate (or their equivalent) have been released as CO_2. Only one molecule of ATP is produced directly by phosphorylation of a substrate with each turn of the cycle. Thus, so far in glycolysis and the citric acid cycle, the energy of one glucose molecule has resulted in the formation of only four ATPs. However, to maintain their highly ordered state, most cells need to expend much more energy than these four ATPs can provide. How, then, is most of the ATP produced?

Most Of The ATP Is Produced By The Electron Transport System And Chemiosmotic Phosphorylation

Now we will consider the fate of all the hydrogens removed from the fuel molecule during glycolysis, acetyl CoA formation, and the reactions of the citric acid cycle. These hydrogens were transferred to the hydrogen acceptors—NAD^+ or FAD. What becomes of them?

Electron Transport Provides Energy For Pumping Protons Across The Inner Mitochondrial Membrane

The electron transport system is a chain of electron acceptors embedded in the inner membrane of the mitochondrion. Hydrogens are passed from NADH to **flavin mononucleotide (FMN),** the first acceptor in the chain (Fig. 8–8). FMN is an electron carrier related to FAD. As hydrogens are transferred from one to another of the electron acceptor molecules, the hydrogen protons become separated from their electrons. Hydrogens, or their electrons, are passed along the chain of acceptors in a series of redox reactions.

When the hydrogen protons (H^+) separate from their electrons, they are released into the surrounding medium. The electrons entering the electron transport system have a relatively high energy content. As they pass along the chain of electron acceptors they lose much of their energy. Some of this energy is used to pump the protons across the inner membrane of the mitochondrion from the matrix to the intermembrane space. The proton pumps result in a much greater concentration of protons in the intermembrane space than in the matrix. An electric gradient, as well as a concentration gradient of hydrogen ions, is established (sometimes referred to as an electrochemical gradient). The energy in these gradients provides the energy for ATP synthesis (Fig. 8–9).

The electron acceptors in the chain include FMN, **ubiquinone (Q),** and a group of closely related proteins, the **cytochromes.** A central atom of iron in the

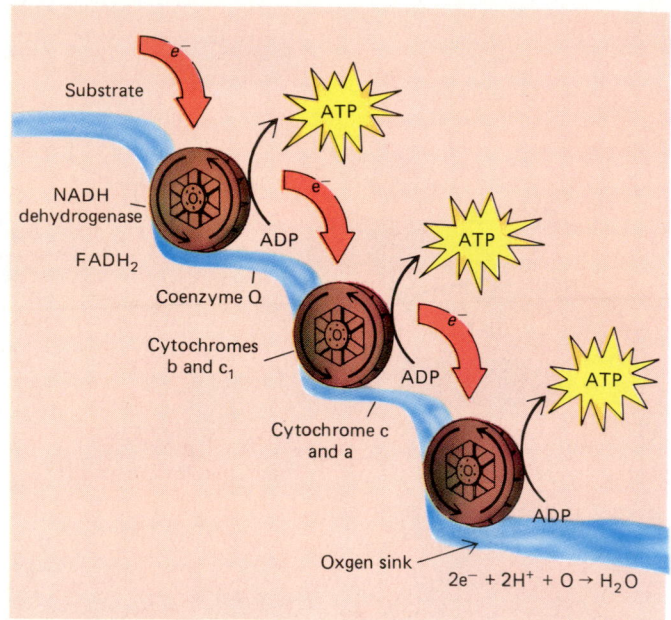

FIGURE 8–8 Electron transport may be compared to a stream of water (electrons) with three waterfalls. The flow of water (electrons) drives water wheels (the proton gradient). There are three sites in the electron transport system where ATP is produced. The electrons end up in the pond at the bottom of the waterfalls (the electron sink of oxygen) where they unite with protons and oxygen to yield H_2O. Electron transport from NADH to oxygen is strongly exergonic, releasing about 53 kilocalories/mole. If released all at once, most of this energy would be lost as heat. Instead, the energy is released slowly in a series of steps, as shown here. The energy released is used to transport protons across the inner mitochondrial membrane to the intermembrane space. The membrane potential established across the inner membrane is the source of the energy needed to synthesize ATP. For each pair of hydrogens that enters this pathway, a maximum of three ATP molecules are produced.

cytochrome molecule combines with the electrons from hydrogen atoms. Cytochrome molecules accept only the electron from the hydrogen, rather than the entire atom. Each of the several types of cytochromes holds electrons at slightly different energy levels. Electrons are passed along from one cytochrome to the next in the chain, losing energy as they go. Finally, the last cytochrome in the chain, cytochrome a_3, passes the two electrons on to molecular oxygen. Simultaneously, the (now low-energy) electrons reunite with protons (H^+) in the surrounding medium, forming hydrogen, and the hydrogen and oxygen unite chemically, producing water.

Oxygen is the final hydrogen acceptor in the electron transport system, which explains why we require oxygen. What happens when cells are deprived of oxygen? When no oxygen is available to accept the hydro-

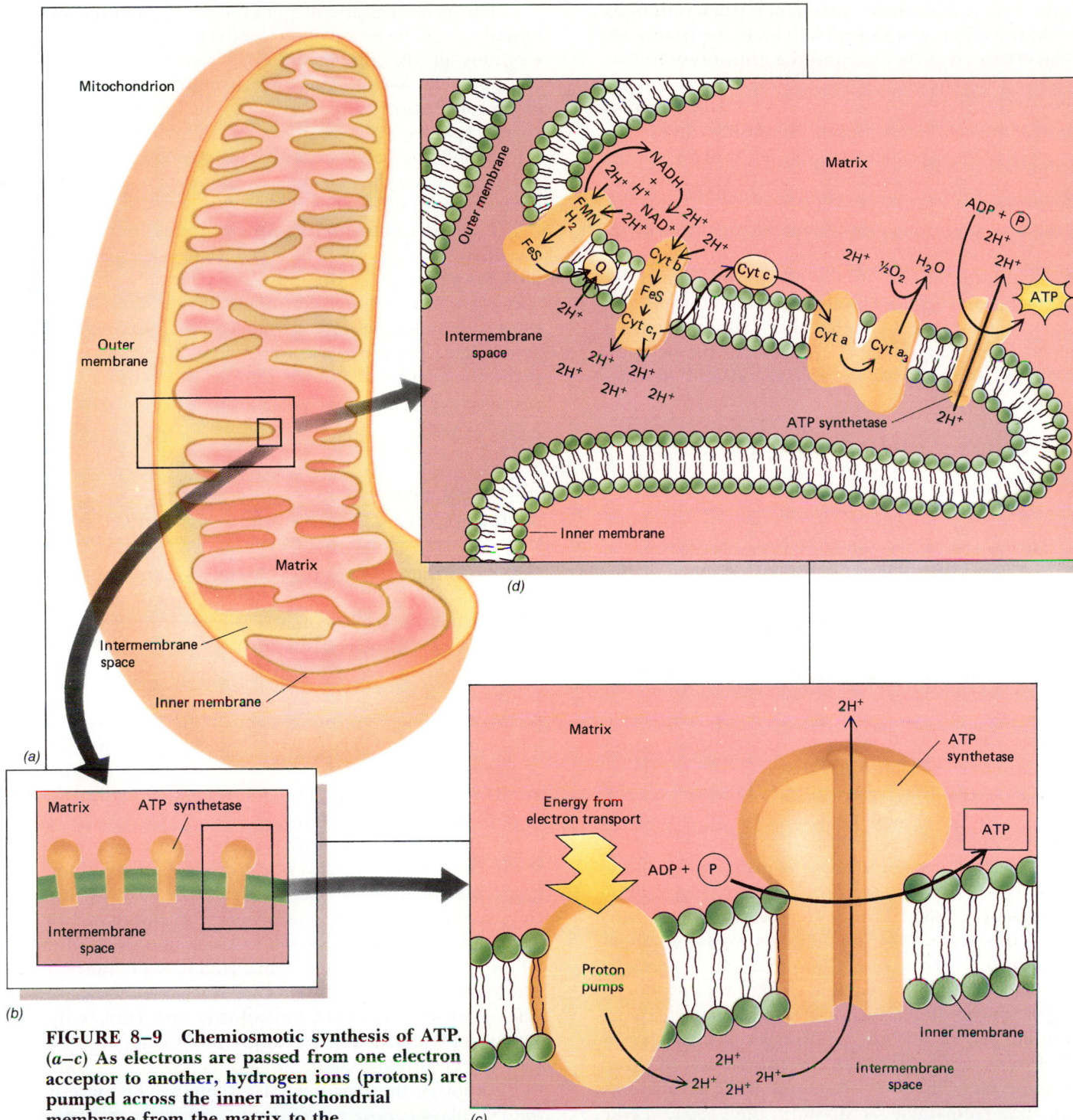

FIGURE 8–9   Chemiosmotic synthesis of ATP. (*a–c*) As electrons are passed from one electron acceptor to another, hydrogen ions (protons) are pumped across the inner mitochondrial membrane from the matrix to the intermembrane space. The protons can flow back into the matrix only through special channels within the enzyme ATP synthetase. The energy released as the protons move down the energy gradient is used to synthesize ATP. (*d*) A more detailed view of electron transport and chemiosmotic synthesis of ATP. The electron transport chain in the inner mitochondrial membrane is a proton pump. The electron acceptors in the membrane are located in three main complexes. FMN, which oxidizes NADH, is located in the first complex. The cytochrome b–c_1 complex consists of two cytochromes and some additional electron acceptors. The cytochrome oxidase complex includes cytochromes a and a_3. Ubiquinone (Q) and cytochrome c are mobile carriers that transfer electrons between the complexes. At three sites in the electron transport chain, energy released during electron transport is used to pump protons across the inner membrane. The protons are prevented from diffusing back into the matrix through the inner membrane everywhere except through special channels in ATP synthetase in the membrane. The flow of the protons through the ATP synthetase generates ATP at the expense of the free energy released as the protons pass from a region of high concentration (outside) to a region of lower concentration (inside).

gen, the last cytochrome in the chain is stuck with it. As a result, the other acceptor molecules in the chain cannot pass their electrons on, and the entire system may be blocked all the way back to NADH. No further ATPs can be produced by way of the electron transport system. Most cells of complex organisms cannot live long without oxygen because the amount of energy they can produce in its absence is insufficient to sustain life processes.

Lack of oxygen is not the only factor that may interfere with the electron transport system. Some poisons, including cyanide, inhibit the normal activity of the cytochrome system. Cyanide combines electrons tightly to cytochrome a_3 so that they cannot be passed along to molecular oxygen. This blocks further passage of electrons through the chain, and ATP production halts.

Electron Transport And ATP Synthesis Are Coupled By A Proton Gradient

As electrons are transferred along the acceptors in the electron transport chain, they are passed down an energy hill. Sufficient energy is released at three points on this hill to transport protons across the inner mitochondrial membrane, and ultimately to synthesize ATP. The passage of each pair of electrons from NADH to oxygen is thought to yield three ATPs.

The flow of electrons is tightly coupled to ATP synthesis and generally does not occur unless phosphorylation of ADP can proceed also. This, in a sense, prevents waste, for electrons will not flow unless ATP can be produced. When electron flow is uncoupled from phosphorylation, there is no production of ATP and the energy of the electrons is dissipated as heat. Because the phosphorylation of ADP to form ATP is coupled with the oxidation of electron transport components, this process is referred to as **oxidative phosphorylation.**

It has long been known that oxidative phosphorylation occurs in the mitochondria and many experiments had shown that the transfer of electrons from NADH to oxygen resulted in the production of three ATP molecules. However, just how these ATPs were synthesized remained a mystery. In 1961, Peter Mitchel proposed the chemiosmotic model for which he was awarded the Nobel Prize in 1978. Mitchell proposed that electron transport and ATP synthesis are coupled by a **proton gradient** which is established across the inner mitochondrial membrane. According to this model, the stepwise transfer of electrons from NADH or $FADH_2$ through the electron carriers to oxygen results in the pumping of protons across the inner mitochondrial membrane into the space between the inner and outer mitochondrial membranes.

The proton gradient generates a membrane potential across the inner mitochondrial membrane. The medium in the intermembrane space is positively charged (Fig. 8–9d). The difference in concentration of protons between the matrix and the intermembrane space represents potential energy. This potential energy results in part from the difference in electrical charge between the two sides of the membrane and in part from the difference in pH.

The inner mitochondrial membrane is impermeable to the passage of protons. The protons can flow back to the matrix of the mitochondrion only through special protein channels in the membrane. In this case the protein channels are formed by the enzyme **ATP synthetase.** As the protons move down the energy gradient, the energy released is used by ATP synthetase to produce ATP.

■ CELLULAR RESPIRATION IS REGULATED BY THE AMOUNT OF ADP AND PHOSPHATE AVAILABLE

Cellular respiration requires a steady input of nutrient fuel molecules and oxygen. Under normal conditions these materials are adequately provided and do not affect the rate of respiration. Instead, the rate of cellular respiration is regulated by the amount of ADP and phosphate available. In a resting muscle cell, for example, ATP synthesis continues until all the ADP has been converted to ATP. Then, when there are no more acceptors of phosphate, phosphorylation stops. Because electron flow is tightly coupled to phosphorylation, the flow of electrons also stops.

When an energy-requiring process like muscle contraction occurs, ATP is split to yield ADP and inorganic phosphate plus energy. The ADP formed can then accept phosphate and energy to become ATP once again. Oxidative phosphorylation continues until all the ADP has again been converted to ATP. Because phosphorylation is tightly coupled to electron flow, the cell possesses a system of control that can regulate the rate of ATP production and adjust it to the momentary rate of energy utilization.

Although the reactions of glycolysis take place in the cytoplasm, those of the citric acid cycle and electron transport occur within the mitochondria. The outer membrane of the mitochondrion engages in active transport and is thought to regulate the intake of materials. Most respiratory enzymes are associated with the inner mitochondrial membrane. It is thought that these enzymes must be exactly located with respect to one another so that hydrogens and electrons may be passed from one to another in the multitudinous and still somewhat bewildering metamorphoses of cellular respiration. No assembly line yet built can match the

TABLE 8–2
The Energy Yield From The Complete Oxidation of Glucose

(1) Net ATP profit from glycolysis		2 ATP* (Substrate level)
Also from glycolysis:	2 NADH $\longrightarrow$	4–6 ATP†
(2) 2 pyruvate to 2 acetyl CoA	2 NADH $\longrightarrow$	6 ATP
(3) 2 acetyl CoA through critic acid cycle		2 ATP (Substrate level)
	6 NADH $\longrightarrow$	18 ATP
	2 FADH$_2$ $\longrightarrow$	4 ATP
(4) Total ATP profit		36–38 ATP

*These are the only 2 ATPs that can be generated anaerobically; production of all other ATPs depends on the presence of oxygen.
†Some energy may be expended to transport NADH across the mitochondrial membrane.

mitochondrial disassembly line for speed and efficiency. What a pity we cannot see it as it must be, vibrating with thousands of elaborate transformations every second, pouring forth streams of ATP, water, and carbon dioxide as it labors along with thousands like it in every one of the trillions of cells of the body.

THE ENERGY YIELD FROM ONE MOLECULE OF GLUCOSE MAY AMOUNT TO 38 ATPs

Just how much ATP can be produced from the complete breakdown of one molecule of glucose is currently a matter of controversy. The net gain of ATP from glycolysis is two ATP per glucose molecule. Two additional ATPs are produced from the substrate-level phosphorylation of the fuel molecule during the citric acid cycle.

Each pair of hydrogens passed through the electron transport chain is thought to provide sufficient energy to produce up to three ATP molecules. The hydrogens transferred from FADH$_2$ provide enough energy for two ATPs. As indicated in Table 8–2, the complete aerobic catabolism of one molecule of glucose is thought to produce a maximum total of 38 ATPs. Recently, some researchers have challenged this figure and have suggested that the actual number of ATPs produced is substantially less.

OTHER NUTRIENTS CAN BE USED AS ENERGY SOURCES

Many organisms depend on nutrients other than glucose, or in addition to glucose, as sources of energy. Humans and many other animals generally obtain more of their energy by oxidizing fatty acids than by oxidizing glucose. Each gram of triacylglycerol contains more than twice as many kilocalories as a gram of glucose or amino acids. Fatty acids are broken down into two-carbon acetyl groups, which combine with coenzyme A and then enter the citric acid cycle (Fig. 8–10).

Amino acids are catabolized by reactions in which the amino group is first removed, in a process called **deamination.** The amino group is converted to urea and excreted. The carbon chain is converted to pyruvate, acetyl CoA, or some other metabolic intermediate that can enter the citric acid cycle.

WHEN OXYGEN IS NOT AVAILABLE MANY CELLS UTILIZE ANAEROBIC PATHWAYS

As we have discussed, in aerobic cellular respiration electrons are removed from fuel molecules during glycolysis and the citric acid cycle and are transferred through a sequence of electron acceptors, the electron transport chain. The final electron acceptor is molecular oxygen. Bacteria and some other types of organisms that inhabit the soil or stagnant ponds where oxygen is in short supply may engage in anaerobic respiration. In **anaerobic respiration** oxygen is not used as an electron acceptor. Instead an inorganic compound, such as nitrate (NO_3) or sulfate (SO_4), serves as the final acceptor of electrons. The end products in anaerobic respiration are inorganic substances (Fig. 8–11).

Certain bacteria adapted to anaerobic conditions utilize an anaerobic pathway referred to as **fermentation.** The end products of fermentation are organic compounds. Under conditions of insufficient oxygen human muscle cells can temporarily utilize a type of fermentation (lactic acid fermentation). Both anaerobic respiration and fermentation depend upon the reactions of glycolysis. Recall that the net profit of two

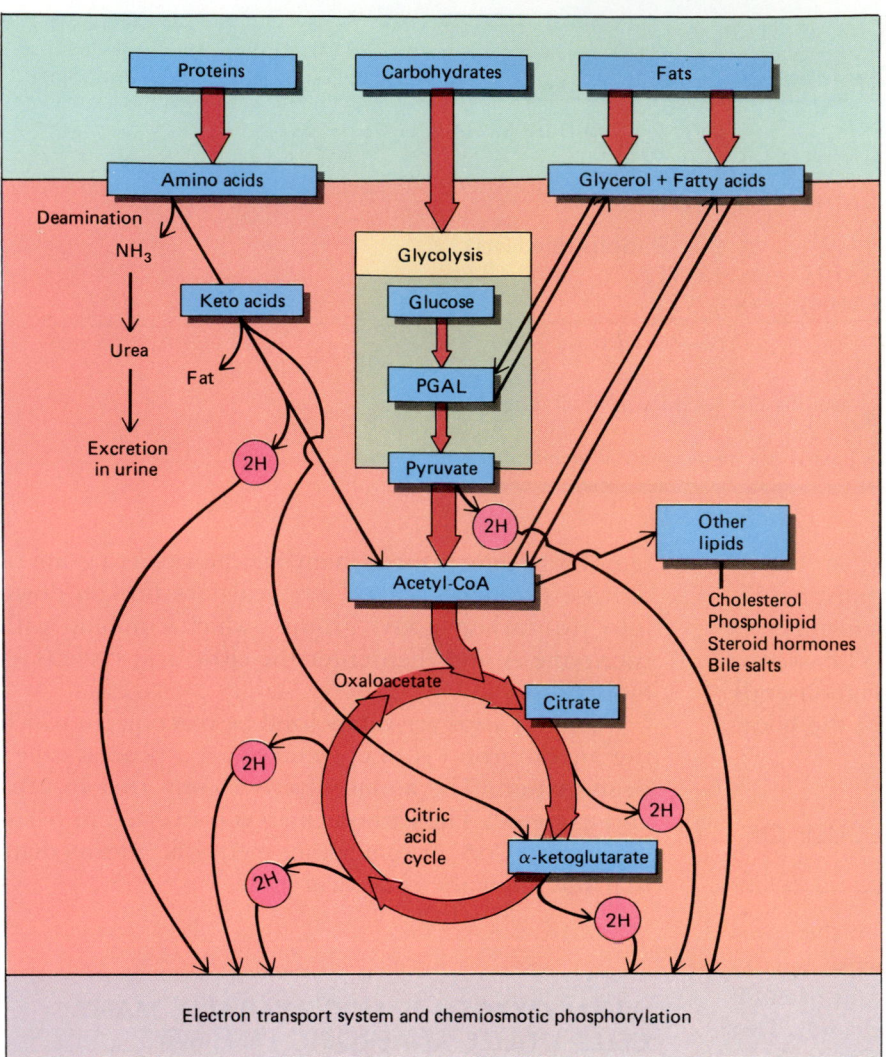

FIGURE 8–10 Carbohydrates, proteins, and fats are all sources of energy for the cell. When these compounds are catabolized, their molecular subunits can be converted to metabolic intermediates that enter glycolysis or the citric acid cycle. This diagram is greatly simplified and illustrates only a few of the principal pathways.

ATPs produced during glycolysis does not require the presence of oxygen.

Most organisms, including the majority of plants and animals, can survive only in an environment that provides sufficient oxygen to support aerobic respiration. These organisms are strict **aerobes.** Anaerobic bacteria that do not require oxygen are referred to as **anaerobes.** Some strict anaerobes are actually poisoned by oxygen. Most versatile are yeasts and certain bacteria that carry on aerobic respiration when oxygen is available but can shift to anaerobic respiration or fermentation when oxygen is in short supply (Fig. 8–12). These organisms are known as **facultative anaerobes.**

In Fermentation The Final Electron Acceptor Is An Organic Compound

In all three pathways—aerobic cellular respiration, anaerobic respiration, and fermentation—glucose or other nutrients are oxidized and hydrogens from them

are transferred to NAD^+. The NAD^+ become reduced to NADH. What happens to these hydrogens is different in each of the three pathways.

In fermentation the final acceptor of electrons from NADH is an organic compound (rather than an inorganic one as in anaerobic respiration). Two common types of fermentation are alcohol fermentation and lactic acid fermentation.

Yeast cells carry on **alcohol fermentation.** When deprived of oxygen, yeast cells split carbon dioxide off from pyruvate to form a compound called **acetaldehyde.** Hydrogen from NADH + H^+ is transferred to the acetaldehyde to produce ethyl alcohol, or drinking alcohol (Fig. 8–13). Such anaerobic reactions are the basis for the production of beer, wine, and other alcoholic beverages. Yeast cells are also used in the baking industry to produce the carbon dioxide that causes dough to rise.

Certain fungi and bacteria and animal muscle cells carry on **lactic acid (lactate) fermentation.** In this pathway, hydrogens removed from the fuel molecule

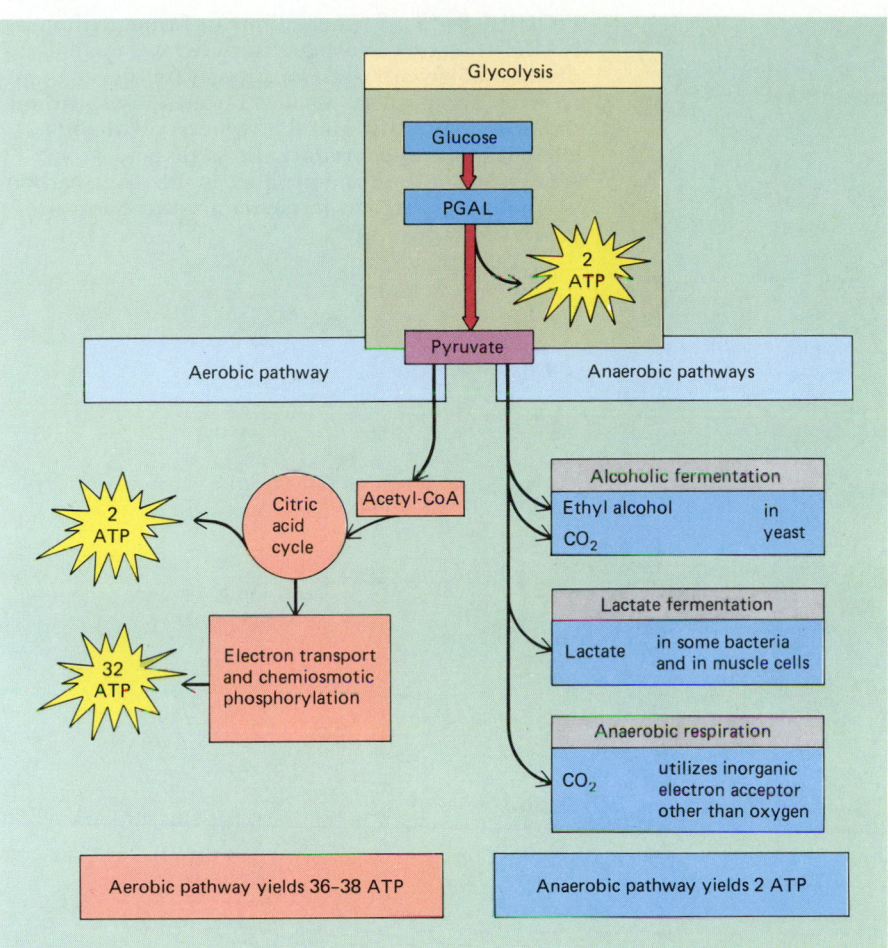

FIGURE 8–11 Comparison of anaerobic with aerobic pathways. Bacteria that inhabit oxygen-poor environments have evolved anaerobic pathways. When oxygen is not available, fermentation occurs in certain bacteria, yeast, and muscle cells.

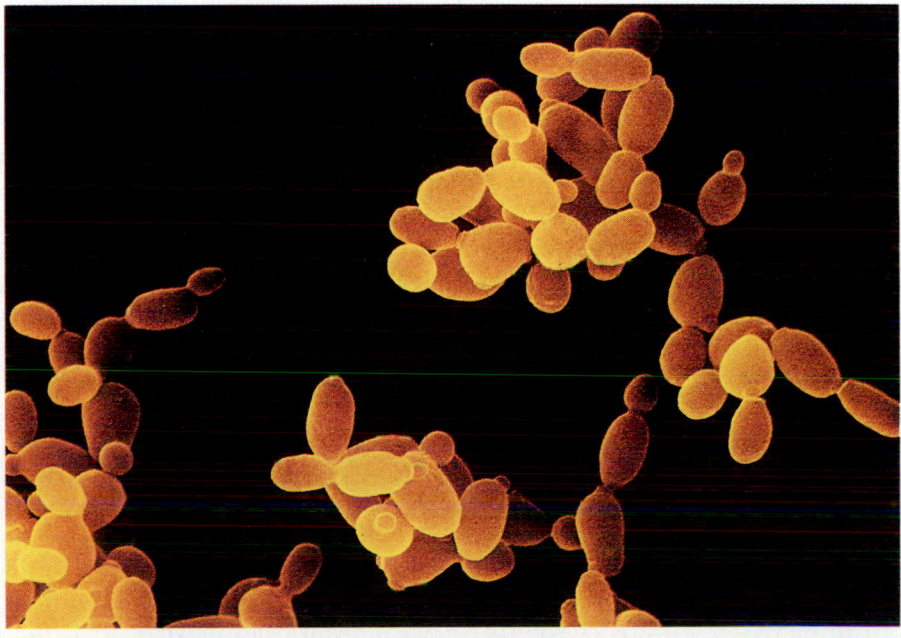

FIGURE 8–12 Yeast cells carry on alcoholic fermentation. (David Phillips/ Visuals Unlimited)

and accepted by NAD^+ during glycolysis are transferred to the pyruvate molecule. When pyruvate is reduced by the addition of hydrogen it becomes **lactate,** the ionic form of lactic acid.

$$Pyruvate + 2H \longrightarrow Lactate$$

Lactate is produced when bacteria sour milk or ferment cabbage to form sauerkraut. In the dairy industry, lactic acid fermentation is used to produce cheese

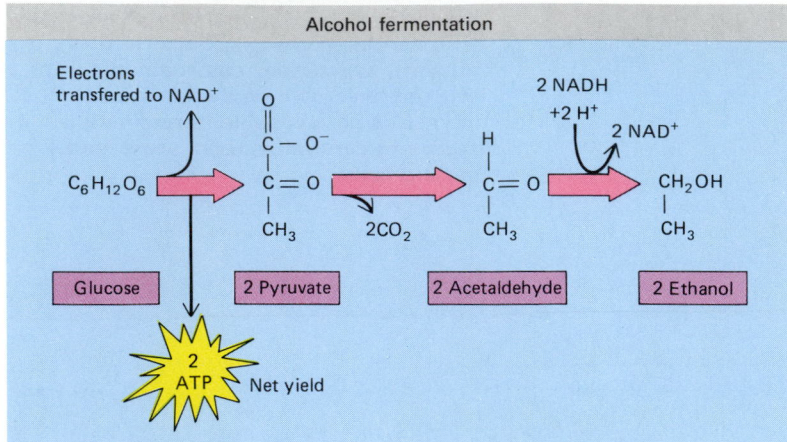

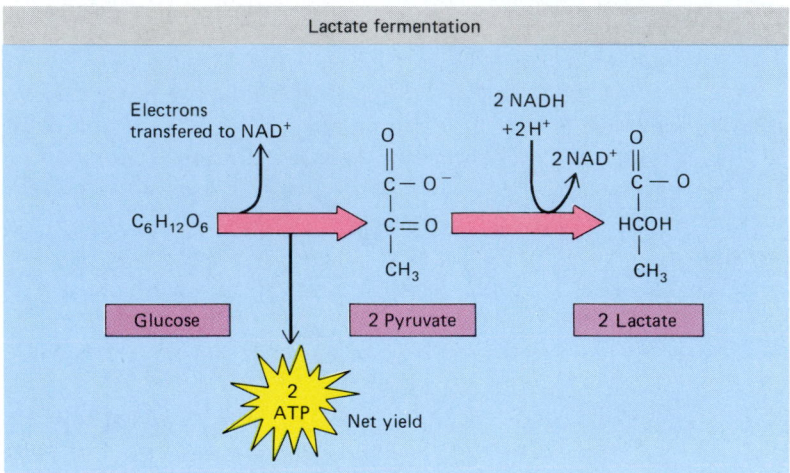

FIGURE 8–13 Fermentation. In fermentation, NADH transfers hydrogen to pyruvate, the end product of glycolysis. Thus, pyruvate serves as an electron acceptor. In alcohol fermentation, carbon dioxide is split off, and the two-carbon alcohol ethanol is the end product. In lactic acid fermentation, the final product is the three-carbon compound lactate. In fermentation there is a net gain of only two ATPs.

and yogurt. Lactic acid fermentation also occurs during muscle activity in the muscle cells of humans and other complex animals.

During strenuous physical activity, such as running, the amount of oxygen delivered to the muscle cells may be insufficient to keep pace with the rapid rate of fuel oxidation. Not all of the hydrogen atoms accepted by NAD$^+$ can be processed in the usual manner through the electron transport chain because there is a shortage of oxygen. In this situation muscle cells can shift from aerobic respiration to lactic acid fermentation. The hydrogens are transferred to pyruvate to form lactate. As lactate accumulates in muscle cells, it contributes to **muscle fatigue.** Gradually, the lactate is transported by the blood to the liver. By running faster than the circulatory system can supply oxygen to the muscles, the athlete incurs an *"oxygen debt"* that must be repaid when the exertion is over. Lactate acidifies the blood, which stimulates respiration, indirectly resulting in rapid breathing. The increased oxygen intake is necessary so that a portion of the lactate can be oxidized, converting it back to pyruvate. This process takes place in the liver cells.

Anaerobic Metabolism Is Inefficient

Anaerobic metabolism is very inefficient compared to aerobic respiration because the fuel is only partially oxidized. Alcohol, the end product of fermentation by yeast cells, can be burned and can even be used as automobile fuel. Obviously, it contains a great deal of energy that the yeast cells were unable to extract using anaerobic methods. Lactate, a three-carbon compound, contains even more energy than the two-carbon alcohol. In contrast, during aerobic respiration all available energy is removed because fuel molecules are completely oxidized.

A net profit of only two ATPs can be produced anaerobically from one molecule of glucose, compared with up to 38 when oxygen is available (Table 8–2). The two ATPs produced during glycolysis represent only about 5% of the total energy in a molecule of glucose. About 55% of the energy in a glucose molecule can be captured by using aerobic pathways. The rest is lost as heat. This level of efficiency is at least double that of the finest machines we can make.

In humans and other endothermic (warm-blooded) animals, body temperatures stay the same regardless of changes in the temperature of the external environment. Some of the heat produced during respiration and other metabolic activities is used to maintain this constant body temperature.

The inefficiency of anaerobic metabolism necessitates a large supply of fuel. By rapidly degrading many fuel molecules a cell can compensate somewhat for the small amount of energy that can be gained from each. To perform the same amount of work, an anaerobic cell must consume up to 20 times as much glucose or other carbohydrate as a cell metabolizing aerobically. Skeletal muscle cells, which often metabolize anaerobically for short periods, must therefore store large quantities of glucose in the form of glycogen.

■ CHAPTER SUMMARY

I. Cells use three different catabolic pathways to extract free energy from nutrients: aerobic cellular respiration, anaerobic respiration, and fermentation.

II. During aerobic cellular respiration a fuel molecule, such as glucose, is oxidized, forming carbon dioxide and water with the release of energy.

III. Aerobic cellular respiration is a redox process in which hydrogen is transferred from glucose to oxygen.

IV. The chemical reactions of cellular respiration can be grouped into four phases: glycolysis, formation of acetyl coenzyme A, the citric acid cycle, and the electron transport system with its associated chemiosmotic ATP synthesis.

V. During glycolysis a molecule of glucose is degraded, forming two molecules of pyruvate.
  A. A net profit of two ATP molecules is gained during glycolysis.
  B. Four hydrogen atoms are removed from the fuel molecule.

VI. In aerobic cellular respiration, pyruvate molecules are combined with coenzyme A, producing acetyl coenzyme A. Carbon dioxide is released during this process.

VII. The four carbons remaining from the original glucose molecule are present in the two acetyl coenzyme A molecules. Acetyl CoA enters the citric acid cycle by combining with a four-carbon compound oxaloacetate, forming citrate, a six-carbon compound.
  A. With two turns of the citric acid cycle the two acetyl CoAs are completely degraded (two carbon-oxygen segments are released as carbon dioxide).
  B. Hydrogens are transferred to NAD^+ or FAD. Only one ATP is produced directly at the substrate level with each turn of the cycle.

VIII. Hydrogen atoms (or their electrons) removed from fuel molecules are transferred from one electron acceptor to another down a chain of acceptor molecules that make up the electron transport system. The final acceptor in the chain is molecular oxygen, which combines with the hydrogen to form water.
  A. According to the chemiosmotic theory, energy liberated in the electron transport chain is used to establish a proton gradient across the inner mitochondrial membrane.
  B. Protons flow back through the membrane through a channel within the enzyme ATP synthetase; energy released in this process is used to synthesize ATP.

IX. In the electron transport system phosphorylation is tightly coupled to electron flow. The rate of ATP synthesis depends upon the availability of ADP and phosphate.

X. Fatty acids and amino acids can also be used as fuel; they are broken down to metabolic intermediates that can enter the citric acid cycle.

XI. Organisms that inhabit oxygen-poor environments use anaerobic pathways for capturing energy.
  A. In anaerobic respiration an inorganic compound such as nitrate serves as the final electron acceptor.
  B. Fermentation is an anaerobic pathway in which electrons from NADH are accepted by an organic compound.
    1. Yeast cells carry on alcohol fermentation in which ethyl alcohol is the final product.
    2. Certain fungi and bacteria, as well as animal muscle cells, carry on lactic acid fermentation in which hydrogen atoms are added to pyruvate to form lactate.
    3. In fermentation there is a net gain of only 2 ATPs per glucose molecule, compared with as many as 38 ATPs per glucose in aerobic respiration.

POST-TEST

1. The process of splitting larger molecules to smaller ones is an aspect of metabolism called _____.

2. An aerobic pathway requires molecular _____.

3. A chemical process during which a substance gains electrons is called _____.

4. The pathway through which glucose is converted to pyruvate is referred to as _____.

5. The reactions of glycolysis take place within the _____, whereas the citric acid cycle takes place within the _____.

6. The hydrogen atoms removed from each PGAL during glycolysis immediately combine with _____ forming _____.

7. When oxygen is present, pyruvate is converted to the two-carbon compound _____.

8. During the citric acid cycle the fuel molecule is completely oxidized; the products are _____, _____, and _____.

9. Acetyl CoA reacts with oxaloacetate to form _____.

10. As electrons are transferred from one acceptor to another in the electron transport chain, they lose much of their _____.

11. The final electron acceptor in the electron transport system is _____.

12. When protons move down an energy gradient in cellular respiration energy is released and used to synthesize _____.

13. The flow of electrons in the electron transport chain is tightly coupled to _____.

14. Yeasts and bacteria that can shift to anaerobic respiration or fermentation when oxygen is in short supply are called facultative _____.

15. The anaerobic process by which alcohol or lactate is produced as a product of glycolysis is referred to as _____.

16. When deprived of oxygen yeast cells transfer hydrogens to pyruvate; the product is _____.

17. During strenuous muscle activity, pyruvate in muscle cells may accept hydrogen; this forms _____ _____.

18. A net profit of only _____ ATPs can be produced anaerobically from one molecule of glucose, compared with _____ which may be produced during aerobic respiration.

19. Anaerobic metabolism is inefficient because the fuel molecule is only partially _____.

REVIEW QUESTIONS

1. What is the specific role of oxygen in the cell? What happens when cells are deprived of oxygen?

2. What aspect of cellular respiration is explained by the chemiosmotic model? How does a proton gradient contribute to ATP synthesis?

3. Justify referring to the mitochondria as the power plants of the cell. Be specific.

4. ATPs are consumed during the first steps of glycolysis. Explain why this occurs.

5. What are the products of glycolysis?

6. In what form is the fuel molecule when it enters the citric acid cycle? What are the products of the citric acid cycle?

7. Trace the fate of hydrogens removed from the fuel molecule during glycolysis when oxygen is present.

8. Trace the fate of hydrogens removed from the fuel molecule when sufficient oxygen is not available in muscle cells.

9. Draw a mitochondrion and indicate the locations of (a) enzymes of the electron transport system and (b) the site of the proton gradient that drives ATP production.

10. Calculate how much energy is made available to the cell by the operation of glycolysis, the citric acid cycle, and the electron transport system; estimate the efficiency of each phase and of the overall process of cellular respiration.

11. Explain the roles in cellular respiration of (a) NAD^+, (b) cytochromes, and (c) oxidative phosphorylation.

12. Some bacteria and fungi live in oxygen-poor environments. How do they obtain energy?

RECOMMENDED READINGS

Alberts, B., et al. *Molecular Biology of the Cell.* Second ed. New York, Garland Publishing, Inc., 1989. Provides an excellent treatment of energy conversions in cells.

Dickerson, E. "Cytochrome c and the Evolution of Energy Metabolism," *Scientific American*, March 1980, 136–153.

Hinckle, P.C., and R.E. McCarty. "How Cells Make ATP," *Scientific American*, March 1978, 104–123. An interesting presentation of the chemiosmotic theory and how it may explain both photosynthesis and oxidative phosphorylation.

Holtzman, E., and A.B. Novikoff. *Cells and Organelles*, 3rd ed. Philadelphia, Saunders College Publishing, 1984. An integrated approach to the structural, biochemical, and physiological aspects of the cell.

Shulman, R.G. "NMR Spectroscopy of Living Cells," *Scientific American*, January 1983, 86–93. Spectroscopy makes it possible to study metabolic processes in intact living cells.

9
Capturing Energy: Photosynthesis

OUTLINE

I. Light exhibits properties of both waves and particles
II. Chlorophyll is a pigment that absorbs light
III. Plants use light energy to make sugar
   A. In light-dependent reactions, light energy is used to make the high-energy compounds, ATP and NADPH + H$^+$
      1. Photosystems I and II are light-harvesting units of chlorophyll molecules
      2. The transport of protons across thylakoid membranes is coupled with the production of ATP
   B. The fixation of carbon dioxide into organic compounds like sugar occurs during light-independent reactions
Focus on photosynthetic efficiency

LEARNING OBJECTIVES

After you have studied this chapter you should be able to:
1. Write a summary reaction for photosynthesis, explaining the origin and fate of each substance involved.
2. Describe the internal structure of a chloroplast.
3. Summarize the events of the light-dependent reactions of photosynthesis, including the role of light in the activation of chlorophyll.
4. Describe how proton gradients allow the formation of ATP according to the chemiosmotic theory.
5. Summarize the events of the light-independent reactions of photosynthesis.

A lupine in the Sierra Nevadas uses solar energy to photosynthesize. (David Cavagnaro)

It is a sweltering July day. The sun beats relentlessly down on acre after acre of corn in a Maryland field (Fig. 9–1). Plants stand motionless as far as the eye can see. The living world is a monotonous green, stretching from one rolling hill to another.

Although it appears that little is taking place in the field, an amazing activity is occurring in every green cell of every corn plant. These plants are converting the sun's energy into a usable chemical form by **photosynthesis.** The chemicals produced by photosynthesis are carbohydrates formed from the simple raw materials, water and carbon dioxide.

The carbohydrates synthesized by photosynthesis are important to the plant for two reasons. First, they are the plant's energy source. That is, they can be broken down by respiration to release energy needed for the plant's life processes. Second, these carbohydrates can be modified in a number of ways to form many different types of biologically important molecules. The proteins, nucleic acids, and lipids that are essential to the plant are largely formed from carbohydrates produced by photosynthesis.

Washington, D.C., is close to the rolling hills of Maryland farms. This large, metropolitan area is filled with millions of people concerned with daily activities that are far-removed from the farm. Yet the existence of all the humans in the nation's capital is as dependent on photosynthesis as are the plants in the corn field. Humans obtain all their food either directly or indirectly from plants. The oxygen we breathe is continually replenished by photosynthesis. And the coal and oil we use to power our technological society represent the products of photosynthesis that occurred millions of years ago. With few exceptions, all life depends on the energy-transforming abilities of photosynthetic organisms.

Plants, algae, and cyanobacteria (or blue-green algae) are producers that obtain their energy from the sun (Fig. 9–2). Each year these remarkable organisms produce more than 200 billion tons of food. The chemical energy stored in this food fuels the chemical reactions that sustain life. Consumers obtain their energy by consuming plants and algae or by eating other consumers that have fed on plants and algae (Fig. 9–3). Almost all life, therefore, is ultimately dependent upon photosynthesis. The only exceptions are a few types of bacteria and certain organisms living in thermal vents in deep sea trenches that metabolize sulfur for their energy requirements.

Since life on our planet depends upon light, it seems appropriate to begin this chapter by discussing the nature of light and how it permits photosynthesis to occur.

KEY CONCEPTS

☐ Almost all living organisms are directly or indirectly dependent on photosynthesis for energy.

☐ Only photosynthetic organisms can convert radiant energy into the chemical energy of food.

☐ In photosynthesis, plants, algae, and cyanobacteria use CO_2 and H_2O to make sugars.

☐ LIGHT EXHIBITS PROPERTIES OF BOTH WAVES AND PARTICLES

Light is a very small portion of a vast, continuous spectrum of radiation, the electromagnetic spectrum (Fig. 9–4). All radiations in this spectrum behave as though they travel in waves. A **wavelength** is the distance from one wave peak to the next. At one end of the spectrum are gamma rays with very short wavelengths (measured in nanometers). At the other end of the spectrum are low-frequency radio waves, with wavelengths so long that they are measured in meters. Within the spectrum of visible light (390 to 760 nanometers), violet has the shortest wavelength and red the longest. Ultraviolet radiation, which is invisible to the human eye, has a still shorter range of wavelengths, and infrared, also invisible, a still longer one.

FIGURE 9–1 Corn and other photosynthetic organisms are able to convert radiant energy to the chemical energy of organic molecules. (Larry Lefever from Grant Heilman)

(a)

(b)

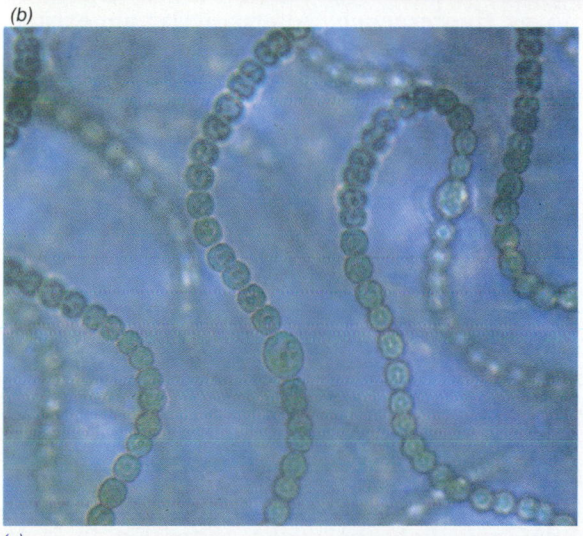

(c)

(a)

(b)

FIGURE 9–3 **The vast majority of living things obtain their energy either directly or indirectly from photosynthesis.** (*a*) **Herbivores are animals that eat plants. The koala, a marsupial from Australia, eats eucalyptus leaves.** (*b*) **Predators are animals that eat herbivores. A Chinese praying mantis (*Tenodera aridifolia sinensis*) consumes insect prey.** (*a* G.R. Roberts; *b* Dwight R. Kuhn)

FIGURE 9–2 **Photosynthetic organisms include** (*a*) **plants (Ponderosa pine, *Pinus ponderosa*), (*b*) algae (*Macrocystis pyrifera*), and (*c*) cyanobacteria (*Nostoc sp*). Plants are primarily terrestrial, while algae are primarily aquatic. Algae may be either microscopic or large seaweeds. Cyanobacteria are prokaryotic organisms that photosynthesize like plants and algae. In addition, there are some photosynthetic bacteria that trap the light's energy in processes different from those of the organisms shown.** (*a, c* Runk/Schoenberger from Grant Heilman; *b* D. Gotshall/Visuals Unlimited)

Light behaves as though it is composed not only of waves but also of discrete energy packets. These particles of energy are called **photons.** The amount of energy in a photon depends on the wavelength of light. The shorter the wavelength, the more energy light has, and the longer the wavelength, the lower the energy (Fig. 9–5). In other words, the energy of the photon is inversely proportional to the wavelength.

Why does photosynthesis depend upon visible light rather than upon some other wavelength of radiation? One reason may be that most of the radiation reaching our planet from the sun is within this portion of the electromagnetic spectrum. Another consideration is that only radiation within the visible light por-

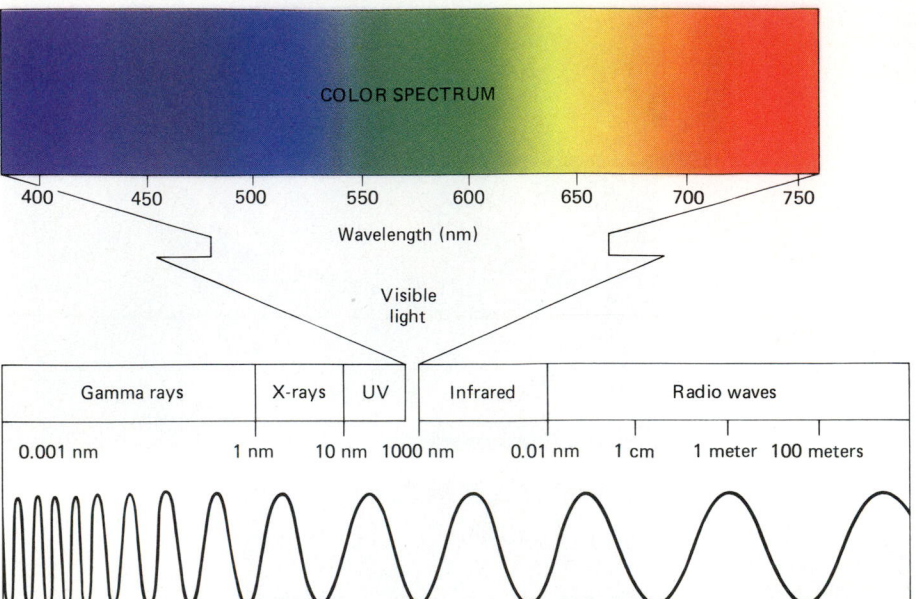

COLOR SPECTRUM

FIGURE 9–4 Visible light is only a portion of the electromagnetic spectrum. Photosynthesis uses light in the visible region.

FIGURE 9–5 Ultraviolet radiation, which is invisible to the human eye, has shorter wavelengths than visible light. It also has more energy per photon than visible light. This energy is so great that it can be destructive to living tissue, as seen by this bad case of sunburn. (Bob Daemmrich/The Image Works)

tion of the spectrum excites certain types of biological molecules, moving electrons into higher energy levels. Wavelengths of radiation that are longer than visible light do not possess enough energy to excite biological molecules. Wavelengths shorter than visible light possess so much energy that they disrupt biological molecules by breaking chemical bonds.

Photons interact with atoms in a variety of ways, but all the ways depend on the arrangement of electrons in the atoms. As you should recall, an atom consists of an atomic nucleus surrounded by one or more energy levels containing electrons. In the hydrogen atom there is only a single electron that occupies the first energy level. The lowest energy state an atom pos-

sesses is called the **ground state,** but energy can be added to an electron so that it will attain a higher energy level. When an electron is raised to a higher energy level than its ground level, the atom is said to be **excited.**

When an electron is raised to a higher energy state it may soon return to its ground level (Fig. 9–6). In this case energy is usually dissipated as heat or as light of a longer wavelength. This type of light production is referred to as **fluorescence.** Alternatively, an excited electron may be lost, leaving the atom with a net positive charge. In this instance the electron may be accepted by a reducing agent (Chapter 7). This is what occurs in photosynthesis.

■ CHLOROPHYLL IS A PIGMENT THAT ABSORBS LIGHT

Chlorophyll is the principal pigment used in photosynthesis. A **pigment** may be defined as any substance that absorbs light. Pigments, which are colored, do not absorb different wavelengths (i.e., colors) of light in the same amounts. Have you ever wondered why most plants are green? The reason is that their leaves reflect most of the green light that strikes them (Fig. 9–7). If green light is reflected, most of it is not being absorbed or used. Chlorophyll, the main pigment plants use in photosynthesis, absorbs light primarily in the blue, violet, and red regions of the spectrum (Fig. 9–8). Actually, there are several kinds of chlorophyll. The most important are chlorophyll *a* and chlorophyll *b*. Plants also have accessory photosynthetic pigments, such as **carotenoids,** that absorb different wavelengths of light.

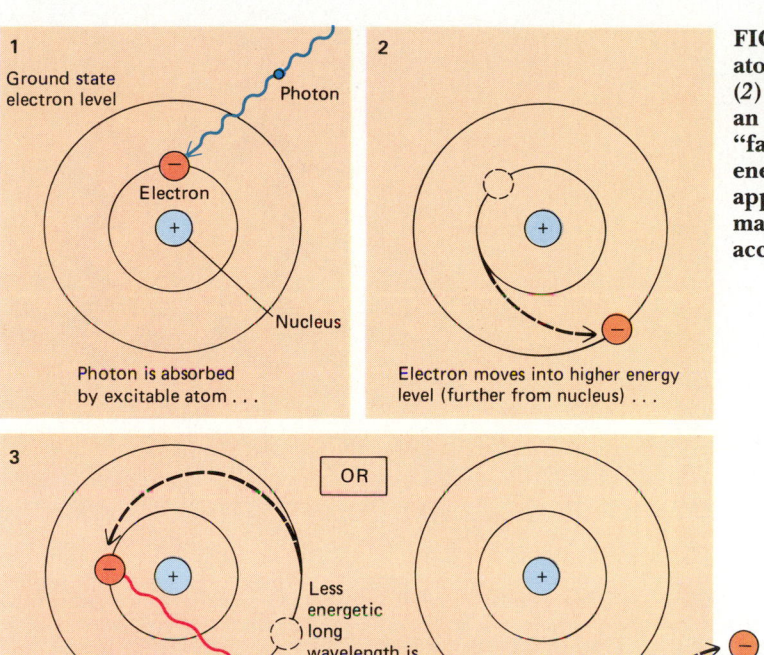

FIGURE 9–6 (*1*) **A photon of light energy strikes an atom or the molecule of which the atom is a part.** (*2*) **The energy of the photon may push the electron to an orbit farther from the nucleus.** (*3*) **If the electron "falls" back to the next lower energy level, a less energetic photon is re-emitted. Alternatively, if the appropriate electron acceptors are available, the electron may leave the atom. In photosynthesis a chain of such acceptors captures the energy of the electron.**

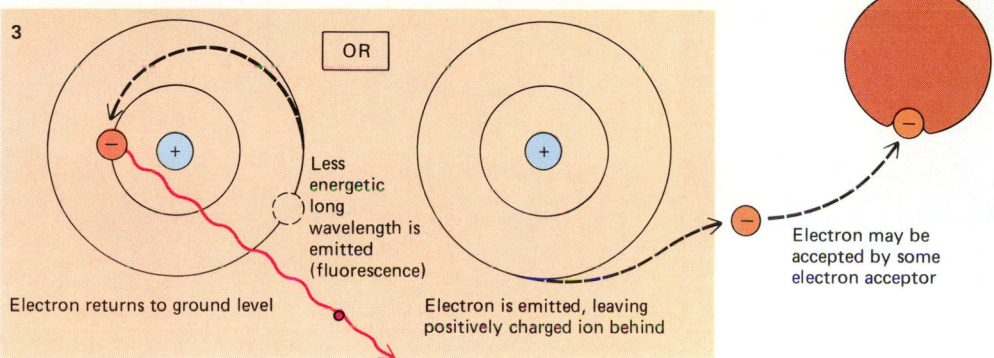

FIGURE 9–7 **Leaves appear green because when white light strikes a leaf, most of the colors comprising the light are absorbed by chlorophyll. However, chlorophyll does not absorb green light strongly; hence, green light is reflected off the leaf to your eye.**

Chlorophyll and other photosynthetic pigments are located within the membranes of **thylakoids,** tiny flattened sacs within cells capable of carrying on photosynthesis. In prokaryotes that photosynthesize, thylakoids often occur as extensions of the plasma membrane and may be arranged around the periphery of the cell. In photosynthetic eukaryotes, thylakoids are found within **chloroplasts** (Chapter 6, Window on the Plant Cell).

Let us examine a chloroplast and its component thylakoids (Fig. 9–10). Surrounded by two membranes, the chloroplast has an interior packed with stacks of thylakoids. These stacks are referred to as **grana.** Each granum looks something like a stack of coins, and each "coin" is a thylakoid (Fig. 9–11). Some thylakoid membranes extend from one granum to another. The region of the chloroplast surrounding the thylakoids is called the **stroma.**

◼ PLANTS USE LIGHT ENERGY TO MAKE SUGAR

The principal raw materials for photosynthesis are water and carbon dioxide. Using the energy that chlorophyll molecules absorb from sunlight, water is split, releasing oxygen and hydrogen (Fig. 9–12). The hydrogen joins with carbon dioxide to form carbohydrates. Although photosynthesis is a complex process composed of many steps, it may be summarized as follows:

Each chlorophyll molecule has a magnesium atom located in the center of a ring structure called a **porphyrin ring** (Fig. 9–9). The magnesium is the part of the molecule that is actually excited by light. The long **phytol** tail of the molecule is a long chain of carbon and hydrogen atoms that give the otherwise flat chlorophyll molecule a shape somewhat like a long-handled skillet.

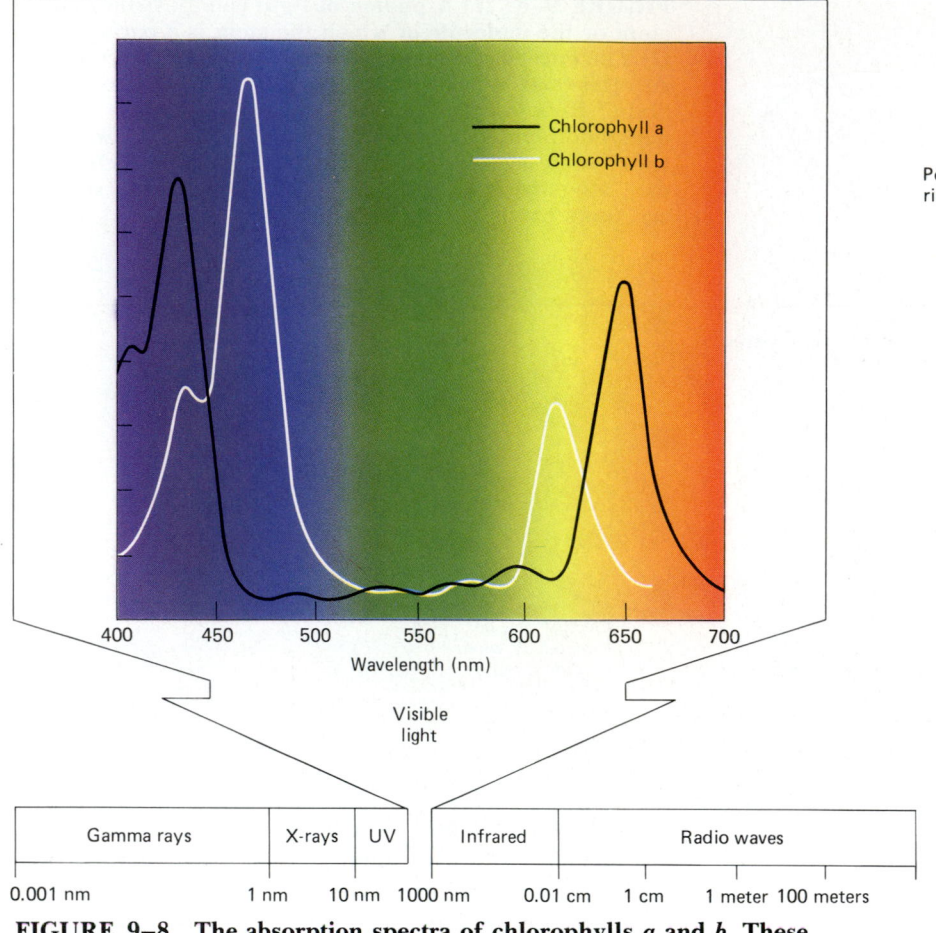

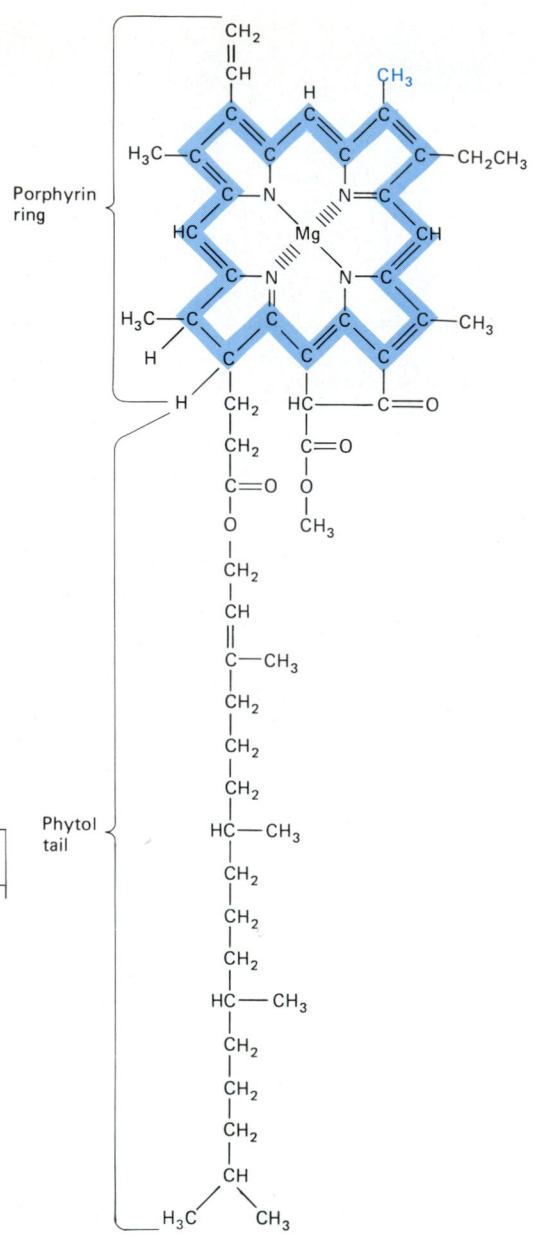

FIGURE 9–8   The absorption spectra of chlorophylls _a_ and _b_. These graphs were obtained by exposing the two types of chlorophyll to different wavelengths of light and measuring how much light was absorbed at each wavelength. Both types of chlorophyll absorb blue and red light strongly and do not absorb green light very strongly. Slight differences in the structures of chlorophylls _a_ and _b_ account for the differences in their absorption spectra.

$$6CO_2 + 12H_2O \xrightarrow[\text{Chlorophyll}]{\text{Light}} C_6H_{12}O_6 + 6O_2 + 6H_2O$$

This equation describes what happens during photosynthesis but not how it happens. Although a detailed examination of photosynthesis is beyond the scope of a beginning biology course, we will describe the principal reactions. For convenience we can divide the reactions of photosynthesis into two parts, the light-dependent and the light-independent reactions.

In Light-Dependent Reactions, Light Energy Is Used To Make The High-Energy Compounds, ATP And NADPH + H$^+$

The light-dependent reactions can take place only in the presence of light. During this phase of photosynthesis several important events occur.

FIGURE 9–9   The chlorophyll molecule is composed of two parts: (_1_) a complex ring structure, called a porphyrin ring, with a magnesium molecule nestled in its center; and (_2_) a long chain of carbons, called a phytol chain. The difference between chlorophylls _a_ and _b_ is slight. At the location on the porphyrin ring where chlorophyll _a_ has a methyl group (blue in the drawing), chlorophyll _b_ has a different group (called an aldehyde group).

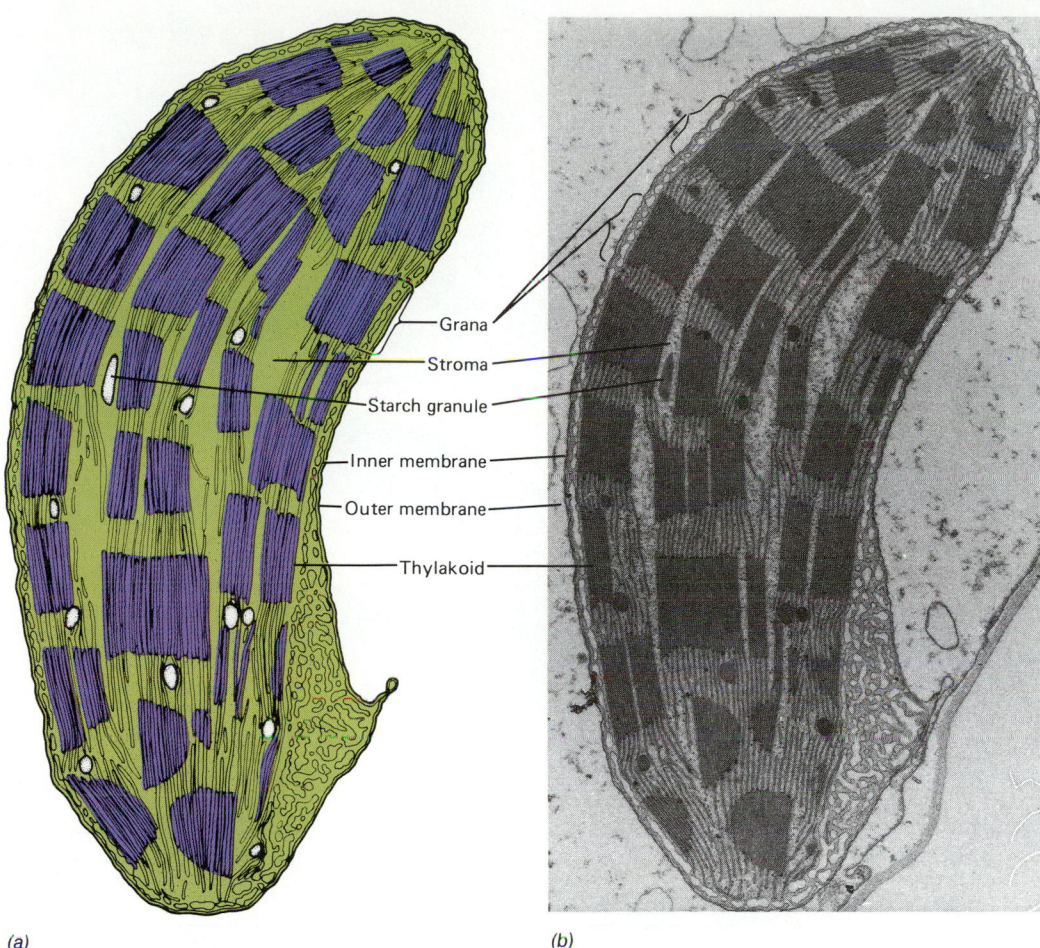

(a)   (b)

FIGURE 9–10   Diagram (a) and electron micrograph (b) of a chloroplast from a leaf cell of a timothy plant, *Phleum pratense*. The interior of a chloroplast is filled with stacks of membranes, called grana, which are interconnected by a network of membranes. The fluid matrix in which the grana are embedded is called the stroma. (L.K. Shumway, Photo Researchers, Inc.)

1. Chlorophyll absorbs light energy, which triggers a flow of excited, or energized, electrons from the chlorophyll molecule.

2. Some of the energy of the energized chlorophyll electrons is transformed to chemical energy and used to make ATP.

3. Some of the light energy trapped by the chlorophyll is used to split water, in a process known as **photolysis.** Oxygen from the water is released. Some of the oxygen is used by the plant for cellular respiration, but most of it is released into the atmosphere.

4. Hydrogen from the water combines with the hydrogen carrier molecule $NADP^+$, forming $NADPH + H^+$ (reduced $NADP^+$). Here again, electrical energy is converted to chemical energy.

We may summarize the light-dependent reactions as follows:

$$12H_2O + 12NADP^+ + 18ADP + 18P_i \longrightarrow$$
$$6O_2 + 12NADPH + 12H^+ + 18ATP$$

Thus, in the light-dependent reactions the energy from sunlight is used to make adenosine triphosphate (ATP) and to reduce $NADP^+$. Some of the captured energy of sunlight is temporarily stored within these two "high-energy" compounds. Note that carbon dioxide is not used in the light-dependent reactions of photosynthesis, nor is sugar produced.

Photosystems I And II Are Light-Harvesting Units of Chlorophyll Molecules

According to the scientific evidence currently available, chlorophyll molecules and associated electron acceptors are physically organized into units called **photosystems.** There are two types of photosystems, each containing up to 400 molecules of chlorophyll. Photosystem I contains a reactive pigment (probably a special form of chlorophyll *a*) known as P700 because it absorbs light with a wavelength of 700 nanometers very strongly. Photosystem II utilizes a pigment, P680,

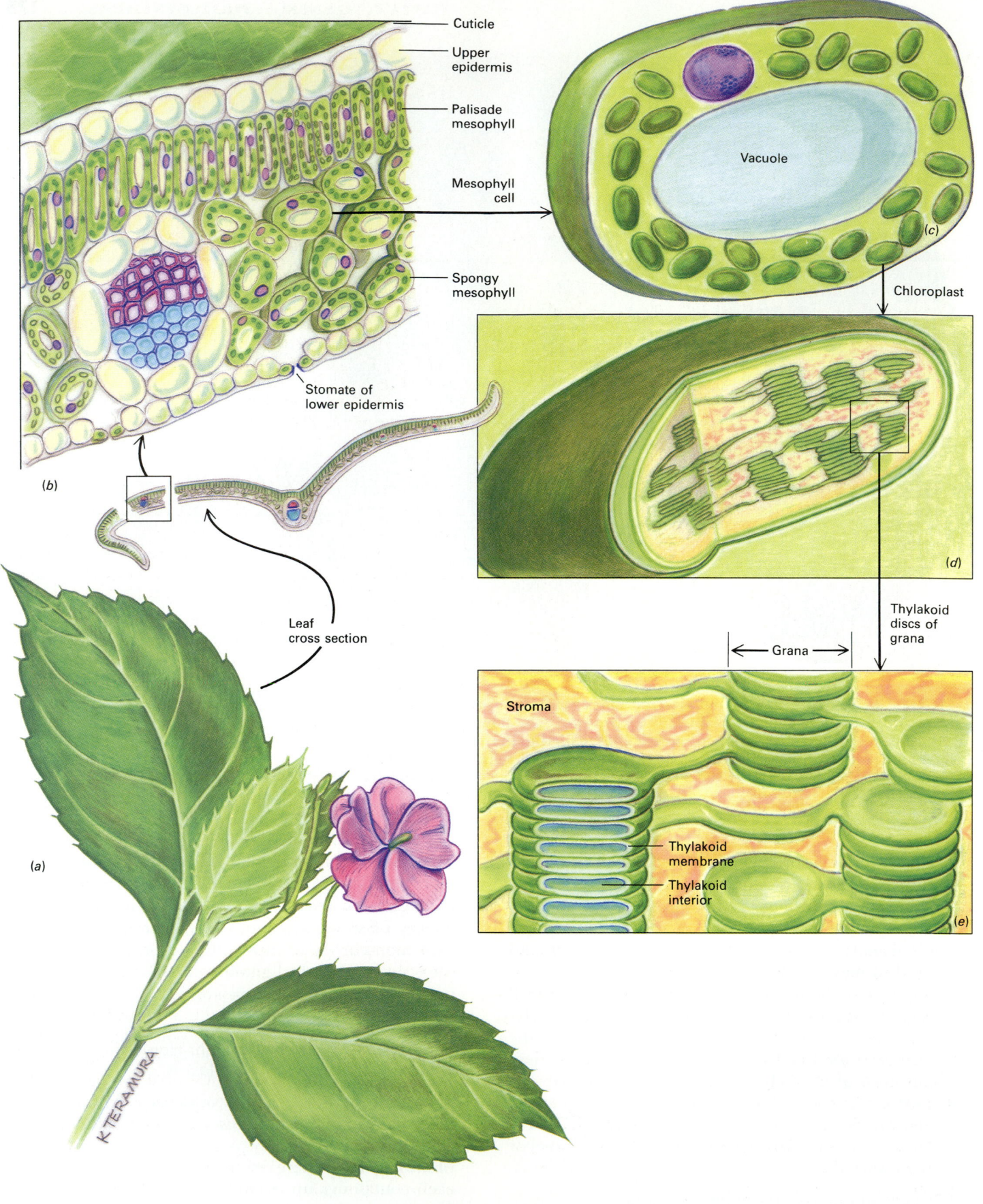

Cuticle

Upper epidermis

Palisade mesophyll

Mesophyll cell

Spongy mesophyll

Stomate of lower epidermis

(b)

Leaf cross section

(a)

Vacuole

(c)

Chloroplast

(d)

Thylakoid discs of grana

Grana

Stroma

Thylakoid membrane

Thylakoid interior

(e)

K. TERAMURA

◀ **FIGURE 9–11 Photosynthesis from different perspectives.** (*a*) Photosynthesis occurs in the green tissues of the plant. (*b*) The cross section of a leaf reveals a structure marvelously adapted for photosynthesis. The middle portion of the leaf, the mesophyll, is the photosynthetic tissue. Carbon dioxide enters the leaf through stomates, and water is carried to the mesophyll in veins. (*c*) A typical mesophyll cell contains numerous chloroplasts. (*d*) Each chloroplast is surrounded by a double membrane. Within the chloroplast, membranous thylakoids are stacked to form grana. The fluid-filled matrix surrounding the grana is the stroma. (*e*) A close-up of the interior of the chloroplast. Chlorophyll is located in the thylakoid membranes. The thylakoids are involved in the light-dependent reactions of photosynthesis, while the light-independent reactions of photosynthesis take place in the stroma.

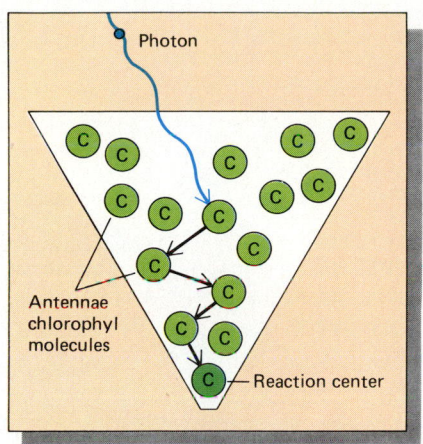

FIGURE 9–13 A photosystem. The many chlorophyll molecules (c) in the unit are excited by photons and transfer their excitation energy to the specially positioned chlorophyll molecule at the reaction center.

FIGURE 9–12 On sunny days the oxygen released by aquatic plants may sometimes be visible as bubbles in the water. This plant (*Elodea*) is actively carrying on photosynthesis, as evidenced by the oxygen bubbles. (E.R. Degginger)

whose absorption maximum is at a wavelength of 680 nanometers.

All chlorophyll molecules of a photosystem apparently serve as antennae to gather solar energy (Fig. 9–13). When they absorb light energy, it is passed from one chlorophyll molecule to another until it reaches the special P700 or P680 pigment molecule, referred to as the **reaction center.** Only this molecule is able to give up its energized electron to an electron acceptor compound.

In photosystem I the energized electron is transferred to several electron acceptors and is finally accepted by NADP+. When NADP+ accepts electrons, the electrons unite with protons present in the chloroplast to form hydrogen, so the reduced form of NADP+ is NADPH + H+. Electrons are restored to photosystem I from photosystem II.

Like photosystem I, photosystem II is activated by a photon and gives up an electron to a chain of electron acceptors. The final electron acceptor is the P700 molecule in photosystem I. The electrons that leave

photosystem II are replaced by electrons from the hydrogen atoms in water. When P680 absorbs light energy, it becomes positively charged and exerts a strong pull on the electrons in water molecules. Water is split (**photolysis**) into its components: protons (H+), electrons, and oxygen. The protons from water are transferred to NADP+, which explains how the NADP+ is converted to NADPH + H+. The oxygen split from the water is released into the atmosphere.

The electron emitted from photosystem II passes from one acceptor to another through a chain of easily oxidized and reduced compounds. As electrons are transferred along this chain of electron acceptors, they become less and less energized. Some of the energy released is used to establish a proton gradient, which leads to the synthesis of ATP (see next section). An electron emitted from photosystem II is eventually donated to photosystem I.

The process just described is known as **noncyclic photophosphorylation** (Fig. 9–14). It is called photophosphorylation because electrons obtain energy from photons of light and then contribute that energy to the phosphorylation of adenosine diphosphate (ADP), producing ATP. In other words, in photophosphorylation the energy of light is used to make ATP. It is noncyclic because there is a one-way flow of electrons from water to NADP+; i.e., $H_2O \rightarrow$ photosystem II $\rightarrow$ photosystem I $\rightarrow$ NADP+. For every two electrons that enter the pathway of noncyclic photophosphorylation, there is an energy yield of two ATP molecules and one NADPH + H+.

The light-dependent reactions of photosynthesis also include **cyclic photophosphorylation,** which is a "shortcut" version of noncyclic photophosphorylation. Only photosystem I is involved in cyclic photophosphorylation. In this pathway the electrons originate from photosystem I and eventually are returned to the

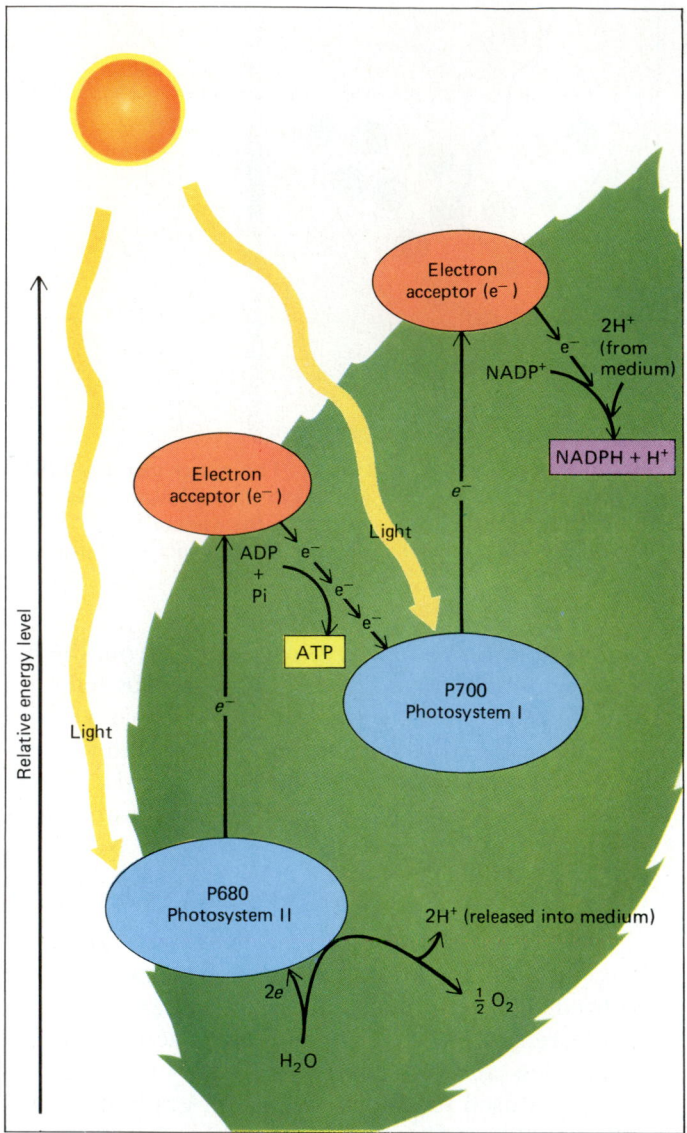

FIGURE 9—14   Light-dependent reactions (noncyclic photophosphorylation). Water is split, producing oxygen, hydrogen ions (protons), and electrons. The flow of energized electrons through photosystems II and I is one way, from the electrons of hydrogen (when water is split) to the formation of NADPH + H⁺. ATP formation is also connected to the electron flow, as the energy released from the energized electrons is used to establish a proton gradient that produces ATP.

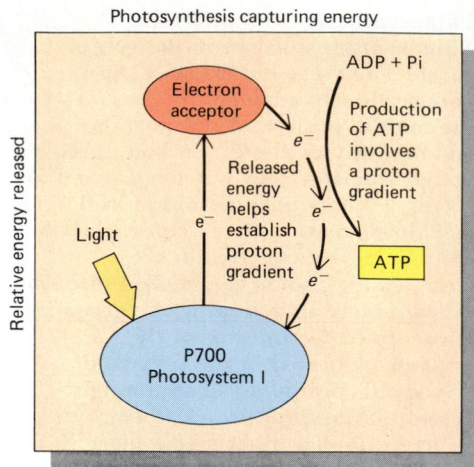

FIGURE 9—15   Cyclic photophosphorylation. Note that only photosystem I is involved. Photolysis does not occur, so neither oxygen nor NADPH + H⁺ is produced.

same photosystem, i.e., they cycle back to photosystem I. For every two electrons that enter the pathway, one ATP molecule is produced. NADPH + H⁺ is not produced, which means that there is a far lower energy yield. Also, water is not split in cyclic photophosphorylation, so oxygen is not produced in this process (Fig. 9–15).

Cyclic photophosphorylation may not occur under normal conditions in photosynthesizing plants. Biologists think that this process was used by ancient bacteria to produce ATP from light energy. A reaction path-

way analogous to cyclic photophosphorylation in plants is present in modern photosynthetic bacteria.

The Transport Of Protons Across Thylakoid Membranes Is Coupled With The Production Of ATP

The photosystems and electron acceptors involved in the light-dependent reactions of photosynthesis are embedded in the thylakoid membrane. As energy is released from electrons traveling through the chain of acceptors, it is used to pump protons (i.e., H⁺ ions) from the stroma across the thylakoid membrane into the center of the thylakoid. Thus, H⁺ ions accumulate within the thylakoids, lowering the pH of the thylakoid interior and making it more acidic. The concentration of positively charged protons inside the thylakoid also creates a difference in electric charge between both sides of the thylakoid membrane. As discussed in Chapter 8, differences in concentration and electric charge, known as concentration gradients and electric gradients, possess energy that can be used to form ATP.

In accordance with the general principles of diffusion, one might expect the highly concentrated hydrogen ions inside the thylakoid to diffuse out. However, they are prevented from doing so because the thylakoid membrane is impermeable to hydrogen ions except at certain points bridged by an enzyme called **ATP synthetase.** This protein extends across the thylakoid membrane, projecting from the membrane surface both inside and outside. It forms a channel through which protons *can* leak out of the thylakoids. As the protons pass through the complex, energy is released, and this energy is tapped by ATP synthetase to form ATP. As you know from Chapter 8, the coupling of

FOCUS ON   Photosynthetic Efficiency

Many plants, including certain agriculturally important crops like soybeans, wheat, and potatoes, do not yield as much carbohydrate from photosynthesis as might be expected. This reduction in yield is especially significant during very hot spells in summer. The reason is an unusual series of reactions involving ribulose bisphosphate, RuBP. You will recall that an enzyme is responsible for CO_2 fixation by attaching CO_2 to RuBP in the Calvin cycle. Under certain conditions, this enzyme binds RuBP to O_2 instead of CO_2. When this occurs, some of the intermediates involved in the Calvin cycle are degraded to CO_2 and H_2O. This process is called **photorespiration** because (1) it occurs during the daylight; (2) it requires oxygen like respiration; and (3) it produces CO_2 and H_2O, like respiration. Unlike normal cellular respiration, however, energy in the form of ATP is not produced. Photorespiration reduces photosynthetic efficiency because it removes some of the intermediates that are used in the Calvin cycle. As a result, CO_2 fixation does not occur as rapidly. The reason plants photorespire is unknown. From our viewpoint, the process is wasteful, particularly when it reduces the yields of important crop plants.

Photorespiration occurs when the concentration of CO_2 is low and the concentration of O_2 is high in the chloroplasts. This would be expected to happen on hot, dry days because these conditions cause water stress in plants. As a result, plants close their stomata, tiny pores in the leaf surface through which gas exchange occurs. The stomata are closed to help the plant conserve water. Once the stomata close, photosynthesis in the chloroplasts rapidly uses up the CO_2 remaining in the leaf. Also, O_2 produced in photosynthesis accumulates in the chloroplasts.

Not all plants have a reduction in photosynthetic efficiency due to photorespiration. Many plants with a tropical origin, including crabgrass, corn, and sugarcane, have evolved mechanisms to bypass photorespiration. One example is C_4 photosynthesis. C_4 plants get their name because the first detectable carbohydrate formed by CO_2 fixation is a 4-carbon compound rather than the 3-carbon compound produced in the Calvin cycle. This 4-carbon compound is then rapidly transported to special cells surrounding the veins of the leaf called **bundle sheath cells**. There the CO_2 is removed from the 4-carbon molecule

Small crabgrass, *Digitaria ischaemum*, a C_4 plant. (Grant Heilman)

and fixed by the Calvin cycle reactions. Photorespiration is negligible in C_4 plants because the concentration of CO_2 in bundle sheath cells is always high. Some scientists are attempting to transfer the C_4 pathway to crops like soybeans and wheat. If this is accomplished, these plants will be able to produce a lot more carbohydrate during hot weather.

ATP synthesis to a hydrogen ion (proton) gradient is called **chemiosmosis** (Fig. 9–16).

The Fixation Of Carbon Dioxide Into Organic Compounds Like Sugar Occurs During Light-Independent Reactions

During the light-dependent reactions of photosynthesis, water is split and light energy is utilized to produce two high-energy molecules, ATP and NADPH + H+. Although these high-energy molecules provide a quick source of energy and reducing power for various metabolic functions of organisms, neither is useful for long-term storage of chemical energy. Cells possess very limited amounts of the precursors from which they are made, so large quantities of either ATP or

NADPH + H+ cannot accumulate. In the light-independent reactions of photosynthesis, sugar is produced from carbon dioxide. This sugar is a source of energy for the cell and, unlike ATP and NADPH + H+, it can be produced in large quantities and stored for future use.

Although the light-independent reactions do not *require* light, they do depend upon the products of the light-dependent reactions. The energy of the ATP and NADPH + H+ produced during the light-dependent phase of photosynthesis is used to "fix" the carbon dioxide of the air into carbohydrate. In other words, in the light-independent reactions of photosynthesis, the energy in ATP and NADPH + H+ is transferred to the bonds of sugar molecules. The process may be summarized as follows:

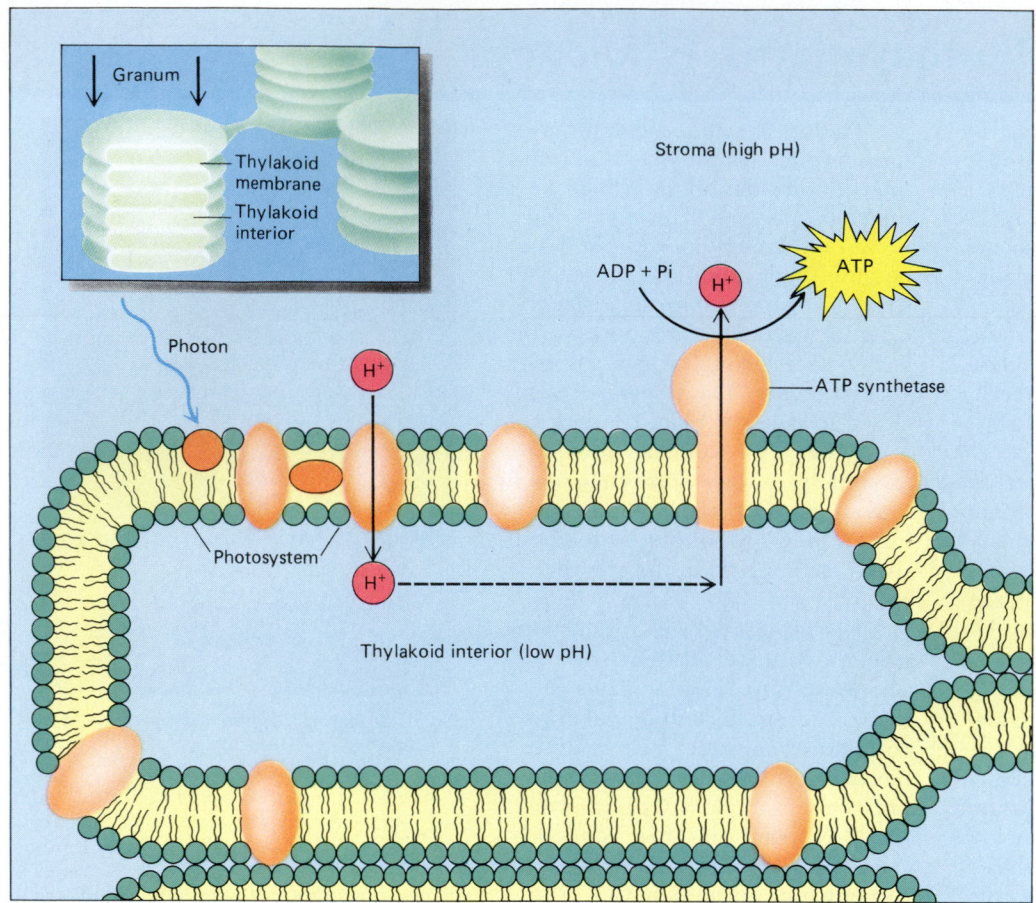

FIGURE 9–16 **The production of ATP by chemiosmosis occurs in the thylakoid membrane. The high concentration of protons in the interior of the thylakoid provides a source of energy (as those protons are transported across the membrane) that is utilized by ATP synthetase to phosphorylate ADP.**

$$6CO_2 + 12NADPH + 12H^+ + 18ATP \longrightarrow$$
$$C_6H_{12}O_6 + 12NADP^+ + 18ADP + 18P_i + 6H_2O$$

The reactions of the light-independent phase proceed by way of a cycle known as the **Calvin cycle.** Melvin Calvin and Andrew Benson at the University of California were able to elucidate the details of this cycle (Fig. 9–17). Dr. Calvin was awarded a Nobel Prize for this significant scientific contribution in 1961.

During the light-independent phase, chemical energy from the ATP and NADPH + H$^+$ produced during the light-dependent phase is transferred to the chemical bonds of carbohydrate molecules. The enzymes for this process are located in the stroma of the chloroplast. Some of the carbohydrate molecules produced during the light-independent phase are used as fuel molecules. Others are used as raw materials to manufacture various types of organic compounds needed by the plant cells, including proteins, lipids, and complex carbohydrates.

As you read the following details of the Calvin cycle, follow the reactions illustrated in Figure 9–18.

The cycle begins with a 5-carbon sugar that has been activated by the addition of a phosphate group, forming **ribulose phosphate.** In the first chemical transformation, ATP from the light-dependent reactions is expended to add a second phosphate to ribulose phosphate. This chemical reaction converts it to **ribulose bisphosphate (RuBP).** A key enzyme, RuBP carboxylase, then combines a molecule of carbon dioxide with RuBP. This is called **CO$_2$ fixation.**

Instantly this 6-carbon molecule splits into two three-carbon molecules called phosphoglycerate (PGA). With the energy from more ATP and hydrogen donated from NADPH + H$^+$, the PGA molecules are converted to **phosphoglyceraldehyde,** known simply as **PGAL.** Keep in mind that the ATP and NADPH + H$^+$ used in this conversion were formed during the light-dependent reactions.

For every six turns of the Calvin cycle, 2 of the 12 PGAL molecules leave the system to be used in carbohydrate synthesis. Each of these three-carbon molecules of PGAL is essentially half a hexose (six-carbon

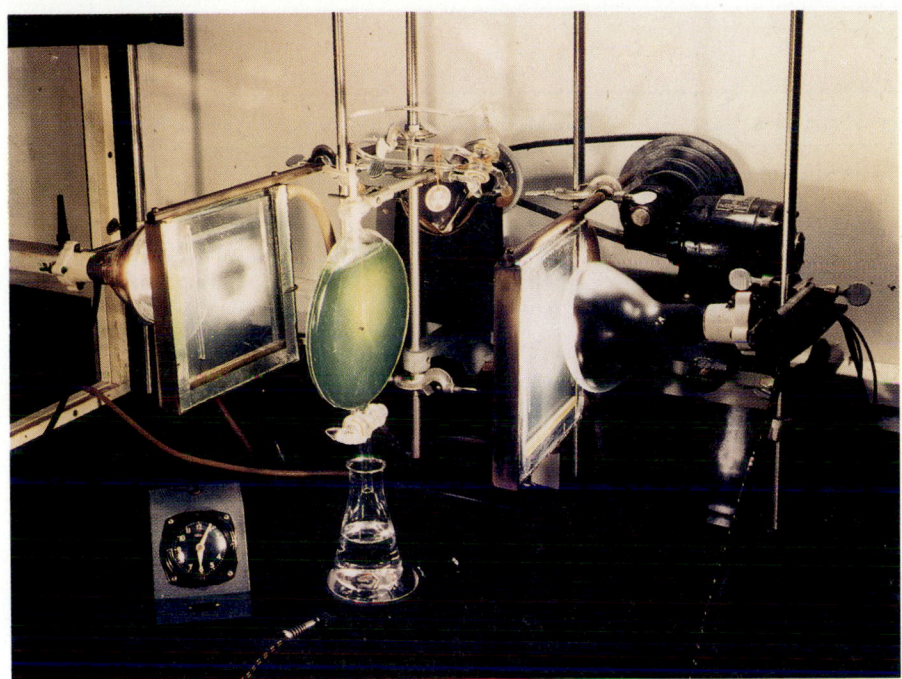

FIGURE 9–17 Calvin's classic experiment that elucidated the steps in the light-independent reactions of photosynthesis, also known as the Calvin cycle. Calvin and his colleagues grew algae in the green "lollipop." $^{14}CO_2$ was bubbled through the algae and they were periodically killed by dumping the "lollipop" contents into a beaker of boiling alcohol. By identifying which compounds contained the ^{14}C at different times, Calvin was able to determine the steps of carbon fixation in photosynthesis. (Melvin Calvin, University of California, Berkeley)

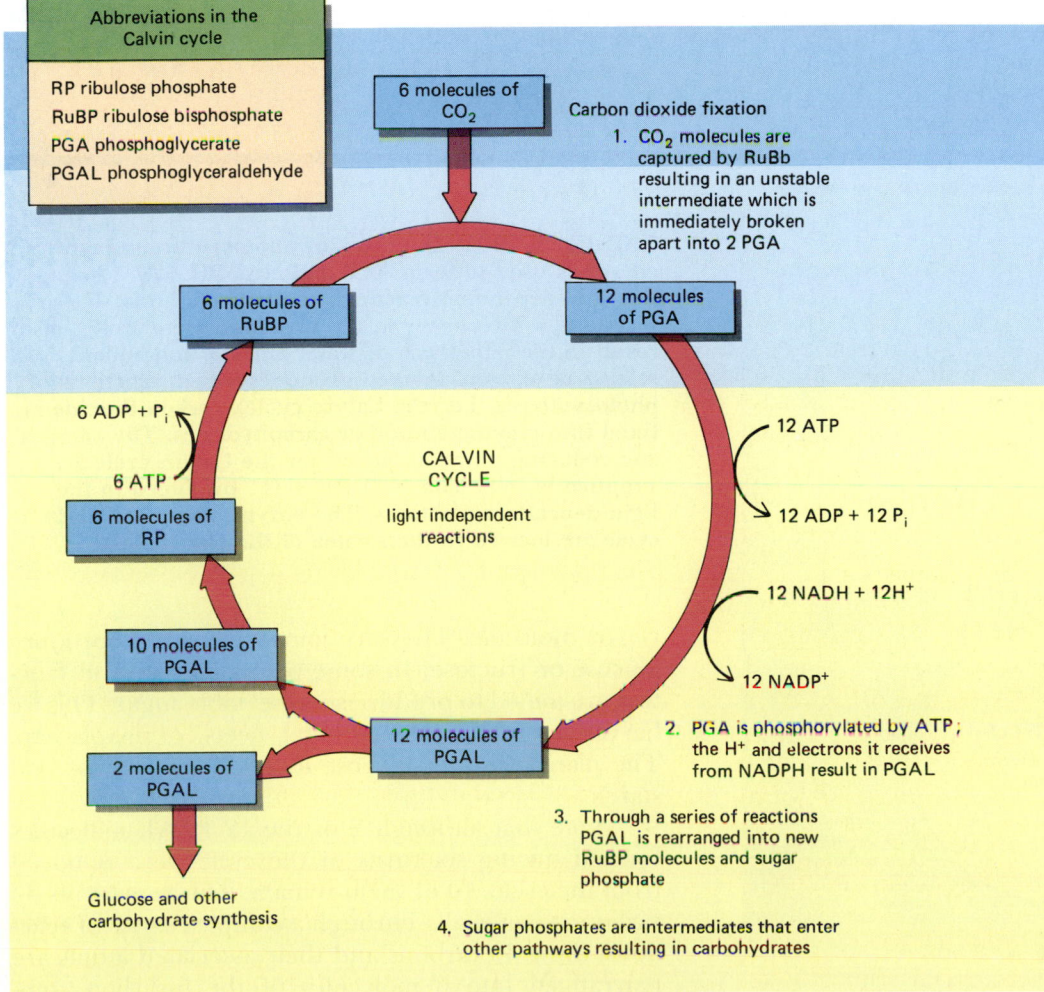

Abbreviations in the Calvin cycle

RP ribulose phosphate
RuBP ribulose bisphosphate
PGA phosphoglycerate
PGAL phosphoglyceraldehyde

6 molecules of CO_2

Carbon dioxide fixation

1. CO_2 molecules are captured by RuBb resulting in an unstable intermediate which is immediately broken apart into 2 PGA

6 molecules of RuBP

12 molecules of PGA

6 ADP + P_i

6 ATP

6 molecules of RP

CALVIN CYCLE

light independent reactions

12 ATP

12 ADP + 12 P_i

12 NADH + 12H^+

12 $NADP^+$

10 molecules of PGAL

2 molecules of PGAL

12 molecules of PGAL

2. PGA is phosphorylated by ATP; the H^+ and electrons it receives from NADPH result in PGAL

3. Through a series of reactions PGAL is rearranged into new RuBP molecules and sugar phosphate

Glucose and other carbohydrate synthesis

4. Sugar phosphates are intermediates that enter other pathways resulting in carbohydrates

FIGURE 9–18 The light-independent reactions of photosynthesis, also known as the Calvin cycle. This diagram depicts six turns of the cycle. For every six turns, six molecules of CO_2 are converted into one molecule of a six-carbon sugar such as glucose. The energy (and reducing power) that drives the Calvin cycle comes from the products of the light-dependent reactions, i.e., ATP and NADPH + H^+.

TABLE 9–1
Summary of Photosynthesis

Reaction series	Summary of process	Needed materials	End products
A. Light-dependent reactions (take place in thylakoid membranes)	Energy from sunlight used to split water, manufacture ATP, and reduce $NADP^+$		
(1) Photochemical reactions	Chlorophyll energized; reaction center gives up energized electron to electron acceptor	Light energy; pigments (chlorophyll)	Electrons
(2) Electron transport	Electrons are transported along chain of electron acceptors in thylakoid membranes; electrons reduce $NADP^+$; splitting of water provides some of H^+ that accumulates inside thylakoids	Electrons, $NADP^+$, H_2O, electron acceptors	$NADPH + H^+$, O_2
(3) Chemiosmosis	H^+ are permitted to move across the thylakoid membrane down proton gradient; they cross membrane through special channels; energy released is used to produce ATP	Proton gradient, $ADP + P_i$	ATP
B. Light-independent reactions (take place in stroma)	Carbon fixation: carbon dioxide is used to make sugar	Ribulose bisphosphate, CO_2, ATP, $NADPH + H^+$, necessary enzymes	Carbohydrates, $ADP + P_i$, $NADP^+$

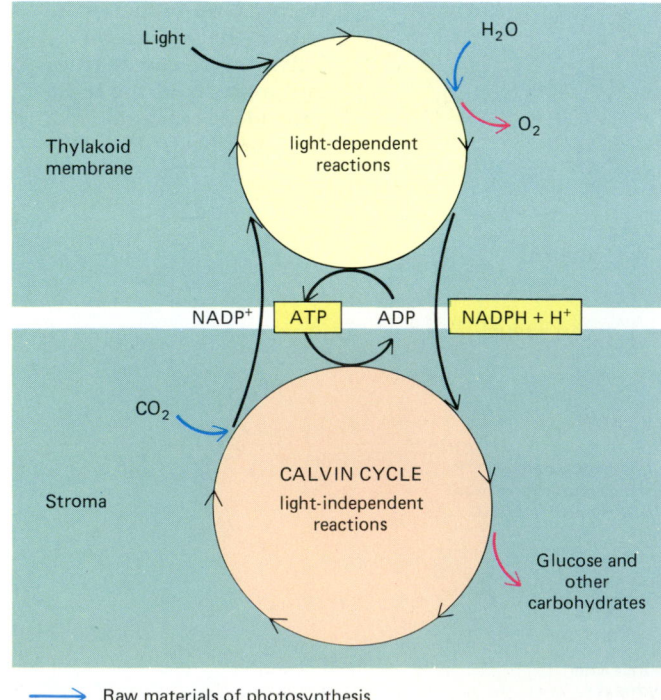

PHOTOSYNTHESIS

FIGURE 9–19 A summary of photosynthesis. Light energy is used to form ATP and $NADPH + H^+$ during the light-dependent reactions of photosynthesis. These reactions, which occur in the thylakoid membranes, also result in the photolysis of water and the subsequent release of oxygen. In the light-independent reactions of photosynthesis (i.e., the Calvin cycle), carbon dioxide is fixed into glucose and other carbohydrates. The energy and reducing power required for the Calvin cycle are supplied by ATP and $NADPH + H^+$ produced in the light-dependent reactions. The enzymes for the Calvin cycle are located in the stroma of the chloroplast.

sugar) molecule. They are joined in pairs to produce glucose or fructose. In some plants glucose and fructose are joined to produce sucrose, table sugar. This we harvest from sugarcane, sugar beets, or maple sap. The plant also uses glucose to produce cellulose and starch.

Note that although 2 of the 12 PGAL molecules formed during six turns of the cycle were removed from the cycle, 10 of them remain. This represents 30 carbon atoms in all. Through a complex series of reactions, these 30 carbons and their associated atoms are rearranged into 6 molecules of the 5-carbon compound, ribulose phosphate, with which we started. Now this same ribulose phosphate is in a position to begin the process of CO_2 fixation and eventual sugar production once again.

In summary, the inputs required for the light-independent reactions are CO_2, NADPH + H$^+$, and ATP (Table 9–1 and Fig. 9–19). The required ATP and NADPH + H$^+$ are provided by the light-dependent reactions of photosynthesis. In the end the CO_2 forms sugar.

CHAPTER SUMMARY

I. During photosynthesis light energy is captured by chlorophyll.
   A. It is used to chemically combine the hydrogen from water with carbon dioxide to produce carbohydrates.
   B. Oxygen is released as a by-product.
II. Light behaves as both a wave and a particle. Its particles of energy, called photons, can excite pigment molecules, such as chlorophyll.
III. Chlorophyll and other photosynthetic pigments are found within the membranes of thylakoids.
   A. In prokaryotes that photosynthesize, thylakoids are found as extensions of the cell membrane.
   B. In eukaryotes thylakoids are found within chloroplasts. The thylakoids are arranged in stacks called grana.
IV. During the light-dependent reactions of photosynthesis electrons of chlorophyll absorb light and become energized. Some of the energy of the energized chlorophyll is used to make ATP, and some is used to split water. Hydrogen from the water is transferred to NADP$^+$, forming NADPH + H$^+$.
   A. Chlorophyll molecules and associated electron acceptors are organized into photosystems. Only the reaction center, a special pigment molecule, actually gives up its energized electron to an electron acceptor compound.
   B. When photosystem I emits an electron, it is ultimately transferred to NADP$^+$.
   C. In noncyclic photophosphorylation the electrons emitted by photosystem II are passed through a chain of electron acceptors. A proton flow is set up that provides the energy for ATP synthesis. Electrons from photosystem II are donated to photosystem I.
      1. As electrons pass through the chain of electron acceptors in photosystem II, protons follow them and accumulate in the thylakoids.
      2. The protons can diffuse out of the thylakoids only through ATP synthetase, a channel in the membrane.
      3. As the protons diffuse through ATP synthetase, energy is released and is used to synthesize ATP.
   D. In cyclic photophosphorylation electrons from photosystem I are eventually returned to the same photosystem; ATP is produced, but no NADPH + H$^+$ is formed.
V. During the light-independent reactions, energy stored within ATP and reduced NADP$^+$ during the light-dependent reactions is used to chemically fix carbon dioxide.
   A. The light-independent reactions proceed via the Calvin cycle.
   B. In the Calvin cycle, carbon dioxide is combined with ribulose phosphate, a five-carbon sugar. With each turn of the cycle, one carbon atom enters the cycle.
   C. Six turns of the cycle result in the synthesis of two molecules of a 3-carbon compound known as PGAL, which combine to produce a molecule of glucose.

POST-TEST

1. The synthesis of food using the energy of light is known as _____ .
2. In photosynthesis light energy is converted to _____ energy in organic compounds.
3. The welfare of the vast majority of animals, including humans, is dependent upon two photosynthetic products, _____ and _____ .
4. Light is composed of particles of energy called _____ .
5. In photosynthesis light is absorbed by _____ .
6. Chlorophyll is located within the membranes of _____ .
7. In photolysis some of the energy captured by chlorophyll is used to _____ _____ .
8. The oxygen liberated by green plants during photosynthesis comes from _____ .

9. Only the reaction center of a photosystem actually gives up its _____.

10. In photophosphorylation the energy from photons is used to add phosphate to _____, producing _____.

11. In _____ photophosphorylation there is a one-way flow of electrons to NADP⁺.

12. In noncyclic photophosphorylation, both ATP and _____ are produced.

13. In cyclic photophosphorylation the electrons that originate from photosystem I are eventually transferred to _____.

14. According to the chemiosmotic theory, energy released from electrons is used to pump _____ across the _____ membrane.

15. As protons pass through ATP synthetase, _____ is synthesized.

16. The inputs for the light-independent reactions are _____, _____, and _____.

17. The enzymes for the light-independent reactions of photosynthesis are located in the _____ of the chloroplast.

18. The light-independent reactions of photosynthesis are also known as the _____ cycle.

19. The process of _____ _____ involves the chemical combination of carbon dioxide with ribulose bisphosphate.

20. _____ turns of the Calvin cycle are required to produce one glucose molecule.

■ REVIEW QUESTIONS

1. What is the role of light in photosynthesis? Explain.
2. Explain the specific role of chlorophyll in photosynthesis.
3. Photosynthetic prokaryotes do not have chloroplasts. How do they manage?
4. Summarize the principal light-dependent reactions of photosynthesis.
5. How is oxygen produced during photosynthesis?

6. Explain how ATP is produced according to the chemiosmotic theory.
7. Summarize the events of the light-independent reactions.
8. Write the overall equation for photosynthesis, including the raw materials and products.
9. It has been said that most of the mass of a tree comes from the air. From what specific substance in air does most of the mass of sugar come?

■ RECOMMENDED READINGS

Hinckle, P.C., and R.E. McCarty. "How Cells Make ATP," *Scientific American*, March 1978, 104–123. An interesting presentation of the chemiosmotic theory and how it may explain both photosynthesis and oxidative phosphorylation.

Miller, K.R. "The Photosynthetic Membrane," *Scientific American*, October 1979, 102–113. Describes the structure of the thylakoid membrane and how it is adapted to convert light energy to chemical energy.

Nakatani, H.Y. "Photosynthesis," *Carolina Biology Reader*, 1988. A clear description of photosynthesis with emphasis on the light-dependent reactions. Burlington, N.C., Carolina Biological Supply Company.

Stryer, L. *Biochemistry*, 2nd ed. San Francisco, W.H. Freeman, 1981. A well-illustrated, readable text that covers the molecules of life and the concepts of cellular energetics from the ground up.

P A R T 4

The Continuity of Life:
Cell Division and Genetics

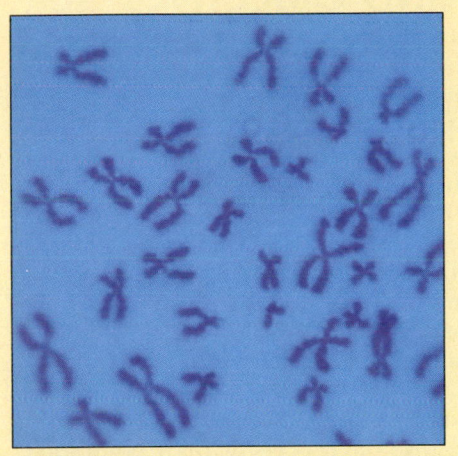
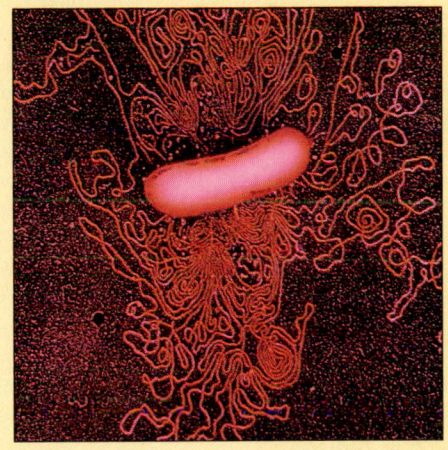
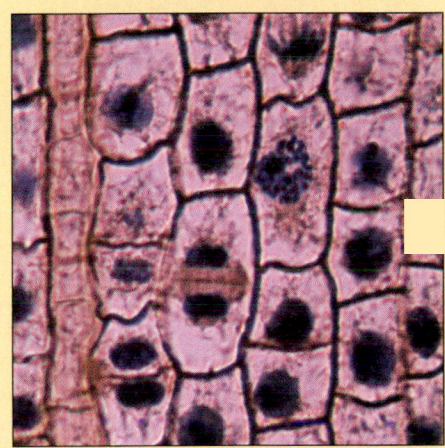
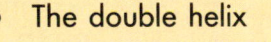

Hereditary information in the nucleus • DNA is composed of nucleotides • The double helix •
Codons and anticodons • Operons in bacteria • Mitosis and meiosis

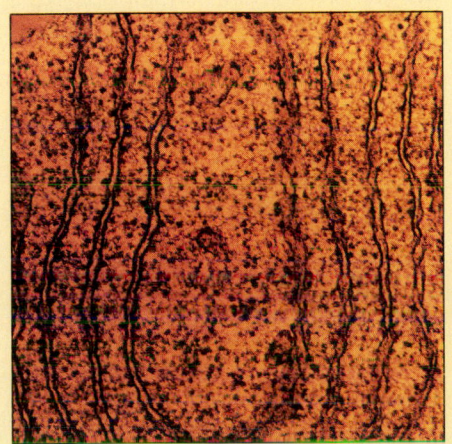

Messenger RNA, transfer RNA, and ribosomal RNA • Multiple alleles cause blood types •
Inheritance and the laws of chance • Monohybrid and dihybrid crosses

10

DNA: The Molecular Basis of Inheritance

OUTLINE

I. The nucleus contains hereditary information
   A. Experiments with the giant *Acetabularia* cell demonstrated the central role of the nucleus
   B. The nucleus controls the cell through a messenger substance
   C. The nucleus alone determines the shape of the *Acetabularia* cell
II. DNA can transfer genetic information from one cell to another
III. Viral nucleic acid contains genetic information to produce new viruses
IV. A variety of methods have uncovered the shape and structure of the DNA molecule
   A. X-ray diffraction provided clues to the shape of the DNA molecule
   B. Watson and Crick demonstrated that DNA is a double helix
   C. DNA is composed of nucleotide subunits
V. DNA replicates semiconservatively

LEARNING OBJECTIVES

After you have studied this chapter you should be able to:

1. Outline the history of the scientific investigations of nucleic acids, giving the major contributions of Hammerling and Brachet, Griffith and Avery, and Watson and Crick.
2. Describe or diagram the basic chemical structure of a nucleic acid strand.
3. Distinguish chemically between DNA and the varieties of RNA.
4. Given the base sequence of one strand of DNA, predict that of a complementary strand of DNA.
5. Summarize the process of DNA replication.

DNA pouring out of *E. coli* bacterium. (K.G. Murti/Visuals Unlimited)

Until recently the deepest secret of life was the mechanism by which organic chemicals came together to form a living organism. A box full of electronic parts, when shaken together, is far more likely to form a color television set than a test tube full of organic molecules is to spontaneously assemble into even the simplest bacterium. According to the cell theory, all cells come from cells (Chapter 5). If so, that bacterium must have inherited the information it needed, first to take shape and then to function, from a parent cell, and, through the parent, from generations upon generations of forebears. Bacteria do not produce amoebas; oak trees do not produce dandelion seeds; nor do people give birth to puppies. It follows that each organism must have within it a set of hereditary instructions characteristic of its species. In turn these instructions can be passed on, making offspring that are of the same species as (and in other ways similar to) the parents. This hereditary information must be copied accurately and delivered to each new cell and to each offspring. Once there, it must express itself by controlling each cell's life and, through the cells, the entire organism.

KEY CONCEPTS

☐ Living things are composed of biological machinery whose plans are stored in their cell's nuclei.

☐ Each eukaryotic cell is controlled by its nucleus.

☐ The cell's nucleus contains DNA, which stores genetic information.

FIGURE 10–1 *Acetabularia crenulata,* the "mermaid's wineglass" seaweed. *Acetabularia* is actually a single cell, utilized in a series of historic grafting studies that demonstrated the control of the cell by its nucleus. (Runk/Schoenberger from Grant Heilman)

☐ DNA is a linear molecule, arranged in the form of a double helix.

☐ DNA is made up of nucleotides. These vary only in the sequence of the bases each contains—adenine, thymine, cytosine, and guanine.

☐ DNA replicates semiconservatively.

☐ THE NUCLEUS CONTAINS HEREDITARY INFORMATION

The blueprint of life must be stored somewhere in each cell, but in what part of the cell? The cell nucleus had long been suspected, because almost all eukaryotic cells have a nucleus and cannot survive long without it (Windows on the Animal and Plant Cells). But how could this be proved?

Experiments With The Giant *Acetabularia* Cell Demonstrated The Central Role Of The Nucleus

In the imaginations of the more romantically inclined biologists, the little seaweed *Acetabularia* resembles a mermaid's wineglass (Fig. 10–1). Less imaginatively, it has been described as a little green toadstool measuring 1 or 2 inches in length. Although it typically grows in tropical seas, it also occurs in some subtropical waters that are both shallow and somewhat rocky. This insignificant underwater organism consists of a single giant cell. A mature *Acetabularia* cell consists of (1) a cup-like cap, (2) a long cylindrical stalk, and (3) a root-like holdfast. The holdfast contains the cell nucleus. There are several species of *Acetabularia*, with caps of different shapes.

If the cap of an *Acetabularia* cell is experimentally removed at the right time in the cell's life cycle, another cap will grow after a few weeks. Such a process, common among lower organisms, is called **regeneration.** This fact attracted the attention of investigators, especially A. Hammerling and J. Brachet, who in the early decades of this century became interested in the possible relationship between the nucleus and the physical characteristics of *Acetabularia*. Because of its great size, they could perform surgery on *Acetabularia* with relative ease. These investigators and their colleagues performed a brilliant series of experiments that in many ways laid the foundation for much of our modern knowledge of the nucleus. In most of these experiments they employed two species of *Acetabularia*, *A. mediterranea*, which has a smooth cap, and *A. crenulata*, with a cap broken up into a series of finger-like projections (Fig. 10–2).

The kind of cap that is regenerated depends upon the species of *Acetabularia* used in the experiment. As

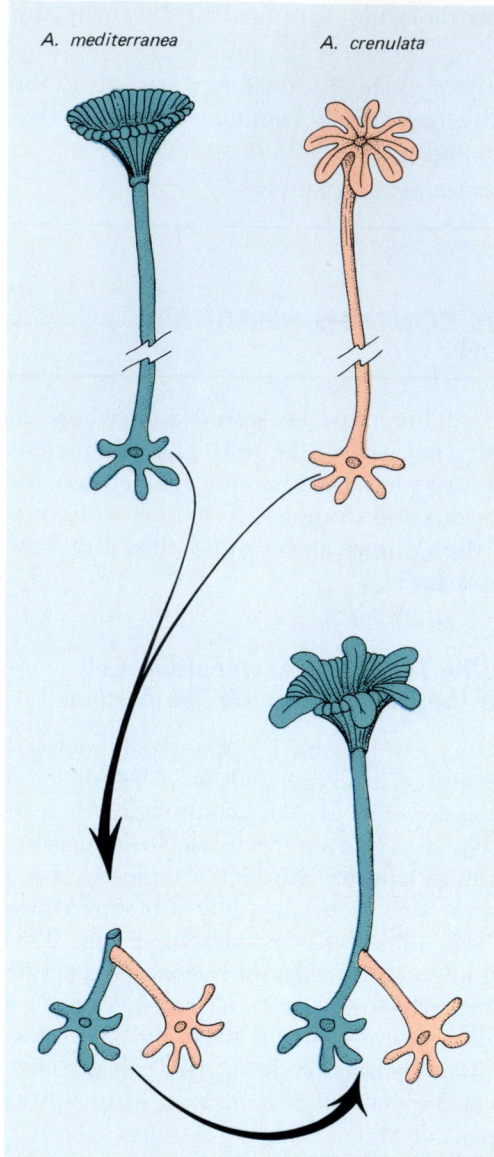

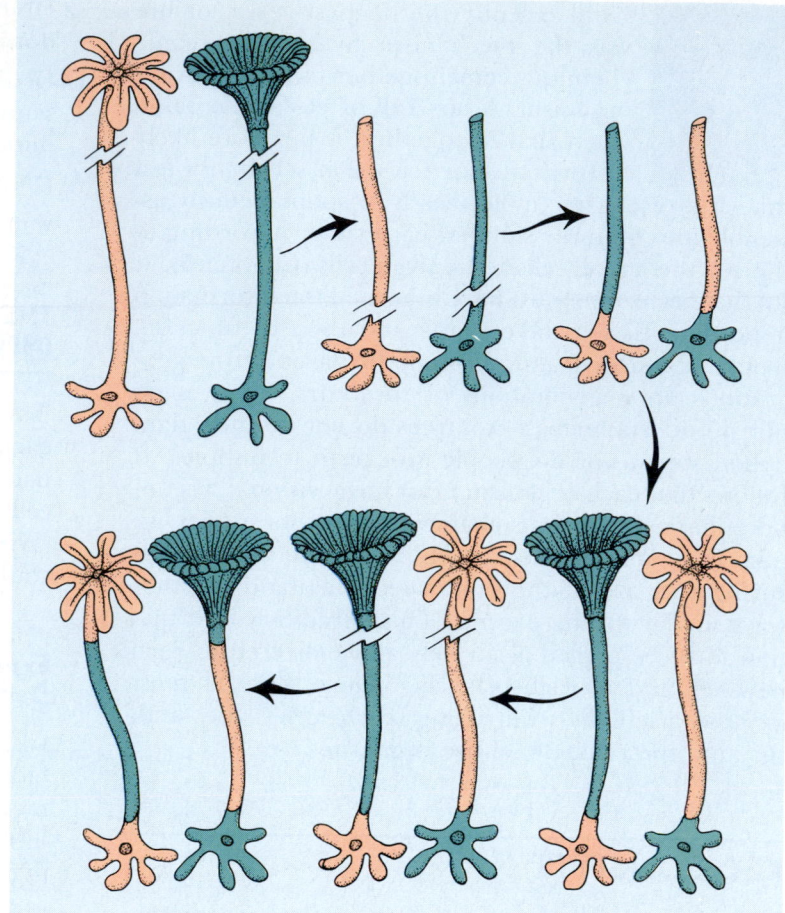

FIGURE 10–3 If the stalks and holdfasts are exchanged, the cap will eventually take the form dictated by the nucleus.

FIGURE 10–2 *Acetabularia crenulata* **has a serrated cap, whereas** *A. mediterranea* **has a smooth cap. If the holdfasts of the two species are grafted together, the resulting cap is intermediate between the shapes which are normal for each species.**

you might expect, *A. crenulata* will ordinarily regenerate a *crenulata* cap, and *A. mediterranea* will regenerate a *mediterranea* cap. But it is possible, if one is very dextrous, to graft two capless organisms of different species together simply by inserting their stalks into one another. After this union, they will regenerate a single cap that has characteristics intermediate between those of the two species involved. Thus, there is evidently something about the lower part of the cell that controls cap shape.

The Nucleus Controls The Cell Through A Messenger Substance

It is also possible to attach a section of one species of *Acetabularia* to the holdfast of another species of *Acetabularia*. First we remove the caps from *A. mediterranea* and *A. crenulata*. Then we sever the stalks from their holdfasts. Finally we exchange the parts. What happens? The caps that regenerate are characteristic not of the species that donated the holdfasts, but of those that donated the stalks. However, if the regenerated caps are removed, the new caps that regenerate will be characteristic of the species that donated the holdfasts (Fig. 10–3). This will continue to be the case no matter how many times the regenerated caps are removed!

From these experiments we may deduce that the holdfast, or something in it, ultimately controls the cell. But what explains the time lag before a new holdfast gains the upper hand? The simplest explanation is that the holdfast produces a temporary messenger substance, which enters the stalk and gains control over the cap. Initially the grafted stalks still contain some of

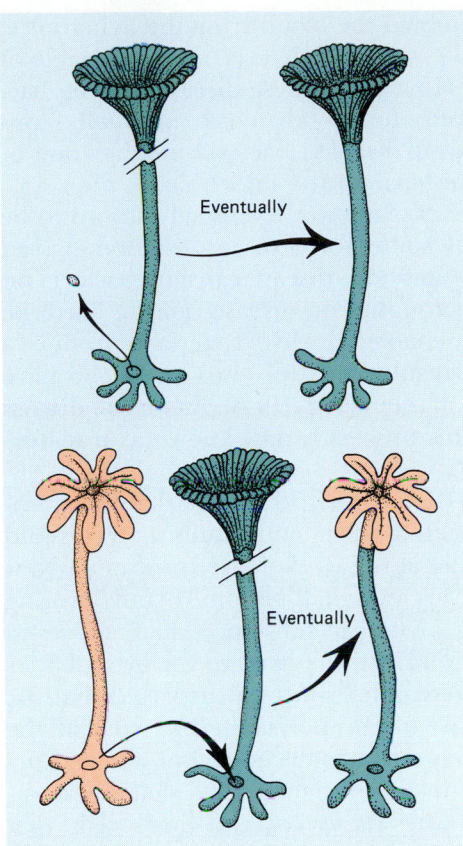

FIGURE 10–4 Closing in on the control of the cell. If nuclei are exchanged, the form of the cap reflects the species of the nucleus.

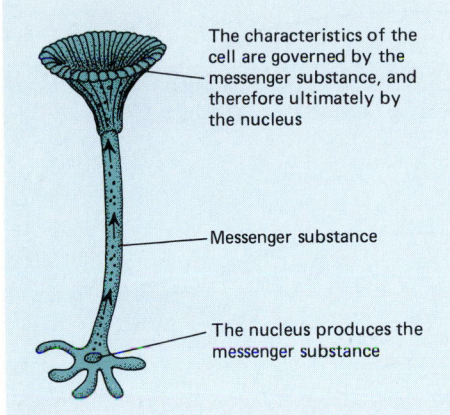

The characteristics of the cell are governed by the messenger substance, and therefore ultimately by the nucleus

Messenger substance

The nucleus produces the messenger substance

FIGURE 10–5 How *Acetabularia* is controlled.

the messenger substance produced by their old holdfasts, so they regenerate caps with the old shape. After this brief delay, the new holdfasts begin sending their messenger substances into the stalks and thus begin directing the stalks to grow caps with the new shapes. (The messenger substance, today known to be messenger ribonucleic acid or mRNA, will be discussed in the next chapter.) But this still leaves us with the question of just what it is about the holdfast that accounts for its authority. An obvious candidate is the nucleus, which these observations seem to indicate also produces the messenger substance.

The Nucleus Alone Determines The Shape Of The *Acetabularia* Cell

If the nucleus is removed and the cap cut off, a new cap will regenerate. *Acetabularia*, however, is usually able to regenerate only once without a nucleus. If the nucleus of an alien species is now inserted and the cap is cut off once again, the new cap will be characteristic of the species that contributed the nucleus (Fig. 10–4).

There is only one reasonable explanation for these observations: The control of the cell exerted by the

holdfast is attributable to the nucleus that is located there. The information *Acetabularia* provided (Fig. 10–5) helped to pave the way for the important findings of other researchers who discovered the roles of the nucleic acids—deoxyribonucleic acid (DNA) and RNA—in heredity and cell function.

■ DNA CAN TRANSFER GENETIC INFORMATION FROM ONE CELL TO ANOTHER

In 1928 the British investigator Frederick Griffith reported the results of what then seemed to be a very eccentric experiment. Bacteriologists had long known that a type of bacteria—*Streptococcus pneumoniae*—causes pneumonia and certain other infections. When grown on special dishes containing a laboratory food substance, these bacteria develop colonies with a smooth, glistening appearance. The appearance results from a polysaccharide coating, or **capsule,** that surrounds each bacterium. Inside the body of a human being or other mammal, this coating protects the microorganisms from the immunological defenses of the host body, allowing the bacteria to stay alive and cause infections.

Researchers also knew that certain weak types of *S. pneumoniae* do not have this polysaccharide coating. Without this protection, they are vulnerable to immunological attack and cannot cause disease. When grown in the laboratory, they form characteristic "rough colonies," since they lack capsules.

Pursuing an investigation with no obvious relationship to DNA, Griffith injected both types of bacteria—smooth, which could cause disease, and rough, which could not—together into mice. The smooth bacteria that he injected had already been killed. Presumably they could not cause disease. Neither should the rough bacteria have been able to produce infection,

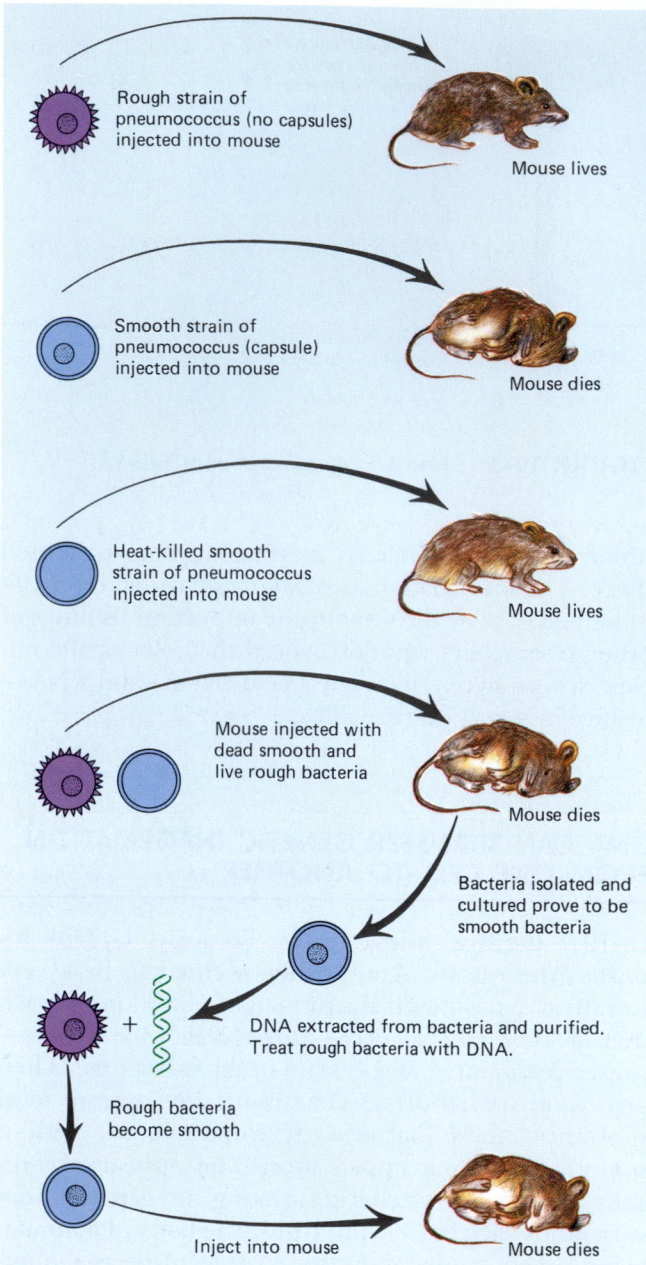

Rough strain of pneumococcus (no capsules) injected into mouse

Mouse lives

Smooth strain of pneumococcus (capsule) injected into mouse

Mouse dies

Heat-killed smooth strain of pneumococcus injected into mouse

Mouse lives

Mouse injected with dead smooth and live rough bacteria

Mouse dies

Bacteria isolated and cultured prove to be smooth bacteria

DNA extracted from bacteria and purified. Treat rough bacteria with DNA.

Rough bacteria become smooth

Inject into mouse

Mouse dies

FIGURE 10–6   The experiments of Frederick Griffith and Oswald Avery. Griffith demonstrated the transfer of genetic information from dead, heat-killed bacteria to living bacteria of a different strain. Although neither the rough strain of bacteria nor the heat-killed smooth strain could kill a mouse, a combination of the two did. Autopsy of the dead mouse showed the presence of living, smooth-strain bacteria. Later, Avery demonstrated the restoration of virulence (that is, the ability to cause disease) to rough bacteria by treating them with DNA from smooth bacteria—proving that it was the DNA that carried the genetic information necessary for the bacterial transformation.

because they lacked a protective coating and would not survive long in the mouse body. The injection as a whole should have been harmless. Instead, many of the mice died.

Griffith examined the blood from the dying mice. In it he found live, smooth bacteria, clearly capable of causing disease. How could these disease-causing bacteria have entered the blood, which originally contained only dead or disabled bacteria, neither one of which could have harmed the mice?

One possible explanation, eventually shown to be correct, was that some form of information passed from the dead, smooth, disease-causing bacteria to some of the live, rough, non-disease-causing bacteria. This information enabled the live bacteria to produce a polysaccharide capsule and thus survive in the mice long enough to kill them (Fig. 10–6). As we will discuss in Chapter 21, this process is now known as **transformation.**

In the early 1940s, Griffith's remarkable experiments were repeated with modifications by Oswald Avery, a researcher at the Rockefeller Institute of New York City. Avery reasoned that some "transforming principle," some chemical substance, had conveyed genetic information from the dead to the living bacteria. He and his research team first suspected that the substance was part of the polysaccharide coat of the bacteria, which was a reasonable idea, but wrong. Further experiments by these scientists revealed that it was something from the *interior* of the dead cells that caused the capsule to develop. They found that they could produce purified samples of DNA from the disease-causing pneumococcal bacteria. These DNA samples were capable of turning non-disease-causing bacteria into bacteria capable of producing vigorous infections. Their revolutionary conclusion was that DNA was the substance that carried genetic information.

■ VIRAL NUCLEIC ACID CONTAINS GENETIC INFORMATION TO PRODUCE NEW VIRUSES

Another class of disease-producing agents, the tiny viruses, actually lack most of the traits of living things; they do not metabolize, grow, or move by themselves. They can, however, reproduce, with the aid of living cellular hosts. This "life-style" is necessarily parasitic, so all viruses probably produce at least some degree of disease in the course of their existence.

All cellular life seems to be susceptible to viral infection, including bacteria. Viruses that attack bacteria are called **bacteriophages,** or **"phages"** for short. They are very simply constructed, consisting of a protein coat surrounding a nucleic acid core. Chance events bring the virus and its host together. The phage then attaches to its host, penetrates the cell's outer coverings, and takes over the bacterium's metabolic machinery. The phage causes this machinery to replicate viral genetic material and thus produces more phages like itself. In other words, the phage forces the bacterium host to follow the genetic directions that the phage injects into it. Eventually, what remains of the host cell bursts open, releasing a swarm of phage particles. Some of these may come in contact with another host cell and then repeat the process.

It was not known before 1952 whether the nucleic acid core or the protein coat (or even both) carried phage information into the bacterial cell, but Alfred Hershey and Martha Chase settled the question in that year. Hershey and Chase worked with *Escherichia coli*, the intestinal bacterium, and a phage (known as "T2") that attacked it. They knew, as you know from reading Chapter 4, that the phage's DNA core contained phosphorus and that it also contained no sulfur. They also knew that the phage's protein coat contained sulfur but little phosphorus. They reasoned that if they could find some sulfur from a phage that had gotten inside a bacterial cell, they would know that the phage's protein coat, not its DNA core, had entered the cell. On the other hand, if they found phosphorus inside the bacterium, they would know that the phage's DNA core, not its protein coat, had entered the cell. Whatever substance entered the cell was responsible for replicating the virus.

Hershey and Chase prepared a batch of phage whose protein coats contained radioactive sulfur (^{35}S). This served as a tracer, allowing them to detect even small quantities of the protein. They also prepared a batch of phage whose DNA core contained radioactive phosphorus (^{32}P). When the phage whose protein was labeled with radioactive sulfur was added to a fresh culture of bacteria, the bacteria did not become radioactive. This showed that the protein did not enter the bacteria. However, when the phage whose DNA was labelled with radioactive phosphorus was added to a bacterial culture, the bacteria did become radioactive. This showed that viral DNA entered the bacteria. Since all possibilities had now been investigated, it was clear that DNA had to be the material that directed the production of new viruses. DNA,[1] not protein, was the genetic material.

■ **A VARIETY OF METHODS HAVE UNCOVERED THE SHAPE AND STRUCTURE OF THE DNA MOLECULE**

Much of the progress in analyzing the structure of macromolecules (very large molecules) has come from the application of a sophisticated form of photography called **x-ray diffraction analysis.** However large a mol-ecule such as DNA may be, until recently one could not reasonably expect to see it in detail. Even the most powerful and sophisticated electron microscopes have unsurpassable limits set by the lengths of the waves of radiation by which they operate.[2]

X-Ray Diffraction Provided Clues To The Shape Of The DNA Molecule

When x-rays of extremely short wavelength are passed through a crystal, they are scattered by the atoms of the crystal somewhat as light rays are scattered by particles of dust or water in a fog. Because of the regularity of the atomic arrangement in a crystal, the scattering of the x-rays is itself regular rather than diffuse. The scattered radiation interferes with itself, producing a characteristic pattern that can be photographed— a kind of molecular fingerprint distinctive for any crystalline substance. The photograph, however, is no more an image of its molecules than a fingerprint is an image of a person. Still, the mathematical analysis of such patterns obtained at different angles of exposure enables the researcher to obtain clues to the shape of the molecule (Fig. 10–7).

Maurice Wilkins and Rosalind Franklin had already employed x-ray diffraction (mostly in the early 1950s) to investigate the structure of DNA when James Watson and Francis Crick became friends at Cambridge University. Watson, who had recently obtained his PhD from Indiana University, wished to gain experience in the rapidly developing new field of molecular biology. Crick was a doctoral candidate in the field of protein structure. Viewing Franklin's observations from their own perspective, Watson and Crick concluded that DNA must be a helical (spiral) molecule. Although inferring the helical structure of DNA was a difficult piece of work, this was not the most challenging part of what they set out to accomplish. The hard part was to demonstrate that the helical shape was consistent with DNA's chemical makeup.

Watson And Crick Demonstrated That DNA Is A Double Helix

It seemed clear that the backbone of the DNA molecule consisted of alternating sugar and phosphate units. Attached somehow to this backbone were four kinds of bases—**adenine, thymine, guanine,** and **cytosine.** Erwin Chargaff of Columbia University had shown that the amount of thymine in a sample of DNA

---

[1] In some viruses the nucleic acid RNA serves to store genetic information, but this is, of course, also a nucleic acid.

[2] Recently developed "tunnelling" microscopes do not have this limitation. Actual images of DNA can now be produced with these microscopes. But tunnelling microscopes were undreamed of in the 1950s when the basic structure of DNA was discovered.

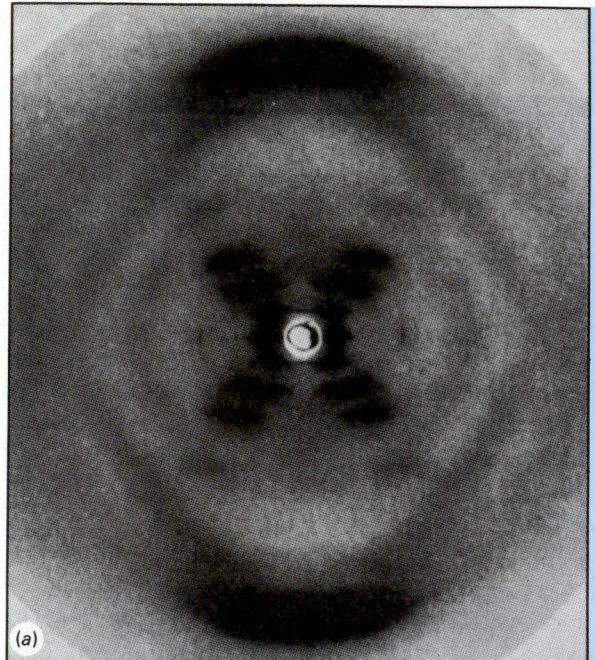

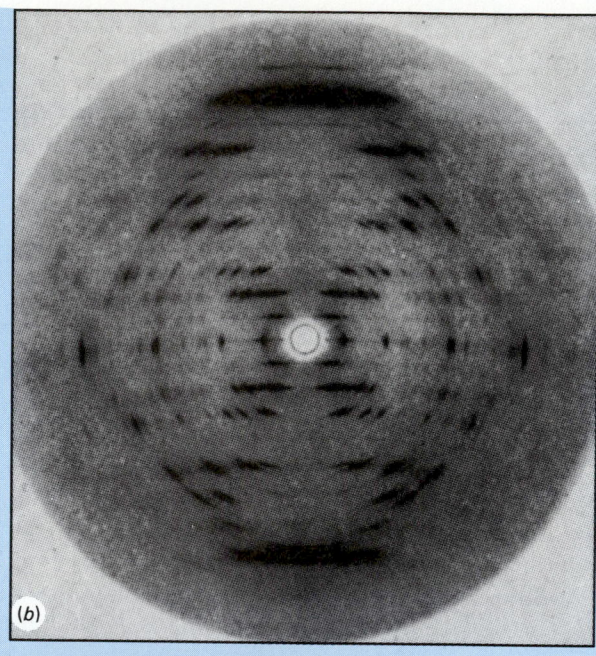

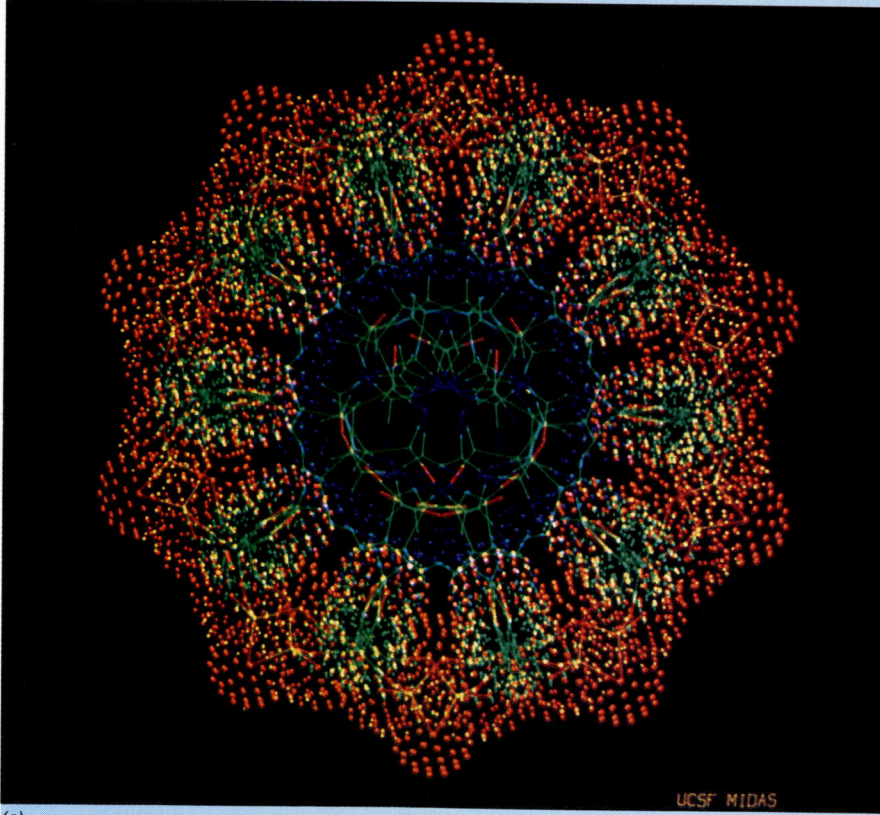

FIGURE 10–7 X-ray diffraction photographs of suitably hydrated fibers of DNA, showing the so-called B configuration. (*a*) Pattern obtained using the sodium salt of DNA. (*b*) Pattern obtained using the lithium salt of DNA. This pattern permits a most thorough analysis of DNA. The diagonal pattern of spots (reflections) stretching from 11 o'clock to 5 o'clock and from 1 o'clock to 7 o'clock provides evidence for the helical structure of DNA. (*c*) A modern computerized reconstruction of the appearance of a DNA molecule if it could be viewed from the top. This detailed picture has been built up from a myriad of clues afforded by indirect evidence like those early x-ray diffraction photographs on which Watson and Crick based so much of their initial thinking. (*a, b* Courtesy of Biophysics Research Unit, Medical Research Council, King's College, London; *c*, Courtesy of Computer Graphics Laboratory, University of California, San Francisco)

was equal to the amount of adenine, and that the amount of cytosine similarly was equal to the amount of guanine. (These equalities are known as Chargaff's rule.) After several false starts, Watson and Crick found that these facts suggested that DNA was not a single, but a *double* helix, held together by bases that connected with one another, every adenine with a thy-

mine, every guanine with a cytosine, producing the equalities that Chargaff had noted.

Watson and Crick next built a model of the DNA molecule as they understood it. This model was built of accurately scaled models of atoms and groups of atoms, with both sizes and bond angles properly proportioned according to those inferred for the actual

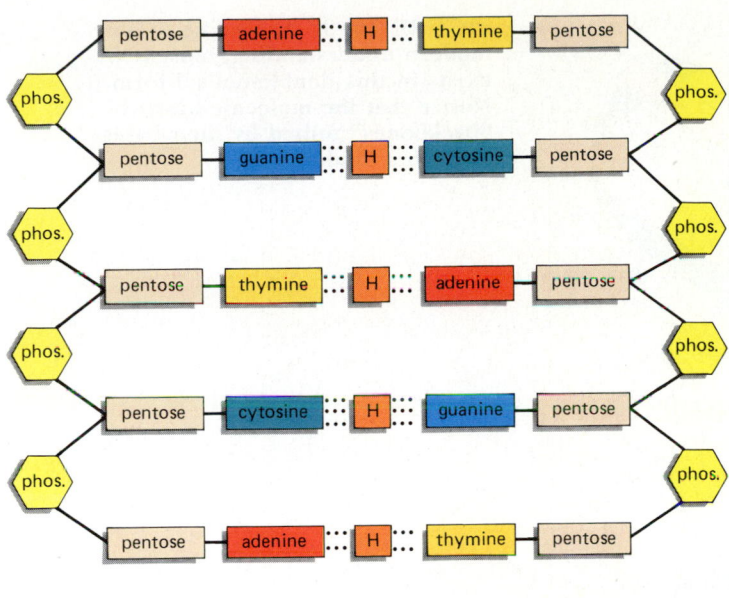

(a)

FIGURE 10–8   Structure of DNA. (a) A portion of the double helix. Its orientation permits the complementary bases to pair. ("H" represents hydrogen bonds.) (b) Note that two sugar-phosphate chains run in opposite directions.

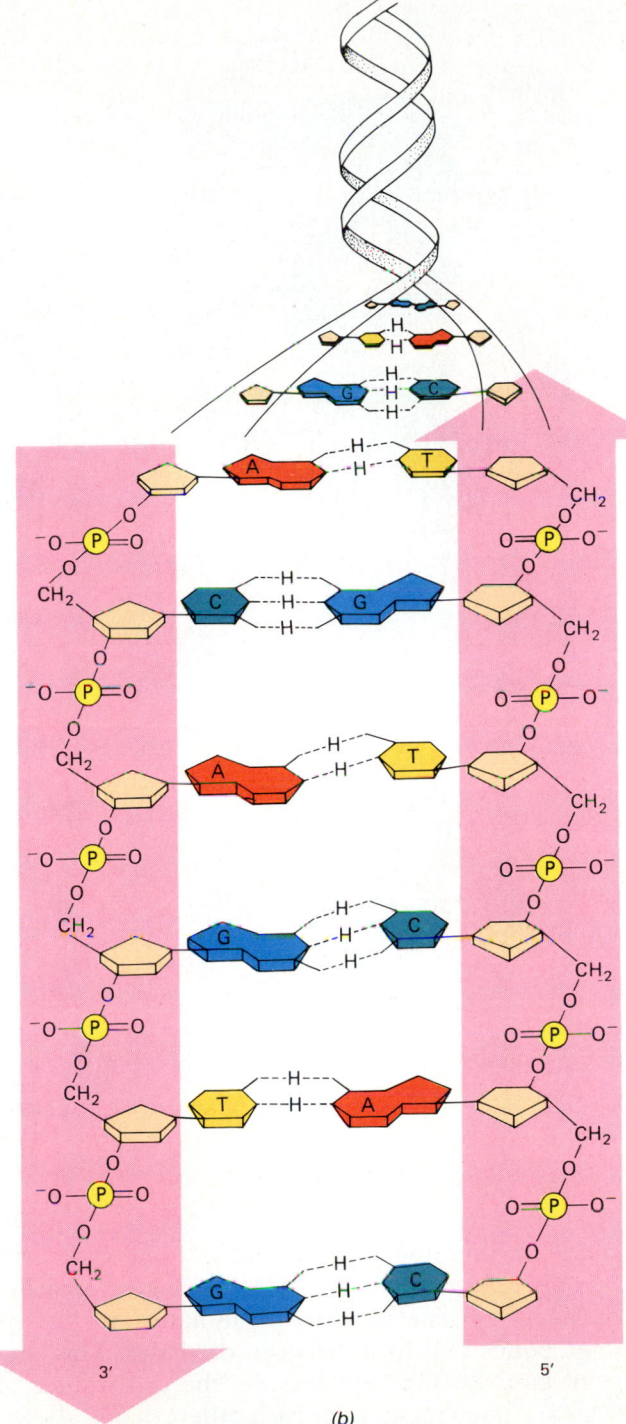

(b)

substances. If everything could be made to fit, their understanding of the DNA molecule would have been shown quite conclusively to be correct. When finally a great double helix stood in their laboratory, they knew they had been right. It was a great moment of discovery, perhaps one of the greatest in the history of science. They had discovered nothing less than the blueprint of life—a molecule whose structure preprogrammed the composition of every component in, and all the functions of, every cell on earth (Fig. 10–7).

DNA Is Composed Of Nucleotide Subunits

DNA is composed of molecular subunits called **nucleotides.** Each nucleotide consists of (1) a five-carbon sugar, **deoxyribose,** (2) a phosphate group, and (3) a nitrogen-containing compound called a **base** (Fig. 10–8). There are actually four bases in DNA: **cytosine (C)** and **thymine (T)** are single-ring **pyrimidine** bases. **Adenine (A)** and **guanine (G)** are double-ring **purine** bases. These bases project like the rungs of a ladder more or less at right angles from the sugar-phosphate backbone of the DNA molecule. Much as the letters forming words, sentences, and books determine the information they carry, the sequence in which these bases appear determines their genetic message. Thus the sequence AGGTCCATCCG bears a different set of information than the sequence TGGAACTAGTCC.

In nature DNA exists in the form of *two* strands, attached to each other by means of their bases as follows:

The ladder-like double strand of DNA is twisted into a double helix, somewhat like a spring or a spiral staircase (Fig. 10–9).

An important fact about the two strands of a double helix is that they are complementary but *not* identical. What that means is that if you know the base sequence of one of the two strands, you can predict the base sequence of the other. This is because a purine base always pairs with a pyrimidine base. More specifically, adenine on one strand normally pairs only with

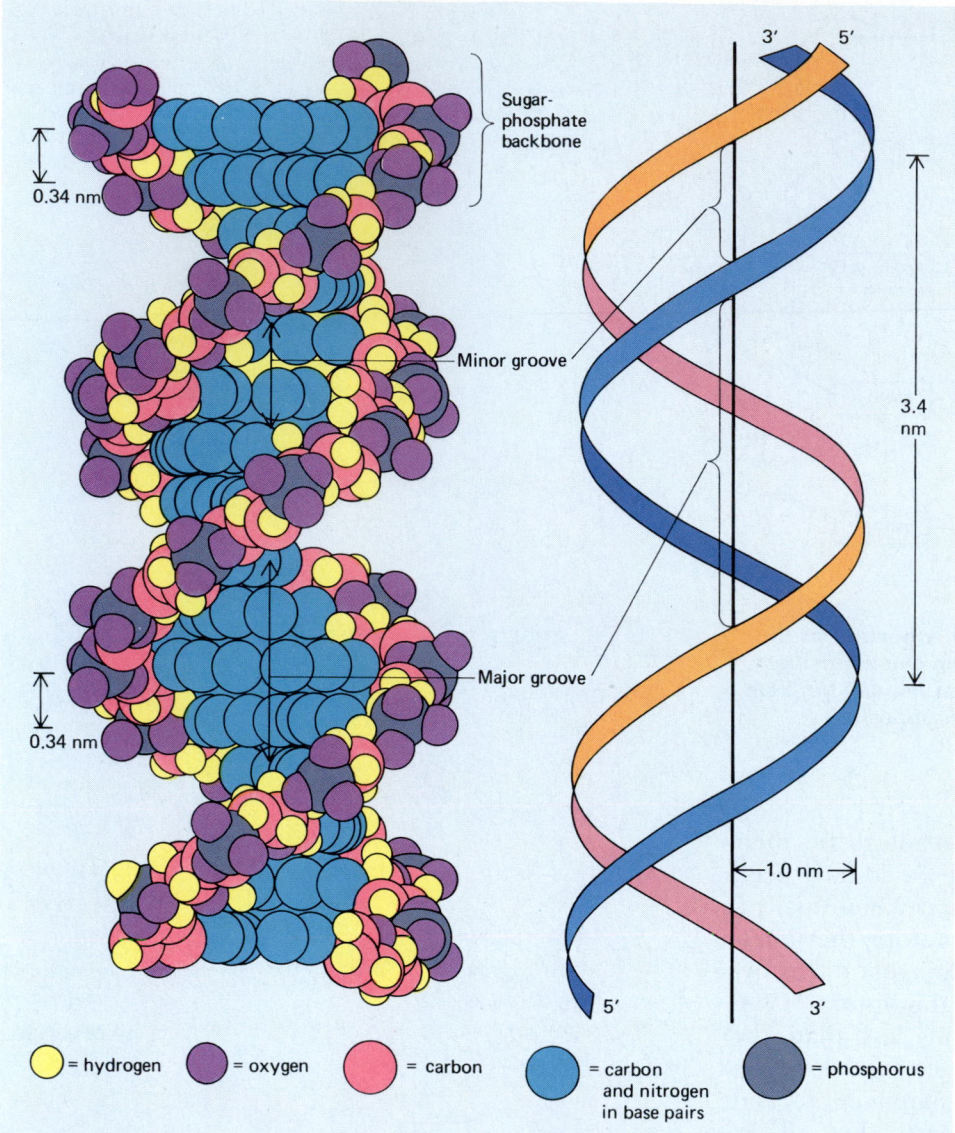

Sugar-phosphate backbone

0.34 nm

Minor groove

Major groove

0.34 nm

3.4 nm

1.0 nm

3'   5'

5'   3'

○ = hydrogen   ● = oxygen   ● = carbon   ● = carbon and nitrogen in base pairs   ● = phosphorus

FIGURE 10–9 According to our modern understanding, most DNA exists in this double-helical form. Notice that the molecule has two "backbones" united by their bases to form a double helix.

thymine on the other strand, and cytosine ordinarily pairs only with guanine. No other pairing relationship is normally possible because of the nature of the hydrogen bonds that form between the bases. This concept of base pairing explains why the two strands of DNA are complementary to each other; that is, the sequence of nucleotides in one chain dictates a complementary sequence of nucleotides in the other. For instance, if a portion of one chain reads AGGCTA then the corresponding part of the other must be TCCGAT. The distance occupied by adenine plus thymine is exactly the same as that occupied by guanine plus cytosine. Because the distance between complementary base pairs is always the same, a double strand of DNA is the same width from one end to the other, regardless of base sequence.

Moreover, the two strands extend in *opposite directions*, as you can see from looking carefully at the orientation of the deoxyribose molecules in Figure 10–8. Notice that each strand has one end labeled "3'" and the other end "5'." This is a designation chemists use to distinguish between the ends of a strand. Notice also that the 3' end of one is paired with the 5' end of the other. Because of this arrangement, the strands are said to be **antiparallel.**

The techniques employed by Watson, Crick, and their colleagues produced an "average" picture of the DNA strand. Their model showed regularly spaced base pairs, and their double helix curved smoothly as it twisted to the right. This type of DNA is known as B-DNA. Today it is understood (Fig. 10–10) that within most DNA helices are regions where coiling varies from the standard. The number of nucleotides per turn can vary, for example.

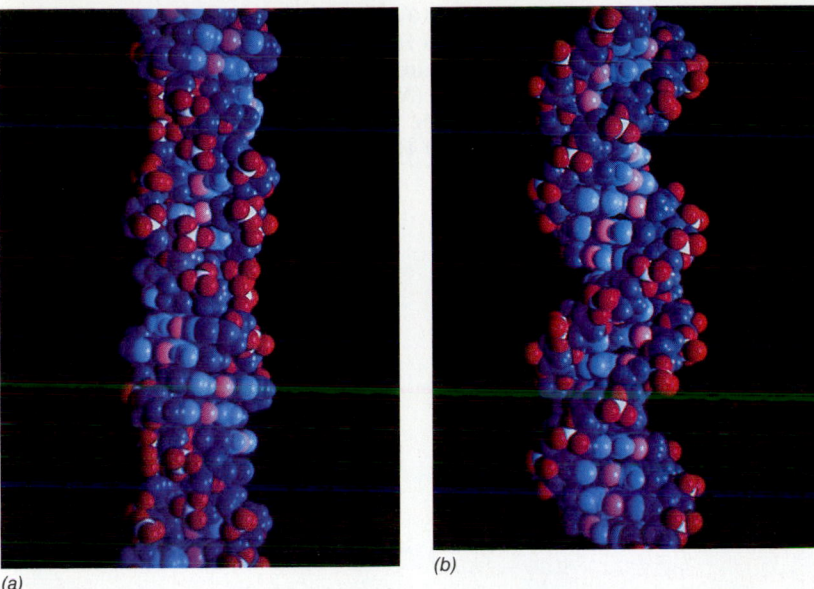

(a)

(b)

FIGURE 10–10 All DNA is not the same. Here "classic" B-DNA is compared with Z-DNA, a form of DNA whose function is as yet unknown. The shape of Z-DNA (*a*) is much less regular than that of B-DNA (*b*). Z-DNA appears to be genetically inactive. (Computer reconstructions by Dr. Richard J. Feldmann)

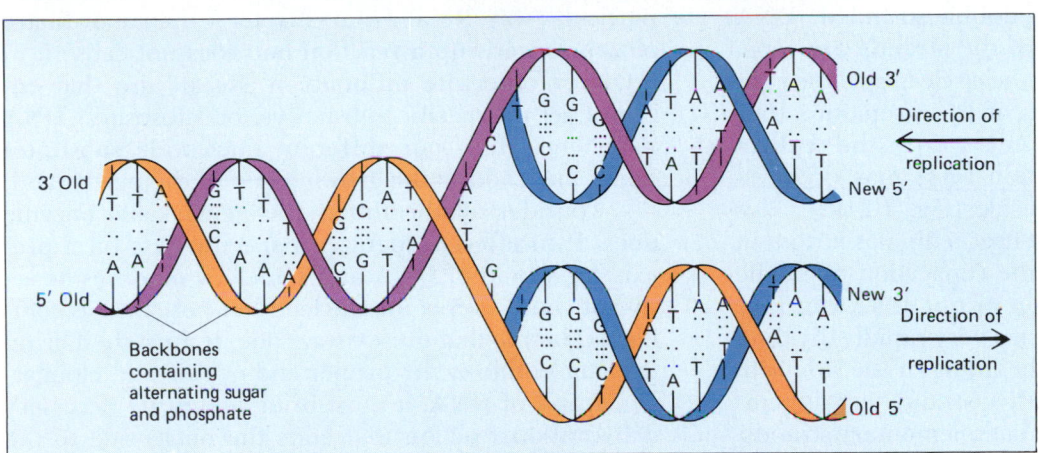

FIGURE 10–11 Mechanism of the replication of DNA. The two ends of each strand of this helix are labelled 3′ or 5′ to establish the direction of replication. The two strands are shown separating. Both are being copied. The strand whose 5′ end is at the left side of the figure is being copied starting from the right (its 3′ end). This corresponds to the 5′ end of the newly synthesized strand. Replication proceeds in the same direction (3′ to 5′) on both templates. In the new strand, as in the old, adenine forms base pairs with thymine, and cytosine forms base pairs with guanine. The new strands grow in the 5′-to-3′ direction because they are antiparallel to their complementary DNA templates.

▪ DNA REPLICATES SEMICONSERVATIVELY

The most distinctive properties of the genetic material are that it carries genetic information and undergoes precise replication. That is, when replicating, DNA makes an exact duplicate of itself. Before a cell divides, the DNA strands in its double helices separate, and each one is used as a template (that is, a kind of model) for a new strand of complementary nucleotides. The result is two double-helix molecules, each identical in

base sequence to the original double helix. Each contains one of the original strands plus a newly synthesized strand (Fig. 10–11). The two original strands are not simply copied and then left behind but are "conserved," that is, kept to be used as part of the two new double helices. Thus, this process is known as **semiconservative replication** (*semiconservative* here meaning "keeping part the same"). Each new double helix contains one old strand and one new strand.

Replication need not begin at the end of a double

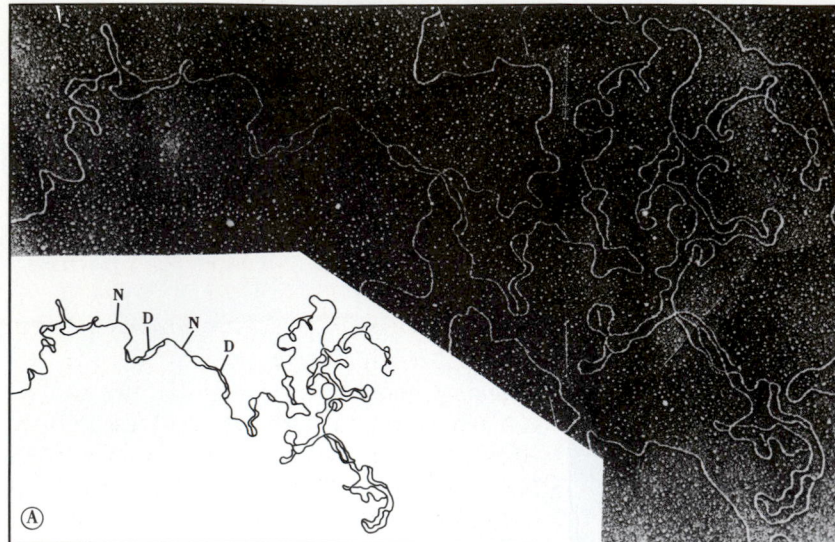

FIGURE 10–12 Electron micrograph of DNA replication in action. Stretches of replicated DNA (*D*) alternate with stretches not yet replicated (*N*). The sketch is a tracing of the photograph. Note the many replication forks. (Courtesy of H.J. Kriegstein and D.S. Hogness)

helix. Enzymes separate the two DNA strands at various sites. (In prokaryotes, double-stranded DNA forms closed circles; so before the strands are separated, other enzymes open up the circle at specific initiation sites.) As the two strands of DNA separate, they form a kind of Y-shaped region known as the **replication fork.** Many such replication forks may occur simultaneously in a DNA molecule (Fig. 10–12). This greatly speeds up the process. Proceeding as it does in many places simultaneously, the replication of all the enormous fund of information in our DNA can take place in as little as 20 minutes in a very rapidly dividing cell such as some of those in the lining of the stomach.

After separating, both DNA strands act as templates for the assembly of new complementary strands. This is the task of the enzyme **DNA polymerase,** which adds new DNA nucleotides to the 3′ end of the new growing strands. Thus, the new strands are said to grow in the "5′-to-3′" direction (notice that because the strands are antiparallel, this is the 3′-to-5′ direction on the old strands). As the new complementary strands are produced, the original double helix separates further, and the newly synthesized half-helices continue to join with the old half-helices until there are two identical double helices of DNA.

Errors in replication do occur, but infrequently. Though the specificity of the base pairing of DNA is necessary for a regular helix to be produced, proper pairing is by no means assured just by the properties of the bases. Improper pairing can and does occur, producing a local distortion in the structure of the helix. The usual fidelity and regularity of the base pairing are owed to DNA polymerase, which is one of the most important enzymes in living things. Proper DNA synthesis could not take place without it. DNA polymerase appears actually to be a complex of enzymes, or one of a closely associated "team." It has a twofold function. Not only must it catalyze the polymerization—chain

formation—of nucleotides, but it must also do this in a particular way. Recall from Chapter 7 that an ordinary enzyme speeds up a reaction but does not cause it to occur or otherwise influence it. Recall also that enzymes act on specific substances, or substrates. DNA polymerase has four different nucleotide substrates that it must add to the growing end of the new strand.

Consider all the things this remarkable enzyme does. It in effect recognizes a specific base on a pre-existing strand of DNA and somehow modifies its active site and accepts the nucleotide whose base is complementary to that pre-existing one. It must then bring that nucleotide to the proper end of the new, elongating strand of DNA, it must bring about the necessary dehydration reaction that adds the nucleotide to the new strand, and next it must release the reaction products. The enzyme then moves one base further on the existing strand and repeats the process all over again. And it continues to do this at a rate of about 200 bases per second for hundreds or thousands of bases every time the cell divides, and occasionally at other times as well! There is even a backup system that will catch most of the few errors that DNA polymerase makes. This backup system is a team of enzymes that snips out erroneous, improperly pairing bases and replaces them with the right ones.

■ CHAPTER SUMMARY

I. The basic plan of all living things and their cells is contained in their DNA.
   A. In eukaryotes, the cell is controlled by its nucleus with the aid of a messenger substance that travels to the cytoplasm.
      1. Experiments performed on *Acetabularia* by Hammerling and Brachet demonstrated the role of the nucleus.

2. If *Acetabularia* stalks are exchanged, the form of the regenerated cap is determined by the holdfast, but only after a delay that implies the existence of a temporary messenger substance.

3. An *Acetabularia* nucleus, when transplanted to a holdfast of a different species, produces a cap characteristic of the nucleus' species.

B. The experiments of Griffith and Avery showed that genetic information is transferred among bacteria by DNA.

C. The genetic information for a virus such as a bacteriophage is found in its nucleic acid.

II. The structure of DNA has been elucidated.

A. X-ray diffraction studies by Rosalind Franklin and others showed that DNA is a helical molecule.

B. James Watson and Francis Crick worked out the basic structure of DNA in the 1950s.

C. DNA is an antiparallel double helix of nucleotides joined by hydrogen bonds between complementary bases.

1. There are four bases: adenine, which pairs with thymine, and cytosine, which pairs with guanine.

2. Given the base sequence of one strand, that of the other can be predicted.

3. Each nucleotide consists of deoxyribose (a sugar), a phosphate group, and a purine or pyrimidine base.

III. DNA duplicates itself by semiconservative replication, in which each of two original strands in a double helix is used in a new double helix.

A. DNA synthesis takes place at replication forks, Y-shaped regions where the two strands of DNA separate and where DNA synthesis occurs on both strands at once.

B. DNA polymerase catalyzes the polymerization of nucleotides, adding the appropriate nucleotide to the strand.

☐ POST-TEST

1. Experiments with *Acetabularia* indicate that the ultimate control of the cell is exercised by the _____.

2. Between them, Griffith and Avery showed that the genetic message was contained within _____.

3. Watson and Crick showed that the shape of the DNA molecule is a _____ _____.

4. Each nucleotide chemically consists of a _____, a _____, and a _____.

5. A sequence of bases AACGGTCA on one strand of DNA would necessitate the sequence _____ in the complementary strand.

6. In the DNA of all known organisms, the amount of adenine equals the amount of _____, and the amount of _____ equals the amount of _____.

7. As the two strands of replicating DNA separate, they form a Y-shaped region known as the _____.

8. The process of DNA replication is described as _____ because the two original strands of DNA are retained in the products.

9. DNA synthesis is catalyzed by the _____.

☐ REVIEW QUESTIONS

1. How did Watson and Crick "prove" their concept of doubly helical DNA?

2. What is *Acetabularia*? What was discovered about the function of the cell nucleus using this organism, and how was this done?

3. How did experiments by Griffith and Avery help show that the transformation of harmless bacteria into virulent ones involves a change in genetic information?

4. How is the molecular structure of DNA uniquely adapted to its function? Explain.

5. What is meant by semiconservative replication?

☐ RECOMMENDED READINGS

Burns, G.W., and P.J. Bottino, *The Science of Genetics,* 6th ed. New York, Macmillan Publishing Company, 1989. An in-depth but readable presentation on the identification of DNA as the genetic material (Chapter 8).

Judson, H.F. *The Eighth Day of Creation.* New York, Simon and Schuster, 1979. Aside from the famous *Double Helix,* this is perhaps the most readable and authoritative history of nucleic acid research from the 1930s through the mid-1970s.

Radman, M., and Wagner, R. "The High Fidelity of DNA Duplication," *Scientific American,* August 1988, 259:40–47.

Watson, J.D. *The Double Helix.* New York, Atheneum, 1968. Not everyone thinks this personal account to be historically accurate, and without doubt it is a highly colored and biased personal history. But there is nothing like it. One feels that, in essence if not in detail, this is how it really was to make a Nobel-Prize-winning discovery.

11

Gene Function: RNA and Protein Synthesis

OUTLINE

I. Proteins are central to an organism's identity
II. The genetic code links gene structure to protein structure
III. RNA encodes and expresses the genetic message stored in DNA
   A. RNA is made by transcription
   B. In RNA processing, introns are removed from the new molecule
IV. In translation, protein is made using the genetic mesage encoded in mRNA
   A. Ribosomal RNA forms part of the ribosome
   B. Transfer RNA carries amino acids to the ribosome
   C. Messenger RNA codons are the instructions the ribosome uses to construct a protein
   D. Special signals direct the ribosome to start and stop adding amino acids
V. Changes can occur in some stages of genetic expression
   A. Mutations are random, permanent changes in DNA
   B. Factors in the environment can produce mutations

LEARNING OBJECTIVES

After you have studied this chapter you should be able to:

1. Define and describe the process of transcription.
2. Define introns and exons. Discuss their occurrence and possible roles.
3. Summarize the sequence of events that occur in translation.
4. Describe the function of transfer RNA, of messenger RNA, and of ribosomal RNA.
5. Distinguish between the processes of initiation, elongation, and termination in protein synthesis.
6. Indicate the significance of genetic proofreading mechanisms.

Rough endoplasmic reticulum with ribosomes. Notice, however, that some ribosomes are not attached to membranes (×121,000). (David M. Phillips/Visuals Unlimited)

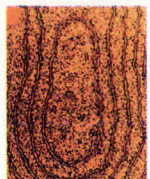

In the last chapter, we learned that deoxyribonucleic acid (DNA) is a double-stranded molecule composed of nucleotide subunits. In this chapter we will see that these subunits make up larger units, still within the DNA molecule, called **genes.** Recall that in Chapter 4 we said that DNA contains the information needed to make all the proteins the cell uses, and that proteins are chains of amino acids joined in specific orders. A gene, generally speaking, is a sequence of DNA nucleotides that represents some piece of information about how or when a particular protein should be synthesized. The term **gene expression** refers to the overall process by which the information in genes is "expressed," or used, to form proteins.

Several types of genes may be involved in making a single protein. In this chapter we focus on the type of gene whose base sequence dictates the order of amino acids in a protein the cell will use. Other types of genes contain information about related aspects of the protein manufacturing process. We will look at some of these in Chapter 12.

KEY CONCEPTS

- ■ DNA directs the activities of an organism by controlling its protein synthesis.

- ■ Ribonucleic acid (RNA), which functions in protein production, is synthesized using DNA as a template. Production of RNA is called transcription.

- ■ There are three types of RNA: mRNA, tRNA, and rRNA. All three are necessary for protein synthesis.

- ■ During translation, the ribosomes manufacture proteins.

■ PROTEINS ARE CENTRAL TO AN ORGANISM'S IDENTITY

Proteins are always present and active, even in cells that are not growing or producing special protein products, such as hormones, that other cells use. At different times in its life cycle, each cell uses different proteins in varying amounts. Proteins function directly in elaborate and subtle ways in such cellular processes as oxygen transport, cell movement, active transport, and facilitated diffusion. Our focus, though, will be on the role of proteins as **enzymes.** Enzymes mediate almost all chemical reactions in the cell. By directing the construction of particular enzymes, the genes control the activities of the cell.

Proteins, you will recall, are large and complex molecules made up of amino acids joined by peptide bonds. A long chain of amino acids is called a **polypeptide;** some proteins consist of more than one polypeptide. The sequence of amino acids in a polypeptide, called the **primary structure of the protein,** is specified directly by the base sequences in the type of genes we will study in this chapter. Proteins also have higher levels of structure, and their resulting shapes are important aspects of their functions. An enzyme without a properly shaped active site, for example, would probably not operate as it should.

How genes ultimately make proteins is the subject of this chapter. Within the typical cell, hundreds of kinds of protein are constantly in production, each molecule requiring approximately 20 minutes to make, with large numbers of protein molecules being produced simultaneously. All this involves not only a multitude of individual protein factories (the ribosomes), but also a whole associated economy of raw material suppliers, subcontractors, and management systems, which might well be thought of as a kind of corporate manufacturing conglomerate—the Ribosome Protein Manufacturing Company. While we tend to think of obvious "fixed" traits such as eye color and hair texture as evidence of genes, such traits are visible ultimately because of protein synthesis. At the same time, other aspects of body structure and function, more difficult to observe directly, are also under the control of genes. The messages in an organism's genes control everything about it by governing its production of protein.

■ THE GENETIC CODE LINKS GENE STRUCTURE TO PROTEIN STRUCTURE

The identification of DNA as the basic genetic material was one of the greatest moments in the history of biology. It may prove to have been one of the most significant events in the history of the human species. It then became clear that hereditary information was stored in the cell in the form of a code—the base sequences in DNA.

In the previous chapter we looked at the way DNA replicates itself, permitting genetic information to be passed on to new cells and new organisms. But the replication of information is only part of the story. Information is of no value if it cannot be put to use. DNA carries and stores genetic information, but how is this information used?

An early clue came from the studies of *Acetabularia* described in Chapter 10, in which a messenger substance was shown to carry information about cell shape from the nucleus to the rest of the cell. In time this substance was shown to be ribonucleic acid, or RNA.

TABLE 11–1
The Genetic Code: The Sequence of Nucleotides in the Triplet Codons of mRNA That Specify a Given Amino Acid

First position (5′ end)	Second position	Third position (3′ end)			
		U	**C**	**A**	**G**
U	U	Phe	Phe	Leu	Leu
	C	Ser	Ser	Ser	Ser
	A	Tyr	Tyr	Terminator	Terminator
	G	Cys	Cys	Terminator	Trp
C	U	Leu	Leu	Leu	Leu
	C	Pro	Pro	Pro	Pro
	A	His	His	Gln	Gln
	G	Arg	Arg	Arg	Arg
A	U	Ile	Ile	Ile	Met
	C	Thr	Thr	Thr	Thr
	A	Asn	Asn	Lys	Lys
	G	Ser	Ser	Arg	Arg
G	U	Val	Val	Val	Val
	C	Ala	Ala	Ala	Ala
	A	Asp	Asp	Glu	Glu
	G	Gly	Gly	Gly	Gly

Each amino acid is usually represented by a three-letter abbreviation: Phe = phenylalanine, Leu = leucine, Ser = serine, Tyr = tyrosine, Cys = cysteine, Pro = proline, His = histadine, Glu = glutamic acid, Arg = arginine, Ile = isoleucine, Met = methionine, Thr = threonine, Asp = aspartic acid, Lys = lysine, Val = valine, Ala = alanine, Gly = glycine, Trp = tryptophan, Gln = glutamine, Asn = asparagine.

Several important types of RNA are involved in transmitting the genetic message from DNA and using that message to make protein.

Twenty different amino acids are commonly found in proteins. These amino acids can be thought of as the 20 "letters" of the protein alphabet. How can DNA, with only four nucleotides in its "alphabet," serve as a code that specifies all 20 amino acids? Single nucleotides could not be the fundamental unit of such a code, for there could be only four kinds of amino acids in any protein. Neither could pairs of nucleotides, since a pool of four nucleotides permits only 16 unique pairs (AA, AC, CA, AT, etc.), specifying only 16 different amino acids.

However, trios of nucleotides could—and do—form such a code, because a pool of four nucleotides permits 64 unique trios (AAA, AAC, CAA, etc.), more than enough to specify 20 amino acids. Each such sequence of three nucleotides—more simply, three bases—is know as a **triplet.** It turns out that these triplets are truly the basic units of genetic information (Table 11–1). With some exceptions, each triplet codes for one amino acid. Some triplets are "synonymous," coding for the same amino acid as another triplet. A few triplets do not code for amino acids but have other roles in protein synthesis.

An important aspect of the genetic code is that it is universal throughout the biological world. The same DNA triplets code for the same amino acids in all living organisms. While protein structure and proportions may vary from one species to another, the relationship of nucleotides to amino acids is the same. As we discuss in Chapter 17, this is considered to be strong evidence that all organisms have evolved from the same early life forms.

■ **RNA ENCODES AND EXPRESSES THE GENETIC MESSAGE STORED IN DNA**

DNA itself does not synthesize proteins. Instead, it directs RNA to synthesize them. RNA differs from DNA in that it contains the sugar ribose rather than deoxyribose. As in DNA, each RNA nucleotide contains one of four bases, although in RNA the base uracil (U) replaces the thymine (T) found in DNA. Uracil behaves chemically like thymine, readily forming base pairs with adenine (Fig. 11–1). Another difference is that RNA molecules usually do not form double helices, instead functioning as single strands that sometimes are elaborately folded.

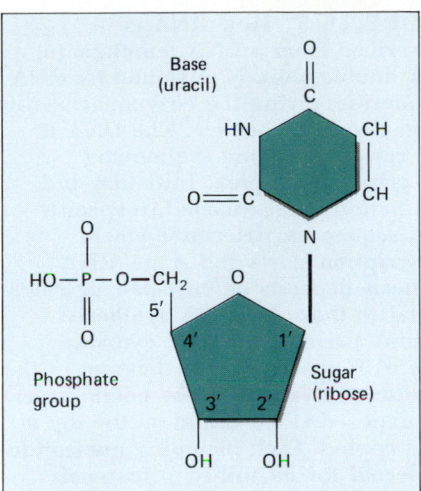

FIGURE 11–1   RNA is composed of four nucleotides, one of which is shown here. The base shown is uracil, but other RNA bases are adenine, guanine, and cytosine. Thymine does not occur in RNA.

Before detailing the steps involved in gene expression, let us look quickly at the overall series of events and the main molecules involved. There are three main types of RNA molecules—**messenger RNA (mRNA), ribosomal RNA (rRNA),** and **transfer RNA (tRNA).** All three are made by DNA, and each has a role in protein production.

Put very simply, the process of protein synthesis consists of two broad steps: (1) **transcription:** a messenger RNA molecule is synthesized from DNA by complementary base pairing; and (2) **translation:** the ribosome assembles amino acids into a polypeptide, guided by the information in the messenger RNA mol-

ecule. Both tRNA and rRNA are also involved in translation. In Step 1, transcription, the genetic message is copied from one nucleotide medium—that of DNA—to another—that of RNA. In Step 2, translation, the genetic message is "translated" from the language of nucleotides into the language of proteins.

In prokaryotic cells (bacteria), which lack a separate nucleus, both transcription and translation occur in the cytoplasm. In eukaryotic cells, in which the DNA is found in a nucleus surrounded by its own membrane, transcription occurs in the nucleus, while translation occurs for the most part outside the nucleus in the cytoplasm.

RNA Is Made By Transcription

Recall that in DNA replication, an existing DNA strand acts as a template, or model, for the synthesis of a new, complementary DNA strand. Something like this happens in transcription (Fig. 11–2). A particular sequence of nucleotides on a DNA strand acts as the template for the synthesis of a complementary RNA strand. For example, where a guanine occurs in the DNA strand, a cytosine pairs with it, becoming part of the new RNA strand. Yet there are several important differences between DNA replication and transcription: Uracil rather than thymine pairs with adenine, and the RNA molecule detaches from the DNA template rather than forming a double helix with it. Although we will focus on the transcription of mRNA, keep in mind that tRNA and rRNA are also synthesized by transcription from special gene sequences in DNA. Thus, we can now define a gene more exactly as

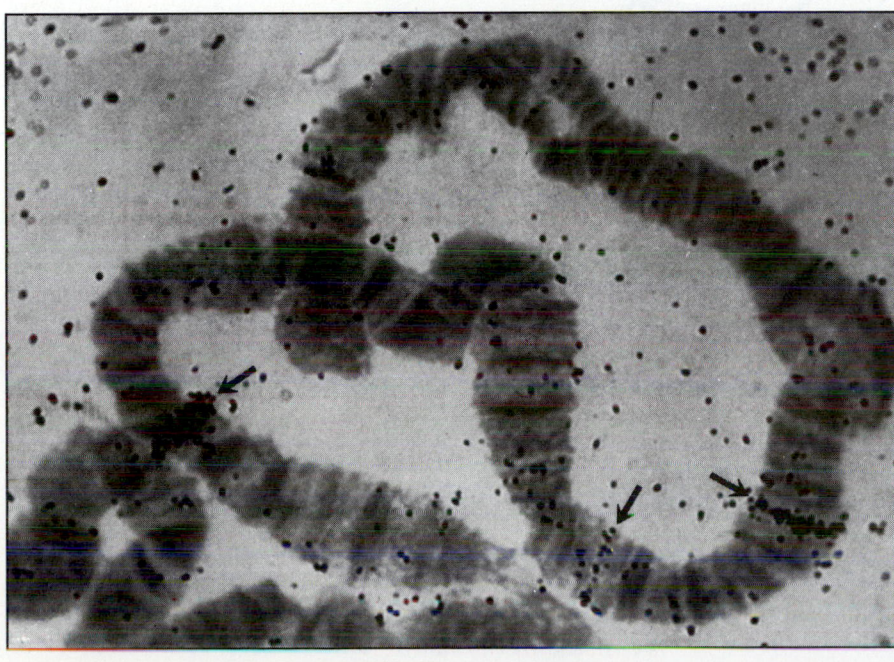

FIGURE 11–2   Production of RNA (transcription) along specific portions (genes) of DNA in a chromosome of the fruitfly *Drosophila melanogaster*. The dark particles are radioactively labeled RNA. Note the three regions along the DNA where the particles are concentrated.
(Courtesy of T. A. Grigliatti)

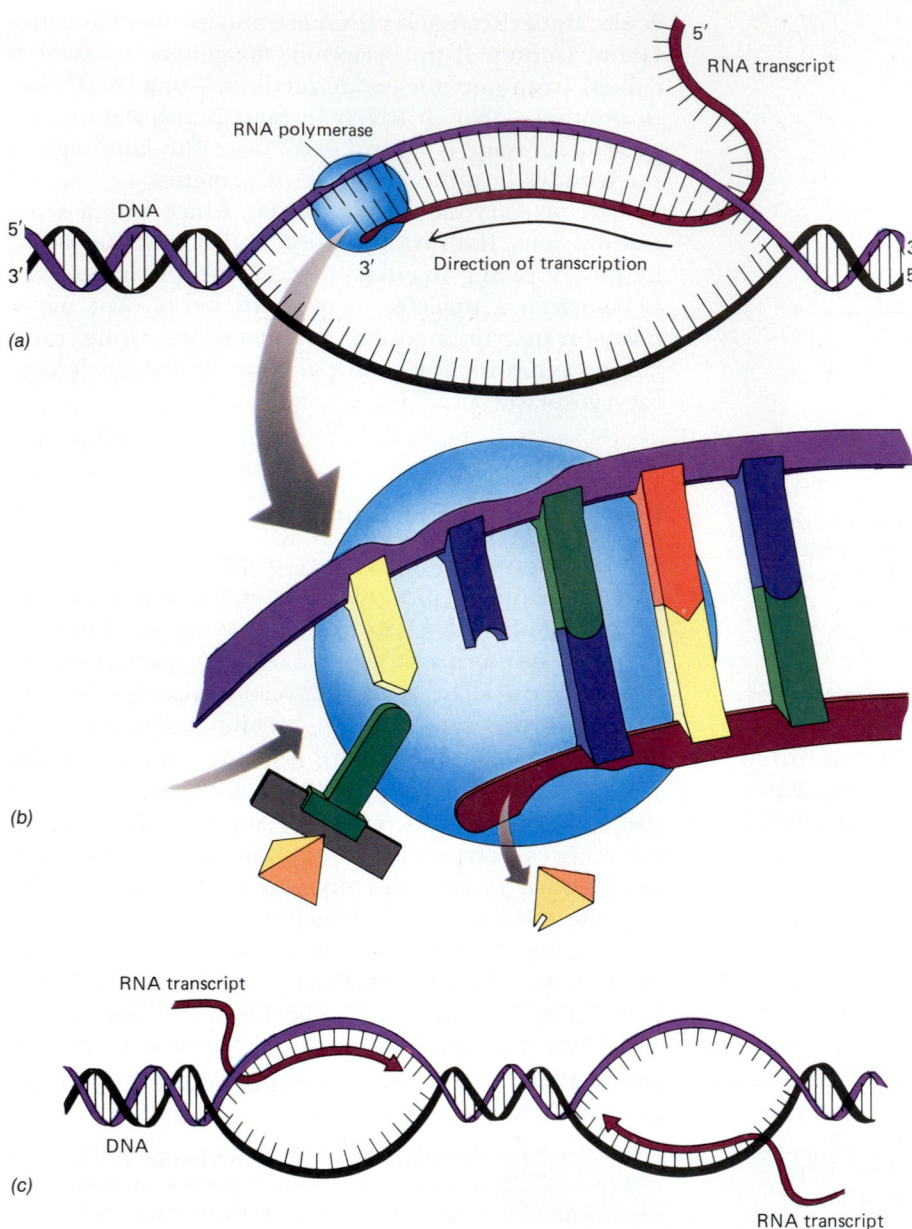

(a)

(b)

(c)

FIGURE 11–3 How RNA is transcribed from a DNA template. (*a*) A DNA double helix is unwound by RNA polymerase, giving the enzyme access to the nucleotide sequence. The DNA is then rewound behind the moving transcription complex. Initiation and termination sites, encoded in specific DNA sequences, determine where transcription starts and stops. RNA synthesis depends on base-pairing rules similar to those for DNA synthesis: Adenine pairs with uracil; cytosine pairs with guanine. (*b*) A diagram of the reaction catalyzed by RNA polymerase. The exposed DNA strand on the top is being copied. Each incoming nucleotide is selected for its ability to base-pair with the DNA template. The nucleotide is then added to the 3' end of the growing RNA chain. (*c*) Only one of the two strands is transcribed for a given gene, but the opposite strand may be transcribed for a neighboring gene.

the sequence of DNA bases needed to specify a single RNA molecule. Let us now look at how a gene produces a molecule of mRNA.

Transcription is controlled by the enzyme **RNA polymerase,** which appears to move along a strand of DNA like a railroad locomotive on a track, constructing a single strand of RNA as it goes (Fig. 11–3). This remarkable enzyme responds to cues built into the base sequence of the DNA. It selects which gene to transcribe, recognizes which of the two paired DNA strands it should copy and identifies where it should begin and end transcription.

Transcription begins when a subunit of RNA polymerase binds to a **promoter** site in the DNA. The promoter consists of a special base pair sequence, indicating that transcription can begin there. Ribonucleo-

tides are matched to complementary bases along the DNA template and added to the 3' end of the growing strand as in DNA replication. Thus, as in DNA replication, the new strand grows in the 5'-to-3' direction. Instead of forming a double helix, however, the mRNA strand separates from the DNA template as it forms. This process continues until the RNA polymerase comes to a specific nucleotide sequence in the DNA that it in effect recognizes as a "stop" signal. The RNA molecule then separates completely from the DNA strand.

In any portion of the double helix, only one of the DNA strands—called the **coding strand,** or plus strand—is the template for an RNA molecule. Which of the two strands serves as the coding strand varies from region to region. One strand may be transcribed

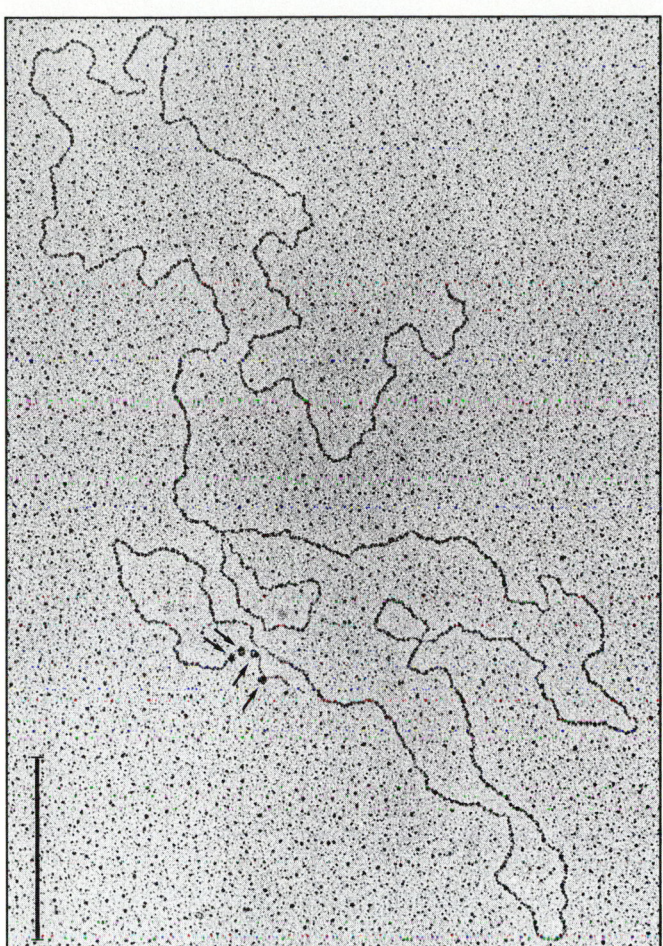

FIGURE 11–4 Electron micrograph of the complex formed by DNA and RNA polymerase, several particles of which are indicated by arrows. The DNA is that of a bacteriophage virus; the RNA polymerase is that of its host cell, the *Escherichia coli* bacterium. If allowed to behave naturally, the RNA polymerase would transcribe the viral DNA, producing viral proteins that might ultimately destroy the host. The bar in the lower left-hand corner of the illustration is about 0.5 micrometer in length and is included for size comparison. (Courtesy of J. M. Sogo and Theodor Koller)

in one region, and the other in another region. The presence of a promoter sequence is what determines whether a particular portion of a DNA strand is able to code (Fig. 11–4). Since the DNA strands are complementary, you can see that a promoter sequence on one strand would have a complementary sequence on the other strand that would not be a promoter sequence.

We said that the genetic code is based on triplets of nucleotide bases in DNA. How is this information encoded in RNA? The sequence of mRNA bases complementary to a triplet in DNA is known as a **codon.** Later, in protein synthesis, the order of the codons in the mRNA will determine the order in which amino acids are added to the polypeptide chain.

In RNA Processing, Introns Are Removed From The New Molecule

In most eukaryotic genes, some very long base sequences are transcribed into mRNA codons that do not go on to participate in the making of a protein. Some of these so-called "interrupted" genes have several such sequences. The complementary codons of these sequences are enzymatically "snipped out" in newly transcribed mRNA molecules. The remaining portions of the mRNA molecule are joined together with great precision, forming the "mature" RNA that directs the construction of a protein. The sequences that are discarded (and their DNA complements) are called **introns;** those that stay and are joined together are called **exons** (Fig. 11–5). Each exon often corresponds to a **domain,** a functional subunit within the protein which will be produced.

Why do many genes possess introns? Intron-exon sequences may play a role in controlling the flow of information from DNA to the ribosomes and in controlling cellular differentiation. The fact that many introns are near-copies of adjacent exons has led to certain evolutionary proposals. For instance, it has been suggested that introns may serve as the raw material from which many new functional genes are constructed, or from which those lost by random changes in bases may be reconstructed. However, it may be that introns lack any function whatever.

☐ IN TRANSLATION, PROTEIN IS MADE USING THE GENETIC MESSAGE ENCODED IN mRNA

Before looking at translation in detail, let us review the entire process briefly. It consists of three steps: initiation, elongation, and termination. (1) **Initiation:** Translation begins when a strand of mRNA forms an initiation complex with the two subunits that make up a ribosome. (2) **Elongation:** The ribosome then "reads out" the codons in the mRNA, using the order of the codons to build a chain of amino acids. As it does this, the ribosome moves down the mRNA strand, usually until it reaches the end. (3) **Termination:** At this point the protein is released and goes to its destination in the cell.

Translation employs all forms of RNA. While mRNA specifies the order of amino acids in the protein, rRNA and its associated enzymes (as part of the ribosome) play several important roles, and tRNA identifies and transports the amino acids themselves.

Ribosomal RNA Forms Part Of The Ribosome

Messenger RNA cannot make proteins by itself. Specialized structures, the ribosomes, take amino acids and build them into a protein according to the instruc-

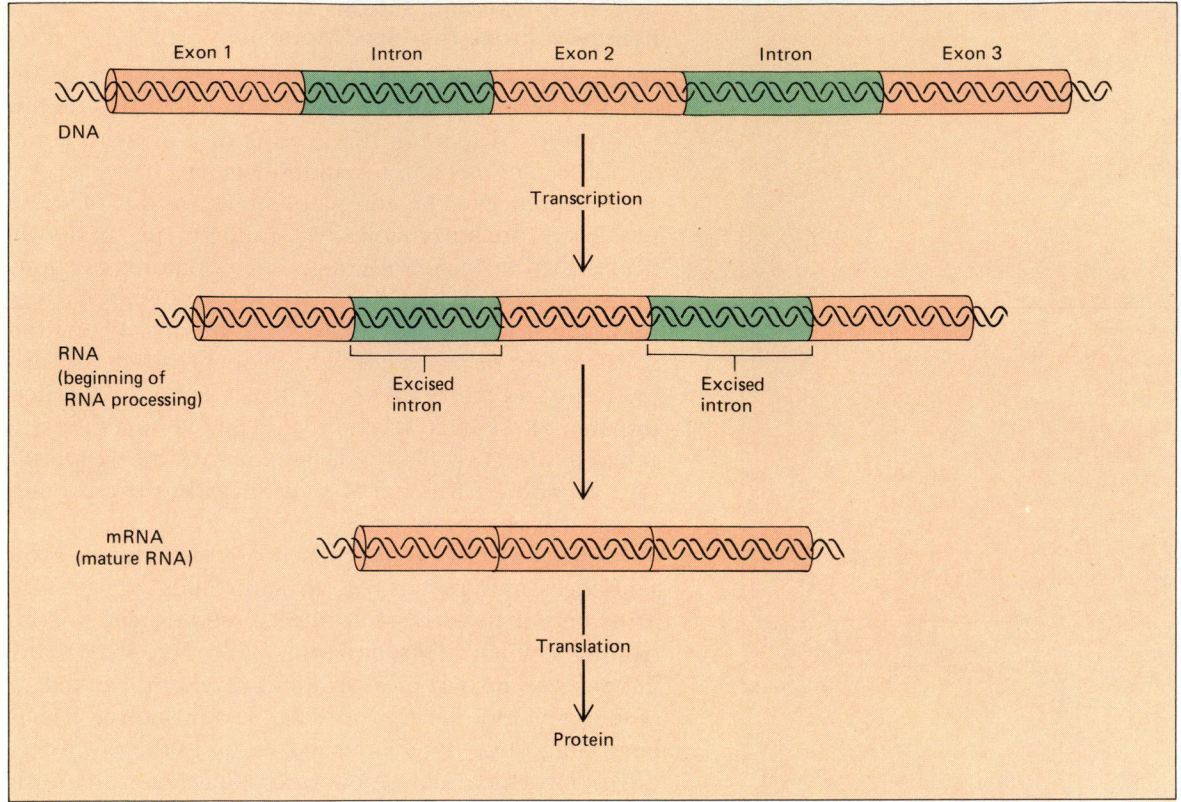

Exon 1   Intron   Exon 2   Intron   Exon 3

DNA

Transcription

RNA
(beginning of
RNA processing)

Excised
intron

Excised
intron

mRNA
(mature RNA)

Translation

Protein

**FIGURE 11–5   Introns are sequences that must be cut out of RNA before
RNA can be translated into a protein. Are the sequences in the DNA that
code for them "genetic junk," or do they have a function?**

tions of the mRNA. A ribosome is composed of two subunits, different in size and weight, which stick together somewhat loosely while translation is actually taking place (Fig. 11–6). The smaller subunit consists of a molecule of RNA and 21 different molecules of protein. The larger subunit comprises two molecules of rRNA and 34 different protein molecules.

A ribosome has three functions. First, it binds to mRNA. Then, it ensures that the genetic message in the mRNA is properly read. Finally, the ribosome catalyzes the formation of peptide bonds in the amino acid chain. The ribosome in effect serves as an enzymatic matchmaker that ensures not only that peptide bonds form but that they form between the correct amino acids in the order specified by mRNA. A variety of active sites and control sites permit the ribosome to carry out this matching process with precision. Most important are the A-site and the P-site (Fig. 11–7). We will look at this aspect of its function in a moment.

Transfer RNA Carries Amino Acids To The Ribosome

The individual amino acids that will go into the protein are first carried to the ribosome by molecules of tRNA. Each tRNA molecule is elaborately folded and capable of forming a temporary chemical bond with a specific

amino acid. Enzymes known as **amino acyl-tRNA synthetases,** one for each kind of amino acid, ensure that the tRNA molecules link up with their specific amino acids. The tRNA molecules thus function as labels or tags. By calling for its tRNA tag, one would obtain the specific amino acid attached to it.

A trio of bases called an **anticodon** projects from the tRNA molecule (Fig. 11–8). The ribosome in effect recognizes the amino acid by recognizing the anticodon of its tRNA carrier. The ribosome recognizes the appropriate anticodon because that anticodon has a base sequence that is complementary to the codon that the ribosome is currently reading on the mRNA strand.

Let us now look at translation in more detail. Because of the complexity of the process, we will start "in the middle," with elongation of a polypeptide chain that has already begun. Later we will return to the initiation process to see how the chain got started.

Messenger RNA Codons Are The Instructions The Ribosome Uses To Construct A Protein

At the point at which we look in on the translation process, we find the amino acid that has been added most recently to the polypeptide chain at the **P-site** of

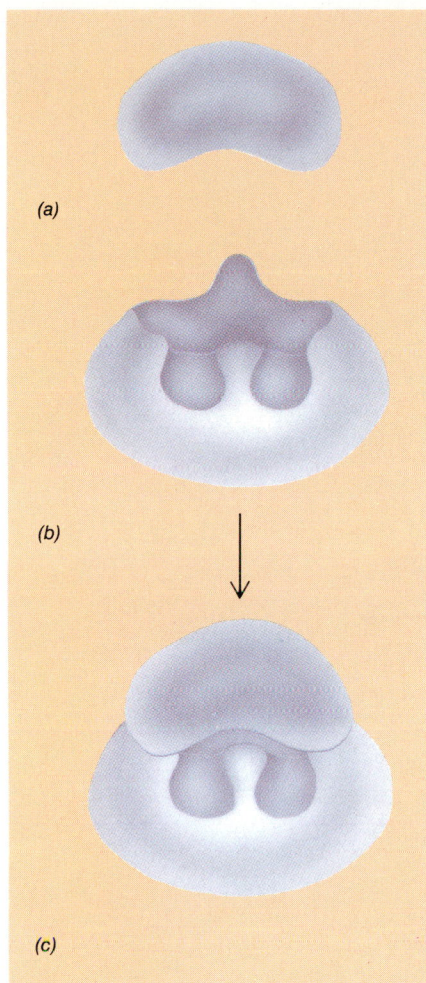

(a)

(b)

(c)

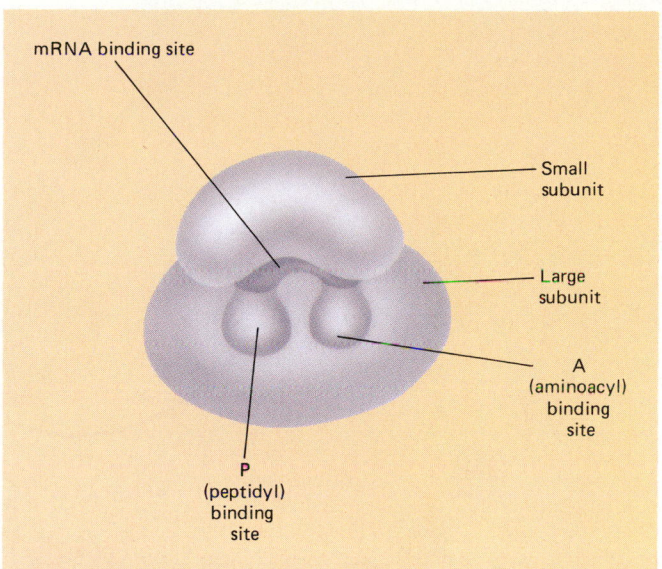

mRNA binding site

Small subunit

Large subunit

A (aminoacyl) binding site

P (peptidyl) binding site

FIGURE 11–7 The binding sites of a ribosome are located mainly on the large subunit. There is a site at which the mRNA binds to the ribosome, together with the A-site and the P-site.

FIGURE 11–6 Portrait of a ribosome. A ribosome is composed of two irregularly shaped subunits, which fit together snugly. The two subunits do move apart slightly during translation.

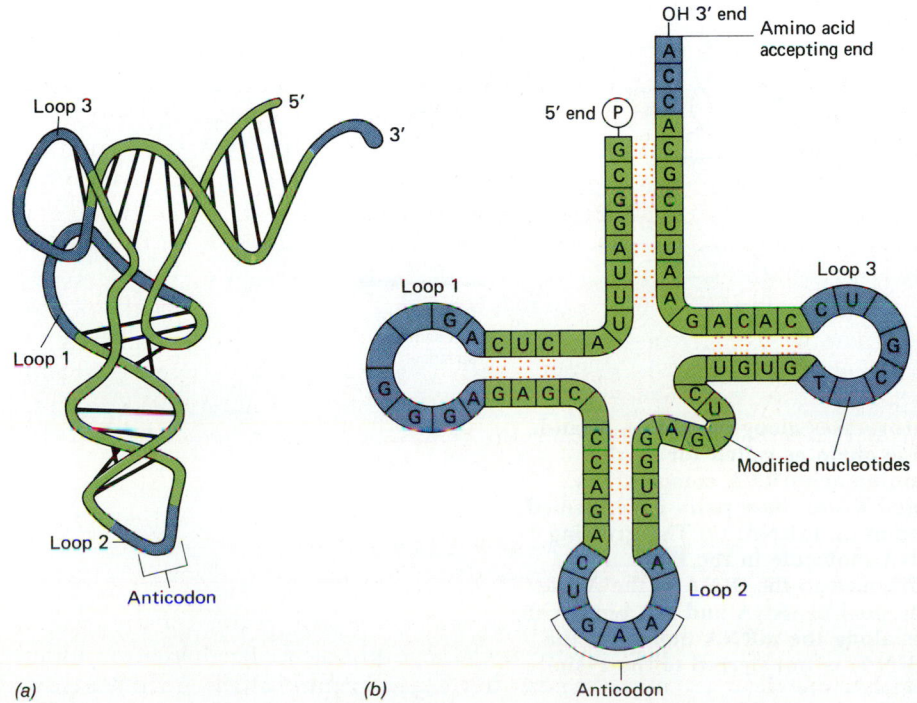

FIGURE 11–8 Two representations of the structure of a typical tRNA molecule; tRNA molecules are the compounds that "read" the genetic code. A diagram of the actual shape of a tRNA molecule is shown in (a). Its three-dimensional shape is determined by intramolecular hydrogen bonds between base-paired regions, which are most clearly observed in the two-dimensional cloverleaf form depicted in (b). One loop contains the triplet anticodon that base pairs with the mRNA codon. The amino acid is attached to the terminal ribose at the 3'—OH end, which has the nucleotide sequence ACC. The pattern of folding permits a constant distance between anticodon and amino acid in all tRNAs that have been studied.

(a)

(b)

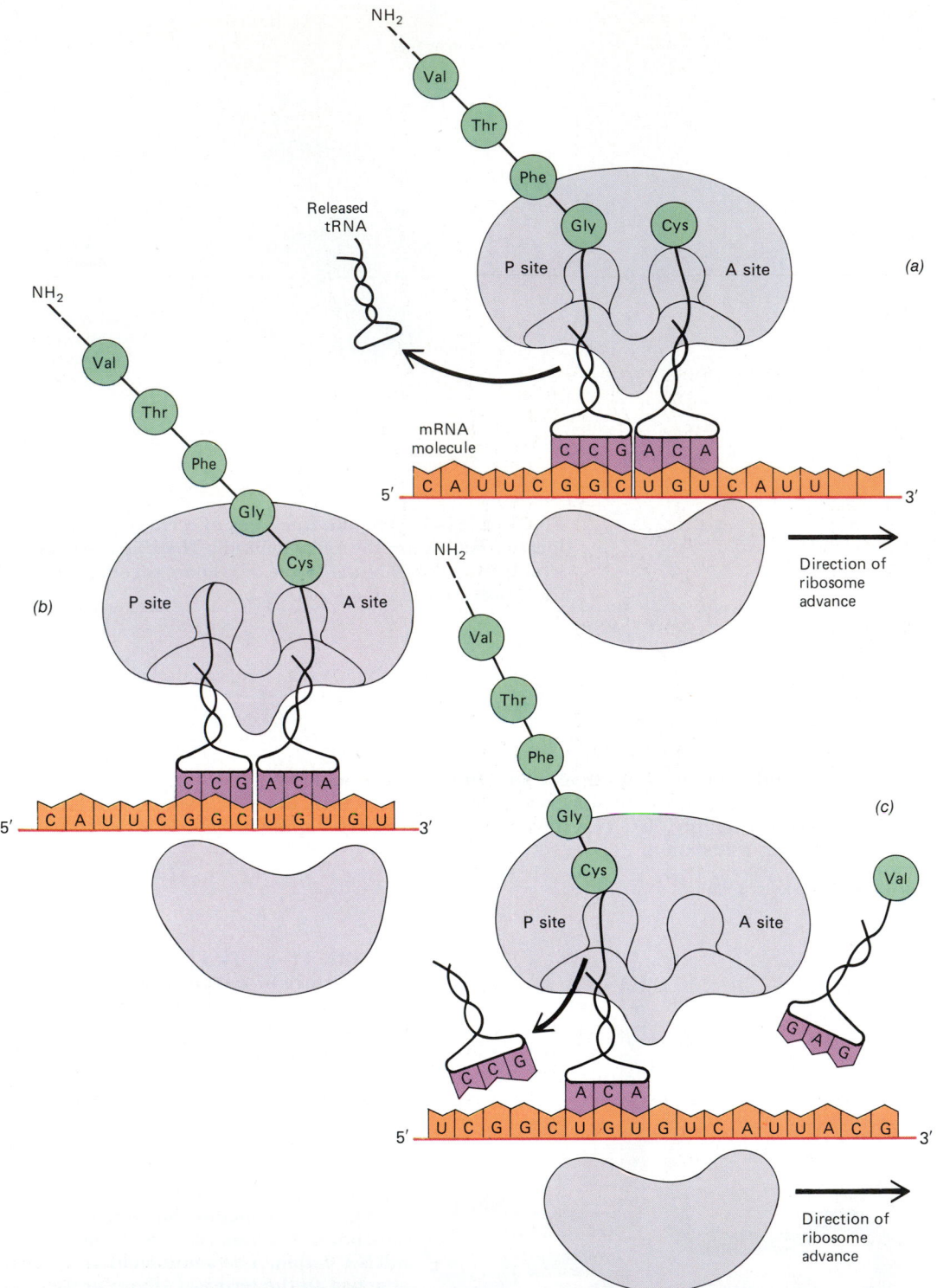

FIGURE 11–9 Translation. A ribosome progresses along an mRNA strand, adding amino acids to a growing polypeptide chain as called for by the sequence of codons in the mRNA. (*a*) **An amino acid-tRNA complex has bound to an empty A-site next to an occupied P-site. Base pairs have formed between the anticodon on tRNA and the codon on mRNA.** (*b*) **The growing polypeptide chain is detached from the tRNA molecule in the P-site and joined by a peptide bond to the amino acid linked to the tRNA at the A-site.** (*c*) **The released tRNA joins the cytoplasmic pool of tRNA and can bind with another amino acid. As the ribosome moves along the mRNA molecule, the growing peptide chain (still attached to a tRNA) is transferred to the P-site.**

the ribosome. Its tRNA molecule is still attached (Fig. 11–9*a*). This amino acid is called the "peptide" amino acid or the "terminal" (end) amino acid. (In fact, it is no longer an amino acid, since it is now part of a larger molecule and thus its properties have been greatly altered. For convenience, however, we will continue to refer to it loosely as an amino acid.)

The **A-site** of the ribosome is where we find the next amino acid to be added to the chain (Fig. 11–9*a*). Exactly which amino acid is added next depends on the mRNA codon that is being "read" at the A-site. The ribosome recognizes a complementary anticodon in a tRNA molecule. It accepts this tRNA molecule and its attached amino acid, which temporarily occupy the A-site together.

Next, a peptide bond forms between the new amino acid in the A-site and the terminal amino acid in the P-site (Fig. 11–9*b*). The tRNA attached to the amino acid in the P-site is released, and the newly incorporated amino acid and its tRNA move from the A-site to the P-site (Fig. 11–9*c*). At the same time the ribosome **translocates,** or moves, along the mRNA strand, so that a new codon can be read at the now-vacant A-site. The process repeats as the next tRNA-amino acid complex arrives.

To sum it up (Fig. 11–9), elongation involves:

1. Acceptance of an amino acid-tRNA complex by the A-site.
2. Establishment of a peptide bond between the A-site amino acid and the P-site (terminal) amino acid.
3. Release of the tRNA formerly occupying the P-site.
4. Translocation of the ribosome down the mRNA strand by one codon.

These steps require energy. While they are taking place, the tRNA-amino acid complex is part of a larger structure called an **elongation complex.** It includes a special protein plus GTP (an energy-charged substance similar to ATP). When the amino acid-tRNA enters the A-site, the GTP in its elongation complex breaks down (much as ATP does), releasing energy that is used in the formation of the peptide bond between the amino acids in the A- and P-sites. Energy is also used in the translocation of the ribosome.

Each strand of mRNA may be used to make multiple copies of a particular protein. A number of ribosomes—as many as 10 or 20—usually bind to a single strand of mRNA, each ribosome independently producing a polypeptide. The entire complex is called a **polyribosome** (Fig. 11–10).

Special Signals Direct The Ribosome To Start And Stop Adding Amino Acids

We have described the way translation proceeds so that you will better understand the special way in which it

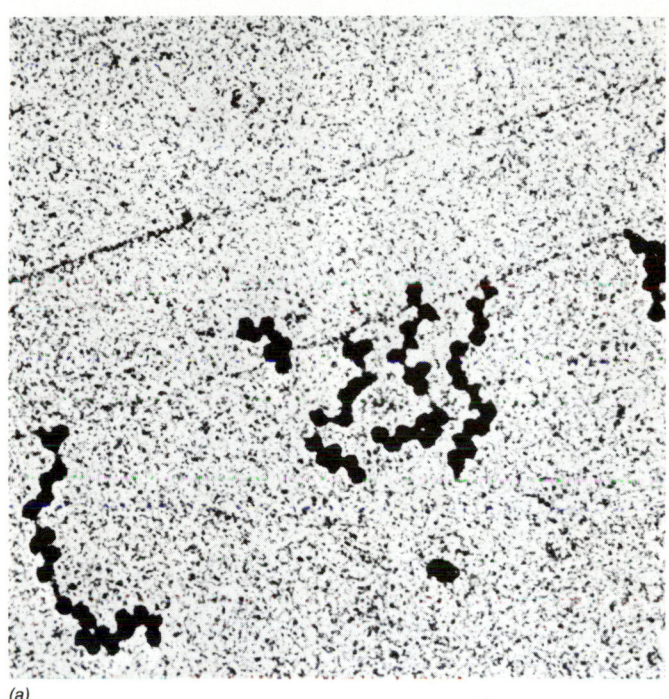

(a)

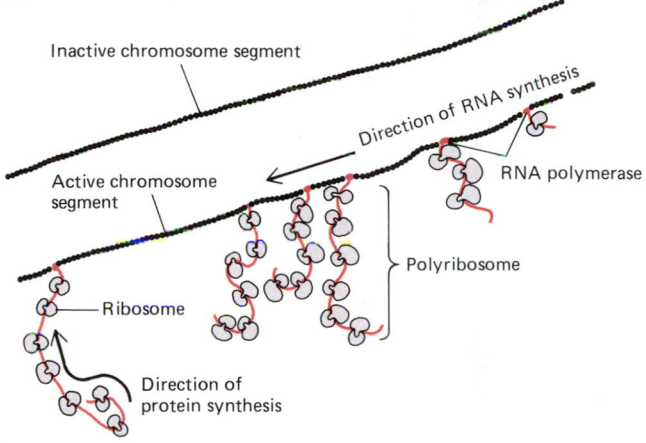

(b)

FIGURE 11–10   Overview of transcription and translation in a bacterial cell. (*a*) Electron micrograph of two strands of DNA, one inactive and the other actively producing mRNA, which is attached to polyribosomes. (*b*) Diagrammatic representation of the processes of transcription and translation. (*a*, Courtesy of O. L. Miller, Jr.; *b*, after Miller, Hamkalu, and Thomas, Jr.)

starts and ends. As with all aspects of translation, initiation and termination are governed by trios of bases.

The "start" signal for translation is the codon AUG on an mRNA strand. The ribosome recognizes the AUG codon as the place to start "reading" the mRNA strand. If you consult Table 11–1, you will see that AUG codes for the amino acid methionine. Thus, proteins ordinarily begin with methionine, although in some proteins the initial methionine is later removed.

For translation to begin at the start codon, the mRNA must bind to the ribosome, and for this to occur, the two ribosomal subunits must first come

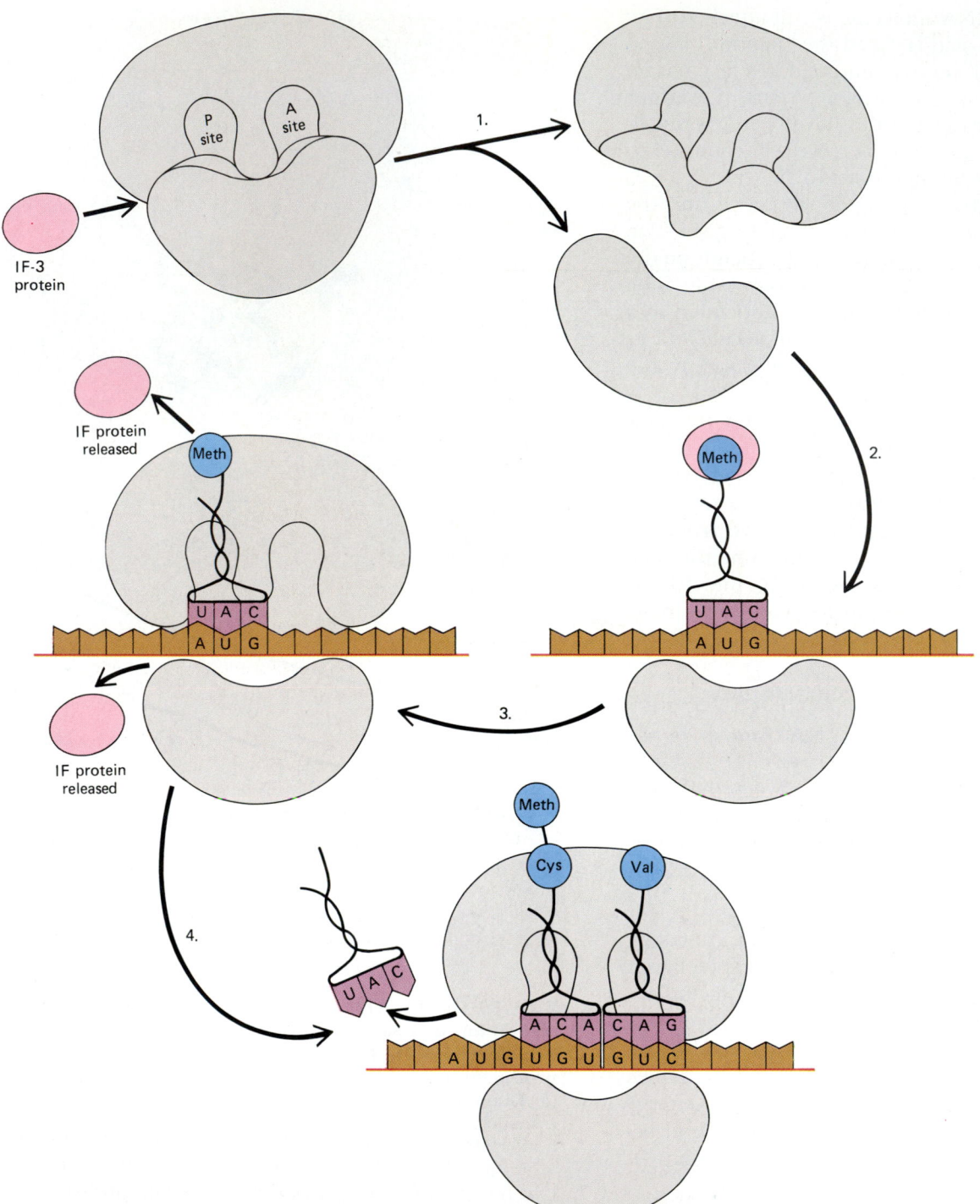

FIGURE 11–11   Initiation of translation. (1) The IF-3 protein dissociates the ribosomal subunits. (2) An initiation complex forms. (3) After the methionine (the first amino acid of the peptide) binds to the P-site, the initiation factor proteins are released. A second amino acid-tRNA complex can now bind to the empty A-site. (4) The process of peptide elongation continues with the addition of new amino acids.

apart (Fig. 11–11). The mRNA strand binds to the small subunit, along with the first amino acid-tRNA complex, carrying the amino acid methionine. This entire assembly is called the **initiation complex.** The large subunit of the ribosome then joins this complex, forming a complete ribosome.

The methionine-tRNA group now occupies the P-site of the completed ribosome. It is in the same position as if it were the terminal amino acid in a chain that had already begun. The elongation of the chain can now proceed, as the ribosome reads the second codon and the second amino acid-tRNA unit arrives at the A-site. This new amino acid will form a peptide bond with methionine and become the new terminal amino acid of the chain.

Termination of protein synthesis is usually accomplished by special stop signals, UGA, UAG, and UAA. These codons do not code for amino acids, but trigger several release factors. These release factors evidently break the bond between the polypeptide and the tRNA occupying the A-site. The ribosome then releases the polypeptide. An instant later the ribosome may release the mRNA strand and then may start translating another one.

CHANGES CAN OCCUR IN SOME STAGES OF GENETIC EXPRESSION

What would happen if one of the bases of a DNA strand were accidentally changed to another base? Or if, during DNA replication, a base that was not complementary to the template base were placed in the newly forming strand? What if an entire sequence of bases from one DNA molecule moved and became part of another portion of the molecule, or of a different DNA molecule? Such a change, or omission or insertion, would be transmitted to all DNA strands replicated from the one in which it originally occurred. Thus, all the descendents of that cell would share its genetic difference. Such a change would also lead to a corresponding change in the complementary mRNA codon and perhaps ultimately to a change in an amino acid added to a protein. The change might not have any result of great importance, but some changes are very consequential. For instance, if the change were to result in what is known as a "nonsense" sequence of bases—that is, a sequence that has no amino acid equivalent—the protein would suffer a break at that point, ending up only a fraction of its proper length. Or the changed sequence could have an amino acid equivalent that, when inserted into the protein, could alter its structure considerably.

Mutations Are Random, Permanent Changes In DNA

Such accidental changes in genetic information do occur and are called **mutations.** Mutations are not necessarily harmful; some are harmless (for example,

FIGURE 11–12 In the mild genetic disorder albinism, as seen in this koala, the lack of genetic information needed to produce the tyrosinase enzyme leads to a melanin deficiency and thence to albinism itself. See also Figure 1–11.

changes in blood type). Indeed, mutations are the raw material for all of evolution and sometimes permit new beneficial traits to arise. These instances are rare, however, because of the already highly adapted nature of living organisms. It is, after all, unlikely that the random removal or replacement of a part will improve a high-fidelity amplifier; such a change would more likely damage it. So, too, a random change in the genetic material is unlikely to improve the function of a cell or an organism (Fig. 11–12).

Some mutations produce disease; in fact, all human genetic diseases must have originated as mutations. The hereditary disease sickle-cell anemia is caused by a mutation that results in a defect in the hemoglobin molecule. Just one inappropriate amino acid is substituted in two of the four polypeptide chains that make up this vital protein. The inappropriate amino acids cause the formation of abnormal chemical linkages between adjacent hemoglobin molecules when oxygen is in short supply. This in turn causes an abnormal crystal-like structure to form in the hemoglobin and to distort the shape of red blood cells. The abnormal red blood cells clog capillaries and then are destroyed by other cells that function to remove dead or damaged cells from the blood. So many red blood cells are destroyed that anemia results.

Every organism ever investigated has been shown to be subject to mutation. The wonder is that we do not observe more mutations than we do. Fortunately, cells possess enzymes whose function is to repair or remove damaged sections of DNA. As a result, most mutations that do occur are not passed on to other cells.

Factors In The Environment Can Produce Mutations

Mutations are caused by a variety of known agents—certain chemicals, for example, including a number of common food additives and some natural agents (such as the aflatoxins produced by some fungi). Ionizing radiation, in the form of x-rays or radioactive fallout, also produces mutations, as does ultraviolet radiation that reaches the earth. Some of these mutations can result in cancer (ultraviolet exposure is strongly linked to skin cancer). If they occur in reproductive cells (eggs and sperm, for example), abnormal genes can be transmitted to future generations.

It appears that under present conditions the effects of most human-produced agents of mutation are small (although they would not be small in the event of a nuclear war or large-scale nuclear accident). Yet the population exposed to them is large. The total number of mutations they produce must therefore be large and is probably increasing.

It is sometimes argued that complete removal of artificial mutagenic substances from the environment would be too expensive to be practical and is probably unnecessary. According to this school of thought, mutagenic agents that remain below certain levels cause no harm. All we need do is to keep the concentration of the mutagen at or below this level, and all will be well.

The biggest objection to this view is that the most diligent searching has failed to disclose the existence of such a minimum threshold level for any known mutagen. What we see instead is called a linear-dose relationship. The smaller the dose, the smaller the effect, yet even the smallest dose produces some effect. Even if levels do exist below which mutagens do no harm, we are totally ignorant of what they are. It seems only prudent to attempt to reduce our exposure to known mutagens as much as possible. At the same time it is imperative that we discover any unknown mutagens that may exist in supposedly harmless substances.

■ CHAPTER SUMMARY

I. Genetic information is stored in trios of nucleotide bases called triplets in DNA, codons in mRNA, and anticodons in tRNA.
   A. The base sequence in a DNA triplet specifies a particular amino acid that will eventually be incorporated into a polypeptide.
   B. In transcription, one DNA strand serves as a template for the synthesis of a complementary mRNA molecule.
   C. The mRNA strand carries the coded genetic message to a ribosome.
II. The mRNA becomes associated with a ribosome, which translates the mRNA, forming a protein.
   A. Translation consists of three steps: initiation, elongation, and termination.
   B. In initiation, a tRNA molecule carrying methionine attaches to the P-site of the ribosome.
   C. In elongation, a tRNA molecule carrying the next amino acid in the chain occupies the A-site of the ribosome. The amino acid forms a peptide bond with methionine and moves into the P-site. Its tRNA molecule is released, and the process repeats. Each tRNA molecule attaches to a specific amino acid. The ribosome recognizes the anticodon of the tRNA molecule, guided by the mRNA codon.
   D. In termination, the mRNA, ribosome, and polypeptide separate.
III. Most eukaryotic genes that code for protein products contain noncoding DNA sequences interspersed among the coding sequences.
   A. The entire gene is transcribed, but the introns, mRNA sequences complementary to the noncoding sequences, are removed by enzymes.
   B. The remaining fragments, called exons, are rejoined, making a mature, functional strand of mRNA.
IV. Thanks to genetic proofreading mechanisms, genetic errors are relatively uncommon. When they do occur, mutations may render specific proteins nonfunctional and thus lead to genetic disease.

■ POST-TEST

1. The three types of RNA are _____, _____, and _____.
2. Information is transferred from DNA to the ribosomes by _____.
3. Each ribosomal subunit consists of molecules of protein and _____.
4. A sequence of three nucleotides in DNA, the triplet, corresponds to the _____ in mRNA.

5. The synthesis of RNA on a DNA template is called _____.

6. Transcription begins when the enzyme _____ _____ binds to a _____.

7. A polyribosome consists of mRNA bound to several _____.

8. The tRNA attached to a new amino acid is accepted by the ribosomal _____.

9. Three steps in translation are _____, _____, and _____.

10. The stretches of noncoding information that are stripped out of mRNA before translation are known as _____.

■ REVIEW QUESTIONS

1. Summarize the complete sequence of information transfer from DNA to finished protein, including the roles of mRNA, tRNA, rRNA, and ribosomes.
2. If DNA directs the synthesis of protein, how can it be said that DNA determines the nature of the cell's chemical constituents other than protein?
3. Contrast DNA replication with mRNA transcription.
4. What is the relationship among triplets, codons, and anti-codons?
5. What is the logical basis of the fact that the unit of the genetic code is the triplet rather than a pair or quadruplet of bases?
6. Match these terms with the appropriate letter in the diagram: translation, replication, transcription.

ⓐ ⟲ DNA —ⓑ→ RNA —ⓒ→ protein

■ RECOMMENDED READINGS

Doolittle, W.F. "The Origin and Functioning of Intervening Sequences in DNA—A Review," *American Naturalist*, 130:915–928, 1987.

Radman, M., and R. Wagner. "The High Fidelity of DNA Duplication," *Scientific American*, August 1988, 40–46.

How the cell proofreads for genetic mistakes.

Watson, J.D., et al. *Molecular Biology of the Gene*, 4th ed. Menlo Park, California, Benjamin Cummings, 1987. The most recent edition of Watson's classic text.

12
Gene Regulation

OUTLINE

I. Gene regulation in prokaryotes primarily involves control of transcription
   A. Operons are units of gene expression in prokaryotes
      1. The *lac* operon is activated by the presence of lactose
      2. The *trp* operon is activated by the absence of tryptophan
   B. The shape of the bacterial chromosome has little or no effect on gene expression
II. There are multiple levels of gene regulation in eukaryotes
   A. Gene control is important in cell growth and differentiation
   B. Most cells of a multicellular eukaryote have the same genes
      1. A plant cell can develop into a functional plant
      2. Some differentiated amphibian cells can form entire organs
      3. The nucleus of a differentiated frog cell can form a tadpole
   C. Chromosome structure is an important part of transcriptional regulation in eukaryotes

LEARNING OBJECTIVES

After you have studied this chapter, you should be able to:

1. Diagram the *lac* operon and describe how it functions.
2. Compare and contrast the *lac* operon with the *trp* operon in *Escherichia coli*.
3. Distinguish between enzyme induction, end-product repression, and end-product inhibition in prokaryotes.
4. Compare the structure of the eukaryotic chromosome with that of the prokaryotic chromosome.
5. Describe the various levels of gene regulation in eukaryotes.
6. Give evidence that some differentiated cells retain the genetic information for the production of other kinds of differentiated tissue, or even the entire organism.

Human female chromosomes (Runk/Schoenberger, from Grant Heilman)

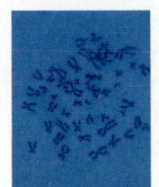 Higher plants and animals may be composed of billions of cells. Many of those cells have specialized structures and functions. We have learned that almost all of those cells have identical genetic blueprints for the entire organism, possibly as many as 100,000 different genes. Clearly, however, not all genes are expressed in all cells all the time. Just as not all books in a library are in use at any one time, so too not all the genes possessed by an organism are being "read" and used all at once. Even a single-celled bacterium, which may possess between 2000 and 4000 genes, does not continuously produce all the proteins its genetic material encodes.

Some genes produce proteins that *are* continuously used. Called **constitutive genes,** they are active throughout the life of the cell. Other genes, however, can be "turned on" or "turned off," that is, directed either to produce their protein product or to stop producing it. These regulated genes are the focus of this chapter. Investigation of the ways in which genes are regulated is one of the most exciting and fast-advancing areas in biology, with great practical and theoretical implications.

KEY CONCEPTS

☐ Many genes are regulated so that their protein products are only formed at certain times.

☐ In prokaryotes, regulated genes are organized into units called operons. Each operon typically encodes for several proteins and controls protein synthesis by regulating transcription.

☐ Cell growth and differentiation in multicellular eukaryotes occur as a result of selective gene expression.

☐ Eukaryotic genes may be regulated at several levels: transcription, messenger ribonucleic acid (mRNA) processing, translation, and post-translational protein processing.

☐ GENE REGULATION IN PROKARYOTES PRIMARILY INVOLVES CONTROL OF TRANSCRIPTION

Prokaryotes such as bacteria are generally single-celled organisms. Under good environmental conditions, they grow rapidly, each cell splitting in half asexually (Chapter 21). They must be able to utilize whatever food sources they encounter during their short lives. Generally, prokaryotes make whatever proteins they need, including enzymes to digest food. However, little protein is made in anticipation of need; that is, bacte-

rial cells do not produce all the enzymes and other proteins they are capable of making just "in case" they need them. Instead, their genes are regulated so that transcription (and ultimately protein synthesis) occurs only when a particular enzyme or other protein is required.

Much of our understanding of gene regulation comes from studies of the colon bacterium, *Escherichia coli.* Its availability and rapid growth when cultured in the laboratory have led to its extensive study. It is probably the best understood organism on the face of the earth.

E. coli live in countless billions in the intestines of mammals. For a short time after the mammal eats, these bacteria are surrounded by the nutrients that the mammal's enzymes have digested. *E. coli* can readily partake of these nutrients without using any digestive enzymes of their own. Their protein production is channeled in other directions. During times when the mammal's intestine is empty of simple, predigested food molecules, however, the bacteria synthesize enzymes to digest unprocessed food materials. They are able to turn on and off the genes that code for many kinds of digestive enzymes.

Our knowledge of the gene-regulating mechanisms in *E. coli* began with the contributions of the French investigators François Jacob and Jacques Monod. Monod began research on *E. coli* in the 1940s during World War II in German-occupied France. In the course of his research, he encountered an oddity in the behavior of *E. coli* bacteria. When fed a mixture of glucose and lactose (a disaccharide found in milk), the bacteria first consumed all the glucose. Then, after a brief pause, they began to consume the lactose. Further investigations showed that in order to consume the lactose, which is composed of two simple sugars (glucose and galactose), the bacteria first had to have the enzyme that breaks down lactose into simpler sugars. *E. coli* that had not been fed lactose ordinarily did not possess this enzyme. However, once they were fed lactose, they started to make the enzyme.

The explanation was some time in coming, but in 1960 Jacob and Monod sent a paper to the *Journal of Molecular Biology* that described how gene regulation in *E. coli* permitted it to produce the lactose-digesting enzyme when it was given lactose as food. They called this mechanism of gene regulation the operon. Since then, similar operons have been described not only in *E. coli* but in other bacteria as well.

Operons Are Units Of Gene Expression In Prokaryotes

It appears that most, if not all, bacterial genes are organized into clusters called **operons.** Operons are composed of four parts: structural genes, a regulator gene, a promoter, and an operator.

Each operon includes several genes, called **structural genes,** that code for the synthesis of a group of enzymes that are all involved in the same function (for example, the digestion of lactose). Transcription of mRNA begins when RNA polymerase recognizes the promoter site on a strand of DNA. The **promoter** is that part of the operon where RNA polymerase binds to begin transcription of the structural genes. The structural genes of an operon all share a single promoter site. When RNA polymerase binds to this site, it transcribes all the structural genes onto one mRNA strand, which may then go on to be translated into the individual proteins.

However, other features of the operon may prevent RNA polymerase from binding to the promoter site and thus may prevent transcription from occurring. Whether the structural genes will be transcribed is under the control of a single **operator.** The operator can switch transcription of the structural genes on or off by allowing or preventing RNA polymerase to interact with the promoter. The operator is regulated by the **regulator gene,** which codes for the production of a protein that can bind to the operator.

The exact mechanisms of gene regulation vary for different operons. We will consider the details of two different operons that have been well-characterized in *E. coli:* the lactose operon and the tryptophan operon.

The *lac* Operon Is Activated By The Presence of Lactose

Let us take as an example the operon in *E. coli* that controls the production of enzymes needed to digest lactose. By convention, this **lactose operon** is called the *lac* operon. It consists of a regulator gene, three structural genes, and operator and promoter sites (Fig. 12–1). The operator site overlaps the promoter site. The structural genes code for three enzymes used in the digestion of lactose: (1) a permease that lets lactose into the cell; (2) the enzyme that breaks down lactose into simpler sugars; and (3) another enzyme that functions in the further digestion of lactose.

However, the regulator gene of the operon codes for a protein called a **repressor protein.** Under normal conditions in *E. coli,* when lactose is *absent,* this repressor protein binds to the operator site of the operon, blocking the promoter site for the structural genes in the process. Now RNA polymerase cannot bind to the promoter site and therefore cannot begin transcribing the structural genes. Thus, when there is no lactose present, the structural genes that code for the enzymes needed to digest it are not even transcribed. The question is, then, when are they transcribed?

As it turns out, when lactose is present, it attaches to the repressor protein, modifying it; as a result, the repressor protein does not bind efficiently to the operator site. Now the promoter site is unblocked, and RNA polymerase may begin transcribing the structural genes. The resulting mRNA may be translated, forming the enzymes involved in the digestion of lactose. This sort of control, in which the presence of a substrate (here, lactose) initiates, or induces, the synthesis of an enzyme, is known as **enzyme induction.** In enzyme induction of the *lac* operon, the presence of the substrate removes the repressor, allowing the structural genes to be expressed in protein synthesis.

Should the lactose once again become scarce, we return to the first scenario. No lactose binds to the repressor. The repressor now attaches to the operator site, preventing further transcription of the structural genes for the enzymes.

What is the advantage of *E. coli*'s ability to repress the formation of these enzymes when lactose is not present? Clearly, it takes energy and raw materials to manufacture those enzymes. For the bacterium that continued to express those genes when their products would not be used, this "unprofitable" diversion of resources might be the margin between life and death.

Yet even with the very small likelihood that any particular *E. coli* will ever be fed lactose, there is an advantage to having the genes for the enzymes that digest it. An *E. coli* bacterium living in the intestine of a mature cow has probably never been exposed to lactose, because the cow probably has not drunk milk for several years, since it was a calf. Nor have a thousand generations of the *E. coli*'s ancestors fed on lactose (*E. coli* can reproduce several times an hour, so a thousand generations could all have lived in the same cow). Yet at some point, its ancestors used the lactose genes, because at some point the cow was a calf feeding on its mother's milk. As the cow grew and stopped drinking milk, the *E. coli* stopped feeding on lactose and stopped expressing the genes for the lactose-digesting enzymes. However, these genes remained a part of the bacterial deoxyribonucleic acid (DNA) and were replicated and passed on through each generation, never once revealing their presence. But when lactose again was present, they were available for use.

The *trp* Operon Is Activated By The Absence Of Tryptophan

In addition to making digestive enzymes, *E. coli* can also make individual amino acids. For example, if given carbon and nitrogen, *E. coli* can manufacture the amino acid tryptophan. The manufacture of tryptophan is a step-by-step pathway. There are five steps, each catalyzed by a specific enzyme. The **tryptophan**

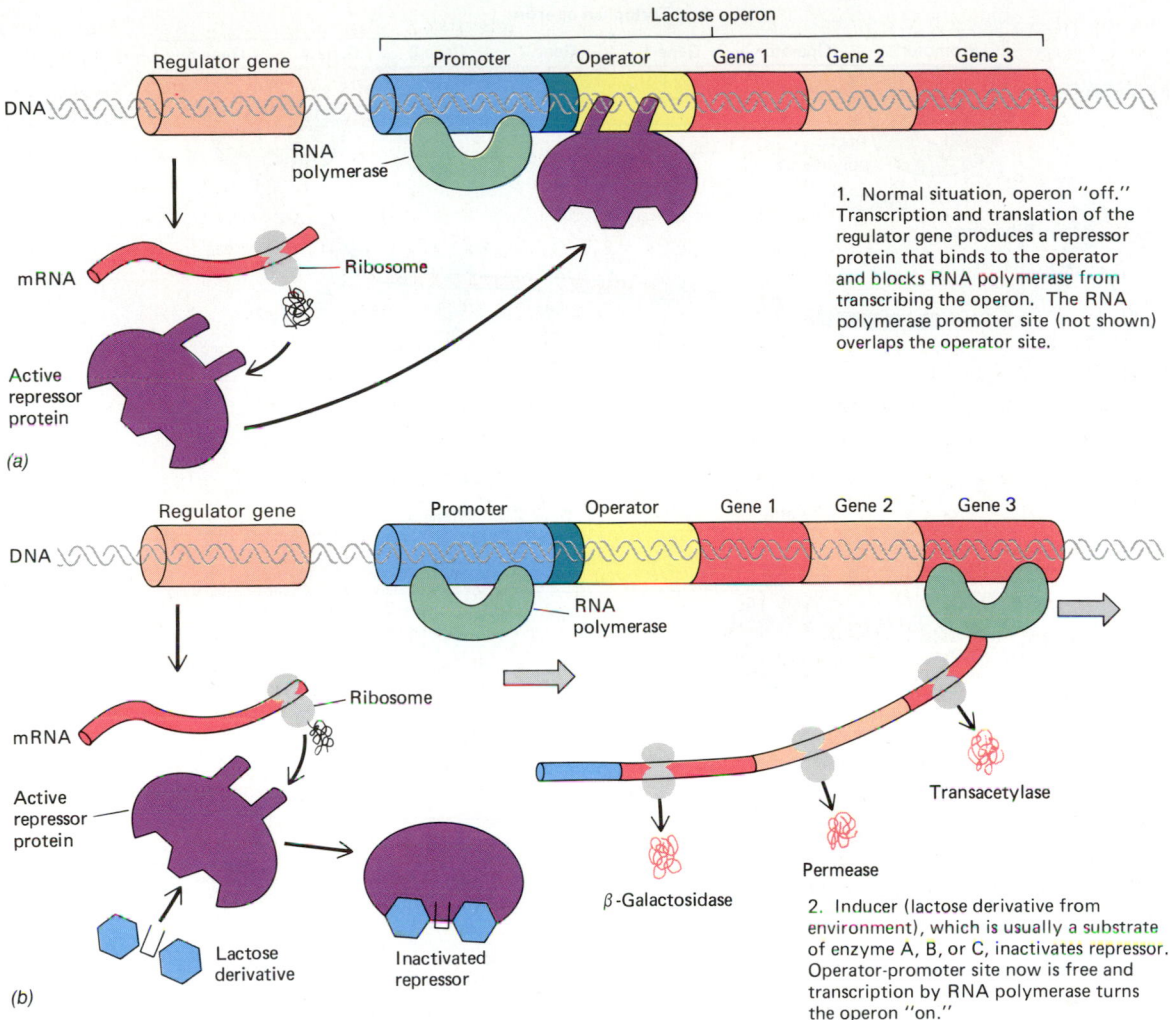

1. Normal situation, operon "off." Transcription and translation of the regulator gene produces a repressor protein that binds to the operator and blocks RNA polymerase from transcribing the operon. The RNA polymerase promoter site (not shown) overlaps the operator site.

2. Inducer (lactose derivative from environment), which is usually a substrate of enzyme A, B, or C, inactivates repressor. Operator-promoter site now is free and transcription by RNA polymerase turns the operon "on."

FIGURE 12–1   The *lac* operon consists of a group of structural genes, a regulator gene, and operator and promoter sites. (*a*) In the absence of lactose, a repressor protein binds to the operator, blocking RNA polymerase from transcribing the structural genes. (*b*) When lactose is present, it binds to the repressor protein. As a result, the repressor cannot bind to the operator, and transcription of the structural genes takes place. The three proteins that are ultimately synthesized enable the bacterium to utilize lactose.

operon, or *trp* operon, in *E. coli* includes five adjacent structural genes that code for the five enzymes needed to produce tryptophan (Fig. 12–2). The operon also includes a regulator gene and operator and promoter sites.

For the tryptophan pathway to proceed, all five enzymes must be present and working. In addition, tryptophan itself must be in low supply. Notice the difference in action between the *trp* operon and the *lac* operon. The *lac* operon is activated by the presence of the substrate (lactose), while the *trp* operon is activated by the absence of the product (tryptophan).

As in the lactose operon, the operator site in the tryptophan operon overlaps the promoter region. The regulator gene codes for a repressor protein that can bind to the operator site and block the promoter site. Unlike the repressor in the lactose operon, however, the repressor protein in the *trp* operon does not by itself stick to the operator site. It adheres only when its shape is slightly modified by the presence of another substance.

Perhaps you have guessed that in this case the substance that modifies the shape of the repressor protein is none other than tryptophan. It acts as a co-repressor, enabling the repressor protein to function in blocking the synthesis of more tryptophan-forming enzymes. This process, in which the presence of the product *prevents* its production at the level of transcription, is

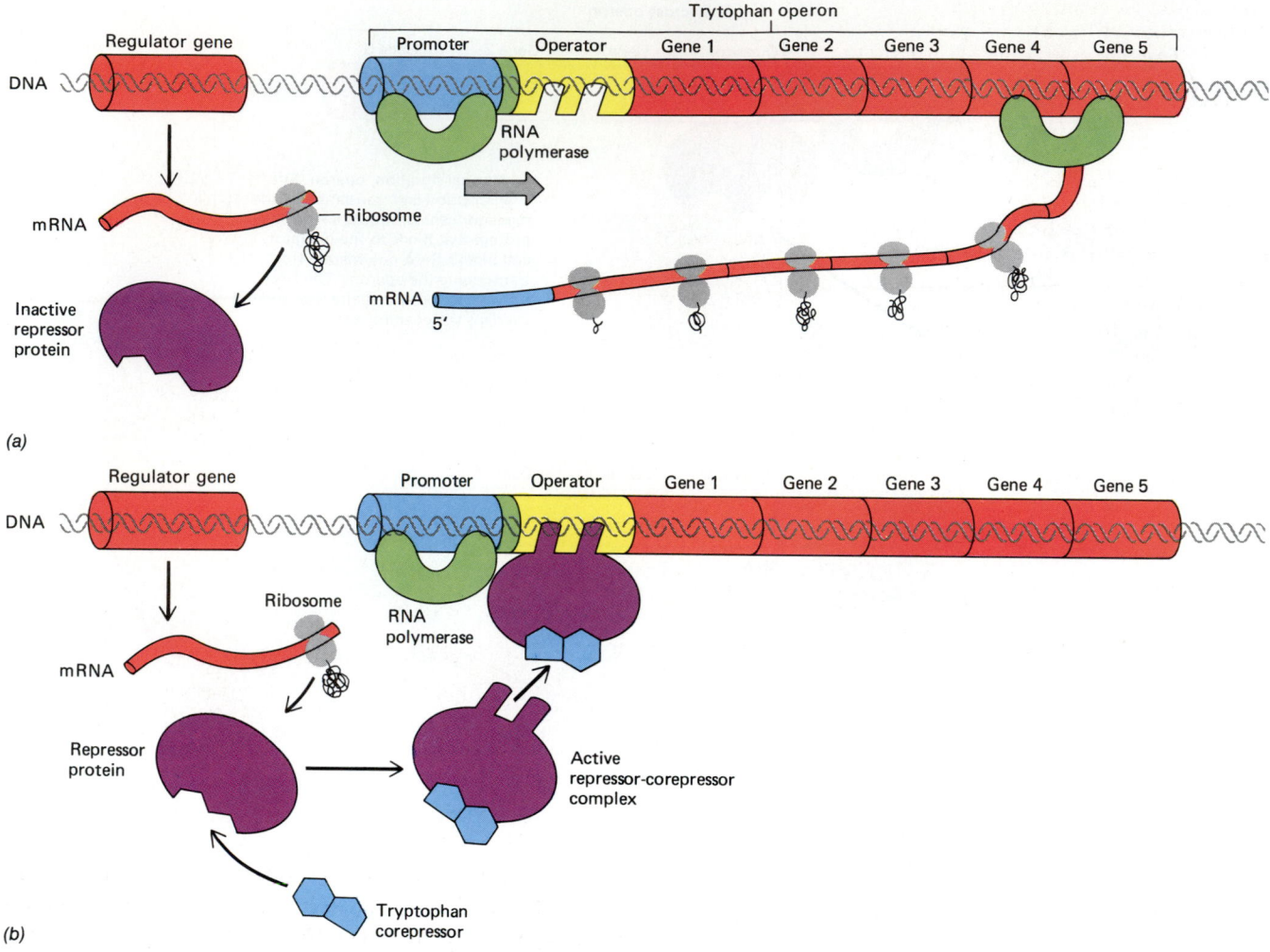

FIGURE 12–2 The *trp* operon of *E. coli*. (*a*) The regulator gene codes for a protein that cannot function by itself. In the absence of tryptophan, the genes for the five enzymes necessary for tryptophan synthesis are transcribed. (*b*) When tryptophan levels are high within the cell, tryptophan attaches to the repressor protein. This activates the repressor protein, which binds to the operator site, preventing transcription of the structural genes.

called **end-product repression.** The end product of a long biosynthetic pathway, here tryptophan, shuts down the pathway before it can even begin.

At the time at which the tryptophan supply becomes great enough to signal repression, however, enough of the five enzymes that form tryptophan may already have been made. As long as all five enzymes exist, the synthesis of tryptophan can still occur. In other words, although end-product repression cuts off the pathway at its source, what about the enzymes that are already at work?

Another method of gene regulation functions in the tryptophan biosynthetic pathway itself. When tryptophan is present in large quantities, tryptophan molecules bind to the first enzyme in the pathway, altering the shape of that enzyme and thereby preventing it from functioning. Since this first step has now been blocked, the entire pathway is blocked and the synthe-

sis of tryptophan does not occur. This type of regulation is called **end-product inhibition.** This is an example of feedback control (Chapter 7).

In summary, when tryptophan is present in *E. coli*, it inhibits its own synthesis in two ways. It prevents the formation of all the enzymes involved in its synthesis by blocking—through co-repression—transcription of their structural genes. It also interferes with the action of the first enzyme needed to synthesize more tryptophan. The end result of these two forms of control is that the bacterium conserves energy and raw materials.

The Shape Of The Bacterial Chromosome Has Little Or No Effect On Gene Expression

The DNA of a prokaryotic cell is found in two forms: as a single, circular molecule called a chromosome, and

FIGURE 12-3 Development of a carrot plant from differentiated root cells. Discs of phloem cells, which are specialized for nutrient transport, were isolated from carrot root tissues. When the cells were cultured in a liquid nutrient medium, clumps of cells developed. These clumps (embryoids) closely resembled plant embryos in their early stages of development. They developed further, forming embryonic roots and shoots. Transferring the embryonic tissue to a solid nutrient medium stimulated the tissues to form plantlets, which could then be grown into mature plants.

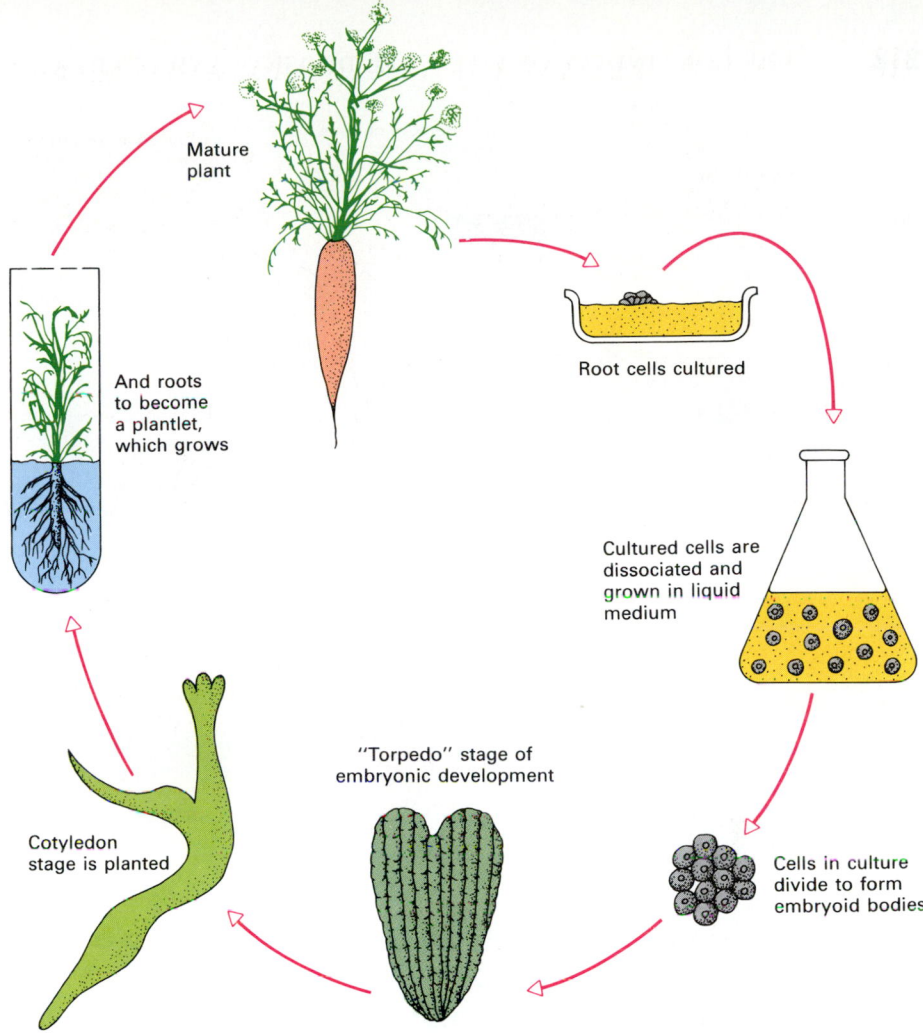

Mature plant

Root cells cultured

Cultured cells are dissociated and grown in liquid medium

And roots to become a plantlet, which grows

"Torpedo" stage of embryonic development

Cotyledon stage is planted

Cells in culture divide to form embryoid bodies

as smaller circlets called plasmids. In both types of circular molecules, the two ends of the double helix join to form a circle. We discuss plasmids in Chapter 16; the genes we have been referring to in this chapter are found in the larger chromosome. The shape of the bacterial chromosome plays little or no role in prokaryotic gene regulation. As we will see, this is not the case in eukaryotic gene regulation.

◼ THERE ARE MULTIPLE LEVELS OF GENE REGULATION IN EUKARYOTES

Molecular biologists, when first developing the operon concept, thought that it would be as important in eukaryotes as it is in prokaryotes. Yet now we believe that operons are not important, and probably do not even occur, in eukaryotes. Consider the greater complexity of eukaryotes, and the precision with which a typical eukaryotic cell plays its role in the life of a multicellular organism. In nine months a human zygote (fertilized egg) develops into a marvel of complexity that makes even the most sophisticated computer look simple. Moreover, that complex living machine is preprogrammed not only to grow, but also to operate, to repair, and eventually to reproduce itself.

In view of such complexity, it is perhaps not surprising that eukaryotes do not employ a single, simple method of gene control, but instead use a combination of methods. The types of control we saw in prokaryotes were all variations of the operon concept; prokaryotic gene expression is controlled largely at the level of transcription. Eukaryotic gene control works at the level of transcription as well. In addition, however, eukaryotes also have other levels of control. Once transcription has occurred, the initial mRNA may be modified while still in the nucleus, in what is known as **mRNA processing.** There is also control of translation, which determines if mRNAs will be translated into polypeptides by the ribosomes. Once polypeptides have been synthesized, they often have to be modified to become active, fully functional proteins. Such posttranslational protein processing represents another level of gene regulation.

Gene Control Is Important In Cell Growth And Differentiation

A multicellular organism grows from a single cell, the fertilized egg, or zygote. In early development, the zygote divides into two, then four, then eight cells, and so on. The cells produced from these early cell divisions are all highly similar. Increasingly, however, they begin to **differentiate,** taking on specific structural characteristics until they become the specialized cells of the mature organism (Fig. 12–3).

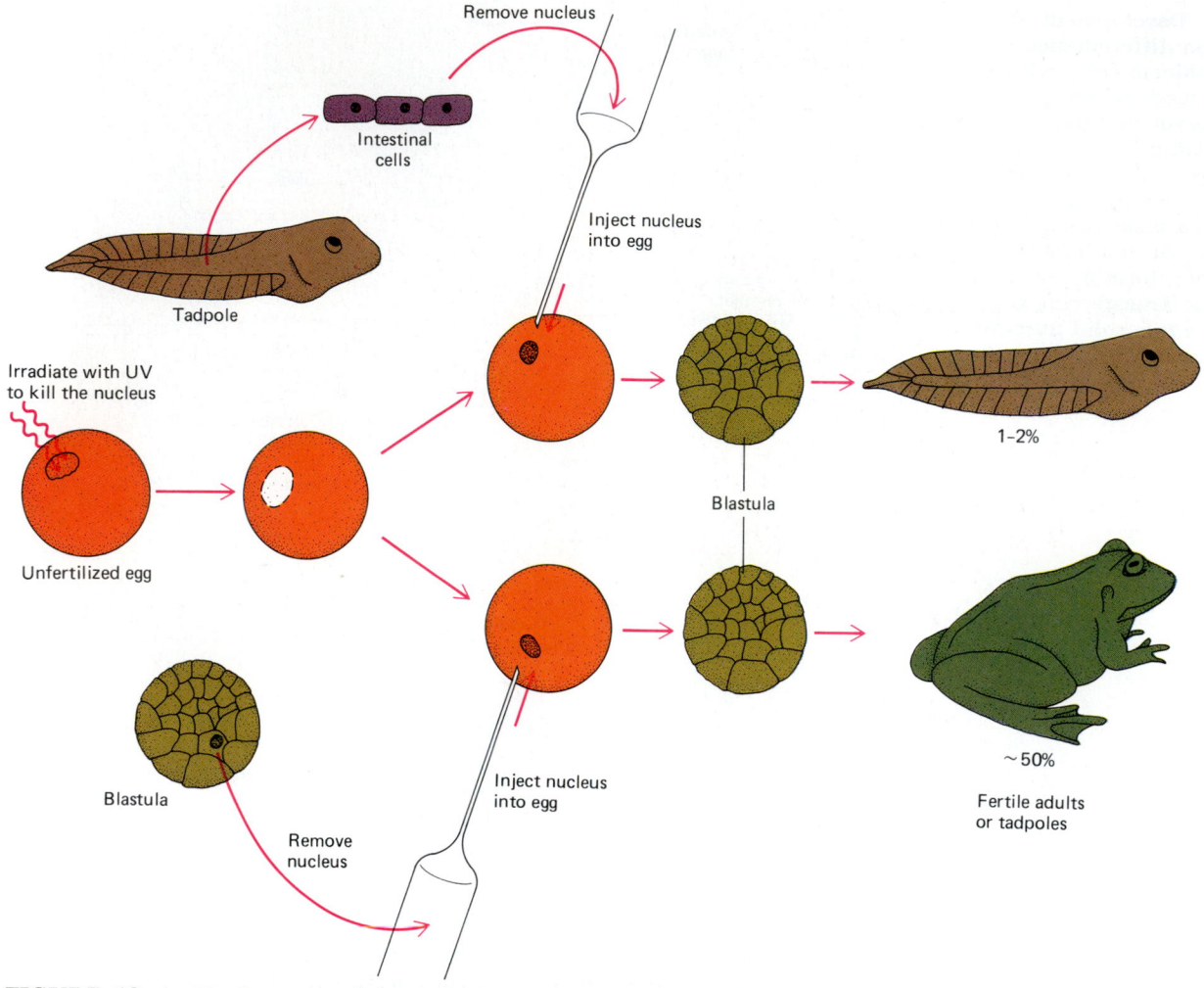

FIGURE 12–4 Nuclear transplantation experiments showed that a nucleus from a differentiated amphibian cell could program development. This was done by injecting it into an egg whose nucleus had been destroyed by ultraviolet radiation. In a small number of trials using nuclei from tadpole intestinal cells, normal development occurred, indicating that the necessary genes to program development to that point were still present.

The orderly development of multicellular organisms is due to the fact that differentiating cells do not produce the same proteins. The distinctive structures and functions of differentiated cells reflect the fact that each type of cell expressed only part of the genetic information it contains, producing some proteins but not others. That is, there is selective gene expression in different tissues and at different times during development. As cells differentiate, they gradually suppress certain genes.

Most Cells Of A Multicellular Eukaryote Have The Same Genes

All the cells in a multicellular eukaryote descended from the same zygote through cell division, in which each cell received a complete set of genetic information from its parent cell. Because of this fact, it is generally accepted that all cells in a multicellular organism must possess the same set of genes. In this regard, even though cells may look and function differently, they are all identical. (There are certain exceptions, such as some white blood cells.) But, can we *prove* that all cells have a complete genetic blueprint for the entire organism?

A Plant Cell Can Develop Into A Functional Plant

We can certainly prove it in the case of higher plants. When you were a child, perhaps you placed a carrot in a container of water. The carrot is a root, that is, a mature organ composed of fully differentiated cells. In

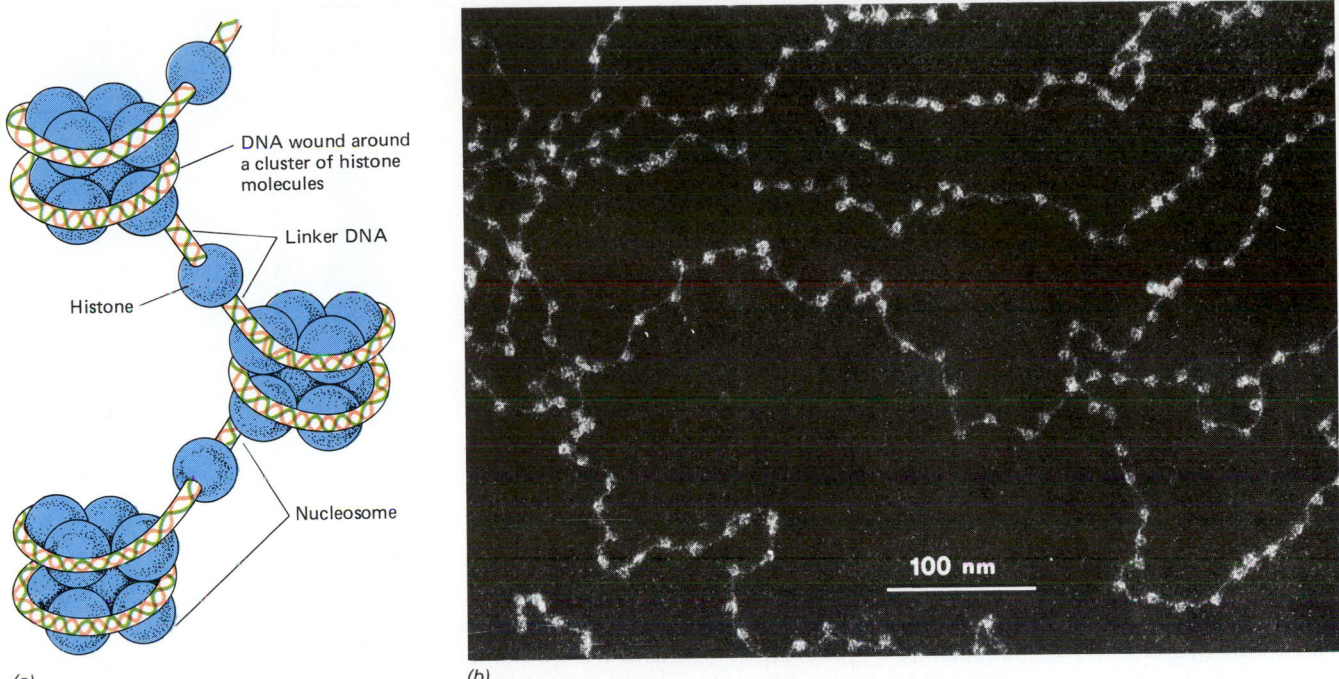

(a)

(b)

DNA wound around
a cluster of histone
molecules

Linker DNA

Histone

Nucleosome

100 nm

FIGURE 12–5   (a) The units of histone surrounded by DNA in the chromosome are called nucleosomes. Here is a model of the structure of a nucleosome. Each nucleosome bead is thought to contain a set of eight histone molecules that form a protein core around which the double-stranded DNA is wound. Another segment of DNA links the nucleosome beads. (b) Electron micrograph of DNA prepared from the nucleus of a chicken red blood cell (×170,000). Note the nucleosome beads. (b D.E. Olins and A.L. Olins)

due course the root sprouted, produced smaller roots, stems, and leaves: a complete new plant, developing not from a seed but from mature, differentiated cells. Today, plant tissue culture techniques are so refined that it is possible to take a single, fully differentiated plant cell and from it grow a mature plant that looks and functions like any other (Fig. 12–3). This property is called **cellular totipotency,** the ability of a cell to develop into any or all the parts of the mature organism.

Some Differentiated Amphibian Cells Can Form Entire Organs

If a salamander limb were to be amputated, and the stump kept moist, muscle cells and other differentiated cells in the stump would lose their distinctiveness and form a **blastema,** or undifferentiated bud. Bone cells, muscle cells, skin cells, and other types of cells used to construct a limb then differentiate from the blastema.

But making a limb grow from an established body is a much smaller task than making the whole organism. Do differentiated cells have all the information they need to produce an *entire* functioning organism? Or have some of the genes in those cells been perma-

nently suppressed, preventing certain types of development from ever occurring?

The Nucleus Of A Differentiated Frog Cell Can Form A Tadpole

If the nucleus of a functional cell from the intestinal lining of a tadpole is injected into a frog egg whose own nucleus has been destroyed, the introduced nucleus will sometimes direct the egg's development (Fig. 12–4). Development may proceed to the tadpole stage, or even to a mature frog. This demonstrates that the nucleus possesses all the information necessary to construct a mature animal. We still do not know how such a nucleus regains the ability to express suppressed genes when placed in a denucleated egg.

Chromosome Structure Is An Important Part Of Transcriptional Regulation In Eukaryotes

In contrast to the simple circular molecules of DNA found in bacteria, the DNA in the nucleus of a eukaryotic cell is in the form of many separate and very long molecules. Each is a double helix, but it does not join ends to form a circle (Fig. 12–5).

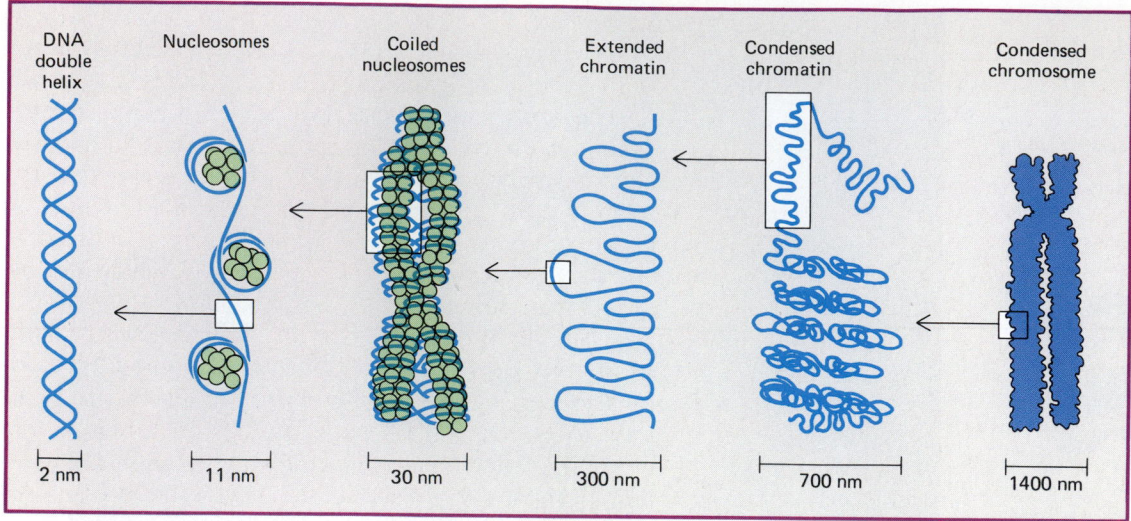

DNA double helix	Nucleosomes	Coiled nucleosomes	Extended chromatin	Condensed chromatin	Condensed chromosome
2 nm	11 nm	30 nm	300 nm	700 nm	1400 nm

FIGURE 12–6 The eukaryotic chromosome has several levels of organization.

Another notable difference is that in eukaryotes the DNA molecules are closely associated with certain proteins, the most notable of which are **histones.** Histone proteins are entwined in regularly arranged, bead-like groups along the DNA molecule. Between each group is a sequence of "linker" DNA. Each "bead" plus the adjacent linker DNA is called a **nucleosome** (Fig. 12–5). Still other histone molecules occur on the DNA molecule between the nucleosomes.

The nucleosomes in turn are very tightly wound into a supercoiled string, which is coiled again, forming a structure known as **chromatin** (Fig. 12–6). This in turn is folded up and looped in three dimensions, forming condensed chromatin. The histone proteins are thought to have a role in determining the pattern of folding. A great deal of DNA, and hence genetic information, can be compacted in this way.

Each separate, entire DNA molecule and its associ-

FIGURE 12–7 (*a*) When eukaryotic DNA is inactive, its nucleosomes are condensed into tight coils. (*b*) We know that active genes are found in decondensed chromatin, which may become so as a result of subtle control signals affecting the histones of the nucleosomes. RNA polymerase would have a much easier time transcribing the information in a decondensed section of chromatin.

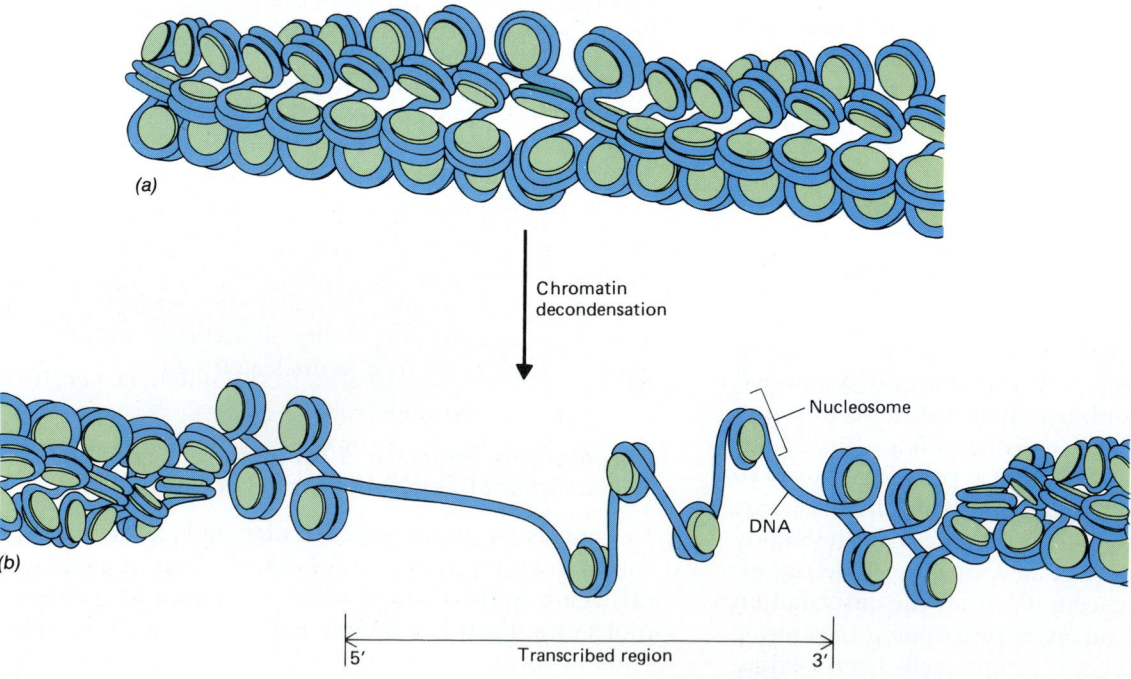

(a)

Chromatin decondensation

Nucleosome

DNA

(b)

5'   Transcribed region   3'

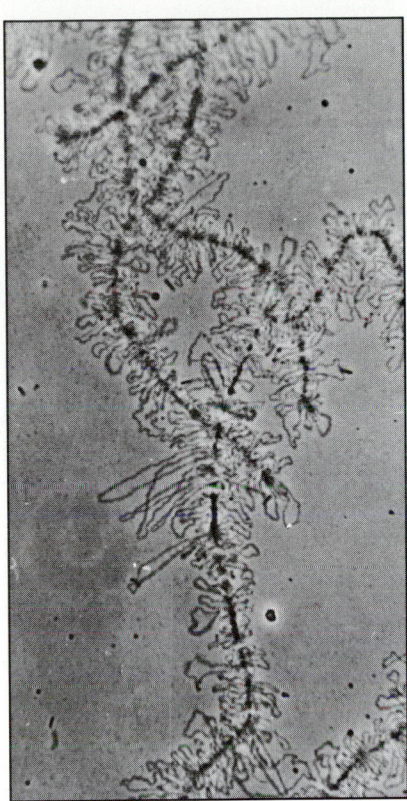

FIGURE 12–8 Photomicrograph of chromosomes from the oocyte of an amphibian (approximately ×1100) showing loops radiating from the central thread. The presence of these loops gave the structures their name, lampbrush chromosomes (lampbrushes were old-fashioned brushes used to clean soot from oil lamps). The loops, which are decondensed chromatin, are sites of intense RNA synthesis. (Dr. Dennis Gould)

ated proteins is called a **chromosome.** Each chromosome carries specific **genes,** precise portions of the cell's genetic information. Genes tend to stay together on the same chromosome, although there are some exceptions to this. During most of the cell's life, all the chromosomes are spread out in the nucleus in the form of chromatin and are indistinguishable from one another. During cell division, as we discuss in the next chapter, the chromatin folds up, forming separate, visible chromosomes.

It is believed that DNA in condensed chromatin cannot be transcribed into RNA. That is, genes in condensed regions cannot be expressed unless they are first "decondensed." Transcription of DNA can only occur if the DNA packaging of the chromosome is relaxed (Fig. 12–7). Because of the way the DNA molecule is coiled, various portions can be relaxed during transcription without the entire huge molecule being uncoiled (Fig. 12–8).

It is thought that transcription is restricted when DNA is tightly associated with histones. It is possible that when a gene is "turned on," other nonhistone proteins in chromosomes may combine with the histones, causing them to loosen their association with DNA. This would permit the transcription of DNA.

We know that the structure of eukaryotic chromosomes plays an important role in the control of transcription. However, a complete understanding of gene regulation at all levels—transcription, mRNA processing, translation, and post-translational protein processing—remains to be determined. Comparing what we know about gene regulation in eukaryotes versus what remains to be discovered is much like comparing the exposed tip of an iceberg to the entire iceberg.

■ **CHAPTER SUMMARY**

I. In prokaryotes, genes are regulated primarily by controlling transcription.
   A. The operon is the unit of gene expression. Each operon consists of: (1) structural genes, which code for proteins used by the cell; (2) a regulator gene, which codes for a repressor protein that binds to (3) the operator site; as a result, the operator site blocks (4) the promoter site where RNA polymerase normally initiates transcription.
   B. In enzyme induction, the presence of a substrate inactivates the repressor protein; therefore, the structural genes are transcribed.
   C. In end-product repression, the presence of a substance (usually the end product of a biosynthetic pathway) signals an operon to stop tran-
   scribing the genes that code for that substance. In end-product inhibition, a substance interferes with an enzyme used to make it.
II. All the genetic information of a zygote is preserved in the mature, differentiated cells of a multicellular organism.
III. Eukaryotic chromosomes are much larger than bacterial chromosomes and are not circular.
   A. Eukaryotic chromosomes are elaborately supercoiled and contain histone and nonhistone proteins.
   B. Eukaryotic genes are regulated in part by changes in the shape of the chromosome, primarily by changes in its degree of condensation.

■ POST-TEST

1. _____ are units of gene expression in prokaryotes, which include structural genes, a regulator gene, and operator and promoter sites.
2. The regulator gene codes for a _____ protein.
3. In the process of _____ _____, the repressor protein is inactivated by the presence of the substrate, resulting in the synthesis of an enzyme.
4. The *lac* operon is activated by the _____ of lactose, while the *trp* operon is activated by the _____ of tryptophan.
5. In end-product _____, the presence of the product prevents its production at the level of transcription.
6. In end-product _____, the presence of the product inactivates the first enzyme in the pathway, preventing further synthesis.
7. The process of change in unspecialized cells that results in their developing the appearance and functions of a specific type of adult cell is known as _____.
8. Cellular _____ is the ability of a cell to develop into specialized parts of a mature organism.
9. An entire tadpole can develop from an egg that has received a _____ from a differentiated intestinal cell.
10. A _____ is a unit of chromatin structure composed of DNA wound around a complex of histone molecules plus a DNA linker region associated with an additional histone molecule.

■ REVIEW QUESTIONS

1. What is an operon? Describe its operation.
2. How are operator and promoter sites similar? How are they different?
3. The histidine operon (*his* operon) in the *Salmonella* bacterium is responsible for the synthesis of the amino acid histidine. It regulates the expression of nine different structural genes involved in histidine synthesis. Based on what you have learned about the *lac* operon and the *trp* operon, do you think the *his* operon would be activated by the presence or absence of histidine? Why?
4. Compare end-product repression, end-product inhibition, and enzyme induction as they occur in bacterial cells. Why are all of these mechanisms advantageous?
5. Give two reasons for the elaborate structure of eukaryotic chromosomes.
6. Explain this phenomenon: when ^3H uracil, which is used in RNA synthesis, is injected into developing fly larvae, radioactivity appears only in certain regions of certain chromosomes.

■ RECOMMENDED READINGS

Alberts, B. et al. *Molecular Biology of the Cell,* 2nd ed. New York, Garland Publishing, 1989. Contains a current and comprehensive review of gene expression in prokaryotes and eukaryotes.

Gurdon, J.B. "The Developmental Capacity of Nuclei Taken from Intestinal Epithelium Cells of Feeding Tadpoles," *J. Embryol. Exp. Morphol.* 10:622–640, 1962. A classic study.

Neidhardt, F.C. et al., eds. Escherichia coli *and* Salmonella typhimurium: *Cellular and Molecular Biology,* Vol. 2. Washington, D.C., American Society for Microbiology, 1987. Includes models of the operon concept.

Steward, F.C., M.O. Mapes, and K. Mears. "Growth and Organized Development of Cultured Cells," *American Journal of Botany* 45:705–713, 1958. A classic study.

13

Chromosomes and Cell Division

OUTLINE

I. All eukaryotes have chromosomes, but chromosome number depends on the species
II. The life cycle of the eukaryotic cell has definite stages
  A. The cell spends most of its life in interphase
  B. Mitosis is the process of nuclear division
    1. Prophase is the first stage of mitosis
    2. In metaphase, chromosomes line up along the equator
    3. In anaphase, chromosomes move to the poles
    4. Telophase ends mitosis
    5. Cytokinesis separates daughter cells
  C. The cell cycle is controllable
III. Some diploid cells produce haploid cells through meiosis
  A. Meiosis differs extensively from mitosis
  B. Meiosis I is the first main stage of meiosis
    1. Chromatin replicates before the first meiotic division
    2. The genes of a chromosome are linked but sometimes cross-over
    3. Homologous chromosomes separate later in meiosis I
  C. In meiosis II, chromatids separate
  D. The cells that result from meiosis mature and form gametes or spores
IV. Sex is determined by chromosome inheritance in animals
V. Chromosome defects can produce inherited disease

LEARNING OBJECTIVES

After you have studied this chapter you should be able to:
1. Identify the stages in the cell cycle and describe the main events of each.
2. Distinguish between haploid and diploid and define homologous chromosomes.
3. Describe the events occurring in each stage of mitosis with emphasis on the behavior of chromosomes.
4. Summarize the significance of meiosis in sexual reproduction.
5. Contrast the events of mitosis and meiosis.
6. Summarize the characteristics of selected chromosomal disorders.

Mitosis in onion root tip. (Runk/Schoenberger from Grant Heilman)

Cells increase in number by dividing. Most cells of a multicellular organism divide frequently, enabling the organism to grow and to repair itself. Those cells that form a functioning part of the organism's body are called body cells, or **somatic cells.** Examples include skin cells, muscle cells, blood cells, and nerve cells.

In most animals, when certain cells found in the reproductive organs divide, the result is not new somatic cells but cells that become or give rise to reproductive cells. These reproductive cells are called **sex cells,** or **gametes.** Examples include the human sperm cell and ovum. Instead of becoming part of the organism that produced them, they form a complete new organism by uniting with one another. If the female gamete, or ovum, unites with the male gamete, or sperm, the result is a "fertilized egg," better called a **zygote,** the first somatic cell of a completely new organism. This type of reproduction, the union of male and female gametes, is called **sexual reproduction.**

The division of the nucleus that results in two identical cells is called **mitosis.** Nuclear division that results in the production of reproductive cells is called **meiosis.** In both forms of cell division, the new cells inherit genetic information, but not in the same way. In mitosis, each new cell inherits an exact, complete copy of its single parent cell's genes. In meiosis, each new reproductive cell inherits only half of its parent cell's genes. The zygote that results from the union of gametes has the proper amount of genetic information because half of it comes from each gamete. In this chapter we look at how cells inherit genes from other cells in cell division; in the next we see how entire organisms inherit genes from other organisms.

The tidy packages of genetic information we call chromosomes not only play a role in the control of eukaryote genes but just as importantly are also involved in the orderly disposition of the hereditary material in cell division. The precision of the chromosome movements is a wonder, and though those movements have been studied for over a century, they are still mysterious in many respects.

KEY CONCEPTS

- During cell division, chromosomes and the genes they contain are distributed to the resulting cells in an orderly fashion.

- Mitosis results in daughter cells with the same number of chromosomes as the parent cell.

- Meiosis reduces the chromosome number by one half, and each resulting cell has exactly one chromosome from each homologous pair.

- Genes tend to be inherited as specific portions of chro-

mosomes, but genes can be exchanged between the members of a homologous pair of chromosomes when crossing-over occurs in meiosis.

ALL EUKARYOTES HAVE CHROMOSOMES, BUT CHROMOSOME NUMBER DEPENDS ON THE SPECIES

Every somatic cell of every organism of a given species contains a characteristic number of chromosomes. Human cells have 46, for example, and cabbage cells have 20. A certain species of roundworm has only 2 chromosomes per cell, and a fern species has been found that has over 1000. The number of chromosomes has no simple relationship to the size or complexity of an organism. Much the same is true of the amount of deoxyribonucleic acid (DNA) occurring in a species; there seems to be no advantage to economizing on information storage capacity in eukaryotes, whether that storage capacity is divided up among many chromosomes or embodied in just a few. (But if we compare viruses, prokaryotes, and eukaryotes with one another, there is much more DNA in prokaryotes than in viruses, and vastly more in eukaryotes than in prokaryotes.)

In somatic cells of complex organisms, each chromosome occurs as a member of a pair. The 46 chromosomes of the human cell (Fig. 13–1), for example, comprise 23 pairs, each consisting of a chromosome that has originated in the female parent (maternal chromosome) and a chromosome that has originated in the male parent (paternal chromosome). Each chromosome pair is different enough in size and shape that biologists can count and identify them. Chromosome shapes and patterns can sometimes be used to infer relationships among species or can be associated with inherited defects.

The members of a given pair of chromosomes are referred to as **homologous chromosomes.** Both members carry information governing the same traits, though this is not necessarily the same information. For example, members of a pair of homologous chromosomes might carry genes that specify part of the hemoglobin molecule, but one might have the information that produces normal hemoglobin structure, and the other might specify the abnormal hemoglobin structure that results in sickle-cell anemia. Homologous chromosomes carry different information because one of each pair originally came from the male gamete and the other came from the female gamete during sexual reproduction.

A cell with chromosomes occurring in pairs is called **diploid,** or **2n.** In human beings, as was mentioned previously, the diploid number is 46 chromo-

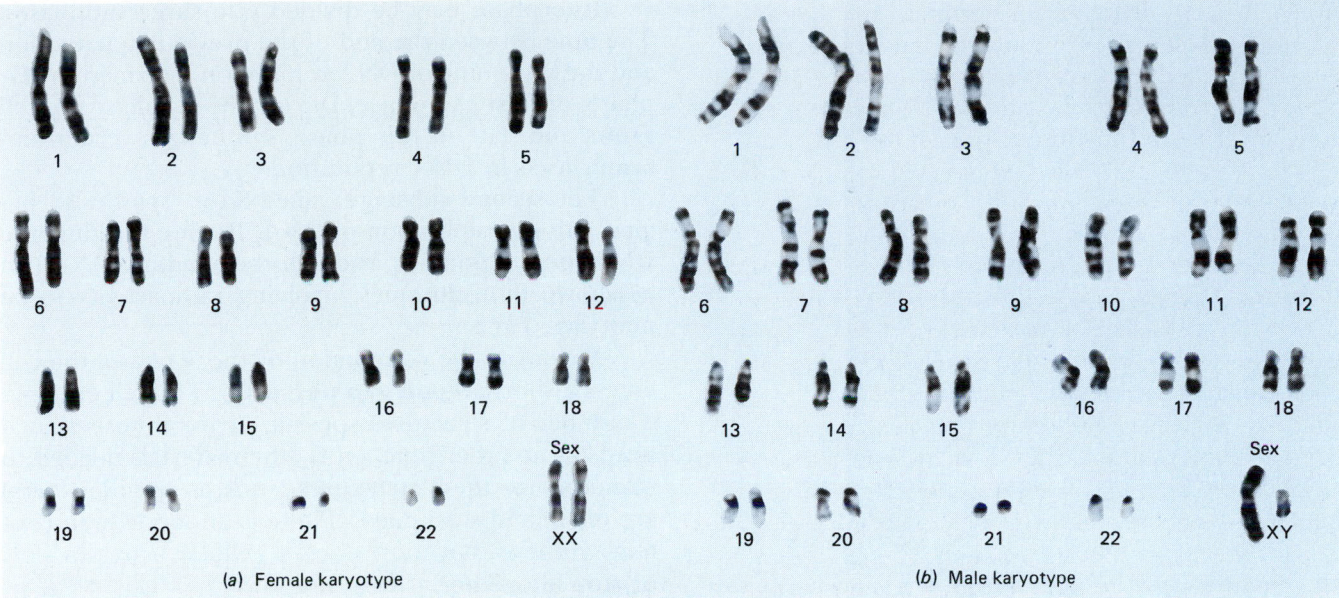

1   2   3   4   5

6   7   8   9   10   11   12

13   14   15   16   17   18

19   20   21   22   Sex
XX

(a) Female karyotype

1   2   3   4   5

6   7   8   9   10   11   12

13   14   15   16   17   18

19   20   21   22   Sex
XY

(b) Male karyotype

FIGURE 13–1   Human chromosomes. Notice the bands on each chromosome. By careful study of such chromosome photographs, known as karyotypes, it is possible to diagnose some hereditary diseases.

somes (23 pairs). The reproductive cells or gametes (eggs and sperm) cannot be diploid, or the zygote (fertilized egg) resulting from their fusion would have at least twice as many chromosomes as it should. The special type of cell division called **meiosis** passes along to each cell only one chromosome of each pair, resulting in a **haploid,** or **n,** sperm or egg. When two such haploid gametes join in fertilization, the normal diploid number of chromosomes is restored.

Plants also can reproduce sexually (that is, by the union of gametes) but in a somewhat different way. Plants and some simple plant-like organisms have a life cycle in which the diploid plant produces **spores** by meiosis. Each spore gives rise to a multicellular organism whose cells *all* have the haploid number of chromosomes. The haploid organism then produces haploid gametes by *mitosis.* When the gametes unite, a new diploid plant is produced. Notice that plant gametes are haploid, just as animal gametes are, but they are not immediate products of meiosis. (The complexities of plant and animal reproduction are more fully discussed in other chapters.)

◼ THE LIFE CYCLE OF THE EUKARYOTIC CELL HAS DEFINITE STAGES

In somatic cells that are capable of dividing, the **cell cycle** is the period from the beginning of one division to the beginning of the next division. The cell cycle may be represented as a circle (Fig. 13–2). The length of time between two successive divisions, represented

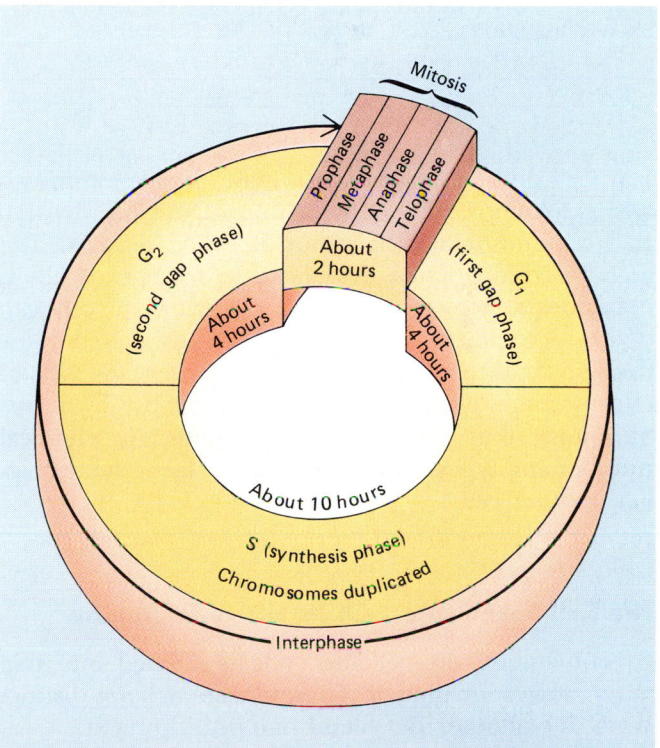

Mitosis

Prophase   Metaphase   Anaphase   Telophase

G_2
(second gap phase)

About
2 hours

About
4 hours

G_1
(first gap phase)

About
4 hours

About 10 hours

S (synthesis phase)
Chromosomes duplicated

Interphase

FIGURE 13–2   The cell cycle. Time relations are illustrative only; actual time varies with the cell type.

by a complete revolution of the circle, is the **generation time, T.** The generation time varies a great deal but is usually about a day in most plant and animal cells.

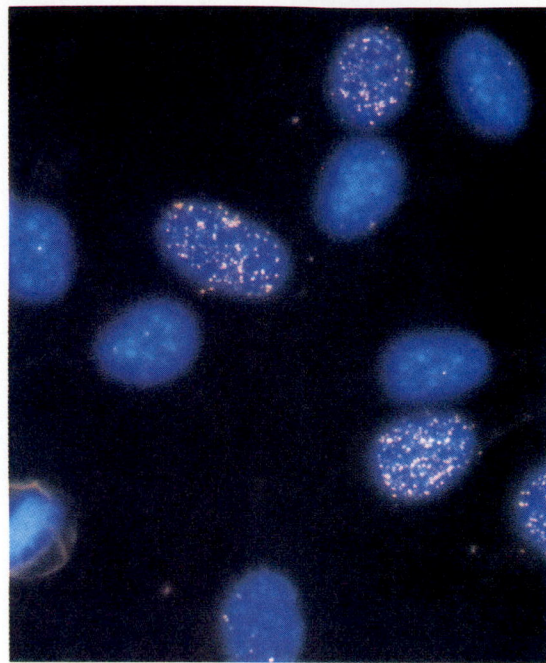

FIGURE 13–3 Photomicrograph of mouse cells in which thymidine (a DNA precursor) has been labeled with tritium (a radioisotope). The bright grains indicate replicated DNA in interphase cells. Only the nuclei are clearly visible. (Jonathan G. Izant)

Cell division involves two main processes, mitosis and cytokinesis. The combination of nuclear division and chromosomal distribution is called **mitosis.** The actual division of the cytoplasm to form two cells is **cytokinesis.** However, before a eukaryotic cell can divide, its nucleus must undergo mitosis, a complex division that precisely distributes a *complete* set of chromosomes to each daughter nucleus. It is by means of mitosis that each new cell contains the identical number and types of chromosomes present in the original mother cell.

The Cell Spends Most Of Its Life In Interphase

By convention the cell life cycle is divided into five major stages, or phases. These phases are not distinct from one another but blend one into the next.

The cell spends most of its life in **interphase,** actively synthesizing materials needed in its growth and maintenance. Because this stage of the cell's life cycle occurs between the phases of successive mitoses, it is called interphase (between phases). In about 1950 it was shown that chromosomes, although not readily visible, undergo replication during interphase. They then condense into visibly separate structures and are distributed to the daughter nuclei during mitosis.

Interphase may be divided into three subphases. The time between the end of the previous cell division and the beginning of DNA replication is termed the G_1 phase, or **first gap phase.** During the G_1 phase the cell grows and, late in this phase, synthesizes certain enzymes used in DNA replication.

The **second substage,** called **S** for "synthesis," begins with the replication of DNA. During this time the DNA content doubles, and half the resulting DNA can be shown by techniques involving radioactivity to be new (Fig. 13–3).

Following the completion of the S phase, the cell enters a short **second gap phase, G_2.** The cell employs a number of structures specialized for mitosis that it needs at no other time. In G_2 the materials needed to manufacture these structures (such as spindle fibers) are made and stockpiled. There is an increase in protein synthesis, but most normal cellular activities slow or stop altogether at this time.

Mitosis Is The Process Of Nuclear Division

The completion of G_2 is marked by the beginning of mitosis, the **M phase.** Mitosis itself may be divided into four stages: **prophase, metaphase, anaphase,** and **telophase.** During mitosis, visible changes associated with cell division take place. Most other cellular activities are suspended during this brief period of the cell's life.

Prophase Is The First Stage Of Mitosis

The first stage of mitosis, **prophase,** begins when the chromatin threads begin to condense and appear as chromosomes (Fig. 13–4). In forming the visible chromosomes of mitosis, the long threads of chromatin are condensed and coiled into much shorter bundles. In this form, the chromosomes can easily move into position and eventually pass into the daughter cells without tangling. Although each human chromosome contains several centimeters of DNA, at mitosis this is condensed into a chromosome that is only 5 to 10 micrometers in length (a 10,000-fold shortening).

As prophase proceeds (Fig. 13–5), the chromosomes become shorter and thicker and are individually visible under the light microscope. The chromosomes have not yet separated completely from their duplicates, and each duplicate is referred to as a **chromatid.** Therefore, each chromosome now consists of two chromatids attached to one another at the centromeric region. The **centromere** may be located about in the middle of the chromosome or at almost any point.

As described in Chapter 5, **centrioles** are hollow cylinders composed of nine triple microtubules. These structures are characteristic of animal cells and some one-celled eukaryotes but are not found in the cells of

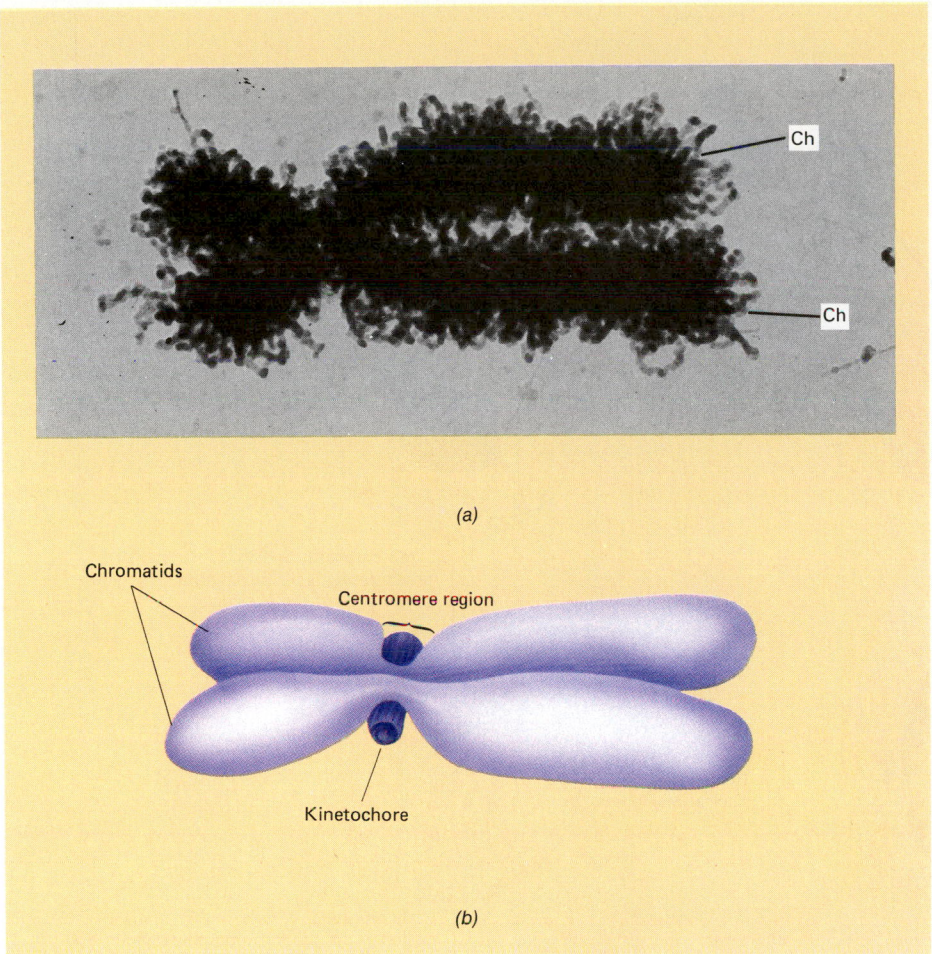

(a)

(b)

FIGURE 13–4   A human chromosome as it appears in an early stage of cell division. *(a)* Photographed at approximately ×50,000, the two members of this pair (each of which is called a chromatid) are held together near their most constricted portions, the centromeres. *(b)* The diagram shows the kinetochore, a protein structure bound to the centromere that serves as a point of attachment for the spindle fibers, which will eventually pull these two identical chromosomes apart. The structure of this chromosome is as tightly wound as a clockspring into a compact package that can be distributed to the daughter cells. *(a,* Courtesy of E. J. Du Praw)

true plants or most fungi. Their exact function in cell division is still not altogether clear after more than a century of careful study. Having duplicated in the S subphase of interphase, the two pairs of centrioles now separate and migrate toward opposite poles of the cell. Clusters of microtubules, composed mainly of protein, extend outward in all directions from nearby the centrioles. These clusters of microtubules are called **asters.** In animal cells microtubules extend from the region surrounding the centrioles and form a **mitotic spindle** (Fig. 13–6). Although plant cells lack centrioles and asters, they do form mitotic spindles. (Window on the Animal Cell).

During prophase the nuclear envelope breaks apart, and the nuclear contents mingle with the cytoplasm. The nucleolus also becomes disorganized in prophase. It is near the end of prophase that the condensed chromosomes attach to the spindle fibers by means of their **kinetochores** (see Fig. 13–4) and they move toward the equator of the cell, midway between the two poles and perpendicular to the axis of the spindle.

In Metaphase, Chromosomes Line Up Along The Equator

The short period during which the chromatids are lined up along the equatorial plane of the cell constitutes **metaphase.** The mitotic spindle is complete and can be seen to be composed of numerous fibers that extend from pole to pole (Figs. 13–4 and 13–5). They end near the centrioles but do not actually touch them.

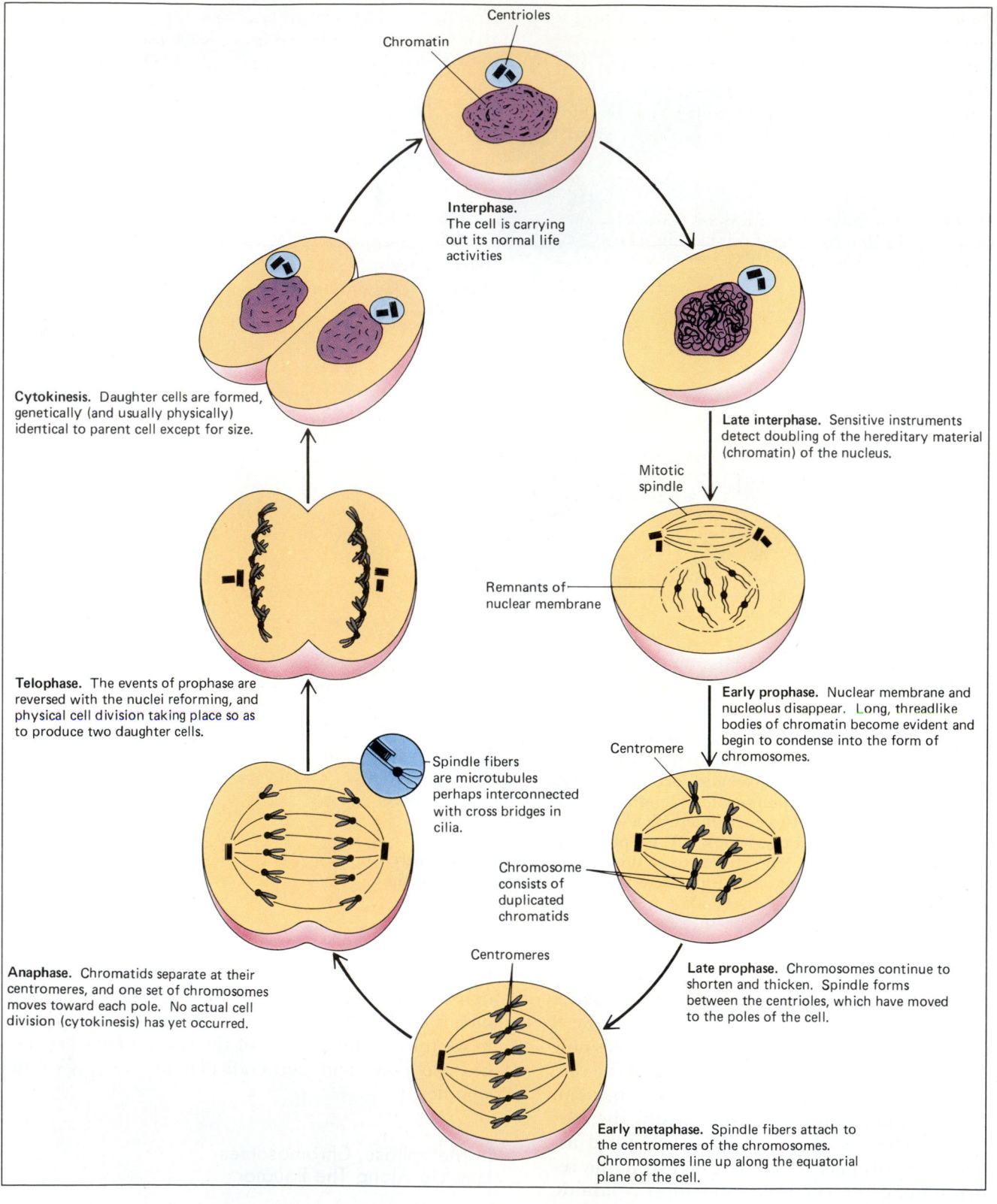

Centrioles

Chromatin

Interphase. The cell is carrying out its normal life activities

Late interphase. Sensitive instruments detect doubling of the hereditary material (chromatin) of the nucleus.

Mitotic spindle

Remnants of nuclear membrane

Early prophase. Nuclear membrane and nucleolus disappear. Long, threadlike bodies of chromatin become evident and begin to condense into the form of chromosomes.

Centromere

Chromosome consists of duplicated chromatids

Late prophase. Chromosomes continue to shorten and thicken. Spindle forms between the centrioles, which have moved to the poles of the cell.

Centromeres

Early metaphase. Spindle fibers attach to the centromeres of the chromosomes. Chromosomes line up along the equatorial plane of the cell.

Spindle fibers are microtubules perhaps interconnected with cross bridges in cilia.

Anaphase. Chromatids separate at their centromeres, and one set of chromosomes moves toward each pole. No actual cell division (cytokinesis) has yet occurred.

Telophase. The events of prophase are reversed with the nuclei reforming, and physical cell division taking place so as to produce two daughter cells.

Cytokinesis. Daughter cells are formed, genetically (and usually physically) identical to parent cell except for size.

As a whole, the spindle is gel-like in consistency and is more viscous than the surrounding cytoplasm.

During metaphase each chromosome is quite condensed and appears thick and discrete. Because metaphase chromosomes can be seen more clearly than those at any other stage, they are typically photo-

◀ **FIGURE 13–5 The stages of the animal cell cycle. Individual steps in the cycle are explained in labels within the figure. The animal cells shown here have a diploid chromosome number of six.**

graphed and studied during this stage, for example, to determine possible chromosome abnormalities and to look for physical similarities and differences in chromosomes for use in evolutionary studies.

In Anaphase, Chromosomes Move To The Poles

Anaphase begins with the centromeres of sister chromatids splitting apart. Each chromatid is now considered to be an independent chromosome. The separated chromosomes now begin to move toward opposite poles. The chromosomes move toward the poles with the centromeres (attached to the spindle fi-

FIGURE 13–6 Interphase and mitosis in human cells grown in culture. These cells are stained with fluorescent dyes. Chromosomes are stained orange, and the microtubules, yellow-green. The rest of the cell is almost invisible. In interphase (*a*) the microtubules are not yet organized into a mitotic spindle. During prophase (*b*) the asters move toward opposite poles of the cell. In metaphase (*c*) the mitotic spindle is well defined, and the chromosomes are lined up along the equatorial plane. In late anaphase (*d*) the complete set of chromosomes is moving toward each end of the cell, and in late telophase (*e*) there is a complete set of chromosomes at each end of the cell. Cytokinesis has begun, evidenced by the split between the groups of microtubules. (Jonathan G. Izant)

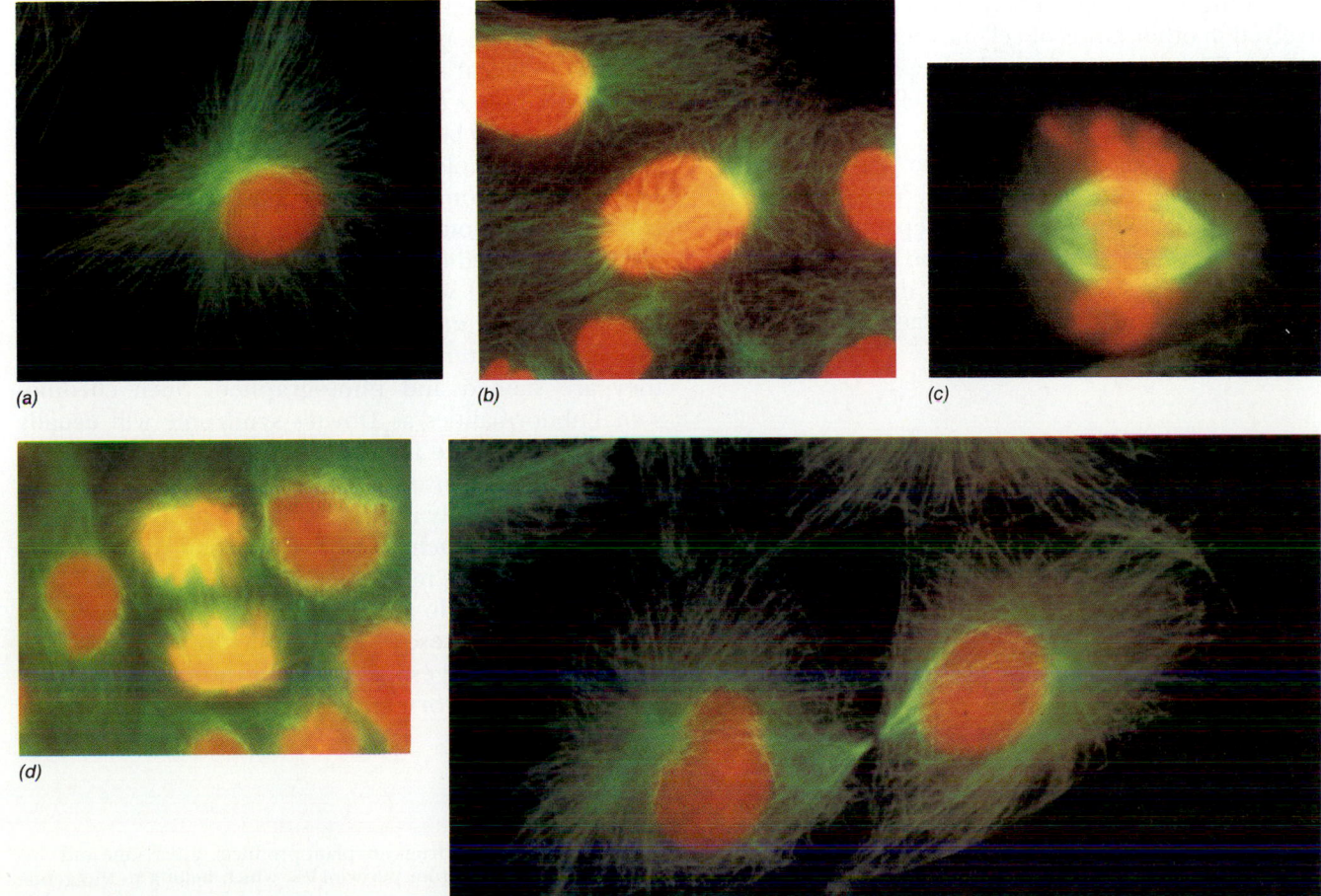

(a)

(b)

(c)

(d)

(e)

bers) leading the way. The two arms of each chromosome trail behind as the spindle fibers seem to pull on the centromeres by their associated kinetochores. Anaphase ends when complete sets of chromosomes have arrived at opposite ends of the cell.

Telophase Ends Mitosis

During the final stage of mitosis, **telophase,** the chromosomes begin to elongate by uncoiling, becoming invisible chromatin threads. A new nuclear envelope forms around each set of chromosomes, produced at least in part from lipid components of the old nuclear membrane. Nucleoli reappear, while spindle fibers disappear.

Cytokinesis Separates Daughter Cells

Cytokinesis usually accompanies mitosis. Cytokinesis generally begins at the conclusion of anaphase and extends through telophase. The division of an animal cell is accomplished by a cleavage furrow that encircles the surface of the cell in the plane of the equator. At the furrow the plasma membrane is pulled inward by a ring of microfilaments composed of contractile proteins, including the widely distributed **actin,** which is also involved in other kinds of cell movement and muscular action. The furrow gradually deepens and separates the cytoplasm into two daughter cells, each with a nucleus.

Although plant cell mitosis is similar to that of animals (Fig. 13–7), the most striking difference is probably the way in which physical cell separation takes place. In plant cells cytoplasmic division occurs by the formation of a **cell plate** between the daughter cells, destined to become the new cell wall that will separate them.

The Cell Cycle Is Controllable

The frequency of mitosis varies greatly not only among cells from different species but among cells from different tissues within a single organism. In any multicellular organism the frequency of mitosis must be closely controlled or growth abnormalities or even tumors may result. The evidence indicates that the frequency of mitosis is basically governed by some kind of intracellular biological clock, though it may also be influenced by such external factors as the temperature of the surroundings.

When conditions are optimal, a bacterium (which is a prokaryote) can divide every 20 minutes. The more complex apparatus of eukaryotes takes longer to operate, but even they can divide every few hours in some

cases. Among the more rapidly dividing cells of human beings are the cells that line the digestive tract, cells in the skin, and the so-called stem cells that give rise to blood cells. These cell types divide rapidly and repeatedly throughout life. In contrast, divisions of the cells that make up the central nervous system usually cease in the first few months of life (although divisions of mature nerve cells have been experimentally obtained under laboratory conditions).

Mitosis can also be affected by certain drugs (Fig. 13–8). Colchicine is a useful drug derived from the autumn crocus plant and can also be employed to block cell division in eukaryotic cells. This substance interferes with the normal function of the mitotic spindle. The chromosomes cannot separate appropriately and move to the opposite ends of the cell. As a result, the cell will end up with an extra set of chromosomes. This can be done not only with somatic cells but also with those destined to become gametes and, in due course, zygotes that can give rise to another generation. This fact is of great commercial importance, especially in connection with the development of new varieties of ornamental and agricultural plants. Plants consisting of cells with extra sets of chromosomes tend to be larger and more vigorous than normal plants and may have such qualities as increased fruit sugar content in tissues or doubled flower petals.

The inhibition of mitosis by such drugs as colchicine has practical application in the diagnosis of chromosomally based disease. Such diagnosis now can be performed even before birth. In one technique, **amniocentesis,** a sample of the fluid surrounding the fetus is withdrawn from the uterus (Fig. 13–9). This fluid contains some sloughed-off cells of the fetus. The cells are then cultured (that is, raised in laboratory containers) and treated with colchicine or a similar drug to arrest mitosis in metaphase. The cells are then treated with distilled water to break them up osmotically. Then they are stained and photographed. Such chromosomal abnormalities as Down's syndrome will usually be evident by some abnormality in the picture.

Some drugs[1] used in cancer therapy block cell division or specifically injure dividing cells. Because cancer cells divide much more rapidly than most normal body cells, they are most affected by these drugs. However, these drugs do affect some types of normal body cells, especially those that multiply rapidly, such as those lining the digestive tract and those that produce the continuously worn layers of the skin and its deriva-

---

[1] Like colchicine, these drugs are plant products. Vincristine and vinblastine are obtained from periwinkles, which belong to the genus *Vinca.* One reason for conserving plant species is the possibility that some of them may well serve as sources of valuable drugs.

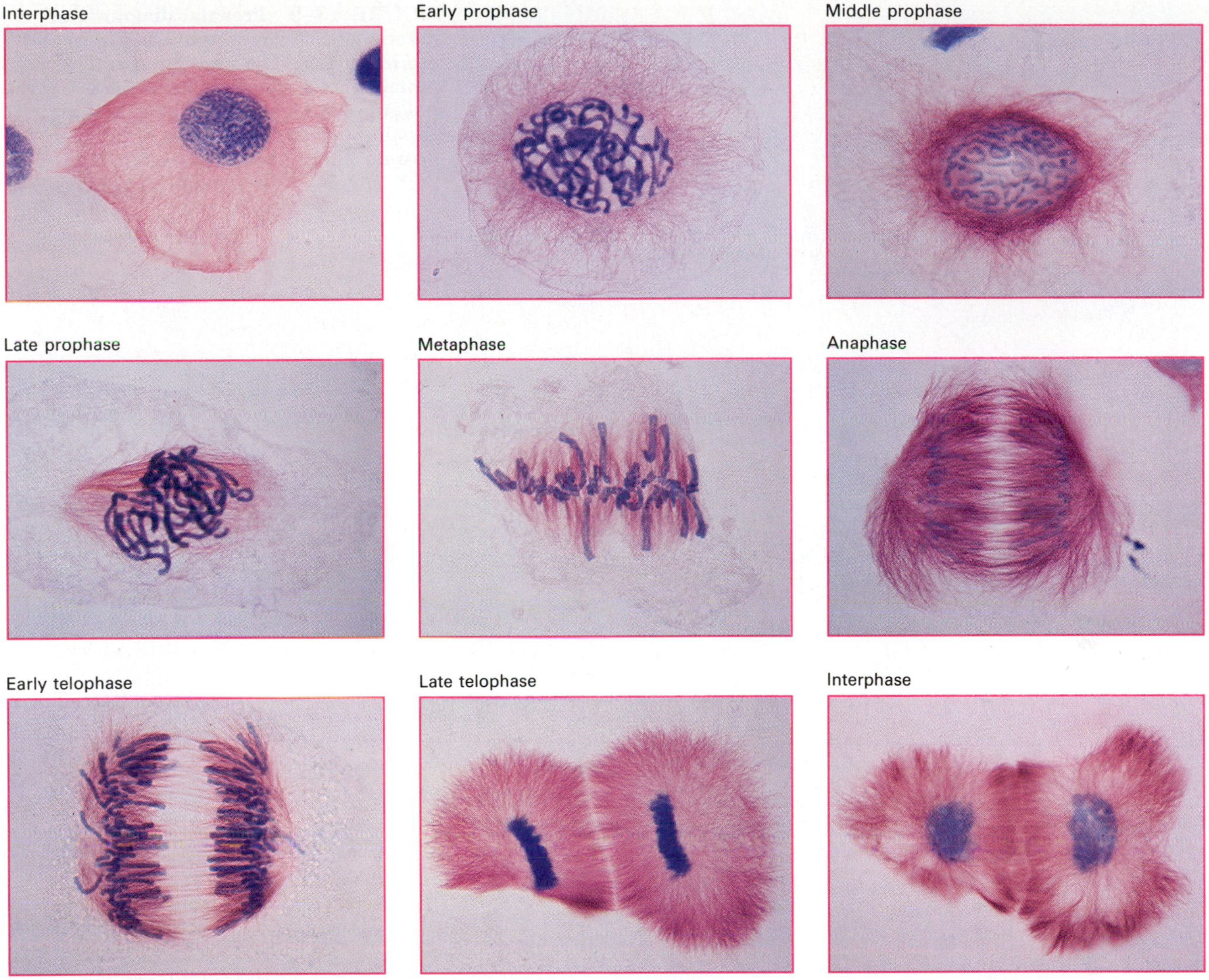

Interphase

Early prophase

Middle prophase

Late prophase

Metaphase

Anaphase

Early telophase

Late telophase

Interphase

FIGURE 13–7   Interphase and the stages of mitosis in plant cells (*Haemanthus*, blood lily) prepared with stains and flattened on slides. (Andrew S. Bajer, University of Oregon)

FIGURE 13–8   When a cell is treated with colchicine, mitosis is arrested in metaphase, when the chromosomes are thickest. The colchicine spreads the chromosomes apart, making them easier to count. (Cytogenetics Laboratory, University of California, San Francisco)

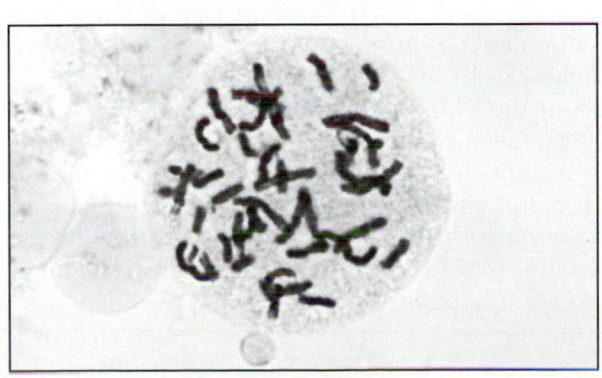

tives such as hair. This is why such drugs may produce side effects, like causing hair loss.

■ SOME DIPLOID CELLS PRODUCE HAPLOID CELLS THROUGH MEIOSIS

The result of mitosis is that each daughter cell receives exactly the same number and kind of chromosomes that the parent cell had. In contrast, in meiosis, a special type of cell division that produces reproductive cells (gametes and some kinds of spores), the number of chromosomes is reduced and each new cell is haploid. Ultimately, meiosis causes the chromosome number of a species to remain the same from generation to generation.

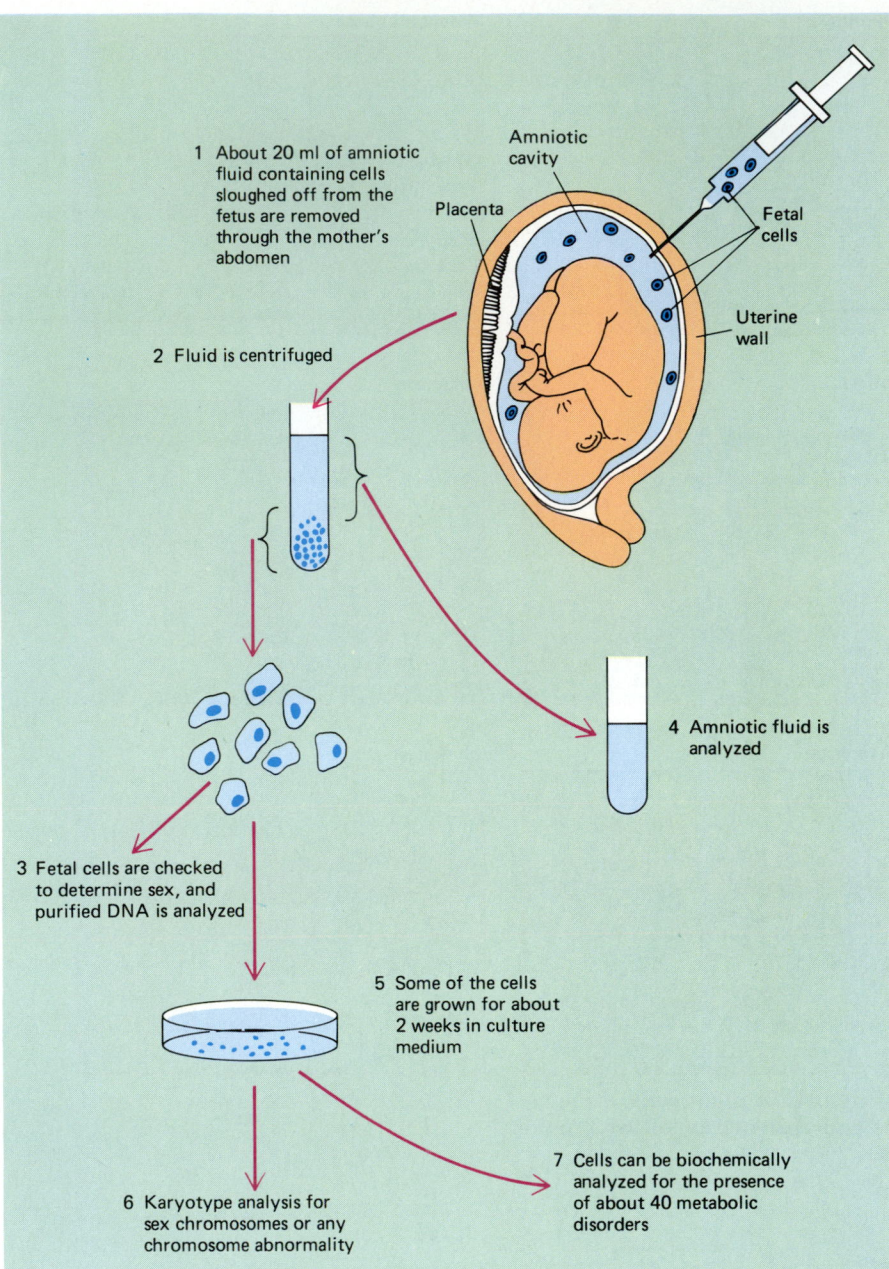

1 About 20 ml of amniotic fluid containing cells sloughed off from the fetus are removed through the mother's abdomen

Amniotic cavity

Placenta

Fetal cells

Uterine wall

2 Fluid is centrifuged

4 Amniotic fluid is analyzed

3 Fetal cells are checked to determine sex, and purified DNA is analyzed

5 Some of the cells are grown for about 2 weeks in culture medium

7 Cells can be biochemically analyzed for the presence of about 40 metabolic disorders

6 Karyotype analysis for sex chromosomes or any chromosome abnormality

FIGURE 13–9 Prenatal diagnosis of genetic and chromosomal disease by amniocentesis. The amniotic fluid contains cells whose chromosomal makeup can be analyzed, thus making possible prenatal diagnosis of chromosomally based disease.

Meiosis *separates the members* of each homologous pair of chromosomes. Any contrasting genetic traits each homologous pair of chromosomes might bear are separated and distributed independently to different gametes or spores. The result is that no two offspring even of the same parents are likely to be exactly alike.

Meiosis Differs Extensively From Mitosis

The events of meiosis are somewhat similar to the events of mitosis, but there are several important differences (Fig. 13–10).

1. In meiosis there are two successive nuclear and cell divisions, potentially producing a total of four cells. In mitosis there is only one nuclear division, and cytokinesis typically occurs only once, yielding two daughter cells.

2. Each of the four cells produced in meiosis contains the haploid number of chromosomes, that is, only one member of each homologous pair. In mitosis each daughter cell contains the diploid number of chromosomes (except in the haploid stage of the plant life cycle, in which haploid cells divide by mitosis to form haploid cells).

3. During meiosis the homologous chromosomes con-

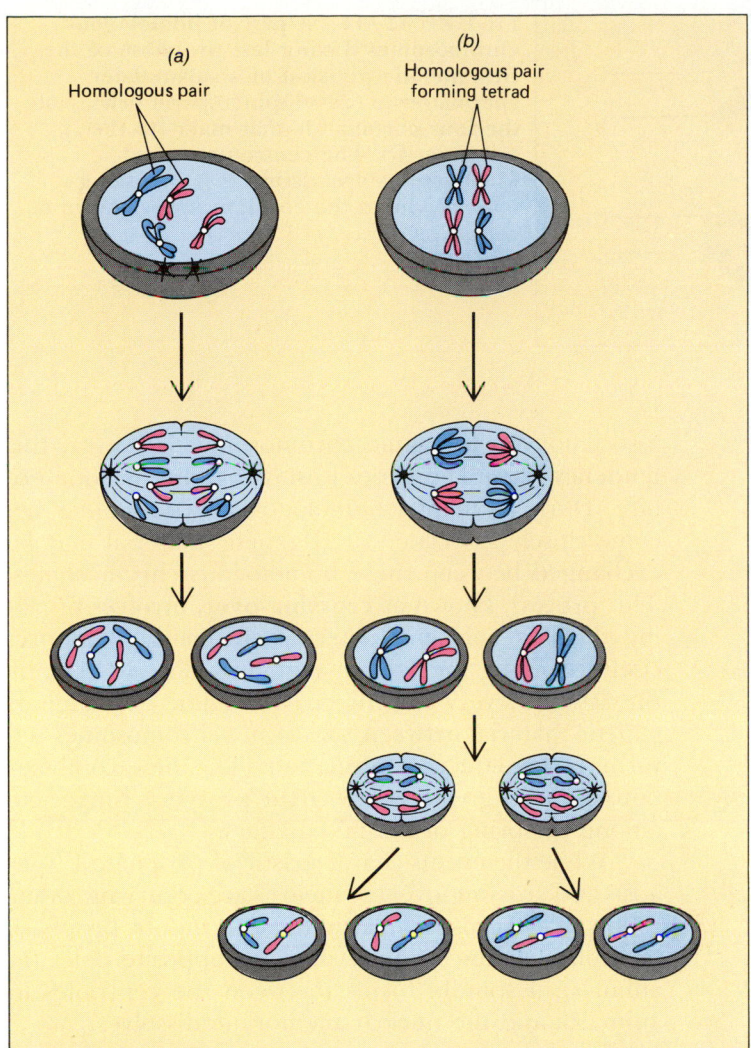

(a) Homologous pair

(b) Homologous pair forming tetrad

FIGURE 13–10 Meiosis compared with mitosis. The diploid number for each cell is four. (*a*) Mitosis. Note that each daughter cell has an identical set of four chromosomes (two pairs), which is the diploid number. (*b*) Meiosis. Two divisions take place, giving rise to four daughter cells. Each daughter cell has only two chromosomes, one of each pair. The chromosomes shown in blue originally came from one parent; those shown in red came from the other parent. Note that in the prophase of the first meiotic division (*top figure*), homologous chromosomes come together, forming tetrads.

taining genetic information from each parent are thoroughly shuffled, and one of each pair is randomly distributed to each new cell; the cells produced possess new combinations of chromosomes, and therefore new combinations of genes. After mitosis the daughter cells contain an identical set of chromosomes, and this chromosome set is identical in every way to that of the mother cell.

4. During meiosis there may be some exchange of parts between homologous chromosomes (called crossing-over) so that even the genes originally located together on one chromosome do not always stay together. Crossing-over further increases the genetic reshuffling of meiosis. In mitosis, there is little opportunity for crossing-over.

Meiosis consists of two cell divisions, logically named the first and second meiotic divisions, or simply **meiosis I** and **meiosis II.** Each division includes stages comparable to the four stages of mitosis, namely prophase, metaphase, anaphase, and telophase.

Meiosis I Is The First Main Stage Of Meiosis

The **first meiotic division** separates the members of each homologous pair of chromosomes and distributes them into daughter cells. These chromosomes had already duplicated, so that each consists of two chromatids. The **second meiotic division** separates the chromatids into individual chromosomes, which then enter separate, haploid daughter cells. Since it is easier to follow these events in an organism that possesses only a few chromosomes, we will discuss a hypothetical organism with a diploid number of only four chromosomes (Fig. 13–10).

Chromatin Replicates Before The First Meiotic Division

As in mitosis, the chromosomes are duplicated during the S phase before meiosis actually begins. Recall that

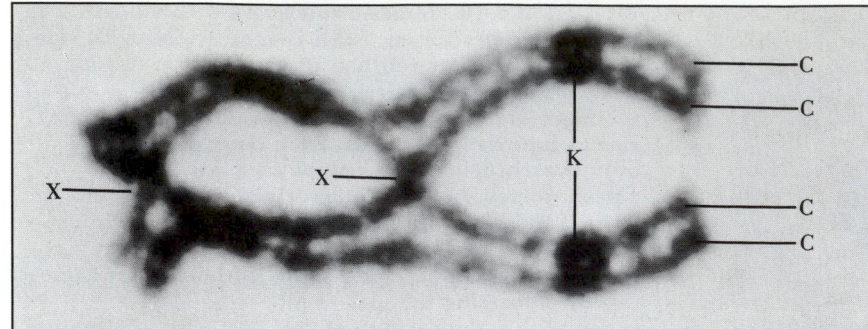

FIGURE 13–11 A pair of homologous chromosomes during late prophase of the first meiotic division of a salamander spermatocyte (developing sperm cell). Note the four chromatids that make up the tetrad (at C). The centromeres and kinetochores are visible at K. Crossing-over produces the configurations shown at each X. (Courtesy of J. Kezer)

when a chromosome is duplicated, it consists for a time of two chromatids joined by their centromeres.

During prophase I (of the first meiotic division), while the chromatids are still elongated and thin, the homologous chromosomes come to lie close together side by side along their entire length—a process called **synapsis.** Since the diploid number in our example is four, at synapsis we would see two homologous pairs. One of each pair, the **maternal chromosomes,** were originally inherited from the organism's mother, whereas the other member of each pair, the **paternal chromosomes,** were contributed by the father. Since each chromosome is replicated (in effect, doubled) at this time and actually consists of *two* chromatids, synapsis results in the coming together of *four* chromatids, forming a complex known as a **tetrad.** The number of tetrads equals the haploid number of chromosomes. In human cells there are 23 tetrads (and a total of 92 chromatids) at this stage.

The Genes Of A Chromosome Are Linked But Sometimes Cross-Over

All the genes located on a particular chromosome are said to be **linked** together and therefore will tend to be inherited together in **linkage groups** that reflect their association on the same chromosome. However, this tendency for linked genes to stay together is not absolute. During synapsis homologous chromosomes become closely associated, and genetic material may be exchanged between these homologous chromosomes. This process, known as **crossing-over,** involves breaking the maternal and paternal chromatids and precisely rejoining the broken segments (Fig. 13–11), by the action of very specific enzymes. The exchange of genetic material between *homologous* chromosomes is a form of **genetic recombination.** The new combinations of genes greatly enhance the prospects for variety among offspring of sexual partners.

While the events characteristic of prophase I (synapsis and crossing-over) of meiosis are occurring, other events that are also similar to those of mitotic prophase take place. The centrioles move to opposite poles (in animals), a spindle forms (between the centrioles in animals), and the nuclear membrane dissolves.

Homologous Chromosomes Separate Later In Meiosis I

The tetrads (paired homologous chromosomes) line up along the equator of the spindle during metaphase I. Both chromatids of one chromosome are

FIGURE 13–12 The stages of meiosis. The diploid number for the cell shown here is four.

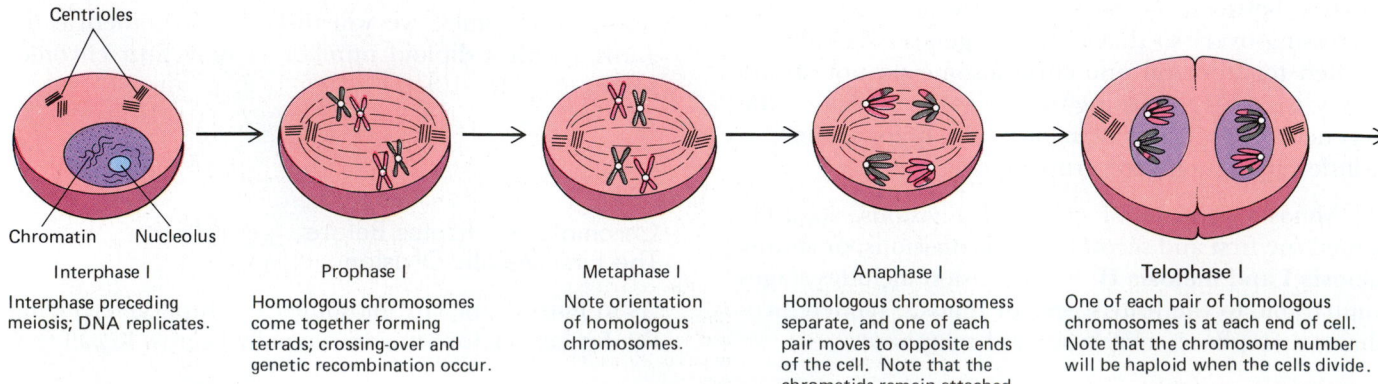

Centrioles

Chromatin   Nucleolus

Interphase I	Prophase I	Metaphase I	Anaphase I	Telophase I
Interphase preceding meiosis; DNA replicates.	Homologous chromosomes come together forming tetrads; crossing-over and genetic recombination occur.	Note orientation of homologous chromosomes.	Homologous chromosomess separate, and one of each pair moves to opposite ends of the cell. Note that the chromatids remain attached at their centromeres.	One of each pair of homologous chromosomes is at each end of cell. Note that the chromosome number will be haploid when the cells divide.

oriented toward the same pole. Their sister chromatids of the homologous chromosome are oriented toward the opposite pole (Fig. 13–12).

During anaphase I the homologous chromosomes of each pair separate, moving toward opposite poles. The chromatids of each chromosome are still united at their centromere regions. In telophase I in this example, there would be two doubled chromosomes (four chromatids) at each pole. During telophase I, the nuclei reorganize, the chromatids begin to elongate, and cytokinesis (cell division) generally takes place. Notice that the haploid number of chromosomes has been established, although each chromosome is double.

In Meiosis II, Chromatids Separate

During the interphase that follows meiosis I, no further DNA or chromosome replication takes place. In most organisms meiotic interphase is very brief; in some organisms it is absent. Since the chromatids do not completely elongate between meiotic divisions, prophase II is also brief. In prophase II there is no pairing of homologous chromosomes (indeed, only one of each pair remains in the cell), and there is no crossing-over.

During metaphase II the chromatids again line up on the equator. Metaphase I and II can be distinguished, because in the first the chromatids are arranged in bundles of four (tetrads), and in the second the chromatids are arranged in groups of two. During anaphase II, the sister centromeres split and the sister chromatids, now complete chromosomes, separate and move to opposite poles. Thus in telophase II there is one of each kind of chromosome at each end of the cell. The haploid condition that existed at telophase I has been maintained but each chromosome is now in an unduplicated state. Nuclear membranes then form, the chromosomes gradually elongate into chromatin, and cytokinesis occurs.

The Cells That Result From Meiosis Mature And Form Gametes Or Spores

The two successive meiotic divisions yield four haploid cells whose nuclei each contain one—and only one—of each kind of chromosome. Each of these cells has a different combination of genes. In the sporophyte generation of plants, meiosis takes place in specialized structures called **sporangia,** or in the case of seed plants, in specialized structures called **cones** or **flowers. Spores** are the reproductive cells formed in plants as a result of meiosis. In animals meiosis typically takes place in specialized reproductive structures, the **gonads,** and results in formation of sperm and eggs by a maturation process that follows or accompanies the events of meiosis. The differences between the development of sperm and that of eggs will be discussed in a later chapter, but for now, we wish to emphasize a fact that has great importance to your understanding of the genetics chapters that follow this one: A gamete normally receives only *one* member of *each* homologous pair of chromosomes.

■ SEX IS DETERMINED BY CHROMOSOME INHERITANCE IN ANIMALS

Sex is an inherited trait. It is important to understand, though, that the mechanism of sex determination found in humans is not necessarily typical of the entire animal kingdom. It is, however, a perfectly good example as long as that reservation is kept in mind.

Up until now, we have created the impression that the two members of each homologous pair of chromosomes are exactly alike in appearance. However, there is a pair of chromosomes in mammals and many other kinds of animals that is unlike all the others in that one of them (the Y chromosome) is smaller than the other (the X chromosome). Since in mammals the Y chromosome occurs only in males, it seems reasonable that it is

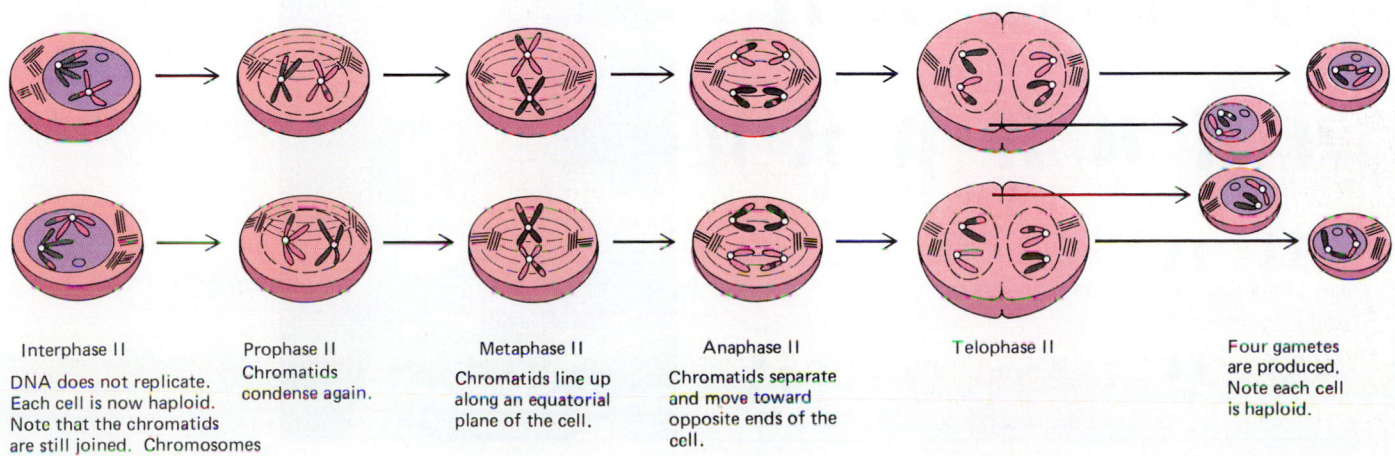

Interphase II
DNA does not replicate. Each cell is now haploid. Note that the chromatids are still joined. Chromosomes do not completely elongate.

Prophase II
Chromatids condense again.

Metaphase II
Chromatids line up along an equatorial plane of the cell.

Anaphase II
Chromatids separate and move toward opposite ends of the cell.

Telophase II

Four gametes are produced. Note each cell is haploid.

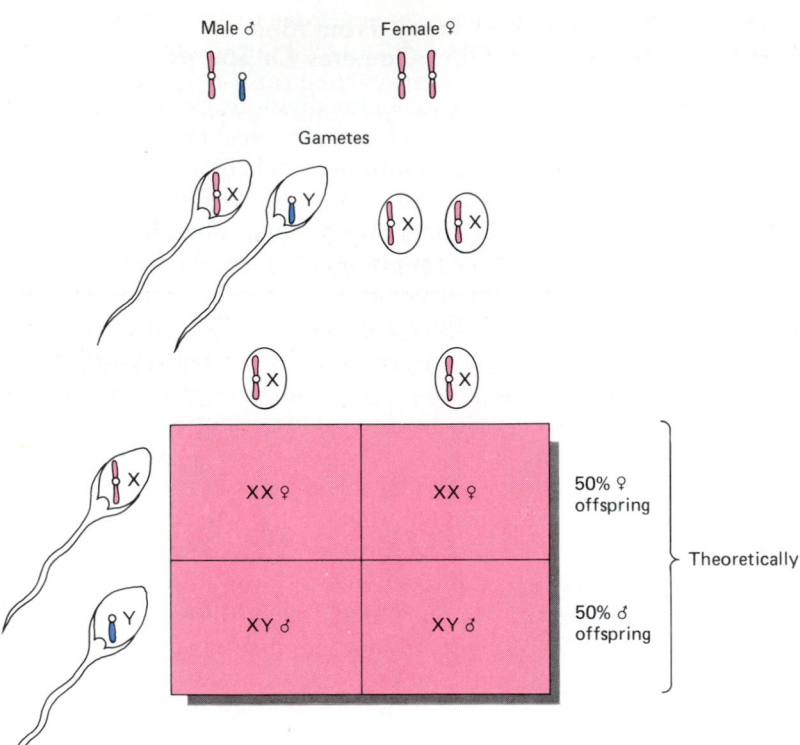

FIGURE 13–13 The inheritance of sex. Sex is determined by the sperm at the moment of fertilization. When the egg is fertilized by an X-bearing sperm, the offspring will be female. When the egg is fertilized by a Y-bearing sperm, the zygote will contain an X and a Y chromosome and the offspring will be male.

this chromosome that makes the difference between the two sexes. The X and the Y chromosomes are known, logically enough, as the **sex chromosomes.** All the remaining chromosomes are called **autosomes.**

In the case of mammals, the Y chromosome bears a single gene that acts as a developmental switch: When present, the embryo becomes male; when ab-

sent, the embryo becomes female. Since an infant's mother is necessarily female, she has two X chromosomes. The father, being male, has an X chromosome and a Y chromosome (Fig. 13–13). Since, as we have emphasized, one gamete receives one chromosome of each pair, all of the mother's eggs will have just one X chromosome. There is no other possibility. The fa-

FIGURE 13–14 Down's syndrome, a disease usually associated with trisomy of the 21st chromosome. (*a*) A karyotype of the chromosomes of a girl with Down's syndrome. Note the extra chromosome 21. (*b*) A 2-year-old boy with Down's syndrome. (Courtesy of Mr. and Mrs. Beny Peretz)

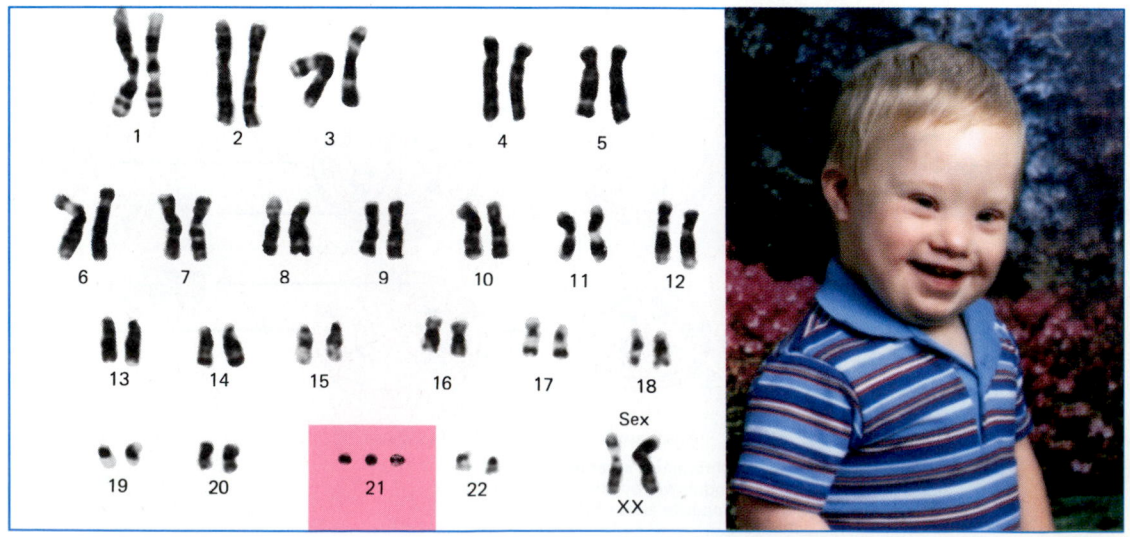

ther's sperm, though, are of two kinds—sperm with an X chromosome and sperm with a Y chromosome. There is a 50% chance that an X-bearing sperm will meet a given egg, with the result that the **zygote,** or fertilized egg, will have two X chromosomes and be female. There is also a 50% chance that a Y-bearing sperm will be the first to reach that egg, thus producing an XY or male zygote.

Since everyone has at least one X chromosome, this chromosome is perfectly free to carry genes that have nothing to do with sex; genes that could, for instance, govern such totally unrelated traits as blood clotting or color vision (discussed more fully in Chapter 15). Less obviously, many sexual traits are doubtless carried by the autosomes, awaiting only the presence

or absence of the Y chromosome to determine whether they are expressed. It is a striking thought that many of one's most explicitly sexual traits may have been inherited from his or her parent of the opposite sex.

■ CHROMOSOME DEFECTS CAN PRODUCE INHERITED DISEASE

Chromosome abnormalities involving one or more extra or missing chromosomes are called **aneuploidies.** When one of the person's homologous sets consists of three chromosomes instead of the normal two, the aneuploidy is called **trisomy** (Fig. 13–14; Table 13–1).

FIGURE 13–15   How nondisjunctions of the X (red) and Y (blue) chromosomes could produce abnormal numbers of these chromosomes in gametes. (a) A nondisjunction in the first meiotic division results in two XY sperm and two sperm with no sex chromosome at all. (b) Second-division nondisjunction of the X chromosome results in one sperm with two X chromosomes, two normal sperm with one Y apiece, and one with no sex chromosome. Nondisjunction of the Y chromosome has the opposite result: one sperm with two Y chromosomes, two with one X, and one with none. Similar events can occur in the formation of eggs. If one of the faulty gametes formed by nondisjunction is involved in fertilization, there could be an offspring with multiple X or Y chromosomes. Alternatively, an individual with a single X chromosome is also possible. Those with a single Y chromosome or no sex chromosome at all do not survive.

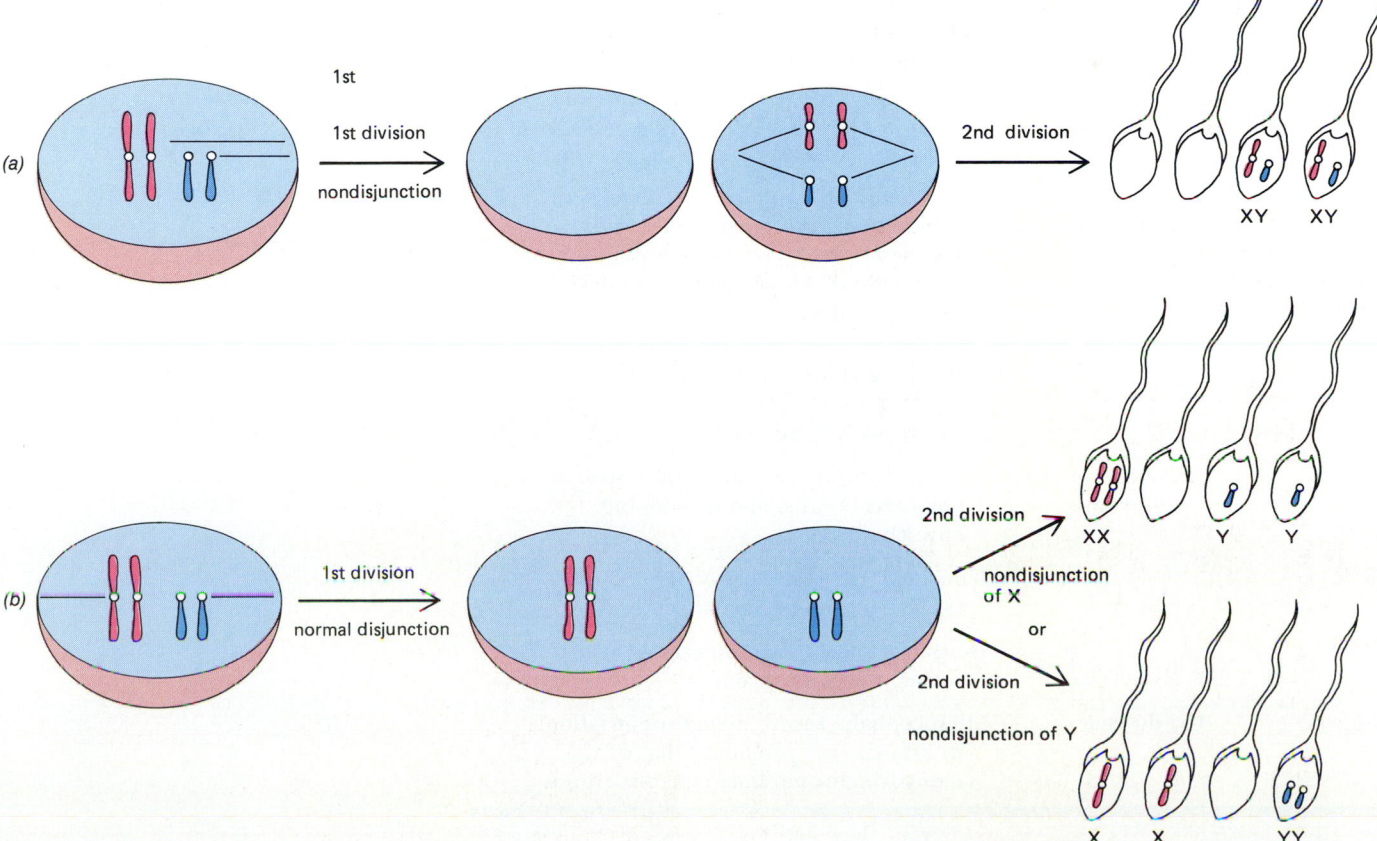

TABLE 13–1
Some Chromosome Abnormalities

Aneuploidy	Common Name	Description
Trisomy 13		Multiple defects, with death by 1–3 months
Trisomy 15		Multiple defects, with death by 1–3 months
Trisomy 18		Ear deformities, heart defects, spasticity, and other damage. Death by 1 year
Trisomy 21	Down's syndrome (mongolism)	Overall frequency is about 1 in 700 live births. True trisomy usually found among children of older (40+) mothers, but translocation resulting in equivalent of trisomy may occur in children of younger women. A 35-year-old mother has 1:200 risk of having Down's syndrome in a child. A 40-year-old mother has 1:50 chance of doing so, and the risk at 44 is 1:20. Epicanthic skin fold, though not same as that in Mongolian race, produces superficial resemblance; hence older name "mongolian idiocy." Varying degrees of mental retardation, usually IQ below 70, though more intelligent exceptions are known. Short stature, protruding furrowed tongue, transverse palmar crease, cardiac deformities common. Patients usually die by age 30–35; 50% die by age 3 or 4, often from leukemia and respiratory infections. Females are fertile if they live to sexual maturity, producing 50% Down's syndrome offspring
Trisomy 22		Similar to Down's syndrome but with more skeletal deformities
XO	Turner's syndrome	Female with short stature, webbed neck, sometimes slight mental retardation. Ovaries degenerate in late embryonic life, leading to rudimentary sexual characteristics. Similar disorders occur sometimes in XX individuals, perhaps resulting from abnormalities of X chromosome inactivation and, very rarely, in XY individuals
XXY	Klinefelter's syndrome	Male with slowly degenerating testes, enlarged breasts, and developing eunuchoidism
XYY		Unusually tall male, heavy acne, some tendency to mild mental retardation
XXX		Despite triploid X chromosomes, usually fertile, normal females
Short 5	Cri-du-chat	Microcephaly, severe mental retardation; in infancy, cry resembles that of cat; defective chromosome is heterozygous

These aneuploidies appear to arise through malfunctions of meiosis, in which chromosomes fail to separate. This is called **nondisjunction.**

In still other cases, a part of a chromosome may break off and attach to another chromosome. The gamete then may have chromosomes that are abnormally short or long. Such **translocations** may be functionally much the same as a trisomy if they result in the presence of too much genetic material in a zygote.

A trisomy might well produce a 50% increase in the activities of the products of most of the involved genes, and recent studies have confirmed this prediction in the case of some genes. So it is little wonder that the body chemistry and development of its possessor could be unbalanced. People who have a trisomy of chromosome 21 (or, rarely, 22) or of the sex chromosomes (Fig. 13–15) may survive into adulthood. Trisomies of the other chromosomes are much rarer because they are much more likely to be fatal. Quite possibly, trisomy of chromosome 21 is less likely to be fatal because this chromosome is among the smallest and probably possesses the fewest genes.

◼ CHAPTER SUMMARY

I. Each somatic cell of every organism of a given species has a characteristic number of chromosomes.

II. The cell cycle is the period from the beginning of one cell division to the beginning of the next division.

  A. During interphase the cell is growing and preparing for the next division. DNA is replicated during interphase.

  B. During mitosis a complete, identical set of chromosomes is distributed to each daughter cell.

   1. During prophase chromatin condenses into chromosomes, the nucleolus breaks down, the nuclear envelope breaks down, and the mitotic spindle begins to form. At the end of prophase, each chromosome is composed of two (sister) chromatids.

   2. During metaphase the chromosomes line up along the equator of the cell.

   3. During anaphase the sister chromatids separate and move toward opposite poles of the cell; each chromatid is now a complete chromosome.

   4. During telophase a nuclear envelope forms around each set of chromosomes, nucleoli reappear, the chromosomes lengthen and become chromatin, the spindle disappears, and cytokinesis generally takes place.

III. Meiosis reduces the chromosome number by one half. It is a required process in the life of all sexually reproducing organisms because it results in haploid gametes.

  A. During prophase I the homologous chromosomes undergo synapsis and crossing-over, resulting in genetic recombination.

  B. The members of each pair of homologous chromosomes separate during anaphase I and are distributed to different daughter cells.

  C. During the second meiotic division the two chromatids of each chromosome separate, and each is distributed to a daughter cell.

  D. As a result of meiosis, four haploid cells are formed.

  E. In sexual reproduction a haploid set of chromosomes from the sperm and a haploid set of chromosomes from the egg unite, forming the diploid zygote.

◼ POST-TEST

1. Except during nuclear division eukaryotic chromosomes appear as long, thin, dark-staining threads called _____.

2. A human somatic cell with 46 chromosomes would be described as _____, or _____.

3. A sperm cell with only one of each pair of chromosomes is described as _____, or _____.

4. The members of a given pair of chromosomes are referred to as _____ chromosomes.

5. The period from the beginning of one cell division to the beginning of the next is known as the _____ _____.

6. In _____ a complete set of chromosomes is distributed to each new daughter nucleus.

7. The actual division of the cytoplasm to form two cells is called _____.

8. During the _____ stage of its life cycle a cell grows and carries out its normal functions.

9. DNA is replicated during the _____ phase of interphase.

10. Sister chromatids are attached at their _____.

11. During _____ of mitosis the chromatids are lined up along the equatorial plane of the cell.

12. During _____ of mitosis the chromatids

separate and begin to move toward opposite poles.

13. In meiosis there are _____ divisions, producing a total of _____ cells.

14. Synapsis results in the formation of complexes of chromosomes known as _____.

15. During _____ segments of DNA of homologous chromatids are broken and exchanged.

16. The exchange of DNA segments between homologous chromatids may result in _____ _____.

17. Down's syndrome is produced by _____ of chromosome 21.

18. A man inherits his X chromosome from his _____.

☐ REVIEW QUESTIONS

1. How does meiosis differ from mitosis?
2. Define the following terms: (a) diploid, (b) haploid, and (c) homologous chromosomes.
3. Draw the stages in meiosis and indicate when and how synapsis and separation of homologous chromosomes occur.
4. Describe the structure of a chromosome. Distinguish between chromatin, chromosome, and chromatid.
5. What is the relationship between genes and chromosomes? What are the functions of genes?
6. Two very different species may have the same diploid chromosome number. How can this be explained?
7. What is an aneuploidy? Give an example.

☐ RECOMMENDED READINGS

Baserga, R. *The Biology of Cell Reproduction.* Cambridge, Massachusetts, Harvard University Press, 1985. An interesting account of mitosis.

Burns, G.W. and P.J. Bottino. *The Science of Genetics,* 6th ed., New York, Macmillan Publishing Company, 1989. Chapter 4, "Cytological Basis of Inheritance," contains straightforward information on both mitosis and meiosis.

Chandley, A.C. "Meiosis in Man," *Trends in Genetics* 4:79–83, 1988.

Patterson, D. "The Causes of Down's Syndrome," *Scientific American,* April 1987, 52–60.

14

Patterns of Inheritance

OUTLINE

I. Gregor Mendel founded the science of genetics in the 19th century

II. Inheritance is governed by laws of chance

III. Genes occur in pairs called alleles and are inherited as parts of chromosomes

    A. The outcome of a simple cross may be predicted by possible combinations of a single pair of alleles

    B. Organisms may have both members of a pair of genes alike, or unlike each other

    C. Some genes are always expressed, but others are expressed only when homozygous

    D. Dominance can be incomplete

IV. Only one allele of each pair goes to each gamete

    A. Genes located on different chromosomes are inherited independently

LEARNING OBJECTIVES

After you have studied this chapter you should be able to:

1. Define the basic terms relating to genetic inheritance, for example, *gene, dominance, recessiveness, codominance, chromosome, homozygous, heterozygous, alleles, homologous, locus, genotype,* and *phenotype.*

2. Relate the inheritance of genetic traits to the behavior of chromosomes in meiosis.

3. Summarize the concepts of homologous chromosomes and allelic genes.

4. Distinguish between homozygous and heterozygous genotypes. Relate genotype to phenotype in terms of dominance.

5. Solve simple problems in genetics involving monohybrid and dihybrid crosses by applying the laws of genetic combination.

Specific combinations of a few genes result in the vast variety of these peppers. (David Cavagnaro)

You are not a duplicate of your mother nor a carbon copy of your father, nor even an exact mixture of the characteristics of both. Yet apart from mutation, all your genetic traits came from the two of them. There must be mechanisms of distribution at work that govern the combinations of genes we receive. Although these mechanisms are governed by chance, their outcome is predictable because the possibilities with which they work are limited and governed by the laws of probability. They are among the best understood of the principles of biology. What mechanisms of inheritance determine whether you are male or female, tall or short, light- or dark-skinned, whether you have type A or type O blood, whether you have normal vision or are colorblind, and countless other of your characteristics? The same principles that can be used to answer those questions can be used to answer others far more grave: What is the likelihood that your son will have hemophilia? Your daughter? Did the hospital give you the right baby? Should you avoid having children because you suffer from a genetic disease? Yet even this is only one kind of application of the principles of genetics. They can also be used to answer questions about breeding a variety of corn which possesses a particular, vital amino acid in its protein, how to increase that protein content, how to impart resistance to plant disease and agricultural pests, and how to increase the number of bushels of corn per acre produced by that variety.

KEY CONCEPTS

☐ Genes combine with one another in accordance with mathematical principles.

☐ Allelic genes govern the same trait; they may be alike or unlike.

☐ In many cases, a gene is expressed only when it and its allele are alike, but in others, it is expressed even when they are unlike. Sometimes alleles are both partially or fully expressed even when they are unlike one another.

☐ Only one allele of a pair normally passes to a gamete.

☐ Since genes are located on chromosomes, they are generally inherited in the same manner as are the chromosomes to which they belong.

☐ GREGOR MENDEL FOUNDED THE SCIENCE OF GENETICS IN THE 19TH CENTURY

The roots of much modern biology, including genetics, extend into the 19th century. The foundation of our modern knowledge of genetics was laid at that time by

FIGURE 14–1   Gregor Mendel, the father of genetics. (V. Orel, Mendelianum of the Moravian museum)

a then obscure Austrian monk, Gregor Mendel (1822–1884), who lived in the town of Brunn, Austria, which is now Brno, Czechoslovakia (Fig. 14–1). "Natural philosophy," as biology was called in the 19th century, was usually not a profession but a hobby; only clergymen, physicians, wealthy people, or others with some leisure time were able to pursue it.

Mendel was active in the natural history society of Brunn and presented his findings in a series of research reports, which were published in the society's journal. His research was revolutionary; despite this, it was almost universally ignored, perhaps because its significance was not understood. Mendel did not gain the recognition rightly due him for more than 30 years. After his death, several biologists independently recognized most of these principles and then rediscovered his papers describing them.

The view of inheritance that had prevailed up to the end of the 19th century was, roughly speaking, a "blending" hypothesis. There was no clear-cut understanding of the special role of the reproductive tissues in inheritance; instead the whole body was thought to participate. This idea is still reflected in common expressions linking inheritance to blood, which of course plays no part in it. So in some vague fashion two parents somehow blended their "blood" to produce an offspring that had a mixture of both "bloods" and ought therefore to have traits intermediate between any contrasting characteristics of its parents.

L.C. Dunn suggests that the significance of Mendel's work was not appreciated because most biologists of the day lacked Mendel's mathematical background and were not used to thinking in statistical terms. We must remember that the behavior of chromosomes in

FIGURE 14–2 A cut-away drawing of a pea flower showing the reproductive parts of the flower. These are the pollen-producing anthers and the stigma, a female part that receives the pollen. Since the petals completely enclose these parts there is little chance of cross-pollination between separate flowers.

meiosis was not worked out until some years after Mendel's studies, so Mendel could have had no clear conception of the physical basis of inheritance. Reading through his original papers (see the Readings list at the end of this chapter) makes it clear that for Mendel the unit of inheritance was a mathematical abstraction whose behavior could be described by equations. At that time, mathematics still was rarely applied to biology. Mendel's mathematical discussions are not the easiest reading even today and probably seemed both eccentric and arcane to most biologists of the time who did run across them.

Mendel's reports were certainly widely distributed. Copies of the journal in which they appeared were regularly sent to 120 learned societies and universities. In addition, 40 reprints of Mendel's most important paper were ordered and sent to some leading biologists of the day. Remarkably and ironically, though Mendel's research would have meant much to Charles Darwin, the page edges of Mendel's paper in Darwin's library were found to be uncut. It had never been read.

Mendel worked with garden peas, which exist in a number of distinct varieties differing in such characteristics as height, flower color, seed coat color, and seed shape. Like other flowering plants, peas produce pollen grains, each of which contains a haploid sperm nucleus. When the pollen grain encounters the appropriate female flower parts, it grows a long tube into them, which eventually reaches the haploid egg. A series of events then occur (which are described in Chapter 23), in which the two haploid gametes fuse and an

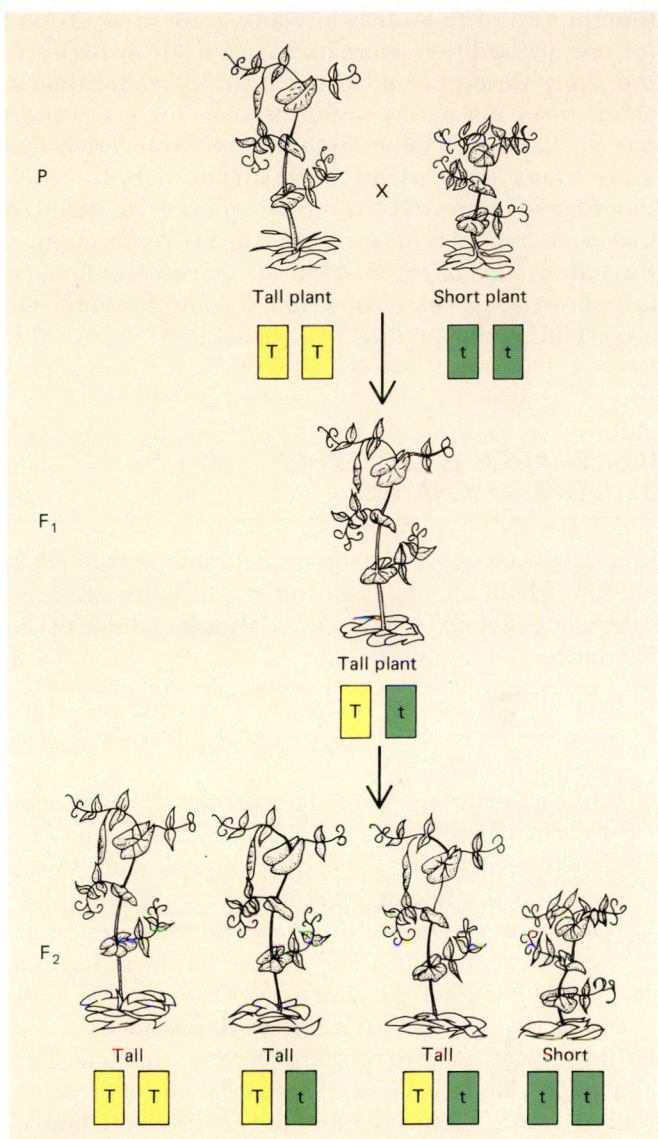

FIGURE 14–3 One of several kinds of crosses carried out by Gregor Mendel. If a tall and short pea plant were crossed, the first (F₁) generation would yield all tall plants. If a member of that generation were self-fertilized, its offspring (the second, or F₂, generation) would include three tall plants for every short one.

embryonic plant develops from the zygote. The seed forms, a little package of nutrients plus the embryo—a miniature plant provided with all that it needs to get a start in life.

Pea plants are normally *self*-fertilized, so simple surgical removal of the male flower parts (Fig. 14–2) makes the plant incapable of being fertilized except by artificial means. In this way, the crossing of varieties can be closely controlled.

In a typical experiment, Mendel crossed a tall-growing variety of pea with a short or bush variety (Fig. 14–3). The offspring were not of some intermediate height, but resembled their tall parent. If, though, two of these hybrid plants were to be crossed (or one allowed to fertilize itself), then "throwbacks" to the short variety would occur in a 3:1 ratio, that is, about three tall plants would be seen for every short one in the second generation. Mendel concluded that some traits can "silently" persist in hybrids, even though not expressed. As a consequence, he deduced that traits must occur as pairs. But his recognition of the importance of the 3:1 ratio in the second generation showed that these pairs combine and recombine in accordance with the laws of probability.

■ INHERITANCE IS GOVERNED BY LAWS OF CHANCE

Mendel's breeding experiments led him to certain conclusions about the mechanisms of heredity, some of which later scholars restated as **Mendel's laws of inheritance:**

1. Inherited traits are transmitted by genes, which occur in pairs. Each member of the pair is now called an **allele.**
2. When gametes are formed in meiosis, the two alleles of each pair separate from one another, and each gamete receives only one allele of each pair. This is known as the **law of segregation.**
3. When two alternative forms of the same gene (i.e., two different alleles) are present in an individual, often only one of the alternatives is expressed. This concept is known as the **law of dominance.**
4. If one considers two or more separate characteristics in a cross, such as flower color and seed coat texture, each characteristic is inherited without relation to the other traits. All possible combinations of independently inherited characteristics thus will occur in the gametes. This is the **law of independent assortment.**

Most of these principles are consequences of the ability of genes to combine and thus result from mathematical laws that govern combinations and probability.

■ GENES OCCUR IN PAIRS CALLED ALLELES AND ARE INHERITED AS PARTS OF CHROMOSOMES

It is remarkable that Mendel inferred all this without ever having observed meiosis or hearing of deoxyribonucleic acid (DNA). In fact, Mendel did not think in terms of genes, although his observations were to lead to the formulation of this concept, and he certainly knew nothing of the role played by chromosomes in

inheritance. He inferred that units of inheritance exist, but could not have said what they are or how they influence an organism's traits.

As we have discussed, a typical gene may now be viewed as a region of DNA containing the information necessary to manufacture a specific polypeptide or protein.[1] This concept can be illustrated with a common genetic abnormality, albinism. One kind of **albino** is a person or animal that lacks the genetic information for the production of the body pigment **melanin.** This pigment is the protein responsible for most of the color of hair, skin, and eyes. Without it an organism would appear completely unpigmented, with white hair and perhaps pink eyes, through which hemoglobin of the blood would show unmasked.

In this kind of albinism it has been shown that all that is lacking is the ability to make **tyrosinase,** an enzyme needed to form melanin from the amino acid tyrosine. Without that enzyme no pigment can be formed. But what is meant by a lack of tyrosinase? It is possible, by immunological techniques, to demonstrate that many tyrosinase-negative persons have at least a version of the tyrosinase enzyme. However, it is not a correct version and is not functional. These albinos lack the genetic information needed to produce functional tyrosinase.

The significance of all this is that the gene "for albinism" does not produce an albino directly. It is a spot of hereditary misinformation in a DNA molecule that results in the production of a defective enzyme. The normal allele of this gene is simply a functional version. If an individual had both an albino gene and a normal skin color gene, the skin would be normally colored due to the presence of the functional enzyme. The **genotype** (genetic makeup) of such a genetically mixed individual would contain an albino gene. One would not know this from appearance, because this gene would not be apparent in the individual, whose **phenotype,** the portion of the genotype that is actually expressed, would be normal.

Recall that two members of a pair of chromosomes are said to be **homologous.** They have genes for similar traits arranged in similar order. The gene for each trait occurs at a particular site in the chromosome called a **locus** (plural, loci). In guinea pigs, for example, if one chromosome of a pair contains a gene for coat color, so will the other chromosome of the pair. The two forms of the gene that govern the same trait and that occupy corresponding loci on homologous chromosomes are known as **alleles** (see Fig. 14–4).

The term allele emphasizes that there are at least two alternative forms of the gene that can occupy a

---

[1] Although some hereditary information (particularly that which is involved in the control of genes) may not code for protein directly, it usually does affect the function of protein-producing genes.

FIGURE 14—4
**Homologous chromosomes
and alleles. Two different
versions of the same gene
can exist in an individual.**

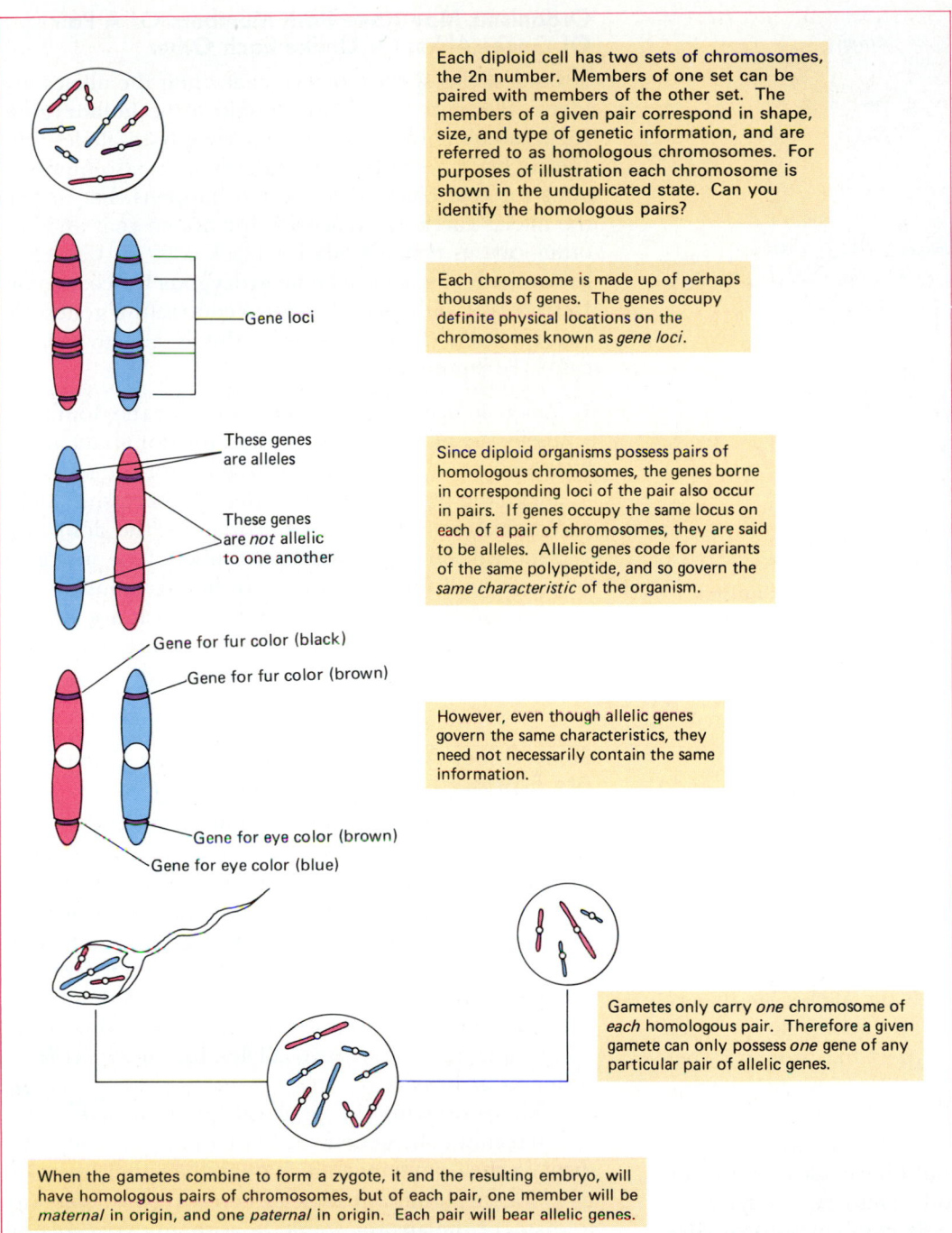

Each diploid cell has two sets of chromosomes, the 2n number. Members of one set can be paired with members of the other set. The members of a given pair correspond in shape, size, and type of genetic information, and are referred to as homologous chromosomes. For purposes of illustration each chromosome is shown in the unduplicated state. Can you identify the homologous pairs?

Gene loci

Each chromosome is made up of perhaps thousands of genes. The genes occupy definite physical locations on the chromosomes known as *gene loci*.

These genes are alleles

These genes are *not* allelic to one another

Since diploid organisms possess pairs of homologous chromosomes, the genes borne in corresponding loci of the pair also occur in pairs. If genes occupy the same locus on each of a pair of chromosomes, they are said to be alleles. Allelic genes code for variants of the same polypeptide, and so govern the *same characteristic* of the organism.

Gene for fur color (black)
Gene for fur color (brown)

However, even though allelic genes govern the same characteristics, they need not necessarily contain the same information.

Gene for eye color (brown)
Gene for eye color (blue)

Gametes only carry *one* chromosome of *each* homologous pair. Therefore a given gamete can only possess *one* gene of any particular pair of allelic genes.

When the gametes combine to form a zygote, it and the resulting embryo, will have homologous pairs of chromosomes, but of each pair, one member will be *maternal* in origin, and one *paternal* in origin. Each pair will bear allelic genes.

specific locus in homologous chromosomes. For bookkeeping purposes, each of these forms can be assigned a letter as its symbol. It is customary to designate the **dominant** allele, the one that always manifests itself, with a capital letter, and the **recessive** allele (the gene that does not express itself in the presence of a dominant allele) with a lowercase letter. Thus, the letter *B* could be used for black coat color and the letter *b* for brown when specifying the alleles that determine coat color in guinea pigs.

The Outcome Of A Simple Cross May Be Predicted By Possible Combinations Of A Single Pair Of Alleles

The usage of genetic terms and some of the basic principles of genetics can be illustrated by considering a simple **monohybrid cross,** that is, a cross between two individuals in which only one heritable trait is being studied. The mating of a genetically pure brown male guinea pig with a genetically pure black female guinea

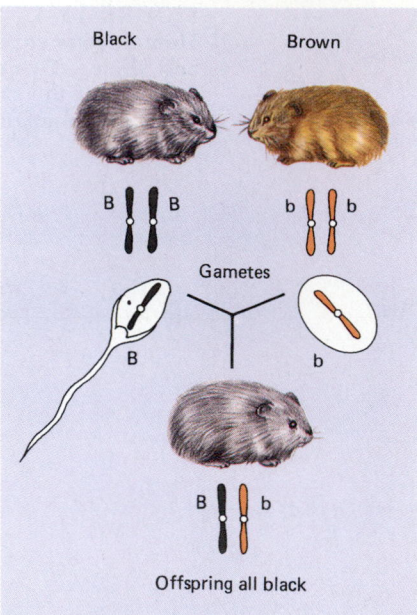

FIGURE 14–5   When a genetically pure black guinea pig is mated with a brown guinea pig, all the offspring are black. (The brown guinea pig has to be genetically pure. If it weren't, it would be black rather than brown.)

Organisms May Have Both Members Of A Pair Of Genes Alike, Or Unlike Each Other

If both alleles specify brown coat, then the alleles are identical and the organism is said to be **homozygous** for coat color. In another guinea pig both alleles may specify black coat. This second animal is also homozygous for coat color. But it often happens that one of the alleles carries instructions for brown coat and the other carries instructions for black coat. In that case, the individual is said to be **heterozygous** for coat color. Despite this, only one of the two contrasting genes will express itself (as we shall see, this is not always the case). To summarize:

1. When an organism is homozygous for the dominant allele, the phenotype will reflect the dominant allele.
2. When the organism is homozygous for the recessive allele, the phenotype will reflect the recessive allele.
3. When the organism is heterozygous, the dominant gene will be expressed in the phenotype just as it would be if the organism were homozygous for the dominant allele. In such cases, one cannot readily tell a heterozygous individual from one that is homozygous for the dominant gene.

Some Genes Are Always Expressed, But Others Are Expressed Only When Homozygous

All this permits us to define "dominant" and "recessive" with precision. When one member of an allelic pair tends to dominate the other completely, so that it alone is expressed in the heterozygous condition, it is said to be **dominant.** If, however, a gene is expressed *only* when homozygous, it is termed **recessive.** In guinea pigs the allele for black coat color is dominant, and the allele for brown coat color is recessive. A guinea pig must have two alleles for brown coat in order to be brown.

There remains the practical question of how we can determine the genetic makeup of an organism that displays the dominant trait. It might be either homozygous for the dominant allele or heterozygous. One way to discover the answer would be with an experimental cross with a recessive organism. Suppose, for instance, that a black guinea pig really is heterozygous. If one were to cross it with a brown one (which could only be homozygous for the recessive brown trait) at least some of the offspring ought to display the recessive trait, that is, be brown (Fig. 14–7). On the other hand, if the unknown animal were homozygous for black coat color, the offspring would all be black.

Dominance Can Be Incomplete

Not every pair of genes consists of a recessive allele and one that can be called dominant. For example, when

pig (that is, one with two black coat color alleles) is illustrated in Figure 14–5. During meiosis in the male, in which sperm cells are formed, the two *bb* alleles separate at anaphase of meiosis I, so each sperm has only one *b* allele. In the formation of ova (egg cells) in the female, the *BB* alleles separate, so each ovum has only one *B* allele. The fertilization of this egg by a *b*-bearing sperm results in an animal with the alleles *Bb*, that is, with one allele for brown coat and one allele for black coat. What color would you expect this animal to be?

Suppose that two black guinea pigs each having alleles for both brown and black coat color were crossed. Because of meiosis half of the gametes produced by each guinea pig would have alleles for black coat color, and the other half would have alleles for brown coat color. The probable combinations of eggs and sperm may be represented in a checkerboard, or **Punnett square,** as illustrated in Figure 14–6. The types of eggs can be represented across the top and the types of sperm indicated along the left side. The squares are filled in with the resulting zygote combinations, so that the letters in each square indicate the genotype of one genetic type of offspring.

The generation with which a particular genetic experiment is begun is called the **parental generation,** or **P.** Offspring of this generation are referred to as the **first filial generation,** or **F₁.** Those resulting when two F₁ individuals are bred constitute the **second filial generation,** or **F₂.**

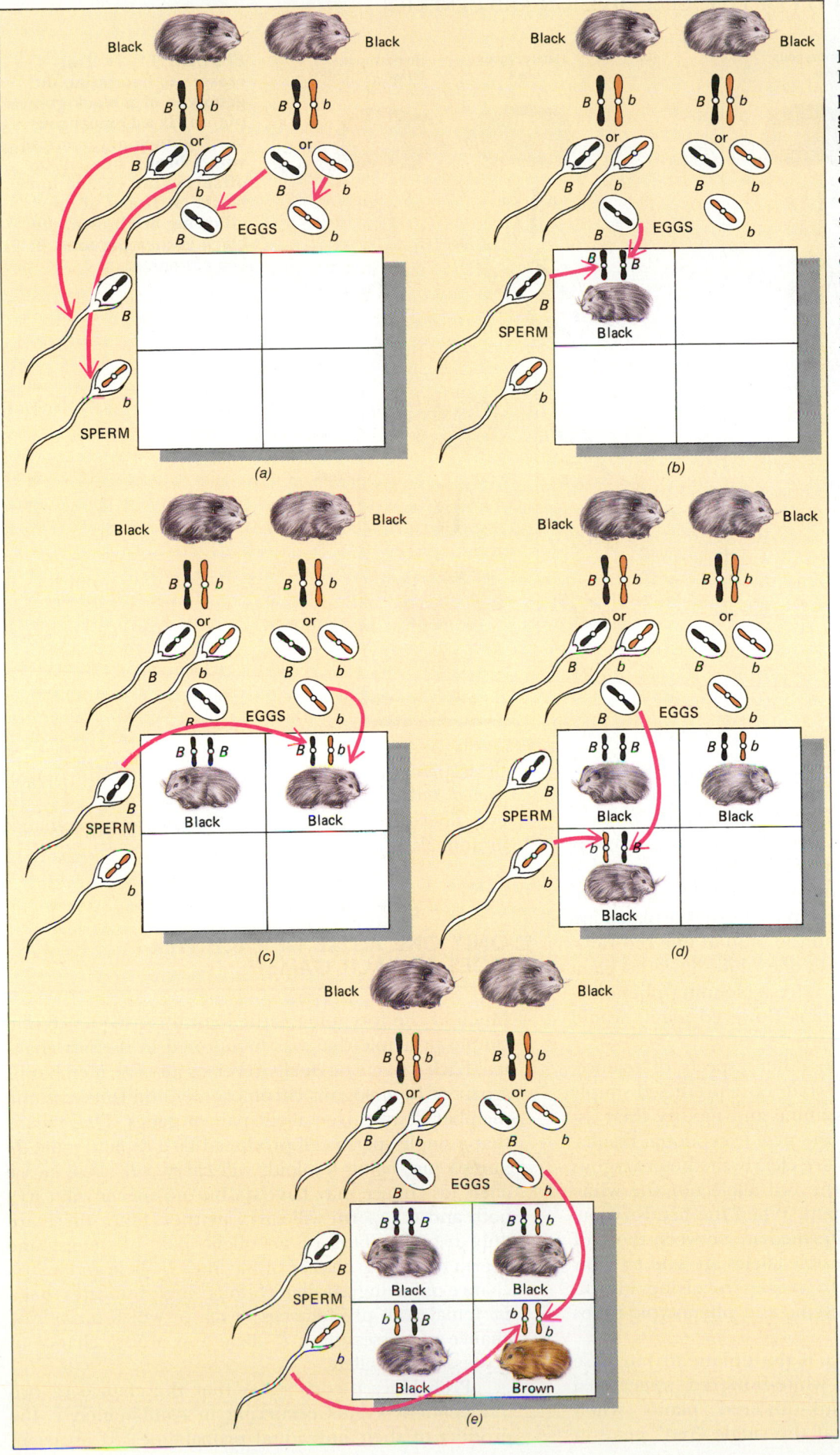

FIGURE 14–6 The Punnett square method of predicting probable genotypes. Note that the letters in each square indicate the genotype of one genetic type of offspring. The reason some combinations of *B* and *b* are written *bB* and others *Bb* is to show that these combinations form in two ways. Usually, though, both are written *Bb*.

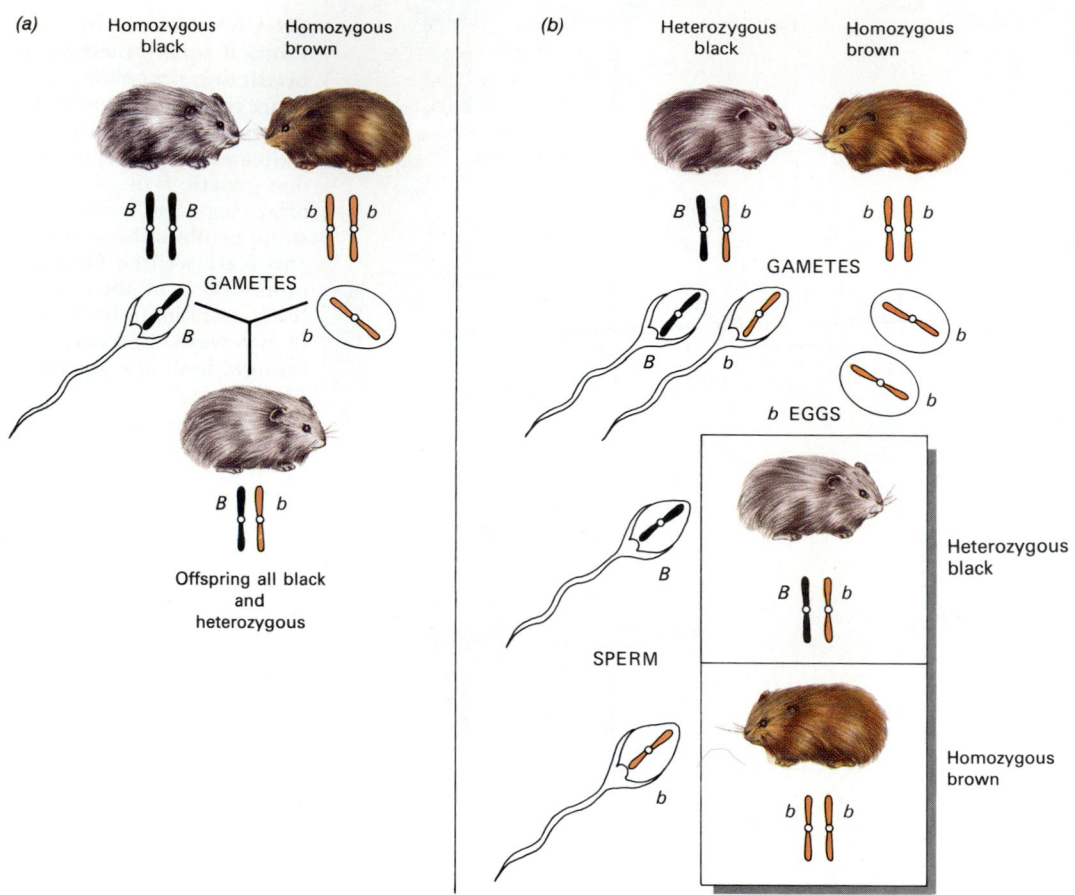

FIGURE 14–7 Test crosses to determine the genotype of a black guinea pig. *(a)* If a homozygous black guinea pig is mated with a brown one, all the offspring are certain to be black. *(b)* If any of the offspring are brown, the black guinea pig must be heterozygous.

red- and white-flowered Japanese four-o'clocks are crossed, the offspring do not have red or white flowers. Instead, all the F_1 offspring have pink flowers. How can we explain that? Does this result in any way prove that Mendel's assumptions that inheritance is not a blending phenomenon are wrong? Quite the contrary, for when two of these pink-flowered plants are crossed, offspring appear in the ratio of one red-flowered to two pink-flowered to one white-flowered plant (Fig. 14–8). If inheritance were a blending phenomenon, all offspring of two pink-flowered plants would be pink.

In this instance, as in other aspects of science, finding results that differ from those predicted simply prompts scientists to reexamine and modify their assumptions to account for the new exceptional results. The pink-flowered plants are clearly the heterozygous individuals, and neither the red allele nor the white allele is completely dominant. When the heterozygote has a phenotype that is intermediate between those of its two parents, the responsible alleles are said to show **incomplete dominance.** In crosses involving incomplete dominance the genotypic and phenotypic ratios are identical.

Incomplete dominance is not unique to Japanese four-o'clocks. Red- and white-flowered sweet pea plants also produce pink-flowered plants when crossed. The reason is that the single "red" gene in

these plants is unable to code for the production of *enough* red pigment to make the petals look red to the eye. There are also examples of incomplete dominance in animals.

■ ONLY ONE ALLELE OF EACH PAIR GOES TO EACH GAMETE

Much of the foregoing is the logical consequence of a simple principle that was mentioned in the last chapter: Each gamete normally receives just one member of a pair of homologous chromosomes and thus just one member of any given allelic pair of genes. That means that a single sperm cell produced by a diploid male Bb guinea pig, being haploid, will carry a B allele *or* a b allele (as chance may have it), but no sperm will carry both and no sperm will carry neither. Since there are only two members of an allelic pair of genes, one sperm has a 50% chance of having a B allele and a 50% chance of having a b allele. Similar logic indicates that a Bb female can produce eggs each of which has a 50% chance of possessing a B allele and a 50% chance of possessing a b allele.

Probability theory states that the chance of two independent events occurring in combination is the product of their individual probabilities. If an event

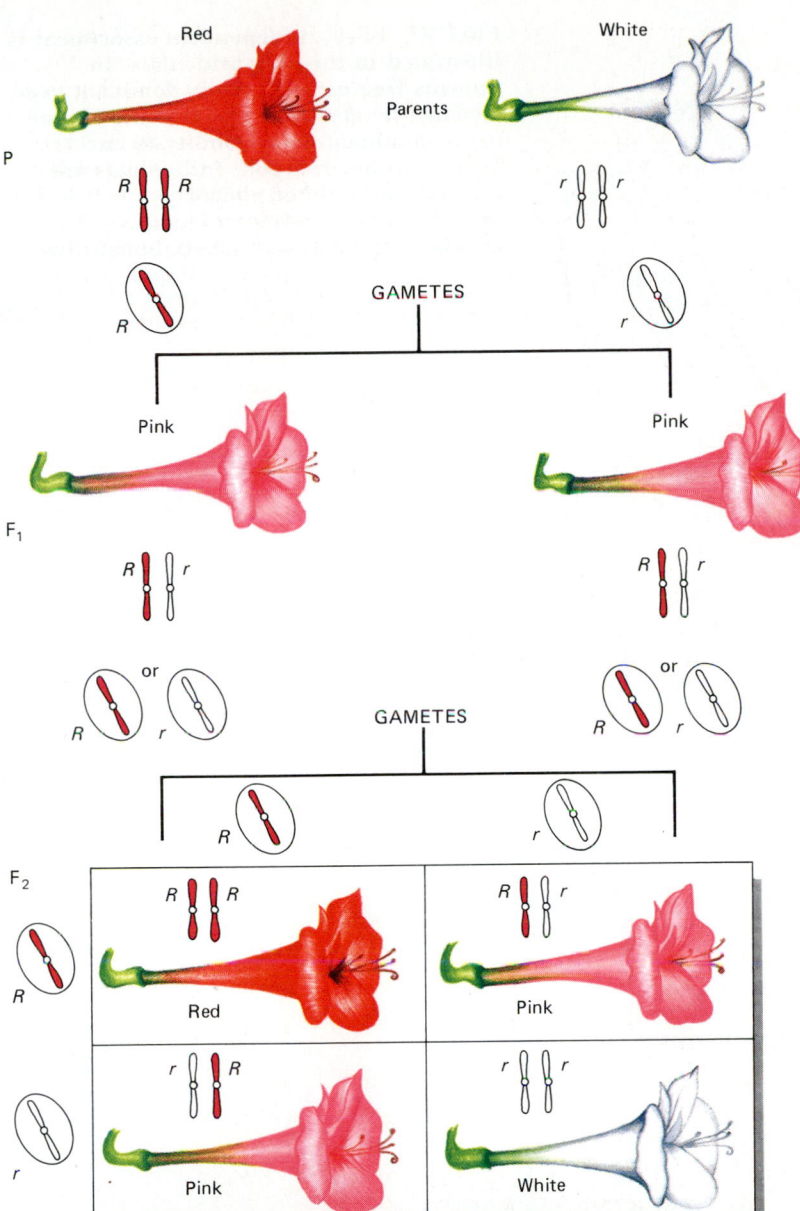

FIGURE 14–8 Incomplete dominance in Japanese four-o'clocks. Red is incompletely dominant to white. A plant with the genotype *Rr* has pink flowers. Note that there is no allele for pink flowers, but only alleles for red and white flowers.

has a 50% chance of occurrence, and another also has a 50% chance of happening, the likelihood of them taking place together is not 100% but 50% times 50%; that is 50% of 50% or 25%. Thus, if sperm are unguided save by chance, the probability of a *b* sperm finding a *b* ovum is 25%, and the chance of a *B* sperm finding a *B* ovum is also 25%. A *B* sperm can be expected to find a *b* egg 25% of the time, and similarly, a *b* sperm will find a *B* egg 25% of the time. But genetically a *Bb* individual is the same as a *bB* individual, so the likelihood of this combination is the total of the probabilities of the two ways in which it can occur, that is, 50%.

Does this mean that out of every four F₂ guinea pigs one is bound to be brown (i.e., *bb*)? Not at all. It is quite likely that a *b* sperm and a *b* egg just won't get together. All the litter might be black, even if both parents are heterozygous. Given a sufficiently large *num-*

ber of instances, though, we should observe very nearly 25% brown offspring of such a cross. Guinea pigs are prolific, but fruit flies, corn, and peas are even more so and for that reason are well suited to genetic studies. One might think people and elephants, with their long generation time and habit of bearing one offspring at a time, would be unsuitable genetic subjects. True, elephants are difficult to work with. With people, however, vast statistical data can be amassed from medical records. As a result, genetic studies of human beings are readily possible, but not, of course, by forced matings between predetermined genotypes. Rather, since a great variety of crosses occur repeatedly in our vast human population, genetic analyses are often carried out by the diligent investigation of medical records followed by statistical summarization of the results of the crosses after they have occurred.

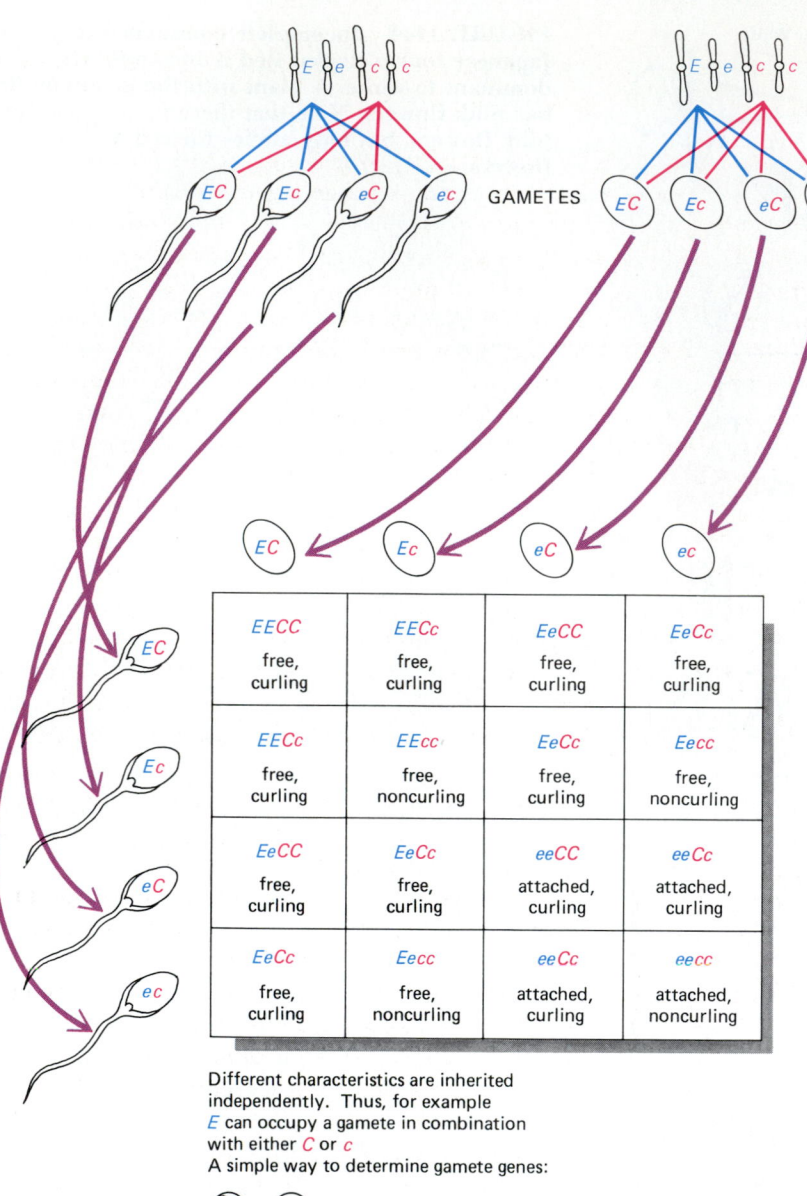

GAMETES

FIGURE 14-9 Independent assortment is illustrated in this dihybrid cross. In humans free earlobes (*E*) is dominant over attached (*e*) and ability to curl the tongue (*C*) is dominant over inability to curl (*c*). When two heterozygous individuals are crossed, the ratio of phenotypes is 9:3:3:1 (9 free, curling; 3 free, noncurling; 3 attached, curling; 1 attached, noncurling).

Different characteristics are inherited independently. Thus, for example *E* can occupy a gamete in combination with either *C* or *c*
A simple way to determine gamete genes:

Combinations: *EC   Ec   eC   ec*

☐ GENES LOCATED ON DIFFERENT CHROMOSOMES ARE INHERITED INDEPENDENTLY

We now know that Mendel's law of independent assortment applies only to traits carried on nonhomologous chromosomes. If different genes are carried on the same chromosome, they will indeed tend to be inherited together. However, if they occur on *different* pairs of homologous chromosomes, they will be inherited independently. To illustrate this principle, let us consider two pairs of human traits. The ability to curl the tongue into a tube is a dominant trait; the absence of this talent is recessive. Attached or absent earlobes is a recessive trait; free earlobes are dominant.

A mating that involves individuals differing in *two* gene pairs is referred to as a **dihybrid cross.** If a man homozygous for tongue curling (*CC*) marries a woman without this trait (*cc*), all their children (*Cc*) may be expected to have this ability, since it is dominant and they are heterozygous for it. Similarly (but quite unrelated), if he has free earlobes (*EE*) and she is homozygous for attached earlobes (*ee*), the children will have free earlobes (*Ee*). One could write the children's genotype for both sets of traits as *CcEe* because the children are heterozygous for both. Since the two sets of traits are due to genes found on separate chromosome pairs, they will be inherited (or *assorted*) independently of each other.

Let us suppose that some of these children were to

marry others who have genotypes identical to their own. As you see by reference to Figure 14–9, four kinds of gametes are possible with respect to these alleles, that is, since each gamete will have one of each pair of alleles, a total of four gametic genotypes is possible, given the genetic makeup of the F_1 generation. If, by chance, the chromosome bearing gene E is sorted during meiosis into the same gamete as the chromosome bearing gene c, the resulting sperm or egg will have the genes Ec. Similarly, if the chromosomes bearing e and C are assorted together, the gamete will have the genes eC. Other gametes would be EC, and still others would be ec. *There are no other possible combinations than these four.* Since allelic genes are always borne on homologous chromosomes that are *separated* from one another in meiosis, a gamete can ordinarily have no more than *one* copy of each allele. Since there are four possible kinds of gametes, working out a cross of that kind will require a Punnett square of 16 cells. Count the phenotypes. If you do so properly, the proportions, or ratios, of phenotypes that you will predict among the offspring will be a 9:3:3:1 ratio.

Obviously, too, genes located on the same chromosome will tend to be inherited together. This tendency is not absolute, however, due to crossing-over (Chapter 13). We shall discuss this further in the next chapter.

The expression and interaction of genes can be much more complex than we are able to indicate in this brief treatment. If *both* alleles are *fully* expressed, they are said to be **codominant.** Human blood types, which we will discuss in the next chapter, provide an example of codominance. Also, a single gene will probably have multiple effects, a quality referred to as **pleiotrophy,** rather than a single effect. Some genes are able to suppress the expressions of other, nonallelic genes. This is referred to as **epistasis.** Finally, even a dominant gene may be expressed only in certain instances, a condition called **incomplete penetrance.**

Albinism illustrates some of these concepts. Whatever skin color genes a person may have, they will have no effect if he or she is an albino. The skin color of an albino of African ancestry will be indistinguishable from that of an albino of Norwegian ancestry. This is an example of **epistasis.** Albinism also produces defects of vision, both by preventing normal eye pigmentation and by producing disorders in the visual pathways of the brain, an example of pleiotropy.

■ CHAPTER SUMMARY

I. The science of genetics was founded by the 19th-century research of Gregor Mendel.
II. The basic unit of heredity is the gene.

    A. Fundamentally, a gene acts by directing the production of a polypeptide and ultimately a protein.
    B. Characteristics of these proteins determine the expressed traits (phenotype) of the organism.
    C. Genes are associated with chromosomes and occur on them. The genes an individual has are its genotype, regardless of how or whether they are manifested.
    D. The behavior of genes in meiosis is similar to the behavior of chromosomes.
    E. The production of most heritable traits is governed by a pair of genes present in each and all body cells. These two forms of a gene in a diploid cell are called alleles.
    F. Allelic genes are carried on homologous chromosomes, and, like them, are paired in meiosis.
    G. Members of a gene pair may be alike (homozygous) or unlike (heterozygous).
    H. When members of a gene pair are unlike, the traits they govern may be determined by one of them if it is clearly dominant. An allele that is expressed only when homozygous is said to be recessive.
    I. Normally, no offspring can receive any more or less than one representative of an allelic pair of genes from each parent.
    J. The member of a pair of chromosomes (and its contained alleles) that a gamete receives is governed entirely by chance.
    K. The inheritance of a particular chromosome has no influence on the inheritance of any other chromosome not homologous to it.
    L. The inheritance of any gene has no influence on the inheritance of any other gene borne on a nonhomologous chromosome.
III. Typically, a monohybrid cross between two heterozygous individuals yields a 3:1 ratio of dominant-to-recessive gene expression in the phenotypes of the offspring. A dihybrid cross between two heterozygous individuals produces a 9:3:3:1 ratio among phenotypes. In cases of incomplete dominance, the ratio of phenotypes is the same as the ratio of genotypes.

■ POST-TEST

1. The modern science of genetics was founded in the _____ century by _____ _____, whose work was unrecognized until its rediscovery after his death.
2. A unit of hereditary information is called a _____, which is somewhat more exactly defined as the hereditary information necessary to produce a particular _____.

3. The expression of a gene in the physical or other characteristics of an organism is its _____.
4. Genes are inherited independently if they occur on non-_____ chromosomes.
5. An organism with both members of an allelic pair of genes alike is said to be _____.
6. A gene that can be expressed even when heterozygous is called _____.
7. If two _____ organisms are crossed, and the trait being studied displays complete dominance, one may expect a 3:1 ratio of _____ among the offspring.
8. Whether a particular sperm finds and fertilizes a given egg is an event that is determined completely by _____.
9. Only _____ gene of each allelic pair will occur in each gamete.

■ REVIEW QUESTIONS

1. What are alleles?
2. What is meant by the terms *homozygous* and *heterozygous?*
3. Why can one receive only one member of a homologous pair of chromosomes from each parent? What would happen if you tried to solve genetic problems without realizing this?
4. Define *dominant, recessive, incompletely dominant.*
5. What usually makes a gene dominant? Why do you suppose that most known genetic defects are recessive?
6. A scientist is studying flower pigment in bean plants. She extracts the pigment from the flowers of two different red flowering plants from a field that also contains white flowering plants. Though both *appeared* equally red to the eye, one of the two plants has significantly less red pigment than the other. Can you propose an explanation?
7. Would the development of genetics in the 20th century have been any different if Mendel had never lived?
8. There is an albinism gene that suppresses the development of any color at all in guinea pigs. It is located on a different chromosome pair from the allelic pair of genes (discussed in the text) that determines whether coat color is black or brown. Suppose a black and a brown guinea pig are crossed, and one-fourth of the offspring are white. How can this be explained, assuming mutation is not the reason?
9. In one study, Gregor Mendel crossed a yellow-seeded with a green-seeded pea plant. The yellow-seeded plant was also tall, while the green-seeded plant was short. The offspring were all yellow-seeded and tall. Assuming independent assortment of these two gene sets, what phenotypes and proportions did he find among the progeny when the hybrid pea plants were allowed to fertilize themselves?
10. If you wanted to determine whether an organism that showed a dominant trait was homozygous or not, you would be best advised to cross the unknown organism with one that is homozygous for the recessive allele rather than to cross it with one known to be heterozygous. Why? (Hint: what phenotype ratios are to be expected among the offspring in the two cases?)
11. In both cattle and horses, reddish coat color is incompletely dominant to white coat color. The heterozygous individuals have roan-colored coats (reddish with interspersed white hairs). If you saw a white mare nursing a roan colt, what would you guess was the coat color of the colt's father? Is there more than one possible answer?

■ RECOMMENDED READINGS

Corcos, A., and F. Monaghan. "Some Myths About Mendel's Experiments," *The American Biology Teacher,* 47:233–236, 1985. Mendel's laws were only implied in his papers. He himself may not have fully appreciated their importance.

Mendel, G. *Experiments in Plant Hybridization,* Edited by J.H. Bennett. English language edition, Edinburgh, Oliver & Boyd, 1965.

Olby, R.C. *The Origins of Mendelism.* New York, Schocken, 1966. Perhaps the most complete account of Mendel and his rediscovery that we are ever likely to get (Mendel's personal papers were destroyed after his death and hardly anyone who knew him was living when his contribution became generally recognized).

15

Some Topics in Human Genetics

OUTLINE

I. The sex of the offspring determines the expression of genes on the X chromosome
II. Genetic linkage can be deduced from the way genes are inherited
III. Some traits are governed by more than one pair of genes
IV. Blood type is an inherited trait
   A. Blood type reflects blood cell surface antigens
   B. The antigens of a blood cell are governed by a set of multiple alleles
   C. The three ABO alleles produce four blood types
   D. The Rh blood factors are potentially hazardous to the newborn
V. Genetic disease is widespread and important
   A. Marriage between blood relatives increases the chance of genetic disease
   B. Some genetic disorders are dominant
   C. Genetics raises ethical questions

LEARNING OBJECTIVES

After you have studied this chapter you should be able to:

1. Describe the inheritance of sex-linked genes.
2. Solve simple problems in genetics involving incomplete dominance, polygenes, and multiple alleles.
3. Recognize a state of genetic linkage and, given an example, be able to solve simple genetic problems involving sex linkage.
4. Describe the inheritance of the Rh and ABO factors and outline the mechanisms of Rh disease.
5. Summarize the characteristics of selected genetic diseases.
6. Define *consanguinity* and give its principal genetic implications.

With new techniques, we may someday prevent human genetic disease. (Myrleen Fergueson/PhotoEdit)

"Our poor family seems persecuted by this awful disease," wrote Queen Victoria (Fig. 15–1) in her journal, "the worst I know." The disease to which Her Majesty referred was hemophilia. It evidently originated with a mutation in one of her X chromosomes, and has been passed on to almost every family of European royalty. It was even a contributory factor in the Russian Revolution. Czar Nicholas had married Princess Alexandra of Germany, a granddaughter of Queen Victoria. The couple had a number of daughters but just one son, Prince Alexis. This heir to the throne had hemophilia. The monk Rasputin, a quack, claimed to be able to preserve his life and thereby gained great influence over the czarina and, through her, over the czar. The resulting public policies are widely believed to have been among the factors that contributed to the Russian Revolution of 1917.

Hemophilia is a disorder of the clotting mechanism of the blood, produced by the absence of a necessary globular protein. If the globulin is artificially provided by injection, the sufferer may live indefinitely, though there is a risk of contracting blood-borne diseases such as hepatitis or acquired immune deficiency syndrome (AIDS). In the absence of this treatment, hemophilia is almost invariably fatal in early youth, though in exceptional cases the patient may reach reproductive age. Typically, death of an untreated hemophiliac is caused by a fatal stroke or bleeding to death, either internally or externally. Should the hemorrhage take place in the brain, death is quick. Long-term sufferers are susceptible to serious arthritis arising from bleeding in joint cavities and other disorders. In developed countries where hemophilia is treated, the picture is not this grim, but in Prince Alexis' day neither the cause nor any effective treatment was known.

A recessive gene carried on the X chromosome produces the disorder. Since males only have one X chromosome, hemophilic disease is much more likely to occur among males than females, but a few cases of homozygous female hemophiliacs are now known. No respecter of rank, hemophilia also occurs in those who are not of royal descent. It should not be confused with the numerous disorders of the clotting mechanism which superficially resemble it but which are not genetic or not associated with the X chromosome.

Royal hemophilia is a striking but not isolated example of the importance of heredity in human affairs. In addition to medical genetics, the practical breeding of plants and animals depends on a knowledge of just what genes we and other organisms have, where they are located, and how they are transmitted. Since genes are information encoded within the DNA of the chromosomes, they are inherited when the chromosomes

FIGURE 15–1   Victoria, Queen of England (1819–1901). Her large family intermarried with most of the royal families of Europe, a circumstance which had the unintended consequence of widely distributing the queen's gene for hemophilic blood disease. (The Bettman Archive, Inc.)

that contain them are inherited. Genes on the same chromosome are said to be linked. We can produce maps of how genes are linked to one another on human and other chromosomes, maps that are of great potential importance in practical and medical genetics. Widespread recognition of the importance of inherited disease has made genetics one of the most rapidly expanding areas of modern medicine.

KEY CONCEPTS

☐ Genes located on the X chromosome exhibit a sex-linked pattern of inheritance.

☐ Genes located on the same chromosome are linked, and this linkage can be determined by their tendency to be inherited together.

☐ The relative locations of linked genes on a chromosome may be inferred from their frequency of crossing-over.

☐ Some traits are governed by several sets of interacting genes.

☐ Some genes occur as multiple alleles, so that more than two varieties of the same gene may occur in a population.

■ THE SEX OF THE OFFSPRING DETERMINES THE EXPRESSION OF GENES ON THE X CHROMOSOME

The Y chromosome is one of the smallest of the human chromosomes and, as far as anyone knows, contains only genes related to male sex. Although the Y chromosome pairs with the X chromosome in meiosis, the Y and X chromosomes are not truly homologous. Thus genes found on the X chromosomes have no known alleles on the Y chromosome. Therefore, in the male any allele that lies on the X chromosome will be expressed whether or not such a gene is dominant or recessive in the XX female. One cannot properly refer to such an X-linked gene in the male as either homozygous or heterozygous. It must be called by a special name, **hemizygous.**

Since no important Y-linked genes are known in men (other than those producing male sex), it is the X chromosomes, present in both men and women, that are being referred to when the term **sex linkage** or **sex-linked genes** is used. Hence **X-linked inheritance** is really a better term. The vast majority of abnormal X-linked genes are recessive. Thus in a female, these recessive genes must be homozygous to be expressed. One practical consequence is that while females may carry recessive X-linked traits in the sense of being heterozygous for them, these traits usually find expression only in their sons. (Remember, a normal son receives his X chromosome from his mother, *never* from his father.)

If an X-linked trait is expressed in a female, the trait must be present on *both* chromosomes and so has been inherited from *both* parents. A color-blind girl, as an example, must have a color-blind father and a mother who is at least heterozygous for color blindness (Fig. 15–2). Such a combination is unusual. Yet a color-blind boy need only have a mother who is heterozygous for the trait. His father can have normal color vision.

Production of brown tooth enamel, another X-linked trait, is dominant over white enamel. It arises from a defect in the development of enamel that produces many tiny discolored cracks in the tooth surface. Suppose for a moment that a woman who is heterozygous for both brown tooth enamel and color blindness marries a man who also bears these traits on his single X chromosome. For our example, assume that in the woman both abnormal traits occur on the same X chromosome (i.e., X^{Bc}). This could produce, as you can see from Figure 15–3, a female in whose chromosomal makeup only one gene for color blindness occurs, along with only one gene for brown tooth enamel (number 1). Since color blindness is recessive, her phenotype does not reflect it. The *dominant* brown enamel, however, is expressed. Offspring number 4, also fe-

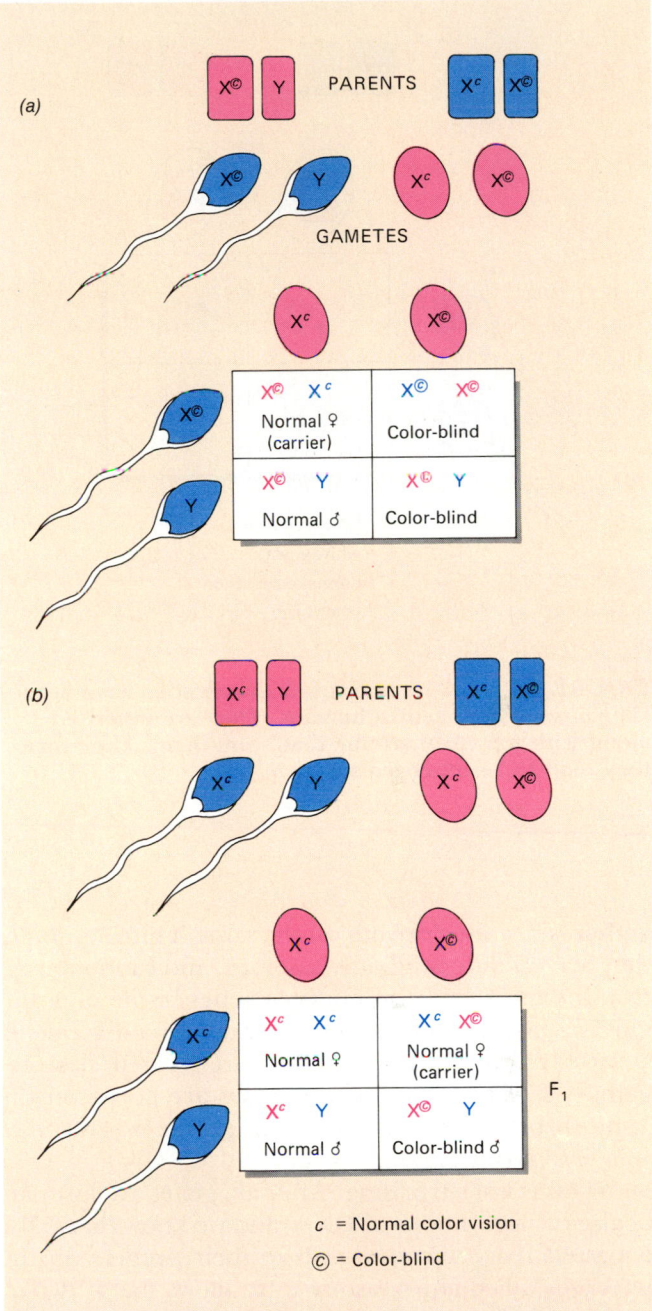

FIGURE 15–2 **Sex-linked genes in humans are usually borne on the X chromosome. Two crosses involving color blindness are shown here. Each circled character represents a gene for color blindness.** *c* **represents the gene for normal vision. In (***a***) the father is color-blind and the mother is not. In (***b***) neither parent is color-blind.**

male, will have brown teeth and will be colorblind. Offspring number 2, a male, will share these shortcomings. Only offspring number 3, again a male, will be phenotypically normal for both traits.

It should be obvious that the genes for color blindness and brown tooth enamel in our example stay to-

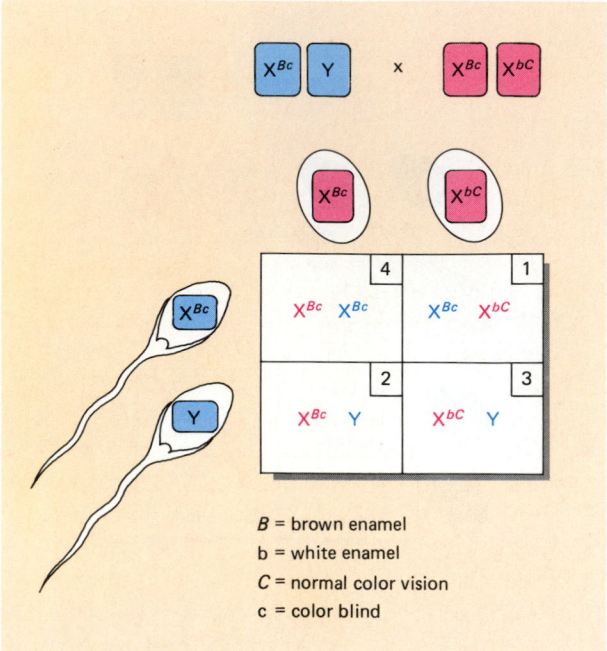

FIGURE 15–3 A number of characteristics have been shown to be X-linked in humans. They are inherited along with the chromosome that bears them. Therefore, they tend to be inherited together.

gether, since they're both on the same X chromosome; they are *not* independently assorted, and cannot be. In fact, the inheritance of multiple genes borne on a single chromosome is very much like that of a *single* gene, at least, as we shall see, if they are located close together. The condition where genes are borne on the same chromosome is termed **linkage.** *Linked genes tend to be inherited together*.

This is not surprising. After all, genes are just particular configurations of DNA, and we know that DNA is passed from ancestral cells to their progeny in the packages called chromosomes. It follows that if two or more genes happen to occur in the same chromosomal package, they will tend to remain together during meiosis and thus will tend to be inherited together. This principle is not limited to the sex chromosomes. Genes also tend to be inherited together if they occur together on a nonsex chromosome, or **autosome.**

Sometimes linked genes are not inherited together. By watching the behavior of chromosomes in meiosis, researchers determined that the explanation of the failure of linked genes to stay together lay in the crossing-over of segments of homologous chromosomes. Recall that when gametes are formed, homologous chromosomes come together during synapsis and crossing-over occurs. As that happens, the chromosomes may exchange parts, as shown in Figure 15–4.

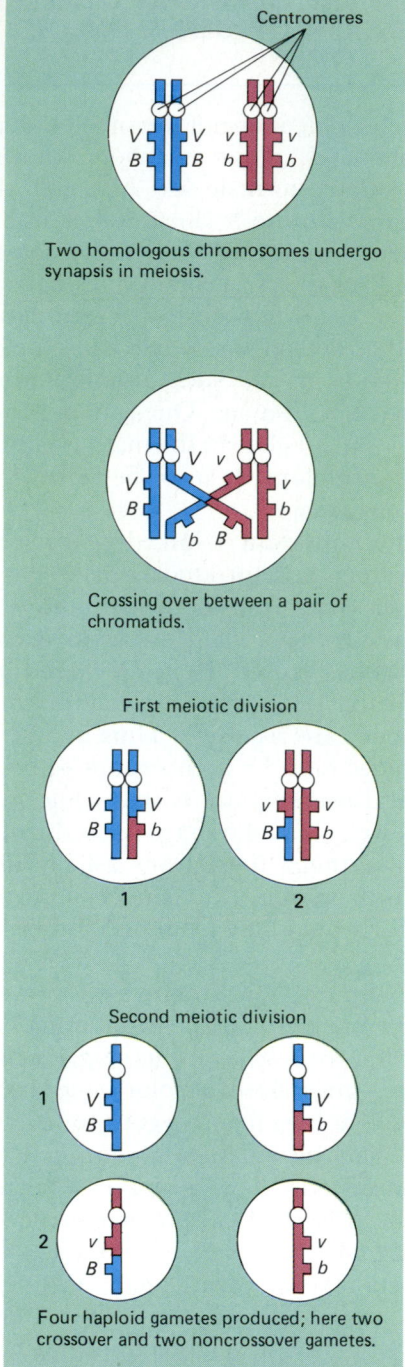

FIGURE 15–4 Crossing-over and genetic recombination. Crossing-over, the exchange of segments between chromatids of homologous chromosomes, permits recombination of genes (*vB* and *Vb*, for example). The farther apart the genes are located on a chromosome, the greater the probability that an exchange of segments between them will occur.

■ GENETIC LINKAGE CAN BE DEDUCED FROM THE WAY GENES ARE INHERITED

By observing a large number of crosses involving linked genes, one can find out how frequently they stay together in their original combination and how frequently they cross-over. If this is done patiently, we can produce maps of chromosomes showing the location of their known genes, for the farther away from one another the loci of different linked genes are on the chromosomes, the more frequently they tend to cross-over. The reason for this is that the portion of two chromosomes where they cross is unlikely to lie exactly in the short segment between two genes that are near one another but likely to occur in the long segment between two distant ones. By counting enough instances of crossing-over, one can estimate the relative distances between linked genes. Using this principle, genetic maps have been constructed that give the order and relative distances of all known genes on them.

An additional mapping technique developed in recent years involves the experimental fusion of human cells with those of mice in tissue culture. For some reason, the resulting "hybrid" cells tend to retain the mouse chromosomes but progressively eject the human chromosomes over the course of several cellular generations. But if just one human chromosome were to remain in such a cell, any human gene products that could be detected would have to result from genes present on that particular chromosome. If a human chromosome fragment remains, the detectable genes must be restricted to that fragment. Finally, if an as yet uncharacterized gene can be shown to be closely linked to a detectable gene, this greatly assists efforts to discover the gene's sequence and the protein it produces. This is important in medical genetics since characterization of the defective protein might suggest medications or other treatments that could counter the effects of a defective protein product of a disease gene.

A great deal of attention is paid to the location of genes. An immense project whose goal is the mapping of the entire human genetic makeup is now under way. This project will probably cost many millions of dollars, untold work, and the combined efforts of many nations. Why is gene mapping so important? For one thing, it can help prove the genetic nature of some disease states. If a disease can be shown to be associated with a nondisease trait known to be genetic, then it would seem that the disease itself must be inherited. Second, one needs to know where something is before it is safe to try to alter it! Precise genetic maps will facilitate the development of advanced genetic engineering techniques for use in the genetic redesign of defective genes (Chapter 16). Third, genetic maps are needed for the development of new varieties of agricultural plants and, increasingly, of animals as well. Last, genetic maps are of importance in keeping track of the evolutionary history of closely related species and populations of animals and plants, including humans. By careful comparison of just where genes lie on the chromosomes of different groups of modern people, it should be possible to learn a great deal about the movements and migrations of our ancestors who lived before such a thing as history was dreamed of.

◼ SOME TRAITS ARE GOVERNED BY MORE THAN ONE PAIR OF GENES

Some traits are governed in their expression by more than one pair of allelic genes known as **polygenes.** Sometimes such polygenes are located on different chromosomes and therefore assort independently. Not many such traits are known, for they are difficult to investigate. It is likely, however, that many human characteristics are inherited in this fashion. One possible instance of polygenic inheritance is human skin color, but height, eye color, and several other traits also seem to be polygenic.

It is well known that if a person of extremely white complexion marries a person of extremely dark complexion, their offspring will be intermediate in color. If that were all there was to it, it would be an obvious case of incomplete dominance. To test that assumption, let us consider the offspring of two such intermediate persons. The assumption of simple incomplete dominance predicts a $1:2:1$ ratio among their offspring. However, what we will observe instead (if their family is sufficiently large, or if we observe offspring in several such families) is a complex distribution not easily expressed by any ratio, ranging from white to very dark, with the majority of intermediate color (Fig. 15–5).

The explanation for this seems to be that human skin color is governed by at least three independent allelic pairs of genes, each pair located on a different chromosome. Thus, the cross dark × white is $AABBCC × aabbcc$. The F_1 offspring can be only $AaBbCc$. What about the F_2? We have worked out some of the genotypes of the F_2 for you in the illustration. It is up to you to determine their phenotypes, which is not difficult to do. (We are not now discussing albinism. Remember that albinism, a complete lack of skin pigment, is an abnormal trait that can occur in any racial group. Albinism behaves as a trait governed by a simple recessive gene.) $AABBCC$ would be very dark indeed. The individual with the genotype $aabbcc$ would be so light that his or her blood would show through the skin. Intermediate degrees of darkness can be predicted by counting the number of capital letters in each genotype.

◼ BLOOD TYPE IS AN INHERITED TRAIT

Although, as the saying has it, all blood is red, there is a surprisingly great diversity of other characteristics, one of which, **blood type,** is of enormous medical and

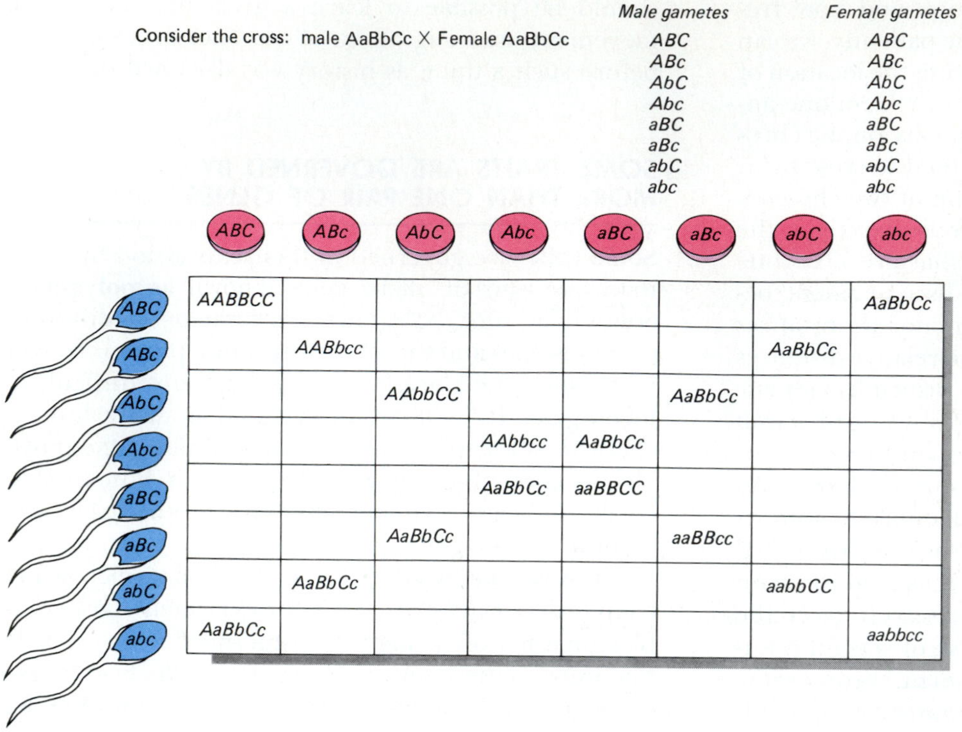

Consider the cross: male AaBbCc X Female AaBbCc

Male gametes
ABC
ABc
AbC
Abc
aBC
aBc
abC
abc

Female gametes
ABC
ABc
AbC
Abc
aBC
aBc
abC
abc

	ABC	ABc	AbC	Abc	aBC	aBc	abC	abc
ABC	AABBCC							AaBbCc
ABc		AABbcc					AaBbCc	
AbC			AAbbCC			AaBbCc		
Abc				AAbbcc	AaBbCc			
aBC				AaBbCc	aaBBCC			
aBc			AaBbCc			aaBBcc		
abC		AaBbCc					aabbCC	
abc	AaBbCc							aabbcc

FIGURE 15–5 Polygenic inheritance. Human skin color is an example of polygenic inheritance, which involves the additive interaction of at least three separate, incompletely dominant sets of genes on three separate chromosomes. We will let *A*, *B*, and *C* represent the genes for dark skin color and *a*, *b*, and *c* represent the light alleles. Skin color can be estimated by counting the number of capital letters in each genotype. See if you can fill in the remaining squares.

legal importance. Through knowledge of blood types, one may transfuse blood safely, sometimes solve paternity cases, unravel tangled disputes about legal inheritance, and help to identify drops of blood left at the scene of a crime.

Blood Type Reflects Blood Cell Surface Antigens

Blood type is determined by particular substances occurring on the surface of red blood cells. These substances, called **antigens,** have the potential for stimulating the production of antibodies. Antibodies (discussed in Chapter 37) are an extremely important part of the immune system with which our bodies confront disease organisms and cancerous cells. Though antigens can belong to different chemical groups, those determining blood type are glycoproteins (Chapter 4).

In effect, the immune system distinguishes between cells that are part of the organism and those that are not, normally attacking alien cells rather than healthy tissues (Chapter 37). This discrimination is made possible by the detection of characteristic antigens on cells. These substances function as a kind of chemical fingerprint, which alerts certain white blood cells to produce **antibodies,** globular proteins possessing a kind of active site similar in some ways to that of an enzyme. The active site of the antibodies fits a particular antigen exactly, but it usually will fit no other antigen. Thus it can attach to the invading disease or-

ganism that stimulated its production but not to normal body cells or even other alien microorganisms. Once tagged with antibodies, the invader's life is forfeit; phagocytic blood cells are attracted to it, or it is destroyed by chemical warfare via a system of enzymes and other proteins called the complement system.

Like those of other cells, the plasma membranes of blood cells possess antigens. In a healthy person there are no antibodies to these antigens, but if the blood of one individual is mixed with that of another, sometimes antibodies are present in the foreign plasma that are capable of reacting with the first person's blood cell's antigens. Such a reaction can produce an abnormal clumping of blood cells called **agglutination,** or their complete breakdown by **hemolysis.**

The Antigens Of A Blood Cell Are Governed By A Set Of Multiple Alleles

Multiple alleles exist when a particular gene, occupying one given locus, exists in more than two forms. Thus an organism can have no more than two of a given set of multiple alleles at a time, but more than two such alleles are known to exist in a population, as in the ABO series of blood types. This series of blood types was the first to be discovered—as a result of early experiments in blood transfusion, which were unpredictably, but all too often, fatal.

Both agglutination and hemolysis reactions can take place in laboratory glassware and presumably also

TABLE 15–1
ABO Blood Types

Phenotype	Genotype	Antigens	Antibodies	How Inherited
A	AA, AO	A	Anti-B	Dominant
B	BB, BO	B	Anti-A	Dominant
AB	AB	A and B	None	Codominant
O	OO	None	Anti-A and B	Recessive

in the body of a living person when red cells come into contact with an antibody that can react with them. For reasons not fully understood, the usual kinds of ABO antibodies develop normally even without known exposure to cells or other sources of antigens that would be capable of provoking them. You can readily understand, though, that a person normally lacks antibodies to the blood type that he or she possesses! However, if blood is transfused into a patient whose plasma contains antibodies hostile to the cells of the donor, damage to these cells is bound to result.[1] Strange to say, we do not understand why these medically inconvenient blood types even exist!

The Three ABO Alleles Produce Four Blood Types

Table 15–1 gives the basic facts of inheritance for both the antibodies and the antigens of the ABO series. Three alleles are known for the ABO series, which by their interaction can produce blood types A, B, AB, and O—depending upon the exact combination of genes in the particular allelic pair possessed by an individual (Fig. 15–6). A normal person has no more than two of the ABO alleles. Type O allele is recessive to all other alleles, and A and B exhibit codominance with respect to each other. Thus two genotypes can give rise to type A and two to type B. AB and O each have only one possible genotype.

See if you can work some practice blood type crosses. What would happen if two people of blood genotype AO and BO were to produce offspring? What blood types could occur in those children?[2]

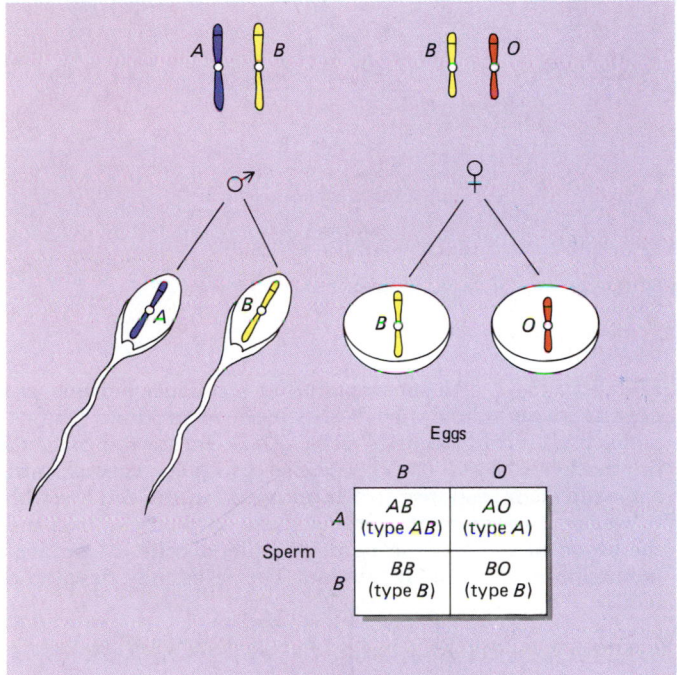

FIGURE 15–6   Multiple alleles for blood types. Three different alleles exist for ABO blood type. In this cross, involving a male with type AB blood and a female with type B blood, each parent produces two kinds of gametes. Four possible genotypes producing three different phenotypes occur among their offspring.

The Rh Blood Factors Are Potentially Hazardous To The Newborn

The Rh series of blood types has been known since the 1940s. There are several somewhat codominant varieties, but all may be thought of as producing either a *positive* or a *negative* phenotype. The positive phenotype exhibits antigens that can provoke dangerous antibodies; the negative phenotype lacks these antigens (although it may have some relatively harmless ones). For our purposes we will simplify the inheritance of the Rh blood types by lumping the several alleles into two groups—*R*, which produces positive phenotypes, and *r*, which produces negative. *R* is dominant. Thus a person with Rh-positive blood has the genotype *RR* or

---

[1] The reverse can also happen. If blood is transfused into a recipient and the donated *plasma* is incompatible with the recipient's cells, trouble may eventually result. Initially, though, such transfusions may be harmless, since the half pint or so of potentially dangerous plasma is rapidly diluted in the recipient's blood volume to harmless levels.

[2] The answer is A, B, AB, and O, resulting from the genotypes AO, BO, AB, and OO.

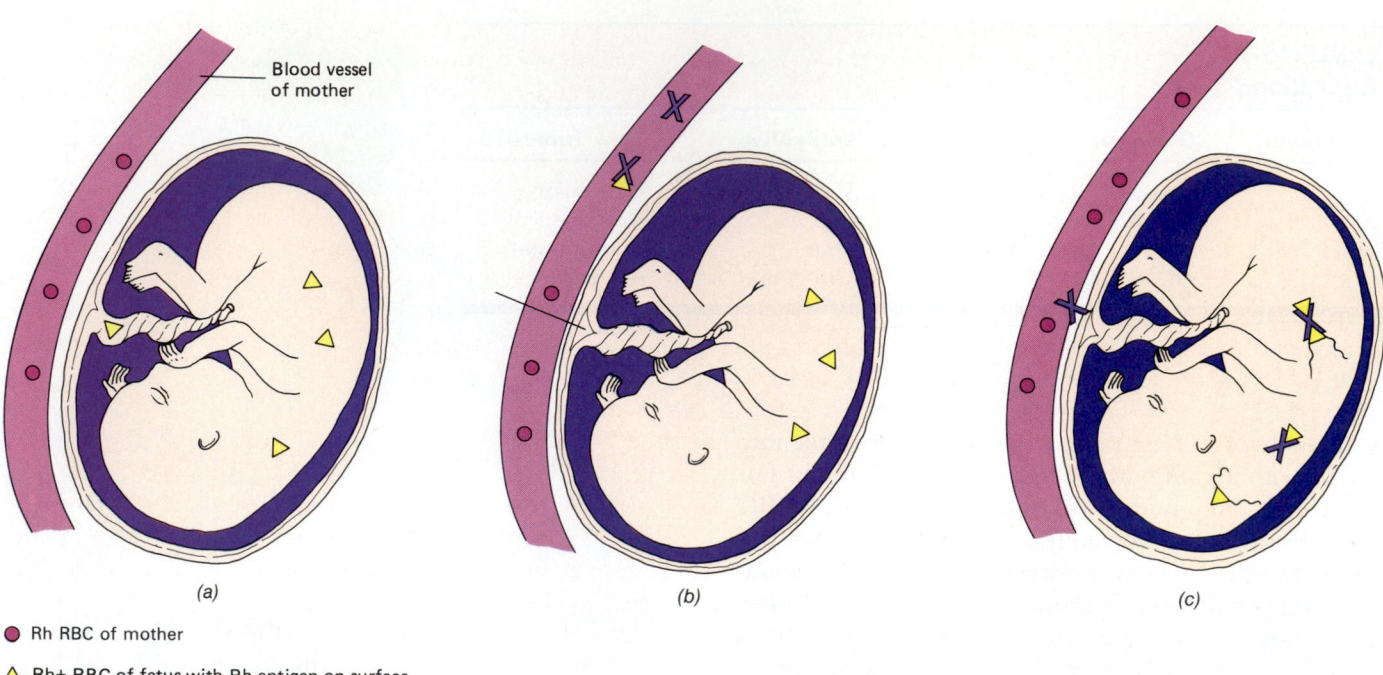

Blood vessel of mother

(a)   (b)   (c)

● Rh RBC of mother

△ Rh+ RBC of fetus with Rh antigen on surface

✗ Anti-Rh antibody made against Rh+ RBC

△ Hemolysis of Rh+ RBC

FIGURE 15–7   Rh incompatibility can cause serious problems when an Rh-negative woman and an Rh-positive man produce Rh-positive offspring. (a) Some Rh⁺ red blood cells (RBCs) leak across the placenta from the fetus into the mother's blood. (b) The mother produces special antibodies (anti-Rh antibodies) in response to Rh antigens on the fetal red blood cells. (c) In a subsequent pregnancy, some of these antibodies cross the placenta and enter the blood of the fetus, causing red blood cells to rupture and release hemoglobin into the circulation. The fetus may develop erythroblastosis fetalis.

Rr. The negative phenotype can only be *rr.* About 85% of the population of the United States has Rh-positive blood.

A person with Rh-positive blood has Rh antigens associated with the blood-cell membranes, and *no* anti-Rh antibodies in the plasma. An Rh-negative person possesses no important Rh antigens on blood-cell membranes and no natural anti-Rh antibodies in the plasma. But if an Rh-negative person were to receive a transfusion of Rh-positive blood, he or she would quickly develop anti-Rh antibodies, which would then hemolyze the cells of the *next* transfusion of Rh-positive blood. The first such transfusion would, however, be relatively harmless.

Although several kinds of maternal-fetal blood type incompatibilities are known, **Rh incompatibility** is probably the most important (Fig. 15–7). If a woman is Rh-negative and her husband is Rh-positive, the fetus may be Rh-positive. A small quantity of blood from the fetus may pass through some defect in the placenta (especially during the birth process) and into the mother's blood, sensitizing her white blood cells,

which then produce antibodies to the Rh antigens. When this woman becomes pregnant again, sensitized white blood cells produce antibodies that may pass through the placenta into the fetal blood and cause clumping of the red blood cells. Notice that although red blood cells cannot ordinarily cross the placenta, antibodies can. The resulting breakdown products of the hemoglobin released into the circulation damage many fetal organs, including the brain. This disease is known as **erythroblastosis fetalis.** In extreme cases so many fetal red blood cells are destroyed that the fetus dies before birth.

When Rh-incompatibility problems are suspected, blood can be exchanged while the baby is still within the mother's uterus, but this is a risky procedure. Rh-negative women are now treated during pregnancy and just after childbirth (or at termination of pregnancy by miscarriage or abortion) with an anti-Rh preparation containing Rh immune globulin. These passively acquired antibodies prevent an active immune response to the red cells, perhaps by clearing the Rh-positive cells from the mother's blood so quickly that her own immune system does not have time to respond. As a result, her white cells never do produce the antibodies that could harm her next baby.

GENETIC DISEASE IS WIDESPREAD AND IMPORTANT

More than 150 human disorders involving enzyme or other protein defects are caused by genetic mutations (see Table 15–2). Most are inherited as autosomal recessive traits. Sickle-cell anemia, phenylketonuria (PKU), cystic fibrosis, and Tay-Sachs disease are well-known diseases that have been linked to gene defects.

Many recessive genetic diseases can now be detected in the heterozygous state by enzyme analysis and other advanced chemical techniques. It is also possible to detect abnormalities in fibroblasts (cells of connective tissue) found in the amniotic fluid, for these fibroblasts originate in the fetus. Thus through genetic screening it can often be determined even before birth whether a fetus will suffer from a genetic disease.

Marriage Between Blood Relatives Increases The Chance Of Genetic Disease

Marriage between relatives is referred to as **consanguinity,** which is derived from two Latin words meaning "related" and "blood" (literally, "shared blood"). It can be shown that the risk of bearing a child with a major congenital abnormality of genetic origin is 70% greater if the parents are, for example, first cousins than if they are unrelated. The death rate for children of such consanguineous matings is between three and four times higher than among a comparison population. Almost half the offspring of brother–sister unions are more or less handicapped by genetic disease.

No mystery enshrouds the increased risk of malformed offspring from consanguineous matings. The problem can be traced to a greater chance of sharing harmful alleles. All of us no doubt shared the same ancestors, hence the same alleles, in the distant past. But with time the pool of human alleles became diversified as a result of mutations and evolutionary processes. Today each of us is genetically unique. Only identical twins have identical alleles. Each of us probably carries a number of harmful *recessive* alleles, which by definition do not express themselves in the heterozygous condition. If you marry a nonrelative, the chance is remote that you will select someone who shares any of the same harmful alleles. There is little chance that any of your offspring will be unfortunate homozygotes, receiving a double dose of harmful recessive alleles. However, if you selected your first cousin as your mate, you would have a common set of grandparents, and so by probability one of every eight of your alleles would be identical. Your offspring would be homozygous for one-sixteenth of all their gene traits. Second cousins share one-thirty-second of their alleles. Thus, there is an increased probability of offspring being homozygous for harmful recessive traits in consanguineous matings. (This reasoning does not apply to dominant genetic defects or those that are sex-linked. Why not?)

Some Genetic Disorders Are Dominant

Although most diseases of genetic origin are recessive, dominant genetic disorders do exist. One possible interpretation is that the protein produced by the normal allele can only be manufactured in useful quantities by two competent alleles. If one allele is absent or nonfunctional, a defect results.

Huntington's disease is a serious hereditary disorder of human beings that is both dominant and lethal. One might think that this disease could only affect one generation at a time and would then immediately die out. It is propagated readily, however, because symptoms do not usually begin until middle age. By that time the sufferer could have produced many children.

Huntington's disease produces mental deterioration, painful paralysis, sensory loss, and ultimately, death. Advances in mapping of deoxyribonucleic acid (DNA) have made it possible to detect this gene even in young people. This at least enables them to make an informed decision as to whether to have children. A person who carries the gene for Huntington's disease and whose mate does not has a 50% chance of producing offspring that will eventually develop this disorder.

Genetics Raises Ethical Questions

Perhaps the major benefit to be gained from a course in biology is biological literacy that may be applied to the exercise of citizenship. There is not much benefit in learning genetics if all that is gained is the ability to draw little squares and estimate ratios. You should apply your knowledge of genetics, even if you have no intention of breeding tropical fish or becoming a genetic engineer.

It is now possible, as we have seen, to diagnose some hereditary disease prenatally by amniocentesis, biopsy of the fetal chorionic membrane, or other means. Having established the existence of the disease, what then? Prospective parents are faced with the al-

TABLE 15–2
Some Important Genetic Disorders*

Name of Disorder	Mode of Inheritance	Description	Treatment, if Any	Comments
Cystic fibrosis (CF)	Autosomal recessive	High level of sweat electrolytes, pulmonary disease, cirrhosis of liver, pancreatic malfunction, and especially, nonsecretion of digestive enzymes; females sometimes reproduce; life expectancy 12–16 years, with some living into 30s and 40s; thick mucus interferes with lung clearance	Symptomatic, with emphasis on digestive enzyme replacement and control of respiratory infections; since CF is now known to result from a defect in cellular sodium transport, a treatment is theoretically possible	CF kills more children than diabetes, rheumatic fever, and poliomyelitis combined; exists in different degrees of severity; commonest in persons of northern European extraction; recent research indicates that CF may become almost routinely detectable prenatally by examination of amniotic fluid
Hemoglobinopathic disease, e.g., sickle-cell anemia (SCA)	Group of autosomal recessive or incompletely dominant traits	Abnormalities of red blood cells caused by presence of certain inappropriate amino acids at crucial locations in hemoglobin molecule	Varies with type of disease. Some, e.g., hereditary methemoglobinemia, may require no treatment. Some can hardly be treated at all; SCA can be treated to some degree	These traits are similar but not allelic; microcytic anemia is commonest in Mediterranean populations, SCA in some black populations; when heterozygous, SCA offers some protection against malaria
Tay-Sachs disease	Autosomal recessive	Abnormal accumulation of lipids due to deficiency of the enzyme hexose-aminidase A; destroys central nervous system in young children		Especially prevalent among Jews of Eastern European ancestry

ternatives of abortion versus crippling medical expenses, worries about the quality of life that the child can be expected to experience, and the strain on the life of the family that must care perpetually for a chronically ill person, perhaps well into his or her adult years. People in this position are faced with difficult decisions.

Increasingly, we are able to detect genetic disease in a heterozygous state, as in the case of sickle-cell anemia. This disorder results from an abnormal form of hemoglobin and can be detected when heterozygous by analysis of the subject's hemoglobin. A heterozygous carrier has a mixture of normal and abnormal hemoglobin, although usually otherwise he or she is normal. Should such knowledge determine who one should marry? If married to another carrier, and therefore likely to produce diseased offspring, should such people produce children? Should a couple who has produced one hemophilic child have another? It is worth noting that anyone, even the healthiest, is probably a heterozygous carrier for several unsuspected genetic diseases.

TABLE 15–2
Continued

Name of Disorder	Mode of Inheritance	Description	Treatment, if Any	Comments
Phenylketonuria (PKU)	Autosomal recessive	Deficiency of liver phenylalanine hydroxylase leads to a depression of the levels of other amino acids, leading in turn to mental deficiency	A low-phenylalanine diet minimizes symptoms; most states have extensive PKU screening programs, in which newborns are tested for excessive blood phenylalanine, or for presence of metabolic products in urine	Since melanin is synthesized from tyrosine, tyrosine deficiency caused by phenylalanine-hydroxylase deficiency results in light coloring
Tyrosinase-negative oculocutaneous albinism (T⁻ albinism)	Autosomal recessive	Absence of pigmentation due to functional absence of tyrosinase; visual acuity 20/200 or less; marked susceptibility to skin cancer	Avoidance of sunlight	Somewhat more common among blacks than whites
Tyrosinase-positive oculocutaneous albinism (T⁺ albinism)	Autosomal recessive	*Reduction* of pigmentation due to malabsorption of tyrosine by body cells; if racial background is heavily pigmented, some pigmentation will survive, though in some cases phenotype is virtually identical with T⁻; pigmentation and visual acuity improve with age		Highest incidence in American Indians, less in blacks, least in whites; hybrid T⁺–T⁻ individuals appear normal

*This table is not complete. For detailed information a book of human genetics should be consulted, such as *The Metabolic Basis of Inherited Disease*, by John B. Stanbury, et al. (New York, McGraw-Hill, 1983). It should also be noted that a number of common diseases, such as diabetes mellitus, though not genetic in the usual sense, appear to have a genetic component influencing susceptibility.

◼ CHAPTER SUMMARY

I. The X chromosome is exceptional in that it has no fully homologous mate in the male. A gene located on the X chromosome of a male will be expressed even if it is recessive.

II. A sex-linked gene is one that is carried on the X chromosome (at least in human genetics). Linked genes occur on the same chromosome, thus being linked together and tend to be inherited together.

A. However, members of homologous chromosome pairs may exchange parts (along with associated genes) when crossing-over occurs in meiosis.

B. The farther apart linked genes are, the more frequently they cross-over.

III. Allelic genes sometimes exist in multiple forms in a population, as in the ABO series of blood types.

Genetically, O is recessive to both A and B, but A and B are codominant with respect to each other.

IV. The Rh series of blood types depend upon antigens that are immunologically and genetically independent of the ABO series.

    A. Anti-Rh antibodies do not occur spontaneously but result from the immunization of an Rh-negative individual to Rh antigens.

    B. Should this occur in the mother of an Rh-positive fetus, the antibodies can cross the placenta, and hemolytic disease of the newborn may result.

V. Genetic disease is widespread among human beings, giving rise to numerous ethical, social and medical problems.

■ POST-TEST

1. Genes are sex-linked (in human beings) if they occur on the _____ chromosome. A father cannot pass such a sex-linked trait to his _____.

2. The ABO series of blood types is a system of _____ alleles in which the genes for types A and B are mutually _____, but both of them are _____ in their expression to O.

3. If a man is color-blind, he received the gene for color blindness from his _____ (mother or father).

4. If two normally pigmented persons produce an albino baby, the parents must have been _____ for this trait.

5. A baby with type AB blood is born into a family whose wife is type B and whose husband is type O. The baby can/cannot be shown to have a different father.

6. A color-blind man is said to be _____ for this trait.

7. Both members of a couple are known to be heterozygous for sickle-cell anemia, a recessive trait. Their chance of producing a healthy baby who is a carrier of this trait is _____ %.

8. A healthy woman has a hemophilic brother. Her husband is healthy and normal. Ultrasound diagnosis determines that the fetus in the next pregnancy is male. There is a _____ % chance that this baby is hemophilic.

■ REVIEW QUESTIONS

1. What are polygenes? Give a possible example.

2. What chromosome or chromosomes bear sex-linked traits?

3. Give an example of genetic screening.

4. Why are female hemophiliacs virtually unknown? What about females with red-green color blindness?

5. Physicians are concerned about possible hemolytic disease of the newborn (erythroblastosis fetalis) when the father is Rh-positive and the mother is Rh-negative. They are not concerned, however, when the father is Rh-negative and the mother is Rh-positive. Why not?

6. Mendel originated the principle of independent assortment on the basis of several crosses involving genes that today are known, in fact, to be linked in some cases. How could linked genes appear to be inherited independently?

7. A color-blind girl has a normally sighted mother. What was the phenotype of her father? the genotype of her mother?

8. Hemophilia is X-linked and recessive. A hemophilic man has a normal son. He plans never to have any other children, but he worries that he may pass this genetic disease on to his grandchildren through this son. Are these fears justified? Why or why not?

9. To be expressed, a recessive genetic disease must be homozygous—if it is autosomal. What relationship does this fact have to consanguineous marriages? Does this apply to sex-linked traits as well?

10. John has type B blood and is Rh-negative. Sue has type AB and also is Rh-negative. Their baby—at least it is supposed to be theirs—has type O and is Rh-negative. If you were their lawyer, what might you advise him, her, or them to do and why?

■ RECOMMENDED READINGS

Langer, A. "Practical Genetics in Office Practice," *Hospital Medicine* 18:109–117, 1982. Excellent introduction to practical medical genetics. Especially useful summary of symptoms of Down's syndrome.

Lewontin, R. *Human Diversity.* San Francisco, W.H. Freeman, 1982. The principles of genetics as applied to human variation.

Roberts, "Race for Cystic Fibrosis Gene Nears End," *Science* 240:282–285, 1988. An illuminating article that will be well worth reading long after the gene is found, both for the light it casts on research techniques and problems of scientific ethics.

16

Genetic Frontiers: Recombinant DNA and Genetic Engineering

OUTLINE

I. Recombinant DNA techniques involve manipulating specific genes in the cells of living organisms, thereby causing them to produce new or unusual substances
   A. First a specific region of DNA is isolated from the rest of the cell's DNA
   B. Recombinant DNA molecules are constructed using DNA from two different sources
   C. The recombinant DNA molecules are transferred into a new host organism's cell
   D. The cells that have taken up the gene of interest are identified
   E. The gene is expressed inside the new cell
   F. Gene insertion in eukaryotes is technically challenging, but feasible

II. Genetic engineering holds great promise for the future
   A. It may be possible to repair human genetic defects
   B. Genetically engineered animals may boost agricultural production
   C. Scientists are using recombinant DNA techniques to improve plants

III. Concern over potential danger caused by genetic engineering has led to the formation of policies on utilizing recombinant DNA techniques

Focus on probing for genetic disease
Focus on molecular genetics and cancer

LEARNING OBJECTIVES

After you have studied this chapter you should be able to:

1. Outline the primary techniques utilized in recombinant DNA experiments.
2. Summarize the problems involved in isolating, identifying, and cloning a single gene.
3. Explain the actions of restriction endonuclease and ligase and their importance in recombinant DNA experiments.
4. Identify the role of vectors in recombinant DNA experiments and give several specific examples of such vectors.
5. Explain the special measures that have been employed to introduce genes experimentally into plant and animal cells.
6. List at least five potential or realized applications of recombinant DNA technology.
7. Summarize the concerns that have been raised about recombinant DNA research and the policies that have been formulated to address these concerns.

Flasks of interferon obtained from bacteria by recombinant DNA techniques. (Photo Researchers/NCI/Science Source)

Techniques of selective breeding have long been used by humans to build a better cow, or honeybee, or corn plant. However, these results come slowly and fall short of producing some of what one might like to accomplish: cow's milk low in cholesterol, for example, or grains of corn with the protein content of soybeans. And selective breeding really leaves us helpless in the face of genetic disease. One cannot help a sufferer of sickle-cell anemia or of cystic fibrosis by simply suggesting that he or she should not have been born.

Using living organisms to produce products is known as **biotechnology.** Humans have been practicing biotechnology for centuries. Selective breeding of plants and animals is an example of biotechnology, as is using yeast to cause bread to rise during baking. However, biotechnology is much in the news today because recent advances in biochemical and genetic techniques promise an unprecedented control of heredity. When people talk about biotechnology today, they may be speaking about it in a narrower sense than the definition given above. The current usage of biotechnology refers to the genetic manipulation of cells so that new strains of organisms with new characteristics can be constructed (Fig. 16–1). It is also known as **genetic engineering.**

There are several important differences between selective breeding that has been practiced by humans for centuries and genetic engineering. For one thing, genetic engineering allows us to tap a much larger gene pool. It is possible to transfer genes from humans to bacteria, from viruses to plants! A second difference is the number of traits we can alter. By traditional breeding methods, for example, it is possible to develop a plant that is resistant to a single pathogen (disease agent). In contrast, resistance to a number of different pathogens can be conferred by genetic engineering. Finally, genetic engineering allows us to develop new strains of organisms in a fraction of the time required by selective breeding.

KEY CONCEPTS

- The deliberate manipulation of genes of different organisms to accomplish human goals is known as biotechnology or genetic engineering.

- Genetic engineering holds great promise for decreasing the amount of time required to produce altered organisms, as compared to selective breeding.

- Genetic engineering allows combinations of genes that are impossible or highly unlikely in nature.

(a)

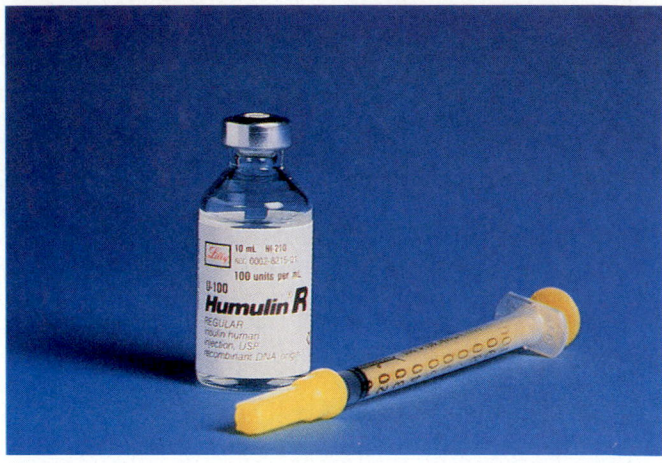

(b)

FIGURE 16–1 In biotechnology, new combinations of genes produce novel traits in organisms. (*a*) In a laboratory test, both tomato plants were subjected to caterpillar infestation. The genetically engineered plant (right) received little damage, while the non-altered plant (left) was almost completely consumed by caterpillars. (*b*) Human insulin for diabetics is now produced by genetically altered bacteria. (*a,* courtesy of Monsanto Co.; *b,* VU/SIU)

■ RECOMBINANT DNA TECHNIQUES INVOLVE MANIPULATING SPECIFIC GENES IN THE CELLS OF LIVING ORGANISMS, THEREBY CAUSING THEM TO PRODUCE NEW OR UNUSUAL SUBSTANCES

Recombinant deoxyribonucleic acid (DNA) techniques permit the formation of new combinations of genes by isolating genes from one organism and introducing them into either a similar or an unrelated organism. By these methods foreign DNA may be inserted not only into the simple single cells of bacteria but also into cells derived from the bodies of complex organisms. The genetic composition may be altered in bacteria so that they might produce substances, such as

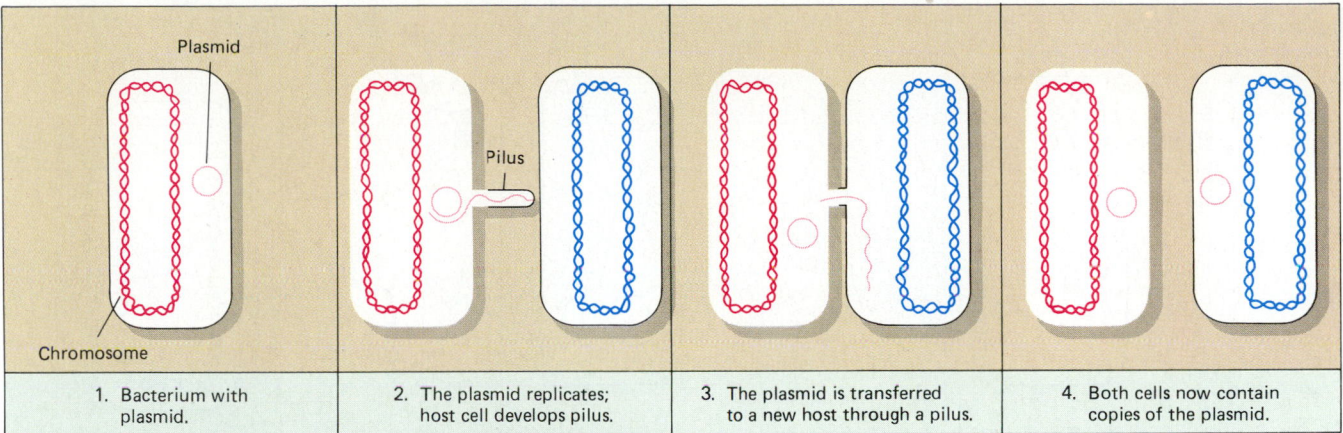

| 1. Bacterium with plasmid. | 2. The plasmid replicates; host cell develops pilus. | 3. The plasmid is transferred to a new host through a pilus. | 4. Both cells now contain copies of the plasmid. |

(a)

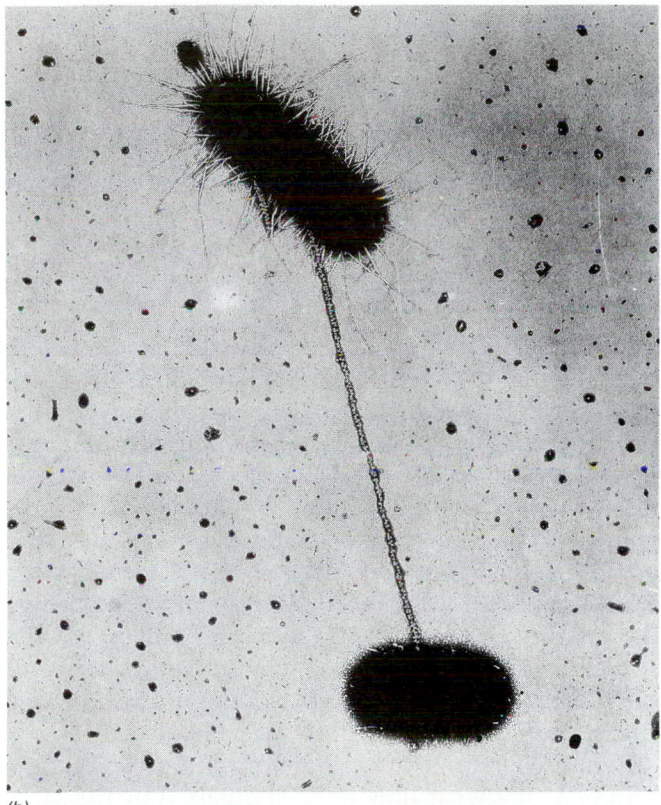

(b)

FIGURE 16–2   Conjugation is a mechanism of natural DNA recombination. (*a*) During conjugation, plasmids, which are small loops of DNA that often contain genes for resistance to antibiotics, can be transferred from one bacterium to another. (*b*) The pilus serves as a conjugation bridge through which DNA passes. (*b*, Charles C. Brinton, Jr., and Judith Carnahan)

insulin or human growth hormone, that they normally cannot produce, or so that they might consume substrates, such as spilled oil, that they normally cannot consume.

Recombinant DNA techniques allow the isolation of genes from the tremendously complex metabolic machinery of eukaryotic cells. The genes can then be inserted into bacteria, where they can be studied far more conveniently. Finally, they can be reinserted into the eukaryotic organism after extensive deliberate modification, possibly after being combined with other genes or parts of genes from other donor organisms.

In a way, genetic engineering is not new. Organisms have mechanisms for exchanging genes, and "experiments" in gene transfer have apparently always occurred in nature. For example, bacterial conjugation results in the transfer of genetic information from one cell to another through a **pilus** (Fig. 16–2). **Transduction,** the transfer of genetic material from one bacterium to another by a bacterial virus, is another example of natural genetic recombination (Fig. 16–3). However, these natural experiments have occurred randomly and certainly not with human goals and desires in view. For deliberate genetic engineering to be effective, we must be able not only to *change* genes but also to change them in a controlled fashion.

How genes operate has been partially understood since the 1950s, but only recently have we developed the technological knowledge necessary to manipulate genes. The greatest problems that must be solved to make genetic engineering practical have only been clearly identified in the last 10 years or so. Among these are locating genes with desired characteristics and isolating them; transferring genes from one organism to another, a process called **vectoring;** and ultimately, making genes to order. The first two of these problems have been resolved fairly well, but work on the third is not so far advanced. Thus far it has been necessary to find a pre-existing living source for the desired gene.

Having a foreign gene expressed in a cell requires several steps. The gene that is introduced to a different cell must be (1) isolated from its original cell; (2) used to construct recombinant DNA; (3) transferred into a

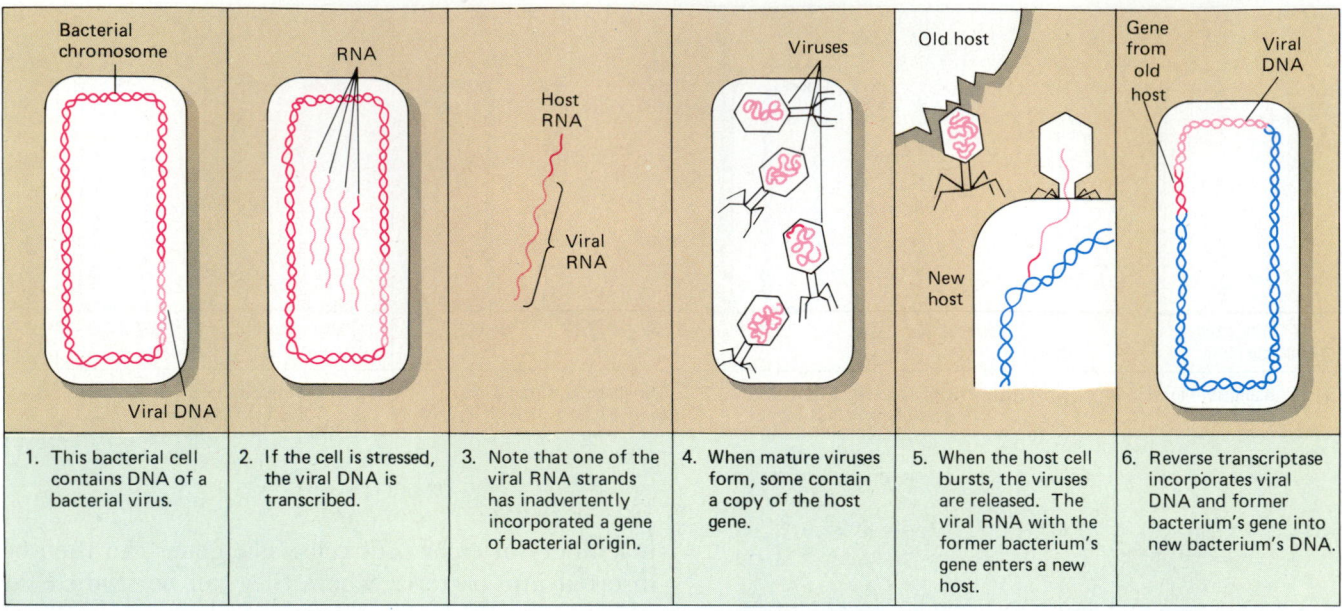

| 1. This bacterial cell contains DNA of a bacterial virus. | 2. If the cell is stressed, the viral DNA is transcribed. | 3. Note that one of the viral RNA strands has inadvertently incorporated a gene of bacterial origin. | 4. When mature viruses form, some contain a copy of the host gene. | 5. When the host cell bursts, the viruses are released. The viral RNA with the former bacterium's gene enters a new host. | 6. Reverse transcriptase incorporates viral DNA and former bacterium's gene into new bacterium's DNA. |

FIGURE 16–3 **In transduction genes are transferred from one bacterial chromosome to another within a virus. There is evidence that genes may sometimes be transferred among different bacterial species by these mechanisms.**

new (i.e., foreign) cell; (4) identified as having been taken up by the cell; and (5) expressed in that cell. We will consider each of these steps in detail.

First A Specific Region Of DNA Is Isolated From The Rest Of The Cell's DNA

To begin the process of genetic engineering, the investigator must isolate the gene of interest. This is done by first breaking up the cell's DNA into more manageable fragments by the use of enzymes known as **restriction endonucleases.** Found in bacteria, these enzymes are used by the bacteria as defense weapons against viral infection. A DNA virus may penetrate a potential host cell but find itself under attack by restriction endonucleases, which seek out certain **recognition sites** in the viral DNA and then chop the alien strand into little, harmless bits. Bacteria produce methyl groups that cover recognition sites on their own DNA. In this way the bacterial DNA is protected from the action of these enzymes; otherwise the bacterium would swiftly chop its own DNA into bits. Eukaryotic DNA possesses recognition sites in random locations throughout its DNA. For that reason restriction endonucleases obtained from bacteria are able to fragment eukaryotic DNA (though under natural conditions they probably never have the opportunity to do so).

The recognition sites for restriction endonucleases are generally four to six base pairs in length. Many of these sequences are **palindromic,** that is, they read the

same in both directions, like the word "radar." An example of a restriction endonuclease is EcoRI (pronounced ECHO-ARE-ONE), so called because it was the first restriction endonuclease identified from *Escherichia coli,* strain R. EcoRI recognizes the DNA base-pair sequence GAATTC and its complementary sequence CTTAAG. Note how the DNA sequence and its complement are palindromic (Fig. 16–4). When EcoRI reacts with this recognition sequence, it cleaves the double-stranded DNA between G and A on *each* strand. This leaves one strand longer than the other on each segment; each segment of DNA is said to have **sticky ends** (Fig. 16–5). DNA segments from entirely different organisms are potentially able to recombine with one another if they've been digested by the same restriction endonuclease. The recognition sites are always the same for each specific endonuclease; no matter what kinds of DNA are cleaved, the recognition sites have the same base sequence. For this reason, their sticky ends are complementary and, in the presence of a "splicing" enzyme, can attach to one another.

Recombinant DNA Molecules Are Constructed Using DNA From Two Different Sources

Restriction endonucleases split the potential donor DNA into many fragments. The next step is to construct recombinant DNA molecules by combining our isolated DNA with DNA from another source. Recall that some bacteria, including *E. coli,* have small acces-

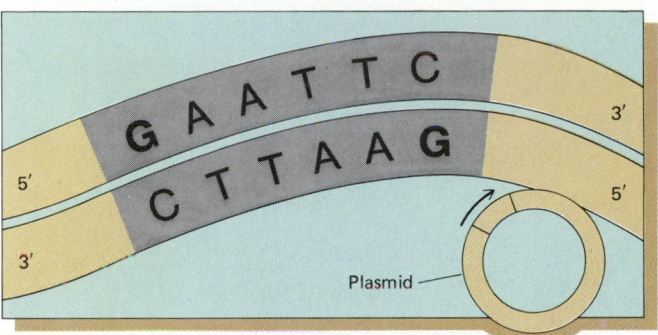

FIGURE 16–4 A portion of a DNA molecule (unwound for simplification) that contains the recognition site for EcoRI. EcoRI, a restriction endonuclease, recognizes the DNA base sequence -GAATTC- and its complimentary base pairs. Note that -GAATTC- and its compliment -CTTAAG- are palindromic: one strand reads the same from left to right as the other strand reads from right to left.

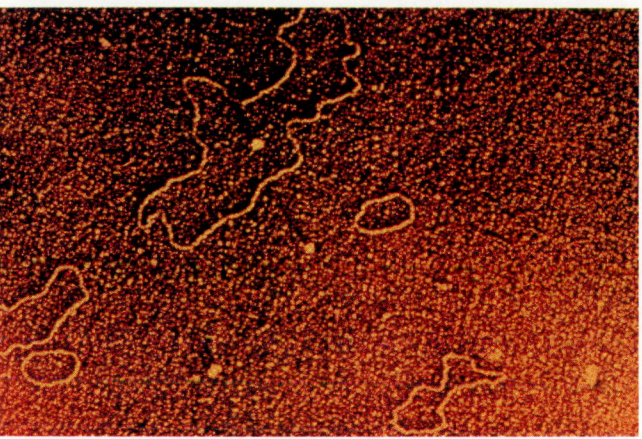

FIGURE 16–6 Electron micrograph of plasmids, circular molecules of bacterial DNA that are often used in genetic engineering to transfer desired genes into bacteria or other organisms. (Thomas Broker/Cold Spring Harbor Laboratory)

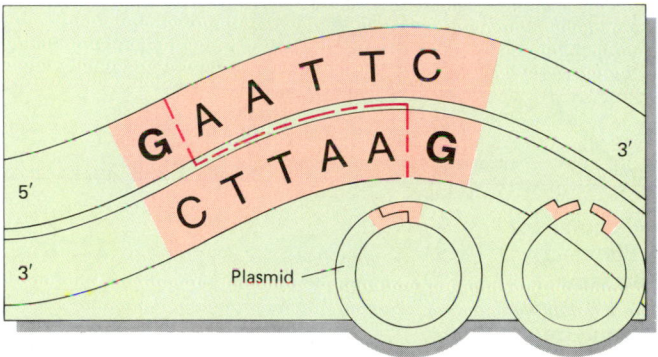

FIGURE 16–5 The cleavage of DNA by the restriction endonuclease EcoRI. This enzyme cleaves bonds within the DNA double helix. The products have short single-strand stubs, i.e., sticky ends.

sory chromosomes known as **plasmids** (Fig. 16–6). When present, many plasmids are replicated along with the main chromosome, and the copies are distributed to daughter cells, as is the chromosome.

Plasmids can be of help to molecular biologists, for they can serve as genetic vectors, carrying desired genes into cells and propagating them. When desired genes are inserted into plasmids, and these altered plasmids are implanted in bacterial hosts, the plasmid and its genes are replicated along with the entire bacterial population descended from that ancestral host cell. Should the gene code for such a protein as human thyroid-stimulating hormone (TSH), that protein accumulates inside the bacterium from which it can later be harvested. Since *E. coli* can host as many as 20 such plasmids, theoretically 20 TSH genes could be accommodated at one time, collectively producing 20 times the amount of the hormone that a single gene might.

To construct recombinant DNA molecules, the investigator treats plasmids from *E. coli* cells with restriction endonuclease (Fig. 16–7). The originally circular plasmids now become linear strands of DNA, with palindromic sticky ends. These are then mixed with sticky DNA fragments from the donor cell. By the use of a splicing enzyme, or **ligase,** the ends of the donor and plasmid DNA are joined. The result is a reconstituted circular plasmid that contains some fragment of donor DNA. In the vast majority of cases this fragment will *not* be the desired one. Yet by chance, a very few can be expected to possess exactly what is desired.

The Recombinant DNA Molecules Are Transferred Into A New Host Organism's Cell

The recombinant DNA molecules (in plasmids) are next allowed to be taken into bacterial cells, thus producing numerous strains of genetically engineered bacteria. The large variety of bacteria with recombinant molecules produced in this way are collectively referred to as a **library,** or gene bank. One of the bacteria with a new DNA combination in this library might contain the target gene that codes for the desired protein. If that one can be picked out from among all the others, it can be allowed to reproduce, thus automatically making many copies of the desired genes that have been artificially implanted in it.

Since the early days of electron microscopy it has been known that some strains of bacteria are able to produce a hair-like structure known as a pilus with which they temporarily attach themselves to other bacteria of the same species. It has been shown that pili are developed only by bacteria that have plasmids (see Fig. 16–2b). During conjugation, two bacteria come to-

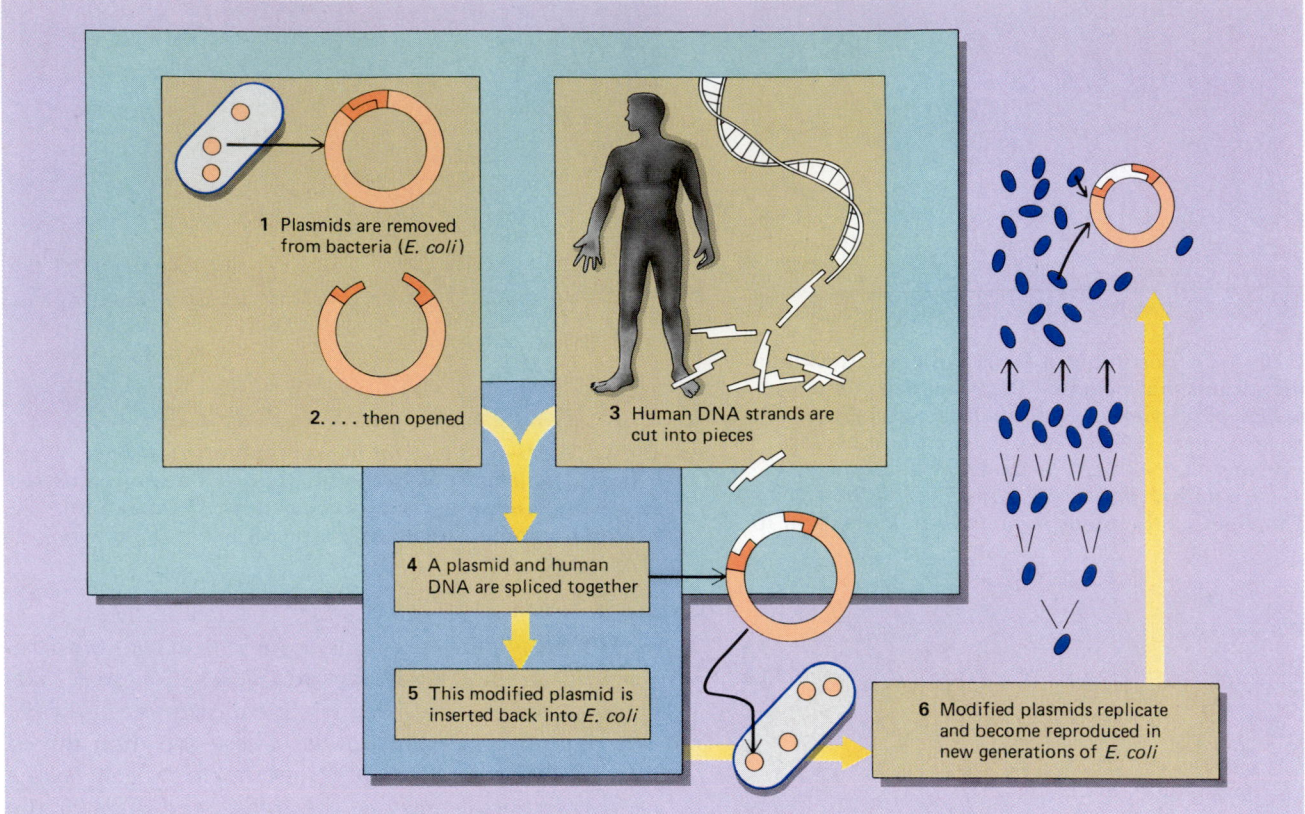

1 Plasmids are removed from bacteria (*E. coli*)

2. . . . then opened

3 Human DNA strands are cut into pieces

4 A plasmid and human DNA are spliced together

5 This modified plasmid is inserted back into *E. coli*

6 Modified plasmids replicate and become reproduced in new generations of *E. coli*

FIGURE 16–7 Using plasmids to construct recombinant DNA molecules for gene cloning. (1) Bacterial plasmids are removed from *E. coli*. (2) The bacterial plasmid is opened with a restriction endonuclease. (3) The same restriction endonuclease is used to cut human DNA into fragments. (4) Ligase is used to splice the human DNA fragment containing the desired gene into the plasmid. The result is recombinant DNA, a hybrid plasmid that contains genes from two different organisms. (5) The hybrid plasmid is inserted into bacterial cells, which replicate it and pass it to their progeny (6), creating a clone of cells capable of producing a human substance.

gether and a plasmid passes through the hollow pilus from one bacterium to another.

To get the patchwork plasmids into host bacteria does not necessarily require a pilus. There are several less natural but more convenient ways, involving, for example, the removal of bacterial cell walls so that the plasmids can penetrate their new hosts easily. Since only a small proportion of the bacteria will take up the plasmids, there has to be a way to determine which bacteria contain the genetically altered plasmids. Fortunately, plasmids often carry genetic information that makes the bacterium resistant to certain antibiotics. If the plasmid that was used has a gene for resistance to a particular antibiotic, then you can grow, or culture, the bacterial population in the presence of that antibiotic. Only bacteria that have taken up the genetically altered plasmid will resist the antibiotic and survive; the rest will conveniently die.

There are several other ways besides using plasmids to implant genes in bacteria. One involves **bacteriophages** (bacterial viruses that are also called

phages), which can actually become part of the main bacterial chromosome and can thus carry new genes into it. In another technique, a $CaCl_2$ solution is used to make bacterial cells take in foreign DNA. With this chemical technique, the DNA enters bacteria without a phage or plasmid vector. Several newer, less involved techniques are also being developed.

The Cells That Have Taken Up The Gene Of Interest Are Identified

A needle could be found in a haystack of any size if one had a magnet or, better, a metal detector. Fortunately, the genetic equivalent of the metal detector is available to us and can be used to find just the DNA base sequence desired among the vast tangle of genes in the cell's chromosomes. This device is called a **genetic probe,** a radioactively labeled segment of messenger ribonucleic acid (mRNA) or artificially synthesized single-stranded DNA complementary to the target gene.

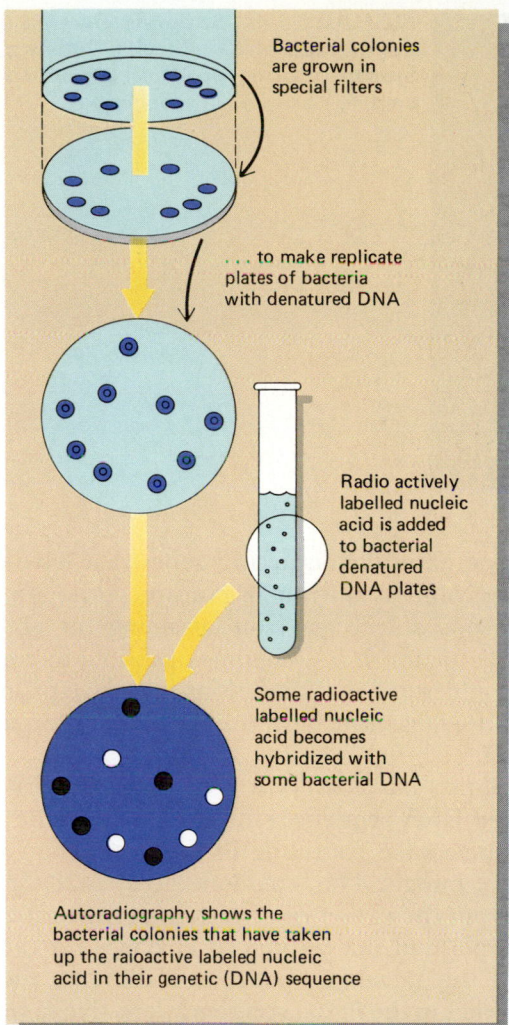

Bacterial colonies are grown in special filters

... to make replicate plates of bacteria with denatured DNA

Radio actively labelled nucleic acid is added to bacterial denatured DNA plates

Some radioactive labelled nucleic acid becomes hybridized with some bacterial DNA

Autoradiography shows the bacterial colonies that have taken up the raioactive labeled nucleic acid in their genetic (DNA) sequence

FIGURE 16–8   Genetic probes can be used to determine which cells have taken up the gene of interest. If mRNA or single-stranded DNA is produced that has a base sequence complimentary to the gene of interest, it will hybridize (form base pairs with that gene). In this technique, the nucleic acid has been radioactively labelled so that it can be visualized by autoradiography. Only the bacterial colonies that have taken up the gene of interest will be labelled.

tains the base sequence needed for production of the protein. If the synthesized mRNA is radioactively labeled, it will identify the sequence of the DNA that is responsible for the production of insulin.

Thus, if after such treatment any radioactive DNA can be detected in that of a particular colony of bacterial cells, this indicates that the bacteria's DNA has hybridized with the radioactive probe. The colonies to which this probe sticks can be visualized by autoradiography (Fig. 16–8). These are the cultures that may be able to produce the desired protein.

The Gene Is Expressed Inside The New Cell

Even though a gene has now been isolated and identified in our example, large-scale propagation of the bacterial strain would not necessarily produce the desired protein. The necessary gene will not be transcribed unless it is associated with a set of regulatory and promoter genes. For example, the lac operon control region has been added to plasmids (Chapter 12).

Suppose that we wanted to engineer bacteria that could produce TSH. After producing recombinant plasmids containing TSH genes and the lac operon control region, and reinserting those plasmids into new bacterial cells, the transcription of the resulting plasmids would be under the control of the lac regulatory region. Therefore, the gene for TSH would be transcribed and translated, producing the thyroid-stimulating hormone, a protein the host bacterium has not produced before.

Although human insulin is now produced commercially by engineered *E. coli* bacteria, the use of *E. coli* for commercial protein production has the drawback that *E. coli* will not normally secrete these products. The insulin accumulates in the cell and must be removed by harsh mechanical disruption (Fig. 16–9). In the future, *Bacillus subtilis* may be used instead, since this bacterium naturally secretes protein into the culture fluid.

To select the cells with the desired genes from the genetic library, the researcher may spread a sample of the bacterial culture on plates. If it is dilute enough, on the average, only one cell falls onto a particular location. When it reproduces it gives rise to a colony of genetically identical cells, which can be propagated further in a liquid medium and then tested for the presence of the desired gene by using a genetic probe.

Suppose we wanted to identify the genes that code for the protein insulin. Because we know the amino acid sequence of insulin, we could synthesize the mRNA that produces it. This mRNA will attach itself to, or **hybridize** with, complementary DNA that con-

Gene Insertion In Eukaryotes Is Technically Challenging, But Feasible

The complexity of eukaryotic cells has, until relatively recently, made altering them by recombinant DNA techniques formidable. For example, it is not enough to simply get the DNA inside the eukaryotic cell. The DNA must be incorporated into a chromosome in order to be expressed. Yeast, a single-celled fungus, represents a simple and convenient eukaryotic model to study recombinant DNA techniques. For one thing, scientists already know a great deal about the biology of yeasts. Also, yeasts offer an advantage over bacteria

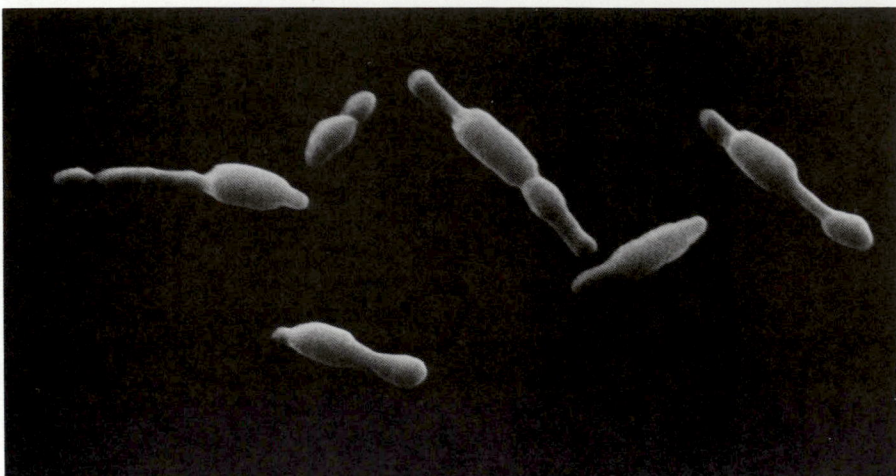

FIGURE 16–9 Scanning electron micrograph of *E. coli* cells "bulging" with human insulin. (Dr. Daniel C. Williams and the Lilly Microscope Laboratory)

in that they will process the proteins they produce into their final form, known as **post-translational processing.** This is important, as many proteins require post-translational processing to become fully functional.

A variety of vectors are used in recombinant DNA techniques involving eukaryotic cells. Although eukaryotic cells don't possess plasmids naturally, bacterial plasmids have been used as vectors in eukaryotes. Another method that has been employed recently to transfer foreign DNA inside cells is **electroporation,** which involves exposing cells to high-voltage electrical pulses. In some way the electricity disrupts the membrane, possibly by tearing a small hole in it. If cells being exposed to electroporation are in a solution containing DNA, some of the cells will take in DNA. There is a high percentage of cell death by this procedure, however. Although electroporation shows promise, there is no current evidence of stable transmission of DNA by this procedure.

A third possible vector for eukaryotic cells is viruses, which are highly efficient at inserting foreign DNA. Viruses that are used as vectors are first altered so they cannot harm the cells they infect. There is a class of RNA viruses, the **retroviruses,** that make DNA copies of themselves by reverse transcription. Sometimes these DNA copies become integrated into the host cell chromosomes, where they are replicated along with host DNA. Any foreign DNA being carried, or vectored, by the virus also becomes part of the host cell's chromosomes.

◼ GENETIC ENGINEERING HOLDS GREAT PROMISE FOR THE FUTURE

One possible future advance in genetic engineering may be the ability to synthesize totally new genes, which are designed in advance to possess exactly the right combination of desired properties. We currently recombine naturally occurring genes that have developed by evolution in living organisms. Although scientists are probably years from attaining the ability to synthesize genes, it is sometimes possible to combine existing genes in new associations that have properties much different from any that the donor organisms ever had.

Scientists have demonstrated the feasibility of introducing DNA sequences into *E. coli* from the bacterium *Pseudomonas*, resulting in the production of the plant dye indigo. This was done accidentally and resulted from the presence of certain genes in the *Pseudomonas* plasmid that produced enzymes for the later parts of the metabolic pathway leading to the dye. Other genes in the *E. coli* recipient were able to provide an earlier part of the chain of intermediates. The really significant aspect of this study may be that the indigo-producing gene was not derived from a plant source but was put together by combining certain *E. coli* genes already present with some that occurred naturally in a *Pseudomonas* plasmid. The equivalent of a eukaryote gene was thus constructed from prokaryote components.

It May Be Possible To Repair Human Genetic Defects

One of the goals of genetic engineering is **gene therapy,** the implantation of genes into eukaryotic cells, in the hope of correcting human genetic disease. The ideal candidate for this type of treatment would be a genetic disorder caused by a single gene defect. For example, adenosine deaminase (ADA) deficiency disease is caused by the body's inability to make a single enzyme, adenosine deaminase. This, in turn, is due to a deficiency in the gene that codes for this enzyme. Without adenosine deaminase, certain white blood cells can't divide and fight infection. Most humans afflicted with this rare genetic disorder die as infants.

FOCUS ON Probing for Genetic Disease

Genetic engineering techniques can be used in the characterization and detection of genetic disease. The earliest studies of this sort were carried out on sickle-cell anemia, since that disease produces a thoroughly characterized protein abnormality. It will soon be evident why that is important.

Sickle-cell hemoglobin differs slightly from normal hemoglobin in its amino acid sequence. That means that the DNA responsible for the disease must also differ from the normal in its base sequence. Knowing the genetic code, we can easily determine what the base sequence of the abnormal DNA must be.

It is possible to make DNA that contains this sequence of bases and it is likely that such a combination is unique in the genome. The artificial DNA is also radioactively labeled, which means, you may recall, that radioactive isotopes are incorporated into some of it. This is now the **probe DNA.**

If the probe DNA is allowed to hybridize with that from a person with sickle-cell anemia, it will unite with his or her DNA at the point where the abnormality lies. Furthermore, even if a person is heterozygous for the disease, the abnormal DNA for which he or she is haploid will hybridize with the probe DNA. This procedure allows us to detect normal carriers of a wide variety of genetic diseases.

Actually, the DNA probe method of detection is not of much use in sickle-cell anemia because the abnormal hemoglobin such people possess can be more simply detected by other methods in either a homozygous or a heterozygous state. However, such abnormal hemoglobin is present only in blood. If we wish to determine whether a fetus will suffer from the disease, we must examine not the blood but the fibroblasts in the amniotic fluid. These tell us nothing by means of the older techniques but can be shown to be heterozygous or homozygous for the disease by the DNA probe method.

An important side benefit of the use of probe DNA is that it allows us to determine on which chromosome the abnormal gene occurs. This is done by hybridizing probe DNA with a chromosome preparation from a person with sickle-cell anemia and observing which of the chromosomes becomes radioactively labeled as a result. Few genetic diseases are as well understood as sickle-cell anemia, but as more and more abnormal proteins are identified and sequenced, an increasing number of hereditary disorders will be detectable in healthy carriers and fetuses.

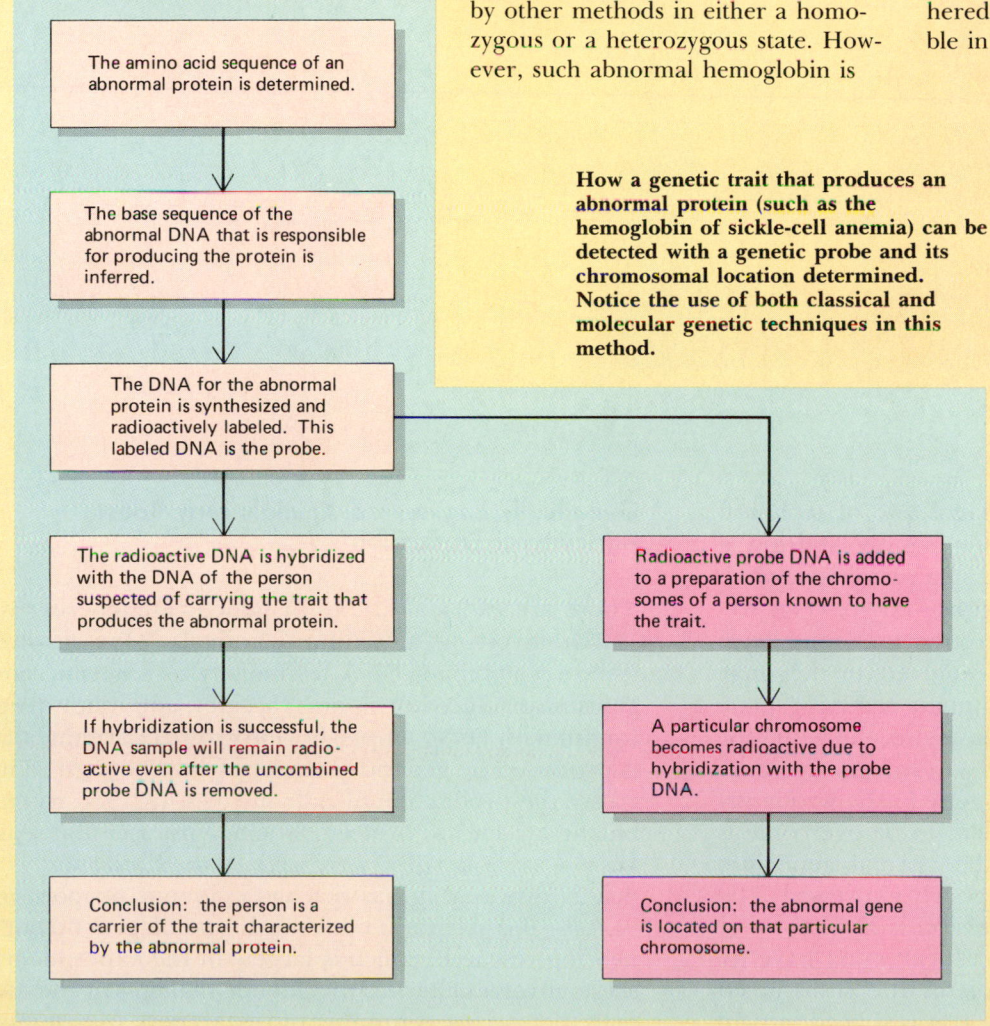

How a genetic trait that produces an abnormal protein (such as the hemoglobin of sickle-cell anemia) can be detected with a genetic probe and its chromosomal location determined. Notice the use of both classical and molecular genetic techniques in this method.

The amino acid sequence of an abnormal protein is determined.

The base sequence of the abnormal DNA that is responsible for producing the protein is inferred.

The DNA for the abnormal protein is synthesized and radioactively labeled. This labeled DNA is the probe.

The radioactive DNA is hybridized with the DNA of the person suspected of carrying the trait that produces the abnormal protein.

If hybridization is successful, the DNA sample will remain radioactive even after the uncombined probe DNA is removed.

Conclusion: the person is a carrier of the trait characterized by the abnormal protein.

Radioactive probe DNA is added to a preparation of the chromosomes of a person known to have the trait.

A particular chromosome becomes radioactive due to hybridization with the probe DNA.

Conclusion: the abnormal gene is located on that particular chromosome.

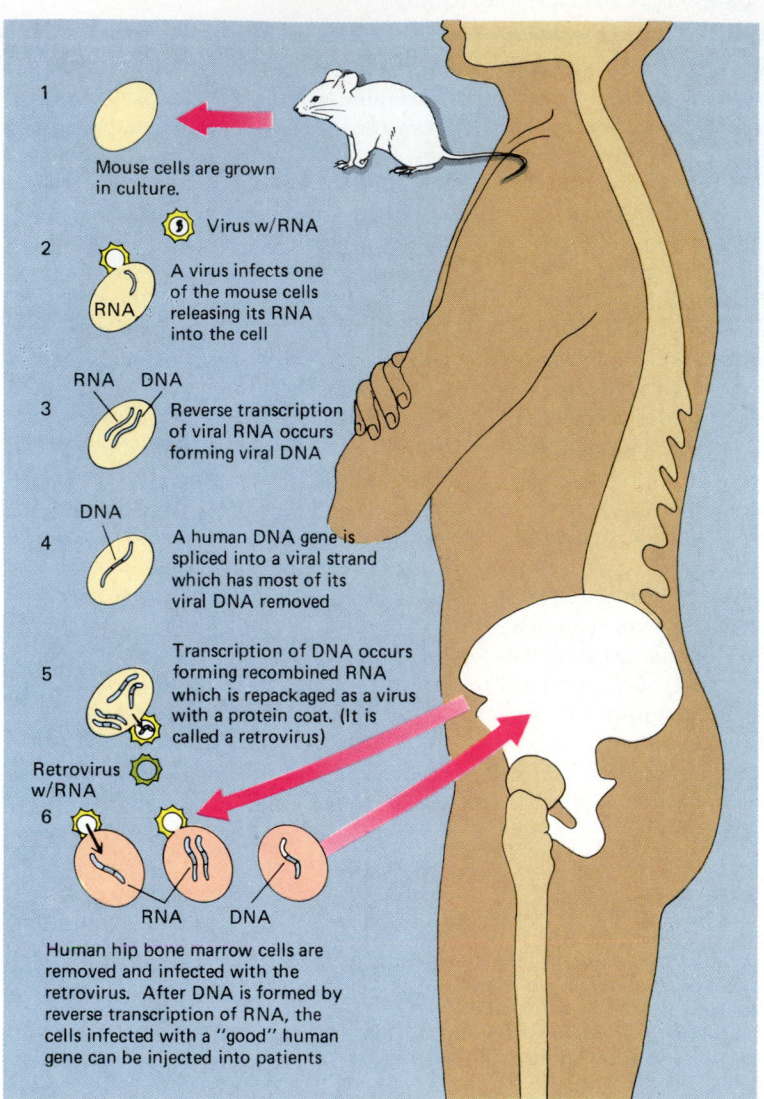

1 Mouse cells are grown in culture.

2 Virus w/RNA — A virus infects one of the mouse cells releasing its RNA into the cell

3 RNA DNA — Reverse transcription of viral RNA occurs forming viral DNA

4 DNA — A human DNA gene is spliced into a viral strand which has most of its viral DNA removed

5 Transcription of DNA occurs forming recombined RNA which is repackaged as a virus with a protein coat. (It is called a retrovirus)

Retrovirus w/RNA

6 RNA DNA

Human hip bone marrow cells are removed and infected with the retrovirus. After DNA is formed by reverse transcription of RNA, the cells infected with a "good" human gene can be injected into patients

FIGURE 16–10   Gene therapy in humans may use a retrovirus, which has RNA instead of DNA, as a vector. This simplified diagram reveals some of the technical complexity involved in gene therapy. Several genetic diseases caused by a single gene defect are candidates for gene therapy.

Gene therapy involves a magnitude of technical problems not encountered when simply splicing a human gene into a bacterium. First, a vector is needed. Retroviruses are usually the vector of choice (Fig. 16–10). As mentioned previously, retroviruses must have their potentially harmful genes removed, or they may cause worse problems in the human receiving gene therapy than he or she already has. Also, the genes for the vector's replication must be removed so the virus can't reproduce and spread the gene to the wrong tissue in the body. Another problem to be overcome is that once the virus transfers the normal gene into human cells, the gene is usually poorly expressed. The exact reason for this poor expression is not known at present, but it may be because we currently have no way to control the final location of the transferred gene in the host's chromosomes. If it is spliced into an inactive portion of a chromosome, it won't be transcribed.

Genetically Engineered Animals May Boost Agricultural Production

There are two main types of genetic engineering research involving agricultural animals. One involves using recombinant DNA technology to have microorganisms produce hormones, vaccines, and other drugs for animals. For example, bovine growth hormone that is produced by genetic engineering can be used to increase the production of milk in cows. In 1988 recombinant vaccines to protect cattle against a viral disease known as rinderpest were reported. Rinderpest is a deadly disease that has reached epidemic proportions in Asia and Africa. The second type of genetic engineering research, which is largely in the experimental stage, involves altering the genetic makeup of the animals themselves. A number of researchers are attempting to develop genetically superior strains of chicken and cattle, for example.

FOCUS ON   Molecular Genetics and Cancer

A cancer cell is an abnormal cell that lacks biological inhibitions. **Cancer** could be considered a genetic disease, transmitted not so much between generations of hosts as between generations of cells. In addition to lacking the normal contact inhibition that stops cell division when its reasonable limits have been reached, cancer cells are immortal. Not that a cancer cell cannot die, but it cannot die of old age. Normal human fibroblasts, for example, can be maintained in tissue culture for perhaps 50 generations before they become enfeebled, yet the HeLa[1] cancer cells isolated from a patient in the 1950s are today alive and well in tissue culture vessels the world over, still going strong after thousands of cell generations. Given the proper

[1] HeLa cells are easily cultured and are used the world over for numerous kinds of research in cell biology.

conditions and care, there seems to be no reason why descendants of today's HeLa cells might not still be multiplying in the year 2100 or beyond.

Recombinant DNA techniques are currently being used in analysis of the processes by which normal cells become transformed into cancer cells. These and other studies indicate that **carcinogenesis** (the development of cancer) in humans and other mammals involves several independent steps. Yet carcinogenesis also seems to be a basic genetic change that produces a cascade of processes, which in turn bring about the multitude of specific properties that a cancer cell must have to become malignant. It appears that cancer cells owe such traits to possession of at least one and probably several genes that are known as **oncogenes.**

In a typical study, a human line of cancer cells from a bladder tumor

was shown to have undergone a mutation in a specific codon (a change from G to T). As a result, the amino acid glycine in position 12 of this gene's protein product was replaced by valine. This one change was shown to be critical to the conversion of the cell's normal gene into an oncogene.

Oncogenes and oncogene-like DNA sequences are widespread among living things. In animals **proto-oncogenes** function in normal cellular growth and embryonic development. When dissociated from their normal regulatory regions, such genes may turn into oncogenes.

Oncogenes that are known to have come from animal cells are found in some viruses, especially retroviruses, which are able to transmit them to new cellular victims. Still other viruses carry oncogene sequences that do not seem to have originated in this way but that also produce cancer. It seems likely, however, that several oncogenes may be necessary to produce cancer. This could explain the observed multistep process of cancer induction required for the production of most experimentally developed cancers. It could also explain the observation that a tendency to cancer is often hereditary.

Imagine, for instance, that each of a patient's body cells contained an oncogene inherited from his parents. A viral infection might implant another, and late in life mitotic malfunction might cause the original gene to be duplicated in one of the patient's continually dividing cell lines. Carcinogenic food additives could produce yet another oncogene in one cell in that line, which might complete its transformation into a cancer cell. Over the course of several years the ancestral cancer cell would refuse to obey normal contact inhibition and would divide without hindrance until an obvious tumor was formed and the cancer diagnosed.

The accumulation of the results of these and other random events would certainly take time. Perhaps that is why most cancers occur with greatest frequency in elderly people.

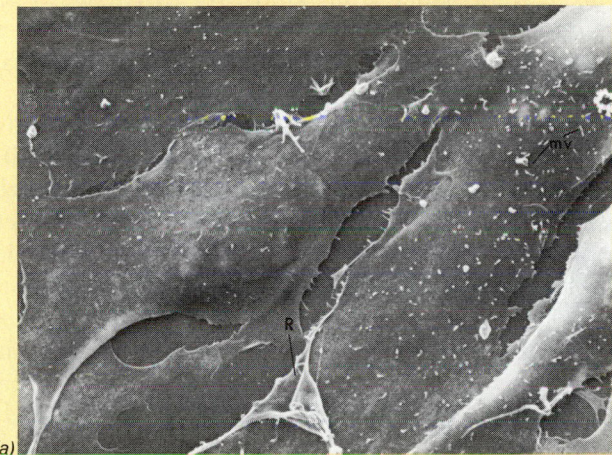

(a)

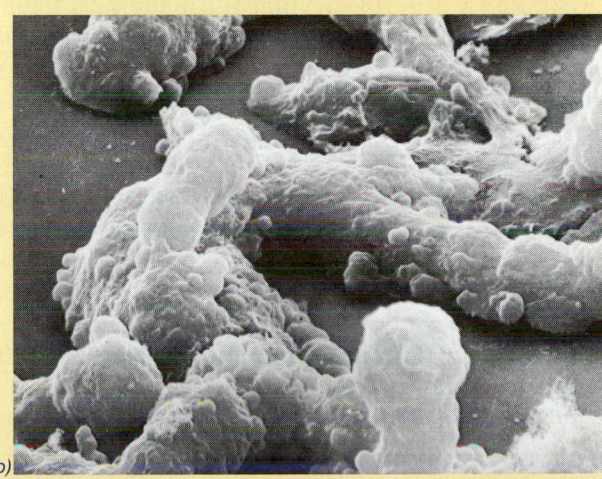

(b)

Scanning electron micrographs of cultured cells. (a) Normal cells from a hamster embryo. A few microvilli (mv) and ruffles (R) are visible. (b) The same cell type after transformation by a cancer-causing virus (approximately ×3145). Note that the cells have blebs, or bubbles, on their surfaces and aggregate to form several layers.
(Courtesy of R.D. Goldman.)

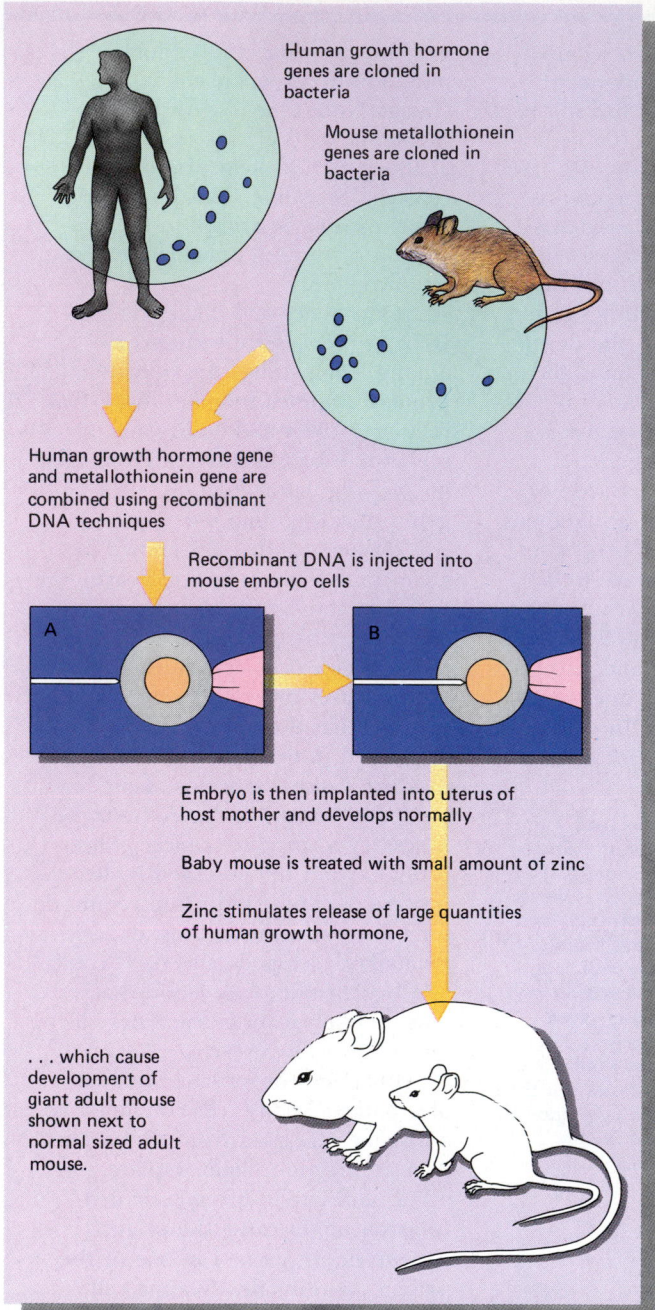

Human growth hormone genes are cloned in bacteria

Mouse metallothionein genes are cloned in bacteria

Human growth hormone gene and metallothionein gene are combined using recombinant DNA techniques

Recombinant DNA is injected into mouse embryo cells

A

B

Embryo is then implanted into uterus of host mother and develops normally

Baby mouse is treated with small amount of zinc

Zinc stimulates release of large quantities of human growth hormone,

. . . which cause development of giant adult mouse shown next to normal sized adult mouse.

FIGURE 16–11   How to make a giant mouse. This experiment demonstrates the enormous potential for incorporating desirable genes into agriculturally important animals using recombinant DNA techniques.

A typical study, using mice as a model, involved transferring human growth hormone genes into mice. This experiment involved a different method of gene insertion, using DNA directly to carry genetic information from one cell to another. There are several ways to do this, one being microinjection of DNA into the recipient cell. In this study the human gene for growth hormone production was isolated from a library of human DNA and transferred to a fertilized mouse egg (Fig. 16–11). In those eggs in which the gene trans-

plant had been successful, growth was enhanced. One mouse grew to more than double the normal size. As might be expected, such mice are also able to transmit their enhanced growth capability genetically to their offspring. One can imagine all sorts of potentially practical applications of similar experiments in such areas as the breeding of domestic meat animals.

Scientists Are Using Recombinant DNA Techniques To Improve Plants

Plants have been selectively bred for centuries, if not millennia. The success of such efforts depends upon the presence of pre-existing genes, either in the variety of plant being selected or in closely related wild or domesticated plants. The desirable traits must also be transferable by crossbreeding. Even primitive varieties of cultivated plants may have certain traits, such as disease resistance, that could be advantageously introduced into varieties more suited to modern needs. However, the rarer varieties of agricultural plants, especially primitive ones, are swiftly becoming extinct. This greatly reduces the size of the potential gene pool from which agricultural researchers may draw. Indeed, just when their genetic potential is being fully realized, wild plants of *all* kinds are threatened with extinction as the last available agricultural land is brought into cultivation to feed the exploding human population.

Recombinant DNA technology allows genes to be introduced into plants from strains or species with which they do not ordinarily interbreed. Much research funding has been made available to plant geneticists because of the economic and humanistic potential of increased plant yields, disease resistance, and more nutritious crops. For example, tomatoes have been genetically engineered to resist the tobacco mosaic virus. Also, in 1988 a gene from soil bacteria was placed in tobacco plants by recombinant DNA techniques. These genetically engineered plants were able to break down a herbicide that normally kills plants. Geneticists working with plants are also at greater liberty to experiment with new techniques than those working with animals, for manipulation of plant genes does not demand the same type of ethical considerations.

Several vectors for the introduction of genes into plant cells have been used. Initially, most genetic introductions were performed by removing plant cell walls to produce naked protoplasts (see Fig. 26–21 from Chapter 26). If two protoplasts are induced to fuse, there is a chance that the resulting cell may give rise to mature, differentiated plants with some of the desirable genetic traits of the "parent" plants. Like the more traditional techniques of sexual propagation, however, it is essentially a hit-or-miss procedure with results that cannot be predicted, let alone guaranteed.

FIGURE 16–12   A crown gall tumor on sunflower. The growth of this tumor is induced by a plasmid carried by *Agrobacterium tumefaciens.* **It is unusual in that the plasmid becomes incorporated into the plant's chromosomes, and the bacterial genes are then expressed by the plant.** (John D. Cunningham/VU)

One example of a successful use of cell fusion involved the white potato. Researchers at the U.S. Department of Agriculture fused cells of wild potato plants with commercial potato cells. From South America, the wild potato plants produced a toxin that repelled insects. The hybrid plant that resulted from the protoplast fusion produced the toxin and repelled the Colorado potato beetle, a devastating pest that costs millions of dollars annually to control with insecticides.

Another technique for introducing plant genes employs the crown gall bacterium, *Agrobacterium tumefaciens,* which produces plant tumors (Fig. 16–12). It does this by introducing a special plasmid into the chromosomes of its host. The plasmid induces abnormal growth by forcing the plant cells to produce abnormal quantities of growth hormone. The plasmid also diverts the metabolism of the host cells to produce substances known as **opines,** simple derivatives of amino and keto acids. These opines are specifically preferred food substances for the bacterium.

Desirable genes can be incorporated into the *Agrobacterium* plasmid for incorporation into a plant host (Fig. 16–13). It is possible to disarm the plasmid so that it does not induce tumor formation. The cells into which such a genetically altered plasmid is introduced are essentially normal except for the genes that the

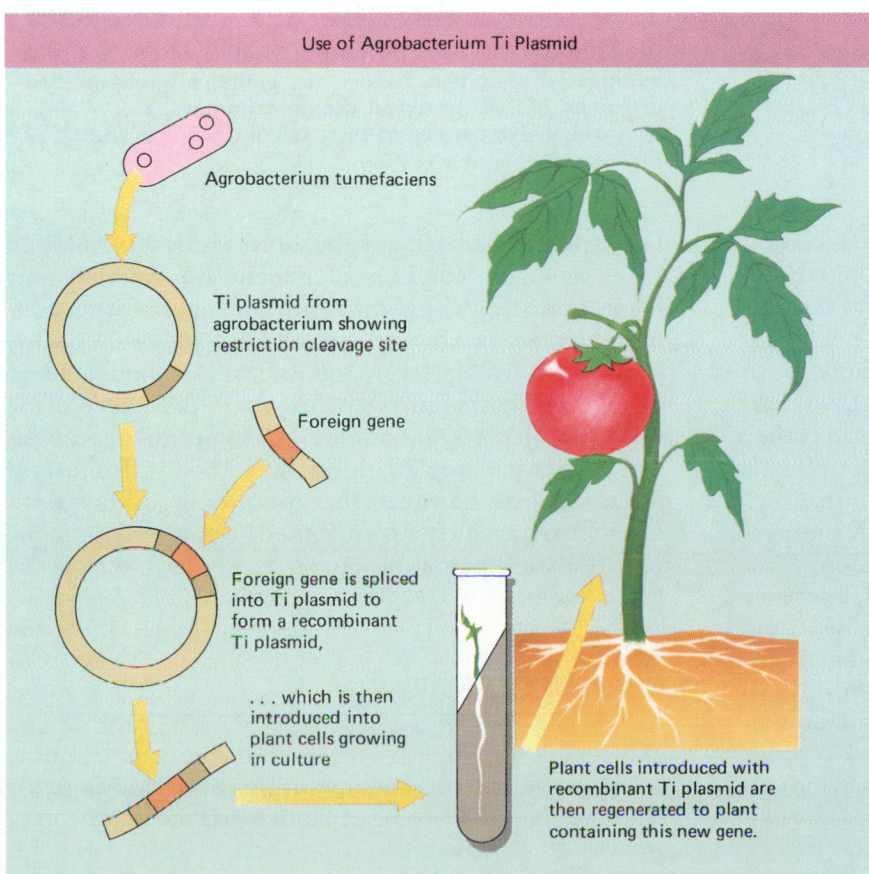

Use of Agrobacterium Ti Plasmid

Agrobacterium tumefaciens

Ti plasmid from agrobacterium showing restriction cleavage site

Foreign gene

Foreign gene is spliced into Ti plasmid to form a recombinant Ti plasmid,

. . . which is then introduced into plant cells growing in culture

Plant cells introduced with recombinant Ti plasmid are then regenerated to plant containing this new gene.

FIGURE 16–13   The Ti plasmid of *Agrobacterium tumefaciens* **can be used to introduce desirable genes from another organism into a plant. Because the T DNA is that portion of the plasmid that gets spliced into plant chromosomes of susceptible plants, it is perfect for transporting foreign genes. Restriction endonuclease and ligase are used to splice the foreign DNA into the Ti plasmid. The recombinant Ti plasmid is inserted into** *Agrobacterium,* **which is then used to infect plant cells in culture. The T DNA, with its foreign gene, is inserted in the plant's chromosome. Genetically engineered plants are produced from the cultured plant cells using plant tissue culture techniques.**

experimenter has inserted. It has been shown that genes placed in the plant chromosomes in this fashion are transmitted sexually via seeds to the next generation, but they can also be propagated asexually if desired.

There are several problems with using *Agrobacterium* plasmids to introduce foreign genes into plants. First, not all plants are susceptible to this bacterium. For example, important crops like corn, wheat, and rice do not get crown gall disease. Only plants that are susceptible to *Agrobacterium* can use its plasmid as a vector, although this difficulty may be overcome in the future. A second problem with this method is that it utilizes plant tissue culture to reconstitute plants after genetic recombination has occurred in plant cells (see Chapter 26). Tissue culture techniques have not been successful for all plants. Therefore, recombinant DNA techniques cannot be used for plants in which tissue culture is a problem.

An additional, very interesting complication of plant genetic engineering is the substantial DNA of the chloroplasts. Chloroplasts are pivotal in photosynthesis, and photosynthesis is the basis of plant productivity. Obviously, it would be useful to develop techniques for changing the portion of the chloroplast genetic information that resides within the chloroplast itself. Methods of chloroplast genetic engineering are currently the focus of intense research interest.

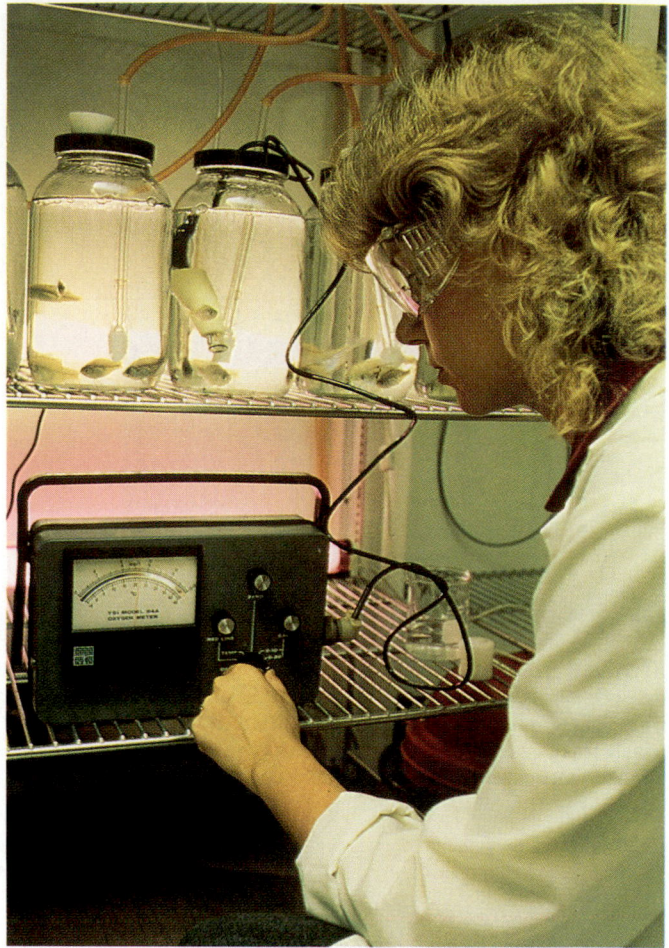

FIGURE 16–14 Determining the safety of a genetically engineered organism. Before any genetically engineered organism is field-tested in the environment, a comprehensive battery of tests is conducted in the laboratory. (Courtesy of Monsanto Co.)

■ CONCERN OVER POTENTIAL DANGER CAUSED BY GENETIC ENGINEERING HAS LED TO THE FORMATION OF POLICIES ON UTILIZING RECOMBINANT DNA TECHNIQUES

While acknowledging the potential uses of recombinant DNA techniques as important and beneficial, many scientists are concerned about potential misuses or unpredictable side effects. There is the possibility that an organism might be produced that would have undesirable ecological or other effects, not by design but by accident. Totally new strains of bacteria or other organisms, with which the world of life has no previous experience, might be similarly difficult to control.

The real problem, however might lie in the permanence of the results of genetic manipulation. Most of what we humans do is quite temporary, however severe the consequences of our actions may be in the short term. Real permanence involves the genetic information of organisms, for *that* is self-propagating in an active fashion that our environmental manipulations cannot match. Only the extinction of plants and animals has had permanent consequences for us to date, and of course that also involves genetic informa-

tion, specifically its loss. Now, however, it is possible to *create* new combinations of genetic information and incorporate them into self-reproducing life forms that, once created, might well persist for geological epochs.

Recent history has failed to bear out these genetic worries. Experiments over the past decade have demonstrated that recombinant DNA experiments can be carried out in complete safety (Fig. 16–14). It must be acknowledged, however, that concern about the safety of DNA experiments contributed to the development of safe experimental design.

■ CHAPTER SUMMARY

I. Genetic engineering refers to the genetic manipulation of cells so that new strains of organisms with new characteristics can be constructed.

A. Recombinant DNA techniques enable investigators to isolate, identify, and manipulate genes from the cells of organisms ranging from bacteria to plants and animals.

1. Many bacteria contain small, accessory, circular DNA molecules called plasmids. These plasmids may be used as vectors, aiding in the transfer of genes from one organism to another.

2. Restriction endonucleases cleave double-stranded DNA at specific sites, breaking DNA into more manageable fragments.

3. Segments of DNA from different sources can be joined by a splicing enzyme, or ligase.

4. If the gene is inserted into a suitable host cell, it may be transcribed and translated, leading to the production of a protein not previously produced by the host organism.

B. Eukaryotic genes can be introduced into eukaryotic cells.

1. Vectoring of new genes into eukaryotic cells has been accomplished with bacterial plasmids, electroporation, cell fusion, and viruses.

2. When such new genes are successfully introduced into nuclear DNA, they are passed on to offspring by sexual reproduction in classic Mendelian fashion.

II. Genetic engineering has great potential.

A. Gene therapy to correct human genetic diseases may soon become a reality.

B. Biotechnology is leading to greater productivity of agricultural animals, through the development of genetically superior animal strains and the production of hormones, vaccines, and other drugs to enhance animal growth.

C. Recombinant DNA techniques are being used to improve crop plants by increasing their productivity, making them more nutritious. and developing disease resistance.

▢ POST-TEST

1. Techniques for manipulating specific genes in the laboratory are termed _____ DNA methods.

2. Transferring genes from one organism to another is known as _____.

3. Small accessory double-helical chromosomes called _____ occur in many bacteria.

4. Plasmids typically contain genes for resistance to _____.

5. The enzymes found in bacteria that cleave DNA at specific places are called _____. They function in the intact bacterial cell as a defense against _____ invasion.

6. A foreign gene can be incorporated into a plasmid that has first been treated with _____ _____ to break it and to produce "sticky" ends.

7. The plasmid (from question 6) can be rejoined by

means of a special enzyme, _____.

8. A _____ _____ is a radioactively labelled segment of mRNA or artificially synthesized single-stranded DNA that is complimentary to a specific target gene.

9. A _____ is a colony of genetically identical cells.

10. Exposing cells to high voltage electrical pulses to assist in the transfer of foreign DNA inside cells is known as _____.

11. _____ _____ is the implantation of genes into eukaryotic cells in the hope of correcting human genetic disease.

12. The _____ _____ disease has provided biologists with a vector capable of introducing recombinant genes into plants. This disease is produced by a _____ that injects a special plasmid into its host cells.

▢ REVIEW QUESTIONS

1. What is genetic engineering? Give some specific examples of its benefits.

2. Describe how genetic information from a eukaryote could be implanted in a bacterium. What is to be gained by this?

3. What is restriction endonuclease? Where is it likely to attack a strand of DNA? What practical application of this is employed in recombinant DNA research?

4. Cigarette smoke has apparently been shown to activate oncogenes. How are oncogenes affected by such muta-

genic chemicals to produce cancer?

5. Why was it necessary to include metallothionein genes in the sequence that implanted functional human growth hormone genes in giant mice?

6. What are some potential ethical problems of genetic engineering? What safeguards could be employed to guard against potential harm that would not excessively hamper practical and theoretical research?

■ RECOMMENDED READINGS

Fox, J.L. "Biotechnology Alfresco," *BioScience* 38:533–537, 1988. A discussion on field-testing of genetically engineered organisms.

Landgren, U., R. Kaiser, C.T. Caskey, and L. Hood. "DNA Diagnostics—Molecular Techniques and Automation," *Science* 242:229–243, 1988. Many aspects of human disease are being elucidated using DNA diagnostics.

Lewis, R. "Genetics Meets Forensics," *BioScience* 39:6–9, 1989. DNA fingerprinting is being used increasingly as evidence in court cases, which are reviewed in this article.

Marx, J.L. "Gene Therapy—So Near and Yet So Far Away," *Science* 232:824, 1986. A statement on some of the technical difficulties yet to be resolved in gene therapy research.

Moody, M.D. "DNA Analysis in Forensic Science," *BioScience* 39:31–36, 1989. Some of the genetic techniques being used in forensic science are explained.

Moses, P.B. "Strange Bedfellows," *BioScience* 37:6, 1987. A discussion of *Agrobacterium*'s use in plant genetic engineering.

Nelson, R.S. et al. "Virus Tolerance, Plant Growth, and Field Performance of Transgenic Tomato Plants Expressing Coat Protein from Tobacco Mosaic Virus," *Biotechnology* 6:403–408, 1988. Resistance to a viral disease has been conferred on tomato plants by genetic engineering.

Stalker, D.M., K.E. McBride, and L.D. Malyj. "Herbicide Resistance in Transgenic Plants Expressing a Bacterial Detoxification Gene," *Science* 242:419–422, 1988. Plants have been developed by genetic engineering techniques to tolerate a herbicide.

PART 5

Evolution

Natural selection • Charles Darwin • Biogeography and evolution • Gene frequencies, gene flow, and gene pools • Fossil evidence • The fossil history of life

Classifying diversity • Population genetics • All life into five kingdoms • Microevolution and macroevolution • Extinction of species • Reproductive isolating mechanisms • Gradualism or punctuated equilibria?

17

Darwin and Natural Selection

OUTLINE

I. The concept of evolution originated before Charles Darwin
II. Darwin gathered much evidence for the theory that organisms evolve by natural selection
III. Many types of scientific evidence support evolution
   A. Comparative anatomy of different organisms reveals evidence of evolution
   B. Developmental biology provides more clues to the evolution of species
   C. Fossils provide still more clues about evolution
   D. Biogeography, the distribution of organisms on the earth, also supports evolution
   E. Comparisons of the biochemistry and molecular biology of different organisms offer evidence of evolution

LEARNING OBJECTIVES

After you have studied this chapter you should be able to:
1. Outline the historical development of the theory of evolution.
2. Explain natural selection as envisioned by Darwin.
3. Compare the types of evidence for evolution that are obtained from the following fields: comparative anatomy, developmental biology, paleontology, biogeography, biochemistry, and molecular biology.
4. Distinguish between compressions, impressions, molds, casts, and petrifactions.
5. Define and give examples of homologous and analogous organs.
6. Relate how scientists make inferences about evolutionary relationships from the sequence of amino acids in specific proteins or the sequence of nucleotides in particular genes in organisms.

Charles Darwin at 30 years of age. (William E. Ferguson)

All the life forms on our planet, from microscopic bacteria to giant blue whales, from tropical tree frogs to desert cacti, evolved from one or a few simple kinds of organisms. This vast diversity of species developed from earlier species by a process Darwin called "descent with modification," or **evolution.** In biology, the term *evolution* means "genetic change in a population of organisms." (It does not refer to the changes that occur to an individual organism within its lifetime: Populations evolve, but individuals do not.)

Another way to describe this process is to say that evolution involves changes in the frequencies with which certain genes occur in a population. The total of all the genes present in a population is known as the **gene pool.** Evolution is a change in the frequencies of genes in the gene pool.

As an example, let us consider the evolution of bacteria that are resistant to antibiotics. The use of antibiotics (drugs intended to harm or kill bacteria and other microorganisms) has increased dramatically since the 1940s (Fig. 17–1). When physicians began to use antibiotics to treat human and animal infections, they believed that these drugs would eliminate diseases caused by bacteria.

This has not taken place, however. When the antibiotic penicillin is used, for example, most of the bacteria present are killed, but some survive. These survivors, because of certain genes they happen to possess, are not killed by the penicillin; they have a genetic resistance to penicillin, and they pass this trait on to their offspring. Once the nonresistant bacteria have died and the penicillin-resistant bacteria have reproduced, the bacterial population has a larger percentage of penicillin-resistant bacteria than before. Since the frequency of the genes that confer resistance has changed in the bacterial population, evolution has occurred; if enough of these types of changes occur over time, a new species might arise. On the practical side, physicians today recognize that they cannot use antibiotics indiscriminately, or we will be unable to control certain diseases as the genetic frequency of resistance to antibiotics increases.

The evolutionary scenario is usually more complex than the example just described. The concept of evolution is the cornerstone of biology, for it enables us to make sense of the tremendous variety in the living world. Biologists no longer question the fact that evolution has occurred. They continue, however, to closely study and actively debate the actual mechanisms that control it.

KEY CONCEPTS

☐ Evolution, one of the basic principles of biology, is a genetic change in a population of organisms.

☐ Charles Darwin proposed that evolution occurs by natural selection.

☐ A vast assemblage of scientific evidence supports evolution.

FIGURE 17–1 **Antibiotic resistance in bacteria. An even coating of bacteria covers the surface of this culture dish except where an antibiotic to which the bacteria are sensitive prevents this growth. Varieties of bacteria that have developed resistance to certain antibiotics are common today.** (Christine Case/Visuals Unlimited)

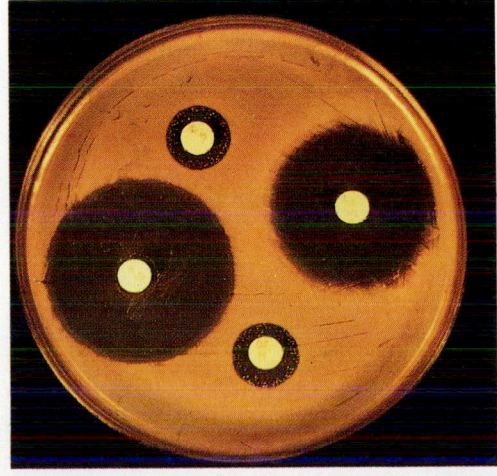

■ THE CONCEPT OF EVOLUTION ORIGINATED BEFORE CHARLES DARWIN

Although Charles Darwin is universally associated with evolution, ideas of evolution predate Darwin by centuries. Aristotle (384–322 B.C.) saw much evidence of design and purpose in nature and arranged all the organisms that he knew of in one "Scale of Nature" that extended from the very simple to the most complex. He visualized living organisms as being imperfect but moving toward a more perfect state. Some have interpreted this concept as the germ of an idea of evolution, but Aristotle is very vague on the nature of this "movement toward the more perfect state" and certainly did not propose any notion of the origin of species.

Long before Darwin, odd fragments resembling bones, teeth, and shells (fossils) had been discovered embedded in rocks. Some of these corresponded to parts of familiar living animals, but others were strangely unlike any known form. Fossils of marine invertebrates were found in rocks high on mountains.

biologists of his time, Lamarck believed that all living things were endowed with a vital force that drove them to evolve toward greater complexity. He also believed that organisms could pass on traits acquired during their lifetimes to their offspring. As an example of this line of reasoning, Lamarck suggested that the long neck of the giraffe evolved when a short-necked ancestor took to browsing on the leaves of trees instead of on grass (Fig. 17–2). Lamarck theorized that the ancestral giraffe, in reaching up, stretched and elongated its neck. Its offspring, inheriting the longer neck, stretched still further. The process, repeated over many generations, supposedly achieved the present long neck.

The proposed mechanism for Lamarckian evolution was an "inner drive" for self improvement. This concept was discredited when the mechanisms of heredity were discovered. Lamarck's contribution to science is important, however, because he was the first to propose that organisms undergo change over time as a result of some natural phenomenon rather than because of divine intervention. It remained for Charles Darwin to discover the mechanism of evolution by natural selection.

FIGURE 17–2   How did the giraffe get its long neck? Lamarck hypothesized that giraffes acquired longer necks by continually stretching into the trees to eat leaves unavailable to other large herbivores and that they passed this characteristic on to their offspring. Although Lamarck's mechanism of evolution was incorrect, he was the first scientist to propose that organisms undergo evolution by natural means. (Walt Anderson/Visuals Unlimited)

Leonardo da Vinci correctly interpreted these finds in the 15th century as the remains of animals that had existed in previous ages but had become extinct.

Jean Baptiste de Lamarck, in his *Philosophie Zoologique* (1809), expressed the most thoroughly considered view of evolution before Darwin. Like most

☐ DARWIN GATHERED MUCH EVIDENCE FOR THE THEORY THAT ORGANISMS EVOLVE BY NATURAL SELECTION

As a young man, Charles Darwin (1809–1882) was appointed naturalist on the ship *Beagle*, which was to

FIGURE 17–3   The voyage of *H.M.S. Beagle*. Observations made during this voyage helped Darwin formulate the concept of evolution by natural selection.

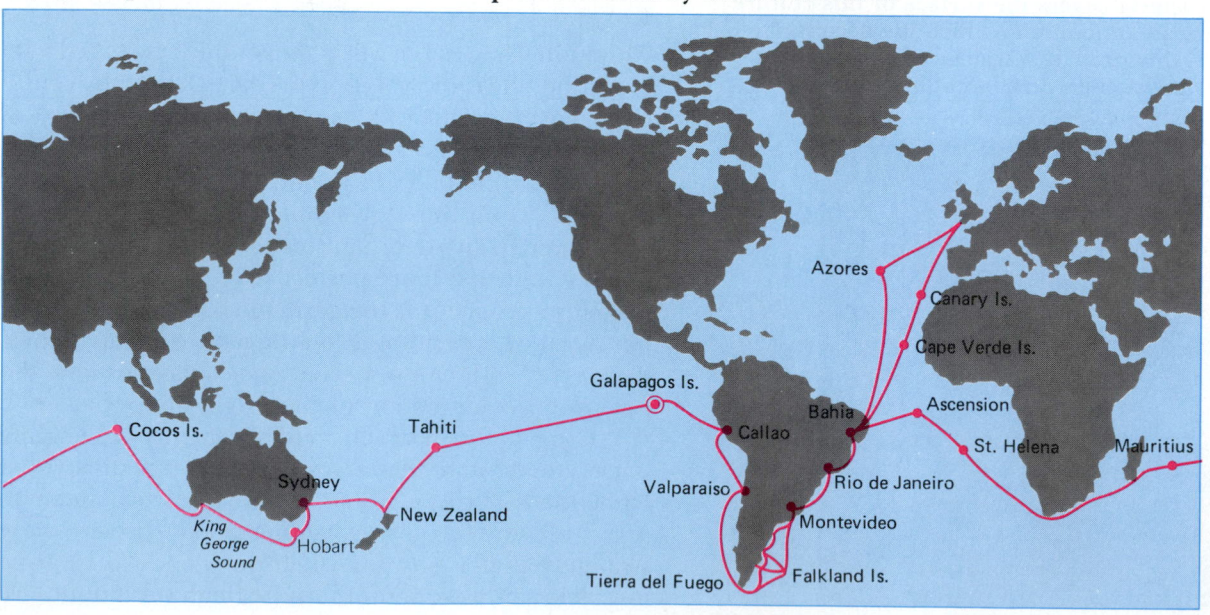

(a)

(b)

(c)

(d)

FIGURE 17–4   Species of the Galapagos Islands. (*a*) A land iguana (*Conolophus pallidus*). This animal is endemic to the Galapagos. (*b*) The webbed feet of the red-footed booby (*Sula sula*) can grasp tree branches. (*c*) A tree cactus (*Opuntia echios*). Other *Opuntia* in the Galapagos are not tree forms. (*d*) The Sally lightfoot crab is a subspecies that is endemic to the Galapagos.   (*a*, David Cavagnaro; *b*, *c*, and *d*, Frans Lanting)

take a 5-year cruise around the world to prepare navigational charts for the British navy (Fig. 17–3). The *Beagle* left Plymouth, England, in 1831 and cruised slowly down the east coast and up the west coast of South America. While other members of the company mapped the coasts and harbors, Darwin had an opportunity to study the animals, plants, fossils, and geological formations of both coastal and inland regions, areas that had not been extensively explored. He collected and catalogued thousands of specimens of plants and animals and kept notes of his observations. He experienced first-hand the tremendous richness of the plants and animals of these regions.

The *Beagle* spent some time at the Galapagos Islands, 600 miles west of Ecuador, where Darwin continued his observations and collections (Figs. 17–4, 17–5). He compared the animals and plants of the Galapagos with those of the South American mainland. He was particularly impressed by their similarities and

wondered why the creatures of the Galapagos should resemble those from South America more than those from Africa, for example. Moreover, although there were similarities between the two groups, there were distinct differences. There were even differences in the birds and reptiles from one island to the next! He pondered these observations and tried to develop an adequate explanation for them.

The general notion in the mid-1800s was that creatures did not change significantly over time, that they looked the same as the day they were created. True, there were some troubling exceptions to this idea. For one thing, breeders could induce a great deal of variation in domesticated plants and animals in just a few generations. They did this by selecting desirable traits and breeding only those individuals that possessed the desired traits, in a procedure known as **artificial selection** (Fig. 17–6).

Also, evidence found in rocks was beginning to

(a)

(b)

FIGURE 17–5 Two species of Darwin's famous Galapagos Island finches. Although these drab, unremarkable birds differ considerably in their adaptations, there seems no doubt that they have a common ancestry. Darwin observed that the various species of Galapagos finches occupied ecological niches filled elsewhere by birds of different families. The South American equivalents of the various Galapagos finches are presumably better adapted to these life-styles than almost all of the finches are, but as chance had it, were never afforded the opportunity to colonize the Galapagos Islands. The likely derivation of such different birds from a common ancestry helped suggest to Darwin that species were not unchanging and that they originated by natural selection. (*a*) Cactus finch, *Geospiza scandens*. The cactus finch may be close to the relatively unspecialized form of the original finch colonists. (*b*) A large ground finch, *Geospiza magnirostra*. This bird has an extremely heavy nutcracker-type bill adapted for eating heavy-walled seeds. (*a*, Jeanne White/Photo Researchers; *b*, Mignel Castro/Photo Researchers)

FIGURE 17–6 A number of common vegetables are members of the same species, *Brassica oleracea*, including broccoli, cauliflower, kale, brussels sprouts, and cabbage. Artificial selection is responsible for the tremendous variation shown within this species. (Raymond Tschoepe)

mountains, valleys, and other physical features of the earth's surface were not created in their present form. Instead, they were formed over long periods of time by the slow geological processes of volcanic activity, uplift, erosion, and glaciation, which still occur today. The slow pace of these geological processes indicated that the Earth was very, very old.

The ideas of Thomas Malthus were another important influence on Darwin. Malthus was a clergyman and economist who noted that populations increase in size geometrically ($2 \rightarrow 4 \rightarrow 8 \rightarrow 16 \rightarrow 32$ and so on) until checked by factors in the environment. In the case of humans, Malthus suggested that wars, famine, and pestilence served as the inevitable and necessary brakes on the growth of populations.

Darwin's years of observing the habits of animals and plants had introduced him to the struggle for existence described by Malthus. It occurred to Darwin that, in this struggle, favorable variations would tend to be preserved and unfavorable ones destroyed. As a result, a population would adapt to the environment and, eventually, new species would arise. Time was all that was required in order for new species to originate, and the geologists of the era, including Lyell, had supplied evidence that the earth was indeed old enough to provide an adequate amount of time.

Darwin had at last obtained a working theory, that of "survival of the fittest." He spent the next 20 years accumulating a tremendous body of facts demonstrating that evolution had occurred and formulating his arguments for natural selection.

As Darwin was pondering his ideas, Alfred Russel Wallace, who was studying the plants and animals of Malaysia and Indonesia, was similarly struck by the diversity of living things and the peculiarities of their distribution. Wallace arrived at a similar conclusion

contradict the accepted views. A number of fossils were discovered that did not have living counterparts. Then too, geological evidence suggested that the earth was far older than previously had been suspected. During the early 19th century, Charles Lyell proposed that

that evolution occurred by natural selection. In 1858, he sent a brief essay to Darwin, who was by then a world-renowned biologist, asking his opinion.

Darwin's friends persuaded him to have Wallace's paper presented along with an abstract of his own views, which he had prepared and circulated to a few friends several years earlier. Both papers were presented in July 1858 in London at a meeting of the Linnaean Society. Darwin's monumental book, *Origin of Species by Means of Natural Selection,* was published in November 1859.

Darwin's mechanism of natural selection consists of four observations about the natural world. (1) **Overproduction:** Each species produces more offspring than will survive to maturity. (2) **Variation:** There is variation among the offspring. It is important to remember that the variation necessary for evolution by natural selection is genetic and can be passed on to offspring. (3) **Competition,** or the "struggle for existence": Organisms compete with one another for the limited resources available to them. (4) **Survival** to reproduce, or "survival of the fittest": Those offspring that possess the most favorable combination of characteristics will be most likely to survive and reproduce. Natural selection results in the increase of "favorable" genes and the decrease of "unfavorable" genes within a population. Over time these changes may be significant enough to cause a new species to arise. In the next few chapters we will examine the role of natural selection in evolution in greater detail.

■ MANY TYPES OF SCIENTIFIC EVIDENCE SUPPORT EVOLUTION

Evolution is now supported by an enormous body of observations and experiments. In this text we can report only a small fraction of this wealth of evidence. Although biologists still do not agree completely on some aspects of the mechanism by which evolutionary changes occur, the concept that evolution has taken place is now well documented. It is consistent with all the information that has been brought to bear upon it.

Any scientific theory should lead to observations or testable predictions that, if not true, would require the theory to be modified or rejected. The concept of evolution is testable. For example, we would predict that fossils of early mammals would not be found in rocks that are as old as rocks containing the fossils of early fish. Instead, we would predict that fossils of mammals would be found in more recent rocks laid down later on top of the rock layers containing the fossil fish. As another example, we would predict that the sequence of amino acids in human hemoglobin would be very similar to that of the chimpanzee, but

less similar to the sequence in the hemoglobin of a horse or a whale. If such predictions proved to be untrue, the theory of evolution would be falsified and would need modification or replacement. However, all findings to date do conform with those predicted by the theory.

Comparative Anatomy Of Different Organisms Reveals Evidence Of Evolution

The study of the structure of any particular organ found in different but related organisms reveals a basic similarity of form that is varied to some extent from one group to another. For example, a bird's wing, a dolphin's front flipper, a bat's wing, and a human arm and hand, although superficially dissimilar, have a very similar arrangement of bones, muscles, and nerves (Fig. 17–7). Each has a single bone, the **humerus,** in the part of the limb nearest the trunk of the body, followed by the **radius** and **ulna,** the two bones of the forearm, then a group of **carpals** in the wrist, and a variable number of digits. This similarity is particularly striking because wings, flippers, and the human arm are used in different ways for different functions, and there is no mechanical need for them to be so similar. Similar arrangements of parts of the forelimb are evident in the ancestral reptiles and amphibians and even in the first fishes to come out of water onto land. Darwin pointed out that such basic structural similarities in organs used in different ways are precisely the outcome we would expect if evolution has taken place. Organs that are similar in form in different organisms owing to a common evolutionary origin are termed **homologous.**

With the acceptance of the theory of evolution, biologists came to realize that the homology of organs is due to their common evolutionary origin. Both bird and bat wings evolved from the forelimb of a common vertebrate ancestor. However, the flying surfaces of their wings are quite different. Feathers grow out from the posterior margin of the wings on the bird, while the flight surface of the bat's wing is essentially a webbed hand. Flight evolved independently in the two groups. Therefore, although both birds and bats use the forelimbs as wings, the limbs are modified in quite different ways.

Organs that are not homologous but simply have similar functions in different organisms are termed **analogous** organs. For example, the lungs of mammals and the trachea of insects are analogous organs that have evolved to meet, in quite different ways, the common problem of obtaining oxygen. The wings of various unrelated flying animals, such as insects and vertebrates, resemble one another superficially. In more fundamental aspects, however, the wings are quite dif-

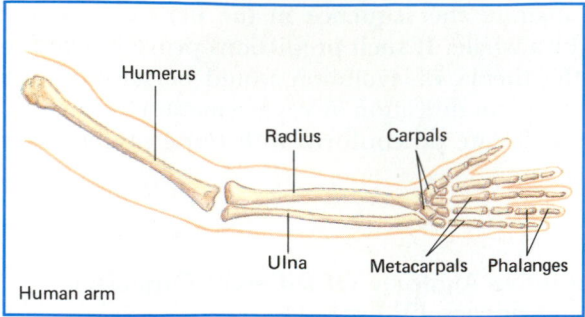

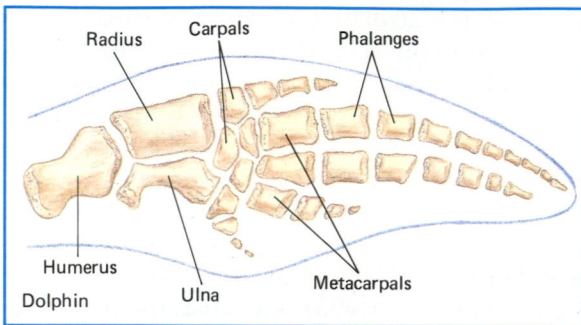

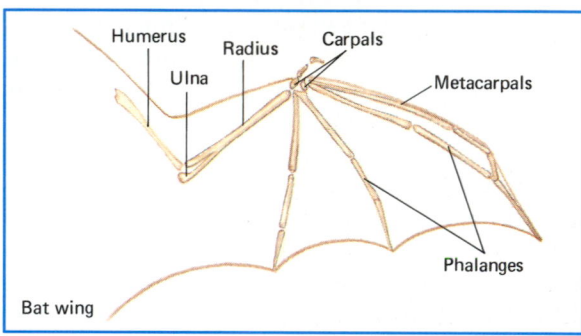

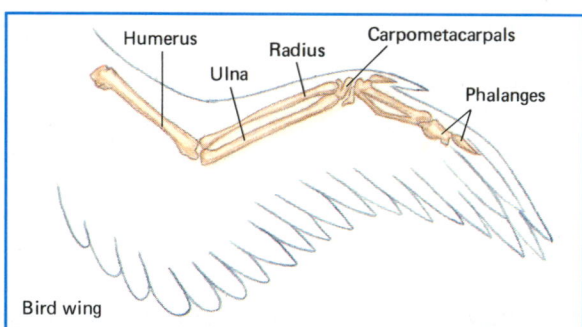

FIGURE 17–7   Homology. The human arm, dolphin flipper, bat wing, and bird wing are homologous because they have a basic, underlying structural similarity.

(a)

(b)

FIGURE 17–8   Analogous structures. The wings of birds and insects, although used for similar functions, have no underlying structural similarity. (*a*, Frans Lanting; *b*, Biophoto Associates)

ferent. Vertebrate wings are modified forelimbs supported by bones, while insect wings are outgrowths of the upper wall of the thorax and are supported by chitinous veins (Fig. 17–8).

The existence of homologous organs is important evidence of evolution. They are used to determine the interrelationships of living organisms. The existence of analogous organs is also crucial proof of evolution and adaptation. Comparisons of organisms with analogous organs indicate they have separate ancestries. Analogous organs are of evolutionary interest because they show how unrelated groups may adapt to common problems as their evolution leads to their convergence in similar environments.

Comparative anatomy also demonstrates the existence of **vestigial organs** (a vestige is a trace of something once present). Many organisms contain such organs or parts of organs that are seemingly nonfunctional and degenerate, often undersized or

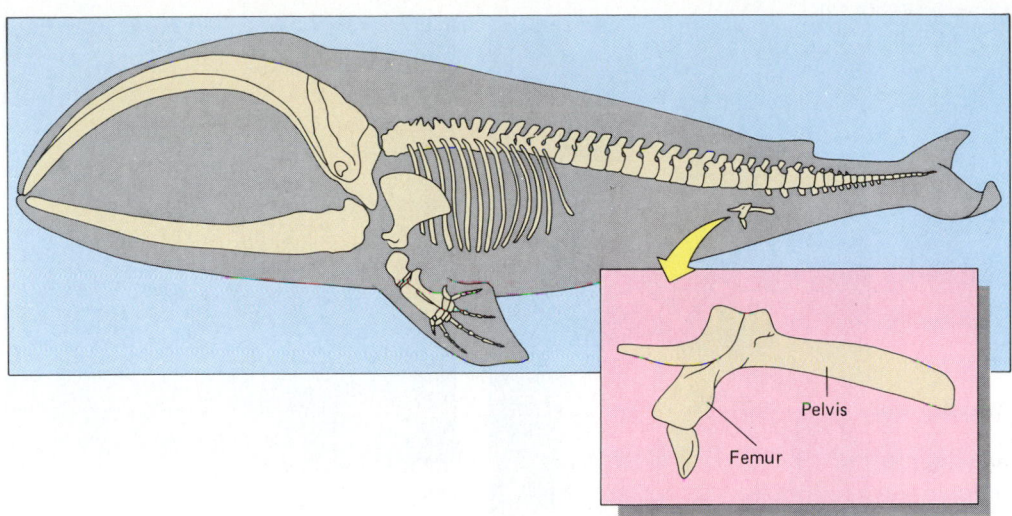

FIGURE 17–9 **The pelvis and femur of a whale are vestigial organs.**

Pelvis

Femur

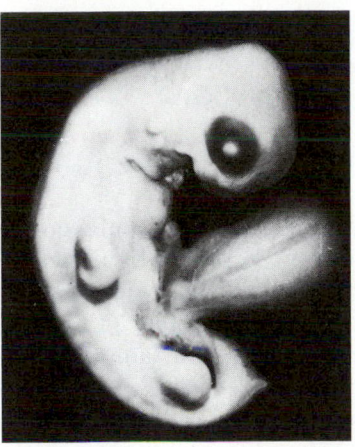

Turtle

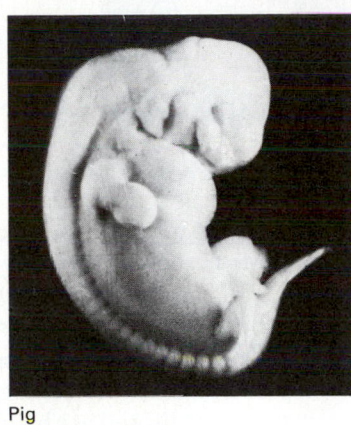

Pig

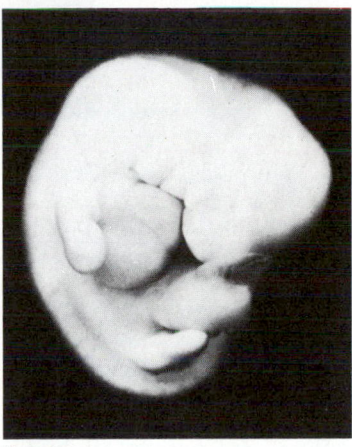

Human

FIGURE 17–10 **The early stages of embryonic development in several vertebrates. Numerous structural similarities are shared in the early stages, including the presence of a tail and gill pouches.** (Robert Pugh)

lacking some essential part. In the human body there are more than 100 structures that have been viewed as vestigial, including the coccyx (fused tail vertebrae), the wisdom teeth, and muscles that move the ears. Whales and pythons have vestigial hind leg bones (Fig. 17–9); wingless birds have vestigial wing bones; many blind, burrowing, or cave-dwelling animals may have vestigial eyes, and so on.

The occasional presence of a vestigial organ is to be expected as a species evolves and adapts to different modes of life. Some organs become much less important for survival and may end up as vestiges. When an organ loses all or most of its function, there is no longer any selective advantage in possessing it. At the same time, selective pressure for getting rid of the vestigial organ altogether is weak, and so the organ tends to remain.

Developmental Biology Provides More Clues To The Evolution Of Species

Embryos of animals that are very different in adult form can be strikingly similar. In fact, it is difficult to distinguish the early embryos of a turtle, mouse, chick, pig, or human (Fig. 17–10). Segmented muscles, gill pouches, a tubular heart without left and right sides, a system of aortic arches in the gill region, and many other features are found in the embryos of all vertebrates. However, none of these features remains in the adults of reptiles, birds, or mammals.

Why are these fish-like features present in the embryos of reptiles, birds, and mammals? Since the higher vertebrates evolved from fish, they share the fish's basic pattern of development. The accumulation of genetic changes since the fish diverged in evolution

(a)

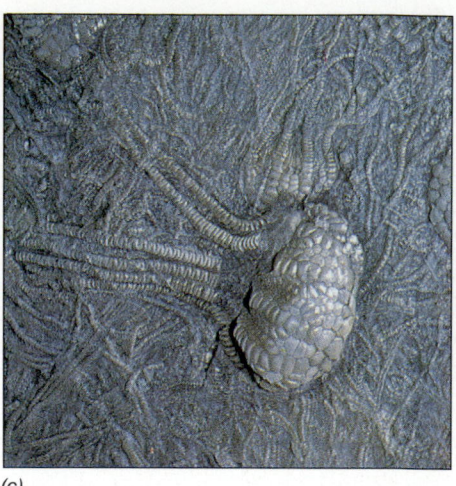

(c)

(b)

FIGURE 17–11   Several types of fossils.
(*a*) **Impression fossil of a fern leaf.**
(*b*) **Petrifaction of wood from the Petrified Forest National Park in Arizona. Cellular details are preserved in the fossils.** (*c*) **Cast fossil of crinoids (ancient echinoderms).**
(*a*, Arthur M. Siegelman; *b*, David Muench, 1988; *c*, E.R. Degginger)

from the higher vertebrates modifies the pattern of development of the higher vertebrate embryos.

Fossils Provide Still More Clues About Evolution

Perhaps the most direct evidence for evolution comes from the sciences of geology and paleontology. **Geology** deals with studies of the earth and its history. **Paleontology** is the science of discovery, identification, and interpretation of fossils.

The term **fossil** (Latin: *fossilis,* something dug up) refers not only to the parts of an animal's or plant's body that may survive, but also to any impression or trace left by previous organisms. If the body part has been trapped in sediments without being completely decomposed, the fossil is known as a **compression.** Some organic material still remains in compressions. If the pressure and heat is great during the formation of rock in which the organism is embedded, all of the organic material may be "vaporized." In this case, all that is left is an **impression** of the original plant or animal (Fig. 17–11).

The most common vertebrate fossils are skeletal parts. From the shapes of the bones and the positions of the bone scars that indicate points of muscle attachment, paleontologists can infer an animal's posture and style of walking, the position and size of its muscles, and the contours of its body. By a careful study of the fossil remains, paleontologists can reconstruct what an animal probably looked like in life.

In some fossils, the original hard parts, or even the soft tissues of the body, may be replaced by minerals. Iron pyrites, silica, and calcium carbonate are some of the common minerals that infiltrate buried tissues. These are known as **petrifactions.** The famous petrified forest of Arizona consists of trees that were buried and infiltrated with minerals (Fig. 17–11). The muscles of a shark more than 350 million years old were so well preserved by this process that details of individual muscle fibers could be observed under the microscope.

Molds and **casts** are produced in a different fashion. Molds are formed by the hardening of the material surrounding the buried organism, followed by the decay and removal of the tissues. The mold may subsequently be filled by minerals that harden to form casts that are replicas of the original structures (Fig. 17–11).

Footprints or trails made in soft mud that later hardened are a common type of fossil. From such remains, the paleontologist can infer something of the structure and locomotion of the animal that made them. Thus, fossils provide a record of animals and plants that lived earlier, some understanding of where and when they lived, and an idea of the kind of environment in which they lived. When enough fossils of organisms of different geological ages have been found, we can trace the lines of evolution that gave rise to those organisms.

Some more recent animal remains have been exceptionally well preserved by being embedded in bogs, tar, amber (ancient tree resin), or ice. The remains of some wooly mammoths deep-frozen in Siberian ice for more than 25,000 years were so well preserved that their deoxyribonucleic acid (DNA) could be analyzed.

The formation and preservation of a fossil require that an organism be buried under conditions that will slow the process of decay. This is most likely to occur if an organism's remains are covered quickly by fine particles of soil suspended in water. The soil particles are deposited as sediment around the animal or plant and cover it. Remains of aquatic organisms may be trapped in bogs, mud flats, sand bars, or deltas. Remains of terrestrial organisms that lived on a flood plain may also be covered by water-borne sediments or, if the organism lived in an arid region, by wind-blown sand. Animals may be trapped in a tar pit as in La Brea in Los Angeles or be covered by volcanic ash as in Pompeii after the eruption of Mount Vesuvius.

Because of the conditions required for preservation, the fossil record is not a random sample of past life. There is bias in the record toward aquatic organisms and those living in the few terrestrial habitats conducive to fossil formation. For example, relatively few fossils of forest animals have been found because plant and animal remains decay very rapidly on the forest floor, before fossils can form. Another reason for bias in the fossil record is that those organisms with hard body parts like bones and shells are more likely to form fossils.

To be interpreted, the sedimentary layers containing fossils must be arranged in chronological order. The layers of sedimentary rock, if they have not been disturbed, occur in the sequence of their deposition, with the more recent layers on top of the older, earlier ones. However, geological events that occur after the rocks have been formed initially may change the relationship of some of the layers.

Geologists can identify specific sedimentary layers of rock by features such as their mineral content, their position in the layers, and by certain key invertebrate fossils, known as **index fossils,** that characterize a specific layer over large geographical areas. With this information, geologists can arrange layers and the fossils they contain in chronological order and identify comparable layers in widely separated localities.

Biogeography, The Distribution Of Organisms On The Earth, Also Supports Evolution

If climate and topography were the only factors determining their distribution, most plant and animal species would be found everywhere that they *could* survive. But this is not the case. Central Africa, for example, has elephants, gorillas, chimpanzees, lions, and antelopes, while Brazil, with a similar climate and environmental conditions, has none of these. South America does have prehensile-tailed monkeys, sloths,

and tapirs, however. The present distribution of organisms seems understandable only in terms of evolution.

The **range** of a given species—that is, the portion of the earth over which it is found—may be only a few square miles or, as with humans, almost the entire planet. In general, closely related species do not have identical ranges, nor are their ranges far apart. They are usually adjacent, but separated by a barrier of some sort, such as a mountain or desert.

As we might expect, then, regions such as Australia and New Zealand, which have been separated from the rest of the continents for a long time (Chapter 20), have plants and animals specific to these areas. Australia has populations of monotremes (egg-laying mammals) and marsupials (mammals that raise young in pouches) found nowhere else. During the Mesozoic era (Chapter 20), Australia was isolated from the rest of the world. Its primitive mammals, therefore, never had any competition from the later-evolving placental mammals (mammals that develop a placenta in the uterus to nourish the unborn young). Placental mammals are thought to have competitively eliminated the monotremes and most of the marsupials everywhere else they may have existed. The original Australian mammals gave rise to a variety of forms that were able to take advantage of the different habitats available (Fig. 17–12).

The kinds of animals and plants found on oceanic islands in general resemble those of the nearest mainland; yet they include some species found nowhere else. Darwin studied the plants and animals of the Cape Verde Islands, some 400 miles west of Dakar, Africa, and of the Galapagos Islands, a comparable distance west of Ecuador. On each island group, the plants and terrestrial animals were unique to that group of islands. However, those of the Cape Verdes Islands resembled African species, while those of the Galapagos resembled South American species.

Darwin concluded that organisms from the neighboring continent migrated, or were carried, to the islands and subsequently evolved into new species. The animals and plants found on oceanic islands are only those that could survive the trip there. Even though there are woodland spots ideally suited for such creatures, there are no frogs or toads on the Galapagos, because neither the animals nor their eggs can survive exposure to sea water. There are no terrestrial mammals either, although there are many bats, as well as land and sea birds. The occurrence of these particular forms—closely related to, yet not identical with, those of the Ecuador coast—suggests strongly that evolution has modified the descendants of the first animals and plants to reach the islands.

The study of the distribution of plants and animals is the science of **biogeography.** One of its basic

Placentals	Marsupials
Lemur (*Lemur*)	Cuscus (*Phalanger*)
Anteater (*Myrmecophaga*)	Anteater (*Myrmecabius*)
Mouse (*Mus*)	Mouse (*Dasycerus*)
Flying squirrel (*Glaucomys*)	Flying phalanger (*Phalanger*)
Wolf (*Canis*)	Tasmanian wolf (*Thylacinus*)
Mole (*Talpa*)	Mole (*Notoryctes*)
Cat (*Felis*)	Cat (*Dasyurus*)

FIGURE 17–12 Parallel adaptations among Australian marsupials and placental mammals found in the rest of the world.

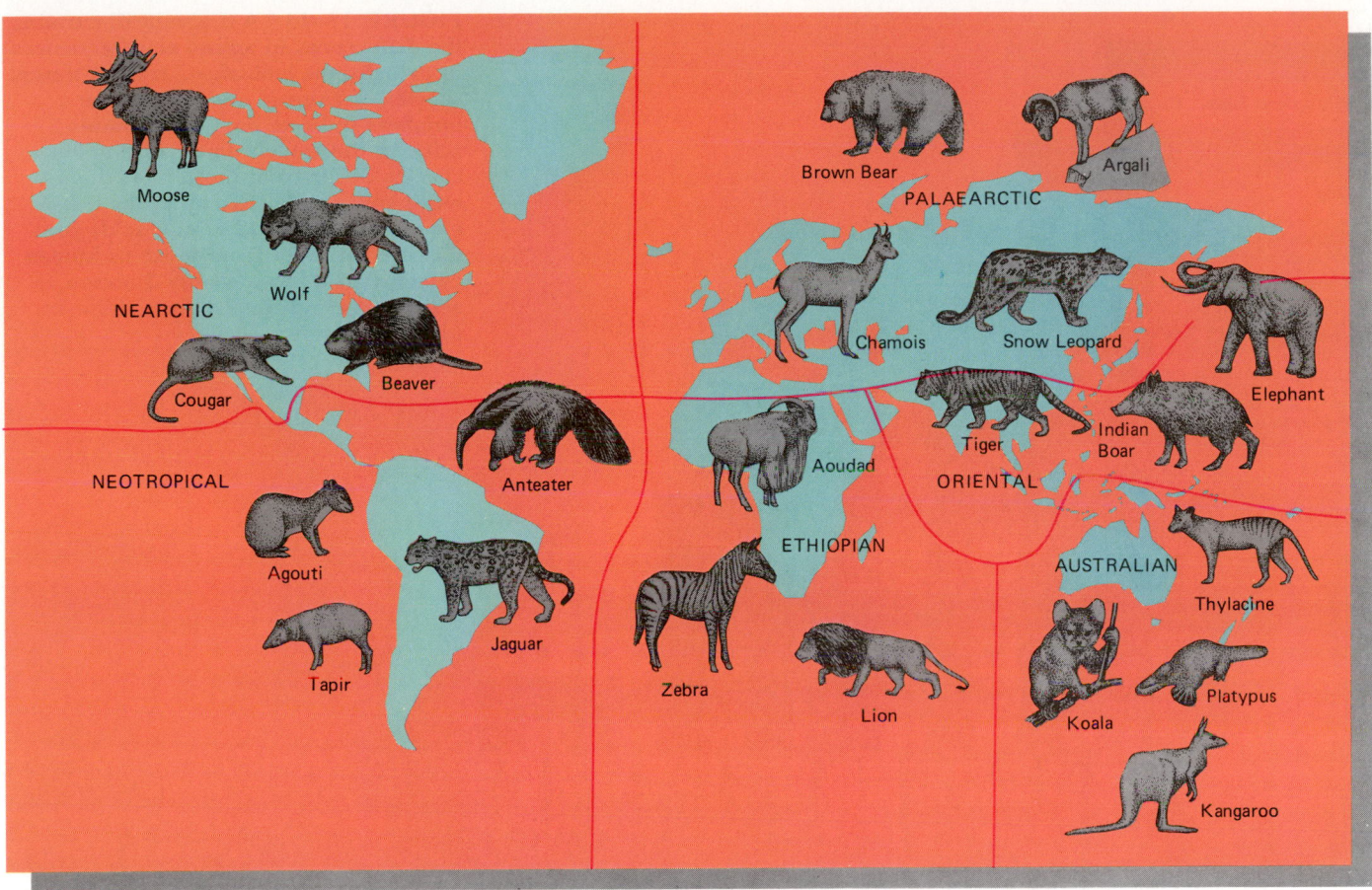

FIGURE 17–13 Biogeographical realms. Large parts of the earth have characteristic animal species and genera, which are different from those in other parts of the world. These biogeographical realms are the direct outcome of the centers of origin of certain species, of their past migrations, and of the barriers they encountered. Kangaroos, for example, occur naturally in Australia but not in the adjacent oriental realm.

tenets is that each species of animal and plant originated only once. The particular place where this occurred is known as the species' **center of origin.** The center of origin is not a single point but the range of the population when the new species was formed. From its center of origin, each species spreads out until halted by a barrier of some kind—physical, such as an ocean or mountain; environmental, such as an unfavorable climate; or ecological, such as the presence of organisms that compete with it for food or shelter (Fig. 17–13).

Comparisons Of The Biochemistry And Molecular Biology Of Different Organisms Offer Evidence Of Evolution

It has become clear that similarities and differences in the biochemistry and molecular biology of organisms closely parallel structural similarities and differences. Indeed, if we established evolutionary relationships based solely on biochemical and molecular features instead of the usual morphological (structural) ones, the end result would be a very similar family tree.

The blood serum of each vertebrate species contains specific proteins, coded for by specific genes. The degree of similarity of these proteins from one species to the next can be determined by laboratory tests. When serum proteins are compared, our closest "blood relations" are found to be the great apes and then, in order, the Old World monkeys, the New World monkeys, and, finally, the tarsioids and lemurs. The biochemical relationships of a variety of organisms tested in this way correlate with and complement evolutionary relationships determined by other means. Cats, dogs, and bears have very similar plasma proteins. Cows, sheep, goats, and deer constitute another group with closely related serum proteins. Similar tests of crustaceans, insects, and mollusks have shown that the forms regarded as being closely related from anatomical or paleontological evidence have comparably similar serum proteins.

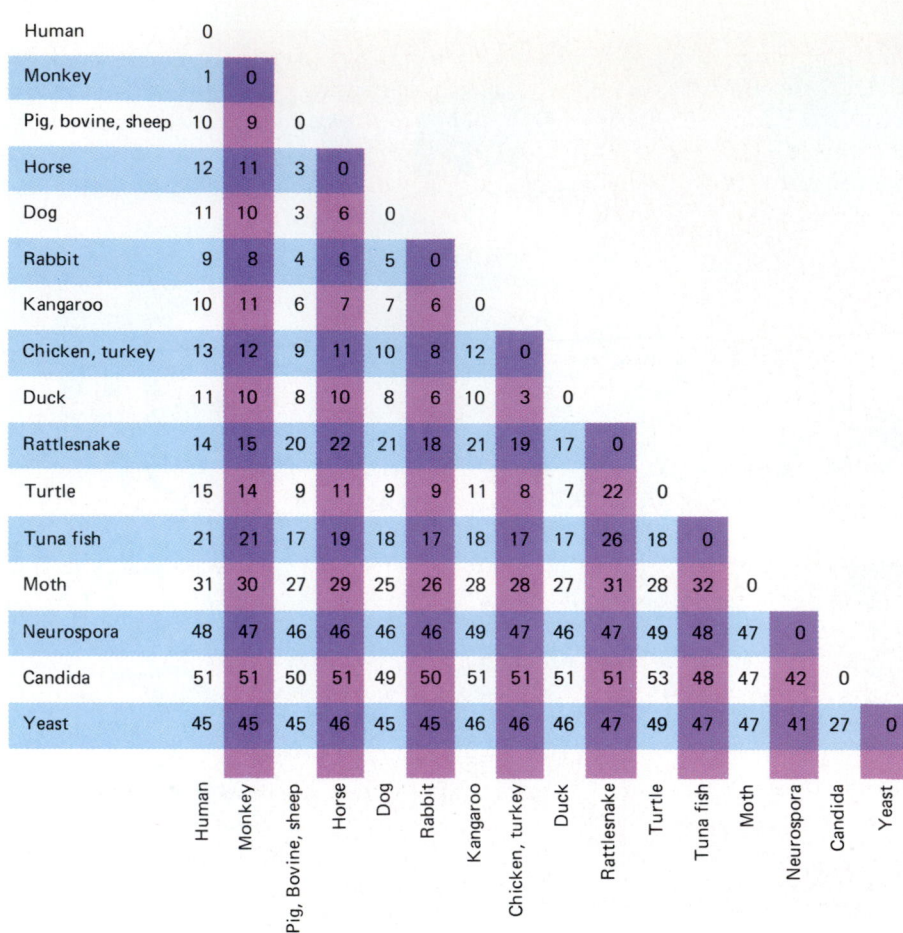

	Human	Monkey	Pig, Bovine, sheep	Horse	Dog	Rabbit	Kangaroo	Chicken, turkey	Duck	Rattlesnake	Turtle	Tuna fish	Moth	Neurospora	Candida	Yeast
Human	0															
Monkey	1	0														
Pig, bovine, sheep	10	9	0													
Horse	12	11	3	0												
Dog	11	10	3	6	0											
Rabbit	9	8	4	6	5	0										
Kangaroo	10	11	6	7	7	6	0									
Chicken, turkey	13	12	9	11	10	8	12	0								
Duck	11	10	8	10	8	6	10	3	0							
Rattlesnake	14	15	20	22	21	18	21	19	17	0						
Turtle	15	14	9	11	9	9	11	8	7	22	0					
Tuna fish	21	21	17	19	18	17	18	17	17	26	18	0				
Moth	31	30	27	29	25	26	28	28	27	31	28	32	0			
Neurospora	48	47	46	46	46	46	49	47	46	47	49	48	47	0		
Candida	51	51	50	51	49	50	51	51	51	51	53	48	47	42	0	
Yeast	45	45	45	46	45	45	46	46	46	47	49	47	47	41	27	0

FIGURE 17–14 A diagram illustrating the differences in amino acid sequences in cytochrome *c* obtained from different species of animals and fungi. The numbers indicate how many amino acids in the cytochrome *c* of a given species differ from those of various other species. For example, the horse cytochrome *c* has 12 amino acids in it that are different from those in human cytochrome *c* but has only 3 amino acids that are different from those in pig cytochrome *c*.

Investigations of the sequence of amino acids in proteins, such as hemoglobin or cytochrome, obtained from different species have revealed great similarities and certain specific differences. The pattern of these differences demonstrates the nature and number of underlying mutations that must have occurred in evolution. Mutations occur, on average, at a certain rate within a given taxonomic group. Comparing the number of alterations in the nucleotide sequence of one organism to those of another, we can estimate the age of a species or higher taxonomic group. The evolutionary relationships of organisms inferred from such biochemical studies parallel the relations inferred earlier on the basis of structural similarities and fossil evidence (Fig. 17–14).

Advances in molecular biology, including the development of methods to determine the sequence of the nucleotide base pairs in DNA, have provided another means of demonstrating evolutionary relationships (Table 17-1). We would expect that species that are believed to be closely related on the basis of other evidence would also have a greater proportion of their DNA nucleotides in common than distantly related species. This has been shown in a number of investigations. The more closely species are believed to be re-

TABLE 17–1		
Differences In Nucleotide Sequences In DNA as Evidence of Phylogenetic Relations		
Species Pairs		**Percentage Differences in Nucleotide Sequences Between Pairs of Species**
Human–chimpanzee		2.5
Human–gibbon		5.1
Human–Old World monkey		9.0
Human–New World monkey		15.8
Human–lemur		42.0

From G.L. Stebbins, *Darwin to DNA, Molecules to Humanity*, San Francisco, W.H. Freeman, 1982.

lated on the basis of other evidence, the greater is the percentage of DNA sequences that they have in common.

Darwin speculated that all forms of life are related through descent with modification from earliest organisms. This speculation has been verified as we have learned more about molecular biology. Even organisms that are very remotely related, such as humans

(*Homo sapiens*) and the bacterium *Escherichia coli,* have some proteins such as cytochrome *c* in common. In the course of the long, independent evolutions of the two organisms, mutations have resulted in substitution of amino acids at various locations in the protein, but the cytochrome *c* molecules of all species are clearly similar in structure and function. Further evidence that all life is related comes from the universality of the genetic code. The genetic code has been passed along essentially unchanged through all the branches of the evolutionary tree since its origin in an extremely early form of life.

☐ CHAPTER SUMMARY

I. Evolution is a genetic change in a population of organisms.

II. Charles Darwin proposed the theory of evolution by natural selection, which is based on four observations.
   A. Overproduction: Each species produces more offspring than will survive to maturity.
   B. Variation: Genetic variation exists among these offspring.
   C. Competition: Organisms compete with one another for the resources needed for life, i.e., food, space, habitat.
   D. Survival to reproduce: The offspring with the most favorable combinations of genetic characters are most likely to survive and reproduce, passing those genetic characters to the next generation.

III. The concept that evolution has taken place is now well documented.
   A. Evidence supporting evolution is derived from comparative anatomy.
      1. Homologous organs have basic structural similarities, even though the organs may be used in different ways. Homologous organs indicate close evolutionary ties between the organisms possessing them.
      2. Analogous organs have similar functions but are not homologous and do not indicate close evolutionary ties.
      3. The occasional presence of a vestigial organ is to be expected as an ancestral species evolves and adapts to different modes of life.
   B. Developmental biology provides evidence for evolution.
      1. The resemblance between embryos of different animals is closer than the resemblance between their adults.
      2. The accumulation of genetic changes since organisms diverged in evolution modifies the pattern of development of higher vertebrate embryos.
   C. Perhaps the most direct evidence for evolution comes from paleontology.
      1. Fossils are remains or traces of ancient animals and plants.
      2. There are several types of fossils: compressions, impressions, molds, casts, and petrifactions.
   D. Biogeography, the distribution of plants and animals, supports evolution.
      1. Areas that have been separated from the rest of the world for a long time have plants and animals specific to those areas.
      2. Each species originated only once, at its center of origin.
      3. From its center of origin, each species spread out until halted by a barrier of some kind.
   E. Biochemistry and molecular biology provide compelling evidence for evolution.
      1. Blood sera of closely related vertebrates are more similar than sera of distantly related vertebrates.
      2. The sequence of amino acids in common proteins such as cytochrome or hemoglobin reveals greater similarities in closely related species.
      3. A greater proportion of the sequence of nucleotides in DNA is identical in closely related organisms.
      4. The universality of the genetic code is further evidence that all life is related.

☐ POST-TEST

1. Genetic change in a population of organisms is known as _____.

2. _____ evolve, but individuals do not.

3. All the genes present in an entire population of a species is its _____ _____.

4. The first to propose that organisms evolve as a result

of some natural phenomenon rather than divine intervention was _____ .

5. Thomas Malthus believed that populations increase in numbers _____ .

6. Inherent in _____'s theory of evolution is the concept that organisms have the potential to produce more offspring than will survive to reproductive maturity.

7. The wings of butterflies and bats, which have similar functions but are quite different in structure, are said to be _____ .

8. An organ that appears to have little or no function, and is smaller than a similar, fully functional equivalent in the organism's ancestor or relatives, is known as a _____ organ.

9. A fossil in which the body part has been trapped in sediments without being completely decomposed is known as a _____ .

10. Geologists can identify specific sedimentary layers of rock by certain key invertebrate fossils, known as _____ _____ .

11. The portion of the earth over which a given species is found is its _____ .

REVIEW QUESTIONS

1. Explain briefly the concept of biological evolution.
2. Propose a mechanism whereby DDT resistance in mosquitos could develop.
3. In what ways does Lamarck's theory of adaptation not agree with present evidence?
4. Consider the giraffe's long neck. Explain how this came about using Lamarck's theory of evolution. Then explain the giraffe using Darwin's mechanism of natural selection.
5. Why is it that only inherited variations are important in the evolutionary process?
6. How can you account for the fact that both Darwin and Wallace independently and almost simultaneously proposed similar theories of evolution by natural selection?
7. List as many vestigial structures in the human body as you can.
8. Distinguish between the different kinds of paleontological evidence used to support evolution.
9. Discuss the factors that might interfere with our obtaining a complete and unbiased picture of life in the past from a study of the fossil record.
10. Explain why marsupials are widespread in Australia and almost nonexistent elsewhere.

RECOMMENDED READINGS

Darwin, C.R. *On the Origin of Species by Means of Natural Selection or the Preservation of the Favored Races in the Struggle for Life*. New York, Cambridge University Press, 1975. A readily obtainable reprint of one of the most important books of all time. Darwin's long essay is still of great significance to modern readers.

Nelson, G., and N. Platnick. *Biogeography*. North Carolina, Carolina Biological Reader, Carolina Biological Supply Company, 1984. A detailed explanation of biogeographical concepts.

Stanley, S.M. *Earth and Life Through Time*. San Francisco, W.H. Freeman, 1985. A presentation of evolution for the general reader.

18
Population Genetics

OUTLINE

I. The Hardy-Weinberg law explains gene frequencies in the gene pool of a population that is not evolving
II. Evolution occurs when there are changes in gene frequencies in a gene pool
   A. Mutation increases variation in the gene pool, thereby causing changes in gene frequencies
   B. Genetic drift causes changes in gene frequencies by random, or chance, events
   C. Gene flow, which changes the amount of variation in the gene pool, is caused by the differential migration of organisms
   D. Natural selection changes gene frequencies in a way that leads to adaptation to the environment
III. Natural selection increases the fitness of a species for the environment in which it lives
   A. Stabilizing selection selects against phenotype extremes and favors phenotypes near the mean, or average
   B. Directional selection results in the change from one phenotype to another
   C. Disruptive selection separates the population into several distinct phenotypes
Focus on Evolution of the Africanized Honeybee

LEARNING OBJECTIVES

After you have studied this chapter, you should be able to:
1. Distinguish between the gene pool of a population and the genotype of an individual.
2. Explain the dependence of evolution upon genetics.
3. State the Hardy-Weinberg law and discuss its significance in population genetics.
4. Explain how each of the following alters the gene frequencies in populations: mutation, genetic drift, gene flow, and natural selection.
5. Distinguish between stabilizing selection, directional selection, and disruptive selection and describe how each plays a role in evolution.

Black-browed albatross chicks. (Frans Lanting)

Evolution is the change in a population of organisms over a period of time. This change is inherited from one generation to the next. Although Darwin recognized that evolution occurred in populations, he did not understand how genetics worked in this regard. One of the most significant advances in biology since Darwin's time has been the discovery of the genetic basis of evolution.

Each species possesses an isolated **gene pool,** which includes all possible alleles at each locus of each chromosome present in the individuals that make up the population (Fig. 18–1). Because most eukaryotic species are diploid, each individual member of a population contains only two alleles at each locus, one on each homologous chromosome, and may be either homozygous or heterozygous at each locus. Therefore, an individual within a population has only some of all the genes present in its gene pool. Moreover, the variation in individuals in a given population indicates that each individual has a different portion of the genes that exist in the gene pool.

If a population is not evolving, frequencies of each allele remain constant from generation to generation. However, changes in allele frequencies over successive generations indicate that evolution has occurred. This type of evolution is sometimes referred to as **microevolution,** because it involves changes taking place *within* a population. Microevolutionary changes are relatively small; they are not large enough to result in the formation of a new species. In this chapter, we examine the factors responsible for microevolution after first considering the genetics of a population that is not evolving.

KEY CONCEPTS

◻ Evolution occurs in populations, not individuals.

◻ Populations exhibit genetic variation, which may be expressed as gene frequencies.

◻ Populations evolve by changes in their gene frequencies.

◻ THE HARDY-WEINBERG LAW EXPLAINS GENE FREQUENCIES IN THE GENE POOL OF A POPULATION THAT IS NOT EVOLVING

Suppose we were doing a study of a population of plants in a field. The environment is assumed to be stable. If we were to count the number of plants with red petals versus white petals, we might find that we have 182 plants with red petals and 18 with white pet-

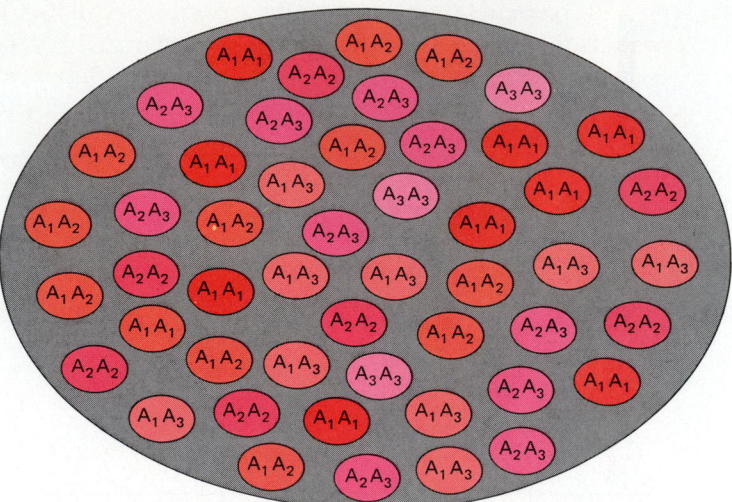

FIGURE 18–1 A gene pool. This drawing shows only one genetic locus (A), with three different alleles possible at that locus (A_1, A_2, A_3). Because each individual (represented by the small circles) is diploid, it will possess only two alleles for each genetic locus.

als, a ratio of 9 to 1. If we came back to the same field the following summer, we would find a population that is essentially the same as the previous one, with roughly nine red-flowered plants to every white-flowered plant. If we did this for a series of generations, we would always get the same result.

The explanation for this stability of populations in successive generations was provided in 1908 by G.H. Hardy, an English mathematician, and W. Weinberg, a German physician, who independently pointed out that in a freely interbreeding population that mates randomly with respect to two allelic genes, the genotypes of the population may be described by the following expression:

$$p^2 + 2\,pq + q^2$$

where

p = the frequency of the dominant allele
q = the frequency of the recessive allele
p^2 = the frequency of the homozygous dominant genotype
$2pq$ = the frequency of the heterozygote
q^2 = the frequency of the homozygous recessive genotype.

Imagine a population of organisms that mate at random with respect to two alleles—*A*, whose frequency is p, and *a*, whose frequency is q. Their combined frequency is 1, so we may write:

$$p + q = 1$$

so

$$p = 1 - q \text{ and } q = 1 - p \qquad (1)$$

A gamete bearing gene *A* may combine with a similar gamete to form an *AA* zygote. Similarly, $A \times a \rightarrow Aa$, $a \times A \rightarrow Aa$, and $a \times a \rightarrow aa$. The frequency of each combination is the product of the frequencies of its component genes, as follows:

	p(A)	q(a)
p(A)	p²(AA)	pq(Aa)
q(a)	pq(Aa)	q²(aa)

When we add these, the total of all genotypes is as follows:

$$p^2 + 2\,pq + q^2 = 1 \qquad (2)$$

where p^2 = frequency of *AA*, 2 pq = frequency of *Aa*, and q^2 = frequency of *aa*.

For the next generation, the present *Aa* organisms will be able to produce *either A* or *a* gametes; that is, half the gametes will be *A* and half *a* from this source. The *AA* organisms will, of course, yield only *A* gametes, and the *aa* organisms will yield only *a* gametes. Thus, the total frequency of all *A* and *a* genes may be expressed by the following equations, where p' stands for the new frequency of *A* and q' for the new frequency of *a* (in the succeeding generation):

$$
\begin{aligned}
p' &= p^2 + 1/2(2\,pq) \qquad (3)\\
&= p^2 + pq\\
&= p^2 + p(1 - p)*\\
&= p^2 + p - p^2\\
&= p
\end{aligned}
$$

$$
\begin{aligned}
q' &= q^2 + 1/2(2\,pq) \qquad (4)\\
&= q^2 + pq\\
&= q^2 + q(1 - q)*\\
&= q^2 + q - q^2\\
&= q
\end{aligned}
$$

This demonstrates that, if left undisturbed, gene frequencies in a randomly mating population do not change from generation to generation, regardless of dominance or recessiveness. Any population in which the distribution of alleles *A* and *a* conforms to the relationship $p^2 + 2\,pq + q^2$, whatever the absolute values for p and q may be, is in genetic equilibrium. No evolution is taking place within that population.

Mendel's laws, as we have seen, describe the frequency of genotypes among offspring of a single mating pair. In contrast, the Hardy-Weinberg law describes the frequencies of alleles in the genotypes of an entire breeding population. The Hardy-Weinberg law shows that, in large populations, sexual reproduction alone will not cause changes in the gene frequencies. Thus, knowledge of the Hardy-Weinberg law is essential for an understanding of the mechanisms of evolutionary change in sexually reproducing populations.

*See Equation (1).

The proportion of alleles in successive generations in a situation of genetic equilibrium will be the same, provided the following conditions exist:

1. **Random mating.** The three types of genotypic individuals (*AA*, *Aa*, *aa*) must not select their mates on the basis of genotype. There must be equal probabilities of mating between genotypes.
2. **No mutations.** In this instance, there must be no mutations of *A* or *a*.
3. **Large size.** The population of individuals must be large enough so that the laws of probability function.
4. **Isolation.** There can be no exchange of genes with other populations that might have different gene frequencies.
5. **No natural selection.** If natural selection is occurring, certain alleles will be favored over others and the gene frequencies will change.

■ EVOLUTION OCCURS WHEN THERE ARE CHANGES IN GENE FREQUENCIES IN A GENE POOL

In studies of populations in nature or in the laboratory, the Hardy-Weinberg law is used to test whether mating is random and whether evolutionary changes are taking place. If the members of a population are mating randomly, and if no other factors are affecting allele frequencies, then the frequencies of the various genotypes will be very close to those calculated with the Hardy-Weinberg formula.

However, the frequencies of the genotypes are often significantly different from those the Hardy-Weinberg law would predict. *Evolution, stated in its simplest terms, represents a departure from the Hardy-Weinberg law of genetic stability.* These changes in the gene pool of a population result from such phenomena as mutations, genetic drift, gene flow, and natural selection. Without one or more of these, genetic frequencies in a large, freely interbreeding population will not change from generation to generation.

Mutation Increases Variation In The Gene Pool, Thereby Causing Changes In Gene Frequencies

Variation is introduced into a gene pool through mutation, which is the source of all new alleles (Fig. 18–2). Mutations result from three types of changes: a change in the nucleotide base pairs of the gene, a rearrangement of genes within chromosomes so that their interactions produce different effects, or a change in the chromosomes.

(a)   (b)

FIGURE 18–2 The fruit fly, *Drosophila melanogaster.* (*a*) A normal fly. (*b*) A mutant with vestigial wings. Hundreds of mutations are known to occur in these little fruit flies, which breed rapidly enough to allow us to easily follow the transmission of such altered traits from generation to generation. In addition to those shown here, mutations affecting eye color, behavior, and many other traits have been described and studied. Since mutations are random changes in genetic material, most of them are bound to be harmful to the organism, as is obviously the case in the *vestigial* mutation shown here. Yet for island-dwelling insects, fully developed wings might well be more of a disadvantage than an advantage, permitting the insect to be too easily blown away from land. Perhaps for this reason flies and other insects that dwell on small oceanic islands frequently have reduced wings or are entirely wingless. (Runk/Schoenberger from Grant Heilman)

Mutations occur randomly and spontaneously. Because most mutations occur in body cells rather than in reproductive cells, they are not inherited. When the organism with the body-cell mutation dies, the mutation dies with it. Some mutations, however, alter the DNA in the reproductive cells. These mutations may or may not affect the offspring because most of the DNA in a cell is "silent" and does not code for specific polypeptides or proteins. If a mutation occurs in the DNA that codes for a polypeptide, it may still have little effect in altering the structure or function of that polypeptide. However, when the polypeptide is altered enough to change how it functions, the mutation is usually harmful.

Most mutations produce small changes in the phenotype that only sophisticated biochemical techniques can detect. By acting against seriously abnormal phenotypes, natural selection eliminates or reduces to low frequencies the major harmful mutations. Mutations causing small changes, even ones with slightly harmful phenotypic effects, have a better chance of being incorporated into the gene pool.

Although mutations do provide evolution's raw material, they do not determine the direction of evolutionary changes. Rather, as unfavorable alleles are continuously weeded out by selective pressures of the environment, the production of new mutations simply keeps up a genetic variability within the population. These mutations, in turn, are usually weeded out and replaced by other mutations.

Genetic Drift Causes Changes In Gene Frequencies By Random, Or Chance, Events

Genetic frequencies can remain constant only if a population is fairly large. Otherwise, random events will tend to cause changes. If a population consists of only a few individuals, predators could destroy the only representatives of a particular genotype and miss the others purely by chance. Such an event would be most unlikely in a large population. It is also a matter of chance whether a particular genotype will be represented in those gametes that do manage to unite in fertilization—a chance that depends upon the random distribution of maternal and paternal chromosomes in meiosis. The production of changes in gene frequency by random events other than mutations is known as **genetic drift** (Fig. 18–3). Genetic drift is a significant factor in small populations.

Even if a gene were of no particular advantage or disadvantage to an organism, its frequency could change by genetic drift. This process may continue until eventually only one of the alternative alleles is present in the population, the others having been eliminated. This is called **fixation.** Genetic drift is a hazard in small populations of organisms with limited gene pools. These organisms are in danger of extinction because the loss of genetic variability tends to reduce the versatility of the organisms, making them more susceptible to stressful changes in their environment.

If a small number of organisms colonize a new habitat, the only genes that will be represented among their descendants will be those few that the founders chanced to possess. Thus, isolated populations may have very different gene frequencies than other populations of the species. The disproportionate effect that a limited number of ancestors exerts on a population is termed the **founder effect** (Fig. 18–4). It can produce great changes in gene frequency even in the absence of natural selection.

In some species, very few individuals survive some critical stage of their life cycle. Among houseflies in northern areas, for example, only a few survive the winter, and they give rise to most of the summer population. In principle, this is similar to the founder effect. Only a few individuals, which perhaps do not truly represent the genetics of the population from which they came, will give rise to the entire future population. As the population again increases in size, the frequencies of many alleles may be quite different from

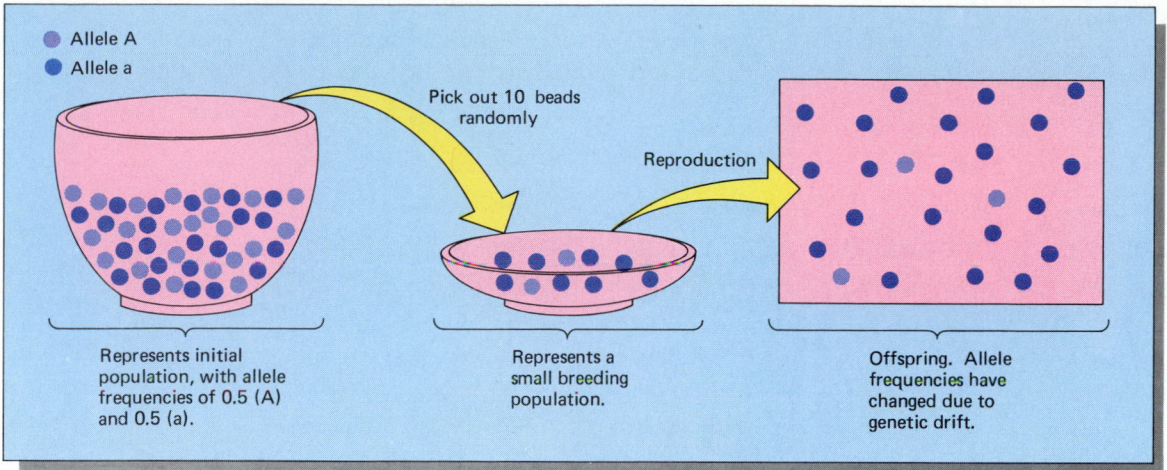

FIGURE 18–3 The smaller the breeding population, the more likely it is that allele frequencies will change. This phenomenon is known as genetic drift.

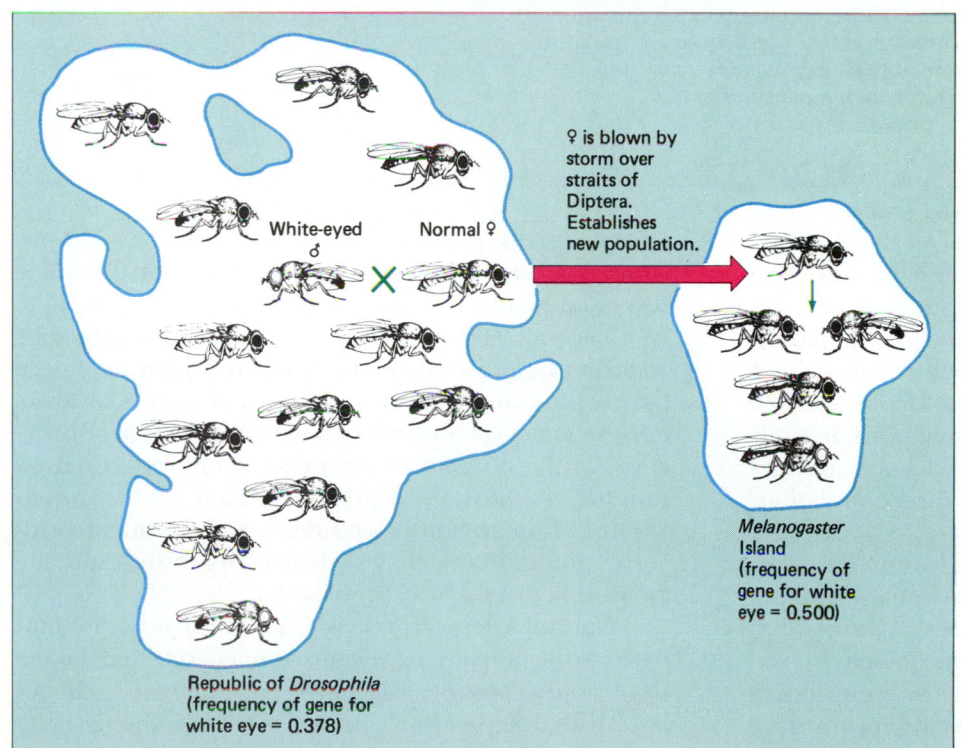

FIGURE 18–4 Founder effect. In this instance, a single female fruit fly is blown to an island. Even though not typical of the population of fruit flies as a whole, her genes and those of her mate will serve as the foundation for the entire gene pool of all fruit flies living on that island. Their gene frequencies will not be the same as those of the original population of fruit flies. In such cases, there is usually a greater range of variation among the ancestral population than among the island group.

those in the population preceding the decline. Because we can think of this phenomenon as a periodic squeezing out of some of the genes in a gene pool in random fashion, it is termed the **bottleneck effect** (Fig. 18–5). The bottleneck effect can change gene frequencies even in the absence of natural selection, usually by greatly reducing the genetic variability of a population. It has been shown, for example, that the genetic variability of African cheetahs is very low. This probably means the cheetah came close to extinction in fairly recent times, perhaps as recently as 3000 years ago.

Gene Flow, Which Changes The Amount Of Variation In The Gene Pool, Is Caused By The Differential Migration Of Organisms

Selective immigration (entering) or emigration (leaving) can change genetic frequencies, because it represents the arrival or departure of organisms that possess one genotype at a greater rate than others in the population. This changes the proportional representation of genotypes in the gene pool.

The migration of individuals causes a correspond-

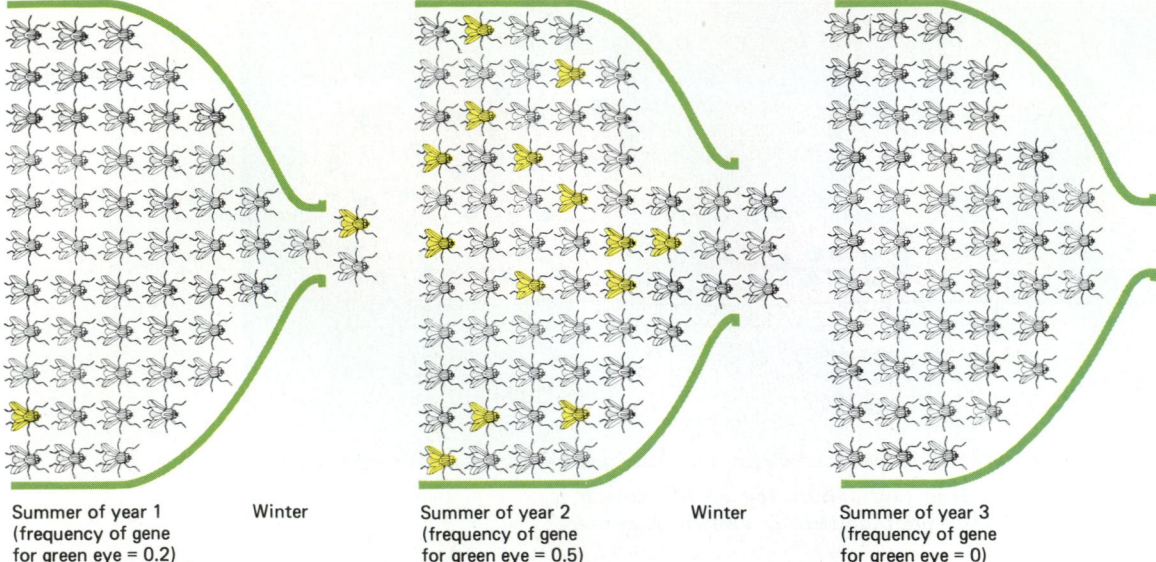

Summer of year 1
(frequency of gene
for green eye = 0.2)　　　Winter

Summer of year 2
(frequency of gene
for green eye = 0.5)　　　Winter

Summer of year 3
(frequency of gene
for green eye = 0)

FIGURE 18–5 **The bottleneck effect. Since only a small population of flies survives the winter, its genotypes, not necessarily resulting from natural selection, determine the genetic frequencies of the entire succeeding summer population. With results similar to the founder effect, descendant populations will have gene frequencies typical of their ancestors, but these may not be typical of the larger populations from which their ancestors came.**

ing movement of alleles, or **gene flow,** that can have significant evolutionary consequences. As alleles "flow" from one population to another, they increase the amount of variability within the population that receives them. If the gene flow between two populations is great enough, these populations will become more similar genetically. Because gene flow reduces the amount of variation between two populations, it tends to counteract the effects of natural selection and genetic drift, both of which cause individual populations to become increasingly distinct.

The amount of migration depends on patterns of breeding and dispersal in a species. Although the migration of certain animals (like birds) is apparent, many migrations are less obvious. Plant pollen, for example, may be carried long distances by wind. Seeds and fruits, which are formed after sexual reproduction has taken place, are often modified for dispersal, in some cases over long distances. Coconut fruits, for example, may be transported hundreds, or even thousands, of miles by ocean currents.

Natural Selection Changes Gene Frequencies In A Way That Leads To Adaptation To The Environment

We have said that genetic drift is random change in gene frequency. We have also said that gene flow and mutation may occur in a given direction, but the direction is unrelated to the nature of the environment.

Only natural selection is adaptive and brings the variability in gene pools into harmony with the environment. It checks the disorganizing effects of the other forces and leads to adaptation.

Natural selection may operate at any number of different times in the life cycle of an organism. There may be nonrandom mating (that is, preferential selection of mates), nonrandom fecundity (that is, differences in the number of offspring produced), or nonrandom survival to reproductive age. Nonrandom survival is particularly common and frequently involves subtle interactions between organisms and the environment in which they live.

Natural selection has two facets, for it both eliminates unfit individuals (negative selection) and favors the fit ones (positive selection). By eliminating the alleles with less favorable traits, selection changes the composition of the gene pool in a favorable direction and increases the probability that the favorable alleles responsible for an adaptation will come together in the same individuals.

To summarize, despite the Hardy-Weinberg law, gene frequencies are always changing. Nonrandom mating seldom prevails, mutation and natural selection never rest, and many collections of organisms in nature are divided into small, more or less isolated gene pools that are separated by geographical barriers. In most instances, then, microevolutionary changes in gene frequencies are inevitable in natural populations of organisms.

FOCUS ON   Evolution of the Africanized Honeybee

Periodically, one reads reports in the news about the migration of Africanized honeybees toward the United States. The movement of these bees is of great concern for several reasons. First, the Africanized honeybee is dangerous because it attacks as a group at the slightest provocation. Also, it is feared that the Africanized honeybee will interbreed or compete with the European honeybee, which is the type of bee raised commercially in this country. This would adversely affect the honeybees' important roles in pollination and honey production.

Africanized honeybees evolved from a small number of African honeybees that were introduced into Brazil in 1956. Because the number of African honeybees was so small, they contained only a fraction of the genes present in the gene pool of African honeybees. Owing to the founder effect, the few genes present in the introduced population formed the gene pool upon which natural selection would act in its new habitat in South America. Natural selection of this limited gene pool resulted in the Africanized honeybee.

Africanized honeybees have spread beyond their original point of origin in Brazil to occupy large areas of Latin America. Wherever they have migrated, decreases in honey production have occurred. Apiculture in countries occupied by the Africanized bees has also changed because the beekeepers must now use a lot of protective equipment.

Will these honey bees eventually be displaced by Africanized bees? (Grant Heilman, from Grant Heilman Photography)

How far will Africanized honeybees go? Will they migrate into the southern United States and stop because of a climatic barrier? Or is it possible that they can expand their range into more northerly areas of North America? Based on research done in Argentina, it is believed that climate will pose no barrier to Africanized honeybees. Consequently, they will probably continue to expand their range.

We can take two different approaches to this problem. We can continue to mount expensive pest-control measures whenever populations of Africanized bees are discovered in the United States. In Kern County, California, for example, one million dollars were spent eradicating Africanized bees in 1985.

Some biologists advocate a different approach—one that utilizes genetics. Instead of fighting their invasion of our land, we can start breeding Africanized honeybees. The African honeybees from which they evolved have a number of desirable traits that are missing in Africanized honeybees. By selectively breeding African honeybees that possess desirable traits with Africanized honeybees, it would be possible to increase their gene pool. Artificial selection might even give rise to a variety of honeybee that is commercially as important as the European honeybee.

NATURAL SELECTION INCREASES THE FITNESS OF A SPECIES FOR THE ENVIRONMENT IN WHICH IT LIVES

A subtle example of natural selection is the development of **protective coloration,** coloration that permits an organism to blend with its surroundings. This protects it from its predators or, in the case of a predator, keeps the prey from noticing it until too late.

Many examples of protective form and coloration readily come to mind. Walking-stick insects resemble twigs so closely that you would never guess they are animals—until they start to walk (Fig. 18–6). The chicks of ground-nesting birds are usually colored to blend in with the surrounding weeds and earth so that they cannot be discerned from a distance (Fig. 18–7). Some katydids resemble leaves not only in color but in the pattern of veins in their wings (Fig. 18–8). Pipefishes have almost perfect camouflage coloration in green eel grass (Fig. 18–9). Evidently, such protective coloration has been preserved and accentuated by means of natural selection.

FIGURE 18-6 Walking sticks resemble twigs when resting on a plant or other object. Most species occur in the tropics, where some of the largest walking sticks reach a length of more than one foot! (Frans Lanting)

FIGURE 18-7 Camouflage conferred by protective coloration. As with many ground-nesting birds, in their natural surroundings the chicks of nighthawks are almost invisible both to us and to sight-hunting predators.

(a)

(b)

FIGURE 18-8 Different katydid species have evolved various forms of protective coloration to blend into their surroundings. (*a*) The wings of this katydid even have venation patterns similar to leaves. (*b*) This leaf-mimicking katydid resembles a dead leaf. Some katydids are mottled to match partly dead leaves; still others are perfectly camouflaged when resting on rainforest tree trunks. (James L. Castner)

FIGURE 18-9 Bay pipefish, *Syngnathus leptorhynchus.* The pipefish is closely related to the sea horse. Most species have thin, narrow bodies from 1 to 18 inches in length. In addition to its protective coloration, its habit of holding its body in positions resembling waving eel grass or algae aids in its camouflage. (Doug Wechsler)

FIGURE 18–10   Batesian mimicry. (*a*) At the right is Jordan's salamander, *Plethodon jordani*, a distasteful species. To its left is *Desmognathus imitator*, a palatable species (at least to its predators). (*b*) Few would want to get close enough to this insect to discover that it is actually a moth. Note the moth-like antennae. (*c*) A genuinely noxious insect, the golden paper wasp. (*a*, E.D. Brodie, Jr., Adelphi University/BPS; *b*, L.E. Gilbert, University of Texas at Austin/BPS; *c*, P.J. Bryant, University of California at Irvine BPS)

Some organisms look very much like other, unrelated organisms. This resemblance is termed **mimicry.** In one type, **Batesian mimicry,** a harmless or edible species resembles one that is dangerous in some way (Fig. 18–10). A harmless moth may look so much like a bee or wasp that even a biologist would hesitate to pick it up. Likewise, many butterflies "mimic" the monarch butterfly. The monarch, having fed while a caterpillar on the poisonous milkweed plant, is toxic to birds. Its imitators look like the monarch but are nonpoisonous. Birds avoid the monarch; they also avoid its imitators. Apparently, natural selection has maintained a resemblance that gives the mimic almost as much protection as the model, for as soon as predators learn to associate the distinctive markings of the model with its undesirable characteristics, they tend to avoid all similarly marked animals.

Müllerian mimicry is a type of mimicry in which different species, all of which are poisonous, harmful, or distasteful, resemble one another. Although their harmfulness protects them as individual species, their similar coloration works as an added advantage. Potential predators can more easily learn their common warning coloration than if each species had its own distinctive pattern (Fig. 18–11).

Protective coloration and mimicry are two examples of adaptive evolution brought about by natural selection. An organism that is well adapted to its environment has an increased chance of surviving and reproducing, passing some of its genes on to the next generation.

We now examine some of the phenomena responsible for adaptive evolution of populations. We have learned that most traits are controlled by several different genes at different loci—that is, polygenes. When traits, such as human height, are under polygenic control, a range of phenotypes occurs, with most of the population located in the median range and fewer at either extreme (Fig. 18–12). This normal distribution forms a standard bell curve. Three main processes of natural selection cause changes in the normal distribution of phenotypes in a population: stabilizing, directional, and disruptive selection. Although we consider each process separately, in nature their influences generally overlap.

Stabilizing Selection Selects Against Phenotype Extremes And Favors Phenotypes Near The Mean, Or Average

Toward the end of the 19th century, an ornithologist named Hermon Bumpus collected a group of sparrows killed in an exceptionally severe snowstorm. He compared these dead birds with those that had survived the storm, using nine characteristics, including wingspread and body weight, as the basis of his comparison.

Bumpus discovered that the dead birds tended to be abnormal; that is, they represented the extreme ends of the normal range of variation in a sparrow population. Bumpus concluded that there is a more or

FIGURE 18–11 Müllerian mimicry. These various lepidopteran species are all unpalatable. (L.E. Gilbert, University of Texas at Austin/BPS)

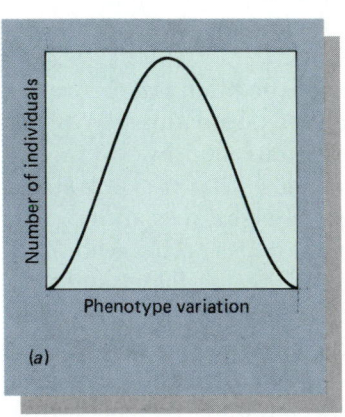

(a)

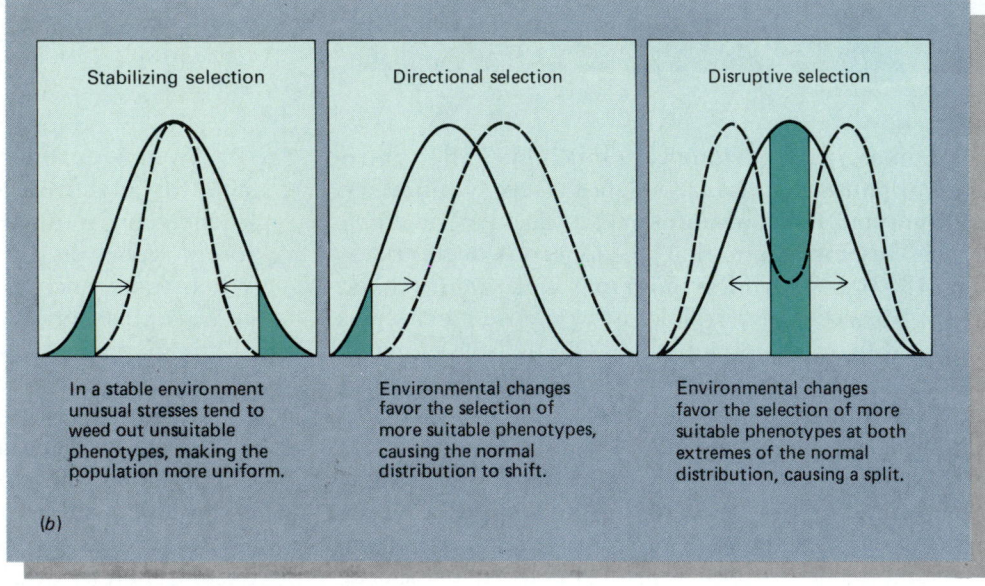

Stabilizing selection	Directional selection	Disruptive selection
In a stable environment unusual stresses tend to weed out unsuitable phenotypes, making the population more uniform.	Environmental changes favor the selection of more suitable phenotypes, causing the normal distribution to shift.	Environmental changes favor the selection of more suitable phenotypes at both extremes of the normal distribution, causing a split.

(b)

FIGURE 18–12 Different types of natural selection. (a) A trait, such as height, that is under polygenic control exhibits a normal distribution of phenotypes. **(b)** As a result of stabilizing selection, the curve is narrower. Directional selection moves the curve in one direction. Disruptive selection results in two or more peaks.

less standard body build suitable for a bird with the life-style of a sparrow. Although extreme deviants from this standard may do well when conditions are not rigorous, unusual stresses, such as a blizzard, periodically tend to weed out the unsuitable phenotypes, along with the genotypes that produced them.

Although Bumpus's data have been subjected to several reanalyses (one cannot duplicate the observations until a comparable blizzard recurs), his basic thesis seems to stand. The phenomenon is known as **stabilizing selection** (Fig. 18–12 b). It is, in a sense,

antievolutionary. It tends to maintain a standard phenotype in a population of organisms. In other words, stabilizing selection is genetically homeostatic as long as the environment of the organism does not undergo long-term change. Ordinarily, natural selection will tend to stabilize the genetic composition of populations. Should the environment change, however, or should the organism find itself able to expand its range into a new kind of environment or ecological niche, then (and only then) natural selection might produce evolutionary changes.

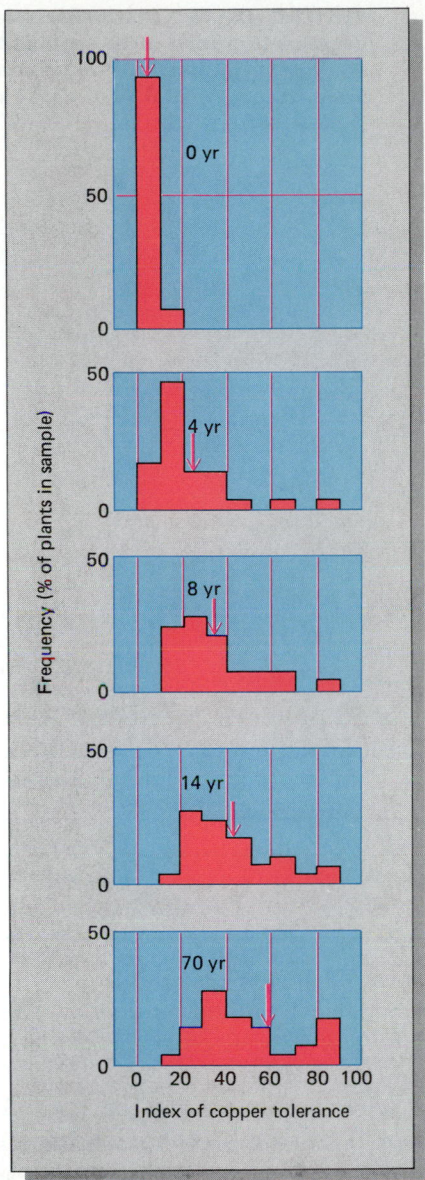

FIGURE 18–13 Directional selection. Year 0 represents grass growing in uncontaminated soil, while the other graphs represent grass growing in soil known to be contaminated by copper for a specified time (4 to 70 years). All plants were grown in a nutrient solution with 0.5 parts per million of copper. Root growth was measured to give an index of copper tolerance; 0 = no growth (complete inhibition) and 100 = maximum growth (no inhibition). The arrows indicate the mean. Note how the arrows progress in one direction with an increase in time. Copper tolerance is under polygenic control.

Directional Selection Results In The Change From One Phenotype To Another

If an environment changes over time, **directional selection** may favor those phenotypes at one of the extremes of the normal distribution (see Fig. 18–12 b).

One phenotype may gradually replace another. Directional selection can only occur, however, if the appropriate allele or alleles the new circumstances favor are already present in the population.

A classic example of directional selection concerns wing color in peppered moths in England (see Focus Box in Chapter 1). Another example of directional selection involves genes for copper tolerance in populations of *Agrostis stolonifera*, which is commonly called bent grass. Individual plants were taken from grass populations growing in areas that had been contaminated by copper from copper smelting in Lancashire, England. The soil in these areas had between 10 and 20 parts per million of copper, a level that is toxic to most plants. The amount of time each area had been contaminated by copper was well documented. These plants, along with control plants taken from areas uncontaminated by copper, were grown in a nutrient solution that contained 0.5 parts per million copper, and root growth was measured. Plants taken from uncontaminated soils barely grew in the experiment, whereas plants from copper-contaminated soil grew progressively better, depending on how long the soil had been known to be contaminated (Fig. 18–13). It is assumed that soils contaminated for a longer period allowed more time for the plants to adapt by directional selection.

Disruptive Selection Separates The Population Into Several Distinct Phenotypes

Sometimes extreme changes in the environment may favor two or more variant phenotypes at the expense of the mean; that is, more than one phenotype may be favored in the new environment, whereas the average phenotype originally present is selected against. **Disruptive selection** is a special type of directional selection in which there is a trend in several directions rather than one (see Fig. 18–12 b).

A clear example of disruptive selection involves Batesian mimicry. In some localities in Africa, there are three different distasteful species of butterfly. Different females of the edible swallowtail butterfly, *Papilio dardanus*, mimic each of the distasteful models (Fig. 18–14). Disruptive selection has favored varieties of the swallowtail that resemble any of the model species. The initial single population has been disrupted into three different populations that differ in their color pattern as each has mimicked a different distasteful model.

We have seen that evolution is a change in gene frequencies in the gene pool of a species population. Natural selection, originally proposed by Charles Darwin, is the most significant factor in changing gene frequencies in populations, whether it involves preserving

FIGURE 18–14 Disruptive selection. Females (*top row*) of the palatable swallowtail butterfly, *Papilio dardanus*, often mimic several distasteful species (*bottom row*). (BPS)

the status quo or favoring trends in one or more directions. It is important to recognize that the forces of natural selection do not cause the development of the "perfect" organism. Natural selection does not develop new phenotypes; rather, it "weeds out" those phenotypes that are less adapted to environmental challenges so that those that are better adapted survive and pass their genes on to their offspring.

CHAPTER SUMMARY

I. Evolution is a change in gene frequencies in the gene pool of a population. Each individual within a population contains only a portion of the genes in the gene pool.
   A. The Hardy-Weinberg law states that gene frequencies in a population tend to remain constant in successive generations unless certain factors are operating.
   B. Gene frequencies may be changed by mutation, genetic drift, gene flow, and natural selection.
      1. The source of new genes in a gene pool is mutation.
      2. Genetic drift is the random change in gene frequencies of a small breeding population. The changes are usually not adaptive.
      3. The migration of individuals between local populations causes a corresponding movement of alleles, or gene flow.
      4. Changes in gene frequencies that lead to adaptation to the environment are caused by natural selection.
II. Natural selection can change the composition of a gene pool in a favorable direction for a particular environment.
   A. Stabilizing selection promotes the status quo for well-adapted populations and favors phenotypes near the mean.
   B. Directional selection causes a shift in the direction of variation due to environmental change, resulting in the change from one phenotype to another.
   C. Disruptive selection separates the population into several phenotypes by favoring two or more phenotype extremes.

POST-TEST

1. The Hardy-Weinberg law demonstrates that regardless of dominance or recessiveness, relative _____ _____ do not change from generation to generation if they are left undisturbed.

2. The genetic variability that is the raw material of evolution is provided by _____.

3. _____ _____ is the production of changes in allele frequencies by random events.

4. In the bottleneck effect, genetic diversity is

_____ (increased or decreased) by a temporary, but extreme, reduction in population size.

5. The impoverished gene pool of Africanized honeybees is due to the _____ effect.

6. Both emigration and immigration cause changes in gene frequencies, a concept known as

_____ _____ .

7. Changes in gene frequencies that lead to adaptation to a particular environment are due to _____

_____ .

8. _____ _____ permits an organism to blend with its surroundings.

9. Stabilizing selection is genetically homeostatic as long as the _____ of an organism does not undergo long-term change.

10. _____ selection of the grass _Agrostis stolonifera_ for tolerance to copper is an indirect consequence of soil pollution.

11. _____ is a type of evolution in which changes in allele frequencies within a population occur over successive generations.

■ REVIEW QUESTIONS

1. In a hypothetical insect, spotted wings (_A_) is dominant over solid wings (_a_). A population of 100 insects was found to have 75 individuals with spotted wings (_AA_ or _Aa_) and 25 with solid wings (_aa_). Determine if the population is in genetic equilibrium.
   a. Calculate q^2, the frequency of individuals having the _aa_ genotype. (Hint to get you started: Divide the number of individuals with solid wings by the total number of insects in the population.)
   b. Based on your calculation in part a, determine the value for q, which is the frequency of _a_.
   c. Calculate the value for p, the frequency of _A_, knowing that $p = 1 - q$.
   d. Now plug your values for p and q into the Hardy-Weinberg equation, $p^2 + 2pq + q^2$. If your final answer equals 1, the insect population is in genetic equilibrium. Is it?

2. Explain the effect of each of the following on genetic variation in a population:
   a. Mutation.
   b. Gene flow.
   c. Genetic drift.

3. If a mutation occurs in a body cell, can it become established in a population? Explain why or why not.

4. We have said that mutations are almost always harmful. Why?

5. Katydids that resemble leaves are an example of protective coloration. How did katydids originally obtain these adaptations? How have these adaptations been preserved?

6. Consider the giraffe. Explain the evolution of the giraffe's long neck by directional selection. Explain how stabilizing selection accounts for the giraffe's long neck today.

■ RECOMMENDED READINGS

Ayala, F.J. _Genetic Variation and Evolution,_ Carolina Biology Reader, 1983. Burlington, N.C., Carolina Biological Supply. An introduction to population genetics.

19
Speciation and Macroevolution

OUTLINE

I. Different species have various mechanisms to achieve reproductive isolation from one another
  A. Prezygotic isolating mechanisms interfere with mating
  B. Postzygotic isolating mechanisms prevent successful reproduction if mating occurs
II. The key to speciation is the development of reproductive isolating mechanisms
  A. Allopatric speciation occurs through the effects of long physical isolation and different selective pressures
  B. Sympatric speciation results from the divergence of two populations in the same physical location
III. Macroevolution involves changes in the *kinds* of species over evolutionary time
IV. Evolution is gradual, occurs in spurts, or is a combination of both processes
V. Extinction of species is an important aspect of evolution
Focus on The Pace of Evolution in Trilobites

LEARNING OBJECTIVES

After you have studied this chapter, you should be able to:
1. Define a species and explain the limitations of your definition.
2. Explain the significance of biological isolating mechanisms.
3. Distinguish between prezygotic and postzygotic isolating mechanisms and give several examples of each.
4. Distinguish between allopatric and sympatric speciation and give an example of each.
5. Define macroevolution and distinguish between microevolution, speciation, and macroevolution.
6. Take either side in a debate on the pace of evolution, by representing the opposing views of gradualism and punctuated equilibria.
7. Define extinction and discuss its biological ramifications.

Eschscholzia mexicana, **Mexican poppy.** (Grant Heilman)

In Chapter 17, we considered the substantial evidence for evolution; in Chapter 18, we examined the mechanisms leading to evolutionary changes within species. We will now consider whether the adaptive changes that occur within populations can ultimately lead to the evolution of new species, the process known as **speciation.**

The concept of distinct groups of living organisms, known as **species,** is not new. However, every definition of exactly what constitutes a species has some sort of limitation. The term *species* means "kind" in Latin. Linnaeus classified plants into separate species based on differences in morphology, or physical form. This method is still used to characterize species, and indeed it is the only method available to characterize fossil species. Morphology alone, however, is not adequate to explain what constitutes a species. For example, dogs come in a wide variety of sizes and shapes, but all dogs are members of the same species. Similarly, broccoli, cabbage, cauliflower, and kohlrabi all belong to the same species.

The discipline of population genetics did much to clarify the concept of species as a group of organisms with a common gene pool. Members of a species interbreed with other members of the same species to produce fertile offspring and do not mate with members of different species. In other words, a species has a gene pool that is isolated from that of other species. This concept, known as **reproductive isolation,** is crucial to our understanding of species. Each species is bound by reproductive barriers that preserve its integrity by preventing genetic mixing with other species. A species is the largest unit in which gene flow is possible.

One of the problems with this definition is that it only applies to sexually reproducing organisms. Organisms that reproduce asexually do not mate in the first place, so we can't think of them in terms of reproductive isolation. For these organisms, the species concept is still valid; they are classified into species on the basis of morphological and biochemical characteristics. Two populations that are widely separated geographically may be so much alike that they are placed in the same species on the basis of their morphology, but it is impossible to test whether they will interbreed in the wild. Organisms that are assigned to different species in the wild may interbreed if they are brought into a zoo, a greenhouse, an aquarium, or the laboratory (Fig. 19–1). Therefore, we usually include in our definition of species that they do not normally interbreed *in the wild.*

KEY CONCEPTS

☐ A species is a group of more or less distinct organisms that are capable of interbreeding with one another in the wild and are reproductively isolated from other species.

☐ When two populations have become reproductively isolated so that they can no longer interbreed, they have evolved into two separate species.

☐ Based on current knowledge, it seems likely that, given enough time, the mechanisms of microevolution lead to speciation and macroevolution.

☐ Extinction is the eventual fate of all species.

FIGURE 19–1 Lions and tigers are recognized as separate species. Although their geographical ranges overlap in parts of Asia, a hybrid between a tiger and a lion has never been found in the wild. However, they have mated when brought together in zoos. (*a,* Animals Animals © 1988, Anthony Bannister; *b,* E.R. Degginger)

(a)

(b)

(a)   (b)   (c)

FIGURE 19–2 Reproductive isolating mechanisms in closely related flycatcher species in North America. Although the flycatchers are nearly identical in appearance and have overlapping ranges, they remain as distinct, reproductively isolated species. They are isolated ecologically because each species is found in a particular habitat within its range during mating. Also, they are isolated behaviorally because each species has its own characteristic song, which serves to identify it to other flycatchers of the same species. (*a*) Least flycatcher, *Empidonax minimus.* (*b*) Acadian flycatcher, *E. virescens.* (*c*) Traill's flycatcher, *E. trailii.* (*a*, Dwight R. Kuhn; *b*, J.R. Woodward/VIREO; *c*, Tom J. Ulrich/Visuals Unlimited)

◻ DIFFERENT SPECIES HAVE VARIOUS MECHANISMS TO ACHIEVE REPRODUCTIVE ISOLATION FROM ONE ANOTHER

A number of mechanisms prevent interbreeding between different species. This preserves the integrity of each species' gene pool because gene flow between species is prevented. Isolating mechanisms that work to restrict the gene flow between species also may be found *within* a species. Each species is composed of local populations, or races, that are separated geographically and/or ecologically. This results in limited genetic exchange between certain populations. Sometimes these populations, in adapting to local conditions, diverge to the point where they become reproductively isolated from the rest of the species. This may lead to the formation of a new species.

Prezygotic Isolating Mechanisms Interfere With Mating

Prezygotic isolating mechanisms prevent fertilization from ever taking place. Because male and female gametes never come into contact, an interspecific (between species) zygote never forms.

Sometimes genetic exchange is prevented between two groups because they reproduce at different times of the day, season, or year. There are many examples of **temporal isolation.** The fruit flies, *Drosophila pseudoobscura* and *Drosophila persimilis,* have ranges that over-lap to a great extent, but they do not interbreed. *Drosophila pseudoobscura* is sexually active in the afternoon and *D. persimilis* in the morning. Similarly, there are two species of sage, *Salvia,* with overlapping ranges in southern California. *Salvia mellifera* (black sage) flowers in early spring, whereas *S. apiana* (white sage) blooms in late spring and early summer.

Although several closely related species may be found in the same geographical area, they usually live and breed in different habitats in that area (Fig. 19–2). This can cause reproductive isolation between the groups. Wood frogs, for example, breed in temporary woodland ponds, whereas bullfrogs breed in larger, permanent bodies of water. They are separated by **ecological isolation.**

Many animal species have distinctive courtship behaviors, so mating between species is prevented by **behavioral isolation.** Courtship is an exchange of signals between a male and a female (Fig. 19–3). A male approaches a female and gives a sign or pattern of signals that may be visual, auditory, or chemical. If the female belongs to the same species, she recognizes the signal and returns her own distinctive signal. Further correct exchanges of signals eventually result in mating. If members of two different species begin courtship, one partner may not recognize one of the signals and may fail to respond. The courtship behavior will stop at that point. Fruit flies, for instance, exhibit a definite, species-specific courting behavior. Part of the behavior is a "love song," a series of buzzes of just the right pitch and rhythm performed by the male. Differ-

FIGURE 19–3 The larger firefly is a female of the genus *Photuris;* being tightly grasped is a male of a different genus, *Photinus.* This embrace will be his last, for she is gnawing through his neck, having lured him from afar by imitating the light signals of a female of his own species. Courtship rituals have evolved among many animals as a mechanism of reproductive isolation, preventing genetic exchange among species. Courtship rituals help ensure that mating takes place only among members of the same species. In this instance, the ritual was imitated not for reproduction but for predation. (James E. Lloyd)

FIGURE 19–4 Hybrid sterility. Mules are interspecific hybrids produced by mating a female horse with a male donkey. Although the mule exhibits valuable characteristics of each of its parents, it is sterile. (Grant Heilman Photography)

ences in "love songs" keep some species of *Drosophila* apart.

Morphological or anatomical differences that inhibit mating between species are known as **mechanical isolation.** Sometimes members of different species will court and attempt copulation, but the structure of their genital organs is incompatible, so successful mating is prevented. The interbreeding of certain insect species is thwarted in this way. Many flowering plants have physical differences in their flower parts that help them maintain their reproductive isolation from one another. The sage plants that were used earlier as an example of temporal isolation also exhibit mechanical isolation. *Salvia mellifera,* which is pollinated by small bees, has a different floral structure than *S. apiana,* which is pollinated by large carpenter bees. The differences in floral structures prevent the insects from cross-pollinating the two species.

Postzygotic Isolating Mechanisms Prevent Successful Reproduction If Mating Occurs

When prezygotic isolating mechanisms fail, as they occasionally do, **postzygotic isolating mechanisms** may come into play. This ensures reproductive failure even though fertilization has taken place.

Generally, the embryonic development of an interspecific hybrid is aborted. Development is a complex process that requires the precise interaction and coordination of many genes. Apparently, the genes from parents belonging to different species do not interact properly in regulating the mechanisms for normal embryonic development. For example, nearly all of the hybrids die in the embryonic stage when the eggs of a bullfrog are fertilized artificially with sperm from a leopard frog. In crosses between different species of *Iris,* the hybrid embryo develops but dies before reaching maturity as a result of breakdown of the endosperm (stored food) in the seed.

If an interspecific hybrid develops successfully, the hybrid may still not be able to reproduce. There are several reasons why this is so. Hybrid animals may exhibit courtship behaviors incompatible with those of either parental species. As a result, they will not mate. More often, the gametes of an interspecific hybrid are abnormal because of problems during meiosis. This is particularly true if the two species have different chromosome numbers. For example, a mule is the offspring of a female horse ($2n = 64$) and a male donkey ($2n = 62$) (Fig. 19–4). This type of union almost always results in sterile offspring ($2n = 63$) because synapsis, pairing of homologous chromosomes during meiosis, cannot occur properly. Many examples of hybrid sterility in plants have been documented. Sometimes the interaction of genes from two species causes a hybrid's anthers to develop improperly. Such male sterility has been found in hybrids between different tobacco (*Nicotiana*) species.

(a)

(b)

FIGURE 19–5 (*a*) The nene (pronounced "nay-nay"), *Branta sandvicensis*, is a goose found in the Hawaiian Islands. It is believed to have evolved from a small population of geese that originated in North America. (*b*) The Canadian goose, *B. canadensis*, is believed to be a close relative of the Hawaiian goose. Although the nene is endangered, strict conservation measures have brought it back from the brink of extinction. (*a*, M.J. Rauzon/VIREO; *b*, R. Villani/VIREO)

■ THE KEY TO SPECIATION IS THE DEVELOPMENT OF REPRODUCTIVE ISOLATING MECHANISMS

We are now ready to consider how entirely new species may arise from previously existing ones. A required step in the evolution of a new species is the reproductive isolation of a population from the rest of the species. When a population is sufficiently different from its ancestral species so that no genetic exchange can occur between them, even if the two populations meet, we say that speciation has occurred. Such a situation is thought to arise in two ways: through allopatric or sympatric speciation.

Allopatric Speciation Occurs Through The Effects Of Long Physical Isolation And Different Selective Pressures

Speciation that occurs when one population becomes geographically separated from the rest of the species and subsequently evolves is known as **allopatric speciation.** Allopatric speciation is believed to be the most common method of speciation and has been the most important mechanism in the evolution of new species of animals.

There are several ways geographic isolation might occur. The earth's surface is in a constant state of change. Rivers change their courses. Glaciers migrate. Mountain ranges form. Land bridges develop, separating previously united aquatic populations. Large lakes diminish into several smaller, geographically separated pools. It is important to recognize that what might be an imposing geographical barrier to one species may be of no consequence to another. Each species has its own methods of dispersal. For example, as a lake sub-

sides into smaller pools, fish are usually unable to cross the land barriers between the pools and so become isolated. Birds, on the other hand, can easily fly from one pool to another.

Alternatively, a small population may migrate and colonize a new area, away from the original species range. This colony would be geographically isolated from its parent species. The Galapagos Islands and the Hawaiian Islands represent examples of geographical areas that were colonized by a few individuals of a few species. From these original colonizers, the distinctive groups of unique species characteristic of each island arose (Fig. 19–5).

Because the population is geographically isolated, there is no interbreeding with the rest of the species and, therefore, no gene flow between it and the parent species. Moreover, the isolated habitat may be different in several ways from the parent species habitat. Climate and soil factors will be distinct, and there will be a different set of biological organisms with which the isolated population must interact. As a result of these differences in habitat, the isolated population will face different selective pressures. Over time, the isolated population will adapt to the new habitat, and its gene pool will diverge from the gene pool of the original species. Eventually, the differences between the two may be so great that they will be unable to interbreed even if their range becomes continuous again.

Many examples of allopatric speciation can be traced to the barriers formed by the glaciations of the Pleistocene Epoch. A western population of the Pleistocene European bear, *Ursus arctos*, was separated from the rest of the species and evolved into the cave bear, *Ursus spelaeus*. The eastern population remained as *Ursus arctos*. This reconstruction has been supported by fossil evidence.

Allopatric speciation may sometimes occur in a relatively short period of time. Early in the 15th century, a small population of rabbits was released on Porto Santo, a small island off the coast of Portugal. There were no rabbits or other competitors and no carnivorous enemies on the island, and the rabbits thrived. By the 19th century, they were markedly different from the ancestral European stock. They were only half as large and had a different color pattern. Moreover, their life-style was different, as they were more nocturnal. Most significant, they could not produce offspring when bred with members of the ancestral European species. Within 400 years, a new species of rabbit had developed.

Lakes and pools of water provide the isolation for allopatric speciation of aquatic organisms that islands provide for terrestrial plants and animals. Large lakes formed by glacial melt at the end of the Pleistocene Epoch in what is now Nevada were populated by one or several species of pupfish. With the gradual demise of the large glacial lakes as a result of glacial retreat and a drier climate, isolated pools were left. Today, there are numerous species of pupfish, but each is restricted to a single water hole.

Sympatric Speciation Results From The Divergence Of Two Populations In The Same Physical Location

Although geographical isolation is an important factor in many cases of evolution, it is not an absolute requirement. When a population forms a new species within the same geographical region as its parent species, **sympatric speciation** has occurred. The divergence of two gene pools in the same geographical range is especially common in plants, although some cases of sympatric speciation in animals have been documented.

We have seen that hybrids formed from the union of two species rarely produce offspring and that these offspring are usually sterile, like the mule. Often sterility is the case because the two parent species have different chromosome numbers. During gamete formation, meiosis occurs to reduce the chromosome number. In order for the chromosomes to be parceled correctly into the gametes, the homologous chromosomes pair during metaphase I. This cannot occur properly in the hybrid offspring of two species because the chromosomes are not homologous. However, *if the 2n chromosome number was doubled before meiosis,* then the pairing of homologous chromosomes could occur. This spontaneous doubling of chromosomes has been documented in both plants and animals. It is not a common occurrence, but neither is it rare. It produces nuclei with multiple sets of chromosomes.

Polyploidy is the possession of more than two sets of chromosomes. When it occurs in conjunction with sexual reproduction between two different species, it is known as **allopolyploidy,** and it can cause the interspecific hybrid to be fertile. This is because the polyploid condition provides the homologous chromosomes for pairing during meiosis. As a result, the gametes may be viable. Allopolyploids can mate with themselves (self-fertilization) or with similar individuals. However, they are reproductively isolated from both parents because the gametes of the allopolyploid have a different number of chromosomes than those of either parent.

Although allopolyploidy is extremely rare in animals, it has been a significant factor in the evolution of the flowering plants. Slightly less than one-half of all flowering plants are believed to be polyploid. Moreover, allopolyploidy provides a mechanism for extremely rapid speciation. A single generation is all that is needed to form a new, reproductively isolated species. Allopolyploidy is believed to explain the rapid appearance of flowering plants in the fossil record and the incredible diversity (more than 250,000 species) in flowering plants today.

Several species of hemp nettle occur in temperate parts of Europe and Asia. One of these, *Galeopsis tetrahit* ($2n = 32$), is a naturally occurring allopolyploid that was formed by the hybridization of two species, *G. pubescens* ($2n = 16$) and *G. speciosa* ($2n = 16$). This speciation, which occurred in nature, was experimentally reproduced in the laboratory. *Galeopsis pubescens* and *G. speciosa* were crossed to produce F_1 hybrids that were mostly sterile. Nevertheless, both F_2 and F_3 generations were formed. In the F_3 hybrid, there was a plant with $2n = 32$, which yielded fertile F_4 offspring. These artificial allopolyploid plants had the same morphology and chromosome number as the naturally occurring *G. tetrahit*. When the experimentally produced plants were crossed with the naturally occurring *G. tetrahit*, a fertile F_1 generation was formed.

A new species of cordgrass was formed in nature by allopolyploidy in the recent past (Fig. 19–6). The parent species were *Spartina maritima*, which is native to Europe and has a diploid chromosome number of 60, and *Spartina alterniflora*, which is native to North America and has a diploid chromosome number of 62. *Spartina alterniflora* was accidentally introduced to Europe around 1860. In approximately 10 years, a new species of cordgrass appeared along the English coast. *Spartina townsendii* was formed by interspecific hybridization of the parent species and had a diploid chromosome number of 62 (61 would have been the expected chromosome number, but apparently one of the parents had a meiotic abnormality.) It was sterile (incapable of sexual reproduction) but reproduced asexually. Shortly after, a vigorous new species of *Spartina* arose by chromosome doubling in *S. townsendii*. The new species, *S. anglica*, was a polyploid with a chromosome number of 124. It spread rapidly and soon was found on both sides of the English channel.

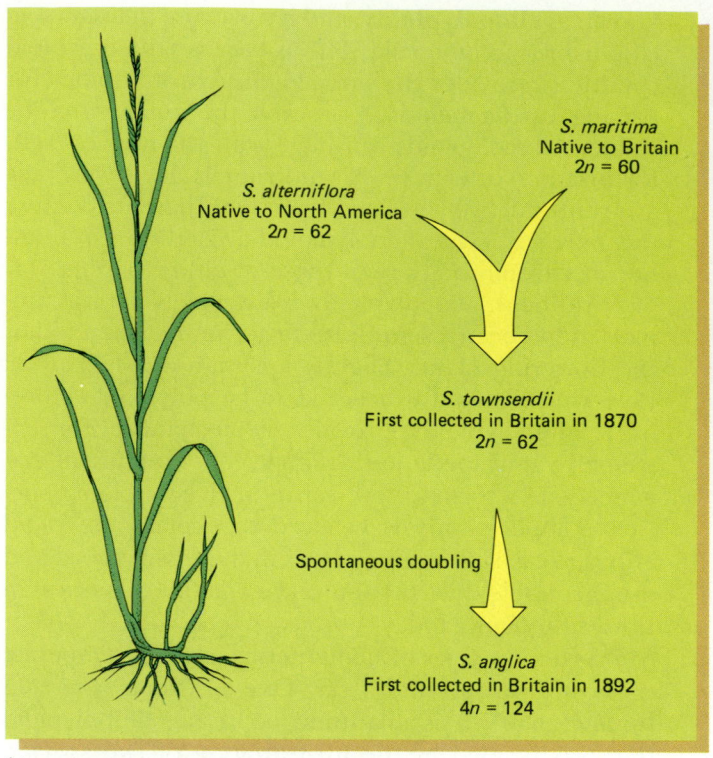

(a)

(b)

FIGURE 19–6 **How the allopolyploid** *Spartina anglica* **was formed. (a) This plant was formed in nature by the interspecific hybridization of two species of cordgrass and the subsequent doubling of the hybrid's chromosomes. (b) Close-up photo of** *S. alterniflora* **growing in a salt marsh.** (*b*, Dwight R. Kuhn)

Within the figure (a):

S. maritima
Native to Britain
2n = 60

S. alterniflora
Native to North America
2n = 62

S. townsendii
First collected in Britain in 1870
2n = 62

Spontaneous doubling

S. anglica
First collected in Britain in 1892
4n = 124

■ MACROEVOLUTION INVOLVES CHANGES IN THE *KINDS* OF SPECIES OVER EVOLUTIONARY TIME

Macroevolution refers to evolution above the species level that might give rise to new genera or higher-level categories of organisms. Because of macroevolution, it follows that all forms of life developed over time from very different and often much simpler ancestors, and all lines of their descent can be traced back to a common ancestral organism.

A change in the basic design of an organism can produce a structure that is unique. Usually these "new" structures, called **preadaptations,** are variations of some structure already in existence; that is, they evolved to fulfill one role, but were changed for another. As examples, bird feathers evolved from reptilian scales, and the mammalian middle ear represents a modified jaw element of reptiles. How do such novel changes occur? Many are probably due to changes in development. Regulatory genes may exert control over hundreds of other genes. Very slight genetic changes in regulatory genes could cause major structural changes. For example, most organisms have **allometric growth,** varied rates of growth for different parts of the body during development (Fig. 19–7). If the rates

FIGURE 19–7 **Different rates of growth for different parts of the body can transform the appearance of an organism. If the tail end of a fish grew faster than the head, the fish would end up looking more like an ocean sunfish than a porcupine fish.**

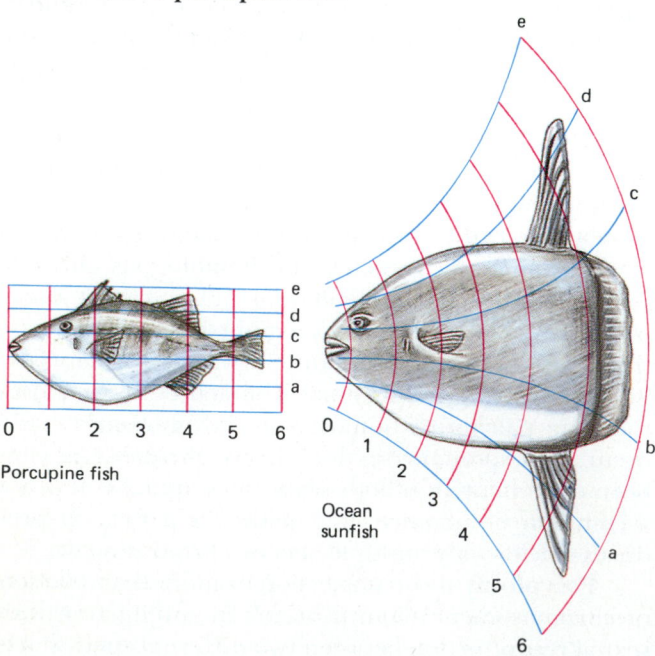

Porcupine fish

Ocean sunfish

FIGURE 19–8 Allometric growth is responsible for the huge antlers of the extinct Irish elk. (Painting by Charles R. Knight, courtesy Dept. of Library Services, American Museum of Natural History)

FIGURE 19–9 Paedomorphosis in salamanders. In most salamanders, the external gills found in the larval stage are not present in the adult. Some salamander species retain this juvenile feature as adults. (Jane Burton/Bruce Coleman, Inc.)

are altered even slightly, drastic changes in the shape of the organism may occur. The incredibly large antlers on the extinct Irish elk, *Megaloceros,* were due to allometric growth (Fig. 19–8). The ancestor of the Irish elk did not possess such exaggerated antlers. Sometimes novel changes are the result of changes in the timing of development. **Paedomorphosis** is the preservation of a juvenile characteristic in an adult. Many salamanders have features as adults that were found only in the immature stages of their ancestors. Retention of gills is an example (Fig. 19–9).

How can evolution explain the *diversity* of species? Every potential habitat or ecological niche represents a source of resources. If the habitat or niche could be exploited, it would afford an advantage to any organism that could adapt to its demands. Because of the constant competition for food and living space, each group of organisms does tend to spread out and occupy as many different habitats and ecological niches as possible. Within each habitat, those individuals with genotypes that produce superior phenotypic adaptations would propagate themselves better than those less well suited to the demands of the environment. Because habitats and ecological niches vary, appropriate adaptations would differ as well, resulting in a variety of physical and behavioral specializations. This process of evolution from a single ancestral species to a variety of forms that occupy somewhat different habitats and ecological niches is termed **adaptive radiation.**

One of the classic examples of adaptive radiation is the evolution of mammals (Fig. 19–10). The earliest known mammal was an insect-eating, five-toed, short-

legged creature that walked with the soles of its feet flat on the ground. Today we see a great variety of mammalian types. These include dogs and deer adapted for a terrestrial life in which running rapidly is important for survival; squirrels and primates adapted for life in the trees; bats equipped for flying; beavers and seals, which maintain an amphibious existence; the completely aquatic whales, porpoises, and sea cows; and the burrowing animals—gophers, moles, and shrews. In each of these, the number and shape of the teeth, the length and number of leg bones, the number and attachment sites of muscles, the thickness and color of fur, the length and shape of the tail, and so on are specifically adapted to the animal's life-style and environment.

Let's return to the example of the Galapagos Islands. Darwin took these islands to be a microcosm of adaptive radiation, represented by the variety of finches present there today (Fig. 17–5). Some of these birds live on the ground and feed on seeds, others feed mainly on cactus, and still others have taken to living in trees and eating insects. These variations in feeding have been accompanied by changes in the size and structure of the beak. This suggested to Darwin that the essence of adaptive radiation is the evolution from a single ancestral form to a variety of different forms, each of which is adapted and specialized in some unique way to survive in a particular habitat (Fig. 19–11).

We have seen that such mechanisms as differential reproduction through natural selection, mutation, genetic drift, and gene flow explain the microevolu-

FIGURE 19–10 Adaptive radiation in mammals. A primitive, insectivorous, shrew-like cohabitant with the dinosaurs is believed to have been the ancestor of all the highly specialized orders of modern mammals.

tionary changes within a population. But can these mechanisms be used to explain the evolution of genera, orders, and higher groupings? Does natural selection account for macroevolution? Can population genetics help explain major evolutionary events?

The concepts of evolution presented in Chapters 17, 18, and 19 represent the **modern synthesis** of evolution. Most biologists believe that the modern synthesis is adequate to explain macroevolution; that is, given enough time, the same processes that lead to speciation will produce new genera, new families, new orders, new classes, and new phyla. It is always possible that other mechanisms will be discovered that have an im-

portant role in macroevolutionary events. If this occurs, evolution will have to be reevaluated in light of the new evidence.

■ EVOLUTION IS GRADUAL, OCCURS IN SPURTS, OR IS A COMBINATION OF BOTH PROCESSES

In studying the fossil record, it becomes apparent that the pace of evolution varies from one group of organisms to another. In some instances, it is fast, and in others, it is slow. It appears that certain major groups have evolved a great deal in a relatively short amount

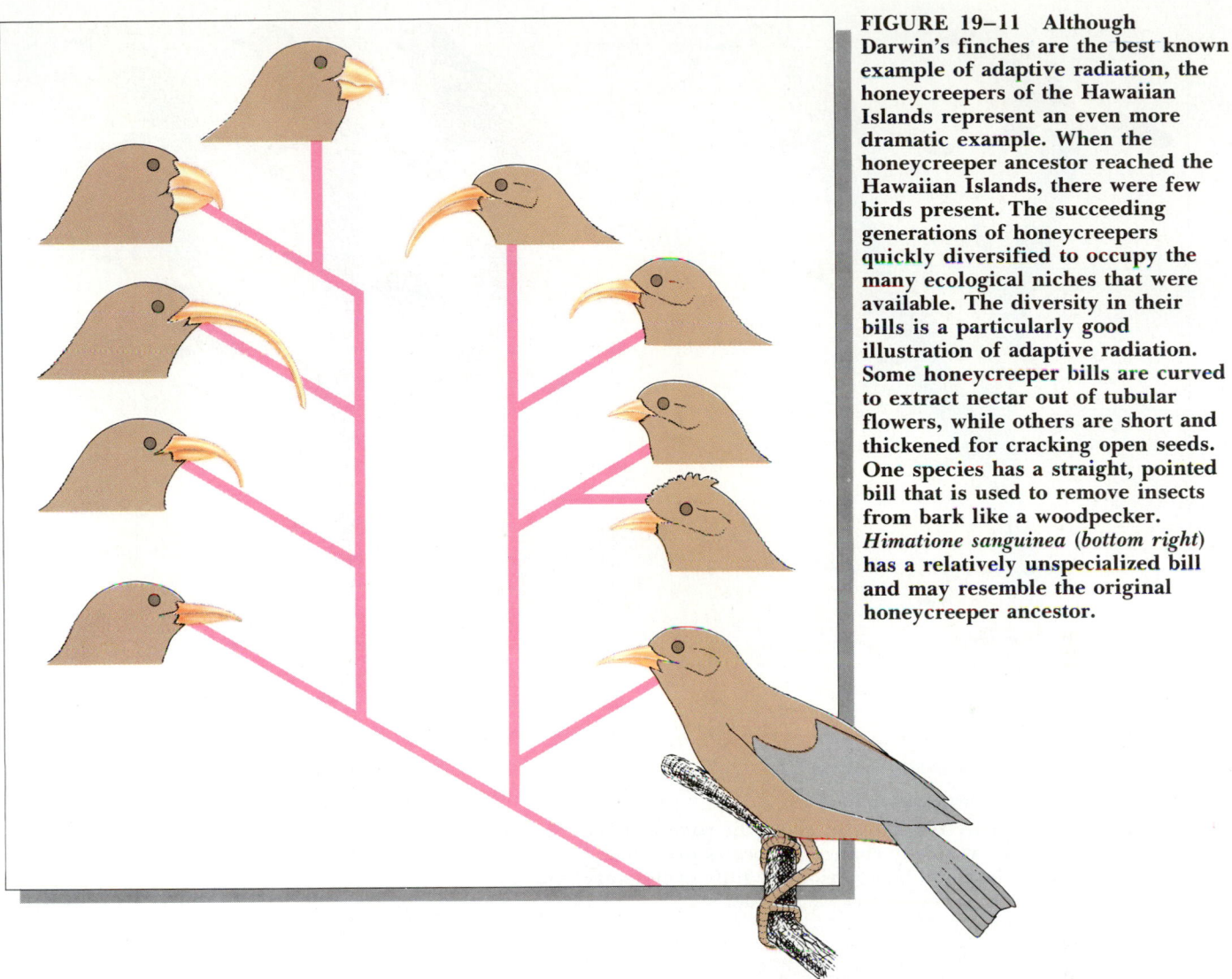

FIGURE 19–11 Although Darwin's finches are the best known example of adaptive radiation, the honeycreepers of the Hawaiian Islands represent an even more dramatic example. When the honeycreeper ancestor reached the Hawaiian Islands, there were few birds present. The succeeding generations of honeycreepers quickly diversified to occupy the many ecological niches that were available. The diversity in their bills is a particularly good illustration of adaptive radiation. Some honeycreeper bills are curved to extract nectar out of tubular flowers, while others are short and thickened for cracking open seeds. One species has a straight, pointed bill that is used to remove insects from bark like a woodpecker. *Himatione sanguinea* (*bottom right*) has a relatively unspecialized bill and may resemble the original honeycreeper ancestor.

of time—for example, mammals and flowering plants. Other organisms appear relatively unchanged, sometimes for millions of years. Mosses do not appear to have evolved much in the past 4 million years, and some organisms like the lungfish appear much the same as in 150-million-year-old fossils. Although the pace of evolution varies, there is no question that it accelerates when there are strong selective forces. A changing, challenging environment causes rapid evolutionary change in populations.

Although evolutionary biologists generally agree that natural selection is the main mechanism responsible for speciation, there is currently much debate on the timing of evolutionary change during a species' existence. Biologists fall into two groups in this debate, and each group has compelling evidence for its position. Sometimes the two groups use the same evidence, interpreting it differently to support opposing ideas. One group supports **gradualism,** a slow and steady accumulation of changes, whereas the other group proposes **punctuated equilibrium,** in which evolution proceeds with periods of inactivity that are followed by very active phases (Fig. 19–12). It is also possible that both may be correct: evolutionary pace may be quite erratic, fast at times and slow at others.

Gradualism represents the traditional approach to evolution. It proposes that populations slowly diverge from one another by the accumulation of adaptive characteristics within a population. These adaptive characteristics accumulate as a result of different selective pressures brought on by the populations living in different environments. If a species evolves by gradualism, there will be a number of intermediate steps, or "missing links." Proponents argue that there are few transitional forms in the fossil record because the fossil record is incomplete. A strong case for gradualism was recently presented in trilobite evolution (see Focus on the Pace of Evolution in Trilobites).

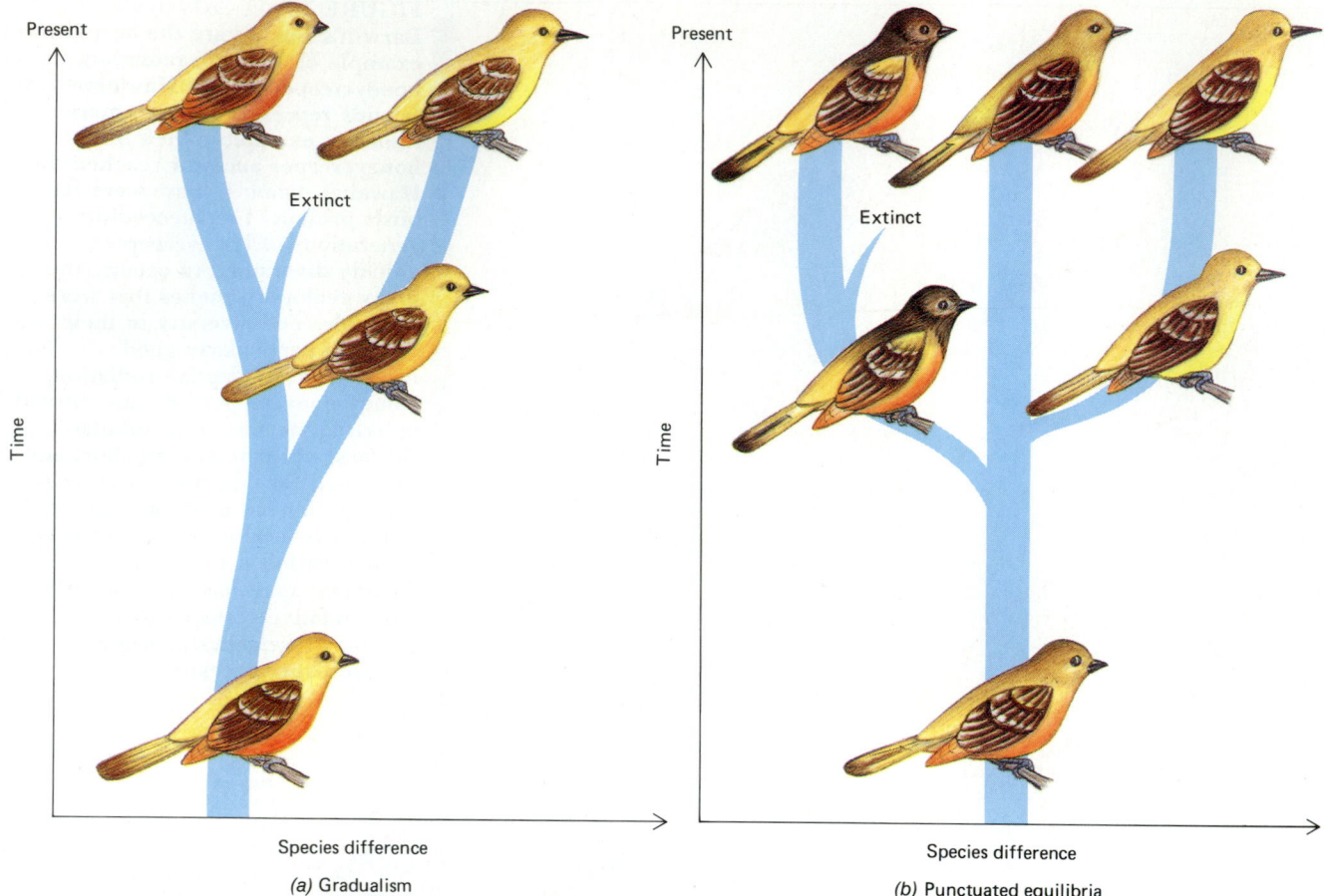

FIGURE 19–12   There are two theories about the pace of evolution. (*a*) **In gradualism there is a slow, steady change in species over time.** (*b*) **In punctuated equilibria there are long periods of little evolutionary change (stasis) followed by short periods of rapid speciation.**

The fact that there is abundant evidence in the fossil record of long periods of **stasis,** or no change in a species, seems to argue against gradualism. However, proponents point out that stasis in fossils is deceptive because fossils do not show all aspects of evolutionary change. Fossils can show changes in external anatomy and skeletal structure, but they cannot reveal such characteristics as internal anatomy, molecular changes, and behavioral changes, which also represent evolution.

Many evolutionary biologists support punctuated equilibrium, whereby evolution of new species normally proceeds in "spurts." These relatively short periods of active evolution produce new species and are followed by long periods of little or no evolutionary change (stasis). Then evolution resumes, new species form, and many old species are out-competed and become extinct. According to punctuated equilibrium, most of a species' existence is spent in stasis, and a very small percentage is spent in active evolutionary change. It is important to realize that a "short" amount of time for speciation may mean thousands of years.

Such a period of time is short when compared to the period of time a species exists. Evolutionary biologists that support punctuated equilibrium point out that sympatric speciation and even allopatric speciation can occur in a relatively short period of time.

The concept of punctuated equilibrium accounts for the abrupt appearance of a new species in the fossil record, with little or no record of intermediate forms; that is, proponents believe there are few transitional forms in the fossil record because there were no transitional forms during speciation. Also, the fossil record shows long periods of little change; this stasis is supported by punctuated equilibrium.

▪ EXTINCTION OF SPECIES IS AN IMPORTANT ASPECT OF EVOLUTION

Extinction, the end of a lineage, occurs when the last individual of a species dies. It is a permanent loss, for once a species is extinct, it can never reappear. Extinc-

FOCUS ON   The Pace of Evolution in Trilobites

One of the criticisms of gradualism made by proponents of punctuated equilibrium is that the fossil record shows scant evidence of a gradual transition during the evolution from one species to another. In other words, there are few intermediate forms that would indicate slow, progressive change.

In 1987 a paper was published in the journal *Nature* about an exhaustive study of approximately 15,000 fossil trilobites from a 3-million-year period. Peter Sheldon studied eight families of the small, invertebrate marine organisms, concentrating on the number of ribs in the exoskeleton of each. He found that each lineage showed gradual change during the 3 million years, with each of the lineages showing a gradual increase in the number of ribs. There was no evidence of a long period of equilibrium (stasis) followed by a brief period of speciation.

The significance of ribs in trilobites, which are extinct, is unknown. One suggestion is that each rib covered a leg. Another is that the addition of ribs provided extra strength. It is also possible that extra ribs had a neutral effect, but were selected for because they were genetically linked to other beneficial traits.

Trilobites. Fossil trilobites showing the area of new ribs. (A.J. Copley/Visuals Unlimited)

Area of new ribs

The publication of Sheldon's work has reignited the discussion about gradualism versus punctuated equilibrium. Some supporters of punctuated equilibrium interpret Sheldon's work quite differently. They say that such minor change as the addition of a few more ribs to the trilobite exoskeleton in 3 million years is equivalent to stasis. Also, they point to recent studies on clam and other marine fossils that clearly support punctuated equilibria.

Some scientists are of the opinion that both types of evolution may occur, that the pace of evolution may be steady and gradual in certain instances and abrupt in others.

---

tions have occurred continually since the origin of life on earth. By one estimate, there is only 1 species living today for every 2000 that have become extinct. Extinction is the eventual fate of all species, in the same way that death is the eventual fate of all living things.

Although extinction does have a negative impact on biological diversity, it has a positive evolutionary aspect. When species become extinct, the ecological niches they occupied become vacant. As a result, those organisms still living evolve and radiate out to fill the unoccupied niches. In other words, the extinct species are replaced by new species.

During the course of life on earth, there appear to have been two types of extinction. The continuous, low-level extinction of species, sometimes called **background extinction,** is one. The second type has occurred five or six times during the earth's history. At

these times, **mass extinctions** of numerous species and higher groupings have taken place (Fig. 19–13). The time periods when each mass extinction occurred may have been for several million years, but that is a relatively short period compared with the history of life on earth. Each period of mass extinction, which appears to have been indiscriminate in its choice of which species survived and which became extinct, was followed by a period of "mass speciation."

The causes of extinction, particularly mass extinction, are not well understood. Both environmental and biological factors seem to be involved. Major changes in the climate could adversely affect plants and animals that are unable to adapt to them. Marine organisms, in particular, are adapted to a very steady, unchanging climate. If the earth's temperature were to decrease overall by just a few degrees, many marine species

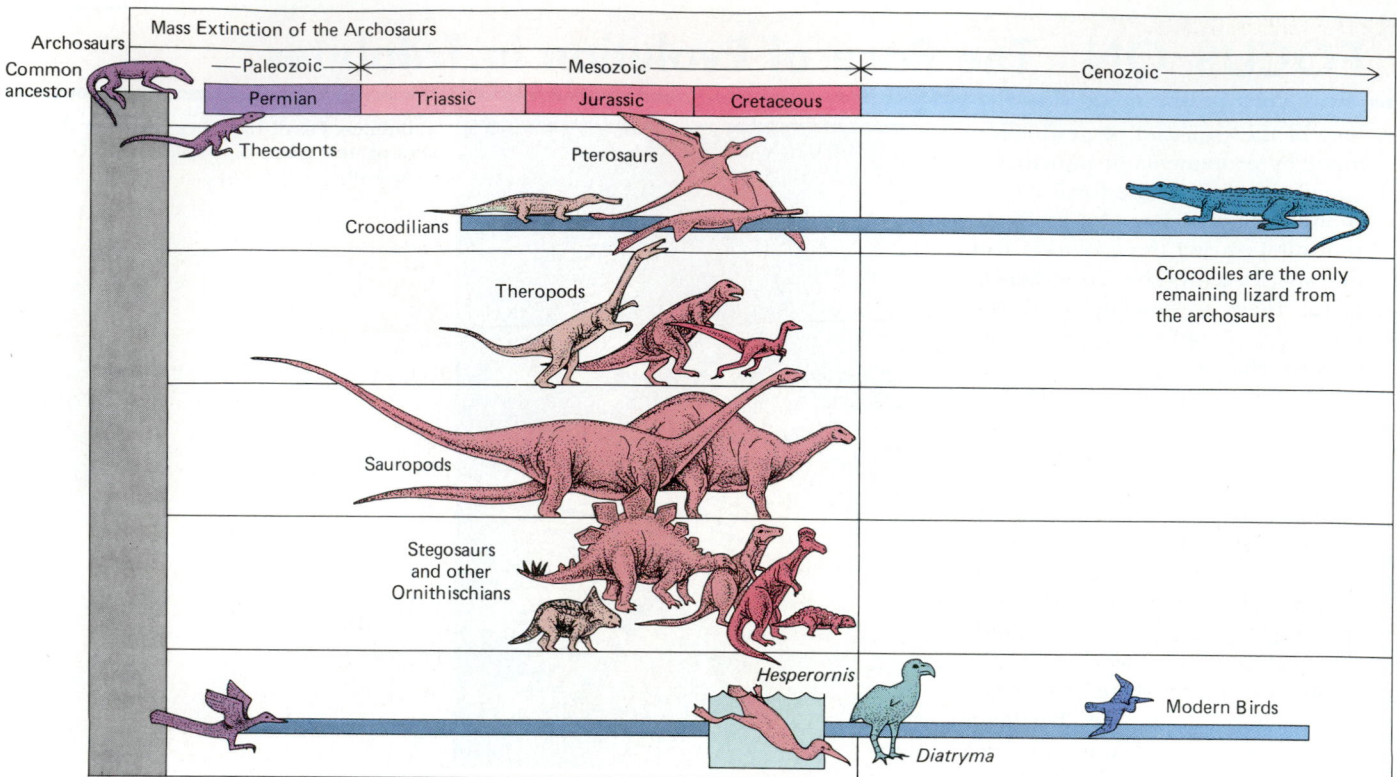

FIGURE 19–13   Mass extinctions have taken place several times in the earth's history. For example, at the end of the Cretaceous Period, which occurred approximately 65 million years ago, a mass extinction of many organisms occurred. At that time, the archosaurs (one of five main groups of reptiles) largely became extinct. The only lines to survive were the crocodiles and birds, both of which are archosaur descendants.

would probably die. Some biologists believe that climatic changes could have been responsible for mass extinctions in the past.

It is also possible that mass extinctions were due to changes in the environment triggered by catastrophes. If the earth was bombarded by a large meteorite, for example, the dust going into the atmosphere upon impact could have blocked much of the sunlight. In addition to killing many plants, this would have lowered the earth's temperature, leading to the death of many marine organisms.

Biological factors can also trigger extinction. When a new species forms, it may be able to outcompete an older species, leading to its demise. Humans have had a profound impact on extinction. The tremendous increase in human population has caused us to spread into areas of the earth that were previously not part of our range. The habitats of many animal and plant species are destroyed whenever humans invade an area. This can result in their extinction (Fig. 19–14). Indeed, some biologists believe the earth has entered the largest period of mass extinction in its entire history, and that this has been triggered by human activity.

FIGURE 19–14   **The dusky seaside sparrow,** *Ammodramus maritimus nigrescens,* **became extinct in 1987, largely due to human destruction of its habitat. In an attempt to preserve some of its gene pool, the last six survivors were successfully crossed with a related seaside sparrow subspecies in the mid-1980s.** (P.W. Sykes, Jr./VIREO)

CHAPTER SUMMARY

I. A species is a group of more or less distinct organisms that have the potential to interbreed with one another in the wild, but not with members of different species.

II. Biological isolating mechanisms restrict the gene flow between species and sometimes between different populations within a species.
   A. Prezygotic isolating mechanisms prevent fertilization from taking place.
      1. When two groups reproduce at different times of the day, season, or year it is known as temporal isolation.
      2. Ecological isolation is caused by differences in habitat between two closely related species living in the same geographical area.
      3. In behavioral isolation, distinctive courtship behaviors prevent mating between species.
      4. Mechanical isolation is due to morphological or anatomical differences in the reproductive structures of plants and animals.
   B. Postzygotic isolating mechanisms assure reproductive failure even when fertilization has taken place.
      1. In hybrid inviability, the hybrid embryo dies at an early stage.
      2. Hybrid sterility prevents hybrid adults from reproducing successfully.

III. Speciation is the evolution of a new species.
   A. Allopatric speciation occurs when one population becomes geographically isolated from the rest of the species and subsequently evolves.
   B. Sympatric speciation does not require geographical isolation and can occur as a result of allopolyploidy.

IV. Macroevolution is the evolution of groups above the species level.
   A. Preadaptations, allometric growth, and paedomorphosis may result in unusual features.
   B. Adaptive radiation is the process of evolution from a single species to a variety of species.
   C. The modern synthesis of evolution is probably adequate to explain macroevolution.

V. The pace of evolutionary change is currently being debated.
   A. According to proponents of gradualism, populations slowly diverge from one another.
   B. According to proponents of punctuated equilibria, evolution proceeds in spurts. Short periods of active evolution are followed by long periods of stasis.

VI. Extinction is the death of a species. Once a species is extinct, it can never reappear.

POST-TEST

1. A _____ is a group of organisms with a common gene pool.
2. In _____ isolation, two closely related species that are found in the same geographical range reproduce at different times of the year.
3. If two different species have reproductive structures that prevent mating, they fail to reproduce owing to _____ isolation.
4. _____ isolating mechanisms interfere with mating, whereas _____ isolating mechanisms prevent successful reproduction if mating has occurred.
5. The most important type of speciation in animal evolution is _____ speciation.
6. An individual that possesses multiple sets of chromosomes, in which one or more of those sets came from a different species, is known as an _____.
7. _____ is evolution of higher taxa (above the species level).
8. A "new" structure that is a variation of some structure already in existence is called a _____.
9. The incredibly large antlers of the extinct Irish elk were due to _____ growth.
10. Several to many species formed from a single ancestral species is known as _____.
11. The fact that the fossil record shows few transitional forms during speciation is used to support _____ _____.
12. _____ extinctions are believed to have occurred during five or six periods of the earth's history.

REVIEW QUESTIONS

1. Give an example of each of the following: temporal isolation, ecological isolation, behavioral isolation, and mechanical isolation.
2. When prezygotic isolating mechanisms fail, postzygotic isolating mechanisms may come into play. Describe two types of postzygotic isolating mechanisms and give an example of each.

3. Give at least five geographical barriers that might lead to allopatric speciation.

4. Why is speciation more likely to occur if the original isolated population is small?

5. Explain how allopolyploidy can cause a new species to form in as little time as one generation.

6. Give an example of each of the following: preadaptation, allometric growth, and paedomorphosis.

7. Does microevolution lead to macroevolution? Why or why not?

8. If you were in a debate and had to support gradualism, what would you say? What would you say if you were supporting punctuated equilibrium?

9. What role does extinction play in evolution?

▢ RECOMMENDED READINGS

Futuyma, D. *Evolutionary Biology,* 2nd ed. Sunderland, MA, Sinauer Associates, 1986. A good in-depth treatment of evolution.

20

The Origin, History, and Diversity of Life

OUTLINE

I. Early earth provided the conditions for chemical evolution
   A. Before cells existed, organic molecules formed on primitive earth
   B. The first cells probably assembled from organic molecules
   C. Eukaryotic cells evolved after prokaryotic cells
II. The fossil record provides us with clues to the history of life
   A. Evidence of living cells may be found in Precambrian times
   B. An incredible diversity of life forms evolved during the Paleozoic Era
   C. The dinosaurs and other reptiles dominated the Mesozoic Era
   D. The Cenozoic Era is known as the Age of Mammals
III. Organisms are classified using the binomial system of nomenclature
IV. Many biologists recognize five kingdoms
   A. Protists do not fit together well in one kingdom
   B. Viruses cannot be assigned to any of the five kingdoms

Focus on Continental Drift

LEARNING OBJECTIVES

After you have studied this chapter, you should be able to:

1. Describe the conditions on early earth.
2. Outline the major steps that are thought to have occurred in the origin of living cells.
3. Briefly describe the geological features and distinguishing plant and animal life for the Precambrian, Paleozoic, Mesozoic, and Cenozoic Eras.
4. Offer at least two justifications for the use of scientific names and classifications of organisms.
5. Arrange the Linnaean categories in hierarchical fashion, from highest to lowest.
6. Determine in which kingdom an organism belongs and summarize the basic characteristics of each kingdom.
7. Define a virus and explain why it does not fit into any of the five kingdoms.

Dawn at a blow hole in the Galapagos Islands.
(Frans Lanting)

We have been concerned with how living things evolved from other living things, but we have not dealt with a fundamental question: How did living cells first appear on earth? Scientists generally accept the hypothesis that the first living cells developed from organic molecules. This process, called **chemical evolution,** involved several stages. First, small organic molecules formed. Over time, they accumulated. Large macromolecules such as proteins and nucleic acids formed from smaller molecules. The macromolecules acted on one another, collecting into more complicated assemblages that could eventually metabolize and replicate. These assemblages developed into cell-like structures that ultimately became the first true cells.

Once the first cells originated, they evolved over millions of years into the rich biological diversity on our planet today. It is believed that life originated on earth only once and that this occurred under environmental circumstances that were quite different from those on earth today. And so, to understand the origin of living cells, we must consider the conditions of early earth.

KEY CONCEPTS

☐ Living cells are thought to have originated from organic molecules.

☐ Geological evidence provides us with much of what we know about the history of life.

☐ Using the binomial system of nomenclature, we assign each species a scientific name composed of two parts.

☐ Living organisms can be classified into five kingdoms.

☐ EARLY EARTH PROVIDED THE CONDITIONS FOR CHEMICAL EVOLUTION

Although we will never be certain of the exact conditions on earth when life arose, scientific evidence from a number of sources can provide us with valuable clues. The atmosphere of early earth, which contained little or no free oxygen, included carbon dioxide (CO_2), water vapor (H_2O), carbon monoxide (CO), hydrogen (H_2), and nitrogen (N_2). The early atmosphere may also have contained some ammonia (NH_3), hydrogen sulfide (H_2S), and methane (CH_4), although these reduced molecules may have been rapidly broken down by ultraviolet radiation from the sun. As the temperature of the earth slowly cooled, water vapor condensed and torrential rains fell, forming the oceans. The falling rain eroded the earth's surface, adding minerals to the oceans, making them "salty."

There are four requirements for chemical evolution. First, life could have evolved only in the absence of free oxygen. Oxygen is very reactive and would have broken down the organic molecules that are a necessary step in the origin of life. The earth's atmosphere was strongly reducing, meaning that any free oxygen would have reacted with other elements. Thus, oxygen would be tied up in compounds. A second requirement for the origin of life was energy. Early earth was a place of high energy, with violent thunderstorms, volcanoes, and intense radiation, including ultraviolet radiation from the sun (Fig. 20–1). The "young" sun probably produced more ultraviolet radiation than it does today, and the earth had no protective ozone layer to block much of this radiation. This radiation provided energy used in the formation of organic molecules from simple compounds. Third, the chemicals that would be the building blocks for chemical evolution must have been present. These included water, dissolved inorganic minerals (present as ions), and the gases present in the early atmosphere. A final requirement would have been time—time for molecules to accumulate and react. The age of the earth provides adequate time for chemical evolution. The earth is approximately 4.6 billion years old, and there is geological evidence that simple life forms appeared 3.5 billion years ago.

Before Cells Existed, Organic Molecules Formed On Primitive Earth

Because organic molecules are the building materials for living organisms, it is reasonable to consider how they might have originated. The concept that simple organic molecules such as sugars, nucleotides, and amino acids could form from simpler raw materials was first hypothesized in the 1920s by two scientists working independently—Oparin, a Russian biochemist, and Haldane, a Scottish physiologist and geneticist. Their hypothesis was tested in the 1950s by Urey and Miller, who designed an apparatus that simulated conditions then believed to be prevalent on early earth (Fig. 20–2). They exposed an atmosphere rich in H_2, CH_4, H_2O, and NH_3 to an electric discharge, which simulated lightning. Their analysis of the chemicals produced in a week revealed that amino acids and other building blocks had formed. We now believe that the earth's early atmosphere was not rich in methane or ammonia, but similar experiments using different combinations of gases have produced a wide variety of organic molecules, including nucleotide bases of RNA and DNA.

Early Earth

N₂   Ammonia   Hydrogen cyanide   Benzene   Fatty acid

H₂O   Hydrogen sulfide   CO₂   Methane   Amino acid   N₂

FIGURE 20–1 Conditions on early earth would have been inhospitable for most of today's life forms. The strongly reducing atmosphere lacked oxygen. Volcanoes erupted, spewing gases that contributed to the atmosphere. Violent thunderstorms produced torrential rainfall that eroded the land.

Oparin envisioned that the organic molecules would, over vast spans of time, accumulate in the shallow seas, as a "sea of organic soup." Under such conditions, he believed that larger organic molecules (polymers) would form by the union of smaller ones (monomers). Based on scientific evidence gathered since Oparin's time, most scientists believe it is more likely that organic polymers formed and accumulated on rock or clay surfaces. Clay is particularly intriguing as a possible site for this first polymerization because it contains zinc and iron ions that might have served as catalysts.

After the first polymers formed, could they have assembled into more complex structures? Scientists have worked with several different **protobionts,** which are assemblages of organic polymers (Fig. 20–3). They have been able to make protobionts that resemble simple life forms in several ways, helping us to envision how complex molecules took that giant leap and became living cells. Protobionts often divide in half after they have "grown." Their interior is chemically different from the external environment, and some of them show the rudiments of metabolism. They are amazingly organized, considering their relatively simple composition.

FIGURE 20–2 Stanley Miller and Harold Urey used this apparatus to replicate what they believed were the conditions of early earth. An electrical spark was produced in the upper right flask to simulate lightning. The gases present in the flask reacted together, forming a number of simple organic compounds, which accumulated in the flask on the lower left. (Courtesy of Dr. Stanley Miller)

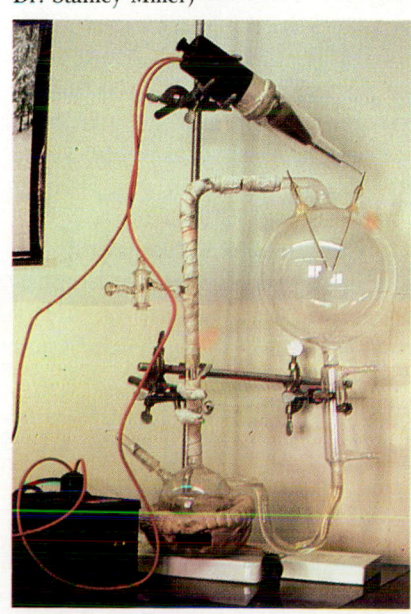

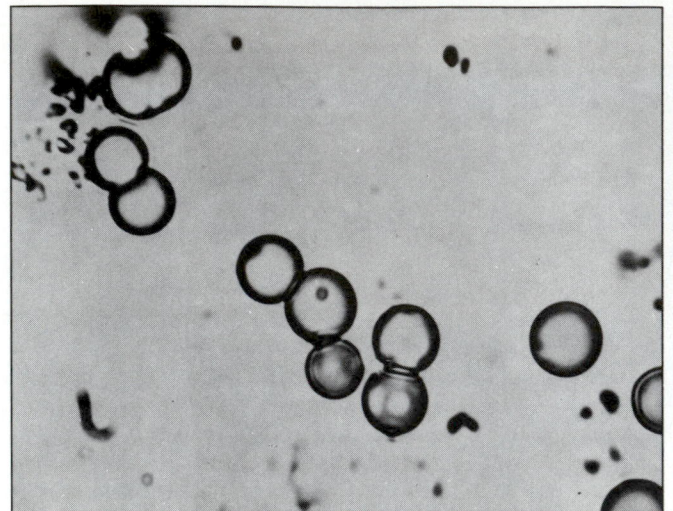

FIGURE 20–3 Proteinoid microspheres, a type of protobiont, are tiny spheres (1–2 μm in diameter) that exhibit some of the properties of life.

FIGURE 20–4 Stromatolites at Shark Bay in Western Australia that are approximately 2000 years old. These formations are composed of mats of cyanobacteria and minerals like calcium carbonate. Some fossil stromatolites are 3.5 billion years old.

The First Cells Probably Assembled From Organic Molecules

Studying protobionts can help us appreciate that relatively simple "pre-cells" can exhibit some of the properties of life; however, it is a major step to go from molecular aggregates such as protobionts to living cells. Fossil evidence indicates that cells were thriving 3.5 billion years ago. The first cells to evolve were prokaryotic. Australian and South African rocks have yielded microscopic fossils of prokaryotic cells 3.4 to 3.5 billion years old. **Stromatolites** are another type of fossil evidence of the earth's earliest cells (Fig. 20–4). These column-like rocks are composed of many minute layers of prokaryotic cells. Stromatolites are found in a number of places in the world, including the Canadian Great Slave Lake and the Gunflint Iron Formations along Lake Superior in the United States. Some of them are extremely ancient. One group in Western Australia, for example, is several billion years old. Living stromatolites are still found in hot springs and in shallow pools of fresh and salt water.

The earliest cells were **anaerobes,** cells that did not need oxygen. Some may have been **heterotrophic,** obtaining the organic molecules they needed for energy from the environment, as opposed to synthesizing them. They probably consumed many types of organic molecules that had formed: sugars, nucleotides, and amino acids, to name a few. By fermenting these organic compounds, they obtained the energy needed to support life. Fermentation is, of course, an anaerobic process.

Before the supply of spontaneously generated organic molecules was exhausted, mutations may have

occurred that gave certain organisms a distinct selective advantage. These cells could obtain energy from sunlight and did not require the energy-rich organic compounds that were in short supply in the environment.

Photosynthesis requires not only light energy, but also a source of hydrogen, which is used to reduce CO_2 when organic molecules are synthesized. Most likely, the first photosynthetic **autotrophs** used the energy of sunlight to split hydrogen-rich molecules like H_2S, releasing sulfur (not oxygen) in the process. Indeed, some types of bacteria still use H_2S as a hydrogen source for photosynthesis. The first photosynthetic autotrophs to split water in order to obtain hydrogen were the **cyanobacteria.** Water was quite abundant on earth, and the selective advantage that splitting water bestowed on them caused the cyanobacteria to thrive. In the process of splitting water, oxygen was released as a gas, O_2. Initially, the oxygen released from photosynthesis oxidized minerals in the ocean and the earth's crust. Over time, oxygen began to accumulate in the oceans and atmosphere.

The increase in atmospheric oxygen had a profound effect on the earth and on life. First, oxygen in the upper atmosphere reacted to form **ozone,** O_3 (Fig. 20–5). Ozone blanketed the earth, preventing much of the sun's ultraviolet radiation from penetrating to the earth's surface. The protective ozone layer enabled living organisms to live closer to the surface in aquatic environments and even on land! Because the energy in ultraviolet radiation had been used to form organic molecules, their synthesis decreased. Some anaerobes

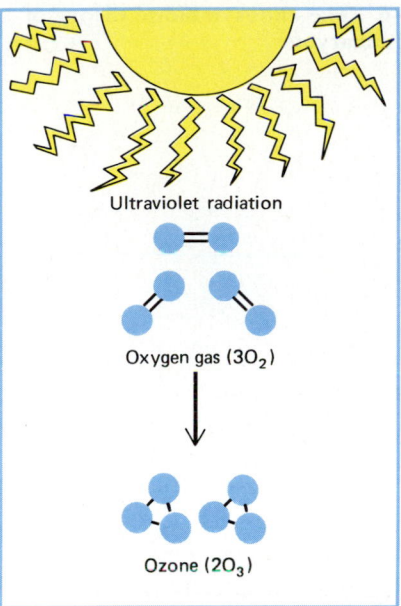

Ultraviolet radiation

Oxygen gas ($3O_2$)

Ozone ($2O_3$)

FIGURE 20–5 Ozone, O_3, is formed in the upper atmosphere when ultraviolet radiation from the sun breaks the double bonds of oxygen molecules.

were poisoned by the oxygen, and many species undoubtedly perished. Some anaerobes, however, evolved ways to neutralize the oxygen so it could not harm them. Some organisms even evolved ways to *use* the oxygen to extract energy from food.

Eukaryotic Cells Evolved After Prokaryotic Cells

Eukaryotes appeared in the fossil record 1.5 billion years ago. How did eukaryotic cells arise from prokaryotes? Recall that eukaryotic cells differ from prokaryotic cells in that they have organelles such as mitochondria and chloroplasts (see Windows on the Plant and Animal Cells). The **endosymbiont theory** suggests that organelles such as mitochondria and chloroplasts may have originated from symbiotic relationships between two prokaryotic organisms (Fig. 20–6). Thus, chloroplasts are thought to have evolved from photosynthetic bacteria that lived inside other prokaryotes. Mitochondria are thought to have evolved from aerobic (oxygen-requiring) bacteria that lived inside other prokaryotes. How did these bacteria come to be **endosymbionts** (organisms that live symbiotically inside a host cell)? Probably they were originally ingested by the host cell but not digested by it. Thus, they survived and reproduced along with the host cell so that future generations of the host also contained endosymbionts. The two organisms developed a mutualistic relationship, and eventually the endosymbiont lost the ability to exist outside its host.

The principle evidence in favor of the endosymbiont theory is that the mitochondria and chloroplasts we see today in eukaryotic cells possess *some* (although not *all*) of their own genetic apparatus. They have their own DNA (as a circular molecule much like that of prokaryotes) and their own ribosomes (which resemble prokaryotic ribosomes rather than eukaryotic ribosomes). They have some of the machinery for protein synthesis, including tRNA molecules, and are able to conduct protein synthesis on a limited scale independently of the nucleus.

In Chapter 21, we will discuss endosymbiotic relationships existing today that help us understand how eukaryotic cells may have arisen. The endosymbiont theory does not completely explain how eukaryotic cells evolved from prokaryotes. It does not explain how the genetic material in the nucleus came to be surrounded by an envelope, for example.

■ THE FOSSIL RECORD PROVIDES US WITH CLUES TO THE HISTORY OF LIFE

The sediments of the earth's crust consist of five major rock strata (layers), each subdivided into minor sublayers, lying one on top of the other. These sheets of rock were formed by the accumulation of mud and sand at the bottom of oceans, seas, and lakes. Each sheet contains certain characteristic fossils that serve to identify deposits made at approximately the same time in different parts of the world.

Geological time has been divided into **eras,** which are subdivided into **periods,** which in turn are composed of **epochs** (Fig. 20–7 and Table 20–1). Between the major eras, and serving to distinguish them, there were widespread geological disturbances that raised or lowered vast regions of the earth's surface and created or eliminated shallow inland seas. These disturbances altered the distribution of sea and land organisms and may have triggered the mass extinction of many life forms. The raising and lowering of portions of the earth's crust result from the slow movements of the enormous plates that compose the crust (see Focus on Continental Drift).

Evidence Of Living Cells May Be Found In Precambrian Times

Signs of **Precambrian** life date from about 3.5 billion years ago. Precambrian times were characterized by widespread volcanic activity and giant upheavals that raised mountains. The heat, pressure, and churning associated with these movements probably destroyed most of whatever fossils may have been formed, but

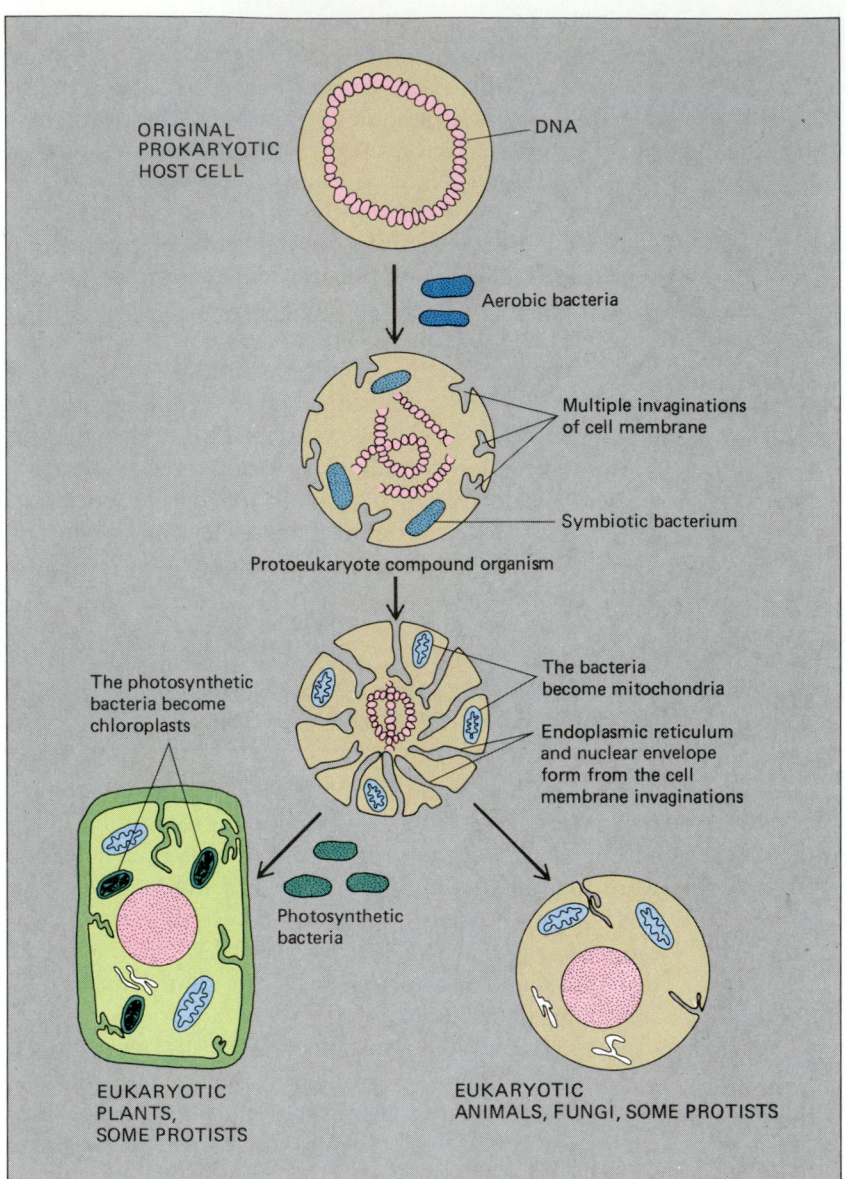

ORIGINAL
PROKARYOTIC
HOST CELL

DNA

Aerobic bacteria

Multiple invaginations
of cell membrane

Symbiotic bacterium

Protoeukaryote compound organism

The photosynthetic
bacteria become
chloroplasts

The bacteria
become mitochondria

Endoplasmic reticulum
and nuclear envelope
form from the cell
membrane invaginations

Photosynthetic
bacteria

EUKARYOTIC
PLANTS,
SOME PROTISTS

EUKARYOTIC
ANIMALS, FUNGI, SOME PROTISTS

FIGURE 20–6 The endosymbiotic theory of the origin of eukaryotes.

some evidence of life still remains. This evidence consists of traces of graphite or pure carbon, which may be the transformed remains of primitive life. These remains are especially abundant in what were the oceans and seas of that era. Fossils of what appear to be cyanobacteria have been recovered from several formations.

The fossils found in the later Precambrian rocks show clear-cut examples of some major groups of fungi, protists, plants, and animals. One source of rich deposits of late Precambrian fossils has been South Australia. The forms of life found there include jellyfish, corals, and segmented worms.

An Incredible Diversity Of Life Forms Evolved During The Paleozoic Era

The oldest subdivision of the **Paleozoic Era,** the **Cambrian Period,** is represented by rocks rich in fossils.

Fossils of all the present-day animal phyla, except the chordates, are present, at least in marine sediments. The sea floor was covered with simple sponges, corals, snails, bivalves, and other organisms.

In the **Ordovician Period,** much of what is now land was covered by shallow seas. Inhabiting the seas were giant **cephalopods,** squid or nautilus-like animals with straight shells 5 to 7 meters long and 30 centimeters in diameter. The first traces of the early vertebrate fish are also found in Ordovician rocks.

Two life forms of great biological significance appeared in the **Silurian Period,** the land plants and the air-breathing animals. The first known land plants resembled ferns. The evolution of terrestrial plants allowed terrestrial animals to evolve. Plants provided food and shelter for the first land animals. The only air-breathing land animals that have been discovered in Silurian rocks resembled scorpions.

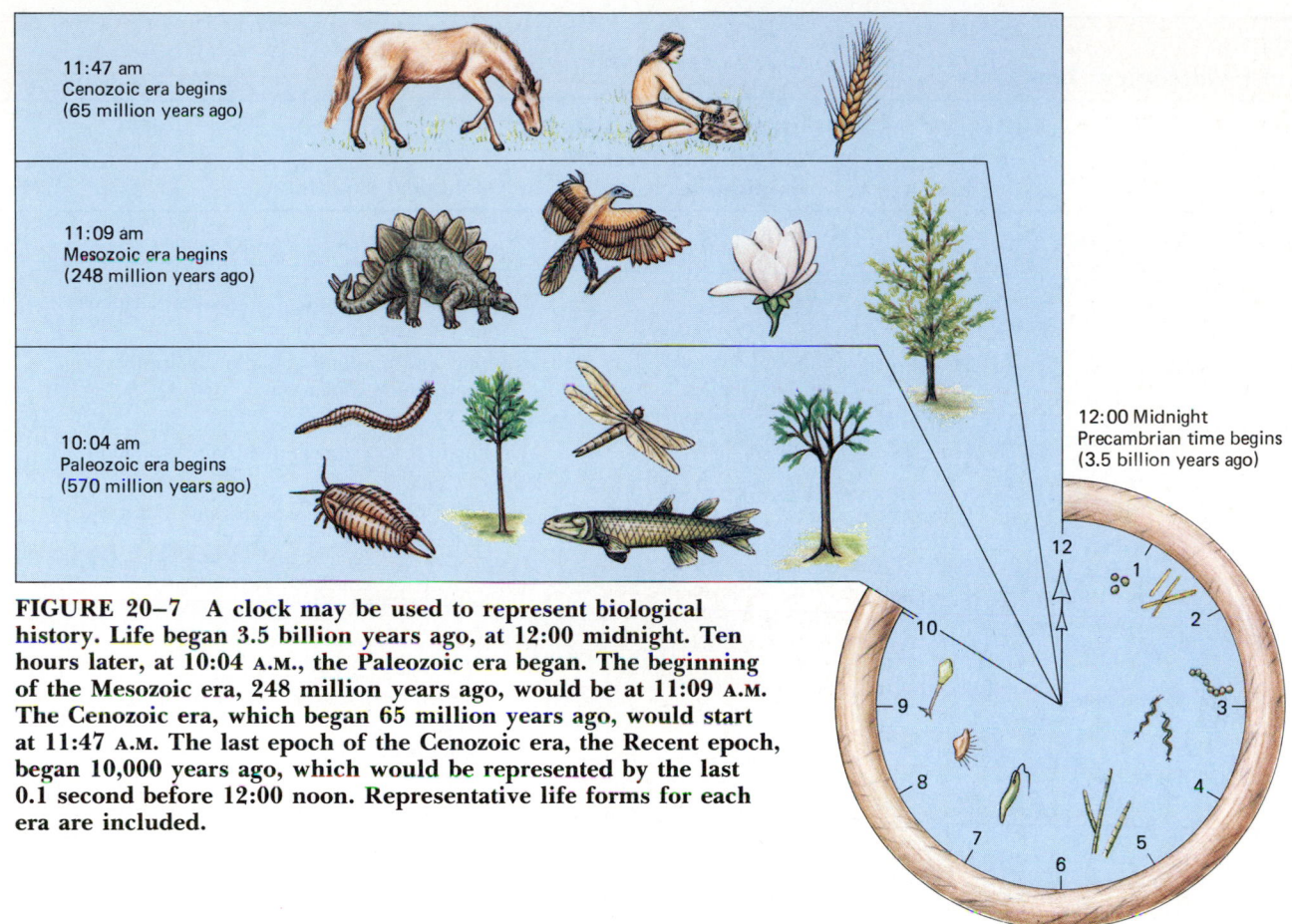

FIGURE 20–7 A clock may be used to represent biological history. Life began 3.5 billion years ago, at 12:00 midnight. Ten hours later, at 10:04 A.M., the Paleozoic era began. The beginning of the Mesozoic era, 248 million years ago, would be at 11:09 A.M. The Cenozoic era, which began 65 million years ago, would start at 11:47 A.M. The last epoch of the Cenozoic era, the Recent epoch, began 10,000 years ago, which would be represented by the last 0.1 second before 12:00 noon. Representative life forms for each era are included.

A great variety of fishes appeared in the **Devonian Period.** In fact, the Devonian is frequently called the "Age of Fishes." More recent Devonian sediments contain fossil remains of salamander-like ancient amphibians as well as wingless insects and millipedes. The early plants diversified during the Devonian Period. Forests of ferns, club mosses, horsetails, and seed ferns flourished.

The **Carboniferous Period** is named for the great swamp forests whose remains persist today as major coal deposits (Fig. 20–8). The land was covered with low swamps filled with horsetails, club mosses, ferns, seed ferns, and gymnosperms. The first reptiles appeared in the Carboniferous Period. Two important groups of winged insects occurred for the first time in the Carboniferous Period—cockroaches and dragonflies.

The final period of the Paleozoic Era, the **Permian Period,** was characterized by great changes in climate and topography. At the end of the Permian Period, mountain ranges formed in North America and Europe. A glaciation (ice sheet), spreading from the Antarctic, covered most of the Southern Hemisphere. Many Paleozoic forms of life may have been unable to adapt to the climatic and geological changes and became extinct. The seed plants became dominant, with the diversification of conifers and the appearance of cycads.

The Dinosaurs And Other Reptiles Dominated The Mesozoic Era

The **Mesozoic Era** began about 248 million years ago and lasted some 183 million years. It is divided into the **Triassic, Jurassic,** and **Cretaceous Periods.** The outstanding feature of the Mesozoic Era was the origin, differentiation, and final extinction of a large variety of reptiles. For this reason, the Mesozoic Era is commonly called the "Age of Reptiles." Of all the reptilian branches, the **dinosaurs** are the most famous (Fig. 20–8). Some of these were among the largest animals that ever lived: *Brontosaurus*, with a length of 21 meters; *Diplodocus*, with a length of 29 meters; and *Brachiosaurus*, with an estimated weight in excess of 50 tons.

Although the reptiles were the dominant animals of the Mesozoic Era, many other important organisms occur in the same formations. Most of the modern orders of insects appeared during that era. Snails and

TABLE 20–1
Life and the Geological Timetable[1]

Era	Period	Epoch	Time from Beginning of Period to Present (millions of years)[2]	Geological Conditions
Cenozoic (Age of Mammals)	Quaternary	Recent	0.01	End of last Ice Age; warmer climate
		Pleistocene	2.0	Four Ice Ages; glaciers in Northern Hemisphere; uplift of Sierras
	Tertiary	Pliocene	5	Uplift and mountain-building; volcanoes; climate much cooler
		Miocene	25	Climate drier, cooler; mountain formation
		Oligocene	38	Rise of Alps and Himalayas; most land low; volcanic activity in Rockies
		Eocene	55	Climate warmer
		Paleocene	65	Climate mild to cool; continental seas disappear
Mesozoic (Age of Reptiles)	Cretaceous		144	Continents separated; formation of Rockies; other continents low; large inland seas and swamps
	Jurassic		213	Climate mild; continents low; inland seas; formation of mountains; continental drift continues
	Triassic		248	Many mountains form; widespread deserts; continental drift begins
Paleozoic (Age of Ancient Life)	Permian		286	Continents merge as Pangaea; glaciers; formation of Appalachians; continents rise
	Carboniferous		360	Lands low; great coal swamps; climate warm and humid; later cooler
	Devonian		408	Glaciers; inland seas
	Silurian		438	Continents mainly flat; flooding
	Ordovician		505	Sea covers continents; climate warm
	Cambrian		570	Climate mild; lands low; oldest rocks with abundant fossils
(Precambrian) Proterozoic			1500	Planet cooled; glaciers; formation of earth's crust; mountains form
Archean			3.5 billion years ago	
Origin of the earth			4.6 billion years ago	
Origin of the universe			15–20 billion years ago	

[1] You may want to study this table starting from the bottom and working your way up through time.
[2] Based on Harland et al. *A Geologic Time Scale*. Cambridge, Cambridge University Press, 1982.

Plants and Microorganisms	Animals
Decline of woody plants; rise of herbaceous plants	Age of *Homo sapiens*
Extinction of many species	Extinction of many large mammals
Development of grasslands; decline of forests	Large carnivores; many grazing mammals; first known human-like primates
Flowering plants continue to diversify	Many forms of mammals evolve
Spread of forests, flowering plants, rise of monocotyledons	Apes evolve; all present mammal families are represented
Gymnosperms and flowering plants dominant	Beginning of Age of Mammals; modern birds
	Evolution of mammals
Rise of flowering plants	Dinosaurs reach peak, then become extinct; toothed birds become extinct; first modern birds; primitive mammals
Gymnosperms common	Large, specialized dinosaurs; first toothed birds; insectivorous marsupials
Gymnosperms dominate	First dinosaurs; egg-laying mammals
Conifers diversify; cycads evolve	Modern insects appear; mammal-like reptiles; extinction of many Paleozoic invertebrates
Forests of ferns, club mosses, horsetails, and gymnosperms	First reptiles; spread of ancient amphibians; many insect forms; ancient sharks abundant
Terrestrial plants well established; first forests; gymnosperms appear; bryophytes appear	Age of Fishes; amphibians appear; wingless insects appear; many trilobites
Vascular plants appear; algae dominant in aquatic environment	Fish evolve; terrestrial arthropods
Marine algae dominant; terrestrial plants first appear	Invertebrates dominant; first fish appear
Algae dominant in aquatic environment	Age of marine invertebrates; most modern phyla represented
Primitive algae and fungi, marine protozoans	Toward end, marine invertebrates
Evidence of first prokaryotic cells	

bivalves increased in number and diversity. Sea urchins reached their peak. Mammals first appeared in the Triassic Period, and birds first appear in Jurassic formations. During the early Triassic Period, the most abundant plants were gymnosperms. By the end of the Cretaceous Period, many flowering plants resembling present-day species had appeared and were the dominant vegetation.

Excellent bird fossils, some even showing the outlines of feathers, have been preserved from the Jurassic Period. *Archaeopteryx* is the classic example (Fig. 20–9). This animal was about the size of a crow, had rather feeble wings, jawbones armed with teeth, and a long reptilian tail covered with feathers. Increasingly, *Archaeopteryx* is interpreted as a representative of a rather rare group of reptiles, one branch of which gave rise to the birds, but not as a bird itself. True birds do occur in the Cretaceous rocks, some of them apparently even older than *Archaeopteryx*.

At the end of the Cretaceous Period, a great many animals and plants abruptly became extinct (Chapter 25). Changes in climate may have been a factor in their demise, although climatic change does not appear to be the sole cause of this mass extinction. Other explanations have been proposed, including the catastrophic collision of the earth with a giant meteorite. There is scientific evidence to support this collision (Fig. 20–10).

The Cenozoic Era Is Known As The Age Of Mammals

With equal justice, the **Cenozoic Era** could be called the "Age of Mammals," the "Age of Birds," the "Age of Insects," or the "Age of Flowering Plants." It is marked by the appearance of all these forms in great variety and numbers of species. It extends from 65 million years ago to the present. The Cenozoic Era is subdivided into two periods, the **Tertiary Period,** encompassing some 63 million years, and the **Quaternary Period,** which covers the last 2 million years.

During the Tertiary Period, grasses, which served as food, and dense forests, which afforded protection from predators, may have been important factors in leading to changes in the mammalian body pattern. Along with the tendency toward increased size, the mammals displayed tendencies toward an increase in the relative size of the brain and toward changes in the teeth and feet.

Evidence of the first known carnivores (meat eaters) appears in early Tertiary formations (Fig. 20–11). They were replaced in the middle Tertiary Period by more modern forms ancestral to the present-day carnivores, such as cats, dogs, bears, and weasels, as well as the web-footed marine carnivores, the seals and walruses.

FIGURE 20–8 *Tyrannosaurus*, the largest of the flesh-eating dinosaurs, reached a length of 15 meters and a height of 6 meters. Its head was as much as 2 meters long and was equipped with many sharp teeth, whose edges were serrated like the blades of steak knives. The long tail was probably used as a counterweight to the immense head.

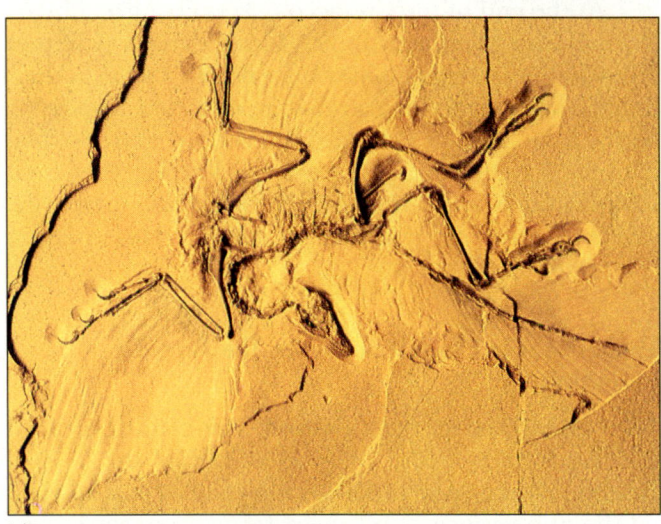

(a)

(b)

FIGURE 20–9 (*a*) Fossil of *Archaeopteryx*, a tailed, toothed, bird-like reptile from the Jurassic period. (*b*) A reconstruction of *Archaeopteryx*. (*a*, John D. Cunningham/Visuals Unlimited)

The Quaternary Period, which extends to the present, has had four "ice ages," or periods of glaciation. At their greatest extent, ice sheets covered nearly 4 million square miles of North America, extending south as far as the Ohio and Missouri Rivers. During these glaciations, enough water was removed from the sea and locked in the ice to lower the sea level by 65 to 100 meters. This created land connections, highways for the dispersal of many land forms. Examples included a land bridge between Siberia and Alaska at the Bering Strait and between England and the European continent.

A considerable number of New World mammals, including the saber-toothed tiger, the mammoth, and the giant ground sloth, became extinct during the Quaternary Period, very possibly as a result of early human hunting. The Quaternary Period has been marked by the extinction of many species of plants, especially woody ones, and the appearance of numerous herbaceous ones.

Today millions of distinguishable kinds of organisms inhabit our planet. To make some order out of this diversity and to communicate knowledge, we need a system of classification. Precise descriptions and designations of organisms are essential to biologists in their work. The science of naming and classifying organisms is known as **taxonomy.**

ORGANISMS ARE CLASSIFIED USING THE BINOMIAL SYSTEM OF NOMENCLATURE

In the 18th century, Carolus Linnaeus, a Swedish botanist and natural historian, simplified the scientific classification of organisms. Before Linnaeus, each organism had a lengthy descriptive name, sometimes composed of ten or more Latinized words! In the Linnaean system, which is also called the **binomial system of nomenclature,** each species is assigned a two-part name (see Chapter 1). The first part of the name designates the **genus** (plural, genera), and the second part designates the **species epithet.** Corn, for example, is assigned to the genus *Zea;* its species epithet is *mays.* The proper scientific name for corn is *Zea mays.* By convention, the genus name is always capitalized, whereas the species epithet is usually not capitalized. Both names are underlined or italicized. The genus name is sometimes used alone to designate all species in the genus, but the species epithet is never used alone; it is always preceded by the genus name.

Scientific names are generally composed from Greek or Latin roots, or from Latinized versions of the names of persons, places, or characteristics. For example, the genus name for the bacterium *Escherichia coli* is based on the name of the scientist Theodor Escherich, whereas the specific epithet *coli* reminds us that *E. coli* live in the colon, or large intestine.

FIGURE 20–10 Sandwiched between the Mesozoic and Cenozoic sediments is a small band of dark clay with a high concentration of iridium. Iridium is rare on earth but abundant in meteorites, leading many to believe that the earth was hit by a large extraterrestrial object at this time. The coin on the iridium-rich layer indicates the size of the layer. (Lawrence Berkeley Laboratory, University of California-Berkeley)

Assigning scientific names permits taxonomy to be a truly international study, for many scientifically important organisms do not have common names, and the names of those that do often vary in different locations and languages. A researcher in Argentina can

FIGURE 20–11 Reconstruction of an ancient carnivorous mammal from the Eocene period. It is eating a tiny horse, *Hyracotherium*, which is thought to have been ancestral to modern horses. (No V/C 1326 (painting by Charles R. Knight), Courtesy Department of Library Sciences, American Museum of Natural History)

FOCUS ON Continental Drift

In 1912, Alfred Wegener, who had noted a similarity between the geographical shapes of South America and Africa, proposed that all the land masses had been joined into one huge supercontinent, which he called Pangaea. He further suggested that Pangaea had subsequently broken apart, and the various land masses had separated, in a process known as **continental drift.** Wegener did not know of any mechanism that could have caused continental drift, so his theory was largely ignored.

In the 1960s, scientific evidence accumulated that provided the explanation for continental drift. The earth's crust is composed of seven large plates (plus a few smaller ones) that float on the mantle. The land masses are situated on some of these plates. As the plates move about, the continents change their relative positions. The movement of the crustal plates is termed **plate tectonics.**

Any area where two plates meet is a site of intense geological activity. Earthquakes and volcanoes are common in this region. Both San Francisco, noted for its earthquakes, and the volcano Mount Saint Helens are situated where two plates meet. If land masses are on the edges of two meeting plates, mountains may be formed. The Himalayas formed when the plate carrying India rammed into the plate carrying Asia. When two plates grind together, one of them is sometimes buried under the other, in a process known as **subduction.** When two plates move apart, a ridge of lava forms between them that continually expands as the plates move farther apart. The Atlantic Ocean is getting larger because of the buildup of lava along the mid-Atlantic ridge, where two plates are separating.

Knowledge that the continents were at one time connected and have since drifted apart is useful in explaining the geographical distribution of plants and animals, or biogeography (Chapter 17). Likewise, continental drift has played a major role in the evolution of different life forms. When Pangaea originally

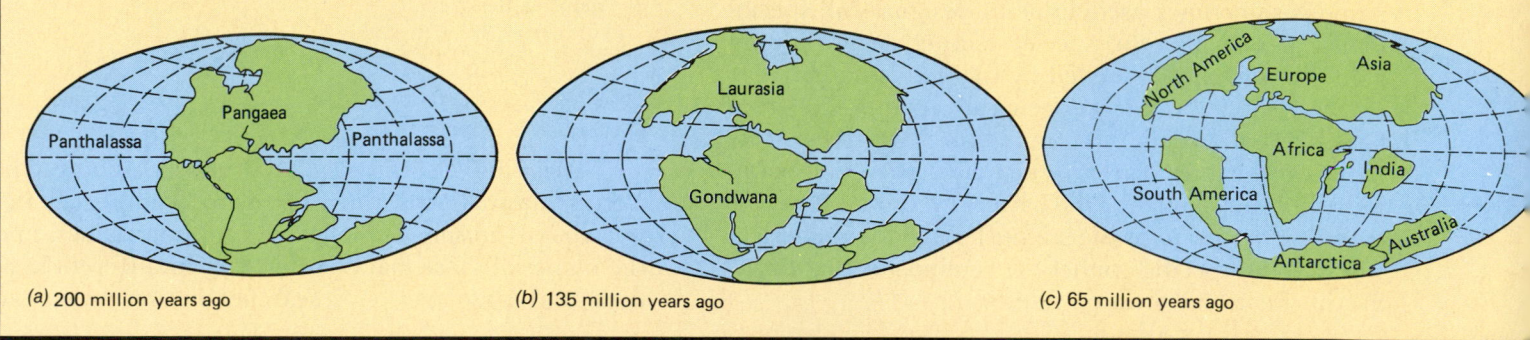

(a) 200 million years ago   (b) 135 million years ago   (c) 65 million years ago

---

know exactly which organisms were used in a published study by an Australian, and if it is possible to obtain them, can repeat or extend the Australian's experiments.

The narrowest category in the Linnaean system is the species, and the broadest is the kingdom. In between there exists a range of categories that form a hierarchy (Fig. 20–12 and Table 20–2). A species is a group of organisms that can interbreed in their natural environment and are reproductively isolated from other organisms (Chapter 19). Closely related species are assigned to the same genus, and closely related genera may be grouped together in a single **family.** Families are grouped into **orders,** orders into **classes,** classes into **phyla,** and phyla into **kingdoms.** In classifying plants, the term **division** is used rather than phylum.

MANY BIOLOGISTS RECOGNIZE FIVE KINGDOMS

For hundreds of years, biologists regarded living things as falling into two broad categories—plants and animals. With the development of microscopes, it became increasingly obvious that many organisms did not fit very well into either the plant or animal kingdom. For instance, the bacteria lack nuclear envelopes and other internal membranes. This difference, which separated them from all other organisms, is far more fundamental than the differences between plants and animals. With our present knowledge, it is difficult to consider bacteria as plants, as was done formerly. Certain organisms—the protist *Euglena*, for example—seem to possess characteristics of both plants and animals. In fact, many single-celled organisms seem to

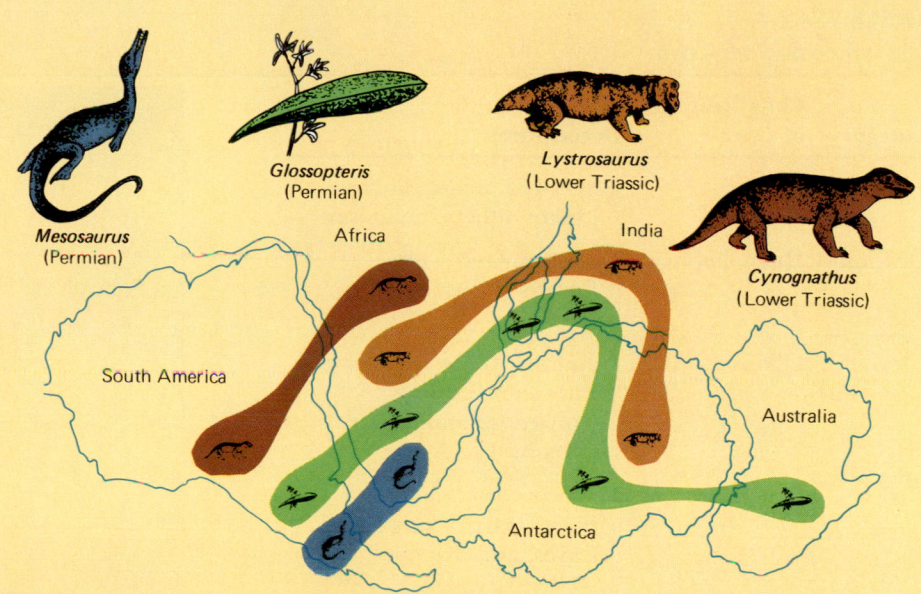

formed, about 250 million years ago, it brought together plants and animals that had evolved separately from one another, leading to competition and possible extinctions. Marine life was adversely affected, largely because, with the continents joined as one large mass, there would have been less coastline. Coastal areas are shallower and, therefore, have high concentrations of marine organisms.

Pangaea separated into several land masses approximately 180 million years ago. As the continents began to drift apart, populations became geographically isolated in different environmental conditions, the ideal setting for evolution.

The distribution of the same fossils of different animal and plant species on four continents suggests that the continents were once joined.

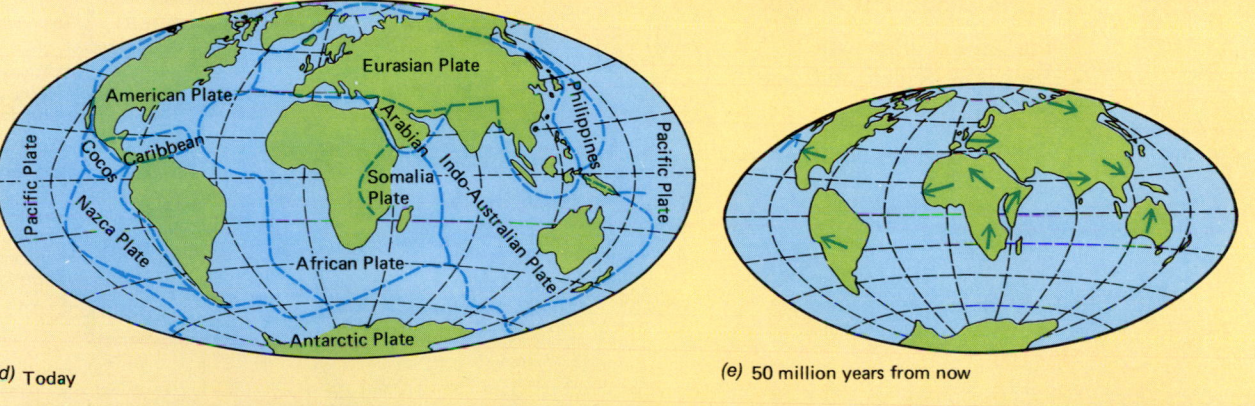

(d) Today

(e) 50 million years from now

FIGURE 20–12 The principal categories used in classifying an organism are arranged in a hierarchy from highest, the kingdom, to lowest, the species. The classification of the cat is shown.

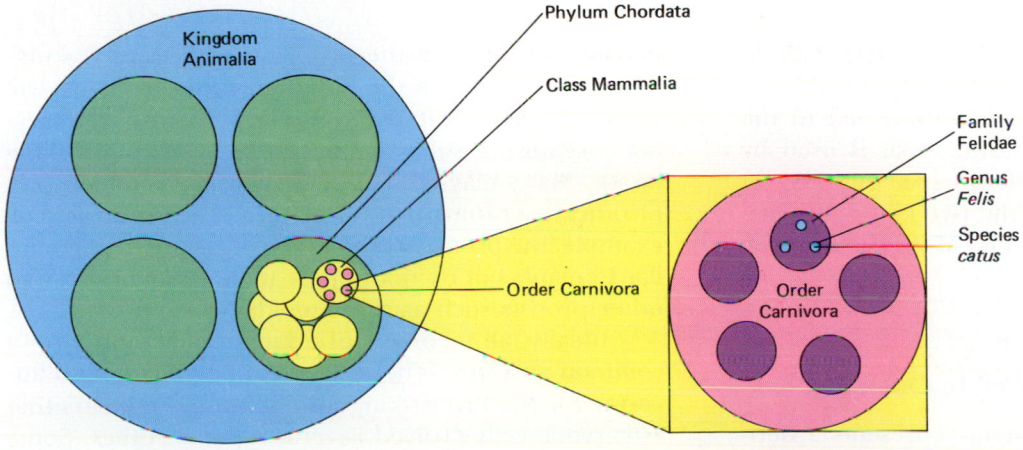

TABLE 20–2
Classification of Corn

Category	Classification of Corn	Description
Kingdom	Plantae	Terrestrial, multicellular, photosynthetic organisms
Division	Magnoliophyta	Vascular plants with flowers, fruits, and seeds
Class	Liliopsida	Monocots: flowering plants with one seed leaf (cotyledon) and flower parts in three's
Order	Cyperales	Monocots with reduced flower parts, elongated leaves, and dry one-seeded fruits
Family	Poaceae	Grasses with hollow stems, fruit and grain, and abundant endosperm in seed
Genus	*Zea*	Tall annual grass with separate female and male flowers
Species	*Zea mays*	Only one species in genus—corn

Classification of Corn, *Zea mays*

KINGDOM Plantae
Terrestrial, multicellular photosynthetic organisms

DIVISION Magnoliophyta
Vascular plants with flowers, fruits, and seeds

CLASS Liliopsida
Monocots: Flowering plants with one seed leaf (cotyledon) and flower parts in threes

ORDER Cyperales
Monocots with reduced flower parts, elongated leaves, and dry 1-seeded fruits

FAMILY Poaceae
Grasses with hollow stems, fruit a grain, and abundant endosperm in seed

GENUS *Zea*
Tall annual grass with separate female and male flowers

SPECIES *Zea mays*
Only one species in genus —corn

have more in common with one another than with either multicellular plants or multicellular animals.

These and other considerations have led to the five-kingdom system of classification that is used by many biologists today (Fig. 20-13). As outlined in Chapter 1 and in Table 20–3, the five kingdoms currently recognized are kingdoms Monera, Protista, Fungi, Plantae, and Animalia.

Protists Do Not Fit Together Well In One Kingdom

Although the five-kingdom system represents a definite improvement over the two-kingdom system, it is not perfect. Most of the problems with the five-kingdom system concern the kingdom Protista. As we will see in Chapter 21, it includes relatively simple eukaryotic organisms such as amoebas, algae, and slime molds. Many may be more closely related to members of other kingdoms than to certain other protists. For example, the protists called green algae are clearly similar to plants but do not appear to be closely related to other protists such as slime molds or red algae.

Ideally, all members of a kingdom should have a common ancestor. There seems to be no common ancestor for the Protist kingdom because it appears that eukaryotic cells evolved several separate times. Some biologists want to resolve this problem by dividing the

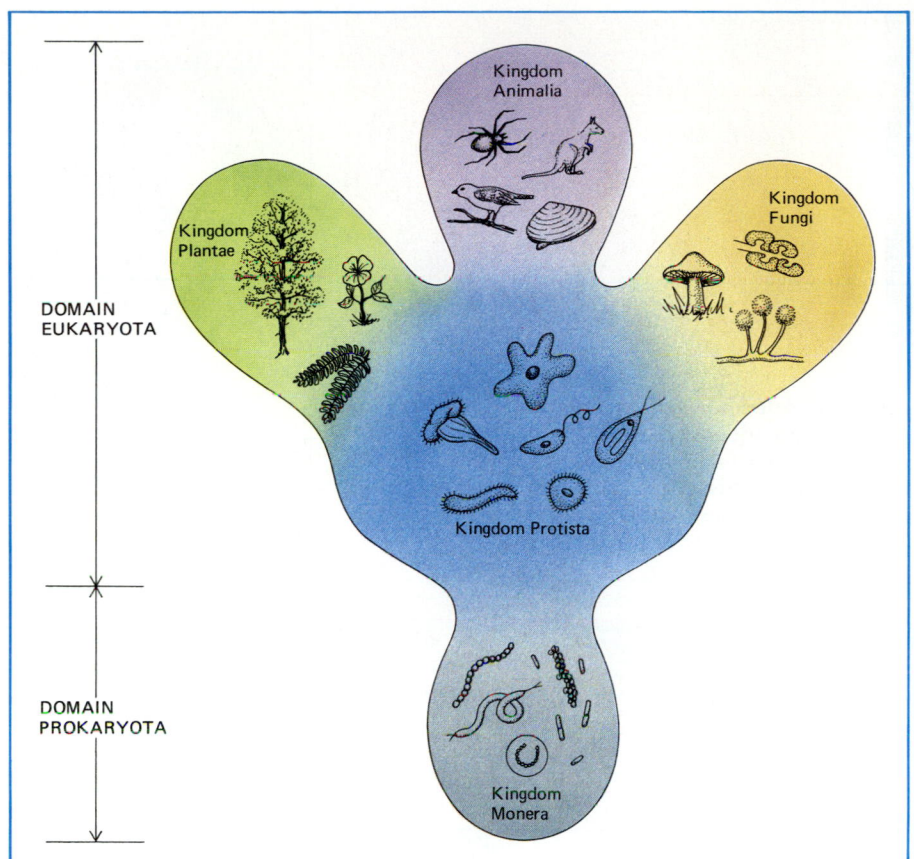

FIGURE 20–13 The five-kingdom system of classification. Protists, which are simple eukaryotic organisms, evolved from ancient prokaryotes (Kingdom Monera). Plants, animals, and fungi arose from different ancestral protists.

Protist kingdom into several additional kingdoms along more natural groupings. However, most biologists would rather deal with the limitations of the five-kingdom system than with additional kingdoms.

Viruses Cannot Be Assigned To Any Of The Five Kingdoms

The five-kingdom system cannot comfortably accommodate the **viruses.** All other life is at least cellular, but viruses are not. Indeed, it is not certain whether they are alive at all in the conventional sense of the term. They are large, infectious macromolecules (Fig. 20–14). Most viruses consist of a nucleic acid core, either DNA or RNA, which is surrounded by a protein coat.

As mentioned in Chapter 10, a bacteriophage, or phage, is a virus that attacks a bacterium. Ordinarily a phage drifts passively until, by chance, it attaches itself to a host bacterial cell by means of a protein coat, and its nucleic acid is injected into the host (Fig. 20–15). Once inside, the viral nucleic acid may take over the metabolic machinery of the host cell, using this machinery to replicate itself many times. When the cell finally bursts open, swarms of viruses are released.

Some biologists have suggested that viruses, because of their simplicity, represent a primitive form of life. However, because all viruses totally depend on living cells to reproduce, other biologists think that viruses are descendants of cellular organisms that have become highly specialized as parasites. Still others hypothesize that viruses were originally fragments of DNA or RNA broken off from the nucleic acids of cellular organisms.

Viruses are no longer considered the simplest form of life. **Viroids** are even smaller and simpler than viruses. Each viroid consists of a very short strand of RNA without any sort of protective coat. To date, viroids have only been found to infect plants. They have been linked to several plant diseases, including potato spindle-tuber disease and a disease that causes stunting of chrysanthemums. A disease-causing agent even smaller than a viroid has apparently been discovered. Tentatively named a **prion,** this infectious particle appears to consist only of a protein. A prion is thought to cause scrapie, a neurological disease of sheep, and may be linked to several rare human diseases of the central nervous system.

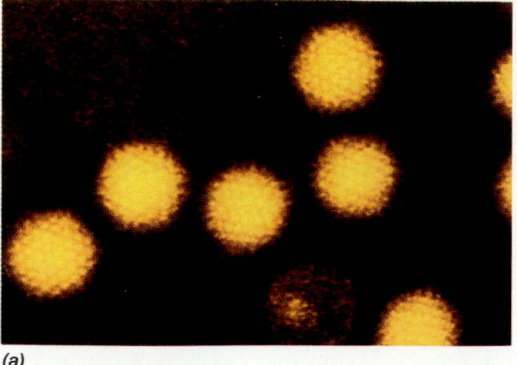

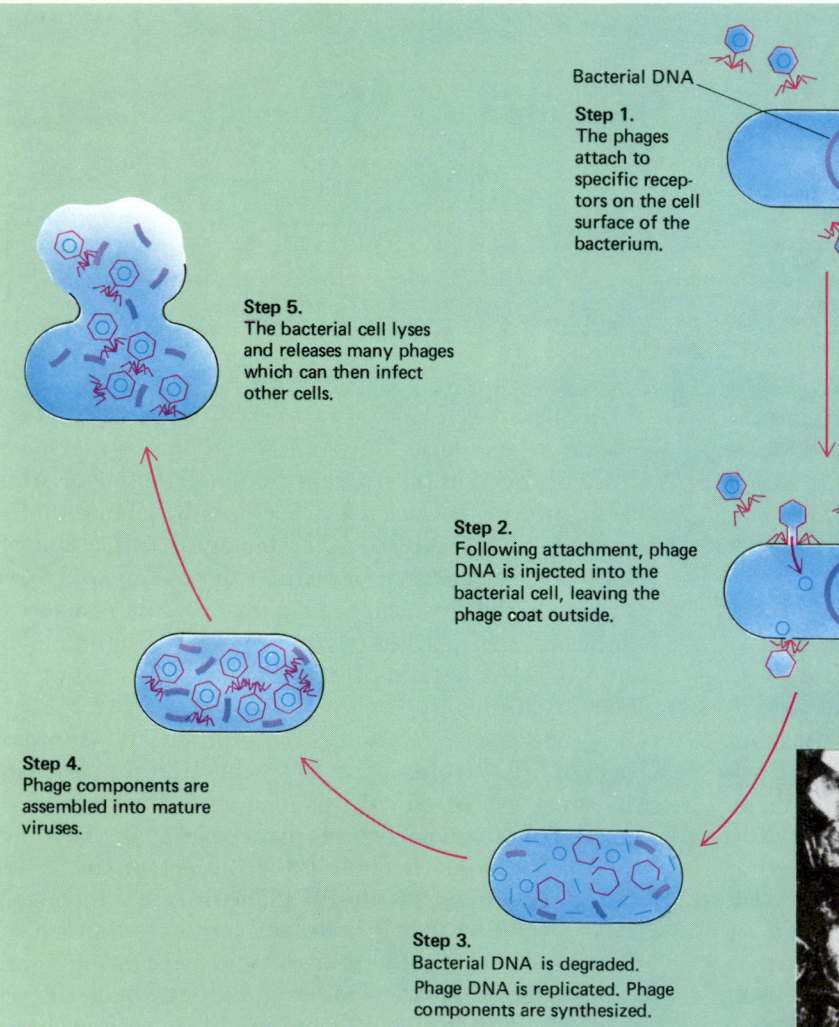

FIGURE 20–14 Several representative viruses. (*a*) **Adenovirus. The common cold is caused by many different viruses, including an adenovirus.** (*b*) **Tobacco mosaic virus. This rigid, elongated virus causes plant disease in a number of plants, including tobacco.** (*c*) **Bacteriophage. This virus, which infects bacterial cells, is a complex combination of helical and polyhedral shapes.** (*a*, Robert Caughey/Visuals Unlimited; *b*, K. G. Murti/Visuals Unlimited; *c*, Courtesy of Dr. Lyle C. Dearden)

FIGURE 20–15 (*a*) **Sequence of bacteriophage (phage) infection of a bacterial cell.** (*b*) **Phages infecting a bacterial cell, *Escherichia coli*. The heads, tails, and base plate of most of the viruses are clearly visible.** (*b*, Courtesy of Dr. Lee D. Simon, Institute for Cancer Research, Philadelphia)

Bacterial DNA

Step 1.
The phages attach to specific receptors on the cell surface of the bacterium.

Step 2.
Following attachment, phage DNA is injected into the bacterial cell, leaving the phage coat outside.

Step 3.
Bacterial DNA is degraded. Phage DNA is replicated. Phage components are synthesized.

Step 4.
Phage components are assembled into mature viruses.

Step 5.
The bacterial cell lyses and releases many phages which can then infect other cells.

(*a*)

(*b*)

TABLE 20-3		
Five Kingdoms: Monera, Protista, Fungi, Plantae, and Animalia		
Kingdom	**Characteristics**	**Ecological Role and Comments**
Monera	Prokaryotes (lack distinct nuclei and other membranous organelles); single-celled; microscopic	
Bacteria	Cell walls composed of peptidoglycan; cells are spherical (cocci), rod-shaped (bacilli), or coiled (spirilla); metabolically varied	Most are decomposers; some parasitic (and pathogenic); some chemosynthetic autotrophs; some photosynthetic; important in recycling nitrogen and other elements; some utilized in industrial processes
Cyanobacteria (blue-green algae)	Specifically adapted for photosynthesis and use water as hydrogen source; chlorophyll and associated enzymes organized into layers in cytoplasm; some can fix nitrogen	Producers; blooms (population explosions) associated with water pollution
Protista	Eukaryotes; mainly unicellular or colonial	
Protozoa	Microscopic; heterotrophic; depend upon diffusion to support their metabolic activities	Important part of zooplankton; near base of many food chains; some are pathogenic
Eukaryotic algae	Photosynthetic; sometimes hard to differentiate from protozoa; some have brown or red pigments in addition to chlorophyll	Very important producers, especially in marine and freshwater ecosystems; phytoplankton
Slime molds	Animal or protozoan characteristics during part of life cycle; fungal traits during remainder	
Fungi	Heterotrophic; absorb nutrients; do not photosynthesize; body composed of threadlike hyphae; hyphae may form tangled masses called mycelia, which infiltrate fungus's food or habitat	Decomposers; some parasites (pathogenic); some used as food; yeast used in making bread and alcoholic beverages; some used to make industrial chemicals or antibiotics; responsible for much spoilage and crop loss
Plantae	Multicellular; complex; adapted for photosynthesis; plants have multicellular reproductive organs; pass through distinct developmental stages and alternation of generations; cell walls of cellulose	Almost entire ecosphere depends upon plants in their role as primary producers; one of most important sources of oxygen in earth's atmosphere
Animalia	Multicellular heterotrophs, many of which exhibit advanced tissue differentiation and complex organ systems; most able to move about by muscular contraction; extremely and quickly responsive to stimuli, with specialized nervous tissue to coordinate responses	Almost sole consuming organisms in biosphere; some specialized as herbivores, carnivores, or detritus feeders

■ CHAPTER SUMMARY

I. Living cells formed from organic molecules by chemical evolution.
   A. The four requirements for chemical evolution are (1) absence of oxygen, (2) energy, (3) chemical building blocks, and (4) sufficient time.
   B. The sequence of events in chemical evolution were (1) the origin and accumulation of small organic molecules, (2) the assembly of macromolecules, (3) the formation of macromolecular assemblages (pre-cells), and (4) the formation of cells.

II. The first cells were anaerobic and prokaryotic.
   A. The oldest cells in the fossil record are 3.4 to 3.5 billion years old.
   B. The evolution of photosynthesis ultimately changed early life because it generated oxygen.
   C. Mitochondria and chloroplasts may have evolved from prokaryote endosymbionts.

III. The earth's history is divided into eras, periods, and epochs.
   A. Cells arose and evolved into different groups of animals, plants, protists, and fungi during the Precambrian.
   B. During the Paleozoic Era, land plants evolved to the gymnosperms, and fish and amphibians flourished.
   C. The Mesozoic Era was characterized by the evolution of flowering plants and reptiles. Insects flourished, and birds and early mammals appeared.
   D. In the Cenozoic Era, which includes the present time, flowering plants and mammals diversified.

IV. In the binomial system of nomenclature, every organism is given a two-part scientific name.
   A. Scientific names designate the genus and species.
   B. The complete hierarchy includes kingdom, phylum (division), class, order, family, genus, and species.
V. Organisms are grouped into five kingdoms.
   A. These kingdoms are Monera, Protista, Fungi, Plantae, and Animalia.
   B. Viruses do not fall into any of the five kingdoms.
      1. Viruses are not cellular.
      2. A typical virus consists of a nucleic-acid core that is surrounded by a protein coat.

■ POST-TEST

1. _____ is a gas that was not part of the earth's atmosphere before living organisms evolved.
2. Energy, chemical building blocks, absence of oxygen, and _____ were the four main requirements for chemical evolution.
3. Although Oparin envisioned life as originating in a "sea of organic soup," it is more likely that it evolved on _____ surfaces.
4. The first autotrophs probably used sunlight to split _____ _____.
5. The _____ theory may explain how certain eukaryotic organelles like chloroplasts evolved.
6. All the present-day animal phyla except the chordates are represented by fossils from the _____ Era.
7. Fossil remnants of dinosaurs are found in rocks from the _____ Era.
8. The most recent era in the earth's history, which includes the present time, is the _____ Era.
9. Earthquakes, subduction, mountain formation, and _____ may occur where two tectonic plates meet.
10. Using the binomial system of nomenclature, the scientific name of each species consists of two parts, the _____ and the _____ epithet.
11. The mold that produces penicillin is *Penicillium notatum*. Its genus is _____.
12. Closely related genera may be grouped together in a single _____.
13. The botanical equivalent of a phylum is a _____.
14. Most viruses consist of two parts, a nucleic-acid core and a protective _____ coat.
15. A _____ is an "organism" even smaller than a virus; it consists only of a very short strand of RNA.

■ REVIEW QUESTIONS

1. If chemical evolution occurred once, why can't it occur again?
2. What are the four requirements for chemical evolution, and why is each essential?
3. Give some of the evidence used to support the endosymbiont theory.
4. Put the following in chronological order with respect to their evolution:
   a. Reptiles, mammals, amphibians, fish.
   b. Flowering plants, ferns, gymnosperms.
5. List from memory the complete hierarchy of Linnaean classification, from kingdom to species.
6. Give a distinguishing characteristic for each of the five kingdoms.
7. Which, if any, properties of life does a virus possess?

■ RECOMMENDED READING

Lewin, R. *Thread of Life*. Washington, D.C., Smithsonian Books, 1982. A visually striking discussion of evolution from the origin of life to the present by one of world's best scientific writers.

P A R T 6

Diversity

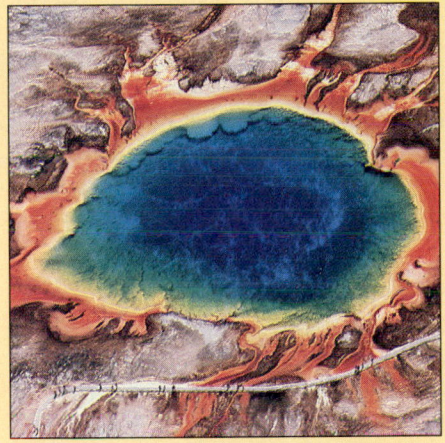

Protozoa, algae, slime molds, and water molds • Decomposers in the ecosystem • Fungi and disease • Invertebrate diversity • Animals of the sea • Symbiosis in lichens • Seeds versus spores

Fish, amphibians, reptiles, birds, and mammals • Adaptable insects • Radial and bilateral symmetry • Flatworms, roundworms, and annelids • The success of the vertebrates • Hominid evolution

21
Monerans and Protists

OUTLINE

I. Monerans are prokaryotes, the simplest and oldest known cells
   A. Prokaryotes differ structurally from eukaryotes
   B. Bacteria differ in their needs for oxygen
   C. Eubacteria have diverse methods of obtaining food
   D. The archaebacteria differ from eubacteria
   E. Bacteria reproduce by fission
   F. Bacteria can exchange DNA
   G. Bacteria survive harsh conditions through dormancy and endospore formation
   H. Studies of bacteria have increased our knowledge of all cells

II. Protists are eukaryotes and are mostly unicellular
   A. Protozoa are animal-like protists
      1. Amoebas and some other protozoa move by forming pseudopods
      2. Flagellates use flagella for locomotion
      3. The ciliates include the paramecia
      4. Sporozoa are parasitic
   B. Algae are plant-like protists
      1. Dinoflagellates have cellulose armor
      2. Diatoms have shells made of silica
      3. Green algae are very much like plants
      4. Red algae are important commercially
      5. Brown algae are the giants among the protists
      6. Euglenoids resemble both plants and animals
   C. Slime molds and water molds are fungus-like protists
      1. The plasmodial slime mold is a mass of naked cytoplasm
      2. Cellular slime molds resemble amoebas
      3. Water molds resemble fungi
   D. The simple protists are considered the earliest eukaryotes

LEARNING OBJECTIVES

After you have studied this chapter you should be able to:
1. Describe the distinguishing characteristics of monerans.
2. Describe the structure of a bacterial cell.
3. Compare the types of bacteria with regard to methods of obtaining food.
4. Summarize the ecological importance of the bacteria and of the cyanobacteria.
5. Characterize the common features of the kingdom Protista.
6. Summarize current theories on the origin of eukaryotic cells and of multicellularity among the protists.
7. Briefly characterize representative protozoa phyla.
8. Briefly characterize representative algae groups.
9. Briefly characterize the slime molds and water molds.

Grand Prismatic Spring, Yellowstone National Park. (Paul Chesley/Photographers, Aspen)

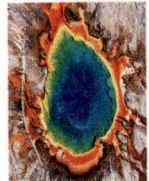

Microscopic life is essential for the continued existence of all living things. Some bacteria keep all life going by "fixing" nitrogen, that is, by combining the nitrogen of the atmosphere with other elements. Other bacteria are decomposers. Without bacteria and fungi, all available carbon, nitrogen, phosphorus, and sulfur would be tied up in the wastes and dead bodies of plants and animals. Life would soon cease because no materials would be available to make new cell parts. Algae are the base of the food chain in aquatic environments.

In this chapter we look at some of the organisms in the kingdoms Monera and Protista. The next chapter, which discusses kingdom Fungi, completes our survey of "simple" organisms.

KEY CONCEPTS

■ The prokaryotes, or monerans, are simple cells with few or no internal membranes.

■ Most bacteria are heterotrophs, but some are photosynthetic. Cyanobacteria practice photosynthesis in much the same fashion as do plants.

■ The protists are simple eukaryotes, with most living as single cells or as colonies.

■ The protists include protozoa, algae, slime molds, and water molds.

■ Protists are widely assumed to have originated from symbiotic unions of prokaryotes. Protists or protist-like ancestors are believed to have given rise to animals, plants, and fungi.

■ MONERANS ARE PROKARYOTES, THE SIMPLEST AND OLDEST KNOWN CELLS

All prokaryotes, the bacteria and cyanobacteria, are assigned to their own kingdom, the **kingdom Monera.** They are much smaller than eukaryotic cells. Most are unicellular organisms, though some form colonies or filaments.

Prokaryotic cells have no nuclear envelope nor the membrane-bounded organelles such as the chloroplasts, mitochondria, endoplasmic reticulum, and Golgi complex that are typical of eukaryotic cells (Fig. 21–1). Prokaryotes do have some structures, such as ribosomes, seen in eukaryotes, but the prokaryotic versions often differ. Though a prokaryote usually has a cell wall surrounding the plasma membrane, for example, its composition differs from that of eukaryotic cell

walls. Bacterial flagella also differ from eukaryotic flagella.

Despite the relative simplicity of prokaryotic cells, the diversity of the microbial world is astonishing. The electron microscope has revealed that what we used to think of as simple plants, the "blue-green algae," are in essence bacteria, what we now call "cyanobacteria." Even more striking, some microorganisms, the **archaebacteria,** have molecular traits that set them apart from all other cells. All monerans that are not archaebacteria are called **eubacteria.**

Prokaryotes Differ Structurally From Eukaryotes

As in eukaryotic cells, the plasma membrane governs the passage of molecules into and out of the bacterial cell. Bacterial plasma membranes, in addition, contain the enzymes needed to operate the electron transport system during cellular respiration (recall that in eukaryotes these enzymes are found in the membranes of mitochondria). In some bacterial cells, the enzymes for cellular respiration are located in the **mesosome,** a complex of internal membranes formed by infoldings of the plasma membrane.

The **cell wall** surrounding the bacterial plasma membrane is a strong, rigid framework that supports the cell, maintains its shape, and probably most important, keeps it from bursting because of osmotic inflow of water. Most bacteria seem adapted to hypotonic surroundings (surroundings in which water is more concentrated outside the cell than inside), but some are able to live in concentrated salt solutions.

The great strength of the bacterial cell wall may be due to a unique macromolecule found only in prokaryotes. **Peptidoglycan** consists of two unusual types of sugar linked with short peptides to form, in effect, a single, giant macromolecule that surrounds the entire plasma membrane. Bacteria rarely survive without walls under natural conditions.

Almost 100 years ago the Danish physician Christian Gram developed a staining procedure we still use to group bacteria. Those that absorb and retain a special violet stain are called **gram-positive.** Those that do not retain the stain are called **gram-negative.** The plasma membrane of a gram-positive bacterium is surrounded by a very thick cell wall consisting primarily of peptidoglycan. The plasma membrane of a gram-negative bacterium is surrounded by a thin peptidoglycan layer and an additional outer layer resembling a second plasma membrane (Fig. 21–2).

The antibiotic penicillin causes bacterial cells to die by interfering with peptidoglycan synthesis, ultimately leaving a fragile cell wall that cannot protect the bacterial cell. Penicillin works most effectively against gram-positive bacteria.

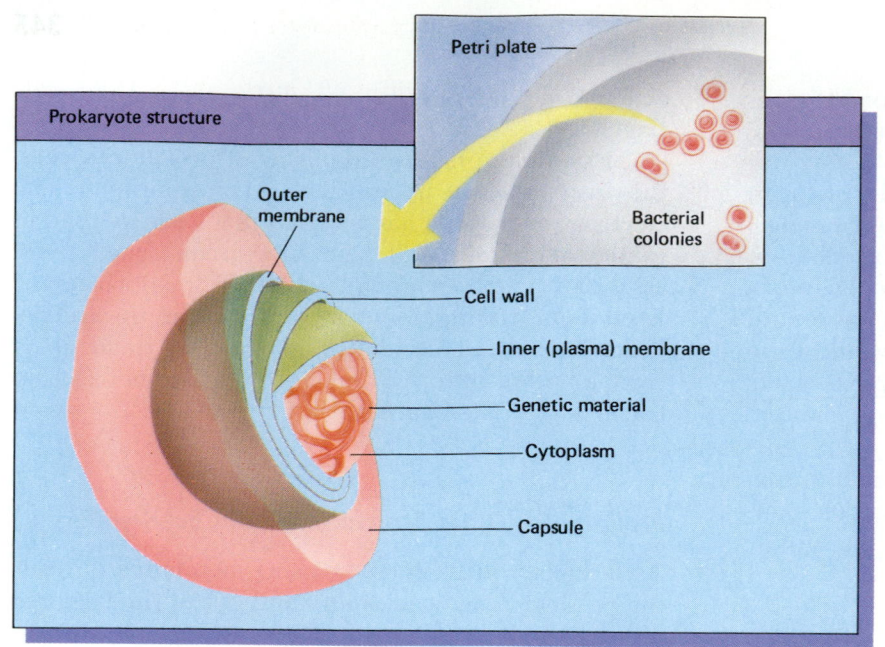

Petri plate

Prokaryote structure

Outer membrane

Cell wall

Inner (plasma) membrane

Genetic material

Cytoplasm

Capsule

Bacterial colonies

FIGURE 21–1 Much simpler in structure than eukaryotes, prokaryotes are mainly distinguished by what they lack. The much-folded DNA molecule (the genetic material) is not surrounded by a nuclear envelope. There are no mitochondria or chloroplasts. The plasma membrane of prokaryotes may be surrounded, however, by several layers of material lacking in eukaryotes. The cell wall maintains turgidity and cellular shape, additional membranes may help to confine protons used in chemiosmosis, and a capsule serves for defense. Shown here is a bacterium, one of millions of microscopic cells found in a single colony in culture.

In bacteria, DNA is found mainly in a single, long circular molecule called a **chromosome** (although histones and other proteins are not associated with the DNA as they are in eukaryotic chromosomes). Stretched out, the bacterial chromosome is about 1000 times longer than the cell itself. Some DNA may also be present in smaller molecules called **plasmids,** which replicate independently of the chromosome. Bacterial plasmids often carry genes that make bacteria resistant to antibiotics.

Many types of bacteria propel themselves with whip-like outgrowths called **flagella** (Fig. 21–3). Bacte-

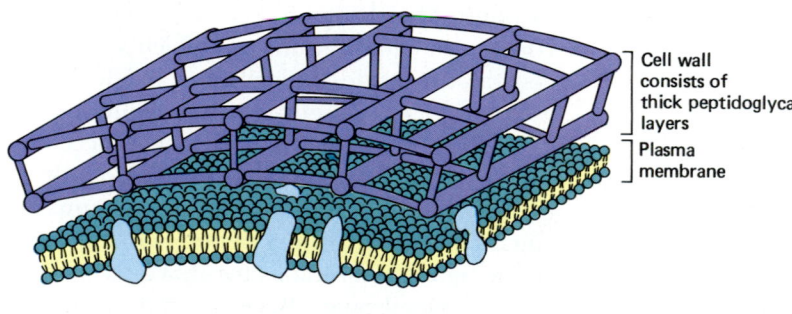

Cell wall consists of thick peptidoglycan layers

Plasma membrane

Gram positive wall

FIGURE 21–2 A simplified, schematic representation of the peptidoglycan layer in gram-positive and gram-negative bacterial cell walls. The gram-negative bacterium has a second lipid bilayer outside the cell wall which resembles the plasma membrane.

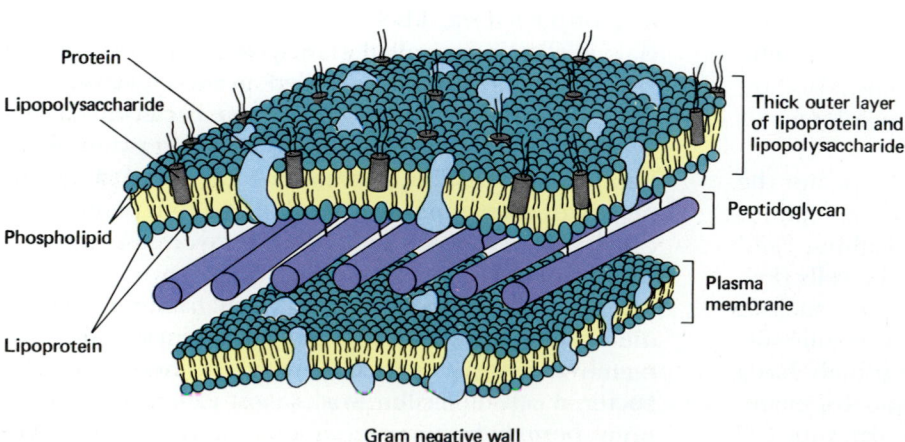

Protein

Lipopolysaccharide

Phospholipid

Lipoprotein

Thick outer layer of lipoprotein and lipopolysaccharide

Peptidoglycan

Plasma membrane

Gram negative wall

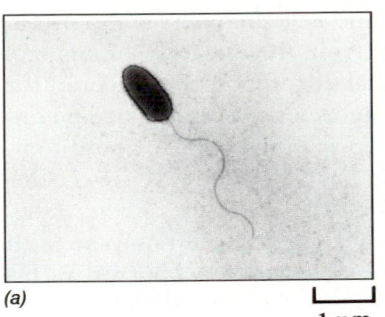 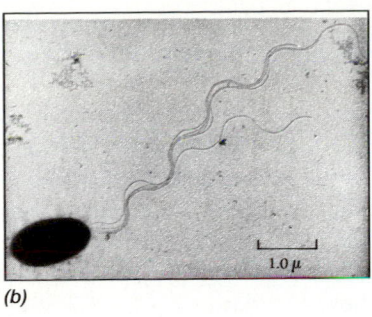 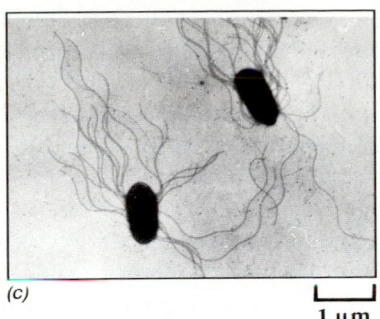

(a)   (b)   1.0 μ   (c)

1 μm   1 μm

FIGURE 21–3   Bacterial flagella. (a) A single flagellum at the end of the bacterium *Psuedomonas aeruginosa* (approximately ×5000). (b) Some bacteria have a tuft of flagella at one end of the cell. (c) In the bacterium *Proteus mirabilis*, flagella project from many surfaces of the cell (approximately ×3500). (a, c, V. Chambers; b, E.S. Boatman)

rial flagella are solid structures made of long protein filaments. (In contrast, recall that eukaryotic flagella consist of two central hollow microtubules surrounded by nine pairs of similar microtubules often connected to other internal tubular structures.) Some species have a single flagellum or a bundle of them at one end of the cell, while in others flagella extend from all around the cell.

Many gram-negative bacteria have hundreds of hair-like extensions known as **pili** (singular, *pilus*) (Fig. 21–4). These structures help the bacteria attach to surfaces such as the cells of host organisms and sometimes function in the transfer of plasmids from one cell to another.

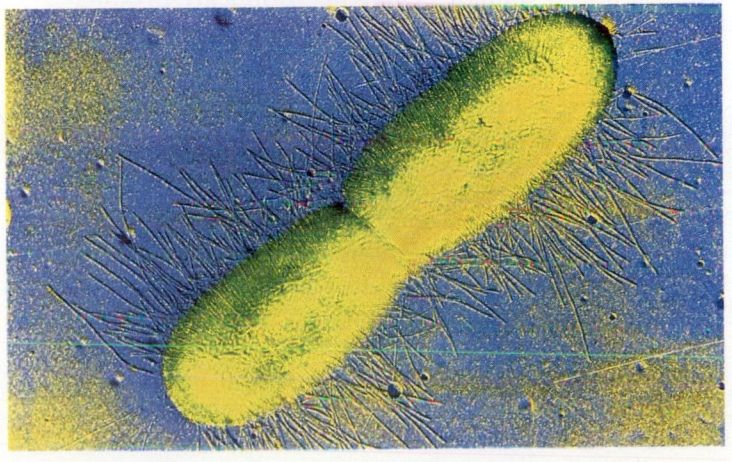

FIGURE 21–4   False color transmission electron micrograph of the bacterium *Escherichia coli*. The fine hair-like structures are probably sex pili, although other kinds of pili are also used by harmful bacteria to attach to host tissues as a preliminary to infection. This bacterium is dividing into two daughter cells (×14,700). (CNRI/Science Photo Library/Photo Researchers, Inc.)

Bacteria Differ In Their Needs For Oxygen

Like most plants and animals, some bacteria can survive only if they have enough oxygen. These are called **aerobic bacteria.** Other bacteria, however, are adapted to environments such as soil or stagnant ponds where oxygen is scarce. These are **anaerobic bacteria,** or **anaerobes.** Recall that some bacteria use an anaerobic pathway called **fermentation;** in lactic acid fermentation, for example, bacteria in milk cause souring. Some anaerobic bacteria (**obligate anaerobes**) are poisoned by oxygen. Others are more versatile, able to use aerobic or anaerobic respiration depending on whether oxygen is present. These are called **facultative anaerobes.**

Eubacteria Have Diverse Methods Of Obtaining Food

Most bacteria are **heterotrophs,** obtaining preformed organic compounds—food—from other organisms. While some of these heterotrophic bacteria are **consumers,** eating other living organisms, most of them are free-living decomposers called **saprobes,** organisms that feed on dead organic matter. Other heterotrophic bacteria are **symbionts,** living in symbiosis (a close relationship with an organism of a different species). These symbionts may be **commensals,** which neither help nor harm their hosts, or **parasites,** which do harm their hosts. Some parasites are **pathogens,** causing disease in plants and animals.

Some bacteria are chemosynthetic autotrophs, or **chemoautotrophs.** They produce their own food from simple inorganic ingredients. Chemosynthetic bacteria absorb carbon dioxide, water, and simple nitrogen compounds from their surroundings. From these they synthesize complex organic substances using energy they obtain by oxidizing ammonia, sulfur or iron compounds, or hydrogen gas. Some of these bacteria play an important role in the nitrogen cycle (Chapter 46).

Several types of bacteria are photosynthetic autotrophs, using energy from sunlight to make food. For example, **cyanobacteria** (formerly known as the "blue-green algae") use a photosynthetic process very similar to that of plants. They use water as a hydrogen source and release oxygen as a by-product. Cyanobacteria live in ponds, lakes, swimming pools, and moist soil, as well as on dead logs and tree bark. Some also live in the oceans, and a few species even inhabit hot springs. They often become the dominant microorganisms in polluted lakes and bays, where in excessive numbers they may do great ecological damage.

All cyanobacteria are microscopic. A few types are unicellular, but most live in large globular colonies or long filaments joined by extracellular materials. Some species show a division of labor among members of a colony. In certain species, some cells are specialized for fixing atmospheric nitrogen, that is, making it available for living organisms, while others are specialized to reproduce; still others may function to attach the colony to objects.

As do other monerans, cyanobacteria lack a nuclear envelope and organelles such as mitochondria and chloroplasts. However, unlike most bacteria, cyanobacteria do have some true internal membranes called **photosynthetic lamellae,** which contain chlorophyll and the enzymes needed for photosynthesis (Fig. 21–5).

Cyanobacteria form symbiotic relationships with many organisms, including protists, fungi, and some plants. Together with fungi they form some kinds of lichens (see Chapter 22). Those that form symbiotic relationships usually lack a cell wall and function quite like chloroplasts within the partner (host) cell, producing food for it.

While cyanobacteria photosynthesize much as plants do, other eubacteria photosynthesize quite differently. These include the green sulfur bacteria, the purple sulfur bacteria, and the green nonsulfur bacteria. The chlorophyll in these bacteria absorbs light most strongly in the near-infrared portion of the light spectrum rather than in the visible light range. This enables them to carry on photosynthesis in light that would appear very dim to human eyes. Also, these bacteria do not produce oxygen during photosynthesis because they do not use water as a hydrogen source. For instance, some of them use hydrogen sulfide instead of water.

The Archaebacteria Differ From Eubacteria

The **archaebacteria** are biochemically very different from eubacteria. One striking difference is the absence of peptidoglycan from their cell walls. Other important differences in proteins and cell chemistry suggest that archaebacteria diverged from the eubacteria long ago. Some of these biochemical differences may account for the fact that many archaebacteria inhabit extreme environments, such as deep-sea hot springs where temperatures may exceed 100°C. Some live in cooler (80°C) hot springs on the earth's surface. Their ability to survive where no other species can compete with them may be what has enabled them to persist to this day.

The archaebacteria include three groups: the **extreme halophiles,** which can live only in very salty environments such as salt ponds; the common **methanogens,** which produce methane from carbon dioxide and hydrogen in the human intestinal tract, sewage, and similar surroundings; and the **thermoacidophiles,** which normally grow in hot, acid environments such as hot sulfur springs. Archaebacteria include one type of photosynthetic bacteria, the purple nonsulfur bacteria.

Bacteria Reproduce By Fission

Bacteria generally reproduce asexually by transverse binary fission, in which the bacterial chromosome replicates, and the cell develops a transverse cell wall and then divides into two new cells. After the chromosome has replicated, the transverse wall forms as an ingrowth of both the plasma membrane and the cell wall. (When the newly formed cell wall does not separate completely into two walls, a chain of bacteria may form.)

With such a simple division process, involving nothing like the precise distribution of replicated chromosomes seen in meiosis or mitosis, bacterial cells can divide with remarkable speed. Some species grown under ideal conditions can divide every 20 minutes. At this rate, if nothing interfered, one bacterium would give rise to some 250,000 bacteria within 6 hours. This explains why the entrance of only a few pathogenic bacteria into a human being can so quickly produce disease.

Bacteria Can Exchange DNA

Although complex sexual reproduction does not occur in monerans, bacteria sometimes do exchange genetic material. Such genetic recombination can take place by transformation, conjugation, or transduction.

In **transformation,** a bacterial cell takes in deoxyribonucleic acid (DNA) fragments released by another bacterium that has broken up. As we saw in Chapter 10, this mechanism has been used experimentally to show that genes can be transferred from one bacterium to another and that DNA is the chemical basis of

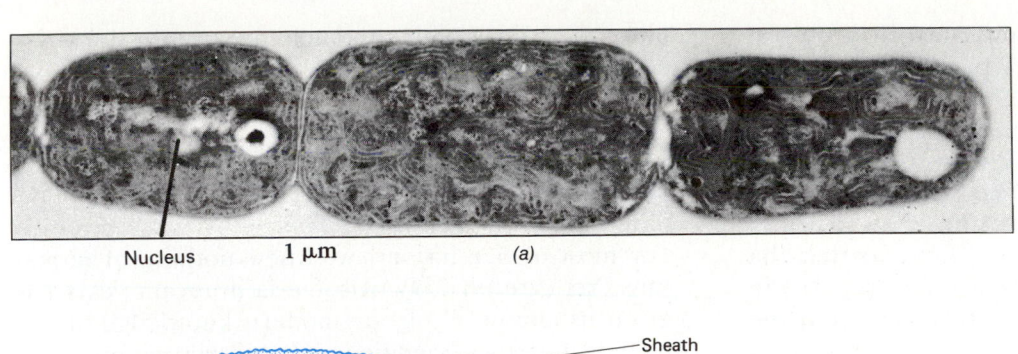

Nucleus 1 μm (a)

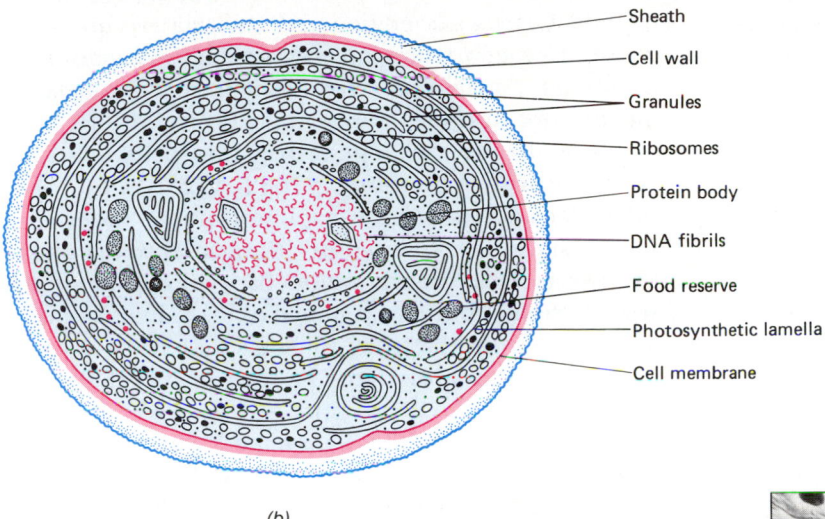

Sheath
Cell wall
Granules
Ribosomes
Protein body
DNA fibrils
Food reserve
Photosynthetic lamella
Cell membrane

(b)

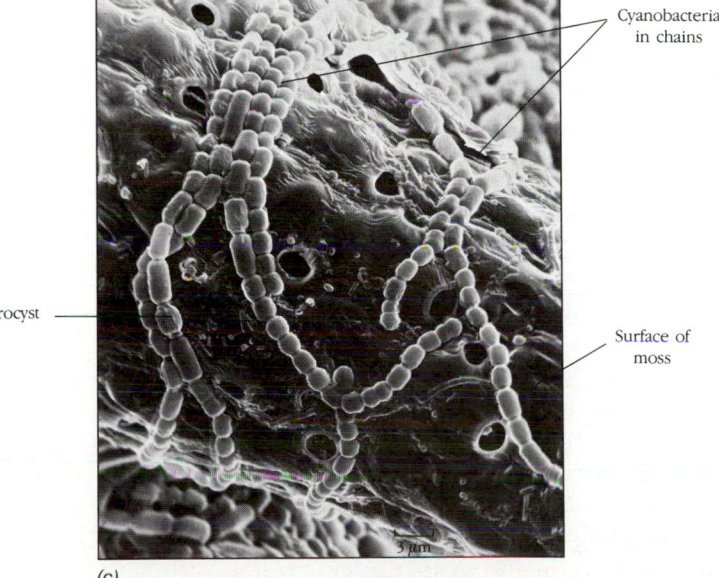

Cyanobacteria in chains

Heterocyst

Surface of moss

(c)

**FIGURE 21–5 Structure of cyanobacteria.
(a) Electron micrograph of *Anabaena*. The pale area
is nucleic acid, but no nuclear envelope is present.
Membranous lamellae function in photosynthesis
somewhat as do the thylakoid membranes in plant
chloroplasts. Gas vacuoles help to keep the
organisms afloat in sunlit layers of water.
(b) Diagram of the structure of a cyanobacterium.
(c) Scanning electron micrograph of cyanobacteria
(probably *A. cylindrica*) on a moss, to whose
nitrogen nutrition it may well make some
contribution.** (*a*, G.B. Chapman; *c*, National Research
Council of Canada)

heredity. Recall that non-disease-causing bacteria were
transformed into disease-causing bacteria by taking in
the genes of dead, disease-causing bacteria.

In **conjugation,** two cells exchange DNA directly.
Two bacterial cells of different mating types come together, and one passes DNA to the other (see Fig. 16–
2). Conjugation has been studied most extensively in
the bacterium *Escherichia coli,* of which there are F^+
strains and F^- strains. F^+ cells are covered with pili and
contain a plasmid known as the F factor, or fertility

factor. When an F^+ cell contacts an F^- cell, the pili
organize into a conjugation tube, called the sex pilus or
the F-pilus, which connects the F^+ cell to the F^- cell.
The F-pilus is hollow, enabling DNA to pass from one
bacterium to the other.

Most strains even of *E. coli* never develop F-pili,
however, and the phenomenon is probably best viewed
as a curiosity of scientific interest with use in genetic
engineering but little significance in nature. Both
transformation and conjugation occur only rarely

among bacteria but have a significant effect on the genetic makeup and evolutionary processes that occur in bacterial populations.

In **transduction,** bacterial genes are carried from one bacterial cell into another within a bacteriophage, a bacterial virus (Fig. 16–3). When a bacteriophage enters a bacterial cell, the DNA of the virus becomes mixed with some of the bacterial DNA, so that the virus now carries some of the bacterial DNA. If the virus infects a second bacterium, the DNA from the first bacterium may become mixed with that of the second bacterium. This new genetic information is then replicated with each new division.

Bacteria Survive Harsh Conditions Through Dormancy And Endospore Formation

When the environment becomes too dry or otherwise hostile, many bacterial cells become dormant. They lose water, shrink slightly, and remain quiescent until water is again available. Other species form **endospores**—modified dormant resting cells—to survive in extremely dry, hot, or frozen environments, or when food is scarce (Fig. 21–6). Endospores can sometimes survive hours of boiling or centuries of freezing. They

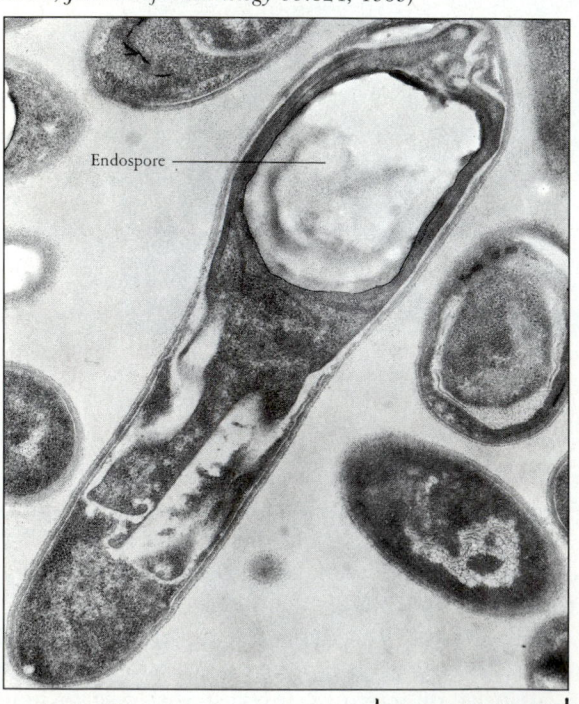

FIGURE 21–6 Endospore within a cell of *Clostridium*. Each cell contains only one endospore, a resistant, dehydrated remnant of cytoplasm. (Santo, L., Hohl H., and Frank H., *Journal of Bacteriology* 99:824, 1969)

Endospore

1 μm

make it very difficult to control some infectious diseases such as anthrax.

Studies Of Bacteria Have Increased Our Knowledge Of All Cells

For most of our history, we knew nothing of monerans. Yet careful study of bacteria in recent years has given us almost all of our modern knowledge of cell biology. Genetically engineered bacteria are now producing formerly scarce materials such as human insulin in commercial quantities, and there is promise of much more to come.

Very likely, the first person to see bacteria was Anton van Leeuwenhoek (1632–1723), who, with hand lenses, curiously examined pond water, sea water, saliva, and many other things. In one letter to the Royal Society of London, his description of the sizes, shapes, and movements of certain organisms leaves no doubt that they were bacteria.

The research of Louis Pasteur in the 1870s and 1880s revealed the importance of bacteria as agents of disease and decay. This stimulated work by Robert Koch, Joseph Lister, and others, and the science of bacteriology blossomed in the late 1800s. Pasteur found that the "diseases" of souring wine and beer were caused by microorganisms that entered from the air and brought about undesirable fermentations, yielding products other than alcohol. By gently heating the grape juice or beer mash to kill the undesirable organisms (a process now known as **pasteurization**), he found he could prevent this souring. As we have seen, Pasteur also convinced the scientific world that spontaneous generation, even of bacteria, did not occur.

Pasteur's work was also the basis for the modern **germ theory** of infectious disease. Robert Koch developed a system (Koch's postulates) for determining whether a disease is caused by living agents. To prove that disease is caused by an organism, one must first isolate the suspected organism and grow it outside the body in artificial culture. Then, when reintroduced into the host, it must produce disease again. Finally, it must be possible to isolate the same organism from the body or secretions of the experimentally infected subject.

▪ PROTISTS ARE EUKARYOTES AND ARE MOSTLY UNICELLULAR

The protists are a vast group of diverse eukaryotes, possibly including as many as 200,000 extant (living) and extinct species. The boundaries and divisions of the kingdom Protista are currently in a state of flux, with some biologists proposing as many as 50 phyla. We will be able to discuss only some typical groups.

Put simply, the **kingdom Protista** includes the protozoa, algae, slime molds, and water molds. **Protozoa** are heterotrophic animal-like protists, which ingest their food. **Algae** are autotrophic plant-like protists, which make their food by photosynthesis. **Slime molds** and **water molds** are heterotrophic fungus-like protists, which absorb their food. These are general categories: Some protists can make their own food at certain times and ingest it at other times.

Chiefly distinguishing protists from bacteria is their larger size and eukaryotic structure. As do all eukaryotic cells, protists have true nuclei, surrounded by a nuclear envelope, which divide by meiosis and mitosis. They also have membrane-bounded organelles such as plastids and mitochondria.

Protists vary tremendously in size, from single-celled protozoa to giant algae 60 meters long. Most protists are microscopic and single-celled; some live in colonies. Some protists are **coenocytic,** that is, multinucleate (having several nuclei), and some are multicellular.

Many protists are free-living, but some form symbiotic or even parasitic associations. Most protists are aquatic, living in oceans, freshwater ponds, lakes, and streams. They make up the **plankton,** the floating microscopic organisms that are the base of the food chain in aquatic ecosystems. Other aquatic protists attach to rocks and other surfaces in the water. The land protists are restricted to damp places such as soil and leaf litter.

The parasitic protists live in the body fluids of plants and animals.

Reproduction in protists is quite varied. All protists can reproduce asexually. Many also reproduce sexually, through meiosis and **syngamy,** the union of gametes. However, most protists do not develop anything comparable to multicellular sex organs, nor do they form embryos.

Protists, as we will see, have various means of locomotion, and most are motile at some point in their life cycle. Protists may move by changing shape, by flexing individual cells, or by waving cilia or flagella; some use several means of locomotion. The cilia and flagella of protists, like those of other eukaryotic cells but unlike bacterial flagella, have a nine-plus-two arrangement of microtubules (i.e., nine outer double microtubules encircling two single microtubules) (Fig. 21–7).

Protozoa Are Animal-Like Protists

The name protozoa was originally given to animal-like organisms that were single-celled. Unicellularity does not imply simplicity, however, and many protozoa are structurally complex. They are animal-like in that most ingest their food.

Amoebas And Some Other Protozoa Move By Forming Pseudopods

Some protozoans, members of the **phylum Sarcomastigophora,** have no definite body shape. Their single cells change form as they move. A typical example is the amoeba, which moves by pushing out temporary cytoplasmic projections called **pseudopods** from the surface of the cell (Fig. 21–8). More cytoplasm flows

FIGURE 21–7 **Electron micrograph of cross section through several cilia showing the 9 + 2 arrangement of the microtubules which is characteristic of eukaryotes.** (W.L. Dentler, University of Kansas/BPS)

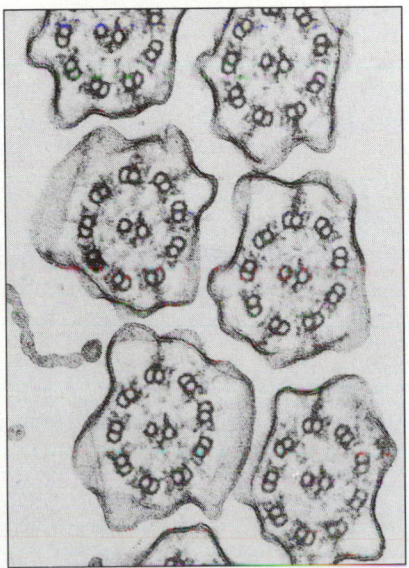

FIGURE 21–8 *Chaos carolinense,* **a giant amoeba ingesting a colonial green alga. Note the pseudopods extending to surround the prey.** (Michael Abbey/Photo Researchers, Inc.)

FIGURE 21–9
Foraminiferans secrete a shell or test. Cytoplasm is extruded through the pores.
(Manfred Kage/Peter Arnold, Inc.)

into the pseudopods, enlarging them until all the cytoplasm has entered and the cell as a whole has moved. Amoebas also use pseudopods to capture food. Two or more pseudopods surround a bit of debris or another microorganism. The food is then enveloped by a **food vacuole** and digested with enzymes added by lysosomes. The amoeba absorbs digested materials from the food vacuole, which gradually shrinks as it empties. The amoeba expels any indigestible remnants, leaving them behind as it moves on. These organisms reproduce asexually by cell division. Sexual reproduction has not been reported in them.

Not all "amoebas" are shapeless. Some members of this group produce calcified shells, or **tests.** For example, the oceans contain trillions of foraminiferans, which secrete chalky, many-chambered shells with pores through which pseudopods can extend (Fig. 21–9). Foraminiferan pseudopods form a sticky, interconnected net that entangles prey. Dead foraminiferans sink to the ocean bottom, where their shells form a gray mud that turns to chalk. With geological uplifting, these chalk formations can become dry land, like the white cliffs of Dover.

Flagellates Use Flagella For Locomotion

Flagellates, also members of the phylum Sarcomastigophora, have spherical or elongate bodies, a single central nucleus, and one or more flagella. Flagellates move rapidly, pulling themselves forward by lashing one or more flagella that are usually located at the front end.

Some flagellates are also amoeboid; they can engulf food by forming pseudopods. Others have a defi-

nite "mouth" or oral groove, a "gullet" or cytopharynx, and specialized organelles for processing food. Some, the zooflagellates, are heterotrophic, obtaining food by ingesting living or dead organisms or by absorbing nutrients from decaying matter.

Flagellates may be free-living or symbionts. Those with the most flagella and the most specialized bodies live in the intestines of termites. Unlike the termites, they have the enzymes needed to digest wood, which

FIGURE 21–10 Rat blood infested with a parasitic flagellate, *Trapanosoma brucei*, visible as dark, wavy bodies among the red blood cells. Similar trypanosomes infest the central nervous system of human beings, causing sleeping sickness. (From Clarkson, Allen B., Jr., and Frederick H. Brohn, *Science* 194:204, 1976; © 1976 by AAAS)

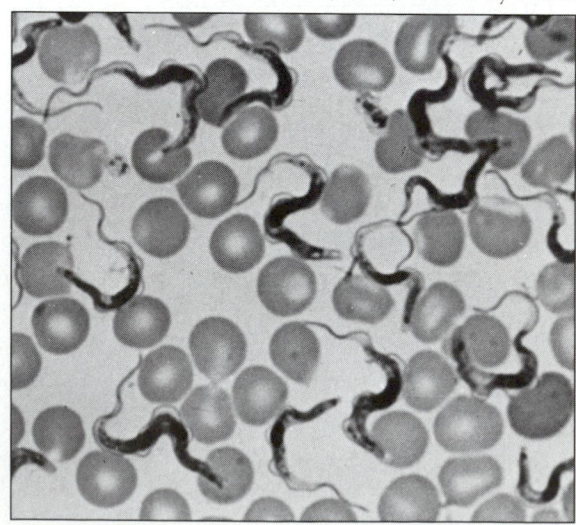

both the flagellates and the termites use as food. Some parasitic flagellates cause disease (Fig. 21–10).

The Ciliates Include The Paramecia

Ciliates are the major members of the protist **phylum Ciliophora.** They have a definite but somewhat flexible shape owing to their flexible outer pellicle. In *Paramecium* several thousand fine cilia extend through pores in the pellicle and permit movement (Fig. 21–11). The cilia beat with an oblique stroke so that the cell revolves as it swims. This ciliary beating is so precisely coordinated that the ciliate can go forward and backward and can turn around.

Most ciliates ingest rather than absorb their food. Near their surface many have numerous small **trichocysts,** organelles that discharge filaments believed to aid in trapping and holding prey or in defense. Although no ciliates are photosynthetic, some have symbiotic algae living in them.

Unlike other protozoans, ciliates have at least two nuclei per cell, often one or more **micronuclei** that function in reproduction, and a larger **macronucleus** that controls cell metabolism and growth. Ciliates may have from two to eight "sexes" called **mating types,** and most are capable of a type of conjugation.

Not all ciliates are motile. Some forms have stalks. Others, such as those of the genus *Stentor,* can swim but more often stay attached to one spot. Strong cilia set up currents in the surrounding water to bring food to them.

Sporozoa Are Parasitic

The **phylum Apicomplexa** is a large group of parasitic protozoa, some causing serious diseases such as malaria in humans. They have no contractile vacuoles nor structures for locomotion; they move by flexing. They often spend parts of their lives in different host species. They are also called **sporozoa** because at some stage in their life many develop into a resistant spore, which is the agent that infects the next host.

Algae Are Plant-Like Protists

The plant-like protists are a diverse group. They range from single-celled microscopic forms to large, multicellular algae (of all the protists, only the algae include truly multicellular forms). Most algae have **gametangia,** specialized cells that form gametes. The plant-like protists known as **euglenoids** resemble animals as well as plants.

Algae, which include dinoflagellates, diatoms, and green, red, and brown algae, are mostly photosynthetic. In addition to chlorophyll *a* and yellow and or-

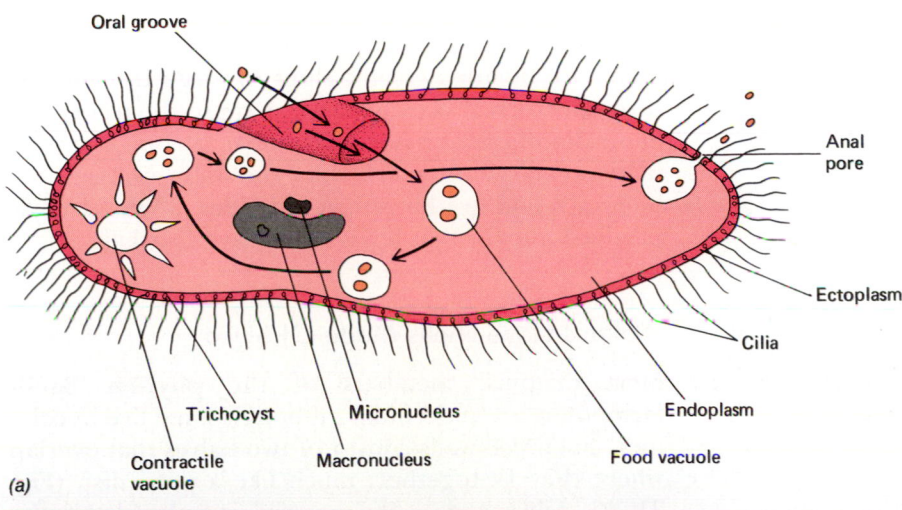

Oral groove

Anal pore

Ectoplasm

Cilia

Endoplasm

Trichocyst   Micronucleus

Contractile vacuole   Macronucleus   Food vacuole

(a)

FIGURE 21–11 Paramecia. (*a*) A diagram of *Paramecium caudatum.* (*b*) A scanning electron micrograph of *Paramecium multilicronucleatum.* Note the oral groove. (*b*, Dr. Eugene Small)

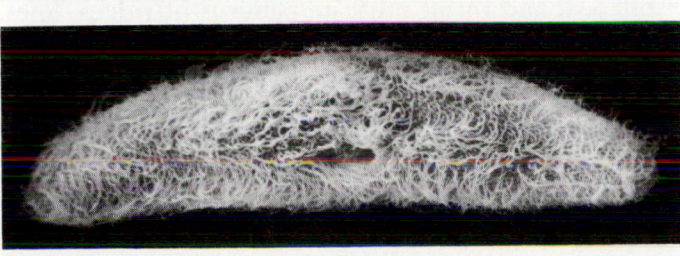

(b)

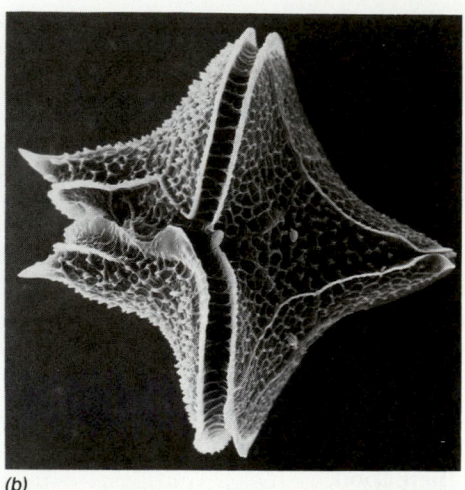

(a)

(b)

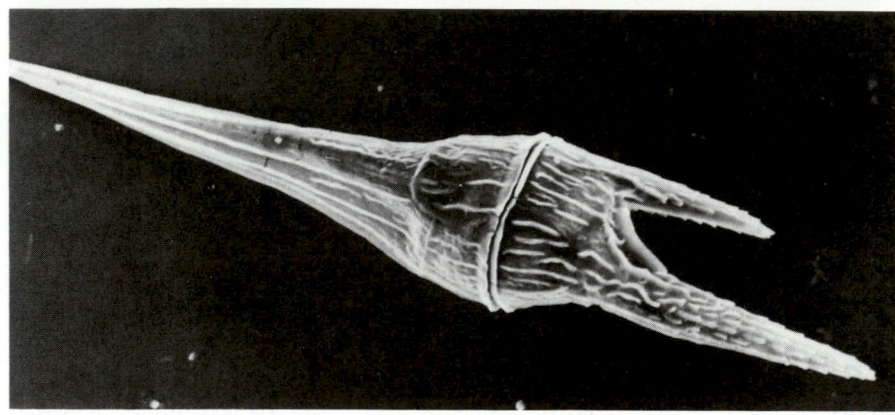

(c)

FIGURE 21–12 Scanning electron micrographs of some dinoflagellates. Note the plates that encase the single-celled body. The two flagella are located in grooves. (*a*) *Gonyaulax.* (*b*) *Protoperidinium.* (*c*) *Ceratium.* (*a, b, c,* John D. Dodge)

ange carotenoids, they have other pigments. Algae are grouped into phyla (also often called **divisions**) largely by their pigments and storage products.

Dinoflagellates Have Cellulose Armor

Among the most unusual protists are the dinoflagellates, members of the **phylum Dinoflagellata.** Most are unicellular, although a few colonial forms exist. A shell of interlocking cellulose plates often covers their cells (Fig. 21–12). Each cell has two flagella, one wrapped like a belt around a transverse groove in the center of the cell. The other flagellum is located in a longitudinal groove perpendicular to the first and projects beyond the cell. The undulation, or wave-like motion, of these flagella propels the cell through the water like a spinning top.

Most dinoflagellates are photosynthetic, possessing chlorophylls *a* and *c* in addition to carotenoids. Their energy storage products are usually oils or polysaccharides. Dinoflagellates are among the most important producers in aquatic ecosystems. Most species are marine (ocean-dwelling rather than freshwater).

Some are bioluminescent; ships or large fish moving through tropical oceans at night cause them to glow. Some dinoflagellates have occasional "blooms," or population explosions, which color the water orange, red, or brown and are known as **red tides.** Some of these produce a toxin that leads to massive fish kills.

Diatoms Have Shells Made Of Silica

Most diatoms, members of the **phylum Bacillariophyta,** are unicellular, although some live in colonies. Their cell walls consist of two halves that overlap where they fit together, much like a petri dish (Fig. 21–13). Silica, a glass-like material, is embedded in the cell wall and forms intricate patterns of ridges, lines, and pores. Diatoms are classified as having either **radial symmetry** (wheel-shaped patterns) or **bilateral** (two-sided) **symmetry.** Most diatoms float as part of plankton. Those that grow on rocks and other surfaces move by gliding, aided by material secreted from a small groove along the shell.

Most diatoms are photosynthetic and contain chlorophylls *a* and *c* as well as carotenoids. Their pigments

FIGURE 21–13 Scanning electron micrograph of several diatoms. The shell is built like a pillbox, with the lid overlapping the base. (Manfred Kage/Peter Arnold, Inc.)

give them a yellow or brown color. They store food reserves as oils or carbohydrates.

Diatoms primarily reproduce asexually by cell division. When a diatom divides, the two halves of the cell wall separate, and each becomes the larger half for a new cell. Therefore, each generation is smaller than the last. When the diatoms are a fraction of their ancestors' size, sexual reproduction occurs, producing gametes that shed their cell walls before uniting to form a zygote. Because the zygote grows substantially before producing a new shell, this process restores the diatom to its ancestral size. After that, the process begins again.

Green Algae Are Very Much Like Plants

If one had to pick a single word to describe the green algae (**phylum Chlorophyta),** it would be "variety." These protists have an amazing number of forms and reproductive methods. Their body structures range from single cells to colonial forms to coenocytic, siphonous algae to multicellular filaments and sheets (Fig. 21–14). Most of the multicellular forms do not have differentiated tissues, however. Most have flagella at least during part of their life, although some are nonmotile. Many green algae are symbionts. Some live in cells of invertebrates; some as part of lichens (see Chapter 22).

The green algae are structurally diverse but biochemically very uniform. They are photosynthetic, with chlorophylls *a* and *b* and carotenoids present in chloroplasts of many varied shapes. Starch is the main food reserve of green algae. Most have cell walls with cellulose, although some lack walls and some are covered with scales. Note that the pigments, storage prod-

ucts, and cell walls of green algae are identical to those of plants. Because of these and other similarities, plants are thought to have evolved from ancestors similar to green algae.

Reproductive methods in the green algae are as varied as form, including both sexual and asexual reproduction. Single-celled green algae may reproduce asexually by cell division; multicellular forms may reproduce by fragmentation. Many green algae produce spores asexually by mitosis. If these spores are flagellated and motile, they are called **zoospores.** Sexual cycles in green algae are often complex, involving the formation of flagellated gametes in single-celled gametangia.

Both aquatic and land-based forms of green algae exist. Aquatic green algae live primarily in fresh water, but there are some marine (ocean) species. Green algae that inhabit the land are restricted to damp soil, tree bark, and other moist places. They are important as the base of the food chain, particularly in freshwater habitats.

Red Algae Are Important Commercially

Most red algae (**phylum Rhodophyta)** are multicellular, typically composed of complex interwoven filaments that are delicate and feathery (Fig. 21–15). Some red algae are flattened sheets of cells. Most of these attach to rocks or other objects by root-like **holdfasts.**

The chloroplasts of red algae contain chlorophyll *a* and carotenoids. They also contain the red pigment phycoerythrin and the blue pigment phycocyanin. The cell walls of red algae often contain mucilaginous (sticky) polysaccharides that are used commercially.

(a)

(b)

(c)

(d)

(e)

FIGURE 21–14   Diversity in the green algae. (*a***)** *Chlamydomonas* **is a biflagellate unicellular organism. (***b***)** **An example of a colonial green alga is** *Volvox.* **A daughter colony can be observed leaving the mother colony through a rent in its wall. (***c***)** **Siphonous green algae, like** *Codium,* **are coenocytic. (***d***)** *Spirogyra* **is a multicellular green alga with a filamentous body form. Note the spiral-shaped chloroplasts. (***e***)** **Some multicellular green algae are sheet-like. The thin, leaf-like form has given** *Ulva* **its common name of "sea lettuce."** (*a*, J. Robert Waaland, University of Washington/BPS; *b*, James Bell/Photo Researchers, Inc.; *c*, Doug Wechsler; *d*, Ed Reschke; *e* Biophoto Associates)

(a)

(b)

FIGURE 21–15   Most red algae are multicellular, many with complex filamentous bodies. (a) *Odonthalia*. (b) *Plumaria*. (a, b, Biophoto Associates)

The red algae are important human food, particularly in the Orient (Fig. 21–16).

The red algae live primarily in warm tropical oceans, though a few freshwater and soil species occur. Some red algae incorporate calcium carbonate into their cell walls from ocean waters. These **coralline red algae** are very important in building "coral" reefs, possibly more important than coral animals.

Brown Algae Are The Giants Among The Protists

All members of the **phylum Phaeophyta,** the brown algae, are multicellular, ranging in length from several centimeters to approximately 60 meters. Their bodies may form tufts, "ropes," or thick flattened branches. They often have gas-filled floats to increase buoyancy.

Kelps, the largest brown algae, appear tough and leathery and differentiate into leaf-like **blades,** stem-like **stipes,** and anchoring **holdfasts,** although these are not the same as the leaves, stems, and roots of plants (Fig. 21–17).

Brown algae contain chlorophylls *a* and *c* and carotenoids in their chloroplasts. A special yellow-brown carotenoid, **fucoxanthin,** is found only in brown algae, dinoflagellates, and diatoms. The main food storage reserve in brown algae is **laminarin,** a carbohydrate.

Like red algae, brown algae are commercially important. **Algin,** a polysaccharide found in their cell walls, may serve to help cement the walls together. Algin is used as a thickening agent in ice cream, marshmallows, and cosmetics. Brown algae are important foods for humans, especially in oriental countries, and they are rich sources of minerals, especially iodine.

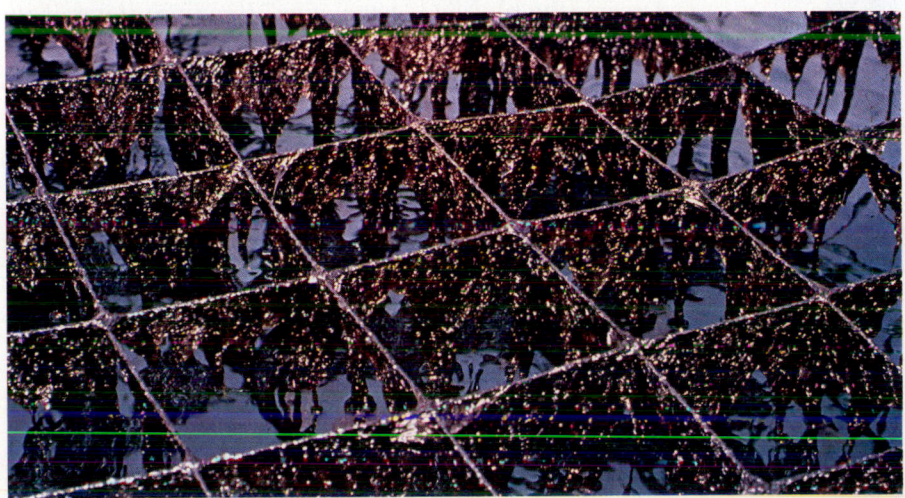

FIGURE 21–16   *Porphyra* (nori), a red alga used as food, is grown in seaweed beds such as this one in Japan. The nets are arranged so the algae are exposed during low tide and submerged during high tide. (Biophoto Associates)

FIGURE 21–17 *Laminaria*, **a typical brown alga. Note the blade, stipe, and holdfast.** (J. Robert Waaland, University of Washington/BPS)

Brown algae are common in cooler ocean waters, especially along rocky coastlines, where they live mainly in the intertidal zone or relatively shallow waters. The kelps form extensive underwater "forests"

FIGURE 21–18   **A kelp bed off the coast of California. These underwater forests are ecologically important, supporting large numbers of aquatic organisms.** (Richard Herrmann)

and are essential in such ecosystems as the primary food producer (Fig. 21–18).

Euglenoids Resemble Both Plants And Animals

All euglenoids, members of the **phylum Euglenophyta,** are unicellular flagellates (Fig. 21–19). They generally have two flagella, one that is long and whip-like and one that is too short to protrude beyond the cell. Euglenoids change shape continually because their outer covering is a flexible, proteinaceous pellicle rather than a rigid cell wall. The euglenoids reproduce asexually by longitudinal cell division. They are not known to reproduce sexually. They store their food as **paramylon,** a polysaccharide.

We include euglenoids in our discussion of plant-like protists because most contain chloroplasts and perform photosynthesis using chlorophylls *a* and *b* and carotenoids. Although these pigments are identical to those of green algae and plants, euglenoids are not thought to be related to either of them but may be close to the sarcomastigophora. Some euglenoids lack chloroplasts and must ingest food.

Slime Molds And Water Molds Are Fungus-Like Protists

Some protists resemble fungi in that they do not perform photosynthesis and have a thread-like body. Unlike fungi, however, many produce flagellated cells and have centrioles; in some, cellulose is a major component of the cell wall.

The Plasmodial Slime Mold Is A Mass Of Naked Cytoplasm

During the so-called "vegetative" (feeding) stage of its life, the plasmodial slime mold **(phylum Myxomycota)** is a mass of naked cytoplasm, often brightly colored, called a **plasmodium** (Fig. 21–20). It has many nuclei but no separate cells. The cytoplasm streams over damp, decaying logs and leaf litter, often forming a network of channels to better cover a large area. Creeping along, it ingests bacteria, yeasts, spores, and decaying matter much as an amoeba does.

When food or water is scarce, the plasmodium crawls to an exposed surface and begins to reproduce. The drying plasmodium usually forms stalked structures of intricate beauty (Fig. 21–21). These **sporangia** form cell walls around each nucleus. Within the sporangia, haploid spores form by meiosis. These spores are very resistant to adverse conditions. When conditions are favorable, they crack open, and a haploid re-

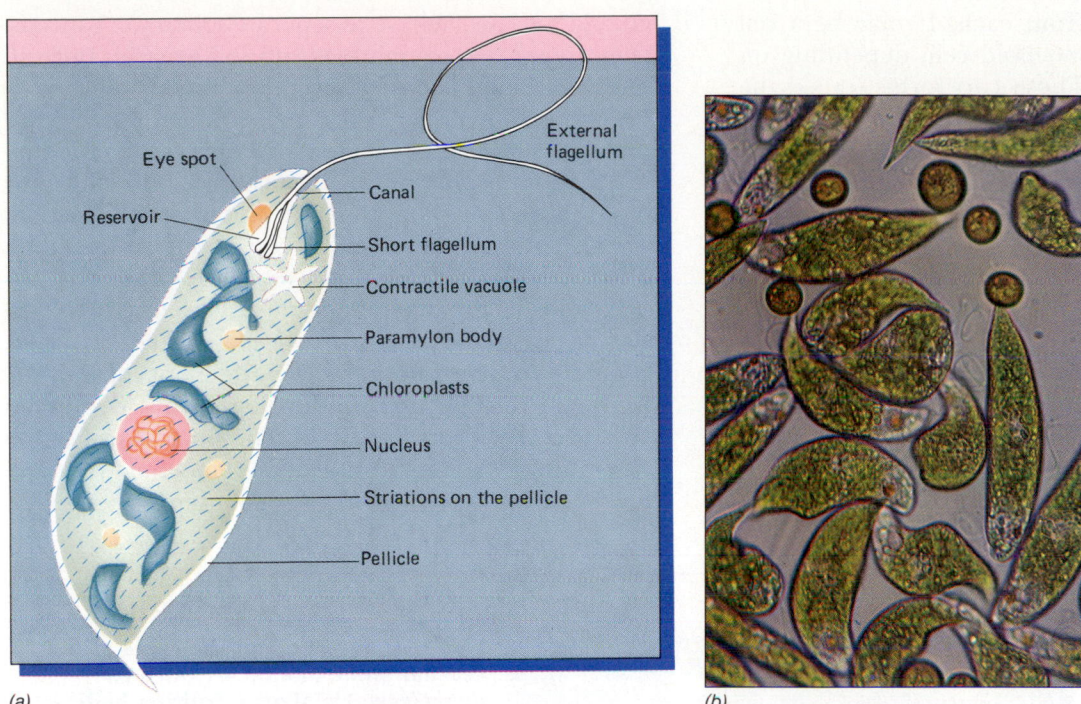

(a)

(b)

FIGURE 21–19   *Euglena.* (*a*) **This protist has both plant-like and animal-like traits. It has at various times been classified in the plant kingdom (with the algae) and in the animal kingdom (when protozoa were considered to be animals). (*b*) Living euglenoids. Note the red eyespots.** (*b*, T.E. Adams)

Labels in figure (a):
Eye spot
Reservoir
External flagellum
Canal
Short flagellum
Contractile vacuole
Paramylon body
Chloroplasts
Nucleus
Striations on the pellicle
Pellicle

FIGURE 21–20   (*a*) **The plasmodium of *Physarum* is colored bright yellow. This naked mass of protoplasm is multinucleate and feeds on bacteria and other microorganisms. (*b*) The reproductive structures of plasmodial slime molds are often stalked sporangia.** (*a*, Patrick W. Grace/Photo Researchers, Inc.; *b*, E.R. Degginger)

(a)

(b)

productive cell emerges from each. It may be a cell with two flagella or an amoeboid cell, depending on how wet conditions are. These two forms, called the **flagellated swarm cell** and the **myxamoeba,** act as gametes. Two of them fuse, and the resulting diploid nucleus of the zygote divides by mitosis. The cytoplasm does not divide, however, so the result is a multinucleate plasmodium.

Cellular Slime Molds Resemble Amoebas

Protists in the **phylum Acrasiomycota** are called cellular slime molds, but they resemble amoebas more than they do the plasmodial slime molds. During their feeding stage, the individual amoeboid cells, each with a haploid nucleus, behave as separate solitary organisms. They creep over rotting logs and soil or swim in fresh water, ingesting bacteria and other food.

When food or water becomes scarce, the cells send out a chemical signal, **cyclic AMP** (cyclic 3′,5′-adenosine monophosphate), which causes them to gather by the hundreds or thousands for asexual reproduction. During this stage the cells creep about as one unit, called a **pseudoplasmodium** or "slug," though each cell retains its plasma membrane and identity. Eventually the slug settles and constructs a stalked structure. The stalk forms from the cells in the front third of the slug. The rear portion of the slug forms a rounded structure at the tip, inside which spores form. Each spore may grow into an individual amoeboid cell, and the cycle repeats (Fig. 21–21). Sexual reproduction has

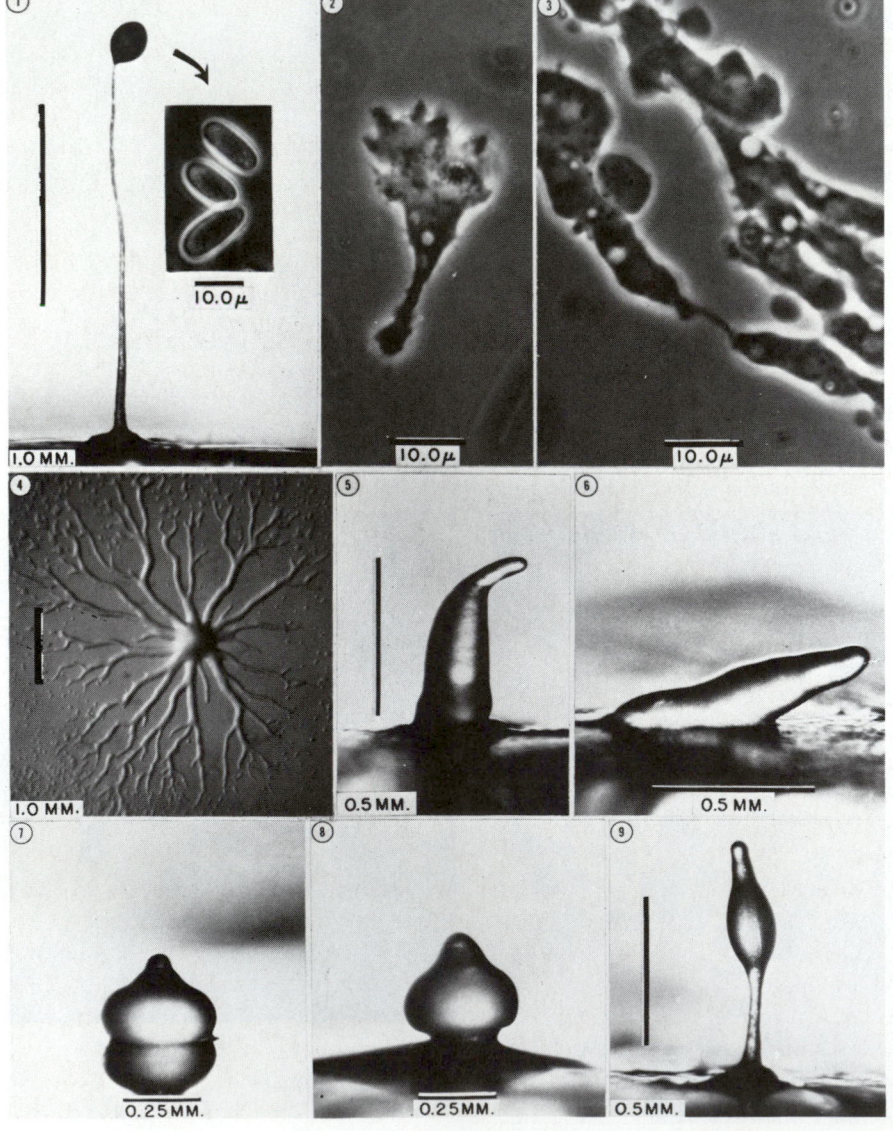

FIGURE 21–21   The life cycle of the cellular slime mold, *Dictyostelium discoideum.* **(1) Mature fruiting body releasing spores. (2) Each spore opens to liberate an amoeba-like, one-celled organism which eats, grows, and reproduces by cell division. (3) After their food supply is depleted, the cells stream together. (4) An aggregation of cells. (5) The aggregation organizes into a slug-shaped, multicellular organism. (6) The "slug" migrates for a period of time before forming a stalked fruiting body (7), (8), and (9). Cells making up the anterior third of the slug differentiate into stalk cells, while those in the posterior two-thirds form the spores.** (Courtesy of Dr. J. Gregg, *The Fungi,* New York, Academic Press)

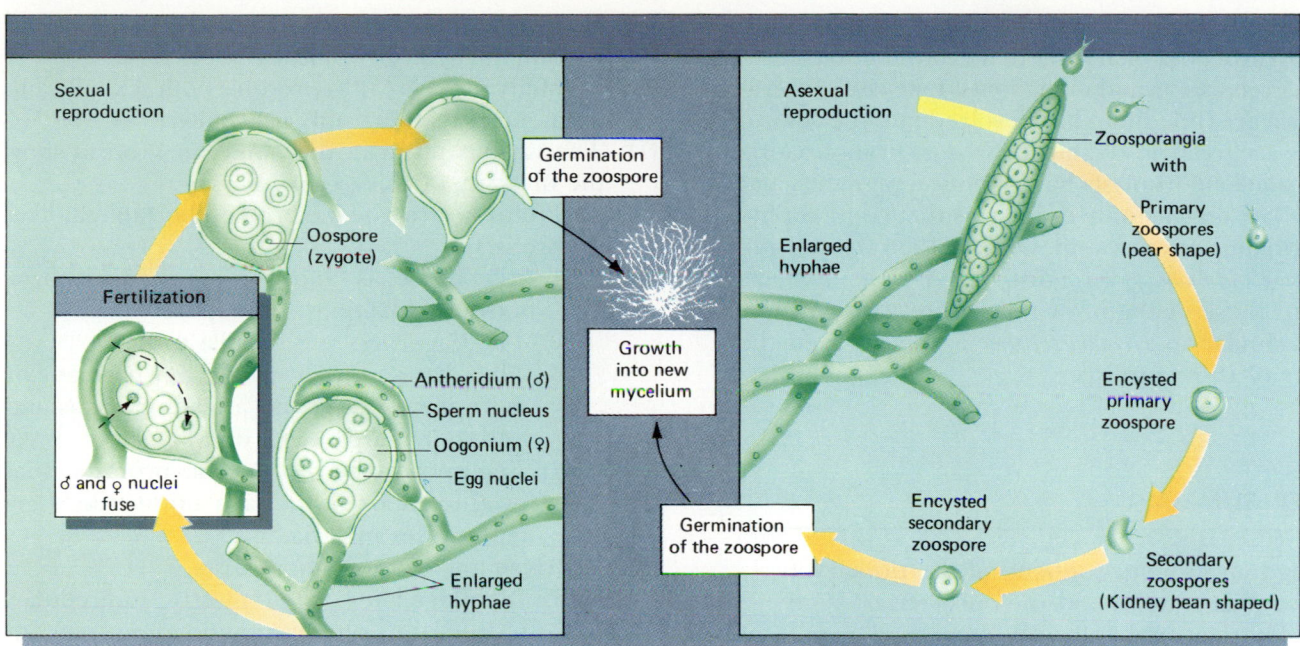

FIGURE 21–22 **The life cycle of a typical water mold. Oomycetes reproduce both asexually and sexually.**

also been observed in cellular slime molds. Most have no flagellated stages.

Water Molds Resemble Fungi

The water molds (**phylum Oomycota**) have a body, called a **mycelium,** which grows over a food source, digesting it and then absorbing the digested nutrients. The thread-like **hyphae** that make up the mycelium are coenocytic; no walls or membranes divide the nuclei. Hence the body is like one giant multinucleate cell. The cell wall may be composed of cellulose (as in plants) or chitin (as in fungi), or both.

When food is plentiful and conditions are good, water molds reproduce asexually (Fig. 21–22). The tip of a hypha swells and a cross wall forms, separating the hyphal tip from the mycelium. In the tip, tiny zoospores with two flagella form, each capable of developing into a new mycelium. When conditions worsen, water molds begin sexual reproduction. After male and female nuclei fuse, an **oospore** develops. The oospore can eventually develop into a new water mold organism. Water molds often spend the winter in this stage.

A water mold caused the infamous Irish potato famine of the 19th century. During several rainy cool summers, the protist multiplied unchecked, causing potatoes (a staple food of Irish farmers at the time) to rot in the fields. Mass starvation and death resulted, and many Irish people emigrated to other countries such as the United States or Australia.

THE SIMPLE PROTISTS ARE CONSIDERED THE EARLIEST EUKARYOTES

The word protist comes from the Greek, meaning "very first." The protists are considered the first eukaryotic cells to exist. There is reason to think that some cell organelles such as mitochondria and chloroplasts may have evolved from prokaryotic cells that lived as endosymbionts in other prokaryotes (see discussion of endosymbiont theory in Chapter 20). For example, chloroplasts are thought to have evolved from photosynthetic prokaryotes that lived inside larger single-celled prokaryotes. Mitochondria may have evolved from aerobic bacteria that lived symbiotically inside other prokaryotes. The symbiosis in each case would have helped both organisms, the endosymbiont and its host.

Most biologists think that the green, red, and brown algae evolved from three independent endosymbiont lineages. For example, the green algae (and euglenoids and plants) possess chlorophylls *a* and *b*. Do any living prokaryotes have the same pigments? In 1975 the bacterium *Prochloron*, which has chlorophylls *a* and *b*, was discovered living as an endosymbiont in certain marine animals. It has been suggested that

Prochloron is a survivor of a now largely extinct group of photosynthetic prokaryotes that evolved into chloroplasts in green algae and euglenoids. Likewise, the cyanobacteria are thought to have evolved into the chloroplasts we see today in red algae, because cyanobacteria and red algal chloroplasts have identical pigments and because they are the only organisms that contain phycoerythrin and phycocyanin. Only in 1983 was a living prokaryote, *Heliobacterium,* discovered with pigments similar to those of the brown algae and diatoms. Organisms similar to *Heliobacterium* may have been the ancestors of these protists.

■ CHAPTER SUMMARY

I. Kingdom Monera contains all the prokaryotes—the archaebacteria and the eubacteria.
   A. The bacterial cell wall, composed mainly of peptidoglycan, supports the cell and keeps it from bursting.
   B. Typically the bacterial cell lacks membranous organelles but may have a mesosome. Bacterial DNA forms a circular chromosome and sometimes smaller plasmids. Bacteria may have flagella and pili.
   C. Most bacteria are heterotrophic. Many form symbiotic relationships with other organisms.
   D. Cyanobacteria occur mainly as colonies or filaments and perform photosynthesis much as plants do. Other photosynthetic bacteria do not release oxygen.
   E. Bacteria reproduce asexually by transverse binary fission.
   F. The formation of endospores permits some bacteria to survive adverse conditions.
II. The kingdom Protista is composed of "simple" eukaryotic organisms.
   A. Most protists are single-celled, though their cell structure is more complex than those of animals or plants.
   B. Most protists are aquatic.
   C. Protists range in size from microscopic single cells to huge multicellular algae.
   D. Protists may obtain their nutrients autotrophically or heterotrophically.
   E. Many protists can reproduce both sexually and asexually, others only asexually.
III. The various phyla in the kingdom Protista show the diversity of this group.
   A. Protozoa are the heterotrophic (animal-like) protists.
      1. The amoebas and some other protozoa move by forming pseudopods.
      2. The flagellates use flagella to move.
      3. The ciliates move by cilia. They have one or more micronuclei and a macronucleus and undergo complex cell division.
      4. The sporozoa are parasitic protozoa that produce spores and are nonmotile. One type causes malaria.
   B. Algae are autotrophic protists.
      1. Dinoflagellates are mostly unicellular, biflagellate, photosynthetic organisms of great ecological importance.
      2. Diatoms are major producers in aquatic ecosystems. These are mostly single-celled with silica in their cell walls.
      3. Green algae vary greatly in size, complexity, and methods of reproduction. They are believed to be the ancestors of plants.
      4. Red algae are mostly multicellular and lack motile cells.
      5. All brown algae are multicellular and some have differentiated tissues.
      6. Euglenoids are single-celled, flagellated, photosynthetic protists.
   C. Slime molds and water molds are fungal-like protists.
      1. The body of the plasmodial slime mold is a multinucleate plasmodium. Slime molds reproduce by forming spores.
      2. The cellular slime molds live as single-celled organisms, aggregating to reproduce.
      3. The water molds have a coenocytic mycelium and reproduce asexually by biflagellate zoospores and sexually by oospores.

■ POST-TEST

1. All of the prokaryotes are assigned to kingdom _____.

2. Prokaryotes lack a _____ _____ and other membrane-bounded _____ characteristic of eukaryotic cells.

3. Ecologically, the cyanobacteria function as _____.

4. Peptidoglycan is found in the bacterial _____ _____.

5. Bacteria that absorb and retain violet stain are known as _____ _____ bacteria.

6. _____ are small molecules of DNA that replicate independently of the main chromosome.

7. Hair-like structures that help some gram-negative bacteria attach to other cells are called _____-_____.

8. Protozoa, algae, slime molds, and water molds are members of the kingdom _____.

9. Cilia and flagella of protists have a nine-plus-two arrangement of _____.

10. Ciliates have a micronucleus that functions in _____ and a macronucleus that functions in _____.

11. Malaria is caused by a parasitic _____.

12. Dinoflagellates are photosynthetic, biflagellate, and often covered by _____ plates.

13. A dinoflagellate bloom is known as a _____ _____.

14. The _____ are photosynthetic protists with cell walls composed of two halves that fit together like a petri dish.

15. Chlorophylls *a* and *b* and carotenoids are found in green algae, _____, and plants.

16. The multicellular bodies of _____ algae are differentiated into blades, stipes, holdfasts, and gas-filled floats.

17. The vegetative stage of slime molds in the phylum Myxomycota is a multinucleate _____.

18. The _____ slime molds behave as single-celled organisms until reproduction, when they aggregate.

■ REVIEW QUESTIONS

1. Imagine that you discover a new microorganism. After careful study you determine that it should be classified in the kingdom Monera, with the cyanobacteria. What characteristics might lead you to such a classification?

2. Identify each of the following and tell why it is important:
   a. peptidoglycan
   b. mesosome
   c. plasmid
   d. pilus
   e. endospore

3. Describe the process of bacterial conjugation and compare it with transduction and transformation.

4. What are archaebacteria, and why are they considered important?

5. What are the characteristics of a typical protist? Why might protists be considered difficult to characterize?

6. How do dinoflagellates sometimes contribute to fish kills?

7. Some biologists still classify the algae as plants. Why could algae be considered plants?

8. Why aren't protozoa considered to be animals in this text?

■ RECOMMENDED READINGS

Donelson, J.E., and M.J. Turner. "How the Trypanosome Changes Its Coat," *Scientific American*, February 1985. An explanation of how the flagellate that causes African sleeping sickness eludes the immune system in humans.

Godson, G.N. "Molecular Approaches to Malaria Vaccines," *Scientific American*, May 1985. A general explanation of the life cycle of the malarial sporozoon.

Lee, J.J., Hutner, S.H., and E.C. Bovee, eds. *Illustrated Guide to the Protozoa*. Lawrence, Kansas, Society of Protozoologists, 1985. Some beautiful line drawings and electron micrographs of most of the protozoa.

Margulis, L. *Symbiosis in Cell Evolution*. San Francisco, W.H. Freeman & Company, 1981. The endosymbiont theory of the origin of eukaryotic cells is presented in depth.

Tortora, G.J., et al. *Microbiology*, 2nd ed. Menlo Park, California, Benjamin/Cummings, 1986.

Woese, C.R. "Archaebacteria," *Scientific American*, June 1981, pp 98–122. First popularization and still an excellent general summary of the main traits and possible significance of the archaebacteria.

22
Fungal Life

OUTLINE

I. Fungi make an important contribution to the ecological balance of our world
II. Most fungi are filamentous in structure
III. Fungi are classified into four divisions
   A. Zygomycetes reproduce sexually by forming zygospores
   B. Ascomycetes (sac fungi) reproduce sexually by forming ascospores
   C. Basidiomycetes (club fungi) reproduce sexually by forming basidiospores
   D. Deuteromycetes (imperfect fungi) do not reproduce sexually
IV. Lichens are dual organisms composed of a fungus and an alga or cyanobacterium
V. Fungi are economically important
   A. Fungi provide food for humans
   B. Fungi produce useful chemicals
   C. Fungi cause many important diseases of plants
   D. Fungi cause certain diseases of animals

LEARNING OBJECTIVES

After you have studied this chapter you should be able to:
1. List the distinguishing characteristics of the kingdom Fungi.
2. Explain the role of fungi as decomposers.
3. Summarize the special ecological roles of lichens and mycorrhizae.
4. Contrast the body of a yeast with that of a mold.
5. Trace the fate of a fungal spore that lands on an appropriate substrate such as an overripe peach, and describe conditions that permit fungal growth.
6. List distinguishing characteristics for each of the divisions of fungi and give examples of each group.
7. Summarize the economic significance of the fungi.
8. Identify several fungal diseases of plants and several human fungal diseases.

Shaggy-mane mushrooms, *Coprinus comatus.*
(Grant Heilman, from Grant Heilman Photography)

Mushrooms, morels, and truffles—delights of the gourmet—have much in common with the black mold that grows on stale bread and the mildew that collects on damp shower curtains. All of these life forms belong to the kingdom Fungi, a diverse group of more than 100,000 known species. Fungi are eukaryotes that possess cell walls, a characteristic that accounted in part for their original classification in the plant kingdom. However, fungal cell walls have a different chemical composition than plant cell walls. Furthermore, fungi lack chlorophyll, a basic characteristic of plants. Fungi are not producers as are plants, but **heterotrophs** that *absorb* their food through the cell wall and plasma membrane. They are nonmotile and reproduce by means of spores, which may be produced sexually or asexually.

KEY CONCEPTS

☐ Although they possess cell walls, fungi are not plants. They neither contain chlorophyll nor produce food by photosynthesis.

☐ Fungi absorb food that they predigest with enzymes.

☐ Fungi play a significant role in the biosphere as decomposers and recyclers.

☐ Fungi are classified into divisions based on their mode of sexual reproduction.

☐ FUNGI MAKE AN IMPORTANT CONTRIBUTION TO THE ECOLOGICAL BALANCE OF OUR WORLD

As are most bacteria, most fungi are decomposers, **saprobes** that absorb nutrients from organic wastes and dead organisms. Instead of taking food inside its body and then digesting it as an animal would, a fungus digests food outside its body by secreting strong digestive enzymes into the food (generally called a substrate) on which it's growing. In this way, complex organic compounds are broken down into simpler compounds that the fungus can absorb. When fungi degrade organic material in this way, carbon, nitrogen, and mineral components of organic compounds are released. These elements can be recycled—that is, used by other organisms as building materials. Without the continuous decomposition performed by fungi and bacteria, essential nutrients would soon become locked up in huge mounds of dead animals, feces, branches, logs, and leaves. These nutrients would be unavailable for use by new generations of organisms.

Although most fungi are saprobes, some form symbiotic relationships. Recall from Chapter 21 that a symbiotic relationship is an intimate relationship between organisms of different species. Some fungi are **parasites,** organisms that live in or on other organisms and that are harmful to their hosts. Parasitic fungi absorb food from the living bodies of their hosts. Such fungi cause disease in humans and other animals, and are the most important disease-causing organisms of plants. Their activities cost billions of dollars in agricultural damage yearly.

Some types of fungi form mutualistic symbiotic relationships with other organisms. In **mutualism,** both symbiotic partners benefit from the relationship. Hence, at the same time a mutualistic fungus absorbs nutrients from its host, it makes some contribution to its host's well-being. **Mycorrhizae** are mutualistic symbiotic relationships between fungi and the roots of plants (Fig. 22–1). Such relationships occur in more than 90% of all families of higher plants. The fungus benefits the plant by decomposing organic material in the soil, thus making minerals available to the plant. The roots supply sugars, amino acids, and some other organic substances that the fungus may use. The importance of mycorrhizae first became evident when horticulturalists observed that orchids do not grow unless an appropriate fungus lives with them. Similarly, it has been shown that many forest trees die from malnutrition when transplanted to nutrient-rich grasslands soils that lack appropriate fungi. When forest soil that contains the appropriate fungi or their spores is added to the soil around these trees, they quickly assume a normal growth pattern.

FIGURE 22–1 Mycorrhizae are mutualistic associations between fungi and the roots of higher plants. An experiment demonstrating that soybeans respond to mycorrhizae: *left,* **a control plant; the other two plants were grown under identical conditions, except that their roots have mycorrhizal associations.** (R. Ronacordi/Visuals Unlimited)

◾ MOST FUNGI ARE FILAMENTOUS IN STRUCTURE

The body structures of fungi vary in complexity, ranging from single-celled yeasts to multicellular molds (a term used loosely to include mildews, rusts and smuts, mushrooms, and many other fungi). In most fungi, the rigid cell wall encasing each cell is composed in part of chitin, which is also a component of the external skeletons of insects and other arthropods. Chitin consists of subunits of a nitrogen-containing sugar. Chitin is far more resistant to breakdown by microbes than is the cellulose that makes up plant cell walls.

Yeasts are unicellular fungi that reproduce asexually mainly by **budding,** in which a small protuberance (bud) grows and eventually separates from the parent cell. Each bud that separates from the mother yeast cell can grow into a new yeast (Fig. 22–2). Yeasts can also reproduce asexually by fission and sexually through spore formation. The yeasts are not classified as a single taxonomic group because many different fungi can be induced to form a yeast stage.

Most fungi are filamentous molds. A mold consists of long, branched threads (or filaments) of cells called **hyphae** (singular, *hypha*). Hyphae form a tangled mass or tissue-like aggregation known as a **mycelium** (plural, *mycelia*). The cobweb-like mold sometimes seen on bread is the mycelia of fungi. What is not seen is the extensive mycelium that grows down into the substance of the bread. The color of the mold usually comes from the reproductive cells called **spores,** which are produced in large numbers on the mycelium. Some hyphae are divided by walls, called **septa** (singular, septum), into individual cells containing one or more nuclei. Other hyphae are **coenocytic,** or undivided by septa, and are something like an elongated, multinucleated giant cell. Septa often contain large pores that permit organelles to flow from cell to cell. Cytoplasm flows within the hypha, providing a system of internal transport (Fig. 22–3).

Fungi grow best in dark, moist habitats, but they are found universally wherever organic material is available. They require moisture to grow, and they can obtain water from the atmosphere as well as from the medium upon which they live. When the environment becomes very dry, fungi survive by going into a resting stage or by producing spores that are resistant to desiccation (drying out). Although the optimum pH for most species is about 5.6, some fungi can tolerate and grow in environments where the pH ranges from 2 to 9. Many fungi are less sensitive to high osmotic pressures than bacteria, and can grow in concentrated salt solutions or sugar solutions such as jelly that discourage or prevent bacterial growth. Fungi may also thrive over a wide temperature range. Even refrigerated food is not immune to fungal invasion.

FIGURE 22–2   Yeasts are unicellular fungi that reproduce asexually mainly by budding. They may also reproduce sexually. (*a*) A cell of the common yeast. The endoplasmic reticulum is more extensive than depicted. Budding of yeast cells is also shown. (*b*) Micrograph of yeast cells, *Saccharomyces cerevisiae*, commonly known as baker's yeast. Note that many of the cells are budding. (John D. Cunningham/Visuals Unlimited)

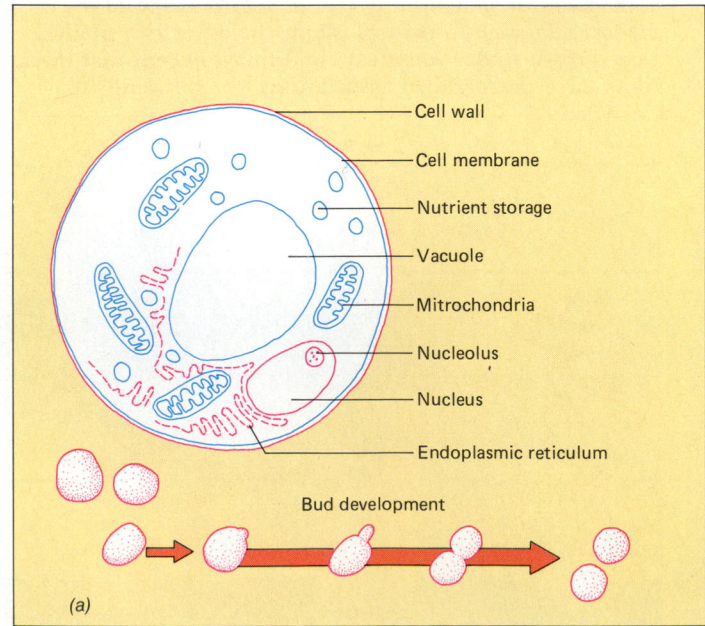

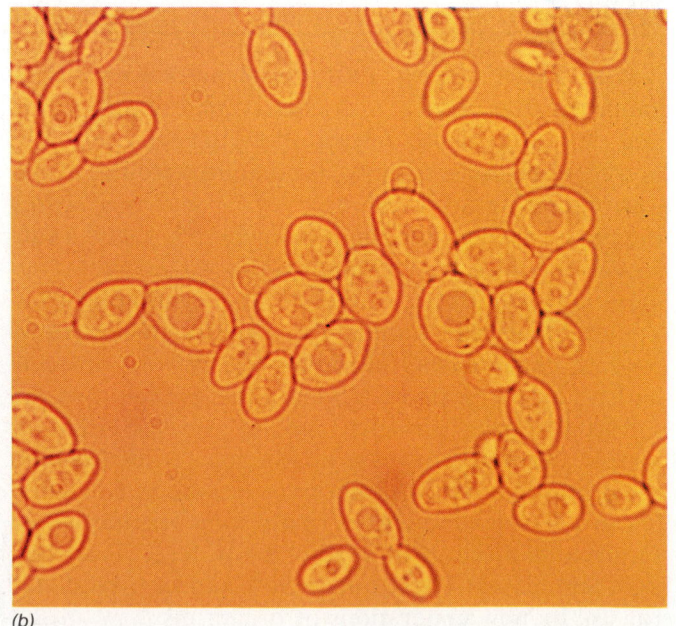

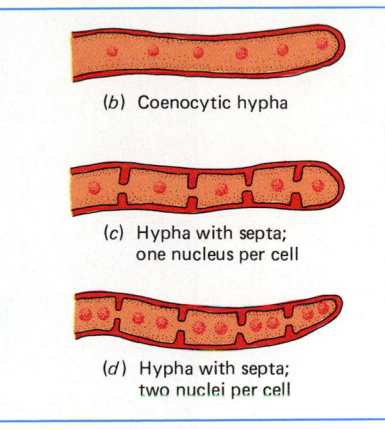

(b) Coenocytic hypha

(c) Hypha with septa; one nucleus per cell

(d) Hypha with septa; two nuclei per cell

(a)

FIGURE 22–3   Molds. (a) The fuzzy appearance of molds is due to their body form, which is a mass of threads called hyphae. (b) Coenocytic hypha. (c) Hypha divided into cells by septa. Each cell has one nucleus. In some fungi the septa are perforated, permitting cytoplasm to stream from one cell to another. (d) Septate hypha in which each cell has two **nuclei.** (a, E.R. Degginger)

Fungal spores are nonmotile reproductive cells dispersed by wind or by animals. Spores are usually produced on aerial hyphae (hyphae that project up into the air). This arrangement permits the spores to be blown by air currents and distributed to new areas. In some fungi, the aerial hyphae form large, complex reproductive structures in which spores are produced. These structures are called **sporocarps** or **fruiting bodies.** The familiar part of a mushroom or toadstool is a large sporocarp. We do not normally see the bulk of the organism, a nearly invisible network of hyphae buried out of sight in the rotting material upon which it grows.

Fungi may produce spores either sexually or asexually. Unlike animal and plant cells, fungal cells usually contain haploid nuclei. In sexual reproduction, fungi often carry out some type of conjugation in which hyphae of two genetically different mating types come together and their nuclei fuse. In certain fungi, the genetically different nuclei do not fuse immediately, but remain separate within the fungal cytoplasm for most of its life. Hyphae that contain two genetically distinct nuclei within each cell are **dikaryotic.** Hyphae that contain only one nucleus per cell are **monokaryotic.**

When a fungal spore contacts an appropriate substrate, perhaps an overripe peach that has fallen to the ground, the spore germinates and begins to grow (Fig. 22–4). A thread-like hypha emerges from the tiny spore. Soon a tangled mat of hyphae infiltrates the

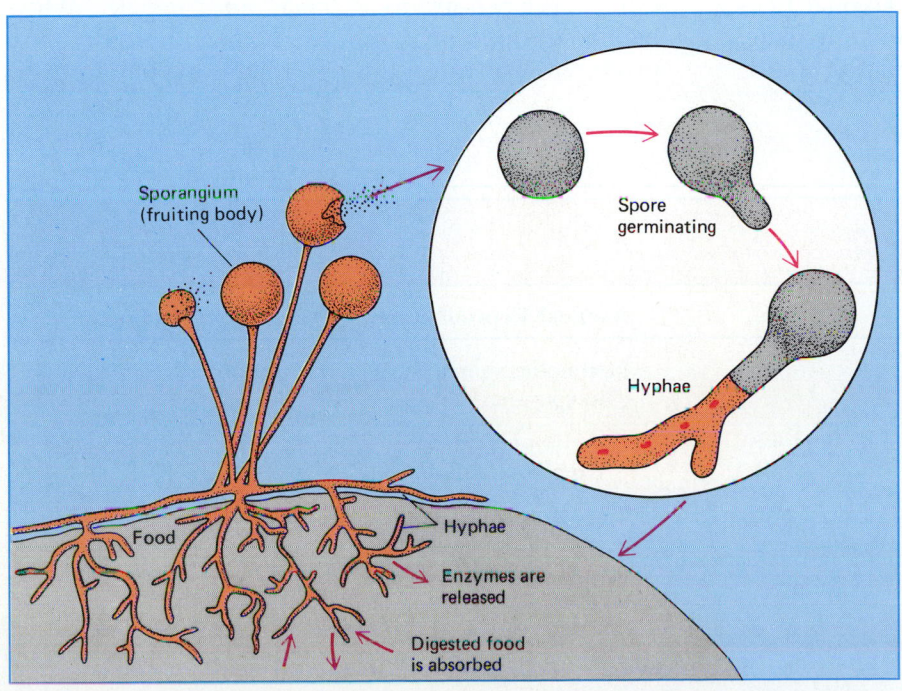

Sporangium (fruiting body)

Spore germinating

Hyphae

Food

Hyphae

Enzymes are released

Digested food is absorbed

FIGURE 22–4   Spore germination and growth of a typical mold.

peach, and other hyphae extend upward into the air. Cells of the hyphae secrete digestive enzymes into the peach, degrading its organic compounds to small molecules that the fungus can absorb.

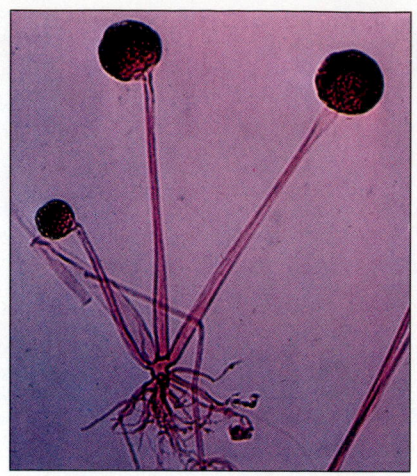

FIGURE 22–5 A photomicrograph of the asexual reproductive structure, a sporangium, of black bread mold, *Rhizopus nigricans.* (Carolina Biological Supply)

■ FUNGI ARE CLASSIFIED INTO FOUR DIVISIONS

The classification of fungi is based mainly on the characteristics of the sexual spores and fruiting bodies. Authorities do not agree on how to classify these diverse organisms, but most assign them to four divisions (equivalent to phyla in animal taxonomy): Zygomycota, Ascomycota, Basidiomycota, and Deuteromycota. Table 22–1 summarizes fungal classification. The fungi were once thought to have evolved from algae, but many biologists now think they can be traced back to unicellular eukaryotes that are extinct. There is evidence that fungi are among the oldest eukaryotes.

Zygomycetes Reproduce Sexually By Forming Zygospores

The members of division Zygomycota are referred to as **zygomycetes.** They produce sexual spores, called **zygospores,** that remain dormant for a time. Their hyphae are coenocytic; that is, they lack septa. Many zygomycetes live in the soil on decaying plant or animal matter. Some are parasites of plants and animals.

One common zygomycete is the black bread mold *Rhizopus nigricans.* Bread becomes moldy when a mold spore falls upon it and then germinates and grows into a tangled mass of threads, the mycelium. Hyphae penetrate the bread and absorb nutrients. Some hyphae, termed **stolons,** grow horizontally with amazing speed;

others, called **rhizoids,** anchor the stolons to the bread. Eventually, certain hyphae grow upward and develop a **sporangium,** or spore sac, at the tip. Clusters of black spherical spores develop within this sac and are released when the delicate sporangium ruptures (Fig. 22–5).

Sexual reproduction occurs when the hyphae of two different mating types (plus and minus) grow into contact with one another (Fig. 22–6). Sexual reproduction can occur only between hyphae of a plus strain and hyphae of a minus strain. This is a sort of physiological sex differentiation. There is no morphological (structural) sex differentiation, so it is not proper to refer to the two strains as "male" and "female." When hyphae of opposite mating types meet, hormones are produced that cause the tips of the hyphae to grow

TABLE 22–1
Divisions of Kingdom Fungi

Division	Common Types	Asexual Reproduction	Sexual Reproduction
Zygomycota	Black bread mold	Nonmotile spores form in sporangium	Zygospores
Ascomycota (sac fungi)	Yeasts, powdery mildews, molds, morels, truffles.	Conidia pinch off from conidiophores	Ascospores
Basidiomycota (club fungi)	Mushrooms, bracket fungi, puffballs, rusts, smuts	Uncommon	Basidiospores
Deuteromycota (imperfect fungi)	Molds	Conidia	Sexual stage not observed

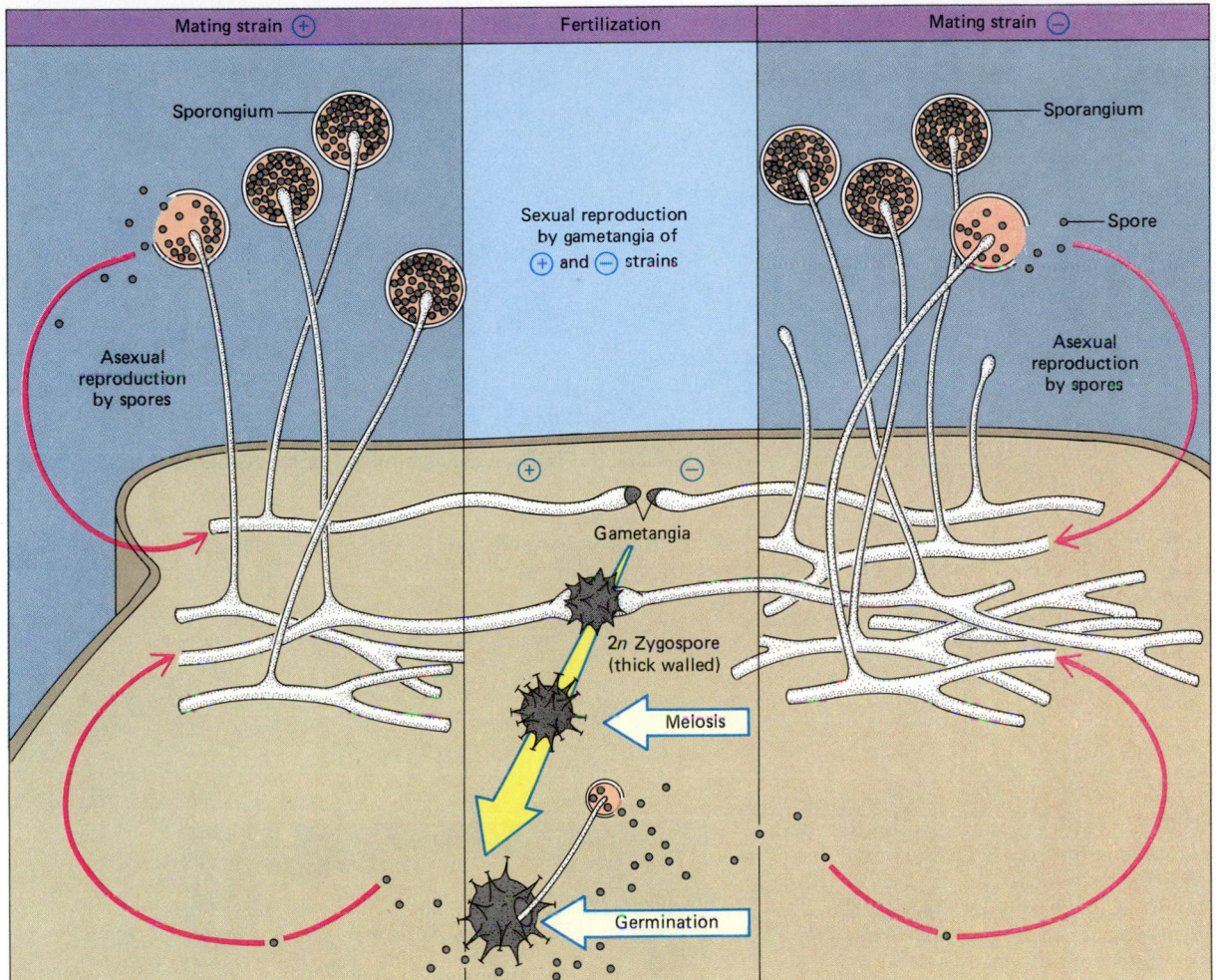

FIGURE 22–6 Life cycle of the black bread mold, *Rhizopus nigricans*. Sexual reproduction takes place only between different mating types.

together. Plus and minus nuclei then fuse to form a diploid nucleus, the zygote. The zygospore develops, providing a thick protective covering around the zygote (Fig. 22–7). The zygospore may lie dormant for several months. Meiosis probably occurs at or just before germination of the zygospore. When the zygospore germinates, an aerial hypha develops with a sporangium at the top. Each spore formed in the sporangium is capable of growing into a new mycelium.

Ascomycetes (Sac Fungi) Reproduce Sexually By Forming Ascospores

Division Ascomycota, the **ascomycetes,** is a large group consisting of about 30,000 described species. The ascomycetes are sometimes referred to as **sac fungi** because their sexual spores are produced in little sacs called **asci** (singular, *ascus*). Their hyphae usually have septa,

but these cross walls are perforated so that cytoplasm can move from one compartment to another.

Ascomycetes vary in complexity from unicellular yeasts to multicellular powdery mildews and cup fungi. They include most of the blue-green, pink, and brown molds that cause food to spoil, and the edible morels and truffles.

In most ascomycetes, asexual reproduction involves production of spores called **conidia** (Fig. 22–8). These spores are pinched off at the tips of certain specialized hyphae known as **conidiophores** (spore bearers). Sometimes called "summer spores," the conidia are a means of rapidly propagating new mycelia. They vary in shape, size, and color in different species; the color of the conidia is what gives the characteristic black, blue, green, pink, or other tint to many of these molds.

Some species of ascomycetes have different mating strains; others have the ability to mate with them-

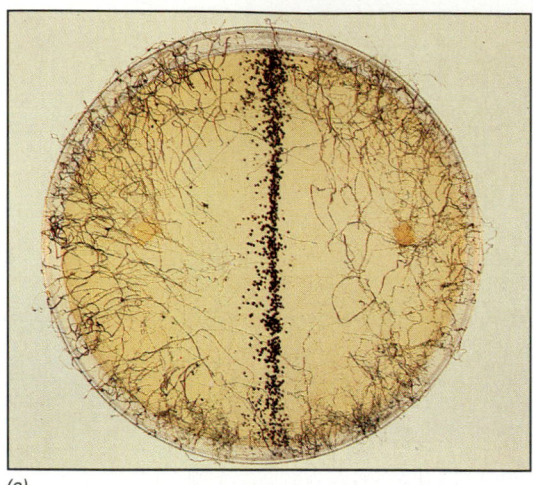

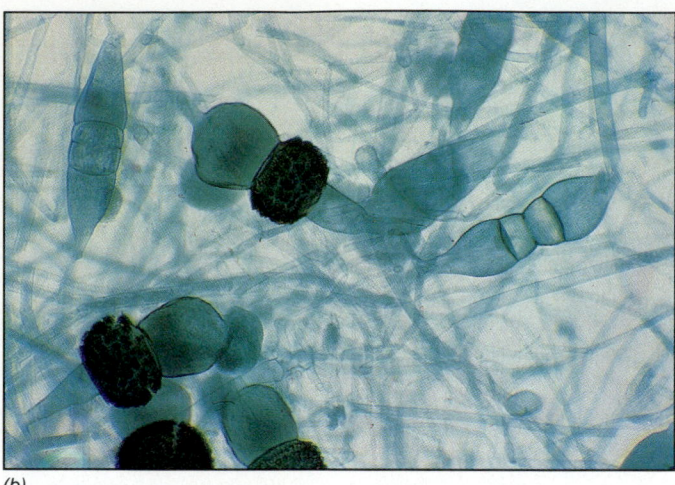

(a)

(b)

FIGURE 22–7 **Sexual reproduction in the zygomycetes.** (*a*) **A Petri dish was inoculated with both + and − strains of** *Rhizopus*. **A line of zygospores formed where the two strains came into contact.** (*b*) **Photomicrograph of zygospores. Note their warty appearance.** (*a*, Runk/Schoenberger, from Grant Heilman; *b*, William E. Schadel © 1988/BPS)

FIGURE 22–8 **Conidia are asexual reproductive cells produced by Ascomycetes and most Deuteromycetes. The arrangement of conidia on conidiophores varies from species to species and is used to help identify these fungi.** (*a*) **Scanning electron micrograph of** *Penicillium* **conidiophores, which resemble paintbrushes.** (*b*) **Scanning electron micrograph of** *Aspergillus* **conidiophore, which produces a tight head of conidia. The taxonomy of both** *Penicillium* **and** *Aspergillus* **is somewhat confusing, as certain species of each genus are classified as Ascomycetes (because they produce a sexual stage), while other species are classified as Deuteromycetes (because they have only been observed to reproduce asexually).** (*a*, Stanley L. Flegler/Visuals Unlimited; *b*, David M. Phillips/Visuals Unlimited)

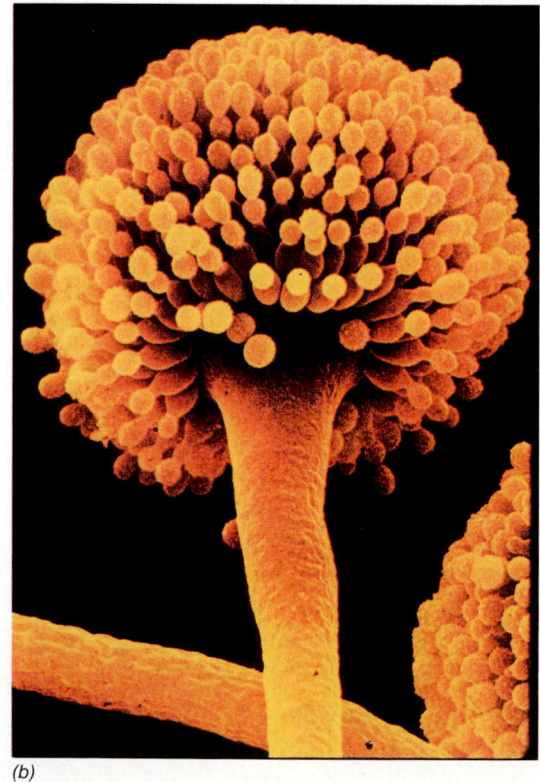

(b)

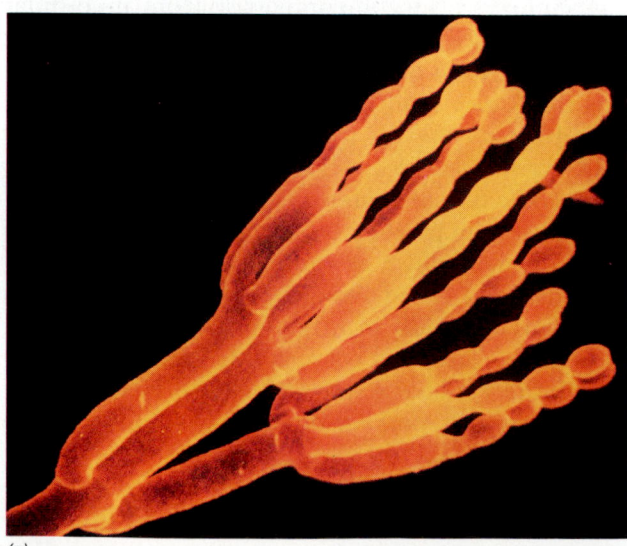

(a)

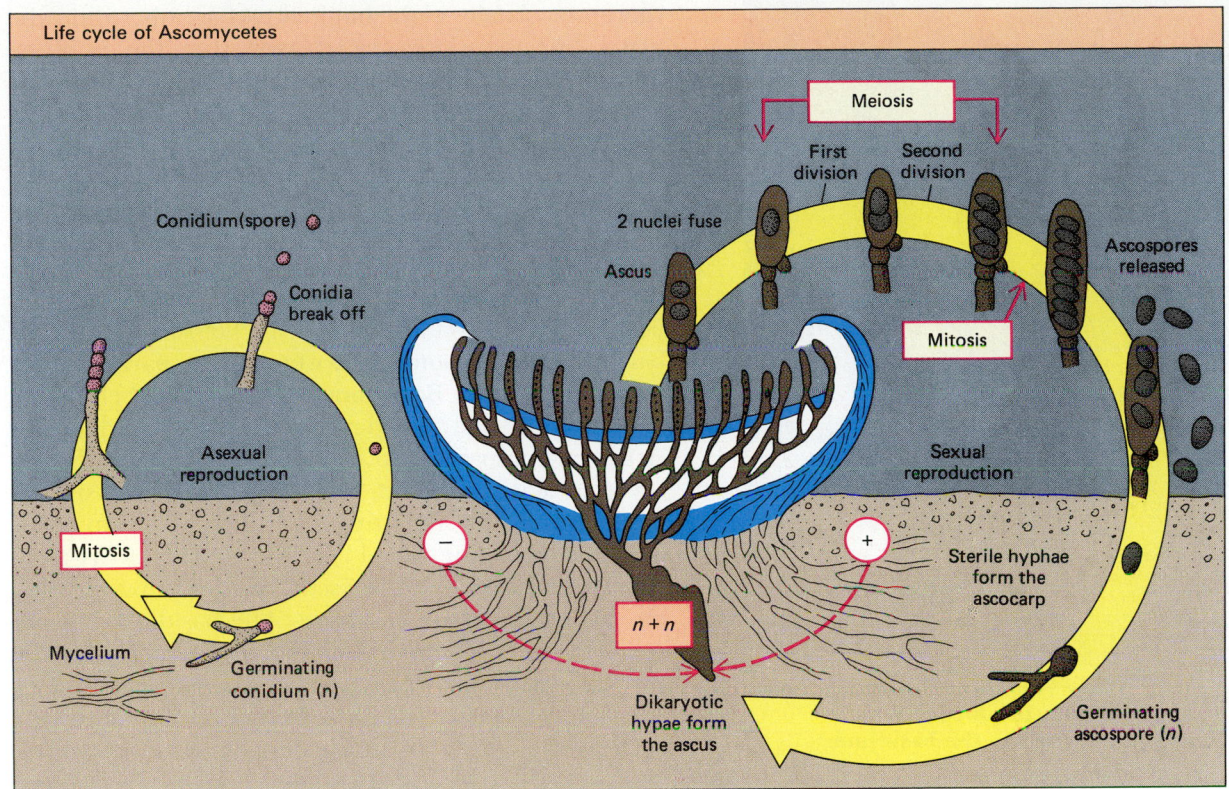

Life cycle of Ascomycetes

Conidium(spore)

Conidia break off

Asexual reproduction

Mitosis

Mycelium

Germinating conidium (n)

2 nuclei fuse

Ascus

Meiosis

First division

Second division

Mitosis

Ascospores released

Sexual reproduction

Sterile hyphae form the ascocarp

Germinating ascospore (n)

Dikaryotic hypae form the ascus

n + n

FIGURE 22–9   Life cycle of an ascomycete. Asexual reproduction involves production of conidia. Sexual reproduction involves the formation of a fruiting body known as an ascocarp in which the sac-like asci develop. Ascospores form within the asci. When an ascus breaks open, ascospores are released and can germinate, each forming a hypha.

selves. Sexual reproduction takes place after two hyphae grow together and their cytoplasm mingles (Fig. 22–9). Within this fused structure, the two nuclei come together, but they do not fuse. New hyphae develop from the fused structure; cells of these hyphae are dikaryotic—that is, each has two nuclei. These hyphae form a fruiting body, known as an **ascocarp,** that is characteristic of the species (Fig. 22–10). This is where the **asci,** sexual spore sacs, develop. Within each cell that will develop into an ascus, the two nuclei fuse and produce a diploid nucleus, the zygote. Each zygote then undergoes meiosis to form four haploid nuclei. This process is usually followed by one mitotic division of each of the four nuclei, resulting in formation of eight haploid nuclei. The haploid nuclei (surrounded by cytoplasm) separate, forming **ascospores,** so that

FIGURE 22–10   The ascocarp of *Peziza coccine*, a cup fungus, is saucer-shaped. Asci line the inner portion of the saucer. (E.R. Degginger)

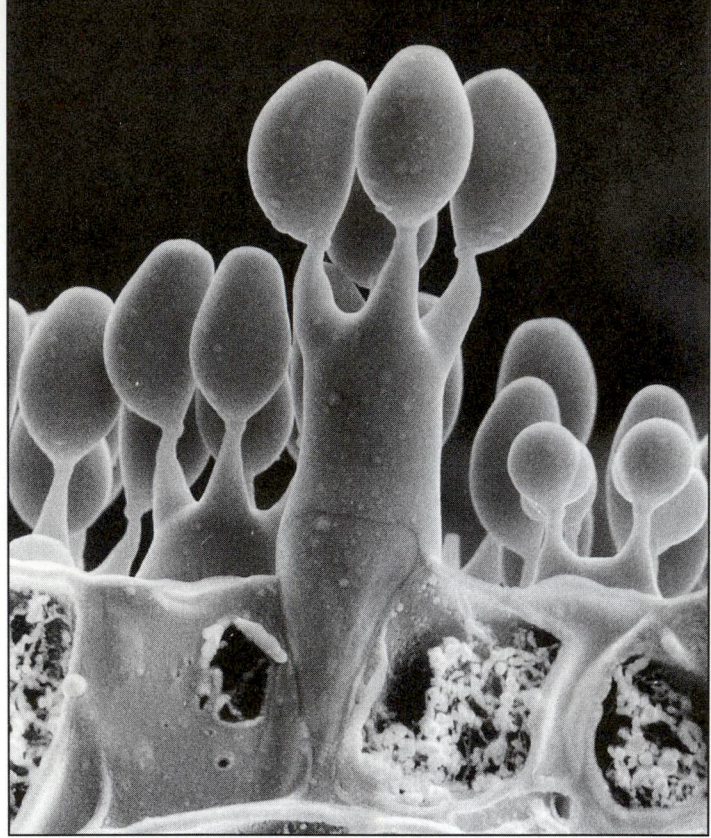

FIGURE 22-11 The basidium produces four basidiospores, which are attached to the basidium. (Biophoto Associates)

there are usually eight haploid ascospores within the ascus. The ascospores are released when the tip of the ascus breaks open.

Basidiomycetes (Club Fungi) Reproduce Sexually By Forming Basidiospores

The 25,000 or more species that make up the division Basidiomycota, or **club fungi,** include the most familiar of the fungi—mushrooms, bracket fungi, and puff balls—as well as some important plant parasites, the rusts and smuts. **Basidiomycetes** derive their name from the fact that they develop a **basidium,** which is a structure comparable in function to the ascus of ascomycetes. Each basidium is an enlarged, club-shaped hyphal cell, at the tip of which develop four **basidiospores** (Fig. 22–11). Note that whereas ascospores develop *within* the ascus, basidiospores develop on the *outside* of the basidium. The basidiospores are released, and when they come in contact with the proper environment, they develop into new mycelia.

The vegetative (feeding) body of a basidiomycete such as the cultivated mushroom, *Agaricus campestris,*

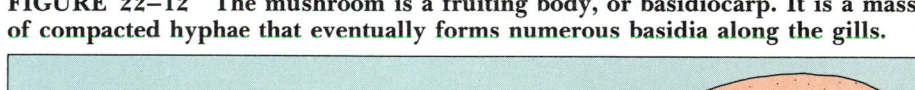

FIGURE 22-12 The mushroom is a fruiting body, or basidiocarp. It is a mass of compacted hyphae that eventually forms numerous basidia along the gills.

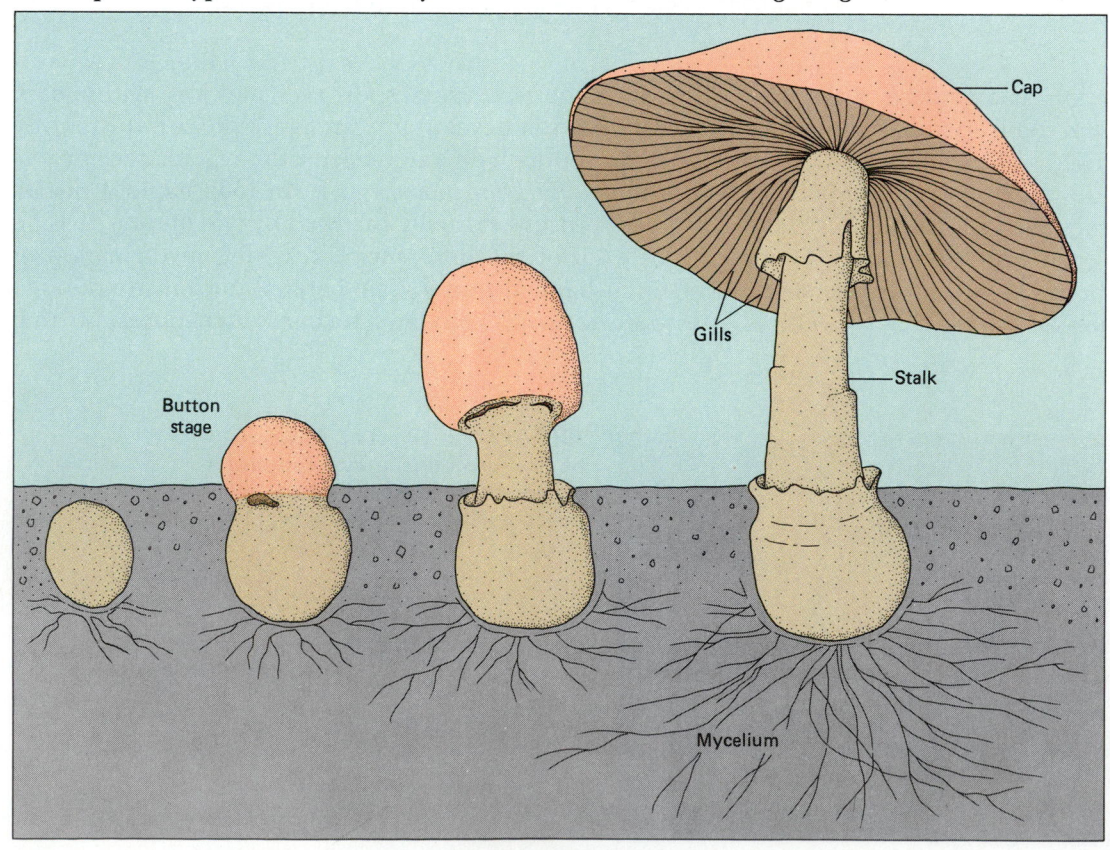

Cap

Gills

Stalk

Button stage

Mycelium

372

consists of a mass of white, branching, thread-like hyphae that occurs mostly below ground. The hyphae are divided into compartments by septa. As in ascomycetes, the septa are perforated and allow cytoplasmic streaming.

Compact masses of hyphae, called buttons, develop along the mycelium. The button grows into the structure we ordinarily call a mushroom, which consists of a stalk and cap (Fig. 22–12). More formally, the mushroom is referred to as a sporocarp, or **basidiocarp** (Fig. 22–13). The lower surface of the cap consists of many thin plates called **gills,** extending radially from the stalk to the edge of the cap. The basidia develop on the surfaces of these gills.

Each individual fungus produces millions of basidiospores, and each basidiospore has the potential, should it happen upon an appropriate environment, to give rise to a new **primary mycelium** (Fig. 22–14). Hyphae of this mycelium consist of monokaryotic cells (cells with a single nucleus). When in the course of its growth such a hypha encounters another hypha of a different mating type, the two hyphae fuse. As in the ascomycetes, however, the two haploid nuclei remain separate. In this way, a **secondary mycelium** with dikaryotic hyphae is produced, in which each cell contains two haploid nuclei. The dikaryotic hyphae of the mycelium grow extensively and eventually form compact masses, which are the mushrooms or basidiocarps. Each basidiocarp actually consists of intertwined hyphae that are matted together. On the gills of the mushroom, the nuclei fuse, forming diploid zygotes. Meiosis then takes place, forming four haploid nuclei.

(a)

(c)

(b)

(d)

FIGURE 22–13 Diversity in basidiocarps. (*a*) Basidia line the gills of the mushroom. (*b*) At maturity, a dried-out puffball will often have a pore through which the basidiospores are discharged in a puff of "smoke." Some puffballs attain remarkable sizes. (*c*) The stinkhorn is also commonly called the devil's penis. Its foul smell attracts flies, which help disperse the slimy mass of basidiospores. (*d*) Bracket fungi, which grow on both dead and living trees, produce shelf-like basidiocarps. Underneath each shelf are many pores. Basidia with basidiospores line the pores. (*a*, Richard H. Gross; *b*, Connie Toops; *c*, Richard D. Poe/Visuals Unlimited; *d*, Ed Reschke)

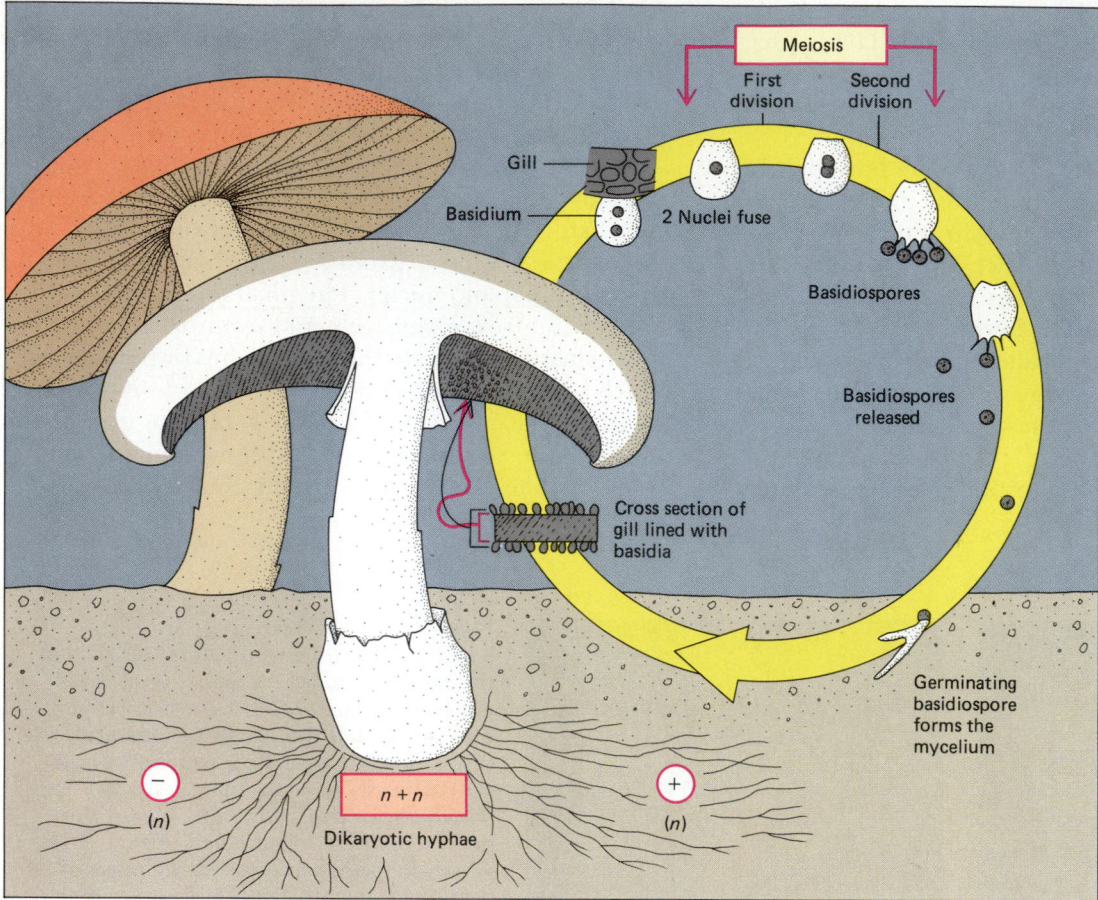

FIGURE 22–14   Life cycle of a mushroom, a typical basidiomycete. A basidiocarp (i.e., a mushroom) develops from the mycelium, a mass of branching threads found underground. On the undersurface of the cap are gills, thin perpendicular plates extending radially from the stalk. Basidia develop on the surface of these gills and produce basidiospores, which are shed. If these spores reach a suitable environment, they give rise to new mycelia.

These nuclei move to the outer edge of the basidium. Finger-like extensions of the basidium develop into which the nuclei and some cytoplasm move; each of these becomes a basidiospore.

Deuteromycetes (Imperfect Fungi) Do Not Reproduce Sexually

About 25,000 species of fungi have been assigned to a group called the **deuteromycetes.** They are also known as "imperfect fungi" because a sexual stage has not been observed during their life cycle. (Should further study reveal a sexual stage, these species will be reassigned to a different division.) Most deuteromycetes reproduce only by means of conidia and so are closely related to the ascomycetes; a few appear to be related to the basidiomycetes.

◻ LICHENS ARE DUAL ORGANISMS COMPOSED OF A FUNGUS AND AN ALGA OR CYANOBACTERIUM

There are some 20,000 species of lichens. Able to tolerate extremes of temperature and moisture, lichens grow everywhere that life can be supported at all, except in polluted, industrial cities. They exist farther north than any plants of the Arctic region and are equally at home in the steaming equatorial jungle. They can grow on tree trunks, mountain peaks, and bare rock (Fig. 22–15). In fact, they are often the first organisms to inhabit rocky areas and play an important role in the formation of soil. Lichens gradually etch tiny cracks in rocks, aiding further disintegration of the rocks by wind and rain.

The "reindeer mosses" of Arctic regions are not mosses but lichens that serve as the main source of

(a) *(b)* *(c)*

FIGURE 22–15 **Lichens vary in their color, shape, and overall appearance.**
**(a) Some lichens grow tightly attached to rocks. It would require a razor blade to
scrape this lichen off its substrate. (b) Many lichens are leaf-like in appearance.
(c) British soldier lichen,** *Cladonia cristatella*, **is branching and shrub-like.** (*a*, M.L.
Dembinsky, Jr.; *b*, Ed Reschke; *c*, Frans Lanting)

food for the reindeer (caribou) of the region. Some
lichens produce colored pigments. One of them, or-
chil, is used to dye woolens, and another, litmus, is
widely used in chemistry laboratories as an acid-base
(pH) indicator.

Although a **lichen** looks like a single organism, it is
actually a symbiotic partnership between a photosynthetic organism and a fungus. The photosynthetic
partner is usually either a green alga or a cyanobacterium, and the fungal partner is most often an ascomycete. In some lichens from tropical regions, the fungal
partner is a basidiomycete. The algae or cyanobacteria
found in lichens are also found as free-living species in
nature, but the fungal components of lichens have
been found only as part of lichens to date.

In the laboratory, the fungal and algal components of a lichen can be isolated and grown separately
in appropriate culture media. The alga grows more
rapidly when separated, whereas the fungus grows
slowly and requires many complex carbohydrates.
Generally, the fungus does not produce fruiting bodies
when separated in this way. The alga and fungus can
be reassembled as a lichen, but only if they are placed
on a culture medium under conditions that cannot
support either of them independently.

What is the nature of this partnership? In the past,
the lichen has been considered a definitive example of
mutualism, a symbiotic relationship that benefits both
species. The photosynthetic partner carries on photosynthesis, producing food for both members of the
lichen; however, it is unclear how the photosynthetic
partner benefits from the relationship. It has been sug-

gested that the photosynthetic partner obtains water
and minerals from the fungus as well as protection,
mainly against drying out; however, there is evidence
that the lichen partnership is not really a case of mutualism but one of controlled parasitism of the photosynthetic partner by the fungus.

Lichens vary greatly in size. Some are almost invisible; others, like the reindeer mosses, may cover miles
of land with a growth that is ankle deep. Growth proceeds slowly; the radius of a lichen may increase by less
than a millimeter each year. Some very large lichens
are thought to be thousands of years old.

Lichens absorb minerals mainly from the air and
from rainwater, but also directly from the surface on
which they grow. They cannot excrete the elements
they absorb, and perhaps for this reason they are very
sensitive to toxic compounds. The absence of lichen
growth has been used to indicate air pollution, especially sulfur dioxide. Absorption of such toxic compounds damages the chlorophyll of the photosynthetic
partner. The return of lichens to an area indicates a
reduction in air pollution in that area.

When a lichen dries out, photosynthesis stops, and
the lichen enters a state of suspended animation in
which it can tolerate severely adverse conditions such
as great extremes of temperature. Lichens reproduce
mainly by asexual means, usually by fragmentation.
Generally, bits of the thallus break off and land on a
suitable surface, where they establish themselves as
new lichens. Some lichens release special dispersal
units containing cells of both partners. In others, the
alga reproduces asexually by mitosis, while the fungus

produces ascospores. The ascospores may be dispersed by wind and find an appropriate algal partner only by chance.

FUNGI ARE ECONOMICALLY IMPORTANT

We have already discussed the vital ecological role of fungi as decomposers. Remember that without these organisms, life on earth eventually would become impossible. However, the same powerful digestive enzymes that enable fungi to decompose wastes and dead organisms also permit them to reduce wood, fiber, and food to their basic components with great efficiency. Thus, various fungi cause incalculable damage to stored goods and building materials each year. Bracket fungi cause enormous losses by bringing about the decay of wood, both in living trees and in stored lumber. The timber these basidiomycetes destroy each year approaches in value that destroyed by forest fires.

Fungi Provide Food For Humans

The ability of yeasts to produce ethyl alcohol and carbon dioxide from glucose by fermentation is of great economic importance. Yeasts used in making wine and beer and in baking are cultivated strains that are carefully maintained to prevent contamination (Fig. 22–16). Wine is produced when yeasts ferment fruit sugars. Beer is made when yeasts ferment grain, usually barley. During the process of making bread, carbon dioxide produced by the yeast becomes trapped in the dough as bubbles, which cause the dough to rise and give leavened bread its light texture. Both the carbon dioxide and the alcohol produced by the yeast are driven off during baking.

The unique flavor of cheeses such as Roquefort and Camembert is produced by the action of members of the ascomycetes genus *Penicillium* (see Fig. 22–16). The mold *P. roquefortii* is found in caves near the French village of Roquefort; only cheeses produced in this area can be called Roquefort cheese. *Aspergillus tamarii* and other fungi are used in the Orient to produce soy sauce by fermenting soybeans. Soy sauce provides other foods with more than its special flavor; it also adds vital amino acids from both the soybeans and the fungi themselves to the low-protein rice diet.

Among the basidiomycetes are some 200 kinds of edible mushrooms and about 70 species of poisonous ones, sometimes called toadstools. Edible mushrooms can be cultivated commercially: more than 60 thousand metric tons are produced each year in the United States alone. Morels, which are gathered and eaten like mushrooms, and truffles, which produce under-

FIGURE 22–16 It is hard to imagine dining without wine or beer, bread, or distinctive cheeses, all of which are produced in part by fungi. Yeasts ferment fruits (wine) or grains (beer), producing ethyl alcohol. That same process produces the carbon dioxide bubbles responsible for making bread rise. The bluish splotches in blue cheese are patches of conidia. (Raymond Tschoepe)

ground fruiting bodies, are ascomycetes (Fig. 22–17). These delights of the gourmet are now being cultivated as mycorrhizae on the roots of tree seedlings.

Edible and poisonous mushrooms can look very much alike and may even belong to the same genus. There is no simple way to distinguish edible from poisonous mushrooms; they must be identified by an expert. Some of the most poisonous mushrooms belong to the genus *Amanita*. Toxic species of this genus have been appropriately called such names as "destroying angel," *Amanita virosa*, and "death angel," *Amanita phalloides* (Fig. 22–18). Ingestion of a single cap can kill a healthy adult human.

Ingestion of certain species of mushrooms causes intoxication and hallucinations. The sacred mushrooms of the Aztecs, *Conocybe* and *Psilocybe*, are still used in religious ceremonies by Central American Indians and others for their hallucinogenic properties. The chemical ingredient psilocybin, chemically related to lysergic acid diethylamide (LSD), is responsible for

(a)

(b)

FIGURE 22–17   Edible ascomycetes. (*a*) *Morchella deliciosa*, commonly called a morel, and (*b*) *Tuber melanosporum*, truffles, are expensive gourmet treats. Both are fruiting bodies known as ascocarps that produce ascospores. Truffles are subterranean ascocarps. Here they are shown entire and sectioned. (*a*, E.R. Degginger; *b*, John D. Cunningham)

the trance-like state and colorful visions experienced by those who eat these mushrooms.

Fungi Produce Useful Chemicals

In 1928, Alexander Fleming noticed that one of his Petri dishes containing staphylococci bacteria was contaminated by mold. The bacteria were not growing in the vicinity of the mold, leading Fleming to the conclusion that the mold was releasing some substance harmful to them. Within a decade of Fleming's discovery, penicillin produced by the ascomycete *Penicillium notatum* was purified and used in treating bacterial infections. Penicillin is still the most widely used and most effective antibiotic. Another fungus, *Penicillium griseofulvicum*, produces the antibiotic griseofulvin, which is used clinically to inhibit the growth of fungi. Cyclosporine, the drug used to suppress immune responses in patients receiving organ transplants, is also derived from a fungus.

Fungi Cause Many Important Diseases Of Plants

Fungi are responsible for many serious plant diseases, including epidemic diseases that spread rapidly and often result in complete crop failure, causing great economic loss and human suffering in some cases. All plants are apparently susceptible to fungal infection. Damage may be localized in certain tissues or struc-

tures of the plant, or the disease may be systemic, affecting the entire plant. Fungus infections may cause stunting of plant structures or of the entire plant; they may cause growths like warts; or they may kill the plant.

A plant may become infected after hyphae enter through pores (stomata) in the leaf or stem or through wounds in the plant body. As the fungal mycelium

FIGURE 22–18   *Amanita phalloides*, the death angel. About 50 g (2 ounces) of this mushroom could kill a 68 kg man. (Hal H. Harrison, from Grant Heilman)

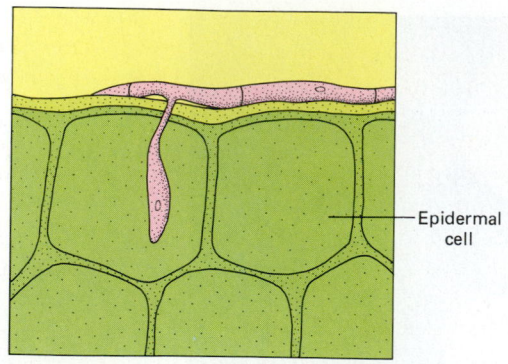

FIGURE 22–19 Haustoria are produced in plant epidermal cells that have been penetrated by the powdery mildew.

grows, it may remain mainly between the plant cells or may penetrate the cells. Parasitic fungi often produce special hyphal branches called **haustoria,** which penetrate the host cells and obtain nourishment from the cytoplasm (Fig. 22–19).

Some important plant diseases caused by ascomycetes are powdery mildews, chestnut blight, Dutch elm disease, apple scab, and brown rot, which attacks cher-

ries, peaches, plums, and apricots (Fig. 22–20). Basidiomycetes include some 700 species of smuts and 6000 species of rusts that attack various plants, including the cereals—corn, wheat, oats, and other grains (Fig. 22–21). Certain deuteromycetes are also important plant pathogens.

Fungi Cause Certain Diseases Of Animals

Although the skin and mucous membranes of healthy animals present effective barriers to fungal penetration, some fungi cause disease in humans and other animals. Some of these cause superficial infections in which the fungi infect only the skin, hair, or nails. Others cause systemic infections, in which fungi infect deep tissues and internal organs and may spread through many regions of the body.

Ringworm and athlete's foot are examples of superficial fungal infections. Candidiasis is an infection of mucous membranes of the mouth or vagina and is among the most common fungal infections. Histoplasmosis is a serious human systemic fungal infection that

FIGURE 22–20 Many ascomycetes are plant pathogens. *Venturia inaequalis* causes apple scab, which infects both fruits and leaves. (Runk/Schoenberger, from Grant Heilman)

FIGURE 22–21 Corn smut on ear of sweet corn. *Ustilago maydis* is the basidiomycete that causes corn smut. (Runk/Schoenberger, from Grant Heilman)

is caused by a fungus that sporulates abundantly in soil containing bird droppings; a person who inhales the spores may then develop the infection. Most pathogenic (disease-causing) fungi cause infections only when the body's immunity is lowered.

CHAPTER SUMMARY

I. Fungi are eukaryotes with cell walls.
  A. Fungi lack chlorophyll, are heterotrophic, and absorb predigested food.
  B. They reproduce by means of spores, which may be produced sexually or asexually.
II. A fungus may be unicellular (the yeast form) or multicellular (the mold form).
  A. The body of a mold consists of long, branched hyphae, which form a mycelium.
    1. In the zygomycetes, the hyphae are coenocytic (undivided by septa).
    2. In other fungi, perforated septa are present.
  B. When a fungal spore comes into contact with an appropriate substrate, it germinates and begins to grow.
III. Fungi are classified into divisions based on their mode of sexual reproduction.
  A. Members of division Zygomycota produce sexual spores called zygospores.
  B. Members of division Ascomycota produce asexual spores called conidia at the tips of conidiophores. Sexual spores called ascospores are produced in asci.
  C. Members of division Basidiomycota produce sexual spores called basidiospores on the outside of a basidium.
  D. The deuteromycetes are the imperfect fungi, species for which a sexual stage has not been observed. Most reproduce asexually by conidia.
IV. A lichen is a symbiotic combination of a fungus and an alga or cyanobacterium in which the fungus benefits from the photosynthetic partner.
V. Fungi function ecologically as decomposers.
  A. Mycorrhizae are symbiotic relationships between fungi and the roots of higher plants.
  B. Lichens play an important role in soil formation.
VI. Fungi are of both positive and negative economic importance.
  A. Mushrooms, morels, and truffles are used as food; yeasts are vital in production of alcoholic beverages and bread; certain fungi are used to produce cheeses and soy sauce.
  B. Fungi are used to make penicillin and other antibiotics.
  C. Fungi cause many plant and animal diseases.

POST-TEST

1. Fungi were originally classified as plants, in part because they possess _____.
2. Fungi reproduce both asexually and sexually by _____.
3. Ecologically, fungi serve as _____.
4. Fungi that obtain their nutrients from living hosts are known as _____.
5. Mycorrhizae are fungi that form mutualistic relationships with the _____ of plants.
6. Yeasts reproduce asexually mainly by _____.
7. A mold consists of thread-like strings of cells called _____, which form a tangled mass called a _____.
8. Some hyphae are divided by walls, called _____, into individual cells.
9. The familiar portion of a mushroom is actually a large _____.
10. Hyphae that contain two genetically distinct nuclei within each cell are _____.
11. *Rhizopus* and other members of the zygomycetes form sexual spores called _____.
12. Sexual reproduction in ascomycetes involves production of _____ within structures called _____.
13. The type of sexual spore produced by a puffball is a _____.
14. In a mushroom, basidia develop on the surface of perpendicular plates called _____.
15. The deuteromycetes are known as imperfect fungi because their _____ stage has not been observed.
16. A _____ is a dual organism that consists of a phototroph and a fungus.
17. _____ are required in both beer and bread production.
18. _____ are special hyphae produced by parasitic fungi that can penetrate host cells and obtain nourishment from the cytoplasm.

■ REVIEW QUESTIONS

1. How does a fungus differ from a plant? What are the distinguishing features of a fungus?
2. What is the ecological importance of each of the following: fungi, lichens, mycorrhizae?
3. How does the body plan of a yeast differ from that of a mold?
4. What is the difference between a hypha and a mycelium? Between an ascus and a basidium?
5. What measures can you suggest to prevent bread from becoming moldy?
6. Briefly describe three important fungal diseases of plants and three fungal diseases of humans.

■ RECOMMENDED READINGS

Ahmadjian, V., and S. Paracer. *Symbiosis: An Introduction to Biological Associations*. Hanover, NH, University Press of New England, 1986. A fascinating description of lichens.

Angier, N. "A Stupid Cell with All the Answers," *Discover* (November): 70–83, 1986. An interesting report on the usefulness of yeasts in biological experiments.

Bessette, A., and W.J. Sundberg. *Mushrooms: A Quick Reference Guide to Mushrooms of North America*. New York, Macmillan Publishing Company, 1987. A beautifully illustrated guide to common fungi.

Kosikowski, F.V. "Cheese," *Scientific American* (May): 88–99, 1985. An absorbing account of how bacteria and fungi are used to make different varieties of cheese.

Vogel, S. "Taming the Wild Morel." *Discover* (May): 58–60, 1988. This brief article gives an introduction to the gourmet's delight, morels.

23
Plant Life

OUTLINE

I. Complex photosynthetic organisms are placed in the kingdom Plantae
II. The mosses and other bryophytes are nonvascular plants
III. Seedless vascular plants include the ferns and their allies
IV. The production of seeds represents a major evolutionary advancement
  A. The gymnosperms are the "naked seed" plants
  B. The flowering plants produce flowers, fruits, and seeds
    1. Sexual reproduction in the flowering plants has a number of novel features
    2. Flowering plants are the most successful plant group on the earth
Focus on Ancient Plants and Coal Formation
Focus on Pollen and Hay Fever

LEARNING OBJECTIVES

After you have studied this chapter you should be able to:

1. Discuss the environmental challenges faced by land plants and relate adaptations they evolved to meet these challenges.
2. Summarize the features that distinguish bryophytes from green algae.
3. Discuss the advancements the ferns and fern allies have over the mosses and liverworts.
4. Diagram a generalized plant life cycle, clearly showing alternation of generations.
5. Compare seeds with spores and discuss the advantages of plants that reproduce primarily by seeds rather than spores.
6. Summarize the features that distinguish gymnosperms from the ferns.
7. Contrast dicots with monocots.
8. Discuss the evolutionary advancements of the angiosperms.

Purple cone flowers. (Carl R. Sams II/ Marvin L. Dembinsky Photography Associates)

The plant Kingdom comprises thousands of different species that live in every conceivable habitat, from the frozen Arctic tundra to lush tropical rain forests. Plants range in size from minute, almost microscopic duckweeds to massive giant sequoias. Although plants exhibit an incredible diversity in size, habit, and form, they are believed to have evolved from common ancestors. There are four major groups of plants living today (Table 23–1). The mosses and other bryophytes lack a vascular, or conducting, system and are therefore restricted in size. The other three groups of plants possess vascular tissues, **xylem** for water and mineral conduction and **phloem** for food conduction. Ferns and their allies are vascular plants that reproduce by spores. The gymnosperms and angiosperms (flowering plants) are vascular plants that reproduce by forming seeds. Gymnosperms are naked seed plants, whose seeds are often produced in a cone, while flowering plants produce seeds enclosed within a fruit.

Land plants are thought to have evolved from ancient aquatic green algae (Fig. 23–1). Evidence supporting this hypothesis includes the fact that the green algae share a number of biochemical and metabolic traits with plants. Both contain the same photosynthetic pigments, chlorophylls *a* and *b* and carotenoids. Also, both store their excess carbohydrates as starch. Cellulose is a major component of the cell walls of both. Finally, certain details of cell division, including

TABLE 23–1
The Plant Kingdom

I Nonvascular plants

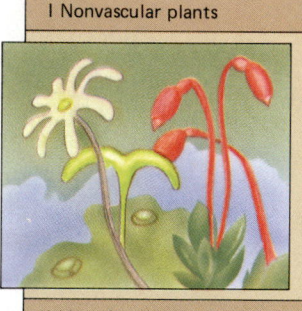

I. Nonvascular plants with a dominant gametophyte generation
      Division Bryophyta (bryophytes)
        Class Bryopsida (mosses)
        Class Hepatopsida (liverworts)
        Class Anthoceropsida (hornworts)

II Vascular plants

II. Vascular plants with a dominant sporophyte generation
  A. Seedless plants
      Division Pterophyta (ferns)
      Division Psilophyta (whisk ferns)
      Division Sphenophyta (horsetails)
      Division Lycophyta (club mosses)
  B. Seed plants
    1. Naked seed plants
      Division Coniferophyta (conifers)
      Division Cycadophyta (cycads)
      Division Ginkgophyta (ginkgo)
      Division Gnetophyta (gnetophytes)
    2. Seeds enclosed within a fruit
      Division Magnoliophyta (flowering plants)
        Class Magnoliopsida (dicots)
        Class Liliopsida (monocots)

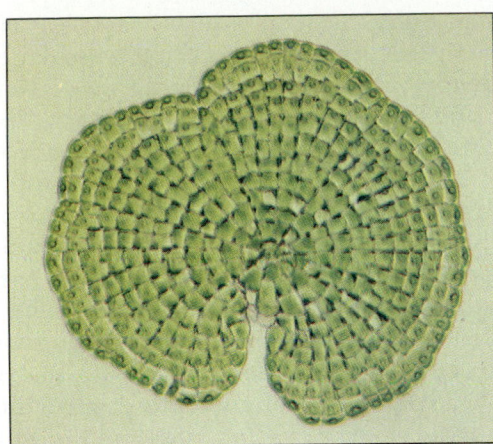

FIGURE 23–1 **The green alga *Coleochaete* resembles a group of green algae which may have been the ancestors of the land plants. It is considered quite primitive because it shares a number of features with its extinct ancestors.** (Courtesy of Dr. Linda Graham, University of Wisconsin, Madison)

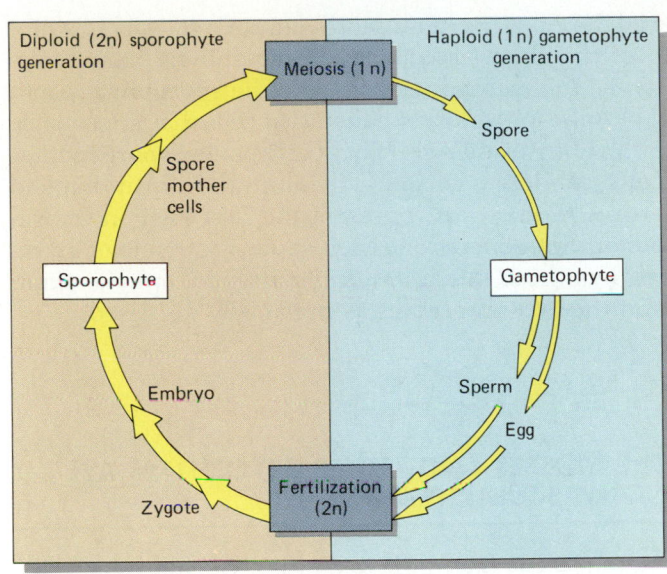

FIGURE 23–2 **The basic plant life cycle. All plants have modifications of this cycle. Note that plants have alternation of generations. They spend part of their life in the haploid (gametophyte) stage and part in the diploid (sporophyte) stage.**

the formation of a cell plate, are shared by plants and many green algae but are found in no other photosynthetic organisms.

KEY CONCEPTS

- A number of adaptations enable terrestrial plants to survive on land.
- The plant life cycle alternates between a haploid, gametophyte stage and a diploid, sporophyte stage.
- Reproduction by seeds is a significant evolutionary advancement that helps explain why seed plants dominate the earth today.

COMPLEX PHOTOSYNTHETIC ORGANISMS ARE PLACED IN THE KINGDOM PLANTAE

Plants use the green pigment chlorophyll to absorb radiant energy, which is then converted to the chemical energy found in carbohydrates. In addition to chlorophylls *a* and *b*, all plants have **carotenoids** (yellow and orange pigments).

One of the most important adaptations plants have in order to survive on land is a waxy covering, the **cuticle,** over their aerial parts. The cuticle is essential for existence on land because it helps prevent evaporation from drying out plant tissues.

Land plants obtain their carbon from the atmosphere as carbon dioxide. This gas must be accessible to the chloroplasts inside green plant cells (see Window on the Plant Cell). However, there is little gas exchange through the waxy cuticle. Instead, gas is exchanged through tiny openings, or **stomata,** in stems and leaves.

Plants have multicellular sex organs, or **gametangia,** each of which possesses a sterile layer of cells surrounding the gametes. Each female organ, the **archegonium,** produces a single egg. Sperm are produced in the male sex organ, the **antheridium.** Presumably, the gametes produced within these organs are protected from desiccation (drying out) by the outer jacket of sterile cells. Algae lack such organs. In plants the fertilized egg develops into a multicellular **embryo** *within* the female gametangium. Thus, during its development the embryo is protected.

Plants have a clearly defined **alternation of generations;** that is, they spend part of their lives in the haploid stage and part in the diploid stage (Fig. 23–2). The haploid portion of the life cycle is called the **gametophyte generation** because it gives rise to gametes by mitosis. The diploid portion of the life cycle is the **sporophyte generation,** which produces **spores** immediately following meiosis.

The gametophytic plant produces antheridia and archegonia. The sperm get to the archegonia in a variety of ways and one sperm fuses with the egg. This process, known as **fertilization,** results in a fertilized egg, or **zygote.**

The zygote is diploid and is the first cell in the sporophyte generation. The zygote divides by mitosis and develops into a multicellular embryo, which is supported and protected by the gametophytic plant. Eventually, the embryo matures into the sporophyte plant. The sporophyte plant has special cells that are capable of dividing by meiosis. These cells, called **spore mother cells,** undergo meiotic division and form haploid spores. The spores represent the first stage in the gametophyte generation. Each spore is capable of growing by mitosis into a multicellular gametophytic plant, and the cycle continues as previously discussed.

■ THE MOSSES AND OTHER BRYOPHYTES ARE NONVASCULAR PLANTS

Aquatic algae have little need for a vascular system because their aquatic environment transports their food and other essential substances. Hence, kelps and other algae can grow quite large. In contrast, land plants that lack vascular tissues, because they have no way to transport water, food, and essential minerals for extensive distances, are restricted in size. Also, since they require a moist environment for reproduction, most bryophytes are found actively growing in moist areas, although some are tolerant of dry areas.

The **bryophytes** are the only nonvascular plants. They are sometimes divided into three classes: the **mosses,** the **liverworts,** and the **hornworts.** These three groups of plants may or may not be closely related. Their life cycles are similar, however.

Moss plants usually live in dense colonies or beds. Each individual plant has tiny root-like structures, or **rhizoids,** that anchor the plant to the soil. Each plant

also has an upright "stem," which bears "leaves." (Because these structures lack specialized vascular tissues, however, they cannot be called true roots, stems, or leaves.) The leafy green moss plant that you are most familiar with is the gametophyte generation of mosses (Fig. 23–3). It bears its gametangia at the top of the plant. Many moss plants have separate sexes; these mosses have male plants that bear antheridia and female plants that bear archegonia. Other mosses produce antheridia and archegonia on the same plant.

In order for fertilization to occur, sperm must fertilize the egg held within the archegonium. The zygote formed as a result of fertilization grows into a multicellular embryo by mitosis and matures into the moss sporophyte. This sporophyte plant grows out of the top of the female gametophyte (Fig. 23–3). Although it is initially green in color and photosynthetic, it becomes a golden brown at maturity and is composed of three main parts: a **foot,** which anchors the sporophyte to the gametophyte; a **seta,** or stalk; and a **capsule,** which contains spore mother cells. These spore mother cells undergo meiosis to form haploid spores. When the spores are mature, the capsule opens by various mechanisms to release the spores. These microscopic cells are carried by wind or rain to other places. If a spore lands in a suitable spot, it germinates and grows into a filamentous thread of green cells called a **protonema.** The protonema, which looks like a filamentous green alga, forms buds. Each bud grows into a leafy green gametophyte plant, and the life cycle continues (Fig. 23–4).

The gametophyte generation is considered the dominant generation in mosses because it is capable of living independently of the sporophyte. By contrast, the sporophyte generation in mosses is at all times at-

(a)

(b)

FIGURE 23–3 Mosses. (a) The leafy green gametophyte plants grow in dense clusters. These are male moss plants with multiple antheridia located at the top of each plant. **(b)** Haircap moss. The moss sporophyte generation grows out of the top of the gametophyte plant. Each sporophyte is attached to and dependent on the gametophyte plant for nourishment. Spores are produced by meiosis within the capsule at the tip of each sporophyte. (a and b, David Cavagnaro)

tached to and dependent on the gametophyte plant. A dominant gametophyte generation is considered to be a primitive characteristic.

Mosses make up an inconspicuous but significant part of their environment. They play an important role in forming soil. Because they grow tightly packed together in dense colonies, they hold the soil in place and help to prevent soil erosion. They provide food for animals, especially birds and mammals. Commercially, the most important mosses are the peat mosses in the genus *Sphagnum*. Peat mosses are particularly beneficial as a soil conditioner; for example, when added to sandy soils, they help to hold and retain mois-

ture in the soil. In some countries peat moss is often collected, dried, and burned for fuel.

The structure of some liverworts is quite different from that of mosses. Their body form is often a flattened, leaf-like **thallus** (plural, *thalli*) which is lobed (Fig. 23–5). They were named liverworts because the lobes of their thalli superficially resemble the lobes of the human liver. In medieval times many people believed in the Doctrine of Signatures, that a "signature" on each plant gave a clue to its use. Because of their resemblance to the human liver, liverworts were considered to be useful in treating liver ailments, a belief that has not been supported by modern medical re-

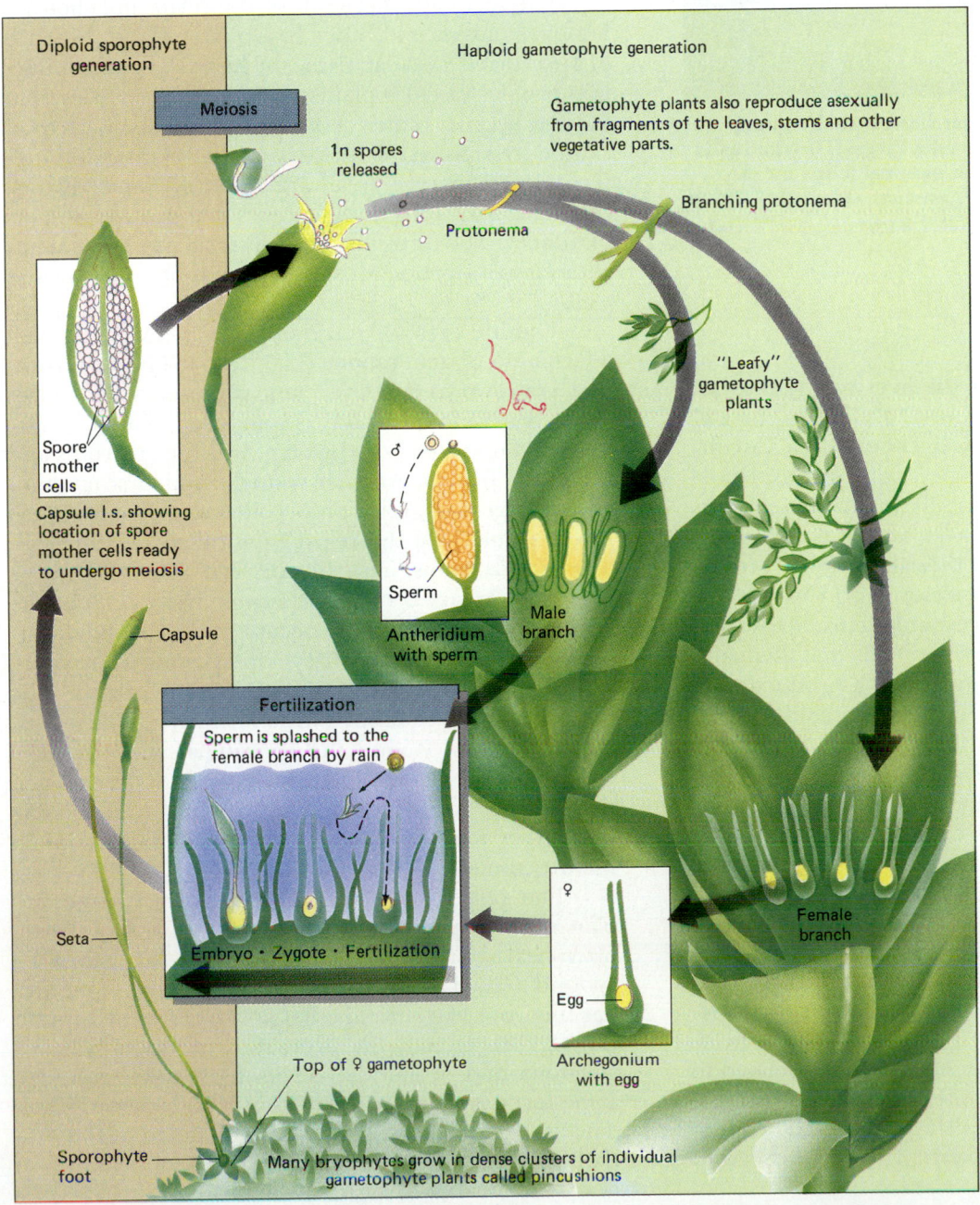

FIGURE 23–4 Alternation of generations in the mosses.

Diploid sporophyte generation

Haploid gametophyte generation

Meiosis

1n spores released

Gametophyte plants also reproduce asexually from fragments of the leaves, stems and other vegetative parts.

Protonema

Branching protonema

Spore mother cells

Capsule l.s. showing location of spore mother cells ready to undergo meiosis

"Leafy" gametophyte plants

Sperm

Antheridium with sperm

Male branch

Capsule

Fertilization

Sperm is splashed to the female branch by rain

Female branch

Seta

Embryo · Zygote · Fertilization

♀

Egg

Archegonium with egg

Sporophyte foot

Top of ♀ gametophyte

Many bryophytes grow in dense clusters of individual gametophyte plants called pincushions

FIGURE 23–5 Many liverworts have a thallus body form which is flattened, ribbon-like lobes. The thallus is the gametophyte plant. Note the gemmae cups on the thallus, which contain gemmae, asexual structures which can grow into new thalli. (Connie Toops)

search. On the underside of the liverwort thallus are root-like rhizoids, which anchor the plant to the soil. As with other bryophytes, the liverworts lack vascular tissue. They are small and generally inconspicuous plants that are restricted largely to damp environments. Many liverworts have a leafy appearance rather than a lobed thallus. Some of these leafy liverworts are superficially very similar to mosses, with rhizoids and structures that resemble leaves and stems.

Liverworts reproduce both sexually and asexually. Their sexual reproduction involves the production of archegonia and antheridia on the haploid thallus. Their life cycle is basically the same as that of the mosses, although some of the structures look quite different. One of the ways that liverworts reproduce asexually is by forming tiny balls of tissue called **gemmae.** These gemmae are borne in a saucer-shaped structure, the **gemmae cup,** directly on the liverwort thallus (Fig. 23–5). Splashing raindrops and small animals aid in the dispersal of gemmae. When gemmae land in a suitable place, each can grow into a liverwort thallus.

The hornworts are a small group of plants whose gametophytes superficially resemble the thalloid liverworts. Hornworts may or may not be closely related to other bryophytes. For example, their cell structure, particularly the presence of a single large chloroplast in each cell, is reminiscent of certain algae. Mosses and liverworts, on the other hand, are like all other plants because each cell has many disk-shaped chloroplasts.

◻ SEEDLESS VASCULAR PLANTS INCLUDE THE FERNS AND THEIR ALLIES

The ferns represent an ancient group of plants that is still successful today. They are especially common in temperate woodlands and tropical rain forests (Fig. 23–6). The three groups of plants that are considered to be allies of the ferns because of similarities in life cycles are the whisk ferns, club mosses (lycopods), and horsetails.

The main advancement of the ferns and their allies over mosses and other bryophytes is the presence of specialized vascular tissues. This system of conduction enables vascular plants to achieve larger sizes than mosses because water, dissolved minerals, and food can be transported to all parts of the plant. Although ferns in temperate areas are relatively small plants, there are tree ferns in the tropics that grow to heights of 60 feet. The ferns and fern allies all have true stems with vascular tissues. Most have true roots and leaves as well.

The life cycle of ferns involves a clearly defined alternation of generations (Fig. 23–7). The fern plant that is grown as a house plant represents the sporophyte generation. Its body is composed of a horizontal underground stem, or **rhizome,** which bears roots and leaves, or **fronds.** As each young frond first emerges from the ground, it is tightly coiled and resembles the top of a violin. For this reason, it is called a **fiddlehead.** The fronds, roots, and rhizome are considered to be true plant organs because each contains vascular tissue. Spore production usually occurs on the fronds. Certain areas on the fronds develop **sporangia,** or spore cases, in which spore mother cells are formed. The sporangia are frequently borne in clusters, called **sori,** on the fronds. Within the sporangia, spore mother cells undergo meiosis to form haploid spores. When these spores are disseminated and land in a suitable place, they may germinate and grow by mitosis into a mature gametophyte plant.

The gametophyte plant of ferns bears no resemblance to the sporophyte generation. It is a tiny, green, often heart-shaped structure that grows flat against the ground. Called a **prothallus,** it lacks vascular tissue and has tiny root-like rhizoids that anchor it to the ground (Fig. 23–8). Usually the prothallus produces both archegonia and antheridia on its underside, although some ferns produce prothalli that have separate sexes. The archegonia are located near the notch of the prothallus, and each contains a single egg. Numerous

(a)   (b)   (c)

FIGURE 23–6   Representative ferns. (a) Developing fronds showing characteristic "fiddlehead" growth pattern of new leaves. (b) Cinnamon fern, showing brown reproductive spikes and green sterile leaves (the spikes bear the sporangia). (c) Tree ferns, *Cyathea arborea*. These plants, which are native to tropical rain forests, are found in New Zealand, South Africa, and South America. (a, Ed Reschke; b, Connie Toops; c, E.R. Degginger)

sperm are produced in the antheridia, which are found scattered among the rhizoids.

Although ferns are considered to be more advanced than mosses because they have vascular tissue, they have retained one primitive requirement: They require water for fertilization. A thin film of water on the ground underneath the prothallus provides the transport medium in which the flagellated sperm swim to the neck of the archegonia. After one of the sperm unites with the egg in an archegonium, a diploid zygote grows by mitosis into a multicellular embryo. As the embryo matures into the sporophyte fern plant, the prothallus withers and dies.

In the fern life cycle the sporophyte generation is considered to be dominant not only because it is larger in size than the gametophyte but also because it persists for an extended period of time whereas the gametophyte dies soon after reproducing. A trend observed in more advanced land plants is that the sporophyte generation becomes increasingly dominant and less dependent on water for fertilization. Most algae have a dominant gametophyte generation, with its corresponding production of flagellated gametes that can swim in the water. For land plants it is more advantageous to produce nonmotile spores than it is to produce flagellated gametes.

The few extant species of whisk ferns are found mainly in the tropics. Although whisk ferns do not closely resemble the ferns, they are considered to be fern allies because of similarities in their life cycles. *Psilotum nudum* is a representative whisk fern (Fig. 23–9).

The horsetails were more important millions of years ago, when they were among the dominant land plants and grew to be the size of trees. The ancient horsetails are still significant to us today because they contributed largely to the vast coal deposits that we are currently using (see Focus on Ancient Plants and Coal Formation). The few extant plants are small but very distinctive (Fig. 23–9). Horsetails often grow in wet, marshy habitats. The hollow, jointed stems are impregnated with silica, which gives them a gritty feeling. In Colonial America they were referred to as scouring rushes and were used to scrub out pots and pans along the stream banks.

Like horsetails, club mosses were dominant plants millions of years ago. Species that are now extinct often attained great size. These large trees were major contributors to the coal deposits on our earth (see Focus on Ancient Plants and Coal Formation). The club mosses today are small, attractive plants most common in woodlands (Fig. 23–9). The plants are evergreen

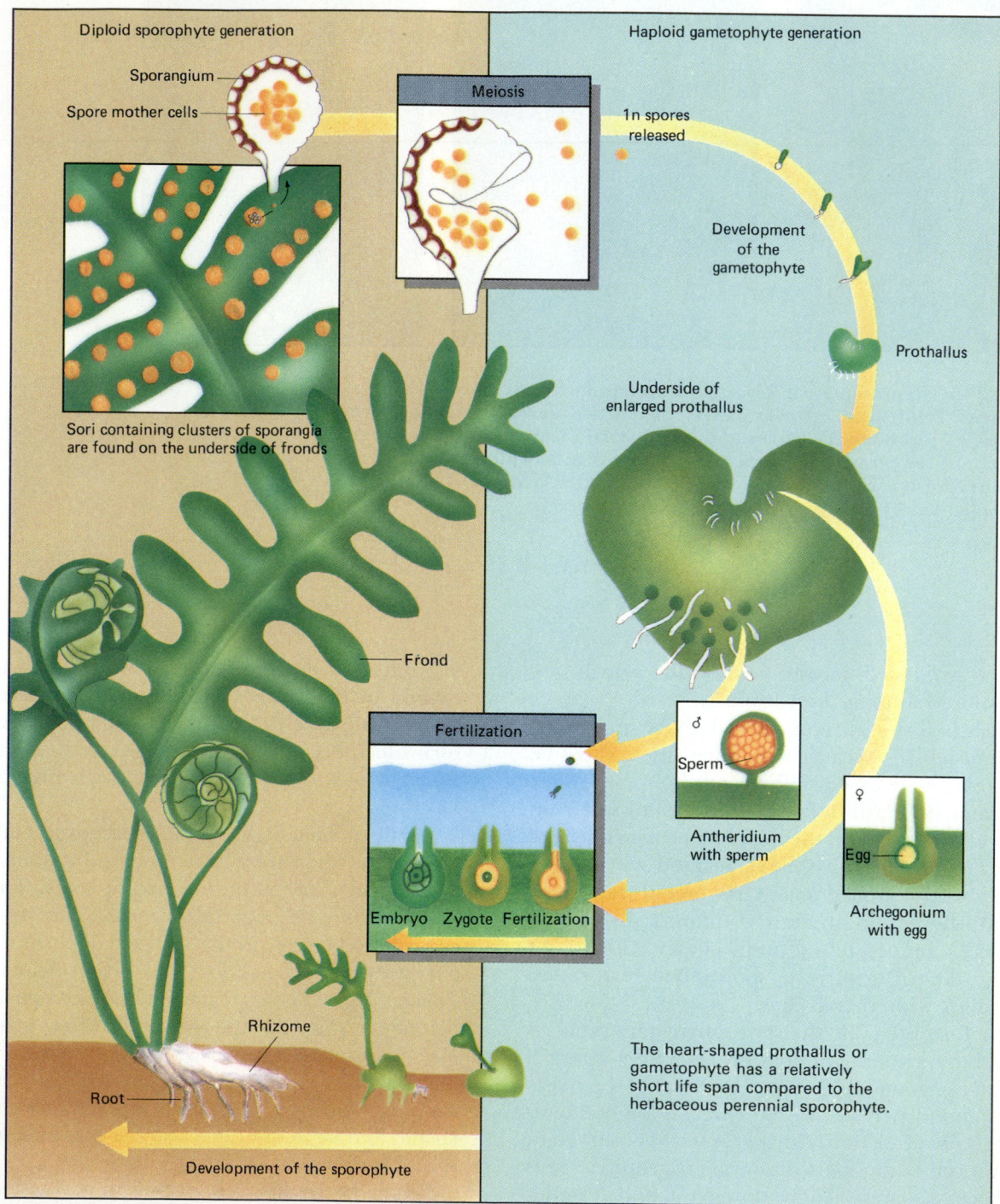

Diploid sporophyte generation

Haploid gametophyte generation

Sporangium

Spore mother cells

Meiosis

1n spores released

Development of the gametophyte

Prothallus

Underside of enlarged prothallus

Sori containing clusters of sporangia are found on the underside of fronds

Frond

Fertilization

Sperm

Antheridium with sperm

♀

Egg

Archegonium with egg

Embryo Zygote Fertilization

The heart-shaped prothallus or gametophyte has a relatively short life span compared to the herbaceous perennial sporophyte.

Rhizome

Root

Development of the sporophyte

FIGURE 23–7 The fern life cycle. Note the clearly defined alternation of generations between the haploid and diploid stages.

FIGURE 23–8 The prothallus is the gametophyte generation of ferns. The dark spots on the prothallus are archegonia. (Runk/Schoenberger, from Grant Heilman)

FIGURE 23–9 Representative fern allies. (*a*) The growth habit of *Psilotum nudum*. The stem is the main organ of photosynthesis since leaves are absent. This ancient-appearing vascular plant bears sporangia directly on the stems. (*b*) *Equisetum telematia*, a horsetail, with a wide distribution in Eurasia, Africa, and North America. It has unbranched, nongreen, nonphotosynthetic "fertile" shoots bearing cone-like structures and separate highly-branched, green, photosynthetic "sterile" shoots. Both types of shoots arise from an underground rhizome. (*c*) Club moss. Although the club mosses may superficially resemble mosses, they are fern allies. The sporophyte plant has reduced, scale-like leaves which are evergreen. Spores are produced in sporangia clustered in a cone-like structure. (*a*, Biophoto Associates; *b*, J. Robert Waaland, University of Washington/BPS; *c*, Dwight R. Kuhn)

(a)

(b)

(c)

FOCUS ON   Ancient Plants and Coal Formation

The industrial society in which we live depends on energy from fossil fuels. One of the most important fuels is coal. Although coal is mined from the earth as a mineral, it is not a mineral like gold or aluminum. Coal is organic, formed from ancient plant material.

Much of the coal we use today was formed from the prehistoric remains of primitive land plants, particularly those of the Carboniferous Period, approximately 300 million years ago. Five main groups of plants contributed to coal formation. Three of them were seedless vascular plants—club mosses, horsetails, and ferns. The other two important groups of coal formers were seed plants—seed ferns (now extinct) and primitive gymnosperms.

It is hard to imagine that the small, relatively inconspicuous club mosses, ferns, and horsetails of today could have been so significant in forming the vast beds of coal in the earth. However, the extinct members of these groups that existed during the Carboniferous Period were giants by comparison and they formed vast forests of trees.

The climate during the Carboniferous Period was warm and mild. Plants could grow year-round because of the favorable weather conditions. The forests of these plants occurred in low-lying areas that were periodically flooded when the sea level rose. When the sea level receded, these plants would become established again.

When these large plants died or were blown over during storms, they were incompletely decomposed because they were covered by the swampy waters. The anaerobic conditions of the water prevented wood-rotting fungi from decomposing the plants, and anaerobic bacteria do not decompose wood rapidly. Over time, the partially decomposed plant material accumulated and consolidated.

Layers of sediment formed over the plant material each time the sea level rose and flooded the low-lying swamps. With time, heat and pressure built up in these accumulated layers and converted the plant material to coal and the sediment layers to sedimentary rock.

Much later, geological upheavals raised the layers of coal and sedimentary rock. For example, coal is found in seams (layers) in the Appalachian Mountains. The various grades of coal (lignite, bituminous, and anthracite) formed as a result of the different temperatures and pressures to which they were exposed.

Some of the plants that grew during the Carboniferous Period included giant ferns, horsetails, and club mosses. The early gymnosperms were also present. Recent work suggests that Carboniferous swamps were more open than depicted here. (No. Geo. 75400C, Field Museum of Natural History, Chicago)

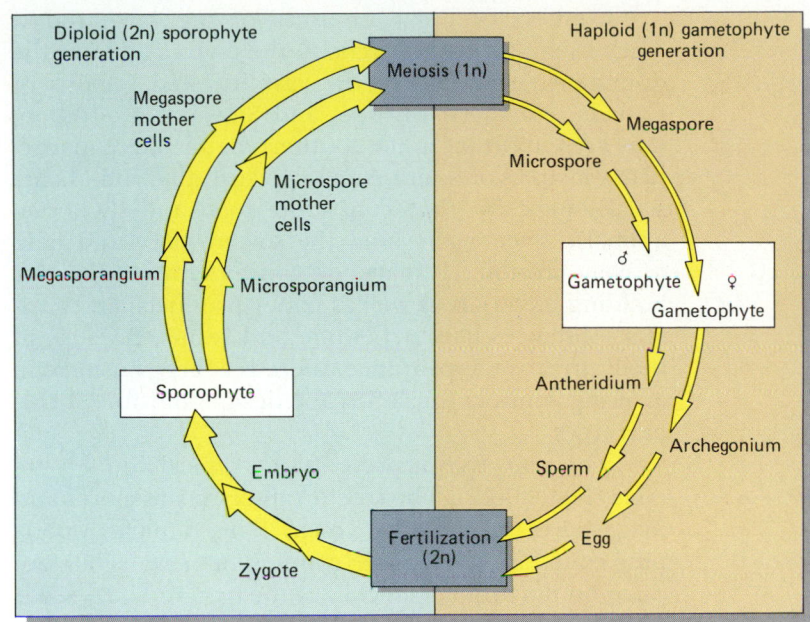

Diploid (2n) sporophyte generation

Haploid (1n) gametophyte generation

Meiosis (1n)

Megaspore mother cells

Megaspore

Microspore

Microspore mother cells

Megasporangium

Microsporangium

♂ Gametophyte   ♀ Gametophyte

Sporophyte

Antheridium

Archegonium

Embryo

Sperm

Egg

Fertilization (2n)

Zygote

FIGURE 23–10   A generalized life cycle for heterosporous plants. These plants produce two types of spores, microspores and megaspores.

and are often fashioned as Christmas wreaths and other holiday decorations. In some areas they are endangered by overuse.

In the alternation of generations that we have considered thus far, plants produced one type of spore as a result of meiosis. This condition, known as **homospory,** is found in the bryophytes, horsetails, whisk ferns, and most ferns and club mosses. However, certain ferns and club mosses are **heterosporous** (Fig. 23–10). They produce two different types of spores as a result of meiosis, **microspores** and **megaspores.** Microspores eventually produce male gametophytes and sperm, while megaspores produce female gametophytes and eggs. **Heterospory** is a significant development in plant evolution because it is found in the two most successful groups of plants on earth today, the gymnosperms and the flowering plants.

■ **THE PRODUCTION OF SEEDS REPRESENTS A MAJOR EVOLUTIONARY ADVANCEMENT**

The primary means of reproduction and dispersal for the most successful plants is by seeds, which develop from the female gametophyte and tissues associated with it. The seed plants comprise the dominant plants in most habitats. Indeed, one could think of the present time as the "Age of Seeds."

It has been said that a single seed may produce more plants than a million spores. Why do seeds confer such an apparent advantage? Seeds are reproductively superior to spores for several reasons. First, plant spores are composed of a single cell, while seeds contain a multicellular, well-developed young plant with embryonic root, stem, and leaves. Second, seeds contain a food supply. After germination, the plant embryo within the seed is nourished by food stored in the seed until it is self-sufficient. Because a spore is a single cell, little food reserve exists for the plant that develops from a spore. Also, seeds are protected by a resistant seed coat.

The two groups of seed plants are the **gymnosperms** and the **angiosperms.** The word *gymnosperm* is adapted from a Greek word meaning "naked seed." These plants produce seeds that are totally exposed or borne on the scales of cones. Pine, spruce, fir, and other conifers are examples of gymnosperms. The Greek from which the term *angiosperm* is derived translates as "seed enclosed in a vessel or case." Angiosperms, or flowering plants, produce their seeds within a fruit. Flowering plants include such diverse plants as corn, oaks, water lilies, cacti, and buttercups.

Both gymnosperms and flowering plants possess vascular tissue, xylem for conduction of water and dissolved minerals and phloem for conduction of food. Both have alternation of generations. The gametophyte generation, however, is significantly reduced in size and is entirely dependent on the sporophyte generation. They both are heterosporous, producing microspores and megaspores.

The Gymnosperms Are The "Naked Seed" Plants

The gymnosperms include some of the most interesting plants in the plant kingdom. A number of record holders are in this group. For example, the world's

FIGURE 23–11 **The majestic sequoias are an example of the division Coniferophyta, gymnosperms that produce their seeds in cones. Some of the sequoias are the world's most massive plants.** (Gary R. Bonner)

largest organism (in terms of sheer bulk) is the General Sherman tree, a giant sequoia. A redwood is the world's tallest tree, measuring almost 380 feet in height. The living organism with the longest life span is the bristlecone pine. One living bristlecone pine in Nevada has been dated by tree ring analysis as 4900 years old.

The gymnosperms are usually classified into four divisions. The largest group is the **conifers** (division Coniferophyta), woody plants that bear their seeds in cones (Fig. 23–11). Two divisions of gymnosperms represent evolutionary remnants of groups that were more significant in the past, the ginkgoes and the cycads. A final group of gymnosperms, the gnetophytes, is a collection of some very unusual plants that share certain advancements not found in the other gymnosperms.

The conifers include pines, spruces, hemlocks, and firs. They are woody trees or shrubs; there are no herbaceous (nonwoody) conifers. They are attractive plants even during winter because most are evergreen. Only a few, such as the larch and the bald cypress, are **deciduous** and shed their leaves, called **needles,** at the end of the growing season. Most conifers are **monoecious,** having separate male and female reproductive parts on the same plant. These reproductive parts are generally borne in **cones.**

Conifers occupy vast areas of the earth today. They range from the Arctic to the tropics and are the dominant vegetation in the vast forested regions of Canada, Northern Europe, and Siberia. In addition, they are important in the southern hemisphere, particularly in areas of South America, Australia, and Malaysia. Ecologically, they contribute food and shelter to animals. Their roots hold the soil in place and help prevent erosion. Humans use conifers for lumber (for building materials as well as paper products) and various substances like turpentine and resins. Because of their attractive appearance, there is a large business in growing conifers for landscape design and for Christmas trees.

The pine tree represents the typical conifer life cycle (Fig. 23–12). The tree produces microspores and megaspores in separate cones. The familiar woody pine cones are the female cones. They are usually located on the upper branches of the tree and bear seeds after fertilization has occurred. Male cones are smaller than female cones, are generally produced on the lower branches in the spring, and are composed of overlapping leaf-like structures called **sporophylls.** At the base of each sporophyll is a sac, or **microsporangium,** which contains numerous microspore mother cells. These cells undergo meiosis to form haploid microspores. Each microspore develops into an extremely reduced male gametophyte generation. The immature male gametophyte is also called a **pollen grain.** The pollen grains are shed from the male cones in great numbers, and some are carried by wind currents to the immature female cones.

The woody bracts of the female cones have megasporangia at their bases. Within each megasporangium, meiosis of a megaspore mother cell produces four haploid megaspores. One of these develops into the female gametophyte, which produces an egg within each of several archegonia. The pollen grain grows a tube that digests its way through the female gametophyte tissue to the egg within the archegonium. Then, a cell within the pollen grain divides to form two nonflagellated sperm (also called sperm nuclei). One of these fuses with the egg to form the zygote, which grows into the young pine embryo, forming part of the seed.

The haploid female gametophyte tissue surrounding the developing embryo becomes the nutritive tissue in the mature seed. The food and embryo are surrounded by a tough, protective seed coat. The mature seed has a papery wing that enables it to be carried by wind currents. When the female cone opens, the seeds are dispersed.

There are several key points to remember about the pine life cycle. The sporophyte generation is dominant, and the gametophyte has decreased in size to microscopic structures in the cones. A major advance-

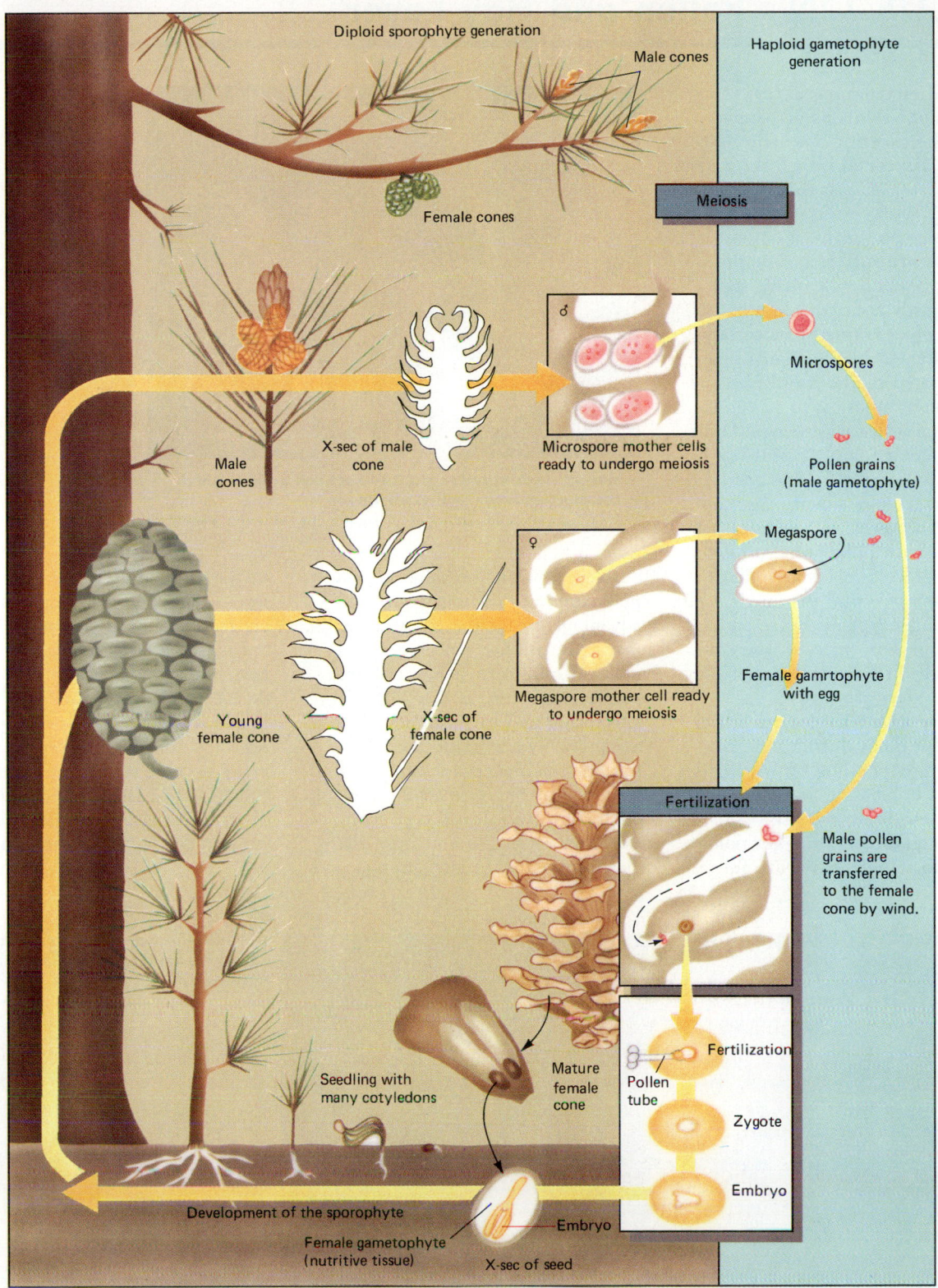

Diploid sporophyte generation

Haploid gametophyte generation

Male cones

Female cones

Meiosis

X-sec of male cone

Microspore mother cells ready to undergo meiosis

Microspores

Male cones

Pollen grains (male gametophyte)

Megaspore

Young female cone

X-sec of female cone

Megaspore mother cell ready to undergo meiosis

Female gamrtophyte with egg

Fertilization

Male pollen grains are transferred to the female cone by wind.

Fertilization

Pollen tube

Seedling with many cotyledons

Mature female cone

Zygote

Embryo

Development of the sporophyte

Female gametophyte (nutritive tissue)

X-sec of seed

Embryo

FIGURE 23–12   The life cycle of pine. One major evolutionary advancement of gymnosperms over the lower vascular plants is their wind-borne pollen. Pines and other gymnosperms are not dependent on water as a transport medium for sperm.

FOCUS ON Pollen and Hay Fever

If you suffer from hay fever, you are not alone. Millions of people endure the sneezing and itchy, watery eyes associated with this condition. Everyone knows that one of the causes of hay fever is pollen, but many blame any plant in bloom when they are suffering. For this reason, roses and goldenrod are often unjustly accused.

Hay fever is caused by certain wind-pollinated plants. Plants that are pollinated by the wind must produce copious amounts of pollen to ensure that at least some of it lands on the stigmas for successful reproduction. In contrast, plants that are pollinated by insects and other animals, and whose large, colorful petals attract such pollinators, do not cause hay fever; relatively little of their pollen gets into the air. Not all wind-pollinated plants cause an allergic reaction. For example, the conifers are all wind-pollinated, yet allergies to conifers are rare.

People with allergies can suffer at different times during the growing season, depending on which plants are pollinating and whether they are sensitized to those plants. In early spring, tree flowers get pollinated before their leaves are fully

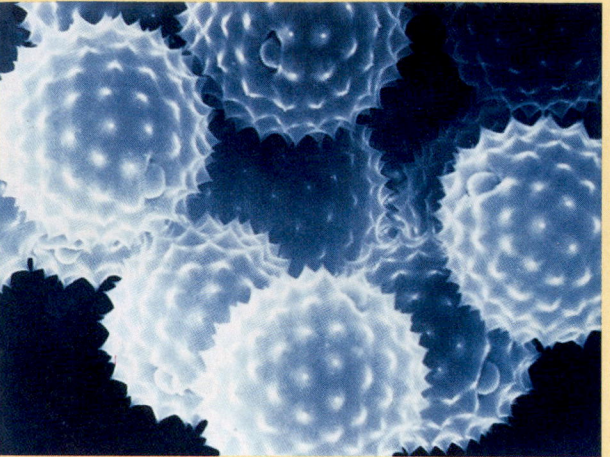

Scanning electron micrograph of ragweed pollen, which is the most common pollen allergen. (Carolina Biological Supply Company)

developed. (Can you think of why pollination at this time would be advantageous for the tree?) Trees that cause allergic reactions in humans include oaks, ashes, walnuts, maples, and elms. If you suffer in late spring and early summer, you are probably allergic to the grasses, for example, bluegrass, timothy, and redtop. It is interesting that most of our major grass crops (e.g., corn, rice, and wheat) do not cause allergies in humans. In late summer and early fall, people are allergic to different plants, depending on their geographical location. In the United States, ragweed is the culprit in the East, while saltbush and Russian thistle are problems in the West.

Most people are born with some resistance to pollen allergies, but many become sensitized by repeated contact. For that reason, a move to a different geographical location often temporarily halts the suffering.

Wind-pollinated plants produce copious amounts of pollen. (Biophoto Associates)

FIGURE 23–13   A cycad growing in South Africa. Cycads are tropical gymnosperms with a palm-like appearance. Note the immense seed cones on this plant. (W.H. Hodge/Peter Arnold, Inc.)

FIGURE 23–14   *Ginkgo biloba*, the ginkgo or maidenhair tree. The unusual leaves of the ginkgo resemble the maidenhair fern, hence its common name. Note the naked, fleshly seeds. (W.H. Hodge/Peter Arnold, Inc.)

ment in the pine life cycle is elimination of water as a transport medium for the sperm. The pollen is carried to the female cones by air currents. Nonflagellated sperm accomplish fertilization by moving through a pollen tube to the egg. Therefore, the gymnosperms are the first land plants whose reproduction is totally adapted for life on land.

The **cycads** were a very important plant group in the prehistoric past. The few remaining extant cycads are tropical plants with a palm-like appearance (Fig. 23–13). Cycad reproduction is similar to pine reproduction. Their ovule and seed structure is most like that of the earliest seeds found in the fossil record.

The **ginkgo,** or maidenhair, tree is native to China, where it has been under cultivation for centuries (Fig. 23–14). It has not been found in the wild, and it is likely that it would have become extinct had it not been cultivated in Chinese monasteries. The ginkgo represents the oldest genus of living trees. Fossil ginkgoes 200 million years old have been discovered that are nearly identical to the modern-day ginkgo. The ginkgo is common in North America today, particularly in cities, because it is somewhat resistant to air pollution.

The **gnetophytes** are an amazingly diverse group of gymnosperms that are divided into three genera (Fig. 23–15). The gnetophytes share a number of fea-

tures that make them clearly more advanced than the rest of the gymnosperms. They have more efficient water-conducting cells, called vessels, in their xylem. Flowering plants have vessels in their xylem, but gymnosperms, with the exception of the gnetophytes, do not. Also, the cone clusters produced by some of the gnetophytes resemble flower clusters.

The Flowering Plants Produce Flowers, Fruits, And Seeds

The **angiosperms,** or flowering plants, are the most successful plants on earth today, surpassing even the gymnosperms in importance. They have adapted to almost every habitat, except Antarctica, and, with over 250,000 species, are the dominant plants (Fig. 23–16). Flowering plants reproduce sexually by forming flowers, fruits, and seeds. They possess vessels in xylem tissue and their fertilization process is unique. Flowering plants are extremely important to humans. Our survival as a species literally depends upon the flowering plants. All of our major food crops are angiosperms, including rice, wheat, and corn. We use angiosperms to provide us with fibers like cotton and

(a)

(b)

FIGURE 23–15   Two examples of gnetophytes. (a) *Ephedra*. Species native to desert areas in the southwestern United States were used by pioneers to make a beverage. A common name for this plant is Mormon tea. (b) The most bizarre gymnosperm in the world is *Welwitschia*, native to an African desert. Although the plant only produces two leaves, the winds tear them so that it appears to have many. (a, John D. Cunningham/Visuals Unlimited; b, No. Bot. 83024c Courtesy, Field Museum of Natural History)

(a)

(b)

(c)

(d)

FIGURE 23–16   Flowering plant diversity. (a) Flowering plants like the water lily are adapted to wet environments. (b) Grass flowers have pendulous stamens. Pollen produced in these stamens gets wind-borne easily. (c) All four floral parts (sepals, petals, stamens, carpels) can be seen in trillium. (d) Cactus plants have a number of adaptations, such as fleshy stems for water storage, that enable them to survive the harsh desert environment. (a, Barbara J. Miller/BPS; b, Dwight R Kuhn; c, Don and Esther Phillips/Tom Stack and Associates; d, E.R. Degginger)

medicines like digitalis and codeine. Woody flowering plants, such as oak, cherry, and walnut, provide us with valuable lumber. Plant products as diverse as rubber, tobacco, coffee, and aromatic oils for perfumes come from flowering plants.

The flowering plants (division Magnoliophyta) are divided into two classes, the **monocots** (class Liliopsida) and **dicots** (class Magnoliopsida) (Table 23–2). Monocots include palms, grasses, orchids, and lilies. Examples of the dicots are oaks, roses, cacti, and sunflowers. The monocots are mostly herbaceous plants with leaves that may be narrow and have parallel veins. The flower parts occur in multiples of three. Monocot seeds have a single **cotyledon** (embryonic seed leaf), and **endosperm** (nutritive tissue) is usually present in the mature seed. Dicots may be herbaceous (for example, the tomato) or woody (for example, the hickory). Their leaves are variable but often broader than monocot leaves, with netted veins. Flower parts occur in multiples of four or five. Two cotyledons are present in seeds of dicots, and endosperm is usually absent in the mature seed.

Sexual Reproduction In The Flowering Plants Has A Number Of Novel Features

The organ of sexual reproduction in the angiosperms is the **flower.** Flowers have four main parts (sepals, petals, stamens, and carpels) that are arranged in whorls (Fig. 23–17). The **sepals** make up the outer-most whorl. They are leaf-like in appearance and often green. The sepals cover and protect the flower parts when the flower is a bud. As the blossom opens from the bud, the sepals fold back to reveal the more conspicuous petals. The **petals** play an important role in assuring that pollination will occur, but they are not directly involved in the fertilization process. Just inside the petals are the **stamens,** the male reproductive part. Each stamen is composed of a thin stalk, the **filament,** which terminates in the **anther,** where pollen is formed. In the center of most flowers are one or more **carpels,** the female reproductive part. The carpel has three sections: the **stigma,** where the pollen lands; the **style,** a neck through which the pollen tube must grow; and the **ovary,** which contains one or more **ovules.** After fertilization of the egg within, each ovule develops into a seed and the ovary develops into the **fruit.**

Like other plants, flowering plants have alternation of generations (Fig. 23–18). The sporophyte generation is clearly dominant, and the gametophyte generation is reduced in size to several cells only. There are no archegonia or antheridia.

Each ovule within the ovary contains a megaspore mother cell which undergoes meiosis, producing four haploid megaspores. One of these develops into the female gametophyte generation, also called the **embryo sac.** The embryo sac contains eight nuclei, including one egg and two **polar nuclei.** The egg and the polar nuclei are involved in fertilization.

FIGURE 23–17 Diagram of a typical flower. This cut away view shows the details of basic floral structure.

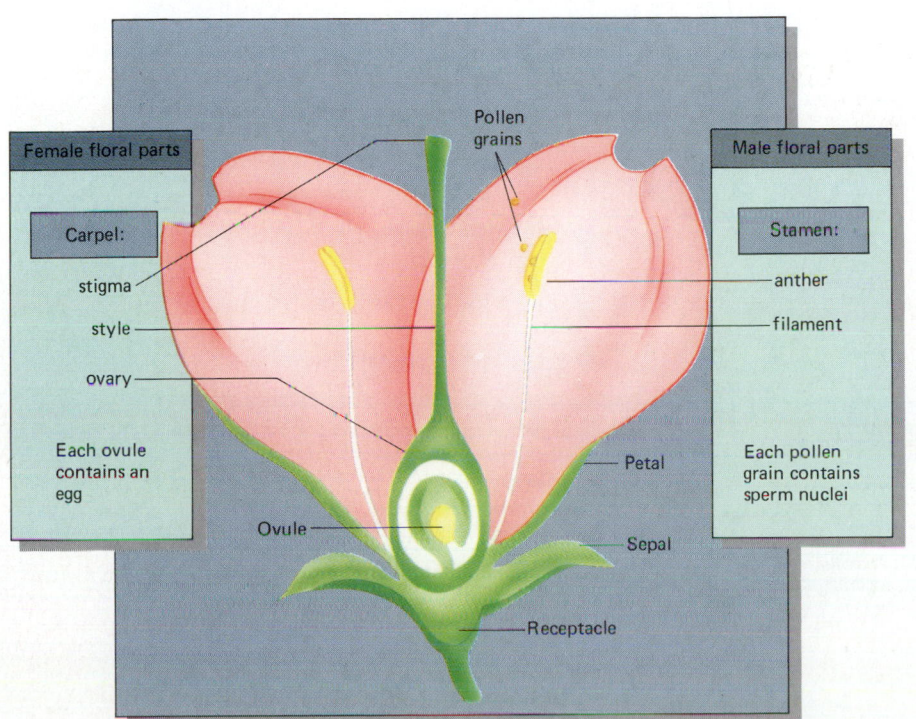

Female floral parts

Carpel:
- stigma
- style
- ovary

Each ovule contains an egg

Pollen grains

Male floral parts

Stamen:
- anther
- filament

Each pollen grain contains sperm nuclei

Petal

Ovule

Sepal

Receptacle

TABLE 23–2
A Generalized Comparison of Monocots and Dicots

Monocots	Dicots
Flowers: floral parts in multiples of 3's	Flowers: floral parts in multiples of 4's or 5's
Seeds: contain one cotyledon	Seeds contains two cotyledons
Stems: vascular bundles are scattered	Stem: organized vascular bundle arrangement
Leaves: long tapering blades with parallel venation. The base of the leaf, the sheath, encircles the stem.	Leaves: broad to narrow leaves with petioles and netted venation.

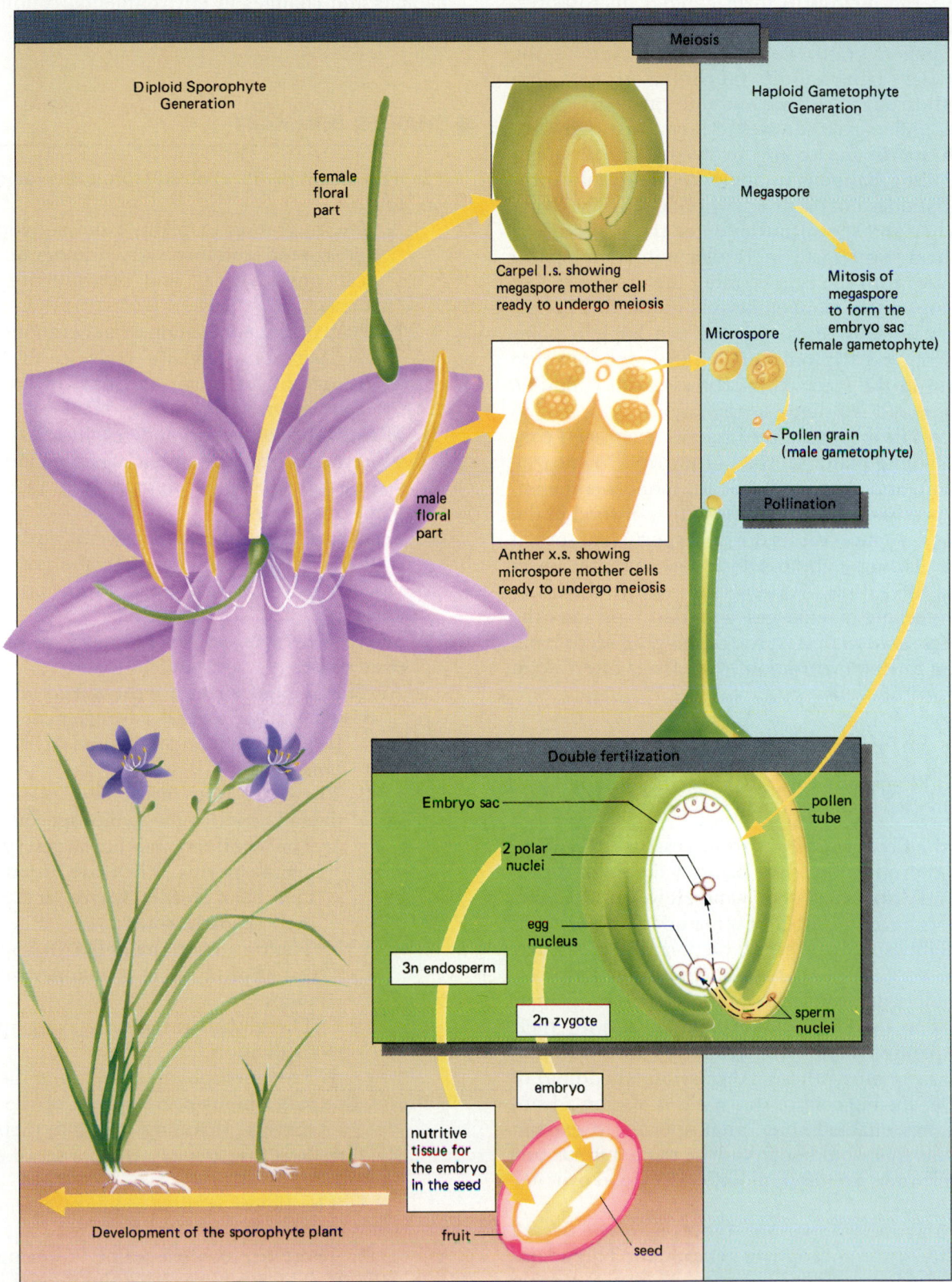

Meiosis

Diploid Sporophyte
Generation

Haploid Gametophyte
Generation

female
floral
part

Carpel l.s. showing
megaspore mother cell
ready to undergo meiosis

Megaspore

Mitosis of
megaspore
to form the
embryo sac
(female gametophyte)

Microspore

male
floral
part

Pollen grain
(male gametophyte)

Anther x.s. showing
microspore mother cells
ready to undergo meiosis

Pollination

Double fertilization

Embryo sac

pollen
tube

2 polar
nuclei

3n endosperm

egg
nucleus

2n zygote

sperm
nuclei

embryo

nutritive
tissue for
the embryo
in the seed

Development of the sporophyte plant

fruit

seed

FIGURE 23–18 Generalized life cycle of a typical flowering plant. The most significant feature is double fertilization, which is found nowhere else in the living world.

The anther contains microspore mother cells which undergo meiosis to form haploid microspores. These microspores each develop into the male gametophyte, a pollen grain. Pollen is transferred to the stigma of the carpel and, if compatible, grows a thin tube down through the style and into the ovary. A cell within the pollen grain divides to form two sperm nuclei. Both sperm nuclei are involved in fertilization.

Something happens in angiosperm sexual reproduction that does not occur anywhere else in the living world. When the sperm nuclei enter the embryo sac, *both* of them participate in fertilization. One sperm nucleus fuses with the egg, forming the diploid zygote that develops into a plant embryo in the seed. The second sperm nucleus fuses with the two haploid polar nuclei, forming a **triploid** (3n) cell that develops into endosperm in the seed. This process is called **double fertilization** and is unique to the angiosperms.

As a result of double fertilization, each seed contains (1) a young plant, (2) food or nutritive tissue (the endosperm), and (3) a seed coat. In monocots the endosperm persists and is the main source of food in the mature seed. In most dicots the endosperm is absorbed by the developing embryo, which subsequently stores food in its cotyledons. As the seed develops, the ovary wall surrounding it enlarges and develops into the fruit. Fruits serve to protect the developing seeds from desiccation during maturation. Also, fruits often aid in the dispersal of seeds.

Flowering Plants Are The Most Successful Plant Group On The Earth

Why are the flowering plants so successful today? Certainly, seed production as the primary means of reproduction and dispersal is significant. However, the flowering plants have a number of other advanced features besides highly successful reproduction. Their vascular tissues are very efficient at conduction; they have vessels in their xylem and sieve tube elements in their phloem. The leaves of flowering plants, with their broad, expanded blades, are structured for maximum efficiency in photosynthesis. **Abscission** (shedding) of these leaves during cold or dry spells is also an advantage that has enabled some angiosperms to expand into habitats that would otherwise be too harsh for survival. Their roots are often modified for food or water storage.

Probably most important, however, is the overall adaptability of the sporophyte generation. This adaptability is evident in the diversity of the flowering plants. The cactus is marvelously adapted for desert environments. The water lily is equally well adapted for wet environments. The flowering plants are successful because they evolve readily, adapting to new habitats and changes in environmental conditions.

■ CHAPTER SUMMARY

I. Terrestrial plants evolved from green algal ancestors.
   A. The migration of plants from aquatic ecosystems to the land involved a number of anatomical, physiological, and reproductive adaptations.
   B. Plants have alternation of generations, spending part of their life cycle in the haploid (gametophyte) stage and part in the diploid (sporophyte) stage.
II. Mosses and other bryophytes have several advancements over the green algae, including possession of a cuticle, stomata, and multicellular gametangia.
III. Ferns and fern allies have several advancements over the bryophytes, including the possession of vascular tissue and a dominant sporophyte generation.
IV. Seeds represent an evolutionary advancement over spores.
   A. The gymnosperms are the naked seed plants.
      1. They have several advancements over the ferns, including production of wind-borne pollen.
      2. There are four divisions of plants, collectively known as the gymnosperms.
         a. The conifers are the largest group of gymnosperms. They are woody plants that bear needle leaves and produce their seeds in cones.
         b. The cycads are palm-like in appearance but reproduce in a manner similar to pines.
         c. The ginkgo is the only living species in its division.
         d. The gnetophytes share a number of advancements over the rest of the gymnosperms, including vessels in their xylem.
   B. The flowering plants produce seeds enclosed within a fruit.
      1. The flower is their organ of sexual reproduction.
      2. There are two classes of flowering plants.
         a. The monocots have floral parts in multiples of three, and their seeds contain one cotyledon. The nutritive tissue in their mature seeds is endosperm.

b. The dicots have floral parts in multiples of four or five, and their seeds contain two cotyledons. The nutritive tissue in their mature seeds is usually in the cotyledons.
3. The flowering plants have several advanced features.

a. Double fertilization, which results in the formation of a zygote and endosperm tissue, is unique to the flowering plants.
b. They possess vessels in their xylem.

■ POST-TEST

1. The bryophytes lack a _____ system and are therefore restricted in size.
2. The waxy layer that covers aerial parts of land plants is the _____.
3. The land plants evolved from the _____ _____.
4. The openings in plants that allow gas exchange for photosynthesis are called _____.
5. The female gametangium, or _____, produces an egg.
6. The fusion of gametes is called _____ and results in a diploid fertilized egg, or _____.
7. Meiosis of spore mother cells results in the formation of _____.
8. Plants have _____ of _____ in which they spend part of their life in the haploid stage and part in the diploid stage.
9. The leafy green moss plant is the _____ generation.
10. Clusters of sporangia, termed _____, are often found on fern fronds.
11. _____ have hollow, jointed stems that are impregnated with silica.

12. _____ are better than spores for reproduction because they have an embryo and food tissue.
13. Most conifers are _____ and have separate male and female reproductive parts on the same plant.
14. Although conifers bear their seeds in cones, they are considered the naked seed plants because their seeds are not enclosed in a _____.
15. The male gametophyte generation of pine is called _____.
16. The nutritive tissue in the pine seed is _____ _____.
17. This class of flowering plants, the _____, includes the palms, grasses, and orchids.
18. The _____ is a nutritive tissue formed as a result of double fertilization.
19. The _____ is composed of the stigma, style, and ovary.
20. After fertilization, the _____ develops into the fruit and the _____ develops into the seed.

■ REVIEW QUESTIONS

1. What is the single most important environmental challenge that land plants face, and what adaptations do they have to meet this challenge?
2. State the advancements that the bryophytes have over the algae.
3. State the advancements that the ferns have over the bryophytes.
4. Why are seeds such a significant evolutionary development?

5. List several ways the conifers are advanced over the ferns.
6. How are the flowering plants different from the gymnosperms?
7. Diagram a flower and label the following parts: stamen, carpel, sepal, petal, anther, filament, stigma, style, ovary, ovule.
8. What are the two classes of flowering plants, and how can one distinguish between them?

■ RECOMMENDED READINGS

Kaufman, P.B., Carlson, T.F., Dayanandan, P., Evans, M.L., Fisher, J.B., Parks, C., and J.R. Wells. *Plants: Their Biology and Importance.* New York, Harper & Row, 1989. An excellent general botany textbook.

K.R. Stern. *Introductory Plant Biology,* 4th ed., Dubuque, Iowa, Wm. C. Brown Publishers, 1988. This reference is written in an enjoyable style. Readers learn botany almost effortlessly.

24
Animal Life: Non-Chordate Invertebrates

OUTLINE

I. Animals are multicellular heterotrophs
   A. Most animal groups inhabit salt water
   B. Animals can be grouped according to body structure or pattern of development
   C. We can describe the animal body using standard directions and body planes

II. Sponges have specialized cells but no true tissues

III. Cnidarians (hydras, jellyfish, and corals) have tissue layers but no true organs
   A. Cnidarians have two body plans: a polyp and a medusa form
   B. The Hydra has a solitary, carnivorous life-style

IV. Flatworms have three tissue layers, a head region, and organs
   A. Planarians are free-living carnivores
   B. Flukes have suckers and other adaptations for their parasitic life-style
   C. Tapeworms are strikingly specialized for their parasitic life-style

V. Proboscis worms are the simplest animals to have organ systems

VI. The roundworms (nematodes) have organ systems but lack circulatory structures

VII. Most mollusks are covered by a shell and have an open circulatory system
   A. There are four principal classes of mollusks
   B. The clam has all the organ systems typical of complex animals

VIII. The annelids have segmented bodies

IX. Arthropods have jointed appendages and an exoskeleton of chitin
   A. The arthropods can be assigned to three subphyla
   B. The insects have adapted to almost every available ecological niche
   C. The grasshopper has a representative insect life-style

X. The echinoderms are spiny-skinned animals of the sea

Focus on Adaptations of Squids and Octopods

LEARNING OBJECTIVES

After you have studied this chapter, you should be able to:

1. Develop a definition of an animal using the characteristics common to animals.
2. Justify classification and proposed relationships of the animal phyla on the basis of (a) symmetry, (b) type of body cavity, (c) pattern of embryonic development (that is, protostomes and deuterostomes).
3. Classify a given animal in the appropriate phylum and class.
4. Contrast the animal phyla based on their distinguishing characteristics.
5. Describe the body plan and life-style of one member of each phylum.
6. Trace the life cycle of each parasitic worm described in the chapter, including tapeworm and *Ascaris*. Identify adaptations that contribute to the success of these parasites.
7. Identify factors contributing to the great biological success of the insects.

Strawberry anemones (*Corynactis californica*). (Gary Milburn, Tom Stack & Associates).

More than a million species of animals have been described, and perhaps several million more remain to be identified. Most members of the animal kingdom are classified in about 35 phyla (see Table 24–1 at end of chapter). The animals most familiar to us—dogs, birds, fish, frogs, snakes—are vertebrates, or animals with backbones. Vertebrates, however, account for only about 5% of the species of the animal kingdom. The majority of animals are the less familiar invertebrates, which are animals without backbones. The invertebrates include such diverse forms as corals, worms, and butterflies. In this chapter, we first discuss some general characteristics of all animals. Then we examine distinguishing characteristics used in classifying animals. Finally, we describe each of the major invertebrate phyla. The chordates, which include some invertebrates, as well as all of the vertebrates, are covered in the following chapter.

KEY CONCEPTS

☐ Animals are multicellular heterotrophs. Most are capable of locomotion, can respond rapidly to stimuli, and reproduce sexually.

☐ In very simple, small animals, life processes such as gas exchange, circulation of materials, and waste disposal take place by diffusion. In larger, more complex animals, specialized structures and mechanisms have evolved to handle these life processes.

■ ANIMALS ARE MULTICELLULAR HETEROTROPHS

We can easily identify a lion as an animal and a rose bush as a plant, but we may mistake an aquatic animal that lives attached to a rock or dock for a plant. There are so many diverse animal forms that exceptions can be found to almost any definition of an animal (Fig. 24–1). Still, there are some characteristics that describe most animals:

1. All animals are multicellular eukaryotic heterotrophs. In contrast to the fungi, most are consumers, ingesting their food first and then digesting it inside the body, usually within a digestive system. As consumers, animals depend on producers for their raw materials, energy, and oxygen. They depend on decomposers to recycle nutrients.
2. The cells that make up the animal body exhibit a division of labor. In all but the simplest animals, cells are organized to form tissues, and tissues are organized to form organs. In most animal phyla, specialized organ systems carry on specific functions.
3. Most animals are capable of locomotion at some time during their life cycle; however, some animals (the sponges, for example) are sessile (firmly attached to an object or surface) as adults.
4. Most animals have well-developed sense organs and nervous systems and can respond rapidly to stimuli in their environment.
5. Most animals reproduce sexually, with large, nonmotile eggs and small, flagellated sperm. Sperm and egg unite to form a fertilized egg, or zygote, which goes through a series of embryonic stages before

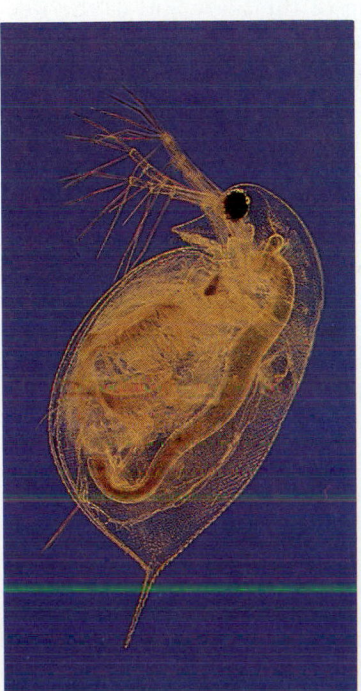

FIGURE 24–1 Despite their diversity, most members of the animal kingdom share several distinctive traits. (a) The nearly transparent body of this freshwater crustacean, a water flea (*Simocephalus vetulus*), shows a complex set of organ systems. (b) As heterotrophs, all animals must feed either on producers or on other animals that eat producers. The goldenrod crab spider shown here with its prey is an example of an animal that eats other consumers. The butterfly relies directly on producers for food. (c) Some animals are easily mistaken for plants. This fern hydroid from the Gulf of Mexico belongs to Phylum Cnidaria. (a, Herman Eisenbeiss/Photo Researchers, Inc.; b, John Gerlach/Visuals Unlimited; c, E.R. Degginger)

(a)   (b)   (c)

developing into a larva or immature form. Some animals can also reproduce asexually.

Most Animal Groups Inhabit Salt Water

Animals originated in the sea but have radiated to occupy virtually every type of environment on the earth. Of the three principal environments—salt water, fresh water, and land—salt water is the most hospitable. Sea water is isotonic (that is, it has a similar concentration of solute and solvent molecules) to the tissue fluids of most marine animals, so there is little problem in maintaining fluid and salt balance. The buoyancy of sea water supports its inhabitants, so they have less need for skeletal support than terrestrial organisms. The temperature of the sea is relatively constant owing to the large volume of water. Plankton, the protists and tiny animals that are suspended in the water and float with its movement, provide a ready source of food.

Fresh water offers a less constant environment and generally contains less food. Moreover, fresh water is hypotonic to the tissue fluids of most animals, so maintaining fluid and salt balance becomes difficult. For these reasons, far fewer animals make their homes in fresh water than in the sea.

Terrestrial life, or life on dry land, is most difficult. Dehydration is a serious threat because water is continuously lost by evaporation and is often difficult to replace. Only a few animal groups—most notably, insects, spiders (and some other arthropods), and higher vertebrates—have evolved adaptations that permit them to make their homes on land.

Animals Can Be Grouped According To Body Structure Or Pattern Of Development

Most biologists agree that animals evolved from the Protista. Although the relationships among the various animal phyla are a matter of conjecture, there is no scarcity of hypotheses. A few of the more widely held hypotheses are presented in this section.

The animal kingdom may be divided into two main subkingdoms: the **Parazoa,** which are the sponges, and the **Eumetazoa,** which include all other animals. This distinction is made because the sponges are so different that most biologists think they did not give rise to any other animal phylum.

Animals Have Radial or Bilateral Symmetry

Members of the Eumetazoa may be classified on the basis of body symmetry. Two phyla, the cnidarians (jellyfish and relatives) and the ctenophores (comb jellies), exhibit **radial** (wheel-like) **symmetry** and are included in the *branch Radiata.* In radial symmetry, similar structures are regularly arranged as spokes from a central axis. Multiple planes can be drawn through the central axis, each dividing the organism into two mirror images. Radial symmetry is considered an adaptation for a sessile life-style because the animal receives stimuli equally from all directions in the environment.

Most animals are **bilaterally symmetrical** (at least in their larval stages) and belong to the *Bilateria.* A bilaterally symmetrical animal can be divided through only one plane (which passes through the midline of the body) to produce roughly equivalent right and left halves that are mirror images (Fig. 24–2). Bilateral symmetry is considered an adaptation to motility. The front end of the animal generally has a head, where sense organs are concentrated; this end receives most environmental stimuli. The rear end of the animal may be equipped with a tail for swimming, or it may just follow along.

Animals Can Be Grouped According to Type of Body Cavity

A widely held system for grouping animal phyla is based on the presence and type of body cavity. Some animals have no body cavity, whereas others have simple body cavities and still others have a type of body cavity referred to as a true coelom. In order to understand these categories, we must look briefly at the animal's embryonic development. The structures of most animals develop from three embryonic tissue layers, called **germ layers,** that are present in the embryo.

The outer germ layer, called the **ectoderm,** gives rise to the outer covering of the body and to the nervous system (if the animal has one). The inner layer, or **endoderm,** lines the digestive tract. **Mesoderm,** the middle layer, gives rise to most of the other body structures, including the muscles, bones, and circulatory system (when they are present).

In the simplest Eumetazoans (the Cnidaria and flatworms), the body is essentially a double-walled sac surrounding a digestive cavity with a single opening to the outside—the mouth. There is no body cavity, so these animals are referred to as **acoelomates** (*a,* meaning "without"; *coelom,* meaning "cavity") (Fig. 24–3).

Other, more complex animals generally have a **tube-within-a-tube body plan.** Tissue derived from endoderm lines the inner tube—the digestive tract. The digestive tract has an opening at each end—the mouth and the anus. The outer tube is the body wall. It is covered with tissue derived from ectoderm. Beneath the ectoderm, the tube consists of mesoderm. The space between the two tubes is the body cavity. It is called a **pseudocoelom** (false coelom) if mesoderm lines only part of it. Animals with a pseudocoelom are called *pseudocoelomates.*

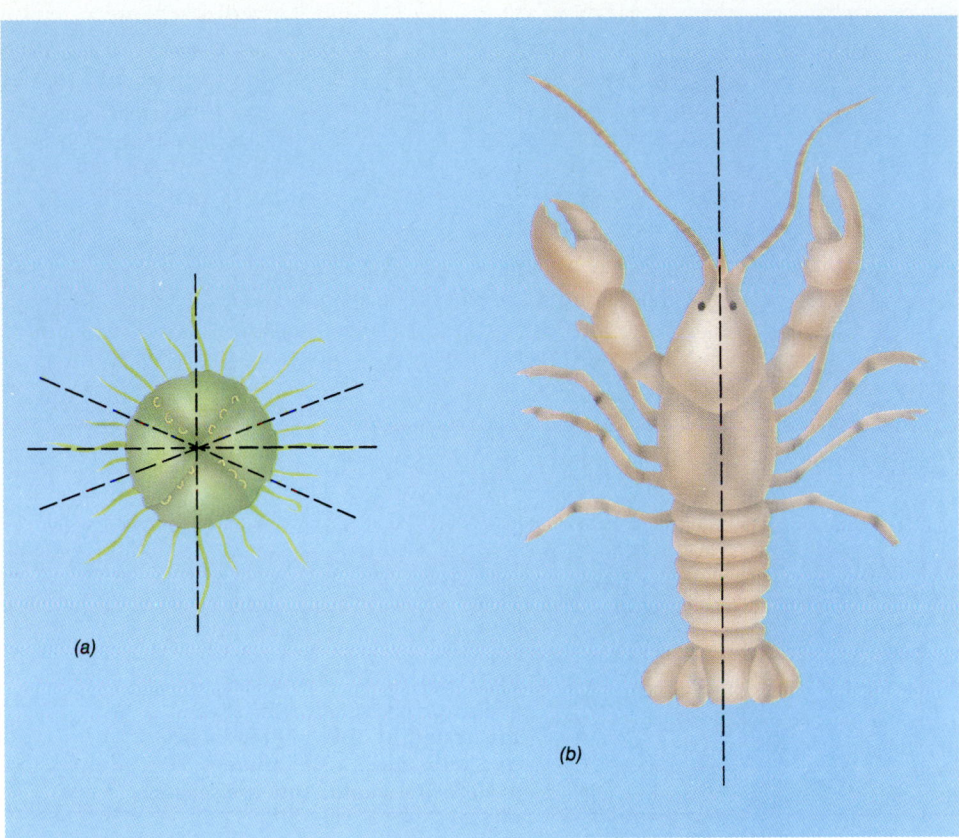

FIGURE 24–2 **Types of body symmetry in animals.** (*a*) In radial symmetry multiple planes can be drawn through the central axis, each dividing the organism into two mirror images. (*b*) Most animals exhibit bilateral symmetry.

In still more complex animals, the body cavity is completely lined with mesoderm because it forms within the mesoderm. Such a body cavity is a true **coelom.** Only animals with true coeloms are referred to as *coelomates*. The tree shown in Figure 24–4 indicates the relationships of the major phyla of animals based on their type of body cavity.

Animals Can Be Grouped Based on Pattern of Embryonic Development

A different family tree is shown in Figure 24–5. This important scheme is based partly on the pattern of embryonic development. According to this view, animals are divided into two groups: the *protostomes* and the *deuterostomes*. These groups reflect two main lines

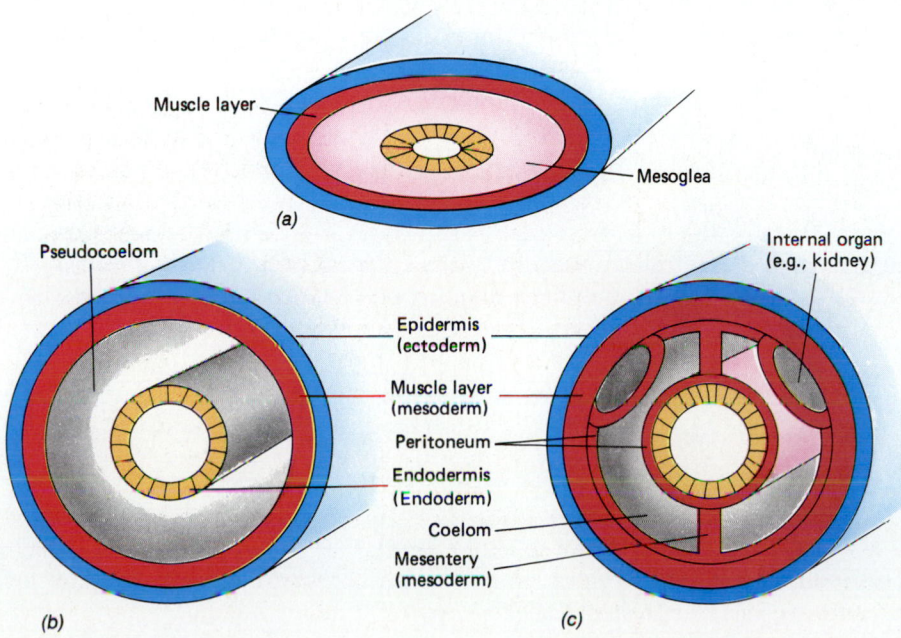

FIGURE 24–3 **Three basic animal body plans are illustrated by these cross sections.** (*a*) An acoelomate animal has no body cavity. (*b*) A pseudocoelomate animal has a body cavity that develops between the mesoderm and endoderm. (*c*) In a coelomate animal the body cavity is completely lined with tissue derived from mesoderm.

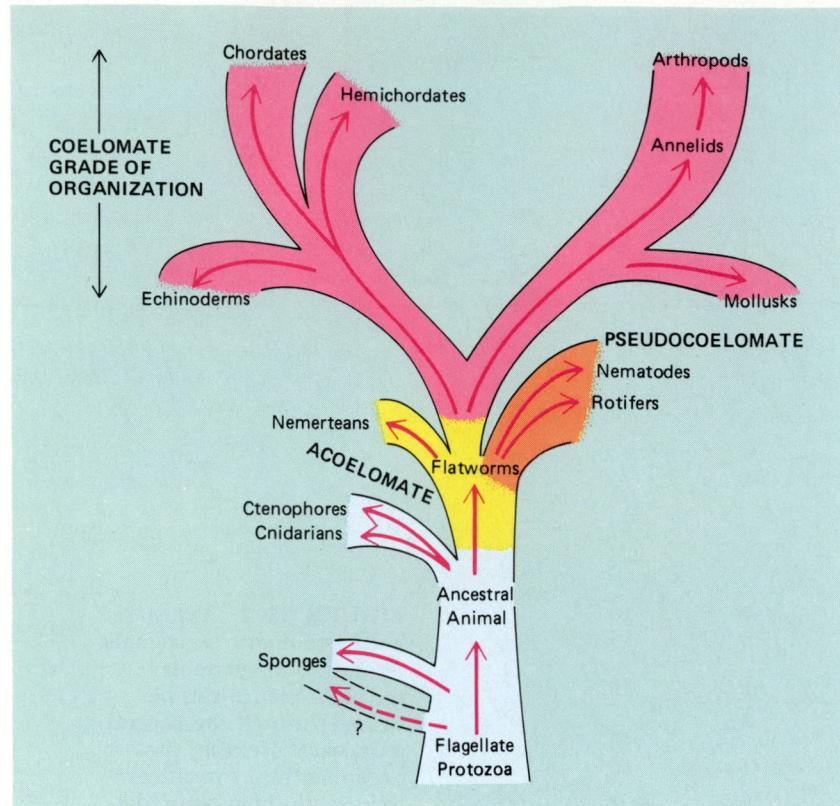

FIGURE 24–4 Proposed evolutionary relationships are illustrated by this phylogenetic tree indicating acoelomate, pseudocoelomate, and coelomate phyla. (After Barnes)

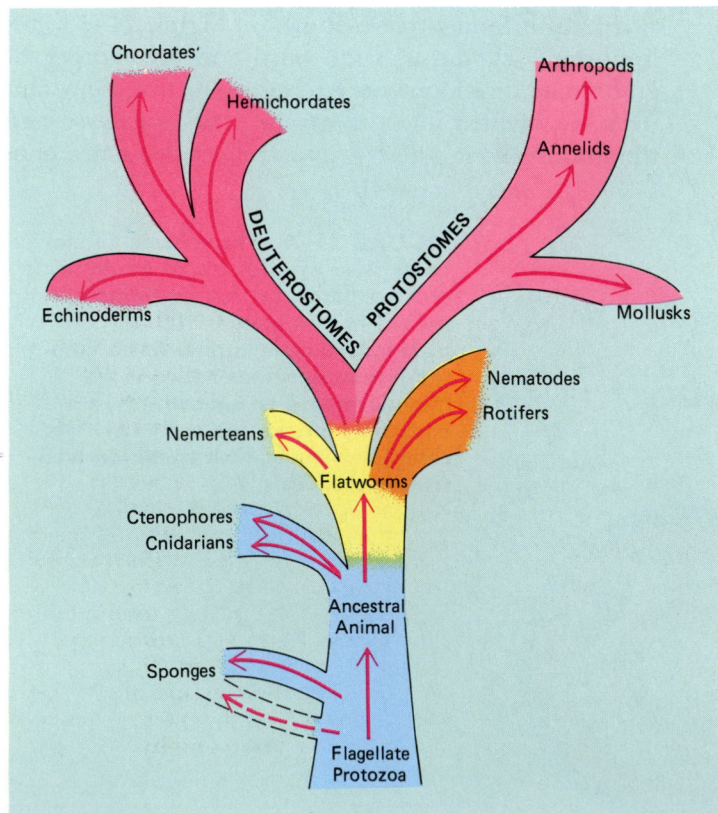

FIGURE 24–5 A phylogenetic tree based on protostome-deuterostome characteristics. (After Barnes)

of evolution. The deuterostomes are thought to have diverged from the main protostome line at a point in evolution considerably after the flatworms.

Early during embryonic development, a group of cells move inward to form an opening called the *blastopore*. In most of the mollusks, annelids, and arthropods, this opening develops into the mouth. These animals are **protostomes** (meaning "first, the mouth"). In echinoderms (for example, sea stars and sea urchins) and chordates (the phylum that includes the vertebrates), the blastopore develops into the anus. The opening that develops into the mouth forms later in development. These animals are the **deuterostomes** ("second, the mouth"). Other differences in the development of protostomes and deuterostomes are the pattern of coelom formation and the pattern of early cell division.

We Can Describe the Animal Body Using Standard Directions and Body Planes

To locate body structures, it is helpful to define some basic terms and directions. The back surface of an animal is its **dorsal** surface; the belly side is its **ventral** surface. **Anterior** means at or toward the front (head-end) of the animal. **Posterior** means at or toward the back (tail-end) of the animal. (In animals that stand on two limbs, as humans do, the term posterior refers to the dorsal surface and anterior to the ventral surface.) A structure is said to be **medial** if it is located toward the midline of the body and **lateral** if it is toward one side of the body; for example, the human ear is lateral to the nose.

The terms **cephalic** and **rostral** (and superior, in humans) refer to the head end of the body; the term

caudal refers to structures closer to the tail. In human anatomy, the term inferior is used to refer to structures located relatively lower in the body.

In a bilaterally symmetrical animal, we can distinguish three planes (flat surfaces that divide the body into specific parts). The **midsagittal** plane (or section) divides the body into equal right and left halves (Fig. 24–6). Any section or plane cut parallel to the midsagittal plane is described simply as a **sagittal** section; it divides the body into unequal right and left parts. A **frontal** (or coronal) section, or plane, divides the body into *dorsal* and *ventral* parts. A **transverse,** or **cross,** section cuts at right angles to the body axis and divides the body into *anterior* and *posterior* parts.

◻ SPONGES HAVE SPECIALIZED CELLS BUT NO TRUE TISSUES

About 5000 species of sponges make up **phylum Porifera.** The name Porifera, which means "to have pores," aptly describes the sponges, which resemble sacs perforated by tiny holes. Sponges occupy aquatic, mainly marine, habitats. Living sponges may be bright red, orange, green, purple, or quite drab. They are generally asymmetrical and may be flat or shaped like fans, balls, or vases (Fig. 24–7).

Sponges are classified on the basis of the type of skeleton they secrete. Members of one class (Calcispongiae) have a chalky skeleton that is composed of calcium carbonate spikes, or **spicules.** Members of a second class (Hexactinellida) are glass sponges with skeletons of six-rayed silica spicules. A third class of sponges (Demospongiae) have a skeleton of spongin (a protein material) fibers. What we recognize as a bath

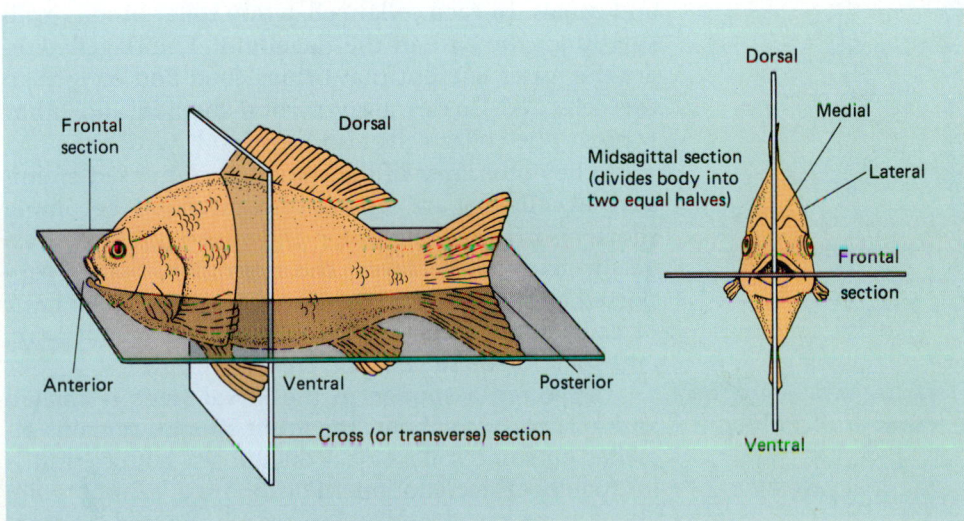

FIGURE 24–6 Body planes and directions. A midsagittal section (a lengthwise vertical cut through the midline) divides the animal into right and left halves. The diagram illustrates various ways in which the body can be sectioned so that its internal structure can be studied. Many cross sections and other types of sections are used in illustrations throughout this book to show relationships among tissues and organs. Some basic directional terms are indicated. Note that terms like lateral and medial are relative, just like the terms east and west in geography.

FIGURE 24–7 Sponges vary widely in size, shape, and color. (a) Tube sponge from the Grand Cayman Islands. (b) View of the open end (osculum) of a sulfur sponge. (a, Tom Stack & Associates; b, H.W. Pratt/BPS)

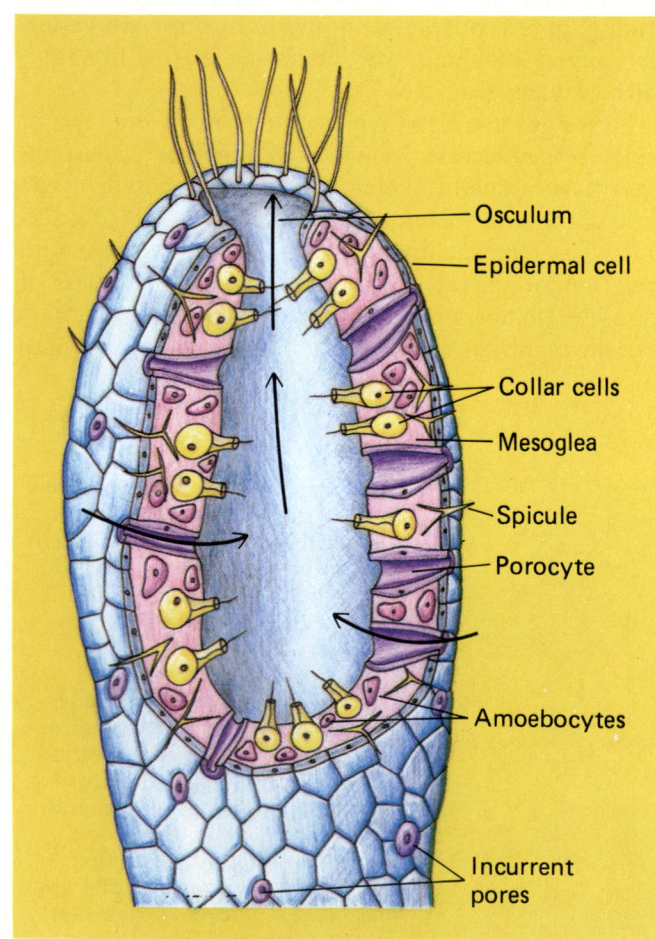

Osculum

Epidermal cell

Collar cells

Mesoglea

Spicule

Porocyte

Amoebocytes

Incurrent pores

FIGURE 24–8 A simple sponge cut open to expose its cellular organization.

sponge is actually a dried spongin skeleton from which all living tissue has been removed.

In a simple sponge, water enters through hundreds of tiny pores, passes into the central cavity, or **spongocoel,** and flows out through the sponge's open end (osculum) (Fig. 24–8). In some types of sponges, the body wall is extensively folded and there are complicated systems of canals.

Although the sponge is a multicellular organism, its cells are so loosely associated that they do not form definite tissues. There is a division of labor, among the several types of cells that make up the sponge, with certain cells specializing in nutrition, support, or reproduction. Flagellated collar cells line the spongocoel and canals. In each collar cell, a tiny collar of microvilli surrounds the base of the flagellum. The flagella create the water current that brings food and oxygen to the cells and carries away carbon dioxide and other wastes. The collar cells also trap food particles.

The outer layer of the sponge is composed of **epidermal cells** that are capable of contraction (becoming shorter). Between the outer and inner cell layers is a gelatin-like middle layer, the **mesoglea,** that is supported by skeletal spicules. The mesoglea is not a layer of cells, but amoeba-like cells do wander about here, and some of these cells secrete the spicules.

Although a sponge in the larval stage is ciliated and able to swim about, the adult sponge remains attached to some solid object on the sea bottom and is incapable of locomotion. The sponge cannot swim about in search of food and instead is adapted for trapping and eating whatever food the sea brings to it. The sponge is a filter feeder. As water circulates through the body, small particles of food are trapped along the sticky collars of the collar cells. Digestion is intracellular, the job of each individual cell, rather than specialized organs. Undigested food is simply eliminated into the water.

Oxygen from the water diffuses throughout the sponge. Gas exchange and waste disposal are carried on by individual cells. Each cell of the sponge body is irritable and can react to stimuli; however, there are no sensory or nerve cells that would enable the animal to react as a whole.

Sponges can reproduce asexually. A small fragment or bud may break free from the parent sponge and give rise to a new sponge. Such fragments may remain to form a colony with the parent sponge. Sponges also reproduce sexually. Most sponges are **hermaphroditic,** meaning that the same individual can produce both egg and sperm. Some of the amoeba-like cells develop into sperm cells, others into egg cells. Hermaphroditic sponges can cross-fertilize with other sponges, however. Fertilization and early development take place within the jelly-like middle layer. Embryos eventually move into the spongocoel and leave the parent along with the stream of outflowing water. After swimming about for a while, the larva finds a solid object, attaches to it, and settles down to a sessile life.

■ CNIDARIANS (HYDRAS, JELLYFISH, AND CORALS) HAVE TISSUE LAYERS BUT NO TRUE ORGANS

Most of the 10,000 or so species of **phylum Cnidaria** (pronounced "nie-dare'-e-a") are marine (sea-dwelling). They are grouped in three classes: (1) **class Hydrozoa,** including the hydras and Portuguese man-of-war; (2) **class Scyphozoa,** the jellyfish; and (3) **class Anthozoa,** the sea anemones and true corals (Fig. 24–9). All the cnidarians have stinging cells called **cnidocytes,** from which they get their name (cnidaria is from a Greek word meaning "sea nettles"). Although many cnidarians live a solitary existence, others group into colonies.

The radially symmetrical cnidarian body is a hollow sac with the mouth and surrounding tentacles located at one end. The mouth leads into the digestive cavity, called the **gastrovascular cavity.** The mouth is the only opening into this cavity and so must serve for both ingestion of food and egestion (expulsion) of wastes.

Much more highly organized than the sponge, the cnidarian has two definite tissue layers. The outer **epidermis** and the inner **gastrodermis** are composed of several types of epidermal cells. These layers are separated by a gelatin-like **mesoglea,** which is not cellular but sometimes contains a few cells.

FIGURE 24–9 Representative cnidarians. (*a*) A member of class Hydrozoa, *Obelia* forms a colony of polyps. (*b*) Portuguese man-of-war (*Physalia physalis*) with a fish it has captured. The Portuguese man-of-war is actually a colony of polyps and medusas. A modified medusa acts as a float for the colony in the form of a gas-filled sac. The trailing tentacles are armed with stinging capsules that can paralyze a fish. (*c*) Polyps from the coral *Montastrea cavernosa* extended for feeding. (*a*, *b*, Runk/Schoenberger from Grant Heilman; *c*, Charles Seaborn)

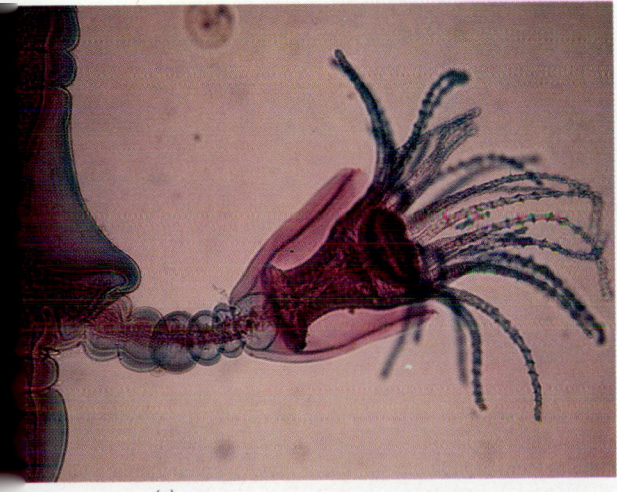

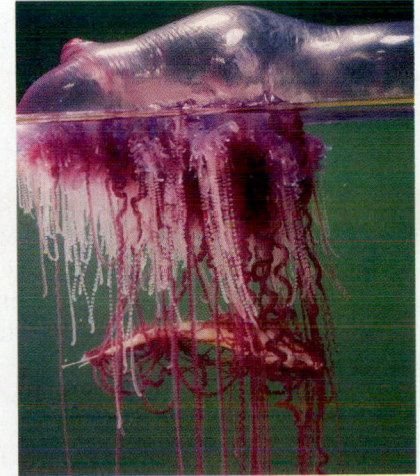

(a)   (b)   (c)

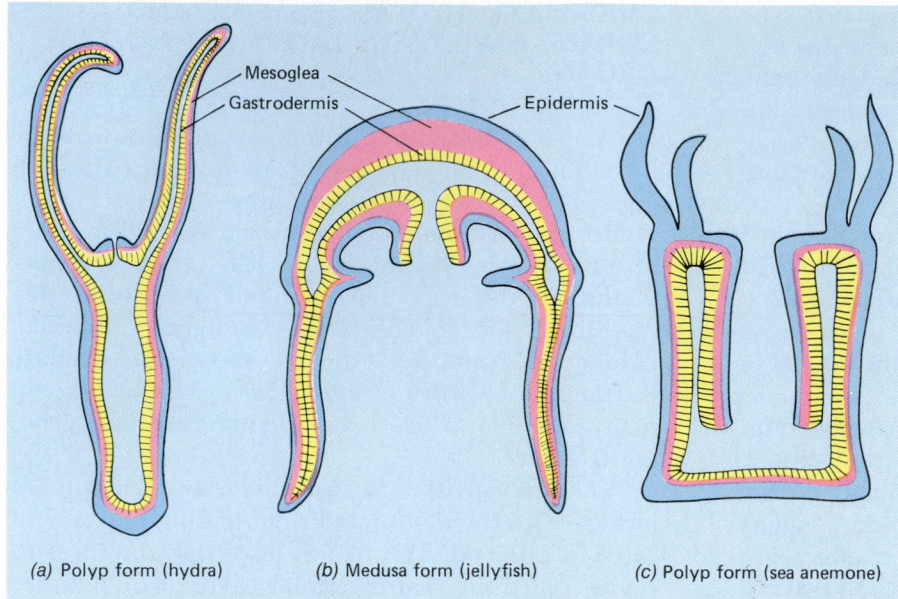

(a) Polyp form (hydra) (b) Medusa form (jellyfish) (c) Polyp form (sea anemone)

FIGURE 24–10 The polyp and medusa body forms characteristic of phylum Cnidaria are basically similar. However, polyps are generally attached to the substrata, whereas medusas are free-swimming. (*a*) The polyp form as seen in *Hydra*. (*b*) The medusa form is basically an upside-down polyp, as shown by inverting a jellyfish. (*c*) In the anthozoan polyp the gastrovascular cavity is characteristically divided into chambers by vertical partitions.

FIGURE 24–11 Life cycle of *Obelia*, a colonial marine hydrozoan. Polyps specialized for reproduction give rise asexually to medusae. The medusae, which are free-swimming, reproduce sexually. The zygote formed develops into a type of larva known as a planula larva. The larva develops into a polyp and gives rise to a new colony. Note the specialization of members of the polyp colony.

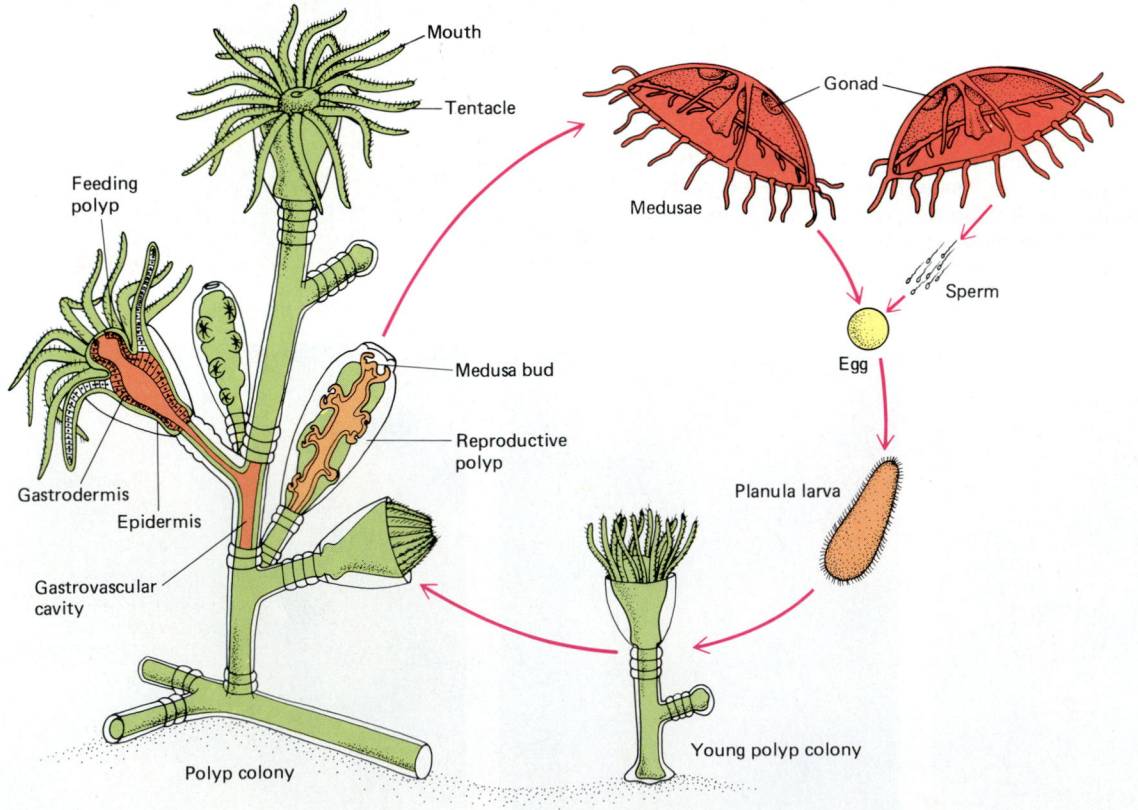

Cnidarians Have Two Body Plans: A Polyp And A Medusa Form

Cnidarians have two body shapes, the **polyp** and the **medusa** (jellyfish) (Fig. 24–10). The polyp form, represented by *Hydra*, resembles an upside-down, slightly elongated jellyfish. Many cnidarian life cycles include both the polyp and the medusa form (Fig. 24–11). Some cnidarian colonies (for example, the Portuguese man-of-war) consist of both polyp and medusa forms.

FIGURE 24–12 *Hydra.* (*a*) **A brown hydra,** *Hydra oligactis,* **capturing a small crustacean. Note the buds on the hydra's body. One bud has already detached as a separate animal.** (*b*) **This hydra is cut longitudinally to reveal its internal structure. Asexual reproduction by budding is represented on the animal's right. Sexual reproduction is represented by the ovary on the animal's left. Male hydras develop testes, which produce sperm.** (*c*) **Cross section of a hydra.** (*a*, Tom Branch, Photo Researchers, Inc.)

The Hydra Has A Solitary, Carnivorous Life-Style

The freshwater hydra, a common cnidarian, is a solitary polyp seldom more than 1 centimeter long (Fig. 24–12). Although capable of locomotion, an adult hydra generally attaches to a rock, twig, or even a leaf and waits for dinner to come along. When a likely prospect happens to brush by one of its tentacles, the hydra's stinging cells respond.

Coiled within each stinging cell is a "thread capsule," or **nematocyst.** Each stinging cell has a small projecting trigger on its outer surface. This responds to touch and to chemicals dissolved in the water (a form of "taste") and causes the nematocyst to fire its thread. Some types of nematocysts have sticky threads that adhere to the prey. Others have long threads that coil around the prey; a third type of nematocyst is tipped with a barb or spine and can inject a protein toxin that paralyzes the prey. The tentacles encircle the

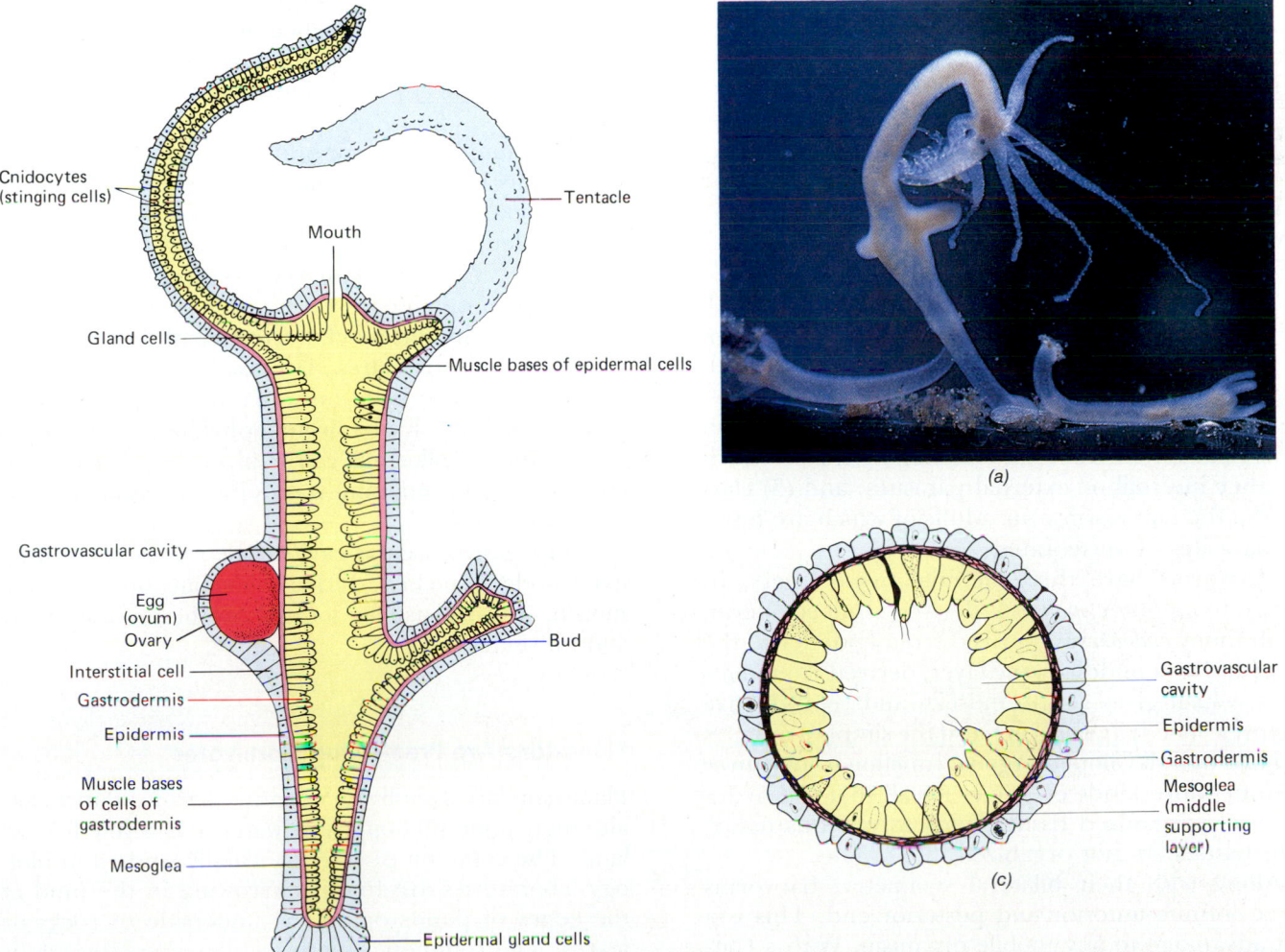

Cnidocytes (stinging cells)

Mouth

Tentacle

Gland cells

Muscle bases of epidermal cells

Gastrovascular cavity

Egg (ovum)

Ovary

Bud

Interstitial cell

Gastrodermis

Epidermis

Muscle bases of cells of gastrodermis

Mesoglea

Epidermal gland cells

(*b*)

(*a*)

Gastrovascular cavity

Epidermis

Gastrodermis

Mesoglea (middle supporting layer)

(*c*)

snared prey and stuff it through the mouth into the gastrovascular cavity, where digestion begins. Partially digested fragments are taken up by pseudopods of the gastrodermis cells. Digestion is completed within food vacuoles.

Gas exchange and waste disposal occur by diffusion, for the body of a hydra is small enough that no cell is far from the surface. The cnidaria are the simplest animals to possess true nerve cells. These cells are simply arranged, forming irregular **nerve nets** that connect sensory cells in the body wall with muscle and gland cells. The coordination achieved is of the simplest sort; there is no grouping of nerve cells forming a brain or nerve cord, and an impulse set up in one part of the body passes in all directions more or less equally, rather than along specific pathways as occurs in more complex animals.

Hydras reproduce asexually by budding during periods when environmental conditions are optimal; however, they differentiate as males and females in the fall or when pond water becomes stagnant. Females develop an *ovary* that produces a single egg, and males form a *testis* that produces sperm. After fertilization, the zygote (fertilized egg) becomes covered with a shell, leaves the parent, and remains within the protective shell throughout the winter.

◼ FLATWORMS HAVE THREE TISSUE LAYERS, A HEAD REGION, AND ORGANS

As their name implies, flatworms are flat, elongated, legless animals. The **phylum Platyhelminthes** (*platy,* meaning "flat"; *helminthes,* meaning "worms") are divided into three classes: (1) class Turbellaria (Fig. 24–13), the free-living flatworms, including the planarian and its relatives; (2) class Trematoda, the flukes, which are either internal or external parasites; and (3) class Cestoda, the tapeworms, the adults of which are intestinal parasites of vertebrates.

Flatworms have three definite tissue layers. In addition to an outer *epidermis*, derived from ectoderm, and an inner *endodermis*, derived from endoderm, the flatworms has a middle tissue layer, derived from mesoderm, which gives rise to muscles and reproductive structures. The flatworms are also the simplest animals that have well-developed **organs,** functional structures of two or more kinds of tissue. Recall that the hydra has loosely organized tissues, such as nervous tissue, but its tissues are not organized into organs.

Along with their bilateral symmetry, flatworms have a definite anterior and posterior end. This is a great advantage to any mobile organism. With a concentration of sense organs in the part of the body that first meets the environment, the animal is able to de-

FIGURE 24–13 A marine turbellarian (*Prostheceraeus bellostriatus*). Turbellarians are voracious predators, feeding primarily on other worms. (Stan Elems/Visuals Unlimited)

tect an enemy quickly enough to escape. The animal is also more likely to see or smell prey quickly enough to capture it. The beginning of **cephalization,** the development of a head, is an important advance in flatworms.

The simple flatworm brain consists of two masses of nervous tissue, called **ganglia,** in the head region. The ganglia are connected to two nerve cords that extend the length of the body. In this *ladder-type arrangement of nerves,* a series of nerves connect the cords like the rungs of a ladder.

Another important characteristic of the flatworms is its excretory structures. Two excretory tubes extend the length of the body and give off branches called **protonephridia.** Each protonephridium ends in a **flame cell,** a collecting cell equipped with cilia that channels fluid containing wastes into the system of excretory tubules.

The gastrovascular cavity, when present, is often extensively branched. It has only one opening, the mouth, which is usually located on the middle of the ventral surface.

Planarians Are Free-Living Carnivores

Planarians are free-living, mainly marine, flatworms, although some inhabit fresh water and a few live on land. The common planarians usually studied in biology laboratories are found burrowing in the mud at the edges of ponds or on the underside of rocks or leaves. The common American planarian, *Dugesia,* is about 15 millimeters long, with what appear to be crossed eyes and distinct "ears" called **auricles** (Fig. 24–

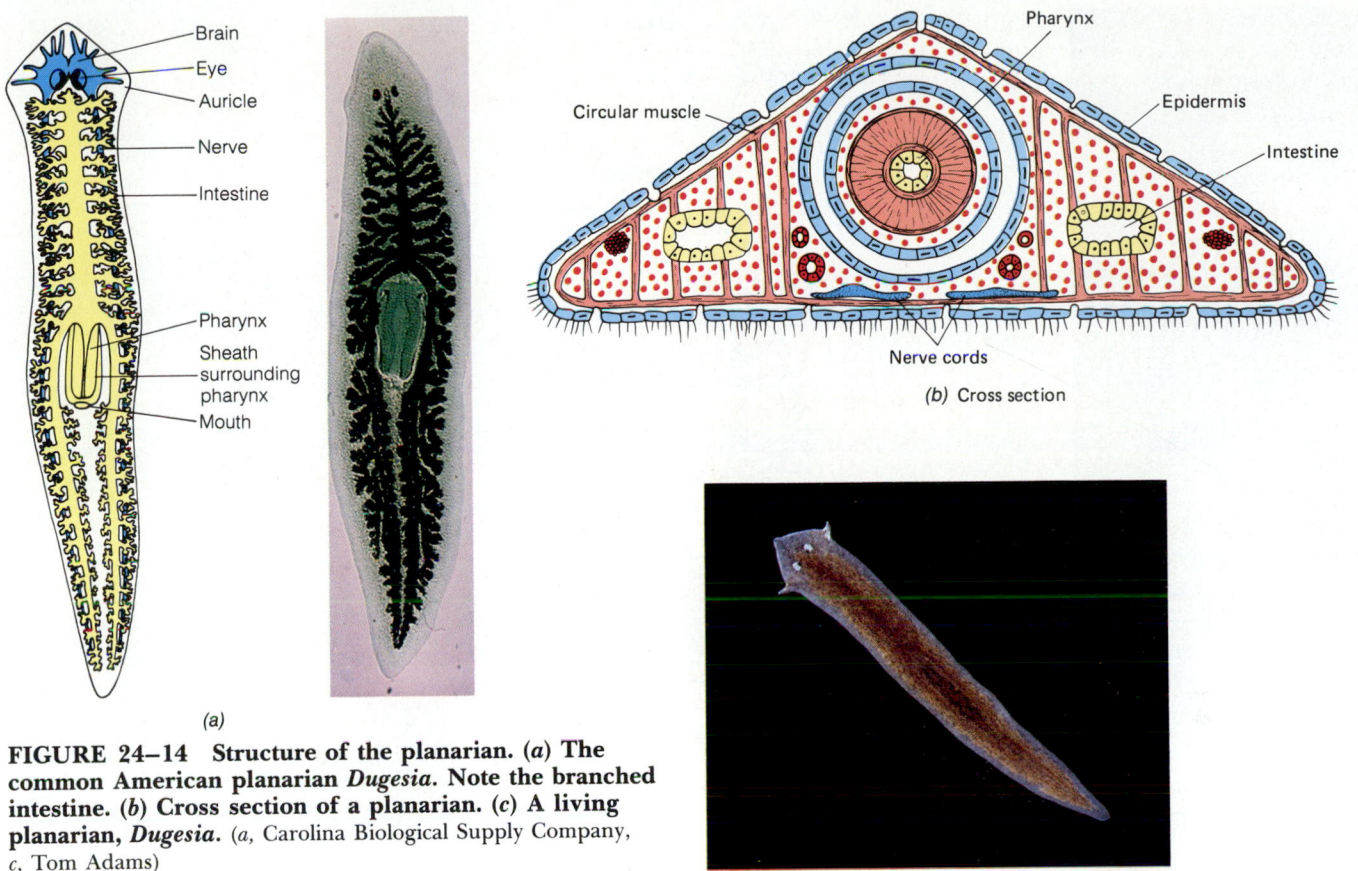

(a)

(b) Cross section

(c)

FIGURE 24–14   Structure of the planarian. (a) The common American planarian *Dugesia*. Note the branched intestine. (b) Cross section of a planarian. (c) A living planarian, *Dugesia*. (a, Carolina Biological Supply Company, c, Tom Adams)

14). The auricles actually serve as organs of smell or, more precisely, **chemoreception.**

Planarians are carnivorous and trap small animals in a mucous secretion. The digestive structures include a single opening (the mouth), a pharynx, and a branched intestine. A planarian can project its pharynx (the first portion of the digestive tube) outward through its mouth, using it like a vacuum cleaner to suck up small pieces of its prey. Extracellular digestion takes place in the intestine by enzymes secreted by gland cells. Digestion is completed after the nutrients have been absorbed into individual cells. Undigested food is eliminated through the mouth. The highly branched gastrovascular cavity helps distribute food to all parts of the body, so that each cell is within range of diffusion.

Because the planarian's body is flattened, gases can reach all the cells by diffusion. There are no specialized respiratory or circulatory structures. Although excretion takes place in part by diffusion, a simple excretory system with flame cells is present.

Planaria can reproduce either asexually or sexually. In asexual reproduction, an individual constricts in the middle and divides into two individuals. Each regenerates its missing parts. Sexually these animals are hermaphroditic. During the warm months of the year, each is equipped with a complete set of male and female organs. Two planaria come together in copulation and exchange sperm cells. Thus, their eggs are cross-fertilized.

Flukes Have Suckers And Other Adaptations For Their Parasitic Life-Style

Although their body plan resembles that of the free-living flatworms, the flukes have evolved specialized adaptations that contribute to their success as parasites. Both blood flukes and liver flukes have one or more suckers for clinging to their host (Fig. 24–15). Flukes also have extremely complex and efficient reproductive organs.

Flukes have complicated life cycles, involving a number of different forms, alternation of sexual and asexual generations, and parasitism on one or more intermediate hosts, such as snails and fishes (Fig. 24–16). The aquatic snails that serve as intermediate hosts thrive in ponds and marshy areas, including rice paddies. When dams are built, the marshy areas created often provide habitats for these snails. Blood flukes of

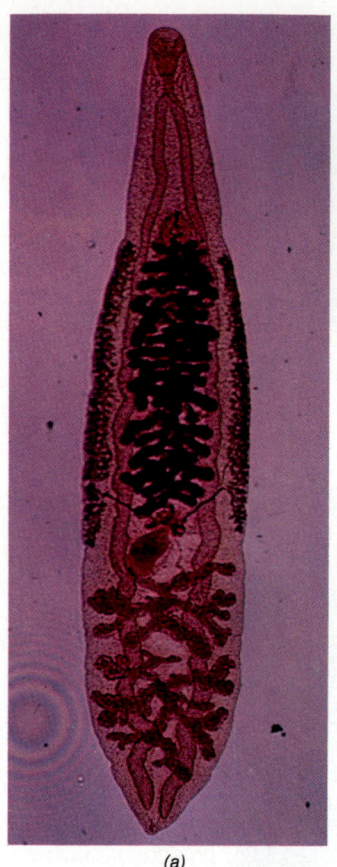

(a)

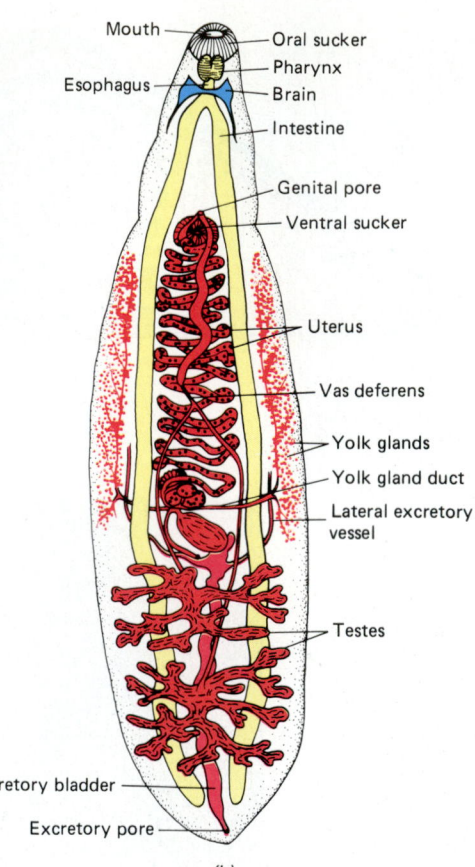

Mouth
Oral sucker
Pharynx
Esophagus
Brain
Intestine

Genital pore
Ventral sucker

Uterus

Vas deferens

Yolk glands
Yolk gland duct
Lateral excretory
vessel

Testes

Excretory bladder

Excretory pore

(b)

**FIGURE 24–15 Structure of a fluke.
(a) The common liver fluke of sheep,
Clonorchis sinensis. (b) Internal
structure of a fluke.** (a, Carolina
Biological Supply Company)

genus *Schistosoma* infect about 200 million people who live in tropical areas.

Tapeworms Are Strikingly Specialized For Their Parasitic Life-Style

Tapeworms are long, flat, ribbon-like animals. As adults, members of the more than 1000 species live as parasites in the intestines of probably every kind of vertebrate, including humans. Sometimes thought of as the most degenerate members (changed from a higher to a lower form) of phylum Platyhelminthes, the tapeworms are actually strikingly specialized for their parasitic mode of life.

Among their many adaptations are suckers and sometimes hooks on the head (scolex), which enable the parasite to stay attached to the host's intestine (Fig. 24–17). Their reproductive adaptations and abilities are extraordinary. The body of the tapeworm consists of a long chain of segments called **proglottids.** Each proglottid is an entire reproductive machine equipped with both male and female organs and containing as many as 100,000 eggs. Because an adult tapeworm may have 2000 segments, its reproductive

potential is staggering. A single tapeworm may produce 600 million eggs in 1 year. Segments farthest from the tapeworm's head contain the ripest eggs and are shed daily, leaving the host's body along with the feces.

Tapeworms lack certain organs. They absorb food directly through their body walls from the host's intestine, and they have no mouths and no digestive systems of their own. Some tapeworms have rather complex life cycles, spending the larval stage within the body of an intermediate host and their adult lives within the body of a different, final host.

Let us consider the life cycle of the beef tapeworm, so named because human beings become infected when they eat poorly cooked beef containing the larva (Fig. 24–18). The microscopic tapeworm larva spends part of its life cycle encysted (enclosed in a little sac or cyst) within the muscle tissue of beef. When a human being ingests infected meat, the digestive juices break down the cyst, releasing the larva. The larva attaches itself to the intestinal lining and within a few weeks matures into an adult tapeworm, growing to a length of perhaps 30 feet. The parasite reproduces sexually within the human intestine and sheds proglottids filled with ripe eggs.

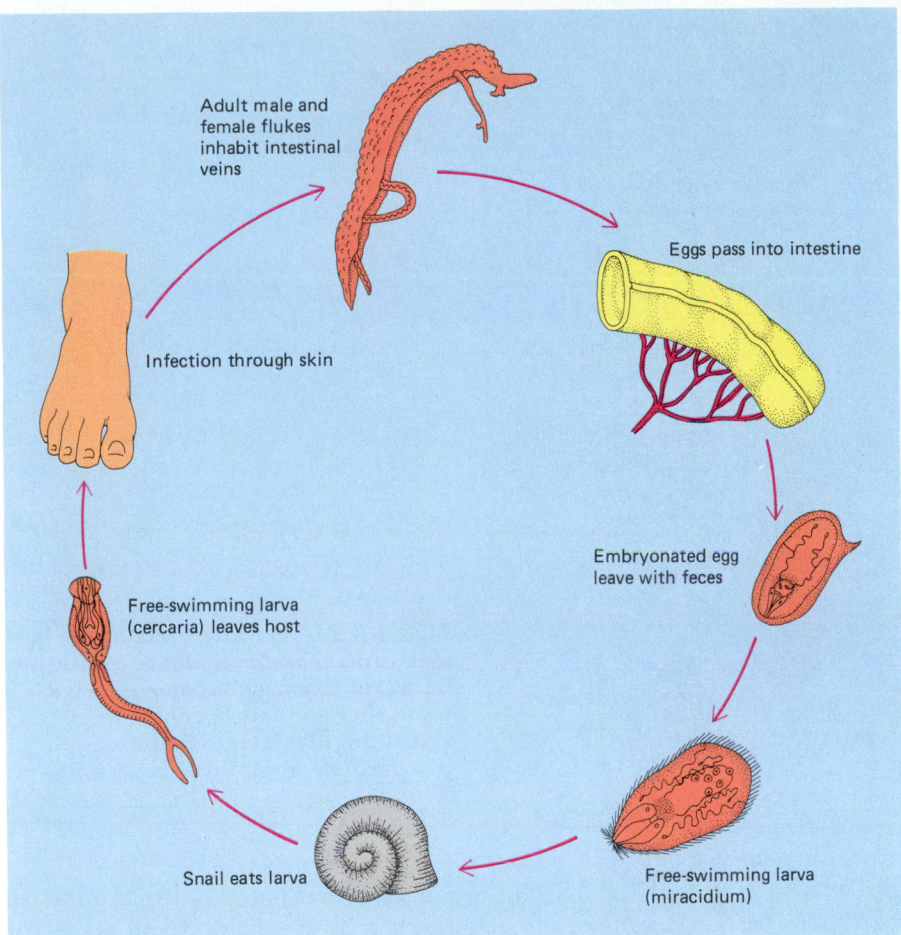

Adult male and
female flukes
inhabit intestinal
veins

Eggs pass into intestine

Infection through skin

Embryonated egg
leave with feces

Free-swimming larva
(cercaria) leaves host

Snail eats larva

Free-swimming larva
(miracidium)

FIGURE 24–16   Life cycle of a blood fluke, a schistosome. Larvae burrow through human skin and find their way to the circulatory system. Sexual reproduction takes place in small blood vessels within the intestine. Some of these vessels break and the eggs are released into the intestinal lumen (cavity). The eggs leave the human body with the feces. If they find their way to fresh water, the eggs hatch, releasing free-swimming larvae called miracidia. To survive, the miracidia must find an intermediate host, an aquatic snail. If successful, they burrow into the tissues of the snail and develop into a form that reproduces asexually. Finally, fork-tailed larvae known as cercaria develop and leave the snail. These larvae swim actively and when they contact a human host they burrow into the skin, completing the cycle.

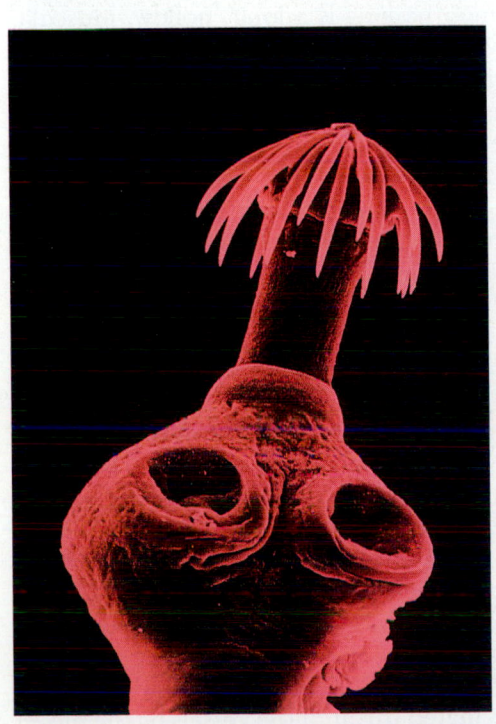

FIGURE 24–17   Some species of tapeworms are armed with powerful hooks that enable these parasites to maintain their attachment to the host. This false-color scanning electron micrograph of the head of the tapeworm *Acanthocirrus retrisrostris* was taken of the larval, encysted form. This tapeworm reaches maturity in the intestines of wading birds that eat barnacles. It then encysts in the body of the barnacle. The piston-like rostellum can be withdrawn into the head or thrust out and buried in the host's tissue. Beneath the rostellum, two of the four powerful suckers are visible (×80 at 35-mm size). (Cath Ellis, Dept. of Zoology, University/Science Photo Library/Photo Researchers, Inc.)

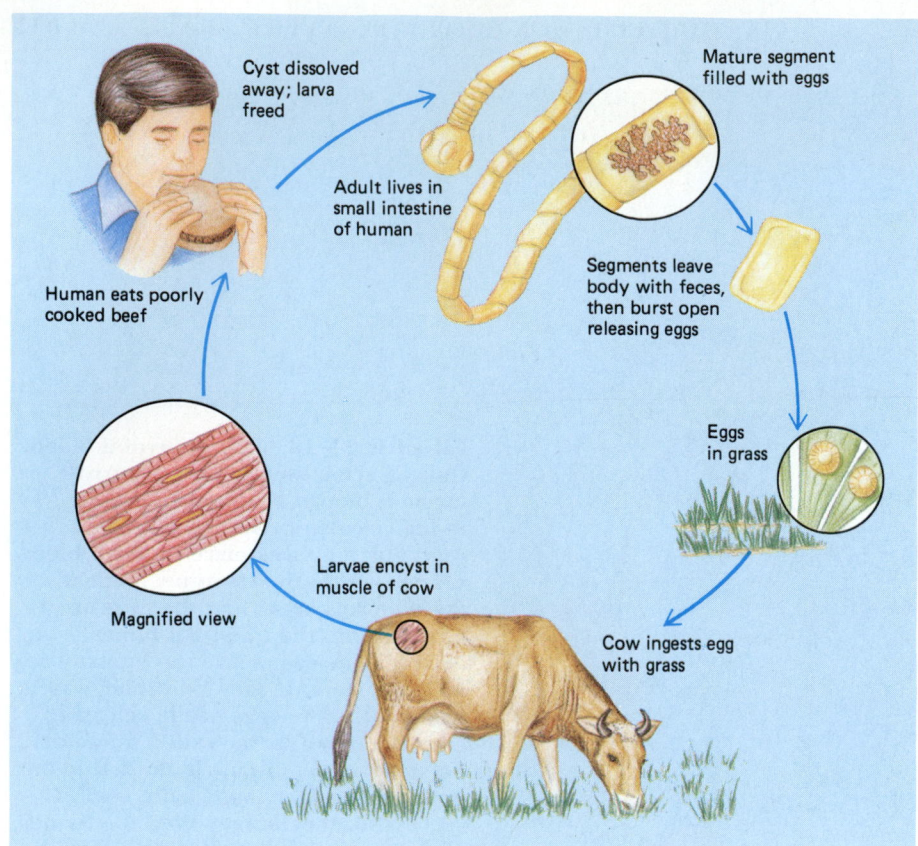

FIGURE 24–18 Life cycle of the beef tapeworm (*Taenia saginata*), a parasitic flatworm. Humans become infected when they eat poorly cooked beef containing the larvae. Cattle are intermediate hosts for this parasite.

Once established within a human host, the tapeworm makes itself at home and may remain there for the rest of its life, as long as 10 years. For the life cycle of the beef tapeworm to continue, its eggs must be ingested by an **intermediate host,** a cow. (This requirement explains why the tapeworm must produce millions of eggs to increase the probability that at least a few will survive.) When a cow eats grass or other food contaminated with infected human feces, the eggs hatch in the cow's intestine. The larvae make their way into muscle, where they encyst. There they await release by a **final host,** perhaps a human eating rare steak.

Two other tapeworms that infect human beings are the pork tapeworm and the fish tapeworm. The pork tapeworm infects us when we eat poorly cooked infected pork, and the fish tapeworm is contracted when we eat raw or poorly cooked infected fish. Like most parasites, tapeworms tend to be species specific; that is, each type can infect only certain species.

■ PROBOSCIS WORMS ARE THE SIMPLEST ANIMALS TO HAVE ORGAN SYSTEMS

Phylum Nemertinea is a relatively small group of animals (about 550 species) that are important to biologists mainly because they are the simplest animals to

have definite **organ systems** (Fig. 24–19). Almost all are marine, although a few inhabit fresh water or damp soil. They have long, narrow bodies, from 5 centimeters to 20 meters long. Some are vivid orange, red, or green, with black or colored stripes. Their most remarkable organ—the **proboscis,** from which they get their common name—is a long, hollow, muscular tube that they eject from the anterior end of the body and use for seizing food or for defense. The proboscis secretes mucus and may be equipped with a barb and with poison-secreting glands.

An important advance of the nemerteans is a *tube-within-a-tube body plan.* The digestive tract is a complete tube with a mouth at one end for taking in food and an anus at the other for eliminating wastes. A digestive tract with two openings is an evolutionary advance over the cnidarians and planarians, whose digestive tracts have a single opening for food intake and waste disposal.

A second advance of the nemerteans is the separation of digestive and circulatory functions. These animals are the most primitive organisms to have a separate circulatory system. It consists simply of muscular tubes (the blood vessels) that extend the length of the body and are connected by transverse vessels. Some of these primitive forms have red blood cells filled with hemoglobin, the same red pigment that transports oxygen in human blood. Nemerteans have no heart; the blood is circulated through the vessels by the move-

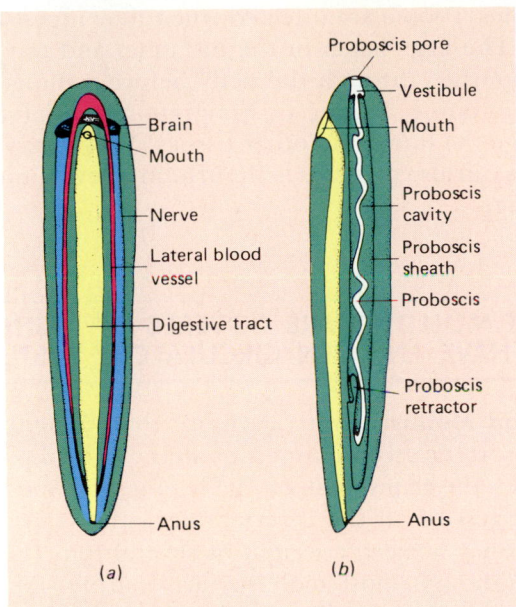

FIGURE 24–19 **Structure of a typical nemertean.** **(a) Dorsal view of the digestive, circulatory, and nervous systems. (b) Lateral view of the digestive tract and proboscis. Note the complete digestive tract that extends from mouth to anus, giving this animal a tube-within-a-tube body plan.**

■ THE ROUNDWORMS (NEMATODES) HAVE ORGAN SYSTEMS BUT LACK CIRCULATORY STRUCTURES

The **nematodes (phylum Nematoda)** are of great ecological importance because they are widely distributed in the soil, the sea, and fresh water, and because they are so numerous. A spadeful of soil may contain more than a million of these mainly microscopic white worms, which thrash around coiling and uncoiling. Although many are free living (Fig. 24–20), others are important parasites in plants and animals. Among the human parasites belonging to phylum Nematoda are the hookworms, the intestinal roundworm *Ascaris,* pinworms, trichina worms, and filaria worms.

The elongate, cylindrical, thread-like nematode body is pointed at both ends and covered with a tough, flexible **cuticle** (Fig. 24–21). Nematodes are the most primitive animals to have a body cavity, a pseudocoelom. It is not a true coelom because it is not completely lined with mesoderm. Like the proboscis worms, the nematodes exhibit bilateral symmetry, a complete digestive tract, three definite tissue layers, and definite organ systems; however, they lack circulatory structures. The sexes are usually separate, and the male is smaller than the female.

Ascaris lumbricoides is an example of a parasitic roundworm. A whitish worm about 25 centimeters long, *Ascaris* is a common intestinal parasite of human beings. The nematode spends its adult life in the human intestine, where it makes its living by sucking in partly digested food. Like the tapeworm, its reproductive output is enormous. The sexes are separate, and copulation takes place within the host. A mature female may lay as many as 200,000 eggs a day.

ments of the body and the contractions of the muscular blood vessels.

Despite the complexity of proboscis worms compared with the flatworms, they are not considered to be ancestral to other, more complex phyla. However, they provide some clue to what the immediate ancestors of the annelids and mollusks may have been like.

FIGURE 24–20 **A free-living nematode among the cyanobacteria** *Oscillatoria,* **its typical food.** (Tom Adams)

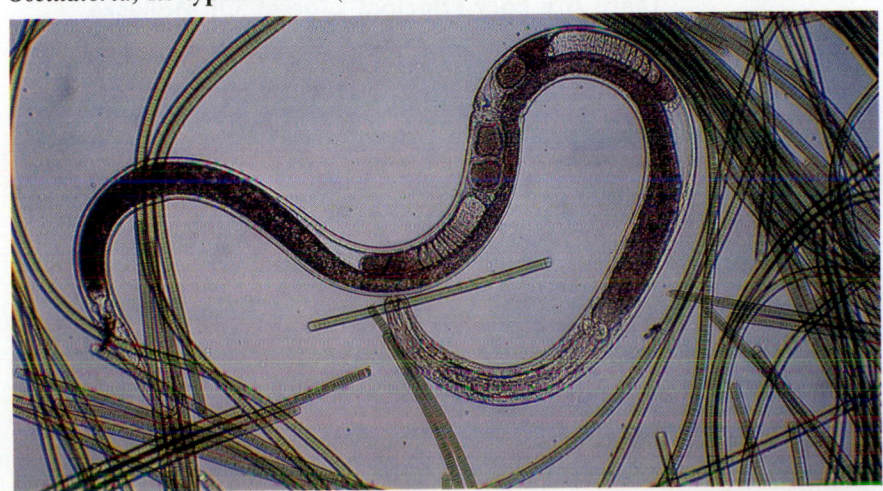

Ascaris eggs leave the human body with the feces and, where sanitation is poor (in most of the world), find their way into the soil. In many parts of the world, human wastes are used as fertilizer, a practice that encourages the survival of *Ascaris* and many other human parasites. People are infected when they ingest ascaris eggs. The eggs hatch in the intestine, and the larvae then journey through the body before settling in the small intestine. During this migration, the larvae can damage the lungs and other tissues. Very heavy infestations can also produce malnutrition or even intestinal blockage.

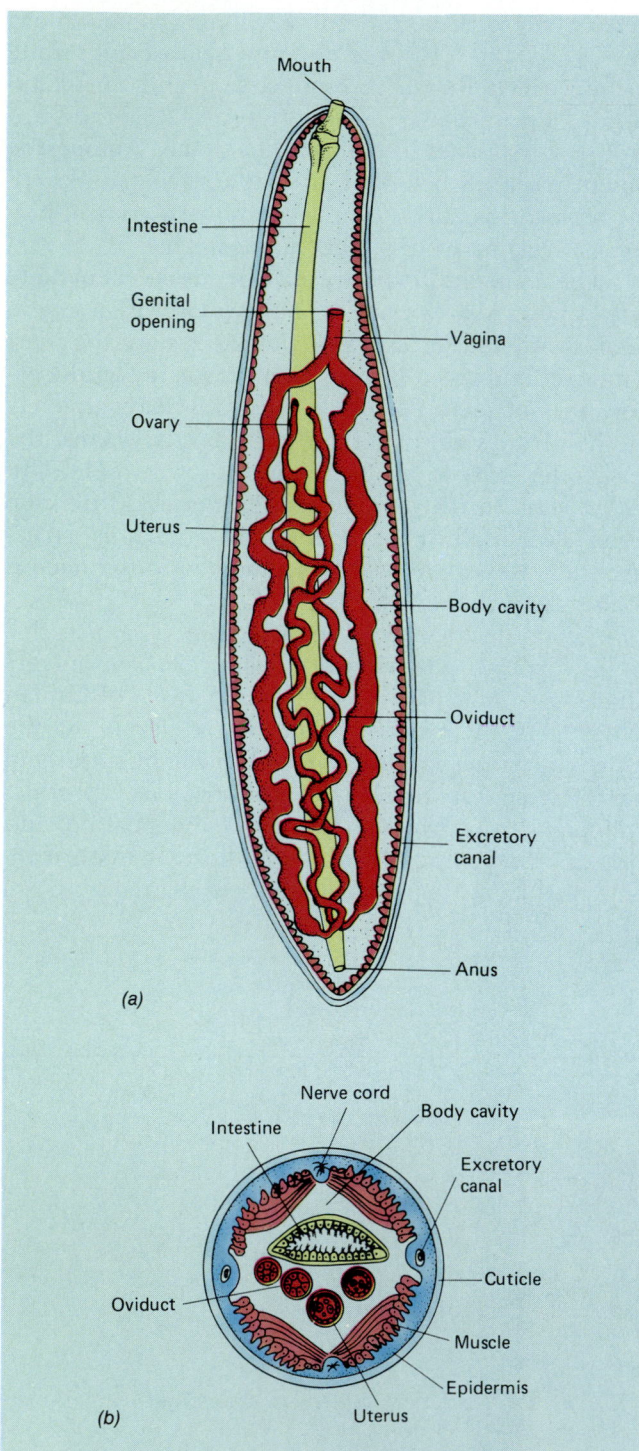

(a)

(b)

FIGURE 24–21   The structure of the roundworm *Ascaris*. (a) Longitudinal section to show internal anatomy. Note the complete digestive tract that extends from mouth to anus. (b) Cross section of *Ascaris*.

☐ MOST MOLLUSKS ARE COVERED BY A SHELL AND HAVE AN OPEN CIRCULATORY SYSTEM

Phylum Mollusca, with its more than 50,000 living species, is the second largest of all the animal phyla. It includes the clams, oysters, snails, slugs, octopods, and the largest of all the invertebrates—the giant squid, which may achieve a weight of several tons (Fig. 24–22). Although most mollusks are marine, there are snails and clams that live in fresh water and many species of snails and slugs that inhabit the land.

Mollusks are soft-bodied animals that are usually covered by a dorsal shell. A broad, flat muscular **foot,** located ventrally, can be used for locomotion. Most of the organs make up a **visceral mass** that is located above the foot. The **mantle,** a heavy fold of tissue, covers the visceral mass and, in most species, contains glands that secrete the shell. The mantle generally overhangs the visceral mass, forming a mantle cavity that often contains a pair of gills. Most mollusks (except clams) have a **radula,** a tongue-like structure with teeth.

Mollusks have a true coelom. Most mollusks have an **open circulatory system,** a system in which blood does not remain within a circuit of blood vessels; instead, blood leaves the blood vessels and flows into a network of large spaces, or sinuses, where it bathes the tissues directly. This network of sinuses makes up the **hemocoel,** or blood cavity.

There Are Four Principal Classes Of Mollusks

Mollusks can be assigned to the following classes:

1. **Class Polyplacophora** consists of the chitons, which are primitive marine animals with segmented shells. Their heads are small, and they lack eyes and tentacles.
2. **Class Gastropoda** includes the snails, slugs, and whelks, animals whose bodies and shells are coiled. A few species, such as the nudibranchs and garden slugs, lack shells. Gastropods have well-developed heads with eyes and one or two pairs of tentacles. Most are marine, but some live in fresh water or on land. The land snails are among the few types of terrestrial invertebrates.

3. **Class Bivalvia** includes the clams and oysters and their relatives. (The clam is discussed in the next section.) Members of this class have two-part shells that are hinged dorsally and ventrally. The head is not distinct, but the foot is usually large and used for burrowing.

4. **Class Cephalopoda** includes the squids and octopods, which are active predators. These animals have large heads with long tentacles and conspicuous eyes. (See "Focus on Adaptations of Squids and Octopods.")

(a)

The Clam Has All The Organ Systems Typical Of Complex Animals

The soft body of the clam, a bivalve, is laterally compressed and completely enclosed by two shells that are hinged dorsally and open ventrally (Fig. 24–23). This arrangement allows the hatchet-shaped foot to protrude ventrally for locomotion. Large, strong muscles attached to the shell enable the animal to open or close its shell.

Openings are present through which water flows into and out of the mantle cavity. Extensions of the mantle, called "siphons," permit the animal to obtain water relatively free of sediment. There is an **incurrent siphon** for water intake and an **excurrent siphon** for water output.

The inner pearly layer of the bivalve shell is secreted in thin sheets by the epithelial cells of the mantle. Composed of calcium carbonate and known as *mother-of-pearl*, it is valued for making jewelry and buttons. Should a bit of foreign matter lodge between the shell and the epithelium, the epithelial cells are stimulated to secrete concentric layers of calcium carbonate around the intruding particle. This is how a pearl is formed.

The clam obtains its food by filtering the sea water brought in over the gills by the siphon. The water is kept in motion by the beating of cilia on the surface of the gills. This stream of water carries food particles that have been trapped in the mucus that has been secreted by the gills to the mouth.

All of the organ systems typical of complex animals are present in the clam. The digestive system is a coiled tube extending from mouth to anus. The open circulatory system consists of a heart that pumps blood into a single blood vessel, which may branch into other vessels. Eventually, blood flows into a network of large spaces that make up the hemocoel. The blood finds its way into vessels that conduct it to the gills, where gas exchange takes place, then back to the heart.

Neurons connect three pairs of ganglia with one another and with the various organs. Sense receptors

(b)

(c)

FIGURE 24–22   There are many beautiful forms of mollusks. (*a*) A flamingo tongue, *Cyphoma gibbosum*, photographed in the Virgin Islands. (*b*) A bay scallop, *Argopecten irradians*, photographed in a sea grass bed in Tampa Bay. (*c*) *Octopus macropus*, a cephalopod, lives in a den among the rocks. (*a* and *b*, Robin Lewis, Coastal Creations; *c*, Jane Burton/Bruce Coleman, Inc.)

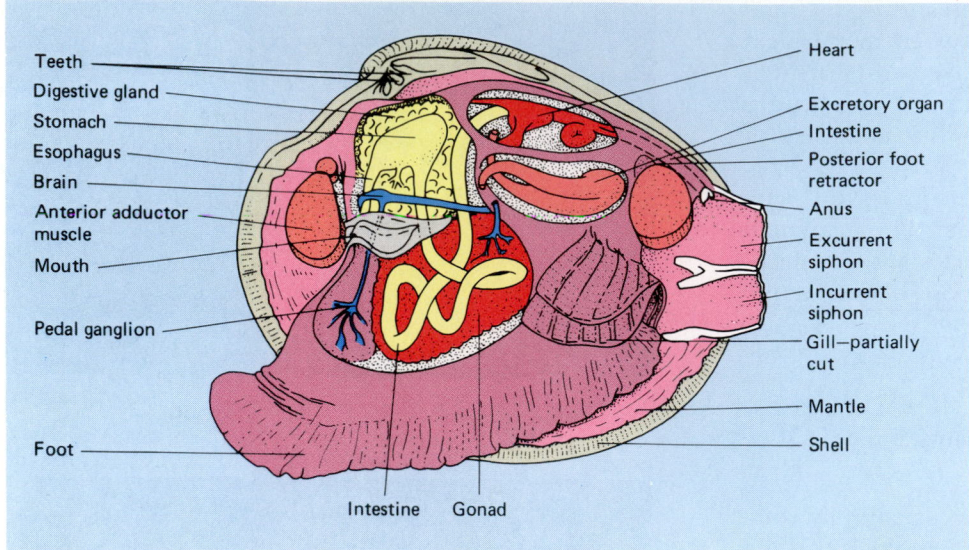

Teeth
Digestive gland
Stomach
Esophagus
Brain
Anterior adductor muscle
Mouth
Pedal ganglion
Foot

Heart
Excretory organ
Intestine
Posterior foot retractor
Anus
Excurrent siphon
Incurrent siphon
Gill—partially cut
Mantle
Shell

Intestine   Gonad

FIGURE 24–23   Internal anatomy of a clam.

include organs of balance and cells sensitive to light and touch.

The sexes are separate in clams. Sperm are usually discharged into the water, and in some bivalves fertilization takes place within the mantle cavity of the female. In these species, the female broods her young within the mantle cavity. Development takes place among the gill filaments. Typically, a free-swimming, ciliated, top-shaped larva, called a **trochophore larva,** develops (Fig. 24–24). The trochophore larva then develops further into a **veliger larva** with shell and foot. Larvae of some freshwater species spend several weeks as parasites on the gills of fishes.

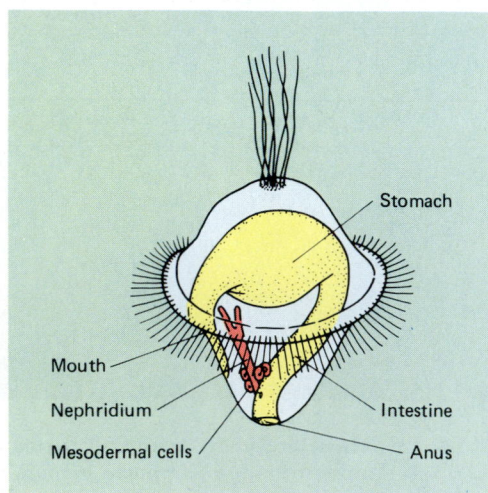

Stomach

Mouth
Nephridium
Mesodermal cells
Intestine
Anus

FIGURE 24–24   Trochophore larva, the first larval stage of many marine mollusks. This type of larva is also characteristic of annelids.

THE ANNELIDS HAVE SEGMENTED BODIES

Phylum Annelida are worms whose bodies are partitioned internally and externally into ring-like segments. This phylum includes the earthworms, leeches, and many marine and freshwater worms. The body wall and many of the internal organs are segmented. The segments are separated from one another by transverse partitions, called **septa.** The bilaterally symmetrical, tubular body may consist of more than 100 segments. Some structures, such as the digestive tract and certain nerves, run the length of the body, passing through successive segments. Other structures are repeated in each segment (Fig. 24–25).

Segmentation is an advantage because not only is the coelom divided into segments, but each segment has its own muscles, enabling the animal to elongate one part of its body while shortening another part. (The annelid's hydrostatic skeleton, which enables it to do this, is discussed in Chapter 32.) In the annelid, the individual segments are almost all alike, but in many segmented animals (arthropods and chordates), different segments and groups of segments are specialized to perform different functions. In some groups, the specialization may be so pronounced that the basic segmentation of the body plan may be obscured.

Bristle-like structures called **setae** aid in locomotion. An annelid has a well-developed coelom, a closed circulatory system, and a complete digestive tract, which consists of a tube extending from mouth to anus. Respiration takes place through the skin or by gills. Typically, a pair of excretory structures called **nephridia** are found in each segment (Fig. 24–26). The nervous system generally consists of a simple brain

FOCUS ON   Adaptations of Squids and Octopods

In contrast to most other mollusks, the cephalopods are active, predatory animals. They are fast-swimming organisms that are adapted for an entirely different life-style than their filter-feeding relatives. Some biologists consider members of this group the most advanced of the invertebrates.

The octopus has no shell, and the shell of the squid is reduced to a small "pen" in the mantle. The cephalopod foot is divided into tentacles, ten in squids, eight in octopods. The tentacles, or arms, surround the central mouth of the large head. Cephalopods have large, well-developed eyes that form images. Although they are not homologous, the eyes appear strikingly like vertebrate eyes and function in much the same way.

The tentacles of squids and octopods are covered with suckers for seizing and holding prey. The mouth is equipped with a radula, something like a belt of teeth, and with two strong, horny beaks used to kill prey and tear it to bits. The mantle is thick and muscular and fitted with a funnel. By filling the mantle cavity with water and ejecting it through the funnel, the animal can attain rapid jet propulsion in the opposite direction.

Besides their speed, many cephalopods have two other important adaptations that enable them to escape from their predators, which include the whales and moray eels. One is the ability to confuse the enemy by rapidly changing colors. By expanding and contracting pigment cells in its skin, the cephalopod can display an impressive variety of mottled colors. Another defense mechanism is the ink sac, which produces a thick black liquid. This is released in a dark cloud when the animal is alarmed. While its enemy pauses, temporarily blinded and confused, the cephalopod easily escapes. The ink has been shown to paralyze the chemical receptors of some predators.

The octopus feeds on crabs and other arthropods, catching and killing them with a poisonous secretion of its salivary glands. During the day, the octopus usually hides among the rocks; in the evening, it emerges to hunt for food. Its motion is incredibly fluid, giving little hint of the considerable strength in its eight arms.

Small octopods survive well in aquaria and have been studied extensively. They have a high degree of intelligence and can make associations among stimuli. Their very adaptable behavior resembles that of the vertebrates more closely than the stereotyped patterns of behavior seen in other invertebrates.

composed of a pair of ganglia and a double ventral nerve cord. A pair of ganglia and lateral nerves are repeated in each segment.

The 10,000 or so species are assigned to three main classes (Fig. 24–27):

1. **Class Polychaeta** (meaning "many hairs") includes marine worms, such as sandworms and tubeworms (Fig. 24–28). These animals swim freely in the sea, burrow in the mud near the shore, or live in tubes formed by cementing bits of shell and sand together with mucus and other secretions from the body wall. Each body segment has a pair of paddle-shaped appendages called **parapodia** that extend laterally and function in locomotion. These fleshy structures bear many stiff setae. Most polychaetes have well-

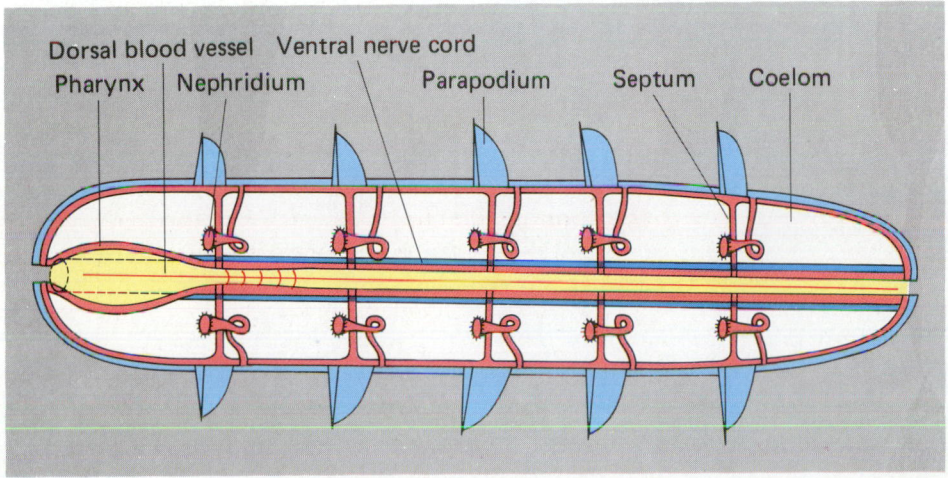

FIGURE 24–25 A generalized annelid. Note that the body is segmented, and there is a serial repetition of body parts.

Dorsal blood vessel   Ventral nerve cord
Pharynx   Nephridium   Parapodium   Septum   Coelom

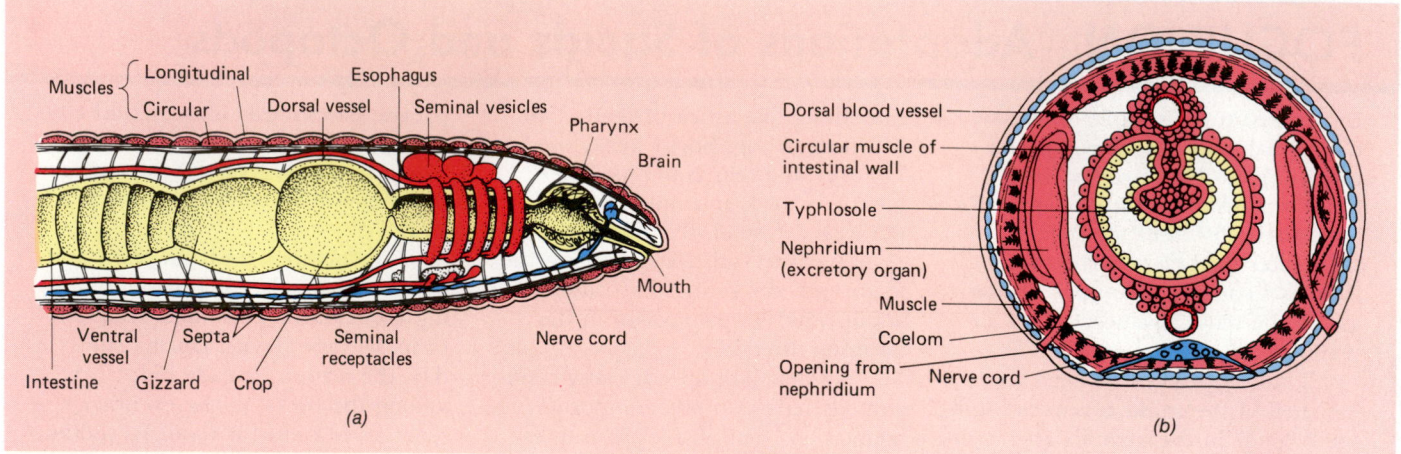

FIGURE 24–26 **Structure of the earthworm:** (*a*) **Sagittal section through the anterior portion of an earthworm.** (*b*) **Cross section.**

developed heads (or prostomia) bearing eyes and antennae. The head may also be equipped with tentacles, bristles, and palps (feelers).

2. **Class Oligochaeta** includes the earthworm and some freshwater worms. These worms lack well-developed heads and have no parapodia. Fewer setae are present than in the polychaetes. Oligochaetes are hermaphroditic.

3. **Class Hirudinea** (Fig. 24–29) comprises the leeches. Most are blood-sucking parasites that inhabit fresh water, although some tropical species are terrestrial and drop on their hosts from foliage. Prominent muscular suckers, present at each end of the body, are used for clinging to the prey. Leeches lack both setae and appendages but have a very complex coelomic cavity.

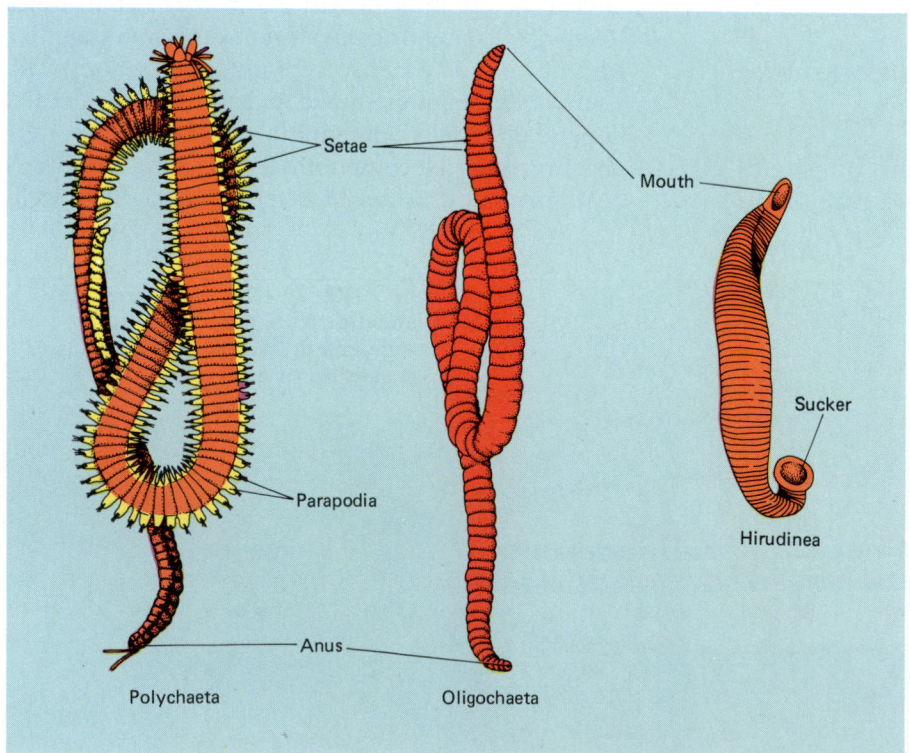

FIGURE 24–27 **Comparison of the classes of the phylum Annelida—Polychaeta, Oligochaeta, and Hirudinea.**

(a)

(b)

FIGURE 24–28 Polychaete annelids. (*a*) *Hermodice carunculata*, a West Indian fireworm (*b*) The Christmas tree worm, *Spirobranchus giganteus*, photographed in a Florida coral reef. (*a*, Charles Seaborn; *b*, James H. Carmichael, Coastal Creations, Inc.)

FIGURE 24–29 The medicinal leech, *Hirudo medicinalis*, was widely used by physicians to absorb "bad blood" from their patients. Today, leeches are being used in some modern surgical procedures because they release an anticoagulant (hirudin) that prevents clotting in the area where they are attached. In the photo shown here leeches are being used to treat hematoma, an accumulation of blood within body tissues that results from injury or disease. The leech attaches its sucker near the site of injury and releases hirudin, which dissolves blood clots. (St. Bartholomew's Hospital/Science Photo Library/Photo Researchers, Inc.)

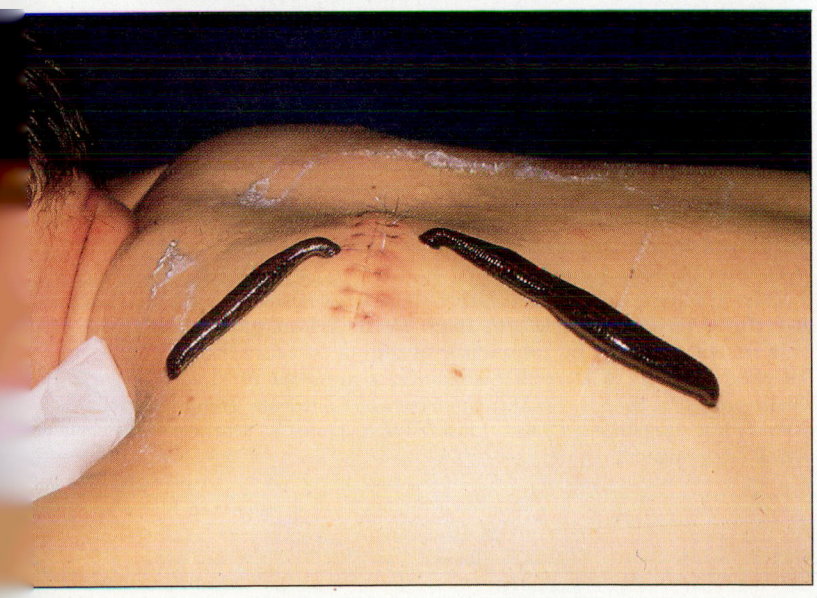

☐ ARTHROPODS HAVE JOINTED APPENDAGES AND AN EXOSKELETON OF CHITIN

Arthropods are the most diverse and biologically successful group of animals. There are more of them (about 800,000 described species), they live in a greater range of habitats, and they can eat a greater variety of food than the members of any other phylum (Fig. 24–30).

Arthropods have paired, jointed appendages from which they get their name (arthropod means "jointed foot"). These appendages function as swimming paddles, walking legs, mouthparts, or accessory reproductive organs that transfer sperm.

Another important arthropod characteristic is the hard, armor-like **exoskeleton** composed of chitin that covers the entire body and appendages. The exoskeleton provides protection against predators and against excessive loss of moisture. It also gives support to the underlying soft tissues. Distinct muscle bundles somewhat comparable to individual vertebrate muscles attach to the inner surface of the exoskeleton. These act upon a system of levers that permit the extension and flexion of parts at the joints. The exoskeleton has certain disadvantages, however. Body movement is somewhat restricted, and as the arthropod grows, it periodically outgrows this nonliving shell. This process of shedding an old shell and growing another larger one

(a)

(b)

(c)

(d)

(e)

FIGURE 24–30 **Arthropods are considered the most successful animals.** (*a*) The Merostomata include only a few closely related species that have survived to the present. Seasonally, horseshoe crabs, *Limulus polyphemus*, return to beaches for mating. (*b*) The vast majority of the arachnids prey on other small invertebrates. In this photograph a garden spider is shown wrapping its prey, a webworm. The worm has already been killed or paralyzed by the spider's venomous bite. (*c*) Most members of class Crustacea are aquatic or semiaquatic. Freshwater crayfish from California. (*d*) Leaf-mimicking katydid (*Pterachroza ocellata*) from Peru. When startled, this insect displays its menacing eyespots to scare would-be predators. (*e*) A striped millipede, a member of class **Chilopoda.** (*a*, Peter J. Bryant, University of California, Irvine/BPS; *b*, E.R. Degginger; *C*, Ken Lucas/BPS; *d*, James L. Castner; *e*, E.R. Degginger)

is known as **molting.** The shed exoskeleton represents a net metabolic loss and molting also leaves the arthropod temporarily vulnerable to predators.

The arthropod body, like that of the annelid, is segmented. In some arthropod classes, however, segments become fused together or lost during development. The bodies of most arthropods are divided into three regions: the *head, thorax,* and *abdomen.* There is an incredible range of variations in body·plan and in the shape of the jointed appendages in the numerous species.

Most of the aquatic arthropods have a system of gills that function in gas exchange. The land forms, in contrast, usually have a system of fine, branching air tubes, called **tracheae,** that conduct air to the internal organs. A few (some spiders and land crabs) have structures comparable to lungs. Arthropods have a variety of well-developed sense organs: complicated eyes, such as the compound eyes of insects; organs of hearing; organs in the antennae sensitive to touch and chemicals; and cells on the surface of the body that are sensitive to touch.

The open circulatory system includes a dorsal, tubular heart that pumps blood into a dorsal artery, and sometimes other arteries. From the arteries, blood flows into large spaces, which collectively make up the hemocoel. Blood in the hemocoel bathes the tissues directly. Eventually blood finds its way back into the heart through openings, referred to as **ostia,** in its walls.

The Arthropods Can Be Assigned To Three Subphyla

There is much disagreement concerning arthropod classification. Here we will use a scheme that divides the phylum into three living subphyla:

1. In **subphylum Chelicerata,** the arthropod body is divided into a cephalothorax (fused head and thorax) and an abdomen. These animals have no antennae and no chewing mandibles, mouthparts that are characteristics of other arthropod subphyla. Instead, the first pair of appendages, located immediately anterior to the mouth, are the **chelicerae,** which are used to manipulate food and pass it to the mouth. The second pair of appendages, called **pedipalps,** are modified to form different functions in various groups. Posterior to the pedipalps are usually four pairs of legs.

Two classes in this subphylum are **class Merostomata** (see Fig. 24–30*a*) and **class Arachnida.** The only living merostomes are the horseshoe crabs. The 60,000 or so species of class Arachnida include the spiders, scorpions, mites, ticks, and harvestmen (daddy longlegs). Arachnids have six pairs of jointed appendages. The first pair are the **chelicerae,** fanglike structures used to penetrate prey and suck out its body fluids. In some arachnids, the chelicerae are used to inject poison into the prey. The second pair of appendages are the **pedipalps,** which are used by spiders to hold and chew food and are modified as sense organs for tasting food in some species. The other four pairs of legs are used for walking. Most arachnids are carnivorous and prey upon insects and other small arthropods (see Fig. 24–30*b*). Many types of arachnids have glands that secrete silk, which is used for making webs.

2. **Subphylum Crustacea** includes the lobsters, crabs, and shrimp. These arthropods have **biramous appendages,** which have two jointed branches at their ends. They also are the only group to have two pairs of **antennae,** sensory organs for touch and taste. Crustaceans are also characterized by **mandibles,** which are located on each side of the ventral mouth and used for biting and grinding food. There are usually five pairs of walking legs. Other appendages may be specialized for swimming, sperm transmission, carrying eggs and young, or sensation (Fig. 24–30*c*).

3. The insects, centipedes, and millipedes are grouped together in **subphylum Uniramia** because they all possess unbranched (uniramous) appendages. They also bear only a single pair of antennae rather than two pairs, as in crustaceans. The insects belong to **class Insecta** (Fig. 24–30*d*). **Class Chilopoda** includes the centipedes (hundred-legged worms), and **class Diplopoda** includes the millipedes (thousand-legged worms) (Fig. 24–30*e*). Both centipedes and millipedes have heads and elongated bodies with many segments, each bearing legs. Each centipede has one pair of legs on each segment, but a millipede has two pairs of legs per segment.

The Insects Have Adapted To Almost Every Available Ecological Niche

With more than 750,000 described species, the insects are the most successful group of animals on our planet in terms of number of species as well as number of individuals and diversity. Some insect species live in fresh water, a few are truly marine, and others inhabit the shore between the tides. Most species, however, are terrestrial; the insects are one of only a few groups of animals that have adapted to life on land.

An insect may be described as an **articulated** (jointed), **tracheated** (having tracheal tubes for gas exchange) **hexapod** (having six feet). The insect body consists of three distinct parts—**head, thorax,** and **abdomen.** Three pairs of legs and usually two pairs of wings emerge from the thorax. One pair of antennae

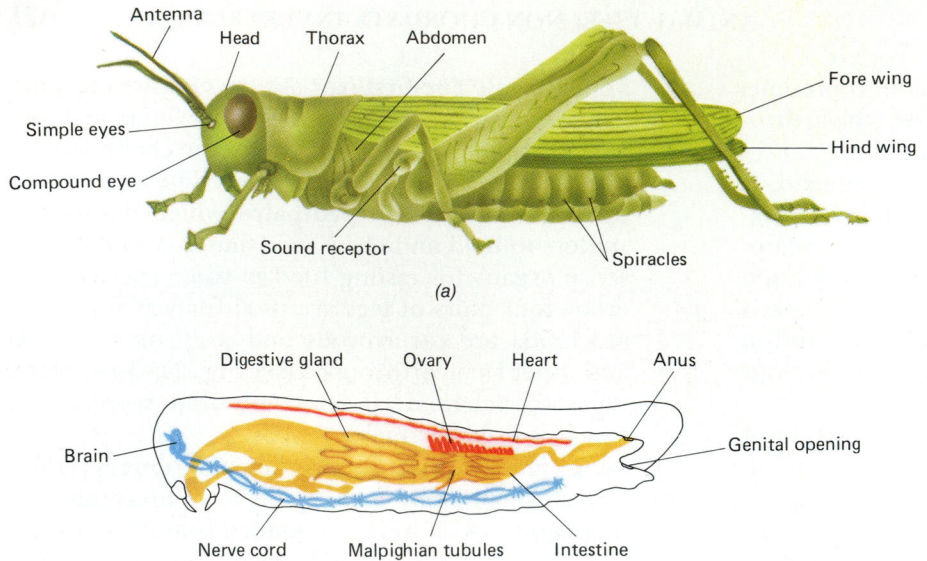

FIGURE 24–31 Structure of the grasshopper. (a) External anatomy. Note the three pairs of articulated legs. (b) Internal anatomy.

protrude from the head (Fig. 24–31). A complex set of mouthparts is present and may be adapted for piercing, chewing, sucking, or lapping. Excretion is accomplished by two or more slender **Malpighian tubules,** which receive metabolic wastes from the blood. After concentrating wastes, the Malpighian tubules discharge them into the intestine.

The sexes are separate in insects and fertilization takes place internally, an important adaptation for land animals. During development, there are several molts. In some orders, there are several developmental stages, called **nymphal stages,** and gradual **metamorphosis** (change in body form) to the adult form. In others, there is a **complete metamorphosis** with four distinct stages in the life cycle: **egg, larva, pupa,** and **adult** (Fig. 24–32).

There are more species of insects than of all other classes of animals combined. What they lack in size, insects make up in sheer numbers. It has been calculated that all the insects in the world would weigh more than all of the remaining land animals on earth. What are the secrets of insect success? One important factor is their body plan, which can be modified and specialized in so many ways that insects have been able to

adapt to an incredible number of life-styles. Insects have filled almost every variety of ecological niche. One of their most important adaptations is their ability to fly. Unlike other invertebrates, which creep slowly along (or under) the ground, the insects fly rapidly through the air. Their wings and their small size facilitate their wide distribution.

The insect body is well protected by the tough exoskeleton, which also helps prevent water loss by evaporation. Other protective mechanisms include mimicry, protective coloration (Chapter 18) and aggressive behavior. Metamorphosis divides the insect life cycle into different stages, a strategy that has the advantage of placing larval forms into their own niches so that in most cases they do not have to compete with adults for food or habitats.

The Grasshopper Has A Representative Insect Life-Style

The structure and activities of the grasshopper can be considered as representative of the insect world (Fig. 24–33). The grasshopper body consists of head,

FIGURE 24–32 **Life cycle of the monarch butterfly.** (a, E.R. Degginger; b–e, Courtesy of Leo Frandzel)

(a) Egg (b) Larva (c) Pupa (early) (d) Pupa (late) (e) Adult

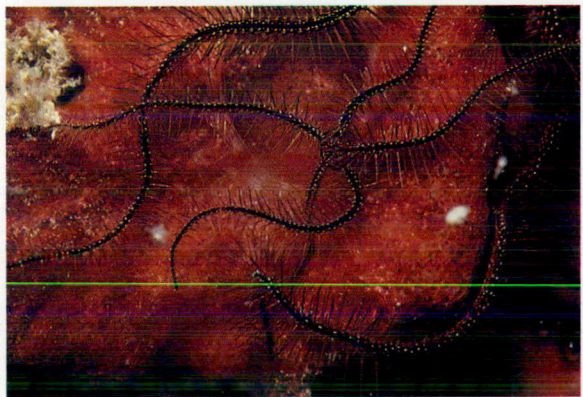

FIGURE 24–33 Some representatives of the five classes of phylum Echinodermata. (*a*) Blue spotted sea urchin, *Astropyga radiata*. (*b*) Pacific Henrichia sea star, *Henricia laeviuscula*, Asteroidea), of the South Pacific. (*c*) A feather star from the Caribbean Sea. (*d*) A sea cucumber, *Thelonota* sp., raising its body to spawn. (*e*) Brittle stars living on the surface of a sponge. (*a*, Brian Parker/Tom Stack and Associates; *b*, Phil Degginger; *c*, Susan Blanchet/M.L. Dembinsky, Jr. Photography Associates; *d*, Peter Scoones/SeaPHot, Ltd; *e*, Charles Seaborn)

thorax, and abdomen. Two pairs of wings and three pairs of legs are present. Its jointed legs provide for much greater ability to maneuver than is possible for an annelid. A complex set of mouthparts is adapted in the grasshopper for chewing. In other insects, mouthparts may be specialized for sucking or lapping.

The leaves ingested by the grasshopper are processed in its complete digestive tract. Absorbed nutrients are distributed by an open circulatory system. A tubular heart pumps blood into the head region through a large artery. The blood then flows freely in the hemocoel, bathing the tissues directly. Eventually the blood finds its way back into the heart chamber and is pumped out again.

Air enters the tracheae through tiny openings, called **spiracles,** that pierce the body wall. The Malpighian tubules remove wastes from the blood in the hemocoel and empty them into the intestine. This excretory system is well adapted for water conservation, and wastes are discharged as rather dry crystals.

The simple brain is continuous with a paired **ventral nerve cord.** Like other insects, the grasshopper has efficient sense organs. Prominent among these are two types of eyes—compound eyes for image formation and simple eyes for light perception. A well-developed endocrine system regulates growth and development. Very complex behavior patterns are known in insects, but these operate on an instinctive or programmed level (see Chapter 44).

The male grasshopper is equipped with a copulatory organ for transferring sperm into the female reproductive tract. The adults usually die soon after the eggs are laid. Grasshoppers do not undergo complete metamorphosis. The young grasshoppers resemble their parents but lack wings. They gradually grow in size and develop into adults. As in other arthropods, the grasshopper periodically outgrows its nonliving exoskeleton. At such times, the skeleton is shed and replaced by a new, larger one.

■ THE ECHINODERMS ARE SPINY-SKINNED ANIMALS OF THE SEA

All of the members of **phylum Echinodermata** inhabit the sea. About 6000 living and 20,000 extinct species have been identified. The living species are divided into five principal classes (Fig. 24–34). **Class Crinoidea** includes the sea lilies and feather stars; **class Asteroidea,** the sea stars; **class Ophiuroidea,** the brittle stars; **class Echinoidea,** the sea urchins and sand dollars; and **class Holothurioidea,** the sea cucumbers.

The echinoderms are in many ways unique in the animal kingdom. Although their larvae exhibit bilateral symmetry, the adults have **pentaradial symmetry.** This means that the body is arranged in five parts around a central disc, where the mouth is located. A thin, ciliated **epidermis** covers the endoskeleton, which consists of small plates (Fig. 24–34). Composed of $CaCO_3$, these plates typically bear spines that project outward. The phylum name Echinodermata, meaning "spiny-skinned," reflects this trait.

A characteristic found only in echinoderms is the **water vascular system,** a network of canals through which sea water circulates. Branches lead to numerous tiny **tube feet,** which extend when filled with fluid and serve in locomotion, obtaining food, and, in some forms, gas exchange. The water vascular system is a hydraulic system. When the echinoderm begins to extend a foot, a rounded muscular sac **(ampulla)** at the upper end contracts forcing water into the tube of the foot. At the bottom of the foot is a suction-type structure that adheres to whatever surface the animal is perched upon. The foot can be withdrawn by contraction of muscle in its walls, which forces water back into the ampulla.

Echinoderms have well-developed coeloms and complete digestive systems, but they have only rudimentary circulatory systems and no specialized excretory structures. There are a variety of respiratory structures in the various classes, including dermal (skin) gills in the sea stars. The nervous system is simple, usually consisting of nerve rings about the mouth with radiating nerves.

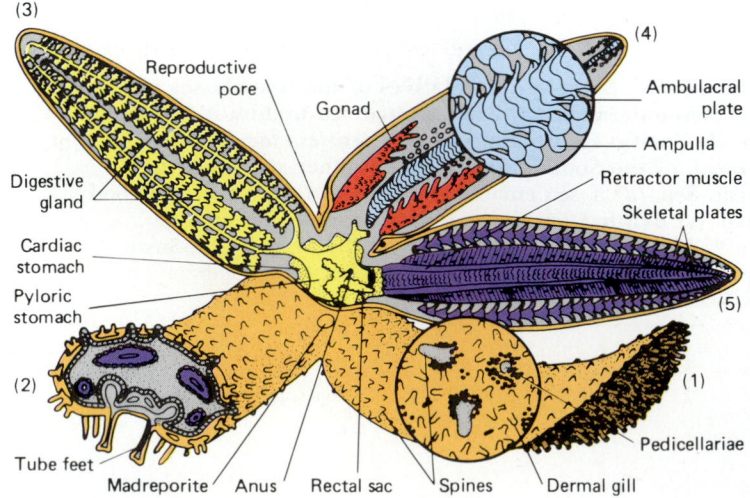

FIGURE 24–34 Structure of the sea star. The sea star *Asterias* viewed from above with the arms in various stages of dissection. (1) Upper surface with a magnified detail showing the features of the surface. The end is turned up to show the tube feet on the lower surface. (2) The arm is shown in cross section. (3) The upper body wall of the arm has been removed. (4) The upper body wall and digestive glands have been removed, and the ampullas and plates are shown in magnified view. (5) All the internal organs have been removed except the retractor muscles, showing the inner surface of the lower body wall.

◾ SUMMARY

I. Animals are euykaryotic, multicellular heterotrophs whose cells exhibit a division of labor. They are generally capable of locomotion at some time during their life cycle, and they generally reproduce sexually.

II. Animals have complex evolutionary relationships.
   A. Most animals exhibit bilateral symmetry; animals in only two phyla have radial symmetry.
   B. Acoelomates are animals that lack a body cavity. Pseudocoelomates have a body cavity derived from the cavity of the embryo. Coelomates have a true coelom lined with mesoderm.
   C. Animals can also be grouped as protostomes or deuterostomes.

III. Phylum Porifera consists of the sponges, the simplest animals.
   A. Sponges are divided into classes on the basis of the type of skeleton they secrete.
   B. The sponge body is a sac perforated by tiny holes through which water enters. As water circulates through the sponge, materials are exchanged by diffusion.

IV. Phylum Cnidaria includes the hydras, jellyfish, and corals.
   A. Cnidarians are characterized by radial symmetry, stinging cells, definite tissue layers, and a nerve net.
   B. Cnidarians have a polyp or medusa body form.

V. Phylum Platyhelminthes includes the planarians, flukes, and tapeworms.
   A. Flatworms have bilateral symmetry, exhibit cephalization, have three definite tissue layers, simple brains, and nervous systems, and other well-developed organs, including protonephridia with flame cells.
   B. The flukes and tapeworms have suckers, complex life cycles, and other adaptations for their parasitic life-styles.

VI. Members of phylum Nemertinea (proboscis worms) have tube-within-a-tube body plans, complete digestive tracts with mouth and anus, and separate circulatory systems.

VII. Phylum Nematoda, the roundworms, includes species of great ecological importance and species parasitic in plants and animals.
   A. Nematodes have pseudocoeloms and complete digestive tracts.
   B. Nematodes parasitic in humans include *Ascaris*, hookworms, trichina worms, and pinworms.

VIII. Phylum Mollusca includes clams, oysters, snails, squids, and octopods.
   A. Mollusks are soft-bodied animals that are usually covered by shells. A mollusk has a ventral foot for locomotion, and a mantle that covers the visceral mass.
   B. Squids and octopods are active predatory animals with large heads, long tentacles, and prominent eyes.

IX. Phylum Annelida, the segmented worms, includes many marine worms, earthworms, and leeches.
   A. Annelids have long bodies that are segmented internally as well as externally. Setae aid in locomotion.
   B. Polychaetes have bristled parapodia that function in locomotion. Leeches lack setae and appendages, but are equipped with suckers.

X. Phylum Arthropoda, the largest phylum, includes horseshoe crabs, spiders, scorpions, mites, crustaceans, insects, millipedes, and centipedes.
   A. Arthropods are animals with jointed appendages and armor-like exoskeletons.
   B. Insects are the most successful group of animals, with more species and greater numbers than any other group.

XI. Phylum Echinodermata is made up of the sea stars, sea urchins, sand dollars, and sea cucumbers.
   A. Echinoderms have spiny skins and exhibit radial symmetry.
   B. Other unique features of the echinoderms are the water vascular system and tube feet.

◾ POST-TEST

1. Animals without backbones are _____.
2. An animal that is divided into roughly right and left halves when sliced down the midline may be described as _____.
3. An acoelomate animal lacks a _____.
4. The two phyla that are deuterostomes are the _____ and the _____.

TABLE 24–1
Comparison of the Principal Animal Phyla

Phylum	Level of organization	Symmetry	Digestion	Circulation	Gas exchange
Porifera (pore bearers) Sponges	Multicellular but tissues loosely arranged	Radial or none	Intracellular	Diffusion	Diffusion
Cnidaria Hydra, jellyfish, coral	Tissues	Radial	Gastrovascular cavity with only one opening; intra- and extra-cellular digestion	Diffusion	Diffusion
Platyhelminthes (flatworms Planarians, flukes, tapeworms	Organs	Bilateral; rudimentary head	Digestive tract with only one opening	Diffusion	Diffusion
Nemertinea Proboscis worms	Organ systems	Bilateral	Complete digestive tract with mouth and anus	At least two pulsating blood vessels; no heart; blood cells with hemoglobin	Diffusion
Nematoda (roundworms) *Ascaris,* hookworms, nematodes	Organ systems	Bilateral	Complete digestive tract with mouth and anus	Diffusion	Diffusion
Mollusca Clams, snails, squids	Organ systems	Bilateral	Complete digestive tract	Open system	Gills and mantle
Annelida (segmented worms) Earthworms, leeches, marine worms	Organ systems	Bilateral	Complete digestive tract	Closed system	Diffusion through moist skin; oxygen circulated by blood
Arthropoda (jointed-footed animals Crustaceans, insects, spiders	Organ systems	Bilateral	Complete digestive tract	Open system	Trachea in insects; gills in crustacea; book lungs or trachea in spider group
Echinodermata (spiny-skinned animals) Sea stars, sea urchins, sand dollars	Organ systems	Embryo bilateral; adult radial	Complete digestive tract	Open system; reduced	Skin gills
Chordata Tunicates, Lancelets, Vertebrates	Organ systems	Bilateral	Complete digestive tract	Closed system; ventral heart	Gills or lungs

TABLE 24–1 (CONTINUED)

Waste disposal	Nervous system	Reproduction	Other characteristics
Diffusion	Irritability of cytoplasm	Asexual by budding; sexual, may be hermaphroditic	Filter feeders; skeleton of chalk, glass, or spongin; larvae swim by cilia; adults incapable of loco-motion
Diffusion	nerve net; no centralization of nerve tissue	Asexual by budding; sexual, separate sexes	Have cnidocytes (stinging cells) along tentacles
Protonephridia; flame cells and ducts	Simple brain; two nerve cords; ladder-type system; simple sense organs	Asexual, by fission; sexual, hermaphroditic, but some cross-fertilize	Three definite tissue layers; no body cavity; many parasitic
Two lateral excretory canals with flame cells	Simple brain; two nerve cords with cross nerves; simple sense organs	Asexual, by fragmentation; sexual, sexes separate	No body cavity; proboscis for defense and capturing prey
Excretory canals	Simple brain; dorsal and ventral nerve cords; simple sense organs	Sexual, sexes separate	Have pseudocoelom; many parasitic
Kidneys	Three pairs of ganglia; simple sense organs	Sexual, sexes separate; fer-tilization in water	Soft-bodied; usually have shell and ventral foot for locomotion
Pair of metanephridia in each segment	Simple brain; ventral nerve cord; simple sense organs	Sexual; hermaphroditic but cross-fertilize	Earthworms till soil
Malpighian tubules in in-sects; antennal (green) glands in crustaceans	Simple brain; ventral nerve cord; well-developed sense organs	Sexual, sexes almost always separate	Hard exoskeleton; most suc-cessful group of animals
Diffusion	Nerve rings; no brain	Sexual, sexes separate	Water vascular system
Kidneys and other organs	Dorsal nerve cord with brain at anterior end	Sexual, sexes separate	(1) Notochord, (2) Hollow tubular, dorsal nerve cord, (3) Pharyngeal grooves

5. Sponges secrete skeletal structures called _____.

6. A hermaphroditic animal can produce _____ and _____.

7. A distinctive feature of the cnidarians is the presence of _____ cells.

21. Humans become infected with beef tapeworms by _____.

22. Malphighian tubules are excretory structures found in _____.

Select the most appropriate match in Column B for each entry in Column A (8 through 20).

Column A

_____ 8. Nerve net
_____ 9. Tube feet
_____ 10. Notochord
_____ 11. Protonephridia with flame cells
_____ 12. Jointed appendages; exoskeleton
_____ 13. Soft-bodied with ventral foot and mantle

_____ 14. Sponge
_____ 15. Snail
_____ 16. Sea star
_____ 17. Tapeworm
_____ 18. Earthworm
_____ 19. Grasshopper
_____ 20. Crayfish

Column B

a. Phylum Cnidaria
b. Phylum Platyhelminthes
c. Phylum Mollusca
d. Phylum Arthropoda
e. Phylum Echinodermata
f. None of the above

◻ REVIEW QUESTIONS

1. Which animals are considered the most successful? Describe some of their characteristics that contribute to this success.
2. What advances do the members of phylum Platyhelminthes exhibit over those animals that belong to phylum Cnidaria? In what ways are they alike?
3. Which group of animals (a) exhibit radial symmetry, (b) have flame cells, (c) have the most primitive brains, (d) are the simplest to have complete digestive tubes with mouth and anus?
4. Distinguish between insects and spiders.
5. Which groups of animals are able to depend upon diffusion for gas exchange? Why?

6. Describe the life cycle (draw a diagram if you can) of the tapeworm.
7. Give two distinguishing characteristics of (a) phylum Annelida, (b) phylum Arthropoda, (c) phylum Mollusca, and (d) phylum Echinodermata.
8. Following generally accepted evolutionary principles, draw a hypothetical ancestral tree of the animal kingdom.
9. What do echinoderms and chordates have in common?
10. Which phyla are acoelomate? Which are coelomate?
11. Locate each of the following in a specific type of animal and give its function: (a) cnidocytes, (b) spicules, (c) nerve net, (d) mantle, (e) setae, (f) tracheae.

◻ READINGS

Barnes, R.D. *Invertebrate Zoology*, 5th ed. Philadelphia, Saunders College Publishing, 1987. This comprehensive textbook discusses the life processes of each invertebrate phylum.

Fingerman, M. *Animal Diversity*, 3rd ed. Philadelphia, Saunders College Publishing, 1981. A very brief summary of both the invertebrates and the vertebrates with emphasis on evolutionary relationships.

25

ANIMAL LIFE: CHORDATES

OUTLINE

I. The tunicates have a protective covering made of cellulose
II. The lancelets may be similar to primitive chordates that gave rise to vertebrates
III. The success of the vertebrates is linked to the evolution of a few key adaptations
  A. The jawless fish are the most primitive vertebrates
  B. The cartilaginous fish include the sharks, rays, and skates
  C. The bony fish are the most numerous vertebrates
    1. The ray-finned fish gave rise to modern bony fish
    2. Descendants of the lobe-finned fish moved onto the land
  D. The amphibians were the first successful land vertebrates
  E. Reptiles were the dominant land animals for almost 200 million years
    1. Reptiles evolved adaptations to life on land
    2. The large reptiles suddenly and mysteriously became extinct
  F. Birds are adapted for flight
  G. Mammals evolved from reptiles
    1. Monotremes are mammals that lay eggs
    2. Marsupials are pouched mammals
    3. Placental mammals complete their embryonic development within the mother's body.
IV. Primates evolved from shrew-like mammals
  A. The fossil record suggests general trends in hominid evolution
    1. The earliest hominids belong to genus *Australopithecus*
    2. *Homo habilis* is the oldest member of genus *Homo*
    3. Numerous fossils of *Homo erectus* have been discovered
    4. *Homo sapiens* appeared about 200,000 years ago
  B. Humans undergo cultural evolution
    1. Development of agriculture resulted in a more dependable food supply
    2. The Industrial Revolution has had profound impact on the ecosphere

LEARNING OBJECTIVES

After you have studied this chapter, you should be able to:

1. Describe the distinguishing characteristics of a chordate, and give examples of members of the three subphyla of chordates.
2. Characterize each of the seven classes of vertebrates, and give examples of each group.
3. Describe the course of vertebrate evolution according to contemporary evolutionary theory.
4. Identify adaptations that enabled vertebrates to succeed on land.
5. Compare the three main groups of living mammals and identify specific mammals that belong to each group.
6. Compare the following hominids: australopithecines, *Homo habilis*, *Homo erectus*, *Homo sapiens*.
7. Describe cultural evolution and its impact on the ecosphere.

Male impala, Nairobi, Kenya. (Stan Osolinski/Marvin L. Dembinsky, Jr., Photography Associates)

Phylum Chordata, the phylum to which we humans belong, is a diverse group of animals that has radiated extensively, filling many types of ecologic niches in fresh water, sea water, and land. There are three subphyla: **subphylum Urochordata** consists of the tunicates, which include the sea squirts, filter-feeding marine animals; **subphylum Cephalochordata** consists of the lancelets, which are small, translucent, fish-like animals; and **subphylum Vertebrata** consists of the vertebrates, which are animals with backbones.

Chordates share three characteristics that distinguish them from all other groups (Fig. 25–1):

1. *All chordates have a notochord during some time in their life cycle.* The notochord is a dorsal longitudinal rod that is firm yet flexible and supports the body.
2. *All chordates have a dorsal tubular nerve cord.* The nerve cord differs from that of other animals not only in its position but in being single rather than double, and hollow rather than solid.
3. *All chordates have pharyngeal gill grooves during some time in their life cycle.* In some chordates, the grooves perforate through the wall of the pharynx and become functional gill slits, but in terrestrial (land) animals, they become modified to form entirely different structures adaptive for life on land.

No clear fossil record of the ancestors of the chordates exists. Biologists think the chordate ancestors were small, soft-bodied animals, perhaps echinoderms. Recall that echinoderms and chordates are both deuterostomes, and that echinoderm larvae are bilaterally symmetrical.

KEY CONCEPTS

◻ Phylum Chordata consists of three subphyla; two of these are invertebrate subphyla (tunicates and lancelets) and the third is subphylum Vertebrata.

◻ At some time in their life cycle, chordates have a notochord, a dorsal, tubular nerve cord, and pharyngeal gill grooves.

◻ Vertebrates include jawless fish, cartilaginous fish, bony fish, amphibians, reptiles, birds, and mammals.

◻ Primates evolved from small, tree-dwelling, shrew-like mammals.

◻ Two significant advances in human cultural evolution were development of agriculture and the Industrial Revolution.

◻ THE TUNICATES HAVE A PROTECTIVE COVERING MADE OF CELLULOSE

The tunicates, which are marine animals, include the sea squirts and their relatives (Fig. 25–2). Tunicates are considered the most successful of the invertebrate chordates. Adults develop a covering, or tunic, that protects them. Interestingly, the tunic is made of cellulose, a polysaccharide found in walls of plant cells.

In their larval stage, tunicates are typically chordate and look like frog tadpoles. They have a pharynx with gill slits, a dorsal nerve cord, and a long muscular tail that contains the notochord. In sea squirts, the larva eventually attaches itself to the sea bottom and loses its tail, notochord, and much of its nervous system. In the adult, only the gill slits suggest that the tunicate is a chordate. Appendicularians, another class of urochordates, retain their ability to swim about and are among the most common members of the zooplankton.

Adult sea squirts are often mistaken for sponges or cnidarians. They are filter feeders that remove plankton from the stream of water that passes through the pharynx. Some species form large colonies in which members may share a common mouth and tunic. Colonial forms often reproduce asexually by budding. Sexual forms are usually hermaphroditic.

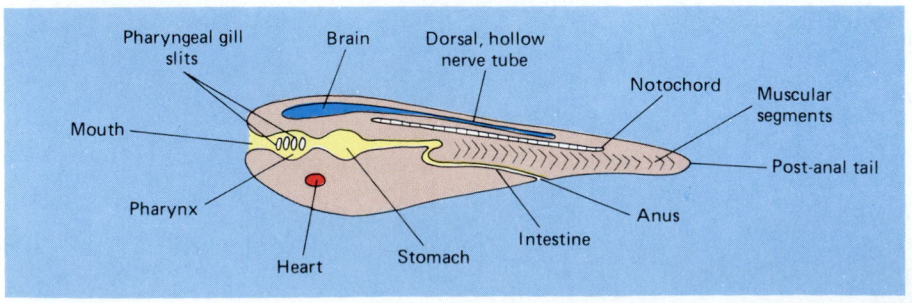

FIGURE 25–1 A generalized chordate illustrating the main chordate characteristics.

FIGURE 25–2   Clear tunicates, *Clavelin picta*, photographed in the Virgin Islands. Tunicates are nonvertebrate chordates. (Courtesy of Robin Lewis, Coastal Creations)

Although they superficially resemble fish, lancelets are far more primitive. They lack paired fins, jaws, a well-defined brain, a heart, and sense organs. Biologists speculate that *Amphioxus* may be similar to the primitive ancestor from which the vertebrates evolved.

☐ THE LANCELETS MAY BE SIMILAR TO PRIMITIVE CHORDATES THAT GAVE RISE TO VERTEBRATES

The chordate characteristics are highly developed in the **lancelets,** commonly referred to as *Amphioxus.* These small, translucent, segmented animals are pointed at both ends (Fig. 25–3). The notochord extends from the tip of the head to the end of the tail. The 25 or so species are widely distributed in shallow seas, either swimming freely or burrowing in the sand near the low-tide line.

☐ THE SUCCESS OF THE VERTEBRATES IS LINKED TO THE EVOLUTION OF A FEW KEY ADAPTATIONS

Vertebrates are distinguished from all other chordates in having a backbone, or **vertebral column,** which forms the chief skeletal axis of the body. This flexible support develops around the notochord and, in most species, largely replaces the notochord during embryonic development. The vertebral column consists of segments called **vertebrae** that are made of cartilage or bone. Dorsal projections of the vertebrae enclose the nerve cord along its length. Anterior to the vertebral column, a **cranium,** or braincase, encloses and protects the brain, the enlarged anterior end of the nerve cord.

The cranium and vertebral column are part of the **endoskeleton.** In contrast to the nonliving exoskeleton of some invertebrates, the endoskeleton is a living tissue that can grow with the animal. In addition to providing support and protection, the endoskeleton provides structures for muscle attachment. Together, skeleton and muscle form a system that permits rapid, efficient movement. Evolution of this system helped

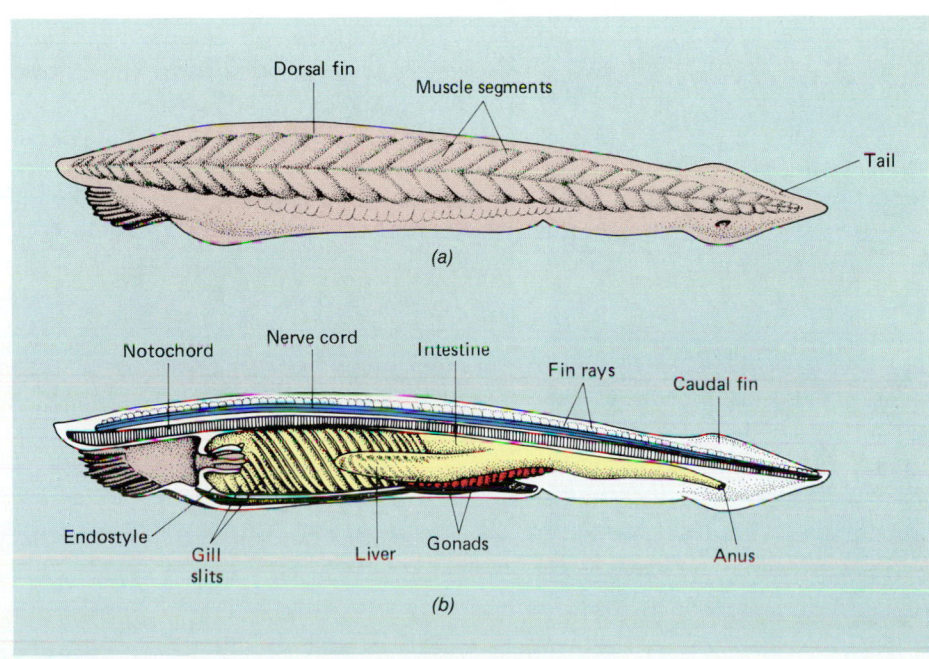

FIGURE 25–3   Amphioxus, a member of the subphylum Cephalochordata. (*a*) External view; (*b*) longitudinal section.

(a)
Dorsal fin
Muscle segments
Tail

(b)
Notochord
Nerve cord
Intestine
Fin rays
Caudal fin
Endostyle
Gill slits
Liver
Gonads
Anus

set the stage for development of the carnivorous life-style typical of many vertebrates.

Two pairs of appendages (pectoral and pelvic) are present in most vertebrates. The fins of fish stabilize them in the water. As vertebrates moved onto the land, jointed appendages evolved that facilitated locomotion.

Recall that in invertebrates there is an evolutionary trend toward cephalization or concentration of nerve cells and sense organs in a definite head. Vertebrate evolution is characterized by *pronounced* cephalization. The brain becomes larger and more elaborate, and its various regions become specialized to perform different functions. Ten or 12 pairs of cranial nerves emerge from the brain and extend to various organs of the body. Vertebrates have well-developed sense organs concentrated in the head: eyes; ears that serve as organs of balance and, in some vertebrates, for hearing as well; and organs of smell and taste.

Vertebrates have a closed circulatory system with a two-, three-, or four-chambered ventral heart. They have paired kidneys that regulate fluid balance. Vertebrates have a complete digestive tract and large digestive glands (liver and pancreas). The sexes are almost always separate.

The vertebrates are less diverse and much less numerous than the insects but rival them in their adaptation to an enormous variety of life-styles. Most exceed the insects in the ability to receive stimuli and react appropriately to them. The 43,000 or so living vertebrate species are assigned to seven classes: class Agnatha, the jawless fish, such as lamprey eels; class Chondrichthyes, the sharks and rays with cartilaginous skeletons; class Osteichthyes, the bony fish; class Amphibia, frogs, toads, and salamanders; class Reptilia, lizards, snakes, turtles, and alligators; class Aves, birds; and class Mammalia, the mammals.

The Jawless Fish Are The Most Primitive Vertebrates

The earliest known fish were small, armored, jawless freshwater fish that lived on the bottom and filtered their food from the water. Today, the only living members of **class Agnatha** are the lamprey eels and hagfishes (Fig. 25–4). Many species of adult lampreys are parasites that live on other fish; they are the only parasitic vertebrates.

Fossil evidence suggests that during the Silurian and Devonian Periods (more than 400 million years ago), one type of jawless fish evolved jaws and paired appendages. These fish shifted from a filter-feeding bottom-dwelling life-style and became active predators. Primitive jawed fish inhabited mainly fresh water. The **placoderms,** a group of primitive jawed fish that

FIGURE 25–4 Class Agnatha. (a) Three lampreys attached to a carp by their suction-cup-type mouths. (b) Suction-cup mouth of adult lamprey. Note the rasp-like teeth. (*a,* Tom Stack/Tom Stack & Associates; *b,* courtesy of Dr. Kiyoko Uehara)

(a)

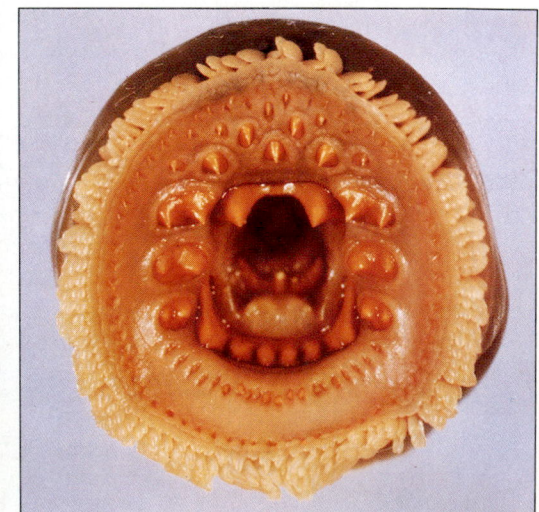

(b)

(a)

(b)

FIGURE 25–5   Chondrichthyes. (*a*) Dorsal view of a skate, *Raja binoculara*. (*b*) Ratfish, *Hydrolagua colliei*. (Charles Seaborn)

lived during the Paleozoic era and are now extinct, are thought to be the ancestors of both the cartilaginous and the bony fish.

The Cartilaginous Fish Include The Sharks, Rays, And Skates

Class Chondrichthyes consists of the cartilaginous fish—sharks, rays, and skates (Fig. 25–5). Almost all marine, these animals have paired jaws and two pairs of fins. The skin contains tooth-like scales known as placoid scales.

Most sharks are streamlined predators that swim actively. The largest sharks, like the largest whales, dine on plankton. The whale shark, which may reach a length of 12 meters, is the largest fish known. Most rays and skates are sluggish, flattened animals that live partly burrowed in the mud. Wave-like movements of its large pectoral fins propel the ray or skate along the bottom as it seeks out a meal of mussels and clams. The stingray has a whip-like tail with a barbed spine at the tip that can inflict a painful wound. The electric ray has muscles modified as electric organs that can discharge enough electricity to stun fairly large fish, as well as human swimmers.

The Bony Fish Are The Most Numerous Vertebrates

The fish most familiar to us are the bony fish of **class Osteichthyes** (Fig. 25–6). Numbering about 20,000 species, the bony fish are the most numerous verte-

brates. They vary greatly in color and shape and range in size from the Philippine goby, which is only about 10 millimeters (0.4 inch) long, to the ocean sunfish, which may reach 907 kilograms (2000 pounds).

The bodies of most bony fish are covered with overlapping bony scales (which develop from the inner layer of the skin). Most bony fish have both median and paired fins, with fin rays made of cartilage. A lateral protective flap of the body wall, the **operculum,** extends posteriorly from the head and covers the gills.

Bony fish generally are **oviparous** (that is, they lay eggs) and fertilize their eggs externally. Fish lay impressive numbers of eggs. The ocean sunfish is said to

FIGURE 25–6   Flame angelfish (*Centropyge loriculus*). (E.R. Degginger)

lay over 300 million! Many species of fish build nests for their eggs and even watch over them. Fish eggs and young offspring often become food for other animals.

The Ray-Finned Fish Gave Rise to Modern Bony Fish

During the Devonian Period, the fish diverged into two major groups: The *ray-finned fish* and the *lobe-finned fish*. The **ray-finned fish** gave rise to most modern bony fish. The ancestors of the ray-finned fish are thought to have had lungs. In them, the lungs became modified as swim bladders, hydrostatic organs that may also store oxygen in some species. The density of the fish body changes when gases are secreted into the swim bladder or absorbed from it. This allows the fish to hover at a given depth.

Descendants of the Lobe-Finned Fish Moved onto the Land

The **lobe-finned fish,** which include the lungfish, are generally thought to have given rise to the land vertebrates (Fig. 25–7). During the Devonian Period, there were frequent seasonal droughts. Ponds dried up, and the lobe-finned fish with lungs adapted for breathing air had a tremendous advantage for survival. Their fleshy lobe fins could support their weight, enabling them to emerge onto dry land and make their way to another pond or stream.

FIGURE 25–7 Ancestors of this lobe-finned fish, a coelacanth, probably gave rise to the amphibians. The paired fins show the basic plan of a jointed series of bones that could evolve into the limbs of a terrestrial vertebrate. Living coelacanths are difficult to observe. Its members (genus *Latimeria*) inhabit deep ocean waters; when brought to the surface, the fish cannot survive the change in atmospheric pressure. (Peter Scoones/Seaphot Ltd.)

The ability to move about, however awkwardly, on dry land also gave these animals the advantage of new food sources—land plants that were already established and land insects that were rapidly evolving. A vertebrate that could survive on land had almost no competition for food. Laying eggs on land away from the many predators in the sea also increased their chances for successful reproduction. Natural selection favored those individuals best adapted for making their way on land, resulting ultimately in the evolution of the amphibians.

Terrestrial (land) vertebrates are referred to as **tetrapods** (four-footed). They include the amphibians, reptiles, birds, and mammals. Most amphibians must return to the water to reproduce, but members of the reptiles, birds, and mammals are fully adapted to life on land.

The Amphibians Were The First Successful Land Vertebrates

The first successful land vertebrates were the **labyrinthodonts,** which were clumsy, salamander-like animals with short necks and heavy, muscular tails. These ancient members of class Amphibia closely resembled their ancestors, the lobe-finned fish; however, they had evolved limbs strong enough to support the weight of the body on land. The labyrinthodonts ranged in size from small, salamander-sized animals to creatures as large as crocodiles. They gave rise to the modern frogs and salamanders and to the earliest reptiles, (the cotylosaurs).

Modern **amphibians** include the frogs, toads, salamanders, and the legless, worm-like caecilians (Fig. 25–8). Although some amphibians are quite successful as land animals and can live in dry environments, most spend at least their early lives in an aquatic environment and most return to the water to reproduce. Eggs and sperm are generally released in water.

The embryos of frogs and toads develop into larvae called **tadpoles.** These larvae have tails and gills and feed on aquatic plants. After a time, the tadpole undergoes metamorphosis. The gills, gill slits, and tail disappear, and the limbs emerge. When these structural modifications are complete, the amphibian can move onto the land.

Adult amphibians do not depend solely on their primitive lungs for the exchange of respiratory gases. Their moist skin, which lacks scales and is richly supplied with blood vessels, also serves as a respiratory surface. Glands within the skin secrete mucus, which helps keep the body moist. The mucus also makes the animal slippery; as a result it can escape from predators. Some amphibians have skin glands that secrete poisonous substances that harm predators.

(a)

(b)

FIGURE 25–8 **Modern amphibians.** (*a*) **Bullfrog** (*Rana catesbeiana*). (*b*) **The red salamander** (*Pseudotriton ruber*) **belongs to the family Plethodontidae. The plethodonts spend their entire lives in fairly moist or humid environments. They have lost their lungs and now rely almost entirely on their moist skin and the membranes that line their mouth and pharynx as organs for gas exchange. The red salamander is common in the eastern United States.** (*a*, Dwight R. Kuhn; *b*, Carolina Biological Supply Company)

The amphibian heart has three chambers—two **atria** (receiving chambers) and a single **ventricle** (a chamber that pumps blood into the arteries) (see Chapter 37). Although there is some mixing of oxygen-rich and oxygen-poor blood, there is a double circuit of blood vessels. One circuit directs blood to the various tissues and organs of the body, while the other circuit conducts blood to the lungs and skin to be recharged with oxygen.

Reptiles Were The Dominant Land Animals For Almost 200 Million Years

Biologists generally agree that the reptiles evolved from the labyrinthodont amphibians during the Carboniferous Period about 300 million years ago. The Mesozoic Era (thought to have ended about 65 million years ago) is known as the Age of Reptiles. During that time, reptiles radiated into an impressive variety of ecologic niches comparable to those of modern mammals. Although some were small, others were the largest, most monstrous animals that ever stalked the earth.

Reptiles Evolved Adaptations to Life on Land

Reptiles are terrestrial animals that do not have to return to water to reproduce. Many adaptations make this life-style possible. The female secretes a protective leathery shell around the egg, which helps prevent the developing embryo from drying out (Fig. 25–9). Fertilization occurs within the body of the female before the shell is added. In this process of **internal fertilization,** the male uses a copulatory organ to transfer sperm into the female reproductive tract.

As the embryo develops within the protective shell, a membrane called the **amnion** forms and surrounds the embryo. The amnion secretes amniotic fluid, providing the embryo with its own private pond. This fluid keeps the embryo moist and also serves as a shock absorber should the egg get bounced about. Like all terrestrial vertebrates, the reptile embryo also has other membranes that protect and support its development.

The reptile body is covered with hard, dry, horny scales that protect the animal from drying out and from predators. This dry skin cannot serve as an organ

FIGURE 25–9 **Hognose snakes hatching. Note the protective leathery shell.** (Zig Leszczynski/Animals Animals)

(a)

(b)

FIGURE 25–10 Modern reptiles. (*a*) **The desert collared lizard,** *Crotaphytus collaris.* (*b*) **Three-toed box turtle.** (*a*, Charles Seaborn; *b*, L. Stone/The Image Bank)

for gas exchange. However, the reptile lungs are better developed for gas exchange than the sac-like lungs of amphibians. Most reptiles have a three-chambered heart, but the ventricle is partly divided by an incomplete partition.

Like fish and amphibians, reptiles lack metabolic mechanisms for regulating body temperature. They are **ectothermic,** which means that their body temperature fluctuates with that of the environment. Some reptiles do have behavioral adaptations that enable them to maintain a body temperature higher than that of the environment. You may have observed lizards basking in the sun. What you may not have known is that the lizard was waiting for its body temperature to rise so that its metabolic rate would increase. Only then could it actively hunt for food. When the body temperature of a reptile is cold, the metabolic rate is low and it is very sluggish (see Fig. 1–5*b*). This probably explains why reptiles are more abundant in warm than in cold climates.

Modern reptiles include the turtles, lizards, snakes, alligators, and crocodiles (see Fig. 25–10). Lizards and snakes are the most common. These animals have rows of scales that overlap like shingles on a roof, forming a protective armor that must be shed periodically. Snakes have a flexible, loosely jointed jaw structure that permits them to swallow animals many times larger than the diameter of their own jaws.

The Large Reptiles Suddenly and Mysteriously Became Extinct

For almost 200 million years, the reptiles were the dominant land animals on earth. Dinosaurs, swimming and flying reptiles, and mammal-like reptiles were abundant and highly successful. Then, quite suddenly during the end of the Cretaceous Period, many of them, including all the dinosaurs, disappeared from the fossil record.

Many theories have been proposed to explain the sudden extinction of the large reptiles. Some blame the newly evolving mammals, which were intelligent and aggressive. The mammals may have successfully competed with the reptiles for food and feasted on reptile eggs. Others think changes in climate may have been responsible for the extinction of the large reptiles. The environment was getting colder at the time they disappeared.

Another theory suggests that about 65 million years ago a large extraterrestrial body, such as a large comet or meteorite, hit or passed near the earth. The impact would have raised a massive cloud of rock particles and dust that blocked out sunlight for several months or even years, resulting in perpetual darkness. Photosynthesis would have come to a halt. Algae and plants would have died out, and both herbivores (plant-eating animals) and carnivores (meat-eating animals) would have declined. Only animals that could survive on decaying vegetation or insects (or on animals that ate these foods) could survive. Mass extinctions occurred not only among reptiles, but also among numerous marine invertebrates.

Gradually, as the dust settled, an iridium-rich layer of clay was deposited, marking the boundary between the Cretaceous and Tertiary Periods. (The element iridium is extremely rare in the earth's crust but is comparatively abundant in extraterrestrial sources. Thus, the presence of an iridium layer is evidence for this theory.) Whatever the cause, by the end of the Mesozoic Era, most species of reptiles had disappeared, leaving only three orders of reptiles. There are many more extinct than living species of reptiles. Long before their decline, however, the reptiles gave rise to both the birds and the mammals.

Birds Are Adapted For Flight

Birds have some characteristics in common with the reptiles. For example, their legs have scales like those of reptiles and they lay eggs. Birds are the only animals with feathers.

About 9000 species of **birds** have been described and classified in 27 orders (Fig. 25–11). Birds live in a wide variety of habitats and can be found on all the continents, most islands, and even the open sea. The largest living birds are the ostriches of Africa, which may be 2 meters (7 feet) tall and weigh 136 kilograms (300 pounds), and the great condors of the Americas, with wingspreads of up to 3 meters (10 feet). The smallest known bird is Helena's hummingbird of Cuba, which is less than 6 centimeters (2.3 inch) long and

weighs less than 4 grams (0.1 ounce). Beautiful and striking colors are found among the birds. Many birds, especially females, are protectively colored by their plumage. During the breeding season, the male usually assumes brighter colors, which help in attracting a mate.

Birds are beautifully adapted for flight. Thought to have evolved from reptilian scales, the feathers of birds are flexible and very strong for their light weight. They protect the body, decrease water loss through the body surface, decrease the loss of body heat, and aid in flying by presenting a plane surface to the air.

The anterior limbs of birds are usually modified for flight; the posterior pair for walking, swimming, or perching. Not all birds fly. Some, such as penguins, have small, flipper-like wings that are used in swimming.

In addition to feathers and wings, birds have many other adaptations for flight. They have compact, streamlined bodies, and the fusion of many bones gives the rigidity needed for flying. Their bones are strong but very light. Many bones are hollow, containing large air spaces. The jaw is light, and instead of teeth, there is a lightweight, horny beak.

FIGURE 25–11 **Modern birds. (a) Male cardinal. (b) Two parrots. (c) Vulturine guineafowl, Africa.** (a, Carl R. Sams II/Marvin Dembinsky, Jr., Photography Associates; b, Stock Imagery; c, Stan Osolinski/Marvin Dembinsky, Jr., Photography Associates)

(a)

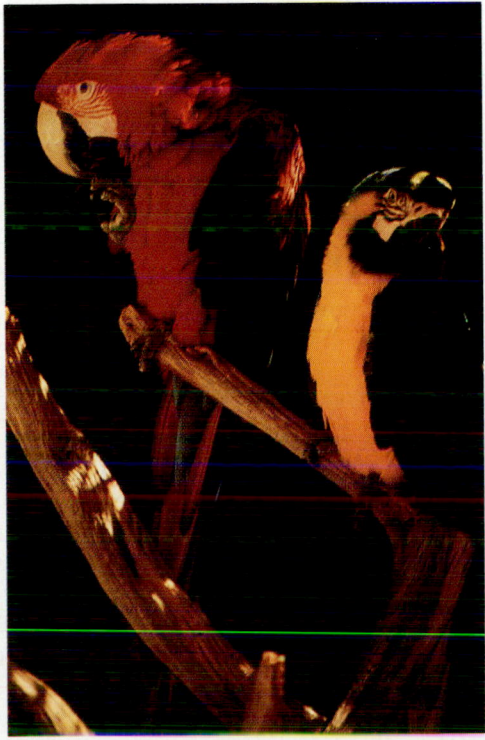
(b)

(c)

The lungs of a bird are very efficient, with thin-walled extensions, called air sacs, that occupy spaces between the internal organs and within certain bones. Birds, like mammals, have four-chambered hearts and a double circuit of blood flow in which oxygen-poor blood is pumped into the lungs, recharged with oxygen, returned to the heart, and pumped out again to the tissues, where it releases oxygen. The very efficient respiratory and circulatory systems deliver sufficient oxygen to the cells to permit a high metabolic rate. This is necessary for the tremendous muscular activity required for flying. Some of the heat generated by metabolic activities is used to maintain a constant body temperature. This ability permits metabolic processes to proceed at constant rates and enables birds to remain active in cold climates. Birds and mammals are the only modern animals that can maintain a constant body temperature internally by means of metabolic mechanisms. They are sometimes called "warm blooded," but **endothermic** is the preferred term.

Mammals Evolved From Reptiles

Mammals, the animals most familiar to us, evolved from a group of reptiles called **therapsids** during the Triassic Period some 200 million years ago. The therapsids were carnivores with differentiated teeth (a mammalian trait) and legs adapted for running (Fig. 25–12). Some of them may have been endothermic, and some may even have had fur. The fossil record indicates that the earliest mammals were small, about the size of a mouse or shrew.

How did the mammals manage to coexist with the reptiles during the 160 million or so years that the reptiles ruled the world? Many adaptations permitted the mammals to compete for a place on the earth, but perhaps most important was that the early mammals specialized in being inconspicuous. They were **arboreal** (tree-dwelling) and **nocturnal** (active at night), searching for food (mainly insects and plant material and

perhaps reptile eggs) while the reptiles slept. The large eye sockets seen in fossil species indicate that these animals had the large eyes characteristic of present-day nocturnal mammals.

As reptiles died out, the mammals began to move into their abandoned territories and ecological niches. During this time, the flowering plants, including many trees, underwent adaptive radiation, providing new habitats, sources of food, and protection from predators. Larger forms and numerous varieties of mammals evolved. During the early Cenozoic Era (perhaps 50 million years ago), the mammals underwent adaptive radiation, becoming widely distributed and adapted to an impressive variety of ecological niches.

Three main lines of mammals had evolved by the end of the Cretaceous Period: (1) the first line is thought to have given rise to the egg-laying **monotremes,** like the duck-billed platypus; (2) the second group were **marsupials,** pouched mammals that were the ancestors of modern-day kangaroos and opossums; and (3) the third group were small, shrew-like mammals that ate insects and lived a nocturnal existence in the trees. All of the modern-day **placental mammals** are thought to have evolved from these shrew-like mammals.

Today mammals inhabit virtually every corner of the earth—on the land, in the water, and even in the air. Their sizes range from the tiny pigmy shrew, weighing less than an ounce, to the blue whale, which may weigh more than 90,000 kilograms (100 tons) and is the largest animal ever known.

Distinguishing mammalian features are the presence of hair, **mammary glands,** which produce milk for the young, and the differentiation of teeth into incisors, canines, premolars, and molars. A muscular diaphragm helps move air in and out of the lungs. Like the birds, but unlike other vertebrate groups, mammals are endotherms, which means that they maintain a constant body temperature. This process is supported by the covering of hair, which serves as insulation, by the four-chambered heart and double circulation, and by the presence of sweat glands. Contributing significantly to the success of the mammals is the nervous system, which is more highly developed than in any other group of animals.

FIGURE 25–12 A mammal-like reptile, *Lycaenops,* **from the late Permian period in South Africa.** (From A.S. Romer and T.S. Parsons)

Monotremes Are Mammals that Lay Eggs

The duck-billed platypus (*Ornithorhynchus*) and the spiny anteater (*Tachyglossus*) are monotremes native to Australia (Fig. 25–13). The females lay eggs, which may be carried in a pouch on the abdomen or kept warm in a nest. When the young hatch, they are nourished with a fluid from specialized glands on the abdomen (primitive mammary glands).

FIGURE 25–13 The spiny anteater, *Tachyglossus*, is a monotreme. (Photograph by Robert Anderson, reprinted with permission of Hubbard Scientific Company)

Marsupials Are Pouched Mammals

Marsupials are pouched mammals such as kangaroos and opossums (Fig. 25–14). Embryos begin their development in the mother's uterus (womb), where they are nourished by yolk and from fluid in the uterus. After a few weeks, still in a very undeveloped stage, the young are born and crawl to the **marsupium** (pouch), where they complete their development. Each of the young attaches itself by its mouth to a mammary gland nipple in the marsupium and is nourished by its mother's milk.

Like the monotremes, the marsupials are found mainly in Australia. The opossum is the only common marsupial in North America, although a few species inhabit South America. At one time, marsupials may have inhabited much of the world but were replaced by the placental mammals. Australia became geographically isolated from the rest of the world before placental mammals reached it, so the marsupials remained the dominant type of mammal on that continent. Their evolution proceeded in many directions and fitted them for many different life-styles, paralleling the evolution of placental mammals elsewhere. Thus, in Australia and adjacent islands, we find marsupials that correspond to our placental wolves, bears, rats, moles, flying squirrels, and even cats (Fig. 17–12). There are also a number of forms without placental counterparts, such as the kangaroo and wallaby.

FIGURE 25–14 Marsupials. (*a*) Koala bears are marsupials specialized to live in trees. Their diet is restricted to leaves of a very few species of eucalyptus. This young koala has outgrown the pouch and will ride on its mother's back until it is old enough to go out on its own. (*b*) The kangaroo is a marsupial native to Australia. Gray kangaroo with joey. (*c*) Young marsupials are born in a very immature state; kangaroo soon after birth. (*d*) The young continue to develop in the safety of the marsupium. (*a*, E.R. Degginger; *b*, Stock Imagery; *c*, Robert Anderson, reprinted with permission of Hubbard Scientific Company; *d*, Robin Lewis, Coastal Creations)

(a)

(b)

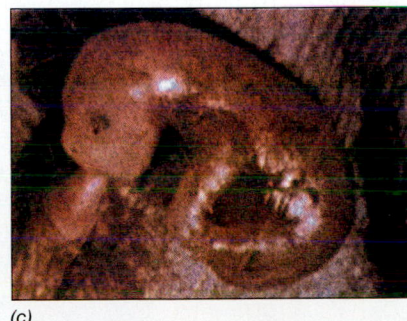

(c)

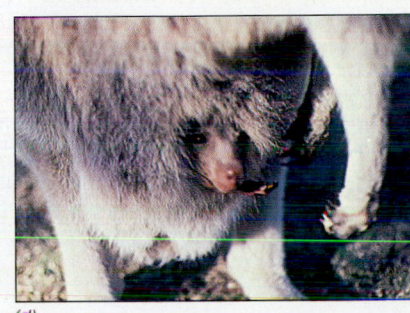

(d)

(a)

(b)

FIGURE 25–15   Placental mammals. (*a*) Raccoon and young in hollow tree trunk. (*b*) A common dolphin, *Delthinus delphis*. The streamlined body and overall fish-like form are adaptations strikingly similar to those possessed by some of the oceanic reptiles of the Mesozoic. Dolphins belong to the order Cetacea. (*a*, Steve Maslowski/Visuals Unlimited; *b*, E.R. Degginger)

FIGURE 25–16   A vertebrate family tree.

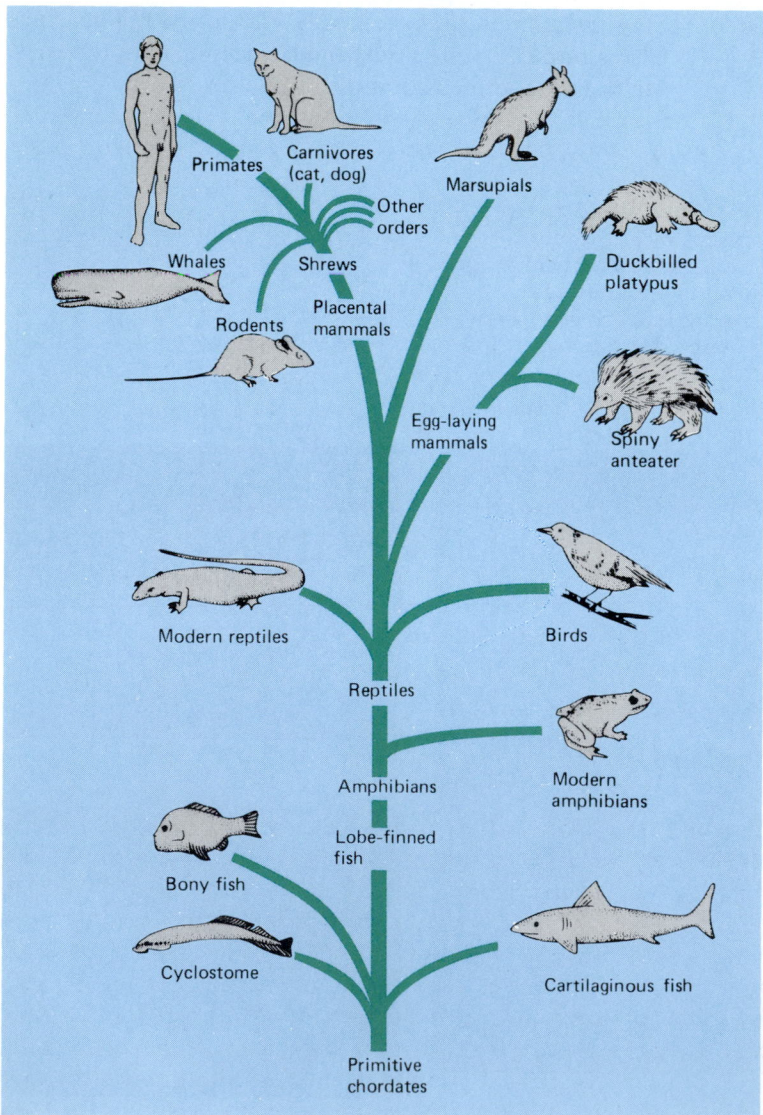

Placental Mammals Complete their Embryonic Development within the Mother's Body

Placental mammals are the animals most familiar to us (Fig. 25–15). In these mammals, tissues of the embryo and of the mother form an organ of exchange, the **placenta.** This organ enables the mother to supply the developing embryo with food and oxygen and to remove its wastes. As a result, the young can remain within the body of the mother until embryonic development is complete. At birth, the offspring of placental mammals are at a more mature stage than those of marsupials. Indeed, among some species, the young can walk around and begin to interact with other members of the group within a few minutes of birth.

There are about 17 living orders of placental mammals. These mammals have adaptations that allow them to fly, swim, and occupy a multitude of terrestrial environments. Their range of adaptations is comparable to those of the reptiles during the Mesozoic Era. Human beings, along with the lemurs, monkeys, and apes, belong to the order Primates. The probable evolutionary relationships of the vertebrates are illustrated in Figure 25–16.

FIGURE 25–17   The common tree shrew most resembles the ancient insectivores that gave rise to the primates. (Warren Garst/Tom Stack & Associates)

■ PRIMATES EVOLVED FROM SHREW-LIKE MAMMALS

For nearly a century after Darwin's *Origin of Species,* fossil evidence of human ancestry was rather sparse and unsatisfactory. Research over the last four or five decades, however, especially in East Africa, has provided us with some reasonable answers to the question, "Where did we come from?" Human ancestry can be traced back some 65 million years to the earliest primates.

The first primates evolved from the arboreal shrew-like mammals that had appeared during the "Age of Reptiles." The living organism that most resembles these ancient mammals is the tree shrew of South America (Fig. 25–17). Because of their ancestry, most primates have adaptations for an arboreal existence even if they do not live on trees. Among the most significant features of primates are the five digits on their hands and feet, four digits plus an **opposable thumb.** This enables primates to grasp objects, such as tree branches. Nails provide a protective covering for the tips of the digits. The fleshy pads at the ends of the digits are sensitive and dexterous.

Another adaptation to life in the trees is long, slender limbs that rotate freely at hips and shoulders. These allow primates full mobility for climbing and searching for food in the treetops. The location of the eyes in front of the head, along with a shortened snout, is an adaptation that provides stereoscopic, or three-dimensional, vision. This is essential for arboreal animals, as an error in depth perception might cause a fatal fall. In addition to sharp sight, hearing is acute in primates, although the sense of smell is relatively poor compared with that of most other mammals.

Primates share several other characteristics, including a complex social behavior. Some biologists think that the learning that was associated with social interactions in primate societies may have been a factor in the evolution of increasingly large, more complex brains. Primate reproduction usually results in only one offspring, which is helpless and requires a long period of nurturing and protection.

There are two suborders in the order **Primates**. The **prosimians** (which means "before apes") include the lemurs, lorises, and tarsiers. The prosimians were the first primates to evolve. The **anthropoids,** comprising the monkeys, apes, and humans, are primates with larger brains, particularly the cerebrum.

The Fossil Record Suggests General Trends In Hominid Evolution

The **hominid** line (humans and their ancestors) separated from the ape line approximately 3.5 or 4 million years ago (see Fig. 25–18). General trends in human evolution are evident from the fossil record, but we do not have enough evidence to make specific conclusions. There are simply too few early hominid fossils, and the ones we have are represented by only a few bones. Moreover, it is impossible to determine many aspects of early hominid biology, appearance, or behavior from fossilized bones. Nevertheless, it is evident

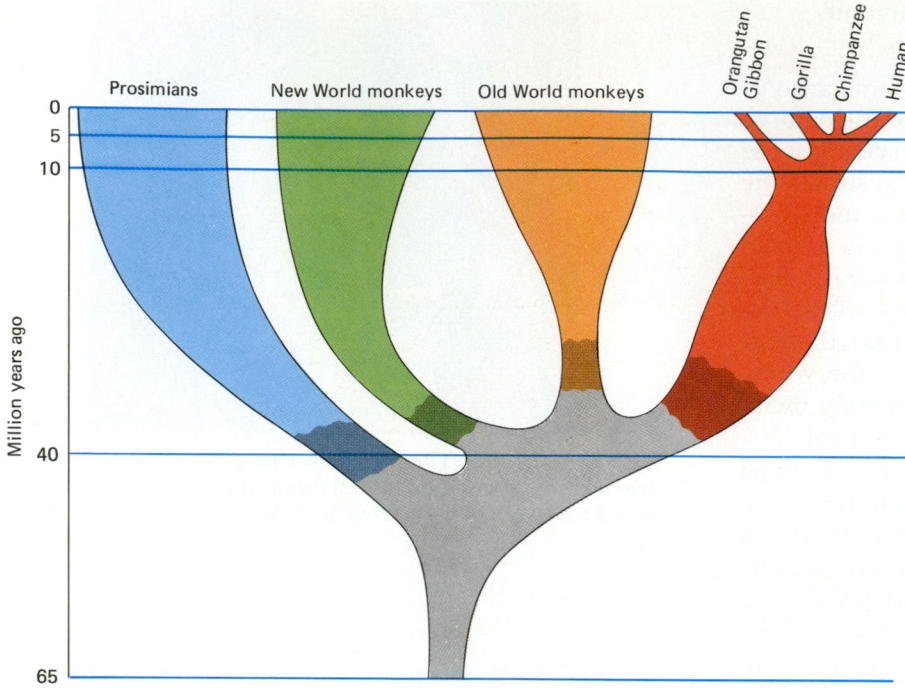

Million years ago

Prosimians   New World monkeys   Old World monkeys   Orangutan   Gibbon   Gorilla   Chimpanzee   Human

FIGURE 25–18 One interpretation of hominid evolution. Although the overall picture of hominid evolution is well-established, details, including precise dates for the divergence of lines, are not firmly fixed.

that early hominids evolved a **bipedal** (two-footed) posture before their brains enlarged. This is an example of **mosaic evolution,** which refers to two traits evolving independently of one another and at different rates. In order to understand the evolutionary progression from the earliest hominids to modern humans, we must examine some of the characteristics of the skeleton and skull.

Compared to an ape's skeleton, the human skeleton possesses distinct features that result in our ability to stand erect and walk on two feet (Fig. 25–19). These differences also reflect the change in habitat for early hominids, from the forest to the ground. The human spine has a greater curvature than an ape's spine, resulting in better balance and weight distribution. The human pelvis is shorter and more rounded, permitting

FIGURE 25–19 Comparison of gorilla and human skeletons. Note the skeletal adaptations for bipedalism in humans.

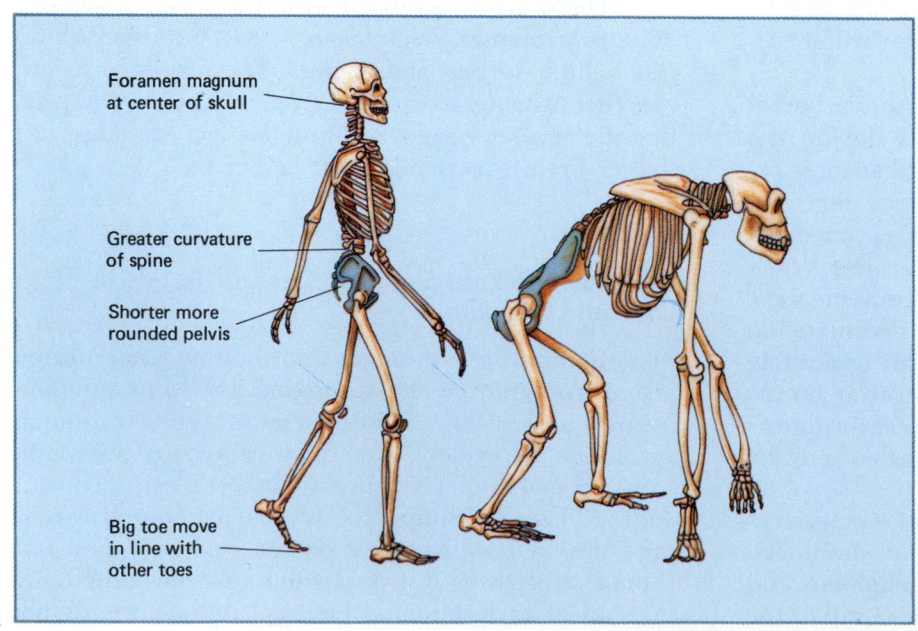

Foramen magnum at center of skull

Greater curvature of spine

Shorter more rounded pelvis

Big toe move in line with other toes

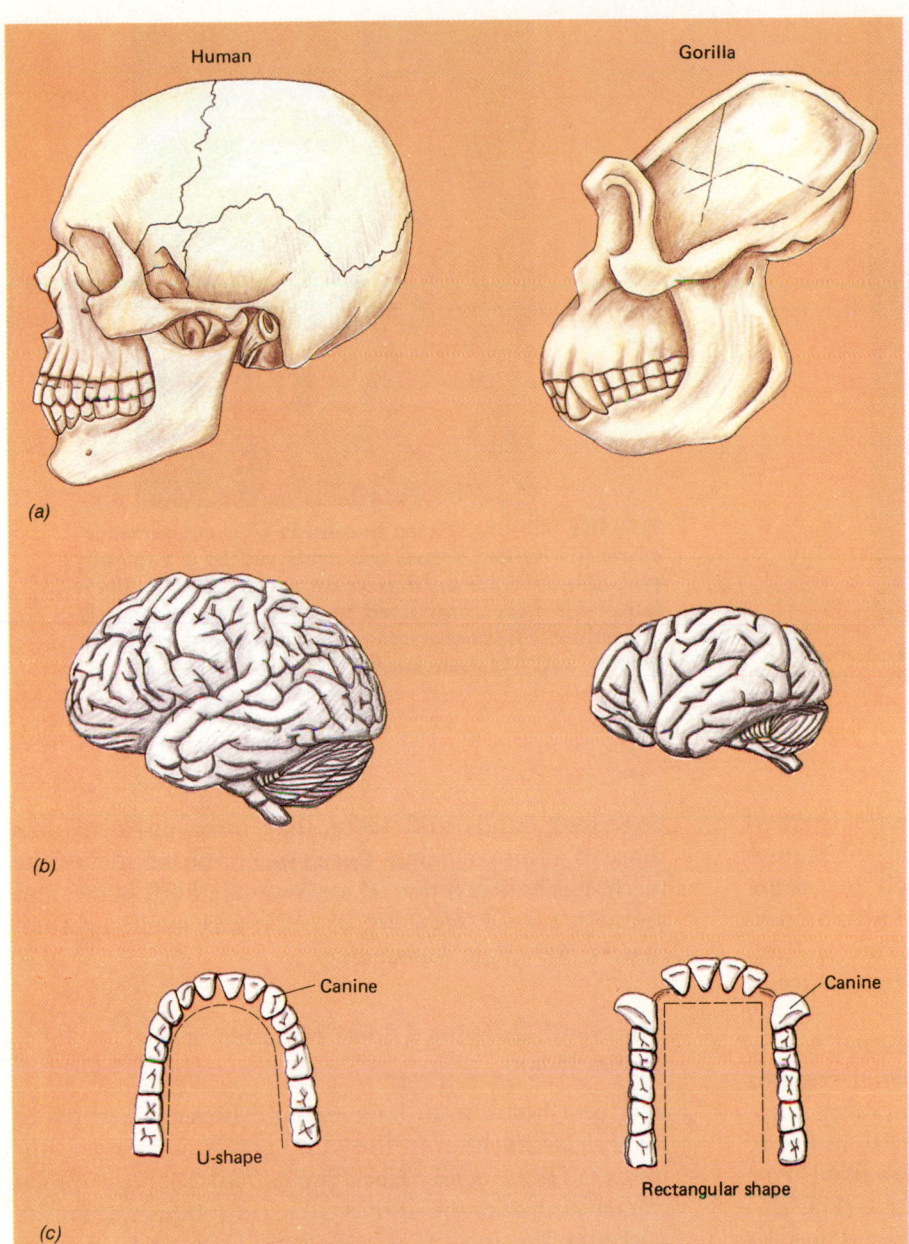

Human

Gorilla

(a)

(b)

Canine

U-shape

(c)

Canine

Rectangular shape

FIGURE 25–20 Comparison of features of the ape and human head. (*a*) The ape skull has pronounced supraorbital ridges. Note how the human skull is flatter in the front and has a more pronounced chin. (*b*) The human brain, particularly the cerebrum, is larger than that of an ape. (*c*) The human jaw is structured so that the teeth are arranged in a U-shape. Human canines are reduced in size compared to ape canines.

a better attachment of the muscles used for upright walking. In apes, the foramen magnum (the opening in the base of the skull where the spinal cord emerges), is located at the back of the skull. In contrast, in humans, the foramen magnum is located in the middle of the bottom of the skull. As a result, the head can be positioned for erect walking. An increase in the length of the legs relative to the arms, and movement of the big toe so it is in line with the rest of the toes were further adaptations in the early hominids for walking on two feet.

Another major trend in human evolution is an increase in the size of the brain relative to the size of the body (Fig. 25–20). In addition, the ape skull possesses prominent bony ridges above the eye sockets, called **supraorbital ridges,** that are lacking in human skulls. Human faces are flatter than ape faces, and their jaws are different. Apes have larger teeth than humans, and their canine teeth are especially large.

The Earliest Hominids Belong to Genus *Australopithecus*

Human evolution first occurred in Africa. The earliest hominids belong to the genus *Australopithecus*, or "Southern ape," and appeared approximately 3.8 mil-

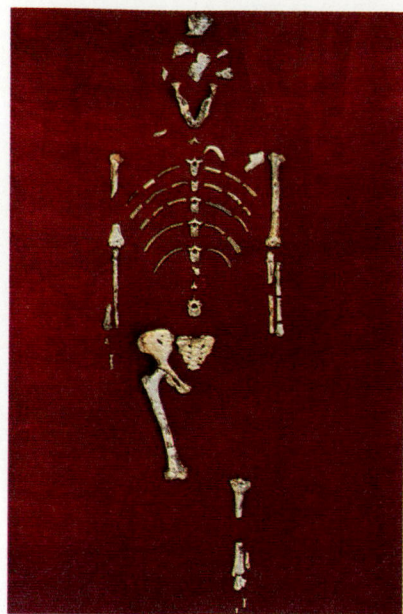

FIGURE 25–21   The skeletal remains of Lucy, a hominid approximately 3.5 million years old. (Institute of Human Origins)

FIGURE 25–22   Three hominids (*Australopithecus afarensis*) walked across ash scattered by a volcanic eruption over 3.6 million years ago in Africa. Their footprints were compacted by a rain shower shortly thereafter. The footprints were discovered by Mary Leakey and her associates.

lion years ago. The actual number of species assigned to this genus is a matter of debate. It is very difficult to decide whether differences in the relatively few skeletal fragments that have been discovered indicate individual variation within a species or separate species. Most biologists recognize between two and four species of australopithecines.

The most ancient hominids are assigned to the species *A. afarensis*. Several fossils of skeletal remains have been discovered, including a remarkably complete skeleton named Lucy (Fig. 25–21). In addition, fossil footprints of three individuals that walked over 3.6 million years ago were discovered in 1976 (Fig. 25–22). The footprints plus pelvis, leg, and foot bones indicate that the development of an upright posture and bipedalism occurred early in human evolution.

Australopithecus afarensis was a small hominid, approximately 3 feet tall. Its face projected forward, and its ape-like skull covered a small brain. The cranial capacity was 450 to 500 cubic centimeters compared to a modern human cranial capacity of 1400 cubic centimeters. Even when differences in body size are taken into account, *A. afarensis* still had a small brain. The number and arrangement of teeth were primitive and included long canines.

Many scientists think *A. afarensis* evolved into the more advanced australopithecine, *A. africanus*, which appeared approximately 3 million years ago. The first *A. africanus* fossil was discovered in South Africa in 1924, and since then a number of others have been found. This rather small hominid walked erect and possessed hands and teeth that were distinctly like those of a human being. Based on characteristics of the teeth, it is believed that *A. africanus* ate both plants and animals. Like *A. afarensis*, its brain was small, approximately 500 cubic centimeters.

Homo habilis is the Oldest Member of Genus *Homo*

The first hominid to have enough human features to justify classification in the same genus as modern humans is *Homo habilis*. *Homo habilis* had a larger cranial capacity, an average of 650 cubic centimeters, than the australopithecines (Fig. 25–23). This early human appeared approximately 1.8 million years ago and persisted for over half a million years. Fossils of *H. habilis* have been found in numerous areas in Africa. These sites contain the first primitive tools, stones that had been chipped to make sharp edges for cutting or scraping. Although other primates occasionally use tools, many consider *H. habilis* the first species to have consciously designed them.

The relationship between the australopithecines and *H. habilis* is not clear. Using physical characteristics as evidence, some biologists believe that the australopithecines were ancestors of *H. habilis*. Others feel that *H. habilis* and *A. africanus* were contemporaries for much of their existence and that *H. habilis* was in a direct line to humans, but *A. africanus* was not (Fig. 25–23). Discoveries of additional fossils may help clarify their relationship.

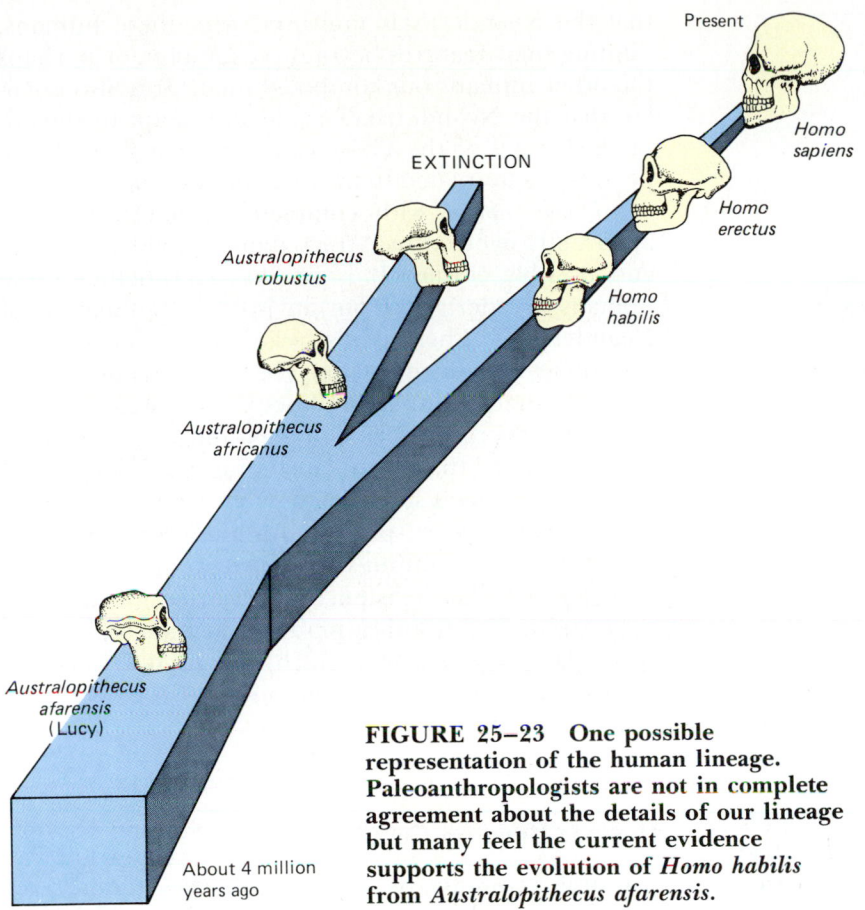

Present

EXTINCTION

Homo sapiens

Homo erectus

Australopithecus robustus

Homo habilis

Australopithecus africanus

Australopithecus afarensis (Lucy)

About 4 million years ago

FIGURE 25–23 One possible representation of the human lineage. Paleoanthropologists are not in complete agreement about the details of our lineage but many feel the current evidence supports the evolution of *Homo habilis* from *Australopithecus afarensis.*

Numerous Fossils of *Homo erectus* Have Been Discovered

There is more agreement on interpretation of the fossils classified as *Homo erectus,* as numerous fossils have been found. *Homo erectus* evolved in Africa as did the other hominids, but migrated into Europe and Asia. For this reason, the oldest fossils of *H. erectus,* 1.5 million years old, are found in Africa, and the later ones are more widely distributed in the Old World. The Peking man and Java man discovered in Asia were later examples of *H. erectus,* which existed until approximately 200,000 years ago.

Homo erectus was bipedal, fully erect, and taller than *H. habilis.* Its skull, although larger, did not possess totally modern features. It retained the heavy supraorbital ridge and projecting face that is characteristic of the apes. During the course of its existence, the *H. erectus* brain got progressively larger.

The increase in mental faculties associated with an increase in brain size enabled these early humans to make more advanced stone tools, including hand axes and other tools that have been interpreted as choppers, borers, and scrapers. Their intelligence enabled them to survive in areas that were cold. *Homo erectus*

wore clothing, built fires, and lived in caves or shelters. It is not known for sure whether they were hunters or scavengers. To date, no weapons have been unearthed at their sites.

Homo sapiens Appeared about 200,000 Years Ago

Humans that have modern enough features to be classified within our species appeared approximately 200,000 years ago. Their brains continued to enlarge, developing from 850 cubic centimeters in earliest individuals to the current cranial capacity of 1400 cubic centimeters.

One of the earliest groups of *H. sapiens* was the **Neanderthals.** They were first discovered in the Neander Valley in Germany, but had a widespread distribution. These early humans had a short, sturdy build. Their brains were slightly larger than modern *H. sapiens,* but their faces still projected slightly, with less pronounced chins and heavy supraorbital ridges (Fig. 25–24).

Neanderthal tools, including spear points, were more sophisticated than those of *H. erectus.* Studies of

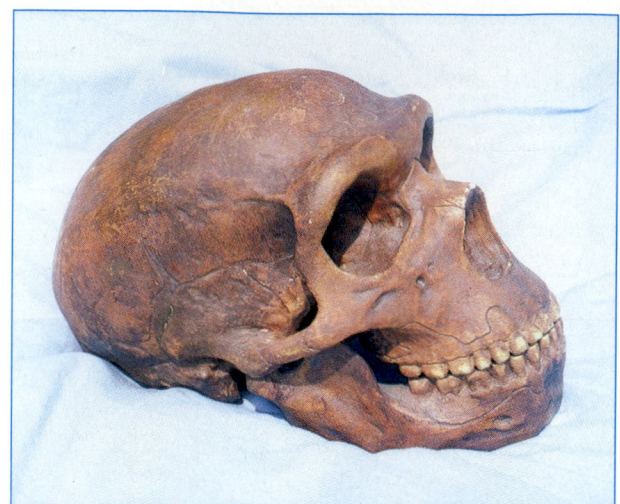

FIGURE 25–24   Neanderthal skull. Note the very heavy supraorbital ridge and the protruding face. The size of the Neanderthal brain was greater than that of modern humans.

Neanderthal sites indicate their culture included hunting for large animals. The existence of skeletons that were old or had healed fractures demonstrates that they cared for the elderly and sick, which is an example of advanced social cooperation. They apparently had rituals, possibly of religious significance, and buried their dead. The presence of food, weapons, and flowers in the graves indicates they had the abstract concept of an afterlife.

The disappearance of the Neanderthals is a mystery. Other groups of *H. sapiens* with more modern features coexisted with the Neanderthals. It is possible that the Neanderthals interbred with these humans, diluting their features beyond recognition, or perhaps the other humans out-competed them. It is also possible that the Neanderthals could not adapt to the climate changes of the Pleistocene and that their disappearance is unrelated to the presence of other humans.

Homo sapiens with completely modern features apparently evolved in Africa about 40,000 years ago, and possibly earlier. It is not known whether these humans are derived from an isolated population of Neanderthals. Their skulls lacked heavy brow ridges and possessed a distinct chin. The **Cro-Magnon** culture in France and Spain exemplifies these humans. Their weapons and tools were complex and often made of materials other than stone, including bone, ivory, and wood. They made stone blades that were very sharp. They developed art, possibly for ritualistic purposes, including cave paintings, engraving, and sculpture (Fig. 25–25). The existence of a variety of complex tools and art is an indication that they may have possessed language abilities, which they used to transmit their culture to younger generations.

Humans Undergo Cultural Evolution

Cultural evolution is the progressive addition of knowledge to the human experience. Human culture is dynamic; it is modified as we obtain new knowledge. Cultural evolution is generally divided into three stages: (1) the development of hunter/food-gatherer societies; (2) the development of agriculture; and (3) the Industrial Revolution.

FIGURE 25–25   The Cro-Magnon people painted animals on cave walls in Europe. These are some of the earliest representations of human art and have been interpreted as having a religious significance, possibly for guaranteeing a successful hunt. (John D. Cunninghan/Visuals Unlimited)

Early humans were nomadic hunter/gatherers who relied on what was available in the environment. As the resources in a given area were exhausted or as the population increased, they would migrate to a different area. These societies required a division of labor and the ability to make tools and weapons, which are needed not only to kill game, but also to scrape hides, dig up roots and tubers, and cook food. Although we are not certain when hunting was incorporated into human society, we do know that it declined in importance approximately 15,000 years ago. This decline may have been due to a decrease in large animals that was triggered in part by a change in climate. A few isolated groups of hunter/gatherer societies survived to the 20th century, including the Mountain Lapps of Scandinavia and the Bushmen of Australia.

Development of Agriculture Resulted in a More Dependable Food Supply

Evidence that humans had begun to cultivate crops approximately 10,000 years ago includes the presence of agricultural tools and plant material at archaeological sites. Agriculture, keeping animals as well as cultivating plants, resulted in a more dependable food supply. It appears from recent archaeological evidence that agriculture arose in several steps. Although there is a lot of variation from one site to another, plant cultivation usually occurred first in combination with hunting. Animal domestication followed at a later period. Agriculture, in turn, often led to more permanent dwellings, as considerable time was invested in growing crops in one area. Often, villages and cities grew up around the farmlands.

Producing food agriculturally was more time consuming than obtaining food by hunting and gathering; however, it was also more productive. In hunter/gath-

erer societies, everyone shares the responsibility of obtaining food. In agricultural societies, fewer people were needed to provide food for everyone, freeing some people to pursue other endeavors, including religion, art, and various crafts.

The Industrial Revolution Has Had Profound Impact on the Ecosphere

Cultural evolution has had a profound effect on human society and on other life forms. The Industrial Revolution, which began in the 18th century, resulted in the concentration of people in urban areas where centers of manufacturing are located. Advances in agriculture encouraged this, as fewer and fewer people were needed to provide food for everyone. The spread of industrialization has increased the demand for natural resources to supply the raw materials for industry. Human population has increased so dramatically that some biologists fear it is approaching the carrying capacity of the earth. As it is, millions of people are malnourished or undernourished. Almost all the arable land on earth is under cultivation.

Cultural evolution has resulted in large-scale disruption and degradation of the environment. Tropical rain forests and other natural environments are being eliminated at a rapid rate. Soil, water, and air pollution now impact on most of the ecosphere. Desertification is increasing as plant cover is removed from marginal, arid lands so they can be cultivated. Many plant and animal species cannot adapt to the rapid changes humans are causing to the environment and are rapidly becoming extinct. The decrease in biological diversity due to extinction is alarming (Fig. 25–26).

On a positive note, we are becoming increasingly aware of the negative impact we have had on the environment, and we have the intelligence to modify our

FIGURE 25–26 This black rhinoceros, *Ngorongoro Crater*, Tanzania, is a member of an endangered species. (E.R. Degginger)

behavior to improve these conditions. By educating younger generations, we can help them develop environmental sensitivity. If we succeed, cultural evolution could be our salvation rather than our destruction.

■ CHAPTER SUMMARY

I. At some time in its life cycle, a chordate has a notochord, a dorsal tubular nerve cord, and pharyngeal gill grooves.

II. Phylum Chordata consists of three subphyla: Urochordata, Cephalochordata, and Vertebrata.

III. Subphylum Urochordata consists of the tunicates, which are sessile, filter-feeding marine animals that have tunics made of cellulose.

IV. Subphylum Cephalochordata consists of the lancelets, which are small, segmented fish-like animals.

V. Subphylum Vertebrata includes animals with a vertebral column, cranium, pronounced cephalization, differentiated brain, muscles attached to an endoskeleton for movement, and two pairs of paired appendages.

VI. The vertebrate classes include jawless fish, cartilaginous fish, bony fish, amphibians, reptiles, birds, and mammals.

A. The jawless fish (class Agnatha) include the lamprey eels and hagfishes.

B. Class Chondrichthyes, the cartilaginous fish, consists of the sharks, rays, and skates.

C. Class Osteichthyes, the bony fish, includes about 20,000 species of freshwater and saltwater fish. Most modern bony fish are ray-finned fish with swim bladders.

D. Modern amphibians include the salamanders, frogs, toads, and worm-like caecilians.

1. Most amphibians return to the water to reproduce. Frog embryos develop into tadpoles, which undergo metamorphosis to become adults.

2. Amphibians use their moist skin as well as lungs for gas exchange. They have a three-chambered heart with systemic and pulmonary circulations, and they have mucous glands in the skin.

E. Class Reptilia includes turtles, lizards, snakes, and alligators.

1. Reptiles are true terrestrial animals.

2. Fertilization is internal. Most reptiles secrete a leathery protective shell around the egg. The embryo develops an amnion and other extraembryonic membranes, which protect it and keep it moist.

3. A reptile has a dry skin with horny scales, lungs with many chambers, and a three-chambered heart.

F. Reptiles dominated the earth during the Mesozoic Era. During the Cretaceous period, most of them, including all of the dinosaurs, became extinct.

G. Birds (class Aves) have many adaptations for flight, including feathers, wings, and light, hollow bones containing air spaces.

1. Birds have a four-chambered heart and very efficient lungs.

2. Birds maintain a constant body temperature and have a high metabolic rate.

H. Mammals have hair, mammary glands, and differentiated teeth and maintain a constant body temperature. They have a highly developed nervous system and a muscular diaphragm.

1. Monotremes, mammals that lay eggs, include the duck-billed platypus and the spiny anteater.

2. Marsupials are pouched mammals such as kangaroos and opossums. The young are born in an immature stage and complete their development in the marsupium.

3. Placental mammals are characterized by an organ of exchange, the placenta, that develops between the embryo and the mother. This organ supplies oxygen and nutrients to the fetus and enables it to complete development within the uterus. There are about 17 living orders of placental mammals.

VII. Primates evolved from small, arboreal, shrew-like mammals.

VIII. The hominid line separated from the ape line approximately 3.5 to 4 million years ago.

A. The earliest hominids belong to the genus *Australopithecus.* The australopithecines walked on two feet, a human feature.

B. *Homo habilis* was an early hominid that had some human features the australopithecines lacked, including a slightly larger brain. *Homo habilis* fashioned tools from stone.

C. *Homo erectus* had a larger brain than *H. habilis,* made more sophisticated tools, and discovered how to use fire.

D. *Homo sapiens* appeared approximately 200,000 years ago.

1. The brain continued to enlarge during their evolution.

2. It is likely that modern *H. sapiens* evolved

from a common African ancestor.
IX. Cultural evolution is the progressive addition of knowledge to the human experience.
  A. It is made possible by an evolutionary in-

crease in brain size in humans.
  B. Two significant advances in cultural evolution were the development of agriculture and the Industrial Revolution.

■ POST-TEST

1. The three distinguishing characteristics of a chordate are a _____, a dorsal, _____ _____, and pharyngeal _____ _____.

2. _____ are sessile, marine chordates often mistaken for sponges.

3. Vertebrates are distinguished from all other animals in having a _____ _____; anterior to this structure, a _____ encloses and protects the brain.

4. Modern fish are thought to have descended from the _____ fish; the lobe-finned fish are credited with being the ancestors of the _____.

5. Vertebrates that have a three-chambered heart and moist skin are the _____.

6. The amnion is an adaptation to _____ life; it secretes a fluid that _____.

7. Reptiles gave rise to the _____ and the _____.

8. Monotremes are mammals that _____ _____; marsupials are distinguished by their _____.

15. Which of the following features is more like humans than apes: (a) large supraorbital ridges, (b) brachiation, (c) foramen magnum located toward back of skull, (d) opposable big toe, (e) small canines?

16. The first hominid to walk erect on two feet was *Homo* _____.

17. The earliest hominid to be placed in the genus *Homo* was *H.* _____.

Match the answer in Column B with the description in Column A; there may be more than one answer for each question.

Column A
_____ 9. Have amnion
_____ 10. Have hair
_____ 11. Have four-chambered heart (two atria and two ventricles)
_____ 12. Body covered with hard, dry, horny scales
_____ 13. Bones contain air spaces; no teeth
_____ 14. Have pharyngeal gill slits at some time in life cycle

Column B
a. Bony fish
b. Amphibians
c. Reptiles
d. Birds
e. Mammals
f. All of the preceding

■ REVIEW QUESTIONS

1. What are the three principal distinguishing characteristics of a chordate? How are these evident in a lancelet? In a human?

2. What characteristics distinguish the vertebrates from the rest of the chordates?

3. How do lampreys and hagfishes differ from other fishes? Of what economic importance are agnathans?

4. Compare the skins of sharks, frogs, snakes, and mammals.

5. Give the location and function of each of the following: (a) swim bladder, (b) placenta, (c) amnion.

6. Give the phylum, subphylum, and class, for each of the following animals: (a) human being, (b) turtle, (c) lamprey eel, (d) lancelet, (e) shark, (f) whale, (g) frog, (h) pelican, (i) bat.

7. Why are monotremes considered to be more primitive than other mammals? Some paleontologists consider

them to be therapsid reptiles rather than mammals. Give arguments for and against this position.

8. Which vertebrate groups maintain a constant body temperature? How do they accomplish this? Why is this advantageous?

9. Which are more specialized animals—birds or mammals? Explain your answer.

10. According to current evolutionary theory, give the significance of each of the following: (a) placoderms, (b) labyrinthodonts, (c) therapsids.

11. Cite one anatomical feature and one behavioral feature that distinguishes each of the following from its immediate ancestor: (a) *Australopithecus afarensis,* (b) *Homo erectus,* (c) *Homo sapiens* (Neanderthal), (d) *Homo sapiens* (modern).

12. How is cultural evolution related to biological evolution? (*Hint:* The evolution of what biological characteristic contributed to cultural evolution?) How has cultural evolution impacted the ecosphere?

■ READINGS

Alldredge, A.L., and L.P. Madin. "Pelagic Tunicates: Unique Herbivores in the Marine Plankton," *Bioscience;* 655–663, September 1982. An account of the ecology and unique adaptations of pelagic tunicates.

del Pino, E.M. "Marsupial Frogs," *Scientific American* 260 (5):1989. Marsupial frogs have a long incubation period within the mother's body, resembling pregnancy in mammals. However, the eggs and embryos of these frogs are bird-like.

Eastman, J.T., and A.L. DeVries. "Antarctic Fishes," *Scientific American* 255 (5):1986. Most species of fish died out when the Antarctic Ocean became icy cold, but fish in the suborder Notothenioidei survive by making biological antifreezes and conserving energy.

Griffiths, M. "The Platypus," *Scientific American* 258 (5):84–91, 1988. Everything you might want to know about this interesting monotreme. The platypus has mechanoreceptors and electroreceptors on its beak for detecting prey.

Klein, R., *The Green World: An Introduction to Plants and People,* 2nd ed. New York, Harper & Row, 1987. Includes a detailed description of the origin of agriculture, along with some of the environmental problems that are the result of cultural evolution.

Lewin, R. *Thread of Life.* Washington, D.C., Smithsonian Books, 1982. Includes a readable presentation of human evolution.

Romer, A.S., and T.S. Parsons. *The Vertebrate Body,* 6th ed. Philadelphia, Saunders College Publishing, 1986. A well-respected, classic textbook that takes a comparative approach to life processes in vertebrates.

Vaughan, T.A. *Mammalogy,* 3rd ed. Philadelphia, Saunders College Publishing, 1986. A systematic approach to an introduction to mammals.

Weaver, K. "The Search for Our Ancestors," *National Geographic:* 560–623, November 1985. The story of human evolution written in a spell-binding fashion.

Welty, J.C. *The Life of Birds,* 4th ed. Philadelphia, Saunders College Publishing, 1988. An introduction to the biology of birds.

PART 7

Plant Structure and Life Processes

Germination of seeds • The potassium ion mechanism • Shedding leaves for survival • Monocots and dicots • Structure/function relationships in leaves

The transport of food in phloem • Pollination before fertilization • Fruit and seed dispersal • Phytochrome and flowering • The five plant hormones

26

Plant Development and Growth

OUTLINE

I. Embryonic development in plants follows an orderly and predictable path
II. A number of external and internal factors affect seed germination
III. Plants exhibit localized growth after seed germination
   A. Primary growth takes place at apical meristems
   B. Secondary growth takes place at lateral meristems
IV. Cells and tissues differentiate in the developing plant body
   A. The ground tissue system is composed of parenchyma, collenchyma, and sclerenchyma
   B. The vascular system is composed of xylem and phloem
   C. Epidermis and periderm comprise the dermal tissue system
V. Roots, stems, leaves, flowers, and fruits make up the plant body
VI. Differentiation in plants is under both genetic and environmental control
Focus on Some Experimental Methods in Embryogenesis

LEARNING OBJECTIVES

After you have studied this chapter you should be able to:

1. Trace the stages in embryo development in flowering plants.
2. Distinguish between primary and secondary growth and between apical and lateral meristems.
3. Distinguish between determinate and indeterminate growth and give two examples of each.
4. Characterize the ground tissue system, the vascular tissue system, and the dermal tissue system of plants.
5. Outline the basic features of leaves, stems, and roots.
6. Relate how plant development is different from development in animals.

Young radish seedlings. (Doug Wechsler)

Plants exhibit an amazing variety of sizes and forms. This complexity is even more marvelous to behold when one considers that each plant, which is composed of millions of specialized cells, develops from a single cell, the fertilized egg. Our task now is to trace this pathway of development in plants, from the zygote, or fertilized egg, to the mature plant body. The exact mechanisms for much of what we describe are unknown at this time and remain to be discovered by future plant biologists.

KEY CONCEPTS

☐ Genetic and environmental factors affect all aspects of plant growth and development.

☐ Plant growth occurs in localized areas of the plant body.

☐ The plant body is composed of three tissue systems: ground, dermal, and vascular.

■ EMBRYONIC DEVELOPMENT IN PLANTS FOLLOWS AN ORDERLY AND PREDICTABLE PATH

Seed plants produce a young plant embryo complete with nutrients in a compact package, the seed (Chapter 23). The seed develops after a zygote, or fertilized egg, forms at fertilization. Mitotic divisions of the zygote to form a multicellular embryo progress in an orderly, predictable fashion that is essentially the same for both dicots and monocots. The following description is for dicot embryonic development.

The two cells formed as a result of the first division of the zygote establish polarity in the embryo (Fig. 26–1). The bottom cell develops into the **suspensor,** a multicellular structure that anchors the embryo and aids in nutrient uptake from the endosperm. The top cell develops into the embryo proper. Initially, the top cell divides to form a chain of cells, the **proembryo.** As mitosis continues, a multicellular sphere of cells develops, the **globular embryo.** Tissue differentiation begins during this stage. When the two cotyledons begin to form, the embryo resembles a heart; this is called the **heart stage.** As the embryo elongates, the **torpedo stage** develops, which continues to grow into the mature embryo. Like all other aspects of plant growth and development, the embryonic stages (proembryo, globular embryo, heart stage, and torpedo stage) are under genetic control. It is possible to culture entire multicellular plants from single cells using plant-tissue culture methods. Under certain conditions, the genes that control the development of the embryo are expressed in plant-tissue culture, and all the stages can be observed in their normal progression. Haploid embryos grown from pollen cells in culture may also go through the embryonic stages, demonstrating that the diploid condition is not a requirement for embryo development.

FIGURE 26–1 Embryonic development in dicots. With the first division of the zygote, polarity is established: The bottom cell develops into the suspensor and the top cell develops into the plant embryo. In the final drawing, the embryo is still immature. In most dicots the endosperm is gone in the mature seed, its reserves having been used for further growth and development of the embryo.

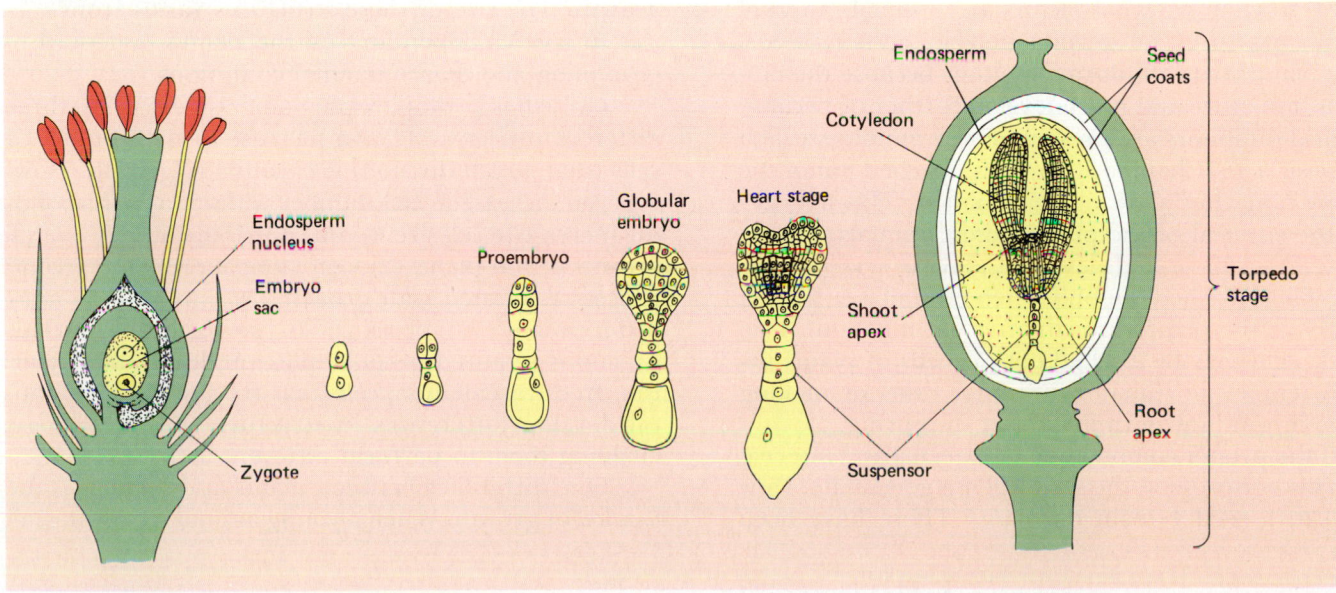

◼ A NUMBER OF EXTERNAL AND INTERNAL FACTORS AFFECT SEED GERMINATION

Mature seeds often will not **germinate** immediately, even if growing conditions are ideal. A number of factors influence whether a seed will germinate. Many of these are environmental factors, including water, oxygen, temperature, and sometimes light requirements. No seed will germinate unless it has imbibed, or absorbed, water. A watery medium in cells is necessary for active metabolism. When a seed germinates, its metabolic machinery is turned on, with numerous materials being synthesized and degraded. Therefore, water is an absolute requirement for germination. Seed germination and growth also require a high amount of energy. Plants have the same aerobic respiratory pathway as animals, so oxygen is usually a requirement for plant development during germination.

Another environmental factor that affects germination is temperature. Each plant species has an optimum temperature for seed germination, although germination will occur over a range of temperatures. For most plants, the optimum germination temperature is 25 to 30°C. Some plant seeds, such as apples, require exposure to prolonged periods of cold before their seeds are able to germinate. Also, certain plants, especially those with tiny seeds, require light for germination. Some of the environmental factors that affect germination ensure the survival of the plant. If plant seeds germinated at extremely low temperatures, the young plants would probably not survive. The requirement of a prolonged cold period ensures that the seed will germinate in the spring rather than the winter. The requirement of light guarantees that a tiny seed will germinate only if it is close to the surface of the soil. If such a seed germinated several inches below the soil surface, it would not have enough food reserves to grow to the surface.

Even when external factors are optimal, internal factors may prevent germination in certain seeds. Many plant seeds are **dormant** either because the embryo is immature and must develop further or because chemical inhibitors are present. These inhibitors, such as abscisic acid (Chapter 30), may be leached out of the seed by rain, thus allowing germination. This may ensure the survival of the plant. Desert annuals, for example, often have high levels of abscisic acid in their seeds that is leached out only when rainfall is sufficient to support the plant's growth after germination. Also, abscisic acid may be leached out of seeds of plants living in temperate climates by winter rains or melting snows, in time for germination in the spring.

If the proper combination of external and internal factors is not present, the seed will not germinate. How long can a seed remain dormant? There have been stories of seeds germinating after thousands of years when archaeologists excavated the tombs of the Pharaohs. These accounts have not been verified. However, a scientific experiment was conducted at Michigan State University starting in 1879, when a variety of seeds were enclosed in jars and buried. Periodically, some of the jars were removed, and an attempt was made to germinate the seeds. As late as 1980, 101 years later, some of the seeds still germinated.

◼ PLANTS EXHIBIT LOCALIZED GROWTH AFTER SEED GERMINATION

The first part of the plant to emerge when the seed germinates is the **radicle,** or embryonic root. As the root grows, it forces its way through the soil, encountering considerable friction. The delicate cells at the tip of the root are protected by a layer of cells known as the **root cap.**

Plant-stem tips are not covered by a cap of cells; other means protect the tips as they grow through the soil to the surface. The stem of the bean seedling is curved over, forming a hook, so the tip is actually pulled up through the soil (Fig. 26–2). Corn and other monocots have a special sheath of cells, the **coleoptile,** surrounding the shoot (Fig. 26–3). The coleoptile pushes through the soil, and the more delicate shoot then grows up through the middle of the coleoptile sheath.

The seedling that emerges from the seed continues to grow into an adult plant. The stems and roots of the plant grow throughout the life of the plant; this is called **indeterminate growth.** Theoretically, these parts of the plant could continue to elongate forever. Leaves and flowers, in contrast, stop growing after reaching a certain size; this is called **determinate growth.** The size of these structures varies from species to species depending on the plant's genetic programming and environmental conditions.

Growth is a complex phenomenon involving three different processes. (1) An increase in the number of cells (that is, cell division) is essential to growth. However, an increase in cell number without a corresponding increase in cell size would contribute little to overall growth of the plant. (2) Cell elongation is the second process associated with growth. (3) Finally, cells must **differentiate,** or specialize, to perform the various functions required in a complex, multicellular organism. In differentiation, cells that are genetically identical develop differences in structure and physiology, enabling them to perform different activities.

One difference between plants and animals is the *location* of growth. When a young animal is growing, all

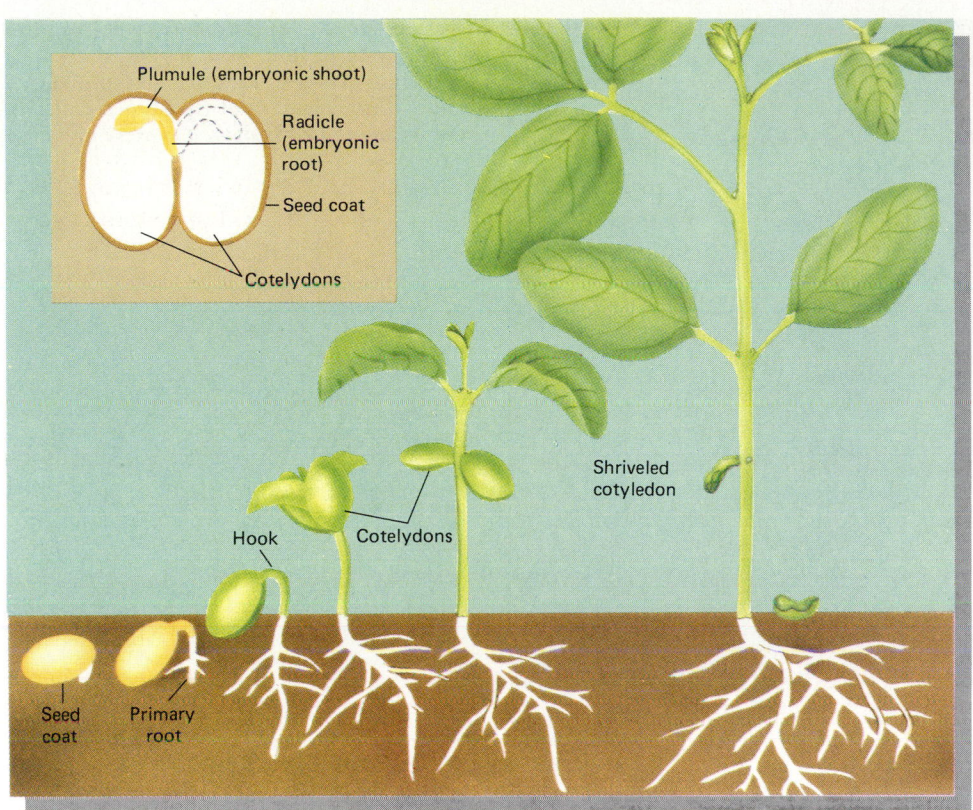

FIGURE 26–2 Seed germination and growth of the young soybean plant, a dicot. Note the hook in the stem of the young seedling that protects the delicate stem tip as it moves up through the soil. Once the shoot has emerged from the soil, the hook straightens. The stem and roots elongate by growth at their tips. As the stored food in the cotyledons is used by the developing plant, the cotyledons shrivel and fall off the stem. At this point the young plant is capable of making its own food by photosynthesis.

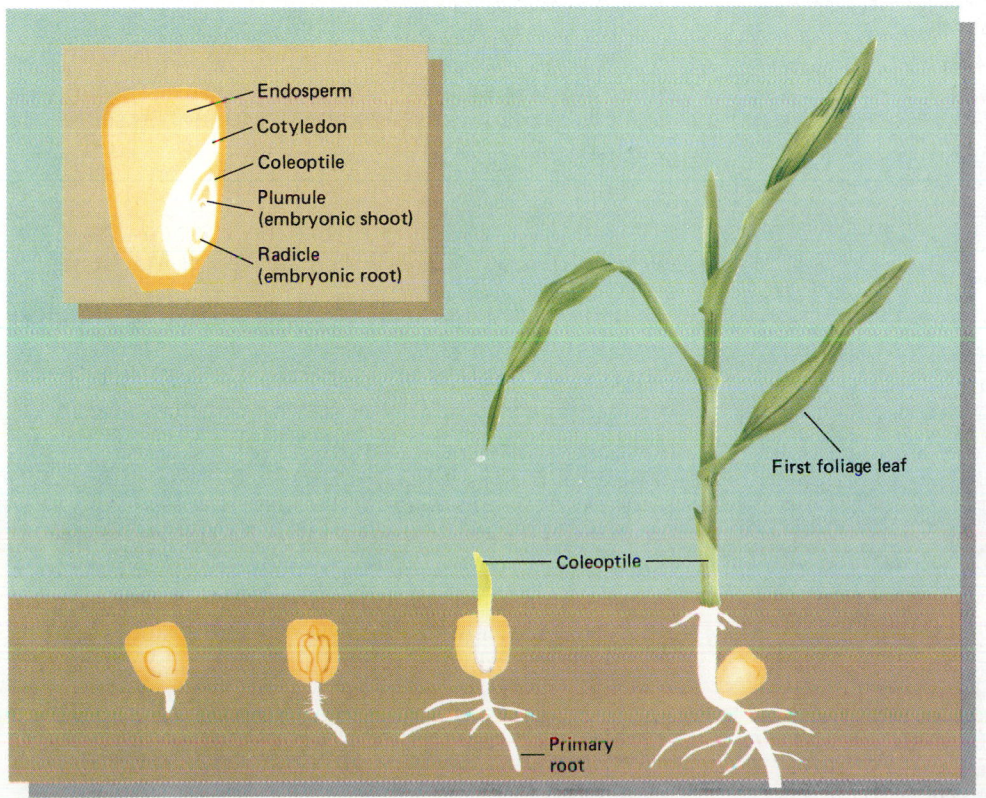

FIGURE 26–3 Seed germination and growth of the young corn plant. Note the coleoptile, a sheath of cells that emerges first from the soil. The delicate shoot tip grows up through the middle of the coleoptile.

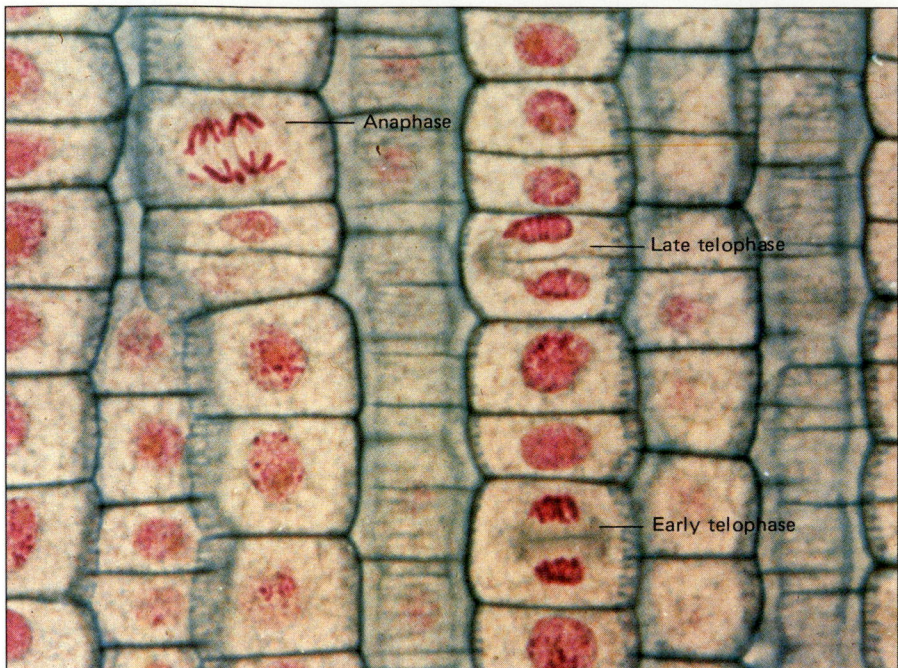

FIGURE 26–4 Meristematic cells retain the ability to divide, giving rise to all other plant tissues. This tissue from the onion (*Allium cepa*) root tip shows cells in several stages of mitosis. (Courtesy of Triarch)

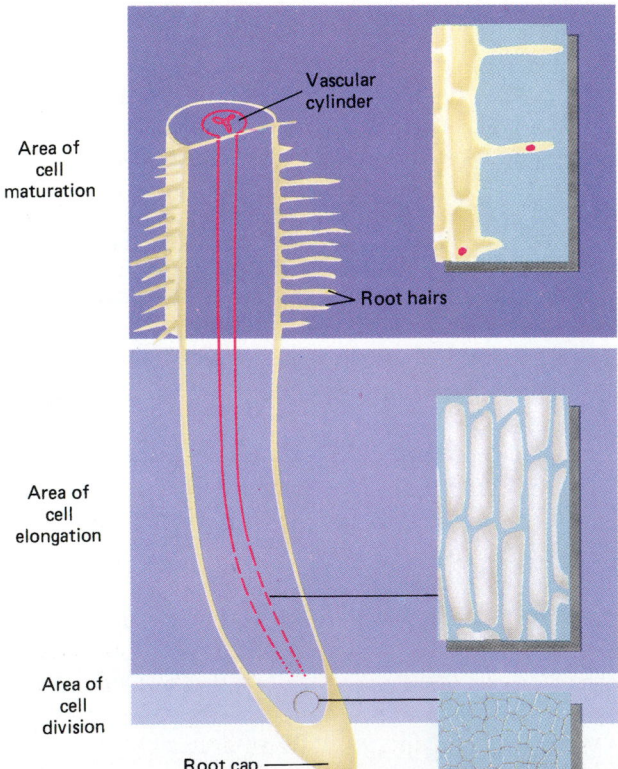

FIGURE 26–5 The root apical meristem. Just behind the root cap is the area of cell division, where mitosis occurs. Further from the tip is the area of cell elongation, where cells enlarge and begin to differentiate. The area of cell maturation has fully mature, differentiated cells. Note the root hairs in this area.

parts of its body grow, although not at the same rate. Not all parts of a plant grow, however. Plant growth is localized into areas, called **meristems,** that are composed of cells that remain unspecialized and retain the ability to divide by mitosis (Fig. 26–4).

There are two kinds of growth in plants. One is **primary growth,** which is an increase in the length of the plant. The other is **secondary growth,** an increase in the girth, or width, of the plant. All plants have primary growth, but only woody plants have secondary growth. A plant with primary growth only is said to be **herbaceous.**

Primary Growth Takes Place At Apical Meristems

Primary growth occurs because of the activity of **apical meristems,** meristematic areas at the tips (*apices;* singular, *apex*) of both stems and roots. As discussed previously, the very end of the root is a protective layer of cells called the root cap (Fig. 26–5); however, directly behind the root cap is the **area of cell division.** A microscopic examination of the area of cell division reveals the meristematic cells. These cells are very small and "boxy" in shape. They remain small because they are continually dividing. Further back from the tip of the root, just behind the area of cell division, the cells are no longer dividing, but they are enlarging. This is the **area of cell elongation.** Some differentiation also occurs here, and immature tissues become evident. The immature tissues continue to develop and differentiate into the mature tissues of the adult plant. Behind the area of cell elongation, further from the root

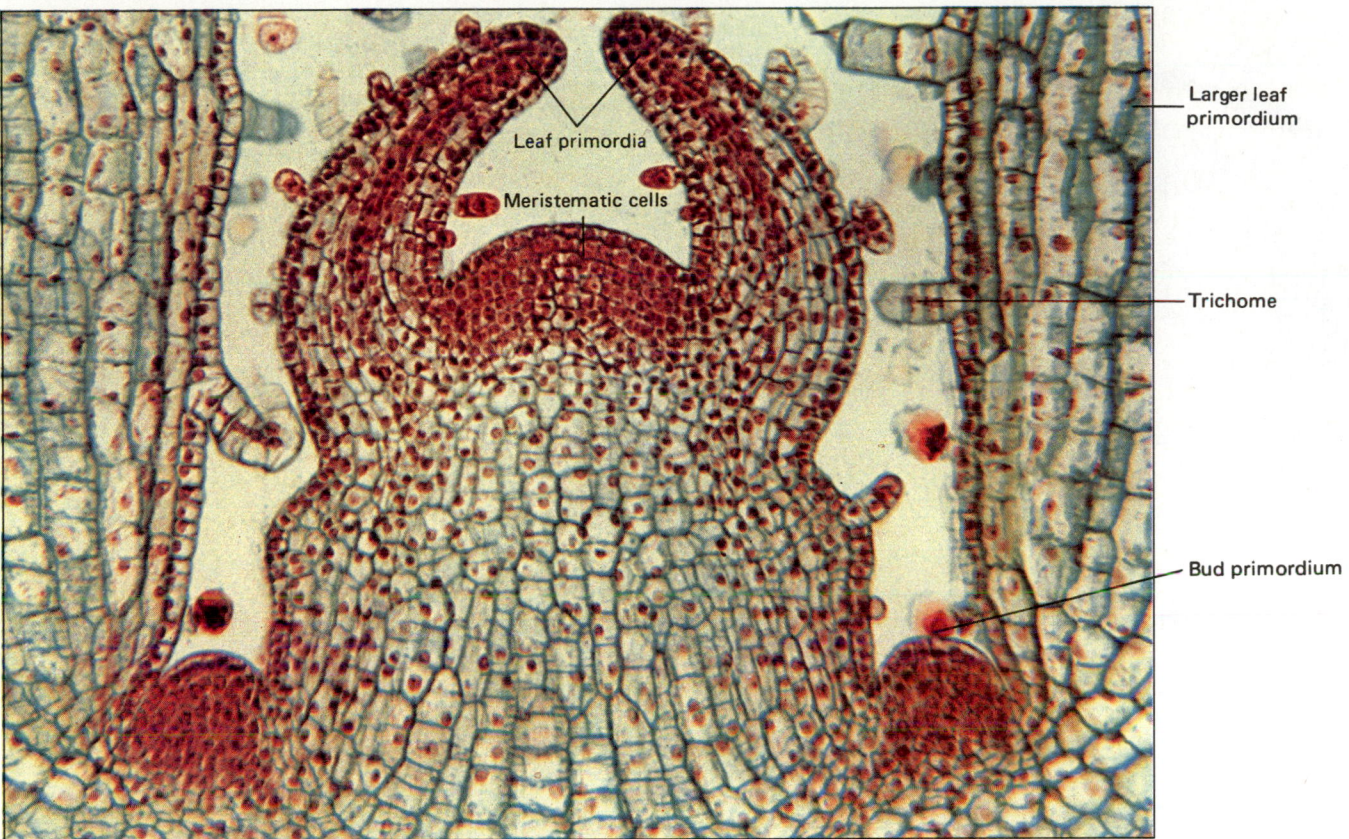

Leaf primordia

Meristematic cells

Larger leaf primordium

Trichome

Bud primordium

FIGURE 26–6   The dicot stem apical meristem. Note the leaf primordia and bud primordia. (Courtesy of Triarch)

tip, the cells have completely differentiated and are fully mature. A number of specialized tissues within the root are evident in this region, called the **area of cell maturation;** for example, root hairs, extensions of epidermal cells, may be observed here.

The apical meristem of the stem is essentially the same in function as that of the root apical meristem, although it looks quite different (Fig. 26–6). **Leaf primordia** (embryonic leaves) and **bud primordia** (embryonic buds) emerge from it. A dome of tiny meristematic cells is located in the center at the very tip of the stem. Further from the tip of the stem, the immature cells elongate and start to differentiate. The immature tissues continue to develop into mature tissues that are located further back from the stem tip. The three areas (cell division, elongation, and maturation) are present in stem tips, although they are not as obvious as in the root.

Secondary Growth Takes Place At Lateral Meristems

Plants with secondary growth have stems and roots that increase in girth. Woody trees and shrubs all have secondary growth in addition to primary growth; that is, these plants increase in length by primary growth and increase in girth by secondary growth. The increase in girth is due to the activity of **lateral meristems,** which are located on the sides of the stem and root. Two lateral meristems are responsible for secondary growth, the **vascular cambium** and the **cork cambium** (Fig. 26–7).

The vascular cambium is a layer of meristematic cells that forms a ring, or cylinder, around the stem and root trunk. It is located exactly between the wood and bark of the plant, and its cells divide to form more wood and more bark (the inner bark, to be more precise). The cork cambium is composed of patches of meristematic cells that are located in the outer bark region. Cells of the cork cambium divide to form the tissues of the outer bark.

What is the advantage of secondary growth and of wood and bark? Secondary growth in plants confers the advantage of a longer life span than most plants without secondary growth have. Individual cells do not live forever. Typical plant cells live approximately 3 years, although there is a lot of individual variation. Plants with only primary growth have no way to replace older tissues in the stem and root. Although their

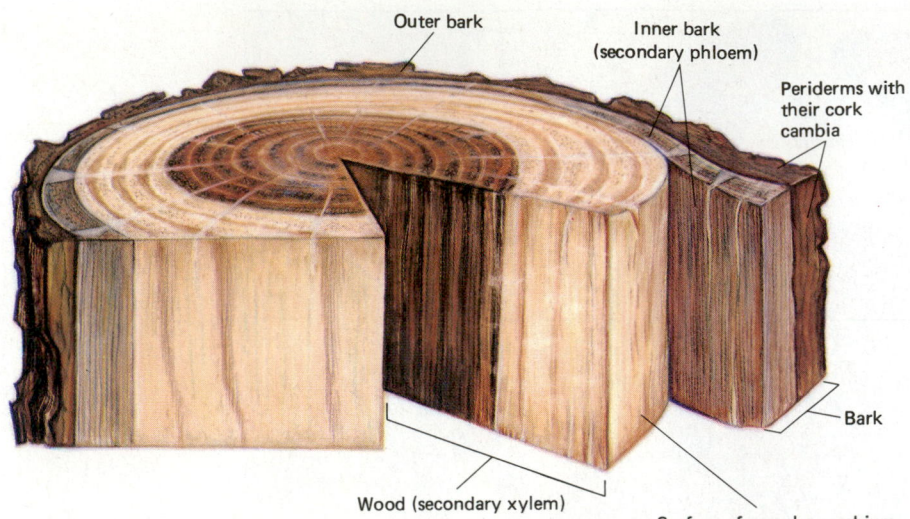

Outer bark

Inner bark
(secondary phloem)

Periderms with
their cork
cambia

Bark

Wood (secondary xylem)

Surface of vascular cambium

FIGURE 26–7 In secondary growth plants increase in girth as a result of the activity of two lateral meristems. The vascular cambium produces secondary vascular tissues, the wood and inner bark. The cork cambium produces the outer bark tissues that replace the epidermis in the secondary plant body.

tips are continually producing new cells, the older parts eventually die. Plants that have secondary growth produce new stem and root tissues to replace the older parts throughout the length of the plant, not just at the tips. Therefore, plants with secondary growth have an extended life span, sometimes for thousands of years!

Plants with secondary growth are **perennial,** living year after year. Plants that are herbaceous and have only primary growth have no persistent (enduring) above-ground parts. Many of them (for example, corn and rice) are **annuals,** and grow, reproduce, and die in one season. Some herbaceous plants (carrots and beets, for example) are **biennials,** and take 2 years to complete growth and reproduction before dying (Fig. 26–8). Herbaceous plants that are perennials (for example, rhubarb and asparagus) live year after year, but die back each winter. Their body parts in the soil do not die but remain dormant during the winter and send out new growth each spring.

CELLS AND TISSUES DIFFERENTIATE IN THE DEVELOPING PLANT BODY

As growth occurs, some cells become specialized, developing into **tissues** that make up the plant body. A tissue is a group of cells that is a structural and functional unit. The tissues of different plant organs are interconnected throughout the plant. All parts of the plant have three tissue systems: dermal, vascular, and ground. The **dermal tissue system** provides a covering for the plant body. The **vascular tissue system** is responsible for conduction of various substances in the plant, including water, dissolved minerals, and food. The rest of the plant body is composed of the **ground tissue system,** which is composed of various cell types with a variety of functions. Some plant tissues are com-

posed of only one cell type (simple tissues), whereas other plant tissues have two or more cell types (complex tissues).

(a)

(b)

FIGURE 26–8 The carrot is a biennial. (a) It grows vegetatively the first year, storing food in its root. (b) The second year, the energy stored in the root is utilized to reproduce. (a and b, G.R. Roberts)

FOCUS ON   Some Experimental Methods in Embryogenesis

During the past 50 years, plant biologists have developed increasingly sophisticated methods to study embryo development in plants. Initially, biologists concentrated on descriptive aspects of embryogenesis, involving light microscopy studies of different stages in the developmental process. Even today, important contributions are being made in descriptive embryology.

Autoradiography is one technique that is used to help determine where specific chemical events are occurring in the embryo during its development. To monitor RNA in developing cells, for example, one can incubate them with radioactively labeled precursors of RNA. After incubation, the cells are stained, sectioned, and placed on slides. The slides are covered with a photographic emulsion and left in the dark. During this time, the radiation emitted from the radioactively labeled RNA exposes the film. After development, the exact location of radioactivity in the cell can be determined.

Analysis of the proteins produced during different stages of embryo development is done in several ways. Polyacrylamide-gel electrophoresis is an effective way to separate individual proteins from one another. In electrophoresis, the proteins are placed into slots cut into a slab of gel. An electric current is applied across the gel for a period of time. Different proteins have different charges that are determined by their amino acid composition. As a result of these charge differences, as well as differences in molecular weight, the proteins will migrate at different rates of speed across the gel. The proteins may be visualized by using a chemical that changes color in the presence of protein or by autoradiography (if the proteins have first been labeled with radioactive amino acids).

Tissue-culture techniques have had a great impact on experimental approaches to embryology. It is

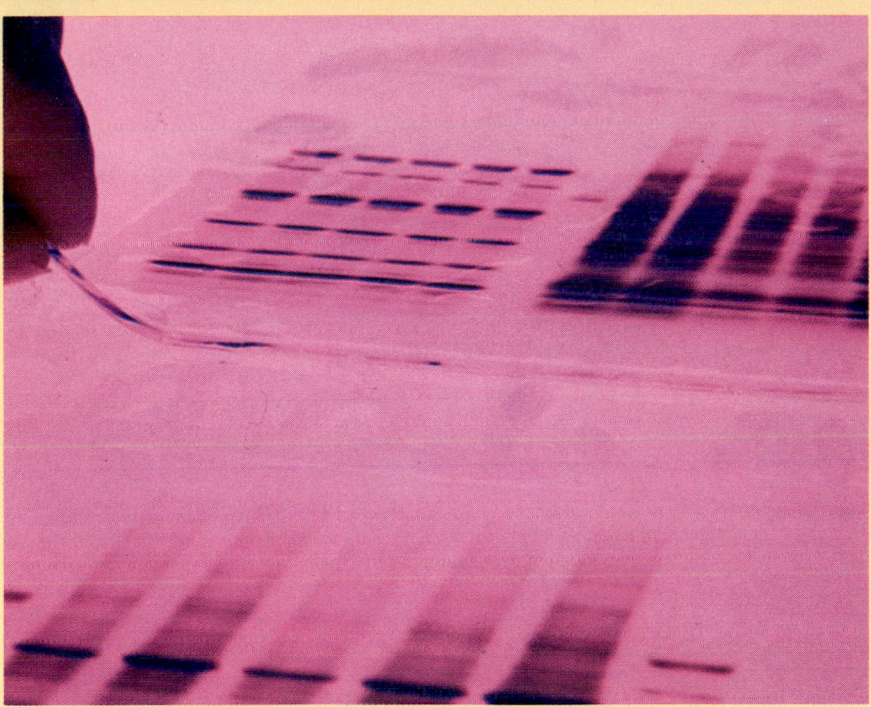

Gel electrophoresis. The separation of the proteins or other large macromolecules can be visualized under ultraviolet radiation if a dye that binds to the organic molecules is added to the gel. The dye fluoresces, giving off pink light when exposed to ultraviolet.

much easier to study certain aspects of embryo development in a tissue-culture system than in a developing seed. For example, the development of haploid embryos from pollen cells provides data that intact diploid embryos within the seed never could. Manipulation of levels of plant hormones in tissue culture has shed some light on their roles in embryology, but many questions remain. For example, it is unclear how the same plant hormones can have different effects on the embryo at different stages.

Gene expression during development is of great interest to plant biologists. But how does one go about studying which genes affect different stages in embryo development? Corn, or maize, is one of the most useful plants to use for several reasons. First, because maize is one of our most important crop plants, it

has been studied a great deal. Much is already known about different aspects of its growth and development as well as its genetics and physiology. Also, there are a number of maize mutants that have abnormalities in their embryonic development. Because most mutants have abnormalities in their endosperm development as well as their embryos, it is relatively easy to identify them by looking for unusual characteristics in the endosperm. It is useful to determine how normal development is interrupted in mutants. For example, it is now known that different genes direct at least some of the development of root and shoot meristems in embryos. This was determined by studying an embryo that has abnormal development in the shoot meristem but normal development in the root meristem.

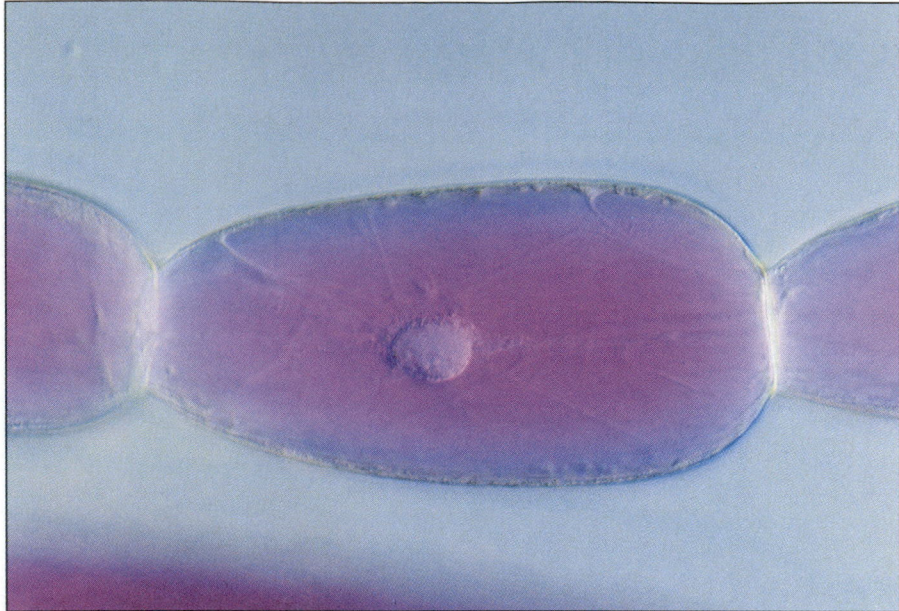

(a)

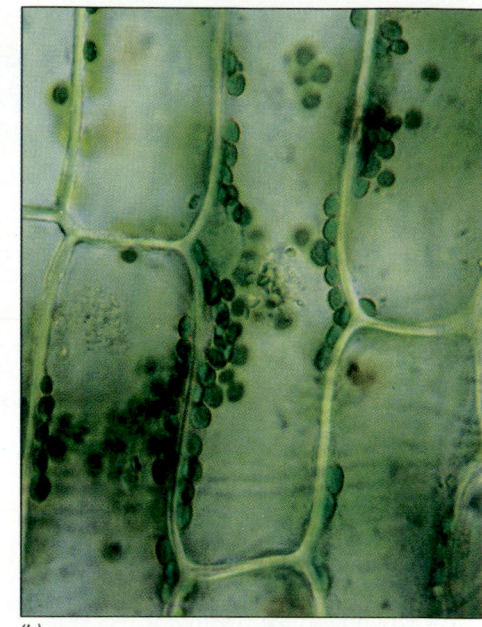

(b)

(c)

FIGURE 26–9 **Parenchyma cells are relatively unspecialized and living at maturity. (a) Parenchyma cells from the stamen hairs of the spiderwort (*Tradescantia virginiana*). The large vacuole contains pigmented material and occupies most of the cell. Note the nucleus and cytoplasmic strands. (b) Some parenchyma cells contain chloroplasts. The primary function of these cells is photosynthesis. (c) Parenchyma cells often function in storage. These parenchyma cells are from the cortex of a *Ranunculus* root. Note the starch grains filling the cells.** (a, Phil Gates, University of Durham/BPS; b and c, Ed Reschke)

The Ground Tissue System Is Composed Of Parenchyma, Collenchyma, And Sclerenchyma

If one were to think of a "typical" plant cell, it would be a **parenchyma** cell. Parenchyma, which is a simple plant tissue, is found throughout the plant body (Fig. 26–9). Parenchyma cells are relatively unspecialized, especially when compared with certain other plant tissues that will be considered in this chapter. Parenchyma cells perform a number of important metabolic functions for plants. They often contain chloroplasts and are responsible for photosynthesis. In addition, parenchyma cells store various materials, including food (visible as starch grains or oil droplets) and salts (visible as crystals). Secretions also may be produced by parenchyma cells. Like all plant cells, each parenchyma cell is enclosed by a cell wall. This wall often contains layers and provides structural support for the plant.

All plant cells have a **primary cell wall.** Many plant cells, as they mature, deposit additional cell wall material *inside* the primary wall (that is, between the primary wall and the plasma membrane). This **secondary cell wall** reinforces the primary wall. Parenchyma cells typically have primary walls only.

Support is a crucial function in plants. One of the simple plant tissues specialized for structural support in plants is **collenchyma** (Fig. 26–10a). Plants lack a skeletal system; their body parts are supported by individual cells, especially strengthening tissues such as collenchyma cells. Their primary walls are unevenly thickened, being especially thick in the corners. Collenchyma is not located throughout the plant. It is found in long strands, often just under the epidermis in stems and along leaf veins. In addition to providing support, collenchyma is an extremely flexible tissue.

A second simple plant tissue specialized for structural support is **sclerenchyma** (Fig. 26–10b and c). The word sclerenchyma is derived from a Greek term meaning "hard." Sclerenchyma cells have both primary and secondary cell walls. Their walls become not only strong, or hard, but extremely thick, so thick that the cell dies. Therefore, at functional maturity, when they are supporting the plant body, sclerenchyma cells are dead. Sclerenchyma may be located in several areas

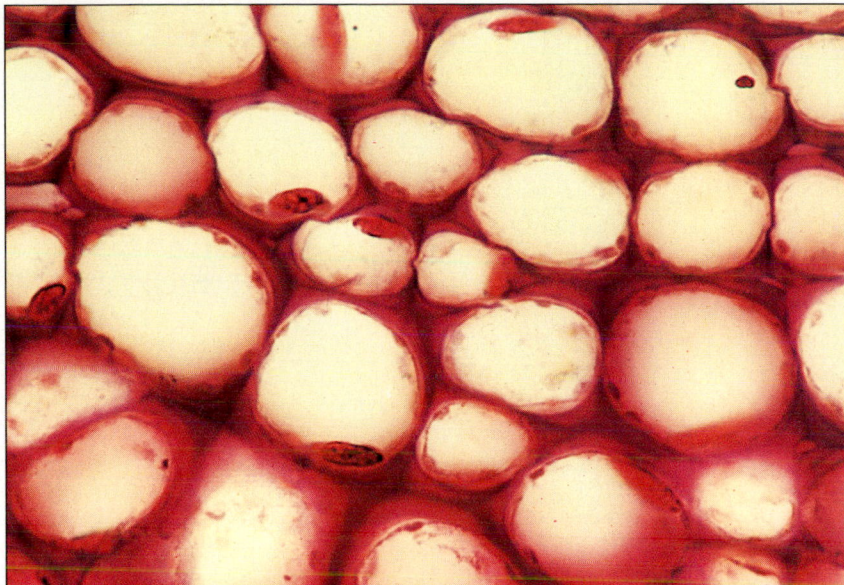

(a)

FIGURE 26–10   **Collenchyma and sclerenchyma are plant tissues specialized for structural support. (a) Collenchyma cells. Note the unevenly thickened cell walls that are especially thick in the corners. Nuclei are evident in several cells, indicating they are living at maturity. (b, c) Sclerenchyma cells produce both primary and secondary walls. The cells walls are extremely thick and hard, providing structural support. The fibers shown here (b) in cross section and (c) in longitudinal section are a type of sclerenchyma. These long, tapering cells are dead at functional maturity.** (a, Biophoto Associates/Photo Researchers, Inc.; b, Dwight R. Kuhn)

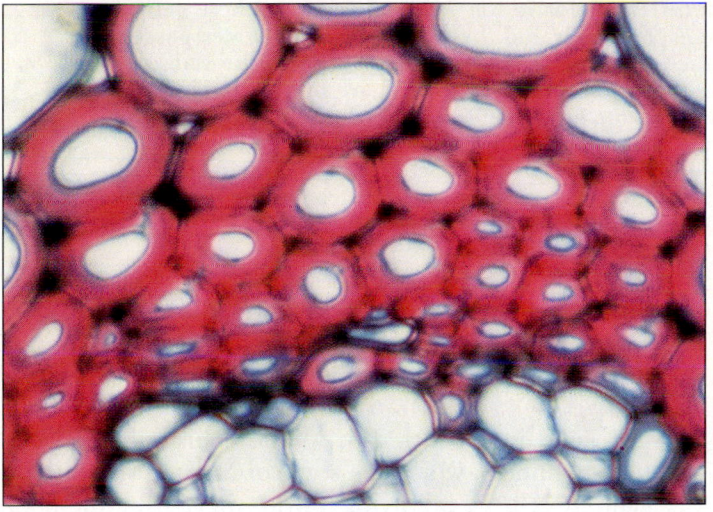

(b)

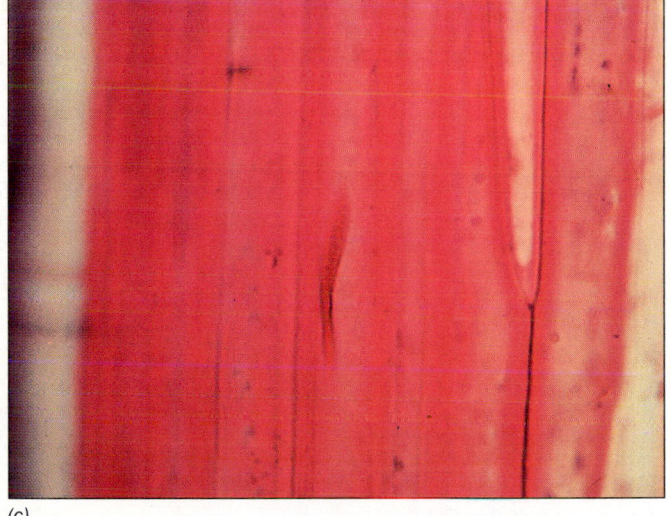

(c)

of the plant. One type of sclerenchyma is **fibers,** which are long, tapered cells that are often located in patches or clumps. In cross section, one can appreciate the thickness of their walls and understand why the cells are dead.

The Vascular System Is Composed Of Xylem And Phloem

The primary function of **xylem** is to conduct water and dissolved minerals from the roots to the stems and leaves. A secondary function of xylem is to provide structural support. Xylem is located throughout the plant body. The xylem of the root is continuous with stem xylem, and stem xylem is continuous with leaf xylem. Xylem is a complex plant tissue that is composed of four different cell types in flowering plants. Two of the four cell types found in xylem are involved

in conduction; these are the **tracheids** and **vessel elements** (Fig. 26–11). In addition to these cells, xylem also contains parenchyma and fibers.

Tracheids and vessel elements are marvelously specialized for conduction. Both are dead at maturity. Tracheids are long, tapering cells that are hollow. Only their cell walls remain. Tracheids are located in patches, or clusters. Water is conducted up through tracheids, passing from one tracheid into another through thin places in the walls, called **pits.** Under the microscope, pits often appear to be holes in the wall, but in this case, appearances are misleading. Vessel elements are considered to be more advanced than tracheids. The cell diameter of vessel elements is wider than tracheids, so they are more efficient at water conduction. These cells are hollow, and their end walls have holes, or perforations. Vessel elements are stacked end-on-end. Water is conducted readily from

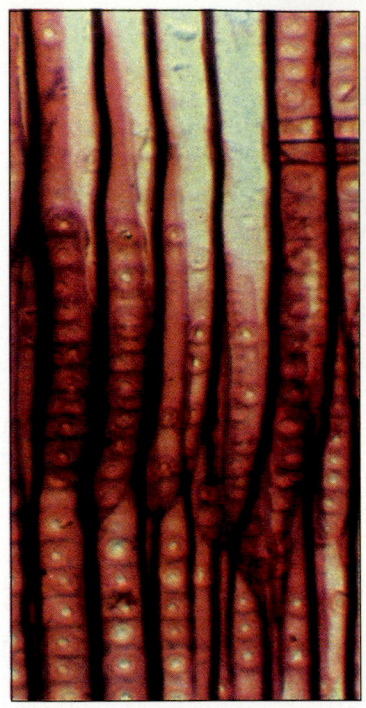

(a)

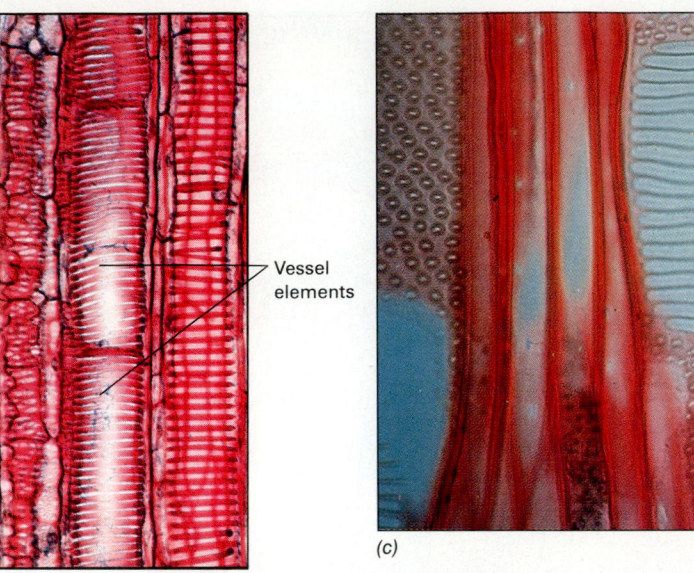

Vessel
elements

(b)

(c)

FIGURE 26–11   Tracheids and vessel elements are the two types of conducting cells in xylem. (*a*) Tracheids in longitudinal section. These cells, which occur in clumps, transport water and dissolved minerals. Water passes readily from tracheid to tracheid through the pits, thin places in the wall. (*b*) Vessel elements in longitudinal section. These cells are more efficient than tracheids in conducting water. Note how they are stacked end-on-end. (*c*) The end walls of vessel elements, called perforation plates, have large holes. Several perforation plates are visible in this photomicrograph. Water readily passes through the perforation plate from one vessel element to the next. (*a*, Courtesy of Triarch; *b*, J. Robert Waaland, University of Washington/BPS)

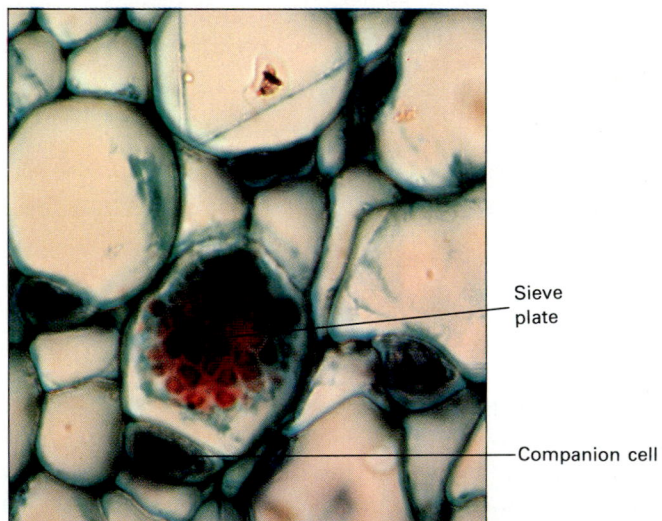

Sieve
plate

Companion cell

FIGURE 26–12   Phloem tissue in cross section. Note the sieve plate, the end wall of the sieve tube member. The smaller cells are companion cells. (J. Robert Waaland, University of Washington/BPS)

one vessel element into the next. A stack of vessel elements is called a **vessel.** Like tracheids, vessel elements also have pits.

Food is conducted throughout the plant by the **phloem** (Fig. 26–12). Like xylem, phloem is a complex tissue. In flowering plants, it is composed of four cell

types, **sieve-tube members, companion cells,** fibers, and parenchyma. Fibers are frequently quite extensive in phloem, providing additional support for the plant body.

Food is conducted in solution through the sieve-tube members, which are some of the most highly specialized cells in the living world. Sieve-tube members are stacked end-on-end, forming sieve tubes. Their end walls, called **sieve plates,** have a series of holes. Cytoplasmic connections extend from one sieve-tube member into the next. Sieve-tube members are living at maturity, but during maturation, many cell parts disintegrate, including the nucleus, vacuole, and ribosomes. Sieve-tube members are among the few eukaryotic cell types that can function without nuclei. One example of such cells in mammals is the red blood cell. However, red blood cells can only function for a very limited period of time (approximately 120 days in humans), presumably because they lack nuclear control. Sieve-tube members typically live for less than a year, although there are notable exceptions. Certain palms have sieve-tube members that have remained alive approximately 100 years! It is not clear how these cells without nuclei can function as long as they do.

Adjacent to each sieve-tube member is a companion cell. The companion cell is a living cell, complete with nucleus and other organelles. There are numerous cytoplasmic connections between companion cells

TABLE 26–1
A Summary of Plant Cell Types

Cell Type		Function	Location
Parenchyma		Secretion, storage, photosynthesis	Throughout the plant body
Collenchyma		Support	Just under stem epidermis; along leaf veins
Sclerenchyma		Support	Throughout the plant body; common in stems and certain leaves
Tracheids		Conduction of water and minerals; also provide support	Xylem
Vessel elements		Conduction of water and minerals; also provide support	Xylem
Sieve-tube members		Conduction of food	Phloem
Companion cells		Aids sieve-tube members in food conduction	Phloem

and sieve-tube members. Although the companion cell does not conduct food, it plays an essential role in phloem transport (Chapter 28).

Epidermis And Periderm Comprise The Dermal Tissue System

The dermal tissue system provides a protective covering over plant parts. In plants with primary growth, the dermal covering is a single layer of cells, the **epidermis** (Fig. 26–13). The epidermis is a complex tissue that is composed of several types of cells. Most of the cells in the epidermis are parenchyma. Their cell walls are thicker toward the outside of the plant, providing protection. In addition, epidermal parenchyma cells generally do not contain chloroplasts. Their transparent nature allows light energy to penetrate into interior tissues of the stem and leaf, where photosynthesis does

FIGURE 26–13 *Tradescantia* leaf epidermis (×350). Note the pink-colored guard cells that form openings for gas exchange. (James Bell/Photo Researchers, Inc.)

occur. One of the greatest threats to the above-ground parts of the plant (that is, stems and leaves) is desiccation (drying out). The epidermal cells secrete a waxy layer, the **cuticle,** over their outer walls. This wax greatly restricts the loss of water from plant surfaces. The root epidermis does not produce a thick cuticle, because roots must be permeable to water in order to absorb it from the soil.

The cuticle prevents water loss through epidermal cells; at the same time, however, it also prevents gases from passing through. In both stems and leaves, photosynthetic tissues are *inside* the epidermis. The gases involved in photosynthesis pass through this barrier through **stomates,** tiny pores formed in the epidermis by two rounded cells, called **guard cells** (see Fig. 26–13). A number of gases pass through the stomates by diffusion, including carbon dioxide, oxygen, and water vapor. Stomates are generally open during the day when photosynthesis is occurring. They close during the night, conserving water in the plant (Chapter 27).

The epidermis also may contain special outgrowths, or hairs, termed **trichomes.** Trichomes have a variety of functions. Root hairs are epidermal cell extensions that increase the surface area of the root that comes into contact with the soil, increasing water absorption (Chapter 28). Plants that can tolerate salty environments may have trichomes specialized for salt removal. Research indicates that the presence of trichomes on the aerial portions of desert plants may increase reflection of light off the plants, thereby cooling the internal tissues and decreasing water loss.

In plants with secondary growth, the epidermis splits apart as the plant increases in girth. The **peri-derm,** a complex tissue, is the functional replacement of the epidermis (Chapter 28). It is several to many cells thick, and it forms the outer bark of the stem and root (Fig. 26–14). Periderm is continually being formed by a lateral meristem located within it, the cork cambium. Cork cambium cells divide, forming cork cells to the outside and cork parenchyma to the inside. Cork cells are dead at maturity. Their walls are heavily coated with a waterproof substance that reduces water loss. Cork parenchyma cells function primarily in storage.

■ ROOTS, STEMS, LEAVES, FLOWERS, AND FRUITS MAKE UP THE PLANT BODY

All the cell and tissue types just discussed are organized into the plant body. Plants have a root system and a shoot system. The root system is generally the below-ground portion. The above-ground portion, the shoot system, is made up of a stem, which bears leaves, flowers, and fruits. Roots, stems, leaves, flowers, and fruits are considered plant organs because each is composed of several different tissues. Some plant tissues are continuous throughout the length of the plant (for example, vascular tissue), whereas others may be localized in certain organs (fibers in stems, for example).

There are several types of root systems (Fig. 26–15). Plants with a **tap root** system have one primary root with smaller roots branching off it. The dandelion root is a good example of a tap root system. The **fibrous root** system has several main roots developing from the end of the stem. Smaller roots branch off these roots. Crabgrass and other grasses have fibrous root systems. Some plants have their roots modified for storage. These **storage roots** may be modified taproots (for example, carrots) or fibrous roots (for example, sweet potatoes).

One way that most stems can be distinguished from roots is that stems bear leaves. The area on the stem where leaves attach is the **node,** and the region of the stem between two successive nodes is the **internode.** Stems have **buds,** the **terminal bud** being the embryonic shoot (apical meristem) at the tip of the stem. When the apical meristem is not actively growing, the terminal bud is covered and protected by **bud scales,** which are modified leaves. Plants also have **lateral buds** that are located in the **axils** of leaves. The axil is the area on the stem directly above where the leaf attaches to the stem. When lateral buds grow, they form stems that bear leaves or flowers.

A woody twig that has shed its leaves can be used to demonstrate stem structures (Fig. 26–16). The terminal bud is covered by bud scales that protect the embryonic tissues (apical meristem) during dormancy.

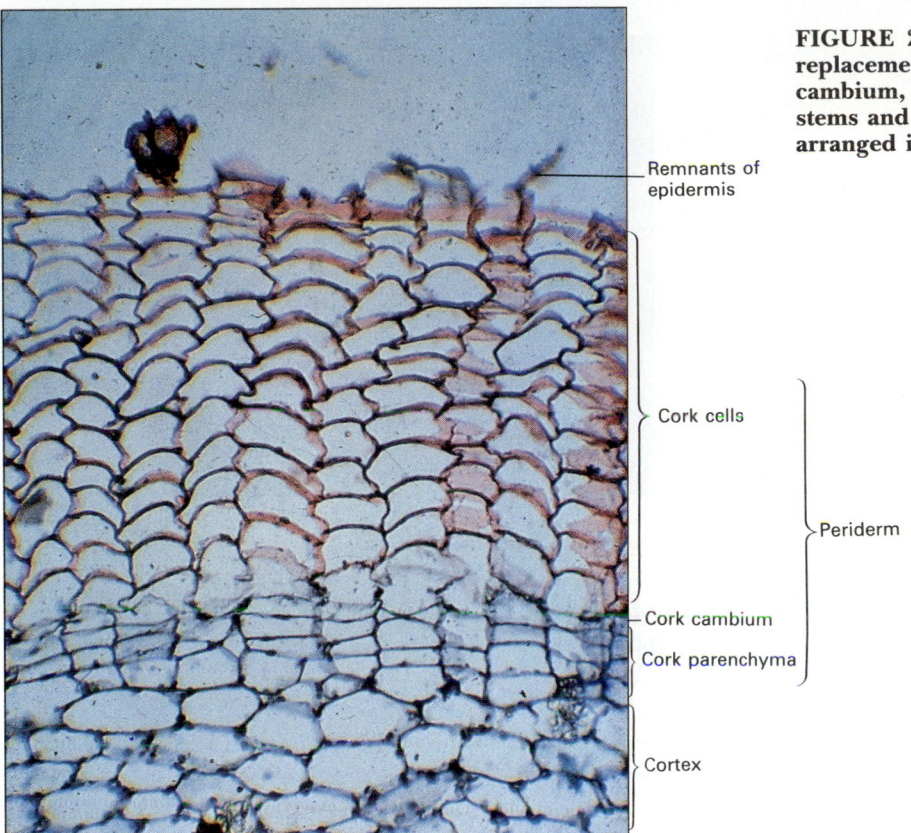

Remnants of epidermis

Cork cells

Periderm

Cork cambium

Cork parenchyma

Cortex

FIGURE 26–14 Periderm is the secondary replacement for epidermis. Formed by the cork cambium, it makes up the outer bark of woody stems and roots. The cells of periderm are always arranged in vertical stacks.

FIGURE 26–15 Root systems in plants. (*a*) The fibrous root system is characteristic of monocots. (*b*) The taproot system is common in many dicots. Both tap roots and fibrous roots may be modified for food storage.

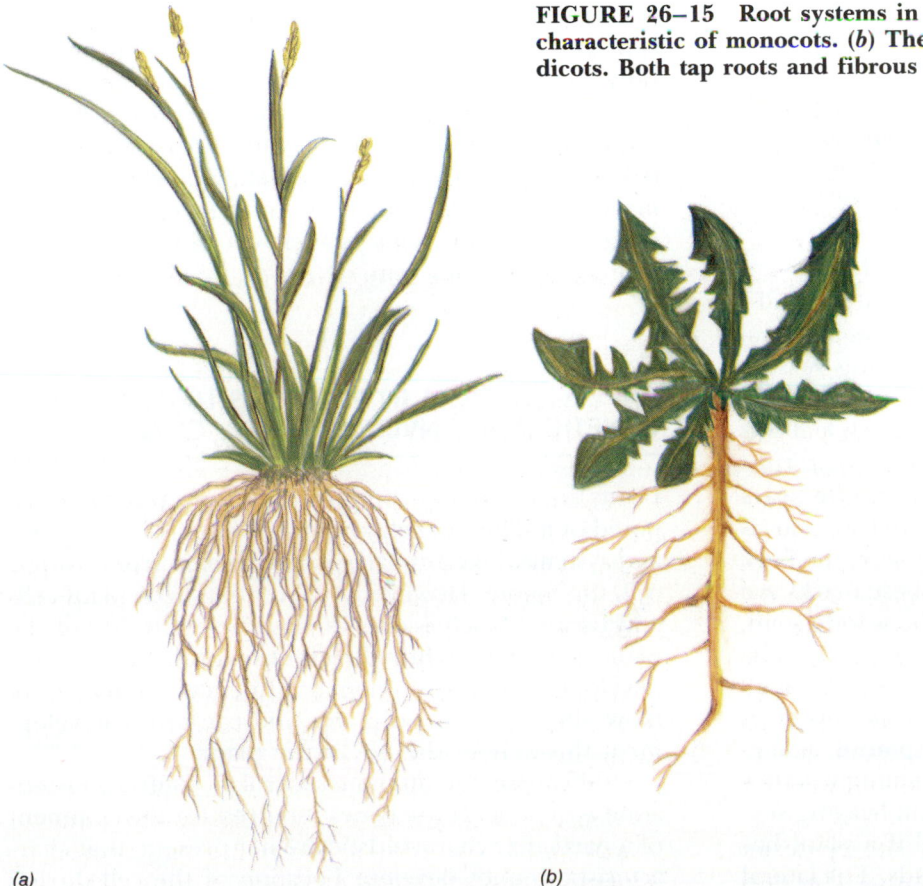

(a)   (b)

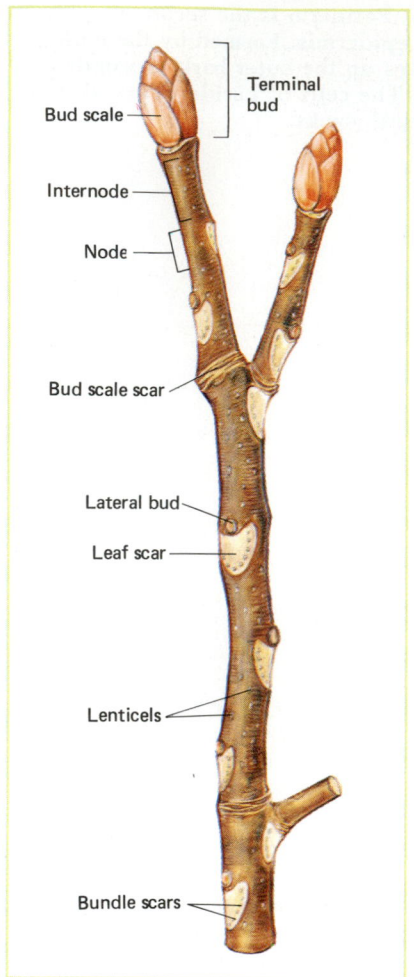

FIGURE 26–16 The external structure of a woody twig of horse chestnut, *Aesculus hippocastanum.* **One can determine the age of a woody twig by counting the number of bud scale scars (don't count side branches). How old is this twig?**

FIGURE 26–17 A simple leaf. Most leaves have a broad, expanded blade and a petiole. (B. Miller/BPS)

When the plant resumes growth, the bud scales fall off, leaving a **bud-scale scar.** Because plants form terminal buds once a year, at the end of the growing season, counting the number of bud-scale scars indicates the age of the twig. A **leaf scar** shows where each leaf was attached to the plant. The vascular tissue that runs from the stem out into the leaf forms **bundle scars** within the leaf scar. Directly above the leaf scar is where the lateral bud may be found. Finally, the bark of a woody twig has tiny marks on it. These marks are the **lenticels,** sites of gas exchange in the woody stem.

Most leaves are composed of two parts. The broad, expanded portion is the **blade,** and the stalk that attaches the blade to the stem is the **petiole** (Fig. 26–17). Leaves may be **simple** or **compound.** Sometimes a beginning student has difficulty telling whether a leaf is compound or really a small stem bearing several simple leaves. One easy way to tell if a plant has compound leaves is to look for lateral buds. The lateral

buds form at the base of the leaf, whether it is simple or compound, never at the base of **leaflets.** Also, compound leaves lie in a single plane, whereas simple leaves are never arranged in one plane on a stem. Leaves can be arranged on a stem in several ways (Fig. 26–18). Plants with **alternate** leaf arrangement have one leaf at each node. In **opposite** leaf arrangement, two leaves occur at each node. In **whorled** leaf arrangement, three or more leaves occur per node. Leaves have a variety of vein patterns (Fig. 26–19). Monocots have **parallel** veins, whereas dicots have **netted** veins. Netted veins can be **palmate,** where several major veins radiate out from one point, or **pinnate,** where the major veins branch off along the entire length of the main vein.

◼ DIFFERENTIATION IN PLANTS IS UNDER BOTH GENETIC AND ENVIRONMENTAL CONTROL

Plants are extremely complex organisms that are composed of millions of cells organized into tissues, organs, and systems. Like animals, plants develop from a single cell, the zygote. However, differentiation of plant cells and tissues, which occurs during the entire life of the plant, is under slightly different controls than animal development. Environmental influences seem to have more direct effects on plants and can affect development throughout the life of the plant.

Of course, the ultimate control of plant differentiation is genetic. If the genes required for development of a particular characteristic are not present, that characteristic cannot develop. Location of the cell during

FIGURE 26–18 Leaf arrangement on the stem may be alternate, opposite, or whorled, depending on the number of leaves at each node.

Alternate   Opposite   Whorled

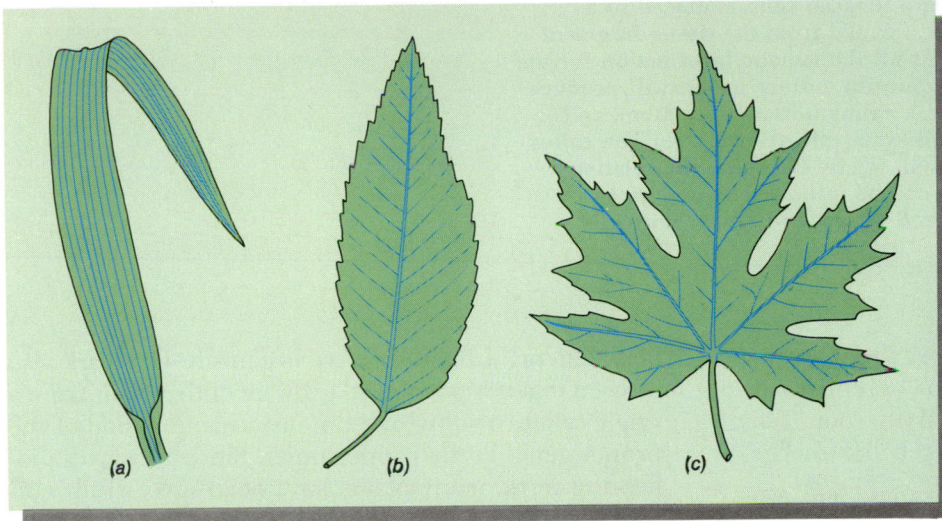

FIGURE 26–19 Vein patterns in leaves. (*a*) Kentucky bluegrass has parallel veins, which are characteristic of monocot leaves. Siberian elm (*b*) and silver maple (*c*) have netted veins, which are characteristic of dicot leaves. Siberian elm is pinnately netted and silver maple is palmately netted.

(a)   (b)   (c)

early development also has a profound effect on what that cell will ultimately become. All this is true for animals as well, but in plants, other nongenetic factors can have a tremendous influence on gene expression.

One nongenetic factor that affects plant growth and development, including differentiation, is the influence of other plant tissues and organs. Much of this control is mediated by **hormones,** substances produced in one part of the plant and transported to another, where they elicit some type of response (Chapter 31). Unlike animal hormones, each plant hormone affects a wide variety of growth responses in the plant throughout its lifetime. These hormones interact with one another in both stimulatory and antagonistic ways.

The external environment is a very important nongenetic factor that affects all aspects of plant growth and development. Environment plays a role in animal growth, but it is a much more profound influence in plants. This should not be surprising when one considers that plants, being sessile, cannot respond to their environment by departing, as animals can. All

aspects of plant growth are intimately connected with environmental cues. In many plants, for example, the initiation of flowering is controlled by differences in day length and darkness that occur with the changing seasons. The environment modifies and, in some cases, controls plant growth and gene expression.

Much of our current understanding of plant cell, tissue, and organ differentiation has come from experimental studies involving cell and tissue cultures. Plant biologists attempted unsuccessfully to grow isolated plant cells in culture beginning in the early 1900s. Initially, plant cells could be kept alive in a chemically defined, sterile medium, but they wouldn't divide. It was discovered that addition of certain natural materials like coconut milk induced the cells to divide in culture. Coconut milk has a complex composition, so the division-inducing substance was not chemically identified for some time (Chapter 31). By the late 1950s, plant cells from a variety of sources could be cultured successfully, dividing to produce a mass of undifferentiated cells, or **callus.**

(a)

(b)

(c)

(d)

Carolina Biological Supply Company

FIGURE 26–20   Propagating tobacco by tissue culture. (*a*) **A fragment of undifferentiated tissue from the center of a tobacco stem is placed in a culture medium. A complete plant can be formed from the tissue fragment because each cell of the fragment contains all the genetic information for the entire organism. Different kinds of hormones in culture media will produce different growth responses.** (*b*) **Placed on a callus initiation medium, cells begin to proliferate and undifferentiated tissue (callus) forms.** (*c*) **The callus produces roots on a root initiation medium.** (*d*) **By changing the relative levels of several plant hormones, shoots can be initiated. Plants grown by tissue culture techniques can be transferred to soil and grown normally.**

In 1958, F.C. Steward, a plant physiologist at Cornell University, succeeded in generating an entire carrot plant from a single cell in the carrot root. This demonstrated conclusively that each cell has the ge-

FIGURE 26–21   The hybrid cell undergoing mitosis was derived by protoplast fusion between soybean (*Glycine max*) and vetch (*Vicia hajastana*). The large chromosomes are derived from the vetch and the small ones from the soybean. (From Constabel, F., et al., *C.R. Acad. Sci.* (Paris) 285:319–322, 1977)

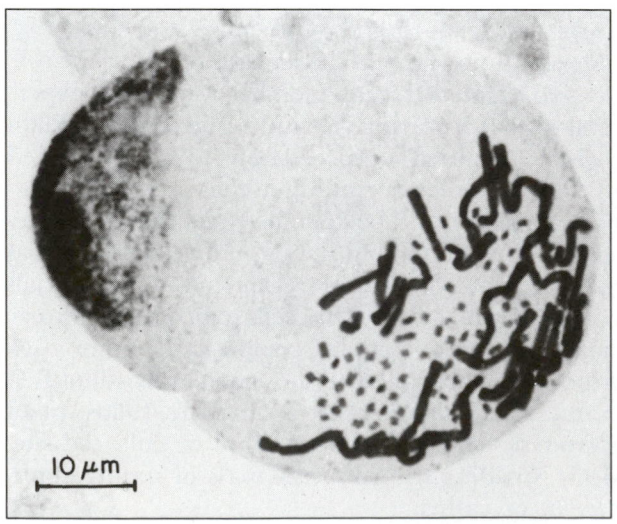

10 μm

netic blueprint for the entire organism. His work also showed that it is possible to grow an entire plant from a single cell, providing one can induce expression of the proper genes at the proper times. Since Steward's pioneering work, many plants have been successfully cultured using a variety of cell sources (Fig. 26–20). Plants have been regenerated from different tissues (for example, pith), organ explants (for example, root apical meristems, young embryos), and individual cells. The generation of plants from haploid pollen cells has proved to be particularly valuable because every gene in a haploid plant is expressed.

It is possible to remove the cell walls from individual plant cells. The resulting cells, called **protoplasts,** can then be induced to fuse with protoplasts from different plants (Fig. 26–21). After fusion, a new plant can sometimes be regenerated from the protoplast. For example, tomato and potato protoplasts have been experimentally induced to fuse, and the hybrid plant (a topato?) subsequently developed. One disadvantage of protoplast fusion is that the results of the fusion cannot be predicted. In the case of the tomato-potato fusion, the regenerated plants had stems like potatoes (no fruits) and below-ground parts like tomatoes (no tubers). More recently, potato protoplasts were fused with protoplasts of wild potatoes. When regenerated, the resulting plants had obtained a desirable trait from the wild ancestor—namely, insect resistance.

Cell- and tissue-culture techniques can obviously be used to help elucidate many fundamental questions involving growth and development in plants. The practical potential of these techniques should also be obvious. Using tissue culture, it is possible to regenerate large numbers of genetically identical plants from cells of a single, genetically superior plant. It is possible to alter genetic composition while in cell culture and then have these changes expressed during regeneration. Research involving cell and tissue cultures is one of the exciting and fruitful areas of biological research today.

◻ CHAPTER SUMMARY

I. Plant embryos develop in the seed in an orderly fashion, from proembryo to globular embryo to heart stage to torpedo stage.
II. Seed germination is affected by a variety of things.
   A. External environmental factors that may affect seed germination include requirements for oxygen, water, temperature, and light.
   B. Internal factors that may affect seed germination include immaturity and the presence of inhibitors.
III. Plant growth is localized in regions, or meristems, and involves cell division, cell elongation, and differentiation.
IV. Plants have two kinds of growth.
   A. Primary growth is an increase in length.
   B. Secondary growth is an increase in girth.
V. Plants are organized into three tissue systems.
   A. The ground tissue system makes up the bulk of the plant body.
      1. Parenchyma tissue is composed of relatively unspecialized, living cells.
      2. Collenchyma tissue is composed of living cells that help support the plant.
      3. Sclerenchyma tissue is composed of dead cells that help support the plant.
   B. The vascular tissue system conducts materials throughout the plant.
      1. Xylem is a complex tissue that conducts water and dissolved minerals.
      2. Phloem is a complex tissue that conducts food.
   C. The dermal tissue system is the outer protective covering of the plant body.
      1. The epidermis covers the primary plant body.
      2. The periderm covers the plant body in plants with secondary growth.
   D. Although separate organs (roots, stems, leaves, flowers, and fruits) exist in the plant, many tissues are integrated throughout the plant body, providing continuity from organ to organ.
VI. Plant development is controlled not only by genetic factors but also by external factors.
   A. Other plant tissues and organs exert a profound influence on plant development.
   B. Many environmental factors determine gene expression and affect plant development.

◻ POST-TEST

1. The _____ is a multicellular structure that anchors the embryo and aids in nutrient uptake from the endosperm.
2. Stems and roots have _____ growth because they grow throughout the life of the plant.
3. The _____ _____ is a protective covering over the root tip.
4. Primary growth, the increase in the length of the plant, is found in localized areas of the plant, the _____ meristems.
5. Stem apical meristems differ from root apical meristems in bearing embryonic structures, _____ _____ , and _____ .
6. The two lateral meristems responsible for secondary growth are the _____ and the cork cambium.
7. Plants that complete their life cycle in 1 year are called _____ , those that complete it in 2 years are _____ , and those that live year after year are _____ .
8. Storage, secretion, and photosynthesis are the functions of _____ .
9. The two tissues specialized for support are _____ and sclerenchyma.
10. Conduction of water and minerals in the xylem occurs in _____ and vessel elements.
11. Conduction of food in the sieve-tube members of the phloem is aided by _____ cells.
12. The outer covering of plants with primary growth is _____ , whereas plants with secondary growth are covered by the _____ .
13. Plants like grasses have a _____ root system.

14. Dormant terminal buds are covered by _____ .

15. Herbaceous stems have _____ for gas exchange, whereas woody stems have _____ .

16. A single leaf composed of several leaflets is said to be _____ .

17. Leaf arrangement on the stem may be alternate, _____ , or whorled.

18. Monocots have _____ leaf venation, and dicots have _____ leaf venation.

19. _____ are substances produced in one part of the plant and transported to another, where they affect growth and development.

20. A mass of cultured cells that have not differentiated is called a _____ .

◼ REVIEW QUESTIONS

1. Put the following stages of embryonic development in order and briefly describe each: torpedo stage, globular stage, proembryo, and heart stage.

2. How is growth in plants different from growth in animals? How is plant development different from animal development?

3. What factors influence the germination of seeds? How are these factors advantageous for plants?

4. What benefit does secondary growth confer on plants (as compared to plants with primary growth only)?

5. Why do you think plant development is so sensitive to nongenetic factors such as the presence of other cells and tissues and environmental influences?

◼ RECOMMENDED READINGS

Raghaven, V. *Embryogenesis in Angiosperms*. Cambridge, Cambridge University Press, 1986. An excellent review of all aspects of embryogenesis in angiosperms, especially good in reviewing experimental methods employed.

Sheridan, W.F., and J.K. Clark. "Maize Embryogeny: A Promising Experimental System," *Trends in Genetics* 3:3–6, 1987. A paper extolling the virtues of corn for embryology studies.

27
Leaf Structure and Function

OUTLINE

I. Epidermis, mesophyll, xylem, and phloem are the major tissues of the leaf
II. There is a relationship between structure and function in leaves
  A. The potassium ion mechanism explains how stomates open and close
  B. Leaves lose water by transpiration and guttation
III. Shedding of leaves allows plants in temperate climates to survive winter
IV. Leaves with functions other than photosynthesis exhibit modifications in structure
Focus on Comparative Plant Anatomy
Focus on Photosynthesis in Desert Plants

LEARNING OBJECTIVES

After you have studied this chapter you should be able to:

1. Identify the major tissues of the leaf.
2. Compare and contrast leaf anatomy of dicot and monocot leaves.
3. Explain how the structure of the leaf is related to its primary function, photosynthesis.
4. Outline the physiological changes that accompany stomatal opening and closing.
5. Define transpiration and list both positive and negative effects it has on the plant.
6. Relate why leaf abscission occurs and what physiological and anatomical changes precede it.
7. List several modified leaves and give the function of each.

Narrow beech fern, *Thelypteris phegopteris*. (Rod Planck/Marvin L. Dembinsky, Jr., Photography Associates)

Leaves are the main photosynthetic organs of the plant. Most leaves are thin and flat, a shape that allows maximum absorption of light energy and efficient internal diffusion of gases. As a result of their arrangement on the plant, they interfere minimally with one another's light supply. The leaves of plants form an intricate green mosaic, bathed in sunlight and atmospheric gases (Fig. 27–1).

KEY CONCEPTS

☐ The primary function of leaves is photosynthesis.

☐ Leaf structure is marvelously adapted for photosynthesis.

☐ Some leaves are modified for functions other than photosynthesis.

FIGURE 27–1 Leaf mosaic. The leaves of this ivy are growing in such a way that each shades the other minimally, promoting the maximal absorption of solar energy for the entire plant. (E.R. Degginger)

▦ EPIDERMIS, MESOPHYLL, XYLEM, AND PHLOEM ARE THE MAJOR TISSUES OF THE LEAF

The leaf is considered a plant organ because it is composed of several different tissues. These tissues are organized in a way that optimizes their main function, photosynthesis. Most leaves of flowering plants have a broad, flattened blade that is very efficient in collecting radiant energy. The blade has top and bottom sides, so

the leaf has two epidermal layers, the **upper epidermis** and the **lower epidermis** (Fig. 27–2). The cells making up this outer covering of the leaf are living parenchyma cells (Chapter 26). They generally lack chloroplasts and are relatively transparent. One interesting feature of leaf epidermal cells is that the cell wall on the outside of the leaf is thicker than on the inside. This may afford the plant additional protection from injury or water loss.

Leaves have such a large surface area exposed to the atmosphere that water loss by evaporation from the surface is unavoidable. The very feature that

FIGURE 27–2 The arrangement of tissues in a typical leaf blade. The blade is covered by an upper and lower epidermis. The photosynthetic tissue, the mesophyll, is often organized into palisades and spongy layers. Veins branch throughout the mesophyll.

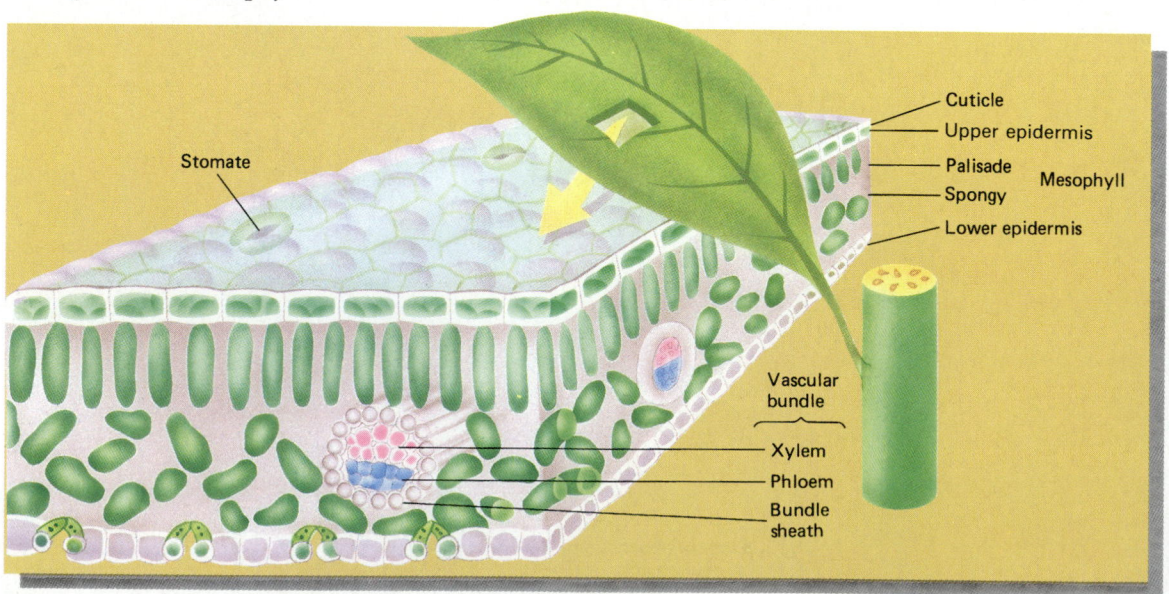

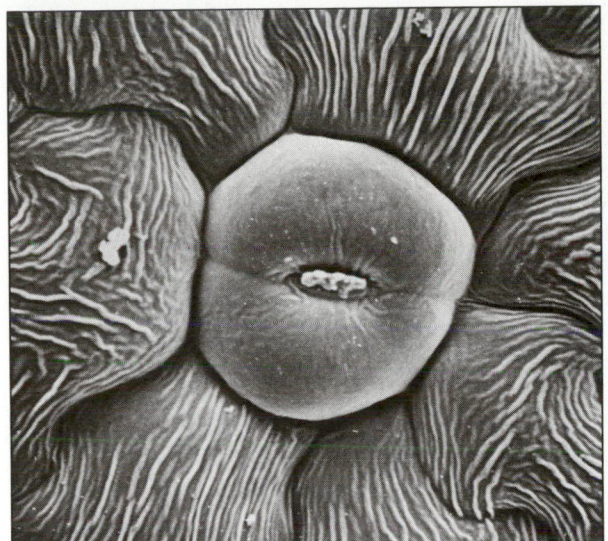

FIGURE 27–3 Scanning electron micrograph (approximately ×4000) of epidermal cells covered by a waxy cuticle, which is evident as ridges over the surface of the cells. The two cells in the center are guard cells. Note that the cuticle is not as obvious on the guard cells. (Biophoto Associates/Photo Researchers, Inc.)

makes a leaf so efficient in collecting the sun's rays is its undoing in regard to water. However, the epidermal cells secrete a waxy layer, the **cuticle,** which serves to reduce water loss (Fig. 27–3). Generally, the upper epidermis has a thicker cuticle than the lower epidermis. The cuticle varies in thickness in different plants.

As one might expect, the leaves of plants living in hot, dry climates have very thick cuticles.

The epidermis of most leaves is covered with various **trichomes** (hairs) (Fig. 27–4). Some leaves have so many trichomes that they feel quite fuzzy. Trichomes frequently aid in the reduction of water loss from the leaf surface by maintaining a layer of moist air next to the leaf. This reduces evaporation from the leaf's surface.

The leaf epidermis is covered with tiny openings, or **stomates,** flanked by specialized cells in the epidermis, the **guard cells** (Fig. 27–5). They are usually the only cells in the epidermis that have chloroplasts. Each pore and the two guard cells that form it are called a **stomatal apparatus.** Stomates are especially numerous on the lower epidermis of terrestrial plants and in many plants are located *only* on the lower surface. This reduces water loss because the stomates on the lower epidermis are shielded from direct sunlight.

The photosynthetic tissue of the leaf, the **mesophyll,** is sandwiched between the upper and lower epidermis. The word *mesophyll* comes from Greek, meaning "the middle of the leaf." Mesophyll cells are parenchyma cells (see Chapter 26) with chloroplasts. They are very loosely arranged with lots of air spaces between them. In dicots the mesophyll is divided into two specific areas. Toward the upper epidermis the cells are stacked into a **palisades** layer, while in the lower portion the cells are more loosely and more irregularly arranged in the **spongy** layer.

FIGURE 27–4 The leaf epidermis is often covered with trichomes that may limit the transpiration of water, discourage herbivores, sting, or perform other functions. (*a*) Stinging nettle, *Urtica ferox,* from New Zealand. The hairs on both leaves and stem will readily break off inside the skin of any animal that brushes against it or attempts to eat it. Irritating substances injected into the skin produce a lively stinging sensation. (*b*) Scanning electron micrograph of a nettle leaf. (*a*, G.R. Roberts; *b*, Biophoto Associates)

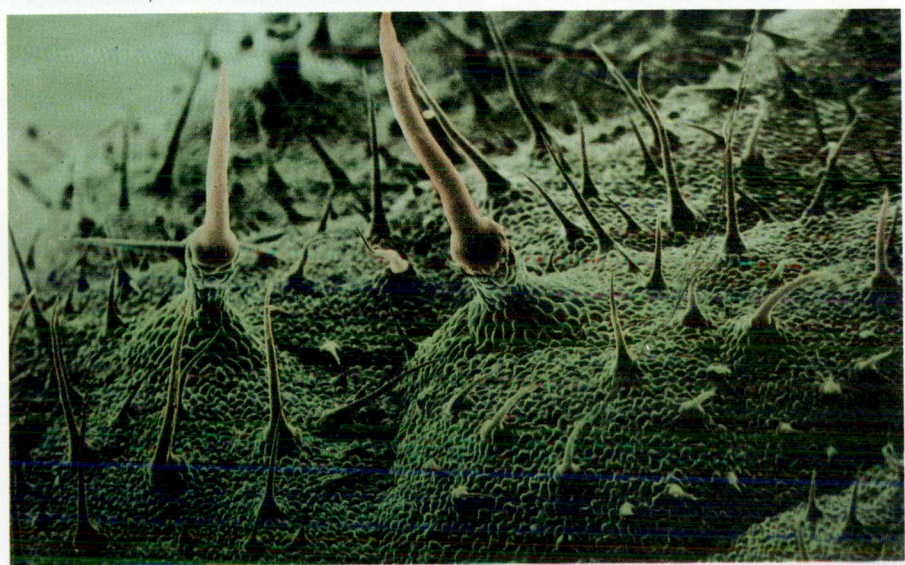

(a)   (b)

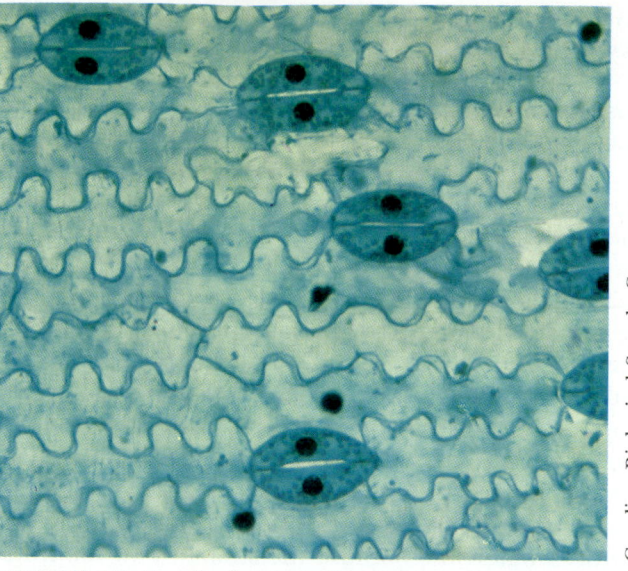

FIGURE 27–5   The lower epidermis of a lily leaf. Note the puzzle-shaped epidermal cells, which are relatively transparent. Each stomatal pore is flanked by two guard cells.

The veins of a leaf extend through the mesophyll. Branching is extensive; no mesophyll cell is very far from a vein. Each vein contains two types of vascular tissue. **Xylem** is located on the upper half of the vein, toward the upper epidermis, while **phloem** is always on the lower side of the vein.

Veins are usually surrounded by one or more layers of nonvascular cells called the **bundle sheath.** Bundle sheaths are composed of parenchyma or sclerenchyma cells (see Chapter 26). Frequently the bundle sheath has columns, or **extensions,** that extend through the mesophyll to the upper and lower epidermis. Bundle sheath extensions may be composed of parenchyma, collenchyma, or sclerenchyma.

Leaf structure differs in monocots and dicots. In dicots, a group that includes plants as diverse as beans and maples, leaves are usually composed of a broad, flattened blade and a petiole. Dicot veins are netted (Fig. 27–6). Monocots like lilies and corn often have long, narrow leaves that wrap around the stem in a sheath, rather than being attached by a petiole. Parallel veins are characteristic of monocot leaves.

The internal anatomy of dicot and monocot leaves is different as well (Fig. 27–7). Dicot leaves typically have two distinct regions in the mesophyll, the pali-

(a)

(b)

FIGURE 27–6   Vein patterns in leaves. (a) *Tibouchina urvilleana,* a dicot with netted veins. (b) Corn, a monocot with parallel veins. (a and b, G.R. Roberts)

FIGURE 27–7   Leaf cross sections. (a) Camellia *(Camellia japonica),* a dicot leaf, has a mesophyll with distinct palisades and spongy sections. (b) Corn *(Zea mays),* a monocot leaf. Note the absence of distinct regions of palisades and spongy mesophyll. Also evident is the evenly spaced parallel venation characteristic of monocots. (a, E.J. Cable/Tom Stack & Associates)

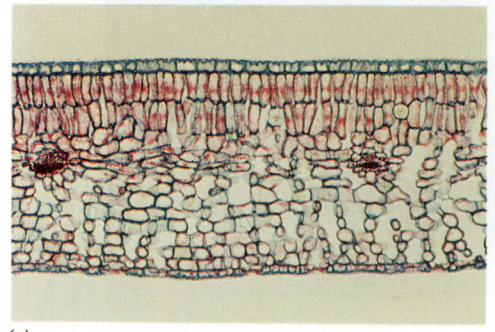
(a)

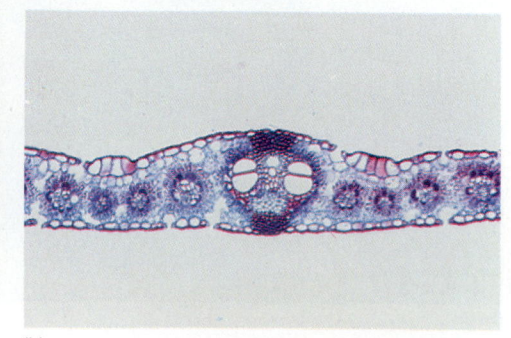
(b)

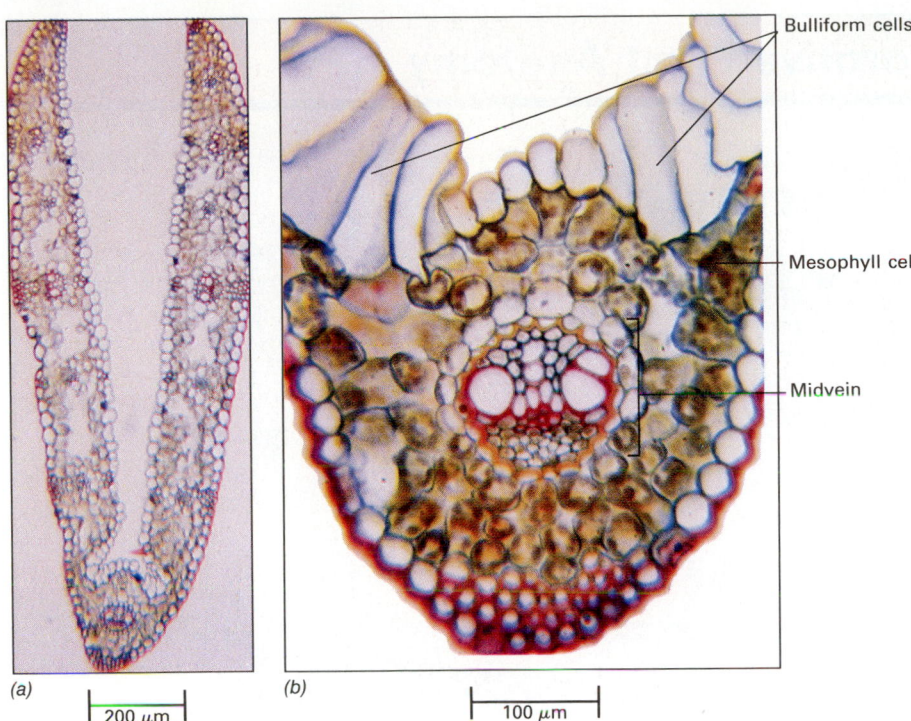

Bulliform cells

Mesophyll cell

Midvein

(a)

(b)

200 μm

100 μm

FIGURE 27–8 Cross section of bluegrass, a monocot. (*a*) The overall view of the leaf. (*b*) Higher magnification of the midvein region showing the bulliform cells. These bulliform cells are partially expanded. When the bulliform cells are fully turgid, the leaf blade is expanded rather than folded up. (*a* and *b*, Jack M. Bostrack)

sades and spongy layers. Mesophyll in many monocot leaves is not differentiated into palisades and spongy tissue. Because dicots have netted veins, a cross section of a dicot blade often shows veins in cross section as well as lengthwise views. In cross section the parallel pattern of monocot veins produces evenly spaced veins, all of uniform size except the midvein.

The upper epidermis of some monocot leaves has large, thin-walled cells, **bulliform cells,** located on ei-

ther side of the midvein (Fig. 27–8). These cells may be involved in the folding of the leaf during drought. When water is plentiful, the bulliform cells are turgid and the leaf is open. When the bulliform cells lose water, the leaf folds inward, reducing transpiration.

There are also differences between the guard cells in monocot and dicot leaves (Fig. 27–9). Dicots have guard cells that are shaped like beans. The epidermal cells surrounding them are not noticeably different

FIGURE 27–9 Variation in guard cells. Dicot guard cells are bean-shaped, while monocot guard cells are thin and are often narrow in the center and thicker at each end. Each monocot guard cell is associated with a subsidiary cell.

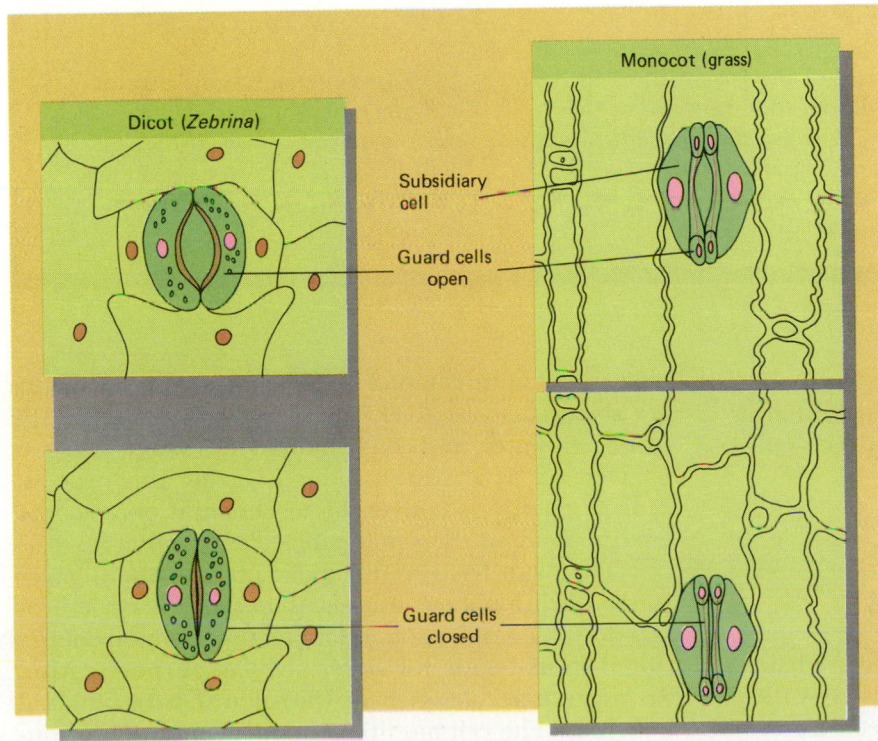

Dicot (*Zebrina*)

Monocot (grass)

Subsidiary cell

Guard cells open

Guard cells closed

FOCUS ON Comparative Plant Anatomy

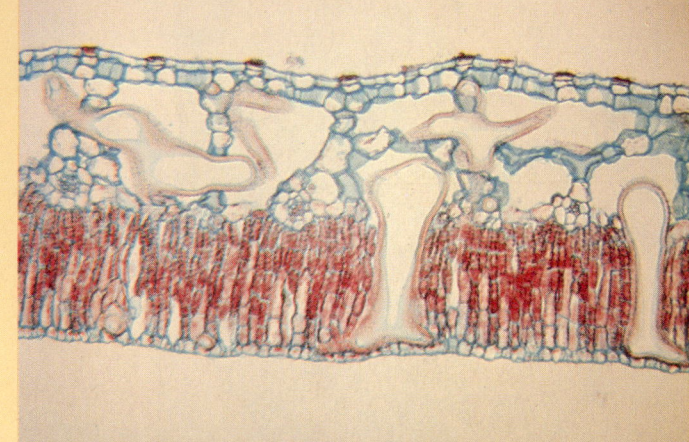

Plant A.

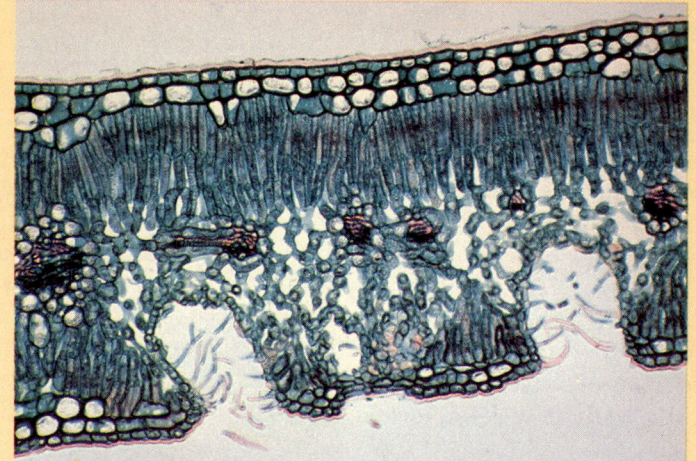

Plant B.

Certain modifications in leaf anatomy are characteristic of plants native to certain habitats, particularly habitats with environmental extremes. Figures A and B are cross sections of leaves from plants growing in entirely different environments. One is a **hydrophyte**, a plant adapted to a very wet habitat, and one is a **xerophyte**, a plant adapted to extremely dry conditions. Let's compare the details of each plant's leaf anatomy to identify the plants.

The epidermis of each plant is quite unusual. You will recall that typical plants have a single layer of cells comprising both the upper and lower epidermis, with most stomates located on the lower epidermis. Plant A has an upper epidermis with many stomates, while its lower epidermis lacks stomates entirely! Its cuticle is negligible. Plant B is even stranger, with a multiple-layered upper epidermis, covered by a thick cuticle. The stomates are located in indentations of the lower epidermis called **stomatal crypts.** A number of trichomes are also apparent in the stomatal crypt.

The mesophyll of each plant also reveals differences, although perhaps not as pronounced as in the epidermis. Both plants have a palisades layer and a spongy layer, although plant A has larger air spaces between its mesophyll cells. The large, red sclereid in the A mesophyll is a support cell to prevent the leaf's collapse since so much of its interior is air space. Support for the interior of the B leaf is provided by bundle-sheath extensions of the veins. The vascular tissue provides a clue on the identity of A and B as well. Plant A has less vascular tissue than plant B.

Based on your knowledge of leaf anatomy, can you identify which plant is xerophytic and which is hydrophytic? Turn to page 488 to see if you're a good comparative anatomist!

from other epidermal cells. Monocot leaves, on the other hand, often have guard cells shaped like dumbbells. Each guard cell is associated with a special epidermal cell called a **subsidiary cell.**

▢ THERE IS A RELATIONSHIP BETWEEN STRUCTURE AND FUNCTION IN LEAVES

The primary function of leaves is to collect radiant energy and convert it to a form that can be used by the plant. This process, **photosynthesis,** has been exam-ined in detail in Chapter 9. In photosynthesis, plants are able to take relatively simple molecules, carbon dioxide and water, and convert them into sugar. Oxygen is given off as a waste product. During this process, radiant energy is converted to chemical energy, the energy bonding the sugar molecules together.

The plant has two metabolic uses for the sugar it forms during photosynthesis. The sugar may be broken down by respiration, releasing the chemical energy stored in its bonds for other cellular purposes. Also, the sugar molecules provide the cell with basic building materials. The cell modifies the sugar molecules, con-

(a)

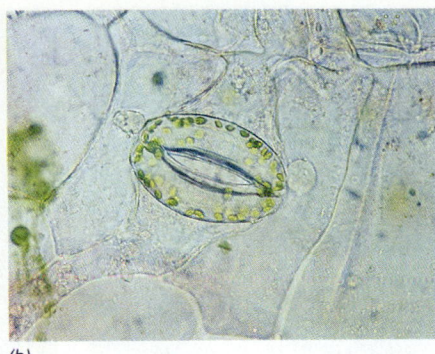
(b)

FIGURE 27–10 Stomatal opening in
***Zebrina:* (a) closed; (b) open.** (a and b,
Dwight R. Kuhn)

verting them into a number of other important compounds.

The transparent epidermis allows light to penetrate to the center of the leaf where the photosynthetic tissue, the mesophyll, is located. Carbon dioxide from the atmosphere diffuses into the leaf's interior through the stomates. Water from the soil is transported to the leaf in the xylem. The loose arrangement of the mesophyll tissue, with its moist cell surfaces and air spaces between cells, allows for rapid diffusion of both carbon dioxide and water into the chloroplasts. Also, the oxygen produced in photosynthesis diffuses rapidly out of the mesophyll cells and passes into the atmosphere through the stomates. Both carbon dioxide and oxygen are dissolved in the water film at the cell surface before going in or out of mesophyll cells. The veins supply water to the photosynthetic tissue (in the xylem) and carry the sugar produced in photosynthesis to other parts of the plant (in the phloem). The bundle sheath and bundle sheath extensions provide additional support to prevent the leaf, which is structurally weak because of the large amount of air space in the mesophyll, from collapsing under its own mass.

The Potassium Ion Mechanism Explains How Stomates Open And Close

Stomates are open during the day when photosynthesis is occurring and closed at night when photosynthesis is shut down (see Focus on Photosynthesis in Desert Plants for an interesting exception). This opening and closing of the stomates is caused by changes in the shape of the guard cells. When water moves into guard cells from surrounding cells, they become turgid (swollen) and bend outward at the center, creating a pore. This occurs because guard cells have thicker walls on the pore side that can't expand much as the cells become turgid. When water leaves the guard cells, they become flaccid and collapse against one another, closing the pore (Fig. 27–10).

The opening and closing of guard cells is triggered by an environmental cue, daylight or darkness. Other environmental factors are also involved, including carbon dioxide concentration. A low concentration of CO_2 within guard cells or in the atmosphere induces the stomates to open even in the dark. Another factor that affects stomatal opening and closing is severe water stress. During prolonged drought, stomates will remain closed even during the day. This mechanism is under hormonal control (Chapter 30). The opening and closing of stomates also appears to be under the control of an internal clock that approximates the 24-hour cycle. Plants placed in darkness continue to open and close their stomates at more or less the same times even in the absence of environmental cues like light and darkness. This internal biological clock is known as a **circadian rhythm.** Other examples of circadian rhythms are given in Chapter 30.

The data from numerous experiments and observations suggest that stomates of plants open and close by the **potassium ion (K^+) mechanism** (Fig. 27–11). The appearance of light triggers an influx of potassium ions into the guard cells from surrounding cells of the epidermis. This movement of potassium, which occurs through ion channels in the membrane by active transport and requires adenosine triphosphate (ATP), has been experimentally measured and verified. The increase of K^+ ions in the guard cells lowers the relative concentration of water in those cells. Therefore, water passes by osmosis into the guard cells from surrounding epidermal cells. This in turn changes the shape of the guard cells, and the pore opens. The guard cells remain electrically neutral during the influx of positively charged K^+ ions because of the concurrent movement of other ions. H^+ ions produced in the guard cells from ionized organic acids are pumped into the surrounding epidermal cells. Also, Cl^- ions are brought into the guard cells with the K^+ ions.

In the late afternoon or early evening the stomates close by a reversal of the process. The K^+ ions are pumped out of the guard cells into the surrounding epidermal cells. Water leaves the guard cells by osmosis, the cells collapse, and the pore closes.

Like other plant responses to light, stomates have varying sensitivities to light of different colors (i.e., wavelengths). The action spectrum for stomatal open-

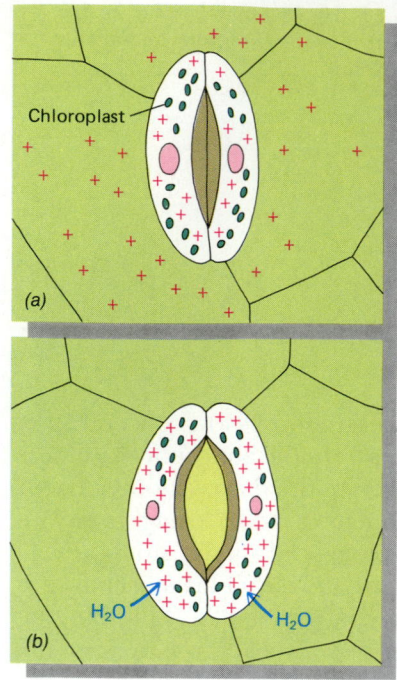

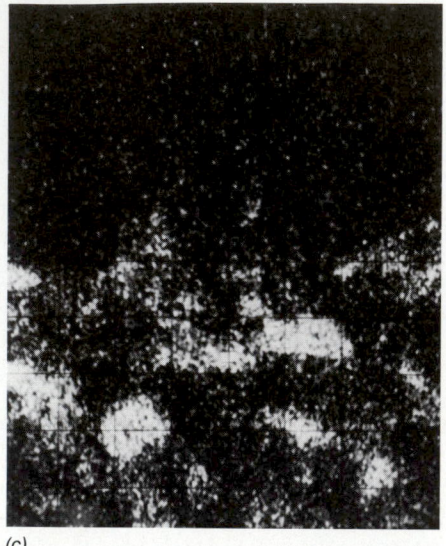

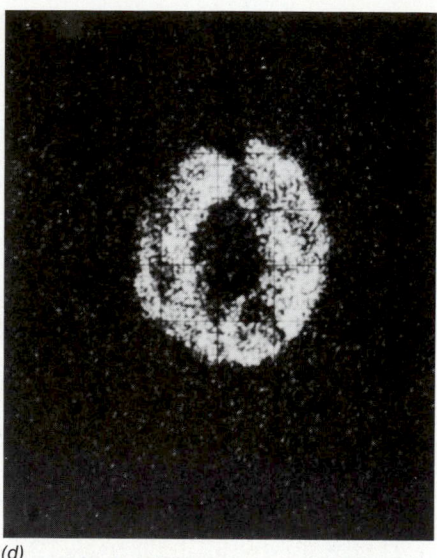

FIGURE 27–11 Movement of K$^+$ ions into and out of the guard cells affects stomatal opening and closing. (*a*) When the stomate is closed, K$^+$ ions are randomly distributed throughout the epidermis. (*b*) With the influx of K$^+$ ions into the guard cells from surrounding epidermal cells, water moves into the guard cells by osmosis. This changes the shape of the guard cells and the pore appears. (*c*) Radiograph of a strip of epidermis with a closed stomate. The white spots indicate K$^+$ ions. Note how they are more or less randomly distributed. (*d*) Radiograph of the same strip of epidermis with the stomate open. Note how the K$^+$ ions are concentrated within the guard cells. (*c* and *d*, Courtesy of Dr. Klaus Raschke, from Humble and Raschke, *Plant Physiology* 48:447–453, 1971, Michigan State University Atomic Energy Plant Research Laboratory)

ing shows greatest activity in the blue and, to a lesser extent, in the red regions. Also, dim blue light induces stomatal opening while dim red light does not. Any physiological response to light must involve a **photoreceptor,** a pigment that absorbs the light before the biological response occurs. These and other data suggest that the photoreceptor for stomatal opening and closing is a yellow flavoprotein located on the tonoplast (vacuolar membrane) in the guard cells. Light also affects stomatal opening indirectly by inducing photosynthesis. This reduces the internal concentration of CO_2 in the leaf, which triggers opening of the stomates.

Leaves Lose Water By Transpiration And Guttation

Despite leaf adaptations, such as the cuticle, that control water loss, approximately 99% of the water a plant absorbs from the soil is lost by evaporation from the leaves and stem. The loss of water vapor from land plants is called **transpiration.** The cuticle is effective in reducing water loss; it is estimated that only 1 to 3% of water lost from a plant passes directly through the cuticle. Most of the water loss occurs through the stomates. The numerous pores, so effective in gas exchange for photosynthesis, are also openings through which water vapor can escape. Also, the loose arrangement of the mesophyll cells provides a large amount of internal air within the leaf in which water can evaporate.

A number of environmental factors influence the transpiration rate. With higher temperatures more water is lost from plant surfaces. Wind also increases the transpiration rate. A high relative humidity decreases transpiration because the air is already saturated, or nearly so, with water vapor. Light increases the transpiration rate, in part because it causes the stomates to open.

Transpiration may seem like a wasteful process. However, there may be some benefits to the large amount of water plants lose by transpiration. First, transpiration, like sweating in animals, has a cooling effect on the plant. When water passes from a liquid state to a vapor, it absorbs a great deal of heat. As the water molecules leave the plant, they carry this heat with them. It is possible that the cooling effect of transpiration prevents overheating of the plant, particularly in direct sunlight. In fact, on a hot summer day the internal temperature of leaves is lower than that of the surrounding air.

A second benefit of transpiration is that it provides the plant with essential minerals. Remember that the water a plant transpires is absorbed from the soil. Soil water is not pure water, but rather a very dilute solution of dissolved mineral salts. The plant uses

FOCUS ON   Photosynthesis in Desert Plants

Plants living in dry, or **xeric,** conditions have a number of special anatomical adaptations that enable them to survive (see Focus on Comparative Plant Anatomy). Many xeric plants have physiological adaptations as well. For example, their stomates may open during the cooler night and close during the hot day to reduce water loss from transpiration. This is in contrast to most plants, which have stomates that are opened during the day and closed at night. But xeric plants that have their stomates closed during the day cannot exchange gases for photosynthesis. You will recall that plants typically fix carbon dioxide during the day, when sunlight is available to produce ATP and NADPH + H$^+$ (Chapter 9).

Many xeric plants have evolved a special photosynthetic pathway, **Crassulacean Acid Metabolism,** or **CAM,** that in effect solves this dilemma. The name comes from the stonecrop plant family, the Crassulaceae, which possesses the CAM pathway, although it has been identified in over 25 different plant families. A number of unrelated plants, including the pineapple and most cacti, have it. It is most common in angiosperms, especially dicots, but occurs in several monocots, *Wel-*

witschia (a gymnosperm), and some ferns.

CAM plants fix CO_2 during the night, when stomates are open, into malic acid (a four-carbon compound), which is stored in the vacuole. During the day, when stomates are closed and gas exchange cannot occur between the plant and the atmosphere, the malic acid is decarboxylated to yield CO_2 again. Now the CO_2 is available *within the leaf tissue* to be fixed into sugar by the usual photosynthetic pathway, the C_3 pathway (Chapter 9).

CAM photosynthesis may sound familiar to you. It is very similar to the C_4 pathway discussed in the Focus Box in Chapter 9. There are important differences, however. You will recall that C_4 plants initially fix carbon dioxide into four-carbon organic acids in leaf mesophyll cells. The acids are later decarboxylated to produce CO_2, which is fixed by the C_3 pathway in the bundle sheath cells. In other words, the C_4 and C_3 pathways occur in *different locations* within the leaf of a C_4 plant. In CAM plants, the initial fixation of CO_2 occurs at night. Decarboxylation of malic acid and subsequent production of sugar from CO_2 by the normal photosynthetic pathway occur

Prickly pear cactus, a typical CAM plant that grows in arid habitats. (Grant Heilman)

during the day. In other words, the C_4 and C_3 pathways occur at *different times* within the same cell of a CAM plant.

The CAM pathway is a very successful adaptation to xeric conditions. CAM plants are able to have gas exchange for photosynthesis *and* significantly reduce water loss by transpiration. Plants with CAM photosynthesis can survive in deserts where neither C_3 nor C_4 plants can.

---

many of these minerals to grow. Some evidence suggests that transpiration enables a plant to take in enough essential minerals and that plants could not satisfy their mineral requirement if they did not transpire.

There is no doubt, however, that under certain circumstances transpiration can be harmful to a plant. On hot summer days, plants frequently lose more water by transpiration than is replaced from the soil. Their cells experience a loss of **turgor** and the plant wilts. If a plant is able to recover overnight, as a result of negligible transpiration (closed stomates) while water is still being absorbed from the soil, the plant is said to have experienced temporary wilting. Most plants recover from temporary wilting with no ill effects. In cases of prolonged drought, the soil may not contain sufficient moisture to permit overnight recov-

ery from wilting. A plant that cannot recover overnight is said to be permanently wilted and will die unless water is supplied immediately. It appears that transpiration is a "mixed blessing" for plants. It has possible benefits and potential hazards but is unavoidable in plants because they have stomates.

There are times when some plants excrete water as a *liquid*. This process, known as **guttation,** occurs when transpiration is negligible and available soil moisture is high. Guttation frequently occurs at night because the stomates are closed, but water continues to move into the roots by osmosis (Chapter 28). Many leaves have special openings through which the water is literally forced out. The early morning droplets of water produced on leaves by guttation are sometimes mistaken for dew, which is water condensation from the air (Fig. 27–12).

FIGURE 27–12   Guttation in wild strawberry (*Fragaria* sp.) plants. (James L. Castner)

■ SHEDDING OF LEAVES ALLOWS PLANTS IN TEMPERATE CLIMATES TO SURVIVE WINTER

In temperate climates, the leaves turn color and **abscise,** or fall off, when winter approaches. Abscission is a complex process that involves many physiological changes in the plant. All of it is orchestrated by changing levels of plant hormones (Chapter 30). What is the advantage of shedding leaves? You have seen that leaves lose a tremendous amount of water by transpiration. During the winter, water is in very short supply. As the ground chills, the roots absorb less water; if

the ground freezes, there can be *no* absorption. If a plant were to maintain its broad leaves during the winter, it would continue to lose water by transpiration but would be unable to replace water by absorption from the soil. Plants have little use for leaves in winter, anyway; in the lower temperatures of winter, the plant's metabolism, including its photosynthetic machinery, slows down a great deal. Shedding leaves is an adaptation to the temporary "drought" and cold of winter.

As autumn approaches, the plant reabsorbs many of the essential minerals located in the leaves. Nitrogen, phosphorus, and possibly potassium move into the woody tissues from the leaves. The level of sugar in the leaves rises. Chlorophyll is broken down, and some of the accessory pigments in the chloroplast, orange **carotenes** and yellow **xanthophylls,** become evident. These pigments were always present in the leaf but were masked by the chlorophyll. In addition, other water-soluble pigments, the red **anthocyanins,** may be synthesized and stored in the vacuole; their function is unknown. The various combinations of these pigments are responsible for the brilliant colors found in autumn landscapes in temperate climates.

The area where the leaf petiole detaches from the stem is structurally different from surrounding tissues (Fig. 27–13). This area, called the **abscission zone,** is composed primarily of thin-walled parenchyma cells and is anatomically weak. There are few fibers. On the stem side of the abscission zone a protective layer of cork cells develops. These cells have a waxy material called **suberin,** which is impermeable to water, impregnated in their walls. As fall approaches, enzymes dissolve the middle lamella, which is the "cement" holding the cells together, in the abscission zone. By this time, there is nothing holding the leaf to the stem but a few xylem cells. A sudden breeze is enough to make the final break, and the leaf detaches. The protective layer remains, sealing off the area and forming the leaf scar.

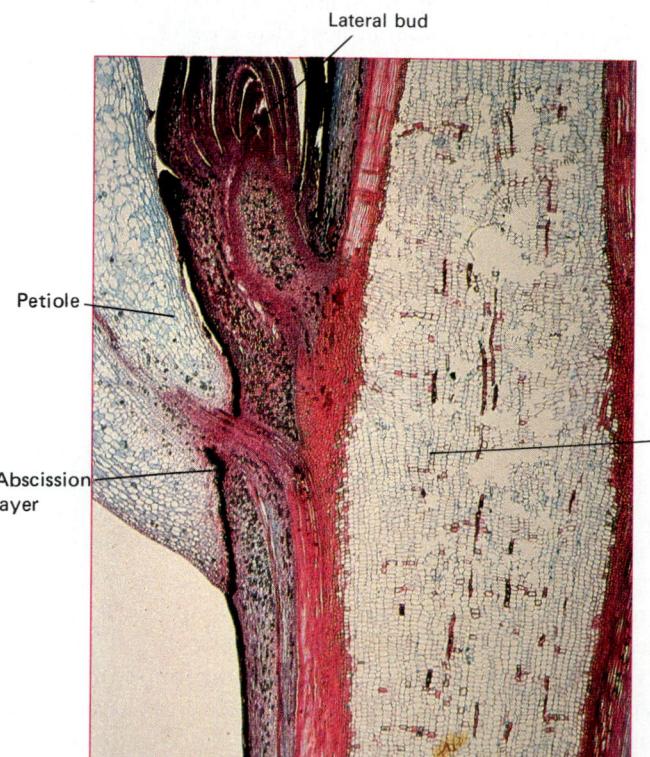

Lateral bud

Petiole

Abscission layer

Stem

FIGURE 27–13   A longitudinal section through a maple branch, showing the base of the petiole. Note the abscission zone where the leaf will abscise from the stem. A lateral bud with its protective bud scales is evident above the petiole. (Biophoto Associates)

FIGURE 27–14   Leaf modifications. (*a*) The leaves of cacti are modified as spines for protection. The stem of a cactus functions both for photosynthesis and water storage. (*b*) Tendrils, which grasp onto objects and aid the plant in climbing, may be modified leaves or stems. These pea tendrils are modified leaves. (*c*) Overlapping bud scales protect buds. Shown here are a terminal bud and several lateral buds. (*d*) The leaves of bulbs such as the onion are fleshy for storage. (*e*) Some plants, such as *Echeveria* sp., have succulent leaves, which are modified for water storage as well as photosynthesis. (*a*, Tom Algire/Tom Stack & Associates; *b*, Dwight R. Kuhn; *c*, William E. Ferguson; *d*, William D. Adams; *e*, Peter F. Zika/Visuals Unlimited)

LEAVES WITH FUNCTIONS OTHER THAN PHOTOSYNTHESIS EXHIBIT MODIFICATIONS IN STRUCTURE

In certain plants, leaves have evolved to perform functions other than photosynthesis (Fig. 27–14). Some plants have leaves specialized for protection. **Spines,** which are hard and pointed, may be found on plants like cacti. (In the cactus, the main organ of photosyn- thesis is the stem rather than the leaf.) Many vines, which are climbing stems that do not support their own weight, have **tendrils.** Tendrils, which are usually spe- cialized leaves, are used for grasping and holding onto other structures. Tendrils keep the vine attached to the structure on which it's growing. The winter buds of a dormant woody plant are covered by protective **bud scales.** Bud scales are modified leaves. They protect the delicate meristematic tissue of the shoot from in-

(a)

(b)

FIGURE 27–15 The pitcher plant has leaves modified to form a pitcher that collects water, drowning its prey. (*a*) Growth habit of a common pitcher plant, *Sarracenia purpurea*. (*b*) A cutaway view of a pitcher reveals accumulated insect bodies and debris. (*a*, Skip Moody/M.L. Dembinsky, Jr., Photography Assoc.)

Carolina Biological Supply Company

jury and desiccation. Leaves may be modified for storage of water or food. For example, the **bulb** is a short stem to which large, fleshy leaves are attached. Onions and tulips form bulbs, which grow underground. Many desert plants have fleshy, succulent leaves for water storage. These leaves are usually green and function for photosynthesis as well.

Some of the most bizarre examples of modified leaves are those of insectivorous plants. These plants have leaves modified to trap animal prey, usually in-

FIGURE 27–16 The modified leaves of a Venus flytrap snap shut on its prey. Venus flytraps are found in bogs in North and South Carolina. (*a*) Growth habit of Venus flytrap, *Dionaea muscipula*, with opened traps. (*b*) A closed trap and insect prey. After enzymes produced by the plant digest as much of the insect as is digestible, the trap will reopen. (*a*, Howard A. Miller, Sr./Photo Researchers, Inc.; *b*, Nuridsany et Perennou/Photo Researchers, Inc.)

(a)

(b)

sects. Most insectivorous plants grow in poor soil that is deficient in certain essential minerals. Insectivorous plants obtain some nutrients by digesting insects and other small animals.

Some insectivorous plants have passive traps. For example, the leaves of the pitcher plant are shaped so that rainwater collects within, forming a reservoir (Fig. 27–15). An insect that is attracted to the pitcher by its scent or nectar may lean over the edge and fall in. It is prevented from crawling out by a row of stiff spines that point downward around the lip of the pitcher. The insect eventually drowns. Enzymes produced by

the plant digest part of the insect's body. This material is absorbed by the plant. In the tropics pitcher plants may be large enough to hold 1 liter or more of water.

The Venus flytrap is an example of an insectivorous plant with active traps (Fig. 27–16). The leaves resemble tiny bear traps. Each side of the leaf has three small hairs located on it. If an insect alights and brushes against two of the hairs, the trap springs shut with amazing rapidity (Chapter 30). After the insect has been digested, the trap reopens and the indigestible remains fall off.

■ CHAPTER SUMMARY

I. Leaf anatomy reflects its main function, photosynthesis.
  A. The transparent epidermis allows light to penetrate into the mesophyll. Stomates in the epidermis are for gas exchange.
  B. The mesophyll has air spaces to allow for rapid diffusion of carbon dioxide and water into the cells. Photosynthesis takes place in the mesophyll.
  C. Veins of a leaf have xylem (to conduct water and essential minerals to the leaf) and phloem (to conduct sugar produced by photosynthesis to the rest of the plant).
II. Monocot and dicot leaves can be distinguished based on their external morphology and their internal anatomy.
III. Stomates open during the day and close at night.
  A. The K^+ ion mechanism explains the opening

and closing process.
  B. A number of factors affect stomatal opening, including light or darkness, CO_2 concentration, and a circadian rhythm within the plant.
IV. Transpiration is the loss of water vapor from plants.
  A. It occurs primarily through the stomates.
  B. The rate of transpiration is affected by environmental factors like temperature and wind.
  C. Transpiration may be both beneficial and harmful to the plant.
V. Guttation is the loss of water as a liquid from plant leaves.
VI. Leaf abscission is a complex process involving physiological and anatomical changes.
VII. Leaves may be modified for functions other than photosynthesis.

■ POST-TEST

1. The _____ is the photosynthetic tissue in the middle of the leaf.
2. Gas exchange occurs through tiny pores formed by _____ _____.
3. The _____ of a leaf vein transports water and dissolved minerals.
4. Stomates are usually more numerous on the _____ epidermis.
5. The leaves of _____ usually have netted venation, distinct palisade and spongy layers in the mesophyll, and bean-shaped guard cells.
6. _____ cells may be involved in folding, or rolling, of monocot leaves during periods of drought.
7. The opening and closing of stomates is due to the movement of _____ ions.
8. When light is involved in a physiological response, the

plant must have a _____ to absorb and perceive the light prior to the physiological response.
9. Most of the water a plant absorbs from the soil is lost by _____.
10. Responses based on an internal biological clock are called _____ _____.
11. The fall of leaves in the autumn is known as leaf _____.
12. The leaf scar is coated with a waterproof material called _____.
13. _____ are leaves that are modified for climbing.
14. Onion bulbs have fleshy leaves that are modified for _____.
15. The leaves of a cactus are modified as _____.

■ REVIEW QUESTIONS

1. Give the general equation for photosynthesis and discuss how the leaf is structured to deliver raw materials and products of photosynthesis.
2. What plant structure is related to both photosynthesis and transpiration? How is it tied to each physiological process?
3. Stomates open in response to light. Could the photoreceptor involved in stomatal opening be chlorophyll? Why or why not?

4. Relate the series of physiological changes that occur in the guard cells during stomatal opening.
5. How does the environment influence transpiration rate? stomatal opening and closing?
6. Why do many woody plants lose their leaves in autumn?
7. Discuss the specialized features of leaves of insectivorous plants.

■ RECOMMENDED READING

Stern, K.R. *Introductory Plant Biology*, 4th ed. Dubuque, Iowa, Wm. C. Brown Publishers, 1988. Includes an interesting discussion on the ecological and human relevance of leaves.

Answer to question on page 480: Plant A, a water lily, is a dicot that exhibits hydrophytic characteristics. It has stomates on its upper epidermis only because the leaf floats on water. Roots and stems, as well as leaves, have large air spaces because gas exchange is more limiting than water for hydrophytic plants. Plant B, oleander, is xerophytic. Xerophytic plants often have a thick cuticle, multiple layered epidermis, and stomatal crypts to reduce transpiration.

28

Stems and Roots

OUTLINE

I. Herbaceous monocots and dicots can be distinguished by the arrangement of vascular bundles in their stems
II. Gymnosperms and certain dicots have stems with secondary growth
III. There are structural differences between primary roots and primary stems
    A. Considerable variation occurs in the internal arrangement of tissues in dicot and monocot roots
    B. There is a relationship between structure and function in primary roots
IV. Gymnosperms and certain dicots have roots with secondary growth
V. Transport in plants occurs in the xylem and phloem
    A. Water and minerals are translocated in the xylem
    B. Sugar is translocated in the phloem
VI. Roots obtain most of the naturally occurring minerals that are found in plant tissues
    A. Sixteen elements are essential for plant growth
    B. Both organic and inorganic fertilizers can replace certain key elements if they are missing from the soil
Focus on Tree-Ring Analysis
Focus on Commercial Hydroponics

LEARNING OBJECTIVES

After you have studied this chapter you should be able to:
1. List several functions of stems and explain how the structure of stems relates to their function.
2. Compare and contrast the structure of herbaceous dicot stems and monocot stems.
3. Trace the pathway of water movement in plants.
4. Distinguish between root pressure and tension-cohesion as mechanisms to explain the rise of water in xylem.
5. Outline the pressure-flow theory of sugar transport in phloem.
6. List several functions of roots.
7. Compare the arrangement of tissues in primary dicot and monocot roots. Give at least one function for each tissue.
8. Outline the criteria an element must satisfy in order to be considered essential for plant growth.
9. List at least five elements that are essential for plant growth and give a physiological role for each in plants.

American beech, *Fagus grandifolia.* (Ed Reschke)

490

The basic plant body has three parts: leaves, roots, and stems. Leaves are the site of photosynthesis, which is the conversion of radiant energy into the chemical energy of sugar (Chapter 27). Roots serve to anchor the plant and aid in absorption of materials from the soil. Stems are usually above-ground structures that link the leaves to the roots.

Stems have three main functions. They support the leaves and reproductive structures. The upright position of most stems, and the arrangement of leaves on them, are adaptations that allow each leaf to absorb maximal light. A second function of stems is internal transport. Stems conduct water and dissolved minerals that roots absorb from the soil to the leaves and other plant structures through the xylem. Stems also conduct the food produced in the leaves by photosynthesis to the roots and other parts of the plant through the phloem. It should be emphasized, however, that stems are not the only plant organ in which conduction takes place. The vascular system is continuous throughout all parts of the plant, and conduction occurs in roots, stems, leaves, and reproductive structures (Fig. 28–1). The third main function of stems is to produce new living tissue. Stems continue to grow throughout the life of the plant. In addition to support, conduction, and production of new stem tissue, a number of stems are modified for asexual reproduction (Chapter 29).

Roots are underground and out of sight, so most people don't appreciate their importance to plants. Roots are essential plant organs. They serve to anchor the plant. They also absorb water and dissolved minerals from the soil, which are then transported throughout the plant. In addition, many roots serve as storage organs. Excess sugars produced in the leaves by photosynthesis are transported through the phloem to the roots, where they are stored until they are used. The roots use some of this sugar for their own respiratory needs, but most of the sugar that is stored in the root is transported via the phloem to other parts of the plant when needed. Some plants, like beets and sweet potatoes, have roots that are enormously swollen with food-storage tissues (Fig. 28–2).

KEY CONCEPTS

- The main function of stems is to support leaves and reproductive structures of plants.

- Roots anchor the plant and absorb water and dissolved minerals from the soil.

- Although there is considerable variation in stem anatomy, all vascular stems have an outer protective covering, some type of ground tissue, and the vascular tissues, xylem and phloem.

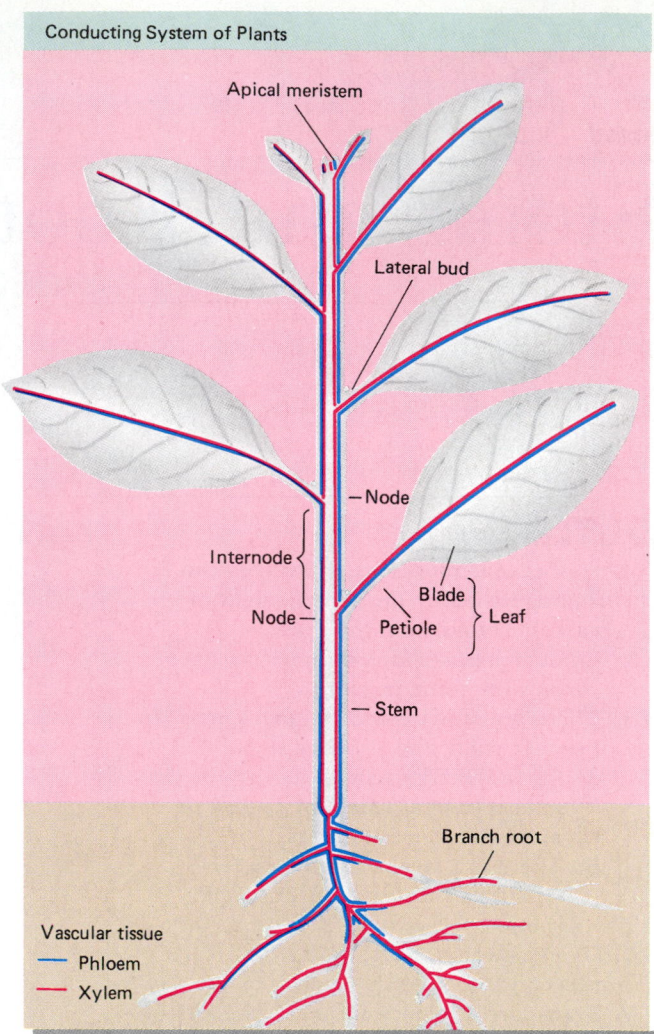

FIGURE 28–1 The plant body of a typical herbaceous plant. Note the continuity of the vascular tissues from one plant organ to the next.

- Roots differ from stems in that roots possess root hairs, root caps, and endodermis and pericycle tissues.

HERBACEOUS MONOCOTS AND DICOTS CAN BE DISTINGUISHED BY THE ARRANGEMENT OF VASCULAR BUNDLES IN THEIR STEMS

In Chapter 26, you learned that plants have two different types of growth. Primary growth is the increase in the length of the plant and occurs at the tips of plants, which are called **apical meristems.** Secondary growth is the increase in the girth of the plant. It is due to the activity of lateral meristems called **cambia** (singular, *cambium*) that are located around the sides of the plant. All plants have primary growth, whereas only some plants have secondary growth. Plant stems with only

FIGURE 28–2   Beets and other root crops are important sources of human food because of the accumulated food that the plant stores in the root. (G.R. Roberts)

primary growth are herbaceous. Although primary stems all have the same basic tissues, the arrangement of tissues in the stem varies considerably. We shall consider primary stem structure in the two groups of flowering plants, the dicots and the monocots.

The sunflower (*Helianthus*) stem is a representative dicot stem with primary growth (Fig. 28–3). Its outer covering is the **epidermis.** Inside the epidermis is a layer that is several cells thick, the **cortex.** The vascular tissue is located in patches that are arranged in a circle. Each patch, or vascular bundle, contains both **xylem** and **phloem.** The xylem is usually located on the inner side of the vascular bundle, whereas the phloem is usually located to the outside. Sandwiched between the xylem and phloem is a single layer of cells, the **vascular cambium.** In sunflowers, there is a cluster of fibers directly outside the phloem, the **phloem fiber cap;** this is not always found in other herbaceous dicots. Although the vascular bundles are arranged in a circle in cross section, it is important to remember that they run as long strands throughout the length of the stem, continuous with the vascular tissues of the root and leaves. The middle of the dicot stem is **pith,** a tissue composed of large thin-walled parenchyma cells. Due to the arrangement of the vascular tissues in bundles, there is no distinct separation of cortex and pith between the bundles. These areas of the stem are usually referred to as **pith rays.**

Monocot stems such as corn (*Zea mays*) are also covered with an epidermis (Fig. 28–4). As in dicot stems, the vascular tissue runs in strands throughout the length of the stem. However, monocot vascular bundles are not arranged in a circle as in dicots. Rather, they are scattered throughout the stem so that distinct areas of cortex and pith are lacking. The tissue in which the vascular tissues are embedded is called **ground tissue** or, sometimes, **ground parenchyma.** Each vascular bundle contains xylem to the inside and

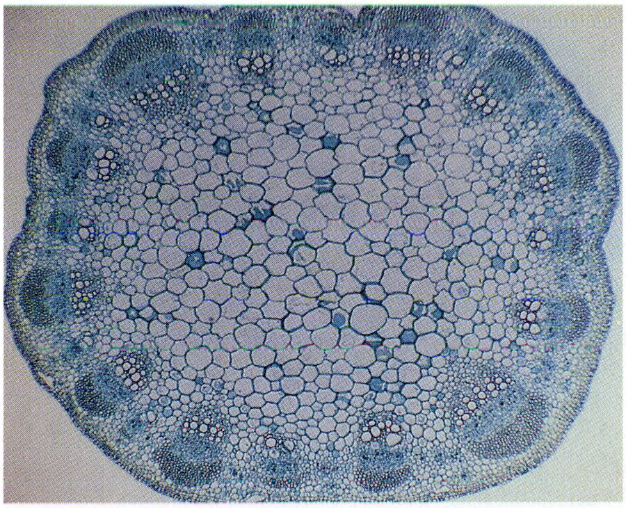

(a)

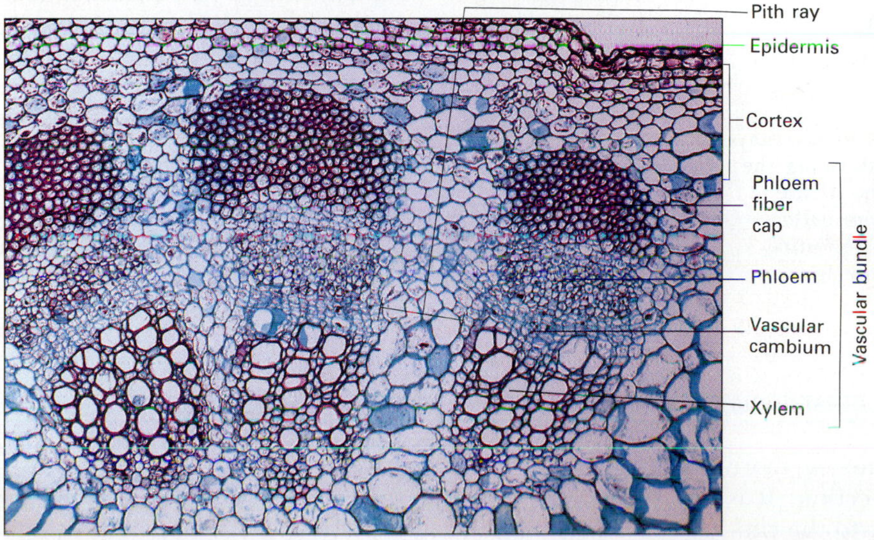

Pith ray

Epidermis

Cortex

Phloem fiber cap

Phloem

Vascular cambium

Xylem

Vascular bundle

FIGURE 28–3   Primary growth in a dicot stem. (a) Cross section of a *Helianthus* (sunflower) stem showing the arrangement of tissues. The vascular bundles are arranged in a circle. (b) Close-up of *Helianthus* vascular bundle. The xylem is to the inside and the phloem to the outside. Each vascular bundle is "capped" by a batch of fibers for additional support. (Ed Reschke)

(b)

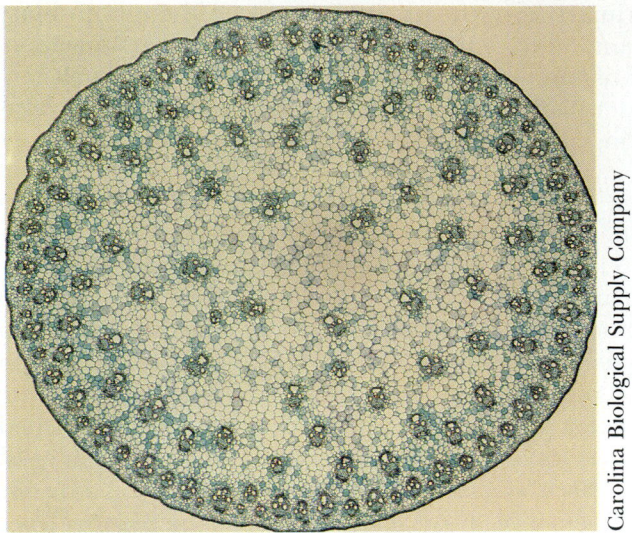

(a)

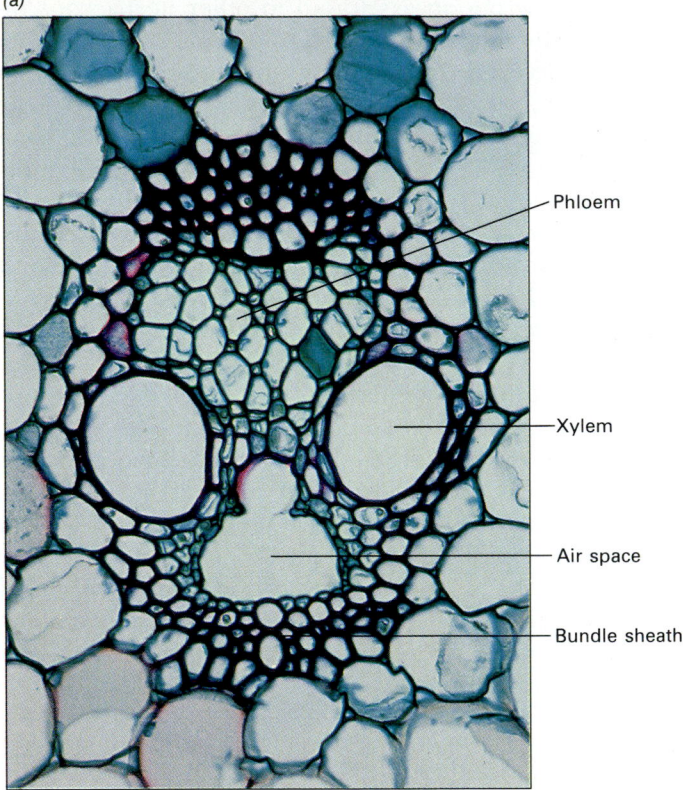

- Phloem

- Xylem

- Air space

- Bundle sheath

(b)

FIGURE 28–4   Arrangement of stem tissues in *Zea mays* (corn), a monocot. (a) Cross section of stem showing the scattered vascular bundles. (b) Close-up of one of the bundles. The air space is where the first xylem cells were formed that are no longer functional. The entire bundle is enclosed in a sheath of sclerenchyma for additional support. (Ed Reschke)

phloem to the outside. There is no vascular cambium in monocot stems.

Each tissue in primary stems has one or more functions. The epidermis serves for protection. It is covered by the cuticle, a waxy layer similar to the cuticle of the leaf epidermis. The stem cuticle reduces water loss from the surface of the stem. The cortex in dicot stems has several functions: photosynthesis, storage, and support. The vascular tissues function not only for conduction but also for support. Xylem transports water and dissolved minerals, and phloem transports food. Fibers may be found in both xylem and phloem, although they are usually more extensive in the phloem. These fibers add considerable strength to the stem body. The vascular cambium is responsible for secondary growth, which is considered later. The pith in the center of dicot stems functions primarily for storage. Although monocots lack a distinct cortex and pith, their ground tissue performs the same kinds of functions.

■ GYMNOSPERMS AND CERTAIN DICOTS HAVE STEMS WITH SECONDARY GROWTH

Secondary growth occurs in plants as a result of the activity of two lateral meristems or cambia (Chapter 26). Cells in the **vascular cambium** divide and produce secondary xylem and secondary phloem, which become the functional replacements for primary xylem and primary phloem. The tissue formed to the inside of the vascular cambium is secondary xylem, commonly known as wood (Fig. 28–5). The tissue formed to the outside of the vascular cambium is secondary phloem, which makes up the inner bark of a woody plant.

The second lateral meristem, the **cork cambium,** divides to produce cork cells and cork parenchyma. The cork cambium and the tissues it produces are collectively referred to as **periderm,** which functions as a replacement for the epidermis. The periderm and any tissues external to it comprise the outer part of the bark.

In flowering plants, only certain dicots have secondary growth. Woody plants such as apple, hickory, and maple are examples of dicots with secondary growth. All gymnosperms have secondary growth; examples include pine, juniper, and spruce. Plants with secondary growth start their development as plants with primary growth; that is, a woody plant, when it begins growth, has primary growth. However, as it continues to increase in length (by primary growth) at the tips of the branches and roots, the older parts of the plant further back from the tips develop secondary tissues.

■ THERE ARE STRUCTURAL DIFFERENCES BETWEEN PRIMARY ROOTS AND PRIMARY STEMS

Primary roots have certain tissues that are also found in primary stems such as epidermis, cortex, xylem, and

Carolina Biological Supply Company

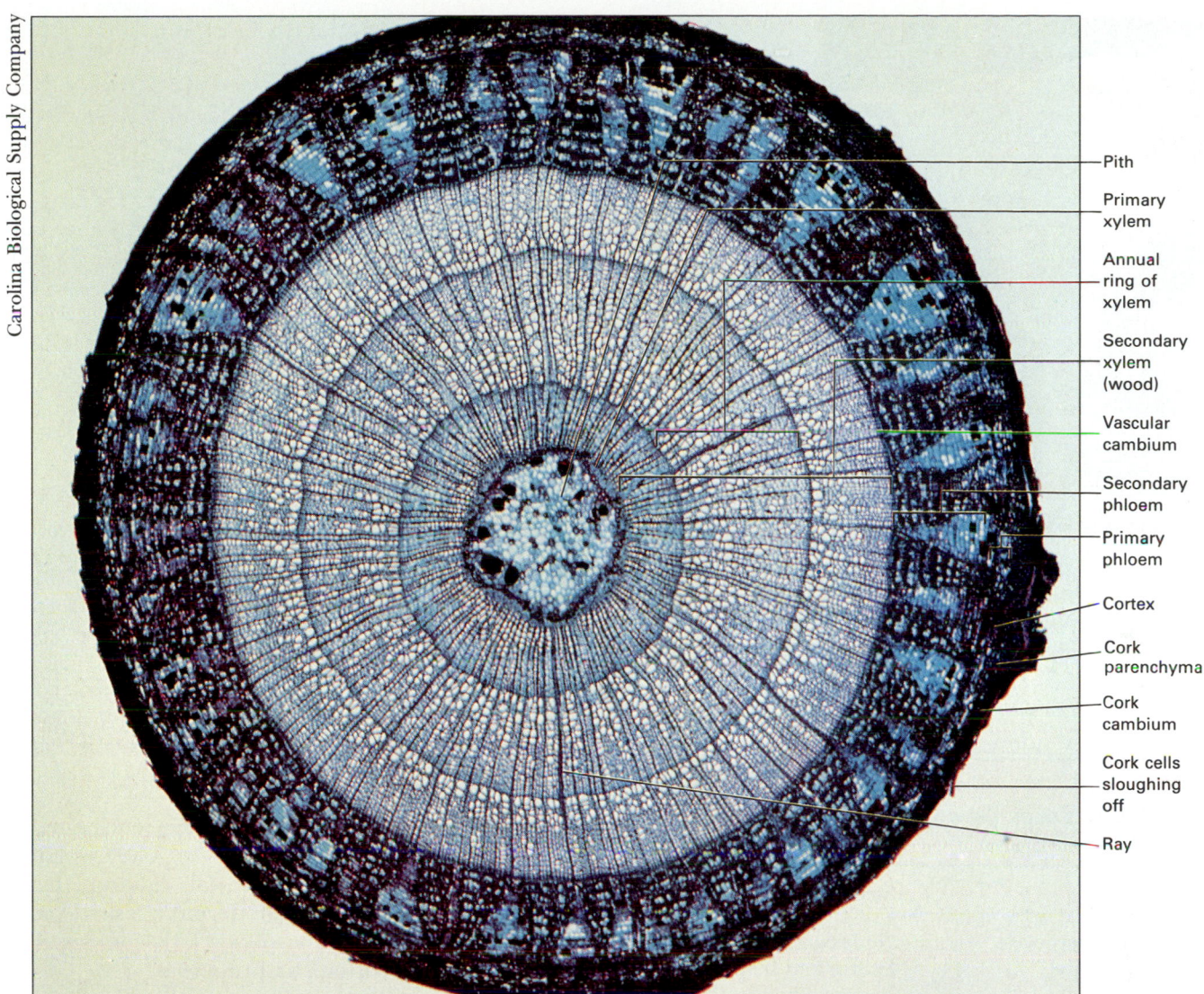

Pith
Primary xylem
Annual ring of xylem
Secondary xylem (wood)
Vascular cambium
Secondary phloem
Primary phloem
Cortex
Cork parenchyma
Cork cambium
Cork cells sloughing off
Ray

FIGURE 28–5 **Cross section of a 3-year-old *Tilia* (basswood) stem showing secondary tissues.**

phloem. In addition, roots have several tissues and structures not found in stems, including a root cap and root hairs. Each root tip is protected by a **root cap,** a layer many cells thick that covers the delicate root apical meristem. As the root grows, the tip encounters a great deal of friction as it pushes through the soil. This friction constantly rubs off root-cap cells, which must continually be replaced by the root apical meristem. **Root hairs** are extensions of epidermal cells located in the area of maturation near the root tip (Chapter 26). They greatly increase the surface area of the root, aiding in absorption (Fig. 28–6). Although stems and leaves may have various types of trichomes, or hairs, they are distinct from root hairs in both structure and function. Table 28–1 summarizes the major differences between dicot stems and roots.

Considerable Variation Occurs In The Internal Arrangement Of Tissues In Dicot And Monocot Roots

We will now examine the structure of typical primary dicot roots and monocot roots. The buttercup *(Ranunculus)* is a representative dicot with primary growth (Fig. 28–7). Like other parts of the plant, dicot roots are covered by a single layer of protective tissue, the **epidermis.** Unlike the epidermis of aerial parts of the plant, however, the root epidermis does not secrete a thick, waxy cuticle, particularly near the root hairs. Such a cuticle would impede the absorption of water from the soil. The root hairs are another modification that enables the root to absorb more water from the soil, as they greatly increase the surface area of the root

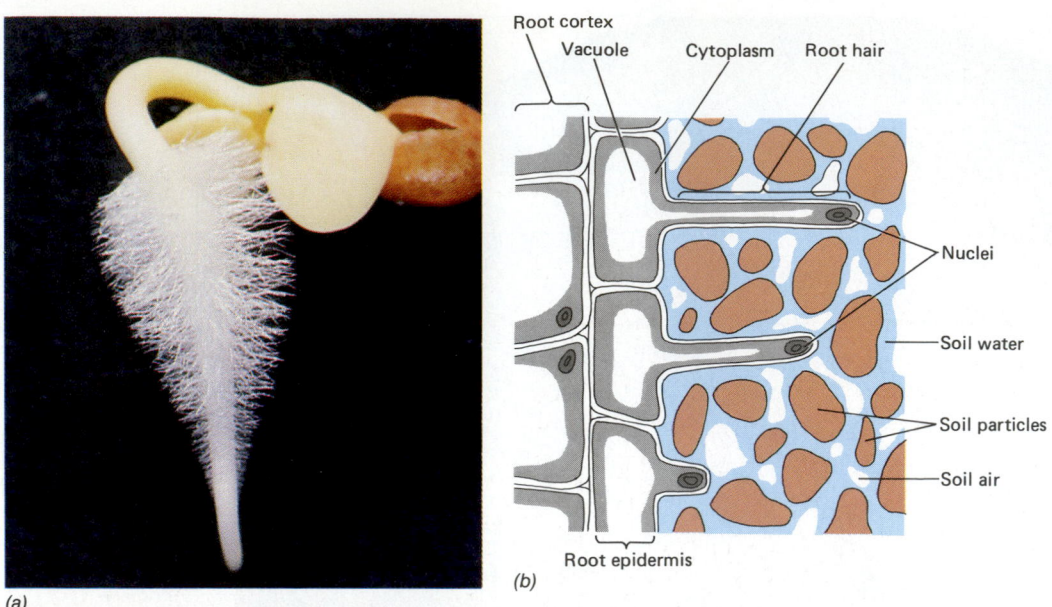

(a)

(b)

FIGURE 28–6 Root hairs increase the surface area of the root in contact with the soil. (*a*) Root hairs on a radish seedling. Each delicate hair is a single cell extension of the root epidermis. (*b*) Stages in root hair development (from bottom to top). The nucleus migrates into the root hair. (*a*, Robert and Linda Mitchell)

TABLE 28–1
General Differences Between Dicot Roots and Stems

Roots	Stems
No nodes or internodes	Nodes and internodes
No leaves or buds	Leaves and buds
Nonphotosynthetic	Photosynthetic
No pith	Pith
No cuticle	Cuticle
Root cap	No cap
Root hairs	No root hairs
Pericycle	No pericycle
Endodermis	Endodermis seldom
Branches form internally from the pericycle	Branches form externally from lateral buds

in contact with the moist soil. Root hairs are short-lived extensions of epidermal cells and never develop into multicellular root branches; branches of roots develop in a different manner.

The root **cortex** is primarily composed of loosely arranged parenchyma (Chapter 26), with lots of intercellular spaces. The inner layer of the cortex, the **endodermis,** is different from the rest of the cortex. Endodermal cells fit snugly against each other. Each has a

special ring called the **Casparian strip** encircling its radial and transverse walls (Fig. 28–8). The Casparian strip contains a fatty, waxy material, **suberin,** that is waterproof. Just inside the endodermis is a single layer of cells, the **pericycle.** The pericycle is composed of cells that retain meristematic properties.

The center of the root is occupied by vascular tissue. The **xylem** is in the middle and often has two, three, four, or more "arms," or extensions. The **phloem** is located between these xylem arms. The **vascular cambium** is sandwiched between the xylem and phloem. With a solid core of vascular tissue in the dicot root, there is no pith.

The tissues in a monocot root are basically the same as in the dicot root, but they are arranged in a slightly different manner in the center part of the root (Fig. 28–9). Starting at the outside of the monocot root, there is epidermis, then cortex, endodermis, and pericycle. However, the vascular tissues inside the pericycle do not form a solid cylinder in the center of the root as in dicots. Instead, phloem and xylem are located in separate patches that are arranged in a circle around the centrally located **pith.** All of these tissues function similarly in monocots as they do in dicots. Because the vast majority of monocots do not have secondary growth, there is no vascular cambium in monocot roots.

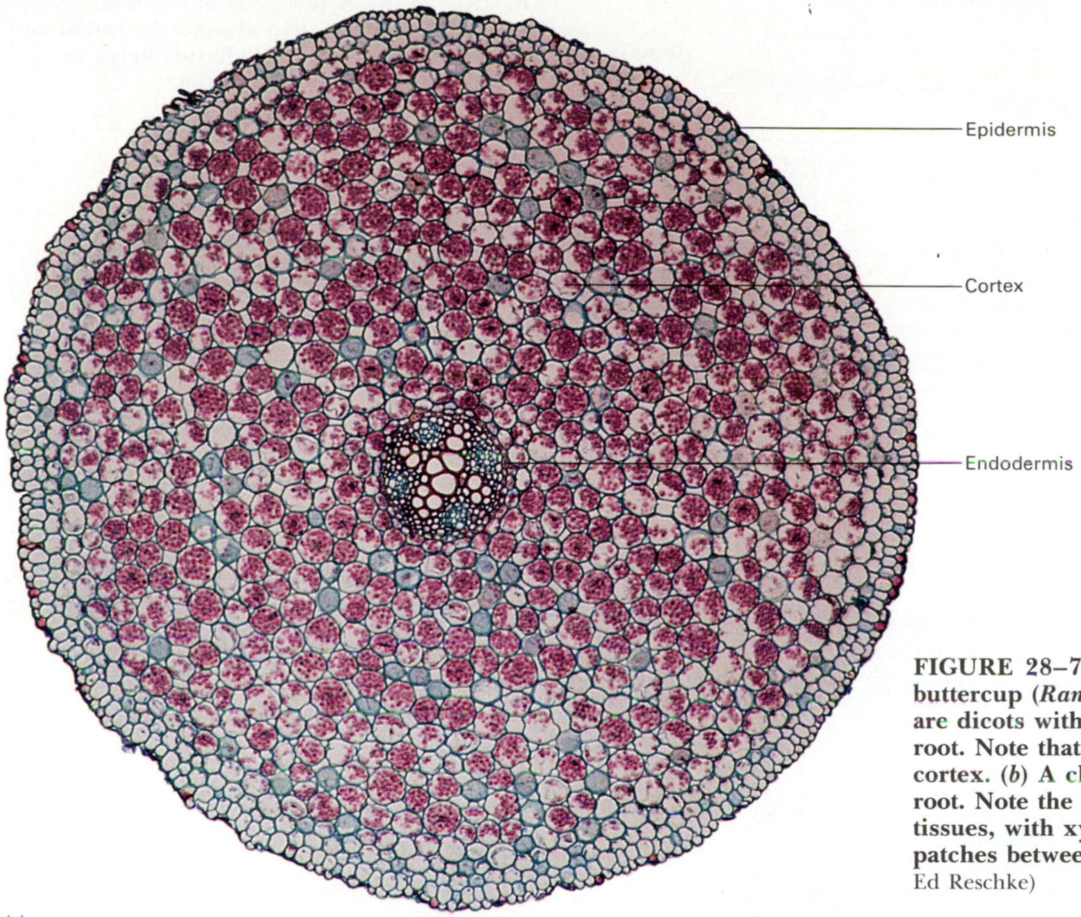

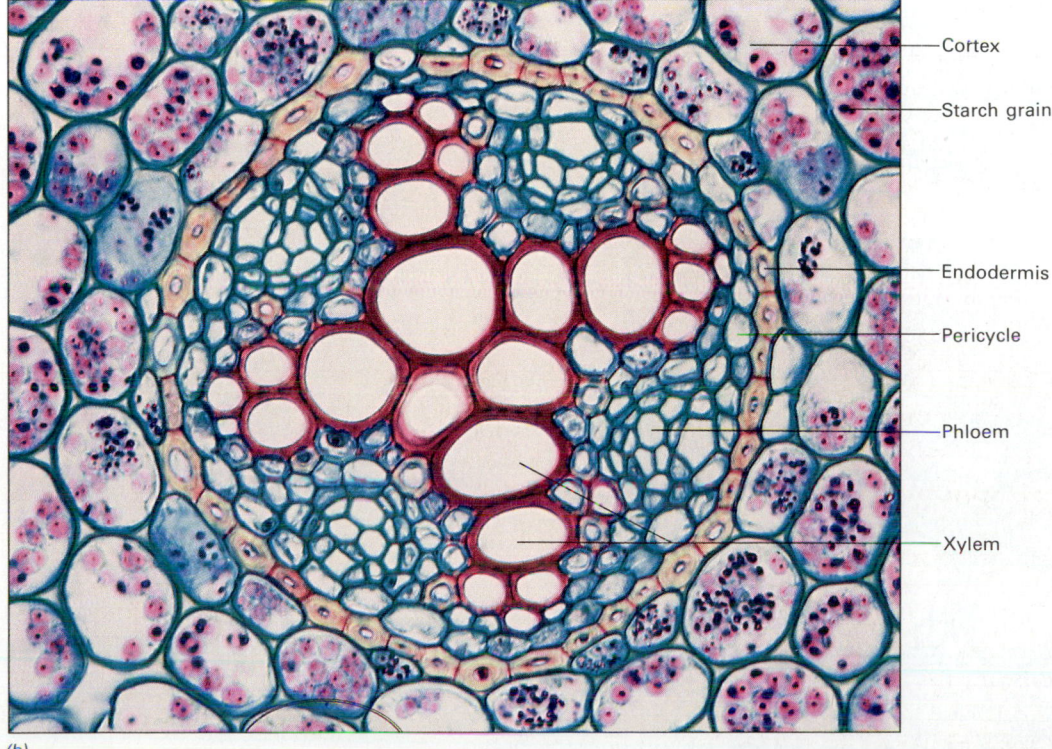

— Epidermis

— Cortex

— Endodermis

(a)

FIGURE 28–7 Cross section of a buttercup (*Ranunculus*) root. Buttercups are dicots with primary growth. (*a*) Entire root. Note that the bulk of the root is the cortex. (*b*) A close up of the center of the root. Note the solid core of vascular tissues, with xylem arms and phloem patches between the xylem arms. (*a* and *b*, Ed Reschke)

Cortex

Starch grain

Endodermis

Pericycle

Phloem

Xylem

(b)

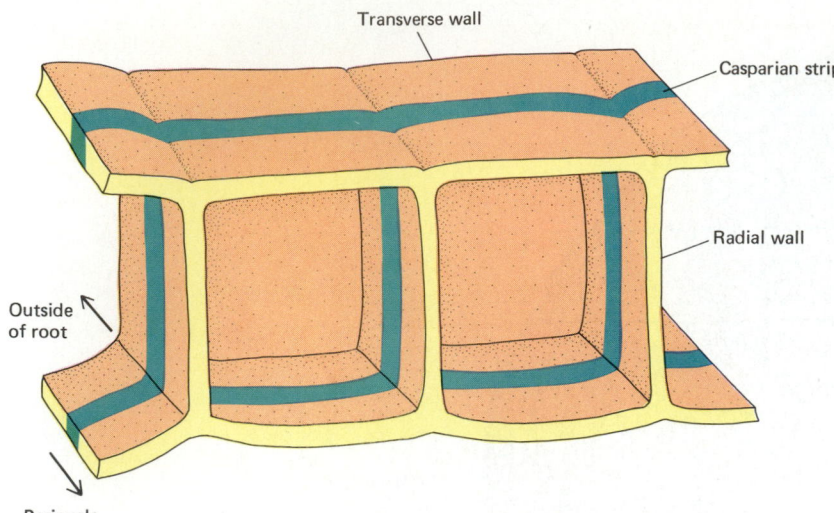

Transverse wall

Casparian strip

Radial wall

Outside of root

Pericycle

FIGURE 28–8 A few cells of the endodermis. Note the Casparian strip around the radial and transverse walls. The endodermis helps to control water uptake by the root.

FIGURE 28–9 Cross section of a greenbriar (*Smilax*) root. Greenbriar exhibits the general characteristics of a monocot root. (*a*) Entire root. (*b*) Close up of a portion of the center of the root, showing the vascular tissues and the pith.

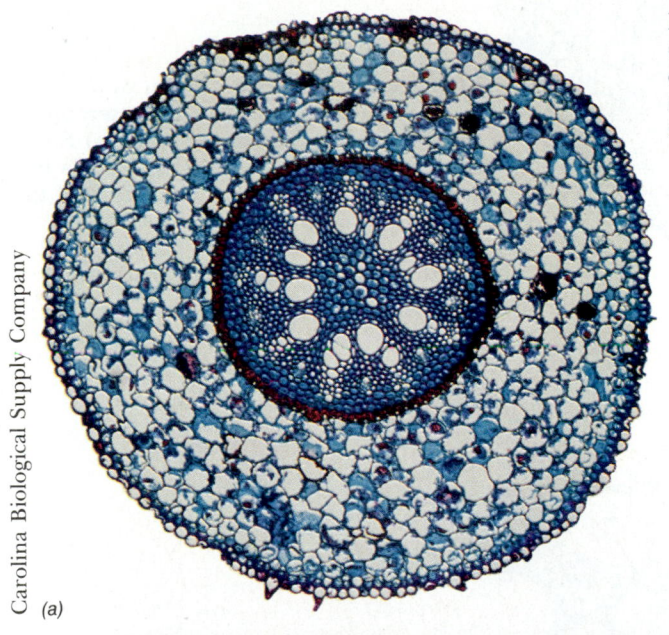

Carolina Biological Supply Company

(a)

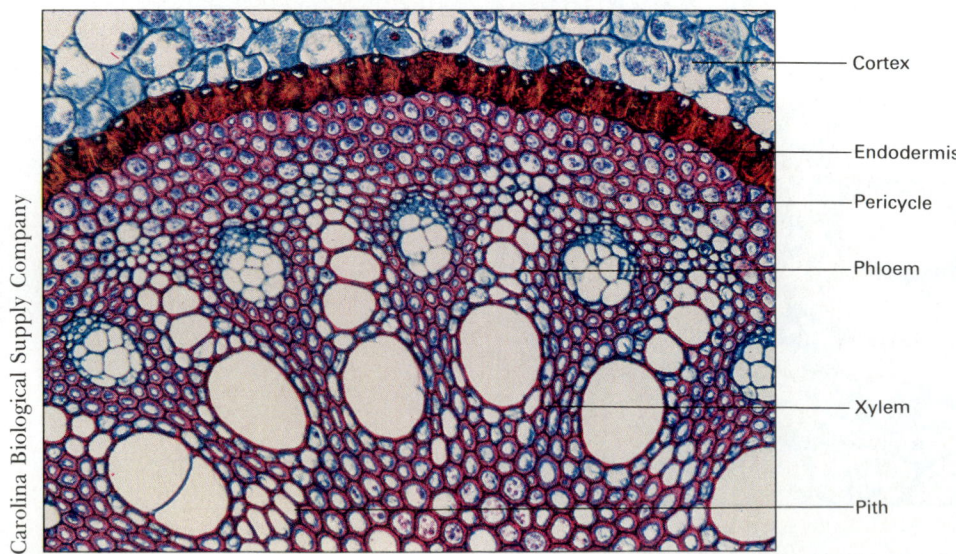

Carolina Biological Supply Company

Cortex

Endodermis

Pericycle

Phloem

Xylem

Pith

(b)

FOCUS ON   Tree-Ring Analysis

In temperate areas, the age of a tree can be determined by counting the number of tree rings. Other useful information can be obtained by analyzing tree rings as well. For example, the size of each ring varies depending on environmental conditions, including precipitation and temperature. Sometimes the variation in tree rings can be attributed to one factor, and similar patterns appear in the rings of many tree species over a large geographical area. For example, trees in the southwestern United States have similar ring patterns due to variation in the amount of annual precipitation. Years with an adequate amount of precipitation produce larger rings of growth, whereas years of drought produce much smaller layers.

It is possible to study the sequence of rings several thousand years back in time. First a master chronology, a complete sample of rings dating back as far as possible, is developed. One starts with an old tree that is currently living. The older rings toward the center of the living tree can be matched with the youngest rings toward the outside of a dead tree or even a piece of wood from a house. By using older and older sections of wood, even those found in prehistoric dwellings, and overlapping their matching ring sequences, one obtains a master chronology of the area.

Tree-ring analysis, or dendrochronology, has been used extensively in several unrelated fields. Astronomers have correlated growth patterns over the years with cycles of sunspots. This work was pioneered by the American astronomer Andrew Douglass during the early years of the 20th century. Tree-ring analysis has been extremely useful in dating prehistoric sites of native Americans in the Southwest. The Cliff Palace in the Mesa Verde National Park, for example, has been dated to 1073 AD. Tree-ring analysis indicates that an extended drought forced the original inhabitants to abandon their home. Climatologists use tree-ring data to study climate patterns in the past. Tree rings are also being analyzed for other disciplines, including ecology (to study changes in plant communities over time), environmental science (to study the effects of pollution), and geology (dating past earthquakes and locating their sources).

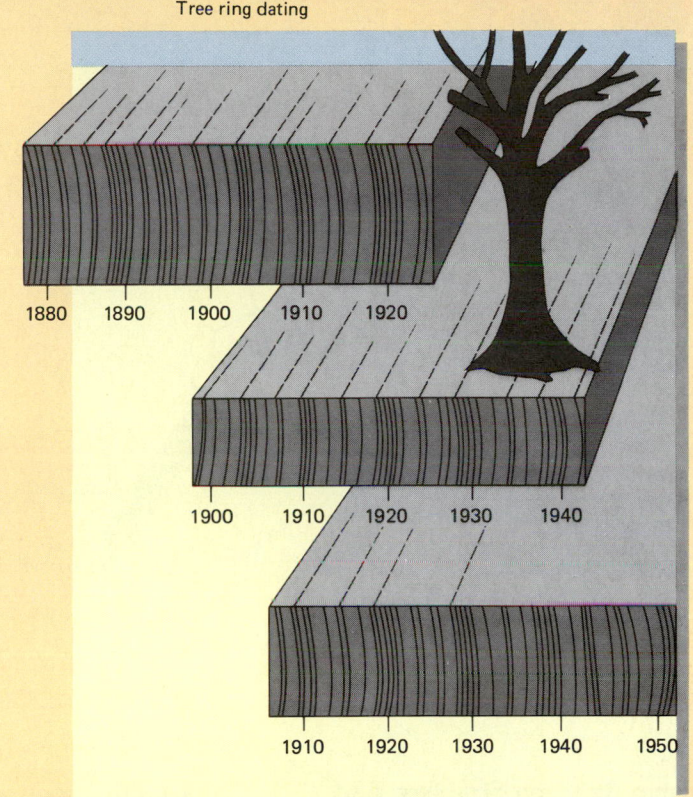

Tree ring dating

Tree ring dating. A master chronology is developed using progressively older pieces of wood from the same geographical area. By matching the rings of a wood sample of unknown age to the master chronology, the age of the sample can be accurately determined.

The Cliff Palace in the Mesa Verde National Park in Colorado has been dated using tree ring analysis. (E.R. Degginger)

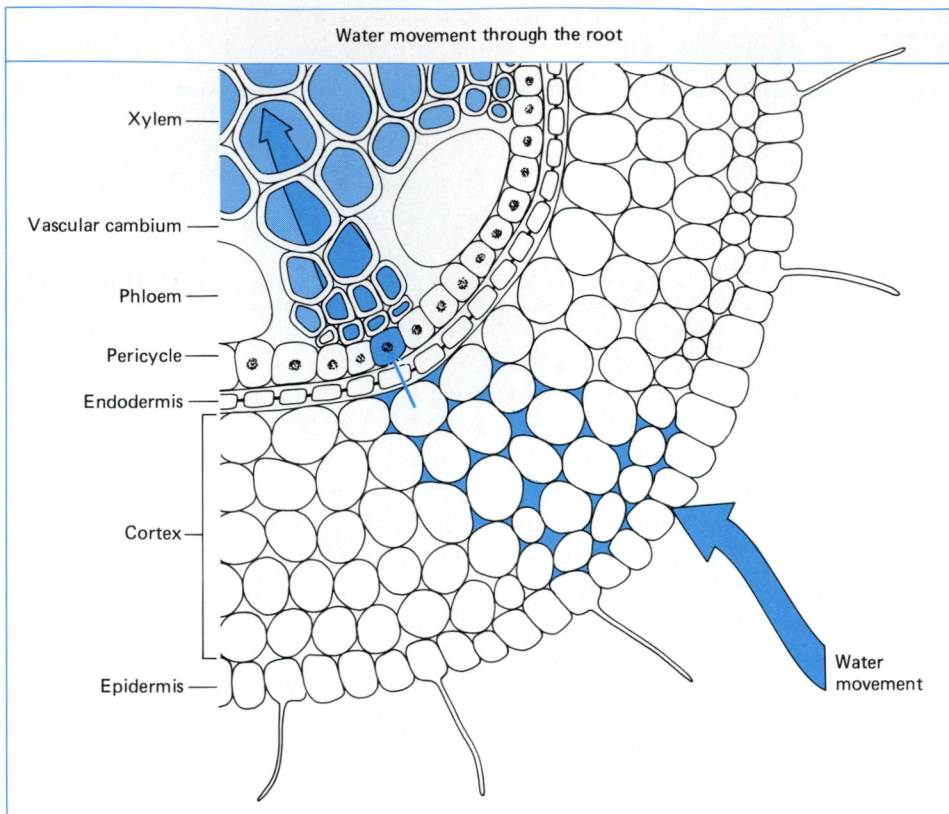

Water movement through the root

Xylem

Vascular cambium

Phloem

Pericycle

Endodermis

Cortex

Epidermis

Water movement

FIGURE 28–10 Most water that enters the root travels along the cell walls and intercellular spaces. When water reaches the endodermis, it must pass through the plasma membrane and enter the cell because the Casparian strip blocks its passage along the wall.

There Is A Relationship Between Structure And Function In Primary Roots

Features of the epidermis, cortex, and endodermis aid in the absorption of water and dissolved minerals from the soil. The lack of a cuticle and presence of root hairs obviously help to increase absorption; however, most of the water that enters the root moves along the cell walls rather than entering the cells (Fig. 28–10). One of the major components of cell walls is cellulose, which readily absorbs water and acts like a wick. (As an example of the absorptive properties of cellulose, consider the cotton ball, which is almost pure cellulose.) When water enters the cortex, it moves along the cell walls and intercellular spaces until it reaches the endodermis. Up to this point, most of the water has never passed through a plasma membrane or entered the cytoplasm of a root cell. The endodermis, with its waterproof Casparian strip running around the four walls that are perpendicular to the surface of the root, blocks passage along the cell wall and in effect forces the water to enter the endodermal cell; that is, water must pass through the plasma membrane of the epidermal cell. For this reason, the endodermis is considered to regulate the movement of water into the root, even though it is an internal tissue and water must pass through other tissues to reach it.

The xylem and phloem of the root have the same functions as they do in the rest of the plant. Phloem conducts food as sucrose to the root where it is stored, or from the root to other parts of the plant where it is used. Water enters the root xylem, often at one of the xylem arms. Up to this point, the pathway of water has been horizontal, from the soil into the center of the root. Upon entering the xylem, water is transported upward into the stem xylem and throughout the rest of the plant.

How roots absorb minerals is under intensive study. The concentrations of various minerals are different in the xylem fluid than in soil water. This indicates that the plant selectively accumulates certain minerals. There is some evidence that minerals may travel through the root tissues from cell to cell, rather than along the cell walls as does water. Dissolved minerals may pass through plasma membranes both passively and actively. In active transport, the mineral ions move against the concentration gradient at a special site in the membrane. Recall that active transport requires the expenditure of energy, usually in the form of ATP (Chapter 6).

The primary function of the root cortex is storage. A microscopic examination of the cells that form the cortex often reveals numerous starch grains. Starch, which is an insoluble carbohydrate composed of glucose units, is the most common form of food stored in plants. Another feature of root cortex, the large intercellular air spaces, allows for aeration of the root as well as a pathway for water uptake.

food is transported in phloem. Translocation in plants does not resemble the movement of materials in animals because nothing *circulates*. The materials being translocated in the xylem travel in one direction only. Although movement in different phloem cells can be in several directions, it cannot be said to circulate.

Water And Minerals Are Translocated In The Xylem

The water and dissolved minerals, which form approximately a 0.1% solution, move within the tracheids and vessel elements, which, you will recall, are hollow, dead cells (Chapter 26). Translocation of water in the xylem is the most rapid of any movement in plants. On a hot summer day, water has been measured moving upward in xylem at 2 feet per minute. The dissolved minerals are carried along passively in the water. The plant does not expend any energy of its own to transport water, which moves as a result of natural physical processes. How does water move to the tops of plants? Clearly, it is either pushed up from the bottom of the plant, or it is pulled up from the top of the plant. Both mechanisms exist in plants.

Root pressure is a phenomenon in which water that enters the roots of a plant is "pushed up" in the xylem toward the top of the plant. In the root-pressure mechanism, water moves into the roots by osmosis. This accumulation of water in root cells creates a "pressure" in the root that forces the water up the xylem. Root pressure is a real phenomenon in plants. Guttation is a manifestation of this process (Chapter 27). However, plant physiologists have measured root pressure and found that it is not strong enough to explain the rise of water to the tops of the tallest trees. Root pressure exerts an influence in smaller plants, but it clearly does not generate sufficient force to cause water to rise hundreds of feet in the tallest plants. Further, root pressure is greatest in the spring, when water is plentiful in the soil. It does not occur to any appreciable extent in summer, yet water transport is highest during hot summer days.

A second possible explanation for the rise of water in plants is that a tension is created at the top of the plant that pulls the water up. This process works much like a person sucking a liquid up a straw. The tension created at the top of the plant is the evaporation pull of transpiration (Chapter 27). This literally pulls the water up through the plant in the narrow "tubes" of xylem.

This pulling of water is only possible as long as there is a solid, unbroken column of water in the xylem. Water has a tendency to form an unbroken column because of the cohesiveness of water molecules. Recall that water molecules are strongly attracted to one another because of hydrogen bonding (Chapter 3). Also, the adhesion of water to the walls of the xylem

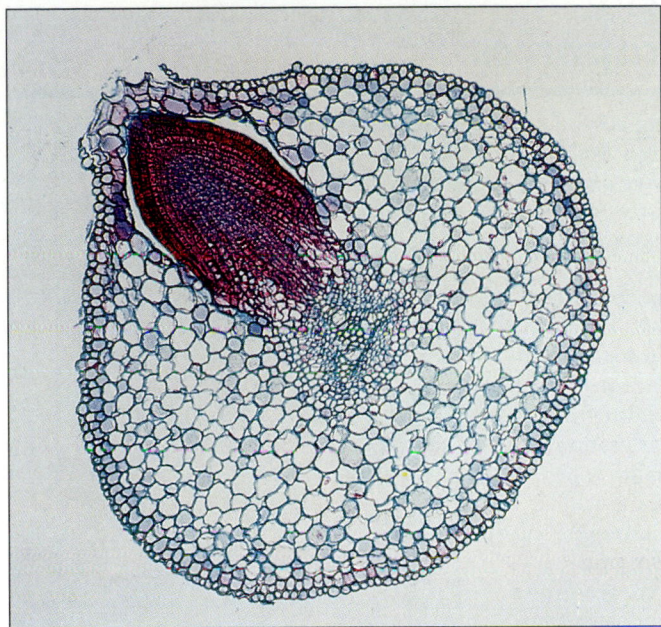

FIGURE 28–11 A multicellular branch root emerges from the root. Branch roots originate at the pericycle. (Ed Reschke)

The pericycle is the origin of multicellular branch roots (lateral roots) (Fig. 28–11). Branch roots originate when a portion of the pericycle, usually at the tip of a xylem arm, becomes meristematic and starts dividing. The branch root that forms from these divisions has all the structures and anatomical features of the root from which it branches. The branch root breaks through several layers of root tissue (the endodermis, cortex, and epidermis) before entering the soil. The pericycle is also involved in secondary growth in plants with woody roots.

GYMNOSPERMS AND CERTAIN DICOTS HAVE ROOTS WITH SECONDARY GROWTH

Plants that produce woody stems also produce woody roots. These plants have both primary and secondary growth. The production of secondary tissues occurs some distance back from the root tip (where primary growth originates). As in the stem, secondary growth in roots is the result of the activity of two lateral meristems, the vascular cambium and the cork cambium. Woody roots have both wood and bark.

TRANSPORT IN PLANTS OCCURS IN THE XYLEM AND PHLOEM

The movement of food, water, and minerals within a multicellular plant is known as **translocation.** Water and minerals are transported in xylem, and dissolved

FOCUS ON   Commercial Hydroponics

Hydroponics, the practice of growing plants in an aerated solution of chemically defined mineral salts, has been used by scientists to determine which elements are essential. Initially, entrepreneurs hailed hydroponics as the scientific way to grow plants in places where soil was poor or unavailable. However, the expenses involved in commercially growing produce for human consumption prevented hydroponics from becoming more than a curiosity. Recent technical improvements have revived the interest in commercial hydroponics.

Hydroponics has great potential in several places. It is being tried experimentally in desert countries in the Middle East, where the soil is too arid to support cultivation and water is unavailable for irrigation. When plants are grown hydroponically in greenhouses, little water is used compared with traditional agriculture. Hydroponics is also being tried in temperate latitudes, particularly for winter crops.

Hydroponics has several advantages. First, it is possible to grow these crops in areas where pathogens and pests are completely absent. This means that the crops are not exposed to pesticides. Also, hydroponics can be used to grow crops near their area of use, saving on transportation costs.

The main disadvantage of hydroponics is the expense. The plants must be supplied with nutrient solution, which must be continually monitored and adjusted. Heating and lighting costs are high. Aeration of the roots has been a major expense, although recent developments like the nutrient-film technique have cut costs considerably. In the nutrient-film technique, the plants are grown in plastic trenches through which a film of nutrient solution is run. In this way, the roots get adequate aeration. The nutrient solution is saved and reused on the plants, which cuts down on water and mineral costs.

Although we will probably never replace traditional agriculture with hydroponics, it has been shown to be a viable alternative in certain situations. As new techniques are developed, hydroponics may become even more common.

Hydroponically grown lettuce. There is no soil in the plastic container, which serves only as a float to keep this and other lettuce plants (shown in rows in the background) from sinking into the nutrient solution (kept covered with white plastic film except for holes in the pots). Although soil is not used, these plants yield excellent salad greens.

---

cells is an important factor in maintaining an unbroken column of water.

To summarize, in the **tension-cohesion mechanism,** a tension is created at the top of the plant by transpiration, the loss of water by evaporation from leaf surfaces. This tension pulls the water in the xylem up to the leaves at the top of the plant. The cohesive and adhesive properties of water enable it to form a solid column in the xylem, which can be pulled. Is tension-cohesion powerful enough to explain the rise of water in the tallest plants? Plant physiologists have calculated that the tension created by transpiration is strong enough to pull water 500 feet in tubes the diameter of xylem vessels. Because the tallest trees on earth are approximately 350 feet high, tension-cohesion easily accounts for their water transport.

Sugar Is Translocated In The Phloem

The sugar produced by photosynthesis is converted into the disaccharide, sucrose, before being loaded into the phloem and transported to the rest of the plant.

Sucrose, or common table sugar, is the predominant form of food carried in phloem. Materials move rapidly in the phloem. Although not as rapid as xylem transport, phloem transport has been measured at approximately 1 inch per minute.

Translocation within the phloem may be up or down in the plant. Sucrose may be transported from its place of manufacture (the leaf) to a place of storage (the root, fruit, or seed). It may also be transported from the leaf or root to actively growing regions like the root or shoot apical meristems, where it would be quickly utilized.

The **pressure-flow hypothesis** best explains what we currently know about translocation in the phloem. Phloem transport occurs as a result of a sugar gradient that is established between the *source,* where the sugar is loaded into the phloem, and the *sink,* where the sugar is removed from the phloem (Fig. 28–12). At the source (for example, a leaf), the dissolved sucrose moves from the mesophyll cells, where it was manufactured, and is actively loaded into the companion cells. The active loading requires ATP energy, which proba-

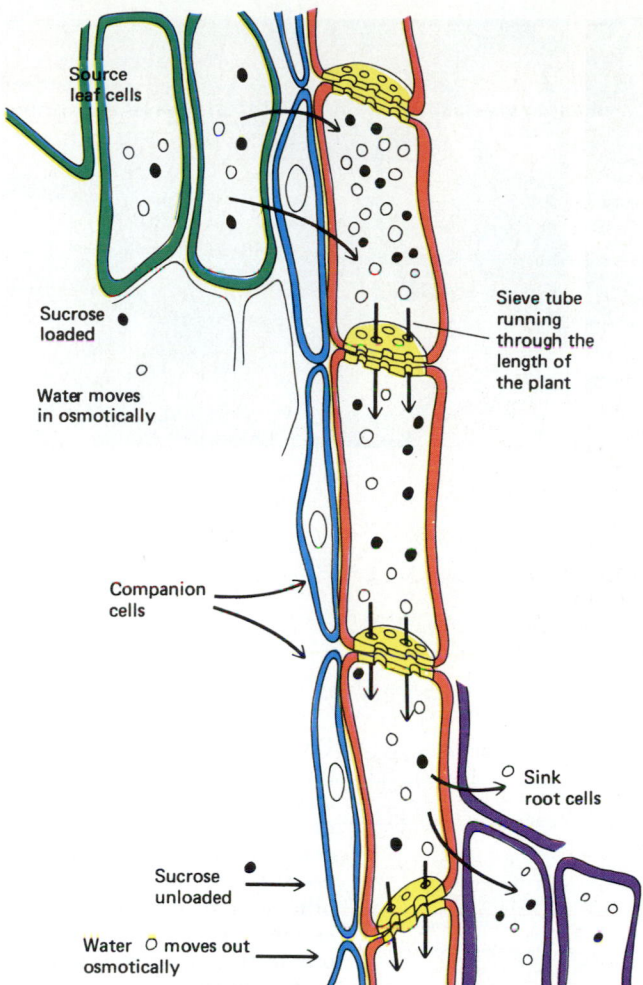

**Source
leaf cells**

**Sucrose
loaded**

**Water moves
in osmotically**

**Sieve tube
running
through the
length of
the plant**

**Companion
cells**

**Sink
root cells**

**Sucrose
unloaded**

**Water moves out
osmotically**

**FIGURE 28–12 Pressure flow mechanism for phloem
transport. Sugar is actively loaded into the sieve tube
element at the source (e.g., the leaf); as a result, water
moves osmotically into the sieve tube element. At the
sink (e.g., the root) the sugar is actively unloaded and
water leaves the sieve tube element by osmosis. The
gradient of sugar from source to sink causes pressure
flow through the sieve tube toward the sink.**

bly works through a proton pump, forming a gradient
of hydrogen ions (Chapter 8). This mechanism is supported by changes in pH that have been observed in
sugar loading. Presumably sugar accompanies the flow
of H^+ ions back across the membrane. Once the sugar
is in the companion cell, it readily moves into the sieve
tube element through the many cytoplasmic connections between the two cells. As a result of the increase
in dissolved sugars in the sieve-tube elements, water
moves by osmosis into the sieve tube, creating a pressure at the source. This pressure pushes the sugar solution through the phloem much as water is forced
through a hose. At its destination, the sink (for example, a root), sugar is actively unloaded from the sieve-
tube elements, with ATP being required. With a loss in
sugar, water moves out of the sieve tubes by osmosis. *It*

*is the sugar gradient (that is, the difference in sugar concen-
trations between the source and the sink) that causes transport
in the phloem.* The actual flow of sugar solution in the
phloem does not require energy; however, both loading and unloading sugar at the source and sink require
the energy of ATP.

Study of phloem transport in plants is difficult.
The cells are under hydrostatic pressure, so cutting
into the phloem to observe it causes the contents of the
sieve-tube element to be sucked against one end wall.
Much useful information about phloem transport has
been obtained using radioactive tracers. Aphids, small
wingless insects that insert their mouthparts into
phloem sieve tubes for feeding, have also been a useful
tool in phloem research (Fig. 28–13). The pressure-
flow mechanism adequately explains current data on
phloem transport. There is still a lot to be learned
about this complex process, however.

ROOTS OBTAIN MOST OF THE NATURALLY OCCURRING MINERALS THAT ARE FOUND IN PLANT TISSUES

It is necessary in our discussion of roots to consider the
minerals absorbed by roots from the soil. There are
more than 90 naturally occurring elements in the
earth. Over 60 of these, including elements as common
as carbon and as rare as gold, have been found in plant
tissues. Not all of these elements are considered essential for plant growth, however. How do biologists determine whether an element is essential? It is impossible to conduct mineral nutrition experiments by
growing plants in soil, which is too complex and contains too many elements. One of the most useful methods is **hydroponics,** growing plants in aerated water
that contains dissolved mineral salts. If a biologist sus-
pects that a particular element is essential for plant
growth, he or she grows the plants in a nutrient solution that contains all known essential elements except
the one in question. If plants grown in the absence of
that element are unable to develop normally or to complete their life cycle, it is an indication that the element
may be essential (Fig. 28–14). Additional criteria are
used to determine whether an element is essential. The
element must be shown to have a direct effect on the
metabolism of the plant. Also, the element must be
demonstrated to be essential for a wide variety of plant
species.

Sixteen Elements Are Essential For Plant Growth

Sixteen elements have been demonstrated to be essential for plant growth. Nine of these are required in
fairly large quantities (greater than 0.05 % dry weight)

(a)

(b)

FIGURE 28–13   **Mature aphid feeding on a lower side of a branch of basswood (*Tilia americana*). The aphid is about 6 mm long. (*a*) The aphid's feeding apparatus (stylet) is inserted into the phloem. (*b*) A microscopic view of *Tilia* phloem, showing that the aphid stylet has penetrated a sieve tube element.** (*a*, Dwight R. Kuhn; *b*, M.H. Zimmerman, *Science* 133:73–79 (Fig. 4), January 13, 1961, © 1988 by the American Associates for the Advancement of Science)

FIGURE 28–14   Tobacco plants illustrating the effects of deficiencies of specific elements. The plant in the center (Ck.) received all the essential elements. The others were supplied with all essential elements except the one indicated on the label (—S = sulfur, —P = phosphorus, —N = nitrogen, —Mg = magnesium, —B = boron, —Ca = calcium, —K = potassium). All plants are the same age and variety. Some of them exhibit chlorosis (breakdown of chlorophyll) and necrosis (death of tissue). (W.R. Robbins, Rutgers University)

and are known as **macronutrients:** carbon, hydrogen, oxygen, nitrogen, phosphorus, potassium, sulfur, calcium, and magnesium. The remaining seven **micronutrients** are needed in trace amounts for normal plant growth and development: iron, boron, manganese, copper, molybdenum, chlorine, and zinc.

Four of the 16 elements—carbon, oxygen, hydrogen, and nitrogen—are obtained from water or gases in the atmosphere. Carbon is taken from carbon dioxide in the atmosphere and incorporated by photosynthesis. Oxygen is obtained from oxygen gas and water. Water also supplies hydrogen to the plant. Plants get their nitrogen from the soil as ions of nitrogen salts, nitrate (NO_3^-), and ammonium (NH_4^+), but nitrogen is "fixed" into that form from nitrogen gas (N_2) by various microorganisms in the soil (Chapter 46). The re-

TABLE 28–2
Functions of Essential Elements in Plants

Element	Major Functions	Specific Role
Carbon	Structural	In carbohydrates, lipids, proteins, and nucleic acids
Hydrogen	Structural	In carbohydrates, lipids, proteins, and nucleic acids
Oxygen	Structural	In carbohydrates, lipids, proteins, and nucleic acids
Nitrogen	Structural	In proteins, nucleic acids, chlorophyll, and certain coenzymes
Phosphorus	Structural	In nucleic acids, phospholipids, and ATP (energy-transfer compound)
Sulfur	Structural	In certain amino acids and vitamins
Calcium	Structural Physiological	In middle lamella of cell walls Role in membrane permeability
Magnesium	Structural Physiological	In chlorophyll Enzyme activator in carbohydrate metabolism
Potassium	Physiological	Osmosis and ionic balance (e.g., opening and closing of stomates); enzyme activator
Chlorine	Physiological	Ionic balance; involved in light reactions of photosynthesis
Iron	Physiological	Part of enzymes involved in photosynthesis and respiration
Manganese	Physiological	Part of enzymes involved in respiration and nitrogen metabolism
Copper	Physiological	Part of enzymes involved in photosynthesis
Zinc	Physiological	Part of enzymes involved in respiration and nitrogen metabolism
Molybdenum	Physiological	Part of enzymes involved in nitrogen metabolism
Boron	Physiological	Exact role unclear; involved in membrane transport and calcium utilization

maining 12 essential elements are obtained from the soil as dissolved mineral ions. Their ultimate source is the parent rock from which the soil was formed.

Some of the roles of the essential elements are summarized in Table 28–2. Carbon, hydrogen, and oxygen are found in all biologically important molecules, including lipids, carbohydrates, nucleic acids, and proteins. Nitrogen is part of proteins, nucleic acids, and chlorophyll. Phosphorus is critical for plants because it is found in nucleic acids, phospholipids (an essential part of membranes), and energy-transfer molecules like ATP. The **middle lamella,** the cementing layer of the plant cell wall, contains calcium. Calcium has also been implicated in a number of

physiological roles in plants, including membrane permeability. Magnesium is part of the chlorophyll molecule. Sulfur is essential because it is found in certain amino acids and vitamins.

Potassium, which plants use in fairly substantial amounts, is not found in a specific compound or group of compounds. Rather, it remains as free K^+ ions in the plant cells. It has a very important role in maintaining the turgidity of cells by inducing osmosis. Its role in the opening and closing of stomates through its effect on osmosis in the guard cells has already been discussed (Chapter 27). Another element that has a role in turgor balance of cells is chlorine. In addition to its osmotic role, the Cl^- ions, which are present in very minute amounts in plants, are essential for photosynthesis. Five of the micronutrients (iron, manganese, copper, zinc, and molybdenum) are involved in various enzymatic reactions, often as enzyme activators. Potassium is also involved in certain enzymatic reactions. The role of boron in plants is unclear. Recent experiments have suggested that boron is involved in membrane transport. It also appears to affect calcium utilization.

Both Organic And Inorganic Fertilizers Can Replace Certain Key Elements If They Are Missing From The Soil

In a balanced ecosystem, the minerals removed from the soil by plants are returned when the plants or the animals that eat them die and decompose; however, the agricultural practices of humans prevent this cycle from occurring. The removal of crops from the land gradually depletes the soil of certain essential elements. Likewise, homeowners mow their lawns and remove the clippings, preventing decomposition and cycling of minerals that were in the grass blades. Plant growth is usually limited by the essential material (water, sunlight, or some essential element) that is in shortest supply. This is sometimes called the **concept of limiting factors.** The three elements that are most often limiting factors for plants are nitrogen, phosphorus, and potassium. In order to sustain productivity of agricultural soils, fertilizers are periodically added to replace those minerals that have become limiting factors.

There are two main types of fertilizer—organic and inorganic. Organic fertilizers come from natural sources such as horse or cow manure and ground corncobs. Green manure, another type of organic fertilizer, is actually a crop that is planted in the soil and deliberately plowed under to decompose rather than being harvested. Frequently, this crop is a plant that has nitrogen fixation in its roots, thereby increasing the amount of nitrogen in the soil. Organic fertilizers have several advantages over inorganic fertilizers. First, they increase the amount of organic material in the soil, which improves the water-holding capacity of the soil. Organic fertilizers also release the minerals they contain gradually, as decomposers break down the organic material. The addition of organic material to the soil changes the biota, or types of living organisms, in the soil. In ways that are not clear to scientists, organic fertilizer added to soil sometimes suppresses microorganisms that could cause plant disease.

Inorganic fertilizer is manufactured. Its exact chemical composition is known. Most inorganic fertilizers contain nitrogen, phosphorus, and potassium. The numbers on fertilizer bags (for example, 10,20,20) tell the relative concentrations of each of the three elements (N,P,K). An advantage of inorganic fertilizers over organic fertilizers is that one can determine precisely how much of which elements are being applied to the soil. By varying the relative concentrations of nitrogen, phosphorus, and potassium, a grower can induce different growth responses in plants. If one were growing a lettuce crop, for example, it would be best to use a fertilizer with a high nitrogen content, because that stimulates vigorous vegetative growth rather than reproduction. However, if one were growing tomatoes and applied a fertilizer with a high nitrogen content, the production of tomatoes would be quite low. Although the plants would grow vigorously, they would form few flowers and, therefore, few tomato fruits. Obviously, there are advantages for both organic and inorganic fertilizers. The chemical elements supplied by each are identical, however. Nitrogen from commercial, inorganic fertilizer is the same as nitrogen from organic fertilizer.

■ CHAPTER SUMMARY

I. Stems function in support, internal transport of materials, and production of new stem tissue.
  A. Stems with primary growth have an epidermis, vascular tissue, and cortex and pith, or ground tissue.
    1. Dicot stems have a distinct cortex and pith. The vascular bundles are arranged in a circle.
    2. Monocot stems have ground tissue instead of a distinct cortex and pith. Their vascular bundles are scattered throughout the ground tissue.
  B. Secondary growth occurs in some dicots and all gymnosperms.

    1. The vascular cambium produces secondary xylem (wood) to the inside and secondary phloem (inner bark) to the outside.
    2. The cork cambium produces cork parenchyma to the inside and cork cells to the outside.
II. Anchorage, absorption, conduction, and storage are the main functions of roots.
  A. Primary roots have an epidermis, cortex, endodermis, pericycle, xylem, and phloem.
    1. The epidermis protects the root.
    2. The cortex contains storage tissue.
    3. The endodermis controls water uptake by the root.

4. The pericycle is the origin of branch roots.
5. The xylem conducts water and dissolved minerals.
6. The phloem conducts food.
B. There are some differences between monocot and dicot roots.
   1. Monocot roots often have a pith.
   2. Dicot roots have a vascular cambium.
III. Water and dissolved minerals move upward in the xylem from the root to the stem and leaves.
   A. Root pressure, caused by the movement of water into the root from the soil, can explain the rise of water in the xylem of small plants.
   B. Tension-cohesion causes the rise of water in even the largest plants.
      1. The evaporation pull of transpiration produces a tension at the top of the plant.
      2. A solid, unbroken column of water is pulled up through the xylem as a result of the cohesive and adhesive properties of water.
IV. Dissolved food is transported in the phloem.
   A. Sucrose is the predominant form of food in the phloem.
   B. Movement of materials in the phloem may be caused by pressure flow.

1. Sugar is actively loaded into the sieve tubes at the source. This requires ATP. As a result of the sugar loading, water moves into the sieve tubes by osmosis.
2. Sugar is actively unloaded from the sieve tubes at the sink. This requires ATP. As a result, water leaves the sieve tubes by osmosis.
3. Transport of materials between the source and sink is driven by mass flow, or the pressure created by the additional water entering the phloem at the source.
V. Sixteen elements are essential for normal plant growth.
   A. Nine elements are macronutrients: carbon, oxygen, hydrogen, nitrogen, potassium, phosphorus, sulfur, magnesium, and calcium.
   B. Seven elements are micronutrients: iron, boron, manganese, copper, zinc, molybdenum, and chlorine.
   C. These elements are part of the structure of biological molecules, important in the ionic balance of cells, and involved in enzyme reactions.

■ POST-TEST

1. Monocots and some dicots have _____ growth, whereas gymnosperms and certain other dicots have _____ growth.
2. Vascular tissue arranged in a circle is characteristic of the primary stems of _____.
3. Primary _____ stems lack a distinct pith and cortex.
4. The two lateral meristems responsible for secondary growth are the _____ _____ and the _____ _____.
5. In plants with secondary growth, the _____ serves as a functional replacement for the epidermis.
6. Botanically speaking, the wood of a tree is _____ _____.
7. The _____ _____ covers the delicate apical meristem of the root tip.
8. The waterproof region around the radial and transverse walls of endodermis cells is the _____ _____.
9. The _____ is the origin of branch roots.
10. The center of a dicot root is _____; the center of a monocot root is _____.
11. _____ is the science of tree-ring analysis.
12. Roots absorb _____ and _____ from the soil.
13. The movement of food, water, and minerals within a plant is called _____.
14. This mechanism of water movement, _____ _____, is not strong enough to explain the rise of water to the tops of the tallest trees.
15. In the tension-cohesion mechanism, the tension is created at the top of the plant by the evaporation pull of _____.
16. The area of the plant where sugar is loaded into the phloem is known as the _____.
17. The pressure created by water entering the phloem at the source transports, or drives, materials from the source to the sink. This phenomenon is known as _____ _____.
18. Minerals may pass through a membrane against the concentration gradient by _____ _____.
19. Growing plants in aerated water with dissolved mineral salts is known as _____.
20. _____ are essential elements required in fairly large quantities.
21. Although more than 60 elements have been found in plant tissues, only _____ of them are essential for plant growth.
22. _____ is an essential element found in phospholipids, nucleic acids, and energy transfer molecules like ATP.
23. _____ and chlorine have a role in maintaining the turgidity of cells.
24. Green manure is an example of a(an) _____ fertilizer.
25. The three elements that most often limit plant growth are potassium, phosphorus, and _____.

■ REVIEW QUESTIONS

1. List at least three functions of stems.
2. Diagram a typical dicot stem with primary growth and label the various tissues. Give at least one function for each of the tissues.
3. Diagram a cross section of a monocot root and label these tissues: epidermis, endodermis, xylem, phloem, cortex, pith, and pericycle.
4. Trace the pathway of water into the root from the soil.
5. Explain the tension-cohesion mechanism of water transport. Make sure you consider both the "tension" and "cohesion" aspects of the mechanism.
6. Describe the pressure-flow mechanism of sugar movement in the phloem, including the activities at the source and sink.
7. Are minerals absorbed in the same manner as water? If not, how are they absorbed by the root?
8. What criteria have biologists used to determine which elements are essential for plant growth?
9. If you were conducting an experiment to determine whether gold was essential for plant growth, what would you use for an experimental control?
10. Give the advantages of both organic and inorganic fertilizers.

■ RECOMMENDED READINGS

Galston, A.W., P.J. Davies, and R.L. Satter. *The Life of the Green Plant,* 3rd ed. Englewood Cliffs, NJ, Prentice-Hall, 1980. Contains an excellent explanation of conduction in both xylem and phloem.

Kaufman, P.B., T.F. Carlson, P. Dayanandan, M.L. Evans, J.B. Fisher, C. Parks, and J.R. Wells. *Plants: Their Biology and Importance.* New York, Harper & Row, 1989. An excellent general botany textbook.

29

Reproduction in Flowering Plants

OUTLINE

I. Asexual reproduction in flowering plants may involve modified stems, leaves, or roots

II. Sexual reproduction in flowering plants involves flowers, fruits, and seeds
- A. Many mechanisms have evolved that accomplish pollination
- B. Fruits are mature, ripened ovaries
- C. Fruit and seed dispersal is highly varied in flowering plants

III. Environmental cues may induce flowering in plants
- A. Flowering may be initiated by light
  1. Phytochrome detects varying periods of day length and darkness
  2. Phytochrome is involved in many plant responses to light besides flowering
- B. Temperature may also affect reproduction

Focus on Localizing Phytochrome in Plant Cells

LEARNING OBJECTIVES

After you have studied this chapter you should be able to:

1. Distinguish between sexual and asexual reproduction.
2. Define each of the following structures: rhizome, tuber, stolon, corm, and bulb.
3. Distinguish between seeds and fruits produced by apomixis and seeds and fruits produced by sexual reproduction.
4. List some generalizations about flowers that are pollinated by insects, by birds, and by wind.
5. Distinguish between simple, aggregate, multiple, and accessory fruits. List and define several types of simple fruits.
6. Cite several different methods of seed and fruit dispersal.
7. Relate how flowering is induced by variations in the amounts of light and darkness, including the role of phytochrome in this process.
8. Relate the effect temperature may have on flowering.

Indian or Chinese lotus, *Nelumbium nelumbo.*
(William E. Ferguson)

508

One of the reasons for the success of flowering plants on earth is their reproductive flexibility. Many flowering plants are able to reproduce both sexually and asexually. Sexual reproduction involves the formation of flowers and, after fertilization, fruits and seeds (Chapter 23). More specifically, in sexual reproduction, there is a fusion of haploid gametes, the egg and the sperm nucleus. The union of these cells, which is called fertilization, occurs within the ovary of the flower. The offspring produced by sexual reproduction exhibit a great deal of individual variation. This is due in part to the independent assortment of chromosomes that occurs during meiosis and to the union of dissimilar gametes, often from two different plants. Sexual reproduction offers several advantages for the plant. It makes it possible for new combinations of genes to occur that might make the plant better suited to its habitat. Also, the fruits and seeds of many plants have various mechanisms for dispersal, allowing the plant to extend its range.

Asexual reproduction in flowering plants does not usually involve the formation of flowers, fruits, and seeds. Instead, vegetative structures (roots, stems, and leaves) produce offspring. In asexual reproduction, there is one parent rather than two, and the offspring are formed by mitosis. This means the offspring are genetically identical to the parent and to each other. Asexual reproduction is ideal for producing large numbers of identical offspring from a parent plant. It is common in plants because asexual reproduction is favored in sessile (nonmoving) organisms living in stable environments. Humans take advantage of the ability of plants to reproduce asexually to produce many plants from a single plant that has a superior combination of genetic characteristics.

KEY CONCEPTS

- In flowering plants, sexual reproduction involves the formation of flowers, fruits, and seeds.
- Flowering plants exhibit tremendous variation in both asexual and sexual reproductive structures.
- Reproductive processes such as pollination and fruit/seed dispersal are highly variable in flowering plants.
- The initiation of sexual reproduction is often under environmental controls (for example, day length and temperature).

ASEXUAL REPRODUCTION IN FLOWERING PLANTS MAY INVOLVE MODIFIED STEMS, LEAVES, OR ROOTS

Flowering plants have evolved many methods of asexual reproduction. Most of these involve modified vegetative parts. In particular, there are a number of structures involved in asexual reproduction that are modified stems. These are rhizomes, tubers, bulbs, corms, and stolons.

The **rhizome** is a horizontal, underground stem (Fig. 29–1). Although rhizomes may resemble roots, they are really stems, as shown by the presence of scale-

FIGURE 29–1 The rhizome is a horizontal, underground stem. New shoots arise from buds that develop along the rhizome.

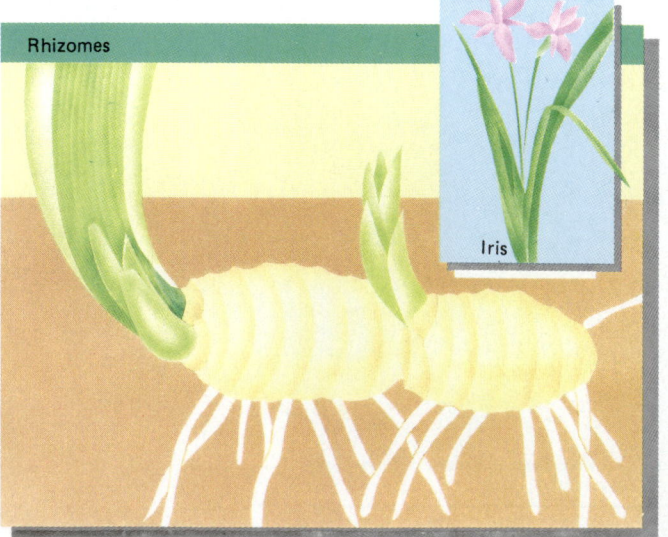

FIGURE 29–2 The potato forms rhizomes, which enlarge at the ends into tubers. Potatoes are seldom grown from seed. Instead, a tuber is cut into pieces, each with an "eye." (G.R. Roberts)

like leaves, buds, nodes, and internodes. Rhizomes frequently branch in different directions. Over time, the old portion of the rhizome dies, eventually dividing the two branches into separate plants. Examples of plants that have rhizomes include the iris and many grasses. Humans propagate plants with rhizomes by dividing or cutting the rhizome into smaller pieces, each with a bud; each piece is capable of growing into an entire plant.

Another underground stem is the **tuber,** which is greatly enlarged, allowing food storage (Fig. 29–2). White potatoes and *Caladium* (Elephant's ear) are examples of plants that produce tubers. The "eyes" of the white potato are actually lateral buds, evidence that the tuber is a modified stem rather than a root. Humans propagate plants with tubers by cutting them into pieces, each with a lateral bud; when planted, each grows into a plant.

A **bulb** is a shortened underground stem to which fleshy storage leaves are attached (Fig. 29–3). Bulbs are globose, or round, and are covered by paper-like bulb scales. They frequently form small daughter bulbs that are initially attached to the mother bulb. Humans separate these daughter bulbs to increase the number of plants, but this process also occurs in nature. The contractile roots (Chapter 28) of some daughter bulbs contract and eventually pull the daughter bulb away from its parent. Lilies, tulips, onions, and daffodils form bulbs.

An underground stem that superficially resembles the bulb is the **corm** (see Fig. 29–3). The storage organ in the corm is the much thickened stem, rather than leaves as in the bulb. The entire corm is stem tissue that is covered with papery scales. These scales are modified leaves that are attached to the corm at nodes. Lateral buds frequently arise on the corm. Plants that produce corms include the crocus, gladiolus, and cyclamen.

Stolons, or runners, are horizontal stems that grow above ground (Fig. 29–4). They are character-

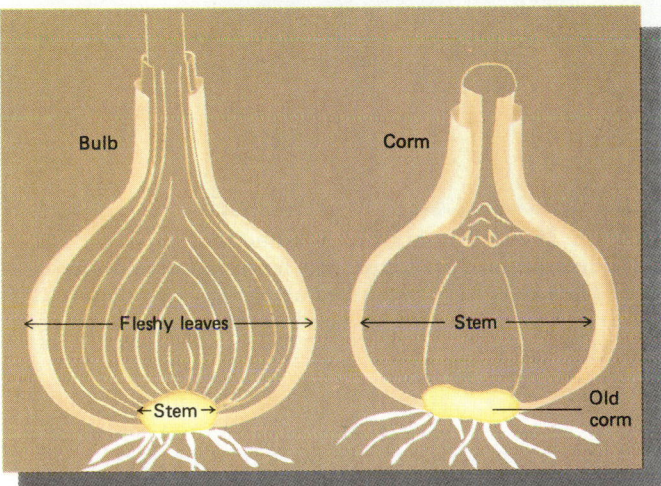

FIGURE 29–3 Bulbs and corms. (*a*) The bulb is an underground stem to which overlapping, fleshy leaves are attached. (*b*) The entire corm is stem tissue, in contrast to the bulb.

FIGURE 29–4 The strawberry reproduces asexually by forming stolons, or runners.

FIGURE 29–5 The "mother of thousands" (*Kalanchoe*) produces young plants along the margins of the leaves. When the young plants attain a certain size, they drop off and root in the ground. (James L. Castner)

FIGURE 29–6 The head of a dandelion. Each individual plumed structure is a fruit that contains a seed. The seeds were produced asexually, by apomixis. The fruits are dispersed by wind currents. (E.R. Degginger)

ized by having long internodes. Adventitious buds develop along the stolon, and each bud gives rise to a new plant. The strawberry is an example of a plant that produces stolons.

Some plants are capable of forming young plants (plantlets) along their leaf margins (Fig. 29–5). *Kalanchoe*, commonly called "mother of thousands," has meristematic tissue in the leaf that gives rise to an individual plant at each "notch" in the leaf. When these plants attain a certain size, they drop to the ground and grow. Some plants have roots that produce **suckers,** which are above-ground stems that develop from adventitious buds on the roots. (Generally, buds originate on stems rather than roots; thus, buds on roots are adventitious.) Each sucker grows roots at the base of the stem. Examples of plants that form root suckers include the black locust, pear, apple, cherry, red raspberry, and blackberry. It is possible to separate the suckers from the parent plant. Some weeds are able to produce considerable numbers of plants by this method. These plants are difficult to control, because pulling the plant out of the soil seldom removes all the roots. In response to wounding, the roots produce more and more adventitious buds, which can be a considerable nuisance.

Sometimes plants produce seeds and fruits without meiosis, fusion of gametes, and the other aspects of sexual reproduction. When this occurs, it is known as **apomixis.** For example, the embryo may develop from a diploid cell in the ovule rather than from a diploid zygote that forms from the union of two haploid gametes. Apomixis is a form of asexual reproduction, even though seeds are produced, because the embryo is genetically identical to the original parent. Example of plants that reproduce asexually by apomixis include the dandelion, citrus plants, blackberry, and certain grasses (Fig. 29–6).

SEXUAL REPRODUCTION IN FLOWERING PLANTS INVOLVES FLOWERS, FRUITS, AND SEEDS

The life cycle of flowering plants, including details about the flower, were considered in Chapter 23. You will recall that meiosis in the male part of the flower (the stamen) results in the formation of haploid pollen grains. The female portion of the flower (the carpel) is also the site of meiosis, with eight haploid nuclei being formed in the embryo sac. Before fertilization can occur, the pollen must be transferred from the stamen to the stigma of the carpel. This transfer of pollen, known as pollination, is accomplished by wind, water, or animals. Double fertilization, a process unique to flowering plants, follows pollination. After fertilization has occurred within the ovule in the ovary, the ovule develops into a seed, and the ovary surrounding it develops into a **fruit** (Fig. 29–7).

Many Mechanisms Have Evolved That Accomplish Pollination

Flowering plants have evolved a number of mechanisms that transfer pollen from the anther of the stamen to the stigma of the carpel, usually from one plant to another. (Although some plants have flowers that pollinate themselves and are **self-pollinated,** most plants have evolved mechanisms to assure **cross-pollination** between separate plants.) Some of these mecha-

(a)   (b)   (c)

(d)   (e)

FIGURE 29–7   Flowering and fruit production in a citrus tree. (a) Flower buds. (b) Orange flowers. Note that several of the flowers have already lost their petals and stamens, revealing the stigma, style, and ovary of the carpel. (c) Maturing ovaries. Fertilization has taken place and the stigma and style have dropped off several ovaries already. (d) Immature fruits. (e) Mature fruits. The orange is a modified berry.

nisms involve animals, including insects, birds, and bats. Biologists who study pollination have developed the concept of **pollination syndromes.** According to this concept, unrelated species of plants that have the same agent of pollination share similar features. Animal-pollinated flowers have various features that attract animals, including showy petals, nectar (a sugary solution that is an attractant for pollinators), pollen (a protein-rich food that pollinators eat), and scent. As the animal moves from flower to flower, it inadvertently carries pollen.

Many insects are involved in pollen transfer. Plants that are pollinated by insects often have blue or yellow petals. The insect eye does not perceive color in the same manner as the human eye. Insects see very well in the blue and yellow range of visible light but do not see red well. Consequently, flowers that are pollinated by insects are not usually red. Insects can also see in the ultraviolet range of the electromagnetic spectrum, an area that is invisible to the human eye. Many flowers have dramatic ultraviolet markings that are invisible to us but direct the insect to the center of the

flower, where the pollen or nectar is located (Fig. 29–8). Insects have a well-developed sense of smell. Many insect-pollinated flowers have a strong scent that is often pleasant, but not always. The carrion plant, which is pollinated by flies, has petals that are dappled with a reddish-brown color (like dried blood) and smells like rotting flesh (Fig. 29–9). Flies move from one flower to another; while looking for a place to deposit their eggs, they accomplish pollination.

Birds such as hummingbirds are important pollinators (Fig. 29–10). Flowers pollinated by birds are usually red, orange, or yellow. Birds see well in this region of visible light. Birds do not have strong sense of smell; consequently, bird-pollinated flowers usually lack much scent.

Bats, which feed at night and do not see very well, are frequent pollinators in the tropics (Fig. 29–11). Bat-pollinated flowers have dusky, dull-colored petals. The flowers of these plants produce a strong scent, usually of fermented fruit. Bats are attracted to the flowers by the scent and lap up the nectar. As they move from flower to flower, pollen is transferred.

(a)

(b)

FIGURE 29–8 Many insect-pollinated flowers have ultraviolet markings that are invisible to humans but very conspicuous to insects. (*a*) A flower as seen by the human eye. (*b*) The same flower viewed with a filter that transmits ultraviolet radiation indicates how the insect eye perceives it. The ultraviolet markings draw attention to the center of the flower where the pollen and nectar are located. (*a* and *b*, Thomas Eisner)

FIGURE 29–9 *Stapelia variegata*. This desert angiosperm is sometimes called the carrion plant due to its coloration and bad smell. It is pollinated by flies. (E.R. Degginger)

FIGURE 29–10 A ruby-throated hummingbird visiting a trumpet vine flower. Birds are important plant pollinators. (Steve Maslowski/Photo Researchers, Inc.)

FIGURE 29–11 A greater short-nosed fruit bat (*Cynopterus sphinx*) visiting a banana plant. The pollen grains on the bat's fur will be carried to the next plant, where cross pollination will occur. Bats are surprisingly important in the pollination of many tropical plants. (Merlin Tuttle/Photo Researchers, Inc.)

Other animals, including snails and small rodents, sometimes pollinate plants.

Some flowering plants have evolved a pollination mechanism that relies on wind rather than animals. Flowering plants that are pollinated by wind produce many, often inconspicuous, flowers (Fig. 29–12). They do not produce large, colorful petals or scent or nectar. Wind pollination is a "hit-or-miss" affair, and the likelihood of pollen landing on the stigma of the same species of flower is slim. However, wind-pollinated plants produce large quantities of pollen. Wind-pollinated plants include grasses, ragweed, and maples.

There are a number of examples of obligate relationships occurring between an animal pollinator and the plant it pollinates. A yucca found in the Southwestern United States, for example, is pollinated by only one species of moth. The female moth lays her eggs in the flower ovary. Neither would be able to reproduce successfully without the other. In this case, if one species were to become extinct, they both would become extinct.

FIGURE 29–12 Corn (*Zea mays*) is wind-pollinated. (*a*) The tassel is a cluster of inconspicuous male flowers. Note how the anthers protrude and dangle in the wind. (*b*) The ear is a cluster of female flowers. The stigma and style of each female flower is a silk. The silk is receptive to corn pollen; when corn pollen lands, a pollen tube grows through the silk to a "kernel" in the ear, where fertilization occurs. (*a* and *b*, G.R. Roberts)

Fruits Are Mature, Ripened Ovaries

There are several types of fruits that vary in structure owing to variations in the structure or arrangement of the flowers from which they were formed. We will consider a few examples of the four basic types of fruits: simple, aggregate, multiple, and accessory fruits (Table 29–1).

Most fruits are simple fruits. A **simple fruit** develops from a single ovary of a single flower. At maturity, simple fruits may be fleshy or dry. Two examples of fleshy fruits are the berry and the drupe. The **berry** is a fleshy fruit that has soft tissues throughout. A tomato is a berry, as are grapes and bananas. A **drupe** is a

TABLE 29–1
Some Types of Fruits[1]

   I. Simple fruit
     A. Fleshy
       1. Berry
       2. Drupe
     B. Dry
       1. Dehiscent
         a. Follicle
         b. Legume
         c. Capsule
       2. Indehiscent
         a. Grain
         b. Achene
  II. Aggregate fruit
 III. Multiple fruit
 IV. Accessory fruit

[1] There are many different fruit types. This table includes only those that are discussed in the text.

simple, fleshy fruit that has a hard, stony "pit" surrounding the seed. Examples of drupes include peaches, plums, and avocados.

Many simple fruits are dry at maturity. These fruits fall into two main categories: dehiscent and indehiscent. **Dehiscent** fruits split open when mature, releasing their seeds (Fig. 29–13). The milkweed **follicle** is an example of a simple, dry, dehiscent fruit that splits open along one seam or suture to release the seeds. The **legume** is a fruit that splits open along two seams or sutures. Pea pods are examples of legumes, as are green beans, although they are harvested before the fruit has dried out and split open.[1] A **capsule** splits open along multiple seams or pores. Poppy and cotton fruits are capsules. Other simple, dry fruits do not split open; these fruits are **indehiscent** (Fig. 29–14). The **grain** is an example of a simple, dry, indehiscent fruit. Each grain contains one seed; the seed coat is fused to the fruit wall, so it appears that the grain is a seed rather than a fruit. Kernels of corn and wheat are actually fruits of this type. The **achene** is a similar fruit in that it is simple, dry, indehiscent, and contains a single seed. However, the seed coat is not fused to the fruit wall in achenes; rather, the seed is attached to the fruit wall at one point only. Therefore, the achene can be separated from its seed. The sunflower fruit is an example of an achene. One can peel off the fruit wall to reveal the seed within.

In addition to simple fruits, there are aggregate, multiple, and accessory fruits (Fig. 29–15). An **aggregate fruit** is formed from a single flower that contains many separate carpels. After fertilization, each ovary

[1] The term *legume* refers to both a fruit type and any member of the pea family (Fabaceae), all of which produce this fruit type.

(b)

(a)

(c)

FIGURE 29–13 Simple, dry, dehiscent fruits. (*a*) The milkweed follicle splits open along one seam. (*b*) The bean fruit is a legume, which splits open along two seams at maturity. (*c*) The capsule splits open along multiple seams or pores at maturity. These iris fruits dehisce along three seams. (*a*, James L. Castner; *b*, David Cavagnaro; *c*, G.R. Roberts)

FIGURE 29–14 Simple, dry, indehiscent fruits. (*a*) The corn fruit is a grain. In grains the fruit wall is fused to the seed coat. (*b*) The sunflower fruit is an achene. The seed coat is not fused to the fruit wall. Therefore, it is possible to peel off the fruit wall, separating it from the seed.

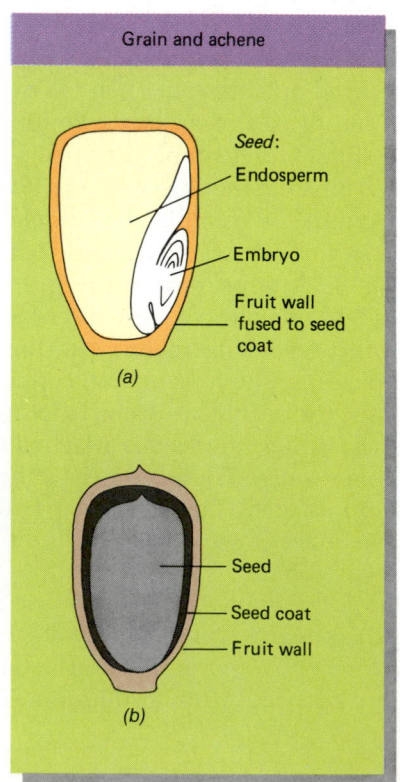

Grain and achene

Seed:
Endosperm
Embryo
Fruit wall fused to seed coat
(a)

Seed
Seed coat
Fruit wall
(b)

from each individual carpel enlarges. As they enlarge, they fuse to form a single fruit. The raspberry and blackberry are examples of aggregate fruits. A third type of fruit is the **multiple fruit,** which is formed from the ovaries of many flowers. Because these flowers grow in close proximity, the ovary from each fuses with nearby ovaries as it develops and enlarges after fertilization. Pineapples and osage oranges are multiple fruits. **Accessory fruits** are the fourth type. They are different from other fruits in that other plant tissues, in addition to ovary tissue, make up the fruit. For example, the edible portion of the strawberry is the red, fleshy **receptacle,** which is the terminal part of the flower stalk. Apples and pears are also accessory fruits; the outer part of each fruit is the enlarged **floral tube** that surrounds the ovary.

Fruit And Seed Dispersal Is Highly Varied In Flowering Plants

Flowering plants have evolved a number of methods of seed dispersal. Dispersal mechanisms for sexually produced offspring increase the probability that some of the variable offspring will find a suitable habitat. As a result, some plants have expanded their range. If the seed travels to an environment that is suitable for growth, it will germinate, and the plant will become established in that habitat. In some cases, the seed is

(a)　　　　　　　　　　(b)　　　　　　　　　　(c)

FIGURE 29–15 Aggregate, multiple, and accessory fruits. (*a*) Raspberries are aggregate fruits developing from a flower that contains many ovaries. (*b*) The pineapple is a multiple fruit, formed from the ovaries of many separate flowers. (*c*) The strawberry is three fruits in one! It is an accessory fruit because the major part of it is tissue other than ovary tissue. It is an aggregate fruit because it develops from a single flower that has many separate carpels. Finally, each speck on the strawberry is a tiny achene that develops from one of the separate carpels. (*a* and *c*, Dwight R. Kuhn; *b*, Phil Degginger)

the actual agent of dispersal; in others, it is the fruit that is dispersed and carries along the enclosed seeds.

Wind is responsible for seed and fruit dispersal in many plants (Fig. 29–16 *a* and *b*). Plants that have winged fruits, such as the maple, are adapted for wind dispersal. Also, light, feathery plumes on the fruit or seed allow it to be transported by wind, often for considerable distances. The dandelion fruit and milkweed seed have this type of adaptation. The tumbleweed is an example of an entire plant being the agent of dispersal. It detaches and blows across the ground; as it bumps along, seeds fall out.

Some plants have evolved special structures that aid in animal dispersal of their seeds and fruits (Fig. 29–16 *c* and *d*). The spines and barbs of the cocklebur and similar fruits catch in the fur of animals. Fleshy, edible fruits are also adapted for animal dispersal. As they are eaten, the seeds are often swallowed. Because of their thick seed coats, they are not digested. Rather, they pass through the animal's digestive tract and are deposited in the animal's feces some distance from the plant. In fact, some seeds will not germinate until they have passed through an animal's digestive tract.

The coconut, with its air spaces and corky floats for buoyancy, is a good example of a fruit that is adapted for dispersal by water (Fig. 29–16 *e*). It can be carried by ocean currents for thousands of miles. When it washes ashore, it germinates and grows into a coconut palm tree. Some fruits are dispersed by neither wind, animals, nor water. These fruits are often explosive, forcibly discharging their seeds. Pressures due to differences in turgor, or to drying out, cause them to burst open suddenly, scattering the seeds for considerable distances.

ENVIRONMENTAL CUES MAY INDUCE FLOWERING IN PLANTS

The initiation of sexual reproduction is often under environmental control, particularly in temperate latitudes. This is important for the plant's survival, because the timing of sexual reproduction is critical to reproductive success. Plants must be able to flower and form fruits and seeds before the onset of winter or a dry season induces dormancy. A number of plants can detect changes in the relative amounts of daylight and darkness that accompany the changing seasons, and flower in response to these changes. These plants vary

FIGURE 29–16 Methods of seed and fruit dispersal. (a) The feathery plumes of a milkweed seed make it buoyant for dispersal by wind. (b) The fruits of sugar maple have wings for wind dispersal. (c) *Acaena* fruits clinging to sheep wool will be carried away from the parent plant. (d) Fleshy fruits are eaten by animals such as this meadow vole. The seeds are frequently swallowed whole and pass unharmed through the animal's digestive tract. (e) Coconuts are adapted for water dispersal. When it washes ashore, the coconut germinates, often thousands of miles from its original home. (*a*, David Cavagnaro; *b* and *e*, James L. Castner; *c*, G.R. Roberts; *d*, Dwight R. Kuhn.)

in their response to the duration and timing of light and dark, but the overall mechanism of detection is the same. Other plants have temperature requirements that induce flower formation.

Flowering May Be Initiated By Light

Photoperiodism is the response of a plant to the relative lengths of daylight and darkness. Flowering is one of several physiological activities that respond to pho-

toperiod in plants. For example, if one were to plant biloxi soybeans at 2-week intervals from early May to August, they would all flower at the same time in September, regardless of size or age (because they would be responding to the same daylength).

Plants can be placed into three main groups on the basis of how photoperiodism affects their flowering: short-day, long-day, and day-neutral plants. **Short-day plants** were initially defined as plants that flower when exposed to some critical day length *or less*. However, further research brought out that the important factor

FOCUS ON   Localizing Phytochrome In Plant Cells

We have seen that phytochrome plays an important role in a variety of physiological functions in plants. This pigment is present in extremely small amounts in plant tissue. Phytochrome can be extracted from plants and detected with sensitive spectrophotometric techniques, but it is impossible to "see" phytochrome in plant cells, even in cells that have been grown in the dark and lack chlorophyll.

We can, however, locate phytochrome in plant tissues and within plant cells by immunochemical techniques. The animal immune system produces various specific substances in response to the presence of foreign proteins. If phytochrome is injected into a rabbit, for example, the rabbit's immune system produces an antibody specific for phytochrome. If this material is removed from the rabbit and purified, it can be added to plant tissues or cells, where it binds to phytochrome. So that researchers can see the binding, they attach one end of the antibody molecule to an enzyme that catalyzes a reaction that produces a colored product in plant cells.

With this technique, it has been possible to localize phytochrome with some success. Within plant cells that have never been exposed to light (and, therefore, contain P_R but no P_{FR}), phytochrome is evenly distributed throughout the cell. After a brief treatment with red light, the phytochrome appears to be more localized, but it is still not associated with the nucleus, chloroplasts, or mitochondria. It has been suggested that it may be associated with the endoplasmic reticulum at this point. After a longer red-light treatment, phytochrome has been found in the nucleus.

Other physiological experiments indicate that phytochrome may also be present in mitochondria and other cellular organelles. Perhaps the various locations of phytochrome within the cell are due to its several different physiological functions. Biologists are trying to determine how phytochrome triggers such diverse growth responses in plants as flowering and seed germination. Information on the cellular location of phytochrome is an important step toward this goal.

The immune system of rabbits is used to help localize phytochrome in plant cells.
(Sharon Cummings/M. L. Dembinsky)

in initiation of flowering in short-day plants is the long, uninterrupted period of darkness rather than the short period of daylight. In other words, short-day plants flower when the night length is equal to or greater than some critical length. Common short-day plants are the chrysanthemum and the poinsettia. Commercial florists manipulate light to produce these flowers at desired times of the year. These plants typically flower in late summer or fall.

Long-day plants were initially defined as being able to flower when the day length is equal to or greater than some critical amount. However, a more accurate definition would be that long-day plants cannot exceed some critical period of darkness. Plants that flower in late spring or summer, such as clover, black-eyed Susan, and lettuce, are long-day plants. The critical day length (or night length) varies from species to species. Two different plants could have the same critical day length, but one could be a short-day plant and the other could be a long-day plant.

Some plants do not initiate flowering in response to changing amounts of daylight and darkness. These **day-neutral plants** have some other type of stimulus, either external or internal, that causes them to flower. The tomato, dandelion, string bean, and pansy are day-neutral plants.

If plants, or any living organism, have a biological response to light, it follows that something in that organism *perceives* the light (that is, a **photoreceptor**). The photoreceptor for photoperiodism is a substance that can detect whether it is daylight or darkness and can cause a physiological change in plants that leads to the initiation of flowering. What is this photoreceptor?

Phytochrome Detects Varying Periods Of Day Length And Darkness

The photoreceptor involved in photoperiodism and a number of other light-initiated physiological responses of plants is a blue-green proteinaceous pigment called **phytochrome.** Phytochrome, which is present in all vascular plants, has two forms. It can readily convert from one form to the other upon absorption of light of specific wavelengths. One form, designated P_R (for red-absorbing phytochrome), absorbs red light (660 nm) strongly. After absorbing red light, the conformation (shape) of the molecule changes to the second form of phytochrome, P_{FR}. This form of phytochrome is so designated because it absorbs red light of longer wavelengths than P_R, described as far-red light (730 nm). When P_{FR} absorbs far-red light, it reverts back to the original form, P_R. The P_{FR} form of phytochrome is less stable than the P_R form and reverts spontaneously, albeit slowly, to P_R in the dark.

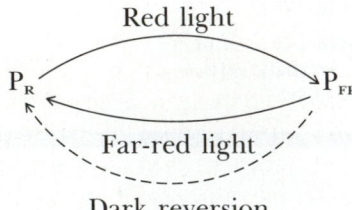

Red light

P_R ⟶ P_{FR}

Far-red light

Dark reversion

But what does a pigment that absorbs red light and far-red light have to do with daylight and darkness? The sun's light is composed of the entire spectrum of visible light in addition to ultraviolet and infra-

red. However, sunlight has more red light than far-red light. As a result, the phytochrome in a plant exposed to the sunlight will be a *mixture* of both P_R and P_{FR}, with P_{FR} predominating. Therefore, the *ratio* of P_R to P_{FR} is critical in determining the physiological response. During the night, the P_{FR} will revert back to P_R.

In short-day plants, the P_{FR} form of phytochrome *inhibits* flowering. In order to flower, these plants need long nights. The long period of darkness allows the P_{FR} to revert back to P_R, so the plant has some minimum time during the 24-hour period with *no* P_{FR} present. This initiates flowering. Biologists have experimented with short-day plants by growing them under a short-day/long-night regime, but interrupting the night with a short burst of red light (Fig. 29–17). Exposure to red light for as brief a period as 10 minutes in the middle of the night will prevent flowering in short-day plants. This effect occurs because the brief exposure to red light converts some of the phytochrome to the P_{FR} form. Therefore, the plant does not have a sufficient period of time at night without any P_{FR}.

The effect that a short period of red light in the middle of the night has on short-day plants is reversible; that is, if a short-day plant is grown under conditions of short days and long nights, with a brief flash of red light followed by a brief flash of far-red light in the middle of the night, that plant will flower. Based on our understanding of the photoreversible nature of phytochrome, this observation is easy to explain. Short-day plants need long nights to allow for dark reversion of P_{FR} to P_R in order to induce flowering. A brief flash of red light in the middle of the night converts P_R into P_{FR}. However, if this is followed by a period of far-red light, the P_{FR} that was formed is converted back into P_R; therefore, flowering occurs.

In long-day plants, the active form of phytochrome, P_{FR}, *induces* flowering. Long-day plants that are exposed to a long-day/short-night regime will flower. The long days cause these plants to produce predominantly P_{FR}. During the short nights, some P_{FR} is changed to P_R, but because the night is short, the plant has little or no time with no P_{FR} present during a 24-hour period; hence, it flowers.

Plant biologists are puzzled by the observation that P_{FR} inhibits flowering in short-day plants and induces flowering in long-day plants. Why different plants respond so differently to P_{FR} is not known at this time. Biologists are also seeking the exact mechanism of phytochrome action; that is, once it has absorbed light and changed into another form, what happens next? Does this somehow trigger the production of a hormone, which then induces flowering? Does phytochrome work by affecting membranes, or enzymes, or gene expression? Or does it work by some combination of these methods? Why does phytochrome have no effect

Long day | Short day | Short day with night interruption

24 hrs | 24 hrs | 24 hrs

Short-day plant

Long-day plant

FIGURE 29–17 Photoperiodic response of a short-day plant (*top row*) and a long-day plant (*bottom row*) to different periods of light and dark. Note that the short-day plant does not flower when exposed to 8 hours of daylight and 16 hours of darkness interrupted with a brief flash of light. This same treatment induces the long-day plant to flower.

on flowering in day-neutral plants? Regardless of its mode of action, the universal presence of phytochrome in vascular plants attests to its importance.

Phytochrome Is Involved In Many Plant Responses To Light Besides Flowering

Phytochrome has been implicated in a number of physiological responses besides flowering. For example, it is involved in the light requirement that some seeds have for germination (Chapter 26). Proof that phytochrome is the photoreceptor for this response is evident in its photoreversibility when exposed to red or far-red light. Seeds with a light requirement must be exposed to red light. Exposure to red light converts P_R to P_{FR}, and germination occurs. If the seeds are exposed to a brief period of red light followed by a brief period of far-red light, however, they will not germinate, because P_{FR} is converted back to P_R, the inactive

form. Experiments on the photoreversible nature of phytochrome have been conducted in which seeds are exposed to alternating forms of light many, many times. Regardless of how many light treatments one gives the seeds, they always respond to the *last* treatment (Fig. 29–18). If the last treatment is red light, the seeds will germinate. If the last treatment is far-red light, the seeds remain dormant.

Other physiological functions under the influence of phytochrome include sleep movements in leaves, shoot dormancy, leaf abscission, and pigment formation in flowers, fruits, and leaves. If phytochrome has been implicated in such diverse physiological responses, then light is required to initiate these responses. The importance of light in various plant functions besides photosynthesis cannot be overemphasized. Timing of daylight and darkness is a key way in which plants "measure" the change in time from one season to the next. This measurement is crucial for survival, particularly in environments where the cli-

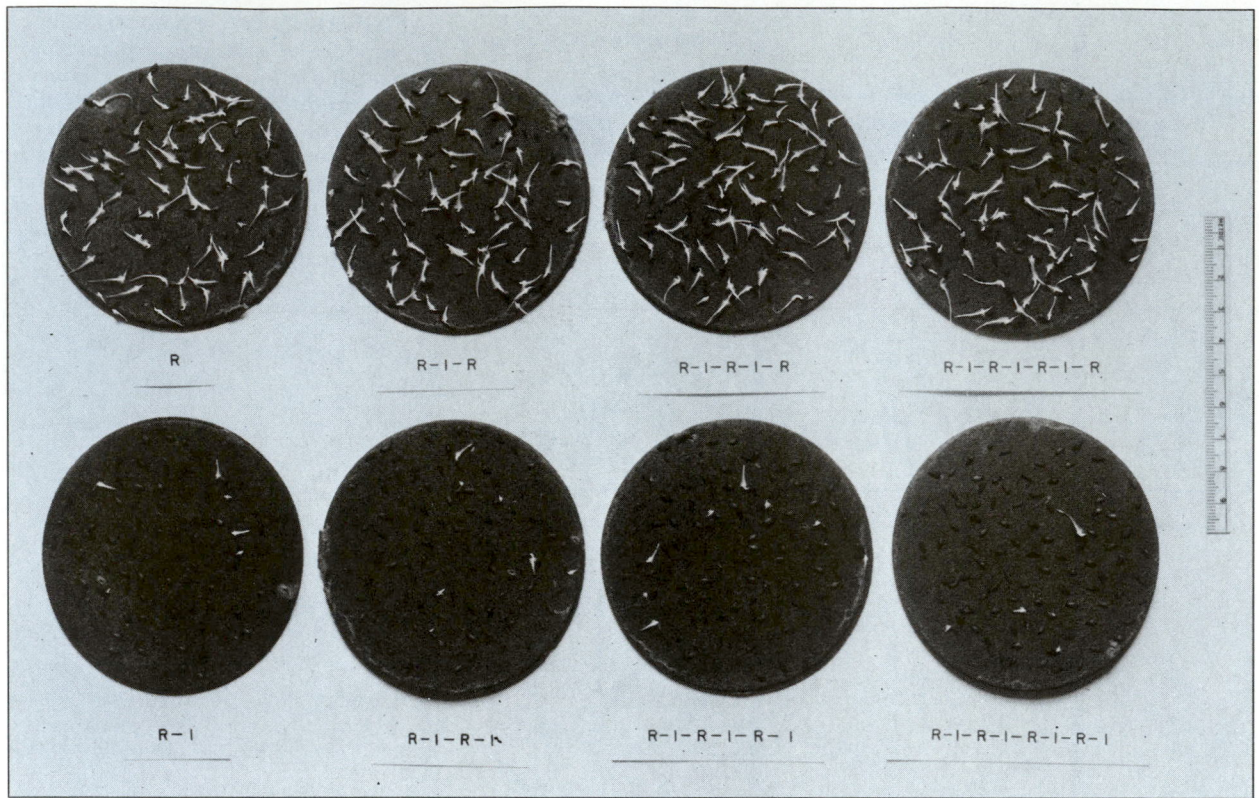

FIGURE 29–18 The control of lettuce seed germination by red (R) and far-red (I) light. Seeds are moistened and then exposed to red light (for 1 minute each exposure) and far-red light (for 4 minutes each exposure) in the sequences indicated. If the last exposure is red light, most of the seeds germinate. If the last exposure is far-red light, they **remain dormant.**
(Borthwick, H.A., et al., *Proceedings of the National Academy of Science USA* 38:662–666, 1952)

mate goes through a regular, annual pattern of favorable and unfavorable seasons.

Temperature May Also Affect Reproduction

In certain plants, the temperature has an effect on flowering. The promotion of flowering by treatment with cold is known as **vernalization.** The part of the plant that must be exposed to cold varies. For some plants, such as most alpine plants, the moist seeds must be exposed to a period of several weeks of cold. For other plants, the young, recently germinated seedlings have a cold requirement. Some plants have an absolute requirement for the cold period and will not flower unless they have been vernalized. Other plants will flower sooner if exposed to low temperatures, but would have flowered at a later time, regardless of temperature. Examples of plants with a cold requirement include biennials like the carrot and winter annuals like winter wheat. Carrots and other biennials grow vegetatively the first year, storing food that is produced by photosynthesis in their roots. After dormancy during the winter, they flower and reproduce during the second year. Carrots that are left in a warm environment and are not exposed to a cold treatment continue vegetative growth indefinitely. They do not initiate sexual reproduction.

The external stimuli that a plant responds to (in this case, temperature) are moderated and influenced by internal conditions within the plant (such as hormone levels). It is possible to eliminate the cold requirement for flowering in biennials by treatment with a certain plant hormone. We will discover in the next chapter that plant hormones affect every aspect of plant growth and development.

■ SUMMARY

I. In asexual reproduction, offspring are genetically identical to the parent.
   A. Rhizomes, tubers, bulbs, corms, and stolons are stems specialized for asexual reproduction.
   B. Some leaves have meristematic tissue along their margins and give rise to small plants asexually.

C. Roots may develop adventitious buds that form suckers, which may produce new plants.

D. Apomixis is the production of seeds and fruits without sexual reproduction.

II. Sexual reproduction involves pollination (transfer of pollen from anther to stigma) and fertilization (fusion of gametes). The flower is the organ in which sexual reproduction occurs. The fruit and seed develop as a result of successful fertilization. The offspring produced by sexual reproduction are genetically variable.

A. Fruits are mature, ripened ovaries.

1. Simple fruits develop from a single ovary of a single flower.

a. Berries and drupes are simple, fleshy fruits.

b. Follicles, legumes, and capsules are simple, dry, dehiscent fruits.

c. Grains and achenes are simple, dry, indehiscent fruits.

2. Aggregate fruits develop from many ovaries within a single flower.

3. Multiple fruits develop from many ovaries of many flowers growing in close proximity.

4. In accessory fruits, the major part of the fruit is tissue other than ovary tissue.

B. Angiosperm seeds and fruits are adapted for various means of dispersal, including wind, animals, and water.

III. Photoperiodism is the response of plants to the duration and timing of light and dark.

A. Flowering is a photoperiodic response, with some plants being short-day plants, long-day plants, or day-neutral plants.

B. The photoreceptor in photoperiodism is a photoreversible pigment, phytochrome.

IV. Vernalization is the promotion of flowering by exposure to cold temperatures.

■ POST-TEST

1. Genetic variability in the offspring is characteristic of _____ reproduction.

2. The _____ is a horizontal, underground stem that is specialized for asexual reproduction.

3. The white potato is an example of an underground stem called a _____.

4. Another name for the runner is _____, an above-ground, horizontal stem.

5. In apomixis, fruits and seeds are produced by _____ means.

6. Plants with blue petals, nectar, and a strong scent are most likely pollinated by _____.

7. Plants with reduced or absent petals, no nectar, no scent, and large quantities of pollen are pollinated by _____.

8. A(n) _____ may be defined as a mature, ripened ovary.

9. Grapes, tomatoes, and bananas are simple fruits known as _____.

10. The peach is an example of a _____.

11. A fruit that splits open to liberate the seeds is said to be _____.

12. A legume is a simple, dry, dehiscent fruit that splits open along _____ seam(s).

13. _____ fruits form from many ovaries of a single flower, whereas _____ fruits develop from many ovaries of many flowers.

14. Apples, strawberries, and pears are examples of _____ fruits.

15. Light, feathery plumes on the seed or fruit signify that it is most likely dispersed by _____.

16. Fleshy, edible fruits are adapted for dispersal by _____.

17. The response of the plant to the relative amounts of daylight and darkness is known as _____.

18. The critical factor in the flowering response of short-day and long-day plants is the amount of _____ (daylight or darkness).

19. The P_{FR} form of pytochrome is produced when _____ light is absorbed.

20. _____ is the promotion of flowering by a cold treatment.

■ REVIEW QUESTIONS

1. Would sexual or asexual reproduction be more beneficial in the following circumstances, and why?

a. Tree in a stable environment.

b. Annual, herbaceous plant in a rapidly changing environment.

c. Plant with an extremely narrow habitat range.

2. Would production of seeds and fruits by apomixis confer any special reproductive advantage that other asexual structures like corms and bulbs lack?

3. Draw pictures to show the kinds of flowers that might form simple, aggregate, multiple, and accessory fruits.

4. Explain the features possessed by fruits and seeds that are dispersed by animals.

5. Will flowering occur in the following situations? Explain

why or why not with respect to phytochrome.

a. A short-day plant is exposed to 15 hours of daylight and 9 hours of darkness.

b. A short-day plant is exposed to 9 hours of daylight and 15 hours of darkness, with a 10-minute flash of red light in the middle of the night.

c. A short-day plant is exposed to 9 hours of daylight and 15 hours of darkness with the following light treatments in the middle of the night: 10 minutes of red light, 10 minutes of far-red light, 10 minutes of red light, 10 minutes of far-red light.

■ RECOMMENDED READINGS

Galston, A.W., P.J. Davies, and R.L. Satter. *The Life of the Green Plant*, 3rd ed. Englewood Cliffs, NJ, Prentice-Hall, 1980. Contains an excellent explanation of phytochrome and photoperiodism.

Kaufman, P.B. et al. *Plants: Their Biology and Importance.* New York, Harper & Row, 1989. A good general reference on plant biology, including reproductive structures.

30

Plant Hormones and Responses

OUTLINE

I. Changes in turgor can induce plant movements
II. A biological clock influences many plant responses
III. A tropism is plant growth in response to an external stimulus
IV. Hormones regulate plant growth and development
   A. Charles Darwin first provided evidence for the existence of auxin
   B. Gibberellins were first discovered in a fungus
   C. Cytokinins promote cell division
   D. Ethylene is the only gaseous plant hormone
   E. Abscisic acid promotes dormancy in higher plants

LEARNING OBJECTIVES

After you have studied this chapter you should be able to:

1. Distinguish between phototropism, geotropism, and thigmotropism.
2. Distinguish between a tropism and a turgor movement.
3. List several different ways each of the following hormones affects plant growth and development: auxin, gibberellin, cytokinin, ethylene, and abscisic acid.
4. Give an example of a physiological response in plants that may be due to varying ratios of several hormones rather than to one specific hormone.
5. Relate which hormones are involved in the following and how their functions interact:
   a. leaf abscission.
   b. seed germination.
   c. apical dominance.

The fall coloring of leaves is under the influence of several plant hormones. (Milton Rand/Tom Stack & Associates)

Because plants lack a nervous system and because most plants are firmly rooted in the ground, it is generally assumed that they are incapable of self-directed movements. However, a variety of growth movements and responses are found in plants. Most of these are very gradual, but some are quite rapid and spectacular, as when the Venus flytrap snaps its leaf shut in less than 0.1 second after being mechanically stimulated (Chapter 27). In many aspects of plant growth and development, environmental cues determine when, whether, and to what extent changes will occur. We have already examined several of these. Water, oxygen, temperature, and light influence the germination of seeds (Chapters 26 and 29). Light affects such diverse physiological responses as photosynthesis (Chapter 9), stomatal opening and closing (Chapter 27), and flowering (Chapter 29). Periods of low temperature may also promote flowering in certain plants (Chapter 29).

ment. The tannins, which are normally stored in the vacuole, impart a bad taste to the tissue, and some researchers have suggested this as a further adaptation to avoid predation.

The closure of the Venus flytrap leaf (Chapter 27) is similar in its mechanism to that of *Mimosa*. An electrical signal, which moves much more rapidly than in *Mimosa*, causes water to move out of certain cells. A movement of ions is also associated with the water movement, but it is not known which ions are involved.

Changes in turgor are also responsible for **solar tracking,** the ability of leaves or flowers to follow the sun's movement across the sky (Fig. 30–2). Frequently, the leaves of these plants are arranged perpendicular to the sun's rays, regardless of the time of day or the sun's position in the sky. This allows for maximal light absorption. Many solar trackers have pulvini at the bases of their petioles. Sunflower, soybean, and cotton are examples of plants that are solar trackers.

KEY CONCEPTS

- Plants, like all living organisms, respond to both external and internal stimuli.
- All aspects of plant growth and development are under the influence of hormones.
- Five different hormones are known to operate in higher plants.
- The functions of plant hormones overlap.

CHANGES IN TURGOR CAN INDUCE PLANT MOVEMENTS

Mimosa pudica, the sensitive plant, dramatically folds its leaves in response to an external stimulus (Fig. 30–1). The stimulus may be mechanical, electrical, chemical, or thermal. It is possible that this unusual behavior protects the plant from predators.

When a *Mimosa* leaf is stimulated by touching, an electrical impulse moves down the leaf to special cells in an organ at the base of the petiole, the **pulvinus.** While we know that in plants such as *Mimosa* electrical signals are a form of intercellular communication, we do not completely understand the actual mechanism of the transmission. When the electrical signal reaches the pulvinus cells, it triggers a loss of turgor in those cells as potassium ions, tannins, and water leave the cells. The sudden change in turgidity causes the leaf move-

A BIOLOGICAL CLOCK INFLUENCES MANY PLANT RESPONSES

Plants, animals, and microorganisms appear to have an internal timer, or biological clock, that approximates a 24-hour cycle. These internal cycles are known as **circadian rhythms** (obtained from Latin words meaning "around" and "day"). Circadian rhythms usually are somewhere between 20 and 30 hours. In nature the rising and setting of the sun resets the clock (to a 24-hour cycle) each day. Phytochrome (Chapter 29) has been implicated as the photoreceptor involved in resetting the biological clock for many plants.

One example of circadian rhythms in plants is the opening and closing of stomates that occurs independently of light and darkness (Chapter 27). Plants placed in continual darkness for extended periods continue to open and close their stomates, maintaining an approximate 24-hour schedule. Another example of circadian rhythms in plants is the sleep movements observed in the common bean and other plants (Fig. 30–3). During the day the leaves are horizontal for optimum light absorption. At night the leaves fold down or up, perpendicular to their daytime orientation. Results from several studies indicate that the plant actually "anticipates" sunrise and sunset, as the movements begin *before* the sun rises or sets. These sleep movements occur independently of the 24-hour cycle in nature. If bean plants are placed in constant darkness or constant light, the movements continue on an approximate 24-hour cycle.

(a)

(b)

FIGURE 30–1   Turgor movements in the "sensitive plant," *Mimosa pudica.* **(a)** *Mimosa pudica* **before being disturbed. (b) The plant several seconds after being touched. Note how the leaves have folded and drooped. (c) (1) The base of the petiole, showing the pulvinus. (2) Section through the pulvinus, showing cells when leaf is undisturbed. (3) Section through the pulvinus, showing loss of turgor that produces the folding of the leaves.** (*a* and *b*, Richard F. Trump/Photo Researchers, Inc.)

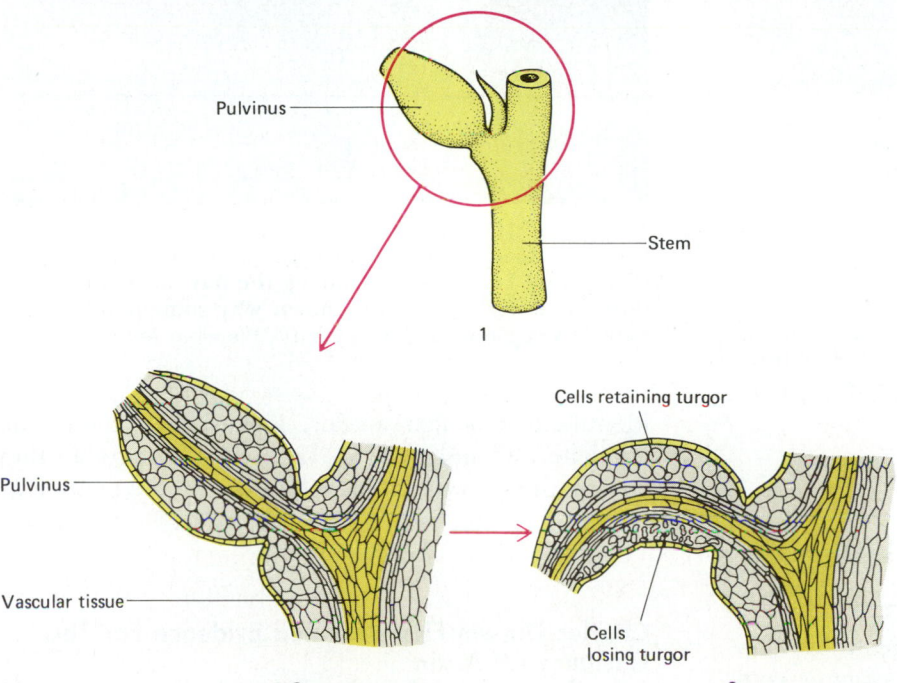
(c)

A TROPISM IS PLANT GROWTH IN RESPONSE TO AN EXTERNAL STIMULUS

Tropisms may be positive or negative depending on whether the plant grows toward (positive) or away from (negative) the stimulus (Fig. 30–4). **Phototropism** is the growth of a plant due to the direction of light. Most stems exhibit positive phototropism and bend toward light. A growth in response to the direction of gravity is **gravitropism** (sometimes called geotropism). Stems generally exhibit negative gravitropism, while roots exhibit positive gravitropism. **Thigmotropism** is growth in response to a mechanical stimulus, such as contact with a solid object. The twin-

FIGURE 30–2 Solar tracking in sunflowers. Note how all the flower heads are oriented in the same direction. The orientation of the plants toward light in solar tracking is due to changes in turgor. (Grant Heilman, from Grant Heilman Photography)

(a)

(b)

FIGURE 30–3 Sleep movements in wood sorrel (*Oxalis*). (*a*) Leaf position during the day. (*b*) Leaf position at night. It is not known why some plants exhibit sleep movements. (*a* and *b*, Biophoto Associates)

ing or curling growth of tendrils is an example of thigmotropism. Tropisms are also caused by other stimuli in the environment such as water, temperature, chemicals, and oxygen.

HORMONES REGULATE PLANT GROWTH AND DEVELOPMENT

In plants, as in animals, **hormones** regulate development and growth. Plant hormones are organic compounds produced in one part of the plant and transported to another part, where they elicit a physiological response. Hormones are effective in extremely small amounts. For that reason their study is very challenging. In plants the study of hormones is even more difficult because, unlike animal hormones, each plant hormone elicits many different responses. Also, the effects of different hormones overlap, so that it is difficult to determine which hormone, if any, is the primary cause of a particular response. To further complicate matters, plant hormones may be stimulatory at one concentration and inhibitory at a different concentration. Five types of plant hormones have been

identified thus far: auxins, gibberellins, cytokinins, ethylene, and abscisic acid (Table 30–1). Together they control the growth and development of the plant at all stages of its life.

Charles Darwin First Provided Evidence For The Existence Of Auxin

Although he is known mostly for developing the concept of natural selection to explain evolution, Charles Darwin was a gifted naturalist who experimented on many plants and animals. Darwin and his son, Francis, were interested in phototropism, the growth of plants toward light. In the 1880s they experimented with newly germinated canary grass seedlings (Fig. 30–5). The first part of the grass seedling to emerge from the soil is the **coleoptile,** the protective sheath that encircles the stem (Chapter 26). When the Darwins exposed coleoptiles to unidirectional light, the coleoptiles bent toward the light. The bending occurred close to, but not at the very tip of, the coleoptile.

(a)

(c)

(b)

FIGURE 30–4 Tropisms demonstrate that plants respond to their environment. (*a*) Growth in the direction of light is known as positive phototropism. (*b*) Stems exhibit negative gravitropism. When placed on its side, this white oak stem responded by growing upward, against the direction of gravity. The response was evident in 15 hours or less. (*c*) The twining motion of a tendril is an example of thigmotropism. Though the *Passiflora* plant is not carnivorous, its tendrils have snared an adult moth, which made the mistake of roosting on them overnight. (*a*, Runk/Schoenberger, from Grant Heilman; *b*, E.R. Degginger; *c*, BPS[9-227])

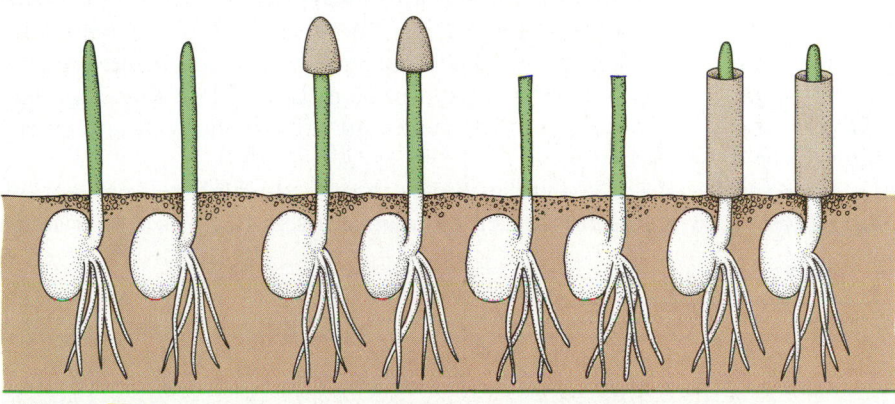

Light rays

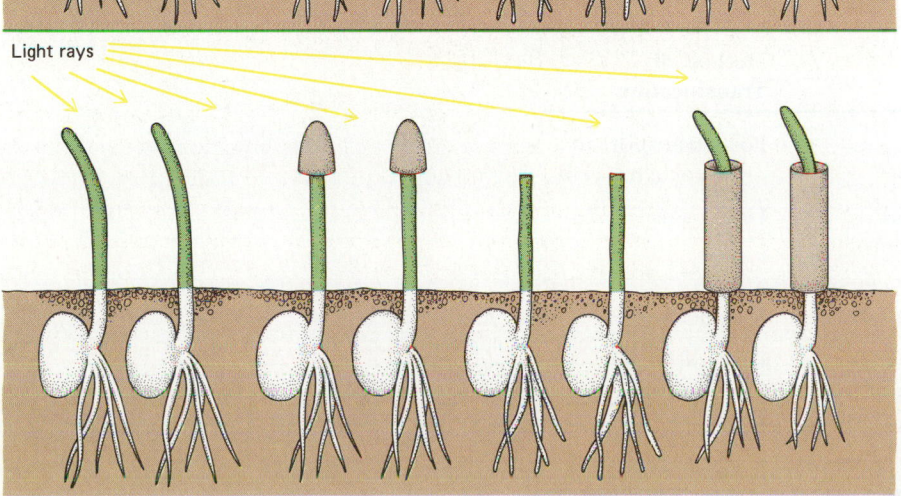

FIGURE 30–5 The Darwins' experiment with coleoptiles of canary grass seedlings. (*Upper row*) Some plants were uncovered, some were covered only at the tip, some had the tip removed, and some were covered everywhere but at the tip. (*Lower row*) After exposure to light coming from one direction, the uncovered plants and the plants with uncovered tips (*far right*) grew toward the light. The plants with covered tips (*center left*) or tips removed (*center right*) did not bend toward light. Darwin and his son concluded that the tip is sensitive to light and produces some "influence" that moves down the plant and causes the bending.

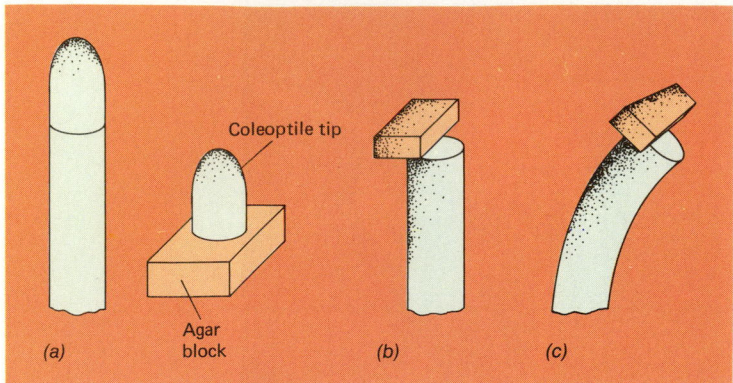

FIGURE 30–6 Frits Went's experiment. (*a*) Coleoptile tips were placed on agar blocks for a period of time. (*b*) The agar block was transferred to a decapitated coleoptile. It was placed off-center and the coleoptile was left in continuous darkness. (*c*) The coleoptile bent, indicating that a chemical had been transferred from the original coleoptile tip to the agar block to the decapitated coleoptile.

The Darwins tried to influence this bending in several ways. For example, they covered the tip of the coleoptile as soon as it emerged from the soil. The plants treated in this manner did not bend! Likewise, bending did not occur when the coleoptile tip was removed (i.e., the coleoptile was decapitated). When the bottom of the coleoptile was shielded from the light, the coleoptile still bent toward light. From these experiments, the Darwins concluded that "some influence is transmitted from the upper to the lower part, causing it to bend." This conclusion fits the definition of a hormone exactly. Thus, Charles Darwin was the first person to produce data suggesting that plants have hormones.

However, it was a number of years before the techniques necessary to extract and identify this substance were available. In the 1920s Frits Went, a young Dutch scientist, isolated the phototropic hormone from oat coleoptiles. He removed the oat coleoptile tips and placed them on tiny blocks of agar for a period of time. When he put an agar block on the side of a decapitated coleoptile in the dark, bending occurred (Fig. 30–6). Went named this substance **auxin** (from the Greek word for "enlarge" or "increase"). The chemical structure of auxin was elucidated by a research team led by Kenneth Thimann at the California Institute of Technology.

Auxins are any of a group of compounds that stimulate phototropic curvature in coleoptiles, stems, and certain other plant structures. The main auxin found in plants is **indoleacetic acid,** or **IAA.** Its structure is similar to that of the amino acid, tryptophan, from which it is synthesized. IAA is synthesized in the shoot apical meristem, young leaves, and seeds. It is not translocated appreciably in either the xylem or the phloem. It moves through the plant within the parenchyma cells, at a rate that is too fast to be accomplished by diffusion. The movement of auxin is called **polar transport** because it is always unidirectional, or polar, from the top of the shoot toward the roots. Polar transport requires energy and is not influenced by gravity. If a section of stem is inverted, the auxin still moves from the stem end toward the root end of the plant.

Auxin promotes growth through cell elongation. Recall that cell elongation occurs in apical meristems

TABLE 30–1
The Five Plant Hormones

Hormone	Site of Production	Method of Translocation
Auxin	Shoot apical meristem, young leaves, seeds	Polar transport in parenchyma cells
Gibberellin	Young leaves, root and shoot apical meristems, embryo in seed	Unknown
Cytokinin	Roots	Xylem
Ethylene	Stem nodes, ripening fruit, senescing tissue	Unknown (diffusion?)
Abscisic acid	Older leaves, root cap, stem	Vascular tissue

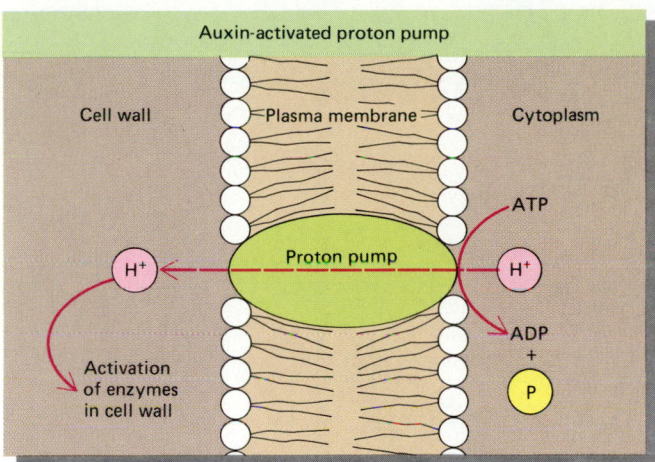

FIGURE 30–7 The acid-growth hypothesis. Auxin activates a proton pump in the plasma membrane. This pumps hydrogen ions out of the cell to the cell wall, changing its pH. The lowered pH of the cell wall activates enzymes that break the cross-links holding the cellulose microfibrils together. The pressure created by increasing turgor then allows the wall to expand.

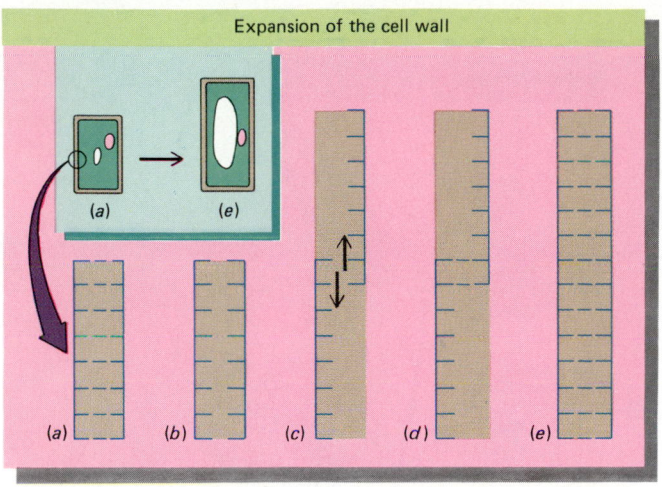

FIGURE 30–8 How wall expansion associated with cell elongation may occur. (*a*) The cell wall is composed of cellulose microfibrils held in place by polysaccharide cross-links. (*b*) Enzymes activated by a lowered pH break the cross-links. (*c*) The pressure created by increasing turgor causes the wall to stretch. (*d*) New polysaccharide cross-links are formed, holding the wall in its new position. (*e*) New cell wall materials are synthesized, completing the expanded wall.

just behind the area of cell division (Chapter 26). Auxin apparently exerts this effect by changing the cell walls so they can expand. According to the **acid-growth hypothesis,** auxin triggers a proton (H$^+$) pump in the plasma membrane (Fig. 30–7). This causes a flow of H$^+$ ions through to the cell wall, acidifying it and activating certain enzymes that break

bonds between cell wall molecules (Fig. 30–8). As a result, the wall becomes flexible and can stretch as water accumulates in the vacuole.

Auxin's effect on cell elongation explains phototropism as well as several other tropisms. When a plant is exposed to a unidirectional source of light, the auxin migrates to the dark side of the stem before moving

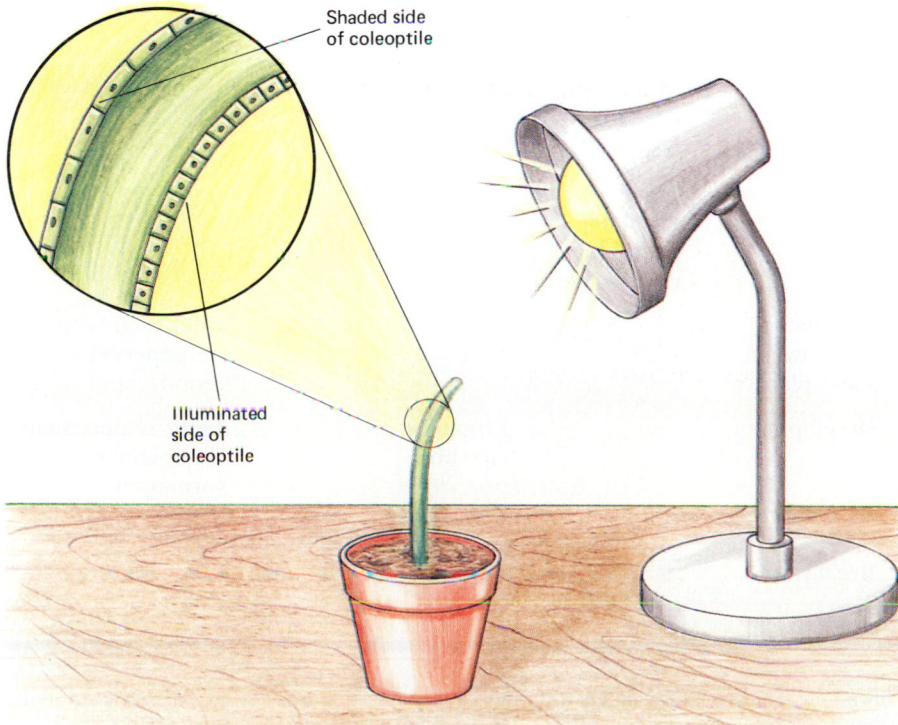

FIGURE 30–9 Phototropism is due to the unequal distribution of auxin. Auxin travels down the side of the stem or coleoptile *away* from the light, causing the cells on the shaded side to elongate. Therefore, the stem or coleoptile bends toward light.

Apical dominance

(a)   (b)

FIGURE 30–10   Auxin inhibits the development of lateral buds. (*a*) When the tip of the plant is intact, the lateral buds do not develop. (*b*) The tip of the plant (source of auxin) has been removed. Because there is no auxin to diffuse down the stem, lateral buds develop into branches.

TABLE 30–2
Some of the Interactions Between Plant Hormones During Various Aspects of Plant Growth

Physiological Activity	Auxin	Gibberellin	Cytokinin	Ethylene	Abscisic Acid	Other Factors for Some Plants
Seed germination		Promotes	?	Promotes	Inhibits	Cold requirement, light requirement
Growth of seedling into mature plant	Cell elongation, organogenesis[1]	Cell division and elongation	Cell division and differentiation, organogenesis[1]	Inhibits cell elongation		
Apical dominance	Inhibits lateral bud development		Promotes lateral bud development	?		
Initiation of reproduction (flowering)		Stimulates flowering in some plants[2]	?			Cold requirement, photoperiod requirement
Fruit development and ripening	Development	Development		Promotes ripening		Light requirement (for pigment formation)
Leaf abscission	Inhibits		Inhibits	Promotes	?	Light requirement
Winter dormancy of plant		Breaks	?		Promotes	Light requirement
Seed dormancy		Breaks	?		Promotes	

[1]In plant tissue culture.
[2]Gibberellin cannot be considered as *the* flowering hormone. There is evidence for a flowering hormone that has not yet been isolated and characterized.

FIGURE 30–11   African violet leaf cuttings.
(*Left***) Cuttings treated with NAA, a synthetic auxin,**
formed many adventitious roots. (*Right***) Cuttings placed**
in water (control) did not form roots in the same time
period. Although African violets are not propagated in
this way commercially, it is easy to see that NAA
stimulates root formation. (Runk/Schoenberger, from Grant
Heilman)

down the stem by polar transport (Fig. 30–9). As a result, the cells on the dark side of the stem elongate
more than the cells on the light side, and the stem
bends. Auxin also influences gravitropism, although
the mechanism is incompletely understood at this time.
To complicate matters, other hormones have been
implicated in gravitropism in addition to IAA. Auxin is
believed to be the hormone involved in thigmotropism
and the other tropisms as well.

Certain plants tend to branch out very little.
Growth in these plants occurs from the apical meristem, rather than from lateral meristems. Such plants
are said to exhibit **apical dominance** (Fig. 30–10). In
plants with strong apical dominance, it appears that
auxin produced in the apical meristem inhibits the

development of lateral buds. When the apical meristem is pinched off, the auxin source is removed and
lateral buds grow into branches.

IAA produced by seeds stimulates the development of the fruit. When auxin is applied to flowers in
which fertilization has not been allowed to occur, the
ovary enlarges and develops into a seedless fruit. Seedless tomatoes have been produced in this manner.
Auxin is not the only hormone involved in fruit development, however (Table 30–2).

A number of manufactured, or synthetic, auxins
have been made that have structures similar to IAA.
Synthetic auxins have several commercial applications.
Naphthalene acetic acid (NAA) is used to stimulate
root development on stem cuttings for asexual propagation, particularly of woody plants (Fig. 30–11). Another synthetic auxin, 2,4-dichlorophenoxyacetic acid
(2,4-D), is used as a selective herbicide, mainly for
broad-leaved plants. It is applied at high concentrations and causes exaggerated growth in some plant
parts and growth inhibition in others. For reasons that
are not understood at the present time, monocots are
less sensitive to the concentration of 2,4-D that is applied. Therefore, an application of 2,4-D to a lawn or a
field of corn will kill the broad-leaved weeds (dicots)
and will not harm the grass or corn (monocots) (Fig.
30–12).

Gibberellins Were First Discovered In A Fungus

In the 1920s a Japanese plant scientist, E. Kurosawa,
was working on a disease of rice in which the young
rice seedlings grow extremely tall and spindly, fall
over, and die. The cause of the disease, called the
"foolish seedling disease of rice," was discovered to be
a fungus, *Gibberella*, that produces a substance, **gibberellin,** which causes the symptoms. Not until after
World War II did scientists in Europe and North

**FIGURE 30–12   The application of 2,4-D to a lawn kills the broad-leaved
dicots but does not harm grass, which is a monocot. Grass (***a***) before and
(***b***) after 2,4-D application. Note the conspicuous absence of weeds.** (*a* and *b*,
Gary Bonner)

(a)

(b)

FIGURE 30–13   **Effect of the continued application of gibberellin on normal and dwarf corn plants.** *From left to right:* **dwarf, untreated; dwarf treated with gibberellin; normal treated with gibberellin; normal, untreated. Note that the dwarf plants respond to gibberellin much more dramatically than the normal plants. Dwarf plants treated with gibberellin resemble normal plants in their growth rate. This dwarf variety is a mutant with a single recessive gene that impairs gibberellin metabolism.)** (Courtesy of B.O. Phinney)

FIGURE 30–14   **Bolting in cabbage. Spectacular stem elongation that is often accompanied by flowering was caused by treatment with gibberellin. (***Left***) Untreated control. (***Right***) Cabbage treated with gibberellin.)** (Courtesy of Sylvan Wittwer)

America learn of the exciting work done by the Japanese. The first gibberellin discovered in plants was isolated from bean seeds in 1960. Gibberellins are involved in many normal functions of plants. In the case of the foolish seedling disease, the symptoms were caused by an abnormally high gibberellin concentration in the plant tissue.

Gibberellins have a complex chemical structure composed of five rings. More than 70 naturally occurring gibberellins have been discovered; they have the same basic structure but differ in number of double bonds and location of chemical groups. These structural differences are important, however. Some gibberellins have pronounced effects on plant growth, while others are inactive. Some of the inactive forms may be precursors to active forms. Gibberellins are produced in the root and stem apical meristems, the young leaves, and the embryo in the seed. It is not known how gibberellins are translocated.

As in the foolish seedling disease of rice, gibberellins promote stem elongation in many plants. When a gibberellin is applied to a plant this elongation may be spectacular, particularly in plants that normally have very short stems. Single-gene dwarf mutants of corn and peas will grow to a normal height when treated

with gibberellins (Fig. 30–13). Gibberellins are also involved in **bolting,** the rapid stem elongation that occurs when many plants initiate flowering (Fig. 30–14). In all these cases, gibberellins cause stem elongation by inducing both cell division and cell elongation. The mechanism of cell elongation appears to be different from that caused by auxin, however.

Gibberellins are involved in several reproductive processes in plants. They stimulate flowering, particularly in long-day plants. In addition, they can substitute for the cold requirement that biennials have before the initiation of flowering (Chapter 29). If gibberellins are applied to biennials during their first year of growth, flowering occurs without the cold period. Gibberellins, like auxins, affect the development of fruits. Commercially, gibberellins are applied to several varieties of grapes to produce larger berries.

Gibberellins are involved in the germination of seeds in many plants. The embryo in the seed pro-

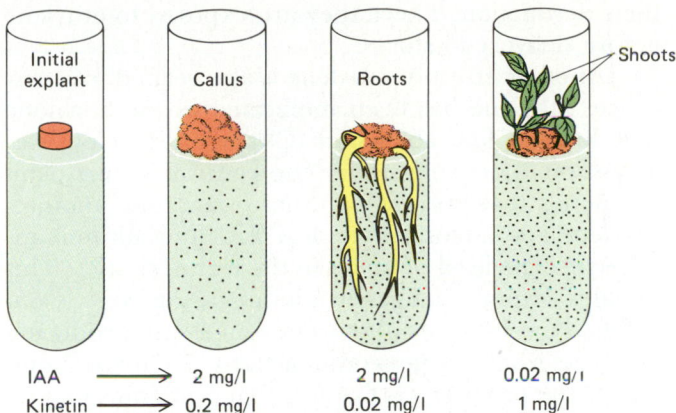

FIGURE 30–15   Growth responses of tobacco tissue culture to auxin and cytokinin. The initial explant is a small piece of sterile tissue from the pith of a tobacco stem, which is placed on a nutrient agar medium as shown at left. After several weeks, the kinds of growth illustrated occur on media supplemented with the indicated levels of auxin and kinetin (a cytokinin). (A.C. Leopold, Boyce Thompson Institute, Ithaca, NY)

FIGURE 30–16   Senescence was delayed in the green leaf by repeated application of cytokinins. Compare this leaf with the rest of the plant, which was not treated with cytokinin.

duces gibberellins that trigger other physiological responses involved in germination. In plants with light or cold requirements for seed germination, artificial application of gibberellins can substitute for the specific environmental requirement. Gibberellins have an important role in the production of enzymes in germinating cereal seeds. The mechanism of action has been studied in detail in germinating barley seeds. The young plant embryo produces gibberellins, which stimulate the seed to synthesize digestive enzymes. These enzymes digest the stored starch in the endosperm, making it available to the young plant as sugar.

Gibberellins appear to be a factor in **hybrid vigor,** the superiority of a hybrid over inbred, genetically uniform plants. For example, inbred strains of corn are not as large or productive as their hybrid offspring. Until recently, the physiological reason for these differences was unknown. In 1988 an analysis of gibberellins in hybrid and inbred strains of corn revealed that hybrid strains have higher levels of gibberellin. When gibberellin was applied to inbred strains of corn, their growth approached that of the hybrid strains. Thus, it appears that production of gibberellin is enhanced in hybrid corn or, alternatively, that gibberellin production is less in inbred corn.

Cytokinins Promote Cell Division

In the 1940s and 1950s a number of researchers were trying to find substances that might induce plant cells to divide in tissue culture (Chapter 26). Skoog and others at the University of Wisconsin discovered that cells would not divide without some substance that was transported in the vascular tissue of plants. This active

substance was also found in coconut milk and autoclaved (super-heated) herring sperm. Finally, in 1956 the active substance was isolated from herring sperm and called a **cytokinin** because it induces cell division, or cytokinesis. In 1963 the first naturally occurring cytokinin was identified from corn and named **zeatin.** Since that time several similar molecules have been extracted from other plants.

Cytokinins have an intriguing structure (Table 30–1). They are similar to the purine base, adenine, which is an important component of nucleic acids. Cytokinins may be found as part of certain transfer-ribonucleic acid (t-RNA) molecules not only in plants, but also in animals and microorganisms. In plants cytokinins are produced in the roots and transported in the xylem to all parts of the plant.

Cytokinins promote cell division and differentiation in intact plants. They are a required ingredient of plant tissue culture media (Chapter 26) and must be present in order to induce mitosis. In tissue culture, cytokinins interact with auxin during **organogenesis,** the formation of plant organs like roots and stems (Fig. 30–15). For example, in tobacco tissue culture a high ratio of cytokinin to auxin induces shoot formation, while a low ratio of cytokinin to auxin induces root formation.

Cytokinins and auxin also interact in the control of apical dominance. Here their relationship is antagonistic, as auxin inhibits the growth of lateral buds while cytokinin promotes their growth. The relationship is reversed in roots, with auxin promoting the growth of branch roots and cytokinins inhibiting it.

One very interesting effect of cytokinins on plant cells is a delay in their **senescence,** or aging (Fig. 30–16). Plant cells, like all living cells, go through a natural aging process. This process is accelerated in cells of plant parts that are cut, such as cut flowers. Cytokinins somehow promote cells to maintain their normal levels of protein and nucleic acids, thus delaying the rapid aging associated with cut plant parts. It is believed that plants must have a continual supply of cytokinins from the roots. Cut flowers, of course, lose their source of cytokinins. Commerically, cytokinins are sprayed on cut flowers to prevent their rapid senescence.

Ethylene Is The Only Gaseous Plant Hormone

The effects of **ethylene** on plants had been noted in the 1800s, long before it was recognized as a natural plant hormone. Before electricity was discovered, a mixture of various gases called coal gas was used to illuminate homes and street lights. People noticed that plants growing near street lights were altered in several ways. Trees shed their leaves early, flowers faded quickly and their petals fell off, and newly sprouted seedlings grew horizontally rather than erect. In 1901 a Russian plant physiologist determined that ethylene was the ingredient in coal gas that caused these effects, but it wasn't until 1934 that scientists demonstrated that ethylene was also produced by plants themselves.

Ethylene is the only plant hormone that is a gas. It is colorless and smells like ether. Ethylene is produced in several places in plants. It is produced in the nodes of stems, in ripening fruits, and in senescing tissues such as leaves.

Many diverse plant processes are influenced by ethylene. Ethylene inhibits cell elongation, promotes the germination of seeds, and is involved in plant responses to wounding or invasion by pathogens. Ethylene has a major role in many aspects of senescence, including the ripening process in fruits. A number of physiological changes occur during fruit ripening. Fruits often change color, as chlorophyll is degraded and other pigments are synthesized. Starch and acids stored in the fruit are converted to sugars, giving the fruit a sweet taste. The fruit cell walls are partly broken down, making the tissue softer. And flavors characteristic of the particular fruit are synthesized. Ethylene triggers these physiological changes. Further, ethylene has a "domino effect." As a fruit ripens, it produces ethylene, which triggers an acceleration of the ripening process. This induces the fruit to produce more ethylene, which further accelerates ripening. The expression "one rotten apple spoils the lot" is true. A rotten apple is one that is overripe. This apple produces large amounts of ethylene, which diffuse and trigger the ripening process in nearby apples. Ethylene is used commercially to promote the uniform ripening of bananas. Bananas are picked while green and shipped to their destination. There they are exposed to ethylene before delivery to stores.

Another effect of ethylene is also related to senescence. Ethylene has been implicated as the hormone that induces leaf abscission (Chapter 27). However, abscission is actually under the control of two plant hormones that are antagonistic toward one another, ethylene and auxin. As the leaf ages and autumn approaches, the level of auxin in the leaf decreases. This initiates several changes in the abscission zone. Concurrently, cells in the abscission zone begin producing ethylene, which triggers other actions. To further complicate the process, it is possible that cytokinins may be involved. Cytokinins, like auxin, decrease in amount as leaf tissue ages.

Exactly how plants recognize the presence of ethylene or respond in so many diverse ways to its presence is unknown. In 1988 a mutant *Arabidopsis* plant was reported that is resistant to ethylene. Subsequent studies of this mutant may provide insight into ethylene's mode of action.

Abscisic Acid Promotes Dormancy In Higher Plants

Abscisic acid was discovered simultaneously in 1963 by two independent research teams. P.F. Wareing in England was working on a hormone that induced bud dormancy in woody plants, and F.T. Addicott in California was working on a hormone that promoted the abscission of cotton fruits. Later, when the structures of both hormones were found to be identical, the hormone was given one name, abscisic acid. This was an unfortunate choice because abscisic acid is primarily involved in dormancy and is not very important in abscission in most plants.

Abscisic acid, or ABA, is a six-carbon ring with a number of substitutions. It is produced in the leaf, root cap, and stem and is transported in the vascular tissue. The level of abscisic acid is also high in seeds and fruits, but it is not clear whether it is synthesized or transported there.

Abscisic acid is sometimes referred to as the "stress hormone," although ethylene also affects plant responses to certain stresses. Abscisic acid promotes change in plant tissues that are "stressed," or exposed to unfavorable conditions. The effect of ABA on plants suffering from water stress is best understood. Abscisic acid increases dramatically in the leaves of plants that are exposed to severe drought conditions. The high level of ABA in the leaves triggers the outflow of potassium ions from the guard cells. This induces water to leave the guard cells by osmosis, and the guard cells collapse (Chapter 27). The closing of stomates in water-stressed plants saves a large amount of water that is normally transpired through the stomates, increasing the plant's likelihood of survival. When water is restored to the plant, the stomates do not open immediately. The level of ABA in the leaf cells must decrease before that can occur.

The onset of winter could also be considered a type of stress on the plant. As winter approaches, woody plants cease growth, and protective coverings of bud scales form over their terminal buds (Chapter 26). Abscisic acid promotes these adaptations. Another winter adaptation that involves abscisic acid is dormancy in seeds. If seeds germinated in the autumn, the delicate seedlings would be killed by the first frost. Many seeds have high levels of ABA in their tissues and are, therefore, unable to germinate. In a corn mutant that is unable to synthesize ABA, the seeds germinate as soon as the embryos are mature, while still attached to the ear (Fig. 30–17).

The evidence that abscisic acid is the only hormone involved in both plant and seed dormancy is not conclusive. The addition of gibberellin reverses the effects of dormancy. In seeds the level of ABA decreases during the winter, while the level of gibberellin increases. Cytokinins have also been implicated. Once again we see that a single physiological activity in plants may be controlled by the interaction of several hormones (Table 30–2). The actual response may be due to changing ratios of hormones instead of the effect of a single hormone.

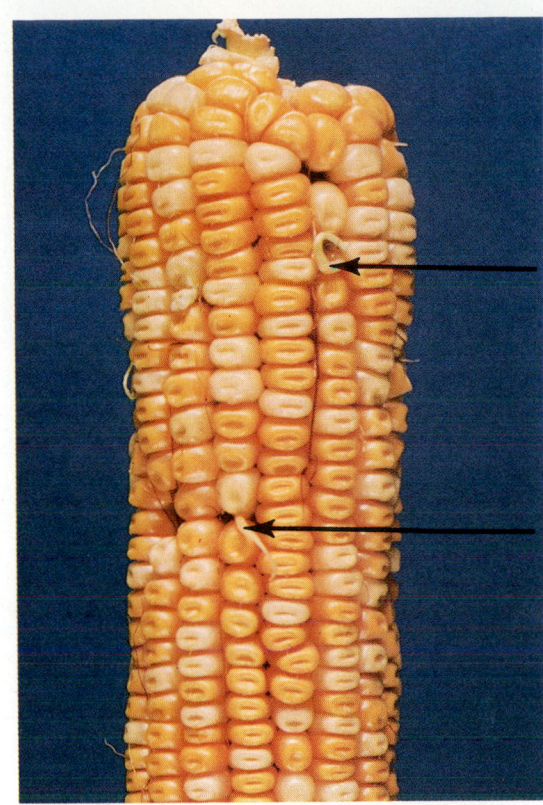

FIGURE 30–17 The lack of formation of abscisic acid can prevent seed dormancy in corn. Some of the white kernels have germinated prematurely to produce white coleoptiles (see arrows). (Courtesy of M.G. Neuffer)

■ SUMMARY

I. Plants respond to external stimuli.
   A. Phototropism is the growth of a plant due to the direction of light.
   B. Gravitropism is the growth of a plant due to the influence of gravity.
   C. Thigmotropism is the growth of a plant in response to contact with a solid object.

II. Some plants respond to external stimuli by changes in turgidity in special cells.
   A. The sensitive plant, *Mimosa pudica*, dramatically folds its leaves in response to various stimuli.
   B. The closure of Venus flytrap leaves is an example of turgor movements.

III. Circadian rhythms, which are regular rhythms in growth or activities of the plant, approximate the 24-hour day and are referred to as the biological clock.

IV. Hormones regulate plant growth and development and are effective in small amounts.

   A. The functions of hormones overlap.
   B. Many effects of hormones may be due to the concentration ratios of several hormones rather than the effect of a single hormone.
   C. There are five plant hormones.
      1. Auxins are involved in cell elongation, phototropism, gravitropism, apical dominance, and fruit development.
      2. Gibberellins are involved in stem elongation, flowering, seed germination, and hybrid vigor.
      3. Cytokinins promote cell division and differentiation, delay senescence, and interact with auxins in apical dominance.
      4. Ethylene affects a wide variety of plant processes, including the ripening of fruits, leaf abscission, and senescence.
      5. Abscisic acid is involved in stomatal closure due to water stress and in bud and seed dormancy.

◼ POST-TEST

1. _____ is the growth of a plant due to the direction of light.
2. Plant roots generally exhibit _____ gravitropism.
3. The twining of tendrils is an example of _____.
4. The _____ is an organ at the base of the petiole in *Mimosa* that can undergo rapid changes in turgor, causing dramatic movements of the leaves.
5. Sleep movements observed in beans are an example of _____ _____.
6. _____ are organic compounds produced in one part of the plant and transported to another, where they cause a positive or negative effect.
7. The movement of auxin, called _____ _____, is unidirectional from the top of the shoot to the roots.

8. A synthetic _____, 2,4-D, is used as a selective herbicide.
9. Research on a disease of rice provided the first clues about _____.
10. _____ are involved in hybrid vigor.
11. _____ interact with auxin during organogenesis in tissue culture.
12. The relationship between cytokinins and auxins in apical dominance is _____.
13. _____ delay senescence, while _____ promotes it.
14. The only plant hormone that is a gas is _____.
15. Abscisic acid promotes the _____ of woody twigs.

◼ REVIEW QUESTIONS

1. Why might some plants have sleep movements?
2. Of what value is dormancy in seeds?
3. How is auxin involved in phototropism?
4. Discuss the various hormones that are involved in each of the following physiological processes: (a) germination of seeds; (b) growth and development of the plant; (c) ripening of fruits; (d) abscission of leaves; (e) dormancy of seeds.

◼ RECOMMENDED READINGS

Galston, A.W., Davies, P.J. and R.L. Satter. *The Life of the Green Plant.* Englewood Cliffs, New Jersey, Prentice-Hall, Inc., 1980. Contains a detailed account of plant hormones.

Kaufman, P.B. et al. *Plants: Their Biology and Importance.* New York, Harper & Row, 1989. Contains a nice account of hormonal regulation in seed plants.

PART 8

Animal Structure and Life Processes

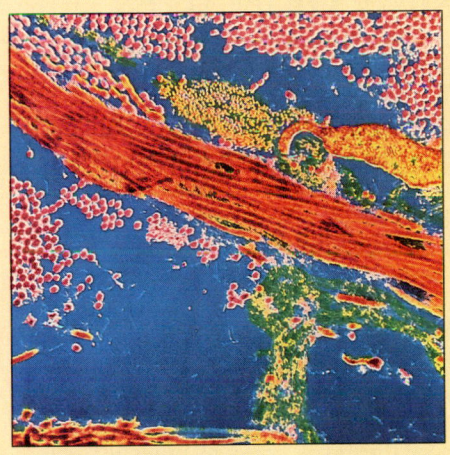  

Information flow through the nervous system • Organization of neural circuits • Neural integration • External and internal skeletons • Evolution of the vertebrate brain

 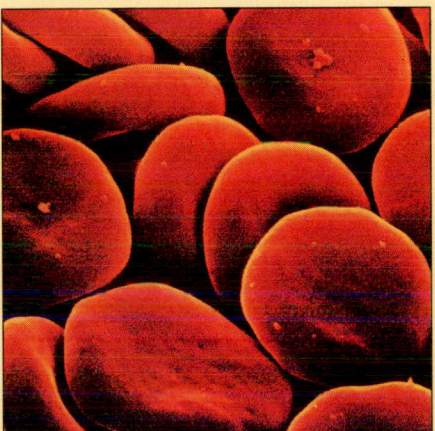

Adaptations for processing food • Gas exchange and transport • Internal defense • Waste disposal and osmoregulation • Male and female reproductive function • Patterns of development

31

ANIMAL TISSUES, ORGANS, AND ORGAN SYSTEMS

OUTLINE

I. Animals are multicellular
II. Tissues of a complex animal are adapted to carry out specific functions
   A. Epithelial tissue covers the body and lines its cavities
   B. Connective tissue joins and supports other body structures
      1. Loose connective tissue is found in the subcutaneous layer; dense connective tissue is found in the skin and tendons
      2. Elastic tissue is found in lungs; reticular tissue supports organs
      3. Adipose tissue stores fat
      4. Cartilage and bone form a supporting skeleton
      5. Blood and lymph are circulating tissues
   C. Muscle tissue contracts, permitting movement
   D. Nervous tissue receives stimuli and transmits information
III. An organ consists of more than one type of tissue and performs one or more functions
IV. Ten organ systems make up the complex animal organism
Focus on Neoplasms: Unwelcome Tissues

LEARNING OBJECTIVES

After you have studied this chapter, you should be able to:
1. Discuss the advantages of multicellularity.
2. Define tissue, organ, and organ systems.
3. Compare the four principal types of animal tissues (epithelial, connective, muscle, and nervous tissues) with respect to general structure and function.
4. List the functions of epithelial tissue, describe the three main shapes of epithelial cells, and describe how these cells can be arranged into tissues.
5. Compare the main types of connective tissue and their functions.
6. Compare the three types of muscle tissue and their functions.
7. Identify the cells that make up neural tissue and give their functions.
8. Distinguish between benign and malignant neoplasms and contrast the cells of a malignant neoplasm with those of normal tissue.
9. Briefly describe the organ systems of a complex animal.

False-color transmission electron micrograph of human connective tissue. (CNRI/Science Photo Library/Photo Researchers, Inc.)

Animals, like plants and most fungi, are **multicellular** (that is, they are composed of many cells). They are not simply colonies or aggregations of similar cells, but are composed of a number of different types of cells, each with a characteristic size, shape, structure, and function. In most animals, cells are organized into tissues, tissues into organs, and organs into organ systems.

KEY CONCEPTS

☐ Animals are multicellular. Advantages of multicellularity include large size, specialization of cells, and diversity of form.

☐ In most animals, cells are organized into tissues, tissues into organs, and organs into organ systems.

☐ Animal tissues may be classified as epithelial, connective, muscle, or nervous tissue. Each type of tissue performs one or more functions.

☐ ANIMALS ARE MULTICELLULAR

Why is a human being or an elephant made up of trillions of cells rather than one giant cell? Cells rarely become large because it is inefficient for them to do so. All materials coming into or out of the cell must pass through the plasma membrane, so the size of the membrane in comparison with the rest of the cell is critical. When the membrane becomes too small in relation to the cell to allow efficient transport of materials, a cell divides to form two cells. This explains why we do not see amoebas as large as whales slithering about. In fact, most one-celled organisms are so tiny that even the larger ones are barely visible to the unaided eye.

In unicellular organisms, cell division results in the production of two new individuals; in multicellular organisms, the two new cells may remain associated, forming a *part* of the organism. One advantage of multicellularity is large size. The number of cells, not their individual size, is responsible for the different sizes of various organisms. The cells of an earthworm and an elephant correspond in size; the elephant is larger because its genes are programmed to provide for a larger number of cells (Fig. 31–1).

Consider how different a sea urchin is from a butterfly, lizard, or tiger. Animals come not only in many different sizes, but also in many different shapes and with many different life styles and patterns of behavior. Diversity is another advantage of multicellularity.

Multicellularity permits specialization. In a one-celled organism, the single cell carries on all life activities. In an organism composed of many cells, there can be a division of labor, with each type of cell performing specific functions. When cells specialize, the organism can become highly proficient at performing a wide variety of activities. Some cells, for example, can become sensory cells and specialize in receiving information about the environment, whereas other cells, which are specialized to contract, enable the organism to escape a predator or capture food. How do cells associate and perform such specialized functions? To answer this, we examine the tissues, organs, and organ systems of complex animals.

☐ TISSUES OF A COMPLEX ANIMAL ARE ADAPTED TO CARRY OUT SPECIFIC FUNCTIONS

Even in the simplest animals, the sponges, there is a division of labor among several cell types. In all other animals, cells are not only specialized but organized, forming tissues. A **tissue** consists of a few types of closely associated cells that are adapted to specific

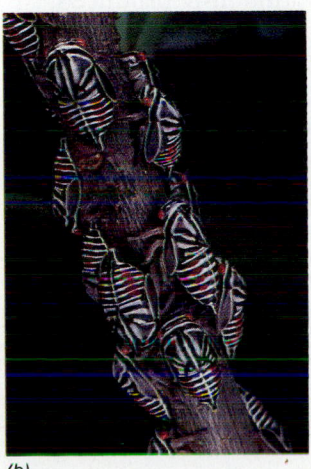

FIGURE 31–1 All animals are multicellular. A large animal such as the bull elephant from western Kenya (*a*) is composed of more cells than any of the much smaller oak treehoppers (*Platycotis vittata*) gathered on a Florida tree (*b*). (*a*, Frans Lanting; *b*, James Castner)

(a)

(b)

functions. Animal tissues may be classified as *epithelial, connective, muscle,* or *nervous tissue.* Each kind of tissue is composed of cells with a characteristic size, shape, and arrangement.

Epithelial Tissue Covers The Body And Lines Its Cavities

The outer layer of skin, the linings of digestive and respiratory tracts, and the lining of kidney tubules are examples of epithelial tissues. **Epithelial tissue** consists of cells that are fitted tightly together, forming a continuous layer or sheet of cells covering a body surface or lining a cavity within the body. One surface of the sheet is attached to the underlying tissue by a *basement membrane* composed of nonliving polysaccharide material (a product of the epithelial cells) and by tiny fibers.

Epithelial tissues may function in protection, absorption, secretion, and sensation. As a covering or lining, epithelial tissue protects the body. The epithelial layer of the skin, called the epidermis, covers the entire body and protects it from mechanical injury, invading bacteria, and excessive water loss. The epithelial tissue lining the digestive tract absorbs nutrients and water into the body. Other epithelial cells may be organized into glands, which are adapted for the secretion of cell products like hormones, enzymes, or sweat.

Everything that enters or leaves the body crosses one or more layers of epithelium. Even food that is taken into the mouth and swallowed is not really inside the body until it is absorbed through the epithelial lining of the digestive tract and enters the blood. The permeability of the various epithelial tissues regulates to a large extent the exchange of substances between the different parts of the body and between the organism and the external environment.

Many epithelial membranes are subjected to continuous wear and tear. As outer cells are sloughed off, they are replaced by new ones from below. Such epithelial tissues generally have a rapid rate of cell division. As a result, new cells are continuously produced, taking the place of those lost.

Several types of epithelial tissue may be distinguished by the number of cell layers, the shape of the cells, and their arrangement (Table 31–1). Epithelium may be **simple,** consisting of one cell layer, or **stratified,** consisting of many layers (as in the outer layer of the skin). A third arrangement is **pseudostratified** epithelium, in which the cells (falsely) appear to be layered but in fact are not. The epithelial cells may be **squamous** (flattened), **cuboidal** (cube-shaped), or **columnar** (elongated like columns). The free surface of the outer cells of the tissue may have specialized structures like cilia or microvilli.

A **gland** consists of one or more epithelial cells that produce and secrete a product, such as mucus, sweat, milk, saliva, hormones, or enzymes (Fig. 31–2). The epithelial tissue lining the cavities and passageways of the body typically contains **goblet cells,** unicellular glands that secrete mucus. The mucus lubricates these surfaces and facilitates the movement of materials.

Connective Tissue Joins And Supports Other Body Structures

Connective tissue binds together and supports the other structures of the body. It protects underlying organs and may store or transport materials. In addition, almost every organ in the body has a supporting framework of connective tissue that supports and cushions its epithelial components.

Unlike epithelium, connective tissue consists of relatively few cells separated by large amounts of **intercellular substance.** This intercellular substance consists of nonliving, thread-like microscopic fibers that are scattered throughout a **matrix,** which is a thin gel composed of polysaccharides. The intercellular substance is secreted by connective tissue cells called **fibroblasts** (Fig. 31–3). Three types of fibers found in connective tissue are collagen, elastic, and reticular fibers. Most numerous are the **collagen fibers** (Fig. 31–4), which are composed of the protein collagen (familiar in its hydrated form as gelatin). The tensile strength (capacity to withstand longitudinal strain or rupture) of these flexible fibers has been compared to that of steel.

Elastic fibers stretch easily and then, like a rubber band, snap back to their normal length when the stress is removed. These fibers are an important component of structures that must stretch. For example, elastic fibers in the walls of a large artery permit the artery to stretch as it fills with blood. Reticular fibers are very fine, branched fibers that form a supporting network within many tissues and organs.

The main types of connective tissue are (1) loose and dense connective tissues, (2) elastic connective tissue, (3) reticular connective tissue, (4) adipose tissue, (5) cartilage, (6) bone, and (7) blood, lymph, and tissues that produce blood cells (Table 31–2). Each of these tissues is exquisitely adapted to the specific functions it performs.

Loose Connective Tissue Is Found In The Subcutaneous Layer; Dense Connective Tissue Is Located In The Skin And Tendons

Loose connective tissue is the most widely distributed connective tissue in the body. It is found as a thin filling between body parts and serves as a reservoir for fluid and salts. Nerves, blood vessels, and muscles are wrapped in this tissue. Together with adipose tissue,

TABLE 31–1
Epithelial Tissues

Tissue name	Main locations	Functions	Description and comments
Simple squamous epithelium Nuclei	Air sacs of lungs, lining of blood vessels	Passage of materials where little or no protection is needed and where diffusion is major form of transport	Cells are flat and arranged as single layer
Simple cuboidal epithelium Nuclei of cuboidal epithelial cells Lumen of tubule	Lining of kidney tubules, gland ducts	Secretion and absorption	Single layer of cells; from the side each cell looks like short cylinder; sometimes have microvilli for absorption
Simple columnar epithelium Goblet cell Nuclei of columnar cells	Lining of much of digestive tract, upper part of respiratory tract	Secretion, especially mucus; absorption, protection, movement of mucous layer	Single layer of columnar cells, often with nuclei located in base of each cell almost in row; sometimes with enclosed secretory vesicles (goblet cells), highly developed Golgi complex, and cilia

continued

TABLE 31–1
Epithelial Tissues *(continued)*

Tissue name	Main locations	Functions	Description and comments
Stratified squamous epithelium 	Skin, mouth lining, vaginal lining	Protection only; little or no absorption or transport of materials; outer layer continuously worn away by friction and replaced from below	Several layers of cells, with only the lower ones columnar and metabolically active; division of lower cells causes older ones to be pushed upward toward surface
Pseudostratified epithelium Approximately ×250 (Ed Reschke)	Some respiratory passages, ducts of many glands	Secretion, protection, movement of mucus	Comparable in many ways to columnar epithelium, except not all cells are of same height; though all cells touch basement membrane, tissue appears stratified; nuclei not in line; may be ciliated, may secrete mucus

loose connective tissue forms the subcutaneous layer, the tissue layer that attaches skin to the muscles and other structures beneath. Loose connective tissue consists of fibers strewn in all directions through a semifluid matrix. Its flexibility permits the parts it connects to move.

Dense connective tissue is very strong and somewhat less flexible than loose connective tissue. Collagen fibers predominate. In *irregular dense connective tissue*, the collagen fibers are arranged in bundles that are distributed in all directions through the tissue. This type of tissue is found in the lower layer (dermis) of the skin. In *regular dense connective tissue*, the collagen bundles are arranged in a definite pattern, making the tissue greatly resistant to stress. Tendons, the cable-like cords that connect muscles to bones, consist of this tissue.

Elastic Tissue Is Found In The Lungs; Reticular Tissue Supports Organs

Structures that expand and then return to their original size, like the walls of large arteries and lung tissue, contain **elastic connective tissue.** This tissue is also

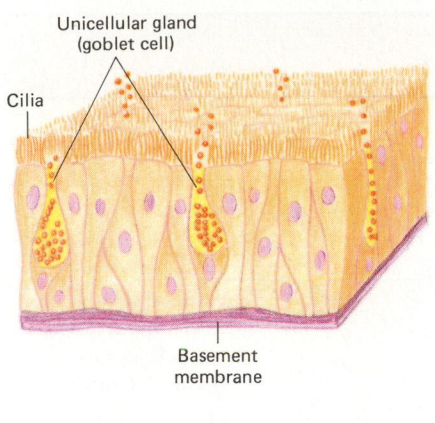

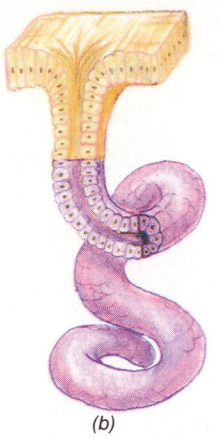

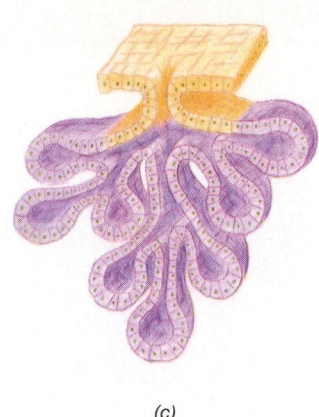

(a)   *(b)*   *(c)*

FIGURE 31–2   A gland consists of one or more epithelial cells. Goblet cells are unicellular glands that secrete mucus. Sweat glands are simple tubular glands with coiled tubes similar to the one shown here. The parotid salivary glands are compound glands like the one shown here.

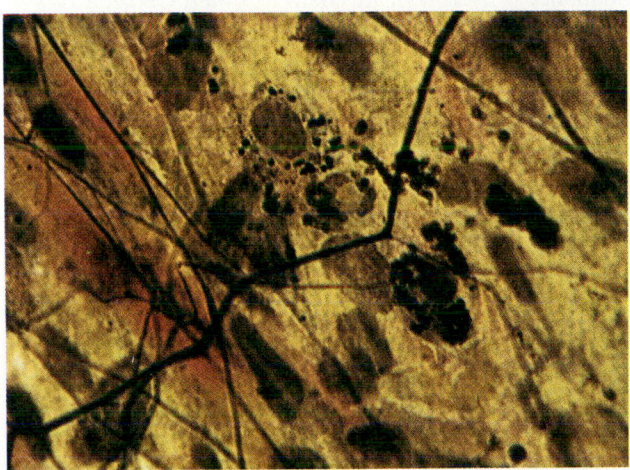

FIGURE 31–3   Loose connective tissue in the mesentery of a rabbit that had an injection with india ink (approximately ×1000). The macrophages have ingested the ink particles. Fibroblasts (the elongated cells) are also present. Note that the collagen is stained pink and the elastic, black. (From Warwick, R., and P.L. Williams, *Gray's Anatomy*, 36th ed., Edinburgh, Churchill Livingstone, 1980)

FIGURE 31–4   Scanning electron micrograph of collagen fibers taken from the tendon of a human biceps muscle. In transmitting forces generated by muscle to bones, the collagen of tendons acts something like an organic rope and is flexible, tough, and strong. (From Jozsa, L. A. Reffy, and J. Balint, National Institute of Traumatology, Budapest, *Acta Histochemica* 74: 1984)

found in ligaments, the bands of tissue that connect bones to one another. Elastic connective tissue consists mainly of bundles of parallel elastic fibers.

Reticular connective tissue is composed mainly of interlacing reticular fibers. It forms a framework that supports many organs, including the liver, spleen, and lymph nodes.

Adipose Tissue Stores Fat

Adipose tissue is rich in cells that store fat and release it when fuel is needed for cellular respiration. It is found in the subcutaneous layer and in tissue that cushions internal organs. An immature fat cell is somewhat star-shaped. As fat droplets accumulate within the cytoplasm, the cell assumes a more rounded appearance (Fig. 31–5). Fat droplets eventually merge with one another until finally a single large drop of fat is present. This large drop occupies most of the volume of the mature fat cell. The cytoplasm and organ-

TABLE 31–2
Connective Tissues

Tissue name	Main locations	Functions	Description and comments
Loose connective tissue Collagen fibers Nuclei of fibroblasts Approximately ×200	Everywhere support must be combined with elasticity, e.g., subcutaneous layer	Support; reservoir for fluid and salts	Fibers produced by fibroblast cells embedded in semifluid matrix and mixed with miscellaneous group of other cells
Dense connective tissue Approximately ×200	Tendons, strong attachments between organs; dermis of skin	Support; transmission of mechanical forces	Bundles of interwoven collagen fibers interdigitated with rows of fibroblast cells
Elastic connective tissue Approximately ×300	Structures that must both expand and return to their original size, such as lung tissue and large arteries; ligaments	Confers elasticity	Branching elastic fibers interspersed with fibroblast
Reticular connective tissue Approximately ×500	Framework of liver, lymph nodes, spleen	Support	Consists of interlacing reticular fibers

TABLE 31–2
Connective Tissues (continued)

Tissue name	Main locations	Functions	Description and comments
Adipose tissue	Subcutaneous layer; pads around certain internal organs	Food storage, insulation, support of such organs as breast, kidneys	Fat cells star-shaped at first; fat droplets accumulate until typical ring-shaped cells are produced
Cartilage Chondrocytes / Lacuna / Intercellular substance	Supporting skeleton in sharks, rays, and some other vertebrates; in other vertebrates forms ends of bones; supporting rings in walls of some respiratory tubes; tip of nose; external ear	Flexible support and reduction of friction in bearing surfaces	Chondrocytes separated from one another by gristly intercellular substance, and occupy little spaces in it
Bone Lacunae / Haversian canal / Matrix Approximately ×150	Most of skeleton in most vertebrates	Support, protection of internal organs, calcium reservoir; skeletal muscles attach to bones	Osteocytes located in lacunae; in compact bone, lacunae arranged in concentric circles about haversian canals
Blood Red blood cells / White blood cell Approximately ×1100	Heart and blood vessels of circulatory system	Transports oxygen, nutrients, wastes, other materials	Consists of cells dispersed in fluid intercellular substance

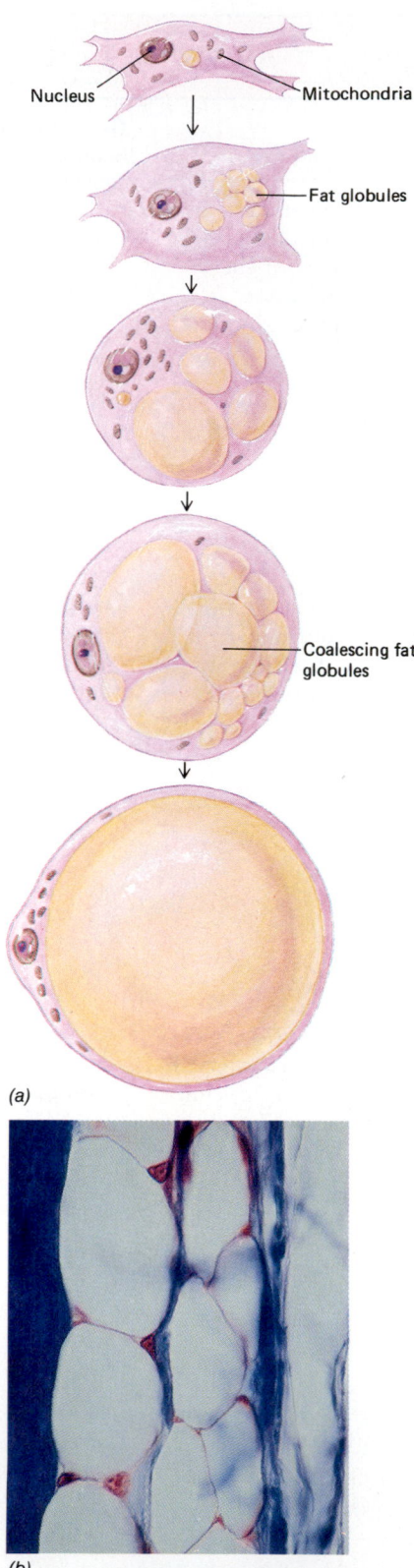

Nucleus

Mitochondria

Fat globules

Coalescing fat globules

(a)

(b)

FIGURE 31-5 Storage of fat in a fat cell. (a) As more and more fat droplets accumulate in the cytoplasm, they coalesce to form a very large globule of fat. Such a fat globule may occupy most of the cell, pushing the cytoplasm and the organelles to the periphery. (b) Photomicrograph of adipose tissue (approximately ×250). (Ed Reschke)

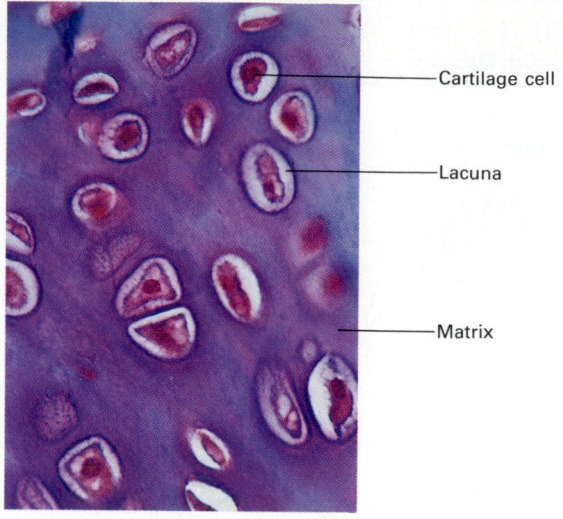

Cartilage cell

Lacuna

Matrix

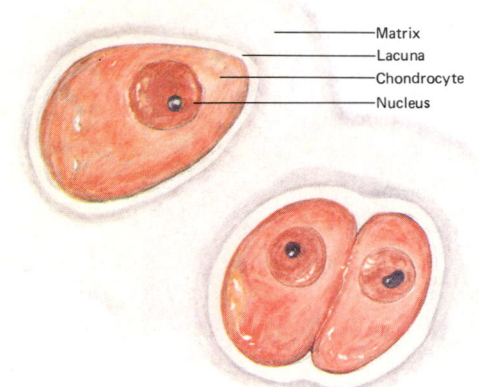

Matrix
Lacuna
Chondrocyte
Nucleus

FIGURE 31-6 Cartilage cells become trapped in small spaces called lacunae. The rubbery matrix contains collagen fibers. (A photo by Ed Reschke)

elles are pushed to the cell edges, where a bulge is typically created by the nucleus. A cross section of such a fat cell looks like a ring with a single stone. In a photomicrograph, adipose tissue looks somewhat like chicken wire. The "wire" represents the rings of cytoplasm, and the large spaces indicate where fat drops existed before they were dissolved by chemicals used to prepare the tissue. The empty spaces may cause the cells to collapse, resulting in a wrinkled appearance.

Cartilage And Bone Form A Supporting Skeleton

The supporting skeleton of vertebrates is composed of cartilage or bone. **Cartilage** is the supporting skeleton in the embryonic stages of all vertebrates. During development, it is largely replaced by bone, except in the sharks and rays. In humans, the supporting structure of the external ear, the supporting rings in the walls of

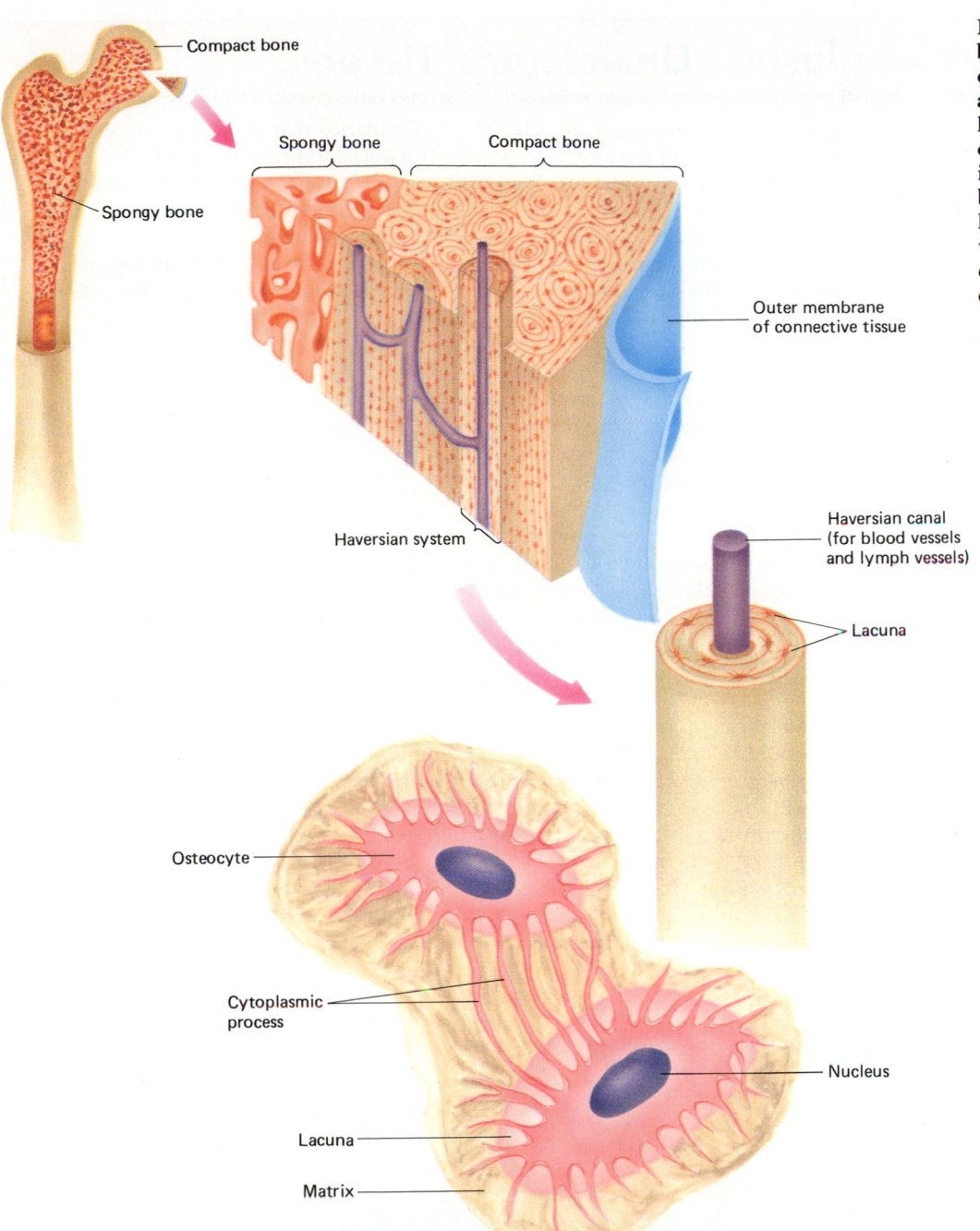

Compact bone
Spongy bone
Spongy bone
Compact bone
Outer membrane of connective tissue
Haversian system
Haversian canal (for blood vessels and lymph vessels)
Lacuna
Osteocyte
Cytoplasmic process
Nucleus
Lacuna
Matrix

FIGURE 31–7 Compact bone is made up of units called osteons. Blood vessels and nerves run through the haversian canal within each osteon. In bone, the matrix is rigid and hard. Bone cells become trapped within lacunae but communicate with one another by way of cytoplasmic processes that extend through tiny canals.

the respiratory passageways, and the tip of the nose are examples of structures composed of cartilage.

Cartilage is firm yet elastic. Cartilage cells, called **chondrocytes,** secrete this hard, rubbery matrix around themselves and also secrete collagen fibers, which become embedded in the matrix and strengthen it. Chondrocytes eventually come to lie singly or in groups of two or four in small cavities called lacunae in the matrix (Fig. 31–6). The cartilage cells in the matrix remain alive. Cartilage tissue lacks nerves, lymph vessels, and blood vessels. Chondrocytes are nourished by diffusion of nutrients and oxygen through the matrix.

Bone is the principal vertebrate skeletal tissue. It is similar to cartilage in that the **osteocytes** (bone cells) that secrete and maintain the matrix are located in lacunae within the matrix (Fig. 31–7). Unlike cartilage, however, bone is a highly vascular tissue (having many blood vessels). Diffusion alone would be too slow to provide nourishment for the osteocytes because the matrix is calcified (hardened). The osteocytes of bone communicate with one another and with capillaries by tiny channels **(canaliculi)** that contain fine extensions of the cells themselves. In compact bone, the osteocytes are arranged around central capillaries in concentric

FOCUS ON Neoplasms: Unwelcome Tissues

A **neoplasm** (new growth), or **tumor,** is an abnormal mass of cells. A **benign** ("kind") tumor tends to grow slowly, and its cells stay together. Because benign tumors form discrete masses that are often surrounded by connective tissue capsules, they can usually be removed surgically. Unless a benign tumor develops in a place where it interferes with the function of a vital organ, it is not lethal.

Cancer is the common name for **malignant** ("wicked") neoplasms. Unlike benign neoplasms, cancer cells invade other tissues, typically spread to new locations, and do not retain the typical structural features of the cells from which they develop. Neoplasms that develop from connective tissues or muscles are referred to as **sarcomas,** whereas those that originate in epithelial tissue are **carcinomas.** Common cancers originating in blood or bone marrow are leukemias, lymphomas, and myelomas.

Cancer is thought to be triggered when the DNA of a cell is mutated by radiation, certain chemicals or irritants, or viruses (see the discussion of oncogenes in Chapter

16). When the transformed cell divides, all the cells derived from it bear the identical mutation. The changes in the DNA affect the plasma membrane and interfere with the cell's control mechanisms. Membrane proteins that normally help regulate cell division and interaction with other cells are replaced by tumor-specific proteins. Two basic defects in behavior that are typical of cancer cells are wild, often rapid, multiplication and abnormal relations with neighboring cells. Unlike normal cells that respect one another's boundaries and form tissues in an orderly, organized manner, cancer cells grow helter skelter upon one another and infiltrate normal tissues. Apparently, they are no longer able to receive or respond appropriately to signals from surrounding cells; communication is lacking.

Studies indicate that many neoplasms grow to only a few millimeters in diameter and then enter a dormant stage, which may last for months or even years. At some point, cells of the neoplasm release a chemical substance that stimulates nearby blood vessels to develop new

capillaries that grow out toward the neoplasm and invest it. Once a blood supply is ensured, the neoplasm grows rapidly and may soon become life-threatening.

Death from cancer almost always results from **metastasis,** a migration of cancer cells through blood or lymph channels to distant parts of the body. Once there, they multiply, forming new malignant neoplasms; these may interfere with the normal function of the tissues being invaded. Cancer often spreads so rapidly and extensively that surgeons are unable to locate all the malignant masses.

Why some persons are more susceptible to cancer than others remains a mystery. Some researchers think that cancer cells arise daily in everyone, but that in most people, the immune system (the system that provides protection from disease organisms and other foreign invaders) is capable of destroying them. According to this theory, cancer is a failure of the immune system. Another suggestion is that people have different levels of tolerance to carcinogens (cancer-producing agents) in

Normal skin (a) compared with cancerous tissue (b). Note the disruption of the normal tissue structure by the invasion of the neoplasm.

Normal epidermis of skin

Cancer tissue

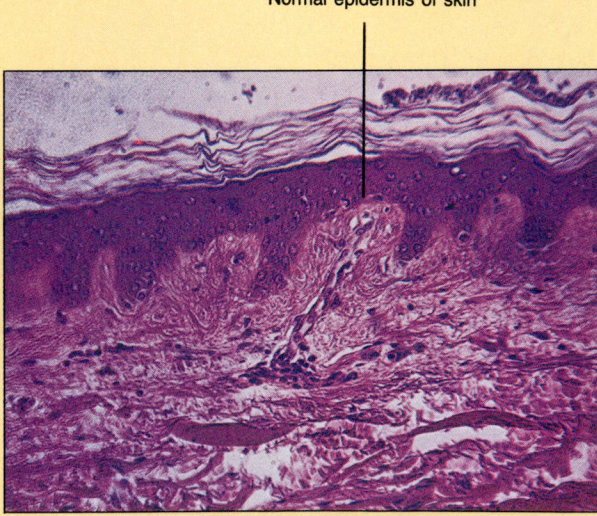

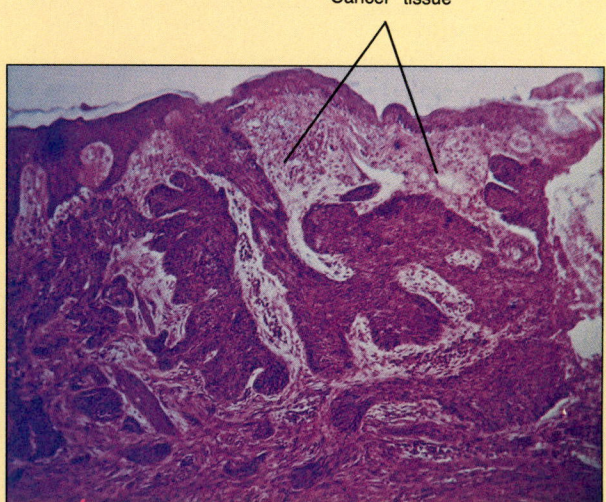

(a) Approximately ×200

(b) Approximately ×150

the environment. More than 80% of cancer cases are thought to be triggered by carcinogens in the environment.

Cancer is the second-highest cause of death in the United States. One in three persons in the United States gets cancer at some time in his or her life, and two out of three cancer patients die as a result of their cancer within 5 years of diagnosis. Currently, the key to survival is early diagnosis and treatment with a combination of surgery, hormonal treatment, radiation therapy, and drugs that suppress mitosis (chemother-

apy). Because cancer is actually an entire family of closely related diseases (there are more than 100 distinct varieties), it is probable that there is no single cure. Most investigators agree, however, that a greater understanding of the control mechanisms and communication systems of cells is necessary before effective cures can be developed.

Risk for cancer can be decreased by following these recommendations:

1. Do not smoke or use tobacco. Smoking is responsible for more than 80% of lung cancer cases.

2. Avoid prolonged exposure to the sun; use sun screen or sun block. Exposure to the sun is responsible for almost all of the 400,000 cases of skin cancer reported each year.

3. Increase the fiber content of your diet and avoid high-fat, smoked, salt-cured, or nitrite-cured foods.

4. Avoid unnecessary exposure to x-rays.

5. Women should examine their breasts each month and obtain annual Pap tests.

---

layers, called **lamellae,** which form spindle-shaped units known as **osteons.** The capillaries, as well as nerves, run through central microscopic channels in the osteons known as **haversian canals.**

Bone also contains large, multinucleated cells called **osteoclasts,** which can dissolve and remove the bony substance, as can the osteocytes themselves. The shape and internal architecture of the bone can gradually change in response to normal growth processes and to physical stress. The calcium salts of bone render the matrix very hard, and the collagen prevents the bony matrix from being overly brittle. Bone is discussed further in Chapter 32.

Blood And Lymph Are Circulating Tissues

Like other connective tissues, **blood** and **lymph** consist of specialized cells dispersed in an intercellular substance. However, these tissues are unique in that the intercellular substance is fluid, and this fluid is not secreted by the blood cells. Blood and lymph are circulating tissues that enable various parts of the body to communicate. These tissues deliver nutrients and oxygen to the cells and take away wastes and carbon dioxide.

Vertebrate blood consists of a fluid component, called **plasma,** in which are suspended **red blood cells, white blood cells,** and **platelets.** The functions of these various components of blood will be discussed in Chapter 36. Blood cells are produced within another connective tissue, the *red bone marrow* found within certain bones.

Muscle Tissue Contracts, Permitting Movement

In most animals, **muscle tissue** is the most abundant tissue. It accounts for nearly two-thirds of the body weight in a human being. Muscle tissue is specialized for contraction and permits a wide range of movement in animals. Because they are long and narrow, muscle cells are referred to as **fibers.** Muscle fibers are usually arranged in layers or bundles surrounded by connective tissue. There are three types of muscle tissue: skeletal, cardiac, and smooth (Table 31–3 and Fig. 31–8).

Skeletal muscle, which is attached to the bones, can be contracted voluntarily. This muscle tissue permits us to walk, run, write, and move the body in other ways. Characterized by a pattern of light and dark stripes, or striations, it is also referred to as **striated muscle.** Each skeletal muscle fiber has several nuclei that lie just under the plasma membrane.

Cardiac muscle, the main tissue of the heart, is a kind of striated muscle that is not under voluntary control. The fibers of cardiac muscle are joined end to end and branch and rejoin, forming complex networks. One or two nuclei are found within each fiber. A characteristic feature of cardiac muscle tissue is the presence of **intercalated discs,** specialized junctions where the fibers join.

The third type of muscle, **smooth muscle,** lacks striations and is involuntary. Each spindle-shaped cell contains only one nucleus. Found within the walls of many organs, smooth muscle is responsible for such internal movements as moving food through the digestive tract.

Nervous Tissue Receives Stimuli And Transmits Information

Nervous tissue receives stimuli and transmits information in the form of nerve impulses. This tissue controls the action of muscles and glands. Although the bulk of nervous tissue is located within the brain and spinal

**TABLE 31–3
The Types of Muscle Tissues**

	Skeletal	Smooth	Cardiac
Location	Attached to skeleton	Walls of stomach, intestines, etc.	Walls of heart
Type of control	Voluntary	Involuntary	Involuntary
Shape of fibers	Elongated, cylindrical, blunt ends	Elongated, spindle-shaped, pointed ends	Elongated, cylindrical fibers that branch and fuse
Striations	Present	Absent	Present
Number of nuclei per fiber	Many	One	One or two
Position of nuclei	Peripheral	Central	Central
Speed of contraction	Most rapid	Slowest	Intermediate
Ability to remain contracted	Least	Greatest	Intermediate

(a) Skeletal muscle fibers   (b) Smooth muscle fibers   (c) Cardiac muscle fibers

cord, bundles of nerve fibers (that is, the nerves) are found in all parts of the body, where they pick up information from the sense organs and return information in the form of "decisions" or "instructions" from the brain and spinal cord.

Nervous tissue consists of nerve cells, or **neurons,** and supporting cells called **glial cells** (Fig. 31–9). A typical neuron consists of a **cell body** containing the nucleus, and elongated extensions of the cytoplasm—the dendrites and axon. **Dendrites** are specialized for receiving impulses, and the single **axon** conducts impulses away from the cell body toward another neuron or a muscle or gland. (Neurons are discussed in Chapter 33.)

◻ AN ORGAN CONSISTS OF MORE THAN ONE TYPE OF TISSUE AND PERFORMS ONE OR MORE FUNCTIONS

Different types of tissues that together perform one or more biological functions constitute an organ. The brain, heart, stomach, and eye are organs, and although you may not think of the skin as an organ, it is the largest one in the body.

Although an organ may be composed primarily of one type of tissue, other types provide support, protection, blood supply, and conduction of nerve impulses. The intestine, for example, is lined with epithelium that secretes digestive enzymes and absorbs nutrients. Layers of muscle make up the bulk of its wall and contract in waves, moving food through the digestive tube. Nervous tissue places the intestine in communication with other parts of the body, such as the brain. Connective tissue supplies the intestine with blood and holds its tissues together; it also holds the tube in place in the body.

◻ TEN ORGAN SYSTEMS MAKE UP THE COMPLEX ANIMAL ORGANISM

Various tissues and organs coordinate their activities, performing a specialized set of functions. Such an organized group of structures is termed an **organ system.** In the human body, as in other complex animals, we can identify ten organ systems, each performing a specific group of activities (Fig. 31–10). Working together, these ten organ systems make up the complex **organism.**

(a)

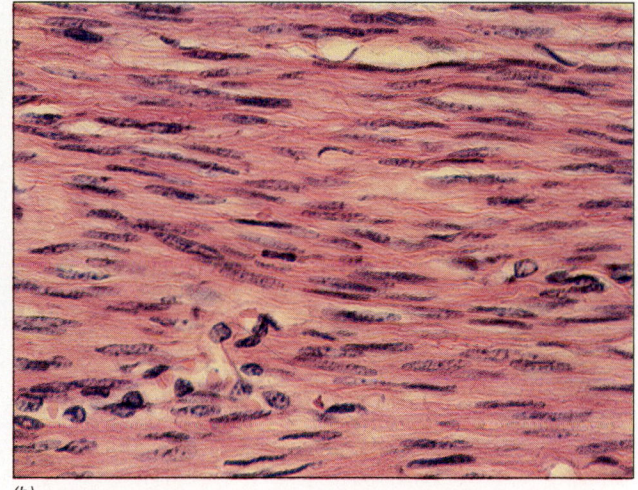

(b)

Nuclei   Intercalated disc

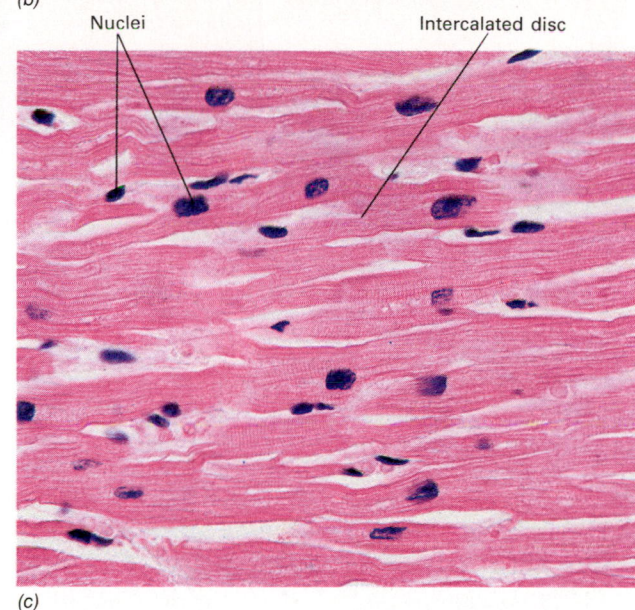

(c)

FIGURE 31–8   Muscle tissue.
(*a*) **Skeletal muscle is striated, voluntary muscle (×1000). (*b*) Smooth muscle tissue lacks striations and is involuntary. This section of human muscle tissue has been magnified about 400 times. (*c*) Cardiac muscle tissue is striated, has branched fibers, and is involuntary. The special junctions between cardiac muscle cells are called intercalated disks.** (*a*, Peter Arnold, Inc.; *b*, Runk/Schoenberger from Grant Heilman)

Cell body of neuron   Neurons   Dendrites

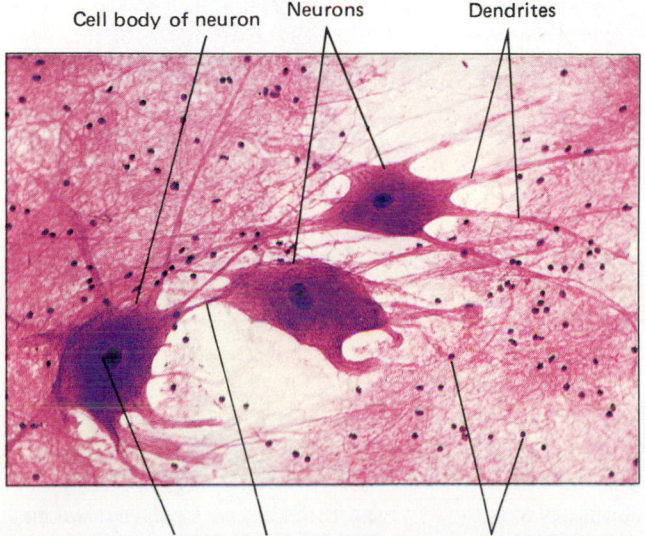

Nucleus   Axon of neuron   Nuclei of glial cells

FIGURE 31–9   Nervous tissue consists of neurons and glial cells (approximately ×50). (Ed Reschke)

The organ systems of complex animals include the integumentary, skeletal, muscle, nervous, circulatory, digestive, respiratory, urinary, endocrine, and reproductive systems. See Table 31–4 for a summary of their principal organs and functions. As an example of an organ system, consider the digestive system. Organs of the digestive system include the mouth, esophagus, stomach, small and large intestines, liver, pancreas, and salivary glands. This system digests food, reducing it to simple molecular components, and then absorbs these nutrients, enabling them to enter the blood for transport to all of the body's cells.

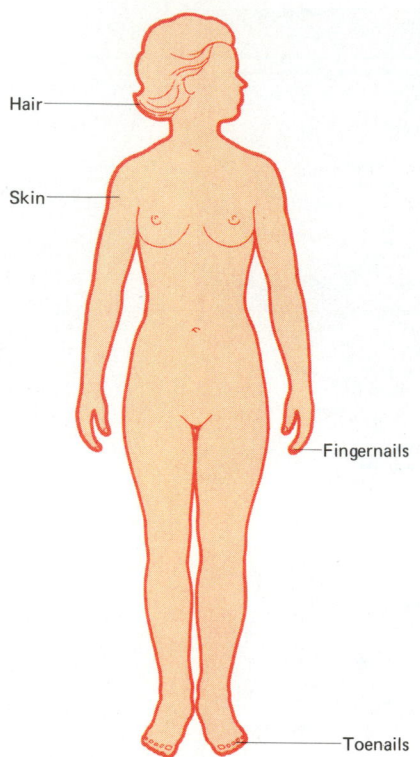

(1) The integumentary system consists of the skin and the structures such as nails and hair that are derived from it. This system protects the body, helps to regulate body temperature, and receives stimuli such as pressure, pain, and temperature.

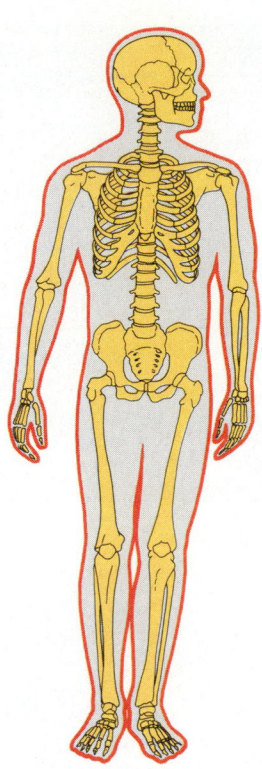

(2) The skeletal system consists of bones and cartilage. This system helps to support and protect the body.

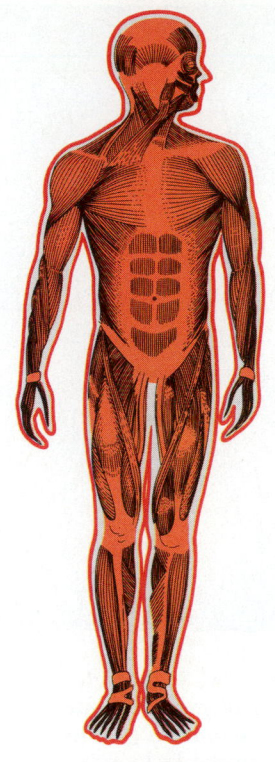

(3) The muscular system consists of the large skeletal muscles that enable us to move, as well as the cardiac muscle of the heart and the smooth muscle of the internal organs.

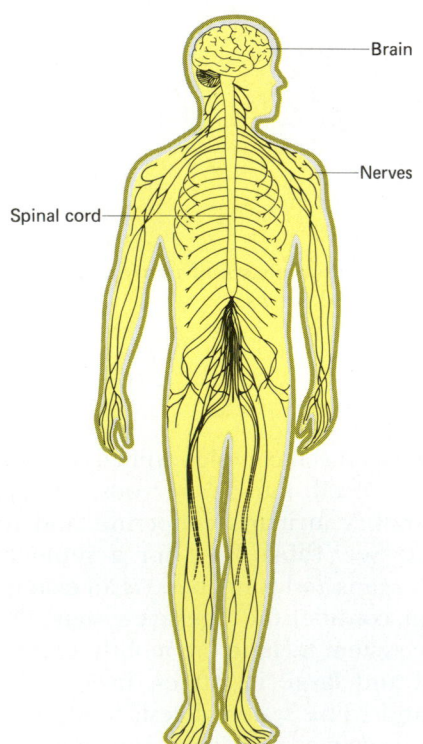

(4) The nervous system consists of the brain, spinal cord, sense organs, and nerves. This is the principal regulatory system.

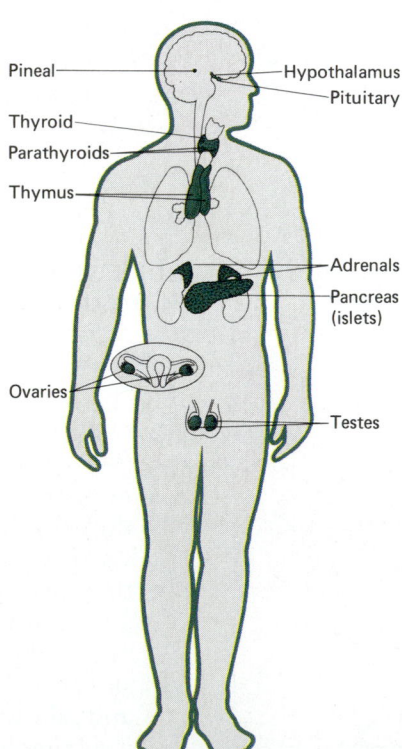

(5) The endocrine system consists of the ductless glands that release hormones. It works with the nervous system in regulating metabolic activities.

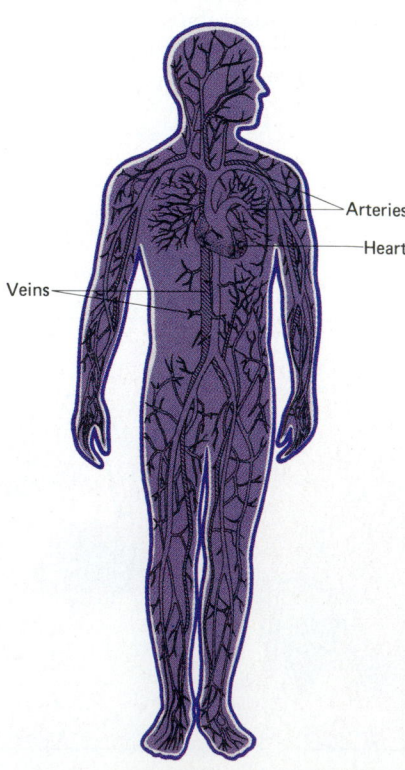

(6a) The circulatory system includes the heart and blood vessels. This system serves as the transportation system of the body.

FIGURE 31–10 The principal organ systems of the human body.

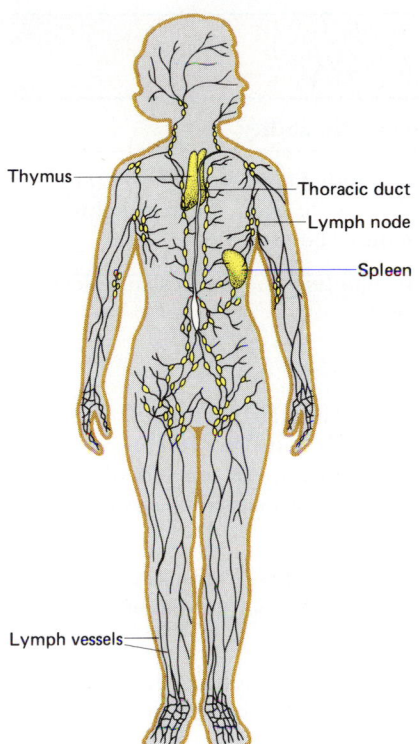

Thymus

Thoracic duct

Lymph node

Spleen

Lymph vessels

(6b) The lymphatic system is a subsystem of the circulatory system; it returns excess tissue fluid to the blood and defends the body against disease.

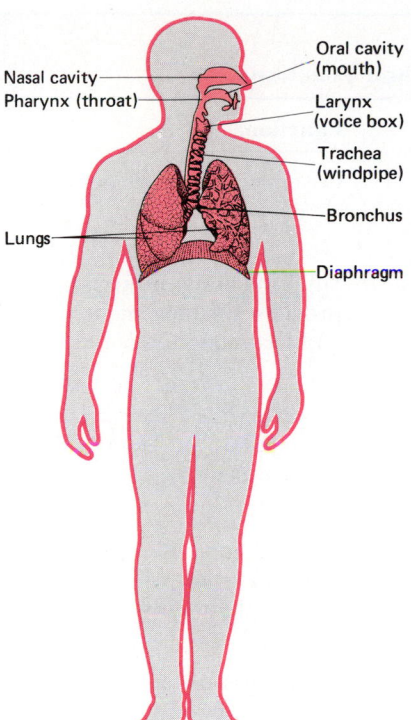

Nasal cavity

Pharynx (throat)

Oral cavity (mouth)

Larynx (voice box)

Trachea (windpipe)

Bronchus

Lungs

Diaphragm

(7) The respiratory system. Consisting of the lungs and air passageways, this system supplies oxygen to the blood and excretes carbon dioxide.

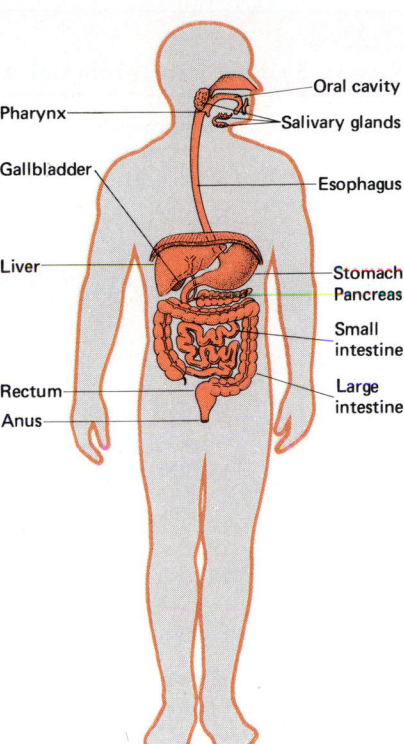

Pharynx

Oral cavity

Salivary glands

Gallbladder

Esophagus

Liver

Stomach

Pancreas

Small intestine

Rectum

Large intestine

Anus

(8) The digestive system consists of the digestive tract and glands that secrete digestive juices into the digestive tract. This system mechanically and enzymatically breaks down food and eliminates wastes.

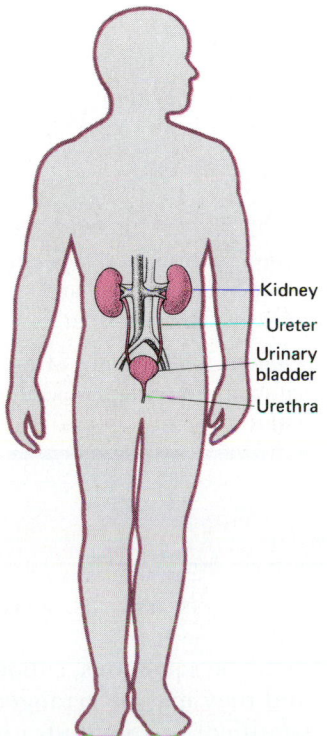

Kidney

Ureter

Urinary bladder

Urethra

(9) The urinary system is the main excretory system of the body, and helps to regulate blood chemistry. The kidneys remove wastes and excess materials from the blood and produce urine.

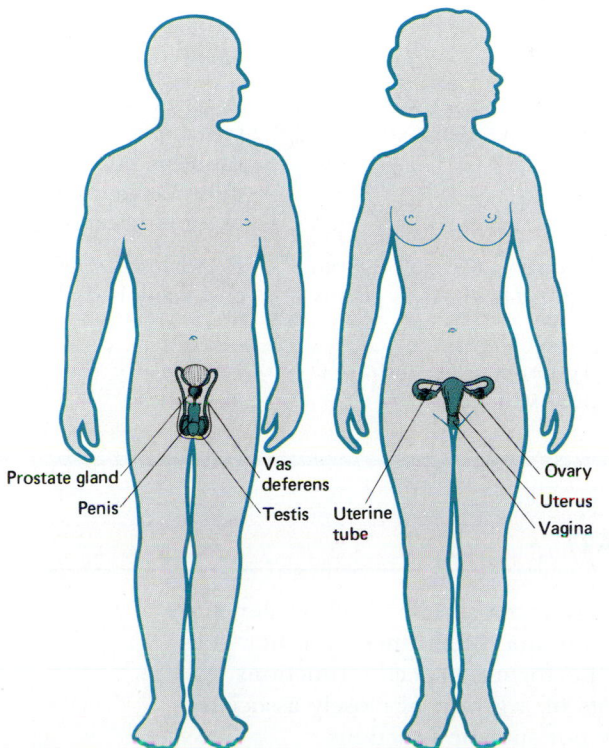

Prostate gland

Penis

Vas deferens

Testis

Uterine tube

Ovary

Uterus

Vagina

(10) Male and female reproductive systems. Each reproductive system consists of gonads and associated structures. The reproductive system maintains the sexual characteristics, and perpetuates the species.

TABLE 31–4
The Organ Systems of a Mammal and Their Functions

System	Components	Functions	Homeostatic ability
Integumentary	Skin, hair, nails, sweat glands	Covers and protects body	Sweat glands help control body temperature; as barrier, the skin helps maintain steady state
Skeletal	Bones, cartilage, ligaments	Supports body, protects, provides for movement and locomotion, calcium depot	Helps maintain constant calcium level in blood
Muscular	Organs mainly of skeletal muscle; cardiac muscle; smooth muscle	Moves parts of skeleton, locomotion; movement of internal materials	Ensures such vital functions as nutrition through body movements; smooth muscle maintains blood pressure; cardiac muscle circulates the blood
Digestive	Mouth, esophagus, stomach, intestines, liver, pancreas	Ingests and digests foods, absorbs them into blood	Maintains adequate supplies of fuel molecules and building materials
Circulatory	Heart, blood vessels, blood; lymph and lymph structures	Transports materials from one part of body to another; defends body against disease	Transports oxygen, nutrients, hormones; removes wastes; maintains water and ionic balance of tissues
Respiratory	Lungs, trachea, and other air passageways	Exchange of gases between blood and external environment	Maintains adequate blood oxygen content and helps regulate blood pH; eliminates carbon dioxide
Urinary	Kidney, bladder, and associated ducts	Eliminates metabolic wastes; removes substances present in excess from blood	Regulates blood chemistry in conjunction with endocrine system
Nervous	Nerves and sense organs, brain and spinal cord	Receives stimuli from external and internal environment, conducts impulses, integrates activities of other systems	Principal regulatory system
Endocrine	Pituitary, adrenal, thyroid, and other ductless glands	Regulates body chemistry and many body functions	In conjunction with nervous system, regulates metabolic activities and blood levels of various substances
Reproductive	Testes, ovaries, and associated structures	Provides for continuation of species	Passes on genetic endowment of individual; maintains secondary sexual characteristics

■ CHAPTER SUMMARY

I. Multicellular organisms can be much larger and more diverse than unicellular ones, and their cells can specialize, performing specific functions.

II. A tissue consists of a group of closely associated cells that carry out specific functions.

    A. Epithelial tissue consists of cells that are fitted tightly together. It covers the body surface and lines its cavities.

1. Epithelial tissue protects, absorbs, secretes, or receives stimuli.

2. Epithelial cells may be squamous, cuboidal, or columnar, and they may be arranged to form simple, stratified, or pseudostratified tissue.

    B. Connective tissue supports and binds together structures of the body. It protects underlying structures and may store or transport materials.

1. Connective tissue consists of relatively few cells separated by intercellular substance, which in turn is composed of fibers scattered through a matrix.
2. The main types of connective tissue are loose and dense, elastic, reticular, adipose, cartilage, bone, blood, lymph, and tissues that produce blood cells.

C. Muscle tissue contracts, enabling an animal to move. The three types of muscle are skeletal, cardiac, and smooth muscle.

D. Nervous tissue receives and transmits stimuli; this tissue controls the action of muscles and glands.

III. Cells of a neoplasm differ from cells of a normal tissue in that they divide in an uncontrolled manner and invade normal tissues.

IV. An organ consists of different types of tissue that work together to perform a particular biological function. The heart, brain, and stomach are examples of organs.

V. Organs and tissues work together, forming organ systems. In complex animals, about ten principal organ systems work together to constitute the total living organism. Among these are the digestive system, the nervous system, and the skeletal system.

■ POST-TEST

1. A group of closely associated cells that carry out specific functions forms a _____ .
2. A _____ consists of epithelial cells that produce and secrete a product.
3. Epithelial cells may be flattened or _____ ; cube shaped or _____ ; or elongated like columns, _____ .

Select the appropriate answer or answers from column B for each entry in column A.

Column A

_____ 4. Protects, absorbs, or secretes
_____ 5. Receives stimuli and transmits impulses
_____ 6. Contracts
_____ 7. Abnormal mass of tissue
_____ 8. Contains fibroblasts
_____ 9. May be simple or stratified
_____ 10. Bone is an example

Column B

a. Muscle tissue
b. Epithelial tissue
c. Connective tissue
d. Nervous tissue
e. Neoplasm

_____ 11. Forms the subcutaneous layer; the most widely distributed connective tissue
_____ 12. Forms a network that supports many organs
_____ 13. Stores fat
_____ 14. Cells found in lacunae
_____ 15. Capillaries run through haversian canals
_____ 16. Contains chondrocytes

a. Bone
b. Adipose tissue
c. Loose connective tissue
d. Reticular connective tissue
e. Cartilage

17. In cardiac muscle, _____ _____ are the junctions between adjoining fibers.
18. Two types of striated muscle are _____ and _____ .
19. Two types of cytoplasmic extensions of neurons are the _____ and _____ .
20. The organ system made up of glands that secrete hormones is the _____ system.
21. The organ system that covers the body is the _____ system.

■ REVIEW QUESTIONS

1. Distinguish among tissues, organs, and organ systems.
2. Imagine that all of the epithelium in a complex animal, such as a human being, suddenly disappeared. What effects would this have on the body and its ability to function?
3. Contrast epithelial tissue with connective tissue.
4. Compare cartilage with bone.
5. What is a gland? Of what tissue is it composed?
6. Compare skeletal, cardiac, and smooth muscle with respect to structure and location in the body.
7. How is a neuron adapted for its specific function?
8. How do the cells of a malignant neoplasm differ from

the cells of a normal tissue? What is metastasis?

9. Locate each of the following: (a) osteocytes, (b) chondrocytes, (c) intercalated discs, (d) fibroblasts, (e) collagen fibers, (f) stratified squamous epithelium.

10. Name the functions of the following: (a) endocrine system, (b) skeletal system, (c) circulatory system, (d) nervous system.

◻ READINGS

Cairns, J. "The Treatment of Diseases and the War against Cancer," *Scientific American:* 51–59, November 1985. An overview of the natural history of some common forms of cancer.

Caplan, A.I. "Cartilage," *Scientific American:* 84–94, October 1984. The basic properties of strength and resilience are explained in terms of the tissues' molecular structure.

Kessel, R.G., and R.H. Kardon. *Tissues and Organs: A Text-Atlas of Scanning Electron Microscopy.* San Francisco, W.H. Freeman Co., 1979. A stunning collection of scanning electron micrographs of tissues and organs.

National Geographic Society Book Service. *The Incredible Machine.* Washington, D.C., National Geographic, 1986. A beautiful and informative introduction to the human body featuring the incredible photographs of Lennart Nilsson.

Solomon, E.P., Schmidt, R., and Adragna, P. *Human Anatomy and Physiology,* 2nd ed. Philadelphia, Saunders College Publishing, 1990. A very readable presentation of human anatomy and physiology that includes a chapter on tissues.

Weinberg, R.A. "A Molecular Basis of Cancer," *Scientific American:* 126–142, November 1983. A discussion of the role of oncogenes in cancer.

See also the readings for Chapter 32.

32

PROTECTION, SUPPORT, AND MOVEMENT: SKIN, SKELETON, AND MUSCLE

Human bone tissue magnified more than 50 times. (Michael Abbey/Photo Researchers, Inc.)

OUTLINE

I. **The animal body is covered and lined with a protective epithelial tissue**
  A. The protective epithelial covering of invertebrates may function in secretion or gas exchange
  B. The vertebrate skin is an important organ system that performs diverse functions
    1. The epidermis is a waterproof protective barrier
    2. The dermis contains sweat glands, hair follicles, blood vessels, and sense organs

II. **Skeletons are important in locomotion, protection, and support of the animal body**
  A. In hydrostatic skeletons, body fluids are used to transmit force
  B. External skeletons of mollusks and arthropods are nonliving shells
    1. The mollusk exoskeleton protects the soft body beneath
    2. The arthropod exoskeleton functions in protection, support, and locomotion
  C. Internal skeletons are living tissue capable of growth
    1. The echinoderm endoskeleton functions in protection and support
    2. The vertebrate endoskeleton is important in protection, support, and locomotion
      a. A typical long bone has a thin shell of compact bone and a filling of spongy bone
      b. Bones develop in the embryo from cartilage or from noncartilage connective tissue and are remodeled throughout life in response to physical stresses
      c. Joints hold bones together and may permit movement

III. **Muscle is the contractile tissue that allows movement in complex animals**
  A. All animals have the ability to move
  B. A vertebrate muscle consists of hundreds of muscle fibers wrapped in connective tissue
  C. Muscle contraction occurs when actin and myosin filaments slide past each other
  D. ATP powers muscle contraction
  E. Skeletal muscle action depends on muscle pairs working antagonistically
  F. Smooth, cardiac, and skeletal muscle are each specialized for particular types of responses

LEARNING OBJECTIVES

After you have studied this chapter, you should be able to:

1. Describe the external epithelium of invertebrates and summarize its functions.
2. Compare the structure and function of vertebrate skin with the external epithelium of invertebrates, and identify the principal derivatives of vertebrate skin.
3. Compare advantages and disadvantages of different types of skeletal systems, including the hydrostatic skeleton, exoskeleton, and endoskeleton.
4. Identify main divisions of the vertebrate skeleton and the bones that make up each division.
5. Describe the structure of a typical long bone.
6. Summarize bone development, differentiating between endochondral and membranous bone development.
7. Describe the gross and microscopic structure of skeletal muscle.
8. List in sequence the events that take place in muscle contraction.
9. Compare the roles of glycogen, creatine phosphate, and ATP in providing energy for muscle contraction.
10. Describe the antagonistic action of muscles.
11. Summarize the functional relationship between skeletal and muscular tissues.

 Some animals run, some jump, and some fly. Others remain rooted to one spot, sweeping their surroundings with tentacles. Many contain internal circulating fluids, pumped by hearts and contained by hollow vessels that maintain their pressure by gentle squeezing. Digestive systems push food along with peristaltic writhings. In all these cases, each action is powered by muscle, a specialized tissue that, however varied its effects, has but one action—contraction.

In many animals, the muscles responsible for locomotion are anchored to the skeleton, which, then serves to transmit forces (Fig. 32–1). The skeleton also supports the body and protects the delicate organs within. Epithelial coverings of invertebrates and skin of vertebrates protect the tissue beneath and may also perform a variety of other functions such as secretion or gas exchange. In this chapter, we discuss skin, skeleton, and muscle—systems that are closely interrelated in function and significance.

KEY CONCEPTS

- The animal body is covered by a protective shield of epithelium that may perform other functions such as secretion, gas exchange, reception of stimuli, or regulation of body temperature.

- Skeletons are important in locomotion and may protect and support the animal body.

- Muscle and skeleton often work together in movement. Muscles, the body's motors, are anchored to the skeleton, which transmits forces.

THE ANIMAL BODY IS COVERED WITH A PROTECTIVE EPITHELIAL TISSUE

Epithelial tissue covers all external and internal surfaces of the animal body. The outer epithelial covering forms a protective shield around the body.

The Protective Epithelial Covering Of Invertebrates May Function In Secretion Or Gas Exchange

In invertebrates, epithelial tissue is simple, rather than stratified, and the external epithelium is generally cuboidal or a low columnar type. In addition to their protective function, the outer epithelial cells may also function as sensory cells that are selectively sensitive to light, chemical stimuli, or mechanical stimuli such as contact or pressure.

FIGURE 32–1 This cheetah is making the most of its powerful muscles as it pursues its prey.

FIGURE 32–2 The foot of the land snail (*Helix* sp.) releases a mucous secretion that produces a slime track through which the animal glides. (Runk/Schoenberger from Grant Heilman)

In many species, the epithelium contains secretory cells that produce a tough, protective cuticle or secrete lubricants or adhesives. In some animals, these cells release odorous secretions that allow communication among members of the species or mark trails. Others produce poisonous secretions that are used in offense or defense. In earthworms, a lubricating, mucous secretion produces a moist slime that aids diffusion of gases across the body wall. This lubricating secretion also reduces friction as the earthworm pushes its way through the soil.

In some species, an epithelial secretion may be limited to a particular region of the body surface. In the gastropod mollusk, for example, a mucous secretion is released from the foot, producing a slime track through which the snail glides (Fig. 32–2). Epithelial cells in the basal region of *Hydra* allows this animal to

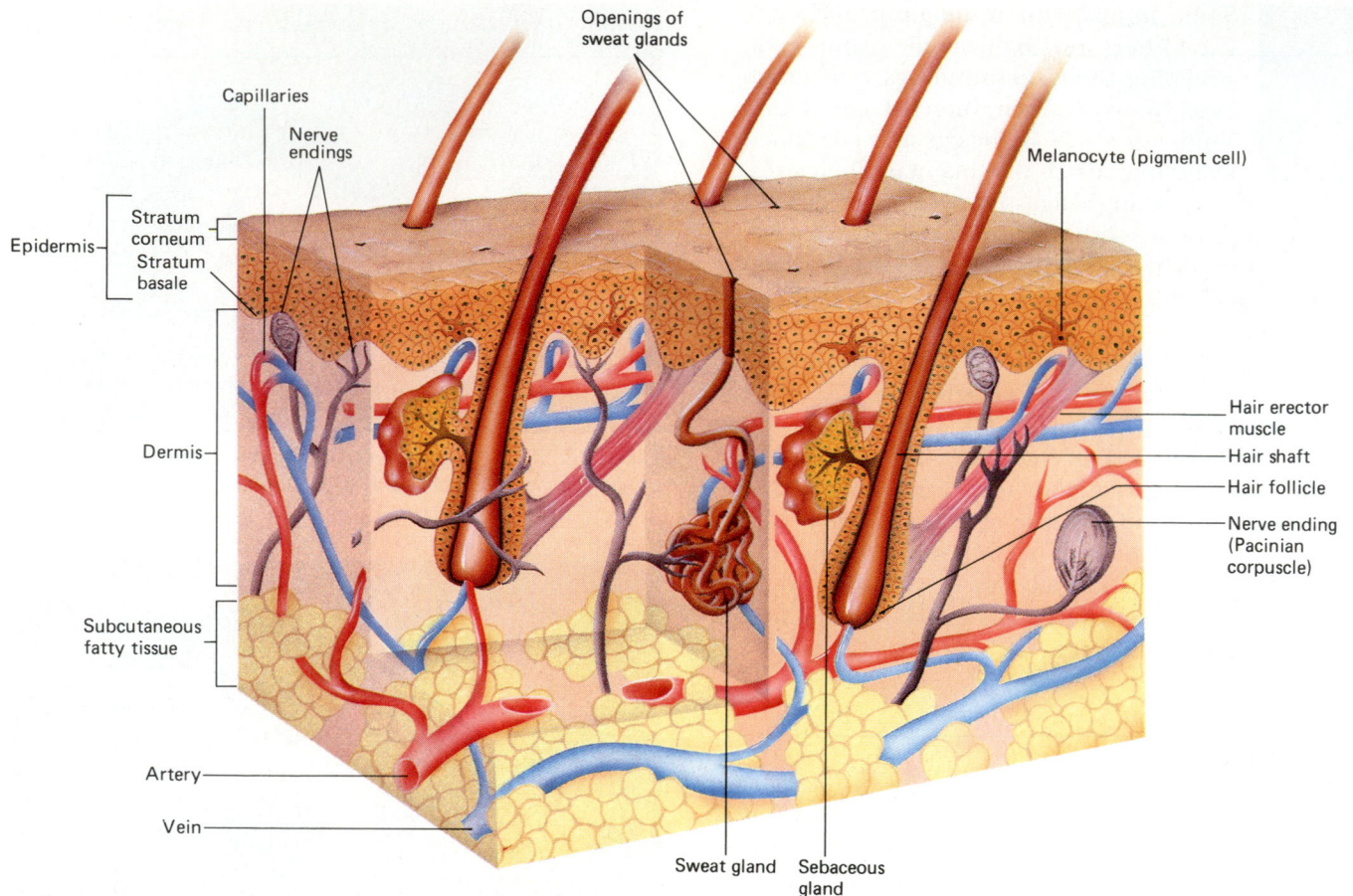

FIGURE 32–3   The structure of mammalian skin.

temporarily cling to some object. Insects are able to walk upside down across a ceiling because of the gummy substance secreted by gland cells in the epithelium of the terminal segments of their legs.

In some insects (weaver ants, for example), epithelial secretions are released as fine threads of considerable strength and are used to construct nests. The spinning glands of spiders develop as invaginations of epidermal cells. Some insects (such as butteflies and moths), synthesize silk from amino acids in their silk-forming glands.

The Vertebrate Skin Is An Important Organ System That Performs Diverse Functions

In many fish, in the African ant-eating pangolin (a mammal), and in some reptiles, the skin develops into a set of scales formidable enough to be considered armor. Even in human beings, the skin has considerable strength. Human skin (Fig. 32–3) includes a variety of structures, including fingernails and toenails, hair of various types, sweat glands, oil (sebaceous) glands, and several types of sensory receptors that permit one to feel pressure, temperature, and pain.

Human skin and the skin of other mammals contain mammary glands that, in females, secrete milk. Oil glands present in human skin empty via short ducts into hair follicles (see Fig. 32–3). They secrete a waxy substance called sebum. In humans, these glands are especially numerous on the face and scalp. The oil secreted keeps the hair moist and pliable and prevents the skin from drying and cracking. (Excessive production of sebum at puberty in response to increased levels of sex hormones fills the glands and follicles, producing the too familiar inflammation called acne.)

In a human being, the skin functions as a thermostatically controlled radiator, regulating the elimination of heat from the body (see Chapter 1). About 2.5 million sweat glands secrete sweat, and its evaporation from the surface of the skin lowers the body temperature.

The skin in some other vertebrates varies considerably from ours. Instead of hairs, birds have feathers, which nevertheless form in a manner comparable to hairs and provide even more effective insulation than fur. Among the ectothermic vertebrates (those whose body temperatures fluctuate with that of the environment), one finds epidermal scales (as in reptiles), naked skin covered with mucus (as in many amphibi-

FIGURE 32–4   Arctic mammals, such as this Pacific walrus, a native of Alaska, have thick layers of subcutaneous fat that serve as insulation. (Nada Pecnik/ Visuals Unlimited)

ans and fish), and skin with bony or tooth-like scales. Some skin, such as that of certain tropical frogs, even has poison glands. Skin and its derivatives are often brilliantly colored in connection with courtship rituals, territorial displays, and various kinds of communication. The human blush pales alongside the spectacular displays of such animals as peacocks.

The Epidermis Is A Waterproof Protective Barrier

The outer layer of skin, the **epidermis,** is the interface between the delicate tissues within the body and the hostile universe. The epidermis consists of several strata (layers) of cells, the lowest of which is the **stratum basale,** and the outermost, the **stratum corneum** (see Fig. 32–2). In the stratum basale, cells continuously divide, and the new cells are pushed upward by yet other cells being produced below them. As the epidermal cells move upward in the skin, they mature. In almost all vertebrates, there are no capillaries in the epidermis, so the maturing cells are progressively deprived of more and more nourishment. They become ever less active metabolically.

As they move upward, epidermal cells manufacture the distinctive skin substance **keratin.** Keratin is a protein that gives the skin considerable mechanical strength and flexibility. Keratin is quite insoluble and serves to waterproof the body-surface. As epidermal cells move through the stratum corneum, the outermost layer of the skin, they die. The cells of this outer layer continuously wear off and must be continuously replaced.

The Dermis Contains Sweat Glands, Hair Follicles, Blood Vessels, And Sense Organs

Underneath the epidermis lies the **dermis** (see Fig. 32–2), which consists of a dense, fibrous connective tissue composed principally of collagen fibers. Collagen imparts strength and flexibility to the skin. The major part of each sweat gland is embedded in the dermis, and the hair follicles reach down into it. The dermis also contains blood vessels, which nourish the skin, and sense organs concerned with touch. Mammalian skin rests on a layer of subcutaneous tissue that is composed mainly of fat that insulates us from unfavorable outside temperature extremes (Fig. 32–4).

■ SKELETONS ARE IMPORTANT IN LOCOMOTION, PROTECTION, AND SUPPORT OF THE ANIMAL BODY

In some of the simplest animals, muscle acts directly on the jelly-like substance of the body itself, or perhaps on a fluid-filled body cavity. In more complex animals, however, a skeleton receives, transmits, and transforms the single movement—contraction—of their muscular tissues into the variety of motions animals use.

In a few instances, this skeleton is internal—plates or shafts of calcium-impregnated tissue. In most cases, however, the skeleton is not a living tissue at all but a lifeless deposit atop the epidermis—a shell, or exoskeleton. In addition to its function in locomotion, the skeleton supports the body and protects the internal organs.

In Hydrostatic Skeletons, Body Fluids Are Used To Transmit Force

Imagine an elongated balloon full of water. If one were to pull on it, it would lengthen, but it would also lengthen if it were squeezed. Conversely, it would shorten if the ends were pushed. In *Hydra* and other cnidarians, cells of the two body layers are capable of contraction, which produces similar changes in shape.

The contractile cells in the outer epidermal layer are arranged longitudinally, whereas the contractile cells of the inner layer (the gastrodermis) are arranged in a circular fashion around the central body axis (Fig. 32–5). These two groups of cells work antagonistically. What one can do, the other can undo. When the epidermal, longitudinal layer contracts, the hydra shortens, and because of the fluid present in its gastrovascular cavity, force is transmitted so that it thickens as well. On the other hand, when the endodermal circular layer contracts, the hydra thins, but its fluid con-

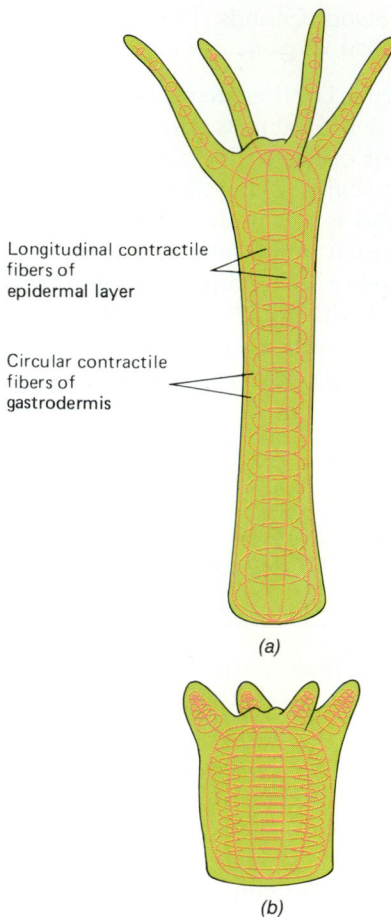

Longitudinal contractile
fibers of
epidermal layer

Circular contractile
fibers of
gastrodermis

(a)

(b)

**FIGURE 32–5 Movement in *Hydra*. The longitudinally
arranged cells are antagonistic to the cells arranged
around the body axis. (*a*) Contraction of the circular
muscles elongates the body. (*b*) Contraction of the
longitudinal muscles shortens the body.**

tents also force it to lengthen. The hydra is mechanically little more than a simple bag of fluid. Its fluid
interior acts as a hydrostatic skeleton, because it transmits force when the contractile cells contract against it.

Hydrostatic skeletons permit only crude mass
movements of the body or its appendages. Precise
movement is difficult because in a fluid, force tends to
be transmitted equally in all directions and, hence,
throughout the entire fluid-filled body of the animal.
It is not easy for the hydra to thicken one part of its
body, for example, while thinning another.

The annelid worms have a more sophisticated
hydrostatic skeleton that permits more versatile body
movements than those of *Hydra*. The body of an earthworm, to take a familiar example, may protrude from
its burrow on damp evenings to feed on bits of decayed
vegetation on the surface. Its posterior end may then
protrude in order to defecate the familiar worm cast-

ings. This practice is not without its hazards, for if the
hungry worm waits too long, an equally hungry early
bird is likely to find it. If the bird is quick enough, the
story ends right there. If not, the giant nerve axons of
the worm's ventral, solid nerve cord swiftly transmit
impulses that stimulate the longitudinal muscles and
inhibit the circular muscles. Abruptly the longitudinal
muscles contract, pulling the body of the worm toward
the safety of its burrow. If the worst occurs and the
bird obtains a firm hold, the worm holds on too, with
its swollen, contracted anterior (front) now fitting the
burrow like a cork in a bottleneck. If the bird releases
its hold, the worm will rapidly crawl down the burrow
to safety. But how?

If the worm is progressing anterior end first, it
protrudes a thinned portion of its body into the burrow ahead. Then, while anchored posteriorly by a
thickened portion of itself, the anterior end swells. The
worm grips the burrow ahead, then releases its posterior grip. By longitudinal muscle contraction, it drags
the whole body toward the anchored anterior end. It
repeats this process again and again (Fig. 32–6).

All this is made possible by the transverse partitions, or **septa,** that subdivide the body cavity of the
worm. These septa isolate portions of the body cavity
and its contained fluid, permitting the hydrostatic skeletons of each segment to be largely independent of one
another. Thus, the contraction of the circular muscle
in the elongating anterior end need not interfere with
the action of the longitudinal muscle in the segments
of the still-anchored posterior.

For animals that do more than drag themselves
along on their bellies, the hydrostatic skeleton is insufficient. Yet some examples of it occur to a small extent
in the higher invertebrates and even in the vertebrates.
Among the mollusks, for example, the feet of bivalves
are extended and anchored by a hydrostatic blood
pressure mechanism not too different from that used
by the earthworm. The multitudinous tube feet of
echinoderms, such as the sea star and sea urchin, are
moved by an ingenious version of the hydrostatic skeleton, and even in man, the penis becomes erect and
stiff because of the turgidity of pressurized blood in its
cavernous spaces.

External Skeletons Of Mollusks And Arthropods Are Nonliving Shells

Although there are others, the two major groups of
animals with external skeletons are the mollusks and
the arthropods. In both the mollusks and the arthropods, the shell is a nonliving product of the cells of the
epidermis, but it functions differently in each phylum.

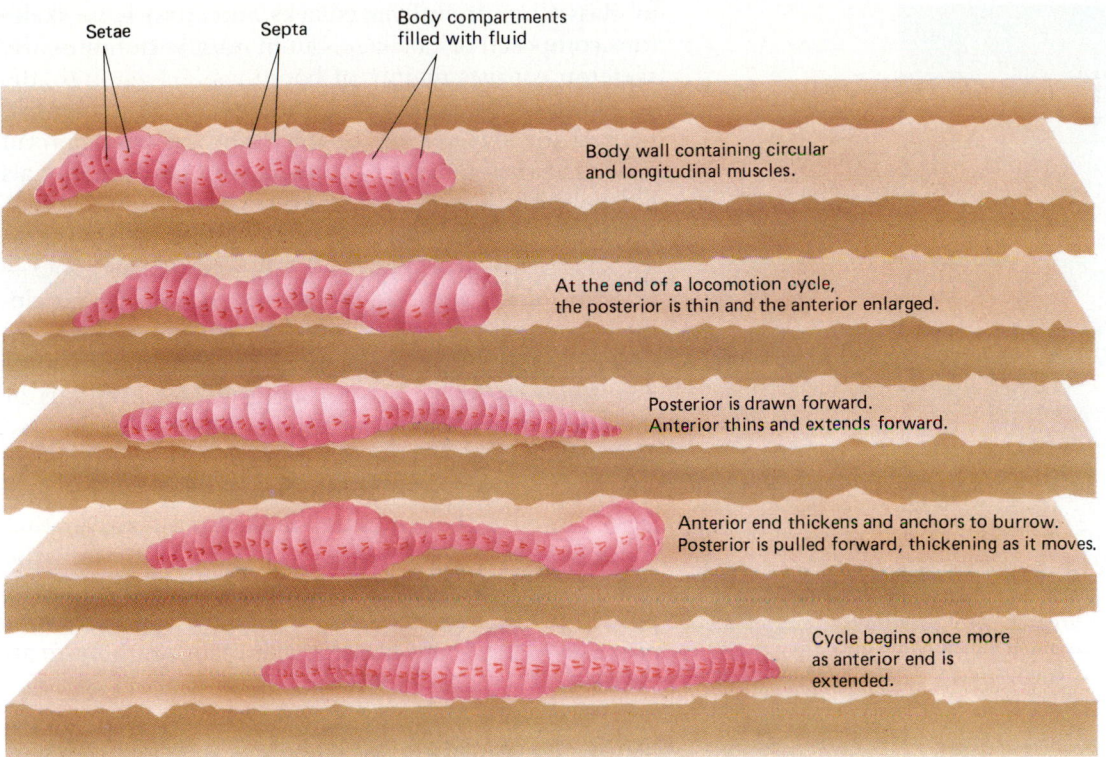

Setae

Septa

Body compartments filled with fluid

Body wall containing circular and longitudinal muscles.

At the end of a locomotion cycle, the posterior is thin and the anterior enlarged.

Posterior is drawn forward. Anterior thins and extends forward.

Anterior end thickens and anchors to burrow. Posterior is pulled forward, thickening as it moves.

Cycle begins once more as anterior end is extended.

FIGURE 32–6   Annelid locomotion. Can you infer the segments in which longitudinal or circular muscle is active in each stage? The worm is aided in anchoring itself by bristle-like setae.

The Mollusk Exoskeleton Protects The Soft Body Beneath

In mollusks, the exoskeleton basically provides protection, with its major muscle attachments serving the skeleton rather than the skeleton serving the muscles. Thus, the common clam has a pair of muscles whose major action is holding the two valves of the shell tightly shut against the onslaughts of sea star and chowder maker.

The Arthropod Exoskeleton Functions In Protection, Support, And Locomotion

The exoskeleton is a major factor in the success of the arthropods. It serves not only to protect and support but also to transmit forces in ways fully comparable to those found in the skeletons of vertebrates. Whereas in mollusks the shell is primarily an emergency retreat, with the bulk of the body exposed at other times, in arthropods the exoskeleton covers every bit of the body. It even extends inward as far as the stomach on one end, and for a considerable distance inward past the anus on the other. Although the arthropod exoskeleton is a continuous one-piece sheath, it varies greatly in thickness and flexibility, with large, thick, inflexible plates separated from one another by thin, flexible joints that are arranged segmentally. Enough joints are provided to make the arthropod's body just as flexible as that of many vertebrates. This exoskeleton is also extensively modified to form specialized tools or weapons, or otherwise adapted to a vast variety of life-styles.

The chief disadvantage of the arthropod exoskeleton is that it is rigid and does not grow, and so does not allow growth of the body beneath. To overcome this disadvantage, arthropods **molt;** that is, cast off their old covering (Fig. 32–7) from time to time.

Internal Skeletons Are Living Tissue Capable Of Growth

Endoskeletons, or internal skeletons, are extensively developed only in the echinoderms and the chordates. Composed of living tissue, the endoskeleton grows in pace with the growth of the animal as a whole, eliminating the need for molting. It also permits the animal to grow, potentially, to great size. Compare the largest land vertebrates—elephants and dinosaurs—to the

FIGURE 32–7 A cicada molting. This dogday harvestman (*Tibicen linnei*) is an immature adult emerging from the nymphal form. (Barry L. Runk from Grant Heilman)

largest land arthropods—beetles only a few inches long. If beetles grew to the size of horses, their external armor would weigh so much that it would probably collapse, or at least prevent the unfortunate animal from moving. So much for the giant insects of the horror movies!

The endoskeleton probably also permits a greater variety of possible motions than does an exoskeleton. In vertebrates, complex motions are produced by a complex interaction of many muscles, but there simply is not room for a great many muscles inside the armor of an arthropod limb. Indeed, in some arthropods, especially the spiders, the hydrostatic action of the body fluid is just as important as the muscles in producing limb movement.

The Echinoderm Endoskeleton Functions In Protection And Support

The echinoderm endoskeleton consists of plates that are composed of nonliving calcium salts embedded in the tissues of the body wall. These form what amounts to an internal shell that provides mainly support and protection. In many echinoderms, spines project from the endoskeleton to the outer surface.

The Vertebrate Endoskeleton Is Important In Protection, Support, And Locomotion

In the vertebrates, the internal skeleton provides protection, support, and transmission of forces. Members of class Chondrichthyes (sharks and rays) have skeletons composed of cartilage, but in most vertebrates, the skeleton consists mainly of bone.

Many vertebrates possess bones that humans lack, such as the skeleton of the gill arches of fish. Careful studies of the embryos of humans and other mammals have shown, however, that a number of elements of the skull originate embryonically in the same way as do the gill arches of fishes. The tiny middle ear bones—malleus, incus, and stapes—are examples of such elements.

The vertebrate skeleton has two main divisions. The **axial skeleton,** located along the central axis of the body, consists of the skull, vertebral column, ribs, and sternum (breastbone). The **appendicular skeleton** consists of the bones of the appendages (arms and legs), as well as the bones that make up the girdles that connect the appendages to the axial skeleton—the shoulder (or pectoral) girdle and most of the hip (or pelvic) girdle (Fig. 32–8). In the following description of the vertebrate skeleton, we will focus on the human skeleton.

The **skull,** the bony framework of the head, consists of the cranial and facial bones. In humans, 8 cranial bones enclose the brain, and 14 bones make up the facial portion of the skull. Several cranial bones that are single in the adult human result from the fusion of two or more bones that were originally separate in the embryo or even in the newborn.

The vertebrate spine, or **vertebral column,** is the main supporting axis below the skull. In humans, the vertebral column supports the body and bears its weight. It consists of 24 **vertebrae** and two fused bones, the *sacrum* and *coccyx*. The regions of the vertebral column are the *cervical* (neck), which consists of 7 vertebrae; the *thoracic* (chest), which consists of 12 vertebrae; the *lumbar* (back), composed of 5 vertebrae; the *sacral* (pelvic), which consists of 5 fused vertebrae; and the *coccygeal*, which also consists of fused vertebrae.

Although they differ in size and shape in different regions of the vertebral column, a typical vertebra consists of a bony central portion, the *centrum*, that bears most of the body weight, and a dorsal ring of bone, the *neural arch*, which surrounds and protects the delicate spinal cord. Vertebrae may also have various projections for the attachment of ribs and muscles and for articulation (joining) with neighboring vertebrae. The first vertebra, the *atlas* (named for the mythical giant of Greek mythology who held the world on his shoulders), has rounded depressions on its upper surface into which fit two projections from the base of the skull.

The *rib cage* is a bony basket formed by the *sternum* (breastbone), thoracic vertebrae, and 12 pairs of ribs. It protects the internal organs of the chest, including the heart and lungs, and supports the chest wall, prevent-

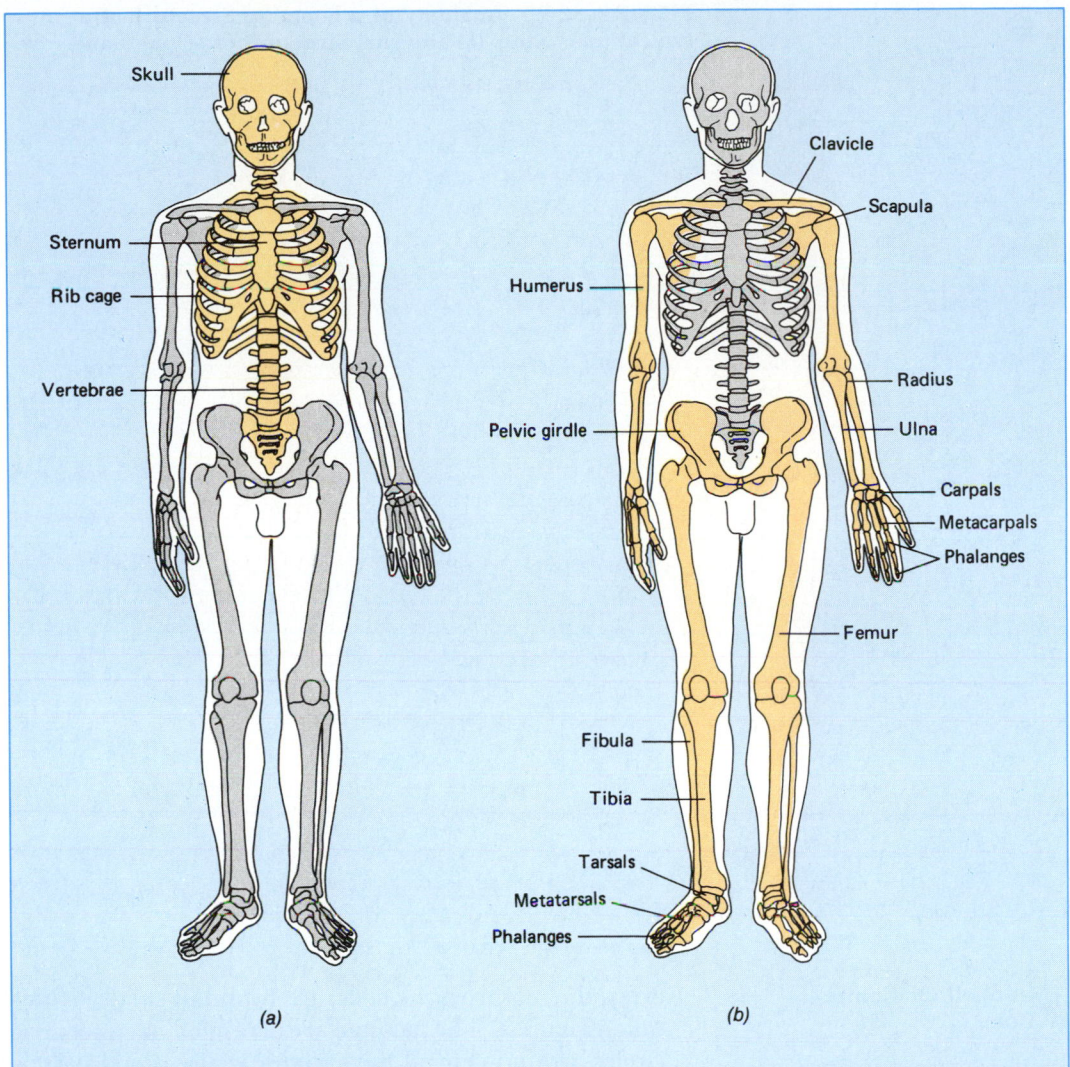

FIGURE 32–8 The human skeleton. (*a*) The bones of the axial skeleton (anterior view). (*b*) The bones of the appendicular skeleton (anterior view).

Labels in (a): Skull, Sternum, Rib cage, Vertebrae

Labels in (b): Clavicle, Scapula, Humerus, Radius, Ulna, Pelvic girdle, Carpals, Metacarpals, Phalanges, Femur, Fibula, Tibia, Tarsals, Metatarsals, Phalanges

ing it from collapsing as the diaphragm contracts with each breath. Each pair of ribs is attached posteriorly to a separate vertebra. Of the 12 pairs of ribs in the human being, the first 7 are attached ventrally to the sternum (breastbone), the next 3 are attached indirectly by cartilages, and the last 2, called "floating ribs," have no attachments to the sternum.

The **pectoral girdle** (shoulder girdle) consists of the two collarbones, or *clavicles,* and the two shoulderblades, or *scapulas*. The **pelvic girdle** consists of a pair of large bones, each composed of three fused hipbones. Whereas the pelvic girdle is securely fused to the vertebral column, the pectoral girdle is loosely and flexibly attached to it by muscles.

Each human limb terminates in five **digits**—the fingers and toes. The more specialized appendages of other vertebrates may be characterized by four digits (as in the pig), three (as in the rhinoceros), two (as in the camel), or one (as in the horse).

Great apes and humans do have a highly specialized feature—the opposable thumb. (Great apes also have an opposable big toe. In humans, although it is not opposable, the big toe is similar enough in structure to the thumb that it can be used as a substitute through surgery.) The opposable thumb can readily be wrapped around objects such as a tree limb in climbing, but it is especially useful in grasping and manipulating objects. It can be opposed to each finger singly, or to all of them collectively. The muscles that move the thumb are almost as powerful as those of all the other fingers put together.

In humans, the upper limbs are not used for locomotion as in other mammals, including the great apes. Our hands have been emancipated. The combination of opposable thumbs and upright posture enables us to use our hands to shape and build, and to effect changes on our environment to a greater extent than any other organism on earth.

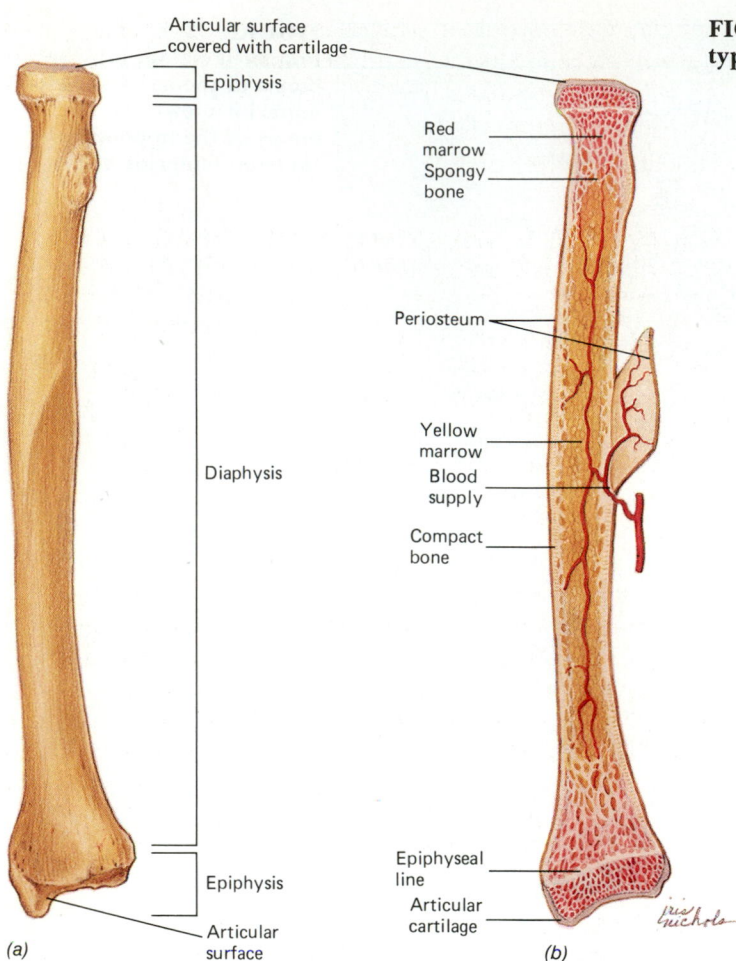

Articular surface covered with cartilage
Epiphysis
Red marrow
Spongy bone
Periosteum
Diaphysis
Yellow marrow
Blood supply
Compact bone
Epiphysis
Epiphyseal line
Articular cartilage
Articular surface

(a)　　(b)

FIGURE 32–9　Anatomy of a bone. (*a*) Structure of a typical long bone. (*b*) Internal structure of a long bone.

A Typical Long Bone Has A Thin Shell Of Compact Bone And A Filling Of Spongy Bone

The radius, one of the two bones of the forearm, is a typical long bone (Fig. 32–9; see also Fig. 31–7). It has numerous muscle attachments. As a result of the way the attachments are arranged, the bone operates as a lever that amplifies the motion the muscles generate.

Like other bones, the radius is covered by a connective tissue membrane, the **periosteum,** which is capable of laying down fresh layers of bone and thus increasing the diameter of the bone. The main shaft of a long bone is known as its **diaphysis.** The expanded ends of the bone are called **epiphyses.** In children, a disc of cartilage, the **metaphysis,** is found between the epiphyses and the diaphysis. The metaphyses are growth centers that disappear at maturity, becoming vague **epiphyseal lines.** Within a long bone, there is a central **marrow cavity** that is filled with a fatty connective tissue known as yellow bone marrow. The marrow cavity is lined with a thin membrane, the **endosteum.**

The radius has a thin outer shell of **compact bone,** which is very dense and hard and is found primarily near the surfaces of a bone, where great strength is an advantage. Recall from Chapter 31 that compact bone consists of interlocking, spindle-shaped units called **osteons,** or **haversian systems.** Within an osteon, osteo-cytes, the mature bone cells, are found in small cavities called **lacunae.** The lacunae are arranged in concentric circles around central **haversian canals.** Blood vessels that nourish the bone tissue pass through the haversian canals. Thread-like extensions of the cytoplasm of the osteocytes extend through narrow channels called **canaliculi.** These cellular extensions connect the osteocytes (see Fig. 31–7).

Interior to the thin shell of compact bone is a filling of **spongy bone,** which despite its loose structure provides most of the mechanical strength of the bone. Spongy bone consists of a meshwork of thin strands of bone. The spaces within the spongy bone are filled with bone marrow. Spongy bone is found within the epiphyses and lining the marrow cavity. At its joint surfaces, the outer layer of a bone consists of articular cartilage that serves as a low-friction bearing for the joints.

Bones Develop In The Embryo From Cartilage Or From Noncartilage Connective Tissue And Are Remodeled Throughout Life In Response To Physical Stresses

During fetal development, bones form in two ways. Long bones, such as the radius, develop from cartilage replicas, a process called **endochondral** bone development. The flat bones of the skull, the irregular vertebrae, and some other bones develop from a noncartilage connective tissue scaffold; this is known as **membranous** bone development.

Osteoblasts are bone-building cells. They secrete the protein collagen, which forms the strong, elastic fibers of bone. A complex calcium phosphate called **apatite** is present in the tissue fluid. This compound automatically crystallizes around the collagen fibers, forming the hard matrix of bone. The mineral component imparts rigidity to bone, whereas the collagen provides flexibility. As the matrix forms around the osteoblasts, they become isolated within the lacunae. The trapped osteoblasts are referred to as osteocytes.

Bones are modeled during growth and remodeled continuously throughout life in response to physical stresses and other changing demands. As muscles develop in response to physical activity, the bones to which they are attached thicken and become stronger. As bones grow, bone tissue is removed from the interior, especially from the walls of the marrow cavity. This process keeps bones from getting too heavy.

Osteoclasts are the cells that break down bone, a process referred to as bone resorption. These bone-breaker cells are very large cells that move about secreting enzymes that digest bone. Osteoclasts and osteoblasts work side by side, in effect shaping bones and forming the precise grain needed in the finished bone. It is believed that most bone is completely made over as many as ten times during the course of an average lifetime.

Joints Hold Bones Together And May Permit Movement

Joints hold bones together and many of them permit flexibility and movement. One way to classify joints is according to the degree of movement they allow. The sutures found between bones of the skull are **immovable joints.** In a suture, the bones are held together by a thin layer of dense fibrous connective tissue. This tissue may be replaced by bone in the adult. **Slightly movable joints** are found between bodies of vertebrae. These joints, which are made of cartilage, help absorb shock.

The most common joints are **freely movable joints.** They are enclosed by a joint capsule composed of connective tissue. The joint capsule is lined with a membrane that secretes a lubricant, **synovial fluid.** Generally, the joint capsule is reinforced by ligaments, bands of fibrous connective tissue that connect the bones and also limit movement at the joint.

Joints wear down with time and use. In osteoarthritis, a common joint disorder, cartilage repair does not keep up with degeneration and the articular carti-

lage wears out. In rheumatoid arthritis, the synovial membrane thickens and becomes inflamed. Synovial fluid accumulates, causing pressure and pain, and the joints become stiff.

■ MUSCLE IS THE CONTRACTILE TISSUE THAT ALLOWS MOVEMENT IN COMPLEX ANIMALS

The muscles serve as motors of the body, generating mechanical forces and motion. Muscles make possible locomotion, manipulation of objects, circulation of blood, the propulsion of food through the digestive tract and many other movements. The three types of muscle—skeletal, smooth, and cardiac—were described in Chapter 31.

All eukaryotic cells contain the contractile protein **actin.** This protein is the major component of microfilaments and is thought to be important in many cell processes such as amoeboid movement, attachment of cells to an object or surface, and contraction of the cleavage furrow in telophase cells. The contractile protein **myosin** is functionally associated with actin in most cells. These contractile proteins are most highly organized in muscle cells, generally referred to as **muscle fibers** because they are elongated.

All Animals Have The Ability To Move

The humblest of animals has the ability to move. The simple cnidarian *Hydra* has only two layers of cells, both consisting of epithelial tissue, but these cells are muscular as well as epithelial; their elongated bases contain contractile strands of actin. In some other cnidarians, distinct muscle fibers are present, and in some instances, these are grouped together in conspicuous bands of muscle. In flatworms, muscle occurs as a specialized tissue that is organized into definite layers.

Bivalve mollusks have both smooth and striated muscle. The smooth muscle, which is capable of slow, sustained contraction, can be used to keep the two shells tightly closed for long periods, even days. The mollusk's striated muscle, which contracts rapidly, shuts the shells quickly when the mollusk is threatened. Arthropod muscles are typically striated like vertebrate skeletal muscle.

A Vertebrate Muscle Consists Of Hundreds Of Muscle Fibers Wrapped In Connective Tissue

In vertebrates, skeletal muscle fibers are organized into bundles that are wrapped by connective tissue. Each of these muscles is an organ. The biceps in your arms, for example, consists of thousands of individual muscle fibers and their connective tissue coverings.

FIGURE 32–10   Skeletal muscle structure. (*a* and *b*) **The structure of a skeletal muscle cell. Notice the eccentric placement of the nuclei, just under the sarcolemma.** (*c*) **Light photomicrograph showing striations (approximately × 200).** (*d*) **Electron micrograph (approximately × 30,000). Note that striations persist at this much higher magnification. Black line indicates 1 micrometer.** *GLY*, glycogen; *MY*, myosin filaments; *ACT*, actin filaments; *M*, mitochondria; *TS*, transverse tubule or T-system; *A, H, I*, and *Z*, the zones and bands in the muscle tissue.

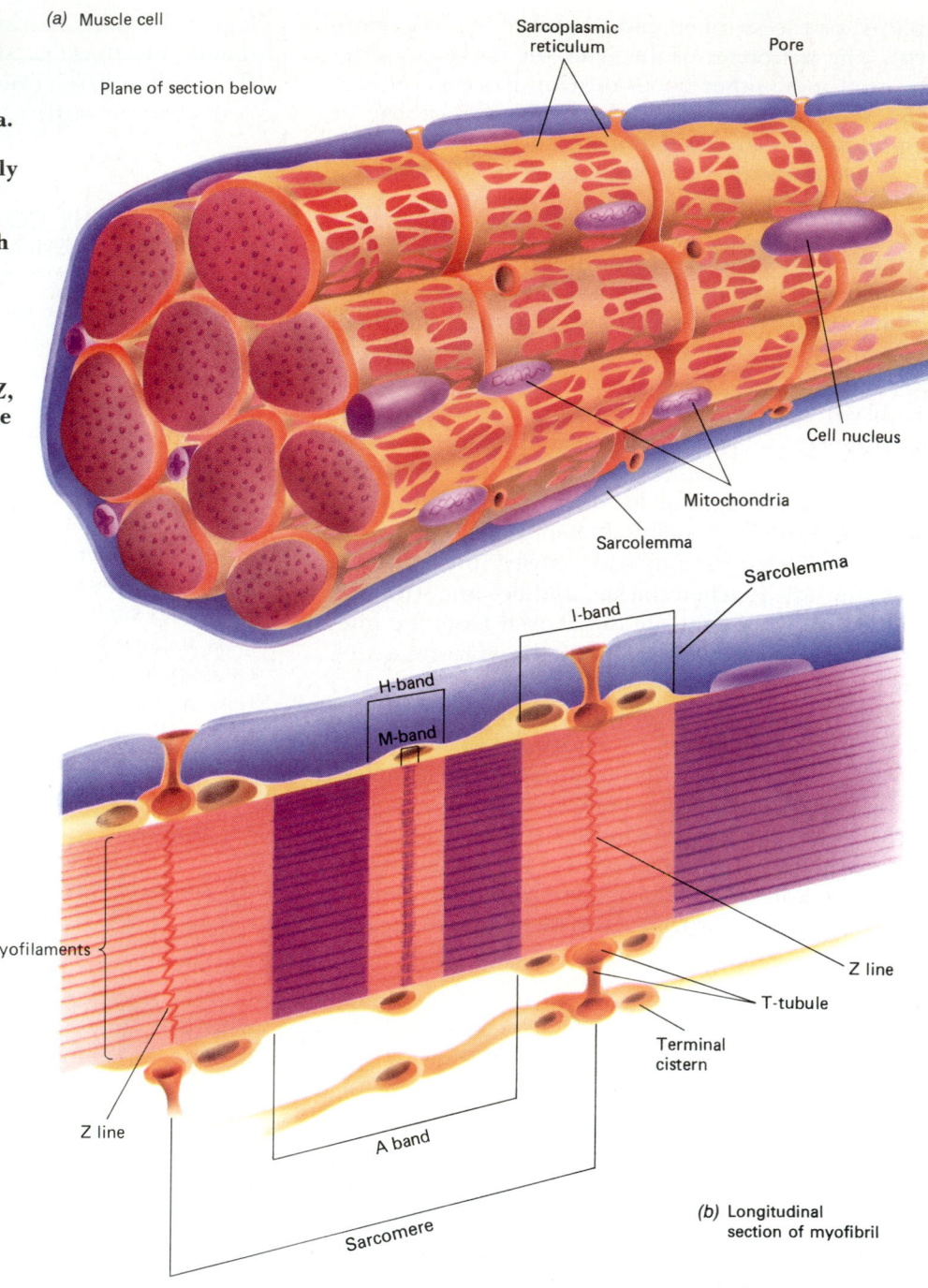

(*a*) Muscle cell

Plane of section below

Sarcoplasmic reticulum

Pore

Cell nucleus

Mitochondria

Sarcolemma

Sarcolemma

I-band

H-band

M-band

Myofilaments

Z line

Z line

T-tubule

Terminal cistern

A band

Sarcomere

(*b*) Longitudinal section of myofibril

Recall that each striated muscle fiber is a spindle-shaped cell with many nuclei (Fig. 32–10). The plasma membrane, known in a muscle cell as the **sarcolemma,** has multiple inward extensions that form a set of **T tubules** (transverse tubules). The cytoplasm of a muscle fiber is referred to as **sarcoplasm,** and the endoplasmic reticulum as **sarcoplasmic reticulum.**

Thread-like structures called **myofibrils** run lengthwise through the muscle fiber. The myofibrils are composed of two types of even tinier structures, the **myofilaments.** The thick myofilaments, called **my-**osin filaments, consist mainly of the protein myosin, whereas the thin **actin filaments** consist of the protein actin. Myosin and actin filaments are both arranged lengthwise in the muscle fibers and overlap. Their overlapping produces a pattern of bands, microscopic or striations, that is characteristic of striated muscle. A **sarcomere** is a functional unit of contraction containing myosin and actin filaments. Sarcomeres are joined at their ends by an interweaving of filaments called the Z line. A single muscle cell has tens to hundreds of sarcomeres arranged end to end.

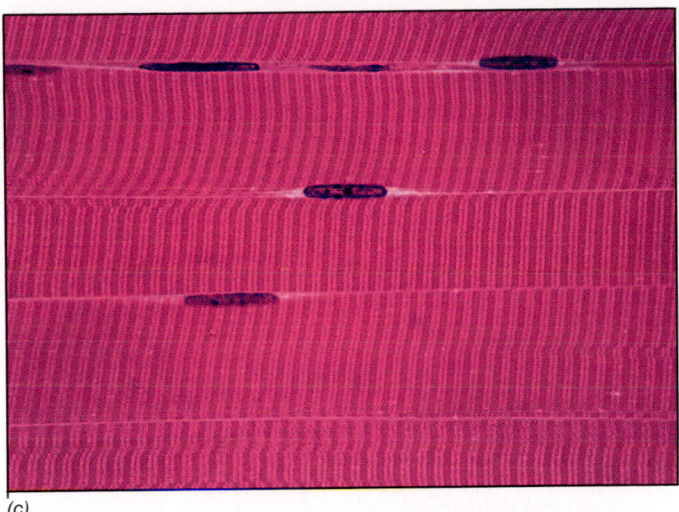

(c)

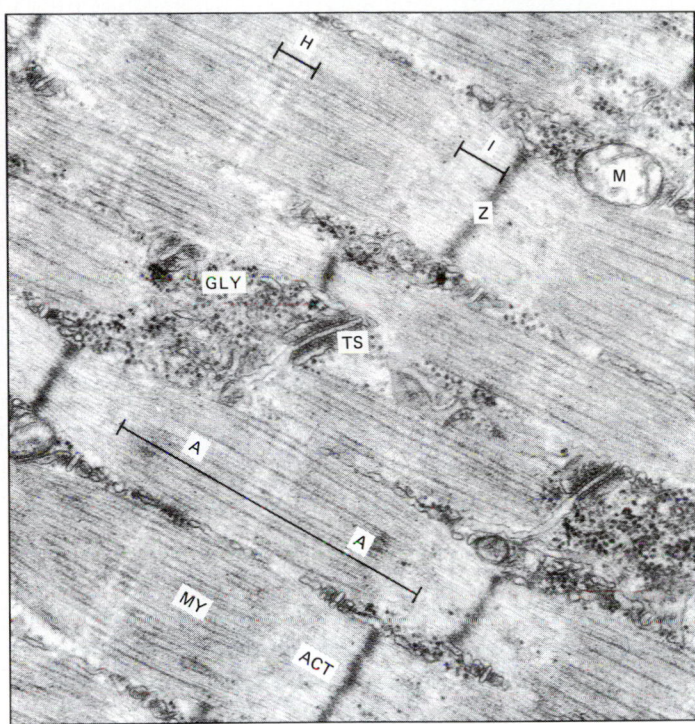

(d)

Muscle Contraction Occurs When Actin And Myosin Filaments Slide Past Each Other

The typical pull of a muscle results from the shortening of its fibers, which in turn results from the actin and myosin filaments actively pulling themselves past and between one another (Fig. 32–11). Each myosin filament consists of about 200 molecules of the protein myosin in a parallel arrangement. A rounded head extends from each rod-shaped myosin molecule. The head of the myosin molecule bears a binding site that is complementary to binding sites on the actin filament. Each actin filament contains 300 to 400 rounded actin molecules that are arranged in two chains. During muscle contraction, the actin filaments are pulled inward between the myosin filaments. As this occurs, the muscle shortens.

Each sarcomere can contract independently. When many sarcomeres contract together, they produce the contraction of the muscle as a whole. We can summarize the process of muscle contraction as follows (Fig. 32–12):

1. When a nerve impulse passes down a motor neuron (a nerve cell that controls muscle) and arrives at the junction (myoneural cleft) between the neuron and muscle, the neuron releases the compound acetylcholine (Fig. 32–13).
2. Acetylcholine diffuses across the myoneural cleft between the neuron and the muscle fiber and combines with receptors on the surface of the muscle fiber. Excess acetylcholine is broken down by the enzyme **cholinesterase.**
3. Acetylcholine causes the sarcolemma to undergo an electrical change called **depolarization,** which we will study in more detail as it occurs in nerve cells. Depolarization may initiate an electrical current that spreads over the sarcolemma. The electrical current generated is known as an **action potential,** or **impulse.**

4. The action potential spreads through the T tubules and stimulates protein channels in the sarcoplasmic reticulum to open, allowing calcium ions to move out of storage and flow into the sarcoplasm.
5. The calcium induces a process that uncovers binding sites on the actin filaments. Myosin heads are now able to attach to them.
6. Myosin combines with ATP, then splits ATP. Energy from the ATP energizes the myosin heads, which then attach to the binding sites on the actin filaments. When attached to the actin filaments, the heads of the myosin molecules are referred to as **cross bridges** because they bridge the gaps between myosin and actin filaments.
7. After the myosin head attaches to the actin filament, it swings 45 degrees (flexes) and releases the ADP. This swinging motion is the power stroke that pulls the actin filament along. The myosin heads pick up another ATP and release their hold on the first set of binding sites. Energized once again, the myosin heads reach for the second set of binding sets. The process is repeated with the third, and so on (Fig. 32–14). This series of stepping motions actively pulls the myosin and actin filaments past one another. Thus, the muscle shortens when the cross bridges attach, move 45 degrees, detach, and then reattach farther along the actin filament.
8. Relaxation of the muscle occurs when calcium is pumped back into the sarcoplasmic reticulum. And all of these events happens in milliseconds.

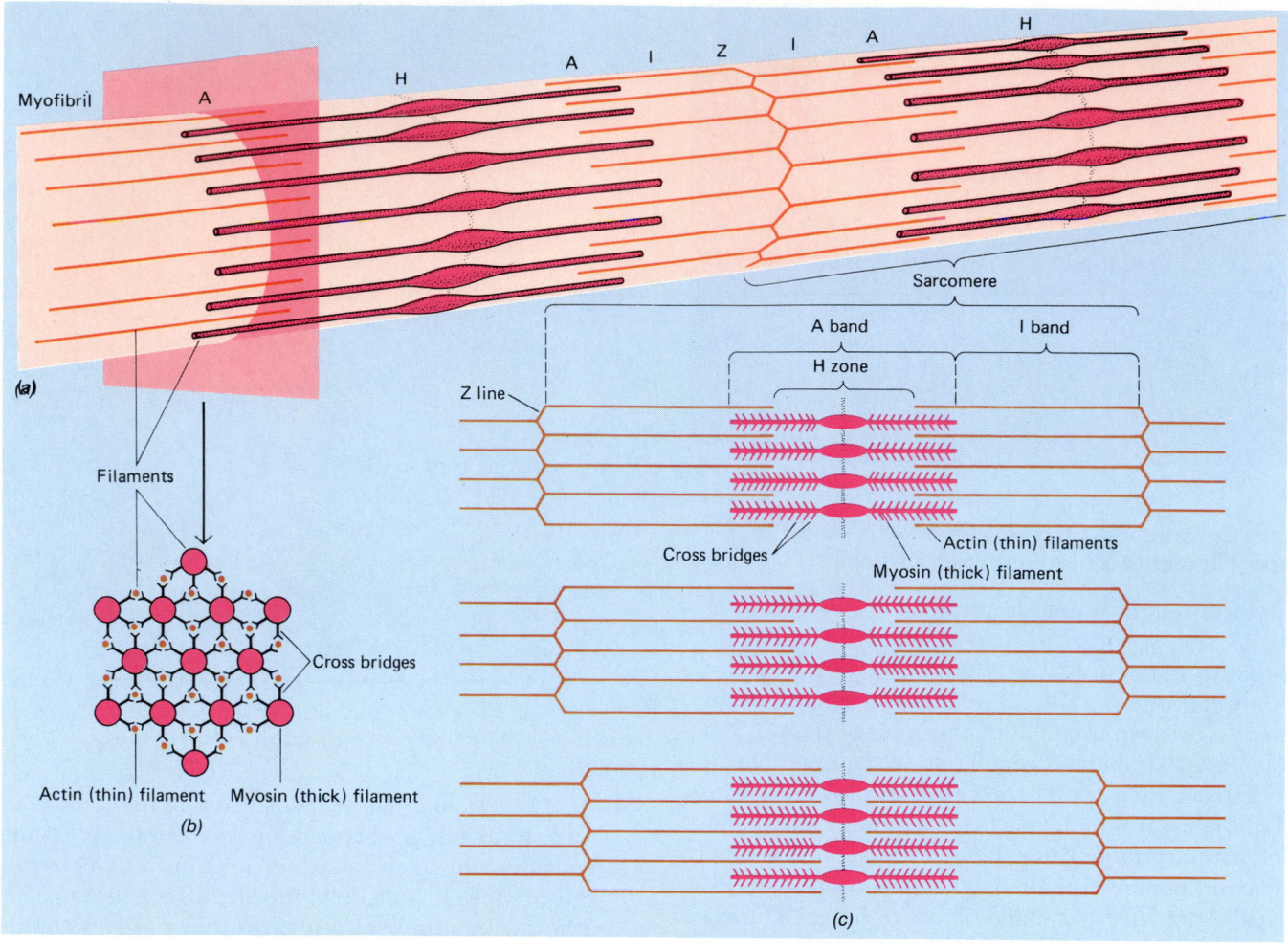

FIGURE 32–11   A myofibril stripped of the accompanying membranes. The Z lines mark the ends of the sarcomeres. (b) Cross section of myofibril shown in (a). (c) Filaments slide past each other during contraction. Notice the way the filaments overlap. It is the regular pattern of overlapping filaments that gives rise to the striated appearance of skeletal and cardiac muscle. In the top drawing of c, the myofibril is relaxed. In the middle drawing, the filaments have slid toward each other, increasing the amount of overlap and shortening the muscle cell by shortening its sarcomeres. At bottom, maximum contraction has occurred; the sarcomere has shortened considerably. Letters represent zones along the myofibril.

Even when we are not moving, our muscles are in a state of partial contraction known as **muscle tone.** Stimulated by messages from nerve cells, some muscle fibers are contracted at any given moment. Muscle tone helps keep muscles prepared for action. When the motor nerve to a muscle is cut, the muscle becomes limp (completely relaxed) and weak.

ATP Powers Muscle Contraction

Muscle cells are often called upon to perform strenuously, and must be provided with large amounts of energy. ATP is the immediate source of energy necessary for muscle contraction. It is necessary both for the pull exerted by the cross bridges and for their release

from each active site, as they engage, hand-over-hand fashion, in their tug of war on the actin filaments. *Rigor mortis*, the temporary but very marked muscular rigidity that occurs after death, results from ATP depletion following the cessation of cellular respiration that occurs at death.[1]

---

[1] Rigor mortis does not persist indefinitely, however, for the entire contractile apparatus of the muscles degenerates eventually, restoring pliability. The phenomenon is temperature-dependent, so given the prevailing temperature, a police officer can estimate the time of death of a cadaver from its degree of rigor mortis. Rigor mortis is not by itself muscular contraction; it only tends to freeze the corpse in its position at the time of death. Thus, tales of corpses sitting, pointing to their murderers, and otherwise carrying on posthumously may be entertaining but have no factual basis.

FIGURE 32–12   **Summary of the events of muscular contraction.**

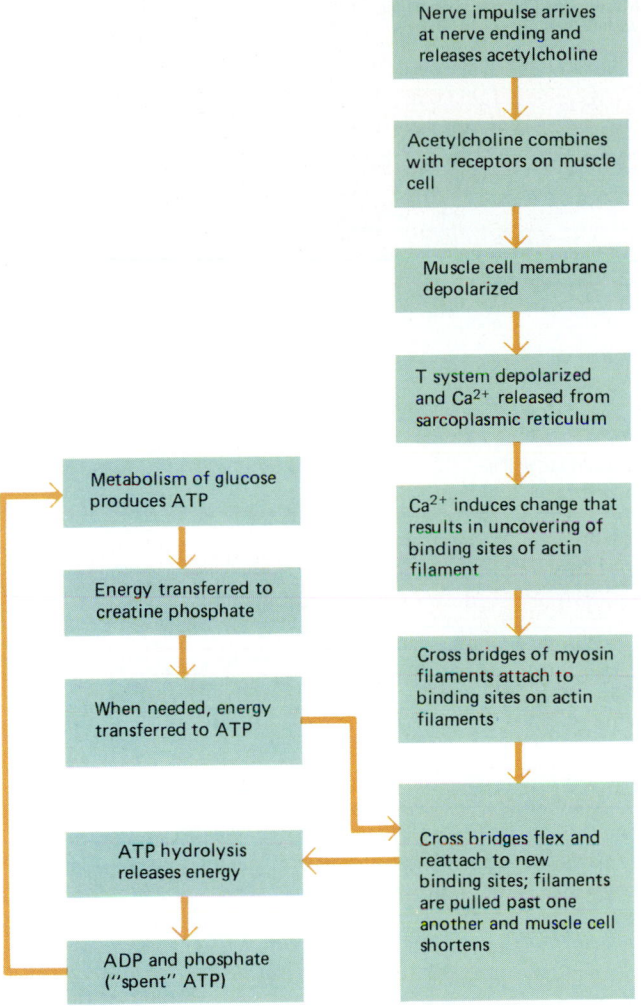

Nerve impulse arrives at nerve ending and releases acetylcholine

Acetylcholine combines with receptors on muscle cell

Muscle cell membrane depolarized

T system depolarized and Ca²⁺ released from sarcoplasmic reticulum

Metabolism of glucose produces ATP

Ca²⁺ induces change that results in uncovering of binding sites of actin filament

Energy transferred to creatine phosphate

When needed, energy transferred to ATP

Cross bridges of myosin filaments attach to binding sites on actin filaments

ATP hydrolysis releases energy

Cross bridges flex and reattach to new binding sites; filaments are pulled past one another and muscle cell shortens

ADP and phosphate ("spent" ATP)

Sufficient energy can be stored in the energy-rich bonds of ATP molecules for less than a second of strenuous activity (perhaps ten contractions). In vertebrate muscle fibers, energy is stockpiled in the energy storage compound **creatine phosphate.** The energy stored in creatine phosphate is transferred to ATP as needed. However, the supply of creatine phosphate does not last very long during vigorous exercise. As ATP and creatine phosphate stores are depleted, muscle cells replenish their supply of these high-energy compounds.

Fuel is stored in muscle fibers in the form of glycogen, a large polysaccharide formed from hundreds of glucose units. Stored glycogen is degraded, yielding glucose, which is then broken down in cellular respiration. When sufficient oxygen is available, enough energy is captured from the glucose to produce needed quantities of ATP.

During strenuous exercise, sufficient oxygen may not be available to meet the needs of the rapidly metabolizing muscle cells. Under these conditions, muscle cells are capable of breaking down fuel molecules anaerobically (without oxygen) for short periods of time. Anaerobic metabolism is a very rapid method of generating ATP, but (as discussed in Chapter 8) this process does not yield very much ATP. The depletion of ATP results in weaker contractions and muscle fatigue. The waste product lactic acid is produced during anaerobic

FIGURE 32–13   **Motor units. (a) A motor unit typically includes many more muscle fibers than appear here, averaging about 150 muscle fibers each. Some units, however, have less than a dozen fibers, whereas others have several hundred. (b) Scanning electron micrograph of some of the cells in a motor unit (× 900). Note how the large neuron branches send subdivisions to each cell in the motor unit.**

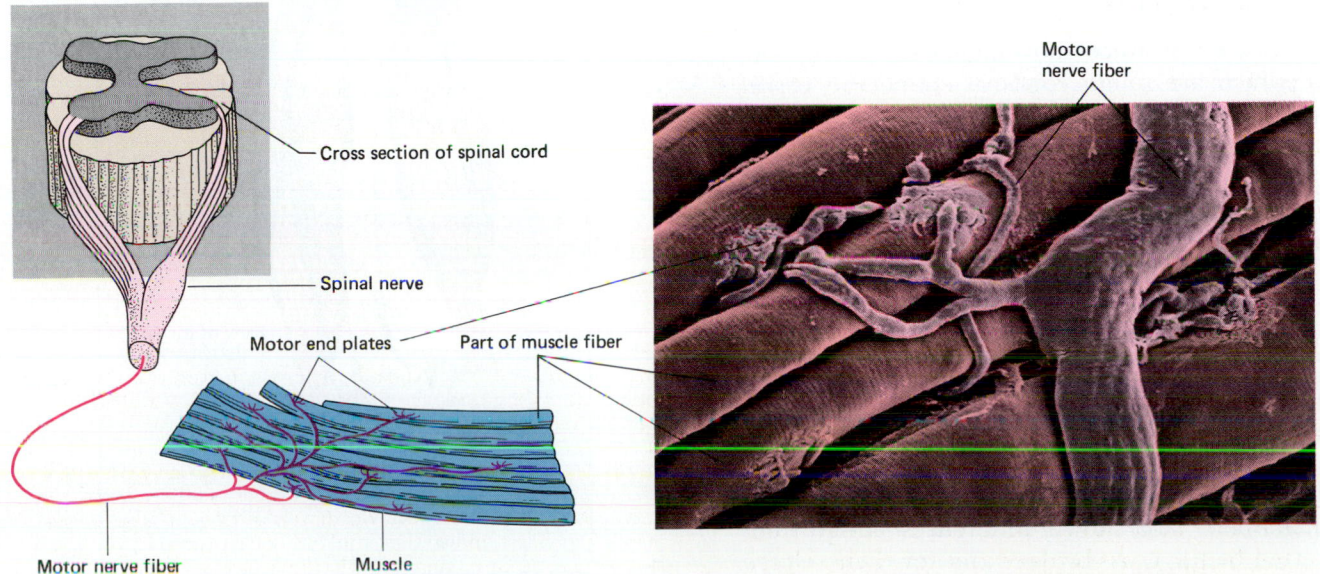

Cross section of spinal cord

Spinal nerve

Motor end plates

Part of muscle fiber

Motor nerve fiber

Muscle

Motor nerve fiber

(a)

(b)

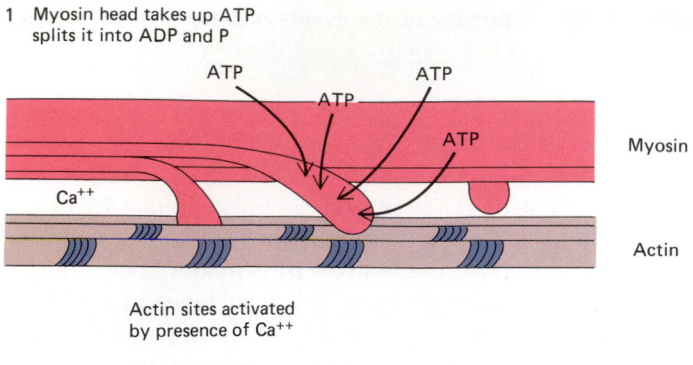

1 Myosin head takes up ATP splits it into ADP and P

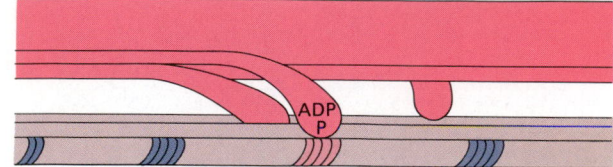

2 Myosin head forms cross-bridge with calcium-activated site on actin

Actin sites activated by presence of Ca++

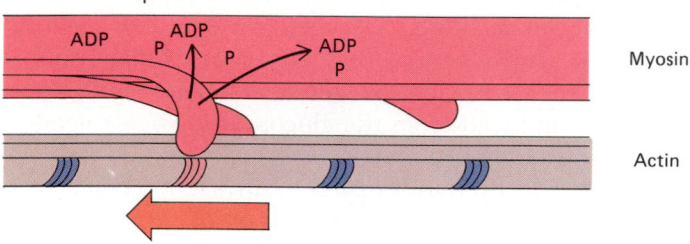

3 Myosin head pulls actin filament, releases its ADP and P

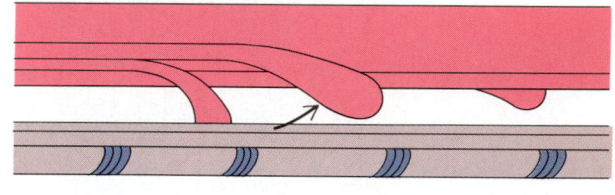

4 Cross-bridge broken

FIGURE 32–14 Simplified diagram of the interaction of actin and myosin in muscle contraction.

FIGURE 32–15 The antagonistic arrangement of the biceps and triceps muscles.

breakdown of glucose. Lactic acid buildup contributes to muscle fatigue. During muscle exertion, an oxygen debt develops that is paid back during the period of rapid breathing that typically follows strenuous exercise.

Skeletal Muscle Action Depends On Muscle Pairs Working Antagonistically

Skeletal muscles produce movements by pulling on tendons, which are tough cords of connective tissue that anchor muscles to bone. Tendons, in turn, pull on bones. Most muscles pass across a joint and are attached to the bones that articulate, or connect, forming the joint. The contracting muscle draws one bone toward or away from the bone with which it articulates.

Muscles can only pull; they cannot push. Muscles act **antagonistically** to one another; the movement produced by one can be reversed by another. The biceps muscle, for example, permits you to flex your arm, whereas the triceps muscle allows you to extend it once again (Fig. 32–15). Thus, the biceps and triceps can work antagonistically to one another. What one does, the other can undo. The superficial muscles of the human body are shown in Figures 32–16 and 32–17.

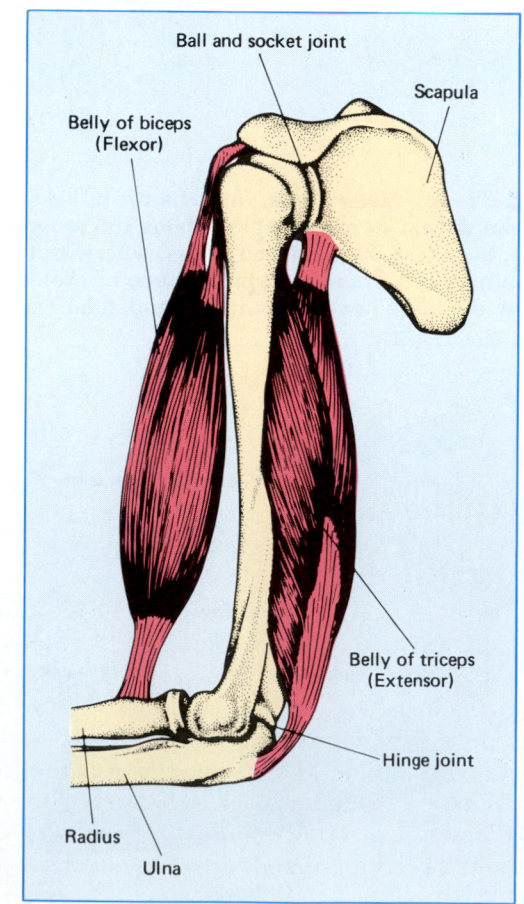

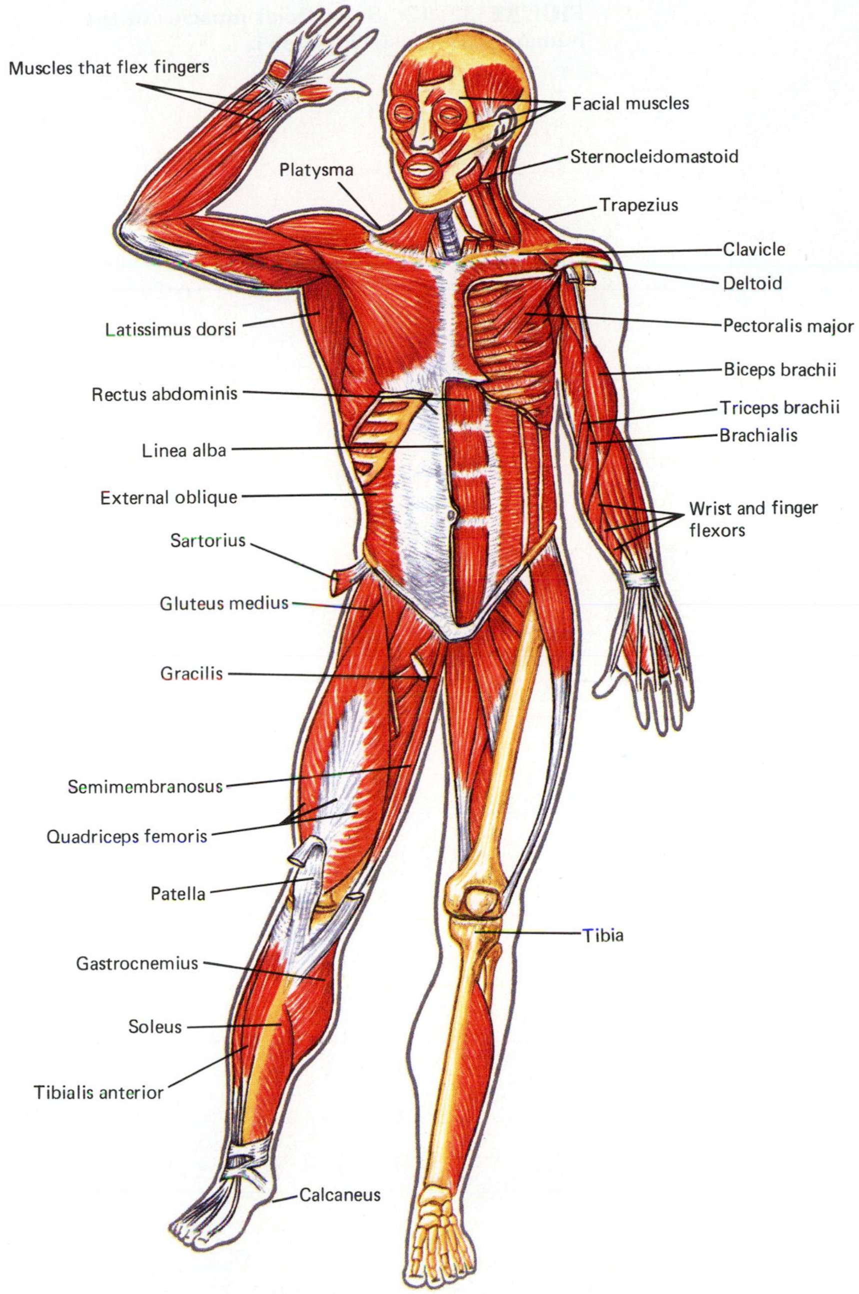

FIGURE 32–16 Superficial muscles of the human body (anterior view).

Muscles that flex fingers
Facial muscles
Platysma
Sternocleidomastoid
Trapezius
Clavicle
Deltoid
Latissimus dorsi
Pectoralis major
Biceps brachii
Rectus abdominis
Triceps brachii
Brachialis
Linea alba
External oblique
Wrist and finger flexors
Sartorius
Gluteus medius
Gracilis
Semimembranosus
Quadriceps femoris
Patella
Tibia
Gastrocnemius
Soleus
Tibialis anterior
Calcaneus

Smooth, Cardiac, And Skeletal Muscle Are Each Specialized For Particular Types Of Responses

The three types of muscle differ in the ways they respond. Smooth muscle often contracts in response to simple stretching, and its contraction tends to be lengthy and sustained. It is well adapted to such functions as the regulation of blood pressure by sustained contraction of the walls of the arterioles. Although smooth muscle contracts slowly, it shortens much more than striated muscle does. Although it is not well suited for running or flying, smooth muscle squeezes superlatively.

Cardiac muscle contracts abruptly and rhythmically, propelling blood with each contraction. Sustained contraction of cardiac muscle would be disastrous! Skeletal muscle, when stimulated by a single brief stimulus, contracts with a quick, single contraction called a **simple twitch.** Ordinarily, simple twitches do not occur except in laboratory experiments. In the normal animal, skeletal muscle receives a series of separate stimuli very close together. These produce not a series of simple twitches, however, but a single, smooth, sustained contraction called **tetanus.** Depending upon the identity and number of muscle cells tetanically contracting, we thread a needle, haul a rope, or swim a lap.

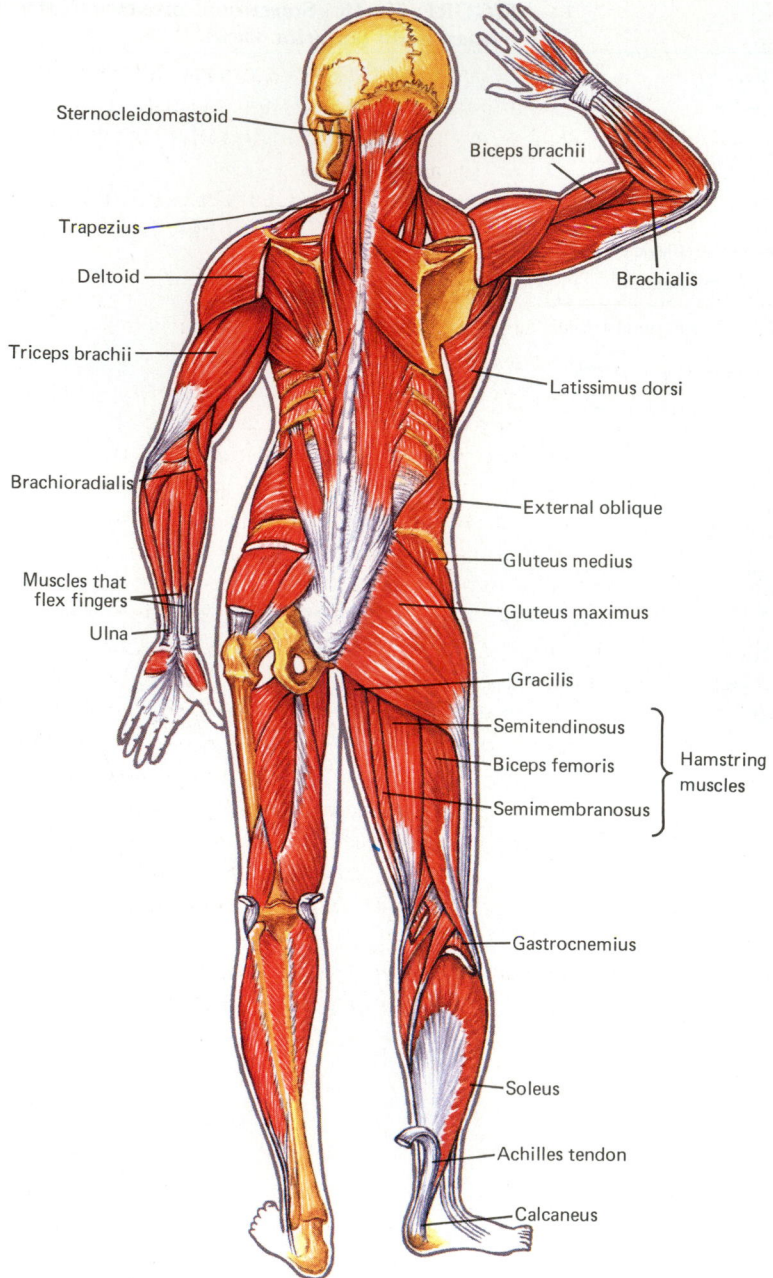

FIGURE 32–17 Superficial muscles of the human body (posterior view).

Sternocleidomastoid

Biceps brachii

Trapezius

Brachialis

Deltoid

Triceps brachii

Latissimus dorsi

Brachioradialis

External oblique

Gluteus medius

Muscles that flex fingers

Gluteus maximus

Ulna

Gracilis

Semitendinosus

Biceps femoris   } Hamstring muscles

Semimembranosus

Gastrocnemius

Soleus

Achilles tendon

Calcaneus

▪ SUMMARY

I. In invertebrates, epithelial tissue may contain secretory cells that produce a protective cuticle, secrete lubricants or adhesives, produce odorous or poisonous secretions, or produce threads for nests or webs. Invertebrate epithelium may be specialized for sensory or respiratory functions.

II. Vertebrate skin protects, may prevent dehydration, may be specialized for secretion or reception of stimuli, and may help regulate body temperature.

   A. In human skin, cells in the stratum basale of the epidermis continuously divide. As they are pushed upward toward the skin surface, these cells mature, produce keratin, and eventually die.

   B. The dermis, which consists of dense, fibrous connective tissue, rests on a layer of subcutaneous tissue that is composed largely of fat.

III. The skeleton transmits mechanical forces that are generated by muscle and also supports and protects the body.

   A. *Hydra* and many other invertebrates have a

hydrostatic skeleton in which fluid transmits forces generated by contractile cells or muscle. In *Hydra,* the circular and longitudinal layers of contractile cells form an antagonistic relationship.

B. Exoskeletons are characteristic of mollusks and arthropods. The arthropod skeleton, composed mainly of chitin, is jointed for flexibility. This nonliving skeleton does not grow; therefore, arthropods molt periodically, shedding their outgrown skeletons and forming new ones.

IV. Endoskeletons, found in echinoderms and chordates, are composed of living tissue and therefore are capable of growth.

A. The vertebrate skeleton consists of an axial portion and an appendicular portion.

B. A typical long bone consists of a thin outer shell of compact bone surrounding the inner spongy bone. Within the long bone is a central marrow cavity.

C. Long bones develop from cartilage replicas; this is endochrondral bone formation. Other bones, such as the flat bones of the skull, develop from a noncartilage connective tissue replica; this is membranous bone development.

V. All animals have the ability to move. Specialized muscle tissue is found in most invertebrate phyla and in all of the vertebrates. As muscle contracts (shortens), it moves body parts by pulling on them.

A. A muscle such as the biceps consists of hundreds of muscle fibers.

B. The striations of skeletal muscle fibers reflect the interdigitations of their actin and myosin filaments. A unit of actin and myosin filaments makes up a sarcomere.

C. During muscle contraction, the actin filaments are pulled inward between the myosin filaments.

1. Muscle contraction begins when a motor neuron releases acetylcholine into the myoneural cleft.

2. The acetylcholine combines with receptors on the surface of the muscle fiber.

3. This results in depolarization of the sarcolemma and initiation of an action potential.

4. The action potential spreads through the T tubules and stimulates calcium release.

5. Calcium initiates a process that uncovers the binding sites of the actin filaments.

6. Myosin heads attach to the binding sites. When attached, they form cross bridges.

7. As the cross bridges flex and reattach to new binding sites, the filaments are pulled past one another and the muscle shortens.

D. ATP is the immediate source of energy for muscle contraction, but muscle tissue has another energy storage compound, creatine phosphate. Glycogen is the fuel stored in muscle fibers.

E. As muscle contracts (shortens), it moves body parts by pulling on them. Muscles act antagonistically to one another.

◼ POST-TEST

1. The vertebrate skin consists of two main layers, the outer _____ and the inner _____.

2. The cells of the stratum _____ of the epidermis are dead and almost waterproof.

3. The protein _____ confers mechanical strength, flexibility, and waterproofing on the skin.

4. _____ skeletons have the principal or even sole function of transmitting muscular force.

5. Arthropods _____ from time to time as they grow, replacing their exoskeletons.

6. The internal skeletons of echinoderms and chordates are known as _____.

7. The radius has a thin outer shell of _____ bone and a spongy filling of _____ bone.

8. Synovial fluid serves as a _____ in _____.

9. The two types of myofilaments in muscle tissue are _____ filaments and _____ filaments.

10. Unscramble this list of the events of muscle contraction and place them in the correct sequence:
a. Calcium release
b. T-system depolarization
c. Acetylcholine release
d. Nerve impulse
e. Uncovering of the binding sites of the actin filaments
f. Cross bridges flex
g. Cross bridges release binding sites

11. Creatine phosphate's function is _____ in the muscle cell.

12. Fuel is stored in muscle cells in the form of the polysaccharide _____; the immediate source of energy for muscle contraction is _____.

■ REVIEW QUESTIONS

1. Compare vertebrate skin with the external epithelum of invertebrates.
2. What properties does keratin confer on human skin?
3. What is a hydrostatic skeleton? Which functions does it perform?
4. How do the septa in the annelid worm contribute to the flexibility of its hydroskeleton?
5. What are the disadvantages of an exoskeleton? Compare the arthropod exoskeleton to the vertebrate endoskeleton.
6. Describe the divisions of the vertebrate skeleton.
7. Draw a typical long bone, such as the radius, and label its parts.
8. Contrast the functions of osteoblasts and osteoclasts. Why is it important that bones be continuously remodeled?
9. Identify three types of joints based on their function.
10. Compare the two types of myofilaments in muscle tissue. What is a sarcomere?
11. Outline the sequence of events that causes a muscle fiber to contract, beginning with the stimulation of its nerve and including cross-bridge action.
12. What is the role of ATP in muscle contraction? What is the function of creatine phosphate? Of glycogen?

■ READINGS

Austin, P.R., et al. "Chitin: New Facets of Research," *Science* 212: 749–753, May 15, 1981. The practical uses of chitin, the most widely distributed animal skeleton carbohydrate, may eventually rival those of cellulose, the major plant skeletal carbohydrate.

Cole, R.P. "Myoglobin Function in Exercising Skeletal Muscle," *Science* 216: 523–525, April 30, 1982. Its function long a mystery, muscle hemoglobin at last yields up some of its secrets. Although the details remain unknown, the substance is shown to be necessary for normal muscular oxygen consumption.

Hadley, N.F. The arthropod cuticle. *Scientific American:* 104–112, July 1986. The author discusses the properties of the arthropod exoskeleton that have contributed significantly to the adaptive success of the arthropods.

Harrington, W.F. *Muscle Contraction.* Burlington, NC, Carolina Biological Supply Company, 1981. Contains hard-to-find and valuable criticisms of the cross-bridge theory of muscle contraction.

Huxley, A. *Reflections on Muscle.* Princeton, Princeton University Press, 1980. The originator of the sliding filament theory of muscular contraction discusses the history and prospects of the scientific understanding of contraction.

Lowenstam, H.A. "Minerals Formed by Organisms," *Science* 211: 1126–1130, March 13, 1981. A fine comparative study of skeletal systems.

Luttgens, K., and K.F. Wells. *Kinesiology: Scientific Basis of Human Motion,* 7th ed. Philadelphia, Saunders College Publishing, 1982. How skeleton, muscles, and nervous system interact to permit and produce the multitude of motions of which the human body is capable.

Morey, E.R. "Spaceflight and Bone Turnover," *Bioscience:* 168–172, 1984. Spaceflight may become practical only when the demineralization of bone that it produces is stopped. This will require an extension of our fundamental knowledge of bone-mineral turnover mechanisms and their control.

Neville, C. *The Biology of the Arthropod Cuticle.* Burlington, NC, Carolina Biological Supply Company. A beautiful short summary.

33

Responsiveness: Neural Control

OUTLINE

I. Neural response includes reception, transmission, integration, and response by muscles or glands

II. The cell types of the nervous system are neurons and glial cells
- A. Glial cells support and protect neurons
- B. A typical neuron consists of a cell body, dendrites, and an axon

III. Neurons convey information by transmitting rapidly moving electrical impulses
- A. The resting potential is the electrical potential difference across the plasma membrane of the neuron
- B. An excitatory stimulus can cause a local depolarization of the membrane
- C. The action potential is a wave of depolarization that moves down the axon
- D. Movement of potassium ions out of the axon repolarizes the membrane
- E. Saltatory conduction occurs in myelinated neurons

IV. Information must be transmitted across synapses
- A. Axons release neurotransmitters that affect postsynaptic neurons
- B. Many neurotransmitters have been identified
- C. Neurons are one-way streets
- D. Myelination and large axon diameter increase rate of impulse transmission

V. Neural integration is the process of averaging all of the incoming information and determining whether an action potential will be generated

VI. The reflex arc is an example of a simple neural pathway

VII. Complex neural pathways are possible because neurons associate in a variety of ways

LEARNING OBJECTIVES

After you have studied this chapter you should be able to:

1. Compare the functions of glial cells and neurons.
2. Draw a neuron. Label each part and give its function.
3. Summarize the process by which an impulse is transmitted along a neuron.
4. Contrast a local potential with an action potential.
5. Summarize the process of synaptic transmission.
6. Describe the factors that determine whether a neuron will transmit an impulse, and summarize their interaction.
7. Draw a reflex pathway consisting of three neurons, label each structure, and indicate the direction of information flow.

Parakeet (*Perruches ondulees*) in flight. (Gérard Lacz/ Peter Arnold, Inc.)

The ability of an organism to survive and to maintain its steady state depends largely upon the effectiveness with which it can respond to changes in its internal or external environment. Changes within the body or in the outside world that can be detected by an organism are called **stimuli.** In simple animals with simple nervous systems, the range and types of responses to stimuli are very limited and somewhat fixed. In complex animals varied and sophisticated responses to stimuli are possible because responsiveness is controlled by two highly specialized systems—the nervous and the endocrine systems.

The nervous system permits very rapid response, whereas the endocrine system responds relatively slowly. As will be discussed in Chapter 41, the endocrine system provides long-lasting chemical regulation. The nervous system receives information, transmits messages, sorts out and interprets incoming data, and then issues appropriate commands so that responses are coordinated and homeostatic.

KEY CONCEPTS

- Neurons are specialized to receive information and to produce and transmit rapid electrical signals; these signals convey information to other neurons or trigger responses in muscle or glands.

- In most animals information flow through the nervous system requires reception, transmission of information to a control organ such as brain or spinal cord, integration of the information, transmission of the "decision," and response by a muscle or gland.

NEURAL RESPONSE INCLUDES RECEPTION, TRANSMISSION, INTEGRATION, AND RESPONSE BY MUSCLES OR GLANDS

Thousands of stimuli bombard an organism each day. Appropriate response to a stimulus involves four processes: *reception, transmission, integration,* and *response* by muscles or glands (Fig. 33–1). **Reception** is the process of detecting a stimulus; it is the job of specialized sense organs as well as of neurons themselves. **Transmission** is the process of sending messages along neurons, from one neuron to another, or from a neuron to a muscle or gland. **Integration** is the process of sorting and interpreting incoming information and determining the appropriate mode of response. In vertebrates, integration is primarily the function of the **central ner-**

vous system (CNS)—the brain and spinal cord. The actual **response** is carried out by **effectors,** the muscles and glands.

THE CELL TYPES OF THE NERVOUS SYSTEM ARE NEURONS AND GLIAL CELLS

The key unit of the nervous system in complex animals is the nerve cell, or **neuron,** which is specialized to receive and send information. The neuron works by producing and transmitting rapid electrical signals called **nerve impulses.** A second cell type unique to the nervous system is the **glial cell.** These cells provide support for the neurons.

Glial Cells Support And Protect Neurons

Some glial cells envelop neurons and form insulating sheaths about them. Others are phagocytic and serve to remove debris from the nervous tissue. A third type of glial cell lines the cavities of the brain and spinal cord. **Schwann cells,** supporting cells found outside the central nervous system, form sheaths about some neurons. Sometimes glial cells are referred to collectively as the **neuroglia,** which literally means "nerve glue."

A Typical Neuron Consists Of A Cell Body, Dendrites, And An Axon

Highly specialized to receive and transmit messages in the form of electrical impulses, the neuron is distinguished from all other cells by its long cytoplasmic processes. We will examine the structure of a common type of neuron, the **multipolar neuron** (Fig. 33–2).

The largest portion of the neuron, the **cell body,** contains the bulk of the cytoplasm, the nucleus, and most of the other organelles. Two types of cytoplasmic extensions project from the cell body. These are the dendrites and the long, single axon.

Dendrites are typically short, highly branched fibers specialized to receive nerve impulses and send them to the cell body. The cell body integrates incoming signals and can also receive impulses directly.

Although microscopic in diameter, an **axon** may be 3 feet or more in length. The axon conducts nerve impulses from the cell body to another neuron or to a muscle or gland. At its end the axon branches, forming **axon terminals** that end in tiny structures called **synaptic knobs.** These structures release neurotransmitters, chemicals that transmit signals from one neuron

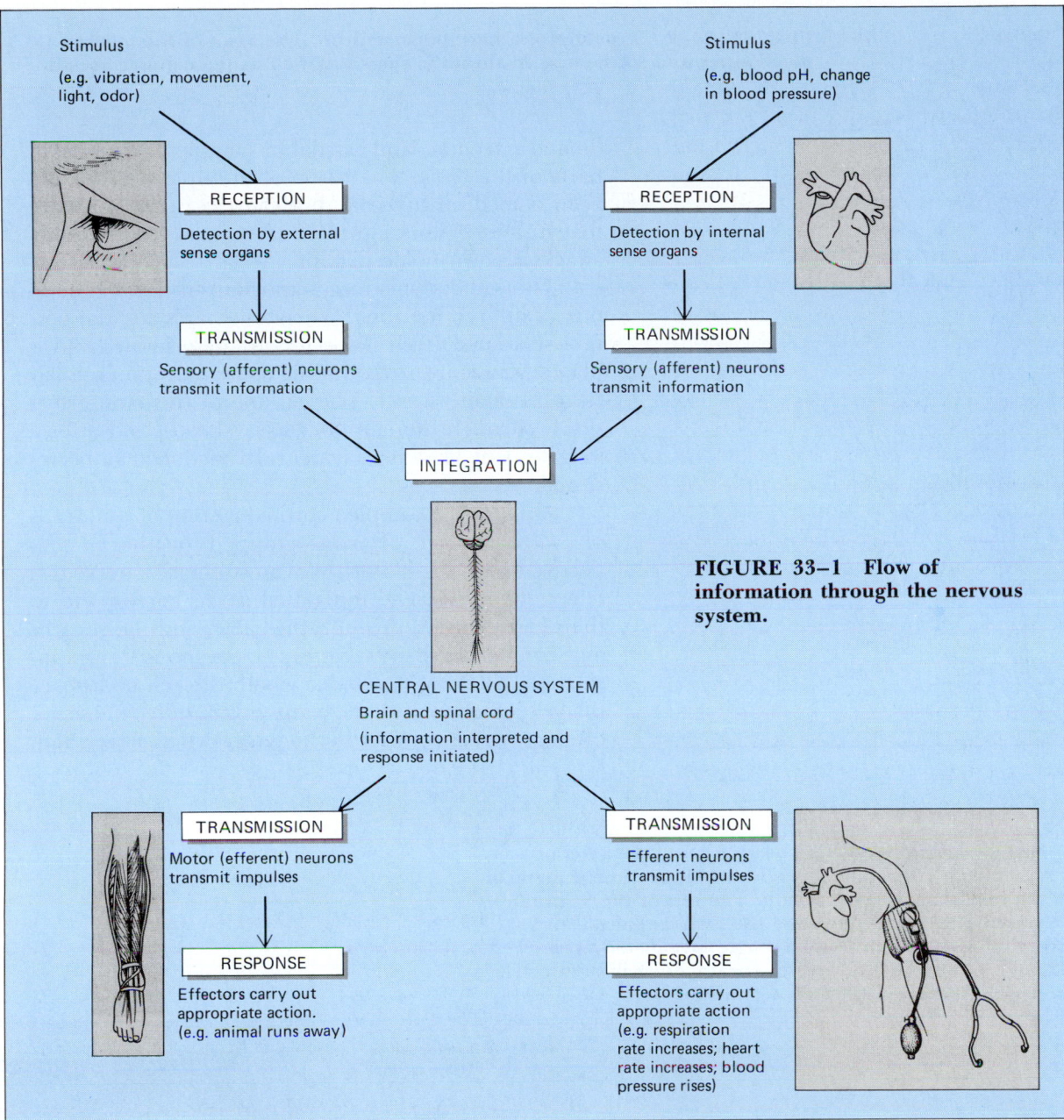

Stimulus

(e.g. vibration, movement, light, odor)

RECEPTION

Detection by external sense organs

TRANSMISSION

Sensory (afferent) neurons transmit information

Stimulus

(e.g. blood pH, change in blood pressure)

RECEPTION

Detection by internal sense organs

TRANSMISSION

Sensory (afferent) neurons transmit information

INTEGRATION

FIGURE 33–1   Flow of information through the nervous system.

CENTRAL NERVOUS SYSTEM

Brain and spinal cord

(information interpreted and response initiated)

TRANSMISSION

Motor (efferent) neurons transmit impulses

RESPONSE

Effectors carry out appropriate action. (e.g. animal runs away)

TRANSMISSION

Efferent neurons transmit impulses

RESPONSE

Effectors carry out appropriate action (e.g. respiration rate increases; heart rate increases; blood pressure rises)

to another. Along its course an axon can give off branches.

Axons of many neurons outside the central nervous system have two coverings—an outer **cellular sheath,** or *neurilemma,* and an inner **myelin sheath** (Fig. 33–3). Both sheaths are formed by Schwann cells. The cellular sheath is formed by Schwann cells that line up along the axon. The myelin sheath, which lies between the axon and the cellular sheath, is formed when the Schwann cell winds its plasma membrane about the axon several times.

The plasma membrane of the Schwann cell contains **myelin,** a white, lipid-rich substance. Myelin is an excellent insulator, and its presence influences transmission of nerve impulses. Between successive

Schwann cells, gaps called **nodes of Ranvier** occur in the myelin sheath. At these points the axon is not insulated with myelin.

Almost all axons more than 2 micrometers in diameter are myelinated, that is, they have myelin sheaths. Those of smaller diameter are generally unmyelinated. In the brain and spinal cord, myelin sheaths are formed by certain glial cells, but cellular sheaths are not present.

In **multiple sclerosis,** a neurological disease that affects about 300,000 people in the United States alone, patches of myelin deteriorate at irregular intervals along the length of the neurons and are replaced by scar tissue. This damage interferes with conduction of neural impulses, and the victim suffers loss of coor-

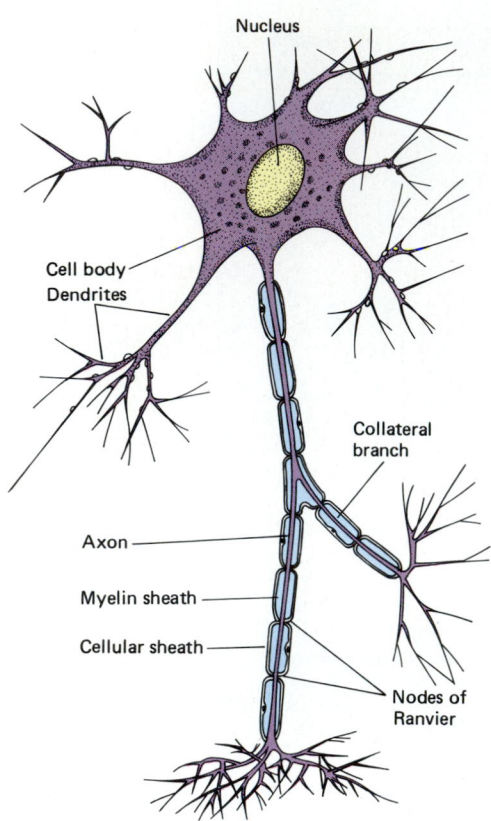

FIGURE 33–2   Structure of a multipolar neuron. The axon of this neuron is myelinated, and so the myelin sheath is shown as well as the cellular sheath.

dination, tremor, and partial or complete paralysis of parts of the body. The cause of multiple sclerosis has been a medical mystery, but there is some evidence that it is an autoimmune disease, in which the body attacks its own tissue (Chapter 37).

The cellular sheath is important in the regeneration of injured neurons. When an axon is cut, the portion separated from the cell body deteriorates and is phagocytized by surrounding cells, but the cellular sheath remains intact. The cut end of the axon grows slowly through the empty cellular sheath, and after a long time at least partial neural function may be restored.

A **nerve** is a complex cord consisting of hundreds or even thousands of axons wrapped together in connective tissue (Fig. 33–4). We can compare a nerve to a telephone cable. The individual axons correspond to the wires that run through the cable, and the sheaths and connective-tissue coverings correspond to the insulation. You might wonder where the cell bodies are that are attached to all the axons in a bundle. These are often grouped together in a mass known as a **ganglion.**

FIGURE 33–3   (a) Formation of the myelin sheath around the axon of a peripheral neuron. A Schwann cell wraps its cell membrane around the axon many times to form the insulating myelin sheath. The rest of the Schwann cell remains outside the myelin sheath, forming the cellular sheath. (b) Electron micrograph of a section through a single myelinated axon (approximately ×42,500). *AX*, axon; *MS*, myelin sheath; *SC*, Schwann cell. (b, Mary B. Bunge, Washington University School of Medicine/BPS)

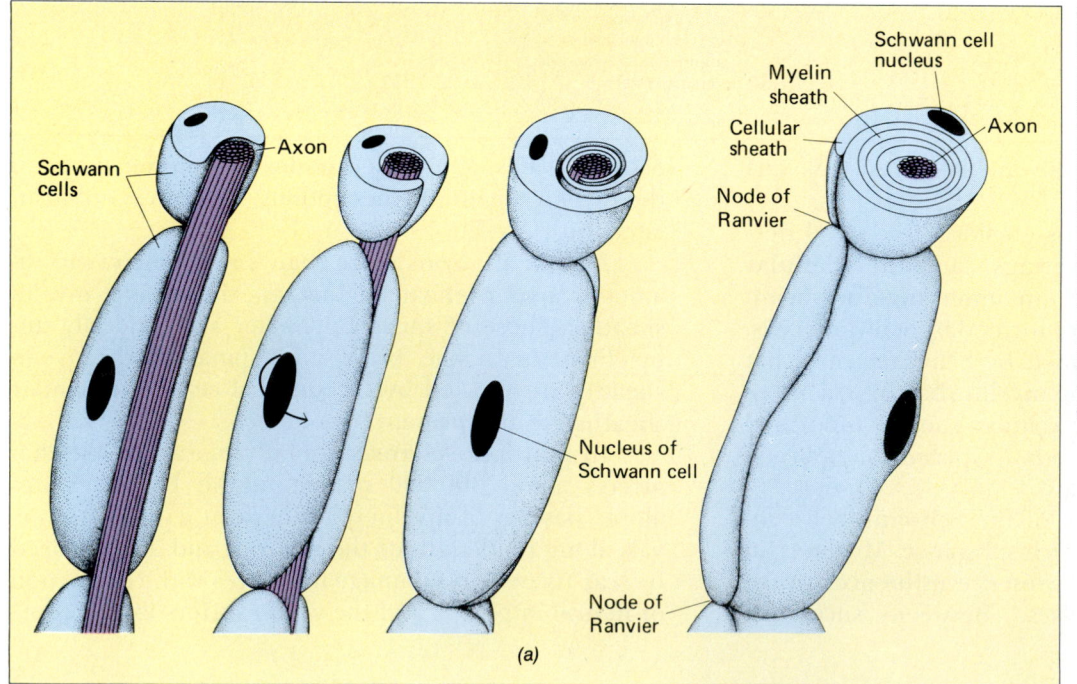

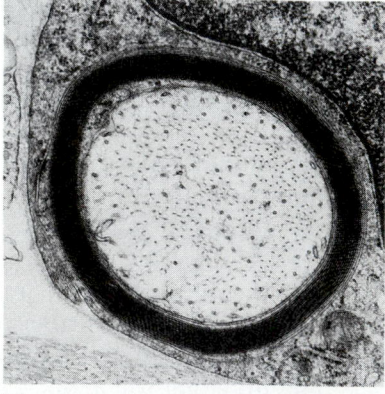

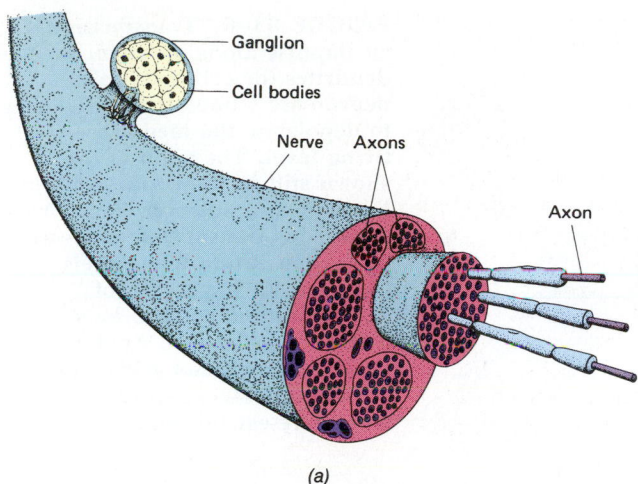

(a)

FIGURE 33–4 (*a*) Structure of a nerve and a ganglion. A nerve consists of bundles of axons held together by connective tissue. The cell bodies belonging to these axons are grouped together in a ganglion. (*b*) The optic nerve of a developing human embryo, shown in cross section ×25,000). In the center of this electron micrograph is a large glia cell, whose nucleus is very prominent. It is surrounded by many axons of the optic nerve, visible as a mass of small circles. (Courtesy of Dr. R.R. Sturrock, University of Dundee, Scotland, *Journal of Anatomy* 139:1984)

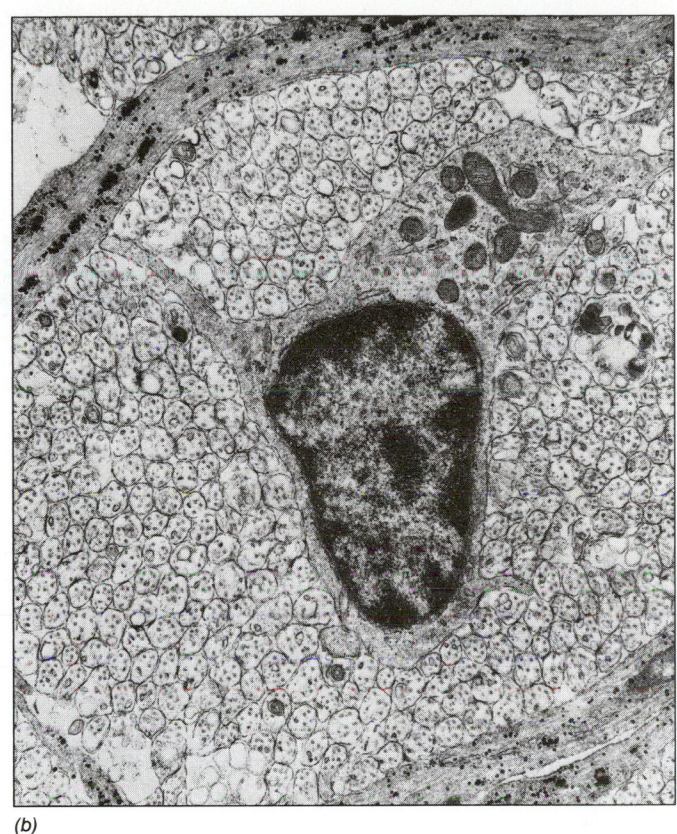

(b)

NEURONS CONVEY INFORMATION BY TRANSMITTING RAPIDLY MOVING ELECTRICAL IMPULSES

When a neuron receives a stimulus that is strong enough, its axon fires a **nerve impulse.** This is an electrical current that travels rapidly down the axon into the synaptic knobs. This electrical impulse, once begun, is self-propagating. For a nerve impulse to begin, however, the plasma membrane has to maintain what is called a resting potential.

The Resting Potential Is The Electrical Potential Difference Across The Plasma Membrane Of The Neuron

In a resting neuron—that is, an unstimulated one that is not transmitting an impulse—the inner surface of the plasma membrane has a negative charge compared with the tissue fluid surrounding it (Fig. 33–5). The plasma membrane is said to be **polarized** (i.e., one side, or pole, has a different charge than the other side). When electric charges are separated in this way, there

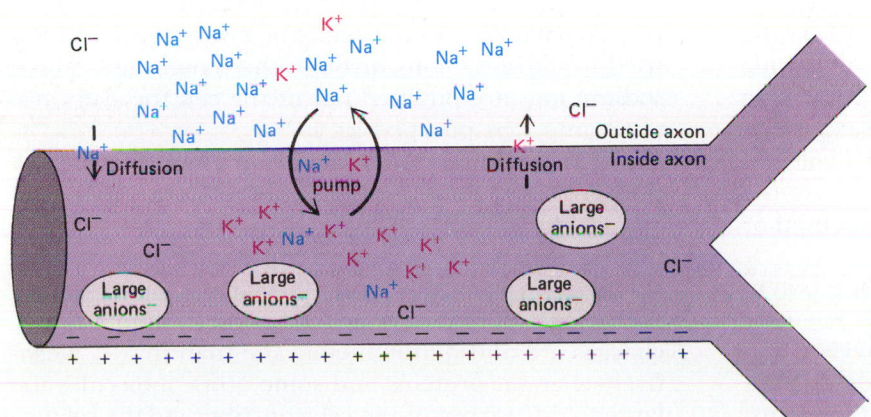

FIGURE 33–5 Segment of an axon of a resting (nonconducting) neuron. Sodium is actively pumped out of the cell and potassium is pumped in. Sodium is unable to diffuse back to any extent, but potassium does diffuse out along its concentration gradient. Because of the unequal distribution of ions, the inside of the axon is negatively charged compared with the outer tissue fluid. The presence of negatively charged proteins and other large anions in the cell contributes to this polarity.

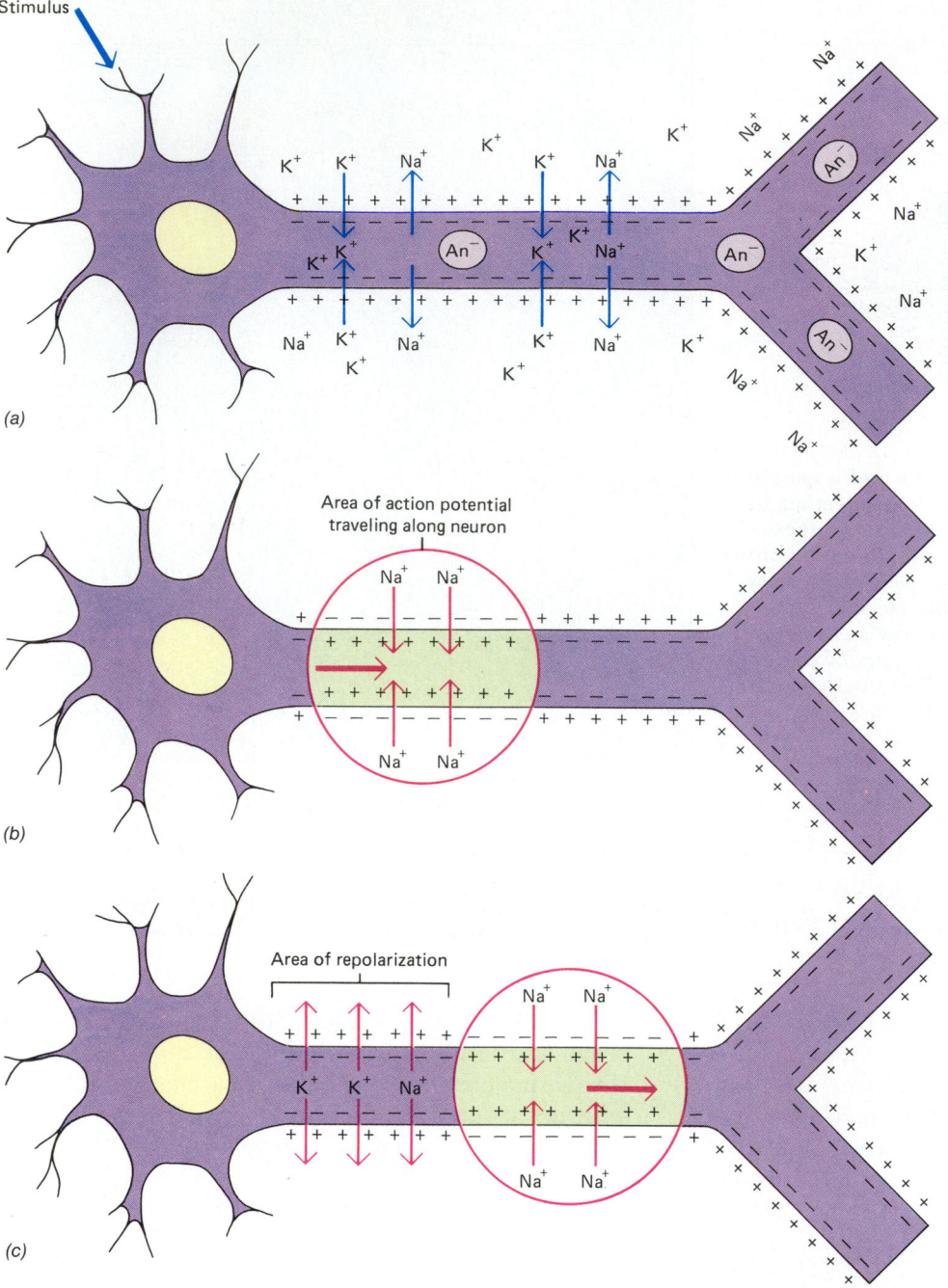

FIGURE 33–6 Transmission of an impulse along an axon. (a) The dendrites (or cell body) of a neuron are stimulated sufficiently to depolarize the membrane to firing level. The axon in (a) is shown still in the resting state and has a resting potential. (b, c) An impulse is transmitted as a wave of depolarization that travels down the axon. At the region of depolarization, sodium ions diffuse into the cell. As the impulse passes along from one region to another, polarity is quickly re-established.

is an electrical **potential energy difference** across the membrane. Should the charges be permitted to come together, they have the *potential* of doing work. The amount of work they could perform may be expressed in millivolts (a millivolt is one-thousandth of a volt). In resting neurons, the potential difference across the plasma membrane is called the **membrane potential** or **resting potential.**

How does the membrane reach its resting potential? The plasma membrane of the neuron has very efficient **sodium-potassium pumps** that actively transport sodium ions (Na$^+$), which each have a single positive charge, out of the cell. At the same time they transport potassium ions (K$^+$), which also each have a single

positive charge, into the cell. The energy from ATP is used to move the ions through the membrane. Three sodium ions are pumped out of the cell for every two potassium ions pumped in.

Both sodium ions and potassium ions diffuse back through the membrane along their concentration gradients. However, the membrane is more permeable to potassium than to sodium. Large amounts of potassium ions leak back out, but only small amounts of sodium ions leak back in. As a result, more positive charges are outside the membrane than inside. Negatively charged proteins and some other molecules too large to diffuse out of the cell contribute to the relative negative charge along the inside of the plasma membrane. At a certain point the positive charge outside the membrane becomes so high that it prevents any more potassium ions from leaking out. At this point the neuron has reached its resting potential. The resting potential of a neuron amounts to about 70 millivolts, written as −70 millivolts (because the inner surface of the plasma membrane is negatively charged compared to the tissue fluid).

An Excitatory Stimulus Can Cause A Local Depolarization Of The Membrane

Any stimulus (electrical, chemical, or mechanical) that makes the neuron more permeable to sodium or potassium at a certain place may change the resting potential. Such local disturbances in the membrane are called **local potentials,** or **graded potentials,** because they vary in magnitude depending upon the magnitude of the stimulus.

The plasma membrane contains sodium and potassium channels, formed by proteins that are embedded in the membrane. **Excitatory stimuli** open these sodium channels, permitting sodium ions to rush into the cell. The passage of sodium ions into the cell briefly causes the membrane potential to become less negative; this is known as **depolarization.**

Depolarization can cause a flow of electric current. The greater the change in potential, the greater is the flow of current. Such a local current flow fades out after flowing only a few millimeters. As we will see, however, local potentials can add up, resulting in transmission of an impulse.

The Action Potential Is A Wave Of Depolarization That Moves Down The Axon

The membrane of a neuron can depolarize as much as 15 mV (which changes the resting potential to about −55 mV), and still no nerve impulse will be transmitted. When the depolarization reaches about −55 mV, the membrane reaches a critical point called the **threshold level,** or firing level. At this point the resulting depolarization is self-propagating; that is, it spreads down the axon as a **wave of depolarization** without fading. The electrochemical gradient established by this spreading wave of depolarization is called a **nerve impulse** or **action potential** (Fig. 33–6).

When threshold level is reached, an almost explosive action occurs as the action potential is produced. The neuron membrane quickly reaches zero potential and even overshoots to about +35 mV so that there is a momentary reversal in polarity. That is, the inside of the neuron has a positive charge compared to the surrounding fluid. The sharp rise and fall of the action potential is called a **spike.** Figure 33–7 illustrates an action potential.

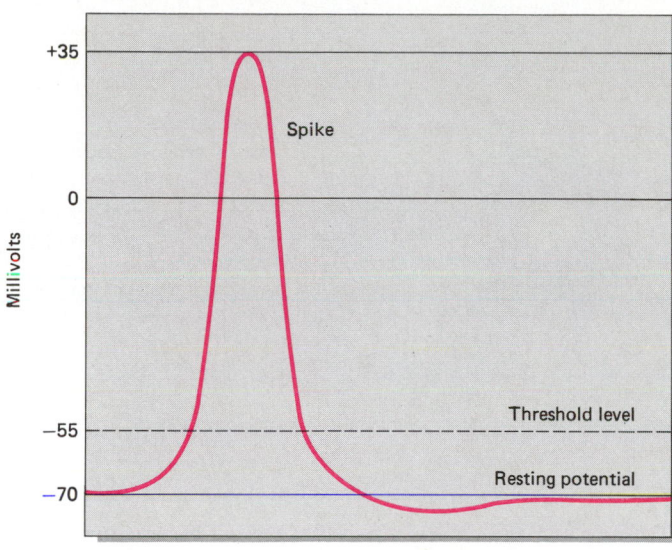

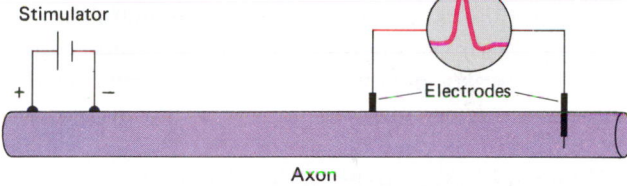

FIGURE 33–7 An action potential recorded with one electrode inside the cell and one just outside the plasma membrane. When the axon depolarizes to about −55 millivolts, an action potential is generated. (The numerical values are included as representative examples. These values may vary for different nerve cells.)

The action potential is an electric current strong enough to cause the resting potential to collapse in the adjacent area of the membrane. The impulse moves along the axon at a constant velocity and amplitude for each type of neuron. Conduction of a neural impulse is somewhat analogous to setting up a path of gunpowder and lighting one end with a match. By igniting the powder particles ahead of it, the flame moves steadily from one end of the trail of gunpowder to the other. Here the analogy breaks down: There is no way of restoring the fuse to its original condition after it has burned, but the nerve cell does restore itself.

The neuron is said to obey an **all-or-none law** since there are no variations in intensity of the action potential. Either the neuron fires completely, or it does not fire at all.

Movement Of Potassium Ions Out Of The Axon Repolarizes The Membrane

As the wave of depolarization moves along the axon, the normal polarized state is quickly reestablished behind it. By the time the action potential moves a few millimeters along the axon, the membrane over which

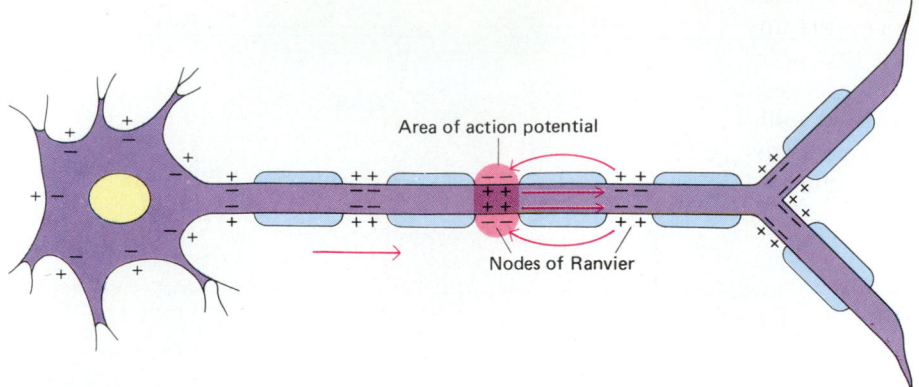

Area of action potential

Nodes of Ranvier

it has just passed begins to *repolarize*. The sodium channels close, so that no sodium can enter the cell. Once sodium channels have closed after the action potential has passed, they cannot reopen while the membrane is still depolarized. During this time the neuron is said to be in a **refractory period.** It cannot transmit another action potential, because sodium channels cannot open to let sodium into the cell.

When the action potential passes, however, as sodium channels close, potassium channels open. Potassium ions can then leave the cell. The resulting reduction of positive ions inside the cell results in **repolarization.** That is, the inside of the cell becomes more negative than the outside. However, the cell interior is still not as negatively charged as it is at the resting potential. The return to the resting potential requires more time, because it requires that the sodium-potassium pumps begin actively transporting excess sodium out of the neuron. Still, despite the limits imposed by their refractory periods, neurons can transmit several hundred impulses per second!

Saltatory Conduction Occurs In Myelinated Neurons

The smooth, progressive impulse transmission just described occurs in neurons that lack myelin sheaths. In myelinated neurons, the myelin insulates the axon except at the nodes of Ranvier. The plasma membrane makes direct contact with the interstitial fluid only at the nodes. As a result the action potential skips along the axon from one node of Ranvier to the next (Fig. 33–8). The ion activity at the node depolarizes the next node down the axon. This type of transmission is known as **saltatory conduction** (from a Latin word meaning to jump).

Saltatory conduction is more rapid and requires less energy than continuous transmission. Only the nodes depolarize, so fewer sodium and potassium ions

are displaced, and the neuron does not have to work as hard to reestablish resting conditions each time an impulse is transmitted.

■ INFORMATION MUST BE TRANSMITTED ACROSS SYNAPSES

When an action potential is transmitted down an axon, it eventually reaches the end of the axon. Here, a small gap, the **synapse,** separates the axon from another neuron or from muscle fibers or gland cells. A neuron that transmits an impulse to a synapse is a **presynaptic neuron,** whereas a neuron that transmits an impulse away from the synapse is a **postsynaptic neuron.** The same neuron may be postsynaptic with respect to one synapse and presynaptic relative to another.

Two types of synapses are **electrical synapses** and **chemical synapses.** In an electrical synapse, the action potential can be transmitted directly to another cell by means of cell-to-cell connections called **gap junctions.** Electrical synapses allow very rapid transmission of impulses and so are especially important in rapid responses such as escape responses. Because chemical synapses are more common, we will focus on them.

In chemical synapses the synaptic cleft, the gap between the two cells, is more than 20 nm wide (still less than a millionth of an inch). Since depolarization is a property of the neuron membrane, when the impulse reaches the end of the axon, it does not jump the gap. An entirely different mechanism conducts the message across the synaptic cleft to the next neuron in the sequence. Chemical compounds called **neurotransmitters** conduct the neural message across the synapse. These chemical messengers diffuse across the synapse from synaptic knobs at the end of the axon. They bind to specific receptors on the plasma membranes of the postsynaptic neurons. Let us look at this process in more detail.

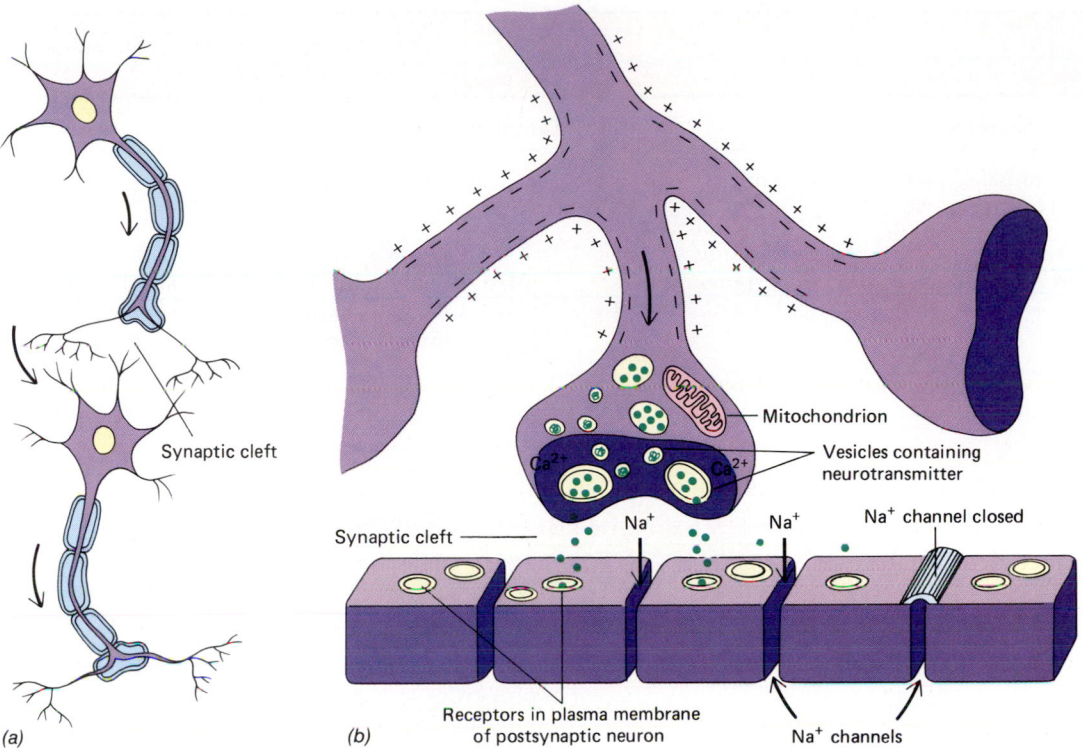

FIGURE 33–9 Transmission of an impulse between neurons or from a neuron to an effector. (*a*) **In most synapses the wave of depolarization is unable to jump across the synaptic cleft between the two neurons.** (*b*) **The problem is solved by the release of neurotransmitter from vesicles within the synaptic knobs of the axon. The neurotransmitter diffuses across the synaptic cleft and may combine with receptors in the membrane of the postsynaptic neuron. This may trigger an impulse in the postsynaptic neuron. It is thought that when neurotransmitter combines with the postsynaptic receptors, hypothetical sodium gates open, permitting sodium to rush into the axon.** (*c*) **Electron micrograph of a synaptic knob and cleft (approximately ×125,000). Two synaptic vesicles (*circled area*) are merging with the cell membrane of the knob and discharging neurotransmitter into the cleft.** *SV*, **synaptic vesicles;** *SC*, **synaptic cleft; neuron above cleft, muscle below.** (Courtesy of Dr. John Heuser)

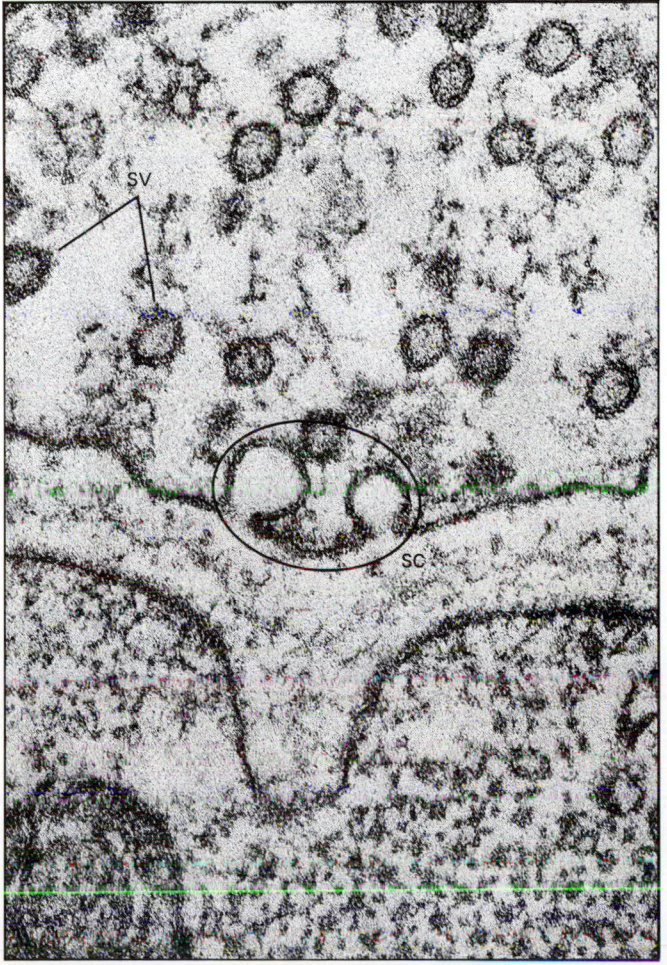

Axons Release Neurotransmitters That Affect Postsynaptic Neurons

Neurotransmitters are continuously synthesized in the synaptic knobs. After being formed, they are stored in synaptic vesicles (little sacs) in the cytoplasm (Fig. 33–9). Each synaptic vesicle contains only one type of neurotransmitter, though the same neuron can produce two or more different neurotransmitters.

Each time an action potential travels down an axon to the synaptic knob, calcium channels in the membrane open, permitting calcium ions to pass into the axon terminal. The calcium ions induce several hundred vesicles to fuse with the membrane and release their neurotransmitters into the synaptic cleft.

TABLE 33–1
Some Neurotransmitters

Substance	Where Secreted	Comments
Acetylcholine	Nerve-muscle junctions; autonomic system[1]; parts of brain	Inactivated by cholinesterase
Norepinephrine	Autonomic system; reticular activating system and other areas of brain and spinal cord	Inactivated slowly by monoamine oxidase (MAO); mainly inactivated by reabsorption by vesicles in the synaptic knob; norepinephrine level in brain affects mood
Dopamine	Limbic system; cerebral cortex; basal ganglia; hypothalamus	Thought to affect motor function; may be involved in schizophrenia[2]; amount reduced in Parkinson's disease
Serotonin	Limbic system; hypothalamus; cerebellum; spinal cord	May play role in sleep; LSD antagonizes serotonin; thought to be inhibitory
GABA (gamma-aminobutyric acid)	Spinal cord, cerebral cortex, cerebellum	Acts as inhibitor; may play role in pain perception
Endorphins	CNS and pituitary gland	Neuropeptides that have morphine-like properties and suppress pain; may help regulate cell growth; linked to learning and memory
Enkephalins	Brain and digestive tract	Neuropeptides that inhibit pain impulses by inhibiting release of substance P; bind to same receptors in brain as morphine
Substance P	Brain and spinal cord, sensory nerves, intestine	Transmits pain impulses from pain receptors into CNS

[1]These and other structures listed in this table will be discussed in Chapter 34.
[2]Studies suggest that the brains of schizophrenics have more dopamine receptors than those of nonschizophrenics.

The neurotransmitters diffuse across the synaptic cleft. Some may be taken up by specific receptors on the dendrites or cell bodies of other neurons. Excess neurotransmitters are reabsorbed into the synaptic vesicles or inactivated by enzymes.

Once they diffuse into the synaptic cleft, neurotransmitters have different effects on other neurons across the synapse. Certain neurotransmitters may cause sodium channels to open in the membrane of another neuron. The resulting influx of sodium ions depolarizes the membrane of this neuron, bringing it closer to firing. As a result, this neuron is more likely to set off an action potential, thus continuing to transmit the "message." This type of synapse is called an **excitatory synapse.** A "message" transmitted by one neuron is picked up, in effect, by the next neuron in the sequence.

Other neurotransmitters, instead of causing sodium channels to open, may cause potassium channels to open in the membrane of a neuron across a synapse. Potassium ions then flow out of this neuron. With additional positive charges outside the membrane, the interior of the neuron becomes even more negative relative to the surrounding fluid. This is called **hyperpolarization.** When such neurotransmitters hyperpolarize the membrane of a neuron across the synapse, an action potential is less likely to arise in the neuron. This type of synapse is called an **inhibitory synapse.**

Many Neurotransmitters Have Been Identified

More than 60 different substances are now known (or suspected) to be neurotransmitters or chemicals that modify the effects of neurotransmitters. Many types of neurons secrete two or even three different types of neurotransmitters. Furthermore, a postsynaptic neuron can have receptors for more than one type of neurotransmitter. Some of its receptors may be excitatory and some inhibitory.

The two neurotransmitters that have been investigated most extensively are acetylcholine and norepinephrine. **Acetylcholine** triggers muscle contraction. It is released not only from motor neurons (neurons that control skeletal muscle) but also by neurons in some parts of the nervous system including the brain. Cells that release acetylcholine are referred to as **cholinergic neurons.** Acetylcholine has an excitatory effect on skeletal muscle but an inhibitory effect on cardiac muscle. Whether a neurotransmitter excites or inhibits is at least in part a property of the postsynaptic receptors with which it combines. After acetylcholine is released into a synaptic cleft and combines with receptors on the postsynaptic neuron, the remaining molecules must be removed to prevent repeated stimulation of the muscle. The enzyme **cholinesterase** catalyzes the breakdown of acetylcholine.

Norepinephrine is released by neurons of the sympathetic nervous system (discussed in Chapter 34) as well as by many neurons in the brain and spinal cord. Neurons that release norepinephrine are called **adrenergic neurons.** Norepinephrine and the neurotransmitters epinephrine and dopamine belong to a class of compounds known as **catecholamines.** After their release from synaptic knobs, catecholamines are removed mainly by re-uptake into the synaptic vesicles. Some are inactivated by enzymes such as monoamine oxidase (MAO). Catecholamines affect mood, and certain drugs such as antidepressants and amphetamines modify mood by altering the levels of these substances in the brain.

Neurons Are One-Way Streets

Within the central nervous system (brain and spinal cord) nerve impulses are usually transmitted from the axon of one neuron to the dendrite (or cell body) of the next. Because neurotransmitter is found only within the synaptic knobs of the axons, neurons function as one-way streets, transmitting from dendrite to cell body to axon and then across the synaptic cleft to the dendrite or cell body of the next neuron in the sequence.

Myelination And Large Axon Diameter Increase Rate Of Impulse Transmission

Compared with the speed of an electric current or the speed of light, a nerve impulse travels rather slowly. The speed of a nerve impulse varies from less than 1 meter to more than 120 meters (398 feet) per second, depending upon the type of neuron. However, these speeds are more than fast enough to allow rapid responses in humans and other animals. What factors affect speed of transmission? In general, the greater the diameter of an axon, the greater its speed of conduction. Moreover, the heavier the myelin sheath, the faster the nerve impulse travels. The largest neurons seem also to be the most heavily myelinated.

■ NEURAL INTEGRATION IS THE PROCESS OF AVERAGING ALL OF THE INCOMING INFORMATION AND DETERMINING WHETHER AN ACTION POTENTIAL WILL BE GENERATED

Each neuron may synapse with hundreds of other neurons. Indeed, as much as 40% of a neuron's dendritic surface may be covered by synaptic knobs of neurons communicating with it (Fig. 33–10). **Neural integration** is the process of adding and subtracting excitatory and inhibitory incoming signals and determining whether an action potential will be generated in the postsynaptic neuron. Each neuron functions like a little computer, averaging the hundreds of messages that continually bombard its dendrites and cell body. More than 90% of the body's neurons are located in the central nervous system, so that is where most neural integration takes place. For this reason, the brain and spinal cord are responsible for making the "decisions."

The process of integration is mechanical and is carried out on the molecular level. Some neurotransmitters bring the neuron closer to firing, whereas others make the neuron less likely to fire. Thus one neurotransmitter may cancel the effect of another neurotransmitter. After these molecular tabulations are carried out, the type and amount of neurotransmitter that predominates determines the result. If sufficient excitatory neurotransmitter is present, the neuron will be stimulated and a message will be transmitted.

■ THE REFLEX ARC IS AN EXAMPLE OF A SIMPLE NEURAL PATHWAY

Neurons are organized into specific **pathways,** or **neural circuits.** Generally, neurons are arranged so that the axon of one neuron in the circuit forms junctions with the dendrites of the next neuron in the circuit. One of the simplest examples of a neural pathway is the **reflex arc** (Fig. 33–11).

A **reflex action** is a relatively fixed reaction pattern to a simple stimulus. The response is predictable and automatic, not requiring conscious thought. Many of the activities of the body such as breathing are regulated by reflex actions.

Although most reflex actions are much more complex, let us consider a **withdrawal reflex,** in which a neural circuit consisting of only three neurons is needed to carry out a response to a stimulus (Fig. 33–11). Suppose you touch a hot stove. Almost instantly, and before you are consciously aware of the situation, you jerk your hand away from this unpleasant stimulus. But in this brief instant a message has been carried from pain receptors in the skin to the spinal cord by a

(a)

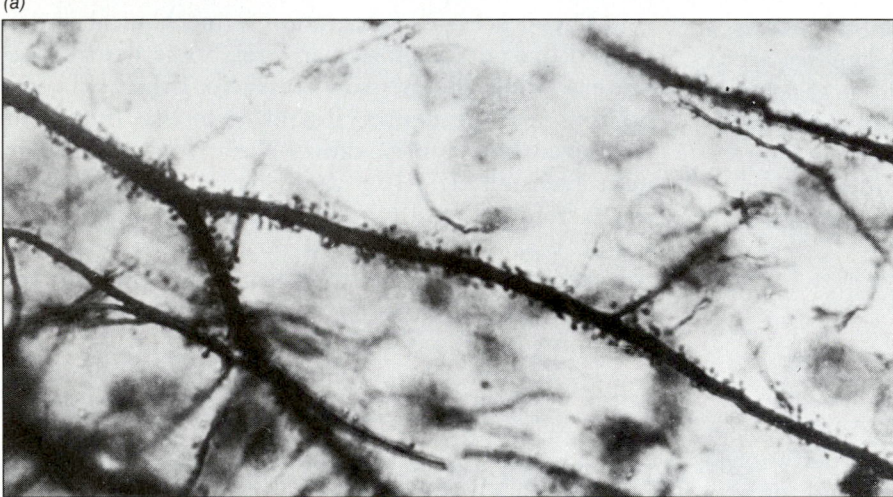

(b)

FIGURE 33–10 (*a*) Neurons in the brain of a squirrel monkey, stained by a recently developed technique. Only a few of the very numerous cells are actually stained, or it would not be possible to make any of them out clearly. (*b*) Close-up of an axon of one of the neurons shown in (*a*). Notice the numerous lollipop-shaped structures attached to the side of the axon. Each of these is actually a synapse with another neuron. This should help you to appreciate the tremendous complexity of the interconnections of nerve cells in the central nervous system. (Courtesy of Dr. Fernando E. D'Amelio, *Stain Technology,* 58:1983)

sensory neuron. In the tissue of the spinal cord the message is transmitted from the sensory neuron to an **association neuron.** Finally, the message is transmitted to an appropriate **motor neuron,** which conducts the message to groups of muscles that respond by contract-

ing and pulling the hand from the stove. Actually, many neurons located in sensory, association, and motor nerves participate in such a reaction, and complicated switching is involved. We move our hands *up* from a hot stove but *down* from a hot light bulb. Gener-

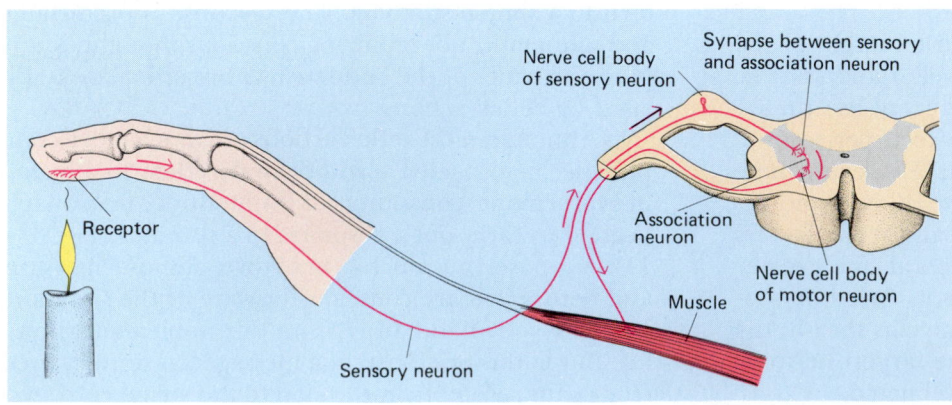

FIGURE 33–11 The withdrawal reflex involves a chain of three neurons. A sensory neuron transmits the message from the receptor to the central nervous system, where it synapses with an association neuron. Then an appropriate motor neuron (shown in red) transmits an impulse to the muscles that move the hand away from the flame (the response).

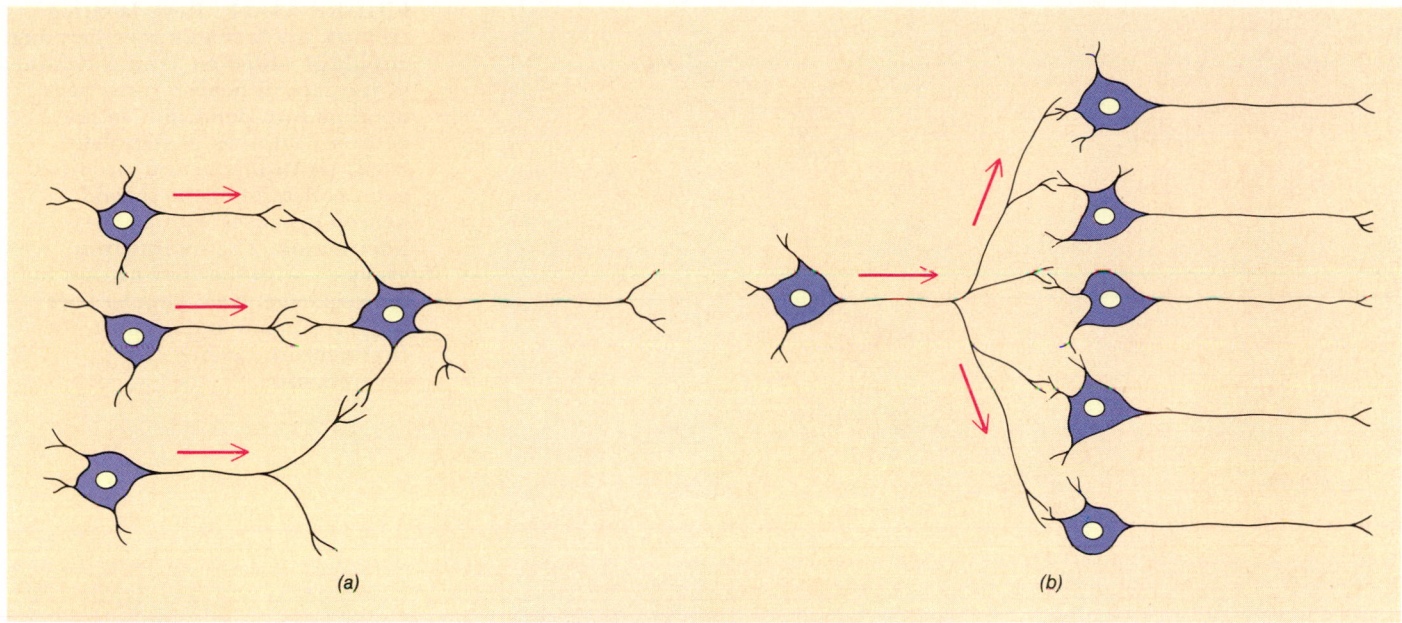

FIGURE 33–12   Organization of neural circuits. (*a*) Convergence of neural input. Several presynaptic neurons synapse with one postsynaptic neuron. This organization in a neural circuit permits one neuron to receive signals from many sources. (*b*) Divergence of neural output. A single presynaptic neuron synapses with several postsynaptic neurons. This organization allows one neuron to communicate with many others.

ally, we are not even consciously aware that all these responding muscles exist.

Quite probably, at the same time that the association neuron sends a message out along a motor neuron, it also sends one up the spinal cord to the conscious areas of the brain. As you withdraw your hand from the hot stove, you become aware of what has happened and feel the pain. This awareness, however, is a feature apart from the reflex response.

☐ COMPLEX NEURAL PATHWAYS ARE POSSIBLE BECAUSE NEURONS ASSOCIATE IN A VARIETY OF WAYS

Within a neural pathway many presynaptic neurons may converge upon a single postsynaptic neuron. In **convergence,** the postsynaptic neuron is controlled by signals coming from two or more presynaptic neurons (Fig. 33–12*a*). An association neuron in the spinal

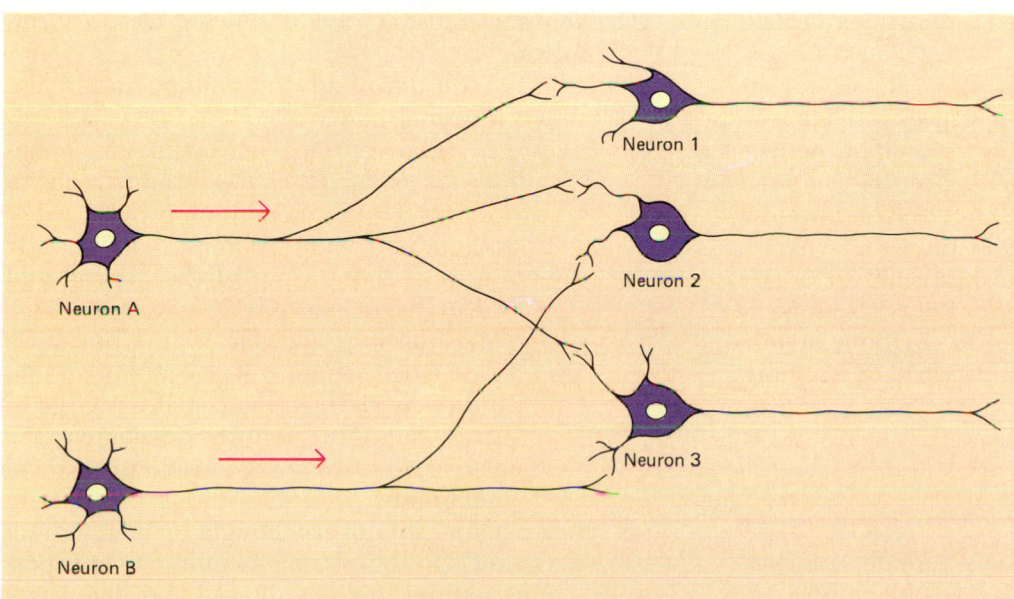

Neuron 1

Neuron 2

Neuron A

Neuron 3

Neuron B

FIGURE 33–13   Facilitation. Neither neuron *A* nor neuron *B* can itself fire neuron *2* or *3*. However, stimulation by either *A* or *B* does depolarize the neuron toward threshold level (if the stimulation is excitatory). This facilitates the postsynaptic neuron, so if another presynaptic neuron stimulates it, the threshold level may be reached and an action potential generated.

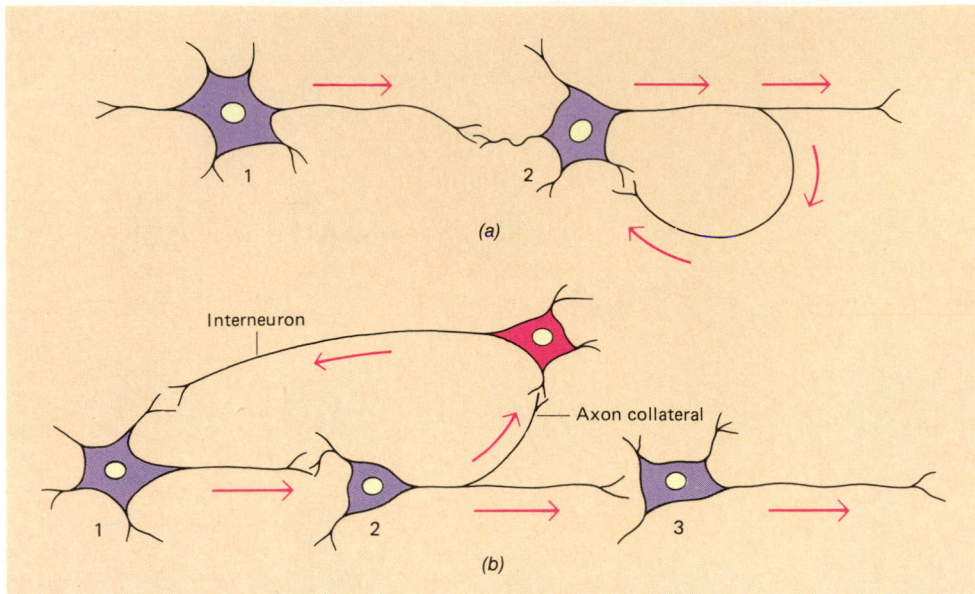

FIGURE 33–14   Reverberating circuits. (a) A simple reverberating circuit in which an axon collateral of the second neuron turns back upon its own dendrites, so the neuron continues to stimulate itself. (b) In this neural circuit an axon collateral of the second neuron synapses with an interneuron. The interneuron synapses with the first neuron in the sequence. New impulses are triggered again and again in the first neuron, causing reverberation.

cord, for instance, may receive converging information from sensory neurons entering the cord, from neurons originating at other levels of the spinal cord, and even from neurons bringing information from the brain. Information from all of these converging neurons is integrated before an action potential is generated in the association neuron and an appropriate motor neuron stimulated.

In **divergence** a single presynaptic neuron stimulates many postsynaptic neurons (Fig. 33–12b). Each presynaptic neuron may synapse with up to 25,000 or more postsynaptic neurons. In **facilitation** the neuron is brought close to threshold level by stimulation from various presynaptic neurons but is not yet at the threshold level. The neuron can be easily excited by further stimulation. Figure 33–13 illustrates facilitation.

The **reverberating circuit** is a neural pathway arranged so that a neuron branch synapses with an association neuron (Fig. 33–14). The association neuron synapses with a neuron in the sequence that can send new impulses again through the circuit. New impulses can be generated again and again until the synapses fatigue (from depletion of neurotransmitter) or are stopped by some sort of inhibition. Reverberating circuits are thought to be important in rhythmic breathing, in maintaining alertness, and perhaps in short-term memory.

■ CHAPTER SUMMARY

I. Neural function depends on reception, transmission, integration, and response by muscles or glands.

II. Glial cells and neurons are the two types of cells characteristic of neural tissue.
   A. Glial cells are supporting cells.
   B. Neurons are specialized to receive stimuli and transmit impulses.
   C. A typical multipolar neuron consists of a cell body from which project many branched dendrites and a single long axon, which ends in axon terminals.
   D. Outside the central nervous system axons are surrounded by a neurilemma, and many axons are also enveloped by a myelin sheath.
   E. A nerve consists of hundreds of axons wrapped in connective tissue; a ganglion is a mass of cell bodies.

III. Neurons transmit rapidly moving electrical impulses.
   A. A neuron that is not transmitting an impulse has a resting potential.
      1. The inner surface of the plasma membrane is negatively charged compared with the outside; the membrane is polarized.
      2. Sodium-potassium pumps continuously transport sodium out of the neuron and transport potassium in.
      3. Potassium ions are able to leak out more readily than sodium ions are able to leak in.
      4. Negatively charged molecules too large to diffuse out of the neuron contribute to the relative negative charge inside the plasma membrane.
   B. Excitatory stimuli are thought to open sodium channels in the plasma membrane. This permits sodium to enter the cell and depolarize the membrane.

C. When the extent of depolarization reaches threshold level, an action potential may be generated.
   1. The action potential is a wave of depolarization that moves down the axon.
   2. The action potential obeys an all-or-none law.
   3. As the action potential moves down the axon, repolarization occurs quickly behind it.
   4. Saltatory conduction takes place in myelinated neurons.
   5. The largest, most heavily myelinated neurons conduct impulses most rapidly.
IV. Synaptic transmission generally depends upon release of a neurotransmitter from vesicles in the synaptic knobs of the presynaptic neuron.
   A. Neurotransmitter diffuses across the synaptic cleft and combines with receptors on the post-synaptic neuron.
   B. Neurons that release the neurotransmitter acetylcholine are known as cholinergic neurons. Those that release norepinephrine are adrenergic.
V. Neural integration is the process of adding and subtracting incoming signals and determining whether to fire an impulse or not.
VI. A reflex arc is a simple neural pathway.
   A. A withdrawal reflex requires a sequence of only three neurons: a sensory neuron, an association neuron, and a motor neuron.
   B. Reflex action is predictable and automatic; it does not require conscious thought.
VII. Complex neural pathways are possible because of such neuron associations as convergence, divergence, and facilitation.

■ POST-TEST

1. The process of receiving a stimulus is called _____.
2. The actual response is carried out by effectors, _____ and _____.
3. The supporting cells of the nervous system are called _____ cells; cells specialized to transmit impulses are _____.
4. The nucleus of a neuron is found within the _____ _____.
5. The _____ transmits impulses from the cell body to the synapse.
6. The _____ _____ is important in the regeneration of injured neurons.
7. A _____ consists of a mass of cell bodies.
8. Sodium is actively transported out of a resting neuron by _____ _____.
9. Any stimulus that increases the neuron's permeability to sodium to threshold level may result in the transmission of an _____ _____.
10. During the _____ _____ a neuron cannot transmit another impulse.
11. The junction between two neurons is called a _____.
12. Synaptic knobs release _____.
13. A _____ action is a predictable, automatic response to a simple stimulus.
14. Impulses are transmitted from sense organs to the central nervous system by _____ neurons.
15. In _____ the postsynaptic neuron is controlled by signals from several presynaptic neurons.

16. Label the following diagram.

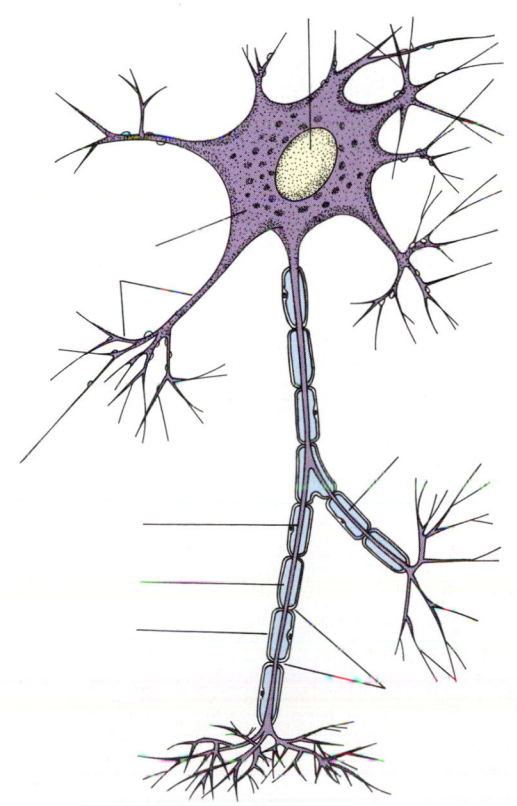

■ REVIEW QUESTIONS

1. Imagine that you are driving down the street when suddenly a child darts in front of your car. What sequence of events must take place within your nervous system before you can slam on the brake?

2. Give the functions of (a) myelin, (b) ganglia, (c) glial cells, (d) dendrites, (e) an axon.
3. What is meant by the resting potential of a neuron? How does the sodium-potassium pump contribute to the resting potential?
4. Contrast the resting potential with an action potential. What is responsible for generation of an action potential?
5. How does the all-or-none law affect neural action?
6. How is neural function affected by the presence of too much calcium? Too little calcium?
7. Contrast transmission along an axon with transmission across a chemical synapse.
8. Give the functions of (a) acetylcholine, (b) cholinesterase, (c) norepinephrine.
9. Imagine that you have just burned your finger by touching a hot pot. Draw a diagram to illustrate the reflex action that would occur. Label each structure and indicate the direction of information flow.
10. Contrast convergence and divergence. What is their general significance?

▣ RECOMMENDED READINGS

Bloom, F.E.: "Neuropeptides," *Scientific American* 245(4), October 1981. An account of the discovery and actions of neuropeptides, which help regulate bodily activities, in some cases acting as both neurotransmitters and hormones.

Fine, A.: "Transplantation in the central nervous system," *Scientific American* 255(2):52–58, 1986. Neurons transplanted from embryos can establish functional connections in the adult brain and spinal cord.

Gottlieb, D.I.: "GABAergic neurons," *Scientific American* 258(2):82–89, 1988. GABA is an inhibitory neurotransmitter in the brains of all mammals.

Llinas, R.R.: "Calcium in Synaptic Transmission," *Scientific American* 247(4), October 1982. The role of calcium is studied in the giant synapse of a squid.

Morell, P., and W.T. Norton: "Myelin," *Scientific American* 242(5), May 1980. A description of myelin and its functions.

Schwartz, J.H.: "The Transport of Substances in Nerve Cells," *Scientific American* 242(4), April 1980. A discussion of the movement of substances long distances between the cell body and the neuron endings.

Wurtman, R.J.: "Nutrients that modify brain function," *Scientific American* 246(4), April 1982. Increasing the level of precursors of neurotransmitters in the blood amplifies signals from some nerve cells. They may eventually be used clinically.

34

Responsiveness: Nervous Systems

OUTLINE

I. All invertebrates except sponges have nervous systems
  A. Some invertebrates have nerve nets or radial nervous systems
  B. Most invertebrates have bilateral nervous systems
II. Key features of the vertebrate nervous system are the hollow, dorsal nerve cord and well-developed brain
III. The evolution of the vertebrate brain is marked by increasing complexity especially of the cerebrum and cerebellum
  A. The hindbrain develops into medulla, pons, and cerebellum
  B. The midbrain is most prominent in fish and amphibians
  C. The forebrain gives rise to the thalamus, hypothalamus, and cerebrum
IV. The human central nervous system is the most complex biological mechanism known
  A. The spinal cord transmits impulses to and from the brain and controls many reflex activities
  B. The largest, most prominent part of the human brain is the cerebrum
  C. The limbic system affects emotional aspects of behavior
  D. The reticular activating system is an arousal system
  E. When signals from the RAS slow, a person may fall asleep
  F. Learning involves the storage of information in the nervous system and its retrieval on demand
  G. Experience affects the brain
V. The peripheral nervous system includes somatic and autonomic systems
  A. The somatic system helps the body adjust to the external environment
  B. The autonomic system helps maintain homeostasis in response to changes in the internal environment
VI. Many mood drugs change the levels of neurotransmitters in the brain
Focus on dopamine and motor function
Focus on cerebral dominance
Focus on alcohol abuse
Focus on crack cocaine

LEARNING OBJECTIVES

After you have studied this chapter you should be able to:
1. Contrast nerve nets and radial nervous systems with bilateral nervous systems.
2. Compare the vertebrate nervous system with a bilateral invertebrate nervous system.
3. Trace the development of the principal regions of the vertebrate brain from the forebrain, midbrain, and hindbrain.
4. Compare the relative sizes and functions of principal regions of the brain in fish, amphibians, reptiles, birds, and mammals.
5. Describe the functions and structure of the spinal cord.
6. Locate the following parts of the human brain and give the functions of each: medulla, pons, midbrain, thalamus, hypothalamus, cerebellum, and cerebrum.
7. Compare the reticular activating system with the limbic system.
8. Contrast REM and non-REM sleep.
9. Describe how we perceive sensation, including pain.
10. Review current theories of learning and memory.
11. Cite experimental evidence linking environmental stimuli with changes in the brain and with learning ability.
12. Compare the somatic system with the autonomic system.
13. Contrast the sympathetic and parasympathetic divisions of the autonomic system.
14. Discuss the biological actions and effects on mood of the following types of drugs: alcohol, barbiturates, antianxiety drugs, antipsychotic drugs, opiates, stimulants, hallucinogens, and marijuana.

The nervous systems of these fish (coy) respond to food in a feeding frenzy (Richard H. Gross)

An organism's type of nervous system is closely linked with its life-style. The simple, sluggish nervous system of *Hydra* is adequate for an animal that remains rooted in one spot waiting for dinner to brush by its tentacles. With its more sophisticated nervous system, a frog can hop about in search of food and eject its tongue with lightning speed to capture a passing fly. However, neither the *Hydra* nor the frog is able to solve algebra problems or learn about its own physiology. The range of possible responses depends in large part on the number of neurons and how they are organized in an animal's nervous system.

KEY CONCEPTS

☐ An animal's range of possible responses depends on the complexity of its nervous system; as animal groups evolved, nervous systems became increasingly complex.

☐ The vertebrate nervous system consists of the central nervous system (CNS) and the peripheral nervous system (PNS).

☐ The CNS (brain and spinal cord) integrates incoming information and determines appropriate responses.

☐ The PNS (receptors, afferent nerves, and efferent nerves) transmits information from receptors to the CNS and transmits "decisions" from the CNS to the muscles and glands that must respond.

☐ ALL INVERTEBRATES EXCEPT SPONGES HAVE NERVOUS SYSTEMS

There is no nervous system in the sponge. Whatever responses it makes are due to the basic irritability of its individual cells. Among other invertebrates there are two main types of nervous systems—nerve nets and bilateral nervous systems.

Some Invertebrates Have Nerve Nets Or Radial Nervous Systems

The simplest organized nervous tissue is the **nerve net** found in *Hydra* and other cnidarians (Fig. 34–1). In a nerve net the nerve cells are scattered throughout the body. No central control organ and no definite pathways are present. Impulses are transmitted in all directions, becoming less intense as they spread from the

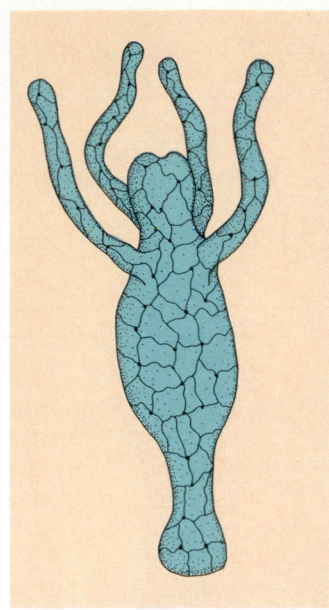

FIGURE 34–1 The nerve net of *Hydra* and other cnidarians is the simplest organized nervous tissue. No central organ and no definite neural pathways are present.

region of initial stimulation. If the stimulus is strong, the message will spread to more neurons of the net than if it is weak.

Since it produces responses that involve the body as a whole, or large parts of it at the same time, such a diffuse pattern of transmission is adequate in a radially symmetrical animal with sluggish locomotion. Responses in cnidarians are limited to discharge of stinging structures (nematocysts) and contractions that permit the movements associated with locomotion and feeding. The nerve net is adaptive in that the cnidarian can respond to dinner approaching from any direction. Many cnidarians have two nerve nets. One is a slow-transmitting system that coordinates movement of the tentacles, while the other is faster and is used to coordinate swimming.

The somewhat more sophisticated nervous system of the echinoderm consists of a nerve ring that surrounds the mouth from which a large radial nerve extends into each arm. These nerves coordinate movement of the animal. In sea stars a nerve net mediates the responses of the dermal gills to tactile stimulation (touch).

Most Invertebrates Have Bilateral Nervous Systems

In bilaterally symmetrical animals the nervous system is usually more complex than in radially symmetrical animals. A bilateral form of symmetry usually reflects a more active way of life, with rapid and sophisticated responses to the environment. The following trends in the evolution of nervous systems can be identified:

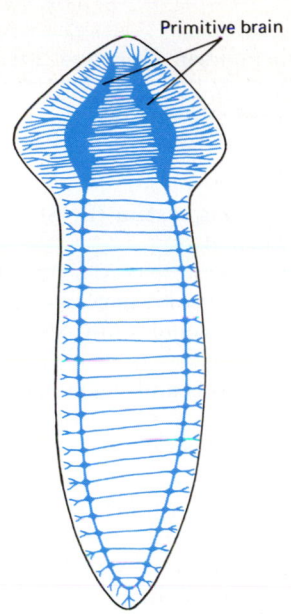

FIGURE 34–2 Planarian flatworms have a ladder-type nervous system. Cerebral ganglia in the head region serve as a simple brain and, to some extent, control the rest of the nervous system.

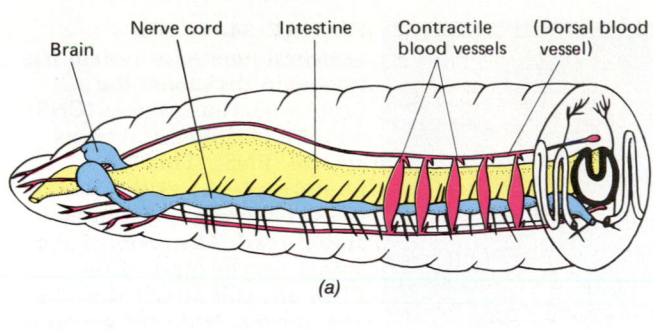

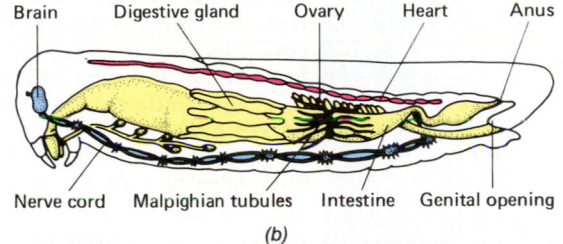

FIGURE 34–3 Annelids and arthropods have bilateral nervous systems. (*a*) The nervous system of the earthworm is typical of those found in other annelids. The cell bodies of the neurons are located in ganglia found in each body segment. They are connected by the ventral nerve cord. (*b*) In the insect nervous system the cerebral ganglia serve as a simple brain. Two ventral nerve cords are present.

1. Increased number of nerve cells.
2. Concentration of nerve cells forming thick cords or masses of tissue, which become nerves, nerve cords, ganglia, and brain.
3. Specialization of function. For example, various parts of the central nervous system may be specialized to perform specific functions, so distinct structural and functional regions can be identified.
4. Increased number of association neurons and more complex synaptic contacts. This permits much greater integration of incoming messages, provides a greater range of responses, and allows far more precision in responses.
5. Cephalization, or formation of a head with a brain. When an animal has a head, the principal sense organs such as eyes, ears, olfactory (sense of smell) receptors, and taste receptors tend to be concentrated there. A bilaterally symmetrical animal generally moves forward, head first. With sense organs concentrated at the front of the body, the animal can detect an enemy quickly enough to escape or sense food in time to capture it. Response can be more rapid if these sense organs are linked by short pathways to decision-making nerve cells nearby. Therefore, nerve cells are also usually concentrated in the head region, constituting a definite brain.

In flatworms, concentrations of nerve cells in the head region form **cerebral ganglia** (Fig. 34–2). These serve as a primitive "brain" and exert some measure of control over the rest of the nervous system. Two ventral longitudinal nerve cords extend from the ganglia to the posterior end of the body. The nerve cords are connected by a series of transverse nerves which also connect the brain with the eyespots. This arrangement is aptly referred to as a **ladder-type nervous system.**

In annelids and arthropods there is also typically a pair of ventrally located longitudinal nerve cords (Fig. 34–3). The cell bodies of the nerve cells are concentrated in pairs of ganglia located in *each* body segment. Afferent and efferent neurons are located in lateral nerves that link the ganglia with muscles and other body structures. In some arthropods specific functional regions have been identified in the cerebral ganglia.

In mollusks there are typically at least three pairs of ganglia; each pair has specific functions. In cephalopods, such as the octopus, nerve cells tend to be concentrated in a central region. All the ganglia are massed in a ring that surrounds the esophagus. The ring contains about 168 million nerve cell bodies. With this complex brain, it is no wonder that the octopus is capable of considerable learning and can be taught quite complex tasks. In fact, the octopus is considered to be among the most intelligent invertebrates.

FIGURE 34–4 The vertebrate nervous system has two main divisions: the central nervous system (CNS) and the peripheral nervous system (PNS). This photograph shows the human CNS consisting of brain and spinal cord. The roots of the spinal nerves (part of the PNS) are still attached to the spinal cord. Note the group of nerves that extend caudally from the lower region of the cord. Because they resemble a horse's tail, they are referred to as the cauda equina. These nerves have been left undisturbed on the right but have been fanned out on the left. (Dissection by Dr. M.C.E. Hutchinson, Department of Anatomy, Guy's Hospital Medical School, London; from Williams and Warwick, eds., *Gray's Anatomy*)

TABLE 34–1
Divisions of the Human Nervous System

I. **Central nervous system (CNS)**
   A. Brain
   B. Spinal cord

II. **Peripheral nervous system (PNS)**
   A. Somatic portion
      1. Receptors
      2. Afferent (sensory) nerves—transmit information from receptors to CNS
      3. Efferent nerves—transmit information from CNS to glands and involuntary muscle in organs
   B. Autonomic portion
      1. Receptors
      2. Afferent (sensory) nerves—transmit information from receptors in internal organs to CNS
      3. Efferent nerves—transmit information from CNS to internal organs
         a. Sympathetic nerves—generally stimulate activity that results in mobilization of energy (e.g., speeds heartbeat)
         b. Parasympathetic nerves—action results in energy conservation or restoration (e.g., slows heartbeat)

KEY FEATURES OF THE VERTEBRATE NERVOUS SYSTEM ARE THE HOLLOW, DORSAL NERVE CORD AND WELL-DEVELOPED BRAIN

The vertebrate nervous system has two main divisions: the central nervous system (CNS) and the peripheral nervous system (PNS; Fig. 34–4). The **CNS** consists of a complex tubular brain that is continuous with the single, dorsal, tubular spinal cord. Serving as central control, these organs integrate incoming information and determine appropriate responses.

The **PNS** is made up of the sensory receptors (e.g., touch, auditory, and visual receptors) and the nerves, which are the communication lines. Various parts of the body are linked to the brain by cranial nerves and to the spinal cord by spinal nerves. Afferent neurons in these nerves continuously inform the CNS of changing conditions. Then efferent neurons transmit the "decisions" of the CNS to appropriate muscles and glands, which make the adjustments needed to preserve homeostasis.

For convenience we subdivide the PNS into **somatic** and **autonomic** portions. Most of the receptors

and nerves concerned with changes in the external environment are somatic; those that regulate the internal environment are autonomic. Both systems have **sensory (afferent) nerves,** which transmit messages from receptors to the CNS, and **motor (efferent) nerves,** which transmit information back from the CNS to the structures that must respond. In the autonomic system there are two kinds of efferent pathways—**sympathetic** and **parasympathetic** nerves (see Table 34–1).

THE EVOLUTION OF THE VERTEBRATE BRAIN IS MARKED BY INCREASING COMPLEXITY ESPECIALLY OF THE CEREBRUM AND CEREBELLUM

In the early vertebrate embryo, the brain and spinal cord differentiate from a single tube of tissue, the **neural tube.** Anteriorly, the tube expands and develops into the brain. Posteriorly, the tube becomes the spinal cord. Brain and spinal cord remain continuous and their cavities communicate. As the brain begins to differentiate, three bulges become visible—the hindbrain, midbrain, and forebrain (Fig. 34–5). All vertebrates, from fish to mammals, have the same basic brain structure. Different parts of the brain may be

FOCUS ON   Dopamine and Motor Function

The neurotransmitter dopamine plays an important role in motor function. Its function became known through an interesting series of somewhat unrelated events. During the mid-1950s the drug reserpine became popular as a major tranquilizer used for mental patients. Then, in 1959, investigators noticed that some patients taking reserpine developed extrapyramidal symptoms, such as muscle rigidity and tremor. These symptoms were very similar to those seen in patients with Parkinson's disease, a disorder in which movement is shaky and difficult. Victims of Parkinson's disease suffer from tremors even when they are not attempting to move. This observation led to studies that showed that the drug reserpine greatly reduces the amount of dopamine within the caudate nucleus and putamen (two of the basal ganglia within the white matter of the cerebrum). Investigators then discovered that patients with Parkinson's disease have only about 50% of the normal amount of dopamine in their basal ganglia.

Attempts to administer dopamine to these patients were not successful because dopamine cannot penetrate the blood-brain barrier. However, a substance known as L-dopa, from which dopamine is synthesized in the body, does penetrate the blood-brain barrier and has dramatically relieved the symptoms of Parkinson's disease in most patients.

It has been shown that dopamine is released by neurons that extend from the substantia nigra (an area in the midbrain) to the basal ganglia. Dopamine is thought to inhibit neurons that produce acetylcholine. When dopamine is absent or present in too small a quantity, the acetylcholine causes overactivity of certain neurons, producing the motor symptoms of Parkinson's disease.

Even in healthy persons, the aging process causes changes in motor abilities. Body movements and even reflexes slow, and movement becomes more difficult. Recent studies suggest that these changes may be due to dopamine depletion, and that treatment with L-dopa may be helpful.

Too much dopamine sometimes causes schizophrenic symptoms. In fact, it has been suggested that people suffering from schizophrenia may have too many dopamine receptors in their brains and thus too much dopamine activity.

---

specialized in the various vertebrate classes, and there is a trend toward increasing complexity, especially of the cerebrum and cerebellum.

The Hindbrain Develops Into The Medulla, Pons, And Cerebellum

The **hindbrain** gives rise to the medulla, pons, and cerebellum. The medulla is continuous with the spinal cord. Together the medulla, pons, midbrain, thalamus, and hypothalamus make up the **brain stem,** the elongated portion of the brain that looks like a stalk holding up the large cerebrum. The **medulla** consists largely of nerve tracts that connect the spinal cord with various parts of the brain. In complex vertebrates, the medulla contains discrete nuclei that are **vital centers** regulating respiration, heart rate, and blood pressure. Other reflex centers in the medulla regulate such activities as swallowing, coughing, and vomiting.

The size and shape of the **cerebellum** vary greatly among the vertebrate classes (Fig. 34–6). The cerebellum coordinates muscle activity and is responsible for muscle tone, posture, and equilibrium. Development of the cerebellum is correlated roughly with the extent and complexity of muscular activity. In some fish, birds, and mammals the cerebellum is highly devel-

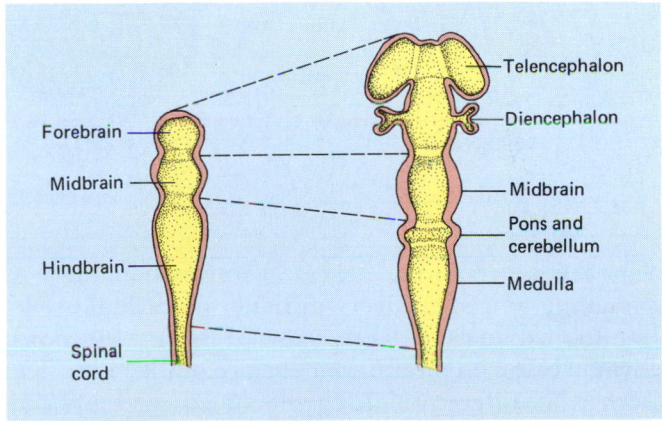

FIGURE 34–5   Early in the development of the vertebrate embryo, the anterior end of the neural tube differentiates into the forebrain, midbrain, and hindbrain. These primary divisions subdivide and then give rise to specific structures of the adult brain.

oped, whereas it tends to be small in amphibians and reptiles.

Injury or removal of the cerebellum results in impaired muscle coordination. A bird without a cerebellum is unable to fly, and its wings thrash about jerkily. When the human cerebellum is injured by a blow or by disease, muscular movements are uncoordinated.

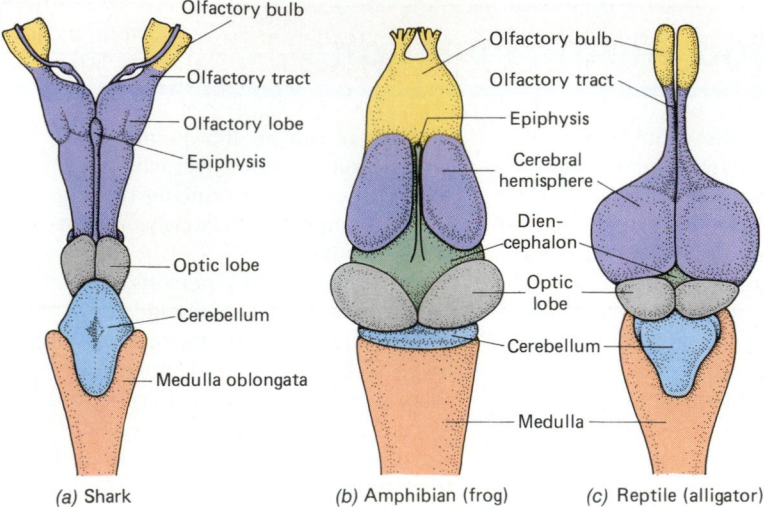

(a) Shark   (b) Amphibian (frog)   (c) Reptile (alligator)

FIGURE 34–6 Comparison of the brains of members of vertebrate classes shows basic similarities and evolutionary trends. Note that different parts of the brain may be specialized in the various groups. For example, the large olfactory lobes in the shark brain (a) are essential to this predator's highly developed sense of smell. During the course of evolution, the cerebrum and cerebellum have become larger and more complex. In the mammal (e) the cerebrum is the most prominent part of the brain.

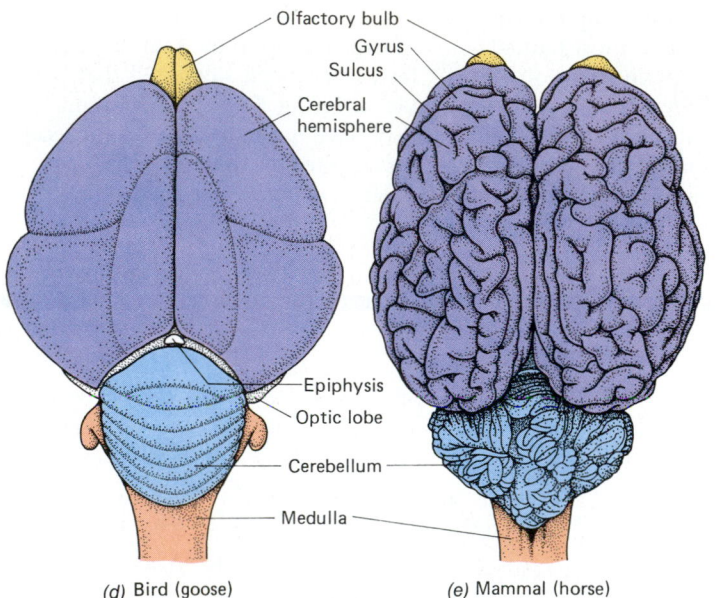

(d) Bird (goose)   (e) Mammal (horse)

Any activity requiring delicate coordination, such as threading a needle, is very difficult, if not impossible.

In mammals, a large mass of fibers, the **pons,** forms a bulge on the anterior surface of the brain. The pons is a bridge connecting the spinal cord and medulla with upper parts of the brain. It contains nuclei that relay impulses from the cerebrum to the cerebellum and also contains centers that help regulate respiration.

The Midbrain Is Most Prominent In Fish And Amphibians

In fish and amphibians the midbrain is the most prominent part of the brain. In these animals the midbrain is the main association area. It receives incoming sensory information, integrates it, and sends decisions to appropriate motor nerves. The dorsal portion of the midbrain is differentiated to some degree in these

lower vertebrates. For example, the **optic lobes,** specialized for visual interpretations, are part of the midbrain.

In reptiles, birds, and mammals, many of the functions of the optic lobes are assumed by the cerebrum. In mammals, the midbrain consists of the **superior colliculi,** which are centers for visual reflexes such as pupil constriction, and the **inferior colliculi,** which are centers for certain auditory reflexes. The mammalian midbrain contains the **red nucleus,** a center that integrates information regarding muscle tone and posture.

The Forebrain Gives Rise To The Thalamus, Hypothalamus, And Cerebrum

The forebrain differentiates into two subdivisions—the diencephalon and telencephalon. The **diencephalon** contains the thalamus and hypothalamus. In all vertebrate classes the **thalamus** is a relay center for motor and sensory messages. In mammals all sensory messages (except those from the olfactory receptors) are delivered to the thalamus before being relayed to the sensory areas of the cerebrum.

The **hypothalamus,** which lies below the thalamus, contains olfactory centers and is the principal integration center for the regulation of the internal organs. The hypothalamus links the nervous and endocrine systems. In fact, the pituitary gland (an important endocrine gland) extends downward from the hypothalamus. Releasing hormones produced by the hypothalamus regulate the secretion of several hormones produced by the anterior lobe of the pituitary gland.

In reptiles, birds, and mammals, body temperature is controlled by a "thermostat" in the hypothalamus. The hypothalamus also regulates appetite and water balance and is involved in emotional and sexual responses. In mammals, the hypothalamus provides input to centers in the medulla and spinal cord that regulate activities such as heart rate, respiration, and digestion.

The **telencephalon** differentiates to form the cerebrum, and, in most vertebrate groups, the **olfactory bulbs.** The olfactory bulbs receive information about odors; the chemical sense of smell is the dominant sense in most aquatic and terrestrial vertebrates. In fact, much of brain development in vertebrates appears to be focused upon the integration of olfactory information. In fish and amphibians, the olfactory bulbs are well developed and the cerebrum is almost entirely devoted to the integration of olfactory information.

In most vertebrates the cerebrum is divided into right and left hemispheres. Most of the cerebrum is made of **white matter** consisting mainly of axons connecting various parts of the brain. In mammals and most reptiles there is a layer of **gray matter,** the **cerebral cortex,** that makes up the outer portion of the cerebrum. Certain reptiles possess a different type of cortex, not found in lower vertebrates, known as the **neopallium;** it serves as an association area, a region that links sensory and motor functions and is responsible for higher functions such as learning. In mammals, the neopallium is extensive, making up the bulk of the cerebrum. In mammalian embryonic development, the cerebrum expands and grows backwards, covering many of the other brain structures. Thus, the cerebrum is the most prominent part of the mammalian brain.

In mammals the cerebrum is responsible for many of the functions that are performed by other parts of the brain in lower vertebrates. It has many complex association functions lacking in reptiles, amphibians, and fish. In small or simple mammals, the cerebral cortex may be smooth. However, in large, complex mammals, the surface area is greatly expanded by numerous folds called **convolutions** (or gyri; singular, gyrus). The furrows between them are called *sulci* (singular, sulcus) when shallow and *fissures* when deep.

■ THE HUMAN CENTRAL NERVOUS SYSTEM IS THE MOST COMPLEX BIOLOGICAL MECHANISM KNOWN

The soft, fragile brain and spinal cord are well protected. Encased within bone, they are covered by three layers of connective tissue, the **meninges.** A special shock-absorbing fluid, **cerebrospinal fluid,** cushions the brain and spinal cord against mechanical injury.

The Spinal Cord Transmits Impulses To And From The Brain And Controls Many Reflex Activities

The **spinal cord** is a hollow cylinder that emerges from the base of the brain and extends downward to about the level of the waist (Fig. 34–4). Its two functions are to (1) control many reflex activities of the body and (2) transmit messages back and forth to the brain through its **ascending** and **descending nerve tracts.** Each tract is a large bundle of axons. As axons pass down through the brain and spinal cord, some of them cross over from one side of the cord (or brain) to the other. For this reason the right side of the brain mainly controls the left side of the body, and the left side of the brain mainly controls the right side of the body.

When examined in cross section, the spinal cord is seen to have a small central canal surrounded by a butterfly-shaped area of gray matter (Fig. 34–7). Outside the gray matter the cord is composed of white matter. **Gray matter** consists mainly of large masses of cell bodies and dendrites of the neurons present within the cord. **White matter** is composed of the myelinated axons of the tracts within the cord.

The Largest, Most Prominent Part Of The Human Brain Is The Cerebrum

Although computers have been designed along similar principles and have been likened to it, even the most intricate computer does not begin to rival the complexity of the human brain. A soft, wrinkled mass of tissue weighing about 1.4 kilograms (3 pounds), the human brain contains about 25 billion neurons. Each neuron is functionally connected to as many as 100,000 others, and there may be as many as 10^{14} synapses. No wonder scientists have barely begun to unravel some of the tangled neural circuits that govern human physiology and behavior!

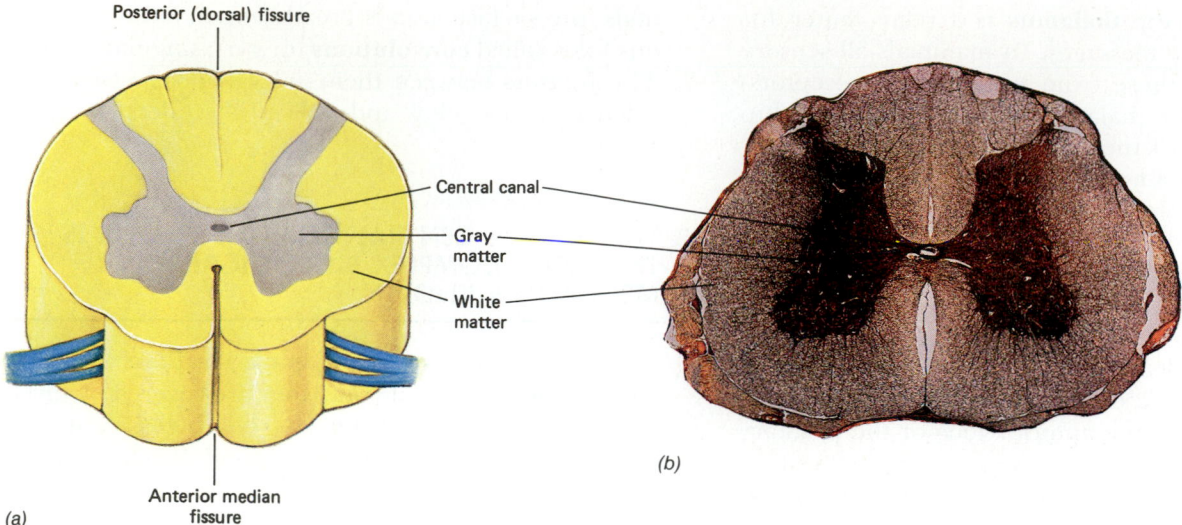

Posterior (dorsal) fissure

Central canal

Gray matter

White matter

Anterior median fissure

(a)

(b)

FIGURE 34–7 The spinal cord consists of gray matter and white matter. (a) Cross section through the spinal cord. (b) Photomicrograph of a cross section through the spinal cord (approximately ×25). (Manfred Kage/Peter Arnold, Inc.)

Brain cells require a continuous supply of oxygen and glucose. Although the brain accounts for only about 2% of the body weight, it receives about 20% of the blood pumped by the heart each minute and consumes about 20% of the oxygen used by the body. The brain is so dependent upon its blood supply that when it is deprived of blood, the person loses consciousness after about 5 seconds, and irreversible damage occurs within a few minutes. In fact, the most common cause of brain damage is stroke (cerebrovascular accident), in which a portion of the brain is deprived of its blood supply (often because a blood vessel has been blocked by a blood clot).

The structure and functions of the main divisions of the brain are summarized in Table 34–2. The brain is illustrated in Figures 34–8 and 34–9.

As in other mammals, the largest, most prominent part of the human brain is the cerebrum. The cerebrum is divided into right and left cerebral hemispheres (see Focus on Cerebral Dominance). Functionally, the human cerebral cortex is divided into three areas: (1) the **sensory areas,** which receive incoming signals from the sense organs; (2) the **motor areas,** which control voluntary movement; and (3) the **association areas,** which link the sensory and motor areas and are responsible for thought, learning, intelligence, language abilities, memory, judgment, and personality.

The size of the motor area in the brain that controls a given part of the body is proportional to the precision and intricacy of the movements of that body part. Thus, a relatively large portion of motor area is devoted to controlling hand movements and facial

expression. There is a similar relationship between the parts of the sensory area and the region of the skin from which it receives impulses.

The cerebrum is divided into lobes (Fig. 34–8). The posterior occipital lobes contain the visual centers. Centers for hearing are located in the lateral temporal lobes. The central sulcus, a fissure that crosses the top of each hemisphere from medial to lateral edge, separates the frontal lobes from the parietal lobes. The primary motor areas, located in the frontal lobes, control skeletal muscle. The parietal lobes are responsible for sensations of heat, cold, touch, and pressure that result from stimulation of sense organs in the skin. The white matter of the cerebrum lies beneath the cerebral cortex. Nerve fibers of the white matter connect the cortical areas with one another and with other parts of the nervous system. A large band of white matter, the **corpus callosum,** connects the right and left hemispheres.

The Limbic System Affects Emotional Aspects Of Behavior

The **limbic system** is an action system of the brain. It consists of certain structures of the cerebrum, thalamus, and hypothalamus and has connections with many other parts of the CNS. The limbic system affects the emotional aspects of behavior, including motivation and feelings of pleasure and punishment. This system also affects memory, sexual behavior, biological rhythms, and autonomic responses. Stimulation of certain areas of the limbic system in an experimental animal results in increased general activity and may cause

TABLE 34-2
The Brain

Structure	Description	Function
Brain stem		
Medulla	Continuous with spinal cord; primarily made up of nerves passing from spinal cord to rest of brain	Contains vital centers (clusters of neuron cell bodies) that control heartbeat, respiration, and blood pressure; contains centers that control swallowing, coughing, vomiting
Pons	Forms bulge on anterior surface of brain stem	Connects various parts of brain with one another; contains respiratory center
Midbrain	Just above pons; largest part of brain in lower vertebrates; in human beings most of its functions are assumed by cerebrum	Center for visual and auditory reflexes (e.g., pupil reflex, blinking, adjusting ear to volume of sound)
Thalamus	At top of brain stem	Main sensory relay center for conducting information between spinal cord and cerebrum. Neurons in thalamus sort and interpret all incoming sensory information (except olfaction) before relaying messages to appropriate neurons in cerebrum
Hypothalamus	Just below thalamus; pituitary gland is connected to hypothalamus by stalk of neural tissue	Contains centers for control of body temperature, appetite, fat metabolism, and certain emotions; regulates pituitary gland; link between "mind" (cerebrum) and "body" (physiological mechanisms)
Cerebellum	Second largest division of brain	Reflex center for muscular coordination and refinement of movements; when injured, performance of voluntary movements is uncoordinated and clumsy
Cerebrum	Largest, most prominent part of human brain; more than 70% of brain's cells located here; longitudinal fissure divides cerebrum into right and left hemispheres, each divided by shallow sulci (furrows) into six lobes: frontal, parietal, temporal, insular, occipital, and limbic	Center of intellect, memory, consciousness, and language; also controls sensation and motor functions
Cerebral cortex (outer gray matter)	Arranged into convolutions (folds) that increase surface area; functionally, cerebral cortex is divided into:	
	1. Motor cortex	Controls movement of voluntary muscles
	2. Sensory cortex	Receives incoming information from eyes, ears, pressure and touch receptors, etc.
	3. Association cortex	Site of intellect, memory, language, and emotion; interprets incoming sensory information
White matter	White matter within cortex consists of myelinated axons of neurons that connect various regions of brain; these axons are arranged into bundles (tracts)	Connects: 1. Neurons within same hemisphere 2. Right and left hemispheres 3. Cerebrum with other parts of brain and spinal cord

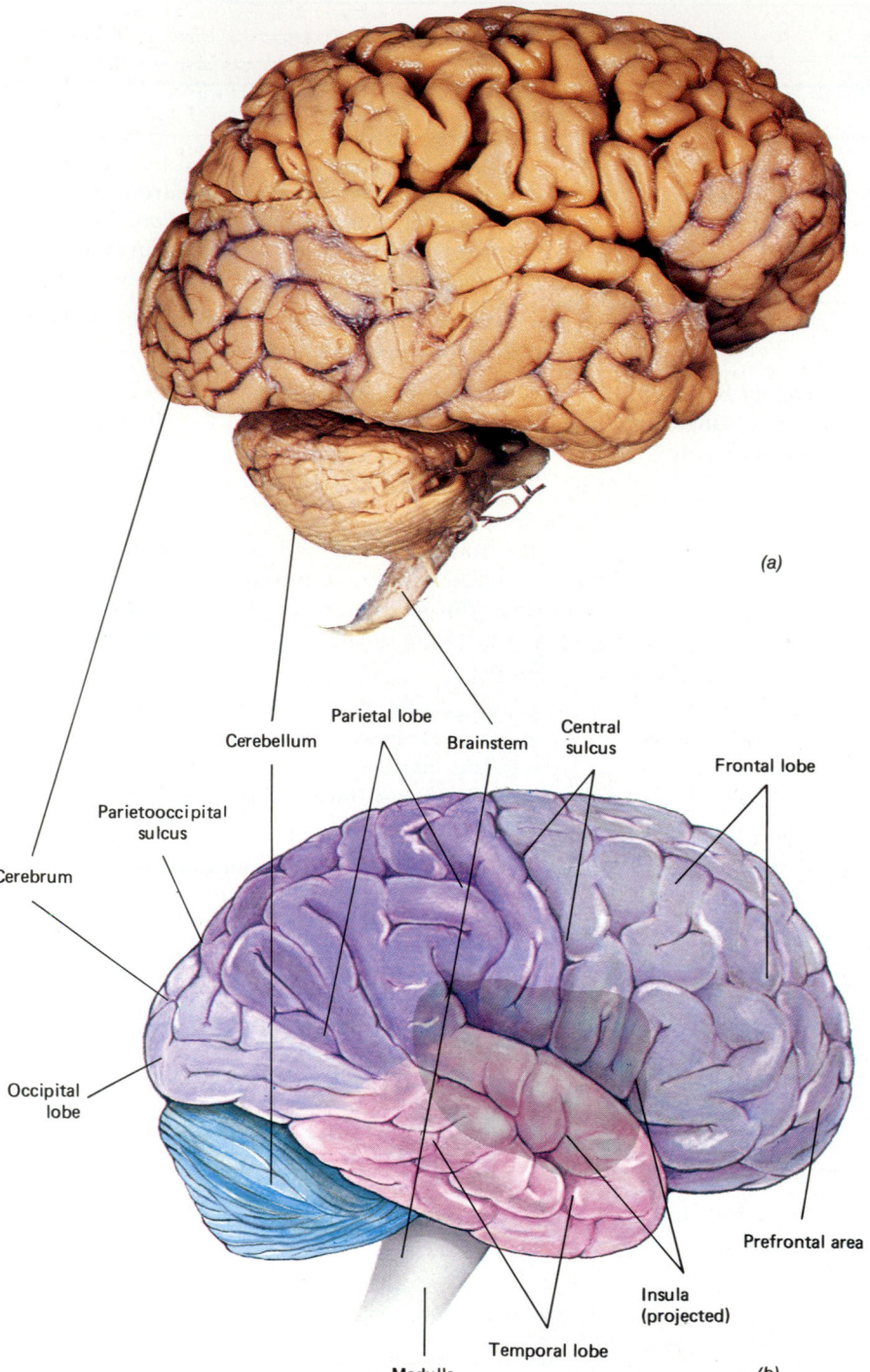

FIGURE 34–8 (*a*) Photograph of the human brain, lateral view. Note that the cerebrum covers part of the brain stem. (*b*) Lateral view of the human brain showing the lobes of the cerebrum. Part of the brain has been made transparent so that the underlying insular lobe can be located. (*a*, Fred Hossler/Visuals Unlimited)

(a)

Parietal lobe

Cerebellum   Brainstem   Central sulcus

Frontal lobe

Parietooccipital sulcus

Cerebrum

Occipital lobe

Prefrontal area

Insula (projected)

Temporal lobe

Medulla   (b)

fighting behavior or what appears to be extreme rage.

When an electrode is implanted in the so-called reward center of the limbic system, a rat will press a lever that stimulates this area as many as 15,000 times per hour. Stimulation of this area is apparently so rewarding that an animal will forgo food and drink and may continue to press the lever until it drops from exhaustion. When an electrode is implanted in the punishment center of the limbic system, an experimental animal quickly learns to press a lever to *avoid* stimula-

tion. The reward and punishment centers are important in influencing motivation and behavior.

The Reticular Activating System Is An Arousal System

Sometimes called the arousal system, the **reticular activating system (RAS)** is a complex pathway of neurons in the brain stem. It receives messages from neurons in

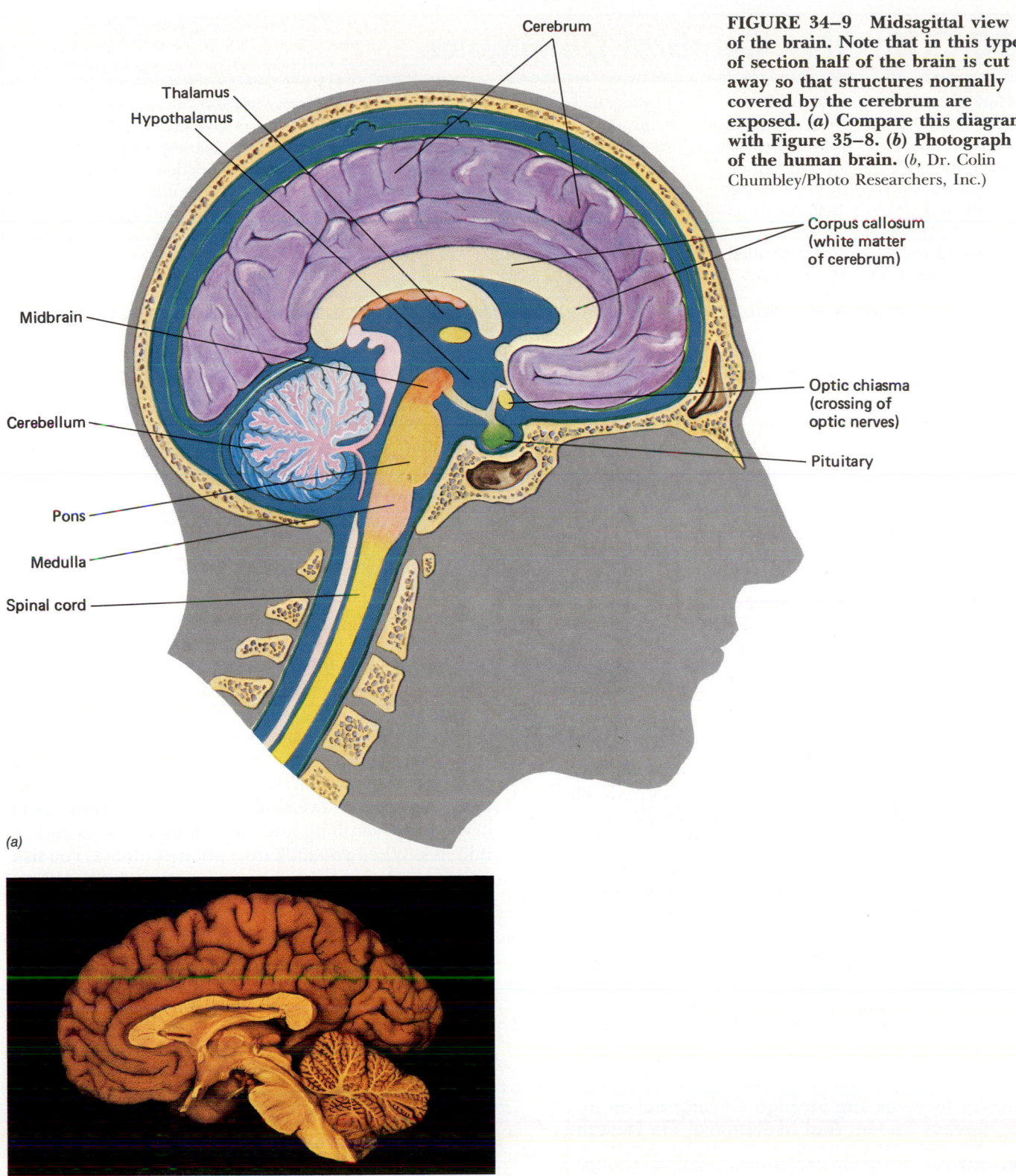

Cerebrum

Thalamus

Hypothalamus

Midbrain

Cerebellum

Pons

Medulla

Spinal cord

Corpus callosum
(white matter
of cerebrum)

Optic chiasma
(crossing of
optic nerves)

Pituitary

FIGURE 34–9 Midsagittal view of the brain. Note that in this type of section half of the brain is cut away so that structures normally covered by the cerebrum are exposed. (a) Compare this diagram with Figure 35–8. (b) Photograph of the human brain. (b, Dr. Colin Chumbley/Photo Researchers, Inc.)

(a)

(b)

the spinal cord and from many other parts of the nervous system and communicates with the cerebral cortex by complex circuits. The RAS is ultimately responsible for maintaining consciousness, and the extent of its activity determines the state of alertness. When the RAS bombards the cerebral cortex with stimuli, you feel alert and are able to focus your attention on specific thoughts. When its activity slows, you begin to feel sleepy. Sometimes when you are listening to a boring lecture, the RAS becomes habituated to the monoto-

FOCUS ON Cerebral Dominance

How many left-handed people do you know? Probably not very many, for 90% of us are right-handed. The remaining 10% are left-handed or ambidextrous. In right-handed people the left cerebral hemisphere is more highly developed for the motor functions related to handedness. In about 98% of adults (re-gardless of handedness) the left hemisphere is also dominant for language abilities, including the ability to speak, read, learn mathematics, and perform all other intellectual functions associated with language.

Until recently the left cerebral hemisphere was thought to be dominant in all respects in most people.

Now investigators think that the two hemispheres actually complement one another. The right hemisphere specializes in functions such as recognizing faces, identifying objects on the basis of shape, and appreciating and recognizing music and form. Some have suggested that creative abilities reside therein.

nous repetition of the professor's voice. As its signals become progressively weaker, the cerebrum may lapse into sleep. Should the RAS be severely damaged, as might happen in an accident, the unfortunate victim may pass into a deep, permanent coma.

When Signals From The RAS Slow, A Person May Fall Asleep

Sleep is a state of unconsciousness during which there is decreased electrical activity in the cerebral cortex and from which a person can be aroused. When the RAS slows the process of relaying incoming information to the cerebrum, we lose awareness of stimuli in our surroundings and go to sleep. Although we tend to stay awake in the presence of attention-holding stimuli, there is a limit beyond which sleep is inevitable.

Two main stages of sleep are recognized: **non-REM** and **REM sleep.** The letters REM are an acronym for rapid eye movements. During non-REM sleep, metabolic rate decreases, breathing slows, and blood pressure decreases. Every hour and a half or so during the night this relaxed-sleep pattern shifts to REM sleep. In REM sleep the individual dreams, electrical activity greatly increases, and beneath the closed eyelids the eyes move rapidly. Everyone dreams, though some people do not recall their nocturnal adventures.

Learning Involves The Storage Of Information In The Nervous System And Its Retrieval On Demand

Learning is a relatively long-lasting adaptive change in behavior resulting from experience. It is a modification of behavior that cannot be accounted for by sensory adaptation, central excitatory states, biological rhythms, or maturation. Laboratory experiments have shown that members of every animal phylum can learn, and learning is adaptive for a wide variety of environmental situations.

Despite extensive research, the mechanisms by which an animal thinks, learns, and remembers are still poorly understood. The human brain differs most dramatically from the brains of other animals by the remarkable development of the association areas within its cerebral cortex. Neurons within these areas form highly complex pathways. Damage to association areas can prevent a person from thinking logically, even though he or she may still be able to hear or even read.

To learn, the brain must be able to (1) focus attention on specific stimuli, whether they be the words on this printed page or an angry wasp buzzing overhead, (2) compare incoming sensory stimuli with stimuli it has encountered before, and (3) store information. **Memory** is the ability to recall stored information.

Just how the brain stores information and retrieves the memory on command has been the subject of much speculation. According to current theory, there are several levels of memory. Short-term memory involves recalling information for a few seconds or minutes. When you look up a phone number, you usually remember it only long enough to dial. Should you need the same number the next day, you would have to look it up again. One theory of short-term memory suggests that it is based on reverberating neural circuits. In a reverberating circuit a neural message is fed back into the circuit by way of an axon collateral (branch) so that a new impulse is sent through the circuit. A memory circuit may continue to reverberate for several minutes until it fatigues or until new signals are received that interfere with the old.

When you select a bit of information for long-term storage, the brain is thought to rehearse the material and then store it in association with similar memories. According to one theory, some physical or chemical change takes place in the synaptic knobs or postsynaptic neurons that permanently facilitates the transmission of impulses within a newly established circuit. Perhaps specific neurons become more sensitive to neurotransmitter. Each time a memory is stored, a new neural pathway is facilitated. Whatever the exact mechanism, memory almost certainly depends on a sequence of molecular events.

Several minutes are required for a memory to become consolidated within long-term memory. Should a person suffer a brain concussion or undergo electroshock therapy, memory of what transpired immediately prior to the incident may be completely lost. When certain parts of the limbic system are injured or removed, a person can recall information stored in the past but is no longer able to store new information.

Retrieval of information stored in long-term memory is of considerable interest—especially to students. Some researchers believe that once information is deposited in long-term storage, it remains in the brain permanently. The only problem is finding the information when we need it. When we seem to forget a particular bit of information, it is because we have not efficiently stored it or searched for it.

Experience Affects The Brain

An expanding array of multicolored and multishaped crib mobiles, educational toys for babies, and emphasis on early childhood education reflect the view that the environment plays an important role in shaping development and academic potential. Studies indicate that the way parents relate to their children during the first 18 months of life is critical in establishing personality and potential for future academic achievement. Parents who interact warmly with their young children and who stimulate their intellectual curiosity from the earliest months of life help to ensure future social and academic success. The structure and biochemistry of the brain can be changed by environmental enrichment.

Studies on experimental animals confirm that environmental experience results in physical and chemical changes in brain structure. Several studies have been performed in which one group of rats is provided with a stimulating environment and given the opportunity to learn while other groups are deprived of stimulation and social interaction. After some months the animals are sacrificed and their brain structures are compared. Those exposed to enriched environments exhibit an increased concentration of synaptic contact. Some investigators have reported that the cerebral cortex actually becomes thicker and heavier. Characteristic biochemical changes also take place. Other experiments have indicated that animals reared in a complex environment may be able to process and remember information more quickly than animals not provided with such advantages.

During early life there are apparently certain critical or sensitive periods of nervous system development that are influenced by environmental stimuli. For example, when the eyes of young mice first open, large numbers of dendritic spines (sites where synapses form) develop on neurons in the visual cortex. When the animals are kept in the dark and deprived of visual stimuli, fewer dendritic spines form. If later in life the mice are exposed to light, some new dendritic spines form, but never as many as develop in a mouse reared in a normal environment.

Early environmental stimulation can also promote development of motor areas in the brain. For example, young rats encouraged to exercise show increased development of the cerebellum. Such studies linking the development of the brain with environmental experience support the concept that early stimulation is important for sensory, motor, and intellectual development.

■ THE PERIPHERAL NERVOUS SYSTEM INCLUDES SOMATIC AND AUTONOMIC SYSTEMS

The **peripheral nervous system (PNS)** consists of all the receptors, the nerves that link the receptors with the CNS, and the nerves that link the CNS with the muscles and glands (the effectors). The portion of the PNS that helps the body respond to changes in the external environment is the somatic system; the nerves and receptors that maintain homeostasis despite internal changes make up the autonomic nervous system.

The Somatic System Helps The Body Adjust To The External Environment

The **somatic nervous system** includes those receptors that react to changes in the external environment, the sensory neurons that keep the CNS informed of those changes, and the motor neurons that adjust the positions of the skeletal muscles, maintaining the body's well-being. In mammals, 12 pairs of **cranial nerves** emerge from the brain itself. These nerves transmit information regarding the senses of smell, sight, hearing, and taste from the special sensory receptors, and information from the general sensory receptors, especially in the head region. The cranial nerves also bring orders from the CNS to the voluntary muscles that control movements of the eyes, face, mouth, tongue, pharynx, and larynx.

In humans 31 pairs of spinal nerves emerge from the spinal cord (Fig. 34–10). Named for the general region of the vertebral column from which they originate, there are 8 pairs of cervical spinal nerves, 12 pairs of thoracic, 5 pairs of lumbar, 5 pairs of sacral, and 1 pair of coccygeal spinal nerves.

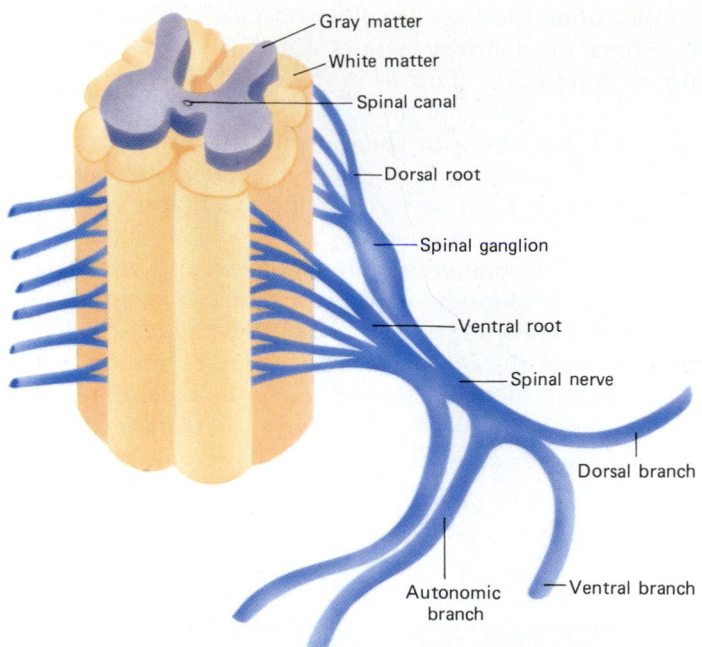

FIGURE 34–10 Dorsal and ventral roots emerge from the spinal cord and join to form a spinal nerve. The spinal nerve divides into several branches.

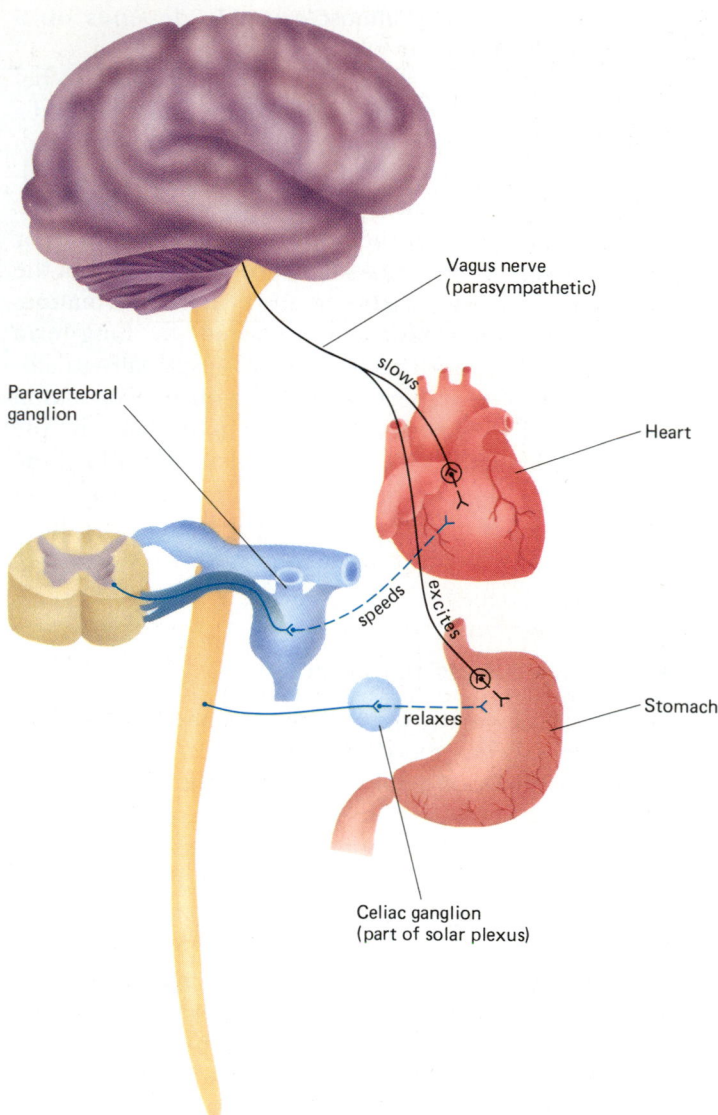

FIGURE 34–11 Dual innervation of the heart and stomach by sympathetic and parasympathetic nerves. Sympathetic nerves are shown in blue.

Each spinal nerve has a **dorsal root** and a **ventral root.** The dorsal root consists of sensory (afferent) fibers, which transmit information from the sensory receptors to the spinal cord. Just before the dorsal root joins with the cord, it is marked by a swelling, the **spinal ganglion,** which consists of the cell bodies of the sensory neurons. The ventral root consists of the motor (efferent) fibers leaving the cord en route to the muscles and glands. Cell bodies of the motor neurons are located within the gray matter of the cord.

The Autonomic System Helps Maintain Homeostasis In Response To Changes In The Internal Environment

The **autonomic system** helps to maintain homeostasis in the internal environment. For instance, it regulates the rate of the heartbeat and maintains a constant body temperature. The autonomic system works automatically and without voluntary input. Its effectors are smooth muscle, cardiac muscle, and glands. Like the somatic system, it is functionally organized into reflex pathways. Receptors within the viscera (internal organs) relay information via afferent nerves to the CNS, the information is integrated at various levels, and the decision is transmitted along efferent nerves to the appropriate muscles or glands.

Information from the various organs is transmitted to the CNS by afferent neurons. Some of these are located within cranial or spinal nerves. These afferent neurons are thought to synapse with association neu-

rons within the CNS. They bring information about blood pressure, respiration, heartbeat, contractions of the digestive tract, and other visceral activities.

The efferent portion of the autonomic system is subdivided into **sympathetic** and **parasympathetic systems.** In general, the sympathetic nerves operate to stimulate organs and to mobilize energy, especially in response to stress, whereas the parasympathetic nerves influence organs to conserve and restore energy, particularly when one is engaged in quiet, calm activities, such as studying biology. Many organs are innervated by both types of nerves, which act upon the organ in a complementary way (Fig. 34–11). For example, the heart rate is slowed by impulses from its parasympathetic nerve fibers and speeded up by messages from its sympathetic nerve supply.

FOCUS ON Alcohol Abuse

Alcohol abuse is responsible for more than 100,000 deaths each year and costs our society more than $100 billion annually. According to pollster Louis Harris, there are 28 million alcoholics in the United States and about one in three homes include someone with a serious drinking problem. Alcohol abuse is not limited to adults. About 4.6 million adolescents or nearly one of every three high school students experience negative consequences from alcohol use, including difficulty with parents, poor performance at school, and breaking the law. In 1986 alcohol was still the drug most widely abused by youth in the United States.

Alcohol abuse results in physiological, psychological, and social impairment for the abuser and also has serious negative consequences for family, friends, and society.

Alcohol abuse has been linked to:

- More than 50% of all traffic fatalities
- More than 50% of violent crimes
- More than 50% of suicides
- More than 60% of cases of child abuse and spouse abuse
- The birth of 15,000 babies with

serious birth defects born every year because their mothers drank alcohol excessively during pregnancy

- Breast cancer: Recent studies suggest that as little as three drinks per week increase risk of breast cancer by 50%.
- Greater risk of liver disease and brain impairment for women than males (clinical effects are seen in women consuming about half the alcohol consumed by men).

Alcohol accumulates in the blood because absorption occurs more rapidly than oxidation and excretion. Alcohol causes depression of the CNS. Effects of various blood alcohol levels are: 50 mg/dL causes sedation; 50–150 mg/dL results in loss of coordination; 150–200 mg/dL produces intoxication; 300–400 mg/dL may result in unconsciousness. Blood levels greater than 500 mg/dL may be fatal. Alcohol is oxidized to carbon dioxide and water. About 10% is excreted unchanged in expired air, urine, and sweat. Blood alcohol levels are usually estimated from the amount present in expired air.

Alcohol commonly causes cirrhosis of the liver, peripheral nerve de-

generation, brain damage, and cardiac damage accompanied by arrhythmias. Gastritis and damage to the pancreas are also common.

Tolerance in those who drink excessive amounts of alcohol occurs because cells of the CNS adapt to the presence of the drug. Physical dependence develops along with tolerance, and withdrawal may result in serious physiological derangements that can lead to death.

According to one study children see more than 100,000 beer commercials on television before they are legally old enough to drink. Congress is currently discussing legislation that would force the alcoholic beverage industry to place labels on its products warning that alcohol can cause mental retardation and other birth defects and impairs the ability to drive or operate machinery. However, there is strong opposition from the alcohol industry, which spends $2 billion annually on advertising.

Treatment for alcohol problems includes various forms of psychotherapy, including relapse prevention, a type of cognitive-behavioral modification. The group support offered by Alcoholics Anonymous (AA) has proved effective for many struggling with alcohol abuse.

MANY MOOD DRUGS CHANGE THE LEVELS OF NEUROTRANSMITTERS IN THE BRAIN

Drugs help us fall asleep; drugs help us stay awake. Drugs help us relax, forget our problems, feel like part of the group. Drugs "expand the mind." Some are purchased freely over the counter; some are prescribed by physicians; others are bought from illegal sources. Of all prescribed drugs, some 25% are for psychological conditions. Each year physicians write more than 200 million prescriptions for sedatives, tranquilizers, and "pep pills." More and more people are attempting to control their moods with an ever-growing variety of stronger and stronger drugs. We are a pill-popping society and may someday be remembered as "the drugged generation."

Why is this emphasis on drugs undesirable? One reason is that no drug is completely harmless. The body is a delicately balanced machine, and even a mild

drug is a chemical trespasser. Even aspirin has been shown to cause noteworthy side effects such as gastric bleeding. But besides the individual matter of doing harm to one's own body, the use and abuse of drugs have become an important social issue. Drugs affect not only individual users but also their families, their employers, and even strangers with whom they share the highway or from whom they steal to support their habit.

Many mood drugs act by altering the levels of neurotransmitters within the brain. For example, when excessive amounts of norepinephrine are released in the RAS, we feel stimulated and energetic. Abnormally low concentrations of this neurotransmitter reduce anxiety.

Habitual use of almost all mood drugs may result in **psychological dependence,** in which the user becomes emotionally dependent upon the drug. When deprived of it, the user craves the feeling of **euphoria**

TABLE 34-3
Effects of Some Commonly Used Drugs

Name of Drug	Effect on Mood	Actions on Body	Dangers Associated with Abuse
Barbiturates (e.g., Nembutal, Seconal)	Sedative-hypnotic[1]; "downers"	Inhibit impulse conduction in RAS: depress CNS, skeletal muscle, and heart; depress respiration; lower blood pressure; cause decrease in REM sleep	Tolerance, physical dependence, death from overdose, especially in combination with alcohol
Methaqualone (e.g., Quaalude, Sopor)	Hypnotic	Depresses CNS; depresses certain polysynaptic spinal reflexes	Tolerance, physical dependence, convulsions, death
Meprobamate (e.g., Equanil, Miltown; "minor tranquilizers")	Antianxiety drug[2]; induces calmness	Causes decrease in REM sleep; relaxes skeletal muscle; depresses CNS	Tolerance, physical dependence; coma and death from overdose
Valium, Librium ("mild tranquilizers")	Reduce anxiety	May reduce rate of impulse firing in limbic system; relax skeletal muscle	Minor EEG abnormalities with chronic use; very large doses cause physical dependence
Phenothiazines (chlorpromazine; "major tranquilizers")	Antipsychotic; highly effective in controlling symptoms of psychotic patients	Affect levels of catecholamines in brain (block dopamine receptors, inhibit uptake of norepinephrine, dopamine, and serotonin); depress neurons in RAS and basal ganglia	Prolonged intake may result in Parkinson-like symptoms
Antidepressant drugs (e.g., Elavil)	Elevate mood; relieve depression	Block uptake of norepinephrine, so more is available to stimulate nervous system	Central and peripheral neurological disturbances; uncoordination; interfere with normal cardiovascular function
Alcohol	Euphoria; relaxation; release of inhibitions	Depresses CNS; impairs vision, coordination, judgment; lengthens reaction time	Physical dependence; damage to pancreas; liver cirrhosis; possible brain damage
Narcotic analgesics (e.g., morphine, heroin)	Euphoria; reduction of pain	Depress CNS; depress reflexes; constrict pupils; impair coordination. Block release of substance P from pain-transmitting neurons.	Tolerance; physical dependence; convulsions; death from overdose
Cocaine	Euphoria; excitation followed by depression	CNS stimulation followed by depression; autonomic stimulation; dilates pupils; local anesthesia; inhibits re-uptake of norepinephrine	Mental impairment; convulsions; hallucinations; unconsciousness; death from overdose
Amphetamines (e.g., Dexedrine)	Euphoria; stimulant; hyperactivity; "uppers," "pep pills"	Stimulate release of dopamine and norepinephrine; block re-uptake of norepinephrine and dopamine into neurons; inhibit monoamine oxidase (MAO); enhance flow of impulses in RAS; increase heart rate; raise blood pressure; dilate pupils	Tolerance; possible physical dependence; hallucinations; death from overdose
Caffeine	Increases mental alertness; decreases fatigue and drowsiness	Acts on cerebral cortex; relaxes smooth muscle; stimulates cardiac and skeletal muscle; increases urine volume (diuretic effect)	Very large doses stimulate centers in the medulla (may slow the heart); toxic doses may cause convulsions
Nicotine	Psychological effect of lessening tension	Stimulates sympathetic nervous system; combines with receptors in postsynaptic neurons of autonomic	Tolerance; physical dependence; stimulates development of atherosclerosis

Name of Drug	Effect on Mood	Actions on Body	Dangers Associated with Abuse
		system; effect similar to that of acetylcholine, but large amounts result in blocking transmission; stimulates synthesis of lipid in arterial wall	
LSD (lysergic acid diethylamide)	Overexcitation; sensory distortions; hallucinations	Alters levels of transmitters in brain (may inhibit serotonin and increase norepinephrine); potent CNS stimulator; dilates pupils sometimes unequally; increases heart rate; raises blood pressure	Irrational behavior
Marijuana	Euphoria	Impairs coordination; impairs depth perception and alters sense of timing; inflames eyes; causes peripheral vasodilation; exact mode of action unknown	In large doses, sensory distortions, hallucinations; evidence of lowered sperm counts and testosterone (male hormone) levels

[1]Sedatives reduce anxiety; hypnotics induce sleep.
[2]Antianxiety drugs reduce anxiety but are less likely to cause drowsiness than the more potent sedative-hypnotics.

(well-being) that the drug induces. Some drugs induce **tolerance** when they are taken continually for several weeks. This means that increasingly larger amounts are required to obtain the desired effect. Tolerance occurs when the liver cells are stimulated to produce larger quantities of the enzymes that break down the drug. Use of some drugs, such as alcohol, barbiturates, or heroin, results in **physical dependence,** or **addiction,** in which physiological changes take place. When the drug is withheld, the addict suffers physical illness and characteristic withdrawal symptoms.

Some commonly used and abused drugs and their effects are listed and described in Table 34–3. Some of the classes of drugs commonly abused are alcohol, stimulants, depressants, cocaine, opiates, hallucinogens, and marijuana.

FOCUS ON   Crack Cocaine

The majority of persons seeking treatment for drug abuse are now crack cocaine addicts. Cocaine use by teenagers alone has increased about 400% during the past 10 years, involving an estimated 2 million youngsters. Crack is a very concentrated and extremely powerful form of cocaine—5 to 10 times as addictive as other forms of cocaine. This drug is produced in illegal, makeshift labs by converting powdered cocaine into small "rocks" which are up to 80% pure cocaine. Crack is smoked in pipes or in tobacco or marijuana cigarettes.

Use of crack results in an intense, brief high beginning in 4 to 6 seconds and lasting for 5 to 7 minutes. Physiologically, crack stimulates a massive release of catecholamine neurotransmitters (norepinephrine and dopamine) in the brain and is thought to block reuptake. Excitation of the sympathetic nervous system occurs and users report experiencing feelings of self-confidence, power, and euphoria. As the neurotransmitters are depleted, the high is followed by a "crash," a period of deep depression. The abuser experiences an intense craving for another crack "hit" in order to get more stimulation.

Some abusers spend days smoking crack without stopping to eat or sleep. Although a vial of rocks can be obtained for about $20, many abusers develop habits that cost hundreds of dollars a week. Supporting an expensive drug habit leads many abusers to prostitution, drug dealing, and other forms of crime.

Cocaine addicts report problems with memory, fatigue, depression, insomnia, paranoia, loss of sexual drive, violent behavior, and attempts at suicide. Crack can cause respiratory problems, brain seizures, cardiac arrest, and elevation of blood pressure leading to stroke. Many users have suffered fatal reactions to impurities in the drug or have died as a result of accidents related to drug use.

■ CHAPTER SUMMARY

I. Among invertebrates, nerve nets and radial nervous systems are typical of radially symmetrical animals, and bilateral nervous systems are characteristic of bilaterally symmetrical animals.
  A. A nerve net consists of nerve cells scattered throughout the body; no CNS is present. Response in these animals is generally slow and imprecise.
  B. Echinoderms typically have a nerve ring and nerves that extend into various parts of the body.
  C. In a bilateral nervous system there is a concentration of nerve cells to form nerves, nerve cords, ganglia, and (in complex animals) a brain. There is also an increase in numbers of neurons, especially of the association neurons. This permits greater precision and a wider range of responses.

II. The vertebrate nervous system consists of central nervous system and peripheral nervous system.

III. In the vertebrate embryo, the neural tube gives rise to brain and spinal cord. The anterior end of the tube differentiates into forebrain, midbrain, and hindbrain.
  A. The hindbrain develops into the medulla, cerebellum, and pons.
    1. In complex vertebrates, the medulla contains vital centers regulating respiration, heart rate, and blood pressure.
    2. The cerebellum is highly developed in fish, birds, and mammals—animals that tend to engage in complex movement.
    3. In mammals, the pons serves as a bridge connecting spinal cord and medulla with upper parts of the brain.
  B. In fish and amphibians the midbrain is the main association area linking sensory input and motor output. In reptiles, birds, and mammals, the midbrain serves as a center for visual and auditory reflexes.
  C. The forebrain differentiates to form the diencephalon and telencephalon.
    1. The diencephalon gives rise to the thalamus and hypothalamus.
      a. The thalamus is a relay center for motor and sensory information.
      b. The hypothalamus controls autonomic functions, links nervous and endocrine systems, and controls temperature, appetite, and fluid balance.
    2. The telencephalon develops into the cerebrum and olfactory bulbs. In fish and amphibians, the cerebrum mainly integrates incoming sensory information.

IV. The vertebrate central nervous system consists of brain and spinal cord.
  A. The spinal cord consists of ascending tracts, which transmit information to the brain, and descending tracts, which transmit information from the brain. Its gray matter contains many nuclei, which serve as reflex centers.
  B. The brain consists of the cerebrum, cerebellum, and brain stem. The brain stem includes the medulla, pons, midbrain, thalamus, and hypothalamus.
  C. The cerebral cortex contains motor areas, which control voluntary movement; sensory areas, which receive incoming sensory information; and association areas, which link sensory and motor areas and are also responsible for learning, language, thought, and judgment.
  D. The limbic system affects the emotional aspects of behavior, motivation, sexual behavior, autonomic responses, and biological rhythms.
  E. The reticular activating system is responsible for maintaining consciousness.
  F. Metabolic rate slows during non-REM sleep. REM sleep is characterized by rapid eye movements and dreaming.
  G. Short-term memory may depend upon reverberating circuits in the brain. Mechanisms of long-term memory are not understood.
  H. Experience can cause physical and chemical changes in the brain.

V. The peripheral nervous system consists of sensory receptors and nerves, including the cranial and spinal nerves and their branches.
  A. The somatic system helps the body to adjust to the external environment.
  B. The autonomic system regulates the internal activities of the body.
    1. The sympathetic system enables the body to respond to stressful situations.
    2. The parasympathetic system influences organs to conserve and restore energy.

VI. Many drugs alter mood by increasing or decreasing the concentrations of specific neurotransmitters within the brain.

■ POST-TEST

1. The simplest organized nervous tissue is the
   _____ _____ which is found in
   Hydra.
2. Afferent nerves conduct impulses toward
   _____ .
3. In a planarian flatworm the _____ serve as
   a primitive brain.
4. The vertebrate central nervous system consists of the
   _____ and _____ _____ .
5. The most prominent part of the amphibian brain is
   the _____ .
6. In fish and amphibian brains, the well-developed
   _____ bulbs receive information about
   _____ .
7. The neopallium serves as an _____ area.
8. The function of the cerebrospinal fluid is to
   _____ .
9. The largest, most prominent part of the mammalian
   brain is the _____ .
10. Voluntary movement is controlled by
    _____ in the cerebral cortex.
11. The reticular activating system (RAS) is responsible
    for _____ .

For each group, select the most appropriate answer from
Column B for the description given in Column A.

Column A
12. Coordinates and refines muscular movement
13. Contains center that controls heartbeat
14. Regulates pituitary gland
15. Receives incoming signals from sense organs
16. Interprets incoming motor information

17. The efferent portion of the autonomic system is sub-
    divided into the _____ and
    _____ systems.
18. An example of _____ is the need for a
    user to increase the dose of a drug to obtain the de-
    sired effect; this often occurs after several weeks of
    taking the same drug.
19. Label the diagram.

Column B
a. hypothalamus
b. sensory cortex
c. cerebellum
d. medulla
e. none of the above

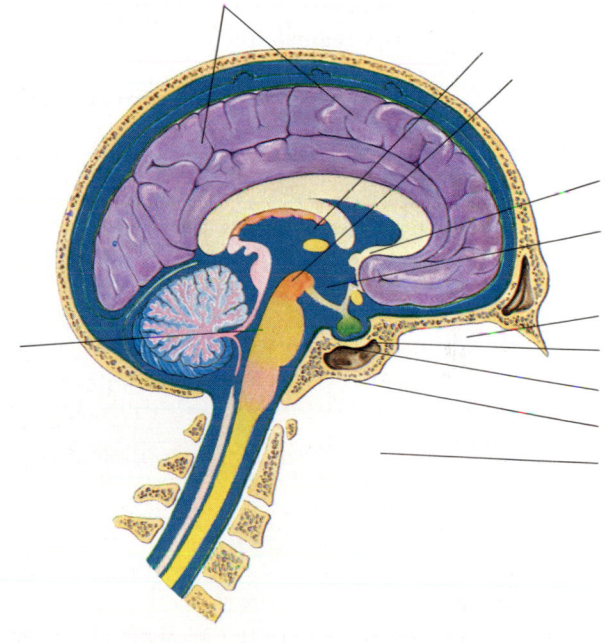

■ REVIEW QUESTIONS

1. Compare the nervous system of a hydra with that of a
   planarian flatworm.
2. Compare the flatworm nervous system with that of a vertebrate.
3. What general trends can you identify in the evolution of
   the vertebrate brain?
4. Compare the fish brain and the mammalian brain.

5. What structures protect the brain and spinal cord?
6. What are the functions of the spinal cord?
7. Identify the part of the brain most closely associated with
   each of the following: (a) regulation of body temperature, (b) regulation of heart rate, (c) link between nervous and endocrine systems, (d) interpretation of incoming sensory messages, (e) coordination of movements.

8. Imagine that you have just become a parent. What kind of things could you do to ensure development of your child's academic abilities?
9. What is the RAS? How does it function?
10. In what way does electrical activity of the brain reflect a person's state of consciousness? Of what use might it be to learn to control one's brain wave patterns?
11. What factors influence pain perception?
12. Contrast somatic and autonomic systems.
13. Contrast sympathetic and parasympathetic systems.
14. Describe how these drugs affect the CNS: (a) alcohol, (b) Dexedrine or other amphetamines, (c) barbiturates, (d) antipsychotic drugs, (e) hallucinogens.

■ RECOMMENDED READINGS

Aoki, C. and Siekevitz, P.: "Plasticity in Brain Development," *Scientific American* 259(6), 1988. Experience and use affect and restructure the developing brain.

Black, I.B., et al: "Biochemistry of Information Storage in the Nervous System," *Science* 256:1263–1268, 1987. Use of molecular approaches has defined new mechanisms that store information in the mammalian nervous system.

Constantine-Paton, M., and M. Law: "The Development of Maps and Stripes in the Brain," *Scientific American* 247 (6): 62–70, 1982. An account of studies of the visual system and its organization.

Goldstein, G.W., and A.L. Betz: "The Blood-Brain Barrier," *Scientific American* 255 (3):74–83, 1986. A discussion of how brain capillaries regulate the entrance of materials into the brain.

Kandel, E.R., and J.H. Schwartz: "Molecular Biology of Learning: Modulation of Transmitter Release," *Science* 218, October 29, 1982. A review that focuses on the biochemistry of learning in a marine mollusk.

Mishkin, M., and T. Appenzeller: "The Anatomy of Memory," *Scientific American* 256 (6):80–89, 1987. Deep structures in the brain may interact with perceptual pathways in outer layers of the brain to transform sensory stimuli into memories.

Morrison, A.R.: "A Window on the Sleeping Brain," *Scientific American* 248 (4):94–101, 1983. A study of REM sleep and accompanying paralysis.

Wurtman, R.J.: "Nutrients That Modify Brain Function," *Scientific American* 246 (4):50–59, 1982. An exciting discussion of the effect of such nutrients as choline, tryptophane, and tyrosine on brain function.

35
Sense Organs

OUTLINE

I. Different types of sense organs respond to different types of energy
II. Sense organs work by producing receptor potentials
III. Sensation depends on transmission of a "coded" message
IV. Receptors adapt to stimuli
V. Mechanoreceptors respond to touch, pressure, gravity, stretch, or movement
  A. Touch receptors are located in the skin
  B. Proprioceptors help coordinate muscle movement
  C. Lateral line organs supplement vision in fish
  D. Many invertebrates have gravity receptors called statocysts
  E. The labyrinth of the vertebrate ear is an organ of equilibrium
  F. Auditory receptors are located in the cochlea
VI. Chemoreceptors detect taste and smell
  A. Taste buds are the organs of taste in humans
  B. The olfactory epithelium is responsible for the sense of smell
VII. Thermoreceptors are sensitive to heat
VIII. Electroreceptors detect electrical currents in water
IX. Photoreceptors use pigments to absorb light
  A. Eyespots, simple eyes, and compound eyes are found among invertebrates
  B. Vertebrate eyes form sharp images
    1. The retina is the light-sensitive part of the eye
    2. A chemical change in rhodopsin leads to transmission of an action potential

Focus on pain perception
Focus on defects in vision

LEARNING OBJECTIVES

After you have studied this chapter you should be able to:

1. Distinguish among the five types of receptors that are classified according to the types of energy to which they respond. Give examples of specific sense organs of each type.
2. Summarize how a sense organ functions, including definitions of energy transduction, receptor potential, and adaptation in your answer.
3. Describe the following mechanoreceptors: tactile receptors, proprioceptors, lateral line organs, and statocysts.
4. Compare the function of the saccule and utricle with that of the semicircular canals in maintaining equilibrium.
5. Trace the path taken by sound waves through the structures of the ear, and explain how the organ of Corti is able to function as an auditory receptor.
6. Describe the receptors of taste and smell.
7. Relate the presence of thermoreceptors to the life-style of animals that have them.
8. Contrast simple eyes, compound eyes, and the vertebrate eye.
9. Label the structures of the vertebrate eye on a diagram, and give the functions of each of the accessory structures.
10. Compare the two types of photoreceptors in the human retina.

Cottontail rabbit. (Sharon Cummings/Marvin Dembinsky, Jr., Photography Associates)

Sense organs link animals with the outside world and enable them to receive information about their external and internal environment. Most familiar to us are the complex sense organs located in the head—the eyes, ears, nose, and taste buds. However, many other types of sense organs are known. The kinds of sense organs an animal has determine just how it perceives the world. We humans live in a world of rich colors, multiple shapes, and varied sounds. But we cannot hear the high-pitched whistles audible to dogs and cats or the ultrasonic echoes by which bats navigate (Fig. 35–1). Nor do we ordinarily recognize our friends by their distinctive odors. And although vision is our dominant and most refined sense, we are blind to the ultraviolet hues that light up the world for insects.

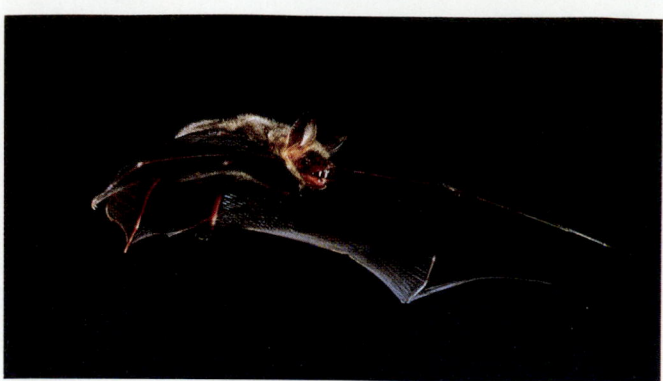

FIGURE 35–1 Large mouse-eared bat *(Grand Murin myotis myotis)*. Bats navigate by ultrasonic echoes inaudible to the human ear. They emit high-pitched clicking sounds and use the echos to locate objects. (Gerard Lacz/Peter Arnold, Inc.)

KEY CONCEPTS

☐ Sense organs detect an event in the environment by absorbing energy; converting that energy into electrical energy; and producing a receptor potential that may result in an action potential that transmits information to the central nervous system (CNS).

☐ Mechanoreceptors detect touch, pressure, movement, or gravity; chemoreceptors detect specific chemical compounds; thermoreceptors detect heat; electroreceptors detect electrical energy; photoreceptors detect light energy.

☐ DIFFERENT TYPES OF SENSE ORGANS RESPOND TO DIFFERENT TYPES OF ENERGY

A sense organ consists of one or more **receptor cells** and sometimes accessory cells. Receptor cells may be either neuron endings or specialized cells that are in close contact with neurons. Human taste buds, for example, are modified epithelial cells connected to one or more neurons. Sense organs initiate nerve impulses in response to specific changes in their environment.

Sense organs can be classified according to the type of energy to which they respond (Table 35–1). **Mechanoreceptors** respond to mechanical energy—touch, pressure, gravity, stretching, or movement. **Chemoreceptors** respond to certain chemical compounds, while **photoreceptors** detect light energy. **Thermoreceptors** respond to heat or cold. Some fish have **electroreceptors,** which detect electrical energy.

Traditionally, mammals are said to have five senses: touch, smell, taste, sight, and hearing. Balance is now also recognized as a sense, and touch is viewed as a compound sense that involves detection of pressure, pain, and temperature. In this chapter we will also consider receptors that enable us to sense muscle tension and joint position.

Sense organs can also be classified according to the location of the stimuli affecting them. **Exteroceptors** receive stimuli from the outside environment, enabling an animal to know and explore the world, search for food, find and attract a mate, find shelter, recognize friends, detect enemies, and even learn. **Proprioceptors** are sense organs within muscles, tendons, and joints that enable the animal to perceive the position of its arms, legs, head, and other body parts, along with the orientation of its body as a whole. With the help of our proprioceptors we humans can get dressed or eat in the dark.

Interoceptors are sense organs *within* body organs that detect changes in pH, osmotic pressure, body temperature, and the chemical composition of the blood. We are usually not conscious of messages sent to the CNS by these receptors as they work continuously, maintaining homeostasis. We become aware of their activity when they enable us to perceive such diverse internal conditions as thirst, hunger, nausea, pain, and orgasm. Interoceptors will be described in later chapters with discussions of blood pressure, temperature regulation, respiration, and other specific body functions.

☐ SENSE ORGANS WORK BY PRODUCING RECEPTOR POTENTIALS

Receptor cells absorb energy, **transduce** (convert) that energy into electrical energy, and produce a receptor potential (Fig. 35–2). In its capacity as a detector or

TABLE 35–1
Classification of Receptors by Stimuli to Which They Respond

Type of Receptor	Examples	Effective Stimulus
Mechanoreceptors	Tactile receptors 　Pacinian corpuscles 　Meissner's corpuscles	Touch, pressure
	Proprioceptors	Movement and body position
	Muscle spindles	Detect muscle contraction
	Golgi tendon organs	Stretch of a tendon
	Joint receptors	Detect movement in ligaments
	Lateral line organs in fish	Waves, currents in the water
	Statocysts in invertebrates	Gravity
	Labyrinth of vertebrate ear	
	Saccule and utricle	Gravity; linear acceleration
	Semicircular canals	Angular acceleration
	Cochlea	Pressure waves (sound)
Chemoreceptors	Taste buds, olfactory epithelium	Specific chemical compounds
Thermoreceptors	Temperature receptors in blood-sucking insects and ticks; pit organs of pit vipers; nerve endings and receptors in skin and tongue of many animals	Heat
Electroreceptors	Organs in skin of some fish	Electrical currents in water
Photoreceptors	Eyespots: ommatidia of arthropods; retina of vertebrates	Light energy

sensor, a receptor receives a small amount of energy from the environment. Each kind of receptor is especially sensitive to one particular form of energy. For example, pigment molecules in photoreceptor cells absorb light energy. Temperature receptors respond to radiant energy transferred by radiation, conduction, or convection. Taste buds and olfactory cells detect the change in energy accompanying the binding of specific molecules to their chemical receptors. Electricity is detected by the energy of electrons.

Receptor cells are remarkably sensitive to *appropriate* stimuli. The photoreceptors (rods and cones) of the human eye, for example, are stimulated by an extremely faint beam of light, whereas only a very strong light can stimulate the optic nerve directly. The negligible amount of vinegar that can be tasted, or the amount of vanilla that can be smelled, would have no effect if applied directly to a nerve fiber.

The various kinds of environmental energy act as triggers for the receptor cells stimulating them to perform biological work. These relationships are best understood by a simple sense organ, the tactile hair of an insect. This hair plus its associated cells constitute a complete sense organ, but only the neuron at the base of the hair is a receptor cell. The dendrite of the neuron is attached to the base of the hair near the socket, and the axon passes directly to the CNS without synapsing.

In its unstimulated state, the neuron maintains a *resting potential*. Recall (Chapter 34) that the resting potential results from the potential difference between the inside and outside of the neuron. This potential difference exists because the ionic compositions of the fluids on each side of the selectively permeable plasma membrane are different. The difference is maintained by sodium-potassium pumps powered by adenosine triphosphate (ATP). When the hair is touched (a mechanical stimulus), its shaft moves in the socket, causing the dendrite to change shape. This change in shape represents mechanical energy and increases the permeability of the neuron membrane to ions. As a result the potential difference between the two sides of the

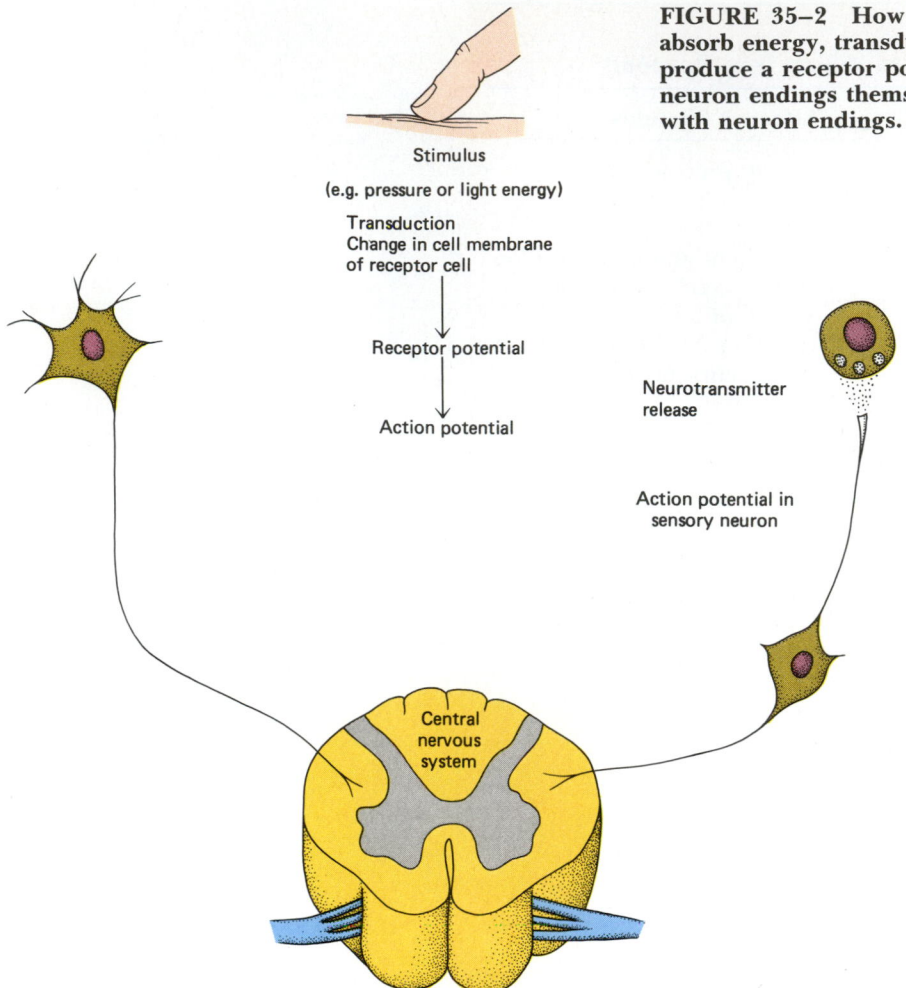

FIGURE 35–2 How a sense organ works. All receptor cells absorb energy, transduce that energy into electrical energy, and produce a receptor potential. Receptor cells may be either neuron endings themselves or specialized cells in close contact with neuron endings. Both types are shown in the diagram.

membrane changes. If the potential difference increases, the cell becomes hyperpolarized. If it decreases or disappears, the cell becomes depolarized.

The depolarization caused by a stimulus is the **receptor potential.** It is a graded response that spreads relatively slowly down the dendrite, fading as it goes. When the depolarization reaches the axon, the threshold level may be reached and an action potential is then generated. The action potential travels along the axon to the central nervous system.

The receptor thus performs all the essential functions of a sense organ: (1) It detects an event in the environment (a force acting on the hair) by absorbing energy; (2) it converts the energy of the stimulus into electrical energy; and (3) it produces a receptor potential, which may result in an action potential that transmits the information to the CNS. This is how all receptors operate.

In invertebrates, the axon of a sensory neuron generally extends all the way to the CNS without synapsing. This means that the message generated at the periphery arrives at the CNS unaltered. However, in many vertebrate sense organs interneurons are interposed between the receptor and the CNS. Because there are many synaptic connections, the original message may be altered and may lose or gain some of its information.

SENSATION DEPENDS ON TRANSMISSION OF A "CODED" MESSAGE

All action potentials are qualitatively the same. Light of the wavelength 400 nanometers (blue), sugar molecules (sweet), and sound waves of 440 hertz (A above middle C) all cause transmission of similar action potentials. How does the organism know whether it is seeing, hearing, or smelling? Our ability to differentiate stimuli depends on both the sense organ itself and on the brain. We can discriminate red from green, a sweet taste from a light breeze, or red from cold because cells of each sense organ are connected to specific neurons in particular parts of the brain. Since a receptor normally responds to only one category of stimuli (e.g., light), a message arriving in the CNS from a particular

FOCUS ON   Pain Perception

Pain is a protective mechanism that signals an organism to react to remove a damaging stimulus. Pain receptors are dendrites of certain sensory neurons found in almost every tissue in the human body. An excess of any type of stimulus—pressure, heat, cold—stimulates pain receptors.

When stimulated, pain receptors signal the spinal cord by way of sensory neurons. The message is transmitted to the opposite side of the spinal cord and then sent upward to the thalamus, where pain reception begins (see the figure). From there impulses are sent into the parietal lobes of the cerebrum. At that time the individual becomes fully aware of the pain and can assess the situation. How threatening is the stimulus? How intense is the pain? What can be done about it? From the thalamus, messages are also sent to the limbic system, where the emotional aspects of the discomfort are addressed.

Pain can be initiated or facilitated at many levels. How intense one's perception of pain is depends upon the particular situation and how one has learned to deal with pain. A child with a bruised knee may emotionally heighten the feeling of pain, whereas a professional fighter may virtually ignore a long series of well-delivered blows.

The brain locates pain on the basis of past experience. Generally pain at the body surface is accurately projected back to the injured area. For example, when you step on a nail, the pain is perceived by the brain and then subjectively projected back to the injured foot, so that you feel pain at the site of puncture. Artificial stimulation of the leg nerves may produce a sensation of pain even though the foot is untouched. In fact, when the foot is amputated, a patient may feel **phantom pain** from the missing limb. Phantom pain occurs because when the severed nerve is stimulated and sends a message to the brain, the brain "remembers" the nerve as it originally was—connected to the missing limb.

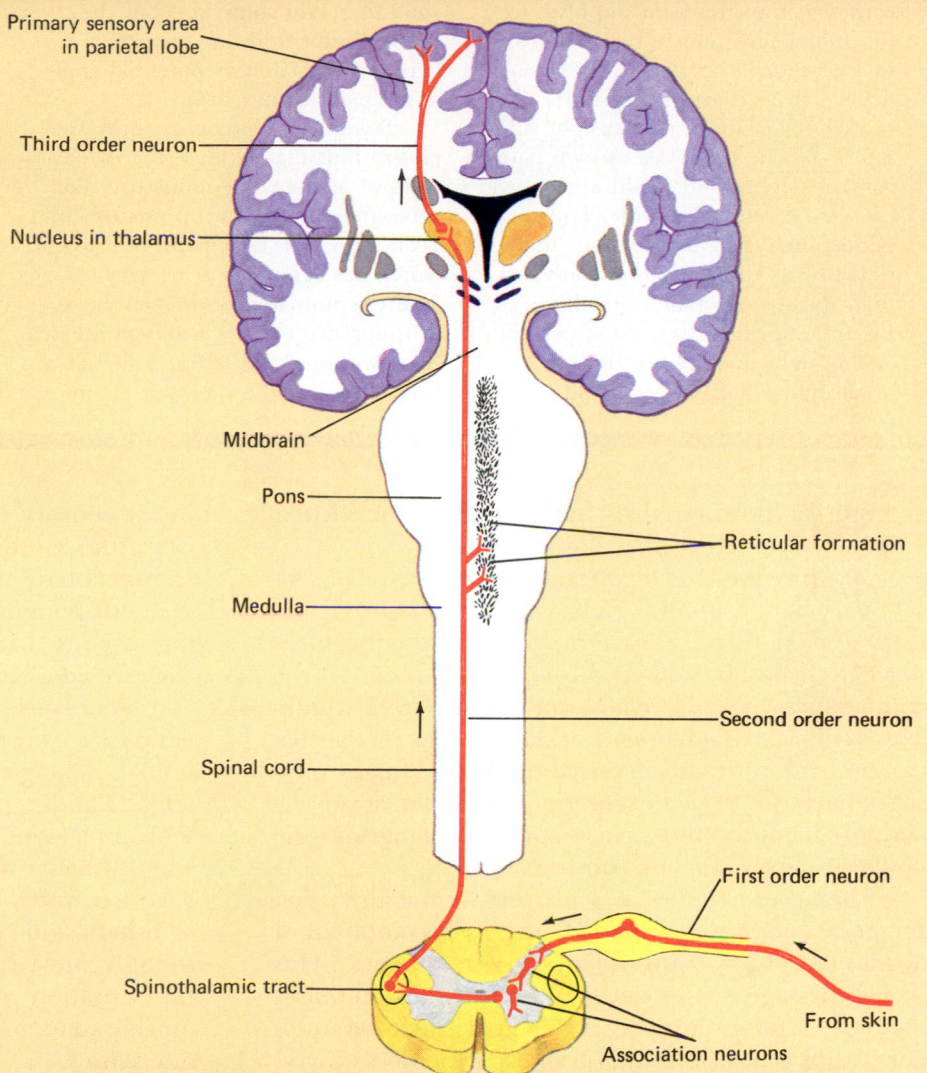

Most internal organs are poorly supplied with pain receptors. For this reason pain from internal structures is often difficult to locate. In fact, pain is often not projected back to the organ that is stimulated. Instead the pain is *referred* to an area just under the skin that may be some distance from the organ involved. The area to which the pain is referred generally is connected to nerve fibers from the same level of the spinal cord as the organ involved. A person with angina who feels heart pain in his left arm is experiencing **referred pain.** The pain originates in the heart as a result of ischemia (insufficient blood in the blood vessels of the heart muscle). However, the pain is actually felt in the arm. One explanation is that neurons from both the heart and the arm converge upon the same neurons in the central nervous system. The brain interprets the incoming message as coming from the body surface because somatic pain is far more common than pain from internal organs; the brain acts on the basis of its past experience. When pain is felt both at the site of the distress and as a referred pain, it may seem to spread, or *radiate*, from the organ to the superficial area.

The physiology of pain is not completely understood. A peptide

(continues on p. 618)

known as **substance P** functions as a neurotransmitter (or perhaps as a modulator of neural activity) in neurons that transmit pain impulses to the spinal cord and brain. Opiates, such as morphine, are analgesic drugs (drugs that relieve pain). They work by blocking the release of substance P. The body has its own pain control system. The brain and pituitary gland release peptides known as **endorphins** (for "endogenous morphine-like") that are more powerful than the strong opiate morphine. Like the opiate drugs, these peptides are thought to suppress the release of substance P from pain-transmit-ting neurons. The neurotransmitter GABA (gamma-amino-butyric acid) is also thought to inhibit release of substance P in some areas of the brain. Endorphins are currently being investigated as potential analgesic (pain-killing) drugs.

Some neurobiologists think that endorphins may explain the mechanism of action of acupuncture. For thousands of years acupuncture has been used to relieve pain, but how it works has remained a mystery. There is now some evidence that acupuncture needles stimulate nerves deep within the muscles, which in turn stimulate the pituitary gland and parts of the brain to release endorphins.

Various clinical methods have been developed for relieving pain. Stimulation of the skin over the painful area with electrodes has been successful in some patients. This procedure is called transcutaneous electrical nerve stimulation. In a few patients, electrodes have been implanted in appropriate areas of the brain so that the patient can stimulate the brain at will. This procedure is thought to relieve pain by stimulating the release of endorphins.

---

receptor is interpreted as meaning that a particular stimulus occurred (e.g., a flash of color).

Interpretation of the message and the type of sensation depends upon which association neurons receive the message. Sensation, when it occurs, takes place in the brain. Rods and cones do not see; only the combination of rods, cones, and centers in the brain see. Artificial stimulation of brain centers (e.g., electrical stimulation) results in sensation. Many sensory messages never give rise to sensations at all. For example, certain chemoreceptors sense internal changes in the body but never stir our consciousness.

Upon stimulation, a sense organ initiates what might be considered a "coded" message, composed of action potentials transmitted by nerve fibers. This coded message is later decoded in the brain. Impulses from the sense organ may differ in (1) the total number of fibers transmitting, (2) the specific fibers carrying action potentials, (3) the total number of action potentials passing over a given fiber, and (4) the frequency of the action potentials passing over a given fiber. For example, the intensity of the stimulus (e.g., how soft or loud) is conveyed by the number of neurons transmitting action potentials and by the frequency of action potentials transmitted by each neuron. Just how the sense organ initiates different codes and how the brain analyzes and interprets them to produce various sensations are not completely understood.

in the sensory neuron decreases. This may occur because the sensory neuron becomes less responsive to stimulation or because the receptor produces a smaller receptor potential, or for both reasons. This diminishing response to a continued, constant stimulus is called **sensory adaptation.**

Some receptors, such as those for pain or cold, adapt so slowly that they continue to trigger action potentials as long as the stimulus persists. Other receptors adapt rapidly, permitting an animal to ignore persistent unpleasant or unimportant stimuli. For example, when you first pull on a pair of tight jeans your pressure receptors let you know that you are being squished and you may feel uncomfortable. Soon, though, these receptors adapt, and you hardly notice the sensation of the tight fit. In the same way, we quickly adapt to odors that at first smell seem to assault our senses.

■ MECHANORECEPTORS RESPOND TO TOUCH, PRESSURE, GRAVITY, STRETCH, OR MOVEMENT

Mechanoreceptors respond to touch, pressure, gravity, stretching, or movement. They are activated when they are mechanically deformed, that is, pushed or pulled so that they change shape.

Touch Receptors Are Located In The Skin

The simplest **mechanoreceptors** are free nerve endings in the skin that are stimulated by objects contacting the body surface. Thousands of more specialized touch receptors are also located in the skin (Fig. 35–3). Some (Merkel's discs and Meissner's corpuscles) are

■ RECEPTORS ADAPT TO STIMULI

Many receptors do not continue to respond at their initial rate, even if the stimulus remains unabated in intensity. With time, the frequency of action potentials

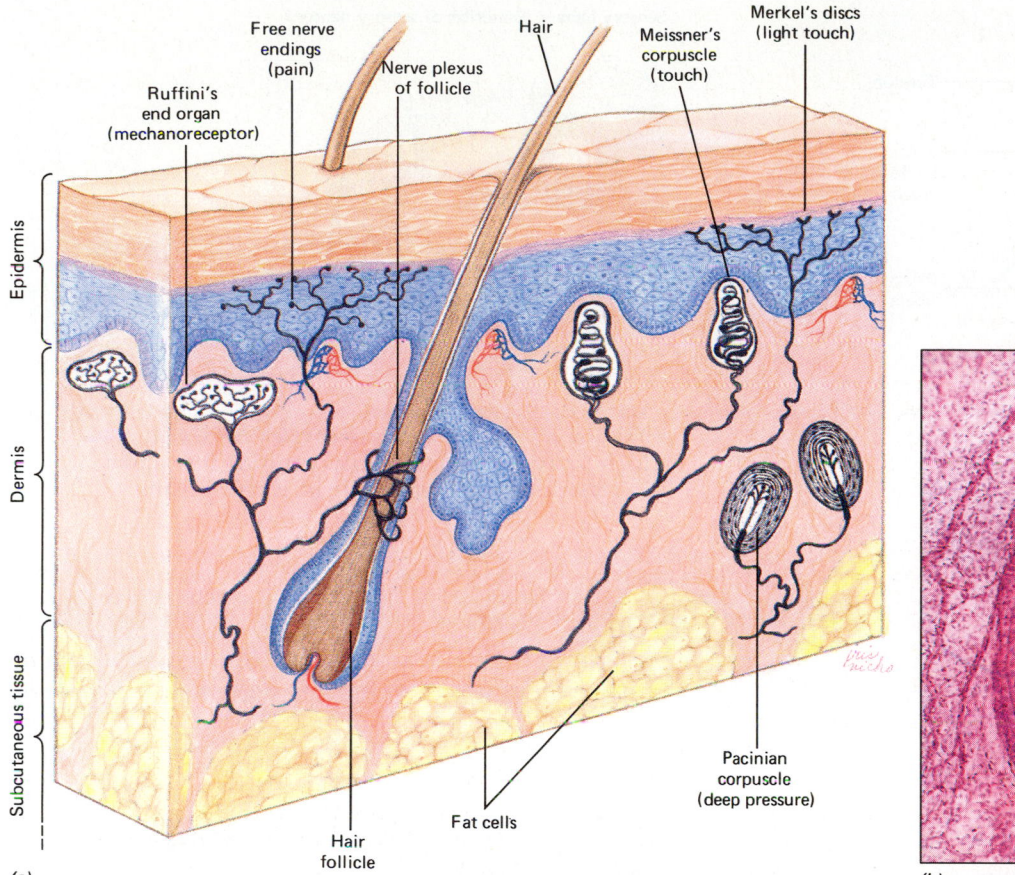

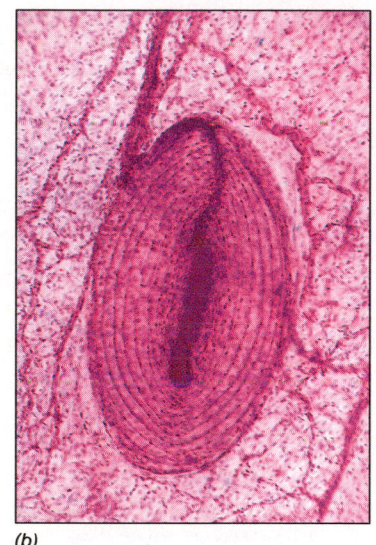

(a)

(b)

FIGURE 35–3   Sense organs within the human skin. (*a*) **Diagrammatic section through the human skin showing the types of sense organs present. The free nerve endings respond to pain; Krause's corpuscles are thought to respond to hot and cold stimuli; tactile hairs, Merkel's disks, Ruffini's end organs, and Meissner's corpuscles respond to touch; pacinian corpuscles respond to deep pressure.** (*b*) **Pacinian corpuscle.**

sensitive to light touch. Others, like the **Pacinian corpuscles,** are especially sensitive to deep pressure.

In the Pacinian corpuscle, a bare nerve ending is surrounded by layers of connective tissue. Compression causes displacement of the connective tissue, which in turn deforms the axon. The Pacinian corpuscle responds to rapid movement of the tissue.

In many invertebrates as well as vertebrates, tactile (touch) receptors lie at the base of a hair or bristle. They are stimulated indirectly when the hair is bent or displaced. Such tactile hairs are involved in orientation to gravity, in postural orientation, and in the reception of vibrations in air and water, as well as in contacts with other objects.

Proprioceptors Help Coordinate Muscle Movement

Proprioceptors are receptors sensitive to changes in movement, tension, and position in muscles and joints.

Their continuous reports to the CNS help ensure that muscle movement will be properly coordinated. Vertebrates have three main types of proprioceptors: **muscle spindles,** which detect muscle movement (Fig. 35–4); **Golgi tendon organs,** which are sensitive to stretch in the tendons that attach muscle to bone; and **joint receptors,** which detect movement in ligaments.

By means of these sense organs we can, even with our eyes closed, perform manual acts such as dressing or playing the piano. Impulses from the proprioceptors are also extremely important in ensuring the harmonious contraction of different muscles involved in a single movement; without such receptors, complicated, skillful acts would be impossible. Impulses from these organs are also important in maintaining balance. Proprioceptors, which were discovered a little more than 100 years ago, are probably more numerous and more continuously active than any of the other sense organs, although we are less aware of them than of most of the others.

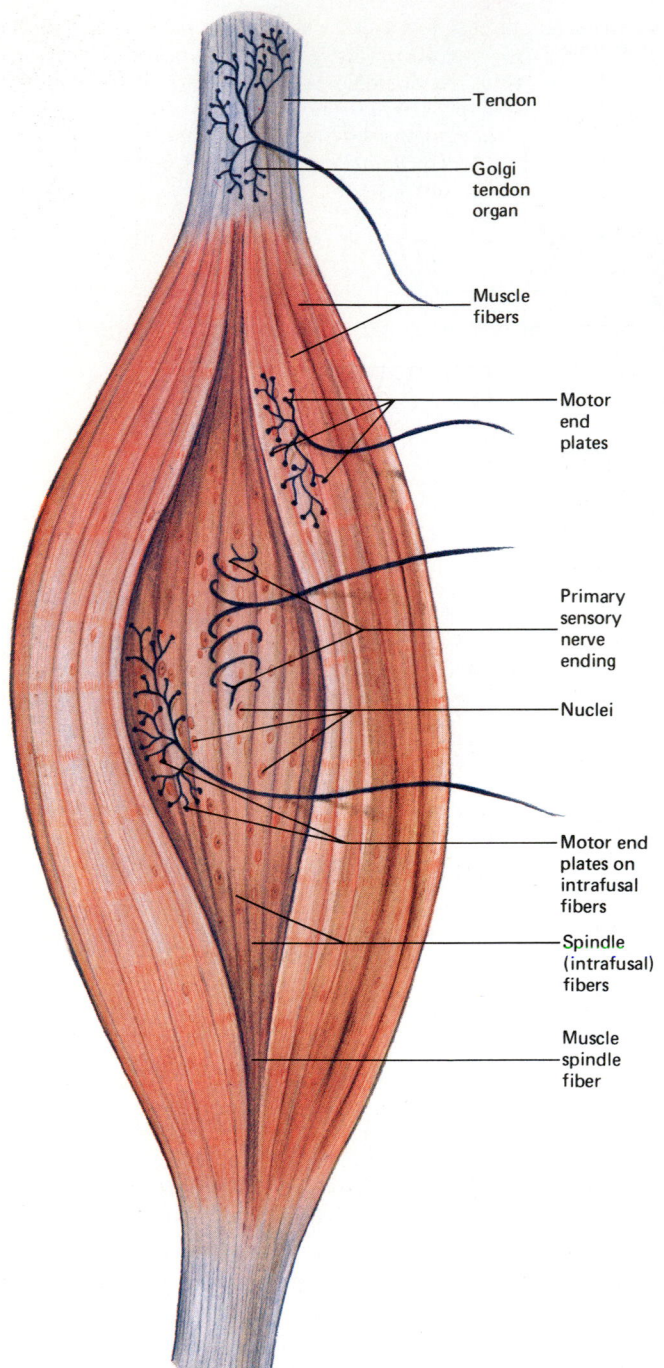

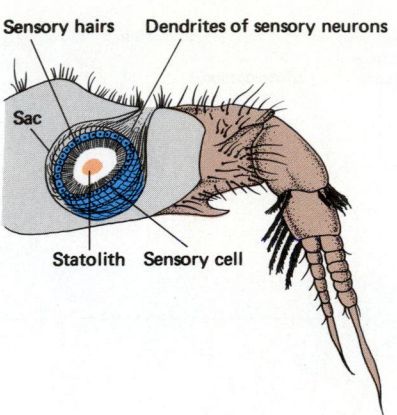

FIGURE 35–5 Many invertebrates have statocysts, which serve as sensors of gravitational force. Shown here is a statocyst in the antennule of a crustacean.

FIGURE 35–4 A muscle spindle and Golgi tendon organ. Muscle spindles detect muscle movement; Golgi tendon organs determine stretch in tendons.

thought to supplement vision by informing the fish of obstacles in its way or of moving objects such as prey or enemies.

Many Invertebrates Have Gravity Receptors Called Statocysts

An organism must have a way of sensing its own orientation with respect to gravity. When displaced from its normal position, an animal quickly adjusts the body to reassume it. This rapid adjustment is possible because receptors continually send information regarding the position and movements of the body to the CNS.

Many invertebrates have specialized sense organs called **statocysts** that serve as gravity receptors. A statocyst is basically an infolding of the epidermis lined with receptor cells that have hairs (Fig. 35–5). The cavity contains a **statolith** (sometimes more than one), which is a tiny granule of loose sand grains or calcium carbonate. Normally the particles are pulled downward by gravity and stimulate the hair cells. When the body shifts position, the statolith shifts, also causing the hair cells to bend. This mechanical displacement results in receptor potentials and action potentials that inform the CNS of the change in position. Because different hair cells fire with each change of position, the animal "knows" where down is and can correct its position if necessary.

Lateral Line Organs Supplement Vision In Fish

Lateral line organs are mechanoreceptors found in fish and in aquatic and larval amphibians. Typically this sense organ consists of a long canal running the length of the body. The canals are lined with receptor cells that are thought to respond to waves, currents, or disturbances in the water. The lateral line organ is

The Labyrinth Of The Vertebrate Ear Is An Organ Of Equilibrium

When we think of the ear, we think of hearing. However, fish do not use their ears for hearing. In fact, in all vertebrates the basic function of the ear is to help

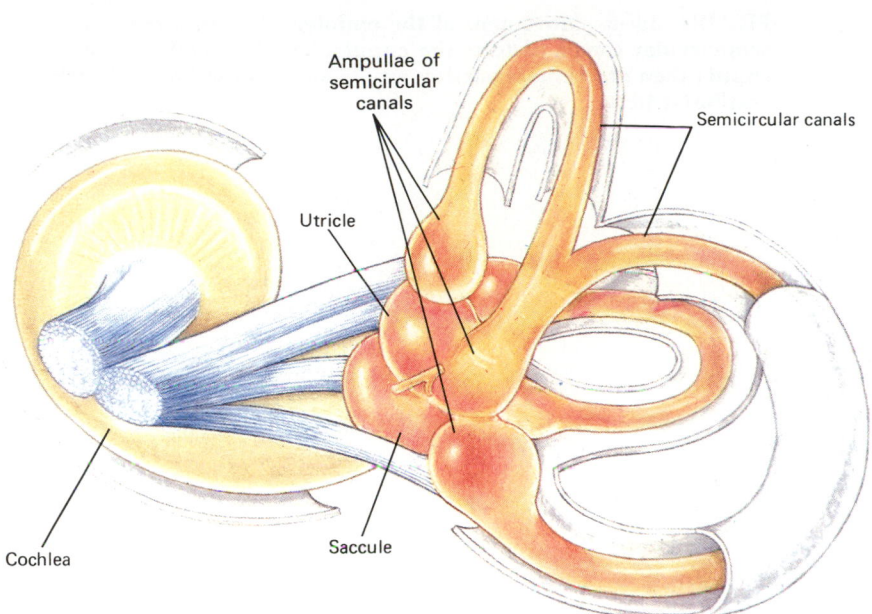

Ampullae of semicircular canals

Semicircular canals

Utricle

Saccule

Cochlea

FIGURE 35–6 The human inner ear with the membranous labyrinth exposed. Because this is a posterior view the utricle and saccule can be seen. Note that the membranous labyrinth is shown in color only within the semicircular canals and is not colored in the cochlea.

maintain equilibrium. The state of balance or adjustment between opposing forces that enables an organism to maintain its orientation is known as **equilibrium.** Typically the ear also contains gravity receptors. Although many vertebrates do not have outer or middle ears, all of them have inner ears.

The inner ear consists of a group of interconnected canals and sacs, referred to as the **labyrinth.** In jawed vertebrates the labyrinth consists of two sac-like chambers, the **saccule** and **utricle,** and three **semicircular canals,** as well as a snail-shaped structure known as the **cochlea.** Collectively, the saccule, utricle, and semicircular canals are referred to as the **vestibular apparatus** (Fig. 35–6). This structure is responsible for the sense of equilibrium.

The saccule and utricle house gravity detectors in the form of small crystals of calcium carbonate called **otoliths** (Fig. 35–7). The sensory cells consist of groups of hair cells surrounded at their tips by a jelly-like mass, the **cupula.** The receptor cells in the saccule and utricle lie in different planes. Normally, the pull of gravity causes the otoliths to press against particular hair cells, stimulating them to initiate impulses. These impulses travel to the brain by way of sensory nerve fibers at their bases. When the head is tilted, or in linear acceleration (change in speed when the body is moving in a straight line), the otoliths press upon the hairs of other cells and stimulate them. This enables the animal to perceive its position relative to the ground regardless of what position the head is in at the time.

Information about **angular acceleration** (turning movements) is furnished by the three semicircular canals. Each of these is connected with the utricle and lies in a plane at right angles to the other two. Each canal is a hollow ring filled with fluid called **endolymph.** At

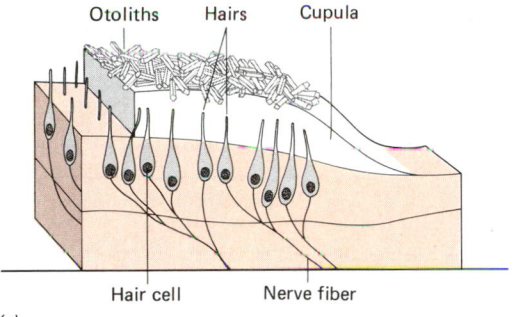

Otoliths Hairs Cupula

Hair cell Nerve fiber

(a)

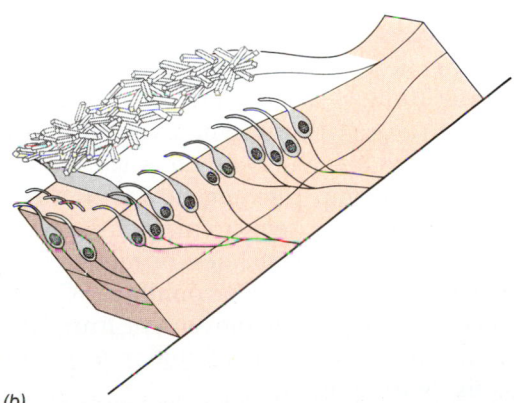

(b)

FIGURE 35–7 Receptors concerned with the position of the body relative to the ground (gravity) are located in the saccule and utricle. Compare the positions of the otoliths and hairs in (*a*) with those in (*b*). Changes in head position cause the force of gravity to distort the cupula, which in turn distorts the hairs of the hair cells; the hair cells respond by sending impulses down the vestibular nerve (part of the auditory nerve) to the brain.

one of the openings of each canal into the utricle is a small, bulb-like enlargement, the **ampulla.** Within each ampulla is a clump of hair cells called a **crista,** similar

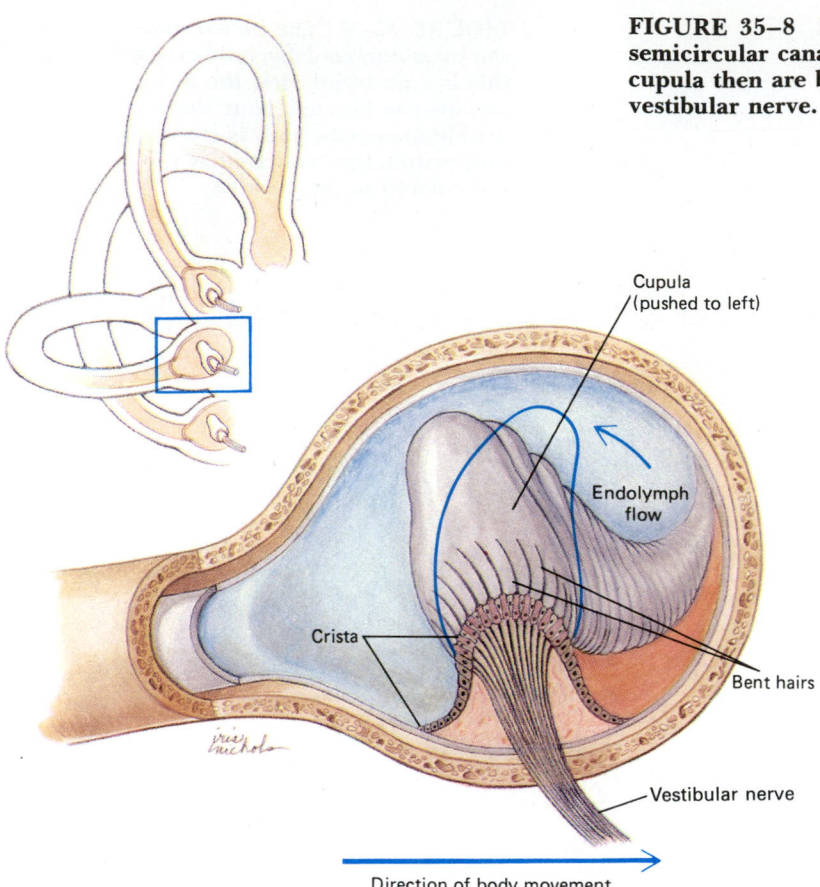

Cupula
(pushed to left)

Endolymph
flow

Crista

Bent hairs

Vestibular nerve

Direction of body movement

to those in the utricle and saccule, but lacking otoliths. These receptor cells are stimulated by movements of the endolymph in the canals (Fig. 35–8). Since the three canals are located in three different planes, a movement of the head in any direction stimulates movement of the fluid in at least one of the canals.

We humans are used to movements in the horizontal plane but not to movements such as the motion of an elevator or of a ship pitching in a rough sea. These motions stimulate the semicircular canals in an unusual way and may cause sea sickness or motion sickness, with their resulting nausea or vomiting. When a person so affected lies down, the movement stimulates the semicircular canals in a more familiar way, and nausea is less likely to occur.

Auditory Receptors Are Located In The Cochlea

Many arthropods and most vertebrates have sound receptors, but for many of them hearing does not seem to be an important sense. Hearing is important for tetrapods, however, and both birds and mammals have a highly developed sense of hearing. Their auditory receptors, located in the cochlea of the inner ear, are mechanoreceptor hair cells that detect pressure waves.

The cochlea is a spiral tube that resembles a snail's shell (Fig. 35–9). If the cochlea were uncoiled, as in Figure 35–10, it would be seen to consist of three canals separated from each other by thin membranes and coming almost to a point at the apex. Two of these canals, the **vestibular canal** (also known as the scala vestibuli) and the **tympanic canal** (scala tympani), are connected with one another at the apex of the cochlea and are filled with fluid. The middle canal, the **cochlear duct** (scala media), is filled with endolymph and contains the actual auditory receptor, the **organ of Corti.**

Each organ of Corti contains about 24,000 hair cells arranged in five rows extending the entire length of the coiled cochlea. Each cell has hair-like projections that extend into the cochlear duct. These cells rest upon the **basilar membrane,** which separates the cochlea from the tympanic canal. Overhanging the hair cells is the **tectorial membrane,** attached along one edge to the membrane on which the hair cells rest, with its other edge free. When stimulated by changes in pressure transmitted through the endolymph, the hair cells initiate impulses in the fibers of the cochlear (auditory) nerve.

In terrestrial vertebrates, accessory structures in the outer and middle ear change sound waves in air to

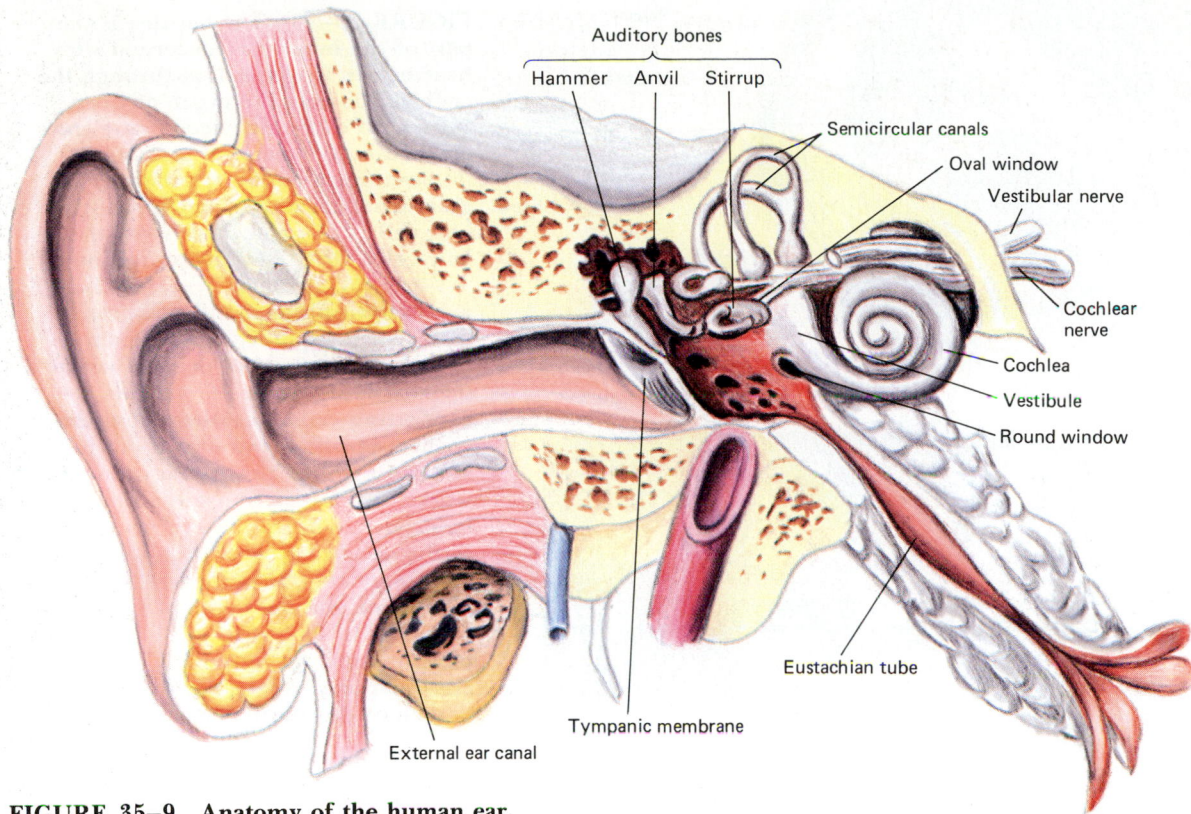

Auditory bones
Hammer   Anvil   Stirrup
Semicircular canals
Oval window
Vestibular nerve
Cochlear nerve
Cochlea
Vestibule
Round window
Eustachian tube
Tympanic membrane
External ear canal

FIGURE 35–9   Anatomy of the human ear.

pressure waves in the cochlear fluid. In the human ear, for example, sound waves pass through the **external auditory meatus** (the canal of the outer ear) and set the **eardrum** (the tympanic membrane separating outer ear and middle ear) vibrating. These vibrations are transmitted across the middle ear by three tiny bones, the **hammer, anvil,** and **stirrup** (so called because of their shapes). The vibrations pass through the oval window, a small opening leading from the middle to the inner ear, and pass to the fluid in the vestibular canal.

Since liquids cannot be compressed, the oval window could not cause movement of the fluid in the vestibular duct unless there were an escape valve for the pressure. This is provided by the **round window** at the end of the tympanic canal. The pressure wave presses upon the membranes separating the three ducts, is transmitted to the tympanic canal, and causes the round window to bulge. The movements of the basilar membrane produced by these pulsations are believed to rub the hair cells of the organs of Corti against the overlying tectorial membrane. This stimulation of the hair cells initiates nerve impulses in the dendrites of the cochlear nerve lying at the base of each hair cell.

When the ear is subjected to intense sound, the organ of Corti may be injured. Members of rock bands and workers subjected to loud, high-pitched noises over a period of years frequently become deaf to high tones because the cells near the base of the organ of Corti become injured.

◼ CHEMORECEPTORS DETECT TASTE AND SMELL

Two highly sensitive types of chemoreceptors are the senses of **taste** and **smell** (olfaction). One of the most thoroughly studied taste receptors is the taste hair of the fly. The terminal segments of the legs and the mouth parts of flies, moths, butterflies, and a number of other insects have very sensitive hairs. In the fly, each one of these contains four taste receptors and a tactile receptor. One taste receptor is more-or-less specific to sugars, one to water, and two to salts.

Taste Buds Are The Organs Of Taste In Humans

The organs of taste in humans and other mammals are bud-like structures known as **taste buds,** which are located in the mouth, mainly on the surface of the tongue, where they are found mainly in tiny elevations, or papillae (Fig. 35–11). There is a rapid turnover of taste bud cells; every 10 to 30 hours the cells are completely replaced.

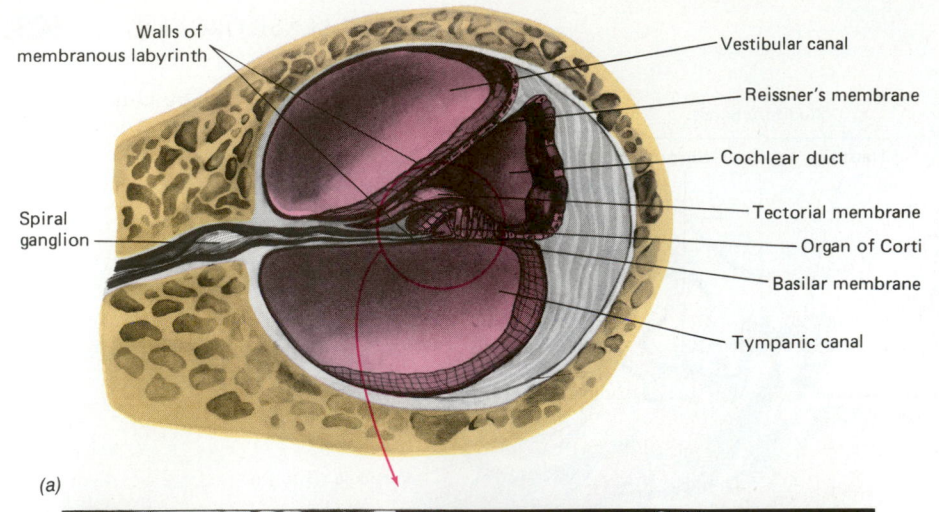

Walls of membranous labyrinth

Spiral ganglion

Vestibular canal

Reissner's membrane

Cochlear duct

Tectorial membrane

Organ of Corti

Basilar membrane

Tympanic canal

(a)

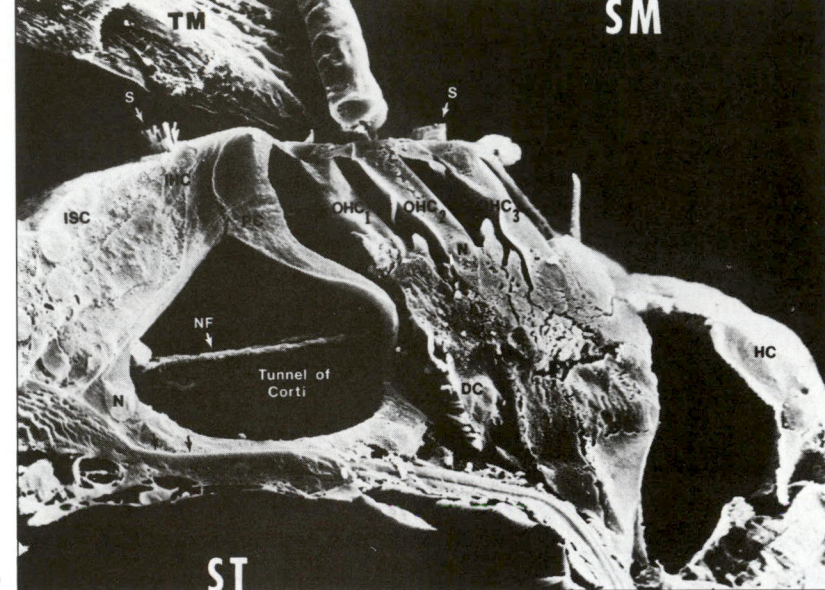

(b)

FIGURE 35-10 The cochlea is the part of the inner ear concerned with hearing. (*a*) Cross section through the cochlea showing the organ of Corti resting on the basilar membrane and covered by the tectorial membrane. (*b*) Scanning electron micrograph of guinea pig organ of Corti showing inner hair cells, *IHC*, and three rows of outer hair cells, *OHC 1–3* (magnification ×1,790). (*N*, nucleus; *S*, stereocilia; *TM*, tectorial membrane; *SM*, scala media (cochlear duct); *ST*, scala tympani (tympanic duct); *black arrows*, basilar membrane; *HC, DC, PC*, various supporting cells.) (*c*) Diagram of the cochlea uncoiled and drawn out in a straight line. (*d*) How the organ of Corti works. Vibrations transmitted by the hammer, anvil, and stirrup set the fluid in the vestibular canal in motion; these vibrations are transmitted to the basilar membrane and the organ of Corti. The hair cells of the organ of Corti are the receptor cells for hearing. The hair cells are innervated by the cochlear nerve, a branch of the auditory nerve.

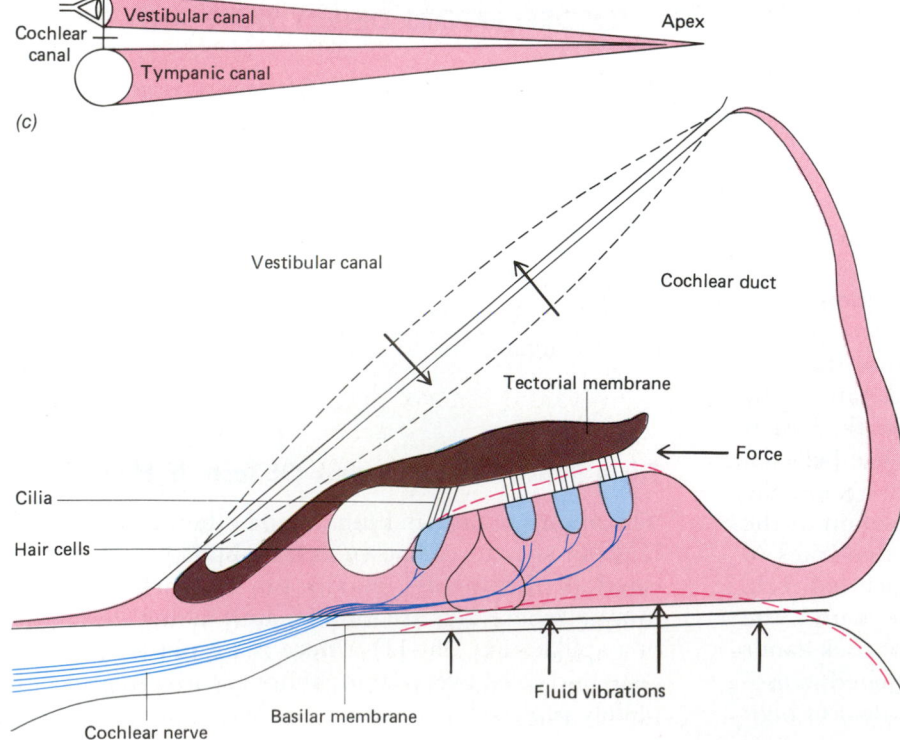

Cochlear canal

Vestibular canal

Tympanic canal

Apex

(c)

Vestibular canal

Cochlear duct

Tectorial membrane

Force

Cilia

Hair cells

Fluid vibrations

Cochlear nerve

Basilar membrane

Tympanic canal

(d)

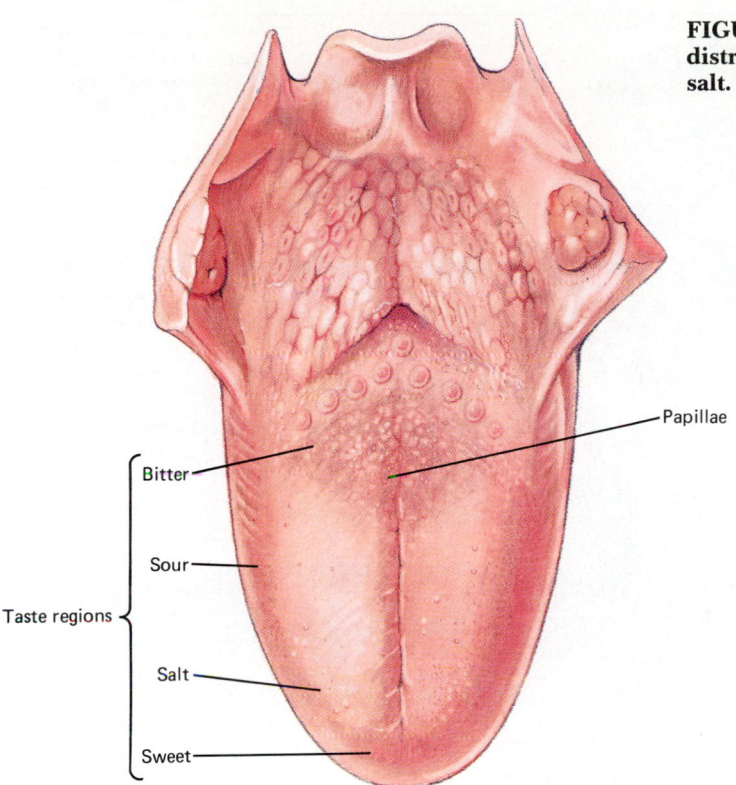

Papillae

Bitter

Sour

Taste regions {

Salt

Sweet

FIGURE 35–11   The surface of the tongue showing distribution of taste buds sensitive to sweet, bitter, sour, and salt.

A taste bud is an oval body that consists of an epithelial capsule containing several taste receptors. Each taste receptor is an epithelial cell with a border of microvilli at its free surface (Figs. 35–12). A hair-like projection extends to the external surface of the taste bud through an opening called the **taste pore.**

Traditionally four basic tastes are recognized: sweet, sour, salty, and bitter. Flavor depends on these four basic tastes in combination with smell, texture, and temperature. When you have a cold and your nose is stopped up, food seems to have little "taste."

The Olfactory Epithelium Is Responsible For The Sense Of Smell

In terrestrial vertebrates, olfaction occurs in the nasal epithelium. In humans the **olfactory epithelium** is located in the roof of the nasal cavity. This epithelium contains about 20 million specialized olfactory cells with axons that extend upward as the fibers of the olfactory nerves. The end of each olfactory cell bears several olfactory hairs that are thought to react to odors (chemicals) in the air.

■ THERMORECEPTORS ARE SENSITIVE TO HEAT

Mosquitos, ticks, and other blood-sucking arthropods use heat receptors to locate endothermic (warm-blooded) animals. At least two types of snakes, pit vipers and boas, use thermoreceptors to locate their prey. Pits in the heads of pit vipers can detect heat generated by a small animal more than a meter away (Fig. 35–13). In mammals, free nerve endings in the skin and tongue detect temperature changes. Thermoreceptors in the hypothalamus detect internal changes in temperature and receive and integrate information from thermoreceptors on the body surface. The hypothalamus then initiates homeostatic mechanisms which we become aware of as sweating or shivering. Prior to these responses, other more subtle responses (e.g., rerouting blood) have probably taken place. The end result of these responses is a constant body temperature.

■ ELECTRORECEPTORS DETECT ELECTRICAL CURRENTS IN WATER

Electrical organs are found in a few species of rays and bony fishes. Some of these animals have electroreceptors on the body that are linked with the neurons supplying the lateral line organs. These receptors are used to detect electrical currents in the water. In species that produce a weak current, the electrical organs help in orientation. This is particularly useful in murky water where visibility and olfaction are poor. The fish is informed of an object in the surrounding water when the electrical conductivity of the object distorts the flow-pattern of the electrical current. The electric eel and some other animals with electrical organs can deliver a powerful shock that stuns prey or enemies.

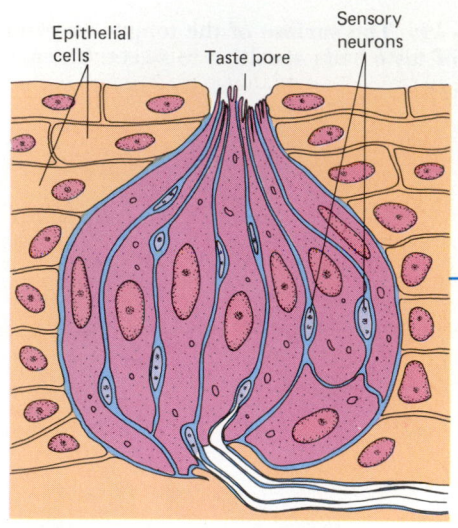

(a)

FIGURE 35–12 **Each taste bud consists of an epithelial capsule containing several taste receptors. (*a*) Diagram of a taste bud. (*b*) Electron micrograph of papillae of the tongue containing taste buds. The taste buds are the oval bodies (approximately ×400).**

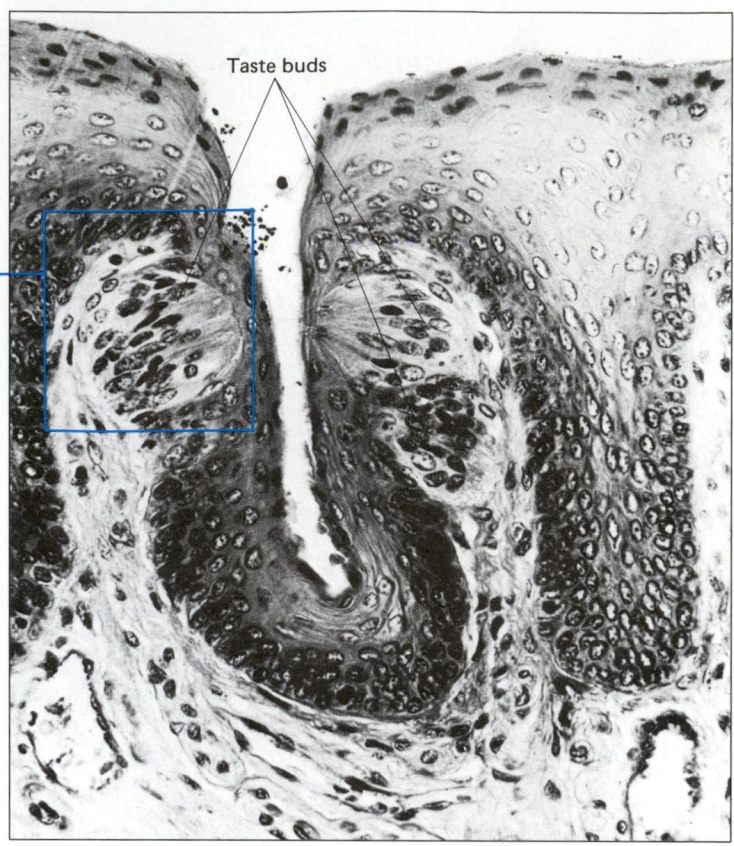

(b)

■ PHOTORECEPTORS USE PIGMENTS TO ABSORB LIGHT

Most animals have photoreceptors that use pigments to absorb light energy. **Rhodopsins** are the photosensitive pigments found in the eyes of cephalopod mollusks, arthropods, and vertebrates. Light energy striking a light-sensitive receptor cell triggers chemical changes in the pigment molecules. As a result the receptor cell may transmit a nerve impulse.

Eyespots, Simple Eyes, And Compound Eyes Are Found Among Invertebrates

The simplest light-sensitive organs in animals are found in certain cnidarians and in flatworms (Fig. 35–14). Their eyespots, or **ocelli,** detect light but do not see objects. Eyespots are often bowl-shaped clusters of light-sensitive cells within the epidermis. They may detect the direction of the source of light and distinguish light intensity.

Effective image formation, called **vision,** requires a more complex **eye,** usually with a lens. A lens is a structure that concentrates light on a group of photoreceptors. As lens systems evolved, the photoreceptors were able to form images. Two fundamentally differ-

FIGURE 35–13 **A golden eye-lash viper showing a pit organ, a sensory structure located between each eye and nostril. The pit organ can detect the heat from a warmblooded animal up to a distance of 1 to 2 meters.**

ent types of eyes evolved: The camera eye of some mollusks (squids and octopuses) and vertebrates, and the compound eye of the arthropods. (Vertebrate and mollusk eyes are analogous structures; that is, they evolved independently of one another and have similar functions but different structures.)

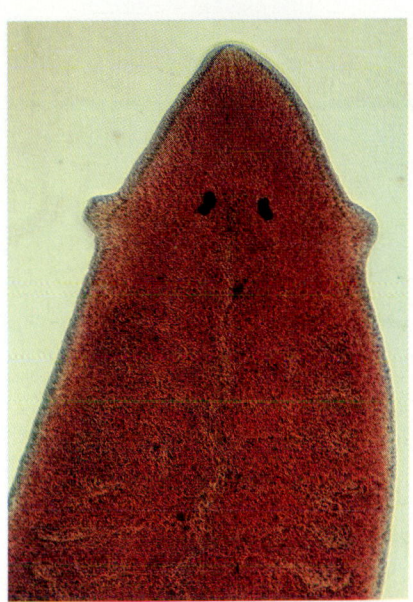

(a)

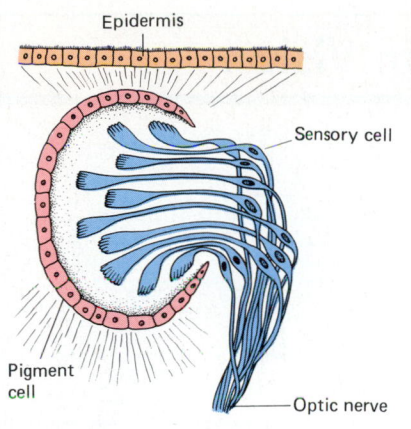

Epidermis

Sensory cell

Pigment cell

Optic nerve

(b) Ocellus of planarian worm

**FIGURE 35–14 Simple invertebrate eyes.
(a) Planarian worm showing eyespots.
(b) Eyespot of planarian worm.**

(a)

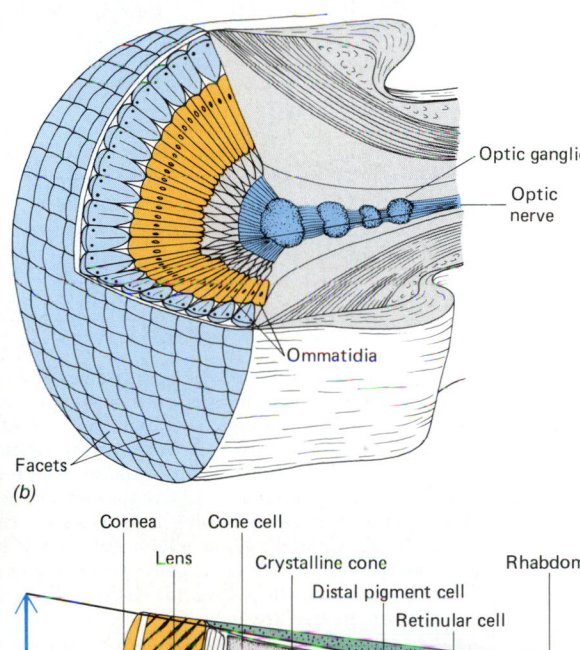

Optic ganglion

Optic nerve

Ommatidia

Facets

(b)

Cornea

Lens

Cone cell

Crystalline cone

Distal pigment cell

Retinular cell

Rhabdome

Proximal pigment cell

(c)

**FIGURE 35–15 (a) The
horsefly (*Tabanus* sp.) has
prominent compound eyes.
(b) Structure of the compound
eye showing several adjacent
ommatidia. This type of eye
registers changes in light and
shade so that the animal can
detect movement. (c) Structure
of an ommatidium, a unit
from the compound eye of an
arthropod. The rhabdome is
the light-sensitive core of the
ommatidium.**

Compound eyes are found in crustaceans and insects. Not only do these eyes look different from vertebrate eyes, but they also see differently. The surface of a compound eye appears faceted, having many faces, like a diamond (Fig. 35–15). Each facet is the convex cornea of one of its visual units, called an **ommatidium.** The number of ommatidia varies in different species, from just a few in the eye of certain crustaceans to as many as 28,000 in the eye of a dragonfly.

Each ommatidium has a transparent covering, the **cornea,** and a lens that focuses light onto a receptor, forming a small inverted image. Each receptor receives light from only a small portion of the visual field. In this way it samples the average light intensity from that area. All the ommatidia together produce a composite image, a **mosaic** picture.

Although the compound eye forms relatively coarse images, a single ommatidium can detect any

FOCUS ON Defects in Vision

The most common defects of the human eye are **nearsightedness** (myopia), **farsightedness** (hypermetropia), and **astigmatism.** In the normal eye, as shown in part *a* of the figure, the shape of the eyeball is such that the retina is the proper distance behind the lens for the light rays to converge in the fovea. In a nearsighted eye, illustrated in part *b* of the figure, the eyeball is too long and the retina is too far from the lens. The light rays converge at a point in front of the retina and are again diverging when they reach it, resulting in a blurred image. In a farsighted eye, the eyeball is too short and the retina too close to the lens (part *c* of the figure). Light rays strike the retina before they have converged, again resulting in a blurred image. Concave lenses correct for the nearsighted condition by bringing the light rays to a focus at a point farther back, and convex lenses correct for the farsighted condition by causing the light rays to converge farther forward.

In astigmatism the cornea is curved unequally in different planes, so that the light rays in one plane are focused at a different point from those in another plane, as shown in part *d* of the figure. To correct for astigmatism, lenses must be ground unequally to compensate for the unequal curvature of the cornea.

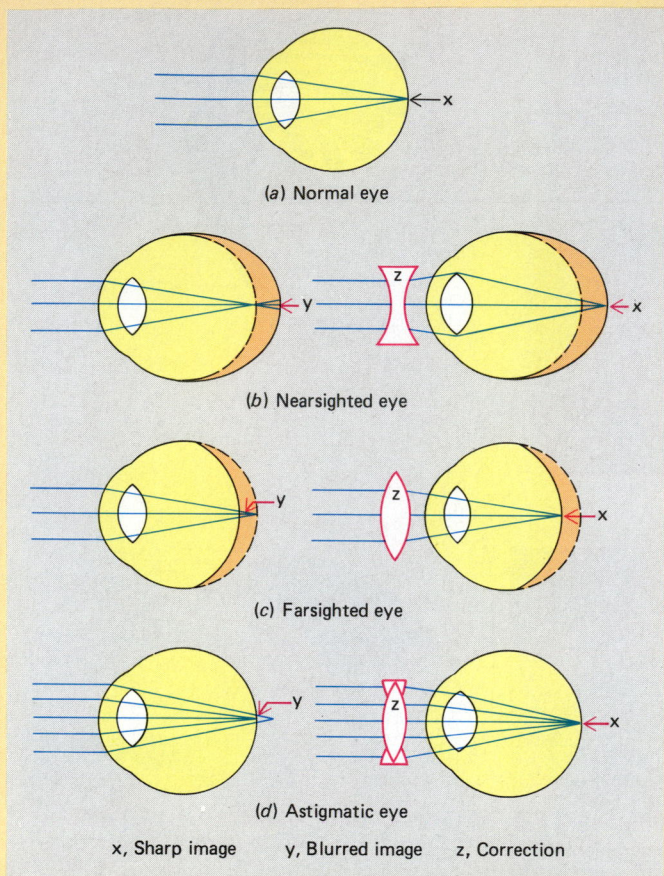

(a) Normal eye

(b) Nearsighted eye

(c) Farsighted eye

(d) Astigmatic eye

x, Sharp image   y, Blurred image   z, Correction

Common abnormalities of the eye that result in defects in vision. (*a*) Normal eye, in which parallel rays coming from a point in space are focused as a point on the retina. (*b*) Nearsighted eye, in which the eyeball is elongated so that parallel light rays are brought to a focus in front of the retina (on dotted line, which represents the position of the retina in a normal eye) and so form a blurred image on the retina. This is corrected by placing a concave lens in front of the eye, which diverges the light rays, making it possible for the eye to focus these rays on the retina. (*c*) Farsighted eye, in which the eyeball is shortened and light rays are focused behind the retina. A convex lens converges the light rays so that the eye focuses them onto the retina. (*d*) Astigmatic eye, in which light rays passing through one part of the eye are focused on the retina, while light rays passing through another area of the lens are not focused on the retina. This is a result of the unequal curvature of the lens, or cornea. A cylindrical lens corrects this by bending light rays going through only certain parts of the eye.

movement of prey or enemy. The compound eye is well adapted to the arthropod's way of life in two other respects. It is sensitive to different wavelengths of light, ranging from the red into the ultraviolet (UV), and it is able to analyze the **plane of polarization** of light. Because an insect can see UV radiation, its world of color is much different from ours. Since different flowers reflect UV to different degrees, two flowers

that appear identically colored to us may appear strikingly different to insects. We can appreciate how the world appears to an insect with UV vision by viewing the landscape through a television camera with a UV-transmitting lens (Fig. 35–16).

A sky that appears equally blue to us in all regions reveals quite different patterns to an insect. The plane of polarization of the light is not the same in all parts of

(a)

(b)

FIGURE 35–16 Insects see ultraviolet light and their world appears differently colored from ours. (*a*) *Nedelia trilobata* appears to be uniformly yellow to the human eye. (*b*) When viewed with UV film, the flowers have darker areas that represent light-absorbing centers.

(a)

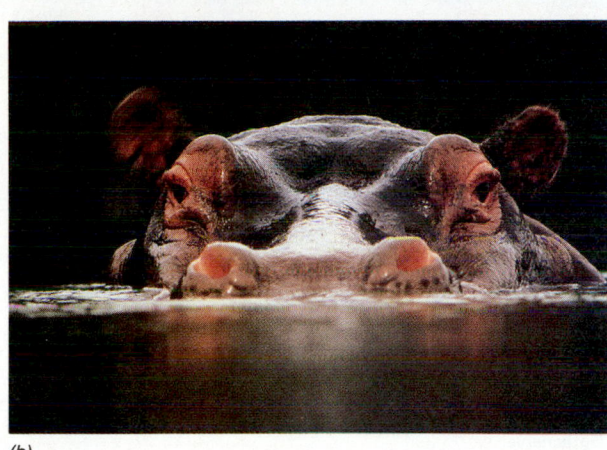

(b)

(c)

FIGURE 35–17 The location of the eyes varies in different vertebrates, resulting in differences in vision. (*a*) The eyes of this white-tail doe are positioned laterally, enabling the animal to see on both sides; even while grazing, it can spot a predator approaching from behind. (*b*) The orbits (bony cavities that contain the eyeballs) of the hippopotamus are elevated, enabling the animal to see even when most of its head is under water. (*c*) Like many other nocturnal animals, the Eastern screech owl (*Otus asio*) has large eyes positioned at the front of its head. The resulting binocular vision permits it to judge distances. (*a*, Skip Moody/Marvin L. Dembinsky, Jr., Photography Associates; *b*, Frans Lanting; *c*, Barry L. Runk, from Grant Heilman)

the sky, and the insect's eye can detect the differences. Honey bees and some other arthropods use this ability as a navigational aid.

Vertebrate Eyes Form Sharp Images

The position of the eyes in the head of humans and certain other higher vertebrates permits both of them to be focused on the same object (Fig. 35–17). This **binocular vision** is an important factor in judging distance and depth.

The vertebrate eye can be compared to a camera. An adjustable lens can be focused for different distances and a diaphragm, the **iris,** regulates the size of the light opening, the **pupil** (Fig. 35–18). The iris is a ring of smooth muscle that appears as blue, green, or

brown, depending on the amount and nature of pigment present. The **retina** corresponds to the light-sensitive film used in a camera. Next to the retina is the **choroid layer,** a sheet of cells filled with black pigment that absorbs extra light and prevents internally reflected light from blurring the image. (Cameras are also black on the inside.)

The outer coat of the eyeball, called the **sclera,** is a tough, opaque, curved sheet of connective tissue that protects the inner structures and helps to maintain the rigidity of the eyeball. On the front surface of the eye this sheet becomes the thinner, transparent **cornea,** through which light enters.

The lens is a transparent, elastic ball located just behind the iris. It bends the light rays coming in and brings them to a focus on the retina. The lens is aided by the curved surface of the cornea and by the refrac-

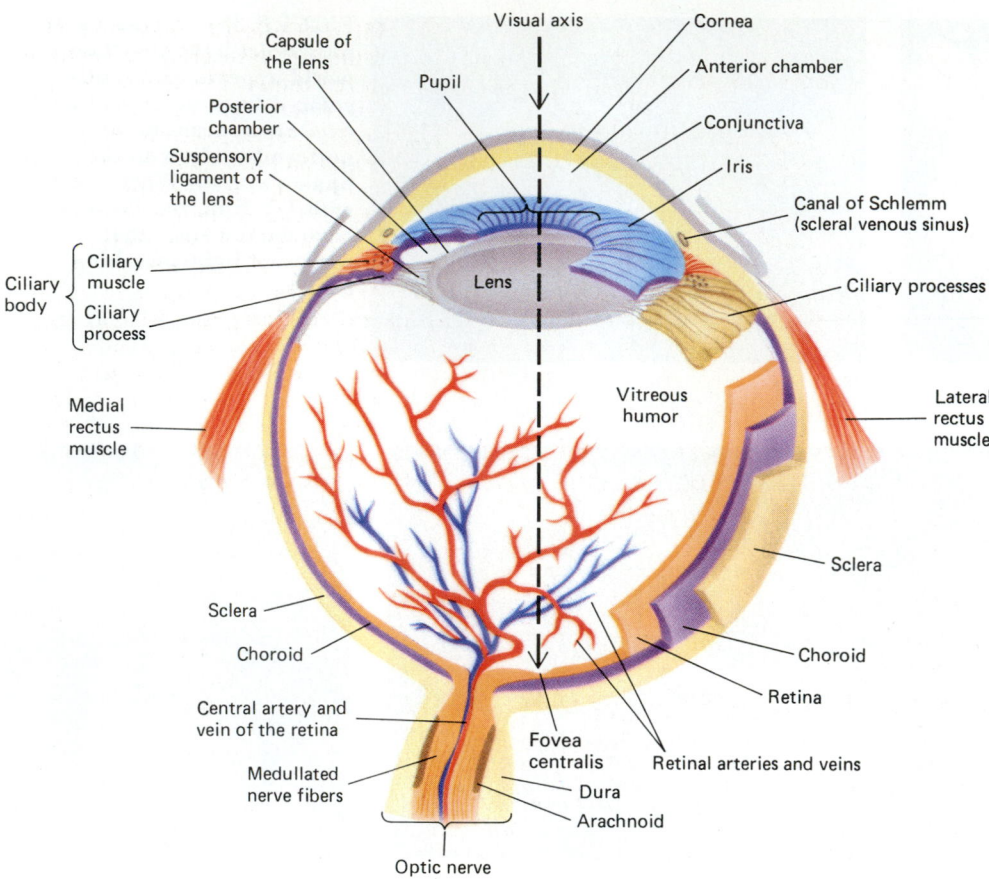

FIGURE 35–18 Structure of the human eye.

tive capabilities (ability to bend light rays) of the liquids inside the eyeball. The cavity between the cornea and the lens is filled with a watery substance, the **aqueous fluid.** The larger chamber between the lens and the retina is filled with a more viscous fluid, the **vitreous body.** Both fluids are important in maintaining the shape of the eyeball by providing an internal, fluid pressure.

The eye has the power of **accommodation,** meaning it can change focus for near or far vision by changing the curvature of the lens. This is made possible by the stretching and relaxing of the lens by the **ciliary muscle.** As people grow older, the lens enlarges and becomes less elastic and thereby less able to accommodate for near vision. When this occurs, eyeglasses with one portion ground for distance vision and one portion ground for near vision (bifocals) may be worn to accomplish what the eye can no longer do. Loss of transparency of the lens is the condition known as a cataract.

The Retina Is The Light-Sensitive Part Of The Eye

The light-sensitive part of the vertebrate eye is the retina, which lines the posterior two-thirds of the eyeball covering the choroid. The retina contains photorecep-

tor cells called, according to their shape, **rods** and **cones** (Fig. 35–19). The 125 million or so rods function in dim light, allowing us to detect shape and movement. Rods are not sensitive to colors. About 6.5 million cones are responsible for bright-light vision, for the perception of fine detail, and for color vision. The cones permit color vision by being differentially sensitive to different frequencies (colors) of light. The cones are concentrated in the **fovea,** a small, depressed area in the center of the retina. The fovea is the region of sharpest vision. In addition to rods and cones, the retina contains many sensory and connector neurons and their axons.

At a point in the back of the eye, the individual axons of the sensory neurons unite to form the **optic nerve,** which then passes out of the eyeball. Here there are no rods and cones. This area is called the "blind spot," since images falling on it cannot be perceived.

A Chemical Change In Rhodopsin Leads To Transmission Of An Action Potential

Rhodopsin, found in the rod cells, and some very closely related pigments in the cone cells, are responsible for the ability to see. Rhodopsin consists of opsin, a large protein, chemically joined with **retinal,** which is

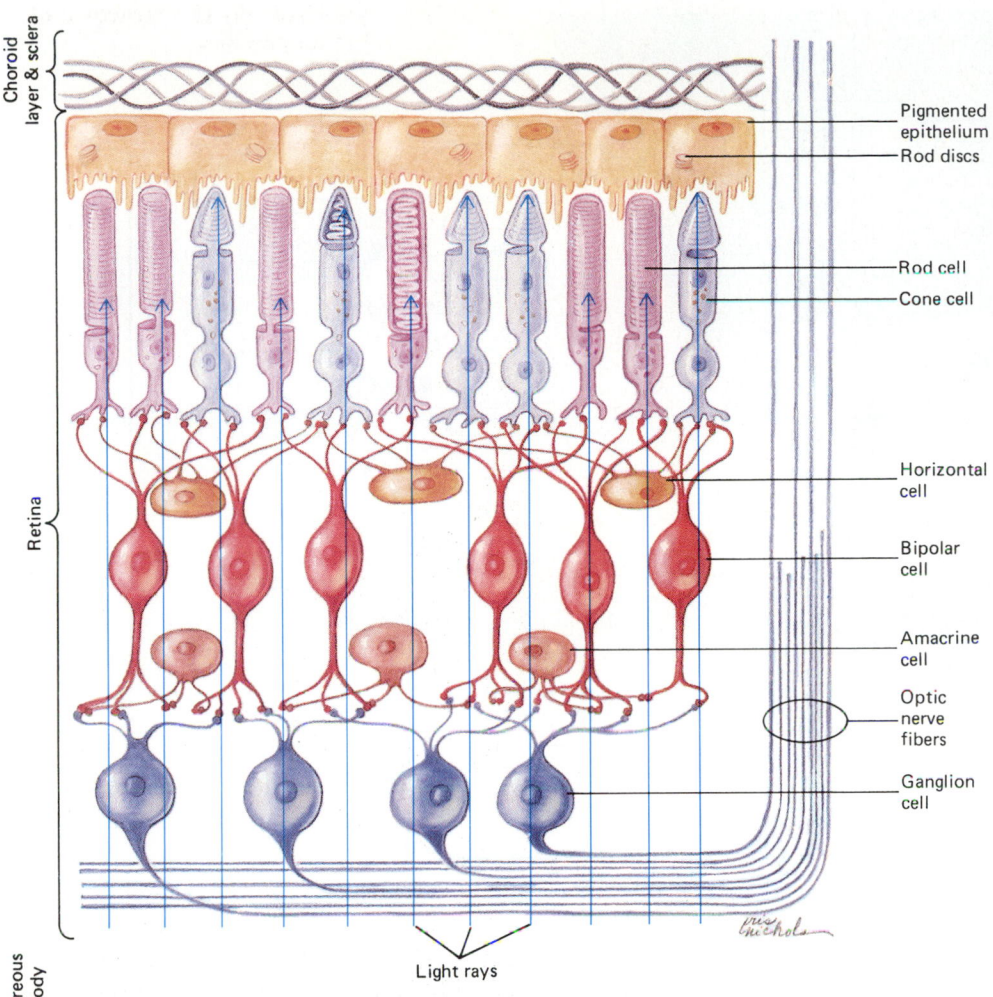

Choroid layer & sclera

Retina

Vitreous body

Pigmented epithelium

Rod discs

Rod cell

Cone cell

Horizontal cell

Bipolar cell

Amacrine cell

Optic nerve fibers

Ganglion cell

Light rays

(a)

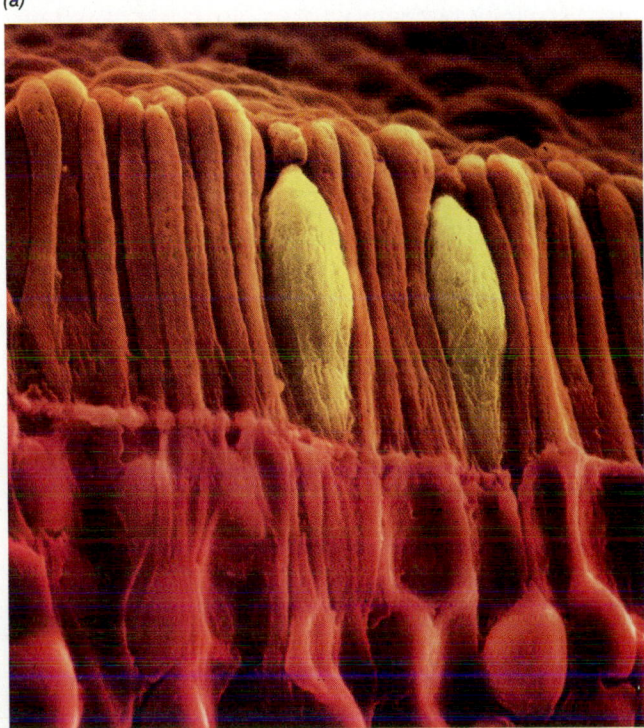

(b)

FIGURE 35–19 (*opposite*) **The retina.** (*a*) **Neuronal connections in the retina. The elaborate interconnections among the various layers of cells allow them to interact and to influence one another in a number of ways.** (*b*) **Rods (red elongated structures) and two cones (shorter, thicker, yellow structures) make up the top row in this micrograph (magnification × 10,000). The elongated rods permit us to see shape and movement, whereas the shorter cones allow us to view our world in color.** (*b*, Lennart Nilsson, from *The Incredible Machine*, p. 279)

made from vitamin A. Two isomers of retinal exist: the 11-*cis* form and the all-*trans* form.

When light strikes rhodopsin, it transforms **11-*cis* retinal** to **all-*trans* retinal.** This change in shape causes rhodopsin to break down into its components, opsin and retinal. During this process rhodopsin is converted into a series of intermediate compounds. One of these intermediates is **metarhodopsin II,** which is thought to be the key compound leading to the generation of action potentials and ultimately transmission of neural impulses to the brain. The all-*trans* retinal is converted back to the 11-*cis* form by an enzyme. Then the retinal combines with opsin to produce rhodopsin once again. This sequence of reactions is known as the *visual cycle* (Fig. 35–20).

Just how visual images are processed is not certain. The size, intensity, and location of light stimuli determine initial processing in the retina. The pattern of neuron firing in the retina appears to be very important. The optic nerves are thought to transmit information to the brain by way of compound, encoded signals.

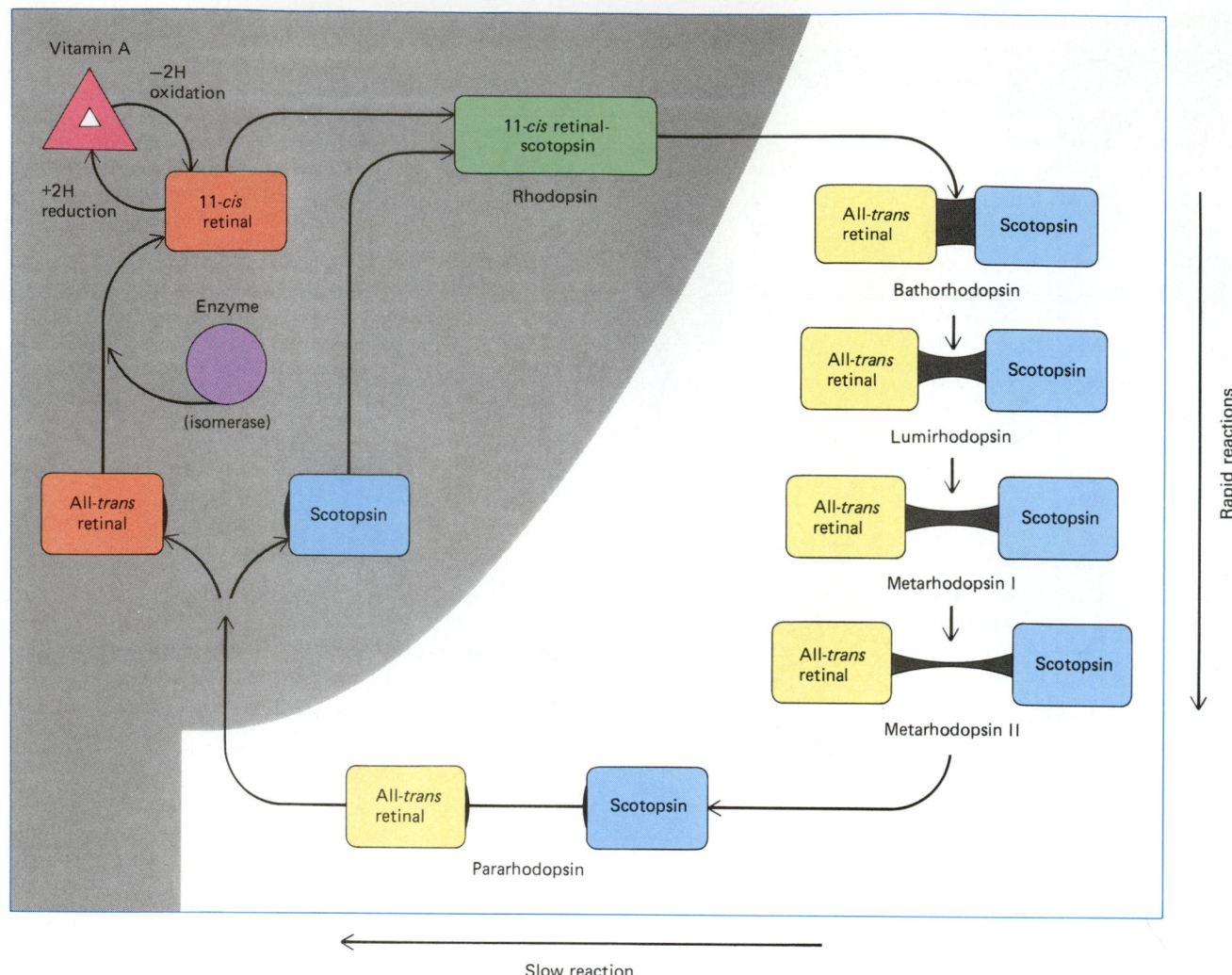

FIGURE 35–20 The visual cycle. When light strikes rhodopsin, it breaks down, depolarizing the rod cell which contains it. This produces an impulse.

■ CHAPTER SUMMARY

I. A sense organ is a receptor specialized to respond to specific energy stimuli in the environment. Sense organs are composed of receptor cells and accessory cells. Receptor cells may be neuron endings or specialized cells in close contact with neurons.

II. Sense organs can be classified on the basis of the type of energy to which they respond; there are mechanoreceptors, chemoreceptors, photoreceptors, thermoreceptors, and electroreceptors.

III. Exteroceptors are sense organs that receive information from the outside world. Proprioceptors are sense organs within muscles, tendons, and joints that enable the animal to perceive orientation of the body and position of its parts. Interoceptors are sense organs within internal body organs.

IV. Receptor cells absorb energy, transduce that energy into electrical energy, and produce receptor potentials.

V. Impulses from a sense organ may differ in the number of neurons transmitting, which particular neurons are firing, the total number of action potentials transmitted by a given neuron, and the frequency of the action potentials transmitted by a given neuron.

VI. Adaptation of a receptor to a continuous stimulus results in diminished perception.

VII. Mechanoreceptors respond to touch, pressure, gravity, stretch, or movement.
  A. The tactile receptors in the skin are mechanoreceptors that respond to mechanical displacement of hairs or of the receptor cells themselves.

B. Muscle spindles, Golgi tendon organs, and joint receptors are proprioceptors that respond continuously to tension and movement in muscles and joints.

C. Lateral line organs supplement vision in fish by informing the animal of moving objects or objects in its path.

D. Statocysts are gravity receptors found in many invertebrates.
   1. When the position of the statolith within the statocyst changes, hairs of receptor cells are bent.
   2. Messages sent to the CNS inform an animal of which hairs have been stimulated; from this the animal can determine where "down" is and can correct for any abnormal orientation.

E. The saccule and utricle of the vertebrate ear contain otoliths that change position when the head is tilted or when the body is moving forward. Hair cells stimulated by the otoliths send impulses to the brain, enabling the animal to perceive the direction of gravity.

F. The semicircular canals of the vertebrate ear inform the brain about turning movements. Their hair cells are stimulated by movements of the endolymph.

G. The organ of Corti within the cochlea is the auditory receptor in birds and mammals.
   1. Sound waves pass through the external auditory meatus, cause the eardrum to vibrate, and are transmitted through the middle ear by the hammer, anvil, and stirrup.
   2. Vibrations pass through the oval window to fluid within the vestibular duct. Pressure waves press upon the membranes separating the three ducts of the cochlea.

   3. Movements of the basilar membrane rub the hair cells of the organ of Corti against the overlying tectorial membrane, thus stimulating them.
   4. Nerve impulses are initiated in the dendrites of the auditory neurons lying at the base of each hair cell.

VIII. Chemoreceptors detect specific chemical compounds, permitting taste and smell.
   A. Taste receptors are specialized epithelial cells located in taste buds.
   B. The olfactory epithelium contains specialized olfactory cells with axons that extend upward as fibers of the olfactory nerves.

IX. Thermoreceptors provide endothermic animals with information about body temperature. In some animals they are used to locate prey.

X. Electroreceptors detect electrical currents in water.

XI. Photoreceptors use pigments to absorb light energy.
   A. Eyespots detect light but do not form images effectively.
   B. The compound eye of arthropods consists of ommatidia, which collectively form a mosaic image.
   C. In the vertebrate eye, light enters through the cornea, is focused by the lens, and sensed as an image by the retina. The iris regulates the amount of light that can enter.
   D. When light strikes rhodopsin in the rod cells, a chemical change in retinal occurs that breaks down the rhodopsin, triggering depolarization of the rod cell.
   E. The rods form images in black and white, whereas the cones are responsible for color vision.

■ POST-TEST

1. A sense organ consists of one or more _____ cells.

2. Exteroceptors are sense organs that _____; proprioceptors enable an animal to perceive _____, along with the _____ of the body as a whole.

3. _____ detect light energy; _____ respond to touch, gravity, or movement.

4. Receptor cells absorb _____, transduce this energy into _____ energy, and produce a _____.

5. The diminishing response of a receptor to a continued, constant stimulus is called _____.

6. Statocysts serve as _____ receptors; their action depends upon mechanical displacement of receptor cell hairs by change in position of a _____.

7. The lateral line organ of fishes is thought to supplement _____.

8. Three main types of vertebrate proprioceptors are _____ _____, which detect muscle movement; _____ _____ organs, which determine stretch in tendons; and _____ receptors, which detect movement in ligaments.

9. The basic function of the vertebrate ear is to help maintain _____.

10. The inner ear consists of interconnected canals and sacs called the _____; in jawed vertebrates this structure consists of two sac-like chambers, the _____ and _____, and three _____ canals, as well as the cochlea.

11. The rocks in your head (within the saccule and utricle) called _____ are actually _____ detectors.

12. Each semicircular canal is filled with fluid called _____; at one of the openings of each canal into the utricle is a small enlargement, the _____.

13. The cochlea, located in the _____ ear, contains mechanoreceptor hair cells that detect _____ waves.

14. The actual auditory receptor is the organ of _____; it is located within the _____ duct.

15. The senses of taste and smell depend upon _____.

16. The photosensitive pigments in the eyes of vertebrates and arthropods are _____.

Select the most appropriate answer in Column B for each description in Column A.

Column A		*Column B*
_____**17.** Light-sensitive part of human eye		**a.** Ommatidium
_____**18.** Regulates size of pupil		**b.** Cones
_____**19.** Perceive color		**c.** Retina
_____**20.** Region of sharpest vision		**d.** Iris
_____**21.** Visual unit in compound eye		**e.** Fovea

22. Label the diagram below. (Refer to Fig. 35–18 as necessary.)

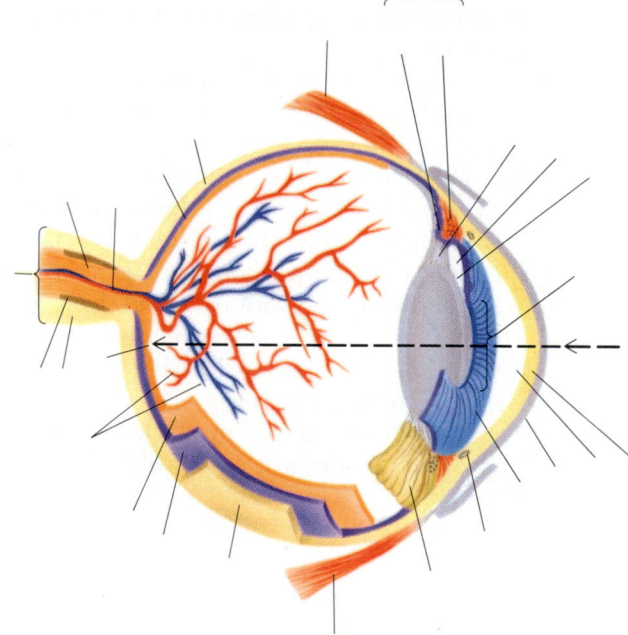

■ REVIEW QUESTIONS

1. Imagine that as you walk into a room you are met by an offensive odor. After a few minutes you hardly notice the smell. Explain.

2. If all neurons transmit the same type of message, how do we know the difference between sound and light? How are we able to distinguish between an intense pain and a mild one?

3. What do statocysts, lateral line organs, and semicircular canals have in common?

4. What is a proprioceptor? What is its function in the mammalian body?

5. Discuss the mechanism by which the sensory cells of the ear are stimulated by sound waves.

6. What are the functions of rods and cones? How are they distributed in the retina?

7. Discuss the mechanism by which photoreceptors are stimulated by light. What is the function of rhodopsin?

8. Contrast the function of the insect's compound eye with that of the vertebrate eye.

■ RECOMMENDED READINGS

Abu-Mostafa, Y.S., and D. Psaltis: "Optical Neural Computers," *Scientific American* 256(3):88–94, 1987. The arrangement of neurons in the brain can be used as a model for building a computer that can solve problems, such as recognizing patterns, that involve memorizing all possible solutions.

Hudspeth, A.J.: "The Hair Cells of the Inner Ear," *Scientific American* 248(1):54–64, 1983. A description of the mechanism by which hair cells in the inner ear respond to convey information about acoustic tones and acceleration. Some of this author's conclusions are controversial.

Levine, J.S., and E.F. MacNichol, Jr.: "Color Vision in Fishes," *Scientific American* 247(2) February 1982, 140–149. A discussion of retinal pigments in fishes and their relationship to the evolution of the eye.

Miller, J.A.: "Colorful Views of Vision," *Science News* Vol. 120, October 3, 1981. An account of a technique for mapping components of the visual system.

Newman, E.A., and P.H. Hartline: "The Infrared Vision of Snakes," *Scientific American* 247(3) March 1982, 116–124. Snakes of two families can detect and localize sources of infrared radiation.

Poggio, T., and C. Koch: "Synapses That Compute Motion," *Scientific American* 256(5):46–52, 1987. Studies of cells in the eye that interpret movement may help clarify mechanisms involved in other neural processes.

Schnapf, J.L., and D.A. Baylor: "How Photoreceptor Cells Respond to Light," *Scientific American* 256(4):40–47, 1987. How a single photoreceptor cell in the eye registers the absorption of a single photon.

Treisman, A.: "Features and Objects in Visual Processing," *Scientific American* 255(5):114B–125, 1986. We perceive meaningful wholes visually by automatically extracting features from a scene and assembling them into objects.

Vaughan, C.: "A New View of Vision," *Science News* 134:58–60, 1988. Discussion of a new theory about how the brain processes visual signals.

36

Internal Transport

OUTLINE

I. Some invertebrates have no circulatory system
II. Some invertebrates have an open circulatory system
III. Many invertebrates have a closed circulatory system
IV. Vertebrates have a closed circulatory system adapted to carry out a variety of functions
V. Vertebrate blood consists of plasma, blood cells, and platelets
  A. Plasma is the fluid component of blood
  B. Red blood cells transport oxygen
  C. White blood cells defend the body against disease organisms
  D. Platelets function in blood clotting
VI. Vertebrates have three main types of blood vessels
VII. The evolution of the vertebrate heart culminated in a four-chambered heart and double-circuit circulation
  A. The fish heart has a single atrium and single ventricle
  B. Amphibians have a three-chambered heart
  C. The reptilian heart consists of two atria and two ventricles
  D. Birds and mammals have a four-chambered heart
  E. The structure of the human heart is marvelously adapted for pumping blood
    1. Each heartbeat is initiated by a pacemaker
    2. Heart rate is regulated by the nervous system
    3. Two main heart sounds can be distinguished
VIII. Arterial pulse results from the alternate expansion and recoil of an artery
IX. Blood pressure depends on blood flow and resistance to blood flow
  A. Blood pressure is highest in arteries
  B. Blood pressure is carefully regulated

LEARNING OBJECTIVES

After you have studied this chapter you should be able to:

1. Compare internal transport in animals that lack a circulatory system, animals with an open circulatory system, and animals with a closed circulatory system.
2. Relate structural adaptations of the vertebrate circulatory system to each function it performs.
3. Compare the structure and functions of red blood cells, white blood cells, and platelets.
4. Compare the structure and function of the different types of blood vessels, including arteries, arterioles, capillaries, and veins.
5. Trace the evolution of the vertebrate heart from fish to mammal.
6. Describe the structure and function of the parts of the heart and label them on a diagram.
7. Summarize how the heart works; include a description of the heartbeat, neural regulation of heart rate, and the sounds produced by the heart.
8. Explain the physiological basis of arterial pulse and tell how pulse is measured.
9. Identify factors that determine blood pressure.
10. Compare blood pressure in different types of blood vessels and summarize how arterial blood pressure is regulated.
11. Trace a drop of blood through the pulmonary and systemic circulations, naming in proper sequence each structure through which it passes.
12. Identify the risk factors of atherosclerosis, trace the progress of the disorder, and summarize its possible complications (including angina pectoris and myocardial infarction).
13. List the functions of the lymphatic system, and describe how the system operates to maintain fluid balance.

SEM of red blood cells (David M. Phillips/Visuals Unlimited.)

OUTLINE (Continued)

X. **In mammals blood is pumped through a pulmonary and a systemic circulation**
   A. **The pulmonary circulation functions to oxygenate the blood**
   B. **The systemic circulation delivers blood to all of the tissues**
      1. **The coronary circulation delivers blood to the heart**
      2. **The hepatic portal system delivers nutrients to the liver**

XI. **The lymphatic system is an accessory circulatory system**
   A. **The lymphatic system consists of lymphatic vessels and lymph tissue**
   B. **The lymphatic system plays an important role in fluid homeostasis**

Focus on the electrical activity of the heart
Focus on cardiovascular disease

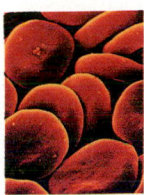

Each cell of the animal body requires a constant supply of oxygen and nutrients and constantly generates metabolic wastes. Very small organisms have few cells and each is in close contact with the surrounding environment. In such organisms simple diffusion may adequately distribute oxygen, nourish each cell, and dispose of waste products. However, gases diffuse about 10,000 times more slowly through tissues than through air. As long as an organism is less than 1 millimeter thick, diffusion may be adequate, but if an organism is to grow any larger, it needs a more effective mechanism for transporting materials to all of its cells.

In complex animals internal transport of nutrients, oxygen, and wastes is accomplished by specialized structures that make up the circulatory system. Most circulatory systems consist of (1) blood, a fluid connective tissue consisting of cells and cell fragments dispersed in fluid, (2) a pumping device, generally a heart, and (3) a system of blood vessels through which the blood is pumped. An efficient circulatory system enables large animals to carry on rapid metabolism and thus allows a more active life-style.

KEY CONCEPTS

☐ Animals that are more than a few cells thick or that have an active life-style have a circulatory system that transports oxygen and nutrients to all of its cells.

☐ Some invertebrates (arthropods and mollusks) have an open circulatory system in which blood flows into a hemocoel and bathes the tissues directly.

☐ Other invertebrates and vertebrates have a closed circulatory system in which blood is pumped through a continuous circuit of blood vessels.

☐ Vertebrate blood consists of a fluid called plasma which contains red blood cells, white blood cells, and platelets.

☐ The vertebrate heart is a muscular pump that has evolved from the single atrium and ventricle of fish to the four-chambered (two atria and two ventricles) heart of birds and mammals.

SOME INVERTEBRATES HAVE NO CIRCULATORY SYSTEM

Many invertebrates lack a circulatory system. In them internal transport depends on diffusion and on other body systems. Cnidarians (e.g., jellyfish and hydras) have a laid-back life-style. Their success does not depend on a high metabolic rate. Diffusion through the thin body wall supplies sufficient oxygen to support their activities and carries wastes from the body.

Cnidarians and many flatworms combine digestive and some internal transport functions in the gastrovascular cavity (Fig. 36–1). Movement of the animal's body, as it stretches and contracts, stirs up the contents of the cavity and helps distribute nutrients to all parts of the body.

The flattened body of the flatworms permits effective gas exchange by diffusion. Metabolic rate tends to be higher than in cnidarians, and this allows a more active life-style. The branching excretory system of planarians provides for internal transport of wastes that are then expelled from the body.

SOME INVERTEBRATES HAVE AN OPEN CIRCULATORY SYSTEM

Arthropods and most mollusks have an **open circulatory system,** in which the heart pumps blood into vessels that have open ends (Fig. 36–2). Blood spills out of them, filling large spaces that make up the **hemocoel** (blood cavity), bathing the cells of the body directly.

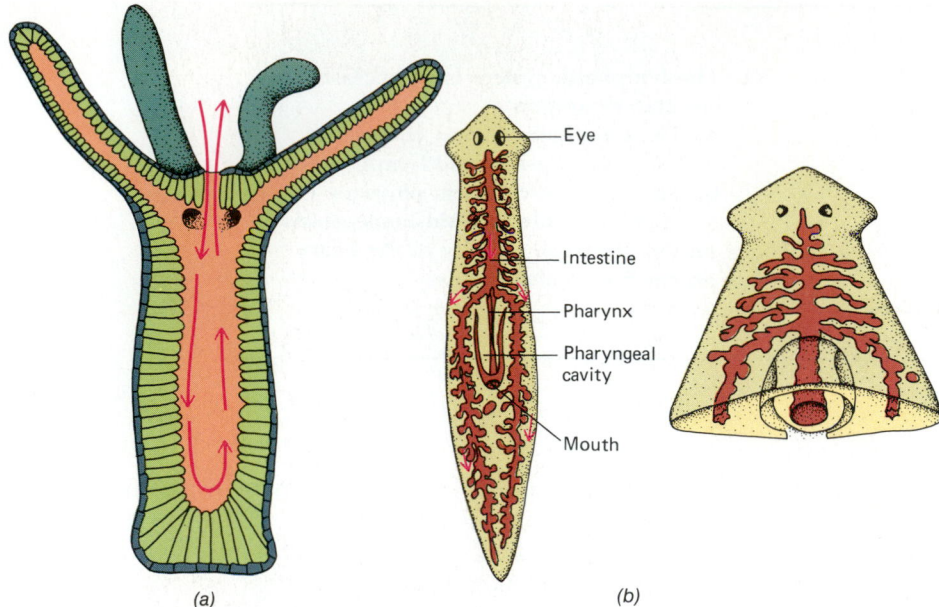

FIGURE 36–1 Some invertebrates lack a circulatory system. (*a*) In *Hydra* and other cnidarians, the gastrovascular cavity helps circulate nutrients to all parts of the body. (*b*) In planarian flatworms, the branched intestine circulates food to all regions of the body.

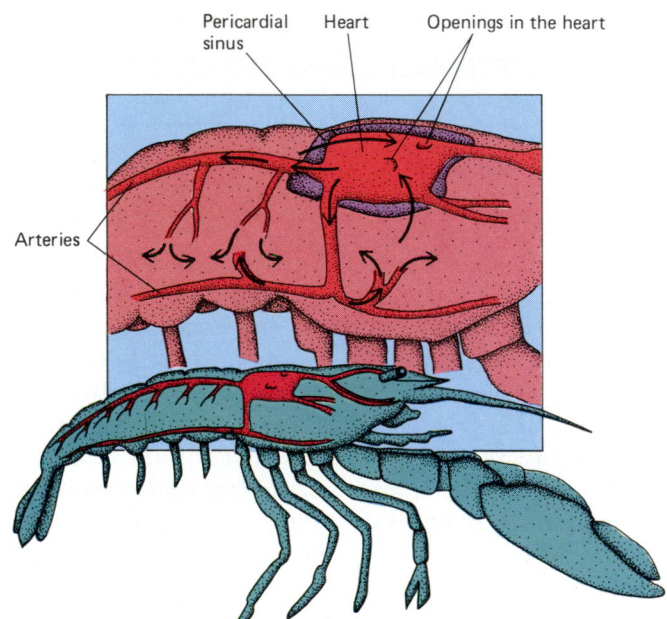

FIGURE 36–2 The crayfish, like other arthropods, has an open circulatory system. Lateral view.

Blood re-enters the circulatory system through openings in the heart (in arthropods) or through open-ended vessels that lead to the gills (in mollusks).

An open circulatory system does not supply oxygen to cells very rapidly even though the blood of some of these invertebrates contains a pigment (hemocyanin) that transports oxygen. An open circulatory system cannot provide enough oxygen to maintain the active life-style of the insects, for example. Indeed, insect blood mainly distributes nutrients and hormones. Oxygen is delivered directly to the cells by a system of air (tracheal) tubes.

☐ MANY INVERTEBRATES HAVE A CLOSED CIRCULATORY SYSTEM

Annelids, some mollusks, and echinoderms are among the invertebrates that have a **closed circulatory system** (Fig. 36–3). In them, blood flows through a continuous circuit of blood vessels. The walls of the smallest blood

FIGURE 36–3 The earthworm has a closed circulatory system. Five pairs of contractile blood vessels deliver blood from the dorsal vessel to the ventral vessel.

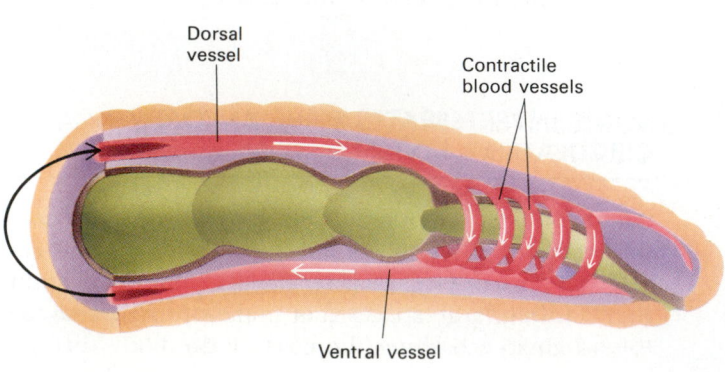

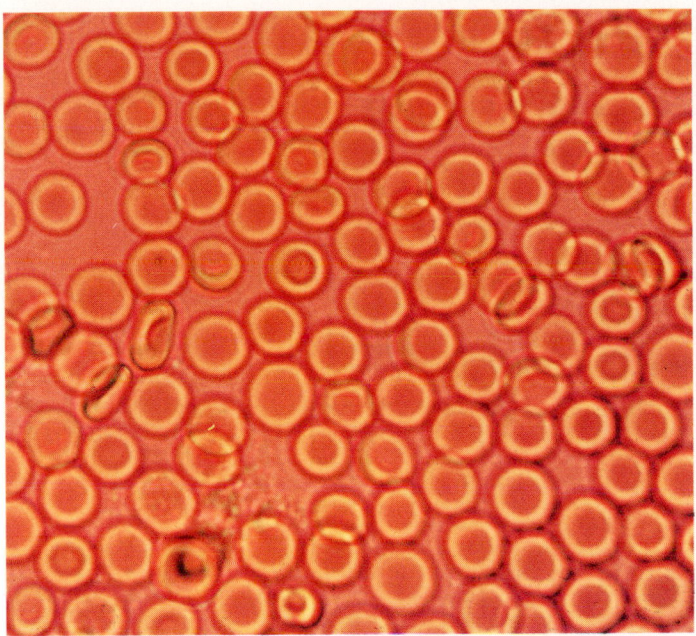

(a)

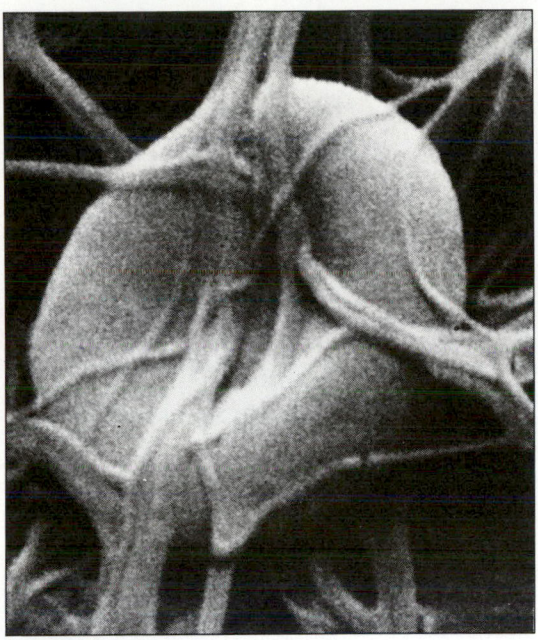

(b)

FIGURE 36–4   Blood. (a) Photomicrograph of blood showing red blood cells (approximately ×450). (b) A scanning electron micrograph of a red blood cell. (a, Runk/Schoenberger, from Grant Heilman; b, from Bernstein, E., *Science* 173, 1971; © 1971, the American Association for the Advancement of Science)

vessels are thin enough to permit diffusion of gases, nutrients, and wastes between blood in the vessels and the extracellular fluid that bathes the cells.

Annelids (e.g., earthworms) have no heart. Contractions of certain blood vessels, as well as contractions of the body wall muscles, circulate the blood. Earthworms possess hemoglobin, the same red pigment that transports oxygen in vertebrate blood. In earthworms, hemoglobin is not present inside red blood cells but is dissolved in the blood plasma.

Although most mollusks have an open circulatory system, the fast-moving cephalopods (squid, octopus) have a closed system made even more effective by the presence of extra "hearts" at the base of the gills that speed the passage of blood through the gills.

The circulatory system of the sea cucumbers is the most highly developed system of any of the echinoderms. It appears to transport both nutrients and oxygen. Invertebrate chordates typically have a closed circulatory system with a ventral heart.

▪ VERTEBRATES HAVE A CLOSED CIRCULATORY SYSTEM ADAPTED TO CARRY OUT A VARIETY OF FUNCTIONS

Vertebrates have a closed circulatory system. The system consists of a ventral heart, blood vessels, blood, lymph, lymph vessels, and some associated organs,

such as the thymus, spleen, and liver. The tiniest blood vessels, **capillaries,** have very thin walls that permit exchange of materials between blood and extracellular fluid.

The circulatory system

1. Transports nutrients from the digestive system and from storage depots to each cell of the body.
2. Transports oxygen from respiratory structures (gills, lungs) to the cells of the body.
3. Transports metabolic wastes from each cell to organs that excrete them.
4. Transports hormones from endocrine glands to target tissues.
5. Helps maintain fluid balance.
6. Defends the body against invading microorganisms.
7. Helps to distribute metabolic heat within the body and to maintain normal body temperature in endothermic (warm-blooded) animals.

▪ VERTEBRATE BLOOD CONSISTS OF PLASMA, BLOOD CELLS, AND PLATELETS

In vertebrates, **blood** consists of a pale yellowish fluid, known as **plasma,** in which red blood cells, white blood cells, and platelets are suspended (Fig. 36–5). An adult man weighing about 70 kilograms (154 pounds) possesses about 5.6 liters (6 quarts) of blood.

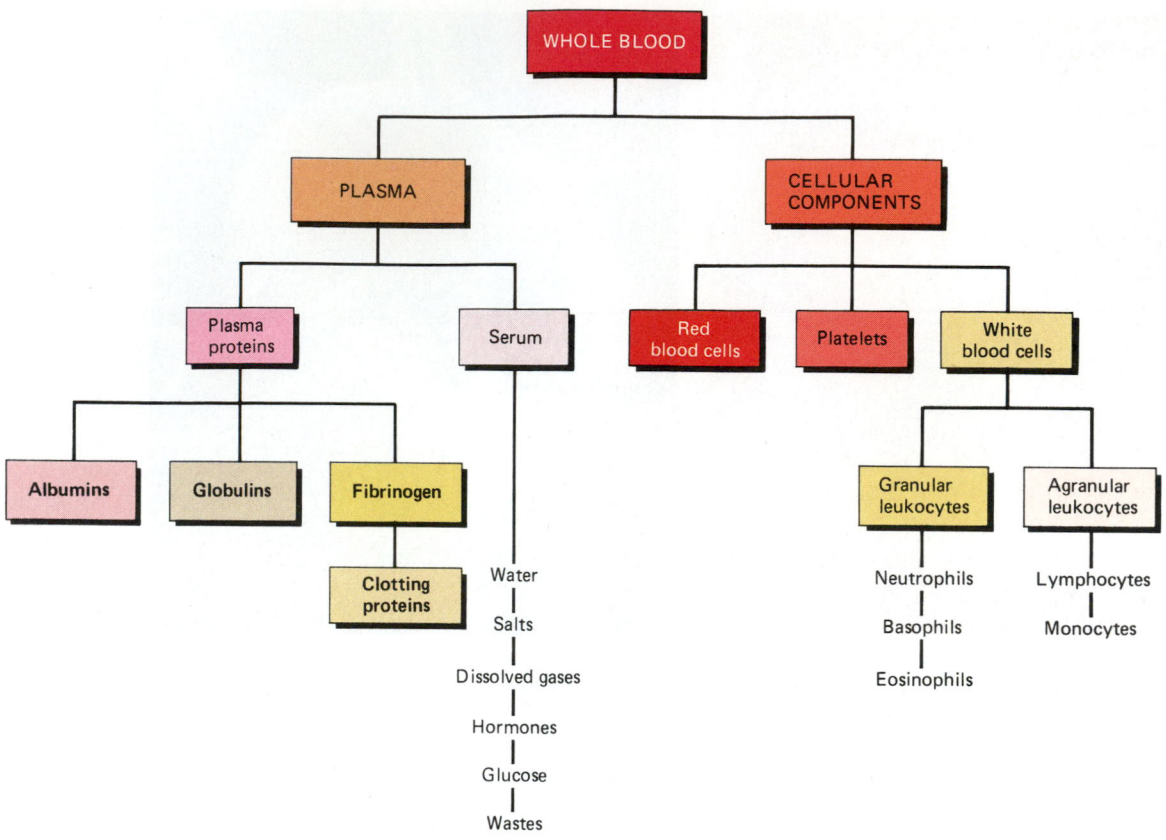

FIGURE 36-5 **Components of whole blood.**

Plasma Is The Fluid Component Of Blood

Plasma is mostly water (92%) but also contains plasma proteins, a sprinkling of salts, and a vast array of materials being transported, such as nutrients, dissolved gases, metabolic wastes, hormones, and even contaminants, such as drugs. Plasma proteins may be divided into three groups, or fractions—the albumins, globulins, and fibrinogen. One of the main functions of these plasma proteins is to maintain an appropriate blood volume. As blood flows through the capillaries, some plasma seeps through the capillary walls and passes into the tissues. However, large protein molecules have difficulty passing through the capillary walls, so most of them remain in the blood. There they exert an osmotic pressure, which helps pull plasma back into the blood.

Certain **albumins** transport substances such as specific hormones, keeping them bound in the blood until needed. One group of **globulins,** the gamma globulins, are antibodies, substances that provide immunity against invading disease organisms. Alpha and beta globulins transport lipids, hormones, iron, and some other substances. **Fibrinogen** and several other plasma proteins are involved in the clotting process. When the proteins involved in blood clotting have been removed from the plasma, the remaining liquid is called **serum.**

Lipoproteins in the blood transport triacylglycerols (triglycerides) and cholesterol. **High-density lipoprotein (HDL)** is thought to prevent cholesterol from clogging the walls of arteries. **Low-density lipoprotein (LDL)** contains a large amount of cholesterol and has been associated with deposit of cholesterol in the arterial wall (atherosclerosis).

Red Blood Cells Transport Oxygen

Each of us has about 30 trillion (3×10^{13}) **red blood cells,** or **erythrocytes,** suspended in our plasma. These cells are so tiny that about 3000 of them lined up end-to-end would measure only 1 inch. Red blood cells are exquisitely adapted for transporting oxygen.

Red blood cells develop inside certain bones in special tissue called **red bone marrow.** As each cell differentiates, it manufactures great quantities of **hemoglobin,** the pigment that gives blood its characteristic red color. Each hemoglobin molecule contains four atoms of iron, each of which can combine with one molecule of oxygen. Thus, each hemoglobin molecule can combine chemically with four molecules of oxygen, forming the compound **oxyhemoglobin.** It is in this molecular form that oxygen is transported throughout the body. Red blood cells also transport carbon dioxide.

The mature red blood cell is a tiny sac of hemoglobin, shaped with a considerable surface area for gas exchange. In mammals the mature red blood cell lacks a nucleus and carries on only limited cellular functions. It has a short life span—about 120 days. As blood circulates through the liver and spleen, worn-out red blood cells are removed from circulation and destroyed. Their hemoglobin molecules are taken apart, and some of the components, such as iron, are sent back to the red bone marrow for reuse. About 2.4 million red blood cells are destroyed each second; these are immediately replaced by new red blood cells from the bone marrow.

A deficiency of hemoglobin, usually accompanied by a reduced number of red blood cells, is called **anemia.** With less hemoglobin, less oxygen is transported, so the body cells do not receive enough oxygen. An anemic person complains of never having enough energy—the "tired-blood" syndrome.

Three general causes of anemia are (1) loss of blood due to hemorrhage or internal bleeding, (2) decreased production of hemoglobin or red blood cells as in iron deficiency anemia, and (3) increased rate of red blood cell destruction (the **hemolytic anemias** such as sickle cell anemia).

White Blood Cells Defend The Body Against Disease Organisms

White blood cells, or **leukocytes,** defend the body against invading bacteria and other foreign substances. These cells are able to leave the blood, squeezing out through the walls of the capillaries. White blood cells are capable of independent locomotion similar to that of an amoeba. They wander about through the tissues of the body, destroying invading microorganisms or worn-out cells.

There are two main types of white blood cells; both have their origin in red bone marrow. **Agranular leukocytes** lack specific granules in their cytoplasm. **Granular leukocytes** have specific granules. Two kinds of agranular leukocytes are lymphocytes and monocytes (Fig. 36–6). **Monocytes** leave the circulation and complete their development in the tissues. There, they increase to about five times their original size and become **macrophages,** the giant scavenger cells of the body. Macrophages have voracious appetites for bacteria, dead cells, and other matter littering the tissues. The role of lymphocytes in immune responses will be discussed in Chapter 37.

Three types of **granular leukocytes** are neutrophils, eosinophils, and basophils. **Neutrophils** are especially adept at seeking out and ingesting bacteria. **Eosinophils** are thought to play a part in allergic reactions. **Basophils** contain large amounts of **histamine,** a chemical released in injured or infected tissues and in

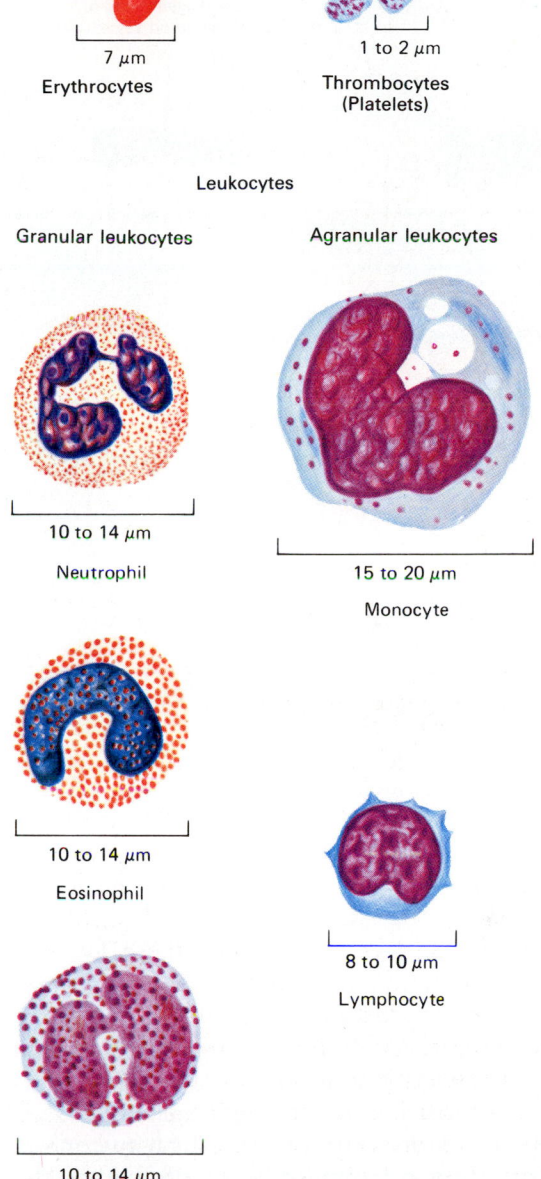

FIGURE 36–6 Principal varieties of blood cells in the circulating blood.

allergic reactions. Basophils also contain **heparin,** an anticlotting chemical that may be important in preventing inappropriate clotting within blood vessels.

Normally, there are about 7000 white blood cells per cubic millimeter of human blood. In contrast, there may be as many as 5 million red blood cells per cubic millimeter, so there are almost 700 red cells to every white cell. Measuring the number of white blood cells is a useful diagnostic tool because the white cell count becomes elevated in bacterial infections and

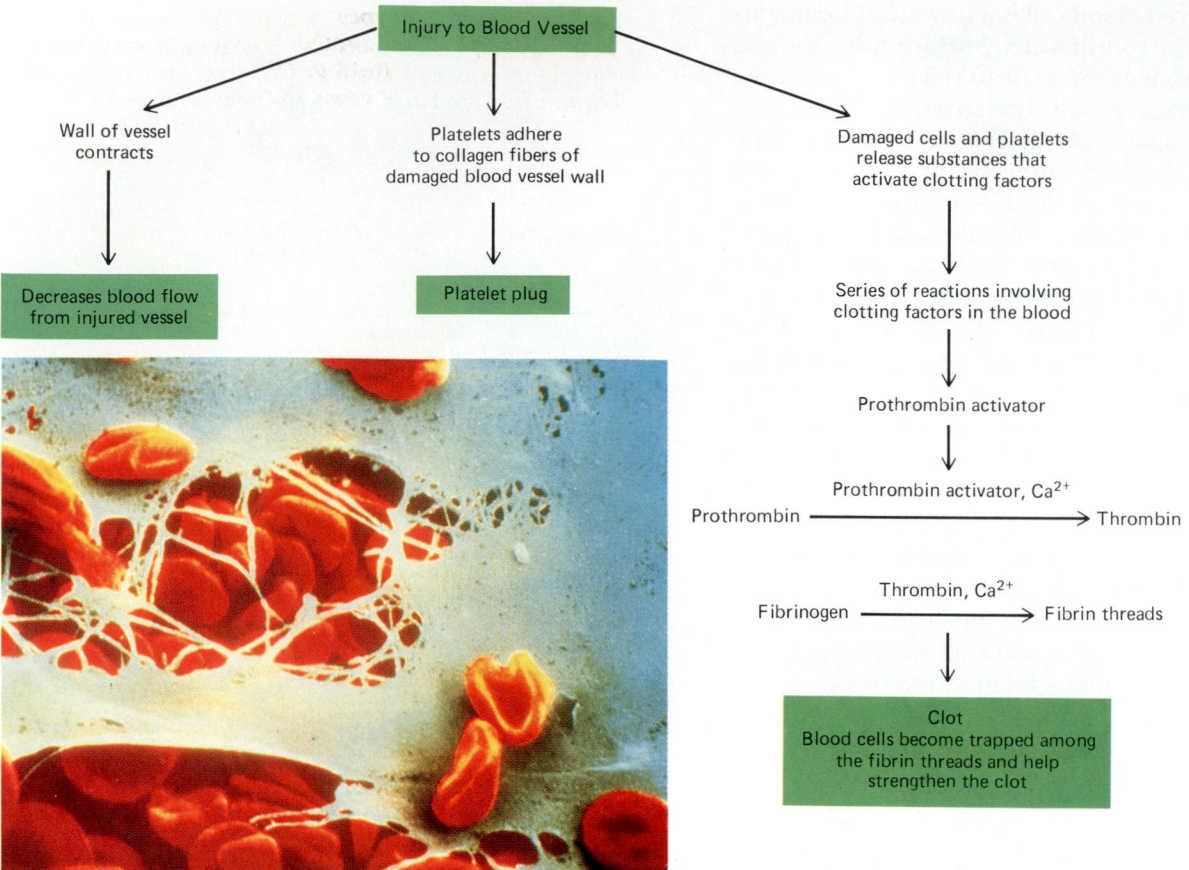

FIGURE 36–7   Overview of blood clotting. Scanning electron micrograph of part of a blood clot, showing red blood cells enmeshed in a network of fibrin. (Lennart Nillson/Boehringer Ingelheim International, GmbH.)

in certain other disorders. On the other hand, lowered white cell counts often accompany viral infections, rheumatoid arthritis, and a few other conditions. This explains why physicians often perform white blood cell counts before prescribing antibiotics, for these drugs are effective against bacteria but not against viruses.

Leukemia is a form of cancer in which any one of the kinds of white cells multiply rapidly within the bone marrow. Many of these cells do not mature. Their large numbers crowd out developing red blood cells and platelets, leading to anemia and impaired clotting. A common cause of death from leukemia is internal bleeding, especially in the brain. Another frequent cause of death is infection because, although there may be a dramatic rise in the white cell count, the cells are immature and abnormal and unable to defend the body against disease organisms. Although no cure for leukemia has been discovered, radiation treatment and therapy with antimitotic drugs can induce partial or complete remissions that have lasted as long as 15 years in some patients.

Platelets Function In Blood Clotting

Platelets are small fragments of cytoplasm that separate from certain large cells (megakaryocytes) in the bone marrow. When the wall of a blood vessel is injured, as when you cut your finger, platelets seal the break by adhering to the wall in large numbers. A complex series of chemical reactions produces tiny fibers that reinforce the platelets, forming a strong clot (Fig. 36–7). This process may be summarized as follows:

$$\text{Prothrombin} \xrightarrow{\text{Several clotting factors, calcium}} \text{Thrombin}$$
(a plasma protein)   (active form of prothrombin)

$$\text{Fibrinogen} \xrightarrow{\text{Thrombin}} \text{Fibrin}$$

Prothrombin is a globulin manufactured in the liver. Vitamin K is necessary for its production. Fibrin consists of tiny threads that serve to reinforce the platelet clot and to entrap red and white blood cells. (The red blood cell in Fig. 36–7 is enmeshed in the fibrin of a

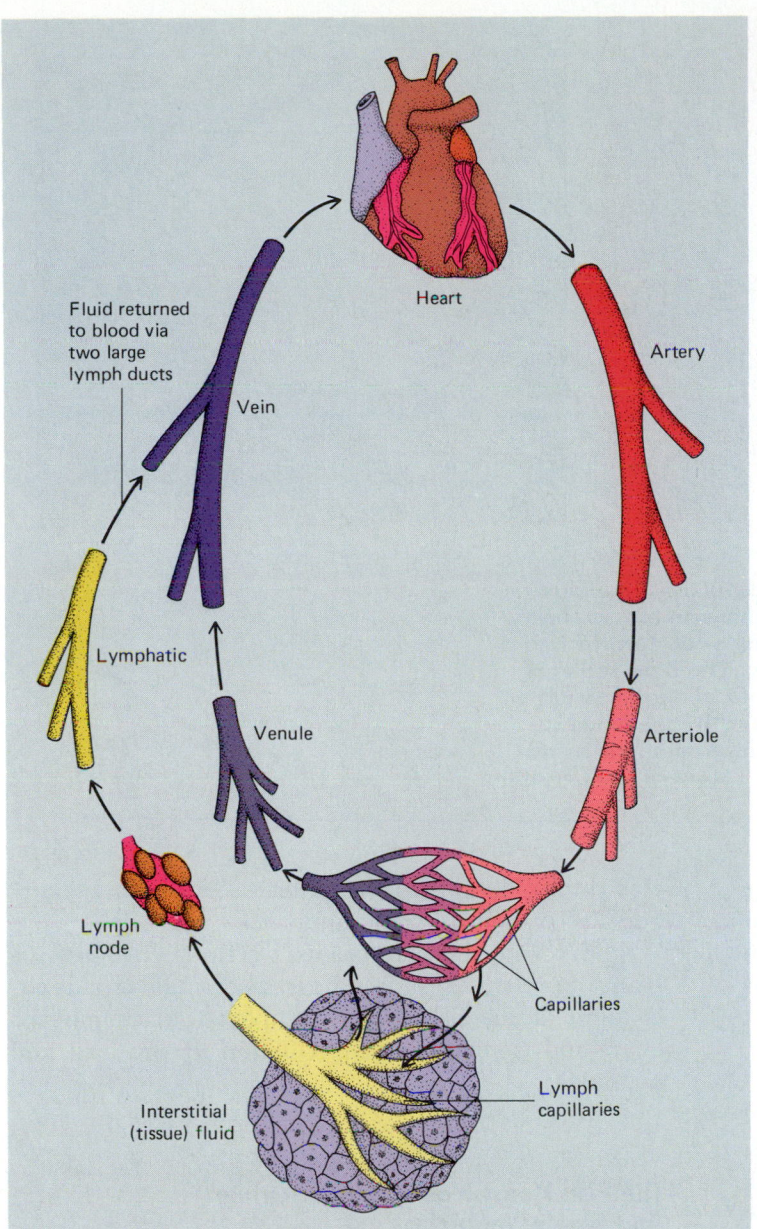

FIGURE 36–8 Types of blood vessels and their relationship to one another. Lymphatic vessels return excess interstitial fluid to the blood by way of ducts that lead into large veins in the shoulder region.

Fluid returned to blood via two large lymph ducts

Vein

Heart

Artery

Lymphatic

Venule

Arteriole

Lymph node

Capillaries

Interstitial (tissue) fluid

Lymph capillaries

developing clot.) In people with hemophilia ("bleeder's disease") one of the clotting factors is absent due to a genetic mutation.

■ VERTEBRATES HAVE THREE MAIN TYPES OF BLOOD VESSELS

The vertebrate circulatory system includes three main types of blood vessels: arteries, capillaries, and veins (Fig. 36–8). An **artery** carries blood away from the heart, toward other tissues. When an artery enters an organ, it divides into many smaller branches called **arterioles.** The arterioles deliver blood into the micro-

scopic **capillaries.** After coursing through an organ, capillaries eventually merge to form **veins** that transport the blood back toward the heart.

The walls of arteries and veins are thick, which prevents gases and nutrients from passing through them. Materials are exchanged between the blood and tissue fluid bathing the cells through the capillary walls, which are only one cell thick (Fig. 36–9). Capillary networks in the body are so extensive that at least one of these tiny vessels is located close to almost every cell in the body. The total length of all capillaries in the body has been estimated as more than 60,000 miles.

Smooth muscle in the arteriole wall can constrict **(vasoconstriction)** or relax **(vasodilatation),** changing the radius of the arteriole. Such changes help maintain

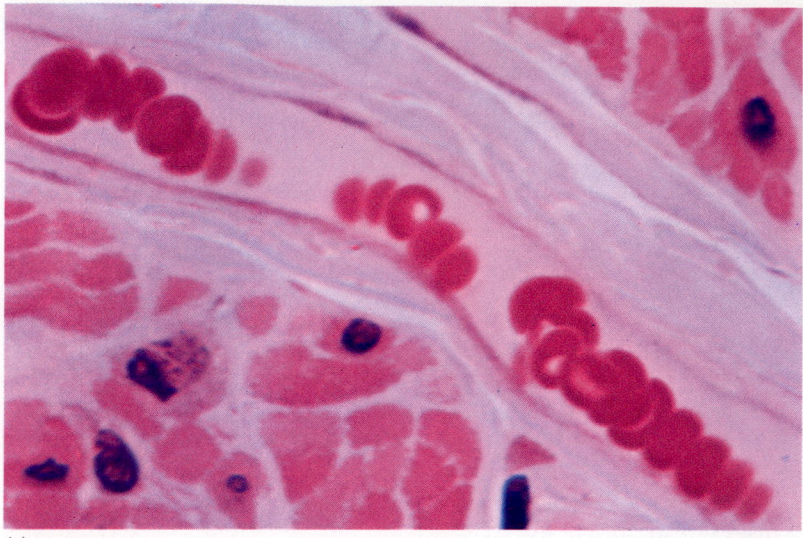

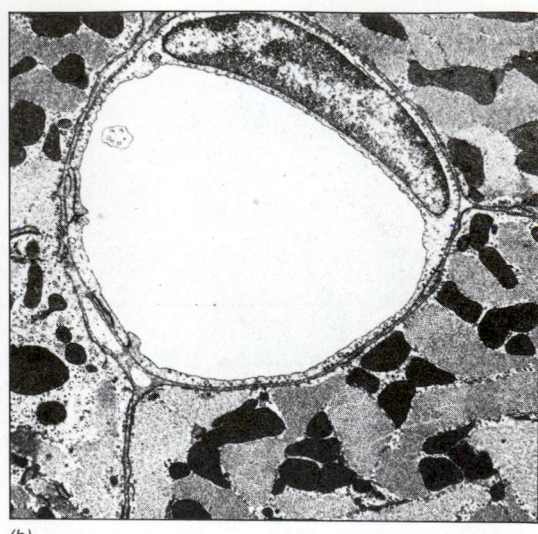

(a)

(b)

FIGURE 36–9 (*a*) **Red blood cells must pass through capillaries in almost single file. (*b*) Cross section of capillary in the papillary muscle of a rat heart. Notice that the wall of the capillary is composed of a single cell formed into a tube. A series of such cells makes up the entire capillary. The light-and-dark pattern surrounding the capillary consists of cross sections of cardiac muscle cells. The heart is much too thick and active to be supplied by the blood it contains; it therefore requires its own system of capillaries to nourish it and provide it with oxygen.** (*a*, Ed Reschke; *b*, Courtesy of Torsten Mattfeldt and Gerhard Mall, University of Heidelberg, *Cardiovascular Research* 17, 1983)

appropriate blood pressure and can help control the volume of blood passing to a particular tissue. Changes in blood flow are regulated by the nervous system in response to the metabolic needs of the tissue, as well as by the demands of the body as a whole. For example, when a tissue is metabolizing rapidly, it needs a greater supply of nutrients and oxygen. During exercise arterioles within the muscles dilate, increasing by more than tenfold the amount of blood flowing to the muscle cells.

If all the blood vessels were dilated at the same time, there would not be sufficient blood to fill them completely. Normally the liver, kidneys, and brain receive the lion's share of the blood. However, if an emergency suddenly occurred requiring you to take rapid action, your blood would be rerouted quickly in favor of your heart and muscles. This would enable you to act effectively. At such a time the digestive system and kidneys can do with a bit less blood, for they are not critical in responding to the crisis.

▪ THE EVOLUTION OF THE VERTEBRATE HEART CULMINATED IN A FOUR-CHAMBERED HEART AND DOUBLE-CIRCUIT CIRCULATION

The vertebrate heart consists of one or two chambers called atria and one or two chambers called ventricles. **Atria** receive blood returning from the tissues; **ventri-**

cles pump blood into the arteries. Additional chambers are present in some animals.

Surveying the vertebrates, we find a progression from the relatively simple heart and single-circuit circulation of the fish to the complex four-chambered heart and double-circuit circulation of the bird and mammal.

The Fish Heart Has A Single Atrium And Single Ventricle

Because it has only one atrium and one ventricle, the fish heart can be described as two-chambered. Actually, two accessory chambers are present. A thin-walled **sinus venosus** receives blood returning from the tissues and pumps it into the atrium (Fig. 36–10*a*). The atrium then contracts, sending blood into the ventricle. The ventricle in turn pumps the blood into an elastic **conus arteriosus,** which does not contract. These four compartments are separated by valves, which prevent blood from flowing backward. From the conus, blood flows into a large artery, the ventral aorta, which branches to distribute blood to the gills. Because blood must pass through the capillaries of the gills before flowing to the other tissues of the body, blood pressure is low through most of the system. This single-circuit, low-pressure circulatory system permits only a low rate of metabolism in fish.

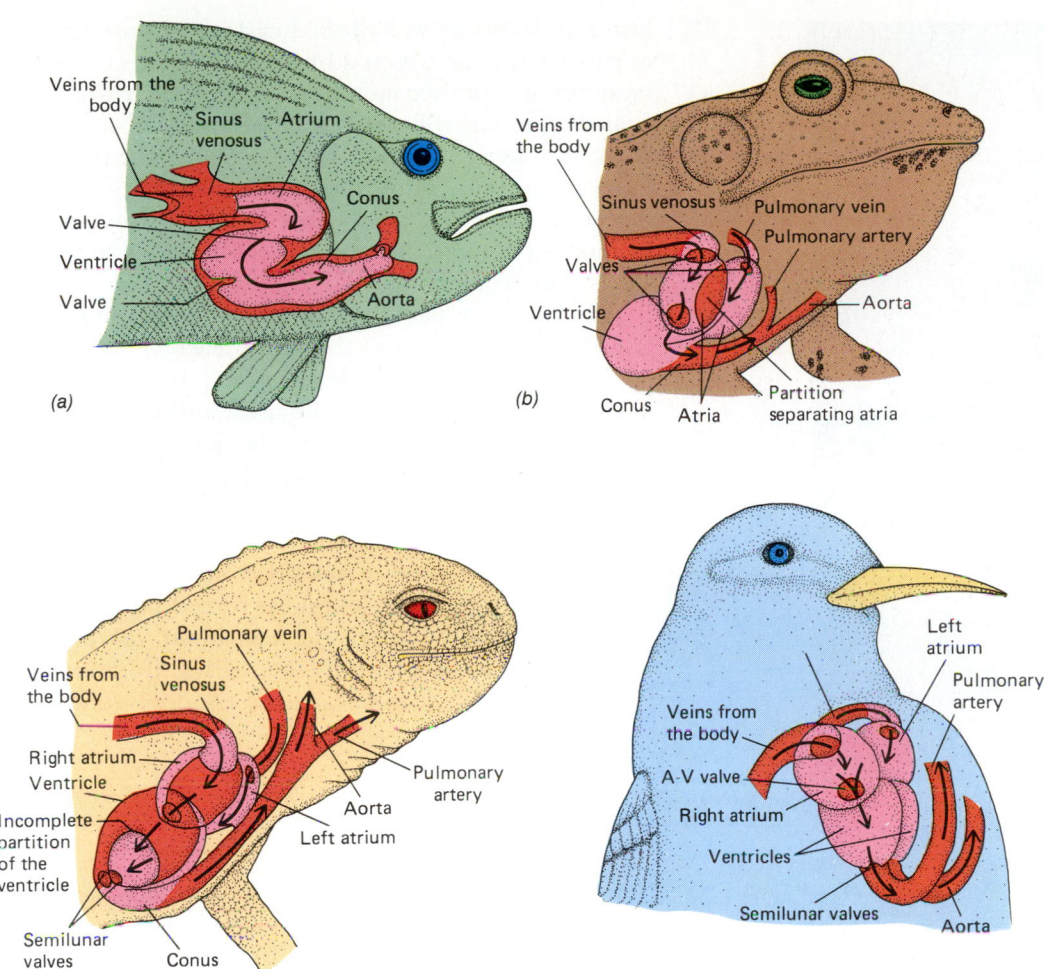

FIGURE 36–10 The evolution of the vertebrate heart. (*a*) The fish heart includes one atrium and one ventricle. (*b*) The amphibian heart consists of two atria and one ventricle. (*c*) The reptilian heart has two atria and two ventricles, but the wall separating the ventricles is incomplete so that blood from the right and left chambers mixes to some extent. (*d*) Birds and mammals have two atria and two ventricles. A complete wall separates right and left sides of the heart so that blood rich in oxygen is kept separate from oxygen-poor blood.

Amphibians Have A Three-Chambered Heart

The three-chambered amphibian heart consists of two atria and a ventricle (Fig. 36–10*b*). A thin-walled sinus venosus collects blood returning from the veins and pumps it into the right atrium. Blood returning from the lungs passes directly into the left atrium. Both atria pump into the single ventricle. In the frog heart, oxygen-rich and oxygen-poor blood are kept somewhat separate. Oxygen-poor blood is pumped out of the ventricle first and passes into the tubular conus arteriosus, which has a spiral fold that helps keep the blood separate. Much of the oxygen-poor blood is directed to the lungs and skin, where it can pick up oxygen from the environment. Oxygen-rich blood is delivered into arteries that conduct it to the various tissues of the body.

The Reptilian Heart Consists Of Two Atria And Two Ventricles

Although the reptilian heart consists of two atria and two ventricles, the wall between the ventricles is incom-

plete, so some mixing of oxygen-rich and oxygen-poor blood occurs (Fig. 36–10*c*). Mixing is minimized by the timing of contractions of the left and right sides of the heart and by pressure differences. In the crocodile, the wall between the ventricles is complete so that the heart consists of four separate chambers.

Birds And Mammals Have A Four-Chambered Heart

The hearts of birds and mammals have completely separate right and left sides (Fig. 36–10*d*). The wall between the ventricles is complete, preventing the mixture of oxygen-rich blood in the left side with oxygen-poor blood in the right side. The conus has split and become the base of the aorta and pulmonary artery. No sinus venosus is present as a separate chamber, although a vestige remains as the sinoatrial node (the pacemaker).

Complete separation of right and left sides of the heart makes it necessary for blood to pass through the heart twice each time it makes a tour of the body. As a

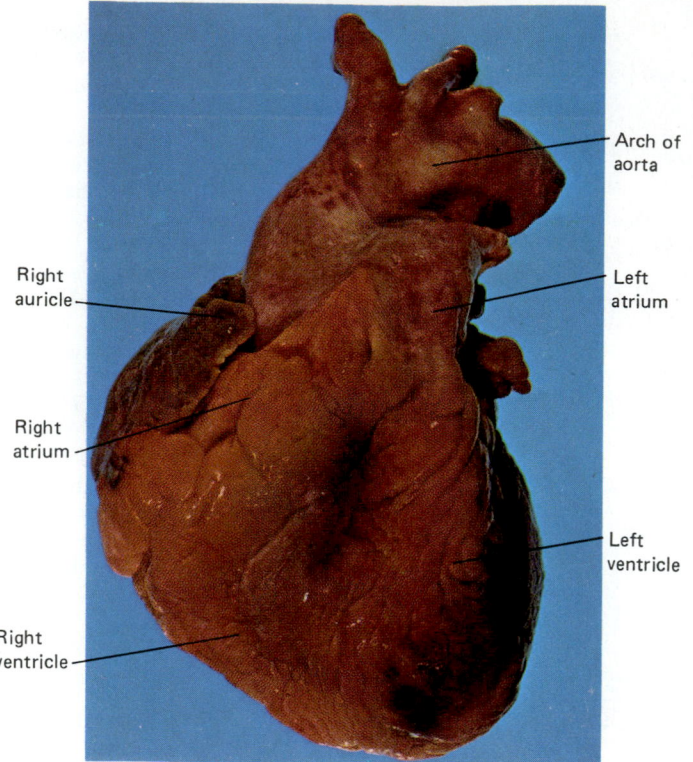

Right auricle

Arch of aorta

Left atrium

Right atrium

Left ventricle

Right ventricle

FIGURE 36–11 Photograph of a human heart, anterior external view. (Courtesy of Phil Horne, Stanford University School of Medicine)

result, it is possible to maintain higher blood pressures and materials are delivered to the tissues rapidly and efficiently. Because the blood of birds and mammals contains more oxygen per unit volume and circulates more rapidly than in other vertebrates, the tissues of the body receive more oxygen. A higher metabolic rate can be maintained, and the endothermic (warm-blooded) condition is possible. Birds and mammals can maintain a constant, high body temperature even in cold surroundings.

The Structure Of The Human Heart Is Marvelously Adapted For Pumping Blood

Not much bigger than a fist and weighing less than a pound, the human **heart** (Fig. 36–11) is a remarkable organ that beats about 2.5 billion times in an average lifetime, pumping about 300 million liters (80 million gallons) of blood. It can vary its output from 5 to 35 liters (5 to 37 quarts) of blood per minute in accordance with the body's needs.

The human heart is a hollow, muscular organ consisting of four chambers (Figs. 36–11 and 36–12). Each side of the heart has an **atrium** (which receives blood from veins) and a **ventricle** (which pumps blood

into arteries). The wall of the heart is composed mainly of cardiac muscle covered by a tough connective-tissue membrane. Another membrane, the **pericardium,** surrounds the entire heart but is separated from it by a space, the **pericardial cavity.** Normally, a thin film of lubricating fluid within the pericardial cavity moistens the contracting surfaces. This fluid reduces friction, facilitating smooth movement of the heart as it contracts and relaxes.

To prevent backflow of blood, a flaplike **valve** guards the entrance and the exit of each ventricle (Fig. 36–13). When the ventricles contract, the **atrioventricular (AV) valves** between the atria and the ventricles close. As a result, blood does not flow backward into the atria. The AV valve between the left atrium and ventricle is known as the **mitral valve.** The AV valve between the right atrium and ventricle is called the **tricuspid valve** because it has three cusps, or flaps. As blood leaves the ventricles and is forced into the great arteries leaving the heart, the **semilunar valves** close, preventing backflow into the ventricles. Both types of valves open and close because of the hydraulic pressure of the blood.

Each Heartbeat Is Initiated By A Pacemaker

You may have watched horror films in which a heart separated from the body of its owner continues to beat spookily. Script writers of such films may have actually rooted their fantasies in fact, for when removed from the body, the heart will continue to beat for many hours if bathed in an appropriate nutritive fluid. This is possible because the heart has its own specialized conduction system and can beat independently of its nerve supply.

Each heartbeat begins in a node of specialized cardiac muscle called the **pacemaker (sinoatrial node,** or simply **SA node),** located in the posterior wall of the right atrium. From the pacemaker, impulses are transmitted through the muscle fibers of the atria, causing them to contract. One group of atrial muscle fibers transmits the muscle impulse directly to a second node (the **atrioventricular node,** or **AV node),** located in the wall between the atria. From this node impulses sweep through specialized fibers to all parts of the ventricles, producing contraction.

Cardiac muscle fibers are separated at their ends by dense bands called **intercalated discs** (Fig. 36–14). Each intercalated disc is actually a tight junction between two cells. This junction is of great significance because it offers very little resistance to the passage of a muscle impulse. It allows an impulse to pass across the disc, so the entire mass of atrial or ventricular muscle tends to contract in response to the impulse as if it were one giant cell.

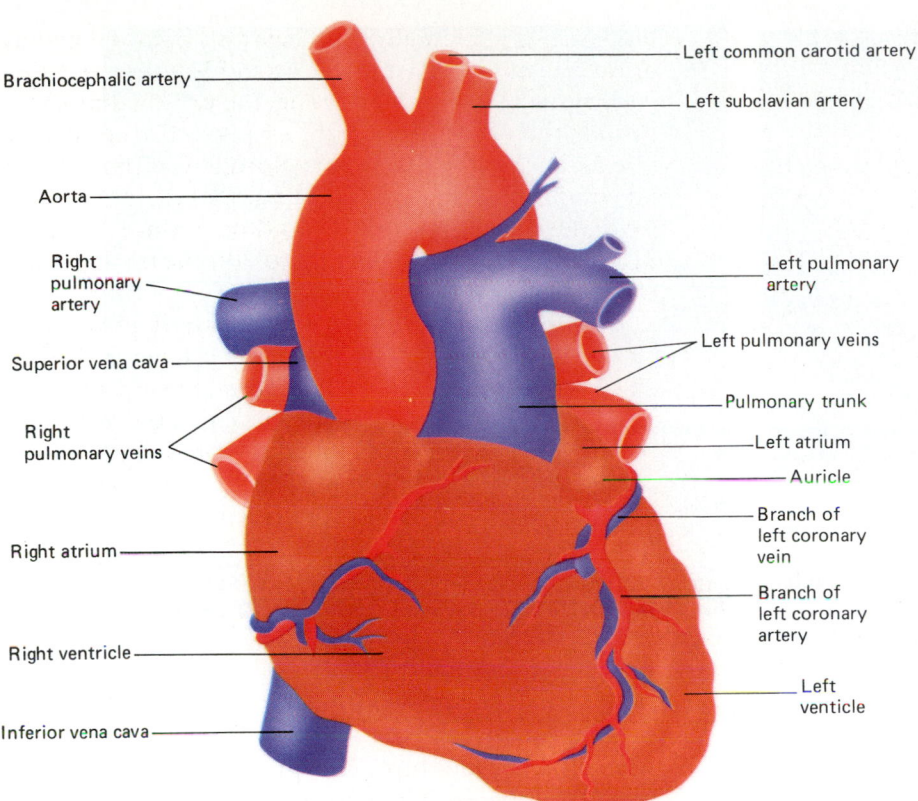

Brachiocephalic artery

Aorta

Right pulmonary artery

Superior vena cava

Right pulmonary veins

Right atrium

Right ventricle

Inferior vena cava

Left common carotid artery

Left subclavian artery

Left pulmonary artery

Left pulmonary veins

Pulmonary trunk

Left atrium

Auricle

Branch of left coronary vein

Branch of left coronary artery

Left venticle

FIGURE 36–12 Diagram of the heart, anterior view. Note the coronary blood vessels that bring blood to and from the heart muscle itself.

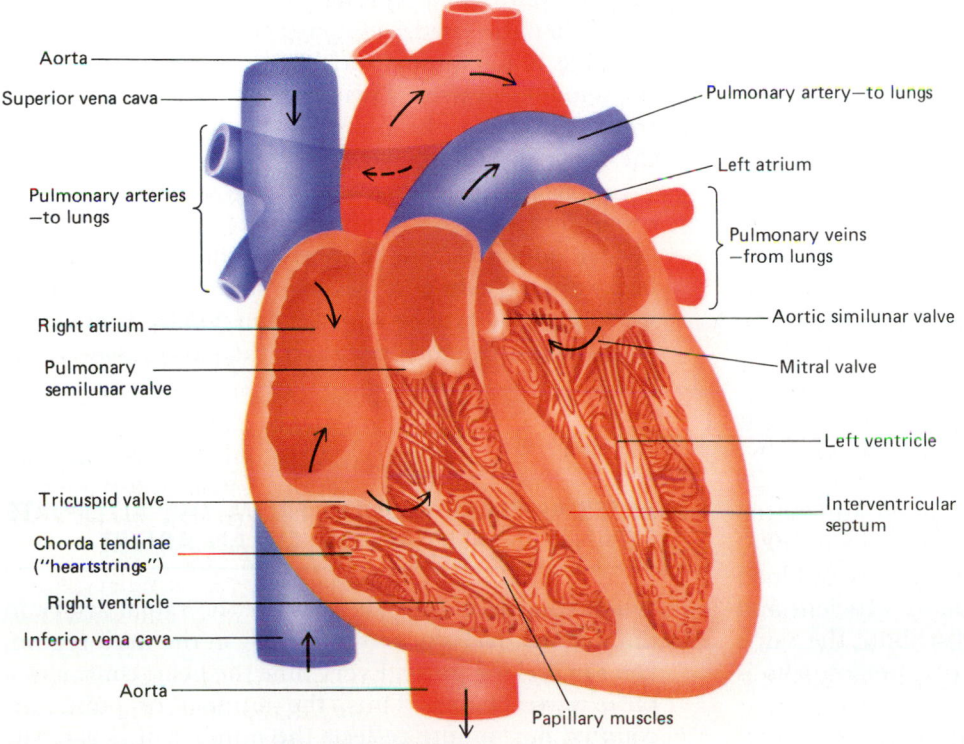

Aorta

Superior vena cava

Pulmonary arteries —to lungs

Right atrium

Pulmonary semilunar valve

Tricuspid valve

Chorda tendinae ("heartstrings")

Right ventricle

Inferior vena cava

Aorta

Pulmonary artery—to lungs

Left atrium

Pulmonary veins —from lungs

Aortic similunar valve

Mitral valve

Left ventricle

Interventricular septum

Papillary muscles

FIGURE 36–13 Section through the human heart showing chambers, valves, and connecting blood vessels.

Heart Rate Is Regulated By The Nervous System

Although the heart is capable of beating rhythmically on its own, it cannot by itself effectively change the strength and rate of contraction to meet the changing needs of the body. Recall from Chapter 34 that this kind of control is the function of the autonomic nervous system. Under conditions of stress, sympathetic nerves can increase strength of contraction as much as

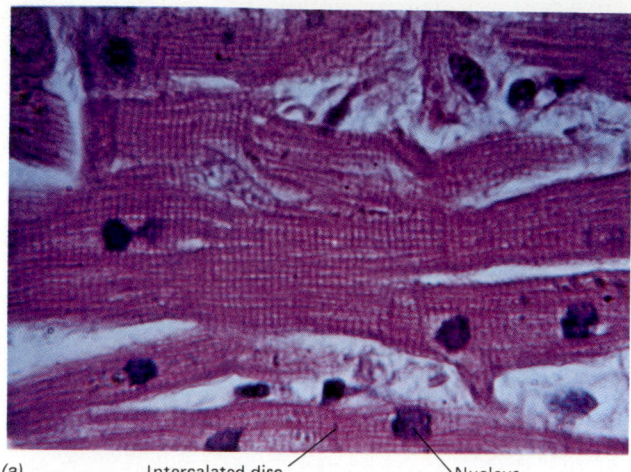

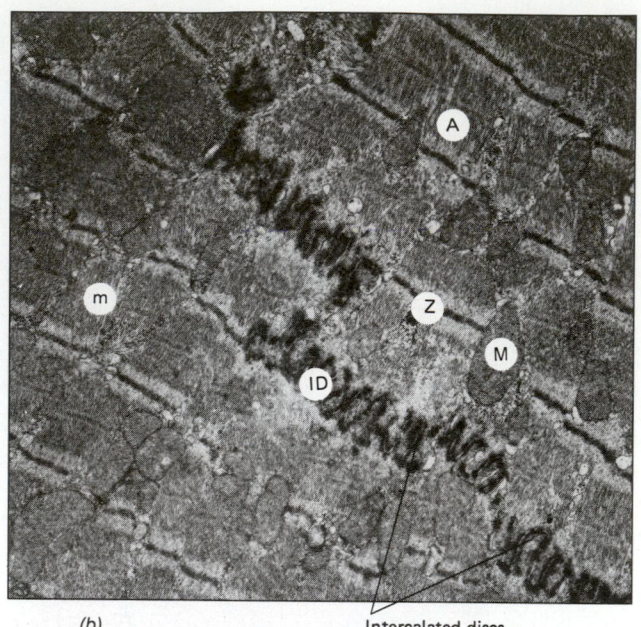

(a)   Intercalated disc   Nucleus

(b)   Intercalated discs

FIGURE 36–14   Cardiac muscle. *(a)* **Cardiac muscle as seen with the light microscope (approximately ×400).** *(b)* **An electron micrograph of cardiac muscle. A, a band; Z, z line; M, mitochondrion; ID, intercalated disc.** *(b, courtesy of Lyle C. Dearden)*

100%. Under more placid conditions, the vagus nerve, a parasympathetic nerve, slows the heart. It is the balance between sympathetic and parasympathetic stimulation that determines heart rate, and this balance is determined by the central nervous system.

The endocrine system also plays a part in regulating heartbeat. When the body is under stress, the hormones epinephrine and norepinephrine, released from the adrenal glands, stimulate the force and rate of the heartbeat.

The normal heart rate is about 70 beats per minute, and the **cardiac output,** the volume of blood pumped by one ventricle, is about 5 liters (5 quarts) per minute. This amount is approximately equal to the total volume of blood in the body. Cardiac output depends primarily on **venous return,** the volume of blood delivered to the heart by the veins. During vigorous exercise the heart may beat as many as 200 times per minute and its output may increase to about 28 liters (30 quarts) per minute. In a trained athlete the heart actually enlarges (in extreme cases up to 50%) and is capable of pumping a greater quantity of blood per beat. An athlete's heart is thus more efficient and does not have to beat as often to distribute the same quantity of blood as does the heart of a person who is not in good physical condition.

Two Main Heart Sounds Can Be Distinguished

A physician listening to the heartbeat with a stethoscope can distinguish two principal sounds, which occur in repeating rhythm. These sounds, usually described as a "lub-dup," are produced each time the heart valves close. The first sound, the "lub," is caused by the closing of the AV valves and marks the beginning of ventricular **systole,** the phase of the heart's cycle when the ventricles contract. The "dup" sound is heard as a quick snap and is caused by the closing of the semilunar valves. This marks the beginning of ventricular **diastole,** the phase of the heart's cycle when the ventricles relax.

A **heart murmur** is a common type of abnormal sound that sometimes indicates a valve disorder. When a valve does not close properly, some blood may flow backward, creating a hissing sound. Characteristic murmurs may also be heard when a valve is enlarged with scar tissue and is rough, so the passageway is narrowed.

■ ARTERIAL PULSE RESULTS FROM THE ALTERNATE EXPANSION AND RECOIL OF AN ARTERY

When you place your finger over the radial artery in the wrist or over the carotid artery in the neck region, you can feel the pulse. Every time the heart contracts, a pulse wave begins. Thus, the number of pulsations counted per minute reflects the number of heartbeats per minute.

Pulse is the alternate expansion and recoil of an artery. Each time the left ventricle pumps blood into the aorta (the great blood vessel conducting blood away from the heart), the elastic wall of the aorta expands. This expansion moves down the aorta and its

FOCUS ON   The Electrical Activity Of The Heart

As each wave of contraction spreads through the heart, electrical currents spread into the tissues surrounding the heart and onto the body surface. This electrical activity can be recorded by placing electrodes on the body surface on opposite sides of the heart. The **electrocardiograph** is the machine used to amplify and record the electrical activity, and the record produced is called an **electrocardiogram (ECG** or **EKG).** In intensive-care units and operating rooms an oscilloscope is often used instead of an electrocardiograph. The oscilloscope continuously monitors the heart, displaying a moving beam of electrons on a screen.

An ECG begins with a **P wave,** which represents the spread of an impulse through the atria just before atrial contraction. Then a **QRS complex** appears, reflecting the spread of an impulse through the ventricles just before they contract. As the ventricles recover, currents generated are reflected upon the graph as a **T wave.** The heart then repeats its pattern of electrical impulses, generating a new P wave, QRS complex, and T wave.

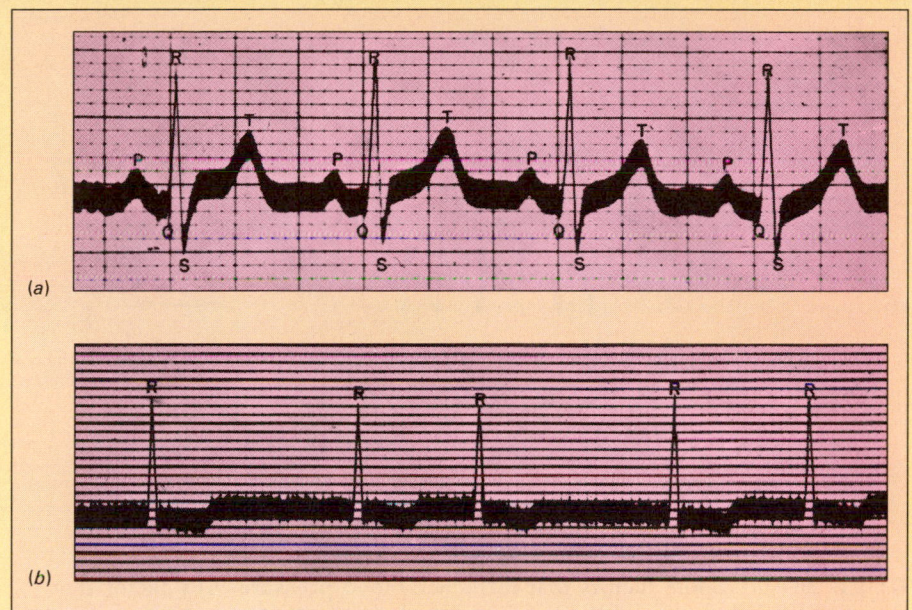

Electrocardiograms. (*a*) Tracing from a normal heart. The P wave corresponds to the contraction of the atria, the QRS complex to the contraction of the ventricle, and the T wave to the relaxation of the ventricle. (*b*) Tracing from a patient with atrial fibrillation. The individual muscle fibers of the atrium twitch rapidly and independently. There is no regular atrial contraction and no P wave. The ventricles beat independently and irregularly, causing the QRS wave to appear at irregular intervals. (Courtesy of Dr. Lewis Dexter and the Peter Bent Brigham Hospital, Boston, Mass.)

Abnormalities in the ECG indicate disorders in the heart or its rhythm. One class of disorders which can be diagnosed with the help of the ECG is **heart block.** In this condition transmission of an impulse is delayed or blocked at some point in the conduction system. Artificial pacemakers can be implanted in patients with severe heart block. A pacemaker is implanted beneath the skin, and its electrodes are connected to the heart. This device provides continuous rhythmic impulses that avoid the block and drive the heartbeat.

---

branches in a wave. As soon as the wave has passed, the elastic arterial wall snaps back to its normal size.

■ BLOOD PRESSURE DEPENDS ON BLOOD FLOW AND RESISTANCE TO BLOOD FLOW

Blood pressure is the force exerted by the blood against the inner walls of the blood vessels. It is determined by the blood flow and the resistance to that flow (Fig. 36–15). Blood flow depends directly upon the pumping action of the heart. When cardiac output increases, blood flow increases, causing a rise in blood pressure. When cardiac output decreases, blood flow decreases, causing a fall in blood pressure. The volume of blood flowing through the system also affects blood pressure. When blood volume is reduced by hemorrhage or chronic bleeding, the blood pressure drops. On the other hand, an increase in blood volume results in an increase in blood pressure. For example, a high dietary salt intake causes water retention. This results in an increase in blood volume and leads to higher blood pressure.

Blood flow is impeded by resistance; when the resistance to flow increases, blood pressure rises. **Peripheral resistance** is the resistance to blood flow caused by the viscosity of the blood and by friction between the blood and the wall of the blood vessel. The length and diameter of a blood vessel determine the surface area of the vessel in contact with the blood. The length of a blood vessel does not change, but the diameter,

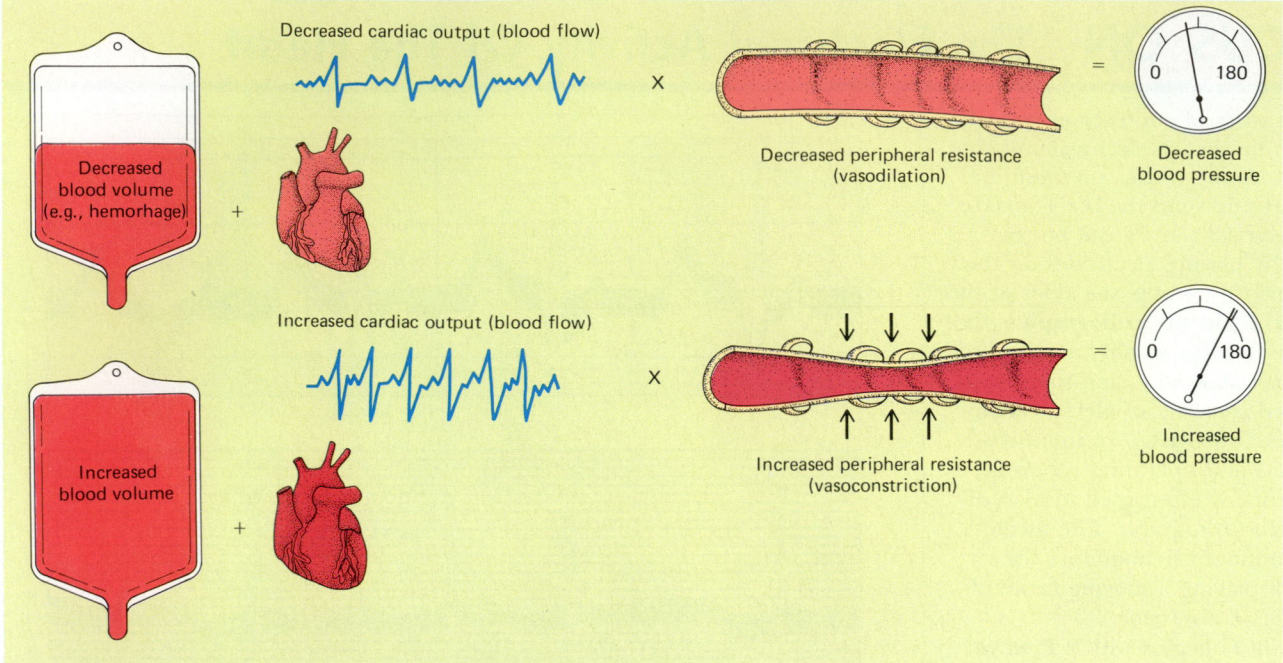

FIGURE 36–15 Some factors that influence blood pressure. Any factor that increases blood flow or resistance raises blood pressure.

especially of an arteriole, does. Even a small change in the diameter of a blood vessel causes a big change in blood pressure. For example, if the radius of a blood vessel is doubled, the resistance is reduced to one-sixteenth of its former value, and the blood flow increases 16-fold.

Blood pressure in arteries rises during systole and falls during diastole. A blood pressure reading is expressed as systolic pressure over diastolic pressure. Normal blood pressure of a young adult would approximate 120/80 (read as "120 over 80" and measured in millimeters of mercury, or mm Hg). When the diastolic pressure consistently reads over 95 mm Hg, the patient may be suffering from **hypertension** (high blood pressure). Hypertension is one of the most common cardiovascular disorders in the United States. Heredity, obesity, stress, and high dietary salt intake are thought to be important factors in the development of hypertension. This disorder places a heavy burden upon the heart, which must pump harder against greater blood pressure.

Blood Pressure Is Highest In Arteries

As you might imagine, blood pressure is greatest in the arteries and lessens as blood flows through the capillaries. By the time blood reaches the veins, its pressure is very low. When the body is in an upright position, gravity offers a great deal of resistance to blood flow

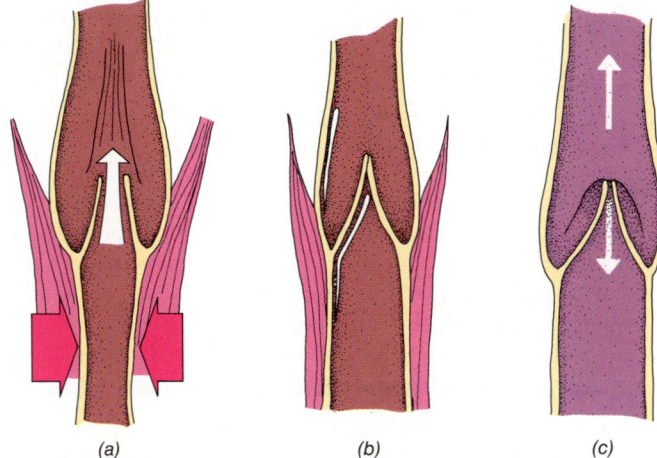

(a)   (b)   (c)

FIGURE 36–16 Valves prevent backflow of blood in veins. (a) When blood flows up toward the heart, the valves are pushed open. (b) When blood begins to fall downward in the vein, it fills the cup-like flaps of the valve. The flaps are forced together, preventing backflow. (c) The weak walls of a varicose vein are easily distended by blood. As a result the flaps of the valves do not meet, and blood can move backward.

through the veins. It is really quite remarkable that blood in the feet manages to make its way back up to the heart. Much of the success of this journey may be attributed to flap-like valves within the veins. These valves prevent the blood from flowing (or falling) backwards. Blood is pushed along through the veins by the pressure of the blood behind it and by compression of veins by muscular movement that occurs when we move about (Fig. 36–16).

In people whose jobs require that they stand for long periods each day, blood accumulates in the veins of the legs. Excessive pooling of the blood stretches the veins, so that the cusps of their valves no longer meet and thus do not close properly. This may lead to **varicose veins,** especially in those who are obese or who have inherited weak vein walls. A varicose vein is dilated, tortuous, and elongated. Varicose veins occur most often in the superficial veins (those close to the surface), especially those in the legs. These veins have the least external support and are subjected to the greatest increases in pressure. **Hemorrhoids,** which are varicose veins in the anal region, occur when venous pressure in that region is constantly elevated, as in chronic constipation (because of straining) and during pregnancy (because of pressure of the enlarged uterus).

When one stands perfectly still for a long period of time (as when a soldier stands at attention), blood pools in the veins. Pressure increases in the capillaries because the veins are not able to accept more blood from them. This causes large amounts of plasma to leave the circulation through the thin walls of the capillaries. Within a few minutes so much blood volume can be lost from the circulation that arterial blood pressure may fall drastically. Blood supply to the brain is diminished, and sometimes fainting occurs. Less dramatic changes in blood pressure occur each time you get up from a horizontal position.

Blood Pressure Is Carefully Regulated

Several complex homeostatic mechanisms interact to maintain normal blood pressure. Tiny pressure receptors called **baroreceptors,** located in certain arteries, are sensitive to changes in blood pressure. When stimulated, the baroreceptors send messages to centers in the medulla of the brain. Nerves then signal the heart to slow down or speed up. Other nerves send messages to arterioles and veins, causing them to dilate or constrict. For example, when blood pressure decreases as you get out of bed in the morning, the heart rate increases slightly and blood vessels are constricted, so blood pressure increases. These neural reflexes act continuously to maintain a steady state of blood pressure.

Hormones are also involved in regulation of blood pressure. The **angiotensins** are a group of hormones that are powerful vasoconstrictors (they cause blood vessels to constrict strongly). When blood pressure is low, the kidneys release the hormone **renin,** which stimulates the formation of angiotensins from a plasma protein. The kidneys also help maintain blood pres-

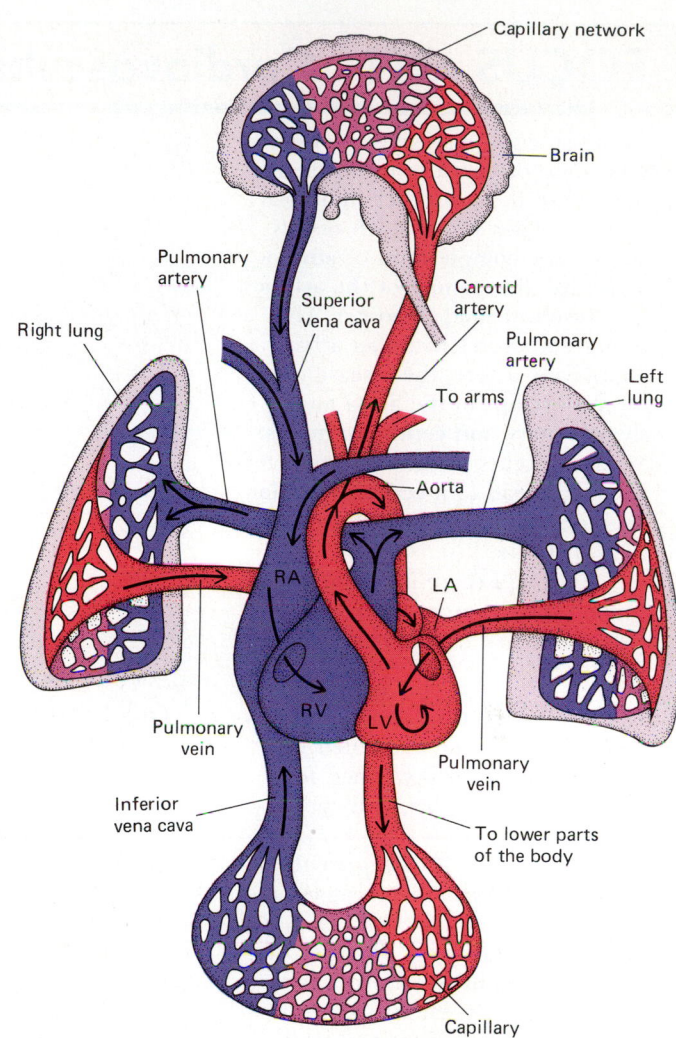

FIGURE 36–17 Highly simplified diagram showing the pattern of circulation through the systemic and pulmonary circuits. Red represents oxygen-rich blood; blue represents oxygen-poor blood.

sure indirectly by influencing blood volume. In response to hormones, the kidneys vary the rate of excretion of salts and water, thereby increasing or decreasing the volume of the blood plasma.

■ IN MAMMALS BLOOD IS PUMPED THROUGH A PULMONARY AND A SYSTEMIC CIRCULATION

In humans and other mammals blood flows through a continuous network of blood vessels that form a double circuit. The **pulmonary circulation** connects the heart with the lungs and functions to oxygenate the blood. After being charged with oxygen in the pulmonary circulation, blood is pumped into the **systemic circulation,** which connects the heart with all the tissues of the body. This general pattern of circulation is illustrated in Figure 36–17.

FOCUS ON Cardiovascular Disease

Cardiovascular disease is the number one cause of death in the United States and in most other industrial societies. Most often death results from some complication of **athero-sclerosis**[1] (hardening of the arteries as a result of lipid deposits). Although atherosclerosis can affect almost any artery, the disease most often develops in the aorta and in the coronary and cerebral arteries. When it occurs in the cerebral arteries, it can lead to a **cerebrovascular accident (CVA),** commonly referred to as a stroke.

Although there is apparently no single cause of atherosclerosis, several major risk factors have been identified:

1. Elevated levels of cholesterol in the blood, often associated with diets rich in total calories, total fats, saturated fats, and cholesterol.
2. Hypertension: The higher the blood pressure, the greater the risk.
3. Cigarette smoking: The risk of developing atherosclerosis is two to six times greater in smokers than in nonsmokers and is directly proportional to the number of cigarettes smoked daily.
4. Diabetes mellitus, an endocrine disorder in which glucose is not metabolized normally.

The risk of developing atherosclerosis also increases with age. Estrogen hormones are thought to offer some protection in women until after menopause, when the concentration of these hormones decreases. Other suggested risk factors that are currently being studied are obesity, hereditary predisposition, lack of exercise, stress and behavior patterns, and dietary factors, such as excessive intake of salt or refined sugar.

In atherosclerosis, lipids are deposited in the smooth muscle cells of the arterial wall. Cells in the arterial wall proliferate and the inner lining thickens. More lipid, especially cholesterol from low-density lipopro-

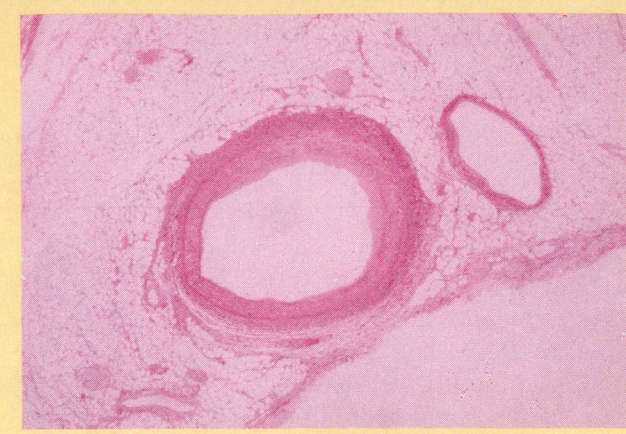
(a)

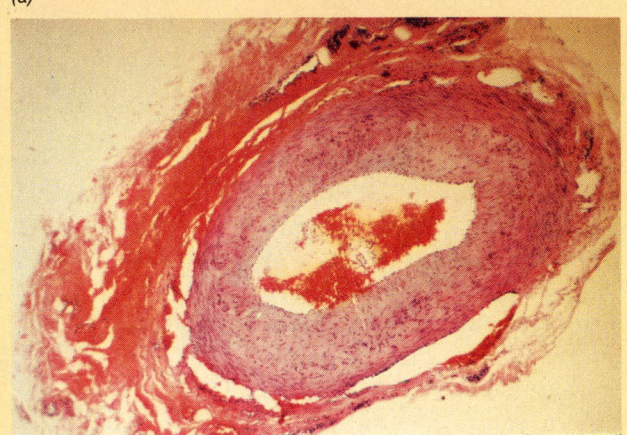
(b)

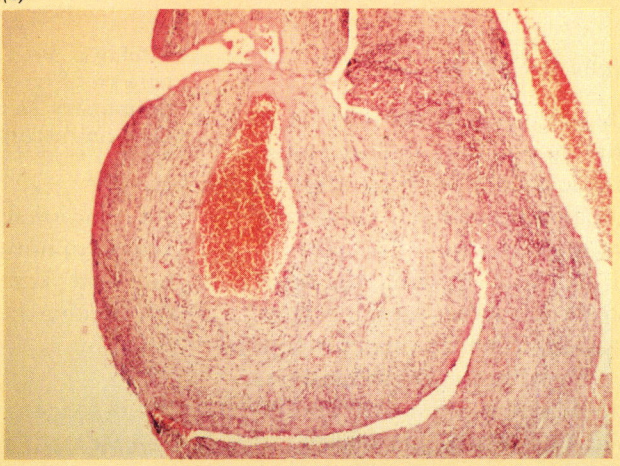
(c)

Progression of atherosclerosis. Cross sections through three arteries showing changes that take place in atherosclerosis. (a) Early stage of atherosclerosis. Inner lining has thickened slightly at lower left. (b) Pronounced changes have taken place in this artery. Note the marked thickening of the wall. (c) This artery is almost completely blocked with atherosclerotic plaque. (a, Martin M. Rotker/Phototake, NYC; b and c, M. Cubberly/Phototake, NYC)

teins, accumulates in the wall. Eventually calcium is deposited there, contributing to the slow formation of hard plaque. As the plaque develops, arteries lose their ability to stretch when they fill with blood, and they become progressively occluded (blocked), as shown in the figure. As the artery narrows, less blood can pass through to reach the tissues served by that vessel and the tissue may become **ischemic** (lacking in blood). Under these conditions the tissue is deprived of an adequate oxygen supply.

When a coronary artery becomes narrowed, **ischemic heart disease** can occur. Sufficient oxygen may reach the heart tissue during normal activity, but the increased need for oxygen during exercise or emotional stress results in the pain known as **angina pectoris.** People with this condition often carry nitroglycerin pills with them for use during an attack. This drug dilates veins so the

venous return is reduced. Cardiac output is lowered, so the heart is not working so hard and requires less oxygen. Nitroglycerin also dilates the coronary arteries slightly, allowing more blood to reach the heart muscle.

Myocardial infarction (MI) is the very serious, often fatal, form of ischemic heart disease that often results from a sudden decrease in coronary blood supply. The portion of cardiac muscle deprived of oxygen dies within a few minutes and is then referred to as an **infarct.** The term myocardial infarction is used as a synonym for heart attack. MI is the leading cause of death and disability in the United States. Just what triggers the sudden decrease in blood supply that causes MI is a matter of some debate. It is thought that in some cases an episode of ischemia triggers a fatal arrhythmia, such as **ventricular fibrillation,** a condition in which the ventricles

contract very rapidly without actually pumping blood. In other cases, a **thrombus** (clot) may form in a diseased coronary artery. Because the arterial wall is roughened, platelets may adhere to it and initiate clotting.

If a thrombus blocks a sizable branch of a coronary artery, blood flow to a portion of heart muscle is impeded or completely halted. This condition is referred to as a **coronary occlusion.** If the coronary occlusion prevents blood flow to a large region of cardiac muscle, the heart may stop beating—that is, **cardiac arrest** may occur—and death can follow within moments. If only a small region of the heart is affected, however, the heart may continue to function. Cells in the region deprived of oxygen die and are replaced by scar tissue.

[1] Atherosclerosis is the most common form of arteriosclerosis, any disorder in which arteries lose their elasticity.

The Pulmonary Circulation Functions To Oxygenate The Blood

Blood from the tissues returns to the right atrium of the heart partially depleted of its oxygen supply but loaded with carbon dioxide wastes. This oxygen-poor blood is directed to the lungs where oxygen diffuses into the blood and carbon dioxide diffuses out. The right ventricle pumps blood into the pulmonary circulation (Fig. 36–17). The **pulmonary arteries** carry blood to the lungs. These are the only arteries in the body that carry oxygen-poor blood. In the lungs the pulmonary arteries branch into smaller and smaller vessels which finally lead into an extensive network of **pulmonary capillaries** that course throughout the lung tissue. As blood circulates through this capillary network, gases are exchanged. **Pulmonary veins,** the only veins in the body to carry oxygen-rich blood, return blood to the left atrium of the heart.

In summary, blood flows through the vessels of the pulmonary circulation in the following sequence:

Right atrium → right ventricle → pulmonary artery → pulmonary capillaries (in lung) → pulmonary veins → left atrium.

The Systemic Circulation Delivers Blood To All Of The Tissues

Blood returning from the pulmonary circulation enters the left atrium of the heart, then passes into the

left ventricle. From there it is pumped into the largest artery of the body, the **aorta.** The aorta divides into arterial branches that carry blood to all regions of the body, including the heart muscle itself. Some of the principal branches include the **carotid arteries** to the brain, the **subclavian arteries** to the shoulder and arm region, the **mesenteric arteries** to the intestine, the **renal arteries** to the kidneys, and the **iliac arteries** to the legs (Fig. 36–18). Each of these branches into smaller and smaller arteries that bring blood to the capillary networks within each tissue or organ.

Blood returning from the brain is carried back toward the heart by the **jugular veins.** Blood from the shoulders and arms drains into the **subclavian veins.** These veins and others bringing blood from the upper portion of the body merge, forming the **superior vena cava,** a very large vein that empties blood into the right atrium. **Renal veins** from the kidneys, **iliac veins** from the lower limbs, **hepatic veins** from the liver, and other veins returning blood from the lower regions of the body empty blood into the **inferior vena cava,** which returns blood to the right atrium.

As an example of blood circulation through the systemic system, let us trace a drop of blood from the heart to the brain and back:

Left atrium → left ventricle → aorta → carotid artery → capillaries in brain → jugular vein → superior vena cava → right atrium → right ventricle → into pulmonary circulation.

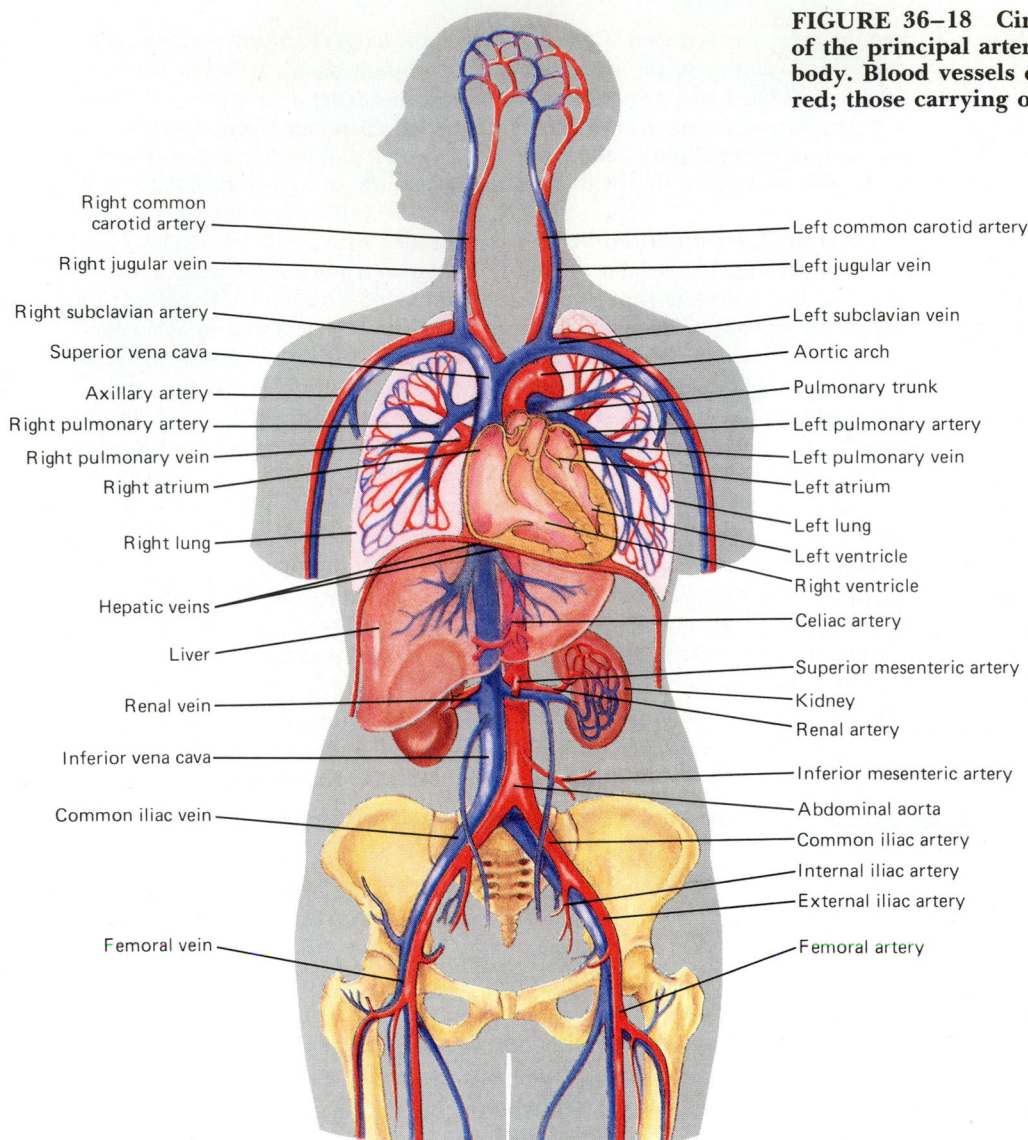

FIGURE 36–18 Circulation of blood through some of the principal arteries and veins of the human body. Blood vessels carrying oxygen-rich blood are red; those carrying oxygen-depleted blood are blue.

Right common carotid artery

Right jugular vein

Right subclavian artery

Superior vena cava

Axillary artery

Right pulmonary artery

Right pulmonary vein

Right atrium

Right lung

Hepatic veins

Liver

Renal vein

Inferior vena cava

Common iliac vein

Femoral vein

Left common carotid artery

Left jugular vein

Left subclavian vein

Aortic arch

Pulmonary trunk

Left pulmonary artery

Left pulmonary vein

Left atrium

Left lung

Left ventricle

Right ventricle

Celiac artery

Superior mesenteric artery

Kidney

Renal artery

Inferior mesenteric artery

Abdominal aorta

Common iliac artery

Internal iliac artery

External iliac artery

Femoral artery

The Coronary Circulation Delivers Blood To The Heart

The heart is an actively metabolizing organ that requires a large and continuous supply of nutrients and oxygen. Blood flowing through its chambers cannot serve these needs because the heart wall is far too thick to permit effective diffusion. Fortunately, the heart is equipped with its own complex of blood vessels, the **coronary circulation** (see Fig. 37–9*b*).

Two **coronary arteries** branch off from the aorta just as it leaves the heart. These arteries give rise to an extensive system of blood vessels within the heart tissue. Most of the coronary capillaries empty into veins that join, forming a large vein, the **coronary sinus,** which empties into the right atrium. Blockage of the coronary arteries is a principal cause of heart disease (see Focus on Cardiovascular Disease).

The Hepatic Portal System Delivers Nutrients To The Liver

Generally, blood travels from artery to capillary to vein. An exception to this sequence is the **hepatic portal system,** which transports nutrients from the intestine to the liver. Blood reaches the intestine via the mesenteric artery and enters capillaries in the intestinal villi, where it receives nutrients. Capillaries from the intestinal villi merge to form the **superior mesenteric vein,** which empties into the **hepatic portal vein.** This vein does not deliver blood to the heart but to the liver.

In the liver the hepatic portal vein branches into a vast network of tiny vessels called **sinusoids.** (Sinusoids are exchange vessels similar to capillaries.) As the blood courses through these sinusoids, the liver cells absorb and process the excess nutrients present. Liver capillaries eventually merge, forming the hepatic veins, which in turn empty into the inferior vena cava. Thus, in the hepatic portal system we find this unusual sequence of blood vessels:

Capillaries → vein → sinusoids → vein.

Note that there is an extra set of exchange vessels. The liver is also supplied with oxygen-rich blood by a hepatic artery, for the oxygen content of the blood in the hepatic portal vein is not sufficient to sustain the liver cells.

■ THE LYMPHATIC SYSTEM IS AN ACCESSORY CIRCULATORY SYSTEM

The lymphatic system is an accessory circulatory system that is connected with blood circulation. Its three principal functions are to (1) collect and return tissue fluid to the blood, (2) defend the body against disease organisms, and (3) absorb lipids from the digestive system. Here we will focus on the first function; the other two are discussed in Chapters 37 and 39, respectively.

The Lymphatic System Consists Of Lymphatic Vessels And Lymph Tissue

The lymphatic system has tiny "dead-end" capillaries that extend into almost all tissues of the body (Fig. 36–19). Tissue fluid enters the lymph capillaries and is then referred to as **lymph.** Lymph capillaries convey the lymph to larger vessels called **lymph veins** (or **lymphatics**).

At strategic locations lymph veins enter **lymph nodes,** small organized masses of lymph tissue. Lymph nodes have two main functions: (1) they filter the lymph as it slowly passes through, and (2) they produce **lymphocytes,** white blood cells important in immune responses. Lymph nodes (sometimes called **lymph glands**) are most numerous in the neck region, under the arms, in the groin region, and in the chest and abdomen. Lymph nodes in an infected area enlarge conspicuously and may be felt as hard little knots below the skin.

Lymph veins that leave the lymph nodes conduct lymph toward the shoulder region. Eventually lymph veins empty their contents into the subclavian veins by way of the thoracic and right **lymphatic ducts.**

Tonsils are masses of lymph tissue under the lining of the oral cavity and throat. (The pharyngeal tonsils in back of the nose are called **adenoids** when they are enlarged.) Tonsils help protect the respiratory system from infection by destroying bacteria and other foreign matter that enter the body through the mouth or nose. Unfortunately, tonsils are sometimes over-

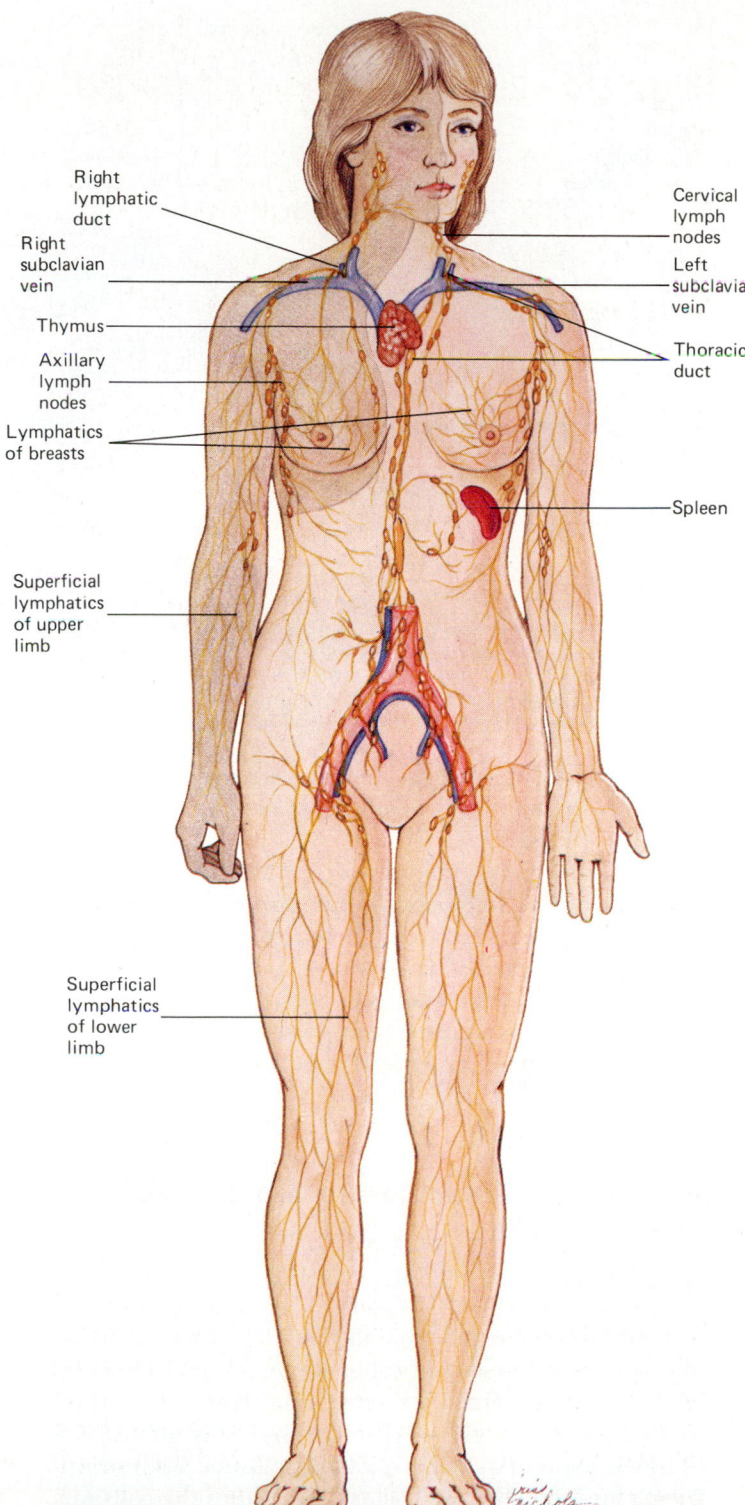

FIGURE 36–19 The lymphatic system. Note that while the lymphatic vessels extend into most tissues of the body, the lymph nodes are clustered in certain regions. The right lymphatic duct drains lymph from the upper right quadrant of the body. The thoracic duct drains lymph from other regions of the body.

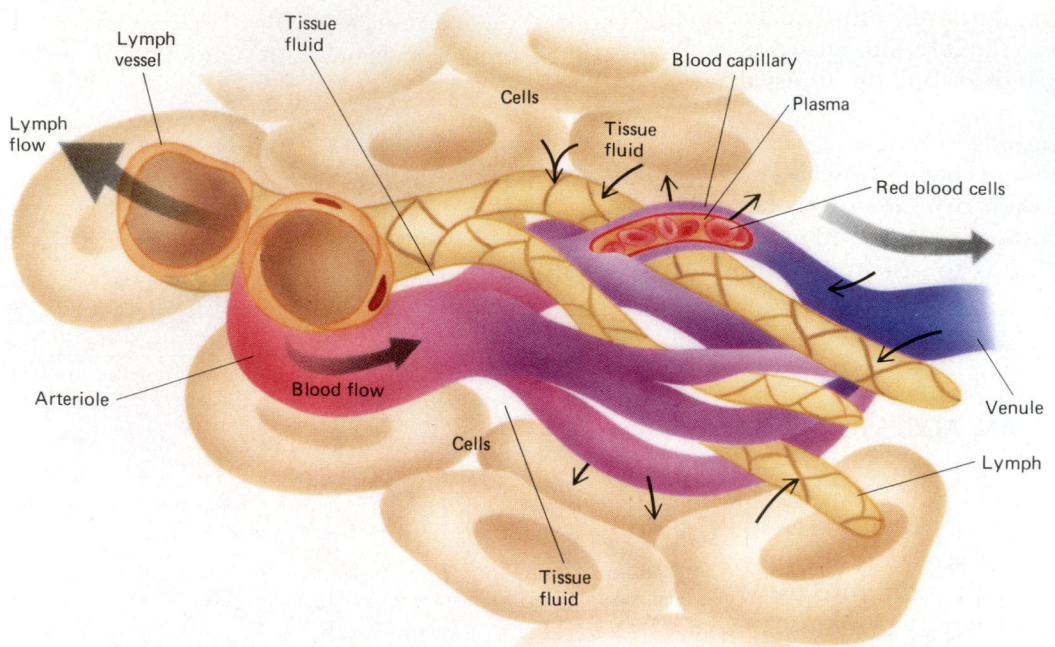

FIGURE 36–20 The relation of lymph capillaries to blood capillaries and tissue cells. Note that blood capillaries are connected to vessels at both ends, whereas lymph capillaries, shown in yellow, are dead-end streets. The arrows indicate direction of flow.

come by invading germs, become the site of frequent infection themselves, and then become prime targets for surgical removal.

Some nonmammalian vertebrates, such as the frog, have lymph "hearts" that pulsate and squeeze lymph along. In mammals, however, the walls of the lymph vessels themselves pulsate, pushing the lymph through the vessels. When muscles contract or arteries pulsate, pressure on the lymph vessels enhances lymph flow. Valves within the lymph vessels prevent backflow.

The Lymphatic System Plays An Important Role In Fluid Homeostasis

As blood enters a capillary network, it is under sufficiently high pressure that some of the plasma is forced out through the capillary wall (Fig. 36–20). Recall that plasma that has escaped from the blood circulation is known as **tissue fluid,** or **interstitial fluid.** This fluid contains few red cells or platelets and only about 25% as much protein as the circulating plasma. Rich in nutrients and oxygen, the tissue fluid bathes the cells in a nourishing sea of needed materials. It also receives wastes from the cells; wastes eventually find their way into the blood.

At the venous ends of the capillaries some tissue fluid moves back into the blood. This is because blood pressure is much lower there, and osmotic pressure of the blood tends to pull fluid back into the vessels. However, not as much fluid returns to the blood as escaped, and plasma proteins do not effectively re-enter the

FIGURE 36–21 Lymphatic drainage is blocked in the limbs of this individual because of a parasitic infection known as filariasis. The condition characterized by such swollen limbs is elephantiasis. (From Markell, E.K., and M. Voge. *Medical Parasitology,* 5th ed., Philadelphia, W.B. Saunders Company, 1981)

blood through the capillary walls. These problems would cause a serious fluid imbalance if it were not for the lymphatic system. This system makes a vital contribution to fluid homeostasis by collecting the excess fluid (amounting to about 10% of the tissue fluid) and protein and returning them to the blood.

When for any reason lymph vessels become blocked, tissue fluid accumulates in the affected area, causing **edema** (swelling). Obstruction of lymph flow may result from injury, inflammation, parasitic infection, or surgery (Fig. 36–21). For example, when a breast is removed (radical mastectomy) because of cancer, lymph nodes in the region under the arm are surgically removed in an effort to prevent the spread of cancer cells. The patient's arm may swell tremendously because of the disrupted lymph circulation. However, new lymph vessels usually develop within a few months, and the swelling slowly subsides.

CHAPTER SUMMARY

I. Small, simple invertebrates, such as sponges, cnidarians, and flatworms, depend on diffusion for internal transport. More complex invertebrates require a specialized circulatory system.

II. Arthropods and most mollusks have an open circulatory system, in which blood flows into a hemocoel, bathing the tissues directly.

III. Other invertebrates have a closed circulatory system, in which blood flows through a continuous circuit of blood vessels.

IV. The vertebrate circulatory system is a closed system that transports nutrients, oxygen, wastes, and hormones; it helps maintain fluid balance and body temperature; and it defends the body against disease.

V. Vertebrate blood consists of liquid plasma in which are suspended red blood cells, white blood cells, and platelets.
   A. Plasma consists of water, salts, substances in transport, and three types of proteins: albumins, globulins, and fibrinogen.
   B. Red blood cells transport oxygen and carbon dioxide.
   C. Lymphocytes and monocytes are agranular white blood cells; neutrophils, eosinophils, and basophils are granular white cells. White blood cells can leave the blood and wander through the tissues.
   D. Platelets patch damaged blood vessels and release substances essential for blood clotting.

VI. Arteries carry blood away from the heart; veins convey blood back toward the heart. Capillaries are tiny, thin-walled vessels through which materials are exchanged between blood and tissues.

VII. The vertebrate heart is a muscular pump that has evolved from the single atrium and ventricle of fish to the four-chambered heart of birds and mammals.
   A. The four-chambered hearts of birds and mammals have completely separate right and left sides which separate oxygen-rich and oxygen-depleted blood.
   B. The heart is enclosed by a pericardium and is equipped with valves that prevent backflow of blood.
   C. Although the heart has its own conduction system and can beat independently of its nerve supply, the heart rate is regulated by sympathetic and parasympathetic nerves.

VIII. Arterial pulse is the alternate expansion and recoil of an artery that occurs each time the left ventricle pumps blood into the aorta.

IX. Blood pressure is determined by blood flow and resistance to flow.
   A. Blood pressure is greatest in arteries and lowest in veins.
   B. Neural and hormonal mechanisms act continuously to maintain normal blood pressure.

X. In humans and other mammals, blood flows through a double circuit—the pulmonary circulation and the systemic circulation.
   A. The right atrium receives oxygen-depleted blood from the tissues and pumps it into the right ventricle. The right ventricle pumps blood to the lungs through the pulmonary circuit.
   B. Blood from the lungs returns to the heart via the pulmonary veins to the left atrium and is pumped into the aorta by the left ventricle. The aorta sends arterial branches into all parts of the systemic circulation.
      1. In the coronary circulation blood is transported to the cells of the heart itself.
      2. In the hepatic portal system a vein gives rise to an extensive network of exchange vessels within the tissues of the liver.

XI. Atherosclerosis is the basis of most cardiovascular disease. It can lead to ischemic heart disease or cerebrovascular accidents.

XII. The lymphatic system, a subsystem of the circulatory system, collects tissue fluid and returns it to the blood.

POST-TEST

1. In a(n) _____ circulatory system the heart pumps blood into a hemocoel.

2. Small invertebrates, such as sponges, depend upon _____ for internal transport.

3. The fluid component of blood is _____.

Select the most appropriate answers from column B for each item in column A. The same answer may be used more than once.

Column A
4. Transport oxygen
5. Seek out and ingest bacteria
6. Become macrophages
7. Initiate clotting
8. Contain hemoglobin
9. Agranular leukocyte

Column B
a. platelets
b. monocytes
c. red blood cells
d. neutrophils
e. eosinophils

10. A deficiency in hemoglobin is referred to as
_____.

11. During clotting, fibrinogen is converted to fibrin by
the action of _____.

Select the most appropriate answers from column B for each item in column A.

Column A
12. Conducts blood toward heart
13. Helps regulate blood pressure
14. Exchange vessel
15. Largest blood vessel
16. Equipped with valves

Column B
a. capillary
b. artery
c. vein
d. arteriole
e. aorta

17. The force exerted by the blood against the inner walls
of the blood vessels is known as _____
_____.

18. The pulmonary vein delivers blood to the
_____.

19. The hepatic portal vein delivers blood to the
_____.

20. Blood pressure is sensed by _____ within
certain arteries.
21. The _____ valve is located between the left
atrium and ventricle.
22. The angiotensins are a group of powerful
_____.

23. When a tissue is ischemic, it lacks sufficient
_____.

24. In atherosclerosis, the _____ wall thickens
and may block the passage of _____.
25. Lymph is produced when _____ fluid enters vessels.
26. Label the figure. If you need help, see Figure 36–12.

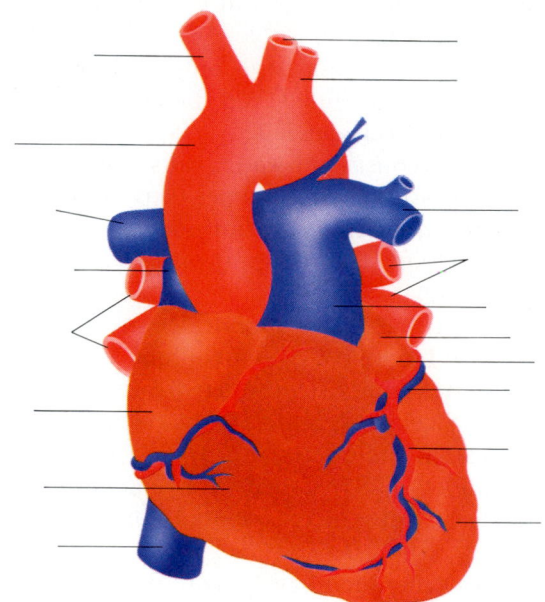

■ REVIEW QUESTIONS

1. Give the functions of (a) red blood cells, (b) plasma proteins, (c) platelets, (d) monocytes, (e) macrophages,
   (f) neutrophils.
2. Where do red blood cells originate? How are they destroyed?
3. What are three general causes of anemia?
4. How are (a) arterioles and (b) capillaries structurally
   adapted for carrying out their specific function?
5. Draw a diagram of the heart and label its chambers,
   valves, and the principal blood vessels that enter and exit
   from it.

6. How is the heartbeat initiated? Regulated?
7. How does blood manage to flow against gravity through
   veins in the legs on its route back to the heart?
8. Give an example of a normal blood pressure reading and
   of one from an individual with hypertension.
9. Trace a drop of blood from (a) superior vena cava to
   aorta, (b) brain to kidney, (c) intestine to lung.
10. How does the lymphatic system help maintain fluid bal-
    ance?
11. What is the relationship among plasma, interstitial (tis-
    sue) fluid, and lymph?

■ RECOMMENDED READINGS

Bodde, T.: "Coping in Space: The Body's Answer to Zero Gravity," *Bioscience* 32(4), April 1982. An interesting discussion of the physiological changes experienced by astronauts.

Cantin, M., and J. Genest: "The heart as an endocrine gland," *Scientific American* 254(2), February 1986. In addition to its pumping function, the heart produces a hormone that helps regulate blood pressure and volume.

Doolittle, R.F.: "Fibrinogen and Fibrin," *Scientific American* 245(6):126–135, 1981. A description of the process of blood clot formation and of clot breakdown.

Levy, R.I., and J. Moskowitz: "Cardiovascular Research: Decades of Progress, a Decade of Promise," *Science* 217:121–126, 1982. A review of current techniques for dealing with cardiovascular disease and a discussion of the decrease in mortality due to these diseases.

Stallones, R.A.: "The Rise and Fall of Ischemic Heart Disease," *Scientific American* 243(5):53–59, 1980. An interesting statistical analysis of the decline of U.S. death rates from heart disease since the 1960s.

37

Internal Defense

OUTLINE

I. Internal defense depends on the ability to distinguish between self and nonself
II. Invertebrates have internal defense mechanisms that are mainly nonspecific
III. Vertebrates can launch both nonspecific and specific immune responses
   A. Nonspecific defense mechanisms include mechanical and chemical barriers against pathogens
      1. Interferons help defend the body against viral infection
      2. Inflammation is a protective mechanism
      3. Phagocytes destroy pathogens
   B. Specific defense mechanisms include antibody-mediated immunity and cell-mediated immunity
      1. Cells of the immune system include lymphocytes and phagocytes
      2. The thymus "instructs" T cells and produces hormones
      3. The major histocompatibility complex enables the vertebrate to distinguish self from nonself
      4. Antibody-mediated immunity is a chemical warfare mechanism
         a. A typical antibody is a Y-shaped molecule consisting of four polypeptide chains
         b. There are five classes of antibodies
         c. The binding of antibody to antigen activates other defense mechanisms
      5. Cell-mediated immunity provides cellular warriors
   C. A secondary immune response is more rapid than a primary response
   D. Active immunity follows exposure to antigens
   E. Passive immunity is borrowed immunity
   F. Normally the body effectively defends itself against cancer
   G. Graft rejection is an immune response against transplanted tissue
   H. Certain sites in the body are immunologically privileged
   I. In an autoimmune disease the body attacks its own tissues
   J. Allergic reactions are inappropriate immune responses

Focus on AIDS

LEARNING OBJECTIVES

After you have studied this chapter you should be able to:
1. Compare in general terms the types of internal defense mechanisms in invertebrates and vertebrates.
2. Distinguish between specific and nonspecific defense mechanisms.
3. Describe the physiological changes and clinical symptoms associated with inflammation, and summarize the role of inflammation in the defense of the body.
4. Describe the process of phagocytosis.
5. Contrast T and B lymphocytes with respect to life cycle and function.
6. Cite the functions of the thymus in immune mechanisms.
7. Define the terms antigen and antibody, and describe how antigens stimulate immune responses.
8. Draw the basic structure of an antibody.
9. Describe the mechanisms of antibody-mediated immunity, including the effects of antigen-antibody complexes upon pathogens; include a discussion of the complement system.
10. Describe the mechanisms of cell-mediated immunity, including development of memory cells.
11. Contrast a secondary with a primary immune response.
12. Contrast active and passive immunity, giving examples of each.
13. Summarize the theory of immunosurveillance, and describe how the body destroys cancer cells.
14. Summarize the immunological basis of graft rejection, and explain how the effects of graft rejection can be minimized.
15. Describe the immunological basis of autoimmune diseases, give two examples, and list possible causes.
16. Explain the immunological basis of allergy, and briefly describe the events that occur during (a) a hayfever response and (b) systemic anaphylaxis.

A large cancer cell (green) is attacked by one T lymphocyte while others stand ready to help. (Lennart Nillson/Boehringer Ingelheim International GmbH)

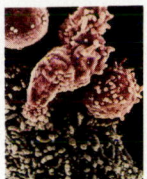

All animals have internal defense mechanisms that provide protection against foreign agents. **Nonspecific defense mechanisms** are directed against a multitude of foreign agents including pathogens (disease-causing organism). These mechanisms prevent pathogens from entering the body and act rapidly to destroy foreign agents that do penetrate the outer defenses. Phagocytosis of invading bacteria is an example of a nonspecific defense mechanism.

Specific defense mechanisms are very precise and highly effective. Their weapons are tailor-made to combat specific macromolecules associated with each foreign agent. Specific defense mechanisms are collectively referred to as **immune responses.** The term immune is derived from a Latin word meaning *safe.* Immune responses can be directed to the particular type of pathogen that infects the body. One of the body's most important specific defense mechanisms is the production of **antibodies,** highly specific proteins that help destroy pathogens. **Immunology,** the study of specific defense mechanisms, is one of the most exciting fields of medical research today.

KEY CONCEPTS

☐ Internal defense depends upon the ability of an organism to distinguish between self and nonself.

☐ Most invertebrates have only nonspecific defense mechanisms such as phagocytosis; vertebrates have both nonspecific and specific defense mechanisms.

☐ Internal responses include cell-mediated immunity and antibody-mediated immunity.

☐ INTERNAL DEFENSE DEPENDS UPON THE ABILITY TO DISTINGUISH BETWEEN SELF AND NONSELF

Internal defense depends upon the ability of an organism to distinguish between itself and foreign cells and large molecules. Such recognition is possible because organisms are biochemically unique. Cells have surface macromolecules (proteins or large carbohydrates) that are different from the surface macromolecules on the cells of other species or even other members of the same species. An organism "knows" its own macromolecules and "recognizes" those of other organisms as foreign.

A single bacterium may have from 10 to more than 1000 distinct macromolecules on its surface.

When a bacterium invades an animal its distinctive surface macromolecules stimulate the animal's defense mechanisms. A substance capable of stimulating an immune response is called an **antigen.**

☐ INVERTEBRATES HAVE INTERNAL DEFENSE MECHANISMS THAT ARE MAINLY NONSPECIFIC

All invertebrate species that have been studied demonstrate the ability to distinguish between self and nonself. However, most invertebrates are able to make only nonspecific immune responses, such as phagocytosis and the inflammatory response.

Sponge cells possess specific glycoproteins on their surfaces that enable them to distinguish between self and nonself. When cells of two different species of sponges are mixed together, they reaggregate according to species. Cnidarians also have this ability and can reject grafted tissue and destroy foreign tissue.

Coelomate invertebrates have wandering amoeba-like phagocytes that effectively engulf and destroy bacteria and other foreign matter. Any particle too large to be phagocytized is walled off or encapsulated by the phagocytes. Coelomate invertebrates also have nonspecific substances in the hemolymph that kill bacteria, inactivate cilia in some pathogens, and cause some foreign cells to clump or agglutinate. In mollusks these hemolymph substances enhance phagocytosis by the phagocytes.

Echinoderms and tunicates are the simplest animals known to possess differentiated white blood cells that perform immune functions. Tunicates also have nodules of lymphatic tissue. Certain annelids (e.g., earthworms) and cnidarians (e.g., corals) are thought to possess specific immune mechanisms and immunological memory. In them and in some echinoderms and simple chordates, the body appears to remember antigens for a short period of time and can respond to them more effectively in a second encounter.

☐ VERTEBRATES CAN LAUNCH BOTH NONSPECIFIC AND SPECIFIC IMMUNE RESPONSES

Vertebrates possess many of the basic mechanisms present in invertebrates and also have more sophisticated defense mechanisms (Fig. 37–1). These are made possible by the development of a specialized lymphatic system. In the discussion that follows we will focus on the human immune system, with references to those of other vertebrates.

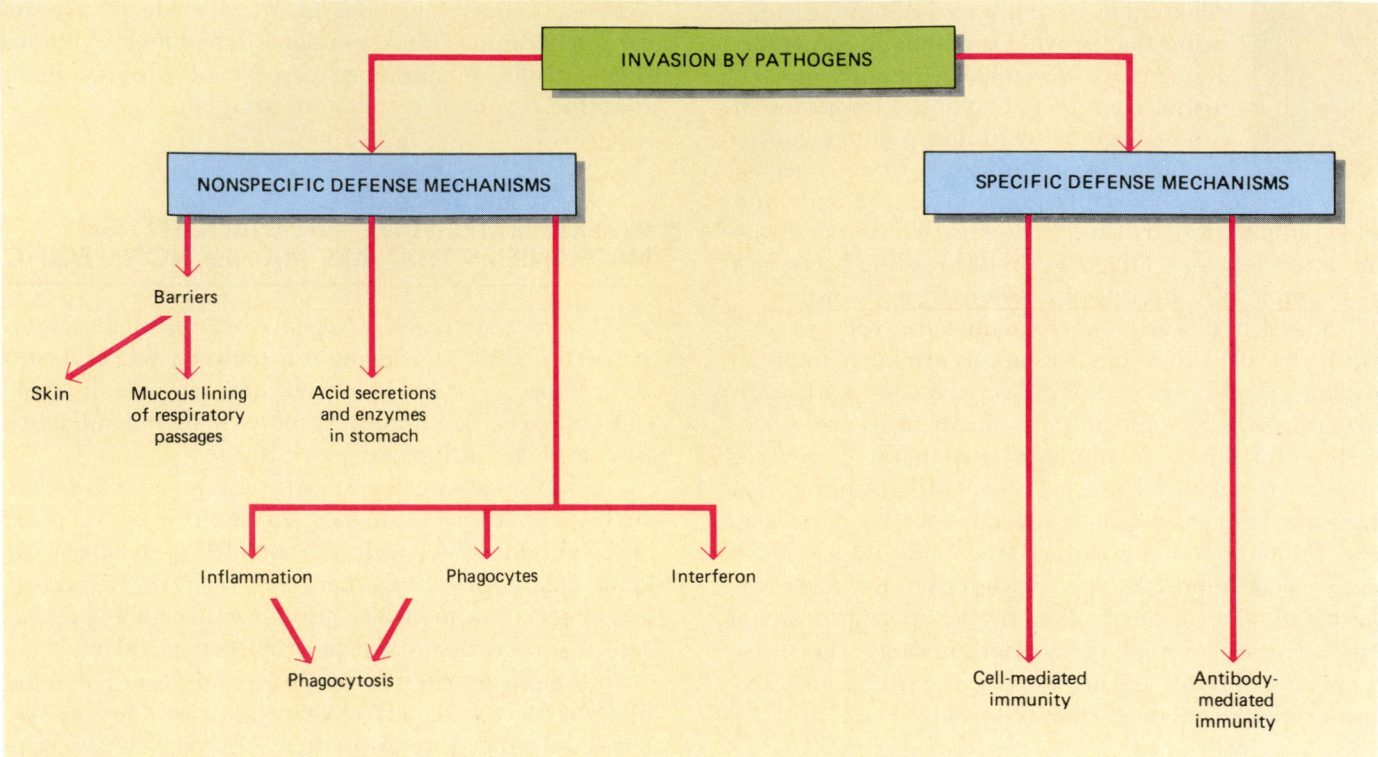

FIGURE 37–1 Summary of nonspecific and specific defense mechanisms. Nonspecific mechanisms prevent entrance of a great many pathogens and act rapidly to destroy those that manage to traverse the barriers. Specific defense mechanisms take longer to mobilize but are highly effective in destroying invaders.

Nonspecific Defense Mechanisms Include Mechanical And Chemical Barriers Against Pathogens

The outer covering of an animal is its first line of defense against pathogens. This covering is often more than just a mechanical barrier. For example, the human skin is populated by millions of harmless microorganisms, sometimes referred to as the normal flora. These organisms normally live in harmony with their host. Their presence appears to inhibit the multiplication of potentially harmful organisms that happen to land on the skin. Such invaders compete unsuccessfully with the resident microorganisms for essential nutrients. Other nonspecific defenses of the skin include sweat and sebum which contain chemicals that destroy certain kinds of bacteria.

Microorganisms that enter the vertebrate body with food are usually destroyed by the acid secretions and enzymes of the stomach. Pathogens that enter the body with inhaled air may be filtered out by hairs in the nose or trapped in the sticky mucous lining of the respiratory passageways. There, they may be destroyed by phagocytes. Should pathogens invade the tissues, other nonspecific defense mechanisms are activated.

Interferons Help Defend The Body Against Viral Infection

When infected by viruses or other intracellular parasites (some types of bacteria, fungi, and protozoa), certain cells respond by secreting proteins called **interferons.** These proteins stimulate other cells to produce antiviral proteins, which prevent the cell from manufacturing macromolecules required by the virus. The virus particles produced in cells exposed to interferon do not infect cells very effectively. Interferons also stimulate **natural killer (NK) cells.** These cells recognize body cells that have been altered by viruses and kill them quickly.

Drug companies have invested millions of dollars trying to develop an inexpensive, effective method of producing human interferon. Successes have been achieved using recombinant DNA techniques. Studies have established that interferon is useful in treating some viral infections and might be helpful in treating certain forms of cancer.

Inflammation Is A Protective Mechanism

When pathogens invade tissues, they trigger an **inflammatory response** (Fig. 37–2). Blood vessels in the

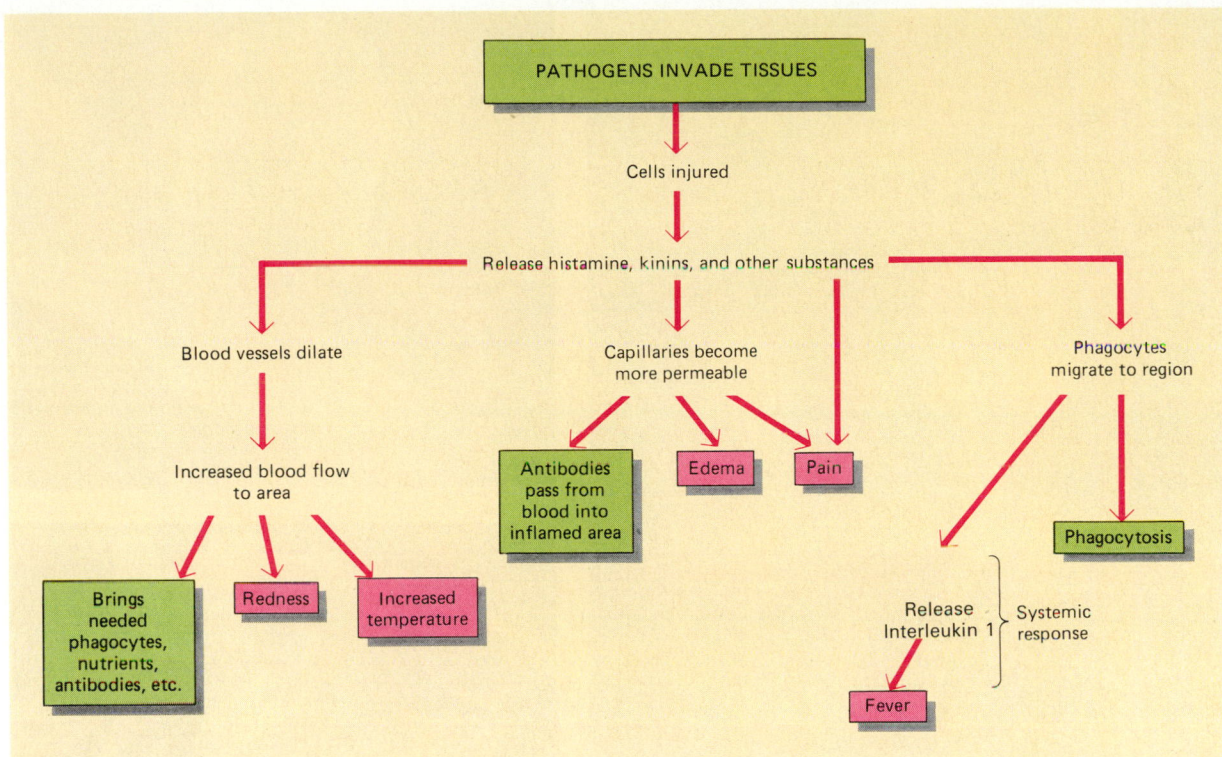

FIGURE 37–2 Inflammation is a mechanism by which protective immune mechanisms can be localized at a region where infection occurs. It is a vital process that permits phagocytic cells, antibodies, and other needed compounds to enter the tissue where microbial invasion is taking place.

affected area dilate, increasing blood flow to the infected region. The increased blood flow makes the skin look red and feel warm. Capillaries in the inflamed area become more permeable, allowing more fluid to leave the circulation and enter the tissues. As the volume of interstitial fluid increases, **edema** (swelling) occurs. The edema (and also certain substances released by the injured cells) causes the pain that we associate with inflammation. Thus, the signs of inflammation are **redness, heat, edema,** and **pain.**

The increased blood flow that occurs during inflammation brings great numbers of phagocytic cells (first, neutrophils and, later, macrophages; see Chapter 36) to the infected area. The increased permeability of the blood vessels allows needed gamma globulins, which serve as antibodies, to leave the circulation and enter the tissues. As fluid leaves the circulation, it also brings with it needed oxygen and nutrients.

Although inflammation is often a local response limited to one area, sometimes the entire body is involved. A peptide called interleukin 1, released by macrophages, can reset the body's thermostat in the hypothalamus. Prostaglandins are also involved in this resetting process. The resulting rise in body temperature known as **fever** is a common clinical symptom of widespread inflammatory response. Fever interferes

with viral activity and decreases circulating iron. When microorganisms have difficulty obtaining iron, they are at a metabolic disadvantage.

Phagocytes Destroy Pathogens

One of the main functions of inflammation appears to be increased phagocytosis. Two types of phagocytes in vertebrates are neutrophils and macrophages. A phagocyte ingests a bacterium or other invading microorganism by flowing (amoeboid fashion) around it and engulfing it. As it ingests the microorganism, the phagocyte wraps it within membrane pinched off from its plasma membrane. The vesicle containing the bacterium is called a **phagosome.** One or more lysosomes adhere to the phagosome membrane and fuse with it. The lysosome releases potent digestive enzymes onto the captured bacterium, and the phagosome membrane releases hydrogen peroxide onto the invader. These substances destroy the bacterium, breaking down its macromolecules to small, harmless compounds that the phagocyte can either use or release.

A neutrophil can phagocytize 20 or so bacteria before it becomes inactivated (perhaps by leaking lysosomal enzymes) and dies. A macrophage can phagocytize about 100 bacteria during its life-span. Can bacte-

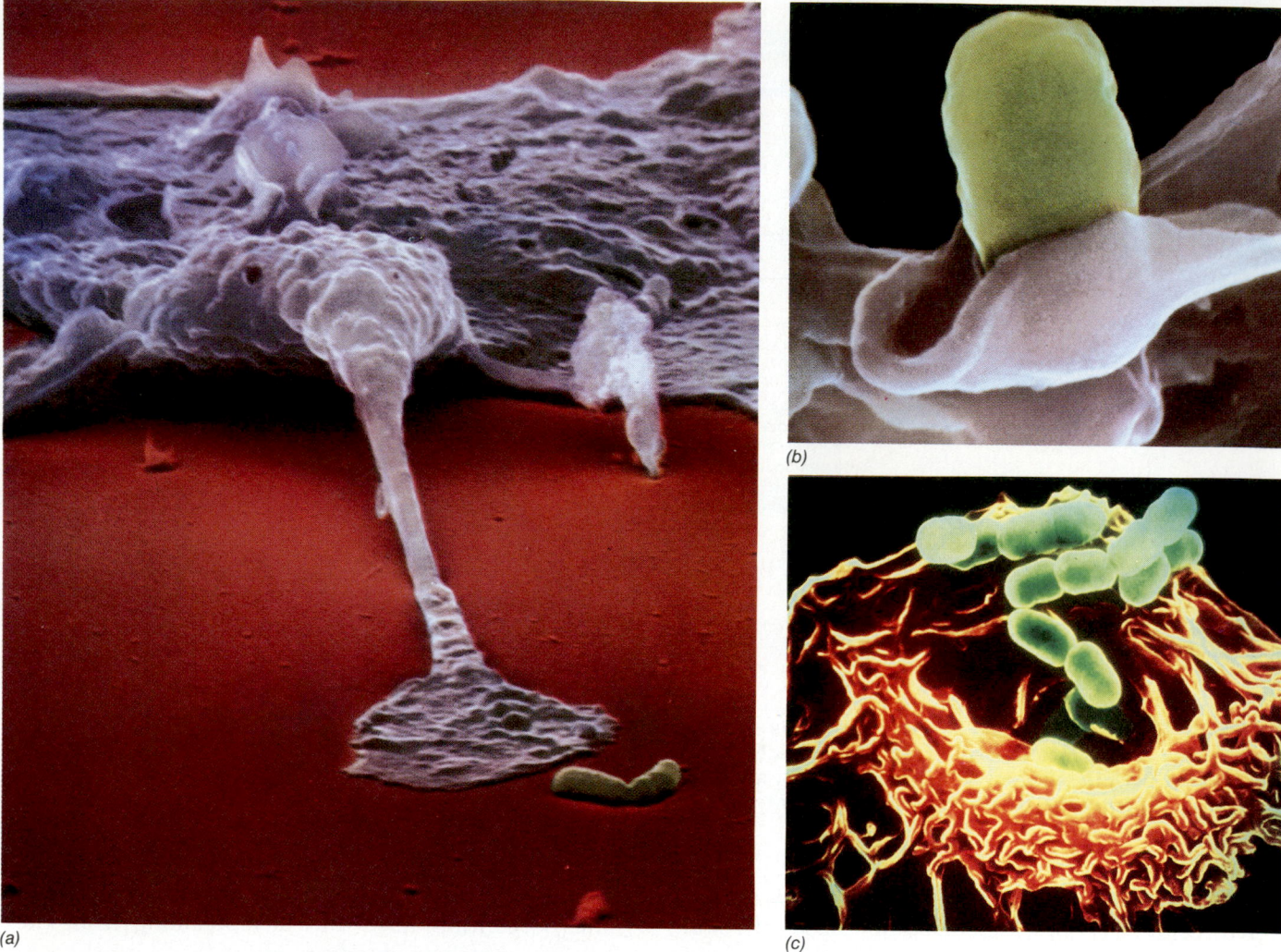

(a)

(b)

(c)

FIGURE 37–3 The macrophage is an incredibly efficient warrier. (*a*) A macrophage (colored gray) extends a pseudopod toward an invading *Escherichia coli* bacterium (green) that is already multiplying. (*b*) The bacterium is trapped within the engulfing pseudopod. (*c*) The macrophage sucks the trapped bacteria in along with its own cell membrane. The macrophage's cell membrane will seal over them and powerful lysosomal enzymes will destroy them. (From Lennart Nillson, INCREDIBLE MACHINE, p. 170, Copyright Boehringer Ingelheim International GmbH)

ria counteract the phagocyte's attack? Certain bacteria release enzymes that destroy the membranes of the phagocyte lysosomes. The powerful lysosomal enzymes then spill out into the cytoplasm and may destroy the phagocyte. Other bacteria have cell walls or capsules that resist the action of lysosomal enzymes.

Some macrophages wander through the tissue, phagocytizing foreign matter and, when appropriate, release antiviral agents (Fig. 37–3). Others stay in one place and destroy bacteria that pass by. For example, air sacs in the lungs contain large numbers of macrophages that destroy foreign matter entering with inhaled air.

Specific Defense Mechanisms Include Antibody-Mediated Immunity And Cell-Mediated Immunity

While nonspecific defense mechanisms destroy pathogens and prevent the spread of infection, specific defense mechanisms are being mobilized. Several days are required to activate specific immune responses, but once in gear, these mechanisms are extremely effective. There are two main types of specific immunity: **antibody-mediated immunity,** in which lymphocytes produce specific antibodies designed to destroy the pathogen, and **cell-mediated immunity,** in which lymphocytes attack the invading pathogen directly. Specific immunity depends on several types of cells.

FIGURE 37–4 Origin and functions of T and B lymphocytes.

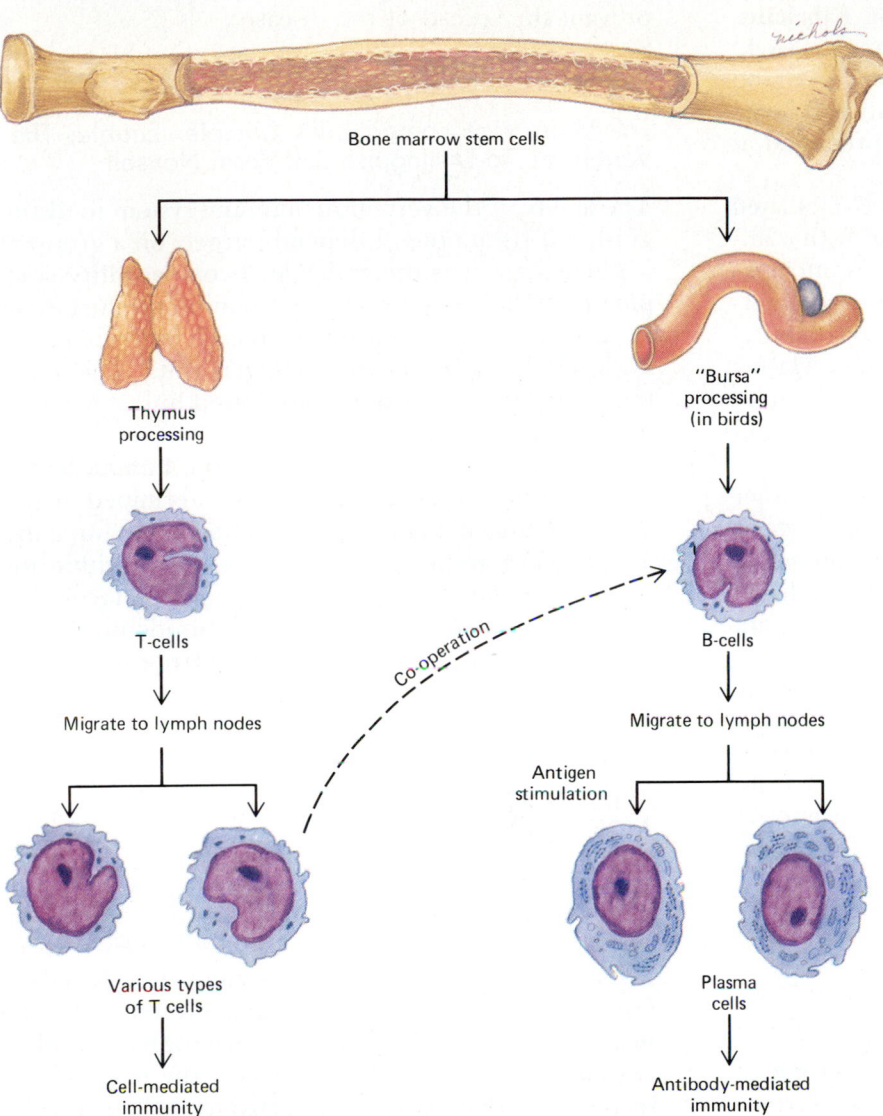

Bone marrow stem cells

Thymus processing

T-cells

Migrate to lymph nodes

Co-operation

Various types of T cells

Cell-mediated immunity

"Bursa" processing (in birds)

B-cells

Migrate to lymph nodes

Antigen stimulation

Plasma cells

Antibody-mediated immunity

Cells Of The Immune System Include Lymphocytes And Phagocytes

Immune responses depend on two main groups of white blood cells: lymphocytes and phagocytes. The trillion or so lymphocytes are the main warriors in specific immune responses. Lymphocytes are stationed strategically in the lymphatic tissue throughout the body. Two main types of lymphocytes are **T lymphocytes,** or **T cells,** and **B lymphocytes,** or **B cells.** A third type of lymphocyte includes the NK cell mentioned earlier, which kills virally infected cells and tumor cells.

T cells originate from stem cells in the bone marrow (Fig. 37–4). On their way to the lymph tissues, the future T cells stop off in the thymus gland for process-

ing. The *T* in T lymphocytes stands for thymus-derived; somehow the thymus gland influences the differentiation of lymphocytes capable of immunological response. T lymphocytes are responsible for cellular immunity.

Three main types, or subsets, of T cells have been identified:

1. **Cytotoxic T cells,** sometimes called killer T cells, recognize cells with foreign antigens on their surfaces and destroy these cells.

2. **Helper T cells** stimulate other types of T cells as well as B cells to respond to antigens.

3. **Suppressor T cells** inhibit the activity of helper T cells and cytotoxic T cells.

As with T cells, millions of B cells are produced in the bone marrow. In birds B cells are processed in a lymphatic organ known as the **bursa of Fabricius.** (The *B* in B lymphocytes refers to bursa-derived.) Other vertebrates do not have a bursa, however, and an equivalent organ has not yet been identified. Some immunologists think that B cells may be processed as they form in the bone marrow.

B cells are responsible for antibody-mediated immunity. Each B cell is specialized to bind with a different type of antigen. When a B cell comes into contact with the specific type of antigen to which it is targeted, it divides rapidly, forming a clone of identical cells. These B cells develop into **plasma cells.** Plasma cells have an extensive, highly developed rough endoplasmic reticulum, a structural adaptation for the synthesis of antibodies.

When a macrophage ingests a bacterium or other pathogen, it digests most, but not all, of the pathogen's antigens. A few molecules of the antigen remain intact and are displayed on the surface of the macrophage. This displayed antigen is necessary to stimulate antigen-sensitive lymphocytes.

Macrophages also secrete substances that promote and help regulate immune responses. When macrophages are stimulated by bacteria they secrete interleukin 1, which helps activate B cells and helper T cells. Interleukin 1 also promotes a general response to injury. It causes fever, mobilizes neutrophils, and activates other mechanisms that defend the body.

The Thymus "Instructs" T Cells And Produces Hormones

Present in all vertebrates, the **thymus gland** has at least two functions. First, in some unknown way, the thymus confers immunological competence upon T cells. Within the thymus these cells develop the ability to differentiate into cells that can respond to specific antigens. This "instruction" within the thymus is thought to take place just before birth and during the first few months of postnatal life. When the thymus is removed from an animal before this processing takes place, the animal is not able to develop cellular immunity. If the thymus is removed after that time, cellular immunity is not seriously impaired.

The second function of the thymus is that of an endocrine gland. It secretes several hormones, including one known as **thymosin.** Although not much is known about these hormones, thymosin is thought to stimulate T cells after they leave the thymus, causing them to complete their differentiation and become immunologically active. Thymosin has been used clinically in patients who have poorly developed thymus glands. It is also being tested as a modifier of biological response in patients with certain types of cancer; stimulating cellular immunity in such patients may help to prevent the spread of the disease.

The Major Histocompatibility Complex Enables The Vertebrate To Distinguish Self From Nonself

The ability of the vertebrate immune system to distinguish self from nonself depends largely on a group of antigens known as the **major histocompatibility complex (MHC).** These proteins present on the surface of every cell are slightly different in each individual, giving each one a biochemical "fingerprint." The genes that code for these proteins are found linked together on one chromosome.

In humans the MHC is called the **human leukocyte antigen (HLA) group.** HLA is determined by five different linked genes (chromosome 6 in humans). These genes are all polymorphic—that is, within the population there are multiple alleles at each locus. Tissues from the same individual or from identical twins have the same HLA alleles and thus identical HLA antigens.

Antibody-Mediated Immunity Is A Chemical Warfare Mechanism

B cells are responsible for antibody-mediated immunity (also called humoral immunity). Antibodies on the surface of B cells serve as receptors. Only the variety of B cell with a matching receptor—**competent B cells**—can bind with a particular antigen presented by the macrophage (Fig. 37–5). When a macrophage displaying antigen contacts a helper T cell, the macrophage secretes interleukin 1, which activates helper T cells.

Activated helper T cells detect B cells that have bound to antigen on the macrophage and bind to the same antigen. A helper T cell recognizes an antigen only if it is part of a complex with a molecule that the T cell recognizes as self. In humans this self-molecule is HLA. The helper T cell binds to both the antigen and the HLA attached to the macrophage. Then the activated helper T cells secrete interleukins which are a group of **lymphokines,** substances secreted by lymphocytes. Interleukins are soluble compounds that activate competent B cells.

Once activated, B cells increase in size. Then they divide by mitosis, each giving rise to a sizable clone of identical cells. Some of these B cells mature into **plasma cells** that produce antibody. Unlike T cells, most plasma cells do not leave the lymph nodes. Only the antibodies they secrete pass out of the lymph tissues and make their way via the lymph and blood to the infected area.

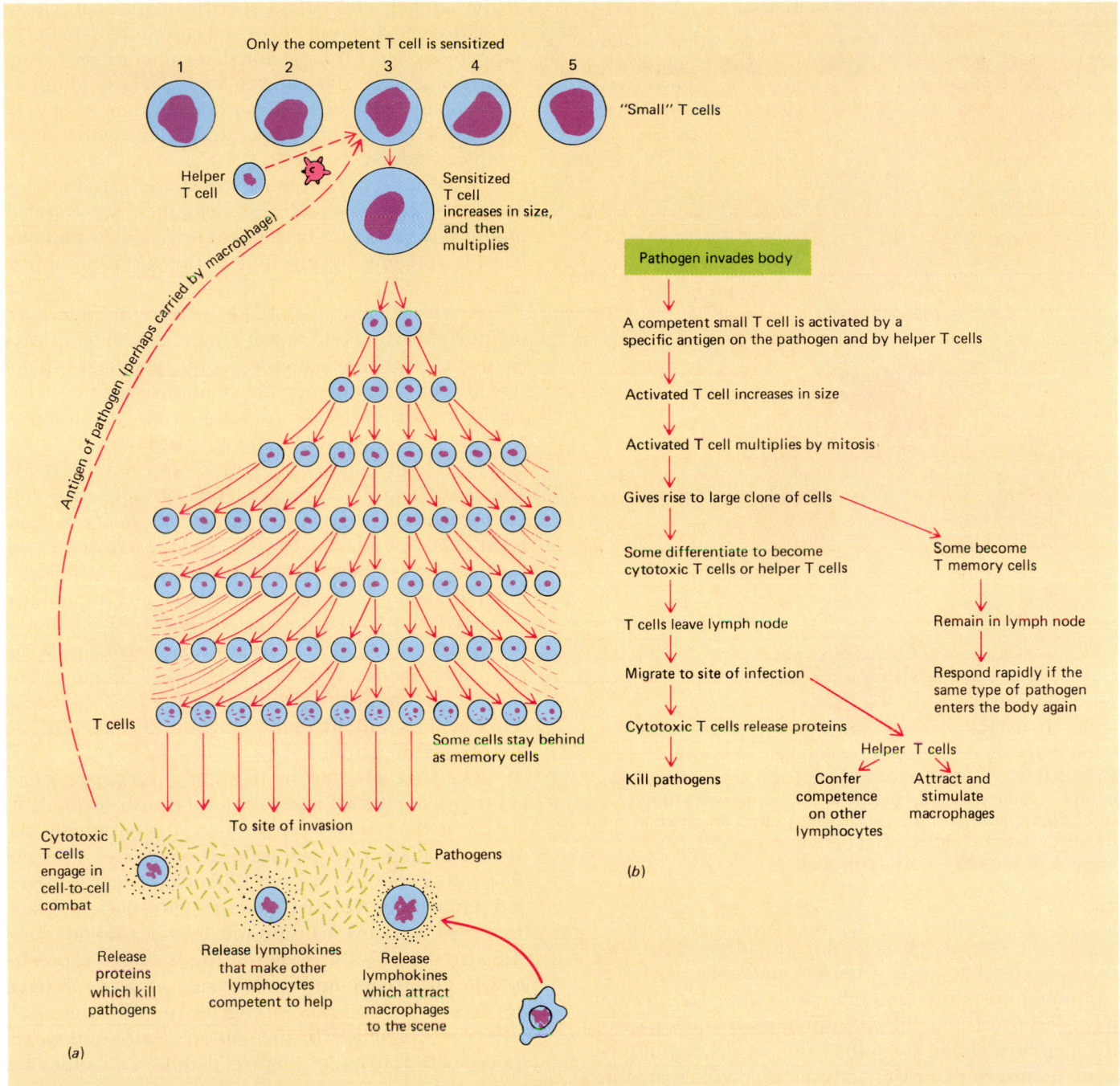

Only the competent T cell is sensitized

"Small" T cells

Helper T cell

Antigen of pathogen (perhaps carried by macrophage)

Sensitized T cell increases in size, and then multiplies

T cells

Some cells stay behind as memory cells

To site of invasion

Cytotoxic T cells engage in cell-to-cell combat

Release proteins which kill pathogens

Release lymphokines that make other lymphocytes competent to help

Release lymphokines which attract macrophages to the scene

Pathogens

(a)

Pathogen invades body

A competent small T cell is activated by a specific antigen on the pathogen and by helper T cells

Activated T cell increases in size

Activated T cell multiplies by mitosis

Gives rise to large clone of cells

Some differentiate to become cytotoxic T cells or helper T cells

Some become T memory cells

T cells leave lymph node

Remain in lymph node

Migrate to site of infection

Respond rapidly if the same type of pathogen enters the body again

Cytotoxic T cells release proteins

Helper T cells

Kill pathogens

Confer competence on other lymphocytes

Attract and stimulate macrophages

(b)

FIGURE 37–5 Antibody-mediated immunity. When a macrophage presents an antigen to a competent B cell and a helper T cell releases appropriate interleukins, the B cell becomes sensitized. Once activated in this way, the competent B lymphocyte multiplies, producing a large clone of cells. Many of these differentiate and become plasma cells, which secrete antibodies. The plasma cells remain in the lymph tissues, but the antibodies are transported to the site of infection by the blood or lymph. Antigen-antibody complexes form that directly inactivate some pathogens and also turn on the complement system. Some of the B lymphocytes become memory cells that persist and continue to secrete small amounts of antibody for years after the infection is over.

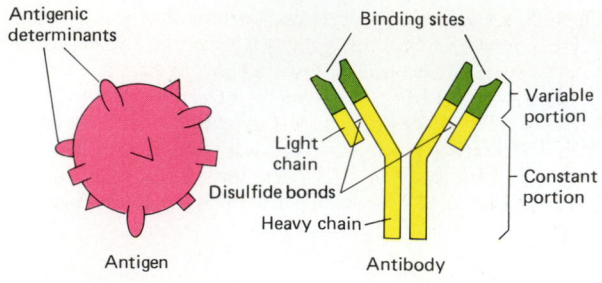

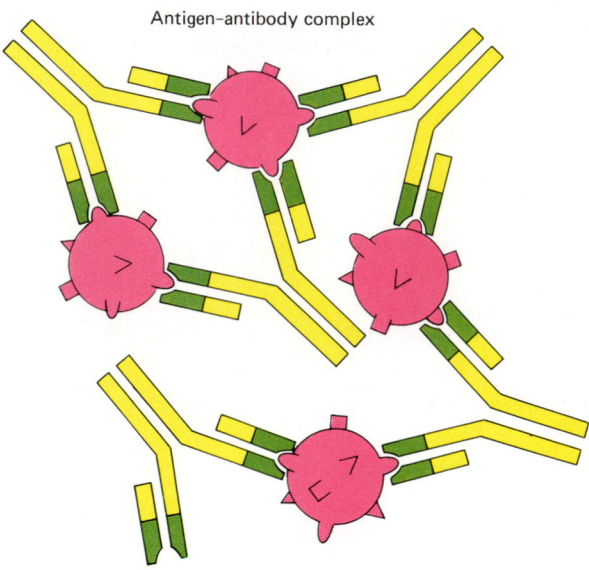

FIGURE 37–6 **Antigen, antibody, and antigen-antibody complex. The antibody molecule is composed of two light chains and two heavy chains, joined together by disulfide bonds. The constant (C) and variable (V) regions of the chains are indicated.**

Some activated B cells do not differentiate into plasma cells, but instead become **memory cells** that continue to produce antibodies long after an infection has been overcome. These antibodies become part of the gamma globulin fraction of the plasma. Should the same pathogen enter the body again, these circulating antibodies are immediately available to mark it for destruction. At the same time memory cells quickly divide to produce new clones of the appropriate plasma cells.

A Typical Antibody Is A Y-Shaped Molecule Consisting Of Four Polypeptide Chains

Antibodies, known more formally as **immunoglobulins,** or **Ig,** are highly specific proteins produced in response to specific antigens. The function of an antibody is to bind to an antigen. How does an antibody "recognize" a particular antigen? In a protein antigen specific sequences of amino acids constitute **antigenic determinants** (Fig. 37–6). These amino acids give part

of the antigen molecule a specific shape that can be recognized by an antibody or cell receptor. Usually, an antigen has 5 to 10 antigenic determinants on its surface. Some have 200 or even more. These antigenic determinants may differ from one another, so several different kinds of antibodies can combine with a single complex antigen.

Some substances found in dust and certain drugs are too small to be antigenic; yet they do stimulate immune responses. These substances, called **haptens,** become antigenic by attaching to the surface of a protein.

A typical immunoglobulin consists of four polypeptide chains: two identical long chains called heavy chains and two identical short chains called light chains (Fig. 37–6). The polypeptide chains are held together and their configurations are stabilized by disulfide (—S—S—) linkages and by noncovalent bonds.

Each light chain is made up of approximately 214 amino acids, and each heavy chain of more than 400. Each chain has a constant segment, a junctional segment, and a variable segment. In the **constant segment,** or **C region,** the amino acid sequence is the same in all antibodies of a particular class. The amino acid sequence of the **junctional segment,** or **J region,** is somewhat variable. The **variable segment,** or **V region,** has a highly variable amino acid sequence. The variable region of the antibody protrudes from the B cell and the constant region anchors the molecule to the plasma membrane.

The V region may be thought of as the part of a key that is unique for a specific antigen (the lock). At its V region the antibody folds three-dimensionally, assuming a shape that enables it to recognize and combine with a specific antigen. When they meet, antigen and antibody fit together *somewhat* like a lock and key; they must fit in just the right way for the antibody to be effective (Fig. 37–7). However, the fit is not as precise as with an enzyme and its substrate. A given antigen can bind with different strengths, or **affinities,** with different antibodies. In the course of an immune response better, stronger (higher-affinity) antibodies are generated.

A typical antibody is a Y-shaped molecule that contains two binding sites, enabling the antibody to combine with two antigen molecules. This permits formation of **antigen-antibody complexes.**

There Are Five Classes Of Antibodies

Antibodies are grouped in five classes according to their structure. At the constant end, the heavy chains of an antibody have amino acid sequences characteristic of the particular antibody class. Using the abbreviation Ig for immunoglobulin, the classes are designated IgG, IgM, IgA, IgD, and IgE. In the simpler verte-

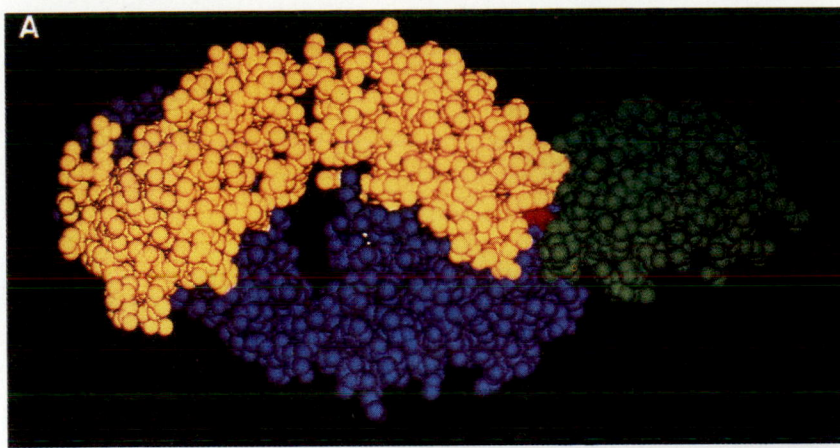

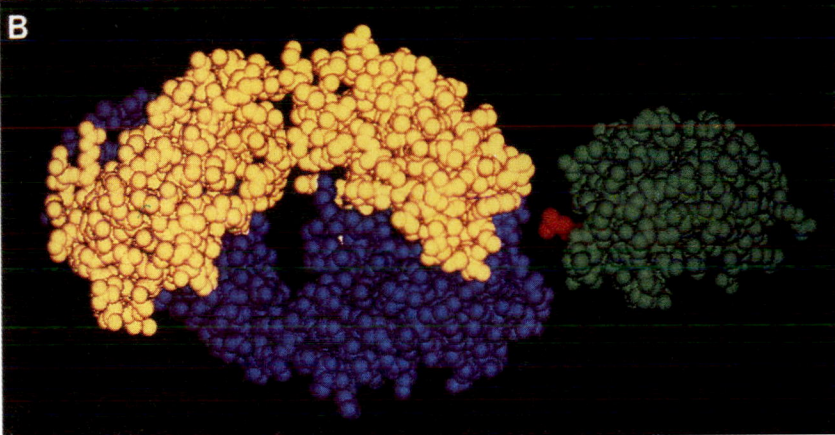

FIGURE 37–7 An antigen-antibody complex. The antigen lysozyme is shown in green. The heavy chain of the antibody is shown in blue, and the light chain in yellow. (*a*) The antigenic determinant, shown in red, fits into a groove in the antibody molecule. (*b*) The antigen-antibody complex has been pulled apart. Note how they fit each other.

brates, only IgM is present. In the amphibians, IgM and IgG are characteristic. These immunoglobulin classes and also IgA are found in birds, and all five classes are produced in humans.

The classes of antibodies have different functions determined by their C regions. For example, IgA is responsible for immunity on body surfaces; IgE mediates allergic responses. We can speculate that as more complex animals evolved, it was advantageous to have a variety of antibody classes with specialized functions.

In humans about 75% of the antibodies in the blood belong to the **IgG** group; these are part of the gamma globulin fraction of the plasma. IgG contributes to immunity against many blood-borne pathogens, including bacteria, viruses, and some fungi. IgG, along with the IgM antibodies, stimulate macrophages and activate the complement system (discussed in the following section). IgM is highly effective against viruses, and characterizes blood type.

IgA is the principal type of antibody found in body secretions, such as mucous secretions of the nose, respiratory passageways, digestive tract, and tears, saliva, and vaginal secretions. IgA is strategically located to defend against infections by inhaled or ingested pathogens. The function of the **IgD** type of antibody is

not known. **IgE,** the mediator of allergic responses, is discussed in a later section.

The Binding of Antibody to Antigen Activates Other Defense Mechanisms

Antibodies mark a pathogen as foreign by combining with an antigen on its surface. Often several antibodies bind with several such antigens, creating a mass of clumped antigen-antibody complex. The combination of antigen and antibody activates several defense mechanisms:

1. The antigen-antibody complex may inactivate the pathogen or a toxin (harmful substance) it secretes. For example, when an antibody attaches to the surface of a virus, the virus may lose its ability to attach to a host cell.
2. The antigen-antibody complex stimulates phagocytosis of the pathogen by macrophages and neutrophils.
3. Antibodies of the IgG and IgM groups work mainly through the complement system. This system consists of about 11 proteins present in plasma and other body fluids. Normally, complement proteins are inactive, but an antigen-antibody complex stim-

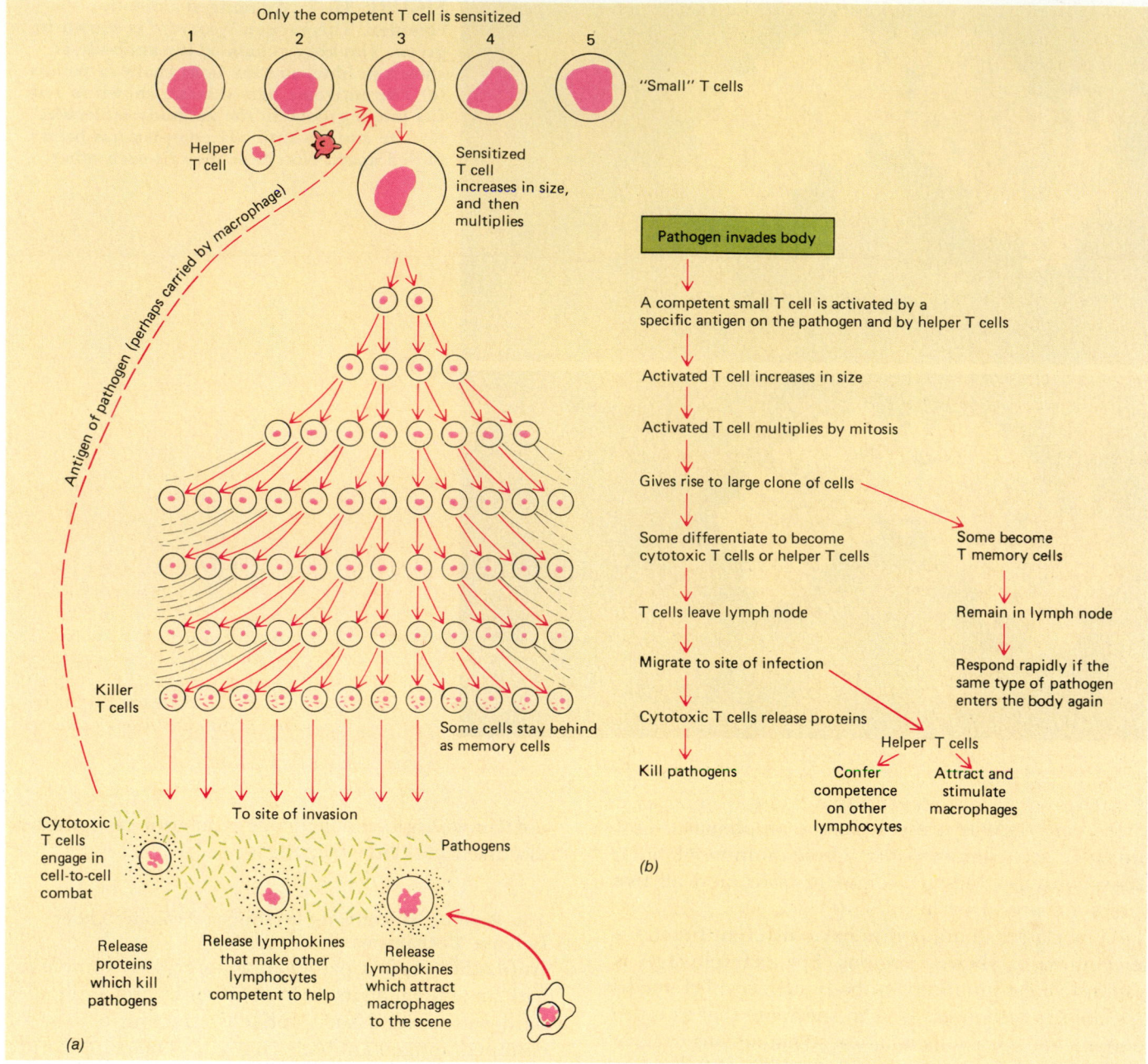

Only the competent T cell is sensitized

1   2   3   4   5   "Small" T cells

Helper T cell

Antigen of pathogen (perhaps carried by macrophage)

Sensitized T cell increases in size, and then multiplies

Killer T cells

Some cells stay behind as memory cells

Cytotoxic T cells engage in cell-to-cell combat

To site of invasion

Pathogens

Release proteins which kill pathogens

Release lymphokines that make other lymphocytes competent to help

Release lymphokines which attract macrophages to the scene

(a)

Pathogen invades body

A competent small T cell is activated by a specific antigen on the pathogen and by helper T cells

Activated T cell increases in size

Activated T cell multiplies by mitosis

Gives rise to large clone of cells

Some differentiate to become cytotoxic T cells or helper T cells

Some become T memory cells

T cells leave lymph node

Remain in lymph node

Migrate to site of infection

Respond rapidly if the same type of pathogen enters the body again

Cytotoxic T cells release proteins

Kill pathogens

Helper T cells

Confer competence on other lymphocytes

Attract and stimulate macrophages

(b)

FIGURE 37–8   Cell-mediated immunity. When activated by an antigen and by a helper T cell, a competent T cell gives rise to a large clone of cells. Many of these differentiate to become cytotoxic T cells that migrate to the site of infection. There, they release proteins that destroy invading pathogens.

ulates a series of reactions that activate the system. The antibody is said to "fix" complement. Proteins of the complement system then work to destroy pathogens. Some complement proteins digest portions of the pathogen cell. Others coat the pathogens; this apparently stimulates macrophages and neutrophils to phagocytize the pathogens. Complement proteins also increase the extent of inflammation.

Complement proteins are not specific. They act against any antigen, provided they are activated by antigen-

antibody complex. After antibodies identify the pathogen very specifically, complement proteins *complement* their action by destroying the pathogens.

Cell-Mediated Immunity Provides Cellular Warriors

The T cells and macrophages are responsible for cell-mediated immunity (Fig. 37–8). These cells are especially effective in attacking viruses, fungi, and the types of bacteria that live within body cells. How do the T cells know which cells to attack? Once a pathogen invades a body cell, macromolecules on the surface of that host cell may be altered. The immune system no longer recognizes the altered cell as "self," so T cells attack and destroy it. Cytotoxic T cells also destroy cancer cells, and, unfortunately, the cells of transplanted organs.

As in antibody-mediated immunity, activated helper T cells are necessary for a response. T cells have **T-cell receptors** on their surfaces that are capable of reacting with antigens on the surfaces of invading cells or altered host cells. There are millions of different T cells, each with specific types of T-cell receptors. Each type is capable of recognizing a specific foreign antigen. As with B cells, only the variety of lymphocyte able to react to the specific antigen presented—that is, the competent lymphocyte—becomes activated. However, a T cell cannot recognize an antigen if it is presented alone. The antigen must be presented to the T cell as part of a complex with a HLA that the T cell recognizes as "self."

Once stimulated, competent T cells increase in size and give rise to a sizable clone of cytotoxic T cells and memory cells. Cytotoxic T cells make up the cellular infantry; they leave the lymph nodes and make their way to the infected area. These killer cells can destroy a target cell within seconds after contact. Then the T cell disengages itself from its victim cell and seeks out a new target cell.

T cells and macrophages at the site of infection secrete interleukins, interferons, and a variety of other substances that help regulate immune function. Some interleukins confer competence upon other lymphocytes in the area, increasing the ranks of cytotoxic T cells. Other interleukins enhance the inflammatory reaction, attracting great numbers of macrophages to the site of infection. Interleukins also stimulate macrophages, making them more active and effective at destroying pathogens.

Suppressor T cells are stimulated by antigen. These cells help regulate both T cells and B cells. Suppressor T cells multiply more slowly than cytotoxic T cells, so more than a week generally elapses before they suppress an immune response.

A Secondary Immune Response Is More Rapid Than A Primary Response

The first exposure to an antigen stimulates a **primary response.** Injection of an antigen into an animal causes specific antibodies to appear in the blood plasma in 3 to 14 days. After injection of the antigen there is a brief latent period, during which the antigen is recognized and appropriate lymphocytes begin to form clones. Then there is a logarithmic phase, during which the antibody concentration rises logarithmically for several days until it reaches a peak (Fig. 37–9). IgM is the prin-

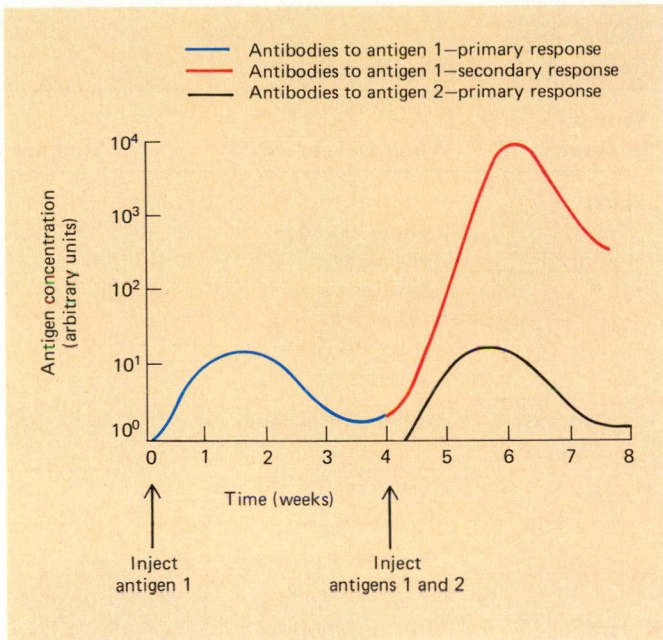

FIGURE 37–9 **Primary and secondary responses of antibody formation to successive doses of antigens. Antigen 1 was injected at day 0 and immune response was assessed by measuring antibody levels to the antigen. At week 4, the primary response had subsided. Antigen 1 was injected again along with a new protein, antigen 2. Note that the secondary response to antigen 1 was greater and more rapid than the primary response. A primary response was made to the newly encountered antigen 2.**

cipal antibody synthesized. Finally, there is a decline phase, during which the antibody concentration decreases to a very low level.

A second injection of the same antigen, even years later, evokes a **secondary response** (Fig. 37–9). Because memory cells bearing antibodies to that antigen persist throughout an individual's life, the secondary response is generally a much more rapid response with a shorter latent period. Much less antigen is necessary to stimulate a secondary response than a primary response. More antibodies are produced than in a primary response, and the decline phase is slower. In a secondary response the predominant antibody is IgG. The affinity (strength of fit) of the antibody also increases following secondary exposure.

The body's ability to launch a rapid, effective response during a second encounter with an antigen explains why we do not usually suffer from the same disease several times. Most persons get measles or chicken pox, for example, only once. When exposed a second time, the immune system destroys the pathogens before they have time to establish themselves and cause symptoms of the disease. Booster shots of vaccine are given in order to elicit a secondary response, reinforcing the immunological memory of the disease-producing antigens.

TABLE 37–1
Active And Passive Immunity

Type of Immunity	When Developed	Memory Cells	Duration of Immunity
Active			
Naturally induced	Pathogens enter the body through natural encounter, e.g., infected person sneezes	Yes	Many years
Artificially induced	After immunization	Yes	Many years
Passive			
Naturally induced	After transfer of antibodies from mother to developing baby	No	Few months
Artificially induced	After injection with gamma globulin	No	Few months

You may wonder, then, how a person can get flu or a cold more than once. Unfortunately, there are many varieties of these diseases, each caused by a virus with slightly different antigens. To further complicate matters, viruses mutate often (a survival mechanism for them), which may result in changes in their surface antigens. Even a slight change may prevent recognition by memory cells. Because the immune system is so specific, each different antigen is treated by the body as a new immunological challenge.

Active Immunity Follows Exposure To Antigens

We have been considering **active immunity,** immunity developed following exposure to antigens. After you have had measles as a young child, for example, you develop immunity that protects you from contracting measles again. Active immunity can be either *naturally* or *artificially* induced (Table 37–1). If someone with measles sneezes near you and you contract the disease, you develop naturally acquired immunity. Active immunity can also be artificially induced by **immunization,** that is, by injection of a vaccine—a suspension of dead or weakened disease organisms. In this case, the body launches an immune response against the antigens contained in the vaccine and develops memory cells so that future encounters with the same pathogen will be dealt with swiftly.

Effective vaccines can be prepared in a number of ways. A virus may be attenuated (weakened) by successive passage through cells of nonhuman hosts. In the process, mutations occur that may result in the pathogen becoming adapted to the nonhuman host, so that it can no longer cause disease in humans. This is how polio, smallpox, and measles vaccines are produced. Whooping cough and typhoid fever vaccines are made from pathogens that, even though they have been killed, still have the necessary antigens to stimulate an immune response. Tetanus and botulism vaccines are made from toxins secreted by the respective pathogens. The toxin is altered so that it no longer can destroy tissues, but its antigens are still intact. When any of these vaccines is introduced into the body, the immune system actively develops clones, produces antibodies, and develops memory cells.

Passive Immunity Is Borrowed Immunity

In **passive immunity,** an individual is given antibodies actively produced by another organism. The serum or gamma globulin containing these antibodies can be obtained from humans or animals. Animal sera are less desirable because its nonhuman proteins can themselves act as antigens, stimulating an immune response that may result in a clinical illness termed *serum sickness.*

Passive immunity is borrowed immunity, and its effects are not lasting. It is used to boost the body's defense temporarily against a particular disease. For example, during the Vietnam War, in areas where hepatitis was widespread, soldiers received injections of gamma globulin containing antibodies specific to the hepatitis pathogen. Such injections of gamma globulin offer protection for only a few months. Because the body has not actively launched an immune response, it has no memory cells and cannot produce antibodies specific to the pathogen. Once the injected antibodies wear out, the immunity disappears.

Pregnant women confer natural passive immunity upon their developing babies by manufacturing antibodies for them. These maternal antibodies, of the IgG class, pass through the placenta (the organ of exchange between mother and developing child) and provide the fetus and newborn infant with a defense system until its own immune system matures. Babies who are breast-fed continue to receive immunoglobulins, particularly IgA, in their milk. These immunoglobulins provide considerable immunity to the pathogens responsible for gastrointestinal infection and perhaps to other pathogens as well.

Normally The Body Effectively Defends Itself Against Cancer

Some immunologists think that a few normal cells are transformed into cancer cells every day in each of us in response to viruses, hormones, radiation, or carcinogens in the environment. Because they are abnormal cells, some of their surface proteins are different from those of normal body cells. Such proteins act as antigens, stimulating an immune response. According to the **theory of immune surveillance,** the body's immune system destroys these abnormal cells whenever they arise. Only when these mechanisms fail do abnormal cells divide rapidly, resulting in cancer.

Every component of the immune system helps defend against cancer cells. Tumor cells exhibit abnormal surface antigens which induce both cellular and antibody-mediated responses. T cells produce interleukins which attract macrophages and NK cells and activate them. The T cells also produce interferons which have an antitumor effect. The macrophages themselves produce factors including tumor necrosis factor (TNF) which inhibit tumor growth.

Cytotoxic T cells, macrophages, and natural killer cells attack cancer cells (Fig. 37–10). Natural killer cells are capable of killing tumor cells or virally infected cells upon first exposure to the foreign antigen. Patients with advanced cancer are thought to have lower natural killer cell activity than healthy persons.

What prevents killer T cells, macrophages, and natural killer cells from effectively destroying cancer cells in some persons? The immune system cells may fail to recognize the cancer cells as foreign, or they may recognize them but be unable to destroy them. Sometimes the presence of cancer cells stimulates B cells to produce IgG antibodies that combine with antigens on the surfaces of the cancer cells. These **blocking anti-bodies** may block the T cells so that they are unable to adhere to the surface of the cancer cells and destroy them. For some unknown reason, the blocking antibodies are not able to activate the complement system which would destroy the cancer cells. Paradoxically, the presence of antibodies in this case is harmful.

An exciting approach in cancer research involves the production of **monoclonal antibodies.** In this procedure, mice or rabbits receive injections of antigens from human cancer cells. After the mice have produced antibodies to the cancer cells, their spleens are removed, and cells containing the antibodies are extracted from this tissue. These cells are fused with cancer cells from other mice. Because of the apparently unlimited ability of cancer cells to divide, these fused hybrid cells will continue to divide indefinitely. Researchers select hybrid cells that are manufacturing the specific antibodies needed and then clone them in a separate cell culture. Cells of this clone produce large amounts of the specific antibodies needed—hence the name monoclonal antibodies. Such antibodies can be injected into the very same cancer patients whose cancer cells were used to stimulate their production and are highly specific for destroying the cancer cells. (Monoclonal antibodies specific for a single antigenic determinant can now be produced.) In trial studies such antibodies are being tagged with toxic drugs that are then delivered specifically to the cancer cells.

Graft Rejection Is An Immune Response Against Transplanted Tissue

Skin can be successfully transplanted from one part of the body to another or between identical twins. However, when skin is taken from one individual and grafted onto the body of another, the skin graft is rejected and it sloughs off. Why?

Recall that tissues from the same individual or from identical twins have the identical HLA alleles and thus the same HLA antigens. Such tissues are compatible. Because its HLA antigens are the same, tissue transplanted from one location to another in the same individual is called an **autograft.**

Because there are several possible varieties of each of the HLA genes (multiple alleles), it is difficult to find identical matches among strangers. When a tissue or organ is taken from a donor and transplanted to the body of an unrelated host, several of the HLA antigens are likely to be different. Such a graft made between members of the same species is called a **homograft.** The host's immune system regards the graft as foreign and launches an effective immune response called **graft rejection.** T lymphocytes attack the transplanted tissue and can destroy it within a few days (Fig. 37–11).

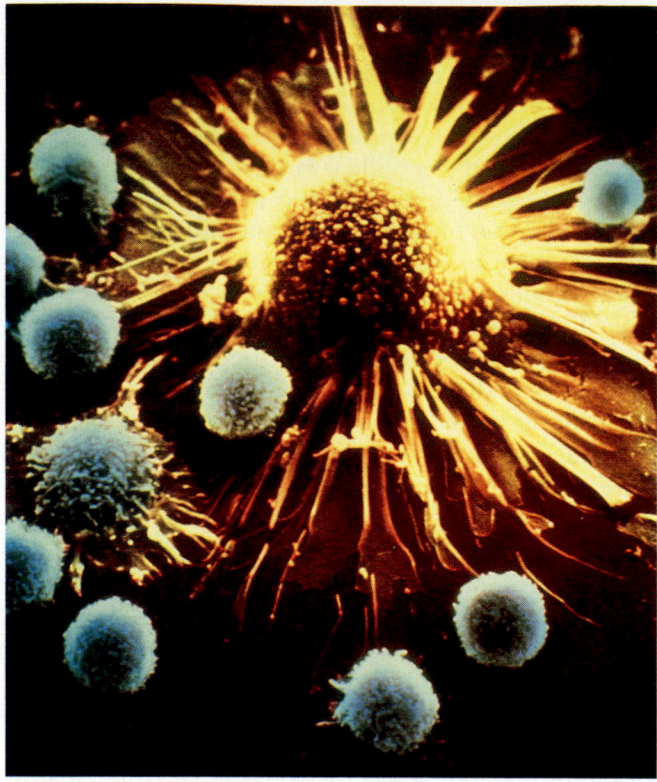

(a)

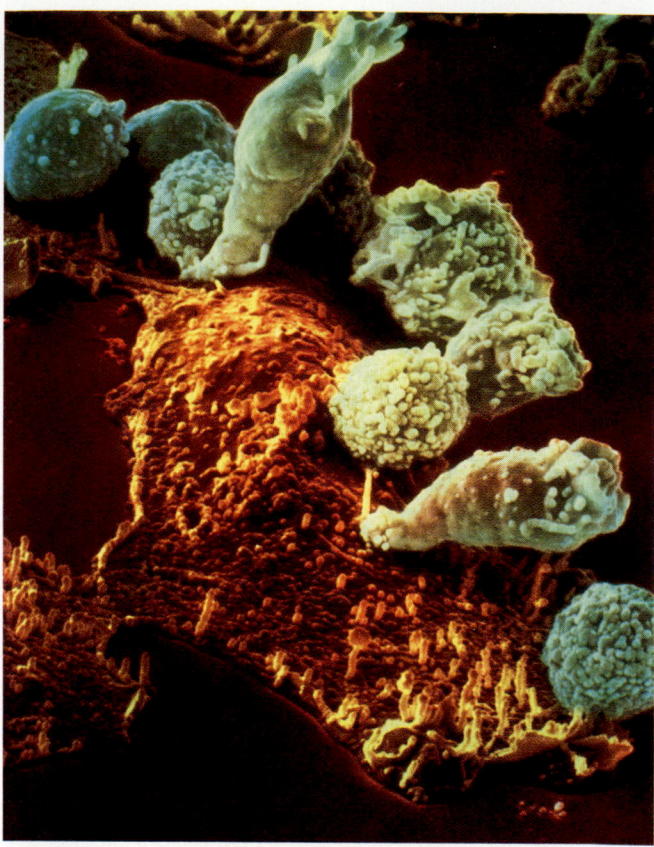

(b)

FIGURE 37–10 **Cytotoxic T cells defend the body against cancer cells.** (*a*) An army of cytotoxic T cells surround a large cancer cell. The T cells recognize the cancer cell as nonself because it displays altered antigens on its surface. (*b*) Some of the cytotoxic T cells elongate as they chemically attack the cancer cell, breaking down its cell membrane. (*c*) The cancer cell has been destroyed; only a collapsed fibrous cytoskeleton remains.

Before transplants are performed, tissues from the patient and from potential donors must be typed and matched as well as possible. Cell typing is somewhat similar to blood typing but more complex. The first obstacle is obtaining the typing sera. For ABO blood typing the sera is readily attainable because A-type individuals naturally have anti-B antibodies, B-type individuals have anti-A antibodies, and O-type individuals have anti-A and anti-B antibodies. In contrast, in order to obtain antibodies directed to antigens of the MHC, an individual must be actively immunized to the foreign proteins. Since some of these antigens are found on all nucleated cells, multiparous females (women who have experienced multiple births) inadvertently get immunized to the histocompatibility antigens of their offspring. Similar immunization occurs in patients undergoing multiple blood transfusions. The antibodies from these serum sources are purified, and the specific antibodies generated are used to tissue type.

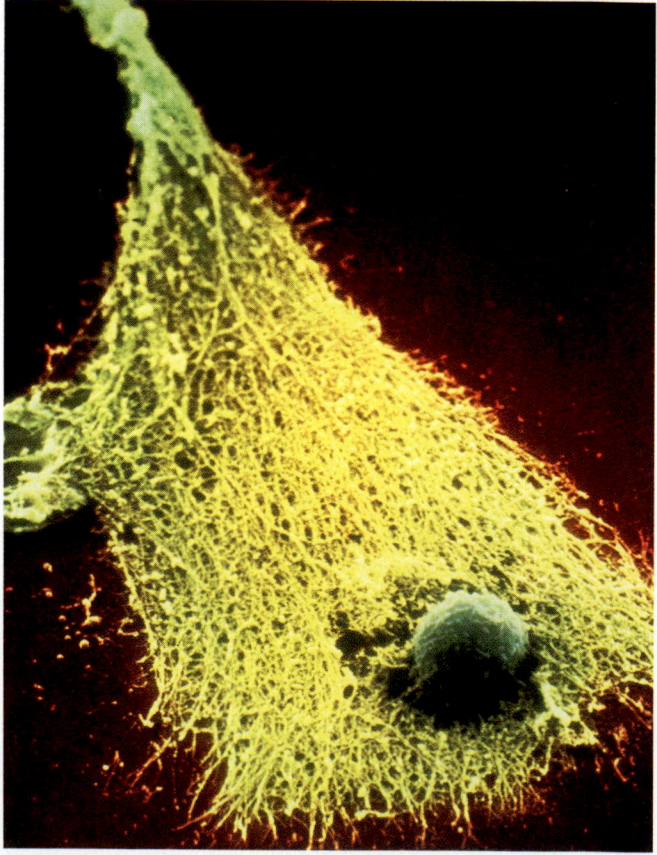

(c)

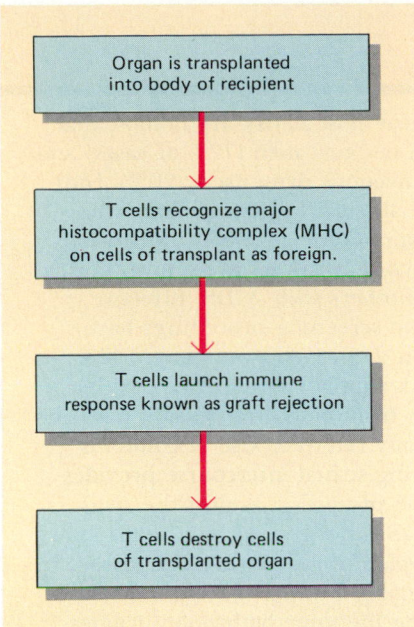

FIGURE 37–11   Graft rejection.

Serological typing is often made by a lymphocyte proliferation test which takes about 5 days. Thus, the results of a tissue match may not be known until after the organ has been transplanted. The information is still useful, however, because it gives the physician an idea of how serious the graft rejection may be and how to treat it. If all five of the HLA group of antigens are matched, the graft has about a 95% chance of surviving the first year. Unfortunately, not many persons are lucky enough to have an identical twin to supply spare parts, so perfect matches are difficult to find. Furthermore, some parts such as the heart cannot be spared. Most organs to be transplanted, therefore, are removed from unrelated donors, often from dying patients or from those who have just died.

To try to prevent graft rejection in less compatible matches, physicians use drugs and x-rays to destroy T lymphocytes. Most of these methods do not kill T lymphocytes selectively, however, so all types of lymphocytes are indiscriminately destroyed. Unfortunately, lymphocyte destruction suppresses not only graft rejection but other immune responses as well, so that many transplant patients succumb to pneumonia or other infections. In immunosuppressed patients there is also an increased incidence of certain types of tumor growths.

Cyclosporin A, an antibiotic extracted from fungi, appears to suppress T cells that have been activated by antigens on the graft but has little effect on B cells. Thus, the graft is not rejected and the patient can still resist infection. Virtually all organ graft recipients today are treated with cyclosporin A because of its significant effect on organ graft survival.

Certain Sites In The Body Are Immunologically Privileged

There are a few immunologically privileged locations in the body in which foreign tissues are accepted by a host. The brain is one such area. Cornea transplants are highly successful because the cornea has almost no blood or lymphatic vessels associated with it and so is out of reach of most lymphocytes. Furthermore, antigens in the corneal graft probably would not find their way into the circulatory system, and so would not stimulate an immune response. The uterus may be an immunologically privileged site. There the human fetus is able to develop its own biochemical identity in safety.

In An Autoimmune Disease The Body Attacks Its Own Tissues

Sometimes self-tolerance appears to break down and the body reacts immunologically against its own tissues, causing an **autoimmune disease.** Some of the diseases that result from such failures in self-tolerance are rheumatoid arthritis, multiple sclerosis, systemic lupus erythematosus (SLE), juvenile insulin-dependent diabetes, and perhaps infectious mononucleosis.

Myasthenia gravis is an autoimmune disease in which a circulating antibody combines with acetylcholine receptors in the motor endplates. This interaction blocks the receptors and can damage or even destroy them. Affected persons experience muscle weakness and respiratory muscles may be affected to a life-threatening extent.

What causes the production of abnormal antibodies in myasthenia gravis and in other autoimmune diseases? No one really knows. Some investigators have suggested genetic predisposition, perhaps involving HLA types; others speculate that prior damage to the tissue is involved. Some studies suggest that a previous viral infection in the involved tissue stimulated the body to manufacture antibodies against the infected cells. Then, after the virus has been destroyed, the body continues to manufacture harmful antibodies capable of attacking the body cells—even though they are no longer infected. A combination of these factors may be responsible.

Allergic Reactions Are Inappropriate Immune Responses

Allergy is a state of altered immune response that is harmful to the body. Allergic persons have a tendency to manufacture antibodies against mild antigens, called **allergens,** that do not stimulate a response in nonaller-

FOCUS ON AIDS

First recognized in 1981, **acquired immunodeficiency syndrome (AIDS)** is a deadly disease that is spreading through the population at an alarming rate. By mid-1988 more than 66,000 cases of AIDS had been reported in the United States; the U.S. Public Health Service estimates that by 1992 there will be about 365,000 AIDS cases. Worldwide, at least 5 million people were infected by the AIDS virus by 1989, and more than a million new cases of AIDS are expected within the next 5 years.

AIDS results from infection with a retrovirus identified as **human immunodeficiency virus (HIV).** The virus rapidly infects helper T cells, resulting in irreversible defects in immunity. Recall that helper T cells stimulate the proliferation of B and T cells, stimulate macrophages, and perform other functions that enhance the immune response. When the helper T cell population is depressed, the ability to resist infection is severely impaired and AIDS victims die within several months to about 5 years of rare forms of cancer, pneumonia, and other opportunistic infections that pose little threat to individuals with fully functioning immune systems.

About 90% of patients dying of AIDS develop a neuropsychological disorder known as AIDS dementia complex which results from direct infection of the central nervous system by the retrovirus. AIDS dementia complex is characterized by progressive cognitive, motor, and behavioral dysfunction that typically ends in coma and death.

Current evidence indicates that AIDS is transmitted mainly by semen during sexual intercourse with an infected person or by direct exposure to blood or blood products.

Those most at risk are homosexual and bisexual men (72% of cases) and intravenous drug users (20%). Individuals such as hemophiliacs who require frequent blood transfusions are also at risk, as are infants born to mothers with AIDS. Effective blood-screening procedures have been developed to safeguard blood bank supplies so that risk of infection from blood transfusion has been greatly reduced. Use of condoms during sexual intercourse provides some protection against the virus. AIDS is not spread by casual contact. People do not contract the disease by hugging, kissing, sharing a drink, or using the same bathroom facilities. Close friends and family members who live with AIDS patients are not more likely to get the disease.

HIV virus particles (blue) that cause AIDS attack a helper T cell (white). HIV seriously impairs the immune system by rapidly destroying helper T cells. Exposure to semen or blood that contains HIV can lead to AIDS-related complex (ARC) or to AIDS. Many factors apparently determine whether a person exposed to AIDS virus will develop the disease. The risk increases with multiple exposures.

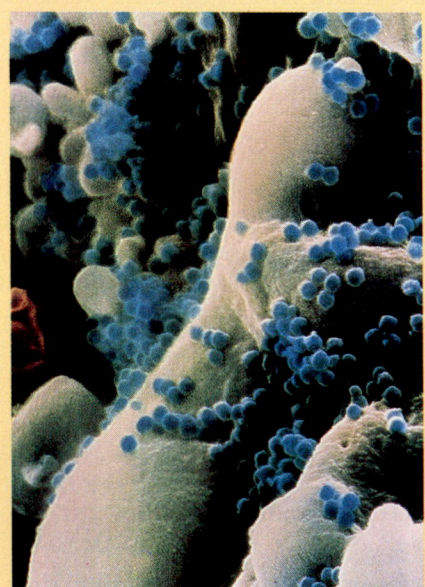

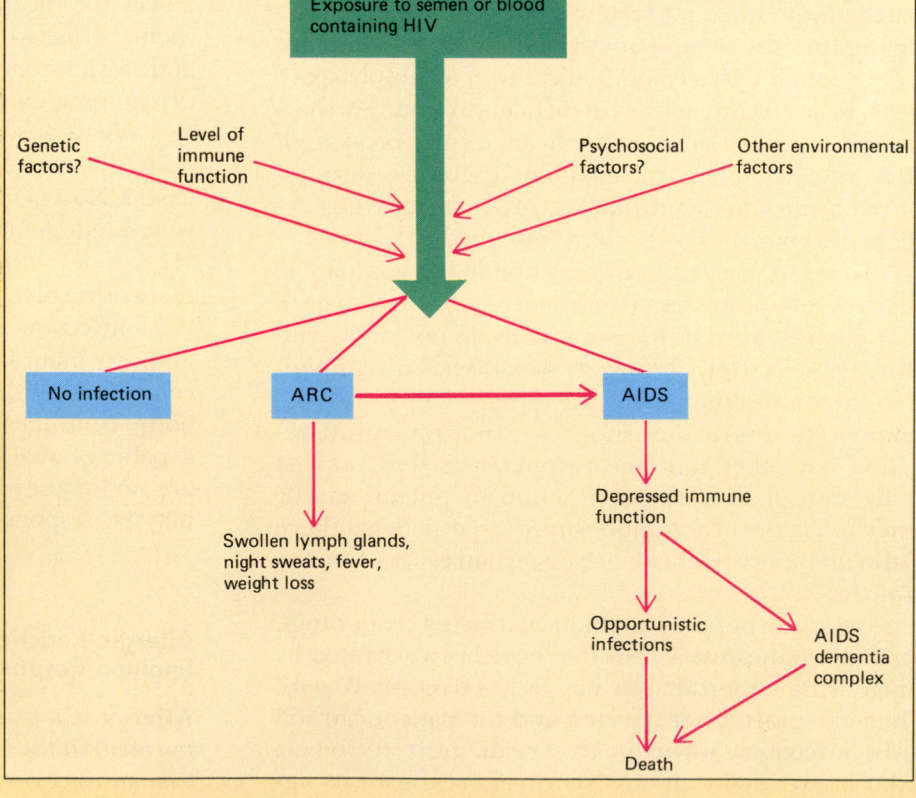

More than 2 million persons in the United States are estimated to have antibody to HIV but have no symptoms of the disease. Possibly not everyone exposed to HIV actually contracts the disease. AIDS may be an opportunistic infection that causes disease in individuals with inadequate immune function. Susceptibility to HIV may depend on a combination of genetic, environmental, and psychosocial factors. The latter include personality variables and coping styles that influence susceptibility to environmental stressors.

Many persons exposed to HIV develop the **AIDS-related complex (ARC);** its symptoms include night sweats, fever, swollen lymph glands, and weight loss. Patients with ARC may eventually develop AIDS, but the extent of that risk is not known.

Research laboratories throughout the world are searching for drugs that will successfully combat the AIDS virus. Because the AIDS virus often infects the central nervous system, an effective drug will have to cross the blood-brain barrier. One drug currently being tested, AZT (azidothymidine), blocks HIV replication. AZT blocks the action of reverse transcriptase, the enzyme needed by the retrovirus for incorporation into the host cell's DNA. Recent evidence suggests that AZT prolongs the onset of AIDS symptoms. Unfortunately, the virus has already developed strains that are resistant to AZT. Research directed at developing a vaccine against AIDS has not yet been successful because the retrovirus mutates rapidly, giving rise to many strains. Thus, a vaccine

that is effective today may not be effective tomorrow.

While immunologists work to develop a successful vaccine, massive educational programs are being developed aimed at slowing the spread of AIDS. Spreading the word that having multiple sexual partners increases the risk for AIDS and teaching sexually active individuals the importance of "safe" sex may help slow the epidemic. Some have suggested that public health facilities offer free condoms to those who are sexually active and free hypodermic needles to those addicted to drugs. The cost of these measures would be far less than the cost of medical care for increasing numbers of AIDS patients and the toll in human suffering.

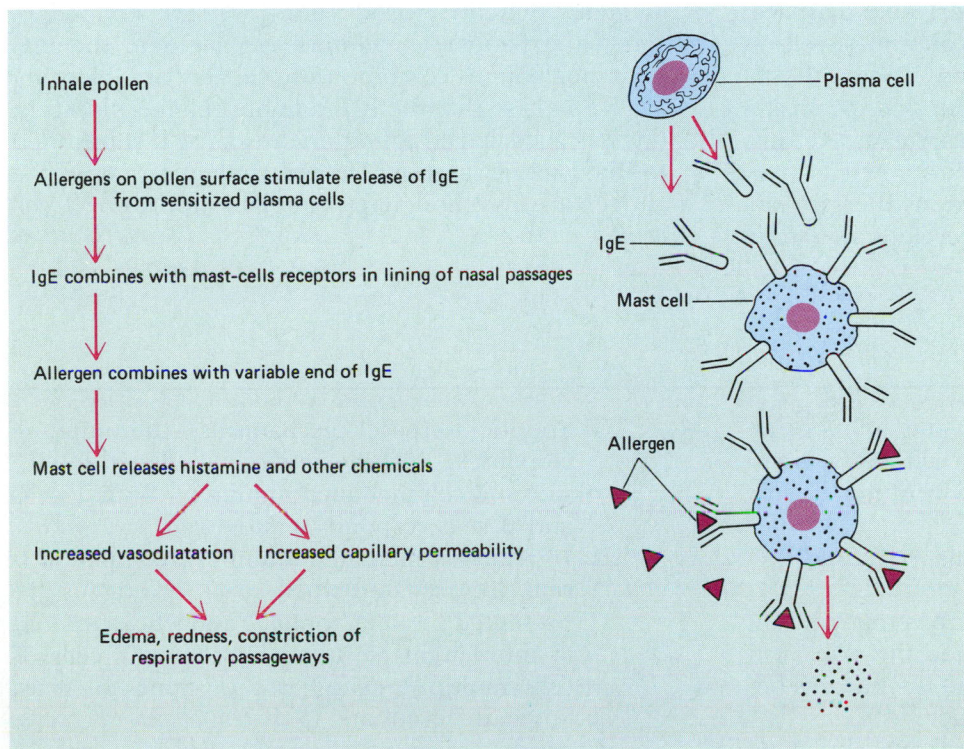

FIGURE 37–12 A common type of allergic response.

gic individuals. In many kinds of allergic reactions, distinctive IgE immunoglobulins are produced. About 15% of the population of the United States is plagued by an allergic disorder such as allergic asthma or hayfever. A tendency to these disorders appears to be inherited.

Let us examine a common allergic reaction—a hayfever response to ragweed pollen (Fig. 37–12). When an allergic person inhales the microscopic pollen, allergens stimulate the release of IgE from sensitized plasma cells in the nasal passages. The IgE attaches to receptors on the membranes of mast cells, large connective tissue cells filled with distinctive granules. Each mast cell has thousands of receptors to which the IgE may attach. Each IgE molecule attaches to a mast cell receptor by its C region end, leaving the V region end of the immunoglobulin free to combine with the ragweed pollen allergen.

When the allergen combines with IgE antibody, the mast cell rapidly releases its granules. When exposed to extracellular fluid, the granules release histamine, serotonin, and other chemicals that cause inflammation. These substances cause blood vessels to dilate and capillary permeability to increase, leading to edema and redness. Such physiological responses cause the victims' nasal passages to become swollen and irritated. Their noses run, they sneeze, their eyes water, and they feel generally uncomfortable.

In **allergic asthma,** an allergen-IgE response occurs in the bronchioles of the lungs. Mast cells release slow-reacting substance of anaphylaxis (SRS-A), which causes smooth muscle to constrict. The airways in the lungs sometimes constrict for several hours, making breathing difficult.

Certain foods or drugs act as allergens in some persons, causing a reaction in the walls of the gastrointestinal tract that leads to discomfort and diarrhea. The allergen may be absorbed and cause mast cells to release granules elsewhere in the body. When the allergen-IgE reaction takes place in the skin, the histamine released by mast cells causes the swollen red welts known as **hives.**

Systemic anaphylaxis is a dangerous allergic reaction that can occur when a person develops an allergy to a specific drug such as penicillin or to compounds in the venom injected by a stinging insect. Within minutes after the substance enters the body, a widespread allergic reaction takes place. Large amounts of histamine are released into the circulation, causing extreme vasodilatation and permeability. So much plasma may be lost from the blood that circulatory shock and death can occur within a few minutes.

The symptoms of allergic reactions are often treated with **antihistamines,** drugs that block the effects of histamines. These drugs compete for the same receptor sites on cells targeted by histamine. When the antihistamine combines with the receptor, it prevents the histamine from combining and thus prevents its harmful effects. Antihistamines are useful clinically in relieving the symptoms of hives and hayfever. They are not completely effective, however, because mast cells release substances other than histamines that also cause allergic symptoms.

In serious allergic disorders patients are sometimes given **desensitization therapy.** Very small amounts of the very antigen to which they are allergic are either injected or administered orally in the form of drops daily over a period of months or years. The antigen stimulates production of IgG antibodies. When the patient encounters the allergen, the IgG immunoglobulins combine with the allergen, blocking its receptors so that the IgE cannot combine with it. In this way a less harmful immune response is substituted for the allergic reaction. Desensitizing injections of the antigen are also thought to stimulate suppressor T cell activity.

■ CHAPTER SUMMARY

I. Internal defense depends upon the ability of an organism to distinguish between self and nonself.

II. Most invertebrates are capable only of nonspecific responses such as phagocytosis.

III. Vertebrates can launch both nonspecific and specific responses.

  A. Nonspecific defense mechanisms that prevent entrance of pathogens include the skin, acid secretions in the stomach, and the mucous lining of the respiratory passageways.

  B. Should pathogens succeed in breaking through the first line of defense, other nonspecific defense mechanisms are activated to destroy the invading pathogens.

   1. When pathogens invade tissues, they trigger an inflammatory response, which brings needed phagocytic cells and antibodies to the infected area.

   2. Neutrophils and macrophages phagocytize and destroy bacteria.

  C. Specific defense mechanisms—the immune responses—include antibody-mediated immunity and cell-mediated immunity. Both T cells and B cells respond to antigens.

  D. In antibody-mediated immunity, competent B cells are activated when specific antigens are presented by a macrophage and when exposed to interleukins secreted by helper T cells. B cells multiply, giving rise to clones of cells. Some differentiate to become plasma cells, which secrete specific antibodies.

  E. Antibodies are highly specific proteins called immunoglobulins. They are produced in response to specific antigens. Antibodies are grouped in five classes according to their structure.

  F. An antibody combines with a specific antigen to form an antigen-antibody complex, which may inactivate the pathogen, stimulate phagocytosis, or activate the complement system.

The complement system increases the inflammatory response and phagocytosis; some complement proteins digest portions of the pathogen cell.

G. In cell-mediated immunity, specific T cells are activated by the presence of specific antigens and by helper T cells; these activated T cells multiply, giving rise to a clone of cells.
1. Some T cells differentiate to become cytotoxic T cells, which migrate to the site of infection and chemically destroy pathogens.
2. Some sensitized T cells remain in the lymph nodes as memory cells; others become helper T cells or suppressor T cells.

H. Second exposure to an antigen evokes a secondary immune response, which is more rapid and more intense than the primary response.

I. Active immunity develops as a result of exposure to antigens; it may occur naturally after recovery from a disease or may be artificially induced by immunization. Passive immunity develops when an individual receives injections of antibodies produced from another person or animal, and is temporary.

J. According to the theory of immune surveillance, the immune system destroys abnormal cells whenever they arise; diseases such as cancer develop when this immune mechanism fails to operate effectively.

K. Transplanted tissues possess protein markers known as major histocompatibility complex that stimulate graft rejection, an immune response launched mainly by T cells that destroys the transplant.

L. In autoimmune diseases the body reacts immunologically against its own tissues.

M. In an allergic response, an allergen can stimulate production of IgE antibody, which combines with the receptors on mast cells; the mast cells then release histamine and other substances, causing inflammation and other symptoms of allergy.

POST-TEST

1. An antigen is a substance capable of stimulating a(n) _____.
2. Specific proteins produced in response to specific antigens are called _____.
3. When infected by viruses, some cells respond by producing proteins called _____.
4. The clinical characteristics of inflammation are _____, _____, _____, and _____.
5. T lymphocytes are thought to originate in the _____ _____; they are processed in the _____ and then proliferate in the _____ tissues.
6. _____ T cells inhibit immune responses.
7. When the body is invaded by the same pathogen a second time, the immune response can be launched more rapidly owing to the presence of _____ cells.
8. The cells that produce antibodies are _____ cells.
9. An antigenic determinant gives the antigen molecule a specific configuration that can be "recognized" by an _____.
10. The _____ confers immunological competence upon T cells.
11. The complement system is activated when an _____ - _____ complex is formed.
12. After a _____ ingests a pathogen, it may display a few molecules of antigen on its surface.
13. Although artificially induced, immunization is a form of _____ immunity.
14. An individual receiving injections of antibodies produced by another organism is receiving _____ immunity.
15. In humans the major histocompatibility complex is called the _____ group.
16. An autograft consists of tissue transplanted from _____.
17. In graft rejection the host launches an effective _____ against _____ tissue.
18. Cornea transplants are highly successful because the cornea is an immunologically _____ site.
19. An _____ is a mild antigen that does not stimulate a response in an individual who is not _____.
20. In a typical allergic reaction mast cells secrete _____ and other compounds that cause _____.

REVIEW QUESTIONS

1. How does the body distinguish between self and nonself? Are invertebrates capable of making this distinction?
2. Contrast specific and nonspecific defense mechanisms. Which type confronts invading pathogens immediately? How do the two systems work together?
3. How does inflammation help to restore homeostasis?

4. Give two specific ways in which cell-mediated and antibody-mediated immune responses are similar and two ways in which they are different.

5. Describe three ways in which antibodies work to destroy pathogens.

6. John is immunized against measles. Jack contracts measles from a playmate in nursery school before his mother gets around to having him immunized. Compare the immune responses in each child. Five years later, John and Jack are playing together when Judy, who is coming down with measles, sneezes on both of them. Compare the response in Jack and John.

7. Why is passive immunity temporary?

8. What is graft rejection? What is the immunological basis for it?

9. List the immunological events that take place in a common type of allergic reaction such as hayfever.

10. Explain the theory of immune surveillance. What happens when immune surveillance fails?

11. What is an autoimmune disease? Give two examples.

12. What public policy decisions would you recommend that might help slow the spread of AIDS?

▪ RECOMMENDED READINGS

Bolotin, C.: "Drug as Hero," *Science* 85:68–71, 1985. The story of cyclosporin, a drug that selectively inhibits the rejection of transplanted organs.

Buisseret, P.D.: "Allergy," *Scientific American* 249(2):86–95, 1982. A discussion of the cellular and biochemical changes that occur during an allergic response.

Cohen, I.R.: "The Self, the World and Autoimmunity," *Scientific American* 258 (4):52–60, 1988. Self-recognition is important for health as well as for certain diseases.

Darnell, J., H. Lodish, and D. Baltimore: *Molecular Cell Biology.* New York, Scientific American Books, 1986. Chapter 24 clearly presents the basics of immunology.

Edelson, R.L., and J.M. Fink: "The Immunologic Function of Skin," *Scientific American* 252 (6):46–53, 1985. Specialized cells in the skin play interacting roles in the response to foreign invaders.

Gallo, R.C.: "The AIDS Virus," *Scientific American* 256 (1): 46–56, 1987. An overview of the discovery of the AIDS virus and a description of the pathogen.

Jaroff, L.: "Stop That Germ," *TIME* May 23, 1988, 56–64. An excellent summary of immune function, with a focus on recent discoveries in immunology.

Kennedy, R.C., J.L. Melnick, and G.R. Dressman: "Antiidiotypes and Immunity," *Scientific American,* 255 (1): 48–52, 1986.

Laurence, J.: "The Immune System in AIDS," *Scientific American,* December 1985. A discussion of the effects of AIDS on the immune system and of how we may be able to stop this disease.

Lerner, R.A.: "Synthetic Vaccines," *Scientific American,* 250 (2): 66–74, 1983. A report on experiments on the preparation of synthetic vaccines.

Marrack, P., and J. Kappler: "The T Cell and Its Receptor," *Scientific American,* 254 (2), February 1986. The surface proteins on T cells are a vital component of cell-mediated immunity.

Marx, J.L.: "Antibodies: Getting Their Genes Together," *Science,* 212:1015–1017, 1981. A brief review of studies on the rearrangement of genes to provide diverse antibodies.

Marx, J.L.: "Monoclonal Antibodies in Cancer," *Science,* 216: 283–285, 1982. An interesting account of the clinical use of monoclonal antibodies.

Old, L.J.: "Tumor Necrosis Factor," *Scientific American,* 258 (5): 59–75, 1988. This protein, first identified because of its anticancer activity, has been found to help regulate inflammation and immunity.

Tizard, I.R.: *Immunology: An Introduction,* 2nd ed. Philadelphia, Saunders College Publishing, 1988. A basic introduction to immunology.

Tonegawa, S.: "The Molecules of the Immune System," *Scientific American,* 253 (4), October 1985. A summary of the molecules and processes in immune response.

Weiss, R.: "Improving the AIDS Test," *Science News,* April 2, 1988, 218–221. Genetic engineering is being used for AIDS-antibody testing.

Science, Vol. 238, November 20, 1987 issue. Frontiers in Biology: Immunology. Six articles in this issue are devoted to new developments in immunology, including articles on the T cell receptor and on development of the antibody repertoire.

Scientific American, 259 (3), October 1988. A single-topic issue including ten articles on AIDS.

38

Gas Exchange

OUTLINE

I. Respiratory structures are adapted for gas exchange in air or water
II. Animals have evolved several different adaptations for gas exchange
  A. The body surface may be adapted for gas exchange
  B. Tracheal tubes are an adaptation for gas exchange in arthropods
  C. Many aquatic animals exchange gases through gills
  D. Terrestrial vertebrates exchange gases through lungs
III. The human respiratory system is typical of air-breathing vertebrates
  A. The airway conducts air into the lungs
  B. Gas exchange occurs in the lungs
  C. Ventilation is accomplished by breathing
  D. Breathing is regulated by respiratory centers in the brain
  E. Gas exchange takes place in the air sacs
  F. Oxygen is transported in combination with hemoglobin
  G. Carbon dioxide is transported mainly as bicarbonate ions
  H. Physiological adaptation to changes in pressure takes time
IV. The effect of breathing dirty air is respiratory insult
  A. A variety of defense mechanisms protect the lungs
  B. Continued respiratory insult leads to respiratory disease
Focus on choking
Focus on cardiopulmonary resuscitation (CPR)
Focus on facts about smoking

LEARNING OBJECTIVES

After you have studied this chapter you should be able to:

1. Compare the advantages and disadvantages of gas exchange in air with those in water.
2. Compare various adaptations for gas exchange, including tracheal tubes, gills, lungs, and the body surface.
3. Trace the route traveled by a breath of air through the human respiratory system from nose to air sacs and, finally, to recipient cells.
4. Summarize the mechanics of breathing and regulation of breathing.
5. Compare the composition of exhaled air with that of inhaled air, and describe the exchange of oxygen and carbon dioxide in the lungs and tissues.
6. Summarize the mechanisms by which oxygen and carbon dioxide are transported in the blood.
7. Describe the defense mechanisms that protect the lungs and the effects of breathing polluted air on the respiratory system.

Gills of fish. (Fred Hossler/Visuals Unlimited)

682

Gas exchange between the animal and the environment is known as **respiration.** During **organismic respiration** oxygen from the environment is taken up by the animal and delivered to its individual cells, and carbon dioxide is released into the environment. Gases move in and out of cells by diffusion. The oxygen supplied to the cells by organismic respiration is used in cellular respiration. Recall from Chapter 8 that **cellular respiration** is the complex series of reactions by which cells break down fuel molecules (e.g., glucose), releasing carbon dioxide and energy. In this process oxygen is present as the final electron acceptor in the electron transport system.

KEY CONCEPTS

- Gases move in and out of cells by diffusion.
- Adaptations for gas exchange include the body surface, gills, tracheal tubes, and lungs.
- In humans and other mammals the respiratory system consists of a series of air passageways that branch into smaller and smaller tubes, ending in the alveoli within the lungs. Gas exchange takes place through the thin walls of the alveoli.
- In humans and other mammals oxygen diffuses from the alveoli into the blood in the pulmonary capillaries and is transported to the cells in the form of oxyhemoglobin. Carbon dioxide, transported mainly in the form of bicarbonate ions, diffuses from blood in the pulmonary capillaries into the alveoli and is expired.

RESPIRATORY STRUCTURES ARE ADAPTED FOR GAS EXCHANGE IN AIR OR WATER

Animals are specifically adapted to exchange gases in air or water. Some respiratory structures, like tracheal tubes and lungs, are best adapted for gas exchange in air, whereas others, like gills, function best in water. However, even in structures adapted for gas exchange in air, the actual exchange of gases takes place across a moist surface.

In one way a watery medium is optimal for gas exchange because gas molecules must be dissolved in water to pass through plasma membranes. However, water has a much greater density and viscosity (resistance to flow) than air, so a large animal must expend more energy to move water over its respiratory surface than to move air. A fish uses up to 20% of its total energy expenditure to perform the muscular work needed to move water over its gills. An air-breather expends much less energy, only 1% or 2% of the total, to move air in and out of its lungs.

Gas exchange in air has certain advantages over gas exchange in water. Compared with water, air contains far more oxygen. Oxygen also diffuses much faster through air than through water. Moreover, air is not salty, so air-breathers do not have to cope with the diffusion of ions into their body fluids along with their oxygen. They have an easier time maintaining homeostasis with respect to ion composition.

On the other hand, organisms that respire in air struggle continuously with water loss. Air dwellers must have adaptations to help them avoid drying out, and their respiratory surfaces must be kept moist so oxygen and carbon dioxide can pass through the plasma membranes. In addition to having fairly impermeable skin, the lungs of air-breathing vertebrates are located deep within the body, not exposed like gills. This arrangement prevents excessive loss of water from the respiratory surface. Air must pass through a long sequence of passageways before reaching the blood-rich, wet respiratory surfaces of the lung, and expired air must again pass through these passageways before leaving the body. The lungs are thus partially protected from the drying effects of air.

ANIMALS HAVE EVOLVED SEVERAL DIFFERENT ADAPTATIONS FOR GAS EXCHANGE

Gases move in and out of cells by diffusion. The air or water supplying the oxygen must be continuously renewed so that as soon as oxygen is used up, more will be available. For this reason animals carry on **ventilation,** that is, they actively move their air or water supply over their respiratory surfaces. Sponges do this by setting up a current of water through the channels of their bodies by means of flagella; most fish gulp water and then actively pass it over their gills; terrestrial vertebrates breathe air.

Gas exchange is a fairly simple process in small, aquatic animals, such as sponges, hydras, and flatworms. Dissolved oxygen from the surrounding water diffuses directly into the cells, and carbon dioxide diffuses out of the cells and into the water. In large, complex animals, cells deep within the body cannot efficiently exchange gases directly with the environment because oxygen cannot diffuse rapidly enough through tissues to reach all the cells. Thus, specialized respiratory structures are required.

Specialized respiratory structures have thin walls that facilitate diffusion. The respiratory surface must be kept moist so that oxygen and carbon dioxide can be dissolved in water. And they are generally richly sup-

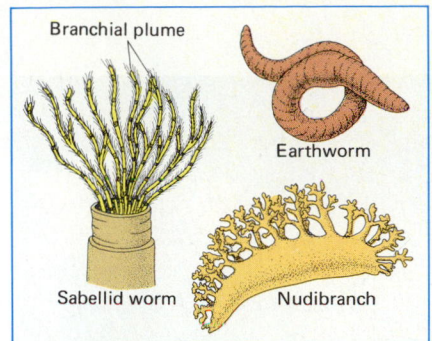

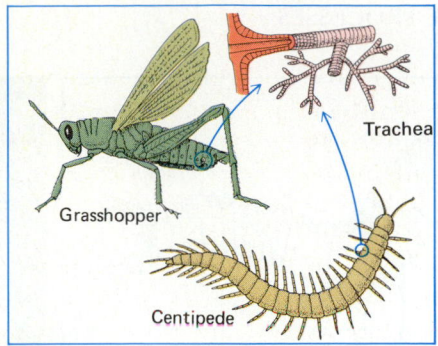

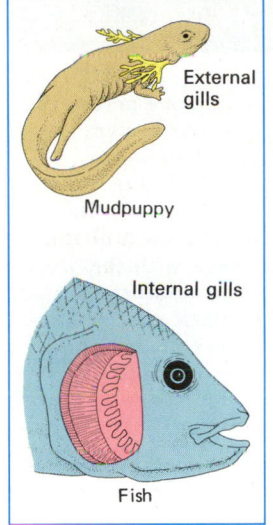

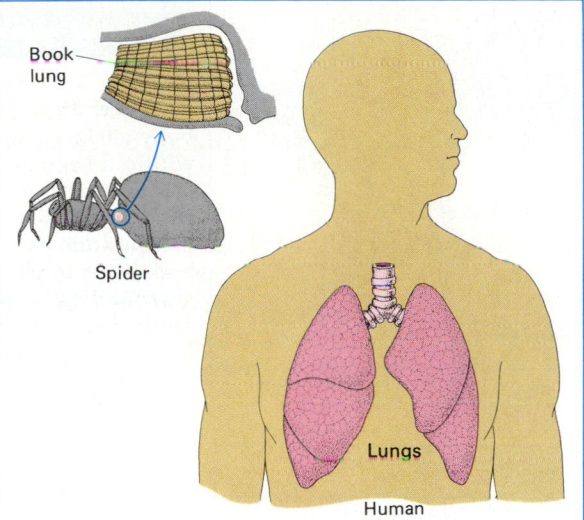

FIGURE 38–1   Principal types of respiratory structures found in animals.

plied with blood vessels allowing transport of respiratory gases. Four main types of respiratory surfaces used by animals are the body surface, tracheal tubes, gills, and lungs (Fig. 38–1).

The Body Surface May Be Adapted For Gas Exchange

Gas exchange occurs through the entire body surface in some animals, including some mollusks, most annelids, small arthropods, and a few vertebrates (Fig. 38–2). All of these animals are small with a high surface-to-volume ratio. They also have a low metabolic rate so smaller quantities of oxygen are used per cell. In aquatic animals the body surface is kept moist by the surrounding water. In terrestrial animals, the body secretes fluids that keep its surface moist.

How does an animal such as the earthworm exchange gases across its body surface? Gland cells in the epidermis secrete mucus, which keeps the body surface moist and also protects it. Oxygen, present in air pockets in the loose soil that the earthworm inhabits, dissolves in the mucus. Then, the oxygen diffuses through the body wall. Oxygen diffuses into blood circulating in a network of capillaries just beneath the outer cell layer. Carbon dioxide is transported by the

blood to the body surface, from which it diffuses out into the environment.

Tracheal Tubes Are An Adaptation For Gas Exchange In Arthropods

In insects and some other arthropods, the respiratory system consists of a network of **tracheal tubes.** Air enters the tracheal tubes through a series of up to 20 tiny openings called **spiracles** along the body surface (Fig. 38–3). In large or active insects air moves in and out of the spiracles by movements of the body or by rhythmic movements of the tracheal tubes.

Once inside the body, the air passes through the branching tracheal tubes, which extend to all parts of the body. The tracheal tubes terminate in microscopic, fluid-filled tracheoles. Gases are exchanged between the body cells and the fluid in these tracheoles.

Many Aquatic Animals Exchange Gases Through Gills

Gills are respiratory structures found mainly in aquatic animals. Gills are supported in water but tend to collapse in air. They are moist, thin structures that ex-

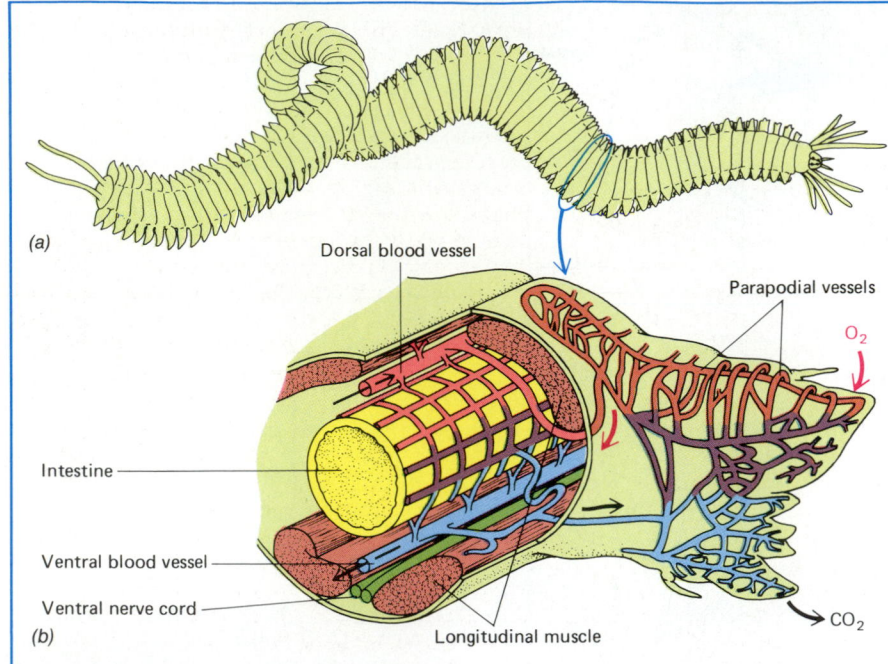

(a)

Dorsal blood vessel

Parapodial vessels

Intestine

O_2

Ventral blood vessel

Ventral nerve cord

CO_2

(b)

Longitudinal muscle

(c)

FIGURE 38–2   Gas exchange across the body surface. (a, b) Vascular system within a segment of the clam worm, *Nereis virens*. Arrows indicate the direction of blood flow. The limb-like parapodium acts as an extension of the body wall in gas exchange with the surrounding water. (c) In nudibranch mollusks, such as this black speckled sea lemon, gas exchange takes place across the body wall, as well as through the gills. (c, Chuck Davis)

(a)

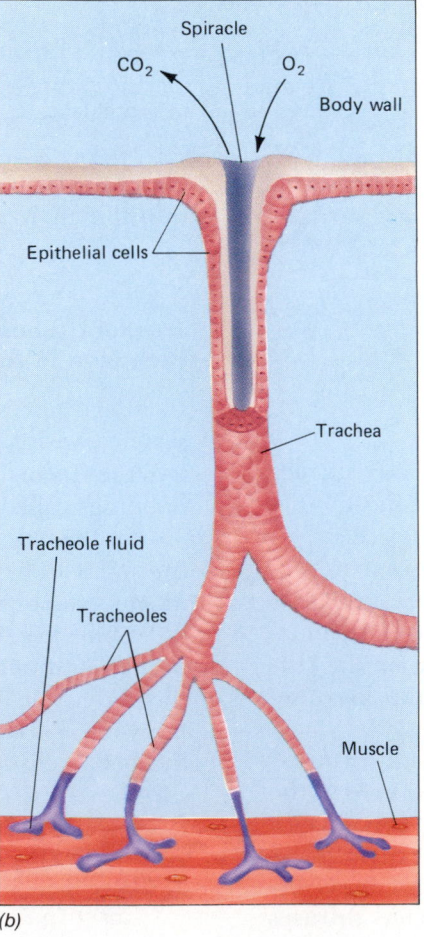

Spiracle

CO_2 O_2

Body wall

Epithelial cells

Trachea

Tracheole fluid

Tracheoles

Muscle

(b)

FIGURE 38–3   Tracheal tubes. (a) A scanning electron micrograph of a mole cricket trachea (approximately ×1300). The corrugations are a very long spiral around the tube, which may strengthen the tracheal wall somewhat as does the spring within the plastic hoses of many vacuum cleaners and hair dryers. The tracheal wall is composed of chitin. (b) Each tracheal tube and its branches conduct oxygen to the body cells of the insect. (a, Courtesy of Dr. James L. Nation and *Stain Technology*, Vol. 58, 1983)

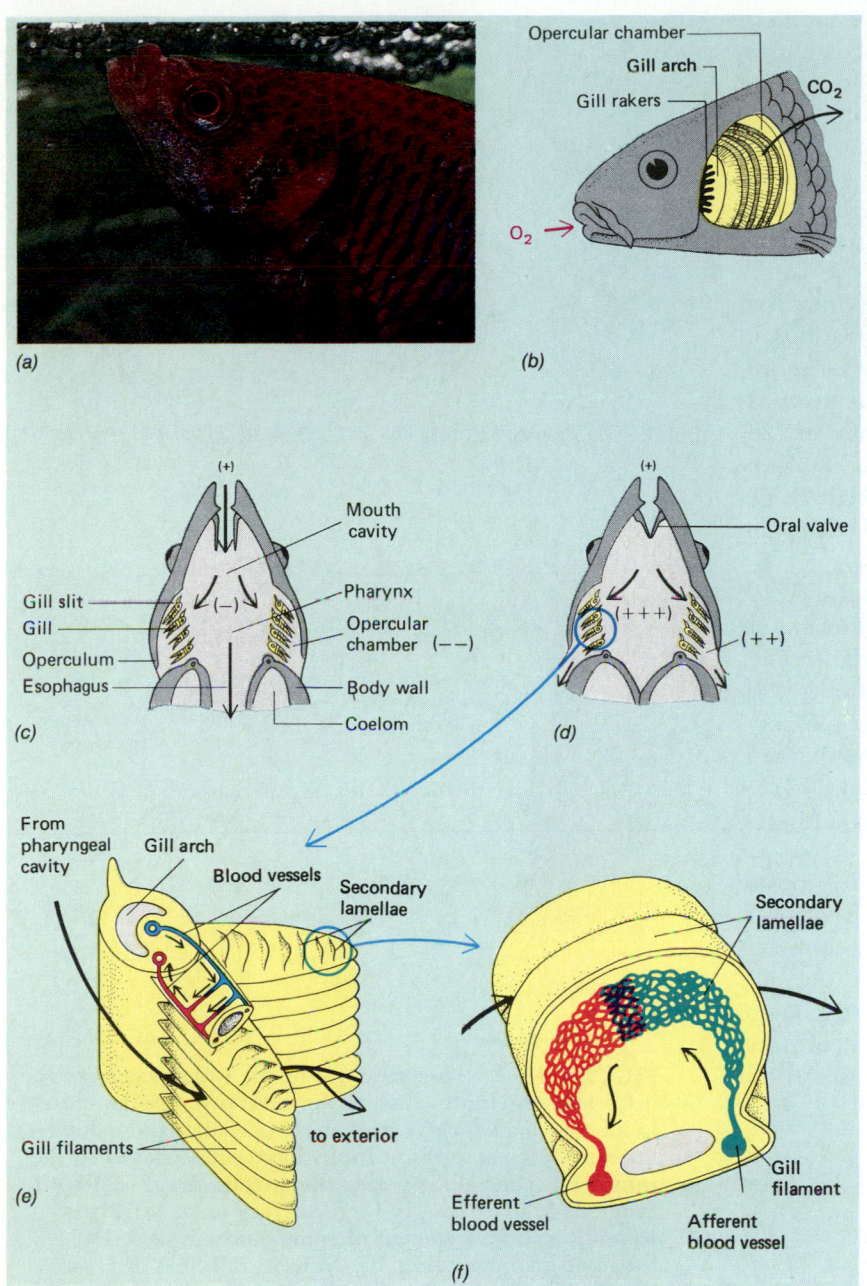

FIGURE 38–4 How the fish gill works. (*a*) A fish's gills are located under a plate, the operculum, which has been removed in this side view. They occupy the opercular chamber and form the lateral wall of the pharyngeal cavity. (*b*) The fish respires by movements of its mouth and pharyngeal cavity. Initially, in inhalation, it enlarges its pharyngeal cavity, seen here from the top. This produces a negative (−) pressure in the pharyngeal cavity, and water enters through the mouth. (*c*) When the fish exhales, it closes the mouth and the oral valve behind the lips so that water is forced between the gills and out. Muscular action produces the high positive (+++) pressure in the pharyngeal cavity that is necessary to force the water between the gills. (*d*) Each gill consists of a cartilaginous gill arch to which two rows of leaf-like gill filaments are attached. Blood circulates through the gill filaments as water passes among them. (*e*) Each gill filament has many even smaller extensions called secondary lamellae. These contain capillaries full of initially deoxygenated blood. The blood flows through the capillaries in a direction *opposite* to that taken by the water. In this example of countercurrent exchange, the blood is charged with oxygen very efficiently.

tend from the body surface. In many animals the outer surface of the gills is exposed to water, and the inner side is in close contact with networks of blood vessels.

Sea stars and sea urchins have simple **dermal gills,** which project from the body wall. Their ciliated epidermal cells ventilate the gills by beating a stream of water over them. Gases are exchanged through the gills between the water and the coelomic fluid inside the body.

In mollusks the gills are folded, providing a large surface for respiration. In bivalve mollusks (e.g., clams) and simple chordates the gills are also adapted for trapping and sorting food. Rhythmic beating of cilia draws water over the gill area, and food is filtered

out of the water at the same time as gases are exchanged. In mollusks gas exchange also takes place through the mantle.

In chordates the gills are usually located internally. A series of slits perforate the pharynx, and the gills are located along the edges of these gill slits. In bony fish, the fragile gills are protected by an external bony plate, the **operculum.** Movements of the operculum help to pump oxygenated water in through the mouth. The water flows over the gills and then exits through the gill slits.

Each gill in the bony fish consists of many **filaments,** which provide an extensive surface for gas exchange (Fig. 38–4). The filaments extend out into the water that continuously flows over them. A capillary network delivers blood to the gill filaments, facilitating diffusion of oxygen and carbon dioxide between blood and water. The very impressive efficiency of this system depends on the flow of blood in a direction opposite to the movement of the water. This arrangement, referred to as a **countercurrent exchange system,** maximizes the difference in oxygen concentration between blood and water.

If blood and water flowed in the *same* direction, the difference between the oxygen concentrations in blood (low) and water (high) would be very large initially and very small at the end. The oxygen concentration in the water would decrease as the oxygen concentration in the blood increased. When the concentrations of oxygen in the two fluids became equal, net diffusion of oxygen would stop. A great deal of oxygen would remain in the water.

In the countercurrent exchange system, however, blood low in oxygen comes in contact with water that is partly oxygen-depleted. Then, as the blood becomes more and more oxygen-rich, it comes in contact with water with a progressively higher concentration of oxygen. In this way a high rate of diffusion is maintained. As a result a very high percentage (more than 80%) of the available oxygen in the water diffuses into the blood.

Oxygen and carbon dioxide diffuse in opposite directions at the same time. This is because oxygen is more concentrated outside the gills than within, but carbon dioxide is more concentrated inside the gills than outside. Thus the same countercurrent exchange mechanism that results in efficient inflow of oxygen also results in equally efficient outflow of carbon dioxide.

Terrestrial Vertebrates Exchange Gases Through Lungs

Lungs are respiratory structures that develop as ingrowths of the body surface or from the wall of a body cavity, such as the pharynx. Arachnids and some small mollusks (particularly terrestrial snails and slugs) have lungs that depend almost entirely on diffusion for gas exchange. Larger mollusks and vertebrates with lungs have some means of forcefully changing the air that is in contact with the lung surface, that is, of ventilating the lung.

The **book lungs** of some spiders are enclosed in an inpocketing of the abdominal wall. These lungs consist of a series of parallel, thin plates (like the pages of a book) filled with blood. The plates are separated by air spaces that are connected to the outside environment through a spiracle.

Not all fish breathe exclusively by gills. African lungfish use lungs as well as gills. Amphibians depend

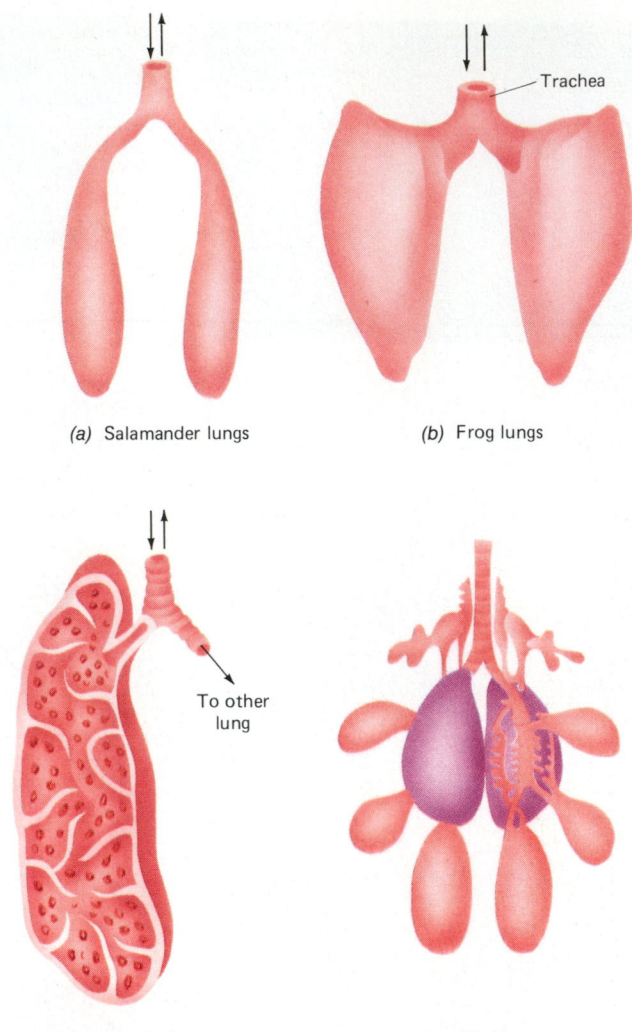

FIGURE 38–5 **The structure of the lungs varies in the different vertebrate classes. Note the progressive increase in surface area for gas exchange. (*a*) Salamander lungs are simple sacs. (*b*) Frog lungs have small ridges in the lung wall that help increase the surface area. (*c*) Reptile lungs have increasingly large surface area. (*d*) Birds have an elaborate system of lungs and air sacs. The lungs of mammals (Fig. 38–6) have millions of air sacs that increase the surface available for gas exchange.**

mainly on their body surface for gas exchange, but they do have simple lungs. The lungs of salamanders are two long, simple sacs, covered on the outside by capillaries (Fig. 38–5). Frogs and toads have ridges containing connective tissue on the inside of the lung, which increase the respiratory surface somewhat.

In reptiles, birds, and mammals, lungs are the principal respiratory surface. The lungs of most reptiles are rather simple, with only some folding that increases the surface for gas exchange. In some lizards and in turtles and crocodiles, the lungs have many subdivisions that increase the surface area for gas exchange.

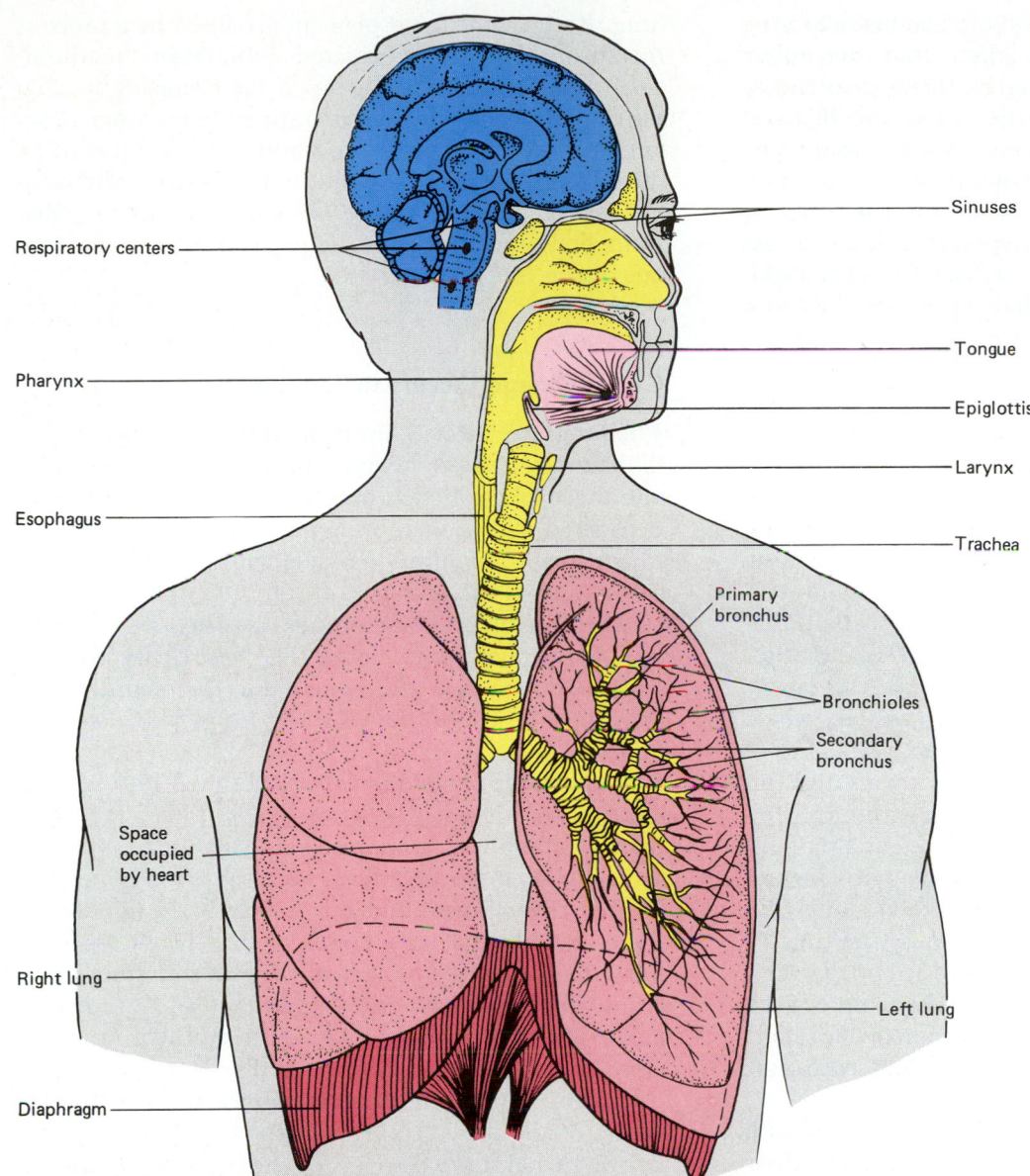

Respiratory centers

Pharynx

Esophagus

Space occupied by heart

Right lung

Diaphragm

Sinuses

Tongue

Epiglottis

Larynx

Trachea

Primary bronchus

Bronchioles

Secondary bronchus

Left lung

FIGURE 38–6 The human respiratory system. The paired lungs are located in the thoracic cavity. The muscular diaphragm forms the floor of the thoracic cavity, separating it from the abdominal cavity below. An internal view of one lung illustrates its extensive system of air passageways. The microscopic alveoli are shown in later figures.

Birds are very active animals with high metabolic rates. They require large amounts of oxygen, and they have highly effective respiratory systems. In birds the lungs have developed several extensions (usually nine) called **air sacs,** which reach into all parts of the body and even penetrate into some of the bones. The respiratory system is arranged so that air flows in one direction through the lungs and is renewed during each inspiration. The lungs have tiny, thin-walled ducts, the **parabronchi,** which are open at both ends. Gas exchange takes place across the walls of these ducts. The direction of blood flow in the lungs is opposite that of air flow through the parabronchi. This countercurrent flow increases the amount of oxygen that enters the lung.

The lungs of mammals are very complex and have an enormous surface area. In the following sections we will examine the human respiratory system in some detail.

THE HUMAN RESPIRATORY SYSTEM IS TYPICAL OF AIR-BREATHING VERTEBRATES

The respiratory system in humans and other air-breathing vertebrates consists of a series of tubes through which air passes on its journey from the nostrils to the air sacs of the lungs and back. A breath of air enters the body through the **nose,** flows through the twin compartments of the nasal cavity to the **pharynx** (throat region), through the **larynx** (voice box), and into the **trachea,** or windpipe (Fig. 38–6). From the trachea the stream of air divides, first into two streams as the trachea branches into the **bronchi** (one bronchus enters each lung), then into the many branches of the bronchi. These branches give rise to thousands of **bronchioles.** As these, too, divide again and then again, the breath becomes a thousand tiny breezes that finally reach the microscopic air sacs, or **alveoli.** From them oxygen diffuses into the blood of the multitude of capillaries enveloping each air sac. At the same time carbon dioxide diffuses from the capillaries into the air sacs and is exhaled through the respiratory tubes.

The Airway Conducts Air Into The Lungs

Air finds its way into the pharynx whether one breathes through the nose or mouth. Nose-breathing is more desirable because as air passes through the nose, it is filtered, moistened, and brought to body temperature. The nostrils are fringed with coarse hair that helps to prevent the entrance of small insects and other foreign matter.

The nostrils open into the **nasal cavities,** which are lined with moist, ciliated epithelium. The lining of the nose has a rich blood supply that warms and humidifies the incoming air. Mucous cells within the epithelium produce more than a pint of mucus a day. Inhaled dirt, bacteria, and other foreign particles are trapped in the layer of mucus and pushed along with the stream of mucus toward the throat by the cilia. In this way foreign particles are delivered to the digestive system, which is far more capable of disposing of such materials than the delicate lungs. A person normally swallows more than a pint of nasal mucus each day, more if he or she has an allergy or infection.

The back of the nasal cavities is continuous with the throat region, or pharynx. An opening in the floor of the pharynx leads into the larynx, sometimes called the "Adam's apple." Because the larynx contains the vocal cords, it is also referred to as the voice box. Cartilage embedded in its wall prevents the larynx from collapsing and makes it hard to the touch when felt through the neck.

During swallowing, a flap of tissue, the **epiglottis,** automatically closes off the larynx from the esophagus so that neither food nor liquid can enter the lower airway. Should this defense mechanism fail and foreign matter come in contact with the sensitive larynx, a **cough reflex** is initiated, expelling the material from the respiratory system. Despite these mechanisms, choking sometimes occurs (see Focus on Choking).

From the larynx air passes into the **trachea.** Like the larynx, the trachea is kept from collapsing by C-shaped rings of cartilage in its wall. The trachea divides into two branches, the bronchi, one going to each lung. Both trachea and bronchi are lined by a mucous membrane containing ciliated cells. Many medium-sized particles that have escaped the cleansing mechanisms of nose and larynx are trapped here. Mucus containing these particles is constantly beaten upward by the cilia to the pharynx, where it is periodically swallowed. This mechanism, functioning as a cilia-propelled mucus elevator, helps keep foreign material out of the lungs.

Gas Exchange Occurs In The Lungs

The lungs are large, paired, spongy organs occupying the thoracic (chest) cavity. The right lung is divided into three lobes, the left lung into two lobes. Each lung is covered with a membrane, the **pleural membrane,** which forms a continuous sac enclosing the lung and continuing as the lining of the chest cavity. The space between the pleural membranes covering the lung and the pleural membrane lining the chest cavity is called the **pleural cavity.** A film of fluid in the pleural cavity provides lubrication between the lungs and the chest wall.

Inside the lungs the bronchi branch into smaller and smaller airways, the bronchioles. There are more than a million tiny bronchioles in each lung, and each leads into a cluster of tiny air sacs, the alveoli (Fig. 38–7). The alveoli are lined by an extremely thin single layer of epithelial cells. Gases diffuse freely through the wall of the alveolus and into the surrounding capillaries. Thus, only two thin membranes separate the air in the alveolus from the blood: the epithelium of the alveolar wall and the capillary wall.

Because the lung consists largely of air tubes and elastic tissue, it is a spongy, elastic organ with a very large internal surface area for gas exchange. In normal adults the surface area of the lung is estimated as approximately that of a tennis court.

Ventilation Is Accomplished By Breathing

Breathing is the mechanical process of moving air from the environment into the lungs and of expelling air from the lungs. Inhaling air is referred to as **inspiration;** exhaling air is **expiration.** A resting adult breathes about 12 times each minute.

The chest cavity is closed so no air can enter except through the trachea. It has a muscular floor, the **diaphragm,** and two lateral **pleural cavities,** each containing a lung. During inspiration, the chest cavity is expanded by the contraction of the diaphragm, which moves the diaphragm downward, and by the contraction of the rib muscles, which moves the ribs upward.

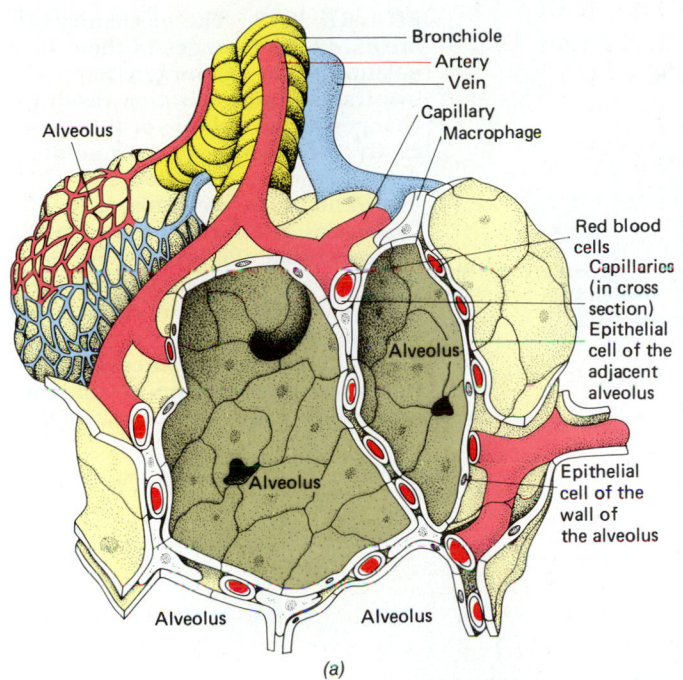

Bronchiole
Artery
Vein
Capillary
Macrophage
Alveolus
Red blood cells
Capillaries (in cross section)
Epithelial cell of the adjacent alveolus
Alveolus
Alveolus
Epithelial cell of the wall of the alveolus
Alveolus   Alveolus

(a)

FIGURE 38–7 Gas exchange takes place across the thin wall of the alveolus. (*a*) Structure of the alveolus. Note that the alveolar wall consists of extremely thin squamous epithelium. The alveoli share their walls, and between the walls of the alveoli lie extensive capillary networks. (*b*) A cross section through one of these capillaries (approximately ×11,450). The nucleus of the endothelial cell that makes up the capillary wall is visible at the bottom of the photograph. Large, dark structures within the capillary are red blood cells. (*c*) An enlargement (approximately ×48,100) of a portion of a capillary similar to that shown in (*b*). The dark structure extending through the capillary is a portion of a red blood cell. The wall of the alveolus is visible just above the wall of the capillary. Notice the very short distance oxygen need travel to get from the air within the alveolus to the red blood cell in which it is transported to the body tissues. (*b* and *c*. Courtesy of Drs. Peter Gehr, Marianne Bachofen, and Ewald R. Wiebel)

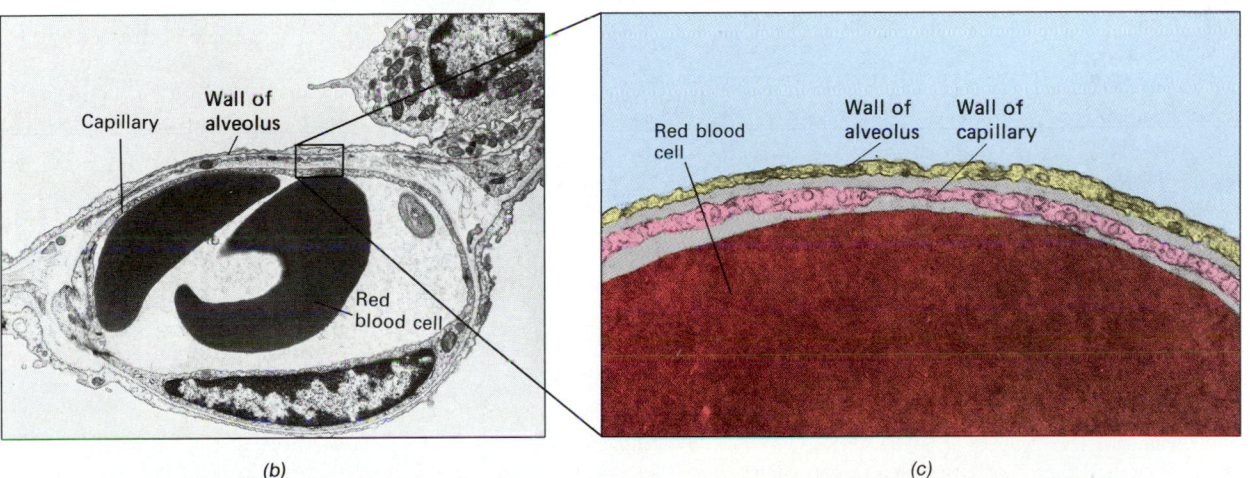

Capillary   Wall of alveolus
Red blood cell

(b)

Red blood cell   Wall of alveolus   Wall of capillary

(c)

These contractions increase the circumference of the chest cavity (Fig. 38–8). As the chest expands, the film of fluid on the pleura pulls the membranous walls of the lungs outward along with the chest walls. This increases the space within each lung. The air in the lungs now has more space in which to move about, and the pressure of the air in the lungs falls below the pressure of the air outside the body. As a result, air from the outside rushes in through the respiratory passageways and fills the lungs until the two pressures are again equal.

Expiration occurs when the diaphragm and rib muscles relax. The volume of the chest cavity decreases, increasing the pressure in the lungs. The millions of distended air sacs deflate, expelling the air that was inhaled. The pressure returns to normal and the lung is ready for another change of air.

Breathing Is Regulated By Respiratory Centers In The Brain

The amount of oxygen used by the body varies with different levels of activity. When you are engaged in a strenuous game of racketball, for example, you require more oxygen than when reading quietly. Breathing is controlled by respiratory centers (groups of specialized neurons) in the brain that are indirectly sensitive to increases in the amount of carbon dioxide in the blood (Fig. 38–6). Groups of neurons in the medulla regulate the basic rhythm of respiration; one group regulates inspiration; a second group is concerned with expiration. Respiratory centers in the pons can stimulate or inhibit the medullary respiratory center.

Chemoreceptors in the medulla are sensitive to concentrations of carbon dioxide in the blood. Recall that an increase in carbon dioxide lowers the pH.

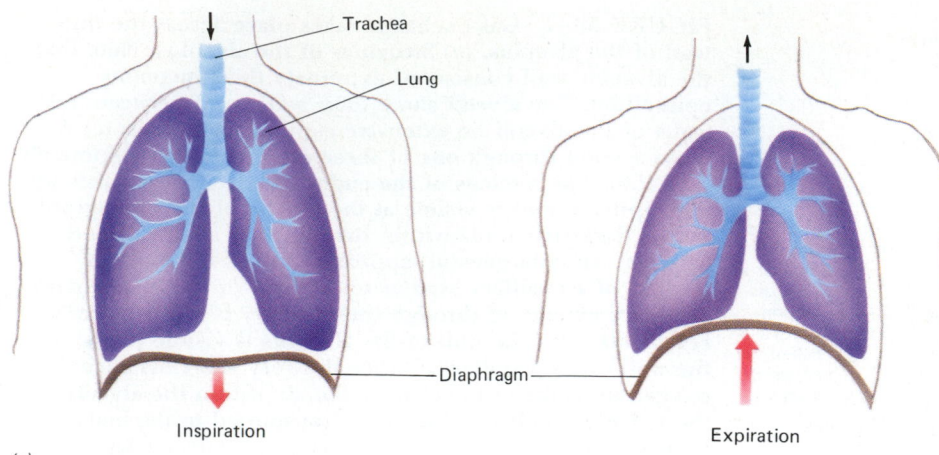

(a)

FIGURE 38–8   **The mechanics of breathing. (a) Changes in the position of the diaphragm in inspiration and expiration result in changes in the volume of the chest cavity. (b) Changes in position of the rib cage in expiration and inspiration. The elevation of the front ends of the ribs by the chest muscles causes an increase in the front-to-back dimension of the chest and a corresponding increase in the volume of the chest cavity. When the volume of the chest cavity increases, a corresponding amount of air moves into the lungs.**

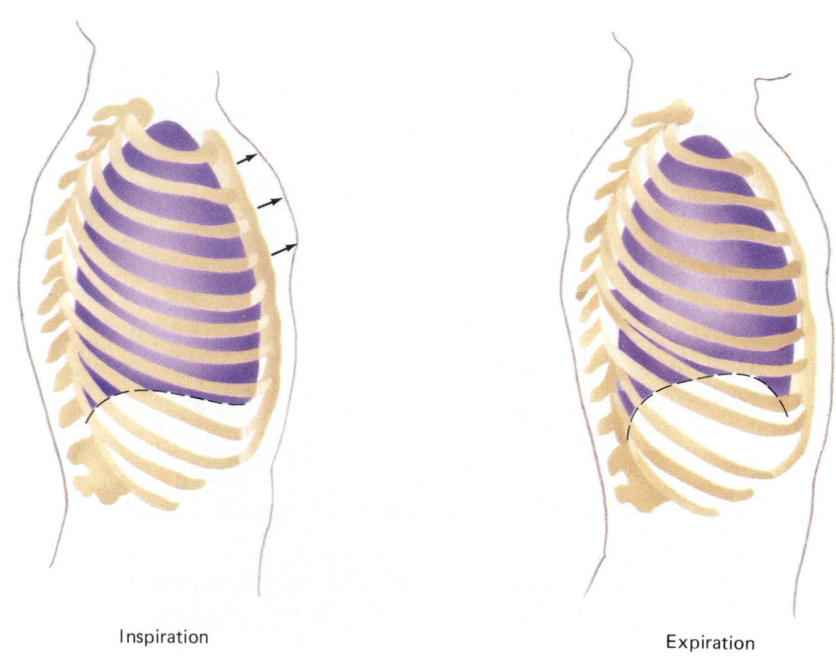

(b)

These receptors respond to a decrease in pH and cause an increase in the rate of respiration. Chemoreceptors in the carotid arteries and aorta are also sensitive to changes in carbon dioxide concentration. These receptors are sensitive to large decreases in oxygen concentration as well. When stimulated these chemoreceptors send neural impulses to the inspiratory center and respiration increases.

Nerve impulses from the inspiratory center are delivered to the diaphragm by the phrenic nerves and to the intercostal muscles by the intercostal nerves. The impulses stimulate contraction of the diaphragm and chest muscles.

During exercise greater amounts of carbon dioxide are produced. The carbon dioxide stimulates the respiratory centers to produce more rapid and more forceful breathing. In this way more oxygen is brought in to meet the body's increased need.

Individuals who have stopped breathing because of drowning, smoke inhalation, electric shock, or cardiac arrest can sometimes be sustained by mouth-to-mouth resuscitation until their own breathing reflexes can be initiated again. Cardiopulmonary resuscitation (CPR) is a method for aiding victims who have suffered respiratory or cardiac arrest, or both. For an overview of the procedure, see Focus on Cardiopulmonary Resuscitation (CPR).

Gas Exchange Takes Place In The Air Sacs

The respiratory system delivers oxygen to the air sacs, but if oxygen were to remain in the lungs, all the other body cells would soon die. The vital link between air sac and body cell is the circulatory system. Each air sac serves as a tiny depot from which oxygen is loaded into

FOCUS ON   Cardiopulmonary Resuscitation (CPR)

Cardiopulmonary resuscitation, or **CPR,** is a method for aiding victims of accidents or heart attacks who have suffered cardiac arrest and respiratory arrest. It should not be used if the victim has a pulse or is able to breathe. It must be started immediately, because irreversible brain damage may occur within about 3 minutes of respiratory arrest. Here are its ABCs:

Airway   Clear airway by extending victim's neck. This is sometimes sufficient to permit breathing to begin again.

Breathing   Use mouth-to-mouth resuscitation.

Circulation   Attempt to restore circulation by using external cardiac compression.

The procedure for CPR may be summarized as follows:

I. Establish the unresponsiveness of the victim.
II. Procedure for mouth-to-mouth resuscitation:
   1. Place the victim on his or her back on a firm surface.
   2. Clear the throat and mouth, and tilt the head back so that the chin points outward. Make sure that the tongue is not blocking the airway. Pull the tongue forward if necessary.
   3. Pinch the nostrils shut and forcefully exhale into the victim's mouth. Be careful, especially in children, not to overinflate the lungs.
   4. Remove your mouth and listen for air rushing out of the lungs.
   5. Repeat about 12 times per minute. Do not interrupt for more than 5 seconds.
III. Procedure for external cardiac compression:
   1. Place the heel of one hand on the lower third of the victim's breastbone. Keep your fingertips lifted off the chest. (In infants, two fingers should be used for cardiac compression; in children, use only the heel of the hand.)
   2. Place the heel of the other hand at a right angle to and on top of the first hand.
   3. Apply firm pressure downward so that the breastbone moves about 4 to 5 cm (1.6 to 2 in) toward the spine. Downward pressure must be about 5.4 to 9 kg (12 to 20 lb) with adults (less with children). Excessive pressure can fracture the sternum or ribs, resulting in punctured lungs or a lacerated liver. This rhythmic pressure can often keep blood moving through the heart and great vessels of the thoracic cavity in sufficient quantities to sustain life.
   4. Relax your hands between compressions to allow the chest to expand.
   5. Repeat at the rate of at least 60 compressions per minute. (For infants or young children, 80 to 100 compressions per minute are appropriate.) Fifteen compressions should be applied, then two breaths, in a ratio of 15:2.

---

blood brought close to the alveolar air by capillaries (Fig. 38–9).

Oxygen molecules diffuse from the air sacs into the blood because the air sacs contain a greater concentration of oxygen than does blood entering the pulmonary capillaries. On the other hand, carbon dioxide moves from the blood, where it is more concentrated, to the air sacs, where it is less concentrated. Each gas diffuses through the cells lining the alveoli and the cells lining the capillaries.

Table 38–1 shows the percentages of oxygen and carbon dioxide present in exhaled air compared with inhaled air. Because carbon dioxide is produced during cellular respiration, there is more of this gas—100 times as much—entering the alveoli from the blood than there is in air inhaled from the environment.

The movement of gases between air sacs and blood is not completely efficient. Not every molecule of inhaled oxygen actually finds its way into the blood, and not every molecule of carbon dioxide is removed from the blood. Part of the reason for this is that newly inhaled air must mix with air from the last inhalation, so the carbon dioxide content of alveolar air is always higher than that of the atmosphere, and its oxygen content is lower. The amount of exchange that takes place, however, is obviously sufficient to support the metabolic well-being of the body. The lungs of some marine mammals that remain underwater for considerable periods of time are more efficient at removing oxygen from inhaled air.

Oxygen Is Transported In Combination With Hemoglobin

When oxygen diffuses into the blood, it forms a weak chemical bond with hemoglobin molecules in the red blood cells, forming **oxyhemoglobin.** Each hemoglobin molecule can combine with, and thus transport, four molecules of oxygen. Since the chemical bond formed between the oxygen and the hemoglobin is weak, the reaction is readily reversible. In the pulmonary capillaries:

Oxygen + Hemoglobin $\longrightarrow$ Oxyhemoglobin

When oxyhemoglobin reaches body cells low in oxygen, the reverse reaction occurs:

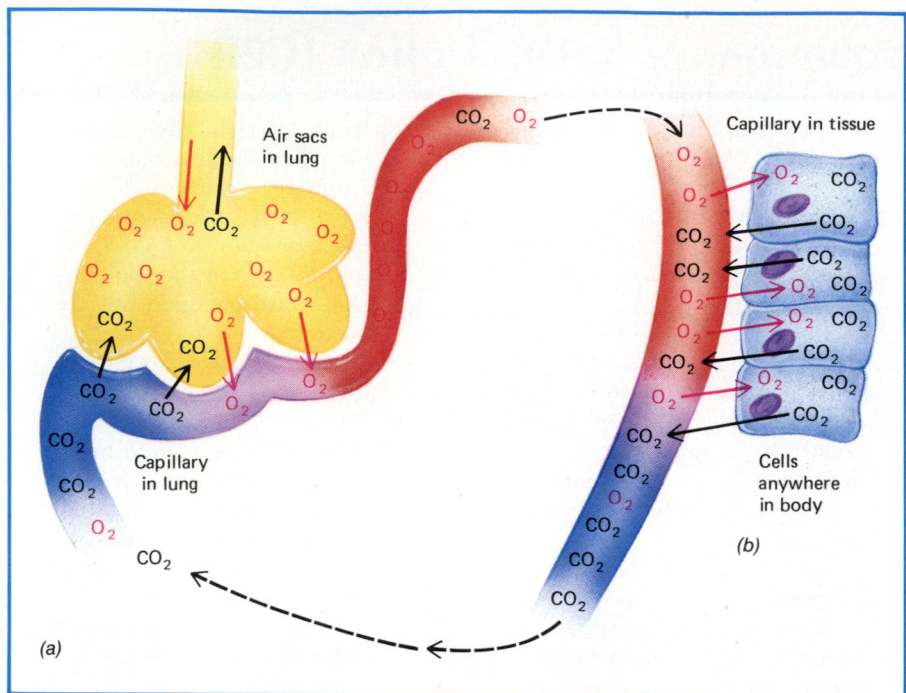

FIGURE 38–9   Gas exchange.
(*a*) **Exchange of gases between air sacs and capillaries in the lung. The concentration of oxygen is greater in the air sacs than in the capillaries, so oxygen moves from the air sacs into the blood. Carbon dioxide is more concentrated in the blood than in the air sacs, so it moves out of the capillaries and into the air sacs. (*b*) Exchange of gases between capillary and body cells. Here, oxygen is more concentrated in the blood, so it moves out of the capillary and into the cells. Carbon dioxide is more concentrated in the cells, and so it diffuses out of the cells and moves into the blood.**

TABLE 38–1 Composition of Inhaled Air Compared with That of Exhaled Air			
	% Oxygen (O_2)	% Carbon Dioxide (CO_2)	% Nitrogen (N_2)
Inhaled air (atmospheric air)	20.9	0.04	79
Exhaled air (alveolar air)	14.0	5.60	79

As indicated, the body uses up about one-third of the inhaled oxygen. The amount of CO_2 increases more than 100-fold because it is produced during cellular respiration.

Oxyhemoglobin $\longrightarrow$ Oxygen + Hemoglobin

The oxygen released diffuses from the capillaries into the cells.

Carbon Dioxide Is Transported Mainly As Bicarbonate Ions

As blood flows through the capillary networks of an organ, carbon dioxide moves out of the cells where it has accumulated and into the blood, where it is less concentrated. Carbon dioxide is transported in the blood in three ways. About 20% is carried by the hemoglobin molecule, and 7% is transported in the plasma as carbon dioxide itself. Most of the carbon dioxide is dissolved in the plasma as **bicarbonate ions** (HCO_3^-).

Carbon dioxide combines with water in the blood to form carbonic acid. This reaction is catalyzed by the enzyme carbonic anhydrase. The carbonic acid dissociates, forming hydrogen ions and bicarbonate ions.

$$CO_2 + H_2O \xrightarrow{\text{Carbonic anhydrase}} H_2CO_3 \longrightarrow H^+ + HCO_3^-$$

Carbon   Water   Carbonic   Hydrogen   Biocarbonate
dioxide   acid   ion   ion

Hemoglobin buffers the hydrogen ions produced so that pH change is minimized.

When it reaches the lungs, carbon dioxide diffuses out of the blood and into the alveoli. Most but by no means all of the transported carbon dioxide leaves the blood and eventually the respiratory system. The human body is adjusted to function at an internal pH of about 7.4 that results from a balance of substances in

FOCUS ON   Choking

Choking kills an estimated 8,000 to 10,000 people per year in the United States. Many of these have long suffered from some degree of paralysis or other malfunction of the muscles involved in swallowing, often without consciously realizing it. Swallowing is a very complex process in which the mouth, pharynx, esophagus, and vocal cords must be coordinated with great precision. Functional muscular disorders of this mechanism can originate in a variety of ways—as birth defects, for example, or from brain tumors or vascular accidents involving the swallowing center of the brain stem.

Choking is more likely to occur in restaurants, where social interactions and unfamiliar surroundings are likely to distract a person's attention from swallowing and where alcohol is more likely to be taken with the meal. A large number of choking victims have a substantial blood alcohol content upon autopsy, which suggests the possibility that in them a marginally effective swallowing reflex has been further and fatally compromised by the effects of alcohol on the brain.

Anyone who begins to choke and gasp during a meal should be asked if he or she can speak. If not, as indicated by shaking the head or other gestures, the person is probably suffering a laryngeal obstruction rather than a coronary heart attack. If you are present during such an episode, be aware that you can take certain steps that can save the person's life: As a first step, deliver a strong blow to the victim's back with the open hand. If this fails, stand behind the victim, bring your arms around the person's waist, and clasp your hands just about the beltline.

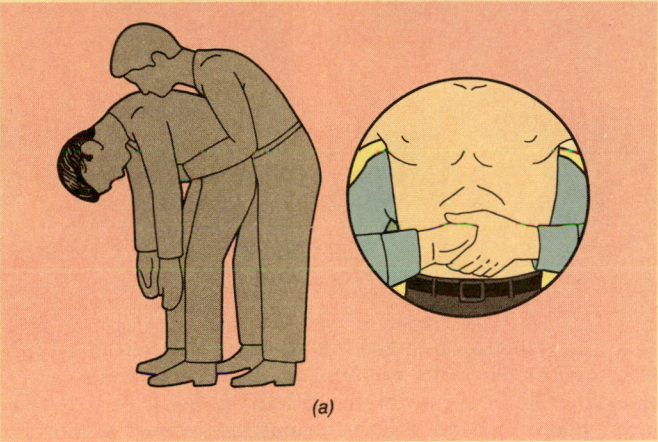

(a)

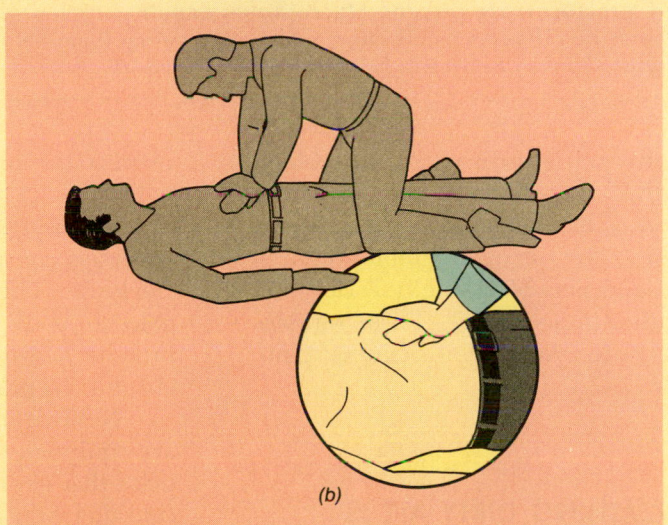

(b)

Your thumbs should be facing inward against the victim's body. Then squeeze abruptly and strongly in an upward direction. In most instances the residual air in the lungs will pop the obstruction out like a cork from a bottle. This is called the abdominal thrust, or **Heimlich maneuver.**

If the Heimlich maneuver must be performed with the victim lying down, place the victim face up. Kneel astride the victim's hips and, with one of your hands on top of the other, place the heel of your bottom hand on the abdomen slightly above the navel but below the rib cage. Press into the victim's abdomen with a quick upward thrust. This may be repeated if necessary. If there is no response within 15 or 20 seconds, it may be necessary to start cardiopulmonary resuscitation, or CPR. (See Focus on Cardiopulmonary Resuscitation [CPR].)

the blood, one of which is a moderate amount of carbon dioxide. If too much is retained because of respiratory insufficiency, the blood pH becomes abnormally low (acidosis) due to excess carbonic acid formation. If hyperventilation blows off too much carbon dioxide, an equally undesirable state, alkalosis, results from abnormally low concentrations of carbonic acid in the blood.

Physiological Adaptation To Changes In Pressure Takes Time

Scuba divers who return to the surface too quickly, or pilots who ascend to over 35,000 feet too rapidly, may suffer from **decompression sickness** (also called the bends). While submerged, a diver breathes gases under high pressure, and because of this pressure, ex-

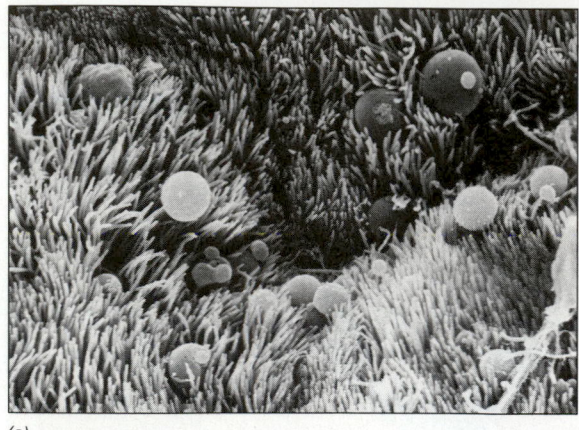
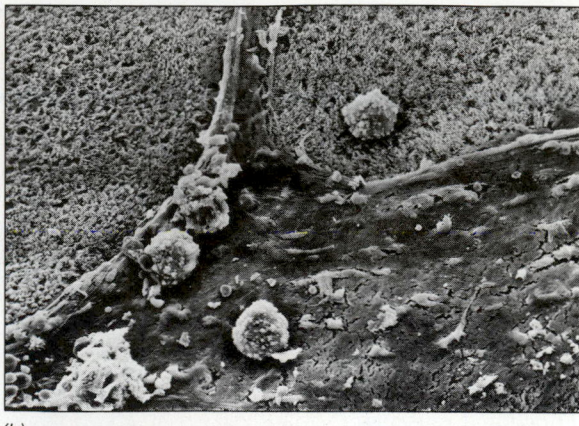

(a)   *(b)*

FIGURE 38–10 (*a*) Scanning electron micrograph showing a carpet of cilia in the lining of a small bronchus. Note the globules of mucus that have emerged from goblet (gland) cells in the epithelial lining. (*b*) View of irritated bronchial lining at a lower magnification. Excess mucus has formed a pool (bottom of photograph) overlaying part of the ciliary carpet. Four macrophages (rounded structures) are present. Macrophages engulf particles and then carry them onto the cilia-mucus elevator, which may sweep them out of the respiratory system. (Courtesy of Dr. R.J. Pack, University of London)

cessive amounts of nitrogen gas dissolve in the blood and tissues. As the diver ascends to a lower pressure, the dissolved nitrogen comes out of solution. If he or she surfaces too rapidly, tiny nitrogen bubbles form in the blood and tissues and may block the flow of blood in capillaries and cause other damage. These bubbles cause the symptoms of decompression sickness: pain, paralysis, even death.

■ THE EFFECT OF BREATHING DIRTY AIR IS RESPIRATORY INSULT

We breathe about 20,000 times each day, inhaling about 35 pounds of air—six times more than the food and drink we consume. Most of us breathe dirty urban air laden with particulates, carbon monoxide, and other harmful substances that are damaging to the respiratory system.

A Variety Of Defense Mechanisms Protect The Lungs

Several defense mechanisms help protect the delicate lungs from the harmful substances we breathe. The hair around the nostrils, the ciliated mucous lining in the nose and pharynx, and the cilia-mucus elevator serve to trap foreign particles in inspired air (Fig. 38–10). One of the body's most rapid defense responses to breathing dirty air is **bronchial constriction.** In this

process the bronchial tubes narrow. As a result, inhaled particles are more likely to land on their sticky mucous lining. Unfortunately, when the bronchial passageways constrict, less air can pass through them to the lungs. This decreases the amount of oxygen available to body cells. Chain smokers and those who breathe heavily polluted air may remain in a state of chronic bronchial constriction.

Neither the smallest bronchioles nor the alveoli are equipped with mucus or ciliated cells. Foreign particles that get through other respiratory defenses and find their way into the alveoli may be engulfed by macrophages. The macrophages may then accumulate in the lymph tissue of the lungs. Lung tissue of chronic smokers and those who work in dirty fossil-fuel–burning industries contains large blackened areas where carbon particles have been deposited.

Continued Respiratory Insult Leads To Respiratory Disease

Continued insult to the respiratory system results in disease. Chronic bronchitis and emphysema are **chronic obstructive pulmonary diseases (COPD)** that have been linked to smoking and breathing polluted air. More than 75% of patients with **chronic bronchitis** have a history of heavy cigarette smoking (See Focus on Facts About Smoking). In chronic bronchitis, irritation from inhaled pollutants causes the bronchial tubes to secrete too much mucus. Ciliated cells, damaged by the pollutants, cannot effectively clear the mucus and

FOCUS ON   Facts About Smoking

- The life of a 30-year-old who smokes 15 cigarettes a day is shortened by an average of more than 5 years.
- If you smoke more than one pack per day, you are about 20 times more likely to develop lung cancer than a nonsmoker. According to the American Cancer Society, cigarette smoking causes more than 75% of all lung cancer deaths.
- If you smoke, you are more likely to develop atherosclerosis, and you double your chances of dying from cardiovascular disease.
- If you smoke, you are 20 times more likely to develop chronic bronchitis and emphysema than a nonsmoker.
- If you smoke, you are seven times more likely to develop peptic ulcers (especially malignant ulcers) than a nonsmoker.
- If you smoke, you have about 5% less oxygen circulating in your blood (because carbon monoxide binds to hemoglobin) than a nonsmoker.
- If you smoke when you are pregnant, your baby will weigh about 6 ounces less at birth, and there is double the risk of miscarriage, stillbirth, and infant death than if you did not smoke.
- Workers who smoke one or more packs of cigarettes per day are absent from their jobs because of illness 33% more often than nonsmokers.
- Risks increase with the number of cigarettes smoked, inhaling, smoking down to a short stub, and use

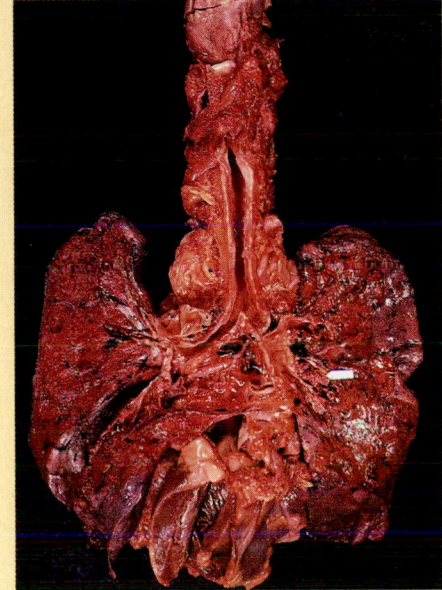 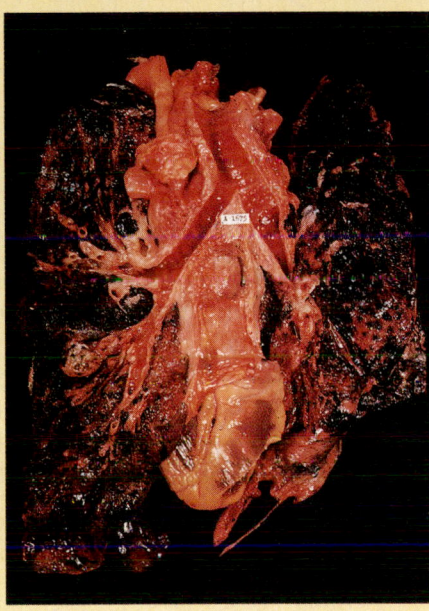

(a)   (b)

A diseased lung compared to a healthy lung. (a) Normal human lungs and major bronchi. (b) Human lungs and heart showing effects of cigarette smoking.

of nonfilter or high-tar, high-nicotine cigarettes. Cigar and pipe smokers have lower risks than cigarette smokers because they do not inhale as much. Cigarette smokers who switch to cigars and continue to inhale actually increase their risks.
- Nonsmokers confined in living rooms, offices, automobiles, or other places with smokers are adversely affected by the smoke. For example, when parents of infants smoke, the infant has double the risk of contracting pneumonia or bronchitis in its first year of life.
- When smokers quit smoking, their risk of dying from chronic pulmonary disease, cardiovascular disease, or cancer decreases. (Precise changes in risk figures depend upon the number of years the person smoked, the number of cigarettes smoked per day, the age of starting to smoke, and the number of years since quitting.)
- If everyone in the United States stopped smoking, more than 300,000 lives would be saved each year.

---

trapped particles from the airways. The body resorts to coughing in an attempt to clear the airways. The bronchioles become constricted and inflamed, and the patient is short of breath.

Victims of chronic bronchitis often develop **pulmonary emphysema,** a disease most common in cigarette smokers. In this disorder alveoli lose their elasticity and walls between adjacent alveoli are destroyed. Surface area of the lung is so reduced that gas ex-

change is seriously impaired. Air is not expelled effectively and stale air accumulates in the lungs. The emphysema victim struggles for every breath and still the body does not get enough oxygen. To compensate, the right ventricle of the heart pumps harder and becomes enlarged. Emphysema patients frequently die of heart failure.

Cigarette smoking is also the main cause of lung cancer. More than 10 of the compounds in the tar of tobacco smoke have been shown to cause cancer. These carcinogenic substances irritate the cells lining the respiratory passages and alter their metabolic balance.

Normal cells are transformed into cancer cells, which multiply rapidly and invade surrounding tissues.

■ CHAPTER SUMMARY

I. Gas exchange in air has certain advantages over gas exchange in water: Air contains more oxygen than water, oxygen diffuses more rapidly through air than through water, and less energy is required for ventilation. However, air-dwellers must have adaptations that prevent drying out.

II. In very small animals gas exchange depends upon diffusion between the environment and each cell. Larger animals have specific adaptations to ensure that oxygen reaches each cell of the body. These gas exchange structures include the body surface, gills, tracheal tubes, and lungs.
   A. Many invertebrates and some vertebrates exchange gases through their skin.
   B. In insects and some other arthropods, the respiratory system consists of a network of tracheal tubes that extend to all parts of the body.
   C. Gills are moist, thin extensions of the body surface that occur mainly in aquatic animals.
   D. Large mollusks and terrestrial vertebrates have lungs, respiratory structures that develop as ingrowths of the body surface or from the wall of the pharynx.

III. In humans respiration is accomplished by a system of air passageways that branch into smaller and smaller tubes, ending finally in the alveoli within the lungs.
   A. In the nasal cavities air is filtered, brought to body temperature, and humidified.
   B. From the nasal cavities air passes through the pharynx and into the larynx. The larynx helps prevent the entrance of foreign material into the lungs by initiating a cough reflex when touched by foreign matter.
   C. From the larynx inhaled air passes into the trachea and then into the right or left bronchus.

   D. Within the lungs the bronchi branch into an extensive system of bronchioles, which eventually terminate in the millions of tiny alveoli, through which gas exchange takes place with the blood.
   E. Breathing is the mechanical process of moving air back and forth between the environment and the air sacs of the lungs.
      1. When the diaphragm and rib muscles contract, expanding the chest, air rushes into the lungs.
      2. When these muscles relax, pressure in the lung increases and air is expired.
   F. Breathing is normally regulated by respiratory centers in the brain. These centers are sensitive to the acidity of the blood, which is directly influenced by the amount of carbon dioxide in the blood.
   G. Oxygen diffuses from the air sacs into the blood while carbon dioxide diffuses from the blood into the air sacs.
   H. Oxygen is transported to the body cells in the form of oxyhemoglobin. As oxygen is needed by the cells, the oxyhemoglobin dissociates, and oxygen diffuses from the blood into the cells.
   I. Carbon dioxide is transported mainly as bicarbonate ions in the blood.
   J. The respiratory system defends itself against dirt and other foreign matter by trapping particles on mucous surfaces, the cilia-mucus elevator, bronchial constriction, and phagocytosis by macrophages. Inhaling polluted air or smoking cigarettes can wear down these defense mechanisms and cause serious damage to the respiratory system.

■ POST-TEST

1. Specialized respiratory structures must have thin walls so that _____ can easily occur; they must be _____ so that gases can be dissolved; and they must be richly supplied with _____ to ensure transport of gases.

2. In insects air enters a network of _____

tubes through openings called _____.

3. Respiratory structures that develop from the wall of a body cavity, such as the pharynx, or as ingrowths of the body surface are called _____.

4. In birds the lungs have several extensions referred to as _____ _____.

5. In the mammalian respiratory system, inhaled air passing through the larynx would next enter the _____.

6. In the mammalian respiratory system, gas exchange takes place through the thin walls of the _____.

7. In mammals the floor of the thoracic cavity is formed by the _____.

Select the most appropriate answer in column B for each description in column A.

Column A

_____ 8. Seals off larynx during swallowing
_____ 9. Cavities in bones of skull
_____ 10. Initiates cough reflex
_____ 11. Covers lung
_____ 12. Structure through which gas exchange actually takes place

Column B

a. sinuses
b. larynx
c. alveoli
d. pleural membrane
e. epiglottis

13. _____ can result when a diver surfaces too rapidly and nitrogen bubbles form in the blood and tissues.

14. Bronchial constriction is one of the body's most rapid responses to _____.

15. In _____, the alveolar walls break down so that several air sacs join to form larger, less elastic alveoli.

16. The main cause of lung cancer is _____.

17. Label the diagram to the right.

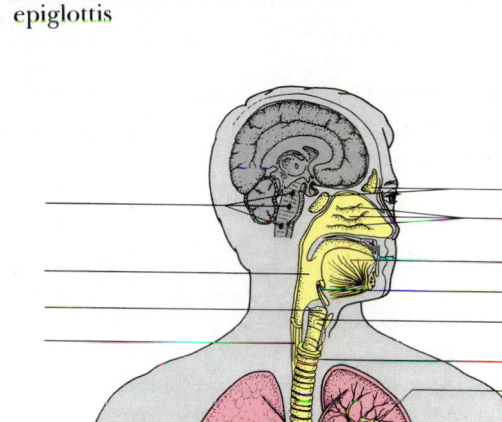

■ REVIEW QUESTIONS

1. What adaptations for gas exchange are found in fish? In insects? In terrestrial vertebrates? Why is diffusion alone an inadequate means of gas exchange in large animals?
2. Why are lungs more suited for an air-breathing vertebrate and gills more effective in a fish?
3. Trace a breath of inhaled air from nose to alveoli, listing each structure through which it must pass.
4. What are the advantages of having millions of alveoli rather than a pair of simple, balloon-like lungs?
5. Describe the protective mechanisms of the respiratory system, including the cilia-mucus elevator and bronchial constriction.
6. Describe the processes of inspiration and expiration.
7. How is breathing regulated?
8. What role does diffusion play in gas exchange in humans?
9. How is oxygen transported in the blood of humans?
10. In what way does the composition of inhaled air differ from that of exhaled air? Why?
11. Summarize the health effects of smoking.

■ RECOMMENDED READINGS

Feder, M.E., and W.W. Bruggren: "Skin Breathing Vertebrates," *Scientific American* 253(2):126–142, 1985. An interesting account of gas exchange through the body surface in some vertebrates.

Kolata, G.: "Cell Biology Yields Clues to Lung Cancer," *Science* 218:38–39, 1982. By studying lung cancer cells in culture, researchers are gaining insights into how the cells multiply and how they can be prevented from multiplying.

Zapol, W.M.: "Diving Adaptations of the Weddell Seal," *Scientific American* 256(6):100–105, 1987. Physiological adaptations permit the seal to swim deeper and hold its breath longer than most other mammals.

39

Processing Food

OUTLINE

I. In the animal kingdom we find all kinds of dinner jackets

II. Some invertebrates have incomplete digestive systems

III. Most invertebrates and all vertebrates have complete digestive systems

IV. The human digestive system has highly specialized structures for processing food
   A. Food begins its journey inside the mouth
   B. The pharynx and esophagus conduct food to the stomach
   C. Food is mechanically and enzymatically digested in the stomach
   D. Most enzymatic digestion takes place inside the small intestine
   E. The liver secretes bile which mechanically digests fats
   F. The pancreas secretes digestive enzymes
   G. Nerves and hormones regulate digestion
   H. Absorption takes place mainly through the villi of the small intestine
   I. The large intestine eliminates wastes

V. Adequate amounts of required nutrients are necessary to support metabolic processes
   A. Carbohydrates are a major energy source in the human diet
      1. Carbohydrates are commonly ingested as starch, cellulose, and sucrose
      2. Glucose is used as fuel by the cells
   B. Lipids are used as an energy source and to make needed biological molecules
      1. Most lipids are ingested as triglycerides
      2. Dietary triglycerides and cholesterol have been associated with heart disease
      3. Fat is stored in adipose tissue

LEARNING OBJECTIVES

After you have studied this chapter you should be able to:

1. Compare adaptations that herbivores, carnivores, and omnivores have evolved for their particular modes of nutrition.

2. Compare how an animal with an incomplete digestive system (for example, a *Hydra*) and an animal with a complete digestive system, such as a vertebrate, process food with respect to ingestion, digestion, absorption, and elimination.

3. Identify on a diagram or model each of the structures of the human digestive system described in this chapter, and give the function of each structure.

4. Trace the pathway traveled by an ingested meal, describing each of the changes that takes place en route.

5. Summarize the functions of the accessory digestive glands of terrestrial vertebrates.

6. Trace the step-by-step digestion of carbohydrate, protein, and lipid.

7. Draw and label a diagram of an intestinal villus, and explain how its structure is so suitably adapted to its function.

8. Identify commonly ingested carbohydrates and trace the fate of glucose after its absorption.

9. Trace the fate of lipids after they are absorbed from the intestine and summarize the relationship between lipid intake and coronary heart disease.

10. Trace the fate of proteins in the body.

11. Distinguish between water-soluble and fat-soluble vitamins, and describe the effects of specific vitamin deficiencies.

12. Summarize the role of minerals as essential nutrients, giving three specific examples.

Whitetail doe eating leaves from tree. (Carl R. Sams II/Marvin L. Dembinsky, Jr., Photography Associates)

OUTLINE

C. Proteins serve as enzymes and are essential structural components of cells
  1. Essential amino acids must be ingested in the diet
  2. Amino acids are deaminated, then used as fuel or stored as fat
D. Vitamins are organic compounds essential for normal metabolism
E. Minerals are inorganic nutrients required by cells
F. Energy metabolism is balanced when energy input equals energy output
G. Obesity is a serious nutritional problem
H. Malnutrition can cause serious health problems
Focus on peptic ulcers
Focus on vegetarian diets

LEARNING OBJECTIVES

13. Contrast basal metabolic rate with total metabolic rate.
14. Write the basic energy equation for maintaining body weight and describe the consequences of altering it in either direction.
15. Summarize the causes of obesity and its treatment.
16. In general terms, describe the problem of world food supply relative to world population, and describe the effects of malnutrition.
17. Summarize the difficulties encountered in obtaining adequate amounts of amino acids in a vegetarian diet, and describe how a nutritionally balanced vegetarian diet could be planned.

Nutrients are the substances present in food that are needed by an organism as an energy source to run the machinery of the body, as ingredients to make compounds for metabolic processes, and as building blocks to permit growth and repair of tissues. Obtaining nutrients is of such vital importance that both individual organisms and ecosystems are designed around the central theme of **nutrition,** the process of taking in and assimilating food. An organism's body plan, as well as its life-style, is adapted to its particular mode of obtaining food.

All animals are heterotrophs, organisms that must obtain their energy and nourishment from the organic molecules manufactured by other organisms. Most animals have a digestive system that processes the food they eat. Food processing may be divided into several steps: *ingestion, digestion, absorption,* and *elimination.*

After foods are selected and obtained, they are ingested, that is, taken into the body. **Ingestion** generally includes taking the food into the mouth and swallowing it. Because animals eat the macromolecules tailor-made by and for other organisms, they must break down these molecules and refashion them for their own needs. We cannot incorporate the proteins in steak directly into our own muscles, for example. The body **digests** the steak, mechanically breaking down the large bites of meat into smaller ones and then enzymatically hydrolyzing (breaking down with the addition of water) the proteins into their component amino acids. The amino acids can then be **absorbed** and **transported** to the muscle cells, which arrange these components into human muscle proteins.

Most animals are equipped with digestive tracts. In them, absorption is the passage of nutrients through the cells lining the digestive tract and into the blood or other body fluids. Nutrients are distributed to all parts of the body and used for metabolic activities within each cell. Food that is not digested and absorbed is discharged from the body, in a process called **egestion** in simple animals and **elimination** in more complex animals.

KEY CONCEPTS

◻ Nutrition is so vital that an animal's body plan and life-style are adapted to its particular mode of obtaining food.

◻ Processing food involves ingestion, digestion, absorption, and elimination of wastes.

◻ In animals with a complete digestive tract, various regions of the digestive tract are specialized for carrying on food processing functions.

◻ With only slight variation, all animals require the same basic nutrients: minerals, vitamins, carbohydrates, lipids, and proteins.

◻ Carbohydrates, lipids, and proteins can all be used as energy sources. Eating too much of any of these nutrients can result in weight gain. Eating too few nutrients or an unbalanced diet can result in malnutrition and death.

◻ The basic energy equation for maintaining body weight is:

Energy (calorie) input = energy output.

(a)

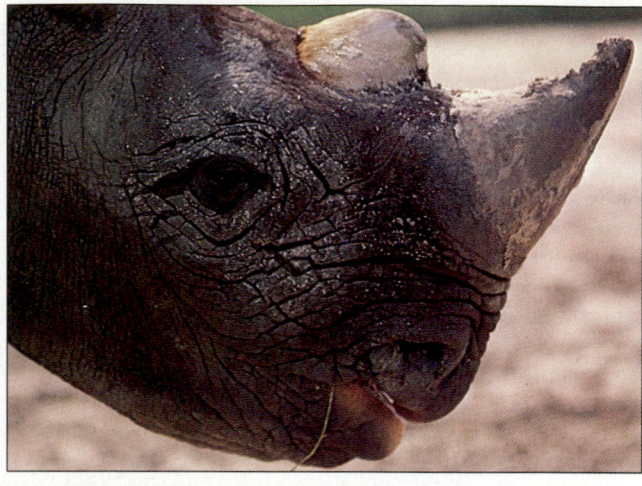

(b)

FIGURE 39–1 **Adaptations of herbivores. (*a*) An acorn weevil. The impressively long "snout" of this little beetle is used both for feeding and to make a hole in the acorn through which an egg is deposited. When it has hatched, the larva feeds on the contents of the acorn seed. (*b*) The rhinoceros can use its horn to uproot and overturn small trees and bushes; it then eats the leaves. Members of some species use their lips to break off grass.** (*a*, Darwin Dale, Photo Researchers Inc.; *b*, courtesy of Busch Gardens, Tampa)

◼ IN THE ANIMAL KINGDOM WE FIND ALL KINDS OF DINNER JACKETS

Consider how any organism processes food and you will see that it has a built-in dinner jacket. Consider the sharp teeth and claws of the lion, for example, as well as its long, quick legs. These structural adaptations enable it to hunt and kill other animals.

Some animals are **herbivores,** or primary consumers, which eat exclusively, or mainly, plant materials (Fig. 39–1). Because animals cannot digest the cellulose of plant cell walls, herbivores have evolved many adaptations for obtaining nutrients from the plant material they eat. For example, vertebrate herbivores generally have a specialized section of the digestive tract in which live bacteria capable of digesting cellulose. In the cud-chewing ruminants (cattle, sheep, deer), the stomach is divided into four chambers. Bacteria inhabiting the first two chambers digest cellulose. Food that is not sufficiently chewed, called cud, is regurgitated back up into the animal's mouth and chewed again.

Many herbivores eat great quantities of food. Grasshoppers, locusts, elephants, and cattle, for example, all spend a major part of their lives eating. Most of what they eat is not efficiently digested and moves out of the body, almost unchanged, as waste. However, by eating large enough quantities, these herbivores digest and absorb sufficient material to provide the nourishment necessary to sustain their life processes.

Herbivores are sometimes eaten by flesh-eating **carnivores,** which may also eat one another. Carni-

vores (secondary and higher-level consumers in ecosystems) are adapted for capturing and killing prey. The fast-moving tongue of the frog captures many a fly, and the long, quick legs and sharp teeth of the cheetah enable it to catch and kill gazelles. Some carnivores seize their victim and swallow it alive and whole (Fig. 39–2). Others paralyze, crush, or shred their prey before ingesting it. Carnivorous mammals have well-developed canine teeth for stabbing during combat. The digestive juice of the stomach breaks down proteins, and because meat is more easily digested than plant food, their digestive tracts are shorter than those of herbivores.

Omnivores, such as bears and humans, include both plant materials and meat in their diet. Earthworms take in large amounts of soil containing both animal and plant material. The blue whale, the largest animal, is a filter feeder that swims through the sea, straining out tiny plants and animals as it swims. Omnivores often possess adaptations that permit them to distinguish among a wide range of smells and tastes. This ability enables them to select a variety of foods.

◼ SOME INVERTEBRATES HAVE INCOMPLETE DIGESTIVE SYSTEMS

Some invertebrates, such as sponges, have no digestive system at all and others have an **incomplete digestive system** with only a single opening. Sponges obtain food by filtering microscopic organisms from the sur-

(a)

(b)

FIGURE 39–2   Adaptations of carnivores. (*a*) With lightning speed the Burmese python strikes at its prey, then suffocates it before consuming it whole. (*b*) With its wide field of vision and fast reflexes, the Chinese praying mantis (*Tendora aridifolia sinensis*) is an able carnivore. (*a*, Courtesy of Mical Solomon and Trudi Segal; *b*, Dwight R. Kuhn)

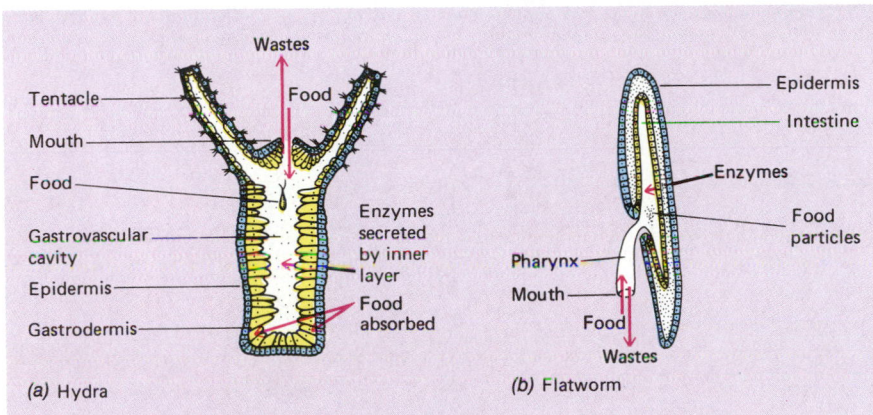
(a) Hydra   (b) Flatworm

FIGURE 39–3   Food processing in invertebrates with an incomplete digestive system. The hydra (*a*) and the flatworm (*b*) have a digestive tract with a single opening that serves as both mouth and anus.

rounding water. Individual cells phagocytize the food particles, and digestion is intracellular within food vacuoles. Wastes are egested into the water that continuously circulates through the sponge body.

Cnidarians, such as hydras and jellyfish, capture small aquatic animals with the help of their cnidocytes and tentacles (Fig. 39–3*a*). The mouth opens into a large gastrovascular cavity lined by cells that secrete enzymes which break down proteins. During digestion within the gastrovascular cavity, proteins are enzymatically split to polypeptides. Digestion continues intracellularly within food vacuoles, and the digested nutrients are passed to other cells by diffusion. Large undigested food particles are ejected through the mouth by contraction of the body.

Free-living flatworms (e.g., planaria) begin to digest their prey even before ingesting it. They extend the pharynx out through their mouth and secrete digestive enzymes onto the prey (Fig. 39–3*b*). When ingested, the food enters the branched intestine. Extracellular digestion proceeds as intestinal cells secrete

digestive enzymes. Partly digested food fragments are then phagocytized by cells of the intestinal lining, and digestion is completed intracellularly within food vacuoles. As in cnidarians, the flatworm digestive system has only one opening, so undigested wastes are eliminated through the mouth.

◻ MOST INVERTEBRATES AND ALL VERTEBRATES HAVE COMPLETE DIGESTIVE SYSTEMS

Most other invertebrates, and all vertebrates, have a **complete digestive system**—the digestive tract is a complete tube with two openings (Fig. 39–4). Food enters through the mouth, and undigested food is eliminated through the anus. Waves of muscular contractions push the food in one direction, so that more food can be taken in while previously eaten food is being digested and absorbed farther down the tract.

Pharynx
Crop
Intestine
Food absorbed
Food
Mouth
Esophagus
Gizzard
Enzymes
Wastes
Anus

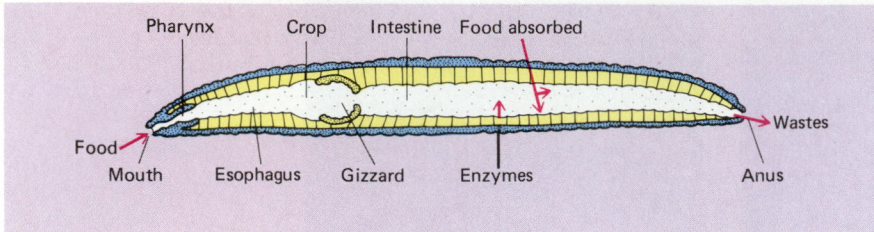

FIGURE 39–4 The earthworm, like most complex animals, has a complete digestive tract extending from mouth at one end of the body to anus at the other end. Various regions of the digestive tract are specialized to perform different food processing functions.

FIGURE 39–5 The human digestive system. Note the complete digestive tract, a long, coiled tube extending from mouth to anus. Locate the three types of accessory glands.

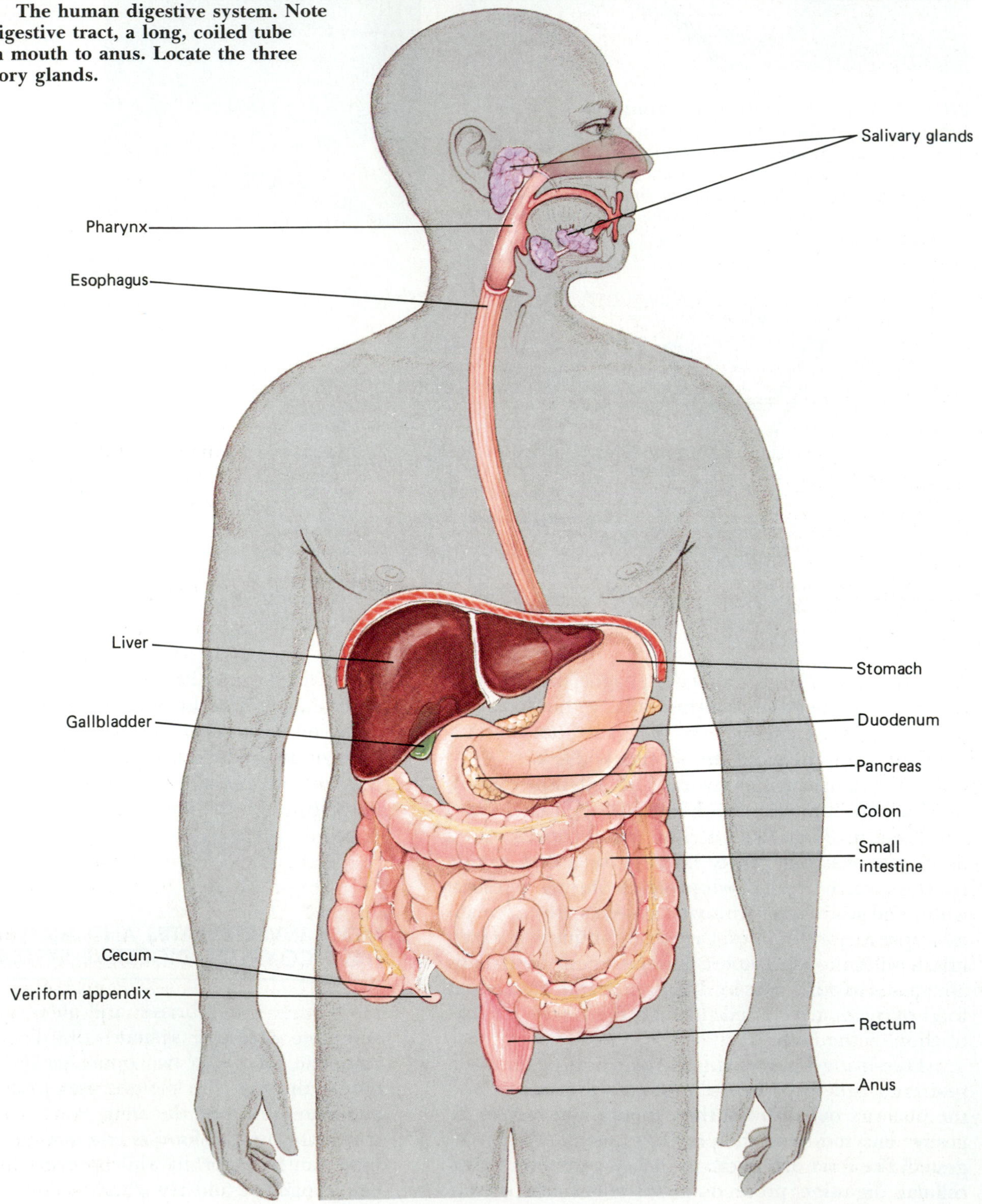

Salivary glands
Pharynx
Esophagus
Liver
Gallbladder
Stomach
Duodenum
Pancreas
Colon
Small intestine
Cecum
Veriform appendix
Rectum
Anus

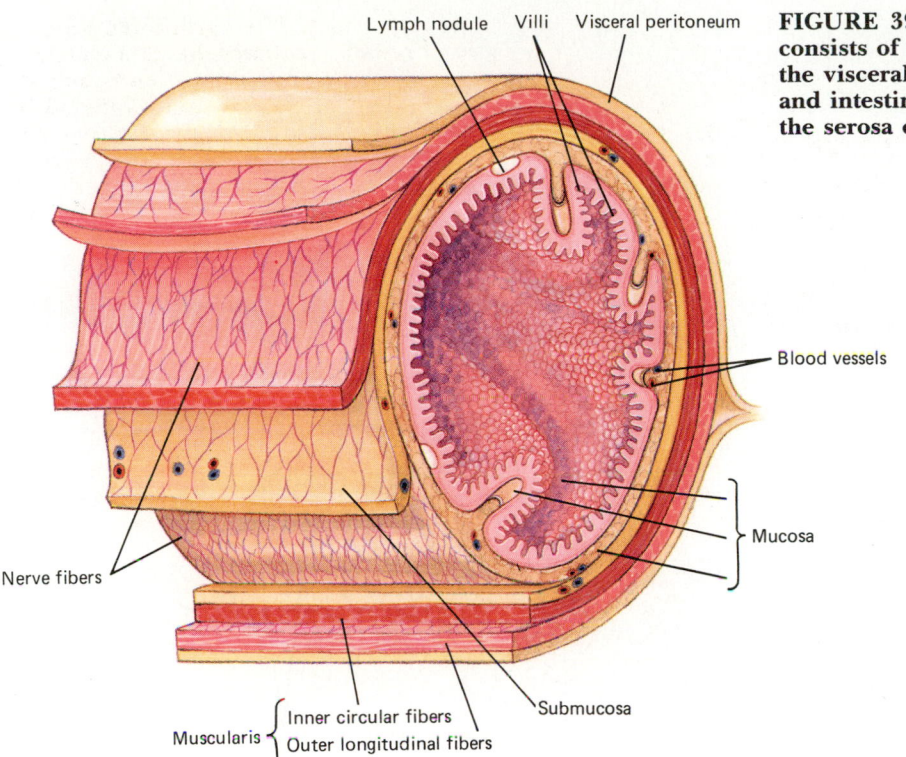

Lymph nodule   Villi   Visceral peritoneum

Blood vessels

Mucosa

Nerve fibers

Muscularis { Inner circular fibers
            Outer longitudinal fibers

Submucosa

FIGURE 39–6 The wall of the digestive tract consists of four layers. The outer layer is known as the visceral peritoneum in the wall of the stomach and intestine. (Above the stomach it is referred to as the serosa or adventitia.)

In a complete digestive tract, various regions of the tube are specialized to perform specific functions. In the vertebrate digestive tract, food successively passes through the following specialized regions: mouth, pharynx (throat), esophagus, stomach, small intestine, large intestine, and anus. All vertebrates have accessory glands that secrete digestive juices into the digestive tract. These include the liver, the pancreas, and in terrestrial vertebrates, the salivary glands.

■ THE HUMAN DIGESTIVE SYSTEM HAS HIGHLY SPECIALIZED STRUCTURES FOR PROCESSING FOOD

In the human digestive system, various regions of the digestive tract have highly specialized structures and functions (Fig. 39–5). The wall of the digestive tract is composed of four layers. Although they vary somewhat in structure in various regions, the layers are basically similar throughout the digestive tract (Fig. 39–6). The **mucosa,** a layer of epithelial tissue and underlying layer of connective tissue, lines the lumen (inner space) of the digestive tract. Surrounding the mucosa is the **submucosa,** a connective tissue layer rich in blood vessels, lymphatic vessels, and nerves. Surrounding the submucosa is a **muscle layer,** consisting of two sublayers of smooth muscle. In the inner sublayer, the muscle fibers are arranged circularly around the digestive tube, whereas in the outer sublayer the muscle fi-

bers are arranged longitudinally. The outer connective tissue coat of the digestive tract is the **adventitia.** Below the level of the diaphragm, the adventitia becomes the **visceral peritoneum.**

Food Begins Its Journey Inside The Mouth

Imagine that you have just taken a big bite of a hamburger. The mouth is specialized for ingestion and for beginning the digestive process. Mechanical digestion begins as you bite, grind, and chew the meat and bun with your teeth. Unlike the simple, pointed teeth of fish, amphibians, and reptiles, the teeth of mammals vary in size and shape and are specialized to perform specific functions. The chisel-shaped **incisors** are used for biting, while the long, pointed **canines** are adapted for stabbing and tearing food. The flattened surfaces of the **premolars** and **molars** are specialized for crushing and grinding food.

Each tooth is covered by **enamel,** the hardest substance in the body. Most of the tooth consists of **dentin,** which resembles bone in composition and hardness. Beneath the dentin is the **pulp cavity,** a soft connective tissue containing blood and lymph vessels and nerves.

While the food is being mechanically disassembled by the teeth, it is also moistened by saliva. Some of its molecules dissolve, enabling you to taste it. (Taste buds, which are located on the tongue and other surfaces of the mouth, were discussed in Chapter 35.) Three pairs of **salivary glands** secrete about a liter of

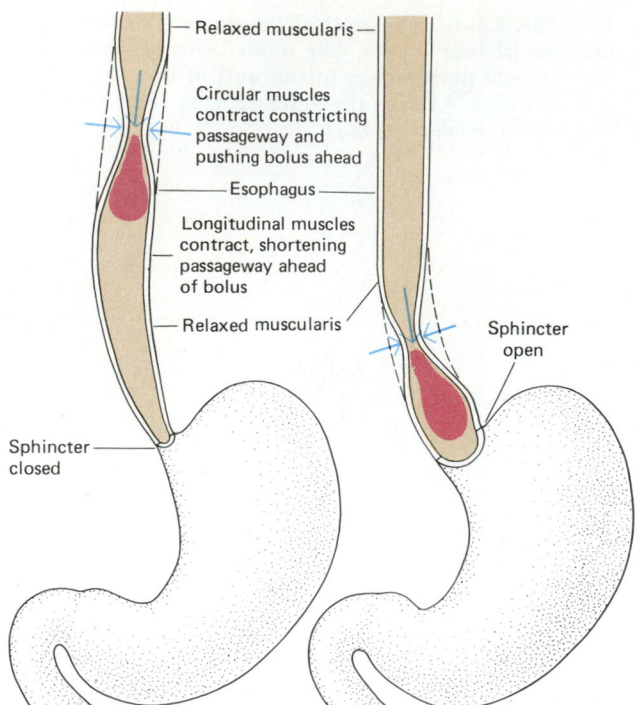

FIGURE 39–7 Peristalsis. Food is moved through the digestive tract by waves of muscular contraction known as peristalsis.

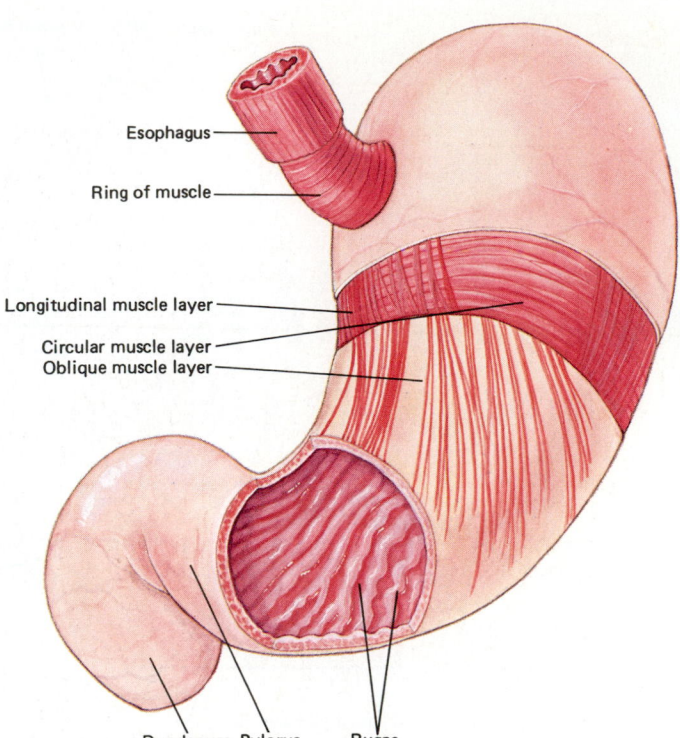

FIGURE 39–8 Structure of the stomach. From the esophagus, food enters the stomach, where it is mechanically and enzymatically digested.

saliva into the mouth cavity each day. Saliva contains an enzyme, **salivary amylase,** which initiates the digestion of carbohydrates.

The Pharynx And Esophagus Conduct Food To The Stomach

After the bite of food has been chewed and fashioned into a lump called a **bolus,** it is swallowed, that is, moved through the **pharynx** and into the **esophagus.** The pharynx, or throat, is a muscular tube that serves as the hallway of the respiratory system as well as the digestive system. During swallowing, the opening to the airway is closed by a small flap of tissue, the **epiglottis.**

Waves of muscular contraction, called **peristaltic contractions,** sweep the bolus through the pharynx and esophagus toward the stomach (Fig. 39–7). Circular muscle fibers in the wall of the esophagus contract around the top of the bolus, pushing it downward. Almost at the same time, longitudinal muscles around the bottom of the bolus and below it contract, shortening the tube.

When the body is in an upright position, gravity helps to move the food through the esophagus, which is about 25 centimeters (10 inches) long, but gravity is not necessary. Astronauts are able to eat in its absence, and even if you are standing on your head, food will reach the stomach.

Food Is Mechanically And Enzymatically Digested In The Stomach

The entrance to the stomach is normally closed by a ring of muscle at the lower end of the esophagus. When a peristaltic wave passes down the esophagus, the muscle relaxes, permitting the bolus to enter the **stomach,** a large muscular organ (Fig. 39–8). When empty, the stomach is collapsed and shaped almost like a hotdog. Folds of the stomach wall called **rugae** give the inner lining a wrinkled appearance. As more and more food enters the stomach, the rugae gradually iron out, stretching the capacity of the stomach to more than a quart (about a liter).

The stomach is lined with simple columnar epithelium that secretes large amounts of mucus. Tiny pits mark the entrances to the millions of gastric glands, which extend deep into the stomach wall (Fig. 39–9). **Parietal cells** in the gastric glands secrete hydrochloric acid and a substance known as **intrinsic factor,** which is needed for adequate absorption of vitamin B. Chief cells in the gastric glands secrete **pepsinogen,** the inactive form of the enzyme **pepsin.** When pepsinogen comes in contact with the acidic gastric juice in the stomach, it is converted to pepsin, the main digestive enzyme of the stomach. Pepsin hydrolyzes proteins, reducing them to polypeptides. (Also see the Focus on Peptic Ulcers.)

FOCUS ON   Peptic Ulcers

One of the wonders of physiology is that gastric juice does not normally digest the stomach wall itself. Several protective mechanisms prevent this from happening. Cells of the gastric mucosa secrete an alkaline mucus that coats the stomach wall and also neutralizes the acidity of the gastric juice along the lining. In addition, the epithelial cells of the lining fit tightly together, preventing gastric juice from leaking between them and onto the tissue beneath. Should some of the epithelial cells be damaged, they are quickly replaced. In fact, the lifespan of an epithelial cell in the gastric mucosa is only about 3 days. About a half million of these cells are shed and replaced every minute.

Still, these mechanisms sometimes malfunction or prove inadequate, and a small bit of the stomach lining is digested, leaving an open sore or **peptic ulcer.** Substances such as alcohol and aspirin reduce the resistance of the stomach mucosa to digestion by gastric juice. Peptic ulcers occur more often in the duodenum than in the stomach. They also sometimes occur in the lower part of the esophagus.

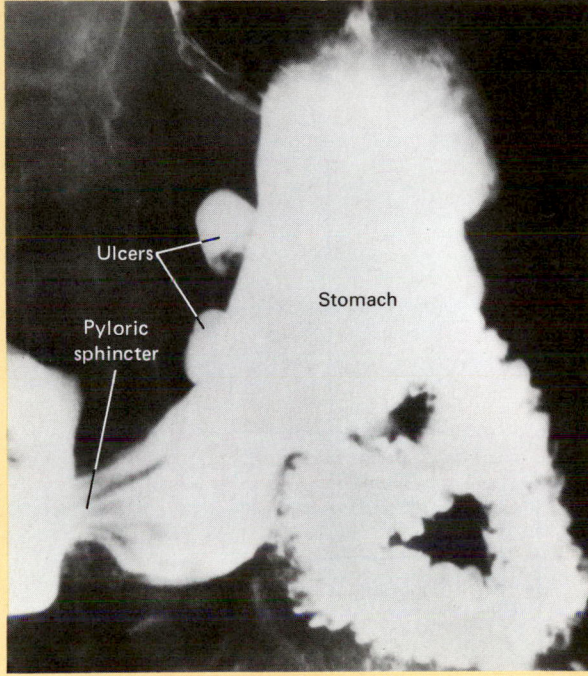

An x-ray of an ulcer in the wall of the stomach. The stomach and intestine have been filled with a contrast medium, making them appear white. This fluid also fills the cavities of the ulcers. (Courtesy of Dr. Jon Ehringer)

Peptic ulcers may bleed, leading to anemia. If the ulcer extends into the muscle layer, blood vessels may also be damaged, resulting in hemorrhage. A **perforated ulcer** is one that extends all the way through the wall of the stomach or other affected organ. The opening created may allow bacteria and food to pass through to the peritoneum, leading to peritonitis and shock. Perforation is the main cause of death from ulcers.

What changes have occurred in our bite of hamburger during its stay in the stomach? The stomach has churned and chemically degraded the food so that it has assumed the consistency of thick soup and is referred to as **chyme.** Protein digestion has begun, and much of the hamburger protein has been degraded to polypeptides. Digestion of the starch in the bun to small polysaccharides and maltose continued until salivary amylase was inactivated by the acidic pH of the stomach. After 3 or 4 hours of digestion in the stomach, peristaltic waves propel a few milliliters of chyme at a time through the stomach exit, the **pylorus,** and into the small intestine.

Most Enzymatic Digestion Takes Place Inside The Small Intestine

Digestion of food is completed in the **small intestine** and nutrients are absorbed through its wall. The small intestine has three regions: the **duodenum,** the **jejunum,** and the **ileum.** Most chemical digestion takes place in the duodenum (the first portion of the small intestine), not in the stomach, as is commonly believed. Bile from the liver and enzymes from the pancreas are released into the duodenum and act upon the chyme. Then enzymes produced by the epithelial cells lining the duodenum catalyze the final steps in the digestion of the major types of nutrients (Table 39–1).

The lining of the small intestine appears velvety because of millions of tiny finger-like projections in the lining, the intestinal **villi** (Figs. 39–10 and 39–11). The villi (singular, villus) increase the surface area of the small intestine for digestion and absorption of nutrients. The intestinal surface is further expanded by thousands of **microvilli,** folds of cytoplasm on the exposed borders of the epithelial cells. About 600 microvilli protrude from the surface of each cell, giving the epithelial lining a fuzzy appearance when viewed with the electron microscope. This fuzzy surface is referred to as a brush border.

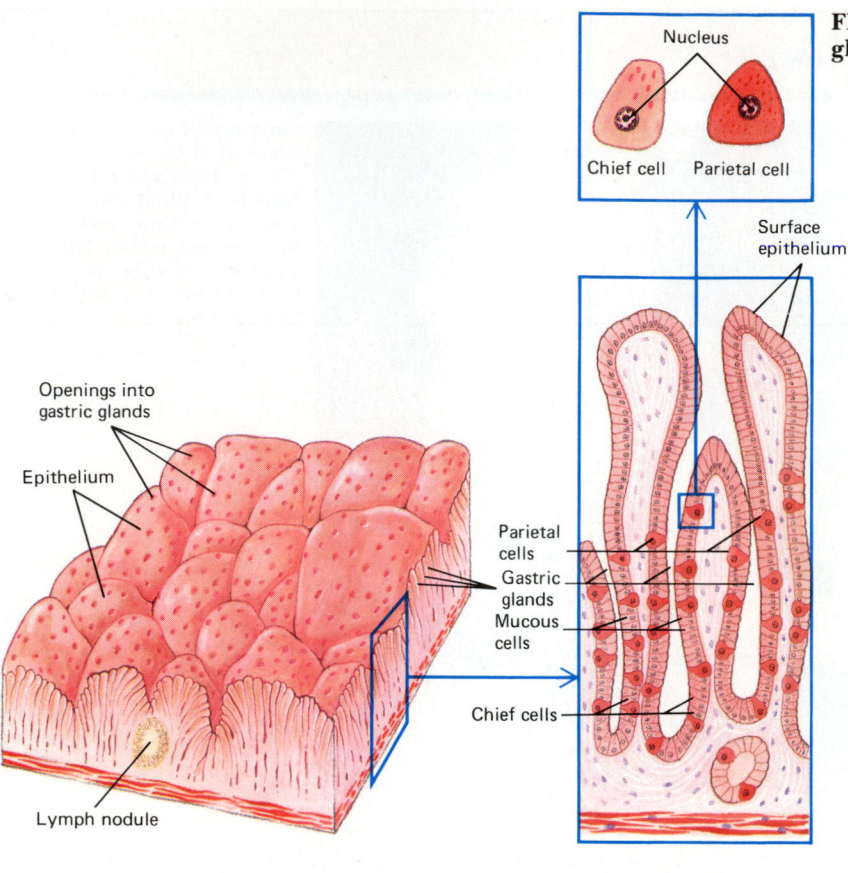

Nucleus

Chief cell   Parietal cell

Surface
epithelium

Openings into
gastric glands

Epithelium

Parietal
cells

Gastric
glands

Mucous
cells

Chief cells

Lymph nodule

Gastric mucosa

Gastric glands

FIGURE 39–9   The stomach lining and gastric glands.

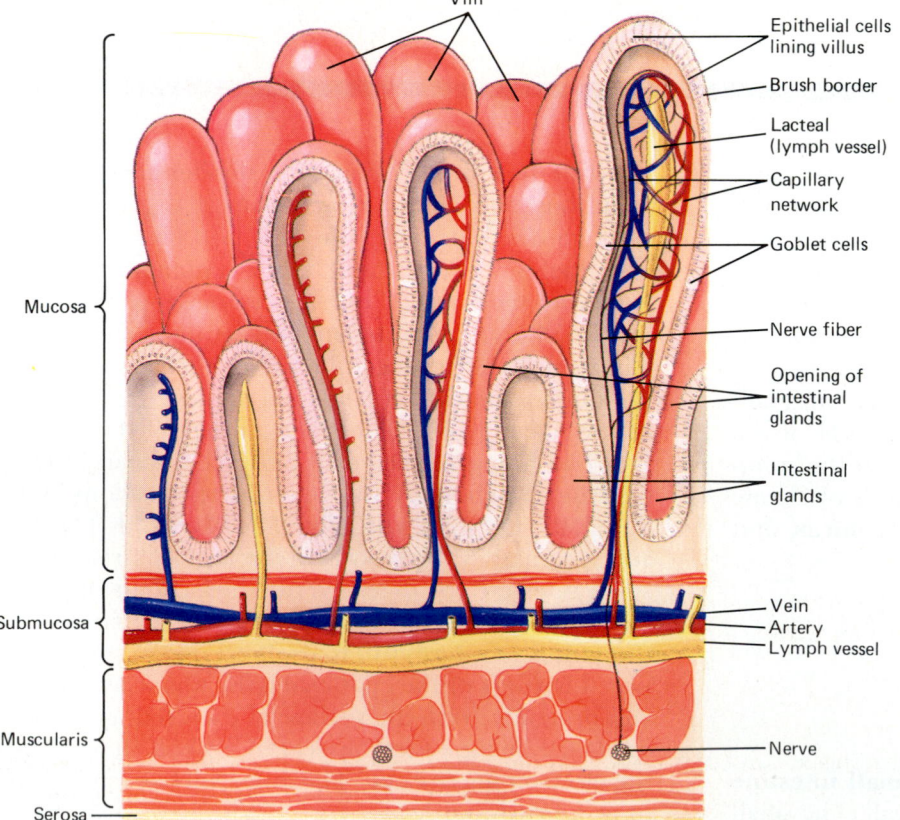

Villi

Epithelial cells
lining villus

Brush border

Lacteal
(lymph vessel)

Capillary
network

Goblet cells

Nerve fiber

Opening of
intestinal
glands

Intestinal
glands

Vein
Artery
Lymph vessel

Nerve

Mucosa

Submucosa

Muscularis

Serosa

**FIGURE 39–10   Diagram of the wall of
the human intestine showing the villi.
Some of the villi have been opened to
show the blood and lymph vessels
within.**

TABLE 39–1
Summary of Digestion

Location	Source of Enzyme	Digestive Process*

Carbohydrate digestion

Mouth — Salivary glands — Polysaccharides (e.g., starch) $\xrightarrow{\text{salivary amylase}}$ Maltose + Small polysaccharides

Stomach — — Action continues until salivary amylase is inactivated by acidic pH

Small intestine — Pancreas — Undigested polysaccharides and small polysaccharides $\xrightarrow{\text{pancreatic amylase}}$ Maltose

Intestine — Disaccharides hydrolyzed to monosaccharides as follows:

Maltose (malt sugar) $\xrightarrow{\text{maltase}}$ Glucose + Glucose

Sucrose (table sugar) $\xrightarrow{\text{sucrase}}$ Glucose + Fructose

Lactose (milk sugar) $\xrightarrow{\text{lactase}}$ Glucose + Galactose

Protein digestion

Stomach — Stomach (gastric glands) — Protein $\xrightarrow{\text{pepsin}}$ Polypeptides

Small intestine — Pancreas — Polypeptides
A—A—A—A—A
|
A—A—A—A—A
$\xrightarrow[\text{chymotrypsin}]{\text{trypsin,}}$ Tripeptides + Dipeptides
A—A—A   A—A

Dipeptides
A—A
$\xrightarrow{\text{carboxypeptidase}}$ Free amino acids
A   A
A

Small intestine — Tripeptides + Dipeptides
A—A—A   A—A
$\xrightarrow{\text{peptidases}}$ Free amino acids
A   A
A
A   A
A

Lipid digestion

Small intestine — Liver — Glob of fat $\xrightarrow{\text{bile salts}}$ Emulsified fat (individual triacylglycerols)

Pancreas — Triacylglycerol $\xrightarrow{\text{lipase}}$ Fatty acids + Glycerol

*◯ = monosaccharide; ⌐ = triacylglycerol; Ɛ = glycerol; ～ = fatty acid; A = amino acid units or, when standing alone, a free amino acid.

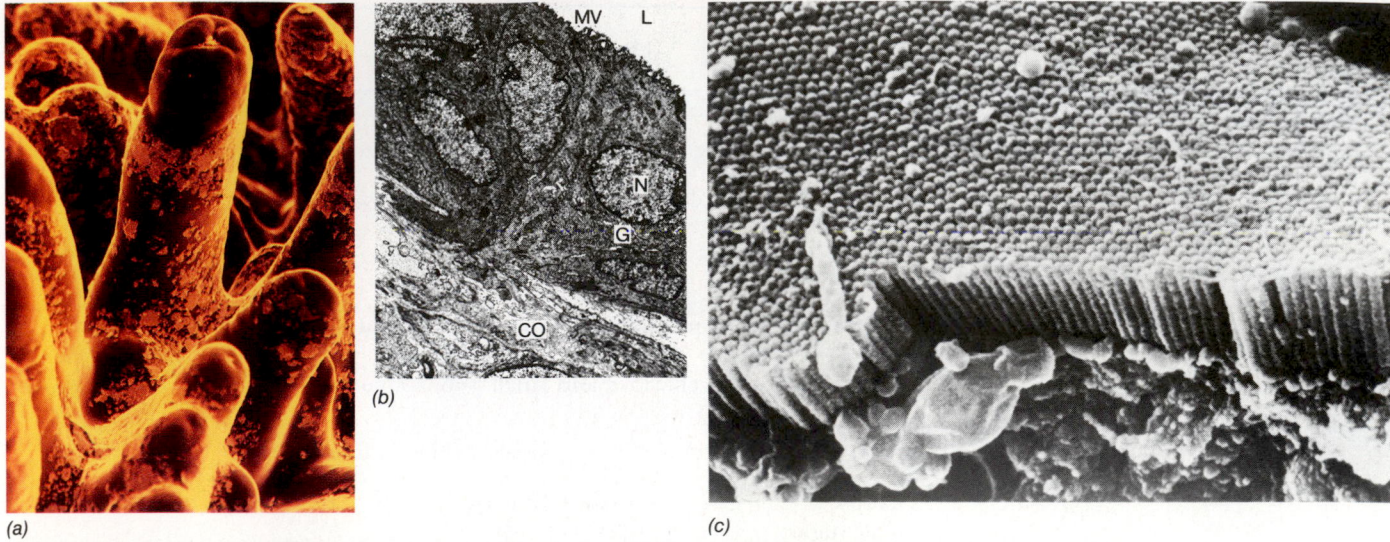

FIGURE 39–11   The intestinal lining. (*a*) Scanning electron micrograph (approximately ×1200) of villi in the small intestine. (*b*) Electron micrograph (approximately ×8000) of epithelial cells lining the small intestine, showing the microvilli (MV). L, lumen; N, nucleus of an epithelial cell; G, Golgi complex; CO, collagen. (*c*) Scanning electron micrograph (approximately ×14,000) of the surface of an epithelial cell from the lining of the small intestine showing microvilli. The epithelium has been cut vertically to allow the microvilli to be viewed from the side as well as from above. (*a*, David Phillips/Visuals Unlimited; *b*, courtesy of Dr. Lyle C. Dearden; *c*, courtesy of J.D. Hoskins, W.G. Henk, and Y.Z. Abdelbaki, from *American Journal of Veterinary Research*, 43:10)

If the intestinal lining were smooth like the inside of a water pipe, food would zip right through the intestine and many valuable nutrients would not be absorbed. Folds in the wall of the intestine, the villi, and microvilli together increase the surface area of the small intestine by about 600 times. If we could unfold and spread out the lining of the small intestine of an adult human, its surface would approximate the size of a tennis court.

The Liver Secretes Bile Which Mechanically Digests Fats

Just under the diaphragm lies the **liver,** the largest and also one of the most complex organs in the body (Fig. 39–12). A single liver cell can carry on more than 500 separate metabolic activities. The liver's food processing functions include the following:

1. Secretes **bile,** which is important in the mechanical digestion of fats.
2. Helps maintain homeostasis by removing nutrients from the blood.
3. Converts excess glucose to glycogen and stores it.
4. Converts excess amino acids to fatty acids and urea.
5. Stores iron and certain vitamins.
6. Detoxifies alcohol and many drugs and poisons that enter the body.

Bile consists of water, bile salts, bile pigments, cholesterol, salts, and lecithin (a phospholipid). Bile pro-

duced in the liver is stored in the pear-shaped **gallbladder.** The gallbladder concentrates the bile and releases it into the duodenum as needed. Bile mechanically digests fats by a detergent-like action in which it decreases the surface tension of fat particles. This permits the fat molecules to disperse so they can be worked on by lipases (fat-digesting enzymes). This dispersion of fat globules by bile is called **emulsification.** Bile does not digest any food material chemically, for it contains no digestive enzymes.

The Pancreas Secretes Digestive Enzymes

The **pancreas** is an elongated gland that secretes both digestive enzymes and hormones that help regulate the level of glucose in the blood. Among its enzymes are (1) **trypsin** and **chymotrypsin,** which digest polypeptides to dipeptides; (2) **pancreatic lipase,** which degrades neutral fats; (3) **pancreatic amylase,** which breaks down almost all types of carbohydrates, except cellulose, to disaccharides; and (4) **ribonuclease** and **deoxyribonuclease,** which split the nucleic acids ribonucleic acid (RNA) and deoxyribonucleic acid (DNA) to free nucleotides. Digestion of carbohydrates, protein, and lipid is summarized in Table 39–1.

Nerves And Hormones Regulate Digestion

Most digestive enzymes are produced only when food is present in the digestive tract. Salivary gland secretion is controlled entirely by the nervous system, but

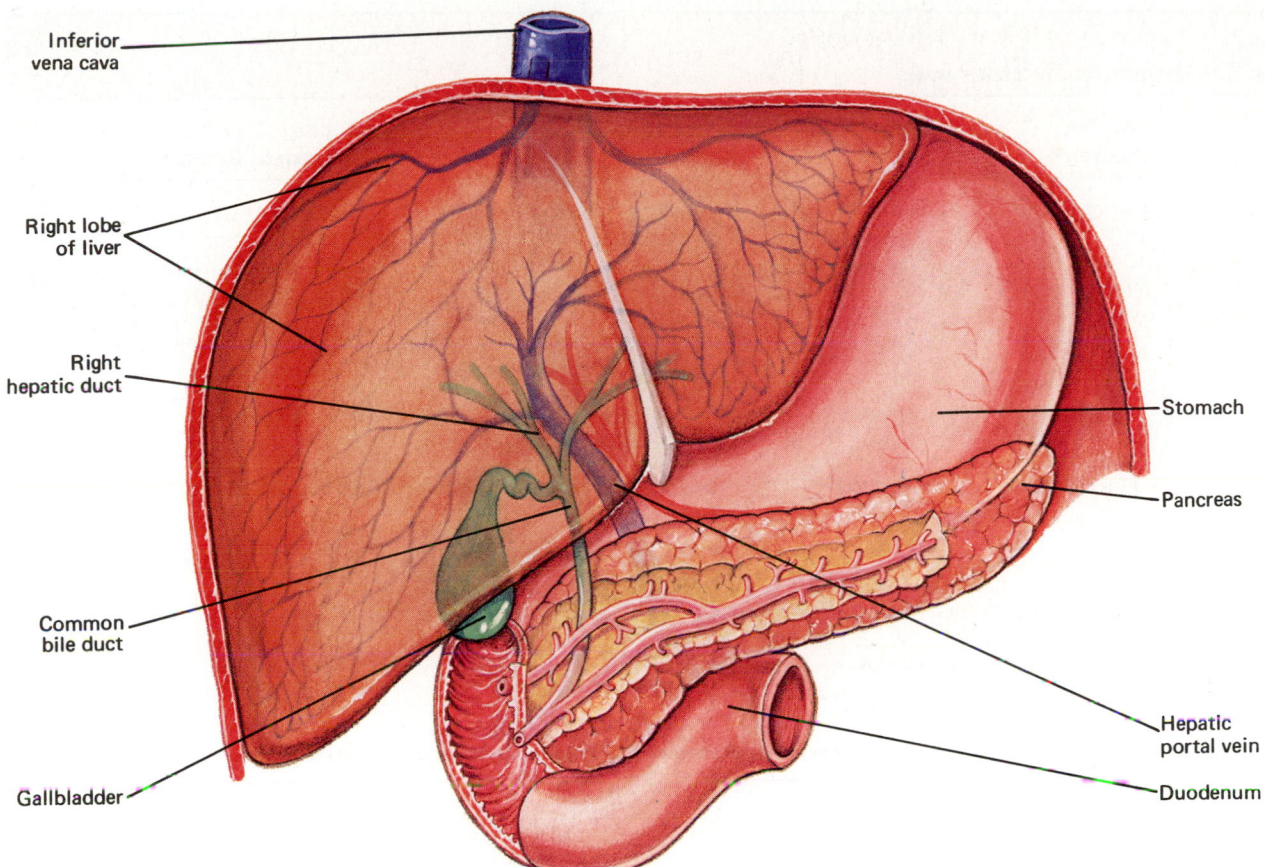

Inferior
vena cava

Right lobe
of liver

Right
hepatic duct

Common
bile duct

Gallbladder

Stomach

Pancreas

Hepatic
portal vein

Duodenum

**FIGURE 39–12   Structure of the liver and pancreas. Note the gallbladder
and ducts.**

secretion of other digestive juices is regulated by both
nerves and hormones (Table 39–2). As an example,
consider the secretion of gastric juice. Seeing, smelling,
tasting, or even thinking about food causes the brain to
send neural messages to the glands in the stomach,
stimulating them to secrete. In addition, when food
distends the stomach, it stimulates glands in the stomach wall to release the hormone **gastrin.** Gastrin is absorbed into the blood and transported to the gastric
glands, where it stimulates release of gastric juice.

Absorption Takes Place Mainly Through The Villi Of The Small Intestine

Only a few substances—water, simple sugars, salts, alcohol, and certain drugs—are composed of molecules
small enough to be absorbed through the wall of the
stomach. Absorption of nutrients is primarily the job
of the intestinal villi. As illustrated in Figure 39–10,
the wall of a villus consists of a single layer of epithelial
cells. Inside each villus is a network of capillaries and a
central lymph vessel called a **lacteal.** To reach the
blood (or lymph), a nutrient molecule must pass
through an epithelial cell of the intestinal lining and

through a cell lining the blood or lymph vessel. Glucose and amino acids cannot diffuse through the intestinal lining and are absorbed by active transport. Absorption of these nutrients is coupled with the active
transport of sodium (Chapter 6). Lipids enter the
lymph system by diffusion and are transported to the
upper trunk region, where the lymph fluid and its contents enter the blood.

The Large Intestine Eliminates Wastes

Indigestible material, such as the cellulose of plant
foods, along with unabsorbed chyme, passes into the
large intestine. This organ, though only about
1.3 meters long (about 4 feet), is called "large" because
its diameter is greater than that of the small intestine.
The small intestine joins the large intestine about
7 centimeters (2.8 inches) from the end of the large
intestine, thereby creating a blind pouch, the **cecum.**
The **vermiform appendix** projects from the end of the
cecum. (Appendicitis is an inflammation of the appendix.) The functions of the cecum and appendix in
human beings are not known, and they are generally
considered vestigial organs, perhaps important in the

TABLE 39–2
Regulation Of Digestion By Hormones

Hormone	Where Secreted	Target Tissue	Actions	Factors That Stimulate Release
Gastrin	Stomach	Stomach (gastric glands)	Stimulates gastric glands to secrete pepsinogen	Distention of stomach by food; certain substances, such as partially digested proteins and caffeine
Secretin	Duodenum	Pancreas	Stimulates release of alkaline component of pancreatic juice	Acidic chyme acting on duodenum
		Liver	Increase rate of bile secretion	
Cholecystokinin (CCK)	Duodenum	Pancreas	Stimulates release of digestive enzymes	Presence of fatty acids and partially digested proteins in duodenum
		Gallbladder	Stimulates contraction and emptying	
Gastric inhibitory peptide	Duodenum	Stomach	Decreases motor activity of stomach and thus slows stomach's emptying	Presence of fat or carbohydrate in duodenum

vegetarian past of the human species. Herbivores such as rabbits have a large, functional cecum containing bacteria that digest cellulose.

From the cecum to the **rectum** (the last portion of the large intestine) the large intestine is known as the **colon.** The regions of the large intestine are the cecum, ascending colon, transverse colon, descending colon, sigmoid colon, rectum, and anus, the opening for the elimination of wastes.

As the chyme passes slowly through the large intestine, water and sodium are absorbed from it, and it gradually assumes the consistency of normal feces. Bacteria inhabiting the large intestine enjoy the last remnants of the meal and return the favor by producing vitamin K and certain B vitamins that can be absorbed and utilized.

A distinction should be made between elimination and excretion. *Elimination* is the process of getting rid of digestive wastes, materials that have never left the digestive tract and did not participate in metabolic activities. *Excretion* refers to the process of getting rid of metabolic wastes, and in mammals is mainly the function of the kidneys. The large intestine, however, does excrete bile pigments.

When chyme passes through the intestine too rapidly, **defecation** (expulsion of feces) becomes more frequent and the feces are watery. This condition, called diarrhea, may be caused by anxiety, certain foods, or by certain disease organisms that irritate the intestinal lining. Prolonged diarrhea results in loss of water and salts, leading to dehydration, a serious condition, especially in infants.

Constipation results when chyme passes through the intestine too slowly. Because more water than usual is removed from the chyme, the feces may be hard and dry. Constipation is often caused by a diet containing insufficient fiber.

Cancer of the colon is one of the most common causes of cancer deaths in the United States (Fig. 39–13). Research indicates that this type of cancer may be related to diet, for the disease is more common in people whose diets are very low in fiber. It has been suggested that less fiber results in less frequent defecation, allowing prolonged contact between the mucous membrane of the colon and such carcinogens as nitrites (used as preservatives) in foods.

▢ ADEQUATE AMOUNTS OF REQUIRED NUTRIENTS ARE NECESSARY TO SUPPORT METABOLIC PROCESSES

From about 20 chemical elements, which are absorbed in the form of simple compounds and salts, plants are able to produce all the different kinds of organic molecules they need. Animals are not such sophisticated chemists. Although they require approximately the

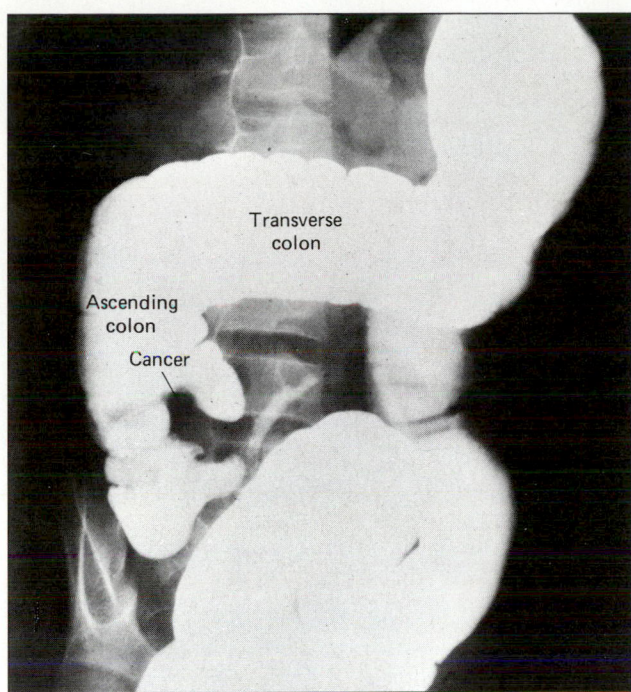

FIGURE 39-13 X-ray view of the large intestine of a patient with cancer of the colon. The lumen of the large intestine has been filled with a suspension of barium sulfate, which makes irregularities in the wall visible. The cancer is evident as a mass that projects into the lumen.

same 20 chemical elements, these elements must already be chemically combined in the form of about 40 chemical substances, many of them organic compounds. With only slight variation, all animals require the same basic nutrients—carbohydrates, lipids, proteins, vitamins, and minerals. Although not considered a nutrient in a strict sense, water is a necessary dietary component. Sufficient fluid must be ingested to replace fluid lost in urine, sweat, feces, and breath.

Our discussion will focus on human nutritional needs and on the metabolic fate of these nutrients. Adequate nutrition is an important global concern. In a 1988 report issued by then U.S. Surgeon General C. Everett Koop, "diseases of dietary excess and imbalance" were cited as among the leading causes of death in the United States. Based on studies by more than 2000 scientists around the world, Koop recommended that Americans change their eating habits to improve health and prevent disease.

Adequate amounts of essential nutrients are necessary for metabolic processes. Recall that metabolism refers to all of the chemical processes that take place in the body. Metabolic processes include anabolism and catabolism. Anabolism refers to synthetic processes; catabolism includes breakdown processes such as biological oxidation.

Once nutrients are absorbed from the digestive tract, they are transported by the blood. Surplus nutri-ents are taken up by the liver cells, where they are either stored or converted into other materials. Under normal circumstances blood leaving the liver carries sufficient nutrients to meet the requirements of all the cells of the body. The blood has been appropriately described as a traveling smorgasbord from which each cell selects whatever nutrients it needs to carry on its metabolic processes.

Carbohydrates Are A Major Energy Source In The Human Diet

Sugars and starches are the principal sources of energy in the ordinary human diet. However, they are not considered essential nutrients, because the body can obtain energy from a mixture of proteins and fats. In the average American diet carbohydrates provide about 50% of the Calories (Cal) ingested daily. Nutritionists measure the energy value of food in Calories per gram of food. (A Calorie, spelled with a capital C, is actually a kilocalorie. It is defined as the amount of heat required to raise the temperature of a kilogram of water from 15°C to 16°C. The physical unit of heat— the calorie [spelled with a lower case c], used by chemists—is 1000 times smaller.)

Carbohydrates Are Commonly Ingested As Starch, Cellulose, And Sucrose

Most carbohydrates are ingested in the form of starch and cellulose, both polysaccharides. (You may want to review the discussion of carbohydrates in Chapter 4.) Nutritionists refer to polysaccharides as complex carbohydrates. Foods rich in complex carbohydrates include rice, potatoes, corn, and other cereal grains. These are the least expensive foods, and for this reason the proportion of carbohydrate in a family's diet often reflects economic status. Very poor people subsist on diets that are almost exclusively carbohydrate, while the more affluent enjoy the more expensive protein-rich foods, such as meat and dairy products.

Fiber is mainly a complex mixture of cellulose and other indigestible carbohydrates. The American diet is low in fiber due to low intake of fruit and vegetables and use of refined flour. Increasing fiber in the diet may decrease the risk of cancer of the colon. Fiber may also stimulate the feeling of being satisfied with the amount of food intake (satiety), and thus be useful in treating obesity. In his 1988 report on nutrition, then U.S. Surgeon General C. Everett Koop suggested that Americans increase their consumption of complex carbohydrates and fiber by eating more fruits, vegetables, and whole grains. He strongly urged Americans to reduce their dietary intake of animal products which are high in fat.

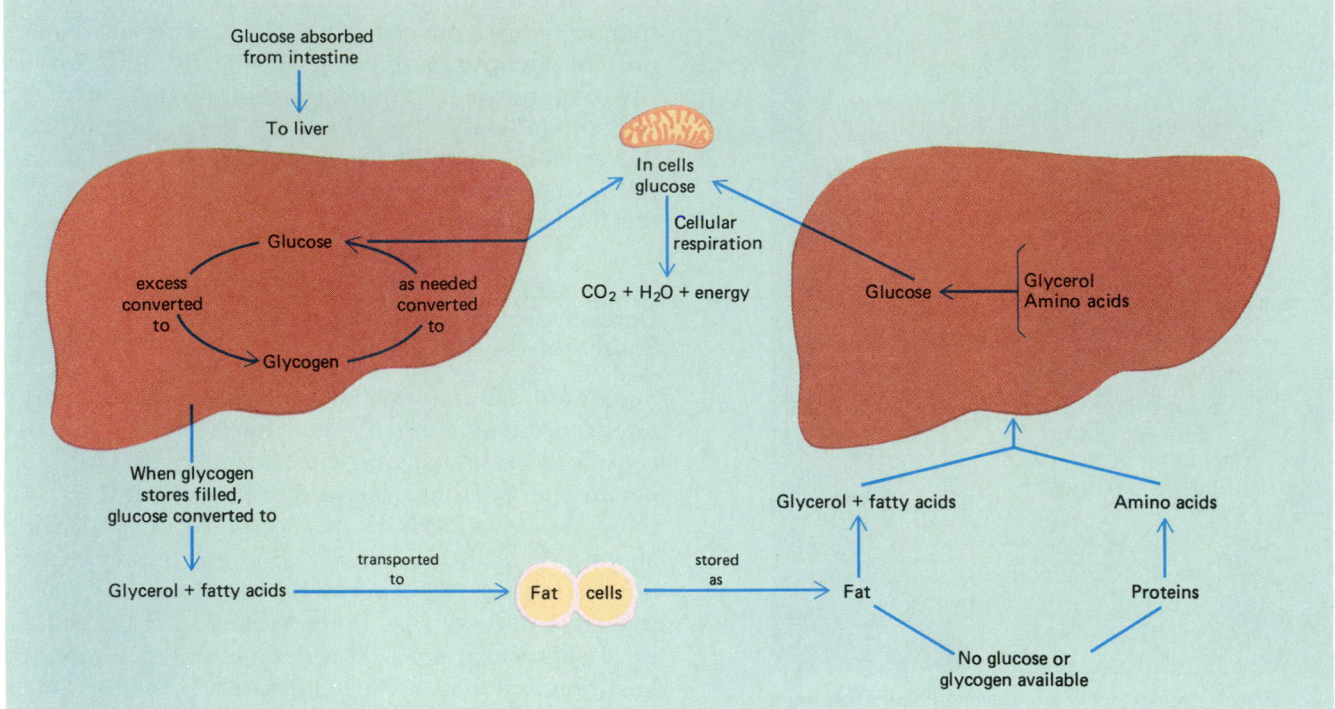

FIGURE 39–14   The fate of glucose in the body. The liver plays a central role in maintaining an appropriate level of glucose in the blood.

In affluent societies about 25% of the carbohydrate intake (more in children) is in the form of the disaccharide sucrose—cane or beet sugar. Sucrose is the so-called refined sugar put in coffee and desserts. Other important sugars are lactose, the sugar in milk, and fructose, found in fruits.

Glucose Is Used As Fuel By The Cells

Monosaccharides are the product of carbohydrate digestion. In the liver the various monosaccharides are converted to glucose. An important function of the liver is to help regulate the concentration of glucose in the blood (blood sugar level). Cells of the body require that a constant supply of glucose be delivered by the blood. Brain cells are especially dependent because they are unable to store glucose themselves. If deprived of an energy source for even a few minutes, they cease to function. After a meal or rich dessert, when there is an excess of glucose in the blood, the liver cells remove and store it as glycogen. Between meals, when the glucose level begins to fall, the liver cells slowly disassemble glycogen and release glucose back into the blood. In this way the liver maintains a rather steady glucose level in the blood (Fig. 39–14).

The normal blood glucose content while fasting is about 90 milligrams per 100 milliliters of blood. After a carbohydrate-rich meal the level may increase briefly to about 140 milligrams per 100 milliliters. If the liver did not remove the excess, the level would rise to more than three times normal after a carbohydrate-rich meal and then fall disastrously between meals or during the night.

The amount of glycogen stored in the liver is sufficient to maintain the blood glucose level for several hours. After the glycogen is used up, liver cells convert amino acids and the glycerol portions of fat to glucose (Fig. 39–15). Several hormones influence the various processes by which the liver regulates blood glucose level; these will be discussed in Chapter 41.

When an excess of carbohydrate-rich food is eaten, the liver cells may become fully packed with glycogen and still have more glucose coming in. In this situation liver cells convert excess glucose to fatty acids and glycerol, which solves the excess glucose problem. However, these compounds are then synthesized into triglycerides (triacylglycerols) and sent to the fat depots of the body for storage.

Lipids Are Used As An Energy Source And To Make Needed Biological Molecules

Cells use ingested lipids as fuel, as components of cell membranes, and to make lipid compounds, such as steroid hormones and bile salts. Lipid accounts for about 40% of the Calories in the average American diet. In

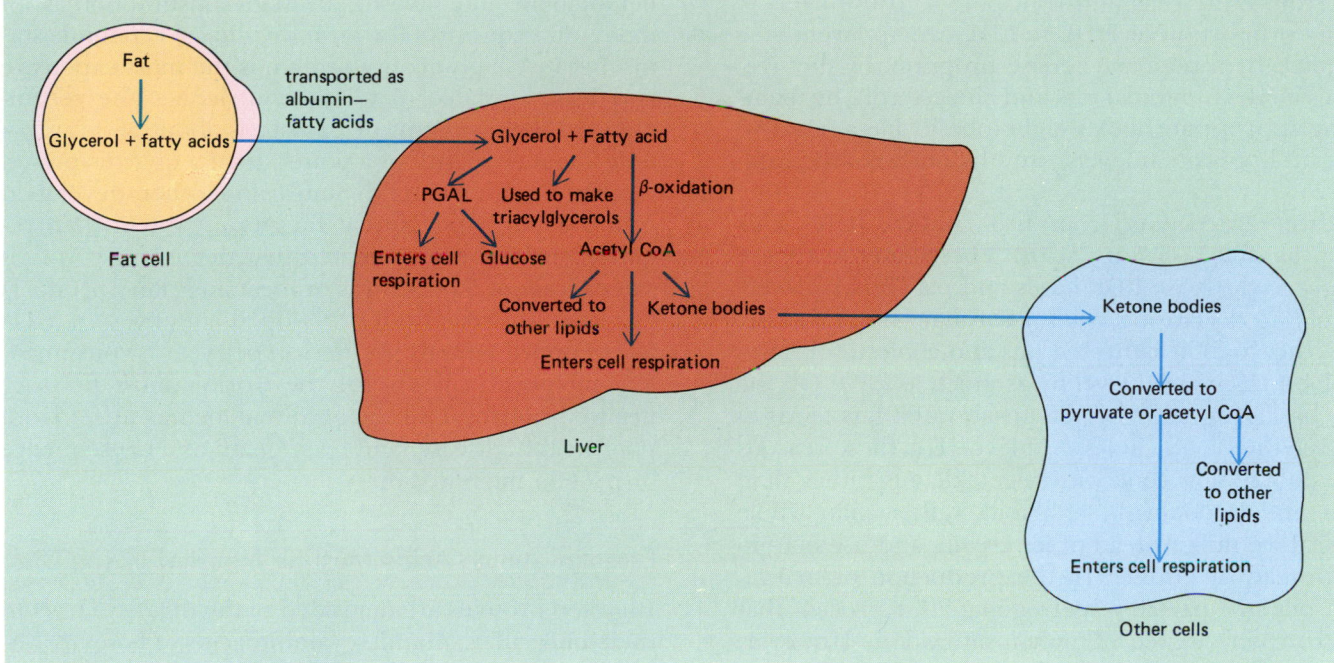

FIGURE 39–15 Overview of lipid metabolism.

poor countries this percentage falls to less than 10%, because most lipid-rich foods—meats, eggs, and dairy products—are relatively expensive.

Three polyunsaturated fatty acids (linoleic, linolenic, and arachidonic acids) are essential fatty acids that must be obtained in the human diet. Given these and sufficient nonlipid nutrients, the body can make all of the lipid compounds (including fats, cholesterol, phospholipids, and prostaglandins) that it needs. For this reason and because lipids are widespread in foods, dietary deficiency of lipids is uncommon.

Most Lipids Are Ingested As Triglycerides

About 98% of lipids in the diet are ingested in the form of triglycerides (triacylglycerols). (Recall from Chapter 4 that a triglyceride is a glycerol molecule chemically combined with three fatty acids; Fig. 4–6.) Triglycerides may be saturated, that is, fully loaded with hydrogens. They may be mono-unsaturated (containing one double bond in the carbon chain of a fatty acid, so two more hydrogen atoms can be added), or polyunsaturated (containing two or more double bonds, so four or more hydrogen atoms can be added).

Generally, animal foods are rich in both saturated fats and cholesterol, and plant foods contain unsaturated fats and no cholesterol. Commonly used polyunsaturated vegetable oils are corn, soya, cottonseed, and safflower oils. Olive and peanut oils contain large amounts of mono-unsaturated fats. Butter contains mainly saturated fats.

The average American diet provides about 700 milligrams of cholesterol each day, whereas only about 300 milligrams is recommended. Cholesterol sources are egg yolks, butter, and meat. The body is not dependent upon dietary sources for cholesterol because it is able to synthesize cholesterol from other nutrients. In fact, dietary intake of saturated fats can increase cholesterol level markedly.

Dietary Triglycerides And Cholesterol Have Been Associated With Heart Disease

Lipids have been the focus of much research because of their role in atherosclerosis, a progressive disease in which the arteries become occluded with fatty material. As discussed in Chapter 36, atherosclerosis leads to circulatory impairment and heart disease.

Cholesterol and triglycerides are not transported free in the blood plasma but are bound to proteins and transported as macromolecular complexes called **lipoproteins.** About 70% of plasma cholesterol is transported on **low-density lipoproteins (LDLs).** The remaining cholesterol is transported mainly on **high-density lipoproteins (HDLs).** High levels of LDL have been correlated with increased risk for coronary artery and heart disease. LDL is thought to pick up cholesterol and deposit it in body cells; under some conditions the cholesterol is deposited in the smooth muscle cells of the arterial wall. When LDL levels are high, HDL may play a protective role and decrease risk for coronary heart disease. HDL may collect cholesterol from body cells and transport it to the liver. A healthy proportion of HDL to LDL can apparently be promoted by a regular exercise program, by diet (reducing intake of animal fats and cholesterol), by maintaining appropriate body weight (obesity has a negative effect on lipoprotein levels), and by not smoking cigarettes.

Unsaturated fats from fish contain fatty acids known as omega-3 fatty acids. These fatty acids are thought to decrease LDL levels and play other protective roles in decreasing risk for coronary heart disease.

A diet high in saturated fats and cholesterol raises the blood cholesterol level by as much as 25%. On the other hand, ingestion of polyunsaturated fats tends to decrease the blood cholesterol level. For these reasons many people now cook with vegetable oils rather than butter and lard, drink skim milk rather than whole milk, eat ice milk instead of ice cream, and use margarine instead of butter. (In the production of margarines, oils are partially hydrogenated, a process that reduces their degree of polyunsaturation. However, this process can be controlled so that soft margarine is produced which contains a large proportion of polyunsaturated fats.)

Fat Is Stored In Adipose Tissue

Although they are absorbed into the lymphatic system, fats eventually enter the blood. From the blood, fats are taken up by the adipose tissues and stored. When needed, stored fats are hydrolyzed to fatty acids and released into the blood. Before these fatty acids can be used by cells as fuel, they must be broken down into smaller compounds and combined with coenzyme A to form molecules of acetyl coenzyme A (Fig. 39–15). This transformation is accomplished in the liver (by a process known as beta oxidation).

For transport to the cells, acetyl coenzyme A is converted into one of three types of ketone bodies. Normally, the level of ketone bodies in the blood is low, but in certain abnormal conditions, such as starvation or diabetes mellitus, fat metabolism is tremendously increased. Ketone bodies are then produced so rapidly that their level in the blood becomes excessive and causes the blood to become too acidic. Such disruption of normal pH balance can lead to death. (How Eskimos, who live on diets extremely high in fat, manage to maintain acid-base homeostasis is somewhat of a mystery.)

Proteins Serve As Enzymes And Are Essential Structural Components Of Cells

Proteins are critical nutrients because they are essential building blocks of cells. Proteins also serve as enzymes and are used to make many needed substances such as hemoglobin and myosin. Protein consumption is an index of a country's (or an individual's) economic status, because high-quality protein is the most expensive and least available of all the nutrients. The recommended daily amount of protein is about 56 grams—only about an eighth of a pound (half a quarter-pound burger). In the United States and other developed countries most people eat far more protein than required. It has been estimated that the average American eats about 300 pounds of meat and dairy products per year. In some underdeveloped countries an average of only 2 pounds per person per year is consumed. Protein poverty is one of the world's most pressing health problems; millions of human beings suffer from poor health, disease, and even death as a consequence of protein malnutrition.

Essential Amino Acids Must Be Ingested In The Diet

Ingested proteins are degraded in the digestive tract to their molecular subunits—amino acids. These are absorbed and used by the cells to make the types of proteins needed. Of the 20 or so amino acids important in nutrition, the body is able to make several by rearranging the atoms of certain organic acids. About eight of the amino acids (nine in children) cannot be synthesized by human body cells at all, or at least not in sufficient quantity to meet the body's needs. These, which must be provided in the diet, are referred to as **essential amino acids.**

Not all ingested proteins contain the same kinds or quantities of amino acids, and many proteins lack some of the essential amino acids. Complete proteins, those that contain the most appropriate distribution of amino acids for human nutrition, are found in eggs, milk, meat, and fish. Some foods, such as gelatin or soybeans, contain a high proportion of protein but do not contain all the essential amino acids, or they do not contain them in proper nutritional proportions. Most plant proteins are deficient in one or more essential amino acids (usually lysine, tryptophan, or threonine).

Most human beings depend upon cereal grains as their staple food—usually rice, wheat, or corn (see Focus on Vegetarian Diets). None of these foods provides an adequate proportion of total amino acids or adequate distribution of essential amino acids, especially not for growing children. In some underdeveloped countries, starchy crops, such as sweet potatoes or cassava, are the principal food. Total protein content of these foods is less than 2%, far below minimum needs.

Amino Acids Are Deaminated, Then Used As Fuel Or Stored As Fat

Amino acids circulating in the blood are removed as needed by individual body cells and used primarily for

FOCUS ON Vegetarian Diets

Most of the world's population depends almost entirely upon plants, especially cereal grains—usually rice, wheat, or corn—as the staple food. None of these foods contain adequate amounts of all of the essential amino acids. Besides being deficient in some of the essential amino acids, plant foods contain a lower percentage of protein than do animal foods. Meat contains about 25% protein, whereas even the new high-yield grains contain only 5% to 13%. What protein is available in plant food is also less digestible than that in animal foods. Because most of the protein is encased within indigestible cellulose cell walls, much of it passes right through the digestive tract.

Despite these potential nutritional problems, more and more people are turning to vegetarian diets. Meats are becoming increasingly expensive because they are ecologically expensive to produce. About 21 kg of protein in grain, for example, is required to produce just 1 kg of beef protein. If the human population of our planet continues to expand at a much greater rate than food production, more grain will be diverted for human food and less for animal feed. The price of meat will continue to soar and may become unaffordable for many of us.

Can a vegetarian diet be nutritionally balanced? With an awareness of the special nutritional problems associated with a vegetarian diet (especially in growing children), they can be overcome. The most important rule is to select foods that complement each other. This requires knowledge of which amino acids are deficient in each kind of food. Since the body cannot store amino acids, all of the essential amino acids must be ingested at the same meal. For example, if rice is eaten for dinner, and beans for lunch the next day, the body will not have all of the essential amino acids that are needed at the same time to manufacture proteins. If beans and rice are eaten together, however, all of the needed amino acids will be provided, because one food provides what the other lacks. Similarly, if dairy products are not excluded from the vegetarian diet, then macaroni can be paired with cheese, or cereal with milk, and all the essential amino acids will be obtained.

---

synthesis of proteins. Excess amino acids are removed from circulation by the liver. In the liver cells these are deaminated; that is, the amine group is removed. Deamination forms ammonia, which is toxic at high concentrations. The ammonia is converted to urea, which is excreted from the body.

The remaining carbon chain of the amino acid (called a keto acid) may be converted into carbohydrate or lipid and used as fuel or stored (Fig. 39–16). Thus, even people who eat high-protein diets can gain weight if they eat too much. Figure 39–17 summarizes the interrelationships of carbohydrate, protein, and lipid metabolism.

Vitamins Are Organic Compounds Essential For Normal Metabolism

Vitamins are organic compounds required by the body for biochemical processes. Many function as components of coenzymes (see Chapter 7). Very small amounts are required in comparison with other dietary constituents. Vitamins may be divided into two main groups. **Fat-soluble vitamins** are those that can be dissolved in fat and include vitamins A, D, E, and K. **Water-soluble vitamins** are the B and C vitamins. Table 39–3 gives the sources, functions, and consequences of deficiency for most of the vitamins (also see Fig. 39–18).

While millions of people gulp down a variety of vitamin pills daily with great fervor, most nutritionists contend that people would not require vitamin supplements and would be healthier if they would eat a nutritionally balanced diet. Proponents of vitamin supplements argue that most of us do not eat a balanced diet and therefore are likely to suffer from vitamin deficiencies.

Debates also rage over the advisability of taking large amounts of certain specific vitamins, such as vitamin C to prevent colds or vitamin E to protect against vascular disease. To date there is no compelling evidence to support claims that massive quantities of any vitamin are beneficial. We do not even understand all the biochemical roles played by vitamins or the interactions between various vitamins and other nutrients. We do know that large overdoses of vitamins, like vitamin deficiency, can be harmful. Moderate overdoses of the B and C vitamins are excreted in the urine, but surpluses of the fat-soluble vitamins are not easily excreted and can accumulate to harmful levels.

Minerals Are Inorganic Nutrients Required By Cells

Minerals are inorganic nutrients generally ingested in the form of salts dissolved in food and water. Essential minerals required in amounts of a gram or more daily include sodium, chlorine, potassium, magnesium, calcium, sulfur, phosphorus, and fluorine (Table 39–4). Several others, such as copper, iodine, and cobalt, are required in milligram or microgram amounts and so are known as **trace elements.**

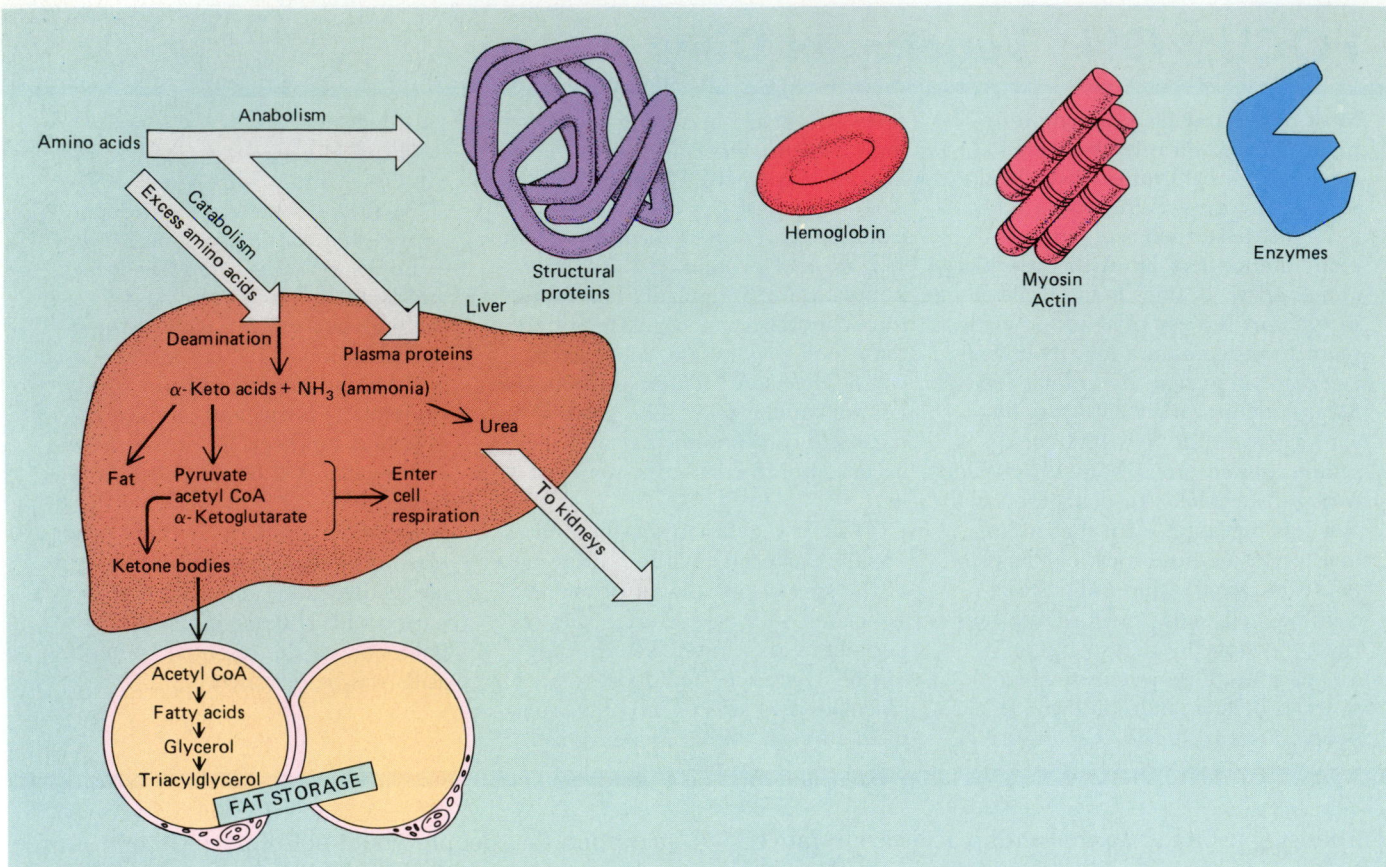

FIGURE 39–16   Overview of protein metabolism.

Minerals are needed as components of all body tissues and fluids. Salt content (about 0.9%) is vital in maintaining the fluid balance of the body, and since salts are lost from the body daily in sweat, urine, and feces, they must be replaced by dietary intake. Sodium chloride (common table salt) is the salt needed in largest quantity in blood and other body fluids. A deficiency of this salt results in dehydration.

Phosphorus and calcium are important structural components of bones and teeth. Phosphorus is also a component of nucleic acids, adenosine triphosphate (ATP), creatine phosphate, and several other biologically important compounds. Calcium is required for blood clotting, muscle contraction, and transmission of nerve impulses.

Iron, the mineral most likely to be deficient in the diet, is an essential part of the cytochrome components of the electron transport system and an essential part of the hemoglobin molecule. When iron intake is inadequate, the body cannot synthesize enough hemoglobin, and as a result the capacity of the blood to transport oxygen to the cells is reduced. People with this condition, known as iron-deficiency anemia, lack energy and are easily fatigued.

Energy Metabolism Is Balanced When Energy Input Equals Energy Output

The amount of energy (heat) liberated by the body during metabolism is a measure of the **metabolic rate;** much of the energy expended by the body is ultimately converted to heat. Metabolic rate may be expressed either in Calories of heat energy expended per hour per day or as a percentage above or below a standard normal level.

The **basal metabolic rate (BMR)** is the rate at which the body releases heat as a result of breaking down fuel molecules. BMR is the body's basic cost of metabolic living, that is, the rate of energy used during resting conditions. An individual's **total metabolic rate** is the sum of his or her BMR and the energy used to carry on all daily activities. An athlete or a laborer has a greater total metabolic rate than a teacher or executive who does not exercise regularly.

An average-sized man who does not engage in any exercise program and who sits at a desk all day expends about 2000 Calories daily. If the food he eats each day contains about 2000 Calories, he will be in a state of energy balance; that is, his energy input will

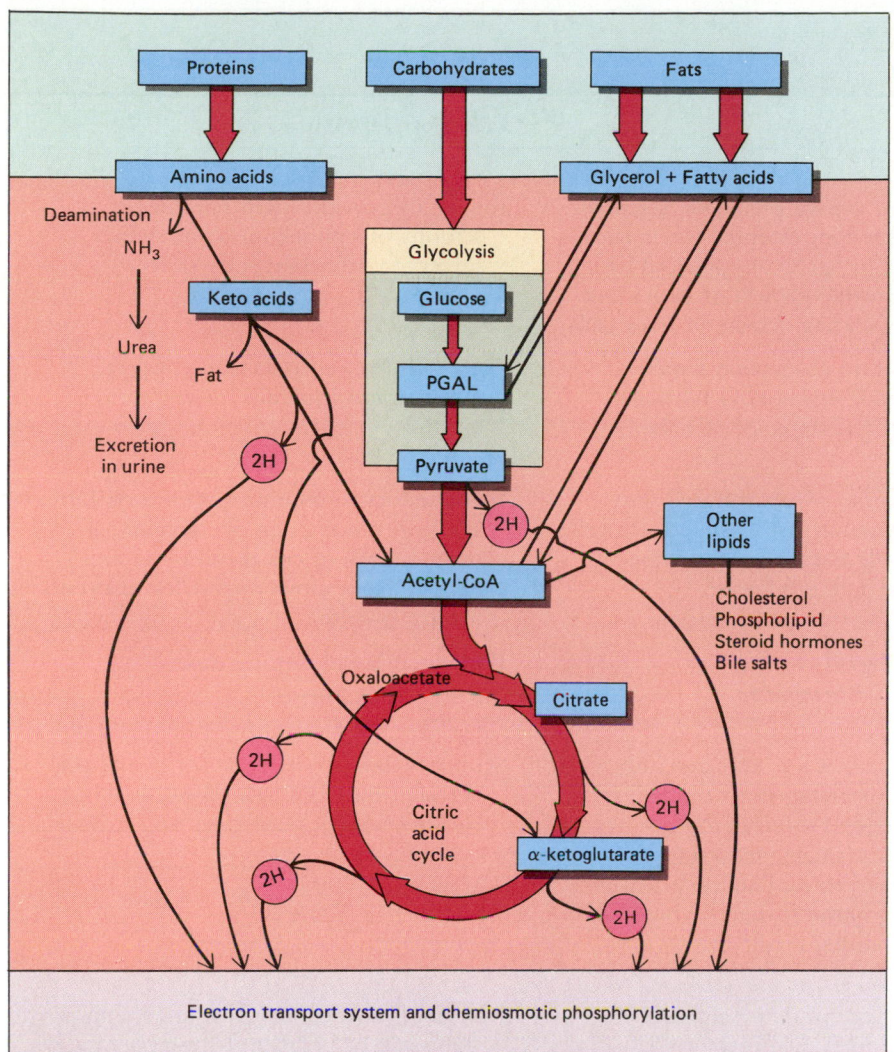

FIGURE 39–17 Overview of the integrated metabolism of carbohydrates, fats, and proteins. This diagram is greatly simplified and illustrates only a few of the principal pathways.

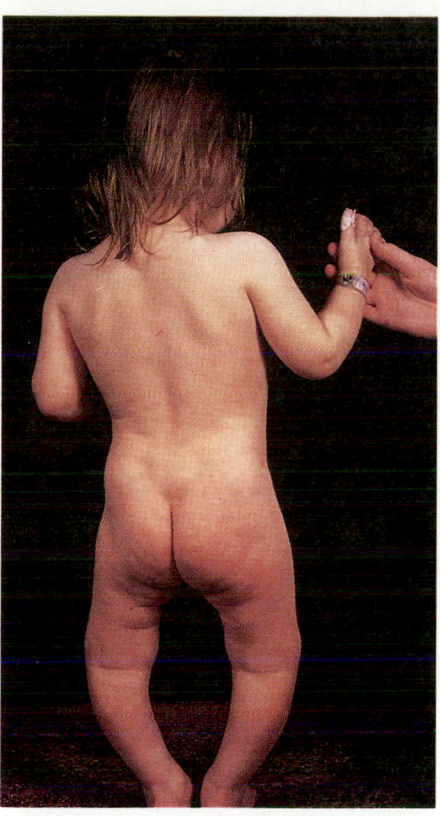

FIGURE 39–18 Child with rickets, a condition that results from deficiency of vitamin D during childhood. This deficiency decreases the body's ability to absorb and use calcium and phosphorus and produces soft, malformed bones. Note the bowed legs. (Biophoto Associates)

equal his energy output. This is an extremely important concept, for when

$$\text{Energy (Calorie) Input} = \text{Energy Output}$$

body weight remains constant. When energy output is greater than energy input, stored fat is burned and body weight decreases. On the other hand, people gain weight when they take in more energy (Calories) in food than they expend in daily activity—in other words, when

$$\text{Energy (Calorie) Input} > \text{Energy Output}$$

Obesity Is A Serious Nutritional Problem

Obesity, the excess accumulation of body fat, is a serious form of malnutrition, and in our affluent society it has become a problem of epidemic proportions. An overweight person places an extra burden upon the heart and is susceptible to heart disease and other ailments. Obese persons generally die at a younger age than people of normal weight. According to insurance statistics, men who are 20% or more overweight bear a 43% greater risk of dying from heart disease, a 53% greater risk of dying from cerebral hemorrhage, and a 133% greater risk of dying as a result of diabetes, compared with men of normal weight. A man who is 20% overweight is 30% more likely to die before retirement age than if his weight were normal. Yet one-third of our working population is 25% or more overweight.

Obesity can result from an increase in the size of fat cells or from an increase in the number of fat cells, or both. The number of fat cells in the adult is apparently determined mainly by the amount of fat stored during infancy and childhood. When babies or small

TABLE 39–3
The Vitamins

Vitamins and U.S. RDA*	Actions	Effect of Deficiency
Fat-soluble		
A 5000 IU†	Component of retinal pigments, essential for normal vision; essential for normal growth and integrity of epithelial tissue; promotes normal growth of bones and teeth by regulating activity of bone cells	Failure of growth; night blindness; atrophy of epithelium; epithelium subject to infection; scaly skin
D 400 IU	Promotes calcium absorption from digestive tract; essential to normal growth and maintenance of bone	Bone deformities; rickets in children; osteomalacia in adults
E 30 IU	Inhibits oxidation of unsaturated fatty acids and vitamin A that help form cell and organelle membranes; precise biochemical role not known	Increased catabolism of unsaturated fatty acids, so not enough are available for maintenance of cell membranes and other membranous organelles; prevents normal growth
K probably about 1 mg	Essential for blood clotting	Prolonged blood-clotting time
Water-soluble		
C (ascorbic acid) 60 mg	Needed for synthesis of collagen and other intercellular substances; formation of bone matrix and tooth dentin, intercellular cement; needed for metabolism of several amino acids; may help body withstand injury from burns and bacterial toxins	Scurvy (wounds heal very slowly and scars become weak and split open; capillaries become fragile; bone does not grow or heal properly)
B-complex vitamins		
Thiamine (B_1) 1.5 mg	Acts as coenzyme in many enzyme systems; important in carbohydrate and amino acid metabolism	Beriberi (weakened heart muscle, enlarged right side of heart, nervous system and digestive tract disorders)
Riboflavin (B_2) 1.7 mg	Used to make coenzymes (e.g., FAD) essential in cellular respiration	Dermatitis; inflammation and cracking at corners of mouth; mental depression
Niacin (nicotinic acid) 20 mg	Component of important coenzymes (NAD† and NADP†) essential to cellular respiration	Pellagra (dermatitis, diarrhea, mental symptoms, muscular weakness, fatigue)
Pyridoxine (B_6) 2 mg	Coenzyme needed for amino acid synthesis and protein metabolism	Dermatitis; digestive tract disturbances; convulsions
Pantothenic acid 10 mg	Constituent of coenzyme A (important in cellular metabolism)	Deficiency extremely rare
Folic acid 0.4 mg	Coenzyme needed for reactions involved in nucleic acid synthesis and for maturation of red blood cells	A type of anemia
Biotin 0.3 mg	Coenzyme needed for carbon dioxide fixation	Deficiency unknown
B_{12} 6 mg	Coenzyme important in nucleic acid metabolism	Pernicious anemia

*RDA: the recommended dietary allowance, established by the Food and Nutrition Board of the National Research Council, to maintain good nutrition for healthy persons.
†International Unit: the amount that produces a specific biological effect and is internationally accepted as a measure of the activity of the substance.

Sources	Comments
Liver, fish-liver oils, egg, yellow and green vegetables	Can be formed from provitamin carotene (a yellow or red pigment); sometimes called anti-infection vitamin because it helps maintain epithelial membranes; excessive amounts harmful
Liver, fish-liver oils, egg yolk, fortified milk, butter, margarine	Two types: D_2 (calciferol), a synthetic form, and D_3, formed by action of ultraviolet rays from sun upon a cholesterol compound in the skin; excessive amounts harmful
Oils made from cereals, seeds, liver, eggs, fish	
Normally supplied by intestinal bacteria; green leafy vegetables	Antibiotics may kill bacteria, then supplements are needed in surgical patients
Citrus fruits, strawberries, tomatoes	Possible role in preventing common cold or in the development of acquired immunity(?); harmful in very excessive dose
Liver, yeast, cereals, meat, green leafy vegetables	Deficiency common in alcoholics
Liver, cheese, milk, eggs, green leafy vegetables	
Liver, meat, fish, cereals, legumes, whole-grain and enriched breads	
Liver, meat, cereals, legumes	
Widespread in foods	
Produced by intestinal bacteria; liver, cereals, dark-green leafy vegetables	
Produced by intestinal bacteria; liver, chocolate, egg yolk	
Liver, meat, fish	Contains cobalt; intrinsic factor secreted by gastric mucosa needed for absorption

TABLE 39–4
Some Important Minerals And Their Functions

Mineral	Functions	Comments
Calcium	Component of bones and teeth; essential for normal blood clotting; needed for normal muscle and nerve function	Bones serve as calcium reservoir; sources: milk and other dairy products, green leafy vegetables
Phosphorus	As calcium phosphate, an important structural component of bone; essential in energy transfer and storage (component of ATP) and in many other metabolic processes; component of DNA and RNA	Performs more functions than any other mineral; antacids can impair absorption
Sulfur	Component of many proteins (e.g., insulin), essential for normal metabolic activity	Sources: high-protein foods, such as meat, fish, legumes, nuts
Potassium	Principal positive ion within cells; influences muscle contraction and nerve excitability	Occurs in many foods
Sodium	Principal positive ion (cation) in interstitial fluid; important in fluid balance; essential for conduction of nerve impulses	Occurs naturally in foods; sodium chloride (table salt) added as seasoning; too much is ingested in average American diet; excessive amounts may lead to high blood pressure
Chlorine	Principal negative ion (anion) of interstitial fluid; important in fluid balance and acid-base balance	Occurs naturally in foods; ingested as sodium chloride
Copper	Component of enzyme needed for melanin synthesis; component of many other enzymes; essential for hemoglobin synthesis	Sources: liver, eggs, fish, whole wheat flour, beans
Iodine	Component of thyroid hormones (hormones that stimulate metabolic rate)	Sources: seafoods, iodized salt, vegetables grown in iodine-rich soil; deficiency results in goiter (abnormal enlargement of thyroid gland)
Cobalt	As component of vitamin B_{12}, essential for red blood cell production	Sources: meat, dairy products; strict vegetarians may become deficient in this mineral
Manganese	Necessary to activate arginase, an enzyme essential for urea formation; activates many other enzymes	Sources: whole-grain cereals, egg yolks, green vegetables; poorly absorbed from intestine
Magnesium	Appropriate balance between magnesium and calcium ions needed for normal muscle and nerve function; component of many coenzymes	Occurs in many foods
Iron	Component of hemoglobin, myoglobin, important respiratory enzymes (cytochromes), and other enzymes essential to oxygen transport and cellular respiration	Mineral most likely to be deficient in diet; deficiency results in anemia; sources: meat (especially liver), nuts, egg yolk, legumes
Fluorine	Component of bones and teeth; makes teeth resistant to decay	Where it does not occur naturally, fluorine may be added to municipal water supplies (fluoridation); excess causes tooth mottling
Zinc	Component of at least 70 enzymes; component of some peptidases, and thus important in protein digestion; may be important in wound healing	Occurs in many foods

children are overfed, abnormally large numbers of fat cells are formed. Later in life these fat cells may be fully stocked with excess lipids or may be shrunken, but they are always there. People with such increased numbers of fat cells are thought to be more susceptible to obesity than those with normal numbers.

Humans regulate their body weight around a specific set point that represents their characteristic weight. In obesity the set point is thought to be elevated so that the body regulates around the abnormally high weight. When an obese person restricts calorie intake, the body attempts to keep its weight at the familiar set point. It does this by lowering its metabolic rate and becoming more efficient in digesting and absorbing food. This explains why losing weight is so difficult and why dieters must struggle to maintain weight loss once it is achieved.

Exercise has been shown to be an important component of any weight loss program because exercise raises metabolic rate. When an obese person loses weight and maintains the loss for a period of from 6 months to 2 years, the body is thought to lower its set point. Then maintaining the weight loss becomes less difficult.

Most overweight people overeat due to a combination of poor eating habits and psychological factors. Whatever the underlying causes, overeating is the only way to become obese. Although water retention does increase body weight, it does not affect fat storage; water excesses can be diminished faster and more easily than fat excess. It has been estimated that for every 9.3 Calories of excess food taken into the body, 1 gram of fat is stored. (An excess of about 140 Calories per day for a month will result in gaining 1 pound.)

Because so many people are overweight, dieting has generated a multimillion-dollar industry embracing diet foods, formulas, pills, books, clubs, slenderizing devices, and even surgical procedures such as gastric stapling or insertion of plastic bubbles inside the stomach. Unfortunately, there is no magic cure for obesity. The only sure (and healthful) way to lose weight is to adjust food intake to meet energy needs. Energy intake must be less than energy output. Then the body will have to draw on its fat stores for the missing calories, and as the fat is mobilized and burned, body weight decreases. This can best be accomplished by a combination of increased exercise and decreased caloric intake (1000 to 1500 Calories daily for the mildly obese). Most nutritionists agree that the best reducing diet is a well-balanced one that provides the bulk of the calories in the form of complex carbohydrates.

Malnutrition Can Cause Serious Health Problems

While millions of people eat too much, countless others do not have enough to eat or do not eat a balanced diet. Individuals suffering from malnutrition cannot function efficiently because they are weak, easily fatigued, and highly susceptible to infection. Essential amino acids, iron, calcium, and vitamin A are commonly deficient nutrients. An estimated quarter of a million children become permanently blind every year because their diets are deficient in vitamin A.

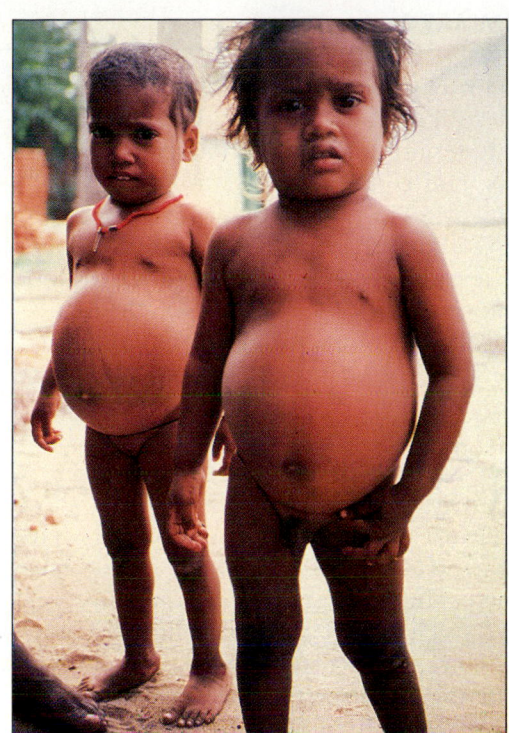

FIGURE 39–19   Children suffering from kwashiorkor, a disease caused by severe protein deficiency. Note the characteristic swollen belly, which results from fluid imbalance. (United Nations, Food and Agricultural Organization photo by P. Pittet)

Of all the required nutrients, essential amino acids are the ones most often deficient in the diet. Millions of people suffer from poor health and a lowered resistance to disease because of protein deficiency. Children's physical and mental development are retarded when the essential building blocks of cells are not provided in the diet. Because their bodies cannot manufacture antibodies (which are proteins) and cells needed to fight infection, common childhood diseases, such as measles, whooping cough, and chicken pox, are often fatal in children suffering from protein malnutrition.

In young children, severe protein malnutrition results in the condition known as **kwashiorkor.** The term, an African word that means "first-second," refers to the situation in which a first child is displaced from its mother's breast when a younger sibling is born. The older child is placed on a diet of starchy cereal or cassava that is deficient in protein. Growth becomes stunted, muscles are wasted, edema develops (as displayed by a swollen belly), the child becomes apathetic and anemic, and metabolism is impaired (Fig. 39–19). Without essential amino acids the digestive enzymes themselves cannot be manufactured, so that eventually what little protein is ingested cannot be digested. Dehydration and diarrhea develop, often leading to death.

■ CHAPTER SUMMARY

I. An organism's body plan and life-style are adapted to its mode of nutrition. Processing food includes ingestion, digestion, absorption of nutrients, and elimination of wastes.
  A. Herbivores eat mainly plant materials; they eat great quantities of food, and many house microorganisms that digest cellulose in their digestive tracts.
  B. Carnivores are meat-eaters; they have adaptations for capturing and killing their prey.
  C. Omnivores can distinguish among a wide range of smells and tastes; some are filter feeders; others are predators.

II. In the simplest invertebrates, the sponges, there is no digestive system; digestion is carried on intracellularly. Cnidarians and flatworms have incomplete digestive systems with only one opening, which serves as both mouth and anus.

III. In more complex invertebrates and in all vertebrates the digestive tract is a complete digestive system, a tube with an opening at each end.

IV. The wall of the human digestive tract consists of four layers: mucosa, submucosa, muscle, and adventitia.
  A. Mechanical digestion and enzymatic digestion of carbohydrates begin in the mouth.
  B. As food is swallowed, it is propelled through the pharynx and esophagus. A bolus of food is moved along through the digestive tract by peristaltic action.
  C. In the stomach, food is mechanically digested by vigorous churning and proteins are enzymatically digested by the action of pepsin in the gastric juice.
  D. Most enzymatic digestion takes place in the duodenum, which receives secretions from the liver and pancreas and produces several digestive enzymes of its own.
  E. The liver produces bile, which emulsifies fats.
  F. The pancreas releases enzymes that digest protein, lipid, and carbohydrate, as well as RNA and DNA.
  G. Activities of the digestive system are regulated by both nerves and hormones.
  H. Most nutrients are absorbed through the thin walls of the intestinal villi.
  I. The large intestine is responsible for the elimination of undigested wastes. It also incubates bacteria that produce vitamin K and certain B vitamins.

V. For a balanced diet human beings and other animals require carbohydrates, lipids, proteins, vitamins, and minerals.
  A. Most carbohydrates are ingested in the form of polysaccharides—starch and cellulose.
    1. Carbohydrates are used primarily as fuel.
    2. Glucose concentration in the blood is carefully regulated; glucose is stored as glycogen and can also be converted to fat.
  B. Lipids are used as fuel, as components of cell membranes, and to synthesize steroid hormones and other lipid substances.
    1. Most lipids are ingested in the form of triglycerides.
    2. Cholesterol is transported on low-density lipoproteins; high levels of LDL are associated with increased risk for heart disease.
    3. Fatty acids are converted to molecules of acetyl coenzyme A and used as fuel. Excess fatty acids are stored as fat.
  C. Proteins serve as enzymes and are essential structural components of cells.
    1. The best distribution of essential amino acids is found in the complete proteins of animal foods.
    2. Excess amino acids are deaminated by liver cells. Amine groups are converted to urea and excreted in urine, and the remaining keto acids are converted to carbohydrate and used as fuel or converted to lipid and stored in fat cells.
  D. Vitamins are required for biochemical processes; many serve as components of coenzymes.
    1. Vitamins are fat-soluble (A, D, E, K) or water-soluble (B complex and C).
    2. Deficiency of each vitamin results in specific physiological consequences.
  E. Minerals are inorganic nutrients ingested as salts dissolved in food and water.

VI. Basal metabolic rate is the body's cost of metabolic living.
  A. Total metabolic rate is the BMR plus the energy used to carry on daily activities.
  B. When energy (Calorie) input equals energy output, body weight remains constant.

VII. Obesity is a serious nutritional problem in which an excess amount of fat accumulates in the adipose tissues.
  A. A person gains weight by taking in more energy, in the form of Calories, than is expended in activity.
  B. Weight may be lost by expending more energy than is taken in. The needed energy is obtained by mobilizing fat and using it as fuel.

VIII. Millions of people suffer from malnutrition. Essential amino acids are the nutrients most often deficient in the diet.

■ POST-TEST

1. The process of taking food into the body is called _____.

2. _____ consists of mechanically and enzymatically breaking down food into molecules small enough to be absorbed.

3. _____ is the process of getting rid of undigested and unabsorbed food.

4. An animal which spends a lot of its time eating and which houses bacteria that digest cellulose in its digestive tract is a _____.

5. Carnivorous mammals have well-developed _____ teeth.

6. The most characteristic feature shared by the cnidarian and flatworm digestive system is that the system is _____.

7. In a complete digestive system the digestive tract has _____ openings.

8. Salivary _____ is an enzyme that initiates the digestion of carbohydrates.

9. A mammalian tooth consists mainly of _____.

Select the most appropriate term in column B for each description in column A. Answers may be used more than once or not at all.

	Column A	Column B
_____ 10.	Protein digestion begins here	**a.** duodenum
_____ 11.	Incubates bacteria	**b.** stomach
_____ 12.	Receives secretions from pancreas	**c.** liver
_____ 13.	Secretes bile	**d.** large intestine
_____ 14.	Converts food to chyme	**e.** none of the above
_____ 15.	Conducts food to stomach	

16. The surface area of the stomach is increased by the presence of _____.

17. Absorption takes place through finger-like projections in the lining of the small intestine called _____.

18. Food leaving the stomach next enters the _____.

19. An open sore in the wall of the stomach or duodenum is called an _____.

20. The function of the gallbladder is to _____.

21. Inorganic nutrients generally ingested as dissolved salts are _____.

22. Vitamins function as _____.

Match the nutrients in column B with the descriptions in column A.

	Column A	Column B
_____ 23.	Needed for hemoglobin synthesis	**a.** carbohydrate
_____ 24.	Used as fuel molecule	**b.** vitamin D
_____ 25.	Deficiency results in goiter	**c.** iron
_____ 26.	Water-soluble vitamin	**d.** iodine
_____ 27.	Deficiency results in rickets	**e.** none of the above

28. Lipids are ingested in the form of _____.

29. In the digestive tract proteins are degraded to _____; most carbohydrates are degraded to _____.

30. Glucose is stored in the liver as _____.

31. In the liver cells excess _____ are deaminated.

32. The _____ _____ is the body's rate of energy use during resting conditions.

33. When energy input is greater than energy output, _____.

34. Kwashiorkor is a disease caused by extreme _____ deficiency.

35. Label the diagram. For correct labeling, see Figure 39–5.

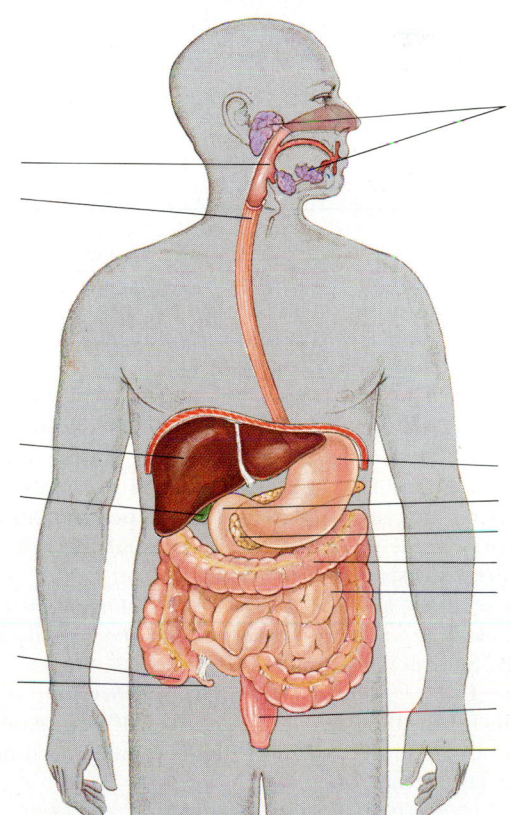

■ REVIEW QUESTIONS

1. How are herbivores and carnivores each adapted for their particular nutritional life-style?
2. Compare the advantages of intracellular and extracellular digestion.
3. How are digestive structures and methods of processing food in sponges, hydras, and flatworms, adapted to each group's life-style? Give specific examples.
4. Why must food be digested?
5. Trace a bite of food through the digestive tract, listing each structure through which it passes.
6. What mechanisms prevent gastric juice from digesting the wall of the stomach? What happens when they fail?
7. Give the functions of the three types of vertebrate accessory glands that secrete digestive juices. Identify their secretions.
8. The inner lining of the digestive tract is not smooth like the inside of a water pipe. Why is this advantageous? What structures increase its surface area?
9. Summarize the step-by-step digestion of (a) carbohydrates, (b) lipids, (c) proteins.
10. What happens to ingested cellulose? Why?
11. Draw and label an intestinal villus.
12. How does the absorption of fat differ from the absorption of glucose?
13. What is the adaptive advantage of specialization of different regions in a complete digestive tract?
14. List the nutrients that must be included in a balanced diet.
15. Why, specifically, are each of the following essential? (a) iron, (b) calcium, (c) iodine, (d) vitamin A, (e) vitamin K, (f) essential amino acids?
16. Draw a diagram to illustrate the fate of carbohydrates in the body.
17. Describe the fate of absorbed amino acids.
18. Describe the fate of absorbed fat.
19. Write an equation to describe energy balance and tell what happens when the equation is altered in either direction.
20. What is the most effective way to treat obesity?
21. Summarize the relationship between diet and heart disease.
22. What are some of the difficulties in planning a nutritionally balanced vegetarian diet?

■ RECOMMENDED READINGS

Beddington, J.R., and R.M. May: "The Harvesting of Interacting Species in a Natural Ecosystem," *Scientific American* 247(5), November 1982, 62–69. Harvesting a biological resource such as krill affects the whales and other animals that normally feed upon it.

Brown, M.S., and J.L. Goldstein: "How LDL Receptors Influence Cholesterol and Atherosclerosis," *Scientific American* 251(5), November 1984, 58–66. Many Americans have too few LDL receptors, which normally remove particles carrying cholesterol from the circulation. Absence of these receptors puts individuals at high risk for atherosclerosis and heart attacks.

Hartbarger, J.C., and N.J. Hartbarger: *Eating for the Eighties: A Complete Guide to Vegetarian Nutrition.* Philadelphia, W.B. Saunders Company, 1981.

Hinman, C.W.: "Potential New Crops," *Scientific American* 255(1), July 1986, 33–37. A number of new plants show promise as sources of food; some are approaching commercial production.

Krause, M.B., and L.K. Mahan: *Food, Nutrition, and Diet Therapy*, 7th ed. Philadelphia, W.B. Saunders Company, 1984. A comprehensive discussion of the science of nutrition and its application to the maintenance of health.

Kretchmer, N., and W. van B. Robertson: *Human Nutrition.* San Francisco, W.H. Freeman Company, 1978. An interesting collection of articles from *Scientific American* covering all levels of nutrition.

Moog, F.: "The Lining of the Small Intestine," *Scientific American* 245(5), November 1981, 154–176. The cells lining the small intestine are covered by a membrane that actively digests foods and speeds nutrients into the blood.

Raloff, J.: "Do You Know Your HDL?" *Science News*, 136(11): 171–173, 1989. An overview of the controversy surrounding the HDL level as a predictor of coronary risk.

Swaminathan, M.S.: "Rice," *Scientific American* 250(1):80–93, 1984. This member of the grass family is one of three on which the human species largely subsists.

40

Osmoregulation and Disposal of Metabolic Wastes

OUTLINE

I. Excretory systems help maintain homeostasis
II. The principal metabolic waste products are water, carbon dioxide, and nitrogenous wastes
III. Invertebrates have solved problems of osmoregulation and metabolic waste disposal in a number of ways
  A. Nephridial organs are tubules specialized for osmoregulation and/or excretion in some invertebrates
  B. Antennal glands are important in osmoregulation in crustaceans
  C. Malpighian tubules are an important adaptation for conserving water in insects
IV. The kidney is the key vertebrate organ of osmoregulation and excretion
  A. Osmoregulation is a continuous challenge for aquatic vertebrates
  B. The mammalian kidney is vital in maintaining homeostasis
    1. The kidneys, urinary bladder, and their ducts make up the urinary system
    2. The nephron is the functional unit of the kidney
    3. Urine is produced by filtration, reabsorption, and secretion
    4. Filtration is not a selective process
    5. Reabsorption is highly selective
    6. Urine concentration depends on a countercurrent mechanism
    7. Urine volume is regulated by the hormone ADH
    8. Sodium reabsorption is regulated by the hormone aldosterone
    9. Urine is composed of water, nitrogenous wastes, and salts

Focus on kidney disease, dialysis, and transplant

LEARNING OBJECTIVES

After you have studied this chapter you should be able to:

1. Relate the principal functions of excretory systems to specific osmoregulatory challenges posed by various environments.
2. Compare the advantages of excreting ammonia, uric acid, or urea.
3. Compare nephridial organs, antennal glands, and Malpighian tubules as osmoregulatory organs.
4. Relate the function of the vertebrate kidney to the success of vertebrates in a wide variety of habitats.
5. Compare the adaptations that freshwater fish have evolved to solve their problems of osmoregulation with those of a marine bony fish.
6. Describe adaptations that have solved osmoregulatory problems in sharks, marine mammals, and marine birds.
7. Label on a diagram the organs of the mammalian urinary system and give the functions of each.
8. Label on a diagram the principal parts of the nephron, including circulatory structures, and give the functions of each structure.
9. Trace a drop of filtrate from Bowman's capsule to its release from the body as urine.
10. Summarize the importance of countercurrent exchange in the process of urine formation.
11. Describe the regulatory effects of ADH and aldosterone.

Scanning electron micrograph of a glomerulus.
(Lennart Nilsson/Boehringer Ingelheim Internat'l GmbH, from *The Incredible Machine*, p. 139)

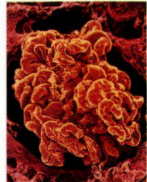

The cells of small marine animals are in direct contact with the surrounding sea water. They obtain food and oxygen directly from the water and release waste products into it. Larger aquatic animals and most terrestrial animals have their own internal sea—the blood and tissue fluid—that bathes their cells and transports nutrients, gases, and waste products. The composition of the body fluids must be kept within limits that the animal can tolerate.

Two processes for maintaining homeostasis of fluids are osmoregulation and excretion of metabolic wastes. **Osmoregulation** is the process by which an animal regulates the volume and composition of its body fluids. **Excretion** is the process of ridding the body of metabolic wastes. Animals have evolved efficient excretory systems to handle these processes. Excretory systems maintain homeostasis by selectively adjusting the concentration of salts and other substances in blood and other body fluids.

KEY CONCEPTS

- The composition of body fluids must be maintained within limits that the animal can tolerate.
- Osmoregulation and excretion are homeostatic mechanisms for maintaining body fluids in a steady state.
- Many excretory organs, including the vertebrate kidney, function by nonselectively filtering body fluids (for example, blood) and then adjusting the concentration of the filtrate by the processes of reabsorption and secretion.

■ EXCRETORY SYSTEMS HELP MAINTAIN HOMEOSTASIS

Typically, an excretory system helps maintain homeostasis in three ways:

1. It carries on osmoregulation.
2. It excretes metabolic wastes.
3. It regulates the concentrations of most of the components of body fluids.

To carry out these functions, an excretory system collects fluid, generally from the blood or interstitial fluid. It then adjusts the composition of this fluid by selectively returning needed substances to the body fluid. The adjusted excretory product (urine, for example), containing excess or potentially toxic substances, is released from the body.

■ THE PRINCIPAL METABOLIC WASTE PRODUCTS ARE WATER, CARBON DIOXIDE, AND NITROGENOUS WASTES

As cells carry on metabolic activities, waste products are generated. If allowed to accumulate, metabolic wastes would eventually reach toxic concentrations and threaten the homeostasis of the animal. For that reason, they must be *excreted* from the body.

The principal metabolic waste products in most animals are water, carbon dioxide, and nitrogenous (nitrogen-containing) wastes. Carbon dioxide is mainly excreted by respiratory structures (Chapter 38); excretory organs, such as kidneys, remove and excrete most of the water and nitrogenous wastes.

Nitrogenous wastes include ammonia, uric acid, and urea. Recall that amino acids and nucleic acids contain nitrogen. During the breakdown of amino acids the nitrogen-containing amino group is removed (deamination) and converted to **ammonia.** However, ammonia is highly toxic. Some aquatic animals excrete it into the surrounding water before it can build up to toxic concentrations in their tissues, and a few terrestrial animals vent it directly to the air. But in many organisms, ourselves included, ammonia is converted to some less toxic nitrogenous waste, such as urea or uric acid.

Uric acid is produced from ammonia and also when nucleotides from nucleic acids are broken down. Uric acid forms crystals and can be excreted as a crystalline paste with little fluid loss. For this reason uric acid excretion is an important adaptation for conserving water in many terrestrial animals such as insects, certain reptiles, and birds. In birds, the absence of a urinary bladder and the frequent excretion of uric acid as part of the feces contribute to the light body weight essential for flight. Because it is not toxic, excretion of uric acid is an adaptive advantage for species whose young begin their development enclosed in eggs.

Urea is the principal nitrogenous waste product of amphibians and mammals. It is produced by attaching a molecule of ammonia to a small carbon compound. Because it is less toxic than ammonia, urea can accumulate in higher concentrations without damaging tissues. Thus, urea can be excreted in concentrated form.

■ INVERTEBRATES HAVE SOLVED PROBLEMS OF OSMOREGULATION AND METABOLIC WASTE DISPOSAL IN A NUMBER OF WAYS

The body fluids of most marine invertebrates are in osmotic equilibrium with the surrounding seawater. These animals are known as *osmotic conformers* because

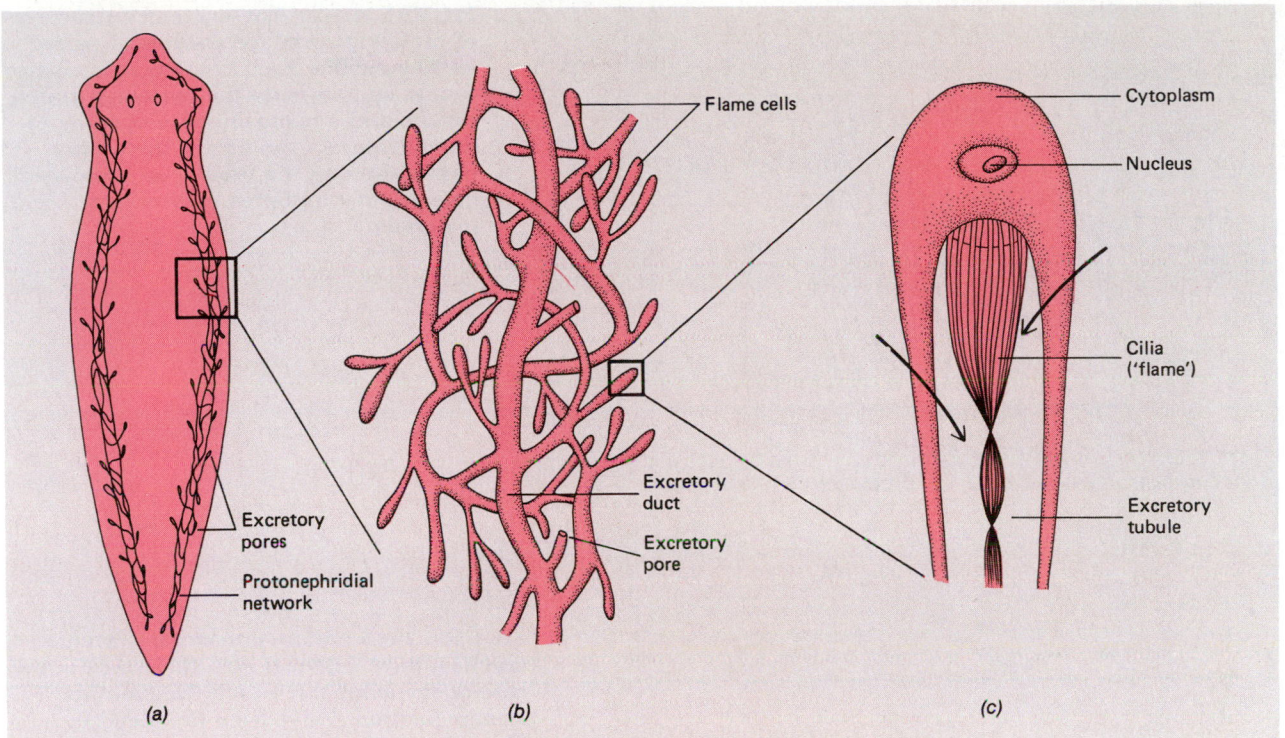

FIGURE 40–1 **Nephridial organs are a common type of excretory structure in invertebrates.** (*a*) **The excretory system of a typical flatworm consists of protonephridia.** (*b*) **Wastes collected by flame cells pass through excretory ducts and out of the animal through excretory pores.** (*c*) **A single flame cell.**

the concentration of their body fluids varies along with the changes in the seawater. Fortunately, the sea is a stable environment and its salt concentration does not vary much over a short period of time.

Marine sponges and cnidarians need no specialized excretory structures. Their wastes pass by diffusion from the intracellular fluid to the external environment. Unlike animals with specialized excretory systems, they expend little or no energy in excretion; energy is provided by the water currents that sweep by or even through them. When changes in water currents or stagnation do occur, aquatic environments, such as coral reefs, are especially prone to damage from the accumulation of metabolic wastes.

Coastal habitats and brackish water such as estuaries are much less stable environments than the open sea. Salt concentrations change frequently with shifting tides. Animals that dwell in these habitats are *osmotic regulators.* In a coastal environment where fresh water enters the sea, the water may have a lower salt concentration than the body fluids of the animal. Water osmotically moves into the animal body, and salt diffuses out. An animal adapted to this environment has excretory structures that actively remove the excess water. Many also have cells in their gills that remove salts from the surrounding water and transport them into the body fluids.

Terrestrial animals have a higher fluid concentration than the air surrounding them. They tend to lose water by evaporation from the body surface and from respiratory surfaces. They may also lose water as body wastes are excreted. Adaptation to life on land has required the evolution of structures and processes that conserve water.

Nephridial Organs Are Tubules Specialized For Osmoregulation And/Or Excretion In Some Invertebrates

Nephridial organs are a common type of excretory structure in invertebrates. They consist of simple or branching tubes that usually open to the outside of the body through nephridial pores. Flatworms are the simplest animals with specialized excretory organs. Although metabolic wastes pass by diffusion through their body surface, these animals also have osmoregulatory nephridial organs, which consist of tubules with enlarged blind ends containing cilia. These organs, known as **protonephridia,** have **flame cells** with brushes of cilia. The constant motion of these cilia reminded early biologists of flickering flames (Fig. 40–1). A system of branching excretory ducts connects the protonephridia with the outside. The flame cells lie in

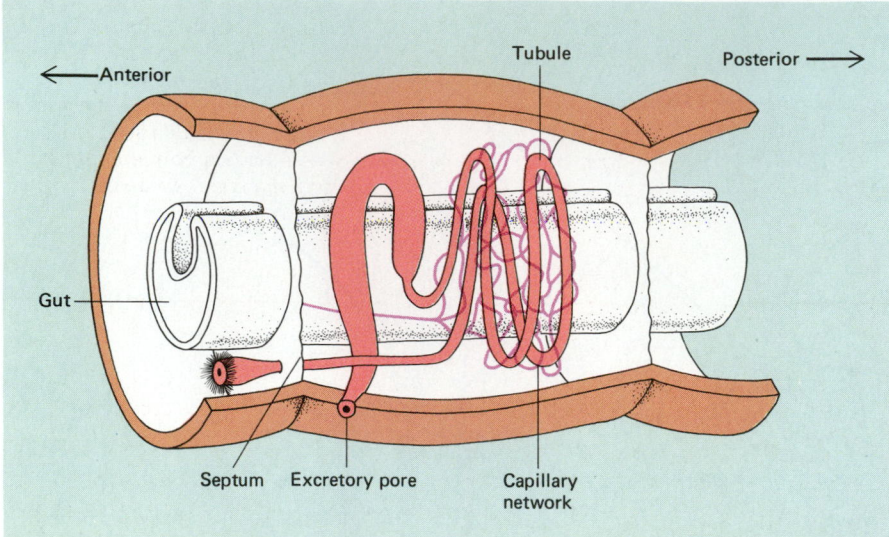

FIGURE 40–2 The excretory structures of the earthworm are a series of paired metanephridia. Each consists of a ciliated funnel opening into the coelom, a coiled tubule, and a nephridiopore opening to the outside. As fluid from the coelom passes through the tubule, an associated network of capillaries reabsorbs usable materials.

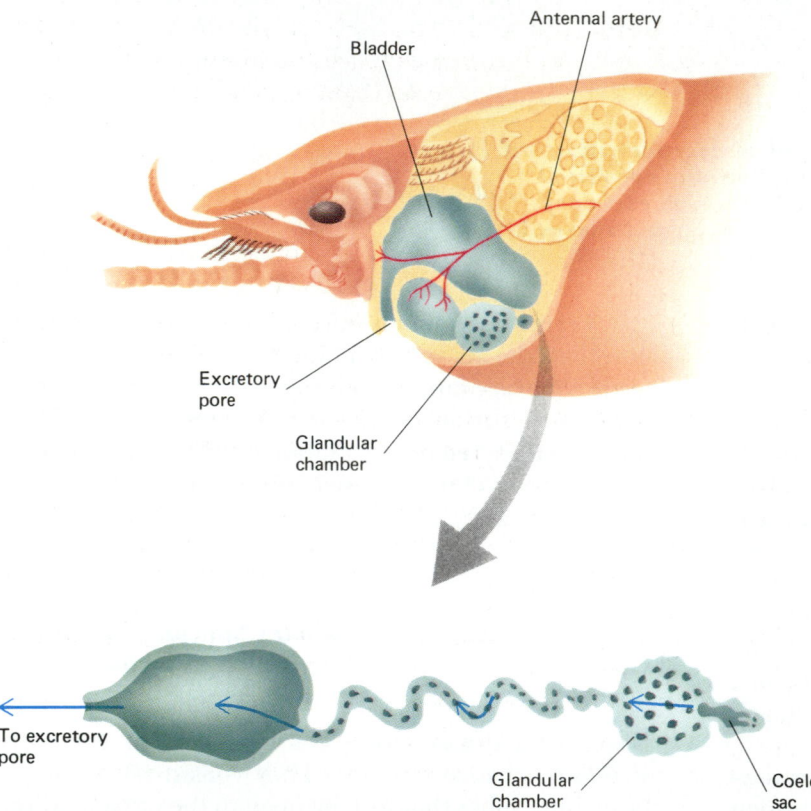

FIGURE 40–3 Many crustaceans, such as crayfish, have antennal glands, located at the base of the antennae, which function in osmoregulation. Fluid from the blood is filtered into the coelomic sac, and its composition is adjusted as it passes through the excretory ducts. The exit duct may be enlarged to form a bladder.

the fluid that bathes the body cells; fluid enters the flame cells, passes through the tubules and excretory ducts, and leaves the body through excretory pores.

Annelids and mollusks have nephridial organs called **metanephridia.** A pair of metanephridia is present in each segment of the earthworm. The metanephridium is a tubule open at both ends. The inner end opens into the coelom as a ciliated funnel (Fig. 40–2). The outer end opens to the outside

through an excretory pore. Around each tubule is a network of capillaries.

Fluid from the coelom (body cavity) passes into the tubule, bringing with it whatever it contains—glucose, salts, wastes. As the fluid moves through the tubule, needed materials, such as water or glucose, are reabsorbed by the capillaries, leaving the wastes behind. In this way urine is produced that contains concentrated wastes.

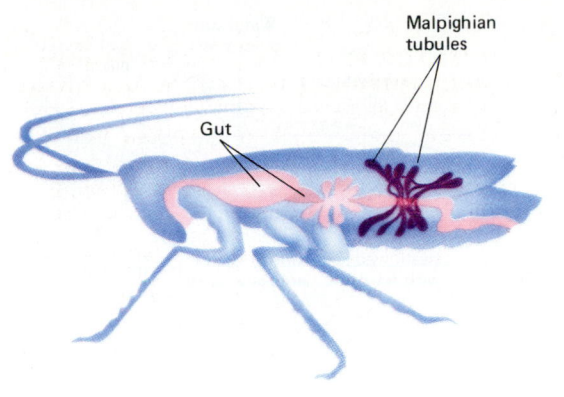

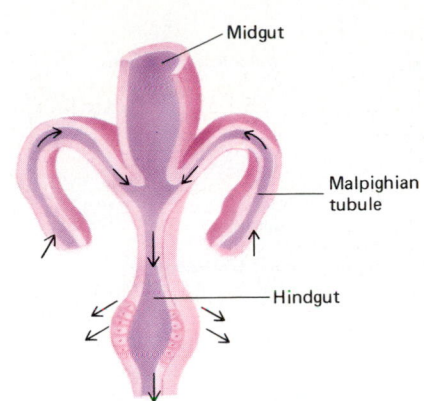

FIGURE 40–4 The slender Malpighian tubules of insects have blind ends that lie in the hemocoel. Their cells transfer fluid from the blood to the cavity of the tubule. Uric acid is concentrated and excreted as a semisolid paste.

Antennal Glands Are Important In Osmoregulation In Crustaceans

Antennal glands, also called **green glands,** are the principal excretory organs of crustaceans (Fig. 40–3). A pair of these structures is located in the head, often at the base of the antennae.

Fluid from the blood is filtered into the antennal gland, and its composition is adjusted as it passes through an excretory tubule. Needed materials are reabsorbed into the blood. Wastes can also be actively secreted from the blood into the filtrate within the excretory organ.

Malpighian Tubules Are An Important Adaptation For Conserving Water In Insects

The excretory system of insects and spiders consists of **Malpighian tubules** (Fig. 40–4). There may be two to several hundred tubules, depending upon the species. Malpighian tubules have blind ends that lie in the body cavity (hemocoel), bathed in blood. Their cells transfer salts and wastes by diffusion or active transport from the blood to the cavity of the tubule. The Malpighian tubules empty into the intestine. Water and some salts are reabsorbed into the blood by specialized rectal glands. Uric acid, the major waste product, is excreted as a semidry paste with a minimum of water. Malpighian tubules help to conserve the insect's body fluids and for this reason have contributed significantly to the success of the insects in terrestrial environments.

■ THE KIDNEY IS THE KEY VERTEBRATE ORGAN OF OSMOREGULATION AND EXCRETION

Vertebrates live successfully in a wide range of habitats—in fresh water, sea, tidal regions, and on land, even in extreme environments such as deserts. In response to the requirements of these environments, vertebrates have evolved adaptations for regulating their

salt and water content and for excreting wastes. An extreme example is the desert-dwelling kangaroo rat, which must carefully conserve water. It obtains most of its water metabolically by oxidizing food, and its kidneys are so efficient that it loses little fluid as urine.

The main osmoregulatory and excretory organ in vertebrates is the kidney. Typically the vertebrate kidney functions by a combination of three processes: *filtration, reabsorption,* and *secretion.* Blood plasma is *filtered* nonselectively, so the initial filtrate that enters the tubules of the kidney contains all the substances present in the blood except blood cells, platelets, and large compounds, such as proteins. As the filtrate passes through the coiled tubules of the kidney, needed materials, such as glucose, amino acids, salts, and water, are selectively *reabsorbed* into the blood. Some substances are actively *secreted* from the blood into the filtrate, in a process that moves materials in the opposite direction from reabsorption. Through reabsorption and secretion, the composition of the filtrate is slowly adjusted, and the urine that is finally excreted consists of metabolic waste products and excess water and salts.

In most vertebrates, not only the kidneys but also the skin, lungs or gills, and digestive system function to some extent to maintain fluid balance and dispose of metabolic wastes. Some reptiles and marine birds have salt glands in the head that excrete salt that enters the body with the seawater they drink. By removing excess water, salts, and other potentially toxic waste materials, all these organs help to maintain homeostasis.

Osmoregulation Is A Continuous Challenge For Aquatic Vertebrates

As fish began to move into freshwater habitats about 400 million years ago, a major evolutionary challenge was the development of mechanisms for effective osmoregulation. Freshwater animals are immersed in fluid with a lower salt concentration than that of their body fluid. Yet because they maintain a salt concentration that is higher than (hyperosmotic to) their sur-

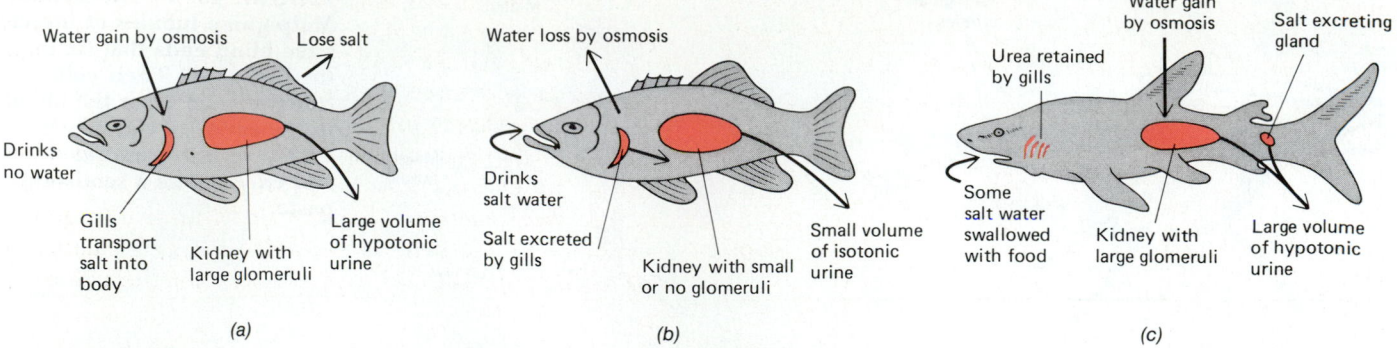

FIGURE 40–5   Problems of osmoregulation in marine and freshwater fish.
(*a*) In freshwater fish, water continuously enters the body by osmosis. To compensate, they produce large amounts of urine. A freshwater fish loses salt to its surroundings and must absorb salt from the water through the gills. (*b*) Marine fish continuously lose water to their surroundings and gain salt. They compensate by drinking large amounts of seawater, producing only a small amount of urine, and excreting excess salt from the gills. (*c*) Sharks retain urea, which establishes a salt concentration in body fluids slightly higher than that of the surrounding seawater. As a result they tend to gain water by osmosis. Their kidneys excrete the excess water.

roundings, water passes into them osmotically; they are in constant danger of becoming waterlogged. Freshwater fish are covered by scales and a mucous secretion that retards the passage of water into the body. However, water does enter through the gills. The kidneys of these fish have become adapted to filter out excess water, and they excrete a large amount of dilute urine (Fig. 40–5).

Water entry, though, is only part of the problem of osmoregulation in freshwater fish. These animals also tend to lose salts to the surrounding fresh water. To compensate, special cells in the gills have evolved that actively transport salt (mainly sodium chloride) from the water into the body.

Most amphibians are at least semiaquatic, and their mechanisms of osmoregulation are similar to those of freshwater fish. They produce a large amount of dilute urine. Through its urine and skin a frog can lose an amount of water equivalent to one-third of its body weight in a day. Active transport of salt inward by special cells in the skin compensates for salt loss through skin and urine.

Freshwater fish have adapted very successfully to their aquatic habitats. One of their chief adaptations was the evolution of body fluids more dilute than seawater. Modern vertebrates have a salt concentration in their body fluids about one-third that of seawater. Thus, when some freshwater fish returned to the sea about 200 million years ago, their blood and body fluids were less salty than (hypo-osmotic to) their surroundings. They tended to lose water osmotically and to take in salt.

To compensate, many marine bony fish drink seawater to replace their water loss. They retain the water

and excrete salt by the action of specialized cells in their gills. Very little urine is excreted by the kidneys, and the nephrons (microscopic units of the kidney) have only small (or no) capillary tufts (glomeruli) that filter the blood in other vertebrates.

Marine chondrichthyes (sharks and rays) have a different set of osmoregulatory adaptations that allows them to tolerate the salt concentrations of their environment. These animals accumulate and tolerate urea. Their tissues are adapted to function at concentrations of urea that would be toxic to most other animals. The high concentration of urea makes the salt concentration of body fluids slightly higher than (hyperosmotic to) seawater. This results in a net inflow of water into their bodies. Their well-developed kidneys excrete a large volume of urine. Excess salt is excreted by the kidneys and, in many species, by a rectal gland.

Whales, dolphins, and other marine mammals ingest seawater along with their food. Their kidneys produce a very concentrated urine, much more salty than seawater. This is an important physiological adaptation, especially for marine carnivores (Fig. 40–6). The high-protein diet of these animals results in production of large amounts of urea, which must be excreted in the urine or, in some cases, by special accessory salt glands.

The Mammalian Kidney Is Vital In Maintaining Homeostasis

In mammals as in other terrestrial vertebrates, the kidneys, skin, lungs, and digestive system all play important roles in osmoregulation and waste disposal

Most of the deamination of amino acids takes place in the liver, which is also the site of production of both urea and uric acid. In addition, most of the bile pigments produced by the breakdown of red blood cells are normally excreted by the liver into the intestine and then pass out of the body with the feces. However, the kidneys are the principal excretory organs and are responsible for excretion of most nitrogenous wastes and for helping to maintain fluid balance by adjusting the salt and water content of the urine.

The Kidneys, Urinary Bladder, And Their Ducts Make Up The Urinary System

The mammalian **urinary system** consists of the kidneys, the urinary bladder, and associated ducts. The overall structure of the human urinary system is shown in Figure 40–8. Located just below the diaphragm in the "small" of the back, the kidneys look like a pair of giant, dark-red lima beans, each about the size of a fist. The outer portion of the kidney is called the **renal cortex;** the inner portion, the **renal medulla** (Fig. 40–9).

FIGURE 40–6 **Marine mammals ingest seawater along with their food. Their kidneys produce a very concentrated urine, much more salty than seawater. Killer whale (*Orcinus orca*).** (Ken Lucas/BPS [5-816])

(Fig. 40–7). Most carbon dioxide and a great deal of water are excreted by the lungs. Though primarily concerned with the regulation of body temperature, the sweat glands excrete 5–10% of all metabolic wastes.

FIGURE 40–7 **In terrestrial vertebrates several body systems work together in osmoregulation and waste disposal. (*a*) In terrestrial mammals the kidney filters the blood and then conserves water by reabsorbing most of the filtrate. (*b*) Nitrogenous wastes are produced by the liver and transported to the kidneys by the circulatory system. All cells produce carbon dioxide and water during cellular respiration. In many terrestrial vertebrates, the kidneys, lungs, skin, and digestive system all participate in the disposal of metabolic wastes.**

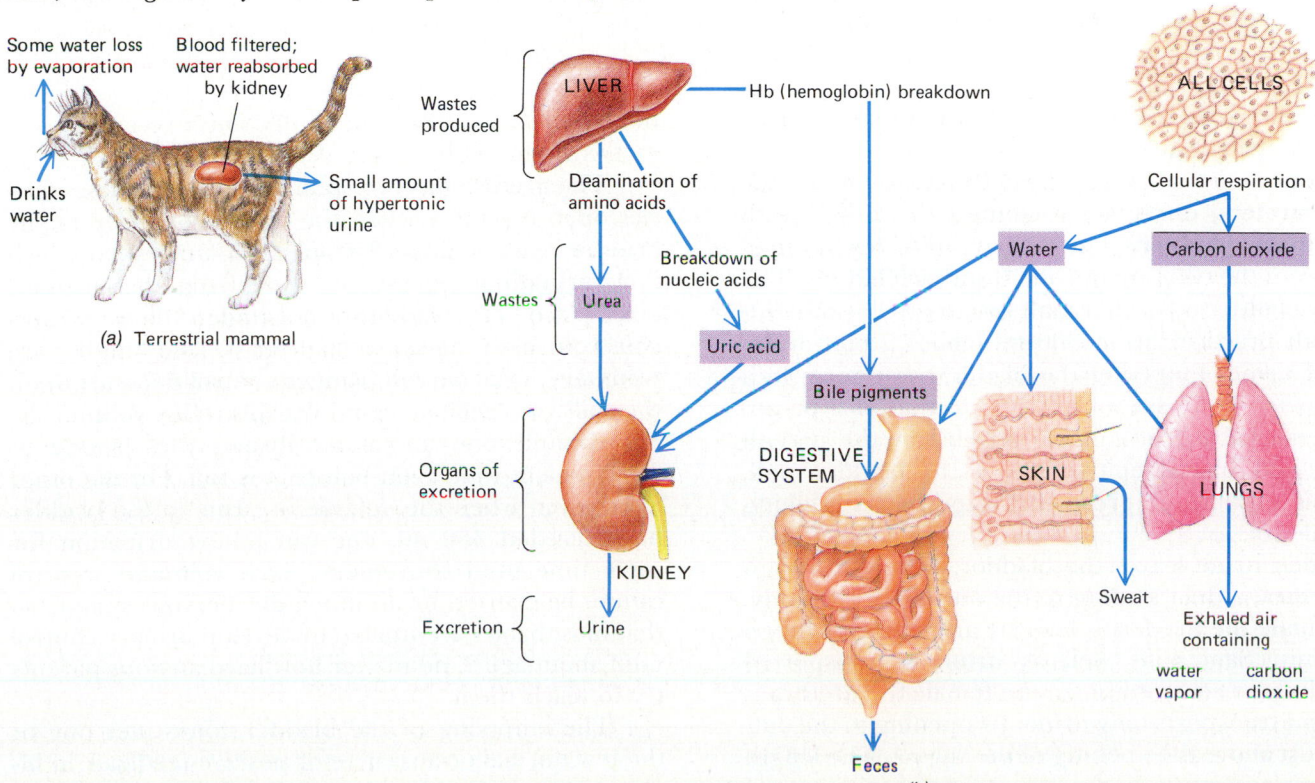

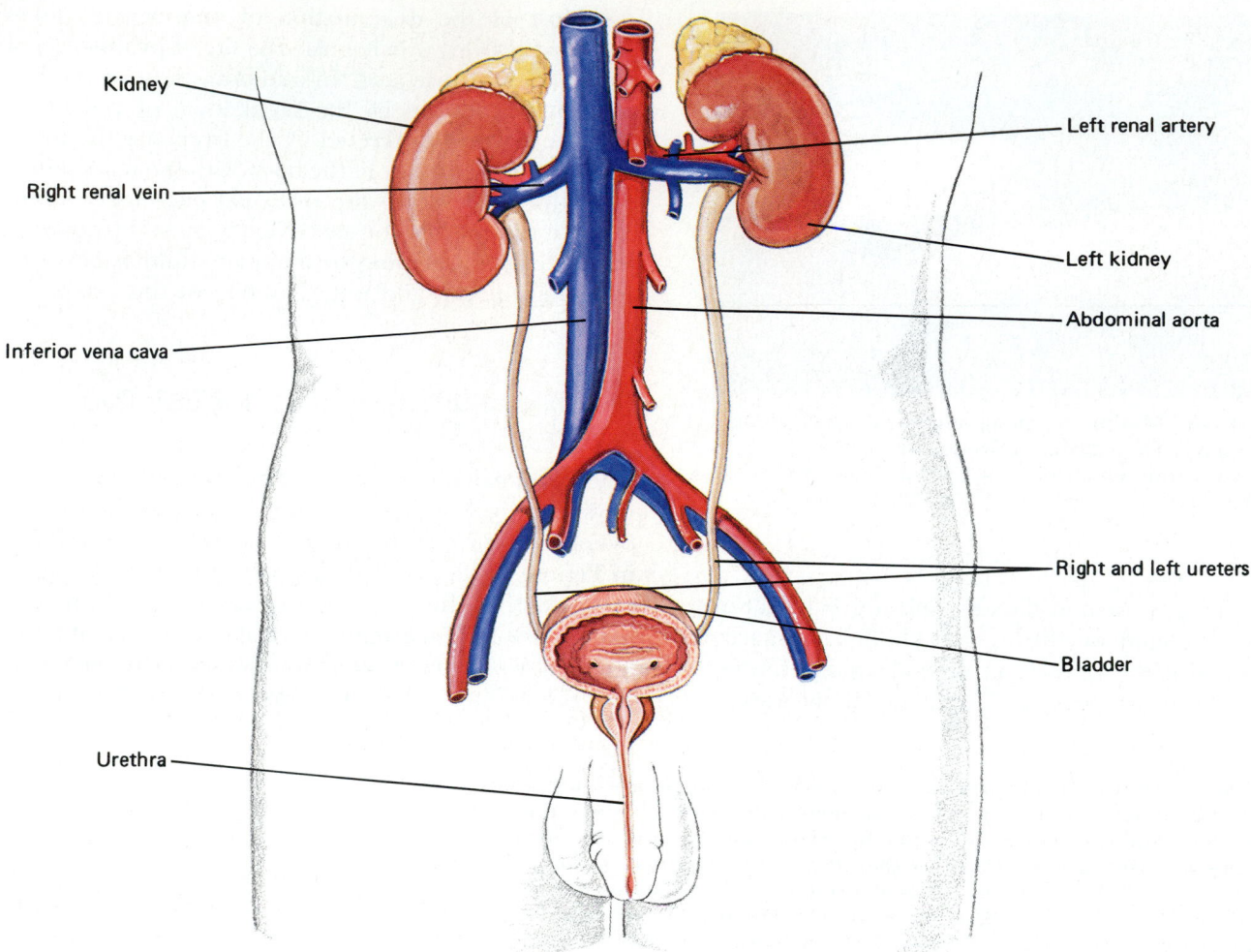

FIGURE 40–8   The human male urinary system.

As urine is produced, it flows into the **renal pelvis,** a funnel-shaped chamber.

From the renal pelvis, urine flows into one of the paired **ureters,** ducts which connect the kidney with the **urinary bladder.** The urinary bladder receives urine from the two ureters, one from each kidney. The urinary bladder is a remarkable organ capable of holding (with practice) up to 800 ml (about a pint and a half) of urine. Emptying the bladder changes it in a moment from the size of a melon to that of a pecan. This remarkable feat is made possible by the smooth muscle and special epithelium of the bladder wall, which is capable of great shrinkage and stretching (Fig. 40–10).

When urine leaves the bladder, it flows through the **urethra,** a duct leading to the outside of the body. In the male the urethra is lengthy and passes through the penis. Semen as well as urine is transported through the male urethra. In the female the urethra is short and transports only urine. Its opening to the outside is just above the opening of the vagina. The length of the male urethra discourages bacterial invasions of

the bladder, and thus such infections are more common in females than in males.

The urethra has two sphincter valves. These muscles open by reflex when the volume of urine in the bladder reaches about 350 ml (10.5 ounces), at which time **urination,** release of urine from the bladder, takes place. The bladder is not under voluntary nervous control in the sense that our skeletal muscles are voluntary; what we call bladder control depends upon the ability to facilitate or inhibit this reflex voluntarily. For example, one can voluntarily empty the bladder at a convenient time even before it is full. On the other hand, even when the volume of urine in the bladder has exceeded 350 ml, one can inhibit urination for some time until convenient. Such voluntary control cannot be exerted by an immature nervous system, so that most babies are unable to develop urinary control until about age 2, no matter how hard anxious parents try to teach them.

The emptying of the bladder constitutes one of the few normal occurrences of positive feedback in biology. One might think that when urination begins, the

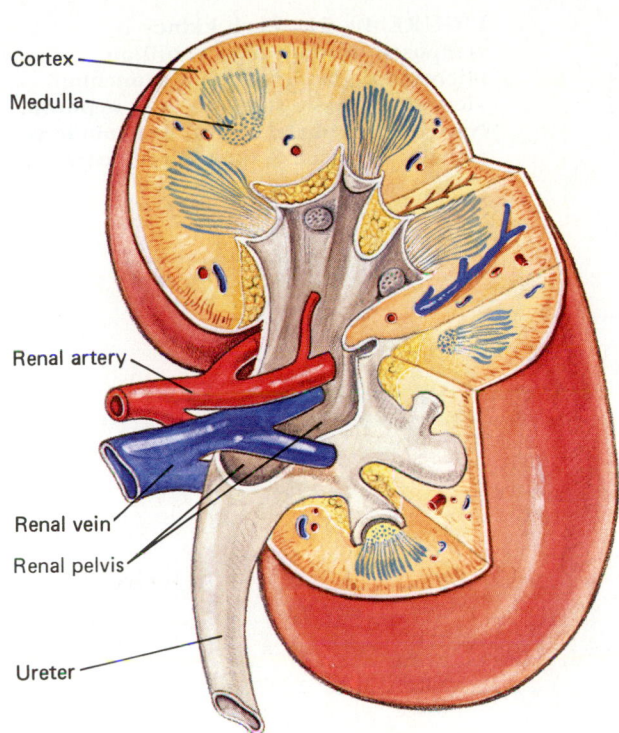

FIGURE 40–9   The human kidney.

FIGURE 40–10   Scanning electron micrograph of the lining of the urinary bladder illustrating the special transitional epithelium and folds resulting from the contraction of the underlying smooth muscle. As urine accumulates in the bladder, the folds stretch out. (From Kessel, R.G., and R.H. Kardon, *Tissues and Organs: A Text-Atlas of Scanning Electron Microscopy,* San Francisco, W.H. Freeman, 1979)

pressure within the bladder would swiftly fall below the threshold necessary to trigger the urination reflex so that urination would promptly stop, leaving the bladder mostly full. Yet the healthy bladder empties completely. The explanation is that the bladder contracts so strongly that the pressure within it actually rises once urination has begun, which stimulates the urination reflex even more strongly. That is why it is so difficult to stop urination once it has begun.

The Nephron Is The Functional Unit Of The Kidney

Each kidney consists of more than a million functional units called **nephrons.** Each nephron consists of a cup-like **Bowman's capsule** connected to a long, partially coiled **renal tubule** (Fig. 40–11). Positioned within the cup-shaped Bowman's capsule is a cluster of capillaries known as the **glomerulus.** Three main regions of the renal tubule are the **proximal tubule,** which conducts the filtrate from Bowman's capsule; the **loop of Henle,** an elongated, hairpin-shaped portion; and the **distal tubule,** which conducts the filtrate to a **collecting duct.**

Blood is delivered to the kidney by the renal artery. Small branches of the renal artery give rise to afferent arterioles. An **afferent arteriole** conducts blood into the capillaries that make up each glomerulus. As blood flows through the glomerulus, some of the plasma is forced into Bowman's capsule.

You may recall that in the usual circulatory pattern capillaries deliver blood into veins. Circulation in the kidneys is an exception in that blood flowing from the glomerular capillaries next passes into an **efferent arteriole,** so called because it conducts blood *away* from the glomerulus. The efferent arteriole delivers blood to a second capillary network, the **peritubular capillaries** that surround the renal tubule. The first set of capillaries, those of the glomerulus, provides the blood to be filtered; the second set receives materials returned to the blood by the tubule. Blood from the peritubular capillaries enters small veins which eventually lead to the renal vein.

Urine Is Produced By Filtration, Reabsorption, And Secretion

Blood circulating through the glomerulus is filtered by Bowman's capsule. Blood cells, platelets, proteins, and other large solutes do not normally leave the blood. However, filtration is not a selective process and the filtrate contains plasma and whatever small solutes, such as glucose, salts, and urea, are present in the plasma.

The filtrate flows into the renal tubule, where reabsorption occurs. Most of the water and useful solutes are returned to the blood, while excess water, salts, and waste materials are retained in the filtrate (Fig. 40–12).

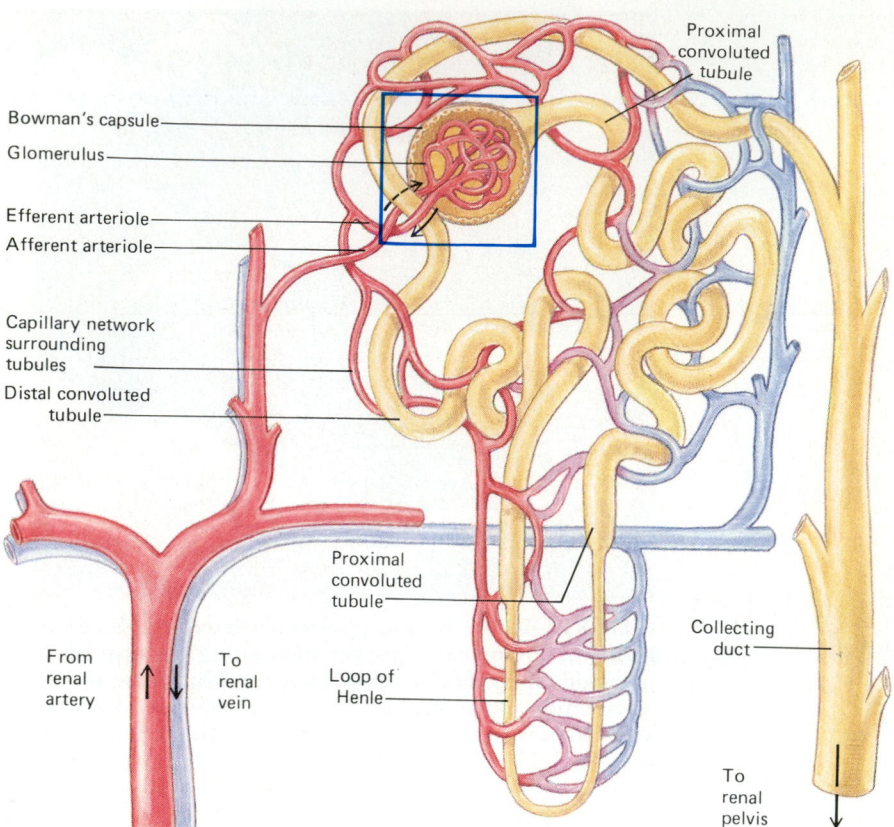

Proximal convoluted tubule

Bowman's capsule

Glomerulus

Efferent arteriole

Afferent arteriole

Capillary network surrounding tubules

Distal convoluted tubule

From renal artery

To renal vein

Proximal convoluted tubule

Loop of Henle

Collecting duct

To renal pelvis

FIGURE 40–11 Each kidney is composed of more than a million microscopic nephrons. Diagrammatic view of the basic structure of a nephron. Note that the distal convoluted tubule is actually adjacent to the afferent arteriole.

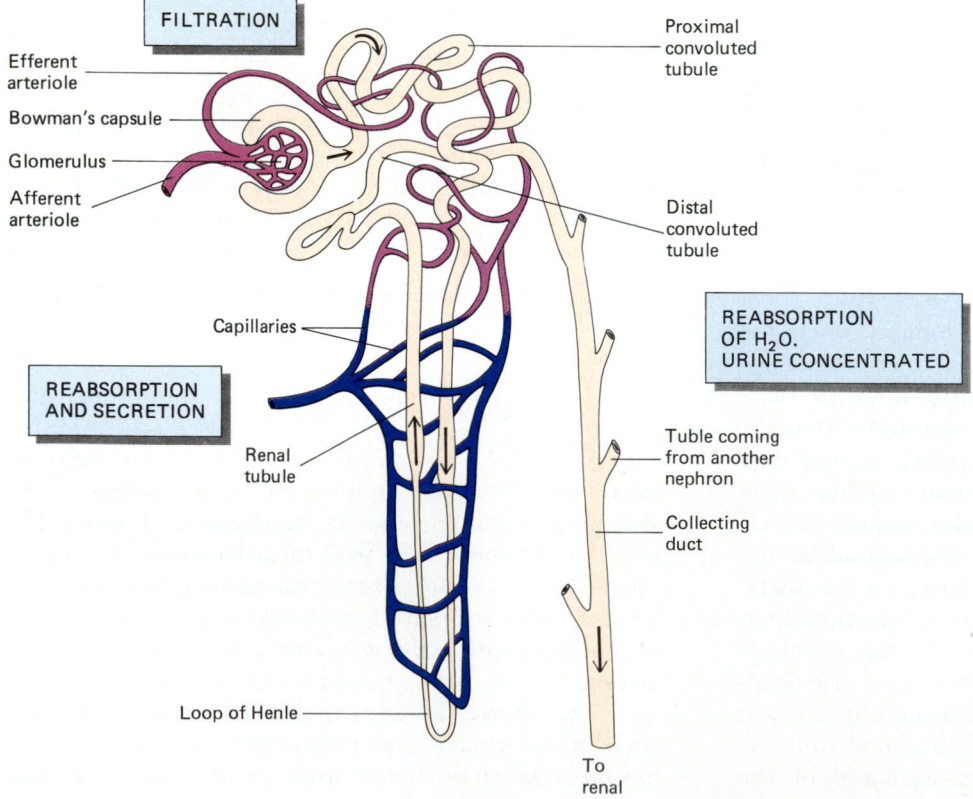

FILTRATION

Efferent arteriole

Bowman's capsule

Glomerulus

Afferent arteriole

Proximal convoluted tubule

Distal convoluted tubule

Capillaries

REABSORPTION AND SECRETION

Renal tubule

REABSORPTION OF H_2O. URINE CONCENTRATED

Tuble coming from another nephron

Collecting duct

Loop of Henle

To renal pelvis

FIGURE 40–12 Urine is produced by filtration, reabsorption, and secretion.

FOCUS ON   Kidney Disease, Dialysis, and Transplant

Kidney disease ranks fourth among major diseases in the United States. Kidney function can be impaired by infections, poisoning caused by substances such as mercury or carbon tetrachloride, lesions, tumors, kidney stone formation, shock, or many circulatory diseases. One of the most common kidney diseases is *glomerulonephritis*, which is actually a large number of related chronic diseases in which the glomeruli are damaged. The damage is thought to result from an autoimmune response.

In chronic kidney disease there is a progressive loss of renal function, which may eventually reach the stage of **kidney failure.** In kidney failure there is a decrease in the glomerular filtration rate and the kidneys are unable to maintain homeostasis of the blood. Homeostatic balance of water, sodium, potassium, calcium, and other salts is no longer possible, and nitrogenous wastes are not excreted. Retention of water causes edema, and as the concentration of hydrogen ions increases, acidosis develops. Nitrogenous wastes accumulate in the blood and tissues, causing a condition referred to as **uremia.** If untreated, the acidosis and uremia can cause coma and eventually death. Chronic kidney failure can be treated by kidney dialysis or by kidney transplant.

Kidney Dialysis Is Used to Treat Patients with Kidney Failure

Dialysis is the process of separating solutes in a solution by diffusion across a semipermeable membrane. A kidney dialysis machine can be used to restore appropriate solute balance to a patient whose kidneys are not functioning.

In extracorporeal dialysis, a plastic tube is surgically inserted into both an artery and a vein in the patient's arm or leg. These tubes can then be connected to a circuit of plastic tubing from a dialysis machine. The patient's blood flows through the tubing, which is immersed in a solution containing most of the normal blood plasma constituents in their normal proportions. The walls of the plastic tubing constitute a semipermeable membrane. Since the dialysis fluid contains no wastes, nitrogenous wastes such as urea pass from the patient's blood through minute pores in the tubing and into the surrounding solution. As the blood circulates repeatedly through the tubing in the machine, dialysis continues, eventually adjusting most of the values of the patient's blood chemistry to normal ranges. Although much improved by recent engineering advances, machine dialysis is very expensive ($20,000 to $30,000 per year per patient), clumsy, and inconvenient and may produce serious side effects

such as osteoporosis (a disorder characterized by loss of calcium from the bones).

A different dialysis technique, continuous ambulatory peritoneal dialysis (CAPD), makes use of the fact that the peritoneum (the lining of the abdominal cavity) is a differentially permeable membrane. A plastic bag containing dialysis fluid is attached to the patient's abdominal cavity, and the fluid is allowed to run into the abdominal cavity. After about 30 minutes, the fluid is withdrawn into the bag and discarded. This process is repeated about three times each day. This type of dialysis is much more convenient but poses the threat of peritonitis, should bacteria enter the body cavity with the dialysis fluid.

Kidney Transplant Is More Effective Than Dialysis

Long-term use of dialysis is not as desirable for the patient as would be a functioning kidney. With a successful kidney transplant a patient can live a more normal life with far less long-term expense. At present more than two-thirds of kidney transplants are successful for several years, although physicians must routinely treat the problems of graft rejection (discussed in Chapter 37). Several recipients of kidney transplants have survived for more than 20 years.

Some substances, especially potassium, hydrogen, and ammonium ions, are actively secreted from the blood into the filtrate. Some drugs, such as penicillin, are also removed from the blood by secretion. The process of secretion is more highly developed in marine bony fish, reptiles, and birds than in mammals. The filtrate passes from the renal tubule into a larger **collecting duct** that eventually empties into the renal pelvis.

Filtration Is Not A Selective Process

Blood flows through the glomerular capillaries under high pressure, forcing more than 10% of the plasma

out of the capillaries and into Bowman's capsule (Fig. 40–13). This process of filtration is somewhat similar to the mechanism whereby tissue fluid is formed as blood flows through other capillary networks in the body. However, much more plasma is filtered in the kidney. Several factors contribute to this process. First, the hydrostatic pressure in the glomerular capillaries is higher compared to other capillaries. The high pressure is due in part to the high resistance to flow presented by the efferent arteriole, which is smaller in diameter than the afferent arteriole. Another factor that contributes to the large amount of filtrate is the large surface area for filtration provided by the highly coiled glomerular capillaries. A third fac-

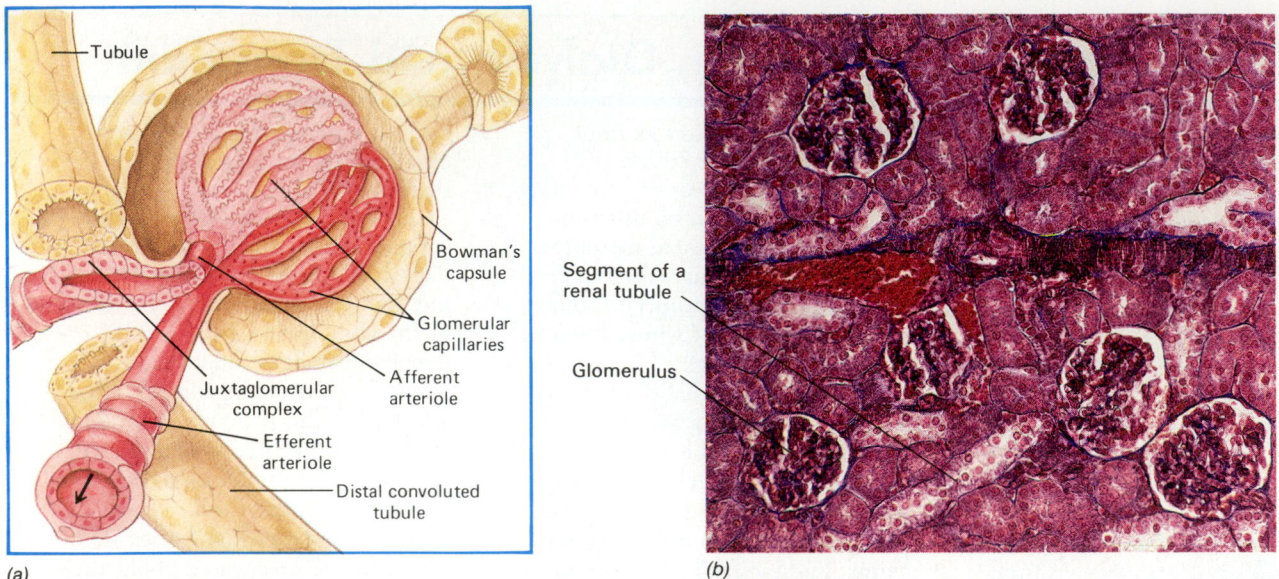

(a)

(b)

FIGURE 40–13 Blood is filtered from the glomerular capillaries into Bowman's capsule and then passes into the proximal tubule. (*a*) Magnified view of the glomerulus and Bowman's capsule. (*b*) Several glomeruli (large circular structures) can be seen in this photomicrograph of kidney tissue. The relationships among the various structures of the kidney are not clear in this photomicrograph because it is a slice through the tissue. For this reason only *segments* of some renal tubules are visible. Note the cuboidal epithelial tissue lining the tubules. (Manfred Kage/Peter Arnold, Inc.)

tor is the great permeability of the glomerular capillaries. These vessels are far more porous than typical capillaries.

Almost 25% of the blood pumped by the heart is delivered to the kidneys each minute, so that every 4 minutes the kidneys receive a volume of blood equal to the total volume of blood in the body. Every 24 hours about 180 liters (about 45 gallons) of filtrate are produced. Common sense tells us that no one could excrete urine at the rate of 45 gallons per day: Within a few minutes dehydration would become a life-threatening problem.

Reabsorption Is Highly Selective

Fortunately, about 99% of the filtrate is reabsorbed into the blood, leaving only about 1.5 liters to be excreted as urine. Reabsorption is highly selective. Wastes, surplus salts, and water are left in the renal tubules, but glucose, amino acids, vitamins, and other useful materials are reabsorbed into the blood. This is accomplished by a combination of active transport, diffusion, and osmosis.

Normally, substances that are useful to the body are completely reabsorbed from the tubules. However, if a large excess of a particular substance is present in

the blood, the tubules may not be able to return all of it. The maximum concentration of a specific substance in the blood at which complete reabsorption can take place is termed the **renal threshold** for that substance. When a substance exceeds its renal threshold, the portion not reabsorbed is excreted in the urine. Some substances, such as urea, have very low thresholds, so even when present in small concentrations, not much is reabsorbed. Other substances, such as glucose, amino acids, and hormones, have high renal thresholds and are normally completely reabsorbed. The threshold values of all these substances are so arranged that the tubules not only largely cleanse the blood of wastes but also regulate each component of the internal chemical environment of the body. Every day the tubules reabsorb more than 40 gallons of water, 2.5 pounds of salt, and about 0.5 pound of glucose. Most of this has been reabsorbed many times over.

What happens if a substance in the blood exceeds its threshold value? An important example of this occurs in the condition **diabetes mellitus.** Because of an insufficiency of the hormone insulin, a diabetic suffers from impaired carbohydrate metabolism. Glucose accumulates in the blood instead of being efficiently absorbed and utilized by the cells. The concentration of glucose filtered into the nephron exceeds the renal threshold, and the amount of the glucose in excess of

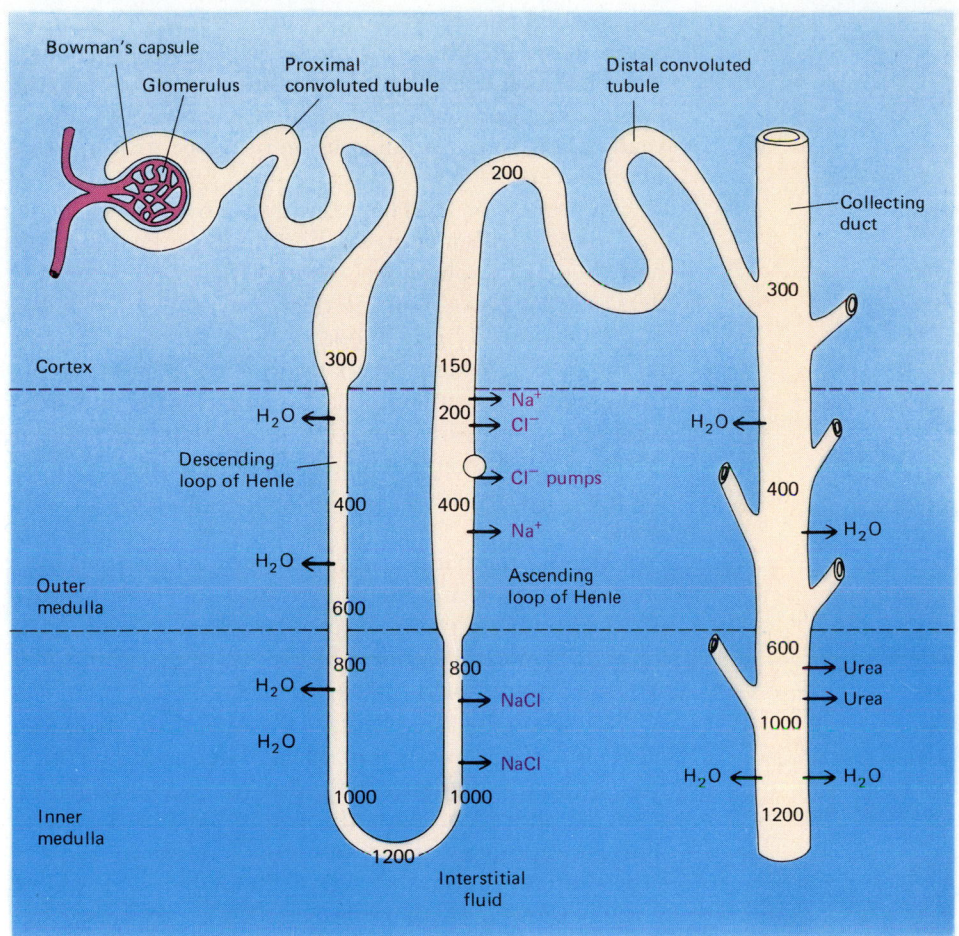

Bowman's capsule
Glomerulus
Proximal convoluted tubule
Distal convoluted tubule

Collecting duct

Cortex

Descending loop of Henle

Outer medulla

Ascending loop of Henle

Inner medulla

Interstitial fluid

200

300

300   150
H₂O ←   200 → Na⁺
                  → Cl⁻
400   Cl⁻ pumps
             400 → Na⁺
H₂O ←
600   800
800   → NaCl
H₂O ←
H₂O   → NaCl
1000   1000
    1200

H₂O ←

H₂O → H₂O

→ Urea
→ Urea

H₂O ← → H₂O

300
400
600
1000
1200

FIGURE 40–14 A countercurrent mechanism in the loops of Henle help maintain a high salt concentration in the interstitial fluid in the medulla of the kidney. Water passes out of the descending loop of Henle, leaving a more concentrated filtrate inside. In the ascending loop, salt moves out. The saltier the tissue fluid becomes, the more water moves out of the descending loop. This leaves a concentrated filtrate inside so more salt passes out. Note that this is a positive feedback system. Water from the collecting ducts moves out into this hypertonic tissue fluid.

the renal threshold is excreted in the urine. The presence of glucose in the urine is evidence of this disorder, as is an increased total production of urine.

Urine Concentration Depends On A Countercurrent Mechanism

The kidneys produce either a concentrated or dilute urine, depending on the needs of the body and the volume of fluid ingested. This variability depends in part on the salt concentration gradient in the tissue fluid surrounding the nephrons in the medulla. The salt concentration is maintained by salt reabsorption from various regions of the renal tubule and by a *countercurrent mechanism.*

Sodium ions are actively transported out of the proximal tubule, and water follows osmotically. The walls of the descending portion of the loop of Henle are relatively permeable to water but relatively impermeable to sodium and urea. There is a high concentration of sodium in the tissue fluid, so as the filtrate passes down the loop of Henle, water moves out by osmosis. This leaves a more concentrated filtrate inside the loop of Henle (Fig. 40–14).

At the turn of the loop of Henle, the walls become more permeable to salt and less permeable to water. As the concentrated filtrate moves up the ascending portion of the loop of Henle, salt moves out into the tissue fluid. Higher in the ascending part of the loop of Henle, pumps are present that transport chloride into the tissue fluid. The negatively charged chloride ions attract positively charged sodium ions, which follow them out of the tubule.

Because water passes out of the descending portion of the loop of Henle, there is a high salt concentration at the bottom of the loop. However, since salt (but not water) is removed in the ascending loop, by the time the filtrate reaches the distal tubule, it is isotonic to blood or even hypotonic to it. As the filtrate moves down the collecting duct, water continues to pass osmotically into the tissue fluid where it is collected by blood vessels.

Note the *counterflow* of fluid through the two limbs of the loop of Henle. Filtrate passing down through the descending region of the loop is flowing in a direction opposite to filtrate moving upward through the ascending loop. The filtrate is concentrated as it moves down the descending loop and diluted as it moves up the ascending loop. This **countercurrent mechanism**

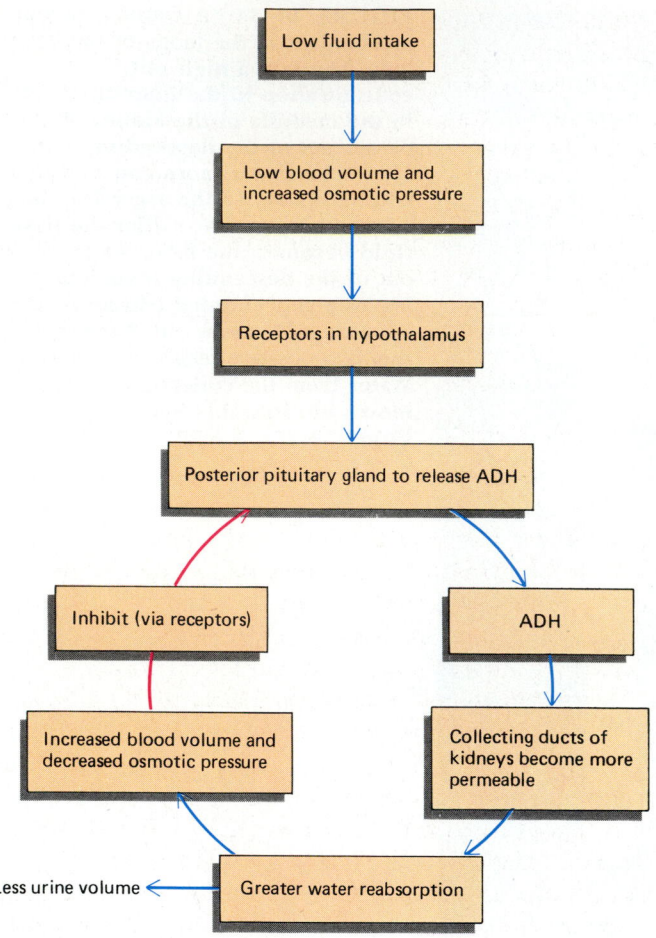

FIGURE 40–15 Regulation of urine volume reflects the blood volume and osmotic pressure. When the body is dehydrated, the hormone ADH increases the permeability of the collecting ducts to water. More water is reabsorbed, and only a small volume of concentrated urine is produced.

helps maintain a high salt concentration in the tissue fluid surrounding the nephrons in the medulla. The hypertonic fluid draws water osmotically from the filtrate in the collecting ducts.

The collecting ducts are routed so that they pass through the zone of very salty tissue fluid on their way to the renal pelvis. So much water may leave the collecting ducts that highly concentrated urine can be produced.

Urine Volume Is Regulated By The Hormone ADH

The amount of urine produced depends upon the body's need to retain or rid itself of water. We have seen that salt reabsorption and the countercurrent mechanism in the loops of Henle establish a very salty tissue fluid that draws water osmotically from the collecting ducts. Permeability of the collecting ducts to water is regulated by **antidiuretic hormone (ADH).**

When the body needs to conserve water, ADH is released from the posterior pituitary gland (Fig. 40–15). This hormone acts on the collecting ducts, making them more permeable to water so that more water is reabsorbed, and a small volume of concentrated urine is produced.

Secretion of ADH is stimulated by special receptors in the hypothalamus. When fluid intake is low, the body begins to dehydrate, causing the blood volume to decrease. As blood volume decreases, the *concentration* of salts dissolved in the blood becomes greater, causing an increase in osmotic pressure. Receptors in the hypothalamus are sensitive to this osmotic change and stimulate the posterior lobe of the pituitary to release ADH. A thirst center in the hypothalamus also responds to dehydration, stimulating an increase in fluid intake.

When one drinks a great deal of water, the blood becomes diluted and its osmotic pressure falls. Release of ADH by the pituitary gland decreases, lessening the amount of water reabsorbed from the collecting ducts. A large volume of dilute urine is produced.

Occasionally, the pituitary gland malfunctions and does not produce sufficient ADH. The resulting condition is termed **diabetes insipidus** (not to be confused with the more common disorder, diabetes mellitus). This condition can also result from a developed insensitivity of the kidney to ADH. In diabetes insipidus, water is not efficiently reabsorbed from the ducts, and therefore a large volume of urine is produced. A person with severe diabetes insipidus may excrete up to 25 quarts of urine each day, a serious loss of water to the body. The affected individual becomes dehydrated and must drink almost continually to offset fluid loss. Diabetes insipidus can often be controlled by injections of ADH or with an ADH nasal spray.

Sodium Reabsorption Is Regulated By The Hormone Aldosterone

Sodium is the most abundant extracellular ion, accounting for about 90% of all positive ions outside cells. Concentration of sodium is carefully regulated by the hormone **aldosterone,** secreted by the cortex of the adrenal glands. This hormone stimulates the distal tubules and collecting ducts to increase sodium reabsorption.

Aldosterone secretion can be stimulated by a decrease in blood pressure. When blood pressure falls, cells of the **juxtaglomerular apparatus** secrete the enzyme renin and activate the renin-angiotensin pathway. The juxtaglomerular apparatus is a small region of cells where the distal tubule contacts the afferent arteriole (Fig. 40–13*a*). Renin acts on a plasma protein, converting it to angiotensin, which stimulates aldosterone secretion.

Urine Is Composed Of Water, Nitrogenous Wastes, And Salts

By the time the filtrate reaches the renal pelvis, its composition has been precisely adjusted. Useful materials have been returned to the blood by reabsorption. Wastes and excess materials that entered by filtration or secretion have been retained by the tubules. The adjusted filtrate, called **urine,** is composed of about 96% water, 2.5% nitrogenous wastes (primarily urea), 1.5% salts, and traces of other substances, such as bile pigments, which may contribute to the characteristic color and odor. Healthy urine is sterile and has been used to wash battlefield wounds where clean water is not available. However, urine swiftly decomposes when exposed to bacterial action, forming ammonia and other products. It is the ammonia that produces the diaper rash of infants.

■ CHAPTER SUMMARY

I. Excretory systems help maintain homeostasis by osmoregulation, excretion of metabolic wastes, and regulation of concentration of body fluid components.

II. The principal waste products of animal metabolism are water, carbon dioxide, and nitrogenous wastes, including ammonia, urea, and uric acid.

III. Invertebrate mechanisms of osmoregulation and waste disposal are diverse and are adapted to the body plan and life-style of an organism.
  A. Marine sponges and cnidarians have no specialized excretory organs.
  B. Flatworms have nephridial organs characterized by protonephridia with flame cells.
  C. Annelids possess coelomic metanephridia.
  D. Crustaceans employ antennal glands for osmoregulation.
  E. Malpighian tubules contribute to insect success on land by conserving water.

IV. The vertebrate kidney is a key organ for maintaining homeostasis of body fluids and disposal of metabolic wastes.
  A. The kidneys produce urine, which then passes through the ureters to the urinary bladder for storage. During urination the urine passes through the urethra to the outside of the body.
  B. Each nephron consists of Bowman's capsule, a cluster of capillaries called a glomerulus, and a long, coiled renal tubule.
  C. Urine formation is accomplished by filtration of plasma, reabsorption of needed materials, and secretion of a few substances such as potassium and hydrogen into the renal tubule.
    1. Plasma filters out of the glomerular capillaries and into Bowman's capsule. Because filtration is mostly a nonselective process, both needed materials, such as glucose, and wastes become part of the filtrate.
    2. About 99% of the filtrate is reabsorbed from the renal tubules into the blood; this is a highly selective process that returns usable materials to the blood but leaves wastes and excessive quantities of other substances to be excreted in the urine.
    3. In secretion certain substances and drugs are actively transported into the renal tubule to become part of the urine.
    4. The tissue fluid surrounding the nephrons has a high salt concentration due to a countercurrent mechanism in the loops of Henle.
    5. Urine volume is regulated by the hormone ADH, which is released by the posterior lobe of the pituitary gland in response to an increase in osmotic pressure of the blood (caused by dehydration). ADH increases the permeability of the collecting ducts. As a result more water is reabsorbed and only a small volume of urine is produced.
    6. Urine consists of water, nitrogenous wastes, salts, and excesses of other substances.

■ POST-TEST

1. The process of removing metabolic wastes from the body is called _____ .

2. _____ is the ability of an organism to regulate its fluid content.

3. The principal nitrogenous waste product of insects and birds is _____ .

4. The principal nitrogenous waste product of amphibians and mammals is _____ .

5. Flatworms have excretory structures called _____ , which are characterized by _____ cells.

6. Earthworms have _____ in each of their body segments.

7. The principal excretory organs of crustaceans are
_____ _____.

8. The excretory structures of insects are
_____ _____.

9. The vertebrate kidney consists of functional units called _____.

Select the most appropriate answer from column B for each description in column A.

Column A
10. outer portion of human kidney
11. delivers urine to bladder
12. part of kidney that receives urine from collecting ducts

Column B
a. cortex
b. medulla
c. ureter
d. urethra
e. renal pelvis

Column A
13. site of filtration
14. site of reabsorption
15. delivers urine to outside of body

Column B
a. urethra
b. renal tubules
c. renal pelvis
d. Bowman's capsule
e. ureter

16. The glomerulus consists of a tuft of _____, which project into _____ _____.
17. Blood is delivered to the glomerulus by the _____ and leaves the glomerulus in the _____.
18. Fluid that leaves the glomerular capillaries and enters Bowman's capsule is called _____.
19. When a substance exceeds its renal threshold, the portion not reabsorbed is _____.
20. The countercurrent mechanism helps maintain a high _____ concentration in the interstitial fluid.
21. Antidiuretic hormone (ADH) increases the permeability of the _____ _____ so that more water is _____ and the volume of urine is _____.
   (increased or decreased)
22. Label the diagram at the right.

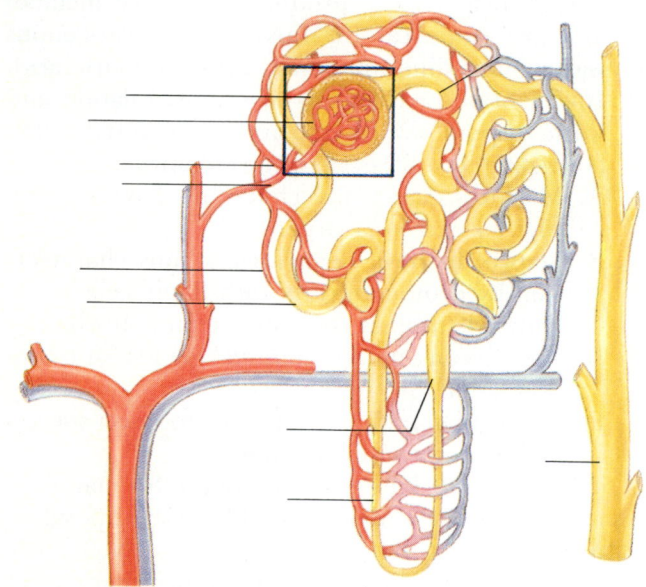

■ REVIEW QUESTIONS

1. Compare osmoregulation in flatworms and insects.
2. The number of protonephridia in a planarian is adjusted to the salinity of the environment. Planaria inhabiting slightly salty water develop fewer protonephridia, but the number quickly increases when the concentration of salt in the environment is lowered. Can you explain why?
3. What type of osmoregulatory problem is faced by marine fish? By freshwater fish? How are these problems met in each case? Do human beings ever have osmoregulatory problems? Explain.
4. Name the structure in the mammalian body that is associated with each of the following: (1) urea formation, (2) urine formation, (3) temporary storage of urine, (4) conduction of urine out of the body.
5. Draw a diagram of a nephron and label its parts.

6. Which part of the nephron is associated with the following: (1) filtration, (2) reabsorption, (3) secretion.
7. Contrast reabsorption and secretion.
8. List the sequence of blood vessels through which a drop of blood passes as it makes its way from renal artery to renal vein.
9. Why is glucose normally not present in urine? Why is it present in diabetes mellitus? Why do you suppose diabetics experience an increased output of urine?
10. How is urine volume regulated? Explain. Why must victims of untreated diabetes insipidus drink great quantities of water?
11. How does the countercurrent mechanism in the loops of Henle help concentrate urine?

■ RECOMMENDED READINGS

Dantzler, W.H.: "Renal Adaptations of Desert Vertebrates," *Bioscience* 32(2):108–112, 1982. Adaptations in renal physiology are integrated with other physiological and behavioral mechanisms for desert survival.

Heatwole, H.: "Adaptations of Marine Snakes," *American Scientist,* Sept.–Oct. 1978, 594–604. Among the adaptations discussed are those that permit several groups of snakes to inhabit the marine environment by maintaining fluid balance.

Pollie, R.: "Comprehending Kidney Disease," *Science News,* October 2, 1982, 218. A simple theory is offered to explain a complex syndrome.

Smith, H.W.: *From Fish to Philosopher,* 2nd ed., Boston, Little, Brown, & Co., 1961. A fascinating summary of the evolution of the vertebrate kidney.

Solomon, E.P., Schmidt, R., and P. Adragna: *Human Anatomy and Physiology.* Philadelphia, Saunders College Publishing, 1990. Chapters 27 and 28 focus on the human urinary system and fluid balance.

41
Endocrine Regulation

OUTLINE

I. Hormone secretion is regulated by negative feedback mechanisms
II. Hormones combine with specific receptor proteins in target cells
   A. Some hormones activate genes
   B. Some hormones work through second messengers
III. In invertebrates hormones regulate growth, metabolism, reproduction, molting, and pigmentation
IV. In vertebrates hormones help regulate growth, reproduction, and many aspects of metabolism
   A. Endocrine disorders may involve hyposecretion or hypersecretion
   B. Nervous regulation and endocrine regulation are integrated by the hypothalamus
   C. The posterior lobe of the pituitary gland releases two hormones
   D. The anterior lobe of the pituitary gland regulates growth and several other endocrine glands
   E. Thyroid hormones stimulate metabolic rate
   F. The parathyroid glands regulate calcium concentration
   G. The islets of the pancreas regulate glucose concentration
   H. The adrenal glands help the body adapt to stress
   I. Prostaglandins are released by many types of mammalian cells
   J. Many other hormones are known

LEARNING OBJECTIVES

After you have studied this chapter you should be able to:

1. Define the terms *hormone* and *endocrine gland* and distinguish between endocrine and exocrine glands.
2. Compare the mechanisms of action of steroid and protein-type hormones. (Include the role of second messengers, such as cyclic adenosine monophosphate.)
3. Summarize the role of hormones in invertebrates.
4. Identify the principal vertebrate endocrine glands and locate them in the body. (Consult Fig. 41–5.)
5. Summarize the regulation of endocrine glands by negative-feedback mechanisms and relate the concept of negative feedback to the specific hormones discussed.
6. Justify the description of the hypothalamus as the link between nervous and endocrine systems and describe the mechanisms by which the hypothalamus exerts its control.
7. Compare the functions of the posterior and anterior lobes of the pituitary; identify their hormones and describe the actions of these hormones.
8. Describe the actions of growth hormone on growth and metabolism and contrast the consequences of hyposecretion and hypersecretion.
9. Describe the actions of the thyroid hormones, their regulation, and the thyroid disorders discussed in this chapter.
10. Contrast the actions of insulin and glucagon and describe the disorders associated with malfunction of the islets of the pancreas.
11. Describe the actions of the mineralocorticoid and glucocorticoid hormones and describe the effects of malfunction of the adrenal cortex.
12. Summarize the role of the adrenal glands in helping the body to adapt to stress.

Protective coloration in the ghost crab, *Ocypode quadrata*. (Connie Toops)

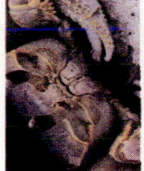

The **endocrine system** is a diverse collection of glands and tissues that secrete chemical messengers responsible for the control of many body processes. These chemical messengers, called **hormones,** have specific regulatory effects on the activities of some of the other tissues in the body. The term *hormone* is derived from a Greek word meaning "to excite." Hormones do indeed excite their target tissues, usually by stimulating a change in some metabolic activity. The study of endocrine activity, **endocrinology,** is an exciting field of medical research.

Endocrine glands produce hormones and secrete them into the surrounding tissue fluid from where they diffuse into capillaries. Hormones are then transported throughout the body by the blood but elicit responses only in their **target tissues.** The target tissue may be another endocrine gland or it may be an entirely different type of organ, such as a bone or the kidney. Often the target tissue is located far from the endocrine gland. Figure 41–1 illustrates the difference between endocrine glands and **exocrine glands** (such as sweat glands and gastric glands), which release their secretions into ducts.

In recent years it has been discovered that specialized cells in the digestive tract and in some other organs, such as the heart and kidneys, also release hormones. The scope of endocrinology has been broadened to include the study of chemical messengers produced by cells that are widely distributed in the body, rather than by single, discrete organs. Chemically, hormones either are lipids (usually steroids) or belong to the protein family (proteins, peptides, and derivatives of amino acids).

KEY CONCEPTS

☐ Endocrine glands produce and secrete hormones, chemical messengers that help regulate the activities of various organs, thereby maintaining homeostasis.

☐ Endocrine activity is regulated by negative-feedback control mechanisms.

☐ Hormones are transported to all the cells of the body by the blood; only target cells have appropriate receptors that combine with the hormone and produce a response.

☐ In mammals the nervous and endocrine systems are linked by the hypothalamus, which regulates the activity of the pituitary gland.

☐ HORMONE SECRETION IS REGULATED BY NEGATIVE-FEEDBACK MECHANISMS

How does an endocrine gland "know" how much hormone to release at any given moment? Hormone secretion is regulated internally by **negative-feedback control mechanisms.** Information regarding the hormone level or the strength of its effect is fed back to the gland, which then responds in a homeostatic manner. The parathyroid glands located in the neck, which regulate calcium level in the blood, provide a good example of negative feedback.

The parathyroid hormone causes the level of calcium in the blood to rise. A low level of calcium in the blood signals the parathyroid glands to release more

FIGURE 41–1 Comparison of (a) an exocrine with (b) an endocrine gland. The secretion of an exocrine gland passes through a duct to reach its final destination. For example, sweat passes through the duct of a sweat gland to reach the surface of the skin. The hormone of an endocrine gland is released into the surrounding tissue fluid and diffuses into the blood, which transports it to its target tissue.

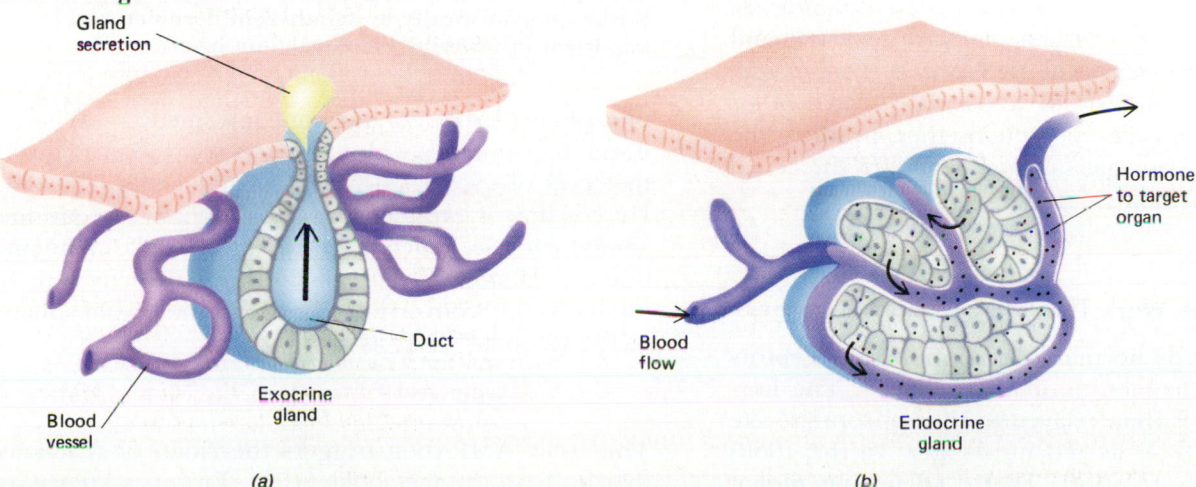

hormone (Fig. 41–2). But when the calcium level rises beyond normal limits, the parathyroid glands are inhibited and slow their output of hormone. Both responses are negative-feedback mechanisms, since in both cases the effects are opposite (negative) to the stimulus (that is, more calcium leads to less hormone). Negative feedback is the basis of hormone regulation. As you will see, variations of this theme abound.

☐ HORMONES COMBINE WITH SPECIFIC RECEPTOR PROTEINS IN TARGET CELLS

In vertebrates, most endocrine glands secrete at least small amounts of their hormones continuously. Although present in minute amounts, more than 30 different hormones may be circulating in the blood at all times.

A hormone may pass through many tissues seemingly "unnoticed" until it reaches its target tissue. How does the target tissue "recognize" its hormone? Specific receptor proteins in the target tissues bind the hormone. This is a highly specific process. The receptor site is like a lock, and the hormones are like different keys. Only the hormone that fits the lock—the specific receptor—can influence the metabolic machinery of the cell. Once a hormone is taken up by a particular tissue, how does it influence the activity of the cells?

Some Hormones Activate Genes

Steroid hormones and thyroid hormones (small polypeptides) are relatively small, lipid-soluble molecules that easily pass through the plasma membrane of the target cell. They move through the cytoplasm and into the nucleus (Fig. 41–3). Specific protein receptors in the nucleus combine with the hormone to form a hormone-receptor complex. This complex then combines with a protein associated with the deoxyribonucleic acid (DNA). As a result certain genes are activated, and messenger ribonucleic acid (mRNA) is synthesized, leading to the synthesis of a particular protein or group of proteins. These proteins then produce the structural or physiological changes that are the actual effect of the hormone.

Some Hormones Work Through Second Messengers

Many protein-type hormones combine with receptors on the plasma membrane of a target cell. The hormonal message is then relayed to the appropriate site within the cell by a second messenger. In the 1960s Earl Sutherland identified **cyclic adenosine mono-**

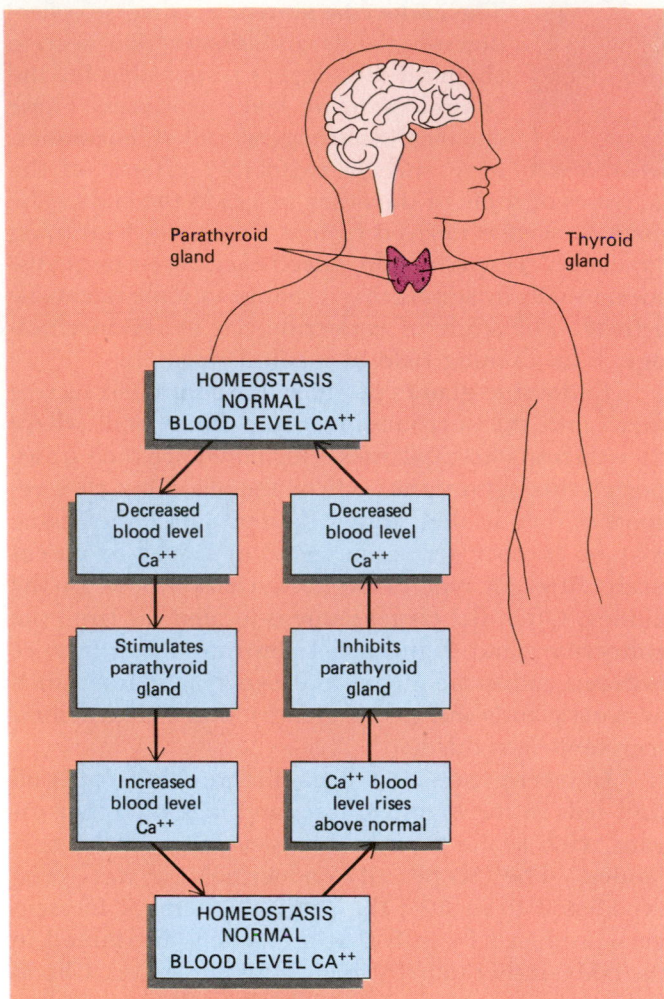

FIGURE 41–2   Regulation of hormone secretion by negative feedback. When the calcium concentration in the blood falls below normal, the parathyroid glands are stimulated to secrete more parathyroid hormone. This hormone acts to increase the calcium concentration in the blood, thus restoring homeostasis. Should the calcium concentration exceed normal, the parathyroid glands are inhibited and slow their secretion of hormone. This diagram has been greatly simplified; calcitonin, a hormone secreted by the thyroid gland, works antagonistically to parathyroid hormone and is important in lowering blood calcium levels.

phosphate (AMP) as a hormone intermediary. When a hormone combines with a receptor on the target cell, the level of cyclic AMP is increased within the cell. Here is how it happens (Fig. 41–4). First, the enzyme **adenyl cyclase,** which is attached to the plasma membrane of almost all cells in the body, is activated. It catalyzes the conversion of adenosine triphosphate (ATP) to cyclic AMP, as follows:

$$\text{ATP} \xrightarrow{\text{Adenyl cyclase}} \text{Cyclic AMP}$$

The cyclic AMP then triggers the chain of reactions that leads to the metabolic effect. Enzymes known as

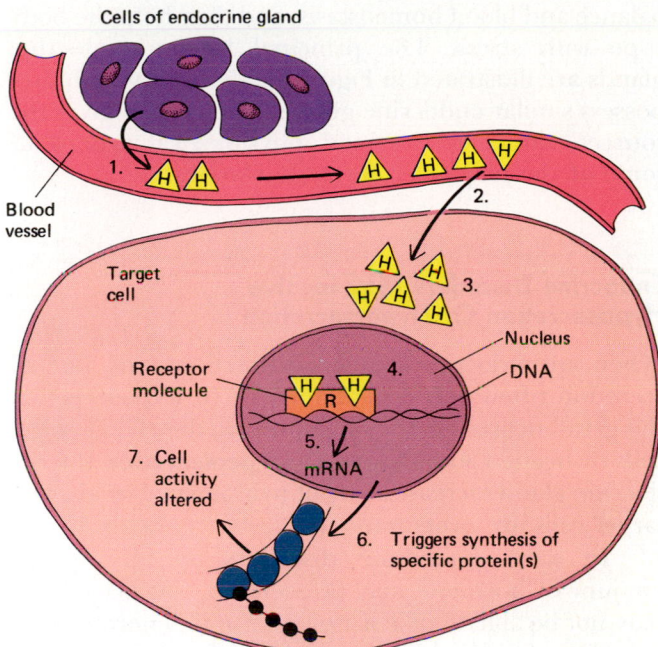

FIGURE 41–3 **Activation of genes by steroid hormones.** Steroid hormones are small lipid-soluble molecules that pass freely through the cell membrane. The hormone combines with receptors within the nucleus of a target cell. The steroid hormone-receptor complex combines with a protein associated with the DNA. This activates specific genes, leading to the synthesis of mRNA coding for specific proteins. The proteins cause the response recognized as the hormone's action.

protein kinases may be activated. These enzymes can add a phosphate group from ATP to a protein. This phosphorylation either activates or inhibits the activity of the protein, which may itself be an enzyme. Membrane permeability may be affected; protein synthesis or secretion may be stimulated.

The particular action initiated by cyclic AMP depends upon the specific kinds of enzyme systems present in the cell. This explains how the same hormone can promote different responses in different cell types. When the target tissue is another endocrine gland, cyclic AMP regulates the release of its hormones.

Several types of hormones may be involved in regulating the metabolic activities of a particular type of cell. In fact, most hormones produce a synergistic effect in which the presence of one hormone may enhance the effects of another.

■ IN INVERTEBRATES HORMONES REGULATE GROWTH, METABOLISM, REPRODUCTION, MOLTING, AND PIGMENTATION

Among invertebrates, hormones are secreted mainly by neurons rather than by endocrine glands. These **neurohormones** help regulate such processes as regeneration in hydras, flatworms, and annelids; molting

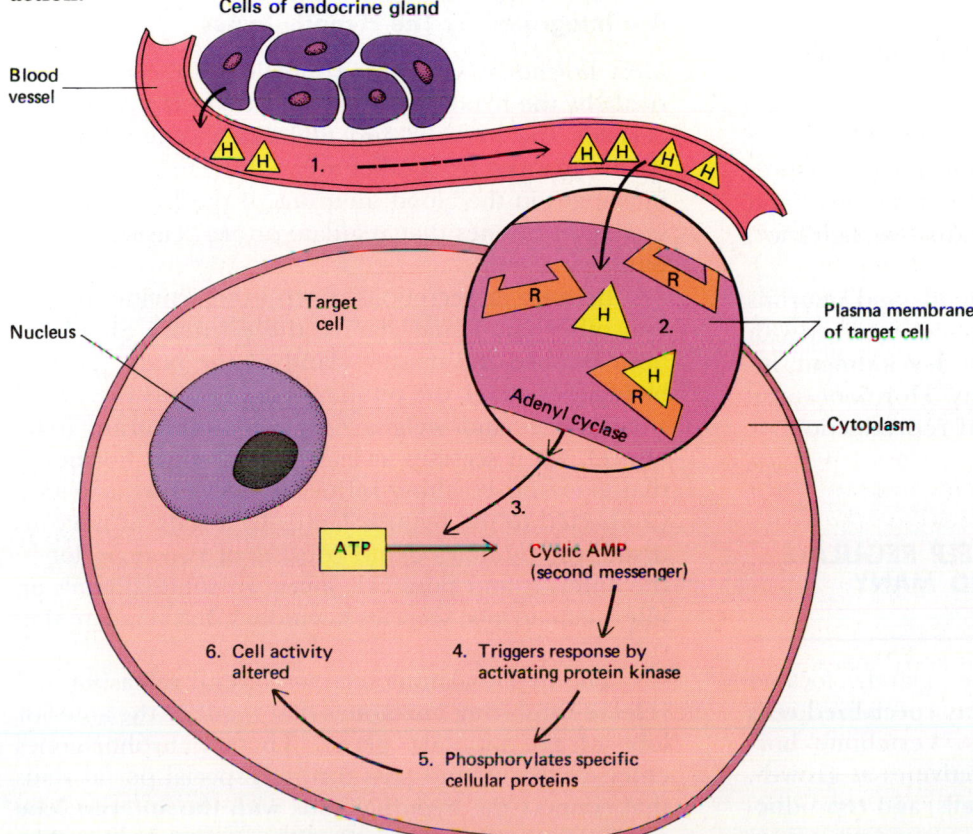

FIGURE 41–4 **Second-messenger mechanism of hormone action.** Many hormones combine with receptors in the cell **membrane** of target cells. The hormone-receptor complex stimulates an enzyme, adenyl cyclase, which catalyzes the conversion of ATP to cyclic AMP, a second messenger. Cyclic AMP then triggers the chain of events leading to the response. A protein kinase may be activated; this enzyme can add a phosphate group to another enzyme, thereby activating or inhibiting its activity.

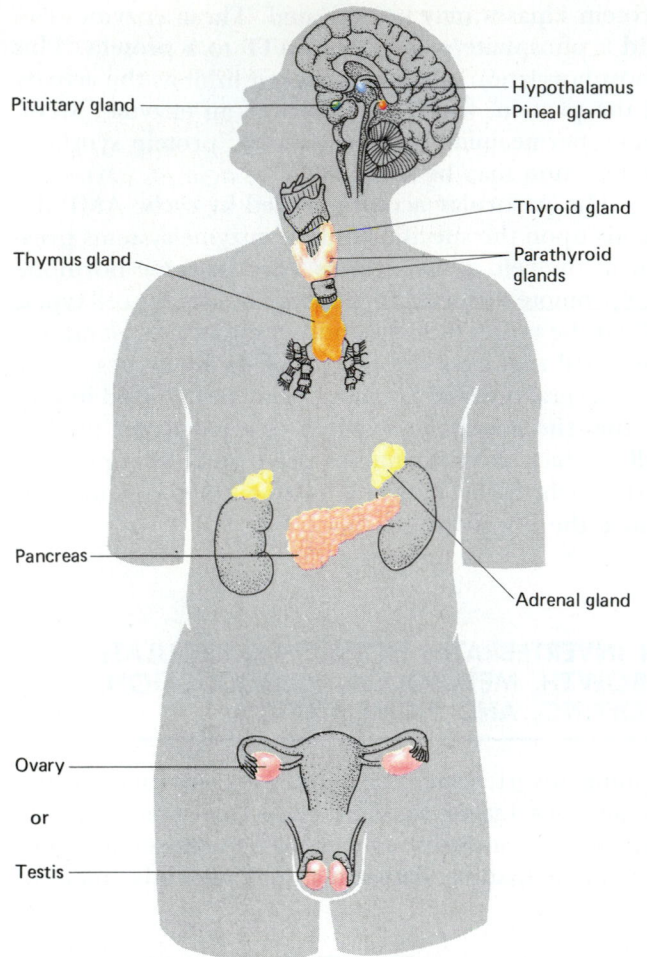

Pituitary gland

Hypothalamus
Pineal gland

Thyroid gland

Thymus gland

Parathyroid glands

Pancreas

Adrenal gland

Ovary

or

Testis

FIGURE 41–5 **Location of the principal endocrine glands of the human male and female.**

and metamorphosis (change in body form, Chapter 24) in insects; color changes in crustaceans; and growth, gamete production, reproductive behavior, and metabolism in other groups.

Insects have both endocrine glands and neurons that secrete hormones. Their hormones interact with one another to regulate growth and development, including molting and morphogenesis. Hormones also help regulate insect metabolism and reproduction.

■ IN VERTEBRATES HORMONES HELP REGULATE GROWTH, REPRODUCTION, AND MANY ASPECTS OF METABOLISM

In vertebrates discrete endocrine glands located throughout the body, as well as many specialized cells in other tissues, secrete hormones. Vertebrate hormones help regulate such diverse activities as growth, metabolic rate, use of nutrients by cells, and reproduction. They are largely responsible for regulating fluid

balance and blood homeostasis, and they help the body cope with stress. The principal human endocrine glands are illustrated in Figure 41–5. Most vertebrates possess similar endocrine glands. Table 41–1 gives the sources, target tissues, and physiological actions of some of the major vertebrate hormones.

Endocrine Disorders May Involve Hyposecretion Or Hypersecretion

When an endocrine gland malfunctions, the rate of secretion often becomes abnormal. In **hyposecretion** the gland decreases hormone output, depriving target cells of needed stimulation. In **hypersecretion** the endocrine gland increases output to abnormal levels, and target cells are overstimulated.

In some endocrine disorders an appropriate amount of hormone may be secreted, but target cells may not be able to take it up and use it. There may be insufficient numbers of receptors or the receptors may not function properly. Any of these malfunctions leads to predictable metabolic abnormalities, with accompanying clinical symptoms (Table 41–2). We will describe some of these disorders as we discuss specific endocrine glands.

Nervous Regulation And Endocrine Regulation Are Integrated By The Hypothalamus

Most hormonal activity is controlled directly or indirectly by the hypothalamus. Thus there is a direct link between the nervous and endocrine systems. In response to input from other areas of the brain and from hormones in the blood, neurons of the hypothalamus secrete hormones that regulate or are released by the pituitary gland.

Because its secretions control the activities of several other endocrine glands, the **pituitary gland** has been dubbed the master gland of the body. Truly a biological marvel, the pituitary gland is only the size of a large pea and weighs only about 0.5 gram (0.02 ounce), yet it secretes at least nine distinct hormones that exert far-reaching influence over body activities. Connected to the hypothalamus by a stalk of nervous tissue, the pituitary gland consists of two main lobes, the anterior and posterior lobes. In some animals an intermediate lobe secretes hormones that regulate skin color.

The hypothalamus secretes several **releasing** and **release-inhibiting hormones** that regulate the anterior lobe of the pituitary gland. These neurohormones enter capillaries and pass through special portal veins that connect the hypothalamus with the anterior lobe of the pituitary (Fig. 41–6). (These portal veins, like

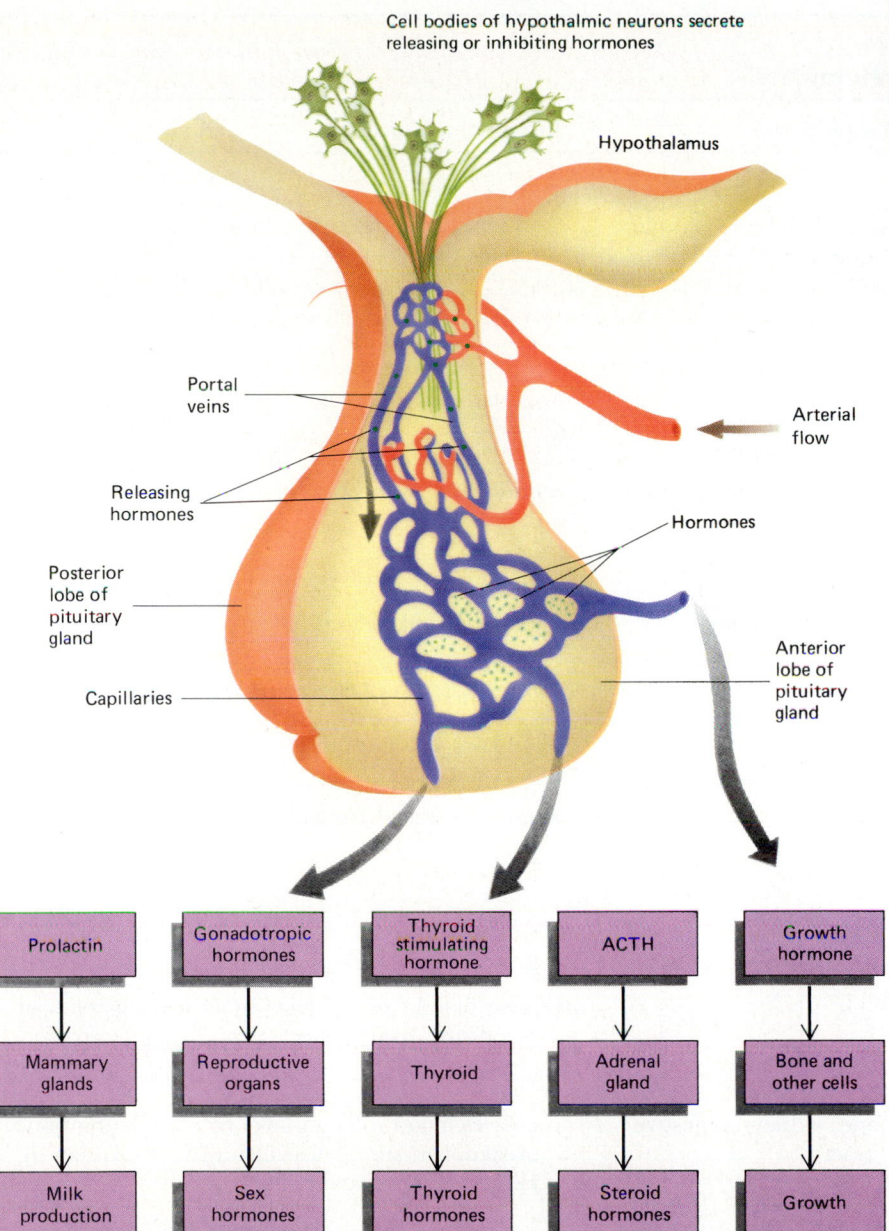

Cell bodies of hypothalmic neurons secrete releasing or inhibiting hormones

Hypothalamus

Portal veins

Releasing hormones

Posterior lobe of pituitary gland

Capillaries

Arterial flow

Hormones

Anterior lobe of pituitary gland

Prolactin	Gonadotropic hormones	Thyroid stimulating hormone	ACTH	Growth hormone
Mammary glands	Reproductive organs	Thyroid	Adrenal gland	Bone and other cells
Milk production	Sex hormones	Thyroid hormones	Steroid hormones	Growth

FIGURE 41–6 The hypothalamus secretes several specific releasing and release-inhibiting hormones, which reach the anterior lobe of the pituitary gland by way of portal veins. Each releasing hormone stimulates the synthesis of a particular hormone by the cells of the anterior lobe. (*R*, releasing hormone; *H*, hormone.)

the hepatic portal vein, do not deliver blood to a larger vein directly but connect two sets of capillaries.) Within the anterior lobe of the pituitary, the portal veins divide into a second set of capillaries. The hormones pass through the walls of these capillaries into the tissue of the anterior lobe where they regulate production and secretion of pituitary hormones.

The Posterior Lobe Of The Pituitary Gland Releases Two Hormones

Two peptide hormones, oxytocin and antidiuretic hormone (ADH; discussed in Chapter 40) are secreted by the **posterior lobe of the pituitary gland.** These hormones are actually produced by specialized nerve cells in the hypothalamus. They reach the posterior lobe of the pituitary by flowing through axons that connect the hypothalamus with the posterior pituitary (Fig. 41–7). Enclosed within tiny vesicles, the hormones pass slowly down the axons of these nerve cells. The axons extend through the pituitary stalk and into the posterior lobe. Hormone accumulates in the axon endings until the neuron is stimulated; then it is released and diffuses into surrounding capillaries.

In a female **oxytocin** levels rise toward the end of pregnancy, stimulating the strong contractions of the uterus needed to expel the baby. Oxytocin is sometimes administered clinically (under the name Pitocin) to initiate or speed labor. After birth, when an infant sucks at its mother's breast, sensory neurons are stimulated and signal the hypothalamus to release oxytocin.

TABLE 41–1
Principal Endocrine Glands and Their Hormones*

Endocrine gland and hormone	Target tissue	Principal actions
Hypothalamus		
Releasing and inhibiting hormones	Anterior lobe of pituitary gland	Stimulates or inhibits secretion of specific hormones
Hypothalamus (production) and posterior lobe of pituitary (storage and release)		
Oxytocin	Uterus	Stimulates contraction
	Mammary glands	Stimulates ejection of milk into ducts
Antidiuretic hormone (ADH)	Kidneys (collecting ducts)	Stimulates reabsorption of water; conserves water
Anterior lobe of pituitary		
Growth hormone (GH)	General	Stimulates growth by promoting protein synthesis
Prolactin	Mammary glands	Stimulates milk secretion
Thyroid-stimulating hormone (TSH)	Thyroid gland	Stimulates secretion of thyroid hormones; stimulates increase in size of thyroid gland
Adrenocorticotropic hormone (ACTH)	Adrenal cortex	Stimulates secretion of adrenal cortical hormones
Gonadotropic hormones (FSH, LH)	Gonads	Stimulates gonad function
Thyroid gland		
Thyroxine (T_4) and triiodothyronine (T_3)	General	Stimulate metabolic rate; essential to normal growth and development
Calcitonin	Bone	Lowers blood calcium level by inhibiting removal of calcium from bone
Parathyroid glands		
Parathyroid hormone	Bone, kidneys, digestive tract	Increases blood calcium level by stimulating bone breakdown; stimulates calcium reabsorption in kidneys; activates vitamin D
Islets of pancreas		
Insulin	General	Lowers blood glucose level by facilitating glucose uptake and utilization by cells; stimulates glycogenesis; stimulates fat storage and protein synthesis
Glucagon	Liver, adipose tissue	Raises blood glucose level by stimulating glycogenolysis and gluconeogenesis; mobilizes fat
Adrenal medulla		
Epinephrine and norepinephrine	Skeletal muscle, cardiac muscle, blood vessels, liver, adipose tissue	Help body cope with stress; increase heart rate, blood pressure, metabolic rate; reroute blood; mobilize fat; raise blood sugar level
Adrenal cortex		
Mineralocorticoids (Aldosterone)	Kidney tubules	Maintain sodium and phosphate balance
Glucocorticoids (Cortisol)	General	Help body adapt to long-term stress; raise blood glucose level; mobilize fat

TABLE 41–1
Principal Endocrine Glands and Their Hormones* (continued)

Endocrine gland and hormone	Target tissue	Principal actions
Ovary†		
Estrogens	General	Stimulate development of secondary sex characteristics
	Reproductive structures	Stimulate growth of sex organs at puberty; promotes monthly preparation of uterus for pregnancy
Progesterone	Uterus	Completes preparation of uterus for pregnancy
	Breasts	Stimulates development
Testis		
Testosterone	General	Stimulates development of secondary sex characteristics and growth spurt at puberty
	Reproductive structures	Stimulates development of sex organs; stimulates spermatogenesis
Pineal gland		
Melatonin	Gonads, pigment cells, other tissues	Influences reproductive processes in hamsters and other animals; pigmentation in some vertebrates; may control biorhythms in some animals; may help control onset of puberty in humans

*The digestive hormones are described in Chapter 39, and the reproductive hormones will be discussed in Chapter 42.

TABLE 41–2
Consequences of Endocrine Malfunction

Hormone	Hyposecretion	Hypersecretion
Growth hormone	Pituitary dwarf	Gigantism if malfunction occurs in childhood; acromegaly in adult
Thyroid hormones	Cretinism (in children); myxedema, a condition of pronounced adult hypothyroidism (BMR is reduced by about 40%; patient feels tired all of the time and may be mentally slow); goiter, enlargement of the thyroid gland	Graves' disease; goiter
Parathyroid hormone	Spontaneous discharge of nerves; spasms; tetany; death	Weak, brittle bones; kidney stones
Insulin	Diabetes mellitus	Hypoglycemia
Adrenocortical hormones	Addison's disease (body cannot synthesize sufficient glucose by gluconeogenesis; patient is unable to cope with stress; sodium loss in urine may lead to shock)	Cushing's disease (edema gives face a full-moon appearance; fat is deposited about trunk; blood glucose level rises; immune responses are depressed)

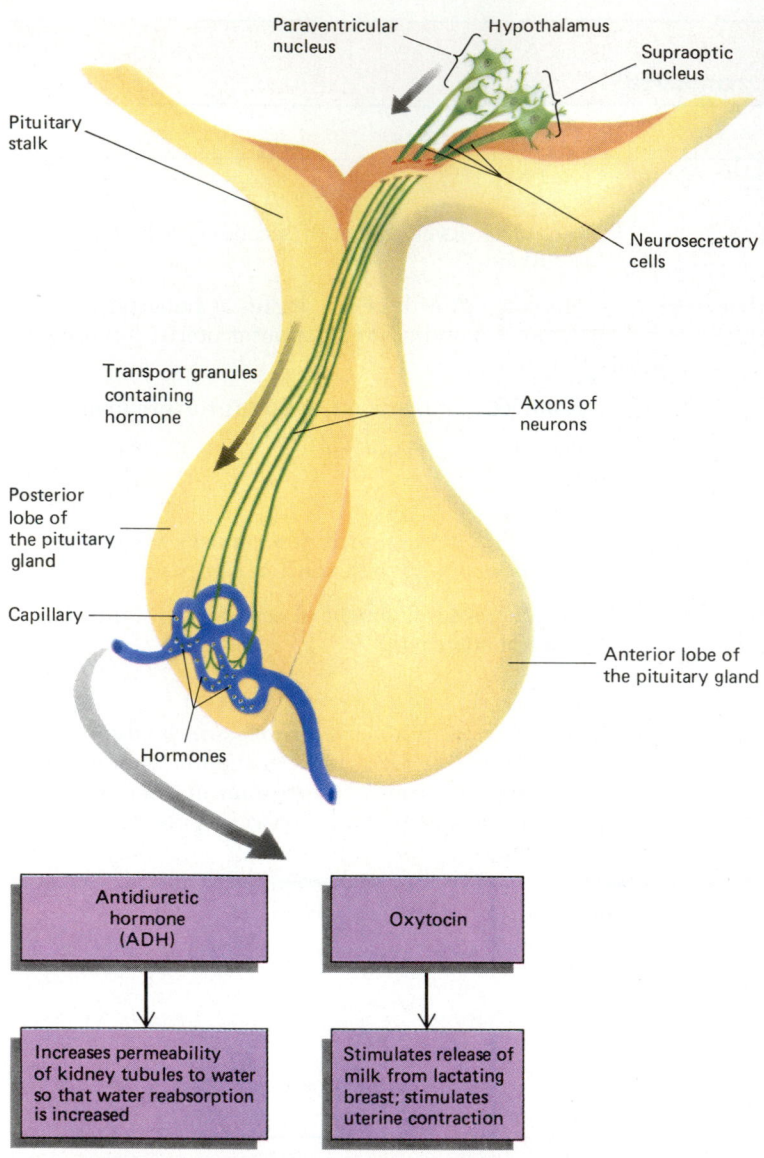

Paraventricular nucleus

Hypothalamus

Supraoptic nucleus

Pituitary stalk

Neurosecretory cells

Transport granules containing hormone

Axons of neurons

Posterior lobe of the pituitary gland

Capillary

Anterior lobe of the pituitary gland

Hormones

Antidiuretic hormone (ADH)

Oxytocin

Increases permeability of kidney tubules to water so that water reabsorption is increased

Stimulates release of milk from lactating breast; stimulates uterine contraction

FIGURE 41–7 The hormones secreted by the posterior lobe of the pituitary are actually manufactured in cells of the hypothalamus. The axons of these neurons extend down into the posterior lobe of the pituitary. The hormones are packaged in granules that flow through these axons and are stored in their ends. The hormone is secreted into the interstitial fluid as needed and then transported by the blood.

The hormone stimulates contraction of cells surrounding the milk glands so that milk is let down into the ducts, from which it can be sucked by the infant. Because oxytocin also stimulates the uterus to contract, breast feeding promotes rapid recovery of the uterus to nonpregnant size. Males have about the same amount of oxytocin circulating in their blood as females, but its function in them is unknown.

The Anterior Lobe Of The Pituitary Gland Regulates Growth And Several Other Endocrine Glands

The **anterior lobe of the pituitary** secretes growth hormone, prolactin, and several **tropic hormones**—hormones which stimulate other endocrine glands (Fig. 41–8). **Prolactin** is the hormone that stimulates the cells of the mammary glands to secrete milk during

lactation. Release of each of the anterior pituitary hormones is in some way regulated by a separate releasing hormone from the hypothalamus and sometimes also by a release-inhibiting hormone, produced in the hypothalamus. When we speak of the pituitary as being stimulated or inhibited, it should be understood that certain receptors in the hypothalamus are generally affected first. They in turn control the pituitary.

Growth Hormone Stimulates Protein Synthesis

Small children measure themselves periodically against their parents, eagerly awaiting that time when they, too, will be "big." Whether one will be tall or short depends upon many factors, including genes, diet, and hormonal balance.

Growth hormone (also called somatotropin) stimulates overall body growth by increasing uptake of

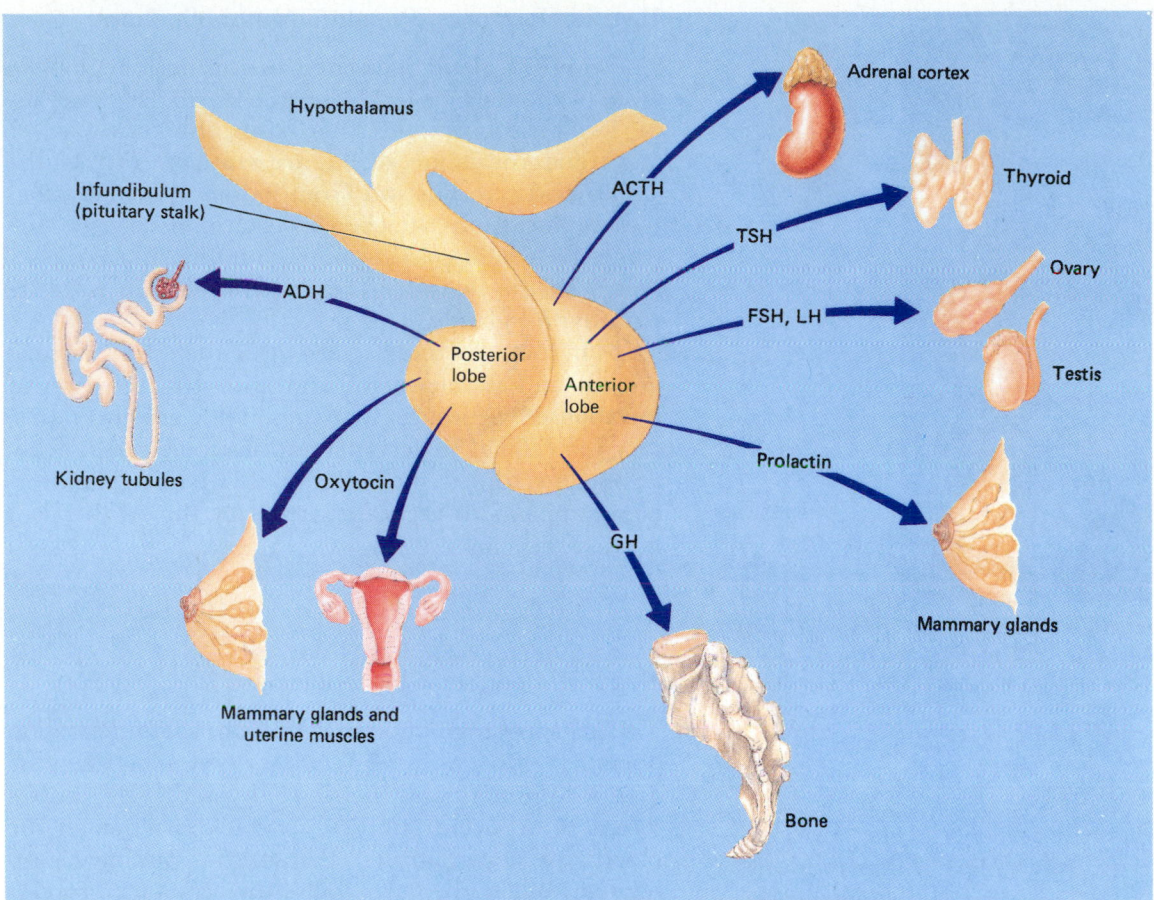

FIGURE 41–8 The pituitary gland is suspended from the hypothalamus by a stalk of neural tissue. The hormones secreted by the anterior and posterior lobes of the pituitary gland and the target tissues they act upon are shown.

amino acids by the cells and by stimulating protein synthesis. Growth hormone promotes mobilization of fat from adipose tissues, raising the level of free fatty acids in the blood. Fatty acids become available for cells to use as fuel, an action that conserves proteins. How does this help to promote growth?

Fat mobilization by growth hormone is also important during fasting or when a person is under prolonged stress. In both of these situations, the blood sugar level is low.

Growth Is Affected By Many Factors

Secretion of growth hormone from the pituitary is regulated both by a growth-hormone-releasing hormone and a growth-hormone-inhibiting hormone released by the hypothalamus. A high level of growth hormone in the blood signals the hypothalamus to secrete the inhibiting hormone. As a result, the pituitary releases less growth hormone. A low level of growth hormone in the blood causes the hypothalamus to secrete the releasing hormone. This hormone stimulates the pituitary gland to release more growth hormone. Many

other factors stimulate secretion of growth hormone, including low blood sugar, increased amino acid concentration in the blood, and stress.

You may recall your parents telling you to get plenty of sleep and exercise so that you would grow properly. Recent studies support these age-old notions. Secretion of growth hormone does increase during exercise, probably because rapid metabolism by muscle cells lowers blood sugar level. Growth hormone secretion also increases during non-REM sleep.

Children who get lots of nurturing also have an advantage. Growth may be retarded in children who are deprived of emotional attention and support even when their physical needs for food and shelter are amply met. Cuddling, playing, and other forms of nurture are apparently essential to normal development. Some emotionally deprived children exhibit psychosocial dwarfism. Their abnormal sleep patterns may be related to decreased secretion of growth hormone.

Other hormones also influence growth. Thyroid hormones appear to be necessary for normal growth-hormone secretion and function. Sex hormone must be present for the growth spurt associated with pu-

FIGURE 41–9 The world's tallest woman, Sandy Allen, with her family. Ms. Allen is 2.22 meters, or 7 ft, 7 1/4 inches tall. Her gigantism is due to an excess of growth hormone. (Bettina Cirone/Photo Researchers, Inc.)

berty to occur. However, the presence of sex hormone eventually causes the growth centers within the long bones to fuse to the shafts, so further increase in height is impossible even when growth hormone is present.

Pituitary Malfunction May Result In Abnormal Growth

Have you ever wondered why midgets failed to grow normally? They are probably **pituitary dwarfs**—individuals whose pituitary glands did not produce sufficient growth hormone during childhood. Though miniature, a pituitary dwarf has normal intelligence and is usually well proportioned. If the growth centers in the long bones are still active when this condition is diagnosed, it can be treated clinically by injection with growth hormone, which can now be synthesized commercially. Growth problems may also result from the malfunction of other mechanisms, such as the regulating hormones from the hypothalamus. Abnormally tall individuals develop when the anterior pituitary secretes excessive amounts of growth hormone during childhood. This condition is referred to as **gigantism** (Fig. 41–9).

If pituitary malfunction leads to hypersecretion of growth hormone during adulthood, the individual cannot grow taller. Connective tissue proliferates, and bones in the hands, feet, and face may increase in diameter. This condition is known as **acromegaly,** meaning "large extremities".

Thyroid Hormones Stimulate Metabolic Rate

The **thyroid gland** is located in the neck region, in front of the trachea and below the larynx (see Fig. 41–5). Two of its hormones, **thyroxine,** also known as T_4, and triiodothyronine, or T_3, are synthesized from the amino acid tyrosine and from iodine. Thyroxine has four iodine atoms attached to each molecule; T_3 has three. We will discuss calcitonin, another hormone secreted by the thyroid gland, in conjunction with the parathyroid glands.

Thyroid hormones are essential for normal growth and development and stimulate the rate of metabolism in most body tissues. They are also necessary for cellular differentiation. Tadpoles cannot develop into adult frogs without thyroxine; This hormone appears to regulate selectively the synthesis of needed proteins.

Thyroid Secretion Is Regulated By Negative Feedback Mechanisms

The principal regulation of thyroid hormone secretion depends upon a negative feedback system between the anterior pituitary and the thyroid gland (Fig. 41–10). When the concentration of thyroid hormone in the blood rises above normal, the anterior pituitary is inhibited. When the level falls, the pituitary secretes more **thyroid-stimulating hormone (TSH).** The TSH acts by way of cyclic AMP to promote synthesis and secretion of thyroid hormones and also to promote increased size of the gland itself.

Too much thyroid hormone in the blood also affects the hypothalamus, inhibiting secretion of TSH-releasing hormone. However, the hypothalamus is thought to exert its regulatory effects primarily in certain stressful situations, such as extreme weather change. Exposure to very cold weather may stimulate the hypothalamus to increase secretion of TSH-releasing hormone, thereby raising body temperature through increased metabolic heat production.

Malfunction Of The Thyroid Gland Leads To Specific Disorders

Extreme hypothyroidism during infancy and childhood results in low metabolic rate and retarded mental and physical development, a condition known as **cretinism.** When diagnosed early enough and treated by administration of thyroid hormones, the effects of cretinism can be prevented.

An adult who feels like sleeping all the time, has little energy, and is mentally slow or confused may also be suffering from hypothyroidism. When there is almost no thyroid function, the basal metabolic rate is

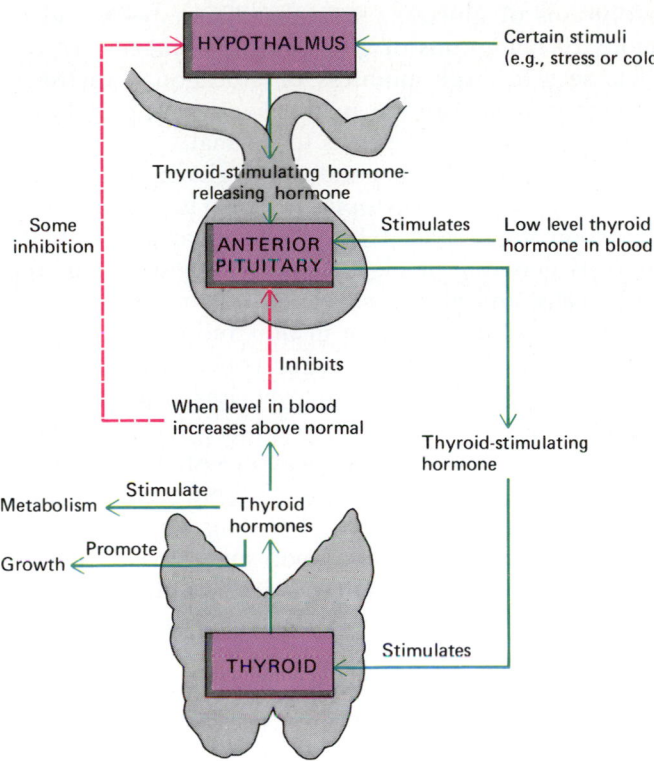

FIGURE 41–10   Regulation of thyroid hormone secretion. Dotted lines indicate inhibition.

FIGURE 41–11   This woman has developed goiter (enlarged thyroid gland) from a deficiency of iodine in her diet. (John Paul Kay/Peter Arnold)

reduced by about 40% and the patient develops the condition called **myxedema,** characterized by a slowing down of physical and mental activity. Hypothyroidism can be treated by using thyroid pills to replace missing hormones.

Hyperthyroidism does not cause abnormal growth but does increase metabolic rate by 60% or even more. This increase in metabolism results in swift use of nutrients, causing the individual to be hungry and to eat more. But this is not sufficient to meet the demands of the rapidly metabolizing cells, so individuals with this condition often lose weight. They also tend to be nervous, irritable, and emotionally unstable.

Any abnormal enlargement of the thyroid gland is termed a **goiter** and may be associated with either hyposecretion or hypersecretion (Fig. 41–11). One cause is dietary iodine deficiency. Without iodine the gland cannot make thyroid hormones, so their concentration in the blood decreases. In compensation the anterior pituitary secretes large amounts of TSH. The thyroid gland enlarges, sometimes to gigantic proportions. However, enlargement of the gland cannot increase production of the hormones, because the needed ingredient is still missing. Thanks to iodized salt, goiter is no longer common in the United States. In other parts of the world, however, hundreds of thousands still suffer from this easily preventable disorder.

The Parathyroid Glands Regulate Calcium Concentration

The **parathyroid glands** are embedded in the connective tissue surrounding the thyroid gland. These glands secrete **parathyroid hormone,** which regulates the calcium level of the blood and tissue fluid. Parathyroid hormone stimulates release of calcium from bones and calcium reabsorption from the kidney tubules. It also activates vitamin D, which then increases the amount of calcium absorbed from the intestine.

Calcitonin, secreted by the thyroid gland, works antagonistically to parathyroid hormone. When the concentration of calcium rises above homeostatic levels, calcitonin is released and rapidly inhibits removal of calcium from bone.

The Islets Of The Pancreas Regulate Glucose Concentration

Besides secreting digestive enzymes (Chapter 39), the pancreas serves as an important endocrine gland. Its hormones, insulin and glucagon, are secreted by cells that form little clusters, or islets, dispersed throughout the pancreas. These islets, first described by the German histologist Paul Langerhans, are called the **islets of Langerhans** (Fig. 41–12). About a million islets are present in the human pancreas. They are composed of **beta cells,** which secrete **insulin,** and **alpha cells,** which secrete **glucagon.**

Insulin Lowers The Concentration Of Glucose In The Blood

Insulin stimulates cells, especially skeletal muscle and fat cells, to take up glucose from the blood. Once glu-

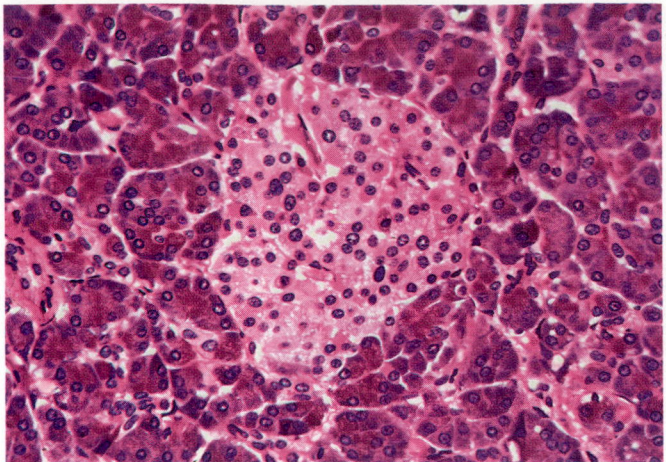

FIGURE 41–12 **Photomicrograph of human pancreas showing the islet of Langerhans.** (Ed Reschke)

cose enters muscle cells, it is either used immediately as fuel or stored as glycogen (in a process called glycogenesis). Insulin activity results in lowering the glucose level in the blood. It also influences fat and protein metabolism. Insulin reduces the use of fatty acids as fuel and instead stimulates their storage in adipose tissue. In a similar manner it inhibits the use of amino acids as fuel, thus promoting protein synthesis.

Glucagon Raises The Concentration Of Glucose In The Blood

Glucagon acts antagonistically to insulin. Its principal effect is to raise blood sugar level. It does this by stimulating liver cells to convert glycogen to glucose (in a process called glycogenolysis) and by stimulating liver cells to make glucose from other metabolites (gluconeogenesis). Note that these actions are opposite to those of insulin. Glucagon mobilizes fatty acids and amino acids as well as glucose. Glucagon is thought to be secreted also by certain cells in the wall of the stomach and duodenum.

Insulin And Glucagon Secretion Is Regulated By Glucose Concentration

Secretion of insulin and glucagon is directly controlled by the concentration of glucose in the blood (Fig. 41–13). After a meal, when the blood glucose level rises as a result of intestinal absorption, beta cells are stimulated to increase insulin secretion. Then, as the cells remove glucose from the blood, decreasing its concentration, insulin secretion decreases accordingly.

When one has not eaten for several hours, the concentration of glucose in the blood begins to fall. When it falls from its normal fasting level of about 90

milligrams of glucose per 100 milliliters of blood to about 70 milligrams of glucose, the alpha cells of the islets secrete large amounts of glucagon. Glucose is mobilized from storage in the liver cells, and blood sugar concentration returns to normal.

The alpha cells respond to the glucose concentration within their own cytoplasm, which is a reflection of the blood sugar level. When blood sugar level is high, there is generally a high level of glucose within the alpha cells, and glucagon secretion is inhibited.

It should be clear that insulin and glucagon work oppositely to keep blood sugar concentration within normal limits. When glucose level rises, insulin release brings it back to normal; when it falls, glucagon acts to raise it again. The insulin-glucagon system is a powerful, fast-acting mechanism for keeping blood sugar level normal. Can you think of reasons why it is important to maintain a constant blood sugar level? One important reason is that brain cells are completely dependent upon a continuous supply of glucose. Brain cells ordinarily are unable to use other nutrients as fuel. As we will discuss, several other hormones also affect blood sugar concentration.

Diabetes Mellitus Is A Serious Disorder Of Carbohydrate Metabolism

The principal disorder associated with pancreatic hormones is **diabetes mellitus.** Of the estimated 10 million diabetics in the United States, about 40,000 die each year as a result of this disorder, making it the third most common cause of death. Diabetes is a leading cause of blindness, kidney disorders, disease of small blood vessels, gangrene of the limbs, and various other malfunctions.

More than 90% of diabetes mellitus cases are non-insulin-dependent, often referred to as Type II. (This type was formerly known as maturity-onset diabetes.) Type II diabetes develops gradually, usually in overweight persons over age 30. In many patients with type II diabetes, the pancreatic islets secrete enough insulin, but receptors on target cells cannot bind it.

Insulin-dependent diabetes, referred to as Type I (and formerly known as juvenile-onset diabetes) usually develops before age 30. In Type I diabetes there is a marked decrease in the number of beta cells in the pancreas, resulting in insulin deficiency. Daily insulin injections are needed to correct the carbohydrate imbalance that results. Type I diabetes is thought to be an autoimmune disease in which antibodies mark the beta cells for destruction. Patients with Type I diabetes have a shortened life expectancy because atherosclerotic disease develops as a result of impaired lipid metabolism (see Chapter 36).

Similar metabolic disturbances occur in both types of diabetes mellitus:

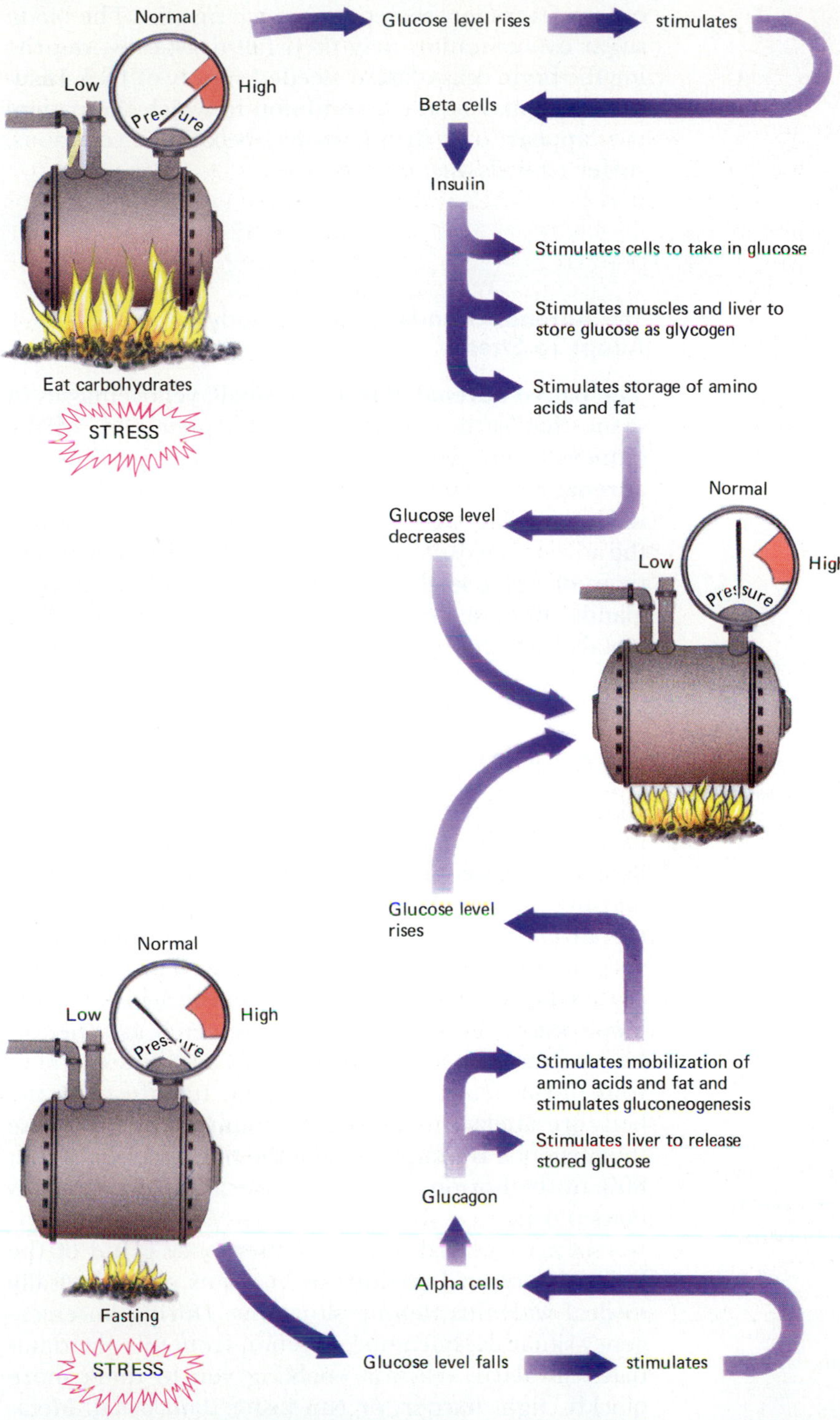

FIGURE 41–13 Regulation of glucose concentration in the blood by insulin and glucagon.

Normal — Low — High — Pressure

Glucose level rises → stimulates

Beta cells

Insulin

Stimulates cells to take in glucose

Stimulates muscles and liver to store glucose as glycogen

Stimulates storage of amino acids and fat

Eat carbohydrates
STRESS

Glucose level decreases

Normal — Low — High — Pressure

Glucose level rises

Stimulates mobilization of amino acids and fat and stimulates gluconeogenesis

Stimulates liver to release stored glucose

Glucagon

Normal — Low — High — Pressure

Alpha cells

Fasting
STRESS

Glucose level falls → stimulates

1. **DECREASED USE OF GLUCOSE** In diabetics, cells dependent on and lacking insulin can take in only about 25% of the glucose they require for fuel. Glucose accumulates in the blood, causing **hyperglycemia** (an abnormally high concentration of glucose in the blood). Instead of the normal fasting level of 90 milligrams per 100 milliliters, the level may reach from 300 to more than 1000 milligrams.

The concentration of glucose is so high in the untreated diabetic that it exceeds the renal thresh-

old: The tubules in the kidneys are unable to return all the glucose in the filtrate to the blood. As a result, glucose is excreted in the urine. The presence of glucose in the urine is a simple screening test for diabetes.

2. **INCREASED FAT MOBILIZATION** Despite the large quantities of glucose in the blood, most cells cannot use it and must turn to other sources of fuel. The absence of insulin promotes the mobilization of fat stores, providing nutrients for cellular respira-

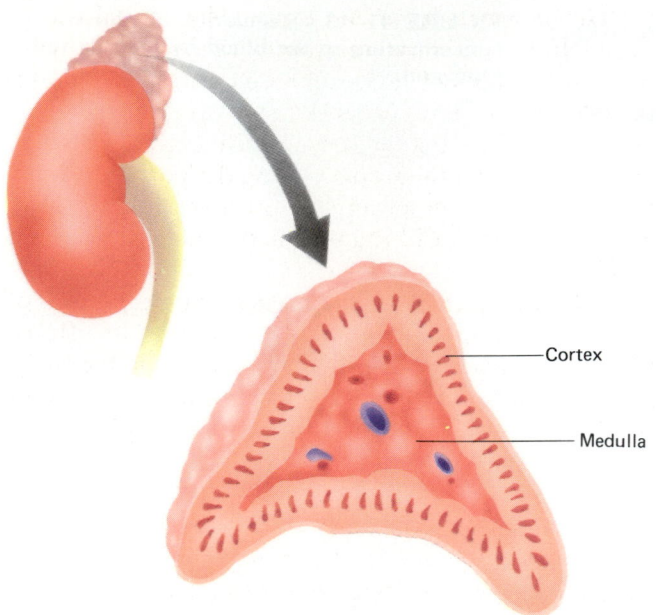

FIGURE 41–14 The adrenal glands.

tion. But unfortunately, the blood lipid level may reach five times the normal level, leading to development of atherosclerosis. Also, the increased fat metabolism increases the formation of ketone bodies. These build up in the blood, causing **ketosis,** a condition in which the body fluids and blood become too acidic. If severe, ketosis can lead to coma and death.

When ketone level in the blood rises, ketones appear in the urine, another clinical indication of diabetes. Because of osmotic pressure, when ketone bodies and glucose are excreted in the urine, they take water with them, so that urine volume increases. The resulting dehydration causes the diabetic to feel continually thirsty.

3. **INCREASED PROTEIN USE** Lack of insulin also results in increased protein breakdown relative to protein synthesis, so the untreated diabetic becomes thin and emaciated.

In Hypoglycemia The Glucose Concentration Is Too Low

Hypoglycemia, low blood glucose concentration, is sometimes seen in people who later develop diabetes. It may be an overreaction by the islets to glucose challenge. Too much insulin is secreted in response to carbohydrate ingestion. About 3 hours after a meal the blood sugar concentration falls below normal, making the individual feel very drowsy. If this reaction is severe enough, the patient may become uncoordinated or even unconscious.

Serious hypoglycemia can develop if diabetics receive injections of too much insulin or if the islets, be-

cause of a tumor, secrete too much insulin. The blood sugar concentration may then fall drastically, depriving the brain cells of their needed supply of fuel. **Insulin shock** may result, a condition in which the patient may appear to be drunk or may become unconscious, suffer convulsions, or even die.

The Adrenal Glands Help The Body Adapt To Stress

The paired **adrenal glands** are small, yellow masses of tissue that lie in contact with the upper ends of the kidneys. Each gland consists of a central portion, the **adrenal medulla,** and a larger outer section, the **adrenal cortex** (Fig. 41–14). Although joined anatomically, the adrenal medulla and cortex develop from different types of tissue in the embryo and function as distinct glands. Both secrete hormones that help to regulate metabolism, and both help the body deal with stress.

The Adrenal Medulla Initiates An Alarm Reaction

The adrenal medulla develops from neural tissue, and its secretion is controlled by sympathetic nerves. Two hormones, **epinephrine** (sometimes called adrenaline) and **norepinephrine** (noradrenaline), are secreted by the adrenal medulla. Chemically, these hormones are very similar; they belong to the chemical group known as **catecholamines** (derived from amino acids). Norepinephrine is the same substance secreted as a neurotransmitter by sympathetic neurons and by some neurons in the central nervous system. Its effects on the body are similar but last about ten times longer because the hormone is removed from the blood slowly. About 80% of the hormone output of the adrenal medulla is epinephrine.

Often referred to as the emergency gland of the body, the adrenal medulla prepares us physiologically to deal with threatening situations. During an emergency situation, hormone secretion from this gland initiates an alarm reaction enabling you to think more quickly, fight harder, or run faster than usual. Metabolic rate increases as much as 100%.

The adrenal medulla hormones cause blood to be rerouted to those organs essential for emergency action. Blood vessels going to the brain, muscles, and heart are dilated, while those to the skin and kidneys are constricted. Constriction of blood vessels serving the skin has the added advantage of decreasing blood loss in case of hemorrhage (and explains the sudden paling that comes with fear or rage). At the same time, the heart beats faster and thresholds in the reticular activating system of the brain are lowered, so you become more alert. Strength of muscle contraction increases. The adrenal medullary hormones also raise fatty acid and glucose levels in the blood, ensuring needed fuel for extra energy.

Under normal conditions both epinephrine and norepinephrine are secreted continuously in small amounts. Their secretion is under nervous control. When anxiety is aroused, messages are sent from the brain through sympathetic nerves to the adrenal medulla. Acetylcholine released by these neurons triggers release of epinephrine and norepinephrine.

The Adrenal Cortex Helps The Body Deal With Chronic Stress

All the hormones of the **adrenal cortex** are steroids synthesized from cholesterol, which in turn is made from acetyl coenzyme A. Recall that steroids are a chemical group classified with the lipids (see Chapter 4). Three types of hormones are produced by the adrenal cortex: (1) sex hormones, (2) mineralocorticoids, and (3) glucocorticoids.

Very small amounts of both **androgens** (hormones that have masculinizing effects) and **estrogens** (the female hormones) are secreted by the adrenal cortex in both sexes. Normally the amounts of these hormones released are so small that they have little physiological effect. However, if tumors develop in the adrenal cortex large quantities of the sex hormones, especially of the androgens, may be secreted.

Mineralocorticoid hormones help regulate salt balance. The principal mineralocorticoid is **aldosterone.** These hormones help the kidneys to reabsorb sodium and excrete potassium, thereby helping to maintain sodium and potassium balance in the body and an appropriate blood pressure. Aldosterone also helps regulate the phosphate concentration in the body. When the adrenal glands do not produce enough aldosterone, large amounts of sodium are excreted in the urine. Water leaves the body with the sodium (because of osmotic pressure), and the blood volume may be so markedly reduced that the patient dies of low blood pressure.

Cortisol, also called hydrocortisone, accounts for about 95% of the **glucocorticoid** activity of the adrenal cortex. The principal action of cortisol is to stimulate liver cells to produce glucose from other nutrients. Cortisol helps provide nutrients for glucose production by stimulating transport of amino acids into liver cells and by promoting fat mobilization so that fatty acids are available for conversion to glucose. Large amounts of glucose and glycogen are produced in the liver, and the blood glucose level rises. Cortisol helps ensure adequate fuel supplies for the cells when the body is under stress. Thus the adrenal cortex provides an important backup system for the adrenal medulla.

Glucocorticoids are used clinically to reduce inflammation in allergic reactions, infections, arthritis, and certain types of cancer. These hormones help stabilize lysosome membranes so that they do not destroy tissues with their potent enzymes. Glucocorticoids also reduce inflammation by decreasing the permeability of capillary membranes, thereby reducing swelling. They reduce the effects of histamine and so are used to treat allergic symptoms.

When used in large amounts over long periods of time, glucocorticoids can cause serious side effects. They decrease the number of lymphocytes in the body, reducing the patient's ability to fight infections. Other side effects include ulcers, hypertension, diabetes mellitus, and atherosclerosis.

Almost any type of stress stimulates the hypothalamus to secrete **corticotropin-releasing factor, CRF**. This hormone stimulates the anterior pituitary to secrete **adrenocorticotropic hormone, ACTH.** ACTH regulates glucocorticoid secretion (as well as aldosterone secretion). ACTH is so potent that it can result in up to a 20-fold increase in cortisol secretion within minutes. When the body is not under stress, high levels of cortisol in the blood inhibit both the hypothalamus and the pituitary.

Abnormally large amounts of glucocorticoids, whether due to disease or drugs, result in **Cushing's disease.** In this condition, fat is mobilized from the lower part of the body and deposited about the trunk. Edema gives the patient's face a full-moon appearance. Blood sugar level rises to as much as 50% above normal, causing adrenal diabetes. If this condition persists for several months, the beta cells in the pancreas can "burn out" from secreting excessive amounts of insulin. This can result in permanent diabetes mellitus. Reduction in protein synthesis causes weakness and decreases immune responses, so the patient often dies of infection.

Destruction of the adrenal cortex and the resulting decrease in aldosterone and cortisol secretion cause **Addison's disease.** Reduction in cortisol prevents the body from regulating the concentration of glucose in the blood because it cannot synthesize enough glucose. The cortisol-deficient patient also loses the ability to cope with stress. If cortisol levels are significantly depressed, even the stress of mild infections can cause death.

The Adrenal Medulla And Cortex Work Together To Help The Body Cope With Stress

Stressors, whether in the form of noise, infection, or even the anxiety of taking a test for which one is not fully prepared, arouse the adrenal glands to action. The brain signals the adrenal medulla rapidly via neural connections to release epinephrine and norepi-

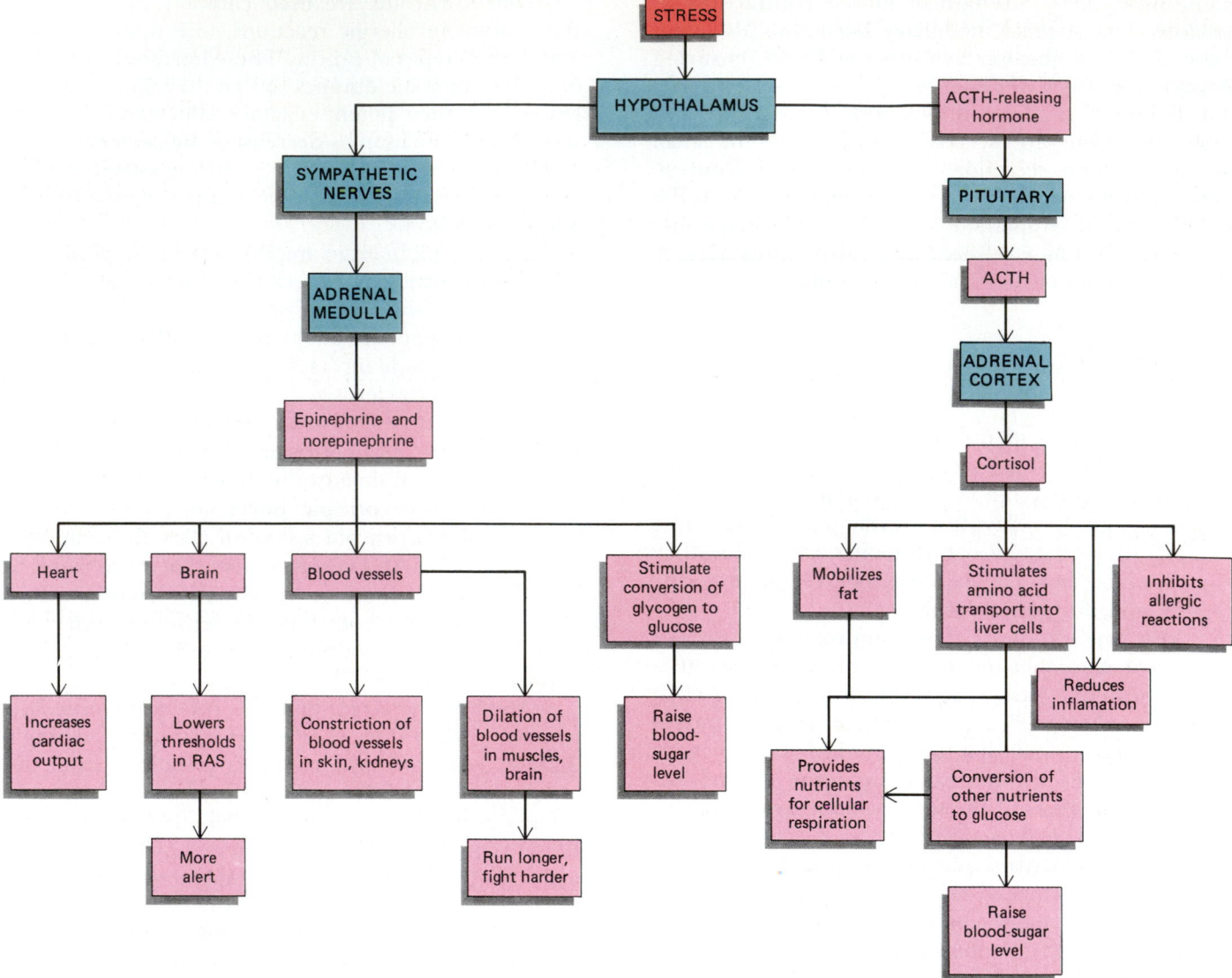

FIGURE 41–15 Some effects of stress.

nephrine, hormones that prepare the body for fight or flight. The hypothalamus also signals the anterior pituitary hormonally to secrete ACTH, which increases cortisol secretion. The cortisol adjusts metabolism to meet the increased demands of the stressful situation (Fig. 41–15).

Some forms of stress are short-lived. We react to the situation and quickly resolve it. Other stressors may last for days, weeks, or even years—a chronic disease, for example, or an unhappy marriage or job situation. General anxiety and tension are examples of nonspecific stress.

Chronic stress is harmful because of the side effects of long-term elevated levels of hormones such as cortisol. Though glucocorticoids help reduce inflammation, they also interfere with normal immune responses. These hormones raise blood pressure, which may contribute to heart disease. They increase levels of fat in the blood, which may promote atherosclerosis.

When animals receive injections of large amounts of glucocorticoids, such disease states are induced, and similar effects are seen when large doses are administered clinically to human patients. Among the diseases linked to excessive amounts of adrenocortical hormones are ulcers, high blood pressure, atherosclerosis, and arthritis.

Prostaglandins Are Released By Many Types Of Mammalian Cells

Prostaglandins are a group of hormones released by many different tissues in the body, including the lungs, liver, and digestive tract. These hormones act in cells in the immediate area in which they are released. Prostaglandins may help regulate the action of other hormones by stimulating or inhibiting cyclic AMP formation.

Prostaglandins influence a variety of metabolic activities. They dilate the bronchial passageways, inhibit gastric secretion, stimulate contraction of the uterus, and affect nerve function, blood pressure, and metabolism. Those synthesized in the temperature-regulating center of the hypothalamus can cause fever. Drugs such as aspirin and acetaminophen (Tylenol) reduce fever and relieve pain by inhibiting prostaglandin synthesis. Prostaglandins are used clinically to initiate labor and induce abortion, and their use as a birth-control drug is being investigated.

Many Other Hormones Are Known

Many other tissues of the body secrete hormones. In Chapter 39 we described hormones released by the digestive tract. The **thymus gland** releases a hormone (thymosin) that plays a role in immune responses, and the kidneys release hormones, one of which (renin) helps regulate blood pressure. The **pineal gland,** located in the brain, produces a hormone called **melatonin,** which may affect reproduction. In Chapter 42 we will discuss the principal reproductive hormones.

■ CHAPTER SUMMARY

I. The endocrine system consists of endocrine glands and tissues that secrete hormones; this system helps regulate many aspects of metabolism, growth, and reproduction.

II. Hormones are transported to their target tissues via the blood.
   A. Some hormones combine with receptors on the plasma membrane of their target cells and act by way of a second messenger, such as cyclic AMP.
   B. Steroid hormones combine with receptor proteins within the target cell; the hormone-receptor complex may stimulate a particular gene to initiate protein synthesis.
   C. Hormone secretion is self-regulated by negative-feedback control mechanisms.

III. In invertebrates hormones influence growth, reproduction, molting, morphogenesis, and pigmentation.

IV. In vertebrates hormones help regulate growth, reproduction, salt and fluid balance, and many aspects of metabolism.

V. Nervous and endocrine system regulation is integrated in the hypothalamus, which regulates the activity of the pituitary gland.
   A. The hormones oxytocin and ADH are produced by the hypothalamus and released by the posterior lobe of the pituitary.
   B. Secretion of anterior pituitary hormones is regulated by releasing and release-inhibiting hormones secreted by the hypothalamus.
      1. Growth hormone stimulates body growth by promoting protein synthesis.
      2. Malfunctions in growth hormone secretion can lead to pituitary dwarfism and gigantism.

VI. Thyroid hormones stimulate the rate of metabolism.
   A. Regulation of thyroid secretion depends mainly upon a feedback system between the anterior pituitary and the thyroid gland.
   B. Hyposecretion of thyroxine during childhood may lead to cretinism; during adulthood it may result in myxedema. Goiter may develop from hyposecretion or hypersecretion.

VII. The parathyroid glands regulate calcium level.

VIII. The islets of the pancreas secrete insulin and glucagon.
   A. Insulin stimulates cells to take up glucose from the blood and so lowers blood sugar concentration.
   B. Glucagon raises blood glucose concentration by stimulating conversion of glycogen to glucose and production of glucose from other nutrients.
   C. Insulin and glucagon secretion are regulated directly by blood glucose levels.
   D. In diabetes mellitus insulin deficiency results in decreased utilization of glucose, increased fat mobilization, and increased protein utilization.

IX. The adrenal glands secrete hormones that help the body cope with stress.
   A. The adrenal medulla, sometimes referred to as the "emergency gland," secretes epinephrine and norepinephrine.
   B. The adrenal cortex secretes sex hormones; mineralocorticoids, such as aldosterone, which increases the rate of sodium reabsorption and potassium excretion by the kidneys; and glucocorticoids, such as cortisol, which promotes gluconeogenesis.
   C. The hormones of the adrenal medulla help the body respond to stress by increasing heart rate, metabolic rate, and strength of muscle contraction and by causing blood to be rerouted to those organs needed for fight or flight. The adrenal cortex acts as a backup system, ensuring adequate fuel supplies for the rapidly metabolizing cells.

X. Prostaglandins may help regulate hormone action by regulating formation of cyclic AMP.

■ POST-TEST

1. Endocrine glands lack _____; they release their secretions into the surrounding tissue fluid and they are transported by the _____.
2. Endocrine glands produce chemical messengers called
_____.
3. A second messenger important in the action of many

hormones is _____.
4. The _____ serves as a link between the nervous and endocrine systems.
5. Hormones bind to specific receptor proteins in their _____ tissues.

Select the most appropriate term in column B for each description in column A.

Column A
6. located in neck region
7. secrete insulin
8. sometimes called "emergency gland"
9. regulates other endocrine glands via tropic hormones
10. secretes glucocorticoids

Column B
a. anterior lobe of pituitary
b. adrenal medulla
c. thyroid
d. islets of pancreas
e. none of the above

Column A
11. stimulates rate of metabolism
12. stimulates release of calcium from bone
13. helps maintain sodium balance
14. raise blood sugar level
15. causes heart to beat faster and blood to be rerouted

Column B
a. thyroid hormones
b. glucagon
c. aldosterone
d. epinephrine
e. none of the above

16. Cretinism is caused by _____-secretion of _____ during childhood.
17. An abnormal enlargement of the thyroid gland is a _____.
18. The principal action of _____ is to pro-

mote production of glucose from other nutrients.
19. Calcitonin is secreted by the _____ gland.
20. In untreated _____, glucose utilization is decreased and ketosis occurs.

■ REVIEW QUESTIONS

1. What is a hormone? What are some of the important functions of hormones?
2. How are hormones transported? How do they "recognize" their target tissues? What is the role of cyclic AMP in hormone action?
3. How do steroid hormones influence the activity of cells?
4. Why is the hypothalamus considered the link between the nervous and the endocrine systems? Explain.
5. Describe the actions of (1) prolactin, (2) oxytocin, (3) thyroid-stimulating hormone.
6. Draw a diagram to illustrate the regulation of thyroid hormone secretion by the anterior pituitary gland.
7. Explain the hormonal basis for (1) acromegaly, (2) pitui-

tary dwarfism, (3) cretinism, (4) hypoglycemia, (5) Cushing's disease.
8. Explain the antagonistic actions of insulin and glucagon in regulating blood glucose level. What other hormones studied in this chapter affect blood glucose level?
9. Describe several physiological disturbances that result from diabetes mellitus.
10. What are the actions of epinephrine and norepinephrine?
11. How is the adrenal medulla regulated?
12. What three types of hormones are released by the adrenal cortex, and what are the actions of each type?
13. Explain how the adrenal glands help the body deal with stress.
14. What are prostaglandins? What are their functions?

■ RECOMMENDED READINGS

Bloom, F.: "Neuropeptides," *Scientific American* October 1981. Neuropeptides are sometimes considered hormones, sometimes neurotransmitters.

Cantin, M., and J. Genest: "The Heart as an Endocrine Gland," *Scientific American* 255(5): 76–81, 1986. A recently discovered hormone that helps regulate blood pressure and volume is secreted by the atria.

Crews, D.: "The Hormonal Control of Behavior in a Lizard,"

Scientific American August 1979. Sexual behavior is controlled by the interaction of the brain and hormones from the gonads.

Edwards, D.D.: "Diabetes Autoimmunity Seen, Stopped," *Science News*, 132(7), 1987. Reports of two clinical trials suggest that early use of an immune system suppressor could stop destruction of insulin-producing cells in diabetics.

42

Reproduction

OUTLINE

I. Asexual reproduction is common among some animal groups
II. Sexual reproduction is the most common type of animal reproduction
III. Animals have evolved interesting reproductive variations
    A. Metagenesis is characteristic of some animal groups
    B. In parthenogenesis there is no fertilization
    C. In hermaphroditism one individual produces sperm and eggs
IV. Human reproduction: The male provides sperm
    A. The testes produce sperm
    B. A series of ducts transport sperm
    C. Semen is produced by the accessory glands
    D. The penis transfers sperm to the female
    E. Reproductive hormones promote sperm production and maintain masculinity
V. Human reproduction: The female produces ova and incubates the embryo
    A. The ovaries produce ova and sex hormones
    B. The uterine tubes transport the ovum
    C. The uterus incubates the embryo
    D. The vagina receives sperm
    E. External genital structures are the vulva
    F. The breasts function in lactation
    G. The menstrual cycle is regulated by hormones
VI. Sexual response involves physiological changes
VII. Fertilization is the fusion of sperm and egg to produce a zygote
VIII. Infertility is the inability to achieve conception
IX. Birth control methods allow individuals to choose
    A. Oral contraceptives prevent ovulation
    B. Usage of the intrauterine device (IUD) has declined
    C. Other common contraceptive methods include the diaphragm and condom
    D. Sterilization renders an individual incapable of producing offspring
X. There are three types of abortion
XI. Sexually transmitted diseases are spread by sexual contact
Focus on breast cancer
Focus on novel origins

LEARNING OBJECTIVES

After you have studied this chapter you should be able to:

1. Compare asexual and sexual reproduction, giving two examples of asexual reproduction.
2. Compare adaptive advantages of the following reproductive styles: metagenesis, parthenogenesis, and hermaphroditism.
3. Trace the passage of sperm cells through the male reproductive system from their origin in the seminiferous tubules until they leave the body in the semen.
4. Label the structures of the male reproductive system on a diagram, and describe the functions of each.
5. Label the structures of the female reproductive system on a diagram, and describe the functions of each.
6. Trace the development of an ovum and its passage through the female reproductive system until it is fertilized.
7. Describe the actions of testosterone and of the gonadotropic hormones in the male.
8. Describe the hormonal regulation of the menstrual cycle, and identify the time of important events of the cycle such as ovulation and menstruation.
9. Identify the physiological changes that occur during sexual response in male and female.
10. Describe the process of human fertilization.
11. Compare the methods of birth control in Table 42–3 with respect to mode of action, effectiveness, advantages, and disadvantages.
12. Identify common sexually transmitted diseases, and describe their symptoms, effects, and treatment.

Scanning electron micrograph of human fertilization. (David M. Phillips/Visuals Unlimited)

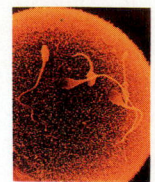

The unique capacity of deoxyribonucleic acid (DNA) to replicate itself permits cells, and the multicellular organisms they compose, to reproduce. In simple animals, fission or budding of a single parent produces identical offspring. More complex animals have male and female sexes, which produce sperm and eggs. The sperm contributes genes coding for some of the male parent's traits and the egg contributes genes coding for some of the female's traits. When sperm and egg unite, a new organism develops that is similar to both parents but not identical to either one. In some animals reproduction is an incredibly complex structural, functional, and behavioral process. In this chapter we will summarize the major features of animal reproductive processes and then focus on human reproduction.

KEY CONCEPTS

☐ In asexual reproduction a single parent splits, buds, or fragments, giving rise to two or more offspring that are identical to the parent.

☐ In sexual reproduction two types of gametes fuse to form a zygote, which develops into an individual that is a unique combination of the genes contributed by each parent.

☐ The reproductive role of the male mammal is to produce sperm cells and deliver them into the female reproductive tract.

☐ The reproductive role of the female mammal is to produce ova, receive sperm, incubate and nourish the developing embryo, and nourish the newborn.

☐ In both sexes, hormones regulate production of gametes. In females, hormones also maintain pregnancy and regulate lactation (milk production) for nourishment of the newborn.

☐ ASEXUAL REPRODUCTION IS COMMON AMONG SOME ANIMAL GROUPS

In **asexual reproduction** a single parent splits, buds, or fragments to give rise to two or more offspring that have hereditary traits identical with those of the parent. Sponges and cnidarians can reproduce by **budding,** in which a small part of the parent's body separates from the rest and develops into a new individual (Fig. 42–1). It may split away from the parent and establish an independent existence or it may remain attached and become a more-or-less independent member of a colony.

FIGURE 42–1 *Hydra* **is one of many organisms that reproduces asexually by budding. A part of the body grows outward and then separates and develops into a new individual. The portion of the parent body that buds is not specialized exclusively for performing a reproductive function.** (Richard Campbell/BPS [28-014])

Salamanders, lizards, sea stars, and crabs can grow a new tail, leg, arm, or certain other organs if the original one is lost. Oyster farmers learned long ago that when they tried to kill sea stars by chopping them in half and throwing the pieces back into the sea, the number of sea stars preying on the oyster bed doubled! A sea star can, in fact, regenerate an entire new individual from a single arm. In some species, this ability to regenerate a part has become a method of reproduction known as **fragmentation.** The body of the parent may break into several pieces; each piece then regenerates the missing parts and develops into a whole animal. Fragmentation is common among flatworms.

☐ SEXUAL REPRODUCTION IS THE MOST COMMON TYPE OF ANIMAL REPRODUCTION

Most animals reproduce by sexual reproduction, which generally involves two parents—a male, which produces sperm, and a female, which produces eggs (ova). The egg is typically large and nonmotile, with a store

(a)

(b)

FIGURE 42–2 Two types of fertilization. (*a*) **External fertilization is illustrated by these mating food frogs. Most amphibians return to water to mate. The female lays a mass of eggs, while the male mounts her and deposits his sperm in the water. (*b*) Mammals practice internal fertilization. These mating lions were photographed in Kenya.** (*a*, Dwight Kuhn; *b*, Joe McDonald/Tom Stack Associates)

of nutrients that supports the development of the embryo. The sperm is usually small and motile, adapted to propel itself by beating its long, whip-like tail. When a sperm cell fertilizes (fuses with) an egg cell a fertilized egg, or **zygote,** forms.

Sexual reproduction has the biological advantage of promoting genetic variety among the members of a species, because the offspring is the product of a particular combination of genes contributed by both parents, rather than a genetic copy of a single individual. By making possible the genetic recombination of the inherited traits of two parents, sexual reproduction gives rise to offspring that may be better able to survive than either parent. Advantageous combinations of genes can spread rapidly through a population that reproduces sexually. This permits the rapid and effective spread of adaptations that enable a species to survive in an ever-changing environment.

In **internal fertilization,** the gametes (sperm and egg) fuse inside the body, whereas in **external fertilization** gametes meet outside the body (Fig. 42–2). Most aquatic animals practice external fertilization. Mating partners usually release eggs and sperm into the water simultaneously. Many gametes are lost—to predators, for example—but so many are released that a sufficient number of sperm and egg cells meet and unite, perpetuating the species.

In internal fertilization, matters are left less to chance. The male generally delivers sperm cells directly into the body of the female. Her moist tissues provide the watery medium required for movement of sperm. Most terrestrial animals, as well as a few fish and some other aquatic animals, practice internal fertilization.

ANIMALS HAVE EVOLVED INTERESTING REPRODUCTIVE VARIATIONS

Animals show tremendous diversity in their methods of sexual reproduction. Even members of the same class may differ markedly in their reproductive processes.

Metagenesis Is Characteristic Of Some Animal Groups

Some animals exhibit **metagenesis,** in which the species has both asexual and sexual generations. For example, in the hydrozoan *Obelia* a polyp generation gives rise asexually by budding to a generation of medusas. The motile medusas produce gametes and reproduce sexually, giving rise to a new generation of polyps. Both types of generations consist of diploid organisms.

In Parthenogenesis There Is No Fertilization

Parthenogenesis (virgin development) is a form of reproduction in which an unfertilized egg develops into an adult animal. Parthenogenesis is common among some mollusks, some crustaceans, insects, especially honeybees and wasps, and some reptiles. Some species of arthropods (and even some vertebrates—lizards) consist entirely of females that reproduce in this way. More commonly, parthenogenesis occurs for several generations, after which males develop, produce sperm, and mate with the females to fertilize their

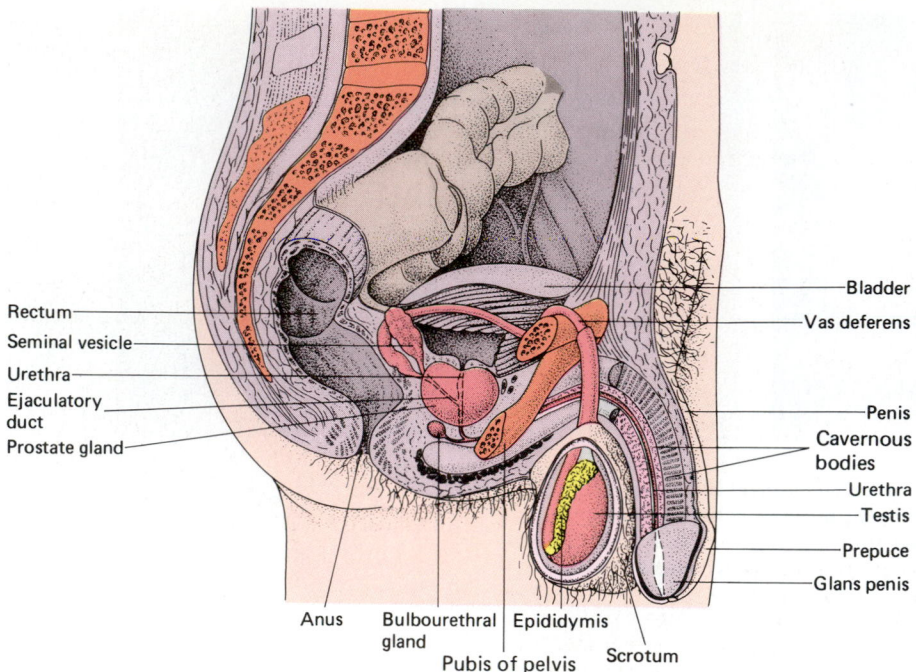

FIGURE 42–3   Anatomy of the human male reproductive system. The scrotum, penis, and pelvic region are shown in sagittal section to illustrate their internal structures.

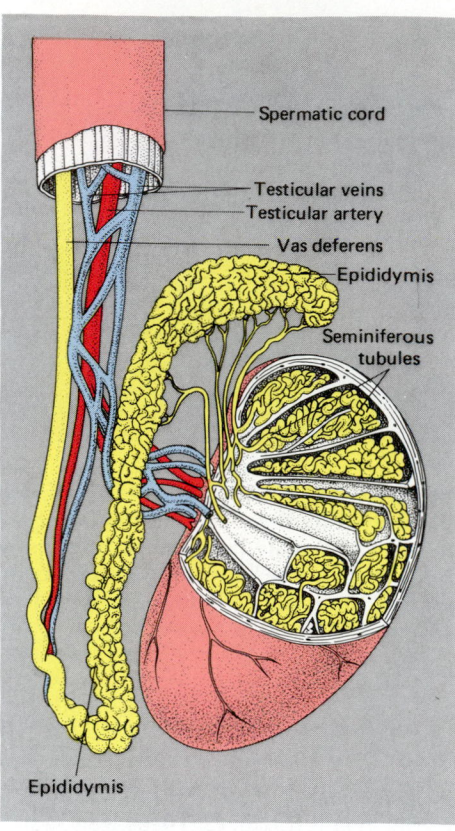

FIGURE 42–4   Structure of the testis, epididymis, and spermatic cord. The testis is shown in sagittal section to illustrate the arrangement of the seminiferous tubules.

eggs. In some species, parthenogenesis is advantageous in maintaining the social order; in others, it appears to be an adaptation for survival in times of stress or when there is a serious decrease in population.

A special form of parthenogenesis occurs in honeybees. The queen honeybee receives sperm from a male during the "nuptial flight." The sperm she receives are stored in a little pouch separated from her genital tract by a muscular valve. As the queen lays eggs, she can either open this valve, permitting the sperm to escape and fertilize the eggs, or keep the valve closed, so that the eggs develop without fertilization. Generally, fertilization occurs in the fall, and the fertilized eggs are quiescent during the winter. The fertilized eggs become females (queens and workers); the unfertilized eggs become males (drones). Some species of wasps alternately produce a parthenogenetic generation and a generation developing from fertilized eggs.

In Hermaphroditism One Individual Produces Sperm And Eggs

Sexual reproduction almost always involves two individuals. **Hermaphroditism** is an important exception to this generalization. In hermaphroditism, a single individual produces both eggs and sperm. A few hermaphrodites, such as the parasitic tapeworms, are ca-

pable of self-fertilization. For the tapeworm, self-fertilization is an adaptation permitting its solitary life-style. Reproduction is possible even when only a single tapeworm infects a host.

Most hermaphrodites do not reproduce by self-fertilization. Rather, as in earthworms, two animals copulate, and each inseminates the other. In some hermaphroditic species, self-fertilization is prevented by the development of testes and ovaries at different times. In the clam *Mercenaria mercenaria*, 98% of the population are males when they first reach maturity; later in their lives they produce eggs and become females. A few continue to produce only sperm and function as males. In the American oyster *Crassostrea virginica*, sperm are mainly produced when the animal first matures. The next year the animal produces eggs, and so on, in a regular annual alternation of sexes.

Certain fish exhibit a somewhat similar reproductive pattern in which sex is related to dominance. In one species of wrasse, the dominant fish is always male, and he lords his position over a harem of females. If he is removed or dies, one of the remaining female fish

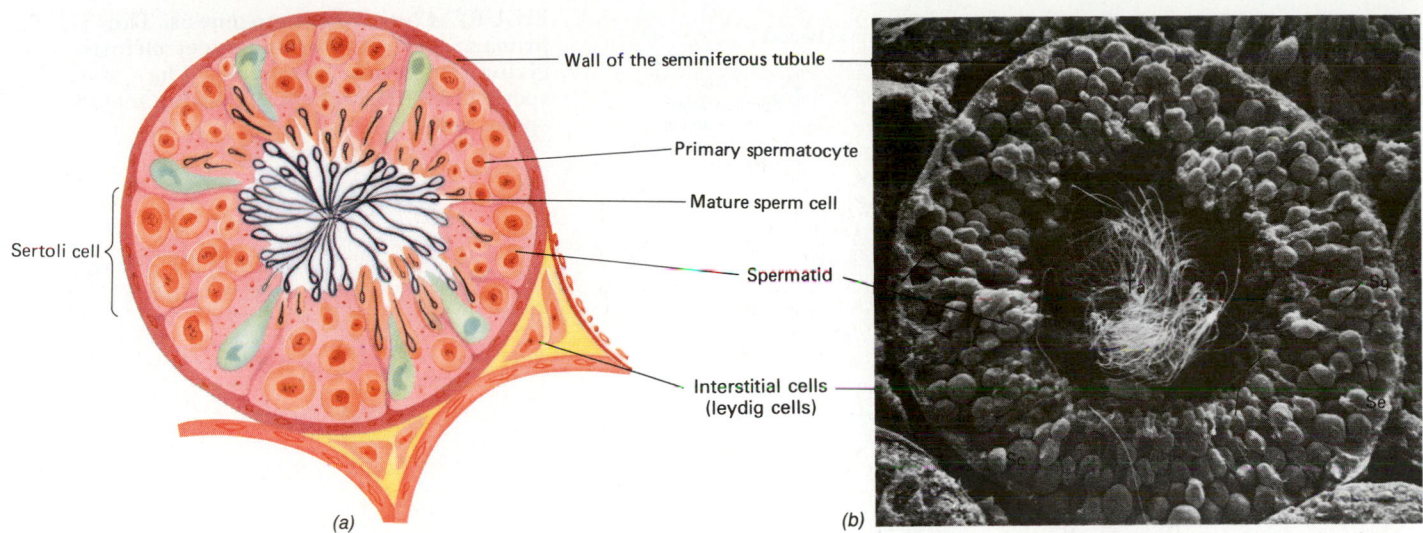

Wall of the seminiferous tubule

Primary spermatocyte

Mature sperm cell

Sertoli cell

Spermatid

Interstitial cells
(leydig cells)

(a)   (b)

FIGURE 42–5   Structure of a seminiferous tubule showing developing sperm cells in various stages of spermatogenesis. (*a*) Identify the sequence of sperm cell differentiation. Note the Sertoli cells and the interstitial cells. (*b*) A scanning electron micrograph of a transverse section through a seminiferous tubule (×580). *Se*, Sertoli cell; *Sc*, primary spermatocyte; *Sg*, spermatogonium. (Kessel, R.G. and R.H. Kardon, *Tissues and Organs: A Text-Atlas of Scanning Electron Microscopy.* San Francisco, W.H. Freeman & Co., 1979)

reverses sex and becomes the new patriarch. In another species of wrasse, the opposite situation exists: The dominant fish is always female, and if she is removed one of the males changes sex and becomes dominant.

HUMAN REPRODUCTION: THE MALE PROVIDES SPERM

The reproductive role of the human male, as in other mammals, is to produce sperm cells and deliver them into the female reproductive tract. When a sperm combines with an egg, it contributes half the genes to the new offspring and determines its sex. The male reproductive system is illustrated in Figure 42–3. Male structures include the testes (which produce sperm and the hormone testosterone), the scrotum (which contains the testes), conducting tubes (which transport sperm from the testes to the outside of the body), accessory glands (which produce semen), and the penis (copulatory organ).

The Testes Produce Sperm

In humans and other vertebrates, **spermatogenesis,** the process of sperm cell production, occurs in the paired male gonads, or **testes.** This process takes place within the walls of a vast tangle of hollow tubules, the **seminiferous tubules,** within each testis (Fig. 42–4). The seminiferous tubules are partially lined with prim-

itive, undifferentiated stem cells called **spermatogonia.** These cells give rise to sperm cells (Fig. 42–5).

During embryonic development and childhood, the spermatogonia divide by mitosis, producing more spermatogonia. At adolescence, some spermatogonia continue to divide by mitosis, but about half of them enlarge and become **primary spermatocytes.** These cells undergo meiosis. (You may want to review the discussion of meiosis in Chapter 13.) In many animals gamete production occurs only in the spring or fall, but humans have no special breeding season. In the human adult male, spermatogenesis proceeds continuously and millions of sperm are produced each day.

Each primary spermatocyte undergoes a first meiotic division producing **secondary spermatocytes** (Figure 42–6). In the second meiotic division each secondary spermatocyte gives rise to two **spermatids.** Four spermatids are produced from the original primary spermatocyte. The spermatid is a fairly large cell with much more cytoplasm than is present in a mature sperm.

The haploid spermatid differentiates into a mature sperm by a rather complicated process. The nucleus shrinks, and part of the Golgi complex becomes the **acrosome,** which produces enzymes that help the sperm penetrate the egg (Fig. 42–7). The two centrioles nestle in a small depression on the nuclear surface. One of these, the distal centriole, gives rise to the flagellum of the sperm. Although somewhat longer than most, the sperm flagellum is typical of eukaryote flagella with the usual 9 + 2 arrangement of microtubules. Finally, most of the cytoplasm is discarded and is phag-

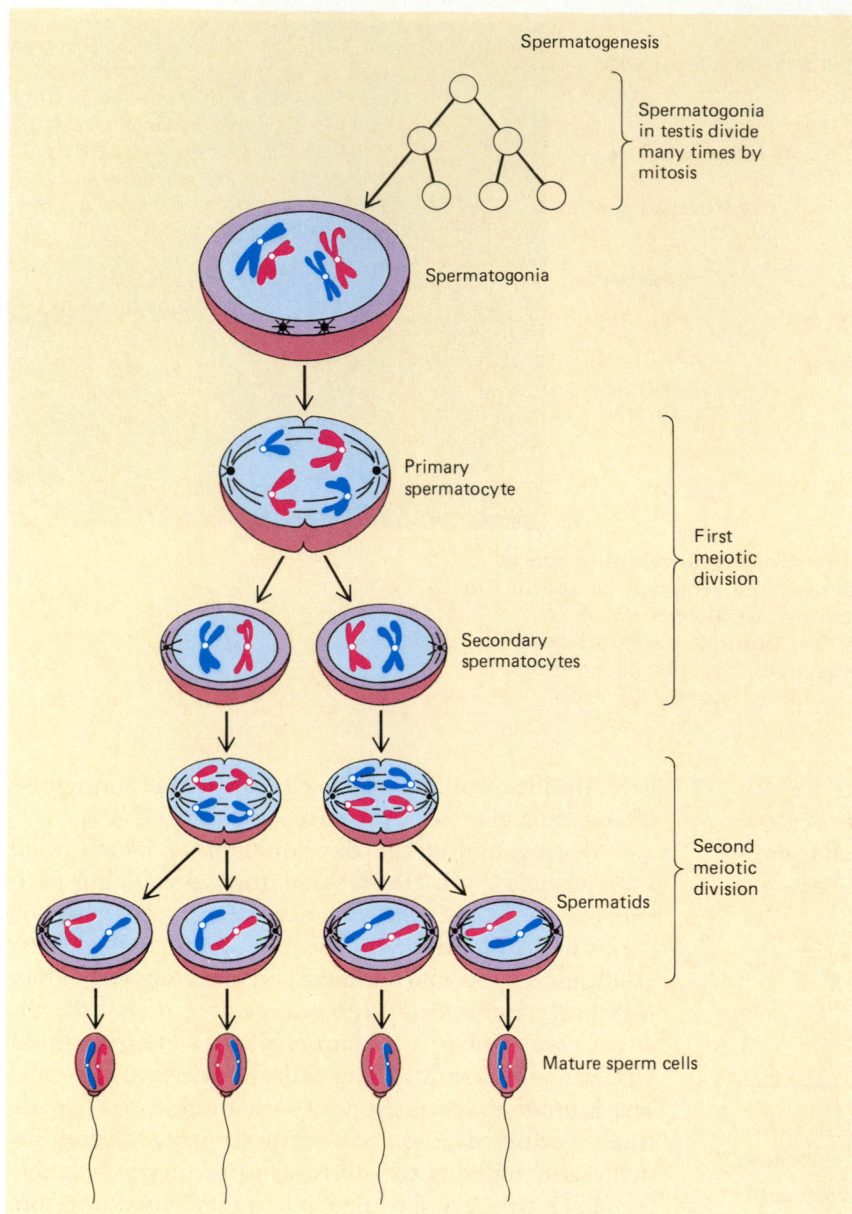

FIGURE 42–6 Spermatogenesis. The primary spermatocyte undergoes meiosis, giving rise to four spermatids. The spermatids differentiate, becoming mature sperm cells.

ocytized by the large nutritive Sertoli cells present within the seminiferous tubules.

In humans sperm cells cannot develop at body temperature. Although the testes develop within the abdominal cavity of the male embryo, about 2 months before birth they descend into the **scrotum,** a skin-covered sac suspended from the groin that serves as a cooling unit, maintaining sperm at about 2°C (about 96°F) below body temperature. In rare cases where the testes do not descend, the seminiferous tubules eventually degenerate and the male becomes **sterile** (because he cannot produce sperm cells).

The scrotum is an outpocketing of the pelvic cavity and is connected to it by the **inguinal canals.** As they descend, the testes pull their blood vessels, nerves, and conducting tubes after them. These structures, encased by muscle and by layers of connective tissue,

make up the **spermatic cord.** The inguinal region is a weak place in the abdominal wall. Straining the abdominal muscles by lifting heavy objects sometimes results in tearing the inguinal tissue. A loop of intestine can then bulge into the scrotum through such a tear. This condition is called an **inguinal hernia.**

A Series Of Ducts Transport Sperm

Sperm cells leave the seminiferous tubules of the testes through a series of small tubules (vasa efferentia) that pass from the testis and empty into a larger tube, the epididymis. The **epididymis** of each testis is a complexly coiled tube in which sperm complete their maturation and are stored.

From each epididymis sperm pass into a sperm duct, the **vas deferens** (plural, vasa deferentia), which

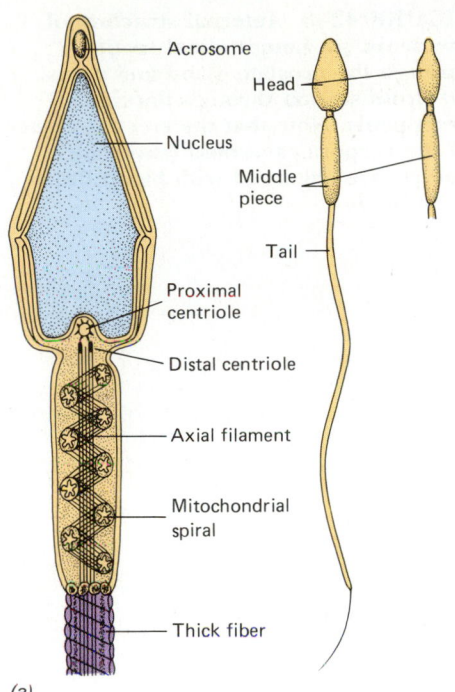

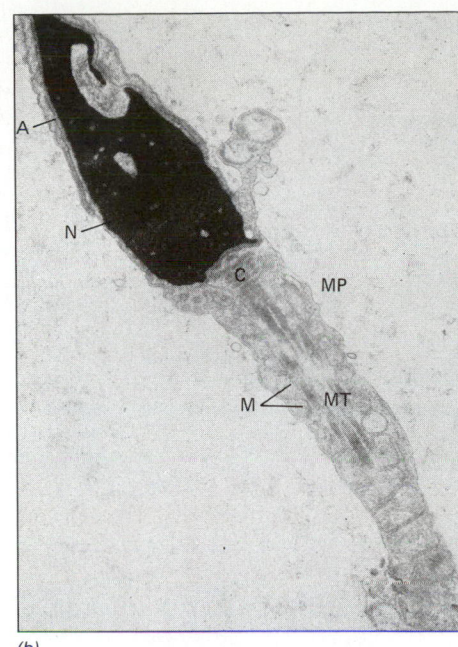

(a)

(b)

FIGURE 42–7 Sperm cell structure. (*a*) A mammalian sperm. (*Left*) A head and middle piece. Structures shown would be visible through an electron microscope. (*Middle and right diagrams*) Top and side views of a sperm. (*b*) Electron micrograph of a human sperm cell (approximately ×37,500). A, acrosome; N, nucleus; MP, midpiece; M, mitochondria; MT, microtubules; C, centrioles.   (*b*, Dr. Lyle C. Dearden)

Labels in diagram (a): Acrosome, Head, Nucleus, Middle piece, Tail, Proximal centriole, Distal centriole, Axial filament, Mitochondrial spiral, Thick fiber

extends from the scrotum through the inguinal canal and into the pelvic cavity. Each vas deferens empties into a short **ejaculatory duct,** which passes through the prostate gland and then opens into the urethra. The single **urethra,** which at different times conducts either urine or semen, passes through the penis to the outside of the body.

Semen Is Produced By The Accessory Glands

As sperm are transported through the conducting tubes, they are mixed with secretions from the accessory glands. About 3.5 ml of **semen** is ejaculated during sexual climax. Semen consists of about 400 million sperm cells suspended in the secretions of the seminal vesicles, prostate gland, bulbourethral glands, and small glands in the walls of the ducts.

The paired **seminal vesicles** empty into the vasa deferentia (Fig. 42–3). The single **prostate gland** releases its alkaline secretion into the urethra. During sexual arousal the paired **bulbourethral glands** release a few drops of alkaline fluid, which can neutralize the acidity of the urethra and aid in lubrication.

The Penis Transfers Sperm To The Female

The **penis** is an erectile copulatory organ adapted to deliver sperm into the female reproductive tract. It consists of a long **shaft** that enlarges to form an expanded tip, the **glans.** Part of the loose-fitting skin of the penis folds down and covers the proximal portion

of the glans, forming a cuff called the **prepuce,** or **foreskin.** In the operation termed **circumcision** (commonly performed on male babies for either hygienic or religious reasons), the foreskin is removed.

Under the skin, the penis consists of three parallel columns of **erectile tissue,** sometimes called the **cavernous bodies** (corpora cavernosa) (Fig. 42–8). One of these columns surrounds the portion of the urethra that passes through the penis. Erectile tissue consists of large blood vessels called venous sinusoids. When the male is sexually stimulated, nerve impulses cause the arteries of the penis to dilate. Blood rushes into the vessels of the erectile tissue, causing the tissue to swell. This compresses veins that conduct blood away from the penis, slowing the outflow of blood through them. Thus more blood enters the penis than can leave, causing the erectile tissue to become further engorged with blood. The penis thus becomes erect, that is, longer, larger in circumference, and firm. Though the human penis does not contain any bone, penis bones do occur in some other mammals, such as bats.

Reproductive Hormones Promote Sperm Production And Maintain Masculinity

At about age 10, the hypothalamus begins to mature in its function of regulating sex hormones. It secretes **gonadotropin-releasing hormone (GnRH),** which stimulates the anterior pituitary to secrete the gonadotropic hormones **follicle-stimulating hormone (FSH)** and **luteinizing hormone (LH).** FSH stimulates devel-

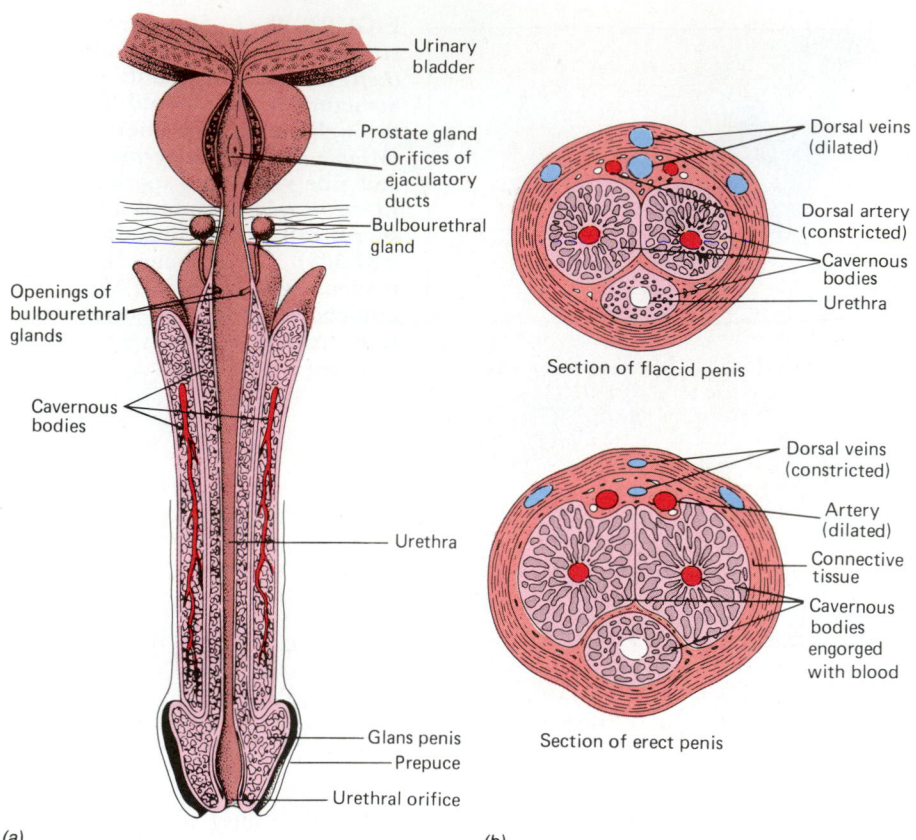

Urinary
bladder

Prostate gland

Orifices of
ejaculatory
ducts

Bulbourethral
gland

Openings of
bulbourethral
glands

Cavernous
bodies

Urethra

Glans penis

Prepuce

Urethral orifice

(a)

Dorsal veins
(dilated)

Dorsal artery
(constricted)

Cavernous
bodies

Urethra

Section of flaccid penis

Dorsal veins
(constricted)

Artery
(dilated)

Connective
tissue

Cavernous
bodies
engorged
with blood

Section of erect penis

(b)

**FIGURE 42–8   Internal structure of
the penis. (a) Longitudinal section
through the prostate gland and penis.
(b) Cross section through flaccid and
erect penis. Note that the erectile tissues
of the corpora cavernosa (cavernous
bodies) are engorged with blood in the
erect penis.**

opment of the seminiferous tubules and may promote
spermatogenesis (Table 42–1). LH stimulates the **interstitial cells,** which lie between the tubules in the
testes, to secrete the hormone **testosterone.**

Testosterone is responsible for the adolescent
growth spurt, which occurs at about age 13 years. This
hormone stimulates growth of the male reproductive
organs and is also responsible for the secondary sexual
characteristics that develop at puberty. The beard begins to grow, and pubic and axillary hair appears.
Vocal cords increase in length and thickness, causing
the voice to deepen, and muscle development is stimulated.

What happens when testosterone is absent? If the
testes are removed—in a procedure known as **castration**—before puberty, the male is deprived of testosterone and becomes a eunuch. He retains child-like sex
organs and does not develop secondary sexual characteristics. If castration occurs after puberty, increased
secretion of male hormone by the adrenal glands helps
maintain masculinity.

■ HUMAN REPRODUCTION: THE FEMALE PRODUCES OVA AND INCUBATES THE EMBRYO

The female reproductive system produces ova (eggs),
receives the penis and sperm released from it during

sexual intercourse, houses and nourishes the embryo
during prenatal development, and produces milk for
the young (lactation). These processes are regulated
and coordinated by the interaction of hormones secreted by the hypothalamus, the anterior lobe of the
pituitary gland, and by the ovaries.

Principal organs of the female reproductive system are the ovaries (which produce ova and hormones), uterine tubes (which transport ova and are the
site of fertilization), uterus (the "incubator" for the
developing child), vagina (which receives the penis and
serves as part of the birth canal), vulva (external genital
structures), and breasts (which nourish the young)
(Figs. 42–9, 42–10).

The Ovaries Produce Ova And Sex Hormones

Like the male gonads, the female gonads, or **ovaries,**
produce both gametes and sex hormones. About the
size and shape of large almonds, the ovaries are located
close to the lateral walls of the pelvic cavity. The ovaries are held in position by several connective tissue
ligaments. Each ovary is covered with a single layer of
epithelium. Internally the ovary consists mainly of connective tissue (called stroma), through which are scattered **ova** in various stages of maturation (see Fig. 42–
11). The process of ovum formation is called **oogenesis.**

TABLE 42–1
Principal Male Reproductive Hormones

Endocrine Gland and Hormones	Principal Target Tissue	Principal Actions
Hypothalamus		
Gonadotropin-releasing hormone (GnRH)	Anterior pituitary	Stimulates release of FSH and LH
Anterior pituitary		
Follicle-stimulating hormone (FSH)	Testes	Stimulates development of seminiferous tubules; may stimulate spermatogenesis
Luteinizing hormone (LH); also called interstitial cell–stimulating hormone (ICSH)	Testes	Stimulates interstitial cells to secrete testosterone
Testes		
Testosterone	General	*Before birth:* stimulates development of primary sex organs and descent of testes into scrotum *At puberty:* responsible for growth spurt; stimulates development of reproductive structures and secondary sex characteristics (male body build, growth of beard, deep voice, etc.) *In adult:* responsible for maintaining secondary sex characteristics; stimulates spermatogenesis

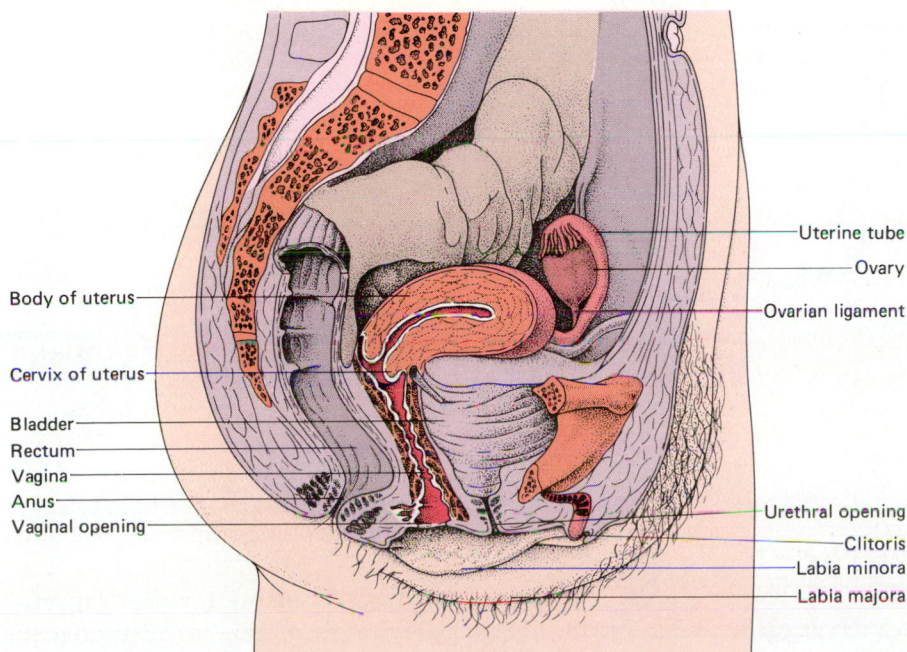

FIGURE 42–9 Midsagittal section of female pelvis showing reproductive organs. Note the position of the uterus relative to the vagina.

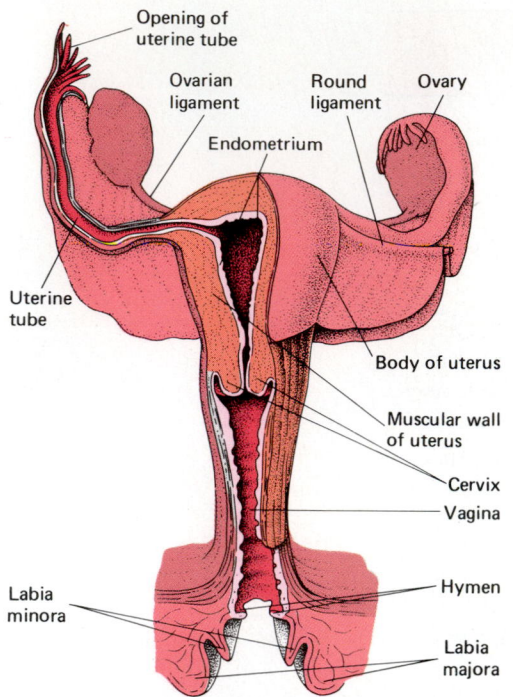

FIGURE 42–10 **Anterior view of female reproductive system. Some organs have been cut open to expose the internal structure. The ligaments help to hold the reproductive organs in place.**

Before birth hundreds of thousands of stem cells, termed **oogonia,** are present in the ovaries.

During prenatal development the oogonia increase in size and become **primary oocytes.** By the time of birth they are in the prophase of the first meiotic division. At this stage they enter a resting phase which lasts throughout childhood and into adult life. The developing ovum and the cluster of cells surrounding it together constitute a **follicle.** All of a female's gametes are produced during embryonic development; no new oogonia arise after birth.

With the onset of puberty a few of the follicles develop each month in response to FSH secreted by the anterior pituitary gland. As the follicle grows, the primary oocyte completes its first meiotic division, producing two cells that are very different in size (Fig. 42–12). The smaller one, the **first polar body,** may later divide, forming two polar bodies, but these eventually disintegrate. The larger cell, the **secondary oocyte,** proceeds to the second meiotic division, but halts in metaphase until it is fertilized. When meiosis does continue, the second meiotic division gives rise to a single ovum and a second polar body. The polar bodies are small and apparently serve to dispose of unneeded chromosomes with a minimum amount of cytoplasm.

FIGURE 42–11 **Microscopic structure of the ovary. Follicles in various stages of development are scattered throughout the ovary.**

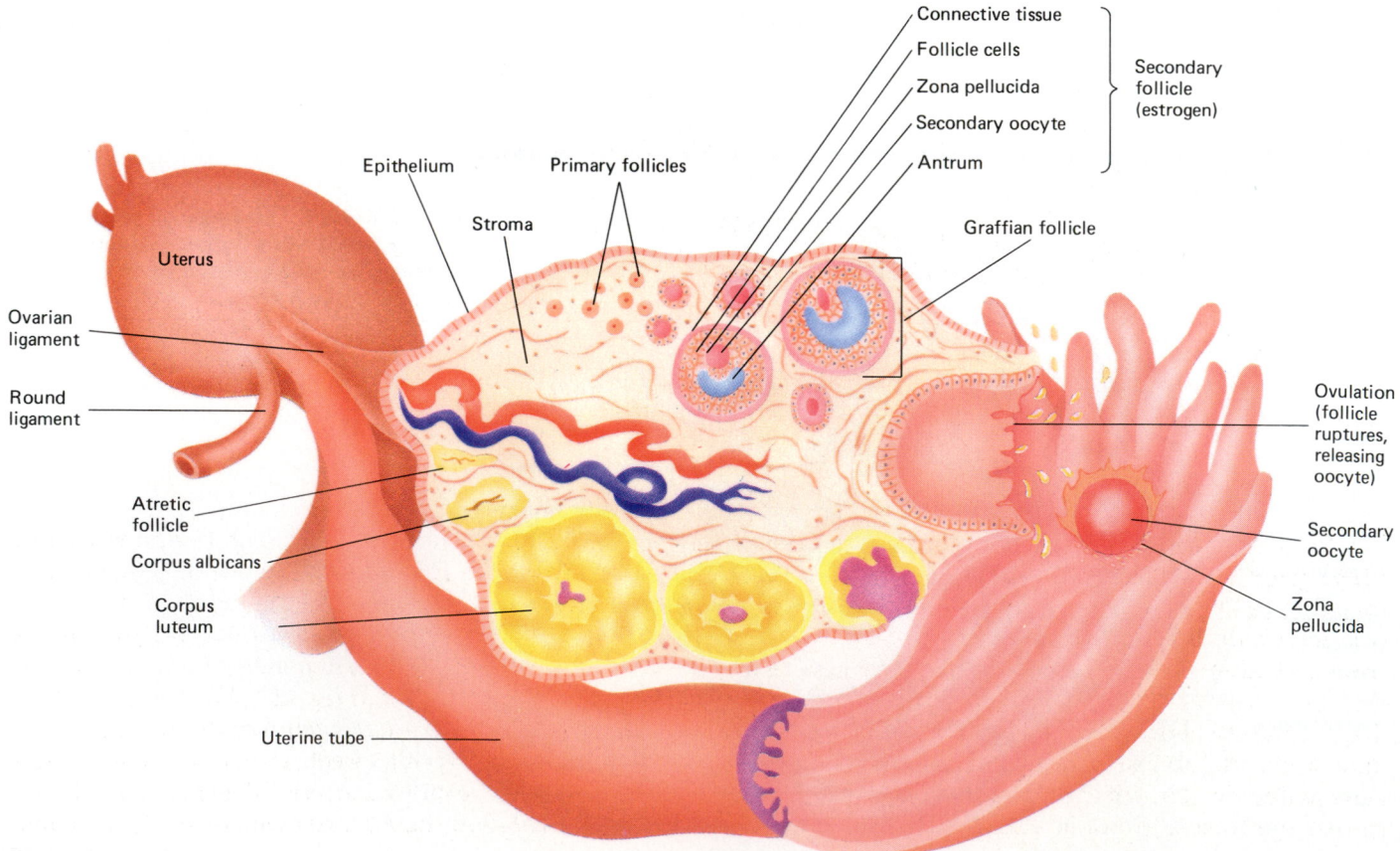

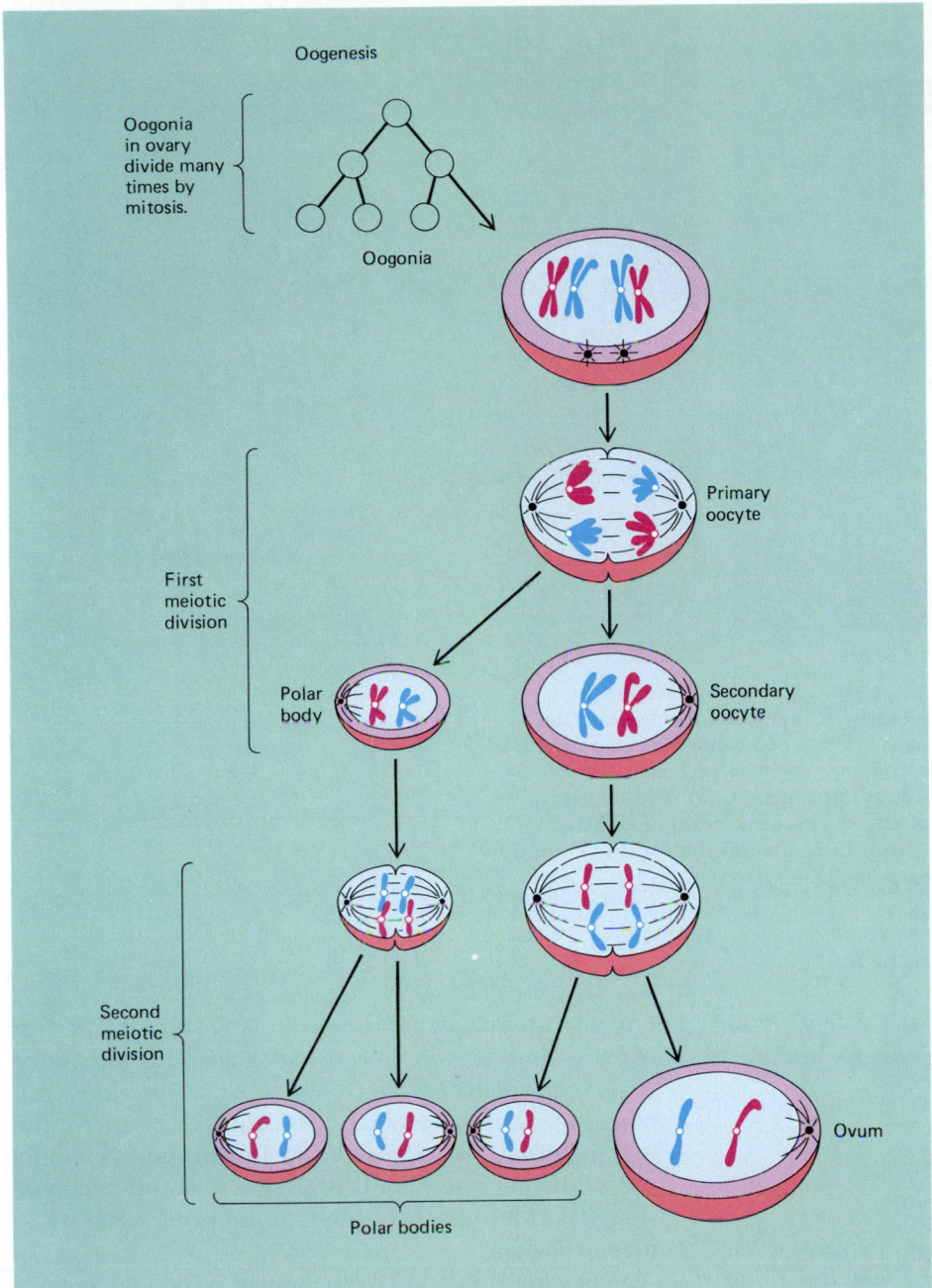

FIGURE 42-12 **Oogenesis. Only one functional ovum is produced from each primary oocyte. The other cells produced are polar bodies that degenerate.**

Recall that in the male large numbers of sperm are needed and in spermatogenesis each primary spermatocyte gives rise to four sperm. In contrast, only a few ova are needed during the reproductive life of a female and, accordingly, each primary oocyte generates only one ovum.

As an oocyte develops, it becomes separated from its surrounding follicle cells by a thick membrane, the **zona pellucida.** The follicle cells themselves proliferate so that the follicle grows in size. As the follicle develops, follicle cells secrete fluid, which collects in a space

created between them (Fig. 42–13a). Connective tissue surrounding the follicle cells contains cells that secrete the steroid sex hormone estradiol.

As a follicle matures, it moves closer to the surface of the ovary, eventually resembling a fluid-filled bulge on the ovarian surface (Fig. 42–13b). Generally only one follicle fully matures each month. Several others may develop for about a week, then deteriorate. These remain in the ovary as atretic (degenerate) follicles.

At **ovulation** the mature ovum (actually a secondary oocyte) is ejected through the wall of the ovary and

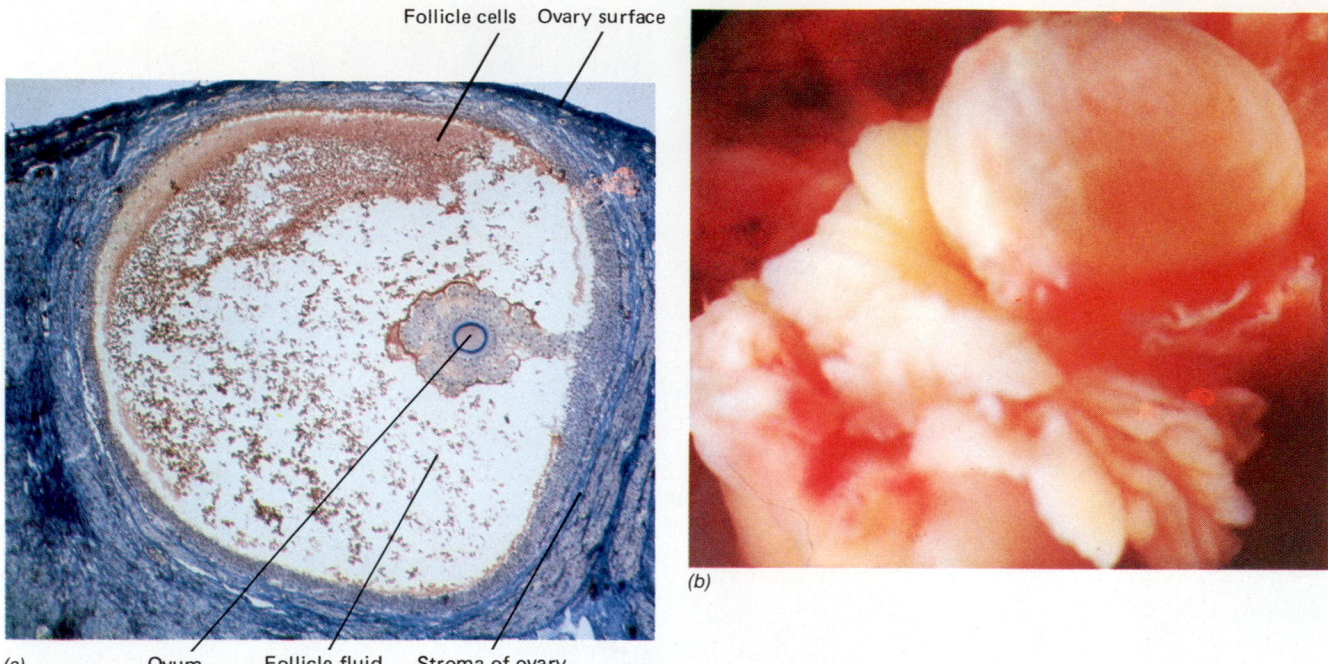

Follicle cells   Ovary surface

(a)   Ovum   Follicle fluid   Stroma of ovary

(b)

FIGURE 42–13   Follicle development. (*a*) **A stained section through a developing follicle. The ovum is surrounded by a layer of follicle cells that will be released along with it. These follicle cells become the corona radiata, a layer that acts as a barrier to sperm cells and may help ensure that the egg is fertilized by only one of the many sperm that approach it.** (*b*) **The mature follicle develops just under the surface of the ovary, producing a tightly stretched, fluid-filled bump (shown here from the outside) that will eventually burst, releasing the ovum.**   (*a*, Biophoto Associates/Photo Researchers, Inc.; *b*, Petit Format/Photo Researchers, Inc.)

into the pelvic cavity. The portion of the follicle that remains in the ovary develops into the **corpus luteum,** a temporary endocrine gland.

The Uterine Tubes Transport The Ovum

Almost immediately after ovulation, the ovum passes into the funnel-shaped opening of the **uterine tube** (also called fallopian tube or oviduct). Peristaltic contractions of the muscular wall of the uterine tube and beating of the cilia in its lining help to move the ovum along toward the uterus. Fertilization takes place within the uterine tube. If fertilization does not occur, the ovum degenerates there.

The Uterus Incubates The Embryo

The uterine tubes empty into the upper corners of the pear-shaped uterus (see Fig. 42–10). About the size of a fist, the **uterus** (or womb) occupies a central position in the pelvic cavity. This organ has thick walls of smooth muscle and a mucous lining, the **endometrium,** which thickens each month in preparation for

possible pregnancy. If an ovum is fertilized, the tiny embryo finds its way into the uterus and is implanted in the endometrium. Here it grows and develops, sustained by nutrients and oxygen delivered by surrounding maternal blood vessels. If fertilization does not occur during the monthly cycle, the endometrium sloughs off and is discharged, in the process known as **menstruation.**

The lower portion of the uterus, called the **cervix,** projects slightly into the vagina. The cervix is a common site of cancer in women. Detection is usually possible by the routine Papanicolaou test (Pap smear) in which a few cells are scraped from the cervix during a regular gynecological examination and studied microscopically. When cervical cancer is detected at very early stages of malignancy, the patient can be cured.

The Vagina Receives Sperm

The **vagina** is an elastic, muscular tube that extends from the uterus to the exterior of the body. The vagina serves as a receptacle for sperm during sexual intercourse and as part of the birth canal when development of the fetus is complete (Fig. 42–10).

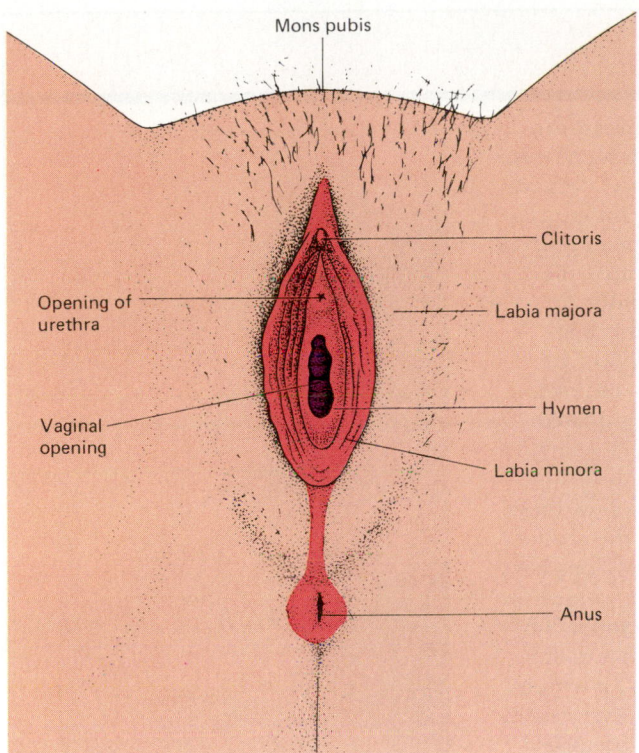

FIGURE 42–14   The vulva, the external genital structures of the female.

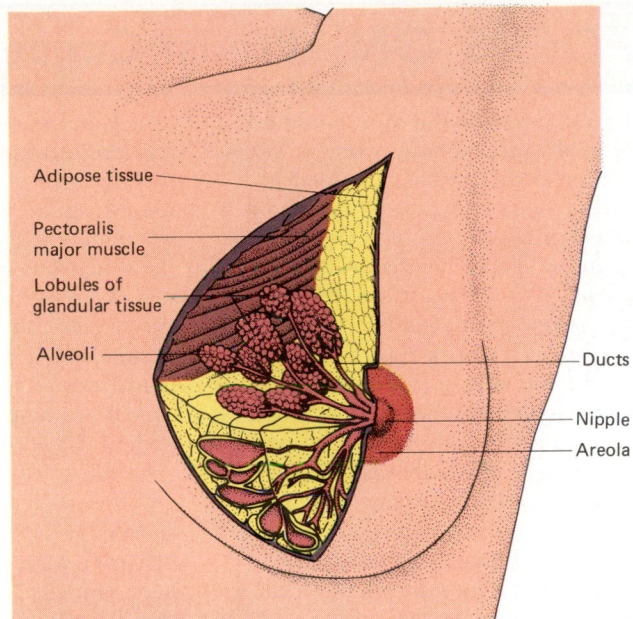

FIGURE 42–15   The mature human female breast.

External Genital Structures Are The Vulva

The external female sex organs, collectively known as the **vulva,** include lip-like folds, the **labia minora,** which surround the opening to the vagina and the opening to the urethra (Fig. 42–14). External to the delicate labia minora are the thicker, hair-covered **labia majora.** Anteriorly the labia minora merge to form the prepuce of the **clitoris,** a very small erectile structure comparable to the male glans penis. Like the penis, the clitoris contains erectile tissue that becomes engorged with blood during sexual excitement. Rich in nerve endings, the clitoris serves as a center of sexual sensation in the female.

The **mons pubis** is the mound of fatty tissue just above the clitoris at the junction of the thighs and torso. At puberty it becomes covered by coarse pubic hair. The **hymen** is a thin ring of tissue that may partially block the entrance to the vagina. It is often ruptured during a woman's first sexual intercourse. However, the hymen may be destroyed by strenuous exercise during childhood or by the use of tampons inserted into the vagina to absorb the menstrual flow.

The Breasts Function In Lactation

The breasts, which contain the mammary glands, overlie the pectoral muscles and are attached to them by connective tissue. Fibrous bands of tissue called **ligaments of Cooper** firmly connect the breasts to the skin. Each breast is composed of 15 to 20 lobes of glandular tissue, further subdivided into lobules made of connective tissue in which gland cells are embedded. The secretory cells are arranged in little grape-like clusters called alveoli (Fig. 42–15). Ducts from each cluster unite to form a single duct from each lobe, so that there are 15 to 20 tiny openings on the surface of each nipple. The amount of adipose tissue around the lobe of the glandular tissue determines the size of the breasts and accounts for their soft consistency. The size of the breasts does not affect their capacity to produce milk. The breasts are the most common site of cancer in women (see Focus on Breast Cancer).

Lactation is the production of milk for the nourishment of the young. During pregnancy, high concentrations of estrogens and progesterone produced by the corpus luteum and by the placenta stimulate the glands and ducts of the breast to develop, resulting in increased breast size. For the first couple of days after childbirth, the mammary glands produce a fluid called **colostrum,** which contains protein and lactose but little fat. After birth the hormone prolactin (secreted by the anterior lobe of the pituitary gland) stimulates milk production, and usually by the third day after delivery milk itself is produced. When the infant suckles at the breast, a reflex action in the mother results in release of prolactin and oxytocin from the pituitary gland. Oxytocin stimulates cells surrounding the alveoli to contract so that the alveoli are compressed. This forces milk from the alveoli into the ducts; the milk can then be sucked out of the breast by the baby.

Breastfeeding promotes recovery of the uterus, because the oxytocin released during breastfeeding

FOCUS ON Breast Cancer

Breast cancer is the most common type of cancer among women. Breast cancer incidence has increased in recent years and now strikes about 1 in every 10 women; it is the leading cause of cancer deaths in women. About 50% of breast cancers begin in the upper outer quadrant of the breast (see figure). As a malignant tumor grows, it may adhere to the deep tissue of the chest wall. Sometimes it extends to the skin, causing dimpling. Eventually the cancer spreads to the lymphatic system. About two-thirds of breast cancers have metastasized (spread) to the lymph nodes by the time the cancer is first diagnosed.

Mastectomy (surgical removal of the breast) and **radiation treatment** are common methods of treating breast cancer. **Chemotherapy** is especially useful in preventing metastasis, especially in premenopausal patients. A recent development in cancer treatment is the use of **biological response modifiers** such as interferons, interleukins, and monoclonal antibodies.

About one-third of breast cancers are estrogen-dependent; that is, their growth depends upon circulating estrogens. Removing the ovaries in patients with these tumors relieves the symptoms and may cause the disease to regress for months or even years. When diagnosis and treatment begin early, 80% of patients survive for 5 years and 62% for 10 years or longer. Untreated patients have only a 20% 5-year survival rate.

Because early detection of these cancers greatly increases the chances of cure and survival, campaigns have been launched to educate women on the importance of self-examination. **Mammography,** a soft-tissue radiological study of the breast, is a technique helpful in detecting very small lesions that might not be identified by palpation. Lesions show on an x-ray plate as areas of increased density. In **xeromammography** the x-ray image is produced on paper rather than on film. This method requires less radiation and provides excellent detail.

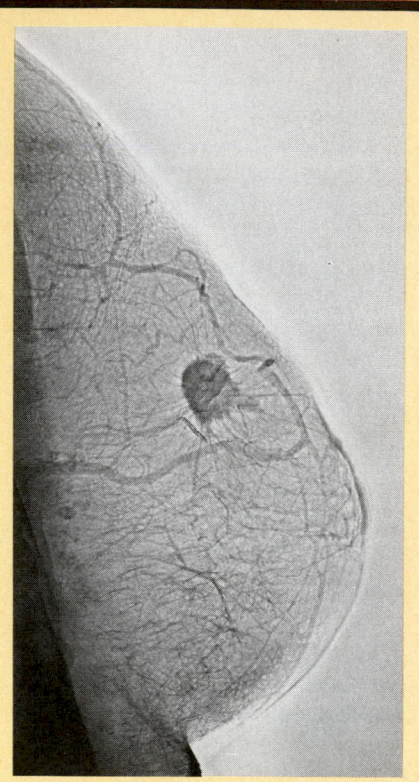

stimulates the uterus to contract to nonpregnant size. Breastfeeding a baby offers many advantages, including promoting a close bond between mother and child. Breast milk is tailored to the nutritional needs of the human infant. Furthermore, breastfed babies receive antibodies in the colostrum and mother's milk that are thought to play a protective role, resulting in a lower incidence of infantile diarrhea and even of respiratory infection during the second 6 months of life.

The Menstrual Cycle Is Regulated By Hormones

As a female approaches puberty the anterior pituitary secretes the gonadotropic hormones FSH and LH, which signal the ovaries to become active. Interaction of FSH and LH with estrogen and progesterone from the ovaries regulates the **menstrual cycle,** the monthly chain of events that prepares the body for possible pregnancy. The menstrual cycle runs its course every month from puberty until **menopause,** the end of a woman's reproductive (though not sexually active) life.

Although there is wide variation, a typical menstrual cycle is 28 days long (Fig. 42–16). The 1st day of the cycle is marked by **menstruation,** the monthly discharge through the vagina of blood and tissue from the endometrium. Ovulation occurs on about the 14th day of the cycle.

During the **menstrual phase** of the cycle, which lasts about 5 days, the pituitary gland releases FSH, which stimulates a few follicles to develop in the ovary. Cells in the connective tissue surrounding the follicle cells secrete the steroid hormone β-estradiol during the **preovulatory phase** of the menstrual cycle. (Estradiol is the principal type of estrogen found in humans; estrogens are a group of closely related 18-carbon steroid hormones.)

Estradiol stimulates growth of the endometrium, which thickens and develops new blood vessels and glands. The sharp rise in the concentration of estradiol in the blood stimulates the anterior lobe of the pituitary to secrete LH. Together, LH and FSH stimulate ovulation. LH then stimulates the portion of the follicle that remains in the ovary after the ovum has been ejected to develop into a corpus luteum.

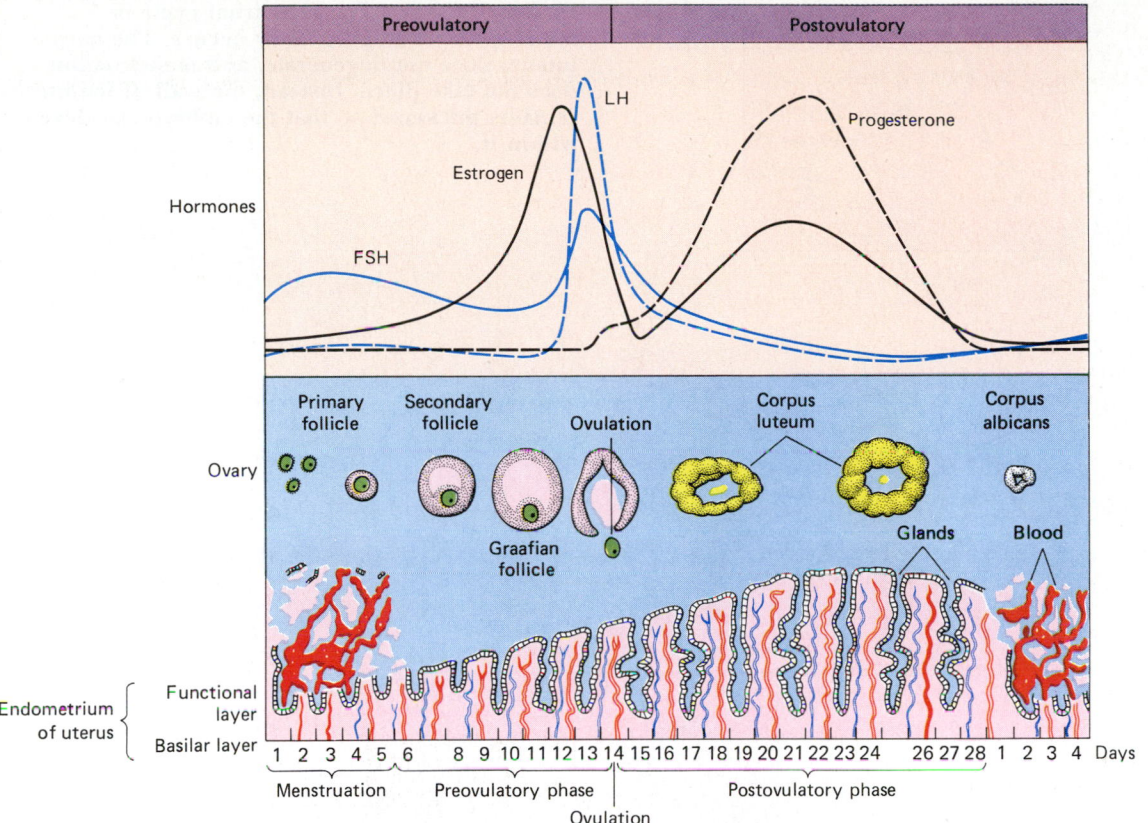

FIGURE 42–16   The menstrual cycle. The events that take place within the pituitary, ovary, and uterus are precisely synchronized. When fertilization does not occur, the cycle repeats itself about every 28 days. Compare this illustration with Figure 42–17.

The corpus luteum produces both estradiol and **progesterone** during the **postovulatory phase.** These hormones stimulate the uterus to continue its preparation for pregnancy. Progesterone stimulates tiny glands in the endometrium to secrete a fluid rich in nutrients. If the ovum is fertilized, this nutritive fluid nourishes the early embryo when it arrives in the uterus on about its fourth day of development (Fig. 42–17). On about the seventh day after fertilization the embryo begins to implant in the thick endometrium. Membranes that develop around the embryo secrete **human chorionic gonadotropin (hCG),** a hormone that signals the mother's corpus luteum to continue to function.

If the ovum is not fertilized, the corpus luteum begins to degenerate and the concentrations of progesterone and estradiol in the blood fall dramatically. Small arteries in the endometrium constrict, reducing the oxygen supply. As cells die and damaged arteries rupture and bleed, menstruation begins once again.

Table 42–2 lists the actions of the pituitary and ovarian reproductive hormones. Note that, like testosterone in the male, estradiol is responsible for the growth of the sex organs at puberty, for body growth, and for the development of secondary sexual charac-

teristics. In the female these include breast development, broadening of the pelvis, and the characteristic development and distribution of muscle and fat responsible for the shape of the female body.

Some testosterone is produced in the female, mainly in the adrenal cortex. This hormone is largely responsible for the adolescent growth spurt and for development of pubic hair and hair under the arms in the female (as well as in the male). In the male some estradiol is produced in the testes.

Hormones Of The Hypothalamus, Anterior Pituitary, And Ovaries Interact To Regulate The Menstrual Cycle

Details of the hormonal interaction that regulates the menstrual cycle are not fully known. GnRH released from the hypothalamus stimulates the anterior pituitary to release FSH and LH (Fig. 42–18). FSH stimulates the early maturation of the follicles in the ovary, and FSH and LH together stimulate the final maturation of a follicle. The rise in estradiol secreted by the developing follicles causes the burst in LH secretion that stimulates ovulation and formation of the corpus luteum.

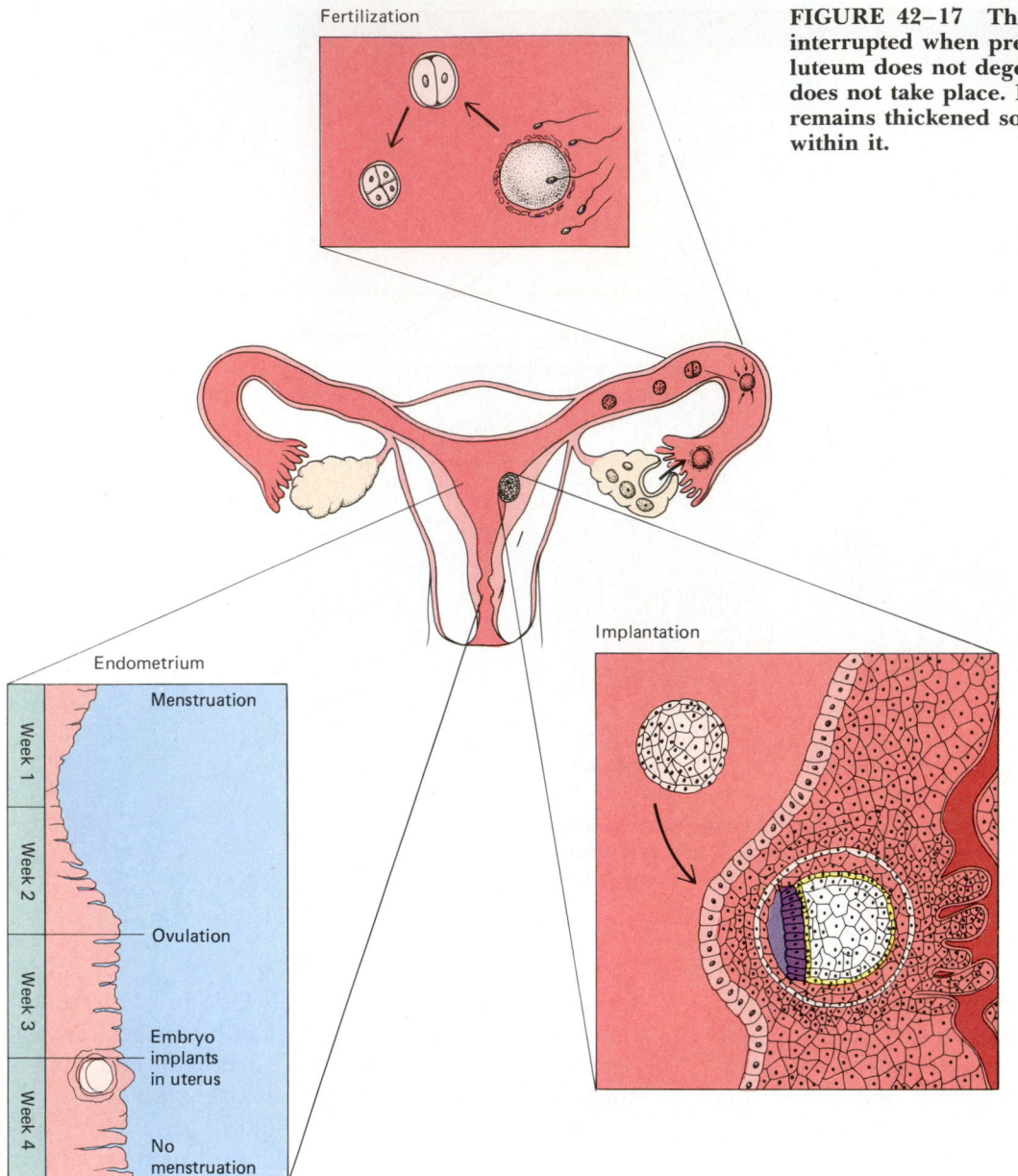

Fertilization

FIGURE 42–17   The menstrual cycle is interrupted when pregnancy occurs. The corpus luteum does not degenerate, and menstruation does not take place. Instead, the wall of the uterus remains thickened so that the embryo can develop within it.

Implantation

Endometrium

Menstruation

Week 1

Week 2

Ovulation

Week 3

Embryo implants in uterus

Week 4

No menstruation

The high constant levels of estradiol and progesterone maintained by the corpus luteum suppress FSH and LH secretion. These negative feedback effects of estradiol and progesterone are thought to act directly on the pituitary gland. Once the corpus luteum begins to degenerate, estradiol and progesterone levels fall, and the secretion of FSH and LH then increase once again.

Estradiol-progesterone imbalance has been suggested as the basis of **premenstrual syndrome (PMS),** a condition experienced by some women several hours to 10 days before menstruation and ending a few hours after onset of menstruation. Symptoms include fatigue, anxiety, depression, irritability, headache, edema, and skin eruptions.

In Menopause The Ovaries Secrete Less Estradiol And Progesterone

As a woman ages, the ovaries become less responsive to FSH and LH. The number of primary follicles in the ovary decreases, resulting in lowered production of estradiol. Thus by about age 50 the menstrual cycle becomes irregular and eventually halts.

Decreased estradiol affects the temperature-regulating center in the hypothalamus and sometimes leads to hot flashes and episodes of sweating. Estradiol deficiency may also contribute to feelings of depression and headaches experienced by some women during the onset of menopause. In menopausal women there is some atrophy of the ovaries, vagina, breasts, and other reproductive structures.

TABLE 42–2
Principal Female Reproductive Hormones

Endocrine Gland and Hormones	Principal Target Tissue	Principal Actions
Hypothalamus		
Gonadotropin-releasing hormone (GnRH)	Anterior pituitary	Stimulates release of FSH and LH
Anterior pituitary		
Follicle-stimulating hormone (FSH)	Ovary	Stimulates development of follicles; with LH, stimulates secretion of estrogen and ovulation
Luteinizing hormone (LH)	Ovary	Stimulates ovulation and development of corpus luteum
Prolactin	Breast	Stimulates milk production (after breast has been prepared by estrogen and progesterone)
Ovary		
Estradiol	General	Growth of sex organs at puberty; development of secondary sex characteristics (breast development, broadening of pelvis, distribution of fat and muscle)
	Reproductive structures	Maturation; monthly preparation of the endometrium for pregnancy; makes cervical mucus thinner and more alkaline
Progesterone	Uterus	Completes preparation of endometrium for pregnancy
	Breast	Stimulates development

Despite these physical changes, menopause does not usually affect a woman's interest or participation in sex. Symptoms of menopause can be alleviated clinically by replacing estrogens.

SEXUAL RESPONSE INVOLVES PHYSIOLOGICAL CHANGES

Human reproduction is accomplished sexually by the union of ovum and sperm. During copulation, also called **coitus** or sexual intercourse in humans, the male deposits semen into the upper end of the vagina. The complex structures of the male and female reproductive systems and the complex physiological, endocrine, and psychological phenomena associated with sexual activity are adaptations that promote the successful union of sperm and ovum and the subsequent development and nurturing of the resulting embryo.

Sexual stimulation results in two basic physiological responses: (1) increased blood flow (vasocongestion) to reproductive structures and certain other tissues such as the skin and (2) increased muscle tension. During vasocongestion, erectile tissues within the penis and clitoris, as well as in other areas of the body, become engorged with blood.

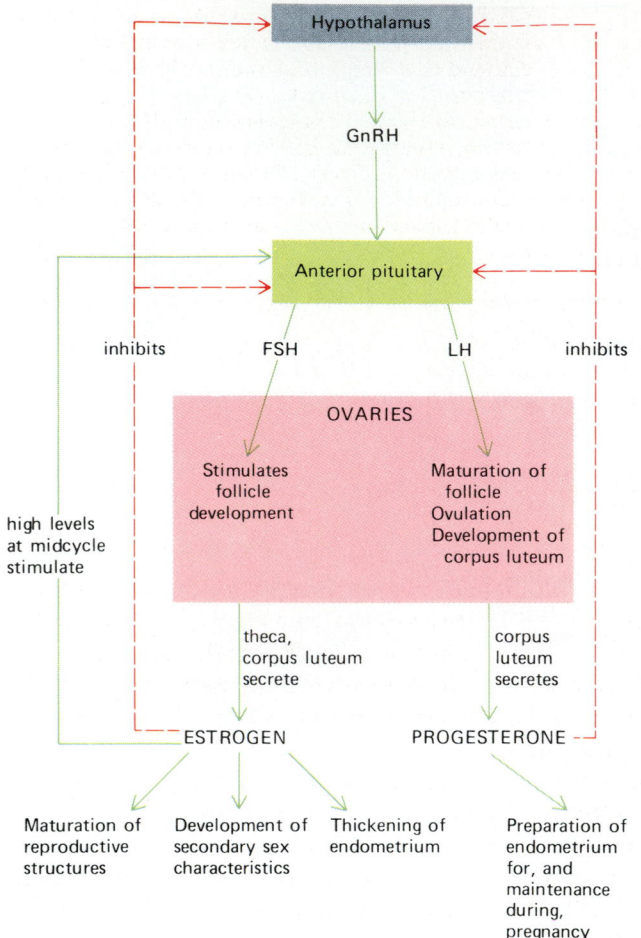

FIGURE 42–18 Hormones that regulate the menstrual cycle. Red arrows indicate inhibition.

vasocongested so that the vaginal entrance becomes somewhat constricted. In this narrowed state the outer one third of the vagina is referred to as the orgasmic platform. In the male the penis increases in circumference. In both sexes, blood pressure increases and heart rate and breathing are accelerated.

Coitus may be initiated during the excitement phase. During coitus the penis is moved back and forth in the vagina by movements referred to as pelvic thrusts. Physical and psychological sensations resulting from this friction (and from the entire intimate experience between the partners) may lead to orgasm, the climax of sexual excitement.

Although lasting only a few seconds, **orgasm** is the phase of maximum sexual tension and its release. In both sexes orgasm is marked by rhythmic contractions of the muscles of the pelvic floor and pelvic reproductive structures. Heart rate and respiration more than double, and blood pressure rises markedly both just before and during orgasm.

In the male, orgasm is marked by a sensation that **ejaculation** of semen is inevitable, followed by emission of semen. Contractions of the vas deferens propel sperm into the ejaculatory ducts. The accessory glands contract, adding their secretions; then, contractions of the ejaculatory ducts, urethra, and certain muscles of the pelvic floor eject the semen from the penis. Muscular contractions continue at about 0.8-second intervals for several seconds. After the first few contractions, their intensity decreases, and they become less regular and less frequent.

In the female, stimulation of the clitoris is important in heightening the sexual excitement that leads to orgasm. Sexual climax is marked by rhythmic contractions of the pelvic muscles and the orgasmic platform, starting at approximately 0.8-second intervals and recurring 5 to 12 times. After the first 3 to 6 contractions their intensity decreases, and the time interval between them increases. No fluid ejaculation accompanies orgasm in the female.

Orgasm is followed by the **resolution phase,** a state of well-being, during which muscle relaxation and detumescence (the subsiding of swelling) restore the body to its normal state. In most males, there is a refractory period during which physiological response to sexual stimulation does not occur. Duration of the refractory period varies in different individuals and also in different situations. Many women are able to respond to further sexual stimuli immediately and may reach orgasm multiple times.

Sexual inhibition may occur at any phase of the sexual response cycle and may result in sexual dysfunction. Psychological and biological factors may contribute to sexual dysfunction. For example, erection of the penis is necessary for effective coitus. Chronic inability to sustain an erection is termed **erectile dysfunction**

Sexual response includes four phases: sexual desire, excitement, orgasm, and resolution. The desire to have sexual activity may be motivated by fantasies about sexual activity or by thinking about sex. This anticipation can lead to sexual excitement. As an individual becomes psychologically and physically stimulated he or she experiences a sense of sexual pleasure. The excitement phase involves vasocongestion and increased muscle tension. Before the penis can enter the vagina and function in coitus it must be erect; accordingly, penile erection is the first male response to sexual excitement. In the female, vaginal lubrication is the first response to effective sexual stimulation. During the excitement phase, the vagina lengthens and expands in preparation for receiving the penis; the clitoris and breasts become vasocongested, and the nipples become erect.

If erotic stimulation continues, sexual excitement heightens. Vasocongestion and muscle tension increase markedly. In the female the inner two thirds of the vagina continues to expand and lengthen. The walls of the outer one third of the vagina become greatly

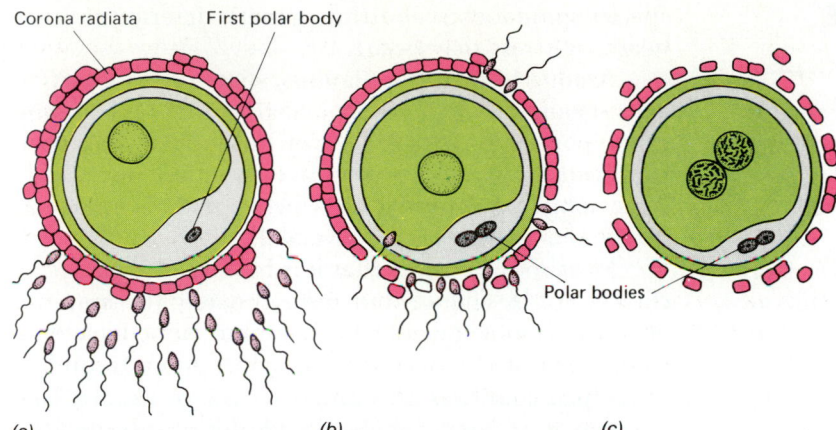

Corona radiata   First polar body

Polar bodies

(a)   (b)   (c)

FIGURE 42–19 Fertilization. (*a*) Each sperm is thought to release a small amount of enzyme that helps to disperse the follicle cells surrounding the ovum. (*b*) After a sperm cell enters the ovum, the ovum completes its second meiotic division, producing an ovum and a polar body. (*c*) Pronuclei of sperm and ovum combine, producing a zygote with the diploid number of chromosomes. (*d*) A scanning electron micrograph of sperm cell fertilizing a hamster ovum. (*d*, David Phillips/Visuals Unlimited)

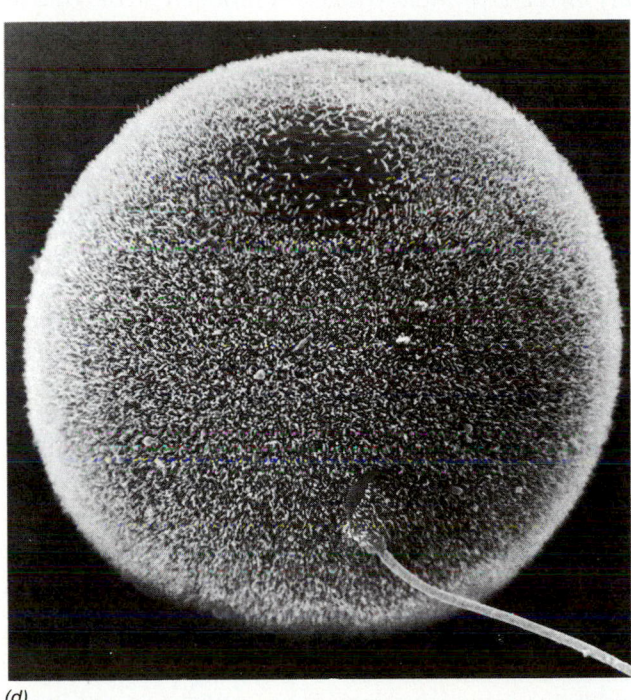

(d)

tility is probably most important in approaching and fertilizing the ovum.

If only one sperm is needed to fertilize an ovum, why are millions involved in each act of coitus? For one thing, sperm movement is undirected, so that many lose their way. Others die as a result of unfavorable pH or phagocytosis by leukocytes and macrophages in the female tract. Only a few thousand succeed in traversing the correct uterine tube and reaching the vicinity of the ovum. Additionally, large numbers of sperm may be necessary to penetrate the covering of follicle cells (the corona radiata) that surrounds the ovum. Each sperm is thought to release small amounts of enzymes from its acrosome that help break down the cement-like substance holding the follicle cells together.

Once one sperm has entered the ovum, there is a rapid electrical change, followed by a slower chemical change in the plasma membrane of the ovum, that prevents the entrance of other sperm. As the fertilizing sperm enters the ovum, it usually loses its tail (Fig. 42–19). Sperm entry stimulates the ovum to complete its second meiotic division. The head of the haploid sperm then swells to form the **male pronucleus** and fuses with the female pronucleus, forming the diploid nucleus of the zygote. The process of fertilization is described in more detail in Chapter 43.

After ejaculation into the female reproductive tract, sperm remain alive and retain their ability to fertilize an ovum for about 48 hours. The ovum itself remains fertile for only about 24 hours after ovulation. Thus, there are only about three days during each menstrual cycle (days 12 to 15 in a very regular 28-day cycle) when sexual intercourse is likely to result in fertilization.

In view of the many factors working against fertilization, it may seem remarkable that it ever occurs! Yet the frequency of coitus and the large number of sperm deposited at each ejaculation enable the human species not only to maintain itself but to increase its numbers at an alarming rate.

(formerly referred to as impotence); in men so affected, the disorder is often associated with psychological issues.

◼ FERTILIZATION IS THE FUSION OF SPERM AND EGG TO PRODUCE A ZYGOTE

In the process of **fertilization** sperm and ovum fuse, producing a zygote. Fertilization and the subsequent establishment of pregnancy together are referred to as **conception.** When conditions in the vagina and cervix are favorable, sperm begin to arrive at the site of fertilization in the upper uterine tube within 5 minutes after ejaculation. Contractions of the uterus and uterine tubes help transport the sperm. The sperms' own mo-

■ INFERTILITY IS THE INABILITY TO ACHIEVE CONCEPTION

Infertility is the inability of a couple to achieve conception after not using contraception for at least 1 year. About 15% of married couples in the United States are affected by infertility. About 30% of cases involve both male and female factors.

A major cause of male infertility is **sterility,** lack of sperm production. Men with fewer than 20 million sperm per milliliter of semen are usually considered to be sterile. When a couple's attempts to produce a child are unsuccessful, a sperm count and analysis may be performed in a clinical laboratory. Sometimes semen is found to contain large numbers of abnormal sperm or occasionally none at all. In about one-fourth of the cases of mumps in adult males, the testes become inflamed; in some of these cases the spermatogonia die, resulting in permanent sterility.

Low sperm counts have been linked to chronic marijuana use, alcohol abuse, and cigarette smoking, and studies show that men who smoke tobacco are more likely than nonsmokers to produce abnormal sperm. Exposure to chemicals such as DDT and PCBs may also result in low sperm count and sterility.

One cause of female infertility is scarring of the uterine tubes. Inflammation of the uterine tubes, sometimes caused by gonorrhea, may result in scarring, which blocks the tubes so that the ovum can no longer pass to the uterus. Sometimes partial constriction of the uterine tube results in tubal pregnancy, in which the embryo begins to develop in the wall of the uterine tube because it cannot progress to the uterus. Uterine tubes are not adapted to bear the burden of a developing embryo; thus, the uterine tube and the embryo it contains must be surgically removed before it ruptures and endangers the life of the mother.

Women with blocked uterine tubes usually can produce ova and can incubate an embryo. However, they need clinical assistance in getting the ovum from the ovary to the uterus. The ovum can be removed from the ovary, fertilized with the husband's sperm in laboratory glassware (in vitro fertilization), and then placed in the woman's uterus, where it may develop normally (see Focus on Novel Origins).

■ BIRTH CONTROL METHODS ALLOW INDIVIDUALS TO CHOOSE

Population sizes of various animal species are limited under natural conditions by a variety of mechanisms, even apart from such factors as disease and food shortage. When conditions are crowded, animals may control their population size by parental neglect, feeding only the strong offspring, or by cannibalism. In some species spontaneous abortion, genetic deterioration, or death by stress may occur.

Among human populations, some use of abortion as a means of birth control has been found among every population studied. Even infanticide has been practiced, especially in primitive societies. Infanticide is no longer sanctioned anywhere, and abortion has become increasingly controversial.

Most couples agree that it is best to have babies by choice, not by chance, but unfortunately the majority of couples who engage in sexual intercourse have only vague notions of how to prevent conception. In underdeveloped countries an estimated 88% of women lack the means to limit family size. Studies indicate that many of these women would use modern birth control methods if they were available and if someone showed them how.

More than 1 million teenagers in the United States become pregnant every year, and thousands of girls aged 14 or younger have babies each year. Yet only one in five sexually active teenagers consistently uses contraception. In addition, the AIDS epidemic poses an increasing risk. The number of unwanted babies could be decreased (and risk of sexually transmitted disease also lessened) if individuals old enough to produce babies knew where they came from and how to prevent conception (and lessen risk of disease). Because parents often neglect to provide such information, it is important that sex education be included in the public school curriculum. One may hope that, in addition, the family and other social and religious institutions will increasingly accept the responsibility of teaching not only about sex itself but also the social and ethical dimensions of sexuality.

When a sexually active woman uses no form of birth control, her chances of becoming pregnant during the course of a year are about 90%. Any method for deliberately separating sexual intercourse from production of babies is considered **contraception** (literally, "against conception"). Since ancient times humans have searched for effective contraceptive methods. Modern science has developed a variety of contraceptives with a high percentage of reliability, but the ideal contraceptive has not yet been devised. Some of the more common methods of birth control are described below and in Table 42–3. (Also see Fig. 42–20.) Note that intrauterine devices (IUDs) as well as some types of oral contraceptives may not actually prevent fertilization; they probably destroy the embryo or prevent its implantation in the wall of the uterus.

Oral Contraceptives Prevent Ovulation

More than 80 million women worldwide use **oral contraceptives** (more than 8 million in the United States alone). The most common preparations consist of a

TABLE 42–3
Birth Control Methods

Method	Failure Rate*	Mode of Action	Advantages	Disadvantages
Oral contraceptives	0.3; 5	Prevents ovulation; may also affect endometrium and cervical mucus and prevent implantation	Highly effective; sexual freedom; regular menstrual cycle	Minor discomfort in some women; possible thromboembolism; hypertension, heart disease in some users; possible increased risk of infertility; should not be used by women who smoke
Intrauterine device (IUD)	1; 5	Not known; probably stimulates inflammatory response	Provides continuous protection; highly effective	Cramps; increased menstrual flow; spontaneous expulsion; prescribed less frequently due to increased risk of pelvic inflammatory disease and infertility; usage has declined due to side effects
Spermicides (sponges, foams, jellies, creams)	3; 20	Chemically kill sperm	No side effects (?) Vaginal sponges are effective in vagina for up to 24 hours after insertion. Sponges also act as physical barriers to sperm cells.	Some evidence linking spermicides to birth defects
Contraceptive diaphragm (with jelly)**	3; 14	Diaphragm mechanically blocks entrance to cervix; jelly is spermicidal	No side effects	Must be prescribed (and fitted) by physician; must be inserted prior to coitus and left in place for several hours after intercourse
Condom	2.6; 10	Mechanical; prevents sperm from entering vagina	No side effects; some protection against STD including AIDS	Interruption of foreplay to fit; slightly decreased sensation for male; could break
Rhythm†	13; 21	Abstinence during fertile period	No side effects (?)	Not very reliable
Douche	40	Flush semen from vagina	No side effects	Not reliable; sperm beyond reach of douche in seconds
Withdrawal (coitus interruptus)	9; 22	Male withdraws penis from vagina prior to ejaculation.	No side effects	Not reliable; contrary to powerful drives present when an orgasm is approached; sperm present in fluid secreted before ejaculation may be sufficient for conception
Sterilization Tubal ligation	0.04	Prevents ovum from leaving uterine tube	Most reliable method	Often not reversible
Vasectomy	0.15	Prevents sperm from leaving scrotum	Most reliable method	Often not reversible
Chance (no contraception)	About 90			

*The lower figure is the failure rate of the method; the higher figure is the rate of method failure plus failure of the user to utilize the method correctly. Based on number of failures per 100 women who use the method per year in the United States.
**Failure rate is lower when used together with spermicidal foam.
†There are several variations of the rhythm method. For those who use the calendar method alone, the failure rate is about 35. However, by taking the body temperature daily and keeping careful records (temperature rises after ovulation), the failure rate can be reduced. Also, by keeping a daily record of the type of vaginal secretion, changes in cervical mucus can be noted and used to determine time of ovulation. This type of rhythm contraception is also slightly more effective. When women use the temperature or mucus method and have intercourse *only* more than 48 hours *after* ovulation, the failure rate can be reduced to about 7.

FOCUS ON Novel Origins

About 10,000 children born each year are products of **artificial insemination.** Usually this procedure is sought when the male partner of a couple desiring a child is sterile or carries a genetic defect. Although the sperm donor remains anonymous to the couple involved, his genetic qualifications are screened by physicians.

In vitro fertilization is a technique by which an ovum is removed from a woman's ovary, fertilized in a test tube, and then reimplanted in her uterus. Such a procedure may be attempted if a woman's fallopian tubes are blocked or if they have been surgically removed. With the help of this technique a healthy baby was born in England in 1978 to a couple who had tried unsuccessfully for several years to have a child. Since that time, thousands of children have been conceived within laboratory glassware.

Another novel procedure is **host mothering.** In this procedure, a tiny embryo is removed from its natural mother and implanted into a female substitute. The foster mother can support the developing embryo either until birth or temporarily until it is implanted again into the original mother or into another host. This technique has already proved useful to animal breeders. For example, embryos from prize sheep can be temporarily implanted into rabbits for easy shipping by air, and then reimplanted into a foster mother sheep, perhaps of inferior quality. Host mothering also has the advantage of allowing an animal of superior quality to produce more offspring than would be naturally possible. In one recent series of experiments mouse embryos were frozen for up to 8 days and then successfully transplanted into host mothers. Host mothering may someday be popular with women who can produce embryos but are unable to carry them to term.

Someday society may have to deal with **cloning** (not yet a reality in the case of humans). In this process the nucleus would be removed from an ovum and replaced with the nucleus of a cell from a person who wished to produce a human copy of himself or herself. Theoretically, any cell nucleus could be used, even a white blood cell nucleus. The fertilized ovum would then be placed into a human uterus for incubation; the resulting baby would be an identical, though younger, twin to the individual whose nucleus was used.

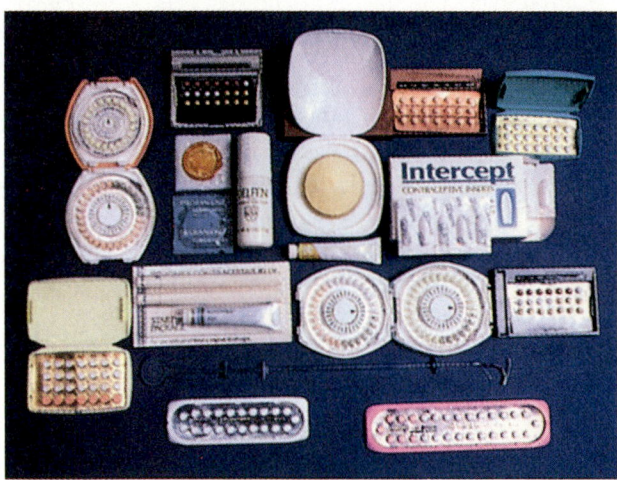

FIGURE 42–20 Some commonly used contraceptive devices. (Ray Ellis/Photo Researchers, Inc.)

combination of progestin (synthetic progesterone) and synthetic estrogen. (Natural hormones are destroyed by the liver almost immediately, but the synthetic ones somehow resist destruction.) Starting on the 5th day of the menstrual cycle, a woman takes one pill each day for about 3 weeks. She then stops taking the pill, and menstruation begins about 3 days later. Ovulation does not occur if she resumes the pills on the 5th day of her cycle. When taken correctly, these pills are about 99.9% effective in preventing pregnancy.

Most oral contraceptives prevent pregnancy by preventing ovulation. When postovulatory levels of ovarian hormones are maintained in the blood, the pituitary gland is inhibited and does not produce the surge of FSH and LH that stimulates ovulation. The chief advantage of oral contraceptives is their high rate of effectiveness.

Studies suggest that women over age 35 who smoke or have other risk factors, such as untreated hypertension, should not take oral contraceptives. Women in this category have an increased risk of death from circulatory diseases such as stroke and myocardial infarction. Nonsmokers can take oral contraceptives safely until age 35 with no increased risk of cardiovascular disease. Oral contraceptives result in death in about 3 per 100,000 users. This compares favorably with the death rate of about 9 per 100,000 pregnancies (Table 42–4).

Usage Of The Intrauterine Device (IUD) Has Declined

The **intrauterine device (IUD)** is a small plastic loop or coil that must be inserted into the lumen of the uterus by a medical professional (Fig. 42–20). Once in place, some types of IUD can be left in the uterus indefinitely or until the woman wishes to conceive. Newer types of IUDs are about 99% effective.

TABLE 42–4
Deaths in the United States from Pregnancy and Childbirth and From Various Birth Control Methods

	Death Rate per 100,000
Pregnancy and childbirth	9
Oral contraception	3
IUD	0.5
Legal abortions—first trimester	1.9
Legal abortions—after first trimester	12.5
Illegal abortion performed by medically untrained individuals	About 100

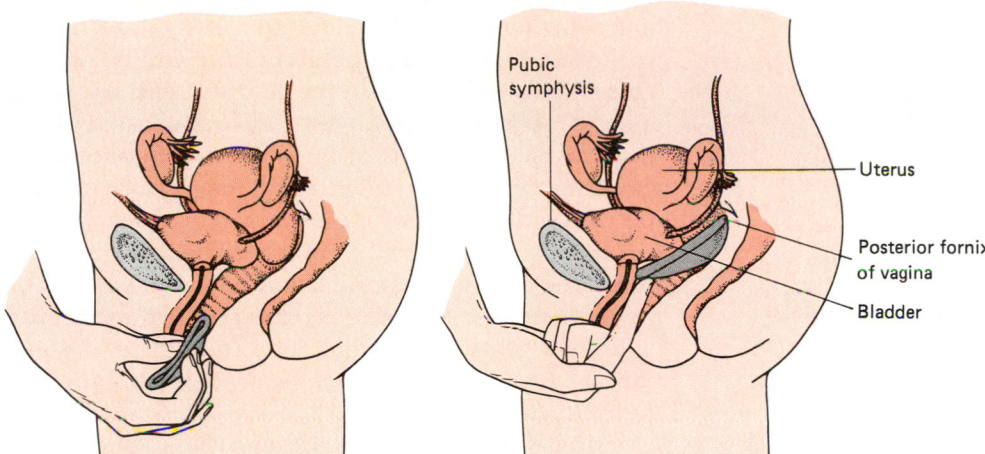

FIGURE 42–21 **Insertion of a contraceptive diaphragm.**

(Labels in figure: Pubic symphysis, Uterus, Posterior fornix of vagina, Bladder)

The mode of action of the IUD is not well understood, but it is thought that the IUD sets up a minor local inflammation in the uterus. Breakdown products of phagocytes are toxic to sperm (and to the embryo, should fertilization occur). Increased levels of antibodies in those who use IUDs support the idea that an immunological mechanism may be involved in its contraceptive action.

Because of side effects such as pelvic inflammatory disease, IUDs are being prescribed less frequently. In fact, the only IUD still marketed in the United States is a T-shaped, progesterone-releasing device that must be re-inserted each year.

Other Common Contraceptive Methods Include The Diaphragm And Condom

Other common contraceptive methods are listed and compared in Table 42–3. The **contraceptive diaphragm** mechanically blocks the passage of sperm from the vagina into the cervix. It is covered with spermicidal jelly or cream and inserted just prior to coitus (Fig. 42–21).

The **condom** is also a mechanical method of birth control. The most commonly used male contraceptive device, the condom provides a barrier that contains the semen so that sperm cannot enter the female tract. The condom is the only contraceptive that provides some protection against AIDS and other sexually transmitted diseases.

Sterilization Renders An Individual Incapable Of Producing Offspring

Aside from total abstinence, sterilization is the only foolproof method of contraception. Sterilization is currently the most popular contraceptive method for couples in which the wife is over age 30. About 75% of sterilization operations are currently performed on males.

Male Sterilization Is Performed By Vasectomy

An estimated 1 million **vasectomies** are performed each year in the United States. Using a local anesthetic, a small incision is made on each side of the scrotum.

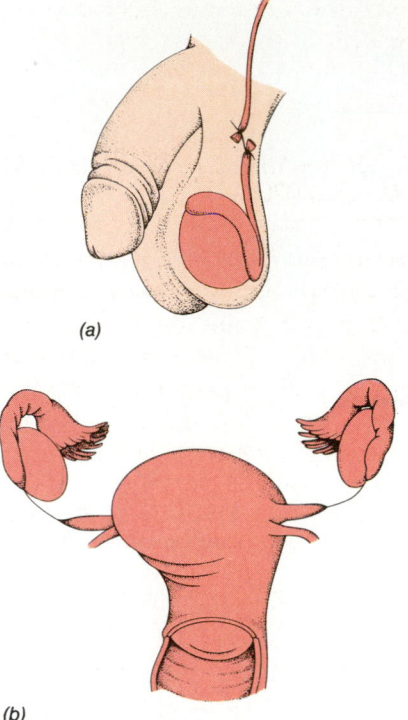

(a)

(b)

FIGURE 42–22 Sterilization. (a) In vasectomy, the vas deferens (sperm duct) on each side is cut and tied. (b) In tubal ligation, each uterine tube is cut and tied so that ovum and sperm can no longer meet.

Then each vas deferens is cut and its ends are tied or clipped so that they cannot grow back together (Fig. 42–22a).

Since testosterone secretion and transport are not affected, a vasectomy in no way affects masculinity. Sperm continue to be produced, though at a much slower rate, and are destroyed by macrophages in the testes. No change in the amount of semen ejaculated is noticed, because sperm account for very little of the semen volume.

In a study of more than 1000 men who had vasectomies, 99% said they had no regrets, and 73% claimed an increase in sexual pleasure, probably because anxiety about pregnancy was erased. By surgically reuniting the ends of the vasa deferentia, surgeons can successfully reverse sterilization in about 50% of attempts made. Apparently, some sterilized men eventually develop antibodies against their own sperm and become sterile.

An alternative to reversible vasectomy is the storage of frozen sperm in sperm banks. If the male should decide to father another child after he has been sterilized, he simply "withdraws" his sperm for use in artificially inseminating his wife. Sperm banks are currently being established throughout the United States. Not much is known yet about the effects of long-term sperm storage, but there may be an increased risk of genetic defects.

Female Sterilization Is By Tubal Ligation

Several techniques are in current use for prevention of ova transport. Most of them involve **tubal ligation,** cutting and tying the uterine tubes (Fig. 42–22b). This can be done through the vagina but is usually performed through an abdominal incision. As in the male, hormone balance and sexual performance are not affected.

■ THERE ARE THREE TYPES OF ABORTION

Abortion is the termination of pregnancy resulting in the death of the embryo or fetus. An estimated 40 million abortions are deliberately performed each year worldwide (more than a million in the United States). Three kinds of abortions may be distinguished: spontaneous abortions, therapeutic abortions, and those undertaken as a means of birth control. **Spontaneous abortions** (popularly called miscarriages) occur without intervention and often are a biological mechanism for destroying a defective embryo. **Therapeutic abortions** are performed in order to maintain the health of the mother, or when there is reason to suspect that the embryo is grossly abnormal. The third type of abortion, the type performed as a means of birth control, is the most controversial.

Most first-trimester abortions (those performed during the first 3 months of pregnancy) and some later ones are performed using a suction method. After the cervix has been dilated, a suction aspirator is inserted in the uterus, and the embryo and other products of conception are evacuated. In pregnancies of more than 12 weeks, the method most commonly used is dilation and evacuation ("D & E"). The cervix is dilated, forceps are used to remove the fetus, and suction is used to aspirate the endometrium.

When abortion is performed during the first trimester by skilled medical personnel, the mortality is about 1.9 per 100,000. After the first trimester this rate rises to 12.5 per 100,000 (Table 42–4). The death rate from illegal abortions performed by medically untrained individuals is about 100 per 100,000. In contrast to these figures, the death rate from pregnancy and childbirth is about 9 per 100,000.

The issue of abortion remains a matter of controversy. Abortions have always been available to those who could afford to travel to places where they were legal. The poor woman took her chances with the unqualified, often unsanitary, but local and less expensive, practitioner. Thousands of women have died as a result of such illegal abortions.

In January 1973 the U.S. Supreme Court ruled that during the first 3 months of pregnancy the decision to have an abortion rests entirely with the pregnant woman and her physician and that the state could not interfere. After the first 3 months the state could regulate abortion procedure in order to protect maternal health. A state could prohibit abortion during the last 10 weeks of pregnancy, when the fetus could survive outside the uterus. The Supreme Court's decision came after several states had liberalized restrictive abortion laws.

A concerted drive by anti-abortion groups, often called "right-to-life" societies, was mounted in reaction to the relaxation of legislative guidelines regarding abortion. Such groups see abortion as a moral issue affecting the embryo, which they consider to be a human being with legal rights. In 1989 the Supreme Court returned the right to regulate abortion to the states. This decision has inspired heated debate and the future of therapeutic and contraceptive abortion remains unclear at this time.

■ SEXUALLY TRANSMITTED DISEASES ARE SPREAD BY SEXUAL CONTACT

Sexually transmitted diseases (STD), also called venereal diseases (VD), are, next to the common cold, the most prevalent communicable diseases in the world. The World Health Organization has estimated that more than 250 million people are infected each year with gonorrhea and more than 50 million with syphilis. Some common sexually transmitted diseases are listed and described in Table 42–5 on page 786. AIDS is discussed in Chapter 38 in the Focus on AIDS.

■ CHAPTER SUMMARY

I. In asexual reproduction, a single parent endows its offspring with a set of genes identical with its own. In sexual reproduction, each of two parents contributes a gamete containing half of the offspring's genetic endowment.

II. Unusual variations of reproduction among various animals include metagenesis, parthenogenesis, and hermaphroditism.

III. The human male reproductive system includes the testes, which produce sperm and testosterone; a series of conducting tubes; accessory glands; and the penis.

 A. The testes, housed in the scrotum, contain the seminiferous tubules, where the sperm are produced, and the interstitial cells, which secrete testosterone.

 B. Sperm complete their maturation and are stored in the epididymis and may also be stored in the vas deferens.

 C. During ejaculation sperm pass from the vas deferens to the ejaculatory duct, and then into the urethra, which passes through the penis.

 D. Semen contains about 400 million sperm suspended in the secretions of the seminal vesicles and prostate gland.

 E. The penis consists of three columns of erectile tissue; when this tissue becomes engorged with blood, the penis becomes erect.

 F. The gonadotropic hormones FSH and LH stimulate sperm production and testosterone secretion. Testosterone is responsible for establishing and maintaining primary and secondary sex characteristics in the male.

IV. The female reproductive system includes the uterus, ovaries, which produce ova and hormones, the uterine tubes, vagina, vulva, and breasts.

 A. After ovulation the ovum enters the uterine tube, where it may be fertilized.

 B. The uterus serves as an incubator for the developing embryo.

 C. The vagina is the lower part of the birth canal; it receives the penis during coitus.

 D. The clitoris is the center of sexual sensation in the female.

 E. The first day of menstrual bleeding marks the first day of the menstrual cycle. Ovulation occurs at about day 14 in a typical 28-day menstrual cycle. Events of the menstrual cycle are coordinated by the gonadotropic and ovarian hormones.

 1. FSH stimulates follicle development; FSH and LH together stimulate ovulation; LH promotes development of the corpus luteum.

 2. The developing follicles release estradiol, which stimulates development of the endometrium and is responsible for the secondary female sex characteristics.

 3. The corpus luteum secretes progesterone, which stimulates final preparation of the uterus for possible pregnancy.

V. Vasocongestion and increased muscle tension are two basic physiological responses to sexual stimulation. The phases of sexual response include sexual desire, excitement, orgasm, and resolution.

TABLE 42–5
Some Common Sexually Transmitted Diseases

Disease and Causative Organism	Course of disease	Treatment
Gonorrhea (*Neisseria gonorrhoeae*, a gonococcus bacterium)	Infection by sexual contact. Bacterial toxin may produce redness and swelling at infection site. Symptoms in males: painful urination and discharge of pus from penis. In about 60% of infected women no symptoms occur in initial stages. Can spread to epididymis (in males) or uterine tubes and ovaries (in females), causing sterility. Can cause widespread pelvic or other infection, plus damage to heart valves, meninges (outer coverings of brain and spinal cord), and joints.	Penicillin, or other antibiotic if penicillin-resistant strain involved.
Syphilis (*Treponema pallidum*, a spirochete bacterium)	Bacteria enter body through defect in skin near site of infection, spread throughout body by lymphatic and circulatory routes. Primary chancre (a small, painless ulcer) forms at site of initial infection; heals in about a month. Highly infectious at this stage. Secondary stage follows, in which a widespread rash and influenza-like symptoms may occur. Scaly lesions may occur that teem with bacteria and are highly infectious. Latent stage that follows can last 20 years. Eventually, lesions called gummae may occur, consuming parts of the body surface or damaging liver, bone, or spleen. Serious brain damage may occur. Death results in 5–10% of cases.	Penicillin. Sensitive blood tests can detect antibodies and hence infection. About one-third of cases recover spontaneously.
Genital herpes (herpes simplex type 2 virus)	Tiny, painful blisters appear on genitals; may develop into ulcers. Influenza-like symptoms may occur. Recurs periodically. Threat to fetus or newborn infant. May predispose to cervical cancer in females.	No effective cure. Some drugs may shorten outbreaks or reduce severity of symptoms.
Pelvic inflammatory disease (PID; usually chlamydial bacteria)	Generalized infection of reproductive organs and pelvic cavity; usually chronic and difficult to treat. May lead to sterility (more than 15% of cases). PID now most common STD in the U.S.	Antibiotics, surgical removal of affected organs
Trichomoniasis (a protozoan)	Symptoms include itching, discharge, soreness. Can be contracted from dirty toilet seats and towels.	Drugs
Yeast infections (genital candidiasis; yeasts)	Irritation, soreness, discharge; especially common in females.	Drugs
Nongonococcal urethritis (NGU)	Inflammation of urethra frequently caused by chlamydia bacteria; urination may become difficult.	Tetracycline

VI. Fertilization is the fusion of egg and sperm to form a zygote.

VII. Effective methods of birth control include oral contraceptives, intrauterine devices, condoms, contraceptive diaphragms, and sterilization.

VIII. Most first-trimester abortions are performed using a suction method.

IX. Important types of sexually transmitted diseases are gonorrhea, syphilis, genital herpes, pelvic inflammatory disease, and AIDS.

■ POST-TEST

1. The type of reproduction in which an animal divides into several pieces and then each piece develops into an entire new animal is called _____.

2. In metagenesis there is an alternation of

   _____.

3. Parthenogenesis is a type of reproduction in which an unfertilized egg _____.

4. An individual that can produce both eggs and sperm is described as _____.

5. A sex cell (either egg or sperm) is properly called a

   _____ ; a fertilized egg is a

   _____.

6. An adult who is unable to parent offspring is said to be _____.

For each group, select the most appropriate answer from Column B for each description in Column A.

Column A
7. Sperm produced here
8. Produce testosterone
9. Columns of erectile tissue
10. Secretes alkaline fluid into urethra
11. Sac that holds the testes

Column B
a. Seminiferous tubules
b. Prostate gland
c. Interstitial cells in testes
d. Cavernous bodies
e. None of the above

Column A
12. Produces gametes
13. Thickens each month in preparation for pregnancy
14. Lower portion of uterus
15. Fertilization takes place here

Column B
a. Uterine tube
b. Ovary
c. Cervix
d. Endometrium
e. None of the above

Column A
16. Produces FSH
17. Produces progesterone
18. Center of sexual sensation
19. Extends from uterus to exterior of body

Column B
a. Hymen
b. Corpus luteum
c. Anterior lobe of pituitary
d. Clitoris
e. None of the above

Column A
20. Responsible for secondary sexual characteristics in female
21. Responsible for secondary sexual characteristics in male
22. Produced by pituitary
23. Stimulates glands in endometrium to develop

Column B
a. Testosterone
b. Estradiol
c. LH
d. Progesterone
e. None of the above

Column A
24. Prevents ovulation
25. Prevents sperm from entering vagina
26. Procedure in which vas deferens is severed
27. Blocks passage of sperm from vagina into uterus

Column B
a. IUD
b. Contraceptive diaphragm
c. Oral contraceptive
d. Condom
e. None of the above

28. A _____ abortion is performed in order to maintain the mother's health or when the embryo is thought to be grossly abnormal.

29. Tubal ligation is a common method of

   _____ _____.

30. The menstrual cycle runs its course every month from puberty until _____.

31. Label the following diagrams. (Refer to Figs. 42–3 and 42–10 in the text as necessary.)

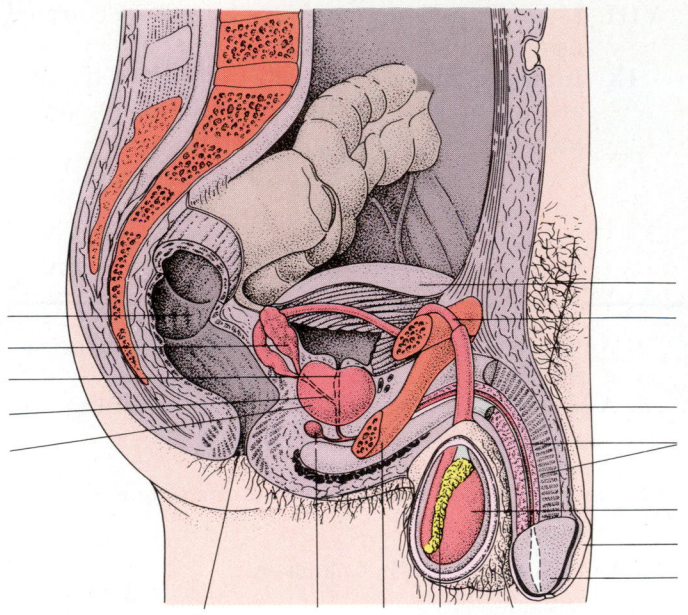

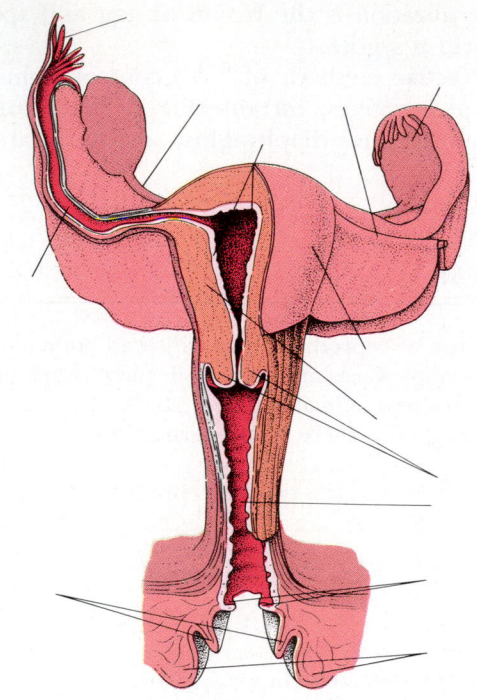

REVIEW QUESTIONS

1. Compare asexual with sexual reproduction, and give specific examples of asexual reproduction.
2. What are the advantages of the following methods of reproduction: metagenesis and parthenogenesis? Give an example of each.
3. Can you think of any biological advantages of parthenogenesis? of hermaphroditism? Explain.
4. Compare the functions of ovaries and testes.
5. Trace the passage of sperm from a seminiferous tubule through the male reproductive system until it leaves the male body during ejaculation. Assuming that ejaculation takes place within the vagina, trace the journey of the sperm until it meets the ovum.
6. What are the actions of testosterone? of estradiol? of progesterone?

7. What is the function of the corpus luteum? Which hormone stimulates its development?
8. What are the actions of FSH and LH in the female?
9. Why are so many sperm produced in the male and so few ova produced in the female?
10. Which methods of birth control are most effective? least effective?
11. Draw a diagram of the principal events of the menstrual cycle, including ovulation and menstruation. Indicate on which days of the cycle sexual intercourse would most likely result in pregnancy.
12. Distinguish among the following terms: erectile dysfunction, infertility, sterility, and castration.

RECOMMENDED READINGS

Masters, W.H., Johnson, V.E., and Kolodny, R.C., Eds. *Human Sexuality*. Boston: Little, Brown and Co., 1985.

Solomon, E.P., Schmidt, R.R., and Adragna, P. *Human Anatomy and Physiology*. Philadelphia: Saunders College Publishing, 1990. This book has an excellent chapter on human reproduction.

Wassarman, P.M. "Fertilization in mammals," *Scientific American*, *259*, (6) December, 1988. A glycoprotein governs many of the events of fertilization including the process by which the fertilized egg prevents other sperm from entering.

43

Development

OUTLINE

I. Development is a balanced combination of several processes
II. Fertilization restores the diploid number of chromosomes
    A. The first steps in fertilization are contact and recognition
    B. Sperm entry is regulated
    C. Sperm and egg pronuclei fuse
    D. Fertilization activates the egg
III. The zygote contains instructions for producing a complete individual
IV. During cleavage the zygote divides, forming many cells
    A. Cleavage provides building blocks for development
    B. The amount of yolk determines the pattern of cleavage
V. The germ layers form during gastrulation
    A. Each germ layer has a specific fate
    B. The pattern of gastrulation varies somewhat
VI. Organogenesis begins with the development of the nervous system
VII. Developmental processes are carefully regulated
    A. Cytoplasmic factors influence the course of development
    B. Cells can influence each other with inducers
    C. Nongenetic and genetic factors interact
VIII. Extraembryonic membranes and placenta protect and nourish the embryo
    A. The chorion and amnion enclose the embryo
    B. The allantois functions in waste disposal
    C. The yolk sac encloses the yolk
    D. The placenta is an organ of exchange
IX. Human prenatal development requires about 266 days

LEARNING OBJECTIVES

After you have studied this chapter you should be able to:

1. Relate the preformation theory and the theory of epigenesis to current concepts of development.
2. Compare the roles of mitosis, growth, morphogenesis, and cellular differentiation in the development of an organism.
3. Summarize the functions of fertilization and describe the four processes involved.
4. Trace the early development of the embryo from zygote through cleavage, morula, blastula, gastrula, and early organogenesis.
5. Compare the fate of each of the germ layers.
6. Discuss genetic and nongenetic factors that interact to regulate development, relating experiments discussed in this chapter.
7. Compare the functions of the extraembryonic membranes and placenta.
8. Describe the general course of human development.
9. Distinguish among the three stages of labor in the birth process.
10. Contrast postnatal with prenatal life, describing several adaptations that the neonate must make in order to live independently.
11. List specific steps that a pregnant woman can take to promote the well-being of her developing child and describe how the embryo can be affected by nutrients, drugs, cigarette smoking, pathogens, and ionizing radiation.
12. Describe the stages of the human life cycle.
13. Identify anatomical and physiological changes that occur with aging and discuss current theories of aging.

Human embryo at 40 days of development.
(Carolina Biological Supply)

OUTLINE (Continued)

A. Development begins in the uterine tube
B. Twins develop when cells separate
C. Implantation begins on about the seventh day
D. Organ development begins during the first trimester
E. The fetus continues to develop during the second and third trimesters

F. The birth process may be divided into three stages
G. The neonate must adapt to its new environment
X. Environmental factors affect the embryo
XI. The human life cycle extends from conception to death
XII. The aging process is marked by a decrease in homeostatic response to stress

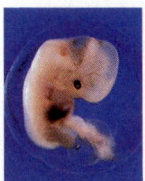 **Development** includes all of the changes that take place during the entire life of an animal from conception to death. In this chapter we focus mainly on development of the embryo, but we also briefly discuss growth and maturation of the animal after birth and the aging process.

Many scientists of the 17th century thought that the human egg cell contained a completely formed, though miniature, human being. They believed that all its parts were already there, so that the embryo had only to grow in size. This concept is known as the **preformation theory.**

By the end of the 17th century two competing groups of preformationists emerged. One group, the ovists, thought that the preformed human resided within the egg; the opposing group, the spermists, were certain that the "little man" was housed in the sperm. Using their crude microscopes some investigators even imagined that they could see a completely formed tiny human being within the head of the sperm (Fig. 43–1).

Some scientists carried the theory to an extreme form, arguing that every woman contained within her body a miniature of every individual who would ever descend from her. Her children, grandchildren, great grandchildren, and so on were thought to be preformed, each within the reproductive cells of the other. Some investigators of that time even computed mathematically how many generations could fit, one within the other's gametes. They concluded that when all these generations had lived and died, the human species would end. Jan Swammerdam, a renowned preformationist, felt that this concept explained original sin. He wrote, "In nature there is no generation, but only propagation, the growth of parts. Thus, original sin is explained, for all men were contained in the organs of Adam and Eve. When their stock of eggs is finished the human race will cease to be."

The idea of preformation was not restricted in its application to human beings alone. All plant and animal species were included. For almost 200 years this

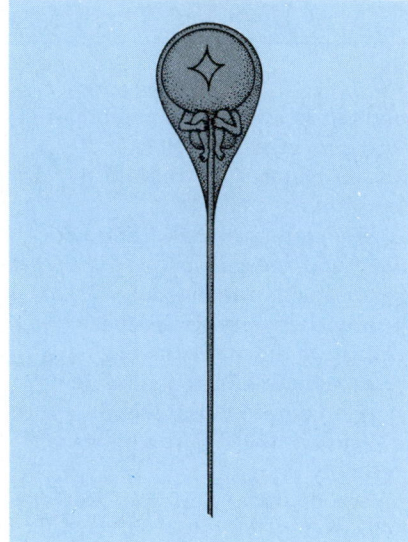

FIGURE 43–1 The preformed "little man" within the sperm as visualized by 17th-century scientists. (After Hartseeker's drawing from "Essay de Dioptrique," Paris, 1694)

theory was seriously debated by scientists and philosophers.

An opposing view, the **theory of epigenesis,** gained experimental support as better techniques for investigation were developed. This theory held that the embryo develops from a formless zygote and that the structures of the body take shape in an orderly sequence, developing their characteristic forms only as they emerge.

Today we know that development is largely epigenetic. No microscopic organism waits preformed in either gamete. Development proceeds from one cell to trillions, from a formless mass of cells to an intricate, highly specialized and organized animal. However, a spark of truth can be found in the preformationist view. Although the "little man" (or woman) itself is not to be found within the zygote, its blueprint *is* there, precisely encoded in the form of chemical specifications within the deoxyribonucleic acid (DNA) of the genes.

KEY CONCEPTS

☐ Embryonic development involves division of the fertilized egg into many cells, movement of these cells to form specific structures, growth, and differentiation of cells to form specialized tissues and organs.

☐ Early development proceeds through the following stages: zygote, cleavage, morula, blastula, gastrula, organogenesis.

☐ Development is regulated by the interaction of genes with cytoplasmic factors, inducers, and environmental factors.

■ DEVELOPMENT IS A BALANCED COMBINATION OF SEVERAL PROCESSES

How does a single-celled, microscopic, unspecialized zygote (fertilized egg) give rise to the blood, bones, brain, and all the other structures of a complex animal? As we will see, development is a balanced combination of several processes: mitosis and cytokinesis, growth, morphogenesis (the cell movements that result in form), and cellular differentiation.

The single-celled zygote undergoes mitosis and cytokinesis, forming two cells, and then each of these cells divides, giving rise to four cells. This process is repeated again and again, producing the trillions of cells of the adult animal. Growth occurs by both an increase in the number of cells and in the size of these cells. An orderly pattern of cellular proliferation and growth provides the cellular building blocks of the organism and results in size increase. But these processes alone would produce only a formless heap of cells.

Cells must arrange themselves into specific structures and appropriate body forms. The precise and complicated cellular movements that bring about the form of a multicellular animal with its intricate pattern of tissues and organs are termed **morphogenesis.**

Not only must cells be arranged into specific structures, but they must also perform varied and specialized functions. Cells are specialized as well as organized. In order to function in different fashions, body structures must be made up of different components. During early development, cells begin to become different from one another, becoming biochemically and structurally specialized and performing specific tasks. More than 200 distinct types of cells can be found in the adult vertebrate body. The process by which cells become specialized is known as **cellular differentiation.**

As you read the following sections on development, bear in mind that mitosis and cytokinesis,

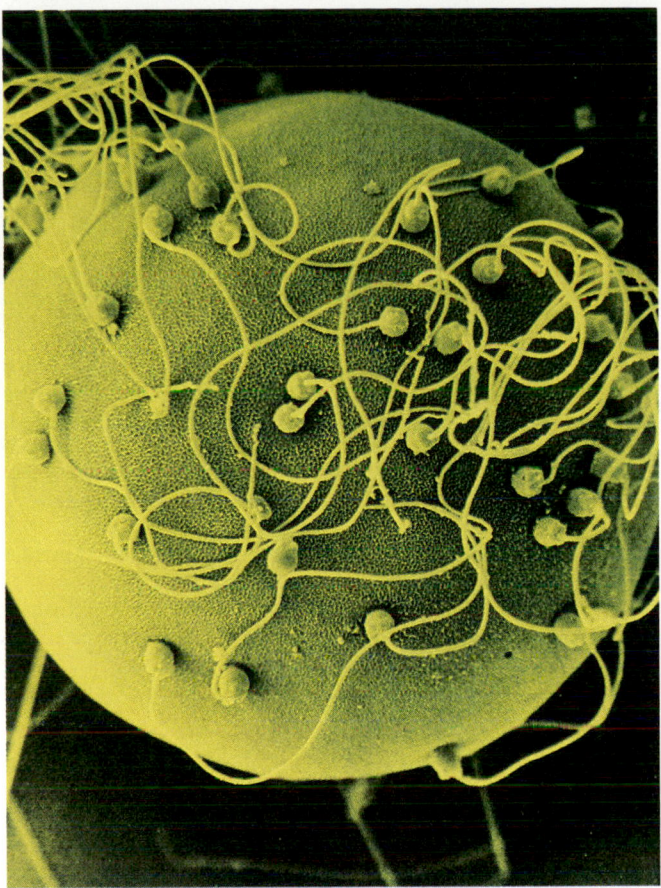

FIGURE 43–2   Sperm swarm around an egg of a surf clam in this scanning electron micrograph (approximately ×2100). (David M. Phillips/Visuals Unlimited)

growth, morphogenesis, and cellular differentiation are intimately interrelated. The pattern of early development is basically similar for all animals.

■ FERTILIZATION RESTORES THE DIPLOID NUMBER OF CHROMOSOMES

Fertilization is the union of a sperm and an ovum to produce a zygote, or fertilized egg (Fig. 43–2). Fertilization serves three functions: (1) The diploid number of chromosomes is restored as the sperm contributes its haploid set of chromosomes to the haploid set in the ovum; (2) in mammals and many other animals, sex of the offspring is determined; (3) the needed stimulation is provided to initiate the reactions in the egg that permit development to take place.

Fertilization involves four steps. First, the sperm contacts the egg and recognition occurs. Second, the sperm enters the egg. Third, the sperm and egg nuclei fuse. Finally, the egg is activated and development begins.

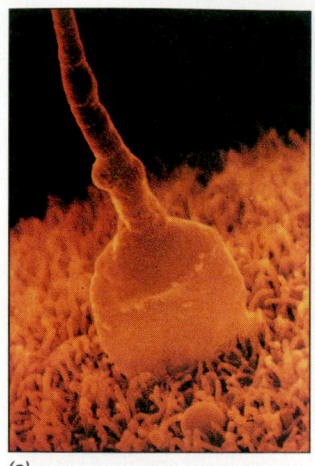 

(a) (b)

FIGURE 43–3 Fertilization. (*a*) **When the first sperm makes contact with the egg's plasma membrane, microvilli elongate and surround the head of the sperm. This fertilization cone draws the sperm into the cytoplasm of the egg. (*b*) This computer reconstruction shows the fusion of sperm and egg cell membranes that marks the instant of fertilization.**
(*a*, David M. Phillips/Visuals Unlimited; *b*, Courtesy of Drs. Gerald and Heide Schatten)

The First Steps In Fertilization Are Contact And Recognition

The egg is surrounded by a very thin **vitelline membrane** and, outside of this, by a thick glycoprotein layer called the **jelly coat** (zona pellucida in mammals). In a few species (e.g., some cnidarians) the egg (or a surrounding structure) secretes a substance that attracts sperm of the same species. However, in most species chemical attraction of sperm by the egg has not been shown, and the gametes meet largely by chance.

The activation of sperm by the egg, the **acrosome reaction,** has been studied most in marine invertebrates. When the sperm contacts the jelly coat surrounding the egg, the acrosome, a structure at the head of the sperm, releases enzymes that digest a path through the jelly coat to the vitelline membrane of the egg. Actin molecules join, forming actin filaments, which permit the acrosome to extend outward and contact the vitelline membrane of the egg. Recognition that the sperm and egg are of the same species occurs at this time.

If the species are the same, a species-specific protein known as **bindin,** located on the acrosome, adheres to a species-specific bindin receptor located on the vitelline membrane.

Sperm Entry Is Regulated

Once acrosome-vitelline fusion occurs, enzymes dissolve a bit of the vitelline membrane in the area of the sperm head. The plasma membrane of the egg is covered with microvilli. Several microvilli elongate to surround the head of the sperm, forming a **fertilization cone.** The sperm is then drawn into the egg by contraction of the fertilization cone. As this occurs, the plasma membranes of sperm and egg fuse (Fig. 43–3).

As soon as one sperm enters the egg two reactions occur that prevent additional sperm from entering. Within a few seconds the so-called fast block to polyspermy occurs. This incomplete block involves an electrical change in the plasma membrane of the egg. Ion channels in the plasma membrane open, permitting sodium ions to pass into the cell. The depolarization that occurs prevents other sperm from fusing with the plasma membrane.

A second mechanism preventing entrance of more than one sperm, referred to as the slow block to polyspermy, is the **cortical reaction.** Depolarization of the egg plasma membrane results in calcium ion release from thousands of cortical granules present beneath the membrane. The cortical granules then release enzymes into the region between the plasma membrane and vitelline membrane. Some of these enzymes dissolve the protein that links the vitelline membrane to the plasma membrane. Other substances released by the cortical granules result in osmotic passage of water into the space between the plasma membrane and vitelline membrane. This elevates the vitelline membrane, and it then becomes the fertilization membrane, a hardened membrane that prevents entry of sperm. This slow block requires one to several minutes, but it is a complete block. (In mammals, a fertilization membrane does not form, but the enzymes released alter the sperm receptors in the zona pellucida so that no additional sperm can bind to them.) In some species polyspermy does occur, but only the first sperm that enters fertilizes the egg.

Sperm And Egg Pronuclei Fuse

After the sperm nucleus enters the egg through the fertilization cone, it is thought to be guided toward the egg nucleus by a system of microtubules that form within the egg. The sperm nucleus swells, forming the male pronucleus, and the nucleus of the ovum becomes the female pronucleus. The haploid pronuclei then fuse, forming the diploid nucleus of the zygote.

Fertilization Activates The Egg

Release of calcium ions into the egg cytoplasm is necessary for the cortical reaction and also triggers metabolic changes within the cell. Oxygen use by the cell increases as certain compounds within the egg cell are oxidized. The pH within the cell increases, bringing about many metabolic changes. Within a few minutes after sperm entry there is a burst of protein synthesis.

In some species sperm penetration initiates rearrangement of the cytoplasm. For example, in the amphibian egg some of the superficial cytoplasm containing dark granules shifts, exposing the underlying lighter-colored cytoplasm. This cytoplasm that appears gray is called the gray crescent; it marks the region where gastrulation, an early developmental process, begins in the amphibian embryo.

In some species the egg can be artificially activated without sperm penetration by swabbing it with blood and pricking the plasma membrane with a needle, by calcium injection, or by diverse other treatments. These eggs develop into embryos parthenogenetically (that is, without being fertilized).

THE ZYGOTE CONTAINS INSTRUCTIONS FOR PRODUCING A COMPLETE INDIVIDUAL

Although it appears to be a relatively simple cell, the **zygote,** or fertilized egg, contains all of the genetic information necessary to produce a complete individual. This simple cell has the potential to give rise to all the diverse cell types of the mature animal. The sperm cell is quite tiny in comparison with the egg, so the bulk of the zygote cytoplasm comes from the ovum. However, the zygote nucleus contains the chromosomes contributed equally by both sperm and egg.

The eggs of many kinds of animals contain yolk—a metabolically inert mixture of proteins, phospholipids, and fats—that serves as food for the developing embryo. The amount and distribution of yolk vary among different animal groups. Whereas bird eggs contain great quantities of yolk, those of mammals and some invertebrates have only small amounts. In eggs with large amounts of yolk, it is concentrated at one pole (end) of the cell. The pole of the cell that contains the largest amount of yolk is known as the **vegetal pole;** the opposite, more metabolically active pole is the **animal pole.** In yolk-poor eggs the yolk is uniformly dispersed throughout the cytoplasm; such eggs are described as **isolecithal.** However, using other criteria, animal and vegetal poles can be distinguished even in yolk-poor eggs.

DURING CLEAVAGE THE ZYGOTE DIVIDES, FORMING MANY CELLS

Shortly after fertilization the zygote undergoes a series of rapid cell divisions collectively referred to as **cleavage.** First, the zygote undergoes mitosis and divides to

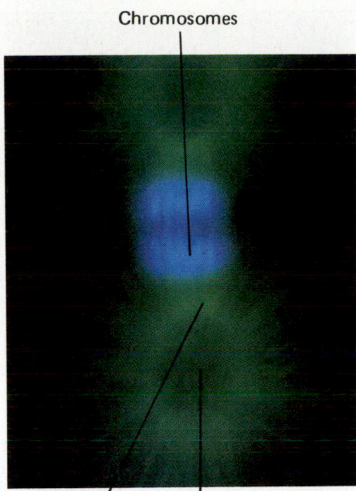

FIGURE 43–4   The first mitotic division of the new organism. The centrioles are derived from the sperm, but the microtubules (in green) of the mitotic spindle are part of the egg's dowry. The brilliantly blue-fluorescing chromosomes were provided by both. (Courtesy of Drs. Gerald and Heide Schatten)

form a two-cell embryo (Fig. 43–4). Then, each of these cells undergoes mitosis and divides, bringing the number of cells to four (Fig. 43–5). Repeated divisions continue to increase the number of cells, or **blastomeres,** making up the embryo. At about the 32-cell stage, the embryo is referred to as a **morula.** The cells of the morula continue to multiply, eventually forming a hollow ball of several hundred cells, the **blastula.** The cells of the blastula surround a fluid-filled cavity, the **blastocoel.**

Cleavage Provides Building Blocks For Development

During cleavage the cells do not grow in size, so the mass of cells produced is not larger than the original zygote. The principal effect of early cleavage, then, is that the zygote is partitioned into many small cells that serve as basic building units. Their small size allows the cells to move about with relative ease, arranging themselves into patterns. Each cell moves by amoeboid motion, probably guided by the proteins of its own cell coat and those of other cell surfaces. Specific properties conferred upon plasma membranes by these surface proteins are important in helping cells to "recognize" one another and therefore in determining which ones will adhere to form a tissue.

The Amount Of Yolk Determines The Pattern Of Cleavage

The amount of yolk in the egg affects the pattern of cleavage. In the isolecithal eggs characteristic of most

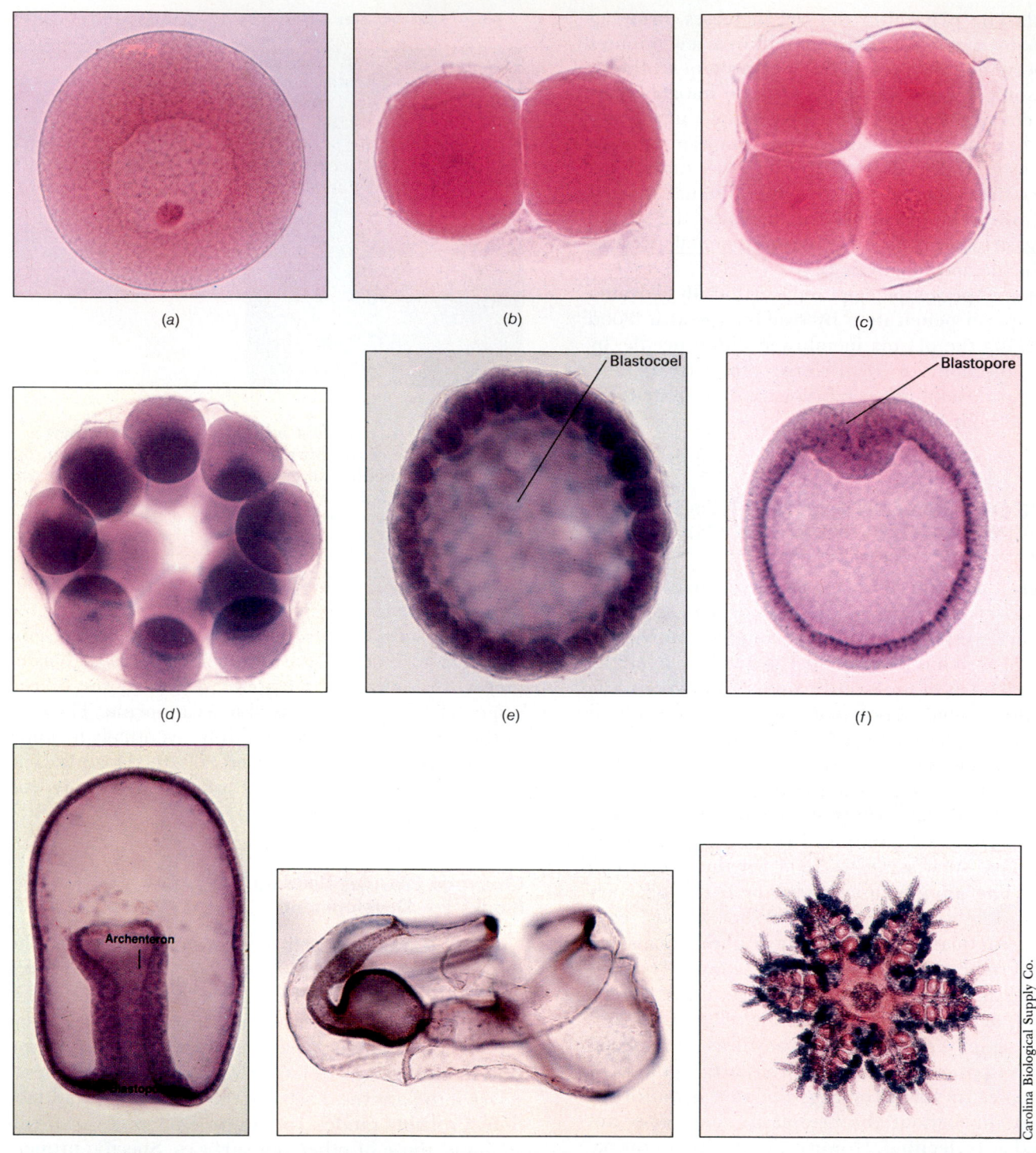

FIGURE 43–5   Development of a seastar. (*a*) Unfertilized seastar egg.
(*b*) Two-cell stage. (*c*) Top view of four-cell stage. (*d*) Sixteen-cell stage.
(*e*) Cross section through 64-cell blastula. (*f*) Section through early gastrula.
(*g*) Section through middle gastrula. (*h*) Seastar larva. (*i*) Young seastar. In this
type of cleavage the entire egg becomes partitioned into cells. The blastopore
is the opening into the inner cavity, the archenteron. All views are side views
with the animal pole at the top except (*c*) and (*i*), which are top views.

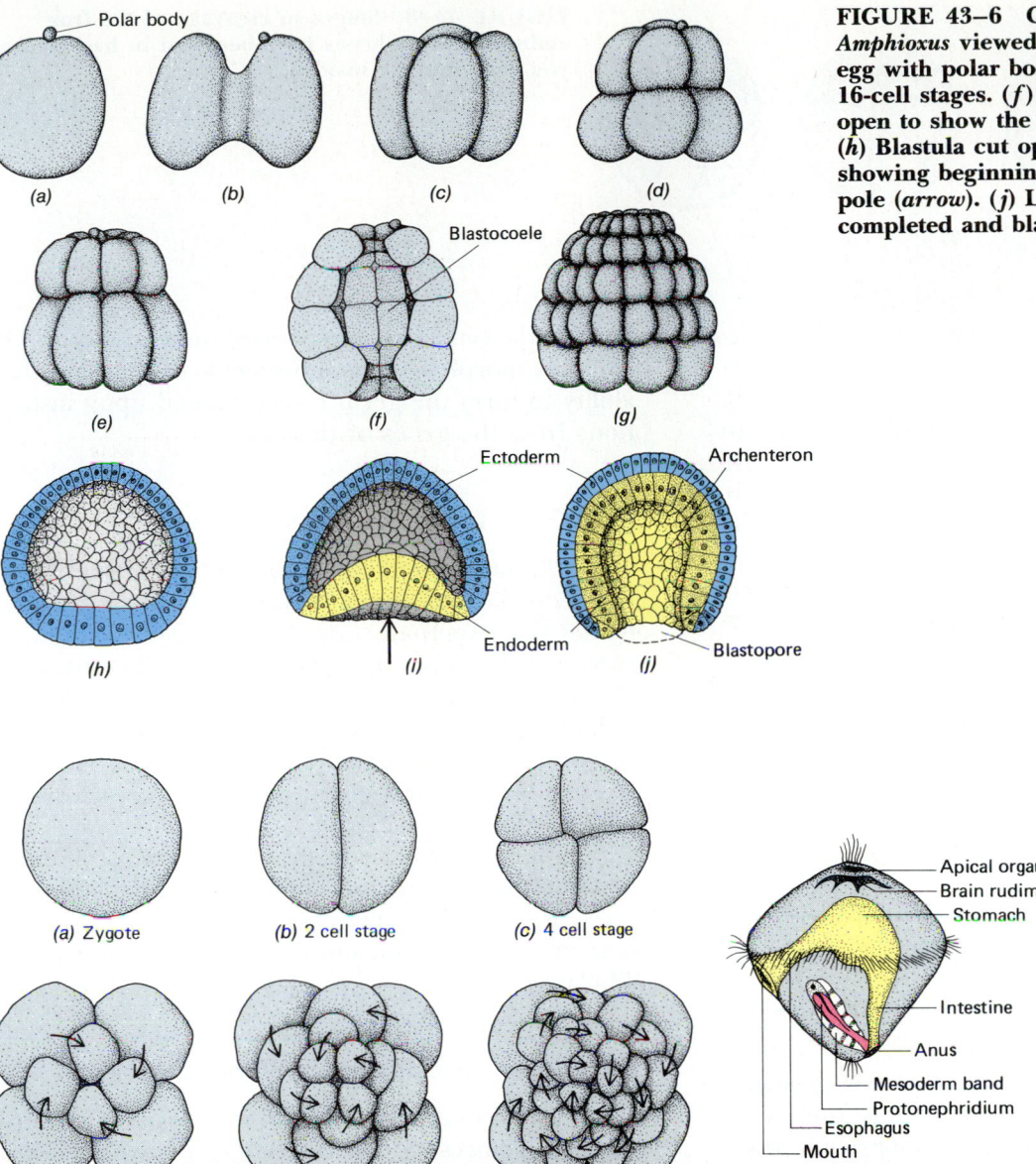

Polar body

(a) (b) (c) (d)

Blastocoele

(e) (f) (g)

Ectoderm   Archenteron

Endoderm   Blastopore

(h) (i) (j)

FIGURE 43–6 Cleavage and gastrulation in *Amphioxus* viewed from the side. (*a*) Mature egg with polar body. (*b, e*) Two-, 4-, 8-, and 16-cell stages. (*f*) Embryo at 32-cell stage cut open to show the blastocoele. (*g*) Blastula. (*h*) Blastula cut open. (*i*) Early gastrula showing beginning of invagination at vegetal pole (*arrow*). (*j*) Late gastrula. Invagination is completed and blastopore has formed.

(a) Zygote   (b) 2 cell stage   (c) 4 cell stage

(d) 8 cell stage   (e) 16 cell stage   (f) 32 cell stage

Apical organ
Brain rudiment
Stomach

Intestine

Anus

Mesoderm band
Protonephridium
Esophagus
Mouth

(g) Trochophore larva

FIGURE 43–7 Development in annelids. (*a*) through (*f*) are top views of the animal pole. The successive cleavage divisions occur in a spiral pattern as indicated. (*g*) A typical trochophore larva. The upper half of the trochophore develops into the extreme anterior end of the adult worm; the rest of the adult body develops from the lower half.

invertebrates and simple chordates there are relatively small amounts of uniformly distributed yolk. During cleavage the entire egg divides, producing cells that are roughly the same size. (This type of cleavage is described as holoblastic.)

Cleavage of isolecithal eggs can be radial or spiral. Radial cleavage is characteristic of deuterostomes (echinoderms and chordates; animals in which the anus develops from the blastopore; see Chapter 24). In radial cleavage, the first division passes through both animal and vegetal poles and splits the egg into two equal cells. The second cleavage division passes through both poles of the egg at right angles to the first and separates the two cells into four equal cells. The third division is horizontal, at right angles to the other two, and separates the four blastomeres into eight—four above and four below the line of cleavage. Radial cleavage is illustrated in the sea star and in *Amphioxus* (Figs. 43–5, 43–6).

Some protostomes (animals in which the mouth develops from the blastopore, e.g., annelids) have a pattern of early cell division known as **spiral cleavage** (Fig. 43–7). In spiral cleavage, the early cell divisions after the first two are oblique to the polar axis, result-

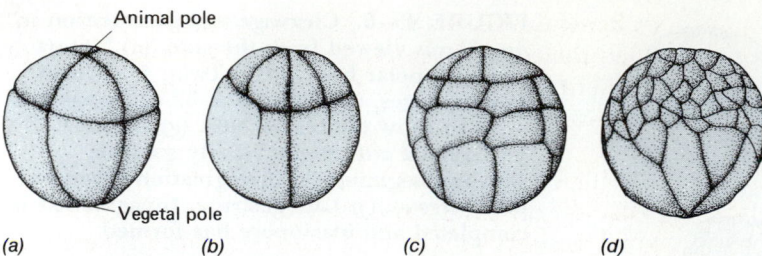

Animal pole

Vegetal pole

(a)　　　(b)　　　(c)　　　(d)

FIGURE 43–8 Stages in cleavage of the frog embryo. The embryos have been cut in half so that you can view the insides.

ing in a spiral arrangement of cells; each cell is located between the two cells below it.

The telolecithal eggs of bony fishes and amphibians contain a large amount of yolk concentrated toward the vegetal pole. The cleavage divisions in the vegetal hemisphere are slowed by the presence of the inert yolk. As a result the blastula consists of many small cells in the animal hemisphere and fewer, but larger, cells in the vegetal hemisphere (Fig. 43–8). For this reason the lower wall is much thicker than the upper one and the blastocoel is displaced upward.

The telolecithal eggs of reptiles, birds, and some fishes have a very large amount of yolk. The small amount of cytoplasm is concentrated at the animal pole. In such eggs, cell division takes place only in the **blastodisc,** the small disc of cytoplasm at the animal pole. (This type of cleavage is meroblastic.)

☐ THE GERM LAYERS FORM DURING GASTRULATION

The process by which the blastocyst becomes a three-layered embryo, called a **gastrula,** is termed **gastrulation.** During gastrulation, the cells arrange themselves into three distinct **germ layers,** or embryonic tissue layers—the ectoderm, mesoderm, and endoderm. The complex morphogenetic movements of cells and their ability to form the germ layers depend upon instructions from the genes of the embryo.

Each Germ Layer Has A Specific Fate

The cells lining the cavity of the embryo (archenteron) make up the **endoderm.** The endoderm gives rise to tissues that eventually line the digestive tract and organs that develop as outgrowths of the digestive tract (including the liver, pancreas, and lungs). The outer wall of the gastrula consists of the germ layer known as **ectoderm.** The ectoderm eventually forms the outer layer of the skin and gives rise to the nervous system and sense organs.

A third layer of cells, the **mesoderm,** proliferates between the ectoderm and endoderm. The mesoderm gives rise to the skeletal tissue, muscle, circulatory system, excretory system, and reproductive system (Table 43–1).

The Pattern Of Gastrulation Varies Somewhat

Gastrulation in the sea star and in *Amphioxus* are illustrated in Figures 43–5 and 43–6, respectively. Gastru-

TABLE 43–1 **Fate of the Germ Layers**		
GASTRULA	**Ectoderm**	Nervous system and sense organs Outer layer of skin (epidermis) and its associated structures (nails, hair, etc.) Pituitary gland
	Mesoderm	Skeleton (bone and cartilage) Muscles Circulatory system Excretory system Reproductive system Inner layer of skin (dermis) Outer layers of digestive tube and of structures that develop from it, such as respiratory system
	Endoderm	Lining of digestive tube and of structures that develop from it, such as respiratory system

lation begins when the blastoderm at the vegetal pole flattens and then bends inward. The cells of a section of the vegetal wall of the blastula move inward in a process referred to as invagination. The invaginated wall eventually meets the opposite wall, obliterating the original blastocoel. We can roughly demonstrate this process by pushing inward on the wall of a deflated rubber ball until it rests against the opposite wall. In a similar way the embryo is converted into a double-walled, cup-shaped structure. The cavity of the cup communicates with the exterior on the side that was originally the vegetal pole of the embryo. The internal wall lines the newly formed cavity, the **archenteron** (primitive gut). The opening of the archenteron to the exterior is the **blastopore,** which becomes the anus.

In the amphibian, the large yolk-laden cells in the vegetal half of the blastula obstruct the inward movement at the vegetal pole. Instead, cells from the animal pole move down toward the yolk-rich cells and then inward and away from the yolk-rich cells, forming the dorsal lip of the blastopore (Fig. 43–9). As the process continues, the blastopore becomes ring-shaped as cells lateral, and then ventral, to the blastopore become involved in the same movements. The yolk-filled cells of the vegetal hemisphere remain as a yolk plug filling the space enclosed by the lips of the blastopore. The rim of the blastopore continues to contract and eventually completely covers the yolk plug. The archenteron forms as a cavity leading from the groove on the surface of the embryo into the interior and is lined on all sides by cells that have involuted from the surface. At first, the archenteron is a narrow slit, but it gradually expands at the anterior end, encroaching on the blastocoel, which is eventually obliterated. Although the details differ somewhat, gastrulation in the bird is basically similar to amphibian gastrulation.

ORGANOGENESIS BEGINS WITH THE DEVELOPMENT OF THE NERVOUS SYSTEM

The process of organ formation is called **organogenesis.** In the vertebrate embryo, the brain and spinal cord are among the first organs to develop (Fig. 43–10). First, the notochord, the flexible skeletal axis in all chordate embryos, grows forward along the length of the embryo as a cylindrical rod of cells. (Later, the notochord will be replaced by the vertebral column.) The developing notochord induces (stimulates) the overlying ectoderm to thicken, forming the **neural plate.** Central cells of the neural plate move downward, forming a depression, the **neural groove;** the cells flanking the groove on each side form **neural folds.** Continued movements of their cells bring the folds closer together until they meet and fuse, forming the **neural tube.** In this process, the neural tube comes to

lie beneath the surface. The ectoderm overlying it will form the outer layer of skin. The anterior portion of the neural tube grows and differentiates into the brain; posterior to the brain, the tube develops into the spinal cord.

The anterior part of the neural tube becomes much larger than the posterior part and continues to grow rapidly. At the same time the neural tube bends down at the anterior end of the embryonic disc. The forebrain, midbrain, and hindbrain differentiate. Then, the forebrain begins to grow outward on either side, forming the beginnings of the cerebral hemispheres.

The various motor nerves grow out of the developing brain and spinal cord, but the sensory nerves have a separate origin. When the neural folds fuse, forming the neural tube, bits of nervous tissue known as the **neural crest** arise from the approximate region of neural fold fusion on each side of the tube. These crest cells migrate downward from their original position and form the dorsal root ganglia of the spinal nerves and the postganglionic sympathetic neurons. From sensory cells in the dorsal root ganglia, dendrites grow out to the sense organs and axons grow in to the spinal cord. Other structures that develop from neural crest cells are the adrenal medulla of each adrenal gland, parts of certain sense organs, teeth, nearly all pigment-forming cells in the body, and the cranial bones that encase the brain.

DEVELOPMENTAL PROCESSES ARE CAREFULLY REGULATED

One of the important unsolved problems of modern biology is the nature of the mechanisms that regulate developmental processes. How does each organ emerge at the appropriate time and in the proper spatial relations to other structures? How can the zygote, a single cell, give rise to the many different types of cells that make up the adult animal—cells that differ so widely in their structure, functions, and chemical properties? In Chapter 12, we examined some factors that appear to regulate gene expression. Even though all the cells of an animal have an identical set of genes with the same instructions, differential gene activity occurs in the cells of different tissues so that different proteins are manufactured in different types of cells. Factors outside the nucleus can also influence development.

Cytoplasmic Factors Influence The Course Of Development

The initial influence upon differentiation is the distribution of materials in the cytoplasm of the zygote itself. Because the zygote cytoplasm is not homogeneous, the

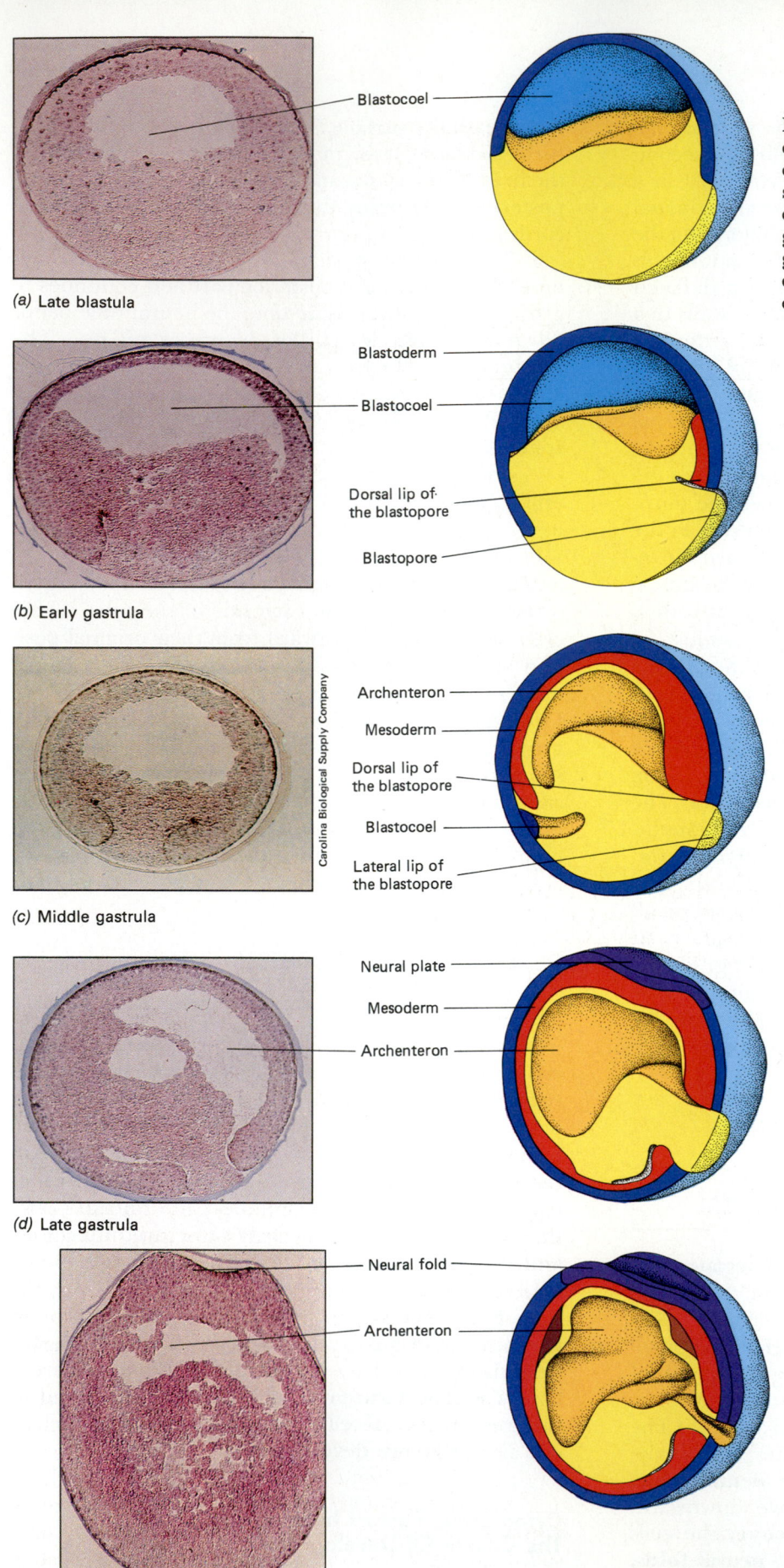

(a) Late blastula

Blastocoel

(b) Early gastrula

Blastoderm

Blastocoel

Dorsal lip of the blastopore

Blastopore

Carolina Biological Supply Company

(c) Middle gastrula

Archenteron

Mesoderm

Dorsal lip of the blastopore

Blastocoel

Lateral lip of the blastopore

(d) Late gastrula

Neural plate

Mesoderm

Archenteron

(e) Early development of the nervous system

Neural fold

Archenteron

FIGURE 43–9 Stages in the development of a frog embryo. The diagrams show the embryo cut in half so you can view the insides of the embryo. (*a*) Late blastula. (*b*) Early gastrula. (*c*) Middle gastrula. (*d*) Late gastrula. (*e*) Nervous system development begins with the formation of the neural plate.

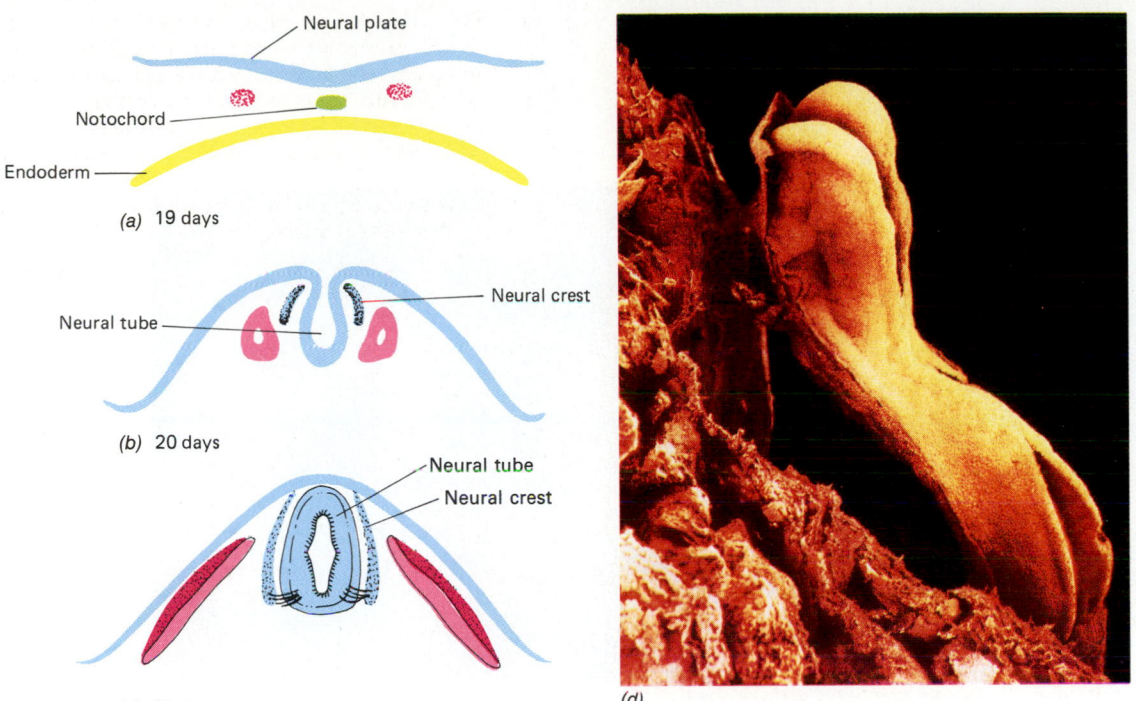

Neural plate

Notochord

Endoderm

(a) 19 days

Neural tube

Neural crest

(b) 20 days

Neural tube

Neural crest

(c) 26 days

(d)

FIGURE 43–10   Cross sections of human embryos at successively later stages, illustrating the early development of the nervous system. The neural crest cells form the doral root ganglion and the postganglionic sympathetic neurons. (a) Approximately 19 days. The neural plate has indented to form a shallow groove flanked by neural folds so that it forms a shallow groove. (b) Approximately 20 days. The neural folds approach one another and upon fusing together will form the neural tube. (c) Approximately 26 days. The neural tube has now formed and will give rise to the brain at the anterior end of the embryo and the spinal cord posteriorly. (d) Photograph of a 20-day-old human embryo shows the developing nervous system. (*d*, Lennart Nilsson, © Boehringer Ingelheim Internat'l GmbH, from *The Incredible Machine*, p. 25)

cytoplasm portioned out to each new cell during division may be different. Such differences have been shown to influence the course of development (Fig. 43–11).

Cells Can Influence Each Other With Inducers

The position of a cell in relation to its neighbors may be critical in determining its fate. One cell, or a group of cells, may release a substance that diffuses out among neighboring cells. This chemical will be most concentrated near the cell that produced it and less concentrated farther away. Such a concentration gradient could provide information to the cells regarding their position in relation to the group. This information could trigger the biochemical events that would determine differentiation.

A chemical that determines the differentiation of nearby cell groups is called an **inducer.** Let us look at a specific example, the development of the nervous system. In the early 1920s, the biologist Hans Spemann performed a series of experiments that helped to explain why ectoderm cells in a specific region of the embryo develop into nerve tissue, while other ectoderm cells give rise to the outer layer of the skin.

Using early frog gastrulas, Spemann transplanted a tiny bit of ectoderm from the region that would normally develop into neural tube to the belly region of the embryo. The transplanted pre-neural tube ectoderm no longer formed neural tissue; instead, it joined with its new cellular neighbors, forming outer skin. Likewise, when belly ectoderm was transplanted to the pre-neural tube region, its fate was determined by its new surroundings—it contributed to the formation of neural tube.

When these same experiments were performed on older gastrulas a few hours later in development, the results were entirely different. Then, when a piece of pre-neural tube tissue was transplanted into the belly region, it differentiated into neural tissue, even though this tissue is out of place in the midst of the belly skin. And, as might be expected, transplanted belly ectoderm cells proceeded to form skin cells even in the midst of the developing neural tube.

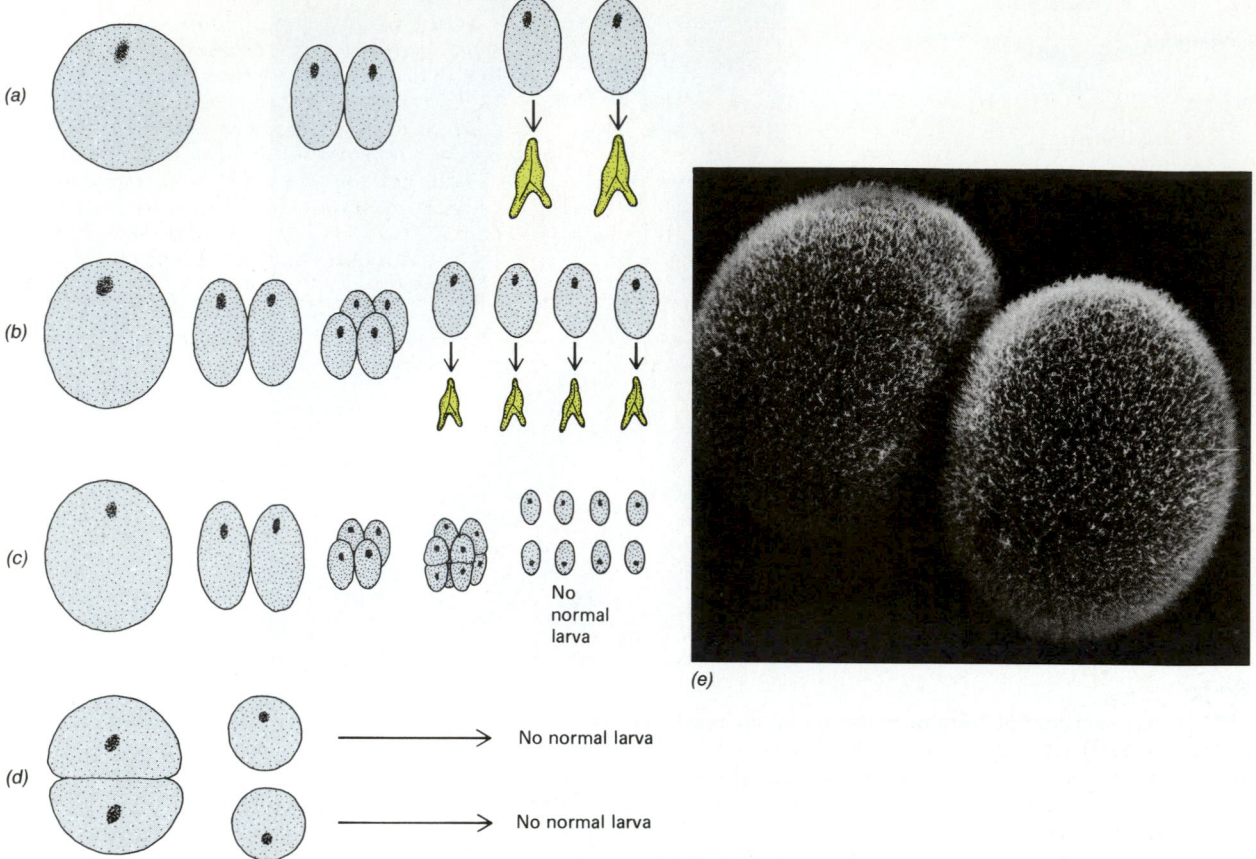

FIGURE 43–11 Experiments that illustrate the influence of cytoplasmic factors upon development. (*a*) When the first two cells of a sea urchin embryo are separated from one another, each cell is able to give rise to a normal larva. (*b*) If each of the first four cells of the sea urchin embryo are separated from one another, four normal larvae develop. (*c*) When the cells of the eight-cell stage are separated from one another, not even one is able to develop into a normal larva. (*d*) If an experimenter causes the plane of cleavage of the first division to extend horizontally, instead of from pole to pole of the zygote, neither cell of the two-cell stage is able to develop into a normal larva. Apparently, certain cytoplasmic substances necessary for normal development are heterogeneously distributed in the zygote so that neither the animal nor the vegetal half has all of the needed ingredients. (*e*) Scanning electron micrograph of the first cleavage division of the egg of a sea urchin (×1300). (*e*, Courtesy of Dr. Everett Anderson)

Apparently something happened between the early and late gastrula stages that accounted for the difference in results. Spemann solved this mystery by showing that the notochord cells lying just beneath the pre-neural tube ectoderm are responsible for the change. The notochord cells induce the ectoderm cells to form neural tube. By producing and secreting a chemical substance, the notochord cells communicate with the ectoderm cells, stimulating them to differentiate into neural tissue.

In the early gastrula the pre-neural tube cells have not yet been subjected to the inducer, so their fate has not yet been determined. Once these cells have been influenced by the inducer later in gastrulation, their fate is decided and cannot easily be altered. When a piece of notochord tissue is transplanted beneath the

belly ectoderm of an early gastrula, not only a second neural tube, but often an almost complete secondary embryo forms in this region. These and thousands of similar experiments have provided evidence that development depends upon chemical communication among cells and the resulting differential activity of their genes.

Nongenetic And Genetic Factors Interact

Although nongenetic influences are important in determining differentiation, the genes exercise ultimate control. Consider a specific experimental example. When ectoderm from the mouth region of an early frog embryo is transplanted to the mouth region of a salamander, the surrounding salamander tissue in-

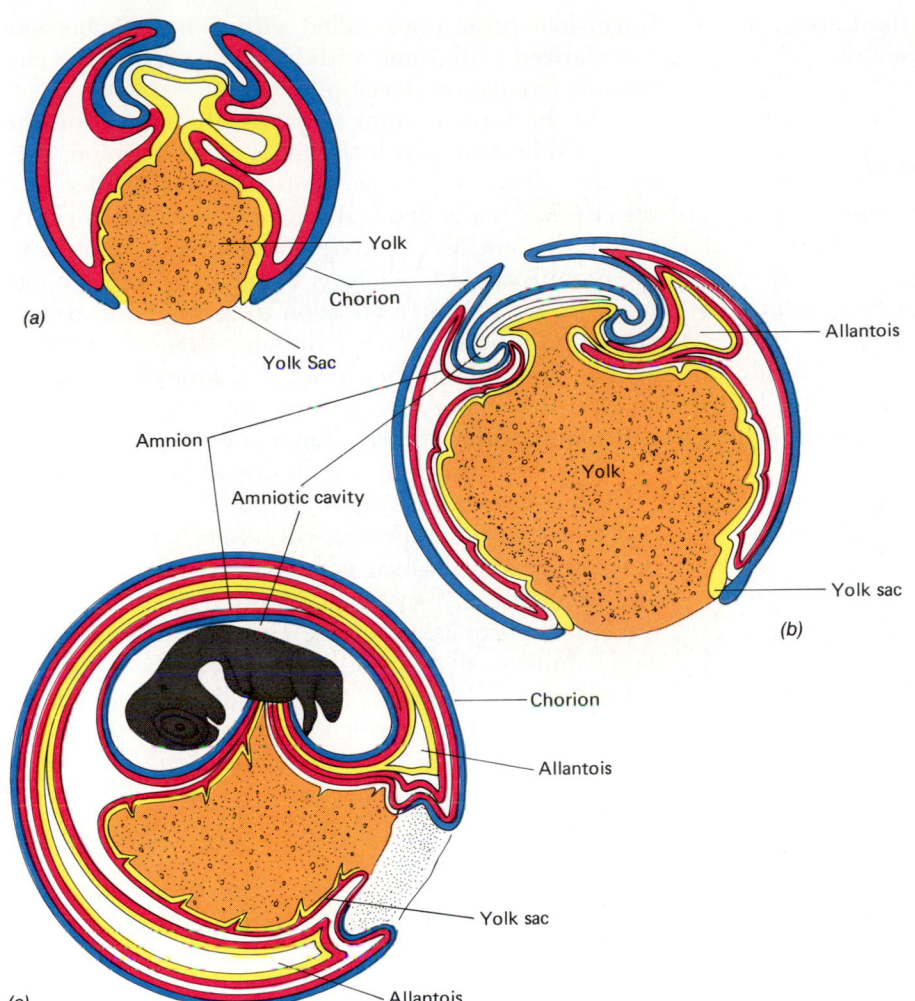

(a)

Yolk

Chorion

Yolk Sac

Amnion

Amniotic cavity

(b)

Allantois

Yolk

Yolk sac

(c)

Chorion

Allantois

Yolk sac

Allantois

FIGURE 43–12 Development of extraembryonic membranes. (*a, b, c*) Successive stages in the development of the extraembryonic membranes of the chick. Each of the membranes develops from a combination of two germ layers. The chorion and amnion are formed from lateral folds of the ectoderm and mesoderm that extend over the embryo and fuse. The allantois develops from an outpocketing of the hindgut. Consisting of endoderm and mesoderm, the allantois is an elongated sac. The yolk sac also consists of endoderm and mesoderm.

duces the transplanted cells to develop into mouth structures. However, the mouth that forms is not the characteristic mouth of a salamander. Instead it bears the horny rows of teeth and horny jaws of a frog. Why? As you might guess, the frog cells are not *competent* to form structures specific to salamanders.

Cellular differentiation is an expression of changes in the activity of specific genes, and genetic activity in turn is influenced by a variety of factors within and outside the cell. Although all cells of an organism possess the same set of genes, differential gene activity results in variations in chemistry, behavior, and structure among cells. Through this process, an embryo comes to be composed of more than 200 types of cells, each exquisitely designed to perform specific functions.

■ EXTRAEMBRYONIC MEMBRANES AND PLACENTA PROTECT AND NOURISH THE EMBRYO

All terrestrial vertebrates have four **extraembryonic membranes:** the chorion, allantois, yolk sac, and amnion (Fig. 43–12). Although they develop from the

germ layers, these membranes are not part of the embryo proper and are discarded at birth. The extraembryonic membranes are adaptations to the challenges of embryonic development on land. During development they protect the embryo, prevent it from drying out, and help in obtaining food and oxygen and eliminating wastes.

The Chorion And Amnion Enclose The Embryo

The **chorion** and **amnion** develop together as folds of the body wall. These membranes grow around the embryo, meeting and fusing above it. Eventually the chorion and amnion enclose the entire embryo (Fig. 43–12). In the eggs of reptiles and birds, the chorion remains in contact with the inner surface of the shell. In mammals it lies next to the cells of the uterine wall.

The space between the embryo and the amnion, known as the amniotic cavity, becomes filled with clear, watery amniotic fluid secreted by the membrane. Embryos of terrestrial vertebrates develop within this pool of fluid. The amniotic fluid prevents desiccation (drying) of the embryo and acts as a protective cushion that absorbs shocks and prevents the amniotic membrane from sticking to the embryo. Amniotic fluid also permits the embryo a certain freedom of motion.

The Allantois Functions In Waste Disposal

The **allantois** is an outgrowth of the developing digestive tract. In reptiles and birds, it serves as a depot for nitrogenous wastes and so becomes quite large. The products of nitrogen metabolism are excreted as uric acid by the kidney of the developing embryo. The poorly soluble uric acid is deposited as crystals in the cavity of the allantois and is discarded along with the allantois when the young animal hatches out of the egg shell.

The allantois fuses with the chorion to form the **chorioallantoic membrane,** which is rich in blood vessels. In the chick embryo, blood in these vessels provides oxygen and receives carbon dioxide and other wastes from the embryo. Gases are exchanged through the shell. In the human, the allantois is small and nonfunctional, except that its blood vessels contribute to the formation of umbilical vessels joining the embryo to the placenta. When the chick hatches or the child is born, most of the allantois, like the other extraembryonic membranes, is discarded. However, the base of the allantois, the portion originally connected to the digestive tract, is converted into part of the urinary bladder.

The Yolk Sac Encloses The Yolk

Like the allantois, the **yolk sac** forms as an outpocketing of the developing digestive tract. In vertebrates with yolk-rich eggs, the yolk sac encloses the yolk, slowly digests it, and makes it available to the embryo. Even in vertebrate embryos with little or no yolk, a yolk sac forms and is evident between the second and sixth weeks of development. Its walls serve as temporary centers for the formation of blood cells.

The Placenta Is An Organ Of Exchange

In humans and other placental mammals, the placenta is the organ of exchange between mother and fetus, providing nutrients and oxygen for the fetus, and removing its wastes, which the mother then excretes (Fig. 43–15). In addition, the placenta is an endocrine organ which secretes hormones that maintain pregnancy.

The placenta develops from both the chorion of the embryo and the uterine tissue of the mother. In early development the chorion grows rapidly, invading the endometrium (lining of the uterus) and forming finger-like projections called **villi.** The villi become vascularized (infiltrated with blood vessels) as the embryonic circulation develops.

As the human embryo grows, the region on the ventral side from which the folds of the amnion, yolk sac, and allantois grew becomes relatively smaller, and the edges of the amniotic folds come together, forming a tube that encloses the other membranes. This tube, the **umbilical cord,** connects the embryo with the placenta (Fig. 43–15). In addition to the yolk sac and allantois, the umbilical cord contains the two umbilical arteries and the umbilical vein. The **umbilical arteries** connect the embryo with a vast network of capillaries developing within the villi; blood from the villi returns to the embryo through the **umbilical vein.**

The placenta consists of the portion of the chorion that develops villi, together with the uterine tissue underlying the villi, which contains maternal capillaries and small pools of maternal blood. The blood of the fetus in the capillaries of the chorionic villi comes in close contact with the mother's blood in the tissues between the villi. However, they are always separated by a membrane through which substances may diffuse or be actively transported. *Maternal and fetal blood do not normally mix in the placenta or any other place.* What problems might arise if the blood did mix?

Several hormones are produced by the placenta. From the time the embryo first begins to implant itself, its trophoblastic cells release **human chorionic gonadotropin (hCG),** which signals the corpus luteum that pregnancy has begun. In response, the corpus luteum increases in size and releases large amounts of progesterone and estrogen, which in turn stimulate continued development of endometrium and placenta. Without hCG, the corpus luteum would degenerate and the embryo would be aborted and flushed out with the menstrual flow. In such a case the woman would probably not even know that she had been pregnant. When the corpus luteum is removed before about the 11th week of pregnancy, the embryo is spontaneously aborted. After that time, however, the placenta itself produces enough progesterone and estrogens to maintain pregnancy.

■ HUMAN PRENATAL DEVELOPMENT REQUIRES ABOUT 266 DAYS

Human prenatal (before birth) development is similar to development in other placental mammals. The human **gestation period,** the duration of pregnancy, averages 280 days from the time of the mother's last menstrual period to the birth of the baby, or 266 days (about 9 months) from the time of conception.

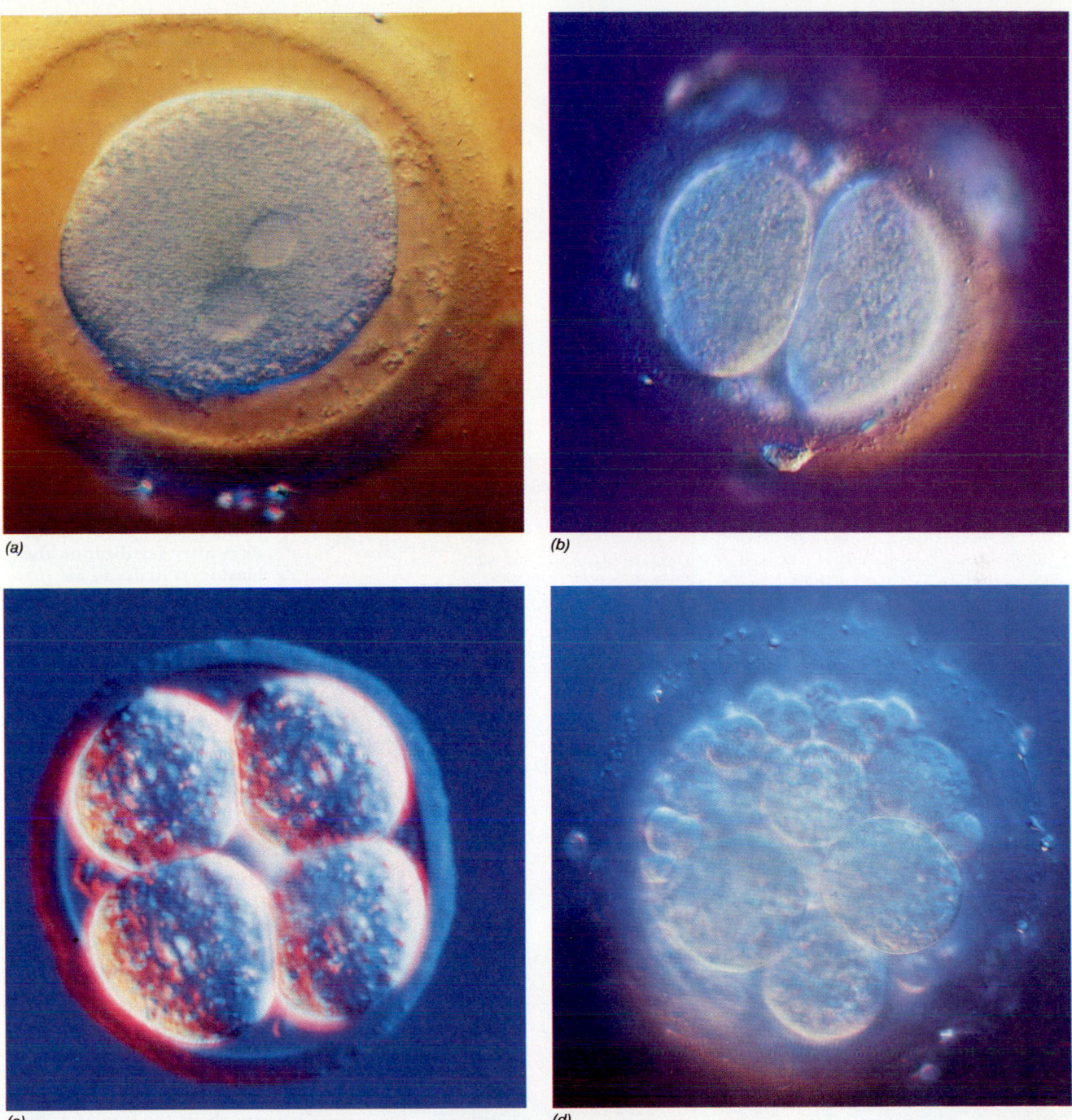

(a)

(b)

(c)

(d)

FIGURE 43–13   Early human development (approximately ×250).
(*a*) **Human zygote. This single cell contains the genetic instructions for**
producing a complete human being. (*b*) **Two-cell stage.** (*c*) **Eight-cell stage.**
(*d*) **Cleavage continues, giving rise to a cluster of cells called the morula.**
(Lennart Nilsson, from *Being Born*, pp. 14, 15, 17)

Development Begins In The Uterine Tube

By about 24 hours after fertilization, the human zygote has completely divided to become a two-cell embryo. Each of the cells of the two-cell embryo undergoes mitotic division, bringing the number of cells to four. Repeated divisions continue as the embryo is pushed along the uterine tube by ciliary action and muscular contraction. By the time the embryo reaches the uterus, on about the fifth day of development, it is in the morula stage (Figs. 43–13; 43–14).

When the embryo enters the uterus, the membrane that has surrounded it, the zona pellucida, is dis-

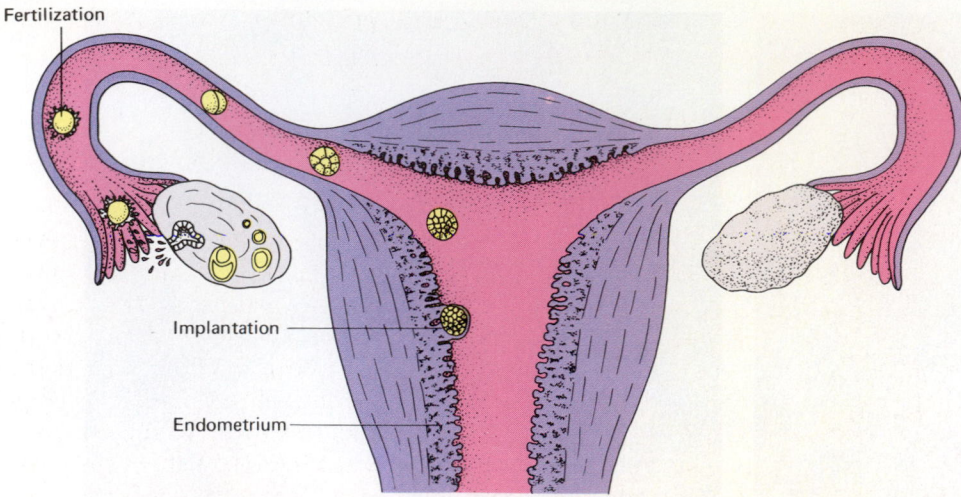

FIGURE 43–14 Cleavage takes place as the embryo is moved along through the uterine tube to the uterus.

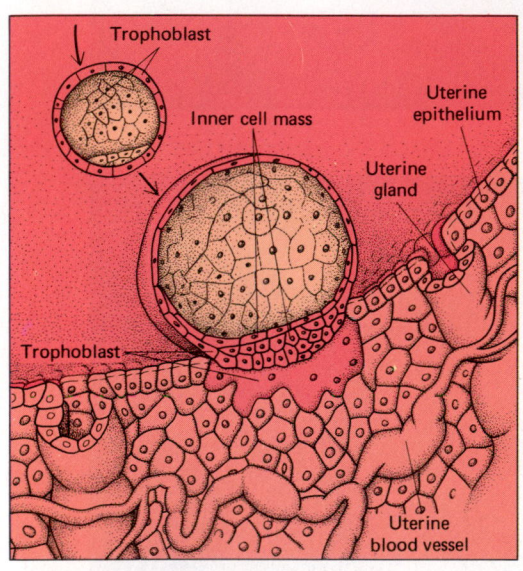

(a)

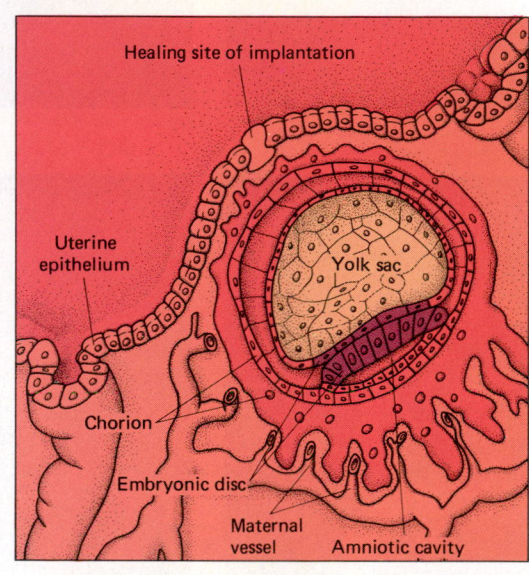

(b)

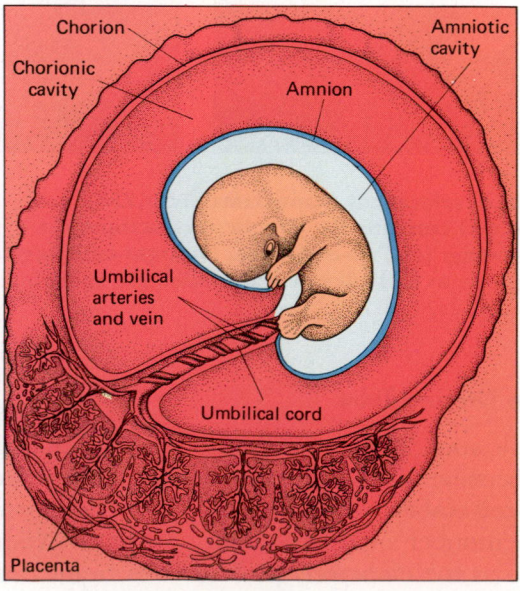

(c)

(d)

FIGURE 43–15 Implantation and development of the early human embryo. (a) About 7 days after fertilization the blastocyst drifts to an appropriate site along the uterine wall and begins to implant itself. The cells of the trophoblast proliferate and invade the endometrium. (b) About 10 days after fertilization the chorion has formed from the trophoblast. (c) By 25 days intimate relationships have been established between the embryo and the maternal blood vessels. Oxygen and nutrients from the maternal blood are now satisfying the embryo's needs. Note the specialized region of the chorion that will soon become the placenta. The embryonic stalk will become part of the umbilical cord. (d) At about 45 days the embryo and its membranes together are about the size of a Ping-Pong ball, and the mother still may be unaware of her pregnancy. The amnion filled with amniotic fluid surrounds and cushions the embryo. The yolk sac has been incorporated into the umbilical cord. Blood circulation has been established through the umbilical cord to the placenta.

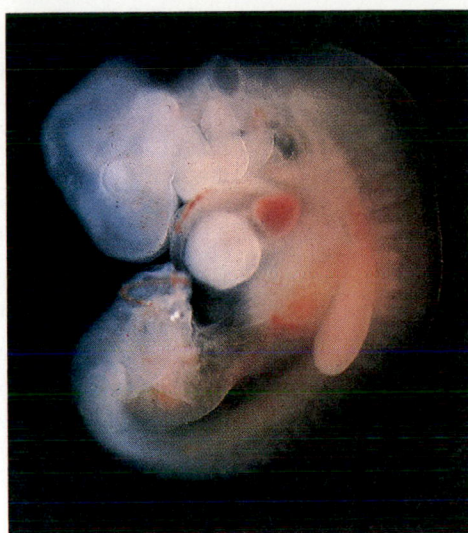

FIGURE 43–16 Photograph of developing human embryo. Human embryo at 29 days, about 7 mm (0.3 inch) long. Note the slender tail and developing limb buds. The tail will regress during later development. The heart can be seen below the head near the mouth of the embryo. Branchial arches appear as "double chins." (Lennart Nilsson, from *A Child Is Born*, Dell Publishing Co., Inc.)

solved. Now the embryo is bathed in a nutritive fluid secreted by the glands of the uterus. Nourished in this manner, the embryo continues its development for two or three days, floating free in the uterine cavity. During this period its cells arrange themselves, forming a blastula, which in mammals is called a **blastocyst** (Fig. 43–15). The outer layer of cells, the **trophoblast,** eventually forms the chorion and amnion that surround the embryo. A little cluster of cells, the **inner cell mass,** projects into the cavity of the blastocyst. These cells give rise to the embryo proper (Fig. 43–15).

Twins Develop When Cells Separate

Occasionally, the cells of the two-cell embryo may separate and each cell may develop into a complete organism. More often the inner cell mass may subdivide, forming two separate groups of cells, and each develops independently. Since each cell has an identical set of genes, the individuals formed are exactly alike—**identical twins.** Very rarely, the two inner cell masses are not completely separated and give rise to **conjoined** (Siamese) **twins. Fraternal twins** develop when a woman ovulates two eggs and each is fertilized by a different sperm. Each zygote has its own distinctive set of genes, so the individuals produced are not identical; they are only as genetically similar as any pair of siblings. Triplets (and other multiple births) may similarly be either identical or fraternal. In the United States twins are born once in about 88 births, triplets once in

88^2 (or 7744), and quadruplets once in 88^3 (or 681,472).

Implantation Begins On About The Seventh Day

On about the seventh day of development the embryo begins to **implant** in the endometrium of the uterus (Fig. 43–15a). The trophoblast cells in contact with the uterine lining secrete enzymes that erode an area just large enough to accommodate the tiny embryo. Slowly the embryo works its way down into the underlying connective and vascular tissues. The opening through which the blastocyst enters the lining of the uterus is closed, first by a blood clot, and eventually by the overgrowth of regenerated epithelial cells. All further development of the embryo takes place *within* the endometrium of the uterus.

During implantation, enzymes destroy some tiny maternal capillaries in the wall of the uterus. Blood from these capillaries comes in direct contact with the trophoblast of the embryo, temporarily providing a rich source of nutrition.

Organ Development Begins During The First Trimester

During the second and third weeks of development, gastrulation occurs. Next, the notochord begins to form and then induces formation of the neural plate. The neural tube develops and the forebrain, midbrain, and hindbrain are all established by the fifth week of development. A week or so later, the forebrain begins to grow outward, forming the rudiments of the cerebral hemispheres.

During the first month of development, the heart begins to take shape and to beat—about 60 times each minute (Fig. 43–16). In the pharyngeal region, the pharyngeal pouches, branchial grooves, and branchial arches form. In the floor of the pharynx at the level of the fourth pharyngeal pouches a tube of cells grows downward, forming the primordial trachea, which gives rise to the lung buds. The digestive system also gives rise to outgrowths that will develop into the liver, gallbladder, and pancreas. Near the end of the first month, the limb buds begin to differentiate; these structures give rise to arms and legs.

All of the organs continue to develop during the second month (Fig. 43–17). A thin tail becomes prominent during the fifth week but fails to keep pace with the rapid growth of the rest of the body and so becomes inconspicuous by the end of the second month. Muscles develop, and the embryo is capable of some movement (Table 43–2). The brain begins to send impulses that regulate the functions of some organs,

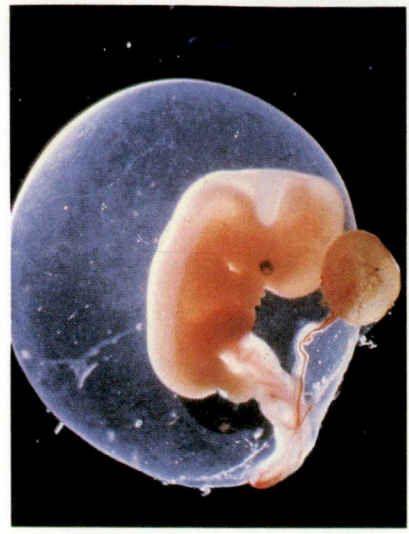

(a)

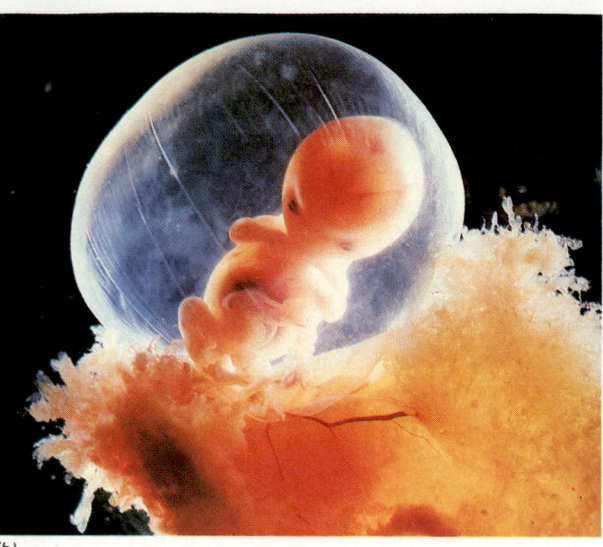

(b)

FIGURE 43–17
Development during the second month. (*a*) Human embryo at 5 1/2 weeks, 1 cm (0.4 inch) long. Limb buds have lengthened and the eyes have become prominent. (*b*) In its 7th week of development, the embryo is 2 cm (0.8 inch) long. The dark red object inside the embryo is the liver. (*a*, Guigoz/Petit Format/Photo Researchers, Inc.; *b*, Lennart Nilsson, from *A Child Is Born*, Dell Publishing Co., Inc.)

TABLE 43–2
Some Important Developmental Events in the Human Embryo

Time from Fertilization	Event
24 hours	Embryo reaches two-cell stage
3 days	Morula reaches uterus
7 days	Blastocyst begins to implant
2.5 weeks	Notochord and neural plate are formed; tissue that will give rise to heart is differentiating; blood cells are forming in yolk sac and chorion
3.5 weeks	Neural tube forming; primordial eye and ear visible; pharyngeal pouches forming; liver bud differentiating; respiratory system and thyroid gland just beginning to develop; heart tubes fuse, bend, and begin to beat; blood vessels are laid down
4 weeks	Limb buds appear; three primary vesicles of brain formed
2 months	Muscles differentiating; embryo capable of movement; gonad distinguishable as testis or ovary; bones begin to ossify; central cortex differentiating; principal blood vessels assume final positions
3 months	Sex can be determined by external inspection; notochord degenerates; lymph glands develop
4 months	Face begins to look human; lobes of cerebrum differentiate; eye, ear, and nose look more "normal"
Third trimester	Lanugo appears, then later is shed; neurons become myelinated; tremendous growth of body
266 days (from conception)	Birth

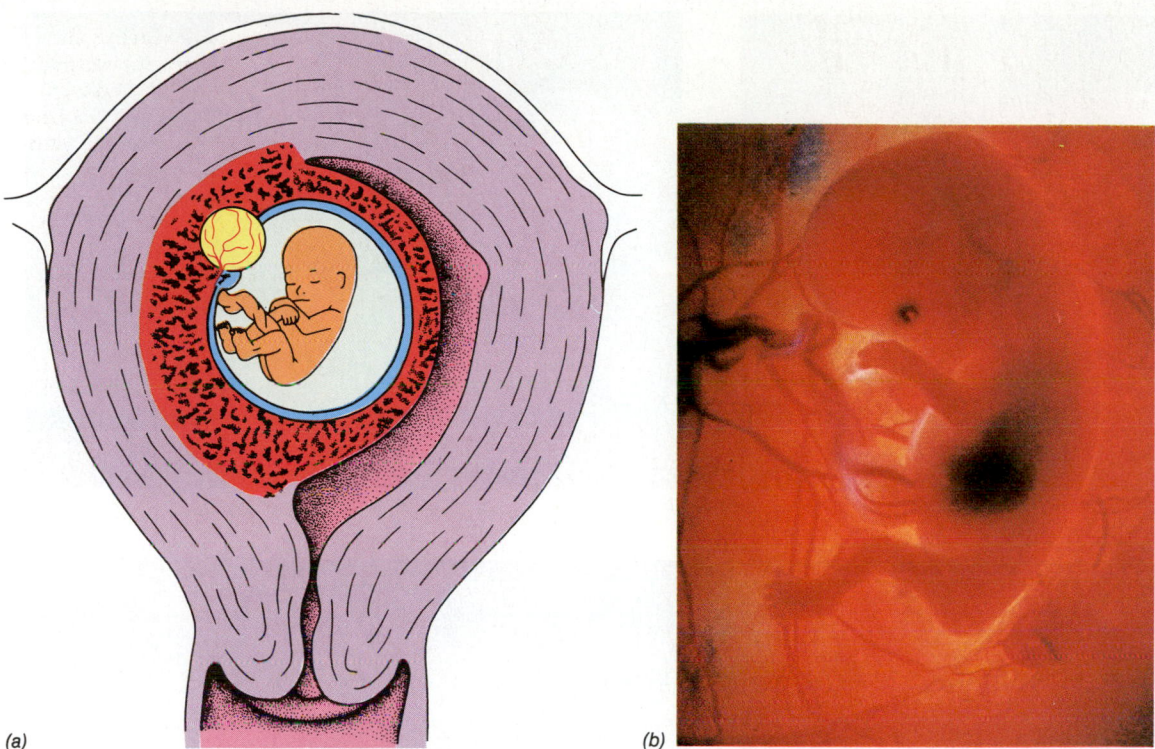

(a)   (b)

FIGURE 43–18   The human fetus at 10 weeks of development. (*a*) Note the position in the uterine wall. (*b*) Photograph at 10 weeks. (Nestle/Petit Format/Photo Researchers, Inc.)

and a few simple reflexes are evident. After the first two months of development the embryo is referred to as a **fetus.**

By the end of the **first trimester** (first three months of development), the fetus is recognizably human (Fig. 43–18). The external genital structures have differentiated, indicating the sex of the fetus. Ears and eyes approach their final positions. Some of the skeleton becomes distinct, and the notochord degenerates. The fetus performs breathing movements, pumping amniotic fluid into and out of its lungs, and even carries on sucking movements. By the end of the third month, the fetus is almost 56 mm (about 2.2 in.) long and weighs about 14 g (0.5 oz).

The Fetus Continues To Develop During The Second And Third Trimesters

During the second trimester (see Fig. 43–21) the fetal heart (now beating about 150 times per minute) can be heard with a stethoscope. The fetus moves freely through the amniotic cavity; during the fifth month, the mother usually becomes aware of fetal movements ("quickening").

The fetus grows rapidly, during the final trimester, and final differentiation of tissues and organs occurs. At the beginning of the sixth month the skin has a wrinkled appearance, perhaps because it is growing faster than the underlying connective tissue. If born prematurely at this age the fetus attempts to breathe and is able to move and cry but almost always dies because its brain is not sufficiently developed to sustain vital functions such as rhythmic breathing and regulation of body temperature.

During the seventh month the cerebrum grows rapidly and develops convolutions. The grasp and sucking reflexes are evident, and the fetus may suck its thumb. Most of the body is covered by a downy hair called **lanugo,** which is usually shed before birth. Occasionally the lanugo is not shed until a few days after birth.

During the last months of prenatal life, a protective cream-like substance, the **vernix,** covers the skin, and hair begins to grow on the scalp. At birth the average full-term baby weighs about 3000 g (7 lb) and measures about 52 cm (20 in.) in total length (or about 35 cm from crown to rump).

The Birth Process May Be Divided Into Three Stages

The factors that initiate the process of birth, or **parturition,** after the period of gestation is complete are not well understood. Childbirth begins with a long series of

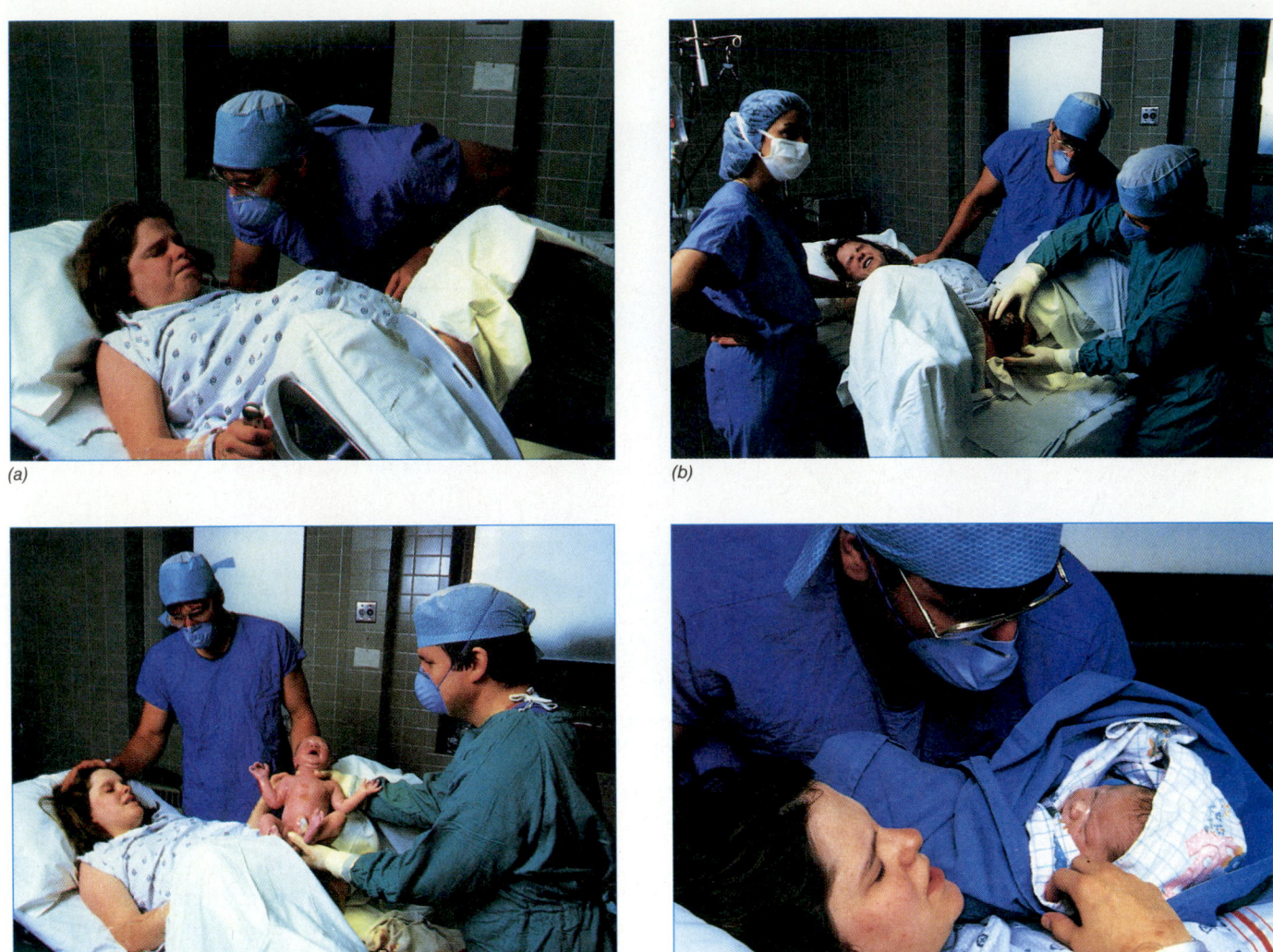

FIGURE 43–19   Birth of a baby. In about 95% of all human births the baby descends through the cervix and vagina in the head-down position. (*a*) The mother bears down hard with her abdominal muscles, helping to push the baby out. When the head fully appears, the physician or midwife can gently grasp it and guide the baby's entrance into the outside world. (*b*) Once the head has emerged, the rest of the body usually follows readily. The physician gently aspirates the mouth and pharynx to clear the upper airway of any amniotic fluid, mucus, or blood. At this time the neonate usually takes its first breath. (*c*) and (*d*) The baby, still attached to the placenta by its umbilical cord, is presented to its mother. During the third stage of labor the placenta is delivered. (SIU/Peter Arnold, Inc.)

involuntary contractions of the uterus, experienced as the contractions of **labor.**

Labor may be divided into three stages. During the **first stage,** which typically lasts about 12 hours, the contractions of the uterus move the fetus down toward the cervix, causing the cervix to dilate (open). The cervix also becomes effaced, losing its normal shape and flattening so that the fetal head can pass through. During the first stage of labor the amnion usually ruptures, releasing about a liter of amniotic fluid, which flows out through the vagina.

During the **second stage,** which normally lasts between 20 minutes and an hour, the fetus passes through the cervix and vagina and is born, or "delivered" (Fig. 43–19). With each uterine contraction the woman holds her breath and bears down so that the fetus is expelled from the uterus by the combined forces of uterine contractions plus the contractions of the muscles of the abdominal wall.

After the baby is born, the contractions of the uterus squeeze much of the fetal blood from the placenta back into the infant. After the pulsations in the

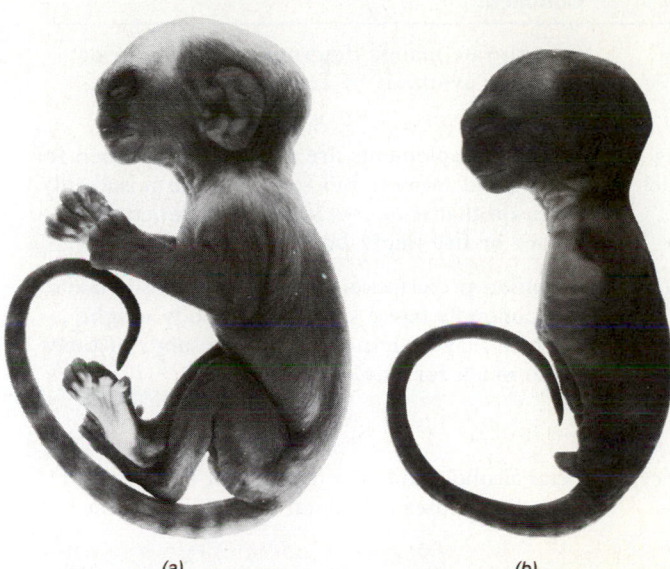

(a)   (b)

FIGURE 43–20   Thalidomide administered to the marmoset (Callithrix jacchus) produces a pattern of developmental defects similar to those found in humans. (a) Control marmoset fetus obtained from an untreated mother on day 125 of gestation. (b) Fetus (same age as control) of marmoset treated with 25 mg/kg thalidomide from days 38 to 52 of gestation. The drug suppresses limb formation, perhaps by interfering with the function of cholinergic nerves. (Courtesy of Dr. W. G. McBride and P. H. Yardy, Foundation 41; from *Development, Growth and Differentiation,* 25[4]:361–373, 1983)

umbilical cord cease, the cord is tied and cut, severing the child from the mother. The stump of the cord gradually shrivels until nothing remains but the scar, the **navel.**

During the **third stage** of labor, which lasts 10 or 15 minutes after the birth of the child, the placenta and the fetal membranes are loosened from the lining of the uterus by another series of contractions and expelled. At this stage they are called collectively the **afterbirth.**

During labor an obstetrician may administer drugs such as oxytocin or prostaglandins to increase the contractions of the uterus or may assist with special forceps or other techniques. In some women, the aperture between the pelvic bones through which the vagina passes is too small to permit the passage of the baby, so the child must be delivered by **cesarean section,** an operation in which an incision is made in the abdominal wall and uterus.

The Neonate Must Adapt To Its New Environment

Great changes take place within a short time after a baby is born. During prenatal life, the fetus received both food and oxygen from the mother through the placenta. Now the newborn's own digestive and respi-

ratory systems must function. Correlated with these changes are several major changes in the circulatory system.

Normally the **neonate** (newborn infant) begins to breathe within a few seconds of birth and cries within a half minute. If anesthetics have been given to the mother, however, the fetus may also have been anesthetized, and breathing and other activities may be depressed. Some infants may not begin breathing until several minutes have passed. This is one of the reasons for the current trend toward natural childbirth, in which as little medication as possible is used.

The neonate's first breath is thought to be initiated by the accumulation of carbon dioxide in the blood after the umbilical cord is cut. This stimulates the respiratory centers in the medulla. The resulting expansion of the lungs enlarges its blood vessels (which previously were partially collapsed); blood from the right ventricle flows in increasing amounts through the pulmonary vessels. During fetal life blood bypassed the lungs by flowing through an arterial duct connecting the pulmonary artery and aorta.

■ ENVIRONMENTAL FACTORS AFFECT THE EMBRYO

We all know that growth and development of babies are influenced by the food they eat, the air they breathe, the disease organisms that infect them, and the chemicals or drugs to which they are exposed. It is less obvious that *prenatal* development is also affected by these environmental influences. Life before birth is even more sensitive to environmental changes than it is in the fully formed baby. Anything that circulates in the maternal blood—nutrients, drugs, pathogens, or even gases—may find their way into the blood of the fetus. Drugs and other agents that can interfere with normal development are called **teratogens** (Fig. 43–20). Table 43–3 describes some of the environmental influences upon development.

About 5% of all babies born alive, or 175,000 babies per year, have a significant defect. Such birth defects account for about 15% of deaths among newborns. Recall from the genetics chapters that birth defects may be caused by genetic as well as environmental factors, or by a combination of the two. In this section we will focus on some environmental factors that affect the well-being of the embryo.

Timing is important. Substances or conditions from the outside environment that intrude upon the developing embryo may cause significant damage at one period in development, yet appear to be harmless during a later stage. Each developing structure has a

TABLE 43–3 Environmental Influences on the Embryo

Factor	Example and effect	Comment
Nutrition	Severe protein malnutrition doubles number of defects; fewer brain cells are produced, and learning ability may be permanently affected	Growth rate mainly determined by rate of net protein synthesis by embryo's cells
Excessive amounts of vitamins	Vitamin D essential, but excessive amounts may result in form of mental retardation; too much vitamins A and K may also be harmful	Vitamin supplements are normally prescribed for pregnant women, but some women mistakenly reason that if one vitamin pill is beneficial, four or five might be even better
Drugs	Many drugs affect development of fetus: Even aspirin has been shown to inhibit growth of human fetal cells (especially kidney cells) cultured in laboratory; it may also inhibit prostaglandins, which are concentrated in growing tissue	Common prescription drugs are generally taken in amounts based on mother's body weight, which may be hundreds or thousands of times too much for tiny embryo
Alcohol	When a woman drinks heavily during pregnancy, the baby may be born with fetal alcohol syndrome—that is, deformed and mentally and physically retarded; low birth weight and structural abnormalities have been associated with as little as two drinks a day; some cases of hyperactivity and learning disabilities may be caused by alcohol intake of a pregnant mother	Fetal alcohol syndrome is thought to be one of leading causes of mental retardation in the U.S.
Cocaine	Premature birth; retarded development; severe cases may be mentally retarded, have heart defects and other medical problems	Thousands of cocaine-addicted babies are being born to mothers who use cocaine during pregnancy
Heroin	High mortality rate and high prematurity rate	Infants that survive are born addicted and must be treated for weeks or months
Thalidomide	Thalidomide, marketed as mild sedative, was responsible for more than 7000 grossly deformed babies born in the late 1950s in 20 countries; principal defect was **phocomelia,** a condition in which babies are born with extremely short limbs, often with no fingers or toes	This drug interferes with cellular metabolism; most hazardous when taken during 4th to 6th weeks, when limbs are developing
Cigarette smoking	Cigarette smoking reduces the amount of oxygen available to the fetus because some of maternal hemoglobin is combined with carbon monoxide; may slow growth and can cause subtle forms of damage; in extreme form carbon monoxide poisoning causes such gross defects as hydrocephaly	Mothers who smoke deliver babies with lower-than-average birth weights and have higher incidence of spontaneous abortions, stillbirths, and neonatal deaths; studies also indicate possible link between maternal smoking and slower intellectual development in offspring
Pathogens	Rubella (German measles) virus crosses placenta and infects embryo; interferes with normal metabolism and cell movements; causes syndrome that involves blinding cataracts, deafness, heart malformations, and mental retardation; risk is greatest (about 50%) when rubella is contracted during first month of pregnancy; risk declines with each succeeding month	Rubella epidemic in the United States in 1963–65 resulted in about 20,000 fetal deaths and 30,000 infants born with gross defects
	Syphilis is transmitted to fetus in about 40% of infected women; fetus may die or be born with defects and congenital syphilis	Pregnant women are routinely tested for syphilis during prenatal examinations
Ionizing radiation	When mother is subjected to x-rays or other forms of radiation during pregnancy, infant has higher risk of birth defects and leukemia	Radiation was one of earliest teratogens to be recognized

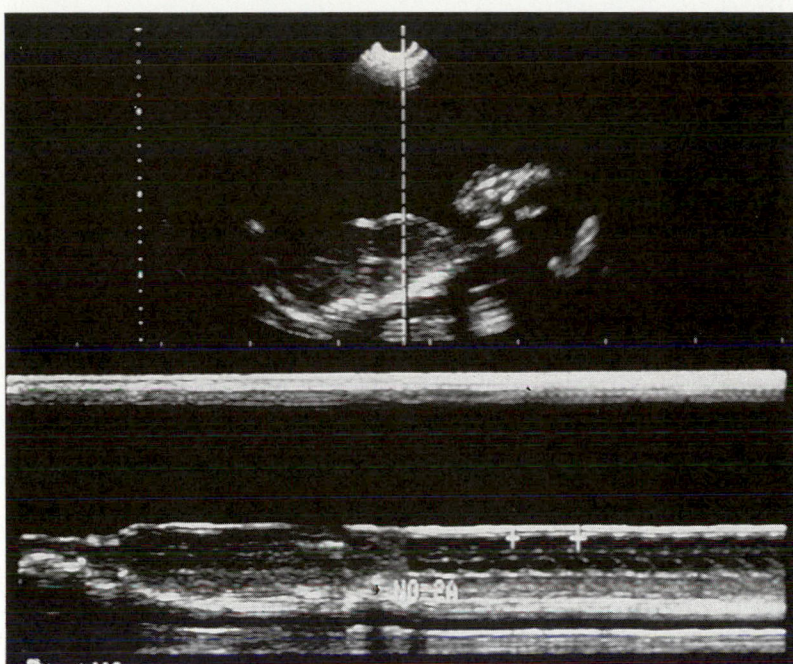

FIGURE 43–21 Sonogram of a fetus. The fetus has been visualized using ultrasound techniques. Such photographs are valuable to the physician in diagnosing multiple births and birth defects. (© Siemens-Lutheran Hospital/Peter Arnold, Inc.)

critical period during which it is most susceptible to unfavorable conditions. Generally this critical period occurs early in the development of the structure, when interference with cell movements or divisions may prevent formation of normal shape or size, resulting in permanent malformation. Since most structures form during the first three months of embryonic life, the embryo is most susceptible to environmental factors during this early period. During a substantial portion of this time the mother may not even realize that she is pregnant and may therefore take no special precautions to minimize potentially dangerous influences.

Recent advances in medicine have enabled physicians to diagnose some defects while the embryo is in the uterus. In some cases treatment is possible before birth. **Amniocentesis,** discussed in Chapter 13, is one technique used to detect certain defects. Figure 43–21 is a **sonogram,** a photograph taken of the embryo by using ultrasound. Such previews are helpful in diagnosing defects and also in determining the position of the fetus and whether a multiple birth is pending.

■ THE HUMAN LIFE CYCLE EXTENDS FROM CONCEPTION TO DEATH

Development begins at conception and continues through the stages of the human life cycle until death (see Table 43–4). We have examined briefly the development of the embryo and fetus, the birth process, and the adjustments it requires of the neonate. The **neonatal period** is usually considered to extend from birth to the end of the first month of extrauterine life. **Infancy** follows the neonatal period and lasts until the rapidly developing infant can assume an erect posture (i.e., can walk), usually between 10 and 14 months of age. Some regard infancy as extending to the end of the first 2 years. **Childhood,** also a period of rapid growth and development, continues from infancy until adolescence.

Adolescence is the time of development between puberty and adulthood. During adolescence a young person experiences the physical and physiological changes that result in physical and reproductive maturity. This is also a time of profound psychological development, as young people make adjustments that help prepare them to assume the responsibilities of adulthood.

■ THE AGING PROCESS IS MARKED BY A DECREASE IN HOMEOSTATIC RESPONSE TO STRESS

Since development in its broadest sense includes any biological change with time, it also includes those changes that result in the decreased functional capacities of the older, mature organism, the changes commonly called **aging.** The declining capacities of the various systems in the human body, though most apparent in the elderly, may begin much earlier in life, during childhood, or even during prenatal life. The newborn female has only 400,000 oocytes remaining of the 4 million she had three months earlier in fetal life!

TABLE 43–4
Stages in the Human Life Cycle

Stage	Time period	Characteristics
Embryo	Conception to end of 8th week of prenatal development	Development proceeds from single-celled zygote to embryo that is about 30 mm long, weighs 1 g, and has rudiments of all its organs
Fetus	Beginning of 9th week of prenatal development to birth	Period of rapid growth, morphogenesis, and cellular differentiation, changing tiny parasite to physiologically independent organism
Neonate	Birth to 4 weeks of age	Neonate must make vital physiological adjustments to independent life: It must now process its own food, excrete its wastes, obtain oxygen, and make appropriate circulatory changes
Infant	End of 4th week to 2 years of age (sometimes ability to walk is considered end of infancy)	Rapid growth; deciduous teeth begin to erupt; nervous system develops (myelinization), making coordinated activities possible; language skills begin to develop
Child	Two years to puberty	Rapid growth; deciduous teeth erupt, are slowly shed and replaced by permanent teeth; development of muscular coordination; development of language skills and other intellectual abilities
Adolescent	Puberty (approx. ages 11–14) to adult	Growth spurt; primary and secondary sexual characteristics develop; development of motor skills; development of intellectual abilities; psychological changes as adolescent approaches adulthood
Young adult	End of adolescence (approx. age 20) to about age 40	Peak of physical development reached; individual assumes adult responsibilities that may include marriage, fulfilling reproductive potential, and establishing career; after age 30, physiological changes associated with aging begin
Middle-aged adult	Age 40 to about age 65	Physiological aging continues, leading to menopause in women and physical changes associated with aging in both sexes (e.g., graying hair, decline in athletic abilities, wrinkling skin); this is period of adjustment for many as they begin to face their own mortality
Old adult	Age 65 to death	Period of senescence (growing old); physiological aging continues; maintaining homeostasis is more difficult when body is challenged by stress; death often results from failure of cardiovascular or immune system

The aging process is far from uniform among different individuals or in various parts of the body. Various systems of the body generally decline at different times and rates. On the average, a 75-year-old man has lost 64% of his taste buds, 44% of the renal glomeruli, and 37% of the axons in his spinal nerves that he had at age 30. His nerve impulses are propagated at a rate 10% slower, the blood supply to his brain is 20% less, his glomerular filtration rate has decreased 31%, and the vital capacity of his lungs has declined 44%. The aging process is also marked by a progressive decrease in the body's homeostatic ability to respond to stress.

Although relatively little is known about the aging process itself, this is now an active field of scientific investigation. Although marked improvements in medicine and public health have led to survival of a larger fraction of the total human population to an advanced age, there has been no concomitant increase in the maximum life expectancy for either men or women.

Cells that differentiate and stop dividing appear to be more subject to the changes of aging than are those that continue to divide throughout life. Nerve and muscle cells, which lose the capacity for cell division at an earlier age, show a decline in their respective functional capacities at an earlier age than do tissues such as liver and spleen, which retain the capacity to undergo cell division.

Several theories have been advanced regarding the nature of the aging process—that it is affected by hormonal changes; that it involves the development of autoimmune responses (immune responses against certain components of the organism's own body that result in destruction of those components by antibodies); that it involves the accumulation of specific waste products within the cell (the "clinker" theory); that it involves changes in the molecular structure of macromolecules such as collagen (an increased cross-linkage between the helical chains); that there is a decrease in

the elastic properties of connective tissues owing to an accumulation of calcium, which results in stiffening of the joints and hardening of the arteries; that it results from the peroxidation of certain lipids by free radicals; or that cells are destroyed by hydrolases released by the breaking of lysosomes.

Other current theories suggest that aging involves the accumulation of somatic mutations caused by continued exposure to cosmic radiation and x-radiation, mutations that decrease the ability of the cell to carry out its normal functions at the normal rate. In all likelihood, aging is both part of, and due to the same kinds of, developmental processes that bring about the increasing functional capacities of the various systems of the body during earlier development. The processes may be part of the program of timed development built into the genome. Like other developmental processes, aging may be accelerated by certain environmental influences and may occur at different rates in different individuals because of inherited differences. For example, there is some experimental evidence that aging, at least in rats, can be delayed by dietary means, specifically, by caloric restriction. Thin rats, by and large, live longer than fat rats! For now, however, genetic predisposition may be the best guarantee of a long life.

CHAPTER SUMMARY

I. Development proceeds as a balanced combination of cell proliferation, growth, morphogenesis, and cellular differentiation.
II. Fertilization involves four processes: contact and recognition; regulation of sperm entry; fusion of sperm and egg nuclei; activation of the egg.
III. The zygote undergoes cleavage, forming a morula and then a blastula.
   A. The main effect of cleavage is to partition the zygote into many small cells.
   B. Cleavage in the isolecithal eggs of most invertebrates and simple chordates involves division of the entire egg to form cells that are about equal in size.
   C. In bony fish and amphibians a concentration of yolk at the vegetal pole slows cleavage so that only a few large cells form there, compared to a large number of smaller cells at the animal pole.
IV. During gastrulation, the ectoderm, mesoderm, and endoderm form; each of these embryonic tissues gives rise to specific adult structures.
V. Organogenesis is the process of organ development. The developing notochord induces nervous system development. The brain and spinal cord develop from the neural tube.
VI. Development is regulated by the interaction of

genes with cytoplasmic factors, organizers, and environmental factors.
VII. In terrestrial vertebrates, four extraembryonic membranes—chorion, amnion, allantois, and yolk sac—protect the embryo and help in obtaining food and oxygen and in eliminating wastes.
   A. The amnion is a fluid-filled sac that surrounds the embryo and keeps it moist; it also acts as a shock-absorber.
   B. In placental mammals, the embryonic chorion and maternal tissue give rise to the placenta, the organ of exchange between mother and developing child.
VIII. Early human development follows a fairly typical vertebrate pattern.
   A. Cleavage takes place as the embryo is moved toward the uterus.
   B. In the uterus, the embryo develops into a blastocyst and implants itself in the endometrium.
   C. After the first 2 months of development, the embryo is referred to as a fetus.
   D. Parturition takes place after about 280 days from the time of the mother's last menstrual period. During the first stage of labor the cervix becomes dilated and effaced; during

the second stage the baby is delivered; and during the third stage the placenta is delivered.

IX. By controlling environmental factors such as alcohol and drug intake, a pregnant woman can help ensure the well-being of her unborn child.

X. The human life cycle includes: embryo, fetus, neonate, infant, child, adolescent, young adult, middle age, and old age.

XI. The aging process is marked by a decrease in homeostatic response to stress.

■ POST-TEST

1. Movement of cells to form a tube such as the neural tube is an example of _____; specialization of cells to form neurons or some other cell type is called _____ _____.
2. When the sperm contacts the jelly coat around the egg, the _____ releases enzymes that digest a path to the _____ envelope of the egg.
3. The sperm is drawn into the egg by contraction of the _____ _____.
4. The fast block to polyspermy involves an _____ change in the plasma membrane of the egg; the slow block to polyspermy is the _____ reaction.
5. The rapid series of mitoses that converts the zygote to a morula is referred to as _____.
6. The process by which the blastula becomes a three-layered embryo is called _____.
7. The germ layer that gives rise to the nervous system is the _____; the germ layer that gives rise to the lining of the digestive tract is the _____.
8. The notochord induces the overlying ectoderm to

form the _____ _____.
9. The neural tube develops into the _____ and _____ _____.
10. The _____ _____ prevents the embryo from drying out and acts as a shock-absorber.
11. In humans the _____ is the organ of exchange between mother and fetus.
12. The cluster of cells that projects into the cavity of the blastocyst is the _____ _____ _____; it gives rise to the _____.
13. On about the seventh day of development the human embryo begins to _____ in the _____.
14. After the first 2 months of development, the human embryo is referred to as a _____.
15. The duration of pregnancy is known as the _____ period.
16. The term neonate refers to the _____.
17. Teratogens are agents that can interfere with normal _____.

■ REVIEW QUESTIONS

1. Contrast the preformation theory with the theory of epigenesis, and relate these theories to current concepts of development.
2. How do the mechanisms of fertilization insure both quality (fertilization by a sperm of the same species), and quantity (fertilization by only one sperm) control?
3. Trace the development of a sea star (or *Amphioxus*) embryo from zygote to gastrula; draw and label diagrams to illustrate your description.
4. Contrast cleavage and gastrulation in the sea star (or *Amphioxus*) and amphibian.
5. Trace some developmental process such as the formation of the neural tube, and explain how growth, morphogenesis, and cellular differentiation are an integral part of the process.

6. Give examples of adult structures that develop from each of the germ layers.
7. Why do terrestrial vertebrate embryos develop an amnion? What are its functions?
8. What is the adaptive value of developing a placenta?
9. Describe human blastocyst formation and implantation.
10. What kinds of adaptations must the neonate make immediately after birth?
11. What are some of the nongenetic factors that influence development?
12. What steps can the pregnant woman take to help ensure the safety and well-being of her developing child?
13. Describe some of the changes that take place during the aging process.

■ RECOMMENDED READINGS

Holiday, R., "A different kind of inheritance," *Scientific American,* Vol. 260 (6), June 1989. DNA methylation may be an important mechanism by which patterns of gene activity are passed from one generation of cells to another during development.

Hynes, R.O., "Fibronectins," *Scientific American,* Vol. 254, No. 6, June 1986. These adhesive proteins serve as organizers in development.

Lagercrantz, H., and Slotkin, T. A. "The stress of being born," *Scientific American,* Vol. 254, No. 4, April 1986. The stress hormones released during birth can be important to the neonate's survival.

Wasserman, P.M., "Fertilization in mammals," *Scientific American,* Vol. 259, No. 6, December 1988. A glycoprotein governs many of the events of fertilization.

P A R T 9

Behavior and Ecology

Biological rhythms and behavioral cycles • Behavioral modification by learning • Instincts • Migrating animals • Defending territories • Sexual behavior • Flexible societies of vertebrates • Kin selection

Communities and ecosystems • Producers, consumers, and decomposers • Population ecology • The flow of energy through the ecosystem • Pollution • Climate and precipitation • Biomes of the land • Freshwater and marine habitats • Agriculture and the environment • Humans in the biosphere

44
Animal Behavior

OUTLINE

I. Behavior fits life-style
II. Behavior is adaptive
   A. Biological rhythms anticipate environmental changes
     1. A variety of behavioral cycles occur among organisms
     2. Biological rhythms are controlled by an internal clock
III. Behavior capacity is inherited
IV. Behavior is modified by learning
   A. In classical conditioning, a reflex becomes associated with a new stimulus
   B. In instrumental conditioning, spontaneous behavior is reinforced
   C. Some innate behaviors may be perfected by instrumental conditioning
   D. Imprinting is a form of learning that occurs only during a critical period
   E. Habituation enables the organism to ignore irrelevant stimuli
   F. Insight learning adapts recalled events to new situations
   G. Learning abilities are biased
V. Instincts are unlearned behavior patterns
   A. Innate behavior is often triggered by specific stimuli
   B. Sign stimuli can trigger inappropriate behavior
   C. Behavior develops
   D. Migration is triggered by environmental changes
   E. Migrating animals must navigate correctly

LEARNING OBJECTIVES

After you have studied this chapter, you should be able to:

1. Support the theses that behavior is (a) adaptive, (b) homeostatic, and (c) flexible.
2. Cite examples of biological rhythms and suggest some of the mechanisms known or thought to be responsible for them.
3. Using appropriate examples, summarize the role of sign stimuli (releasers) in the expression of simple and complex programmed behavior.
4. Summarize the contributions of heredity, environment, and maturation to behavior.
5. Define at least three distinguishable kinds of learning.
6. Compare learning ability with innate behavior as adaptational systems and give at least one example of their interaction.
7. Discuss the adaptive significance of imprinting.
8. Postulate biological advantages for migration.

Ringtail lemurs. (Frans Lanting/Minden Pictures)

In an anatomy course, you would study the structure of the human body in considerable detail, but the multitude of facts you would learn about the physical machinery of the organism would mean little if you did not also learn its function. The powerful muscles of the forearm, for instance, are attached to long tendons resembling bicycle gearshift cables that attach to the bones of the hand, especially those of the fingers. There is another level of significance beyond this. Chimpanzees also have hands that are not too different from our own, but why are they different at all? The differences mostly relate to the different ways in which the hands are *used;* yet the similarities also permit similar uses. People can climb trees, but not as well as chimpanzees. Chimpanzees, like people, can manipulate objects and even use tools, but the chimpanzee has yet to be born that can build a house. Structure, function, and behavior are all parts of the total constellation of adaptations that define an organism and equip it for survival. In this chapter, we consider how behavior fits into the total life of the organism so as to enable it to live and to pass its heritage on to those that follow it.

KEY CONCEPTS

☐ Each organism has a distinctive set of adaptive behaviors that fit its life-style.

☐ Behavior often alters to suit changes in the environment before they take place. Such timing results from the action of biological clocks, which are usually kept running on time by environmental cues.

☐ The capacity for behavior is inherited. Sometimes that capacity is so narrowly circumscribed that this amounts to the inheritance of the behavior itself.

☐ Much inherited behavior is extensively modifiable by experience. Learning involves persistent changes in behavior that result from experiences.

☐ Several types of learning are recognized: passive classical conditioning, active operant conditioning, insight learning, and imprinting. Imprinting differs strikingly from the other forms of learning in that no reinforcing stimulus is required.

☐ Learning abilities tend to be biased in accordance with the requirements of the organism's life-style.

☐ Instincts or fixed action patterns are narrowly circumscribed patterns of behavior that are indeed inherited, although they can usually be modified by learning.

☐ Sign stimuli trigger the release of much innate behavior.

■ BEHAVIOR FITS LIFE-STYLE

Suppose that your instructor were to arm you with a hypodermic syringe full of poison and demand that you find a particular type of insect (that you have never seen and that can fight back) and that you inject the ganglia of its nervous system (about which you have been taught nothing) with just enough poison to paralyze your victim, but not enough to kill it. You would be hard put to accomplish these tasks, but a solitary wasp no longer than the first joint of your thumb does it all with elegance and surgical precision, and without instruction.

The bee-killer wasp *Philanthus* captures bees, stings them, and places the paralyzed insects in burrows excavated in the sand. She then lays an egg on her victims, which are devoured alive by the larva that hatches from that egg. From time to time, the *Philanthus* returns to her hidden nest to reprovision it until the larva becomes a hibernating pupa in the fall. Her offspring will repeat this, doing it to perfection without ever having seen it done.

When a *Philanthus* covers a nest with sand, she takes precise bearings on the location of the burrow before flying off again to hunt. There is no way in which knowledge of the location of the burrow could be genetically programmed in the wasp. How to dig it, how to cover it, how to kill the bees—these behaviors appear to be genetically programmed. Because a burrow can be dug only in a suitable spot, however, its location must be learned after it is dug. This was determined by the Dutch investigator Niko Tinbergen.

Tinbergen surrounded the wasp's burrow with a circle of pine cones, on which the wasp took her bearings (Fig. 44–1). Before she returned with another bee, Tinbergen moved the circles of pine cones. The wasp could not find her burrow because the cones no longer surrounded it. Only when the experimenter restored the cones to their original location could the wasp find her burrow.

Behavior refers to the responses of an organism to signals from its environment. Notice how efficiently the wasp carried out a complex, although largely genetically programmed, sequence of behaviors. Very little of her behavior had to be learned. In contrast, the existence of complex programmed behavior is hard to demonstrate in human beings. We owe the complexity of our behavior to a *generalized* ability to learn. The *Philanthus* wasp's intelligence is as narrowly specialized as her stinger.

Much of what organisms do can be analyzed in terms of specific behavior patterns that occur in response to stimuli (changes) in the environment. A dog may wag its tail, a bird may sing, a butterfly may release a volatile sex attractant, an athlete may apply a deodorant. Behavior is just as diverse as biological

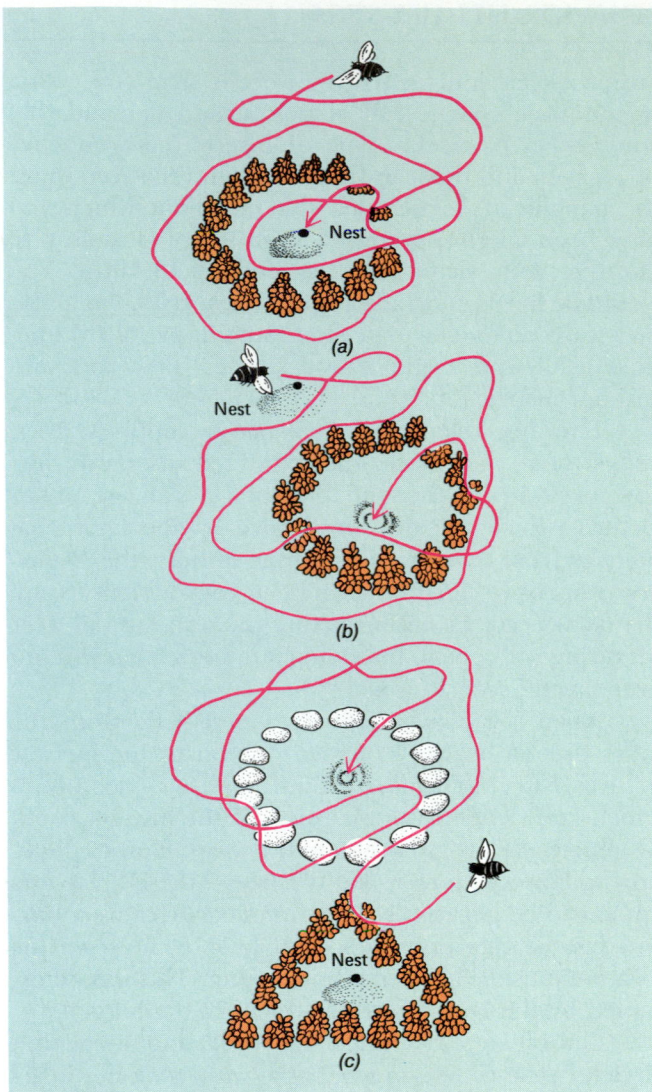

**FIGURE 44–1 Tinbergen's sand wasp experiment.
When the ring of pine cones is moved from position (a)
to position (b), the *Philanthus* wasp behaves as if her nest
were still located at the center. She has therefore learned
its position in relation to the cones. That it is the
arrangement of the cones rather than the cones
themselves that the wasp responds to is shown by the
substitution of a ring of stones for cones in (c). The
learning ability of *Philanthus* is quite limited but is
adequate for situations that normally arise in
nature.** (After Tinbergen)

structure and is just as characteristic of a given species
as its structure and biochemistry.

■ BEHAVIOR IS ADAPTIVE

We study behavior in the laboratory or under controlled conditions out of doors, always keeping in mind
that we cannot isolate what an animal does from the

way in which it lives. **Ethology,** founded by Niko Tinbergen and Konrad Lorenz in the 1920s and 1930s, is
the study of behavior in natural environments from
the point of view of adaptation. Thus, a particular behavior may help an organism obtain food or water,
acquire and maintain territory in which to live, protect
itself, or reproduce. Certain behavioral responses may
even lead to the death of the individual but while increasing the chance that copies of its genes will survive
through the enhanced production or survival of the
offspring.

Behavior does tend to be homeostatic for the individual organism as well as adaptive in the evolutionary
sense. The body of a homeothermic organism has a
collection of physiological responses that help to keep
body temperature constant. For example, a human
may shiver to generate more heat or perspire when too
hot, but most people will not be content merely to
shiver or to sweat! One can don or remove clothing,
light a fire, or turn on an air conditioner.

Many animals, however, can regulate body temperature only by changing their behavior. Lizards, for
example, may warm their bodies by basking in the sunlight. To absorb the maximum amount of heat, the lizard places its body at right angles to the sun's rays,
puffs itself up and spreads out all body membranes
(Fig. 44–2). If the lizard becomes too warm, it may first
orient the body parallel to the rays of sunlight, decreasing the area exposed directly to sunlight. It may also
retract its body membranes and shrink its body as
much as possible. If these measures fail, the lizard will
seek shade, spreading out all body membranes and the
body itself to the maximum to radiate excess heat.

Biological Rhythms Anticipate Environmental Changes

It is often to an organism's advantage to synchronize
its metabolic processes and behavior with the cyclic
changes in the external environment, so that its behavior can *anticipate* these regular changes. The little fiddler crabs (Fig. 44–3) of marine beaches often emerge
from their burrows at low tide to engage in social activities such as territorial disputes. They must return to
their burrows *before* the tide returns or they would
probably be washed away. How do the crabs "know"
that high tide is about to occur? They cannot consult
tide tables!

One might think that clues present in the seashore
environment tip off the crabs. When the crabs are isolated in the laboratory from any known stimulus that
could relate to time and tide, however, their characteristic behavioral rhythms persist (although as we shall
see, in many such cases the periodic behavior of isolated animals may gradually drift away from its original starting and stopping times).

FIGURE 44–2   Behavioral thermoregulation in a lizard. The cold lizard (a) lies at right angles to the sunlight, puffing up its body to increase the surface area available for heat absorption. When too warm (b), the lizard orients itself parallel to the sun's rays and deflates its body. Eventually (c), it seeks shade. (Illustration concept, courtesy of Pam Godfrey)

A Variety Of Behavioral Cycles Occur Among Organisms

Biological rhythms occur throughout the living world. There are not only daily rhythms, but monthly cycles and annual rhythms as well. These rhythms do not require such obvious triggers as physiological imbalance or external cues to follow a regular cycle.

FIGURE 44–3   Fiddler crabs run about on the surface of the sand at low tide but must return to their burrows before the tide returns. The enlarged claw of this crab shows it to be male. (John Shaw/Tom Stack & Associates)

In animals, periods of activity and sleep, feeding and drinking, body temperature, and many other processes also have a cycle 24 hours long. Some animals are **diurnal,** exhibiting their greatest activity during the day, whereas others are **nocturnal,** being most active during the hours of darkness. Still others are **crepuscular,** having their greatest activity during the twilight hours, at dawn, or both. As in the case of the fiddler crabs, there are ecological reasons for these adaptations. If an animal's food is most plentiful in the early morning, for example, its cycle of activity must be regulated so that it becomes active shortly before dawn. Even in human beings, physiological processes seem to follow an intrinsic rhythm. Human body temperature, for example, follows a typical daily curve, as does the heart rate of rats.

Some biological rhythms of animals reflect the **lunar** (moon) **cycle.** The most striking rhythms are those in marine organisms that are tuned to the changes in the tides due to the phases of the moon. For instance, a combination of tidal, lunar, and annual rhythms governs the reproductive behavior of the grunion, a small fish of the Pacific coast of the United States. The grunion swarms from April through June on those three or four nights when the highest tide of the year occurs. At precisely the high point of the tide, the fish squirm onto the beach, deposit eggs and sperm in the sand, and return to the sea in the next wave. By the time the next tide reaches that portion of the beach 15 days later, the young fish have hatched and are ready to enter the sea.

Biological Rhythms Are Controlled By An Internal Clock

The too familiar American cockroach (*Periplaneta americana*) needs no protection to survive. During the day, one may rarely see one and thus imagine that the exterminator has indeed disposed of them. Nightfall brings disillusionment as the agile creatures scamper for cover when the lights are turned on. If we can't get rid of them, we may as well learn from them.

When we record the movements of a roach placed in a circular treadwheel cage, we find that not only is it most active at night, but that its high-activity period begins and ends at the same time each night.

If, however, we enclose the cockroach in a light-tight enclosure in which temperature and humidity are kept constant, and every other known environmental variation eliminated, it continues to exhibit its characteristic daily rhythm of activity (Fig. 44–4). Significantly, however, this activity begins just a little bit later each day and varies slightly from its former length. If the experiment were to continue long enough, the activity cycle of the imprisoned insect would continue to be rhythmic, but could eventually become the opposite of that of its brethren running free on the laboratory floor.

This kind of experiment demonstrates that the activity is rhythmic and has a period length that is **circadian** (meaning "approximately one day"). It suggests that the organisms have a **biological clock** which is precisely adjusted or reset by environmental cues.

If the biological clock needs periodic resetting by the environment, why not simply use environmental cues to trigger daily activities? Remember, organisms may need to anticipate environmental changes that may not be preceded by any detectable signs. If set periodically, the biological clock can be used for a variety of such timing functions throughout the day.

Current evidence seems to indicate that most organisms have no single biological clock. Instead, the interaction of a number of biochemical processes (possibly involving cellular membranes) might be responsible for governing behavioral and physiological rhythms. The pineal gland is thought to play a role in the timing system of rats, birds, and some other vertebrates. Regions of the hypothalamus have been shown to be a part of the biological clock in mammals. Although some parts of an organism may coordinate or dominate the function of the biological clock, it is likely that there is a clock in every cell.

Doubtless, the biological clock has a genetic basis. Normal fruit flies, *Drosophila*, have a clock that has a free-running period of 24.2 hours. (The **free-running period** is the length of the clock's repetitive cycle when the animals are isolated from environmental cycles and kept under constant conditions. It is thought to indicate the rate at which the clock is running.) Mutant fruit flies have been discovered with free-running periods of 19 and 28 hours. Each mutation has been traced to the same locus on the X chromosome.

■ BEHAVIOR CAPACITY IS INHERITED

It is clear from this example, however, that it is not so much behavior itself but the capacity for behavior that is inherited. If those same fruit flies are kept in a more normal environment, their activity cycles are much the same as those of other fruit flies.

This is seen also in the far more complex behavior of honeybees. The size and structure of the nervous system of a honeybee permit only a limited range of behaviors, yet these insects are not automatons. Within those limits, the complex bee society can respond flexibly to changes in the environment in a multitude of ways that we are only just beginning to understand. At the same time, a honeybee is not a little, furry person, no matter how much we are tempted to compare its society with ours. The tiny bee brain could never learn all that is involved in being a good citizen of the hive; bees are hatched knowing the instructions for its society in genetically inherited and preprogrammed form.

Some of the most striking evidence for this has come from the study of disease control among bees. Bees of the Van Scoy strain are very susceptible to a serious epidemic disease known as American foulbrood, which kills immature bees as they develop in brood cells of the honeycomb. Bees of the Brown

FIGURE 44–4 A record of the daily activity of a cockroach. The arrow denotes when all environmental cues to the time of day were eliminated.

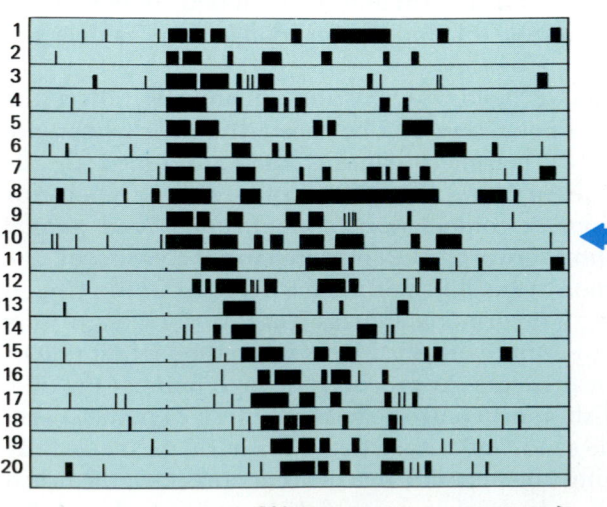

■ Active periods

← 24 hours →

strain are less susceptible because workers remove dead larvae from their waxen cells and discard them. This **hygienic behavior** has two components: (1) the removal of the wax cap of the cell, and (2) the removal of the dead larva. The Van Scoy bees leave the dead larvae to rot. Appropriately, this contrasting behavior is called **unhygienic.** The existence of these two va-

riety-specific traits implied that hygienic behavior in honeybees was under genetic control (see Fig. 44–5).

In 1964, W.C. Rothenbuhler investigated the genetic basis of the behavior. Rothenbuhler crossed hygienic and unhygienic bees. The F_1 generation was unhygienic, so the hygienic trait was evidently recessive. Backcrossing these hybrid unhygienic bees with hygi-

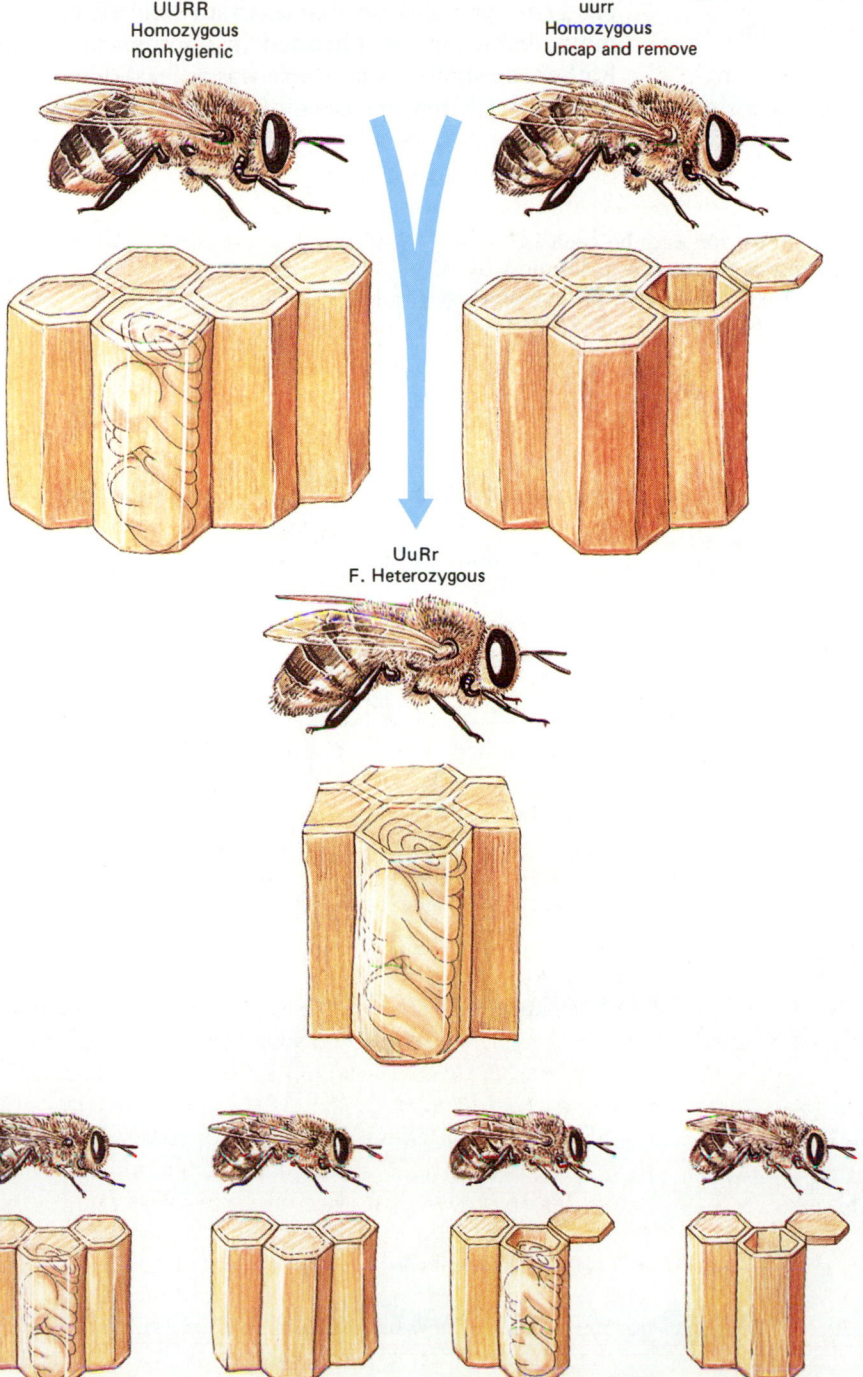

FIGURE 44–5 Inheritance of hygienic behavior in honeybees. Bees uncap and remove dead larvae only if they have the correct genotype. See the text for an explanation.

UURR
Homozygous
nonhygienic

uurr
Homozygous
Uncap and remove

UuRr
F. Heterozygous

UuRr
Nonhygienic

Uurr
Remove but not
uncap

uuRr
Uncap but not
remove

uurr
Uncap and remove

enic ones, he obtained four behavioral phenotypes in approximately equal proportions:

1. Worker bees that would neither uncap the cells of dead larvae nor remove the corpses, even if the experimenter opened the cells for them.
2. Workers that both uncapped the cells and removed larvae.
3. Workers that opened the caps of cells but left the larvae untouched.
4. Workers that did not uncap cells but removed larvae if the cells were uncapped by the experimenter.

These results can be explained easily by the hypothesis that the ability to uncap is controlled by a single pair of genes (Fig. 44–5), and the ability to remove is controlled by an independently assorted pair. Notice that because the total behavior is sequential, the first part (the uncapping) triggers the second part (removal); however, the appropriate neural pathways can develop in the nervous systems of the insects only if functional copies of the genes are present and unsuppressed by abnormal alleles.

Rothenbuhler did not demonstrate that the total uncapping behavior was completely specified by a single gene, nor did he demonstrate that all of the removal behavior was encoded in another single gene. Rather, he showed that there was at least one key gene necessary for the presence of each of these behaviors.

FIGURE 44–6 Evidence for the inheritance of behavior may be seen in lovebirds. (*a*) The peach-faced lovebird carries nesting material to the nest by tucking it among its rump feathers. (*b*) Fischer's lovebird carries the material in its bill, a strip at a time. (*c*) Hybrids appear confused and try to carry material both ways.

(a)

(b)

(c)

Removing that gene was like removing a necessary cog in a machine. Even though the machine becomes non-functional, most of it remains.

One might think that little or no behavior of the more complex and large-brained vertebrates, such as birds or mammals, is inherited. It has been shown, however, that, in at least a few cases, some of the behavior of birds and mammals is indeed very largely predetermined by their genes.

There are several species of the commonly kept cagebird *Agapornis,* better known as the lovebird (an exact English translation of its scientific name), for example, that differ not only in appearance but in behavior as well. Both species build nests of strips of vegetation, but in captivity they will accept newspaper instead (Fig. 44–6). The birds carry this material to the nests they are building in different ways. Fischer's lovebird *(A. fischeri)* carries the strips of paper to its nest in its bill, one at a time. A peach-faced lovebird *(A. roseicollis),* however, can tuck several such strips among its tail and rump feathers at one time. Thus, it can make fewer trips.

William Dilger cross-bred the peach-faced lovebird and Fischer's lovebird. The hybrids that Dilger bred could tuck the strips in among their feathers, but so ineptly that the strips fell out; they never could successfully build a nest in this manner. Eventually, however, they learned to carry the strips in their beaks, but never completely abandoned their futile attempts to tuck them in among their feathers. Whether these birds learned how to build nests by a trial-and-error kind of operant conditioning, or whether they were able, in effect, to choose between two inherited behaviors is not known. What does stand out is that in lovebirds, the method of transportation of nest-building materials is genetically inherited but somewhat flexible.

BEHAVIOR IS MODIFIED BY LEARNING

Both innate and learned behavior govern the life of an organism. Some innate behavior appears to be functional from the moment that the neural circuitry is in place and does not seem to be modified by environmental factors. The orb-weaving spider, for example, builds its first web complete in every detail, and repeatedly builds webs in the same manner throughout its life. Other behavior is totally learned, and still other behavior begins with an inherited framework that experience can modify.

In Classical Conditioning A Reflex Becomes Associated With A New Stimulus

In a type of learning called classical conditioning, an association is formed between some normal body function and a stimulus. Ivan Pavlov, a Russian physiologist who worked early in this century, discovered that when a bell was rung at the same time an experimental dog was fed, an association formed between the normally irrelevant stimulus of the bell and the secretion of digestive juice. Eventually (Fig. 44–7), when the bell was rung by itself, digestive juice was secreted. Pavlov called the physiologically meaningful stimulus (food, in this case) the **unconditioned stimulus.** The normally irrelevant stimulus that became a substitute for it (the bell) was the **conditioned stimulus.** Because a dog does not normally salivate at the sound of a bell, the association was clearly a learned one. It could also be forgotten. If the bell no longer rang at mealtimes, the dog would eventually cease responding to it. Pavlov called this **extinction.**

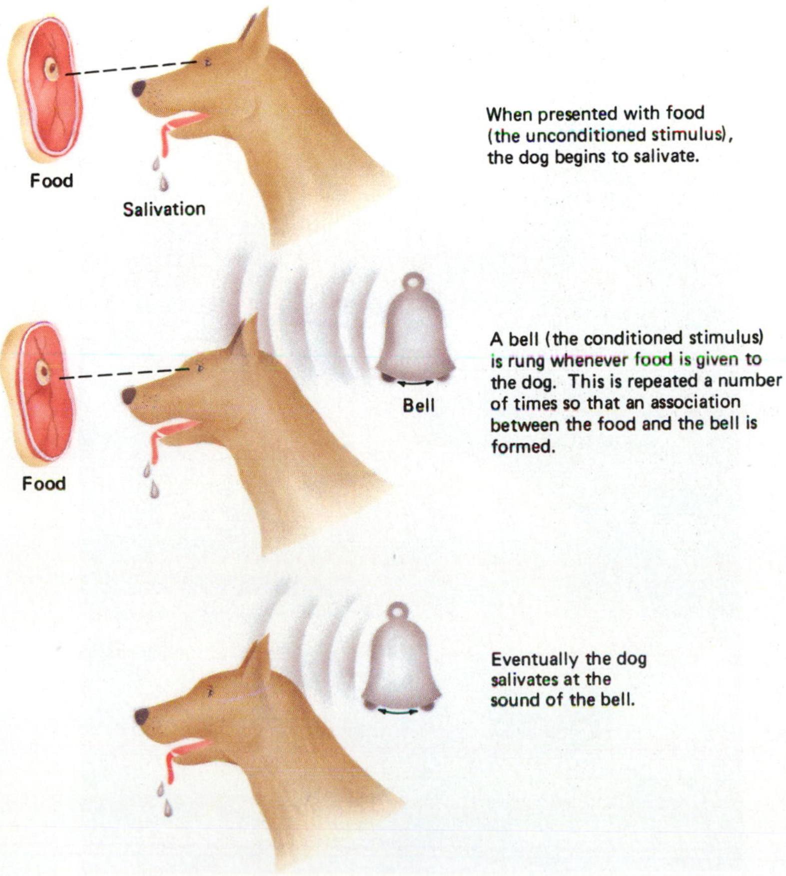

Food

Salivation

When presented with food (the unconditioned stimulus), the dog begins to salivate.

Food

Bell

A bell (the conditioned stimulus) is rung whenever food is given to the dog. This is repeated a number of times so that an association between the food and the bell is formed.

Eventually the dog salivates at the sound of the bell.

FIGURE 44–7 Pavlov's experiment. When the bell was rung, the dog salivated.

In Instrumental Conditioning, Spontaneous Behavior Is Reinforced

In another form of conditioning, **instrumental conditioning** or **operant conditioning,** the subject must do something in order to gain a reward (**positive reinforcement**) or avoid a punishment or deprivation (**negative reinforcement**). In a typical experiment, a rat is placed in a cage containing a moveable bar. When the bar is pressed, a pellet of food is delivered to the rat. This is not known to the rat when the moveable bar is first introduced, but sooner or later some random action moves the bar, and a food pellet rolls down a chute. Eventually, the rat learns the association and presses on the bar whenever it is hungry (Fig. 44–8).

Endless variations of this technique have been concocted. A pigeon might be trained to peck at a lighted circle to obtain food, or a chimpanzee might learn to perform some task in order to get tokens that can be exchanged for food, or a human schoolchild might learn to stay quietly in his seat in school to obtain some reward.

Instrumental conditioning is probably the way that animals learn to perform complex tasks like walking or perfecting feeding skills. Humans can also use it to get an animal to tell us something about the way it perceives the world. If one can train it to distinguish between red and green circles, for instance, then obviously it can see those colors.

Some Innate Behaviors May Be Perfected By Instrumental Conditioning

Instrumental conditioning may even play a large role in the development of behaviors that appear to be entirely innate. A classic example is the feeding behavior of gull chicks. Herring gull chicks peck the beaks of the parents, which regurgitate partially digested food for them. The chicks are attracted by two stimuli: a red spot on the parent's beak and the beak's shape and

FIGURE 44–8   Pigeon in a Skinner cage. Pioneered by the psychologist B. F. Skinner, this apparatus automatically rewards an experimental animal that learns to perform a desired action. In this version, the pigeon presses on a bar to obtain a reward. (Animals Animals © G. I. Bernard)

FIGURE 44–9   A pelican chick begging food from a parent. (H. Cruickshank/VIREO)

downward movement. Like the rat's chance pressing of the bar, this behavior is sufficiently functional to get the chicks their first meal, but they waste much energy in pecking. Some pecks are off target, failing to reach the parent's beak, and are therefore not rewarded. The begging behavior becomes more efficient over time, however (Fig. 44–9). Thus, a behavior that might appear to be entirely instinctive is perfected by learning.

Imprinting Is A Form Of Learning That Occurs Only During A Critical Period

Anyone who has watched a mother duck with a swarm of ducklings must have wondered how she can keep track of such a horde of almost identical little creatures, tumbling about in the weeds and grass, let alone tell them from those belonging to another hen (Fig. 44–10). Although she is capable of recognizing her offspring to an extent, basically it is they that have the responsibility of keeping track of her, which is a far simpler chore. The survival of the duckling requires an extremely rapid establishment of the behavioral bond between it and its parent. This bond, which is usually formed within a few hours of birth (or hatching), forms by a type of learning known as **imprinting.** An early investigator of imprinting, ethologist Konrad Lorenz, discovered that a newly hatched bird may imprint on a human, or even an inanimate object, if its parent is not present. Although the behavior itself is genetically determined, the bird learns the object.

Among many kinds of birds, especially ducks and geese, the older embryos are able to exchange calls with their nest mates and parents right through the porous eggshell. When they hatch, at least one parent is normally on hand, emitting the characteristic vocalizations with which the hatchlings are already familiar. If the parent moves, the chicks will follow it. The movement plus the vocalizations produce imprinting, in which, during a brief critical period after hatching, the chicks learn the appearance of the parent. Other types of learning are not restricted to such a critical period. Furthermore, imprinting takes place without any external reinforcement and is resistant to negative reinforcement. It seems, therefore, to occupy a special category.

Imprinting establishes the bond between mother and offspring among many mammals, as well as among birds. In many species, the mother also establishes a bond with her offspring during a critical period. The mother in some species of hoofed mammals, such as sheep, will accept her offspring for only a few hours after its birth. If they are kept apart past that time, the young are thereafter rejected. Normally, this behavior enables the mother to distinguish her own offspring from those of others, evidently by olfactory cues. There is no proof, incidentally, that this kind of bonding occurs among human beings.

Habituation Enables The Organism To Ignore Irrelevant Stimuli

Habituation is a form of learning in which an organism learns to ignore a repeated or continuing irrelevant stimulus. In Figure 44–11, we see the familiar but extraordinary aggregation of pigeons in a city street.

FIGURE 44–10 The formation of parent-offspring bonds. Through imprinting, some young animals follow the first moving object they encounter. Usually, the object is their mother, although it is possible experimentally to imprint many such infants upon unnatural objects. (J. H. Dick/VIREO)

FIGURE 44–11 In the form of learning known as habituation, an animal's unlearned response to constant or repeated stimulation wanes, as is the case with these pigeons, which are unperturbed by the presence of humans. (Janet Goldwater)

These birds have learned by repeated harmless encounters that human beings are no more dangerous to them than cows are to crows, and behave accordingly. This is to their advantage. The pigeon that was intolerant of people would never get enough to eat.

Insight Learning Adapts Recalled Events To New Situations

The most complex learning is **insight learning,** the ability to remember past experiences that may involve different stimuli and to adapt these recalled events to solve a new problem. Insight learning is most easily demonstrated in primates (Fig. 44–12). A dog can be placed in a blind alley that it must *circumvent* in order to reach a reward. The difficulty of the problem appears to be that the animal must move *away* from the reward in order to get *to* it. At first the dog typically flings itself at the barrier nearest the food. Eventually, by trial and error, the frustrated dog may find its way around the barrier and reach the reward (it will do this immediately the next time). A baboon placed in the same kind of situation, however, is likely to see the solution instantly. Primates appear to be especially good at insight learning, but a broad range of animals seem to have this ability to some degree.

Learning Abilities Are Biased

Learning capabilities are usually biased in that some things are learned more easily than others. One does not really need to teach a baby bird to fly, for instance. Learning language comes very naturally to us. A child will learn the speech of the people who raise it even if no one deliberately instructs it. In general, learning biases reflect the specialized mode of life of an animal. What is most important is most easily learned. The same rat that may have taken a dozen trials to perfect the artificial task of pushing a lever to get an immedi-

ate reward learns in a *single* trial to avoid a food that has made it ill as long as 6 hours after the food was eaten. Those who poison rats to get rid of them can readily appreciate the adaptive value of this learning talent to the rat. Such quick aversive learning forms the basis of warning coloration, which is found in many poisonous insects and brilliantly colored, but distasteful, bird eggs. Once made ill by such an egg, the predators learn to avoid them.

FIGURE 44–12 Insight learning, and in this case, simple tool use. Confronted with the problem of reaching food hanging from the ceiling, the chimpanzee stacks boxes until it can climb and reach the food. Many other examples of apparent insight are known from the behavior of these animals.

Learning appears to depend upon changes in the readiness of individual neurons to form circuit relationships with one another and to transmit impulses in specific directions. For an organism to learn, its neurons must have a large number of potential interactions with one another; hence, there must be many interactions, not just the few required by stereotyped preprogrammed behavior, such as the sand wasp's "professional" ability to sting its victims. Because innate behavior is really a consequence of the biophysical properties of individual neurons and of their interconnections, the more narrowly stereotyped a system of behavior is, the more obvious is its genetic control. Yet without the necessary pre-existence of the proper neural circuitry, even learned behavior would be impossible. Moreover, the kind of learned behavior that the organism typically and most easily develops also depends upon the layout of that circuitry.

INSTINCTS ARE UNLEARNED BEHAVIOR PATTERNS

The term **instinct** refers to innate behavior that need not be learned. Biologists today understand that much behavior falls into this category, but almost from its beginnings, the scientific study of animal behavior has had to contend with controversy surrounding not only

FIGURE 44–13 Suckling reflex in newborn human. (Erika Stone/Peter Arnold, Inc.)

the role of instinct in behavior but also the very validity of the concept itself.

The concept of an instinct is often vague to many people, who often confuse it with intuitive responses such as might be reflected in the phrase "I took an instinctive liking to her." Being unable to pinpoint the basis for a feeling or action, however, does not make it instinctive. In the words of a fortune cookie slip, "ignorance never settles a question." To prove that a behavior is truly innate, one must exclude the possibility that it is learned. This is usually very difficult. In fact, as in the case of the feeding "instinct" of the herring gull chick, there is a growing body of evidence that the dichotomy between instinctive and learned behaviors is largely artificial, for in many cases a combination of learning with instinct produces the behavior that an animal exhibits.

The classic instinct is a pattern of behavior that the animal is predisposed to perform at birth, although it may manifest itself at a later time. An infant, for instance, is born knowing how to swallow or it would never survive the first week or two of its life, even though most of the muscles involved in swallowing are striated "voluntary" muscle. As the infant's blood-sugar content falls and other physiological changes occur, it experiences a period of **appetite** or tension that motivates it to seek food. The way in which it does this is to emit loud noises in a frequency range to which its mother is probably instinctively predisposed to respond. When the infant is held in the proper position (Fig. 44–13), it begins to experience some sensation, probably the taste of milk, that triggers an **innate releasing mechanism** to nurse, consummating the instinctive pattern of behavior that culminates with swallowing. When internal senses report that it has had enough, **satiation** occurs, and the infant ceases to nurse.

In many animals, innate behaviors are far more elaborate than this. When an innate pattern of behavior can be described exactly, and is characteristic of a species, it is known as a **fixed-action pattern**. Such fixed-action patterns govern a wide range of activities such as courtship, nest building, ritualized combat, spider-web spinning, feeding habits, and much more in a wide variety of organisms. They are usually triggered by a specific stimulus, begin in a stereotyped manner, and proceed through a predictable series of stages.

Innate Behavior Is Often Triggered By Specific Stimuli

Some kinds of behavior are triggered or inhibited by a restricted sensory stimulus, a **sign stimulus,** which is also called a **releaser.** A sign stimulus represents a re-

(a)

(b)

FIGURE 44–14 A sign stimulus. (a) The moth, when threatened by a predatory bird, abruptly exposes two "eyespots" on its lower wings (b). These resemble the eyes of an owl or other carnivore and apparently trigger an innate avoidance response in the insectivorous bird. Notice that the entire owl need not be present to trigger this behavior—just the specific eyespot stimulus. (P. William Davis)

stricted aspect of the total sensory stimulus characteristic of some situation that is adaptively significant to the organism. Sign stimuli often serve as triggers for complex, stereotyped fixed-action patterns of behavior, such as mate selection, territoriality, or the recognition of danger. When quick action is essential, as in escaping from a predator, a danger *sign* is more useful than a detailed description of the danger.

Stimuli utilized as alarm signs, whether they are sights or sounds, are usually simple and contrast sharply to the environment. Small birds typically show an immediate flight reaction to animals with large eyes, and understandably so, considering that their major predators, owls and hawks, have large eyes (Fig. 44–14). They will also respond to a faked pair of eyes on the wings of some species of moth by breaking off their attack, which is an obvious advantage to the moth! Birds will respond in exactly the same way to false eyes on the rear end of a species of frog or, for that matter, to a drawing of eyes on a piece of cardboard. The significant thing is that the bird is responding not to an owl, but to a single, narrow sensory cue that is usually characteristic of an owl.

Sign Stimuli Can Trigger Inappropriate Behavior

From the standpoint of the bird, flight behavior triggered by harmless and edible potential prey is clearly not adaptive. If the concept of the sign stimulus as an isolated sensory trigger is valid, it should be possible to use sign stimuli experimentally to produce other maladaptive behavior. Oystercatchers recognize their eggs on the basis of their round shape and characteristically speckled appearance. If an oystercatcher's egg rolls out of the nest, the bird will retrieve the egg by pushing the egg back into the nest with its bill (Fig. 44–15). If, however, an absurdly large model of the egg is provided, the oystercatcher will turn from retrieving its own eggs to incubate the worthless model. In fact, the oystercatcher will neglect to brood its own eggs in favor of a model so huge that the bird cannot sit on it and keeps sliding off! An intense sign stimulus that produces an abnormally augmented response is known as a **supranormal stimulus.**

A bird's basic approach to chick rearing requires many sign stimuli before they are grown. Two important sign stimuli are the appearance of the egg, which triggers incubation, and the markings on the inside of a nestling's mouth, which become visible when the chick gapes and which stimulate the parent to feed it. A number of species of birds take advantage of these sign stimuli, depositing their eggs in the nests of other birds for them to raise. Cowbirds and cuckoos are familiar examples of such **nest parasites.**

A female cuckoo seeks out the nest of a habitual host species and lays just one egg in it, usually removing one of the host's eggs at the same time. The cuckoo egg, although small for the size of its real parent, is usually larger than that of the host species and similarly colored, so the host incubates it. The cuckoo chick hatches before those of the host bird and pushes the unhatched eggs or any fledglings belonging to the host parents out of the nest. The host parents ignore the ejected eggs or chicks even though they are plain to see, but the cuckoo chick gapes more widely and frequently than their own chicks would. Responding to these supranormal stimuli, the foster parents throw themselves into the task of feeding their gluttonous and fast-growing fosterling, which consumes at least as much food all by itself as all of their own fledglings together would have eaten (Fig. 44–16).

FIGURE 44–15 Sometimes birds such as this oystercatcher will choose an artificial egg to incubate rather than their own if the artificial egg is larger. Stimuli of this type are known as supranormal stimuli.

FIGURE 44–16 The stimulus for feeding behavior is not always species-specific, as evidenced by this yellow warbler that is feeding a young cowbird instead of one of her own young. (E. R. Degginger)

Behavior Develops

Behavior involves all body systems, but it is primarily influenced by the coordinating mechanisms of the body (that is, of the nervous and endocrine systems). The capacity for behavior is therefore subject to whatever genetic characteristics govern the development and *range of function* of these systems. One may think of a continuous scale of behaviors, ranging from the most rigidly programmed, genetically inherited types through those that are somewhat modifiable, to those that, although containing a genetic component, are extensively developed through experience.

Before an organism can exhibit any pattern of behavior, it must be physiologically ready to produce the behavior. Breeding behavior does not ordinarily occur among birds or most mammals unless steroid sex hormones are present in their blood at certain concentrations. A human baby cannot walk unless its reflex and muscular development permit it to walk. These states of physiological readiness are themselves produced by a continuous interaction with the environment. The level of sex hormones in a bird's blood may be determined by seasonal variations in day length. The baby's muscles develop in response to exercise. Without the trial and error involved in learning how to walk, walking would be retarded.

Perhaps the best example of such interaction between readiness and environment is afforded by the white-crowned sparrow, which exhibits considerable regional variation in its song.[1] This bird, even if kept in isolation, eventually will sing a very poorly developed but recognizable white-crowned sparrow song. If it is allowed to grow up under the care of its parents for the first 3 months of life, however, when it matures, it will sing in the local "dialect" characteristic of its parents or foster parents. If such learning does not take place in those 3 months, it never will, and if the sparrow consorts with birds of other species after the 3-month period, it will not learn their songs. It appears that the white-crowned sparrow is hatched with a rough model or pattern of its song built in. The regional details are filled in by learning.

Migration Is Triggered By Environmental Changes

The dramatic seasonal migrations of birds have long excited human curiosity, but it is only recently that we have understood even their rudiments. Until Renaissance times, for instance, when many migratory birds in Europe seemed to disappear in winter, people thought they were hibernating in such unlikely places as the muddy bottoms of ponds. Since then, widespread human travel and communication over long distances have led to our modern understanding that, even without any obvious immediate motivation (such as hunger), many animals regularly travel long distances to breed or just to feed for certain seasons of the year in a different location. Some migrations involve astonishing feats of endurance and navigation. Ruby-throated hummingbirds cross the vast reaches of the Gulf of Mexico twice each year, and the sooty tern travels across the entire South Atlantic from Africa to reach its tiny island breeding grounds south of Florida.

Although there is no reason to think that migration is purposeful, it can seem carefully planned (Fig. 44–17). Birds may feed heavily weeks before those

---

[1] Different species of birds vary greatly in the extent to which their songs are learned or innate. Among cowbirds, for example, which rely on birds of other species to raise their young, the song is *entirely innate*. Some song birds, on the other hand, are able to *learn* hundreds of songs in a single breeding season.

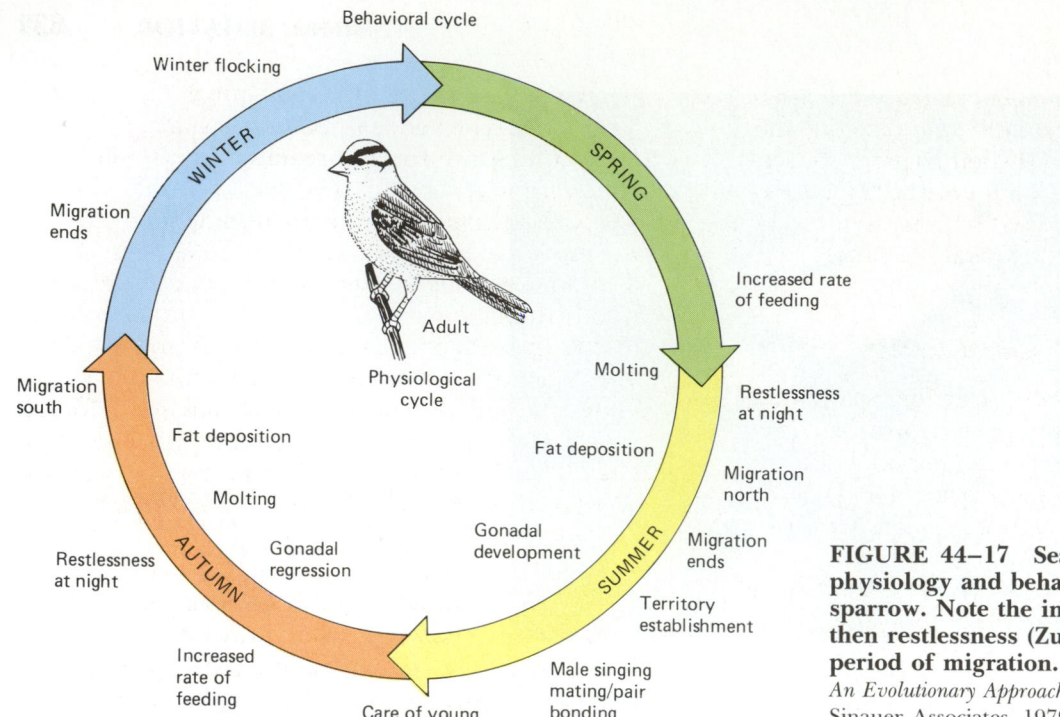

FIGURE 44–17 Seasonal changes in the physiology and behavior of the white-crowned sparrow. Note the increased rate of feeding and then restlessness (Zugenruhe) that precedes each period of migration. (From Alcock, J., *Animal Behavior: An Evolutionary Approach,* 2nd ed., Sunderland, Mass., Sinauer Associates, 1979)

food reserves will be needed and often fly south even *before* the weather turns cold or food becomes scarce. Salmon swim into fresh water toward the end of their life cycles. Monarch butterflies fly southward, and the *next generation* of butterflies flies north in the spring.

The propensity for this behavior must be inherited and maintained by natural selection. Migration, in other words, must be a specific adaptation in the life-

styles of many organisms. The adaptive significance of migratory behavior is not always clear, however. It is obviously to the advantage of birds to fly south in the winter or, because of excellent seasonal food supply, to fly north in the spring, but why certain eels migrate to the Sargasso Sea to spawn is a mystery.

The behavioral trigger that sets off migratory behavior varies. Some animals migrate upon maturation;

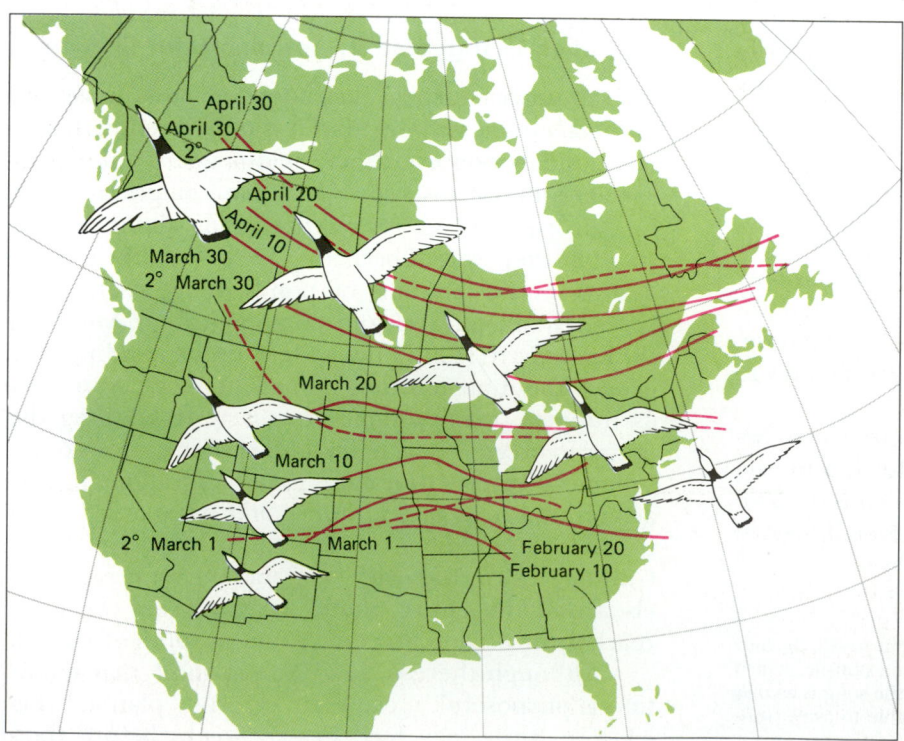

FIGURE 44–18 The northward migration of the Canada goose keeps pace with the arrival of spring in different parts of the North American continent. The geese are shown following lines that connect different points of the map according to a mean temperature of 2°C, or 35°F. (Modified after Lincoln)

in others, explicit environmental cues trigger the process. In migratory birds, for example, the pineal gland senses changes in day length. It then triggers characteristic restless behavior called **Zugunruhe,** or migratory restlessness. The bird shows an increased readiness to fly and flies for longer periods of time.

Migrating Animals Must Navigate Correctly

The *direction* of travel is also obviously important, and this raises the general problem of animal navigation. Birds appear to navigate by a combination of celestial (sun- and star-related) and geographic and climatic cues (Fig. 44–18). Honeybees and some birds are sensitive to the earth's magnetic field.

One example of how animal navigation is studied must suffice for now, but more examples will be considered in Chapter 45. Working in the 1950s, Franz and Eleonore Sauer hand reared a number of whitethroats, a species of small European warbler. This ruled out any possibility that the parents had transmitted any information to their offspring. When (and only when) the birds could see the star patterns of the night sky, they attempted to fly in the normal direction of migration for this species, a direction that they had had no opportunity to learn. When the birds were brought into a planetarium and the night sky of a different locale was simulated on the planetarium dome, they attempted to fly in a direction that would have taken them to their normal wintering grounds from that locality. The conclusion seemed inescapable: although the direction of migration was unlearned, the birds were able to find it by means of celestial navigation.

■ SUMMARY

I. Behavior consists of the responses of an organism to signals from its environment.

II. Ethology is the scientific study of behavior under natural conditions from the point of view of adaptation.

   A. Behavior tends to be adaptive.

   B. Behavior tends to be homeostatic.

III. It is adaptive for an organism's metabolic processes and behavior to be synchronized with the cyclical changes in the environment.

   A. Some biological rhythms reflect the lunar cycle or the changes in tides due to phases of the moon.

   B. In many species, physiological processes and activity follow circadian rhythms.

   C. No single biological clock has been found. Biological rhythms are thought to be regulated by both internal and external factors.

IV. Before an organism can show any pattern of behavior, it must be physiologically ready to produce the behavior.

V. Learning is a change in behavior resulting from experience. The simpler forms of learning are conditioning, both classical and operant, and habituation. Insight learning is particularly characteristic of the more "intelligent" animals; it involves the ability to see through a problem. Imprinting is a unique form of learning that establishes a parent-offspring bond during a critical period early in life.

VI. In some birds, the need to migrate and the direction of migration appear to be genetically programmed, but how to navigate may be learned.

VII. Innate behavior is unlearned and may be triggered by a specific unlearned sign stimulus, or releaser.

VIII. Innate behavior is genetic; learned behaviors develop as a result of experience. The actual development of behavior is generally a product of a complex interaction between heredity and environment. Virtually all behavior is modifiable to some extent, and virtually all behavior possesses some genetic component or predisposition.

■ POST-TEST

1. Behavior may be defined as responses of an organism to _____.

2. _____ is the study of behavior in natural environments from the point of view of adaptation.

3. The movement of flatworms toward a piece of raw meat is a _____.

4. A biological rhythm with approximately a 24-hour cycle is a _____ _____.

5. Animals that are most active at dawn or twilight are described as _____.

6. Another name for a sign stimulus is a _____.

7. _____ behavior is mainly genetic; _____ behavior develops as a result of experience.

8. _____ is a form of learning in which a young animal forms a strong attachment to an individual (usually its parent) within a few hours of birth.

9. The term Zugunruhe refers to _____ _____.

10. Secretion of saliva by a student when the noon bell rings is an example of _____ conditioning.

☐ REVIEW QUESTIONS

1. In what ways are the behaviors of *Philanthus*, the bee-killer wasp, adaptive?
2. What is imprinting and why is it considered a form of learning? What is its significance as an adaptation?
3. How does the response of a gull chick to feeding develop? Which components of the parent-chick interaction are learned and which innate? How do the innate components ensure uniformity of the learned responses?
4. Sensitivity to the earth's magnetic field has been observed in birds, bees, and bacteria. What possible reason would bacteria have to be sensitive to this stimulus? (*Hint:* There is a downward component to the earth's magnetism.)
5. Why is it adaptive for some species to be diurnal but others nocturnal or crepuscular (dusk-loving)?
6. How does physiological readiness affect instinctual behavior? How does it affect learned behavior?
7. When Konrad Lorenz kept a greylag goose isolated from other geese for the first week of its life, the goose persisted in following human beings about in preference to other geese. How could this behavior be explained?

☐ RECOMMENDED READINGS

Crews, D., and W.R. Garstka. "The Ecological Physiology of a Garter Snake," *Scientific American*, November 1982. An account of the physiological and behavioral adaptations of the red-sided garter snake to its harsh environment.

Dilger, W.C. "The Behavior of Lovebirds," *Scientific American*, January 1962. The genetics of nest-material handling in these birds.

Goodenough, J., G.F. Williams, and J.D. Palmer. "The Biological Clock: Regulating the Pulse of Life," *The Science Teacher*: 31–34, May 1977.

Gwinner, E. "Internal Rhythms and Bird Migration," *Scientific American*, April 1986. Update on the topic of how birds "know" when to migrate.

Matthews, G.V.T. *Orientation and Position-Finding by Birds*. Burlington, NC, Carolina Biological Supply Company, 1974. Although written before the magnetic senses of animals were understood, this is a good summary of research on the use of visual cues by birds in navigation.

Nicolai, J. "Mimicry in Parasitic Birds," *Scientific American*, January 1962. A more complete discussion of the complexities of brood parasitism. How closely should a parasite's egg resemble its host's egg?

Palmer, J. "Biological Clocks of the Tidal Zone," *Scientific American*, February 1975. Many marine organisms besides fiddler crabs must synchronize their behavior with the tides.

Von Frisch, K. *Animal Architecture*. New York, Harcourt Brace Jovanovich, 1974. The construction projects of animals, from spider webs to anthills, not forgetting birds' nests.

Wilson, M.F. *Vertebrate Natural History*. Philadelphia, Saunders College Publishing, 1984. Ecology text with much valuable ethological information.

Wursig, B. "Dolphins," *Scientific American*. March 1979. An interesting description of dolphin behavior and learning ability.

45
Social Behavior

OUTLINE

I. Sociality requires communication
   A. Animals communicate in a wide variety of ways
   B. Pheromones are chemical signs used in communication
II. Animals often arrange themselves in dominance hierarchies
   A. Dominance hierarchies suppress aggression
   B. Dominance results from many causes
III. Many animals defend territory
IV. Sexual behavior is usually social
   A. Pair bonds establish reproductive cooperation
   B. Many organisms care for their young
V. Play is often practice behavior
VI. Highly organized societies occur among insects and vertebrates
   A. The social insects include some hymenopterans and termites
   B. Vertebrate societies tend to be relatively flexible
VII. Kin selection could produce altruistic behavior
VIII. Sociobiology attempts to explain altruism by kin selection

LEARNING OBJECTIVES

After you have studied this chapter, you should be able to:

1. Give a description of an animal society, and identify the cooperative result of actions of the organisms, the suppression of aggression, and the modes of communication the animals employ.
2. Present the concept of a dominance hierarchy, giving at least one example, and propose a possible adaptive significance and social function for it.
3. Distinguish between home range and territory and give three theories about the adaptive significance of territoriality.
4. Discuss the adaptive value of courtship behavior and describe a pair bond.
5. Compare the society of a social insect with human society.
6. Define kin selection and summarize its proposed role in the maintenance of insect and other animal societies.
7. Summarize the emphases of sociobiology.

Japanese monkey, or Nihor Saru. (Stock Imagery, 1990)

In a pioneering work, *The Social Life of Animals,* W.C. Allee[1] showed that many animals are far more resistant to noxious environments when they live in groups than when they are alone. Schools of fish are less vulnerable to predators than single fish because large numbers tend to confuse their predators, and some are able to repel predators cooperatively. Some kinds of fish have elaborate evasive maneuvers that can work only when many fish perform them together. Flocks of birds may be able to find food better than single individuals.

By cooperation, some insects are able to construct elaborate nests and raise young by mass-production methods. A pack of wolves and a pride of lions have greater success in hunting than the individual wolves or lions would have if hunting alone. Animals that are hunted may be better able to detect or discourage predators when some individuals in the group are always on watch, and may be able to drive off predators by collective action. It seems clear that social behavior offers definite benefits (Fig. 45–1) that increase the chances of the propagation of the genes that produce it.

The mere presence of more than one individual does not mean that the behavior is social. Many factors of the physical environment bring animals together in **aggregations,** but whatever interaction they experience may be circumstantial. A light shining in the dark is a stimulus that causes large numbers of moths to gather around it. The high humidity under a log may attract aggregations of wood lice. Although it may be adaptive for these organisms to aggregate, their behavior is not truly social unless the presence of some members of the species can be shown to attract others.

Ethologists generally define **social behavior** as adaptive **conspecific** (among members of the same species) interactions, in which individuals modify one another's behavior. Many species that engage in social behavior form societies. A **society** is a group of individuals belonging to the same species that actively cooperates. A hive of bees, a flock of birds, a pack of wolves, and a school of fish are examples of societies. Some societies are loosely organized, whereas others have a complex structure. A well-organized society exhibits cooperation and division of labor among animals of different sexes, age groups, or castes. A complex system of communication reinforces the organization of the society. The members of a society tend to remain together and to resist attempts by outsiders to enter the group.

[1] Allee, W.C. *The Social Life of Animals.* New York, W.W. Norton, 1938.

FIGURE 45–1   Nesting sea birds, like these Falkland Island shags, often defend the territory directly surrounding their nests. This results in a regular spacing of the nests. (Frans Lanting/Minden Pictures)

KEY CONCEPTS

☐ Social interaction is widespread throughout the animal kingdom and probably produces a variety of advantages for organisms that are capable of it.

☐ Social behavior does require interaction and therefore an inhibition of aggression or defense behaviors. It also requires some system of communication.

☐ Aggression is minimized by the use of dominance hierarchies and by territoriality.

☐ Much social behavior centers on reproduction.

☐ Elaborate societies occur among the termites and hymenopteran insects. The behavioral basis of these societies is genetic, although they may employ limited language interactions.

☐ More flexible societies occur among vertebrates, especially mammals. The most elaborate vertebrate society is that of human beings. Its basis is largely cultural, involving the symbolic transmission of culture from one generation to another.

☐ Kin selection may explain the development of altruistic behavior in general and almost certainly underlies the origin of hymenopteran sociality.

■ SOCIALITY REQUIRES COMMUNICATION

The ability to communicate is an essential ingredient of social behavior, for only by exchanging mutually recognizable signals can one animal influence the behavior of another (Fig. 45–2). **Communication** occurs when an animal performs an act that changes the behavior of another organism. Communication may aid in finding food, as in the elaborate dances of the bees. It may hold a group together, warn a group of danger, indicate social status, ask for or indicate willingness to provide care, identify members of the same species, or indicate sexual maturity.

Animals Communicate In A Wide Variety Of Ways

Animal communication differs significantly from most human communication in that it is not symbolic. As you read or listen, words convey information to your mind. The words themselves are not the information; they stand for it. The relationship between the word "sun" and the sun itself is an artificial one and a learned one; a person who could read only Japanese would not recognize it. Yet the sun exists no matter what language we use. The symbol is a substitute that can be manipulated as the sun itself cannot.

Releasers (sign stimuli) used in animal communication are not true symbols. True symbols are learned; releasers are not. Moreover, a releaser is usually an immediate part of an animal's surroundings, but words have no necessary direct link to the information they symbolize. We can say or think "sun" when there is no sun in sight. We can even use abstract symbols, such as some of those employed in mathematics, to stand for things no one has ever seen.

Methods of animal communication are extremely varied. The singing of birds is an obvious example of auditory communication; it serves to announce the presence of a territorial male. Some animals communicate by scent rather than sound. Antelopes rub the secretions of facial glands on conspicuous objects in their vicinity (Fig. 45–3). Dogs mark territory by frequent urination. Certain fish, the gymnotids, use electric pulses for navigation and communication (Fig. 45–4), including territorial threat, in a fashion similar to bird vocalization. As E.O. Wilson has said, "The fish, in effect, sing electrical songs."

Pheromones Are Chemical Signs Used In Communication

Pheromones are chemical signals that convey information between members of a species. They are a simple, widespread means of communication. Many types of messages can be conveyed by pheromones. Most pheromones act as releasers that elicit a very specific, immediate, but transitory type of behavior. Others act as primers that trigger hormonal activities that may result in slow, but long-lasting, responses. Some pheromones may act in both ways.

An advantage to pheromone communication is that little energy must be expended to synthesize the simple, but distinctive, organic compounds involved. Conspecific individuals have receptors that are attuned to the molecular configuration of the pheromone; other species usually ignore it. Pheromones are effective in the dark, they can pass around obstacles, and they last for several hours or longer. Major disadvantages of pheromone communication are slow transmission and limited information content. Some animals compensate for the latter disadvantage by secreting different pheromones with different meanings.

FIGURE 45–2 A male hylid frog of Costa Rica calling to locate a mate. (L.E. Gilbert, University of Texas at Austin/BPS)

FIGURE 45–3 Pronghorn antelope marking territory by applying scent from its facial glands on any convenient object. (Harry Engels, Bruce Coleman Inc.)

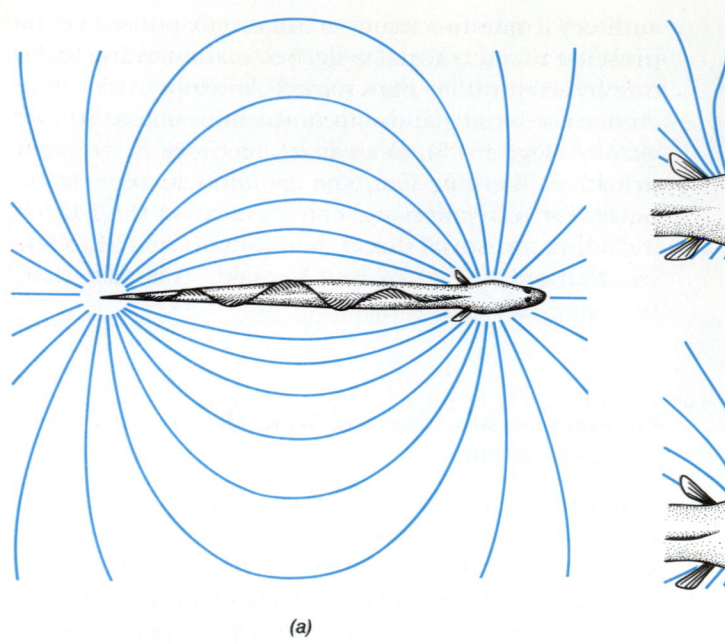

(a)

(b)

(c)

FIGURE 45–4 Certain fish navigate and communicate through electrical signals. In *Gymnarchus,* an electrical field is radiated from the head to the tail (*a*). Objects that conduct less (*b*) or more electricity (*c*) than the surrounding water distort the electrical field. Special sense organs along the sides of the fish help it use this information to navigate, intimidate territorial rivals, and locate mates.

Pheromones are important in attracting the opposite sex and in sex recognition in many species. Many female insects produce pheromones that attract males of the appropriate species. We have taken advantage of some sex-attractant pheromones to help control such pests as gypsy moths by luring the males to traps baited with synthetic versions of the female pheromone.

Some aspects of the sexual cycle of vertebrates are affected by pheromones. When the odor of a male mouse is introduced among a group of females, the estrous cycles of the female mice become synchronized. In some species of mice, the odor of a strange male, a sign of high population density, causes a newly impregnated female to abort. Among humans, it appears that some unconsciously perceived body odor is capable of synchronizing the menstrual cycles of women who associate closely (for instance, college roommates or cellmates in prison). As we shall see, pheromones much more strictly govern the reproduction of many social insects.

ANIMALS OFTEN ARRANGE THEMSELVES IN DOMINANCE HIERARCHIES

In the spring, a paper-wasp nest may be founded cooperatively by females that have survived their winter hibernation. During the early course of construction, a series of squabbles among the females takes place in which the combatants bite one another's bodies or legs and (rarely) sting. Finally, one of the young potential queens gets the upper hand over all the rest, and

thereafter she is hardly ever challenged. The queen spends more and more time on the nest, and less and less time out foraging for herself. She takes the food she needs from the others as they return. If they do not like the setup, they can leave; some do.

The queen then begins to take an interest in raising a family—her family. Because she is almost always at hand, she is able to prevent other wasps from laying eggs in the brood cells by rushing at them, jaws agape. At the same time, she cannot be stopped from laying all the eggs she wants, because she has already demonstrated that she cannot be successfully challenged.

A careful analysis of this aggressive behavior discloses that the queen can bite any other wasp without serious fear of retaliation. There is usually another wasp, however, that can bite any wasp she chooses (other than the queen) without fear of retaliation. Thus, although the queen can bite any wasp in the nest, the other wasps are not equal in their relationships with one another. One can arrange the wasps into a definite **dominance hierarchy,** an arrangement of status that regulates aggressive behavior within the society:

Queen > Wasp A > Wasp B . . . Wasp J > Wasp K

Dominance Hierarchies Suppress Aggression

Once a dominance hierarchy is established, little or no time is wasted in fighting. Subordinate wasps, upon challenge, generally exhibit submissive poses that inhibit the aggressive behavior of the queen toward them. Consequently, few or no colony members are

FIGURE 45–5 **Social animals use many signals to convey messages relating to social dominance. This baboon bares his teeth and screams in an unmistakable show of aggression.** (Gerald Lacz/Peter Arnold, Inc.)

lost through wounds sustained in fighting one another (Fig. 45–5).

Dominance Results From Many Causes

In some animals, dominance is a simple function of aggressiveness, which is itself often influenced directly by sex hormones. Among chickens, the rooster is the most dominant; as with most vertebrates, the hormone testosterone increases the aggressiveness of chickens. If a hen receives testosterone injections, her place in the dominance hierarchy shifts upward. If her ovaries are removed, the reverse takes place. Recent tests on rhesus monkeys have shown that when males are dominant, their testosterone levels are much higher than when they have been defeated. Not only can estrogen sometimes reduce dominance and testosterone increase dominance, but dominance may even increase testosterone. It is not always easy to determine cause and effect.

In many species, males and females have separate dominance systems, but in many monogamous animals, especially birds, the female takes on the dominance status of her mate by virtue of their relationship. This is not always the case, however. Like many fish,

some coral reef fish (labrids) are capable of sex reversal. What is odd is that the most dominant individual is always male, and the remaining fish within his territory are always female. If the male dies or is removed, the most dominant female will become the new male. Should anything new happen to "him," the next ranking female will become the new sultan of the harem. Still other fish exhibit the reverse behavior. In these fish, the most dominant fish becomes a female.

■ MANY ANIMALS DEFEND TERRITORY

Virtually all animals, and even some plants, maintain a minimum personal distance from their neighbors, as one can observe in the even spacing among the members of a flock of birds resting on a telephone line. Most animals have a geographical area that they seldom or never leave. Such an area is called a **home range** (Fig. 45–6). Because the animal has the opportunity to become familiar with everything in that range, it has an advantage over both its predators and its prey in negotiating cover and finding food. Some, but not all, animals defend a portion of the home range against other individuals of the same species and even against individuals of other species. Such a defended area is called a **territory.** The tendency to defend such a territory is known as **territoriality.**

Territoriality is easily studied in birds. Typically, the male chooses a territory at the beginning of the breeding season. This behavior results from high concentrations of sex hormones in the blood. The males of adjacent territories fight until territorial boundaries

FIGURE 45–6 **A coral reef has many secluded areas in which a territorial animal can establish a home range. Among the most territorial of coral reef fish is the moray eel, pictured here, which will attack any animal (including a human diver) that comes too close to its shelter.** (Charles Seaborn)

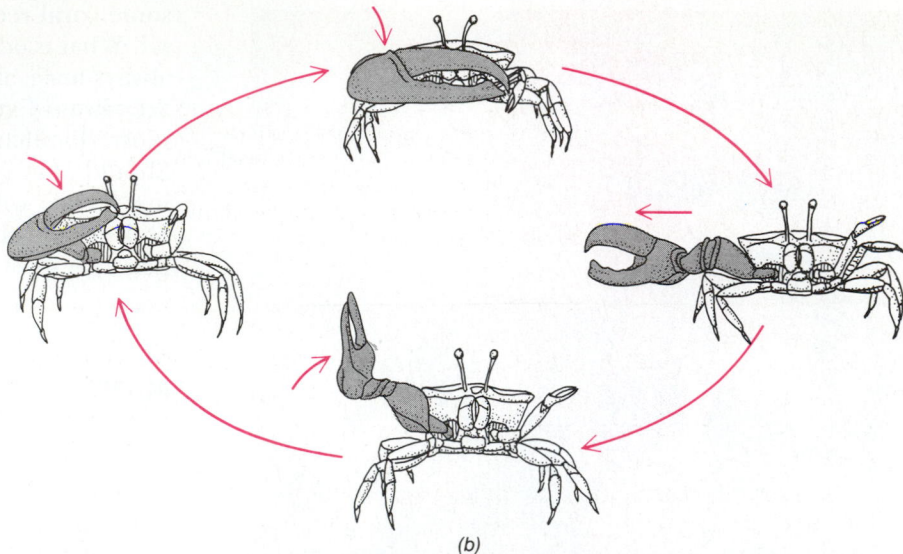

(b)

FIGURE 45–7   Courtship displays. (a) King penguins in courtship. (b) Courtship signals by male fiddler crabs are specific to each species. This particular sequence of the motion of the large right claw is characteristic of the species _Uca lactea_. (_a_, Frans Lanting/Minden Pictures; _b_, after Crane, _J Zoologica_ 42:69–82, 1957)

(a)

become fixed. Generally, the dominance of a male varies directly with his nearness to the center of his territory. Thus, close to "home," he is a lion. When invading some other bird's territory, he is likely to be a lamb. The interplay of dominance values among territorial males eventually produces a neutral line at which neither is dominant. That line is the territorial boundary. Bird songs announce the existence of a territory and often serve as a substitute for violence. Furthermore, they announce to eligible females that a propertied male resides in the territory. Typically, male birds take up a conspicuous station, sing, and sometimes display striking patterns of coloration to their neighbors, rivals, and sometimes their mates (Fig. 45–7).

Territoriality among animals may be adaptive in that it tends to reduce conflict among members of the same species, control population growth, and ensure the most efficient use of environmental resources by encouraging dispersion and thus spacing individuals more or less evenly throughout a habitat. Usually, territorial behavior is related to the specific life-style of the organism that displays it and to whatever aspect of its ecology is most critical to its reproductive success. For instance, sea birds may range over hundreds of square miles of open water but exhibit territorial behavior that is restricted to nesting sites on a rock or island, their resource that is in the shortest supply and for which competition is keenest. The adaptive "reason" for territoriality is not always readily evident, however.

■ SEXUAL BEHAVIOR IS USUALLY SOCIAL

The minimum social contact and, for some species of animals (for example, many species of spiders), the only social contact, is the sex act. Fertilization and perhaps the rearing of young are, for some animals, the only forms of social behavior (Fig. 45–8). Let us consider the sex act as a basic example of social conduct, for the elements to which it can be reduced are also the least common denominators of most social behavior.

The sex act is obviously adaptive in that it is usually necessary for reproduction, at least in animals. It requires _cooperation_, the _temporary suppression of aggressive behavior_, and a _system of communication_. Among some jumping spiders, for example, mating is preceded by a ritual courtship on the part of the male, the effect of which is to produce temporary paralysis in the female. While she is thus enthralled, the male inseminates her. Should she recover before he makes his escape, he becomes the main course at his own wedding feast. Even so, he would thus be able to make the ultimate material contribution to the eggs the female will presently produce and, therefore, to the perpetuation of his genes. She would otherwise have to bear the metabolic burden of their production all by herself.

Because the female usually chooses the mate, selection has favored those male characteristics that make a male most attractive. Success of a male in dominance encounters with other males indicates his quality to the female. The victorious male courts the female.

FIGURE 45–8 Male and female jumping spiders, *Phidippus audax,* **in the courtship behavior that precedes mating. The male performs an elaborate dance that inhibits the female's natural aggression toward him, allowing him to get close enough to inseminate her.** (James H. Carmichael, Bruce Coleman, Inc.)

An important function of courtship is to ensure that the male is a member of the same species, but it also provides the female further opportunity to evaluate the quality of the male. Courtship may also be necessary as a signal to trigger nest building or ovulation. Courtship rituals may be long and complex. The first display of the male releases a counter behavior of a conspecific female. This, in turn, releases additional male behavior, and so on until the pair are ready for copulation.

Courtship, then, is often strenuous and dangerous, especially for males. Tremendous energy seems to be wasted in the fights that occur among rams, bull seals, bull moose, and many other mammals and even some birds. As if the fighting weren't enough, more energy may be wasted in wallowing, roaring, leaping about, and other male acrobatics. It is often suggested that these behaviors are a kind of test of male fitness. The male with the greatest endurance has the most, or possibly the only, opportunity to mate and to propagate his genes. As for the female, the physiological stress of egg production, pregnancy, and/or lactation will test her stamina in due course.

Pair Bonds Establish Reproductive Cooperation

A **pair bond** is a stable relationship between animals of the opposite sex that ensures cooperative behavior in mating and the rearing of the young (Fig. 45–9). In some species, a newly arrived female is initially treated as a rival male. Then, through the use of instinctive appeasement postures and gestures by both male and female, the initial hostility is dissipated and mating takes place. Such sexual-appeasement behavior may be very elaborate and gives rise to mating dances in some birds. Often, courtship displays continue in modified form throughout the reproductive association of mates and sometimes, especially in birds, they may persist for life. The releaser mechanisms involved in the establishment and maintenance of the pair bond are often remarkably detailed. Such cues enable courtship rituals to function as behavioral genetic isolating mechanisms among species.

Many Organisms Care For Their Young

Care of the young is an additional component of successful reproduction in many species; it, too, requires a parental investment (Fig. 45–10). The benefit of parental care is the increased likelihood that the offspring will survive, but the cost is a reduction in the number of offspring that can be produced. Because of the time spent carrying the developing embryo, the female has more to lose than the male if the young do not develop. Thus, females are more likely than males to brood eggs and young, and usually the females invest more in parental care.

Investing time and effort in care of the young is usually less advantageous to a male (assuming that the female can handle the job by herself), for time spent in parenting is time lost from inseminating other females. Even worse, it may not be certain who fathered the offspring. Raising some other male's offspring is a definite genetic disadvantage (remember the cuckoo?),

FIGURE 45–9 A pair of nesting albatrosses. In many species pair bonds are maintained by grooming or other displays of affection. (E.R. Degginger)

(a)

(b)

(c)

FIGURE 45–10 Examples of parental investment.
(a) Cougars and black bears are normally mortal enemies
and actively avoid each other. This confrontation was
initiated when the cougar intruded in the area where a
female bear was raising her cubs. (b) A chinstrap
penguin regurgitating food for her young. Such an
investment of time and energy on the part of the parent
does not benefit the parent directly. It does help ensure
the transmission of the parent's genes into succeeding
generations. (c) A baby baboon rides on its mother's back
during early infancy and comes to inherit some of her
social status. (a, E.R. Degginger; b, P.R. Ehrlich, Stanford
University/BPS; c, courtesy of Busch Gardens)

which is probably why male lions kill the cubs of
former harem masters whose position they have
usurped. In some situations, however, it may be to the
male's advantage to help rear his own young or even
those of a genetic relative. Receptive females may be
scarce, and gathering sufficient food may require
more effort than one parent can provide. In some habitats, the young may need protection against predators
and sometimes against cannibalistic males of the same
species.

■ PLAY IS OFTEN PRACTICE BEHAVIOR

Play is an important aspect of the development of behavior in many species, especially young mammals. It
serves as a means of practicing adult patterns of behavior and perfecting means of escape, prey killing, and
even sexual conduct. In true play, the behavior may
not be actually consummated. Thus, a kitten pounces
upon a dead leaf but of course does not kill it, even
though the kitten administers a typical carnivore neck
bite. When playing with a littermate, the same kitten
may practice the disemboweling stroke with its hind
claws (Fig. 45–11), but the littermate is not intention-
ally injured in the process.

■ HIGHLY ORGANIZED SOCIETIES OCCUR AMONG INSECTS AND VERTEBRATES

Some animal societies exhibit elaborate and complex
patterns of social interactions. In these societies, there
is considerable division of labor that is not directly connected with the care of the young.

The Social Insects Include Some Hymenopterans And Termites

Although many insects cooperate socially, such as tent
caterpillars, which spin a communal nest, the most
elaborate insect societies are found among the bees,
ants, wasps, and termites. The first three of these all
belong, not coincidentally, to the order Hymenoptera.
Insect societies are held together by an elaborate system of sign stimuli that is keyed to social interaction; as
a result, they tend to be quite rigid. In addition to
other modes of communication, the social insects secrete pheromones that accomplish such tasks as sup-
pressing the ovaries of worker honeybees or alerting
an ant hill to the presence of an enemy (an "alarm substance" is given off from a special abdominal gland of
an excited worker).

FIGURE 45–11 Young racoons playfully sparring. Play is behavior that is not consummated and often serves as a means of practicing behavior that will be used in earnest in later life, possibly in hunting, fighting for territory, or competing for mates. (E.R. Degginger)

The social organization of honeybees has been studied more extensively than that of any social insect. A honeybee society generally consists of a single adult queen, up to 80,000 worker bees (all female), and, at certain times, a few males called drones that fertilize newly developed queens. The queen's job is reproduction; she deposits about 1000 fertilized eggs per day in the wax cells of a comb.

Division of labor in the bee society is mostly determined by age. The youngest worker bees serve as nurse bees. They have special glands on the head that secrete **royal jelly,** which is essential for the nutrition of all larval bees (its name results from the belief that, in larger quantities, royal jelly helps to produce queens). After about a week as nurse bees, workers begin to produce wax and build and maintain the wax cells. Older workers are foragers, bringing home the vital nectar and pollen. Most worker bees die at the ripe old age of 42—days, that is.

Behavioral cues tip off the bees when there is a labor shortage in any category. If there are too many larvae for the nurse bees, foragers will pitch in for the duration of the emergency, redeveloping royal-jelly glands, if need be.

The composition of a bee society is controlled by an anti-queen pheromone secreted by the queen. It acts as a releaser, inhibiting the workers from raising a new queen, and it also has a primer effect, for it inhibits the development of the ovaries in the workers (Fig. 45–12). If the queen dies, or if the colony becomes so large that the inhibiting effect of the pheromone is dissipated, the workers begin to feed some larvae the special food that promotes their development into new queens.

The most sophisticated known mode of communication among bees is a stereotyped series of body movements known as a **dance.** If a worker runs across a rich source of nectar, it can communicate this fact to the other bees within the hive by dancing on a vertical comb surface. If the food supply is nearby, the bee performs a **round dance** (Fig. 45–13), which generally excites the other bees and causes them to fly about in all directions (but within a certain distance from the hive) till they have found the nectar. If the source is distant, however, the bee performs a **waggle dance.** This "step" has a figure-eight configuration. As the bee treads the long axis of the figure eight, she emits a series of distinctive sounds and waggles her abdomen from side to side. The angle between the long axis and the force of gravity is the same as the angle between

FIGURE 45–12 A queen honeybee inspecting wax cells. Note the numerous workers that surround her. They constantly lick secretions from the queen bee. These secretions are transmitted throughout the hive and act to suppress the activity of the workers' ovaries. (Treat Davidson/Photo Researchers, Inc.)

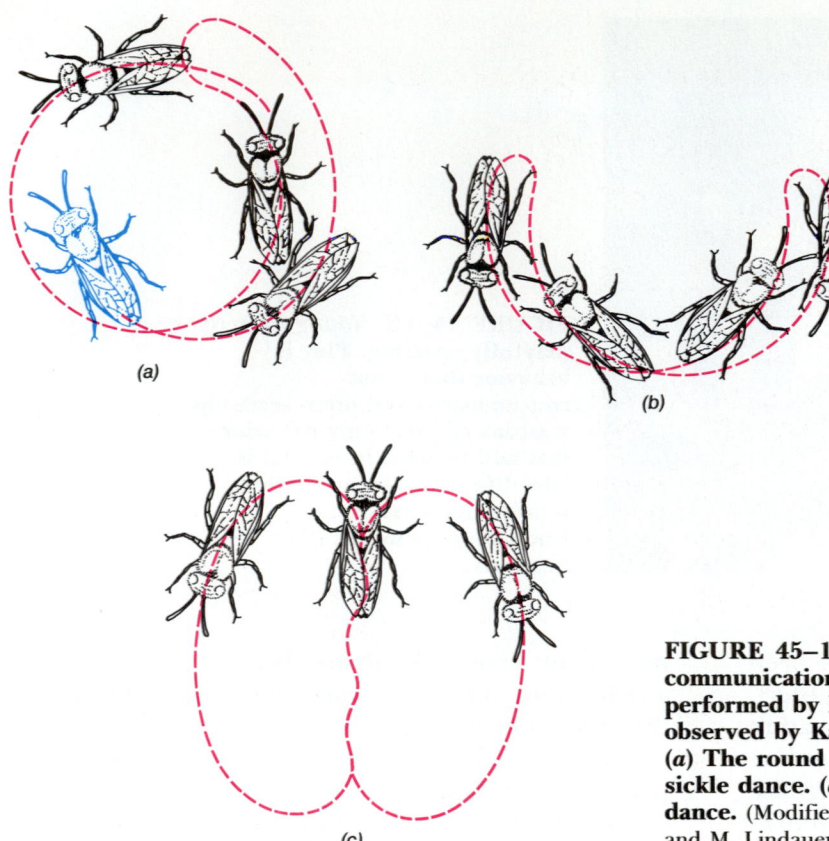

FIGURE 45–13 **Three kinds of communication dances performed by honeybees, as observed by Karl von Frisch. (*a*) The round dance. (*b*) The sickle dance. (*c*) The waggle dance.** (Modified from K. von Frisch and M. Lindauer)

the sun, the hive, and the shortest flight direction to the food source (Fig. 45–14).

Although rudimentary social behavior is widespread among insects, the only elaborate societies with extensive division of labor occur among the Hymenoptera (bees, wasps and ants) and the termites. The question is not so much why other insects are not social as it is why those particular insects *are* social. We cannot answer this question with assurance for the termites, but the basis of Hymenoptera sociality appears to be chromosomal.

Hymenoptera employ a system of sex determination in which males are haploid; thus, males develop from unfertilized eggs. Reproductive females store sperm cells from previous matings in a seminal receptacle. If they permit a sperm cell to contact the egg as it is laid, the resulting insect is female; otherwise, it is male. Because a haploid male drone produces sperm that also has a haploid set, each one of those sperm cells has *all* his chromosomes. Because the queen stores this sperm for much, or even all, of the queen's lifetime, the worker bees of a hive are more closely related than sisters born of a diploid father would be. Indeed, they have three-quarters of their genes in common (they will share half of the queen's chromosomes and all of the drone's). As a consequence, they are more closely related to one another than they would be to

their own offspring, if they could have any. Because new queens will also be their sisters, they are actually more likely to be able to pass on copies of their genes to the next generation by raising these than they would by producing their own offspring. A worker bee's offspring would have only half of its genes in common with its worker mother.

Vertebrate Societies Tend To Be Relatively Flexible

Simpler, at least by first impression, but also more flexible than those of insects, vertebrate societies are very common. There is great variation among vertebrate societies. Hyenas, wolves, red deer, prairie dogs, baboons, and people represent a great range of behaviors, yet a few points of similarity do stand out—with exceptions.

Among vertebrate societies, we find a far greater range and plasticity of potential behavior than among insect societies. Vertebrate societies are far less rigid and much more adaptable to changing needs. Vertebrate societies usually contain nothing comparable to the physically and behaviorally specialized castes of termites or ants. What is more, except for human beings, individual members of vertebrate societies are not as specialized in their tasks as are the social insects. The

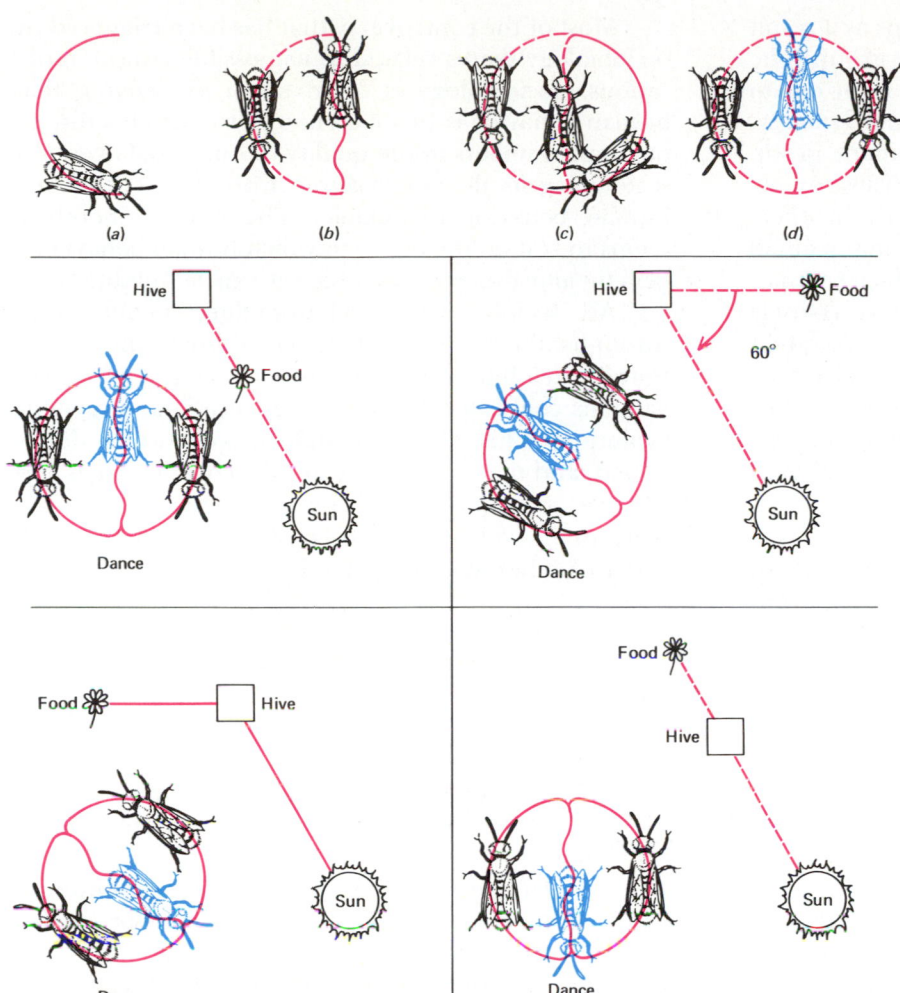

(a)   (b)   (c)   (d)

Hive   Food   Sun   Dance

Hive   Food   60°   Sun   Dance

Food   Hive   Sun   Dance

Food   Hive   Sun   Dance

FIGURE 45–14 Indication of direction by the waggle dance.

very behavioral plasticity of vertebrates makes possible the symbolic transmission of culture, and this opens up whole new worlds inaccessible to the social insects. The only vertebrate society, however, that is based to any great extent on the symbolic transmission of culture is human society.

KIN SELECTION COULD PRODUCE ALTRUISTIC BEHAVIOR

Altruistic behavior, in which one individual appears to act in such a way as to benefit others rather than itself, is frequently seen in the more complex social groups (Fig. 45–15). Biologists Watts and Stokes observed a particularly clear case of altruistic behavior in the mating of wild turkeys. Several groups of males, each of which has an internal dominance hierarchy, gather in a special mating territory and go through their displays of tail spreading, wing dragging, and gobbling in front of females who come to the area to copulate. One

FIGURE 45–15 Prairie dog. Low-ranking members of this social rodent group act as sentries. Sentries place their own lives in danger by exposing themselves outside their burrows. However, in this way they protect their siblings and by so doing ensure that the genes they share in common will be perpetuated in the population. This is a classic example of kin selection. (Tina Waisman)

group attains dominance over other groups as a result of cooperation among the males within the group. The dominant member of the dominant group is the one to copulate most frequently with the females. Seemingly, the males that helped establish the dominant group but have low status within it gain nothing. Close analysis has shown that members of a group are brothers from the same brood. Because they share many genes with the successful male, they are indirectly perpetuating many of their genes. In this case, altruism is closely related to **kin selection,** vicarious gene propagation among closely related individuals, such as we have seen among the hymenopteran social insects. The nonbreeding turkeys, by promoting the breeding success of a kinsman, propagate copies of many of their own genes.

Among some different birds (Florida jays and others), nonreproducing individuals aid in the rearing of the young. Nests tended by these additional helpers as well as parents produce more young than nests with the same number of eggs overseen only by parents. The nonreproducing helpers, siblings of the parents, are apparently increasing their own biological success by ensuring the successful propagation of their genes via their siblings. It has also been suggested that if an organism in a habitat whose carrying capacity is at its limit cannot readily obtain its own territory, it can nevertheless pass on copies of a substantial portion of its genes by aiding an established pair of relatives that do possess a territory.

SOCIOBIOLOGY ATTEMPTS TO EXPLAIN ALTRUISM BY KIN SELECTION

Sociobiology is the school of ethology that focuses upon the evolution of behavior through natural selection. It represents a synthesis of population genetics, evolution, and ethology. Like many biologists of the past (such as Darwin), Edward O. Wilson and other sociobiologists emphasize the animal roots of human behavior, but they have attempted to infuse their discipline with population genetics, with particular emphasis on the effect of kin selection on patterns of inheritance. Many of the concepts discussed in this chapter, such as altruism and paternal investment in care of the young, are based on contributions made by sociobiologists.

For the sociobiologist, the organism and its adaptations, including its behavior, are ways its genes have of making more copies of themselves. The cells and tissues of the body support the functions of the reproductive system. The reproductive system's job is to transmit genetic information to succeeding generations.

Most of the controversy that has been triggered by sociobiology seems related to its possible ethical implications. Sociobiology is often taken as denying that human behavior is flexible enough to permit substantial improvements in the quality of our social lives. Yet sociobiologists do not disagree with their critics that human behavior is flexible. The debate therefore seems to rest on the *degree* to which human behavior is genetic and the *extent* to which it can be modified.

As sociobiologists acknowledge, people can, through culture, change their way of life far more profoundly in a few years than a hive of bees or a troop of baboons could in hundreds of generations of genetic evolution. This ability is indeed genetically determined, and that is a very great gift. How we use it and what we accomplish with it is not a gift but a responsibility upon which our own well-being and the well-being of other species depend.

SUMMARY

I. Social behavior is adaptive conspecific interaction. A society is a group of individuals of the same species that cooperate in an adaptive manner.
   A. In a society, there is a means of communication, cooperation, division of labor, and a tendency to stay together.
   B. Animals form societies because it is adaptive for them to do so.
II. Animal communication involves the transmission of signals but does not utilize (as far as is known) symbolic language in the human sense.
   A. Pheromones are chemical signals that convey information between members of a species.
III. Dominance hierarchies result in the suppression of aggressive behavior.
IV. Organisms often inhabit a home range, from which they seldom or never depart. This range, or some portion of it, may be defended from members of the same (or occasionally different) species.
   A. Defended areas are called *territories,* and the defensive behavior is *territoriality.*
   B. Often, territorial defense is carried out by display behavior rather than actual fighting.
V. Courtship behavior ensures that the male is a member of the same species, and it permits the female to assess the quality of the male.
   A. A pair bond is a stable relationship between a male and a female that ensures cooperative behavior in mating and rearing the young.
   B. Parental care increases the probability that the offspring will survive. A high investment

in parenting is often less advantageous to the male than to the female.

VI. Play gives the young animal a chance to practice adult patterns of behavior.

VII. Insect societies depend upon releasers and so tend to be rigid, with the role of the individual narrowly defined.

VIII. Vertebrate societies are far less rigid than insect ones. Although innate behavior is important, generally the role of the individual is learned.

    A. Human society is by far the most complex of all vertebrate societies.

IX. In altruistic behavior, one individual appears to behave in such a way as to benefit others rather than itself.

    A. Altruism may be closely related to kin selection.

    B. Kin selection may account for the evolution of complex societies of social insects in which only a few members reproduce.

X. Sociobiology is a school of ethology that focuses on the evolution of behavior through natural selection.

☐ POST-TEST

1. Adaptive interactions among members of a population are referred to as _____ _____.

2. A _____ is a group of individuals belonging to the same species that cooperate in an adaptive manner and have a means of communicating with one another.

3. An important difference between human and animal communication is that animal communication is not generally _____.

4. _____ are chemical signals that convey information between members of a species.

5. An arrangement of members of a population by status is called a _____ _____.

6. The geographical area that members of a population seldom leave is the _____.

7. Territoriality tends to reduce _____ and control _____ growth.

8. A _____ _____ is a stable relationship between animals of the opposite sex that en-

sures cooperative behavior in mating and rearing the young.

9. In a beehive, the youngest bees serve as _____ bees; they secrete _____ _____ on which larvae feed.

10. The extensive behavioral repertoire of the bee is almost entirely _____ (innate or learned).

11. Human society differs from other animal societies in that it depends mainly on the transmission of _____.

12. In _____ behavior, one individual appears to act to benefit others rather than itself.

13. _____ selection favors the indirect perpetuation of an animal's genes by a relative.

14. According to sociobiology, an organism and its adaptations are ways that its genes have of _____.

☐ REVIEW QUESTIONS

1. What distinguishes an organized society from a mere aggregation of organisms? Cite an example of an organized society, and describe characteristics that qualify the society as organized.

2. How many similarities between the transmission of information by symbolic language and by heredity can you think of? How many differences?

3. Contrast the "language" of bees with human language.

4. How does an organism learn its place in a dominance

hierarchy? What determines this place? What are the advantages of a dominance hierarchy?

5. What is territoriality? What functions does it seem to serve?

6. What is sociobiology?

7. What is kin selection? How is kin selection used by sociobiologists to explain the evolution of altruistic behavior?

8. Do animals play just for the fun of it?

9. What are some advantages of courtship rituals?

☐ RECOMMENDED READINGS

Bonner, J.T. *The Evolution of Culture in Animals.* Princeton, Princeton University Press, 1980. The title tells the contents but fails to convey the charm of this scholarly little book.

Franks, N.R. and B. Hölldobler, "Sexual competition during colony reproduction in army ants," *Biological Journal of the Linnean Society* 30, 1987.

Goodenough, J. *Animal Communication.* Burlington, NC, Carolina Biological Supply Company, 1984. A perfect textbook supplement that discusses the details of animal communication far more extensively than space permits in this chapter.

Gould, S.J. "The Guano Ring," *Natural History,* January 1982. An easily studied and interpreted example of territoriality.

Grady, R.M. and J.L. Hoogland, "Why do male black-tailed prairie dogs (*Cynomys ludovicianus*) give a mating call?" *Animal Behavior* 34, 1986.

Heinrich, B. "The Regulation of Temperature in the Honey-bee Swarm," *Scientific American,* June 1981. A discussion of thermo-regulation in a swarm of bees.

Hoogland, J.L. "Nepotism in prairie dogs (*Cynomys ludovicianus*) varies with competition but not with kinship," *Animal Behavior* 34, 1986.

Ligon, J.D., and S.H. Ligon. "The Cooperative Breeding Behavior of the Green Woodhoopoe," *Scientific American,* July 1982. Kin selection and altruism in a vertebrate society.

Page, R.E., Robinson, G.E. and M.K. Fondrk, "Genetic specialists, kin recognition and nepotism in honey-bee colonies," *Nature* 338, 1989.

Partridge, B.L. "The Structure and Function of Fish Schools," *Scientific American,* June 1982. Schooling confuses predators and is coordinated by lateral line sensory input.

Sherman, P.W., "Mate guarding as paternity insurance in Idaho ground squirrels," *Nature* 338, 1989.

46

Communities and Population Ecology

OUTLINE

I. Organisms live together in communities and ecosystems
II. Organisms play specific roles in energy transfer
   A. Producers make food from simple chemicals
   B. Consumers obtain food from other organisms
   C. Decomposers recycle materials from corpses and wastes
III. Substances are recycled within communities
   A. Carbon is recycled via carbon dioxide
   B. Bacteria are essential to the nitrogen cycle
   C. The phosphorus cycle is determined by insolubility and a lack of gaseous compounds
IV. Organisms are adapted to characteristic ecological niches
V. Populations of organisms make up communities
   A. Organisms vary in the number of young they can produce
   B. Life spans vary among different species
   C. A population tends to grow
   D. Some limiting factors reduce population regardless of size
   E. Some limiting factors dynamically regulate population growth
VI. Energy and substances travel through food chains and webs
   A. Food relationships are best described as webs
   B. Food pyramids reflect the biomass, energy, and numeric relationships of trophic levels
   C. Ecosystems vary in productivity
VII. Communities vary in species diversity
VIII. Communities develop
   A. Sometimes communities must start with a lifeless habitat
   B. Communities sometimes develop where a predecessor community already exists
   C. Why does succession occur?
   D. Why is the climax community stable?
Focus on Microcosms

LEARNING OBJECTIVES

1. Define community and ecosystem. Give the salient characteristics of each, and give an example of each.
2. Characterize producers, consumers, and decomposers, and give the function of each category of organism in a community.
3. Summarize the concept of limiting factors, and describe their relationship to the ecological niche.
4. Give the main factors that produce population change.
5. Define environmental resistance and give its role in determining population growth and size.
6. Summarize the role of limiting factors in environmental resistance.
7. Distinguish the effects of density-dependent limiting factors from those of density-independent limiting factors.
8. Give examples of density-dependent limiting factors.
9. Define K-strategies and R-strategies, and give an example of each.
10. Define ecological niche and give examples.
11. Summarize the carbon, nitrogen, and phosphorus cycles.
12. Summarize the concept of a food chain and compare it with that of a food web.
13. Describe a typical food pyramid with emphasis on the biomass to be expected at different trophic levels.
14. Give reasons for the reduction of biomass with trophic level, and summarize at least two practical consequences of this reduction.
15. Summarize the main determinants of community diversity.
16. Outline the concept of ecological succession.
17. Contrast autogenic with allogenic succession.

Hippopotamus in Virunga National Park, Zaire. (Robert E. Ford/BPS [129–789])

 As we near the end of this book, we come to the point where what you have learned in biology should begin to come together. Until now we have looked at details—how chlorophyll absorbs light energy, how the citric acid cycle powers cells, how DNA codes genetic information, and how the queen helps to keep a beehive in business. What does it all mean? How does it fit together?

The bee depends on the chlorophyll, without which there could be no nectar. Often, the chlorophyll depends on the bee, without which many plants could not be pollinated. Both depend upon the intricate chemistry whereby certain soil bacteria bring atmospheric nitrogen into chemical combination. All depend upon wind, rain, and sunlight. There is another kind of influence that enters more and more into all of these age-old relationships. A single species, *Homo sapiens,* now generates forces of a geological order of magnitude, forces that reach increasingly into the most remote crannies of the biosphere. As we begin our study of ecology, we must consider them all.

KEY CONCEPTS

☐ No organism exists in solitude; all occur in communities of interacting organisms.

☐ All or almost all organisms play one of three main roles in community life: producer, consumer, or decomposer.

☐ The adaptations of an organism interact with its environment and produce its distinctive life-style or ecological niche.

☐ Competition from other organisms constrains the actual ecological niche to narrower dimensions than it theoretically could assume.

☐ Population growth is fundamentally constrained by birth rate and death rate.

☐ Both death rate and birth rate can be affected by limiting factors in the environment—essentials that are in short or sometimes in excessive supply.

☐ Density-independent limiting factors do not maintain population at any normative value, yet they are not altogether unrelated to population density.

☐ Density-dependent limiting factors dynamically regulate population.

☐ All vital substances are recycled within and between communities, but some more readily than others. The least readily recycled substances are most likely to be limiting factors.

☐ Although energy cannot be recycled within ecosystems, it is transferred through food chains and webs and is dispersed in the process of transfer.

☐ The progressive loss of energy in food chains gives rise to pyramids of biomass. The terminal organisms of food chains usually have the least biomass of any trophic level. The primary producers have the greatest biomass.

☐ Although the reasons that ecological succession occurs are not fully understood and appear to vary, communities replace one another in orderly fashion until a final stable (or very slowly changing) state is attained.

☐ ORGANISMS LIVE TOGETHER IN COMMUNITIES AND ECOSYSTEMS

For the biologist, a **community** is an association of organisms of different species living together with some degree of interdependence (Fig. 46–1). Thus, you, your dog, and the fleas on your dog are all members of the same community! One might add cockroaches, silverfish, dandelions, grass, maple trees, and much more to the list.

Organisms form integrated communities of varying sizes. Because communities are rarely completely isolated, they really do interact with one another, even when that interaction is not readily apparent. Furthermore, we can think of communities as being nested within one another like Chinese boxes; there are communities within communities. A forest is a community, but so is a rotting log in that forest, containing fungi, insects, and even mice. So, too, is the association of microorganisms living within the gut of a termite in that log.

Living things also have a nonliving environment that is essential to their lives. Food, minerals, air, water, and sunlight are just as much a part of a honeybee's life as the flowers that it pollinates and from which it takes nectar. The nonliving environment as well as the living communities it contains make up an **ecosystem.** Although an ecosystem typically is fairly self-sufficient, like a community, it lacks precise boundaries. Thus, the fate of a far-off tropical rain forest may profoundly influence weather patterns in a temperate zone, and the pesticides deposited on a prairie may affect the invertebrates and birds of a distant river delta.

☐ ORGANISMS PLAY SPECIFIC ROLES IN ENERGY TRANSFER

We place the organisms of a community into certain categories: producer, consumer, and decomposer. Most stable communities contain representatives of all three categories. All interact extensively with one another (Fig. 46–2).

(a)

(b)

(c)

FIGURE 46–1 Each organism is a member of a community, often specifically adapted for its life in a given community. (*a*) The water requirements of the African "living stone" plant are minimal; it is specifically adapted for life in a dry habitat. (*b*) How different are the needs of water lilies! Their autecology excludes them from the desert communities inhabited by "living stones," although they might be capable of some degree of interaction in a desert oasis. (*c*) This complex coral reef community contains many subcommunities and is itself part of the marine habitat. (*a* and *b*, P. William Davis; *c*, Robin Lewis, Coastal Creations)

Producers Make Food From Simple Chemicals

Producers manufacture complex organic molecules from simple inorganic substances, usually using the energy of sunlight to do so. In other words, they are autotrophs. By incorporating their chemical output into their own bodies, producers turn their bodies or body parts into a food resource for other organisms. Plants and plant-like protists are the most significant producers, but in some places (such as deep-sea hot springs), other organisms are more important.

Consumers Obtain Food From Other Organisms

Animals, animal-like protists, and a few predatory bacteria and fungi comprise the consumers. **Consumers** are organisms that use the bodies of other organisms as a source of food energy and body-building materials. Herbivores, the plant eaters, are **primary consumers. Secondary consumers** include meat eaters, the **predators** or **carnivores,** which consume only other animals. Still others, the **omnivores,** consume a variety of organisms, whether plant or animal. Many animals, however, do not fit readily into one of these categories.

Parasitism is one of the most effective of such special life-style adaptations. Although a parasite, like a predator, attacks a living victim (called a **host**), it does not immediately (and may never) kill the host organism. Some parasite-like predators do gradually kill the host. These are **parasitoids,** such as certain wasp larvae that infest spiders, caterpillars, or other prey for extended periods (Fig. 46–3). The parasitoid adaptation is actually widespread among insects. Some parasitoids may have hyperparasitoids that, in turn, live on them! It is estimated that fully 25% of all animal species are parasitoids, most of them insects.

Finally, we come to the scavengers, also called **detritus feeders.** In aquatic habitats, these scavengers often burrow in bottom muck, consuming the organic matter collected there. Earthworms are a terrestrial equivalent, as are termites and many fly larvae.

Decomposers Recycle Materials From Corpses And Wastes

Both decomposers and detritus feeders liberate materials that are tied up in wastes and the corpses of dead organisms. These materials then become available for

(a)

(b)

(c)

(d)

(e)

FIGURE 46–2 Producers, consumers, and decomposers. (*a*) The trees and epiphytic vegetation of the tropical rain forest account for most of the primary production of food within this ecosystem. (*b*) The three-toed sloth is a primary consumer; it eats mostly leaves. The low energy content of this fare may account for the sloth's proverbial sluggishness . (*c*) Sloths are a common prey for the Harpy eagle, a secondary consumer or predator. (*d*) Another style of secondary consumption is exemplified by the king vulture, a scavenger or detritus feeder. (*e*) Finally, decomposers such as this cup fungus reduce all living things to their mineral consituents, plus carbon dioxide and water. (*a*, Frans Lanting; *b*, Animals Animals © 1990 Mickey Gibson; *c*, Steven Holt/VIREO; *d*, William J. Weber/Visuals Unlimited; *e*, James L. Castner)

FIGURE 46–3   An adult fly lays her eggs on a gypsy moth caterpillar. The eggs will hatch into parasitoid larvae that will eat the caterpillar alive. (Grant Heilman/Grant Heilman Photography)

re-use by primary producers and, in due course, for consumers. Thus, the decomposers are necessary to the survival of any community. Without them, not only would corpses and waste products accumulate indefinitely, but elements such as carbon, nitrogen, and phosphorus contained in various organisms would become permanently locked up. Many bacteria and fungi are decomposers.

◼ SUBSTANCES ARE RECYCLED WITHIN COMMUNITIES

The earth and its biosphere are a closed system. Short of a cosmic catastrophe, the materials utilized by organisms cannot be lost, although they can end up in locations that are outside the reach of living things.

Carbon Is Recycled Via Carbon Dioxide

Because organic life is based upon the properties of the carbon atom, carbon must be available to living things. Much carbon enters the atmosphere in the form of carbon dioxide emitted from industrial combustion and volcanoes. By photosynthesis, plants and cyanobacteria remove carbon dioxide from the air and **fix** it—that is, incorporate it into complex chemical compounds (Fig. 46–4). These compounds may then be used for fuel, by the producer that first made them or by a consumer or decomposer that has the opportunity. In the process, carbon dioxide is liberated once again.

Carbon also can be carried out of the reach of living things. Vast coal beds have formed from the bodies of ancient trees that did not decay fully before they were buried. A significant long-term loss of carbon

dioxide occurs when it is incorporated into the calcium carbonate shells of marine organisms, which can form sea-bed deposits thousands of feet thick. Almost the only way such carbon can re-enter the atmosphere is by volcanic action.

Although biological carbon fixation and geological deposit should keep the atmospheric carbon dioxide content nearly constant indefinitely, these processes require thousands or even millions of years to be effective. This means that substantial emissions of carbon dioxide by increased volcanic action or human activity could cause carbon dioxide to accumulate in the atmosphere, which would cause the earth's temperature to rise disastrously and produce great climatic changes that would be permanent for most human purposes.

Bacteria Are Essential To The Nitrogen Cycle

We and other consumers ultimately depend upon photosynthetic autotrophs for chemically combined or **fixed nitrogen** that is utilized in the form of proteins and nucleic acids. Plants depend upon both chemosynthetic autotrophs and decomposers (which liberate nitrogen from proteins and wastes in the form of ammonia) for recycling the nitrogen upon which their survival depends (Fig. 46–5). Atmospheric nitrogen can be fixed by high temperatures that cause it to combine with oxygen to form nitrogen oxides. These can combine with water to form nitrate. Although combustion, volcanic action, and lightning discharges are important sources of fixed nitrogen, most nitrogen fixation is carried out by **nitrogen-fixing microorganisms,** which must employ the enzyme **nitrogenase** to do so.

Most plants require nitrogen in the form of nitrate (NO_3^-) or ammonium (NH_2), mainly to manufacture amino and nucleic acids. Nitrate is produced by bacte-

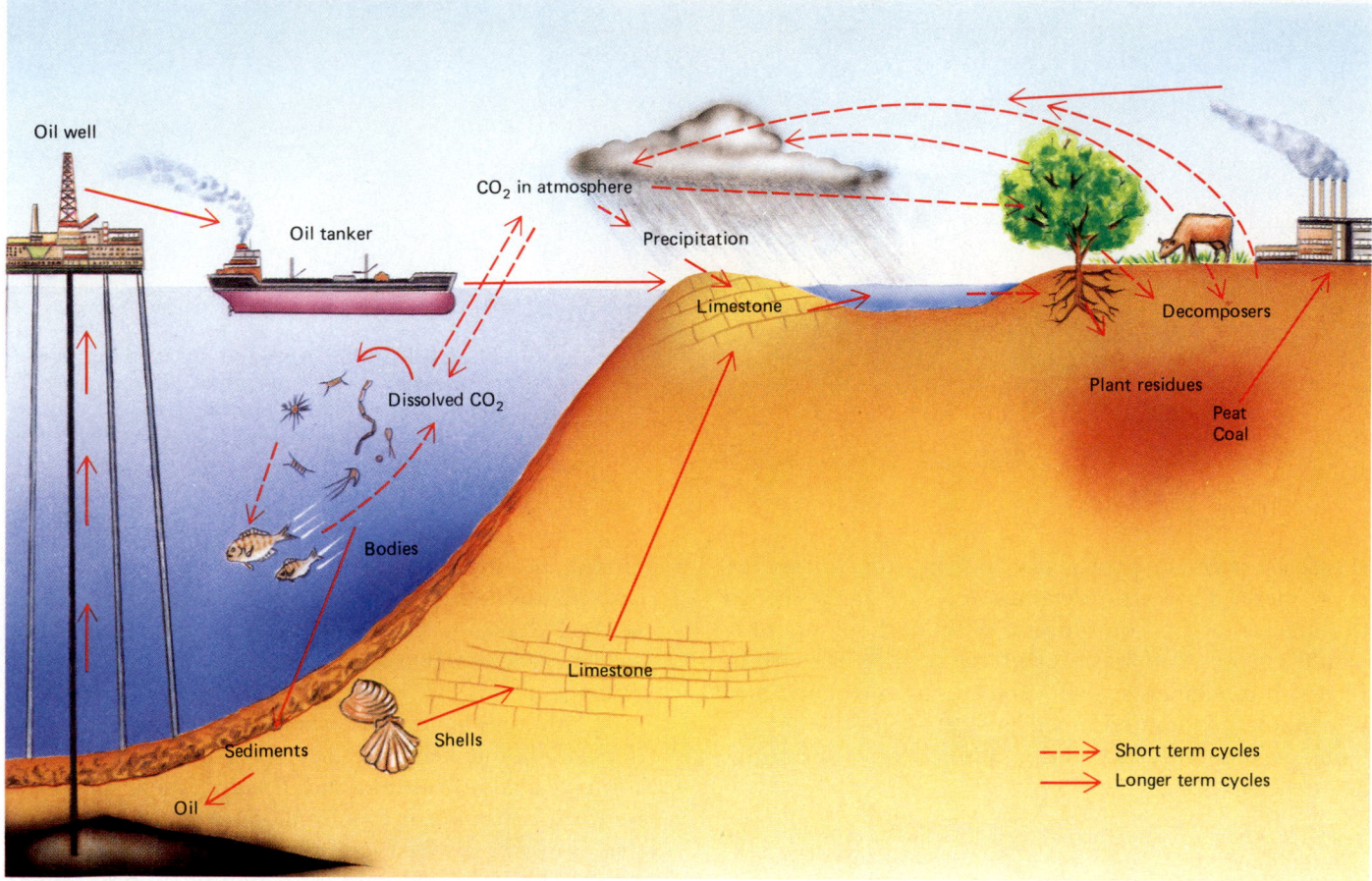

FIGURE 46–4   Simplified diagram of the carbon cycle. The dashed arrows indicate rapid processes; the solid arrows denote relatively slow ones, for example the geological uplift that carries limestone to the surface where its carbonate dissolves in water in the form of bicarbonate ion and is utilized by aquatic plants as a carbon source.

ria known as **nitrifying** bacteria, which oxidize nitrite (NO_2^-) to nitrate (NO_3^-). Nitrite, in its turn, is produced from ammonia by another form of nitrifying bacteria that oxidize ammonia as *their* source of energy. Ammonia is a common product of the decay of protein and other nitrogen-containing materials in corpses and wastes.

Not all of the nitrogen compounds released by decay become available to plants. **Denitrifying bacteria** are anaerobic bacteria that use nitrate to produce free nitrogen. They reverse the action of the nitrifying and nitrogen-fixing bacteria, preventing the direct recycling of a large portion of the nitrogen liberated by decay.

Nitrogen deficits can be made good by nitrogen-fixing microorganisms, cyanobacteria, and certain other bacteria. The nitrogenase that these microorganisms employ can only function in the absence of oxygen, so the enzyme must be insulated from oxygen by some means. The most important terrestrial nitrogen-fixing microorganisms live in special swellings, or **nodules,** on the roots of leguminous plants such as beans or peas (Fig. 46–6) and some other woody plants.

In aquatic habitats, cyanobacteria do much of the nitrogen fixation. Filamentous cyanobacteria have **heterocyst cells,** which fix nitrogen and which are not themselves photosynthetic. (If they were, the oxygen they generated would inactivate their nitrogenase.) Some water ferns have cavities in which such cyanobacteria live, somewhat as nitrogen-fixing bacteria live in root nodules.

The Phosphorus Cycle Is Determined By Insolubility And A Lack Of Gaseous Compounds

As water runs over rocks, it gradually wears away the surface and carries off a variety of minerals, the most important of which are phosphates. Because the most common phosphorus minerals are quite insoluble, they

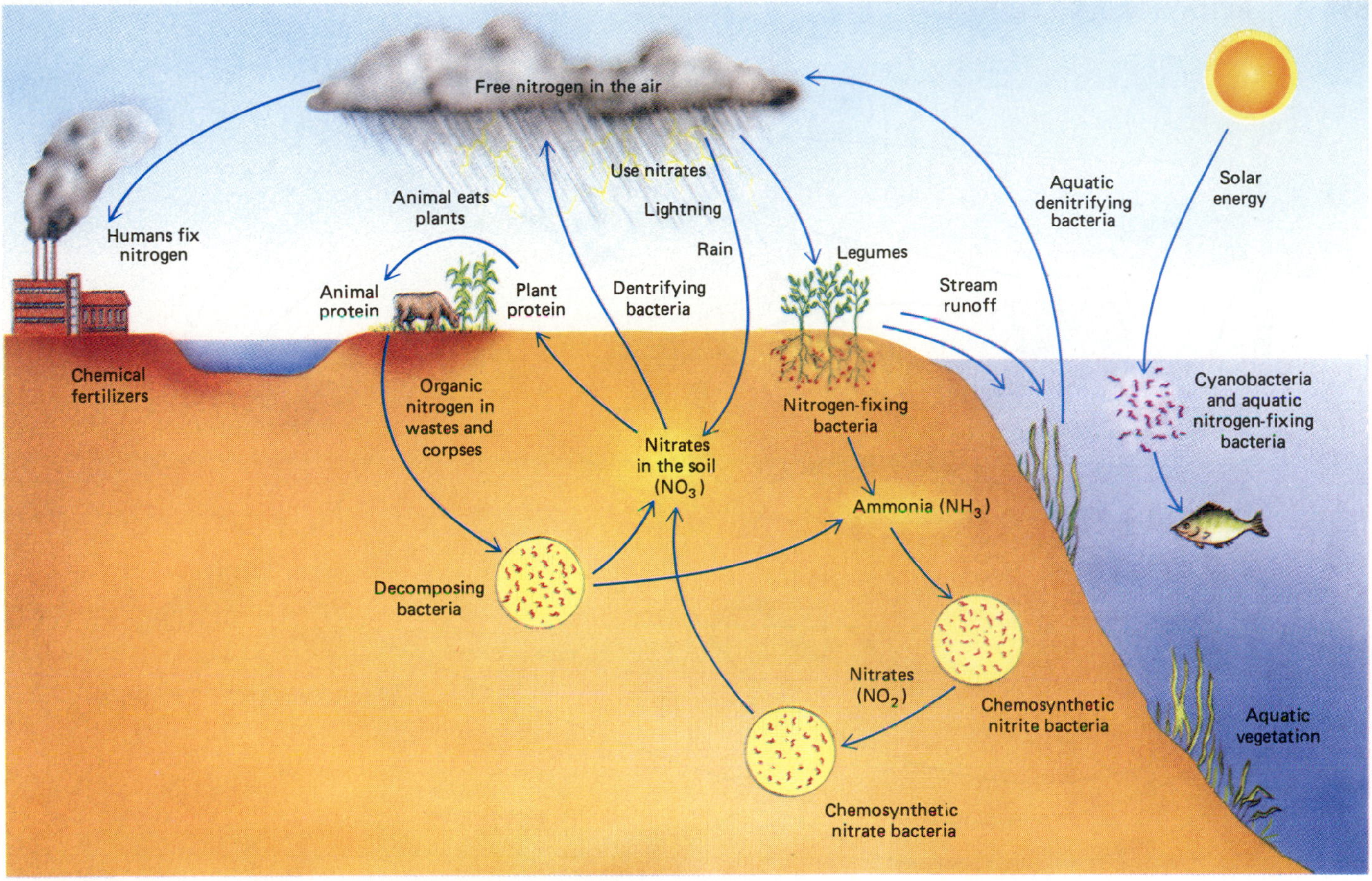

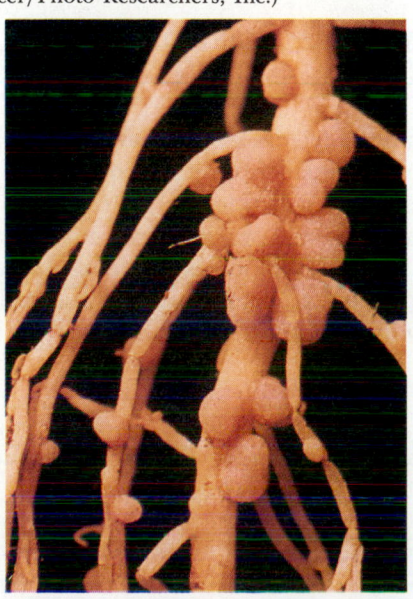

FIGURE 46–5 The nitrogen cycle, including the effects of human activities. We have deliberately *not* simplified this diagram so as to convey some impression of the multitude of pathways whereby this vital substance travels throughout the biosphere. (Adapted from National Research Council, *Nitrates: An Environmental Assessment,* National Academy Press, 1978)

FIGURE 46–6 Root nodules of a pea plant. Mutualistic *Rhizobium* bacteria live in these nodules, living on energy derived from sugars provided by their legume host. The bacteria efficiently fix nitrogen, some of which is utilized by the host plant. The ultimate death and decay of both partners enriches the soil with the nitrogen they have brought into chemical combination. (Hugh Spencer/Photo Researchers, Inc.)

tend to be deposited permanently on the sea floor (Fig. 46–7). However, geological processes of uplift may some day expose these sediments on new land surfaces from which they can once again be eroded.

Phosphorus enters aquatic communities through absorption by algae and plants, which are in turn consumed by large organisms and by microorganisms and other plankton. These are then eaten by various fin and shell fish, which may also feed upon one another. Phosphate-rich body parts, especially bones, eventually end up on the bottom. A small amount of phosphorus is deposited in the droppings of sea birds at their nesting and roosting grounds.

Phosphorus in the soil of plant communities is taken in by plant roots in the form of inorganic phosphates such as magnesium phosphate. Animals obtain most of their phosphate as inorganic or organic compounds in the food they eat, which is ultimately derived from plants.

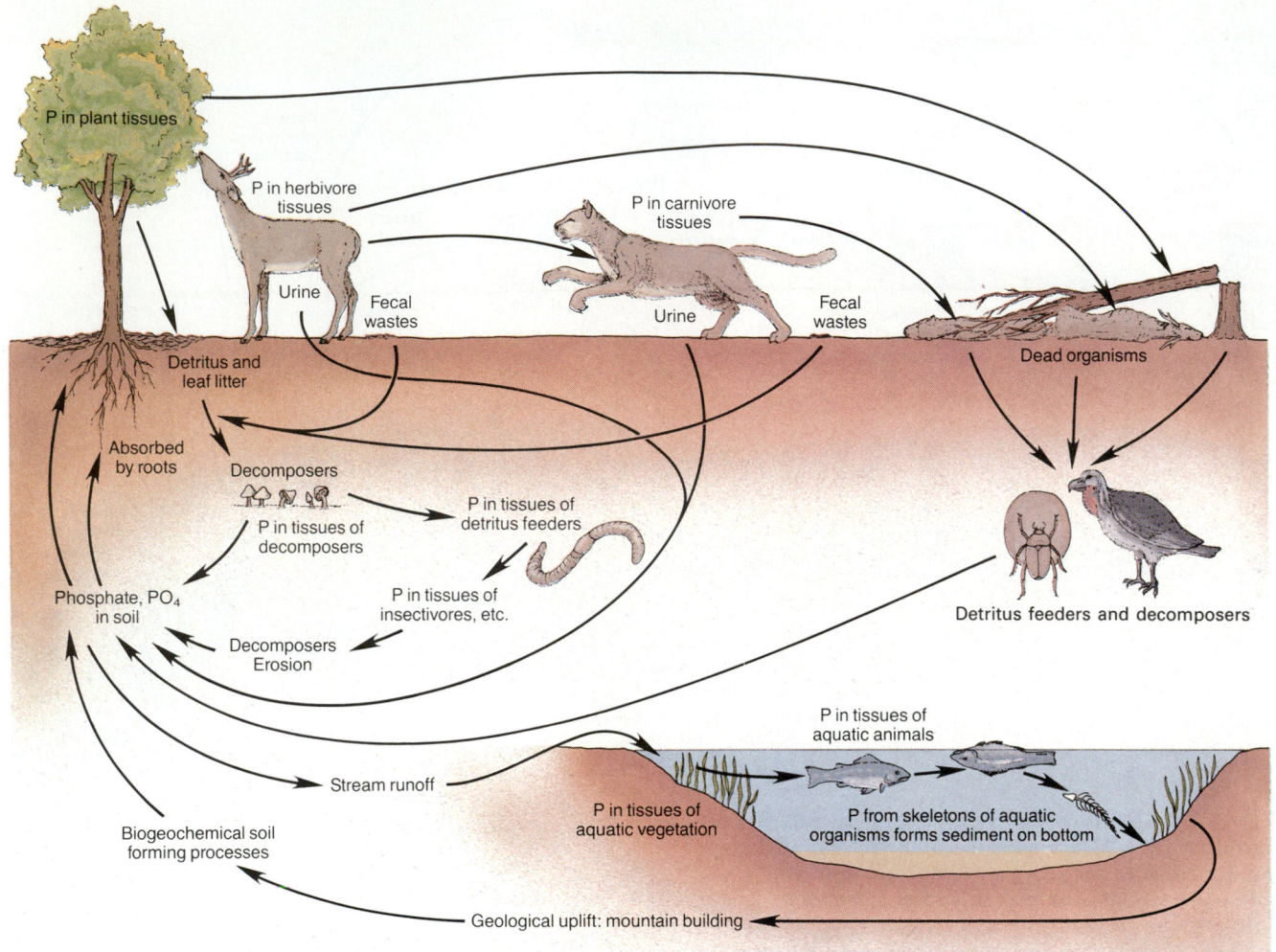

FIGURE 46–7   The phosphorus cycle. Recycling of phosphorus is rendered difficult by the fact that no biologically important compound or state of phosphorus is gaseous.

◻ ORGANISMS ARE ADAPTED TO CHARACTERISTIC ECOLOGICAL NICHES

Every organism has an **ecological niche,** which is the totality of its adaptations, its use of resources, and the life-style to which it is fitted (Fig. 46–8). An ecological niche is an organism's way of life. Thus, the ecological niche of a dog flea is not merely the dog, but the flea's special version of a parasitic way of life, its life history (much of which is spent away from the dog), and much else.

An organism usually has a wide range of potential resources available to it. The *potential* ecological niche of an organism is called its **fundamental niche.** However, an organism at any one time usually is able to use only some of these resources. Various factors, such as competition with other species, may exclude it from

part of its fundamental niche. Thus, the life-style that an organism actually pursues and the resouces that it actually uses comprise its **realized niche.** An organism is usually capable of utilizing much more of its environment's resources than it actually does use.

The little Carolina anole lizard (Fig. 46–9), for example, was once widespread in its native home, Florida. The habitat portion of its fundamental niche included both trunks and crowns of trees, exterior house walls, and much else. When a related species of Cuban anole was introduced to the area, however, the Carolina anoles seemed to disappear. Careful investigation disclosed that the Carolina anoles were now confined largely to the foliated crowns of trees, where they were less easily seen. Their realized niche had become much smaller as a result of competition from the Cuban anole. Because all natural communities consist of nu-

(a) Cape May warbler

(b) Bay-breasted warbler

(c) Blackburnian warbler

(d) Black-throated green warbler

(e) Myrtle warbler

FIGURE 46–8   Some songbird ecological niches. Each of these species of *Dendroica*, common warblers, spends most of its feeding time in different portions of the trees it frequents and also consumes somewhat differing insect food. The colored regions indicate where each species spends at least half its foraging time. (After MacArthur)

FIGURE 46–9   (a) Green Carolina anole. (b) Brown Cuban anole. (*a*, Runk/Schoenberger, from Grant Heilman; *b*, Steven G. Maka)

(a)

(b)

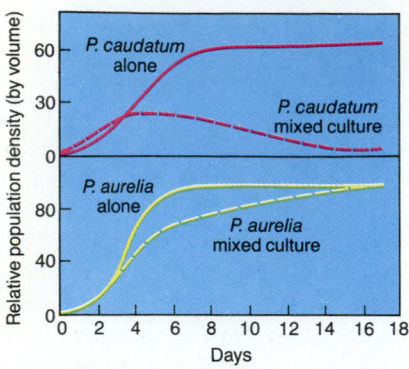

FIGURE 46-10 Competition among *Paramecium*. The solid-line curves show how each species of *Paramecium*'s population grows in a single-species environment; the dotted curves show how they grow when in competition with each other. (After G.F. Gause, *Science*, vol. 79, 1934)

merous species, many of which compete to some extent, the complex interaction among them produces the realized niche of each.

What if there is little difference in the realized niche of two species? The Russian biologist A. F. Gause grew two similar species of *Paramecium*, *P. aurelia* and *P. caudatum*, together (Fig. 46–10). Only *P. aurelia* thrived; *P. caudatum* dwindled and eventually died out. "As a result of competition," wrote Gause, "two similar species scarcely occupy similar niches, but displace each other in such a manner that each takes possession of certain peculiar kinds of food and modes of life in which it has an advantage over its competitor."

These experiments gave rise to the principle of **competitive exclusion:** Evolutionary forces tend to drive the ecological niches of ecologically similar organisms apart, with the result that they acquire differing clusters of adaptations. This is called **niche differentiation.**

■ POPULATIONS OF ORGANISMS MAKE UP COMMUNITIES

A **population** is a group of individuals of the same species that occupies a particular area. Population size is governed by principles of population growth, dispersal, and limitation.

Organisms Vary In The Number Of Young They Can Produce

Two things determine population growth: biotic potential and mortality. The maximum rate at which a population could increase under ideal conditions is its

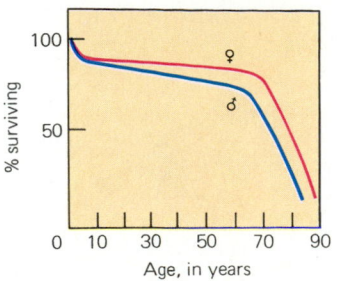

FIGURE 46–11 A human survivorship curve, based on 1980 census data.

biotic potential. This is rarely the rate at which it *does* grow. Many factors can determine the biotic potential of a population. Biotic potential depends not only upon the number of offspring produced at a time, but how frequently they are produced, and on how long the reproductive life of the organism may be.

Life Spans Vary Among Different Species

The number of offspring produced is the birth rate, or **natality,** of a population. No matter how many offspring are produced, however, their death rate, or **mortality,** determines how many will remain. As an example, let us consider the survivorship of human beings in the United States in the early 1980s (Fig. 46–11). As the graph in the figure shows, there was an early hazardous time of life for newborn infants and young children, especially males. This resulted principally from premature birth, congenital defects, and undeveloped immunity to infectious disease. Once this initial period was safely passed, however, few people died until well into their thirties. The mortality rate increased substantially in middle age, with a great acceleration in the fifties and thereafter. In the sample

FIGURE 46–12 Survivorship curves for a herring gull population. Data collected from Kent Island, Maine, 1934–1939, as reported by Paynter. Baby gulls were banded to establish identity.

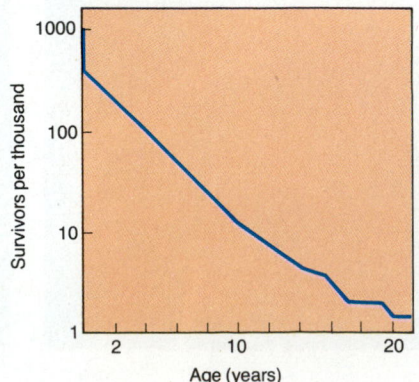

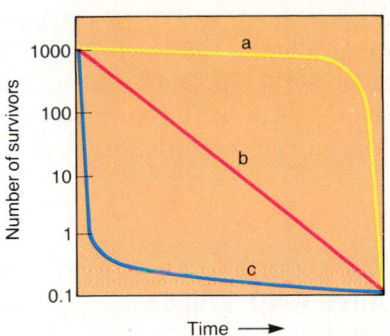

FIGURE 46–13 Three theoretical survivorship curves describing two extremes in life span strategy (*a*) and (*c*) and a third intermediate state (*b*).

studied, no one lived longer than 109 years and the average life span was 68.5 years for males and 73.2 years for females. Because it shows the number of organisms that survive to reach a particular age, this figure is an example of a **survivorship curve.**

Human beings are long-lived organisms whose young have a high likelihood of survival. This curve reflects an almost complete freedom from predation, excellent care of the young, and a combination of medical and social measures conducive to the reduction of infant mortality, although not necessarily to the reduction of mortality at later ages. Although similar survivorship curves have been found in some large and formidable animals, such a curve is not characteristic of most species.

Another type of survivorship is exhibited by the herring gull (Fig. 46–12). Notice that the vast preponderance of mortality occurs almost immediately despite the protection and care given to the chicks by the parent birds. Much of this reflects predation or attack and destruction by other herring gulls, but some occurs as a result of such factors as inclement weather, infectious disease, or starvation following the death of the parent.

Once the chicks become fully fledged and independent, their survivorship increases dramatically. Notice, however, that mortality occurs at about the same rate (as reflected in the slope of the graph) throughout the remaining life of the birds. This constancy probably results from essentially random events that cause death year after year with little age bias. As a result, few or no herring gulls in nature die from the degenerative diseases of "old age" that cause death in most humans. Partly for this reason the maximum ages observed for animals kept in protected captivity are not necessarily indicative of what happens in a natural state.

Based on wide experience with a wide variety of organisms, ecologists can construct a family of survivorship curves that typify (Fig. 46–13) the ideal survivorships of organisms in which mortality is greatest among the young, mortality is spread evenly across all age groups, and mortality is greatest in old age. The survivorship of most organisms can be compared to these curves.

A Population Tends To Grow

When a population's growth is rapid, a graph of its increase resembles the letter J (Fig. 46–14). This shape is caused by an initial **lag phase,** in which growth is slow, followed by a **logarithmic phase** of very rapid growth, in which population growth may *temporarily* approach biotic potential. Only a balance between birth rate and death rate can stop this growth. The growth of any population is usually slowed and then stopped by a reduced birth rate or an increased death rate, so that the graph begins to resemble a slanted letter S. The totality of all the factors (including shortage of food or suitable nesting sites, increased incidence of disease, lack of places to hide from predators, and much else) that work against population growth is called **environmental resistance.**

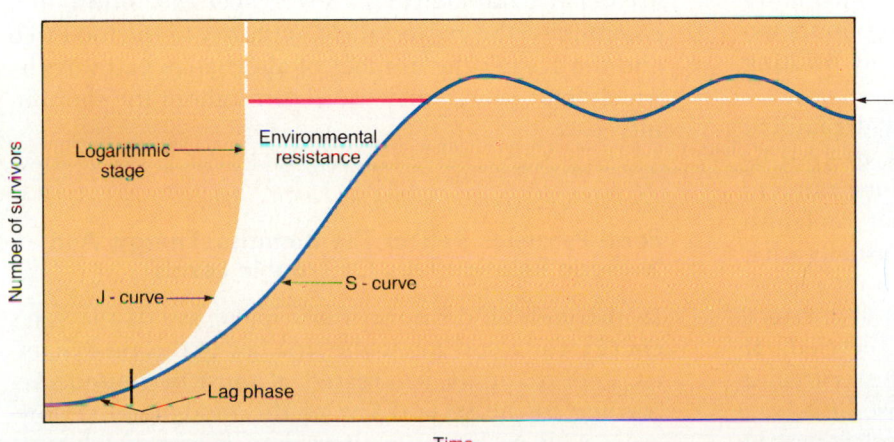

FIGURE 46–14 A typical, simple growth curve such as can be observed initially in a bacterial culture. The J-shaped curve of population growth changes to an S-shaped curve as the result of increasing environmental limitations.

Environmental resistance often causes populations to stabilize somewhere around the highest point they attain, but not always. We sometimes see a **population crash,** which is an abrupt decline from high population density to very low density. Such an abrupt change can even result in local extinction. When all of the vital resources are consumed, all of the organisms—to the last individual—die.

Some Limiting Factors Reduce Population Regardless Of Size

Random events that are hostile to the life requirements of organisms can reduce their population and serve as **density-independent limiting factors.** These factors have about the same effect regardless of population size, and often affect population density in unpredictable ways. A very severe blizzard, a hurricane, a collision with an asteroid—all might cause extreme and irregular reductions in populations. In fact, high-density populations can be more vulnerable to such disasters owing to a shortage of refuges or other resources. If the limiting factor is severe and recurrent, it can periodically halt the otherwise unchecked growth of a population. For example, in many temperate and arctic environments, mosquitoes are able to produce several generations per summer, achieving very high population densities by the end of the season. The coming of winter halts their growth. Not a single adult mosquito survives it; the entire population must grow afresh the next summer from the few eggs and hibernating larvae that do survive.

Some Limiting Factors Dynamically Regulate Population Growth

Although populations of some organisms do fluctuate violently, this is not usually the case. A steady-state population is produced by the presence of limiting factors that act like population "thermostats." Such limiting factors increase in effectiveness as the population increases and are called **density-dependent limiting factors.**

Resource depletion is often a density-dependent factor. Populations do not usually totally destroy or waste the *in*organic resources of their environment because of the recycling mechanisms that replenish them. Thus, a population's resources are usually what we would call **renewable resources.** Yet even renewable resources are limited by their initial abundance and the rapidity with which they are recycled. If human beings consume or pollute water faster than it is recycled, they will effectively lose the use of it.

Competition is another factor that increases at high population densities. Eventually, it may reach the point where many members of the population fail to obtain the minimum of whatever resource is in shortest supply. This raises the mortality rate and inhibits further population growth. If the competition arises from another species, it can lead to extinction.

■ ENERGY AND SUBSTANCES TRAVEL THROUGH FOOD CHAINS AND WEBS

Erasmus Darwin (Charles Darwin's grandfather) is said to have been fond of repeating the phrase "eat or be eaten." Whatever the social truths of the saying may be, to be biologically true it must be amended to "eat *and* be eaten"! Virtually everything eats plants, directly or indirectly. The herbivores or omnivores that feed upon them directly are themselves eaten by predators or fed upon by parasites, who in turn may be eaten by other predators and parasites. When any of the preceding die or produce waste products, their corpses or wastes may be eaten by detritus feeders, scavengers, or decomposers, all of which will also be eaten, for all of them will ultimately die.

Food Relationships Are Best Described As Webs

A simple list of the eaters and the eaten describes a **food chain.** Because few organisms other than parasites eat just one kind of other organism, there are rather few such highly restricted food chains in the world of life. More typically, the flow of energy and materials through communities takes place in accordance with a range of choices of food. In a community of reasonable complexity, numerous alternative pathways are possible. One could best express these as a diagram resembling a web.

Thus, a **food web** is a more realistic model of the flow of energy and materials through the numerous pathways in a typical community (Fig. 46–15). The greater the number of different species residing in a community, the greater the complexity of the food web and the greater the number of alternative pathways by which food energy may travel throughout the community.

Food Pyramids Reflect The Biomass, Energy, And Numeric Relationships Of Trophic Levels

In the end, all creatures must make use of the sun's energy, for according to the first law of thermodynamics, energy cannot be created; it must be reused. According to the second law of thermodynamics, however, with each reuse, there is a substantial loss of energy. Only a small amount of the energy (food)

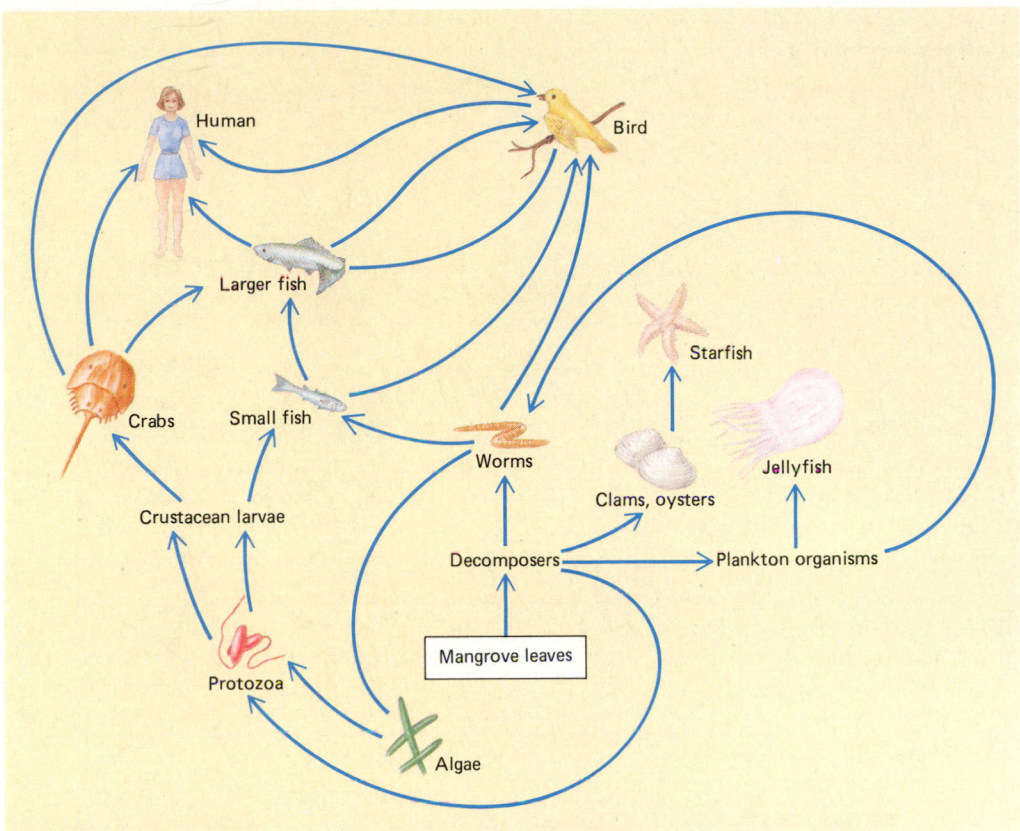

FIGURE 46–15   A greatly simplified diagram of a marine food web. Arrows point from each organism to its consumer or consumers. Inclusion of alternative food choices would greatly increase the interconnections.

available to a group of consumers can be used for body building. This is reflected in the **biomass,** which is the total mass of a particular population.

Organisms may be classified roughly according to how closely their feeding habits and adaptations place them to the community autotrophs. The **primary producers** (which produce all the food in the community in the first place) constitute the first **trophic level** (*trophic* means "feeding") and usually possess the highest biomass and the highest numbers of all trophic levels. At the second trophic level are the primary consumers (the herbivores); secondary consumers (that is, carnivores) are at the third trophic level. It is rare to find more than four trophic levels, at least in terrestrial communities, because each successive level is usually much reduced in both biomass and numbers as compared to its predecessor. A diagram that reflects these relationships graphically resembles a pyramid (Fig. 46–16), and is known as a **food pyramid.**

FIGURE 46–16   Trophic levels; a trophic pyramid, much idealized. Numbers are illustrative only and are not intended to be exact.

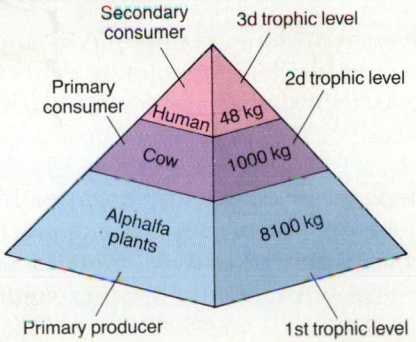

Ecosystems Vary In Productivity

Throughout the world, productivity varies partly because one ecosystem differs from the next in terms of the supply of needed materials (Fig. 46–17). In almost all cases, the amount of food available to all trophic levels is ultimately determined by how well the plants fix carbon dioxide, for that partly determines how well they grow. The productivity of a given community, depending as it does upon that of its plants, varies not only because of raw materials but also in accordance with such factors as how much solar energy it receives, the presence of climatic or other limiting factors, and

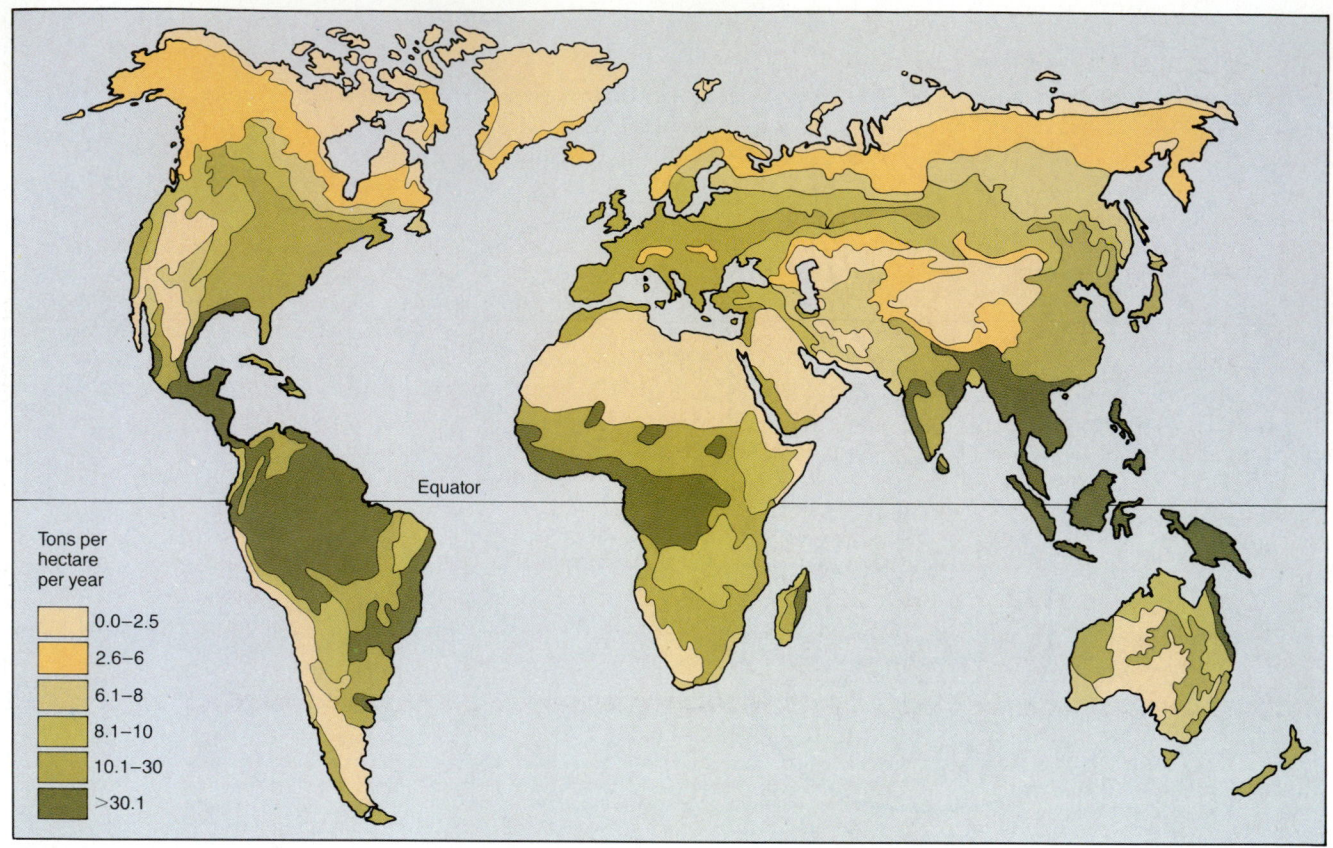

FIGURE 46–17   Worldwide terrestrial productivity. Temperature, rainfall, and insolation determine the productivity of land ecosystems. Wet tropical regions rank the highest, with deserts and arctic regions the lowest.

human intervention. In order to produce growth, however, carbon compounds must remain in the plant body, which is not always the case. Like animals, plants respire oxygen, release energy from food compounds for their own metabolic use, and release carbon dioxide back into the atmosphere. A plant usually keeps 25% of the carbon that it fixes.

The **gross primary productivity** of an ecosystem is the rate at which organic matter is produced during photosynthesis. The **net primary productivity** is the rate at which this organic matter is actually incorpo-

rated into plant bodies so as to produce growth. Only the net primary productivity is available for the nutrition of heterotrophs, and only a portion of this is actually consumed. Only part of *this*, in turn, is digested and metabolized, and most of it is immediately used as a source of energy to run life processes. Thus, the amount of food available to a particular group of organisms, as reflected in their biomass, is progressively reduced in each successive level of the food chain (Fig. 46–18).

FIGURE 46–18   A pyramid of biomass for a given area of temperate grassland.

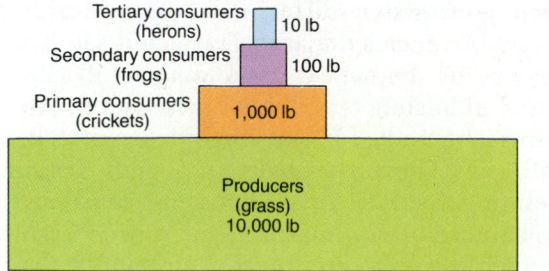

Tertiary consumers (herons) 10 lb
Secondary consumers (frogs) 100 lb
Primary consumers (crickets) 1,000 lb
Producers (grass) 10,000 lb

■ COMMUNITIES VARY IN SPECIES DIVERSITY

Species diversity within a community is important; as species become locally or absolutely extinct, diversity decreases. As alien species are imported, diversity may either increase or decrease. What makes these differences?

No single factor determines diversity, but we have identified some of the factors that interact to do so. First, diversity is related to the abundance of potential ecological niches. An already complex community of-

FOCUS ON   Microcosms

A favorite illustration of ecosystems has always been a balanced aquarium—that is, an aquarium containing fish and plants, along with decomposing bacteria. If properly set up, it should be possible for the inhabitants of such an aquarium to survive indefinitely, even if it is totally sealed off from the outside world. Unfortunately, whenever this is actually tried, everything usually dies in short order. This result is of more than academic interest due to the development of space flight. Spacecraft sent on journeys lasting years cannot be expected to carry all the food and oxygen they will need. It is obvious that they should be balanced terrariums, growing their own

food, recycling their own wastes, and producing their own oxygen by photosynthesis. In the Soviet Bios experiments, however, in which humans were sealed inside completely closed systems similar to projected spacecraft, the walls became covered with green slime and the humans contracted severe diarrhea, a situation not to be desired halfway to Mars.

In 1977, Dr. Joe Hanson of NASA was able to develop stable, sealed ecosystems containing shrimp and algae. Engineering Research Associates of Tucson, Arizona, was able to develop mass-produced versions of Hanson's systems that are now commercially available. By extensive experiment, they were able to

develop a controlled mixture of as much as 100 species of organisms that worked well (most of these were microorganisms). In addition to shrimp and algae, for example, it was also necessary to include *Nitrosomonas* bacteria to convert the toxic ammonia excreted by shrimp into nitrite. Nitrite, also toxic, was in turn converted to nitrate by *Nitrobacter*. The nitrate can be utilized as a nitrogen source by the algae. The entire system, called a **microcosm,** is supplied inside a sealed glass sphere resembling a paperweight. These decorative little ecosystems may tell us much about the management of spacecraft, including our spaceship earth.

---

fers a greater variety of potential ecological niches than a simple one.

Second, diversity is inversely related to isolation. Islands tend to be much less diverse than ecologically similar continental areas. This is due partly to the difficulty that many species may have in reaching the island. Many species of organisms also may become locally extinct (and cannot be readily replaced) as a result of random events.

Third, diversity is sometimes related to stress and extreme environmental conditions. In such a case, only the few species capable of resisting extreme conditions or specifically adapted to them will be present. Thus, the diversity of a polluted stream bed is low compared with that of nearby undisturbed areas.

Fourth, diversity is often higher in the margins of distinctive habitats than in their centers, because the margins contain all or most of the ecological niches of the habitats that they border. This is known as the **edge effect.**

Fifth, diversity is reduced when any one species of organism enjoys a decided position of dominance within a community by appropriating a disproportionate share of available resources, thus crowding out many other species.

Finally, diversity is greatly affected by biotic history. An area recently vacated by glaciers, for instance, will have a low diversity because few species will have had a chance to enter it and become established. A

long-established stable area (such as a geologically ancient desert) might have high diversity even if it is a poor habitat in other respects.

■ COMMUNITIES DEVELOP

Communities of organisms do not spring into existence full blown but develop gradually through a series of stages until they reach a state of maturity. The process of community development is called **succession.**

Sometimes Communities Must Start With A Lifeless Habitat

Primary succession begins in a habitat that is devoid of life, such as a bare rock surface (for instance, cooled volcanic lava (Fig. 46–19), rock scraped clean by glacial action, or even the wall of an abandoned building). Although the details may vary, in such a succession one might first observe a community of lichens, then mosses and drought-resistant ferns, and, finally, tough grasses and herbs. These might be replaced by low shrubs (if sufficient soil has accumulated), and then by forest trees in several distinct stages. Eventually, a relatively stable climax community is formed that undergoes little or no further change.

FIGURE 46–19 Primary succession. This view shows small plants growing on recently cooled volcanic lava in Hawaii. (Charles Seaborn)

Communities Sometimes Develop Where A Predecessor Community Already Exists

Secondary succession takes place in a habitat already substantially modified by a pre-existing community. A moderate forest fire, the systematic logging of a forest, and an abandoned old field are common examples. In secondary succession, a community might go through a shrubby grassland stage at about the midpoint of its progress toward a climax community of mature forest. In other instances, such as in the case of a forest fire, a habitat is produced that is unlike any that occurs in primary succession.

Some organisms are specifically adapted to the early stages of secondary succession. For instance, some plants, such as fireweed (Fig. 46–20) or certain conifers, reproduce and grow especially well right after fires. Very long-lived seeds of such species as dog fennel may also persist for tens to hundreds of years in the soil, only to sprout when and if conditions favor their growth. This is sometimes seen even in urban areas when an old building or pavement is removed from the ground and an almost pure stand of the persistent species sprouts suddenly from the newly exposed soil.

FIGURE 46–20 Secondary succession in the boreal forest. The northern forests of Canada, Siberia, and Alaska are fundamentally coniferous communities, but when they are disturbed by logging or fire, one of the first plants to grow is fireweed (a) followed by shrubs such as alders. The midsuccession trees (b) are usually hardwoods, particularly aspens such as the brilliant yellow trees shown here. In time, conifers will displace the aspens. (a, David Muench 1990; b Tina Waisman)

(a)

(b)

Why Does Succession Occur?

Why does succession take place? In some cases, succession occurs mostly as a result of the influence of physical and chemical changes originating outside the community, such as when a pond fills with silt and eventually becomes land.

In other cases, succession occurs because of the way organisms modify their own habitat and community. There are two main explanations of this type of succession that are not mutually exclusive. According to the first explanation, each community prepares the way for its successor. The other, more recent idea holds that successor communities actively damage or destroy the species occurring in their preceding stages by causing changes in the environment that cannot be tolerated by their predecessors. It is also true that random events may influence succession, especially in determining just which species happen to arrive and become established at a particular stage of development of the ecosystem.

Certain trends often can be made out during the course of succession. In the first place, community productivity often goes up. Secondly, biomass tends to increase. Thirdly, species diversity usually increases for most of the course of the succession. (In the end, diversity may decrease owing to the competitive dominance of a few or even one dominant species.)

Why Is The Climax Community Stable?

Most communities seem to reach a permanent stable stage, forming a **climax community** that does not appear to have a successor. As the concept was originally proposed, the climax community was thought to be a stage of maturity, comparable to the adulthood of an organism, and that all communities in a given area, growing under similar conditions of climate and soil, would be much the same. The members of these climax communities were believed to be the species best adapted to life under the prevailing conditions. The climax community was a kind of optimum assemblage of organisms—the one best ecosystem, as it were. It has become ever more clear that the makeup of communities of all stages result from historical events, chance happenings, invasions of foreign organisms, and much else. It is not even certain that the *concept* of the climax community is a valid one; various disasters occurring within the community (for instance, fire or violent storms) may produce every imaginable successional stage at the same time in scattered spots within it. A more important objection is that even climax communities may well continue to change, but at a rate too slow for us to notice.

■ SUMMARY

I. A biological community consists of a group of organisms that interact and live together. A reasonably complete and independent community comprises an ecosystem.

II. The major community roles are those of producer, consumer, and decomposer.
   A. Producers are the photosynthetic organisms that are the basis of most food chains.
   B. Consumers are almost exclusively animals. They feed upon other organisms.
   C. Decomposers recycle the components of corpses and wastes by feeding on them.

III. All vital substances (although not energy) are continually recycled through ecosystems, becoming continually available to new generations of organisms.
   A. Carbon is converted to carbon dioxide by the metabolic processes of most organisms, but it is reincorporated into complex organic compounds (fixed) by photosynthesis. The rate of entry of carbon into the biosphere is roughly balanced by its loss as insoluble compounds on ocean beds.
   B. Although the atmosphere is mostly nitrogen, the high energy cost of fixing this substance in chemical combination makes it a potential limiting factor. Although nitrogen compounds are liberated from organic substances by decay, these compounds are degraded by some microorganisms. The imbalance is redressed by symbiotic and free-living nitrogen-fixing microorganisms, principally bacteria and cyanobacteria.
   C. Because it has no biologically important gaseous compounds, phosphorus is not readily recycled but tends to be deposited in sea beds, from which it is only slowly and unreliably recycled by geologic processes.

IV. Organisms play particular parts in ecosystems. The distinctive life-style and adaptational package of an organism is its ecological niche.
   A. Organisms usually are able to exploit more resources and play a broader role in the life of their community than they actually do. They are constrained by the competition of other organisms.

V. Population growth results from the difference between the rate of arrival and the rate of departure of organisms. Usually the most significant forms of arrival and departure are birth and death.
   A. Biotic potential is the theoretical maximum rate with which a population can grow.

B. Mortality is the death rate of organisms in a population. Those that survive comprise the actual population.

C. Environmental limitations prevent unbridled population growth. The totality of all such limitations constitutes environmental resistance.

D. Limiting factors are the specific inhibitors of population growth. Generally, a limiting factor is an essential that is in potentially short supply.

1. Density-independent limiting factors are conditions or events that depress population without tending to stabilize it.

2. Density-dependent limiting factors are most effective at high population densities. For this reason, they tend to stabilize population size or density.

a. Resource depletion is a common limiting factor. Examples include a lack of food or nesting sites needed for continued population growth.

b. With resource depletion, competition for whatever resources remain intensifies and becomes a limiting factor itself.

VI. The trophic relationships of a community may be expressed as food chains or, more realistically, as food webs that show the multitude of alternative pathways that energy may take among the consumers and decomposers of a community.

A. Gross primary productivity of an ecosystem is the rate at which organic matter is produced during photosynthesis. Net primary productivity expresses the rate at which some of this matter is incorporated into plant bodies. Net primary productivity is less than gross primary productivity because of the losses resulting from plant metabolism.

B. Food pyramids express the progressive reduction in biomass and numbers of organisms found in successive trophic levels.

VII. Communities succeed one another in orderly fashion.

A. Primary succession occurs in a habitat that is initially devoid of life.

B. Secondary succession may go through some of the same stages as primary, but begins with a pre-existing community (for example, an abandoned pasture).

C. Some successions occur as a result of the actions of the organisms within the changing community. Other successions result primarily from the action of outside influences. It is increasingly believed that community composition results from an individualistic, almost random action on the part of its member species.

POST-TEST

1. Ecologically speaking, mold would be classified as a _____ .

2. Maggots, however, would be considered to be _____ _____ .

3. _____ usually do not kill the organism on which they feed.

4. The _____ _____ is the totality of an organism's adaptations, use of resources, and the life-style to which it is fitted.

5. The _____ _____ is the ideal ecological niche that an organism could potentially occupy.

6. The interaction of limiting factors plus competition from other species help to determine an organisms _____ _____ .

7. "Complete competitors cannot coexist" is a statement of the principle of _____ _____ .

8. Carbon dioxide originally enters the atmosphere mainly by _____ action.

9. Nitrite is oxidized to nitrate by _____ bacteria.

10. Most nitrogen fixation is performed by microorganisms, especially _____ and _____ .

11. Community diversity tends to _____ in later stages of succession.

12. For most species, the highest mortality occurs _____ in life.

REVIEW QUESTIONS

1. Can you distinguish a community from an ecosystem? Could one speak of a deciduous forest as a community? What about a rotting log within that forest? Which would be best referred to as an ecosystem?

2. How might one distinguish between a decomposer and a detritus feeder?

3. What is a parasitoid?

4. What is an ecological niche? A fundamental niche? Why is a realized niche usually "narrower" than a fundamental niche?

5. What is biotic potential? Environmental resistance?

6. Who was A. F. Gause and what important ecological concept did he originate?

7. Describe the carbon and phosphorus cycles. What are their fundamental differences?

8. Why are nitrogen-fixing microorganisms essential to the continuance of life on the earth?

9. Why is the concept of a food web preferable to that of a food chain in most cases?

10. The Florida everglades have been transected by a number of highways that are difficult for the Florida "panther" to cross. However, there is still ample habitat for the panthers that remain. Despite this, the number of panthers continues to decrease. Propose an explanation related to pyramids of biomass (of course, there are other explanations).

11. Suppose that it is proposed that the number of ring-neck pheasants available to hunters should be increased by (a) killing as many foxes (thought to prey upon the birds) as possible and (b) by feeding the birds in the winter. In your opinion, would either approach be successful, or just successful up to a point? Why, or why not?

■ RECOMMENDED READINGS

Begon, M., J.L. Harper, and R. Colin Townsend. *Ecology: Individuals, Populations, and Communities.* Sunderland, MA, Sinauer Associates, 1986. A population-oriented (but clear and understandable) introduction to general ecology.

Crawley, M.J., ed. *Plant Ecology.* Oxford, Blackwell Scientific Publications, 1986. Plant ecology is almost the basis of all community ecology, and here is a thoroughly modern discussion of plant ecology from an up-to-date ecological viewpoint.

Krebs, C. *Ecology: The Experimental Analysis of Distribution and Abundance,* 3rd ed. New York, Harper & Row, 1985. A community-oriented approach to ecology.

Mitsch, W.J., and J.G. Gosselink. *Wetlands.* Van Nostrand Reinhold, 1986. A theoretical and practical analysis of this important ecosystem type with a view to its conservation.

47

The Ecosphere

OUTLINE

I. Life is a thin film on the surface of the earth
   A. The sun warms the earth
      1. The earth's temperature depends on heat balance
      2. Local energy input is governed largely by the angle of lightfall
   B. Atmospheric circulation is driven by solar input
      1. Atmospheric heat is transferred mainly by convection
   C. Air pollution affects global patterns of heat transfer
      1. Atmospheric heat is retained by carbon dioxide
      2. The endangered ozone layer protects the earth from ultraviolet radiation
   D. The Coriolis effect perturbs the movement of air and water masses
II. Water dominates the surface of the earth
   A. Water has a very high specific heat
   B. Water changes its phase readily and is continuously recycled by the hydrologic cycle
   C. Vast oceanic currents redistribute marine water masses
III. Climate and precipitation result from patterns of air and water movement
   A. Microclimates can differ markedly from overall climate
Focus on the Last Winter

LEARNING OBJECTIVES

After you have studied this chapter, you should be able to:
1. List the main determinants of the earth's climate.
2. Summarize the factors that comprise the planetary heat budget.
3. Discuss the role of convection and the Coriolis effect in the production of global air and water flow patterns.
4. Summarize the hydrologic cycle.
5. Give the main causes of precipitation.
6. Summarize the nature of the ecological threat posed by the atmospheric accumulation of carbon dioxide and methane as well as the threat resulting from the depletion of the stratospheric ozone layer.

Dawn in Deadhorse Point State Park, Utah. (David Muench 1990)

Hanging in space, almost completely isolated from all but sunlight, our planet has often been compared to a vast spaceship, one whose life-support system is life itself. The living things aboard this great spacecraft produce oxygen, cleanse its air, adjust its gases, transfer energy, and recycle waste products with great efficiency. Yet none of these things would be possible without the nonliving environment of the world, the ducts and conduits of our spaceship earth. Because the science of ecology deals with the nonliving environment as well as the living, it is fitting to call this, the greatest of all systems of life, the **ecosphere** (Fig. 47–1).

KEY CONCEPTS

☐ Although the mass of living things is insignificant compared with that of the earth or even the atmosphere, living things profoundly influence the nature of the atmosphere and hydrosphere and are in turn completely dependent upon it.

☐ The temperature of the earth's surface depends on the balance between the absorption of solar radiation and the re-radiation of heat into outer space.

☐ Seasonal and local inputs of solar radiation vary in accordance with the angle of the sun's rays with respect to the earth's surface, which is in turn governed both by the geometry of the planet and the inclination of its axis.

☐ Atmospheric and hydrospheric circulation are basically convective, driven by heat but influenced by Coriolis forces.

☐ Some forms of air pollution are of global importance, especially the so-called "greenhouse effect" and the destruction of the stratospheric ozone layer.

☐ Global water circulation results from wind forces, evaporation, and precipitation.

☐ Geographic patterns of air and water circulation produce climate.

☐ LIFE IS A THIN FILM ON THE SURFACE OF THE EARTH

The earth is composed of the **biosphere,** which is the totality of its living inhabitants; the **hydrosphere,** which is its supply of water (both frozen and liquid); the gaseous **atmosphere;** and, bulking by far the greatest, its **lithosphere,** which is composed of rocks and their derivatives. The nonliving environment is influenced by the living biosphere, but all of the constituents of the biosphere, hydrosphere, and atmosphere probably came from the solid crust of the planet, the lithosphere. If you look at a classroom globe of the earth, you can easily form an idea of the proportions of these components. The atmosphere is proportionately as thick as the finishing coat of lacquer on that sphere; the hydrosphere is comparable to the paint; living things collectively amount to about as much as the dust particles and bacteria on that globe—if the janitor is moderately conscientious, that is.

Just the same, the earth would be a vastly different place without living things. The influence of the living biosphere on the nonliving environment can be illustrated by the producers. Without their photosynthesis, earth's surface temperatures would range in hundreds of degrees Celsius, there would be practically no oxygen in the atmosphere, and possibly no water (Fig. 47–2). The composition of the surface of the lithosphere also would be profoundly different from what we know.

FIGURE 47–1   The ecosphere. No significant loss or gain of matter occurs on the earth as a whole. All materials utilized by living things must be continuously recycled within this vast, closed system. (NASA)

A = Atmosphere   H = Hydrosphere
B = Biosphere   L = Lithosphere

FIGURE 47–2 The earth's ecosphere consists of four interdependent subsystems: atmosphere, hydrosphere, lithosphere, and biosphere.

Yet life depends on the lithosphere. Mineral substances used by living things, such as phosphorus or iodine, obviously originated in the lithosphere, and so, less obviously, does the global supply of carbon dioxide and water. These substances are continually emitted by volcanos. Ultimately, the oxygen that primary producers make from water is a derivative of those substances and, hence, of the lithosphere. Minerals are continuously rendered soluble by living things, leached from the lithosphere by percolating water, incorporated temporarily into the bodies of organisms, and often deposited as their remains in sedimentary rocks in the ocean depths. From there, they may be uplifted to form continents or subducted into the earth's interior. Vast deposits of phosphate rock, calcium carbonate, and silica represent eons of accumulated skeletons of once-living things. Spottily recycled by geological pro-

cesses to the earth's surface, these minerals may slowly reenter the biosphere.

The Sun Warms The Earth

In the sun's core, highly compressed, heated hydrogen atoms are forced to fuse with one another, producing helium. Every second, an estimated 596 million metric tons of hydrogen become 540 metric tons of helium. The difference, some 44 million metric tons of mass, does not simply disappear but is converted into energy, released mainly as heat and light. Virtually all of this energy is emitted to outer space in the form of radiation—light, infrared radiation, and other varieties. An infinitesimal portion of this energy strikes the earth's atmosphere, and a minute part of that tiny trickle operates the biosphere. Fusion energy, sought as the power basis of future civilization, is really the oldest of all our energy sources, for it has powered the biosphere from the most ancient times. Although almost all living things ultimately operate on solar power, solar energy also directly influences the nonliving environment, particularly the hydrosphere and atmosphere.

The "antennae" that collect solar radiation to power the life processes of the earth's organisms are the mostly green leaves and photosynthetic tissues of plants and certain microorganisms. Only the sunlight actually reaching photosynthetic tissues can be used for photosynthesis; thus, very little of the sunlight falling on large areas of mountainous regions, extreme deserts, and similar places nearly void of plant life is transformed into biotic energy.

Even when photosynthetic organisms are present, they seldom cover an area completely. By far most of the light that is absorbed simply heats the rocky surface, the water, and the atmosphere. Even that which is absorbed by primary producers is eventually released as heat by one metabolic process or another (unless the organisms incorporating it are geologically buried to form coal or oil deposits). Almost all delays in the flow of heat energy are temporary. All of the absorbed heat, plus whatever is released by nuclear reactions in the depths of the earth, is ultimately radiated away.

The slightest imbalance between absorption and output has far-reaching consequences, raising or lowering the surface and atmospheric temperature of the globe until a new equilibrium is attained. The endless fluctuations of what we call normal climate depend upon that balance, as do droughts, floods, and ice ages.

The Earth's Temperature Depends On Heat Balance

In the daytime, 23% of the solar radiation that falls upon the earth is immediately reflected by clouds, and another 3% by surfaces, especially snow, ice, and oceans. Yet another 14% is lost by immediate re-radiation of light in the form of infrared energy. The remaining 60% is absorbed but is eventually lost by radiation of long-wave infrared energy into space, not only during the day but at night. Notice that these figures add up to 100%. If they did not, the temperature of the earth would fall or rise until energy output exactly equaled input once again. In fact, the blanketing atmosphere keeps the earth's surface much warmer than it would be without one; relatively minor changes in the proportions of atmospheric gases could cause substantial further warming.

These figures are planetary averages and may vary substantially because of local conditions. High clouds (even of dust or soot), for example, increase energy reflection. Low clouds and even water vapor increase energy absorption, as do carbon dioxide and methane. The most significant local variation, however, is produced by a combination of the earth's roughly spherical shape and the oblique inclination of its axis of rotation. These factors produce great variation in **insolation,** the energy delivered by sunlight.

Local Energy Input Is Governed Largely By The Angle Of Lightfall

The fall of light on the face of the earth is determined by two main factors: distance from the sun and the inclination of the earth's axis. Although distance from the sun varies during the year, its effect upon climate and season is tiny compared with that produced by the inclination of the earth's axis.

Because this inclination is always the same (23.5 degrees), during half of the year the Northern Hemisphere is inclined *toward* the sun; during the other half, it is inclined *away* from the sun (Fig. 47–3). Naturally, the orientation of the Southern Hemisphere is just the opposite at these times (Fig. 47–4). The principal difference this inclination makes is seen in the angles at which the sun's rays strike the earth at any one time. Owing to the size of the sun's radiating surface, rays of light emitting from it are approximately parallel to one another. On the average, they fall most nearly in a vertical fashion near the Equator but more and more obliquely as one approaches the poles. This means that the energy content of the light is spread out over a larger area of surface near the poles, where the rays are almost parallel to the surface of the planet. This reduces the light's ability to heat the surface. Also, rays of light entering the atmosphere obliquely near the poles must usually pass through more air than those falling near the Equator, allowing more of their energy to be reflected out to space. Even if the axis of the earth were not inclined at all, these factors would be sufficient to produce great climatic variations at different latitudes. Climatic differences between the various parts of the earth are fundamentally determined by the spherical shape of the earth, which, in turn, determines the angle with which the sunlight strikes the earth at different latitudes.

Figure 47–4 displays the relationship of the earth's surface to the sun's rays in the month of June. Notice that at this time they actually fall vertically on the surface of the earth at the latitude of the Tropic of Cancer. Six months later, in December, the diagram shows that they will fall vertically on the Tropic of Capricorn instead. Thus, each of these nonequatorial latitudes experiences tropical sunshine once a year. By contrast, the seasons in equatorial regions are likely to be distinguished by precipitation, so that wet and dry, rather than warm and cold, seasons are the rule. At sea level, these areas have an endless summer. The temperate and polar zones have pronounced cool and warm seasons, opposite from one another in the Northern and Southern Hemispheres.

Atmospheric Circulation Is Driven By Solar Input

The action of wind and water affects the atmosphere and hydrosphere by making climate much less extreme over the surface of the earth. The transfer of heat

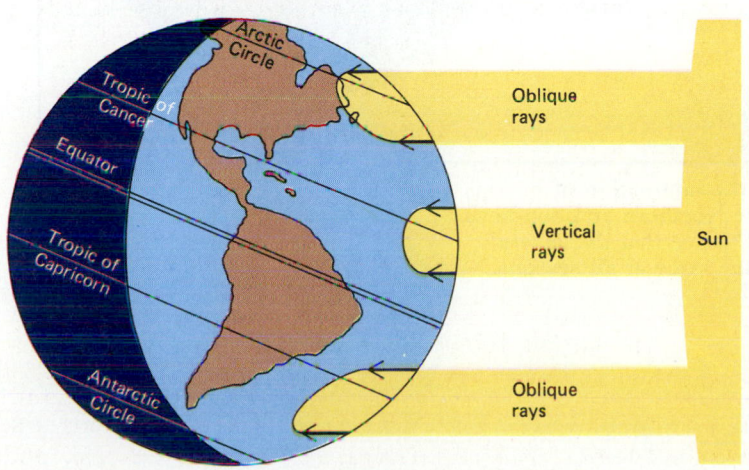

FIGURE 47–3 The amount of energy per unit area that the earth's surface receives is determined by the angle at which the sun's rays strike the earth. This angle varies from place to place due to the earth's spherical shape and the inclination of its axis. In this diagram, which represents the situation in the month of June, solar radiation strikes the earth perpendicularly in the Northern Hemisphere but very obliquely in the Southern Hemisphere. This difference produces summer conditions in the Northern Hemisphere and winter in the Southern Hemisphere, both at the same time.

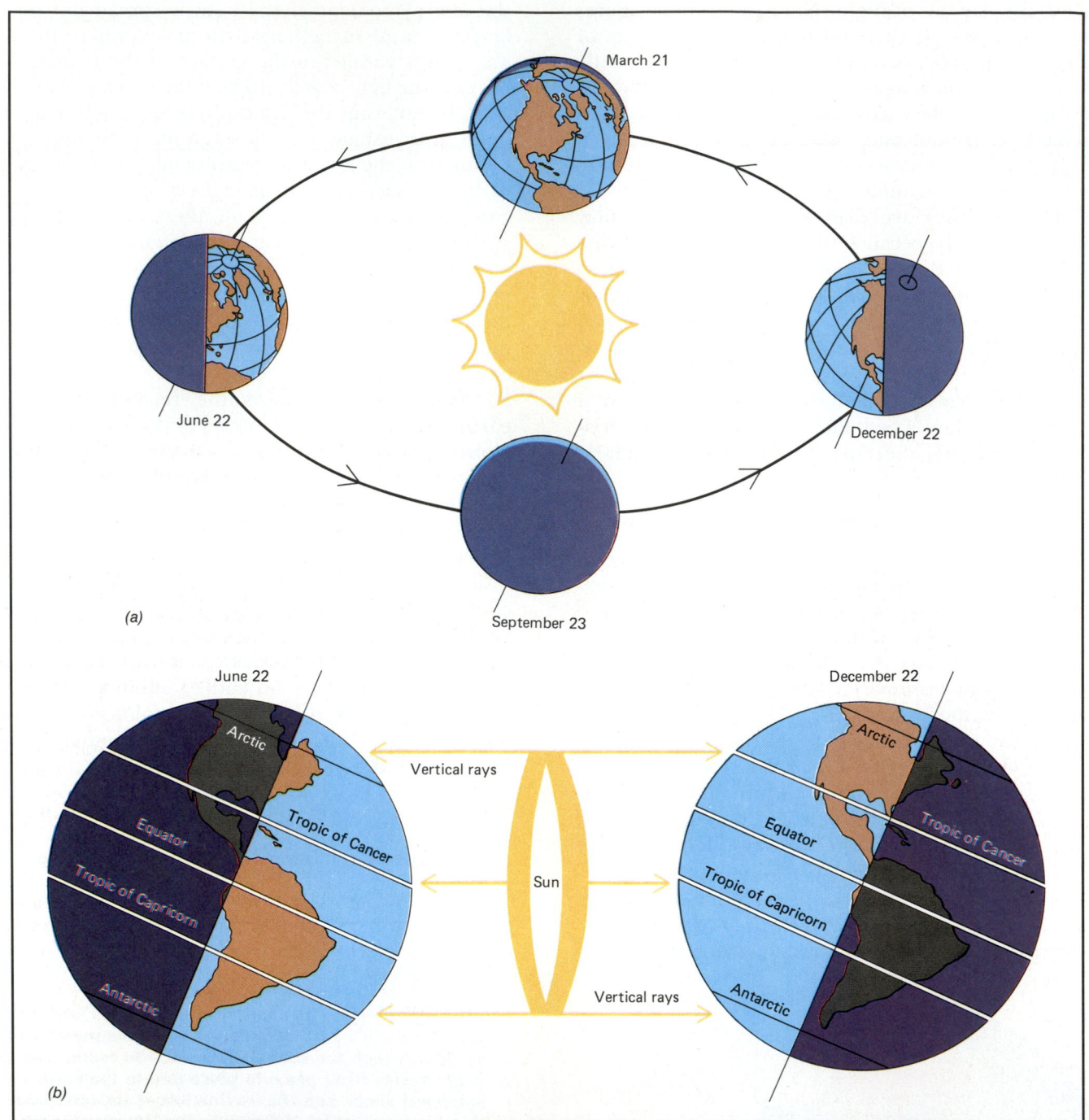

FIGURE 47–4 As the earth travels around the sun, the inclination of its axis remains the same. As a result, the sun's rays strike the Northern and Southern Hemispheres obliquely at different times of the year. Notice that the absolute angle of the sun's rays at the equator is more nearly the same at all seasons.

through rock is very slow; rock is almost a heat insulator. The poles would be colder and the Equator far hotter if heat were conducted between the two regions only by rock. Both air and water are also poor conductors of heat, but because they are fluid, they can transfer heat by their motion. Such heat transfer by circulation is usually called **convection.**

Atmospheric Heat Is Transferred Mainly By Convection

In large measure, differences in temperature drive the circulation of the atmosphere. The heated surface of the Equator heats the air that contacts it, causing this air to expand and, because it is lighter, to rise. As it rises, it cools and sinks again. Much of it recirculates almost immediately to the same areas it has left, but a large part of the remainder flows poleward, where eventually it is chilled. Similar local convection and poleward flow occur at higher latitudes as well (Fig. 47–5). As air cools by contact with the polar ground and ocean, it sinks to flow toward the Equator, generally beneath the sheets of warm air flowing poleward at the same time. Also, seasonal heating of the polar regions and midlatitudes prevents the stagnation of air, so that constant fluxes of air masses do battle over the temperate zones. This produces much more variable weather than is seen at the Equator. Most important, this constant flux of air transfers heat from the Equator toward the poles and, in returning, cools the land over which it passes. This constant turnover does not equalize temperatures over the surface of the earth, but it does moderate them. The atmosphere circulates upward from the Equator and, for the most part, returns to the Equator without substantial detour, although it is dryer.

Air Pollution Affects Global Patterns Of Heat Transfer

The forms of pollution that are the most difficult to control are those that affect very large areas. Recently, we have come to appreciate that some forms of air pollution, whose sources are very widespread, are almost

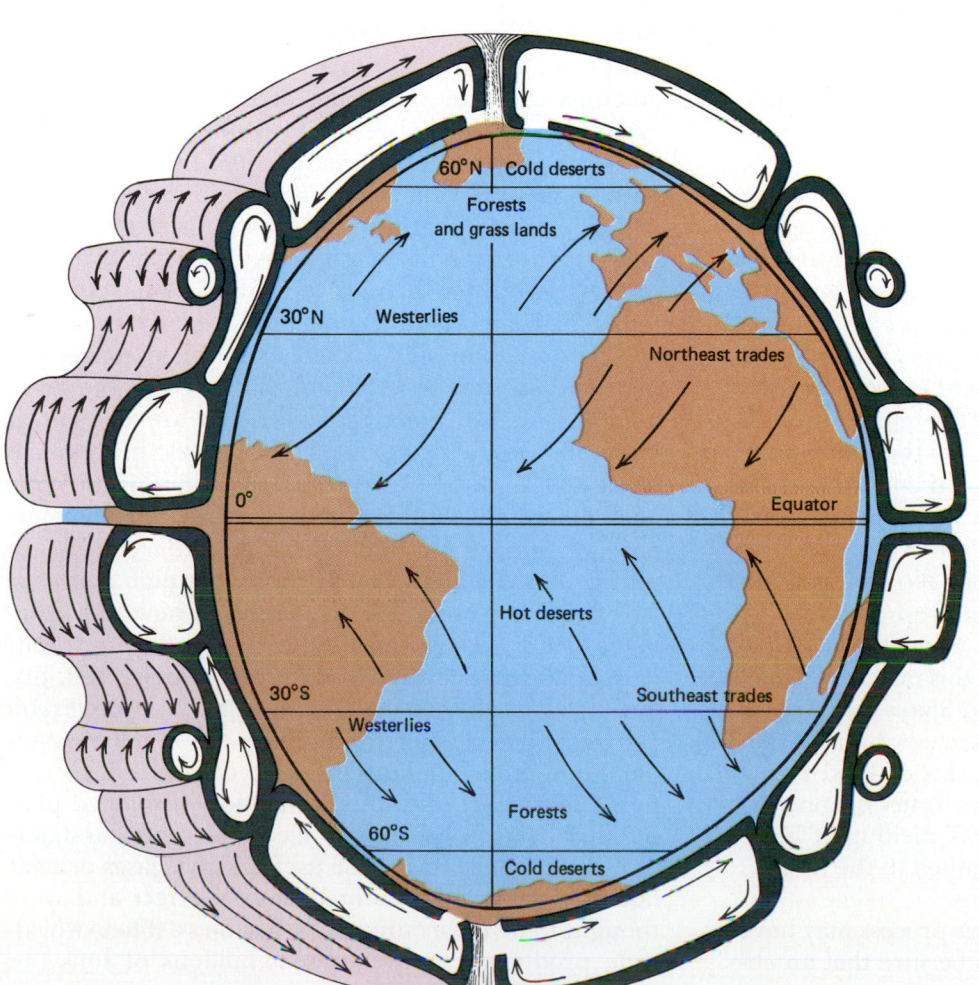

FIGURE 47–5 Because the sun heats the earth's surface unequally, convective forces create atmospheric movement. The greatest solar energy input occurs at the Equator, heating air most strongly in that area. The air rises and travels poleward but is cooled in the process so that much of it descends again around 30 degrees latitude in both hemispheres. The rising air produces a prevailing high-pressure area in most equatorial regions. At higher latitudes the patterns of movement are more complex. For example, the small circles at 50 degrees represent the storm-generating jet streams. Heavy rains occur where air rises and cools, especially in the tropics. Where the dried air descends, deserts may occur. Polar deserts have little precipitation but may be less extreme than temperate or tropical deserts due to the low rate of evaporation that prevails there.

certain to have important effects on the climate and atmosphere of the entire earth.

Atmospheric Heat Is Retained By Carbon Dioxide

The carbon dioxide level of the atmosphere may or may not have fluctuated in past geological time, but it is certainly increasing in our day at the unprecedented rate of some 15 billion tons per year. Increased rates of combustion resulting from industrialization and generation of electricity now return fossil carbon to the atmosphere in large quantities. The clearing of formerly forested areas (especially in the tropics), with the accompanying burning and decay processes that are no longer compensated for by the primary production of their vegetation, may contribute even more carbon dioxide to the atmosphere.

Relatively small changes in carbon dioxide content could produce temperature changes all over the earth. Biological and geological processes remove carbon dioxide from the atmosphere much more slowly than human activities now release it, so that several thousands of years would be required for a return to pre–Industrial Revolution levels, even if the increased rate of carbon dioxide production were to be curtailed immediately.

Carbon dioxide will certainly further accentuate the heat-retaining properties of the earth's atmosphere, the so-called "greenhouse effect." Other gases, such as methane and chlorofluorocarbons, which are artificially produced or whose production is accentuated by human activities, can also be expected to contribute. The warming effect of these gases is partly counteracted by the cooling effect of smoke and suspended particulate matter, which are also produced by many of the same processes that emit carbon dioxide. However, most experts believe that the temperature of the earth will rise several degrees by the end of the next century. This is likely to change the pattern of air and ocean currents so as to shift the climatic zones poleward, making the areas of the world that are most agriculturally productive today both too hot and too dry for high production of current crop plants.

Northerly areas may become agriculturally productive in compensation; however, this may be of small comfort to countries like the United States and Argentina, little of whose land may lie in the new breadbasket zones (the development of new strains of food plants or even new food plants may compensate to some extent). It is also likely that widespread melting of polar ice masses may raise the sea level, much to the distress of heavily populated coastal regions.

There is some evidence that this process may have already begun, but it is difficult to be sure that an abnormally hot, dry summer, for example, is not just a normal fluctuation in weather that has nothing to do with long-term climatic trends. There is no reasonable doubt, however, that the greenhouse effect will be far advanced by the end of the 21st century or before.

The Endangered Ozone Layer Protects The Earth From Ultraviolet Radiation

A dangerous pollutant at sea level, stratospheric ozone (O_3), provides vital protection from damaging ultraviolet radiation for most living things on earth. Almost everyone with a light complexion has at some time experienced a painful sunburn resulting from the small amount of ultraviolet radiation that does succeed in penetrating the atmosphere. Unfortunately for the sun worshippers among us, damage to the skin done by ultraviolet radiation does not stop there. Like many other forms of radiation, ultraviolet radiation has the ability to damage the nuclear DNA of skin cells, producing local mutations that can lead to skin cancer in later life. In small organisms (such as most plankton), even the shallow penetration of ultraviolet radiation can produce mutations that can be passed on to their descendants or, if exposure is severe, can kill them. Because the eyes of insects and some other invertebrates are sensitive to ultraviolet radiation, excessive amounts could be confusing or disorienting to such organisms as honeybees. Ultraviolet radiation also damages plant tissues; increases could have bad effects both on natural plant communities and on agriculture.

The reason that ozone forms in the stratosphere and not (normally) at sea level is the exposure of atmospheric oxygen to ultraviolet radiation. The diatomic oxygen (O_2) that we are used to breathing becomes triatomic (O_3) when exposed to sources of high energy such as electrical discharge or the large amounts of ultraviolet radiation that are present at high altitudes. This normal stratospheric process can be slowed or stopped, however, by oxides of nitrogen and compounds of chlorine. In the past, this interference was apparently minimal, because little chlorine or nitrogen oxides are normally present at high altitudes.

Beginning in the 1930s, a family of new and seemingly inert refrigerants, the chlorofluorocarbons (CFCs), were discovered and became widely used, displacing older alternatives such as sulfur dioxide. Freon refrigerant used in air conditioners is a familiar example. Soon other applications were discovered, so that now CFCs are used in the production of foamed plastics, fire control, cleaning of electronic parts, and aerosol-can propellants. These useful heavy gases or near liquids are almost without biological effect and were thought to be environmentally harmless. Their worldwide production now stands at millions of tons per year.

Within the past ten years, evidence has accumulated that, in the upper atmosphere, these otherwise stable compounds are degraded by high-energy ultraviolet radiation. The resulting chlorine and chlorine compounds interfere with ozone formation. Each year, during the Southern Hemisphere's winter, a large "hole" appears in the stratospheric ozone layer above the South Pole. This is almost certainly not a normal occurrence. Ominously, the hole has deepened each year since its discovery and has persisted longer into the spring and summer. A thinning of the ozone layer over the regions of the North Pole has also been detected. There is obvious danger that holes in the ozone layer may soon spread into the skies of the inhabited parts of our planet.

The ozone-layer hole has presented the world with a classic environmental dilemma. Assuming that it does indeed represent a serious danger and that it is truly caused by CFC production, it is of global concern, not only because of its potential effects, but in its cause as well. How can the nations of the world be persuaded to limit or phase out CFC production? Fortunately, the developed nations seem convinced that this should be done, and even some of the major manufacturers of these compounds believe that the danger is real.

A proposed CFC treaty, drafted in 1987, went into effect in 1989. Under present provisions, nations that sign it agree to first freeze and then to cut CFC production by 50% by mid 1998. It is hoped that, in the meantime, environmentally superior alternative substances will be developed. Some substitutes are known already but, unfortunately, cannot be used in most existing refrigeration and air-conditioning systems. They are also more expensive to manufacture, which has led to resistance to their use on the part of many third-world nations.

The Coriolis Effect Perturbs The Movement Of Air And Water Masses

The earth's rotation also contributes indirectly to air movement (and to water movement as well). A person standing exactly at the North Pole would hardly become dizzy despite the rotation of the earth, because he or she would rotate only once a day. Now imagine hiking (or sledding) a kilometer south of the pole. Even wearing snowshoes, you could easily keep up with the earth's rotation should you desire to do so.

By similar thinking, a sedentary Soviet citizen of Leningrad is moving rotationally at a speed of 830 kilometers per hour (kmph) (Fig. 47-6), while a Kampalan farmer in Africa is zipping along at 1660 kmph owing to the fact that he is sitting on the equator and therefore must traverse a much larger circle in the course of a day.

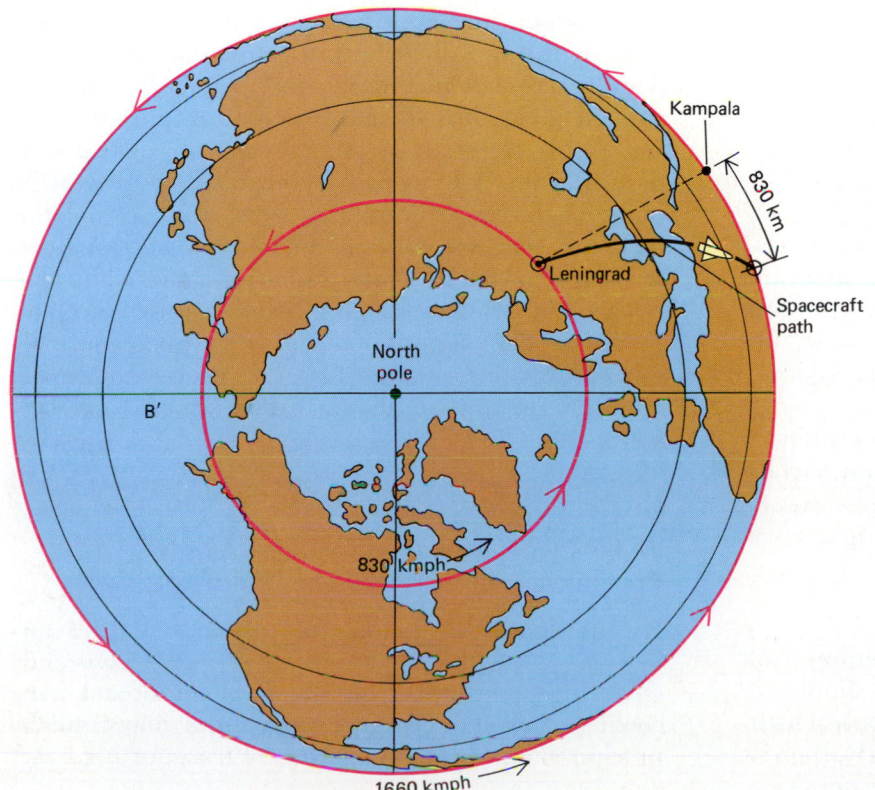

FIGURE 47-6 The Coriolis effect causes the path of a spacecraft to curve with respect to the earth's surface. The absolute speed of the earth's rotation varies with the distance from the poles. Because the entire globe must turn around once each day, a point on the Equator must travel much further during that time than a point near one of the poles.

If a spacecraft were to orbit from Leningrad to Kampala at such a speed that it would take an hour to arrive, its path would simultaneously be affected by another speed, an eastward velocity of 830 kmph imparted by the earth's rotation *at the latitude of Leningrad.* As the craft traveled, the eastward speed of the earth beneath it would steadily increase while the missile's speed remained the same. Because the velocity difference between Kampala and Leningrad is 830 kmph, the ship would land 830 kilometers *west* of Kampala! Less spectacular but nevertheless significant deviations of this sort can be detected even in the case of long-range artillery fire. A missile or artillery shell travels in a straight line but curves with respect to the surface of the earth because the earth is rotating below it.

This effect, called the **Coriolis effect,** is a deflection of the path of objects traveling north or south caused by the rotation of the earth. It is most marked for massive objects like artillery shells, but it influences substances as light as air if the distances traveled are long enough and if their north-south speed is low. The poleward movement of air masses fulfills these criteria. Thus, as a consequence of the Coriolis effect, air masses curve in an eastward direction as they travel poleward, and a prevailing wind from the west can be observed in many areas, especially the tropics. On the other hand, the cooled polar air flows back toward the Equator in a westward direction. These flows interact with the shape of the land and the distribution of oceans and currents to produce consistent weather trends that, in turn, produce the great climatic zones of the earth.

■ WATER DOMINATES THE SURFACE OF THE EARTH

Not only does water comprise most of the contents of living things, but its properties determine many of the conditions on the earth that are favorable for life. Water, most of which resides in the oceans, covers 70% of the earth's surface (Fig. 47–7). The sheer quantity of water gives this remarkable substance broad ecological significance, a significance that is heightened by its remarkable heat and density relationships, its surface tension, and its ability to act as a solvent.

Water Has A Very High Specific Heat

If energy is added to a substance, the temperature will rise, but temperature and energy are not simply related. An example will make this clear. Suppose a bathtub contains 80 liters of water at 60°C. Clearly, there is

more heat energy in the bathtub than in a cup of water (0.2 liter) at the same temperature. In fact, there is far more heat energy in that tub at 60°C than in a cup of boiling water at 100°C. Consider a different aspect of this matter: Which will cool down faster—a 1-liter balloon full of air at 60°C or a 1-liter cannonball of iron at 60°C? Obviously, the balloon of air will cool faster. The two substances differ in the amount of heat energy, even though their volumes are the same.

We may express all this more precisely using the concept of **specific heat,** which is the energy needed to raise or lower a unit of mass of a given substance by 1°C. It takes much less heat to raise the temperature of air than of iron. The surprising fact is, however, that more heat is needed to raise the temperature of water than of iron! Water has a remarkably high specific heat, mostly due to extensive hydrogen bonding among its molecules (see Chapter 3). Consequently, water increases or decreases in temperature more slowly than almost any substance (rock, for instance, and especially air), even if all three are exposed to the same amount of insolation. As a result, water moderates the climate of adjacent land areas, and aquatic habitats experience less temperature variation than land habitats do.

Local variations of temperature common on the earth can easily change water to and from a liquid, solid, or evaporated state. Unlike almost all other substances, water does not have its greatest density as a solid, but as a liquid at 4°C. Below that temperature, it expands and becomes *less* dense. This is why ice floats. When water freezes, it releases a great deal of heat. This does not ordinarily raise its temperature, but it does help to slow further freezing. Similarly, water absorbs a great deal of heat when it evaporates; conversely, it releases a great deal of heat when it condenses again. This release of heat energy drives the immense forces of tropical storms, which, from this point of view, are colossal atmospheric-heat engines, but it is also vital to more ordinary weather. The heat released by water upon condensation raises the temperature of the air with which water vapor is usually mixed. This air, being warmed, rises even higher in the atmosphere, thus inducing turbulence and creating storms.

Water Changes Its Phase Readily And Is Continuously Recycled By The Hydrologic Cycle

Because the average temperatures of the earth's surface favor the evaporation of water, water constantly enters the atmosphere, mostly from the oceans. Any cooling of air that contains water vapor may result in precipitation. This has the effect of transporting water

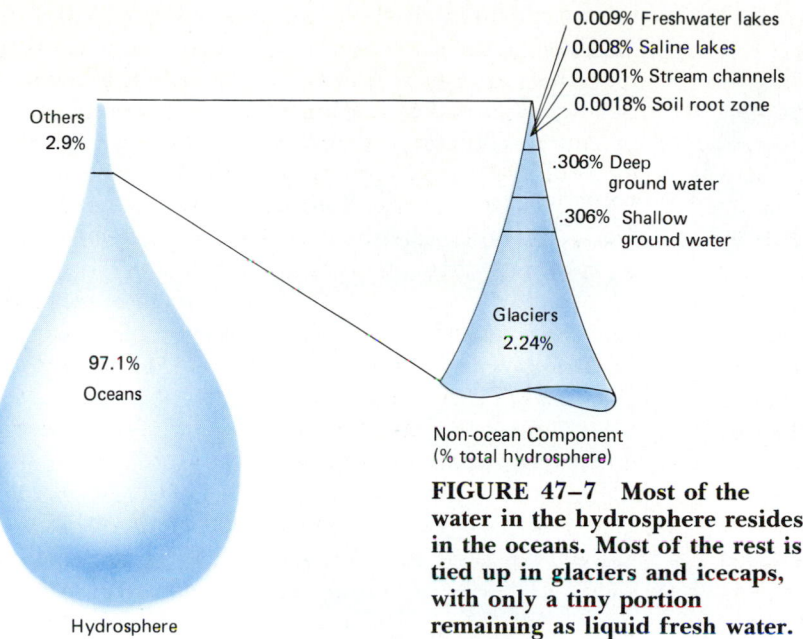

0.009% Freshwater lakes
0.008% Saline lakes
0.0001% Stream channels
0.0018% Soil root zone

.306% Deep
ground water

.306% Shallow
ground water

Others
2.9%

Glaciers
2.24%

97.1%
Oceans

Non-ocean Component
(% total hydrosphere)

Hydrosphere

FIGURE 47–7   Most of the water in the hydrosphere resides in the oceans. Most of the rest is tied up in glaciers and icecaps, with only a tiny portion remaining as liquid fresh water.

inland. Rising moist-air masses carry water vapor steadily upward. As the air enters regions of decreasing pressure, it expands and therefore cools.

The amount of water vapor that air can contain at any given temperature is its **saturation value.** The lower the temperature, the lower the saturation value. Whether cooled by this or some other mechanism, if the air is sufficiently humid, eventually the temperature may fall as low as the **dew point,** the temperature at which water will condense from the air. Especially if accompanied by turbulence and condensation nuclei such as dust, this cooling produces **precipitation**—that is, snow, hail, or rain.[1] Most precipitation occurs over the oceans, but that which is carried onto the land sustains almost all life on land. The amount of precipitation, the rate of its removal by evaporation or runoff, and its seasonal distribution profoundly influence the distribution of terrestrial life. Eventually, water returns to the ocean, although it may be delayed for months in snowfields or for millenia in glaciers. The global recycling of water is called the **hydrologic cycle** (Fig. 47–8).

Worldwide, about a billion tons of water evaporate and another billion are precipitated every minute. Imbalances in this equation produce such disasters as ice ages. It is sobering to realize that the causes of ice ages are not well understood. They may depend upon such seemingly minor variations as a few more or less sunspots than normal, a bit more or less carbon dioxide in the atmosphere, or changes in the number of soot and dust particles. It is even more sobering to realize that through our industrial civilization, we have produced important changes in some of these parameters without really knowing what effect those changes may have or, until recently, without realizing that they may have any effect at all.

Vast Oceanic Currents Redistribute Marine Water Masses

Almost everyone has seen waves raised by the action of wind on a lake or ocean. Given time to act, persistent prevailing winds can produce the deep mass movements of ocean water known as currents.

The oceans are not uniformly distributed over the globe (Fig. 47–9). There is clearly more water in the Southern Hemisphere than in the Northern Hemisphere. Even more important, the circumpolar flow of water in the Southern Hemisphere is almost unimpeded by land masses. In the Northern Hemisphere, circumpolar movement also takes place, but in a westerly direction. In reality, these movements are part of larger, circular current patterns that occur in the Pacific and Atlantic Oceans and that move clockwise in the Northern Hemisphere and counterclockwise in the Southern (Fig. 47–10). The ultimate explanation for these circular current patterns (called **gyres**) is the

---

[1]Although not technically precipitation, dew forms from atmospheric moisture and may sometimes (in foggy coastal deserts or some mountainous areas) locally contribute more water to the community than precipitation does.

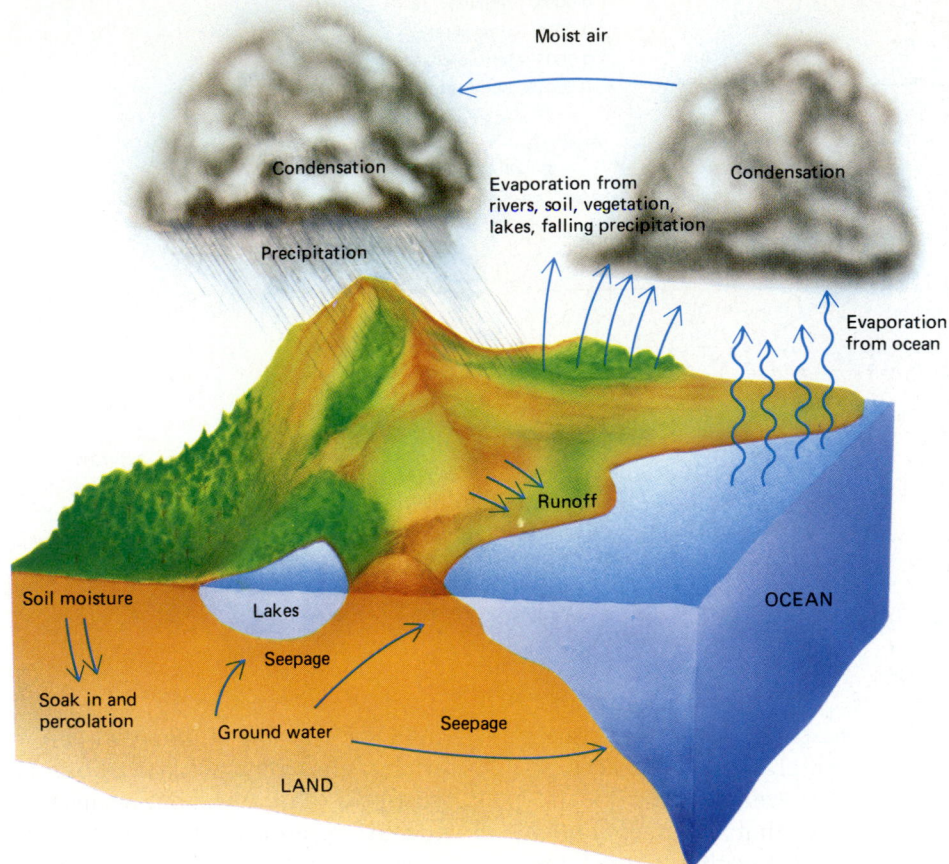

FIGURE 47–8 The hydrologic cycle. Water vapor carries precipitation over land and sea, with temporary stays in freshwater areas and the bodies of living things. It all returns to the sea, to be recycled yet again.

FIGURE 47–9 The Northern and Southern Hemispheres have greatly differing proportions of land and water, with far more water occurring in the Southern Hemisphere. Ocean currents are more free to flow longitudinally in the Southern Hemisphere.

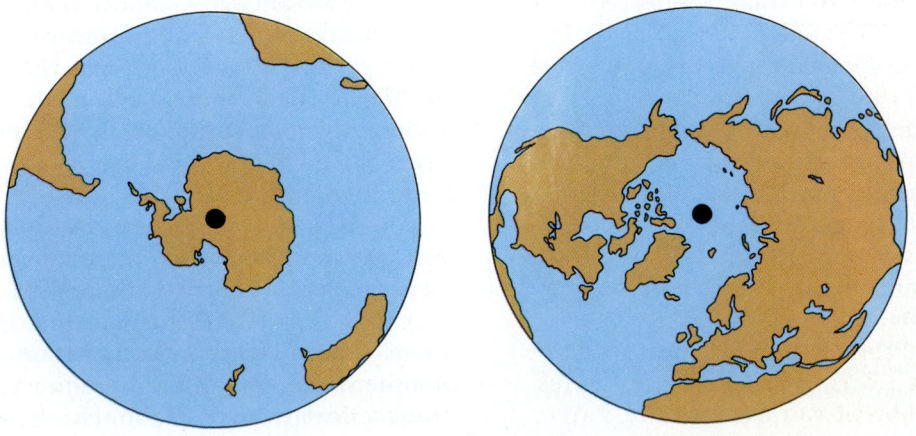

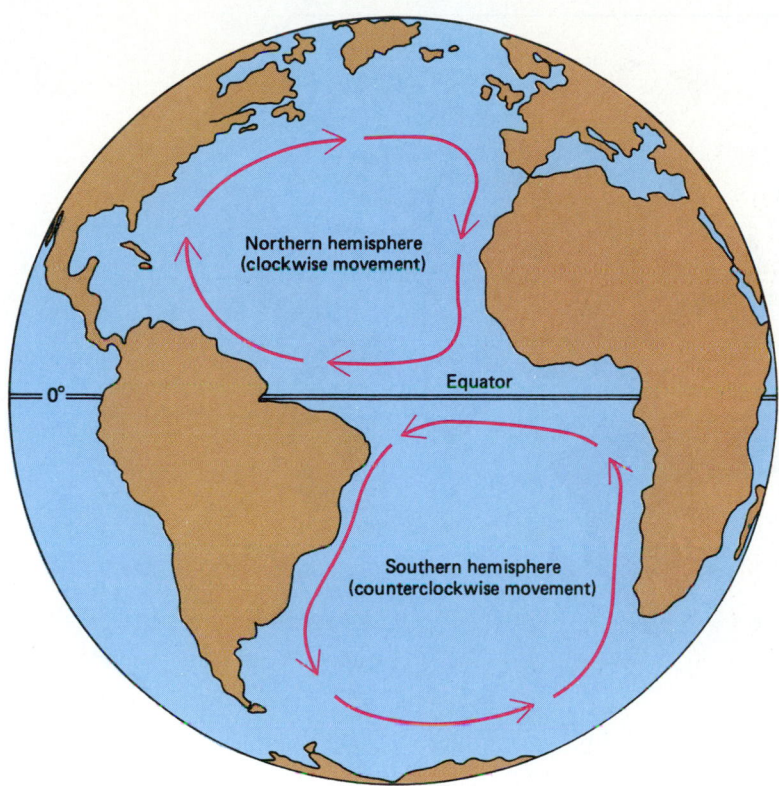

FIGURE 47–10 The basic pattern of the main ocean currents results in part from the action of winds and in part from the Coriolis effect. The main ocean current flow—clockwise in the Northern Hemisphere and counterclockwise in the Southern Hemisphere—results partly from the Coriolis effect, which also causes the difference in direction.

Coriolis movement of the prevailing winds that drive them.

The great Humboldt current, for instance, drives northward along the west coast of South America. This cold current cools the South American continent almost to the Equator. The Gulf Stream, on the other hand, carries warm Caribbean waters northeast, transferring heat and minerals and greatly moderating the coastal European climate.

CLIMATE AND PRECIPITATION RESULT FROM PATTERNS OF AIR AND WATER MOVEMENT

Average temperature, temperature extremes, precipitation, the seasonal distribution of precipitation, day length, and season length are biologically the most important dimensions of climate. Combinations of these climatic factors produce the biomes such as tundra, desert, rain forest, and prairie.

Differences in precipitation are subtly determined and depend upon a combination of all other climatic factors. Rain falls heavily in some areas of the tropics, resulting mainly from the equatorial upwelling of moisture-laden air. High ocean-surface temperatures encourage the evaporation of vast quantities of water,

and prevailing winds blow great packets of the resulting moist air over land masses. Convective heating of the air by the heavily insolated ground causes heavy and frequent storms. The air, driven high by the heat of condensation, eventually returns to earth on either side of the equator between the Tropics of Cancer and Capricorn. By then, most of its moisture has been squeezed from it, and the dry air returns to the Equator. This makes little biological difference over the ocean, but the lack of moisture in this returning air produces some of the great tropical deserts, such as the Sahara Desert (Fig. 47–11).

Air also dries from long journeys over land masses. Near the windward coasts of continents, rainfall may be heavy, but not all the precipitation runs off the ground surfaces; some of it reevaporates into the air. Nevertheless, there is a continuing deficit, so that in the temperate zones, continental interiors are usually dry because they are so far from any oceans that could moisten the air that passes over them.

Moisture is also removed from air by topographic uplift of air masses. If, as occurs on the North American west coast, prevailing winds blow onto a mountain range, the air cools as it rises and moisture condenses out of the air, especially on the windward slopes of the mountains. Downwind, a low-precipitation area, or **rain shadow,** develops. The rain shadow often creates

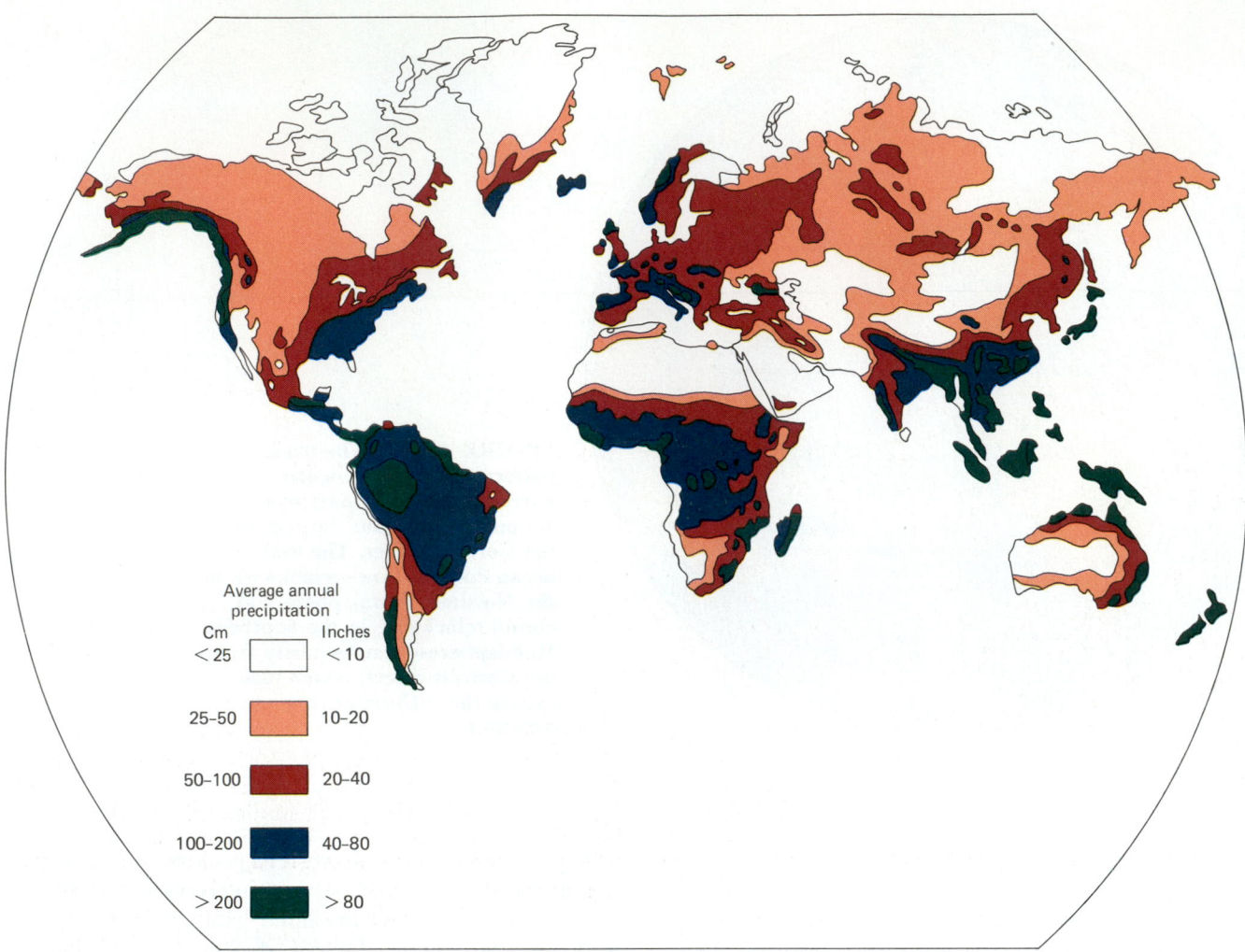

FIGURE 47–11   Hemispheric map of average precipitation. Notice the light precipitation at northern latitudes, in the centers of many continents, and to the leeward of mountain ranges.

Average annual precipitation

Cm	Inches
< 25	< 10
25–50	10–20
50–100	20–40
100–200	40–80
> 200	> 80

deserts, Thus, some of the regional differences in worldwide precipitation result from the drying of air as it is returned to more equatorial areas, some from long travel over continents and some from cooling produced by mountainous regions.

Microclimates Can Differ Markedly From Overall Climate

There are many exceptions to the overall climatic conditions in any habitat. Differences in elevation, in the steepness and directional orientation of slopes, or in exposure to prevailing winds may produce local variations in climate known as **microclimates,** and these may be quite different from the general climate, the **macroclimate,** of their overall surroundings. It is really the microclimate that is most important to an organism, for the microclimate of its habitat is the climate it actually experiences and with which it must cope. Sometimes it is possible for animals to migrate to a local area of favorable microclimate (Fig. 47–12) or even to substantially modify their own microclimate to dimensions more favorable to their existence. For instance, the temperature within a forest is usually lower and the relative humidity higher than that outside the forest, and the temperature and humidity beneath the litter of the forest floor can be expected to differ still more. By burrowing, desert-dwelling organisms survive in climatic conditions that would kill them in minutes on the surface. The cooler daytime microclimate in their burrows permits them to survive until night, when conditions are more favorable and allow them to leave their retreats to forage or hunt.

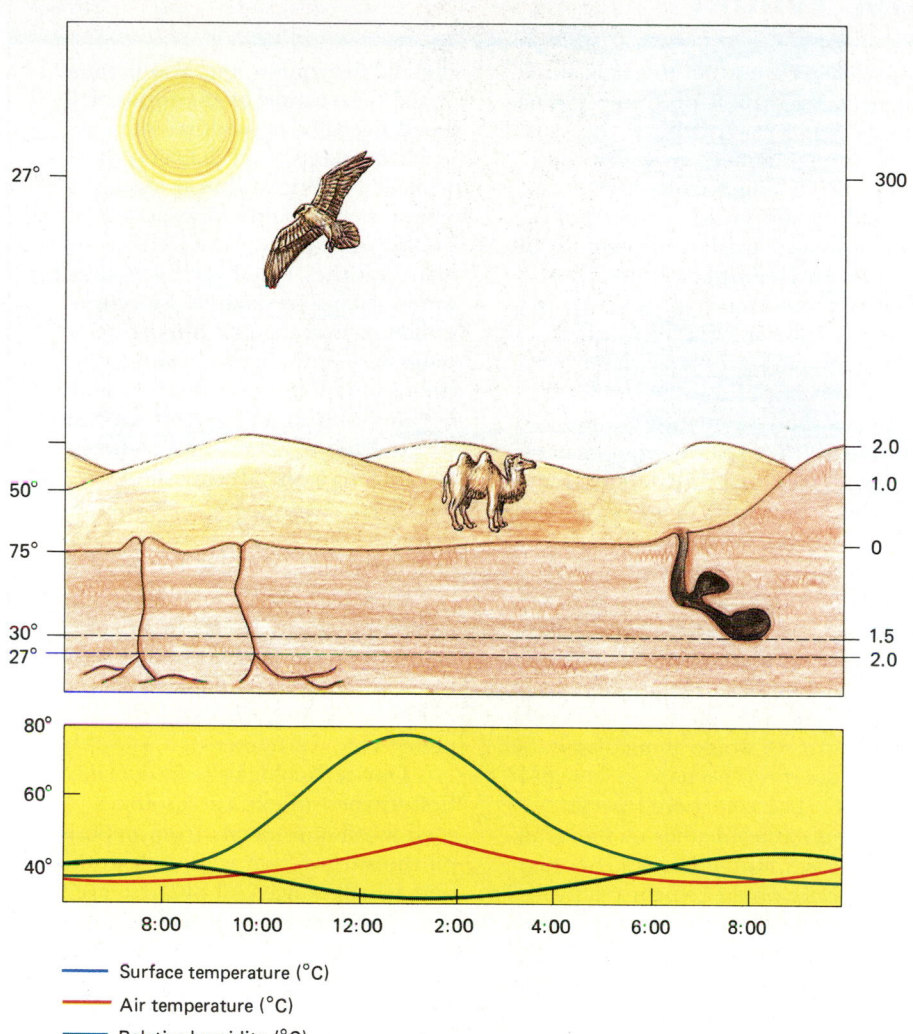

- Surface temperature (°C)
- Air temperature (°C)
- Relative humidity (°C)

FIGURE 47-12 Substantial vertical differences in microclimate in a desert habitat. The sunlit surface is a dangerous 75°C, but no more than a meter above it the temperature is fully 25°C lower. Soaring birds live at an equable 27°C with little energy expenditure, and so do burrowing organisms with even less.

FIGURE 47-13 A cushion of moss campion. These little plants grow in tundra and alpine habitats, producing an internal habitat of still air and moderate temperatures despite the wind and cold outside. (Tina Waisman)

Perhaps the most striking example of microclimate is produced by such alpine plants as moss campion (Fig. 47–13). These little angiosperms grow in a low, ovoid cushion whose streamlined shape encourages the drying alpine wind to pass over them. At the same time, they are able to absorb light and solar heat, which is retained in the still air within. Because wind cannot blow away accumulating litter and dead-plant parts within the cushion, these accumulate, allowing the retention of as much as a teaspoonful of water that the plants otherwise would not have. In the daytime, the temperature of the interior of one of these cushion plants can be several degrees warmer than that of its immediate surroundings. Small animals such as insects may shelter in these miniature warm forests, away from the dry, ultraviolet-radiation-suffused world of cold and wind outside.

There has never been any doubt in the minds of informed people that nuclear war would be unimaginably disastrous, but for many years little attention was paid to the probable ecological consequences of nuclear exchange. Scientists assembled at a 2-day conference in the fall of 1983 agreed that the consequences of nuclear war would produce the most serious ecological damage possible. In addition to immediate injuries and death, it would produce a horrendous climax to the environmental despoliation of our civilization.

The biological concentration of such isotopes as strontium-90 has been studied since the 1950s, and it is obvious that even a limited nuclear exchange would produce widespread environmental pollution by radioactive fallout. It has also been clear for some time that much of the protective atmospheric ozone layer would be destroyed by nuclear-generated nitrogen oxides. What is new is an appreciation of what the smoke, soot, and pulverized earth produced by nuclear explosions would do to the terrestrial climate.

Widespread dust has entered the earth's atmosphere before. There is some evidence (and more speculation) that dust stirred up by asteroidal bombardment of the earth may have been responsible for widespread extinctions of life that we observe in the fossil record. Even within historic times, large volcanic eruptions have placed enough ash in the atmosphere to produce marked, although temporary, climatic changes. The 1815 eruption of the Indonesian volcano Tambora ejected some 25 cubic miles of debris, much of which did not fall to earth immediately. In 1816, it produced a disastrous year for agriculture, the "year without a summer," in which there were three killing frosts during the New England growing season and widespread hardship throughout the Northern Hemisphere. Studies of the effects of forest fires in the summer of 1988 showed a lowering of daily temperatures near these fires of as much as 15° to 20°C. Large scale as these events were, they would be dwarfed by the effects of even a "moderate" nuclear exchange in which only 20% or so of the available weapons were utilized.

The dust and soot produced could be expected to obscure sunlight almost completely over the entire Northern Hemisphere and probably the Southern Hemisphere as well. Even in summer, prevailing temperatures would immediately fall below freezing, perhaps as low as 15° to 20°C. The cold would persist, freezing natural bodies of water, in many cases, to the bottom. Because the oceans would remain relatively warm for a considerable time, the resulting marked temperature difference between land and sea would produce storms of unparalleled violence. Darkness and cold lasting for months would cause most animals and plants to die, and most of them would probably become extinct. This would be especially true of tropical life, but even temperate-zone species would be decimated, particularly if the exchange occurred in summer.

Although some ecological recovery could be expected within a few years, conventional agriculture would be impossible, not only because of climatic disruption, but also because of the destruction of the industrial-based needs of agriculture such as fertilizer plants, fuel delivery, and availability of agricultural machinery and spare parts. Starving people would hunt down any surviving animals that they could but eventually would themselves starve. Moreover, radiation sickness and other disease states, brought about by widespread chemical pollution resulting from the burning of synthetic material, could be expected to weaken even the survivors. Survivors, at least in the Northern Hemisphere, would face extreme cold, water shortages, lack of food and fuel, heavy burdens of radiation and pollutants, disease, and psychological stress—all in twilight or in darkness. It is likely that most tropical plants and animals would be rendered extinct, as would most temperate-zone vertebrates.

Our technologically based society first learned to extract resources such as oil and metals from deposits on the surface that were rich in the desired substance. All of these are now depleted, so industry must employ advanced methods to remove metals from low-grade ores, drill deep wells in the ocean beds, and so on. The technology needed to reestablish industry would be a casualty of a disastrous nuclear war; therefore, there is no reason to think that anything resembling our present industrial civilization could be reestablished in the ecologically impoverished world inhabited by any survivors. It would be of no consequence whether their nations won, lost, or even participated in the warfare that led to this worldwide tragedy.

◼ CHAPTER SUMMARY

I. The unique planetary environment of the earth makes life possible.
   A. Sunlight is the major (almost the sole) source of energy available to the biosphere.
      1. The heat budget of the earth balances heat loss by radiation against heat gain by absorption. Its climate depends in large measure on the set of conditions under which heat loss and gain are equal.
   2. Local climate, however, is determined largely by the local solar energy input on the surface. Most of the time, this is greatest and most constant in equatorial regions, whose climate therefore is hotter and less

variable than that of temperate and polar areas.

II. Global climate is moderated by heat transfer via air and ocean.

   A. Atmospheric heat transfer by convection from Equator to poles produces both a poleward movement of warm air and a countervailing motion of cool air. These meet in the temperate zone, producing unstable weather.

   B. The Coriolis effect diverts air and water currents longitudinally. The Coriolis effect results from the rotation of the earth.

III. Water is a virtually unique substance that transfers heat efficiently, but whose high specific heat resists changes in temperature, thus moderating climate.

   A. The hydrologic cycle results from the evaporation of water from both land and sea and its subsequent precipitation when the air that contains it is cooled.

   B. Oceanic currents result from prevailing winds. They redistribute mineral nutrients globally and often profoundly affect local climate.

IV. Precipitation is greatest where air heavily saturated with water vapor has a chance to cool to the dew point. Accordingly, humid areas tend to be located where air passes over the ocean and then is forced upward by mountainous topography or where such air is otherwise cooled. Deserts develop in the rain shadows of mountain ranges or in continental interiors.

V. Microclimate is the local variation of climate that organisms actually experience. By choosing favorable microclimates, or even by creating them, organisms can often live in areas whose macroclimate is hostile to life.

■ POST-TEST

1. The single largest major component of the earth is the _____.

2. The energy that operates the world of life originates almost exclusively in _____.

3. The prevailing temperature of the earth is determined by the heat balance of the earth—that is, the relationship between gain from the sun and loss by _____.

4. The warmth and constancy of equatorial climates result mostly from the direct _____ with which sunlight falls upon their surface.

5. Heat is transmitted from the Equator to the poles by _____ of atmosphere and hydrosphere.

6. The high _____ _____ of water moderates climates of areas adjacent to large bodies of water.

7. The _____ _____, which results from an interaction between motion, inertia, and the rotation of the earth, displaces the paths of atmospheric and oceanic currents to the east and west.

8. Precipitation will probably occur when the temperature of humid atmosphere falls below its _____ _____.

9. Mountain ranges may produce downwind arid _____ _____.

10. The hollow stems of certain arctic plants are able to produce a temperature that is as much as 10°C higher than that of the outside air. Thus, their tissues are exposed to a favorable _____.

■ REVIEW QUESTIONS

1. What proportion of the sun's light that strikes the earth's surface is employed by plants for photosynthesis?

2. What determines the temperature of the earth? How might this balance be disturbed?

3. What basic forces determine the circulation of the earth's atmosphere? Can you describe its general directions?

4. Give the properties of water that are of the greatest significance in its role in determining the earth's climate.

5. Summarize the hydrologic cycle.

6. What forces produce the main oceanic currents? What is their role in determining the climate of the earth?

7. What conditions produce precipitation? What are some of the factors that produce areas of precipitation extremes such as rain forests and deserts?

8. What is meant by microclimate? Give some examples of how living things seek or produce favorable microclimates in which to live.

■ RECOMMENDED READINGS

Foster, R.J. *Physical Geology.* Columbus, OH, Charles E. Merrill, 1983. Introduction to the geological processes that determine the basic structure of the earth upon which organisms must live.

Gabler, R.E., et al. *Essentials of Physical Geography,* 3d ed. Philadelphia, Saunders College Publishing, 1987. A synthesis of geology, economics, and ecology.

Harwell, M.A. *Nuclear Winter.* New York, Springer Verlag, 1984. As thorough a summary as can be made without actually trying the experiment.

Lutgens, F.K., and E.J. Tarbuck. *The Atmosphere: An Introduction to Meteorology.* Englewood Cliffs, NJ, Prentice-Hall, 1986. The operation and causes of climate and weather.

Turk, J., and A. Turk. *Environmental Science,* 4th ed. Philadelphia, Saunders College Publishing, 1988. Ecology with a human emphasis. Includes an extensive discussion of the natural resources essential to civilization and the geologic processes that produce them or influence them.

48
Life Zones

OUTLINE

I. Organisms have unique geographic distributions
II. Major land life zones are called biomes
  A. Tundra is the northernmost land biome
  B. The boreal forest is dominated by conifers
  C. Temperate coniferous forests occur in western North America
  D. Temperate biomes vary with precipitation
    1. Deciduous forests develop where winters are pronounced and rainfall is high
    2. Temperate grasslands occur in areas of moderate precipitation
    3. Several types of deserts occur where precipitation is sparse
  E. Tropical life zones also vary with precipitation
    1. Tropical rain forests occur where rainfall and temperature are high
    2. Savannah is a tropical grassland with some trees
III. Aquatic life zones occupy most of the earth's surface
  A. Aquatic organisms fall into three ecological categories
  B. Freshwater habitats include streams, lakes, and ponds
  C. The marine zone is the largest aquatic life zone
    1. Marine and freshwater environments meet in salt marshes and estuaries
    2. The intertidal zone is harsh but productive
    3. The subtidal zone is protected from wave action
    4. The shallow continental shelf surrounds all continents
    5. Most of the ocean bottom is too deep for photosynthesis
  D. Plankton forms the basis of most marine food chains
  E. Succession occurs in aquatic habitats
IV. Life zones interact
Focus on Coral Reefs

LEARNING OBJECTIVES

After you have studied this chapter, you should be able to:

1. Briefly describe the principal land biomes of the earth, citing the climatic factors that influence their characteristics.
2. Describe the ways in which aquatic habitats differ most significantly from land habitats and the factors that contribute to these differences.
3. Describe such marine life zones as the estuarine, intertidal, neritic (continental shelf), planktonic, and abyssal zones, summarizing (where applicable) their food-web relationships, their principal environment constraints, and their general ecological significance.
4. Describe thermal and nutrient layering in an aquatic habitat.

Moist desert, Arizona. (Grant Heilman/Grant Heilman Photography)

884

One would hardly expect to find a penguin living in Florida or a hippopotamus in Anchorage, Alaska—at least, not outside of a zoo! Yet we are not surprised to find humans, along with dogs, cats, and spiders, in both places. It is obvious that organisms are not uniformly distributed throughout the earth, but what governs their distribution?

KEY CONCEPTS

☐ Organisms are not distributed uniformly on land or in the ocean; they have a characteristic geographical distribution.

☐ Temperature, precipitation, and sunlight (in other words, climate) determine the distribution of communities of organisms on the land. The northernmost zones seem mainly to be influenced by temperature. The distribution of temperate and tropical organisms is, in addition, greatly influenced by variations in precipitation.

☐ Tundra is dominated by very low, slow-growing vegetation; boreal forest, by coniferous evergreens; temperate forest, usually by deciduous vegetation; prairie, by grasses; and desert, by widely spaced plants with marked adaptations for water conservation.

☐ Aquatic organisms include the free-floating and mostly microscopic plankton, the bottom-dwelling benthos, and the free-swimming nekton.

☐ Intertidal and near-shore oceanic communities are the most productive, but much of the ocean is an unproductive aquatic desert.

☐ Intertidal organisms are strikingly adapted to the stringent demands of this harsh habitat, which include daily exposure to the air and heavy wave action. Abyssal organisms are almost entirely heterotrophic, comprising food chains based on dead organisms that fall from lighted near-surface waters. The subtidal and continental-shelf zones support most marine organisms.

☐ ORGANISMS HAVE UNIQUE GEOGRAPHIC DISTRIBUTIONS

Where and how organisms live is determined by their adaptations and the history of their species. Characteristic, geographically limited assemblages of organisms make up the major geographical life zones of the earth, called **biomes.**

The bigger the differences among these zones, the bigger the differences among the creatures that inhabit them. Probably the greatest contrast between the

areas where plants and animals can live exists between the aquatic and land environments. This contrast reflects the substantial differences in the physical and chemical properties of water and air, which naturally determine the basic and distinctive adaptations of the organisms that live in each habitat.

Thermoregulation, to take one example, is easier for an animal living in air because air does not cool the body as efficiently as water does. Also, when the surroundings are hot, the evaporation of water can be used to cool a land organism.

Skeletal support is another example. Aquatic organisms require less skeletal support than land dwellers because of the greater density of water. Organisms such as jellyfish that have no rigid skeletons could not conceivably live outside of the water. Because of the support water affords, aquatic organisms can grow to a large size. Consider the largest known insects, which are only several inches long. Yet the largest known aquatic arthropods, certain Japanese crabs, span almost 12 feet.

Water can carry nutrients directly to all parts of an aquatic plant or alga; algae do not usually have vascular tissues. Because water also absorbs light heavily in comparison to air, this limits photosynthesis to the shallows. Finally, water (especially warm water) can dissolve only very small amounts of oxygen, so that water-breathing organisms must suffice with far less oxygen than land creatures.

We can find many ways in which geography groups organisms both on land and in the water. There are great differences between the grasslands and the forests, or the deep sea and the shallows. In all of these cases, features of the physical environment such as rainfall or light penetration create most of the differences.

☐ MAJOR LAND LIFE ZONES ARE CALLED BIOMES

The distribution of terrestrial communities is governed fundamentally by climate. Climate is shaped not only by temperature, but also by precipitation and, to some degree, the intensity and availability of light. Climate, in turn, affects soil development, another important governing factor.

Biogeographical realms are regions that are composed of entire continents or large parts of a continent separated by major geographical barriers and characterized by the presence of certain types of animals and plants, such as the nearctic realm of the North American continent. Within these biogeographical realms are **biomes,** which are large, relatively distinct community units that are also characterized by certain kinds of plants and animals (Fig. 48–1). The boundaries of biomes are established by a complex interaction of cli-

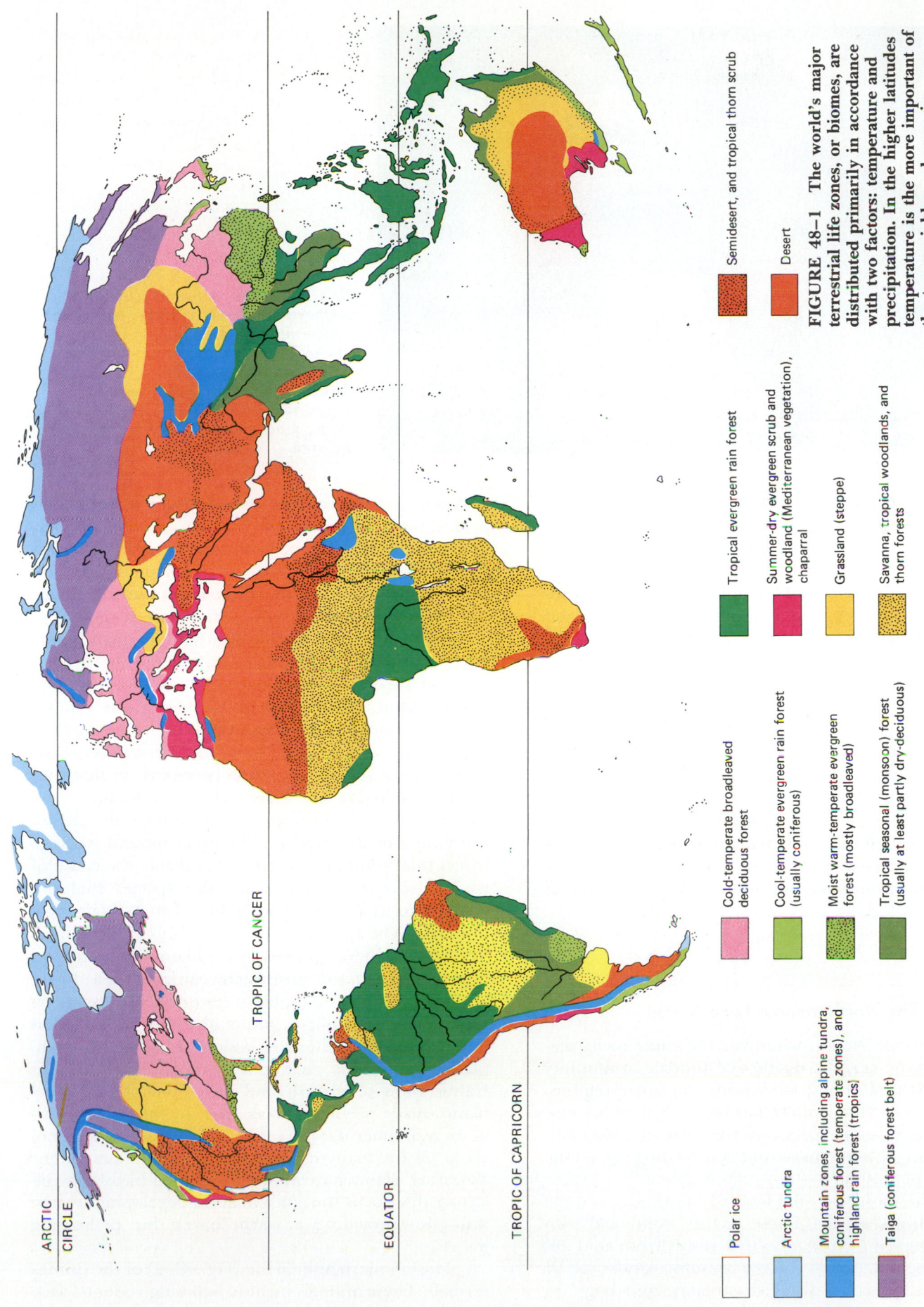

ARCTIC CIRCLE

TROPIC OF CANCER

EQUATOR

TROPIC OF CAPRICORN

Polar ice

Arctic tundra

Mountain zones, including alpine tundra, coniferous forest (temperate zones), and highland rain forest (tropics)

Taiga (coniferous forest belt)

Cold-temperate broadleaved deciduous forest

Cool-temperate evergreen rain forest (usually coniferous)

Moist warm-temperate evergreen forest (mostly broadleaved)

Tropical seasonal (monsoon) forest (usually at least partly dry-deciduous)

Tropical evergreen rain forest

Summer-dry evergreen scrub and woodland (Mediterranean vegetation), chaparral

Grassland (steppe)

Savanna, tropical woodlands, and thorn forests

Semidesert, and tropical thorn scrub

Desert

FIGURE 48–1   The world's major terrestrial life zones, or biomes, are distributed primarily in accordance with two factors: temperature and precipitation. In the higher latitudes temperature is the more important of the two variables, whereas in temperate and tropical zones precipitation is a significant determinant of community composition.

885

FIGURE 48–2   Altitude zonation at a northern temperate latitude. Notice the coniferous forest eventually giving way to bare rock and arctic-type ice caps. This photo was taken in Glacier Park, Montana; the valley and the jagged faces of the rock were formed by the movement of glaciers through the area. (E.R. Degginger)

mate, physical factors, and other conditions produced by their living inhabitants.

The distribution of land life zones is complicated by the altitudinal distribution of organisms in mountainous areas. The base of a tropical mountain, for instance, might be clothed in rain forest. Higher and cooler altitudes might possess deciduous trees reminiscent of those in northern deciduous woodlands. Above that, one might find evergreen forests, and above the evergreens, a kind of tundra resembling that of the arctic. At the very top, a permanent ice or snow cap might be found, similar to the nearly lifeless polar land areas.

Communities found in areas of hilly or mountainous topography (Fig. 48–2) reflect not only the latitude, but the altitude and the direction faced by the various slopes, because this determines how directly the sun shines on them, and sometimes affects rainfall also.

Tundra Is The Northernmost Land Biome

In the extreme North, wherever the snow melts seasonally, there exists a distinctive **tundra** community (Fig. 48–3) that almost surrounds the polar regions. (The Southern Hemisphere has no equivalent because it has no land—only water—in the corresponding latitudes, although some moss and lichens do grow on the Antarctic peninsula.)

For the most part, the land of the tundra is quite flat and drains slowly. These factors, combined with the low rates of evaporation that result from cold annual temperatures, produce a swampy landscape of broad, shallow lakes, sluggish streams, and bogs. In

addition, the tundra has a layer (varying in depth and in thickness) of permanently frozen ground, the **permafrost,** which interferes with subterranean drainage. Frozen mammals, especially extinct mammoths, have been unearthed from the permafrost with the meat still edible—at least by dogs—after tens of thousands of years. Permafrost can be spotty in its distribution. Where substantial amounts of unfrozen soil overlie it, forested areas are more likely to be present. Although there is little precipitation over much of the tundra, low temperatures, flat topography, and the permafrost layer conspire to keep the area wet. Some parts of the tundra, however, have so little precipitation that they are essentially arctic deserts with very little life.

A very short growing season affects the life of everything in the tundra. The great natural stress to which this biome is subjected is probably the cause of its low species diversity, but the species that are adapted to its extreme conditions often exist in great numbers. The tundra is dominated by reindeer moss (actually a lichen), grasses, grass-like sedges, and annual plants that must grow each year from seed. There are no readily recognizable trees or shrubs except in very sheltered localities, although dwarf willows and other dwarf trees do grow widely. The need to complete an entire life cycle in a span of weeks produces frantic rates of growth and development in annual plants, made possible in large part by the great length of each summer day. In many places, the sun does not set at all for many days in midsummer. Many arctic flowering plants have flowers designed like solar collectors that focus the sun's heat on developing ovules, thus also providing a warm haven for pollinating insects.

Most modern animal life (Fig. 48–4) of the tundra is small. These animals include lemming rodents, wea-

(a)

FIGURE 48–3 (*a*) **An example of tundra vegetation in Alaska. (*b*) The alpine tundra vegetation of the Western United States is not the same as the northern tundra and is adapted to somewhat different conditions, although it often contains some of the same species.** (*a*, Sharon Cummings/ Marvin L. Dembinsky, Jr., Photography Associates; *b*, David Muench 1990)

(b)

sels, arctic foxes, snowshoe hares, ptarmigan, snowy owls, hawks, and the like. The immense musk-oxen, caribou, and reindeer are exceptions. There are no reptiles or amphibians in this zone. One might also expect few or no insects, but insects are able to survive over winter as eggs or pupae, and some species occur in great numbers. Mosquitoes are particularly numerous and offensive.

Tundra soils tend to be geologically young because most of them have only recently been covered by glaciers. For this reason, they are usually poor in nutrients, so much so that distinctly different vegetation

FIGURE 48–4 **The tundra biome contains a variety of species, although fewer than more southerly habitats. Like the trees of temperate deciduous forests, many of the low plants of the tundra, such as bearberry (*a*), change their foliage colors. The brilliant hues probably attract birds and mammals, which disperse the seeds by eating the fruit. Wildlife of the tundra includes (*b*) the snowy owl, which like most owls is a nocturnal predator, and (*c*) the arctic fox. The color of this animal varies according to the season. During the long winter its coat is almost entirely white.** (*a*, B.J. O'Donnell/BPS [44-825]; *b*, Louis Campbell; *c*, David Cavagnaro)

(a)

(b)

(c)

often grows in the vicinity of animal roosts and dens and their associated droppings. There may be little organic litter on the surface, for although the low temperatures prevailing over most of the year inhibit decay, there is also relatively little primary production.

Although an incomplete equivalent of the tundra biome occurs on a few mountains (see Fig. 48–3), the thin, dry air, the high levels of ultraviolet-radiation exposure, the higher levels of precipitation (especially snowfall), and the usual absence of a permafrost layer make these mountain "tundras" not fully comparable to their sea-level counterpart. Also, their species diversity tends to be even lower than that of the true tundra.

One reason for this particularly low diversity is that mountaintop tundras are almost like islands, often separated from one another and the true tundra by hundreds of miles of other types of habitat. In fact, a mountaintop tundra usually contains species of the true tundra only if it actually *was* a part of the tundra during past glacial times.

All tundra regenerates very slowly after disturbance. Even casual use by hikers can be enough to destroy mountaintop tundra. As a result of oil exploration and military use, large portions of the arctic tundra have suffered long-lasting damage that is likely to last hundreds of years.

The Boreal Forest Is Dominated By Conifers

Like the tundra, the northern boreal forest biome is also circumpolar, dominated by gymnosperms (Fig. 48–5) but with a soil that has a much thicker organic layer than the tundra. Permafrost is usually absent here, or lies deep beneath the surface. This biome has an acidic, mineral-poor soil that is characterized by a deep layer of partly decomposed pine and spruce needles (or larch, in the harsher areas). The growing

FIGURE 48–5 Taiga. (E.R. Degginger)

season of the boreal forest is somewhat longer than that of the tundra. Deciduous angiosperms such as aspen or birch may form striking assemblages in burned or logged-over areas of the boreal forest, but, overall, conifers clearly dominate. The probable reason is that the growing season is so short that photosynthesis is needed year round. Yet water cannot be absorbed by roots from frozen soil, or even from cold soil (approximately 5°C), and the soil is cold or frozen for most of the year.

Conifers have many drought-resistant adaptations (such as their needle-like leaves with minimal surface area for water loss) that enable them to withstand the water shortage of the northern winter months while permitting them to photosynthesize.

The animals of the boreal forest include some larger species such as caribou (which migrate into the area from the tundra in winter), wolves, bears, and moose. However, most of the animals are medium-sized to small, such as rodents, rabbits, and fur-bearing predators, like lynx, sable, and mink. Most species of birds are seasonally abundant but migrate away in the winter. Insects are abundant, but amphibians and reptiles are scarce, except in the southern extensions.

Temperate Coniferous Forests Occur In Western North America

The boreal forest extends quite far south on the western coast of North America, reaching as far as northern California. For the trees, the extensive drought that follows the rains every year is physiologically much like the water shortage caused by cold in the northern winter. Like the tundra, the boreal forest has a mountain equivalent often found at the middle altitudes, and some mature coniferous forests occur in the North American Southwest.

The upper limit of tree growth on mountains is known as the **timberline.** Usually it is located just below a tundra area. Here a stunted, miniature forest, the **krummholz** (Fig. 48–6), may occur. The krummholz is usually composed of intricately entangled dwarfed conifers that resemble bonsai trees. Below the krummholz, large trees may occur but are often distorted into a characteristic **flagged** shape in which the branches are mostly confined to the downwind side of the trunk. This results from the accumulation of ice on the upwind side, which kills the branches.

Temperate Biomes Vary With Precipitation

Below the boreal forest, biomes do not extend evenly around the world because of the influence of precipitation. In temperate latitudes, rainfall varies greatly with

FIGURE 48–6 Krummholz vegetation on Pennsylvania Mountain in the Mosquito range of the Colorado Rockies. (Connie Toops)

out having the opportunity to be recharged with fresh moisture. The climate of the North American continent, however, is dominated by rain shadows cast by mountain ranges, especially in the West.

As prevailing westerly winds push against the bases of the Sierra Nevada and Cascade Mountains, masses of moist air from the Pacific Ocean are forced upward, subjected to cooling, and precipitate much of their moisture, as discussed in the preceding chapter. Thus, the westerly slopes of the mountains are so well watered that a kind of rain forest develops. Considerable rain falls in the upper reaches of the eastern slopes also, but by the time the air has sunk back to lower altitudes, most of the available moisture has been wrung from it and its relative humidity is very low indeed.

Deciduous Forests Develop Where Winters Are Pronounced And Rainfall Is High

Where temperate-zone precipitation ranges from about 70 to 150 centimeters annually, **temperate deciduous forests** tend to develop. Those of the northeastern and middle eastern areas of the United States are familiar examples. These habitats are dominated by broad-leaved hardwood trees that lose their leaves every year (Fig. 48–8), but broad-leaved evergreen trees make up an increasing proportion of the trees in this life zone as one proceeds southward.

Typically, the soil of a deciduous forest consists of a deep, clay-rich lower layer and a topsoil rich in partly decayed humus. Leaf fall, dead trees, and dead roots all contribute to the humus content of the soil, but humus is perishable. As organic materials in the

longitude, and produces differences in community composition. In general, continental interiors tend to be dry, but this results from a variety of causes. Permanent high-pressure areas, such as those over the Sahara Desert, may nudge moist air masses aside. Air passing over a large land mass also may dry out with-

FIGURE 48–7 Seasonal changes in a temperate deciduous forest, (a) Dense, green hardwood foliage during summer. (b) Color changes in foliage during fall. (a and b, Doug Wechsler)

(a)

(b)

FIGURE 48–8   A broad-leaved evergreen subtropical forest, near Miami, Florida.

FIGURE 48–9   Western grassland of the North American plains. (David Muench 1990)

humus decay further, mineral ions are released from it. If they are not immediately absorbed by the roots of the living trees, these ions are washed into the clay, where they may be retained.

The deciduous woodland originally contained such large mammals as puma, wolves, deer, bison, bears, and other species that are now extinct, plus many small mammals and birds. Both reptiles and amphibians abounded, together with a denser and more varied insect life than exists today.

Temperate Grasslands Occur In Areas Of Moderate Precipitation

The North American midwest is an excellent example of a temperate grassland (Fig. 48–9). Here there are few trees, except for those that grow near water-courses, but grass grows in great profusion in the thick soil once deposited by glaciers and wind. Summers are hot, winters are cold, and rainfall is often uncertain.

As rainfall decreases with longitude, minerals are found to accumulate more and more in a marked layer just below the topsoil instead of washing out of the soil. The soil originally had considerable humus content and a well-developed layer of topsoil, for many prairie grasses are similar to deciduous plants in that the above-ground portions of the plants die off each winter, while the roots survive underground. The leaves and stems then decay to form humus. Also, many prairie grasses are sod-forming grasses. Their roots form a continuous underground mat like those we see in

lawns surrounding residences and other buildings, only much thicker and tougher.

Several Types Of Deserts Occur Where Precipitation Is Sparse

Deserts are arid areas with sparse or nonexistent plant life. The low water content of the desert atmosphere leads to temperature extremes of heat and cold, so that a temperature can sometimes change by as much as 100°F in a single day. Deserts vary greatly, depending upon the amount of precipitation they receive. Desert soil has a low organic content and often a high mineral content. Saline desert soils can actually be toxic to plants.

Desert animals tend to be small and hard to find (Fig. 48–10), remaining undercover or returning to shelter periodically during the heat of the day. At night, they come out to forage. In addition to specialized insects, there are many specialized desert reptiles: lizards, tortoises, and snakes, especially venomous snakes such as the American sidewinder rattlesnake and similarly adapted African and Asian vipers. The moister deserts may have a few specialized amphibians, especially toads, if water is locally available for larval development.

Desert mammals include such rodents as the American kangaroo rat, which does not have to drink water but can subsist solely on the water content of its food plus water generated by its oxidative metabolism. In American deserts, there are also jackrabbits, and in

(a)

(b)

(c)

(d)

FIGURE 48–10 **Inhabitants of hot deserts are often strikingly adapted to the demands of their environment. (a) The moister deserts of North America frequently contain large cacti such as this giant saguaro. (b) Desert cottontail. Especially large ears may function as heat radiators. (c) Gopher snake. A number of desert snakes travel by "sidewinding" because of the uncertain purchase afforded by sand. (d) Nearly leafless Joshua tree, a member of the Lily family.** (a, E.R. Degginger; b, Animals Animals © 1990 Charles Palek; c, Bob McKeever/Tom Stack & Associates; d, David Muench 1990)

Australian deserts, ecologically equivalent kangaroos. Carnivores like the African fennec fox and some raptorial birds, especially owls, live on the rodents and rabbits. A few larger herbivores, such as antelopes, may also be found.

Desert plant adaptations are, if anything, even more striking than those of the animals. On the whole, perennial desert plants tend to have reduced leaves or even none. Still others shed their leaves for most of the year. Desert plants are noted for **allelopathy,** an adaptation in which toxic substances secreted by roots or found in shed leaves inhibit the establishment of competing plants nearby. Many desert plants have defensive spines, thorns, or toxins to resist grazing.

Tropical Life Zones Also Vary With Precipitation

Tropical and subtropical biomes (Fig. 48–11) are at least as varied as temperate ones, and like temperate life zones, they are determined mainly by precipitation. Thus, there are not only tropical forests, but also grasslands and deserts. In the tropics, the seasonal distribution of rainfall is especially important. There are grasslands that would be rain forests except that almost all of their rainfall occurs during 2 months of the year. Lush vegetation could scarcely persist for 10 months unwatered, especially where high temperatures encourage evaporation.

Tropical Rain Forests Occur Where Rainfall And Temperature Are High

Of all the life zones in the world, the coral reef and the tropical rain forest are unexcelled in species diversity and variety (Fig. 48–12). No one species dominates the tropical rain forest; one could travel for a quarter mile without encountering two members of the same species of tree, especially where the community is not a climax.

Despite what you may have seen in Tarzan movies, the vegetation of tropical rain forests is not dense at ground level, except near stream banks and in areas that are recovering from lumbering, fire, or agricultural use. The continuous canopy of leaves overhead produces a dark habitat with an extremely humid microclimate. Much of the precipitation in the tropical

(a)

(b)

(c)

FIGURE 48–11   The tropical rain forest. (*a, b*) Tropical rain forest trees typically possess elaborate systems of buttress roots that support them in the shallow, often wet soil. In very high rainfall areas, their leaves may have elongated "drip tips" which drain water off rapidly, thus discouraging the growth of tiny epiphytes, called epiphylls, which would otherwise develop on the leaves and cut off some sunlight from them. (*c*) Saving the investment of materials in skeletal tissues of the trunk, a strangler fig climbs up the trunk of a palm tree which may eventually be killed. (*b*, Frans Lanting)

rain forest is locally recycled water that comes from the transpiration of the forest's own trees. If trees are removed over a large area, that may so reduce the rainfall that the original forest is unlikely to be re-established.

A fully developed rain forest has at least three distinct levels, or **stories,** of vegetation. The topmost story consists of the crowns of occasional very tall trees. It is exposed entirely to the sunlight. The middle story forms a continuous canopy of leaves that lets very little sunlight through to support the plants of the sparse understory. The understory itself consists of both plants specialized for life there and the seedlings of taller trees.

All stories of vegetation support extensive epiphytic communities of smaller plants that grow in crotches, on bark, or even on the leaves of their hosts (see Fig. 48–12). The epiphytes are not parasitic in the usual sense, but their numbers can become so great as to break branches or interfere with photosynthesis. Most of the turnover of the mineral nutrients in the rain forest probably occurs among epiphytes and the growing leaves of the forest canopy, where most animal biomass is concentrated, rather than on the forest floor. Because little light penetrates to the understory, many of the plants livings there are adapted to climb upon already established host trees rather than investing their meager photosynthetic resources in the dead cellulose tissues of their own trunks. Tropical vines as

thick as a man's thigh abound. This adaptation can be seen in extreme form in several species of **strangler tree** (see Fig. 48–11). The strangler fig, for instance, overgrows the trunk of the host, eventually killing it.

The trees of the tropical rain forest are usually evergreen (although not coniferous), but there are exceptions. Their roots are often shallow, forming a mat 2 or 3 feet thick on the surface of the soil that misses hardly a particle of mineral nutrient released from leaves and litter by decay processes. Despite their poor anchorage, swollen bases or "flying buttress" trunk bases hold the trees upright and aid in the extensive distribution of the shallow roots.

Because the temperature is high year round in the rain forest, decay organisms, ably assisted by detritus-feeding ants and termites, decompose organic litter before it can become humus. At the same time, the heavy rainfall washes nutrients rapidly from the soil, but highly developed mycorrhizae extract nutrients from decomposing material. They are then transferred to the roots of living plants before they have a chance to enter the soil. The absorptive mechanisms of the rain forest community are so efficient that the run-off water often has a lower mineral content than the rain that falls there.

Rain-forest animals include the most abundant and varied insect, reptile, and amphibian fauna on the face of the earth (see Fig. 48–12). Birds, too, are varied and often brilliantly colored. Ants are especially prom-

(a)

(b)

(c)

(d)

FIGURE 48–12   Some animals of the rain forest. (*a*) An arrow-poison frog from Columbia. The bright colors of this frog are an aposematic trait which warn that it is inedible. Human beings, however, utilize the frog's poison to coat the tips of the arrows they use for hunting. (*b*) Three-toed sloth. Since these animals hang from branches, the hairs of their coats point toward their backs to encourage rainwater to run off them. (*c*) Orangutans. (*d*) Tropical termite nest.
(*a*, Edmund D. Brodie, Jr., Adelphi University/BPS; *b*, Kevin Schafer/Tom Stack & Associates; *c*, Brian Parker/Tom Stack & Associates; *d*, David Cavagnaro)

inent. Many mammals live in trees (sloths, monkeys), although some large ground-dwelling mammals, including even elephants, are found there.

Explosive human population growth in tropical countries may spell the doom of most or all rain forests by the end of the century. It is believed that many rain-forest organisms will be rendered extinct in this way before they have even been scientifically described. A great many useful drugs, such as quinine (a tree product) or vinblastine (a chemotherapeutic drug used in cancer treatment that occurs in the tropical periwinkle *Vinca*), have been discovered among the exotic chemical compounds used by tropical plants as chemical de-

fenses against grazers and parasites. One wonders how many other potentially useful species of plant and animal life have already vanished. The total ecological impact of rain-forest destruction is unknown at present, but it is likely that the burning or decay of felled trees will contribute substantially to the atmospheric content of carbon dioxide while removing one of the largest areas of carbon-dioxide fixation by photosynthesis.

The local impact of rain-forest destruction is immediate and substantial. Once the protective mat of roots has been removed from the soil, erosion is likely to become an even more severe problem than in tem-

FIGURE 48–13 The savanna biome of Eastern Africa. These grasslands formerly supported large herds of grazing animals and their predators, which are now swiftly vanishing under pressure from pastoral and agricultural land use. (Donald Marshall Collection, WIV International Programs)

perate climates and often will be accompanied by mineral loss due to leaching. Because the mineral content of the rain-forest community resides mostly in its vegetation, rain-forest soils tend to be poor. Even if left alone, areas of destruction may not become mature forest again if the soil is excessively depleted. Moreover, if a larger area is destroyed, the rather inefficiently dispersed seeds of the forest species may not make their way to a large part of it.

Savannah Is A Tropical Grassland With Some Trees

The **savannah** (or **veld**) life zone is a tropical grassland or very open woodland, depending on one's viewpoint (Fig. 48–13). Widely dispersed trees such as *Acacia*, which are bristling with thorns that protect them against herbivores, grow amid long grasses. The greatest herbivore biomass in the modern world occurs in

(a)

FIGURE 48–14 Inhabitants of the savanna. (*a*) Cape buffalo grazing in Uganda. (*b*) Lions killing a zebra. In the open spaces of the savanna (shown here during the dry season), hunting in pairs or in groups is advantageous in tiring or cornering swift-moving prey. (*c*) What is not consumed by the hunters is left for the scavengers, which often double as predators themselves. Here hyenas compete with buzzards, which are strikingly adapted as scavengers. Their long, almost bald necks allow them to reach inside carcasses and make a last effort to pick the body clean. (*a* and *c*, E.R. Degginger; *b*, Animals Animals © 1990 David C. Fritts)

(b)

(c)

the African savannah. Here live the great herds of antelope, giraffe, zebra, and the like. Large predators, such as lions and hyenas, kill and scavenge the herbivores. Savannah is produced naturally either by low rainfall or by sharply seasonal rainfall. In areas of seasonally varying rainfall, the herds and their predators (Fig. 48–14) may migrate annually, much as caribou migrate between tundra and boreal forest in the north.

Tropical grasslands are being rapidly converted to range for cattle and other animals. Severe overgrazing in places has converted marginal savannah to actual desert. On the other hand, lumbering and grazing have converted much bordering rain forest to savannah. In both cases, large amounts of original wildlife habitat are being subjected to probably irreversible destruction.

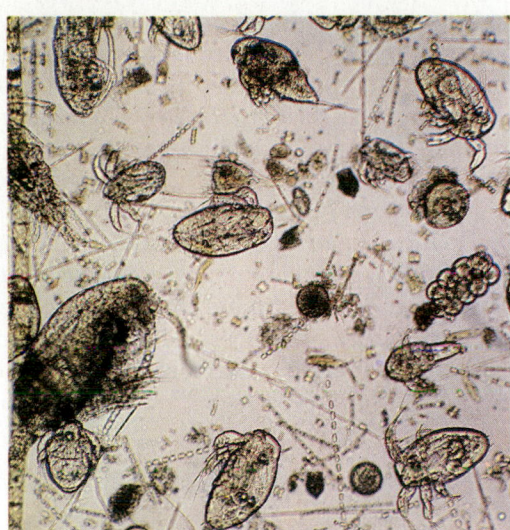

FIGURE 48–15   Mixed marine zooplankton, mainly crustaceans such as copepods. (Runk Schoenberger, from Grant Heilman Photography)

■ AQUATIC LIFE ZONES OCCUPY MOST OF THE EARTH'S SURFACE

The world's surface is mostly water, frozen or liquid, and of the portion that is liquid, most is salt water. In water (as on land), mineral nutrients collectively represent a significant limiting factor. In water, however, temperature is a somewhat less important factor. In most land areas, there is no great shortage of light, the floor of the rain forest being the most important exception. But this is not true in aquatic life zones. Water greatly interferes with the penetration of light, so floating photosynthesizers must remain near the surface, and vegetation attached to the bottom can grow only in the shallowest zones.

The most fundamental division in aquatic ecology is probably between fresh and salt water. The physiological adaptations required to permit osmoregulation in freshwater animals is a major strain on an organism's resources. Some animals, such as echinoderms, are never found in fresh water at all. Such organisms, adapted to a very narrow range of salinity, are known as **stenohaline;** those that can live in a variety of salinities are **euryhaline.** Some fish and crustaceans are euryhaline as adults but stenohaline as larvae. Others, such as salmon and eels, live in fresh water during part of their life cycles and salt during another part, migrating between the two.

Aquatic Organisms Fall Into Three Ecological Categories

The aquatic habitat contains three main ecological categories of organisms: free-floating **plankton,** strongly swimming **nekton,** and bottom-dwelling **benthos.** All categories are constrained by the small amount of light

penetrating the water. This limits productivity to the **euphotic zones,** which are upper or shallow waters in which photosynthesis is practical. The benthos may be further subdivided into **intertidal, subtidal,** and very deep-dwelling **abyssal** communities. Of these, only the abyssal community is without its own primary producers and subsists on the leftovers that wash or fall into it from the lighted areas of the water.

The plankton (Figs. 48–15 and 48–16) forms the basis for most aquatic food chains. It consists of organisms (usually small or microscopic) that are relatively feeble swimmers and thus, for the most part, are carried about at the mercy of currents and waves. They

FIGURE 48–16   A baleen whale siphoning the surface of the open ocean for plankton. The bird nearby has been attracted by fish stirred up near the surface of the water by the motion of the whale. (David W. Hamilton/The Image Bank)

are unable to swim far horizontally, but some species are capable of large daily vertical migrations and occupy different strata of water at different times of the day or sometimes at different seasons. Plankton may be subdivided into two major categories: the phytoplankton and the zooplankton. The **phytoplankton** are primary producers. Phytoplankton are cyanobacteria and free-floating algae of several types (diatoms usually predominate). It is believed that the very smallest of the phytoplankton, the **nannoplankton,** are responsible for most of the primary production of the seas. The **zooplankton** include planktonic protists and animals, including the larval stages of many organisms that are large as adults.

Freshwater Habitats Include Streams, Lakes, And Ponds

Even temporary ponds have a typical assemblage of inhabitants: protists and algae, mosquito larvae, toad tadpoles, fairy shrimp, which hatch from eggs that lie dormant during dry spells in the bottom, and water strider insects, which live atop the surface film while the water lasts (they can fly away when it dries up). The kinds of organisms found in streams vary greatly, depending mostly on the strength of the current. In fast streams, the water is often cold and has a high oxygen content, and the inhabitants may have adaptations

such as suckers that keep them from being swept away. Large, slow-moving streams ecologically resemble lakes, and it is on lake habitats that we will focus our attention.

A typical lake shore is inhabited by partly submerged vegetation, such as cattails and water lilies (Fig. 48–17). The lake shore plus several other concentric communities of deeper-dwelling, completely submerged plants constitute the **littoral zone,** the shallow water area around the margins of a lake, pond, or watercourse. It is the most highly productive zone of the lake. A shallow lake may consist entirely of a littoral zone. Algae, particularly filamentous algae and diatoms, may exceed the biomass of the higher plants in the littoral zone. The littoral zone contains frogs and their tadpoles, turtles, annelid worms, crayfish and other crustaceans, insect larvae, and many fish (see Fig. 48–17). Here, too, at least in the quieter areas, one finds surface-film dwellers, sometimes collectively called the **neuston,** such as water striders and whirligig beetles.

The deeper **limnetic zone** of open water contains sparse life, but larger fish spend most of their time here, although they may visit the littoral zone to feed and breed. Owing to its depth, less vegetation grows in this zone.

The deepest zone, the **profundal zone,** does not get enough light for photosynthesis. Primary producers cannot prosper here because it is below their **com-**

FIGURE 48–17 A small freshwater pond provides an example of both an ecosystem and the littoral zone. These great tropical Victoria lilies support a rich array of invertebrates which are consumed by one another, by larger invertebrates, and by fishes. (*b*) Bass spawning ground. Each circular crater is a bass "nest." These fish use the shallows of the littoral zone for breeding and would disappear with the destruction of this zone—for example, by lakeside "development." (*a*, Richard H. Gross; *b*, SELBYPIC)

(a)   (b)

pensation point, the point at which the products of photosynthesis balance the rate of catabolism. Nevertheless, much food drifts into this zone from adjacent zones or falls into it from the lighted waters above. When dead plants and animals reach the profundal zone, decay bacteria liberate the minerals their bodies contain, but these minerals cannot be effectively recycled because primary producers do not exist there. Consequently, the profundal habitat tends to be both rich in minerals and anaerobic, with few forms of higher life.

This marked layering is accentuated by **thermal stratification** characteristic of large lakes, especially in temperate zones. In thermal stratification, cool water remains at the bottom in the summer, being separated from the warmer water above by a marked and abrupt temperature transition, the **thermocline.** In temperate-zone lakes, falling temperatures in the autumn cause the lake waters to mix. This occurs because as the surface water cools its density increases; thus, it displaces the less dense warmer water beneath, which then rises to the surface, where it cools in its turn. As this process of cooling and sinking continues, the lake eventually is churned to a uniform temperature throughout. Hereafter, even light winds will produce further mixing, since density differences no longer exist. This phenomenon is called the **fall turnover.**

The sudden presence of large amounts of nutritive mineral ions in surface waters encourages high algal populations to develop, which form temporary **blooms** then, and again in the spring. The **spring turnover** occurs as the ice melts and the surface water reaches 4°C (its temperature of greatest density). During the summer, thermal stratification again occurs.

The Marine Zone Is The Largest Aquatic Life Zone

Most of the world's biomass resides in the oceans; however, the oceans are by no means uniformly productive. The "limitless resources of the sea" are anything but limitless, requiring as much careful conservation as those of the land if we are to avoid total ecological disaster in the near future.

Marine And Freshwater Environments Meet In Salt Marshes And Estuaries

Where the sea meets the land, one of several kinds of ecosystems may exist: a rocky shore, a sandy beach, an intertidal mud flat, or a tidal estuary containing salt marshes (Fig. 48–18). An **estuary** is a coastal body of water, such as a bay or large river mouth partly surrounded by land, with access to the open sea and a

FIGURE 48–18   The Chesapeake bay is a major estuary surrounded by minor, tributary estuaries. The Potomac river enters the main estuary at the bottom of the picture. Such environments are extensively threatened by coastal development, overfishing, and many kinds of pollution. (NASA)

(a)

(b)

FIGURE 48–19   The estuarine environment.
(*a*) Brigantine salt marsh estuary in New Jersey. (*b*) The glasswort, *Salicornia virginia*, is an estuary plant that grows just above the high tide mark in marshy areas. It is adapted to salty soil conditions by its ability to accumulate and tolerate comparable concentrations of salt in its own tissues. (*a*, Dwight R. Kuhn; *b*, G.R. Roberts)

large supply of fresh water from rivers. It usually contains **salt marshes,** which are very shallow or swampy areas filled with mud and dominated by grasses (Fig. 48–19). Its salinity fluctuates between that of sea water and that of fresh water. Many estuaries undergo marked daily or seasonal variations in temperature, salinity, and other physical properties. To survive there, organisms must have a wide range of tolerance to these changes.

The waters of estuaries are among the most fertile in the world, often being much more productive than the adjacent sea or the fresh water up the river. This high productivity results from (1) the action of the tides, which promotes a rapid circulation of nutrients and help remove waste products, (2) the flow of nutrients into the estuarine ecosystem from the land drained by rivers and creeks that run into the estuary, and (3) the presence of many kinds of plants, which provide an extensive photosynthetic carpet and whose roots and stems also mechanically trap much potential food material. As leaves and plants die, they decay, forming the basis of many detritus food chains. Most commercially important fin and shell fish spend their larval stages in estuaries among the protecting roots and tangle of decaying stems.

Estuaries and salt marshes have often appeared to uninformed people to be worthless. In consequence, they have been used as dumps for the castoffs of industrial civilization and have become severely polluted. More recently, they have been "filled" with dredged bottom material to form artificial land for residential and industrial development. Much of the total produc-

tivity of the marine environment has been lost in this way, but there are signs that further loss may slow owing to legal restrictions on coastal development.

The Intertidal Zone Is Harsh But Productive

Marine zones are profoundly influenced by tides and currents. The gravitational pulls of both sun and moon produce two tides a day throughout the ocean, but the height of those tides depends on the time of year and the shape of the shoreline. The area between low and high tide is the **intertidal zone** (Figs. 48–20 and 48–21).

The high levels of light and nutrients, together with an abundance of oxygen in the intertidal zone, produce many potentially excellent habitats. Yet it is a very stressful environment. If an intertidal beach is sandy, the inhabitants must contend with a constantly shifting environment that threatens to engulf them and gives them scant protection against wave action (see Fig. 48–21). Consequently, most sand-dwelling organisms are continuous and active burrowers. They can, however, follow the tides up and down the beach, and so do not usually have any notable adaptations to drying and exposure.

On the other hand, a rocky shore (see Fig. 48–20) provides a multitude of fine anchorage sites but is exposed to wave action when submerged and to drying (to say nothing of seasonal heating and freezing) when exposed to the air. A typical resident of a rocky shore will have some way of sealing in moisture (perhaps by

(*text continues on page 902*)

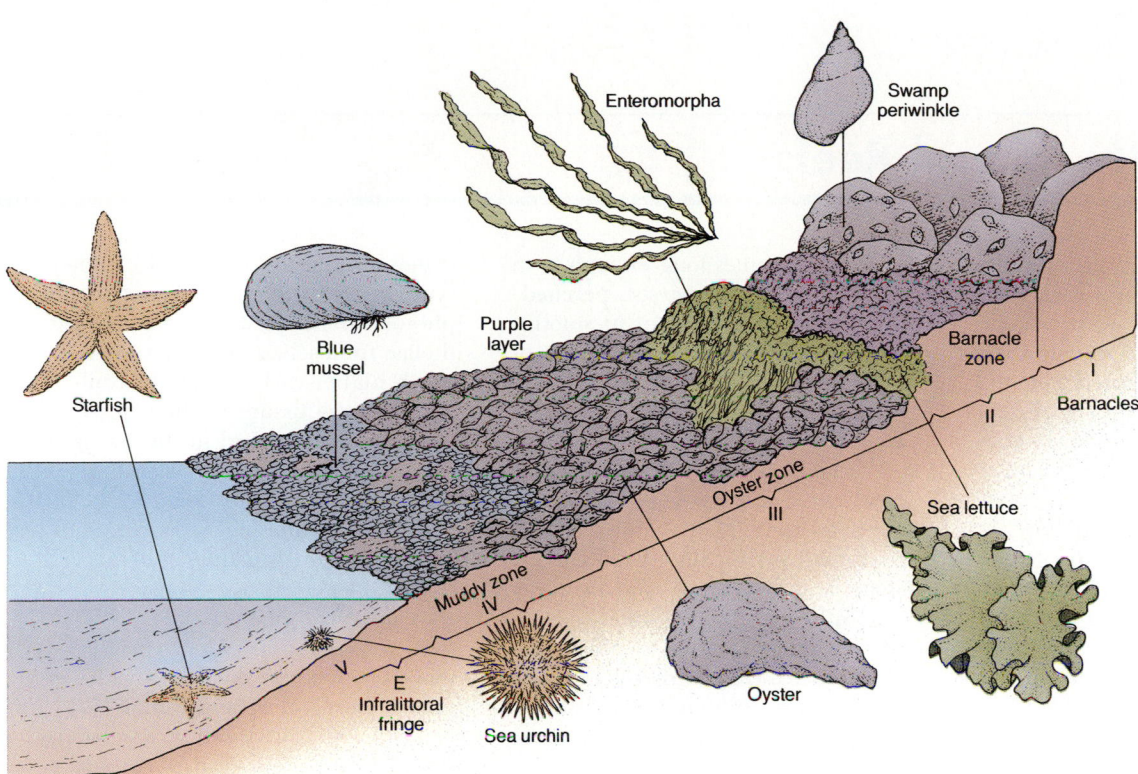

FIGURE 48–20 Zonation along a rocky shore of the mid-Atlantic North American coastline. (I) Bare rock with some black (that is, brown) algae and swamp periwinkle. (II) Barnacle zone. (III) Oyster zone, which also includes sea lettuce and purple (red) algae. (IV) Muddy zone with mussel beds. (V) Infralittoral zone with sea stars and other typical animals.

FIGURE 48–21 Life zones on a sandy beach. The gradation, less marked than that of a rocky shore, is nevertheless evident. (I) Supratidal zone: ghost crabs and sand fleas. (II) Flat beach zone: ghost shrimp, bristle worms, clams. (III) Intertidal zone: clams, lugworms, sand "crabs" which follow the retreating or advancing waters). (IV) Subtidal zone.

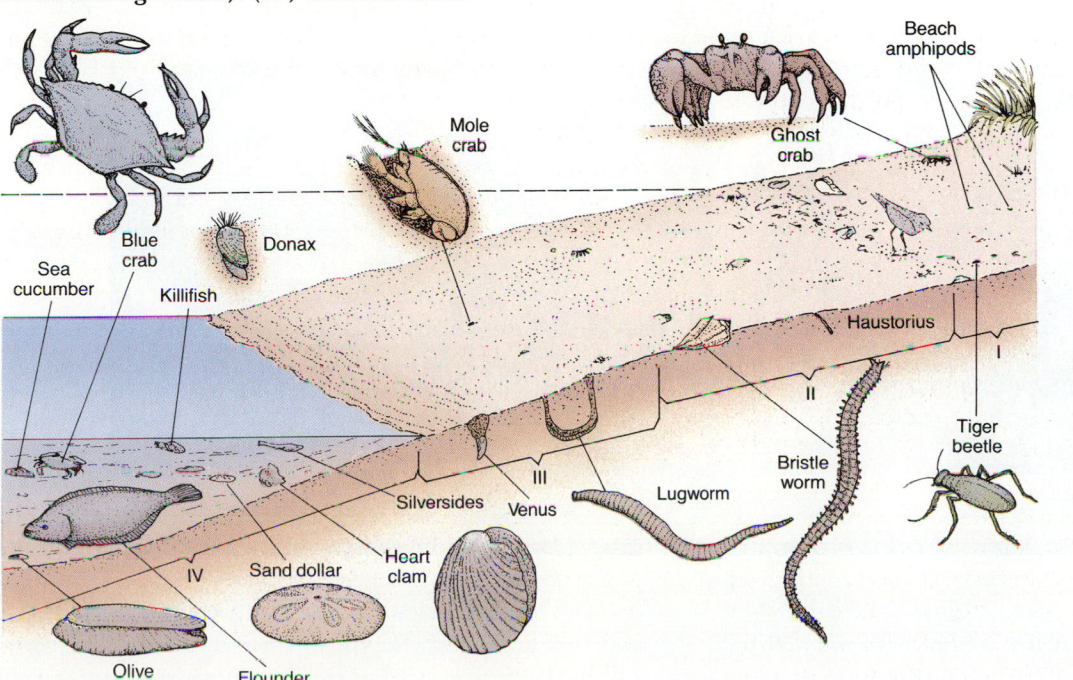

FOCUS ON   Coral Reefs

The most wonderfully varied marine environment is the coral reef. Such reefs are confined to warm waters (although coral is not), much as most rain forests occur in warm climates. Like the rain forest, they are highly productive and highly diverse environments, often occurring in nutrient-poor habitats.

Coral reefs start in shallow water—for example, on the sides of a volcanic island near the surface. Wave and weather action eventually erode the island to water level, but the coral does not erode in the same way. Being alive, it is capable of growing in compensation. The result can be a circular reef of coral, called an **atoll,** surrounding a lagoon of quiet water. Sometimes the bottom sinks, or the ocean level rises slowly enough for the growth of the coral organisms to keep up with it. If this process continues long enough, a living coral reef may persist, perched upon layer after layer of its smothered forebears. Sometimes coral reefs form barriers, comparable to offshore sand bars, as in the Great Barrier Reef of Australia.

The upper living portions of coral reefs must grow in the euphotic zone. Many coral reefs are composed principally of red coralline algae, and even coral animals, usually anthozoan coelenterates, have hordes of intracellular symbiotic dinoflagellates **(zooxanthellae)** living in their tissues. Although species of coral without zooxanthellae do exist, only species with zooxanthellae can build reefs. Other reef-dwelling organisms, such as the giant clam, *Tridacna,* may also have symbiotic algae. Not only do the coral organisms **(polyps)** capture food with stinging tentacles, but thanks to their symbiotic partners, they benefit from photosynthesis as well. (The zooxanthellae themselves benefit from a much higher carbon-dioxide availability in the tissues of their hosts than they would find in the warm water outside.) A coral polyp is almost the marine equivalent of a lichen, and some biologists consider corals "honorary plants," for their tissues may contain even more zooxanthellae than animal cells. If they are viewed as plants, their primary productivity compares with that of rain forests and estuaries.

The shallow location also means that the reef is exposed to turbulent wave action. The disadvantage of this is that organisms are easily damaged by wave action and may be stranded occasionally at low tide. The advantage, however, is that the

The complexity and productivity of the coral reef rivals or exceeds that of any terrestrial environment. In many ways the coral reef is the aquatic equivalent of the tropical rain forest. (a) Pink sea fan coral growing at right. (b) Giant clam. (c) Glassy sweepers. (d) Clownfish and sea anemone to whose sting the fish is immune. (e) Plumeworm with brain coral. (f) Queen angelfish. (a, Al Grotell; b, Phil Degginger; c, Jeffrey L. Rotman; d and e, Susan Blanchet/M.L. Dembinsky, Jr., Photography Associates; e, E.R. Degginger)

(a)

(b)

surf provides filter feeders with a cost-free energy input, much as is the case in the intertidal littoral zone.

The waters in which coral reefs are found are often poor in nutrients. Parts of the South Pacific Ocean, for example, tend to be unproductive for this reason. How coral reefs, productive as they are, overcome this disadvantage is not fully known; however, efficient local recycling mechanisms (similar to those of rain forests) appear to play a part, as do, it is suspected, the zooxanthellae.

Coral reefs contain hundreds of species of fishes and invertebrates that occur nowhere else. Many are brilliantly colored or poisonous or possess other striking and bizarre adaptations. The multitude of ecological niches and relationships occurring there seems comparable only to the tropical rain forest among terrestrial ecosystems. As in the rain forest, however, competition is intense, particularly competition for light and space to grow.

For instance, corals of the genus *Podillopora* possess a striking defense mechanism that protects them from coral browsers such as the crown-of-thorns starfish. Tiny commensal crabs live among the branches of the coral. When a starfish crawls onto the coral, the crabs swarm over it, nipping off its tube feet and eventually killing it if it does not promptly retreat!

(c)

(d)

(e)

(f)

FIGURE 48–22 Intertidal life zonation. Several clearly defined areas can be seen in this photograph of a rocky beach at low tide in the vicinity of Botany Beach, British Columbia. The barnacles covering the higher rocks show up white, and the mussels, on lower rocks, are dark blue or black. Pink tones are contributed by coralline algae, with green or brown (kelp) algae in the very lowest areas. (J. Robert Waaland, University of Washington/BPS)

closing its shell, if it has one), plus powerful mechanisms for anchoring itself to the rocks (Fig. 48–22). Mussels, for example, have horny, thread-like anchors, and barnacles have a special cement gland. Algae of the intertidal zones of rocky shores usually have thick, gummy polysaccharide coats that dry out slowly when exposed, and flexible bodies that are not easily broken by wave action. Some members of this community hide in burrows or crevices at low tide, and some small semiterrestrial crustaceans run about the splash line, following it up and down the beach (Fig. 48–23).

The Subtidal Zone Is Protected From Wave Action

The **subtidal zone** is below the lowest tide but still shallow enough for vigorous photosynthesis (Fig. 48–24). Largely protected from wave action, this area supports a variety of echinoderms (such as sea stars), fish, burrowing worms, eelgrass, and the like. Shore birds find the subtidal and intertidal zones rich hunting and fishing grounds. (The subtidal zone can be considered the shallowest portion of the neritic zone, which we discuss next.)

(a)

(c)

(b)

FIGURE 48–23 Adaptations to life in the intertidal zone. (*a*) **The seaweeds clinging to these rocks have very little hard structure to their bodies yet have strong holdfasts that enable them to remain in place against the onslaught of the tides.** (*b*) **These Bahamian snails shelter from the wave action by clinging tightly to a cavity in the rocks. Their shells are so shaped as to offer minimal resistance to the rushing seawater.** (*c*) **Life underwater in a tidal pool: dog whelk, rock barnacle,** *Obelia*, **rockweed, sea anemone, and blue mussel.** (*a*, Animals Animals © 1990 Ted Levin; *b*, Charles Seaborn; *c*, Runk Schoenberger from Grant Heilman Photography)

FIGURE 48–24 This offshore caye (or sand island) is surrounded by an area of shallow water. Notice that submerged vegetation grows thickly only in the shallows. That is because light penetrates water rather poorly; the deeper the water, the sparser is the bottom vegetation.

FIGURE 48–25 A deep-sea fish. Most fish that live at great ocean depths have weak or vestigial eyes. Many have luminous organs enabling them to locate one another (it is thought) for mating or social display. Inside the nearly transparent body of this particular fish are the remains of small crustaceans that have probably drifted down from the productive near-surface waters of the ocean.

The Shallow Continental Shelf Surrounds All Continents

Nekton and larger benthic organisms are mostly confined to the shallower **neritic** waters, which are less than 200 feet deep, because that is where their food is. Not only is there considerable vegetation on the bottom, but as you can see from Figure 48–24, this is also where the chlorophyll content of the water (reflecting the phytoplankton that is there) is high, although not as high as in an estuary. That means that little commercial fishing is worthwhile more than a hundred miles offshore, a fact that is significant in determining a nation's "territorial waters."

Most Of The Ocean Bottom Is Too Deep For Photosynthesis

Some 88% of the ocean is more than a mile deep, far below where any algae can grow and, therefore, for the most part, an aquatic desert. Most of the life that exists under the tremendous pressures and darkness of the abyss depends upon whatever food filters into its habitat from the upper lighted regions. The principal exceptions are the deep-sea oases located near abyssal hot springs (**thermal vents**).

Animals of the abyss are strikingly adapted to darkness and scarcity of food (Fig. 48–25). Many have illuminated organs, enabling them to see one another for mating. Some abyssal squid even emit a luminous cloud of ink that may distract predators. A great many abyssal organisms are predators (there is little other choice!) and live in dispersed populations. Some are remarkably adapted killing machines that are prepared to take quick and maximum advantage of any rare opportunity to feed. Their scattered population suggests that it might be hard for them to find mates, but studies to confirm this must await the development of ways to observe the ocean bottom over long periods. One species of deep-sea anglerfish hangs onto her mate once she has found him, for he permanently attaches to her body and grows into it.

Until recently, the deeper parts of the oceans seemed safe from ecological damage, but the development of deep-sea mining and fishing techniques and certain kinds of waste disposal could easily threaten this zone.

Plankton Forms The Basis Of Most Marine Food Chains

Marine plankton are mostly small or microscopic forms: tiny crustaceans, diatoms and other protists, jellyfish, swimming tunicates, ocean sunfish, arrow worms, and the larval forms of many bottom-dwelling organisms. One can see from this that "plankton" is not a taxonomic classification. Plankton are at the base of most marine food chains, but not all plankton eaters are small. The blue whale, perhaps the largest animal ever to have lived, eats plankton directly, straining it out of the water through the whalebone in its mouth.

The marine benthos include detritus feeders (such as many worms and bivalve mollusks), many crustaceans (such as crabs, lobsters, and some shrimp), sessile creatures (such as sea anemones, corals, and tunicates), and even some fish (such as flatfish, rays, and moray eels). Algae equipped with holdfasts are also part of the shallow-water benthos, but they do not extend past the euphotic zone.

Marine nekton rely ultimately on plankton and sometimes on benthic organisms as well, for it is with these that the food chains of most fish (including most commercially valuable fish) begin. Thus, the best fishing tends to be near land, in the relatively shallow and well-fertilized continental shelf areas.

Succession Occurs In Aquatic Habitats

Salt marshes and tidal flats experience ecologic succession. Freshwater aquatic habitats also are prone to successional changes.

Geologically young lakes and watercourses are said to be **oligotrophic,** meaning "little food," because they usually have little plant life and therefore little of the primary production needed for food. Fish and other aquatic animals that live in oligotrophic habitats typically require low temperatures and high oxygen concentrations. Sometimes they obtain much of their food from sources outside the aquatic habitat such as insects that fall into the water. The total biomass of an oligotrophic aquatic community is usually low, but species diversity is high and many of its species (for instance trout) are highly valued.

Eutrophication is the typical successional process that occurs in geologically aging aquatic habitats. A shallow, warm lake or watercourse, whose biomass and waters are high in mineral nutrients and biomass, is said to be eutrophic. The reason that a geologically older lake gradually becomes eutrophic is that silt and nutrients accumulate in it from nearby watersheds. The term eutrophic means "good food"; but whether the food is truly good or not, it does become more plentiful. The water warms and the species composition changes to include fish such as bass or carp. Since community productivity improves, more fish are available, but of fewer species.

The accelerated release of minerals from pollutants produces an exaggerated form of eutrophication called **cultural eutrophication.** Cultural eutrophication can change the characteristic form of submerged and shoreline vegetation undesirably, producing swampy tangles of growth through which fish can hardly navigate. Dense blooms of cyanobacteria can occur, crowding out normal and more desirable algae and protists. At night or on cloudy days, the plant life consumes more oxygen than it produces. Finally, the time inevitably comes when some event (such as low winter temperature) kills off a large part of the huge plant biomass. Its decay then reduces the oxygen content of the water so much that the lake cannot support most aquatic animal life, and massive fish kills result.

◼ LIFE ZONES INTERACT

Not one of the biomes or life zones we have discussed exists in isolation. All interact. The details of the interactions can be unexpected. When parts of the Amazon rain forest flood annually, fish leave the stream beds and range widely over the forest floor, where they have been shown to play a role in dispersing the seeds of many species of land parts. At another extreme, the waters of the polar zones may be much more productive than the surrounding land. In Antarctica, for instance, there is hardly any terrestrial community of organisms, but there are many seabirds, seals, and other semiaquatic air breathers. These are supported exclusively by the ocean. Their waste products, cast-off feathers, and the like, if deposited on land, support what lichens and insects may occur there.

Inhabitants of various biomes may interact over wide distances, even global distances in the case of migratory birds and fish. Migratory birds commonly spend critical parts of their life cycles in entirely different countries, which can make their conservation difficult. It does little good, for instance, to protect a songbird in one country if the inhabitants of the next put it in the cooking pot as soon as it lands in their neighborhood. It does little good to bemoan the loss of the rain forests if we, in the developed countries, consume excessive amounts of beef, encouraging the destruction of rain forests so that cattle can be raised to meet our demand. Even if we do care, because the nature of the connections are subtle, we may not realize the impact of our consumptive life-styles upon those same rain forests or acknowledge our own complicity in their destruction.

It is this kind of large-scale interaction that makes ecology difficult for many people to grasp or apply, because it places the consequences of actions far away and out of sight. Because the ocean depths are out of sight, we destroy their life with sewage sludge. Because tropical rain forests are far away, we do not experience their loss, or realize the inevitable impact this will have upon the lives of all human beings.

◼ CHAPTER SUMMARY

I. Living things are restricted in their occurrence to habitats to which their adaptations suit them.

II. Living things make up certain characteristic geographic assemblages. Such terrestrial life zones are known as biomes.

A. Characterized by a permanently frozen layer of subsoil, the permafrost, the tundra is the northernmost of the terrestrial biomes. It possesses low-growing vegetation that is adapted to cold, boggy conditions, permafrost and a very short growing season.

B. The boreal-forests lie south of the tundra and stretch across both North America and Eurasia. The boreal forest is dominated by coniferous trees.

C. Temperate coniferous forests exist where a pronounced dry season gives the evergreen needle-leaved trees a competitive edge. Some of these habitats have high seasonal rainfall and can even qualify as temperate rain forest.

D. Temperate communities are extremely varied, including deciduous forest, prairie, and desert regions. Differences in precipitation produce the various temperate communities.

1. Deciduous forests are dominated by broad-leaved trees that for the most part lose all their leaves seasonally. The soil is well differentiated into several horizons.

2. Moderate precipitation produces the temperate grasslands. Typically, these possess a deep, mineral-rich soil and are well suited to the growing of grain crops.

3. Deserts have low rates of precipitation and possess communities whose organisms have noteworthy water-conserving adaptations. Desert soils are often saline.

E. Tropical biomes include the tropical rain forest, the savannah, and deserts, plus intermediate communities. Most of the differences among tropical life zones result from variations in precipitation.

1. Tropical rain forests are produced by very high rainfall that is evenly distributed throughout the year. They have high species diversity and are noted for three stories of forest foliage, many epiphytes, and often a lateritic soil.

2. Many tropical grasslands are similar to open forests, with scattered trees interspersed with grassy areas. These are called savannah or veld.

III. Aquatic life zones differ from terrestrial largely because of the differences in physical properties of air and water. One of the most striking limiting factors in aquatic environments is the availability of light for photosynthesis. Oxygen also is in short supply. However, water's density provides support for the tissues and bodies of aquatic organisms, which reduces their need for skeletal structures.

A. Freshwater communities are differentiated on the basis of water depth. The marginal littoral zone contains emergent vegetation and heavy growths of algae. The limnetic zone is intermediate in depth, and the deepest profundal zone holds little life.

B. The main marine habitats are similar in principle to those of fresh water but are modified by tidal action.

1. Aquatic life is ecologically divided into the plankton, nekton, and benthos. In large measure, the free-floating microscopic plankton support aquatic communities. Owing to the availability of mineral nutrients near continental margins, shallower waters are the most productive.

2. Marine life zones include estuaries, intertidal habitats, subtidal and continental shelf zones, and the abyss.

a. The very productive estuary community receives a high input of nutrients from the adjacent land and serves as an important nursery area for the young stages of many aquatic organisms.

b. Organisms of the intertidal zone possess adaptations that enable them to resist wave action. Because oxygen, light, and food are readily available, intertidal communities are abundant and productive.

c. The subtidal and continental shelf zones are characterized by rich plankton and benthic vegetation.

d. Coral reef–building organisms employ symbiotic photosynthetic organisms to produce food directly for the use of the coral polyps. The species diversity of coral reefs is unexcelled in aquatic habitats.

e. The abyssal zone is without primary producers (except in the case of hot-spring oasis communities), so that its animal inhabitants are exclusively predators or scavengers subsisting on input from the areas of the ocean that are capable of primary production.

■ POST-TEST

1. The tundra typically has a _____ rate of precipitation, a short growing season, and a permanently frozen underground layer of _____.
2. _____ are the dominant vegetation of the boreal forest.
3. In the temperate zone, the deciding factor in producing forest versus grassland is usually _____.
4. In temperate deciduous woodlands, minerals leached from decomposing humus may accumulate in a layer of _____.
5. Species diversity is exceptionally _____ in tropical rain forest and marine _____ _____ habitats.
6. The _____ is a tropical habitat in which widely spaced trees are interspersed with grassland.
7. Compared with terrestrial habitats, aquatic environments are less variable in _____ and have less available _____ and _____.
8. Because they are shallow and abundantly provided with mineral nutrients, _____ are highly productive habitats that shelter larval stages of many marine organisms.
9. Temperate-zone lakes are thermally stratified, with warm and cold layers separated by a transitional _____. This stratification is lost at the time of the _____ _____.
10. Partly submerged vegetation grows in the _____ zone of freshwater lakes and streams.
11. Organisms living in a sandy beach usually escape wave action by _____.
12. The lightless _____ community is almost entirely heterotrophic, living on an input of dead organisms from other marine habitats.

■ REVIEW QUESTIONS

1. What factors produce the major terrestrial biomes?
2. List the terrestrial biomes and give the location or locations of each.
3. What physical and chemical properties of water are most important in determining the adaptations of the organisms found in aquatic habitats?
4. What are the major freshwater life zones found in a large pond or lake?
5. What is a thermocline? What is its ecological significance?
6. What are plankton? What is their role in aquatic ecology?
7. How do the inhabitants of a rocky beach differ from those of a sandy beach? What characteristic adaptations tend to occur in the two habitats?
8. What is meant by "compensation point"? What is its significance to benthic communities?
9. Describe the ecological structure of a coral reef.

■ RECOMMENDED READINGS

Also consult the general ecology works listed at the end of Chapter 49.

Cole, G.A. *Textbook of Limnology*, 3d ed. St. Louis, C.V. Mosby, 1983. Ecology of fresh waters.

Forsyth, A., and K. Miyata. *Tropical Nature*. New York, Charles Scribner's Sons, 1984. Beautifully written introduction to tropical rain forests. At once scholarly and anecdotal.

Mitsch, W.J., and J.G. Gosselink. *Wetlands*. New York, Van Nostrand Reinhold, 1986. Complete discussion of every aspect of freshwater and coastal wetland ecology.

Moss, S., and L. deLeiris. *Natural History of the Antarctic Peninsula*. New York, Columbia University Press, 1988. A rich marine habitat but an impoverished terrestrial one. Fascinating description of an outpost of life.

Teal, J., and M. Teal. *Life and Death of the Salt Marsh*. Boston, Little, Brown, 1969. Popular, very readable, and personal summary of coastal wetland ecology with a plea for wetlands conservation.

Zwinger, A.H., and B. Willard. *Land Above the Trees: A Guide to American Alpine Tundra*. New York, Harper & Row, 1986 (originally published 1972). Unique and beautifully written guide to the tundra most accessible to Americans.

49

Human Ecology

OUTLINE

I. Agriculture produces artificial communities
   A. Agricultural communities have unique characteristics
   B. New systems of agriculture may permit increased production
   C. Forestry is similar to agriculture in many ways
   D. Pesticides are employed for high-productivity agriculture despite the ecological damage that they produce
      1. Pesticides are broadly toxic, especially to predators
      2. DDT and other chlorinated hydrocarbons are highly persistent and subject to biological magnification
      3. Pesticide use can be reduced
      4. Alternatives to pesticides are possible
II. Pollution results from improper waste disposal
   A. Water pollution is both a threat to public health and an ecological issue
   B. Much water pollution robs aquatic habitats of oxygen
   C. Air pollution results mainly from combustion
      1. Air pollution is a public-health threat and an ecological menace that also produces other kinds of widespread economic damage.
      2. Air pollution is worsened by air stagnation and inversion
      3. Air pollution can be controlled
   D. Radioactive pollutants are long lived
   E. Solid waste destroys wetlands and wildlife and can produce other kinds of pollution
III. Energy is the most basic need of technological civilization
IV. Wild plants and animals are now becoming extinct at an unprecedented rate
V. Human population growth consumes resources and produces pollution
   A. The human population is growing exponentially
   B. Overpopulation is not just a third-world problem
VI. What is to be done?

OBJECTIVES

After you have studied this chapter, you should be able to:

1. Review the development and impact of modern human life-styles upon the ecosystems of the earth.
2. Contrast an agricultural community with a typical natural community, relating the differences to ecological instability.
3. Summarize the direct and indirect ecological impact of modern agriculture and forestry practices and outline new or proposed methods of agriculture.
4. Describe two problems associated with pesticide use and two problems specifically associated with the use of persistent pesticides (chlorinated hydrocarbons).
5. Summarize the principal alternatives to chemical biocides in agricultural pest control and briefly discuss the advantages and disadvantages of each.
6. Summarize the sources of water pollution and describe their ecological effects.
7. Discuss the principal ecological effects and climatic implications of air pollution, together with ways of controlling these effects.
8. Describe the principal methods of solid-waste disposal and discuss the relevance of recycling to waste disposal.
9. Discuss nuclear power, fusion power, and solar power as energy options.
10. Summarize the process of extinction, listing factors that contribute to the decline and extinction of endangered species and providing an example of each.
11. Relate human overpopulation to specific environmental problems and explain how humans can temporarily expand the carrying capacity of their habitat.

Traditional agriculture on an Amish farm in Pennsylvania. (Grant Heilman, from Grant Heilman Photography)

Human beings exist in a web of ecological relationships much as do other animals and plants. Yet human ecology is so unusual that it merits special study, not merely because we ourselves are human beings but because of the great effect the human species has upon the rest of the biosphere. Although one rarely speaks of a species as a geological or ecological disaster, that is the category into which *Homo sapiens* seems to fit. Although other organisms, from ants to beavers, have substantial ecological impact, none is so destructive as we are, and indeed no multicellular organism in the known history of life has had a greater impact on the planet than we have had. In a few short generations, we have utterly transformed the face of the earth and greatly accelerated the extinction of organisms to a rate unparalleled since the demise of the dinosaurs. Two of our unique cultural adaptations have brought the planet to the point of crisis: agriculture and industry. Through these we have been able to greatly distort the natural paths by which energy flows through the biosphere so that they serve a single end: the production of human biomass.

KEY CONCEPTS

☐ Agriculture and industrial development permit the human species to have a great impact on the biosphere. By allowing dense human populations to develop and through attendant pollution and depletion of resources, we have brought ourselves and our planet to the verge of environmental disaster.

☐ Agricultural communities require input of energy, pesticides, and fertilizer that natural communities do not. Chemical cycles are incomplete within them, and a substantial part of their primary production is consumed at remote locations.

☐ A combination of highly consumptive and wasteful life-style with rapid human population growth is generally responsible for environmental degradation.

☐ Numerous cultural and economic forces contribute to pollution—for example, the distribution of some of the costs of business to the general public or to other political jurisdictions via environmental degradation.

☐ Most endangerment of wildlife appears to stem from destruction of habitats.

☐ AGRICULTURE PRODUCES ARTIFICIAL COMMUNITIES

Agriculture is an original and highly distinctive innovation of the human species (Fig. 49–1), but much of the damage that we have done to the environment has been done by agriculture. Technological advances in agriculture have accelerated this impact. Practices such as scientific crop rotation, use of chemical fertilizers and pesticides, mechanical tillage, and large, mechanically driven irrigation systems did not exist 200 years ago. What has made most of these practices widespread and practical is the ready availability of fossil-fuel energy. This energy chemically fixes nitrogen in fertilizer factories, turns the wheels of tractors, and mines water from the ground. Indeed, widespread agriculture itself is an innovation. The most ancient peoples did not employ it and thus apparently had little ecological impact. Today, no animal species can even be compared to *Homo sapiens* in ecological importance.

Agricultural Communities Have Unique Characteristics

Agriculture may be viewed as the establishment of artificial, greatly simplified communities of crop organisms that yield a product that is directly or indirectly convertible into human biomass. Such communities are unstable and very simple. They cannot perpetuate themselves without human intervention, support little wildlife, and recycle nutrients poorly. Corn, for example, has no effective seed release and dispersal mechanism and thus can reproduce only with human aid. Often crops and livestock can resist their natural enemies only with human assistance.

In a natural state, plants that fall prey to particular pests grow interspersed with others that a particular pest species will not eat. Although pests are natural experts at locating their food plants, this dispersal does make it harder for the pests to find them. Moreover, wild plants are usually genetically diverse and often possess natural pest-control adaptations, such as poisonous alkaloids. Breeders of crop plants usually eliminate these traits, making the varieties they produce much more susceptible to insect attack and thus much more dependent on human protection.

Furthermore, at any one time, agriculture tends to be dominated by the latest varieties of crop plants, so that a serious threat to a single variety might wipe out an entire major crop for one or more years. (This happened several years ago with a dominant variety of corn that was uniquely susceptible to a new strain of fungal disease. It is happening now to some kinds of coconut palms.)

In the unnatural communities of modern agriculture, therefore, unnatural pest-control methods have become necessary. Inevitably, these methods have substantial ecological impact, often extending far outside the boundaries of the areas where they are applied. Human management also greatly reduces competition among species. Farmers go to great lengths (Fig. 49–2)

(a)

(b)

(c)

FIGURE 49–1   Primitive slash-and-burn agriculture.
(*a*) The forest trees are felled and (*b*) then burned,
releasing mineral nutrients into the soil. (*c*) A clearing
becomes a village, which is abandoned when the soil has
become depleted, so that the rain forest regenerates.
Archaeological evidence indicates that most agriculture,
including that of early Europeans, was originally of this
type. (*a*, Hajit Kaur; *b*, Courtesy of Don Draper; *c*, Courtesy of
Ben Watson)

(a)

(b)

**FIGURE 49–2   Mechanized versus traditional
agriculture.** (*a*) The extraordinary productivity of
Western agriculture today is due to mechanization and
chemical treatment, most of it connected with weed and
pest suppression. However, modern farming methods
have ecological side effects, including accelerated soil
erosion and pollution of groundwater and surface runoff
alike with fertilizers and pesticides. (*b*) Although less
productive per acre, traditional agriculture may better
preserve those acres for the use of future generations.
The future of agricultural science may lie in the
improvement of such methods; indeed, we may yet see
them used again in the developed world. (*a*, USDA; *b*,
Frans Lanting)

to suppress undesired species (that is, weeds and much
wildlife, together with invertebrate pests) that eat or
compete with cultivated varieties or interfere with their
harvest.

Farmland is not a suitable habitat for most of the
wild species that formerly occupied it. This realization
has led to the development of game parks and preserves in many countries. With enlightened management, agricultural lands can support modest populations of wild animals. Understandably, however, many
farmers feel that they cannot spare the space that
otherwise could be employed more profitably, and
many also view patches of natural habitat as breeding
grounds from which weeds and pests can readily
spread.

New Systems Of Agriculture May Permit Increased Production

Famine and malnutrition are among the gravest global problems of our time, as the news media regularly remind us. American agriculture, although by no means problem free, often has seemed exempt from disastrous crop losses and continues to be one of the most productive agricultural systems anywhere. However, the costs have been substantial in erosion, wildland destruction, and energy consumption. These problems are even more acute in the underdeveloped world.

Perhaps agriculture needs to be reinvented. Many "high technology" alternatives, such as hydroponic farming, have been proposed and put into limited practice, but on the whole, these have proved to be even more expensive and energy consuming than conventional agriculture. "Green revolution" agriculture employs improved varieties of crop plants, usually along with mechanical tillage and heavy use of pesti-cides and fertilizers. There is no doubt that it has greatly increased food production in the less developed nations, but there is doubt whether this increase can keep pace with the burgeoning population.

Forestry Is Similar To Agriculture In Many Ways

The deforestation of all countries has proceeded with alarming speed since the Industrial Revolution and, in some localities, since ancient times. This deforestation resulted not only from simply cutting down trees, but also from heavy grazing by sheep and goats, which prevented the growth of seedlings. More recently, firewood and wood charcoal fueled the early Industrial Revolution. To this day, the greatest use of forest products worldwide is as fuel. Furthermore, since the 19th century, forests have been extensively cut to provide wood for construction timber and paper pulp. Such practices continue today, especially in the tropics (Fig. 49–3), but even in parts of the United States.

FIGURE 49–3 **Destruction of tropical rain forests for subsistence logging in Ecuador (a) and Malaysia (b). Forests, such as these in Costa Rica, are also cleared for charcoal production (c) and to create pastureland (d).** (a–d, Bill Gabriel)

these substances are quite toxic, and they sometimes accumulated in the soil in amounts that inhibited even plant growth. Pests also became resistant to them (a case of natural selection), so that over time they lost their effectiveness. These problems led to some use of natural plant insecticides, such as nicotine (derived from tobacco leaves) and the pyrethins (derived from certain chrysanthemum-like plants).

The era of second-generation pesticides began about 1940 with the discovery of DDT. Pesticides of this type are synthetic organic compounds, most of which can be classified into three groups: (1) DDT and related chlorinated hydrocarbons (Chlordane, Dieldrin, Mirex), (2) organophosphates (Malathion, Parathion), and (3) carbamates (Sevin, Temik). These highly effective organic pesticides have been developed and put into widespread use by everyone from farmers and health departments to suburban homeowners.

The chlorinated hydrocarbons interfere with nerve action by antagonizing the sodium pumps in nerve cell membranes. Most other organic pesticides are anticholinesterases, which render the enzyme cholinesterase incompetent. Anticholinesterases are, in general, more toxic than chlorinated hydrocarbons. Humans are less susceptible to the action of both classes of pesticides than insects, partly because of their large body size and partly because of differences in physiology. Nevertheless, serious accidental poisonings, especially with potent anticholinesterase pesticides, occur regularly.

FIGURE 49–4   A white spruce seedling nursery. After the trees reach a certain height, they are transferred to another location where they are planted in rows like a crop, periodically thinned, and weed trees are removed. Little wildlife inhabits such farms. (T. Kitchin/Tom Stack & Associates)

Many nations have embarked on extensive reforestation programs; however, the trees planted often are not native and do not fit in well with the local species. Additionally, these replanted trees are usually managed as a crop, so ecologically this **silviculture,** as it is called, resembles agriculture. Tree farms are not forest communities; usually they are single-species aggregations (Fig. 49–4) with little wildlife and without species typical of climax communities. They must be managed like any other crop.

Pesticides Are Employed For High-Productivity Agriculture Despite The Ecological Damage That They Produce

The first-generation pesticides—inorganic chemicals such as sulfur, lead, arsenic, and mercury—were used to repel or kill pests for hundreds of years. Many of

Pesticides Are Broadly Toxic, Especially To Predators

Most widely used pesticides are broad-spectrum poisons that kill nonpest species as well as pests. Another major drawback of pesticides is that many insects become resistant to them. Although initially the pesticide may cause a rapid decline in the pest population, resistant mutants gradually replace the susceptible insects. By 1980, more than 200 DDT-resistant pests were known. Often, a pesticide-resistant pest population evolves rapidly, whereas the natural predators of these insects do not become resistant. This occurs partly because predators reproduce more slowly than pests, partly because there are fewer of them, and partly because of biological magnification (Fig. 49–5).

With their natural enemies devastated, the pest population increases in numbers and may become a greater threat than before the pesticides were used. Also, previously unobtrusive insects, whose populations were held in check by the now absent predators, may now become economically significant pests for the first time.

(a)

(c)

(b)

FIGURE 49–5 Biocides, necessary to control insect pests (a), become concentrated as they are passed along the food chain. When predators consume poisoned insects, the concentration of pesticides increases. (b) Ultimate predators, such as hawks, bald eagles, peregrine falcons, and pelicans that feed on contaminated birds and fish are prime victims. (c) Pesticides interfere with egg production and embryonic development of these predatory birds, causing thin-shelled, fragile eggs that rarely hatch in the wild, as in the case of these brown pelican eggs. (a, USDA; b and c, Frans Lanting/ Minden Pictures)

DDT And Other Chlorinated Hydrocarbons Are Highly Persistent And Subject To Biological Magnification

The organophosphate and carbamate pesticides are **biodegradable**—that is, they decompose several weeks or months after being sprayed. DDT and other chlorinated hydrocarbons are **persistent pesticides.** They are not readily biodegradable. An additional drawback of chlorinated hydrocarbon pesticides is that they do not remain confined to the areas where they are sprayed. Food chains, air movements, and, to a lesser extent, water currents distribute persistent pesticides globally, so there is now probably no organism on land or in the sea whose tissues are completely free from them. Traces of DDT have been found even in the fat of penguins in Antarctica, thousands of miles from the nearest place where DDT was deliberately employed. Because DDT is more soluble in body fats than in water, it tends to become concentrated in body fat, and progressively more concentrated as it passes through the food chain. The most predatory creatures thus absorb the most DDT. DDT and other chlorinated hy-drocarbon pesticides produce eggshell thinning in predatory birds that has resulted in the near extinction of some species.

Pesticide Use Can Be Reduced

Partly because DDT does pose a serious threat to ecosystems, its use in the United States was banned in 1973. In the early 1980s, however, it was still being manufactured in the United States for export to many other countries! Other countries are still manufacturing and using it. Several closely related chlorinated hydrocarbon pesticides have also been banned, but others continue to be used widely. A number of international organizations officially sanction the use of DDT because of its usefulness in checking the populations of mosquitoes that spread malaria, especially in the underdeveloped world. Even without considering some of the deeper issues involved in such use, however, it is clear that DDT is becoming steadily less effective as a mosquito pesticide as these insects develop increasing resistance to its action.

Although the problems connected with pesticide use are reduced somewhat when biodegradable compounds such as the carbamates or organophosphates are substituted, there is no known pesticide currently in widespread use that is free from undesirable ecological consequences. The carbamate Sevin, for instance, is almost harmless to mammals but is instant death for bees, and bees are vital for pollination of food and wild plants. Because all pesticides kill other species besides the pests that they are intended to destroy, in the end we can solve pesticide problems by not using them. Yet this seems less and less likely to happen.

Alternatives To Pesticides Are Possible

There are some alternatives to the use of pesticides short of completely abandoning crops to the insect world. Among these alternatives are the use of natural enemies, the release of sterile males, and such practices as careful crop rotation to break pest life cycles by depriving them regularly of their preferred food plant.

The sterile-male technique has been employed against the screw-worm fly, an important pest of the cattle industry (Fig. 49–6). Millions of screw-worm flies were artificially reared, then sterilized by exposure to ionizing radiation. When released in huge numbers, these sterile flies successfully competed with the relatively few fertile ones for mates. The female screw-worm fly mates only once in her lifetime. As a result of many unproductive matings, the screw-worm fly population dramatically declined, and, in some areas, especially islands, this pest has been completely eradicated. In places that are not geographically isolated, however, fertile flies constantly re-enter from adjoining areas, making eradication impossible. Furthermore, in some populations of screw-worm flies, the females are able to detect and discriminate against the sterile males.

Third-generation pesticides may hold the greatest promise. These are synthetic versions of naturally occurring pest pheromones or hormones that, if applied carefully, can disrupt the behavior (particularly reproductive behavior) or physiology of the target organism without damaging other harmless or desirable organisms.

Biological control agents of various types can sometimes suppress the growth of pest populations, but they are most likely to be successful when the pest organism has been introduced to a habitat to which it is not native. Often, in such a case, the pest has no natural enemies in the new habitat. *If* those enemies were among the chief limiting factors on the multiplication of the pest organism in its original habitat, introducing the missing natural enemies may control the pest. Of course, the natural enemies themselves must first be investigated and shown to be harmless before being used.

FIGURE 49–6 Biological control, a superior alternative to the use of biocides against screw-worm flies, which cause tremendous damage in livestock. The male flies are sterilized and then released to mate with wild females. Since the females mate only once, they will never reproduce. (USDA)

■ POLLUTION RESULTS FROM IMPROPER WASTE DISPOSAL

Waste may be defined as any product of our civilization that is usually discarded rather than used, or a formerly useful product that is no longer used for its original purpose or for any other. The disposal of wastes has become an acute problem, not only because of the quantity of such wastes, but also because of their kind. Many are not readily degraded by natural mechanisms, even in small quantities. Careless waste disposal is practiced because it is advantageous to the polluter, forcing others to bear the costs, economic and otherwise.

Pollution is a reduction in the quality of the environment by the addition of materials (or conditions such as heat) normally absent or present only in small quantities. Polluting costs us a great deal, so why does it exist? Most of the reasons are economic. Here are some of them:

1. By avoiding the costs of waste disposal, the polluter compels society to bear either the costs of the pollution or the costs of correcting it.
2. By passing on the cost of clean-up to future generations (as in the case of dumping toxic chemicals), polluters avoid having to pay for it themselves.
3. By providing convenience of use through disposable packaging, manufacturers increase sales of many goods while compelling the public to bear the costs, ecological and otherwise, of that disposal.

4. By deliberately designing goods to wear out or become unstylish, and by otherwise encouraging wasteful consumption, we increase the flow of goods and services through the economy, but also increase the waste materials to be discarded.

Water Pollution Is Both A Threat To Public Health And An Ecological Issue

Water pollution was one of the first varieties of pollution to attract widespread concern and efforts at correction, mainly because of its obvious role in the transmission of disease by drinking water. This has led to an effort to separate drinking water from that used for waste disposal or, failing this, to purify drinking water. However, sewage or other organic wastes in the remaining water reduce the oxygen content of natural waters and promote cultural eutrophication.

At present, the leading sources of water pollution are industrial waste, municipal sewage (which actually may be largely composed of industrial waste), and agriculture. Industry accounts for most water pollution in the United States. Usually far more concentrated than municipal sewage, industrial waste produces up to 12 times more pollution per gallon of effluent (outflow) than municipal wastes do. What they lack in concentration, however, municipal wastes tend to make up in volume, particularly taking into account street drainage and surface runoff after rain and snow. Agriculture contributes pollutants in the form of silt from erosion, pesticides and fertilizers in runoff from fields, and manure from feedlot and barnyard wastes.

Organic wastes provide a rich source of nutrients for decay bacteria and fungi. Hence, feces, blood from slaughterhouses, oxygen-demanding wastes from paper mills, and peelings from vegetable-processing plants (among many other things) stimulate the growth of bacteria whose metabolism rapidly removes oxygen from the water. Industrial wastes may also contain large amounts of sediment, chemically combined nitrogen, phosphorus, carbon dioxide, methane, hydrogen sulfide, and smaller amounts of miscellaneous chemicals, heavy-metal ions, and even pesticides.

Much Water Pollution Robs Aquatic Habitats Of Oxygen

A polluted environment is a demanding one. Not surprisingly, few organisms can tolerate it. Yet those able to exist in it often attain astronomical numbers and very large biomass.

When organic wastes are dumped into a stream, a predictable sequence of events occurs (Fig. 49–7). Near the source of pollution, surprisingly, numbers of fish and other organisms may persist because the organic wastes have not yet had time to decay and the dissolved oxygen level is high. In severe instances, somewhat farther away, conditions worsen as bacteria and fungi that degrade organic sewage use up so much oxygen that anaerobic conditions prevail. Still farther away, the polluted water has begun to purify itself: Organic materials start to disappear, oxygen diffuses into the water from the air, and, except for cultural eutrophication, something like a normal community is established.

Air Pollution Results Mainly From Combustion

Air pollution is produced principally by combustion, either in automobile engines or in industrial and power-generation plants. It is responsible for photochemical smog and the extremely damaging acid rain. Although control technologies exist, they are, as in the case of water pollution, insufficiently employed.

The United States alone dumps more than 220 million tons of pollutants into the air annually. The pollutants that account for most air pollution are carbon monoxide, sulfur oxides, nitrogen oxides, hydrocarbons, and particulates. Their relative contributions to air pollution are shown in Figure 49–8. These pollutants interact to form secondary pollutants and smog and also produce the acid chemicals (chiefly nitric and sulfuric acids or ammonium salts of these acids) that are responsible for acid precipitation.

Air Pollution Is A Public-Health Threat And An Ecological Menace That Also Produces Other Kinds Of Widespread Economic Damage

Plant damage from air pollution occurs mostly in the photosynthetic tissue (mesophyll) of the leaf (Fig. 49–9). The surfaces of mesophyll cells, which are moist to facilitate gas exchange in photosynthesis, are vulnerable to attack by toxic substances in the air. Fluoride particulates, photochemical smog (generated from automobile emissions by a chemically complex atmospheric process), sulfur dioxide, and even very fine soot can kill vegetation. Some species of plants are more susceptible to certain pollutants than other species. At a concentration of less than one part per million, the common pollutant nitrogen dioxide reduces the growth of tomato plants by 30%. Photosynthesis is reduced by 60% when smog concentrations are 0.25 part per million. Such decreases in photosynthesis also slow the flow of resins under the bark of trees, rendering the trees susceptible to plant disease and insect pests.

Acid precipitation damages both land and aquatic communities in several ways. In the first place, rain-

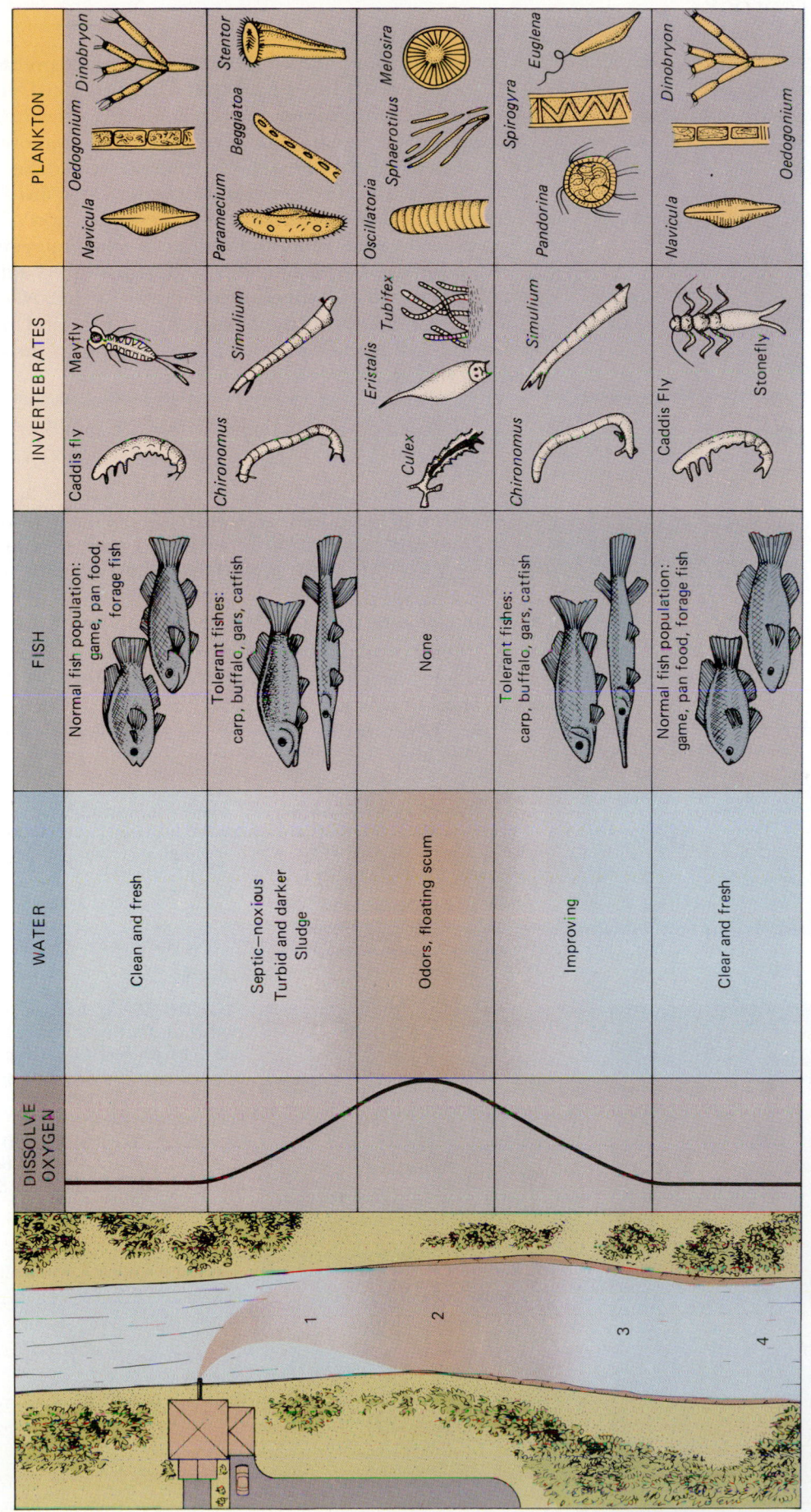

FIGURE 49–7 Zonation of pollution ecology in a stream receiving large amounts of untreated sewage. As the amount of oxygen dissolved in the water decreases, fish disappear. Only organisms able to obtain oxygen from the surface, tolerate low oxygen tensions, or respire anaerobically can survive. When the sewage has all been decomposed by bacteria, the species of plants and animals present in the stream return approximately to normal, although eutrophic changes may persist. (After Odum)

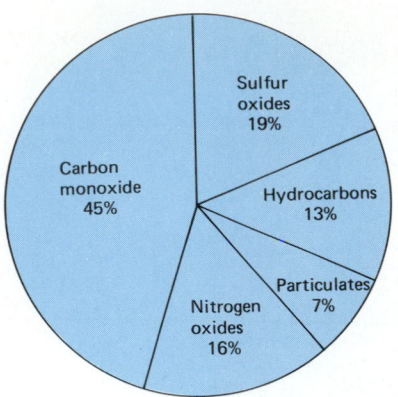

FIGURE 49–8 **The average make-up of air pollution.**

(a)

water or snow of low pH produces aquatic pollution in lakes and streams whose water has little buffering capacity[1] (see Chapter 3) that is often sufficient to destroy most aquatic life. Even extremely remote areas (such as inland China) are seriously affected. Secondly, the high nitrate content of this water can contribute to cultural eutrophication. Acid precipitation directly damages plant roots and indirectly weakens vegetation by nitrogen overnutrition (too much fixed nitrogen may actually do the most damage). At the same time, other mineral nutrients such as calcium and magnesium are leached out of the soil and lost to the plants.

Air Pollution Is Worsened By Air Stagnation And Inversion

Atmospheric inversion is the cause of most acute air-pollution episodes (Fig. 49–10). In such cases, weather conditions form a lid of warm air above cooler, polluted air. Although inversions do not actually increase pollution, they seal the pollutants below and prevent them from being dispersed. In Los Angeles County, where conditions favor them, inversion layers form when warm air moves in from the deserts to the east and lies over the mountains that surround Los Angeles. Beneath this warm air is a layer of cooler air that

[1] "Soft" water of low mineral content, which often occurs where stream and lake beds are composed of granite, has little buffering capacity. "Hard" water, which has high concentrations of calcium and magnesium, has greater buffering capacity.

(b)

(c)

FIGURE 49–9 **Effects of air pollution (sulfur dioxide) on white birch leaf (*a*). Effects of acid rain on (*b*) German pine forest and (*c*) ornamental architecture of the Field Museum, Chicago.** (*a*, USDA; *b*, Animals Animals © 1990 Robert Maier; *c*, Gary Milburn/Tom Stack & Associates)

(a)   (b)

FIGURE 49–10   The meteorology of air pollution: two views of Boise, Idaho. (a) In a normal pattern of air flow, warm air close to the ground rises, carrying with it most atmospheric pollutants. (b) A thermal inversion. A lid of warm air prevents the circulation of air from below, trapping pollution. (a and b, William H. Mullins)

has moved in from the sea. Such inversions also form, although less regularly, in virtually every area of the country. Simple air stagnation, which can occur anywhere, is almost as bad, for then pollution is not dispersed by normal wind action.

It is that very wind action, although helpful in removing polluted air, that unfortunately results in acid rain possibly hundreds of miles from where the pollutants originated. It is common for acid rain to be caused by air pollution in other nations, and this has caused international disputes between such countries as Great Britain and Sweden, and the United States and Canada.

Air Pollution Can Be Controlled

Air pollution can be controlled in a variety of ways, at least in theory. Automobiles are probably the largest source of air pollution nationwide. This has led to the design of engines that are not only more fuel economical, but often less polluting as well. Measures that increase the efficiency of combustion also reduce the amounts of unburned fuel and carbon monoxide that are voided to the air. Catalytic afterburners remove most of the harmful gaseous products from the exhaust effluent. Unfortunately, however, there is no practical way to remove nitrogen oxides from automobile exhaust.

We could alleviate industrial air pollution by requiring the use of low-sulfur fuel, but this is not an entirely practical alternative, especially when coal is used. Sulfur and nitrogen oxides can be removed from stack gases by washing the effluent with water and then treating the water waste that results. It is possible to produce both nitric and sulfuric acids from this waste in commercial quantities, and some day this may be widely practiced, at least if acid-rain control treaties or legislation force the widespread use of such equipment. The most visible industrial pollutant—soot—has long been removed from the effluent gas of some industries by filtering it out mechanically or electrically, using centrifugal (cyclone) separators, or electrostatic precipitators.

Radioactive Pollutants Are Long Lived

Radioactive pollution poses a unique danger owing to the extreme longevity of the pollutants, whose half-lives can run into hundreds of years. In addition, these pollutants often chemically resemble harmless or even vital substances. Thus, the radioactive substances can pass extensively through food chains, where they are deposited easily in tissues, sometimes becoming concentrated by biological magnification. In addition to nuclear weapons, some manufacturing and industrial processes release radioactive isotopes, as do accidents involving nuclear power plants.

Atmospheric radioactive pollution has been partially alleviated by a 1963 treaty that prohibits atmospheric tests of nuclear weapons by the handful of nations that have agreed to its terms. Weapons tests are not the only ground of concern, however. The recent Chernobyl nuclear power plant disaster in the Soviet Union has *by itself* emitted a substantial fraction of the total amount of radioactive material ever released into the earth's atmosphere. In the event of even a limited nuclear war, many radioactive substances would find their way into the food chain, producing a harvest of disease for generations after the acute deaths from the nuclear exchange.

Even during peace, widespread generation of nuclear wastes by power plants could have a similar effect, particularly in the event of accidents encountered in transporting those wastes. To this must be added the potential hazards that could be created if a nuclear power plant were to be destroyed by design flaws, carelessness, sabotage, or warfare.

Many plants that produce nuclear-weapon components in the United States alone have been found to be amazingly irresponsible in observing their obligation to protect the environment from their ordinary discharges and wastes, and to be so lax in their safety and security measures that accident or terrorist attack could easily produce disaster. Some have had to be closed altogether; of these plants, a number are in such hopeless condition that they probably will remain closed indefinitely.

Take one example of a typical radioactive pollutant that occurs in fallout from both weapons and power plant disasters. The radioactive isotope of strontium is chemically similar to calcium and travels in ecosystems much as calcium does. Like calcium, it becomes incorporated into bones and teeth and can produce such damage as bone cancer. It has a half-life of 28 years—that is, in 28 years, half of it will remain, the other half having decayed into nonradioactive products. In another 28 years, the remaining half will not have declined to zero but will have been halved again, so one might say the quarter-life of radioactive strontium is 56 years. That is a long time to carry such a substance in one's bones.

This is not idle speculation. In the arctic, reindeer "moss" greatly concentrates radioactive strontium, which then is further concentrated (somewhat like persistent pesticides) when reindeer eat the lichens. When reindeer herders such as Laplanders take milk from the reindeer, the radioactive strontium is still further concentrated and may end up in the body of a child. Less marked but similar biological magnification takes place in the grass-cow-human food chain.

Solid Waste Destroys Wetlands And Wildlife And Can Produce Other Kinds Of Pollution

Each of us accounts for about 3.6 kilograms (almost 10 pounds) of garbage, trash, and other solid waste *per day*, and the amount is rising steadily. Until quite recently, there was little thought of doing anything more with solid wastes than picking them up from one place and putting them down in another, but we are rapidly running out of rugs under which to sweep the debris of society.

Today's American consumer discards large quantities of paper (newspapers, paper bags, cups, plates, cartons, and other packaging materials) and substantial amounts of edible food scraps, as well as styrofoam containers, nonreturnable glass and plastic bottles, steel and aluminum cans, and wood and garden refuse, little of which is recycled. In addition to these municipal wastes, agricultural activities generate more than 1.8 billion metric tons (a metric ton is about 2205 pounds) of wastes (mainly manure) each year; by no means is all or even most of this waste recycled. Mining and industrial wastes add to the problem.

Although some rural communities still dispose of their solid wastes in open dumps and many coastal cities practice ocean dumping, the two principal means of solid-waste disposal today are the sanitary landfill and incineration. The old-fashioned dump, with its swarming flies, roaches, and rats, is a manifest public-health hazard. Dumps, if they catch fire, also can be dangerous sources of air pollution; once ablaze, their fires can be difficult to extinguish. Finally, the garbage they contain represents a kind of food subsidy supporting huge populations of scavengers such as gulls. The gulls and rats associated with dumps can do direct ecological damage by preying on other plants and animals for miles around.

Ocean dumping is, in some ways, worse than that practiced on land. Many of the materials dumped out of sight of the shoreline are synthetic, and many of these, especially plastics, are not degraded rapidly or at all by decomposer organisms. By their persistence in the marine environment, they damage many organisms, usually by entangling them or damaging their digestive systems.

The sanitary landfill is a slight improvement over a dump. Daily burying of garbage is its principal advantage. After waste is dumped in a sanitary landfill, it may be further compacted by bulldozers. Each day a layer of soil is pushed over the garbage to discourage flies and rats. With the sanitary-landfill method, abandoned strip mines have been filled and eventually reclaimed, and artificial mountains have been constructed for skiing in the Midwestern plains.

Not all land *should* be filled, however. Marshes and lakes, for example, are ecologically vital wetlands, yet they often are favorite sites for landfill. Landfills, like dumps, can pollute groundwater as contaminants leak into the ground. This is a particularly acute problem where highly toxic industrial wastes have been buried, but even ordinary municipal waste is a significant source of groundwater pollution. Another disadvantage of sanitary landfills is the limited potential uses the land has once it is filled. If a landfill is used as a building site, settling may cause walls and foundations to crack. Methane gas resulting from anaerobic decomposition in the depths of the fill may seep into buildings and constitute an explosion hazard. For these reasons, filled land is best used for parks and other recreational purposes.

Many communities burn their garbage rather than dump it. In a modern incinerator, the trash is burned in a carefully engineered furnace, but some air pollution still results. The heat produced by the fire may be used to boil water and generate steam that can be sold for industrial use.

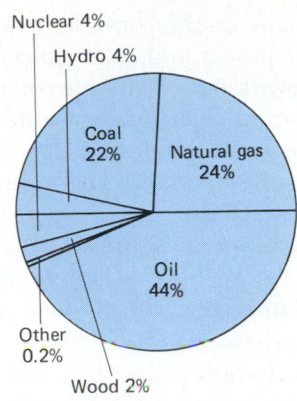

FIGURE 49–12 **The mixture of energy sources currently utilized in the United States consists mostly of fossil fuels.**

ENERGY IS THE MOST BASIC NEED OF TECHNOLOGICAL CIVILIZATION

Human agriculture and industry are dependent on energy. Our technological civilization would be impossible without a constant large energy input. Thus far, that energy has been obtained principally from burning fossil fuels. This depletes the supply of such fuels, increases carbon dioxide in the atmosphere, and may prevent fuel use for important purposes in the future. Of all substitute power-generation technologies, solar power (Fig. 49–11) may have the greatest potential.

Just as there could be no life on earth without the energy of the sun, there could be no modern society without the energy harnessed by human beings. One may credibly make the case that increases in energy procurement and production, from human and animal muscle power, through wind and water power, to modern electricity generation, have produced most of the increase of our population. Yet as population has

grown and technology has expanded, it has become increasingly apparent that we have an energy problem. Simply stated, the problem results from too many people consuming too many goods that require energy input for use or for manufacture. The effect is a rapid depletion of oil, natural gas, and even coal resources.

New technologies are being developed that may enable us to partially replace traditional energy sources with new ones, but presently we still depend on fossil fuels (mainly petroleum) for about 90% of our energy needs (Fig. 49–12), including large amounts used in food processing, shipment, and production. These fuels also produce almost 100% of our air pollution and, as we have seen, may give rise to the extremely significant atmospheric greenhouse effect. When they are gone, moreover, they will be unavailable for higher uses such as the production of commercial chemicals.

FIGURE 49–11 **"Solar One," a solar electrical plant in the Mojave Desert of California. Over 1800 heliostats (the banked mirrors surrounding the tower base) focus the sun's heat on a boiler at the top of the central tower. Thus far, solar power has proved most practical for low-power uses such as water heating and small-scale electricity generation, but in some locales it might prove usable for power production on an industrial scale.** (Courtesy of Southern California Edison Co.)

Two important energy options now being developed are nuclear power and solar power. Thirty years ago, nuclear experts hailed the age of nuclear power with the claim that nuclear power would be "too cheap to meter." These words could not have been less true. With changes in the economy and massive cost overruns, nuclear power has become very expensive. Questions about environmental impact, availability of uranium ore, and safety have seriously threatened the nuclear power industry. (Only two nuclear plants planned for construction in the United States since 1978 have not been later canceled, and these two are unlikely ever to be completed.) Mishaps (by no means all of them minor) occur repeatedly in nuclear plants, mainly due to equipment problems or failures. The problem of safe disposal of nuclear wastes (most of which will still retain half of their original radioactivity after 10,000 years) has never been solved satisfactorily.

The nuclear power plants currently in use are fission plants. A controlled chain reaction is established in some suitable material, usually a uranium isotope or plutonium. The heat that is produced boils water (or, in the future, perhaps some other fluid), producing steam that drives a turbine and generator. It is probable that fission-type nuclear power plants will continue to account for a small percentage (perhaps 5%) of the world energy budget.

Fusion-type nuclear power depends on the fusion of atoms of light elements such as deuterium, tritium, and lithium to produce the heat. Nuclear fusion may prove less harmful than fission, but the technology is still in its developmental stages, and no one knows when or even if this energy option will become a reality.

Many experts consider solar power a safe and viable alternative to nuclear power. The energy is free and abundant, it causes no air pollution and few environmental hazards, and the technology promises to be less expensive than alternatives. Solar energy can be used in buildings through passive designs that maximize the use of natural sunlight and through the installation of solar collectors, which trap and store heat.

Technology is available for converting sunlight into direct-current electricity; thus far it is expensive, although there has been great recent progress in this area. Photovoltaic cells, tiny cells similar to the silicon semiconductor chips used in calculators, can convert up to 20% of the sunlight striking their surface into electricity. Mass-production techniques may greatly reduce the cost of photovoltaic cells in the future, making their household use affordable. Even if (as seems likely) this technology is not adequate for industrial use, it has the potential to greatly reduce the demand for fossil-fueled electric power from residences, leaving more to be used by industry and stretching fossil-fuel supplies.

Other energy options include water, wind, tide, geothermal, and ocean thermal power. With the exception of water power, these sources are not used on large scales at present and are not expected to contribute more than about 1% of the annual energy budget by the year 2000. Energy experts project that we will still be tightly locked into fossil-fuel use for our energy needs as we enter the 21st century. In their 800-page report on the energy situation from 1985 to 2010, the Committee on Nuclear and Alternative Energy Systems concluded that the highest priority of the United States' national energy policy should be directed toward avoiding waste and, in general, reducing the growth of energy *demand*. This approach would minimize both the need for new generating technologies and the pollution produced by existing ones.

■ WILD PLANTS AND ANIMALS ARE NOW BECOMING EXTINCT AT AN UNPRECEDENTED RATE

By its unparalleled use of the earth's resources, the human species has caused extinction of other species in numbers comparable to the great extinctions of the geological past. Most of these have resulted directly or indirectly from habitat destruction.

Although extinctions have always occurred, human beings have raised the rate at which they occur perhaps a thousand-fold. The species most vulnerable to extinction appear to be large, predatory, and migratory animals, such as the polar bear. However, extinction also pursues animals with other combinations of characteristics: those that require large tracts of wilderness or solitude, live in very specialized or restricted habitats, compete with human beings in any respect, or yield economically valuable products. Extinctions also result from importing alien species against which an endangered organism has no effective defense. Plants are more vulnerable than animals, for they cannot actively evade threats and are often restricted to special habitats or small geographic areas. More than 19,000 species of plants are known today to be endangered, rare, or otherwise threatened. This is almost certainly but a fraction of the true number.

Habitat destruction seems to account for most extinctions and endangerments of species. The extinction of the passenger pigeon, for example, apparently resulted from the destruction of the beech forest where it nested as much as from the market hunting that is usually blamed. Another nearly extinct species, the ivory-billed woodpecker, requires large virgin tracts of cypress forest. Today little of that habitat is left, and only a recently discovered remnant population of this once-common bird persists in a remote area of Cuba.

It follows that the best hope of preserving any species is to preserve its habitat, together with the eco-

(a)

(b)

(c)

FIGURE 49–13 Some endangered species: (*a*) *Eschscholzia ramosa,* (*b*) **Himalayan snow leopards,** (*c*) **golden toad of the Monteverde cloud forest preserve in Costa Rica, and** (*d*) **giant panda.** (*a*, John H. Trager/Visuals Unlimited; *b*, E.R. Degginger; *c*, Doug Wechsler; *d*, Animals Animals © 1990 Fran Allan)

(d)

system of which that is a part. Preserving endangered species in zoos and similar facilities is commendable as a last resort but is unlikely to succeed for many species. In the first place, confined animals must adapt to an artificial environment, becoming, in effect, semi-domesticated. It is by no means certain that they or their descendants would ever be competent under natural conditions if and when they could be reintroduced to their original habitat. Moreover, inbreeding and genetic drift among small, confined populations with a limited gene pool may endanger them by genetic disease, as has already been demonstrated in some zoo animals and small wild populations of endangered species.

Why should we care? There are at least four reasons: (1) endangered species may be potentially useful, (2) endangered species may play an important but unsuspected ecological role, (3) the world will be a duller place without many endangered species, and (4) we

have a moral obligation to preserve our fellow creatures, each a work of biological art whose duplication is far beyond us.

Many of our modern drugs are products of living things—the antibiotics, for example. Many, like quinine, were discovered in tropical higher plants. Potential medicinal agents have been discovered in such unlikely species as blowfish, sea anemones, venomous snakes, and tropical yams. Surely we may yet discover that species now considered "useless" have potential for domestication. When that time comes, many of these species, especially organisms from tropical rain forests, will almost certainly already be extinct (Fig. 49–13).

We are now becoming aware that the tropical rain forests play a major role in reducing the carbon-dioxide content of the atmosphere, which so worries us in connection with the atmospheric "greenhouse effect." Programs of reforestation are being discussed as substitutes for the forests that have been destroyed. Supposing that new forests are indeed planted, including some in the tropics, they will grow on degraded soil and will be much less diverse than the ecosystems they replace.

Finally, do we have the moral right to destroy the world of life? The irreversible extinction of species, the least specimen of which is far more complex and sophisticated than the most advanced computer, is vandalism on a scale comparable to using the Mona Lisa as a barroom dartboard.

HUMAN POPULATION GROWTH CONSUMES RESOURCES AND PRODUCES POLLUTION

The appropriate human population density is one that can be sustained in reasonable comfort, in stable balance with the resources of our habitat. Limiting factors can be expected to stop human population growth one day. It is likely that this limitation will be accompanied by widespread and intense human misery connected with such events as starvation and warfare. Because population growth cannot be sustained indefinitely, humane methods of limiting it are desirable, not only to avoid some apocalyptic future, but to help prevent further environmental deterioration in the present.

More people need more food, clothing, houses, schools, roads, automobiles, television sets, energy, and the material goods our society holds so dear. Each of these can be translated into increased pressure on the environment. For example, the need for more food requires the use of more pesticides and more chemical fertilizers, which result in damage to the soil and increased land and water pollution. Production of chemical fertilizers requires the use of more petroleum, contributing to energy shortages. At the same time, more people need housing, stores, schools, and roads, so more land is taken out of agricultural production, leaving farmers to try to raise more food on less land.

The one-third of the world's population that lives in the developed countries consumes 85% of the

earth's resources and is responsible for most of the stress placed on the environment. The other two-thirds of humanity exert much less pressure on the earth's resources, but not by choice. Despite their understandable aspirations to consume at our level, it has been estimated that the maximum world population that could be supported at the United States' level of affluence is less than 1 billion. The environmental impact of enriching even this billion to current United States' norms would involve increased industrial pollution, increased erosion of agricultural land, further depletion of natural resources, and much more. Such a goal is totally unrealistic, however, because at the current growth rate, world population will reach 8 billion by the year 2100! Almost certainly, most human beings living then will subsist in unprecedented conditions of poverty.

Overpopulation is thus one of the most pressing problems of our time—a problem from which we in the United States are not insulated. Aside from the moral issues involved, the National Security Council has described population increases around the world as a threat to our national security. How much greater a threat it must be to the third-world nations in which population growth is now the greatest, and to their neighbors.

The Human Population Is Growing Exponentially

Beginning perhaps 14,000 years ago up through the Middle Ages, the human species was essentially in its lag phase of population growth. Disease and food shortages served as powerful environmental resistance. About the time of the Industrial Revolution, however, our population entered the logarithmic phase of its growth, and it is now increasing by about 200,000 persons per day. In some nations, human population currently is doubling every 15 years, or even more rapidly.

The demographic events of the past 250 years are shown in Figure 49–14. Notice that among developed countries the birth rate has declined over the long run.

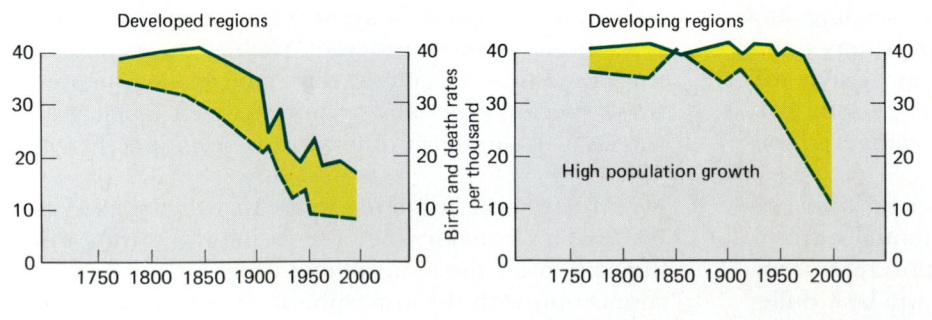

FIGURE 49–14 Birth and death rates in the developed and underdeveloped world for the past 250 years.

The death rate has also declined, perhaps mostly due to the control of infectious disease. Because the birth rate and death rate have mostly marched in step, the population of such countries as Sweden has increased only modestly; in some countries, such as West Germany, it even appears to be in a state of slight decline. The birth rate in less-developed countries has also fallen dramatically, but modern medicine has brought mortality (still high by the standards of the developed nations) down even faster, so that their population has increased greatly despite the decline in the birth rate.

Current and past population growth affects age distribution in a nation's population. In the graph of the United States' population shown in Figure 49–15, notice that there are very few individuals in the uppermost age ranges. Mostly because of the post-World War II baby boom, there is a bulge in the population profile at around the late teens to early forties.

Sweden, shown in the second graph, has a stable population history. Owing to widespread use of birth-control methods and ready availability of excellent medical care for the entire population, about as many Swedes are born each year as die. Thus, Sweden has achieved zero population growth. However, the average Swede lives to quite an advanced age, so that the numbers of people in all age ranges are about the same up to about age 70.

Mexico, shown in the third graph, typifies heavy population growth in a third-world nation. In such a nation, children have always been perceived as an economic asset—that is, they are extra hands to work subsistence farms and, eventually, providers of care to aged parents. Until recently, many children died early

from infectious disease, providing powerful economic motivation for widespread resistance to population control.

But here there is a conflict of economic interests between individuals and their society. In all three graphs, we have stippled the age ranges in which people consume resources without (on the whole) making substantial economic contributions. A demand for schooling, pediatric care, nursery care, and much else is generated by the children of Mexico. Yet the resources of Mexican society are insufficient to meet this demand. Moreover, the growth rate of the Mexican population causes the need for all of society's services to double every 15 years. Even an affluent country like Sweden or the United States could not readily double all its roads, schools, hospitals, sewage-disposal plants, fire departments, and apartment buildings every 15 years. If it did, think of the ecological consequences! However, the developing countries cannot even run fast enough to stay where they are.

What has produced the demographic differences between the developed and less-developed nations? Evidently, the citizens of the developed nations ceased some time ago to regard children as an economic asset. The slowing of population growth that resulted is called the **demographic transition.** In the United States, for example, it costs in excess of $60,000 to raise a child to economic independence (not counting the cost of a college education), almost none of which is returned to the parent. Upwardly mobile people therefore tend to limit their family size. Most developed countries went through a difficult time of social transition in the 19th and early 20th centuries, when they

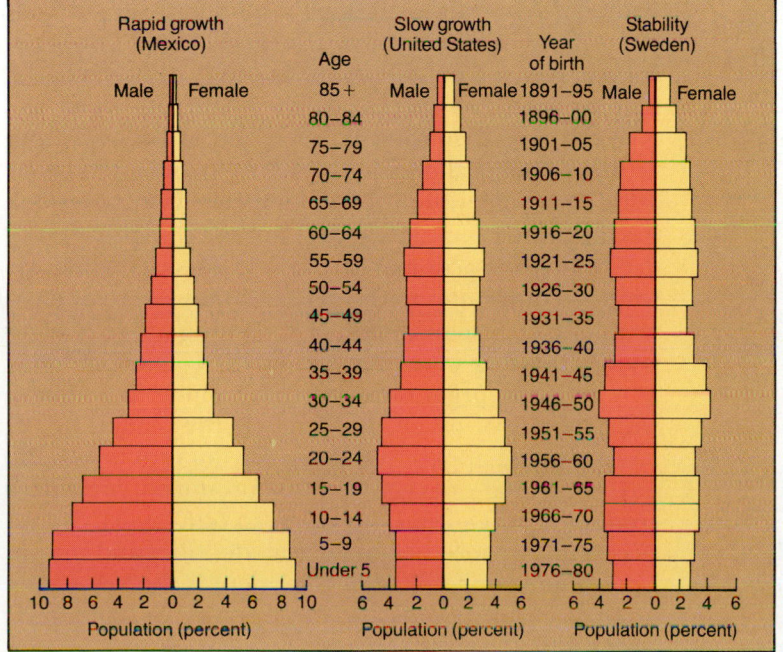

FIGURE 49–15 Population age distribution in three countries.

accumulated the capital necessary to establish the in-dustrial economies in which children are now no longer an economic asset. Not so the less-developed nations. With capital consumed as fast as it is generated by the urgent needs of their populations, they are not likely to achieve industrial economies or undergo the kind of demographic transition that has placed the developed countries in their present position.

Overpopulation Is Not Just A Third-World Problem

One may readily obtain agreement in the United States that other countries, *especially less-developed nations,* are overpopulated. A glance at Figure 49–16 should help to dispel this misunderstanding. Parts of Western Europe and North America are as densely populated as any part of India or China! As suburban sprawl stretches out, merging one city into another, as farms are sold and the cost of food rises, as resources dwindle and we experience fuel and other shortages, as water pollution and air pollution produce increasing blight, many individuals no longer need to be convinced that overpopulation is a present reality in developed nations too. Only their relatively high level of affluence and ability to import needed goods and materials maintains the high standard of living that tends to obscure their overcrowding.

In the long run, the ideal population for a nation, continent, or planet is one that can be stably sustained. It is clear that our resources will not indefinitely sustain the population of even lightly peopled countries at their *present* population densities. We have reached these unprecedented population densities by means of a temporary expansion of the carrying capacity of the habitat by technological means of consuming nonre-newable resources and our passing some of the costs to the ecosystem. Because nonrenewable resources will dwindle and finally cease, and because the ability of the ecosystem to absorb pollution is limited, expansion cannot continue indefinitely.

Ideally, population control will be attained through planning, not through violence or disease and their attendant suffering. However, although many environmental problems are rooted in human over-population, their severity is linked to our choice of life-style. Many ecologically harmful practices that are both wasteful and irrational could be altered. What seems to have been lacking is the will to do so.

☐ WHAT IS TO BE DONE?

Although environmental deterioration is global both in cause and effect, it has often proved very difficult or impossible to implement conservation internationally. The third-world nations where the forces of destruc-tion are now most spectacularly at work have often proved the most recalcitrant, for instance, in the pro-tection of endangered species, the conservation of rain forests, the use of less harmful pesticides, or limitation of the production of chlorofluorocarbons. The indus-trialized nations, however, have precious little to boast about, as they often produce environmental destruc-tion that extends far beyond their own borders. What hope we may have of reversing the flow of events prob-ably centers on the developed nations. Can they set a good example in the management of their own re-sources or in the safe disposal of their own wastes? Can they change import and other policies so as to avoid contributing to the downfall of others?

The developed nations should also lead the way in the development of an economy of thrift. Products should be designed for economical re-use or recycling, which should help both the problem of the depletion of resources and that of pollution by wastes. The devel-oped nations are also in the best position to produce more new varieties of crops suited to the tropics by conventional plant breeding and genetic engineering. If the greenhouse effect proves a reality, they will probably need them themselves! The developed na-tions can also spend money wisely, using their great economic power to encourage the preservation of threatened species and even ecosystems within and outside of their political boundaries.

Above all, we must avoid an attitude of hopeless-ness. To be sure, we will never be able to reverse the damage completely, but we should do what we can; even little bits of the world are worth saving. It is likely that you now have completed the last biology course you will ever take, but your biological education ought not to stop here. As a responsible and educated citizen, you should continue to be alert not only to environ-mental threats but to all the burgeoning issues of biol-ogy. We should all remain informed and involved. Joining environmental organizations, for instance, will help to keep you current on the changing issues, will give you a voice in their policies, and will allow you to influence political decisions, to purchase land to be set aside for conservation, and much else. Don't overlook the little things—after all, they are the things easiest to do. So small an action as throwing trash out of your car window contributes to the general atmosphere of envi-ronmental irresponsibility; just planting a tree may do good outlasting your lifetime.

As living organisms, we share much in common with the fate of other life forms on the planet. It is at last becoming apparent that we are not immune to the environmental damage we have produced. We differ from other organisms, however, in our capacity to re-flect on and probe our biological identities and to con-sider the meaning of life. This talent is the key to any hope of ensuring our, and the ecosphere's, continued survival.

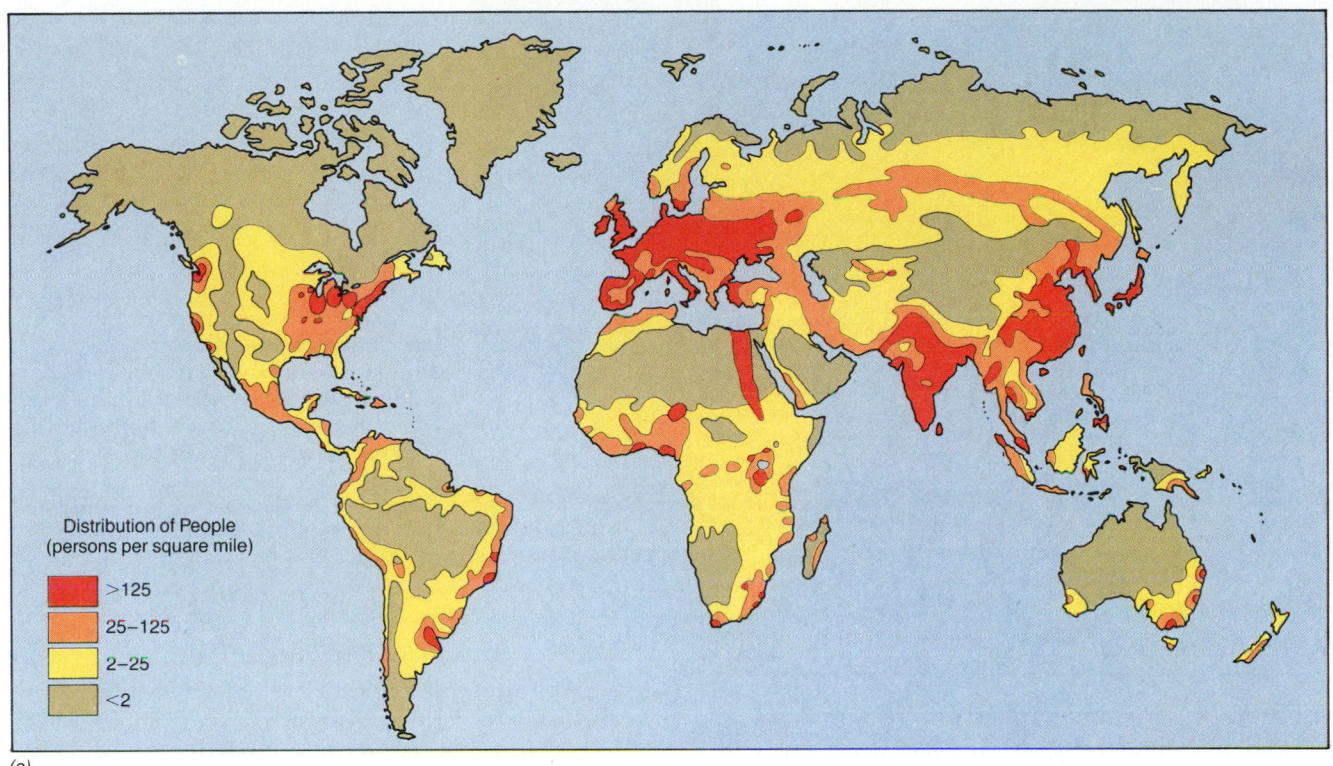

(a)

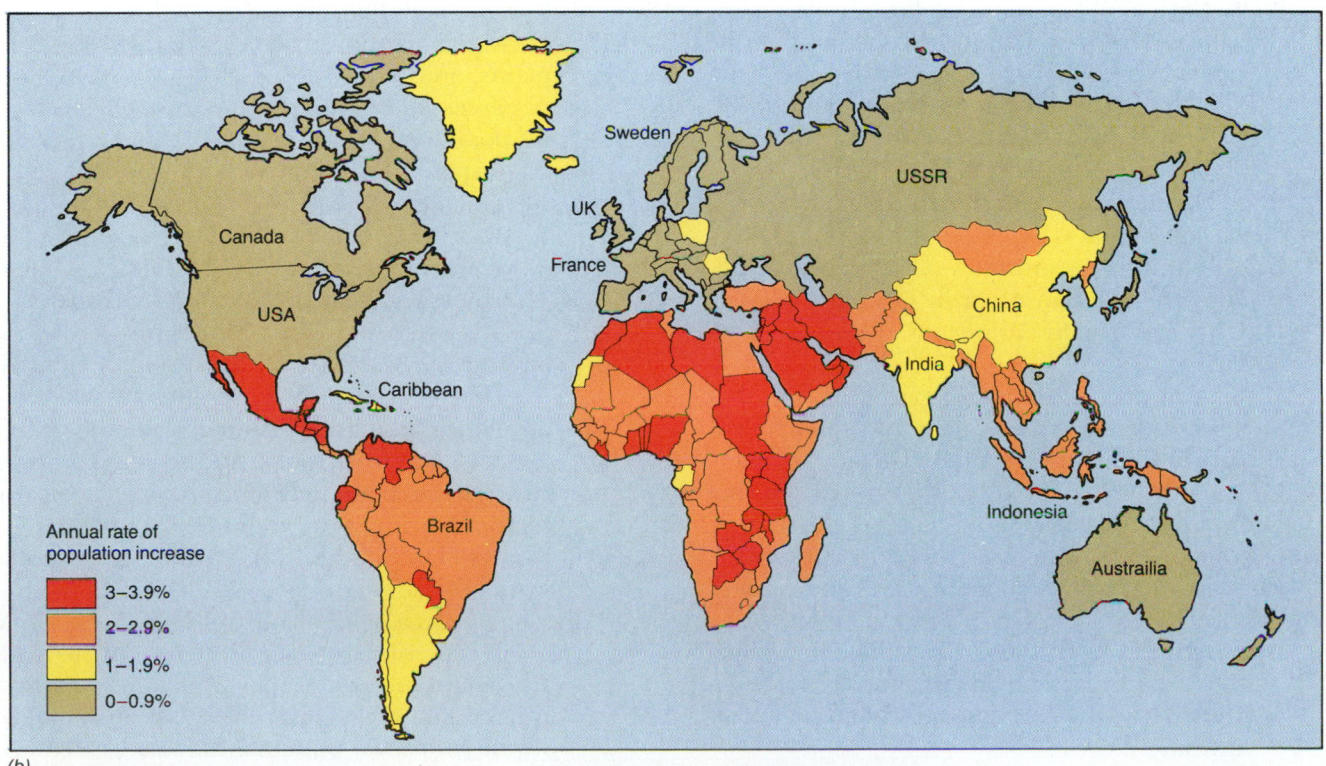

(b)

FIGURE 49–16   (*a*) **Although a great deal of the world seems sparsely populated, heavy population densities occur in East Asia, South Asia, Europe, and parts of North America.** (*b*) **Population** *growth* **rates, however, are low in most developed nations.**

FIGURE 49–17 Overpopulation produces pressures that extend even into remote areas. As Brazilian population expands into the Amazon wilderness, greatly aided by the trans-Amazon highway shown here, the greatest of tropical rain forest ecosystems will be lost not only to Brazil but to the rest of humanity as well. (Earth Sciences © 1990 Dr. Nigel Smith)

☐ CHAPTER SUMMARY

I. Modern agriculture requires substantial energy and technologic input and has great environmental impact.
   A. In agriculture, people replace diverse natural communities with those consisting of one or two specially bred species.
      1. These artificial communities are unstable, simple, and usually lacking in some important ecological mineral cycles.
      2. Some wildlife can coexist with agricultural communities.
   B. Unnatural communities of modern agriculture are prone to attack by pests and disease. This is typically combated by using pesticides.
   C. Silviculture resembles agriculture in that single-species tree farms are cultivated, rather than forest communities.

II. Most pesticides are broad-spectrum poisons that kill nonpest species as well as pests. Many pest populations become resistant to pesticides, whereas predator populations are often eradicated.
   A. Persistent pesticides are not biodegradable. They become widely distributed and then concentrated in predators by biological magnification.
   B. Alternatives to pesticide use exist and should be employed where possible.

III. Pollution exists when wastes degrade the environment's suitability for any of the organisms that would normally inhabit it.

IV. Severe aquatic pollution by organic substances produces a characteristic zonation. Eventually, much of the pollution is rendered harmless by natural processes, but only after a portion of the habitat has been degraded.

V. Various pollutants are deposited in the air and elsewhere by motor vehicles, industries, power plants, and other sources.
   A. Trees and other plants are damaged by air pollution. Widespread acid rain, caused by air pollution, menaces both land and aquatic habitats.
   B. An inversion layer may develop when a layer of warm air acts as a lid, sealing pollutants beneath so that they cannot be dispersed.

VI. Radioactive materials cannot easily be disposed of, and they persist and are passed extensively through food chains. At present, most such radioactive pollution has resulted from atmospheric tests of nuclear weapons, and a bit less from nuclear power plant accidents. The ecological effects of nuclear war would be catastrophic.

VII. Sanitary landfills and incineration are two main ways of solid-waste disposal.

VIII. The nuclear power industry has been threatened by serious questions regarding its safety and environmental impact. Solar power is considered a safe alternative for the future but one whose industrial potential may be limited.

IX. Species may become endangered or extinct by a variety of processes—overexploitation, environmental pollution, or the introduction of exotic species—but habitat destruction is the most important.

X. Increases in population translate into increased needs for food, shelter, clothing, material goods, and educational, medical, and social services. These increases result in additional environmental stress.

    A. By our technology, we have temporarily increased the carrying capacity of the earth for human beings. This expansion depends on consumption of nonrenewable resources and the passage of some of the costs to the environment in the form of pollution.

    B. A nation may be considered overpopulated if its total population is too great to be sustained permanently by its resources. By this criterion, in the author's view, virtually all areas of the modern world are overpopulated at the present time.

■ POST-TEST

1. Agricultural communities are generally simpler/more complex than natural communities.
2. Pesticides that are _____ decompose a few weeks or months after they have been applied.
3. DDT and other chlorinated hydrocarbon biocides are soluble/insoluble in water and very soluble/insoluble in body fat.
4. Owing to _____ _____, persistent pesticides, such as DDT, often tend to accumulate in the _____ levels of trophic pyramids.
5. _____ is a reduction in the quality of the environment by the addition of materials not normally found there or harmful quantities of otherwise harmless substances.
6. In the area of most recent pollution of a stream, the organic content is high/low, but the dissolved level is still high.
7. In an atmospheric inversion, a layer of _____ air forms a lid above one of _____ polluted air.
8. In human beings, radioactive strontium becomes incorporated into _____ and _____.
9. The bulk of extinctions now occurring appear to be due to _____ _____.

■ REVIEW QUESTIONS

1. What are some of the reasons that agricultural communities are unstable compared with natural communities?
2. Why are pest populations more likely to become resistant to pesticides than predator populations?
3. Give some economic and social reasons for pollution.
4. What disadvantages do you see to the use of natural enemies of pest organisms in place of pesticides?
5. What happens to the communities within aquatic ecosystems in extreme instances of pollution by organic wastes?
6. What is an inversion layer?
7. Why are many persistent pesticides and radioactive substances subject to biological magnification? Explain this process.
8. What kinds of animals are most prone to extinction by human agency? Is there reason to be concerned about this, or is it something we can live with?
9. Which contributes more, in your opinion, to our ecological crisis—overpopulation or wasteful consumption? Show how each contributes to the major ecological problems outlined in this chapter.
10. When fossil fuels are exhausted, which means of power generation do you think will be most widely employed? What might its drawbacks be? Its effect on our standard of living?

■ RECOMMENDED READINGS

It is virtually impossible to provide an up-to-date set of readings covering human modification of the environment owing to the constant accumulation of new studies and discoveries. It is suggested that the student look in recent issues of *Science, Scientific American, National Geographic Magazine,* as well as respectable newspapers and news magazines, and the publications of conservation societies such as the Sierra Club, the Nature Conservancy, or the Audubon Society.

Alternative Agriculture, by the National Research Council, Washington, D.C., National Academy Press, 1989. A review, along with case studies, of alternative farming methods in modern production agriculture.

Biodiversity, edited by E.O. Wilson. Washington, D.C. National Academy Press, 1988. A detailed account of the importance and decline of biological diversity. Solutions are also examined.

Biological Pest Control, edited by N.W. Hussey and N. Scopes,

New York, Cornell University Press, 1985. A compilation of successful control of pests through an integrated program of pest control. Enough detail is provided to enable readers to start biological pest control programs themselves.

Brown, M.H., *The Toxic Cloud: The Poisoning of America's Air.* New York, Harper and Row, 1987. The gravity of air pollution in the United States is examined in a clear, compelling fashion.

Conservation for the Twenty-First Century, edited by D. Western and M. Pearl. New York, Oxford University Press, 1989. An agenda for conservation action is presented by experts from many different disciplines.

Flavon, C. "How many Chernobyls?" *Worldwatch* 1:1, 1988. An examination of the risks of nuclear power.

Groundwater Protection, by The National Groundwater Policy Forum and The Conservation Foundation. Washington, D.C., The Conservation Foundation, 1987. An analysis of groundwater pollution, including its causes and control.

Houghton, R.A., and G.M. Woodwell. "Global Climatic Change," *Scientific American* 260(4):36–44, 1989. A current analysis of historical data on temperature and atmospheric carbon-dioxide levels leads the authors to predict dramatically rising temperatures.

Merrick, T.W., "World Population in Transition," *Population Bulletin* 41:2. Washington D.C., Population Reference Bureau, Inc., 1988. Although the human population has been increasing exponentially, there is evidence that the rate of increase may be slowing.

Reisner, M., "The Rise and Fall and Rise of Energy Conservation," *The Amicus Journal* 9:2, 1987. An article that traces the political rollercoaster ride of energy conservation in the United States.

Scientific American 260(9), 1989. The entire September 1989 issue is devoted to the impact of human society upon the ecosphere.

State of the World: *1989.* Edited by L.R. Brown et al. New York, W.W. Norton, 1989. Scrutiny of such environmental issues as land degradation, ozone depletion in the upper atmosphere, and world food prospects.

Appendix

◼ UNDERSTANDING BIOLOGICAL TERMS

Your task of mastering new terms will be greatly simplified if you learn to dissect each new word. Many terms can be divided into a prefix, the part of the word that precedes the main root, the word root itself, and often a suffix, a word ending that may add to or modify the meaning of the root. As you progress in your study of biology, you will learn to recognize the more common prefixes, word roots, and suffixes. Such recognition will help you analyze new terms so that you can more readily determine their meaning and will also help you remember them.

Prefixes

a-, ab- from, away, apart (abduct, lead away, move away from the midline of the body)

a-, an-, un- less, lack, not (asymmetrical, not symmetrical)

ad- (also **af-, ag-, an-, ap-**) to, toward (adduct, move toward the midline of the body)

allo- different (allometric growth, different rates of growth for different parts of the body during development)

ambi- both sides (ambidextrous, able to use either hand)

andro- a man (androecium, the male portion of a flower)

anis- unequal (anisogamy, sexual reproduction in which the gametes are different in size)

ante- forward, before (anteflexion, bending forward)

anti- against (antibody, proteins that have the capacity to react against foreign substances in the body)

auto- self (autotrophic, organisms that manufacture their own food)

bi- two (biennial, a plant that takes two years to complete its life cycle)

bio- life (biology, the study of life)

circum-, circ- around (circumcision, a cutting around)

co-, con- with, together (congenital, existing with or before birth)

contra- against (contraception, against conception)

cyt- cell (cytology, the study of cells)

di- two (disaccharide, a compound made of two sugar molecules chemically combined)

dis- apart (dissect, cut apart)

ecto- outside (ectoplasm, outer layer of cytoplasm)

end-, endo- within, inner (endoplasmic reticulum, a network of membranes found within the cytoplasm)

epi- on, upon (epidermis, upon the dermis)

ex-, e-, ef- out from, out of (extension, a straightening out)

extra- outside, beyond (extraembryonic membrane, a membrane that encircles and protects the embryo)

gravi- heavy (gravitropism, growth of a plant in response to gravity)

hemi- half (cerebral hemisphere, lateral half of the cerebrum)

hetero- other, different (heterozygous, unlike members of a gene pair)

homo-, hom- same (homologous, corresponding in structure; homozygous, identical members of a gene pair)

hyper- excessive, above normal (hypersecretion, excessive secretion)

hypo- under, below, deficient (hypotonic, a solution whose osmotic pressure is less than that of an isotonic solution)

in-, im- not (incomplete flower, a flower that does not have one or more of the four main parts)

inter- between, among (interstitial, situated between parts)

intra- within (intracellular, within the cell)

iso- equal, like (isotonic, equal strength)

macro- large (macronucleus, a large, polyploid nucleus found in ciliates)

mal- bad, abnormal (malnutrition, poor nutrition)

mega- large, great (megakaryocyte, giant cell of bone marrow)

meso- middle (mesoderm, middle tissue layer of the animal embryo)

meta- after, beyond (metaphase, the stage of mitosis after prophase)

micro- small (microscope, instrument for viewing small objects)

mono- one (monocot, a group of flowering plants with one cotyledon, or seed leaf, in the seed)

oligo- small, few, scant (oligotrophic lake, a lake deficient in nutrients and organisms)

oo- egg (oocyte, developing egg cell)

paedo- a child (paedomorphosis, the preservation of a juvenile characteristic in an adult)

para- near, beside, beyond (paracentral, near the center)

peri- around (pericardial membrane, membrane that surrounds the heart)

photo- light (phototropism, growth of a plant in response to the direction of light)

poly- many, much, multiple, complex (polysaccharide, a carbohydrate composed of many simple sugars)

post- after, behind (postnatal, after birth)

pre- before (prenatal, before birth)

pseudo- false (pseudopod, a temporary protrusion of a cell, i.e., "false foot")

retro- backward (retroperitoneal, located behind the peritoneum)

semi- half (semilunar, half-moon)

sub- under (subcutaneous tissue, tissue immediately under the skin)

super-, supra- above (suprarenal, above the kidney)

sym- with, together (sympatric speciation, evolution of a new species within the same geographical region as the parent species)

syn- with, together (syndrome, a group of symptoms which occur together and characterize a disease)

trans- across, beyond (transport, carry across)

Suffixes

-able, -ible able (viable, able to live)

-ad used in anatomy to form adverbs of direction (cephalad, toward the head)

-asis, -asia, -esis condition or state of (hemostasis, stopping of bleeding)

-cide kill, destroy (biocide, substance that kills living things)

-emia condition of blood (anemia, a blood condition in which there is a lack of red blood cells)

-gen something produced or generated or something that produces or generates (pathogen, something that can cause disease)

-gram record, write (electrocardiogram, a record of the electrical activity of the heart)

-graph record, write (electrocardiograph, an instrument for recording the electrical activity of the heart)

-ic adjective-forming suffix which means *of* or *pertaining to* (ophthalmic, of or pertaining to the eye)

-itis inflammation of (appendicitis, inflammation of the appendix)

-logy study or science of (cytology, study of cells)

-oid like, in the form of (thyroid, in the form of a shield)

-oma tumor (carcinoma, a malignant tumor)

-osis indicates disease (psychosis, a mental disease)

-pathy disease (dermopathy, disease of the skin)

-phyll leaf (mesophyll, the middle tissue of the leaf)

-scope instrument for viewing or observing (microscope, instrument for viewing small objects)

Some Common Word Roots

abscis cut off (abscission, the falling off of leaves or other plant parts)

angi, angio vessel (angiosperm, plants that produce seeds enclosed within a fruit or "vessel")

apic tip, apex (apical meristem, area of cell division located at the tips of plant stems and roots)

arthr joint (arthropods, invertebrate animals with jointed legs and segmented bodies)

aux grow, enlarge (auxin, a plant hormone involved in growth and development)

bi, bio life (biology, study of life)

blast a formative cell, germ layer (osteoblast, cell that gives rise to bone cells)

brachi arm (brachial artery, blood vessel that supplies the arm)

bry grow, swell (embryo, an organism in the early stages of development)

cardi heart (cardiac, pertaining to the heart)

carot carrot (carotene, a yellow, orange, or red pigment in plants)

cephal head (cephalad, toward the head)

cerebr brain (cerebral, pertaining to the brain)

cervic, cervix neck (cervical, pertaining to the neck)

chlor green (chlorophyll, a green pigment found in plants)

chondr cartilage (chondrocyte, a cartilage cell)

chrom color (chromosome, deeply staining body in nucleus)

cili small hair (cilium, a short, fine cytoplasmic hair projecting from the surface of a cell)

coleo a sheath (coleoptile, a protective sheath that encircles the stem in grass seeds)

conjug joined together (conjugation, a sexual phenomenon in certain protists)

cran skull (cranial, pertaining to the skull)

cyt cell (cytology, study of cells)

decid falling off (deciduous, a plant that sheds its leaves at the end of the growing season)

dehis split (dehiscent fruit, a fruit that splits open at maturity)

derm skin (dermatology, study of the skin)

ecol dwelling, house (ecology, the study of organisms in relation to their environment, i.e., "their house")

enter intestine (enterobacteria, bacteria that inhabit humans, particularly in their digestive tract)

evol to unroll (evolution, descent of complex organisms from simpler ancestors)

fil a thread (filament, the thin stalk of the stamen in flowers)

gamet a wife or husband (gametangium, the part of a plant or protist that produces reproductive cells)

gastr stomach (gastrointestinal tract, the digestive tract)

gen generate, produce (gene, a hereditary factor)

glyc, glyco sweet, sugar (glycogen, storage form of glucose)

gon seed (gonad, an organ that produces gametes)

gutt a drop (guttation, loss of water as liquid drops from plants)

gymn naked (gymnosperm, a plant that produces seeds that are not enclosed within a fruit, i.e., "naked")

hem blood (hemoglobin, the pigment of red blood cells)

hepat liver (hepatic, of or pertaining to the liver)

hist tissue (histology, study of tissues)

hom, homeo same, unchanging, steady (homeostasis, reaching a steady state)

hydr water (hydrolysis, a breakdown reaction involving water)

leuk white (leukocyte, white blood cell)

menin membrane (meninges, the three membranes that envelop the brain and spinal cord)

morph form (morphogenesis, development of body form)

my, myo muscle (myocardium, muscle layer of the heart)

myc a fungus (mycelium, the vegetative body of fungi)

nephr kidney (nephron, microscopic unit of the kidney)

neur, nerv nerve (neuromuscular, involving both the nerves and muscles)

occiput back part of the head (occipital, back region of the head)

ost bone (osteology, study of bones)

path disease (pathologist, one who studies disease processes)

ped, pod foot (biped, organism with two feet)

pell skin (pellicle, a flexible covering over the body of certain protists)

phag eat (phagocytosis, process by which certain cells ingest particles and foreign matter)

phil love (hydrophilic, a substance that attracts water)

phloe bark of a tree (phloem, food-conducting tissue in plants which corresponds to bark in woody plants)

phyt plant (xerophyte, a plant adapted to xeric, or dry, conditions)

plankt wandering (plankton, microscopic aquatic protists that float or drift passively)

rhiz root (rhizome, a horizontal, underground stem that superficially resembles a root)

scler hard (sclerenchyma, cells that provide strength and support in the plant body)

sipho a tube (siphonous, a type of body form found in certain algae that is tubular)

som body (chromosome, deeply staining body in the nucleus)

sor heap (sorus, a cluster or "heap" of sporangia in the ferns)

spor seed (spore, a reproductive cell that gives rise to individual offspring in plants and protists)

stom a mouth (stomate, a small pore ["mouth"] in the epidermis of plants)

thigm a touch (thigmotropism, plant growth in response to touch)

thromb clot (thrombus, a clot within a blood vessel)

tropi turn (thigmotropism, growth of a plant in response to contact with a solid object, as when a tendril "turns" or wraps around a wire fence)

visc pertaining to an internal organ or body cavity (viscera, internal organs)

xanth yellow (xanthophyll, a yellowish pigment found in plants)

xyl wood (xylem, water-conducting tissue in plants, the "wood" of woody plants)

zoo an animal (zoology, the science of animals)

Glossary

abdomen (ab′doh-men) (1) In mammals the region of the body between the diaphragm and the rim of the pelvis; (2) in arthropods, the posterior-most major division of the body.

abiogenesis (ab″ee-oh-jen′e-sis) The spontaneous generation of life; the origin of living things from inanimate objects.

abscisic acid (ab-sis′ik) A plant hormone involved in responses to stress and in dormancy.

abscission (ab-sizh′en) The normal (usually seasonal) falling off of leaves or other plant parts, such as fruits or flowers.

abscission zone The area at the base of the petiole where the leaf will break away from the stem.

absorption spectrum A measure of the amount of energy at specific wavelengths that has been absorbed as light passes through a substance. Each type of molecule has a characteristic absorption spectrum.

accessory fruit A fruit composed primarily of tissue other than ovary tissue. Apples and pears are accessory fruits.

acclimatization (a-kly′muh-tih-zay′shun) Gradual physiological changes in an organism in response to slow, relatively long-lasting changes in the environment.

acetylcholine (ah″see-til-koh′leen) A compound of choline and acetic acid; the neurotransmitter employed by cholinergic nerves.

acetyl Co A (ah-see′til) A key intermediate compound in metabolism; consists of an acetyl group covalently bonded to coenzyme A.

achene (a-keen′) A simple, dry, indehiscent fruit with one seed. The fruit wall is separate from the seed coat. Sunflower fruits are achenes.

acid A substance that releases hydrogen ions (protons) in water. Acids have a sour taste, turn blue litmus paper red, and unite with bases to form salts.

acoelomate organisms (a-seel′oh-mate) Organisms that lack a body cavity (coelom).

actin (ak′tin) A protein component of muscle fiber that, together with the protein myosin, is responsible for the ability of muscles to contract.

action potential The electrical activity developed in a muscle or nerve cell during activity; a neural impulse.

activation energy The energy required to initiate a chemical reaction.

active site Area of an enzyme that accepts a substrate and catalyzes its breakdown or its reaction with another substrate.

active transport Energy-requiring transport of a molecule across a membrane from a region of low concentration to a region of high concentration.

adaptation The ability of an organism to adjust to its environment.

adaptive radiation The evolution from an unspecialized ancestor of several to many species.

adenine (ad′eh-neen) A purine (nitrogenous base) that is a component of nucleic acids and of nucleotides important in energy transfer—adenosine triphosphate (ATP), adenosine diphosphate (ADP), and adenylic acid (AMP).

adenosine triphosphate (ATP) (a-den′oh-seen) An organic compound containing adenine, ribose, and three phosphate groups; of prime importance for energy transfers in biological systems.

adrenal glands (ah-dree′nul) Paired endocrine glands, one located just superior to each kidney.

adventitious root (ad″ven-tish′us) A root that arises in an unusual position on the plant.

aerobic (air-oh′bik) Growing or metabolizing only in the presence of molecular oxygen.

afferent (af′fur-ent) Structure which conducts fluid or impulses toward an organ or structure, e.g., afferent neurons conduct impulses to the central nervous system.

aggregate fruit A fruit that develops from a single flower with many separate carpels, such as a raspberry.

agnathans (ag-na′thanz) Jawless fishes; class of vertebrates, including lampreys, hagfishes, and many extinct forms.

algae (al′gee) Single-celled or simple multicellular photosynthetic organisms; important producers.

alkali (al′kuh-lie) A substance which, when dissolved in water, produces a pH greater than 7; also called a base.

allantois (a-lan′toe-iss) One of the extraembryonic membranes of reptiles, birds, and mammals; a pouch growing out of the posterior part of the digestive system.

alleles (al-leels′) Genes governing variations of the same characteristic that occupy corresponding positions on homologous chromosomes; alternative forms of a single gene.

allometric growth Varied rates of growth for different parts of the body during development.

allopatric speciation (al-oh-pa′trik) Speciation that occurs when one population becomes geographically separated from the rest of the species and subsequently evolves.

allopolyploid (al″oh-pol′ee-ploid) A polyploid formed by joining one or more sets of chromosomes from each of two different species.

allosteric site (al-oh-steer′ik) A site located on an enzyme molecule which enables a substance, other than the normal substrate, to bind to the molecule, and to change the shape of the molecule and the activity of the enzyme.

allozyme (al′loh-zime) One of two or more slightly different versions of the same enzyme. The presence of allozymes may confer a selective advantage on the organism that possesses them.

alternation of generations A type of life cycle characteristic of plants. Plants spend part of their life in the haploid (gametophyte) stage and part in the diploid (sporophyte) stage.

alveolus (al-vee′o-lus)   (1) An air sac of the lung through which gas exchange with the blood takes place; (2) a sac-like unit of some glands.

amino acid (uh-mee′no)   An organic compound containing an amino group (—NH₂) and a carboxyl group (—COOH). Amino acids may be linked together to form the peptide chains of protein molecules.

aminoacyl tRNA (uh-mee″no-as′seel)   Molecule consisting of an amino acid covalently linked to a tRNA.

amniocentesis (am″nee-oh-sen-tee′sis)   Sampling of the amniotic fluid surrounding a fetus in order to obtain information about its development and genetic makeup.

amnion (am′nee-on)   An extraembryonic membrane that forms a fluid-filled sac for the protection of the developing embryo.

anabolism (an-ab′oh-lizm)   Chemical reaction in which simpler substances are combined to form more complex substances, resulting in the storage of energy, the production of new cellular materials, and growth.

anaerobic (an″air-oh′bik)   Growing or metabolizing in the absence of molecular oxygen.

analogous   Similar in function or appearance but not in origin or development.

anaphase (an′uh-faze)   The stage of mitosis between metaphase and telophase in which the chromatids separate and move toward opposite ends of the cell.

androecium (an-dree′see-um)   The male portion of a flower.

aneuploidy (an′-you-ploy-dee)   Any chromosomal aberration in which there are either extra or missing copies of certain chromosomes.

angiosperm (an′jee-oh-sperm″)   The traditional name for plants having flowers, fruits, and seeds.

anion (an′eye-on)   A negatively charged ion such as Cl⁻.

anisogamy (an″eye-sog′uh-mee)   Reproductive process involving motile gametes of similar form but dissimilar size as in certain algae.

annelids (an′eh-lids)   Segmented worms with true coeloms, such as earthworms.

annual plant   A plant that completes its entire life cycle in one growing season.

annual ring   A layer of wood (secondary xylem) formed in woody plants growing in temperate areas. One layer is usually formed per year.

anther (an′thur)   The part of the stamen in flowers that produces microspores and, ultimately, pollen.

antheridium (an″thur-id′ee-im)   The male gametangium in certain land plants. Antheridia produce sperm.

anthocyanins (an′tho-sigh′ah-ninz)   A class of pigments of blue, red, and violet flowers.

antibiotic (an″ty-by-ot′ik)   Substance produced by microorganisms that has the capacity, in dilute solutions, to inhibit the growth of or destroy bacteria and other microorganisms; used largely in the treatment of infectious diseases in humans, animals, and plants.

antibodies (an′-tih-bod″ees)   Protein compounds (immunoglobulins) produced by plasma cells in response to specific antigens and having the capacity to react against the antigens.

anticodon (an″ty-koh″don)   A sequence of three nucleotides in transfer RNA that is complementary to, and combines with, the three-nucleotide codon on messenger RNA, thus helping to specify the addition of a particular amino acid to the end of a growing peptide.

antigen (an′tih-jen)   Any substance capable of stimulating an immune response; usually a protein or large carbohydrate that is foreign to the body.

anus (ay′nus)   The distal end and outlet of the digestive tract.

aorta (ay-or′tah)   The largest and main systemic artery of the body; arises from the left ventricle and branches to distribute blood to all parts of the body; main artery leaving the heart in vertebrates.

apical dominance (ape′ih-kl)   The inhibition of lateral buds by the apical meristem.

apical meristem (mehr′ih-stem)   An area of cell division located at the tips of plant stems and roots. Apical meristems produce primary tissues.

apomixis (ap″uh-mix′us)   A type of reproduction in which fruits and seeds are formed asexually, such as in dandelions.

arachnids (ah-rack′nids)   Eight-legged arthropods such as scorpions and spiders.

archaebacteria (ar″kuh-bak-teer′ee-uh)   Biochemically unique bacteria thought to be similar to the ancient ancestors of all bacteria.

archegonium (ar″ke-go′nee-um)   The female organ of certain land plants (such as mosses) in which an egg is produced.

arteries   Thick-walled blood vessels that carry blood away from the heart and toward the body organs.

arteriole (ar-teer′ee-ole)   A very small artery.

arthropod (ar′throh-pod)   An invertebrate, such as an insect or crustacean, that has jointed legs.

artificial selection   Selection by humans of traits that are desirable in other plants or animals and breeding of only those individuals that possess the desired traits.

asexual reproduction   Reproduction in which only one parent participates and in which the genetic makeup of parent and offspring is the same.

atom   The smallest quantity of an element that can retain the chemical properties of that element; composed of an atomic nucleus containing protons and neutrons, together with electrons that circle the nucleus in specific orbitals.

atrium (of the heart) (ay′tree-um)   The chamber on each side of the heart that receives blood from the veins.

autonomic nervous system (aw-tuh-nom′ik)   The portion of the peripheral nervous system that controls the visceral functions of the body, e.g., regulates smooth muscle, cardiac muscle, and glands, thereby helping to maintain homeostasis.

autosome (aw′toh-sohm)   A chromosome other than the sex (X and Y) chromosomes.

autotrophy (aw′-toh-troh″fee)   Obtaining organic molecules by synthesizing them from inorganic material.

auxin (awk′sin)   A plant hormone involved in various aspects of plant growth and development.

axon (ax′on)   The long, tubular extension of the neuron that transmits nerve impulses away from the cell body.

bacillus (ba-sil′us)   A rod-shaped bacterium.

background extinction   The continuous, low-level extinction of species that has been evident throughout much of the history of life. Compare with mass extinction.

bacteria (bak-teer′ee-uh)   Unicellular, prokaryotic microorganisms belonging to kingdom Monera. Most are decomposers, but some are parasites or autotrophs.

bacteriophage (bak-teer′ee-oh-fayj)   Virus that can infect a bacterium (literally, "bacteria eater").

basal body (bay′sl)   Structure that is similar to a centriole in its arrangement of microtubules and other components and which is involved in the organization and anchorage of a cilium or flagellum.

basal metabolic rate   The amount of energy expended by the body just to keep alive, when no food is being digested and no voluntary muscular work is being done.

base   A compound that releases hydroxyl ions (OH—) when dissolved in water; turns red litmus paper blue.

basidium (ba-sid′ee-um)   The club-like spore-producing organ of certain fungi.

Batesian mimicry (bate′see-un mim′ih-kree)   The resemblance of a harmless or palatable species to one that is dangerous, unpalatable, or poisonous.

behavioral isolation   A prezygotic isolating mechanism in which gamete exchange between two groups is prevented because each group possesses its own characteristic courtship behavior.

benthos (ben′thos)   The flora and fauna of the bottom of oceans or lakes.

berry   A simple, fleshy fruit in which the fruit wall is soft throughout. Tomatoes, bananas, and grapes are berries.

bicuspid (by-kus′pid)   Having two points or cusps (as bicuspid teeth) or two flaps (as the mitral valve of the heart).

biennial (by-en′ee-ul)   A plant that takes 2 years (i.e., two growing seasons) to complete its life cycle.

binary fission (by′nare-ee fish′un)   Equal division of a cell or organism into two; usually a variety of asexual reproduction.

binomial nomenclature (by-nome′ee-ul)   System of naming organisms by the combination of the names of genus and species.

biogeography   The study of the distribution of organisms.

biological clocks   Means by which activities of plants or animals are adapted to regularly recurring changes.

biological oxidation   Process in which electrons removed from an atom or molecule are transferred through the electron transport system of the mitochondrion.

biomass   The total weight of a group of organisms in a particular community.

biome (by′ohm)   A large, easily differentiated community unit arising as a result of complex interactions of climate, other physical factors, and biotic factors.

biosphere   The entire zone of air, land, and water at the surface of the earth that is occupied by living things.

biotic potential   Inherent power of a population to increase in numbers when the age ratio is stable and all environmental conditions are optimal.

bipedal   Walking on two feet.

blastopore (blas′toh-pore)   Primitive opening into the body cavity of an early embryo which may become the mouth or anus of the adult organism.

blastocyst (blas′toh-sist)   The blastula stage in the development of the mammalian embryo; a spherical mass consisting of a single layer of cells, the trophoblast, from which a small cluster, the inner cell mass, projects into a central cavity.

blastula (blas′tew-lah)   Usually a spherical structure produced by cleavage of a fertilized ovum; consists of a single layer of cells surrounding a fluid-filled cavity.

B lymphocyte (lim′foh-site)   A type of white blood cell responsible for antibody-mediated immunity. When stimulated, B lymphocytes differentiate to become plasma cells that produce antibodies; also called B cells.

bottleneck   Genetic drift that may result from a sudden decrease in a population.

brachiate   To swing, arm to arm, from one branch to another.

brachiopods (bray′kee-oh-pods)   Marine organisms possessing a pair of shells and, internally, a pair of coiled arms with ciliated tentacles.

bronchiole (bronk′ee-ole)   Tiny air duct of the lung that branches from a bronchus; divides to form air sacs (alveoli).

bronchus (bronk′us); pl. **bronchi** (bronk′eye)   One of the branches of the trachea and its immediate branches within the lung.

bryophytes (bry′oh-fites)   Members of the plant kingdom comprising mosses, liverworts, and hornworts.

bryozoans (bry″uh-zoh′uns)   Minute water animals that usually form fixed, branching, moss-like colonies; sometimes mistaken for seaweed; other species form colonies that appear as thin, lacy encrustations on rocks.

bud   An undeveloped shoot that can develop into flowers, stems, or leaves. Buds may be terminal (at the tip of the stem) or lateral (on the side of the stem).

budding   Asexual reproduction in which a small part of the parent's body separates from the rest and develops into a new individual, eventually either taking up an independent existence or becoming a more or less independent member of the colony.

bud primordium (pry-mor′dee-um)   An embryonic bud, evident in the stem apical meristem.

bud scale   A modified leaf that covers and protects winter buds.

buffers   Substances in a solution that tend to lessen the change in hydrogen ion concentration (pH) that otherwise would be produced by adding acids or bases.

bulb   A globose, fleshy, underground bud. A bulb is a short stem with fleshy leaves, e.g., onion.

bulliform cell (bool-ee′form) A large, thin-walled cell found in the epidermis of some monocots; aids in the folding and unfolding of leaves associated with periods of drought.

bundle sheath A ring of cells surrounding the vascular bundle in dicot and monocot leaves.

callus (kal′us) Undifferentiated tissue in plant tissue culture.

calorie (kal′oh-ree) A unit of heat. The calorie used in the study of metabolism is the large Calorie or kilocalorie and is defined as the amount of heat required to raise the temperature of a kilogram of water one degree Celsius.

Calvin cycle Cyclic series of reactions occurring in the light-independent phase of photosynthesis that fixes carbon dioxide and produces glucose.

calyx (kay′liks) The collective term for the sepals of a flower.

cambium (kam′bee-um) A layer of meristematic tissue that produces lateral (secondary) growth in plants.

capillaries (kap′i-lare-eez) Microscopic blood vessels occurring in the tissues which permit exchange of materials between tissues and blood. Lymph capillaries also drain away excess tissue fluid.

capsule (1)The portion of the moss sporophyte that contains spores; (2) a simple, dry, dehiscent fruit that opens along many seams or pores to release seeds, such as cotton fruits.

carbohydrate Compound containing carbon, hydrogen, and oxygen, in the approximate ratio of $1C:2H:1O$, such as sugars, starches, and cellulose.

carboxyl (kar-bok′sil) A monovalent radical characteristic of organic acids; group of atoms in a molecule arranged as —COOH.

carcinogen (kar′sin-oh-jen, kar-sin′oh-gen) A substance that causes cancer or accelerates its development.

cardiac (kar′dee-ak) Pertaining to the heart.

cardiac muscle Distinctive involuntary but striated type of muscle occurring only in the vertebrate heart.

carnivore (kar′ni-vor) An animal that primarily eats flesh.

carotene (kare′oh-teen) Yellow to orange-red pigments found in carrots, sweet potatoes, leafy vegetables, etc.; can be converted in the animal body to vitamin A.

carotenoid A carotene or xanthophyll.

carpel (kar′pul) The female structure in flowers that bears ovules.

carrying capacity The ability of an environment to support a population of organisms.

Casparian strip (kas-pare′ee-un) A band of waterproof material around the walls of the cells of the root endodermis.

cast Fossil formed by a mold being filled in by minerals that harden.

catalyst (kat′al-list) A substance that regulates the speed at which a chemical reaction occurs without affecting the end point of the reaction and without being used up as a result of the reaction.

cation (kat′eye-un) An ion bearing a positive charge.

cell The basic structural and functional unit of the body, consisting of a complex of organelles bounded by a unit membrane, and usually microscopic in size.

cell cycle Cyclic series of events in the life of a dividing eukaryotic cell consisting of M (mitosis), cytokinesis, and the stages of interphase, which are the G_1 (first gap), S (DNA synthesis), and G_2 (second gap) phases.

cell plate Forming cell wall that separates the two daughter cells produced by plant mitosis.

cellulose (sel′yoo-lohs) A complex polysaccharide that composes the cell walls of plants.

center of origin The particular place where a species originated or evolved.

centriole (sen′tree-ohl) One of a pair of small, dark-staining organelles lying near the nucleus in the cytoplasm of animal cells.

centromere (sen′tro-meer) Specialized constricted region of a chromatid that serves as the site of spindle fiber attachment during cell division; sister chromatids are joined in the vicinity of their centromeres.

cerebellum (ser-eh-bel′um) The deeply convoluted subdivision of the brain lying beneath the cerebrum which is concerned with the coordination of muscular movements; second largest part of the brain.

cerebral cortex (ser-ee′brul kor′tex) The outer layer of the cerebrum composed of grey matter and consisting mainly of nerve cell bodies.

cerebrum (ser-ee′brum) Largest subdivision of the brain; functions as the center for learning, voluntary movement, and interpretation of sensation.

chaparral (shap″uh-ral′) Distinctive vegetation type characteristic of certain areas with Mediterranean climates; dominated by drought-resistant shrubs and small trees with small leaves.

chemical evolution The origin of life from nonliving matter.

chemoautotrophs (kee″moh-aw′toh-trofes) Autotrophic organisms that depend on the energy of pre-existing chemical compounds in their environment rather than the energy of sunlight.

chemoreceptor (kee″moh-ree-sep′tor) A sense organ or sensory cell that responds to chemical stimuli.

chemotropism (kee″moh-tro′pizm) A growth response to a chemical stimulus.

chiasma (ky-az′muh); pl. **chiasmata** A site in a tetrad where homologous (non-sister) chromatids have undergone exchange by breakage and rejoining.

chitin (ky′tin) An insoluble, horny protein-polysaccharide that forms the exoskeleton of arthropods and the cell walls of many fungi.

chlorophyll (klor′oh-fil) Light-trapping green pigments found in most photosynthetic organisms.

chloroplast (klor′oh-plast) A chlorophyll-bearing intracellular organelle of plant cells; site of photosynthesis.

chordates (kor′dates) Phylum of animals which possess, at some time in their lives, a cartilaginous dorsal skeletal structure called a notochord.

chorion (kor′ee-on) An extraembryonic membrane in reptiles, birds, and mammals that forms an outer cover around the embryo, and in mammals contributes to the formation of the placenta.

chromatid (kroh′mah-tid)   One of the two halves of a replicated chromosome.

chromatin (kroh′mah-tin)   The complex of DNA and protein which makes up eukaryotic chromosomes.

chromomere (kro′moh-meer)   One of a linear series of bead-like structures composing a chromosome.

chromosomes (kro′moh-soms)   Filamentous or rod-shaped bodies in the cell nucleus that contain the hereditary units, the genes.

chrysalis (krih′suh-lis)   Pupa of a butterfly or moth; immobile stage of life cycle between larval (caterpillar) form and adult winged insect.

cilium (sil′ee-um); pl. **cilia**   One of many short hair-like structures that project from the surface of some cells and are used for locomotion or movement of materials across the cell surface; structurally like flagella, including a cylinder of 9 doublet microtubules and 2 central single microtubules, all covered by a plasma membrane.

circadian rhythm (sir-kay′dee-un)   An internal rhythm that approximates the 24-hour day. Circadian rhythms are found in plants, animals, and other organisms.

citric acid cycle   Aerobic series of chemical reactions in which fuel molecules are completely degraded to carbon dioxide and water with the release of metabolic energy; also known as the Krebs cycle.

cladogenesis (klad″oh-jen′eh-sis)   Diversifying speciation in which one or more new species is derived from the parent species, which continues to exist.

class   A taxonomic grouping of related, similar orders of living organisms.

classical conditioning   A type of learning in which a normal response to a stimulus becomes associated with a new stimulus, after which the new stimulus elicits the response.

cleavage   Series of cell divisions that converts the zygote to a multicellular blastula.

climax community   The final, stable, and mature community in a successional series.

cline   Continuous series of differences in structure or function exhibited by the members of a species along a line extending from one part of their range to another.

clone   A population of cells descended by mitotic division from a single ancestral cell, or a population of genetically identical organisms asexually propagated from a single ancestor.

cnidarians (nye-dare′ee-uns)   Phylum of animals possessing stinging cells called cnidoblasts; characterized by a single body opening, two tissue layers, and radial symmetry, such as the *Hydra*.

coacervate (koh-as′sir-vate)   A protobiont formed from relatively complex mixtures of polypeptides, nucleic acids, and polysaccharides.

coccus (kok′us)   A spherical bacterial cell, usually less than 1 micrometer in diameter.

codominance (koh″dom′in-ints)   Condition in which both alleles of a locus are expressed in a heterozygote.

codon (koh′don)   A trio of mRNA bases that specifies an amino acid or a signal to terminate the polypeptide.

coelom (see′lum)   Body cavity which forms within the mesoderm and is lined by mesoderm.

coenzyme (koh-en′zime)   An organic, nonprotein substance that serves as a cofactor; participates in the reaction by donating or accepting some reactant; loosely bound to enzyme. Most of the vitamins function as coenzymes.

cofactor   A nonprotein substance needed by an enzyme for normal action; cofactors are metal ions; others are coenzymes.

coleoptile (kol-ee-op′tile)   A protective sheath that encloses the stem in grasses.

collagen (kol′ah-gen)   Protein in connective tissue fibers that is converted to gelatin by boiling.

collenchyma (kol-en′kih-mah)   Living cells with moderately but unevenly thickened walls. Collenchyma cells help support the primary plant body.

commensal (kum-men′sul)   An organism that lives in intimate association with another species without harming or benefiting its host.

community   An assemblage of populations that live in a defined area or habitat, which can either be very large or quite small. The organisms constituting the community interact in various ways with one another.

companion cell   A cell in the phloem of plants that is responsible for loading and unloading sugar into the sieve tube member for conduction.

competitive inhibition   Interference with enzyme action by an abnormal substrate that is not permanently bound to the active site and that competes with the normal substrate for the site.

complete flower   A flower that possesses all four main parts: sepals, petals, stamens, and carpels.

compression   A fossil in which the organism was trapped in sediments without being completely decomposed.

cone   (1) In botany, a reproductive structure in many gymnosperms that produces either microspores or megaspores. (2) In zoology, the conical photoreceptive cell of the retina that is particularly sensitive to bright light, and, by distinguishing light of various wavelengths, mediates color vision.

conifers (kon′ih-furs)   Gymnosperms bearing needle-like leaves.

conjugation   A sexual phenomenon in certain protists that involves exchange or fusion of a cell with another cell. The term is also used sometimes for DNA exchange in bacteria.

connective tissue   Vertebrate tissue consisting mostly of a matrix composed of cell products in which the cells are embedded, e.g., bone.

contractile root   A root that contracts and pulls the plant deeper into the soil.

contractile vacuole (vak′yoo-ohl)   A vacuole that expands, filling with water, and periodically contracts, ejecting the water from the cell.

convection   The transfer of heat within liquids and gases that results from currents arising as a result of heat-induced local differences in density.

convergent evolution The independent evolution of similar structures that carry on similar functions, in two or more organisms of widely different, unrelated ancestry.

cork Cells produced by the cork cambium. Cork is dead at maturity and functions for protection.

cork cambium (kam′bee-um) Lateral meristem in plants that produces cork cells and cork parenchyma. Cork cambium and the tissues it produces make up the outer bark on a woody plant.

cork parenchyma (par-en′kih-mah) One or more layers of parenchyma cells produced by the cork cambium.

corm A short, thickened underground stem specialized for food storage and asexual reproduction, as exists in the gladiolus.

corolla (kor-ohl′ah) A collective term for the petals of a flower.

corpus luteum (loo-tee′um) The yellow endocrine tissue in the ovary that develops from the ruptured follicle after ovulation; secretes progesterone and estrogen.

cortex (kor′tex) The outer layer of an organ; in plants, the tissue beneath the epidermis in many nonwoody plants.

cotyledon (kot″i-lee′dun) The seed leaf of the embryo of a plant, which may contain food stored for germination.

covalent bond Chemical bond involving one or more shared pairs of electrons.

cristae (kris′tee) Shelf-like folds of the inner membrane of a mitochondrion.

crossing over The breaking and rejoining of homologous (non-sister) chromatids during early meiotic prophase I, resulting in an exchange of genetic material.

ctenophores (teen′oh-forz) Marine animals (comb jellies) whose bodies consist of two layers of cells enclosing a gelatinous mass. The outer surface is covered with eight rows of cilia resembling combs, by which the animal moves through the water.

cultural evolution The progressive addition of knowledge to the human experience.

cuticle (kew′tih-kl) A waxy covering over the epidermis of the above-ground portion of land plants that reduces water loss from plant surfaces.

cyanobacteria (sy-an′oh-bak-teer′ee-uh) Prokaryotic photosynthetic microorganisms that possess chlorophyll and produce oxygen by the photolysis of water. Formerly known as blue-green algae.

cycads (sih′kads) Members of a class of woody seed plants (the Cycadophyta) which live mainly in tropical and semitropical regions and have either short, tuberous, underground stems, or erect, cylindrical above-ground stems.

cyclic AMP (cAMP) A form of adenosine monophosphate in which the phosphate is part of a ring-shaped structure; acts as a regulatory molecule and second messenger in organisms ranging from bacteria to humans.

cyclosis (sy-kloh′sis) The streaming motion of the living material in cells.

cytochromes (sy′toh-kromz) The iron-containing heme proteins of the electron transport system that are alternately oxidized and reduced in biological oxidation.

cytokinesis (sy″toh-kih-nee′-sis) Stage of cell division in which the cytoplasm is divided to form two daughter cells.

cytokinin (sy″toh-kih′nin) A plant hormone involved in various aspects of plant growth and development.

cytoplasm (sy′toh-plazm) General cellular contents exclusive of the nucleus.

cytosine See nucleic acid.

cytoskeleton Internal structure of microfilaments, intermediate filaments, and microtubules that gives shape and mechanical strength to cells.

day-neutral plant A plant that does not flower in response to variations in day length which occur with changing seasons.

deamination (dee-am-ih-nay′shun) Removal of an amino group ($—NH_2$) from an amino acid or other organic compound.

deciduous (de-sid′yoo-us) Falling off at a certain time or stage of growth, such as some leaves, antlers, and the wings of some insects.

decomposers Microorganisms of decay.

dehiscent fruit (dih-his′sent) A simple, dry fruit that splits open along one or more seams at maturity; compare with indehiscent fruit.

dehydrogenation (dee-hy″dro-jen-ay′shun) A form of oxidation in which hydrogen atoms are removed from a molecule.

deme An interbreeding population within a species.

denature (dee-nay′ture) To alter the physical properties and three-dimensional structure of a protein, nucleic acid, or other macromolecule by mild treatment that does not break the primary structure but which does destroy its activity.

dendrite (den′drite) A short branch of a neuron that receives and conducts nerve impulses toward the cell body.

denitrifying bacteria (dee-ny′tri-fy-ing) Bacteria that convert ammonia to nitrogen gas.

deoxyribose The 5-carbon sugar found in DNA.

depolarization (dee-pol″ar-ih-zay′shun) Change in electric charge of a plasma membrane that produces the action potential.

dermis (dur′mis) The layer of dense connective tissue beneath the epidermis in the skin of vertebrates.

desmosomes (dez′moh-somz) Button-like plaques, present on the two opposing cell surfaces and separated by the intercellular space, that serve to hold the cells together.

detritus feeders (deh-try′tus) Organisms, other than microorganisms, that feed on dead, decaying organisms or their fragments.

deuterostome (doo′ter-oh-stome) A division of coelomate animals that includes the echinoderms and chordates; characterized by radial cleavage and development of the anus from the blastopore.

dicotyledon (dy-kot-ih-lee′dun) A flowering plant with embryos having two seed leaves, or cotyledons; also known as a dicot.

differentiation Development toward a more mature state; a process changing a relatively unspecialized cell to a more specialized cell.

diffusion The net movement of molecules from a region of high concentration to one of lower concentration, brought about by their kinetic energy.

dihybrid cross (dy-hy′brid) A genetic cross which takes into account the behavior of two distinct pairs of genes.

dikaryotic cells (dy-kare-ee-ot′ik) Cells having two nuclei.

dimorphic (dy-mor′fik) Having two forms, as in the alternation of generations of some plants.

dinoflagellates (dy″noh-flaj′eh-lates) Single-celled algae, surrounded by a shell of thick interlocking cellulose plates.

dioecious (dy-ee′shus) Having male and female reproductive structures on separate plants; compare with monoecious.

diploid (dip′loid) A chromosome number twice that found in gametes; containing two sets of chromosomes.

directional selection The gradual replacement of one phenotype with another due to environmental change which favors those phenotypes at one of the extremes of the normal distribution.

disaccharide (dy-sak′ah-ride) A two-unit sugar, such as sucrose, which consists of two hexose subunits.

disruptive selection A special type of directional selection in which changes in the environment favor two or more variant phenotypes at the expense of the mean.

distal Remote; farther from the main body axis.

diurnal (dy-ur′nl) Active during the daytime.

DNA Deoxyribonucleic acid; present in chromosomes; contains genetic information coded in specific sequences of its constituent nucleotides.

DNA replication The synthesis of DNA by using a complementary strand of DNA as a template.

dominant allele (al-leel′) Gene allele which is always expressed when it is present, regardless of whether it is homozygous or heterozygous.

dorsal (dor′sl) Pertaining to the back of an animal.

double fertilization A process in the angiosperm reproductive cycle where there are two fertilizations: one results in the formation of a young plant; the second results in the formation of the endosperm.

drupe (droop) A simple, fleshy fruit in which the inner wall of the fruit is hard and stony. Peaches and cherries are drupes.

echinoderms (eh-kine′oh-derms) Spiny-skinned marine animals such as starfish, sea urchins, and sea cucumbers.

ecological isolation A prezygotic isolating mechanism in which gamete exchange is prevented between two groups that are located in the same geographical area because they live and reproduce in two different ecological habitats.

ecological niche The status of an organism within a community or ecosystem; depends on the organism's structural adaptations, physiological responses, and behavior.

ecology (ee-kol′uh-jee) The study of the interrelations between living things and their environment, both physical and biotic.

ecosystem (ee′koh-sis-tem) A natural unit of living and nonliving parts that interact to produce a stable system in which the exchange of materials between living and nonliving is recycled.

ecotone (ee′koh-tone) A fairly broad transition region between adjacent biomes; contains some organisms from each of the two biomes plus some that are characteristic of, and perhaps restricted to, the ecotone.

ectoderm (ek′toh-derm) The outer of the three embryonic germ layers of the gastrula; gives rise to the skin and nervous system.

ectotherm Having a body temperature that fluctuates with that of the environment; "cold-blooded."

edaphic factors (ee-daf′ik) Factors in the soil that influence the distribution and numbers of plants and animals.

effector A muscle or gland that contracts or secretes in direct response to nerve impulses.

efferent (ef′fur-ent) Pertaining to a structure that leads away from another structure or organ, such as the efferent arteriole of the kidney nephron.

electrochemical potential The potential energy possessed by a system in which there is a difference in electrical charge, as well as a concentration gradient of ions across a membrane.

electron A negatively charged subatomic particle located at some distance from the atomic nucleus.

electron transport system A series of chemical reactions during which hydrogens or their electrons are passed along from one acceptor molecule to another, with the release of energy.

element Chemically, one of the 100 or so types of matter, natural or man-made, composed of atoms, all of which have the same number of protons in the atomic nucleus and the same number of electrons circling in orbits.

embryo (em′bree-oh) A young organism before it emerges from the egg, seed, or body of its mother; the developing human organism until the end of the second month, after which it is referred to as a fetus.

embryo sac The female gametophyte generation in angiosperms.

endergonic reactions (end″er-gon′ik) Nonspontaneous reactions requiring a net input of free energy.

endocrine glands (en′doh-crin) Glands that secrete products into the blood or tissue fluid instead of into ducts.

endocytosis (en″doh-sy-toh′sis) The active transport of substances into a cell by the formation of invaginated regions of the plasma membrane which pinch off and become cytoplasmic vesicles.

endoderm (en′doh-derm) The inner germ layer of the gastrula lining the archenteron; becomes the digestive tract and its outgrowths—the liver, lungs, and pancreas.

endodermis (en″doh-der′mis) The innermost layer of the cortex in the plant root. Endodermis cells have a Casparian strip running around radial and transverse walls.

endogenous (en-doj'eh-nus)   Produced within the body, or due to internal causes.

endometrium (en″doh-mee'tree-um)   Uterine lining.

endoplasmic reticulum (en″doh-plaz'mik reh-tik'yoo-lum) **(ER)**   Structure composed of numerous internal membranes within eukaryotic cells.

endorphins (en-dor'finz)   Polypeptide transmitter substance of certain brain and visceral neurons whose action is mimicked by opiate alkaloids.

endoskeleton (en″doh-skel'eh-ton)   Bony and cartilaginous supporting structures within the body that provide support from within.

endosperm (en'doh-sperm)   The nutritive tissue that is found at some point in all angiosperm seeds.

endosymbiont hypothesis (en″doh-sim'bee-ont)   Idea that certain organelles such as mitochondria and chloroplasts originated as symbiotic prokaryotes that lived inside larger cells.

endotherm (en'doh-therm)   An animal that uses metabolic energy to maintain a constant body temperature despite variations in environmental temperature.

entropy (en'trop-ee)   Disorderliness; a quantitative measure of randomness or disorder, symbolized by S.

environmental resistance   The sum of the physical and biological factors that prevent a species from reproducing at its maximum rate.

enzyme (en'zime)   An organic catalyst produced within a living organism that accelerates specific chemical reactions.

epicotyl (ep'ih-kot″il)   The part of the axis of a plant embryo or seedling above the point of attachment of the cotyledons.

epidermis (ep-ih-dur'mis)   An outer layer of cells covering the body of plants and animals; functions primarily for protection.

epiphyte (ep'ih-fite)   A plant that is self-nourishing but that grows upon another plant, using it for position and support.

epistasis (ep″ih-sta'sis)   Condition in which certain alleles at one locus can alter the expression of alleles at a different locus.

epithelial tissue (ep-ih-theel'ee-al)   The type of tissue that covers body surfaces, lines body cavities, and forms glands; also called epithelium.

epoch   Geological time that is a subdivision of a period.

era   In geology, one of five main divisions of geological time. Eras are divided into periods.

erythrocyte (er-eeth'roh-site)   Vertebrate red blood cell.

esophagus (ee-sof'ah-gus)   The muscular tube extending from the pharynx to the stomach.

estrus (es'trus)   The recurrent period of sexual receptivity occurring around ovulation in female mammals having estrous cycles.

ethology (ee-thol'oh-jee)   The study of the whole range of animal behavior under natural conditions.

ethylene (eth'ih-leen)   A plant hormone involved in various aspects of plant senescence.

eubacteria (yoo'bak-teer″ee-ah)   Bacteria other than the archaebacteria.

euchromatin (yoo-kroh'-mah-tin)   Loosely coiled chromatin which is transcribed into mRNA.

euglenoids (yoo-glee'noids)   A group of unicellular protists that includes *Euglena*.

eukaryote (yoo″kare'ee-ote)   Organism whose cells possess organelles surrounded by membranes.

eutrophication (yoo″troh-fih-kay'shun)   Process of progressive nutritional enrichment of bodies of water by natural processes or pollution.

evolution   Genetic change in a population of organisms.

excretion (ek-skree'shun)   The discharge from the body of a waste product of metabolism (not to be confused with the elimination of undigested food materials).

exergonic (ex-er-gon'ik)   A reaction characterized by the release of energy.

exocrine glands (ex'oh-crin)   Glands that excrete their products through ducts in the epithelium onto the surface of the skin, such as sweat glands.

exocytosis (ex″oh-sy-toh'sis)   Export of materials from the cell by fusion of cytoplasmic vesicles with the plasma membrane.

exogenous (ek-sodj'eh-nus)   Due to, or produced by, an external cause; not arising within the body.

exon   (1) A protein-coding region of a eukaryotic gene; (2) the RNA transcribed from such a region (see intron).

exoskeleton (ex″oh-skel'eh-ton)   An external skeleton, such as the shell of mollusks or outer covering of arthropods; provides protection and sites of attachment for muscles.

extinction   The death of a species. Extinction occurs when the last individual of a species dies.

F_1 generation (first filial)   The first generation of hybrid offspring resulting from a genetic cross.

F_2 generation (second filial)   The offspring of the F_1 generation.

facilitated diffusion   The passage of ions and molecules, bound to specific carrier proteins, across a cell membrane down their concentration gradients. Facilitated diffusion does not require a special source of energy.

family   In classification, the taxonomic grouping of related, similar genera.

fatty acid   One of many organic acids largely composed of long chains of carbon and hydrogen atoms that occur as components of fat.

feedback control   System in which the accumulation of the product of a reaction leads to a decrease in its rate of production, or a deficiency of the product leads to an increase in its rate of production.

fermentation   Anaerobic respiration that utilizes organic compounds both as electron donors and acceptors.

fertilization   Union of male and female gametes.

fiber   (1) In plants, a type of sclerenchyma. Fibers are long, tapered cells with thick walls. (2) In animals, an elongated cell such as a muscle or nerve cell.

fibrous root system   A root system in plants that has several main roots without a dominant root.

filament   In botany, the thin stalk of the stamen.

fission (fish'un)   Process of asexual reproduction in

which an organism divides into two approximately equal parts.

flagellum (flah-jel′um) Long, whip-like, movable structure of cells that is used in locomotion or in moving fluid past the tissue of which the cell is a part. Eukaryote flagellae are composed of two single microtubules surrounded by nine double microtubules, but prokaryote flagellae are filaments rotated by special structures located in the plasma membrane and cell wall.

fluid mosaic model The modern picture of the plasma membrane (and other cell membranes) in which protein molecules float in a phospholipid bilayer.

follicle (fol′i-kl) (1) A simple, dry, dehiscent fruit that splits open along one seam to liberate the seeds; (2) a small sac of cells in the mammalian ovary that contains a maturing egg; (3) the pocket in the skin from which a hair grows.

food chain A sequence of organisms through which energy is transferred from its ultimate source in a plant; each organism eats the preceding and is eaten by the following member of the sequence.

food web Complex feeding relationships in a community of organisms.

foot In botany, the basal portion of the moss sporophyte that serves to anchor it to the gametophyte.

fossil Parts of an ancient organism or traces left by previous life.

founder effect Genetic drift that results from a small population colonizing a new area.

frameshift mutation Mutation that results when nucleotides are inserted into or deleted from the DNA in numbers that are not multiples of three, thereby altering the reading frame so that all of the codons downstream from the mutation will be changed.

frond A leaf of a fern.

fruit In angiosperms, a mature, ripened ovary. Fruits contain seeds and usually serve for seed protection and dispersal.

fucoxanthin (few″koh-zan′thin) The brown pigment found in diatoms, brown algae, and dinoflagellates.

fungus; pl. fungi A eukaryote with cell walls that is incapable of photosynthesis. Most fungi are decomposers; a few are parasitic.

G_1 phase Gap phase in interphase of the cell cycle before DNA synthesis begins.

gametangium (gam″uh-tan′gee-um) Special multicellular or unicellular structure of plants, protists, and fungi in which gametes are formed. In fungi, these form at the tips of hyphae of opposite mating types.

gamete (gam′eet) A cell that functions in sexual reproduction; an egg or sperm whose union, in sexual reproduction, initiates the development of a new individual.

gametic isolation (gam-ee′tik) A prezygotic isolating mechanism in which gamete exchange between two groups cannot occur because of chemical differences between the gametes of two different species.

gametophyte (gam-ee′toh-fite) The haploid or gamete-producing stage in the life cycle of a plant.

ganglion (gang′glee-on) A knot-like mass of the cell bodies of neurons located outside the central nervous system.

gap junction Structure consisting of specialized regions of the plasma membranes of two adjacent cells containing numerous pores that allow passage of certain molecules and ions between them.

gastrula (gas′troo-lah) Early stage of embryonic development during which the three germ layers form.

gemma (jem′mah) Vegetative bud in bryophytes that develops asexually into a new plant.

gene A discrete unit of hereditary information that usually specifies a polypeptide. It consists of DNA and is part of the chromosomes.

gene flow The movement of alleles between local populations, or demes, due to the migration of individuals. Gene flow can have significant evolutionary consequences.

gene pool All the genes present in a species population.

genetic code Code consisting of groups of three bases in mRNA which specifies individual amino acids or translation start and stop signals.

genetic drift A random change in gene frequency in a small breeding population.

genome (jee′nome) A complete set of hereditary factors contained in one haploid assortment of chromosomes.

genotype (jeen′oh-type, jen′oh-type) The complete genetic makeup of an organism.

genus (jee′nus) A rank in taxonomic classification above the species.

germ cells Cells within the body that give rise to gametes.

germ layer Primitive embryonic tissue layer; endoderm, mesoderm, or ectoderm.

germ line In animals, the line of cells that will ultimately undergo meiosis to form gametes.

germination The resumption of growth in seeds or spores.

gibberellin (jib″ur-el′lin) A plant hormone involved in various aspects of plant growth and development.

gill The respiratory organ of aquatic animals, usually a thin-walled projection from the body surface or from some part of the digestive tract.

gland Body structure specialized for secretion.

glomerulus (glom-air′yoo-lus) A tuft of minute blood vessels: specifically, the knot of capillaries at the proximal end of a kidney tubule.

glucagon (gloo′kah-gahn) A pancreatic hormone that stimulates glycogenolysis, increasing the concentration of glucose in the blood.

glycogen (gly′koh-jen) A polysaccharide formed from glucose and stored primarily in liver and (to a lesser extent) muscle tissue; the major carbohydrate stored in animal cells.

glycolysis (gly-kol′ih-sis) The metabolic conversion of glucose into pyruvate with the production of ATP. Glycolysis is a metabolic pathway present in all living cells.

glycoprotein (gly′koh-proh′teen) A protein with covalently attached carbohydrates.

glyoxysome (gly-ox′ih-sohm) Microbody that contains a

large array of enzymes; prominent in the fatty endosperm tissue of germinating plant seeds.

Golgi complex (goal′jee); also called **Golgi body** or **Golgi apparatus**   One of numerous cell organelles found in the cytoplasm of eukaryotic cells. Its membranes modify and sort products of the endoplasmic reticulum.

gonad (goh′nad)   A gamete-producing gland; an ovary or testis.

gradualism   The idea that evolutionary change of a species is due to a slow, steady accumulation of changes over time.

grain   A simple, dry, indehiscent fruit in which the fruit wall is fused to the seed coat, making it impossible to separate the fruit from the seed. Corn kernels are grains.

grana (gran′ah)   Small membranous structures within chloroplasts that contain chlorophyll and are the site of the light-dependent reactions of photosynthesis.

gravitropism (grav″ih-troh′pizm)   Growth of a plant in response to gravity.

greenhouse effect   The warming of the earth resulting from the retention of atmospheric heat caused by the action of certain gases, especially carbon dioxide.

ground meristem   A primary meristem (embryonic tissue) that gives rise to cortex and pith in plants.

guanine (gwan′een)   See nucleic acid.

guard cell   A cell in the epidermis of plant stems and leaves. Two guard cells form a pore for gas exchange, collectively called a stomate.

guttation (gut-tay′shun)   The appearance of water droplets on leaves, forced out through leaf pores by root pressure.

gymnosperms (jim′noh-sperms)   Seed plants in which the seeds are not enclosed in an ovary; gymnosperms frequently bear their seeds in cones.

gynoecium (ji-nee′see-um)   The female portion (all the carpels) of a flower.

habituation (hab-it″yoo-ay′shun)   The process by which organisms become accustomed to a stimulus and cease to respond to it.

haploid (hap′loyd)   The chromosome number characteristic of gametes or spores; half the diploid number. In plants, the chromosome number of body cells of the gametophyte generation.

Hardy-Weinberg law   The principle that, regardless of dominance or recessiveness, the relative frequencies of allelic genes do not change from generation to generation.

hemizygous genes (hem″ih-zy′gus)   Genes carried on a heterokaryotic sex chromosome.

hemoglobin (hee′-moh-gloh″bin)   The red, iron-containing protein pigment of erythrocytes that transports oxygen and carbon dioxide and aids in regulation of pH.

hemophilia (hee″moh-feel′-ee-ah)   "Bleeder's disease"; hereditary disease in which blood does not clot properly; caused by a globulin deficiency affecting the production of thromboplastin.

herbaceous (er-bay′shus)   A plant with soft tissues only.

herbivore (erb′i-vore)   Animals which only consume plants or algae for their nutritional requirements.

hermaphrodite (her-maf′roh-dite)   An organism which can produce both male and female gametes.

heterochromatin (het″ur-oh-kroh′mah-tin)   Highly coiled and compacted chromatin in an inactive state.

heterocysts (het′ur-oh-sists″)   Nitrogenase-containing cells of certain cyanobacteria.

heterogamy (het″ur-og′ah-mee)   Reproduction involving the union of two gametes that differ in size and structure, such as the egg and sperm.

heterospory (het″ur-os′pur-ee)   Production of two types of spores in plants, microspores and megaspores.

heterothallic (het-ur-oh-thal′ik)   Pertaining to an organism having two mating types; only by combining a plus and a minus strain can sexual reproduction occur.

heterotrophs (het′ur-oh-trofes)   Organisms that cannot synthesize their own food from inorganic materials and therefore must live either at the expense of autotrophs or upon decaying matter.

heterozygote advantage   A phenomenon in which the heterozygous condition confers some special advantage on an individual that either homozygous condition does not, i.e., *Aa* has a higher degree of fitness than *AA* or *aa*.

hibernation   The dormant state of decreased metabolism in which certain animals pass the winter.

histones (his′tones)   Small, positively charged (basic) proteins in the nucleus which bind to the negatively charged DNA to form the nucleosomes.

holdfast   The structure for attachment to solid surfaces found in multicellular algae.

homeobox (home′ee-oh-box)   Specific DNA sequence found in many of the animal genes that are involved in controlling the development of the body plan.

homeostasis (home″ee-oh-stay′sis)   The balanced internal environment of the body; the automatic tendency of an organism to maintain such a steady state.

homeotherm (home′ee-oh-therm)   See endotherm.

homeotic gene (home″ee-ot′ik)   A gene that controls the formation of specific structures during development. Such genes were originally identified through mutants in which one body part was substituted for another.

hominid (hah′min-id)   Any of a group of both extinct and living humans.

hominoids   The apes and hominids.

homologous chromosomes (hom-ol′ah-gus)   Chromosomes that are similar in morphology and genetic constitution. In humans there are 23 pairs of homologous chromosomes, each containing one member from the mother and one member from the father.

homology (hom-ol′oh-jee)   Similarity in basic structural plan and development, which is assumed to reflect a common genetic ancestry.

homospory (hoh-mos′pur-ee)   Production of one type of spore in plants. The spore gives rise to a bisexual gametophyte.

homozygous (hoh″moh-zy′gus)   Possessing a pair of identical alleles.

hormone   A chemical messenger produced by an endocrine gland or by certain cells. In animals, hormones are usually transported in the blood and regulate some aspect of metabolism.

hybrid breakdown   A postzygotic isolating mechanism in which, although the interspecific hybrid is fertile and produces a second (F_2) generation, the F_2 has defects that prevent it from successfully reproducing.

hybrid inviability   A postzygotic isolating mechanism in which the embryonic development of an interspecific hybrid is aborted.

hybrid sterility   A postzygotic isolating mechanism in which the hybrid cannot reproduce successfully.

hydrocarbons   Organic compounds composed solely of hydrogen and carbon.

hydrogen bond   A weak bond between the hydrogen of one molecule and oxygen (usually) of another molecule; of primary importance in the structure of nucleic acids and proteins.

hydrolysis (hy-drol'ih-sis)   The splitting of a compound into parts by the addition of water between certain of its bonds, the hydroxyl group being incorporated into one fragment, and the hydrogen atom into the other.

hydrophilic   Attracted to water.

hydrophobic   In biology, the property of repelling water molecules.

hydroxyl (hy-drok'sil)   An ion or chemical group with the formula —OH.

hypertonic   Having a greater concentration of solute molecules and a lower concentration of solvent (water) molecules, and hence an osmotic pressure greater than that of the solution with which it is compared.

hypha (hy'fah)   One of the filaments composing the mycelium of a fungus.

hypocotyl (hy'poh-kah"tl)   The part of the axis of a plant embryo or seedling below the point of attachment of the cotyledons.

hypothalamus (hy-poh-thal'uh-mus)   Part of the brain that functions in regulating the pituitary gland, the autonomic system, emotional responses, body temperature, water balance, and appetite; located below the thalamus.

hypotonic   Having an osmotic pressure or solute content less that that of some standard of comparison.

impression   Fossil in which pressure and heat have destroyed all traces of organic material.

imprinting   A form of rapid learning by which a young bird or mammal forms a strong social attachment to an individual (usually a parent) or object within a few hours after hatching or birth.

inbreeding   Mating of genetically similar individuals. Homozygosity increases with each successive generation of inbreeding.

incomplete dominance   Condition in which neither member of a pair of contrasting alleles is completely expressed when the other is present.

incomplete flower   A flower lacking one or more of the four main parts: sepals, petals, stamens, and/or carpels.

indehiscent fruit (in"dih-his'ent)   A simple, dry fruit that does not split open at maturity; compare with dehiscent fruit.

independent assortment   The mutually independent inheritance of genes that are located on separate chromosome pairs.

index fossil   Certain key invertebrate fossils that are found in the same sedimentary layers in different geographical areas. Index fossils help geologists identify comparable layers in widely separate locations.

induced fit   Hypothesis that the active site of certain enzymes becomes conformed to the shape of the substrate molecule.

induction   Condition in which development of one group of cells is influenced by another group of cells.

inflammation   The response of body tissues to injury or infection, characterized clinically by heat, swelling, redness, and pain, and physiologically by increased vasodilation and capillary permeability.

innate behaviors   Behaviors that are inherited and typical of the species.

innate releasing mechanism   The hypothetical neural circuits that enable an animal to perceive sign stimuli and respond to them with appropriate muscular contractions.

insight learning   The intuitive solution to a problem that has not previously been attempted.

instinct   A genetically determined pattern of behavior or responses that is not based on the individual's previous experience.

integral proteins   Proteins that span or penetrate the lipid bilayer of cellular membranes.

integration   In biology, the process of sorting and interpreting neural impulses leading to an appropriate response.

integumentary system (in-teg"yoo-men'tur-ee)   The body's covering, including the skin and its nails, glands, hair, and other associated structures.

interferon (in"tur-feer'on)   A protein produced by animal cells when challenged by a virus. Important in immune responses, it prevents viral reproduction and enables cells of the same species to resist a variety of viruses.

intermediate filaments   Cytoplasmic fibers that are part of the cytoskeletal network and intermediate in size between microtubules and microfilaments.

interneuron (in"tur-noor'on)   A nerve cell that carries impulses to other nerve cells and is not directly associated with either an effector or a sense receptor.

internode   The section of a stem between two nodes.

interphase   The period in the life cycle of a cell in which there is no visible mitotic division; period between mitotic divisions.

interstitial fluid   The fluid that occupies the spaces between the cells of a tissue; tissue fluid.

intron   A non-protein-coding region of a eukaryotic gene and also of the RNA transcribed from such a region. Introns do not appear in mature mRNA (see exon).

invagination (in-vaj"ih-nay'shun)   The infolding of one part within another, specifically a process of gastrulation in which one region folds in to form a double-layered cup.

inversion, chromosomal   Turning a segment of a chromosome end for end and attaching it to the same chromosome.

ion   An atom or a group of atoms bearing an electric charge, either positive (cation) or negative (anion).

isogamy (eye-sog'ah-mee)   Reproduction resulting from

the union of two gametes that are similar in size and structure.

isomer (eye′soh-mur) One of two or more chemical compounds having the same chemical formula but a different structural formula, such as glucose and fructose.

isotonic (eye″soh-ton′ik) Having identical concentrations of solute and solvent molecules, and hence the same osmotic pressure as the solution with which it is compared.

isotope (eye′suh-tope) An alternate form of an element with a different number of neutrons but the same number of protons and electrons.

karyotype (kare′ee-oh-type) The chromosomal constitution of an individual. Representations of the karyotype are generally prepared by photographing the chromosomes and arranging the homologous pairs according to size and centromere position.

keratin (kare′ah-tin) A horny, water-insoluble protein found in the epidermis of vertebrates and in nails, feathers, hair, and horns.

kilocalorie See calorie.

kinesis (kih-nee′sis) The activity of an organism in response to a stimulus; the direction of the response is not controlled by the direction of the stimulus (in contrast to a taxis).

kinetochore (kin-eh′toh-kore) Portion of the chromosome centromere to which mitotic spindle fibers attach.

kingdoms The broadest category of classification commonly used.

Krebs cycle See citric acid cycle.

lamella (lah-mel′ah) A thin leaf or plate, as of bone.

larva An immature free-living form in the life history of some animals in which it may be unlike the parent.

larynx (lare′inks) The organ at the upper end of the trachea that contains the vocal cords.

latent learning Learning in which an animal stores (without apparent reward) information about its environment that can later influence its behavior.

lateral meristem An area of cell division located on the side of the plant. There are two lateral meristems, the vascular cambium and the cork cambium.

leaf primordium (pry-mor′dee-um) An embryonic leaf, evident in the stem apical meristem.

learning A change in the behavior of an animal that results from experiences during its lifetime.

legume (leg′yoom) A simple, dry, dehiscent fruit that splits open along two seams to release its seeds.

lenticels (len′tih-sels) Masses of cells that rupture the epidermis and form porous swellings in stems, facilitating the exchange of gases.

leucoplasts (loo′koh-plasts) Colorless plastids that act as centers for the storage of materials in the cytoplasm of certain kinds of plant cells.

lichens (ly′kenz) Compound organisms composed of symbiotic algae and fungi.

ligament (lig′uh-ment) A connective tissue cable or strap that connects bones to each other or holds other organs in place.

lignin (lig′nin) The substance responsible for the hard, woody nature of plant stems and roots.

linkage The tendency for a group of genes located on the same chromosome to be inherited together in successive generations.

lipase (lip′ase) Fat-digesting enzyme.

lipid Any of a group of organic compounds which are insoluble in water but soluble in fat solvents; lipids serve as a storage form of fuel and an important component of cell membranes.

lithosphere (lith′oh-sfeer) Portion of the earth that is composed of rock.

littoral (lit′or-ul) The region of shallow water near the shore between the high and low tide marks.

locus The particular point on the chromosome at which the gene for a given trait occurs.

long-day plant A plant that flowers in response to long days (and short nights); compare with short-day plant.

lumen (loo′men) The cavity or channel within a tube or tubular organ, such as a blood vessel or the digestive tract.

lymph (limf) The colorless fluid that is derived from blood plasma and resembles it closely in composition; contains white cells.

lymph nodes A mass of lymph tissue surrounded by a connective tissue capsule; manufactures lymphocytes and filters lymph.

lymphocyte (limf′oh-site) White blood cell with nongranular cytoplasm that is responsible for immune responses.

lysis (ly′sis) The process of disintegration of a cell or some other structure.

lysosome (ly′soh-some) Intracellular organelles present in many animal cells; contain a variety of hydrolytic enzymes which act when the lysosome ruptures or fuses with another vesicle. Lysosomes function in development and in phagocytosis.

macroevolution (mak″roh-eh-voh-loo′shun) Large-scale evolutionary change; evolutionary change involving higher taxa, such as genera and orders, i.e., above the level of species.

macromolecule Very large molecules such as a protein or nucleic acid.

macronutrient An essential element that is required in fairly large amounts for normal growth.

mantle Thin outside layer of mollusk body that is usually responsible for the production of the shell.

marsupials (mar-soo′pee-uls) A subclass of mammals, the Metatheria, characterized by the possession of an abdominal pouch in which the young are carried for some time after being born in a very undeveloped condition.

mass extinction The extinction of numerous species and higher taxa during a relatively short period of geological time. Compare with background extinction.

mechanical isolation A prezygotic isolating mechanism in which gamete exchange between two groups is prevented by morphological or anatomical differences between them.

medusa A jellyfish; a stage in the life cycle of certain cnidarians in which the organism is free-swimming and umbrella-shaped.

megaphyll (meg′uh-fil) A leaf that contains multiple vascular strands. Megaphylls are found in ferns, gymnosperms, and angiosperms. Compare with microphyll.

megasporangium (meg"ah-spor-an'jee-um) A spore sac containing megaspores.

megaspore (meg'ah-spor) The haploid spore in heterosporous plants that gives rise to a female gametophyte.

megaspore mother cell A diploid cell in the megasporangium that undergoes meiosis to form megaspores.

meiosis (my-oh'sis) Division of the cell nucleus that produces haploid cells; produces gametes in animals and spores in plants.

melanin (mel'ah-nin) A dark brown to black pigment common in the integument of many animals and sometimes found in other organisms.

menopause The period (usually from 45 to 55 years of age) in women when the recurring menstrual cycle ceases.

menstruation (men-stroo-ay'shun) The monthly discharge of blood and degenerated uterine lining in the human female; marks the beginning of each menstrual cycle.

meristem (mer'ih-stem) A localized area of mitosis and growth in the plant body.

mesoderm (mez'oh-derm) The middle layer of the three basic tissue layers that develop in the early embryo; gives rise to connective tissue, muscle, bone, blood vessels, kidneys, and many other structures; lies between the ectoderm and the endoderm.

mesophyll (mez'oh-fil) Photosynthetic cells in the interior of a leaf.

mesophytes (mez'oh-fites) Common land plants that live in a climate with an average amount of moisture.

messenger RNA (mRNA) RNA that has been transcribed from DNA that specifies the amino acid sequence of a protein in eukaryotes and prokaryotes.

metabolism The sum of all the chemical processes that occur within a cell or organism; the transformations by which energy and matter are made available for use by the organism.

metamorphosis (met"ah-mor'fuh-sis) An abrupt transition from one developmental stage to another, such as from a larva to an adult.

metaphase (met'ah-faze) The middle stage of mitosis and meiosis during which the chromosomes line up in the equatorial plate.

metastasis (me-tas'tuh-sis) The transfer of a disease such as cancer from one organ or part of the body to another not directly connected to it.

microbodies Membrane-bounded eukaryote cellular structures containing enzymes.

microevolution Changes in gene frequencies that occur within a population over successive generations.

microfilaments Tiny rod-like structures with contractile properties that make up part of the internal skeletal framework of the cell. Microfilaments are composed of actin.

micronutrient An essential element that is required in trace amounts for normal growth.

microphyll A leaf that contains one vascular strand; microphylls are found in horsetails and club mosses.

Compare with megaphyll.

microsporangia (my"kroh-spor-an'jee-ah) Small "pollen" sacs that contain microspore mother cells which divide by meiosis to form microspores.

microspore The haploid spore in heterosporous plants that gives rise to a male gametophyte.

microspore mother cell A diploid cell in the microsporangium that undergoes meiosis to form microspores.

microtubules (my-kroh-too'bewls) Hollow, cytoplasmic cylinders, composed mainly of tubulin protein, that compose such organelles as flagella and centrioles and serve as a skeletal component of the cell.

microvilli (my-kroh-vil'ee) Minute projections of the cell membrane which increase the surface area of the cell; found mainly in cells concerned with absorption or secretion, such as those lining the intestine or kidney tubules.

mimicry (mim'ik-ree) An adaptation for survival in which an organism resembles some other living or nonliving object.

mitochondria (my"toh-kon'dree-ah) Spherical or elongated intracellular organelles that contain the electron transport system and enzymes for the citric acid cycle. Sometimes referred to as the powerhouses of the cell.

mitosis (my-toh'sis) Division of the cell nucleus resulting in the distribution of a complete set of chromosomes to each daughter cell; cytokinesis (division of the cytoplasm) usually occurs during the telophase stage of mitosis. Mitosis consists of four phases: prophase, metaphase, anaphase, and telophase.

mitotic spindle (my-tot'ik) Structure consisting mainly of microtubules that provides the framework for chromosome movement during cell division.

mold 1. Fossil formed by the hardening of material surrounding a buried organism, followed by decay and removal of the tissues. 2. Multicellular fungi including the mildews, rusts, and mushrooms.

mole The amount of a chemical compound whose mass in grams is equivalent to its molecular weight, the sum of the atomic weights of its constituent atoms.

molecule The smallest particle of a covalently bonded element or compound that has the composition and properties of a larger part of the substance.

molting The shedding and replacement of an outer covering such as hair, feathers, or exoskeleton.

monerans (moh-nair'unz) The simplest prokaryotic microorganisms, the bacteria and cyanobacteria; forms lacking true nuclei or plastids and in which sexual reproduction is rare or absent.

monocot (mon'oh-kot) One of the two classes of angiosperms, or flowering plants. Monocots get their name (*mono* = one) from having one cotyledon in the seed.

monoecious (mon-ee'shus) Having separate male and female reproductive parts on the same plant; compare with dioecious.

monomer (mon'oh-mer) A simple molecule of a compound of relatively low molecular weight which can be linked with others to form a polymer.

monosaccharide (mon-oh-sak′ah-ride)   A simple hexose sugar; one that cannot be degraded by hydrolysis to a simpler sugar.

monotremes (mon′oh-treems)   Egg-laying mammals such as the duck-billed platypus of Australia.

morphogenesis (mor-foh-jen′eh-sis)   The development of the form and structures of the body and its parts by precise movements of its cells.

mosaic evolution   Two traits within a species that evolved independently of one another and at different rates.

Müllerian mimicry   Mimicry of one species by another, where both species are dangerous to some predator.

multiple alleles (al′leels)   Three or more alternative allelic genes which govern the same trait, as in the ABO series of blood types.

multiple fruit   A fruit that develops from many ovaries of many separate flowers to form one fruit. Pineapples are multiple fruits.

muscle   An organ which by contraction produces movement.

mutagen (mew′tah-jen)   Any agent that is capable of producing mutations.

mutation   A change in the nucleotide base pairs of a gene, or a rearrangement of genes within chromosomes so that their interactions produce different effects; a change in the chromosomes themselves.

mutualism   An association whereby two organisms of different species each gain from being together and are often unable to survive separately.

mycelium (my-seel′ee-um)   The vegetative body of fungi and certain protists (water molds); consists of a branched network of hyphae.

mycorrhizae (my″kor-rye′zee)   Mutualistic associations of fungi and plant roots that aid in the plant's absorption of materials from the soil.

myosin (my′oh-sin)   A protein which, together with actin, is responsible for muscle contraction.

nekton (nek′ton)   Collective term for free-swimming aquatic animals that are essentially independent of water movements such as currents.

nematocyst (nem-at′oh-cyst)   A minute stinging structure found within cnidocytes (stinging cells) in cnidarians; used for anchorage, defense, and capturing prey.

nephridium (neh-frid′ee-um)   The excretory organ of the earthworm and other annelids which consists of a ciliated funnel, opening into the next anterior coelomic cavity and connected by a tube to the outside of the body.

nephron (nef′ron)   The functional, microscopic unit of the vertebrate kidney.

neritic zone (ner-ih′tik)   Zone of shallow water adjacent to the seacoast.

nerve   A large bundle of axons (or dendrites) wrapped in connective tissue that conveys impulses between the central nervous system and some other part of the body.

net primary productivity   Total carbon fixation by a plant community, minus the carbon lost by respiration.

neuroglia (noor-og′lee-ah)   Connecting and supporting cells in the central nervous system.

neuron (noor′on)   A nerve cell; a conducting cell of the nervous system which typically consists of a cell body, dendrites, and an axon.

neutrons (noo′tronz)   Electrically uncharged particles of matter existing along with protons in the atomic nucleus.

nitrogen fixation   The ability of certain microorganisms to bring atmospheric nitrogen into chemical combination.

node   The area on a plant stem where the leaves attach.

noncompetitive inhibitor   A substance that permanently destroys the ability of an enzyme to function.

nondisjunction   Abnormal separation of homologous chromosomes or sister chromatids caused by their failure to disjoin (move apart) properly during cell division.

nonpolar covalent bond   Chemical bond in which electrons are shared equally among the participating atoms and which does not produce any electrical charge within the molecule.

notochord (no′toe-kord)   The flexible, longitudinal rod in the anteroposterior axis that serves as an internal skeleton in the embryos of all chordates and in the adults of some.

nuclear envelope   The double membrane system that surrounds the cell nucleus of eukaryotes.

nuclear pores   Structures in the nuclear envelope that appear to allow passage of certain molecules between the cytoplasm and the nucleus.

nucleic acid (noo-klay′ik)   DNA or RNA. A polymer composed of nucleotides that contain the purine bases adenine and guanine and/or the pyrimidine bases cytosine, thymine, and uracil.

nucleolus (new-klee′-oh-lus)   Specialized structure in the nucleus formed from regions of several chromosomes; site of ribosome synthesis.

nucleosome (new′klee-oh-sohm)   Repeating unit of chromatin structure consisting of a length of DNA wound around a complex of eight histone molecules (two of each of four different types) plus a DNA linker region associated with a fifth histone protein.

nucleotide (noo′klee-oh-tide)   A molecule composed of a phosphate group, a five-carbon sugar (ribose or deoxyribose) and a nitrogenous base (purine or pyrimidine); one of the subunits of nucleic acids.

nucleus (new′klee-us)   (1) That portion of an atom that contains the protons and neutrons; the core; (2) a cellular organelle containing DNA and serving as the control center of the cell; (3) a mass of nerve cell bodies in the central nervous system.

nutrients   The chemical substances in food that are used by the body as components for synthesizing needed materials and for fuel.

nymph   A juvenile insect that often resembles the adult stage and that will become an adult without an intervening pupal stage.

obligate anaerobe (ob′lih-gate an′air-obe)   An anaerobic organism that is killed by oxygen.

omnivore (om′nih-vore)   An animal which consumes both plant and animal material.

oncogene (on'koh-jeen) Any of a number of genes that usually play an essential role in cell growth or division and that cause the formation of a cancer cell when mutated; also known as cellular oncogenes.

ontogeny (on-toj'uh-nee) The complete developmental history of the individual organism.

oogamy (oh-og'ah-me) The fertilization of a large non-motile female gamete by a small, motile male gamete.

oogenesis (oh"oh-jen'eh-sis) Production of female gametes (eggs).

operant (instrumental) conditioning A type of learning in which an animal is rewarded or punished for performing a behavior it discovers by chance.

operator site The site on DNA to which repressor molecules are bound, thereby inhibiting the synthesis of mRNA by the genes in the adjacent operon; adjacent to the structural genes in the operon.

operon (op'er-on) In prokaryotes, a group of structural genes that are transcribed as a single message plus their associated regulatory elements. An operon is controlled by a single repressor.

optimal foraging The theory that animals feed in a manner that maximizes benefits and/or minimizes costs.

orbital Any one of the permissible patterns of motion of an electron in an atom or molecule.

order In taxonomic classification, a group of related, similar families.

organ A differentiated part of the body made up of tissues and adapted to perform a specific function or group of functions, such as the heart or liver.

organelle One of the specialized structures within the cell, such as the mitochondria, Golgi complex, ribosomes, or contractile vacuole.

organogenesis The development of organs.

osmoregulation (oz"moh-reg-yoo-lay'shun) The active regulation of the osmotic pressure of body fluids so that they do not become excessively dilute or excessively concentrated.

osmosis (oz-moh-sis) Diffusion of water (the principal solvent in biological systems) through a selectively permeable membrane from a region of higher concentration of water to a region of lower concentration of water.

osmotic pressure A measure of the solute concentration of a solution.

osteocyte (os'tee-oh-site) A mature bone cell; an osteoblast that has become embedded within the bone matrix and occupies a lacuna.

osteon (os'tee-on) Spindle-shaped unit of bone composed of concentric layers of osteocytes; Haversian system of bone.

ovary (oh'var-ee) (1) In animals, one of the paired female gonads; responsible for producing eggs and sex hormones; (2) in flowering plants, the base of the carpel that contains ovules. Ovaries develop into fruits after fertilization.

oviduct (oh'vih-dukt) Tube that carries ova from the ovary to the uterus, cloaca, or body exterior.

oviparous (oh-vip'ur-us) A type of development in which the young hatch from eggs laid outside the mother's body.

ovoviviparous (oh"voh-vih-vip'ur-us) A type of development in which the young hatch from eggs incubated within the mother's body.

ovule (ov'yool) The part which develops into the seed after fertilization.

ovum The female gamete, or egg.

oxidation The loss of electrons, or in organic chemistry, the loss of hydrogen from a compound.

oxidative phosphorylation (fos"for-ih-lay'shun) The production of ATP using energy derived from the transfer of electrons in the electron transport system of the mitochondria.

paedomorphosis (pee"doh-mor-foh'sis) Evolutionary retention of characteristics in the adult juvenile.

palisade mesophyll The vertically stacked photosynthetic cells near the upper epidermis in dicot leaves.

paramylon (par"uh-my'lon) Carbohydrate storage compound present in euglenoids; chemically distinct from both starch and glycogen.

parapodia (par"uh-poh'dee-ah) Paired, thickly bristled paddles extending laterally from each segment of polychaete worms.

parasitism An intimate relationship between organisms of two different species in which one benefits and the other is harmed.

parasympathetic A division of the autonomic nervous system concerned primarily with the control of the internal organs; functions to conserve or restore energy.

parenchyma (par-en'kih-mah) Plant cells that are relatively unspecialized, are thin-walled, may contain chlorophyll, and are typically rather loosely packed; they function in photosynthesis and in the storage of nutrients.

parthenogenesis (par"theh-noh-jen'eh-sis) The development of an unfertilized egg into an adult organism; common among honey bees, wasps, and certain other arthropods.

pathogen (path'oh-gen) An organism capable of producing disease.

pelagic (pel-aj'ik) An organism that inhabits open water, as in mid-ocean.

pellicle (pel'ih-kl) A flexible, proteinaceous covering over the body of certain protists.

penis The male sexual organ of copulation in reptiles and mammals.

peptide (pep'tide) A compound consisting of two or more amino acids.

peptide bond A distinctive covalent carbon-to-nitrogen bond that links amino acids in peptides and proteins.

peptidoglycan (pep"tid-oh-gly'kan) A modified protein or peptide possessing an attached carbohydrate.

perennial (pur-en'ee-ul) A plant that grows year after year. Perennials may be woody or herbaceous.

pericycle (pehr'eh-sy'kl) A layer of meristematic cells in roots that gives rise to branch roots.

periderm (pehr'ih-durm) Layers of cells covering the surface of woody stems and roots (i.e., the outer bark). Anatomically, the periderm is composed of cork cells, cork cambium, and cork parenchyma, along with traces of primary tissues.

period In geology, geological time that is a subdivision of an era. Each period is divided into epochs.

peripheral nervous system The nerves and receptors that lie outside the central nervous system.

peristalsis (pehr″ih-stal′sis) Powerful, rhythmic waves of muscular contraction and relaxation in the walls of hollow tubular organs, such as the ureter or parts of the digestive tract, that serve to move the contents through the tube.

peroxisomes (pehr-ox′ih-somz) Lysosome-like vesicles containing enzymes that produce or degrade hydrogen peroxide.

petals The colored cluster of modified leaves that constitute the next-to-outermost portion of a flower.

petiole (pet′ee-ohl) The part of a leaf that attaches to a stem.

petrifaction A fossil in which the soft tissues of the organism are replaced by minerals such as iron pyrites, silica, and calcium.

pH The negative logarithm of the hydrogen ion concentration by which the degree of acidity or alkalinity of a fluid may be expressed.

phagocytosis (fag″oh-sy-toh′sis) Literally, "cell-eating"; a type of endocytosis by which certain cells engulf food particles, microorganisms, foreign matter, or other cells.

pharynx (far′inks) That part of the digestive tract from which the gill pouches or slits develop; in higher vertebrates it is bounded anteriorly by the mouth and nasal cavities and posteriorly by the esophagus and larynx; the throat region in humans.

phenotype (fee′noh-type) The physical or chemical expression of an organism's genes (see also genotype).

pheromone (feer′oh-mone) A substance secreted to the external environment by one organism that influences the development or behavior of other members of the same species.

phloem (floh′em) Vascular tissue that conducts food in plants.

phospholipids (fos″foh-lip′idz) Lipids similar to triglycerides in which a phosphorus-containing group occurs in place of one of the fatty acids. Phospholipids compose most of the plasma and internal membranes of cells.

phosphorylation (fos″for-ih-lay′shun) The introduction of a phosphate group into an organic molecule.

photic zone (foh′tik) Zone of aquatic habitats lying near enough to the surface that sufficient light is present for photosynthesis to take place.

photolysis (foh-tol′uh-sis) The photochemical splitting of water in the light-dependent reactions of photosynthesis.

photon (foh′ton) A particle of electromagnetic radiation; one quantum of radiant energy.

photoperiodism (foh″toh-peer′ee-od-izm) The physiological response of animals and plants to variations of light and darkness.

photophosphorylation (foh″toh-fos-for-ih-lay′shun) The production of ATP in photosynthesis.

photoreceptor (foh′toh-ree-sep′tor) (1) A sense organ specialized to detect light; (2) a pigment that absorbs light before triggering a physiological response.

photorespiration (foh″toh-res-pur-ay′shun) The production of carbon dioxide and consumption of oxygen during photosynthesis at high light intensities by C-3 plants.

photosynthesis (foh″toh-sin′thuh-sis) The production of organic materials, especially glucose, from carbon dioxide and water using the energy of light. Photosynthesis is practiced by plants, some protists, and several kinds of bacteria.

photosystem A group of chlorophyll and other molecules located in the thylakoid membrane (in photoautotrophic eukaryotes) which emits electrons in response to light.

phototropism (foh″toh-troh′pizm) The growth response of an organism to the direction of light.

phycocyanin (fy″koh-sy′ah-nin) A blue chromoprotein found in cyanobacteria.

phylogeny (fy-loj′en-ee) The complete evolutionary history of a group of organisms.

phylum (fy′lum) A taxonomic grouping of related, similar classes; a category beneath the kingdom and above the class. Phyla are used in classifying animals; divisions are used for plants.

phytochrome (fy′toh-krome) A blue-green, proteinaceous pigment that is the photoreceptor for a wide variety of physiological responses, including initiation of flowering in certain plants.

phytoplankton (fy″toh-plank′tun) Microscopic floating algae and plants which are distributed throughout oceans or lakes; autotrophic plankton.

pili (pil′ee); sing. **pilus** Hair or hair-like structures; especially, the external hair-like filaments of bacteria.

pinocytosis (pin″oh-sy-toh′sis) Cell-drinking; the engulfing and absorption of droplets of liquids by cells.

pith Large, thin-walled parenchyma cells found as the innermost tissue in many plants.

placenta (plah-sen′tah) The partly fetal and partly maternal organ whereby materials are exchanged between fetus and mother in the uterus of placental mammals.

plankton Free-floating, mainly microscopic aquatic organisms found in the upper layers of the water; includes phytoplankton, which are photosynthetic organisms, and zooplankton, which are heterotrophic organisms.

plasma cells Cells that secrete antibodies; differentiated B lymphocytes.

plasma membrane A living, functional part of the cell through which all nutrients entering the cell and all waste products or secretions leaving it must pass; the surface membrane of the cell which acts as a selective barrier to passage of molecules and ions into the cell.

plasmids (plaz′midz) Small circular DNA molecules that carry genes separate from the main bacterial chromosome.

plasmodium (plaz-moh′dee-um) (1) Multinucleate, ameboid mass of living matter that constitutes the vegetative phase of the life cycle of slime molds; (2) a single-celled organism that reproduces by spore formation and causes malaria.

plasmolysis (plaz-mol'ih-sis)   The shrinkage of cytoplasm and the pulling away of the plasma membrane from the cell wall when a plant cell (or other walled cell) loses water, usually after being placed in a hypertonic environment.

plastids (plas'tidz)   A family of membrane-bounded organelles occurring in photosynthetic eukaryote cells; examples are chloroplasts and amyloplasts.

platelets (playt'lets)   Cell fragments in the blood that function in clotting.

pleiotropic gene (ply"oh-troh'pik)   A gene that affects a number of different characteristics in a given individual.

ploidy (ploy'dee)   Relating to the number of sets of chromosomes in a cell.

plumule (ploom'yool)   The embryonic shoot of a seed plant.

poikilotherm (poi"kil-oh-thurm)   See ectotherm.

point mutation   Mutation involving a single nucleotide change in DNA.

polar covalent bond   A chemical bond established by electron sharing which produces some difference in the charge of the ends of the molecule.

polar nucleus   One of two haploid cells that fuse with a sperm nucleus during double fertilization in angiosperms.

polar transport   The unidirectional movement of the plant hormone, auxin, from the stem tip to the roots.

pollen   The male gametophytes of seed plants which produce haploid nuclei capable of fertilization.

pollination   In seed plants, the transfer of pollen from the male to the female part of the plant.

polygenes (pol'ee-jeens")   Two or more pairs of genes that affect the same trait in an additive fashion.

polymer (pol'ih-mer)   A molecule built up from repeating units of the same general type, such as a protein, nucleic acid, or polysaccharide.

polymorphism (pol"ee-mor'fizm)   (1) The existence of two or more phenotypically different individuals within the same species; (2) the presence of more than one allele for a given locus.

polypeptide   A chain of many amino acids linked by peptide bonds.

polyploidy (pol'ee-ploy"dee)   Possession of more than two sets of chromosomes per nucleus.

polyps (pol'ips)   Hydra-like animals; the sessile stage of the life cycle of certain cnidarians.

polyribosomes   A complex consisting of a number of ribosomes attached to an mRNA molecule during translation; also known as polysomes.

polysaccharide (pol-ee-sak'ah-ride)   A carbohydrate consisting of many monosaccharide units; examples are starch, glycogen, and cellulose.

polyunsaturated fat (pol"ee-un-sat'yur-ay-ted)   A fat containing fatty acids that have double bonds and are not fully saturated with hydrogen.

postzygotic isolating mechanism (post'zy-got'ik)   A mechanism that restricts gene flow between species and ensures reproductive failure even though fertilization has taken place.

potassium (K⁺) ion mechanism   Mechanism by which plants open and close their stomates. The influx of potassium ions into the guard cells causes water to move in by osmosis, changing the shape of the guard cells and opening the pore.

preadaptation   An evolutionary change in an existing biological structure that enables it to have a different function.

predation   Relationship in which a species kills and devours other animals.

prehensile (pree"hen'sil)   Adapted for grasping by wrapping around an object, as in a prehensile tail.

prezygotic isolating mechanism (pree"zy-got'ik)   A mechanism that restricts gene flow between species by preventing mating from taking place.

primary growth   An increase in the length of a plant. This growth occurs at the tips of the stems and roots due to the activity of apical meristems.

primary succession   Ecological succession taking place in an environment which has not previously supported a community.

producers   Organisms, such as plants, that produce food materials from simple inorganic substances.

prokaryote (pro-kar'ee-ote)   Organisms that lack membrane-bounded nuclei and other membrane-bounded organelles; the bacteria and cyanobacteria.

promoter   A recognition signal encoded in DNA that functions to initiate transcription.

prophase   The first stage in mitosis, during which the chromatin threads condense, distinct chromosomes become evident, and a spindle forms.

protective coloration   The coloring of an organism so that it blends into its surroundings in such a way that it is difficult to see.

protein   A large, complex organic compound composed of chemically linked amino acid subunits; contains carbon, hydrogen, oxygen, nitrogen, and sulfur; proteins are the principal structural constituents of cells.

protist (proh'tist)   One of a vast assemblage of eukaryotic organisms, primarily single-celled or simple multicellular, mostly aquatic.

protobiont (proh"toh-by'ont)   Assemblages of organic polymers that spontaneously form during certain conditions. Protobionts may have been involved in chemical evolution.

proton   A basic physical particle present in the nuclei of all atoms that has a positive electrical charge and a mass of 1; a hydrogen ion consists of a single proton.

protoplast   Plant, fungal, or bacterial cell without its cell wall. Protoplasts are produced by enzymatically digesting the cell wall.

protostome (proh'toh-stome)   Major division of the animal kingdom in which the blastopore develops into the mouth, and the anus forms secondarily; includes the annelids, arthropods, and mollusks.

protozoa (proh"toh-zoh'a)   Single-celled, animal-like protists, including amoebas, ciliates, flagellates, and sporozoans.

proximal   Relatively near the body center.

pseudocoelom (soo″doh-see′lom) A body cavity between the mesoderm and endoderm; derived from the blastocoele.

pseudoplasmodium The aggregation of cells for reproduction in cellular slime molds.

pseudopod A temporary extension of an amoeboid cell, which the cell uses for feeding and locomotion.

punctuated equilibrium The concept that evolution proceeds with periods of inactivity (i.e., periods of little or no change within a species) followed by very active phases, so that major adaptations or clusters of adaptations appear suddenly in the fossil record.

pupa (pew′pah) A stage in the development of an insect, between the larva and the adult; a form that neither moves nor feeds and may be in a cocoon.

purines (pure′eenz) Nitrogenous bases with carbon and nitrogen atoms in two interlocking rings; components of nucleic acids, ATP, NAD$^+$, and other biologically active substances. Examples are adenine and guanine.

pyrimidines (pyr-im′ih-deenz) Nitrogenous bases composed of a single ring of carbon and nitrogen atoms; components of nucleic acids. Examples are thymine, cytosine, uracil.

quadrupedal Walking on all fours.

quantum A unit of radiant energy; the amount of energy emitted or absorbed by atoms or molecules.

radicle (rad′ih-kl) The embryonic root of a seed plant.

range The portion of the earth in which a particular species occurs.

receptacle In botany, the end of a flower stalk where the floral parts are attached.

receptor (1) A specialized sensory neural structure that is excited by a specific type of stimulus; (2) a site on the cell surface specialized to combine with a specific substance such as a hormone or transmitter substance.

recessive genes Genes not expressed in the heterozygous state.

recombinant DNA Any DNA molecule made by combining genes from different organisms.

redirected behaviors Innate, stereotyped behaviors that are directed toward a substitute object.

redox reactions (ree′dox) Chemical reactions in which one substance is oxidized and another reduced; involves the transfer of one or more electrons from one reactant to another.

reduction In chemistry, the gain of electrons by a substance or the chemical addition of hydrogen; the opposite of oxidation.

reflex An inborn, automatic, involuntary response to a given stimulus which is determined by the anatomic relations of the involved neurons; generally functions to restore homeostasis.

regulator genes Special genes that provide codes for the synthesis of repressor or activator proteins.

releaser A stimulus that triggers an unlearned behavior; a communication signal between members of a species.

repressor The protein substance produced by a regulator gene that represses protein synthesis in a specific gene.

reproductive isolation The reproductive barriers that prevent a species from interbreeding with another species. As a result, each species' gene pool is isolated from other species.

respiration (1) Cellular respiration is the process by which cells conserve the energy of food molecules in biologically useful forms, such as ATP; (2) organismic respiration is the act or function of gas exchange.

resting potential The membrane potential (difference in electrical charge) of an inactive neuron (about −70 millivolts).

retrovirus (ret′roh-vy″rus) An RNA virus that produces a DNA intermediate in its host cell.

reverse transcriptase Enzyme produced by retroviruses to enable the transcription of DNA from the viral RNA in the host cell.

rhizoids (ry′zoids) Colorless, hair-like absorptive filaments analogous to roots that extend from the base of the stem of mosses, liverworts, and fern prothallia.

rhizome (ry′zome) A horizontal underground stem that gives rise to above-ground leaves.

ribonucleic acid (RNA) A family of single-stranded nucleic acids that function mainly in protein synthesis.

ribosomes (ry′boh-sohms) Organelles that are part of the protein synthesis machinery; consist of a larger and a smaller subunit, each composed of ribosomal RNA (rRNA) and ribosomal proteins.

RNA polymerase Family of enzymes that catalyze the synthesis of RNA molecules from DNA templates.

ribozyme (ry′boh-zime) A molecule of RNA that has catalytic ability.

rickettsia (rih-ket′see-uh) A type of disease organism intermediate in size and complexity between a virus and a bacterium; parasitic within cells of insects and ticks; transmitted to humans by the bite of an infected insect or tick.

ritualization The modification of a behavior pattern, through evolution, to serve a communicative function.

root cap A covering of cells over the root tip that protects the delicate meristematic tissue directly behind it.

root hair An extension of an epidermal cell in roots. Root hairs increase the absorptive capacity of roots.

rough ER Major division of the endoplasmic reticulum that contains ribosomes and functions in protein synthesis; compare with smooth ER.

saprobic nutrition (sap-roh′bik) A type of heterotrophic nutrition in which organisms absorb their required nutrients from nonliving organic material.

sarcolemma (sar″koh-lem′mah) The muscle cell plasma membrane.

sarcomere (sar′koh-meer) A segment of a striated muscle cell located between adjacent Z-lines that serves as a unit of contraction.

sarcoplasmic reticulum System of vesicles in a skeletal or cardiac muscle cell which surrounds the myofibrils and releases calcium in muscle contraction; a modified endoplasmic reticulum.

saturated fatty acid A fatty acid with no double bonds

between adjacent carbon atoms. It is completely saturated with hydrogen.

savannah   A tropical or subtropical grassland containing scattered trees.

sclerenchyma (skler-en′kim-uh)   Cells that provide strength and support in the plant body. Schlerenchyma cells are dead at maturity and have extremely thick walls.

secondary growth   An increase in the width of a plant due to the activity of lateral meristems (vascular cambium and cork cambium).

seed   A plant reproductive body that is composed of a young, multicellular plant and nutritive tissue (food).

selectively permeable membrane   A membrane that allows some substances to cross it more easily than others. Biological membranes are generally permeable to water but restrict the passage of many solutes.

semen   Fluid composed of sperm suspended in various glandular secretions that is ejaculated from the penis during orgasm.

senescence   The aging process.

sensitization   An increased response by an animal to a stimulus that has been presented before.

sepals (see′puls)   The outermost parts of a flower, usually leaf-like in appearance, that protect the flower as a bud.

sessile (ses′sile)   Permanently attached to one location. Coral animals, for example, are sessile.

sex-linked genes   Genes borne on a sex chromosome. In mammals almost all sex-linked genes are borne on the X-chromosome.

sexual dimorphism (dy-mor′fizm)   Difference in body proportions, coloring, or other characteristics in the two sexes of a species.

short-day plant   A plant that flowers in response to short days and long nights; compare with long-day plant.

sieve tube member   The cell that conducts food in the phloem of plants.

signal sequence   A sequence of amino acids in a protein that indicates that it is to be translocated across the endoplasmic reticulum.

sign stimulus   Any stimulus that elicits an innate response in an animal.

simple fruit   A fruit that develops from a single ovary of a single flower.

siphonous (sy′fun-us)   A type of body form that is tubular and coenocytic; found in certain algae.

skeletal muscle   Voluntary or striated muscle of vertebrates, so called because it usually is directly or indirectly attached to some part of the skeleton.

smooth ER   Portion of the endoplasmic reticulum that has no ribosomes and thus appears smooth; produces steroids and other lipids; compare with rough ER.

sodium-potassium pump   Cellular active transport mechanism that transports sodium out of, and potassium into, cells.

solute (sol′yoot)   The dissolved substance in a solution.

solvent   A liquid substance, such as water, in which other materials may be dissolved.

somatic cell   A cell of the body not involved in sexual reproduction.

somatic nervous system   That part of the nervous system that keeps the body in adjustment with the external environment; includes the sensory receptors on the body surface and within the muscles, and the nerves that link them with the central nervous system.

sorus (soh′rus)   A cluster of sporangia (in the ferns).

speciation   Evolution of a new species.

species   A group of organisms with similar structural and functional characteristics that in nature breed only with each other and have a close common ancestry; a group of organisms with a common gene pool.

specific heat   The amount of heat required to raise 1 gram of a substance 1°C.

sperm   The motile, haploid male reproductive cell of animals and some plants and protists; spermatozoa.

spermatogenesis (spur″mah-toh-jen′eh-sis)   The production of sperm by meiosis.

spermatozoa (spur-mah-toh-zoh′uh)   Mature sperm cells.

sperm nucleus   A nonflagellated male reproductive cell. Sperm nuclei are produced by angiosperms and gymnosperms.

S-phase   Phase in interphase of cell cycle during which DNA and other chromosomal components are synthesized.

sphincter (sfink′tur)   A group of circularly arranged muscle fibers, the contractions of which close an opening, such as the pyloric sphincter at the end of the stomach.

spindle   The intracellular apparatus, composed of microtubules, that separates chromosomes in cell division of eukaryotes.

spine   A leaf that is modified for protection, such as a cactus spine; compare with thorn.

spiracle (speer′ih-kl)   An opening for gas exchange, such as the opening on the body surface of a trachea in insects.

spongy mesophyll (mes′oh-phil)   The irregularly arranged, photosynthetic tissue closest to the lower epidermis in leaves.

sporangium (spor-ran′jee-um)   A spore case, found in plants and certain protists and fungi.

spore   A reproductive cell that gives rise to individual offspring in plants, algae, fungi, and certain protozoa.

spore mother cell   A diploid cell that undergoes meiosis to form haploid spores in plants.

sporophyll (spor′oh-fil)   A leaf-like structure that bears spores.

sporophyte (spor′oh-fite)   The diploid portion of a plant life cycle which produces spores by meiosis.

stabilizing selection   Natural selection that acts against extreme phenotypes and favors intermediate variants; associated with a population well adapted to its environment.

stamen (stay′men)   The male part of flowers that produces pollen.

steroids (steer′oids)   Complex molecules containing carbon atoms arranged in four interlocking rings, three of which contain six carbon atoms each and the fourth of which contains five; the male and female sex hormones and the adrenal cortical hormones of vertebrates are examples, as is the insect hormone ecdysone. Steroids are chemical derivatives of cholesterol.

stigma That portion of the carpel where the pollen lands prior to fertilization.

stimulus A physical or chemical change in the internal or external environment of an organism potentially capable of provoking a response.

stolon An above-ground, horizontal stem with long internodes. Stolons often form buds that develop into separate plants.

stomate Small pore flanked by specialized cells (i.e., guard cells) which are located in the epidermis of land plants; stomata allow for gas exchange necessary for photosynthesis.

strobilus (stroh′bil-us) A cone-like structure that bears sporangia.

stroma The matrix of the chloroplast which surrounds the grana.

stromatolite (stroh-mat′oh-lite) A column-like rock that is composed of many minute layers of prokaryotic cells, usually cyanobacteria. Some stromatolites are over 3 billion years old.

style The neck connecting the stigma to the ovary of a carpel.

suberin (soo′ber-in) A waterproof material found in plants which occurs in the covering of leaf scars, in cork cells, and in the Casparian strip of endodermal cells.

substrate A substance on which an enzyme acts; a reactant in an enzymatically catalyzed reaction.

supraorbital ridges (soop″rah-or′bit-ul) Prominent bony ridges above the eye sockets. Ape skulls have prominent supraorbital ridges.

suspensor In plant embryo development, a multicellular structure that anchors the embryo and aids in nutrient absorption from the endosperm.

symbiosis (sim-bee-oh′sis) An intimate relationship between two or more organisms of different species.

sympathetic nervous system A subdivision of the autonomic nervous system; its general effect is to mobilize energy, especially during stress situations; prepares the body for fight-or-flight response.

sympatric speciation (sim-pat′rik) The evolution of a new species within the same geographical region as the parent species.

synapse (sin′aps) The junction between two neurons or between a neuron and an effector.

synapsis (sin-ap′sis) The pairing of homologous chromosomes during Prophase I of meiosis.

syngamy (sin′gah-mee) Sexual reproduction; the union of the gametes in fertilization.

tap root A root system in plants that has one main root with smaller roots branching off it.

taxis (tak′sis) An orientation movement of a motile organism's response to a stimulus.

taxonomy (tax-on′ah-mee) The science of naming, describing, and classifying organisms.

telophase (teel′oh-faze or tel′oh-faze) The last stage of mitosis and meiosis when, having reached the poles, the chromosomes become decondensed and a nuclear envelope forms around each group.

temporal isolation A prezygotic isolating mechanism in which genetic exchange is prevented between two groups because they reproduce at different times of the day, season, or year.

tendon A connective tissue structure that joins a muscle to another muscle, or a muscle to a bone. Tendons transmit the force generated by a muscle.

tendril A leaf or stem that is modified for holding or attaching onto objects.

termination codon Any codon in mRNA that does not code for an amino acid (UAA, UAG, and UGA). This stops the translation of a peptide at that point. Also called stop codon.

territoriality Behavior pattern in which one organism (usually a male) delineates a territory of its own and defends it against intrusion by other members of the same species and sex.

tertiary structure (tur′she-air″ee) The three-dimensional shape of a protein that forms spontaneously as a result of interactions of side chains.

testis (tes′tis) The male gonad that produces spermatozoa; in humans and certain other mammals the testes are situated in the scrotal sac.

tetrad A group of four homologous chromatids formed during synapsis in the first meiotic prophase.

therapsids (ther-ap′sids) A group of mammal-like reptiles of the Permian period; gave rise to the mammals.

thermodynamics (thurm″oh-dy-nam′iks) Principles governing heat or energy transfer.

thigmotropism (thig″moh-troh′pizm) Plant growth in response to contact with a solid object, such as plant tendrils.

thorax (1) The upper body of vertebrates. (2) The second major division of the arthropod body.

thorn A stem that is modified for protection; compare with spine.

threshold The value at which a stimulus just produces a sensation, is just appreciable, or comes just within the limits of perception.

thylakoids (thy′lah-koids) Stacks of flat membranous sacs inside the chloroplast where light energy is converted into ATP and NADPH used in carbohydrate synthesis.

thymine (thy′meen) See nucleic acid.

tissue A group of closely associated, similar cells that work together to carry out specific functions.

tonus (toh′nus) The continuous partial contraction of muscle.

trachea (tray′kee-uh) (1) Principal thoracic air duct of terrestrial vertebrates; (2) one of the microscopic air ducts ramifying throughout the body of most terrestrial arthropods and some terrestrial mollusks.

tracheids (tray′kee-idz) A type of water-conducting cell in the xylem of plants.

transcription The synthesis of RNA from a DNA template.

transduction   The transfer of a genetic fragment from one cell to another, e.g., from one bacterium to another, by a virus.

transfer RNA (tRNA)   A form of RNA composed of about 70 nucleotides which serves in the synthesis of proteins. An amino acid is bound to a specific kind of tRNA and then arranged in order by the complementary pairing of the nucleotide triplet (codon) in mRNA and the triplet anticodon of tRNA.

transformation   (1) The incorporation of genetic material by a cell that causes a change in its phenotype; (2) the conversion of a normal cell to a malignant cell.

translation   Conversion of information provided by mRNA into a specific sequence of amino acids in the production of a polypeptide chain; the information in the mRNA is translated into a certain kind of protein.

translocation   (1) The movement of materials (water, dissolved minerals, dissolved food) in the vascular tissues of a plant; (2) chromosome abnormality in which part of one chromosome has become attached to another.

transmission, neural   Conduction of a neural impulse along a neuron or from one neuron to another.

transpiration   Evaporation of water from the leaves of a plant; aids in drawing water up the stem.

transposon (tranz-poze′on)   A DNA segment that is capable of moving from one chromosome to another or to different sites within the same chromosome.

triacylglycerol (try-as″il-glis′-er-ol)   The most common form of fat consisting of three fatty acid chains chemically linked with a glycerol. Also called triglyceride.

trichome (trik′ome)   A hair or other appendage growing out from the epidermis of plants.

trilobite (try′loh-bite)   Marine arthropods of the Paleozoic era characterized by two dorsal longitudinal furrows that separated the body into three lobes.

triplet   A sequence of three nucleotides that serve as the basic unit of genetic information, usually signifying the identity and position of an amino acid unit in a protein.

trophic level   The distance of an organism in a food chain from the primary producers of a community.

tropism (troh′pizm)   A growth response in plants that is elicited by an external stimulus.

tubers   Thickened underground stems that are adapted for food storage; found in plants such as the white potato.

tubulin (toob′yoo-lin)   The protein dimers from which microtubules are constructed by cells.

tumor   Mass of tissue that is growing in an uncontrolled manner; a neoplasm.

turgor pressure (tur′gor)   Hydrostatic pressure that develops within a walled cell, such as a plant cell, when the osmotic pressure of the cell's contents is greater than the osmotic pressure of the surrounding fluid.

ubiquinone (yoo-bik′kwin-ohn)   Coenzyme Q, a component of the electron transport system which can take up and release electrons.

uracil (yur′ah-sil)   See nucleic acid.

urea (yur-ee′ah)   The principal nitrogenous excretory product of mammals; one of the water-soluble end products of protein metabolism.

uric acid (yoor′ik)   The principal nitrogenous excretory product of insects, birds, and reptiles; a relatively insoluble end product of protein metabolism; also occurs in mammals as an end product of purine metabolism.

uterus (yoo′tur-us)   The womb; the hollow, muscular organ of the female reproductive tract in which the fetus undergoes development.

vacuole (vak′yoo-ole)   A cavity enclosed by a membrane and found within the cytoplasm; may function in storage, digestion, or in water elimination.

vagina   The elastic, muscular tube extending from the cervix to its orifice that receives the penis during sexual intercourse and serves as the birth canal.

valence   The number of electrons that an atom can donate, accept, or share in the formation of chemical bonds.

vascular cambium   A lateral meristem in plants that produces secondary xylem (wood) and secondary phloem (inner bark).

vector   (1) Nucleic acid molecule such as a plasmid that transfers genetic information; (2) agent that transfers a parasite or a virus from one organism to another.

ventral   Referring to the belly aspect of an animal's body.

vernalization   Promotion of flowering in certain plants by exposing them to a cold period.

vertebrates   Chordates that possess a bony vertebral column; fish, amphibians, reptiles, birds, and mammals.

vesicle (ves′ih-kl)   Any small sac, especially a small spherical membrane-bounded compartment within the cytoplasm.

vessel element   A type of water-conducting cell in the xylem of plants.

villus; pl. villi   A minute, elongated projection from the surface of a membrane, e.g., villi of the mucosa of the small intestine.

viroids (vy′roids)   Tiny, naked viruses consisting only of nucleic acid.

virus   A tiny pathogen composed of a core of nucleic acid usually encased in protein and capable of infecting living cells. A virus is characterized by total dependence upon a living host.

vitamin   An organic substance necessary in small amounts for the normal metabolic functioning of a given organism.

viviparous (vih-vip′er-us)   Bearing living young that develop within the body of the mother.

xylem (zy′lem)   Vascular tissue that conducts water and dissolved minerals in certain plants.

zooplankton (zoh″oh-plank′tun)   The nonphotosynthetic organisms present in plankton.

zoospore (zoh′oh-spore)   A flagellated motile spore produced asexually.

zygote (zy′gote)   The diploid (2n) cell that results from the union of two haploid gametes; a fertilized egg.

Post-test Answers

Chapter 1

1. Study of life 2. metabolism 3. homeostasis 4. DNA 5. adaptations 6. cells 7. organs 8. ecosystems 9. ecology 10. photosynthesis 11. water, energy 12. decomposers 13. mutations 14. selection 15. Protista 16. species

Chapter 2

1. adequate evidence 2. specific conclusions 3. premises 4. no more 5. general—unifying 6. induction 7. tentative 8. test (verification) 9. theory 10. observations 11. public

Chapter 3

1. atom 2. neutrons 3. orbitals 4. It consists of 2 carbon atoms, 6 hydrogen atoms, and 1 oxygen atom 5. electron donors 6. covalent 7. reduction 8. acid 9. basic (alkaline) 10. buffer 11. capillary action; adhesive 12. hydrogen

Chapter 4

1. c 2. d 3. b 4. e 5. c 6. d 7. a 8. b 9. amino acids 10. amino acid sequence in its polypeptide chains 11. Cellulose 12. glycogen

Chapter 5

1. resolving power 2. ribosomes 3. smooth ER 4. Golgi complex 5. mitochondria 6. Chloroplasts 7. microtubules 8. Microfilaments 9. nucleus; nuclear envelope 10. chromatin; chromosomes 11. genes 12. cell wall 13. vacuole 14. nuclear membrane; membranous 15. function it performs

Chapter 6

1. selectively permeable 2. hydrophilic; hydrophobic 3. hydrophobic tails 4. Microvilli 5. cellulose 6. Desmosomes 7. gap junctions 8. diffusion 9. osmosis 10. hypertonic 11. isotonic 12. exocytosis 13. phagocytosis 14. dissolved materials (solutes) 15. active transport 16. contractile vacuoles

Chapter 7

1. Energy 2. potential energy 3. thermodynamics 4. free 5. Endergonic 6. equilibrium 7. three 8. energy; ATP 9. enzyme 10. activation energy 11. ase 12. coenzyme 13. noncompetitive

Chapter 8

1. catabolism 2. oxygen 3. reduction 4. glycolysis 5. cytoplasm; mitochondria 6. NAD^+; NADH-H 7. acetyl coenzyme A 8. carbon dioxide, water, hydrogen 9. citrate (citric acid) 10. energy 11. molecular oxygen 12. ATP 13. ATP synthesis 14. anaerobes 15. fermentation 16. ethyl alcohol 17. lactic acid (lactate) 18. 2; up to 38 19. oxidized

Chapter 9

1. photosynthesis 2. chemical 3. food (or sugar or carbohydrate), oxygen (any order) 4. photons 5. chlorophyll 6. thylakoids 7. split water 8. water 9. electron 10. ADP, ATP 11. noncyclic 12. NADPH + H$^+$ 13. photosystem I 14. H$^+$, thylakoid 15. ATP 16. CO_2, ATP, NADPH + H$^+$ (any order) 17. stroma 18. Calvin 19. CO_2 fixation 20. Six

Chapter 10

1. nucleus 2. DNA 3. double helix 4. sugar, phosphate, base. (any order) 5. TTGCCAGT 6. thymine. guanine. cytosine. (last two answers—any order) 7. replication fork 8. semiconservative 9. DNA polymerase

Chapter 11

1. messenger RNA, ribosomal RNA, transfer RNA (any order) 2. mRNA 3. rRNA 4. codon 5. transcription 6. RNA polymerase; promoter 7. ribosomes 8. A-site 9. initiation, elongation, termination 10. introns

Chapter 12

1. Operons 2. repressor 3. enzyme induction 4. presence, absence 5. repression 6. inhibition 7. differentiation 8. totipotency 9. nucleus 10. nucleosome

Chapter 13

1. chromatin 2. diploid, 2n 3. haploid, n 4. homologous 5. cell cycle 6. mitosis 7. cytokinesis 8. interphase 9. synthesis (S) 10. centromeres 11. metaphase 12. anaphase 13. two; four 14. tetrads 15. crossing-over 16. genetic recombination 17. trisomy 18. mother

Chapter 14

1. nineteenth; Gregor Mendel 2. gene; protein (or polypeptide) 3. phenotype 4. homologous 5. homozygous 6. dominant 7. heterozygous; phenotypes 8. chance 9. one

Chapter 15

1. X; son 2. multiple; codominant; dominant 3. mother 4. heterozygous 5. can 6. hemizygous 7. 50 8. 50

Chapter 16

1. recombinant 2. vectoring 3. plasmids 4. antibiotics 5. restriction endonucleases, viral 6. restriction endonucleases 7. ligase 8. genetic probe 9. clone 10. electroporation 11. gene therapy 12. crown gall, bacterium

Chapter 17

1. evolution 2. populations 3. gene pool 4. Lamarck 5. geometrically 6. Darwin's 7. analogous 8. vestigial 9. compression 10. index fossils 11. range

Chapter 18

1. gene frequencies 2. mutation 3. genetic drift 4. decreased 5. founder 6. gene flow 7. natural selection 8. protective coloration 9. environment 10. directional 11. microevolution

Chapter 19

1. species 2. temporal 3. mechanical 4. pre-zygotic, post-zygotic 5. allopatric 6. allopolyploid 7. macroevolution 8. preadaptation 9. allometric 10. adaptive radiation 11. punctuated equilibria 12. mass

Chapter 20

1. oxygen 2. time 3. clay (or rock) 4. hydrogen sulfide
5. endosymbiont 6. Precambrian 7. Mesozoic 8. Cenozoic
9. volcanoes 10. genus, species 11. *Penicillium* 12. family
13. division 14. protein 15. viroid

Chapter 21

1. Monera 2. nuclear envelope; organelles 3. producers 4. cell
wall 5. gram-positive 6. Plasmids 7. F-pili 8. Protista 9. microtubules 10. reproduction; metabolism (or growth) 11. sporo-
zoan 12. cellulose 13. red tide 14. diatoms 15. euglenoids
16. brown 17. plasmodium 18. cellular

Chapter 22

1. cell walls 2. spores 3. decomposers 4. parasites 5. roots
6. budding 7. hyphae, mycelium 8. septa 9. sporocarp (or
basidiocarp or fruiting body) 10. dikaryotic 11. zygospores
12. ascospores, asci 13. basidiospore 14. gills 15. sexual
16. lichen 17. yeasts 18. haustoria

Chapter 23

1. vascular 2. cuticle 3. green algae 4. stomates (or stomata)
5. archegonium 6. fertilization, zygote 7. spores 8. alternation, generations 9. gametophyte 10. sori 11. horsetails
12. seeds 13. monoecious 14. fruit (or ovary) 15. pollen
16. female gametophyte 17. monocots 18. endosperm 19. carpel 20. ovary, ovule

Chapter 24

1. invertebrates 2. bilaterally symmetrical 3. body cavity
4. echinoderms, chordates 5. spicules 6. eggs, sperm 7. stinging (cnidocytes) 8. a 9. e 10. f 11. b 12. d 13. c 14. f
15. c 16. e 17. b 18. f 19. d 20. d 21. eating poorly
cooked beef 22. insects

Chapter 25

1. notochord; nerve cord; gill slits (grooves) 2. Tunicates
3. vertebral column; cranium 4. ray-finned; amphibians 5. amphibians 6. terrestrial; protects the embryo (and provides a
water environment) 7. birds, mammals 8. lay eggs; pouches
(marsupia) 9. c, d, e 10. e 11. d, e 12. c 13. d 14. f 15. e
16. *erectus* 17. *habilis*

Chapter 26

1. suspensor 2. indeterminate 3. root cap 4. apical 5. leaf
primordia, bud primordia (any order) 6. vascular cambium
7. annuals, biennials, perennials 8. parenchyma 9. collenchyma
10. tracheids 11. companion 12. epidermis, periderm (or cork)
13. fibrous 14. bud scales 15. stomates, lenticels 16. compound 17. opposite 18. parallel, netted 19. hormones
20. callus

Chapter 27

1. mesophyll 2. guard cells 3. xylem 4. lower 5. dicots
6. bulliform 7. potassium 8. photoreceptor 9. transpiration
10. circadian rhythms 11. abscission 12. suberin 13. tendrils
14. storage 15. spines

Chapter 28

1. primary, secondary 2. dicots 3. monocot 4. vascular cambium, cork cambium (any order) 5. periderm 6. secondary
xylem 7. root cap 8. Casparian strip 9. pericycle 10. xylem,

pith 11. dendrochronology 12. water, minerals (any order)
13. translocation 14. root pressure 15. transpiration
16. source 17. pressure flow 18. active transport 19. hydroponics 20. Macronutrients 21. sixteen 22. phosphorus
23. potassium 24. organic 25. nitrogen

Chapter 29

1. sexual 2. rhizome 3. tuber 4. stolon 5. asexual 6. insects
7. wind 8. fruit 9. berries 10. drupe 11. dehiscent 12. two
13. aggregate, multiple 14. accessory 15. wind 16. animals
17. photoperiodism 18. darkness 19. red 20. vernalization

Chapter 30

1. phototropism 2. positive 3. thigmotropism 4. pulvinus
5. circadian rhythms 6. hormones 7. polar transport 8. auxin
9. gibberellins 10. gibberellins 11. cytokinins 12. antagonistic
13. cytokinins, abscisic acid 14. ethylene 15. dormancy

Chapter 31

1. tissue 2. gland 3. squamous; cuboidal; columnar 4. b 5. d
6. a 7. e 8. c 9. b 10. c 11. c 12. d 13. b 14. e 15. a
16. e 17. intercalated disks 18. skeletal, cardiac 19. axon, dendrites 20. endocrine 21. integumentary

Chapter 32

1. epidermis; dermis 2. corneum 3. keratin 4. Hydrostatic
5. molt 6. endoskeleton 7. compact; spongy 8. lubricant;
joints 9. myosin, actin 10. d, c, b, a, e, f, g 11. energy storage
12. glycogen; ATP

Chapter 33

1. reception 2. muscles, glands 3. glial (neuroglia); neurons
4. cell body 5. axon 6. cellular sheath 7. ganglion 8. sodium
pumps 9. action potential (neural impulse) 10. refractory period 11. synapse 12. neurotransmitter 13. reflex 14. afferent
(sensory) 15. convergence

Chapter 34

1. nerve net 2. the central nervous system (CNS) 3. cerebral
ganglia 4. brain, spinal cord 5. midbrain 6. olfactory; odors
7. association 8. protect the brain and spinal cord 9. cerebrum
10. motor areas (motor cortex) 11. maintaining consciousness
12. c 13. d 14. a 15. b 16. e 17. sympathetic, parasympathetic 18. tolerance

Chapter 35

1. receptor 2. receive stimuli from the outside world; position of
body parts; orientation 3. Photoreceptors; mechanoreceptors
4. energy; electrical; receptor potential 5. adaptation 6. gravity;
statolith 7. vision 8. muscle spindles; Golgi tendon; joint
9. equilibrium 10. labyrinth; saccule, utricle; semicircular
11. otoliths; gravity 12. endolymph; ampulla 13. inner; sound
14. Corti; cochlear 15. chemoreceptors 16. rhodopsins 17. c
18. d 19. b 20. e 21. a

Chapter 36

1. open 2. diffusion 3. plasma 4. c 5. d 6. b 7. a 8. c
9. b 10. anemia 11. thrombin 12. c 13. d 14. a 15. e
16. c 17. blood pressure 18. left atrium 19. liver 20. baro-
ceptors 21. mitral 22. vasoconstrictors 23. oxygen 24. arterial; blood 25. interstitial (tissue)

Chapter 37

1. immune response 2. antibodies (or immunoglobulins) 3. interferons 4. redness, heat, swelling, pain 5. bone marrow; thymus; lymph 6. Suppressor 7. memory 8. plasma cells (differentiated B cells) 9. antibody 10. thymus 11. antigen-antibody 12. macrophage 13. active 14. passive 15. HLA 16. one location to another in the same organism 17. immune response; transplanted 18. privileged 19. allergen; allergic 20. histamine; inflammation.

Chapter 38

1. diffusion; moist; capillaries (blood vessels) 2. tracheal; spiracles 3. lungs 4. air sacs 5. trachea 6. alveoli (air sacs) 7. diaphragm 8. e 9. a 10. b 11. d 12. c 13. Decompression sickness 14. inhaling dirty air 15. emphysema 16. cigarette smoking

Chapter 39

1. ingestion 2. Digestion 3. Elimination 4. herbivore 5. canine 6. incomplete 7. two 8. amylase 9. dentin 10. b 11. d 12. a 13. c 14. b 15. e 16. rugae 17. villi 18. duodenum 19. ulcer (peptic ulcer) 20. store bile 21. minerals 22. components of coenzymes 23. c 24. a 25. d 26. e 27. b 28. triglycerides 29. amino acids; glucose 30. glycogen 31. amino acids 32. basal metabolic rate 33. there is an increase in body weight 34. protein

Chapter 40

1. excretion 2. osmoregulation 3. uric acid 4. urea 5. protonephridia; flame cells 6. metanephridia 7. Antennal (green) glands 8. malpighian tubules 9. nephrons 10. a 11. c 12. e 13. d 14. b 15. a 16. capillaries; Bowman's capsule 17. afferent arteriole; efferent arteriole 18. filtrate 19. excreted in the urine 20. salt 21. collecting ducts; reabsorbed; decreased

Chapter 41

1. ducts; blood 2. hormones 3. cyclic AMP 4. hypothalamus 5. target 6. c 7. d 8. b 9. a 10. e 11. a 12. e 13. c 14. b 15. d 16. hypo- (too little); thyroid hormones 17. goiter 18. cortisol (glucocorticoids) 19. parathyroid 20. diabetes

Chapter 42

1. fragmentation 2. sexual and asexual generations 3. develops into a new individual 4. hermaphroditic 5. gamete; zygote 6. sterile 7. a 8. c 9. d 10. b 11. e 12. b 13. d 14. c

15. a 16. c 17. b 18. d 19. e 20. b 21. a 22. c 23. d 24. c 25. d 26. e 27. b 28. therapeutic 29. female sterilization 30. menopause

Chapter 43

1. morphogenesis; cellular differentiation 2. acrosome; vitelline 3. fertilization cone 4. electrical; cortical 5. cleavage 6. gastrulation 7. ectoderm; endoderm 8. neural plate 9. brain, spinal cord 10. amnion (amniotic fluid) 11. placenta 12. inner cell mass; embryo 13. implant; uterus (endometrium) 14. fetus 15. gestation 16. newborn infant 17. development

Chapter 44

1. signals from the environment 2. Ethology 3. positive chemotaxis 4. circadian rhythm 5. crepuscular 6. releaser 7. Instinctive 8. imprinting 9. migratory restlessness 10. classical

Chapter 45

1. social behavior 2. society 3. symbolic 4. Pheromones 5. dominance hierarchy 6. home range 7. conflict; population 8. pair bond 9. nurse; royal jelly 10. innate 11. culture 12. altruistic 13. Kin 14. making more copies of themselves

Chapter 46

1. decomposer 2. detritus feeders 3. parasites 4. ecological niche 5. fundamental niche 6. realized niche 7. competitive exclusion 8. volcanic 9. nitrifying 10. bacterial—cyanobacteria 11. increase 12. early.

Chapter 47

1. lithosphere 2. the sun 3. radiation 4. angle 5. convection 6. specific heat 7. coriolis effect 8. dew point 9. rain shadows 10. microclimate

Chapter 48

1. low—permafrost 2. coniferous evergreens 3. precipitation 4. clay 5. high—coral reef 6. savannah (veldt) 7. temperature—oxygen—light 8. estuaries 9. thermocline—fall overturn 10. littoral 11. burrowing 12. abyssal

Chapter 49

1. simpler 2. biodegradable 3. insoluble—soluble 4. biological magnification upper 5. pollution 6. high—oxygen 7. warm—cooler 8. teeth—bones 9. habitat loss

Index

Page numbers in *italics* indicate illustrations, page numbers followed by "f" indicate footnotes, page numbers followed by "t" indicate tables.

A

Abortion, 784–785
Abscisic acid, 534–535, *535*
Abscission, 400
Abscission zone, of stems, 484, *484*
Absorption, of food, definition of, 699
    in small intestine, 709
Abyssal zone, of aquatic habitat, 895
Accommodation, in vertebrate eye, 630
Acetabularia cell, giant, role of nucleus in, *187*, 187–188, *188*
Acetaldehyde, 164
Acetyl CoA. *See* Acetyl coenzyme A
Acetyl coenzyme A, 155, 158–159
Acetylcholine, 569, 586, 586t
Achenes, 513
Acid(s), 55
Acid-base neutralization, 56
Acid-growth hypothesis, auxin and, 529, *529*
Acid rain, with air pollution, 57, 914, *916*
Acoelomates, 404, *405*
Acorn weevil, *700*
Acquired immunodeficiency syndrome (AIDS), 676–677
Acromegaly, 752
Acrosome, of sperm, 765, *767*
Acrosome reaction, 792
Actin, 112, 230
    in muscle fiber, 567–569, *568–569*, *572*
Action potential, definition of, 583, *583*
    movement of, 584, *584*
    in muscle contraction, 569
Activation energy, 144
Active immunity, antigen exposure and, 672
Active transport, 125–126, *127*
Adaptation, 10, *10*
    behavior as, 818–820
Adaptative radiation, 317, *318*, *319*
Addicott, F.T., 534
Addiction, to mood drugs, 609
Addison's disease, 757
Adenine, 142, 191, 193
Adenoids, lymph tissue of, 655
Adenosine deaminase deficiency disease, 272

Adenosine diphosphate (ADP), 143
    in regulation of cellular respiration, 162–163
Adenosine monophosphate, 143
Adenosine triphosphate (ATP), 75, 125–126, *127*, 142–143, *142–143*, 153
    light energy to make, 174–179
    synthesis of, 160, *161*
        chemiosmotic, 155
Adenyl cyclase, as hormone intermediary, 744
ADH. *See* Antidiuretic hormone (ADH)
Adipose tissue, 543, *544*, 545t
    fat storage in, 714
Adolescence, in human life cycle, 811, 812t
Adrenal gland, *756*, 756–758
    hormones of, 748t
Adrenergic neurons, function of, 587
Adrenocortical hormones, malfunction of, consequences of, 749t
Adrenocorticotropic hormone (ACTH), action of, 748t
    adrenal secretion of, 757
    hypothalamic production of, *747*
Adulthood, in human life cycle, 812t
Adventitia, of digestive tract, 703, *703*
Aerobes, 164
Aerobic respiration, 153
Afferent arteriole, of nephron, 733, *734*
Afferent nerves, 596
Afterbirth, 809
Agapornis, nesting behavior of, *822*, 823
Agglutination, 258
Aggregation, as social behavior, 834
Aggression, suppression of, dominance hierarchies and, 836–837, *837*
Aging, 811–813
Agnatha, 436
Agranular leukocytes, 641, *641*
Agriculture, artificial communities in, 908–913, *909*
    development of, 451
    mechanized, increased production with, 910
    traditional vs, *909*

AIDS, 676–677
Air, inhaled vs exhaled, composition of, 692t
    movement of, Coriolis effect and, *873*, 873–874
        effects of, 877–879
Air pollution, 914, 916–917, *916–917*
    effects of breathing, 694–696
    and global heat transfer, 871–873
Air pressure, changes in, physiological adaptation to, 693–694
Air sacs, in birds, 687
    gas exchange in, 690–691, *692*
Airway, establishment of, in cardiopulmonary resuscitation, 691
    function of, 688
Alarm signs, as innate behavior, 828
Albatross, courtship and bonding of, *838–839*
Albinism, *109*, 244, 251, 257
    tyrosinase-positive oculocutaneous, 263t
Albumins, in plasma, 640
Alcohol, abuse of, 607
    effects of, 608t
        on embryo, 810t
    fermentation of, 164, *165*
Aldosterone, 738, 757
Algae, 351, 353–358, *899*
Algin, 357
All-*trans* retinal, in rhodopsin, 631
Allantois, function of, *801*, 802
Allee, W.C., 834
Allele(s), 244–245, *245*
    ABO and, 259
    codominant, 251
    combinations of, 245–246
    dominant, 245, 246
    and genetics, 248–249
    inheritance of, 250–251
    multiple, blood cell antigens and, 258
    recessive, 245
Allelopathy, 891
Allergens, definition of, 675
Allergic reactions, as immune responses, 675, 677–678
Allopatric speciation, *314*, 314–315
Allopolyploidy, 315, *316*
Allosteric site, 147
Alpha cells, of pancreas, 753

Alpine plants, microclimate of, 879, *879*
Altitude zonation, 886, *886*
Altruistic behavior, kin selection as, 843, *843*
Alveoli, *687*, 688
Amino acids, 71–73, *72*, 111
  deaminated, 714–715, *716*
  in proteins, 200
Amino acyl-tRNA synthetases, 204
Ammonia, as metabolic waste product, 726
Amniocentesis, 230, *232*, 811
Amnion, 439
  function of, 801, *801*
Amoebas, 351, 352
Amphetamines, effects of, 608t
Amphibian(s), 438–439, *439*
  brain of, *598*
  heart of, 645, *645*
Amphibian cells, organ formation and, *218*, 219
Amphioxus, 435, *435*
Ampulla, of semicircular canal, 621
Anaerobes, 164, 347
  as early cells, 328
Anaerobic metabolism, inefficiency of, 166–167
Anaerobic pathways, 153, *153*
  aerobic vs, *165*
  utilization by cells, 163–167
Anaerobic respiration, 163, *165*
Analgesics, narcotic, effects of, 608t
Analogous structures, 287–288, *288*
Anaphylaxis, as allergic reaction, 678
Androgens, adrenal secretion of, 757
Anemia, 641
Aneuploidies, 237
Angina pectoris, atherosclerosis and, 653
Angiosperms, 392, *396*, 396–397
Angiotensins, in blood pressure regulation, 651
Angular acceleration, perception of, 621–622, *622*
Animal(s), 19
  agricultural, genetic engineering and, 274
  behavior of, 816–832. *See also* Behavior
  body of, standard directions and body planes in, 407
  characteristics of, 340t, *403*, 403–404
  classes of, comparison of, 430–431t
  extinction of, 920–922
  grouping of, 404–407, *406*
  multicellular, 539
  non-chordate invertebrate, 402–432

Animal(s), *(Continued)*
  pollination of flowers by, 511
  salt water inhabitants, 404
  symmetry of, 404, *405*
Animal pole, of zygote, 793, 796
Animal tissues, 538–556
Anions, 50
Annelid(s), 420–422, *421*
  characteristics of, 430–431t
  hydrostatic skeleton of, for movement, 562, *563*
Anteater, 442, *443*
Antelope, territorial marking by, *835*
Antennae, 425
Antennal glands, in crustacean osmoregulation, *728*, 729
Anther, 397, 400
Antheridium, 383
Anthropoids, 445
Antibiotics, 283, *283*
Antibody(ies), 258, 259, *668*
  binding to antigens, 669–670
  blocking, 673
  classes of, 668–669
  concentration of, increase in, *671*
  in immune responses, 661
  monoclonal, 673
  structure of, 668
Antibody-mediated immunity, 664–670, *665*, *667*
Anticodon, 204
Antidepressants, effects of, 608t
Antidiuretic hormone (ADH), action of, 748t
  pituitary gland release of, 747, *750*
  urine volume regulation by, 738
Antigen(s), *668*
  affinities of, 668
  blood cell, 258
  in immune response, 661, 672
Antigen-antibody complex, 668, *668*, *669*
Antigenic determinants, 668
Antihistamines, 678
Antiparallel, 194
Antiviral agents, 664
Anvil, of middle ear, 623, *623*
Aorta, 653–655, *654*
Apatite, in bone formation, 567
Aphids, 501, *502*
Apical dominance, of plants, 530, *531*, 533
Apomixis, 510
Appendicular skeleton, 564, *565*
Appendix, 709
Appetite, as instinct, 827
Aquatic habitats, 895–904
  freshwater, 896–897
  succession in, 904

Aquatic organisms, ecological categories of, 895–896
Aqueous fluid, of vertebrate eye, 630
Arachnida, *424*, 425
Archaebacteria, 348
Archaeopteryx, 333, *334*, 441
Archean era, 332–333t
Archegonium, 383
Archenteron, of gastrula, 797, *798*
Arctic fox, *887*
Aristotle, 25, 283
Arrow-poison frog, *893*
Arterial pulse, 648–649
Arterioles, in vertebrates, 643, *643*
Artery(ies), 643, *643*
  blood pressure in, 650
  pulse in, 648
Arthropods, 423–428
  appendages of, 423
  characteristics of, 430–431t
  classification of, 425
  gas exchange in, 683, *683*, *684*
  molting by, 423–425
  organs of, 425
  skeleton of, 423, 563
Artificial insemination, 782
Artificial selection, 285, *286*
Ascaris, 417–418
Asci, 369
Ascocarp, 371, *371*
Ascomycetes, 369–372, *371*, *378*, *378*
Ascospores, formation of, 369–372
Asexual reproduction, 9, *9*, 762
Association area, of cerebral cortex, 600
Association neuron, 588
Asters, 227
Asthma, 678
Astigmatism, 628, *628*
Atherosclerosis, 652, *652*
  smoking and, 695
Athlete's foot, 378
Atmosphere, 867, *868*
  circulation of, 869, 871
  heat retention by, 872
  heat transfer by, 871, *871*
Atoll, coral, 900
Atom(s), 12
  combining of, 46
  definition of, 43
  electron configuration of, 44–45, *45*
  excited, 172
  formation of bonds by, chemical, 48–52
    ionic, 50–51
  ground state of, 172
  sharing of electrons by, 48
Atomic number, 45
ATP, in muscle contraction, 570
ATP synthetase, 155–156, 162, 178

Atria, 439, 644
  human, 646, *646*, *647*
Atrioventricular node, 646
Atrioventricular valves, human, 646, *646*, *647*
Auditory meatus, external, 623
Auditory receptors, 622–623
Auricles, 412–413, *413*
Australian marsupials, placental mammals vs, 291, *292*
Australopithecus, 39
  hominids and, 447–448, *448*
Autograft, definition of, 673
Autoimmune disease, 675
Autonomic nervous system, 596, 606
Autoradiography, embryogenesis and, 463
Autosomes, 236, 256
Autotrophs, 14–15, 328
Auxin(s), 526–531, *527*
  in apical dominance, 533
  in fruit development, 531
  functions of, 528–529, *529*
  during organogenesis, 533
  phototropism and, *529*, 529–531
  polar transport of, 528
Avery, Donald, experiments of, 190, *190*
Axial skeleton, 564, *565*
Axon(s), 578
  neurotransmitter release by, 585–586
Azidothymidine (AZT), for AIDS, 677

B
B cells (lymphocytes), *665*, 665–666
B-DNA, 194, *195*
Bacon, Roger, 25, *25*
Bacteria, 345–347
  characteristics of, 340t
  DNA in, 346, 348–350
  dormancy and endospore formation by, 350
  nitrifying, 852
  plasma membrane, cell wall of, 345, *346*
  polysaccharide coating on, 189
  reproduction by fission, 348
  studies of, 350
  transformation and, 190, *190*
Bacterial cell, transcription and translation in, *207*
Bacteriophage(s), 190–191, 270
  action of, 339, *341*
Bacteriorhodopsin, *114*
Barbiturates, effects of, 608t
Bare rock, in marine habitat, *899*
Barnacle zone, in marine habitat, *899*
Baroreceptors, 651
Basal metabolic rate, definition of, 716
Base(s), 55–56, 193

Basement membrane, of epithelial tissue, 540
Basidiocarp, 373, *373*
Basidiomycetes, 372–374, 376, 378, *378*
Basidiospores, *372*, 372–374
Basidium, 372
Basilar membrane, in organ of Corti, 622
Bass, spawning ground of, *896*
Batesian mimicry, 305, *305*
Bats, pollination by, 511–512
Bearberry, *887*
Behavior, animal, 816–832
  capacity for, inheritance of, 820–823
  cycles of, 819
  definition of, 817
  development of, 829
  emotional aspects of, limbic system and, 600, 602
  evolution of, natural selection and, 844
  innate, instrumental conditioning of, *824*, 824–825
  stimuli triggering, 827–828
  and life style, 817–818
  modification by learning, 823–827
  sign stimulus for, 828
  social, 833–846, *834*
  thermoregulation of, in lizards, 818, *819*
  unlearned patterns of, instincts as, 827–831
Bends, with air pressure changes, 693–694
Benthos, 895
Berry, 513
Beta cells, of pancreas, 753
Bias, experimenter, conclusions and, 30
Bicarbonate ions, 57–58
  blood carbon dioxide transport by, 692–693
Biceps muscles, 572, *572*
Bile, secretion of, 708, *709*
Bindin, and acrosome reaction, 792
Binocular vision, in vertebrates, 629, *629*
Binomial system of nomenclature of organisms, 17, 335
Biocides, effects of, *912*
Biodegradability, of pesticides, 912
Biogenesis, 25, 28, 31
Biogeography, 291–293, *293*
Biological clock, 820
  effect on plant responses, 524, 526
Biological control, of screw-worm flies, *913*

Biological magnification, of pesticides, 911, *912*
Biological organization, hierarchy of, 12–16, *13*, *14*
Biological response modifiers, for breast cancer, 774
Biological rhythms, 818–820
Biologists, concerns of, 3, *3*
Biology, 3, 3–4
Biomass, 859, *860*
Biomes, 884, *885*
Biopolymers, 65, 65f
Biosphere, 14, 867, *868*
Biotechnology, 266, *266*
Biotic potential, 856
Biramous appendages, *424*, 425
Bird(s), adaptation for flight, 441
  brain of, *598*
  heart of, 645, *645*
  lungs of, *686*
  orders of, 441
  pollination by, 511, *512*
  songs of, innate vs learned, 829f
Birth, stages of, 807–809
Birth control, 780–784
  deaths from, 783t
  methods of, 781t, *782*
Birth rate(s), 922
  species differences in, 856
Bivalvia, 418
Bladder wall, epithelium of, *733*
Blastema, 219
Blastocoel, production of, 793
Blastocyst, formation of, *804*, 805
Blastodisc, of telolecithal eggs, 796
Blastomeres, production of, 793
Blastopore, of gastrula, 797, *798*
Blastula, production of, 793
Blocking antibodies, 673
Blood, 545t
  composition of, 549, *640*
    vertebrate, *639*, 639–643
  oxygenation of, *651*, 653
Blood clotting, platelets in, 642, *642*
Blood pressure, 649–651, *650*
Blood type(s), 257–261, *259*, 259t
Blood vessels, of dermis, *560*, 561
  injury to, circulatory response to, *642*
  types of, *643*, 643–644
Blooms, in freshwater habitats, 897
Body surface, gas exchange by, 683, *684*
Body temperature, regulation of, 6, 7
Bohr model(s), of electron configurations, 45, *45*, *46*
Bolting, and gibberellins, 532, *532*
Bond(s), chemical, 48–52
  covalent, *49*, 49–50
Bone(s), 545t, 547, *547*
  formation of, 566–567

Book lungs, *683*, 686–687
Boreal forest, 888
Boron, in plants, 504
Bottleneck effect, gene pool and, 301, *302*
Bowman's capsule, of kidney, 733, *734*, *736*
Brain, respiratory centers of, 689–690
    size of, human evolution and, 447, *447*
        structure of, 600, 601t, *602*, *603*
        vertebrate, evolution of, 596–599, *597*
Brain coral, *901*
Brain stem, 597, 601t
Branchial plume, respiration by, *683*
Breast(s), function of, *773*, 773–774
Breast cancer, 774, *774*
Breastfeeding, 773–774
Breathing, 688–691
Breeding, selective, 266
Bronchi, *687*, 688
Bronchial constriction, lung protection by, 694
Bronchioles, *687*, 688
Bronchitis, 694–695
Brown algae, 357–358, *358*
Bryophytes, 384
Bud scales, 485–486
Budding, in asexual reproduction, 762, *762*
Buffalo, *894*
Buffers, 57–58
Bulbourethral glands, 767
Bulbs, 486, 509, *509*
Bumpus, Hermon, 305–306
Bundle sheath cells, 179
Burmese python, *701*
Bursa of Fabricius, in birds, 666
Buttercup, roots of, 493–494, *495*
Buttons, 373
Buzzard, *894*

C
C region, of polypeptide chain, 668
Caffeine, effects of, 608t
Calcitonin, 748t, 753
Calcium, body need for, 716
    function of, 720t
    parathyroid regulation of, *744*, 753
Calorie, 54
Calvin, Melvin, 180
Calvin cycle, 179–183, *181*
Cambium(a), 461, 490–492, *493*, 494
    cork, 461, *462*
Cambrian period, 330, 332–333t
Canada goose, northward migration of, *830*
Canaliculi, of bone, 547, 566, *566*

Cancer, body defenses against, 673
    of colon, 710, *711*
    definition of, 548
    molecular genetics and, 274–275
    prevention of, recommendations for, 549
Cancer cells, 274
Candidiasis, 378
Canines, 703
Capillary(ies), 639
    blood flow in, *644*
    gas exchange in, 691
    structure of, *644*
    in vertebrates, 643, *643*
Capsule, surrounding bacterium, 189
Carbohydrate(s), 61–66, 170
    classification of, 61
    digestion of, 707t
    as energy source, 711–712
    interrelationship with protein and fat, *717*
    metabolism of, in diabetes mellitus, 754–756
Carbon, atom of, 49–50
    recycling of, 851, *852*
Carbon dioxide, atmospheric heat retention by, 872
    blood transport of, 692–693
    in carbon recycling, 851, *852*
    exchange of, in air sacs, 691
    fixation of, 180, *852*
        in organic compounds, 179–183
    fossil fuels and, 20
    as metabolic waste product, 726
Carbonic acid, 57–58
Carboniferous period, 331, 332–333t, *334*
Carcinogenesis, steps in, 274
Carcinoma, 548
Cardiac arrest, 653
Cardiac muscle, 549, 550t, *551*
    microscopic appearance of, *648*
    specialization of, 573
Cardiac output, 648
Cardiopulmonary resuscitation (CPR), 691
Cardiovascular disease, 652–653
Carnivores, 700, *701*
    as consumers, 849
    first evidence of, 333, *335*
    marine, osmoregulation in, 730, *731*
Carolina anole lizard, 854, *855*
Carotenoids, 172, 383
Carotid arteries, 653–655, *654*
Carpals, 287
Carpels, 397
Cartilage, 545t, 546, *546*
Cartilaginous fish, 437, *437*
Casparian strip, 494, *496*

Castration, 768
Casts, production of, 290
Catabolic pathways, oxidation-reduction reactions and, 153–155
Catalase, 144–145
Catalyst, 144
Catecholamines, 587, 756
Cations, 50, 50t
Cats, domestic, classification of, 19, 19t
Cavernous bodies, of penis, 767
Cecum, 709
Cell(s), 4–5, 12, *14*
    amphibian, organ formation and, *218*, 219
    assembly of, from organic molecules, 328–329
    diploid, production of haploid cells by, 231–235
    division of, chromosomes and, 223–240
    colchicine and, 230, *231*
    processes in, 226
    haploid, production by diploid cells, 231–235
    in interphase, 226
    joining of, 116–117
    movement as property of, 6–8, *8*
    of multicellular eukaryote, genes of, 218–219
    nucleus of, 187–189, *189*
    plasma membrane and, 108
    red. *See* Red blood cells
    sex, 224
    somatic, 224
    structural components of, proteins as, 714–715
    surface area of, microvilli and, 112, *116*
    utilization of anaerobic pathways by, 163–167
    walls of, 113–116, *116*
Cell body, of neuron, 578
Cell cycle, *225*, 225–231
Cell differentiation, 805, *806*, 807
    in development, 791
    gene control in, *217*, 217–218
    inducer and, 799
Cell-mediated immunity, 664–668, *665*, *667*, *670*, 670–678
Cell membranes. *See* Membrane(s)
Cell plate, 230
Cell typing, in organ transplantation, 674
Cellular respiration, 5, 15, *156*, 157t, 682
    phases of, *155*, 155–162
    as redox process, 154–155
    regulation of, 162–163
Cellular sheath, of neuron, 579, *580*

Cellular totipotency, 219
Cellulose, 65–66, 113–114, 498
    in carbohydrates, 711–712
    structure of, 67
Cenozoic era, 332–333t, 333–335
Center of origin of species, 293, *293*
Central nervous system, human, complexity of, 599–605
    response of, 578, *579*
Centrioles, 226–227
Centromere, 226
Cephalization, 412
    of invertebrates, 595
Cephalopods, of Ordovician period, 330
Cerebellum, 597, *597*, 601t
Cerebral cortex, 599–600, 601t
Cerebral dominance, 604
Cerebral ganglia, of invertebrates, 595, *595*
Cerebrospinal fluid, of human central nervous system, 599
Cerebrovascular accident (CVA), atherosclerosis and, 652
Cerebrum, 599–600, 601t, *602*
Cervix, uterine, cancer of, 772
Cesarean section, 809
Chameleon, *8*
Chargaff, Erwin, 191
Chargaff's rule, 191–192
Chase, Martha, 191
Chelicerae, 425
Chemical(s), inorganic, as pecticides, 911
    provided by fungus, 377
Chemical bonds, 48–52
Chemical compounds, 46–48
Chemical evolution, conditions for, 326–329
Chemical principles, 42–59
Chemical reactions, 139–140
    catalysts and, 144
    types of, 140–141, *140–141*
Chemical synapses, 584
Chemiosmosis, 178–179, *180*
Chemoautotrophs, 347
Chemoreceptors, 413
    stimuli of, 615t
    in taste and smell perception, 623, 625
Chemotherapy, for breast cancer, 774
Chesapeake bay, as estuary, *897*
Chest wall motion, with breathing, *690*
Childbirth, deaths from, 783t
    stages of, 807–809
Childhood, in human life cycle, 811, 812t
Chilopoda, *424*, 425

Chitin, 66
Chlorinated hydrocarbons, biological magnification of, 912
Chlorine, function in body, 720t
    in plants, 503–504
Chlorofluorocarbons, effects on ozone layer, 872
Chlorophyll, 172–173, *173–175*
Chloroplast(s), 173, *175*, 278
    evolution of, 329
    grana in, 173, *176–177*
Choking, 693
Cholecystokinin, 710t
Cholesterol, association with heart disease, 713–714
    and atherosclerosis, 652
Cholinergic neurons, 587
Cholinesterase, 569, 587
Chondrichthyes, 437
    osmoregulation in, 730, *730*
Chondrocytes, of cartilage, 547
Chordates, 433–454
    characteristics of, 430–431t, 434, *434*
Chorioallantoic membrane, function of, *801*, 802
Chorion, function of, 801, *801*
Chorionic gonadotropin, human, in menstrual cycle, 775
    placental release of, 802
Chorionic villi, function of, 802
Choroid layer, of vertebrate eye, 629, *630*
Chromatid(s), 226, 235
Chromatin, 220, *220*, 233–234
Chromosome(s), 223–240, 346
    abnormalities of, 238t
    bacterial, shape of, gene expression and, 216–217
    crossing-over of, 256, *256*
    defects of, inherited disease in, 237, *237*
    definition of, 220–221
    genes of, linked, 234
    homologous, 224, 244
      separation of, 234
    human, 224, 225, 227
    inheritance of, 250–251
    genes as part of, 244–248
    lampbush, 221
    linkage and, 256–257
    locus of, 244
    mapping of, genetic linkage and, 256–257
    maternal, 234
    nondisjunction of, 237, *239*
    nonsex, 256
    number of, in eukaryotes, 224
    paternal, 234
    sex, 235–237, *236*

Chromosome(s), *(Continued)*
    structure of, eukaryotes and, 219–221, *220*
    translocations of, 237
    X, sex-linked genes on, *255*, 255–256
Chronic obstructive pulmonary disease (COPD), 694
Chyme, and digestion, 705
Chymotrypsin, secretion of, 708
Ciliary muscle, in vertebrate eye, 630
Ciliated mucus, of lungs, 694, *694*
Ciliates, 353, *353*
Circadian rhythms, 481, 524, *526*, 820
Circulation, double-circuit, 644–648
    establishment of, in cardiopulmonary resuscitation, 691
    in mammals, pattern of, 651, *651*, 653–655
    in vertebrates, *645*
Circulatory system, *552*, 554t
    functions of, 639
    in invertebrates, 637–639, *638*
    in vertebrates, 639
Circumcision, 767
11-*cis* retinal, in rhodopsin, 631
Citrate, 159
Citric acid cycle, 155, *159*, 159–160
    entrance of fatty acids into, 163, *164*
Civilization, need for energy in, 919–920
Clam, 419–420, *420*
Classes, 19
Classical conditioning, 823, *823*
Clavicle, 565, *565*
Cleavage, patterns of, 793–796, *795*, *796*
Climate, causes of, 877–879
    differences in, reasons for, 869
    and distribution of terrestrial communities, 884
Climax community, stability of, 863
Clitoris, 773, *773*
Cloning, 782
Closed system, 135
Clownfish, *901*
Club mosses, 387–391, *389*
Cnidarians, 409–412
    body plans of, *410–411*, 411
    characteristics of, 430–431t
    classification of, 409, *409*
    digestive system of, *701*
    hydrostatic skeleton of, 561, *562*
    internal defense mechanisms of, 661
Coal, formation of, plants and, 390
Cobalt, function in body, 720t
Cocaine, crack, 609
    effects of, 608t
      on embryo, 810t

Cochlea, *621*, 621–623, *624*
Cockroach, daily rhythm of, 820, *820*
Coding strand, 202–203
Codon, 203
Coelom, 404, 405
Coenocytic, 366, *367*
Coenzymes, 146, *146*
Cofactors, and enzymes, 146, *146*
Coitus, 777
Coitus interruptus, 781t
Colchicine, mitosis and, 230, *231*
Coleomates, internal defense mechanisms of, 661
Coleoptile, 526, *527–528*, 528
Collagen fibers, 540, *543*
Collecting duct, of kidney, 733, *734*, 735
Collenchyma, of plants, 464, *465*
Colon, cancer of, 710, *711*
Color blindness, inheritance of, *255*, 255–256
Coloration, protective, 303, *304*
Colostrum, production of, 773
Columnar epithelium, simple, 541t
Combustion, and air pollution, 914, 916–917
Commensals, 347
Communication, necessity for, in sociality, *835*, 835–836
Community(ies), artificial, agriculture as, 908–913, *909*
    composition of, 856–858
    development of, 861–863
    organisms living in, 848, *849*
    and population ecology, 847–865
    species diversity in, 860–861
Compact bone, of long bones, 566, *566*
Compensation point, in freshwater habitats, 896–897
Competition, 287
    effect on population growth, 858
Competitive exclusion, definition of, 856
Competitive inhibition, 148, *149*
Complement system, antibodies in, 669
Compound(s), chemical formulas of, 46–48
Compound eye, structure of, 627, *627*
Concentration gradient, 120
Conception, 779, *779*
Conditioned stimulus, 823, *823*
Condom, for birth control, 781t, *782*, 783
Cone(s), 235, 392
Cone cells, of retina, 630, *631*
Conidia, 369, *370*
Conifers, *392*, 392–396, *393*
    monoecious, 392

Conjugation, in bacteria, 349–350
    DNA recombination and, 267, *267*
Connective tissue, functions of, 540–549
    types of, 540–549, 544t
        dense, 542
        elastic, 542–543
        loose, 540–542
        reticular, 543
Consanguinity, 261
Conspecific interactions, as social behavior, 834
Constipation, causes of, 710
Consumers, 347
    dependence on producers in ecosystem, *15*, 15–16
    food consumption by, 849
Continental drift, 336–337, *336–337*
Continental shelf, 903
Contraception, 780–784
Contractile vacuole, 124–125, *125*
Control, experimental, 26
Convection, atmospheric heat transfer by, 871, *871*
Convergence, of neurons, 589, *589*
Convolutions, of cerebrum, 599
COPD, 694
Copper, function in body, 720t
Coral reefs, 900–901, *900–901*
    as home range of Moray eel, *837*
Coriolis effect, *873*, 873–874
    on oceanic currents, 877
Cork, cabrium, 461, *462*
Corm, 509, *509*
Corn, classification of, 19t, *338*, 338t
    stems of, 491–492, *492*
Cornea, of compound eye, 627, *627*
    of vertebrate eye, 629, *630*
Coronary circulation, 653–654
Corpus callosum, 600
Corpus luteum, 772
Cortex, of plant stem, 491
    root, 494
Cortical reaction, polyspermy prevention by, 792
Cortisol, adrenal secretion of, 757
Cotyledon, 397
Cough reflex, 688
Countercurrent exchange system, of respiration, 686
Countercurrent mechanism, of urine concentration, 737–738
Courtship behavior, 838–839, *838–839*
Covalent bonds, *49*, 49–50
CPR, 691
Crack cocaine, 609
Cranial nerves, 605, *606*
Cranium, of vertebrates, 435

Crassulacean acid metabolism, 483
Creatine phosphate, in muscle contraction, 571
Cretaceous period, 331, 332–333t
Cretinism, 752
Crick, Francis, 191–193
Crista, of semicircular canal, 621
Cro-Magnon culture, 450, *450*
Cross bridges, in muscle contraction, 569
Crossing-over of chromosomes, 256, *256*
Crown gall tumor, in plant improvement, 277, *277*
Crustaceans, 425, *728*
Cuboidal epithelium, simple, 541t
Cuckoo, as nest parasite, 828, *829*
Cultural eutrophication, of waters, 904
Cupula, of ear labyrinth, 621, *621*
Cushing's disease, 757
Cuticle, of leaf epidermis, 477, *477*
    of nematodes, 417
    of plants, 383
Cyanobacteria, 345, 348
    characteristics of, 340t
    as early cells, 328
    and nitrogen fixation, 852
Cycads, 396, *396*
Cyclases, 76
Cyclic AMP, 76, 360, 744
Cyclic nucleotides, 76
Cyclic photophosphorylation, 177–178, *178*
Cyclosporine, 377
    in organ transplantation, 675
Cystic fibrosis, 262t
Cytochromes, 154, 160
Cytokinesis, 226, 230, 791
Cytokinins, 533–534
Cytoplasm, 12
Cytosine, 191, 193
Cytoskeleton, membrane connections of, 112, *115*
Cytotoxic T cells, 665

D
da Vinci, Leonardo, 284
Dance, of honeybees, 841
Darwin, Charles, 10–11
    and natural selection, 282–296
    voyage of, *284*, 284–285
    Francis, 526
Dawson, Charles, 38
DDT, biological magnification of, 912
Deamination, 163
Death rates, *922*
    species differences in, 856
Deciduous forests, development of, *889*, 889–890

Decomposers, 15, 347, 365
    break down of wastes by, 16
    as consumers, 849, 851
    interdependence with producers and consumers, 16, *16*
Decompression sickness, 693–694
Deduction, 25, 28
Deductive reasoning, 29t
Deep-sea fish, *903*
Defecation, definition of, 710
Defense mechanisms, specific vs nonspecific, 661
Delivery, stages of, 808, *808*
Demographic transition, in population growth, 923
Dendrites, 550, 578
Dendrochronology, 497
Denitrifying bacteria, 852
Density, effects on population growth, 858
Dentin, of teeth, 703
Deoxyribonuclease, secretion of, 708
Deoxyribonucleic acid. *See* DNA
Deoxyribose, 61, 193
Depolarization, 569, 583
Dermal gills, *683*, 685
Dermis, structure of, *560*, 561
Desensitization therapy, for allergic reaction, 678
Desert(s), microclimates of, *879*
    tropical, cause of, 877
    types of, 890–891
Desert animals, adaptations of, 890–891, *891*
Desert plants, adaptations of, 891, *891*
    photosynthesis in, 483
Desmosomes, 117, *119*
Detritus feeders, as consumers, 849
Deuteromycetes, 374
Deuterostomes, 405, 407
Development, 789–815
    cytoplasmic factors in, 797, 799, *800*
    definition of, 790
    prenatal, human, 802–809
    processes of, 791
    as property of life, 5
    regulation of, 797–801
Devonian period, 331, 332–333t
Dew point, of air, 875
Dextrose. *See* Glucose
Diabetes insipidus, 738
Diabetes mellitus, 736–737
    and atherosclerosis, 652
    carbohydrate metabolism in, 754–756
    investigation of, 33–34
    metabolic disturbances in, 754–756
Diacylglycerol, 68

Dialysis, 735
Diaphragm, contraceptive, for birth control, 781t, *782*, 783
    insertion of, *783*
    function of, 688
    motion of, with breathing, *690*
Diaphysis, of long bones, 566, *566*
Diarrhea, causes of, 710
Diastole, 648
Diatoms, 354–355
2,4-Dichlorophenoxyacetic acid, 531, *531*
Dicots, 397
    stems of, 490–492
    tissues of, 493–494
Diencephalon, of forebrain, 598
Differentiation, factors in, genetic vs nongenetic, 800–801
Diffusion, 120–125, *121*
    facilitated, 121–122, *121–122*
Digestion, 707t
    definition of, 699
    enzymatic, in small intestine, 705, 708
    hormonal regulation of, 710t
    in stomach, 704–705
Digestive juice, secretion of, regulation of, 708–709, 710t
Digestive system(s), *553*, 554t
    of invertebrates, 700–701, *702*, 703
    wall of, 703, *703*
Digits, of vertebrates, 565, *565*
Dihybrid cross, *250*, 250–251
Dinoflagellates, 354, *354*
Dinosaurs, 331, *334*
    original concepts of, *35*, 35–36
Dipeptide, 71, *71*
Diploid, 224–225
Diplopoda, *424*, 425
Directional selection, 307, *307*
Disaccharides, 64–65, *65*
Disease(s), defense against, white blood cells and, 641–642
    inherited, genetics and, 254
Disruptive selection, 307, *308*
Distal tubule, of nephron, 733, *734*
Divergence, of neurons, 589, *589*
Diversity, multicellularity and, 539
Divisions, 19
DNA, 8–9, 75, 256
    B-DNA, 194, *195*
    in bacteria, 346
    changes in, mutations and, 209–210
    cleavage of, 268, *269*
    coding strand of, 202–203
    isolation of, in genetic engineering, 267–268

DNA, *(Continued)*
    molecular basis of inheritance, 186–187
    molecule of, 191–194, *194*
    mutation of, and cancer, 548
    nucleotide subunits in, *193*, 193–194, 294t
    recombinant, 265, *267*, 267–270, *270*
    replication of, 195–196, *195–196*
    structure of, 193, *193*
    transfer of genetic information by, 8–9, 189–190
    viral, recognition sites in, 267–268, *269*
    Z-DNA, *195*
DNA polymerase, DNA replication and, 196
DNA probe, 273
DNA sequence, palindromic, 268, *269*
Domain, 203
Dominance, causes of, 837
    genetic, 246
    hierarchies of, 836–837, *837*
    incomplete, 246–249
    law of, 244
Dopamine, 586t, 597
Dorsal root, of spinal nerve, 606, *606*
Douche, for birth control, 781t
Douglass, Andrew, 497
Down's syndrome, *237*
Drug(s), dividing cells and, 230, 230f
    effects of, 608–609t
        on embryo, 810t
Drupe, 513
Duck-billed platypus, 442, *443*
Dumps, effects of, 918
Duodenum, 705
Dynamic equilibrium, 141

E
Ear, human, 623, *623*
Earth, early, conditions for chemical evolution in, 326–329, *327*
    inclination on axis, *870*
    life on, 867–874
    temperature of, heat balance and, 868–869
Earthworm, digestive system of, *702*
    excretory system of, *728*
    nervous system of, *595*
Echinoderms, 428, *428*
    characteristics of, 430–431t
    endoskeleton of, 564
Ecological level of organization, 14, *14*
Ecological niches, in adaptation of organisms, 854–856, *855*
Ecology, human, 907–928
Economy of thrift, development of, 924

Ecosphere, 14, 866–882, *867*
  and Industrial Revolution, *451*, 451–452
  society and, 20
  subsystems of, 867, *868*
Ecosystem(s), 14
  definition of, 848
  productivity of, 859–860, *860*
  survival through diversity, 16
Ectoderm, 404
Edema, in inflammatory response, 663, *663*
  lymphatic system and, 656
Edge effect, in species diversity, 861
Efferent arteriole, of nephron, 733, *734*
Efferent nerves, 596
Egestion, definition of, 699
Egg, activation of, fertilization and, 792–793
  fertilization and implantation of, *804*
  sperm surrounding, *791*
Einstein, Albert, 32–33
Ejaculation, 778
Ejaculatory duct, 767
Electrical synapses, 584
Electrocardiography, 649, *649*
Electrolytes, 56
Electromagnetic spectrum, 170, *172*
Electron(s), 44–46, *45–46*, 48
Electron acceptors, 154, *154*
  in fermentation, 164–166
Electron clouds, 45, *46*
Electron microscope, 36, *37*
Electron shells, 45
Electron transport, and ATP synthesis, 155
    proton gradient in coupling of, *160*, 160–162
Electronegativity, 50
Electroporation, 272
Electroreceptors, 615t, 625
Elements, 43–46
  chemical formulas of, 46–48
  and compounds, interaction of, 48
  in living organisms, 44t
Elephantiasis, *656*
Elimination, definition of, 699, 710
Elongation, 203
Elongation complex, 207
Embryo, development of, 803, *803–804*, 806t
    in vertebrates, 289, *289*
  environmental factors affecting, 809, 810t, 811
  in human life cycle, 812t
  incubation of, 772
Embryo sac, 397
Embryogenesis, experimental methods in, 463

Emphysema, smoking and, 695
Emulsification, definition of, 708
Enamel, of teeth, 703
End-product repression, 215–216
Endangered species, 921, *921*
Endochondral bone, development of, 567
Endocrine disorders, 746
Endocrine gland(s), exocrine gland vs, 743, *743*
  hormones of, 748–749t
  location of, *746*
  malfunction of, consequences of, 749t
Endocrine regulation, 742–760
  integration with nervous regulation, 746–747
Endocrine system, *552*, 554t, 743
Endocrinology, 743
Endocytosis, 127–128, *128*, *130*
Endodermis, 494, 498
Endolymph, of semicircular canal, 621
Endometrium, of uterus, 772
Endorphins, 586t, 618
Endoskeleton, 563
Endosperm, 397
Endosteum, of long bones, 566, *566*
Endosymbiont theory, of evolution, 329, *330*
Energy, 133–184
  biological work and, 135–137
  civilization's need for, 919–920
  definition of, 135
  extraction from food molecules, 153, *153*
  forms of, 135–137
  in glycolysis, 156–158
  heat, 136–137
  kinetic, 136, *137*
  law of conservation of, 137–138, *138*
  levels of, in atoms, 45
  of life, 134–151, *135*
  light, in electromagnetic spectrum, 170, *172*
    to make ATP and NADPH and H⁺, 174–179
    of plants to make sugar, 173–183
    waves and particles of, properties of, 170–172
  local, of earth, factors in, 869
  metabolism of, balance of, 716–717
  one-way flow of, 135
  potential, 136, *137*
  radiant, conversion of, to chemical energy, *170*
  release of, pathways of, 152–168
  release of, by falling object, *154*
  sources of, 163, *919*

Energy, *(Continued)*
  transfer of, role of organisms in, 848–851, *850*
  transformations of, in metabolic reactions, 139–141
  yield of, from complete oxidation of glucose, 163, 163t
Enkephalins, 586t
Entropy, *138*, 138–139
Environment, adaptation to, 10
  changes in gene frequencies and, 302
  somatic nervous system and, 605–606
  changes in, species evolution in response to, *11*, 11–12
    triggering of migration by, 829–831, *830*
  in control of flowering plants, 515–520
  deterioration of, prevention of, 924
  plant differentiation and, 470
Environmental resistance, to population growth, 857, *857*
Enzyme(s), 143–148, *145–146*, *148–149*
  digestive, secretion of, 708
  induction, 214
Eosinophils, 641, *641*
Epidermis, of plant, 467–468, *468*, 476, *476–477*
  of plant stem, 491, 493
  structure of, *560*, 561
Epididymis, *764*, 766
Epigenesis, theory of, 790
Epiglottis, 688, 704
Epinephrine, 748t, 756
Epiphysis, of long bones, 566, *566*
Epiphytes, 892, *892*
Epistasis, 251
Epithelial cells, of intestine, *708*
Epithelial tissue, body covering by, 559–561
Epithelial tissues, 540
  types of, 541–542t
Epochs, of geological time, 329, *331*
Equilibrium, 48
  dynamic, 141
  labyrinth of ear and, 620–622
Eras, of geological time, 329, *331*
Erectile dysfunction, 778
Erectile tissue, of penis, 767
Erythroblastosis fetalis, 260
Erythrocytes, oxygen transport by, *639*, 640–641
Escherichia coli, 213
Esophagus, function of, 704
Essential fatty acids, 713
Estradiol, action of, 777t
Estrogens, 748t, 757

Estuary, 897, *897–898*
Ethics, of science, 38
Ethology, 818
Ethylene, effects on plants, 534
Eubacteria, 345, 347
Euglenoids, 353, 358, *359*
Eukaryotes, 345
    chromosome structure and, 219–220, *220, 224*
        gene insertion in, 271–272
        gene regulation in, 217–221
        origin of, *330*
Eukaryotic algae, characteristics of, 340t
Eukaryotic cell, evolution of, 329
    life cycle of, *225,* 225–231
Euphoria, with mood drugs, 607
Euphotic zones, of aquatic habitat, 895
Euryhaline organisms, 895
Eutrophication, of waters, 904
Evidence, interpretation of, 29–30, *30*
Evolution, 10, 12, 282–342
    adaptive, 305
    of Africanized honeybee, 303
    changes in gene pool and, 299–302
    chemical, conditions for, 326–329
    definition of, 283, 298
    developmental biology and, *289,* 289–290
    environmental changes and, *11,* 11–12
    extinction of species in, 320–322
    gradualism in, 319
    by natural selection, 10–11
    pace of, 318–321
    punctuated, 319, *320*
    scientific evidence of, 287–295
        biochemistry and molecular biology, 293–295
        biogeography, 291–293
        comparative antomy, 287–289
        fossils, *290,* 290–291
    stasis in, 320
    theory of, 10–12
Excitatory stimuli, of nerve membrane, 583
Excretion, definition of, 710, 726
Exocrine gland, endocrine gland vs, *743, 743*
Exocytosis, 126, *127, 130*
Exons, 203, *204*
Exoskeleton, 423
Experience, effect on brain, 605
Experiments, example of, 33–34
Expiration, in breathing, 688
External gills, *683,* 685
Exteroceptors, 614
Extinction, background, 321
    in classical conditioning, 823, *823*
    mass, 321–322, *322*

Extinction, *(Continued)*
    species, 320–322
Extraembryonic membranes, function of, *801,* 801–802
Eye, 626, *630*
Eyespots, 626, *627*

F
Facilitation, of neurons, 589, *590*
Facultative anaerobes, 164, *165*
Fall turnover, in freshwater habitats, 897
Family, 19
Farsightedness, 628, *628*
Fat(s), digestion of, 708, *709*
    increased mobilization in diabetes mellitus, 755–756
    interrelationship with carbohydrate and protein, *717*
    neutral, 68
    storage in body, 714
Fat cell, *546*
Fatty acid(s), in citric acid cycle, 163, *164*
    definition of, 68
    essential, 713
    saturation of, 68
Feedback control, 147
Fermentation, 163–166, *165–166*
Ferns, 386–387, *387–389*
Fertilization, 383, *776, 779, 804*
    definition of, 779
    double, 400
    in flower, 508
    functions of, *791,* 791–793
    internal, 439
    in sexual reproduction, 763, *763*
    steps in, 792
Fertilization cone, 792
Fertilizers, soil, 504
Fetus, development of, 807
    in human life cycle, 812t
Fever, in inflammatory response, 663, *663*
Fiber(s), in carbohydrates, 711–712
    of muscle tissue, 549
Fibrinogen, in plasma, 640
Fibroblasts, 540, *543*
Fiddlehead, 386
Fiddler crabs, biological rhythms of, 818, *819*
    courtship behavior of, *838*
Filaments, 397
    of gills, 685, *685*
Filariasis, lymphatic blockage in, *656*
Filtration, in kidney, 729, 735–736
Finches, of Galapagos Islands, Darwin and, *286,* 317
Fireweed, in secondary succession, 862, *862*

Fish, 436–438, *436–438*
    electrical signals of, 835, *836*
    heart of, 644, *645*
    osmoregulation in, *730*
Fixation, and genetic drift, 300
Fixed-action pattern, instinct as, 827
Flagella, 346–347, *347*
Flagellated swarm cell, 360
Flagellates, *352,* 352–353
Flagged trees, of coniferous forests, 888
Flame cells, 412
    of protonephridia, 727, *727*
Flat beach zone, *899*
Flatworms, *412,* 412–416
    characteristics of, 430–431t
    excretory system of, 412, 727
Flavin adenine dinucleotide (FAD), 154
Flavin mononucleotide (FMN), 160, *160*
Fleming, Alexander, 31–32, *32,* 32f, *377*
Flies, pollination by, 511, *512*
    Redi's experiment using, 26, *26*
Flower(s), 235
    parts of, 397, *397*
        reproductive, *243*
    pollination of, 510–511, *512*
Flowering plants, 396–400, 507–520. *See also* Plant(s), flowering
Fluid homeostasis, lymphatic system in, 656–657
Flukes, 413–414, *414, 415*
Fluorescence, 172, *173*
Fluorine, function in body, 720t
Fog, forests and, *57*
Follicle, development of, in female reproduction, 771, *772*
Follicle-stimulating hormone (FSH), in female, 777t
    and sperm production, 767, 769t
Food chains, energy travel through, 858–860, *859*
Food processing, 698–724
Food pyramids, 858–859, *859*
Food supply, development of agriculture and, 451
Food webs, energy travel through, 858–860, *859*
Forebrain, divisions of, 598–599
Foreskin, of penis, 767
Forestry, as agriculture, 910–911
Formula, chemical, 46
    structural, 48
Fossil(s), 283–284, *290,* 290–291
    history of life in, 329–335
    radioisotopes in study of, *47*
Founder effect, gene frequencies and, 300, *301*
Fovea, of retina, 630

Fragmentation, in asexual reproduction, 762
Franklin, Rosalind, 191
Free-running period, definition of, 820
Freeze-fracture, of lipid bilayers, 113
Freshwater fish, osmoregulation in, *730*
Freshwater habitat, *896*, 896–897
Frog, cleavage and gastrulation in, *796*
   lungs of, *686*
   tadpole formation in, *218*, 219
Fronds, 386
Fructose, 61
Fruit(s), dispersal of, 514–515, *516*
   ethylene and, 534
   of flowering plant, 397, 510, *511*
   types of, 513–515, *513–515*
Fruiting bodies, 367
Fucoxanthin, 357
Fuel molecules, breakdown of, 153, 159–160
Fundamental niches, 854
Fungus(i), 19, 364–380
   characteristics of, 340t
   classification of, 368t, 368–374
   contributions of, 365
   and disease, 377–379, *378*
   economic importance of, 376–379
   reproduction of, 369–372
   structure of, 366

G
GABA, function of, 586t
Galactosamine, 66
Galapagos Islands, species of, *285*, *286*
Gallbladder, function of, 708
Gametangia, 353, 383
Gametes, 224
Gametophyte generation, 383
Ganglion(a), 412
   of nerve, 580
Gap junctions, 117, *118*, 584
Gap phase, 226
Gas, noble, 48
Gas exchange, 681–697
   adaptation for, 682–687
   in air sacs, 690–691, *692*
   epithelial tissue and, 559–560
   in lungs, 688
Gastric glands, 704, *706*
Gastric inhibitory peptide, 710t
Gastrin, 709, 710t
Gastropodia, 418
Gastrovascular cavity, in invertebrates, 637, *638*
Gastrulation, germ layer formation during, *796*, 796–797, *798*
Gemmae cup, 386, *386*

Gene(s), 221. *See also* Genetic(s)
   crossing-over of, 234
   desirable, introduction into plant, *277*, 277–278
   in DNA molecule, 199
   flow of, in differential migration of organisms, 301–302
   foreign, expression of, 267
   frequencies of, 300–302, *301*
   hormone activation of, 744, *745*
   inheritance of, 244–248
   insertion of, in eukaryotes, 271–272
   linkage groups of, 234
   linked, 234, 256
   recombinant, 234, 256, *256*
   regulation, 214
      in eukaryotes, 217–221
      in prokaryotes, 213–217
   sex-linked, on X chromosome, *255*, 255–256
   structure of, 199–200, 214
   types of, 246, 255
   vectoring of, 267
Gene bank, 269
Gene expression, 199, 246
   changes in, 209–210
   embryogenesis and, 463
   in new cell, 271
Gene function, 198–211
Gene pool, 283, 298, *298*, 311
   bottleneck effect and, 301, *302*
   gene frequency changes in, 299–302
   variation in, mutation increases, 299–300, *300*
Gene therapy, 272–274, *274*
Generation(s), 246
Generation time, 225
Genetic(s). *See also* Gene(s)
   alleles and, 248–249
   ethical questions and, 261–262
   human, 253–264
   inherited diseases and, 254
   molecular, and cancer, 274–275
   plant differentiation and, 470–471
   population, 297–309
   science of, Mendel and, 242
Genetic analysis, 249
Genetic code, 199–200, 200t
Genetic defects, human, repair of, 272–274, *274*
Genetic disease(s), probing for, 273
Genetic disorders, 261–262, 262–263t
   chromosome defects in, 237, *237*
Genetic drift, changes in gene frequencies and, 300–301, *301*
Genetic engineering, 265–266, *266*
   determination of safety of, 278, *278*

Genetic engineering, (*Continued*)
   future of, 272–273
Genetic information, in cell nucleus, 187–189
   transfer of, by DNA, 189–190
      through pilus, 267, *267*
   in viral nucleic acid, 190–191
Genetic mapping, 256–257
Genetic probe(s), 270–271, *271*
Genetic recombination, 234, 256, *256*
Genital herpes, 786t
Genitalia, female, 773, *773*
Genotype(s), 244
   Punnett square for predicting, 246, *247*
   test crosses to determine, 246, *248*
Genus, 17
Genus species epithet, 335
Geological time, divisions of, 329, *331*, 332–333t
Geology, 290
Germ layers, 404
   formation of, 796–797
Gestation period, in humans, 802
Giant clam, *900*
Gibberellins, 531–533, *532*
Gigantism, pituitary gland malfunction and, 752, *752*
Gills, for gas exchange, 683, *683*, 685–686
Ginkgo, 396, *396*
Giraffes, adaptation to environment, *11*, 11–12
Giraffe(s), Lamarck and, 284
Gland, definition of, 540, *543*
Glans, of penis, 767
Glasswort, *898*
Glassy sweepers, *901*
Glial cells, 550, 578
Globulins, in plasma, 640
Glomerulus, of kidney, 733, *734*, *736*
Glucagon, 748t, 753–754
Glucocorticoids, 748t, 757
Glucosamine, 66
Glucose, 61
   aerobic catabolism of, 153
   concentration of, regulation of, 753–756, *755*
   forms of, 61, *64*
   liver production of, 712, *712–713*
   oxidation of, energy yield from, 163, 163t
   use in body, 712, *712*
      in diabetes mellitus, 755
Glyceraldehyde-3-phosphate (PGAL), 157
Glycerol, 68, *69*
Glycine, 71
Glycocalyx, 112
Glycogen, 65, 66
   in muscle contraction, 571

Glycogen, (Continued)
  use in body, 712, 712
Glycolysis, 155–158, 158
Glycoproteins, 112, 115
Glycosidic bond, 64
Gnetophytes, 396, 397
Goblet cells, 540
Goiter, 753, 753
Golgi tendon organs, 619, 620
Gonadotropic hormones, 747, 748t
Gonadotropin-releasing hormone
  (GnRH), in female, 777t
    and sperm production, 767, 769t
Gonads, 235
Gonorrhea, 786t
Graded potential, of nerve membrane,
  583
Graft rejection, 673, 675
Grain, 513
Gram, Christian, 345
Granular leukocytes, 641, 641
Grasshopper, lifestyle as representative
  of insects, 426–428, 427
Grassland, temperate, 890, 890
Gravitropism, 525
Gravity receptors, in invertebrates,
  620, 620
Gray matter, of cerebrum, 599
Green algae, 355, 356
Green glands, in crustacean osmo-
  regulation, 728, 729
"Greenhouse effect", 872
Griffith, Frederick, 189–190, 190
Ground state, of atom, 172
Groundwater pollution, from landfills,
  918
Growth, abnormal, pituitary gland
  malfunction and, 752
    allometric, 316–317, 316–317
    factors affecting, 751–752
    internal skeleton and, 563–567
    as property of life, 4–5, 5
    regulation of, in invertebrates,
      745–746
        in vertebrates, 746–759
Growth hormone, action of, 748t
    human, transfer into mice, 276,
      276
    hypothalamic production of, 747
    malfunction of, consequences of,
      749t
Guanine, 191, 193
Guttation, 483, 484
Gymnarchus, electrical signals of, 835,
  836
Gymnosperms, 391–396, 392
    of boreal forest, 888
    roots of, secondary growth of,
      499
    stems of, 492
Gyres, of marine water, 875, 876, 877

H
Habitat destruction, and extinction,
  920
Habituation, as learning, 825–826,
  826
Hagfishes, 436, 436
Hair follicles, 560, 560–561
Halophiles, extreme, 348
Hammer, of middle ear, 623, 623
Haploid, 225
Haptens, as antigens, 668
"Hard" water, buffering capacity of,
  916f
Hardy-Weinberg law, 298–299
Haustoria, 378, 378
Haversian canals, of bone, 549, 566,
  566
Hay fever, 394
Heart, development of, 805, 805
    electrical activity of, 649, 649
    evolution of, 644–648
    human, structure of, 646–647
    innervation of, 606
Heart block, 649
Heart disease, 653, 713–714
Heart murmur, 648
Heart rate regulation, 647–648
Heart sounds, 648
Heat, as form of energy, 136–137
    in inflammatory response, 663,
      663
    specific, 54
Heimlich maneuver, 693
HeLa cancer cells, 274
Helper T cells, 665
Hemocoel, in invertebrates, 637–638,
  638
Hemoglobin, 640, 691–692
Hemoglobinopathic disease, 262t
Hemolysis, 258
Hemolytic anemia, 641
Hemophilia, 254
Hemorrhoids, 651
Heparin, in basophils, 641
Hepatic portal system, circulation in,
  653–655, 654
Herbivores, 700, 700
Hermaphoditism, 764–765
Heroin, effects on embryo, 810t
Herring gull, survivorship curve of,
  856, 857
Hershey, Alfred, 191
Heterocyst cells, of cyanobacteria, ni-
  trogen fixation by, 852
Heterospory, 391
Heterotrophs, 15, 347, 365
    as early cells, 328
    multicellular, 403–407
High-density lipoproteins (HDLs), 713
Hindbrain, divisions of, 597, 597
Hirudinea, 422, 423

Histones, 219, 220
Histoplasmosis, 378–379
HIV, 676
Hives, as allergic reaction, 678
Holdfasts, 355, 357
Home range, of animal territory, 837
Homeostasis, 5–6
    autonomic nervous system and,
      606
    excretory systems in maintenance
      of, 726
    fluid, lymphatic system in, 656–
      657
    mammalian kidney and, 730–739
Hominids, 445–448, 446–448
Homo erectus, 38, 449
Homo habilis, 448, 449
Homo sapiens, 449–450, 450
Homograft, definition of, 673
Homology, 287, 288
Homospory, 391
Honeybee, Africanized, evolution of,
  303
    social organization of, 841
Hormone(s), definition of, 743
    of endocrine glands, 748–749t
    plant, 523–535, 528t, 530t
        differentiation and, 471
    production by thymus, 666
    reproductive, female, 775–776,
      777t, 778
        male, 767–768, 769t
    second-messenger mechanism of,
      745
    secretion of, regulation of, 743–
      744
Hornworts, 384, 386
Horsetails, 387
Host, parasite and, 849
Host mothering, 782
Human(s), classification of, 19t
    cultural evolution of, 450–451,
      451
    digestive system of, 702
    lungs of, 687
    survivorship curve of, 856, 856–
      857
Human chorionic gonadotropin
  (hCG), 775, 802
Human ecology, 907–928
Human immunodeficiency virus
  (HIV), 676
Human leukocyte antigen (HLA)
  group, 666
Humerus, 287
Huntingdon's disease, 261
Hybrid sterility, 313, 313
Hybrid vigor, gibberellins and, 533
Hydra, 411, 411–412
    digestive system of, 701
    hydrostatic skeleton of, 561, 562

Hydrogen, transfer of, 154
Hydrogen bonds, 52, *52*
Hydrogen ions, water and, 55–56
Hydrologic cycle, 874–875, *876*
Hydrolysis, 64–65, 142
Hydroponics, 500–501, *502*
Hydrosphere, 867, *868*
Hydrostatic skeleton, 561, *562*
Hyena, *894*
Hygienic behavior, of Van Scoy bees, 820–822, *821*
Hylid frog, *835*
Hymen, 773, *773*
Hymenopterans, as social insects, 840–842
Hyperglycemia, in diabetes mellitus, 755
Hypersecretion, of endocrine gland, 746
Hypertension, 650
    and atherosclerosis, 652
Hyphae, 366–368, *367*
Hypoglycemia, 756
Hyposecretion, of endocrine gland, 746
Hypothalamus, of forebrain, 599
    function of, 601t
    hormones of, *747*, 748t
        menstrual cycle regulation by, 775–776
Hypothesis(es), 32
    historical, *34–35*, 34–36
Hypothyroidism, 752

I
Ice, hydrogen bonding in, 54–55, *55*
Ice ages, 334
Ileum, 705
Iliac blood vessels, 653–655, *654*
Immune responses, 661, 671
Immune surveillance, theory of, 673
Immunity, active vs passive, 672
Immunization, 672
Immunoglobulins, 668
Immunology, 661
Implantation, of egg, *776*, *804*, 805
Impotence, 779
Imprinting, 825, *825*
Impulse transmission, rate of, 587
Incinerators, effects of, 919
Incisors, 703
Independent assortment, law of, 244
Indoleacetic acid, 528
Inducer, function of, 799
Induction, 25, 29
    scientific method to check results of, 30–36
Inductive reasoning, 29, 29t
Industrial Revolution, ecosphere and, *451*, 451–452

Infancy, in human life cycle, 811, 812t
Inferior colliculi, in mammals, 598
Inferior vena cava, 653–655, *654*
Infertility, 780
Inflammatory response, 662–663, *663*
Infralittoral zone, in marine habitat, *899*
Ingestion, definition of, 699
Inguinal hernia, in male, 766
Inheritance, molecular basis of, 186–197
    patterns of, 241–252
    of skin color, 257, *258*
    X-linked, 255
Inhibition, end-product, 216
Inhibitors, of enzymes, 148, *149*
Initiation, 203
Initiation complex, 208
Innate behavior, instrumental conditioning of, *824*, 824–825
    stimuli triggering, 827–828
Inorganic compounds, 53–58
Insect(s), *424*, 425–428, *426–427*
    excretory system of, *729*
    nervous system of, *595*
    pollination by, 511, *512*
Insight learning, 826, *826*
Insolation, and temperature of earth, 869
Inspiration, in breathing, 688
Instincts, 827–831
Instrumental conditioning, 824, *824*
Insulin, 748t, 749t, 753–754
Integration, in neural response, 578, *579*
Integumentary system, *552*, 554t
Intercalated discs, of heart, 646
    of muscle, 549
Intercellular substance, of connective tissue, 540
Intercourse, sexual, 777
Interferons, in viral infections, 662
Internal defense, 660–680
Internal transport, 636–659
Interoceptors, 614
Interphase, 226, *226*
Interstitial fluid, lymphatic system recirculation of, 656
Intertidal zone, 895, 898, *899*, 902, *902*
Intestine, lining of, *708*
    wall of, *706*
Intrauterine device, for birth control, 781t, *782*, 782–783
Intrinsic factor, and digestion, 704
Introns, 203, *204*
Invertebrates, absence of circulatory system in, 637
    animals included in, 403
    digestive systems of, 700–701, *702*, 703

Invertebrates *(Continued)*
    internal defense mechanisms of, 661
    metabolic wastes disposal in, 726–729
    nervous systems of, 594–595
    non-chordate, 402–432
    osmoregulation in, 726–729
Iodine, function in body, 720t
Ions, 50–51, *51*
    biologically important, 50t
Iris, of vertebrate eye, 629, *630*
Iron, function in body, 716, 720t
Ischemia, cardiac, atherosclerosis and, 653
Islets of Langerhans, 753, *754*
Isolation, behavioral, 312–313, *313*
    reproductive, *311*, 311–314
Isolecithal eggs, 793
Isomers, 48, 61
Isotopes, 47

J
J region, of polypeptide chain, 668
Jacob, Francois, 213
Jejunum, 705
Jelly coat, of egg, 792
Joblot, Louis, 27
Joint(s), function of, 567
Joint receptors, as proprioceptors, 619, *620*
Jugular veins, 653–655, *654*
Jurrasic period, 331, 332–333t
Juxtaglomerular apparatus, in blood pressure control, 738

K
Kalanchoe, 510
Kangaroos, 443
Keratin, of epidermis, *560*, 561
Kidney, 729–739
    functions of, 729
    in homeostasis, 730–739
    structure of, 731–732, *733–734*
    water conservation by, *731*
Kidney disease, 735
Kilocalorie (kcal), 137
Kin selection, *843*, 843–844
Kinetochores, 227, *227*
Kingdoms, 19
    of biological life, 336–340, 340t
Koch, Robert, 350
Krebs cycle, *159*, 159–160
Krummholz, of coniferous forests, 888, *889*
Kurosawa, E., 531
Kwashiorkor, 721, *721*

L
Labia, of female genitalia, 773, *773*
Labor, stages of, 807–809

Labyrinth, of vertebrate ear, as organ of equilibrium, 620–622
Lactate, 165
Lactation, breasts and, 773–774
Lactic acid, fermentation of, 164–166
   and muscle fatigue, 572
Lactose, 64
   Lac operons and, 214, *215*
Lacunae, of long bones, 566, *566*
Ladder-type nervous system, of invertebrates, 595, *595*
Lag phase, of population growth, 857, *857*
Lakes, 896–897
Lamarck, Jean Baptiste de, 284
Lamella(e), of bone, 503, 549
Lamprey eels, 436, *436*
Lancelets, 435, *435*
Landfills, effects of, 918
Lanugo hair, in fetus, 807
Large intestine, 709–710
Larynx, 687, *687*
Last winter, 880
Lateral line organs, in fish, 620
Law(s), of rationality, 28
   scientific, 36
Leaf(leaves), 470, *470–471*
   bundle sheath of, 478
   dicot, 478–479, *478–479*
   epidermis of, 477, *478*, 479, *479*
   functions of, nonphotosynthetic, *485–486*, 485–487
   mesophyll of, 477
   monocot, *478–479*, 478–480
   shedding of, winter survival and, 484
   structure and function of, 475–487, *476*
   trichromes of, 468, 477, *477*
   veins of, 478, *478*
Learning, ability for, 826–827
   behavior modification by, 823–827
   process of, 604–605
Leeuwenhoek, Anton van, 350
Legume, 513
Lens, of vertebrate eye, 629, *630*
Leukemia, 642
Leukocytes, 641–642
Librium, effects of, 608t
Lichens, 374–376, *375*
Life, definition of, 4
   organization of, 1–78
   origin of, 31, *31*, 325–342
Life cycle, human, stages of, 811, 812t
Life spans, species differences in, 856–857
Life style, behavior and, 817–818
Life zones, 883–906
   aquatic, 895–904
   interaction of, 904

Life zones, *(Continued)*
   terrestrial, 884–895
Ligaments of Cooper, of breasts, 773
Ligase, in recombinant DNA construction, 269
Light, energy, 170–179. *See also* Energy, light
   polarized, perception of, 628, *629*
Light-independent reactions, 179
Lightfall, angle of, and local solar energy, 869, *869*
Lignin, 114
Limbic system, function of, 600, 602
Limiting factors, for plant life, 504
Limnetic zone, of freshwater habitats, 896
Linnaeus, Carolus, 17
Lions, *894*
Lipid(s), 66–68
   digestion of, 707t
   as energy source, 712–714
   metabolism of, *713*
Lipid bilayer(s), of membranes, 109–110, 113
Lipoproteins, 713
   in blood, 640
Lithosphere, 867–868, *868*
Litorral zone, of freshwater habitats, 896
Liver, function of, 708, *709*
   glucose regulation by, 712, *712*
   role in digestion, 707t
Liverwort, 384, *385*, 385–386
Lizards, behavioral thermoregulation in, 818, *819*
Lobe-finned fish, 438, *438*
Local potential, of nerve membrane, 583
Logarithmic phase, of population growth, 857, *857*
Long bones, structure of, 566, *566*
Loop of Henle, 733, *734*
   sodium concentration in, 737, *737*
Lovebird, nesting behavior of, *822*, 823
Low-density lipoproteins (LDLs), 713
LSD, effects of, 609t
Luminous organs, of deep-sea fish, 903, *903*
Lunar cycles, of behavior, 819
Lung(s), defense mechanisms of, 694, *694*
   respiration by, *683*
   structure of, *689*
   types of, *686*
   in vertebrates, 686–687
Lung cancer, smoking and, 695
Luteinizing hormone (LH), in female, 777t
   and sperm production, 767, 769t
Lyell, Charles, 286

Lymph, 549, 655
Lymph cells, origin of, *665*
Lymph flow, obstruction of, *656*
Lymphatic system, *553*, 554t, *655*, 655–656
   circulation in, *655*, 655–657
   relation to circulatory system, *656*
Lymphocytes, 655
Lymphokines, in antibody-mediated immunity, 666

M
Macroevolution, 316–318
Macromolecules, 65
Macronutrients, 502
Macrophages, 641
Maidenhair, 396, *396*
Major histocompatibility complex, in immune system, 666
Male pronucleus, in conception, 779
Malnutrition, 721
Malpighian tubules, in insects, 729, *729*
Malthus, Thomas, 286
Maltose, 64, *65*
Mammal(s), age of, 333–335
   brain of, *598*
   distinguishing features of, 442
   evolution of, 442
   heart of, 645, *645*
   placental, 442, *444*, 445
Mammary glands, 442
Mammography, 774, *774*
Mandibles, 425
Manganese, function in body, 720t
Manure, green, 504
Marijuana, effects of, 609t
Marine fish, osmoregulation in, *730*
Marine food web, *859*
Marine water, redistribution of, 875–877, *876*
Marine zone, 897–903
Marrow cavity, of long bones, 566, *566*
Marsupials, 442, 443, *443*
Mass extinction, of dinosaurs, 333
Mass number, 45
Mastectomy, for breast cancer, 774
Matrix, in bone formation, 567
   of connective tissue, 540
Matter, definition of, 135
Mechanoreceptors, definition of, 618
   responses of, 618–623
   stimuli of, 615t
Medulla, 597, *597*, 601t
Megasporangium, 392
Megaspores, 391
Meiosis, 224–225, 231–235, *233–235*, 246, 391
Melanin, 244
Melatonin, 748t, 759

Membrane(s), 107–132
  bilayers of, 109–110
    ingestion and ejection by, 128, 129
    splitting of, 113
  fluid mosaic model of, 109, 112
  glycoproteins in, 112, 115
  leaflets of, 109
  mechanism for moving materials through, 117–120t
  outer. See Plasma membrane
  permeability of, 117–120
  phospholipids in, 109, 110–111
  proteins in, 110–112, 114
Membrane potential, of neuron, 582
Membranous bone, 567
Memory, 604–605
Mendel, Gregor, 242–243, 242–244
Mendel's law, 244, 299
Meninges, of human central nervous system, 599
Menopause, 774
  hormone reduction in, 776–777
Menstrual cycle, 774–777, 775
Menstruation, 772
Meprobamate, effects of, 608t
Meristems, 460, 460
  apical, 490
    primary growth at, 460–461, 460–461
    lateral, secondary growth at, 461–462, 462
Merostomata, 424, 425
Mesenteric arteries, 653–655, 654
Mesoderm, 404
Mesophyll, of leaf, 477
Mesosome, 345
Mesozoic era, 331, 332–333t, 333
Messenger RNA (mRNA), 201
  codons, as instructions to ribosomes, 204–209
  processing, 217
Messenger substance, cell control by, 188–189
Metabolic processes, nutrient support of, 710–721
Metabolic rate, 716, 752
Metabolic reactions, energy transformations in, 139–141
Metabolic wastes, disposal of, 725–741
  in invertebrates, 726–729
  by kidney, 731
Metabolism, 5, 6, 135
  regulation of, in invertebrates, 745–746
  in vertebrates, 746–759
Metagenesis, 763
Metanephridia, 728, 728
Metaphysis, of long bones, 566, 566
Metarhodopsin II, in light perception, 631

Metastasis, in cancer, 548
Methanogens, 348
Methaqualone, effects of, 608t
Microclimates, variations in, 878–879
Microcosms, 861
Microevolution, 298
Micronutrients, 502
Microscopes, "tunnelling", 191f
Microspores, 391–392
Microvilli, 112, 116
  of intestine, 705, 708
Midbrain, 598, 601t
Middle ear, structure of, 623, 623
Migration, environmental triggering of, 829–831, 830
  navigation in, 831
Milk secretion, prolactin and, 750
Mimicry, 305
Mimosa pudica, 524, 525
Mineral(s), definition of, 715
  functions of, 720t
  root absorption of, 498, 501–504
  translocation of, 499–500
Mineralocorticoids, 748t, 757
Mitochondria, evolution of, 329
Mitosis, 224, 226
  in development, 791
  drugs affecting, 230, 231
  meiosis vs, 232–233, 233
  phases in, 226–230, 227–228, 231
Mitotic spindle, 227, 229
Mitral valve, of human heart, 646, 646–647
Molars, 703
Molds, 351, 358, 366
  formation of, 290
Molecular genetics, and cancer, 275
Molecules, 12
Mollusks, 418–420, 419
  characteristics of, 430–431t
  external skeleton of, 563
Molting, 563, 564
  hormone regulation of, 745–746
Monera, 18, 19, 344–350
  characteristics of, 340t
Monoclonal antibodies, 673
Monocots, 397
  stems of, 490
  tissues of, 493–494, 496
Monocylglycerol, 68
Monocytes, 641, 641
Monod, Jacques, 213
Monohybrid cross, 245–246, 246
Monosaccharides, 61
Monotremes, 442
Mons pubis, 773, 773
Mood drugs, effect on neurotransmitters, 607, 609
Moray eel, home range of, 837
Morels, 376, 377
Morphogenesis, in development, 791

Mortality, species differences in, 856
Morula, production of, 793
Mosaic picture, with compound eye, 627, 627
Moss(es), of Arctic regions, 374–375
  club, 387–391, 389
  gametophyte generation of, 384, 384
  as nonvascular plants, 384–386
  peat, 385
Moss campion, microclimate of, 879
Moth, peppered, evolution of, 17
Motor area, of cerebral cortex, 600
Motor function, dopamine and, 597
Motor nerves, 596
Motor neuron, 588
Mouth, 703–704, 707t
Movement, as property of cells, 6–8, 8
Mucosa, of digestive tract, 703, 703
Muddy zone, in marine habitat, 899
Mullerian mimicry, 305, 306
Multicellularity, of animals, 539
Multiple sclerosis, 579–580
Multipolar neuron, 578
Muscle(s), 572–573
  and body movement, 567–574
  functions of, 557–576
  for locomotion, 559, 559
Muscle contraction, 569–570, 570–571
Muscle fatigue, 166
Muscle fibers, structure of, 567–568, 568–569
Muscle layer, of digestive tract, 703, 703
Muscle spindles, 619, 620
Muscle tissue, 549, 550t, 551
Muscle tone, 570
Muscular system, 552, 554t
Mushrooms, 373, 373, 376–377, 377
Mutation(s), 11, 209, 209–210
  and gene pool variation, 299–300, 300
Myasthenia gravis, 675
Mycelium, 361, 366, 373, 374
Mycorrhizae, 365, 365
Myelin sheath, of neuron, 579, 580
Myocardial infarction, atherosclerosis and, 653
Myofilaments, of muscle fiber, 568, 568–569
Myosin, 567–569, 568–569, 572
Myxamoeba, 360
Myxedema, with thyroid gland malfunction, 753

N

Nannoplankton, 896
Naphthalenacetic acid, and root development, 531, 531
Narcotic analgesics, effects of, 608t

Nasal cavities, function of, 688
Natality, species differences in, 856
Natural killer (NK) cells, attack on cancer cells, 673, *674*
  in viral infections, 662
Natural selection, changes of gene frequencies, 302
  Darwin's theory of, 10–11, *11*, 282–296
  fitness of species for environment and, 303–308
  types of, 305, *306*
Navel, 809
Navigation, in migrating animals, 831
Neanderthals, 449–450, *450*
Nearsightedness, 628, *628*
Neckton, 895
Needham, John T., 29–30
Needles, 392
Negative-feedback control mechanisms, 743
  in thyroid gland regulation, 752
Negative reinforcement, in instrumental conditioning, 824
Nematocysts, 411
Nematodes, *417*, 417–418, 430–431t
Nemertinea, 416–417, *417*, 430–431t
Neonate, adaptation of, 809
  in human life cycle, 811, 812t
Neopallium, of cerebrum, 599
Neoplasms, 548–549
Nephridia, 420
Nephridial organs, 727–729
Nephron, function of, 733, *734*
Neritic waters, 903
Nerve, definition of, 580, *581*
Nerve cells, interconnections of, *588*
Nerve impulse, 581, 583
  transmission of, 578
Nerve net, 412
  of invertebrates, 594, *594*
Nerve ring, of invertebrates, 594
Nervous regulation, integration with endocrine regulation, 746–747
Nervous system, *552*, 554t
  cell types of, 578–581
  central, 596, *596*, 596t
  development of, 605, 797, *799*
  divisions of, 596t
  heart rate regulation by, 647–648
  of invertebrates, 594–595
  peripheral, 596, *596*, 596t
    divisions of, 605–606
  in responsiveness, 593–612
Nervous tissue, function of, 549–550
Nest parasites, 828, *829*
Neural circuits, 587, *589*
Neural control, in responsiveness, 577–592
Neural integration, 587
Neural response, 578, *579*

Neural structures, development of, 797, *798–799*
Neurilemma, 579, *580*
Neuroglia, of nervous system, 578
Neurohormones, in invertebrates, 745
Neuron(s), 550, 578
  associations of, 589
  function of, 581–584
  impulse transmission between, *585*
  knobs of, 587, *588*
  pathways of, 587
  resting potential of, 581, *581*, 615
  structure of, 578–580, *580*
Neurotransmitters, 584–585, *585*, 586t
  effect of mood drugs on, 607, 609
Neuston, of freshwater habitats, 896
Neutral fats, 68
Neutrons, 44
Neutrophils, 641, *641*
Niche differentiation, 856
Nicotinamide adenine dinucleotide (NAD), 154
Nicotine, effects of, 608t
Nitrifying bacteria, 852
Nitrogen, atom of, 50
  fixation of, 851–852, *853*
Nitrogenous wastes, 726, 739
NK cells, attack on cancer cells, 673, *674*
  in viral infections, 662
Noble gases, 48
Nodes of Ranvier, 579
Nodules, plant, nitrogen fixation in, 852, *853*
Nomenclature, binomial system of, 335
Noncompetitive inhibition, 148, *149*
Noncyclic photophosphorylation, 177, *178*
Non-electrolytes, 56
Nongonococcal urethritis, 786t
Norepinephrine, action of, 748t
  adrenal secretion of, 756
  function of, 586t, 587
Nose, 687, *687*
Novel origins, 782
Nuclear power, 920
Nuclear war, ecological consequences of, 880
Nuclear waste, pollution from, 918
Nucleic acids, structure of, *73*, 75–76, *76*
  viral, genetic information in, 190–191
Nucleosomes, *219*, 220
Nucleotides, 75–76, *76*, 193
  sequences of, 200, *201*
Nucleus, of cells, 12
Nutrients, essential, 68
  ingestion of, and geographic distribution of organisms, 884

Nutrients, *(Continued)*
  as sources of energy, 163
Nutrition, 699
  effects on embryo, 810t

O
Obesity, 717, 719–721
Observation(s), predictions tested by, 32–33, *33*
Ocean dumping, effects of, 918
Oceanic currents, and marine water redistribution, 875–877, *876*
Ocelli, light perception by, 626
Offshore caye, life surrounding, *903*
Olfactory bulbs, 599
Olfactory epithelium, 625
Oligochaeta, 422
Oligotrophic waters, 904
Ommatidium, of compound eye, 627, *627*
Omnivores, 700, 849
Oncogenes, 275
Oocytes, 770
Oogenesis, 768, *770*, *771*
Oogonia, 770
Oospore, 361
Operant conditioning, 824, *824*
Operator, structural genes and, 214
Operculum, 437
  of gills, 685, *685*
Operon(s), 213–214
  Lac, lactose and, 214, *215*
  Trp, tryptophan and, 214–215, *216*
Opines, in plant improvement, 277
Opossums, 443
Opsin, in rhodopsin, 630
Optic nerve, 630
Oral contraceptives, 780, 781t, 782, *782*
Orangutans, *893*
Orders, 19
Ordovician period, 330, 332–333t
Organ(s), 287–289, *289*
  definition of, 550
  development of, 805, *806*, 807
Organ of Corti, 622, *624*
Organ systems, 538–556, *552–553*, 554t
Organelles, 12
Organic compounds, 60–78
  biologically important, 63–64t
  structure of, 61, *62*
  synthetic, as pecticides, 911
Organic molecules, formation of, on primitive earth, 326–327, *327*
Organism(s), adaptation of, ecological niches in, 854–856, *855*
  binomial system of nomenclature of, 17
  classification of, 17, 335–336

Organism(s), *(Continued)*
geographic distributions of, 884
interaction of, *850*
variety of, 16–19
Organismic respiration, 682
Organogenesis, 533, 797
Orgasm, 778
Oribitals, of electrons, 45, *46*
Osmoregulation, 725–741
definition of, 726
in invertebrates, 726–729
in vertebrates, 729–739
aquatic, 729–730, *730*
Osmosis, 122–125, *123*, 123t, *124*
Osmotic conformers, 726
Osmotic regulators, 727
Osteichthyes, 437–438, *438*
Osteoblasts, 567
Osteoclasts, 549, 567
Osteocytes, 547
Osteons, 566, *566*
Otoliths, 621, *621*
Ovary(ies), of flower, 397
function of, 768, 770–772
hormones of, 749t
menstrual cycle regulation by, 775–776
Overpopulation, 922, 924, *925–926*
Overproduction, 287
Overweight, 717, 719–721
Ovulation, 771
prevention of, 780, 782
Ovum(a), maturation of, 768, *770*
transport of, 772
Oxaloacetate, 159
Oxidation, 52–53, 153–154
Oxidation-reduction reactions, 153–155
Oxidative phosphorylation, 162
Oxygen, bacteria needs for, 347
blood transport of, 691–692
depletion with water pollution, 914
plant release of, 173, *177*
Oxygen exchange, in air sacs, 691
Oxyhemoglobin, 640, 691
Oxytocin, 747, 748t, *750*
Oyster zone, in marine habitat, *899*
Oystercatcher, 828, *828*
Ozone, formation of, in early earth, 328, *329*
Ozone layer, endangerment of, 20, 872–873

P

P wave, of electrocardiogram, 649, *649*
Pacemaker, heartbeat and, 646
Pacinian corpuscles, 619, *619*
Paedomorphosis, 317, *317*

Pain, in inflammatory response, 663, *663*
perception of, 617–618
Pair bonds, reproductive cooperation by, 839
Paleontology, 290
Paleozoic era, 330–331, 332–333t
Pancreas, function of, 708
glucose concentration regulation by, 753–756
hormones of, 748t
role in digestion, 707t
Pancreatic enzymes, secretion of, 708
Pantothenic acid, 158
Parabronchi, in birds, 687
Paramecium, 353
competitive exclusion by, 856, *856*
Paramylon, 358
Parapodia, 421
Parasites, 347
Parasympathetic nervous system, 596, 606
Parathyroid gland, calcium regulation by, 753
of monkey, 105, 106
Parathyroid hormone, *744*, 748–749t
Parenchyma, of plants, 464, *464*
Parental investment, 839, *840*
Parietal cells, in gastric glands, 704
Parkinson's disease, 597
Parthenogenesis, 763–764
Parturition, stages of, 807–809
Passive immunity, 672–673
Pasteur, Louis, 27–28, *28*, 31, 350
Pasteurization, 350
Pathogens, 347
defense mechanisms against, 662, *662*
effects on embryo, 810t
inactivation of, 669
phagocytic destruction of, 663–664
Peat mosses, 385
Pectins, 114
Pectoral girdle, 565, *565*
Pedipalps, 425
Peforated ulcer, 705, *705*
Pellicle, 114
Pelvic inflammatory disease, 786t
Penetrance, incomplete, 251
Penicillin, 31–32, 32f, 283, 345
Penis, 767
Pentoses, 61
Pepsin, and digestion, 704
Peptic ulcer, 705, *705*
smoking and, 695
Peptide bonds, 71
Peptidoglycan, 116, 345
Pericardium, of human heart, 646, *646–647*

Pericycle, 494
branch roots and, 499, *499*
Periderm, 492
of plants, 468, *469*
Periods, of geological time, 329, *331*
Periosteum, of long bones, 566, *566*
Peripheral resistance, and blood pressure, 649
Peristalsis, 704, *704*
Peritubular capillaries, of nephron, 733, *734*
Periwinkle, in marine habitat, *899*
Permafrost, 886
Permian period, 331, 332–333t
Pesticides, 911–913, *912*
Pests, suppression of, in mechanized agriculture, *909*
Petals, 397
Petrifactions, 290, *290*
pH buffers, 57–58
pH scale, 56, *56*
Phage(s), 190–191, 270
action of, 339, *341*
Phagocytosis, 127–128, *128*
pathogen destruction by, 663–664
stimulation of, 669
Phagosome, 663
Phalanthus, behavior of, 817
Phantom pain, 617
Pharynx, 687, *687*, 704
Phenothiazines, effects of, 608t
Phenotype, 244
Phenylketonuria, 263t
Pheromones, in animal communication, 835–836
Phloem, 465–466, *466*, 478, 490–492, 494, 498
transport in, 499–501, *501*
Phosphate, regulation of cellular respiration by, 162–163
Phosphoglyceraldehyde (PGAL), 180
Phosphoglycerate, 180
Phospholipids, 68, *70*
in membranes, 109, *110–111*
Phosphorus, body need for, 716, 720t
Phosphorus cycle, 852–853, *854*
Phosphorylation, 143, *143*, 155–156
Photolysis, 175, *177*
Photons, 171, *172*
Photoperiodism, flowering plants and, 516
Photophosphorylation, 177–178, *178*
Photoreceptors, 518
light absorption by, 626–632
stimuli of, 615t
Photorespiration, 179
Photosynthesis, 14–15, 169–184, *170–171*, *182*, 182t, 480–481, 868
absence of, on ocean bottom, 903
dependency of life upon, 170

Photosynthesis, *(Continued)*
in desert plants, 483
efficiency of, 179
Photosynthetic lamellae, 348, *349*
Photosynthetic organisms, in plants, 382–384
Photosystems, 175, 177, *177*
Phototropism, 525
Phyla, 19
animal, comparison of, 430–431t
Phylogenetic relations, nucleotide sequences in DNA and, 294t
Physical dependence, on mood drugs, 609
Phytochrome, and plant responses to light, 517–520, *520*
Phytoplankton, 896
Pigment(s), 172
Pigmentation, hormone regulation in invertebrates, 745–746
Piltdown man, 38–39
Pilus(i), 269–270, 347, *347*
Pine tree, 392
Pineal gland, hormones of, 749t, 759
Pink sea fan coral, *900*
Pinocytosis, 128, *129*
Pit vipers, thermoreceptors of, 625, *626*
Pith rays, 494
Pituitary gland, anterior lobe of, growth regulation by, 750–752
menstrual cycle regulation by, 775–776
function of, 746
hormones of, 748t
target tissues of, *751*
posterior lobe of, hormone secretion by, 747, 750, *750*
Placenta, 445
function of, *801*, 801–802
Placental mammals, 442, *444*, 445
Placoderms, 436–437
Planarians, 412–413
Plankton, 351, 895, *895*
as basis of marine food chain, 903–904
Plant(s), 19, 381–395
absorption by, 498
alternation of generations in, 383, *383*
ancient, and coal formation, 390
annuals, 462, *462*
apical dominance of, 530, *531*, 533
biennials, 462, *462*
body of, 468–470, 490
cells of, 462, 472, *472*
cytokinins in, *533*, 533–534
development of, *217*, 218–219

Plant(s), *(Continued)*
localizing phytochrome in, 517
types of, 467t
characteristics of, 340t
collenchyma of, 464, *465*
comparative antomy of, 468–470
cross-pollination of, 510–511
day-neutral, 518
dermal tissue system of, 467–468, *468–469*
development of, 456–473
differentiation in, 470–473
diseases of, fungus and, 377–378, *378*
essential elements in, functions of, 503t
extinction of, 920–922
fibers of, 465
flowering, 396–400
classes of, 397
environmental cues inducing, 515–520
photoperiodism and, 516
production by, 396–397, *397*
reproduction in, 507–520
asexual, 508–510
sexual, *397–398*, 397–400, 508, 510–515
growth of, 456–473
effect of air pollution on, 914
elements essential for, 501–504
external stimulus of, 525–526, *527*
hormonal regulation of, 526–535
plant hormones in, 530t
herbaceous, 460
hormones of, 523–535, 528t
interactions between, 530t
hydrophyte, 480
insectivorous, *486*, 487
land, 382–383, *383*
leaves of, 470, *471*, *476*
light energy of, to make sugar, 173–183
limiting factors for, 504
localized growth in, after germination, 458–462, *459*
long-day, 517–518
major groups of, 382, 382t
meristems of, 460, *460*
middle lamella of, 503
minerals in, obtained by roots, 501–504
movements of, changes in turgor and, 524
nonvascular, mosses as, 384–386
oxygen released by, 173, *177*
parenchyma of, 464, *464*
perennial, 462, *462*

Plant(s), *(Continued)*
photosynthesis in, 383–384, 480–481, 483
recombinant DNA techniques to improve, 276–278
reproduction of, by apomixis, 510, *510*
temperature and, 520
responses of, biological clock influences on, 524, *526*
roots of. *See* Root(s)
sclerenchyma of, 464–465, *465*
seedless vascular, 386–391
short-day, 516–518
solar tracking by, 524, *526*
sporophyte, 384
stems of. *See* Stem(s)
tension-cohesion mechanism in, 500
tissues in, 462
transpiration and gutation by, 482–483, *484*
transport in, 499–501
vascular system of, 465–467, *466*, *490*, *490*
xeric, 483
xerophyte, 480
Plasma, 639–640
Plasma cells, 666
Plasma membrane, 12, 108–109, *109*, *112*
Plasmids, 216–217, *269*, 269–270, 346
Plasmodesmata, 116–117, *117*
Plasmodium, 358, *359*
Plate tectonics, and continental drift, 336–337, *336–337*
Platelets, 642, *642*
Platyhelminthes. *See* Flatworms
Platypus, duck-billed, 442, *443*
Play, as practice behavior, 840, *841*
Pleiotrophy, 251
Pleural membrane, function of, 688
Plumeworm, *901*
Polar bodies, in female reproduction, 770
Polar nuclei, 397
Polarization, of neuron, 581
Pollen, 392, 394, 400
Pollination, 510–512
Pollution. *See also* Air pollution; Water pollution
definition of, 913
improper waste disposal and, 913–919
population growth and, 922
radioactive, 880, 917–918
Polychaeta, 421–422, *423*
Polygenes, 257
Polymers, 65f
Polypeptides, 71, 199, 668

Polyplacophora, 418
Polyploidy, 315
Polyps, of coral reef, 900
Polyribosome, 207, *207*
Polysaccharides, 65–66
Polyspermy, blocking of, 792
Ponds, 896–897
Pons, 597, *597*, 601t
Population, age distribution of, 923, *923*
    definition of, 856
        growth of, 856–858, *857*
            effects of, 922–924
Population crash, 858
Population ecology, 847–865
Population genetics, 297–309
Porifera. *See* Sponges
Porphyrin ring, 173, *174*
Positive reinforcement, 824
Post-translational processing, 271–272
Postovulatory phase, of menstrual cycle, 775
Postsynaptic neuron, 584
Postzygotic isolating mechanisms, preventing reproduction, 313
Potassium, function in body, 720t
    in plants, 503
Potassium ion(s), in repolarization, 583–584
Potassium ion mechanism, stomates and, 481, *482*
Potential energy difference, in neuron, 582
Pouchet, Felix, studies of, 27
Prairie dog, kin selection in, *843*
Preadaptations, 316
Precambrian era, 329–330, 332–333t
Precipitation, 875, *878*
    and biome distribution, *885*, 888–891
    causes of, 877–879
Predators, as consumers, 849
Prediction(s), 32–34, *33*
Preformation theory, 790, *790*
Pregnancy, deaths from, 783t
    menstrual cycle with, 776
    smoking in, 695
Premenstrual syndrome (PMS), 776
Premolars, 703
Prenatal development, human, 802–809
Preovulatory phase, of menstrual cycle, 774
Prepuce, of penis, 767
Presynaptic neuron, 584
Prezygotic isolating mechanisms, interfering with mating, 312–313
Primary succession, 861, *862*
Primates, evolution of, 445–452
Primitive agriculture, *909*
Prion, 339

Problems, recognition and statement of, 31–32
Proboscis worms, 416–417, *417*
Processing food, 698–724
Producers, in ecosystems, 14–16, *15*
    manufacture of food by, 849
Productivity, of ecosystems, 859–860, *860*
Products of reaction, 48
Profundal zone, of freshwater habitats, 896
Progesterone, action of, 748t, 777t
    in menstrual cycle, 775
Proglottids, 414
Prokaryotes, 213, 345, *346*
    evolution of, 328
    gene regulation in, 213–217
Prolactin, action of, 748t, 777t
    hypothalamic production of, *747*
    pituitary secretion of, 750
Promoter, 214
Promoter site, 202
Pronuclei, of sperm and egg, fusion of, 792
Proprioceptors, 614
    muscle movement coordination by, 619
Prosimians, 445
Prostaglandins, 758
Prostate gland, 767
Protective coloration, 303, *304*
Protein(s), 68–74
    amino acids in, 200
    analysis of, 463
    carrier, 122
    as central to identity of organism, 199
    classification of, 70
    digestion of, 707t
    as enzymes, 199, 714–715
    increased use of, in diabetes mellitus, 756
    interrelationship with carbohydrate and fat, 717
    levels of organization of, 73, *73*
    in membranes, 110–112, *114*
    messenger RNA and, 204–205
    metabolism of, *716*
    as species-specific, 70–71
    structure of, 73, *74*, 199–200
    synthesis of, 715
        growth hormone stimulation of, 750–751
        RNA and, 198–211
        steps in, 201
        termination of, 209
    transport of, 122, 125
Protein kinase, 147, *148*
Protein malnutrition, 721, *721*
Proterozoic era, 332–333t
Prothallus, 386, *389*

Protists, 19, 344, 350–362
    characteristics of, 340t
    classification of, problems with, 338–339
    as earliest eukaryotes, 361
Proto-oncogenes, 275
Protobionts, 327, *328*
Proton(s), 44, 178
Proton gradient, 155, 162
Protonephridia, 412, 727, *727*
Protoplasts, 472, *472*
    fusion of, in plant improvement, 276–277
Protostomes, 405, 407
Protozoa, 340t, 351–353
Proximal tubule, of nephron, 733, *734*
Pseudocoelom, 404
Pseudoplasmodium, *360*, 360–361
Pseudopods, *351*, 351–352
Pseudostratified epithelium, 542t
Psychological dependence, on mood drugs, 607
Publication of scientific findings, 37
Pulmonary circulation, 651, *651*, 653
Pulp cavity, of teeth, 703
Pulvinus, 524
Punnett square, for predicting genotypes, 246, *247*
Pupil, of vertebrate eye, 629, *630*
Purine, 193
Pylorus, 705
Pyrimidine bases, 193
Pyrophosphate group, 143
Pyruvate, 155f, 157–159

Q
QRS complex, of electrocardiogram, 649, *649*
Quantum, 46
Quaternary period, 332–333t, 333
Queen angelfish, *901*

R
Radial cleavage, *794*, 795
Radiation, adaptive, 317, *318–319*
    for breast cancer, 774
    effects on embryo, 810t
    ultraviolet, *172*
Radioactive pollution, 880, 917–918
Radioactivity, 47
Radius, 287
Ragweed pollen, allergic reaction to, 677, *677*
Rain shadow, 877
Range of species, 291
Rat, instrumental conditioning of, 824, *824*
Rationality, law of, 28
Ray(s), 437
Ray-finned fish, 438
Reabsorption, in kidney, 729, 736–737

Reactants, 48
Reaction units, 177
Realized niches, 854
Reasoning, 29, 29t
Reception, in neural response, 578, *579*
Receptor(s), classification by stimuli, 615t
Receptor potentials, 614–616, *616*
Receptor proteins, for hormones, 744–745
Recombinant DNA, 265
  techniques of, 266–272
    to improve plants, 276–278
    policies on utilizing, 278
Rectum, 710
Red algae, 355–357, *357*
Red blood cells, cytoskeleton of, *115,* 122
  oxygen transport by, *639,* 640–641
Red bone marrow, 640
Red nucleus, of mammalian brain, 598
Redi, Francesca, *26,* 26–27
Redness, in inflammatory response, 663, *663*
Redox reactions, 53, 154
Reduction, 53, 154
Referred pain, 617
Reflexes, 587–589, *588*
Refractory period, in action potential, 584
Regeneration, 187
Regulator, 147
Regulator genes, 214
Release-inhibiting hormones, 746, *747*
Releasers, in animal communication, 835
  for innate behavior, 827
Releasing hormones, 746, *747*
Renal blood vessels, 653–655, *654*
Renal cortex, 731, *734*
Renal medulla, 731
Renal pelvis, 732
Renal threshold, 736
Renal tubule, 733, *734*
Renewable resources, effect on population growth, 858
Renin, in blood pressure regulation, 651
Repolarization, potassium ions in, 583–584
Repression, end-product, 215–216
Repressor protein, 214
Reproduction, 8–9, *9–10,* 761–788
  asexual, 9, *9,* 762
  differential, 11
  human, female, 768–777
  hormones in, 777t, *778*
  male, 765–768

Reproduction, *(Continued)*
  regulation of, in invertebrates, 745–746
    in vertebrates, 746–759
  sexual, 9, *9–10,* 224, 762–763
    variations in, 763–765
Reproductive cooperation, by pair bonds, 839
Reproductive isolation, *311–312,* 311–313
Reproductive system, *553,* 554t
  human, female, *769, 770*
    male, *764*
Reptile(s), 439–441
  adaptations of, 439–440, *439–440*
  body temperature of, 440
  brain of, *598*
  evolution of, 299, 442, *442*
  extinction of, 440–441
  heart of, 645, *645*
  lungs of, *686*
Research, scientific, cost of, 37
Resistance, to pesticides, 911
Resources, and population growth, 858, 922
Respiration, 682
  anaerobic, 163, *165*
  cellular. *See* Cellular respiration
  countercurrent exchange system in, 686
Respiratory disease, 694–696
Respiratory surfaces, types of, 673, *683*
Respiratory system, *553,* 554t, 682
  human, *687,* 687–694
Responsiveness, nervous systems in, 593–612
  neural control in, 577–592
Resting potential, of neuron, 581, *581*
Restriction endonucleases, 268
Reticular activating system (RAS), 602–604
Retina, 629–630, *630–631*
Retinal, in rhodopsin, 630
Retroviruses, 272–273, *273*
Reverberating circuit, of neurons, 589, *590*
Rh compatibility, 259–261, *260*
Rhinoceros, *700*
Rhizoids, 368, 384
Rhizomes(s), 386, *508,* 508–509
Rhodopsins, 626, 630–631
Rhythm method, of birth control, 781t
Rib cage, 564, *565*
Ribonuclease, secretion of, 708
Ribonucleic acids (RNA), 75
Ribose, 61, 142
Ribosomal RNA (rRNA), 201, 203–204
Ribosome(s), 71, 203–207, *205–206*
Ribulose phosphates, 180

Rigor mortis, definition of, 570f
Ringworm, 378
RNA, composition of, 200, *201*
  production of, *201–202,* 201–203
  and protein synthesis, 198–211
RNA polymerase, 202–203
Rocky shore, zonation along, *899*
Rod cells, of retina, 630, *631*
Root(s), 468, *469*
  branch, pericycle and, 499, *499*
  food-storage tissues of, 490, *491*
  functions of, 490, 498
  minerals obtained by, 501–504
  and stems, 489–504
    differences between, 492–499, 494t
  structures of, 493, *494*
Root pressure, 499
Round dance, of honeybees, 841, *842*
Round window, of middle ear, 623, *623*
Roundworms, *417,* 417–418
Royal jelly, of honeybees, 841
Rubella, effects on embryo, 810t
Rugae, of stomach, 704
Runners, *509,* 509–510

S
SA node, of heart, 646
Sac fungi, 369
Saccule, of ear labyrinth, 621, *621*
Salamander, lungs of, *686*
Salivary glands, 703, 707t
Salt(s), definition of, 56
  formation of, 56–57
  in urine, 739
Salt marshes, 898, *898*
Saltatory conduction, 584
Sampling, experimental, 30, *30*
Sand island, life surrounding, *903*
Sandy beach, life zones on, *899*
Saprobes, 347, 365
Sarcoma, 548
Sarcoplasm, of muscle fiber, 568, *568–569*
Satiation, as instinct, 827
Saturation value, of air, 875
Savannah, *894,* 894–895
Scavengers, as consumers, 849
School of Athens, by Raphael, *25*
Schwann cells, 578
Science, 23–40
  beginnings of, 25–28
  cost of, 37
  ethical dimensions of, 39
  as human activity, 36–38
  as means of investigation, 24
  and technology, interaction of, 36
  understanding of universe via, 24–28

Scientific method, 30–37
Scientists, 37–38
Scientist(s), 26
Sclera, of vertebrate eye, 629, *630*
Sclerenchyma, of plants, 464–465, *465*
Scrotum, 766
Sea anemone, *901*
Seastar, *794*
Seaweeds, of intertidal zone, *902*
Second messengers, hormone function through, 744, *745*
Secondary response, of immune system, 671
Secondary succession, 862, *862*
Secretin, function in digestion, 710t
Secretion, epithelial tissue and, 559–560
    in kidney, 729
Seed(s), dispersal of, 514–515
    germination of, 458–462, *459*
    production of, 391–400
Segregation, law of, 244
Selection, 306–307, *306–308*
Semen, production of, 767
Semicircular canals, 621, *621*
Semilunar valves, of human heart, 646, *646, 647*
Seminal vesicles, 767
Seminiferous tubule, structure of, 765, *765*
Sensation, basis of, 616, 618
Sense organs, 613–635
    of dermis, *560*, 561
Sensory adaptation, 618
Sensory area, of cerebral cortex, 600
Sensory nerves, 588, 596
Sepals, 397
Septum(a), 366, 420
    in hydrostatic skeleton, 562
Serological typing, in organ transplantation, 675
Serotonin, 586t
Serum, blood, 640
Serum sickness, 672
Setae, 420
Sex, inheritance of, 235–237, *236*
Sex cells, 224
Sex chromosomes, 236
Sex linkage, 255
Sexual behavior, sociality of, *838*, 838–840
Sexual intercourse, 777
Sexual reproduction, 9, *9–10*, 224, 762–763
Sexual response, 777–779
Sexually transmitted diseases, 785, 786t
Sharks, 437
    osmoregulation in, *730*
Shrew, 445, *445*
Sickle-cell anemia, 262, 262t

Sign stimulus(i), in animal communication, 835
    and behavior, 827–828
Silurian period, 330, 332–333t
Silviculture, 911
Sinoatrial node, 646
Sinusoids, of liver, 654
Siphons, of clams, 419
Size, multicellularity and, 539
Skates, 437
Skeletal muscle, 549, 550t, *551*
    specialization of, 573
Skeletal system, *552*, 554t
Skeleton, external, 562–563
    functions of, 557–576
    and geographic distribution of organisms, 884
    internal, 563–567
Skin, color of, polygenic inheritance of, 257, *258*
    functions of, 557–576
    structures of, 560, *560*
Skull, of vertebrates, 564, *565*
Slash-and-burn agriculture, *909*
Sleep, cerebral cortical activity and, 604
Slime molds, 340t, 351, 358–361
Small intestine, 705, 707t, 708–709
Smell, chemoreceptors for, 623, 625
Smoking, and atherosclerosis, 652
    effects on embryo, 810t
    facts about, 695
Smooth muscle, 549, 550t, *551*
    specialization of, 573
Snail(s), of intertidal zone, *902*
    mucous secretion of, 559
Snowy owl, *887*
Social behavior, 833–846, *834*
Social insects, 840–842
Sociality, communication in, *835*, 835–836
Society(ies), among animals, 840–843
    definition of, 834
    and ecosphere, 20
Sociobiology, 844
Sodium, 51
    function in body, 720t
    regulation of, *737*, 737–738
Sodium chloride, function in body, 716
Sodium-potassium pump, 582, *582*
"Soft" water, buffering capacity of, 916f
Solar energy, warming of earth by, 868–869
Solar power, as energy source, *919*, 919–920
Solar tracking, by plants, 524, *526*
Solid waste, effects of, 918–919
Solutions, 123, *123*
Solvent, water as, 53–54

Somatic cells, 224
Somatic nervous system, 596, 605–606
Somatotropin. *See* Growth hormone
Sonogram, 811, *811*
Sori, 386
Spallanzani, Lazaro, 27, *27*
Specialization, multicellularity and, 539
Speciation, 311, *314*, 314–315
Species, 10, 17, 311
    diversity of, 860–861
    extinction of, 320–322
    origin of, 293, *293*
    range of, 291
Specific heat, 874
Sperm, entry into egg, *791–792*, 792
    production of, 765, *766*
    structure of, 767
    transport of, 766–767
Sperm cell differentiation, 765
Spermicides, for birth control, 781t, *782*
Spermatic cord, 766
Spicules, 407
Spike, in action potential, 583
Spinal cord, function of, 599, *600*
    nerves of, 606, *606*
Spines, of plants, 485
Spiracles, 683
    of insects, 428
Spiral cleavage, 795, *795*
Sponges, 407–409, *408*
    characteristics of, 430–431t
Spongocoel, 408, *408*
Spongy bone, of long bones, 566, *566*
Spontaneous abortion, 784
Spontaneous generation, 26–27
Sporangia, 235, 358–360, *360*, 386
Spores, 225, 235, 366–367, 383–384
Sporocarps, 367
Sporophylls, 392
Sporophyte generation, 383
Sporozoa, 353
Spring turnover, in freshwater habitats, 897
Squamous epithelium, 541t, 542t
Stabilizing selection, 306, *306*
Stamens, 397
Starch, 65, *66*
    as energy source, 711–712
Statocysts, in invertebrates, 620, *620*
Stem(s), 468–470, *470*
    abscission zone of, 484, *484*
    functions of, 490
    and roots, 489–504
        differences between, 492, 494t, 499
Stenohaline organisms, 895
Stereoisomers, 61
Sterility, 780
    hybrid, 313, *313*

Sterility, (Continued)
  in male, 766
Sterilization, for birth control, 781t, 783–784, 784
Sternum, 565, 565
Steroids, 68, 71
  gene activation by, 745
Stigma, 397
Stimulus(i), definition of, 578
  response of living things to, 8, 8
  supranormal, and inappropriate behavior, 828, 828
Stirrup, of middle ear, 623, 623
Stolons, 509, 509–510
Stomach, innervation of, 606
  role in digestion, 707t
  structure of, 704, 704
Stomates, 383, 468
  role in photosynthesis, 481, 481
Stories, of tropical rain forests, 892
Strangler tree, 892, 892
Strata, of epidermis, 560, 561
Streams, 896–897
Stress, adrenal gland and, 756–757
  effects of, 758
Striated muscle, 549
Stroke, atherosclerosis and, 652
Stromatolites, 328, 328
Structural formula, 48
Structural genes, 214
Structural isomers, 48
Style, of flower, 397
Subclavian blood vessels, 653–655, 654
Subcutaneous tissue, 560, 561, 561
Suberin, 494
Submucosa, of digestive tract, 703, 703
Substance P, in pain perception, 618
Substrate, 144
Subtidal zone, 895
  of rocky beach, 902, 903
  of sandy beach, 899
Subtropical biomes, 895, 896
  forest, 890
Succession, in aquatic habitats, 904
  in community development, 861
  reason for, 863
Suckers, 510
Sucrose, 64, 65
  in carbohydrates, 711–712
Sugar(s), as energy source, 711
  light energy of plants to make, 173–183
  translocation of, 500–501
Sulfur, function in body, 720t
Sundew, 10
Sunflower, stem of, 491, 491
Superficial muscles, 572, 573–574
Superior colliculi, in mammalian brain, 598

Superior mesenteric vein, 654
Superior vena cava, 653–655, 654
Suppressor T cells, 665
Supratidal zone, of sandy beach, 899
Surrogate motherhood, 782
Survival, 287
Survivorship curve, of humans, 856–857, 856–857
Sweat glands, 6, 560, 561
Syllogisms, 28–29, 29
Symbionts, 347, 352
Sympathetic nervous system, 596, 606
Sympatric speciation, 315
Synapsis(es), 234, 584
  transmission across, 584–587
Synaptic knobs, of neuron, 578
Syngamy, 351
Synovial fluid, in joints, 567
Syphilis, 786t
  effects on embryo, 810t
Systemic circulation, 651, 651, 653–655, 654
Systole, in circulation, 648

T
T cells, activation of, 670
  receptors for, 671
  types of, 665
T cells (lymphocytes), 665, 665
T tubules, of muscle fiber, 568, 568–569
T wave, of electrocardiogram, 649, 649
Tadpoles, 438
Taiga, 888
  secondary succession in, 862, 862
Tapeworms, 414–416, 415–416
Target tissues, in endocrine system, 743
Taste, chemoreceptors for, 623, 625, 626
Taxonomy, 17, 335, 337
Tay-Sachs disease, 262t
Technology, and science, interaction of, 36
Tectorial membrane, in organ of Corti, 622
Teeth, 703
Telencephalon, 599
Telolecithal eggs, 796
Temperate biomes, variation with precipitation, 888–891
Temperature, and biome distribution, 885
  body, regulation of, 6, 7
  increase in, 20
  plant reproduction and, 520
Tension-cohesion mechanism, in plants, 500
Teratogens, 809, 809
Termination, 203

Termites, nest of, 893
  as social insects, 840–842
Territoriality, in animals, 837–838
Tertiary period, of Cenozoic era, 332–333t, 333
Testis(es), hormones of, 749t
  sperm production by, 765
  structure of, 764
Testosterone, 748t, 768, 769t
Tetrads, 234
Tetrapods, 438
Thalamus, of forebrain, 599
  function of, 601t
  pain reception in, 617, 617
Thalidomide, effects on embryo, 810t
Thallus, 385, 385
Theory, 36
Therapeutic abortion, 784
Therapsids, 442, 442
Thermal inversion, air pollution and, 916–917, 917
Thermal stratification, in freshwater habitats, 897
Thermal vents, on ocean bottom, 903
Thermoacidophiles, 348
Thermodynamics, 136
  first law of, 137–138, 138
  second law of, 138–139, 139
Thermoreceptors, and heat perception, 625, 626
  stimuli of, 615t
Thermoregulation, behavioral, in lizards, 818, 819
  and geographic distribution of organisms, 884
Thigmotropism, 525–526
Thimann, Kenneth, 528
Three-toed sloth, 893
Threshold level, of depolarization, 583
Thrift, economy of, development of, 924
Thrombocyte, 641
Thrombus, in coronary artery, 653
Thumb, opposable, of primates, 445
Thylakoids, 173
Thymine, 191, 193
Thymosin, 666
Thymus gland, hormone production by, 666, 759
Thyroid gland, hormones of, 748t
  malfunction of, 749t, 752–753
  and metabolic rate, 752
  secretion of, regulation of, 752
Thyroid-stimulating hormone (TSH), action of, 748t
  hypothalamic production of, 747
  pituitary gland secretion of, 752
Thyroxine, 748t, 752
Tidal pool, life in, 902
Tight junctions, 117, 119
Timberline, of coniferous forests, 888

Tinbergen, sand wasp experiment of, 817, *818*
Tissue(s), 12, *14*
    definition of, 539
Tissue-culture techniques, embryogenesis and, 463
Tissue fluid, lymphatic system recirculation of, 656
Toad, American, eggs of, *11*
Tolerance, to mood drugs, 609
Tongue, taste regions of, *625*
Tonsils, 655
Tooth enamel, brown, inheritance of, 255–256
Touch receptors, 618–619, *619*
Toxins, inactivation of, 669
Trace elements, 45, 715
Trachea, 425, *687*, 687–688
Tracheal tubes, for gas exchange, 683, *683–684*
Traits, genetic control of, 257
Tranquilizers, effects of, 608t
Transcription, RNA synthesis by, *201–202*, 201–203
Transcutaneous electrical stimulation, pain blocking by, 618
Transduction, 267, *268*
    in bacteria, 350
Transfer RNA (tRNA), 201
    structure of, 204, *205*
Transformation, bacterial, 190, *190*, 348–349
Translation, 201
    initiation of, 207–208, *208*
    protein production by, 203–209
    steps in, 203, *206*
Translocation, 499–501, *501*
Transmission, in neural response, 578, *579*
Transpiration, 482
Transportation, of food, definition of, 699
Tree-ring analysis, 497
Triacylglycerol, 68
Triassic period, 331, 332–333t
Tricarboxylic acid (TCA) cycle, *159*, 159–160
Triceps muscles, 572, *572*
Trichocysts, 353
Trichomes, 468, 477, *477*
Trichomoniasis, 786t
Tricuspid valve, of human heart, 646, *646–647*
Triglycerides, association with heart disease, 713–714
Triiodothyronine, 752
Trilobites, evolution of, 321
Trimesters, of pregnancy, 807, *811*
Triplets, 200, 200t
Triploid cell, 400
Trisomy, 237, *237*, 238t

Trochophore larva, 420, *420*
Trophic levels, and biomass, 859
Trophoblast, formation of, *804*, 805
Tropical life, grassland, *894*, 894–895
    rain forest, 891–893
    zones of, variation with precipitation, 891–895, *896*
Tropical rain forests, animals of, *893*
    conditions for, 891–892
    destruction of, 893, *910*
    trees of, 892, *892*
Tropisms, 525–526, *527*
Truth, guarding by scientists, 37–38
Trypsin, secretion of, 708
Tryptophan operon, 214–215, *216*
Tubal ligation, for birth control, 781t, 784, *784*
Tuber, *508*, 509
Tumor, definition of, 548
Tundra, 886–888, *887*
Tunicates, 434, *435*
"Tunnelling" microscopes, 191f
Turgor pressure, 125, *126*
    and plant movement, 524
Twins, development of, 805
Tympanic canal, 622, *624*
Tyrosinase, 244

U
Ubiquinone (Q), 160
Ulna, 287
Ultraviolet light, perception with compound eye, 628, *629*
Ultraviolet radiation, 20, *172*
    effects of, 872
    ozone layer protection from, 872–873
Umbilical cord, function of, 802
Unconditioned stimulus, 823, *823*
Unhygienic behavior, of Van Scoy bees, 820–822, *821*
Uniramia, 425
Universe, understanding of, science and, 24–28
Urea, as metabolic waste product, 726
Uremia, 735
Urethritis, nongonococcal, 786t
Uric acid, as metabolic waste product, 726
Urinary system, *553*, 554t, 731–733
    human, 732, *732*
Urination, definition of, 732
Urine, composition of, 739
    concentration of, 737–738
    production of, 733, 735
    volume of, regulation of, 738
Uterine tube, embryo development in, 803, 805
    ovum transport by, 772
Uterus, function of, 772
Utricle, of ear, 621, *621*

V
V region, of polypeptide chain, 668
Vacuole, food, 352
Vagina, function of, 772
Valence electrons, 48
Valium, effects of, 608t
Valves, of human heart, 646, *646–647*
van der Waals force, 52
Van Scoy bees, behavior of, genetic control of, 820–822, *821*
Variation, 287
Varicose veins, 651
Vas deferens, 766
Vasectomy, 781t, 783–784, *784*
Vasoconstriction, 643
Vasodilatation, 643
Vectoring, of genes, 267
Vegetal pole, of zygote, 793, 796
Vegetarian diets, 715
Veins, valves of, *650*, 650–651
    in vertebrates, 643, *643*
Veld, *894*, 894–895
Veliger larva, 420
Venous return, in cardiac circulation, 648
Ventilation, 682, 688–689, *690*
Ventral nerve cord, of insects, 428
Ventral root, of spinal nerve, 606, *606*
Ventricles, of heart, 439, 644
    human, 646, *646–647*
Ventricular fibrillation, 653
Venus flytrap, *8*, *486*, 487
Vermiform appendix, 709
Vernalization, 520
Vernix, in fetus, 807
Vertebral column, regions of, 564, *565*
Vertebrates, aquatic, osmoregulation in, 729–730, *730*
    brain of, evolution of, 596–599, *597*
    classes of, 436
    digestive systems of, 701, *702*, 703
    endoskeleton of, 435–436, 564–567
    evolution of, 436, 445
    immune responses of, 661–678
    lungs in, 686–687
    muscle of, 567–568
    nervous system of, 596
    osmoregulation in, 729–739
    primitive, 436–437
    skin of, 560–561
    societies of, flexibility of, 842–843
    waste disposal in, 729–739
Vestibular apparatus, of ear, *621*, 621–622, *624*
Vestigial organs, 288–289, *289*
Victoria, Queen of England, *254*
Victoria lilies, *896*
Villi, of intestine, 705, *706*, 708
Vines, tendrils of, 485

Viral infections, interferons in, 662
Viroids, 339
Virus(es), 4f, 190
    classification of, problems with,
        339, *341*
    eukaryotic cells and, 272
    production of new, 191
Visceral peritoneum, of digestive tract,
    703, *703*
Vision, defects in, 628
    requirements for, 626
    rhodopsin in, *632*
Vital centers, of brain, 597
Vitamins, 718–719t
    definition of, 715
    effects on embryo, 810t
Vitelline membrane, of egg, 792
Vitreous body, of vertebrate eye, 630
Vulva, 773, *773*

W

Waggle dance, of honeybees, 841,
    842–843
Wallace, Alfred Russel, 286–287
Warblers, interspecific competition of,
    855
Wareing, P.E., 534
Waste(s), break down of, by decom-
    posers, 16
    definition of, 913
    disposal of, improper, and pollu-
        tion, 913–919
        in vertebrates, 729–739
    elimination of, by large intestine,
        709–710
Water(s), acids and bases in, 55–56
    action of, capillary, 54
        effect on climate, 869
    buffering capacity of, 916f
    cohesive and adhesive forces of,
        54

Water(s), (Continued)
    conservation of, *731, 738*
    distribution in ecosphere, 874–
        877, *875*
    hydrogen bonding in, 54–55, *55*
    as metabolic waste product, 726
    molecules of, 53, *53*
    movement of, Coriolis effect and,
        873, 873–874
        effects of, 877–879
    neutralization of, 58
    phase changes of, 874–875
    properties of, *52*, 53–55
    as solvent, 53–54
    specific heat of, 874
    temperature of, 54–55
    translocation of, 499–500
    in urine, 739
Water molds, 351, 358, 361
Water pollution, 914, *915*
Water strider, *54*
Watson, James, 191–193
Wavelength, 170
Weather, effect of water on, 874
Weeds, suppression of, in mechanized
    agriculture, *909*
Went, Frits, 528
Wetlands, destruction of, solid waste
    and, 918–919
Whale, 895
Whisk ferns, 387, *389*
White blood cells, in defense against
    disease, 641–642
White-crowned sparrow, seasonal
    physiology and behavior of, *830*
White matter, 599, 601t
White spruce seedling nursery, *911*
Wildlife, destruction of, solid waste
    and, 918–919
Wilkins, Maurice, 191
Wind, action of, effect on climate, 869
    pollination by, 512, *513*

Wind, (Continued)
    seed dispersal by, 515, *516*
Wings, 287
Withdrawal, for birth control, 781t
Withdrawal reflex, 587, *588*
Worms, proboscis, 416–417, *417*

X

X-ray diffraction analysis, 191, *192*
Xylem, 465, *466*, 478, 491–492, 494,
    498
    transport in, 499–501

Y

Yeast(s), 366, *366*, 376, *376*
Yeast infections, 786t
Yields, in reaction, 48
Yolk, of zygote, 793
Yolk sac, function of, *801*, 802
Young, caring for, 839–840

Z

Z-DNA, *195*
Zeatin, 533
Zebra, *894*
Zinc, function in body, 720t
Zona pellucida, in female reproduc-
    tion, 771
Zonation, by altitude, 886, *886*
Zooplankton, 895–896
Zoospores, 355
Zooxanthellae, of coral reef, 900
Zugunruhe, of migrating birds, *830*,
    831
Zygomycetes, 368
Zygospores, 368
Zygote, 224, 237, 383–384, 457
    contents of, 793
    division of, 457, *457*, 793–796
    production of, 779, *779*
    in sexual reproduction, 763

The Metric System

Standard Metric Units

Metric Units		Abbreviations
Standard unit of mass	gram	g
Standard unit of length	meter	m
Standard unit of volume	liter	l

Some Common Prefixes

Prefixes		Examples
kilo	1,000	a kilogram is 1,000 grams
centi	0.01	a centimeter is 0.01 meter
milli	0.001	a milliliter is 0.001 liter
micro (μ)	one-millionth	a micrometer is 0.000001 (one-millionth) of a meter
nano (n)	one-billionth	a nanogram is 10^{-9} (one-billionth) of a gram
pico (p)	one-trillionth	a picogram is 10^{-12} (one-trillionth) of a gram

Some Common Units of Length

Unit	Abbreviation	Equivalent
meter	m	approximately 39 in
centimeter	cm	10^{-2} m
millimeter	mm	10^{-3} m
micrometer	μm	10^{-6} m
nanometer	nm	10^{-9} m
angstrom	Å	10^{-10} m

Length conversions

1 in = 2.5 cm	1 mm = 0.039 in
1 ft = 30 cm	1 cm = 0.39 in
1 yd = 0.9 cm	1 m = 39 in
1 mi = 1.6 km	1 m = 1.094 yd
	1 km = 0.6 mi

To convert	Multiply by	To obtain
inches	2.54	centimeters
feet	30	centimeters
centimeters	0.39	inches
millimeters	0.039	inches